Brief Contents

UNIT 1 Organization of the Body

1 The Human Body: An Orientation 1

2 Chemistry Comes Alive 23

3 Cells: The Living Units 60

4 Tissue: The Living Fabric 115

UNIT 2 Covering, Support, and Movement of the Body

5 The Integumentary System 150

6 Bones and Skeletal Tissues 173

7 The Skeleton 199

8 Joints 251

9 Muscles and Muscle Tissue 279

10 The Muscular System 323

UNIT 3 Regulation and Integration of the Body

11 Fundamentals of the Nervous System and Nervous Tissue 390

12 The Central Nervous System 434

13 The Peripheral Nervous System and Reflex Activity 489

14 The Autonomic Nervous System 531

15 The Special Senses 553

16 The Endocrine System 601

UNIT 4 Maintenance of the Body

17 Blood 642

18 The Cardiovascular System: The Heart 670

19 The Cardiovascular System: Blood Vessels 706

20 The Lymphatic System and Lymphoid Organs and Tissues 766

21 The Immune System: Innate and Adaptive Body Defenses 781

22 The Respiratory System 818

23 The Digestive System 868

24 Nutrition, Metabolism, and Energy Balance 926

25 The Urinary System 974

26 Fluid, Electrolyte, and Acid-Base Balance 1012

UNIT 5 Continuity

27 The Reproductive System 1041

28 Pregnancy and Human Development 1091

29 Heredity 1124

Equipping You with 21st-Century Skills to Succeed in A&P *and Beyond...*

The **11th Edition** of Elaine Marieb and Katja Hoehn's best-selling A&P text and media program motivates and supports both novice learners and expert students, more than ever before. Each carefully-paced chapter guides you in advancing from mastering terminology to applying knowledge in clinical scenarios, to practicing the critical thinking and problem-solving skills that are required for entry to nursing, allied health, and exercise science programs.

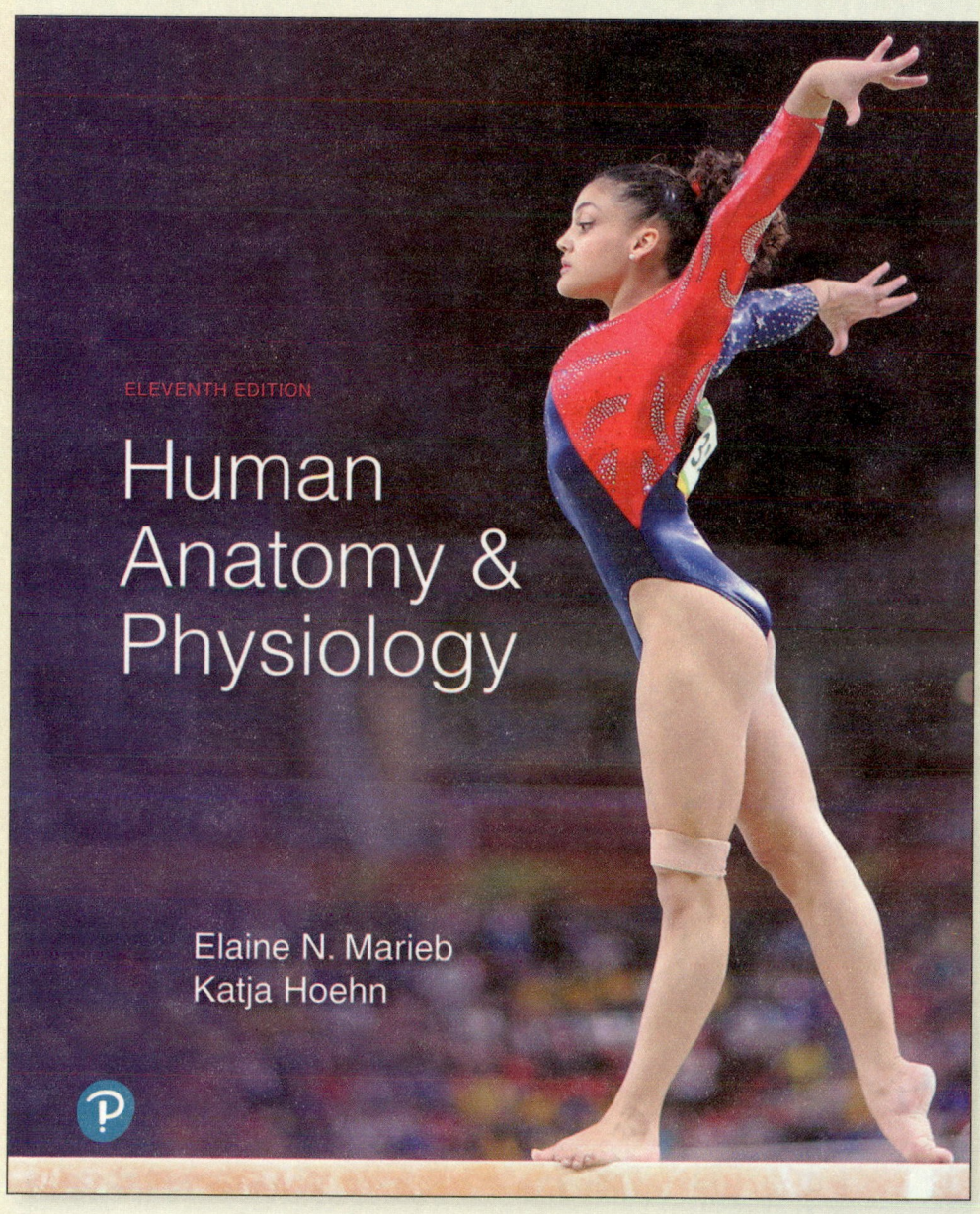

ELEVENTH EDITION

Human Anatomy & Physiology

Elaine N. Marieb
Katja Hoehn

Identify "Big Picture" Concepts
Before Exploring Details

Before you look up details and information within a chapter, read the **Chapter-Opening Roadmap**, which visually groups and organizes "big picture" concepts and shows how they are related. To focus your studying, review the numbered **Key Concept Headings**, **Learning Outcomes**, and summaries.

UNIQUE! Chapter Roadmaps provide a visual overview of the key concepts in the chapter and show how they relate to each other. Each key concept "brick" in the roadmap corresponds to a numbered section within the chapter.

Each numbered section within the chapter begins with a **Key Concept Heading** that helps you quickly grasp the "big idea" of the discussion that follows.

UPDATED! Career Connection Videos feature a health care professional who describes how the chapter content relates to their everyday work. You can access all of the Career Connections videos through an open access web page at https://goo.gl/88srfV.

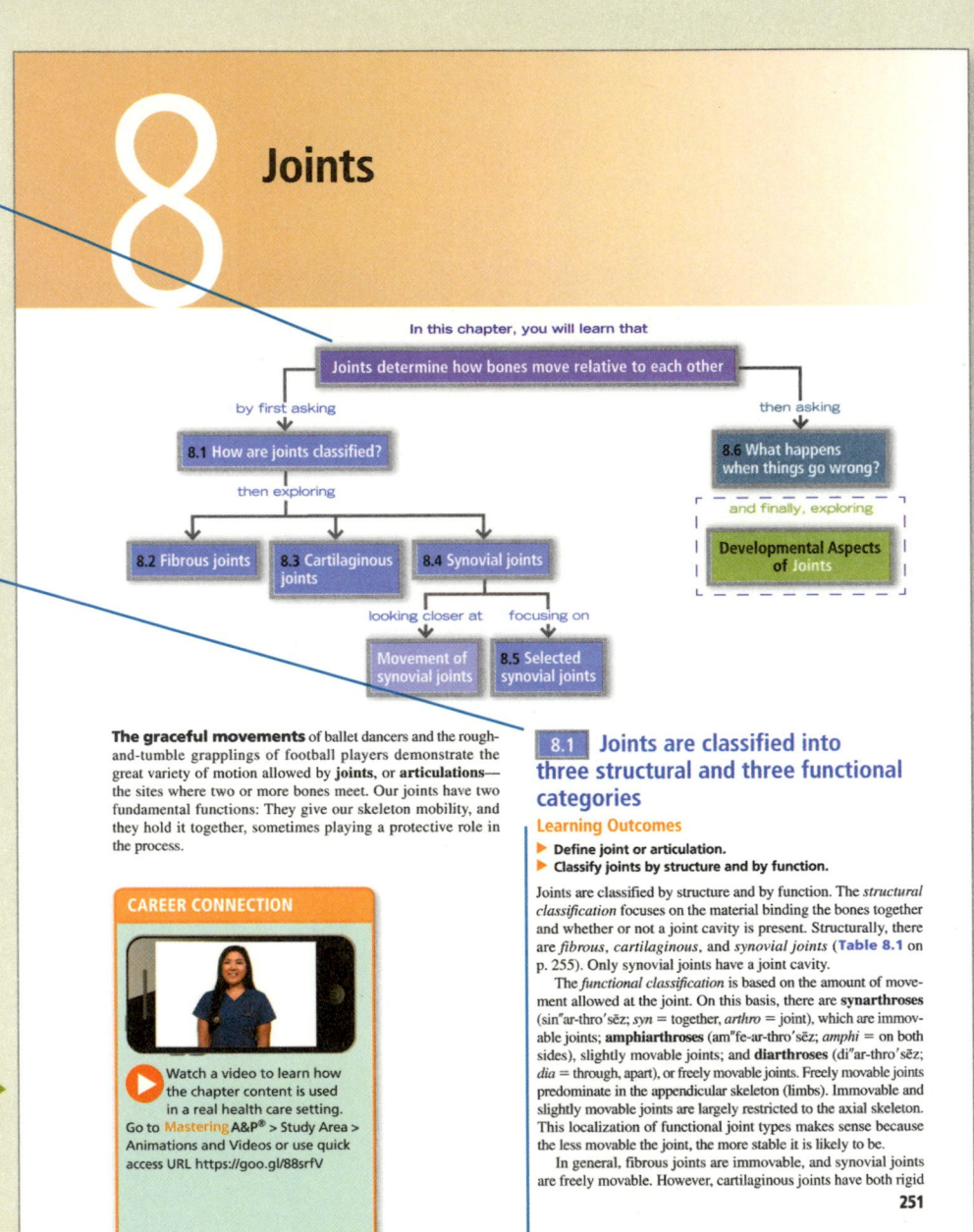

8 Joints

In this chapter, you will learn that

Joints determine how bones move relative to each other

by first asking

8.1 How are joints classified?

then exploring

8.2 Fibrous joints **8.3 Cartilaginous joints** **8.4 Synovial joints**

looking closer at   focusing on

Movement of synovial joints **8.5 Selected synovial joints**

then asking

8.6 What happens when things go wrong?

and finally, exploring

Developmental Aspects of Joints

The graceful movements of ballet dancers and the rough-and-tumble grapplings of football players demonstrate the great variety of motion allowed by **joints**, or **articulations**—the sites where two or more bones meet. Our joints have two fundamental functions: They give our skeleton mobility, and they hold it together, sometimes playing a protective role in the process.

CAREER CONNECTION

▶ Watch a video to learn how the chapter content is used in a real health care setting.
Go to Mastering A&P® > Study Area > Animations and Videos or use quick access URL https://goo.gl/88srfV

8.1 **Joints are classified into three structural and three functional categories**

Learning Outcomes
▶ Define joint or articulation.
▶ Classify joints by structure and by function.

Joints are classified by structure and by function. The *structural classification* focuses on the material binding the bones together and whether or not a joint cavity is present. Structurally, there are *fibrous, cartilaginous,* and *synovial joints* (**Table 8.1** on p. 255). Only synovial joints have a joint cavity.

The *functional classification* is based on the amount of movement allowed at the joint. On this basis, there are **synarthroses** (sin″ar-thro′sĕz; *syn* = together, *arthro* = joint), which are immovable joints; **amphiarthroses** (am″fe-ar-thro′sĕz; *amphi* = on both sides), slightly movable joints; and **diarthroses** (di″ar-thro′sĕz; *dia* = through, apart), or freely movable joints. Freely movable joints predominate in the appendicular skeleton (limbs). Immovable and slightly movable joints are largely restricted to the axial skeleton. This localization of functional joint types makes sense because the less movable the joint, the more stable it is likely to be.

In general, fibrous joints are immovable, and synovial joints are freely movable. However, cartilaginous joints have both rigid

251

See p. 251

Learning Outcomes are presented at the beginning of each chapter section to give you a preview of essential information to study.

Study the Figures as You Read the Text

Anatomy and Physiology is a visual science. To succeed, you need to practice and develop visual literacy skills for understanding and interpreting information. To help you achieve this goal, the text and associated figures are tightly integrated so that you never have to flip pages back and forth to connect visuals with words.

EXPANDED! 6 new Focus Figures (for a total of 26) walk you through complex processes using exceptionally clear, easy-to-follow illustrations with integrated text explanations.

NEW Focus Figures are as follows:
3.1 The Plasma Membrane, *pp. 64–65*
11.4 Postsynaptic Potentials and Their Summation, *pp. 418–419*
16.2 Stress and the Adrenal Gland, *pp. 628–629*
18.2 The Cardiac Cycle, *pp. 694–695*
21.1 An Example of a Primary Immune Response, *pp. 808–809*
28.2 Fetal and Newborn Circulation, *pp. 1108–1109*

See pp. 808–809

NEW! Focus Figure "Mini-Animation" Coaching Activities bring the 6 new Focus Figures to life using short video segments.

Blue text represents the voice of an A&P instructor, highlighting important points to remember.

Activation and Differentiation of B Cells

An immunocompetent but naive B lymphocyte is *activated* when matching antigens bind to its surface receptors and cross-link adjacent receptors together. Antigen binding is quickly followed by receptor-mediated endocytosis of the cross-linked antigen-receptor complexes. As we described previously, this is called *clonal selection* and is fol-

Antigen
Cross-linked receptors

EXPANDED! 31 unique In-Line Figures are strategically placed within the text to visually reinforce the text discussion.

See p. 796

Pace Yourself:
Learn & Review the Basics

EXPANDED! Summary Tables present key information and serve as "one-stop shopping" study tools. 13 new Summary Tables have been added to this edition.

Table 5.1 Summary of Cutaneous Glands

	ECCRINE SWEAT GLANDS	APOCRINE SWEAT GLANDS	SEBACEOUS GLANDS
Functions	• Temperature control • Some antibacterial properties	May act as sexual scent glands	• Lubricate skin and hair • Help prevent water loss • Antibacterial properties
Type of Secretion	Hypotonic filtrate of blood plasma	Filtrate of blood plasma with added proteins and fatty substances	Sebum (an oily secretion)
Method of Secretion	Merocrine (exocytosis)	Merocrine (exocytosis)	Holocrine
Secretion Exits Duct At	Skin surface	Usually upper part of hair follicle; rarely, skin surface	Usually upper part of hair follicle; sometimes, skin surface
Body Location	Everywhere, but especially palms, soles, forehead	Mostly axillary and anogenital regions	Everywhere *except* palms and soles

See p. 162

Sebaceous Glands

The **sebaceous glands** (se-ba′shus; "greasy"), or *oil glands* (Figure 5.9a), are simple branched alveolar glands that are found all over the body except in the thick skin of the palms and soles. They are small on the body trunk and limbs, but quite large on the face, neck, and upper chest. These glands secrete an oily substance called **sebum** (se′bum). The central cells of the alveoli accumulate oily lipids until they become so engorged that they burst, so functionally these glands are *holocrine glands* (◀ p. 126). The accumulated lipids and cell fragments constitute sebum.

NEW! Text Recall icons guide you to review specific pages where a concept was first introduced.

See p. 163

NEW! Building Vocabulary Coaching Activities in Mastering A&P® are a fun way to learn word roots and A&P terminology while building and practicing important language skills.

Apply Your Knowledge to a Range & Variety of Questions

As you build your knowledge and confidence in A&P, practice responding to the more challenging questions—you are likely to encounter similar questions on a test or licensing exam. Your extra effort will pay off at exam time!

NEW! A greater **variety and range of self-assessment questions** have been added to the **Check Your Understanding** sections within each chapter and include **Apply**, **Predict**, **What If?**, **Draw**, and **Make Connections**. Dozens of new **visual questions** ask you to label structures or interpret visual information.

Check Your Understanding

5. How does a nucleus within the brain differ from a nucleus within a neuron?

6. How is a myelin sheath formed in the CNS, and what is its function?

7. What is the structural classification of the neuron shown below? What is its usual functional classification? Name the parts labeled a–d.

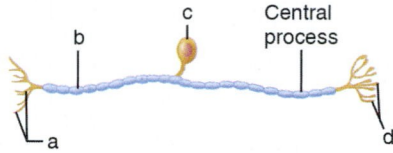

8. **APPLY** Which structural and functional type of neuron is activated first when you burn your finger? Which type is activated last to move your finger away from the source of heat?

9. **MAKE CONNECTIONS** Which part of the neuron is its fiber? How do nerve fibers differ from the fibers of connective tissue (see Chapter 4) and the fibers in muscle (see Chapter 9)?

_____ *For answers, see Answers Appendix.*

See p. 400

NEW! **"Draw" questions** ask you to create visuals that reinforce important concepts by drawing a structure, annotating a figure, or creating a summary table.

3. **DRAW** Create a summary table to help you study the pharynx by comparing and contrasting its three parts. For each part, identify what it conducts (air, food, or both), the type of epithelium found there, and the associated tonsils.

NEW! All of the **End-of-Chapter Review questions** are now organized into 3 levels of difficulty based on **Bloom's Taxonomy categories:**
Level 1: Remember/Understand
Level 2: Apply/Analyze
Level 3: Evaluate/Synthesize

	Conducts	Epithelium	Tonsils
Nasopharynx	Air	Pseudostratified ciliated columnar	Pharyngeal Tubal
Oropharynx	Air and food	Stratified squamous	Palatine Lingual
Laryngopharynx	Air and food	Stratified squamous	—— (none)

See p. 824 and Answers Appendix

Prepare for Your Future Career &
Practice Solving Real-World Problems

The authors of this text, Elaine Marieb and Katja Hoehn, share insights from their own clinical experience to help you prepare for your future career in health care. All clinical examples and applications are signaled with an easy-to-find "Clinical" label.

UPDATED! Homeostatic Imbalance discussions alert you to the consequences of body systems not functioning optimally. Relevant photos have been added to selected discussions for visual reinforcement.

HOMEOSTATIC IMBALANCE 5.6

CLINICAL

Changes in nail appearance can help diagnose certain conditions. For example, yellow-tinged nails may indicate a respiratory or thyroid gland disorder. (Thickened yellow nails are usually due to a fungus infecting the nail.) An outward concavity of the nail (*koilonychia* or "spoon nail," **Figure 5.8**) may signal an iron deficiency. Horizontal lines (Beau's lines) across the nails can be a sign of severe illness that affects the whole body such as uncontrolled diabetes, a heart attack, or cancer chemotherapy.

Figure 5.8 Koilonychia.

See p. 161

UPDATED! Clinical Case Studies are provided at the end of Chapters 5–29 and challenge you to apply your knowledge to realistic clinical scenarios.

CLINICAL CASE STUDY

70-Year-Old Male with Polyuria

Mr. Gutteman, a 70-year-old male, was brought into the ER. He had been sick several days with the flu, and was found confused and barely conscious by his daughter.

Mr. Gutteman is breathing rapidly and has a fever of 39°C (102°F). His skin is dry and flaccid, his mucous membranes are dry, and his eyes are sunken. The physician ordered:

- IV (intravenous) fluid and electrolyte replacement
- Blood and urine tests for presence of glucose and ketones
- Strict I&O [careful measurement of fluid intake (e.g., IV, drinking) and output (e.g., urine)]

NEW! Each Clinical Case Study includes "NCLEX-Style" questions for practice with the kinds of challenge questions that you will eventually encounter on a licensing exam. Practice answering these questions on your own or in collaboration with classmates. Your instructor can also assign **new NCLEX-Style questions in Mastering A&P®**, along with Homeostatic Imbalance questions, Clinical Case Study Coaching Activities, and Nurses Need Physiology Case Studies.

1. **+ NCLEX-STYLE** You would expect high levels of blood glucose and the presence of glucose and ketones in Mr. Gutteman's urine if:
   a. His pancreas is secreting too much insulin
   b. His liver is secreting too little insulin
   c. His pancreas is secreting too little insulin
   d. His liver is secreting too much glucagon

See p. 641

Access the Complete Textbook Online or Offline Using the Pearson eText

You can read your textbook without having to add weight to your bookbag!

The Pearson eText mobile app offers offline access and can be downloaded for most iOS and Android phones and tablets from the Apple App or Google Play stores.

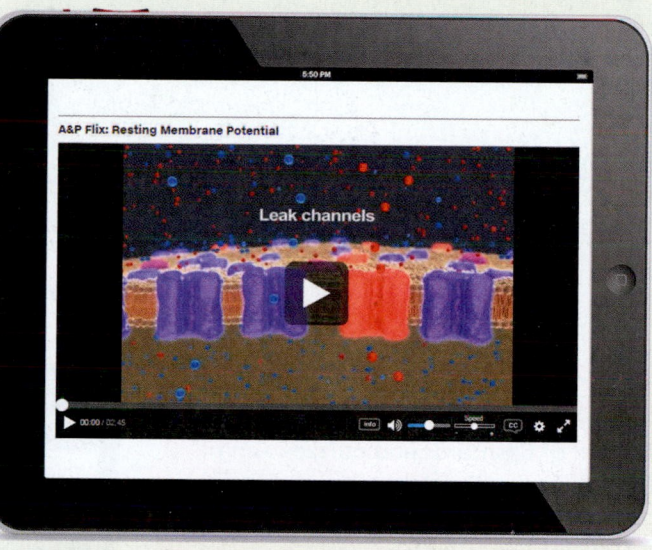

Powerful interactive and customization functions include instructor and student note-taking, highlighting, bookmarking, search, and links to glossary terms. The Marieb/Hoehn eText also includes dozens of embedded videos and animations that bring A&P concepts to life.

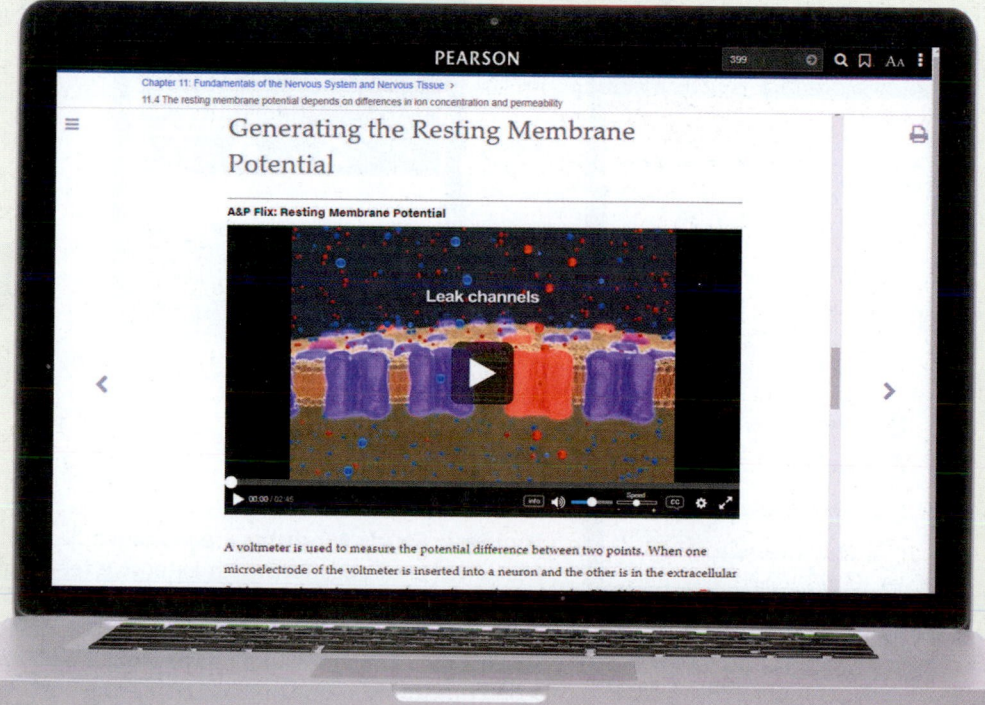

Get Online Practice and Coaching with Mastering A&P®

Mastering A&P® provides tutorials and review questions that you can access before, during, and after class.

EXPANDED! Interactive Physiology 2.0 Coaching Activities teach complex physiology processes using exceptionally clear animations, interactive tutorials, games, and quizzes. IP2 features new graphics, quicker navigation, and a mobile-friendly design. New topics include Generation of an Action Potential and Cardiac Cycle. IP2 and IP animations can be assigned from the Mastering A&P® item library or accessed through the Study Area.

NEW! PAL 3.1 Customizable Flashcards allow you to create a personalized, mobile-friendly deck of flashcards and quizzes using images from Practice Anatomy Lab. Use the checklist to select only those structures covered in your course.

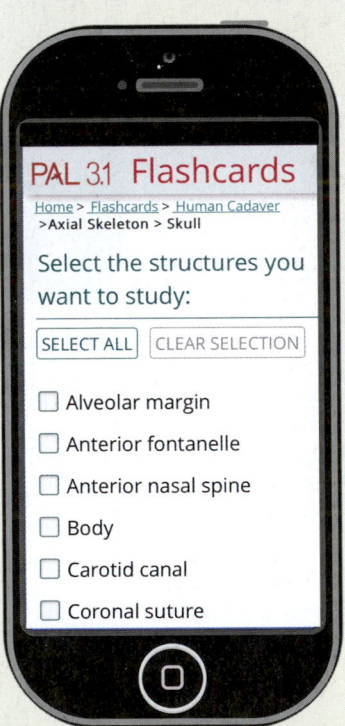

Dynamic Study Modules are manageable, mobile-friendly sets of questions with extensive feedback for you to test, learn, and retest yourself on basic concepts. **NEW!** Instructors can select or deselect specific questions for assignments from more than 3,000 questions, organized by chapter section.

New for Instructors:
Ready-to-Go Teaching Modules

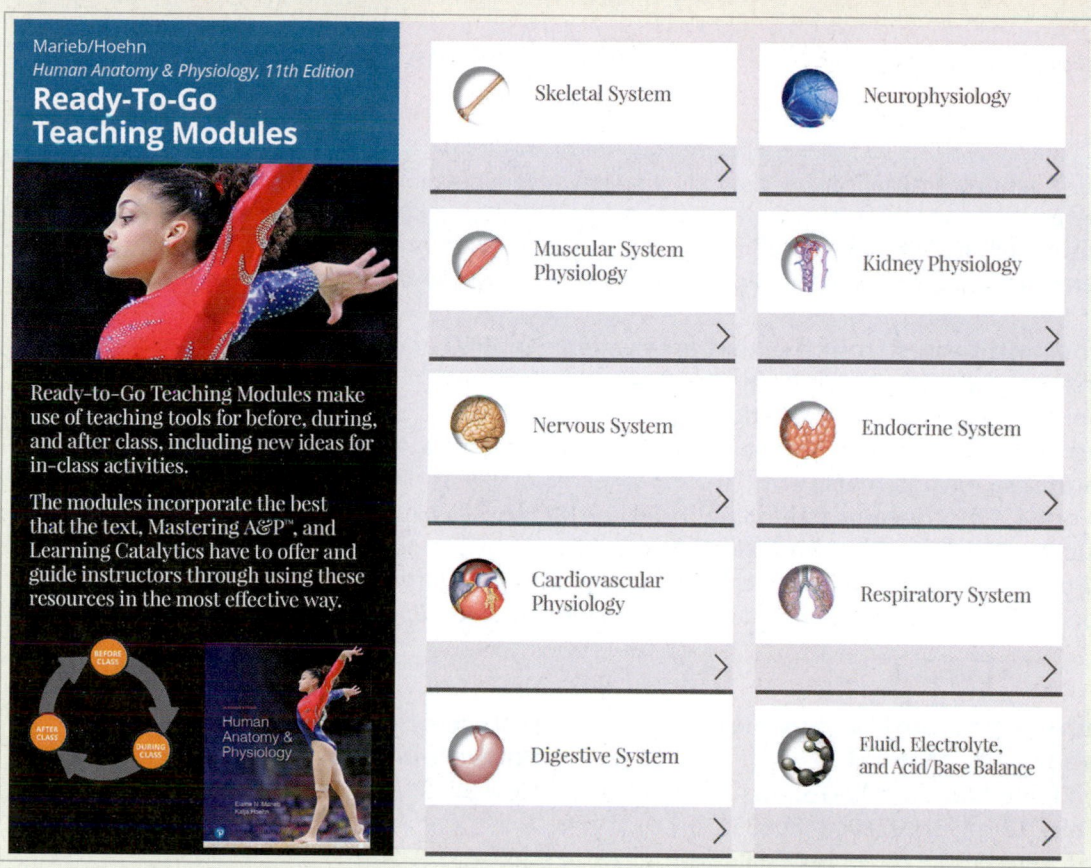

Marieb/Hoehn
Human Anatomy & Physiology, 11th Edition
Ready-To-Go Teaching Modules

Ready-to-Go Teaching Modules make use of teaching tools for before, during, and after class, including new ideas for in-class activities.

The modules incorporate the best that the text, Mastering A&P™, and Learning Catalytics have to offer and guide instructors through using these resources in the most effective way.

Human Anatomy & Physiology

Skeletal System	Neurophysiology
Muscular System Physiology	Kidney Physiology
Nervous System	Endocrine System
Cardiovascular Physiology	Respiratory System
Digestive System	Fluid, Electrolyte, and Acid/Base Balance

Learning Catalytics allows students to use their smartphone, tablet, or laptop to respond to questions in class. Visit learningcatalytics.com to learn more.

Additional Support for Students & Instructors

Mastering A&P® offers thousands of tutorials, activities, and questions that can be assigned for homework and practice. Highlights of new assignment options include:

- **NEW! Building Vocabulary Coaching Activities** give you practice learning and using word roots in context as you learn new A&P terms.
- **NEW! Focus Figure "Mini-Animation" Coaching Activities** bring the 6 new Focus Figures to life and include assessment questions.

- **IMPROVED! Concept Map Coaching Activities** support the concept maps in the text without requiring students to submit their own concept map for grading.
- **NEW! NCLEX-Style Questions** give students practice with the kinds of questions that will eventually appear on a licensing exam.

The Mastering A&P® Instructor Resources Area includes the following downloadable tools for instructors who adopt the Eleventh Edition for their classes:

- **NEW! Ready-to-Go Teaching Modules** provide teaching tools for 10 challenging topics in A&P.
- **Customizable PowerPoint® lecture outlines** include customizable images and provide a springboard for lecture prep.
- **All of the figures, photos, and tables from the text** are available in JPEG and PowerPoint® formats, in labeled and unlabeled versions, and with customizable labels and leader lines.
- **Test bank** provides thousands of customizable questions across Bloom's Taxonomy levels. Each question is tagged to chapter learning outcomes that can also be tracked within Mastering A&P® assessments. Available in Microsoft® Word and TestGen® formats.
- **Animations and videos** bring A&P concepts to life and include A&P Flix 3-D Animations.
- **A comprehensive Instructor Guide to Text and Media**, co-authored by Elaine Marieb and Laura Steele, includes a detailed teaching outline for each chapter, along with a wealth of activities, examples, and analogies that have been thoroughly class-tested with thousands of students.

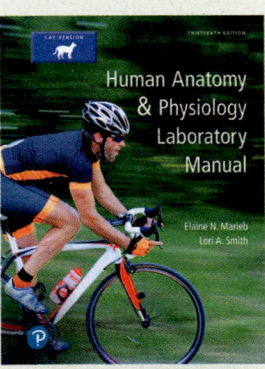

Human Anatomy & Physiology Laboratory Manual
by Elaine Marieb & Lori Smith
Main 12/e ISBN 9780134806358
Cat 13/e ISBN 9780134632339
Fetal Pig 13/e ISBN 9780134806365

Instructor Resource DVD with PowerPoint Lecture Outlines
ISBN 9780134757292

Instructor Guide to Text and Media for Human Anatomy & Physiology 11/e
ISBN 9780134787978

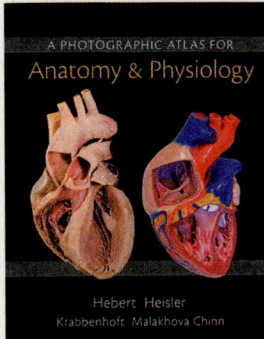

A Photographic Atlas for Anatomy & Physiology
by Nora Hebert, Ruth E. Heisler, et al.
ISBN 9780321869258

Printed Testbank for Human Anatomy & Physiology 11/e
ISBN 9780134788944

Human Anatomy & Physiology

ELEVENTH EDITION

Elaine N. Marieb, R.N., Ph.D.
Holyoke Community College

Katja Hoehn, M.D., Ph.D.
Mount Royal University

Editor-in-Chief: Serina Beauparlant
Senior Courseware Portfolio Manager: Lauren Harp
Managing Producer: Nancy Tabor
Content & Design Manager: Michele Mangelli, Mangelli Productions, LLC
Courseware Director, Content Development: Barbara Yien
Courseware Specialist: Laura Southworth
Courseware Senior Analyst: Barbara Price
Courseware Editorial Assistant: Dapinder Dosanjh
Rich Media Content Producer: Lauren Chen
Mastering Content Developer: Cheryl Chi
Production Supervisor: Karen Gulliver
Copyeditor: Anita Hueftle

Proofreaders: Betsy Dietrich and Martha Ghent
Compositor: Cenveo® Publisher Services
Art Coordinator: Jean Lake
Interior & Cover Designer: Hespenheide Design
Illustrators: Imagineering STA Media Services, Inc.
Rights & Permissions Management: Ben Ferrini
Rights & Permissions Project Manager: Cenveo® Publisher Services, Matt Perry
Photo Researcher: Kristin Piljay
Manufacturing Buyer: Stacey Weinberger
Director of Product Marketing: Allison Rona
Senior Anatomy & Physiology Specialist: Derek Perrigo
Director of Field Marketing: Tim Galligan

Cover photo credit: Tim Clayton/Corbis/Getty Images

Library of Congress Cataloging-in-Publication Data

Names: Marieb, Elaine Nicpon, author. | Hoehn, Katja, author.
Title: Human anatomy & physiology / Elaine N. Marieb, R.N., Ph.D., Holyoke Community College, Katja Hoehn, M.D., Ph.D., Mount Royal University.
Other titles: Human anatomy and physiology
Description: Eleventh edition. | Hoboken, New Jersey : Pearson Education, Inc., [2019] | Includes index.
Identifiers: LCCN 2017044428 | ISBN 9780134580999 (Student edition) | ISBN 0134580990 (Student edition)
Subjects: LCSH: Human physiology. | Human anatomy.
Classification: LCC QP34.5 .M265 2018 | DDC 612—dc23
LC record available at https://lccn.loc.gov/2017044428

1 17

ISBN 10: 0-13-458099-0; ISBN 13: 978-0-13-458099-9 (Student edition)
ISBN 10: 0-13-476299-1; ISBN 13: 978-0-13-476299-9 (Instructor's Review Copy)

www.pearson.com

About the Authors

We dedicate this work to our students both present and past, who always inspire us to "push the envelope."

Elaine N. Marieb

After receiving her Ph.D. in zoology from the University of Massachusetts at Amherst, Elaine N. Marieb joined the faculty of the Biological Science Division of Holyoke Community College. While teaching at Holyoke Community College, where many of her students were pursuing nursing degrees, she developed a desire to better understand the relationship between the scientific study of the human body and the clinical aspects of the nursing practice. To that end, while continuing to teach full time, Dr. Marieb pursued her nursing education, which culminated in a Master of Science degree with a clinical specialization in gerontology from the University of Massachusetts. It is this experience that has informed the development of the unique perspective and accessibility for which her publications are known.

Dr. Marieb has given generously to provide opportunities for students to further their education. She funds the E.N. Marieb Science Research Awards at Mount Holyoke College, which promotes research by undergraduate science majors, and has underwritten renovation of the biology labs in Clapp Laboratory at that college. Dr. Marieb also contributes to the University of Massachusetts at Amherst, where she provided funding for reconstruction and instrumentation of a cutting-edge cytology research laboratory. Recognizing the severe national shortage of nursing faculty, she underwrites the Nursing Scholars of the Future Grant Program at the university.

In 2012 and 2017, Dr. Marieb gave generous philanthropic support to Florida Gulf Coast University as a long-term investment in education, research, and training for healthcare and human services professionals in the local community. In honor of her contributions, the university is now home to the Elaine Nicpon Marieb College of Health and Human Services.

Katja Hoehn

Dr. Katja Hoehn is a professor in the Department of Biology at Mount Royal University in Calgary, Canada. Dr. Hoehn's first love is teaching. Her teaching excellence has been recognized by several awards during her 24 years at Mount Royal University. These include a PanCanadian Educational Technology Faculty Award (1999), a Teaching Excellence Award from the Students' Association of Mount Royal (2001), and the Mount Royal Distinguished Faculty Teaching Award (2004).

Dr. Hoehn received her M.D. (with Distinction) from the University of Saskatchewan, and her Ph.D. in Pharmacology from Dalhousie University. In 1991, the Dalhousie Medical Research Foundation presented her with the Max Forman (Jr.) Prize for excellence in medical research. During her Ph.D. and postdoctoral studies, she also pursued her passion for teaching by presenting guest lectures to first- and second-year medical students at Dalhousie University and at the University of Calgary.

Dr. Hoehn has been a contributor to several books, written numerous research papers in Neuroscience and Pharmacology,

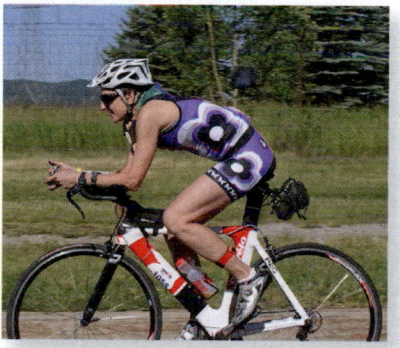

and has co-authored the previous four editions of this textbook. For many years, she has also reviewed and authored electronic media that accompanies Pearson anatomy and physiology books.

Following Dr. Marieb's example, Dr. Hoehn provides financial support for students in the form of a scholarship that she established in 2006 for nursing students at Mount Royal University.

Dr. Hoehn is also actively involved in the Human Anatomy and Physiology Society (HAPS) and is a member of the American Association of Anatomists. When not teaching, she likes to spend time outdoors with her husband and two sons. She also enjoys competing in long-course triathlons, and playing Irish flute down at the local pub.

Preface

Today's students have access to an enormous amount of information about anatomy and physiology. As educators, our biggest challenge is to help students focus on mastering the basic concepts of this field. Providing this firm foundation will help students to become lifelong learners who can critically evaluate new information, connect that information to the foundation they have already established, and apply it in a clinical setting. How can we help students build a strong foundation in anatomy and physiology? We believe that this new edition of our textbook will help learners by building on the strengths of previous editions while using new and innovative ways to help students visualize connections between various concepts.

Unifying Themes

Three unifying themes that have helped to organize and set the tone of this textbook continue to be valid and are retained in this edition. These themes are:

Interrelationships of body organ systems. This theme emphasizes the fact that nearly all regulatory mechanisms have interactions with several organ systems. The respiratory system, for example, cannot carry out its role of gas exchange in the body if there are problems with the cardiovascular system that prevent the normal delivery of blood throughout the body. The System Connections feature and Make Connections questions throughout the book help students connect new information to old information and think of the body as a community of dynamic parts instead of a number of independent units.

Homeostasis. Homeostasis is the normal and most desirable condition of the body. Its loss is always associated with past or present pathology. This theme is not included to emphasize pathological conditions, but rather to illustrate what happens in the body "when things go wrong" and homeostasis is lost. Whenever students see a red balance beam symbol accompanied by an associated clinical topic, their understanding of how the body works to stay in balance is reinforced.

Complementarity of structure and function. This theme encourages students to understand the structure of some body part (ranging from a molecule to an organ) in order to understand the function of that structure. For example, muscle cells can produce movement because they are contractile cells.

New to the Eleventh Edition

New and augmented elements aim to help learners in the following ways.

To help students make connections between new and previously learned material. In order for students to master new concepts, they must link these new concepts with concepts they already understand. In this edition, we help them do this by adding:

- **Text recall icons (◄).** These icons direct the student back to the specific pages where a concept was first introduced.
- **Make Connections questions.** We've added more of this type of question to the Check Your Understanding review questions that follow each module within a chapter. To answer these questions, the student must employ concepts learned previously (most often in previous chapters).
- **New kinds of higher-level questions.** Each chapter now has at least five higher-level questions that require students to think more deeply, pulling together strands from multiple concepts. These questions are clearly identified as APPLY, DRAW, PREDICT, MAKE CONNECTIONS, and WHAT IF? questions.
- **New summary tables.** Students have told us that they want more summary tables. In response, 13 new summary tables (two with illustrations) have been added in order to help students see the big picture.

To enhance students' visual literacy. Anatomy is and has always been taught principally through images. Increasingly, however, physiological data is also represented as images, whether it be molecular interactions or graphical descriptions of processes. Throughout their future health care careers, students will need to be able to understand and interpret information presented visually. In this edition, we help them do this by:

- **Adding new Focus figures.** Focus figures are illustrations that use a "big picture" layout and dramatic art to guide the student through difficult physiological processes in a step-by-step way. Our previous Focus figures have been a hit with both students and instructors. In response to requests for additional Focus figures, we are pleased to present six new two-page features.

- **Adding** DRAW **questions in each chapter.** Students often think that they understand an illustration simply by looking at it, but to truly comprehend an illustration and cement its concepts requires a more active learning approach. For this reason we now include at least one higher-level review question within each chapter that requires a student either to draw an illustration or to add to an existing diagram.

- **Adding questions about illustrations.** To help students practice their visual literacy skills, we have added 47 new Check Your Understanding questions that include an illustration as part of the question. Some of these are as simple as labeling exercises, but many require more advanced interpretation.

- **Updating art to improve its teaching effectiveness.** As always, this is a major part of the revision. Today's students are accustomed to seeing sophisticated photorealistically rendered images. However, many students are not adept at extracting, and thinking critically about, the relevant information contained in such illustrations. With this in mind we continue to refine and update our illustrations as students' needs change, improving their ability to teach important concepts. In many cases we have added blue "instructor's voice" text within the figure to guide a student through it, replacing much of the more remote figure legend. In addition, new photos were painstakingly chosen and labeled to enhance the learning process.

- **Adding new illustrations to existing tables and adding new illustrated tables.** Students find illustrated tables particularly effective because they provide a visual cue that helps them remember a topic. In this edition, we have added illustrations to two tables and added two new illustrated tables.

- **Adding in-line figures.** These are small (less than a half-column wide) illustrations or photos strategically located within the text that discuss the concept they illustrate. This edition now has 31 such in-line figures, most of them newly added.

To help students clinically apply what they have learned

- **Updated Homeostatic Imbalance features.** Many of the Homeostatic Imbalance features have been updated and relevant photos have been added to some. All have been reviewed for accuracy and relevancy. In addition, the updated book design makes these features stand out more clearly.

- **Updated Clinical Case Studies in Chapters 5-29 with added new** + NCLEX-STYLE **questions.** The end-of-chapter review questions, which are now organized into three levels of difficulty based on Bloom's Taxonomy categories, culminate in a clinical case study that allows students to apply some of the concepts they have learned to a clinical scenario. These case studies have been extensively revised and each case study has two questions that are similar in style to those in the NCLEX exam.

- **New clinically relevant photos.** We have added or updated a number of photos that have clinical relevance (procedures, conditions, etc.) that will help students apply what they are reading to real-life situations and to their future careers.

In this edition, certain chapters have received the bulk of our attention and have been more heavily revised. As you can see in the Highlights of New Content (below), these are Chapters 2–4, 9, and 27–29.

As in the previous edition, we have taken painstaking care to ensure that almost all the text and the associated art are covered on the same two-page spread. Although this sounds like a simple goal, it actually takes a great deal of work and has not usually been achieved by other textbooks. We make this effort because it is invaluable to student learning to not have to flip pages back and forth between art and text. Finally, you will notice the appearance of new icons referencing MasteringA&P® interspersed within the text. This guides students to go to the relevant on-line activities to supplement their learning.

Other Highlights of New Content

Chapter 1 The Human Body: An Orientation
- New Figure 1.1 illustrates complementarity of structure and function.
- Updated *A Closer Look* feature on types of medical imaging and added five new photos.
- New Homeostatic Imbalance features about hiatal hernias and about "wrong site surgery."

Chapter 2 Chemistry Comes Alive
- New Homeostatic Imbalance feature about patient's pH predicting outcome of CPR.
- New figures illustrate triglyceride structure (2.16); the difference between saturated and unsaturated fatty acids (2.17); phospholipids (2.18); and protein functions (2.20).
- Revised Figures 2.6 (formation of ionic bonds) and 2.12 (dissociation of salt in water) teach more effectively.
- New summary tables reinforce information about chemical bonds (Table 2.2) and about macromolecules and their monomers and polymers (Table 2.5).

Chapter 3 Cells: The Living Units
- Added Focus Figure 3.1 about the plasma membrane, and reorganized accompanying text.
- Reorganized text about passive membrane transport for improved clarity; updated and reorganized discussion of autophagy and apoptosis.
- Updated information about Tay-Sachs disease.

- New micrographs show micro- and intermediate filaments (Figure 3.20).
- Improved teaching effectiveness of Figures 3.5 (diffusion), 3.17 (processing and distribution of newly synthesized proteins), and 3.30 (stages of transcription).
- New information about telomeres in cancer cells.
- New Homeostatic Imbalance feature about progeria.

Chapter 4 Tissue: The Living Fabric

- New images of cilia show the difference between transmission and scanning electron microscopy (Figure 4.2).
- New in-line figure illustrates apical and basal surfaces of epithelial cells.
- Revised art for epithelial and connective tissue for clarity (Figures 4.4 and 4.11).
- New Figure 4.5 shows how exocrine and endocrine glands differ, and new Figure 4.10 gives an overview of the classification of connective tissue.
- Updated *A Closer Look* feature about cancer.

Chapter 5 The Integumentary System

- New illustrated summary table comparing cutaneous glands (Table 5.1).
- Revised Figures 5.3 and 5.4 for better teaching effectiveness.
- Updated information about skin color and disease states.
- Updated Homeostatic Imbalance features about hirsutism and about hair loss.
- New Homeostatic Imbalance feature about nail changes with disease.
- Updated statistics for and treatment of melanoma, with new photo (Figure 5.11c).

Chapter 6 Bones and Skeletal Tissues

- New summary Table 6.1 compares cartilage and bone tissue.
- New photos of an osteoclast (Figure 6.7); of a femur in longitudinal section to show compact and spongy bone (Figure 6.3); and of a section of a flat bone (skull bone) (Figure 6.4 top).
- Extensive revision of Figure 6.12, which teaches bone growth at epiphyseal plates, including new X ray to show epiphyseal plates, and new photomicrograph of epiphyseal cartilage.
- Updated information about bone remodeling, hormonal regulation of bone growth, and osteoporosis.

Chapter 7 The Skeleton

- New drawings to illustrate the location of the true and false pelves, and the pelvic inlet and outlet (Figure 7.33).
- Updated Homeostatic Imbalance features about pes planus (flat feet) and about developmental dysplasia of the hip.
- New photos of bimalleolar fracture (Figure 7.35) and of cleft lip and palate (Figure 7.39).

Chapter 8 Joints

- New Homeostatic Imbalance feature about shoulder dislocations.
- New Table 8.3 summarizes movements at synovial joints.

- Revised Figure 8.4 (bursae and tendon sheaths).
- Updated *A Closer Look* about prostheses.

Chapter 9 Muscles and Muscle Tissue

- New "Background and Overview" section begins the discussion of the mechanisms of excitation and contraction of skeletal muscle, including a new "big picture" overview in Figure 9.7.
- New introduction to ion channels with art helps students understand skeletal muscle excitation and contraction.
- Reorganized discussions of graded muscle contractions and of smooth muscle, including new Figure 9.24 showing calcium sources for smooth muscle contraction.
- Updated discussion of muscle fatigue.
- Updated Homeostatic Imbalance feature on Duchenne muscular dystrophy.
- Updated *A Closer Look* feature about anabolic steroids.

Chapter 10 The Muscular System

- Revised art about levers for clarity (Figure 10.2 and 10.3).
- New cadaver dissection photos show dissection of muscles of the anterior neck and throat, superficial muscles of the thorax and shoulder in posterior view, and posterior muscles of the thigh and hip (Figures 10.9, 10.14, and 10.21).
- New photos illustrate thumb movements and show torticollis.

Chapter 11 Fundamentals of the Nervous System and Nervous Tissue

- New Focus Figure 11.4 illustrates postsynaptic potentials and their summation.
- Improved teaching effectiveness of Figure 11.12 (coding of action potentials for stimulus intensity) and Figure 11.19 (illustrating a reflex).
- New information about synthetic opiates in *A Closer Look*, with new PET scans showing effects of drug addiction.
- Added new research findings associating synaptic pruning and development of schizophrenia.

Chapter 12 The Central Nervous System

- New Figure 12.26 and revised text teach more effectively about the blood brain barrier.
- New Figure 12.30 shows spinal cord segment location in relation to vertebral column.
- New Table 12.2 summarizes spinal cord cross-sectional anatomy.
- Updated Homeostatic Imbalance features about hypothalamic disorders, cerebral palsy, anencephaly, and spina bifida, and about narcolepsy and insomnia, including new use of orexin receptor antagonists to treat insomnia.
- New type of MRI photo shows fiber tracts in brain and spinal cord.

Chapter 13 The Peripheral Nervous System and Reflex Activity

- New drawings of nerves of cervical, brachial, lumbar, and sacral plexuses show their position in relationship to the vertebrae (and hip bone in some cases) (Figures 13.9–13.12).

- New images illustrating the results of damage to the ulnar and radial nerves.
- New summary table of nerve plexuses (Table 13.7).
- New Homeostatic Imbalance feature and photo about an abnormal plantar reflex (Babinski's sign).
- Redrawn figure illustrating crossed-extensor reflex for improved student understanding.

Chapter 14 The Autonomic Nervous System

- New Figure 14.8 shows sympathetic innervation of the adrenal medulla.
- Clarified section about visceral sensory neurons.
- New photo illustrates Raynaud's disease.
- Revised Figure 14.5 on the sympathetic trunk for better teaching effectiveness.

Chapter 15 The Special Senses

- Revised Figure 15.2 (the lacrimal apparatus) for better teaching effectiveness.
- New photo of fundus of retina (Figure 15.7).

Chapter 16 The Endocrine System

- New Table 16.1 compares the endocrine and nervous systems.
- New Focus Figure 16.2 describes short- and long-term stress responses.
- Figures 16.5 (effects of growth hormone) and 16.9 (synthesis of thyroid hormone) revised for clarity.
- Updated information about diabetes mellitus, Addison's disease, and thyroid deficiency in childhood.

Chapter 17 Blood

- Updated information about anticoagulant medications.
- New photo shows petechiae resulting from thrombocytopenia (Figure 17.16).

Chapter 18 The Cardiovascular System: The Heart

- New Focus Figure 18.2 teaches students how to understand the cardiac cycle, with accompanying text reorganized.
- New photo shows an individual having an ECG (Figure 18.16).

Chapter 19 The Cardiovascular System: Blood Vessels

- New "drinking straw" analogy and art to explain resistance.
- New Figure 19.4 shows the structure of most capillary beds according to current understanding, and new text describes those capillary beds.
- Revised Figure 19.6 on proportions of blood volume throughout the vascular tree for greater teaching effectiveness.
- New illustration of cerebral arterial circle (circle of Willis) (Figure 19.24).

Chapter 20 The Lymphatic System and Lymphoid Organs and Tissues

- New illustrated Table 20.1 summarizes key characteristics of the major lymphoid organs.
- Revised Figure 20.9 with orientation diagrams helps students locate Peyer's patches (aggregated lymphoid nodules).
- Updated information about lymphatic drainage of the CNS.

Chapter 21 The Immune System: Innate and Adaptive Body Defenses

- New Focus Figure 21.1 gives an example of a primary immune response and summarizes innate and adaptive defenses.
- New illustrated Table 21.8 summarizes the components of adaptive immunity and complements the new Focus figure.
- New photo of a macrophage engulfing bacteria.
- Revised Figure 21.4 and text on inflammation, Figure 21.6 on complement activation, and Figure 21.11 on clonal selection of a B cell for greater teaching effectiveness.

Chapter 22 The Respiratory System

- New Figure 22.1 illustrates the four respiratory processes.
- Added section about sleep apnea.
- New scanning electron micrographs of emphysematous and normal lung tissue (Figure 22.22).
- Updated statistics about lung cancer and trends in asthma prevalence.

Chapter 23 The Digestive System

- New Figure 23.25 teaches the enterohepatic circulation of bile salts, and new Figure 23.30 shows the macroscopic anatomy of the small intestine.
- Improved teaching effectiveness of Figure 23.7 (neural reflex pathways in the gastrointestinal tract) and 23.16 (microscopic anatomy of the stomach).
- Added Homeostatic Imbalance features about dry mouth (xerostomia) and about tooth decay in primary teeth.
- Updated Homeostatic Imbalance feature about acute appendicitis to state that surgery is no longer always the first choice of treatment.

Chapter 24 Nutrition, Metabolism, and Energy Balance

- New Figure 24.24 shows the size and composition of various lipoproteins.
- Improved teaching effectiveness of Figure 24.21 (insulin effects during the postabsorptive stage).
- Updated Homeostatic Imbalance features with mechanism of cell death in frostbite, and diet recommendations for individuals with phenylketonuria.
- New information about environmental factors that may contribute to the obesity epidemic in *A Closer Look*.
- Updated nutritional information about lipids, and updated statistics about the prevalence of obesity in adults and children and about the prevalence of diabetes mellitus.

Chapter 25 The Urinary System

- New Figure 25.18 shows the medullary osmotic gradient and interstitial fluid osmolalities in the renal cortex and medulla.
- New Table 25.1 summarizes the regulation of glomerular filtration rate.
- Improved teaching effectiveness of Figures 25.9 (blood vessels of the renal cortex), 25.12 (the filtration membrane), 25.15 (routes for tubular reabsorption), and 25.16 (tubular reabsorption of water and nutrients).

- New pyelogram shows anatomy of kidneys, ureters, and urinary bladder (Figure 25.23).
- Added Homeostatic Imbalance feature about renal trauma.
- Updated Homeostatic Imbalance feature about kidney stones.

Chapter 26 Fluid, Electrolyte, and Acid-Base Balance

- New Figure 26.12 summarizes the body's chemical buffers.
- Improved teaching effectiveness of Figure 26.1 (major fluid compartments of the body), 26.2 (electrolyte composition of blood plasma, interstitial fluid, and intracellular fluid), and 26.7 (disturbances in water balance).
- Clarified definitions of sensible and insensible water loss.

Chapter 27 The Reproductive System

- This chapter has been extensively updated, revised, and reorganized. Almost every figure has been reconceptualized and several new figures have been added. These changes have been made for better teaching effectiveness.
- New opening module now compares male and female reproductive system anatomy and physiology and highlights common features, allowing students to make connections more easily. Homologous structures, patterns of hormone release, and meiosis are included in this section.
- New Figure 27.1 illustrates the basic pattern of interactions along the hypothalamic-pituitary-gonadal (HPG) axis in both males and females.
- The section about meiosis has been extensively rewritten to help increase student understanding. New in-line figures help introduce the basic terminology and some of the concepts before meiosis is discussed in detail.
- A new big-picture overview of meiosis introduces the major events before the details of each step are presented.
- Figures 27.22 (events of oogenesis) and 27.24 (regulation of the ovarian cycle) are extensively revised and updated for increased teaching effectiveness and accuracy.
- New Figure 27.26 depicts the genetic determination of sex.

Chapter 28 Pregnancy and Human Development

- New photo of sperm surrounding an oocyte (Figure 28.2).
- New Figure 28.5 illustrates implantation of a blastocyst.
- New photo of a 22-day embryo illustrates lateral folding (Figure 28.10d).
- Figure 28.12 (neurulation and early mesodermal differentiation) revised for clarity.
- New Focus Figure 28.2 (*Focus on Fetal and Newborn Circulation*) teaches the special features of fetal circulation and changes that occur in this circulation after birth.
- New Table 28.1 summarizes the special structures of the fetal circulation, their functions, and their postnatal structure.
- Updated information about placental hormone secretion and about the hormonal control of the initiation of labor.
- New information about fetal cells that enter the maternal circulation.
- New Homeostatic Imbalance feature about preeclampsia.

Chapter 29 Heredity

- Added Punnett square showing X-linked inheritance.
- Figure 29.1 (preparing a karyotype) and 29.4 (genotype and phenotype probabilities) revised for clarity.
- New photo of a couple with achondroplasia.
- Updated information about small noncoding RNAs.
- It has become increasingly clear that very few benign traits in humans follow a simple dominant-recessive inheritance pattern. Tongue rolling, astigmatism, freckles, dimples, phenylthiocarbamide tasting, widow's peak, and double-jointed thumb were all at one time thought to follow this pattern of inheritance. Closer examination has revealed compelling evidence against each of these. Consequently, the examples throughout the chapter have changed.

Acknowledgments

Producing a new edition of this book is an enormous undertaking. Let us take you through the steps and introduce you to the people behind the scenes that have helped make this book what it is. Every new edition begins with a revision plan. We'd like to thank all of the students and instructors who have provided the feedback (gathered by our editorial team) that forms the basis of this plan. Once this plan was in place, Barbara Price (our text Development Editor) scoured each chapter. This was Barbara's first exposure to the book and her fresh eyes on the text found opportunities to further clarify the presentation. In addition, she noted places where additional chunking of the text (such as bulleted lists) would help the students. Her excellent work has made this text better. We incorporated her ideas, and reviewer feedback, together with our own updates and ideas for reorganization of the text and art. Thanks to Patricia Bowne for contributing to the Clinical Case Studies and Wendy Mercier for reviewing all of the Case Studies. We also very much appreciate the help of Karen Dougherty, who used her expertise as a physician and educator to review all of the Homeostatic Imbalance features and help us revise and update them.

We then laid out each chapter to maintain text-art correlation before passing the manuscript off to Michele Mangelli. Michele wore many different hats during this revision. She was both the Program Manager for the editorial side of things as well as the Goddess of Production. She reviewed the revised manuscript before she sent it to ace copyeditor Anita Hueftle. Anita saved us on many occasions from public embarrassment by finding our spelling and grammar errors, our logical lapses, and various other inconsistencies. We can't thank Anita enough for her meticulous and outstanding work! (Any remaining errors are our fault.)

At the same time the text was in revision, the art program was going through a similar process. This book would not be what it is without the help of Laura Southworth, our superb Art Development Editor. Laura's creativity, attention to detail, and her sense of what will teach well and what won't have helped us immensely. She has worked tirelessly to make our Focus figures and other art even better. Finding good, usable photos is never easy, and we are grateful for the hard work of Kristin Piljay (Photo Researcher). It was also a pleasure to work with Jean Lake again, who expertly juggled the administrative aspects of the art program and kept us all on track. This team ensured that the artists at Imagineering had all the information they needed to produce beautiful final art products.

As the manuscript made the transition from Editorial to Production, Michelle Mangelli (wearing a different hat—this one as the Production and Design Manager) took over again. As head honcho and skilled handler of all aspects of production, everyone answered to her from this point on. Kudos to our excellent production coordinator, Karen Gulliver, who did much of the hands-on handling, routing, and scheduling of the manuscript. We'd also like to thank Martha Ghent (Proofreader), Betsy Dietrich (Art Proofreader), Sallie Steele (Indexer), Alicia Elliot (Project Manager at Imagineering), and Cenveo (Compositor). Izak Paul meticulously read every chapter for scientific accuracy, and we are very grateful for his careful work. Thanks also to Gary Hespenheide for his stunning design work on the cover, chapter opening pages, and the text.

It was a pleasure to work with Lauren Harp, our Acquisitions Editor. Her extensive knowledge of the needs of both faculty and students in anatomy and physiology has helped inform this revision. Her enthusiasm for this book is infectious, her choice for the cover is inspired, and we are delighted to have her on board! Before Lauren became part of the team, Serina Beauparlant, our Editor-in-Chief, stepped up to helm the planning phase of this revision. Fiercely dedicated to making this book and its associated media resources the best teaching tools that they can be, Serina has been invaluable in shaping this revision. We deeply appreciate all she has done for us and this book. Lauren and Serina were competently aided by Editorial Assistant Dapinder Dosanjh (and before her, Nicky Montalvo).

Other members of our team with whom we have less contact but who are nonetheless vital are: Barbara Yien, Director of Content Development, Stacey Weinberger (our Senior Manufacturing Buyer), and Derek Perrigo (our top-notch Marketing Manager). We appreciate the hard work of our media production team headed by Lauren Chen, Lauren Hill, Laura Tommasi, Sarah Young Dualan, and Cheryl Chi, and also wish to thank Eric Leaver for his astute observations on certain figures.

Kudos to our entire team. We feel we have once again prepared a superb textbook. We hope you agree.

Many people reviewed parts of this text—both professors and students, either individually or in focus groups—and we would like to thank them. Input from the following reviewers has contributed to the continued excellence and accuracy of this text and its accompanying MasteringA&P® assignment options, including Interactive Physiology 2.0:

Matthew Abbott, *Des Moines Area Community College*
Emily Allen, *Rowan College at Gloucester County*
Lynne Anderson, *Meridian Community College*

David C. Ansardi, *Calhoun Community College*

Martin W. Asobayire, *Essex Community College*

David Babb, *West Hills College Lemoore*

Yvonne Baptiste-Szymanski, *Niagara County Community College*

Claudia Barreto, *University of New Mexico–Valencia*

Jerry Barton, *Tarrant County College*

Shawn Bearden, *Idaho State University*

Charles Benton, *Madison Area Technical College*

J. Gordon Betts, *Tyler Junior College*

Diana Bourke, *Community College of Allegheny County*

Sherry Bowen, *Indian River State College*

Michael Brady, *Columbia Basin College*

Betsy Brantley, *Valencia College*

Beth Braun, *Truman College*

Carol A. Britson, *University of Mississippi*

C. Steven Cahill, *West Kentucky Community and Technical College*

Christie Campbell, *Ozarks Technical Community College*

Maria C. Carles, *Northern Essex Community College*

Tamyra Carmona, *Cosumnes River College*

Marien Cendon, *Miami Dade College*

Brendon Chastain, *West Kentucky Community Technical College*

Sam Chen, *Moraine Valley Community College*

Alexander G. Cheroske, *Mesa Community College–Red Mountain*

Brandi Childress, *Georgia Perimeter College*

William M. Clark, *Lone Star College–Kingwood*

Joseph Comber, *Villanova University*

Teresa Cowan, *Baker College–Auburn Hills*

Donna Crapanzano, *Stony Brook University*

Maurice M. Culver, *Florida State College at Jacksonville*

Jason Dechant, *University of Pittsburgh*

Smruti A. Desai, *Lone Star College–CyFair*

Karen H. Dougherty, *Hopkinsville Community College*

Sondra Dubowsky, *McLennan Community College*

Karen Dunbar Kareiva, *Ivy Tech Community College*

Kathryn Durham, *Lorain County Community College*

Karen Eastman, *Chattanooga State Community College*

Sharon S. Ellerton, *Queensborough Community College–CUNY*

Paul Emerick, *Monroe Community College*

Elyce Ervin, *University of Toledo*

Martha Eshleman, *Pulaski Technical College*

Colin Everhart, *St. Petersburg Community College*

Brian D. Feige, *Mott Community College*

Michele Finn, *Monroe Community College*

John E. Fishback, *Ozarks Technical Community College*

Maria Florez, *Lone Star College–CyFair*

Reza Forough, *Bellevue College*

Juanita A. Forrester, *Chattahoochee Technical College*

Aaron Fried, *Mohawk Valley Community College*

Dean Furbish, *Wake Technical Community College*

Marie Gabbard, *College of Western Idaho*

Sophia Garcia, *Tarrant County College*

Jane E. Gavin, *University of South Dakota*

Peter Germroth, *Hillsborough Community College*

Emily K. Getty, *Ivy Tech Community College*

Amy Giesecke, *Chattahoochee Technical College*

Anna Gilletly, *Central New Mexico Community College*

Gary Glaser, *Genesee Community College*

Richard Gonzalez-Diaz, *Seminole State College of Florida*

Abigail Goosie, *Walters State Community College*

Pattie S. Green, *Tacoma Community College*

Edwin Griff, *University of Cincinnati*

George G. Hanak, *Pasco-Hernando State College*

Mary Beth Hanlin, *Des Moines Area Community College–Boone*

Heidi Hawkins, *College of Southern Idaho*

Martie Heath-Sinclair, *Hawkeye Community College*

Nora Hebert, *Red Rocks Community College*

Nadia Hedhli, *Hudson County Community College*

D.J. Hennager, *Kirkwood Community College*

Jennifer Hill, *Montgomery College–Takoma Park-Silver Spring*

Shannon K. Hill, *Temple College*

Mark Hollier, *Georgia Perimeter College*

H. Rodney Holmes, *Waubonsee Community College*

Mark J. Hubley, *Prince George's Community College*

Carolyn Huffman, *Wichita Area Technical College*

Julie Huggins, *Arkansas State University*

Jason Hunt, *Brigham Young University–Idaho*

Alexander Ibe, *Weatherford College*

Alexander Imholtz, *Prince George's Community College*

Virginia Irintcheva, *Black Hawk College*

Brian E. Jordan, *C.S. Mott Community College*

Thomas Jordan, *Pima Community College*

Christopher Jung, *University of Alaska Anchorage*

William M. Karkow, *University of Dubuque*

Suzanne Keller, *Indian Hills Community College*

Michael Kielb, *Eastern Michigan University*

Marta Klesath, *North Carolina State University*

Nelson H. Kraus, *University of Indianapolis*

Paul M. Lea IV, *Northern Virginia Community College*

Steven Lewis, *Metropolitan Community College–Penn Valley*

Juanita Limas, *Kirkwood Community College*

Jerri K. Lindsey, *Tarrant County College–Northeast*

Chelsea Loafman, *Central Texas College*

Paul Luyster, *Tarrant County College*

Ken Malachowsky, *Florence-Darlington Technical College*

Theresa Martin, *College of San Mateo*

Nicole Mashburn, *Calhoun Community College*

Abdallah M. Matari, *Hudson County Community College*

Bhavya Mathur, *Chattahoochee Technical College*

Tiffany Beth McFalls-Smith, *Elizabethtown Community and Technical College*
Jennifer Menon, *Johnson County Community College*
Jaime Mergliano, *John Tyler Community College*
Sharon Miles, *Itawamba Community College*
Todd Miller, *Hunter College of CUNY*
Louise Millis, *North Hennepin Community College*
Justin Moore, *American River College*
Christine Morin, *Prince George's Community College*
Qian F. Moss, *Des Moines Area Community College*
Regina Munro, *Chandler-Gilbert Community College*
Necia Nicholas, *Calhoun Community College*
Maria Oehler, *Florida State College–Jacksonville*
Betsy Ott, *Tyler Junior College*
Ellen Ott-Reeves, *Blinn College–Bryan*
Stephen Page, *Community College of Baltimore County & Townson University*
Vikash Patel, *Nevada State College*
Dennis Pearson, *Morton College*
Diane Pelletier, *Green River Community College*
Jessica Petersen, *Pensacola State College*
Jason Pienaar, *University of Alabama*
Becky Pierce, *Delta College*
Gilbert Pitts, *Austin Peay State University*
Renee Prenitzer, *Greenville Technical College*
Fernando Prince, *Laredo Community College*
Sarah A. Pugh, *Shelton State Community College*
Suzanne Pundt, *University of Texas at Tyler*
Rolando J. Ramirez, *The University of Akron*
Wendy Rappazzo, *Harford Community College*
Terrence J. Ravine, *University of South Alabama*
Christine S. Rigsby, *Middle Georgia State University*
Laura H. Ritt, *Burlington County College*
Cynthia Robison, *Wallace Community College*
Susan Rohde, *Triton College*
Brian Sailer, *Central New Mexico Community College*
Sharon Schapel, *Mott Community College*
Mark Schmidt, *Clark State Community College*
Michael W. Sipala, *Bristol Community College*
Amy Skibiel, *Auburn University*
Lori Smith, *American River College–Los Rios*
Kerry Smith, *Oakland Community College–Auburn Hills*
Tom Sobat, *Ivy Tech Community College*
Kay Sourbeer, *Tidewater Community College*
Ashley Spring-Beerensson, *Eastern Florida State College*
Justin R. St. Juliana, *Ivy Tech Community College*
Cindy Stanfield, *University of South Alabama*
Laura Steele, *Ivy Tech Community College–Northeast*
George A. Steer, *Jefferson College of Health Sciences*
Michelle Stettner, *Meridian Community College*
Sherry Stewart, *Navarro College*

Susan E. Tappen, *Central New Mexico Community College*
Dean Thornton, *South Georgia State College*
Rita A. Thrasher, *Pensacola State College*
Brenda Tondi, *George Mason University*
Sheela Vemu, *Waubonsee Community College*
Khursheed Wankadiya, *Central Piedmont Community College*
Chad Wayne, *University of Houston*
Kira L. Wennstrom, *Shoreline Community College*
Shirley A. Whitescarver, *Bluegrass Community and Technical College–KCTCS*
John Whitlock, *Hillsborough Community College*
Patricia Wilhelm, *Johnson and Wales University*
Luann Wilkinson, *Marion Technical College*
Selwyn A. Williams, *Miami Dade College*
Darrellyn Williams, *Pulaski Technical College*
Peggie Williamson, *Central Texas College*
Heather Wilson-Ashworth, *Utah Valley University*
MaryJo A. Witz, *Monroe Community College*
Jackie Wright, *South Plains College*
James Robert Yount, *Brevard Community College*

We would like to acknowledge the following group who reviewed various iterations of the new Focus figures: Matthew Abbott, David Ansardi, Jake Dechant, Karen Dougherty, Peter Germroth, Gary Glaser, Suzanne Keller, Gilbert Pitts, Terry Ravine, Michelle Stettner, and Rita Thrasher.

We would also like to acknowledge the support of Katja's colleagues at Mount Royal University (Trevor Day, Sarah Hewitt, Tracy O'Connor, Sarah Orton, Izak Paul, Lorraine Royal, Karen Sheedy, Kartika Tjandra, and Margot Williams); Department Chairs (Ruth Pickett-Seltner and Melanie Rathburn); and Deans (Jeffrey Goldberg and Jonathan Withey). Thanks also to Katja's husband, Dr. Lawrence Haynes, a fellow physiologist who has worked together with Katja and has been involved in all aspects of this revision. We would like to thank Katja and Larry's sons, Eric and Stefan Haynes, for putting up with their parents through many revisions of this book and for continuing to be an inspiration and a joy.

We really would appreciate hearing from you concerning your opinion—suggestions and constructive criticisms—of this text. It is this type of feedback that will help us in the next revision and underlies the continued improvement of this text.

Elaine Marieb

Elaine Marieb

Katja Hoehn

Katja Hoehn

Elaine N. Marieb and Katja Hoehn
Pearson Education
1301 Sansome Street
San Francisco, CA 94111

Contents

UNIT 1   Organization of the Body

1 The Human Body: An Orientation   1

1.1 Form (anatomy) determines function (physiology)   2

1.2 The body's organization ranges from atoms to the entire organism   4

1.3 What are the requirements for life?   5

1.4 Homeostasis is maintained by negative feedback   9

1.5 Anatomical terms describe body directions, regions, and planes   12

A CLOSER LOOK Medical Imaging: Illuminating the Body   16

1.6 Many internal organs lie in membrane-lined body cavities   17

2 Chemistry Comes Alive   23

PART 1 BASIC CHEMISTRY   24

2.1 Matter is the stuff of the universe and energy moves matter   24

2.2 The properties of an element depend on the structure of its atoms   25

2.3 Atoms bound together form molecules; different molecules can make mixtures   28

2.4 The three types of chemical bonds are ionic, covalent, and hydrogen   31

2.5 Chemical reactions occur when electrons are shared, gained, or lost   35

PART 2 BIOCHEMISTRY   38

2.6 Inorganic compounds include water, salts, and many acids and bases   38

2.7 Organic compounds are made by dehydration synthesis and broken down by hydrolysis   41

2.8 Carbohydrates provide an easily used energy source for the body   43

2.9 Lipids insulate body organs, build cell membranes, and provide stored energy   45

2.10 Proteins are the body's basic structural material and have many vital functions   48

2.11 DNA and RNA store, transmit, and help express genetic information   53

2.12 ATP transfers energy to other compounds   55

3 Cells: The Living Units   60

3.1 Cells are the smallest unit of life   61

PART 1 PLASMA MEMBRANE   63

3.2 The plasma membrane is a double layer of phospholipids with embedded proteins   63

FOCUS FIGURE 3.1 The Plasma Membrane   64

3.3 Passive membrane transport is diffusion of molecules down their concentration gradient   68

3.4 Active membrane transport directly or indirectly uses ATP   73

FOCUS FIGURE 3.2 Primary Active Transport: The Na^+-K^+ Pump   74

3.5 Selective diffusion establishes the membrane potential   79

3.6 Cell adhesion molecules and membrane receptors allow the cell to interact with its environment   81

FOCUS FIGURE 3.3 G Proteins   82

PART 2 THE CYTOPLASM   83

3.7 Cytoplasmic organelles each perform a specialized task   83

3.8 Cilia and microvilli are two main types of cellular extensions   90

PART 3 **NUCLEUS** 91

3.9 The nucleus includes the nuclear envelope, the nucleolus, and chromatin 91

3.10 The cell cycle consists of interphase and a mitotic phase 96

3.11 Messenger RNA carries instructions from DNA for building proteins 98

FOCUS FIGURE 3.4 Mitosis 100

FOCUS FIGURE 3.5 Translation 106

3.12 Autophagy and proteasomes dispose of unneeded organelles and proteins; apoptosis disposes of unneeded cells 108

DEVELOPMENTAL ASPECTS of Cells 109

4 Tissue: The Living Fabric 115

4.1 Tissue samples are fixed, sliced, and stained for microscopy 117

4.2 Epithelial tissue covers body surfaces, lines cavities, and forms glands 117

4.3 Connective tissue is the most abundant and widely distributed tissue in the body 126

4.4 Muscle tissue is responsible for body movement 138

4.5 Nervous tissue is a specialized tissue of the nervous system 140

4.6 The cutaneous membrane is dry; mucous and serous membranes are wet 141

4.7 Tissue repair involves inflammation, organization, and regeneration 142

A CLOSER LOOK Cancer—The Intimate Enemy 144

DEVELOPMENTAL ASPECTS of Tissues 146

UNIT 2   Covering, Support, and Movement of the Body

5 The Integumentary System 150

5.1 The skin consists of two layers: the epidermis and dermis 150

5.2 The epidermis is a keratinized stratified squamous epithelium 152

5.3 The dermis consists of papillary dermis and reticular dermis 154

5.4 Melanin, carotene, and hemoglobin determine skin color 156

5.5 Hair consists of dead, keratinized cells 157

5.6 Nails are scale-like modifications of the epidermis 160

5.7 Sweat glands help control body temperature, and sebaceous glands secrete sebum 161

5.8 First and foremost, the skin is a barrier 163

5.9 Skin cancer and burns are major challenges to the body 165

DEVELOPMENTAL ASPECTS of the Integumentary System 167

SYSTEM CONNECTIONS 168

6 Bones and Skeletal Tissues 173

6.1 Hyaline, elastic, and fibrocartilage help form the skeleton 174

6.2 Bones perform several important functions 175

6.3 Bones are classified by their location and shape 176

6.4 The gross structure of all bones consists of compact bone sandwiching spongy bone 176

6.5 Bones develop either by intramembranous or endochondral ossification 184

6.6 Bone remodeling involves bone deposition and removal 188

6.7 Bone repair involves hematoma and callus formation, and remodeling 190

6.8 Bone disorders result from abnormal bone deposition and resorption 193

DEVELOPMENTAL ASPECTS of Bones 194

SYSTEM CONNECTIONS 196

7 The Skeleton 199

PART 1 **THE AXIAL SKELETON** 199

7.1 The skull consists of 8 cranial bones and 14 facial bones 201

7.2 The vertebral column is a flexible, curved support structure 218

7.3 The thoracic cage is the bony structure of the chest 224

PART 2 **THE APPENDICULAR SKELETON** 227

7.4 Each pectoral girdle consists of a clavicle and a scapula 227

7.5 The upper limb consists of the arm, forearm, and hand 230

7.6 The hip bones attach to the sacrum, forming the pelvic girdle 236

7.7 The lower limb consists of the thigh, leg, and foot 240

DEVELOPMENTAL ASPECTS of the Skeleton 246

8 Joints 251

8.1 Joints are classified into three structural and three functional categories 251

8.2 In fibrous joints, the bones are connected by fibrous tissue 252

8.3 In cartilaginous joints, the bones are connected by cartilage 253

8.4 Synovial joints have a fluid-filled joint cavity 254

FOCUS FIGURE 8.1 Synovial Joints 262

8.5 Five examples illustrate the diversity of synovial joints 264

8.6 Joints are easily damaged by injury, inflammation, and degeneration 272

A CLOSER LOOK Joints: From Knights in Shining Armor to Bionic Humans 274

DEVELOPMENTAL ASPECTS of Joints 275

9 Muscles and Muscle Tissue 279

9.1 There are three types of muscle tissue 280

9.2 A skeletal muscle is made up of muscle fibers, nerves, blood vessels, and connective tissues 281

9.3 Skeletal muscle fibers contain calcium-regulated molecular motors 284

9.4 Motor neurons stimulate skeletal muscle fibers to contract 290

FOCUS FIGURE 9.1 Events at the Neuromuscular Junction 292

FOCUS FIGURE 9.2 Excitation-Contraction Coupling 294

FOCUS FIGURE 9.3 Cross Bridge Cycle 297

9.5 Temporal summation and motor unit recruitment allow smooth, graded skeletal muscle contractions 298

9.6 ATP for muscle contraction is produced aerobically or anaerobically 303

9.7 The force, velocity, and duration of skeletal muscle contractions are determined by a variety of factors 306

9.8 How does skeletal muscle respond to exercise? 309

9.9 Smooth muscle is nonstriated involuntary muscle 310

DEVELOPMENTAL ASPECTS of Muscles 316

A CLOSER LOOK Athletes Looking Good and Doing Better with Anabolic Steroids? 317

SYSTEM CONNECTIONS 318

10 The Muscular System 323

10.1 For any movement, muscles can act in one of three ways 324

10.2 How are skeletal muscles named? 324

FOCUS FIGURE 10.1 Muscle Action 325

10.3 Fascicle arrangements help determine muscle shape and force 326

10.4 Muscles acting with bones form lever systems 327

10.5 A muscle's origin and insertion determine its action 332

Table 10.1 Muscles of the Head, Part I: Facial Expression 333

Table 10.2 Muscles of the Head, Part II: Mastication and Tongue Movement 336

Table 10.3 Muscles of the Anterior Neck and Throat: Swallowing 338

Table 10.4 Muscles of the Neck and Vertebral Column: Head Movements and Trunk Extension 340

Table 10.5 Deep Muscles of the Thorax: Breathing 344

Table 10.6 Muscles of the Abdominal Wall: Trunk Movements and Compression of Abdominal Viscera 346

Table 10.7 Muscles of the Pelvic Floor and Perineum: Support of Abdominopelvic Organs 348

Table 10.8 Superficial Muscles of the Anterior and Posterior Thorax: Movements of the Scapula and Arm 350

Table 10.9 Muscles Crossing the Shoulder Joint: Movements of the Arm (Humerus) 354

Table 10.10 Muscles Crossing the Elbow Joint: Flexion and Extension of the Forearm 357

Table 10.11 Muscles of the Forearm: Movements of the Wrist, Hand, and Fingers 358

Table 10.12 Summary: Actions of Muscles Acting on the Arm, Forearm, and Hand 362

Table 10.13 Intrinsic Muscles of the Hand: Fine Movements of the Fingers 364

Table 10.14 Muscles Crossing the Hip and Knee Joints: Movements of the Thigh and Leg 367

Table 10.15 Muscles of the Leg: Movements of the Ankle and Toes 374

Table 10.16 Intrinsic Muscles of the Foot: Toe Movement and Arch Support 380

Table 10.17 Summary: Actions of Muscles Acting on the Thigh, Leg, and Foot 384

UNIT 3   Regulation and Integration of the Body

11 Fundamentals of the Nervous System and Nervous Tissue   390

11.1 The nervous system receives, integrates, and responds to information   391

11.2 Neuroglia support and maintain neurons   392

11.3 Neurons are the structural units of the nervous system   394

11.4 The resting membrane potential depends on differences in ion concentration and permeability   400

FOCUS FIGURE 11.1 Resting Membrane Potential   402

11.5 Graded potentials are brief, short-distance signals within a neuron   404

11.6 Action potentials are brief, long-distance signals within a neuron   405

FOCUS FIGURE 11.2 Action Potential   406

11.7 Synapses transmit signals between neurons   412

FOCUS FIGURE 11.3 Chemical Synapse   415

11.8 Postsynaptic potentials excite or inhibit the receiving neuron   416

FOCUS FIGURE 11.4 Postsynaptic Potentials and Their Summation   418

11.9 The effect of a neurotransmitter depends on its receptor   420

11.10 Neurons act together, making complex behaviors possible   426

DEVELOPMENTAL ASPECTS of Neurons   428

A CLOSER LOOK Pleasure Me, Pleasure Me!   429

12 The Central Nervous System   434

12.1 Folding during development determines the complex structure of the adult brain   435

12.2 The cerebral hemispheres consist of cortex, white matter, and the basal nuclei   439

12.3 The diencephalon includes the thalamus, hypothalamus, and epithalamus   447

12.4 The brain stem consists of the midbrain, pons, and medulla oblongata   450

12.5 The cerebellum adjusts motor output, ensuring coordination and balance   454

12.6 Functional brain systems span multiple brain structures   456

12.7 The interconnected structures of the brain allow higher mental functions   458

12.8 The brain is protected by bone, meninges, cerebrospinal fluid, and the blood brain barrier   464

12.9 Brain injuries and disorders have devastating consequences   468

12.10 The spinal cord is a reflex center and conduction pathway   470

12.11 Neuronal pathways carry sensory and motor information to and from the brain   476

DEVELOPMENTAL ASPECTS of the Central Nervous System   482

13 The Peripheral Nervous System and Reflex Activity   489

PART 1 **SENSORY RECEPTORS AND SENSATION**   490

13.1 Sensory receptors are activated by changes in the internal or external environment   490

13.2 Receptors, ascending pathways, and cerebral cortex process sensory information   493

PART 2 **TRANSMISSION LINES: NERVES AND THEIR STRUCTURE AND REPAIR**   496

13.3 Nerves are cordlike bundles of axons that conduct sensory and motor impulses   496

13.4 There are 12 pairs of cranial nerves   498

13.5 31 pairs of spinal nerves innervate the body   507

PART 3 **MOTOR ENDINGS AND MOTOR ACTIVITY**   517

13.6 Peripheral motor endings connect nerves to their effectors   517

13.7 There are three levels of motor control   517

PART 4 **REFLEX ACTIVITY**   519

13.8 The reflex arc enables rapid and predictable responses   519

13.9 Spinal reflexes are somatic reflexes mediated by the spinal cord   520

FOCUS FIGURE 13.1 Stretch Reflex   522

DEVELOPMENTAL ASPECTS of the Peripheral Nervous System   526

14 The Autonomic Nervous System   531

14.1 The ANS differs from the somatic nervous system in that it can stimulate or inhibit its effectors   532

14.2 The ANS consists of the parasympathetic and sympathetic divisions   534

14.3 Long preganglionic parasympathetic fibers originate in the craniosacral CNS   536

14.4 Short preganglionic sympathetic fibers originate in the thoracolumbar CNS   538

14.5 Visceral reflex arcs have the same five components as somatic reflex arcs   542

14.6 Acetylcholine and norepinephrine are the major ANS neurotransmitters   543

14.7 The parasympathetic and sympathetic divisions usually produce opposite effects   545

14.8 The hypothalamus oversees ANS activity   547

14.9 Most ANS disorders involve abnormalities in smooth muscle control   548

DEVELOPMENTAL ASPECTS of the ANS   548

SYSTEM CONNECTIONS   550

15 The Special Senses   533

PART 1   THE EYE AND VISION   554

15.1 The eye has three layers, a lens, and humors, and is surrounded by accessory structures   554

15.2 The cornea and lens focus light on the retina   563

15.3 Phototransduction begins when light activates visual pigments in retinal photoreceptors   567

15.4 Visual information from the retina passes through relay nuclei to the visual cortex   573

PART 2   THE CHEMICAL SENSES: SMELL AND TASTE   575

15.5 Airborne chemicals are detected by olfactory receptors in the nose   575

15.6 Dissolved chemicals are detected by receptor cells in taste buds   578

PART 3   THE EAR: HEARING AND BALANCE   580

15.7 The ear has three major areas   580

15.8 Sound is a pressure wave that stimulates mechanosensitive cochlear hair cells   585

15.9 Sound information is processed and relayed through brain stem and thalamic nuclei to the auditory cortex   589

15.10 Hair cells in the maculae and cristae ampullares monitor head position and movement   590

15.11 Ear abnormalities can affect hearing, equilibrium, or both   594

DEVELOPMENTAL ASPECTS of the Special Senses   595

16 The Endocrine System   601

16.1 The endocrine system is one of the body's two major control systems   602

16.2 The chemical structure of a hormone determines how it acts   603

16.3 Hormones act through second messengers or by activating specific genes   603

16.4 Three types of stimuli cause hormone release   607

16.5 Cells respond to a hormone if they have a receptor for that hormone   608

16.6 The hypothalamus controls release of hormones from the pituitary gland in two different ways   609

FOCUS FIGURE 16.1 Hypothalamus and Pituitary Interactions   610

16.7 The thyroid gland controls metabolism   617

16.8 The parathyroid glands are primary regulators of blood calcium levels   621

16.9 The adrenal glands produce hormones involved in electrolyte balance and the stress response   622

16.10 The pineal gland secretes melatonin   627

FOCUS FIGURE 16.2 Stress and the Adrenal Gland   628

16.11 The pancreas, gonads, and most other organs secrete hormones   630

A CLOSER LOOK Sweet Revenge: Taming the Diabetes Monster?   633

DEVELOPMENTAL ASPECTS of the Endocrine System   636

SYSTEM CONNECTIONS   637

UNIT 4   Maintenance of the Body

17 Blood   642

17.1 The functions of blood are transport, regulation, and protection   643

17.2 Blood consists of plasma and formed elements   643

17.3 Erythrocytes play a crucial role in oxygen and carbon dioxide transport   645

17.4 Leukocytes defend the body   651

17.5 Platelets are cell fragments that help stop bleeding   657

17.6 Hemostasis prevents blood loss   657

17.7 Transfusion can replace lost blood   663

17.8 Blood tests give insights into a patient's health   666

DEVELOPMENTAL ASPECTS of Blood   666

18 The Cardiovascular System: The Heart 670

18.1 The heart has four chambers and pumps blood through the pulmonary and systemic circuits 671

18.2 Heart valves make blood flow in one direction 679

18.3 Blood flows from atrium to ventricle, and then to either the lungs or the rest of the body 680

FOCUS FIGURE 18.1 Blood Flow through the Heart 681

18.4 Intercalated discs connect cardiac muscle fibers into a functional syncytium 683

18.5 Pacemaker cells trigger action potentials throughout the heart 686

18.6 The cardiac cycle describes the mechanical events associated with blood flow through the heart 692

FOCUS FIGURE 18.2 The Cardiac Cycle 694

18.7 Stroke volume and heart rate are regulated to alter cardiac output 696

DEVELOPMENTAL ASPECTS of the Heart 700

19 The Cardiovascular System: Blood Vessels 706

PART 1 BLOOD VESSEL STRUCTURE AND FUNCTION 707

19.1 Most blood vessel walls have three layers 709

19.2 Arteries are pressure reservoirs, distributing vessels, or resistance vessels 710

19.3 Capillaries are exchange vessels 710

19.4 Veins are blood reservoirs that return blood toward the heart 712

19.5 Anastomoses are special interconnections between blood vessels 714

PART 2 PHYSIOLOGY OF CIRCULATION 714

19.6 Blood flows from high to low pressure against resistance 714

19.7 Blood pressure decreases as blood flows from arteries through capillaries and into veins 716

19.8 Blood pressure is regulated by short- and long-term controls 718

19.9 Intrinsic and extrinsic controls determine blood flow through tissues 725

19.10 Slow blood flow through capillaries promotes diffusion of nutrients and gases, and bulk flow of fluids 730

FOCUS FIGURE 19.1 Bulk Flow across Capillary Walls 732

PART 3 CIRCULATORY PATHWAYS: BLOOD VESSELS OF THE BODY 734

19.11 The vessels of the systemic circulation transport blood to all body tissues 735

Table 19.3 Pulmonary and Systemic Circulations 736

Table 19.4 The Aorta and Major Arteries of the Systemic Circulation 738

Table 19.5 Arteries of the Head and Neck 740

Table 19.6 Arteries of the Upper Limbs and Thorax 742

Table 19.7 Arteries of the Abdomen 744

Table 19.8 Arteries of the Pelvis and Lower Limbs 748

Table 19.9 The Venae Cavae and the Major Veins of the Systemic Circulation 750

Table 19.10 Veins of the Head and Neck 752

Table 19.11 Veins of the Upper Limbs and Thorax 754

Table 19.12 Veins of the Abdomen 756

Table 19.13 Veins of the Pelvis and Lower Limbs 758

DEVELOPMENTAL ASPECTS of Blood Vessels 759

A CLOSER LOOK Atherosclerosis? Get Out the Cardiovascular Drāno 760

SYSTEM CONNECTIONS 761

20 The Lymphatic System and Lymphoid Organs and Tissues 766

20.1 The lymphatic system includes lymphatic vessels, lymph, and lymph nodes 767

20.2 Lymphoid cells and tissues are found in lymphoid organs and in connective tissue of other organs 770

20.3 Lymph nodes cleanse lymph and house lymphocytes 771

20.4 The spleen removes bloodborne pathogens and aged red blood cells 773

20.5 MALT guards the body's entryways against pathogens 774

20.6 T lymphocytes mature in the thymus 776

DEVELOPMENTAL ASPECTS of the Lymphatic System and Lymphoid Organs and Tissues 776

SYSTEM CONNECTIONS 778

21 The Immune System: Innate and Adaptive Body Defenses 781

PART 1 INNATE DEFENSES 782

21.1 Surface barriers act as the first line of defense to keep invaders out of the body 782

21.2 Innate internal defenses are cells and chemicals that act as the second line of defense 783

PART 2 ADAPTIVE DEFENSES 790

21.3 Antigens are substances that trigger the body's adaptive defenses 791

21.4 B and T lymphocytes and antigen-presenting cells are cells of the adaptive immune response 792

21.5 In humoral immunity, antibodies are produced that target extracellular antigens 796

21.6 Cellular immunity consists of T lymphocytes that direct adaptive immunity or attack cellular targets 801

FOCUS FIGURE 21.1 An Example of a Primary Immune Response 808

21.7 Insufficient or overactive immune responses create problems 811

DEVELOPMENTAL ASPECTS of the Immune System 814

22 The Respiratory System 818

PART 1 FUNCTIONAL ANATOMY 820

22.1 The upper respiratory system warms, humidifies, and filters air 820

22.2 The lower respiratory system consists of conducting and respiratory zone structures 824

22.3 Each multilobed lung occupies its own pleural cavity 833

PART 2 RESPIRATORY PHYSIOLOGY 834

22.4 Volume changes cause pressure changes, which cause air to move 834

22.5 Measuring respiratory volumes, capacities, and flow rates helps us assess ventilation 840

22.6 Gases exchange by diffusion between the blood, lungs, and tissues 842

22.7 Oxygen is transported by hemoglobin, and carbon dioxide is transported in three different ways 847

FOCUS FIGURE 22.1 The Oxygen-Hemoglobin Dissociation Curve 848

22.8 Respiratory centers in the brain stem control breathing with input from chemoreceptors and higher brain centers 853

22.9 Exercise and high altitude bring about respiratory adjustments 857

22.10 Respiratory diseases are major causes of disability and death 858

DEVELOPMENTAL ASPECTS of the Respiratory System 860

SYSTEM CONNECTIONS 862

23 The Digestive System 868

PART 1 OVERVIEW OF THE DIGESTIVE SYSTEM 869

23.1 What major processes occur during digestive system activity? 870

23.2 The GI tract has four layers and is usually surrounded by peritoneum 871

23.3 The GI tract has its own nervous system called the enteric nervous system 874

PART 2 FUNCTIONAL ANATOMY OF THE DIGESTIVE SYSTEM 875

23.4 Ingestion occurs only at the mouth 876

23.5 The pharynx and esophagus move food from the mouth to the stomach 881

23.6 The stomach temporarily stores food and begins protein digestion 884

23.7 The liver secretes bile; the pancreas secretes digestive enzymes 893

23.8 The small intestine is the major site for digestion and absorption 900

23.9 The large intestine absorbs water and eliminates feces 906

PART 3 PHYSIOLOGY OF DIGESTION AND ABSORPTION 912

23.10 Digestion hydrolyzes food into nutrients that are absorbed across the gut epithelium 912

23.11 How is each type of nutrient processed? 912

DEVELOPMENTAL ASPECTS of the Digestive System 918

SYSTEM CONNECTIONS 920

24 Nutrition, Metabolism, and Energy Balance 926

PART 1 NUTRIENTS 927

24.1 Carbohydrates, lipids, and proteins supply energy and are used as building blocks 927

24.2 Most vitamins act as coenzymes; minerals have many roles in the body 931

PART 2 METABOLISM 933

24.3 Metabolism is the sum of all biochemical reactions in the body 934

24.4 Carbohydrate metabolism is the central player in ATP production 936

FOCUS FIGURE 24.1 Oxidative Phosphorylation 941

24.5 Lipid metabolism is key for long-term energy storage and release   946

24.6 Amino acids are used to build proteins or for energy   948

24.7 Energy is stored in the absorptive state and released in the postabsorptive state   949

24.8 The liver metabolizes, stores, and detoxifies   955

A CLOSER LOOK Obesity: Magical Solution Wanted   958

PART 3 ENERGY BALANCE   960

24.9 Neural and hormonal factors regulate food intake   960

24.10 Thyroxine is the major hormone that controls basal metabolic rate   962

24.11 The hypothalamus acts as the body's thermostat   963

DEVELOPMENTAL ASPECTS of Nutrition and Metabolism   968

25 The Urinary System   974

25.1 The kidneys have three distinct regions and a rich blood supply   975

25.2 Nephrons are the functional units of the kidney   978

25.3 Overview: Filtration, absorption, and secretion are the key processes of urine formation   983

25.4 Urine formation, step 1: The glomeruli make filtrate   984

25.5 Urine formation, step 2: Most of the filtrate is reabsorbed into the blood   989

25.6 Urine formation, step 3: Certain substances are secreted into the filtrate   994

25.7 The kidneys create and use an osmotic gradient to regulate urine concentration and volume   995

FOCUS FIGURE 25.1 Medullary Osmotic Gradient   996

25.8 Renal function is evaluated by analyzing blood and urine   1000

25.9 The ureters, bladder, and urethra transport, store, and eliminate urine   1002

DEVELOPMENTAL ASPECTS of the Urinary System   1006

26 Fluid, Electrolyte, and Acid-Base Balance   1012

26.1 Body fluids consist of water and solutes in three main compartments   1013

26.2 Both intake and output of water are regulated   1016

26.3 Sodium, potassium, calcium, and phosphate levels are tightly regulated   1019

26.4 Chemical buffers and respiratory regulation rapidly minimize pH changes   1026

26.5 Renal regulation is a long-term mechanism for controlling acid-base balance   1029

26.6 Abnormalities of acid-base balance are classified as metabolic or respiratory   1033

A CLOSER LOOK Sleuthing: Using Blood Values to Determine the Cause of Acidosis or Alkalosis   1034

DEVELOPMENTAL ASPECTS of Fluid, Electrolyte, and Acid-Base Balance   1035

SYSTEM CONNECTIONS   1036

UNIT 5   Continuity

27 The Reproductive System   1041

27.1 The male and female reproductive systems share common features   1042

PART 1 ANATOMY OF THE MALE REPRODUCTIVE SYSTEM   1047

27.2 The testes are enclosed and protected by the scrotum   1048

27.3 Sperm travel from the testes to the body exterior through a system of ducts   1050

27.4 The penis is the copulatory organ of the male   1050

27.5 The male accessory glands produce the bulk of semen   1052

PART 2 PHYSIOLOGY OF THE MALE REPRODUCTIVE SYSTEM   1053

27.6 The male sexual response includes erection and ejaculation   1053

27.7 Spermatogenesis is the sequence of events that leads to formation of sperm   1054

27.8 Male reproductive function is regulated by hypothalamic, anterior pituitary, and testicular hormones   1059

PART 3 ANATOMY OF THE FEMALE REPRODUCTIVE SYSTEM   1060

27.9 Immature eggs develop in follicles in the ovaries   1061

27.10 The female duct system includes the uterine tubes, uterus, and vagina   1062

27.11 The external genitalia of the female include those structures that lie external to the vagina   1067

27.12 The mammary glands produce milk   1068

PART 4 PHYSIOLOGY OF THE FEMALE REPRODUCTIVE SYSTEM 1069

27.13 Oogenesis is the sequence of events that leads to the formation of ova   1069

27.14 The ovarian cycle consists of the follicular phase and the luteal phase   1073

27.15 Female reproductive function is regulated by hypothalamic, anterior pituitary, and ovarian hormones   1074

27.16 The female sexual response is more diverse and complex than that of males   1078

PART 5 SEXUALLY TRANSMITTED INFECTIONS 1080

27.17 Sexually transmitted infections cause reproductive and other disorders   1080

DEVELOPMENTAL ASPECTS of the Reproductive System   1081

SYSTEM CONNECTIONS   1085

28 Pregnancy and Human Development   1091

28.1 Fertilization combines the sperm and egg chromosomes, forming a zygote   1092

FOCUS FIGURE 28.1 Sperm Penetration and the Blocks to Polyspermy   1094

28.2 Embryonic development begins as the zygote undergoes cleavage and forms a blastocyst en route to the uterus   1097

28.3 Implantation occurs when the embryo burrows into the uterine wall, triggering placenta formation   1098

28.4 Embryonic events include gastrula formation and tissue differentiation, which are followed by rapid growth of the fetus   1102

FOCUS FIGURE 28.2 Fetal and Newborn Circulation   1108

28.5 During pregnancy, the mother undergoes anatomical, physiological, and metabolic changes   1112

28.6 The three stages of labor are the dilation, expulsion, and placental stages   1114

28.7 An infant's extrauterine adjustments include taking the first breath and closure of vascular shunts   1116

28.8 Lactation is milk secretion by the mammary glands in response to prolactin   1116

A CLOSER LOOK Contraception: To Be or Not To Be   1118

28.9 Assisted reproductive technology may help an infertile couple have offspring   1119

29 Heredity   1124

29.1 Genes are the vocabulary of genetics   1125

29.2 Genetic variation results from independent assortment, crossing over, and random fertilization   1126

29.3 Several patterns of inheritance have long been known   1128

29.4 Environmental factors may influence or override gene expression   1131

29.5 Factors other than nuclear DNA sequence can determine inheritance   1131

29.6 Genetic screening is used to detect genetic disorders   1133

Appendices

Answers Appendix   A-1

A   The Metric System   A-18

B   Functional Groups in Organic Molecules   A-20

C   The Amino Acids   A-21

D   Two Important Metabolic Pathways   A-22

E   Periodic Table of the Elements   A-25

F   Reference Values for Selected Blood and Urine Studies   A-26

Glossary   G-1

Photo and Illustration Credits   C-1

Index   I-1

1

The Human Body: An Orientation

In this chapter, you will learn that

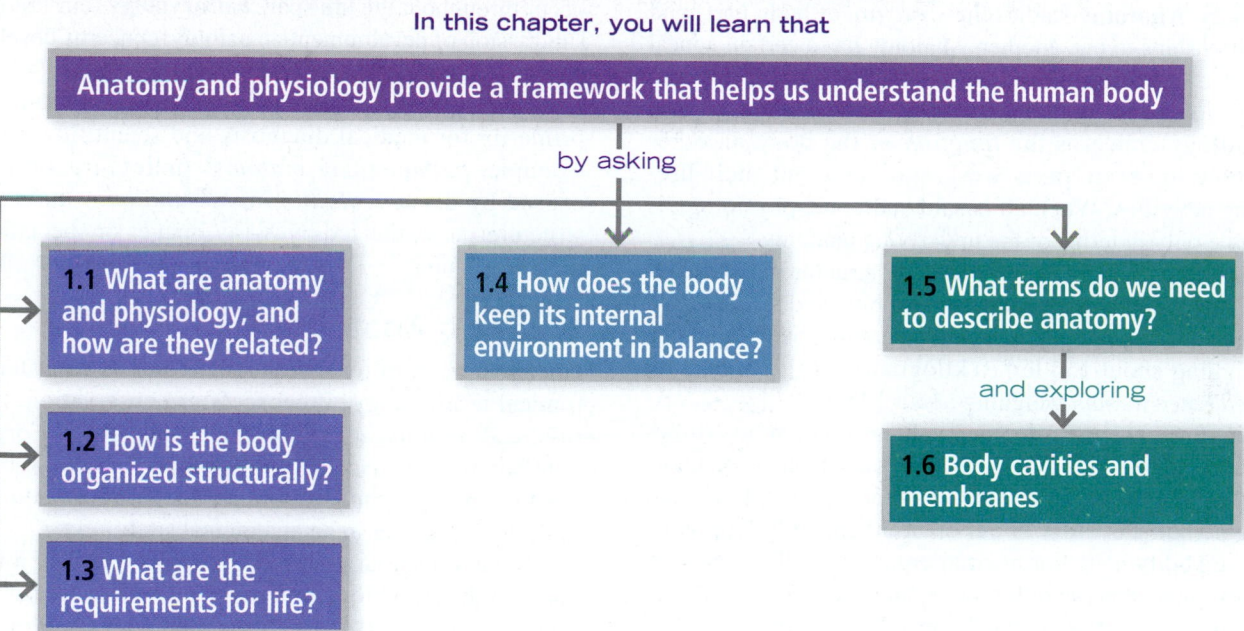

Anatomy and physiology provide a framework that helps us understand the human body

by asking

1.1 What are anatomy and physiology, and how are they related?

1.2 How is the body organized structurally?

1.3 What are the requirements for life?

1.4 How does the body keep its internal environment in balance?

1.5 What terms do we need to describe anatomy?

and exploring

1.6 Body cavities and membranes

Welcome to the study of one of the most fascinating subjects possible—your own body. Such a study is not only highly personal, but timely as well. We get news of some medical advance almost daily. To appreciate emerging discoveries in molecular genetics, to understand new techniques for detecting and treating disease, and to make use of published facts on how to stay healthy, you'll find it helps to learn about the workings of your body. If you are preparing for a career in the health sciences, the study of anatomy and physiology has added rewards because it provides the essential foundation for your clinical experiences.

In this chapter we define and contrast anatomy and physiology and discuss how the human body is organized. Then we review needs and processes common to all living organisms. Three essential concepts—*the complementarity of structure and function, the hierarchy of structural organization,* and *homeostasis*—will unify and form the bedrock for your study of the human body. And finally you'll learn the language of anatomy—terminology that anatomists use to describe the body and its parts.

1.1 Form (anatomy) determines function (physiology)

Learning Outcomes

▶ **Define anatomy and physiology and describe their subdivisions.**

▶ **Explain the principle of complementarity.**

Two complementary branches of science—anatomy and physiology—provide the concepts that help us to understand the human body. **Anatomy** studies the *structure* of body parts and their relationships to one another. Anatomy has a certain appeal because it is concrete. Body structures can be seen, felt, and examined closely. You don't need to imagine what they look like.

Physiology concerns the *function* of the body, in other words, how the body parts work and carry out their life-sustaining activities. When all is said and done, physiology is explainable only in terms of the underlying anatomy.

For simplicity, when we refer to body structures and physiological values (body temperature, heart rate, and the like), we will assume that we are talking about a healthy young *reference man* weighing about 155 lb [70 kilograms (kg)] or a healthy young *reference woman* weighing about 125 lb (57 kg).

Although we use the reference values and common directional and regional terms to refer to all human bodies, you know from observing the faces and body shapes of people around you that we humans differ in our external anatomy. The same kind of variability holds for internal organs as well. In one person, for example, a nerve or blood vessel may be somewhat out of place, or a small muscle may be missing. Nonetheless, well over 90% of all structures present in any human body match the textbook descriptions. We seldom see extreme anatomical variations because they are incompatible with life.

Topics of Anatomy

Anatomy is a broad field with many subdivisions, each providing enough information to be a course in itself. **Gross**, or **macroscopic**, **anatomy** is the study of large body structures visible to the naked eye, such as the heart, lungs, and kidneys. Indeed, the term *anatomy* (from Greek, meaning "to cut apart") relates most closely to gross anatomy because in such studies preserved animals or their organs are dissected (cut up) to be examined.

Gross anatomy can be approached in different ways.

- In **regional anatomy**, all the structures (muscles, bones, blood vessels, nerves, etc.) in a particular region of the body, such as the abdomen or leg, are examined at the same time.

- In **systemic anatomy** (sis-tem′ik),* body structure is studied system by system. For example, when studying the cardiovascular system, you would examine the heart and the blood vessels of the entire body.

- Another subdivision of gross anatomy is **surface anatomy**, the study of internal structures as they relate to the overlying skin surface. You use surface anatomy when you identify the

*For the pronunciation guide rules, see the first page of the glossary in the back of the book.

bulging muscles beneath a bodybuilder's skin, and clinicians use it to locate appropriate blood vessels in which to feel pulses and draw blood.

Microscopic anatomy deals with structures too small to be seen with the naked eye. For most such studies, exceedingly thin slices of body tissues are stained and mounted on glass slides to be examined under the microscope. Subdivisions of microscopic anatomy include **cytology** (si-tol′o-je), which considers the cells of the body, and **histology** (his-tol′o-je), the study of tissues.

Developmental anatomy traces structural changes that occur throughout the life span. **Embryology** (em″bre-ol′o-je), a subdivision of developmental anatomy, concerns developmental changes that occur before birth.

Some highly specialized branches of anatomy are used primarily for medical diagnosis and scientific research. For example, *pathological anatomy* studies structural changes caused by disease. *Radiographic anatomy* studies internal structures as visualized by X-ray images or specialized scanning procedures.

Studying Anatomy

One essential tool for studying anatomy is a mastery of anatomical terminology. Other tools are observation, manipulation, and, in a living person, *palpation* (feeling organs with your hands) and *auscultation* (listening to organ sounds with a stethoscope). A simple example illustrates how some of these tools work together in an anatomical study.

Let's assume that your topic is freely movable joints of the body. In the laboratory, you will be able to *observe* an animal joint, noting how its parts fit together. You can work the joint (*manipulate* it) to determine its range of motion. Using *anatomical terminology*, you can name its parts and describe how they are related so that other students (and your instructor) will have no trouble understanding you. The list of word roots (at the back of the book) and the glossary will help you with this special vocabulary.

Although you will make most of your observations with the naked eye or with the help of a microscope, medical technology has developed a number of sophisticated tools that can peer into the body without disrupting it. See **A Closer Look** on pp. 16–17.

Topics of Physiology

Like anatomy, physiology has many subdivisions. Most of them consider the operation of specific organ systems. For example, **renal physiology** concerns kidney function and urine production. **Neurophysiology** explains the workings of the nervous system. **Cardiovascular physiology** examines the operation of the heart and blood vessels. While anatomy provides us with a static image of the body's architecture, physiology reveals the body's dynamic and animated workings.

Physiology often focuses on events at the cellular or molecular level. This is because the body's abilities depend on those of its individual cells, and cells' abilities ultimately depend on the chemical reactions that go on within them. Physiology also rests on principles of physics, which help to explain electrical

The sharp edges of incisors (structure) make them ideal for cutting like scissors (function).

The flat surfaces of molars (structure) make them ideal for grinding, like a mortar and pestle (function).

Figure 1.1 **Complementarity of structure and function.**

currents, blood pressure, and the way muscles use bones to cause body movements, among other things. We present basic chemical and physical principles in Chapter 2 and throughout the book as needed to explain physiological topics.

Complementarity of Structure and Function

Although it is possible to study anatomy and physiology individually, they are really inseparable because function always reflects structure. That is, what a structure can do depends on its specific form. This key concept is called the **principle of complementarity of structure and function**.

For example, bones can support and protect body organs because they contain hard mineral deposits. Blood flows in one direction through the heart because the heart has valves that prevent backflow. Another example is how the various shapes of our teeth reflect their different actions, as shown in **Figure 1.1**. Throughout this book, we accompany a description of a structure's anatomy with an explanation of its function, and we emphasize structural characteristics contributing to that function.

Check Your Understanding

1. In what way does physiology depend on anatomy?
2. Would you be studying anatomy or physiology if you investigated how muscles shorten? If you explored the location of the lungs in the body?
3. APPLY Use the word root definitions located at the back of this book to define each of the following terms: gastritis, leukocyte, nephropathy.

For answers, see Answers Appendix.

1.2   The body's organization ranges from atoms to the entire organism

Learning Outcomes

▶ **Name the different levels of structural organization that make up the human body, and explain their relationships.**

▶ **List the 11 organ systems of the body, identify their components, and briefly explain the major function(s) of each system.**

The human body has many levels of structural organization (**Figure 1.2**). The simplest level of the structural hierarchy is the **chemical level**, which we study in Chapter 2. At this level, *atoms*, tiny building blocks of matter, combine to form *molecules* such as water and proteins. Molecules, in turn, associate in specific ways to form *organelles* that are the basic components of cells. *Cells* are the smallest units of living things. We examine the **cellular level** in Chapter 3. All cells share some common functions, but individual cells vary widely in size and shape, reflecting their unique functions in the body.

The simplest living creatures are single cells, but in complex organisms such as human beings, the hierarchy continues on to the **tissue level**. *Tissues* are groups of similar cells that have a common function. The four basic tissue types in the human body

Chemical level
Atoms combine to form molecules.

Cellular level
Cells are made up of molecules.

Tissue level
Tissues consist of similar types of cells.

Organ level
Organs are made up of different types of tissues.

Organismal level
The human organism is made up of many organ systems.

Organ system level
Organ systems consist of different organs that work together closely.

Figure 1.2 Levels of structural organization. Components of the cardiovascular system are used to illustrate the levels of structural organization in a human being.

are epithelial tissue, muscle tissue, connective tissue, and nervous tissue.

Each tissue type has a characteristic role in the body, which we explore in Chapter 4. Briefly, epithelial tissue covers the body surface and lines its cavities. Muscle tissue provides movement. Connective tissue supports and protects body organs. Nervous tissue provides a means of rapid internal communication by transmitting electrical impulses.

An *organ* is a discrete structure composed of at least two tissue types (four is more common) that performs a specific function for the body. The liver, the brain, and a blood vessel are very different from the stomach, but each is an organ. You can think of each organ of the body as a specialized functional center responsible for a necessary activity that no other organ can perform.

At the **organ level**, extremely complex functions become possible. Let's take the stomach for an example. Its lining is an epithelium that produces digestive juices. The bulk of its wall is muscle, which churns and mixes stomach contents (food). Its connective tissue reinforces the soft muscular walls. Its nerve fibers increase digestive activity by stimulating the muscle to contract more vigorously and the glands to secrete more digestive juices.

The next level of organization is the **organ system level**. Organs that work together to accomplish a common purpose make up an *organ system*. For example, the heart and blood vessels of the cardiovascular system circulate blood continuously to carry oxygen and nutrients to all body cells. Besides the cardiovascular system, the other organ systems of the body are the integumentary, skeletal, muscular, nervous, endocrine, lymphatic, respiratory, digestive, urinary, and reproductive systems. (Note that the immune system is closely associated with the lymphatic system.) Look ahead to Figure 1.4 on pp. 6–7 for an overview of the 11 organ systems.

The highest level of organization is the *organism*, the living human being. The **organismal level** represents the sum total of all structural levels working together to keep us alive.

Check Your Understanding

4. What level of structural organization is typical of a cytologist's field of study?

5. What is the correct structural order for the following terms: tissue, organism, organ, cell?

For answers, see Answers Appendix.

1.3 What are the requirements for life?

Learning Outcomes

▶ **List the functional characteristics necessary to maintain life in humans.**

▶ **List the survival needs of the body.**

Necessary Life Functions

Now that you know the structural levels of the human body, the question that naturally follows is: What does this highly organized human body do?

Like all complex animals, humans maintain their boundaries, move, respond to environmental changes, take in and digest nutrients, carry out metabolism, dispose of wastes, reproduce themselves, and grow. We will introduce these necessary life functions here and discuss them in more detail in later chapters.

We cannot emphasize too strongly that all body cells are interdependent. This interdependence is due to the fact that humans are multicellular organisms and our vital body functions are parceled out among different organ systems. Organ systems, in turn, work cooperatively to promote the well-being of the entire body. **Figure 1.3** identifies some of the organ systems making major contributions to necessary life functions. Also, as you read this section, check **Figure 1.4** on pp. 6–7 for more detailed descriptions of the body's organ systems.

(Text continues on p. 8.)

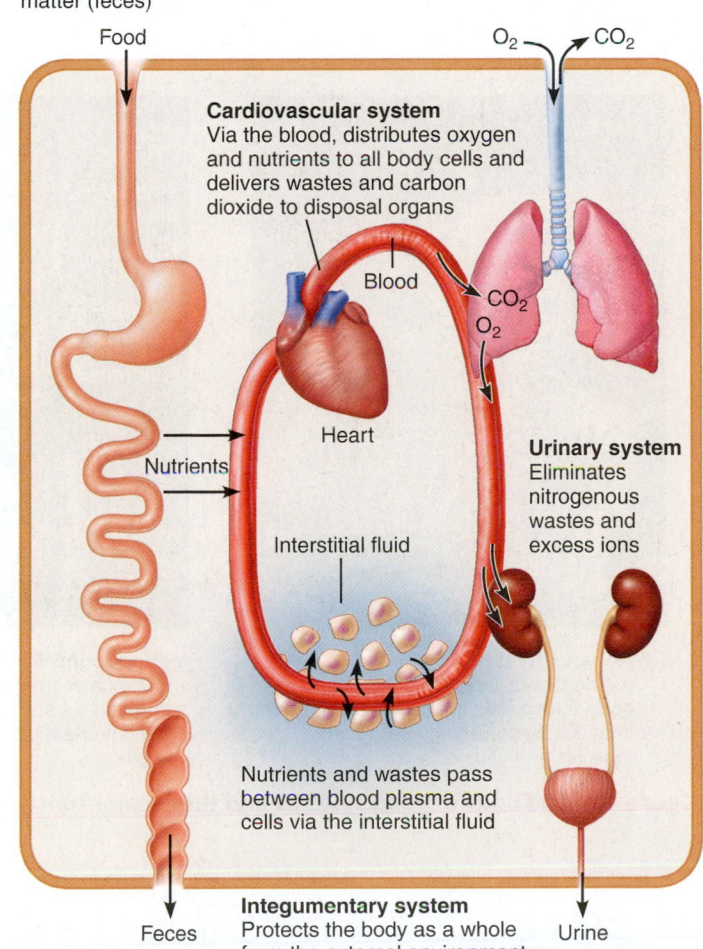

Digestive system
Takes in nutrients, breaks them down, and eliminates unabsorbed matter (feces)

Food

Respiratory system
Takes in oxygen and eliminates carbon dioxide

O_2 CO_2

Cardiovascular system
Via the blood, distributes oxygen and nutrients to all body cells and delivers wastes and carbon dioxide to disposal organs

Blood

CO_2
O_2

Nutrients

Heart

Interstitial fluid

Urinary system
Eliminates nitrogenous wastes and excess ions

Nutrients and wastes pass between blood plasma and cells via the interstitial fluid

Integumentary system
Protects the body as a whole from the external environment

Feces   Urine

Figure 1.3 Examples of interrelationships among body organ systems.

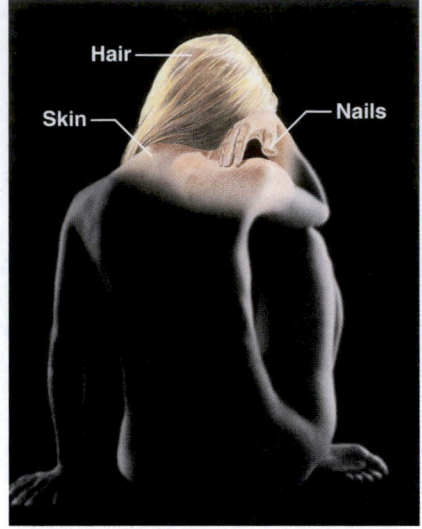

(a) Integumentary System
Forms the external body covering, and protects deeper tissues from injury. Synthesizes vitamin D, and houses cutaneous (pain, pressure, etc.) receptors, and sweat and oil glands.

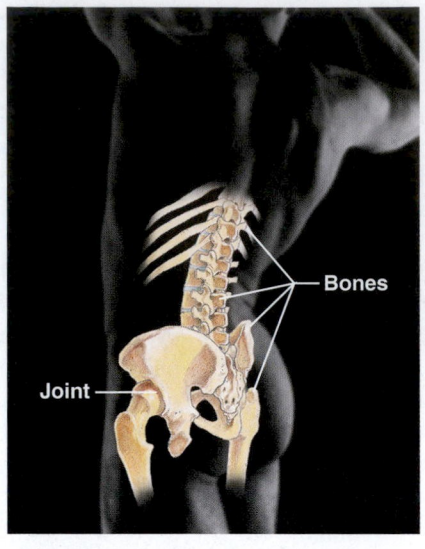

(b) Skeletal System
Protects and supports body organs, and provides a framework the muscles use to cause movement. Blood cells are formed within bones. Bones store minerals.

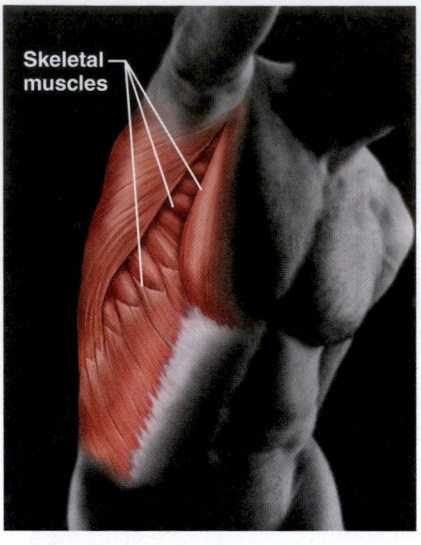

(c) Muscular System
Allows manipulation of the environment, locomotion, and facial expression. Maintains posture, and produces heat.

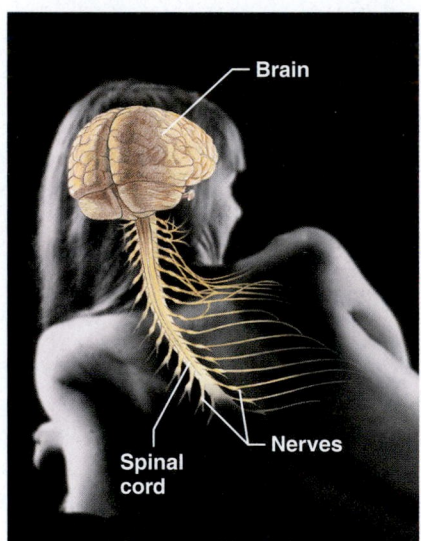

(d) Nervous System
As the fast-acting control system of the body, it responds to internal and external changes by activating appropriate muscles and glands.

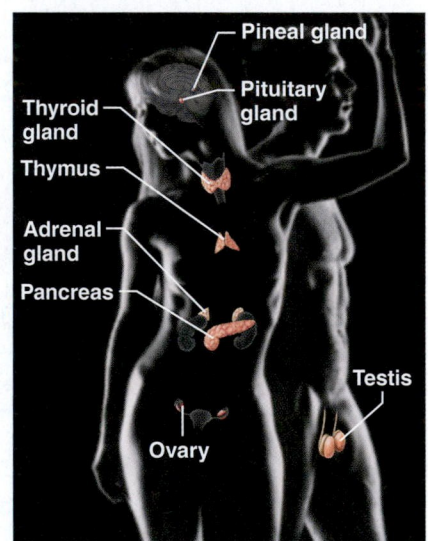

(e) Endocrine System
Glands secrete hormones that regulate processes such as growth, reproduction, and nutrient use (metabolism) by body cells.

(f) Cardiovascular System
Blood vessels transport blood, which carries oxygen, carbon dioxide, nutrients, wastes, etc. The heart pumps blood.

Figure 1.4 **The body's organ systems and their major functions.**

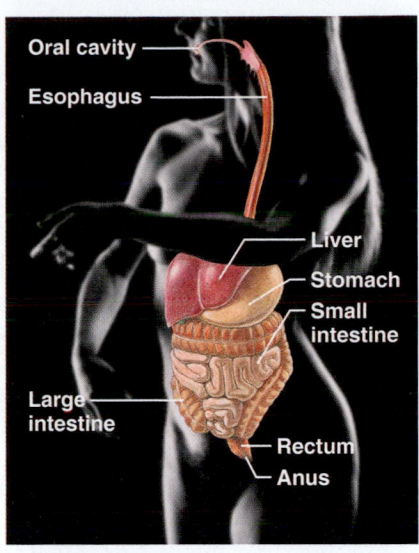

(g) Lymphatic System/Immunity
Picks up fluid leaked from blood vessels and returns it to blood. Disposes of debris in the lymphatic stream. Houses white blood cells (lymphocytes) involved in immunity. The immune response mounts the attack against foreign substances within the body.

(h) Respiratory System
Keeps blood constantly supplied with oxygen and removes carbon dioxide. These exchanges occur through the walls of the air sacs of the lungs.

(i) Digestive System
Breaks down food into absorbable units that enter the blood for distribution to body cells. Indigestible foodstuffs are eliminated as feces.

(j) Urinary System
Eliminates nitrogenous wastes from the body. Regulates water, electrolyte, and acid-base balance of the blood.

(k) Male Reproductive System **(l) Female Reproductive System**
Overall function is production of offspring. Testes produce sperm and male sex hormone, and male ducts and glands aid in delivery of sperm to the female reproductive tract. Ovaries produce eggs and female sex hormones. The remaining female structures serve as sites for fertilization and development of the fetus. Mammary glands of female breasts produce milk to nourish the newborn.

Figure 1.4 *(continued)*

Maintaining Boundaries

Every living organism must **maintain its boundaries** so that its internal environment (its inside) remains distinct from the external environment (its outside). In single-celled organisms, the external boundary is a limiting membrane that encloses its contents and lets in needed substances while restricting entry of potentially damaging or unnecessary substances. Similarly, all body cells are surrounded by a selectively permeable *plasma membrane*.

The plasma membrane separates the *intracellular fluid* inside cells from the *extracellular fluid* outside. Part of the extracellular fluid (blood *plasma*) is enclosed in blood vessels. The remainder, the *interstitial fluid*, surrounds and bathes all of our cells (see Figure 1.3 on p. 5).

Another important boundary, the integumentary system, or skin, encloses the body as a whole (Figure 1.4a). This system protects our internal organs from drying out (a fatal change), infection, and the damaging effects of heat, sunlight, and an unbelievable number of chemicals in the external environment.

Movement

Movement includes the activities promoted by the muscular system, such as propelling ourselves from one place to another by running or swimming, and manipulating the external environment with our nimble fingers (Figure 1.4c). The skeletal system provides the bony framework that the muscles pull on as they work (Figure 1.4b). Movement also occurs when substances such as blood, foodstuffs, and urine are propelled through internal organs of the cardiovascular, digestive, and urinary systems, respectively. On the cellular level, the muscle cell's ability to move by shortening is more precisely called **contractility**.

Responsiveness

Responsiveness, or **excitability**, is the ability to sense changes (stimuli) in the environment and then respond to them. For example, if you cut your hand on broken glass, a *withdrawal reflex* occurs—you involuntarily pull your hand away from the painful stimulus (the broken glass). You don't have to think about it—it just happens! Likewise, when carbon dioxide in your blood rises to dangerously high levels, chemical sensors respond by sending messages to brain centers controlling respiration, and you breathe more rapidly.

Because nerve cells are highly excitable and communicate rapidly with each other via electrical impulses, the nervous system is most involved with responsiveness (Figure 1.4d). However, all body cells are excitable to some extent.

Digestion

Digestion is the breaking down of ingested foodstuffs to simple molecules that can be absorbed into the blood. The nutrient-rich blood is then distributed to all body cells by the cardiovascular system. In a simple, one-celled organism such as an amoeba, the cell itself is the "digestion factory," but in the multicellular human body, the digestive system performs this function for the entire body (Figure 1.4i).

Metabolism

Metabolism (mĕ-tab′o-lizm; "a state of change") is a broad term that includes all chemical reactions that occur within body cells. It includes breaking down substances into simpler building blocks (the process of *catabolism*), synthesizing more complex substances from simpler building blocks (*anabolism*), and using nutrients and oxygen to produce (via *cellular respiration*) ATP, the energy-rich molecules that power cellular activities. Metabolism depends on the digestive and respiratory systems to make nutrients and oxygen available to the blood, and on the cardiovascular system to distribute them throughout the body (Figure 1.4i, h, and f, respectively). Metabolism is regulated largely by hormones secreted by endocrine system glands (Figure 1.4e).

Excretion

Excretion is the process of removing wastes, or *excreta* (ek-skre′tah), from the body. If the body is to operate as we expect it to, it must get rid of nonuseful substances produced during digestion and metabolism.

Several organ systems participate in excretion. For example, the digestive system rids the body of indigestible food residues in feces, and the urinary system disposes of nitrogen-containing metabolic wastes, such as urea, in urine (Figure 1.4i and j). Carbon dioxide, a by-product of cellular respiration, is carried in the blood to the lungs, where it leaves the body in exhaled air (Figure 1.4h).

Reproduction

Reproduction occurs at the cellular and the organismal level. In cellular reproduction, the original cell divides, producing two identical daughter cells that may then be used for body growth or repair. Reproduction of the human organism, or making a whole new person, is the major task of the reproductive system. When a sperm unites with an egg, a fertilized egg forms and develops into a baby within the mother's body. The reproductive system is directly responsible for producing offspring, but its function is exquisitely regulated by hormones of the endocrine system (Figure 1.4e).

Because males produce sperm and females produce eggs (ova), there is a division of labor in reproduction, and the reproductive organs of males and females are different (Figure 1.4k, l). Additionally, the female's reproductive structures provide the site for fertilization of eggs by sperm, and then protect and nurture the developing fetus until birth.

Growth

Growth is an increase in size of a body part or the organism as a whole. It is usually accomplished by increasing the number of cells. However, individual cells also increase in size when not dividing. For true growth to occur, constructive activities must occur at a faster rate than destructive ones.

Survival Needs

The ultimate goal of all body systems is to maintain life. However, life is extraordinarily fragile and requires several factors.

These **survival needs** include nutrients (food), oxygen, water, and appropriate temperature and atmospheric pressure.

- **Nutrients. Nutrients**, taken in via the diet, contain the chemical substances used for energy and cell building. Most plant-derived foods are rich in carbohydrates, vitamins, and minerals, whereas most animal foods are richer in proteins and fats.

  Carbohydrates are the major energy fuel for body cells. Proteins, and to a lesser extent fats, are essential for building cell structures. Fats also provide a reserve of energy-rich fuel. Selected minerals and vitamins are required for the chemical reactions that go on in cells and for oxygen transport in the blood. The mineral calcium helps to make bones hard and is required for blood clotting.

- **Oxygen.** All the nutrients in the world are useless unless **oxygen** is also available. Because the chemical reactions that release energy from foods are *oxidative* reactions that require oxygen, human cells can survive for only a few minutes without oxygen. Approximately 20% of the air we breathe is oxygen. The cooperative efforts of the respiratory and cardiovascular systems make oxygen available to the blood and body cells.

- **Water. Water** accounts for 50–60% of our body weight and is the single most abundant chemical substance in the body. It provides the watery environment necessary for chemical reactions and the fluid base for body secretions and excretions. We obtain water from ingested foods and liquids. We lose it from the body by evaporation from the lungs and skin and in body excretions.

- **Normal body temperature.** If chemical reactions are to continue at life-sustaining rates, **normal body temperature** must be maintained. As body temperature drops below 37°C (98.6°F), metabolic reactions become slower and slower, and finally stop. When body temperature is too high, chemical reactions occur at a frantic pace and body systems stop functioning. At either extreme, death occurs. The activity of the muscular system generates most body heat.

- **Appropriate atmospheric pressure. Atmospheric pressure** is the force that air exerts on the surface of the body. Breathing and gas exchange in the lungs depend on *appropriate* atmospheric pressure. At high altitudes, where atmospheric pressure is lower and the air is thin, gas exchange may be inadequate to support cellular metabolism.

The mere presence of these survival factors is not sufficient to sustain life. They must be present in the proper amounts. Too much and too little may be equally harmful. For example, oxygen is essential, but excessive amounts are toxic to body cells. Similarly, the food we eat must be of high quality and in proper amounts. Otherwise, nutritional disease, obesity, or starvation is likely. Also, while the needs listed here are the most crucial, they do not even begin to encompass all of the body's needs. For example, we can live without gravity if we must, but the quality of life suffers.

Check Your Understanding

6. What separates living beings from nonliving objects?
7. What name is given to all chemical reactions that occur within body cells?
8. The image below shows tissue cells and part of a blood vessel. The cells' nutrients and wastes are exchanged across an important boundary between two fluid compartments. Name the boundary (a) and the fluid in the compartments (b and c). Be specific.

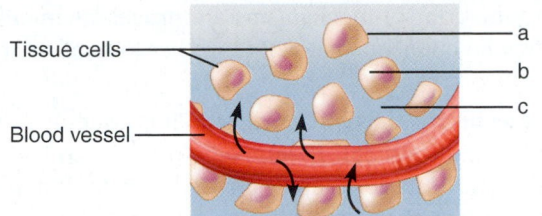

For answers, see Answers Appendix.

1.4 Homeostasis is maintained by negative feedback

Learning Outcomes

▶ **Define homeostasis and explain its significance.**
▶ **Describe how negative and positive feedback maintain body homeostasis.**
▶ **Describe the relationship between homeostatic imbalance and disease.**

When you think about the fact that your body contains trillions of cells in nearly constant activity, and that remarkably little usually goes wrong with it, you begin to appreciate what a marvelous machine your body is. Walter Cannon, an American physiologist of the early twentieth century, spoke of the "wisdom of the body," and he coined the word **homeostasis** (ho″me-o-sta′sis) to describe its ability to maintain relatively stable internal conditions even though the outside world changes continuously.

Although the literal translation of homeostasis is "unchanging," the term does not really mean a static, or unchanging, state. Rather, it indicates a *dynamic* state of equilibrium, or a balance, in which internal conditions vary, but always within relatively narrow limits. In general, the body is in homeostasis when its needs are adequately met and it is functioning smoothly.

Maintaining homeostasis is more complicated than it appears at first glance. Virtually every organ system plays a role in maintaining the constancy of the internal environment. Adequate blood levels of vital nutrients must be continuously present, and heart activity and blood pressure must be constantly monitored and adjusted so that the blood is propelled to all body tissues. Also, wastes must not be allowed to accumulate, and body temperature must be precisely controlled. A wide variety of chemical, thermal, and neural factors act and interact in complex ways—sometimes helping and sometimes hindering the body as it works to maintain its "steady rudder."

Homeostatic Control

Communication within the body is essential for homeostasis. Communication is accomplished chiefly by the nervous and endocrine systems, which use neural electrical impulses or blood-borne hormones, respectively, as information carriers. We cover the details of how these two great regulating systems operate in later chapters, but here we explain the basic characteristics of control systems that promote homeostasis.

The **variable** is the factor or event being regulated. All homeostatic control mechanisms are processes involving at least three components that work together to regulate the variable (**Figure 1.5**).

1. The **receptor** is the first component. It is a sensor that monitors the environment. It responds to *stimuli* (changes) by sending information (input) along the *afferent pathway* to the second component, the *control center*.

2. The **control center** determines the *set point*, which is the level (or range of levels) at which a variable is to be maintained. It analyzes the input it receives by comparing it to the set point and determines the appropriate response. Information (output) then flows from the control center along the *efferent pathway* to the third component, the *effector*. (To help you remember the difference between "afferent" and "efferent," note that information traveling along the afferent pathway *approaches* the control center and efferent information *exits* from the control center.)

3. The **effector** carries out the control center's response to the stimulus. The results of the response then *feed back* to influence the effect of the stimulus, either reducing it so that the whole control process is shut off, or enhancing it so that the whole process continues at an even faster rate.

Negative Feedback Mechanisms

Most homeostatic control mechanisms are **negative feedback mechanisms**. In these systems, the output shuts off the original effect of the stimulus or reduces its intensity. These mechanisms cause the variable to change in a direction *opposite* to that of the initial change, returning it to its "ideal" value.

Let's start with an example of a nonbiological negative feedback system: a home heating system connected to a temperature-sensing thermostat. The thermostat houses both the receptor (thermometer) and the control center. If the thermostat is set at 20°C (68°F), the heating system (effector) is triggered ON when the house temperature drops below that setting. As the furnace produces heat and warms the air, the temperature rises, and when it reaches 20°C or slightly higher, the thermostat triggers the furnace OFF. This process results in a cycling of the furnace between "ON" and "OFF" so that the temperature in the house stays very near the desired temperature. Your body "thermostat," located in a part of your brain called the hypothalamus, operates in a similar fashion (**Figure 1.6**).

Regulation of body temperature is only one of the many ways the nervous system maintains the constancy of the internal environment. Another type of neural control mechanism is seen in the *withdrawal reflex* mentioned earlier, in which the hand is jerked away from a painful stimulus such as broken glass.

The endocrine system is equally important in maintaining homeostasis. A good example of a hormonal negative feedback mechanism is the control of blood sugar (glucose) by insulin. As blood sugar rises, receptors in the body sense this change, and the pancreas (the control center) secretes insulin into the blood. This change in turn prompts body cells to absorb more glucose, removing it from the bloodstream. As blood sugar falls, the stimulus for insulin release ends.

The body's ability to regulate its internal environment is fundamental. All negative feedback mechanisms have the same goal: preventing severe changes within the body. Body temperature and blood sugar are only two of the variables that need to

③ Input: Information sent along afferent pathway to control center.

④ Output: Information sent along efferent pathway to effector.

② Receptor detects change.

① Stimulus produces change in variable.

⑤ Response of effector feeds back to reduce the effect of stimulus and returns variable to homeostatic level.

Control Center

Afferent pathway

Efferent pathway

Receptor

Effector

IMBALANCE

BALANCE

IMBALANCE

Figure 1.5 Interactions among the elements of a homeostatic control system maintain stable internal conditions.

Figure 1.6 **Body temperature is regulated by a negative feedback mechanism.**

be regulated. There are many! Other negative feedback mechanisms regulate heart rate, blood pressure, the rate and depth of breathing, and blood levels of oxygen, carbon dioxide, and minerals.

Positive Feedback Mechanisms

In **positive feedback mechanisms**, the initial response enhances the original stimulus so that further responses are even greater. This feedback mechanism is "positive" because the change that results proceeds in the *same* direction as the initial change, causing the variable to deviate further and further from its original value or range.

In contrast to negative feedback controls, which maintain some physiological function or keep blood chemicals within narrow ranges, positive feedback mechanisms usually control infrequent events that do not require continuous adjustments.

Typically, they set off a linked sequence of events. Once initiated, the results of each reaction feed into the next like a series of waterfalls on a river. Because of these characteristics, positive feedback mechanisms are often referred to as *cascades* (from the Italian word meaning "to fall") that amplify the original stimulus. Two familiar examples are the enhancement of labor contractions during birth and blood clotting.

Chapter 28 describes the positive feedback mechanism in which oxytocin, a hypothalamic hormone, intensifies labor contractions during the birth of a baby (see Figure 28.16, p. 1114). Oxytocin causes the contractions to become both more frequent and more powerful. The increased contractions cause more oxytocin to be released, which causes more contractions, and so on until the baby is born. The birth ends the stimulus for oxytocin release and shuts off the positive feedback mechanism.

① Break or tear occurs in blood vessel wall.

Positive feedback cycle is initiated.

③ Released chemicals attract more platelets.

Positive feedback loop

② Platelets adhere to site and release chemicals.

Feedback cycle ends when plug is formed.

④ Platelet plug is fully formed.

Figure 1.7 A positive feedback mechanism regulates formation of a platelet plug.

Blood clotting is a normal response to a break in the wall of a blood vessel and is an excellent example of an important body function controlled by positive feedback. Once a vessel has been damaged, blood elements called platelets immediately begin to cling to the injured site and release chemicals that attract more platelets. This rapidly growing pileup of platelets temporarily "plugs" the tear and initiates the sequence of events that finally forms a clot (**Figure 1.7**).

Positive feedback mechanisms are likely to race out of control, so they are rarely used to promote the moment-to-moment well-being of the body. Some positive feedback mechanisms, including this one, may have only local effects. For example, blood clotting is accelerated in injured vessels, but does not normally spread to the entire circulation.

Homeostatic Imbalance

Homeostasis is so important that most disease can be regarded as a result of its disturbance, a condition called **homeostatic imbalance**. As we age, our body's control systems become less efficient, and our internal environment becomes less and less stable. These events increase our risk for illness and produce the changes we associate with aging.

Another important source of homeostatic imbalance occurs when the usual negative feedback mechanisms are overwhelmed and destructive positive feedback mechanisms take over. Some instances of heart failure reflect this phenomenon.

Examples of homeostatic imbalance appear throughout this book to enhance your understanding of normal physiological mechanisms. This symbol ⚖ introduces the homeostatic imbalance sections and alerts you to the fact that we are describing an abnormal condition. Each Homeostatic Imbalance section is numbered to correspond with critical thinking questions available in the Study Area of MasteringA&P®—visit the website to find Homeostatic Imbalance questions and other helpful study tools.

Check Your Understanding

9. What process allows us to adjust to either extreme heat or extreme cold?

10. Why is the control system shown in Figure 1.7 called a positive feedback mechanism? What event ends it?

11. **APPLY** When we begin to get dehydrated, we usually get thirsty, which causes us to drink fluids. Is thirst part of a negative or a positive feedback control system? Explain your choice.

For answers, see Answers Appendix.

1.5 Anatomical terms describe body directions, regions, and planes

Learning Outcomes

▶ **Describe the anatomical position.**

▶ **Use correct anatomical terms to describe body directions, regions, and body planes or sections.**

Most of us are naturally curious about our bodies, but our interest sometimes dwindles when we are confronted with the terminology of anatomy and physiology. Let's face it—you can't just pick up an anatomy and physiology book and read it as though it were a novel. Unfortunately, confusion is likely without precise, specialized terminology. To prevent misunderstanding, anatomists use universally accepted terms to identify body structures precisely and with a minimum of words. We present and explain the language of anatomy next.

Anatomical Position and Directional Terms

To describe body parts and position accurately, we need an initial reference point, and we must indicate direction. The anatomical reference point is a standard body position called the **anatomical position**. In the anatomical position, the body is erect with feet slightly apart. This position is easy to remember because it resembles "standing at attention," except that the palms face forward and the thumbs point away from the body. You can see the anatomical position in **Figure 1.8a** and **Table 1.1** (top) on p. 14.

It is essential to understand the anatomical position because most of the directional terms used in this book refer to the body

Cephalic
Frontal
Orbital
Nasal
Buccal (cheek)
Oral
Mental (chin)
Cervical
Thoracic
Sternal
Axillary
Mammary
Abdominal
Umbilical
Pelvic
Inguinal
(groin)
Pubic (genital)

Upper limb
Acromial
Brachial (arm)
Antecubital
Olecranal
Antebrachial
(forearm)
Carpal (wrist)

Manus (hand)
Metacarpal
Palmar
Pollex
Digital

Lower limb
Coxal (hip)
Femoral (thigh)
Patellar
Popliteal
Crural (leg)
Sural (calf)
Fibular or peroneal

Pedal (foot)
Tarsal (ankle)
Calcaneal
Metatarsal
Digital
Plantar
Hallux

Cephalic
Otic
Occipital (back
of head)
Cervical
Back (dorsal)
Scapular
Vertebral
Lumbar
Sacral
Gluteal
Perineal (between
anus and external
genitalia)

Thorax
Abdomen
Back (Dorsum)

(a) Anterior/Ventral

(b) Posterior/Dorsal

Figure 1.8 Regional terms used to designate specific body areas. Common terms are shown in parentheses. **(a)** Anatomical position. **(b)** The heels are raised to show the plantar surface of the foot.

as if it were in this position, regardless of its actual position. Another point to remember is that the terms "right" and "left" refer to those sides of the person or the cadaver being viewed— not those of the observer.

Directional terms allow us to explain where one body structure is in relation to another. For example, we could describe the relationship between the ears and the nose by stating, "The ears are located on each side of the head to the right and left of the nose." Using anatomical terminology, this becomes "The ears are lateral to the nose." Using anatomical terms saves words and is less ambiguous.

Commonly used orientation and directional terms are defined and illustrated in Table 1.1. Many of these terms are also used in everyday conversation, but remember as you study them that their anatomical meanings are very precise.

Regional Terms

The two fundamental divisions of our body are its *axial* and *appendicular* (ap″en-dik′u-lar) parts. The **axial part**, which makes up the main *axis* of our body, includes the head, neck, and trunk. The **appendicular part** consists of the *appendages*, or *limbs*, which are attached to the body's axis. **Regional terms** used to designate specific areas within these major body divisions are indicated in Figure 1.8.

Body Planes and Sections

For anatomical studies, the body is often cut, or *sectioned*, along a flat surface called a *plane*. The most frequently used body planes are *sagittal*, *frontal*, and *transverse* planes, which

Table 1.1 Orientation and Directional Terms

TERM	DEFINITION	EXAMPLE	
Superior (cranial)	Toward the head end or upper part of a structure or the body; above		The head is superior to the abdomen.
Inferior (caudal)	Away from the head end or toward the lower part of a structure or the body; below		The navel is inferior to the chin.
Anterior (ventral)*	Toward or at the front of the body; in front of		The breastbone is anterior to the spine.
Posterior (dorsal)*	Toward or at the back of the body; behind		The heart is posterior to the breastbone.
Medial	Toward or at the midline of the body; on the inner side of		The heart is medial to the arm.
Lateral	Away from the midline of the body; on the outer side of		The arms are lateral to the chest.
Intermediate	Between a more medial and a more lateral structure		The collarbone is intermediate between the breastbone and shoulder.
Proximal	Closer to the origin of the body part or the point of attachment of a limb to the body trunk		The elbow is proximal to the wrist.
Distal	Farther from the origin of a body part or the point of attachment of a limb to the body trunk		The knee is distal to the thigh.
Superficial (external)	Toward or at the body surface		The skin is superficial to the skeletal muscles.
Deep (internal)	Away from the body surface; more internal		The lungs are deep to the skin.

*The terms *ventral* and *anterior* are synonymous in humans, but this is not the case in four-legged animals. *Anterior* refers to the leading portion of the body (abdominal surface in humans, head in a cat), but *ventral* specifically refers to the "belly" of a vertebrate animal, so it is the inferior surface of four-legged animals. Likewise, although the dorsal and posterior surfaces are the same in humans, the term *dorsal* specifically refers to an animal's back (as in the dorsal fin of a shark). Thus, the dorsal surface of four-legged animals is their superior surface.

Figure 1.9 Planes of the body with corresponding magnetic resonance imaging (MRI) scans.

lie at right angles to one another (**Figure 1.9**). A section is named for the plane along which it is cut. Thus, a cut along a sagittal plane produces a sagittal section.

- A **sagittal plane** (saj′ĭ-tal; "arrow") is a vertical plane that divides the body into right and left parts. A sagittal plane that lies exactly in the midline is the **median plane**, or **midsagittal plane** (Figure 1.9a). All other sagittal planes, offset from the midline, are **parasagittal planes** (*para* = near).

- **Frontal planes**, like sagittal planes, lie vertically. Frontal planes, however, divide the body into anterior and posterior parts (Figure 1.9b). A frontal plane is also called a **coronal plane** (kŏ-ro′nal; "crown").

- A **transverse**, or **horizontal**, **plane** runs horizontally from right to left, dividing the body into superior and inferior parts (Figure 1.9c). Of course, many different transverse planes

exist, at every possible level from head to foot. A transverse section is also called a **cross section**.

- **Oblique sections** are cuts made diagonally between the horizontal and the vertical planes. Because oblique sections are often confusing and difficult to interpret, they are seldom used.

Figure 1.9 includes examples of magnetic resonance imaging (MRI) scans that correspond to the three sections shown in the figure. Clinically, the ability to interpret sections made through the body, especially transverse sections, is important. Additionally, certain medical imaging devices (*A Closer Look*, pp. 16–17) produce sectional images rather than three-dimensional images.

It takes practice to determine an object's overall shape from sectioned material. Sectioning the body or an organ along different planes often results in very different views. For example,

Medical Imaging: Illuminating the Body

X rays of a hand and foot

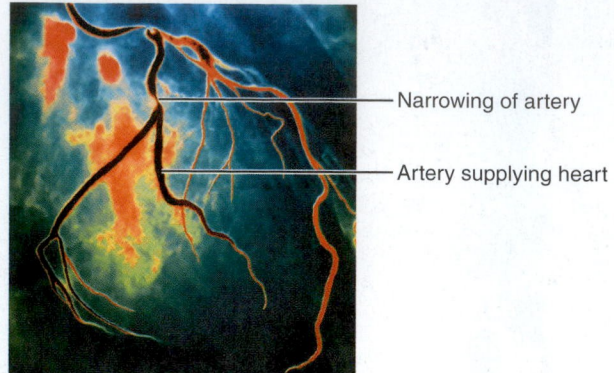

Narrowing of artery

Artery supplying heart

A DSA image of the arteries that supply the heart.

X-Ray Imaging

- Uses X rays (very short-wavelength electromagnetic waves).
- Dense structures (e.g., bones) appear as light areas.
- Hollow air-containing organs (e.g., lungs) and fat appear as dark areas.
- **Used for:** Detecting broken bones; finding breast tumors (mammography); measuring density of bones to screen for osteoporosis (porous bones).
- **Not used for:** Most soft tissue problems except when used with a contrast medium such as barium.
- **Cons:** Radiation exposure; images are two-dimensional.

Digital Subtraction Angiography

- Angiography is visualizing blood vessels (*angi* = vessel) by X ray or CT scan.
- Requires injection of an X-ray-absorbing contrast agent.
- Digitally subtracting images from before and after injection of the contrast agent yields very clear images of blood vessels.
- **Used for:** Detecting blood vessel abnormalities such as blockages in the arteries that supply the heart.
- **Cons:** Time-consuming and expensive. Adverse reactions to the contrast medium can occur.

Right   Left

Liver

Small intestine

Kidney

Muscle

Vertebra

CT scan of the abdomen. By convention, cross sections such as this are shown as if the patient were lying on their back and viewed from the feet toward the head.

Secondary tumor (bone)

Primary tumor (prostate)

CT   PET   Combined

PET scans are used to monitor the spread of cancers. In this case, a CT scan (left) and PET scan (center) are combined to create the final image (right).

Computed Tomography (CT) Scans

- Computerized reconstruction of a series of X-ray images.
- Provide detailed cross-sectional pictures of scanned body regions.
- **Used for:** Images of bone, soft tissues, and blood vessels.
- **Less useful for:** Nervous tissue and joint structure (e.g., knee and shoulder).
- **Cons:** More radiation exposure than X rays—may be a concern if used repeatedly.

Positron Emission Tomography (PET) Scans

- Uses gamma rays that are emitted by radioactively tagged tracer molecules that are injected into the body.
- The radioactive tracer molecule used depends on the reason for doing the scan. Radioactive *glucose* is used to locate cells with the highest metabolic activity, such as cancer cells. *Florbetapir* is used to visualize beta-amyloid plaques in the brain that are associated with Alzheimer's disease.
- **Used for:** Detecting the spread of cancer or monitoring the response to cancer treatment. Sometimes used to help diagnose Alzheimer's disease, and as a research tool to explore brain function.
- **Cons:** Radiation exposure; relatively poor image resolution.

MRI showing a midsagittal section of the head.

Ultrasound image of a fetus in the uterus.

Magnetic Resonance Imaging (MRI)

- Uses powerful magnets and radio waves to image the location of hydrogen atoms in the body (most of which are in water).
- Distinguishes body tissues based on water content. Structures with low water content such as bones are not readily visible.
- Produces high-contrast images of soft tissues, particularly those that are obscured by bone in CT scans.
- Functional MRI (fMRI) tracks blood flow into various parts of the brain.
- **Used for:** Imaging of brain, spinal cord, and nerves to detect abnormalities such as tumors; assessing joint, ligament, cartilage, and other soft tissues; fMRI allows visualization of the activity in specific brain regions.
- **Not used for:** Bones.
- **Cons:** More expensive and much slower than CT scans. Cannot be used in patients with most types of metal implant.

Ultrasound Imaging

- Uses high-frequency sound waves that reflect (echo) off of the body's tissues.
- Does not use X rays, so thought to be safe for imaging a developing fetus.
- Can monitor movement in real time (such as heart valve motion and blood flow through vessels and the heart).
- Inexpensive and easy to use.
- **Used for:** Monitoring a fetus during pregnancy; diagnosing abdominal or pelvic disorders such as gallbladder disease; can detect atherosclerosis (thickening and hardening of the arterial walls) and heart valve disorders.
- **Not used for:** Air-filled structures (such as the lungs) and structures surrounded by bone (such as the brain and spinal cord).
- **Cons:** Images tend to be lower resolution (blurry), although their sharpness is being improved.

a transverse section of the body trunk at the level of the kidneys would show kidney structure in cross section very nicely. A frontal section of the body trunk would show a different view of kidney anatomy, and a midsagittal section would miss the kidneys completely. With experience, you will gradually learn to relate two-dimensional sections to three-dimensional shapes.

Check Your Understanding

12. What is the anatomical position? Why is it important that *you* learn this position?
13. The axillary and acromial regions are both in the general area of the shoulder. Where specifically is each located?
14. What type of cut would separate the brain into anterior and posterior parts?
15. **DRAW** Draw the outline that you would get if you made midsagittal, coronal, and transverse sections of the banana at the right.

For answers, see Answers Appendix.

1.6 Many internal organs lie in membrane-lined body cavities

Learning Outcomes

▶ Locate and name the major body cavities and their subdivisions and associated membranes, and list the major organs contained within them.
▶ Name the four quadrants or nine regions of the abdominopelvic cavity and list the organs they contain.

Anatomy and physiology textbooks typically describe two sets of internal body cavities called the dorsal and ventral body cavities. These cavities are closed to the outside and provide different degrees of protection to the organs within them. Because these two cavities differ in their mode of embryonic development and their lining membranes, the dorsal body cavity is not recognized as such in many anatomical references. However, the idea of two sets of internal body cavities is a useful learning concept and we use it here.

Dorsal Body Cavity

The **dorsal body cavity**, which protects the fragile nervous system organs, has two subdivisions (**Figure 1.10**, gold areas). The **cranial cavity**, in the skull, encases the brain. The **vertebral**, or **spinal**, **cavity**, which runs within the bony vertebral column, encloses the delicate spinal cord. The spinal cord is essentially a continuation of the brain, and the cranial and spinal cavities are continuous with one another. Both the brain and the spinal cord are covered by membranes called meninges.

Ventral Body Cavity

The more anterior and larger of the closed body cavities is the **ventral body cavity** (Figure 1.10, deep-red areas). Like the dorsal cavity, it has two major subdivisions, the *thoracic cavity* and the *abdominopelvic cavity*. The ventral body cavity houses internal organs collectively called the **viscera** (vis′er-ah; *viscus* = an organ in a body cavity), or visceral organs.

The superior subdivision, the **thoracic cavity** (tho-ras′ik), is surrounded by the ribs and muscles of the chest. The thoracic cavity is further subdivided into lateral **pleural cavities** (ploo′ral), each enveloping a lung, and the medial **mediastinum** (me″de-ah-sti′num). The mediastinum contains the **pericardial cavity** (per″ĭ-kar′de-al), which encloses the heart, and it also surrounds the remaining thoracic organs (esophagus, trachea, and others).

The thoracic cavity is separated from the more inferior **abdominopelvic cavity** (ab-dom′ĭ-no-pel′vic) by the diaphragm, a dome-shaped muscle important in breathing. The abdominopelvic cavity, as its name suggests, has two parts. However, these regions are not physically separated by a muscular or membrane wall. Its superior portion, the **abdominal cavity**, contains the stomach, intestines, spleen, liver, and other organs. The inferior part, the **pelvic cavity**, lies in the bony pelvis and contains the urinary bladder, some reproductive organs, and the rectum. The abdominal and pelvic cavities are not aligned with each other. Instead, the bowl-shaped pelvis tips away from the perpendicular as shown in Figure 1.10a.

HOMEOSTATIC IMBALANCE 1.1 — CLINICAL

Each body cavity is uniquely suited to house its contents. Problems arise when a structure strays into a neighboring cavity. A *hiatal hernia* occurs when part of the stomach slides through the diaphragm into the thoracic cavity, allowing stomach acid to cause heartburn (which is actually irritation of the esophagus, not the heart). Severe cases may require surgical repair. ●●●●●●●

Membranes in the Ventral Body Cavity

The walls of the ventral body cavity and the outer surfaces of the organs it contains are covered by a thin, double-layered membrane, the **serosa** (se-ro′sah), or **serous membrane**.

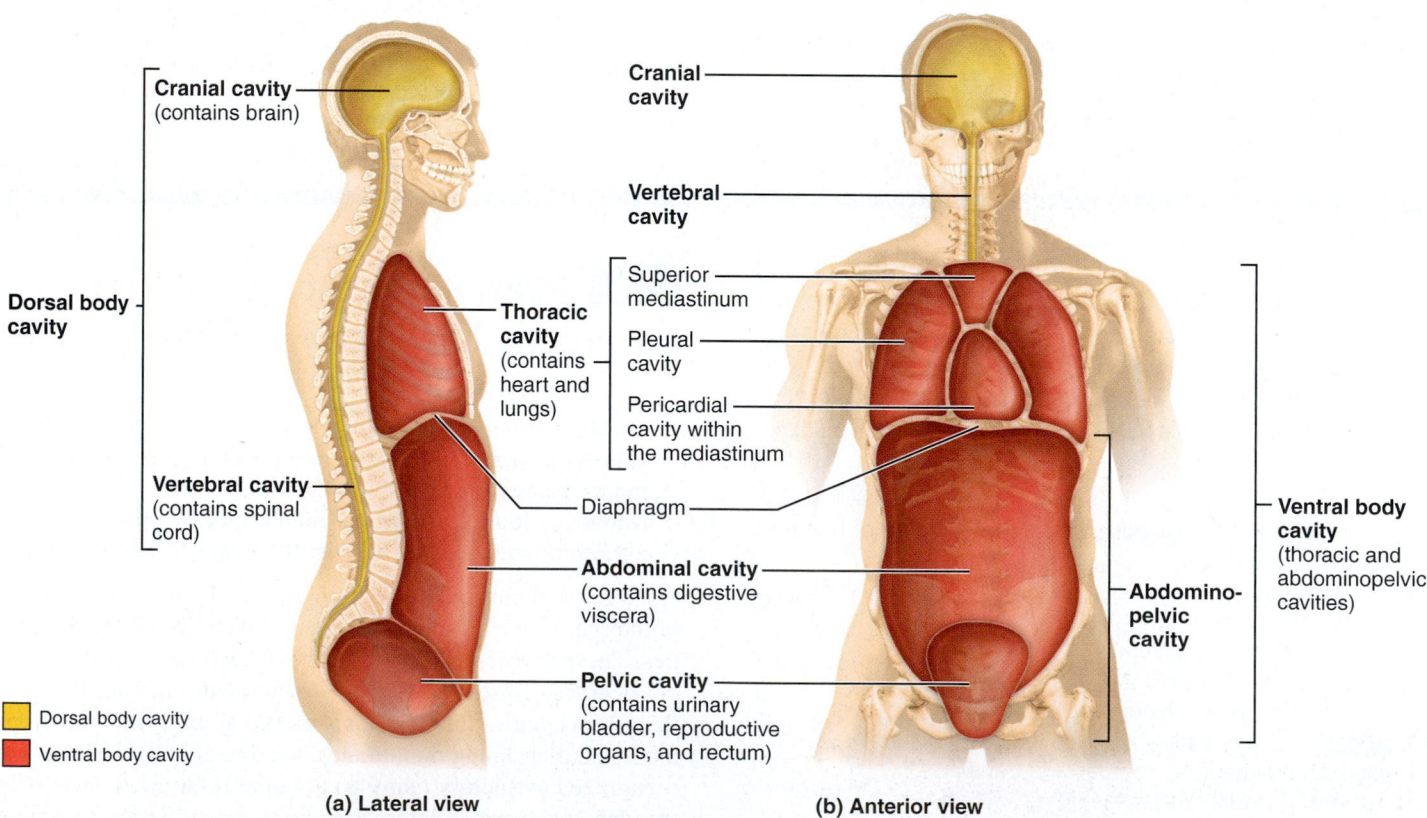

Cranial cavity (contains brain)

Dorsal body cavity

Vertebral cavity (contains spinal cord)

Thoracic cavity (contains heart and lungs)

Diaphragm

Abdominal cavity (contains digestive viscera)

Pelvic cavity (contains urinary bladder, reproductive organs, and rectum)

Cranial cavity

Vertebral cavity

Superior mediastinum

Pleural cavity

Pericardial cavity within the mediastinum

Abdomino-pelvic cavity

Ventral body cavity (thoracic and abdominopelvic cavities)

☐ Dorsal body cavity
☐ Ventral body cavity

(a) Lateral view   **(b) Anterior view**

Figure 1.10 Dorsal and ventral body cavities and their subdivisions.

Outer balloon wall
(comparable to parietal serosa)

Air (comparable to serous cavity)

Inner balloon wall
(comparable to visceral serosa)

(a) A fist thrust into a flaccid balloon demonstrates the relationship between the parietal and visceral serous membrane layers.

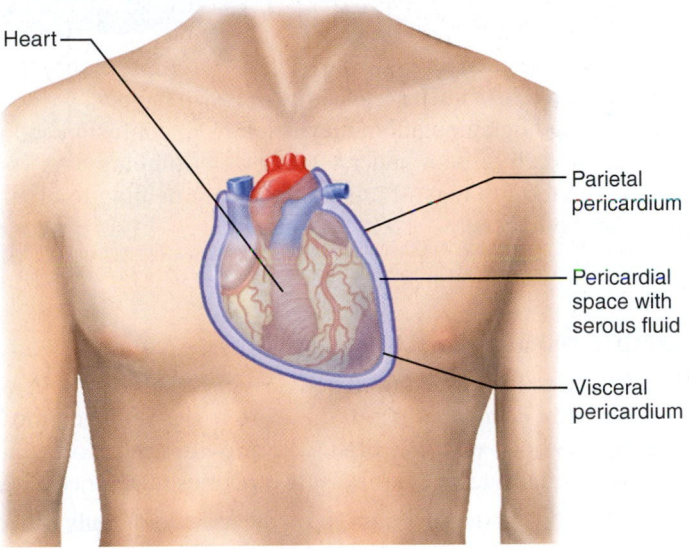

Heart

Parietal
pericardium

Pericardial
space with
serous fluid

Visceral
pericardium

(b) The serosae associated with the heart.

Figure 1.11 Serous membrane relationships.

The part of the membrane lining the cavity walls is called the **parietal serosa** (pah-ri′ĕ-tal; *parie* = wall). It folds in on itself to form the **visceral serosa**, covering the organs in the cavity.

You can visualize the relationship between the serosal layers by pushing your fist into a limp balloon (**Figure 1.11a**). The part of the balloon that clings to your fist can be compared to the visceral serosa clinging to an organ's external surface. The outer wall of the balloon represents the parietal serosa that lines the walls of the cavity. (However, unlike the balloon, the parietal serosa is never exposed but is always fused to the cavity wall.) In the body, the serous membranes are separated not by air but by a thin layer of lubricating fluid, called **serous fluid**, which is secreted by both membranes. Although there is a potential space between the two membranes, the barely present, slitlike *serous cavity* is filled with serous fluid.

The slippery serous fluid allows the organs to slide without friction across the cavity walls and one another as they carry out their routine functions. This freedom of movement is especially important for mobile organs such as the pumping heart and the churning stomach.

The serous membranes are named for the specific cavity and organs with which they are associated. For example, as shown in Figure 1.11b, the *parietal pericardium* lines the pericardial cavity and folds back as the *visceral pericardium*,

which covers the heart. Likewise, the *parietal pleurae* (ploo′re) line the walls of the thoracic cavity, and the *visceral pleurae* cover the lungs. The *parietal peritoneum* (per″ĭ-to-ne′um) is associated with the walls of the abdominopelvic cavity, while the *visceral peritoneum* covers most of the organs within that cavity. (The pleural and peritoneal serosae are illustrated in Figure 4.14c on p. 142.)

HOMEOSTATIC IMBALANCE 1.2   CLINICAL

When serous membranes are inflamed, their normally smooth surfaces become roughened. This roughness causes the membranes to stick together and drag across one another. Excruciating pain results, as anyone who has experienced *pleurisy* (inflammation of the pleurae) or *peritonitis* (inflammation of the peritoneums) knows.

Abdominopelvic Regions and Quadrants

Because the abdominopelvic cavity is large and contains several organs, it helps to divide it into smaller areas for study. Medical personnel usually use a simple scheme to locate the abdominopelvic cavity organs (**Figure 1.12**). In this scheme, a transverse and a median plane pass through the umbilicus at right angles. The four resulting quadrants are named according to their positions from the subject's point of view: the **right upper quadrant (RUQ), left upper quadrant (LUQ), right lower quadrant (RLQ)**, and **left lower quadrant (LLQ)**.

Another division method, used primarily by anatomists, uses two transverse and two parasagittal planes. These planes,

Right upper
quadrant
(RUQ)

Left upper
quadrant
(LUQ)

Right lower
quadrant
(RLQ)

Left lower
quadrant
(LLQ)

Figure 1.12 The four abdominopelvic quadrants. In this scheme, the abdominopelvic cavity is divided into four quadrants by two planes.

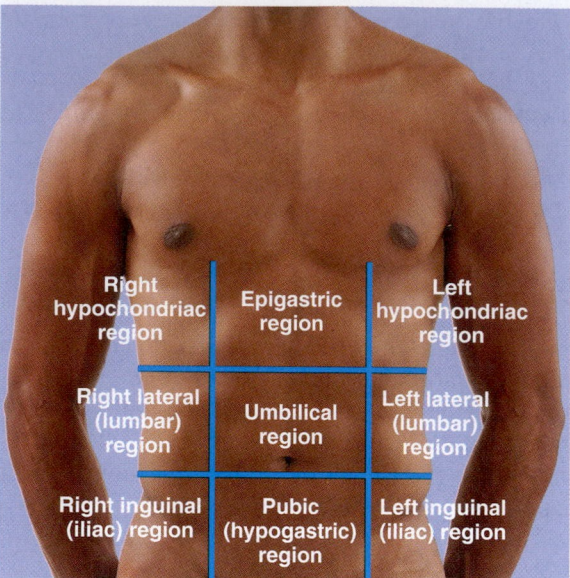

(a) Nine regions delineated by four planes

- The **right** and **left inguinal**, or **iliac**, **regions** (ing'gwĭ-nal) are located lateral to the hypogastric region (*iliac* = superior part of the hip bone).
- The **right** and **left lateral (lumbar) regions** lie lateral to the umbilical region (*lumbus* = loin).
- The **right** and **left hypochondriac regions** lie lateral to the epigastric region and deep to the ribs (*chondro* = cartilage).

HOMEOSTATIC IMBALANCE 1.3   CLINICAL

You may have seen news stories about "wrong site surgery" and wondered how such serious mistakes can happen. Critical errors, including amputation, may result from confusion about right versus left or poor understanding of terminology. As you master the terminology of anatomy, you are helping to eliminate these blunders.

Other Body Cavities

In addition to the large closed body cavities, there are several smaller body cavities. Most of these are in the head and most open to the body exterior. Figure 1.8 provides the terms that will help you locate all but the last two cavities mentioned here.

- **Oral and digestive cavities.** The oral cavity, commonly called the mouth, contains the teeth and tongue. This cavity is part of and continuous with the cavity of the digestive organs, which opens to the body exterior at the anus.
- **Nasal cavity.** Located within and posterior to the nose, the nasal cavity is part of the respiratory system passageways.
- **Orbital cavities.** The orbital cavities (orbits) in the skull house the eyes and present them in an anterior position.
- **Middle ear cavities.** The middle ear cavities in the skull lie just medial to the eardrums. These cavities contain tiny bones that transmit sound vibrations to the hearing receptors in the inner ears.
- **Synovial cavities.** Synovial (sĭ-no've-al) cavities are joint cavities. They are enclosed within fibrous capsules that surround freely movable joints of the body (such as the elbow and knee joints). Like the serous membranes, membranes lining synovial cavities secrete a lubricating fluid that reduces friction as the bones move across one another.

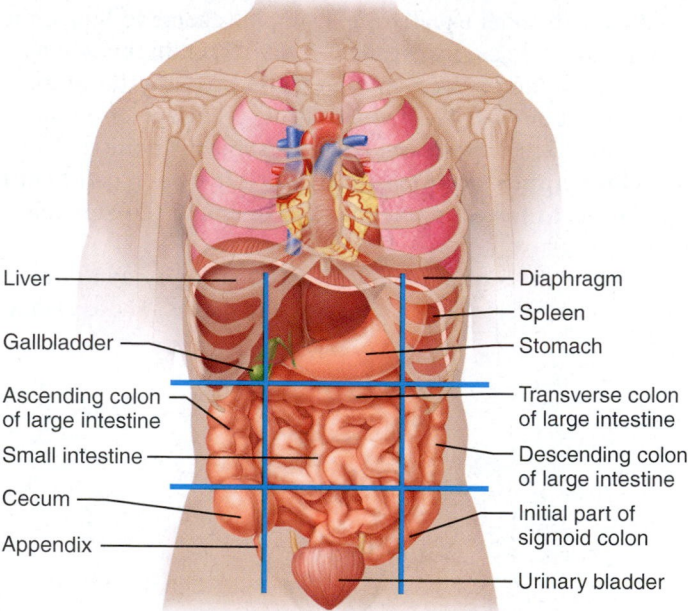

(b) Anterior view of the nine regions showing the superficial organs

Figure 1.13 **The nine abdominopelvic regions.** The superior transverse plane is just inferior to the ribs; the inferior transverse plane is just superior to the hip bones; and the parasagittal planes lie just medial to the nipples.

positioned like a tic-tac-toe grid on the abdomen, divide the cavity into nine regions (**Figure 1.13**):

- The **umbilical region** is the centermost region deep to and surrounding the umbilicus (navel).
- The **epigastric region** is located superior to the umbilical region (*epi* = upon, above; *gastri* = belly).
- The **pubic (hypogastric) region** is located inferior to the umbilical region (*hypo* = below).

Check Your Understanding

16. Of the uterus, small intestine, spinal cord, and heart, which is/are in the dorsal body cavity?
17. **APPLY** When you rub your cold hands together, the friction between them results in heat that warms your hands. Why doesn't warming friction result during movements of the heart, lungs, and digestive organs?
18. **PREDICT** Joe went to the emergency room where he complained of severe pains in the lower right quadrant of his abdomen. What might be his problem?

For answers, see Answers Appendix.

CHAPTER SUMMARY

1.1 Form (anatomy) determines function (physiology) (pp. 2–3)

1. Anatomy is the study of body structures and their relationships. Physiology is the science of how body parts function.

Topics of Anatomy (p. 2)

2. Major subdivisions of anatomy include gross anatomy, microscopic anatomy, and developmental anatomy.

Studying Anatomy (p. 2)

3. Anatomical terminology is essential for studying anatomy.

Topics of Physiology (pp. 2–3)

4. Typically, physiology concerns the functioning of specific organs or organ systems. Examples include cardiovascular physiology, renal physiology, and muscle physiology.
5. Physiology is explained by chemical and physical principles.

Complementarity of Structure and Function (p. 3)

6. Anatomy and physiology are inseparable: What a body can do depends on the unique architecture of its parts. This principle is called the complementarity of structure and function.

1.2 The body's organization ranges from atoms to the entire organism (pp. 4–5)

1. The levels of structural organization of the body, from simplest to most complex, are: chemical, cellular, tissue, organ, organ system, and organismal.
2. The 11 organ systems of the body are the integumentary, skeletal, muscular, nervous, endocrine, cardiovascular, lymphatic, respiratory, digestive, urinary, and reproductive systems. The immune system is a functional system closely associated with the lymphatic system. (For functions of these systems see pp. 6–7.)

1.3 What are the requirements for life? (pp. 5–9)

1. All living organisms carry out certain vital functional activities necessary for life, including maintenance of boundaries, movement, responsiveness, digestion, metabolism, excretion, reproduction, and growth.
2. Survival needs include nutrients, water, oxygen, and appropriate temperature and atmospheric pressure.

1.4 Homeostasis is maintained by negative feedback (pp. 9–12)

1. Homeostasis is a dynamic equilibrium of the internal environment. All body systems contribute to homeostasis, but the nervous and endocrine systems are most important. Homeostasis is necessary for health.

Homeostatic Control (pp. 9–12)

2. Control mechanisms of the body contain at least three elements that work together: receptor(s), control center, and effector(s).

3. Negative feedback mechanisms reduce the effect of the original stimulus, and are essential for maintaining homeostasis. Body temperature, heart rate, blood pressure, breathing rate and depth, and blood levels of glucose and certain ions are regulated by negative feedback mechanisms.
4. Positive feedback mechanisms intensify the initial stimulus, leading to an enhancement of the response. They rarely contribute to homeostasis, but blood clotting and labor contractions are regulated by such mechanisms.

Homeostatic Imbalance (p. 12)

5. With age, the efficiency of negative feedback mechanisms declines. These changes underlie certain disease conditions.

1.5 Anatomical terms describe body directions, regions, and planes (pp. 12–17)

Anatomical Position and Directional Terms (pp. 12–13)

1. In the anatomical position, the body is erect, facing forward, feet slightly apart, arms at sides with palms forward.
2. Directional terms allow body parts to be located precisely. Terms that describe body directions and orientation include: superior/inferior; anterior/posterior; ventral/dorsal; medial/lateral; intermediate; proximal/distal; and superficial/deep.

Regional Terms (p. 13)

3. Regional terms are used to designate specific areas of the body (see Figure 1.8).

Body Planes and Sections (pp. 13–17)

4. The body or its organs may be cut along planes to produce different types of sections. Frequently used planes are sagittal, frontal, and transverse.

1.6 Many internal organs lie in membrane-lined body cavities (pp. 17–20)

1. The body contains two major closed cavities. The dorsal cavity, subdivided into the cranial and spinal cavities, contains the brain and spinal cord. The ventral cavity is subdivided into the thoracic cavity, which houses the heart and lungs, and the abdominopelvic cavity, which contains the liver, digestive organs, and reproductive structures.
2. The walls of the ventral cavity and the surfaces of the organs it contains are covered with thin membranes, the parietal and visceral serosae, respectively. The serosae produce a thin fluid that decreases friction during organ functioning.
3. The abdominopelvic cavity may be divided by four planes into nine abdominopelvic regions (epigastric, umbilical, hypogastric, right and left iliac, right and left lumbar, and right and left hypochondriac), or by two planes into four quadrants. (For boundaries and organs contained, see Figures 1.12 and 1.13.)
4. There are several smaller body cavities. Most of these are in the head and open to the exterior.

REVIEW QUESTIONS

To access additional practice questions using your smartphone, tablet, or computer: MasteringA&P®> Study Area > Practice Tests & Quizzes

(Some multiple choice questions have more than one correct answer. Select the best answer or answers from the choices given.)

Level 1 Remember/Understand

1. The correct sequence of levels forming the structural hierarchy is:
   (a) organ, organ system, cellular, chemical, tissue, organismal
   (b) chemical, cellular, tissue, organismal, organ, organ system
   (c) chemical, cellular, tissue, organ, organ system, organismal
   (d) organismal, organ system, organ, tissue, cellular, chemical
2. The structural and functional unit of life is (a) a cell, (b) an organ, (c) the organism, (d) a molecule.
3. Which of the following is a *major* functional characteristic of all organisms? (a) movement, (b) growth, (c) metabolism, (d) responsiveness, (e) all of these.
4. Two of these organ systems bear the *major* responsibility for ensuring homeostasis of the internal environment. Which two? (a) nervous system, (b) digestive system, (c) cardiovascular system, (d) endocrine system, (e) reproductive system.
5. In (a)–(e), a directional term [e.g., distal in (a)] is followed by terms indicating different body structures or locations (e.g., the elbow/the wrist). In each case, choose the structure or organ that matches the given directional term.
   (a) distal: the elbow/the wrist
   (b) lateral: the hip bone/the umbilicus
   (c) superior: the nose/the chin
   (d) anterior: the toes/the heel
   (e) superficial: the scalp/the skull
6. Which ventral cavity subdivision has no bony protection?
   (a) thoracic cavity, (b) abdominal cavity, (c) pelvic cavity.
7. Terms that apply to the backside of the body in the anatomical position include:
   (a) ventral; anterior
   (b) back; rear
   (c) posterior; dorsal
   (d) medial; lateral
8. According to the principle of complementarity, how does anatomy relate to physiology?
9. Construct a table that lists the 11 systems of the body, names two organs of each system (if appropriate), and describes the overall or major function of each system.
10. List and describe briefly five external factors that must be present or provided to sustain life.
11. Define homeostasis.
12. Define plane and section.
13. Provide the anatomical term that correctly names each of the following body regions: (a) arm, (b) thigh, (c) chest, (d) fingers and toes, (e) anterior aspect of the knee.
14. (a) Make a diagram showing the nine abdominopelvic regions, and name each region. Name two organs (or parts of organs) that could be located in each of the named regions. (b) Make a similar sketch illustrating how the abdominopelvic cavity may be divided into quadrants, and name each quadrant.

Level 2 Apply/Analyze

15. Assume that the body has been sectioned along three planes: (1) a median plane, (2) a frontal plane, and (3) a transverse plane made at the level of each of the organs listed below. Which organs would be visible in only one or two of these three cases? (a) urinary bladder, (b) brain, (c) lungs, (d) kidneys, (e) small intestine, (f) heart.
16. Relate each of the following conditions or statements to either the dorsal body cavity or the ventral body cavity.
    (a) surrounded by the bony skull and the vertebral column
    (b) includes the thoracic and abdominopelvic cavities
    (c) contains the brain and spinal cord
    (d) contains the heart, lungs, and digestive organs
17. Which of the following relationships is *incorrect*?
    (a) visceral peritoneum/outer surface of small intestine
    (b) parietal pericardium/outer surface of heart
    (c) parietal pleura/wall of thoracic cavity
18. Compare and contrast the operation of negative and positive feedback mechanisms in maintaining homeostasis. Provide two examples of variables controlled by negative feedback mechanisms and one example of a process regulated by a positive feedback mechanism.
19. Why is an understanding of the anatomical position important?
20. Use as many directional terms as you can to describe the relationship between the elbow's olecranal region and your palm.
21. At the clinic, Harry was told that blood would be drawn from his antecubital region. What body part was Harry asked to hold out? Later, the nurse came in and gave Harry a shot of penicillin in the area just distal to his acromial region. Did Harry take off his shirt or drop his pants to receive the injection? Before Harry left, the nurse noticed that Harry had a nasty bruise on his gluteal region. What part of his body was black and blue?
22. Calcium levels in Mr. Gallariani's blood are dropping to dangerously low levels. The hormone PTH is released and soon blood calcium levels begin to rise. Shortly after, PTH release slows. Is this an example of a positive or negative feedback mechanism? What is the initial stimulus? What is the result?
23. Mr. Harvey, a computer programmer, has been complaining of numbness and pain in his right hand. The nurse practitioner diagnosed his problem as carpal tunnel syndrome and prescribed use of a splint. Where will Mr. Harvey apply the splint?

Level 3 Evaluate/Synthesize

24. Aiden has been suffering agonizing pain with each breath and has been informed by the physician that he has pleurisy. (a) Specifically, what membranes are involved in this condition? (b) What is their usual role in the body? (c) Explain why Aiden's condition is so painful.
25. A man is behaving abnormally, and his physician suspects that he has a brain tumor. Which of the following medical imaging techniques would best localize the tumor in the man's brain (and why)? Conventional X ray, DSA, PET, ultrasound, MRI.

2 Chemistry Comes Alive

In this chapter, you will learn that

Chemical reactions underlie all physiological processes

by investigating

Part 1 Basic Chemistry

by differentiating between

2.1 Matter and energy

and looking closer at

2.2 Atoms and elements

then asking

2.3 How is matter combined into molecules and mixtures?

and

2.4 What are the three kinds of chemical bonds?

and

2.5 How do chemical reactions form, rearrange, or break bonds?

Part 2 Biochemistry

by asking

2.6 What is the importance of inorganic compounds to the body?

then examining

Organic compounds

by asking

2.7 How are large organic compounds made and broken down?

and looking closer at

2.12 The energy currency, ATP

and looking closer at

2.8 Carbohydrates **2.9 Lipids** **2.10 Proteins** **2.11 Nucleic acids**

Why study chemistry in an anatomy and physiology course? The answer is simple. Your entire body is made up of chemicals, thousands of them, continuously interacting with one another at an incredible pace. Although it is possible to study anatomy without much reference to chemistry, chemical reactions underlie all physiological processes—movement, digestion, the pumping of your heart, and even your thoughts. This chapter presents the basic chemistry and biochemistry (the chemistry of living material) you need to understand body functions.

CAREER CONNECTION

Watch a video to learn how the chapter content is used in a real health care setting. Go to Mastering A&P® > Study Area > Animations and Videos or use quick access URL https://goo.gl/88srfV

BASIC CHEMISTRY

2.1 Matter is the stuff of the universe and energy moves matter

Learning Outcomes

▶ **Differentiate between matter and energy and between potential energy and kinetic energy.**

▶ **Describe the major energy forms.**

Matter

Matter is the "stuff" of the universe. More precisely, **matter** is anything that occupies space and has mass. With some exceptions, it can be seen, smelled, and felt.

We usually consider mass to be the same as weight. However, this statement is not quite accurate. The *mass* of an object is equal to the actual amount of matter in the object, and it remains constant wherever the object is. In contrast, weight varies with gravity. So while your mass is the same at sea level and on a mountaintop, you weigh just slightly less on that mountaintop. The science of chemistry studies the nature of matter, especially how its building blocks are put together and interact.

States of Matter

Matter exists in *solid*, *liquid*, and *gaseous states*. Examples of each state are found in the human body. Solids, like bones and teeth, have a definite shape and volume. Liquids such as blood plasma have a definite volume, but they conform to the shape of their container. Gases have neither a definite shape nor a definite volume. The air we breathe is a gas.

Energy

Compared with matter, energy is less tangible. It has no mass, does not take up space, and we can measure it only by its effects on matter. **Energy** is defined as the capacity to do work, or to put matter into motion. The greater the work done, the more energy is used doing it. A baseball player who has just hit the ball over the fence uses much more energy than a batter who bunts the ball back to the pitcher.

Kinetic versus Potential Energy

Energy exists in two states, and each can be transformed to the other. **Kinetic energy** (ki-net′ik) is energy in action. We see evidence of kinetic energy in the constant movement of the tiniest particles of matter (atoms) as well as in larger objects (a bouncing ball). Kinetic energy does work by moving objects, which in turn can do work by moving or pushing on other objects. For example, a push on a swinging door sets it into motion.

Potential energy is stored energy, that is, inactive energy that has the *potential*, or capability, to do work but is not

presently doing so. The batteries in an unused toy have potential energy, as does water confined behind a dam. Your leg muscles have potential energy when you sit still on the couch. When potential energy is released, it becomes kinetic energy and so is capable of doing work. For example, dammed water becomes a rushing torrent when the dam is opened, and that rushing torrent can move a turbine at a hydroelectric plant and charge a battery.

Matter and energy are inseparable. Matter is the substance, and energy is the mover of the substance. All living things are composed of matter and they all require energy to grow and function. The release and use of energy by living systems gives us the elusive quality we call life. Now let's consider the forms of energy used by the body as it does its work.

Forms of Energy

- **Chemical energy** is the form stored in the bonds of chemical substances. When chemical reactions occur that rearrange the atoms of the chemicals in a certain way, the potential energy is unleashed and becomes kinetic energy, or energy in action.

  For example, some of the energy in the foods you eat is eventually converted into the kinetic energy of your moving arm. However, food fuels cannot be used to energize body activities directly. Instead, some of the food energy is captured temporarily in the bonds of a chemical called *adenosine triphosphate* (*ATP*; ah-den′o-sēn tri″fos′fāt). Later, ATP's bonds are broken and the stored energy is released as needed to do cellular work. Chemical energy in the form of ATP is the most useful form of energy in living systems because it is used to run almost all functional processes.

- **Electrical energy** results from the movement of charged particles. In your home, electrical energy is found in the flow of electrons along the household wiring. In your body, electrical currents are generated when charged particles (*ions*) move along or across cell membranes. The nervous system uses electrical currents, called *nerve impulses* (or *action potentials*), to transmit messages from one part of the body to another. Electrical currents traveling across the heart stimulate it to contract (beat) and pump blood. (This is why a strong electrical shock, which interferes with such currents, can cause death.)

- **Mechanical energy** is energy *directly* involved in moving matter. When you ride a bicycle, your legs provide the mechanical energy that moves the pedals.

- **Radiant energy**, or **electromagnetic radiation** (e-lek″tro-mag-net′ik), is energy that travels in waves. These waves, which vary in length, are collectively called the *electromagnetic spectrum*. They include radio waves, microwaves, infrared waves, visible light, ultraviolet waves, and X rays. Light energy, which stimulates the retinas of our eyes, is important in vision. Ultraviolet waves cause sunburn, but they also stimulate your body to make vitamin D.

Converting Forms of Energy

With few exceptions, energy is easily converted from one form to another. For example, the chemical energy (in gasoline) that powers the motor of a speedboat is converted into the mechanical energy of the whirling propeller that makes the boat skim across the water.

Energy conversions are quite inefficient. Some of the initial energy supply is always "lost" to the environment as heat. It is not really lost because energy cannot be created or destroyed, but that portion given off as heat is at least partly *unusable*. It is easy to demonstrate this principle. Electrical energy is converted into light energy in a lightbulb. But if you touch a lit bulb, you will soon discover that some of the electrical energy is producing heat instead.

Likewise, all energy conversions in the body liberate heat. This heat helps to maintain our relatively high body temperature, which influences body functioning. For example, when matter is heated, the kinetic energy of its particles increases and they begin to move more quickly. The higher the temperature, the faster the body's chemical reactions occur. We will learn more about this later.

Check Your Understanding

1. What form of energy is found in the food we eat?
2. What form of energy is used to transmit messages from one part of the body to another?
3. What type of energy is available when we are still? When we are exercising?

For answers, see Answers Appendix.

2.2 The properties of an element depend on the structure of its atoms

Learning Outcomes

▶ **Define chemical element and list the four elements that form the bulk of body matter.**
▶ **Define atom. List the subatomic particles, and describe their relative masses, charges, and positions in the atom.**
▶ **Define atomic number, atomic mass, atomic weight, isotope, and radioisotope.**

All matter is composed of **elements**, unique substances that cannot be broken down into simpler substances by ordinary chemical methods. Among the well-known elements are oxygen, carbon, gold, silver, copper, and iron.

At present, 118 elements are recognized. Of these, 92 occur in nature. The rest are made artificially in particle accelerator devices.

Four elements—carbon, oxygen, hydrogen, and nitrogen—make up about 96% of body weight, and 20 others are present in the body, some in trace amounts. **Table 2.1** on p. 26 lists those of importance to the body. An oddly shaped checkerboard called the **periodic table** (see Appendix E) provides a listing of the known elements and helps to explain the properties of each element.

Each element is composed of essentially identical particles or building blocks, called **atoms**. The smallest atoms are less than 0.1 nanometer (nm) in diameter, and the largest are only about five times as large. [1 nm = 0.000000001 (or 10^{-9}) meter (m), or 40 billionths of an inch!]

Every element's atoms differ from those of all other elements and give the element its unique physical and chemical properties. *Physical properties* are those we can detect with our senses (such as color and texture) or measure (such as boiling point and freezing point). *Chemical properties* describe the way atoms interact with other atoms (bonding behavior) and account for the facts that iron rusts, animals can digest their food, and so on.

We designate each element by a one- or two-letter chemical shorthand called an **atomic symbol**, usually the first letter(s) of the element's name. For example, C stands for carbon, O for oxygen, and Ca for calcium. In a few cases, the atomic symbol is taken from the Latin name for the element. For example, sodium is indicated by Na, from the Latin word *natrium*.

Structure of Atoms

The word *atom* comes from the Greek word meaning "indivisible." However, we now know that atoms are clusters of even smaller particles called protons, neutrons, and electrons and that even those subatomic particles can be subdivided. Still, the old idea of atomic indivisibility is useful because an atom loses the unique properties of its element when it is split into its subatomic particles.

An atom's subatomic particles differ in mass, electrical charge, and position in the atom. An atom has a central **nucleus** containing protons and neutrons tightly bound together. The nucleus, in turn, is surrounded by orbiting electrons (**Figure 2.1**). **Protons** (p^+) bear a positive electrical charge, and **neutrons** (n^0) are neutral, so the nucleus is positively charged overall. Protons and neutrons are heavy particles and have approximately the same mass, arbitrarily designated as 1 **atomic mass unit** (1 amu). Since all of the heavy subatomic

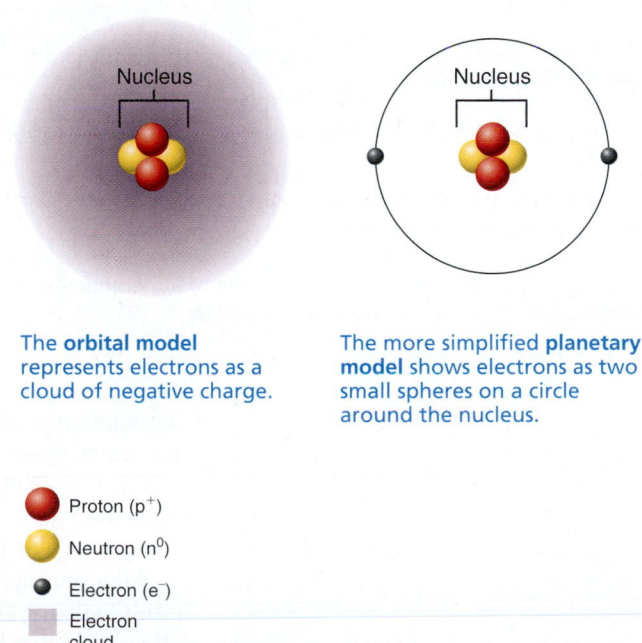

The **orbital model** represents electrons as a cloud of negative charge.

The more simplified **planetary model** shows electrons as two small spheres on a circle around the nucleus.

● Proton (p^+)
● Neutron (n^0)
● Electron (e^-)
▪ Electron cloud

Figure 2.1 Two models of the structure of a helium atom. Helium has 2 protons, 2 neutrons, and 2 electrons ($2p^+$; $2n^0$; $2e^-$).

Table 2.1 Common Elements Composing the Human Body*

ELEMENT	ATOMIC SYMBOL	APPROX. % BODY MASS†	FUNCTIONS
Major (96.1%)			
Oxygen	O	65.0	A component of both organic (carbon-containing) and inorganic (non-carbon-containing) molecules. As a gas, it is needed for the production of cellular energy (ATP).
Carbon	C	18.5	A component of all organic molecules, which include carbohydrates, lipids (fats and oils), proteins, and nucleic acids.
Hydrogen	H	9.5	A component of all organic molecules. As an ion (proton), it influences the pH of body fluids.
Nitrogen	N	3.2	A component of proteins and nucleic acids (genetic material).
Lesser (3.9%)			
Calcium	Ca	1.5	Found as a salt in bones and teeth. Its ionic (Ca^{2+}) form is required for muscle contraction, conduction of nerve impulses, and blood clotting.
Phosphorus	P	1.0	Part of calcium phosphate salts in bones and teeth. Also present in nucleic acids, and as part of ATP and phospholipids.
Potassium	K	0.4	Its ion (K^+) is the major positive ion (cation) in cells. Necessary for conduction of nerve impulses and muscle contraction.
Sulfur	S	0.3	Component of proteins, particularly muscle proteins.
Sodium	Na	0.2	As an ion (Na^+), sodium is the major positive ion found in extracellular fluids (fluids outside of cells). Important for water balance, conduction of nerve impulses, and muscle contraction.
Chlorine	Cl	0.2	Its ion (chloride, Cl^-) is the most abundant negative ion (anion) in extracellular fluids.
Magnesium	Mg	0.1	Present in bone. Also an important cofactor in a number of metabolic reactions.
Iodine	I	0.1	Needed to make functional thyroid hormones.
Iron	Fe	0.1	Component of hemoglobin (which transports oxygen within red blood cells) and some enzymes.
Trace (less than 0.01%)			

Chromium (Cr); cobalt (Co); copper (Cu); fluorine (F); manganese (Mn); molybdenum (Mo); selenium (Se); silicon (Si); tin (Sn); vanadium (V); zinc (Zn)

These elements are referred to as *trace elements* because they are required in very minute amounts; many are found as part of enzymes or are required for enzyme activation.

*A listing of the elements by ascending order of atomic number appears in the periodic table (Appendix E).
†Percentage of "wet" body mass; includes water.

particles are concentrated in the nucleus, the nucleus is fantastically dense, accounting for nearly the entire mass (99.9%) of the atom.

The tiny **electrons** (e^-) bear a negative charge equal in strength to the positive charge of the proton. However, an electron has only about 1/2000 the mass of a proton, and the mass of an electron is usually designated as 0 amu.

All atoms are electrically neutral because the number of protons in an atom is precisely balanced by its number of electrons (the + and − charges will then cancel the effect of each other). For example, hydrogen has one proton and one electron, and iron has 26 protons and 26 electrons. For any atom, the number of protons and electrons is always equal.

The **planetary model** of the atom is a simplified model of atomic structure (Figure 2.1, right). As you can see, it depicts electrons moving around the nucleus in fixed, generally circular orbits. But we can never determine the exact location of electrons at a particular time because they jump around following unknown trajectories. So, instead of speaking of specific orbits, chemists talk about **orbitals**—regions around the nucleus in which a given electron or electron pair is likely to be found most of the time. This more modern **orbital model** is more useful for predicting the chemical behavior of atoms (Figure 2.1, left). The orbital model depicts *probable* regions of greatest electron density by denser shading (this haze is called the *electron cloud*). However, the planetary model is simpler to depict, so we will use that model in most illustrations of atomic structure in this text.

Hydrogen, with just one proton and one electron, is the simplest atom. You can visualize the spatial relationships in the hydrogen atom by imagining it as a sphere enlarged until its diameter equals the length of a football field. In that case, the nucleus could be represented by a lead ball the size of a gumdrop in the exact center of the sphere. Its lone electron could be pictured as a fly buzzing about unpredictably within the sphere. Though not completely accurate, this mental image demonstrates that most of the volume of an atom is empty space, and nearly all of its mass is concentrated in the central nucleus.

Identifying Elements

All protons are alike, regardless of the atom considered. The same is true of all neutrons and all electrons. So what determines the unique properties of each element? The answer is that atoms of different elements are composed of *different numbers* of protons, neutrons, and electrons and this determines the chemical and physical properties of each element.

The simplest and smallest atom, hydrogen, has 1 proton, 1 orbiting electron, and no neutrons (**Figure 2.2**). Next in size is the helium atom, with 2 protons, 2 neutrons, and 2 electrons. Lithium follows with 3 protons, 4 neutrons, and 3 electrons. If we continued this step-by-step progression, we would get a graded series of atoms containing from 1 to 118 protons, an equal number of electrons, and a slightly larger number of neutrons at each step.

All we really need to know about a particular element, however, are its atomic number, mass number, and atomic weight. Taken together, these provide a fairly complete picture of each element.

Atomic Number

The **atomic number** of any atom is equal to the number of protons in its nucleus and is written as a subscript to the left of its atomic symbol. Hydrogen, with one proton, has an atomic number of 1 ($_1$H). Helium, with two protons, has an atomic number of 2 ($_2$He), and so on. The number of protons is always equal to the number of electrons in an atom, so the atomic number *indirectly* tells us the number of electrons in the atom as well. As we will see shortly, this information is important indeed, because electrons determine the chemical behavior of atoms.

Mass Number and Isotopes

The **mass number** of an atom is the sum of the masses of its protons and neutrons. The mass of the electrons is so small that it is ignored. Recall that protons and neutrons have a mass of 1 amu. Hydrogen has only one proton in its nucleus, so its atomic and mass numbers are the same: 1. Helium, with 2 protons and 2 neutrons, has a mass number of 4.

The mass number is usually indicated by a superscript to the left of the atomic symbol. For example, helium is $_2^4$He. This simple notation allows us to deduce the total number and kinds of subatomic particles in any atom because it indicates the number of protons (the atomic number), the number of electrons (equal to the atomic number), and the number of neutrons (mass number minus atomic number). In our example, we can do the subtraction to find that $_2^4$He has two neutrons.

From what we have said so far, it may appear as if each element has one, and only one, type of atom representing it. This is not the case. Nearly all known elements have two or more structural variations called **isotopes** (i'so-tōps; *iso* = same; *topos* = place). Isotopes of an element have the same number of protons and electrons (and so have the same chemical properties), but they differ in the number of neutrons they contain. Earlier, when we said that hydrogen has a mass number of 1, we were speaking of ^{1}H, its most abundant isotope. A few hydrogen atoms have a mass of 2 or 3 amu (atomic mass units),

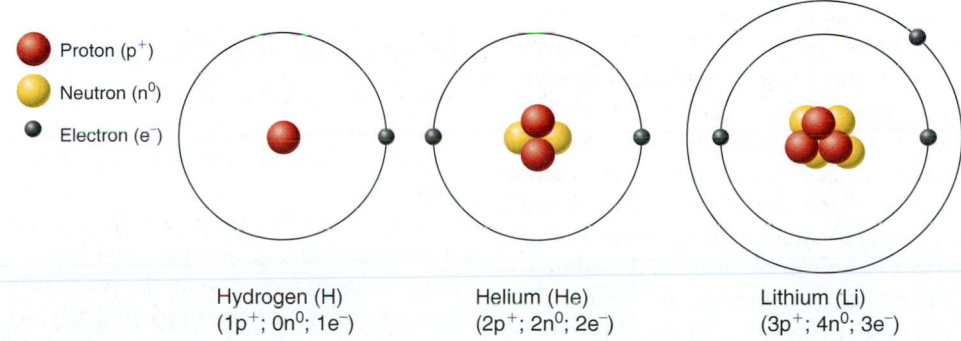

Proton (p$^+$)
Neutron (n^0)
Electron (e$^-$)

Hydrogen (H)	Helium (He)	Lithium (Li)
(1p$^+$; 0n^0; 1e$^-$)	(2p$^+$; 2n^0; 2e$^-$)	(3p$^+$; 4n^0; 3e$^-$)

Figure 2.2 Atomic structure of the three smallest atoms.

Figure 2.3 **Isotopes of hydrogen.**

Hydrogen (^{1}H)
(0 neutrons)

Deuterium (^{2}H)
(1 neutron)

Tritium (^{3}H)
(2 neutrons)

which means that they have 1 proton and, respectively, 1 or 2 neutrons (**Figure 2.3**).

Carbon has several isotopes. The most abundant of these are ^{12}C, ^{13}C, and ^{14}C. Each of the carbon isotopes has six protons (otherwise it would not be carbon), but ^{12}C has six neutrons, ^{13}C has seven, and ^{14}C has eight. Isotopes can also be written with the mass number following the symbol: C-14, for example.

Atomic Weight

You might think that atomic weight should be the same as atomic mass, and this would be so if atomic weight referred to the weight of a single atom. However, **atomic weight** is an average of the weights (mass numbers) of *all* the isotopes of an element, taking into account their relative abundance in nature. As a rule, the atomic weight of an element is approximately equal to the mass number of its most abundant isotope. For example, the atomic weight of hydrogen is 1.008, which reveals that its lightest isotope (^{1}H) is present in much greater amounts in our world than its ^{2}H or ^{3}H forms.

Radioisotopes

The heavier isotopes of many elements are unstable, and their atoms decompose spontaneously into more stable forms. This process of atomic decay is called *radioactivity*, and isotopes that exhibit this behavior are called **radioisotopes** (ra″de-o-i′so-tōps). The disintegration of a radioactive nucleus may be compared to a tiny explosion. It occurs when subatomic *alpha* (α) *particles* (packets of 2p + 2n), *beta* (β) *particles* (electron-like particles), or *gamma* (γ) *rays* (electromagnetic energy) are ejected from the atomic nucleus.

Why does this happen? The answer is complex, but for our purposes, the important point is that the dense nuclear particles are composed of even smaller particles called *quarks* that associate in one way to form protons and in another way to form neutrons. The "glue" that holds these nuclear particles together is weaker in the heavier isotopes. When radioisotopes disintegrate, the element usually transforms to a different element.

Because we can detect radioactivity with scanners, and radioactive isotopes share the same chemistry as their more stable isotopes, radioisotopes are valuable tools for biological research and medicine. Most radioisotopes used in the clinical setting are used for diagnosis, that is, to localize and illuminate damaged or cancerous tissues. For example, iodine-123 is used to determine the size and activity of the thyroid gland and to

detect thyroid cancer. PET scans (described in *A Closer Look* in Chapter 1, pp. 16–17) use radioisotopes to probe the workings of molecules deep within our bodies. All radioisotopes, regardless of the purpose for which they are used, damage living tissue, and they all gradually lose their radioactive behavior. The time required for a radioisotope to lose one-half of its activity is called its *half-life*. The half-lives of radioisotopes vary dramatically from hours to thousands of years.

Alpha emission is easily blocked outside the body, but if absorbed, it causes considerable damage. For this reason, alpha particles from decaying inhaled radon gas are second only to smoking as a cause of lung cancer. (Radon, a gas, results naturally from decay of uranium in the ground.) Gamma emission has the greatest penetrating power. Radium-226, cobalt-60, and certain other radioisotopes that decay by gamma emission are used to destroy localized cancers.

Contrary to what some believe, ionizing radiation does not damage organic molecules directly. Instead, it knocks electrons out of other atoms and sends them flying, like bowling balls smashing through pins all along their path. It is the electron's energy and the unstable molecules left behind that do the damage.

Check Your Understanding

4. What two elements besides H and N make up the bulk of living matter?

5. DRAW Draw a planetary model of an atom with a mass number of 7 and an atomic number of 3.

6. WHAT IF Look at the function of iron in Table 2.1 on p. 26. What would be the effect of an iron deficiency on blood?

For answers, see Answers Appendix.

2.3 Atoms bound together form molecules; different molecules can make mixtures

Learning Outcomes

▶ **Define molecule, and distinguish between a compound and a mixture.**

▶ **Compare solutions, colloids, and suspensions.**

Molecules and Compounds

Most atoms do not exist in the free state, but instead are chemically combined with other atoms. Such a combination of two or more atoms held together by chemical bonds is called a **molecule**.

If two or more atoms of the *same* element combine, the resulting substance is called a *molecule of that element*. When two hydrogen atoms bond, the product is a molecule of hydrogen gas and is written as H_2. Similarly, when two oxygen atoms combine, a molecule of oxygen gas (O_2) is formed. Sulfur atoms commonly combine to form sulfur molecules containing eight sulfur atoms (S_8).

When two or more *different* kinds of atoms bind, they form molecules of a **compound**. Two hydrogen atoms combine with one oxygen atom to form the compound water (H_2O). Four hydrogen atoms combine with one carbon atom to form the compound methane (CH_4). Notice again that molecules of methane and water are compounds, but molecules of hydrogen gas are not, because compounds always contain atoms of at least two different elements.

Compounds are chemically pure, and all of their molecules are identical. So, just as an atom is the smallest particle of an element that still has the properties of the element, a molecule is the smallest particle of a compound that still has the specific characteristics of the compound. This concept is important because the properties of compounds are usually very different from those of the atoms they contain. Water, for example, is very different from the elements hydrogen and oxygen. Indeed, it is next to impossible to tell what atoms are in a compound without analyzing it chemically.

Mixtures

Mixtures are substances composed of two or more components *physically intermixed*. Most matter in nature exists in the form of mixtures, but there are only three basic types: *solutions*, *colloids*, and *suspensions* (**Figure 2.4**).

Solutions

Solutions are homogeneous mixtures of components that may be gases, liquids, or solids. *Homogeneous* means that the mixture has exactly the same composition or makeup throughout—a sample taken from any part of the mixture has the same composition (in terms of the atoms or molecules it contains) as a sample taken from any other part of the mixture. Examples include the air we breathe (a mixture of gases) and seawater (a mixture of salts, which are solids, and water). The substance present in the greatest amount is called the **solvent** (or dissolving medium). Solvents are usually liquids. Substances present in smaller amounts (dissolved in the solvent) are called **solutes**.

Water is the body's chief solvent. Most solutions in the body are *true solutions* containing gases, liquids, or solids dissolved in water. True solutions are usually transparent. Examples are saline solution [table salt (NaCl) and water], a mixture of glucose and water, and mineral water. The solutes of true solutions are very small, usually in the form of individual atoms and molecules. Consequently, they are not visible to the naked eye, do not settle out, and do not scatter light. In other words, if a beam of light is passed through a true solution, you will not see the path of light.

Concentration of Solutions We describe true solutions in terms of their *concentration*, which may be indicated in various ways. Solutions used in a college laboratory or a hospital are often described in terms of the **percent** (parts per 100 parts) of

Solution	Colloid	Suspension
Solute particles are very tiny, do not settle out or scatter light.	Solute particles are larger than in a solution and scatter light; do not settle out.	Solute particles are very large, settle out, and may scatter light.
Solute particles	Solute particles	Solute particles
Example Mineral water	*Example* Jell-O	*Example* Blood
		Plasma / Settled red blood cells / Unsettled Settled

Figure 2.4 The three basic types of mixtures.

the solute in the total solution. This designation always refers to the solute percentage, and unless otherwise noted, water is assumed to be the solvent.

Milligrams per deciliter (mg/dl) is a concentration measurement commonly used to measure the blood concentration of glucose, cholesterol, and so on. (A deciliter is 100 milliliters or 0.1 liter.)

Still another way to express the concentration of a solution is in terms of its **molarity** (mo-lar'ĭ-te), or moles per liter, indicated by *M*. This method is more complicated but much more useful. To understand molarity, you must understand what a mole is. A **mole** of any element or compound is equal to its atomic weight or **molecular weight** (sum of the atomic weights) in grams. This concept is easier than it seems, as illustrated by the following example.

Glucose is $C_6H_{12}O_6$, which indicates that it has 6 carbon atoms, 12 hydrogen atoms, and 6 oxygen atoms. To compute the molecular weight of glucose, you would look up the atomic weight of each of its atoms in the periodic table (see Appendix E) and compute its molecular weight as follows:

Atom	Number of Atoms		Atomic Weight		Total Atomic Weight
C	6	×	12.011	=	72.066
H	12	×	1.008	=	12.096
O	6	×	15.999	=	95.994
					180.156

Then, to make a *one-molar* solution of glucose, you would weigh out 180.156 grams (g) of glucose and add enough water to make 1 liter (L) of solution. In short, a one-molar solution (abbreviated 1.0 *M*) of a chemical substance is the molecular weight of the substance in grams in 1 L (1000 milliliters) of solution.

The beauty of using the mole as the basis of preparing solutions is its precision. One mole of any substance always contains exactly the same number of solute particles, that is, 6.02×10^{23}. This number is called **Avogadro's number** (av″o-gad′rōz). So whether you weigh out 1 mole of glucose (180 g) or 1 mole of water (18 g) or 1 mole of methane (16 g), in each case you will have 6.02×10^{23} molecules of that substance.* This allows almost mind-boggling precision to be achieved.

Because solute concentrations in body fluids tend to be quite low, those values are usually reported in terms of millimoles (m*M*; 1/1000 mole).

Colloids

Colloids (kol′oidz), also called *emulsions*, are *heterogeneous* mixtures, which means that their composition is dissimilar in different areas of the mixture. Colloids often appear translucent or milky and although the solute particles are larger than those in true solutions, they still do not settle out. However, they do scatter light, so the path of a light beam shining through a colloidal mixture is visible.

Colloids have many unique properties, including the ability of some to undergo **sol-gel transformations**, that is, to change reversibly from a fluid (sol) state to a more solid (gel) state. Jell-O, or any gelatin product (Figure 2.4), is a familiar example of a nonliving colloid that changes from a sol to a gel when refrigerated (and that gel will liquefy again if placed in the sun). Cytosol, the semifluid material in living cells, is also a colloid, largely because of its dispersed proteins. Its sol-gel transformations underlie many important cell activities, such as cell division and changes in cell shape.

Suspensions

Suspensions are *heterogeneous* mixtures with large, often visible solutes that tend to settle out. An example of a suspension is a mixture of sand and water. So is blood, in which the living blood cells are suspended in the fluid portion of blood (blood plasma). If left to stand, the suspended cells will settle out unless some means—mixing, shaking, or circulation in the body—keeps them in suspension.

All three types of mixtures are found in both living and nonliving systems. In fact, living material is the most complex mixture of all, since it contains all three kinds of mixtures interacting with one another.

Distinguishing Mixtures from Compounds

Now let's zero in on how to distinguish mixtures and compounds from one another. Mixtures differ from compounds in several important ways:

- The chief difference between mixtures and compounds is that no chemical bonding occurs between the components of a mixture. The properties of atoms and molecules are not changed when they become part of a mixture. Remember they are only physically intermixed.

- Depending on the mixture, its components can be separated by physical means—straining, filtering, evaporation, and so on. Compounds, by contrast, can be separated into their constituent atoms only by chemical means (breaking bonds).

- Some mixtures are homogeneous, whereas others are heterogeneous. A bar of brass is a homogeneous mixture of copper and zinc. However, when brass shavings are swept up from the workshop floor, what you have collected is a heterogeneous mixture of dirt, dust, and metal.

Check Your Understanding

7. What is the meaning of the term "molecule"?
8. Why is sodium chloride (NaCl) considered a compound, but oxygen gas is not?
9. Blood contains a liquid component and living cells. Would it be classified as a compound or a mixture? Why?

                                         For answers, see Answers Appendix.

---

* The important exception to this rule concerns molecules that ionize and break up into charged particles (ions) in water, such as salts, acids, and bases (see p. 39). For example, simple table salt (sodium chloride) breaks up into two types of charged particles. Therefore, in a 1.0 *M* solution of sodium chloride, 2 *moles* of solute particles are actually in solution.

2.4 Three types of chemical bonds are ionic, covalent, and hydrogen

Learning Outcomes

▶ Explain the role of electrons in chemical bonding and in relation to the octet rule.

▶ Differentiate among ionic, covalent, and hydrogen bonds.

▶ Compare and contrast polar and nonpolar compounds.

As noted earlier, when atoms combine with other atoms, they are held together by **chemical bonds**. A chemical bond is not a physical structure like a pair of handcuffs linking two people together. Instead, it is an energy relationship between the electrons of the reacting atoms, and it is made or broken in less than a trillionth of a second.

The Role of Electrons in Chemical Bonding

Electrons forming the electron cloud around the nucleus of an atom occupy regions of space called **electron shells** that consecutively surround the atomic nucleus. The atoms known so far can have electrons in seven shells (numbered 1 to 7 from the nucleus outward), but the actual number of electron shells occupied in a given atom depends on the number of electrons the atom has. Each electron shell contains one or more orbitals. (Recall that *orbitals* are regions around the nucleus in which a given electron is likely to be found most of the time.)

It is important to understand that each electron shell represents a different **energy level**, because this prompts you to think of electrons as particles with a certain amount of potential energy. In general, the terms *electron shell* and *energy level* are used interchangeably.

How much potential energy does an electron have? The answer depends on the energy level that it occupies. The attraction between the positively charged nucleus and negatively charged electrons is greatest when electrons are closest to the nucleus and falls off with increasing distance. This statement explains why electrons farthest from the nucleus (1) have the greatest potential energy (the energy they absorbed to overcome the nuclear attraction and reach the more distant energy levels) and (2) are most likely to interact chemically with other atoms. (They are the least tightly held by their own nucleus and the most easily influenced by other atoms and molecules.)

Each electron shell can hold a specific number of electrons. Shell 1, the shell immediately surrounding the nucleus, accommodates only 2 electrons. Shell 2 holds a maximum of 8, and shell 3 has room for 18. Subsequent shells hold larger and larger numbers of electrons, and the shells tend to be filled with electrons consecutively. For example, shell 1 fills completely before any electrons appear in shell 2.

Which electrons are involved in chemical bonding? When we consider bonding behavior, the only electrons that are important are those in the atom's outermost energy level. Inner electrons usually do not take part in bonding because they are more tightly held by the nucleus.

When the outermost energy level of an atom is filled to capacity or contains 8 electrons, the atom is stable. Such atoms

are *chemically inert*, that is, unreactive. A group of elements called the *noble gases*, which include helium and neon, typify this condition (**Figure 2.5a**). On the other hand, atoms in which the outermost energy level contains fewer than 8 electrons tend to gain, lose, or share electrons with other atoms to achieve stability (Figure 2.5b).

What about atoms that have more than 20 electrons, in which the energy levels beyond shell 2 can contain *more* than 8 electrons? The number of electrons that can participate in bonding is still limited to a total of 8. The term **valence shell** (va'lens) refers to an atom's outermost energy level *or that portion of it* containing the electrons that are chemically reactive. The key to chemical reactivity is the **octet rule** (ok-tet'), or **rule of eights**. Except for shell 1, which is full when it has 2 electrons, atoms

(a) Chemically inert elements

Outermost energy level (valence shell) complete

Helium (He)
($2p^+$; $2n^0$; $2e^-$)

Neon (Ne)
($10p^+$; $10n^0$; $10e^-$)

(b) Chemically reactive elements

Outermost energy level (valence shell) incomplete

Hydrogen (H)
($1p^+$; $0n^0$; $1e^-$)

Carbon (C)
($6p^+$; $6n^0$; $6e^-$)

Oxygen (O)
($8p^+$; $8n^0$; $8e^-$)

Sodium (Na)
($11p^+$; $12n^0$; $11e^-$)

Figure 2.5 Chemically inert and reactive elements. (*Note:* For simplicity, each atomic nucleus is shown as a sphere with the atom's symbol; individual protons and neutrons are not shown.)

2

The lone outer electron of sodium joins the seven outer electrons of chlorine.

The result is a cation (Na⁺) and an anion (Cl⁻), which can form ionic bonds.

Sodium atom (Na)
(1e⁻ in outer shell)

Chlorine atom (Cl)
(7e⁻ in outer shell)

(a)

Sodium ion (Na⁺)   Chloride ion (Cl⁻)

Sodium chloride (NaCl)

(b)

Large numbers of Na⁺ and Cl⁻ ions associate to form salt (NaCl) crystals.

Cl⁻

Na⁺

(c)

Figure 2.6 Formation of ionic bonds.

tend to interact in such a way that they end up having *8 electrons in their valence shell.*

Types of Chemical Bonds

Three major types of chemical bonds—*ionic, covalent,* and *hydrogen bonds*—result from attractive forces between atoms. The key features of these three bond types are summarized in **Table 2.2** on p. 35.

Ionic Bonds

Recall that atoms are electrically neutral. However, electrons can be transferred from one atom to another, and when this happens, the precise balance of $+$ and $-$ charges is lost so that charged particles called **ions** are formed. An **ionic bond** (i-on′ik) is a chemical bond between atoms formed by the transfer of one or more electrons from one atom to the other. The atom that gains one or more electrons is the *electron acceptor*. It acquires a net negative charge and is called an **anion** (an′i-on). The atom that loses electrons is the *electron donor*. It acquires a net positive charge and is called a **cation** (kat′i-on). (To remember this term, think of the "t" in "cation" as a $+$ sign.) Both anions and cations are formed whenever electron transfer between atoms occurs. Because opposite charges attract, these ions tend to stay close together, resulting in an ionic bond.

One example of ionic bonding is the formation of table salt, or sodium chloride (NaCl), by interaction of sodium and chlorine atoms (**Figure 2.6**). Sodium, with an atomic number of 11, has only 1 electron in its valence shell. It would be very difficult to attempt to fill this shell by adding 7 more. However, if this single electron is lost, shell 2 with 8 electrons becomes the valence shell (outermost energy level containing electrons) and is full. By losing the lone electron in its third energy level, sodium achieves stability and becomes a cation (Na⁺). On the other hand, chlorine (atomic number 17) needs only 1 electron to fill its valence shell. By accepting an electron, chlorine achieves stability and becomes an anion.

When sodium and chlorine atoms interact, this is exactly what happens. Sodium donates an electron to chlorine (Figure 2.6a), and the oppositely charged ions created in this exchange attract each other, forming sodium chloride (Figure 2.6b). Ionic bonds are commonly formed between atoms with 1 or 2 valence shell electrons (the metallic elements, such as sodium, calcium, and potassium) and atoms with 7 valence shell electrons (such as chlorine, fluorine, and iodine).

Most ionic compounds fall in the chemical category called *salts*. In the dry state, salts such as sodium chloride do not exist as individual molecules. Instead, they form **crystals**, large arrays of cations and anions held together by ionic bonds (Figure 2.6c).

Sodium chloride is an excellent example of the difference in properties between a compound and its constituent atoms. Sodium is a silvery white metal, and chlorine in its molecular state is a poisonous green gas used to make bleach. However, sodium chloride is a white crystalline solid that we sprinkle on our food.

Covalent Bonds

Electrons do not have to be completely transferred for atoms to achieve stability. Instead, they may be *shared* so that each atom is able to fill its outer electron shell at least part of the time. Electron sharing produces molecules in which the shared electrons occupy a single orbital common to both atoms, which constitutes a **covalent bond** (ko-va′lent).

Hydrogen with its single electron can fill its only shell (shell 1) by sharing a pair of electrons with another atom. When it shares with another hydrogen atom, a molecule of hydrogen

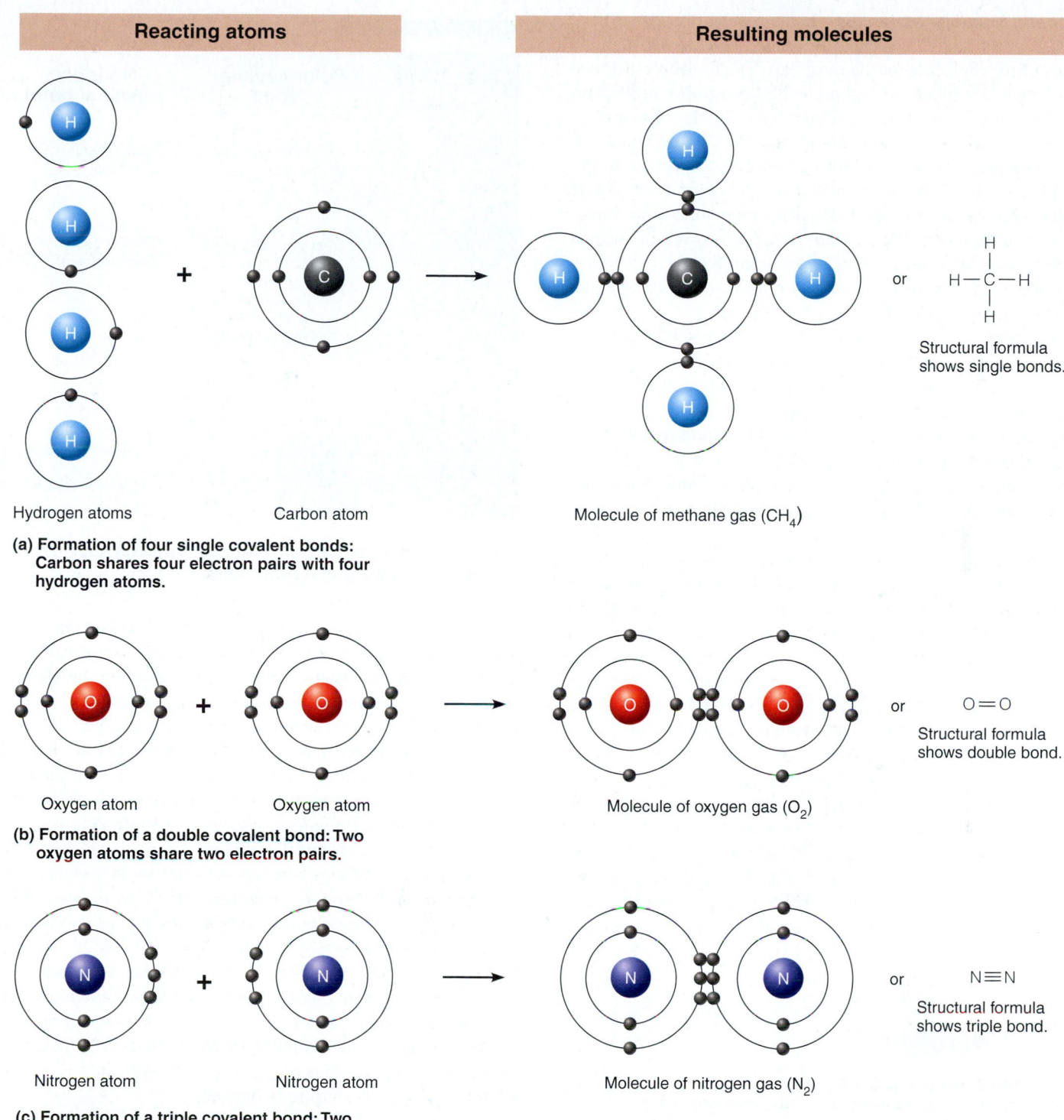

Reacting atoms	Resulting molecules

Hydrogen atoms   Carbon atom   Molecule of methane gas (CH_4)

Structural formula shows single bonds.

(a) Formation of four single covalent bonds:
Carbon shares four electron pairs with four
hydrogen atoms.

Oxygen atom   Oxygen atom   Molecule of oxygen gas (O_2)

Structural formula shows double bond.

(b) Formation of a double covalent bond: Two
oxygen atoms share two electron pairs.

Nitrogen atom   Nitrogen atom   Molecule of nitrogen gas (N_2)

Structural formula shows triple bond.

(c) Formation of a triple covalent bond: Two
nitrogen atoms share three electron pairs.

Figure 2.7 Formation of covalent bonds.

gas is formed. The shared electron pair orbits around the molecule as a whole, satisfying the stability needs of each atom.

Hydrogen can also share an electron pair with different kinds of atoms to form a compound (**Figure 2.7a**). Carbon has 4 electrons in its outermost shell, but needs 8 to achieve stability. Hydrogen has 1 electron, but needs 2. When a methane molecule (CH_4) is formed, carbon shares 4 pairs of electrons with 4 hydrogen atoms (1 pair with each hydrogen). Again, the shared

electrons orbit and "belong to" the whole molecule, ensuring the stability of each atom.

When 2 atoms share 1 pair of electrons, a single covalent bond is formed (indicated by a single line connecting the atoms, such as H—H). In some cases, atoms share two or three electron pairs, resulting in *double* or *triple* covalent bonds (Figure 2.7b and c). (These bonds are indicated by double or triple connecting lines such as O=O or N≡N.)

Polar and Nonpolar Molecules   In the covalent bonds we have discussed, the shared electrons are shared equally between the atoms of the molecule for the most part. The molecules formed are electrically balanced and are called **nonpolar molecules** (because they do not have separate + and − poles of charge).

Such electrical balance is not always the case. When covalent bonds are formed, the resulting molecule always has a specific three-dimensional shape, with the bonds formed at definite angles. A molecule's shape helps determine what other molecules or atoms it can interact with. It may also result in unequal electron pair sharing, creating a **polar molecule**, especially in nonsymmetrical molecules containing atoms with different electron-attracting abilities.

In general, *small* atoms with 6 or 7 valence shell electrons, such as oxygen, nitrogen, and chlorine, are electron-hungry and attract electrons very strongly, a capability called **electronegativity**. On the other hand, most atoms with only one or two valence shell electrons tend to be **electropositive**. In other words, their electron-attracting ability is so low that they usually lose *their* valence shell electrons to other atoms. Potassium and sodium, each with one valence shell electron, are good examples of electropositive atoms.

Carbon dioxide and water illustrate how molecular shape and the relative electron-attracting abilities of atoms determine whether a covalently bonded molecule is nonpolar or polar. In carbon dioxide (CO_2), carbon shares four electron pairs with two oxygen atoms (two pairs are shared with each oxygen). Oxygen is very electronegative and so attracts the shared electrons much more strongly than does carbon. However, because the carbon dioxide molecule is linear and symmetrical (**Figure 2.8a**), the electron-pulling ability of one oxygen atom offsets that of the other, like a standoff between equally strong teams in a game of tug-of-war. As a result, the shared electrons orbit the entire molecule and carbon dioxide is a nonpolar compound.

In contrast, a water molecule (H_2O) is bent, or V shaped (Figure 2.8b). The two electropositive hydrogen atoms are located at the same end of the molecule, and the very

Ionic bond	Polar covalent bond	Nonpolar covalent bond
Complete transfer of electrons	Unequal sharing of electrons	Equal sharing of electrons
Separate ions (charged particles) form	Charge unbalanced among atoms [molecule has slightly positive (δ^+) and slightly negative (δ^-) ends]	Charge balanced among atoms
Na^+ Cl^-	H—O—H δ^+ δ^+ δ^-	$O{=}C{=}O$
Sodium chloride	Water	Carbon dioxide

Figure 2.9 Ionic, polar covalent, and nonpolar covalent bonds compared along a continuum.

electronegative oxygen is at the opposite end. This arrangement allows oxygen to pull the shared electrons toward itself and away from the two hydrogen atoms. In this case, the electron pairs are *not* shared equally, but spend more time in the vicinity of oxygen. Because electrons are negatively charged, the oxygen end of the molecule is slightly more negative (the charge is indicated with a delta and minus as δ^-) and the hydrogen end slightly more positive (indicated by δ^+). Because water has two poles of charge, it is a *polar molecule*, or **dipole** (di'pōl).

Polar molecules orient themselves toward other dipoles or toward charged particles (such as ions and some proteins), and they play essential roles in chemical reactions in body cells. The polarity of water is particularly significant, as you will see later in this chapter.

Different molecules exhibit different degrees of polarity, and we can see a gradual change from ionic to nonpolar covalent bonding as summarized in **Figure 2.9**. Ionic bonds (complete electron transfer) and nonpolar covalent bonds (equal electron sharing) are the extremes of a continuum, with various degrees of unequal electron sharing in between.

Hydrogen Bonds

Unlike the stronger ionic and covalent bonds, hydrogen bonds are more like attractions than true bonds. Hydrogen bonds form when a hydrogen atom, already covalently linked to one electronegative atom (usually nitrogen or oxygen), is attracted by another electron-hungry atom, so that a "bridge" forms between them.

Hydrogen bonding is common between dipoles (such as water molecules) because the slightly negative oxygen atoms of one molecule attract the slightly positive hydrogen atoms of other

(a) Carbon dioxide (CO_2) molecules are linear and symmetrical. They are nonpolar.

(b) V-shaped water (H_2O) molecules have two poles of charge—a slightly more negative oxygen end (δ^-) and a slightly more positive hydrogen end (δ^+).

Figure 2.8 Carbon dioxide and water molecules have different shapes, as illustrated by molecular models.

Table 2.2	Summary: Major Chemical Bond Types	
TYPE	**DESCRIPTION**	**STRENGTH**
Covalent bonds	Sharing of pairs of electrons. May be polar (not equally shared) or nonpolar (equally shared).	Strongest
Ionic bonds	Attraction between two oppositely charged ions.	Intermediate
Hydrogen bonds	Attraction between a hydrogen atom carrying a partial positive charge (δ^+) and an electronegative atom with a slightly negative charge (δ^-).	Weakest

molecules (**Figure 2.10a**). Hydrogen bonding is responsible for the tendency of water molecules to cling together and form films, referred to as *surface tension*. This tendency helps explain why water beads up into spheres when it sits on a hard surface and why water striders can walk on a pond's surface (Figure 2.10b).

Although hydrogen bonds are too weak to bind atoms together to form molecules, they are important *intramolecular bonds* (literally, bonds within molecules), which hold different parts of a single large molecule in a specific three-dimensional shape. Some large biological molecules, such as proteins and DNA, have numerous hydrogen bonds that help maintain and stabilize their structures.

Check Your Understanding

10. What kinds of bonds form between water molecules?

11. Oxygen ($_8$O) and argon ($_{18}$A) are both gases. Oxygen combines readily with other elements, but argon does not. What accounts for this difference?

12. APPLY Assume imaginary compound XY has a polar covalent bond. How does its charge distribution differ from that of XX molecules?

For answers, see Answers Appendix.

2.5 Chemical reactions occur when electrons are shared, gained, or lost

Learning Outcomes

▶ **Define the three major types of chemical reactions: synthesis, decomposition, and exchange. Comment on the nature of oxidation-reduction reactions and their importance.**

▶ **Explain why chemical reactions in the body are often irreversible.**

▶ **Describe factors that affect chemical reaction rates.**

As we noted earlier, all particles of matter are in constant motion because of their kinetic energy. Movement of atoms or molecules in a solid is usually limited to vibration because the particles are united by fairly rigid bonds. But in liquids or gases, particles dart about randomly, sometimes colliding with one another and interacting to undergo chemical reactions. A **chemical reaction** occurs whenever chemical bonds are formed, rearranged, or broken.

(a) **The slightly positive ends (δ^+) of the water molecules become aligned with the slightly negative ends (δ^-) of other water molecules.**

(b) **A water strider can walk on a pond because of the high surface tension of water, a result of the combined strength of its hydrogen bonds.**

Figure 2.10 Hydrogen bonds.

Chemical Equations

We can write chemical reactions in symbolic form as chemical equations. For example, we indicate the joining of two hydrogen atoms to form hydrogen gas as

$$\text{H} + \text{H} \rightarrow \text{H}_2 \text{ (hydrogen gas)}$$

reactants   product

and the combining of four hydrogen atoms and one carbon atom to form methane as

$$4\text{H} + \text{C} \rightarrow \text{CH}_4 \text{ (methane)}$$

reactants   product

Notice that in equations, a number written as a *subscript* indicates that the atoms are joined by chemical bonds. But a number written as a *prefix* denotes the number of *unjoined* atoms or molecules. For example, CH_4 reveals that four hydrogen atoms are bonded together with carbon to form the methane molecule, but 4H signifies four unjoined hydrogen atoms.

A chemical equation is like a sentence describing what happens in a reaction. It contains the following information:

- The **reactants**: The number and kinds of the interacting substances.
- The **products**: The chemical composition of the result of the reaction.
- The *relative proportions*: Balanced equations indicate the relative proportion of each reactant and product.

In the previous equations, the reactants are atoms, as indicated by their atomic symbols (H, C). The product in each case is a molecule, as represented by its **molecular formula** (H_2, CH_4). The equation for the formation of methane may be read in terms of molecules or moles—as *either* "four hydrogen atoms plus one carbon atom yield one molecule of methane" *or* "four moles of hydrogen atoms plus one mole of carbon yield one mole of methane." Using moles is more practical because it is impossible to measure out one atom or one molecule of anything!

Types of Chemical Reactions

Most chemical reactions can be categorized as one of three types: *synthesis*, *decomposition*, or *exchange reactions*.

When atoms or molecules combine to form a larger, more complex molecule, the process is a **synthesis**, or **combination**, reaction. A synthesis reaction always involves bond formation. It can be represented (using arbitrary letters) as

$$A + B \rightarrow AB$$

Synthesis reactions are the basis of constructive, or **anabolic**, activities in body cells, such as joining small molecules called amino acids into large protein molecules (**Figure 2.11a**). Synthesis reactions are conspicuous in rapidly growing tissues.

A **decomposition reaction** occurs when a molecule is broken down into smaller molecules or its constituent atoms:

$$AB \rightarrow A + B$$

Essentially, decomposition reactions are reverse synthesis reactions: Bonds are broken. Decomposition reactions underlie all degradative, or **catabolic**, processes in body cells. For example, the bonds of glycogen molecules are broken to release simpler molecules of glucose (Figure 2.11b).

Exchange, or **displacement**, **reactions** involve both synthesis and decomposition. Bonds are both made and broken. In an exchange reaction, parts of the reactant molecules change partners, so to speak, producing different product molecules:

$$AB + C \rightarrow AC + B \quad \text{and} \quad AB + CD \rightarrow AD + CB$$

An exchange reaction occurs when ATP reacts with glucose and transfers its end phosphate group (indicated by a circled P in Figure 2.11c) to glucose, forming glucose-phosphate. At the same time, the ATP becomes ADP. This important reaction occurs whenever glucose enters a body cell, and it effectively traps the glucose fuel molecule inside the cell.

Another group of important chemical reactions in living systems is **oxidation-reduction reactions**, called **redox reactions**

(a) **Synthesis reactions**	(b) **Decomposition reactions**	(c) **Exchange reactions**
Smaller particles are bonded together to form larger, more complex molecules.	Bonds are broken in larger molecules, resulting in smaller, less complex molecules.	Bonds are both made and broken (also called displacement reactions).
Example Amino acids are joined together to form a protein molecule.	*Example* Glycogen is broken down to release glucose molecules.	*Example* ATP transfers its terminal phosphate group to glucose to form glucose-phosphate.
Amino acid molecules / Protein molecule	Glycogen / Glucose molecules	Adenosine triphosphate (ATP) / Glucose / Adenosine diphosphate (ADP) / Glucose-phosphate

Figure 2.11 **Types of chemical reactions.**

for short. Oxidation-reduction reactions are decomposition reactions in that they are the basis of all reactions in which food fuels are broken down for energy (that is, in which ATP is produced). They are also a special type of exchange reaction because electrons are exchanged between the reactants. The reactant losing the electrons is the *electron donor* and is said to be **oxidized**. The reactant taking up the transferred electrons is the *electron acceptor* and is said to become **reduced**.

Redox reactions also occur when ionic compounds are formed. Recall that in the formation of NaCl (see Figure 2.6), sodium loses an electron to chlorine. Consequently, sodium is oxidized and becomes a sodium ion, and chlorine is reduced and becomes a chloride ion. However, not all oxidation-reduction reactions involve *complete transfer* of electrons—some simply change the pattern of electron sharing in covalent bonds. For example, a substance is oxidized either by losing hydrogen atoms or by combining with oxygen. The common factor in these events is that electrons that formerly "belonged" to the reactant molecule are lost. The electrons are lost either entirely (as when hydrogen is removed and takes its electron with it) or relatively (as the shared electrons spend more time in the vicinity of the very electronegative oxygen atom).

To understand the importance of oxidation-reduction reactions in living systems, take a look at the overall equation for *cellular respiration*, which represents the major pathway by which glucose is broken down for energy in body cells:

$$C_6H_{12}O_6 + 6O_2 \rightarrow 6CO_2 + 6H_2O + ATP$$

glucose   oxygen   carbon   water   cellular
                     dioxide   energy

As you can see, it is an oxidation-reduction reaction. Consider what happens to the hydrogen atoms (and their electrons). Glucose is oxidized to carbon dioxide as it loses hydrogen atoms, and oxygen is reduced to water as it accepts the hydrogen atoms. This reaction is covered in detail in Chapter 24.

Energy Flow in Chemical Reactions

Because all chemical bonds represent stored chemical energy, all chemical reactions ultimately result in net absorption or release of energy. Reactions that release energy are **exergonic reactions**. These reactions yield products with less energy than the initial reactants, along with energy that can be harvested for other uses. With a few exceptions, catabolic and oxidative reactions are exergonic.

In contrast, the products of energy-absorbing, or **endergonic**, reactions contain more potential energy in their chemical bonds than did the reactants. Anabolic reactions are typically endergonic reactions. In the body, endergonic reactions are usually coupled to exergonic reactions. For example, the energy released when fuel molecules are broken down (oxidized) is captured in ATP molecules and then used to synthesize the complex biological molecules the body needs to sustain life.

Reversibility of Chemical Reactions

All chemical reactions are potentially reversible. If chemical bonds can be made, they can be broken, and vice versa.

Reversibility is indicated by a double arrow. When the arrows differ in length, the longer arrow indicates the major direction in which the reaction proceeds:

$$A + B \rightleftharpoons AB$$

In this example, the forward reaction (going to the right) predominates. Over time, the product (AB) accumulates and the reactants (A and B) decrease in amount.

When the arrows are of equal length, as in

$$A + B \rightleftharpoons AB$$

neither the forward reaction nor the reverse reaction is dominant. In other words, for each molecule of product (AB) formed, one product molecule breaks down, releasing the reactants A and B. Such a chemical reaction is said to be in a state of **chemical equilibrium**.

Once chemical equilibrium is reached, there is no further *net change* in the amounts of reactants and products unless more of either are added to the mix. Product molecules are still formed and broken down, but the balance established when equilibrium was reached (such as greater numbers of product molecules) remains unchanged.

Chemical equilibrium is analogous to the admission scheme used by many nightclubs that restrict the number of patrons admitted to comply with safety regulations. To stay within their allowed capacity (for example, 300), once 300 people are inside no one else is admitted until others leave. For instance, when 10 leave, 10 more may go in. So there is a constant turnover but the number of patrons in the club always stays at 300.

Although all chemical reactions are reversible, many biological reactions show so little tendency to go in the reverse direction that they are irreversible for all practical purposes. Chemical reactions that release energy will not go in the opposite direction unless energy is put back into the system. For example, when our cells break down glucose during cellular respiration to yield carbon dioxide and water, some of the energy released is trapped in the bonds of ATP. Because the cells then use ATP's energy for various functions (and more glucose will be along with the next meal), this particular reaction is never reversed in our cells. Furthermore, if a product of a reaction is continuously removed from the reaction site, it is unavailable to take part in the reverse reaction. This situation occurs when the carbon dioxide that is released during glucose breakdown leaves the cell, enters the blood, and is eventually removed from the body by the lungs.

Factors Influencing the Rate of Chemical Reactions

What influences the speed of chemical reactions? For atoms and molecules to react chemically in the first place, they must *collide* with enough force to overcome the repulsion between their electrons. Interactions between valence shell electrons—the basis of bond making and breaking—cannot occur long distance. The force of collisions depends on how fast the particles are moving. Solid, forceful collisions between rapidly moving particles in which valence shells overlap are much more likely to cause reactions than are collisions in which the particles graze each other lightly. The factors that affect the rate of chemical reactions are:

- **Temperature.** Higher temperatures increase the kinetic energy of particles and the force of their collisions, increasing the rate of chemical reactions.

- **Concentration.** High concentrations of reacting particles increase the chances of successful collisions, and reactions progress faster. Lowering concentrations slows reactions. Unless reactants are added or products removed, chemical equilibrium will eventually occur.

- **Particle size.** The smaller the reacting particles, the faster the chemical reaction. Smaller particles move faster than larger ones (at the same temperature) and so collide more frequently and forcefully.

- **Catalysts.** At normal body temperature, most chemical reactions would proceed far too slowly to maintain life were it not for the presence of catalysts. **Catalysts** (kat′ah-lists) are substances that increase the rate of chemical reactions without themselves becoming chemically changed or part of the product. Biological catalysts are called *enzymes* (en′zīmz). Later in this chapter we describe how enzymes work.

Check Your Understanding

13. Which reaction type—synthesis, decomposition, or exchange—occurs when fats are digested in your small intestine?

14. Why are many reactions that occur in living systems irreversible for all intents and purposes?

15. What specific name is given to decomposition reactions in which food fuels are broken down for energy?

For answers, see Answers Appendix.

PART 2

BIOCHEMISTRY

Biochemistry is the study of the chemical composition and reactions of living matter. Chemicals in the body fall into one of two major classes: organic or inorganic compounds. **Organic compounds** contain carbon and are made by living things. All organic compounds are covalently bonded molecules, and many are large.

All other chemicals in the body are considered **inorganic compounds**. These include water, salts, and many acids and bases. Inorganic compounds are generally defined as compounds that lack carbon. You should be aware of a few exceptions to this generalization: Carbon dioxide and carbon monoxide, for example, contain carbon but are considered to be inorganic compounds. Organic and inorganic compounds are equally essential for life.

2.6 Inorganic compounds include water, salts, and many acids and bases

Learning Outcomes

▶ Explain the importance of water and salts to body homeostasis.

▶ Define acid and base, and explain the concept of pH.

Water

Water is the most abundant and important inorganic compound in living material. It makes up 60–80% of the volume of most living cells. What makes water so vital to life? The answer lies in several properties:

- **High heat capacity.** Water has a high heat capacity. In other words, it absorbs and releases large amounts of heat before changing appreciably in temperature itself. This property of water prevents sudden changes in temperature caused by external factors, such as sun or wind exposure, or by internal conditions that release heat rapidly, such as vigorous muscle activity. As part of blood, water redistributes heat among body tissues, ensuring temperature homeostasis.

- **High heat of vaporization.** When water evaporates, or vaporizes, it changes from a liquid to a gas (water vapor). This transformation requires that large amounts of heat be absorbed to break the hydrogen bonds that hold water molecules together. This property is extremely beneficial when we sweat. As sweat (mostly water) evaporates from our skin, large amounts of heat are removed from the body, providing efficient cooling.

- **Polar solvent properties.** Water is an unparalleled solvent. Indeed, it is often called the **universal solvent**. Biochemistry is "wet chemistry." Biological molecules do not react chemically unless they are in solution, and virtually all chemical reactions in the body depend on water's solvent properties.

  Because water molecules are polar, they orient themselves with their slightly negative ends toward the positive ends of the solutes, and vice versa, first attracting the solute molecules, and then surrounding them. This polarity of water explains why ionic compounds and other small reactive molecules (such as acids and bases) *dissociate* in water, their ions separating from each other and becoming evenly scattered in the water, forming true solutions (**Figure 2.12**).

  Water also forms layers of water molecules, called **hydration layers**, around large charged molecules such as proteins, shielding them from the effects of other charged substances in the vicinity and preventing them from settling out of solution. Such protein-water mixtures are *biological colloids*. Blood plasma and cerebrospinal fluid (which surrounds the brain and spinal cord) are colloids.

  Water is the body's major transport medium because it is such an excellent solvent. Nutrients, respiratory gases, and metabolic wastes carried throughout the body are dissolved in blood plasma, and many metabolic wastes are excreted from the body in urine, another watery fluid. Lubricants (e.g., mucus) also use water as their dissolving medium.

- **Reactivity.** Water is an important *reactant* in many chemical reactions. For example, foods are broken down to their building blocks by adding a water molecule to each bond as it is broken. We will discuss such decomposition reactions (called *hydrolysis reactions*) in the next module.

- **Cushioning.** Water is not compressible, but it can flow. These properties allow water to form resilient cushions around certain body organs, helping to protect them from physical trauma. The cerebrospinal fluid surrounding the brain exemplifies water's cushioning role.

Salts

A **salt** is an ionic compound containing cations other than H^+ and anions other than the hydroxyl ion (OH^-). As already noted, when salts are dissolved in water, they dissociate into their component ions (Figure 2.12). For example, sodium sulfate (Na_2SO_4) dissociates into two Na^+ ions and one SO_4^{2-} ion. It dissociates easily because the ions are already formed. All that remains is for water to overcome the attraction between the oppositely charged ions.

All ions are **electrolytes** (e-lek′tro-līts), substances that conduct an electrical current in solution. [Note that groups of atoms that bear an overall charge, such as sulfate (SO_4^{2-}), are called *polyatomic ions*.]

Salts commonly found in the body include NaCl, $CaCO_3$ (calcium carbonate), and KCl (potassium chloride). However, the most plentiful salts are the calcium phosphates that make bones and teeth hard. In their ionized form, salts play vital roles in body function. For instance, the electrolyte properties of sodium and potassium ions are essential for nerve impulse transmission and muscle contraction. Ionic iron forms part of the hemoglobin molecules that transport oxygen within red blood cells, and zinc and copper ions are important to the activity of some enzymes. Other important functions of the elements found in body salts are summarized in Table 2.1 on p. 26.

Water molecule

Negative oxygen regions of polar water molecules are attracted to Na^+ cations.

Positive hydrogen regions of polar water molecules are attracted to Cl^- anions.

Salt crystal

Ions in solution

Figure 2.12 Dissociation of salt in water.

Acids and Bases

Like salts, acids and bases are electrolytes. They ionize and dissociate in water and can then conduct an electrical current.

Acids

Acids have a sour taste, can react with (dissolve) many metals, and "burn" a hole in your rug. But for our purposes the most useful definition of an acid is a substance that releases **hydrogen ions (H^+)** in detectable amounts. Because a hydrogen ion is just a hydrogen nucleus, which consists of a single "naked" proton, acids are also defined as **proton donors**.

When acids dissolve in water, they release hydrogen ions (protons) and anions. It is the concentration of protons that determines the acidity of a solution. For example, hydrochloric acid (HCl), an acid produced by stomach cells that aids digestion, dissociates into a proton and a chloride ion:

$$HCl \rightarrow H^+ + Cl^-$$
proton anion

Other acids found or produced in the body include acetic acid ($HC_2H_3O_2$), which is the acidic portion of vinegar; and carbonic acid (H_2CO_3). The molecular formula for an acid is easy to recognize because the hydrogen is written first.

Bases

Bases have a bitter taste, feel slippery, and are **proton acceptors**—that is, they take up hydrogen ions (H^+) in detectable amounts. Common inorganic bases include the *hydroxides* (hi-drok′sīds), such as magnesium hydroxide (milk of magnesia) and sodium hydroxide (lye). Like acids, hydroxides dissociate when dissolved in water, but in this case **hydroxyl ions (OH^-)** (hi-drok′sil) and cations are liberated. For example, ionization of sodium hydroxide (NaOH) produces a hydroxyl ion and a sodium ion, and the hydroxyl ion then binds to (accepts) a proton present in the solution. This reaction produces water and simultaneously reduces the acidity (hydrogen ion concentration) of the solution:

$$NaOH \rightarrow Na^+ + OH^-$$
cation   hydroxyl
ion

and then

$$OH^- + H^+ \rightarrow H_2O$$
water

Bicarbonate ion (HCO_3^-), an important base in the body, is particularly abundant in blood. **Ammonia (NH_3)**, a common waste product of protein breakdown in the body, is also a base. It has one pair of unshared electrons that strongly attracts protons. By accepting a proton, ammonia becomes an ammonium ion:

$$NH_3 + H^+ \rightarrow NH_4^+$$
ammonium
ion

pH: Acid-Base Concentration

The more hydrogen ions in a solution, the more acidic the solution is. Conversely, the greater the concentration of hydroxyl ions (the lower the concentration of H^+), the more basic, or *alkaline* (al'kuh-līn), the solution becomes. The relative concentration of hydrogen ions in various body fluids is measured in concentration units called **pH units** (pe-āch').

The idea for a pH scale was devised by a Danish biochemist and beer brewer named Sören Sörensen in 1909. He was searching for a convenient means of checking the acidity of his alcoholic product to prevent its spoilage by bacterial action. (Acidic conditions inhibit many bacteria.) The pH scale that resulted is based on the concentration of hydrogen ions in a solution, expressed in terms of moles per liter, or molarity. The pH scale runs from 0 to 14 and is *logarithmic*. In other words, each successive change of one pH unit represents a tenfold change in hydrogen ion concentration (**Figure 2.13**). The pH of a solution is defined as the negative logarithm of the hydrogen ion concentration [H$^+$] in moles per liter, or $-\log$[H$^+$]. (Note that brackets [ ] indicate concentration of a substance.)

At a pH of 7 (at which [H$^+$] is 10^{-7} *M*), the solution is *neutral*—neither acidic nor basic. The number of hydrogen ions exactly equals the number of hydroxyl ions (pH = pOH). Absolutely pure (distilled) water has a pH of 7.

Solutions with a pH below 7 are acidic—the hydrogen ions outnumber the hydroxyl ions. The lower the pH, the more acidic the solution. A solution with a pH of 6 has ten times as many hydrogen ions as a solution with a pH of 7.

Solutions with a pH higher than 7 are alkaline, and the relative concentration of hydrogen ions decreases by a factor of 10 with each higher pH unit. So, solutions with pH values of 8 and 12 have, respectively, 1/10 and 1/100,000 (1/10 × 1/10 × 1/10 × 1/10 × 1/10) as many hydrogen ions as a solution of pH 7.

The approximate pH of several body fluids and of a number of common substances appears in Figure 2.13. Notice that as the hydrogen ion concentration decreases, the hydroxyl ion concentration rises, and vice versa.

HOMEOSTATIC IMBALANCE 2.1   CLINICAL

Many of the life-preserving enzymes in your body work within a very narrow pH range. During cardiopulmonary resuscitation, an arterial pH of 7.0 predicts a poor outcome, and patients presenting with an arterial pH of less than 6.85 rarely survive.

Neutralization

What happens when acids and bases are mixed? They react with each other in displacement reactions to form water and a salt. For example, when hydrochloric acid and sodium hydroxide interact, sodium chloride (a salt) and water are formed.

$$\text{HCl} + \text{NaOH} \rightarrow \text{NaCl} + \text{H}_2\text{O}$$
acid   base   salt   water

This type of reaction is called a **neutralization reaction**, because the joining of H$^+$ and OH$^-$ to form water neutralizes the solution. Although the salt produced is written in molecular form (NaCl), remember that it actually exists as dissociated sodium and chloride ions when dissolved in water.

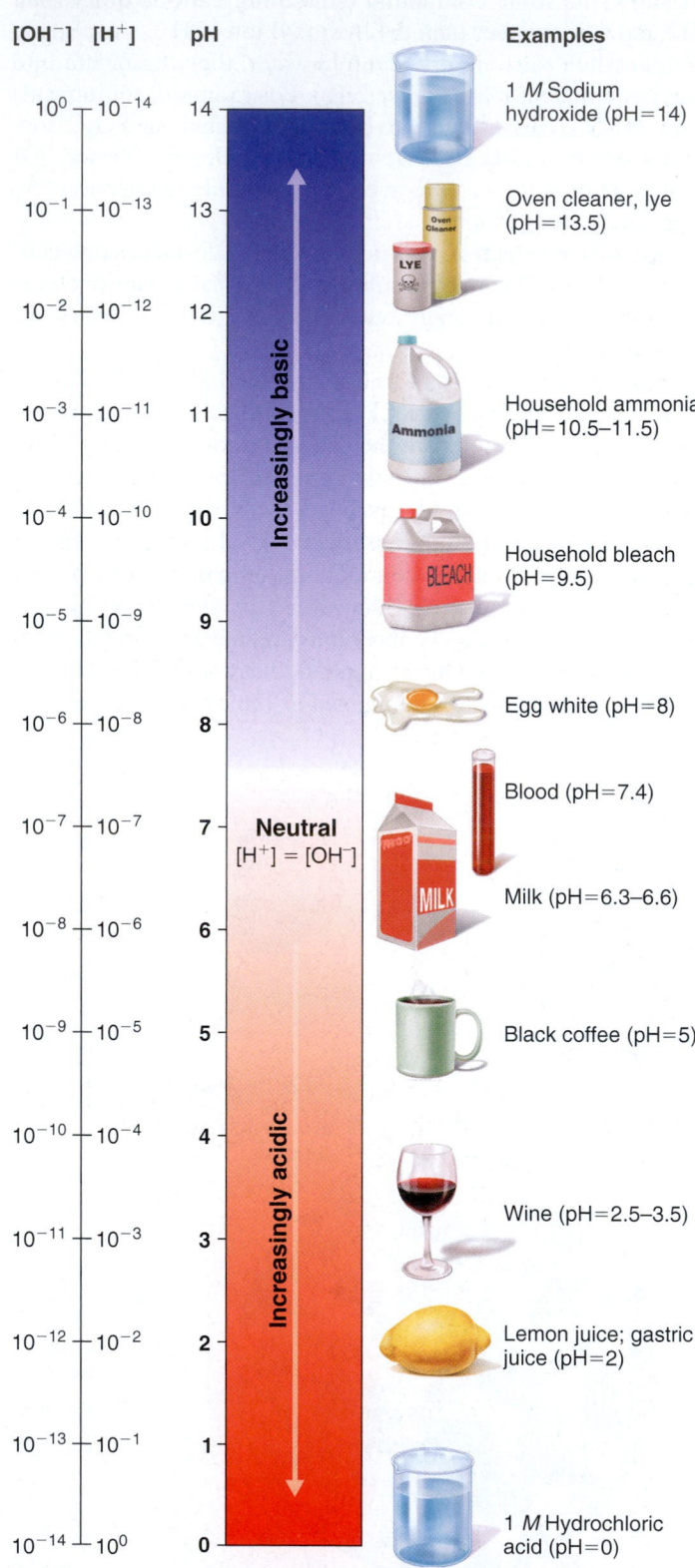

Concentration (moles/liter)

[OH$^-$]	[H$^+$]	pH	Examples
10^0	10^{-14}	14	1 *M* Sodium hydroxide (pH=14)
10^{-1}	10^{-13}	13	Oven cleaner, lye (pH=13.5)
10^{-2}	10^{-12}	12	
10^{-3}	10^{-11}	11	Household ammonia (pH=10.5–11.5)
10^{-4}	10^{-10}	10	
10^{-5}	10^{-9}	9	Household bleach (pH=9.5)
10^{-6}	10^{-8}	8	Egg white (pH=8)
10^{-7}	10^{-7}	7	Blood (pH=7.4)
10^{-8}	10^{-6}	6	Milk (pH=6.3–6.6)
10^{-9}	10^{-5}	5	Black coffee (pH=5)
10^{-10}	10^{-4}	4	
10^{-11}	10^{-3}	3	Wine (pH=2.5–3.5)
10^{-12}	10^{-2}	2	Lemon juice; gastric juice (pH=2)
10^{-13}	10^{-1}	1	
10^{-14}	10^0	0	1 *M* Hydrochloric acid (pH=0)

Increasingly basic

Neutral [H$^+$] = [OH$^-$]

Increasingly acidic

Figure 2.13 The pH scale and pH values of representative substances. The pH scale is based on the number of hydrogen ions in solution. The concentrations of hydrogen ions, [H$^+$], and hydroxyl ions, [OH$^-$], in moles per liter are indicated for each pH value noted. At a pH of 7, [H$^+$] = [OH$^-$] and the solution is neutral.

Buffers

Living cells are extraordinarily sensitive to even slight changes in the pH of the environment. Imagine what would happen to all those hydrogen bonds in biological molecules with large numbers of free H^+ running around. (Can't you just hear those molecules saying, "Why share hydrogen when I can have my own?")

Homeostasis of acid-base balance is carefully regulated by the kidneys and lungs and by chemical systems (proteins and other types of molecules) called **buffers**. Buffers resist abrupt and large swings in the pH of body fluids by releasing hydrogen ions (acting as acids) when the pH begins to rise and by binding hydrogen ions (acting as bases) when the pH drops. Buffers can do this because they consist of a combination of a *weak acid* and a corresponding *weak base*. To comprehend how chemical buffer systems operate, you must thoroughly understand strong and weak acids and bases.

The first important concept is that the acidity of a solution reflects *only* the free hydrogen ions, not those still bound to anions. Consequently, acids that dissociate completely and irreversibly in water are called **strong acids**, because they dramatically change the pH of a solution. Examples are hydrochloric acid and sulfuric acid. If we could count out 100 molecules of hydrochloric acid and place them in 1 milliliter (ml) of water, we could expect to end up with 100 H^+, 100 Cl^-, and no undissociated hydrochloric acid molecules in that solution.

Acids that do not dissociate completely, like carbonic acid (H_2CO_3) and acetic acid (HAc), are **weak acids**. If we were to place 100 molecules of acetic acid in 1 ml of water, the reaction would be something like this:

$$100 \text{ HAc} \rightarrow 90 \text{ HAc} + 10 \text{ H}^+ + 10 \text{ Ac}^-$$

Because only free H^+ determines pH, you can see right away that the acetic acid solution with 10 H^+ is much less acidic than the HCl solution with 100 H^+. Weak acids dissociate in a predictable way, and molecules of the intact acid are in dynamic equilibrium with the dissociated ions. Consequently, the dissociation of acetic acid may also be written as

$$\text{HAc} \rightleftharpoons \text{H}^+ + \text{Ac}^-$$

This viewpoint allows us to see that if H^+ (released by a strong acid) is added to the acetic acid solution, the equilibrium will shift to the left and some H^+ and Ac^- will recombine to form HAc. On the other hand, if a strong base is added and the pH begins to rise, the equilibrium shifts to the right and more HAc molecules dissociate to release H^+. This characteristic of weak acids allows them to play important roles in the chemical buffer systems of the body.

The concept of strong and weak bases is more easily explained. Remember that bases are proton acceptors. **Strong bases** are those, like hydroxides, that dissociate easily in water and quickly tie up H^+. On the other hand, sodium bicarbonate (commonly known as baking soda) ionizes incompletely and reversibly. Because it accepts relatively few protons, its released bicarbonate ion is considered a **weak base**.

Now let's examine how one buffer system helps to maintain pH homeostasis of the blood. Normally, blood pH varies within a very narrow range (7.35 to 7.45). If the pH of blood varies from these limits by more than a few tenths of a unit, it may be fatal. Although there are other chemical blood buffers, the **bicarbonate buffer system** is a major one. In this buffer system, the weak acid is carbonic acid (H_2CO_3). It dissociates reversibly, releasing its corresponding weak base, bicarbonate ions (HCO_3^-), and protons (H^+):

$$\underset{\substack{\text{H}^+ \text{ donor} \\ \text{(weak acid =} \\ \text{carbonic acid)}}}{\text{H}_2\text{CO}_3} \underset{\text{Response to drop in pH}}{\overset{\text{Response to rise in pH}}{\rightleftharpoons}} \underset{\substack{\text{H}^+ \text{ acceptor} \\ \text{(weak base =} \\ \text{bicarbonate)}}}{\text{HCO}_3^-} + \underset{\text{proton}}{\text{H}^+}$$

The chemical equilibrium between carbonic acid and bicarbonate ion resists changes in blood pH by shifting to the right or left as H^+ ions are added to or removed from the blood. As blood pH rises (becomes more alkaline due to the addition of a strong base), the equilibrium shifts to the right, forcing more carbonic acid to dissociate. Similarly, as blood pH begins to drop (becomes more acidic due to the addition of a strong acid), the equilibrium shifts to the left as more bicarbonate ions begin to bind with protons. As you can see, strong bases are replaced by a weak base (bicarbonate ion) and protons released by strong acids are tied up in a weak one (carbonic acid). In either case, the blood pH changes much less than it would in the absence of the buffering system. We discuss acid-base balance and buffers in more detail in Chapter 26.

Check Your Understanding

16. Salts are electrolytes. What does that mean?

17. Which ion is responsible for increased acidity?

18. PREDICT Would you expect a small or large change in pH if you added a strong acid to a strong base? A strong acid to a weak base? Why?

19. MAKE CONNECTIONS We have learned about the complementarity of structure and function as it relates to anatomy and physiology (Chapter 1). See if you can extend your thinking about this principle to a simple molecule, and explain how the structure of a water molecule makes water an excellent solvent.

For answers, see Answers Appendix.

2.7 Organic compounds are made by dehydration synthesis and broken down by hydrolysis

Learning Outcome

▶ **Explain the role of dehydration synthesis and hydrolysis in forming and breaking down organic molecules.**

Molecules unique to living systems—carbohydrates, lipids (fats), proteins, and nucleic acids—all contain carbon. What makes carbon so special that "living" chemistry depends on its presence? To begin with, no other *small* atom is so precisely **electroneutral**. The consequence of its electroneutrality is that carbon never loses or gains electrons. Instead, it always shares

2

them. Furthermore, with four valence shell electrons, carbon forms four covalent bonds with other elements, as well as with other carbon atoms. As a result, carbon can help form long, chainlike molecules (common in fats), ring structures (typical of carbohydrates and steroids), and many other structures that are uniquely suited for specific roles in the body.

Many biological molecules (carbohydrates and proteins for example) are **macromolecules**—large complex molecules containing thousands of atoms. Most macromolecules are **polymers**, which are chainlike molecules made of many smaller, identical or similar subunits (**monomers**).

Monomers are joined together by a process called **dehydration synthesis** (**Figure 2.14**). During dehydration synthesis, a hydrogen atom is removed from one monomer and a hydroxyl group is removed from the monomer it is to be joined with. As a covalent bond unites the monomers, a water molecule is released. This removal of a water molecule at the bond site occurs each time a monomer is added to the growing polymer

chain. The opposite reaction in which molecules are degraded is called **hydrolysis** (hi-drol'ĭ-sis; water splitting). In these reactions, a water molecule is added to each bond that is broken, thereby releasing its building blocks (smaller molecules).

Writing out all of the atoms in chemical structures is cumbersome, so everyone uses shortcuts. Here are two structures. The left structure shows all of the atoms. The structure on the right is the simplified version. It assumes that you know that each corner has a carbon. You will see these simplified structures throughout the chapter.

For the most part, organic molecules are very large molecules, but their interactions with other molecules typically

Figure 2.14 Dehydration synthesis and hydrolysis. Biological molecules are formed from their monomers, or subunits, by dehydration synthesis and broken down to the monomers by hydrolysis reactions. Note the use of simplified structures.

involve only small, reactive parts of their structure called *functional groups* (acid groups, amines, and others). The most important functional groups involved in biochemical reactions are illustrated in Appendix B.

We will explore the four types of macromolecules that form our cells in the next four modules.

Check Your Understanding

20. What is the result of hydrolysis reactions and how are these reactions accomplished?

For answers, see Answers Appendix.

2.8 Carbohydrates provide an easily used energy source for the body

Learning Outcome

▶ Describe the building blocks, general structure, and biological functions of carbohydrates.

Carbohydrates, a group of molecules that includes sugars and starches, represent 1–2% of cell mass. Carbohydrates contain carbon, hydrogen, and oxygen, and generally the hydrogen and oxygen atoms occur in the same 2:1 ratio as in water. This ratio is reflected in the word *carbohydrate* ("hydrated carbon").

A carbohydrate can be classified according to size and solubility as a monosaccharide ("one sugar"), disaccharide ("two sugars"), or polysaccharide ("many sugars"). Monosaccharides are the monomers, or building blocks, of the other carbohydrates. In general, the larger the carbohydrate molecule, the less soluble it is in water.

Monosaccharides

Monosaccharides (mon″o-sak′ah-rīdz), or *simple sugars*, are single-chain or single-ring structures containing from three to seven carbon atoms (**Figure 2.15a**). Usually the carbon, hydrogen, and oxygen atoms occur in the ratio 1:2:1, so a general formula for a monosaccharide is $(CH_2O)_n$, where n is the number of carbons in the sugar. Glucose, for example, has six carbon atoms, and its molecular formula is $C_6H_{12}O_6$. Ribose, with five carbons, is $C_5H_{10}O_5$.

Monosaccharides are named generically according to the number of carbon atoms they contain. Most important in the body are the pentose (five-carbon) and hexose (six-carbon) sugars. The pentose *deoxyribose* (de-ok″sĭ-ri′bōs) is part of DNA, and *glucose*, a hexose, is blood sugar.

Two other hexoses, *galactose* and *fructose*, are **isomers** (ī′so-mers) of glucose. That is, they have the same molecular

(a) Monosaccharides

Monomers of carbohydrates

Example
Hexose sugars (the hexoses shown here are isomers)

Glucose   Fructose   Galactose

Example
Pentose sugars

Deoxyribose   Ribose

(b) Disaccharides

Consist of two linked monosaccharides

Example
Sucrose, maltose, and lactose (these disaccharides are isomers)

Glucose + Fructose → **Sucrose**

Glucose + Glucose → **Maltose**

Galactose + Glucose → **Lactose**

Figure 2.15 Carbohydrate molecules important to the body. *(Figure continues on p. 44 .)*

(c) Polysaccharides

Long chains (polymers) of linked monosaccharides

Example
This polysaccharide is a simplified representation of glycogen, a polysaccharide formed from glucose molecules.

Glycogen

Figure 2.15 *(continued)* **Carbohydrate molecules important to the body.**

formula ($C_6H_{12}O_6$), but their atoms are arranged differently, giving them different chemical properties (Figure 2.15a).

Disaccharides

A **disaccharide** (di-sak′ah-rīd), or *double sugar*, is formed when two monosaccharides are joined by dehydration synthesis (Figure 2.14a, c). In this synthesis reaction, a water molecule is lost as the bond is made, as illustrated by the synthesis of sucrose (soo′krōs):

$$2C_6H_{12}O_6 \rightarrow C_{12}H_{22}O_{11}+H_2O$$

glucose + fructose   sucrose   water

Important disaccharides in the diet are *sucrose* (glucose + fructose), which is cane or table sugar; *lactose* (glucose + galactose), found in milk; and *maltose* (glucose + glucose), also called malt sugar (Figure 2.15b). Disaccharides are too large to be transported through cell membranes, so they must be hydrolyzed to monosaccharides (Figure 2.14b, c) before the digestive tract can absorb them. A water molecule is added as each bond is broken, releasing monosaccharides that can be absorbed from the digestive tract into the blood.

Polysaccharides

Polysaccharides (pol″e-sak′ah-rīdz) are polymers of simple sugars linked together by dehydration synthesis. Because polysaccharides are large, fairly insoluble molecules, they are ideal storage products. Another consequence of their large size is that they lack the sweetness of the simple and double sugars.

Only two polysaccharides are of major importance to the body: starch and glycogen. Both are polymers of glucose. Only their degree of branching differs.

Starch is the storage carbohydrate formed by plants. The number of glucose units composing a starch molecule is high and variable. When we eat starchy foods such as grain products and potatoes, the starch must be digested for its glucose units to be absorbed. We are unable to digest *cellulose*, another polysaccharide found in all plant products. However, it is important in providing the *bulk* (one form of fiber) that helps move feces through the colon.

Glycogen (gli′ko-jen), the storage carbohydrate of animal tissues, is stored primarily in skeletal muscle and liver cells. Like starch, it is highly branched and is a very large molecule (Figure 2.15c). Since each branch can be attacked by an enzyme, many glucose molecules can be released simultaneously when cells need glucose for fuel. Skeletal muscles use their glycogen for themselves, but liver cells use their stored glycogen to maintain blood sugar (glucose) levels. This allows the body's cells to get the fuel they need.

Carbohydrate Functions

The major function of carbohydrates in the body is to provide a ready, easily used source of cellular fuel. Most cells can use only a few types of simple sugars, and glucose is at the top of the cellular "menu." As described in our earlier discussion of oxidation-reduction reactions (p. 37), glucose is broken down and oxidized within cells. During these reactions, electrons are transferred, releasing the bond energy stored in glucose. This energy is used to synthesize ATP. When ATP supplies are sufficient, dietary carbohydrates are converted to glycogen or fat and stored. Those of us who have gained weight from eating too many carbohydrate-rich snacks have personal experience with this conversion process!

Only small amounts of carbohydrates are used for structural purposes. For example, some sugars are found in our genes. Others are attached to the external surfaces of cells where they act as identity molecules to guide cellular interactions.

Check Your Understanding

21. What are the monomers of carbohydrates called? Which monomer is blood sugar?
22. What is the animal form of stored carbohydrate called?

For answers, see Answers Appendix.

2.9 Lipids insulate body organs, build cell membranes, and provide stored energy

Learning Outcome

▶ Describe the building blocks, general structure, and biological functions of lipids.

Lipids are insoluble in water but dissolve readily in other lipids and in organic solvents such as alcohol and ether. Like carbohydrates, all lipids contain carbon, hydrogen, and oxygen, but the proportion of oxygen in lipids is much lower. In addition, phosphorus is found in some of the more complex lipids. Lipids include *triglycerides*, *phospholipids* (fos″fo-lip′idz), *steroids* (stĕ′roidz), and a number of other lipid substances. **Table 2.3** on p. 47 gives the locations and functions of some lipids found in the body.

Triglycerides

Triglycerides (tri-glis′er-īdz) are commonly known as *fats* when solid or *oils* when liquid. Triglycerides are large molecules, often consisting of hundreds of atoms. They provide the body's most efficient and compact form of stored energy, and when they are oxidized, they yield large amounts of energy. A triglyceride is composed of two types of building blocks, **fatty acids** and **glycerol** (glis′er-ol), in a 3:1 ratio of fatty acids to glycerol (**Figure 2.16**). Fatty acids are linear chains of carbon and hydrogen atoms (hydrocarbon chains) with an organic acid group (—COOH) at one end. Glycerol is a modified simple sugar (a sugar alcohol).

Fat synthesis involves attaching three fatty acid chains to a single glycerol molecule by dehydration synthesis. The result is an E-shaped molecule. The glycerol backbone is the same in all triglycerides, but the fatty acid chains vary, resulting in different kinds of fats and oils.

Their hydrocarbon chains make triglycerides nonpolar molecules. Because polar and nonpolar molecules do not interact (oil and water do not mix), digestion and absorption of fats is complicated, and ingested fats and oils must be broken down to their building blocks.

Triglycerides are found mainly beneath the skin, where they insulate the deeper body tissues from heat loss and protect them

(a) One of three dehydration synthesis reactions in the synthesis of a triglyceride.

(b) One triglyceride molecule consists of a glycerol and three fatty acids bound together by dehydration synthesis.

(c) Simplified drawing of the same triglyceride.

Two more dehydration synthesis reactions like the one shown on the left yield a total of 3 molecules of H_2O and one triglyceride molecule.

Figure 2.16 Triglycerides consist of glycerol and three fatty acids. Simplified fatty acid structures are shown in **(c)**.

(a) Saturated fat

At room temperature, the molecules of a saturated fat such as butter are packed closely together, forming a solid.

This simplified structural formula shows that there are **no double bonds** between carbons.

(b) Unsaturated fat

At room temperature, the molecules of an unsaturated fat such as olive oil cannot pack together closely enough to solidify because of the kinks in at least one of their fatty acid chains.

This simplified structural formula shows a **double bond** between one pair of carbons in one of the fatty acids.

Double bond between carbons bends chain.

Figure 2.17 **Saturated and unsaturated fatty acids.**

from mechanical trauma. For example, women are usually more successful English Channel swimmers than men. Their success is due partly to their thicker subcutaneous fatty layer, which helps insulate them from the bitterly cold water of the Channel.

The length of a triglyceride's fatty acid chains and their degree of *saturation* with H atoms determine how solid the molecule is at a given temperature. Fatty acid chains with only single covalent bonds between carbon atoms are referred to as **saturated** (**Figure 2.17a**). Their fatty acid chains are straight and, at room temperature, the molecules of a saturated fat are packed closely together, forming a solid. Fatty acids that contain one or more double bonds between carbon atoms are said to be **unsaturated** (**monounsaturated** or **polyunsaturated**, respectively) (Figure 2.17b). The double bonds cause the fatty acid chains to kink so that they cannot be packed closely enough to solidify. As a result, triglycerides with short fatty acid chains or unsaturated fatty acids are oils (liquid at room temperature) and are typical of plant lipids. Examples include olive and peanut oils (rich in monounsaturated fats) and corn, soybean, and safflower oils, which contain a high percentage of polyunsaturated fatty acids. Longer fatty acid chains and more saturated fatty acids are common in animal fats such as butterfat and the fat of meats, which are solid at room temperature. Of the two types of fatty acids, the unsaturated variety, olive oil for example, is said to be more "heart healthy."

Trans fats, common in many margarines and baked products, are oils that have been solidified by addition of H atoms at sites of carbon double bonds. They increase the risk of heart disease even more than the solid animal fats. Conversely, the

Polar head group (*hydrophilic* = water loving)

Nonpolar tails (*hydrophobic* = water fearing)

Nitrogen-containing group

Phosphate

Glycerol

Fatty acid

Fatty acid

CH_2—$N^+(CH_3)_3$
CH_2
O
$O=P$—O^-
O
CH_2—CH—CH_2

You'll see phospholipids drawn like this throughout this book.

In a watery environment, phospholipids automatically line up with their heads facing the water on either side, and their hydrophobic tails "hiding" away from the water.

Polar head (hydrophilic)

Nonpolar tails (hydrophobic)

(a) Parts of a phospholipid

(b) Chemical structure of a phospholipid

(c) Schematic diagram of a phospholipid

(d) Phospholipids lining up to form a bilayer

Figure 2.18 **Phospholipid structure.** A phospholipid consists of a glycerol backbone with two fatty acids, a phosphate group, and a nitrogen-containing group.

Table 2.3 Representative Lipids Found in the Body

LIPID TYPE	LOCATION/FUNCTION
Triglycerides	
	• Major form of stored energy in the body.
	• Fat deposits (in subcutaneous tissue and around organs) protect and insulate body organs.
Phospholipids	
	• Chief components of cell membranes.
	• Help transport lipids in blood (as part of *lipoproteins*; see below).
Steroids	
Cholesterol	• Component of cell membranes.
	• Starting molecule for synthesis of all body steroids.
Bile salts	• Breakdown products of cholesterol.
	• Released by the liver into the digestive tract, where they help with fat digestion and absorption.
Vitamin D	• Fat-soluble vitamin produced in the skin on exposure to UV radiation.
	• Necessary for normal bone growth and function.
Sex hormones	• Include estrogens and progesterone (female hormones) and testosterone (male hormone).
	• Produced in the gonads.
Corticosteroids (hormones from adrenal cortex)	• Cortisol is a metabolic hormone necessary for maintaining normal blood glucose levels.
	• Aldosterone helps to regulate salt and water balance of the body by targeting the kidneys.
Other Lipoid Substances	
Fat-soluble vitamins	• Vitamins A, D (a steroid, see above), E, and K are fat soluble. See Table 24.2 on p. 932 for details.
Eicosanoids (e.g.,prostaglandins, thromboxanes)	• Derived from fatty acids found in all cell membranes.
	• Prostaglandins have diverse effects, including promoting inflammation, stimulating uterine contractions, regulating blood pressure, and controlling gastrointestinal tract motility.
	• Thromboxanes are powerful vasoconstrictors.
Lipoproteins	• Lipid and protein-based substances that transport fatty acids and cholesterol in the bloodstream.
	• Major varieties are high-density lipoproteins (HDLs) and low-density lipoproteins (LDLs).
Glycolipids	• Components of cell membranes.
	• Carbohydrates attached to lipids determine blood type and play roles in cell recognition and in recognition of foreign substances by immune cells.

omega-3 fatty acids, found naturally in cold-water fish, appear to decrease the risk of heart disease and some inflammatory diseases.

Phospholipids

Phospholipids are modified triglycerides. Specifically, they have two, rather than three, fatty acid chains. The third chain is replaced by a phosphate group (PO_4) with an attached nitrogen-containing group (**Figure 2.18**). The two fatty acid "tails" make one end of the phospholipid molecule nonpolar and **hydrophobic** (*hydro* = water; *phobia* = fear). These nonpolar tails interact only with other nonpolar molecules. The rest of

the molecule, including the phosphate group, form the head—the polar, **hydrophilic** (*phil* = loving) region of the molecule. The head interacts with other polar or charged particles, such as water or ions. Having both hydrophilic and hydrophobic ends makes phospholipids ideally suited to be the chief building material for cellular membranes (Figure 2.18d).

Steroids

Structurally, steroids differ quite a bit from fats and oils. **Steroids** are basically flat molecules made of four interlocking hydrocarbon rings. Like triglycerides, steroids are fat soluble and contain little oxygen. The single most important

Four interlocking hydrocarbon rings form a steroid.

Different steroids have different groups attached to the 4-ring backbone. **Cholesterol** (shown here) is the basis for all steroids formed in the body.

Figure 2.19 Steroid structure.

molecule in our steroid chemistry is *cholesterol* (ko-les'ter-ol) (**Figure 2.19**). We ingest cholesterol in animal products such as eggs, meat, and cheese, and our liver produces some.

Cholesterol has a bad reputation because of its role in atherosclerosis, but it is essential for human life. Cholesterol is found in cell membranes and is the raw material for synthesis of vitamin D, steroid hormones, and bile salts. Although steroid hormones are present in the body in only small quantities, they are vital to homeostasis. Without sex hormones, reproduction would be impossible, and a total lack of the corticosteroids produced by the adrenal glands is fatal.

Eicosanoids

The **eicosanoids** (i-ko'sah-noyds) are diverse lipids chiefly derived from a 20-carbon fatty acid (arachidonic acid) found in all cell membranes. Particularly important are the *prostaglandins*, which play roles in various body processes including blood clotting, regulation of blood pressure, inflammation, and labor contractions. Their synthesis and inflammatory actions are blocked by NSAIDs (nonsteroidal anti-inflammatory drugs; e.g., ibuprofen).

Check Your Understanding

23. How do triglycerides differ from phospholipids in body function and location?

For answers, see Answers Appendix.

2.10 Proteins are the body's basic structural material and have many vital functions

Learning Outcomes

▶ **Describe the four levels of protein structure.**
▶ **Describe enzyme action.**

Protein composes 10–30% of cell mass and is the basic structural material of the body. However, not all proteins are construction materials. Many play vital roles in cell function.

Proteins, which include enzymes (biological catalysts), hemoglobin of the blood, and contractile proteins of muscle, have the most varied functions of any molecules in the body. These and other selected roles of proteins are shown in **Figure 2.20**.

Amino Acids and Peptide Bonds

The building blocks of proteins are molecules called **amino acids**, of which there are 20 common types (see Appendix C). As shown in the amino acid below, all amino acids have two important functional groups: a basic group called an *amine* (ah'mēn) *group* (—NH₂), and an organic *acid group* (—COOH).

Amine group   Acid group

An amino acid may therefore act either as a base (proton acceptor) or an acid (proton donor). All amino acids are identical except for a single group of atoms called their *R group*. It is differences in the R group that make each amino acid chemically unique. All amino acids contain carbon, oxygen, hydrogen, and nitrogen, and two contain sulfur as well.

Proteins are long chains of amino acids joined together by dehydration synthesis, with the acid end of one amino acid linked to the amine end of the next. The resulting bond produces a characteristic arrangement of linked atoms called a **peptide bond** (**Figure 2.21**). Two united amino acids form a *dipeptide*, three a *tripeptide*, and ten or more a **polypeptide**. Although polypeptides containing more than 50 amino acids are called proteins, most proteins are macromolecules containing from 100 to over 10,000 amino acids.

Because each type of amino acid has distinct properties, the sequence in which they are bound together produces proteins that vary widely in both structure and function. We can think of the 20 amino acids as a 20-letter "alphabet" used in specific combinations to form "words" (proteins). Just as a change in one letter can produce a word with an entirely different meaning (flour → floor) or that is nonsensical (flour → flocr), changes in the kinds or positions of amino acids can yield proteins with different functions or proteins that are nonfunctional. Nevertheless, there are thousands of different proteins in the body, each with distinct functional properties, and all constructed from different combinations of the 20 common amino acids.

Structural Levels of Proteins

Proteins can be described in terms of four structural levels: primary, secondary, tertiary, and quaternary. The linear sequence of amino acids composing the polypeptide chain is the *primary structure* of a protein. This structure, which resembles a strand

(a) Structural proteins

Function: Mechanical support.
Example: *Collagen*, found in all connective tissue, is the single most abundant protein in the body. It is responsible for the tensile strength of bones, tendons, and ligaments.

Collagen

(b) Enzyme proteins

Function: Catalysis. Protein enzymes are essential for virtually every biochemical reaction in the body.
Examples: *Disaccharidases* hydrolyze disaccharides, *proteases* hydrolyze proteins, and *oxidases* oxidize food fuels.

Disaccharide

Enzyme

(c) Transport proteins

Function: Moving substances (in blood or across plasma membranes).
Example: *Hemoglobin* transports oxygen in blood. Some plasma membrane proteins transport substances (such as ions) across the plasma membrane.

Hemoglobin

Oxygen

(d) Contractile proteins

Function: Movement.
Example: *Actin* and *myosin* cause muscle cell contraction and function in cell division in all cell types.

Actin

Myosin

(e) Communication proteins

Function: Transmitting signals between cells. Can act as *chemical messengers* (e.g., hormones) or as *receptors* in the plasma membrane.
Example: *Insulin* (a protein) acts at its receptor to regulate blood sugar levels.

Chemical messenger

Receptor

(f) Defensive proteins

Function: Protect against disease.
Example: *Antibodies* released by certain immune cells are specialized proteins that bind and inactivate foreign substances (e.g., bacteria, toxins, viruses).

Virus

Antibodies

Figure 2.20 Examples of protein functions.

Dehydration synthesis: The acid group of one amino acid is bonded to the amine group of the next, with loss of a water molecule.

Amine group Acid group

Amino acid + Amino acid

Peptide bond

Dipeptide

Figure 2.21 Amino acids are linked together by peptide bonds. Peptide bonds are formed by dehydration synthesis and broken by hydrolysis reactions.

Hydrolysis: Peptide bonds linking amino acids together are broken when water is added to the bond.

(a) Primary structure

The sequence of amino acids forms the polypeptide chain.

Amino acid   Amino acid   Amino acid   Amino acid   Amino acid

(b) Secondary structure

The primary chain forms spirals (α-helices) and sheets (β-sheets).

α-Helix: The primary chain is coiled to form a spiral structure, which is stabilized by hydrogen bonds.

β-Sheet: The primary chain "zig-zags," forming a "pleated" sheet. Adjacent strands are held together by hydrogen bonds.

(c) Tertiary structure

Superimposed on secondary structure. α-Helices and/or β-sheets are folded up to form a compact globular molecule held together by intramolecular bonds.

Tertiary structure of transthyretin, a protein that transports the thyroid hormone thyroxine in blood and cerebrospinal fluid.

(d) Quaternary structure

Two or more polypeptide chains, each with its own tertiary structure, combine to form a functional protein.

Quaternary structure of a functional transthyretin molecule. Four identical transthyretin subunits join to form a complex protein.

Figure 2.22 Levels of protein structure.

of amino acid "beads," is the backbone of the protein molecule (**Figure 2.22a**).

Proteins do not normally exist as simple, linear chains of amino acids. Instead, they twist or bend upon themselves to form a more complex *secondary structure*. The most common type of secondary structure is the **alpha (α)-helix**, which resembles a Slinky® toy or a coiled spring (Figure 2.22b). The α-helix is formed by coiling the primary chain. It is stabilized by hydrogen bonds formed between NH and CO groups in amino acids in the primary chain that are about four amino acids apart. Hydrogen bonds in α-helices always link different parts of the *same* chain together.

In another type of secondary structure, the **beta (β)-pleated sheet**, the primary polypeptide chains do not coil, but are linked side by side by hydrogen bonds to form a pleated, ribbonlike structure that resembles an accordion's bellows

(Figure 2.22b). Notice that in this type of secondary structure, the hydrogen bonds may link together *different polypeptide chains* as well as *different parts* of the same chain that has folded back on itself. A single polypeptide chain may exhibit both types of secondary structure at various places along its length.

Many proteins have *tertiary structure* (ter′she-a″re), the next higher level of complexity, which is superimposed on secondary structure and involves the amino acids' R-groups. Tertiary structure is achieved when α-helical or β-pleated regions of the polypeptide chain fold upon one another to produce a compact ball-like, or *globular*, molecule (Figure 2.22c). Hydrophobic R groups are on the inside of the molecule and hydrophilic R groups are on its outside. Their interactions plus those reinforced by covalent and hydrogen bonds help to maintain the unique tertiary shape.

When two or more polypeptide chains aggregate in a regular manner to form a complex protein, the protein has *quaternary structure* (kwah′ter-na″re). The transthyretin molecule with its four identical globular subunits represents this level of structure (Figure 2.22d). (Transthyretin transports thyroid hormone in the blood.)

How do these different levels of structure arise? Although a protein with tertiary or quaternary structure looks a bit like a clump of congealed pasta, the ultimate overall structure of any protein is very specific and is dictated by its primary structure. In other words, the types (hydrophilic versus hydrophobic) and relative positions of amino acids in the protein backbone determine the complex three-dimensional structure of the folded protein. Proteins fold so that hydrophilic amino acids are near the surface and hydrophobic amino acids are buried in the protein's core.

Fibrous and Globular Proteins

The overall structure of a protein determines its biological function. In general, proteins are classified according to their overall appearance and shape as either fibrous or globular.

Fibrous proteins, also known as **structural proteins**, form long strands. Some exhibit only secondary structure, but most have tertiary or even quaternary structure as well. For example, *collagen* (kol′ah-jen) is made of helical tropocollagen molecules packed together side by side to form a strong ropelike structure (Figure 2.20a on p. 49). Fibrous proteins are insoluble in water, and very stable—qualities ideal for providing mechanical support and tensile strength to the body's tissues. Besides structural proteins like collagen (the single most abundant protein in the body), the fibrous proteins include certain contractile proteins (see Figure 2.20d).

Globular proteins, also called **functional proteins**, are compact, spherical proteins that have at least tertiary structure. Some also exhibit quaternary structure. The globular proteins are water-soluble, chemically active molecules, and they play crucial roles in virtually all biological processes. Some (antibodies) help to provide immunity, others (protein-based hormones) regulate growth and development, some are transport proteins, and still others (enzymes) are catalysts that oversee just about every chemical reaction in the body (see Figure 2.20b, c, e, f).

Protein Denaturation

Fibrous proteins are stable, but globular proteins are quite the opposite. The activity of a protein depends on its specific three-dimensional structure. Intramolecular bonds, particularly hydrogen bonds, are important in maintaining that structure. However, hydrogen bonds are fragile and easily broken by many chemical and physical factors, such as excessive acidity or temperature. Although individual proteins vary in their sensitivity to environmental conditions, hydrogen bonds begin to break when the pH drops or the temperature rises above normal (physiological) levels. When this happens, proteins unfold and lose their specific three-dimensional shape. Such a protein is said to be **denatured**.

The disruption is reversible in most cases, and the "scrambled" protein regains its native structure when desirable conditions are restored. However, if the temperature or pH change is so extreme that protein structure is damaged beyond repair, the protein is *irreversibly denatured*. The coagulation of egg white (primarily albumin protein) that occurs when you boil or fry an egg is an example of irreversible protein denaturation. There is no way to restore the white, rubbery protein to its original translucent form.

When globular proteins are denatured, they can no longer perform their physiological roles. For example, an enzyme's function depends on the presence of specific arrangements of atoms where catalytic activity occurs, called **active sites**. Active sites are regions that fit and interact chemically with other molecules of complementary shape and charge. Because atoms contributing to an active site in the folded protein may actually be far apart in the primary chain, disruption of intramolecular bonds separates them and destroys the active site. The globular transport protein hemoglobin becomes totally unable to bind and transport oxygen when blood pH is too acidic, because the structure needed for its function has been destroyed.

We will describe most types of body proteins in conjunction with the organ systems or functional processes to which they are closely related. However, one group of proteins—*enzymes*—is intimately involved in the normal functioning of all cells, so we will consider these incredibly complex molecules here.

Enzymes and Enzyme Activity

Enzymes are globular proteins that act as biological catalysts. *Catalysts* are substances that regulate and accelerate the rate of biochemical reactions but are not used up or changed in those reactions. More specifically, enzymes can be thought of as chemical traffic cops that keep our metabolic pathways flowing. Enzymes cannot force chemical reactions to occur between molecules that would not otherwise react. They can only increase the speed of reaction, and they do so by staggering amounts—from 100,000 to over 1 billion times the rate of an uncatalyzed reaction. Without enzymes, biochemical reactions proceed so slowly that for practical purposes they do not occur at all.

Characteristics of Enzymes

Some enzymes are purely protein. In other cases, the functional enzyme consists of two parts: an **apoenzyme** (the protein portion) and a **cofactor**. Together, these two parts form a **holoenzyme**. Depending on the enzyme, the cofactor may be an ion of a metal element such as copper or iron, or an organic molecule needed to assist the reaction in some way. Most organic cofactors are derived from vitamins (especially the B complex vitamins). In this case the type of cofactor is more precisely called a **coenzyme**.

Each enzyme is chemically specific. Some enzymes control only a single chemical reaction. Others can regulate a small group of related reactions by binding with molecules that differ only slightly. The substance on which an enzyme acts is called a **substrate**. As mentioned earlier, catalytic activity occurs at the *active site*.

The presence of specific enzymes determines not only which reactions will be speeded up, but also which reactions will occur—no enzyme, no reaction. This also means that unwanted or unnecessary chemical reactions do not occur.

Most enzymes are named for the type of reaction they catalyze. *Hydrolases* (hi′druh-lās-es) add water during hydrolysis reactions and *oxidases* (ok′sĭ-dās-es) oxidize reactants by adding oxygen or removing hydrogen. You can recognize most enzyme names by the suffix -*ase*.

In many cases, enzymes are part of cellular membranes. They are often grouped together so that the product of one enzyme-catalyzed reaction becomes the substrate of the neighboring enzyme, and so on. Some enzymes are produced in an inactive form and must be activated in some way before they can function, often by a change in the pH of their surroundings. For example, digestive enzymes produced in the pancreas are activated in the small intestine, where they actually do their work. If they were produced in active form, the pancreas would digest itself.

Sometimes, enzymes are inactivated immediately after they have performed their catalytic function. This is true of enzymes that promote blood clot formation when the wall of a blood vessel is damaged. Once clotting is triggered, those enzymes are inactivated. Otherwise, you would have blood vessels full of solid blood instead of one protective clot. (Eek!)

Enzyme Action

How do enzymes perform their catalytic role? Every chemical reaction requires that a certain amount of energy, called **activation energy**, be absorbed to prime the reaction. The activation energy is needed to alter the bonds of the reactants so that they can be rearranged to become the product. It is present when kinetic energy pushes the reactants to an energy level where their random collisions are forceful enough to ensure interaction. Activation energy is needed regardless of whether the overall reaction is ultimately energy absorbing or energy releasing.

One way to increase kinetic energy is to increase the temperature, but this is not possible in the body because higher temperatures would denature proteins. Enzymes allow reactions to occur at normal body temperature by decreasing the amount of activation energy required (**Figure 2.23**). An enzyme speeds up a reaction by lowering the barrier. Think of a runner slowly climbing a 12-foot wall versus one that is running over hurdles.

Exactly how do enzymes accomplish this remarkable feat? The answer is not fully understood. However, we know that, due to structural and electrostatic factors, they decrease the randomness of reactions by binding to the reacting molecules temporarily and presenting them to each other in the proper position for chemical interaction (bond making or breaking) to occur.

Three basic steps are involved in enzyme action (**Figure 2.24**).

① **Substrate(s) bind to the enzyme's active site, temporarily forming an enzyme-substrate complex.** Substrate binding causes the active site to change shape so that the substrate and the active site fit together precisely, and in an orientation that favors reaction. Although enzymes are specific for particular substrates, other (nonsubstrate) molecules may act as *enzyme inhibitors* if their structure is similar enough to occupy or block the enzyme's active site.

② **The enzyme-substrate complex undergoes internal rearrangements that form the product(s).**

③ **The enzyme releases the product(s) of the reaction.** If the enzyme became part of the product, it would be a reactant and not a catalyst. The enzyme is not changed and returns to its original shape, available to catalyze another reaction.

Because enzymes are unchanged by their catalytic role and can act again and again, cells need only small amounts of each enzyme. Catalysis occurs with incredible speed. Most enzymes can catalyze millions of reactions per minute.

Check Your Understanding

24. What does the name "amino acid" tell you about the structure of this molecule?

Figure 2.23 Enzymes lower the activation energy required for a reaction.

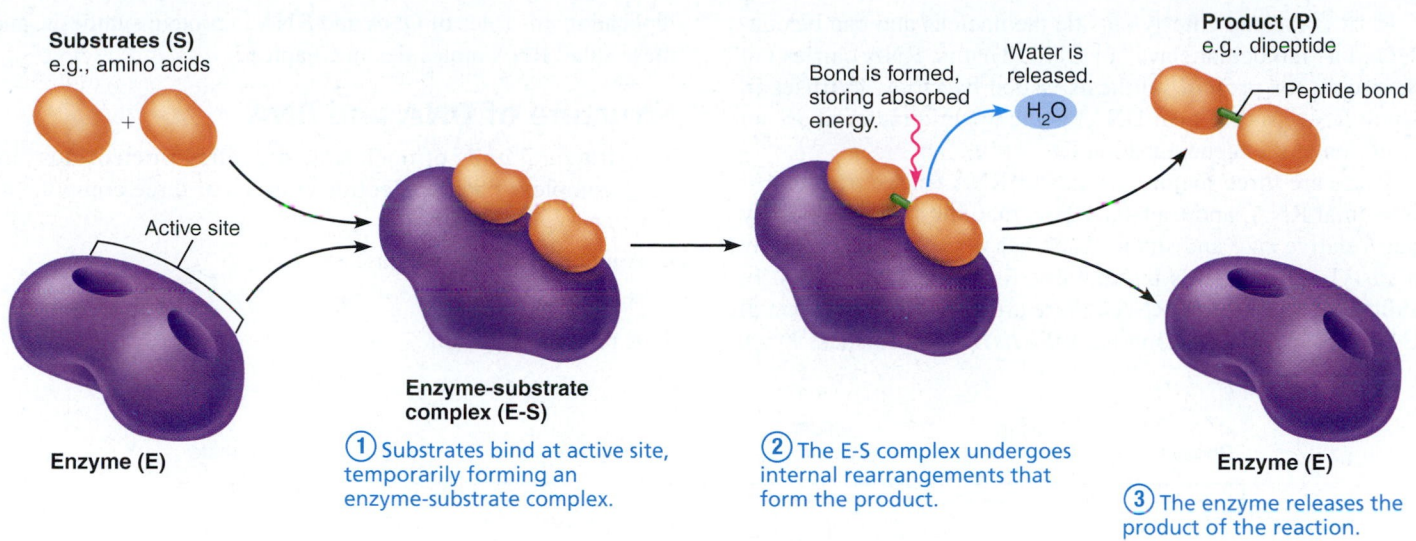

Substrates (S)
e.g., amino acids

Active site

Enzyme (E)

Enzyme-substrate complex (E-S)

① Substrates bind at active site, temporarily forming an enzyme-substrate complex.

Bond is formed, storing absorbed energy.

Water is released.

H_2O

② The E-S complex undergoes internal rearrangements that form the product.

Product (P)
e.g., dipeptide

Peptide bond

Enzyme (E)

③ The enzyme releases the product of the reaction.

Figure 2.24 Mechanism of enzyme action. In this example, the enzyme catalyzes the formation of a dipeptide from specific amino acids. *Summary*: $E + S \rightarrow E\text{-}S \rightarrow P + E$

25. What is the primary structure of proteins?
26. What are the two types of secondary structure in proteins?
27. DRAW Figure 2.23 shows the effect of an enzyme on the activation energy of an exergonic reation. Draw two similar graphs showing the effect of an enzyme on an endergonic reaction (see p. 37 to refresh your memory on these terms).

For answers, see Answers Appendix.

2.11 DNA and RNA store, transmit, and help express genetic information

Learning Outcome

▶ **Compare and contrast DNA and RNA.**

The **nucleic acids** (nu-kle′ic), composed of carbon, oxygen, hydrogen, nitrogen, and phosphorus, are the largest molecules in the body. The nucleic acids include two major classes of molecules, **deoxyribonucleic acid (DNA)** (de-ok″sĭ-ri″bo-nu-kle′ik) and **ribonucleic acid (RNA)**.

Roles of DNA and RNA

DNA and RNA have different roles in the cell, as summarized in **Table 2.4**. Typically, DNA is found in the nucleus (control center) of the cell, where it constitutes the *genetic material*, also called the *genes*, or more recently the *genome*. DNA has two fundamental roles: It replicates (reproduces) itself before a cell divides, ensuring that the genetic information in the descendant cells is identical, and it provides the basic instructions for building every protein in the body. Although we have said that enzymes govern all chemical reactions, remember that enzymes, too, are proteins formed at the direction of DNA.

By providing the information for protein synthesis, DNA determines what type of organism you will be—frog, human, oak tree—and directs your growth and development. It also accounts for your uniqueness. DNA fingerprinting can help solve forensic mysteries (for example, verify one's presence at a crime scene), identify badly burned or mangled bodies after a disaster, and establish or disprove paternity. DNA fingerprinting analyzes tiny samples of DNA taken from blood, semen, or other body tissues and shows the results as a "genetic barcode" that distinguishes each of us from all others.

Table 2.4	Comparison of DNA and RNA	
CHARACTERISTIC	**DNA**	**RNA**
Major cellular site	Nucleus	Cytoplasm (cell area outside the nucleus)
Major functions	Is the genetic material; directs protein synthesis; replicates itself before cell division	Carries out the genetic instructions for protein synthesis
Structure	Double strand coiled into a double helix	Single strand, straight or folded
Sugar	Deoxyribose	Ribose
Bases	Adenine (A), guanine (G), cytosine (C), thymine (T)	Adenine (A), guanine (G), cytosine (C), uracil (U)

RNA is located chiefly outside the nucleus and can be considered a "molecular slave" of DNA. That is, RNA carries out the orders for protein synthesis issued by DNA. [Viruses in which RNA (rather than DNA) is the genetic material are an exception to this generalization.]

There are three major varieties of RNA (messenger RNA, ribosomal RNA, and transfer RNA) that are distinguished by their relative size and shape. Each has a specific role to play in carrying out DNA's instructions for protein synthesis. In addition to these three RNAs, there are several types of small RNA molecules, including *microRNAs*. We will discuss DNA replication, the roles of DNA and RNA in protein synthesis, and these small RNA molecules in Chapter 3.

Structure of DNA and RNA

The structural units of nucleic acids, called **nucleotides**, are quite complex. Each nucleotide consists of three components (**Figure 2.25a**):

- A nitrogen-containing base
- A pentose sugar
- A phosphate group

Figure 2.25 Structure of DNA. (a) Two nucleotides—the structural units of DNA. The nucleotides are linked by hydrogen bonds (shown by dotted lines) between their complementary bases. **(b)** DNA is a *double helix*—a coiled double polymer of nucleotides.

Table 2.5 Summary of Monomers and Polymers of Some Organic Molecules		
CLASS OF ORGANIC MOLECULE*	MONOMERS (BUILDING BLOCKS)	POLYMER
Carbohydrates	Monosaccharides (e.g., glucose)	Polysaccharides
Proteins	Amino acids	Polypeptides or proteins
Nucleic acids	Nucleotides	DNA or RNA

*Lipids are not included in this table because they do not form polymers.

The synthesis of a nucleotide involves the attachment of a base and a phosphate group to the pentose sugar.

Five major varieties of nitrogen-containing bases can contribute to nucleotide structure: **adenine**, abbreviated **A** (ad′ĕ-nēn); **guanine**, **G** (gwah′nēn); **cytosine**, **C** (si′to-sēn); **thymine**, **T** (thi′mēn); and **uracil**, **U** (u′rah-sil). Adenine and guanine are large, two-ring bases (called purines), whereas cytosine, thymine, and uracil are smaller, single-ring bases (called pyrimidines).

DNA is a long, double-stranded polymer—a double chain of nucleotides (Figure 2.25b). The bases in DNA are A, G, C, and T, and its pentose sugar is *deoxyribose* (as reflected in its name). Its two nucleotide chains are held together by hydrogen bonds between the bases, so that a ladderlike molecule is formed. Alternating sugar and phosphate components of each chain form the *backbones* or "uprights" of the "ladder," and the joined bases form the "rungs." The whole molecule is coiled into a spiral staircase–like structure called a **double helix**.

Bonding of the bases is very specific: A always bonds to T, and G always bonds to C. A and T are therefore called **complementary bases**, as are C and G. According to these base-pairing rules, ATGA on one DNA nucleotide strand would necessarily be bonded to TACT (a complementary base sequence) on the other strand.

RNA molecules are single strands of nucleotides. RNA bases include A, G, C, and U (U replaces the T found in DNA), and its sugar is *ribose* instead of deoxyribose.

Table 2.5 offers a quick summary of the monomers and polymers of three types of macromolecules.

Check Your Understanding

28. How do DNA and RNA differ in the bases and sugars they contain?

29. PREDICT Look at Figure 2.25. Which base pair (A–T or C–G) do you think is less likely to come apart? Why?

For answers, see Answers Appendix.

2.12 ATP transfers energy to other compounds

Learning Outcome

▶ Explain the role of ATP in cell metabolism.

Glucose is the most important cellular fuel, but none of the chemical energy contained in its bonds is used directly to power cellular work. Instead, energy released during glucose

catabolism is coupled to the synthesis of **adenosine triphosphate (ATP)**. In other words, some of this energy is captured and stored as small packets of energy in the bonds of ATP. ATP is the primary energy-transferring molecule in cells and it provides a form of energy that is immediately usable by all body cells. It is the dollar of the cell's metabolic economy.

Structurally, ATP is an adenine-containing RNA nucleotide to which two additional phosphate groups have been added (**Figure 2.26**). Chemically, the triphosphate tail of ATP can be compared to a tightly coiled spring ready to uncoil with tremendous energy when the catch is released. Actually, ATP can store energy because its three negatively charged phosphate groups are closely packed and repel each other. When its terminal (third) high-energy phosphate bond is broken (hydrolyzed), the chemical "spring" relaxes and the molecule as a whole becomes more stable.

Cells tap ATP's bond energy during coupled reactions. An enzyme transfers the terminal phosphate group from ATP to another molecule in a process called *phosphorylation*. These newly *phosphorylated* molecules temporarily become more energetic and capable of performing some type of cellular work. In the process of doing that work, they lose the phosphate

Figure 2.26 Structure of ATP (adenosine triphosphate). ATP is an adenine nucleotide to which two additional phosphate groups have been attached during breakdown of food fuels.

group. The amount of energy released and transferred during ATP hydrolysis corresponds closely to that needed to drive most biochemical reactions. As a result, cells are protected from excessive energy release that might be damaging, and energy squandering is kept to a minimum.

Breaking the terminal phosphate bond of ATP yields a molecule with two phosphate groups—*adenosine diphosphate* (*ADP*)—and an inorganic phosphate group, indicated by P_i, accompanied by a transfer of energy:

$$ATP + H_2O \rightleftarrows ADP + P_i + energy$$

As ATP is hydrolyzed to provide energy for cellular needs, ADP accumulates. Breaking the terminal phosphate bond of ADP liberates a similar amount of energy and produces adenosine monophosphate (AMP) (see Figure 2.26).

The cell's ATP supplies are replenished as glucose and other fuel molecules are oxidized and their bond energy is released. The same amount of energy that is liberated when ATP's terminal phosphates are split off must be captured and used to reverse the reaction to reattach phosphates and re-form the energy-transferring phosphate bonds. Without ATP, molecules cannot be made or degraded, cells cannot transport substances across their membranes, muscles cannot shorten to tug on other structures, and life processes cease (**Figure 2.27**).

Check Your Understanding

30. Glucose is an energy-rich molecule. So why do body cells need ATP?

31. What change occurs in ATP when it releases energy?

For answers, see Answers Appendix.

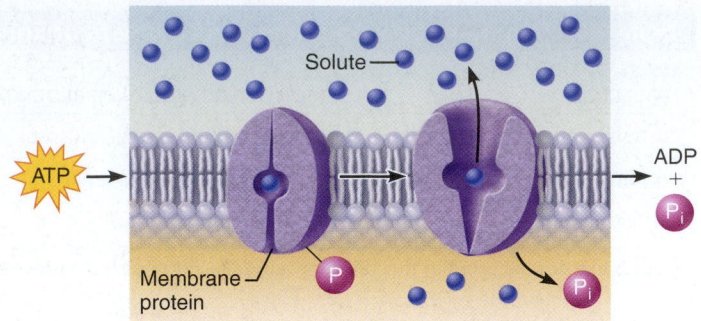

(a) **Transport work:** ATP phosphorylates transport proteins, activating them to transport solutes (ions, for example) across cell membranes.

(b) **Mechanical work:** ATP phosphorylates contractile proteins in muscle cells so the cells can contract (shorten).

(c) **Chemical work:** ATP phosphorylates key reactants, providing energy to drive energy-absorbing chemical reactions.

Figure 2.27 Three examples of cellular work driven by energy from ATP.

RELATED CLINICAL TERMS

Acidosis (as″ĭ-do′sis; *acid* = sour, sharp) A condition of acidity or low pH (below 7.35) of the blood; high hydrogen ion concentration.

Alkalosis (al″kah-lo′sis) A condition of basicity or high pH (above 7.45) of the blood; low hydrogen ion concentration.

Heavy metals Metals with toxic effects on the body, including arsenic, mercury, and lead. Iron, also included in this group, is toxic in high concentrations.

Ionizing radiation Radiation that causes atoms to ionize; for example, gamma rays and X rays.

Radiation sickness Disease resulting from exposure of the body to radioactivity; rapidly dividing cells are most affected—for example, blood-forming cells and the cells lining the digestive tract.

CHAPTER SUMMARY

PART 1

BASIC CHEMISTRY

2.1 Matter is the stuff of the universe and energy moves matter (pp. 24–25)

Matter (p. 24)

1. Matter is anything that takes up space and has mass.

Energy (pp. 24–25)

2. Energy is the capacity to do work or put matter into motion.
3. Energy exists as potential energy (stored energy) and kinetic energy (active or working energy).
4. Forms of energy involved in body functioning are chemical, electrical, radiant, and mechanical. Of these, chemical (bond) energy is most important.
5. Energy may be converted from one form to another, but some energy is always unusable (lost as heat) in such transformations.

2.2 The properties of an element depend on the structure of its atoms (pp. 25–28)

1. Elements are unique substances that cannot be decomposed into simpler substances by ordinary chemical methods. Four elements (carbon, hydrogen, oxygen, and nitrogen) make up 96% of body weight.
2. The building blocks of elements are atoms.

Structure of Atoms (pp. 25–27)

3. Atoms are composed of positively charged protons, negatively charged electrons, and uncharged neutrons. Protons and neutrons are located in the atomic nucleus, constituting essentially the atom's total mass. Electrons are outside the nucleus in the electron shells. In any atom, the number of electrons equals the number of protons.

Identifying Elements (pp. 27–28)

4. Atoms may be identified by their atomic number (p^+) and mass number ($p^+ + n^0$). The notation ^4_2He means that helium (He) has an atomic number of 2 and a mass number of 4.
5. Isotopes of an element differ in the number of neutrons they contain. The atomic weight of any element is approximately equal to the mass number of its most abundant isotope.

Radioisotopes (p. 28)

6. Many heavy isotopes are unstable (radioactive). These so-called radioisotopes decompose to more stable forms by emitting alpha or beta particles or gamma rays. Radioisotopes are useful in medical diagnosis and treatment and in biochemical research.

2.3 Atoms bound together form molecules; different molecules can make mixtures (pp. 28–30)

Molecules and Compounds (pp. 28–29)

1. A molecule is the smallest unit resulting from the chemical bonding of two or more atoms. If the atoms are different, they form a molecule of a compound.

Mixtures (pp. 29–30)

2. Mixtures are physical combinations of solutes in a solvent. Mixture components retain their individual properties.

3. The types of mixtures, in order of increasing solute size, are solutions, colloids, and suspensions.
4. Solution concentrations are typically designated in terms of percent or molarity.

Distinguishing Mixtures from Compounds (p. 30)

5. Compounds are homogeneous; their elements are chemically bonded. Mixtures may be homogeneous or heterogeneous; their components are physically combined and separable.

2.4 Three types of chemical bonds are ionic, covalent, and hydrogen (pp. 31–35)

The Role of Electrons in Chemical Bonding (pp. 31–32)

1. Electrons of an atom occupy areas of space called electron shells or energy levels. Electrons in the shell farthest from the nucleus (valence shell) are most energetic.
2. Chemical bonds are energy relationships between valence shell electrons of the reacting atoms. Atoms with a full valence shell or eight valence shell electrons are chemically unreactive (inert). Those with an incomplete valence shell interact with other atoms to achieve stability.

Types of Chemical Bonds (pp. 32–35)

3. Ionic bonds are formed when valence shell electrons are completely transferred from one atom to another.
4. Covalent bonds are formed when atoms share electron pairs. If the electron pairs are shared equally, the molecule is nonpolar. If they are shared unequally, it is polar (a dipole).
5. Hydrogen bonds are weak bonds formed between one hydrogen atom, already covalently linked to an electronegative atom, and another electronegative atom (such as nitrogen or oxygen). They bind together different molecules (e.g., water molecules) or different parts of the same molecule (as in protein molecules).

2.5 Chemical reactions occur when electrons are shared, gained, or lost (pp. 35–38)

Chemical Equations (pp. 35–36)

1. Chemical reactions involve the formation, breaking, or rearrangement of chemical bonds.

Types of Chemical Reactions (pp. 36–37)

2. Chemical reactions are either anabolic (constructive) or catabolic (destructive). They include synthesis, decomposition, and exchange reactions. Oxidation-reduction reactions may be considered a special type of exchange (or decomposition) reaction.

Energy Flow in Chemical Reactions (p. 37)

3. Bonds are energy relationships and there is a net loss or gain of energy in every chemical reaction.
4. In exergonic reactions, energy is liberated. In endergonic reactions, energy is absorbed.

Reversibility of Chemical Reactions (p. 37)

5. If reaction conditions remain unchanged, all chemical reactions eventually reach a state of chemical equilibrium in which the reaction proceeds in both directions at the same rate.
6. All chemical reactions are theoretically reversible, but many biological reactions go in only one direction because of energy requirements or the removal of reaction products.

Factors Influencing the Rate of Chemical Reactions (pp. 37–38)

7. Chemical reactions occur only when particles collide and valence shell electrons interact.

8. The smaller the reacting particles, the greater their kinetic energy and the faster the reaction rate. Higher temperature or reactant concentration, as well as the presence of catalysts, increases chemical reaction rates.

PART 2
BIOCHEMISTRY

2.6 Inorganic compounds include water, salts, and many acids and bases (pp. 38–41)

1. Most inorganic compounds do not contain carbon. Those found in the body include water, salts, and inorganic acids and bases.

Water (p. 38)

2. Water is the single most abundant compound in the body. It absorbs and releases heat slowly, acts as a universal solvent, participates in chemical reactions, and cushions body organs.

Salts (p. 39)

3. Salts are ionic compounds that dissolve in water and act as electrolytes. Calcium and phosphorus salts contribute to the hardness of bones and teeth. Ions of salts are involved in many physiological processes.

Acids and Bases (pp. 39–41)

4. Acids are proton donors; in water, they ionize and dissociate, releasing hydrogen ions (which account for their properties) and anions.

5. Bases are proton acceptors. The most common inorganic bases are the hydroxides; bicarbonate ion and ammonia are important bases in the body.

6. pH is a measure of hydrogen ion concentration of a solution (in moles per liter). A pH of 7 is neutral; a higher pH is alkaline, and a lower pH is acidic. Normal blood pH is 7.35–7.45. Buffers help to prevent excessive changes in the pH of body fluids.

2.7 Organic compounds are made by dehydration synthesis and broken down by hydrolysis (pp. 41–43)

1. Organic compounds contain carbon. Those found in the body include carbohydrates, lipids, proteins, and nucleic acids, all of which are synthesized by dehydration synthesis and digested by hydrolysis. All of these biological molecules contain C, H, and O. Proteins and nucleic acids also contain N.

2.8 Carbohydrates provide an easily used energy source for the body (pp. 43–45)

1. Carbohydrate building blocks are monosaccharides, the most important of which are hexoses (glucose, fructose, galactose) and pentoses (ribose, deoxyribose).

2. Disaccharides (sucrose, lactose, maltose) and polysaccharides (starch, glycogen) are composed of linked monosaccharide monomers.

3. Carbohydrates, particularly glucose, are the major energy fuel for forming ATP. Excess carbohydrates are stored as glycogen or converted to fat for storage.

2.9 Lipids insulate body organs, build cell membranes, and provide stored energy (pp. 45–48)

1. Lipids dissolve in fats or organic solvents, but not in water.

2. Triglycerides are composed of three fatty acid chains and glycerol. They are found chiefly in fatty tissue where they provide insulation and reserve body fuel. Unsaturated fatty acid chains produce oils. Saturated fatty acids produce solid fats typical of animal fats.

3. Phospholipids are modified phosphorus-containing triglycerides that have polar and nonpolar portions. They are found in all plasma membranes.

4. The steroid cholesterol is found in cell membranes and is the basis of steroid hormones, bile salts, and vitamin D.

2.10 Proteins are the body's basic structural material and have many vital functions (pp. 48–53)

1. The unit of proteins is the amino acid, and 20 common amino acids are found in the body.

2. Many amino acids joined by peptide bonds form a polypeptide. A protein (one or more polypeptides) is distinguished by the number and sequence of amino acids in its chain(s) and by the complexity of its three-dimensional structure.

3. Fibrous proteins, such as keratin and collagen, have secondary (α-helix or β-pleated sheet) and perhaps tertiary and quaternary structure. Fibrous proteins are used as structural materials.

4. Globular proteins achieve tertiary and sometimes quaternary structure and are generally spherical, soluble molecules. Globular proteins (e.g., enzymes, some hormones, antibodies, hemoglobin) perform special functional roles for the cell (e.g., catalysis, molecule transport).

5. Proteins are denatured by extremes of temperature or pH. Denatured globular proteins are unable to perform their usual function.

6. Enzymes are biological catalysts. They increase the rate of chemical reactions by decreasing the amount of activation energy needed. They do this by combining with the reactants and holding them in the proper position to interact. Many enzymes require cofactors to function.

2.11 DNA and RNA store, transmit, and help express genetic information (pp. 53–55)

1. Nucleic acids include deoxyribonucleic acid (DNA) and ribonucleic acid (RNA). DNA specifies protein structure and replicates itself exactly before cell division. RNA is involved in carrying out DNA's instructions for protein synthesis and includes messenger, ribosomal, and transfer RNA.

2. The structural unit of nucleic acids is the nucleotide, which consists of a nitrogenous base (adenine, guanine, cytosine, thymine, or uracil), a sugar (ribose or deoxyribose), and a phosphate group.

3. DNA is a double-stranded helix. It contains deoxyribose and the bases A, G, C, and T.

4. RNA is single stranded. It contains ribose and the bases A, G, C, and U.

2.12 ATP transfers energy to other compounds (pp. 55–56)

1. ATP is the universal energy compound of body cells. Some of the energy liberated by the breakdown of glucose and other food fuels is captured in the bonds of ATP molecules and transferred via coupled reactions to energy-consuming reactions.

REVIEW QUESTIONS

To access additional practice questions using your smartphone, tablet, or computer: MasteringA&P® > Study Area > Practice Tests & Quizzes

(Some multiple choice questions have more than one correct answer. Select the best answer or answers from the choices given.)

Level 1 Remember/Understand

1. Which of the following forms of energy is the stimulus for vision? **(a)** chemical, **(b)** electrical, **(c)** mechanical, **(d)** radiant.
2. All of the following are examples of the four major elements contributing to body mass except **(a)** hydrogen, **(b)** carbon, **(c)** nitrogen, **(d)** sodium, **(e)** oxygen.
3. The subatomic particles responsible for the chemical behavior of atoms are **(a)** electrons, **(b)** ions, **(c)** neutrons, **(d)** protons.
4. Which of the following does *not* describe a mixture? **(a)** properties of its components are retained, **(b)** chemical bonds are formed, **(c)** components can be separated physically, **(d)** includes both heterogeneous and homogeneous examples.
5. In a beaker of water, the water-water bonds can properly be called **(a)** ionic bonds, **(b)** polar covalent bonds, **(c)** nonpolar covalent bonds, **(d)** hydrogen bonds.
6. When a pair of electrons is shared between two atoms, the bond formed is called **(a)** a single covalent bond, **(b)** a double covalent bond, **(c)** a triple covalent bond, **(d)** an ionic bond.
7. Factors that accelerate the rate of chemical reactions include all but **(a)** the presence of catalysts, **(b)** increasing the temperature, **(c)** increasing the particle size, **(d)** increasing the concentration of the reactants.
8. Which of the following molecules is an inorganic molecule? **(a)** sucrose, **(b)** cholesterol, **(c)** collagen, **(d)** sodium chloride.
9. Acids **(a)** release hydroxyl ions when dissolved in water, **(b)** are proton acceptors, **(c)** cause the pH of a solution to rise, **(d)** release protons when dissolved in water.
10. A triglyceride consists of **(a)** glycerol plus three fatty acids, **(b)** a sugar-phosphate backbone to which two amino groups are attached, **(c)** two to several hexoses, **(d)** amino acids that have been thoroughly saturated with hydrogen.
11. The lipid(s) used as the basis of vitamin D, sex hormones, and bile salts is/are **(a)** triglycerides, **(b)** cholesterol, **(c)** phospholipids, **(d)** prostaglandin.
12. Provide the atomic symbol for each of the following elements: **(a)** calcium, **(b)** carbon, **(c)** hydrogen, **(d)** iron, **(e)** nitrogen, **(f)** oxygen, **(g)** potassium, **(h)** sodium.

Level 2 Apply/Analyze

13. A deficiency in this element can be expected to reduce the hemoglobin content of blood: **(a)** Fe, **(b)** I, **(c)** F, **(d)** Ca, **(e)** K.
14. Which of the following covalently bonded molecules are polar?

(a)   (b)   (c)   (d)

15. Identify each reaction as one of the following: **(a)** a synthesis reaction, **(b)** a decomposition reaction, **(c)** an exchange reaction.

    _____ **(1)** $2Hg + O_2 \rightarrow 2HgO$

    _____ **(2)** $HCl + NaOH \rightarrow NaCl + H_2O$

16. A chemist, during the course of an analysis, runs across a chemical composed of carbon, hydrogen, and oxygen in the proportion 1:2:1 and having a six-sided molecular shape. It is probably **(a)** a pentose, **(b)** an amino acid, **(c)** a fatty acid, **(d)** a monosaccharide, **(e)** a nucleic acid.
17. How many moles of aspirin, $C_9H_8O_4$, are in a bottle containing 450 g by weight? (*Note*: The approximate atomic weights of its atoms are C = 12, H = 1, and O = 16.)
18. The following equation, which represents the oxidative breakdown of glucose by body cells, is a reversible reaction.

    $$\text{Glucose} + \text{oxygen} \rightarrow \text{carbon dioxide} + \text{water} + \text{ATP}$$

    (a) How can you indicate that the reaction is reversible?
    (b) How can you indicate that the reaction is in chemical equilibrium?
    (c) Define chemical equilibrium.

Level 3 Evaluate/Synthesize

19. Consider the following information about three atoms:

    $$^{12}_{6}C \qquad ^{13}_{6}C \qquad ^{14}_{6}C$$

    (a) How are they similar to one another? **(b)** How do they differ from one another? **(c)** What are the members of such a group of atoms called? **(d)** Using the planetary model, draw the atomic configuration of $^{12}_{6}C$ showing the relative position and numbers of its subatomic particles.
20. What are hydrogen bonds and how are they important in the body?
21. Differentiate clearly between primary, secondary, and tertiary protein structure.
22. Describe the mechanism of enzyme action.
23. Explain why, if you pour water into a glass very carefully, you can "stack" the water slightly above the rim of the glass.
24. Some antibiotics act by binding to certain essential enzymes in the target bacteria. **(a)** How might these antibiotics influence the chemical reactions controlled by the enzymes? **(b)** What is the anticipated effect on the bacteria? On the person taking the antibiotic prescription?
25. Mrs. Roberts, in a diabetic coma, has just been admitted to Noble Hospital. Her blood pH indicates that she is in severe acidosis (low blood pH), and measures are quickly instituted to bring her blood pH back within normal limits. **(a)** Define pH and note the normal pH of blood. **(b)** Why is severe acidosis a problem?
26. Jason, a 12-year-old boy, was awakened suddenly by a loud crash. As he sat up in bed, straining to listen, his fright was revealed by his rapid breathing (hyperventilation), a breathing pattern effective in ridding the blood of CO_2. At this point, was his blood pH rising or falling?
27. After you eat a protein bar, which chemical reactions introduced in this chapter must occur for the amino acids in the protein bar to be converted into proteins in your body cells?

3 Cells: The Living Units

In this chapter, you will learn that

3.1 Cells are the smallest unit of life

by exploring

Part 1 Plasma Membrane

by asking

3.2 What is the structure of the plasma membrane?

How do substances move across the plasma membrane?

looking closer at

3.3 Passive membrane transport

3.4 Active membrane transport

3.5 How does a cell generate a voltage across its plasma membrane?

3.6 How does the plasma membrane allow the cell to interact with its environment?

Part 2 Cytoplasm

looking closer at

Cytosol

Inclusions

3.7 Cytoplasmic organelles

exploring

3.8 Cellular extensions

Part 3 Nucleus

by investigating

3.9 The structure of the nucleus

and asking

3.10 How does a cell grow and divide?

3.11 What are the roles of DNA and RNA in protein synthesis?

and further asking

3.12 How are cells, organelles, and proteins destroyed?

and finally, exploring

Developmental Aspects of Cells

CAREER CONNECTION

Watch a video to learn how the chapter content is used in a real health care setting. Go to Mastering A&P® > Study Area > Animations and Videos or use quick access URL https://goo.gl/88srfV

Just as bricks and timbers are the structural units of a house, **cells** are the structural units of all living things, from one-celled "generalists" like amoebas to complex multicellular organisms such as humans, dogs, and trees. The human body has 50 to 100 trillion of these tiny building blocks.

This chapter focuses on structures and functions shared by all cells. We address specialized cells and their unique functions in later chapters.

3.1 Cells are the smallest unit of life

Learning Outcomes

▶ **Define cell.**

▶ **Name and describe the composition of extracellular materials.**

▶ **List the three major regions of a generalized cell and their functions.**

Since the late 1800s, cell research has been exceptionally fruitful and has provided us with three concepts collectively known as the **cell theory**:

- **The cell is the smallest unit of life.** When you define the properties of cells, you define the properties of life.

- **All organisms are made of one or more cells.** Cells are the structural and functional building blocks of an organism. Different cell types have different functions within an organism, and the activity of an organism depends on the activities of individual cells and of all of the cells together. According to the *principle of complementarity of structure and function*, the activities of cells are dictated by their shapes, and by the types and relative numbers of the subcellular structures they contain.

- **Cells only arise from other cells.** Most body cells arise by *mitosis*, which we explore later in this chapter. Sperm and ovum (egg) cells arise by a related process called *meiosis* (mi-o′sis), which we describe in Chapter 27.

We will expand on all of these concepts as we progress. Let us begin with the idea that the cell is the smallest living unit. Whatever its form, however it behaves, the cell is the microscopic package that contains all the parts necessary to survive in an ever-changing world. It follows then that loss of homeostasis in cells underlies virtually every disease.

The trillions of cells in the human body include over 250 different cell types that vary greatly in shape, size, and function (**Figure 3.1**). The disc-shaped red blood cells, branching nerve cells, and cubelike cells of kidney tubules are just a few examples of the shapes cells take. Cells also vary in length—ranging from 2 micrometers (1/12,000 of an inch) in the smallest cells to over a meter in the nerve cells that cause you to wiggle your toes. A cell's shape reflects its function. For example, the flat, tilelike epithelial cells that line the inside of your cheek fit closely together, forming a living barrier that protects underlying tissues from bacterial invasion.

(a) Cells that connect body parts, form linings, or transport gases

(b) Cells that move organs and body parts

(c) Cell that stores nutrients

(d) Cell that fights disease

(e) Cell that gathers information and controls body functions

(f) Cell of reproduction

Figure 3.1 Cell diversity. (Note that cells are not drawn to the same scale.)

Chromatin

Nucleolus

Smooth endoplasmic reticulum

Cytoplasm

Mitochondrion

Lysosome

Centrosome matrix

Centrioles

Cytoskeletal elements
• Microtubule
• Intermediate filaments

Nuclear envelope

Nucleus

Plasma membrane

Rough endoplasmic reticulum

Ribosomes

Golgi apparatus

Secretion being released from cell by exocytosis

Peroxisome

Figure 3.2 Structure of the generalized cell. No cell is exactly like this one, but this composite illustrates features common to many human cells. Note that not all of the organelles are drawn to the same scale in this illustration.

Regardless of these differences, all cells have the same basic parts and some common functions. For this reason, we will describe a *generalized*, or *composite*, cell (**Figure 3.2**).

A human cell has three main parts:

- The *plasma membrane*: the outer boundary of the cell, which acts as a selectively permeable barrier.

- The *cytoplasm* (si′to-plazm): the intracellular fluid packed with *organelles*, small structures that perform specific cell functions.

- The *nucleus* (nu′kle-us): an organelle that controls cellular activities. Typically the nucleus lies near the cell's center.

Extracellular Materials

Although we tend to think of the body as collections of cells—and it *is* that—it is impossible to discuss cells and their activities

without saying something about extracellular materials. So—let's do that before going on to details about the generalized cell.

First of all, what are extracellular materials? **Extracellular materials** are substances contributing to body mass that are found outside the cells. The major classes of extracellular materials are:

- **Extracellular fluid. Extracellular fluid (ECF)** includes **interstitial fluid**, blood plasma, and cerebrospinal fluid. ECF dissolves and transports substances in the body. Interstitial fluid is the fluid in tissues that bathes all of our cells, and has endless major roles to play. Like a rich, nutritious "soup," interstitial fluid contains thousands of ingredients, including amino acids, sugars, fatty acids, regulatory substances, and wastes. To remain healthy, each cell must extract from this mix the exact amounts of the substances it needs depending on present conditions.

- **Cellular secretions.** These secretions include substances that aid in digestion (intestinal and gastric fluids) and some that act as lubricants (saliva, mucus, and serous fluids).

- **Extracellular matrix.** The **extracellular matrix** is the most abundant extracellular material. Most body cells are in contact with a jellylike substance composed of proteins and polysaccharides. Secreted by the cells, these molecules self-assemble into an organized mesh in the extracellular space, where they serve as a universal "cell glue" that helps bind body cells together. As described in Chapter 4, the extracellular matrix is particularly abundant in connective tissues—in some cases so abundant that it (rather than living cells) accounts for the bulk of that tissue type. Depending on the structure to be formed, the extracellular matrix in connective tissue ranges from soft to rock-hard.

Check Your Understanding

1. Summarize the three key points of the cell theory.
2. What are the three main parts of a human cell?

For answers, see Answers Appendix.

PART 1

PLASMA MEMBRANE

The flexible **plasma membrane** separates two of the body's major fluid compartments—the **intracellular fluid** within cells and the *extracellular fluid* outside cells. The term *cell membrane* is commonly used as a synonym for plasma membrane, but because nearly all cellular organelles are enclosed in a membrane, in this book we will always refer to the cell's surface, or outer limiting membrane, as the plasma membrane. The plasma membrane is much more than a passive envelope. As you will see, its unique structure allows it to play a dynamic role in cellular activities.

3.2 The plasma membrane is a double layer of phospholipids with embedded proteins

Learning Outcomes

▶ Describe the chemical composition of the plasma membrane and relate it to membrane functions.
▶ Compare the structure and function of tight junctions, desmosomes, and gap junctions.

The **fluid mosaic model** of membrane structure depicts the plasma membrane as an exceedingly thin (7–10 nm) structure composed of a double layer, or bilayer, of lipid molecules with protein molecules "plugged into" or dispersed in it. The proteins, many of which float in the fluid *lipid bilayer*, form a constantly changing mosaic pattern. The model is named for this characteristic. In *Focus on the Plasma Membrane* (**Focus Figure 3.1** on pp. 64–65) we build a plasma membrane one key player at a time. Study this figure carefully before you read about the key players in more detail next.

Membrane Lipids

The lipid bilayer forms the basic "fabric" of the membrane. It is constructed largely of *phospholipids*, with smaller amounts of *cholesterol*.

Phospholipids

Recall from Chapter 2 (◀ **pp. 46–47**) that the polar *hydrophilic* heads of phospholipids (shown schematically at right) are attracted to water—the main constituent of both the intracellular and extracellular fluids. As a result they lie on both the inner and outer surfaces of the membrane. The nonpolar tails of phospholipids, being *hydrophobic*, avoid water and line up in the center of the membrane.

Polar hydrophilic head (includes phosphate group)

Nonpolar hydrophobic tail (fatty acids)

The result is that all plasma membranes, indeed all biological membranes, share a sandwich-like structure: They consist of two parallel sheets of phospholipid molecules lying tail to tail, with their polar heads bathed in water on either side of the membrane. This self-orienting property of phospholipids encourages biological membranes to self-assemble into generally spherical structures (as shown at right) and to reseal themselves when torn.

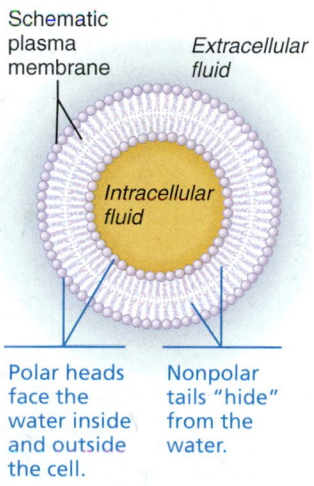

Schematic plasma membrane

Extracellular fluid

Intracellular fluid

Polar heads face the water inside and outside the cell.

Nonpolar tails "hide" from the water.

With a consistency similar to olive oil, the plasma membrane is a dynamic fluid structure in constant flux. Its phospholipids move freely from side to side, parallel to the membrane surface, but because of their self-orienting properties, they rarely flip-flop or move from one half of the bilayer to the other half. The inward-facing and outward-facing surfaces of the plasma membrane differ in the kinds and amounts of lipids they contain, and these variations help to determine local membrane structure and function.

Cholesterol

Some 20% of membrane lipid is cholesterol. Like phospholipids, cholesterol has a polar region (its hydroxyl group) and a nonpolar region (its fused ring system). It wedges its platelike hydrocarbon rings between the phospholipid tails, which stiffens the membrane.

Membrane Proteins

A cell's plasma membrane bristles with proteins that allow it to communicate with its environment. Proteins make up about half of the plasma membrane by mass and are responsible for

The plasma membrane is a phospholipid bilayer with embedded proteins arranged as a fluid mosaic.

Let's build a membrane by adding one key player at a time:

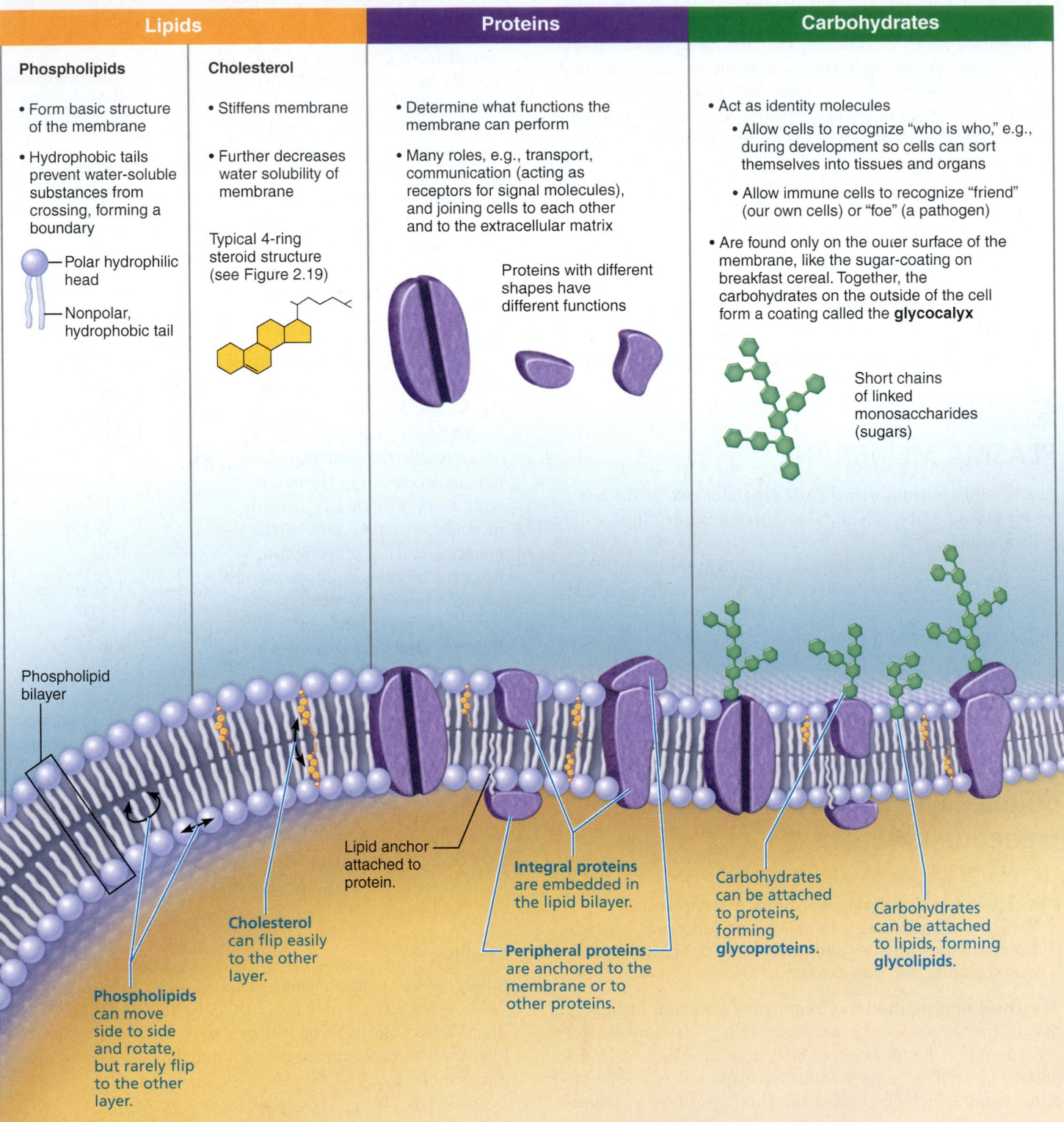

Lipids		Proteins	Carbohydrates

Phospholipids

- Form basic structure of the membrane

- Hydrophobic tails prevent water-soluble substances from crossing, forming a boundary

— Polar hydrophilic head

— Nonpolar, hydrophobic tail

Cholesterol

- Stiffens membrane

- Further decreases water solubility of membrane

Typical 4-ring steroid structure (see Figure 2.19)

- Determine what functions the membrane can perform

- Many roles, e.g., transport, communication (acting as receptors for signal molecules), and joining cells to each other and to the extracellular matrix

Proteins with different shapes have different functions

- Act as identity molecules
  - Allow cells to recognize "who is who," e.g., during development so cells can sort themselves into tissues and organs
  - Allow immune cells to recognize "friend" (our own cells) or "foe" (a pathogen)

- Are found only on the outer surface of the membrane, like the sugar-coating on breakfast cereal. Together, the carbohydrates on the outside of the cell form a coating called the **glycocalyx**

Short chains of linked monosaccharides (sugars)

Phospholipid bilayer

Lipid anchor attached to protein.

Cholesterol can flip easily to the other layer.

Phospholipids can move side to side and rotate, but rarely flip to the other layer.

Integral proteins are embedded in the lipid bilayer.

Peripheral proteins are anchored to the membrane or to other proteins.

Carbohydrates can be attached to proteins, forming **glycoproteins**.

Carbohydrates can be attached to lipids, forming **glycolipids**.

Extracellular fluid
(watery environment
outside cell)

Glycocalyx
(carbohydrates)

Outward-facing layer
of phospholipids

Inward-facing layer
of phospholipids

Filament of
cytoskeleton

Cytoplasm
(watery environment
inside cell)

Functions of the Plasma Membrane:

• **Physical barrier:** Encloses the cell, separating the *cytoplasm* from the *extracellular fluid.*

• **Selective permeability:** Determines which substances enter or exit the cell.

• **Communication:** Plasma membrane proteins interact with specific chemical messengers and relay messages to the cell interior.

• **Cell recognition:** Cell surface carbohydrates allow cells to recognize each other.

most of the specialized membrane functions (**Figure 3.3**). Some membrane proteins drift freely along the membrane surface. Others are "tethered" to intracellular or extracellular structures and are restricted in their movement.

There are two distinct populations of membrane proteins, integral and peripheral (Focus Figure 3.1, pp. 64–65).

Integral Proteins

Integral proteins are firmly inserted into the lipid bilayer. Some protrude from one membrane face only, but most are *transmembrane proteins* that span the entire membrane and protrude on both sides. Whether transmembrane or not, all integral proteins have both hydrophobic and hydrophilic regions. This structural feature allows them to interact with both the nonpolar lipid tails buried in the membrane and the water inside and outside the cell.

Some transmembrane proteins are involved in transport, and form *channels*, or pores. Small, water-soluble molecules or ions can move through these pores, bypassing the lipid part of the membrane. Others act as *carriers* that bind to a substance and then move it through the membrane (Figure 3.3a). Some transmembrane proteins are enzymes (Figure 3.3c). Still others are receptors for hormones or other chemical messengers and relay messages to the cell interior—a process called *signal transduction* (Figure 3.3b).

Peripheral Proteins

Unlike integral proteins, **peripheral proteins** are not embedded in the lipid bilayer. Instead, they either attach loosely to integral proteins or have a hydrophobic region that anchors them into the membrane. Peripheral proteins include a network of filaments that helps support the membrane from its cytoplasmic side (Figure 3.3e). Some peripheral proteins are enzymes. Others are motor proteins involved in mechanical functions, such as changing cell shape during cell division and muscle cell contraction. Still others link cells together.

Membrane Carbohydrates and the Glycocalyx

The extracellular surface (but not the intracellular surface) of the membrane is decorated with short branching carbohydrates. These are attached to most of the membrane's proteins and some of the membrane's lipids that are exposed on the exterior surface. Lipids and proteins with sugars attached are called **glycolipids** (gli″ko-lip′idz) and **glycoproteins**, respectively. Glycolipids have two fatty acid tails (like phospholipids), but a carbohydrate replaces the phosphate head group.

The **glycocalyx** (gli″ko-ka′liks; "sugar covering") consists of the fuzzy, sticky, carbohydrate-rich area at the cell surface created by the sugars of glycoproteins and glycolipids. Your cells are sugar-coated like breakfast cereal. The glycocalyx is further enriched by glycoproteins secreted by the cell.

Because every cell type has a different pattern of sugars in its glycocalyx, the glycocalyx provides identity molecules—highly

(a) Transport

• A protein (left) that spans the membrane may provide a hydrophilic channel across the membrane that is selective for a particular solute.
• Some transport proteins (right) hydrolyze ATP as an energy source to actively pump substances across the membrane.

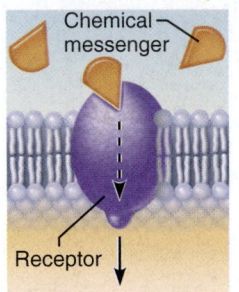

(b) Receptors for signal transduction

• A membrane protein exposed to the outside of the cell may have a binding site that fits the shape of a specific chemical messenger, such as a hormone.
• When bound, the chemical messenger may cause a change in shape in the protein that initiates a chain of chemical reactions in the cell.

(c) Enzymatic activity

• A membrane protein may be an enzyme with its active site exposed to substances in the adjacent solution.
• A team of several enzymes in a membrane may catalyze sequential steps of a metabolic pathway as indicated (left to right) here.

(d) Cell-cell recognition

• Some glycoproteins (proteins bonded to short chains of sugars which help to make up the glycocalyx) serve as identification tags that are specifically recognized by other cells.

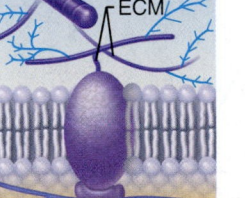

(e) Attachment to the cytoskeleton and extracellular matrix (ECM)

• Elements of the cytoskeleton (cell's internal framework) and the extracellular matrix (fibers and other substances outside the cell) may anchor to membrane proteins.
• Helps maintain cell shape, fixes the location of certain membrane proteins, and plays a role in cell movement.

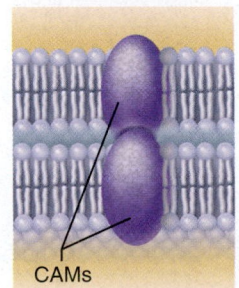

(f) Cell-to-cell joining

• Membrane proteins of adjacent cells may be hooked together in various kinds of intercellular junctions.
• Some membrane proteins (cell adhesion molecules or CAMs) of this group provide temporary binding sites that guide cell migration and other cell-to-cell interactions.

Figure 3.3 Membrane proteins perform many tasks. A single protein may perform a combination of these functions.

specific biological markers by which approaching cells recognize each other (Figure 3.3d). For example, immune system cells use these markers to determine which cells belong in the body and which are foreign.

HOMEOSTATIC IMBALANCE 3.1

CLINICAL

Definite changes occur in the glycocalyx of a cell that is becoming cancerous. In fact, a cancer cell's glycocalyx may change almost continuously, allowing it to keep ahead of immune system recognition mechanisms and avoid destruction. (Cancer is discussed on pp. 144–145.)

Cell Junctions

In many cases, the plasma membranes of adjacent cells are joined together by specialized cell junctions that allow neighboring cells to adhere and sometimes to communicate. These junctions may aid or inhibit movement of molecules between or past cells and also serve to tie cells together into tightly knit communities (**Figure 3.4**).

Let's look at each of the three types of cell junctions.

Tight Junctions

In a **tight junction**, a series of integral protein molecules in the plasma membranes of adjacent cells fuse together like the

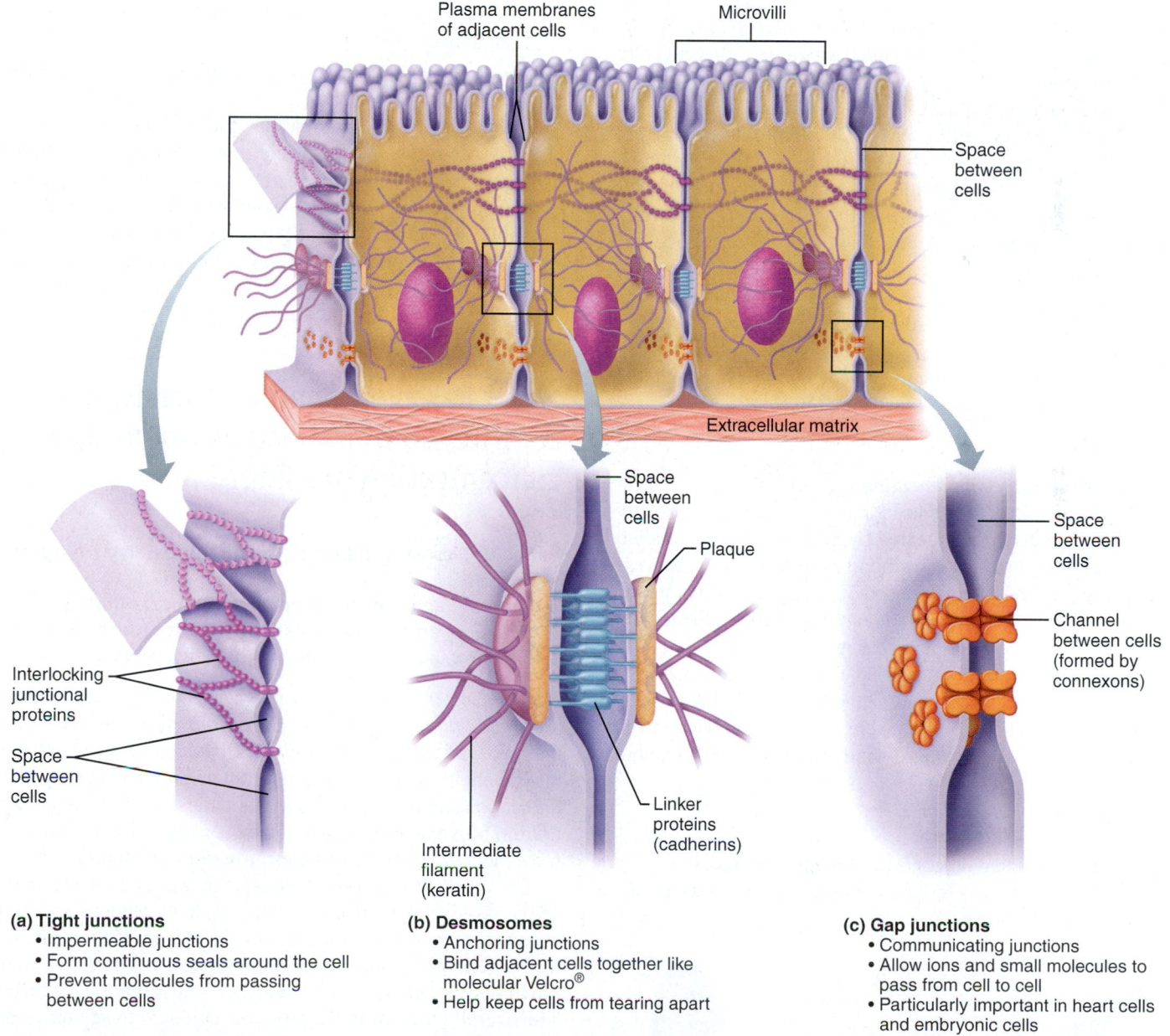

(a) Tight junctions
- Impermeable junctions
- Form continuous seals around the cell
- Prevent molecules from passing between cells

(b) Desmosomes
- Anchoring junctions
- Bind adjacent cells together like molecular Velcro®
- Help keep cells from tearing apart

(c) Gap junctions
- Communicating junctions
- Allow ions and small molecules to pass from cell to cell
- Particularly important in heart cells and embryonic cells

Figure 3.4 Cell junctions. An epithelial cell is shown joined to adjacent cells by three common types of cell junctions. (Note: Except for epithelia, it is unlikely that a single cell will have all three junction types.)

zipper of a Ziploc® bag. This forms an *impermeable junction* that encircles the cell and separates one fluid-filled compartment from another (Figure 3.4a). Tight junctions help prevent molecules from passing through the extracellular space between adjacent cells and restrict the movements of membrane proteins. For example, tight junctions between epithelial cells lining the digestive tract keep digestive enzymes and microorganisms in the intestine from seeping into the bloodstream. (Although called "impermeable" junctions, some tight junctions are leaky and may allow certain ions to pass.)

Desmosomes

Desmosomes (dez′muh-sōmz; "binding bodies") serve as *anchoring junctions*—mechanical couplings scattered like rivets along the sides of adjacent cells to prevent their separation (Figure 3.4b). On the cytoplasmic face of each plasma membrane is a buttonlike thickening called a *plaque*. Adjacent cells are held together by thin linker protein filaments (cadherins) that extend from the plaques and fit together like Velcro® in the intercellular (between cells) space. Thicker keratin filaments (intermediate filaments, which form part of the cytoskeleton) extend from the cytoplasmic side of the plaque across the width of the cell to anchor to the plaque on the cell's opposite side.

In this way, desmosomes bind neighboring cells together into sheets and also contribute to a continuous internal network of strong fibers that act as "guy-wires." These guy-wires distribute tension throughout a cellular sheet and reduce the chance of the sheet tearing when it is subjected to pulling forces. Desmosomes are abundant in tissues subjected to great mechanical stress, such as skin and heart muscle.

Gap Junctions

A **gap junction** is a *communicating junction* between adjacent cells. At gap junctions the adjacent plasma membranes are very close, and the cells are connected by hollow cylinders (called *connexons*) composed of transmembrane proteins. Different types of gap junctions are composed of different transmembrane proteins, and they determine what can pass through them from one cell to its neighbor. Ions, simple sugars, and other small molecules pass through these water-filled channels (Figure 3.4c).

Gap junctions are present in electrically excitable tissues, such as the heart and smooth muscle, where ion passage from cell to cell helps synchronize their electrical activity and contraction.

Check Your Understanding

3. What basic structure do all cellular membranes share?

4. Name the type of each of these molecules and state its role in the plasma membrane.

(a)   (b)   (c)

Table 3.1	Active versus Passive Tranport
PASSIVE TRANSPORT	**ACTIVE TRANSPORT**
No added energy required (uses kinetic energy)	Requires added energy (e.g., ATP)
Substances move from high to low concentration (i.e., "down" their concentration gradient)	Substances can move from low to high concentration (i.e., against their concentration gradient)

5. Which two types of cell junctions would you expect to find between muscle cells of the heart?

6. MAKE CONNECTIONS Phospholipid tails can be saturated or unsaturated (Chapter 2). This is true of phospholipids in plasma membranes as well. Which type—saturated or unsaturated—would make the membrane more fluid? Why?

For answers, see Answers Appendix.

Substances move through the plasma membrane in essentially two ways—passively or actively. In **passive processes**, substances cross the membrane without any energy input from the cell. In **active processes**, the cell provides the metabolic energy (usually ATP) needed to move substances across the membrane. Active and passive transport processes are the topics of the next two modules and are summarized in **Table 3.1**, Table 3.2 on p. 73, and Table 3.3 on p. 78.

3.3 Passive membrane transport is diffusion of molecules down their concentration gradient

Learning Outcomes

▶ **Relate plasma membrane structure to passive transport processes.**

▶ **Compare and contrast simple diffusion, facilitated diffusion, and osmosis relative to substances transported, direction, and mechanism.**

The three types of passive transport across the plasma membrane are *simple diffusion*, *facilitated diffusion*, and *osmosis.** All of these are various types of diffusion. **Diffusion** (dĭ-fu′zhun) is the movement of molecules or ions from an area where they are in higher concentration to an area where they are in lower concentration. Movement from high to low concentration is also called movement down or along a **concentration gradient**.

The driving force for diffusion is the intrinsic kinetic energy of the molecules themselves. The constant random and high-speed motion of molecules and ions (a result of their intrinsic kinetic energy) results in collisions. With each collision, the particles ricochet off one another and change direction. The overall effect of this erratic movement is to scatter or disperse the particles throughout the environment (**Figure 3.5**).

*Some consider filtration as a fourth form of passive transport. However, filtration occurs across capillary walls, not plasma membranes, so we will discuss it in Chapter 19 together with capillary transport.

A **concentration gradient** is a difference in concentration.

Area of low concentration of dye

Area of high concentration of dye

Dye pellet

① The dye pellet dissolves, creating a high concentration of dye around it.

② Random thermal motion (kinetic energy) makes the dye molecules move and collide constantly. As a result, they move down their concentration gradient (from high to low concentration).

③ When the dye is evenly distributed, there is no more *net* movement of dye because there is no more concentration gradient.

Figure 3.5 Diffusion.

The speed of diffusion is influenced by three factors:

- **Concentration.** The greater the difference in concentration of the diffusing molecules or ions between the two areas, the more collisions occur and the faster the particles diffuse.

- **Molecular size.** Smaller molecules diffuse more rapidly.

- **Temperature.** Higher temperature (more kinetic energy) increases the speed of molecular movement and means more rapid diffusion.

In a closed container, diffusion eventually produces a uniform mixture of molecules. In other words, the system reaches equilibrium, with molecules moving equally in all directions (no *net* movement).

Diffusion is immensely important in physiological systems and it occurs rapidly because the distances molecules are moving are very short, perhaps 1/1000 (or less) the thickness of this page! Examples include the movement of ions across cell membranes and the movement of neurotransmitters between two nerve cells.

The plasma membrane is a physical barrier to diffusion because of its hydrophobic core. That is, the membrane is a **selectively**, or **differentially**, **permeable** barrier: It allows some substances to pass while excluding others. For example, it allows nutrients to enter the cell, but keeps many undesirable substances out. At the same time, it keeps valuable cell proteins and other necessary substances in the cell, but allows wastes to exit.

▶ Watch a 3-D animation of this process: MasteringA&P® > Study Area > Animations & Videos > A&P Flix

 HOMEOSTATIC IMBALANCE 3.2

CLINICAL

Selective permeability is a characteristic of healthy, intact cells. When a cell (or its plasma membrane) is severely damaged, the membrane becomes permeable to virtually everything, and substances flow into and out of the cell freely. This phenomenon is evident in patients with severe burns. Precious fluids, proteins, and ions "weep" from the damaged cells.

What determines whether a given substance can cross the plasma membrane? The following characteristics are key:

- *Lipid solubility*. The more lipid soluble, the more readily it will diffuse across.

- *Size*. The smaller the molecule, the more readily it will diffuse across.

In addition, molecules that are not sufficiently small or lipid soluble can still diffuse across if they are assisted by a carrier molecule such as an ion channel or transport protein. The unassisted diffusion of lipid-soluble or very small particles is called *simple diffusion*. Assisted diffusion is known as *facilitated diffusion*. A special name, *osmosis*, is given to the diffusion of a solvent (usually water) through a membrane.

Simple Diffusion

In **simple diffusion**, substances diffuse directly through the lipid bilayer (**Figure 3.6a**). Such substances are usually small nonpolar molecules that readily dissolve in lipids (are lipid soluble). These include gases (such as oxygen and carbon dioxide), steroid hormones, and fatty acids. To reiterate, the two criteria that determine how easily a substance will pass by simple diffusion through a plasma membrane are (1) lipid solubility and (2) size. Simple diffusion is not an all-or-none thing. Some substances diffuse readily and others hardly at all. For example, water is not lipid soluble and you would expect it to be repelled by the hydrophobic lipid tails of the membrane's core. However, its very small size allows very small amounts to move across the lipid bilayer by simple diffusion.

Facilitated Diffusion

Certain molecules, notably glucose and other sugars, some amino acids, and ions are transported passively even though they are unable to pass through the lipid bilayer. Instead they move through the membrane by a passive transport process called **facilitated diffusion** in which the transported substance either (1) binds to *carrier proteins* in the membrane and is ferried across or (2) moves through water-filled *channel proteins*.

- **Carrier-mediated facilitated diffusion. Carriers** are transmembrane proteins that are specific for transporting certain polar molecules or classes of molecules, such as sugars and amino acids, that are too large to pass through membrane channels. Alterations in the shape of the carrier allow it to first

envelop and then release the transported substance, allowing it to bypass the nonpolar regions of the membrane. Essentially, the carrier protein changes shape to move the binding site from one face of the membrane to the other (Figure 3.6b).

Notice that a substance transported by carrier-mediated facilitated diffusion, such as glucose, moves down its concentration gradient (from high concentration to low), just as in simple diffusion. Glucose is normally in higher concentrations in the blood than in the cells, where it is rapidly used for ATP synthesis. So, glucose transport within the body is *typically* unidirectional—into the cells.

Carrier-mediated transport is limited by the number of protein carriers that are available. For example, when all the glucose carriers are "engaged," they are said to be *saturated*, and glucose transport is occurring at its maximum rate.

- **Channel-mediated facilitated diffusion. Channels** are transmembrane proteins that transport substances, usually ions or water, through aqueous channels from one side of the membrane to the other (Figure 3.6c and d). Channels are selective due to pore size and the charges of the amino acids lining the pore. *Leakage channels* are always open and simply allow ions or water to move according to concentration gradients. *Gated channels* are controlled (opened or closed), usually by chemical or electrical signals. Like carriers, many channels can be inhibited by certain molecules, show saturation, and tend to be specific. Substances moving through them also follow the concentration gradient (always moving down the gradient).

Oxygen, water, glucose, and various ions are vitally important to cellular homeostasis. Their passive transport by diffusion

(a) Simple diffusion of lipid-soluble molecules directly through the phospholipid bilayer

(b) Carrier-mediated facilitated diffusion via protein carrier specific for one chemical; binding of solute causes transport protein to change shape

(c) Channel-mediated facilitated diffusion through a channel protein; mostly ions selected on basis of size and charge

(d) Osmosis, diffusion of a solvent such as water through a specific channel protein (aquaporin) or through the lipid bilayer

Figure 3.6 Diffusion through the plasma membrane.

(either simple or facilitated) represents a tremendous saving of cellular energy. Indeed, if these substances had to be transported actively, cell expenditures of ATP would be enormous!

Osmosis

The diffusion of a solvent, such as water, through a selectively permeable membrane is **osmosis** (oz-mo′sis; *osmos* = pushing). Osmosis is extremely important in determining the distribution of water in the various fluid-containing compartments of the body (cells, blood, and so on). In the clinic, you will encounter patients with swelling due to the abnormal accumulation of fluid in their tissues (*edema*, see Figure 19.20 on p. 734). In order to understand the causes and treatments of this condition, you will need to understand osmosis.

As we mentioned earlier, even though water is highly polar, a small amount of it can "sneak through" the plasma membrane by osmosis because of its small size. Water also moves freely and reversibly through water-specific channels constructed by transmembrane proteins called **aquaporins (AQPs)**, which allow single-file diffusion of water molecules. The water-filled aquaporin channels are particularly abundant in red blood cells and in cells involved in water balance such as kidney tubule cells.

Osmosis occurs whenever the water concentration differs on the two sides of a membrane. If distilled water is present on both sides of a selectively permeable membrane, no *net* osmosis occurs, even though water molecules move in both directions through the membrane. If the solute concentration on the two sides of the membrane differs, water concentration differs as well (as solute concentration increases, water concentration decreases).

The extent to which solutes decrease water's concentration depends on the *number*—not the *type*—of solute particles, because one molecule or one ion of solute (typically) displaces one water molecule. The total concentration of all solute particles in a solution is referred to as the solution's **osmolarity** (oz″mo-lar′ĭ-te). When equal volumes of aqueous solutions of different osmolarity are separated by a membrane that is *permeable to all molecules* in the system, net diffusion of both solute and water occurs, each moving down its own concentration gradient. Equilibrium is reached when the water (and solute) concentration on both sides of the membrane is the same (**Figure 3.7a**).

(a) Membrane permeable to both solutes and water

Solution with lower osmolarity (lower solute concentration)

Solution with higher osmolarity (higher solute concentration)

At equilibrium, both sides have the same osmolarity; volume unchanged

Both solute and water molecules move down their concentration gradients.

H₂O

Solute

Freely permeable membrane

Solute molecules (sugar)

(b) Membrane permeable to water, impermeable to solutes

Water, but not solutes can move through the membrane.

At equilibrium, both sides have the same osmolarity; *different* volume (because only water can move)

Water moves by osmosis from an area of higher to lower water concentration (lower to higher solute concentration).

H₂O

Selectively permeable membrane

Solute molecules (sugar)

Figure 3.7 Influence of membrane permeability on diffusion and osmosis.

If we consider the same system, but make the membrane *impermeable to solute particles*, we see quite a different result: Water moves and the volume changes (Figure 3.7b).

The latter situation mimics osmosis across plasma membranes of living cells, with one major difference. In our examples, the volumes of the compartments are infinitely expandable and the effect of pressure exerted by the added weight of the higher fluid column is not considered. What happens in a real cell? That depends on whether the cell is a plant cell (with a rigid cell wall) or an animal cell (with no cell wall).

As water diffuses into living plant cells, the point is finally reached where the **hydrostatic pressure** (the back pressure exerted by water against the cell wall) within the cell is equal to its **osmotic pressure** (the tendency of water to move into the cell by osmosis). At this point, there is no further (net) water entry. As a rule, the higher the amount of nondiffusible, or *nonpenetrating*, solutes in a cell, the higher the osmotic pressure and the greater the hydrostatic pressure must be to resist further net water entry. In our plant cell, hydrostatic pressure is pushing water out, and osmotic pressure is pulling water in; therefore, you could think of the osmotic pressure as an osmotic "suck."

In living animal cells, such major changes in hydrostatic (and osmotic) pressures cannot occur because they lack rigid cell walls. Osmotic imbalances cause animal cells to swell or shrink (due to net water gain or loss) until either (1) the solute concentration is the same on both sides of the plasma membrane, or (2) the membrane stretches to its breaking point.

Tonicity

We have just learned that many solutes, particularly intracellular proteins and selected ions, cannot diffuse through the plasma membrane. Consequently, any change in their concentration alters the water concentration on the two sides of the membrane and results in a net loss or gain of water by the cell.

Tonicity (to-nis'ĭ-te) refers to the ability of a solution to change the shape (or plasma membrane tension) of cells by altering the cells' internal water volume (*tono* = tension).

- **Isotonic** ("the same tonicity") **solutions** have the same concentrations of nonpenetrating solutes as those found in cells (0.9% saline or 5% glucose). Cells exposed to isotonic solutions retain their normal shape, and exhibit no net loss or gain of water (**Figure 3.8a**). As you might expect, the body's extracellular fluids and most intravenous solutions (solutions infused into the body via a vein) are isotonic.

- **Hypertonic solutions** have a higher concentration of nonpenetrating solutes than seen in the cell (for example, a strong saline solution). Cells immersed in hypertonic solutions lose water and shrivel, or *crenate* (kre'nāt) (Figure 3.8b).

- **Hypotonic solutions** are more dilute (contain a lower concentration of nonpenetrating solutes) than cells. Cells placed in a hypotonic solution plump up rapidly as water rushes into them (Figure 3.8c). Distilled water represents the most extreme example of hypotonicity. Because it contains *no* solutes, water continues to enter cells until they finally burst, or *lyse*.

Notice that osmolarity and tonicity are not the same. A solution's osmolarity is based solely on its total solute concentration.* In contrast, its tonicity is based on how the solution affects cell volume, which depends on (1) solute concentration and (2) solute permeability of the plasma membrane.

---

*Osmolarity (Osm) is determined by multiplying molarity (moles per liter, or *M*) by the number of particles resulting from ionization. For example, since NaCl ionizes to $Na^+ + Cl^-$, a 1 *M* solution of NaCl is a 2 Osm solution. For substances that do not ionize (e.g., glucose), molarity and osmolarity are the same. More precisely, the term *osmolality* is used, which is equal to the number of particles mixed into a kilogram of water.

(a) **Isotonic solutions**	(b) **Hypertonic solutions**	(c) **Hypotonic solutions**
Cells retain their normal size and shape in isotonic solutions (same solute/water concentration as inside cells; water moves in and out).	Cells lose water by osmosis and shrink in a hypertonic solution (contains a higher concentration of nonpenetrating solutes than are present inside the cells).	Cells take on water by osmosis until they become bloated and burst (lyse) in a hypotonic solution (contains a lower concentration of nonpenetrating solutes than are present inside cells).

Figure 3.8 The effect of solutions of varying tonicities on living red blood cells.

Table 3.2 Passive Membrane Transport Processes: Diffusion

PROCESS	ENERGY SOURCE	DESCRIPTION	MEMBRANE TRANSPORT PROTEIN REQUIRED	SPECIFIC AND SATURABLE	EXAMPLES
Simple diffusion	Kinetic energy	Net movement of molecules down their concentration gradient (from higher concentration to lower concentration)	No	No (passage depends only on small size and lipid solubility)	Lipids, oxygen, and carbon dioxide
Facilitated diffusion	Kinetic energy	Same as simple diffusion, but the diffusing substance is attached to a membrane carrier protein or moves through a channel protein	Yes	Yes (specificity depends on shape inside transport protein)	Glucose, Na^+, K^+
Osmosis	Kinetic energy	Diffusion of water through a selectively permeable membrane; can occur directly through the lipid bilayer or via membrane channels (aquaporins)	No, except for movement through aquaporins	No, except for movement through aquaporins	Water is the only example

Osmolarity is expressed as osmoles per liter (osmol/L) where 1 osmol is equal to 1 mole of nonionizing molecules. A 0.3 osmol/L solution of NaCl is isotonic because sodium ions are usually prevented from diffusing through the plasma membrane. But if the cell is immersed in a 0.3 osmol/L solution of a penetrating solute, both water and solute will enter the cell. The cell will swell and burst, just as if it had been placed in pure water.

Summary of Passive Membrane Transport

There are two important characteristics of any transport process: *specificity* and *saturability*. Any transport process that depends on a transport protein (such as a carrier or channel) will be *saturable*. This means that there is a *maximum rate* of transport because there are only a limited number of these proteins in the membrane. Like enzymes, transport proteins exhibit a high degree of *specificity*. For example, the carrier for glucose combines specifically with glucose in much the same way that an enzyme binds to its specific substrate.

Simple diffusion and osmosis occurring directly through the plasma membrane are not specific or saturable processes. They are not specific because they don't depend on the shape of the molecule. As long as the molecule can diffuse through the lipid barrier of the membrane, it will pass. Because no proteins are involved, these processes are not saturable. The rate of transport depends only on the size of the concentration gradient—the larger the gradient, the greater the movement.

Table 3.2 summarizes the processes of passive membrane transport.

Check Your Understanding

7. What is the energy source for all types of diffusion?
8. How do the two types of facilitated diffusion differ?
9. **PREDICT** Usually, Na^+ and Cl^- cannot cross the plasma membranes of cells. What would happen to a cell if it suddenly became permeable to both Na^+ and Cl^-?

10. **APPLY** For the two graphs below, which one represents simple diffusion and which represents facilitated diffusion? What is the name of the transport property illustrated by these graphs?

Difference in concentration across membrane
(a)   (b)

For answers, see Answers Appendix.

3.4 Active membrane transport directly or indirectly uses ATP

Learning Outcomes

▶ Differentiate between primary and secondary active transport.
▶ Compare and contrast endocytosis and exocytosis in terms of function and direction.
▶ Compare and contrast pinocytosis, phagocytosis, and receptor-mediated endocytosis.

An *active process* occurs whenever a cell uses energy to move solutes across the membrane. Substances moved actively across the plasma membrane are usually unable to pass in the necessary direction by passive transport processes. The substance may be too large to pass through the channels, incapable of dissolving in the lipid bilayer, or moving against its concentration gradient.

There are two major means of active membrane transport: active transport and vesicular transport.

Primary active transport is the process in which solutes are moved across cell membranes against electrochemical gradients using energy supplied directly by ATP. The action of the Na⁺-K⁺ pump is an important example of primary active transport.

 Watch a 3-D animation of this process: MasteringA&P® > Study Area > Animations & Videos > A&P Flix

Extracellular fluid

Na⁺

Na⁺-K⁺ pump

ATP-binding site

ATP

K⁺

Cytoplasm

① Three cytoplasmic Na⁺ bind to pump protein.

Na⁺ bound

P

ADP

② Na⁺ binding promotes hydrolysis of ATP. The energy released during this reaction phosphorylates the pump.

K⁺ released

ATP

⑥ The pump protein binds ATP and releases K⁺ to the inside, and Na⁺ sites are ready to bind Na⁺ again. The cycle repeats.

Na⁺ released

P

③ Phosphorylation causes the pump to change shape, expelling Na⁺ to the outside.

K⁺ bound

Pi

⑤ K⁺ binding triggers release of the phosphate. The dephosphorylated pump resumes its original conformation.

K⁺

P

④ Two extracellular K⁺ bind to pump.

Active Transport

Like carrier-mediated facilitated diffusion, **active transport** requires transport proteins that combine *specifically* and *reversibly* with the transported substances. However, facilitated diffusion always follows concentration gradients because its driving force is kinetic energy. In contrast, active transporters move solutes, most importantly ions, "uphill" *against* a concentration gradient. To do this work, cells must expend energy.

Active transport processes are distinguished according to their source of energy:

- In *primary active transport*, the energy to do work comes *directly from hydrolysis of ATP* by transport proteins called **pumps**.

- In *secondary active transport*, transport is driven *by energy stored in concentration gradients of ions* created by primary active transport pumps. Secondary active transport systems always move more than one substance at a time using a **cotransport protein**.

Regardless of whether the energy is provided directly (primary active transport) or indirectly (secondary active transport), each membrane pump or cotransporter transports only specific substances. Active transport systems provide a way for the cell to be very specific in cases where substances cannot pass by diffusion. No transporter—no transport.

Primary Active Transport

In **primary active transport**, hydrolysis of ATP results in the *phosphorylation* of the pump. (In other words, the transport protein is energized by the transfer of a phosphate group from ATP.) This step causes the protein to change its shape in such a manner that it pumps the bound solute across the membrane.

Primary active transport systems include calcium and hydrogen pumps, but the most important example of a primary active transport system is the **sodium-potassium pump**, for which the pump protein is an enzyme called **Na$^+$-K$^+$ ATPase**. **Focus Figure 3.2**, *Focus on Primary Active Transport: The Na$^+$-K$^+$ Pump*, describes the operation of this pump. For each molecule of ATP used, the Na$^+$-K$^+$ pump drives three Na$^+$ out of the cell and pumps two K$^+$ back in. As a result the concentration of K$^+$ inside the cell is some 10 times higher than that outside, and the reverse is true of Na$^+$. These ionic concentration differences are essential for excitable cells like muscle and nerve cells to function normally and for all body cells to maintain their normal fluid volume. Because Na$^+$ and K$^+$ leak slowly but continuously through leakage channels in the plasma membrane down their concentration gradient (and cross more rapidly in stimulated muscle and nerve cells), the Na$^+$-K$^+$ pump operates almost continuously.

Earlier we said that solutes diffuse down their concentration gradients. This is true for uncharged solutes, but ions are more complicated. The negatively and positively charged faces of the plasma membrane can help or hinder diffusion of ions driven by a concentration gradient. It is more correct to say that ions diffuse according to **electrochemical gradients**. This recognizes the effect of both electrical and concentration (chemical) forces. The electrochemical gradients maintained by the Na$^+$-K$^+$ pump are crucial for cardiac, skeletal muscle, and neuron function. They also underlie most secondary active transport as we shall see next.

Secondary Active Transport (Cotransport)

Secondary active transport (also called **cotransport**) uses a cotransport protein to couple the "downhill" (down its concentration gradient) movement of one solute to the "uphill" (against its concentration gradient) movement of another solute. The concentration gradient that is the source of energy for secondary active transport is created by primary active transport—in many cases by the Na$^+$-K$^+$ pump. By moving sodium across the plasma membrane against its concentration gradient, the pump stores energy (in the gradient). Then, just as water held back by a dam can do work as it flows downward (to generate electricity, for instance), a substance pumped across a membrane can do work as it leaks back, propelled "downhill" along its concentration gradient.

For example, as sodium moves back into the cell, other substances are "dragged along," or cotransported, by the same protein (**Figure 3.9**). Some sugars, amino acids, and many ions are cotransported via secondary active transport into cells lining the small intestine. Because the

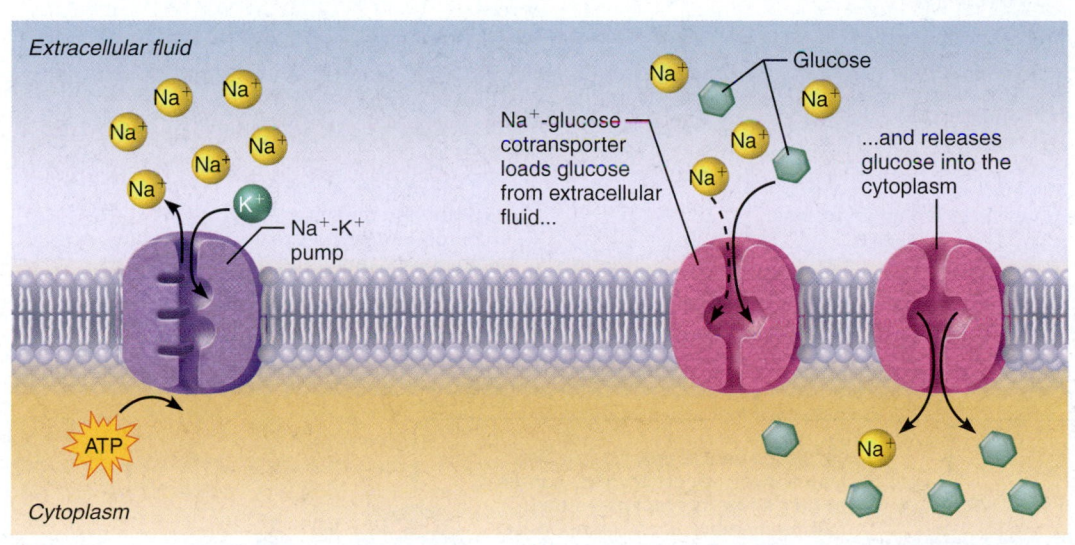

① **Primary active transport**
The ATP-driven Na$^+$-K$^+$ pump stores energy by creating a steep concentration gradient for Na$^+$ entry into the cell.

② **Secondary active transport**
As Na$^+$ diffuses back across the membrane through a membrane cotransporter protein, it drives glucose against its concentration gradient into the cell.

Extracellular fluid

Na$^+$

Na$^+$-K$^+$ pump

K$^+$

ATP

Cytoplasm

Na$^+$-glucose cotransporter loads glucose from extracellular fluid...

Glucose

...and releases glucose into the cytoplasm

Na$^+$

Figure 3.9 Secondary active transport is driven by the concentration gradient created by primary active transport.

energy for this type of transport is the concentration gradient of the ion (in this case Na^+), Na^+ has to be pumped back out of the cell to maintain its concentration gradient.

In a **symport system**, such as the one we just described, the two transported substances move in the same direction (*sym* = same). In an **antiport system** (*anti* = opposite, against), the transported substances "wave to each other" as they cross the membrane in opposite directions. An example of an antiport system is a cotransporter that cells use to regulate their intracellular pH. This cotransporter uses the Na^+ concentration gradient to pump H^+ ions out of the cell.

Vesicular Transport

In **vesicular transport**, fluids containing large particles and macromolecules are transported across cellular membranes inside bubble-like, membranous sacs called *vesicles*. Like active transport, vesicular transport moves substances into the cell (endocytosis) and out of the cell (exocytosis). It is also used for combination processes such as *transcytosis* and *vesicular trafficking*. **Transcytosis** moves substances into, across, and then out of the cell. Transcytosis is common in the endothelial cells lining blood vessels because it provides a quick means to get substances from the blood to the interstitial fluid. **Vesicular trafficking** moves substances from one area (or membranous organelle) in the cell to another. The fleet of vesicles involved in vesicular transport can be thought of as the FedEx® of the cell. Vesicular transport processes are energized by ATP (or in some cases another energy-rich compound, *GTP*—guanosine triphosphate).

Endocytosis

Vesicles provide the main route for bringing bulk solids, most macromolecules, and fluids into a cell (or transporting them across a cell via transcytosis). Many types of endocytosis rely

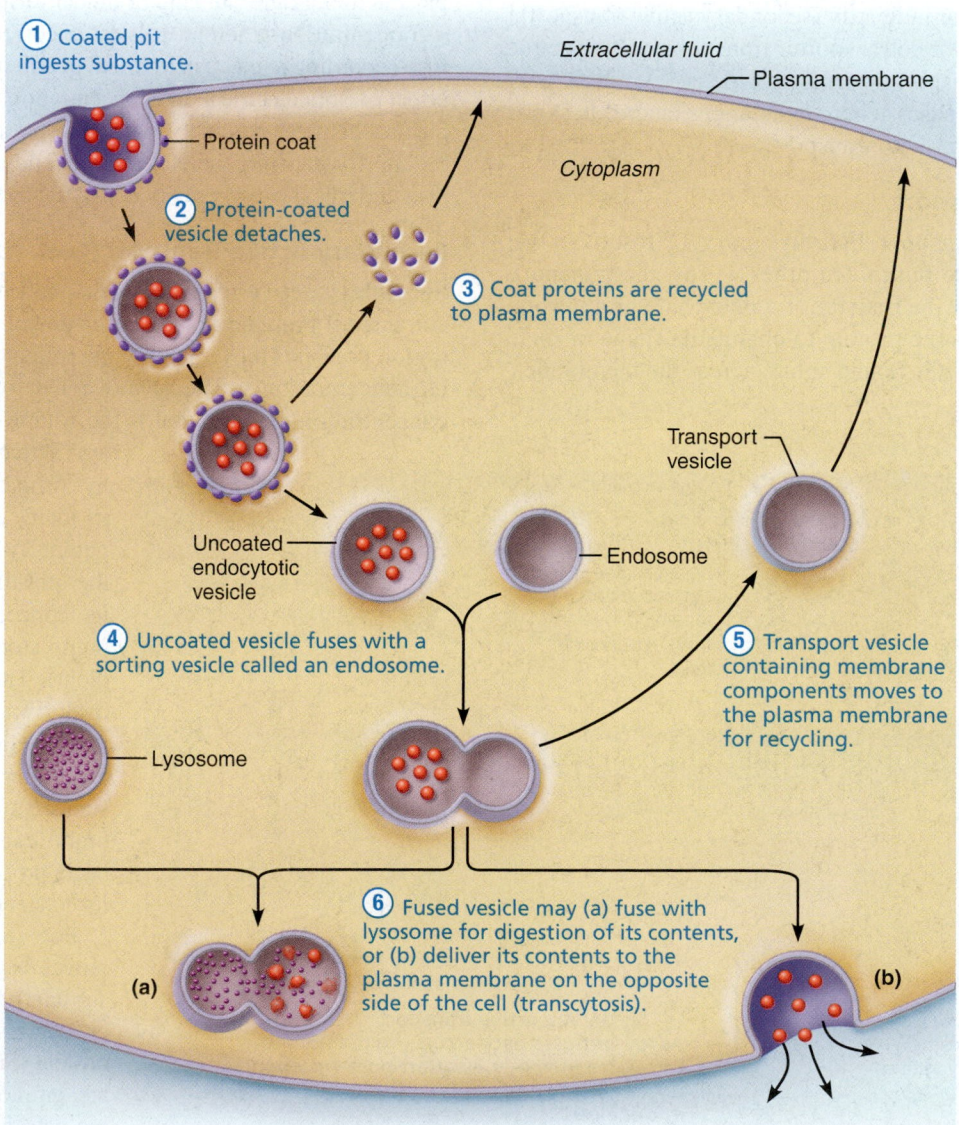

① Coated pit ingests substance.

Extracellular fluid

Plasma membrane

Protein coat

Cytoplasm

② Protein-coated vesicle detaches.

③ Coat proteins are recycled to plasma membrane.

Transport vesicle

Uncoated endocytotic vesicle

Endosome

④ Uncoated vesicle fuses with a sorting vesicle called an endosome.

⑤ Transport vesicle containing membrane components moves to the plasma membrane for recycling.

Lysosome

⑥ Fused vesicle may (a) fuse with lysosome for digestion of its contents, or (b) deliver its contents to the plasma membrane on the opposite side of the cell (transcytosis).

(a)

(b)

Figure 3.10 Events of endocytosis. Note the three possible fates for a vesicle and its contents, shown in ⑤ and ⑥.

on receptors in the membrane to determine the substances to be transported.

Endocytosis begins with a *coated pit*—an infolding of the membrane. Coated pits have a protein coating on the cytoplasmic face that deforms the membrane to produce the vesicle. **Figure 3.10** shows the basic steps in endocytosis and transcytosis.

Three types of endocytosis differ in the type and amount of material taken up and the means of uptake. These are phagocytosis, pinocytosis, and receptor-mediated endocytosis.

- **Phagocytosis.** In **phagocytosis** (fag″o-si-to′sis; "cell eating"), the cell engulfs some relatively large or solid material, such as a clump of bacteria, cell debris, or inanimate particles (asbestos fibers or glass, for example) (**Figure 3.11a**). When a particle binds to receptors on the cell's surface, cytoplasmic extensions called pseudopods (soo′do-pahdz; *pseudo* = false, *pod* = foot) form and flow around the particle. This forms an endocytotic vesicle called a **phagosome** (fag′o-sōm; "eaten body"). In most cases, the phagosome then fuses with a lysosome and its contents are digested.

  In the human body, only cells called *macrophages* and certain white blood cells are "experts" at phagocytosis. Commonly referred to as *phagocytes*, these cells help protect the body by ingesting and disposing of bacteria, other foreign substances, and dead tissue cells. The disposal of dying cells is crucial, because dead cell remnants trigger inflammation in the surrounding area. Most phagocytes move about by **amoeboid motion** (ah-me′boyd; "changing shape"); that is, their cytoplasm flows into temporary extensions that allow them to creep along.

- **Pinocytosis.** In **pinocytosis** ("cell drinking"), also called **fluid-phase endocytosis**, a bit of infolding plasma membrane surrounds a very small volume of extracellular fluid containing dissolved molecules (Figure 3.11b). This droplet enters the cell and fuses with a sorting vesicle called an endosome. Unlike phagocytosis, pinocytosis is a routine activity of most cells, affording them a nonselective way of sampling the extracellular fluid. It is particularly important in cells that absorb nutrients, such as cells that line the intestines.

  As mentioned, bits of the plasma membrane are removed when the membranous sacs are internalized. However, these membranes are recycled back to the plasma membrane by exocytosis as described shortly, so the surface area of the plasma membrane remains remarkably constant.

- **Receptor-mediated endocytosis.** The main mechanism for the *specific* endocytosis and transcytosis of most macromolecules by body cells is **receptor-mediated endocytosis** (Figure 3.11c). This exquisitely selective mechanism allows cells to concentrate material that is present only in small amounts in the extracellular fluid. The receptors for this process are plasma membrane proteins that bind only certain substances. Both the receptors and attached molecules are internalized and then dealt with in one of the ways discussed above. Substances taken up by receptor-mediated endocytosis include enzymes, insulin (and some other hormones), low-density lipoproteins (such as cholesterol attached to a

(a) Phagocytosis
The cell engulfs a large particle by forming a projecting pseudopod ("false foot") around it and enclosing it within a membranous sac called a phagosome. The phagosome combines with a lysosome and its contents are digested. The vesicle has receptors capable of binding to microorganisms or solid particles.

Receptors

Phagosome

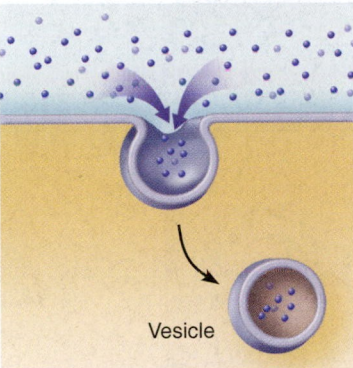

(b) Pinocytosis
The cell "gulps" a drop of extracellular fluid containing solutes into tiny vesicles. No receptors are used, so the process is nonspecific.

Vesicle

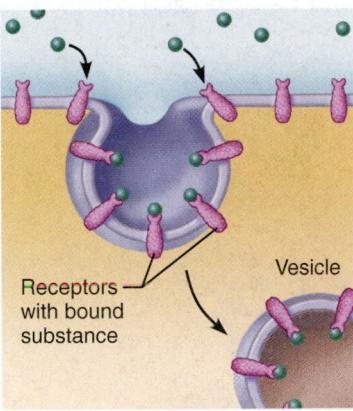

(c) Receptor-mediated endocytosis
Extracellular substances bind to specific receptor proteins, enabling the cell to ingest and concentrate specific substances in protein-coated vesicles. Substances may be released inside the cell or digested in a lysosome.

Receptors with bound substance

Vesicle

Figure 3.11 Comparison of three types of endocytosis.

transport protein), and iron. Unfortunately, flu viruses, diphtheria, and cholera toxins hijack this route to enter our cells.

Exocytosis

Vesicular transport processes that eject substances from the cell interior into the extracellular fluid are called **exocytosis** (ek″so-si-to′sis; "out of the cell"). Exocytosis is typically stimulated by a cell-surface signal such as binding of a hormone to a membrane receptor or a change in membrane voltage. Exocytosis accounts for hormone secretion, neurotransmitter release, mucus secretion, and in some cases, ejection of wastes. The

Table 3.3   Active Membrane Transport Processes

PROCESS	ILLUSTRATION	ENERGY SOURCE	DESCRIPTION	EXAMPLES
Active Transport				
Primary active transport*		ATP	• Transport of substances against a concentration (or electrochemical) gradient. • A pump protein moves substances across the plasma membrane. • Directly uses energy of ATP hydrolysis.	Ions (Na$^+$, K$^+$, H$^+$, Ca^{2+}, and others)
Secondary active transport*		Ion concentration gradient maintained with ATP	• Cotransport (coupled transport) of two solutes across the membrane. • Energy is supplied by the concentration gradient created by primary active transport. • *Symporters* move the transported substances in the same direction; *antiporters* move transported substances in opposite directions across the membrane.	Movement of polar or charged solutes, e.g., amino acids (into cell by symporters); Ca^{2+}, H$^+$ (out of cells via antiporters)
Vesicular Transport				
Endocytosis				
Phagocytosis		ATP	• A large external particle (proteins, bacteria, dead cell debris) is surrounded by a pseudopod ("false foot") and becomes enclosed in a vesicle (phagosome).	Occurs primarily in phagocytes (some white blood cells and macrophages)
Pinocytosis (fluid-phase endocytosis)		ATP	• Plasma membrane sinks beneath an external fluid droplet containing small solutes. Membrane edges fuse, forming a fluid-filled vesicle.	Occurs in most cells; important for taking in dissolved solutes by absorptive cells of the kidney and intestine
Receptor-mediated endocytosis		ATP	• External substance binds to membrane receptors. • Selective endocytosis and transcytosis.	Means of intake of some hormones, cholesterol, iron, and most macromolecules
Vesicular trafficking		ATP	• Vesicles pinch off from organelles and travel to other organelles to deliver their cargo.	Intracellular trafficking between certain organelles, e.g., endoplasmic reticulum and Golgi apparatus
Exocytosis		ATP	• Secretion or ejection of substances from a cell. • The substance is enclosed in a membranous vesicle, which fuses with the plasma membrane and ruptures, releasing the substance to the exterior.	Secretion of neurotransmitters, hormones, mucus, etc.

*Like facilitated diffusion, primary and secondary active transport require a transport protein in the membrane. For this reason, they are both specific and saturable.

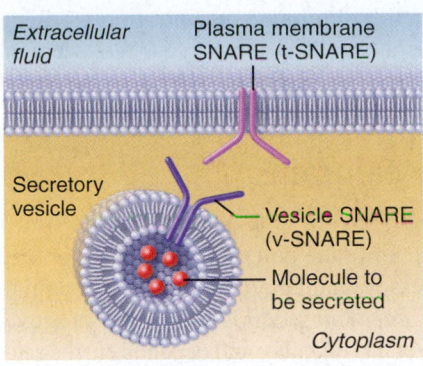

Extracellular fluid

Plasma membrane SNARE (t-SNARE)

Secretory vesicle

Vesicle SNARE (v-SNARE)

Molecule to be secreted

Cytoplasm

(a) The process of exocytosis

① The membrane-bound vesicle migrates to the plasma membrane.

Fused v- and t-SNAREs

② Proteins at the vesicle surface (v-SNAREs) bind with t-SNAREs (plasma membrane proteins).

Fusion pore formed

③ The vesicle and plasma membrane fuse and a pore opens up.

④ Vesicle contents are released to the cell exterior.

(b) Photomicrograph of a secretory vesicle releasing its contents by exocytosis (100,000×)

Figure 3.12 Exocytosis.

substance to be removed from the cell is first enclosed in a protein-coated membranous sac called a *secretory vesicle*. In most cases, the vesicle migrates to the plasma membrane, fuses with it, and then ruptures, spilling the sac contents out of the cell (**Figure 3.12**).

Exocytosis, like other mechanisms in which vesicles are targeted to their destinations, involves a "docking" process in which transmembrane proteins on the vesicles, fancifully called v-SNAREs (*v* for vesicle), recognize certain plasma membrane proteins, called t-SNAREs (*t* for target), and bind with them. This binding causes the membranes to "corkscrew" together and fuse, rearranging the lipid monolayers without mixing them (Figure 3.12a). As described, membrane material added by exocytosis is removed by endocytosis—the reverse process.

Table 3.3 summarizes active membrane transport processes.

Check Your Understanding

11. Does the Na^+-K^+ ATPase pump K^+ into or out of the cell? What is the ratio of Na^+ to K^+ it pumps?

12. As a cell grows, its plasma membrane expands. Does this membrane expansion involve endocytosis or exocytosis?

13. Phagocytic cells gather in the lungs, particularly in the lungs of smokers. What is the connection?

14. Which vesicular transport process allows a cell to take in cholesterol from the extracellular fluid?

For answers, see Answers Appendix.

3.5 Selective diffusion establishes the membrane potential

Learning Outcome

▶ **Define membrane potential and explain how the resting membrane potential is established and maintained.**

As you're now aware, the selective permeability of the plasma membrane can lead to dramatic osmotic flows, but that is not its only consequence. An equally important result is the generation of a **membrane potential**, or voltage, across the membrane. While all body cells have a membrane potential, it is especially important for nerve and muscle cells because they use changes in membrane potential as a form of communication.

A *voltage* is electrical potential energy resulting from the separation of oppositely charged particles. In cells, the oppositely charged particles are ions, and the barrier that keeps them apart is the plasma membrane.

In their resting state, plasma membranes of all body cells exhibit a **resting membrane potential** that typically ranges from -50 to -90 millivolts (mV), depending on cell type. For this reason, all cells are said to be **electrically polarized**. The minus sign before the voltage indicates that the *inside* of the cell is negative compared to its outside. This voltage (or charge separation) exists *only at the membrane*. If we

🔵 Complete an interactive tutorial: MasteringA&P® > Study Area > Interactive Physiology added up all the negative and positive charges in the cytoplasm, we would find that the cell interior is electrically neutral. Likewise, the positive and negative charges in the extracellular fluid balance each other exactly.

So how does the resting membrane potential come about, and how is it maintained? The short answer is that diffusion causes ionic imbalances that polarize the membrane, and active transport processes *maintain* that membrane potential. First, let's look at how diffusion polarizes the membrane.

K^+ Is the Key Player

Many kinds of ions are found both inside cells and in the extracellular fluid, but the resting membrane potential is determined mainly by the concentration gradient of potassium (K^+) and by the differential permeability of the plasma membrane to K^+ and other ions (**Figure 3.13**). K^+ and protein anions predominate inside body cells, and the extracellular fluid contains relatively more Na^+, which is largely balanced by Cl^-. The plasma membrane is somewhat permeable to K^+ because of leakage channels, but it is impermeable to the protein anions. Consequently, as shown in Figure 3.13 ①, K^+ diffuses out of the cell along its concentration gradient but the protein anions are unable to follow, and this loss of positive charges makes the cytoplasmic side of the membrane more negative.

② As more and more K^+ leaves the cell, the negativity of the inner membrane face becomes great enough to attract K^+ back toward and even into the cell. ③ At a membrane voltage of −90 mV, potassium's concentration gradient is exactly balanced by the electrical gradient (membrane potential), and one K^+ enters the cell as one leaves.

In many cells, sodium (Na^+) also contributes to the resting membrane potential. Sodium is strongly attracted to the cell interior by its concentration gradient, bringing the resting membrane potential to −70 mV. However, K^+ still largely determines the resting membrane potential because the membrane is much more permeable to K^+ than to Na^+. Even though the membrane is permeable to Cl^-, in most cells Cl^- does not contribute to the resting membrane potential, because its concentration and electrical gradients exactly balance each other.

We may be tempted to believe that massive flows of K^+ ions are needed to generate the resting potential, but this is not the case. Surprisingly, the number of ions producing the membrane potential is so small that it does not change ion concentrations in any significant way.

In a cell at rest, very few ions cross its plasma membrane. However, Na^+ and K^+ are not at equilibrium and there is some net movement of K^+ out of the cell and of Na^+ into the cell. Na^+ is

strongly pulled into the cell by both its concentration gradient and the interior negative charge. If only passive forces were at work, these ion concentrations would eventually become equal inside and outside the cell.

Active Transport Maintains Electrochemical Gradients

Diffusion establishes the membrane potential. Now let's look at how active transport processes maintain the membrane potential to keep the cell in a *steady state*. This requires that the rate of active transport of Na^+ out of the cell is equal to, and depends on, the rate of Na^+ diffusion into the cell. If more Na^+ enters, more is pumped out. (This is like being in a leaky boat. The more water that comes in, the faster you bail!) The Na^+-K^+ pump couples sodium and potassium transport (see Focus Figure 3.2 on p. 74). The pump maintains both the membrane potential (the charge separation) and the osmotic balance. Indeed, if Na^+ was not continuously removed from cells, so much would accumulate intracellularly that the osmotic gradient would draw water into the cells, causing them to burst.

As we described on p. 75, diffusion of charged particles across the membrane is affected not only by concentration gradients, but by the electrical charge on the inner and outer faces of the membrane. Together these gradients make up the *electrochemical gradient*. The diffusion of K^+ across the plasma membrane is aided by the membrane's greater permeability to it and by the ion's concentration gradient, but the negative charges on the cell interior resist K^+ diffusion. In contrast, a

① K^+ diffuse down their steep concentration gradient (out of the cell) via leakage channels. Loss of K^+ results in a negative charge on the inner plasma membrane face.

② K^+ also move into the cell because they are attracted to the negative charge established on the inner plasma membrane face.

③ A negative membrane potential (−90 mV) is established when the movement of K^+ out of the cell equals K^+ movement into the cell. At this point, the concentration gradient promoting K^+ exit exactly opposes the electrical gradient for K^+ entry.

Extracellular fluid

Potassium leakage channels

Protein anion (unable to follow K^+ through the membrane)

Cytoplasm

Figure 3.13 The key role of K^+ in generating the resting membrane potential. The resting membrane potential is largely determined by K^+ because at rest, the membrane is much more permeable to K^+ than Na^+. The active transport of sodium and potassium ions by the Na^+-K^+ pump maintains the Na^+ and K^+ concentration gradients.

steep electrochemical gradient draws Na$^+$ into the cell, but the membrane's relative impermeability to it limits Na$^+$ diffusion.

The transient opening of gated Na$^+$ and K$^+$ channels in the plasma membrane "upsets" the resting membrane potential. As we describe in later chapters, this is a normal means of activating neurons and muscle cells. We will revisit the resting membrane potential and have a closer look at how it is established in *Focus on the Resting Membrane Potential*, Focus Figure 11.1 on p. 402.

Check Your Understanding

15. What process establishes the resting membrane potential?
16. Is the inside of the plasma membrane negative or positive relative to its outside in a polarized membrane of a resting cell?

For answers, see Answers Appendix.

3.6 Cell adhesion molecules and membrane receptors allow the cell to interact with its environment

Learning Outcomes

▶ Describe the role of cell adhesion molecules in allowing cells to interact with their environment.
▶ List several roles of membrane receptors and that of G protein–coupled receptors.

Cells are biological minifactories and, like other factories, they receive and send orders from and to the outside community. But *how* does a cell interact with its environment, and *what* activates it to carry out its homeostatic functions?

Sometimes cells interact directly with other cells. However, in many cases cells respond to extracellular chemicals, such as hormones and neurotransmitters distributed in body fluids. Cells also interact with extracellular molecules that act as signposts to guide cell migration during development and repair. The most important molecules cells use to interact with their environment fall into two large families—cell adhesion molecules and plasma membrane receptors (see Figure 3.3).

Roles of Cell Adhesion Molecules (CAMs)

Thousands of **cell adhesion molecules (CAMs)** are found on almost every cell in the body. CAMs play key roles in embryonic development and wound repair (situations where cell mobility is important) and in immunity. These sticky glycoproteins (*cadherins* and *integrins*) act as:

- The molecular Velcro® that cells use to anchor themselves to molecules in the extracellular space (see Figure 3.3e on p. 66) and to each other (see Figure 3.3f)

- The "arms" that migrating cells use to haul themselves past one another

- SOS signals sticking out from the blood vessel lining that rally protective white blood cells to a nearby infected or injured area

- Mechanical sensors that transmit information about changes in the extracellular matrix into the cell, bringing about a variety of cellular responses such as cell migration, proliferation (growth by cell division), and specialization

Roles of Plasma Membrane Receptors

A huge and diverse group of integral proteins that serve as binding sites are collectively known as **membrane receptors**. Some function in contact signaling, and others in chemical signaling. Most are glycoproteins. Let's take a look.

Contact Signaling

Contact signaling, in which cells come together and touch, is the means by which cells recognize one another. It is particularly important for normal development and immunity. Some bacteria and other infectious agents use contact signaling to identify their "preferred" target tissues.

Chemical Signaling

Many signals originating outside of the cell are in the form of chemical messengers. *Chemical signaling* is the process in which a **ligand**, the chemical messenger, binds a specific receptor and initiates a response. Ligands include most *neurotransmitters* (nervous system signals), *hormones* (endocrine system signals), and *paracrines* (chemicals that act locally and are rapidly destroyed).

Different cells respond in different ways to the same ligand. Acetylcholine, for instance, stimulates skeletal muscle cells to contract, but inhibits heart muscle. Why do different cells respond so differently? The reason is that a target cell's response depends on the internal machinery that the receptor is linked to, not the specific ligand that binds to it.

Though cell responses to receptor binding vary widely, there is a fundamental similarity: When a ligand binds to a membrane receptor, the receptor's structure changes, and cell proteins are altered in some way. For example, some membrane proteins respond to ligands by becoming activated enzymes, while others common in muscle and nerve cells respond by transiently opening or closing ion channels, which in turn changes the membrane potential of the cell.

G protein–coupled receptors exert their effect *indirectly* through a **G protein**, a regulatory molecule that acts as a middleman or relay to activate (or inactivate) a membrane-bound enzyme or ion channel. This in turn generates one or more intracellular chemical signals, commonly called **second messengers**, which connect plasma membrane events to the internal metabolic machinery of the cell. Two important second messengers are **cyclic AMP** and ionic calcium, both of which typically activate *protein kinase enzymes*. These enzymes transfer phosphate groups from ATP to other proteins, activating a whole series of enzymes that bring about the desired cellular activity. Because a single enzyme can catalyze hundreds of reactions, the amplification effect of such a chain of events is tremendous, much like a video "going viral." *Focus on G Proteins* (**Focus Figure 3.3**, p. 82) describes a G protein signaling system. Take a moment to study this figure carefully because this key signaling pathway is involved in neurotransmission, smell, vision, and hormone action (Chapters 11, 15, and 16).

G proteins act as middlemen or relays between extracellular first messengers and intracellular second messengers that cause responses within the cell.

The sequence described here is like a molecular relay race. Instead of a baton passed from runner to runner, the message (a shape change) is passed from molecule to molecule as it makes its way across the plasma membrane from outside to inside the cell.

Ligand (1st messenger)   Receptor   G protein   Enzyme   2nd messenger

① **Ligand* (1st messenger) binds to the receptor.** The receptor changes shape and activates.

② **The activated receptor binds to a G protein and activates it.** The G protein changes shape (turns "on"), causing it to release GDP and bind GTP (an energy source).

③ **Activated G protein activates (or inactivates) an effector protein by causing its shape to change.**

Extracellular fluid

Effector protein (e.g., an enzyme)

Ligand   Receptor

G protein

GDP   GTP

GTP

GTP

Inactive 2nd messenger

Active 2nd messenger

Activated kinase enzymes

④ **Activated effector enzymes catalyze reactions that produce 2nd messengers in the cell.** (Common 2nd messengers include cyclic AMP and Ca^{2+}.)

⑤ **Second messengers activate other enzymes or ion channels.** Cyclic AMP typically activates protein kinase enzymes.

⑥ **Kinase enzymes activate other enzymes.** Kinase enzymes transfer phosphate groups from ATP to specific proteins and activate a series of other enzymes that trigger various metabolic and structural changes in the cell.

Cascade of cellular responses (The amplification effect is tremendous. Each enzyme catalyzes hundreds of reactions.)

* Ligands include hormones and neurotransmitters.

Intracellular fluid

Check Your Understanding

17. What do you call signaling chemicals that bind to membrane receptors? Which type of membrane receptor is most important in directing intracellular events by promoting formation of second messengers?

For answers, see Answers Appendix.

PART 2

THE CYTOPLASM

Learning Outcome

▶ **Describe the composition of the cytosol.**

Cytoplasm ("cell-forming material"), the cellular material between the plasma membrane and the nucleus, is the site of most cellular activities. Although early microscopists thought that the cytoplasm was a structureless gel, the electron microscope reveals that it consists of three major elements: the *cytosol*, *organelles*, and *inclusions*.

The **cytosol** (si′to-sol) is the viscous, semitransparent fluid in which the other cytoplasmic elements are suspended. It is a complex mixture with properties of both a colloid and a true solution. Dissolved in the cytosol, which is largely water, are proteins, salts, sugars, and a variety of other solutes.

Inclusions are chemical substances that may or may not be present, depending on cell type. Examples include stored nutrients, such as the glycogen granules in liver and muscle cells; lipid droplets in fat cells; and pigment (melanin) granules in certain skin and hair cells.

The **organelles** are the metabolic machinery of the cell. Each type of organelle carries out a specific function for the cell—some synthesize proteins, others generate ATP, and so on.

3.7 Cytoplasmic organelles each perform a specialized task

Learning Outcomes

▶ **Discuss the structure and function of mitochondria.**
▶ **Discuss the structure and function of ribosomes, the endoplasmic reticulum, and the Golgi apparatus, including functional interrelationships among these organelles.**
▶ **Compare the functions of lysosomes and peroxisomes.**
▶ **Name and describe the structure and function of cytoskeletal elements.**

The organelles ("little organs") are specialized cellular compartments or structures, each performing its own job to maintain the life of the cell.

Although certain organelles such as ribosomes and centrioles lack a membrane (are nonmembranous), most organelles are bounded by a membrane similar in composition to the plasma membrane. These membranes enable the *membranous organelles* to maintain an internal environment different from that of the surrounding cytosol. This compartmentalization is crucial to cell functioning. Without it, biochemical activity would be chaotic. Now let's consider what goes on in each of the workshops of our cellular factory.

Mitochondria

Mitochondria (mi″to-kon′dre-ah) are typically threadlike (*mitos* = thread) or lozenge-shaped membranous organelles. In living cells they squirm, elongate, and change shape almost continuously. They are the power plants of a cell, providing most of its ATP supply. The density of mitochondria in a particular cell reflects that cell's energy requirements, and mitochondria generally cluster where the action is. Busy cells like kidney and liver cells have hundreds of mitochondria, whereas relatively inactive cells (such as certain lymphocytes) have just a few.

A mitochondrion is enclosed by *two* membranes, each with the general structure of the plasma membrane (**Figure 3.14**). The *outer membrane* is smooth and featureless, but the *inner membrane* folds inward, forming shelflike **cristae** (kri′ste; "crests") that protrude into the *matrix*, the gel-like substance

(a)

(b)

(c)

Figure 3.14 Mitochondrion. (a) Diagram of a longitudinally sectioned mitochondrion. **(b)** Close-up of a crista showing enzymes (stalked particles). **(c)** Electron micrograph of a mitochondrion (50,000×).

within the mitochondrion. Teams of enzymes, some dissolved in the mitochondrial matrix and others forming part of the crista membrane, break down intermediate products of food fuels (glucose and others) to water and carbon dioxide.

As the metabolites are broken down and oxidized, some of the energy released is captured and used to attach phosphate groups to ADP molecules to form ATP. This multistep mitochondrial process (described in Chapter 24) is called *aerobic cellular respiration* (a-er-o′bik) because it requires oxygen.

Mitochondria are complex organelles: They contain their own DNA, RNA, and ribosomes and are able to reproduce themselves. Mitochondrial genes (some 37 of them) direct the synthesis of 1% of the proteins required for mitochondrial function, and the DNA of the cell's nucleus encodes the other 99%. When cellular requirements for ATP increase, the mitochondria synthesize more cristae or simply pinch in half (a process called *fission*) to increase their number, then grow to their former size.

Intriguingly, mitochondria are similar to bacteria in the purple bacteria phylum, and mitochondrial DNA is bacteria-like. It is widely believed that mitochondria arose from bacteria that invaded the ancient ancestors of plant and animal cells, and that this unique merger gave rise to all complex cells.

Ribosomes

Ribosomes (ri′bo-sōmz) are small, dark-staining granules composed of proteins and a variety of RNAs called *ribosomal RNAs*. Each ribosome has two globular subunits that fit together like the body and cap of an acorn. Ribosomes are sites of protein synthesis, a function we discuss in detail later in this chapter.

Two ribosomal populations divide the chore of protein synthesis:

- *Free ribosomes* float freely in the cytosol. They make soluble proteins that function in the cytosol, as well as those imported into mitochondria and some other organelles.

- *Membrane-bound ribosomes* are attached to membranes, forming a complex called the *rough endoplasmic reticulum* (**Figure 3.15**). They synthesize proteins destined either for incorporation into cell membranes or lysosomes, or for export from the cell.

Ribosomes can switch back and forth between these two functions, attaching to and detaching from the membranes of the endoplasmic reticulum, according to the type of protein they are making at a given time.

Endoplasmic Reticulum (ER)

The **endoplasmic reticulum** (**ER**) (en″do-plaz′mik rĕ-tik′u-lum; "network within the cytoplasm") is an extensive system of interconnected tubes and parallel sacs called **cisterns** (sis′ternz) (Figure 3.15). The cavities of cisterns are filled with fluid. Coiling and twisting through the cytosol, the ER is continuous with the outer nuclear membrane and accounts for about half of the cell's membranes. There are two distinct varieties: rough ER and smooth ER.

(a) Diagrammatic view of smooth and rough ER

Smooth ER

Nuclear envelope

Rough ER

Ribosomes

Cisterns

Nucleus

(b) Electron micrograph of smooth and rough ER (25,000×)

Figure 3.15 The endoplasmic reticulum.

Rough Endoplasmic Reticulum

The external surface of the **rough ER** is studded with ribosomes, which is why it is called "rough" (see Figures 3.2 and 3.15a, b). Proteins assembled on these ribosomes thread their way into the fluid-filled interior of the ER cisterns (as described on p. 105). When complete, the newly made proteins are enclosed in vesicles for their journey to the Golgi apparatus where they undergo further processing.

The rough ER has several functions:

- Its ribosomes manufacture all proteins secreted from cells. For this reason, the rough ER is particularly abundant and well developed in most secretory cells, antibody-producing immune cells, and liver cells, which produce most blood proteins.

- It is also the cell's "membrane factory" where integral proteins and phospholipids that form part of all cellular membranes are manufactured. Within the cisterns, sugar groups are attached to those proteins that will eventually face the extracellular environment. The enzymes that catalyze lipid synthesis have their active sites on the external (cytosolic) face of the ER membrane, where the needed substrates are readily available.

Smooth Endoplasmic Reticulum

The **smooth ER** (see Figures 3.2 and 3.15) is continuous with the rough ER and consists of tubules arranged in a looping network.

Its enzymes (all integral proteins integrated into its membranes) play no role in protein synthesis. Instead, the enzymes catalyze reactions involved with the following tasks:

- Metabolize lipids, synthesize cholesterol and phospholipids, and synthesize the lipid components of lipoproteins (in liver cells)

- Synthesize steroid-based hormones such as sex hormones (for example, testosterone-synthesizing cells of the testes are full of smooth ER)

- Detoxify drugs, certain pesticides, and cancer-causing chemicals (in liver and kidneys)

- Break down stored glycogen to form free glucose (in liver cells especially)

- Store calcium ions in most cell types [skeletal and cardiac muscle cells have an elaborate smooth ER (called the sarcoplasmic reticulum) that stores calcium and releases it as a trigger for contraction]

The amount of smooth ER varies widely between different types of cells. As you can see from the list above, it plays a role in diverse cellular processes, depending on the type of cell.

Golgi Apparatus

The **Golgi apparatus** (gol′je) consists of stacked and flattened membranous sacs, shaped like hollow dinner plates, associated with swarms of tiny membranous vesicles (**Figure 3.16**). The

Cis face—
"receiving" side of
Golgi apparatus

Transport vesicle
from rough ER

Cisterns

New vesicles
forming

Transport
vesicle
from
trans face

Trans face—
"shipping" side of
Golgi apparatus

Secretory vesicle

Newly secreted
proteins

(a) Many vesicles in the process of pinching off from the Golgi apparatus

New vesicles forming

Golgi apparatus

Transport vesicle at
the trans face

(b) Electron micrograph of the Golgi apparatus (90,000×)

Figure 3.16 **Golgi apparatus.**

3

① **Vesicles move from ER to Golgi.** Vesicles containing proteins bud off of the rough ER, migrate, and fuse with membranes of the Golgi apparatus.

② **Proteins are modified.** Inside the Golgi apparatus, some sugar groups are trimmed while others are added. In some cases, phosphate groups are added.

③ **Proteins are distributed.** Proteins are tagged ("addressed"), sorted, and packaged in three different types of vesicles, which bud off the concave *trans* face (the "shipping" side) of the Golgi stack. These vesicles have three possible fates depending on the proteins they carry:

Rough ER

ER membrane   Phagosome

Proteins in cisterns

Vesicle becomes lysosome

Golgi apparatus

Secretory vesicle

Plasma membrane

Extracellular fluid

Secretion by exocytosis

③a **Proteins for secretion.** Proteins destined for export are packaged into **secretory vesicles** that migrate to the plasma membrane and dump their contents by exocytosis.

③b **Proteins destined for membranes.** Vesicles containing lipids and transmembrane proteins destined for the plasma membrane (or other cellular membranes) fuse with and are incorporated into those membranes.

③c **Digestive enzymes.** Inactive digestive enzymes are packaged into vesicles that eventually become lysosomes or fuse with lysosomes.

Figure 3.17 Processing and distribution of newly synthesized proteins.

Golgi apparatus is the principal "traffic director" for cellular proteins. Its major function is to modify, concentrate, and package the proteins and lipids made at the rough ER and destined for export from the cell.

This process is shown in **Figure 3.17**.

Peroxisomes

Resembling small lysosomes, **peroxisomes** (pĕ-roks′ĭ-sōmz; "peroxide bodies") are spherical membranous sacs containing a variety of powerful enzymes, the most important of which are oxidases and catalases.

Oxidases use molecular oxygen (O_2) to detoxify harmful substances, including alcohol and formaldehyde. Their most important function is to neutralize **free radicals**, highly reactive chemicals with unpaired electrons that can scramble the structure of biological molecules. Oxidases convert free radicals to hydrogen peroxide, which is also reactive and dangerous but which the catalases quickly convert to water. Free radicals and hydrogen peroxide are normal by-products of cellular metabolism, but they have devastating effects on cells if allowed to accumulate.

Peroxisomes are especially numerous in liver and kidney cells, which are very active in detoxification. They also play a

role in energy metabolism by breaking down and synthesizing fatty acids.

Some peroxisomes are formed when existing peroxisomes simply pinch in half. However, most new peroxisomes form by budding off of the endoplasmic reticulum via a special ER machinery that differs from that used for vesicles destined for modification in the Golgi apparatus.

Lysosomes

Lysosomes ("disintegrator bodies") are spherical membranous organelles containing activated hydrolytic (digestive) enzymes (**Figure 3.18**). As you might guess, lysosomes are large and abundant in phagocytes, the cells that dispose of invading bacteria and cell debris. Lysosomal enzymes can digest almost all kinds of biological molecules. They work best in acidic conditions and so are called *acid hydrolases*.

The lysosomal membrane is adapted to serve lysosomal functions in two important ways. First, it contains H^+ (proton) "pumps," which are ATPases that gather hydrogen ions from the surrounding cytosol to maintain the organelle's acidic pH. Second, it retains the dangerous lysosomal enzymes (acid hydrolases) while permitting the final products of digestion to escape so that they can be used by the cell or excreted. In this

Lysosomes

Light green areas are
regions where materials
are being digested.

Figure 3.18 Electron micrograph of lysosomes (20,000×).

way, lysosomes provide sites where digestion can proceed *safely* within a cell.

Lysosomes function as a cell's "demolition crew" by:

- Digesting particles taken in by endocytosis, particularly ingested bacteria, viruses, and toxins

- Degrading stressed or dead cells and worn-out or nonfunctional organelles, a process more specifically called autophagy ("self-eating") (discussed further on p. 109)

- Performing metabolic functions, such as glycogen breakdown and release

- Breaking down bone to release calcium ions into the blood

The lysosomal membrane is ordinarily quite stable, but it becomes fragile when the cell is injured or deprived of oxygen and when excessive amounts of vitamin A are present. When lysosomes rupture, the cell digests itself, a process called **autolysis** (aw″tol′ĭ-sis).

**HOMEOSTATIC
IMBALANCE 3.3**   **CLINICAL**

Lysosomes degrade glycogen and certain lipids in the brain at a relatively constant rate. In *Tay-Sachs disease*, a rare neurodegenerative disorder, the lysosomes lack an enzyme needed to break down a specific glycolipid in nerve cell membranes.

These lysosomes swell with undigested lipids, interfering with nervous system functioning. Children with the disease begin to lose developmental milestones at 3 to 6 months. One physical finding is a "cherry red spot" on the macula of the retina at the back of the eye. No cure currently exists, and the disease progresses to blindness, deafness, seizures, and paralysis. Most children with Tay-Sachs die by age 5. ●●●●●●●●

The Endomembrane System

The **endomembrane system** is a system of organelles (most described above) that work together mainly to (1) produce, degrade, store, and export biological molecules, and (2) degrade potentially harmful substances. It includes the ER, Golgi apparatus, secretory vesicles, and lysosomes, as well as the nuclear envelope—that is, all of the membranous elements that are either structurally connected or arise via forming or fusing transport vesicles (**Figure 3.19**). The nuclear envelope is directly connected to the rough and smooth ER (see Figure 3.15). The plasma membrane, though not actually an *endo*-membrane, is also functionally part of this system.

Besides these direct structural relationships, a wide variety of indirect interactions (indicated by arrows in Figure 3.19) occur among the members of the system. Some of the vesicles "born" in the ER migrate to and fuse with the Golgi apparatus or the plasma membrane, and vesicles arising from the Golgi apparatus can become part of the plasma membrane or lysosomes.

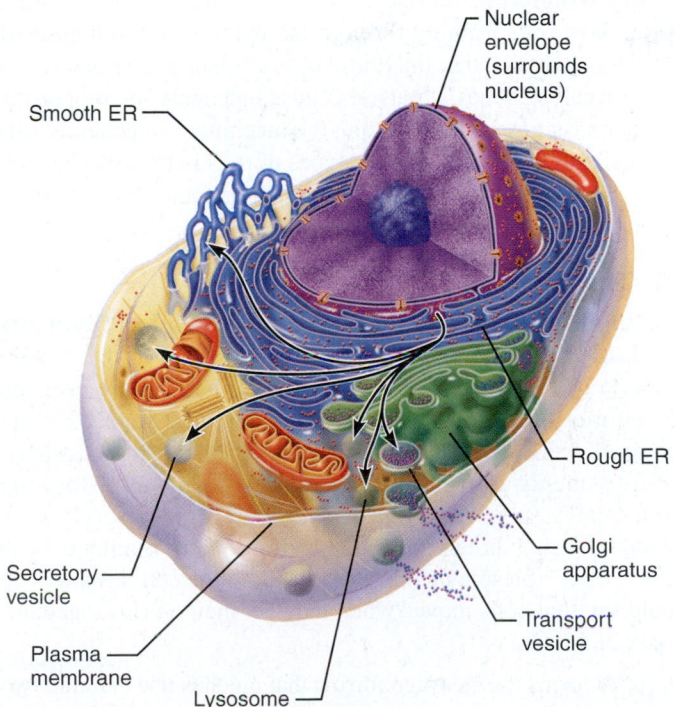

Nuclear envelope (surrounds nucleus)

Smooth ER

Rough ER

Golgi apparatus

Transport vesicle

Secretory vesicle

Plasma membrane

Lysosome

Figure 3.19 The endomembrane system.

(a) **Microfilaments**	(b) **Intermediate filaments**	(c) **Microtubules**
Strands made of spherical protein subunits called actin	Tough, insoluble protein fibers constructed like woven ropes composed of tetramer (4) fibrils	Hollow tubes of spherical protein subunits called tubulin
Actin subunit — 7 nm	Tetramer subunits — 10 nm	Tubulin subunits — 25 nm
Microfilaments form the blue network in this photo.	Intermediate filaments form the lavender network surrounding the pink nuclei in this photo.	Microtubules appear as gold networks surrounding the cells' pink nuclei in this photo.

Figure 3.20 Cytoskeletal elements support the cell and help to generate movement. The photos (bottom) are of fibroblasts treated with fluorescent tags for the structure of interest.

Cytoskeleton

The **cytoskeleton**, literally, "cell skeleton," is an elaborate network of rods running through the cytosol and hundreds of accessory proteins that link these rods to other cell structures. It acts as a cell's "bones," "muscles," and "ligaments" by supporting cellular structures and providing the machinery to generate various cell movements. The three types of rods in the cytoskeleton are *microfilaments*, *intermediate filaments*, and *microtubules*. None of these is membrane covered.

Microfilaments

The thinnest elements of the cytoskeleton, **microfilaments** (mi″kro-fil′ah-ments), are semiflexible strands of the protein *actin* ("ray") (**Figure 3.20a**). Most microfilaments are involved in cell motility (movement) or changes in cell shape. Except in muscle cells, where they are especially abundant and stable, actin filaments are constantly breaking down and re-forming from smaller subunits whenever and wherever their services are needed. As they break down and re-form, actin filaments push or pull on the plasma membrane, changing the cell's shape. You could say that cells move "when they get their act(in) together." For example:

- Actin forms the cleavage furrow that pinches one cell into two during cell division.
- Microfilaments attached to cell adhesion molecules (see Figure 3.3e, f) are responsible for the crawling movements of

amoeboid motion, and for membrane changes that accompany endocytosis and exocytosis.

- In muscle cells, actin filaments interact with another protein, *myosin* (mi′o-sin), to generate contractile forces in a cell.

In nonmoving cells, actin filaments also help maintain cell shape and distribute tension throughout the cell.

Intermediate Filaments

Intermediate filaments are tough, insoluble protein fibers that resemble woven ropes. Made of twisted units of *tetramer* (4) *fibrils*, they have a diameter between those of microfilaments and microtubules (Figure 3.20b). Intermediate filaments are the most stable and permanent of the cytoskeletal elements and strongly resist tension. They attach to desmosomes, and their main job is to act as internal cables to resist pulling forces exerted on the cell. Because their protein composition varies in different cell types, there are numerous names for these cytoskeletal elements—for example, they are called neurofilaments in nerve cells and keratin filaments in epithelial cells.

Microtubules

The elements with the largest diameter, **microtubules** (mi″kro-tu′būlz), are hollow tubes made of spherical protein subunits called *tubulin* (Figure 3.20c). Most microtubules radiate from a small region of cytoplasm near the nucleus called the *centrosome* or *cell center* (see Figure 3.2 on p. 62).

Microtubules are remarkably dynamic organelles, constantly growing out from the centrosome, disassembling, and then reassembling at the same or different sites. The microtubules determine the overall shape of the cell, as well as the distribution of cellular organelles.

Mitochondria, lysosomes, and secretory vesicles attach to the microtubules like ornaments hanging from tree branches. Tiny protein machines called **motor proteins** (*kinesins*, *dyneins*, and others) continually move and reposition the organelles along the microtubules.

Powered by ATP, motor proteins act like train engines moving substances along on the microtubular "railroad tracks." They move "hand over hand" somewhat like an orangutan—gripping, releasing, and then gripping again at a new site further along the microtubule.

Centrosome and Centrioles

As mentioned, microtubules are anchored at one end in an inconspicuous region near the nucleus called the **centrosome** or *cell center*. The centrosome acts as a *microtubule organizing center*. It has few distinguishing marks other than a granular-looking *matrix* that contains paired **centrioles**, small, barrel-shaped organelles oriented at right angles to each other (**Figure 3.21**). The centrosome matrix is best known for generating microtubules and organizing the mitotic spindle in cell division (see *Focus on Mitosis*, Focus Figure 3.4 on pp. 100–101). Each centriole consists of a pinwheel array of nine *triplets* of microtubules, each connected to the next by nontubulin proteins and arranged to form a hollow tube. Centrioles also form the bases of cilia and flagella, our next topics.

Check Your Understanding

18. Compare the functions of lysosomes and peroxisomes.

19. How are microtubules and microfilaments related functionally?

20. Name each of the organelles below. Match the organelle(s) with the applicable statement(s). Answers may be used more than once.

___ (1) Moves organelles within cell using motor proteins
___ (2) Contains its own DNA
___, ___, ___ (3) Has cisterns
___ (4) Major site of ATP synthesis
___ (5) Site of steroid hormone synthesis
___ (6) Has *cis* and *trans* faces

(a)

Centrosome matrix

Centrioles

Microtubules

(b)

Figure 3.21 Centrioles. (a) Three-dimensional view of a centriole pair oriented at right angles, as they are usually seen in the cell. **(b)** An electron micrograph showing a cross section of a centriole (190,000×). Notice that it is composed of nine microtubule triplets.

(a)

(b)

(c)

(d)

(e)

21. APPLY Consider a plasma membrane glycoprotein. Describe the protein's path through the cell to the plasma membrane starting with its synthesis on the ribosome.

For answers, see Answers Appendix.

3.8 Cilia and microvilli are two main types of cellular extensions

Learning Outcomes

▶ Describe the role of centrioles in the formation of cilia and flagella.

▶ Describe how the two main types of cell extensions, cilia and microvilli, differ in structure and function.

Cilia and Flagella

Cilia (sil′e-ah; "eyelashes") are whiplike, motile cellular extensions (**Figure 3.22**) that occur, typically in large numbers, on the exposed surfaces of certain cells. Ciliary action moves substances in one direction across cell surfaces. For example, ciliated cells that line the respiratory tract propel mucus laden with dust particles and bacteria upward away from the lungs.

When a cell is about to form cilia, the centrioles multiply and line up beneath the plasma membrane at the cell's free (exposed) surface. Microtubules then "sprout" from each centriole, forming the ciliary projections by exerting pressure on the plasma membrane.

Flagella (flah-jel′ah) are also projections formed by centrioles, but are substantially longer than cilia. The only flagellated cell in the human body is a sperm, which has one propulsive flagellum, commonly called a tail. Notice that cilia *propel other substances* across a cell's surface, whereas a flagellum *propels the cell itself*.

Centrioles forming the bases of cilia and flagella are referred to as **basal bodies** (ba′sal) (Figure 3.22). The "9 + 2" pattern of microtubules in the cilium or flagellum itself (nine *doublets*, or pairs, of microtubules encircling one central pair) differs slightly from that of a centriole (nine microtubule *triplets*). Additionally, the cilium has flexible "wagon wheels" of cross-linking proteins (purple in Figure 3.22).

Extending from the microtubule doublets are arms composed of the motor protein dynein (green dynein arms in Figure 3.22). The dynein arms produce the movement of the cilium or flagellum. Arms of one doublet grip the adjacent doublet and, powered by ATP, push it up, release, and then grip again. Because the doublets are physically restricted by other proteins, they are forced to bend. The collective bending action of all the doublets causes the cilium to bend.

As a cilium moves, it alternates rhythmically between a propulsive *power stroke*, when it is nearly straight and moves in an arc, and a *recovery stroke*, when it bends and returns to its initial position (**Figure 3.23a**). With these two strokes, the cilium

Outer microtubule doublet

Dynein arms

Central microtubule

Cross-linking protein between outer doublets

Radial spoke

A cross section through the **cilium** shows the "9 + 2" arrangement of microtubles.

Cross-linking proteins between outer doublets

Radial spoke

Plasma membrane

Triplet

A cross section through the **basal body**. The nine outer doublets of a cilium extend into a basal body where each doublet joins another microtubule to form a ring of nine triplets.

The doublets also have attached motor proteins, the **dynein arms.**

The "9 + 2" pattern: Microtubule doublets encircle two central microtubules. Microtubules are held together by cross-linking proteins and radial spokes.

Cilium

Basal body (centriole)

Figure 3.22 Structure of a cilium.

Power, or propulsive, stroke

Recovery stroke, when cilium is returning to its initial position

1 2 3 4 5 6 7

(a) Phases of ciliary motion.

Layer of mucus

Cell surface

(b) Traveling wave created by the activity of many cilia acting together propels mucus across cell surfaces.

Figure 3.23 Ciliary function.

produces a pushing motion in a single direction that repeats some 10 to 20 times per second. The bending of one cilium is quickly followed by the bending of the next and then the next, creating a current at the cell surface that resembles the traveling waves that pass across a field of grass on a windy day (Figure 3.23b).

Microvilli

Microvilli (mi″kro-vil′i; "little shaggy hairs") are tiny, finger-like extensions of the plasma membrane that project from an exposed cell surface (Figure 3.4 top and **Figure 3.24**). They increase the plasma membrane surface area tremendously and are most often found on the surface of absorptive cells such as intestinal and kidney tubule cells. Microvilli have a core of bundled actin filaments that extend into a mat of actin filaments, called the *terminal web*, near the surface of the cell. (Actin is often a contractile protein, but in microvilli and the terminal web it acts as a mechanical "stiffener" that shapes the cell.)

Check Your Understanding

22. The major function of cilia is to move substances across the free cell surface. What is the major role of microvilli?

For answers, see Answers Appendix.

Microvillus

Actin filaments

Terminal web

Figure 3.24 Microvilli.

Practice Histology questions: MasteringA&P® > Study Area > Lab Tools > PAL

PART 3

NUCLEUS

Anything that works, works best when it is controlled. For cells, the control center is the gene-containing **nucleus** (*nucle* = pit, kernel). The nucleus can be compared to a computer, design department, construction boss, and board of directors—all rolled into one. As the genetic library, it contains the instructions needed to build nearly all the body's proteins. Additionally, it dictates the kinds and amounts of proteins to be synthesized at any one time in response to signals acting on the cell.

Most cells have only one nucleus, but some, including skeletal muscle cells, bone destruction cells, and some liver cells, are **multinucleate** (mul″tĭ-nu′kle-āt), that is, they have many nuclei. The presence of more than one nucleus usually signifies that a larger-than-usual cytoplasmic mass must be regulated.

Except for mature red blood cells, whose nuclei are ejected before the cells enter the bloodstream, all of our body cells have nuclei. **Anucleate** (a-nu′kle-āt; *a* = without) cells cannot reproduce and therefore live in the bloodstream for only three to four months before they deteriorate. Without a nucleus, a cell cannot produce mRNA to make proteins, and when its enzymes and cell structures start to break down (as all eventually do), they cannot be replaced.

3.9 The nucleus includes the nuclear envelope, the nucleolus, and chromatin

Learning Outcome

▶ Outline the structure and function of the nuclear envelope, nucleolus, and chromatin.

The nucleus, averaging 5 μm in diameter, is larger than any of the cytoplasmic organelles. Although most often spherical or oval, its shape usually conforms to the shape of the cell. The

Nuclear envelope

Chromatin (condensed)

Nucleolus

Cisterns of rough ER

(a)

Surface of nuclear envelope.

Fracture line of outer membrane

Nuclear pores

Nucleus

Nuclear pore complexes. Each pore is ringed by protein particles.

Nuclear lamina. The netlike lamina composed of intermediate filaments formed by lamins lines the inner surface of the nuclear envelope.

(b)

Figure 3.25 The nucleus. (a) Three-dimensional diagram of the nucleus, showing the continuity of its double membrane with the ER. **(b)** Freeze-fracture transmission electron micrographs (TEMs).

nucleus has three recognizable regions or structures: the *nuclear envelope* (*membrane*), *nucleoli*, and *chromatin* (**Figure 3.25a**).

The Nuclear Envelope

The nucleus is bounded by the **nuclear envelope**, a *double membrane* barrier separated by a fluid-filled space (like the mitochondrial membrane). The outer nuclear membrane is continuous with the rough ER of the cytoplasm and is studded with ribosomes on its external face. The inner nuclear membrane is lined by the *nuclear lamina*, a network of *lamins* (rod-shaped proteins that assemble to form intermediate filaments) that maintains the shape of the nucleus and acts as a scaffold to organize DNA in the nucleus (Figure 3.25b, bottom).

At various points, the nuclear envelope is punctured by **nuclear pores**. An intricate complex of proteins, called a *nuclear pore complex*, lines each pore, forming an aqueous

transport channel and regulating entry and exit of molecules (e.g., mRNAs) and large particles into and out of the nucleus (Figure 3.25b, middle).

Like other cell membranes, the nuclear envelope is selectively permeable, but here substances pass much more freely than elsewhere. Small molecules pass through the relatively large nuclear pore complexes unhindered. Protein molecules imported from the cytoplasm and RNA molecules exported from the nucleus move through the central channel of the pores. This process is energy dependent and guided by soluble transport proteins.

The nuclear envelope encloses a jellylike fluid called *nucleoplasm* (nu′kle-o-plazm) in which other nuclear elements are suspended. Like the cytosol, the nucleoplasm contains dissolved salts, nutrients, and other essential solutes.

Nucleoli

Within the nucleus are **nucleoli** (nu-kle′o-li; "little nuclei"), dark-staining spherical bodies where the ribosomal subunits are assembled. Nucleoli are not membrane bounded. Typically, there are one or two per nucleus, but there may be more. Nucleoli are usually largest in growing cells that are making large amounts of tissue proteins.

Nucleoli are aggregations of all of the components needed to synthesize and assemble ribosomal subunits. They center around the DNA that codes for ribosomal RNA (rRNA). As rRNA molecules are synthesized, they combine with proteins to form the two kinds of ribosomal subunits. (The proteins are manufactured on ribosomes in the cytoplasm and imported into the nucleus.) The finished subunits leave the nucleus through the nuclear pores and enter the cytoplasm, where they join to form functional ribosomes.

Chromatin

Seen through a light microscope, **chromatin** (kro′mah-tin) appears as a fine, unevenly stained network, but special techniques reveal it as a system of bumpy threads weaving through the nucleoplasm. Chromatin is composed of approximately:

- 30% **DNA**, our genetic material
- 60% globular **histone proteins** (his′tōn), which package and regulate the DNA
- 10% RNA chains, newly formed or forming

The fundamental units of chromatin are **nucleosomes** (nu′kle-o-sōmz; "nuclear bodies"), which consist of flattened disc-shaped cores or clusters of eight histone proteins connected like beads on a string by a DNA molecule. The DNA winds (like a ribbon of Velcro®) twice around each nucleosome and continues on to the next cluster via *linker* DNA segments (**Figure 3.26** ① and ②).

Histones provide a physical means for packing the very long DNA molecules (some 2 meters' worth per cell) in a compact, orderly way, but they also play an important role in gene regulation. In a nondividing cell, for example, the presence of methyl groups on histone proteins shuts down the nearby DNA. As another example, addition of acetyl groups to histone exposes different DNA segments, or genes, so that they can dictate the specifications for synthesizing proteins or various RNA species. Such active chromatin segments, referred to as *extended chromatin*, are not usually visible under the light microscope. The generally inactive *condensed chromatin* segments are darker staining and more easily detected. Understandably, the most active body cells have much larger amounts of extended chromatin.

When a cell is preparing to divide, the chromatin threads coil and condense enormously to form short, barlike bodies called **chromosomes** ("colored bodies") (Figure 3.26 ④ and ⑤). Chromosome compactness prevents the delicate chromatin strands from tangling and breaking during the

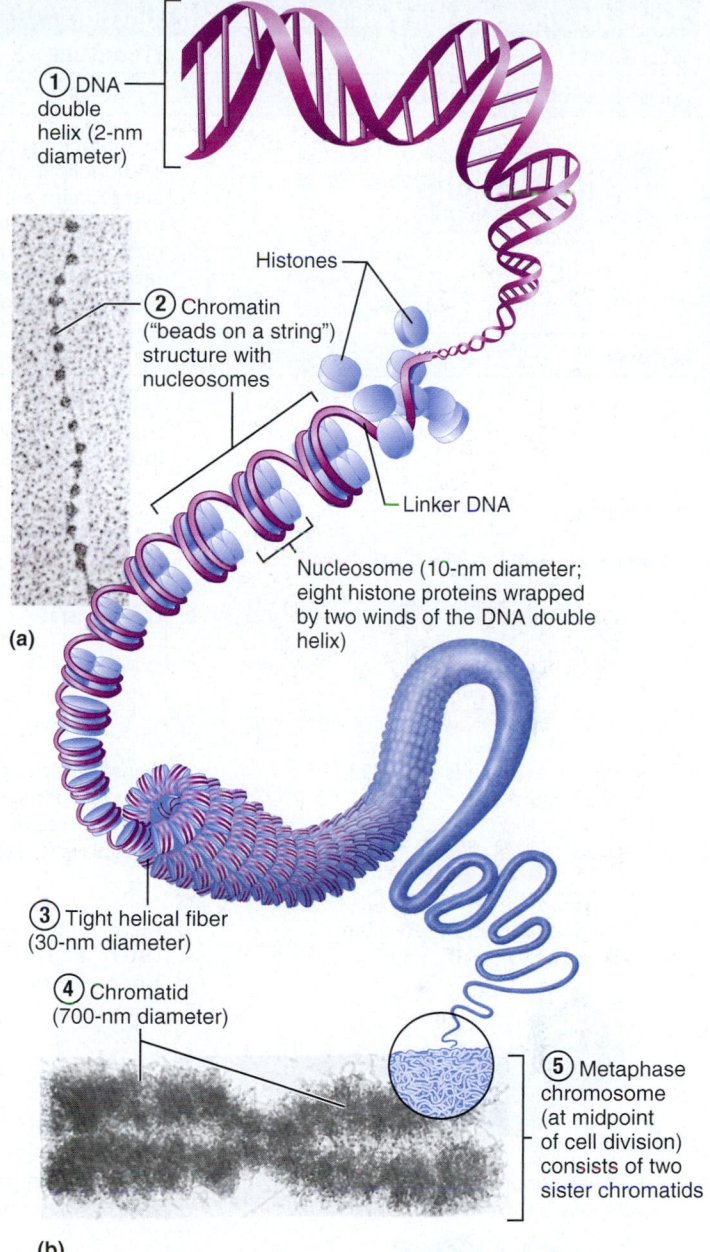

(a)

(1) DNA double helix (2-nm diameter)

Histones

(2) Chromatin ("beads on a string") structure with nucleosomes

Linker DNA

Nucleosome (10-nm diameter; eight histone proteins wrapped by two winds of the DNA double helix)

(3) Tight helical fiber (30-nm diameter)

(4) Chromatid (700-nm diameter)

(5) Metaphase chromosome (at midpoint of cell division) consists of two sister chromatids

(b)

Figure 3.26 Chromatin and chromosome structure.
(a) Electron micrograph of chromatin fiber (125,000×). **(b)** DNA packed in a chromosome. The levels of increasing structural complexity (coiling) are shown in order from the smallest (①–⑤).

movements that occur during cell division. Next, we describe the events of cell division and the functions of DNA.

Table 3.4 beginning on p. 94 summarizes the parts of the cell.

Check Your Understanding

23. If a cell ejects or loses its nucleus, what is its fate and why?

24. What is the role of nucleoli?

25. What is the importance of the histone proteins present in the nucleus?

For answers, see Answers Appendix.

(Text continues on p. 96.)

Table 3.4 Parts of the Cell: Structure and Function

CELL PART*	STRUCTURE	FUNCTIONS
Plasma Membrane (Focus Figure 3.1)		
	Membrane made of a double layer of phospholipids within which cholesterol and proteins are embedded. Proteins may extend entirely through the lipid bilayer or protrude on only one face. Most externally facing proteins and some lipids have attached sugar groups.	Serves as an external cell barrier. Transmembrane proteins act as receptors for chemical messengers (e.g., hormones, neurotransmitters), as transport proteins, and in cell-to-cell recognition. Maintains a resting potential that is essential for functioning of excitable cells.
Cytoplasm		
	Cellular region between the nuclear and plasma membranes. Consists of fluid **cytosol** containing dissolved solutes, **organelles** (the metabolic machinery of the cytoplasm), and **inclusions** (stored nutrients, secretory products, pigment granules).	
Organelles		
• Mitochondria (Figure 3.14)	Rodlike, double-membrane structures; inner membrane folded into projections called cristae.	Site of ATP synthesis; powerhouse of the cell.
• Ribosomes (Figures 3.32, 3.33, Focus Figure 3.5)	Dense particles consisting of two subunits, each composed of ribosomal RNA and protein. Free or attached to rough endoplasmic reticulum.	The sites of protein synthesis.
• Rough endoplasmic reticulum (Figures 3.15, 3.33)	Membranous system enclosing a cavity, the cistern, and coiling through the cytoplasm. Externally studded with ribosomes.	Sugar groups are attached to proteins within the cisterns. Proteins are bound in vesicles for transport to the Golgi apparatus and other sites. External face synthesizes phospholipids.
• Smooth endoplasmic reticulum (Figure 3.15)	Membranous system of sacs and tubules; free of ribosomes.	Site of lipid and steroid (cholesterol) synthesis, lipid metabolism, drug detoxification, and Ca^{2+} storage.
• Golgi apparatus (Figures 3.16, 3.17)	A stack of flattened membranes and associated vesicles close to the nucleus.	Packages, modifies, and segregates proteins for secretion from the cell, inclusion in lysosomes, and incorporation into the plasma membrane. Modifies carbohydrates on proteins.
• Peroxisomes (Figure 3.2)	Membranous sacs containing catalase and oxidase enzymes.	The enzymes detoxify a number of toxic substances. The most important enzyme, catalase, breaks down hydrogen peroxide.

*Individual cellular structures are not drawn to scale.

Table 3.4 *(continued)*		
CELL PART*	STRUCTURE	FUNCTIONS
Cytoplasm		
• Lysosomes (Figure 3.18)	Membranous sacs containing acid hydrolases.	Sites of intracellular digestion.
• Microtubules (Figures 3.20–3.22)	Cylindrical structures made of tubulin proteins.	Support the cell and give it shape. Involved in intracellular and cellular movements. Form centrioles and cilia and flagella, if present.
• Intermediate filaments (Figure 3.20)	Protein fibers; composition varies.	The stable cytoskeletal elements; resist mechanical forces acting on the cell.
• Microfilaments (Figure 3.20)	Fine filaments composed of the protein actin.	Involved in muscle contraction and other types of intracellular movement; help form the cell's cytoskeleton.
• Centrioles (Figure 3.21)	Paired cylindrical bodies, each composed of nine triplets of microtubules.	As part of the centrosome, organize the microtubule network. During mitosis (cell division), form the spindle and asters. As basal bodies, form the bases of cilia and flagella.
Inclusions	Varied; includes stored nutrients such as lipid droplets and glycogen granules, protein crystals, pigment granules.	Storage for nutrients, wastes, and cell products.
Cellular Extensions		
• Cilia (Figures 3.22, 3.23)	Short cell-surface projections; each cilium contains nine pairs of microtubules surrounding a central pair.	Coordinated movement creates a unidirectional current that propels substances across cell surfaces.
• Flagellum	Like a cilium, but longer; only example in humans is the sperm tail.	Propels the cell.
• Microvilli (Figure 3.24)	Tubular extensions of the plasma membrane; contain a bundle of actin filaments.	Increase surface area for absorption.
Nucleus (Figures 3.2, 3.25)		
	Largest organelle. Surrounded by the nuclear envelope; contains fluid nucleoplasm, nucleoli, and chromatin.	Control center of the cell; responsible for transmitting genetic information and providing the instructions for protein synthesis.

➤

Table 3.4   Parts of the Cell: Structure and Function (continued)

CELL PART*	STRUCTURE	FUNCTIONS
Nucleus (continued) (Figures 3.2, 3.25)		
• Nuclear envelope (Figure 3.25)	Double-membrane structure pierced by pores. Outer membrane continuous with the endoplasmic reticulum.	Separates the nucleoplasm from the cytoplasm and regulates passage of substances to and from the nucleus.
• Nucleolus (Figure 3.25)	Dense spherical (non-membrane-bounded) bodies, composed of ribosomal RNA and proteins.	Site of ribosome subunit manufacture.
• Chromatin (Figure 3.26)	Granular, threadlike material composed of DNA and histone proteins.	DNA constitutes the genes.

3.10 The cell cycle consists of interphase and a mitotic phase

Learning Outcomes

▶ **List the phases of the cell cycle and describe the key events of each phase.**

▶ **Describe the process of DNA replication.**

The **cell cycle** is the series of changes a cell goes through from the time it is formed until it reproduces. The outer ring of **Figure 3.27** shows the two major periods of the cell cycle:

• *Interphase* (in green), in which the cell grows and carries on its usual activities

• *Cell division* or the *mitotic phase* (in yellow), during which it divides into two cells

Interphase

Interphase is the period from cell formation to cell division. Early cytologists, unaware of the constant molecular activity in cells and impressed by the obvious movements of cell division, called interphase the resting phase of the cell cycle. However, this image is misleading because during interphase a cell is carrying out all its routine activities and is "resting" only from dividing. Perhaps a more accurate name for this phase would be *metabolic phase* or *growth phase*.

Subphases

In addition to carrying on its life-sustaining activities, an interphase cell prepares for the next cell division. Interphase is divided into G_1, S, and G_2 subphases (the Gs stand for *gaps* before and after the S phase, and S is for *synthetic*). In all three

subphases, the cell grows by producing proteins and organelles, but chromatin is reproduced only during the S subphase.

Let's take a closer look at the subphases:

• G_1 **(gap 1 subphase):** The cell is metabolically active, synthesizing proteins rapidly and growing vigorously (Figure 3.27, light green area). This is the most variable phase in terms of length. G_1 typically lasts several minutes to hours, but it may last for

Figure 3.27 The cell cycle. Important checkpoints at which mitosis may be prevented from occurring are found throughout interphase; the diagram shows two.

days or even years. Cells that permanently stop dividing are said to be in the **G₀ phase**. For most of G_1, virtually no activities directly related to cell division occur. However, as G_1 ends, the centrioles start to replicate in preparation for cell division.

- **S phase:** DNA is replicated, ensuring that the two future cells being created will receive identical copies of the genetic material (Figure 3.27, blue area). New histones are made and assembled into chromatin. One thing is sure, without a proper S phase, there can be no correct mitotic phase. (We will describe DNA replication next.)

- **G₂ (gap 2 subphase):** The final phase of interphase is brief (Figure 3.27, dark green area). Enzymes and other proteins needed for division are synthesized and moved to their proper sites. By the end of G_2, centriole replication (begun in G_1) is complete. At the end of this phase is the G_2/M checkpoint when the cell ensures that all DNA is replicated and damaged DNA has been repaired. If so, the cell is now ready to divide.

Throughout S and G_2, the cell continues to grow and carries on with business as usual.

DNA Replication

Before a cell can divide, its DNA must be replicated exactly, so that identical copies of the cell's genes can be passed to each of the two resulting daughter cells. During the S phase, replication begins simultaneously on several chromatin threads and continues until all the DNA has been replicated.

Human DNA molecules are very long, and replication of a DNA molecule begins at several *origins of replication* along its length. This strategy greatly increases the speed of replication.

Replication involves the following sequence of events (**Figure 3.28**):

1. Uncoiling: Enzymes unwind the DNA molecule, forming a **replication bubble**.
2. Separation: The two DNA strands separate as the hydrogen bonds between base pairs are broken. The point at which the strands unzip is known as the **replication fork**.
3. Assembly: With the old (parental) strands acting as templates, the enzyme **DNA polymerase** positions complementary free nucleotides along the template strands, forming two new strands. Because the polymerases work in one direction only, the two new strands, called leading and lagging strands, are synthesized in opposite directions. Two new (daughter) DNA molecules result from one parental DNA molecule. Since each new molecule consists of one old and one new nucleotide strand, this mechanism is known as **semiconservative replication**.
4. Restoration: Ligase enzymes (not illustrated in Figure 3.28) splice short segments of DNA together, restoring the double helix structure.

Figure 3.28 Replication of DNA: summary. Once the DNA helix is uncoiled, and the hydrogen bonds between its base pairs are broken, each nucleotide strand of the DNA acts as a template for constructing a complementary strand, as illustrated on the right-hand side of the diagram. DNA polymerases work in one direction only, so the two new strands (leading and lagging) are synthesized in opposite directions. (The step in which the RNA primers are formed to start the process is not illustrated, nor are the DNA ligase enzymes that join the DNA fragments on the lagging strand.)

3

The progression from DNA replication to cell division proceeds smoothly when the newly formed DNA is undamaged. If damage occurs, the cycle stops at the G_2/M checkpoint until the DNA repair mechanism has fixed the problem.

During the replication process, histones (made in the cytoplasm and imported into the nucleus) associate with the DNA, completing the formation of two new chromatin strands. The chromatin strands, united by a buttonlike centromere, are held together until the cell enters the anaphase stage of mitotic cell division (see p. 101). They are then distributed to the daughter cells as described next, ensuring that each cell has identical genetic information.

> ▶ Watch a 3-D animation of this process: MasteringA&P® > Study Area > Animations & Videos > A&P Flix

Cell Division

Cell division is essential for body growth and tissue repair. Cells that continually wear away, such as cells of the skin and intestinal lining, reproduce themselves almost continuously. Others, such as liver cells, divide more slowly to maintain the size of the organ they compose but retain the ability to reproduce quickly if the organ is damaged. Most cells of nervous tissue, skeletal muscle, and heart muscle lose their ability to divide when they are fully mature, and repairs are made with scar tissue (a fibrous type of connective tissue).

In most body cells, cell division, which is called the **M (mitotic) phase** of the cell cycle, involves two distinct events (Figure 3.27, yellow area): *mitosis* and *cytokinesis*:

- **Mitosis** (mi-to′sis; *mit* = thread; *osis* = process), the division of the nucleus, is the series of events that parcels out the replicated DNA of the parent cell to two daughter cells. Described as four phases—**prophase**, **metaphase**, **anaphase**, and **telophase**—mitosis is actually a continuous process, with one phase merging smoothly into the next. Its duration varies according to cell type, but in human cells it typically lasts about an hour or less. *Focus on Mitosis* (**Focus Figure 3.4**, pp. 100–101), describes the phases of mitosis in detail.

    A different process of nuclear division called *meiosis* produces sex cells (ova and sperm) with only half the number of genes found in other body cells. We discuss the details of meiosis in Chapter 27. Here we concentrate on mitotic cell division.

- **Cytokinesis** (si-to-kĭ-ne′sis; *kines* = movement), the division of the cytoplasm, begins during late anaphase and is completed after mitosis ends. A *contractile ring* made of actin filaments (Focus Figure 3.4, final phase on p. 101) draws the plasma membrane inward to form a **cleavage furrow** over the center of the cell. The furrow deepens until it pinches the cytoplasmic mass into two parts, yielding two daughter cells. Each is smaller and has less cytoplasm than the parent cell, but is genetically identical to it. The daughter cells then enter the interphase portion of the life cycle until it is their turn to divide.

Control of Cell Division

The cell cycle is regulated by both internal and external factors that are important in determining when cells divide or stop dividing. These factors include:

- The ratio of cell surface area to cell volume. As a cell grows, its volume increases more rapidly than its surface area. Bigger cells require more nutrients, which must pass through proportionately less surface area as the cell grows. This limits how big a cell can be and explains why most cells are microscopic.

- Chemical signals such as growth factors and hormones released by other cells.

- The availability of space (how much room there is to grow). Normal cells stop proliferating when they begin touching, a phenomenon known as contact inhibition.

A number of checkpoints occur throughout interphase. We have already mentioned the G2/M checkpoint. The G_1 checkpoint, the *restriction point* (see Figure 3.27), is another key point at which a stop signal can halt further growth.

In addition, two groups of proteins are crucial to a cell's ability to finish the S phase and enter mitosis: *cyclins* and *cyclin-dependent kinases* (*Cdks*). Cdks are activated (switched on) or deactivated (switched off) by cyclins that function in a regulatory role. Joining of specific Cdk and cyclin proteins initiates enzymatic cascades needed for cell division. At the end of mitosis, enzymes destroy the cyclins and the process begins again.

Check Your Understanding

26. During what phase of the cell cycle is DNA synthesized?

27. What are three events occurring in prophase that are undone in telophase?

28. MAKE CONNECTIONS If a segment of one of the DNA strands being replicated consists of CGAATG, what will be the base sequence of the corresponding DNA strand? (Hint: Use what you learned about complementary bases in Chapter 2.)

For answers, see Answers Appendix.

3.11 Messenger RNA carries instructions from DNA for building proteins

Learning Outcomes

▶ Define gene and genetic code and explain the function of genes.

▶ Name the two phases of protein synthesis and describe the roles of DNA, mRNA, tRNA, and rRNA in each phase.

▶ Contrast triplets, codons, and anticodons.

In addition to directing its own replication, DNA serves as the master blueprint for protein synthesis. Cells also make lipids and carbohydrates, but DNA does not dictate their structure. Historically, DNA is said to specify *only* the structure of protein molecules, including the enzymes that catalyze the synthesis of all classes of biological molecules.

Much of the metabolic machinery of the cell is concerned in some way with protein synthesis. Essentially, cells are miniature protein factories that synthesize the huge variety of proteins that determine the chemical and physical nature of cells—and therefore of the whole body.

Recall from Chapter 2 that proteins are composed of polypeptide chains, which in turn are made up of amino acids (◀ p. 48). For purposes of this discussion, we define a **gene** as a segment of a DNA molecule that carries instructions for creating one polypeptide chain. We humans have about 20,000 protein-encoding genes. (Note, however, that some genes specify the structure of certain varieties of RNA as their final product.)

The four nucleotide bases (A, G, T, and C) are the "letters" of the genetic alphabet, and the information of DNA is found in the sequence of these bases. Each sequence of three bases, called a **triplet**, can be thought of as a "word" that specifies a particular amino acid. For example, the triplet AAA calls for the amino acid phenylalanine, and CCT calls for glycine. The sequence of triplets in each gene forms a "sentence" that tells exactly how a particular polypeptide is to be made: It specifies the number, kinds, and order of amino acids needed to build a particular polypeptide.

Variations in the arrangement of A, T, C, and G allow our cells to make all the different kinds of proteins needed. Even a "small" gene has an estimated 210 base pairs in sequence. The ratio between DNA bases in the gene and amino acids in the polypeptide is 3:1 (because each triplet stands for one amino acid), so we would expect the polypeptide specified by such a gene to contain 70 amino acids.

Our genes consist of long sequences of DNA, only some of which code for protein. The coding regions are called **exons**. Exons are separated from each other by long intervening sequences called **introns**. At first glance, introns may seem wasteful. However, they can act as control elements and they can allow for making different (related) proteins from one gene by omitting or including certain exons. In addition, they provide the raw material for rapid evolution by allowing exons to be swapped from gene to gene, making novel combinations that *might* be useful.

The Role of RNA in Protein Synthesis

By itself, DNA is like a music recording: The information it contains cannot be used without a decoding mechanism (e.g., an MP3 player). Furthermore, most polypeptides are manufactured at ribosomes in the cytoplasm, but in interphase cells, DNA never leaves the nucleus. So, DNA requires not only a decoder, but a messenger as well. The decoding and messenger functions are carried out by RNA, the second type of nucleic acid.

As you learned in Chapter 2, RNA differs from DNA: RNA is single stranded, and it has the sugar ribose instead of deoxyribose, and the base uracil (U) instead of thymine (T) (◀ Table 2.4, p. 53). Three forms of RNA typically act together to carry out DNA's instructions for polypeptide synthesis:

- **Messenger RNA (mRNA)**, relatively long nucleotide strands resembling "half-DNA" molecules (one of the two strands of a DNA molecule coding for protein structure). mRNA carries the coded information to the cytoplasm, where protein synthesis occurs.

- **Ribosomal RNA (rRNA)**, along with proteins, forms the ribosomes, which consist of two subunits—one large and one small. The two subunit types combine to form functional ribosomes, which are the sites of protein synthesis.

- **Transfer RNA (tRNA)**, small, roughly L-shaped molecules that ferry amino acids to the ribosomes. There they decode mRNA's message for amino acid sequence in the polypeptide to be built.

All types of RNA are formed on the DNA in the nucleus in much the same way as DNA replicates itself: The DNA helix separates and one of its strands serves as a template for synthesizing a complementary RNA strand. Once formed, the RNA molecule migrates into the cytoplasm. Its job done, the DNA recoils into its helical, inactive form.

Our 20,000 genes are located on only about 2% of the nuclear DNA, which codes for the synthesis of short-lived mRNA. DNA in the nucleolus (mentioned previously) codes for the synthesis of rRNA, which is long-lived and stable, as is tRNA coded by other DNA sequences. Because rRNA and tRNA do not transport codes for synthesizing other molecules, they are the final products of the genes that code for them. They act together to "translate" the message carried by mRNA.

Polypeptide synthesis involves two major steps:

1. *Transcription*, in which DNA's information is encoded in mRNA
2. *Translation*, in which the information carried by mRNA is decoded and used to assemble polypeptides

Figure 3.29 summarizes the information flow in these two major steps. The figure also indicates the "RNA processing" that removes introns from mRNA before this molecule moves into the cytoplasm.

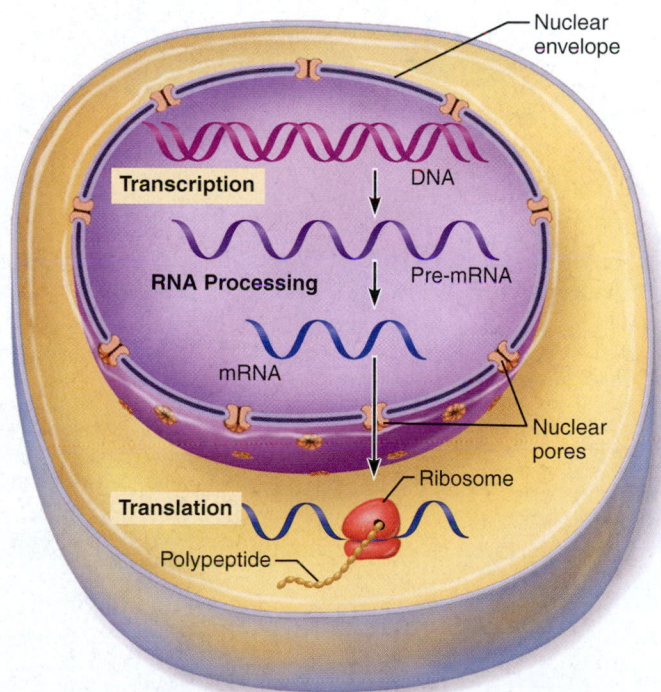

Figure 3.29 Information flow during protein synthesis.
Information flows from the DNA gene through mRNA during transcription, and from mRNA to polypeptide during translation. (Note that mRNA is first synthesized as pre-mRNA, which is processed by enzymes before leaving the nucleus.)

(Text continues on p. 102.)

Mitosis is the process of nuclear division in which the chromosomes are distributed to two daughter nuclei. Together with cytokinesis, it produces two identical daughter cells.

Interphase

Early Prophase

Late Prophase

Centrosomes (each has 2 centrioles)
Plasma membrane
Nucleolus
Nuclear envelope
Chromatin

Early mitotic spindle
Aster
Chromosome consisting of two sister chromatids
Centromere

Spindle pole
Nonkinetochore microtubule
Fragments of nuclear envelope
Kinetochore
Kinetochore microtubule

Interphase

Interphase is the period of a cell's life when it carries out its normal metabolic activities and grows. Interphase is not part of mitosis.

• During interphase, the DNA-containing material is in the form of chromatin. The nuclear envelope and one or more nucleoli are intact and visible.

• There are three distinct periods of interphase: G_1, S, and G_2.

The light micrographs show dividing lung cells from a newt. The chromosomes appear blue and the microtubules green. (The red fibers are intermediate filaments.) The schematic drawings show details not visible in the micrographs. For simplicity, only four chromosomes are drawn.

Prophase—first phase of mitosis

Early Prophase
• The chromatin coils and condenses, forming barlike *chromosomes*.

• Each duplicated chromosome consists of two identical threads, called *sister chromatids*, held together at the *centromere*. (Later when the chromatids separate, each will be a new chromosome.)

• As the chromosomes appear, the nucleoli disappear, and the two centrosomes separate from one another.

• The centrosomes act as focal points for growth of a microtubule assembly called the *mitotic spindle*. As the microtubules lengthen, they propel the centrosomes toward opposite ends (poles) of the cell.

• Microtubule arrays called *asters* ("stars") extend from the centrosome matrix.

Late Prophase
• The nuclear envelope breaks up, allowing the spindle to interact with the chromosomes.

• Some of the growing spindle microtubules attach to *kinetochores* (ki-neʹto-korz), special protein structures at each chromosome's centromere. Such microtubules are called *kinetochore microtubules*.

• The remaining (unattached) spindle microtubules are called *nonkinetochore microtubules*. The microtubules slide past each other, forcing the poles apart.

• The kinetochore microtubules pull on each chromosome from both poles in a tug-of-war that ultimately draws the chromosomes to the center, or equator, of the cell.

▶ Watch a 3D animation of this process: MasteringA&P® >
Study Area > Animations & Videos > A&P Flix

Metaphase	Anaphase	Telophase	Cytokinesis

Spindle

Metaphase plate

Daughter chromosomes

Nuclear envelope forming

Nucleolus forming

Contractile ring at cleavage furrow

Metaphase—second phase of mitosis

• The two centrosomes are at opposite poles of the cell.

• The chromosomes cluster at the midline of the cell, with their centromeres precisely aligned at the spindle *equator*. This imaginary plane midway between the poles is called the *metaphase plate*.

• At the end of metaphase, enzymes that will act to separate the chromatids from each other are triggered.

Anaphase—third phase of mitosis

The shortest phase of mitosis, anaphase begins abruptly as the centromeres of the chromosomes split simultaneously. Each chromatid now becomes a chromosome in its own right.

• The kinetochore microtubules shorten, gradually pulling each chromosome toward the pole it faces.

• At the same time, the nonkinetochore microtubules slide past each other, lengthen, and push the two poles of the cell apart.

• The moving chromosomes look V shaped. The centromeres lead the way, and the chromosomal "arms" dangle behind them.

• Moving and separating the chromosomes is helped by the fact that the chromosomes are short, compact bodies. Diffuse threads of chromatin would trail, tangle, and break, resulting in imprecise "parceling out" to the daughter cells.

Telophase—final phase of mitosis

Telophase begins as soon as chromosomal movement stops. This final phase is like prophase in reverse.

• The identical sets of chromosomes at the opposite poles of the cell begin to uncoil and resume their threadlike chromatin form.

• A new nuclear envelope forms around each chromatin mass, nucleoli reappear within the nuclei, and the spindle breaks down and disappears.

• Mitosis is now ended. The cell, for just a brief period, is binucleate (has two nuclei) and each new nucleus is identical to the original parent nucleus.

Cytokinesis—division of cytoplasm

Cytokinesis begins during late anaphase and continues through and beyond telophase. A contractile ring of actin microfilaments forms the *cleavage furrow* and pinches the cell apart.

Transcription

Transcription software converts a spoken message into a written copy. In other words, information is converted from one form or format to another.

In cells, **transcription** transfers information from a DNA base sequence to the complementary base sequence of an mRNA molecule. The form is different, but the same information is being conveyed. Once the mRNA molecule is made, it detaches and leaves the nucleus via a nuclear pore, and heads for the protein synthesis machinery, the ribosome.

Transcription cannot begin until gene-activating chemicals called *transcription factors* stimulate histones at the gene transcription site to loosen. The transcription factors then bind to the **promoter**, a special DNA sequence that contains the *start point* of the gene to be transcribed. It also specifies which DNA strand is going to serve as the *template strand* (**Figure 3.30** ①). The uncoiled DNA strand not used as a template is called the *coding strand* because it has the same (coded) sequence as the mRNA to be built (except for the U in mRNA in place of T in DNA). Once these preparations are made, **RNA polymerase** (the enzyme that oversees the synthesis of mRNA) can initiate transcription.

Transcription involves three basic phases: initiation, elongation, and termination as shown in Figure 3.30.

Processing of mRNA

Before translation can begin, editing and further processing are needed to clean up the raw mRNA transcript. Because the DNA is transcribed sequentially, the mRNA initially made, called *pre-mRNA*, is still littered with introns. Before the newly formed

① Initiation

RNA polymerase binds to the promoter, pries apart the two DNA strands, and initiates mRNA synthesis at the start point within the promoter region.

RNA polymerase

DNA

Coding strand of gene

Promoter region containing the start point

Template strand of gene

Termination signal

Template strand

mRNA

② Elongation

RNA polymerase moves along the template strand, joining together RNA nucleotides complementary to the DNA. It elongates the mRNA transcript one nucleotide at a time, unwinding the DNA double helix in front and rewinding it behind.

mRNA

③ Termination

mRNA synthesis ends when the polymerase reaches a special nucleotide sequence called a **termination signal**. RNA polymerase and the complete mRNA transcript are released.

Completed mRNA

RNA polymerase

Figure 3.30 Stages of transcription.

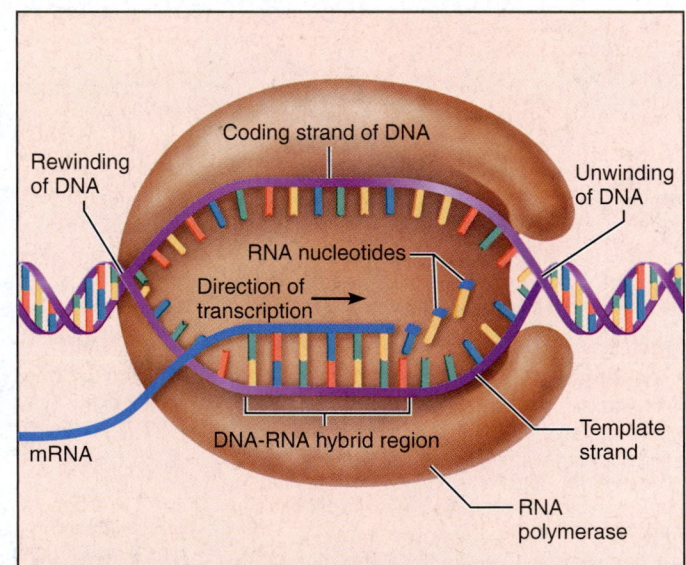

Rewinding of DNA

Coding strand of DNA

Unwinding of DNA

RNA nucleotides

Direction of transcription

Template strand

mRNA

DNA-RNA hybrid region

RNA polymerase

The DNA-RNA hybrid: At any given moment, 16–18 base pairs of DNA are unwound and the most recently made RNA is still hydrogen bonded to the template DNA. This small region is called the DNA-RNA hybrid.

RNA can be used as a messenger, sections corresponding to introns must be removed. Large RNA-protein complexes called *spliceosomes* snip out the introns and splice together the remaining exon-coded sections, producing functional mRNA. Now our messenger is ready to exit the nucleus, enter the cytosol, and participate in the next step of protein synthesis—translation.

Translation

A translator takes a message in one language and restates it in another. In the **translation** step of protein synthesis, the language of nucleic acids (base sequence) is translated into the language of proteins (amino acid sequence).

Genetic Code

The rules by which the base sequence of a gene is translated into an amino acid sequence are called the **genetic code**. For each triplet, or three-base sequence on DNA, the corresponding three-base sequence on mRNA is called a **codon**. Since there are four kinds of RNA (or DNA) nucleotides, there are 4^3, or 64, possible codons. Three of these 64 codons are "stop signs" that call for termination of polypeptide synthesis. All the rest code for amino acids.

Because there are only about 20 amino acids, some are specified by more than one codon. This redundancy in the genetic code helps protect against problems due to transcription (and translation) errors. **Figure 3.31** shows the genetic code and a complete codon list.

Role of tRNA

Translation involves the mRNAs, tRNAs, and rRNAs mentioned above. Before we get into the details of the translation process, let's look at how the structure and function of tRNAs are well matched.

Shaped like a handheld drill, tRNA (shown schematically below) is well suited to its dual function of binding to both an amino acid and an mRNA codon. The amino acid (picked up from the cytoplasmic pool) is bound to one end of tRNA. At the other end is its **anticodon** (an″ti-ko′don), a three-base sequence, which binds to the mRNA codon calling for the amino acid carried by that particular tRNA. Because anticodons form hydrogen bonds with complementary codons, tRNA is the link between the language of nucleic acids and the language of proteins. For example, if the mRNA codon is AUA, which specifies isoleucine (Ile), the tRNAs carrying isoleucine will have the anticodon UAU (as shown above), which can bind to the AUA codons.

There are approximately 45 types of tRNA, each capable of binding with a specific amino acid. The attachment process is controlled by an aminoacyl-tRNA synthetase enzyme and is activated by ATP. Once its amino acid is loaded, the

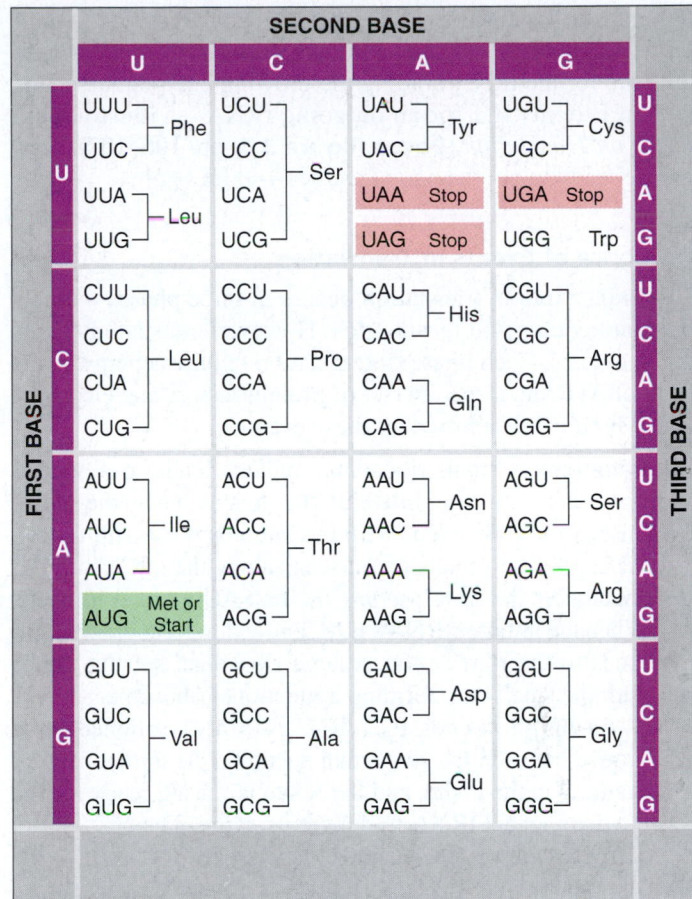

Figure 3.31 The genetic code. The three bases in an mRNA codon are designated as the first, second, and third. Each set of three specifies a particular amino acid, represented here by an abbreviation (see list below). The codon AUG (which specifies the amino acid methionine) is the usual start signal for protein synthesis. The word *stop* indicates the codons that serve as signals to terminate protein synthesis.

Abb.*	Amino acid	Abb.*	Amino acid
Ala	alanine	Leu	leucine
Arg	arginine	Lys	lysine
Asn	asparagine	Met	methionine
Asp	aspartic acid	Phe	phenylalanine
Cys	cysteine	Pro	proline
Glu	glutamic acid	Ser	serine
Gln	glutamine	Thr	threonine
Gly	glycine	Trp	tryptophan
His	histidine	Tyr	tyrosine
Ile	isoleucine	Val	valine

*Abbreviation for the amino acid

tRNA (now called an *aminoacyl-tRNA* because of its amino acid cargo) diffuses to the ribosome, where its amino acid is positioned, as specified by the mRNA codons and described below. The ribosome is more than just a passive attachment site for mRNA and tRNA. Like a vise, the ribosome holds the tRNA and mRNA close together to coordinate the coupling of codons and anticodons. To do its job, the ribosome has one binding site for mRNA and three binding sites for tRNA: an A (aminoacyl) site for an incoming aminoacyl-tRNA, a P (peptidyl) site for the tRNA holding the growing polypeptide chain, and an E (exit) site for an outgoing tRNA, as illustrated in *Focus on Translation* (**Focus Figure 3.5**, pp. 106–107). Now we are ready to put the parts together—so let's go!

Sequence of Events in Translation

Like transcription, translation occurs in three phases—*initiation*, *elongation*, and *termination*. However, translation occurs in the cytosol. Each phase of translation requires energy in the form of ATP and a specific set of protein factors and enzymes. Focus Figure 3.5 summarizes these events.

(1) **Initiation.** A small ribosomal subunit binds to a special methionine-carrying **initiator tRNA**, and then to the "new" mRNA to be decoded. With the initiator tRNA still in tow, the small ribosomal subunit scans along the mRNA until it encounters the *start codon*—the first AUG triplet it meets. When the initiator tRNA's UAC anticodon "recognizes" and binds to the start codon, a large ribosomal subunit unites with the small one, forming a functional ribosome.

As this phase ends, the mRNA is firmly positioned in the groove between the ribosomal subunits, the initiator tRNA is sitting in the P site, and the A site is vacant, ready for the next aminoacyl tRNA to deliver its cargo. The next phase, elongation of the polypeptide, now begins.

(2) **Elongation.** During the three-step cycle of elongation, the ribosome moves along the mRNA in one direction and one amino acid at a time is added to the growing polypeptide (Focus Figure 3.5).

(2a) **Codon recognition.** The incoming aminoacyl-tRNA binds to a complementary codon in the A site of a ribosome.

(2b) **Peptide bond formation.** An enzymatic component in the large ribosomal subunit catalyzes peptide bond formation between the amino acid of the tRNA in the P site and that of the tRNA in the A site.

(2c) **Translocation.** The ribosome translocates, or moves, shifting its position one codon along the mRNA. This shift moves the tRNA in the A site to the P site. The unloaded (vacant) tRNA is transferred to the E site, from which it is released and ready to be recharged with another amino acid from the cytoplasmic pool.

This orderly "musical chairs" process continues: The tRNA in the peptidyl site transfers its polypeptide cargo to the incoming aminoacyl-tRNA, and then the P-site-to-E-site and A-site-to-P-site movements of the tRNAs occur (Focus Figure 3.5). As the ribosome "chugs" along the mRNA "track" and the mRNA is progressively read, the initial portion of the mRNA passes through the ribosome and may attach successively to several other ribosomes, all reading the same message simultaneously and sequentially. This multiple ribosome–mRNA complex, a *polyribosome*, efficiently produces multiple copies of the same protein (**Figure 3.32**).

(3) **Termination.** The mRNA strand is read sequentially until its last codon, the *stop codon* (one UGA, UAA, or UAG) enters the A site. The stop codon is the "period" at the end of the mRNA sentence—it tells the ribosome that translation of that mRNA is finished. As a result, water instead of an amino acid is added to the polypeptide chain. This hydrolyzes (breaks) the bond between the polypeptide and the tRNA in the P site. The completed polypeptide chain is then released from the ribosome, and the ribosome separates into its two subunits (Figure 3.32a). The released protein may undergo processing before it folds into its complex 3-D structure and floats off, ready to work. When the message of the mRNA that directed the formation of the protein is no longer needed, the mRNA is degraded.

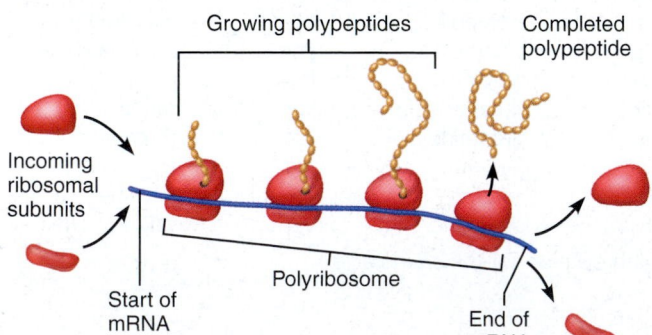

(a) Each polyribosome consists of one strand of mRNA being read by several ribosomes simultaneously. In this diagram, the mRNA is moving to the left and the "oldest" functional ribosome is farthest to the right.

(b) This transmission electron micrograph shows a large polyribosome (400,000×).

Figure 3.32 Polyribosome arrays. Polyribosome arrays allow a single strand of mRNA to be translated into hundreds of the same polypeptide molecules in a short time.

① The SRP directs the mRNA-ribosome complex to the rough ER. There the SRP binds to a receptor site.

② Once attached to the ER, the SRP is released and the growing polypeptide snakes through the ER membrane pore into the cistern.

③ An enzyme clips off the signal sequence. As protein synthesis continues, sugar groups may be added to the protein.

ER signal sequence

Ribosome

mRNA

Signal recognition particle (SRP)

Receptor site

Growing polypeptide

Signal sequence removed

Sugar group

④ In this example, the completed protein is released from the ribosome and folds into its 3-D conformation.

⑤ The protein is enclosed within a protein-coated transport vesicle. The transport vesicles make their way to the Golgi apparatus, where further processing of the proteins occurs (see Figure 3.17).

Released protein

Rough ER cistern

Cytosol

Transport vesicle pinching off

Protein-coated transport vesicle

Figure 3.33 Rough ER processing of proteins. An endoplasmic reticulum (ER) signal sequence in a newly forming protein causes the signal recognition particle (SRP) to direct the mRNA-ribosome complex to the rough ER.

Processing in the Rough ER

As noted earlier in the chapter, ribosomes attach to and detach from the rough ER. When a short "leader" peptide called an **ER signal sequence** is present in a protein being synthesized, the associated ribosome attaches to the membrane of the rough ER. This signal sequence, with its attached cargo of a ribosome and mRNA, is guided to appropriate receptor sites on the ER membrane by a signal recognition particle (SRP), a protein that cycles between the ER and the cytosol. **Figure 3.33** details the subsequent events occurring at the ER.

Summary: From DNA to Proteins

The genetic information of a cell is translated into the production of proteins via a sequence of information transfer that is completely determined by complementary base pairing. The transfer of information goes from DNA base sequence (triplets) to the complementary base sequence of mRNA (codons) and then to the tRNA base

(Text continues on p. 108.)

Translation is the process in which genetic information carried by an mRNA molecule is decoded in the ribosome to form a particular polypeptide. The "translators" are tRNA molecules that can recognize and bind specifically both to an mRNA codon and an amino acid.

Getting Ready
- Making mRNA (transcription)
- Attaching amino acid to tRNA
- tRNAs diffuse to ribosome

Amino acid that corresponds to anticodon

Met

tRNA

The correct amino acid is attached to each species of tRNA by a synthetase enzyme (aminoacyl-tRNA synthetase).

① Initiation:

Initiation occurs when four components combine at the P site:
- A small ribosomal subunit
- An initiator tRNA carrying the amino acid methionine
- The mRNA
- A large ribosomal subunit

Once this is accomplished, the next phase, elongation, begins.

Methionine (amino acid)

Met

Initiator tRNA bearing anticodon

Met

A site

P site

Start codon

Large ribosomal subunit

E site

Small ribosomal subunit

Pre-mRNA

mRNA

Template strand of DNA

Nucleus (site of transcription)

Newly made (and edited) **mRNA** leaves nucleus and travels to a ribosome for decoding.

Cytosol (site of translation)

② Elongation:

Amino acids are added one at a time to the growing peptide chain via a process that has three repeating steps: 2a, 2b, 2c.

Met
Leu — Amino acid corresponding to anticodon
Ile
Pro
GAU — tRNA anticodon
E P A
GGC
AUACCGCUA
Complementary mRNA codon

②a Codon recognition:

The anticodon of an incoming tRNA binds with the complementary mRNA codon (A to U and C to G) in the A site of the ribosome.

②b Peptide bond formation:

The growing polypeptide bound to the tRNA at the P site is transferred to the amino acid carried by the tRNA in the A site. A new peptide bond is formed.

Met
Ile — Growing polypeptide chain
Pro
Leu — New peptide bond
E P A
GGC GAU
AUACCGCUA

②c Translocation:

The entire ribosome translocates, shifting its position one codon along the mRNA.

Released tRNA
GGC
Met
Ile
Pro
Leu
The tRNA that was in the A site is now in the P site.
The unloaded tRNA from the P site is now in the E site. It is released.
E P A
GAU
CCGCUACUC
Direction of ribosome movement
The next codon to be translated is now in the empty A site, ready for step ②a again.

③ Termination:

When a stop codon (UGA, UAA, or UAG) arrives at the A site, elongation ends.

Polypeptide chain
GAC
Release factor
E P A
CCU
CUGGGAUGA
Stop codon
E P A
Release factor triggers the ribosomal subunits to separate, releasing the mRNA and new polypeptide.

3

DNA: DNA base sequence (triplets) of the gene codes for synthesis of a particular polypeptide chain

mRNA: Base sequence (codons) of the transcribed mRNA

tRNA: Consecutive base sequences of tRNA anticodons recognize the mRNA codons calling for the amino acids they transport

Polypeptide: Amino acid sequence of the polypeptide chain

Figure 3.34 Information transfer from DNA to RNA to polypeptide. Notice that the "reading" of the mRNA by tRNA anticodons reestablishes the base (triplet) sequence of the DNA genetic code (except that T is replaced by U).

sequence (anticodons), which is identical to the template DNA sequence except for the substitution of uracil (U) for thymine (T) (**Figure 3.34**).

Other Roles of DNA

The story of DNA doesn't end with the production of proteins encoded by exons. Scientists are finding that DNA also codes for a surprising variety of active RNA species that are not translated into proteins, including the following:

- **MicroRNAs (miRNAs)** are small RNAs that can interfere with and suppress mRNAs made by certain exons, effectively silencing them.

- **Small interfering RNAs (siRNAs)** are like miRNA but originate outside the cell. For example, our cells can use an infecting virus's RNA to create siRNAs that then interfere with viral replication.

Beyond the discussion here and in Chapter 29, we still have much to learn about these versatile RNA species that arise from noncoding DNA and appear to play a role in heredity.

Check Your Understanding

29. How do the A, P, and E ribosomal sites differ functionally during protein synthesis?

30. What is the role of DNA in transcription?

31. **DRAW** Draw a tRNA with the anticodon CAC. What codon does this anticodon bind? Use Figure 3.31 to determine the amino acid this tRNA binds, and draw the amino acid on the tRNA.

For answers, see Answers Appendix.

3.12 **Autophagy and proteasomes dispose of unneeded organelles and proteins; apoptosis disposes of unneeded cells**

Learning Outcomes

▶ **Distinguish between autophagy and apoptosis and indicate their importance.**

▶ **Describe the importance of ubiquitin-dependent degradation of soluble proteins.**

The workings of the cytoplasm are complex and seemingly unending. Without some system to get rid of malfunctioning or obsolete cells and organelles, or to recycle cellular nutrients for reuse within hungry cells, debris would soon shut down normal cell functions. Not to worry—cells use autophagy, proteasomes, and apoptosis to ensure the viability of the organism as a whole.

Autophagy

The process called **autophagy** ("self-eating") sweeps up bits of cytoplasm and unneeded organelles into double-membrane vesicles called *autophagosomes*. They are then delivered to lysosomes for digestion of the contents, which the cell reuses.

Autophagy serves three roles:

- It allows cells to dispose of clumps of unneeded proteins and even cytoplasmic organelles, such as mitochondria, when they are worn out.

- In times of stress, particularly starvation, it allows cells to cannibalize parts of themselves in order to survive.

- It helps restructure cells during development.

There is evidence that problems with autophagy may be related to a wide variety of diseases, including neurodegenerative disorders such as Alzheimer's and Parkinson's diseases.

Ubiquitin-Proteasome Pathway

Another pathway, called the **ubiquitin-proteasome pathway**, handles individual proteins that are misfolded, damaged, or unneeded and must be disposed of. Examples of unneeded proteins include some that are used only in cell division and must be degraded at precise points in the cell cycle, and short-lived transcription factors.

Doomed proteins are marked for destruction (proteolysis) when proteins called **ubiquitins** (u-bĭ′kwĭ-tinz) are attached to them. The tagged proteins are then hydrolyzed to small peptides by **proteasomes**, giant "waste disposal" complexes composed of protein-digesting enzymes. Proteasome activity is critical during starvation when these complexes degrade preexisting proteins to provide amino acids for synthesis of new and needed proteins.

Apoptosis

Apoptosis (ap″o-to′sis; falling away), or **programmed cell death**, rids the body of cells that are programmed to have a limited life span. These include the cells lining the uterus in a menstruating woman, and the webs between the fingers and toes of a developing fetus.

Apoptosis, unlike autophagy, does not use the services of lysosomes as its destructive tool. During the most common pathway of apoptosis:

1. The mitochondrial membranes become permeable in response to internal or external signals.

2. Cytochrome c and other factors leak from the mitochondria into the cytosol, and these factors activate intracellular enzymes called *caspases*.

3. The caspases unleash a torrent of digestive activity within the cell, which initiates apoptosis.

4. The dying cell shrinks into a ball without leaking its contents into the surrounding tissue.

5. The cell then sprouts "eat me" signals, releases a chemical that attracts macrophages, and is phagocytized.

Check Your Understanding

32. What is the importance of ubiquitin in the life of a cell?

33. What is apoptosis and what is its importance in the body?

For answers, see Answers Appendix.

Developmental Aspects of Cells

We all begin life as a single cell, the fertilized egg, and all the cells of our body arise from it. Early in development, cells begin to specialize, some becoming liver cells, some nerve cells, and so on. All our cells carry the same genes, so how can one cell become so different from another? This is a fascinating question.

Cells in various regions of the embryo are exposed to different chemical signals that channel them into specific pathways of development. When the embryo consists of just a few cells, the major signals may be nothing more than slight differences in oxygen and carbon dioxide concentrations between the more superficial and the deeper cells. But as development continues, cells release chemicals that influence development of neighboring cells by triggering processes that switch some genes "off" and others "on." Some genes are active in all cells. For example, genes for rRNA and ATP synthesis are "on" in all cells, but genes for synthesizing the enzymes needed to produce thyroxine are "on" only in cells that are going to be part of the thyroid gland. So, the story of cell specialization lies in the kinds of proteins made and reflects the activation of different genes in different cell types.

Cell specialization leads to *structural* variation—different organelles come to predominate in different cells. For example, muscle cells are full of actin and myosin, and their cytoplasm fills with microfilaments. Liver and phagocytic cells have more lysosomes. The development of specific and distinctive features in cells is called **cell differentiation**.

Cell Destruction and Modified Rates of Cell Division

During early development, cell death and destruction are normal events. Nature takes few chances. More cells than needed are produced, and excesses are eliminated later by apoptosis (see above). Apoptosis is particularly common in the developing nervous system.

Most organs are well formed and functional long before birth, but the body continues to grow and enlarge by forming new cells throughout childhood and adolescence. Once we reach adult size, cell division is important mainly to replace short-lived cells and repair wounds.

During young adulthood, cell numbers remain fairly constant. However, local changes in the rate of cell division are common. For example, when a person is anemic, his or her bone marrow undergoes **hyperplasia** (hi″per-pla′ze-ah), or accelerated growth (*hyper* = over; *plas* = grow), to produce red blood cells at a faster rate. If the anemia is remedied, the excessive marrow activity ceases. **Atrophy** (at′ro-fe), a decrease in size of an organ or body tissue, can result from loss of normal stimulation or from diseases like muscular dystrophy. Muscles that lose their nerve supply atrophy and waste away, and lack of exercise leads to thinned, brittle bones.

Cell Aging

Cell aging is complicated and has many causes. The exact nature of human aging is still a mystery, but there are several current theories on aging:

- The *wear-and-tear theory* holds that the cumulative effect of assaults, such as environmental toxins, leads to accelerated rates of cell death throughout the body.

- The *mitochondrial theory* places the blame on damage caused by free radicals, resulting in diminished energy production by damaged mitochondria.

- The *immune theory* holds that aging results from a progressive weakening of the immune system; the body loses its ability to fight off pathogens or to heal systemic inflammation, which is also associated with aging and risks for chronic diseases.

- The *genetic theory* holds that cell aging is "programmed" into our genes. Telomeres are nonsensical strings of nucleotides that cap the ends of chromosomes, providing protection. Though telomeres carry no genes, they appear to be vital for chromosomal survival. With each cycle of DNA replication, the telomeres get a bit shorter and eventually they become too short for cell survival. Telomerase, an enzyme found in certain specialized cells, has been dubbed the "immortality enzyme" due to its ability to lengthen previously

shortened telomeres. Unfortunately, telomerase is also found in cancer cells where it contributes to their immortality.

HOMEOSTATIC IMBALANCE 3.4   **CLINICAL**

Progeria (pro-jē′re-ah) is a rare disease that mimics aging. It is caused by a defective protein in the nuclear lamina (see p. 92) called *progerin*. The resulting abnormal nucleus makes cells unstable. The disease appears by age 2. Children with the disease display slow growth, thinning hair, brittle bones, arthritis, and severe cardiovascular disease (**Figure 3.35**). Progeria usually results in death by age 20.

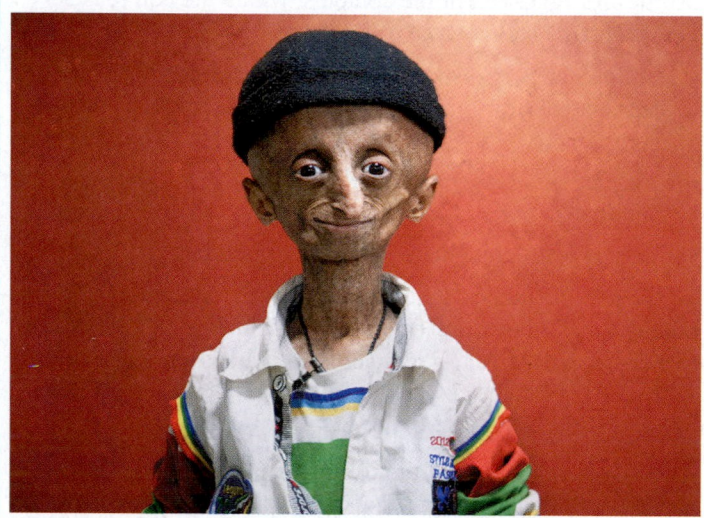

Figure 3.35 A child with progeria.

Scientists have recently found that a drug that stimulates autophagy can help cells clear out progerin, suggesting a possible new treatment for this disease. Healthy, aged individuals also have progerin in their cells, and this may contribute to the normal aging process.

RELATED CLINICAL TERMS

Anaplasia (an′ah-pla′ze-ah; *an* = without, not; *plas* = to grow) Abnormalities in cell structure and loss of differentiation; for example, cancer cells typically lose the appearance of the parent cells and come to resemble undifferentiated or embryonic cells.

Dysplasia (dis-pla′ze-ah; *dys* = abnormal) A change in cell size, shape, or arrangement due to chronic irritation or inflammation (infections, etc.).

Hypertrophy (hi-per′tro-fe) Growth of an organ or tissue due to an increase in the size of its cells. Hypertrophy is a normal response of skeletal muscle cells when they are challenged to lift excessive weight; differs from hyperplasia, which is an increase in size due to an increase in cell number.

Liposomes (lip′o-sōmz) Hollow microscopic sacs formed of phospholipids that can be filled with a variety of drugs. Serve as multipurpose vehicles for drugs, genetic material, and cosmetics.

Mutation A change in DNA base sequence that may lead to incorporation of incorrect amino acids in particular positions in the resulting protein; the affected protein may remain unimpaired or may function abnormally or not at all, leading to disease.

Necrosis (nĕ-kro′sis; *necros* = death; *osis* = process) Death of a cell or group of cells due to injury or disease. Acute injury causes the cells to swell and burst, and induces the inflammatory response. (This is uncontrolled cell death, in contrast to apoptosis described in the text.)

CHAPTER SUMMARY

3.1 Cells are the smallest unit of life (pp. 61–63)

1. All living organisms are composed of cells—the basic structural and functional units of life. Cells vary widely in both shape and size.
2. The principle of complementarity states that the biochemical activity of cells reflects the operation of their organelles.
3. The generalized cell is a concept that typifies all cells. The generalized human cell has three major regions—the nucleus, cytoplasm, and plasma membrane.
4. Extracellular materials are substances found outside the cells. They include body fluids, cellular secretions, and extracellular matrix. Extracellular matrix is particularly abundant in connective tissues.

PART 1

PLASMA MEMBRANE

3.2 The plasma membrane is a double layer of phospholipids with embedded proteins (pp. 63–68)

1. The plasma membrane encloses cell contents, mediates exchanges with the extracellular environment, and plays a role in cellular communication.
2. The plasma membrane is a fluid bilayer of lipids (phospholipids, cholesterol, and glycolipids) within which proteins are inserted.
3. The lipids have both hydrophilic and hydrophobic regions that organize their aggregation and self-repair. The lipids form the structural part of the plasma membrane.
4. Most proteins are integral transmembrane proteins that extend entirely through the membrane. Some are peripheral proteins that are not embedded in the lipid bilayer.
5. Proteins are responsible for most specialized membrane functions: Some are enzymes, some are receptors, and others mediate membrane transport functions.
6. Carbohydrates attached to proteins (forming glycoproteins) and lipids (forming glycolipids) coat the outer surface of the membrane, and help form the glycocalyx. They act as identity molecules.
7. Cell junctions join cells together and may aid or inhibit movement of molecules between or past cells.
8. Tight junctions are impermeable junctions. Desmosomes mechanically couple cells into a functional community. Gap junctions allow joined cells to communicate.
9. The plasma membrane acts as a selectively permeable barrier. Substances move across the plasma membrane by passive processes, which depend on the kinetic energy of molecules, and by active processes, which depend on the use of cellular energy (ATP).

3.3 Passive membrane transport is diffusion of molecules down their concentration gradient (pp. 68–73)

1. Diffusion is the movement of molecules (driven by kinetic energy) down a concentration gradient. Lipid-soluble solutes can diffuse directly through the membrane by dissolving in the lipid.
2. Facilitated diffusion is the passive movement of certain solutes across the membrane either by their binding with a membrane carrier protein or by their moving through a membrane channel. As with other diffusion processes, it is driven by kinetic energy.

Because it requires a transport protein (carrier or channel), it is both saturable and specific.
3. Osmosis is the diffusion of a solvent, such as water, through a selectively permeable membrane. Water diffuses through membrane channels (aquaporins) or directly through the lipid portion of the membrane from a solution of lesser osmolarity (total concentration of all solute particles) to a solution of greater osmolarity.
4. The presence of solutes unable to penetrate the plasma membrane causes the cell to swell or shrink. Net osmosis ceases when the solute concentration on both sides of the plasma membrane reaches equilibrium.
5. Solutions that cause a net loss of water from cells are hypertonic. Those causing net water gain are hypotonic. Those causing neither gain nor loss of water are isotonic.

3.4 Active membrane transport directly or indirectly uses ATP (pp. 73–79)

1. Active transport depends on a pump protein and energy. Substances transported move against concentration or electrical gradients. In primary active transport, such as that provided by the Na^+-K^+ pump, ATP directly provides the energy.
2. In secondary active transport, the energy of a concentration gradient (produced by a primary active transport process) is used to move another substance against its concentration gradient. A cotransport protein uses the energy of one substance going down its concentration gradient to transport another substance against its concentration gradient. Cotransported substances move in either the same (symport) or opposite (antiport) directions across the membrane.
3. Vesicular transport also requires that energy be provided. Endocytosis brings substances into the cell. If the substances are relatively large particles, the process is called phagocytosis. If the substances are dissolved molecules, the process is pinocytosis. Receptor-mediated endocytosis is selective: Engulfed molecules attach to receptors on the membrane before endocytosis occurs. Exocytosis, which uses SNAREs to anchor the vesicles to the plasma membrane, ejects substances (hormones, wastes, secretions) from the cell.
4. Transcytosis is the sequence of endocytosis, vesicular transport through the cell, and exocytosis on the other side.

3.5 Selective diffusion establishes the membrane potential (pp. 79–81)

1. All cells in the resting stage exhibit a voltage across their membrane, called the resting membrane potential. Because of the membrane potential, both concentration and electrical gradients determine the ease of an ion's diffusion.
2. The resting membrane potential is generated by concentration gradients of ions and the differential permeability of the plasma membrane to ions, particularly potassium ions. Sodium is in high extracellular concentration and low intracellular concentration, and the membrane is poorly permeable to it. Potassium is in high concentration in the cell and low concentration in the extracellular fluid. The membrane is more permeable to potassium than to sodium. Protein anions in the cell are too large to cross the membrane.
3. A negative membrane potential is established when the movement of K^+ out of the cell equals K^+ movement into the

cell. Na$^+$ movements across the membrane contribute minimally to establishing the membrane potential. The greater outward diffusion of potassium (than inward diffusion of sodium) leads to a charge separation at the membrane (inside negative). This charge separation is maintained by the operation of the sodium-potassium pump.

 Complete an interactive tutorial: MasteringA&P® >Study Area > Interactive Physiology > Nervous System I: Resting Membrane Potential

3.6 Cell adhesion molecules and membrane receptors allow the cell to interact with its environment (pp. 81–83)

1. Cells interact directly and indirectly with other cells. Indirect interactions involve extracellular chemicals carried in body fluids or forming part of the extracellular matrix.
2. Cell adhesion molecules and membrane receptors are key players in mediating interactions between a cell and its environment.
3. Activated membrane receptors act as catalysts, regulate channels, or, like G protein–coupled receptors, act through second messengers such as cyclic AMP and Ca^{2+}. Ligand binding results in changes in protein structure or function within the targeted cell.

PART 2
THE CYTOPLASM

3.7 Cytoplasmic organelles each perform a specialized task (pp. 83–89)

1. The cytoplasm, the cellular region between the nuclear and plasma membranes, consists of the cytosol (fluid cytoplasmic environment), inclusions (nonliving nutrient stores, pigment granules, crystals, etc.), and cytoplasmic organelles.
2. The cytoplasm is the major functional area of the cell. These functions are mediated by organelles.
3. Mitochondria, organelles limited by a double membrane, are sites of ATP formation. Their internal enzymes carry out the oxidative reactions of cellular respiration.
4. Ribosomes, composed of two subunits containing ribosomal RNA and proteins, are the sites of protein synthesis. They may be free or attached to membranes.
5. The rough endoplasmic reticulum is a ribosome-studded membrane system. Its cisterns act as sites for protein modification. Its external face acts in phospholipid synthesis. Vesicles pinched off from the ER transport the proteins to other cell sites.
6. The smooth endoplasmic reticulum synthesizes lipid and steroid molecules. It also acts in fat metabolism, in drug detoxification, and as a calcium ion depot.
7. The Golgi apparatus is a membranous system close to the nucleus that packages protein secretions for export, packages enzymes into lysosomes for cellular use, and modifies proteins destined to become part of cellular membranes.
8. Peroxisomes are membranous sacs containing oxidase and catalase enzymes that protect the cell from the destructive effects of free radicals and other toxic substances by converting them first to hydrogen peroxide and then water.
9. Lysosomes are membranous sacs of acid hydrolases packaged by the Golgi apparatus. Sites of intracellular digestion, they degrade worn-out organelles and stressed or dead cells, and they release ionic calcium from bone.

10. The endomembrane system incorporates the organelles that work together to produce, degrade, store, and export biological molecules, and to degrade harmful substances. It includes the ER, Golgi apparatus, secretory vesicles, lysosomes, and nuclear envelope, all of which are connected directly or by vesicle formation or fusing.
11. The cytoskeleton includes microfilaments, intermediate filaments, and microtubules. Microfilaments are important in cell motility and changes in cell shape. Intermediate filaments help cells resist mechanical stress and connect other elements. Microtubules organize the cytoskeleton and are important in intracellular transport. Transport and motility functions involve motor proteins.
12. The centrosome is the microtubule organizing center; it organizes the mitotic spindle and contains paired centrioles.

3.8 Cilia and microvilli are two main types of cellular extensions (pp. 90–91)

1. Cilia and flagella are cellular extensions; cilia propel other substances across the cell surface, and flagella propel the cell. Both have a basal body at their base.
2. Microvilli are extensions of the plasma membrane that increase its surface area for absorption.

PART 3
NUCLEUS

3.9 The nucleus includes the nuclear envelope, the nucleolus, and chromatin (pp. 91–96)

1. The nucleus is the control center of the cell. Most cells have a single nucleus. Without a nucleus, a cell cannot divide or synthesize more proteins, and is destined to die.
2. The nucleus is surrounded by the nuclear envelope, a double membrane penetrated by fairly large pores.
3. Nucleoli are nuclear sites of ribosome subunit synthesis.
4. Chromatin is a complex network of slender threads containing histone proteins and DNA. The chromatin units are called nucleosomes. When a cell begins to divide, the chromatin coils and condenses, forming chromosomes.

3.10 The cell cycle consists of interphase and a mitotic phase (pp. 96–98)

1. The cell cycle is the series of changes that a cell goes through from the time it is formed until it divides.
2. Interphase is the nondividing phase of the cell cycle. Interphase consists of G_1, S, and G_2 subphases. During G_1, the cell grows and centriole replication begins. During the S phase, DNA replicates. During G_2, the final preparations for division are made. Many checkpoints occur during interphase at which the cell gets the go-ahead signal to go through mitosis or is prevented from continuing to mitosis.
3. DNA replication occurs before cell division, ensuring that both daughter cells have identical genes. The DNA helix uncoils, and each DNA nucleotide strand acts as a template for the formation of a complementary strand. Base pairing provides the guide for the proper positioning of nucleotides.
4. The semiconservative replication of a DNA molecule produces two DNA molecules identical to the parent molecule, each formed of one "old" and one "new" strand.
5. Cell division, essential for body growth and repair, occurs during the M phase. Cell division consists of two distinct phases: mitosis (nuclear division) and cytokinesis (division of the cytoplasm).

6. Mitosis, consisting of prophase, metaphase, anaphase, and telophase, parcels out the replicated chromosomes to two daughter nuclei, each genetically identical to the parent nucleus. Cytokinesis, which begins late in mitosis, divides the cytoplasmic mass into two parts.

7. Cell division is stimulated by certain chemicals (including growth factors and some hormones) and increasing cell size. Lack of space and inhibitory chemicals deter cell division. Cyclin-Cdk complexes regulate cell division.

3.11   Messenger RNA carries instructions from DNA for building proteins   (pp. 98–108)

1. A gene is defined as a DNA segment that provides the instructions to synthesize one polypeptide chain. Since the major structural materials of the body are proteins, and all enzymes are proteins, this amply covers the synthesis of all biological molecules.

2. The base sequence of exon DNA provides the information for protein structure. Each three-base sequence (triplet) calls for a particular amino acid to be built into a polypeptide chain.

3. The RNA molecules acting in protein synthesis are synthesized on single strands of the DNA template. RNA nucleotides are joined to the template according to base-pairing rules.

4. Instructions for making a polypeptide chain are carried from the DNA to the ribosomes via messenger RNA. Ribosomal RNA forms part of the protein synthesis sites. A transfer RNA ferries each amino acid to the ribosome and binds to a codon on the mRNA strand specifying its amino acid.

5. Protein synthesis involves (a) transcription, synthesis of a complementary mRNA, and (b) translation, "reading" of the mRNA by tRNA and peptide bonding of the amino acids into the polypeptide chain. Ribosomes coordinate translation.

6. Introns and other DNA sequences encode various small RNA species that may interfere with or promote the function of specific genes.

3.12   Autophagy and proteasomes dispose of unneeded organelles and proteins; apoptosis disposes of unneeded cells   (pp. 108–109)

1. Organelles and large protein aggregates are picked up by autophagosomes and delivered to lysosomes for digestion. This process, autophagy, is very important for keeping the cytoplasm free of deteriorating organelles and other debris.

2. Proteins that are damaged or no longer needed are targeted for destruction by attachment of ubiquitin. Proteasomes then degrade these proteins.

3. Apoptosis is programmed cell death. Its function is to dispose of damaged or unnecessary cells.

Developmental Aspects of Cells   (pp. 109–110)

1. The first cell of an organism is the fertilized egg. Early in development, cell specialization begins and reflects differential gene activation.

2. During adulthood, cell numbers remain fairly constant. Cell division occurs primarily to replace lost cells.

3. Cellular aging may reflect chemical insults, progressive disorders of immunity, or a genetically programmed decline in the rate of cell division with age.

REVIEW QUESTIONS

 To access additional practice questions using your smartphone, tablet, or computer: MasteringA&P® > Study Area > Practice Tests & Quizzes

Level 1 Remember/Understand

(Some multiple choice questions have more than one correct answer. Select the best answer or answers from the choices given.)

1. The smallest unit capable of life by itself is (**a**) the organ, (**b**) the organelle, (**c**) the tissue, (**d**) the cell, (**e**) the nucleus.

2. The major types of lipid found in the plasma membranes are (choose two) (**a**) cholesterol, (**b**) triglycerides, (**c**) phospholipids, (**d**) fat-soluble vitamins.

3. Membrane junctions that allow nutrients or ions to flow from cell to cell are (**a**) desmosomes, (**b**) gap junctions, (**c**) tight junctions, (**d**) all of these.

4. Osmosis always involves (**a**) a selectively permeable membrane, (**b**) a difference in solute concentration, (**c**) diffusion, (**d**) active transport, (**e**) a, b, and c.

5. The endocytotic process in which a sampling of particulate matter is engulfed and brought into the cell is called (**a**) phagocytosis, (**b**) pinocytosis, (**c**) exocytosis.

6. The nuclear substance composed of histone proteins and DNA is (**a**) chromatin, (**b**) the nucleolus, (**c**) nucleoplasm, (**d**) nuclear pores.

7. The information sequence that determines the nature of a protein is the (**a**) nucleotide, (**b**) gene, (**c**) triplet, (**d**) codon.

8. The phase of mitosis during which centrioles reach the poles and chromosomes attach to the spindle is (**a**) anaphase, (**b**) metaphase, (**c**) prophase, (**d**) telophase.

9. The RNA synthesized on one of the DNA strands is (**a**) mRNA, (**b**) tRNA, (**c**) rRNA, (**d**) all of these.

10. The RNA species that travels from the nucleus to the cytoplasm carrying the coded message specifying the sequence of amino acids in the protein to be made is (**a**) mRNA, (**b**) tRNA, (**c**) rRNA, (**d**) all of these.

11. Cells lining the trachea have whiplike motile extensions on their free surfaces. What are these extensions, what is their source, and what is their function?

Level 2 Apply/Analyze

12. A physiologist observes that the concentration of sodium inside a cell is decidedly lower than that outside the cell. She also observes that there is a small leakage of sodium into the cell. What cellular process prevents the sodium concentration gradient from disappearing? (**a**) osmosis, (**b**) diffusion, (**c**) primary active transport, (**d**) secondary active transport.

13. If DNA has a sequence of AAA, then a segment of mRNA synthesized on it will have a sequence of (**a**) TTT, (**b**) UUU, (**c**) GGG, (**d**) CCC.

14. A nerve cell and a lymphocyte are presumed to differ in their (**a**) specialized structure, (**b**) suppressed genes and embryonic history, (**c**) genetic information, (**d**) a and b, (**e**) a and c.

15. A pancreas cell makes proteins (enzymes) that it releases to the small intestine. Which of the following best describes the path of these proteins from synthesis to exocytosis at the pancreatic cell's plasma membrane (PM)? (**a**) Golgi → rough ER → PM, (**b**) smooth ER → Golgi → lysosome → PM, (**c**) rough ER → Golgi → PM, (**d**) nucleus → Golgi → PM.

16. Contrast the roles of ER-bound ribosomes with those free in the cytosol.
17. Comment on the role of the sodium-potassium pump in maintaining a cell's resting membrane potential.
18. Differentiate between primary and secondary active transport processes.
19. Two examples of chemotherapeutic drugs (drugs used to treat cancer) and their cellular actions are listed below. Explain why each drug could be fatal to a cell.
    - Vincristine (brand name Oncovin): damages the mitotic spindle
    - Doxorubicin (Adriamycin): binds to DNA and blocks mRNA synthesis
20. In their anatomy lab, many students are exposed to the chemical preservatives phenol, formaldehyde, and alcohol. Our cells break down these toxins very effectively. What cellular organelle is responsible for this?

Level 3 Evaluate/Synthesize

21. Cell division typically yields two daughter cells, each with one nucleus. How is the occasional binucleate condition of liver cells explained?
22. A "red-hot" bacterial infection of the intestinal tract irritates the intestinal cells and interferes with digestion. Such a condition is often accompanied by diarrhea, which causes loss of body water. On the basis of what you have learned about osmotic water flows, explain why diarrhea may occur.
23. Dynein is missing from the cilia and flagella of individuals with a specific inherited disorder. These individuals have severe respiratory problems and, if males, are sterile. What is the structural connection between these two symptoms?
24. Explain why alcoholics are likely to have much more smooth ER than teetotalers.
25. Fresh water is a precious natural resource in Florida and it is said that supplies are dwindling. Desalinizing (removing salt from) ocean water has been recommended as a solution to the problem. Why shouldn't we drink salt water?

4

Tissue: The Living Fabric

In this chapter, you will learn that

Tissues are groups of cells similar in structure that perform a common or related function

by first asking

4.1 How are tissues prepared for microscopy?

and next asking

4.6 How do cutaneous, mucous, and serous membranes differ?

then asking

4.7 How are tissues repaired?

then comparing

4.2 Epithelial tissue

4.3 Connective tissue

4.4 Muscle tissue

4.5 Nervous tissue

and finally, exploring

Developmental Aspects of Tissues

CAREER CONNECTION

Watch a video to learn how the chapter content is used in a real health care setting. Go to **Mastering** A&P® > Study Area > Animations and Videos or use quick access URL https://goo.gl/88srfV

4

Unicellular (one-cell) organisms are independent little creatures. Each cell alone obtains and digests its food, ejects its wastes, and carries out all the other activities necessary to keep itself alive. But in the multicellular human body, cells do not operate independently. Instead, they form tight cell communities that live and work together.

Individual body cells are specialized, with each cell type performing specific functions that help maintain homeostasis and benefit the body as a whole. Cell specialization is obvious: Muscle cells look and act differently from skin cells, which in turn are easy to distinguish from brain cells. Cell specialization allows the body to function in sophisticated ways, but division of labor has certain hazards. When a particular group of cells is indispensable, its injury or loss can disable or even destroy the body.

Tissues (*tissu* = woven) are groups of cells that are similar in structure and perform a common or related function. Four primary tissue types interweave to form the "fabric" of the body. These basic tissues are epithelial, connective, muscle, and nervous tissue.

If we summarized the role of each primary tissue in a single word, we could say that epithelial tissue *covers*, connective tissue *supports*, muscle tissue *moves*, and nervous tissue *controls*. However, these words reveal only a fraction of what each tissue does (**Figure 4.1**).

As we explained in Chapter 1, tissues are organized into organs such as the kidneys and heart. Most organs contain all four tissue types. How these tissues are arranged determines an organ's structure and capabilities. The study of tissues, or **histology**, complements the study of gross anatomy. Together they provide the structural basis for understanding organ physiology.

Nervous tissue: Internal communication
- Brain
- Spinal cord
- Nerves

Muscle tissue: Contracts to cause movement
- Muscles attached to bones (skeletal)
- Muscles of heart (cardiac)
- Muscles of walls of hollow organs (smooth)

Epithelial tissue: Forms boundaries between different environments, protects, secretes, absorbs, filters
- Lining of digestive tract organs and other hollow organs
- Glands (e.g., pancreas)
- Skin surface (epidermis)

Connective tissue: Supports, protects, binds other tissues together
- Bones
- Tendons
- Fat and other soft padding tissue

Figure 4.1 Overview of four basic tissue types: epithelial, connective, muscle, and nervous tissues.

4.1 Tissue samples are fixed, sliced, and stained for microscopy

Learning Outcome

▶ List the steps involved in preparing animal tissue for microscopic viewing.

Microscopy allows us to study tissue structure. Before a specimen can be viewed through a microscope, it must be **fixed** (preserved) and then cut into **sections** (slices) thin enough to transmit light (in light microscopy) or electrons (in electron microscopy). Finally, the specimen must be **stained** to enhance contrast between structures.

In light microscopy, the stains are beautifully colored synthetic dyes, most of which were originally developed by clothing manufacturers. Many dyes consist of negatively charged molecules (acidic stains) or positively charged molecules (basic stains) that bind within the tissue to macromolecules of the opposite charge. Different parts of cells and tissues take up different dyes, distinguishing different anatomical structures.

For *transmission electron microscopy (TEM)*, tissue sections are stained with heavy metal salts. These metals provide contrast by deflecting electrons in the beam to different extents. Electron-microscope images are in shades of gray because color is a property of light, not of electron waves. However, the image may be artificially colored to enhance contrast. Another kind of electron microscopy, *scanning electron microscopy (SEM)*, provides three-dimensional pictures of an unsectioned tissue surface (**Figure 4.2**).

Figure 4.2 Comparison of transmission and scanning electron micrographs. TEM (left) shows a thin section of the cell revealing cilia (tiny hairlike projections of the plasma membrane). SEM (right) shows the surfaces of the cilia. Artificially colored.

Preserved tissue we see under the microscope has been exposed to many procedures that alter its original condition and introduce minor distortions called **artifacts**. For this reason, most structures that we view under a microscope are not exactly like those in living tissue.

Check Your Understanding

1. What is the purpose of fixing tissue for microscopic viewing?
2. What types of stains are used to stain tissues to be viewed with a transmission electron microscope?

For answers, see Answers Appendix.

4.2 Epithelial tissue covers body surfaces, lines cavities, and forms glands

Learning Outcomes

▶ List several structural and functional characteristics of epithelial tissue.
▶ Name, classify, and describe the various types of epithelia, and indicate their chief function(s) and location(s).
▶ Define gland.
▶ Differentiate between exocrine and endocrine glands, and between multicellular and unicellular glands.
▶ Describe how multicellular exocrine glands are classified structurally and functionally.

Epithelial tissue (ep″ĭ-the′le-ul), or an **epithelium** (plural: epithelia), is a sheet of cells that covers a body surface or lines a body cavity (*epithe* = laid on, covering). Two forms occur in the body:

- *Covering and lining epithelium*, which forms the outer layer of the skin; dips into and lines the open cavities of the urogenital, digestive, and respiratory systems; and covers the walls and organs of the closed ventral body cavity

- *Glandular epithelium*, which fashions the glands of the body

Epithelia form boundaries between different environments, and nearly all substances received or given off by the body must pass through an epithelium. For example, the epidermis of the skin lies between the inside and the outside of the body. Epithelium lining the urinary bladder separates underlying cells of the bladder wall from urine.

In its role as a boundary-forming tissue, epithelium accomplishes many functions, including (1) protection, (2) absorption, (3) filtration, (4) excretion, (5) secretion, and (6) sensory reception, all of which will be touched upon later in this chapter.

Special Characteristics of Epithelium

Epithelial tissues have five distinguishing characteristics, discussed below.

Polarity

All epithelia have two surfaces that differ in structure and function. The **apical surface** is not attached to surrounding tissue and is exposed to either the outside of the body or the cavity of an internal organ. The **basal surface** is attached to the underlying connective tissue (see below). For this reason, we say that epithelia exhibit *apical-basal polarity*.

Epithelium

Apical surface:
• Borders open space

Basal surface:
• Next to underlying connective tissue

4

Although some apical surfaces are smooth and slick, most have **microvilli**, fingerlike extensions of the plasma membrane (◄ **Figure 3.24, p. 91**, and see Figures 4.4c and 4.6). Microvilli tremendously increase the exposed surface area. In epithelia that absorb or secrete (export) substances (those lining the intestine or kidney tubules, for instance), the microvilli are often so dense that the cell apices have a fuzzy appearance called a *brush border*. Some epithelia, such as that lining the trachea (windpipe), have motile **cilia** (◄ **p. 90**) that propel substances along their free surface (see Figures 4.2 and 4.4d).

Adjacent to the basal surface of an epithelium is a thin supporting sheet called the **basal lamina** (lam′ĭ-nah; "sheet"). This noncellular, adhesive sheet consists largely of glycoproteins secreted by the epithelial cells plus some fine collagen fibers. The basal lamina acts as a selective filter that determines which molecules diffusing from the underlying connective tissue are allowed to enter the epithelium. The basal lamina also acts as scaffolding along which epithelial cells can migrate to repair a wound.

Specialized Contacts

Except for glandular epithelia (discussed on pp. 123–126), epithelial cells fit closely together to form continuous sheets. The sides of adjacent cells are tied together by *tight junctions* and *desmosomes* (◄ **pp. 67–68**). Recall from Chapter 3 that tight junctions prevent substances from leaking through spaces between cells and that desmosomes keep cells from pulling apart.

Supported by Connective Tissue

All epithelial sheets rest upon and are supported by connective tissue. In between the epithelial and connective tissues is a **basement membrane** (see Figure 4.4b, d–f) that reinforces the epithelial sheet, helps it resist stretching and tearing, and defines the epithelial boundary. The basement membrane consists of two layers: a basal lamina (described above) and a **reticular lamina**. The reticular lamina is deep to the basal lamina. It consists of a layer of extracellular material containing a fine network of collagen protein fibers that "belongs to" the underlying connective tissue.

HOMEOSTATIC IMBALANCE 4.1   **CLINICAL**

An important characteristic of cancerous epithelial cells is their failure to respect the basement membrane boundary, which they penetrate to invade the tissues beneath. ● ● ● ● ● ● ● ●

Avascular but Innervated

Although epithelium is *avascular* (contains no blood vessels), it is *innervated* (supplied by nerve fibers). Epithelial cells are nourished by substances diffusing from blood vessels in the underlying connective tissue.

Regeneration

Epithelium has a high regenerative capacity. Some epithelia are exposed to friction and their surface cells rub off. Others are damaged by hostile substances in the external environment (bacteria, acids, smoke). If and when their apical-basal polarity and lateral contacts are destroyed, epithelial cells begin to reproduce themselves rapidly. As long as epithelial cells receive adequate nutrition, they can replace lost cells by cell division.

Classification of Epithelial Tissue

The name of each epithelium has two parts, just like a first and last name. The first name indicates the number of cell layers present. Based on the number of cell layers, there are simple and stratified epithelia (**Figure 4.3a**).

- **Simple epithelia** consist of a single cell layer. They are typically found where absorption, secretion, and filtration occur and a thin epithelial barrier is desirable.

(a) Classification based on number of cell layers

Apical surface

Basal surface

Simple

Apical surface

Basal surface

Stratified

(b) Classification based on cell shape

Squamous   **Cuboidal**   **Columnar**

Figure 4.3 Classification of epithelia. Note that cell shape influences the shape of the nucleus.

- **Stratified epithelia**, composed of two or more cell layers stacked on top of each other, are common in high-abrasion areas where protection is important, such as the skin surface and the lining of the mouth.

An epithelium's second name describes the shape of its cells. Epithelial cells vary in height, and on that basis, there are three common shapes of epithelial cells (Figure 4.3b):

- **Squamous cells** (skwa′mus) are flattened and scale-like (*squam* = scale).
- **Cuboidal cells** (ku-boi′dahl) are boxlike, approximately as tall as they are wide.
- **Columnar cells** (kŏ-lum′nar) are tall and column shaped.

In each case, the shape of the nucleus conforms to that of the cell. The nucleus of a squamous cell is a flattened disc; that of a cuboidal cell is spherical; and a columnar cell nucleus is elongated from top to bottom and usually located closer to the cell base. Keep nuclear shape in mind when you attempt to identify epithelial types.

Simple epithelia are easy to classify by cell shape because all of the cells usually have the same shape. In stratified epithelia, however, cell shape differs in the different layers. To avoid ambiguity, stratified epithelia are named according to the shape of the cells in the *apical* layer. This naming system will become clearer as we explore the specific epithelial types.

As you read about the epithelial classes, study **Figure 4.4**. Try to pick out the individual cells within each epithelium. This is not always easy, because the boundaries between epithelial cells often are indistinct. Furthermore, the nucleus of a particular cell may or may not be visible, depending on the plane of the cut made to prepare the tissue slides.

Simple Epithelia

The simple epithelia are most concerned with absorption, secretion, and filtration. Because they consist of a single cell layer and are usually very thin, protection is not one of their specialties.

Simple Squamous Epithelium   The cells of a **simple squamous epithelium** are flattened laterally, and their cytoplasm is sparse (Figure 4.4a). In a surface view, the close-fitting cells resemble a tiled floor. When the cells are cut perpendicular to their free surface, they resemble fried eggs seen from the side, with their cytoplasm wisping out from the slightly bulging nucleus.

Thin and often permeable, simple squamous epithelium is found where filtration or the exchange of substances by rapid diffusion is a priority. In the kidneys, it forms part of the filtration membrane that filters blood to make urine. In the lungs, it forms the walls of the air sacs across which gas exchange occurs (Figure 4.4a).

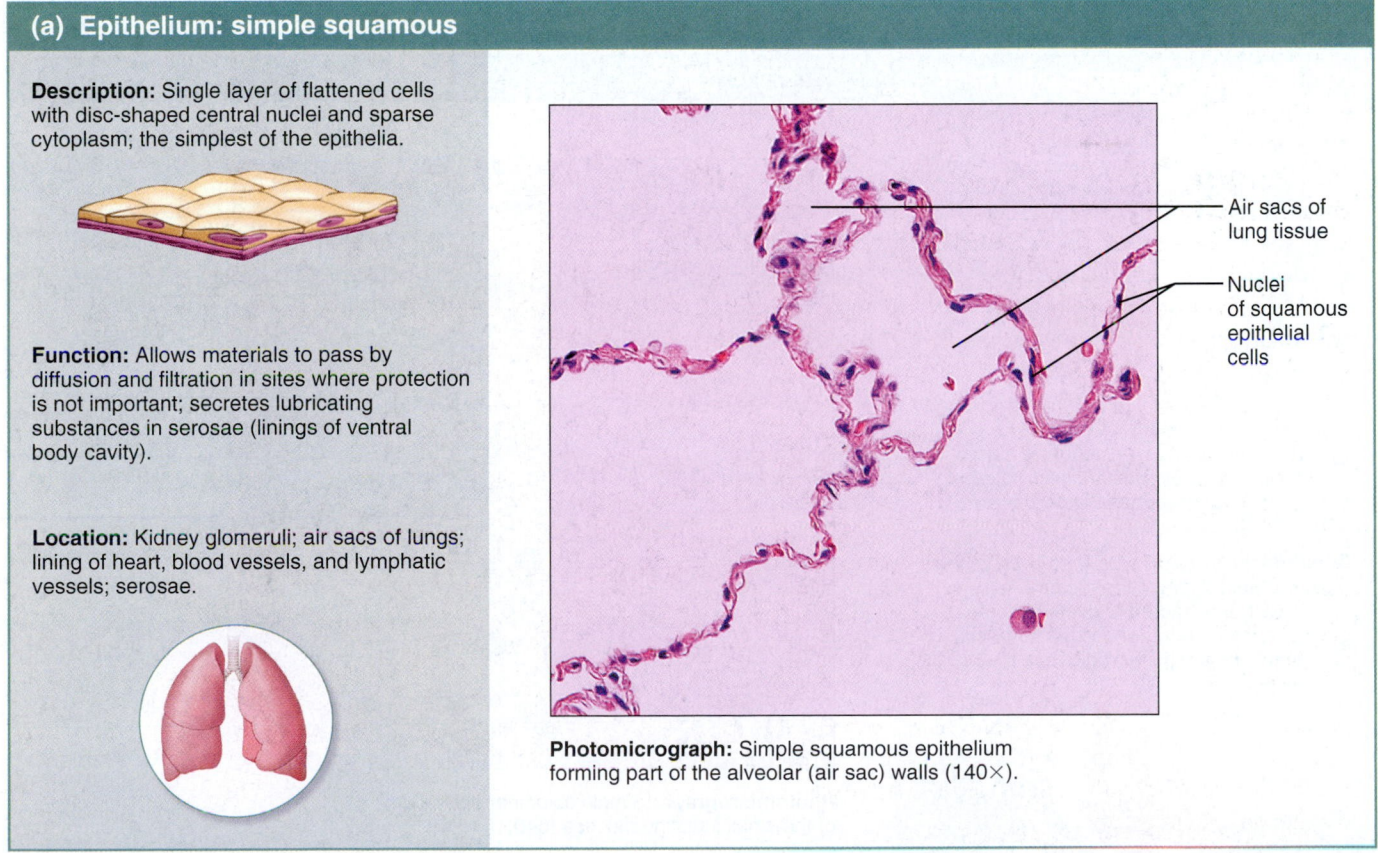

(a) Epithelium: simple squamous

Description: Single layer of flattened cells with disc-shaped central nuclei and sparse cytoplasm; the simplest of the epithelia.

Function: Allows materials to pass by diffusion and filtration in sites where protection is not important; secretes lubricating substances in serosae (linings of ventral body cavity).

Location: Kidney glomeruli; air sacs of lungs; lining of heart, blood vessels, and lymphatic vessels; serosae.

Air sacs of lung tissue

Nuclei of squamous epithelial cells

Photomicrograph: Simple squamous epithelium forming part of the alveolar (air sac) walls (140×).

Figure 4.4 Epithelial tissues.

Practice Histology questions: MasteringA&P® > Study Area > Lab Tools > PAL

(b) Epithelium: simple cuboidal

Description: Single layer of cubelike cells with large, spherical central nuclei.

Function: Secretion and absorption.

Location: Kidney tubules; ducts and secretory portions of small glands; ovary surface.

Simple cuboidal epithelial cells

Nucleus

Basement membrane

Connective tissue

Photomicrograph: Simple cuboidal epithelium in kidney tubules (430×).

(c) Epithelium: simple columnar

Description: Single layer of tall cells with *round* to *oval* nuclei; many cells bear microvilli, some bear cilia; layer may contain mucus-secreting unicellular glands (goblet cells).

Function: Absorption; secretion of mucus, enzymes, and other substances; ciliated type propels mucus (or reproductive cells) by ciliary action.

Location: Nonciliated type lines most of the digestive tract (stomach to rectum), gallbladder, and excretory ducts of some glands; ciliated variety lines small bronchi, uterine tubes, and some regions of the uterus.

Cavity inside intestine

Microvilli (on apical surfaces)

Simple columnar epithelial cells

Mucus of goblet cell

Photomicrograph: Simple columnar epithelium of the small intestine mucosa (640×).

Figure 4.4 *(continued)* **Epithelial tissues.**

(d) Epithelium: pseudostratified columnar

Description: Single layer of cells of differing heights, some not reaching the free surface; nuclei seen at different levels; may contain mucus-secreting cells and bear cilia.

Function: Secrete substances, particularly mucus; propulsion of mucus by ciliary action.

Location: Ciliated variety lines the trachea and most of the upper respiratory tract; nonciliated type in males' sperm-carrying ducts and ducts of large glands.

Trachea

Goblet cell (contains mucus)

Cilia

Pseudo-stratified epithelial layer

Basement membrane

Photomicrograph: Pseudostratified ciliated columnar epithelium lining the human trachea (780×).

Figure 4.4 *(continued)*

Two simple squamous epithelia in the body have special names that reflect their location.

- **Endothelium** (en″do-the′le-um; "inner covering") provides a slick, friction-reducing lining in lymphatic vessels and in all hollow organs of the cardiovascular system (blood vessels and the heart). Capillaries consist exclusively of endothelium, and its exceptional thinness encourages the efficient exchange of nutrients and wastes between the bloodstream and surrounding tissue cells.

- **Mesothelium** (mez″o-the′le-um; "middle covering") is the epithelium found in serous membranes, the membranes lining the ventral body cavity and covering its organs.

Simple Cuboidal Epithelium **Simple cuboidal epithelium** consists of a single layer of cells as tall as they are wide (Figure 4.4b). The generally spherical nuclei stain darkly. Important functions of simple cuboidal epithelium are secretion and absorption. This epithelium forms the walls of the smallest ducts of glands and of many kidney tubules.

Simple Columnar Epithelium **Simple columnar epithelium** is a single layer of tall, closely packed cells, aligned like soldiers in a row (Figure 4.4c). It lines the digestive tract from the stomach through the rectum. Columnar cells are mostly associated with absorption and secretion, and the digestive tract lining has two distinct modifications that make it ideal for that dual function:

- Dense microvilli on the apical surface of absorptive cells

- Tubular glands made primarily of cells that secrete mucus-containing intestinal juice

Additionally, some simple columnar epithelia display cilia on their free surfaces, which help move substances or cells through an internal passageway.

Pseudostratified Columnar Epithelium The cells of **pseudostratified columnar epithelium** (soo″do-strǎ′tĭ-fīd) vary in height (Figure 4.4d). All of its cells rest on the basement membrane, but only the tallest reach the free surface of the epithelium. Because the cell nuclei lie at different levels above the basement membrane, the tissue gives the false (pseudo) impression that several cell layers are present; hence "pseudostratified." The short cells are relatively unspecialized and give rise to the taller cells. This epithelium, like the simple columnar variety, secretes or absorbs substances. A ciliated version containing mucus-secreting goblet cells lines most of the respiratory tract. Here the motile cilia propel sheets of dust-trapping mucus away from the lungs.

4

(e) Epithelium: stratified squamous

Description: Thick epithelium composed of several cell layers; basal cells are cuboidal or columnar and metabolically active; surface cells are flattened (squamous); in the keratinized type, the surface cells are full of keratin and dead; basal cells are active in mitosis and produce the cells of the more superficial layers.

Function: Protects underlying tissues in areas subjected to abrasion.

Location: Nonkeratinized type forms the moist linings of the esophagus, mouth, and vagina; keratinized variety forms the epidermis of the skin, a dry epithelium.

Stratified squamous epithelium

Nuclei

Basement membrane

Connective tissue

Photomicrograph: Stratified squamous epithelium lining the esophagus (285×).

(f) Epithelium: transitional

Description: Resembles both stratified squamous and stratified cuboidal; basal cells cuboidal or columnar; surface cells dome shaped or squamouslike, depending on degree of organ stretch.

Function: Stretches readily, permits stored urine to distend urinary organ.

Location: Lines the ureters, bladder, and part of the urethra.

Transitional epithelium

Basement membrane

Connective tissue

Photomicrograph: Transitional epithelium lining the bladder, relaxed state (360×); note the bulbous, or rounded, appearance of the cells at the surface; these cells flatten and elongate when the bladder fills with urine.

Figure 4.4 *(continued)* **Epithelial tissues.**

Stratified Epithelia

Stratified epithelia contain two or more cell layers. They regenerate from below; that is, the basal cells divide and push apically to replace the older surface cells. Stratified epithelia are considerably more durable than simple epithelia, and protection is their major (but not their only) role.

Stratified Squamous Epithelium   **Stratified squamous epithelium** is the most widespread of the stratified epithelia (Figure 4.4e). Composed of several layers, it is thick and well suited for its protective role in the body. Its free surface cells are squamous, and cells of the deeper layers are cuboidal or columnar. This epithelium is found in areas subjected to wear and tear, and its surface cells are constantly being rubbed away and replaced by division of its basal cells. Because epithelium depends on nutrients diffusing from deeper connective tissue, the epithelial cells farther from the basement membrane are less viable and those at the apical surface are often flattened and may even be dead.

To avoid memorizing all its locations, simply remember that this epithelium forms the external surface of the skin and extends a short distance into every body opening that is directly continuous with the skin. The outer layer, or epidermis, of the skin is *keratinized* (ker′ah-tin″īzd), meaning its surface cells contain *keratin*, a tough protective protein. (We discuss the epidermis in Chapter 5.) The other stratified squamous epithelia of the body are *nonkeratinized*.

Stratified Cuboidal and Columnar Epithelia   **Stratified cuboidal epithelium** is quite rare in the body, mostly found in the ducts of some of the larger glands (sweat glands, mammary glands). It typically has two layers of cuboidal cells.

Stratified columnar epithelium also has a limited distribution in the body. Small amounts are found in the pharynx, the male urethra, and lining some glandular ducts. This epithelium also occurs at transition areas or junctions between two other types of epithelia. Only its apical layer of cells is columnar. Because of their relative scarcity in the body, Figure 4.4 does not illustrate these two stratified epithelia.

Transitional Epithelium   **Transitional epithelium** forms the lining of hollow urinary organs, which stretch as they fill with urine (Figure 4.4f). Cells of its basal layer are cuboidal or columnar. The apical cells vary in appearance, depending on the degree of distension (stretching) of the organ. When the organ is distended with urine, the transitional epithelium appears to thin from about six cell layers to as few as three, and its dome-like apical cells flatten and become squamouslike. The ability of transitional cells to change their shape (undergo "transitions") allows a greater volume of urine to flow through a tubelike organ. In the bladder, it allows more urine to be stored.

Glandular Epithelia

A **gland** consists of one or more cells that make and secrete a particular product. This product, called a **secretion**, is an aqueous (water-based) fluid that usually contains proteins. However, some glands release a lipid- or steroid-rich secretion.

Secretion is an active process. Glandular cells obtain needed substances from the blood and transform them chemically into a product that is then discharged from the cell. Notice that the term *secretion* can refer to both the gland's *product* and the *process* of making and releasing that product.

Glands are classified according to two sets of traits:

- Where they release their product—glands may be *endocrine* ("internally secreting") or *exocrine* ("externally secreting")
- Number of cells—glands may be *unicellular* ("one-celled") or *multicellular* ("many-celled")

Unicellular glands are scattered within epithelial sheets. By contrast, most multicellular epithelial glands form by invagination (inward growth) of an epithelial sheet into the underlying connective tissue (**Figure 4.5**). At least initially, most have *ducts*, tubelike connections to the epithelial sheets.

Endocrine Glands

Because **endocrine glands** lose their ducts during development, they are often called *ductless glands* (Figure 4.5, right). They produce **hormones**, chemical messengers that they secrete by exocytosis directly into the extracellular space. From there the hormones enter the blood or lymphatic fluid and travel to specific target organs. Each hormone prompts its target organ(s) to respond in some characteristic way. For example, hormones

Multicellular epithelial glands form by invagination (inward growth) of an epithelial sheet.

Epithelial sheet

Invaginated epithelial sheet

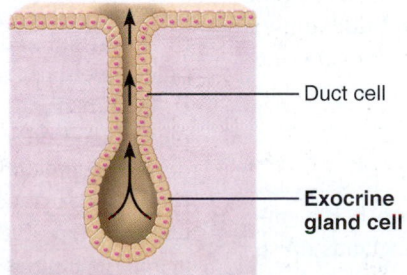

Exocrine glands retain the connecting cells, which form a *duct* that transports secretions to the epithelial surface.

Duct cell

Exocrine gland cell

Endocrine glands lose their ducts during development. They secrete *hormones* into the interstitial fluid. These hormones then enter the blood.

Connecting cells are lost during development

Endocrine gland cell

Blood vessel

Figure 4.5 **Formation of multicellular exocrine and endocrine glands.**

4

(a)   (b)

Figure 4.6 Goblet cell (unicellular exocrine gland). (a) Photomicrograph of a goblet cell in the simple columnar epithelium lining the small intestine (1640×). **(b)** Corresponding diagram. Notice the secretory vesicles and well-developed rough ER and Golgi apparatus.

Practice Histology questions: MasteringA&P® > Study Area > Lab Tools > PAL

produced by certain intestinal cells cause the pancreas to release enzymes that help digest food in the digestive tract.

Endocrine glands are structurally diverse, so one description does not fit all. Most are compact multicellular organs, but some individual hormone-producing cells are scattered in the digestive tract lining (mucosa) and in the brain, giving rise to their collective description as the *diffuse endocrine system*. Endocrine secretions are also varied, ranging from modified amino acids to peptides, glycoproteins, and steroids. Not all endocrine glands arise from epithelial tissue, so we defer their further consideration to Chapter 16.

Exocrine Glands

All **exocrine glands** secrete their products onto body surfaces (skin) or into body cavities. The unicellular glands do so directly (by exocytosis), whereas the multicellular glands do so via an epithelium-walled duct that transports the secretion to the epithelial surface. Exocrine glands are diverse. They include the liver (which secretes bile); the pancreas (which synthesizes digestive enzymes); mucous, sweat, oil, and salivary glands; and many others.

Unicellular Exocrine Glands   The only important examples of **unicellular** (or one-celled) glands are *mucous cells* and *goblet cells*. Unicellular glands are sprinkled in the epithelial linings of the intestinal and respiratory tracts amid columnar cells with other functions (see Figure 4.4c).

In humans, all such glands produce **mucin** (mu′sin), a complex glycoprotein that dissolves in water when secreted. Once dissolved, mucin forms mucus, a slimy coating that protects and lubricates surfaces. In **goblet cells**, accumulating mucin distends the top of the cell, making the cells look like a glass with a stem (which is why it is named "goblet" cell, **Figure 4.6**). This distortion does not occur in **mucous cells**.

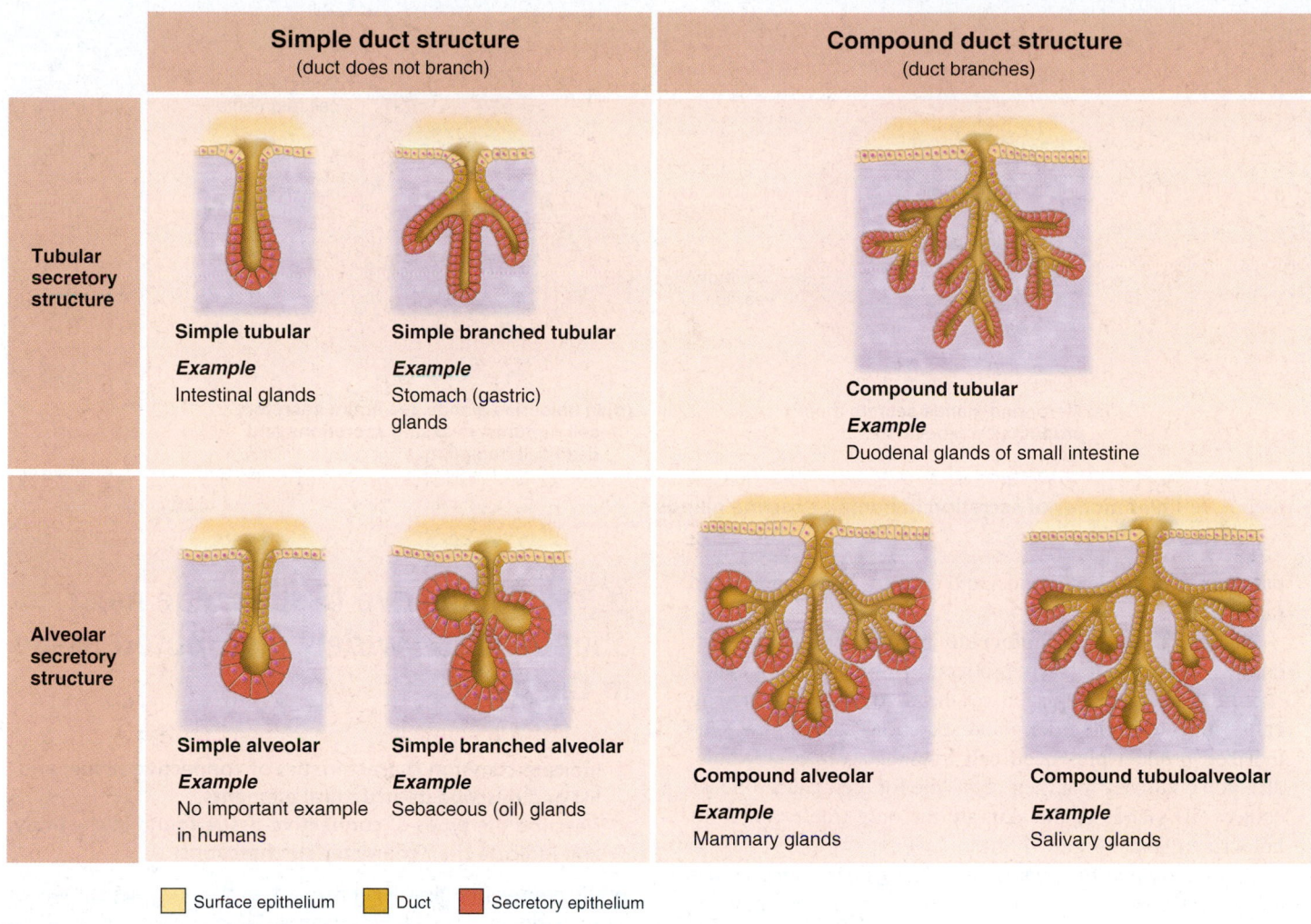

Simple duct structure (duct does not branch)		Compound duct structure (duct branches)	
Tubular secretory structure	**Simple tubular** *Example* Intestinal glands	**Simple branched tubular** *Example* Stomach (gastric) glands	**Compound tubular** *Example* Duodenal glands of small intestine
Alveolar secretory structure	**Simple alveolar** *Example* No important example in humans	**Simple branched alveolar** *Example* Sebaceous (oil) glands	**Compound alveolar** *Example* Mammary glands **Compound tubuloalveolar** *Example* Salivary glands

Surface epithelium   Duct   Secretory epithelium

Figure 4.7 Types of multicellular exocrine glands. Multicellular glands are classified according to duct type (simple or compound) and the structure of their secretory units (tubular, alveolar, or tubuloalveolar).

Multicellular Exocrine Glands Compared to the unicellular glands, **multicellular exocrine glands** are structurally more complex. They have two basic parts: an epithelium-derived *duct* and a *secretory unit* (*acinus*) consisting of secretory cells. In all but the simplest glands, *supportive connective tissue* surrounds the secretory unit, supplies it with blood vessels and nerve fibers, and forms a *fibrous capsule* that extends into the gland and divides it into *lobes*.

Multicellular exocrine glands can be classified by structure and by mode of secretion.

- **Structural classification.** On the basis of their duct structures, multicellular exocrine glands are either simple or compound (**Figure 4.7**). **Simple glands** have an unbranched duct, whereas **compound glands** have a branched duct. The glands are further categorized by their secretory units as (1) **tubular** if the secretory cells form tubes; (2) **alveolar** (al-ve′o-lar) if the secretory cells form small, flasklike sacs (*alveolus* = "small hollow cavity"); or (3) **tubuloalveolar** if they have both types of secretory units. Note that many anatomists use the term **acinar** (as′ĭ-nar; "berrylike") interchangeably with alveolar.

- **Modes of secretion.** Multicellular exocrine glands secrete their products in different ways, so they can also be described functionally as *merocrine*, *holocrine*, or *apocrine* glands. Most are **merocrine glands** (mer′o-krin), which secrete their products by exocytosis as they are produced. The secretory cells are not altered in any way (so think "merely secrete" to remember their mode of secretion). The

Secretory vesicles

(a) Merocrine glands secrete their products by exocytosis.

Secretory cell fragments

(b) In holocrine glands, the entire secretory cell ruptures, releasing secretions and dead cell fragments.

Figure 4.8 Chief modes of secretion in human exocrine glands.

pancreas, most sweat glands, and salivary glands belong to this class (**Figure 4.8a**).

Secretory cells of **holocrine glands** (hol'o-krin), in contrast, accumulate their products within them until they rupture. (They are replaced by the division of underlying cells.) Because holocrine gland secretions include the synthesized product plus dead cell fragments (*holo* = whole, all), you could say that their cells "die for their cause." Sebaceous (oil) glands of the skin are the only true example of holocrine glands (Figure 4.8b).

Although *apocrine glands* (ap'o-krin) are present in other animals, there is some controversy over whether humans have this gland type. Like holocrine glands, apocrine glands accumulate their products, but in this case only just beneath the free surface. Eventually, the apex of the cell pinches off (*apo* = from, off), releasing the secretory granules and a small amount of cytoplasm. The cell repairs its damage and the process repeats again and again. The closest to an example of this process in humans is the release of lipid droplets by lactating mammary glands. However, most histologists classify mammary glands as merocrine glands because this is the means by which milk proteins are secreted.

Check Your Understanding

3. Epithelial tissue is the only tissue type that has polarity, that is, an apical and a basal surface. Why is this important?

4. Stratified epithelia are "built" for protection or to resist abrasion. What are the simple epithelia better at?

5. Some epithelia are pseudostratified. What does this mean?

6. APPLY While examining a multicellular exocrine gland under the microscope, you notice that there are many cells in metaphase. Is this gland a merocrine or holocrine gland? Explain.

7. DRAW Draw three cells lined up to make a simple ciliated columnar epithelium. Add a basement membrane. Label it and the apical and basal surfaces. Indicate which of the four basic tissue types you would find below the basement membrane.

For answers, see Answers Appendix.

4.3 Connective tissue is the most abundant and widely distributed tissue in the body

Learning Outcomes

▶ Indicate common characteristics of connective tissue, and list and describe its structural elements.

▶ Describe the types of connective tissue found in the body, and indicate their characteristic functions.

While **connective tissue** is prevalent in the body, its amount in particular organs varies. For example, skin consists primarily of connective tissue, while the brain contains very little.

There are four main classes of connective tissue and several subclasses (**Table 4.1**). These are *connective tissue proper* (which includes fat and the fibrous tissue of ligaments), *cartilage*, *bone*, and *blood*.

Connective tissue does much more than just *connect* body parts. Its major functions include (1) *binding and supporting*, (2) *protecting*, (3) *insulating*, (4) *storing* reserve fuel, and (5) *transporting* substances within the body. For example, bone and cartilage support and protect body organs by providing the hard underpinnings of the skeleton. Fat insulates and protects body organs and provides a fuel reserve. Blood transports substances inside the body.

Common Characteristics of Connective Tissue

Connective tissues share two characteristics that together set them apart from other primary tissues:

- **Extracellular matrix.** All other primary tissues are composed mainly of cells, but connective tissues consist largely of nonliving **extracellular matrix** (ma'triks; "womb") (◀ p. 63), which separates, often widely, the living cells of the tissue. Because of its matrix, connective tissue can bear weight,

Table 4.1	Comparison of Classes of Connective Tissues				
TISSUE CLASS AND EXAMPLE	**SUBCLASSES**	**CELLS**	**MATRIX**		**GENERAL FEATURES**
Connective Tissue Proper *Dense regular connective tissue*	1. Loose connective tissue • Areolar • Adipose • Reticular 2. Dense connective tissue • Regular • Irregular • Elastic	Fibroblasts Fibrocytes Defense cells Adipocytes	Gel-like ground substance All three fiber types: collagen, reticular, elastic		Six different types; vary in density and types of fibers Functions as a binding tissue Resists mechanical stress, particularly tension Provides reservoir for water and salts Energy (fat) storage
Cartilage *Hyaline cartilage*	1. Hyaline cartilage 2. Elastic cartilage 3. Fibrocartilage	Chondroblasts (found in growing cartilage) Chondrocytes	Gel-like ground substance Fibers: collagen, elastic fibers in some		Resists compression because of the large amounts of water held in the matrix Functions to cushion and support body structures
Bone Tissue *Compact bone*	1. Compact bone 2. Spongy bone (See Chapter 6 for details)	Osteoblasts Osteocytes	Gel-like ground substance calcified with inorganic salts Fibers: collagen		Hard tissue that resists both compression and tension Functions in support
Blood	(See Chapter 17 for details)	Red blood cells (RBCs) or erythrocytes White blood cells (WBCs) or leukocytes Platelets	Plasma No fibers		A fluid tissue Functions to carry O_2, CO_2, nutrients, wastes, and other substances (such as hormones)

withstand great tension, and endure abuses, such as physical trauma and abrasion, that no other tissue can tolerate.

- **Common origin.** All connective tissues arise from *mesenchyme* (an embryonic tissue).

One noticeable difference between different types of connective tissue is how richly they are supplied by blood vessels. Cartilage is avascular. Dense connective tissue is poorly vascularized, and the other types of connective tissue have a rich supply of blood vessels.

Structural Components of Connective Tissue

Connective tissues have three main components: *ground substance*, *fibers*, and *cells* (Table 4.1). Together ground substance and fibers make up the extracellular matrix. (Note that some authors use the term *matrix* to indicate the ground substance only.)

The composition and arrangement of these three components vary tremendously. The result is an amazing diversity of connective tissues, each adapted to perform a specific function in the body. For example, the matrix can be delicate and fragile to form a soft "packing" around an organ, or it can form "ropes" (tendons and ligaments) of incredible strength. Nonetheless, connective tissues have a common structural plan, and we use *areolar connective tissue* (ah-re′o-lar) as our *prototype*, or model (**Figure 4.9** and Figure 4.11a). All other subclasses are simply variants of this plan.

Ground Substance

Ground substance is the unstructured material that fills the space between the cells and contains the fibers. It has three components:

- *Interstitial fluid.* The ground substance consists of large amounts of fluid and functions as a molecular sieve through which nutrients and other dissolved substances can diffuse between the blood capillaries and the cells. The fibers embedded in the ground substance make it less pliable and hinder diffusion somewhat.

- *Cell adhesion proteins.* These proteins serve mainly as a connective tissue glue that allows connective tissue cells to attach to the extracellular matrix.

- *Proteoglycans.* The proteoglycans consist of a protein core to which large polysaccharides called *glycosaminoglycans* (GAGs) (gli″kos-ah-me″no-gli′kanz) are attached. The strand-like GAGs [e.g., *chondroitin sulfate* and *hyaluronic acid* (hi″ah-lu-ron′ik)] stick out from the protein core like the fibers of a bottle brush. The proteoglycans tend to form huge aggregates in which the GAGs intertwine and trap water,

forming a substance that varies from a fluid to a viscous gel. The higher the GAG content, the more viscous the ground substance.

Connective Tissue Fibers

The fibers of connective tissue are proteins that provide support. Three types of fibers are found in connective tissue matrix: collagen, elastic, and reticular fibers. Of these, collagen fibers are by far the strongest and most abundant.

Collagen Fibers These fibers are constructed primarily of the fibrous protein *collagen*. Collagen molecules are secreted into the extracellular space, where they assemble spontaneously into cross-linked fibrils, which in turn are bundled together into the thick collagen fibers seen with a microscope. Because their fibrils cross-link, collagen fibers are extremely tough and provide high tensile strength (the ability to resist being pulled apart) to the matrix. Indeed, stress tests show that collagen fibers are stronger than steel fibers of the same size!

Elastic Fibers Long, thin, elastic fibers form branching networks in the extracellular matrix. These fibers contain a rubber-like protein, *elastin*, that allows them to stretch and recoil like rubber bands. Connective tissue can stretch only so much before its thick, ropelike collagen fibers become taut. Then, when the tension lets up, elastic fibers snap the connective tissue back to its normal length and shape. Elastic fibers are found where greater elasticity is needed, for example, in the skin, lungs, and blood vessel walls.

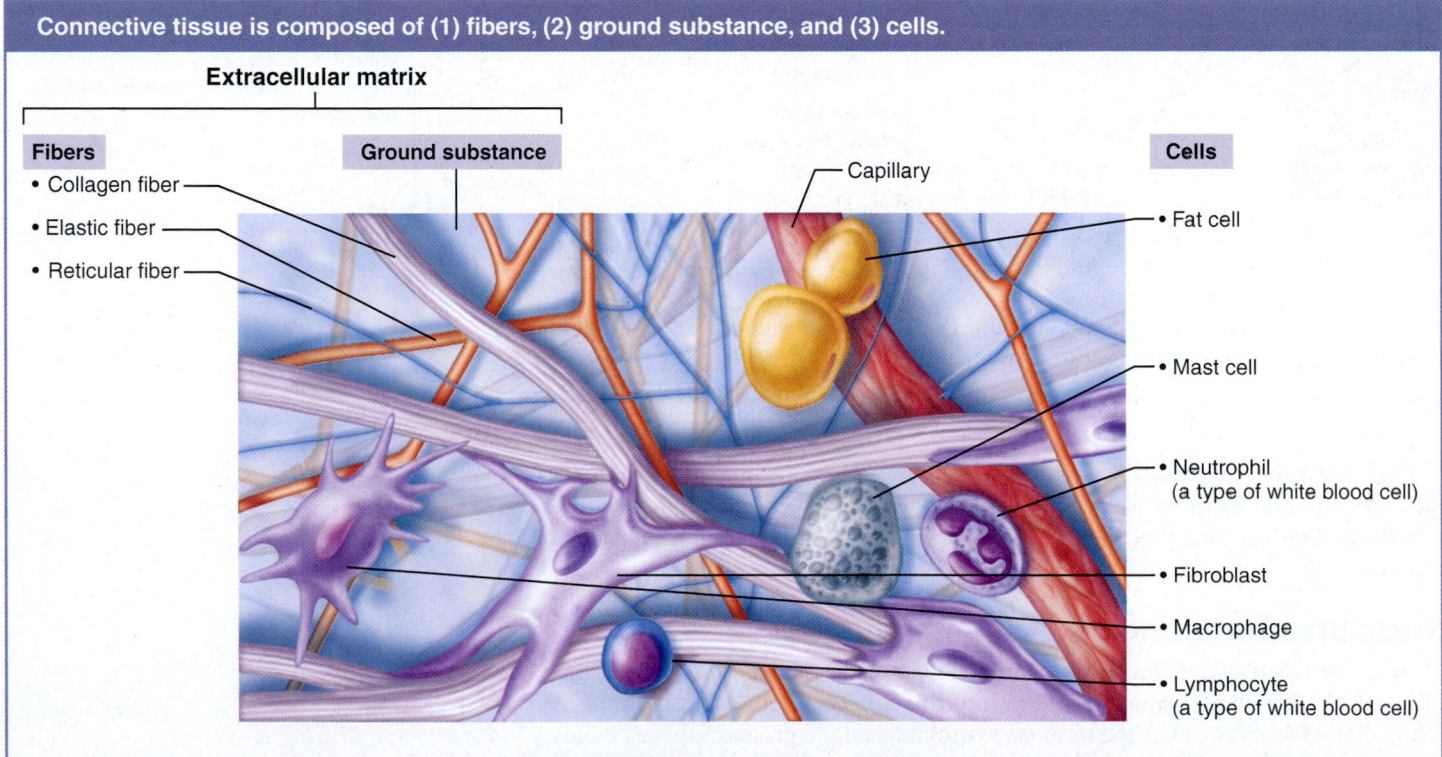

Connective tissue is composed of (1) fibers, (2) ground substance, and (3) cells.

Extracellular matrix

Fibers
- Collagen fiber
- Elastic fiber
- Reticular fiber

Ground substance

Capillary

Cells
- Fat cell
- Mast cell
- Neutrophil (a type of white blood cell)
- Fibroblast
- Macrophage
- Lymphocyte (a type of white blood cell)

Figure 4.9 Areolar connective tissue: A prototype (model) connective tissue. This tissue underlies epithelia and surrounds capillaries. (See Figure 4.11a for a micrograph.)

Reticular Fibers These short, fine fibers are made of a different type of collagen than the more common, thicker collagen fibers. They connect to the coarser collagen fibers, but they branch extensively, forming delicate networks (*reticul* = network) that surround small blood vessels and support the soft tissue of organs. They are particularly abundant where connective tissue is next to other tissue types, for example, in the basement membrane of epithelial tissues, and around capillaries, where they form fuzzy "nets" that allow more stretch than the larger collagen fibers.

Connective Tissue Cells

Each major class of connective tissue has a resident cell type that exists in immature (*-blast*) and mature (*-cyte*) forms. For example:

- **Fibroblasts** in connective tissue proper become **fibrocytes**.
- **Chondroblasts** (kon′dro-blast″) in cartilage become **chondrocytes**.
- **Osteoblasts** (os′te-o-blast″) in bone become **osteocytes**.

The immature blast cells are actively mitotic. Each type of blast cell secretes the ground substance and the fibers characteristic of their particular matrix. Once they synthesize the matrix, the blast cells assume their mature, less active *-cyte* form. The mature cells maintain the health of the matrix. However, if the matrix is injured, they can easily revert to their more active blast form to repair and regenerate the matrix.

Blood, the fourth major class of connective tissue, is an exception to the generalization that we have just made. Its immature blood cell–forming type (once called a hemocytoblast) is called a *hematopoietic stem cell* (hem″ah-to-poy-et′ik). In addition, it is not located in "its" tissue (blood) and does not make the fluid matrix (plasma) of that tissue. Blood formation is considered in Chapter 17.

Connective tissue is also home to an assortment of other cell types, such as:

- **Adipocytes** (ad′ĭ-po-sītz), commonly called *adipose* or *fat cells*, which store energy as fat.
- **White blood cells** (WBCs or leukocytes, including neutrophils, eosinophils, and lymphocytes) and other cell types that are concerned with tissue response to injury.
- **Mast cells**, which typically cluster along blood vessels. These oval cells detect foreign microorganisms (e.g., bacteria, fungi) and initiate local inflammatory responses against them. Mast cell cytoplasm contains secretory granules (*mast* = stuffed full of granules) with chemicals that mediate inflammation, especially in severe allergies. These chemicals include:
  - *Heparin* (hep′ah-rin), an anticoagulant chemical that prevents blood clotting when free in the bloodstream (but in human mast cells it appears to regulate the action of other mast cell chemicals)
  - *Histamine* (his′tah-mēn), a substance that makes capillaries leaky

- *Proteases* (protein-degrading enzymes)
- Other enzymes

- **Macrophages** (mak′ro-fāj″es; *macro* = large; *phago* = eat) are large, irregularly shaped cells that avidly devour a broad variety of foreign materials, ranging from foreign molecules to entire bacteria to dust particles. These "big eaters" also dispose of dead tissue cells, and they are central actors in the immune system. Macrophages are peppered throughout loose connective tissue, bone marrow, and lymphoid tissue. They may be attached to connective tissue fibers (fixed) or may migrate freely through the matrix. Some macrophages have selective appetites. For example, those of the spleen primarily dispose of aging red blood cells, but they will not turn down other "delicacies" that come their way.

Types of Connective Tissue

As noted, all classes of connective tissue consist of living cells surrounded by an extracellular matrix. Their major differences reflect cell type, and the types and relative amounts of fibers, as summarized in Table 4.1 on p. 127.

As mentioned earlier, mature connective tissues arise from a common embryonic tissue called **mesenchyme** (meh′zin-kīm). Mesenchyme has a fluid ground substance containing fine sparse fibers and star-shaped *mesenchymal cells*. It arises during the early weeks of embryonic development and eventually differentiates (specializes) into all other connective tissue cells. However, some mesenchymal cells remain and provide a source of new cells in mature connective tissues.

Figure 4.10 gives an overview of the classification of different types of connective tissue. This will help you orient yourself as you study the various kinds of connective tissue that we will discuss next.

All mature connective tissues except for bone, cartilage, and blood are **connective tissue proper**. Connective tissue proper has two subclasses: **loose connective tissues** (areolar, adipose, and reticular) and **dense connective tissues** (dense regular, dense irregular, and elastic).

Figure 4.10 Overview of types of connective tissue.

4

Connective Tissue Proper—Loose Connective Tissues

Areolar Connective Tissue   **Areolar connective tissue** (**Figure 4.11a**) is the most widely distributed connective tissue in the body, and it serves as a universal packing material between other tissues. It binds body parts together while allowing them to move freely over one another. It is the connective tissue that most epithelia rest on and is present in all mucous membranes as the *lamina propria* (see p. 141).

The functions of areolar connective tissue include:

- Supporting and binding other tissues (the job of the fibers)
- Holding body fluids (the ground substance's role)
- Defending against infection (via the activity of white blood cells and macrophages)
- Storing nutrients as fat in adipocytes (fat cells)

The most obvious structural feature of this tissue is the loose arrangement of its fibers. The rest of the matrix, occupied by ground substance, appears to be empty space when viewed through the microscope, and in fact, the Latin term *areola* means "a small open space." Because of its loose nature, areolar connective tissue provides a reservoir of water and salts for surrounding body tissues, always holding more fluid than there is in the entire bloodstream. Essentially all body cells obtain their nutrients from and release their wastes into this "tissue fluid."

Fibroblasts, flat branching cells that appear spindle shaped (tapered at both ends) in profile, are the predominant cell type. Numerous macrophages are also seen and present a formidable barrier to invading microorganisms. Adipocytes appear singly or in clusters, and occasional mast cells are identified easily by the large, darkly stained cytoplasmic granules that often obscure their nuclei. Other cell types are scattered throughout.

The ground substance of areolar connective tissue is viscous like molasses (because of its high concentration of hyaluronic acid). This may hinder the movement of cells through it. Some white blood cells, which protect the body from disease-causing microorganisms, secrete an enzyme (hyaluronidase) to liquefy the ground substance and ease their passage. (Unhappily, some harmful bacteria have the same ability.)

Adipose (Fat) Tissue   **Adipose tissue** (ad′ĭ-pōs) is similar to areolar tissue in structure and function, but its nutrient-storing ability is much greater. Consequently, *adipocytes* account for 90% of this tissue's mass. The matrix is scanty and the cells are packed closely together, giving a chicken-wire appearance to the tissue. A glistening oil droplet (almost pure triglyceride) occupies most of an adipocyte's volume and displaces the nucleus to one side (Figure 4.11b). Mature adipocytes are among the largest cells in the body. As they take up or release fat, they become plumper or more wrinkled, respectively.

Adipose tissue is richly vascularized, indicating its high metabolic activity. Without the fat stores in our adipose tissue, we could not live for more than a few days without eating. Adipose tissue is certainly abundant: It constitutes 18% of an average person's body weight.

Adipose tissue may develop almost anywhere areolar tissue is plentiful, but it usually accumulates in subcutaneous tissue, where it acts as a shock absorber, as insulation, and as an energy storage site. Because fat is a poor conductor of heat, it helps prevent heat loss from the body. Other sites where fat accumulates are located around the kidneys, behind the eyeballs, and at genetically determined locations such as the abdomen and hips.

The abundant fat beneath the skin serves the general nutrient needs of the entire body, and smaller deposits of fat serve the local nutrient needs of highly active organs. Such deposits occur around the hard-working heart and around lymph nodes (where cells of the immune system are furiously fighting infection), within some muscles, and as individual fat cells in the bone marrow, where new blood cells are produced at a rapid rate. Many of these local deposits are highly enriched in special lipids.

The adipose tissue just described is sometimes called *white fat*, or *white adipose tissue*, to distinguish it from *brown fat*, or *brown adipose tissue*. White fat stores nutrients (mainly for other cells), but brown fat contains abundant mitochondria that use the lipid fuels to generate heat (instead of generating ATP molecules). The bloodstream then carries this heat to warm the rest of the body. The richly vascular brown fat occurs mainly

(a) Connective tissue ⟶ Proper ⟶ Loose ⟶ Areolar

Description: Gel-like matrix with all three fiber types; cells: fibroblasts, macrophages, mast cells, and some white blood cells.

Function: Wraps and cushions organs; its macrophages phagocytize bacteria; plays important role in inflammation; holds and conveys tissue fluid.

Location: Widely distributed under epithelia of body, e.g., forms lamina propria of mucous membranes; packages organs; surrounds capillaries.

Epithelium

Lamina propria

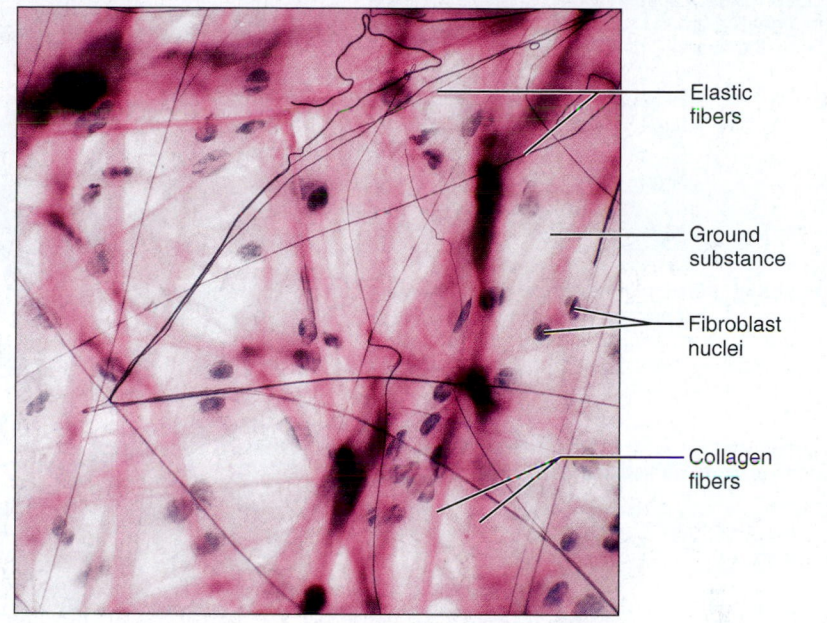

Elastic fibers

Ground substance

Fibroblast nuclei

Collagen fibers

Photomicrograph: Areolar connective tissue, a soft packaging tissue of the body (340×).

(b) Connective tissue ⟶ Proper ⟶ Loose ⟶ Adipose

Description: Matrix as in areolar, but very sparse; closely packed adipocytes, or fat cells, have nucleus pushed to the side by large fat droplet.

Function: Provides reserve food fuel; insulates against heat loss; supports and protects organs.

Location: Under skin in subcutaneous tissue; around kidneys and eyeballs; within abdomen; in breasts.

Adipose tissue

Mammary glands

Nucleus of adipose (fat) cell

Fat droplet

Photomicrograph: Adipose tissue from the subcutaneous layer under the skin (350×).

Figure 4.11   Connective tissues.

Practice Histology questions: MasteringA&P® > Study Area > Lab Tools > PAL

4

(c) Connective tissue ⟶ Proper ⟶ Loose ⟶ Reticular

Description: Loose network of reticular fibers in a gel-like ground substance; reticular cells lie on the fibers.

Function: Fibers form a soft internal skeleton (stroma) that supports other cell types including white blood cells, mast cells, and macrophages.

Location: Lymphoid organs (lymph nodes, bone marrow, and spleen).

Spleen

White blood cell (lymphocyte)

Reticular fibers

Photomicrograph: Dark-staining network of reticular connective tissue fibers forming the internal skeleton of the spleen (350×).

(d) Connective tissue ⟶ Proper ⟶ Dense ⟶ Regular

Description: Primarily parallel collagen fibers; a few elastic fibers; major cell type is the fibroblast.

Function: Attaches muscles to bones or to muscles; attaches bones to bones; withstands great tensile stress when pulling force is applied in one direction.

Location: Tendons, most ligaments, aponeuroses.

Shoulder joint

Ligament

Tendon

Collagen fibers

Nuclei of fibroblasts

Photomicrograph: Dense regular connective tissue from a tendon (430×).

Figure 4.11 *(continued)* **Connective tissues.**

on the back of babies who (as yet) lack the ability to produce body heat by shivering. Small deposits occur in adults, mostly above the collarbones, on the neck and abdomen, and around the spine.

Reticular Connective Tissue **Reticular connective tissue** resembles areolar connective tissue, but the only fibers in its matrix are reticular fibers, which form a delicate network along which fibroblasts called **reticular cells** (Figure 4.11c) are scattered. Although reticular *fibers* are widely distributed in the body, reticular tissue is limited to certain sites. It forms a labyrinth-like **stroma** ("bed" or "mattress"), or internal framework, that can support many free blood cells (mostly lymphocytes) in lymph nodes, the spleen, and bone marrow.

Connective Tissue Proper—Dense Connective Tissues

The three varieties of dense connective tissue are dense regular, dense irregular, and elastic. Since all three have fibers as their prominent element, dense connective tissues are often called **fibrous connective tissues**.

Dense Regular Connective Tissue **Dense regular connective tissue** contains closely packed bundles of collagen fibers running in the same direction, parallel to the direction of pull (Figure 4.11d). This arrangement results in white, flexible structures with great resistance to tension (pulling forces) where the tension is exerted in a single direction. Crowded between the collagen fibers are rows of fibroblasts that continuously manufacture the fibers and a small amount of ground substance.

Collagen fibers are slightly wavy (see Figure 4.11d). This allows the tissue to stretch a little, but once the fibers straighten out, there is no further "give" to this tissue. Unlike our model (areolar) connective tissue, this tissue has few cells other than fibroblasts and is poorly vascularized.

With its enormous tensile strength, dense regular connective tissue forms:

- *Tendons.* Tendons are cords that attach muscles to bones.
- *Aponeuroses.* Aponeuroses (ap″o-nu-ro′sēz) are flat, sheet-like tendons that attach muscles to other muscles or to bones.
- *Ligaments.* Ligaments bind bones together at joints. They contain more elastic fibers than tendons and are slightly more stretchy.

Dense Irregular Connective Tissue **Dense irregular connective tissue** has the same structural elements as the regular variety. However, the bundles of collagen fibers are much thicker and they are arranged irregularly; that is, they run in all directions (Figure 4.11e). This type of tissue forms sheets in body areas where tension is exerted from many different directions. It is found in the skin as the leathery *dermis*, and it forms fibrous joint capsules and the fibrous coverings that surround some organs (kidneys, bones, cartilages, muscles, and nerves).

(e) Connective tissue ⟶ Proper ⟶ Dense ⟶ Irregular

Description: Primarily irregularly arranged collagen fibers; some elastic fibers; fibroblast is the major cell type.

Function: Withstands tension exerted in many directions; provides structural strength.

Location: Fibrous capsules of organs and of joints; dermis of the skin; submucosa of digestive tract.

Shoulder joint

Fibrous joint capsule

Nuclei of fibroblasts

Collagen fibers

Photomicrograph: Dense irregular connective tissue from the fibrous capsule of a joint (430×).

Figure 4.11 *(continued)*

4

(f) Connective tissue ⟶ Proper ⟶ Dense ⟶ Elastic

Description: Dense regular connective tissue containing a high proportion of elastic fibers.

Function: Allows tissue to recoil after stretching; maintains pulsatile flow of blood through arteries; aids passive recoil of lungs following inspiration.

Location: Walls of large arteries; within certain ligaments associated with the vertebral column; within the walls of the bronchial tubes.

Aorta

Heart

Elastic fibers

Photomicrograph: Elastic connective tissue in the wall of the aorta (250×).

Figure 4.11 *(continued)* **Connective tissues.**

Elastic Connective Tissue A few ligaments, such as those connecting adjacent vertebrae, are very elastic. The dense regular connective tissue in those structures is called **elastic connective tissue** (Figure 4.11f). Additionally, many of the larger arteries have stretchy sheets of elastic connective tissue in their walls.

Cartilage

Cartilage (kar′tĭ-lij) stands up to both tension *and* compression. Its qualities are between those of dense connective tissue and bone. It is tough but flexible, providing a resilient rigidity to the structures it supports.

Cartilage lacks nerve fibers and is avascular. It receives its nutrients by diffusion from blood vessels located in the connective tissue layer (perichondrium) surrounding it. Its ground substance is quite firm and contains large amounts of the GAGs chondroitin sulfate and hyaluronic acid, and firmly bound collagen fibers (and in some cases elastic fibers). Cartilage matrix also contains an exceptional amount of tissue fluid. In fact, cartilage is up to 80% water! The movement of tissue fluid in its matrix enables cartilage to rebound after being compressed and also helps to nourish the cartilage cells.

Chondroblasts, the predominant cell type in growing cartilage, produce new matrix until the skeleton stops growing at the end of adolescence. *Chondrocytes* (mature cartilage cells) are typically found in small groups within cavities called *lacunae* (lah-ku′ne; "pits").There are three varieties of cartilage: *hyaline* *cartilage*, *elastic cartilage*, and *fibrocartilage*, each dominated by a particular fiber type.

Hyaline Cartilage **Hyaline cartilage** (hi′ah-līn), or *gristle*, is the most abundant cartilage in the body. Although it contains large numbers of collagen fibers, they are not apparent and the matrix appears glassy (*hyal* = glass, transparent) blue-white when viewed by the unaided eye. Chondrocytes account for only 1–10% of the cartilage volume (Figure 4.11g).

Hyaline cartilage provides firm support with some pliability. It is found in a number of key places:

- It covers the ends of long bones as *articular cartilage*, providing springy pads that absorb compression at joints.

- It supports the tip of the nose.

- It connects the ribs to the sternum.

- It supports most of the respiratory system passages (for example, the trachea).

- It makes up most of the embryonic skeleton before bone forms. Skeletal hyaline cartilage persists during childhood as the *epiphyseal plates* (e″pĭ-fis′e-ul; growth plates), actively growing regions near the ends of long bones.

Elastic Cartilage Histologically, **elastic cartilage** (Figure 4.11h) is nearly identical to hyaline cartilage. However, elastic cartilage has many more elastic fibers. Found where strength and exceptional stretchability are needed, elastic cartilage

(g) Connective tissue ⟶ Cartilage ⟶ Hyaline

Description: Amorphous but firm matrix; collagen fibers form an imperceptible network; chondroblasts produce the matrix and when mature (as chondrocytes) lie in lacunae.

Function: Supports and reinforces; serves as resilient cushion; resists compressive stress.

Location: Forms most of the embryonic skeleton; covers the ends of long bones in joint cavities; forms costal cartilages of the ribs; cartilages of the nose, trachea, and larynx.

Costal cartilages

Chondrocyte in lacuna

Matrix

Photomicrograph: Hyaline cartilage from a costal cartilage of a rib (470×).

(h) Connective tissue ⟶ Cartilage ⟶ Elastic

Description: Similar to hyaline cartilage, but more elastic fibers in matrix.

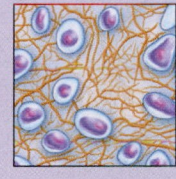

Function: Maintains the shape of a structure while allowing great flexibility.

Location: Supports the external ear (pinna); epiglottis.

Elastic fibers

Chondrocyte in lacuna

Matrix

Photomicrograph: Elastic cartilage from the human ear pinna; forms the flexible skeleton of the ear (510×).

Figure 4.11 *(continued)*

4

(i) Connective tissue ⟶ Cartilage ⟶ Fibrocartilage

Description: Matrix similar to but less firm than that in hyaline cartilage; thick collagen fibers predominate.

Function: Tensile strength allows it to absorb compressive shock.

Location: Intervertebral discs; pubic symphysis; discs of knee joint.

Intervertebral discs

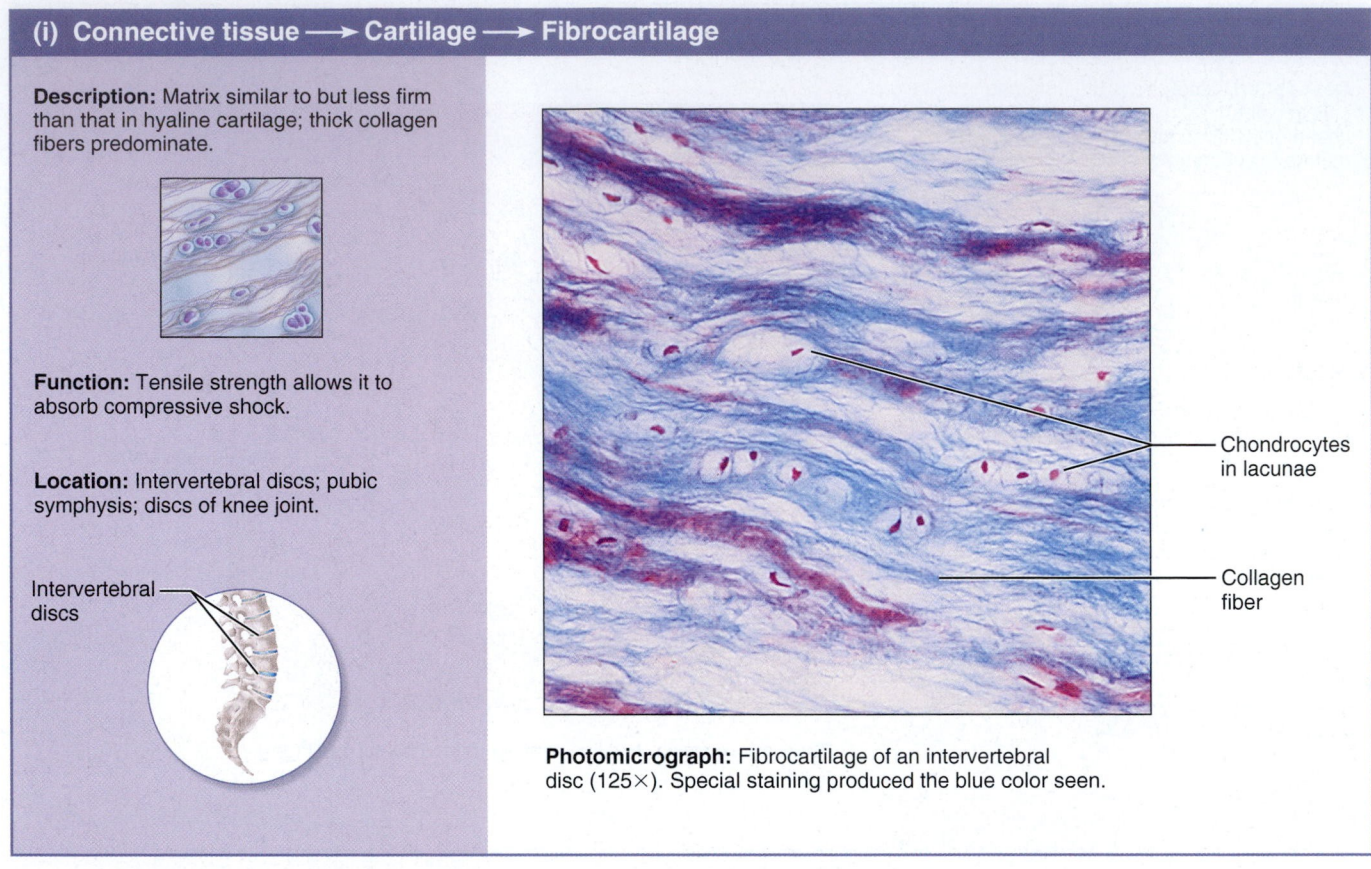

Chondrocytes in lacunae

Collagen fiber

Photomicrograph: Fibrocartilage of an intervertebral disc (125×). Special staining produced the blue color seen.

(j) Connective tissue ⟶ Bone

Description: Hard, calcified matrix containing many collagen fibers; osteocytes lie in lacunae. Very well vascularized.

Function: Supports and protects (by enclosing); provides levers for the muscles to act on; stores calcium and other minerals and fat; marrow inside bones is the site for blood cell formation (hematopoiesis).

Location: Bones.

Central canal

Lacunae

Lamella

Photomicrograph: Cross-sectional view of bone (125×).

Figure 4.11 (*continued*) **Connective tissues.**

(k) Connective tissue ⟶ Blood

Description: Red and white blood cells in a fluid matrix (plasma).

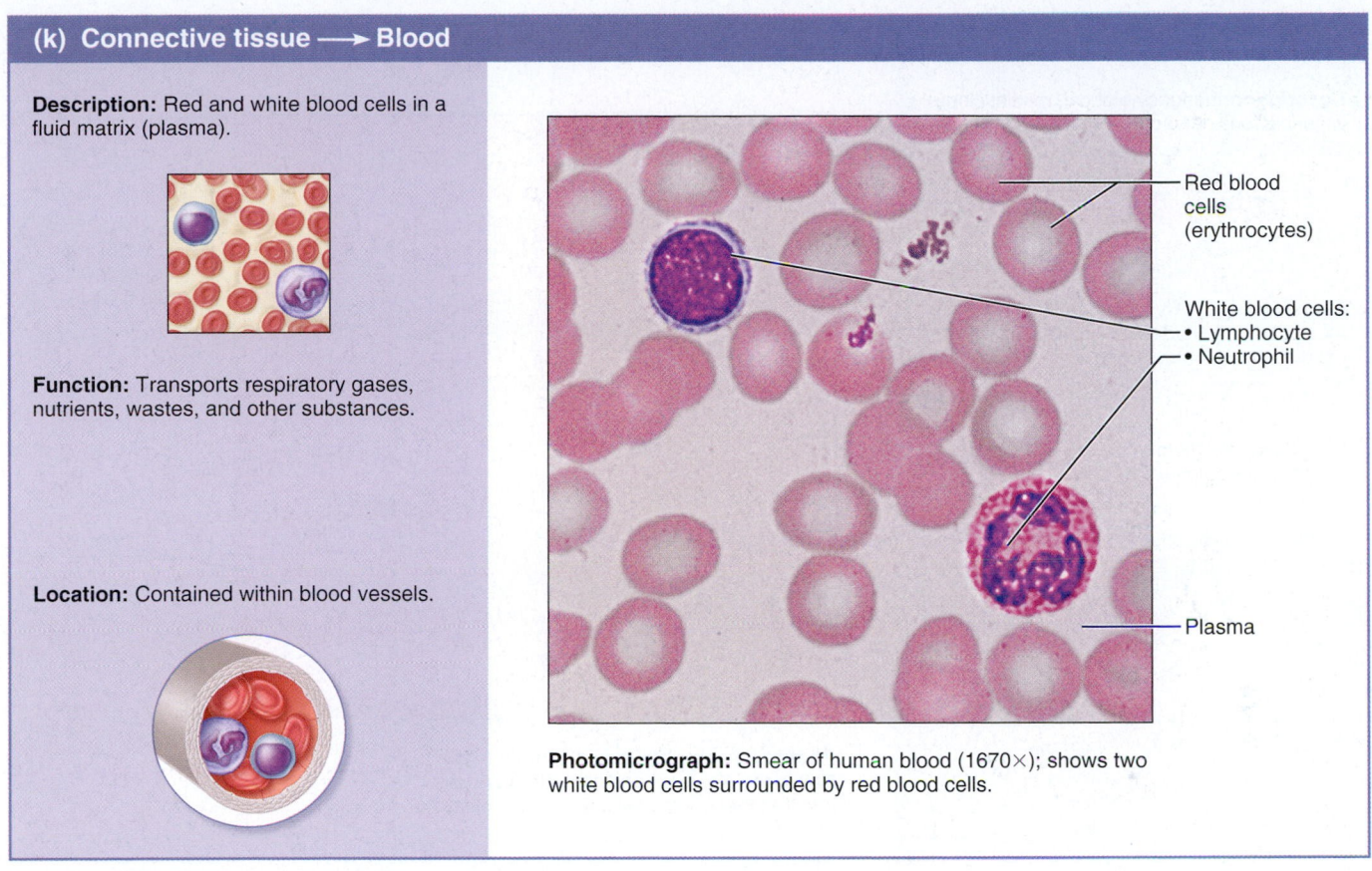

Function: Transports respiratory gases, nutrients, wastes, and other substances.

Location: Contained within blood vessels.

Red blood cells (erythrocytes)

White blood cells:
• Lymphocyte
• Neutrophil

Plasma

Photomicrograph: Smear of human blood (1670×); shows two white blood cells surrounded by red blood cells.

Figure 4.11 *(continued)*

forms the "skeletons" of the external ear and the epiglottis (the flap that covers the opening to the respiratory passageway when we swallow).

Fibrocartilage Structurally, **fibrocartilage** is intermediate between hyaline cartilage and dense regular connective tissues. Its rows of chondrocytes (a cartilage feature) alternate with rows of thick collagen fibers (characteristic of dense regular connective tissue) (Figure 4.11i). Because it resists both compression and tension well, fibrocartilage is found where strong support and the ability to withstand heavy pressure are required: for example, the intervertebral discs (resilient cushions between the bony vertebrae) and the spongy cartilages of the knee.

HOMEOSTATIC IMBALANCE 4.2   CLINICAL

Because cartilage is avascular and aging cartilage cells lose their ability to divide, injured cartilage heals slowly. This phenomenon is excruciatingly familiar to those who have experienced sports injuries. During later life, cartilage tends to calcify or even ossify (become bony). In such cases, the chondrocytes are poorly nourished and die. •••••••

Bone (Osseous Tissue)

Because of its rocklike hardness, **bone,** or **osseous tissue** (os′e-us), has an exceptional ability to support and protect body structures. Bones of the skeleton also provide cavities for storing

fat and synthesizing blood cells. Bone matrix is similar to that of cartilage but is harder and more rigid because, in addition to its more abundant collagen fibers, bone has an added matrix element—inorganic calcium salts (bone salts).

Osteoblasts produce the organic portion of the matrix, and then bone salts are deposited on and between the fibers. Mature bone cells, or *osteocytes*, reside in the lacunae within the matrix they have made (Figure 4.11j). A cross section of bone tissue reveals closely packed structural units called *osteons* that look like tree rings. Osteons are concentric rings of bony matrix (lamellae) surrounding central canals containing the blood vessels and nerves serving the bone. Unlike cartilage, the next firmest connective tissue, bone is well supplied by invading blood vessels.

Blood

Blood, the fluid within blood vessels, is the most atypical connective tissue. It does *not* connect things or give mechanical support. It is classified as a connective tissue because it develops from mesenchyme and consists of *blood cells*, surrounded by a nonliving fluid matrix called *blood plasma* (Figure 4.11k).

The vast majority of blood cells are red blood cells, or erythrocytes, but scattered white blood cells and platelets (needed for blood clotting) are also seen. The "fibers" of blood are soluble protein molecules that precipitate, forming visible fiberlike structures during blood clotting. Blood functions as the transport vehicle for

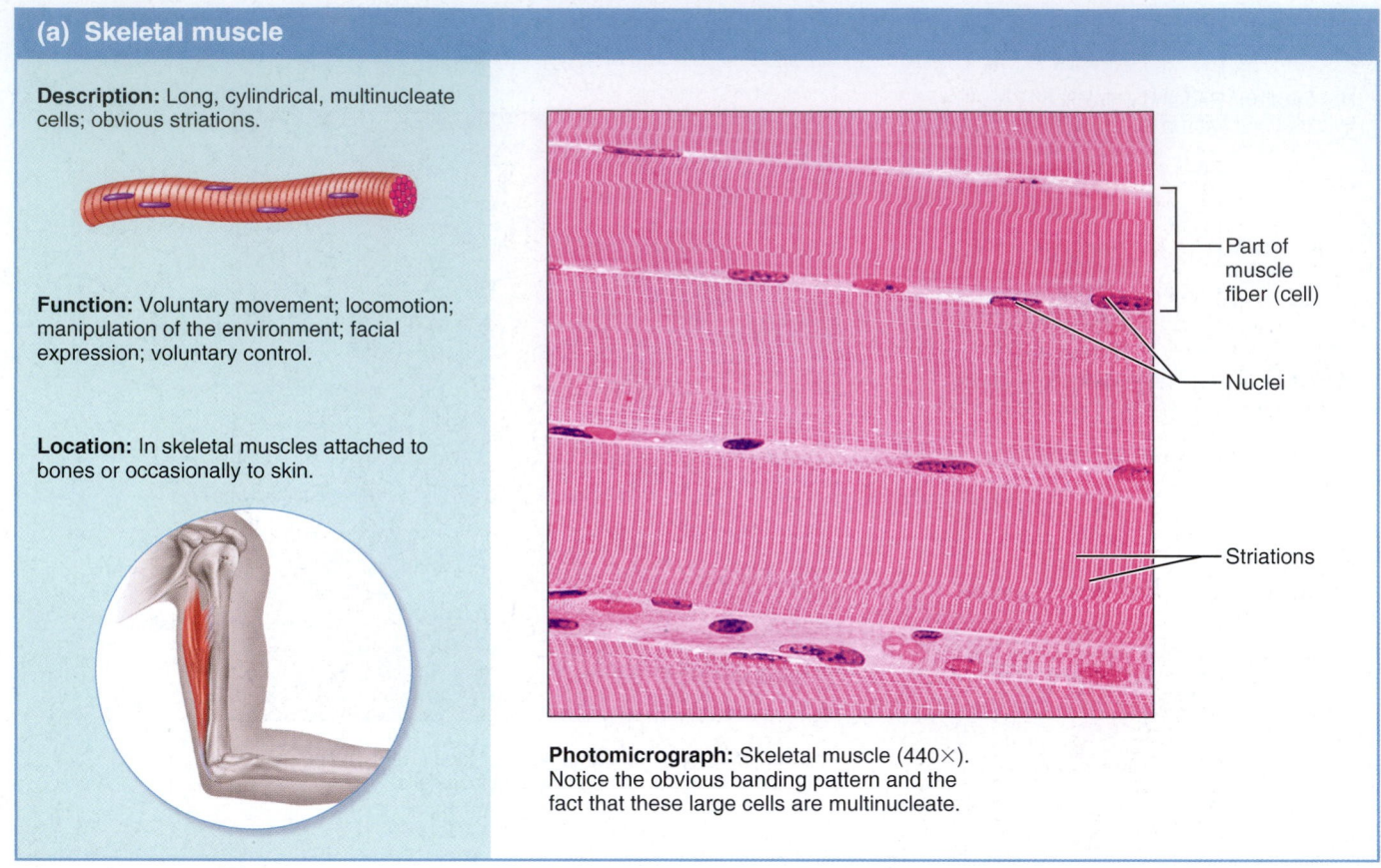

(a) Skeletal muscle

Description: Long, cylindrical, multinucleate cells; obvious striations.

Function: Voluntary movement; locomotion; manipulation of the environment; facial expression; voluntary control.

Location: In skeletal muscles attached to bones or occasionally to skin.

Part of muscle fiber (cell)

Nuclei

Striations

Photomicrograph: Skeletal muscle (440×). Notice the obvious banding pattern and the fact that these large cells are multinucleate.

Figure 4.12 Muscle tissues.

the cardiovascular system, carrying nutrients, wastes, respiratory gases, and many other substances throughout the body.

Check Your Understanding

8. What are four functions of connective tissue?
9. What are the three types of fibers found in connective tissues?
10. Which connective tissue has a soft weblike matrix capable of serving as a fluid reservoir?
11. What type of connective tissue is damaged when you cut your index finger tendon?
12. **MAKE CONNECTIONS** It has been observed that aging cartilage tends to calcify or ossify and its cells die. What survival needs are not being met in these cells and why is this so?

For answers, see Answers Appendix.

4.4 Muscle tissue is responsible for body movement

Learning Outcome

▶ **Compare and contrast the structures and body locations of the three types of muscle tissue.**

Muscle tissues are well-vascularized tissues that are responsible for most types of body movement. As in epithelial tissue, the cells in muscle tissue are all tightly packed together. Muscle cells possess **myofilaments**, elaborate networks of the *actin* and *myosin* filaments that bring about movement or

contraction in all cell types. There are three kinds of muscle tissue: skeletal, cardiac, and smooth. Because skeletal muscle contraction is under our conscious control, skeletal muscle is often referred to as **voluntary muscle**. The other two types are called **involuntary muscle** because we do not consciously control them. We briefly survey all three muscle types here. Skeletal muscle and smooth muscle are described in detail in Chapter 9, and cardiac muscle is discussed in Chapter 18.

Skeletal Muscle

Skeletal muscle tissue is packaged by connective tissue sheets into organs called *skeletal muscles* that are attached to the bones of the skeleton. These muscles form the flesh of the body, and as they contract they pull on bones or skin, causing body movements.

Skeletal muscle cells, also called **muscle fibers**, are long, cylindrical cells that contain many peripherally located nuclei. Their obvious banded, or striated, appearance reflects the precise alignment of their myofilaments (**Figure 4.12**).

Cardiac Muscle

Cardiac muscle is found only in the walls of the heart. Its contractions help propel blood through the blood vessels to all parts of the body. Like skeletal muscle cells, cardiac muscle cells are striated. However, cardiac cells differ structurally in that they are:

(b) Cardiac muscle

Description: Branching, striated, generally uninucleate cells that connect at specialized junctions (intercalated discs).

Function: As it contracts, it propels blood into the circulation; involuntary control.

Location: The walls of the heart.

Intercalated discs

Striations

Nucleus

Photomicrograph: Cardiac muscle (475×); notice the striations, branching of cells, and the intercalated discs.

(c) Smooth muscle

Description: Cells are spindle shaped (tapered on both ends) with central nuclei; no striations; cells arranged closely to form sheets.

Function: Propels substances or objects (foodstuffs, urine, a baby) along internal passageways; involuntary control.

Location: Mostly in the walls of hollow organs.

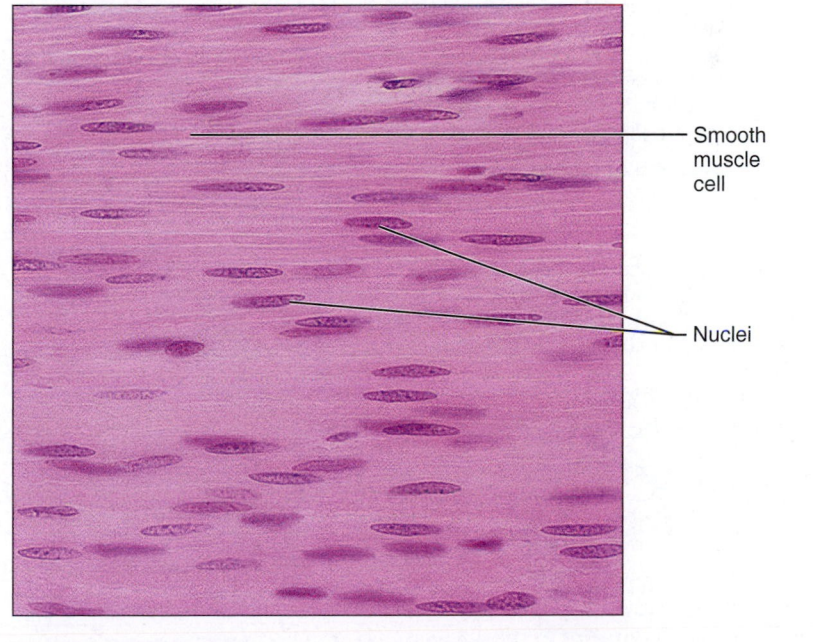

Smooth muscle cell

Nuclei

Photomicrograph: Sheet of smooth muscle from the digestive tract (465×).

Figure 4.12 *(continued)*

Practice Histology questions: MasteringA&P® > Study Area > Lab Tools > PAL

4

- Generally uninucleate (one nucleus) with the nucleus situated centrally
- Branching cells that fit together tightly at unique junctions called **intercalated discs** (in-ter′kah-la″ted) (Figure 4.12b)

Smooth Muscle

Smooth muscle is so named because its cells have no visible striations. Individual smooth muscle cells are spindle shaped (tapered at both ends) and contain one centrally located nucleus (Figure 4.12c). Smooth muscle is found mainly in the walls of hollow organs other than the heart (digestive and urinary tract organs, uterus, and blood vessels). It squeezes substances through these organs by alternately contracting and relaxing.

Check Your Understanding

13. You are looking at muscle tissue through the microscope and you see striped branching cells that connect with one another. What type of muscle are you viewing?

14. Which muscle type(s) is voluntary? Which is injured when you pull a muscle while exercising?

For answers, see Answers Appendix.

4.5 Nervous tissue is a specialized tissue of the nervous system

Learning Outcome

▶ Indicate the general characteristics of nervous tissue.

Nervous tissue is the main component of the nervous system—the brain, spinal cord, and nerves—which regulates and controls body functions. It contains two major cell types: neurons and supporting cells.

Neurons are highly specialized nerve cells that generate and conduct nerve impulses (**Figure 4.13**). Typically, they are branching cells with cytoplasmic extensions or processes that enable them to:

- Respond to stimuli (via processes called *dendrites*)
- Transmit electrical impulses over substantial distances within the body (via processes called *axons*)

Supporting cells (known as glial cells or neuroglia) are nonconducting cells that support, insulate, and protect the delicate neurons. Chapter 11 presents a more complete discussion of nervous tissue.

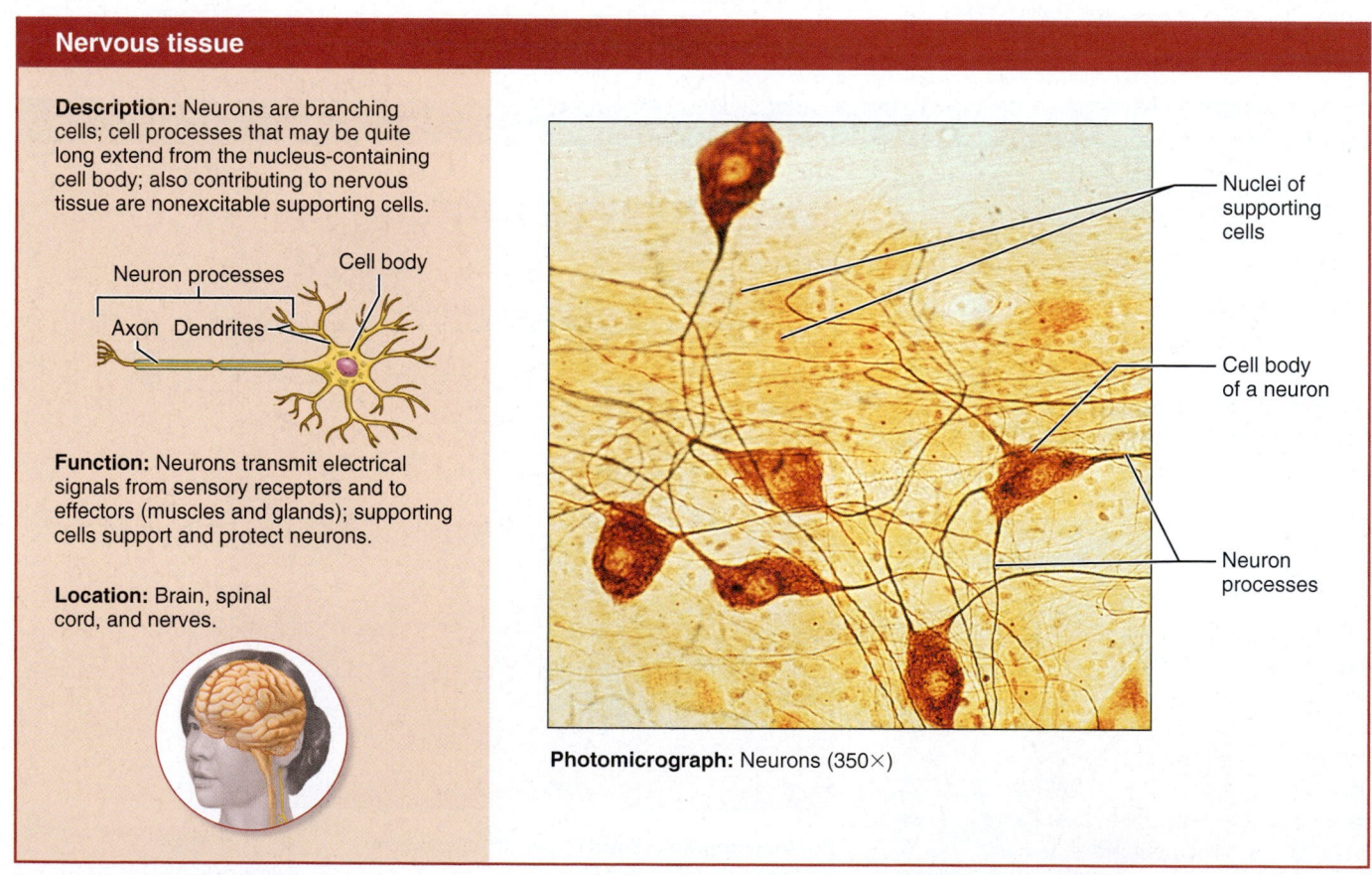

Nervous tissue

Description: Neurons are branching cells; cell processes that may be quite long extend from the nucleus-containing cell body; also contributing to nervous tissue are nonexcitable supporting cells.

Neuron processes
Cell body
Axon   Dendrites

Function: Neurons transmit electrical signals from sensory receptors and to effectors (muscles and glands); supporting cells support and protect neurons.

Location: Brain, spinal cord, and nerves.

Nuclei of supporting cells

Cell body of a neuron

Neuron processes

Photomicrograph: Neurons (350×)

Figure 4.13 Nervous tissue.

Practice Histology questions: MasteringA&P° >
Study Area > Lab Tools > PAL

15. How does the extended length of a neuron's processes aid its function in the body?

_____ *For answers, see Answers Appendix.*

4.6 The cutaneous membrane is dry; mucous and serous membranes are wet

Learning Outcome

▶ Describe the structure and function of cutaneous, mucous, and serous membranes.

Now that we have described all four primary tissues, we can consider the body's covering and lining membranes. These incorporate more than one type of tissue and so are simple organs. There are three types of covering and lining membranes: *cutaneous*, *mucous*, and *serous*. Essentially they all are continuous multicellular sheets composed of at least two primary tissue types: an epithelium bound to an underlying layer of connective tissue proper. Some consider the *synovial membranes* to belong to this group, but they consist only of connective tissue. We will describe them in Chapter 8.

Cutaneous Membrane

The **cutaneous membrane** (ku-ta′ne-us; *cutis* = skin) is your skin (**Figure 4.14a**). It is an organ system consisting of a keratinized stratified squamous epithelium (epidermis) firmly attached to a thick layer of connective tissue (dermis). Unlike other epithelial membranes, the cutaneous membrane is exposed to the air and is a dry membrane. Chapter 5 is devoted to this unique organ system.

Mucous Membranes

Mucous membranes, or **mucosae** (mu-ko′se), line all body cavities that open to the outside of the body, such as the hollow organs of the digestive, respiratory, and urogenital tracts (Figure 4.14b). In all cases, they are "wet," or moist, membranes bathed by secretions or, in the case of the urinary mucosa, urine. Notice that the term *mucosa* refers to the location of the membrane, *not* its cell composition, which varies. However, most mucosae contain either stratified squamous or simple columnar epithelia. The epithelial sheet lies directly over a layer of areolar connective tissue called the **lamina propria** (lam′ĭ-nah pro′pre-ah; "one's own layer"). In some mucosae, the lamina propria rests on a third (deeper) layer of smooth muscle cells.

Mucous membranes are often adapted for absorption and secretion. Although many mucosae secrete mucus, this is not a requirement. The mucosae of both the digestive and respiratory tracts secrete copious amounts of lubricating mucus, but that of the urinary tract does not.

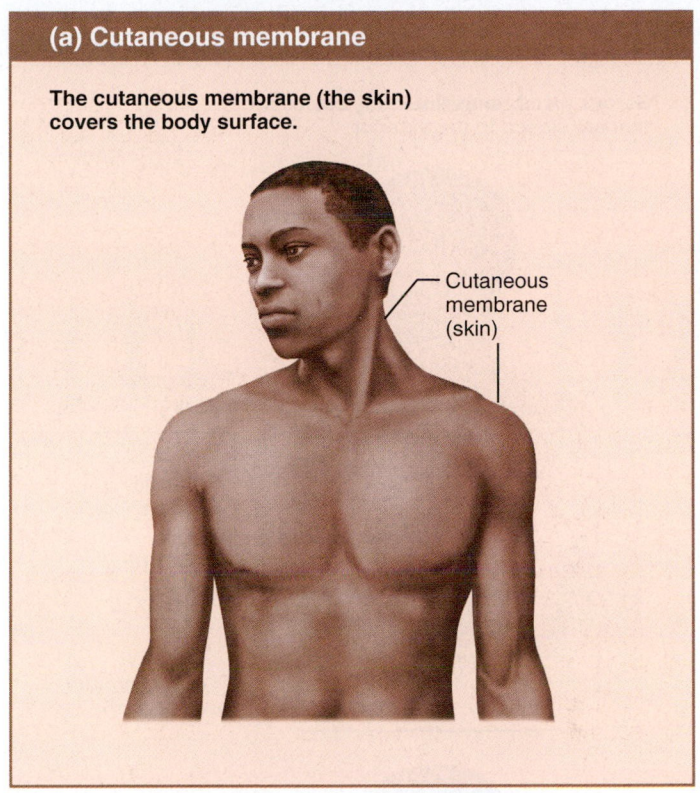

(a) Cutaneous membrane

The cutaneous membrane (the skin) covers the body surface.

Cutaneous membrane (skin)

(b) Mucous membranes (mucosae)

Mucous membranes line body cavities that are open to the exterior.

Mucosa of nasal cavity

Mucosa of mouth

Esophagus lining

Mucosa of lung bronchi

Figure 4.14   Classes of covering and lining membranes.

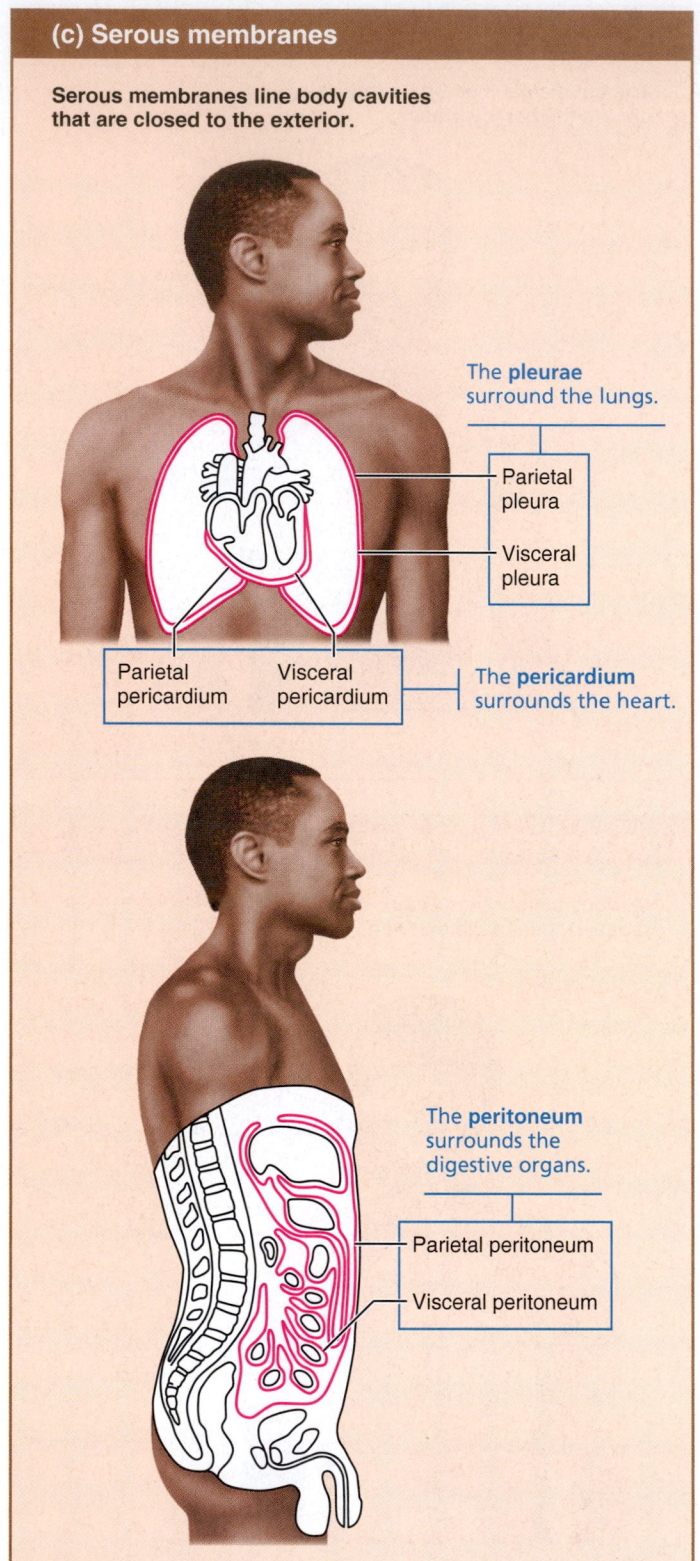

(c) Serous membranes

Serous membranes line body cavities that are closed to the exterior.

The **pleurae** surround the lungs.

Parietal pleura

Visceral pleura

Parietal pericardium   Visceral pericardium

The **pericardium** surrounds the heart.

The **peritoneum** surrounds the digestive organs.

Parietal peritoneum

Visceral peritoneum

Figure 4.14 *(continued)* **Classes of covering and lining membranes.**

Serous Membranes

Serosa membranes, or **serosae)** (se-ro′se) are the moist membranes found in closed ventral body cavities (Figure 4.14c).

Recall that serous membranes have a *visceral layer* and a *parietal layer* separated by serous fluid (◄ **Figure 1.11, p. 19**). Each layer consists of simple squamous epithelium (a mesothelium) resting on a thin layer of areolar connective tissue. The mesothelial cells add the slippery molecule hyaluronic acid to the fluid that filters from the capillaries in the associated connective tissue. The resulting thin, clear *serous fluid* lubricates the facing surfaces of the parietal and visceral layers, so that they slide across each other easily.

The serosae are named according to their location and specific organ associations: The **pleurae** line the thoracic wall and cover the lungs; the **pericardium** encloses the heart; and the **peritoneum** encloses the abdominopelvic viscera.

Check Your Understanding

16. What type of membrane consists of epithelium and connective tissue, and lines body cavities open to the exterior?
17. What type of membrane lines the thoracic walls and covers the lungs, and what is it called?
18. **APPLY** Your classmate, Jennifer, is confused by all the different types of membranes. To help her understand, explain the key differences between the serous and mucous membranes, the plasma membrane (see Chapter 3), and the basement membrane (see p. 118).
19. **MAKE CONNECTIONS** The two layers of serous membranes are held together by serous fluid, which is largely water. Which of the properties of water (Chapter 2) makes these layers "stick" together?

For answers, see Answers Appendix.

4.7 Tissue repair involves inflammation, organization, and regeneration

Learning Outcome

▶ Outline the process of tissue repair involved in normal healing of a superficial wound.

The body has many techniques for protecting itself from uninvited "guests" or injury. Mechanical barriers such as the skin and mucosae, the cilia of epithelial cells lining the respiratory tract, and the strong acid (chemical barrier) produced by stomach glands represent three defenses at the body's external boundaries.

When tissue is injured, these barriers are penetrated. This stimulates the body's inflammatory and immune responses, which wage their battles largely in the connective tissues of the body. The *inflammatory response* is a relatively nonspecific reaction that develops quickly wherever tissues are injured, while other *immune responses* are extremely specific, but take longer to swing into action (detailed discussion in Chapter 21).

Steps of Tissue Repair

Tissue repair requires that cells divide and migrate, activities that are initiated by growth factors (wound hormones) released by injured cells. Repair occurs in two major ways:

Scab

Blood clot in cut skin

Epidermis

Vein

Inflammatory chemicals

Migrating white blood cell

Artery

① Inflammation sets the stage:
- Trauma causes injured tissue cells, mast cells, and others to release inflammatory chemicals.
- Inflammatory chemicals make local blood vessels leaky, allowing white blood cells, fluid, clotting proteins, and other plasma proteins to seep into the injured area.
- Clotting seals off the injured area and prevents bacteria, toxins, and other harmful substances from spreading.
- The surface of the clot exposed to air dries and forms a scab.

Regenerating epithelium

Area of granulation tissue ingrowth

Fibroblast

Macrophage

Budding capillary

② Organization restores the blood supply:
- The clot is replaced by the ingrowth of fragile capillaries that are characteristic of **granulation tissue**. This restores the vascular supply.
- Fibroblasts multiply and produce growth factors and collagen fibers that bridge the gap. When healing is complete, these fibroblasts revert to the resting stage or undergo apoptosis (cell suicide).
- Macrophages phagocytize dead and dying cells and other debris.
- Surface epithelial cells multiply and begin to migrate over the granulation tissue.

Regenerated epithelium

③ Regeneration and fibrosis effect permanent repair:
- The fibrosed area matures and contracts, pulling the margins of the wound together.
- As it regenerates, the epithelium thickens under the scab, which detaches.
- A fully regenerated epithelium with an underlying area of scar tissue results. The scar may be visible as a white line, depending on the severity of the wound.

Fibrosed area

Figure 4.15 **Tissue repair of a nonextensive skin wound: regeneration and fibrosis.**

- **Regeneration** replaces destroyed tissue with the same kind of tissue.
- **Fibrosis** replaces destroyed tissue with **scar tissue**, which is dense connective tissue.

Which of these occurs depends on (1) the type of tissue damaged and (2) the severity of the injury. In skin, the tissue we will use as our example, repair involves both activities as described in detail in **Figure 4.15**.

In the initial part of wound healing, a special tissue, called **granulation tissue**, is formed. Granulation tissue is a delicate

Cancer—The Intimate Enemy

The word cancer elicits dread in everyone. Why does cancer strike some and not others?

Although once perceived as disorganized cell growth, this disease is now known to be a process in which a sequence of many tiny alterations changes a normal cell into a killer.

When cells fail to follow normal controls of cell division and multiply excessively, an abnormal mass of proliferating cells called a **neoplasm** (ne'o-plazm, "new growth") results. Neoplasms are classified as **benign** ("kindly") or **malignant** ("bad"). A benign neoplasm is strictly a local affair. Its cells remain compacted, are often encapsulated, tend to grow slowly, and seldom kill their hosts if removed before they compress vital organs.

In contrast, cancers are malignant neoplasms, nonencapsulated masses that grow relentlessly. Their cells resemble immature cells, and they invade their surroundings. Whereas normal cells die when they lose contact with the surrounding matrix, malignant cells tend to break away from the parent mass—the primary tumor—and travel via blood or lymph to other body organs, where they form *secondary cancer masses*. This capability of traveling to other parts of the body is called **metastasis** (mĕ-tas'tah-sis). Metastasis and invasiveness distinguish cancer cells from the cells of benign neoplasms. Cancer cells consume an exceptional amount of the body's nutrients, leading to weight loss and tissue wasting that contribute to death.

Carcinogenesis

What changes a normal cell into a cancerous one? Factors include radiation, mechanical trauma, certain viral infections, chronic inflammation, many chemicals, and toxins such as those contained in tobacco smoke.

All these factors cause *mutations*—changes in DNA that alter the expression of certain genes. Fortunately, we have several defenses against carcinogens. First, the liver inactivates many carcinogens. Second, cells have intrinsic DNA repair mechanisms to help undo mutations. Third, some immune system cells continuously scan our body cells and eliminate any cancerous cells they find. Furthermore, one mutation isn't enough. It takes several genetic changes to transform (convert) a normal cell into a cancerous cell.

The discovery of **oncogenes** (Greek *onco* = tumor), or cancer-causing genes, provided a clue to the role of genes in cancer. **Proto-oncogenes**, benign forms of oncogenes in normal cells, were discovered later. Proto-oncogenes code for proteins that are essential for cell division, growth, and cellular adhesion, among other things. Many proto-oncogenes have fragile sites that break when exposed to carcinogens, converting them to oncogenes. Oncogenes may also "switch on" dormant genes that allow cells to become invasive and metastasize. Known oncogenes now number over 100.

Like the accelerator of a car, oncogenes accelerate growth. But to get the car to go, you also have to take your foot off the brakes. **Tumor suppressor genes,** another group of genes, act like the brakes. The products of their genes inhibit cell growth and division, so the inactivation of these genes leads to uncontrolled cell growth. Over half of all cancers involve malfunction or loss of *TP53*, just one of the many identified tumor suppressor genes. This is not surprising when you learn that *TP53* codes for a protein that stops cell division in cells whose DNA has been damaged. If the damage cannot be repaired, the protein triggers apoptosis.

Furthermore, although each type of cancer is genetically distinct, human cancers share mutations in a common set of key pathways controlling growth and cell division. Whatever genetic factors are at work, the "seeds" of cancer are in our own genes. Cancer is an intimate enemy indeed.

Let's look at the carcinogenesis of colorectal cancer, one of the best understood human cancers. As with most cancers, colorectal cancer develops gradually. One of the first signs is a polyp (see photo), a small, benign growth consisting of apparently normal mucosa cells. As cell division continues, the polyp enlarges, becoming an adenoma (neoplasm of glandular epithelium). As the mutations accumulate, various tumor suppressor genes are inactivated and oncogenes are mobilized, and the adenoma becomes increasingly abnormal. The final consequence is colon carcinoma, a form of cancer that metastasizes quickly.

Cancer Prevalence

In the U.S. and Canada, almost half of all people develop cancer in their lifetime and a fifth of us will die of it. Cancer can arise from almost any cell type, but the most common cancers originate in the skin, colon, lung, breast, and prostate.

Many cancers are preceded by lumps or other structural changes in tissue—for instance, *leukoplakia*, white patches in the mouth caused by the chronic irritation of ill-fitting dentures. Although these lesions sometimes progress to cancer, in many

---

pink tissue that contains capillaries that grow in from nearby areas and lay down a new capillary bed. Granulation tissue is actually named for these capillaries, which protrude nublike from its surface, giving it a granular appearance. These capillaries are fragile and bleed freely, as we see when someone picks at a scab.

The repair process described in Figure 4.15 follows healing of a wound (cut, scrape, puncture) that breaches an epithelial barrier. In simple *infections* (a pimple or sore throat), healing is solely by regeneration. Only severe (destructive) infections lead to clot formation or scarring.

Regenerative Capacity of Different Tissues

Tissues vary widely in their ability to regenerate. Epithelial tissues, bone, areolar connective tissue, dense irregular connective tissue, and blood-forming tissue regenerate extremely well. Smooth muscle and dense regular connective tissue have a moderate capacity for regeneration, but skeletal muscle and cartilage have a weak regenerative capacity. Cardiac muscle and the nervous tissue in the brain and spinal cord have virtually no *functional* regenerative capacity, and cases they remain stable or even revert to normal if the environmental irritant is removed.

Diagnosis and Staging

Screening tests are vital for early detection. Examples include *mammography* (X-ray examination of breast tissue), examining testicles for lumps, examining the blood for cancer markers, and checking fecal samples for blood.

Unfortunately, most cancers are diagnosed only after symptoms have already appeared. In this case diagnosis is usually by **biopsy**: surgically removing a tissue sample and examining it microscopically for malignant cells. Increasingly, diagnosis includes chemical or genetic analysis of the sample—typing cancer cells by which genes are switched on or off.

Physical and histological examinations, lab tests, and imaging techniques (MRI, CT) can determine the extent of the disease (size of the neoplasm, degree of metastasis, etc.). Then, the cancer is assigned a **stage** from 1 to 4 according to the probability of cure. Stage 1 has the best probability of cure, stage 4 the worst.

Cancer Treatments

Most cancers are removed surgically if possible. Surgery is commonly followed by irradiation and chemotherapy. Chemotherapy is beset with the problem of resistance. Some cancer cells can eject the drugs in tiny bubbles or flattened vesicles dubbed exosomes. These resistant cells multiply, forming new tumors that are also resistant to chemotherapy. Anticancer drugs also have unpleasant side effects—nausea, vomiting, hair loss—because they kill *all* rapidly dividing cells, including normal tissue cells. The anticancer drugs also can affect the brain, producing mental fuzziness and memory loss. X rays also have side effects because, in passing through the body, they destroy healthy tissue as well as cancer cells.

Promising New Therapies

Traditional cancer treatments—"cut, burn, and poison"—are widely recognized as crude and painful. We are now learning how to hone our weapons against this disease. Treatments over the past decades have gradually become more targeted. The amount of tissue removed in surgeries has become smaller and the drugs we use have become more specific. Two treatment approaches are likely to see more widespread use:

- *Interrupting the signaling pathways that fuel cancer growth.* This approach proved its worth in a few specific types of cancer in which all of the cases share mutations in the same signaling pathway. Pioneer drugs such as imatinib dramatically improved outcomes for patients with a specific kind of leukemia. Unfortunately, most types of cancer are more diverse in the pathways that are mutated. Now that we can sequence the DNA of a patient's tumor, we can determine exactly which signaling pathways are disrupted. We can then select appropriate drugs to target those pathways in an individual's particular cancer. Such "personalized medicine" approaches are less likely to damage noncancerous tissues while being more effective at targeting cancerous cells.

- *Cancer immunotherapy.* These therapies range from antibodies that block specific

A polyp in the colon.

receptors on cancer cells, to various methods to activate the immune system to attack and kill cancer cells, to vaccines that block the development of cancer in the first place. This is an area of active research and drug development, with many clinical trials under way.

Other experimental treatments seek to starve cancer cells by cutting off their blood supply, fix defective tumor suppressor genes and oncogenes, destroy cancer cells with viruses, or signal cancer cells to commit suicide by apoptosis.

So far, average survival rates have increased only slightly, but there is much hope for the future. In the meantime, advancements continue to be made in pain management and in reducing the side effects of chemotherapy.

---

they are routinely replaced by scar tissue. However, recent studies show that some unexpected (and highly selective) cellular division occurs in both these tissues after damage, and researchers are investigating ways to encourage regeneration.

In nonregenerating tissues and in exceptionally severe wounds, fibrosis totally replaces the lost tissue. Over a period of months, the fibrous mass shrinks and becomes more and more compact. The resulting scar appears as a pale, often shiny area composed mostly of collagen fibers. Scar tissue is strong, but it lacks the flexibility and elasticity of most normal tissues, and cannot perform the functions of the tissue it has replaced.

HOMEOSTATIC IMBALANCE 4.3　　**CLINICAL**

Scar tissue that forms in the wall of the urinary bladder, heart, or other muscular organ may severely hamper the organ's function. The normal shrinking of the scar reduces the internal volume and may hinder or even block substances from moving through a hollow organ. Scar tissue hampers muscle's ability

Figure 4.16 Embryonic germ layers and the primary tissue types they produce. The three embryonic layers collectively form the very early embryonic body.

to contract and may interfere with its normal excitation by the nervous system. In the heart, these problems may lead to progressive heart failure. In irritated visceral organs, particularly following abdominal surgery, *adhesions* may form as the newly forming scar tissue connects adjacent organs together. Such adhesions can prevent the normal shifting about (churning) of loops of the intestine, dangerously obstructing the flow of foodstuffs. Adhesions can also immobilize joints.   ●●●●●●●

Check Your Understanding

20. What are the three main steps of tissue repair?

21. Why does a deep injury to the skin result in abundant scar tissue?

22. APPLY Three patients are in intensive care because of damage and widespread cell death in three different organs. One patient has brain damage from a stroke, another has had a heart attack, and a third has a damaged liver (a gland) due to a car accident. All of these patients will survive, but only one will regain full function through tissue regeneration in the damaged organ. Which one? Why?

For answers, see Answers Appendix.

Developmental Aspects of Tissues

One of the first events of embryonic development is the formation of the three **primary germ layers**, which lie one atop the next like a three-layered pancake. From superficial to deep, these layers are the **ectoderm**, **mesoderm** (mez′o-derm), and **endoderm** (**Figure 4.16**). These primary germ layers then

specialize to form the four primary tissues—epithelium, nervous tissue, muscle, and connective tissues—that make up all body organs.

By the end of the second month of development, the primary tissues have appeared, and all major organs are in place. In general, tissue cells remain mitotic and produce the rapid growth that occurs before birth. The division of nerve cells, however, stops or nearly stops during the fetal period. After birth, the cells of most other tissues continue to divide until adult body size is achieved. Cellular division then slows greatly, although many tissues retain some ability to regenerate.

In adults, only epithelia and blood-forming tissues are highly mitotic. Some tissues that regenerate through life, such as the glandular cells of the liver, do so through division of their mature (specialized) cells. Others, like the epidermis of the skin and cells lining the intestine, have abundant *stem cells*, relatively undifferentiated cells that divide as necessary to produce new cells.

Given good nutrition, good circulation, and relatively infrequent wounds and infections, our tissues normally function efficiently through youth and middle age. But with increasing age, epithelia thin and are more easily breached. Tissue repair is less efficient, and bone, muscle, and nervous tissues begin to atrophy, particularly when a person is not physically active. These events are due partly to decreased circulatory efficiency, which reduces delivery of nutrients to the tissues, but in some cases, diet is a contributing factor.

Another problem of aging tissues is the likelihood of DNA mutations in the most actively mitotic cells, which increases the risk of cancer (see **A Closer Look** on pp. 144–145).

RELATED CLINICAL TERMS

Adenoma (ad″ĕ-no′mah; *aden* = gland, *oma* = tumor) Any neoplasm of glandular epithelium, benign or malignant. The malignant type is more specifically called adenocarcinoma.

Autopsy (aw′top-se) Examination of the body, its organs, and its tissues after death to determine the cause of death; also called postmortem examination and necropsy.

Carcinoma (kar″sĭ-no′mah; *karkinos* = crab, cancer) Cancer arising in an epithelium; accounts for 90% of human cancers.

Healing by first intention The simplest type of healing; occurs when the edges of the wound are brought together by sutures, staples, or other means used to close surgical incisions. Only small amounts of granulation tissue need be formed.

Healing by second intention The wound edges remain separated, and relatively large amounts of granulation tissue bridge the gap; the manner in which unattended wounds heal. Healing is slower than in wounds in which the edges are brought together, and larger scars result.

Keloid (ke′loid) Abnormal proliferation of connective tissue during healing of skin wounds; results in large mass of scar tissue at the skin surface.

Lesion (le′zhun; "wound") Any injury, wound, or infection that affects tissue over an area of a definite size (as opposed to being widely spread throughout the body).

Marfan's syndrome Genetic disease resulting in abnormalities of connective tissues due to a defect in fibrillin, a protein that is associated with elastin in elastic fibers. Clinical signs include loose-jointedness, long limbs and spiderlike fingers and toes, visual problems, and weakened blood vessels (especially the aorta) due to poor connective tissue reinforcement.

Osteogenesis imperfecta (brittle bone disease) An inherited condition that causes defective collagen production. Because collagen reinforces many body structures including bones, the result is weak bones that break easily. It is not unusual for its victims to have 30 or more fractures during their lifetime. Occurs in 1 out of 20,000 births. Misdiagnosis results in infants coming to the ER with multiple fractures being treated as battered babies.

Pathology (pah-thol′o-je) Scientific study of changes in organs and tissues produced by disease.

Pus A collection of tissue fluid, bacteria, dead and dying tissue cells, white blood cells, and macrophages in an inflamed area.

Sarcoma (sar-ko′mah; *sarkos* = flesh; *oma* = tumor) Cancer arising in the mesenchyme-derived tissues, that is, in connective tissues and muscle.

Scurvy A nutritional deficiency caused by lack of adequate vitamin C needed to synthesize collagen; signs and symptoms include blood vessel disruption, delay in wound healing, weakness of scar tissue, and loosening of teeth.

VAC (vacuum-assisted closure) Innovative healing process for open-skin wounds and skin ulcers. Often induces healing when all other methods fail. Involves covering the wound with a special sponge, and then applying suction through the sponge. In response to the subsequent skin stretching, fibroblasts in the wound form more collagen tissue and new blood vessels proliferate, bringing more blood into the injured area, which also promotes healing.

CHAPTER SUMMARY

Tissues are collections of structurally similar cells with related functions. The four primary tissues are epithelial, connective, muscle, and nervous tissues.

4.1 Tissue samples are fixed, sliced, and stained for microscopy (p. 117)

1. Preparation of tissues for microscopic examination involves cutting thin sections of the tissue and using dyes to stain the tissue. Minor distortions called artifacts can be introduced by the tissue preparation process.

4.2 Epithelial tissue covers body surfaces, lines cavities, and forms glands (pp. 117–126)

1. Epithelial tissue is the covering, lining, and glandular tissue of the body. Its functions include protection, absorption, excretion, filtration, secretion, and sensory reception.

Special Characteristics of Epithelium (pp. 117–118)

2. Epithelial tissues exhibit specialized contacts, polarity, avascularity, support from connective tissue, and high regenerative capacity.

Classification of Epithelial Tissue (pp. 118–123)

3. Epithelium is classified by arrangement as simple (one layer) or stratified (more than one layer) and by cell shape as squamous, cuboidal, or columnar. The terms denoting cell shape and arrangement are combined to describe the epithelium fully.

4. Simple squamous epithelium is a single layer of squamous cells. Highly adapted for filtration and exchange of substances, it forms walls of air sacs of the lungs and lines blood vessels. It contributes to serosae as mesothelium and lines all hollow circulatory system organs as endothelium.

5. Simple cuboidal epithelium, commonly active in secretion and absorption, is found in glands and in kidney tubules.

6. Simple columnar epithelium, specialized for secretion and absorption, consists of a single layer of tall columnar cells that exhibit microvilli and often mucus-producing cells. It lines most of the digestive tract.

7. Pseudostratified columnar epithelium is a simple columnar epithelium that appears stratified. Its ciliated variety, rich in mucus-secreting cells, lines most of the upper respiratory passages.

8. Stratified squamous epithelium is multilayered; cells at the free surface are squamous. It is adapted to resist abrasion. It lines the esophagus and vagina; its keratinized variety forms the skin epidermis.

9. Stratified cuboidal epithelia are rare in the body, and are found chiefly in ducts of large glands. Stratified columnar epithelium has a very limited distribution, found mainly in the male urethra and at transition areas between other epithelial types.

10. Transitional epithelium is a modified stratified squamous epithelium, adapted for responding to stretch. It lines hollow urinary system organs.

Glandular Epithelia   (pp. 123–126)

11. A gland is one or more cells specialized to secrete a product.
12. On the basis of site of product release, glands are classified as exocrine or endocrine. Glands are classified structurally as multicellular or unicellular.
13. Unicellular exocrine glands, typified by goblet cells and mucous cells, are mucus-secreting single-celled glands.
14. Multicellular exocrine glands are classified according to duct structure as simple or compound, and according to the structure of their secretory parts as tubular, alveolar, or tubuloalveolar.
15. Multicellular exocrine glands of humans are classified functionally as merocrine or holocrine.

4.3   Connective tissue is the most abundant and widely distributed tissue in the body   (pp. 126–138)

1. Connective tissue functions include binding and support, protection, insulation, fat storage, and transportation (blood).

Common Characteristics of Connective Tissue   (pp. 126–127)

2. Connective tissues originate from embryonic mesenchyme and have an extensive matrix. Depending on type, a connective tissue may be well vascularized (most), poorly vascularized (dense connective tissue), or avascular (cartilage).

Structural Components of Connective Tissue   (pp. 127–129)

3. The structural elements of all connective tissues are extracellular matrix and cells.
4. The extracellular matrix consists of ground substance and fibers (collagen, elastic, and reticular). It may be fluid, gel-like, or firm.
5. Each connective tissue type has a primary cell type that can exist as a mitotic, matrix-secreting cell (-blast) or as a mature cell (-cyte) responsible for maintaining the matrix. The undifferentiated cell type of connective tissue proper is the fibroblast; that of cartilage is the chondroblast; that of bone is the osteoblast; and that of blood-forming tissue is the hematopoietic stem cell (see Chapter 17).

Types of Connective Tissue   (pp. 129–138)

6. Embryonic connective tissue is called mesenchyme.
7. Connective tissue proper consists of loose and dense varieties. The loose connective tissues are:
   - Areolar: gel-like ground substance; all three fiber types loosely interwoven; a variety of cells; forms the lamina propria and soft packing around body organs; the prototype.
   - Adipose: consists largely of adipocytes; scant matrix; insulates and protects body organs; provides reserve energy fuel. Brown fat is more important for generating body heat.
   - Reticular: finely woven reticular fibers in soft ground substance; the stroma of lymphoid organs including bone marrow.
8. Dense connective tissue proper includes:
   - Dense regular: dense parallel bundles of collagen fibers; few cells, little ground substance; high tensile strength; forms tendons, ligaments, aponeuroses; in cases where this tissue also contains numerous elastic fibers it is called elastic connective tissue.
   - Dense irregular: like regular variety, but fibers are arranged in different planes; resists tension exerted from many different directions; forms the dermis of the skin and organ capsules.
9. Cartilage exists as:
   - Hyaline: firm ground substance containing collagen fibers; resists compression well; found in embryonic skeleton, at articulating surfaces of bones, and trachea; most abundant type.
   - Elastic cartilage: elastic fibers predominate; provides flexible support of the external ear and epiglottis.
   - Fibrocartilage: parallel collagen fibers; resists both tension and compression well; forms intervertebral discs and knee cartilages.
10. Bone (osseous tissue) consists of a hard, collagen-containing matrix embedded with calcium salts; forms the bony skeleton.
11. Blood consists of blood cells in a fluid matrix (plasma).

4.4   Muscle tissue is responsible for body movement   (pp. 138–140)

1. Muscle tissue consists of elongated cells specialized to contract and cause movement.
2. Based on structure and function, the muscle tissues are:
   - Skeletal muscle: attached to and moves the bony skeleton; cells are cylindrical and striated.
   - Cardiac muscle: forms the walls of the heart; pumps blood; cells are branched and striated.
   - Smooth muscle: in the walls of hollow organs; propels substances through the organs; cells are spindle shaped and lack striations.

4.5   Nervous tissue is a specialized tissue of the nervous system   (pp. 140–141)

1. Nervous tissue forms organs of the nervous system. It is composed of neurons and supporting cells.
2. Neurons are branching cells that receive and transmit electrical impulses. They are involved in body regulation. Supporting cells support and protect neurons.

4.6   The cutaneous membrane is dry; mucous and serous membranes are wet   (pp. 141–142)

1. Covering and lining membranes are simple organs, consisting of an epithelium bound to an underlying connective tissue layer. They include mucosae, serosae, and the cutaneous membrane.

4.7   Tissue repair involves inflammation, organization, and regeneration   (pp. 142–146)

1. Inflammation is the body's response to injury. Tissue repair begins during the inflammatory process. It may lead to regeneration, fibrosis, or both.
2. Tissue repair begins with organization, during which the blood clot is replaced by granulation tissue. If the wound is small and the damaged tissue is actively mitotic, the tissue will regenerate and cover the fibrous tissue. When a wound is extensive or the damaged tissue amitotic, it is repaired only by dense connective (scar) tissue.

Developmental Aspects of Tissues   (p. 146)

1. Epithelium arises from all three primary germ layers (ectoderm, mesoderm, endoderm); muscle and connective tissue from mesoderm; and nervous tissue from ectoderm.
2. The decrease in mass and viability seen in most tissues during old age often reflects circulatory deficits or poor nutrition.

REVIEW QUESTIONS

To access additional practice questions using your smartphone, tablet, or computer: MasteringA&P® > Study Area > Practice Tests & Quizzes

Level 1 Remember/Understand

(Some questions have more than one correct answer. Select the best answer or answers from the choices given.)

1. Use the key to classify each of the following described tissue types into one of the four major tissue categories.

   Key: (a) connective tissue   (c) muscle
          (b) epithelium   (d) nervous tissue

   ____ (1) Tissue type composed largely of nonliving extracellular matrix; important in protection and support

   ____ (2) The tissue immediately responsible for body movement

   ____ (3) The tissue that enables us to be aware of the external environment and to react to it

   ____ (4) The tissue that lines body cavities and covers surfaces

2. An epithelium that has several layers, with an apical layer of flattened cells, is called (choose all that apply): (a) ciliated, (b) columnar, (c) stratified, (d) simple, (e) squamous.

3. The gland type that secretes products such as milk, saliva, bile, or sweat through a duct is (a) an endocrine gland, (b) an exocrine gland.

4. The membrane which lines body cavities that open to the exterior is a(n) (a) endothelium, (b) cutaneous membrane, (c) mucous membrane, (d) serous membrane.

5. Scar tissue is a variety of (a) epithelium, (b) connective tissue, (c) muscle tissue, (d) nervous tissue, (e) all of these.

6. Name the primary cell type in connective tissue proper; in cartilage; in bone.

7. Name the two major components of the extracellular matrix and, if applicable, subclasses of each component.

8. Name the specific connective tissue type found in the following body locations: (a) forming the soft packing around organs, (b) supporting the ear pinna, (c) forming "stretchy" ligaments, (d) first connective tissue in the embryo, (e) forming the intervertebral discs, (f) covering the ends of bones at joint surfaces, (g) main component of subcutaneous tissue.

9. What is the function of macrophages?

Level 2 Apply/Analyze

10. Define tissue.

11. Name four important functions of epithelial tissue and provide at least one example of a tissue that exemplifies each function.

12. Provide examples from the body that illustrate four of the major functions of connective tissue.

13. Matrix is extracellular. How does the matrix get to its characteristic position?

14. Differentiate between the roles of neurons and the supporting cells of nervous tissue.

15. Indicate which primary tissue classes derive from each embryonic germ layer.

16. John sustained a severe injury during football practice and is told that he has a torn knee cartilage. Can he expect a quick, uneventful recovery? Explain your response.

17. The epidermis (epithelium of the cutaneous membrane or skin) is a keratinized stratified squamous epithelium. Explain why that epithelium is much better suited for protecting the body's external surface than a mucosa consisting of a simple columnar epithelium would be.

18. Mrs. Delancy went to the local meat market and bought a beef tenderloin (cut from the loin, the region along the steer's caudal vertebral column) and some tripe (cow's stomach). What type of muscle was she preparing to eat in each case?

Level 3 Evaluate/Synthesize

19. Explain the functional classification of multicellular exocrine glands and supply an example for each class.

20. Compare and contrast skeletal, cardiac, and smooth muscle tissue relative to structure, body location, and specific function.

21. Describe the process of tissue repair, making sure you indicate factors that influence this process.

22. In what ways are adipose tissue and bone similar? How are they different?

23. In adults, over 90% of all cancers are either adenomas (adenocarcinomas) or carcinomas. (See Related Clinical Terms for this chapter.) In fact, cancers of the skin, lung, colon, breast, and prostate are all in these categories. Which one of the four basic tissue types gives rise to most cancers? Why do you think this is so?

24. Cindy, an overweight high school student, is overheard telling her friend that she's going to research how she can transform some of her white fat to brown fat. What is her rationale here (assuming it is possible)?

5 The Integumentary System

In this chapter, you will learn that

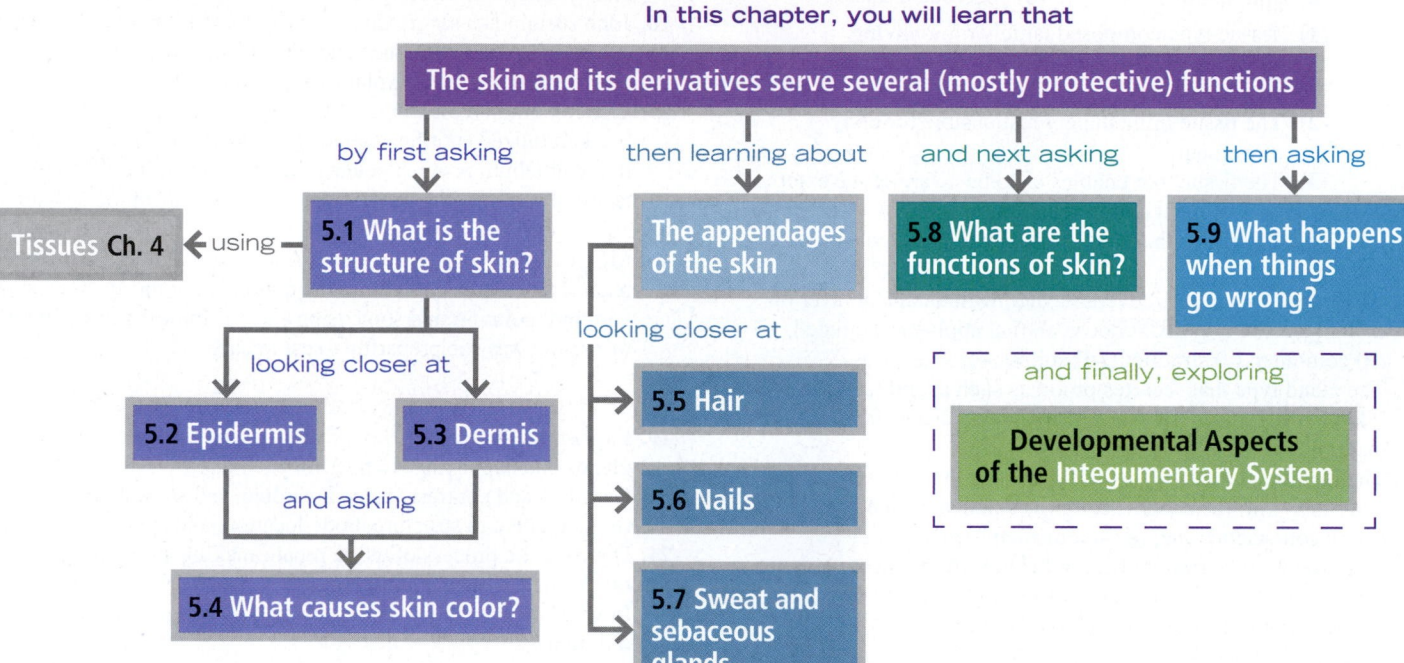

The skin and its derivatives serve several (mostly protective) functions

by first asking → Tissues Ch. 4 ← using → **5.1 What is the structure of skin?**

then learning about → The appendages of the skin

and next asking → **5.8 What are the functions of skin?**

then asking → **5.9 What happens when things go wrong?**

looking closer at:
- 5.2 Epidermis
- 5.3 Dermis

and asking:
- 5.4 What causes skin color?

looking closer at:
- 5.5 Hair
- 5.6 Nails
- 5.7 Sweat and sebaceous glands

and finally, exploring

Developmental Aspects of the Integumentary System

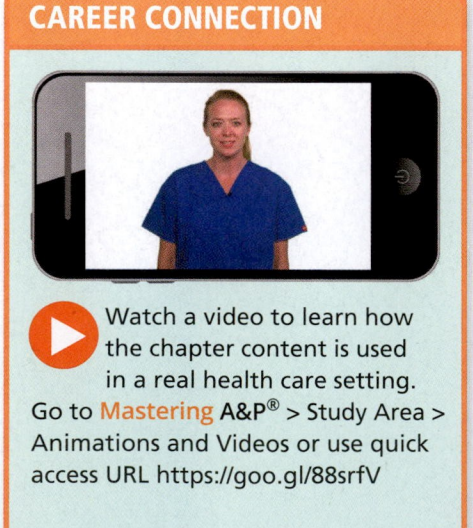
Would you be enticed by an ad for a coat that is waterproof, stretchable, washable, and air-conditioned, that automatically repairs small cuts, rips, and burns? How about one that's guaranteed to last a lifetime? Sounds too good to be true, but you already have such a coat—your skin.

The skin and its appendages (sweat and oil glands, hairs, and nails) make up a complex set of organs that serves several functions, mostly protective. Together, these organs form the **integumentary system** (in-teg″u-men′tar-e).

5.1 The skin consists of two layers: the epidermis and dermis

Learning Outcome

▶ List the two layers of skin and briefly describe subcutaneous tissue.

You may not have thought about it, but your skin is an architectural marvel. It covers the entire body, has a surface area of 1.2 to

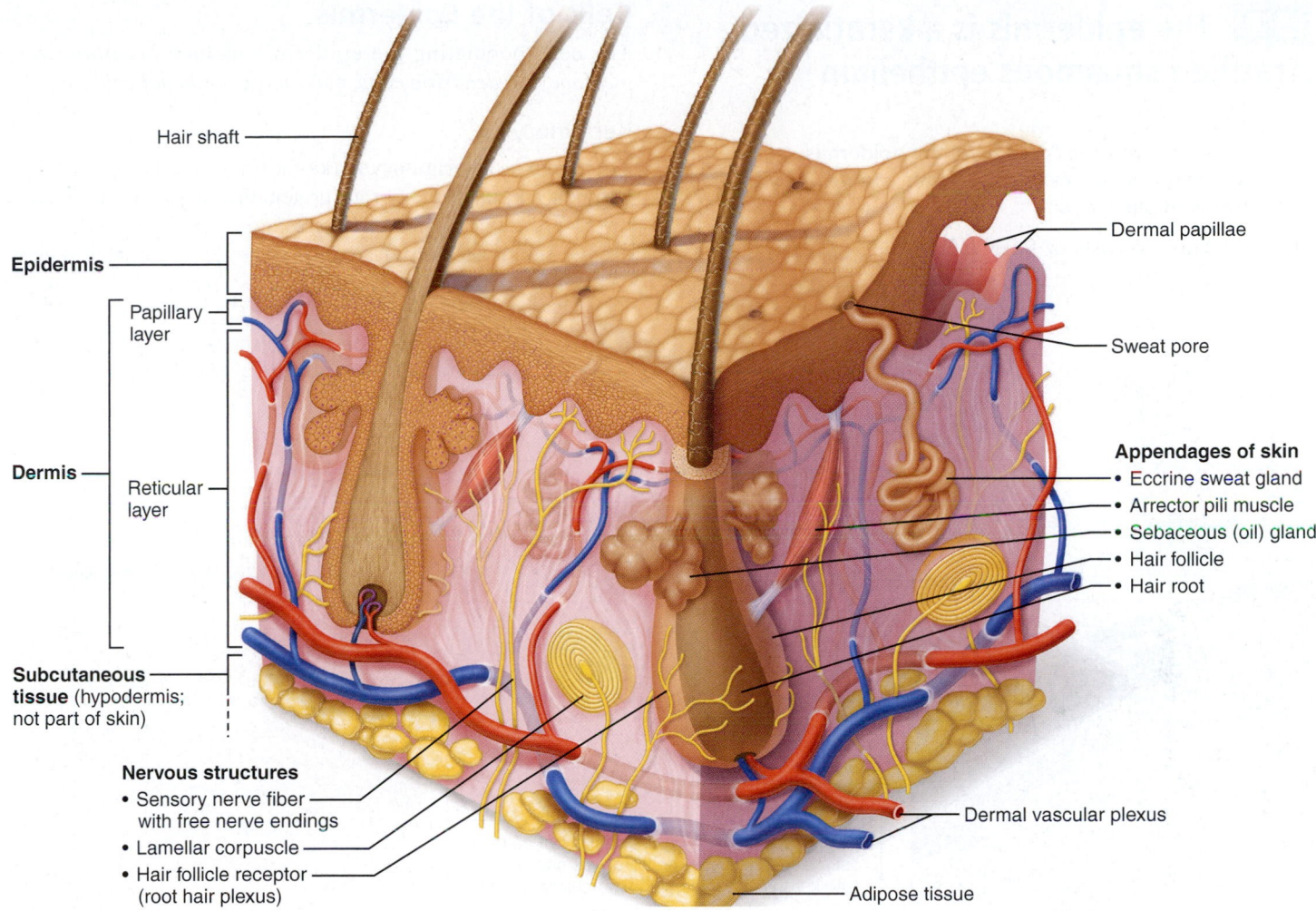

Epidermis

Dermis

    Papillary layer

    Reticular layer

Subcutaneous tissue (hypodermis; not part of skin)

Nervous structures
- Sensory nerve fiber with free nerve endings
- Lamellar corpuscle
- Hair follicle receptor (root hair plexus)

Hair shaft

Dermal papillae

Sweat pore

Appendages of skin
- Eccrine sweat gland
- Arrector pili muscle
- Sebaceous (oil) gland
- Hair follicle
- Hair root

Dermal vascular plexus

Adipose tissue

Figure 5.1 Skin structure. Three-dimensional view of the skin and underlying subcutaneous tissue. The epidermal and dermal layers have been pulled apart at the upper right corner to reveal the dermal papillae.

2.2 square meters, and accounts for about 7% of total body weight in the average adult. Also called the integument ("covering"), the skin serves as much more than a bag for body contents. Pliable yet tough, it takes constant punishment from external agents. Without our skin, we would quickly fall prey to harmful bacteria and perish from water and heat loss.

Varying in thickness from 1.5 to 4.0 millimeters (mm) or more in different parts of the body, the skin is composed of two distinct layers (**Figure 5.1**):

- The *epidermis* (ep″ĭ-der′mis), composed of epithelial cells, is the outermost protective shield of the body (*epi* = upon).
- The underlying *dermis,* making up the bulk of the skin, is a tough, leathery layer composed mostly of dense connective tissue.

Only the dermis is vascularized. Nutrients reach the epidermis by diffusing through the tissue fluid from blood vessels in the dermis.

The **subcutaneous tissue** lies just deep to the skin (Figure 5.1). It is also known as the *hypodermis* or the *superficial fascia* (fash′e-ah; "a bond") because it is superficial to the tough connective tissue wrapping (fascia) of the skeletal muscles. The subcutaneous tissue is *not* part of the skin, but it shares some of the skin's protective functions. It consists mostly of adipose tissue with some areolar connective tissue.

Besides storing fat, the subcutaneous tissue anchors the skin to the underlying structures (mostly to muscles), but loosely enough that the skin can slide relatively freely over those structures. Sliding skin protects us by ensuring that everyday bumps often just glance off our bodies. Because of its fatty composition, the subcutaneous tissue also acts as a shock absorber and an insulator that reduces heat loss.

Check Your Understanding

1. Which layer of the skin—dermis or epidermis—is better nourished?

For answers, see Answers Appendix.

5.2   The epidermis is a keratinized stratified squamous epithelium

Learning Outcomes

▶ Name the tissue type composing the epidermis.
▶ List the major layers of the epidermis and describe the functions of each layer.

The **epidermis** consists of four distinct cell types and four or five distinct layers.

Cells of the Epidermis

The cells populating the epidermis include *keratinocytes*, *melanocytes*, *dendritic cells*, and *tactile epithelial cells*.

Keratinocytes

The chief role of **keratinocytes** (kĕ-rat′ĭ-no-sītz″; "keratin cells") is to produce **keratin**, the fibrous protein that helps give the epidermis its protective properties (Greek *kera* = horn) (**Figure 5.2b**, orange cells). Most epidermal cells are keratinocytes. These

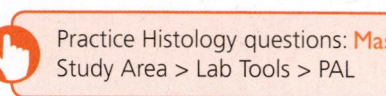

> Practice Histology questions: MasteringA&P® > Study Area > Lab Tools > PAL

Stratum corneum
Most superficial layer; 20–30 layers of dead cells, essentially flat membranous sacs filled with keratin. Glycolipids in extracellular space.

Stratum granulosum
One to five layers of flattened cells, organelles deteriorating; cytoplasm full of lamellar granules (release lipids) and keratohyaline granules.

Stratum spinosum
Several layers of keratinocytes unified by desmosomes. Cells contain thick bundles of intermediate filaments made of pre-keratin.

Stratum basale
Deepest epidermal layer; one row of actively mitotic stem cells; some newly formed cells become part of the more superficial layers. See occasional melanocytes and tactile epithelial cells.

(a)   Dermis

Keratinocytes

Dermis

Melanin granule

Sensory nerve ending

Tactile epithelial cell

Desmosomes   Melanocyte   Dendritic cell

(b)

Figure 5.2 Epidermal cells and layers of the epidermis.
(a) Photomicrograph of the four major epidermal layers in thin skin (200×).
(b) Diagram showing these four layers and the distribution of different cell types. The stratum lucidum, present only in thick skin, is not illustrated here.

keratinocytes are tied together by desmosomes for strength and, in some layers, by tight junctions to hinder movement of water between cells (◀ pp. 67–68).

Keratinocytes arise in the deepest part of the epidermis from a cell layer called the *stratum basale*. These cells undergo almost continuous mitosis in response to epidermal growth factor, a peptide produced by various cells throughout the body. Newly formed keratinocytes are pushed upward by the production of new cells beneath them, all the while making the keratin that eventually fills them. By the time the keratinocytes approach the skin surface, they are dead, scale-like flat sacs completely filled with keratin.

Millions of dead keratinocytes rub off every day, giving us a totally new epidermis every 25 to 45 days. Where the skin experiences friction, both cell production and keratin formation are accelerated. Persistent friction (from a poorly fitting shoe, for example) causes a thickening of the epidermis called a *callus*.

Melanocytes

Melanocytes (mel′ah-no-sītz), the spider-shaped epithelial cells that synthesize the pigment **melanin** (mel′ah-nin; *melan* = black), are found in the deepest layer of the epidermis (Figure 5.2b, gray cell). Melanin is made in membrane-bound granules called *melanosomes* and then transferred through the cell processes (the "spider legs") to nearby keratinocytes. As a result, the basal keratinocytes contain more melanin than do the melanocytes themselves. This melanin clusters on the superficial, or "sunny," side of the keratinocyte nucleus, forming a pigment shield that protects the nucleus from the damaging effects of ultraviolet (UV) radiation in sunlight.

Dendritic Cells

The star-shaped **dendritic cells** arise from bone marrow and migrate to the epidermis. Also called *Langerhans cells* (lahng′er-hanz), they ingest foreign substances and are key activators of our immune system, as described later in this chapter. Their slender processes extend among the surrounding keratinocytes, forming a more or less continuous network (Figure 5.2b, purple cell).

Tactile Epithelial Cells

Occasional **tactile epithelial cells** (or *Merkel cells*) are present at the epidermal-dermal junction. Shaped like a spiky hemisphere (Figure 5.2b, blue cell), each tactile epithelial cell is intimately associated with a disclike sensory nerve ending. The combination functions as a sensory receptor for touch.

Layers of the Epidermis

Variation in epidermal thickness determines if skin is *thick* or *thin*. In **thick skin**, which covers areas subject to abrasion—the palms, fingertips, and soles of the feet—the epidermis consists of five layers, or strata (stra′tah; "bed sheets"). From deep to superficial, these layers are stratum basale, stratum spinosum, stratum granulosum, stratum lucidum, and stratum corneum. In thin skin, which covers the rest of the body, the stratum lucidum appears to be absent and the other strata are thinner (Figure 5.2a, b).

Note that the terms "thick skin" and "thin skin" are really misnomers because they refer to the epidermis only. Indeed, if you include the dermis, the thickest skin is on the upper back.

Stratum Basale (Basal Layer)

The **stratum basale** (stra′tum bah-sa′le), the deepest epidermal layer, is also called the stratum germinativum (jer′mĭ-nă″tiv-um; "germinating layer"). It is attached to the underlying dermis along a wavy borderline that resembles corrugated cardboard. It consists of a single row of stem cells—a continually renewing cell population—representing the youngest keratinocytes. The many mitotic nuclei seen in this layer reflect the rapid division of these cells. Each time one of these basal cells divides, one daughter cell is pushed into the cell layer just above to begin its specialization into a mature keratinocyte. The other daughter cell remains in the basal layer to continue the process of producing new keratinocytes.

Some 10–25% of the cells in the stratum basale are melanocytes, and their branching processes extend among the surrounding cells, reaching well into the more superficial stratum spinosum layer.

Stratum Spinosum (Prickly Layer)

The **stratum spinosum** (spi″no′sum; "prickly") is several cell layers thick. This layer gets its name from the many spinelike extensions of its keratinocytes as seen under a microscope. However, these spines do not exist in living cells: They are artifacts created during tissue preparation because the cells shrink while holding tight at their many desmosomes. This makes them look like tiny versions of the spiked iron balls used in medieval warfare. Cells of the stratum spinosum contain thick bundles of intermediate filaments, which consist of a tension-resisting protein, *pre-keratin*. These intermediate filaments resist tension in the cell and are anchored to the desmosomes. Scattered among the keratinocytes are dendritic cells, which are most abundant in this epidermal layer.

Stratum Granulosum (Granular Layer)

The thin **stratum granulosum** (gran″u-lo′sum) consists of one to five cell layers in which keratinocyte appearance changes drastically, and the process of **keratinization** (in which the cells fill with keratin) begins. These cells flatten, their nuclei and organelles begin to disintegrate, and they accumulate two types of granules.

- The *keratohyaline granules* (ker″ah-to-hi′ah-lin) help to form keratin in the upper layers, as we will see.
- The *lamellar granules* (lam′el-ar; "a small plate") contain a water-resistant glycolipid that is secreted into the extracellular space. Together with tight junctions, the glycolipid plays a major part in slowing water loss across the epidermis.

Proteins within the keratinocytes and lipids deposited outside them make these cells tough and water resistant. This is why the outermost layers of skin are strong and water-tight.

Like all epithelia, the epidermis relies on capillaries in the underlying connective tissue (the dermis in this case) for its nutrients. Above the stratum granulosum, the epidermal cells are too far from the dermal capillaries to survive. In addition, the glycolipids coating their external surfaces cut them off from nutrients. As a result, they die. This is a normal sequence of events.

Stratum Lucidum (Clear Layer)

The **stratum lucidum** (loo'sid-um; "light"), found only in thick skin, is visible through a light microscope as a thin translucent band just above the stratum granulosum. It consists of a few rows of flat, dead keratinocytes. Electron microscopy reveals that its cells are identical to those at the bottom of the next layer, the stratum corneum.

Stratum Corneum (Horny Layer)

An abrupt transition occurs between the nucleated cells of the stratum granulosum and the flattened, anucleate cells of the **stratum corneum** (kor'ne-um). This outermost epidermal layer is a broad zone 20 to 30 cell layers thick that accounts for up to three-quarters of the epidermal thickness (see Figure 5.3). Keratin consists of the pre-keratin intermediate filaments embedded in a "glue" from the keratohyaline granules. Keratin and proteins that accumulate just inside the plasma membrane of cells in this stratum protect the skin against abrasion and penetration. The glycolipid between its cells helps keep this layer nearly waterproof. It is amazing that even a layer of dead cells can still perform so many functions.

The cells of the stratum corneum are shed regularly. These cells are the dandruff shed from the scalp and the flakes that come off dry skin. In a lifetime, the average person sheds 18 kg (40 lb) of these flakes (which act as fodder for the dust mites that inhabit our homes). The saying "Beauty is only skin deep" is especially interesting in light of the fact that when we look at someone, nearly everything we see is dead!

Check Your Understanding

2. While walking barefoot, Jeremy stepped on a rusty nail that penetrated the epidermis. Name the layers the nail pierced from the superficial skin surface to the junction with the dermis.
3. What is the major function of the stratum basale?
4. Why are the desmosomes and tight junctions connecting the keratinocytes so important?
5. **WHAT IF?** Suppose that the dead cells of the stratum corneum did not shed from your body, but rather stuck to it and continued to accumulate. What problems might this cause?

For answers, see Answers Appendix.

5.3 The dermis consists of papillary dermis and reticular dermis

Learning Outcome

▶ **Name the tissue types composing the dermis. List its major layers and describe the functions of each layer.**

The **dermis** (*derm* = skin) is made up of strong, flexible connective tissue. Its cells are typical of those found in any connective tissue proper: fibroblasts, macrophages, and occasional

Figure 5.3 Light micrograph of the dermis. (165×)

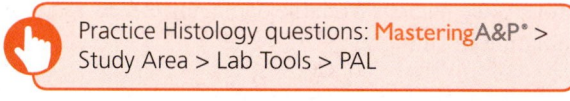

Practice Histology questions: MasteringA&P* > Study Area > Lab Tools > PAL

mast cells and white blood cells (◀ p. 129). Its semifluid matrix, embedded with fibers, binds the entire body together like a body stocking. It is your "hide" and corresponds to animal hides used to make leather.

The dermis has two layers, the papillary and reticular, which lie next to one another along an indistinct boundary (**Figure 5.3**).

Papillary Dermis

The thin, superficial **papillary dermis** (pap'il-er-e) is areolar connective tissue in which fine interlacing collagen and elastic fibers form a loosely woven mat with many small blood vessels. The looseness of this connective tissue allows phagocytes and other defensive cells to wander freely as they patrol the area for bacteria that have penetrated the skin.

Peglike projections from its surface, called **dermal papillae** (pah-pil'e; *papill* = nipple), indent the overlying epidermis (Figures 5.1 and 5.3). Many dermal papillae contain capillary loops. Others house free nerve endings (pain receptors) and touch receptors called *tactile corpuscles* or *Meissner's corpuscles* (mīs'nerz kor'pus-lz). (Note that tactile epithelial cells and tactile corpuscles are different structures.) In thick skin, such as the palms of the hands and soles of the feet, these papillae lie atop larger mounds called *dermal ridges*, which in turn cause the overlying epidermis to form *epidermal ridges* (**Figure 5.4c**).

Collectively, these skin ridges, referred to as **friction ridges**, may enhance our ability to grip certain kinds of surfaces. They also contribute to our sense of touch by enhancing vibrations detected by the large *lamellar corpuscles* (receptors) in the dermis.

Friction ridge patterns are genetically determined and unique to each of us. Because sweat pores open along their crests, our fingertips leave identifying films of sweat called *fingerprints* on almost anything we touch.

Reticular Dermis

The deeper **reticular dermis**, accounting for about 80% of the thickness of the dermis, is coarse, dense irregular connective tissue (Figure 5.3). The network of blood vessels that nourishes this layer, the *dermal vascular plexus*, lies between this layer and the subcutaneous tissue (see Figure 5.1). The extracellular matrix of the reticular dermis contains thick bundles of interlacing collagen fibers. The reticular dermis is named for its network of collagen fibers (*reticulum* = network); the name does not imply any special abundance of reticular fibers.

The collagen fibers run in various planes, but most run parallel to the skin surface. Separations, or less dense regions, between these bundles form **cleavage (tension) lines** in the skin. These externally invisible lines tend to run longitudinally in the skin of the limbs and in circular patterns around the neck and trunk (Figure 5.4a). Cleavage lines are important to surgeons because when an incision is made *parallel* to these lines, the skin gapes less and heals more readily.

The collagen fibers of the dermis give skin strength and resiliency that prevent minor jabs and scrapes from penetrating the dermis. Elastic fibers provide the stretch-recoil properties of skin.

Flexure lines are dermal folds that occur at or near joints, where the dermis is tightly secured to deeper structures. (Notice the deep creases on your palms.) Since the skin cannot slide easily to accommodate joint movement in such regions, the dermis folds and deep skin creases form (Figure 5.4b). Flexure lines are also visible on the wrists, fingers, soles, and toes.

HOMEOSTATIC IMBALANCE 5.1   CLINICAL

Extreme stretching of the skin, such as during pregnancy, can tear the dermis, leaving silvery white scars called *striae* (stri′e;

Cleavage lines
- Represent separations between underlying collagen fiber bundles in the *reticular dermis*.
- Run circularly around the trunk and longitudinally in the limbs.
- Surgical incisions parallel to cleavage lines heal better than those made across them.

Flexure lines form where the dermis is closely attached to the underlying structures.

Friction ridges consist of epidermal ridges that lie on top of dermal ridges.

Sweat duct openings along the crests of friction ridges help make fingerprints.

(a) Cleavage lines

(b) Flexure lines

(c) Friction ridges of fingertip (SEM 12×)

Figure 5.4 Dermal modifications result in characteristic skin markings.

Practice Histology questions: MasteringA&P° > Study Area > Lab Tools > PAL

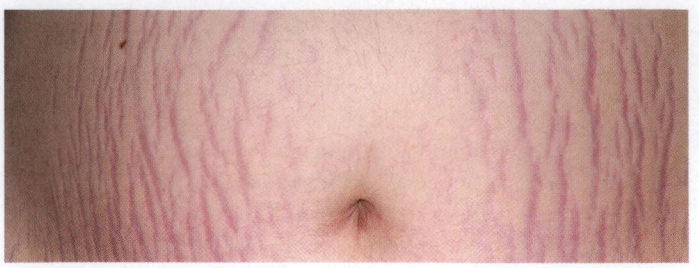

Figure 5.5 Stretch marks (striae).

"streaks"), commonly called *stretch marks* (**Figure 5.5**). Short-term but acute trauma (as from a burn or from repeated friction when wielding a hoe) can cause a *blister*, a fluid-filled pocket that separates the epidermal and dermal layers. ●●●●●●●

Check Your Understanding

6. What type of tissue makes up the papillary and reticular dermis?

7. **APPLY** You have just gotten a paper cut. It is very painful, but it doesn't bleed. Has the cut penetrated into the dermis or just the epidermis?

8. **PREDICT** Stefan's front bike tire slipped on a patch of gravel as he rode to work. He needed five stitches to close the cut above his left eye. The split skin ran along a cleavage line. Is Stefan likely to have a major scar or is the wound likely to heal cleanly? Explain.

9. **DRAW** Draw a wavy line to represent the junction between the dermis and epidermis. Draw and label the layer of epidermal cells next to the dermis. Label the layer of dermis that is next to these epidermal cells and label the projections from this layer that indent the epidermis.

For answers, see Answers Appendix.

5.4 Melanin, carotene, and hemoglobin determine skin color

Learning Outcomes

▶ **Describe the factors that normally contribute to skin color.**

▶ **Briefly describe how changes in skin color may be used as clinical signs of certain disease states.**

Melanin is a polymer made of an amino acid called tyrosine. Its two forms range in color from reddish yellow to brownish black. Melanin synthesis depends on an enzyme in melanocytes called tyrosinase (ti-ro′sĭ-nās). As noted earlier, melanin is transported from melanocytes to the basal keratinocytes. Eventually, lysosomes break down the melanosomes, so melanin pigment is found only in the deeper layers of the epidermis.

Human skin comes in different colors. However, distribution of those colors is not random—populations of darker-skinned people tend to be found nearer the equator (where greater protection from the sun is needed), and those with the lightest skin are found closer to the poles. Since all humans have the same relative number of melanocytes, differences in skin coloring reflect

the kind and amount of melanin made and retained. Melanocytes of black- and brown-skinned people produce many more and darker melanosomes than those of fair-skinned individuals, and their keratinocytes retain it longer. *Freckles* and *pigmented nevi* (moles) are local accumulations of melanin.

When we expose our skin to sunlight, keratinocytes secrete chemicals that stimulate melanocytes. Prolonged sun exposure causes a substantial melanin buildup, which helps protect the DNA of skin cells from UV radiation by absorbing the rays and dissipating the energy as heat. Indeed, the initial signal for speeding up melanin synthesis seems to be a faster repair rate of DNA that has suffered photodamage (*photo* = light). In all but the darkest-skinned people, this defensive response causes skin to darken visibly (tanning occurs).

HOMEOSTATIC IMBALANCE 5.2 — CLINICAL

Despite melanin's protective effects, excessive sun exposure eventually damages the skin. It causes elastic fibers to clump, which results in leathery skin; temporarily depresses the immune system; and can alter the DNA of skin cells, leading to skin cancer. Although darker-skinned people are less likely to get melanoma, the deadliest form of skin cancer, they are not immune and still need to have suspicious skin changes evaluated. Basal cell and squamous cell cancers occur in all skin types.

Many chemicals induce photosensitivity; that is, they increase the skin's sensitivity to UV radiation and can cause an unsightly skin rash. Such substances include some antibiotic and antihistamine drugs, and many chemicals in perfumes and detergents. Small, itchy blisters erupt all over the body. ●●●●●●●

Carotene (kar′o-tēn) is a yellow to orange pigment found in certain plant products such as carrots. It tends to accumulate in the stratum corneum and in the fat of the subcutaneous tissue. In the body, carotene can be converted to vitamin A, a vitamin that is essential for normal vision, as well as for epidermal health.

The pinkish hue of fair skin reflects the crimson color of the oxygenated pigment **hemoglobin** (he′mo-glo″bin) in the red blood cells circulating through the dermal capillaries. Because light-skinned people have only small amounts of melanin in their skin, the epidermis is nearly transparent and allows hemoglobin's color to show through.

HOMEOSTATIC IMBALANCE 5.3 — CLINICAL

Health care providers quickly glean important information about a person's health from the color of their skin. Alterations in skin color provide clues about certain disease states or even emotional states:

● *Blueness, or cyanosis* (si″ah-no′sis; *cyan* = dark blue): When hemoglobin is poorly oxygenated, the skin may take on a bluish-gray tint, especially in the mucous membranes and nail beds. Cyanosis can be a sign of respiratory or cardiovascular problems.

- *Pallor*, or *blanching*: During emotional stress, blood may be diverted from the skin to internal organs, causing the person to look "as white as a sheet." Low blood pressure and *anemia* (decreased ability of blood to carry oxygen) may also cause pallor.

- *Redness*, or *erythema* (er″ĭ-the′mah): Reddened skin may indicate embarrassment (blushing), fever, inflammation, or allergy.

- *Yellowness*: Normally the liver eliminates bilirubin, a waste from red cell breakdown. This yellow pigment accumulates in body tissues when the liver doesn't function properly. The result is *jaundice* (jawn′dis), a yellow discoloration most obvious in the sclerae, or whites, of the eyes. Eating large amounts of yellow vegetables, which contain carotene, may cause golden skin, but the sclerae remain white.

- *Red/purple/green/yellow marks*: Bruises, also called ecchymoses (ek″ĭ-mo′sēs) or hematomas (he″mah-to′mah; "blood swelling"), occur when blood vessels are damaged and leak blood into the surrounding tissue. As the body breaks down the blood, the color of the bruise changes. The size, color, and shape of ecchymoses is important in forensic medicine.

- *Brown or black "necklace,"* or *bruises*: Dark areas in the axillae and around the neck are sometimes mistaken for a lack of cleanliness. These areas of velvety hyperpigmentation may be a sign of insulin resistance and elevated blood glucose.   ●●●●●●●

Check Your Understanding

10. Melanin and carotene are two pigments that contribute to skin color. What is the third and where is it found?

11. Which alteration in skin color may indicate a liver disorder?

For answers, see Answers Appendix.

Along with the skin itself, the integumentary system includes a number of associated structures. These **skin appendages** include hair and hair follicles, nails, sweat glands, and sebaceous (oil) glands. Each plays a unique role in maintaining body homeostasis. Although they all derive from the epithelial cells of the epidermis (i.e., they are *epidermal derivatives*), they all extend into the dermis. We will examine them in the next three modules.

5.5 Hair consists of dead, keratinized cells

Learning Outcomes

▶ **List the parts of a hair follicle and explain the function of each part. Also describe the functional relationship of arrector pili muscles to the hair follicles.**

▶ **Name the regions of a hair and explain the basis of hair color. Describe the distribution, growth, replacement, and changing nature of hair during the life span.**

Millions of hairs are distributed over our entire skin surface except our palms, soles, lips, nipples, and parts of the external genitalia (such as the head of the penis). Although hair helps to keep other mammals warm, our sparse body hair is far less luxuriant and useful. Its main function in humans is to sense insects on the skin before they bite or sting us. Hair on the scalp guards the head against physical trauma, heat loss, and sunlight. Eyelashes shield the eyes, and nose hairs filter large particles like lint and insects from the air we inhale.

Together, hairs and their follicles form complex structural units. In these units, the *hairs* are the long filaments, and the *hair follicles* are tubular invaginations of the epidermis from which the hairs grow.

Structure of a Hair

Hairs, or **pili** (pi′li), are flexible strands produced by hair follicles and consist largely of dead, keratinized cells. The *hard keratin* that dominates hairs and nails has two advantages over the *soft keratin* found in typical epidermal cells: (1) It is tougher and more durable, and (2) its individual cells do not flake off.

Follicle wall
• Peripheral
 connective tissue
 (fibrous) sheath
• Glassy membrane
• Epithelial root sheath
 • External root sheath
 • Internal root sheath

Hair
• Cuticle
• Cortex
• Medulla

(a) Diagram of a cross section of a hair within its follicle

**(b) Photomicrograph of a cross section of a
hair and hair follicle (100×)**

Hair shaft

Arrector
pili

Sebaceous
gland

Hair root

Hair bulb

Follicle wall
• Peripheral
 connective tissue
 (fibrous) sheath
• Glassy membrane
• Epithelial root sheath
 • External root sheath
 • Internal root sheath

Hair root
• Cuticle
• Cortex
• Medulla

Hair matrix

Hair papilla

Melanocyte

Subcutaneous
adipose tissue

**(c) Diagram of a longitudinal view of the expanded hair
bulb of the follicle, which encloses the matrix**

**(d) Photomicrograph of longitudinal view
of the hair bulb in the follicle (150×)**

Figure 5.6 **Structure of a hair and hair follicle.**

The chief regions of a hair are the *root*, the part embedded in the skin, and the *shaft*, the part that projects above the skin's surface (**Figure 5.6**). If the shaft is flat and ribbonlike in cross section, the hair is kinky; if it is oval, the hair is silky and wavy; if it is perfectly round, the hair is straight and tends to be coarse.

A hair has three concentric layers of keratinized cells: the medulla, cortex, and cuticle (Figure 5.6a, b).

• The *medulla* (mĕ-dul′ah; "middle"), its central core, consists of large cells and air spaces. The medulla, the only part of the hair that contains soft keratin, is absent in fine hairs.

- The *cortex*, a bulky layer surrounding the medulla, consists of several layers of flattened cells.

- The outermost **cuticle** is formed from a single layer of cells overlapping one another like shingles on a roof. This arrangement helps separate neighboring hairs so the hair does not mat. (Hair conditioners smooth out the rough surface of the cuticle and make hair look shiny.) The cuticle is the most heavily keratinized part of the hair, providing strength and keeping the inner layers tightly compacted. Because it is subjected to the most abrasion, the cuticle tends to wear away at the tip of the hair shaft. This allows keratin fibrils in the cortex and medulla to frizz, creating "split ends."

Hair pigment is made by melanocytes at the base of the hair follicle and transferred to the cortical cells. Different concentrations of melanin produce hair color from blond to brown to black. Red hair is colored by a related pigment called *pheomelanin*. When melanin production decreases and air bubbles replace melanin in the hair shaft, hair turns gray or white.

Structure of a Hair Follicle

Hair follicles (*folli* = bag) fold down from the epidermal surface into the dermis (and sometimes into the subcutaneous tissue). The deep end of the follicle, located about 4 mm (1/6 in.) below the skin surface, expands to form a **hair bulb** (Figure 5.6c, d). A knot of sensory nerve endings called a **hair follicle receptor**, or **root hair plexus**, wraps around each hair bulb (see Figure 5.1). Bending the hair stimulates these endings. Consequently, our hairs act as sensitive touch receptors.

A *papilla of a hair follicle* (or *hair papilla*) is a dermal papilla that protrudes into the hair bulb. This papilla contains a knot of capillaries that supplies nutrients to the growing hair and signals it to grow. If the hair papilla is destroyed by trauma, the follicle permanently stops producing hair.

The wall of a hair follicle has a dermal component and an epidermal component. These components are described from external to internal (Figure 5.6a–d).

- **Peripheral connective tissue sheath** (*fibrous sheath*). This connective tissue sheath is derived from the dermis. It forms the external layer of the follicle wall.

- **Glassy membrane.** The glassy membrane is at the junction of the fibrous sheath and the epithelial root sheath. It is, in essence, the basement membrane of the follicle epithelium.

- **Epithelial root sheath.** The epithelial root sheath is derived from the epidermis. It has two components: The *external root sheath*, a direct continuation of the epidermis, and the *internal root sheath*, which is derived from the matrix cells (see below).

Hair grows because cells in the bulb of the follicle divide rapidly. These dividing cells make up the **hair matrix** and lie immediately adjacent to the hair papilla. As the matrix produces new hair cells, the older part of the hair is pushed upward, and its fused cells become increasingly keratinized and die. As they divide, hair matrix cells are replenished by stem cells that migrate down to the bulb from a region closer to the skin surface called the *hair bulge* (not shown in Figure 5.6).

Associated with each hair follicle is a bundle of smooth muscle cells called an **arrector pili** (ah-rek′tor pi′li; "raiser of hair") muscle. As you can see in Figure 5.1, most hair follicles approach the skin surface at a slight angle. The arrector pili muscle is attached in such a way that its contraction pulls the hair follicle upright and dimples the skin surface to produce goose bumps in response to cold temperatures or fear. This "hair-raising" response is not very useful to humans, with our short sparse hairs, but it is an important way for other animals to retain heat and protect themselves. Furry animals stay warmer by trapping a layer of insulating air in their fur; and a scared animal with its hair on end looks larger and more formidable to its enemy. The more important role of the arrector pili in humans is that its contractions force sebum out of hair follicles to the skin surface where it acts as a skin lubricant.

Types and Growth of Hair

Hairs can be classified as vellus or terminal. The body hair of children and adult females is pale, fine **vellus hair** (vel′us; *vell* = wool, fleece). The coarser, longer hair of the eyebrows, eyelashes, and scalp is **terminal hair**, which may also be darker.

At puberty, terminal hairs appear in the axillary and pubic regions of both sexes and on the face and chest (and typically the arms and legs) of males. These terminal hairs grow in response to the stimulating effects of *androgens* (male sex hormones of which *testosterone* is the most important).

HOMEOSTATIC IMBALANCE 5.4   CLINICAL

When women develop coarse terminal hairs in a masculine distribution, the condition is called **hirsutism** (her′soot-izm; *hirsut* = hairy). Hirsutism is a cosmetic problem that may signal an underlying medical issue. In *polycystic ovary syndrome*, the ovaries secrete larger amounts of androgens, stimulating hair growth. Less frequently, ovarian or adrenal tumors can secrete androgens. In many cases, no clinical problem is found. ● ● ● ● ● ●

Hairs grow an average of 2 mm per week, although this rate varies widely among body regions as well as with sex and age. Each follicle goes through *growth cycles*. In each cycle, an active growth phase is followed by a resting phase in which the hair matrix is inactive and the follicle shinks somewhat. At the start of each active phase, the newly growing hair pushes out the old hair, which is shed. The life span of hairs varies. In the scalp, the follicles stay active for an average of four years, so individual hairs grow quite long before being shed. The follicles in eyebrows, in contrast, are active for only a few months so the eyebrows never grow very long. Fortunately, the cycles of adjacent hair follicles are not synchronized. For this reason, humans shed only a small percentage of their hairs at any one time.

Hair Thinning and Baldness

Given ideal conditions, hair grows fastest from the teen years to the 40s. When hairs are no longer replaced as quickly as they are shed, the hair thins. By age 60 to 65, both sexes usually experience some degree of balding. Coarse terminal hairs are replaced by vellus hairs, and the hair becomes increasingly wispy.

True baldness is different. The most common type is **male pattern baldness**, a genetically determined, gender-influenced condition that changes the response of the hair follicles to androgens. The hair follicles respond to androgens with shorter and shorter growth cycles. The cycles become so short that many hairs never even emerge from their follicles before being shed, and those that do are fine vellus hairs that look like peach fuzz in the "bald" area. The drugs used to treat male pattern baldness either inhibit the production of androgens or increase blood flow to the skin and hair follicles. These treatments are only partly successful.

HOMEOSTATIC IMBALANCE 5.5

CLINICAL

Abrupt hair thinning may be distressing, and determining its cause can be challenging. *Telogen effluvium* (ef-floo′ve-um), or TE, occurs when too many hair follicles enter the resting phase at the same time. TE often occurs after a "shock to the system," such as surgery, a car crash, hormonal changes with childbirth, crash dieting, or emotional stress.

Hair loss or a change in texture may also signal low thyroid hormone levels. Drugs, including some antidepressants and cancer chemotherapy drugs, may cause hair thinning. Finally, severe dietary deficiencies of protein or minerals may result in thin, brittle hair.

Check Your Understanding

12. Why is having your hair cut painless?
13. Use the image below to answer the following questions. What are the names of the parts of the hair labeled a and c? What are the names of structures b and d and what is the function of each of these structures?

14. **DRAW** Draw a cross section of a hair shaft and label its three concentric regions.

For answers, see Answers Appendix.

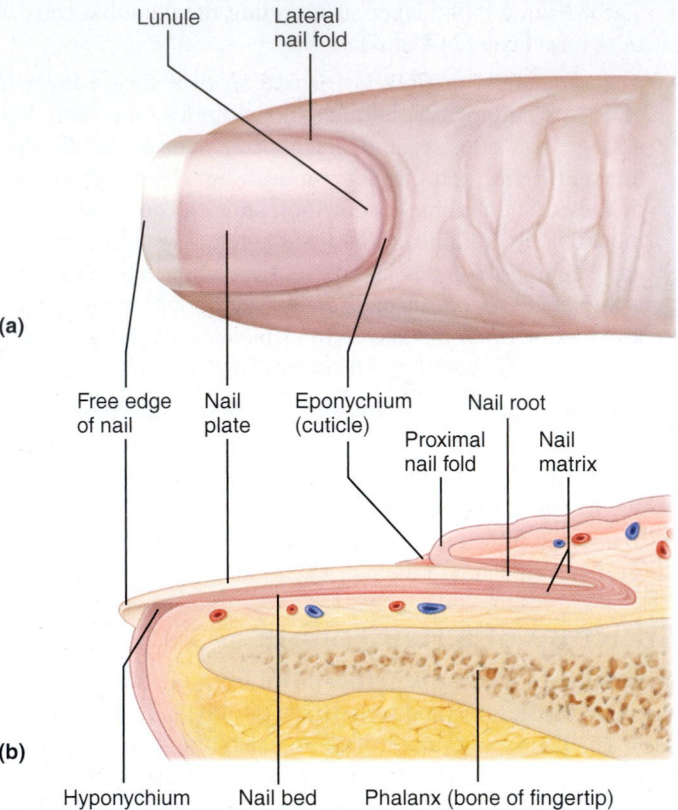

(a)

(b)

Figure 5.7 **Structure of a nail. (a)** Surface view of the distal part of a finger. **(b)** Sagittal section of the fingertip. The nail matrix that forms the nail lies beneath the lunule.

5.6 Nails are scale-like modifications of the epidermis

Learning Outcome

▶ **Describe the structure of nails.**

A **nail** forms a clear protective covering on the dorsal surface of the distal part of a finger or toe (**Figure 5.7**). Nails correspond to the hooves or claws of other animals, and are useful as "tools" to help pick up small objects or scratch an itch.

In contrast to *soft keratin* of the epidermis, nails (like hairs) contain *hard keratin*. Each nail has a proximal *root* (embedded in the skin), a *nail plate* or *body* (visible attached portion), and a *free edge*. The nail rests on a bed of epidermis called the *nail bed*. This bed contains only the deeper layers of the epidermis, because the nail itself corresponds to the superficial keratinized layers. The thickened proximal portion of the nail bed, the **nail matrix**, is responsible for nail growth. As the nail cells produced by the matrix become heavily keratinized, the nail body slides distally over the nail bed.

Nails normally appear pink because of the rich bed of capillaries in the underlying dermis. However, the region that lies over the thick nail matrix appears as a white crescent called the *lunule* (lu′nool; "little moon"). The proximal and lateral borders of the nail are overlapped by skin folds, called **nail folds**. The proximal nail fold projects onto the nail body as the **cuticle** or **eponychium** (ep″o-nik′e-um; "on the nail"). The thickened region beneath the free edge of the nail where dirt and debris tend to accumulate is the **hyponychium** ("below nail"). It secures the free edge of the nail plate at the tip of the finger or toe.

HOMEOSTATIC IMBALANCE 5.6

Changes in nail appearance can help diagnose certain conditions. For example, yellow-tinged nails may indicate a respiratory or thyroid gland disorder. (Thickened yellow nails are usually due to a fungus infecting the nail.) An outward concavity of the nail (*koilonychia* or "spoon nail," **Figure 5.8**) may signal an iron deficiency. Horizontal lines (Beau's lines) across the nails can be a sign of severe illness that affects the whole body such as uncontrolled diabetes, a heart attack, or cancer chemotherapy.

Figure 5.8 Koilonychia.

Check Your Understanding

15. Why is the lunule of a nail white instead of pink like the rest of the nail?
16. Why are nails so hard?

For answers, see Answers Appendix.

5.7 Sweat glands help control body temperature, and sebaceous glands secrete sebum

Learning Outcomes

▶ Compare the structure and locations of sweat and oil glands. Also compare the composition and functions of their secretions.
▶ Compare and contrast eccrine and apocrine glands.

Sweat glands, also called *sudoriferous glands* (su″do-rif′er-us; *sudor* = sweat), are distributed over the entire skin surface except the nipples and parts of the external genitalia. Their number is staggering—up to 3 million per person.

We have two types of sweat glands: eccrine and apocrine. In both types, the secretory cells are associated with *myoepithelial cells*, specialized cells that contract when stimulated by the nervous system. Their contraction forces the sweat into and through the gland's duct system to the skin surface. Sweating is regulated by a branch of the autonomic (involuntary) nervous system.

Eccrine Sweat Glands

Eccrine sweat glands (ek′rin; "secreting"), also called *merocrine sweat glands*, are far more numerous than apocrine sweat glands and are particularly abundant on the palms, soles of the feet, and forehead. Each is a simple, coiled, tubular gland. The secretory part lies coiled in the dermis, and the duct extends to open in a funnel-shaped *pore* (*por* = channel) at the skin surface (**Figure 5.9b**). (These sweat pores are different from the

(a) Sebaceous gland (90×)

(b) Eccrine gland (140×)

Figure 5.9 Cutaneous glands.

Practice Histology questions: MasteringA&P® > Study Area > Lab Tools > PAL

Table 5.1 Summary of Cutaneous Glands

	ECCRINE SWEAT GLANDS	APOCRINE SWEAT GLANDS	SEBACEOUS GLANDS
Functions	• Temperature control • Some antibacterial properties	May act as sexual scent glands	• Lubricate skin and hair • Help prevent water loss • Antibacterial properties
Type of Secretion	Hypotonic filtrate of blood plasma	Filtrate of blood plasma with added proteins and fatty substances	Sebum (an oily secretion)
Method of Secretion	Merocrine (exocytosis)	Merocrine (exocytosis)	Holocrine
Secretion Exits Duct At	Skin surface	Usually upper part of hair follicle; rarely, skin surface	Usually upper part of hair follicle; sometimes, skin surface
Body Location	Everywhere, but especially palms, soles, forehead	Mostly axillary and anogenital regions	Everywhere *except* palms and soles

so-called pores of a person's complexion, which are openings of hair follicles.)

Eccrine gland secretion, commonly called sweat, is a hypotonic filtrate of the blood that passes through the secretory cells of the sweat glands and is released by exocytosis. It is 99% water, with some salts (mostly sodium chloride), traces of metabolic wastes (urea, uric acid, and ammonia), and a microbe-killing peptide called *dermcidin*. Normally, sweat is acidic with a pH between 4 and 6.

Sweating's major role is to prevent the body from overheating. Heat-induced sweating begins on the forehead and spreads inferiorly over the remainder of the body. Emotionally induced sweating—the so-called "cold sweat" brought on by fright or nervousness—begins on the palms, soles, and axillae (armpits) and then spreads to other body areas.

Apocrine Sweat Glands

The approximately 2000 **apocrine sweat glands** (ap′o-krin) are largely confined to the axillary and anogenital areas. In spite of their name, they are merocrine glands, which release their product by exocytosis like the eccrine sweat glands. Larger than eccrine glands, they lie deeper in the dermis or even in the subcutaneous tissue, and their ducts empty into hair follicles.

Apocrine secretion contains the same basic components as true sweat, plus fatty substances and proteins. Consequently, it is viscous and sometimes has a milky or yellowish color. The secretion is odorless, but when bacteria on the skin decompose its organic molecules, it takes on a musky and generally unpleasant odor, the basis of body odor.

Apocrine glands begin functioning at puberty under the influence of the male sex hormones (androgens) and play little role in maintaining a constant body temperature. Their precise function is not yet known. Three lines of evidence suggest that they may be the human equivalent of other animals' sexual scent glands: (1) Sexual foreplay increases their activity; (2) they enlarge and recede with the phases of a woman's menstrual cycle; and (3) behavioral studies show that their secretions may act as pheromones (chemical messengers released by one individual that trigger a response in other members of the same species).

Two important types of modified apocrine glands are:

- **Ceruminous glands. Ceruminous glands** (sĕ-roo′mĭ-nus; *cera* = wax) are modified apocrine glands found in the lining of the external ear canal. Their secretion mixes with sebum produced by nearby sebaceous glands to form a sticky, bitter substance called *cerumen*, or earwax, that is thought to deter insects and block entry of foreign material.

- **Mammary glands. Mammary glands**, another type of specialized sweat gland, secrete milk. Although they are properly part of the integumentary system, we will consider the mammary glands in Chapter 27 with female reproductive organs.

Sebaceous Glands

The **sebaceous glands** (se-ba′shus; "greasy"), or *oil glands* (Figure 5.9a), are simple branched alveolar glands that are found all over the body except in the thick skin of the palms and soles. They are small on the body trunk and limbs, but quite large on the face, neck, and upper chest. These glands secrete an oily substance called **sebum** (se′bum). The central cells of the alveoli accumulate oily lipids until they become so engorged that they burst, so functionally these glands are *holocrine glands* (◄ p. 126). The accumulated lipids and cell fragments constitute sebum.

Most, but not all, sebaceous glands develop as outgrowths of hair follicles and secrete sebum into a hair follicle, or occasionally to a pore on the skin surface. Arrector pili contractions force sebum out of the hair follicles to the skin surface.

Sebum softens and lubricates the hair and skin, prevents hair from becoming brittle, and slows water loss from the skin. Perhaps even more important is its *bactericidal* (bacterium-killing) action. Sebaceous glands increase their activity during puberty under the influence of male sex hormones.

HOMEOSTATIC IMBALANCE 5.7
CLINICAL

Acne is an active inflammation of the sebaceous glands accompanied by "pimples" (pustules or cysts), whiteheads, or blackheads on the skin. If accumulated sebum blocks a sebaceous gland duct, it can cause a whitehead (*closed comedone*) or a blackhead (*open comedone*). Acne is associated with infection by *Propionibacterium acne*. It can range from mild to so severe that it leads to permanent scarring.

Overactive sebaceous glands can cause *seborrhea* (seb″o-re′ah; "fast-flowing sebum"), known as "cradle cap" in infants (**Figure 5.10**). Seborrhea begins on the scalp as pink, raised lesions that gradually become yellow to brown and begin to slough off oily scales.

Figure 5.10 Cradle cap (seborrhea) in a newborn.

Table 5.1 summarizes the three types of cutaneous glands.

Check Your Understanding

17. Which cutaneous glands are associated with hair follicles?
18. When Anthony returned home from a run in 85°F weather, his face was dripping with sweat. Why?
19. What is the difference between heat-induced sweating and a "cold sweat," and which variety of sweat gland is involved?
20. **APPLY** Sebaceous glands are not found in thick skin. Why is their absence in those body regions desirable?

For answers, see Answers Appendix.

5.8 First and foremost, the skin is a barrier

Learning Outcome

▶ **Describe how the skin accomplishes at least five different functions.**

Like the skin of a grape, our skin keeps its contents juicy and whole. The skin and its appendages perform a variety of functions, including protection, body temperature regulation, cutaneous sensation, metabolic functions, blood reservoir, and excretion.

Protection

Given its superficial location, the skin is our most vulnerable organ system, exposed to microorganisms, abrasion, temperature extremes, harmful chemicals, and UV radiation. The skin constitutes at least three types of barriers: chemical, physical, and biological.

Chemical Barriers

Chemical barriers include skin secretions and melanin. Although the skin's surface teems with bacteria, the low pH of skin secretions—the **acid mantle**—retards their multiplication. In addition, dermcidin in sweat and bactericidal substances in sebum kill many bacteria outright. Skin cells also secrete natural antibiotics called *defensins* that literally punch holes in bacteria, making them leak like sieves. Wounded skin releases large quantities of protective peptides called *cathelicidins* that are particularly effective in preventing infection by group A streptococcus bacteria.

As discussed earlier, melanin provides a chemical pigment shield to prevent UV damage to skin cells.

Physical Barriers

The continuity of skin and the hardness of its keratinized cells provide physical barriers. As a physical barrier, the skin is a remarkable compromise. A thicker epidermis would be more impenetrable, but we would pay the price in loss of suppleness and agility.

The outstanding barrier capacity of the skin arises from the structure of its stratum corneum, which has been compared to bricks and mortar. Multiple layers of dead flat cells are the bricks and the glycolipids surrounding them are the mortar. Epidermal continuity works hand in hand with the acid mantle and chemicals in skin secretions to ward off bacterial invasion. The water-resistant glycolipids of the epidermis block most diffusion of water and water-soluble substances between cells, preventing both their loss from and entry into the body through the skin. However, there is a continual small loss of water through the epidermis.

Substances that *do* penetrate the skin in limited amounts include:

- *Lipid-soluble substances*, such as oxygen, carbon dioxide, fat-soluble vitamins (A, D, E, and K), and steroids (estrogens)
- *Oleoresins* (o″le-o-rez′inz) of certain plants, such as poison ivy and poison oak
- *Organic solvents*, such as acetone, dry-cleaning fluid, and paint thinner, which dissolve the cell lipids
- *Salts of heavy metals*, such as lead and mercury
- Selected drugs (nitroglycerine, seasickness medications)
- Drug agents called *penetration enhancers* that help ferry other drugs into the body

HOMEOSTATIC IMBALANCE 5.8   CLINICAL

Organic solvents and heavy metals are devastating to the body and can be lethal. Passage of organic solvents through the skin into the blood can shut down the kidneys and also cause brain damage. Absorption of lead results in anemia and neurological defects. These substances should never be handled with bare hands.

Biological Barriers

Biological barriers include the dendritic cells of the epidermis and macrophages in the dermis:

- Dendritic cells (◄ p. 153) are part of the immune system. They patrol beneath the skin's surface, engulfing any foreign invaders that have penetrated the epidermis. Once they have captured their prey, they leave the skin and migrate to the nearest lymph node. There they display their prey to other immune cells. This triggers an immune response.
- Dermal macrophages constitute a second line of defense. They dispose of viruses and bacteria that manage to penetrate the epidermis. Like dendritic cells, macrophages can initiate an immune response.

Body Temperature Regulation

The body works best when its temperature remains within homeostatic limits. Like car engines, we need to get rid of the heat generated by our internal reactions. As long as the external temperature is lower than body temperature, the skin surface loses heat to the air and to cooler objects in its environment, just as a car radiator loses heat to the air and other nearby parts.

Under normal resting conditions, and as long as the environmental temperature is below 31–32°C (88–90°F), sweat glands secrete about 500 ml (0.5 L) of sweat per day. This routine and unnoticeable sweating is called *insensible perspiration*. When body temperature rises, the nervous system stimulates

dermal blood vessels to dilate and the sweat glands into vigorous secretory activity. On a hot day, sweat becomes noticeable and can account for the loss of up to 12 L (about 3 gallons) of body water in one day. This visible output of sweat is called *sensible perspiration*. Evaporation of sweat from the skin surface dissipates body heat and efficiently cools the body, preventing overheating.

When the external environment is cold, dermal blood vessels constrict. Their constriction causes the warm blood to bypass the skin temporarily and allows skin temperature to drop to that of the external environment. This slows passive heat loss from the body, conserving body heat. We will discuss body temperature regulation in more detail in Chapter 24.

Cutaneous Sensation

The skin is richly supplied with **cutaneous sensory receptors**, which are actually part of the nervous system. The cutaneous receptors are classified as *exteroceptors* (ek″ster-o-sep′torz) because they respond to stimuli arising outside the body. For example:

- Tactile (Meissner's) corpuscles (in the dermal papillae) and tactile epithelial cells with their associated sensory nerve endings allow us to become aware of a caress or the feel of our clothing against our skin.
- Lamellar (also called Pacinian) corpuscles (in the deeper dermis or subcutaneous tissue) alert us to bumps or contacts involving deep pressure.
- Hair follicle receptors report on wind blowing through our hair and a playful tug on a ponytail.
- Free nerve endings that meander throughout the skin sense painful stimuli (irritating chemicals, extreme heat or cold, and others).

Figure 5.1 illustrates all the cutaneous receptors mentioned above with two exceptions: tactile corpuscles (found only in skin that lacks hairs), and tactile epithelial cells (shown in Figure 5.2b). We will discuss these cutaneous receptors in more detail in Chapter 13.

Metabolic Functions

The skin is a chemical factory, fueled in part by the sun's rays. When sunlight bombards the skin, modified cholesterol molecules are converted to a vitamin D precursor. This precursor is transported via the blood to other body areas to be converted to vitamin D, which plays various roles in calcium metabolism. For example, calcium cannot be absorbed from the digestive tract without vitamin D.

Among its other metabolic functions, the epidermis makes chemical conversions that supplement those of the liver. For example, keratinocyte enzymes can:

- "Disarm" many cancer-causing chemicals that penetrate the epidermis
- Activate some steroid hormones—for instance, they can transform cortisone applied to irritated skin into hydrocortisone, a potent anti-inflammatory drug

Skin cells also make several biologically important proteins, including collagenase, an enzyme that aids the natural turnover of collagen (and deters wrinkles).

Blood Reservoir

The dermal vascular supply is extensive and can hold about 5% of the body's entire blood volume. When other body organs, such as vigorously working muscles, need a greater blood supply, the nervous system constricts the dermal blood vessels. This constriction shunts more blood into the general circulation, making it available to the muscles and other body organs.

Excretion

The body eliminates limited amounts of nitrogen-containing wastes (ammonia, urea, and uric acid) in sweat, although most such wastes are excreted in urine. Profuse sweating is an important avenue for water and salt (sodium chloride) loss.

Check Your Understanding

21. What chemicals produced in the skin help provide barriers to bacteria? List at least three and explain how the chemicals are protective.
22. Which epidermal cells play a role in body immunity?
23. How is sunlight important to bone health?
24. **MAKE CONNECTIONS** When blood vessels in the dermis constrict or dilate to help maintain body temperature, which type of muscle tissue that you learned about (in Chapter 4) acts as the effector that causes blood vessel dilation or constriction?

For answers, see Answers Appendix.

5.9 Skin cancer and burns are major challenges to the body   CLINICAL

Learning Outcomes

▶ Summarize the characteristics of the three major types of skin cancers.
▶ Explain why serious burns are life threatening. Describe how to determine the extent of a burn and whether it is a first-, second-, or third-degree burn.

Loss of homeostasis in body cells and organs reveals itself on the skin, sometimes in startling ways. The skin can develop more than 1000 different conditions and ailments. The most common skin disorders are bacterial, viral, or yeast infections (see Related Clinical Terms on pp. 169–170). Less common, but far more damaging to body well-being, are skin cancer and burns, considered next.

Skin Cancer

Skin cancer is the most common kind of cancer—one in five of us will develop skin cancer at some point. The single most important risk factor for skin cancer is exposure to the UV radiation in sunlight and tanning beds, which damages DNA bases. Adjacent pyrimidine bases often respond by fusing, forming lesions called *dimers*. UV radiation also appears to disable a tumor suppressor gene. In limited numbers of cases, frequent irritation of the skin by infections, chemicals, or physical trauma seems to be a predisposing factor.

The three major forms of skin cancer are basal cell carcinoma, squamous cell carcinoma, and melanoma.

Basal Cell Carcinoma

Basal cell carcinoma (kar″sĭ-no′mah), the least malignant and most common, accounts for nearly 80% of cases. Stratum basale cells proliferate, invading the dermis and subcutaneous tissue. The cancer lesions occur most often on sun-exposed areas of the face and appear as shiny, dome-shaped nodules that later develop a central ulcer with a pearly, beaded edge (**Figure 5.11a**). Basal cell carcinoma is relatively slow-growing, and metastasis seldom occurs. Full cure by surgical excision is the rule in 99% of cases.

Squamous Cell Carcinoma

Squamous cell carcinoma, the second most common skin cancer, arises from the keratinocytes of the stratum spinosum. The lesion appears as a scaly reddened papule (small, rounded elevation) that arises most often on the head (scalp, ears, and lower lip), and hands (Figure 5.11b). It tends to grow rapidly and metastasize if not removed. If it is caught early and removed surgically or by radiation therapy, the chance of complete cure is good.

(a) Basal cell carcinoma

(b) Squamous cell carcinoma

(c) Melanoma

Figure 5.11 **Photographs of skin cancers.**

Melanoma

Melanoma (mel″ah-no′mah), cancer of melanocytes, is the most dangerous skin cancer because it is highly metastatic and resistant to chemotherapy. It accounts for only about 1% of skin cancers, but most skin cancer deaths. Its incidence is increasing rapidly. Melanoma can begin wherever there is pigment. Most such cancers appear spontaneously, and about one-third develop from preexisting moles. It usually appears as a spreading brown to black patch (Figure 5.11c) that metastasizes rapidly to surrounding lymph and blood vessels.

The key to surviving melanoma is early detection. Survival rates decline with increasing thickness of the melanoma, degree of involvement of nearby lymph nodes, and extent of metastasis. In advanced cases, surgical treatment is followed by immunotherapy, radiation therapy, or, more recently, targeted gene therapy that has had promising results by shrinking tumors and/or prolonging life.

The American Cancer Society suggests that we regularly examine our skin for new moles or pigmented spots. Apply the **ABCD rule** for recognizing melanoma:

Asymmetry: The two sides of the pigmented spot or mole do not match.

Border irregularity: The borders of the lesion exhibit indentations.

Color: The pigmented spot contains several colors (blacks, browns, tans, and sometimes blues and reds).

Diameter: The spot is larger than 6 mm in diameter (the size of a pencil eraser).

Some experts add an *E*, for *evolution* or *evolving* (changes with time).

Burns

Burns are a devastating threat to the body primarily because of their effects on the skin. A **burn** is tissue damage inflicted by intense heat, electricity, radiation, or certain chemicals, all of which kill cells in the affected areas.

The immediate threat to life resulting from severe burns is a catastrophic loss of body fluids containing proteins and electrolytes. This leads to dehydration and electrolyte imbalance, and then renal failure (kidney shutdown) and circulatory shock (inadequate blood circulation due to reduced blood volume). To save the patient, the lost fluids must be replaced immediately by administration of intravenous (IV) fluids.

Evaluating Burns

Burns are classified according to their severity (depth) as first-, second-, or third-degree burns. In **first-degree burns**, only the epidermis is damaged. Symptoms include localized redness, swelling, and pain. First-degree burns tend to heal in two to three days without special attention. Sunburn is usually a first-degree burn.

Second-degree burns injure the epidermis and the upper region of the dermis. Symptoms mimic those of first-degree burns, but blisters also appear. The burned area is red and

(a) Skin bearing partial-thickness burn (1st- and 2nd-degree burns)

(b) Skin bearing full-thickness burn (3rd-degree burn)

Figure 5.12 Partial-thickness and full-thickness burns.

painful, but skin regeneration occurs with little or no scarring within three to four weeks if care is taken to prevent infection. First- and second-degree burns are considered *partial-thickness burns* (**Figure 5.12a**).

Third-degree burns are *full-thickness burns*, involving the entire thickness of the skin (Figure 5.12b). The burned area appears gray-white, cherry red, or blackened, and initially there is little or no edema. Since the nerve endings have been destroyed, the burned area is not painful. Although skin might eventually regenerate, it is usually impossible to wait that long because of fluid loss and infection. Skin grafting is advised.

In general, burns are considered critical if any of the following conditions exists:

- Over 25% of the body has second-degree burns
- Over 10% of the body has third-degree burns
- There are third-degree burns of the face, hands, or feet (facial burns introduce the possibility of burned respiratory passageways, which can swell and cause suffocation)

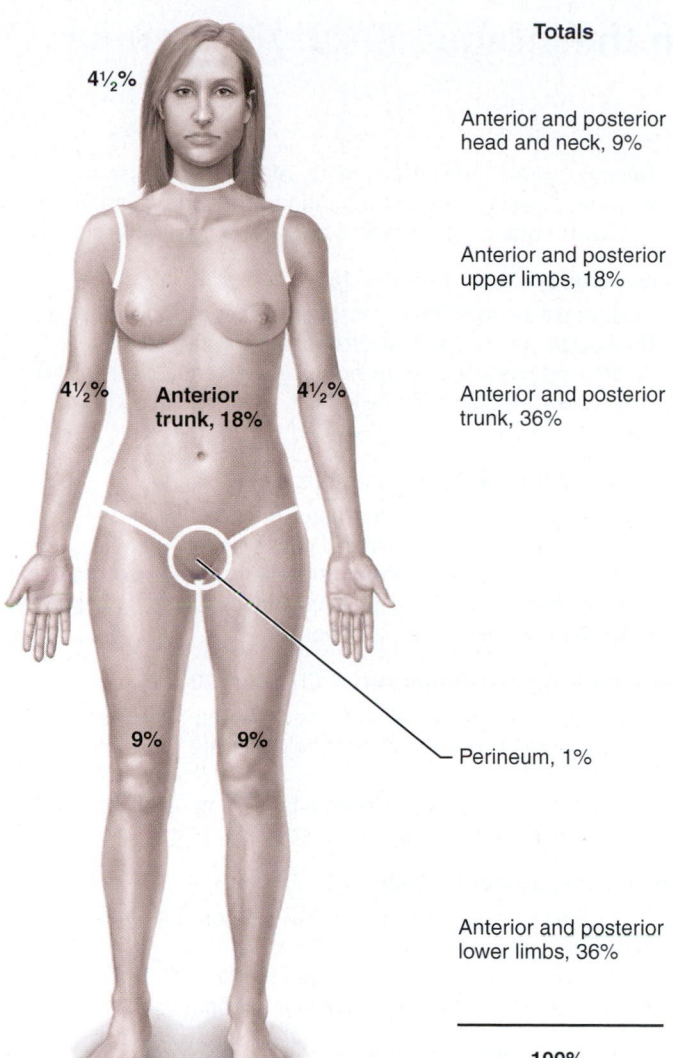

Totals

Anterior and posterior
head and neck, 9%

Anterior and posterior
upper limbs, 18%

Anterior and posterior
trunk, 36%

Perineum, 1%

Anterior and posterior
lower limbs, 36%

4½%

Anterior
trunk, 18%

4½%   4½%

9%   9%

100%

Figure 5.13 Estimating the extent and severity of burns using the rule of nines. Surface area values for the anterior body surface are indicated on the human figure. Total surface area (anterior and posterior body surfaces) for each body region is indicated to the right of the figure.

In adults, the percentage of body surface burned is estimated using the **rule of nines**. This method divides the body into 11 areas, each accounting for 9% of total body area, plus an additional area surrounding the genitals accounting for 1% of body surface area (**Figure 5.13**). The rule of nines is only approximate, so special tables are used when greater accuracy is desired.

Treating Burns

Patients with severe burns need thousands of extra food calories daily to replace lost proteins and allow tissue repair. No one can eat enough food to provide these calories, so burn patients are given supplementary nutrients through gastric tubes and IV lines. Replacing lost fluid by IV hydration is also critical.

After the initial crisis has passed, infection becomes the main threat and sepsis (widespread bacterial infection) is the leading cause of death in burn victims. Bacteria, fungi, and other pathogens easily invade areas where the skin barrier is destroyed, and they multiply rapidly in the nutrient-rich environment of dead tissues. As you might expect, antibiotics play an important role in burn treatment.

Longer-term treatment of full-thickness burns usually involves a skin graft. To prepare a burned area for a skin graft, the *eschar* (es′kar), or burned skin, must first be debrided (removed). Then healthy skin is transplanted to the burned site. If possible, the skin is the patient's own (an *autograft*). Otherwise, there is a good chance that the patient's immune system will reject the foreign skin. Even if the graft "takes," extensive scar tissue often forms—a major problem because it can limit mobility at joints and other critical areas.

Alternatively, synthetic skin—a silicone "epidermis" bound to a spongy "dermal" layer composed of collagen and ground cartilage—can be applied to the debrided area. In time, the patient's own dermal tissue absorbs and replaces the artificial one. Then the silicone sheet is peeled off and replaced with a network of epidermal cells cultured from the patient's own skin.

Check Your Understanding

25. Which type of skin cancer develops from the youngest epidermal cells?

26. What name is given to the rule for recognizing the signs of melanoma?

27. The healing of burns and epidermal regeneration is usually uneventful unless the burn is a third-degree burn. What accounts for this difference?

28. Although the anterior head and face represent only a small percentage of the body surface, burns to this area are often more serious than those to the body trunk. Why?

For answers, see Answers Appendix.

Developmental Aspects of the Integumentary System

The epidermis develops from the embryonic ectoderm (◀ p. 146), and the dermis and subcutaneous tissue develop from mesoderm. By the end of the fourth month of development, the skin is fairly well formed. The epidermis has all its strata, dermal papillae are obvious, fingerprints have developed, and rudimentary epidermal appendages have formed by downward projections of cells from the basal layer. During the fifth and sixth months, the fetus is covered with a downy coat of delicate colorless hairs called the *lanugo coat* (lah-nu′go; "wool"). This hairy cloak is shed by the seventh month, and vellus hairs appear.

From Infancy to Adulthood

When a baby is born, its skin is covered with *vernix caseosa* (ver′niks kă-se-o′sah; "varnish of cheese"), a white, cheesy-looking substance produced by the sebaceous glands that protects the fetus's skin within the water-filled amnion. The newborn's skin is very thin and often has accumulations in the sebaceous glands on the forehead and nose that appear as small white spots called *milia* (mil′e-ah). These normally disappear by the third week after birth.

Homeostatic Interrelationships between the Integumentary System and Other Body Systems

5

Skeletal System   Chapters 6–8
- Skin protects bones; skin synthesizes a vitamin D precursor needed for normal calcium absorption and deposit of bone (calcium) salts, which make bones hard
- Skeletal system provides support for skin

Muscular System   Chapters 9–10
- Skin protects muscles
- Active muscles generate large amounts of heat, which increases blood flow to the skin and may activate sweat glands in skin

Nervous System   Chapters 11–15
- Skin protects nervous system organs; cutaneous sensory receptors for touch, pressure, pain, and temperature located in skin (see Figure 5.1)

- Nervous system regulates diameter of blood vessels in skin; activates sweat glands, contributing to thermoregulation; interprets cutaneous sensation; activates arrector pili muscles

Endocrine System   Chapter 16
- Skin protects endocrine organs; converts some hormones to their active forms; synthesizes a vitamin D precursor
- Androgens produced by the endocrine system activate apocrine and sebaceous glands and are involved in regulating hair growth

Cardiovascular System   Chapters 17–19
- Skin protects cardiovascular organs; prevents fluid loss from body; serves as blood reservoir
- Cardiovascular system transports oxygen and nutrients to skin and removes wastes from skin; provides substances needed by skin glands to make their secretions

Lymphatic System/Immunity   Chapters 20–21
- Skin protects lymphatic organs; prevents pathogen invasion; dendritic cells and macrophages help activate the immune system
- Lymphatic system prevents edema by picking up excessive leaked fluid; immune system protects skin cells

Respiratory System   Chapter 22
- Skin protects respiratory organs; hairs in nose help filter out dust from inhaled air
- Respiratory system furnishes oxygen to skin cells and removes carbon dioxide via gas exchange with blood

Digestive System   Chapter 23
- Skin protects digestive organs; provides vitamin D needed for calcium absorption; performs some of the same chemical conversions as liver cells
- Digestive system provides needed nutrients to the skin

Urinary System   Chapters 25–26
- Skin protects urinary organs; excretes salts and some nitrogenous wastes in sweat
- Kidneys activate vitamin D precursor made by keratinocytes; dispose of nitrogenous wastes of skin metabolism

Reproductive System   Chapter 27
- Skin protects reproductive organs; cutaneous receptors respond to erotic stimuli; highly modified sweat glands (mammary glands) produce milk
- During pregnancy, skin stretches to accommodate growing fetus; changes in skin pigmentation may occur

During infancy and childhood, the skin thickens, and more subcutaneous fat is deposited. Although we all have approximately the same number of sweat glands, the number that function increases in the first two years after birth and is determined by climate. For this reason, people who grow up in hot climates have more active sweat glands than those raised in cooler areas.

During adolescence, the skin and hair become oilier as sebaceous glands increase their activity, and acne may appear. Acne generally subsides in early adulthood, and skin reaches its optimal appearance when we reach our 20s and 30s. Thereafter, the skin starts to show the effects of cumulative environmental assaults (abrasion, wind, sun, chemicals). Scaling and various kinds of skin inflammation, or **dermatitis** (der″mah-ti′tis), become more common.

Aging Skin

As old age approaches, the rate of epidermal cell replacement slows, the skin thins, and its susceptibility to bruises and other injuries increases. The lubricating substances produced by the skin glands that make young skin so soft become deficient. Skin becomes dry and itchy, although people with naturally oily skin seem to postpone this dryness until later in life. Elastic fibers

clump, and collagen fibers become fewer and stiffer. The subcutaneous fat layer diminishes, leading to the intolerance to cold so common in elderly people. Additionally, declining levels of sex hormones result in similar fat distribution in elderly men and women.

The decreasing elasticity of the skin, along with the loss of subcutaneous tissue, inevitably leads to wrinkling. Decreasing numbers of melanocytes and dendritic cells enhance the risk and incidence of skin cancer in this age group. As a rule, redheads and fair-skinned individuals, who have less melanin to begin with, show age-related changes more rapidly than do those with darker skin and hair.

By age 50, the number of active hair follicles has declined markedly and continues to fall, resulting in hair thinning. Hair loses its luster in old age, and the genes responsible for graying and male pattern baldness become active.

Although there is no known way to avoid skin aging, one of the best ways to slow the process is to shield your skin from the sun's rays of both UVA (aging rays) and UVB (rays that burn) light. Aged skin that has been protected from the sun, while it grows thinner and loses some elasticity, still remains unwrinkled and unmarked.

RELATED CLINICAL TERMS

Albinism (al′bĭ-nizm; *alb* = white) Inherited condition in which melanocytes do not synthesize melanin owing to a lack of tyrosinase. An albino's skin is pink, the hair pale or white, and the irises of the eyes unpigmented or nearly so.

Alopecia (al″o-pe′she-ah) Absence of hair from areas of the body where it would normally occur. May be partial or complete (baldness). In *alopecia areata*, the immune system attacks the hair follicles, leading to hair loss.

Boils and carbuncles (kar′bung-klz; "little glowing embers") Inflammation of hair follicles and sebaceous glands in which an infection has spread to the underlying subcutaneous tissue; common on the dorsal neck. A carbuncle is a cluster of boils. A common cause is bacterial infection.

Cold sores (fever blisters) Small fluid-filled blisters that itch or hurt; usually occur around the lips and in the mucosa of the mouth; caused by a herpes simplex infection. The virus localizes in a cutaneous nerve, where it remains dormant until activated by emotional upset, fever, or UV radiation.

Contact dermatitis Itching, redness, and swelling, progressing to blister formation; caused by exposure of the skin to chemicals (e.g., poison ivy oleoresin) that provoke an allergic response in sensitive individuals.

Decubitus ulcer (de-ku′bĭ-tus) Localized breakdown and ulceration of skin due to interference with its blood supply. Usually occurs over a bony prominence, such as the hip or heel, that is subjected to continuous pressure; also called a bedsore.

Dermatology The branch of medicine that studies and treats disorders of the skin.

Eczema (ek′ze-mah) A skin rash characterized by itching, blistering, oozing, and scaling of the skin. A common allergic

reaction in children, but also occurs in adults. Frequent causes include allergic reactions to certain foods (fish, eggs, and others) or to inhaled dust or pollen. Treated by methods used for other allergic disorders.

Epidermolysis bullosa (EB) A group of hereditary disorders characterized by inadequate or faulty synthesis of keratin, collagen, and/or basement membrane "cement" that results in lack of cohesion between layers of the skin and mucosa. A simple touch causes layers to separate and blister. In severe cases fatal blistering occurs in major vital organs. Because the blisters rupture easily, people with EB suffer frequent infections. Treatments are aimed at relieving the symptoms and preventing infection.

Impetigo (im″pĕ-ti′go; *impet* = an attack) Pink, fluid-filled, raised lesions (common around the mouth and nose) that develop a yellow crust and eventually rupture. Caused by staphylococcus infection, it is contagious, and common in school-age children.

Porphyria (por-fer′e-ah; "purple") A group of inherited disorders in which certain enzymes needed to form the heme of hemoglobin of blood are abnormal. Because these enzymes do not work properly, metabolic intermediates of the heme pathway called porphyrins build up, spill into the circulation, and eventually damage various parts of the body. Some porphyrias make the skin very sensitive to sunlight. The skin becomes lesioned and scarred. In some cases, fingers, toes, and nose are disfigured; gums degenerate and teeth become prominent. Believed to be the basis of folklore about vampires.

Psoriasis (so-ri′ah-sis) A chronic autoimmune condition characterized by raised, reddened epidermal patches covered with silvery scales that itch or burn, crack, and sometimes bleed or become infected. When severe, it may be disfiguring and debilitating. Trauma, infection, hormonal changes, or stress often trigger the autoimmune attacks. Cortisone-containing topicals (medications applied to the skin surface) may control mild cases. For more severe cases, self-injected drugs called biologicals and/or phototherapy with UV light in conjunction with chemotherapeutic drugs provides some relief.

Rosacea (ro-za′she-ah) A chronic skin eruption produced by dilated small blood vessels of the face, particularly the nose and cheeks. Papules and acne-like pustules may or may not occur. More common in women. Cause is unknown, but stress, some endocrine disorders, and anything that produces flushing (hot beverages, alcohol, sunlight, etc.) can aggravate this condition.

Vitiligo (vit″ĭ-li′go; *viti* = a vine, winding) The most prevalent skin pigmentation disorder, characterized by a loss of melanocytes and uneven dispersal of melanin, so that unpigmented skin regions (light spots) are surrounded by normally pigmented areas. An autoimmune disorder.

Scleroderma (*scler* = hard) An autoimmune disorder characterized by stiff, hardened skin due to abnormal amounts of collagen in the dermis that severely limit joint movements and facial expressions. A classic sign of the disorder is Raynaud's phenomenon in which the fingers and toes become white and painful because of poor blood flow to those areas. The fibrosis that occurs in systemic cases may affect a variety of organs including the lungs (eventually leading to difficulty breathing) and the kidneys (leading to renal hypertension because of blood vessel constriction and occlusion).

CHAPTER SUMMARY

5.1 The skin consists of two layers: the epidermis and dermis (pp. 150–151)

1. The skin, or integument, is composed of two discrete tissue layers, an outer epidermis and a deeper dermis, resting on subcutaneous tissue.

5.2 The epidermis is a keratinized stratified squamous epithelium (pp. 152–154)

1. The epidermis is an avascular, keratinized sheet of stratified squamous epithelium. Most epidermal cells are keratinocytes. Scattered among the keratinocytes in the deepest epidermal layers are melanocytes, dendritic cells, and tactile epithelial cells.

2. From deep to superficial, the strata, or layers of the epidermis, are the basale, spinosum, granulosum, lucidum, and corneum. The stratum lucidum is absent in thin skin. The mitotically active stratum basale is the source of new cells for epidermal growth. The most superficial layers are increasingly keratinized and less viable.

5.3 The dermis consists of papillary dermis and reticular dermis (pp. 154–156)

1. The dermis, composed mainly of dense irregular connective tissue, is well supplied with blood vessels, lymphatic vessels, and nerves. Cutaneous receptors, glands, and hair follicles reside within the dermis.

2. The more superficial papillary dermis exhibits dermal papillae that protrude into the epidermis above, as well as dermal ridges. Dermal ridges and epidermal ridges together form the friction ridges that produce fingerprints.

3. In the deeper, thicker reticular dermis, the connective tissue fibers are much more densely interwoven. Less dense regions between the collagen bundles produce cleavage, or tension, lines in the skin. Points of tight dermal attachment to the subcutaneous tissue produce dermal folds, or flexure lines.

5.4 Melanin, carotene, and hemoglobin determine skin color (pp. 156–157)

1. Skin color reflects the amount of pigments (melanin and carotene) in the skin and the oxygenation level of hemoglobin in blood.

2. Melanin production is stimulated by exposure to ultraviolet radiation in sunlight. Melanin, produced by melanocytes and transferred to keratinocytes, protects the keratinocyte nuclei from the damaging effects of UV radiation.

3. Skin color is affected by emotional state. Alterations in normal skin color (jaundice, erythema, and others) may indicate certain disease states.

5.5 Hair consists of dead, keratinized cells (pp. 157–160)

1. A hair, produced by a hair follicle, consists of heavily keratinized cells. A typical hair has a central medulla, a cortex, and an outer cuticle, and root and shaft portions. Hair color reflects the amount and kind of melanin present.

2. A hair follicle consists of an inner epithelial root sheath and an outer peripheral connective tissue sheath derived from the dermis. The base of the hair follicle is a hair bulb with a matrix that produces the hair. A hair follicle is richly vascularized and well supplied with nerve fibers. Arrector pili muscles pull the follicles into an upright position, producing goose bumps, and propel sebum to the skin surface when they contract.

3. Except for hairs of the scalp and around the eyes, hairs formed initially are fine vellus hairs; at puberty, under the influence of androgens, coarser, darker terminal hairs appear in the axillae and the genital region.

4. The rate of hair growth varies in different body regions and with sex and age. Differences in life span of hairs account for differences in length on different body regions. Hair thinning reflects factors that lengthen follicular resting periods, and age-related atrophy of hair follicles. In male pattern baldness, genetic predisposition and male sex hormones are determining factors.

5.6 Nails are scale-like modifications of the epidermis (pp. 160–161)

1. A nail covers the dorsum of a finger (or toe) tip. The actively growing region is the nail matrix.

5.7 Sweat glands help control body temperature, and sebaceous glands secrete sebum (pp. 161–163)

1. Eccrine (merocrine) sweat glands, with a few exceptions, are distributed over the entire body surface. Their primary function is thermoregulation. They are simple coiled tubular glands that secrete a salt solution containing small amounts of other solutes. Their ducts usually empty to the skin surface via pores.

2. Apocrine sweat glands, which may function as scent glands, are found primarily in the axillary and anogenital areas. Their secretion is similar to eccrine secretion, but it also contains proteins and fatty substances on which bacteria thrive.

3. Sebaceous glands occur all over the body surface except for the palms and soles. They are simple alveolar glands; their oily holocrine secretion is called sebum. Sebaceous gland ducts usually empty into hair follicles.

4. Sebum lubricates the skin and hair, slows water loss from the skin, and acts as a bactericidal agent. Sebaceous glands increase activity during puberty and are controlled by androgens.

5.8   First and foremost, the skin is a barrier   (pp. 163–165)

1. Protection. The skin protects by chemical barriers (the antibacterial nature of sebum, defensins, cathelicidins, the acid mantle, and the UV shield of melanin), physical barriers (the hardened keratinized and lipid-rich surface), and biological barriers (dendritic cells and macrophages).

2. Body temperature regulation. The skin vasculature and sweat glands, regulated by the nervous system, play an important role in maintaining body temperature homeostasis.

3. Cutaneous sensation. Cutaneous sensory receptors respond to temperature, touch, pressure, and pain stimuli.

4. Metabolic functions. A vitamin D precursor is synthesized from cholesterol by epidermal cells. Skin cells also play a role in some chemical conversions.

5. Blood reservoir. The extensive vascular supply of the dermis allows the skin to act as a blood reservoir.

6. Excretion. Sweat contains small amounts of nitrogenous wastes and plays a minor role in excretion.

5.9   Skin cancer and burns are major challenges to the body   (pp. 165–167)

1. The most common skin disorders result from infections.

2. The most common cause of skin cancer is exposure to ultraviolet radiation.

3. Basal cell carcinoma and squamous cell carcinoma are cured if they are removed before metastasis. Melanoma, a cancer of melanocytes, is less common but more dangerous.

4. In severe burns, the initial threat is loss of protein- and electrolyte-rich body fluids, which may lead to circulatory collapse. The second threat is overwhelming bacterial infection.

5. The extent of a burn may be evaluated by using the rule of nines. The severity of burns is indicated by the terms first degree, second degree, and third degree. Third-degree burns are full-thickness burns that require grafting for successful recovery.

Developmental Aspects of the Integumentary System   (pp. 167, 169)

1. The epidermis develops from embryonic ectoderm; the dermis (and subcutaneous tissue) develops from mesoderm.

2. The fetus exhibits a downy lanugo coat. Fetal sebaceous glands produce vernix caseosa, which helps protect the fetus's skin from its watery environment.

3. A newborn's skin is thin. During childhood the skin thickens and more subcutaneous fat is deposited. At puberty, sebaceous glands increase their activity and terminal hairs appear in greater numbers.

4. In old age, the rate of epidermal cell replacement declines and the skin and hair thin. Skin glands become less active. Loss of collagen and elastic fibers and subcutaneous fat leads to wrinkling. Photodamage is a major cause of skin aging.

REVIEW QUESTIONS

 To access additional practice questions using your smartphone, tablet, or computer: Mastering A&P® > Study Area > Practice Tests & Quizzes

Level 1   Remember/Understand

(Some questions have more than one correct answer. Select the best answer or answers from the choices given.)

1. Which epidermal cell type is most numerous? (a) keratinocyte, (b) melanocyte, (c) dendritic cell, (d) tactile epithelial cell.

2. Which cell functions as part of the immune system? (a) keratinocyte, (b) melanocyte, (c) dendritic cell, (d) tactile epithelial cell.

3. The epidermis provides a physical barrier due largely to the presence of (a) melanin, (b) carotene, (c) collagen, (d) keratin.

4. Skin color is determined by (a) the amount of blood, (b) pigments, (c) oxygenation level of the blood, (d) all of these.

5. The sensations of touch and pressure are picked up by receptors located in (a) the stratum spinosum, (b) the dermis, (c) the subcutaneous tissue, (d) the stratum corneum.

6. Which is not a true statement about the papillary dermis? (a) it is largely areolar connective tissue, (b) it is most responsible for the toughness of the skin, (c) it contains nerve endings that respond to stimuli, (d) it is highly vascular.

7. Which of the following is not an epidermal derivative? (a) hair, (b) sweat gland, (c) sensory receptor, (d) sebaceous gland.

8. An arrector pili muscle (a) is associated with each sweat gland, (b) can cause a hair to stand up straight, (c) enables each hair to be stretched when wet, (d) provides new cells for continued growth of its associated hair.

9. The product of this type of sweat gland includes protein and lipid substances that become odoriferous as a result of bacterial action: (a) apocrine gland, (b) eccrine gland, (c) sebaceous gland, (d) pancreatic gland.

10. Sebum (a) lubricates the surface of the skin and hair, (b) consists of cell fragments and fatty substances, (c) in excess may cause seborrhea, (d) all of these.

11. The rule of nines is helpful clinically in (a) diagnosing skin cancer, (b) estimating the extent of a burn, (c) estimating how serious a cancer is, (d) preventing acne.

12. Distinguish clearly between first-, second-, and third-degree burns.

13. Why does skin wrinkle and what accelerates the wrinkling process?

Level 2   Apply/Analyze

14. Which epidermal cells contain keratohyaline and lamellar granules?

15. Is a bald man really hairless? Explain.

16. Count Dracula, the most famous vampire, rumored to have killed at least 200,000 people, was based on a real person who lived in eastern Europe about 600 years ago. He was indeed a "monster," although he was not a real vampire. The historical Count Dracula may have suffered from which of the following? (Hint: See Related Clinical Terms.) (a) porphyria, (b) EB, (c) halitosis, (d) vitiligo. Explain your answer.

17. On an outline diagram of the human body, mark off various regions according to the rule of nines. What percentage of the total body surface is affected if the skin over the following body parts is burned? (a) the entire posterior trunk and buttocks, (b) an entire lower limb, (c) the entire front of the left upper limb.

18. Dean, a 40-year-old aging beach boy, is complaining to you that although his suntan made him popular when he was young, now his face is all wrinkled, and he has several darkly pigmented moles that are growing rapidly and are as big as large coins. He shows you the moles, and immediately you think "ABCD." What does that mean and why should he be concerned?

19. Patients with third-degree burns demonstrate the loss of vital functions performed by the skin. What are the two most important problems encountered clinically with such patients? Explain each in terms of the absence of skin.

Level 3 Evaluate/Synthesize

20. You go to the beach to swim on an extremely hot, sunny summer afternoon. Describe two ways in which your integumentary system acts to preserve homeostasis during your outing.

21. Describe the process of hair formation, and list several factors that may influence (a) growth cycles and (b) hair texture.

22. Explain each of these familiar phenomena in terms of what you learned in this chapter: (a) pimples, (b) dandruff, (c) greasy hair and "shiny nose," (d) stretch marks from gaining weight, (e) freckles.

23. Why are there no skin cancers that originate from stratum corneum cells?

24. A man got his finger caught in a machine at the factory. The damage was less serious than expected, but the entire nail was torn off his right index finger. The parts lost were the body, root, bed, matrix, and eponychium of the nail. First, define each of these parts. Then, tell if this nail is likely to grow back.

25. Tanya, a 30-year-old resident of a mental hospital, has an abnormal growth of hair on the dorsum of her right index finger. The orderly comments that she gnaws on that finger continuously. What do you think is the relationship between Tanya's gnawing activity and her hairy finger?

26. Mrs. Gaucher received second-degree burns on her abdomen when she dropped a kettle of boiling water. She asked her doctor (worriedly) if she would need a skin graft. What do you think he told her?

CLINICAL CASE STUDY

21-Year-Old Female with Deep Lacerations

While riding her bike to campus, 21-year-old Liliana Rose was struck by a car. Examination in the Emergency Department reveals several injuries. Relative to her integumentary system, the following comments are noted on her chart:

- Epidermal abrasions of the right lateral upper arm and anterior shoulder.
- A deep, 2-cm laceration extending vertically on right lateral cheek and a horizontal 1-cm laceration on the temple.
- Cyanosis is apparent in her nail beds and lips.

The lacerated areas are cleaned, sutured (stitched), and bandaged by the emergency room (ER) personnel.

1. **+ NCLEX-STYLE** Liliana's epidermal layer has been damaged. Which statement best explains the significance of this damage?

   a. It's not significant, because the cells in the epidermis are already dead.

   b. It's significant because cells in the epidermis give rise to all the cells in all the different skin layers.

   c. It's significant because disrupting the epidermis will cause severe, sometimes life-threatening bleeding.

   d. It's significant because the cells of the epidermis protect against evaporative water loss, UV radiation, and infection.

2. Assuming that bacteria have penetrated the dermis in these areas, what remaining skin defenses might act to prevent further bacterial invasion?

3. Liliana is concerned about scarring to her face. The surgeon tells her that the cut extending along her temple should heal with little scarring, but the one extending up her cheek is more likely to scar. Why might this be the case?

4. **+ NCLEX-STYLE** Liliana's cyanosis is clinically significant because:

   a. It indicates that she has less hemoglobin reaching her lips and fingers.

   b. It indicates that her blood is low in oxygen.

   c. It indicates that there are more melanocytes developing in her lips and fingers.

   d. It indicates a buildup of bilirubin in her epidermis.

5. Two nursing students are discussing wound care, especially for abrasions. "The biggest issue is all the stuff the skin secretes!" one of them complains. "It's crazy hard to keep a wound clean, and the patient is losing fluid through the abrasions. Plus all that acid! The last thing a wound needs is acid on it! Why hasn't anyone developed a dressing that can stop skin secretions?"

   The other student disagrees. "I don't think skin would heal as well without its secretions. Doesn't the body usually know what it's doing?"

   What's your opinion, and why?

6

Bones and Skeletal Tissues

In this chapter, you will learn that

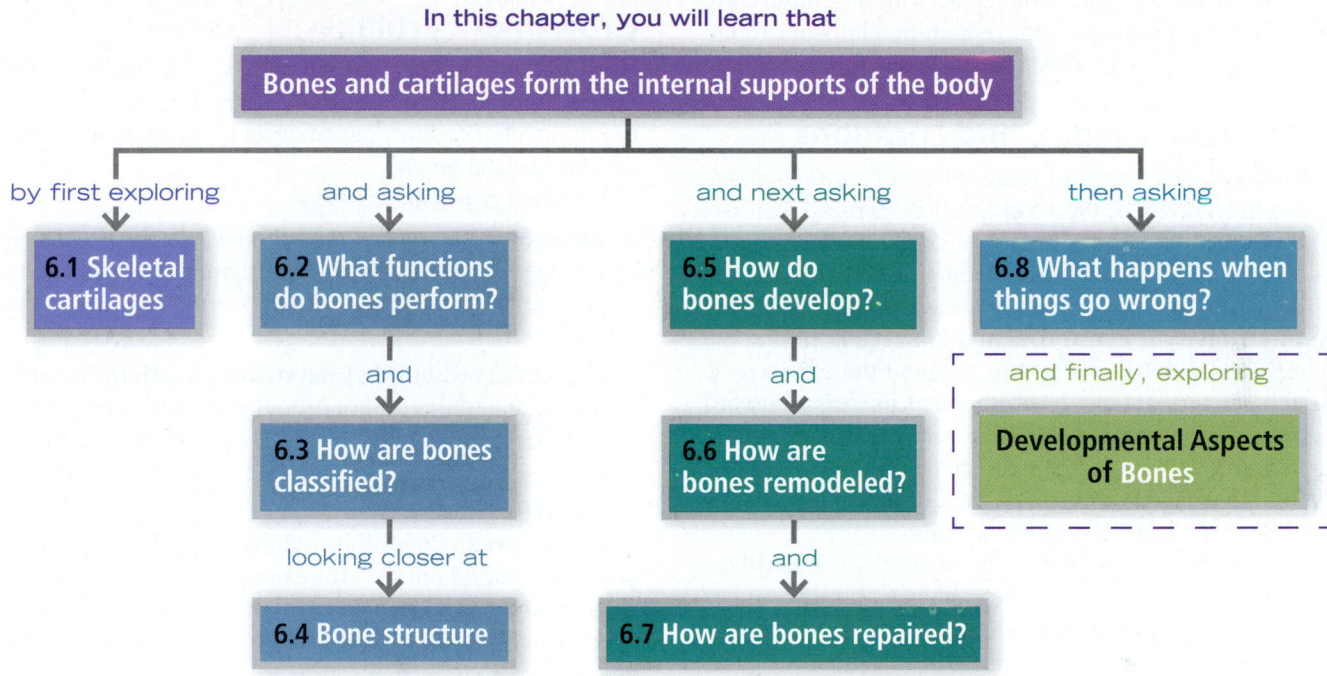

Bones and cartilages form the internal supports of the body

by first exploring

6.1 Skeletal cartilages

and asking

6.2 What functions do bones perform?

and

6.3 How are bones classified?

looking closer at

6.4 Bone structure

and next asking

6.5 How do bones develop?

and

6.6 How are bones remodeled?

and

6.7 How are bones repaired?

then asking

6.8 What happens when things go wrong?

and finally, exploring

Developmental Aspects of Bones

Bones appear to be the most lifeless of body organs, and may even summon images of a graveyard. However, the bones within you are not the dead, dry bones that you may have seen. Rather, they are living, dynamic organs, and bone tissue is formed and remodeled throughout life. Without bones to form our internal supporting skeleton, we would all creep along the ground like slugs, lacking any definite shape or form. Along with its bones, the skeleton contains resilient cartilages, which we briefly discuss at the beginning of this chapter.

Hyaline, elastic, and fibrocartilage help form the skeleton

Learning Outcomes

▶ **Describe the functional properties of the three types of cartilage tissue.**

▶ **Locate the major cartilages of the adult skeleton.**

▶ **Explain how cartilage grows.**

The human skeleton is initially made up of cartilages and fibrous membranes, but bone soon replaces most of these early supports. The few cartilages that remain in adults are found mainly in regions where flexible skeletal tissue is needed.

Basic Structure, Types, and Locations

A **skeletal cartilage** is made of some variety of *cartilage tissue* sculpted to fit its body location and function. Cartilage consists primarily of water. It is very resilient—it has the ability to spring back to its original shape after being compressed.

Cartilage contains no nerves or blood vessels. It is surrounded by a layer of dense irregular connective tissue, the *perichondrium* (per″ĭ-kon′dre-um; "around the cartilage"). The perichondrium acts as reinforcement to resist outward expansion when the cartilage is compressed. The perichondrium also contains blood vessels that nourish the cartilage cells. The thickness of cartilage is limited by the distance nutrients can diffuse through the matrix to reach the cells.

As we described in Chapter 4, the three types of cartilage tissue are hyaline, elastic, and fibrocartilage. All three types have the same basic components—cells called *chondrocytes*, enclosed in small cavities (lacunae) within an *extracellular matrix* containing a jellylike ground substance and fibers.

Hyaline Cartilages

Hyaline cartilages, which look like frosted glass when freshly exposed, provide support with flexibility and resilience. They are the most abundant skeletal cartilages. Their chondrocytes are spherical (◀ **Figure 4.11g, p. 135**), and the only fiber type in their matrix is fine collagen fibers (which are undetectable microscopically). Colored blue in **Figure 6.1**, skeletal hyaline cartilages include:

- *Articular cartilages* (*artic* = joint, point of connection), which cover the ends of most bones at movable joints
- *Costal cartilages*, which connect the ribs to the sternum (breastbone)
- *Respiratory cartilages*, which form the skeleton of the *larynx* (*voice box*) and reinforce other respiratory passageways
- *Nasal cartilages*, which support the external nose

Elastic Cartilages

Elastic cartilages resemble hyaline cartilages (◀ **Figure 4.11h, p. 135**), but they contain more stretchy elastic fibers and so are better able to stand up to repeated bending. They are found in only two skeletal locations (shown in green in Figure 6.1)—the external ear and the epiglottis (the flap that bends to cover the opening of the larynx each time we swallow).

Fibrocartilages

Highly compressible with great tensile strength, **fibrocartilages** consist of roughly parallel rows of chondrocytes alternating with thick collagen fibers (◀ **Figure 4.11i, p. 136**). Fibrocartilages occur in sites that are subjected to both pressure and stretch, such as the padlike cartilages (menisci) of the knee and the discs between vertebrae (colored red in Figure 6.1).

Growth of Cartilage

Unlike bone, which has a hard matrix, cartilage has a flexible matrix that can accommodate mitosis. It is the ideal tissue to use to rapidly lay down the embryonic skeleton and to provide for new skeletal growth.

Cartilage grows in two ways.

- **Appositional growth.** In **appositional growth** (ap″o-zish′un-al), cartilage-forming cells in the surrounding perichondrium secrete new matrix against the external face of the existing cartilage tissue.

- **Interstitial growth.** In **interstitial growth** (in″ter-stish′al), the lacunae-bound chondrocytes divide and secrete new matrix, expanding the cartilage from within. Typically, cartilage growth ends during adolescence when the skeleton stops growing.

Under certain conditions—during normal bone growth in youth and during old age, for example—cartilage can become calcified (hardened due to deposit of calcium salts). Note, however, that calcified cartilage is not bone; cartilage and bone are always distinct tissues.

Table 6.1 compares some of the key characteristics of cartilage and bone (◀ **p. 137**). This overview will help orient you as you discover more about bone in the rest of the chapter.

Check Your Understanding

1. Which type of cartilage is most plentiful in the adult body?
2. What two body structures contain flexible elastic cartilage?
3. Cartilage grows by interstitial growth. What does this mean?

For answers, see Answers Appendix.

Table 6.1	Comparison of Cartilage and Bone
CARTILAGE	**BONE**
Surrounded by perichondrium	Surrounded by periosteum
No blood vessels or nerves (only in perichondrium)	Blood vessels and nerves throughout
Chondrocytes in lacunae	Osteocytes in lacunae
Flexible extracellular matrix	Rigid extracellular matrix (due to inorganic calcium salts)
Extracellular matrix made by chondroblasts	Extracellular matrix (organic part) made by osteoblasts
Appositional growth and interstitial growth	Appositional growth only

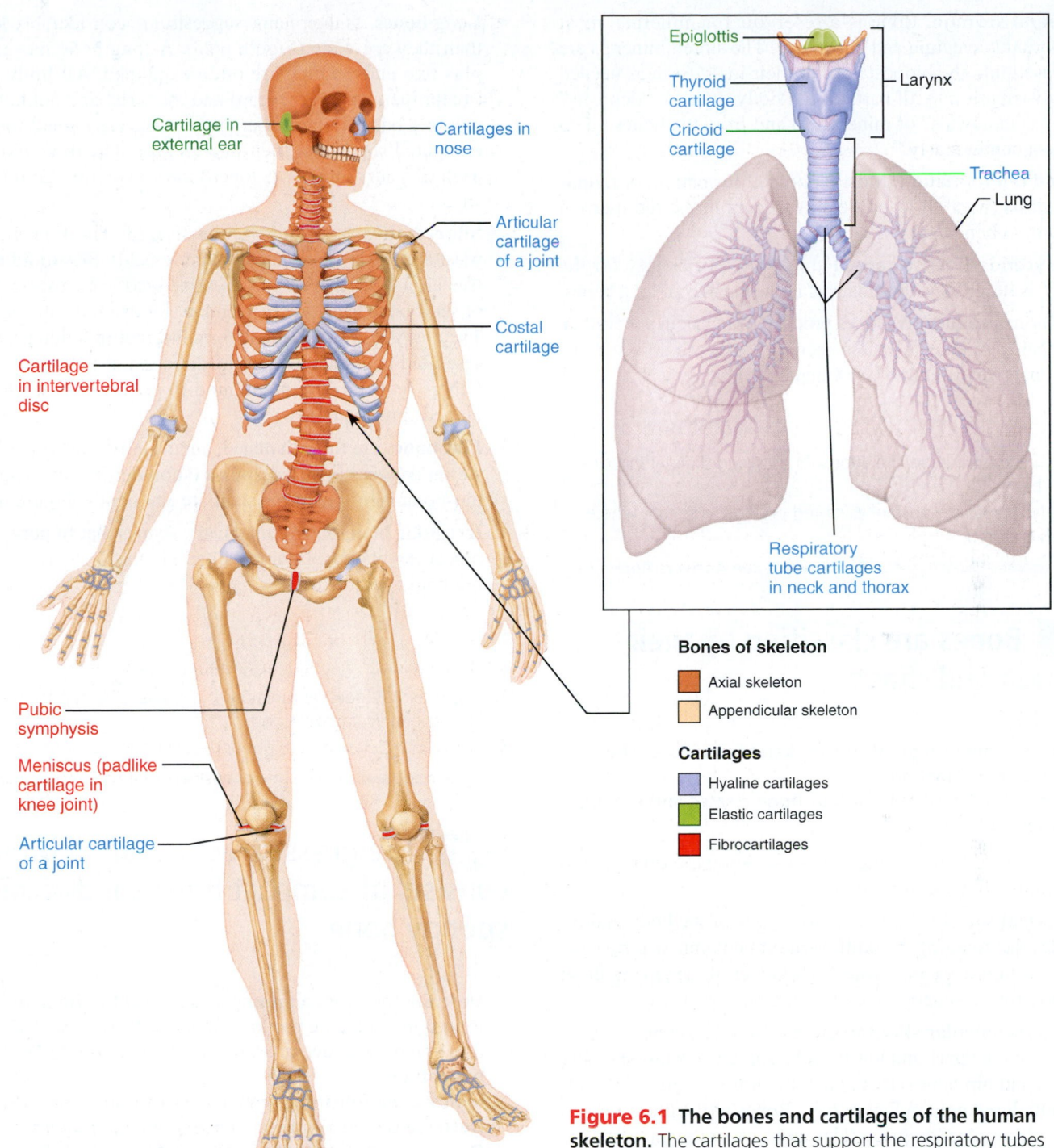

Bones of skeleton

- Axial skeleton
- Appendicular skeleton

Cartilages

- Hyaline cartilages
- Elastic cartilages
- Fibrocartilages

Figure 6.1 The bones and cartilages of the human skeleton. The cartilages that support the respiratory tubes and larynx are drawn separately at the right.

6.2 Bones perform several important functions

Learning Outcome

▶ **Describe the functions of the skeleton and of bone tissue.**

Our bones perform seven important functions:

- **Support.** Bones provide a framework that supports the body and cradles its soft organs. For example, bones of lower limbs act as pillars to support the body trunk when we stand, and the rib cage supports the thoracic wall.

- **Protection.** The fused bones of the skull protect the brain. The vertebrae surround the spinal cord, and the rib cage helps protect the vital organs of the thorax.

- **Anchorage.** Skeletal muscles, which attach to bones by tendons, use bones as levers to move the body and its parts. As a result, we can walk, grasp objects, and breathe. The design of joints determines the types of movement possible.

- **Mineral storage.** Bone is a reservoir for minerals, most importantly calcium and phosphate. The stored minerals are released into the bloodstream in their ionic form as needed for distribution to all parts of the body. Indeed, "deposits" and "withdrawals" of minerals to and from the bones go on almost continuously.

- **Blood cell formation.** Most blood cell formation, or **hematopoiesis** (hem″ah-to-poy-e′sis), occurs in the red marrow of certain bones.

- **Triglyceride (fat) storage.** Fat, a source of energy for the body, is stored as yellow marrow in the cavities of long bones.

- **Hormone production.** Bones produce osteocalcin, a hormone that helps to regulate insulin secretion, glucose homeostasis, and energy expenditure (see Chapter 16).

Check Your Understanding

4. What is the functional relationship between skeletal muscles and bones?
5. List two types of substances stored in bone and state where each is stored.

For answers, see Answers Appendix.

6.3   Bones are classified by their location and shape

Learning Outcomes

▶ **Name the major regions of the skeleton and describe their relative functions.**

▶ **Compare and contrast the four bone classes and provide examples of each class.**

The 206 named bones of the human skeleton are divided into two groups: axial and appendicular.

- The **axial skeleton** forms the long axis of the body and includes the bones of the skull, vertebral column, and rib cage (shown in orange in Figure 6.1). Generally speaking these bones protect, support, or carry other body parts.

- The **appendicular skeleton** (ap″en-dik′u-lar) consists of the bones of the upper and lower limbs and the girdles (shoulder bones and hip bones) that attach the limbs to the axial skeleton (colored gold in Figure 6.1). Bones of the limbs help us move from place to place (locomotion) and manipulate our environment.

Bones come in many sizes and shapes. For example, the pisiform bone of the wrist is the size and shape of a pea, whereas the femur (thigh bone) is nearly 2 feet long in some people and has a large, ball-shaped head. The unique shape of each bone fulfills a particular need. The femur, for example, withstands great pressure, and its hollow-cylinder shape provides maximum strength with minimum weight to accommodate our upright posture.

Bones are classified by their shape as long, short, flat, or irregular (**Figure 6.2**).

- **Long bones**, as their name suggests, are considerably longer than they are wide (Figure 6.2a). A long bone has a shaft plus two ends, which are often expanded. All limb bones except the patella (kneecap) and the wrist and ankle bones are long bones. Notice that these bones are named for their elongated shape, *not* their overall size. The three bones in each of your fingers are long bones, even though they are small.

- **Short bones** are roughly cube shaped. The bones of the wrist and ankle are examples (Figure 6.2d). **Sesamoid bones** (ses′ah-moid; "shaped like a sesame seed") are a special type of short bone that form in a tendon (for example, the patella). They vary in size and number in different individuals. Some sesamoid bones act to alter the direction of pull of a tendon. Others reduce friction and modify pressure on tendons to reduce abrasion or tearing.

- **Flat bones** are thin, flattened, and usually a bit curved. The sternum (breastbone), scapulae (shoulder blades), ribs, and most cranial bones of the skull are flat bones (Figure 6.2c).

- **Irregular bones** have complicated shapes that fit none of the preceding classes. Examples include the vertebrae and the hip bones (Figure 6.2b).

Check Your Understanding

6. What are the components of the axial skeleton?
7. Contrast the general function of the axial skeleton to that of the appendicular skeleton.
8. What bone class do the ribs and skull bones fall into?

For answers, see Answers Appendix.

6.4   The gross structure of all bones consists of compact bone sandwiching spongy bone

Learning Outcomes

▶ **Describe the gross anatomy of a typical flat bone and a long bone. Indicate the locations and functions of red and yellow marrow, articular cartilage, periosteum, and endosteum.**

▶ **Indicate the functional importance of bone markings.**

▶ **Describe the histology of compact and spongy bone.**

▶ **Discuss the chemical composition of bone and the advantages conferred by its organic and inorganic components.**

Because they contain different types of tissue, bones are organs. (Recall that an organ contains several different tissues.) Although bone (osseous) tissue dominates bones, they also contain nervous tissue in their nerves, cartilage in their articular cartilages, dense connective tissue covering their external surface, and muscle and epithelial tissues in their blood vessels. We will consider bone structure at three levels: gross, microscopic, and chemical.

(c) Flat bone (sternum)

(a) Long bone (humerus)

(b) Irregular bone (vertebra), right lateral view

(d) Short bone (talus)

Figure 6.2 **Classification of bones on the basis of shape.**

Gross Anatomy

Compact and Spongy Bone

Every bone has a dense outer layer that looks smooth and solid to the naked eye. This external layer is **compact bone** (**Figures 6.3**, 6.4, and 6.5). Internal to this is **spongy bone** (also called **trabecular bone**), a honeycomb of small needle-like or flat pieces called **trabeculae** (trah-bek′u-le; "little beams"). In living bones the open spaces between trabeculae are filled with red or yellow bone marrow.

Spongy bone has a mesh of bony spines called *trabeculae.*

Compact bone looks smooth and solid.

Figure 6.3 **Compact and spongy bone.** Photo of a sectioned proximal femur (thigh bone).

6

Compact bone

Spongy bone (diploë)

Compact bone

Trabeculae of spongy bone

Figure 6.4 Structure of a flat bone. Flat bones consist of a layer of spongy bone sandwiched between two thin layers of compact bone. (Photomicrograph at bottom, 25×)

Structure of Short, Irregular, and Flat Bones

Short, irregular, and flat bones share a simple pattern: They all consist of thin plates of spongy bone (*diploë*) covered by compact bone. The compact bone is covered outside and inside by connective tissue membranes, respectively the *periosteum* and *endosteum*. However, these bones are not cylindrical and so they have no shaft or expanded ends. They contain bone marrow (between their trabeculae), but no well-defined marrow cavity. Where they form movable joints with their neighbors, hyaline cartilage covers their surfaces.

Figure 6.4 shows a typical flat bone arrangement: compact bone–spongy bone–compact bone, which resembles a stiffened sandwich.

Structure of a Typical Long Bone

With few exceptions, all long bones have the same general structure: a shaft, bone ends, and membranes (**Figure 6.5**).

Diaphysis   A tubular **diaphysis** (di-af′ĭ-sis; *dia* = through, *physis* = growth), or shaft, forms the long axis of the bone. It is constructed of a relatively thick *collar* of compact bone that surrounds a central **medullary cavity** (med′u-lar-e; "middle"), or *marrow cavity*, that contains no bone tissue. Instead, the medullary cavity contains yellow marrow (fat) in adults and so is called the *yellow marrow cavity*. Between the marrow and the compact bone, there is often a thin layer of spongy bone.

Epiphyses   The **epiphyses** (e-pif′ĭ-sēz; singular: epiphysis) are the bone ends (*epi* = upon). An outer shell of compact bone forms the epiphysis exterior and the interior contains spongy bone. A thin layer of articular (hyaline) cartilage covers the joint surface of each epiphysis, cushioning the opposing bone ends during movement and absorbing stress.

Between the diaphysis and each epiphysis of an adult long bone is an **epiphyseal line**, a remnant of the **epiphyseal plate**. The epiphyseal plate, commonly called the *growth plate*, is a disc of hyaline cartilage that grows during childhood to lengthen the bone. The flared portion of the bone where the diaphysis and epiphysis meet, whether it is the epiphyseal plate or line, is called the *metaphysis* (*meta* = between).

Membranes   A glistening white, double-layered membrane called the **periosteum** (per″e-os′te-um; *peri* = around, *osteo* = bone) covers the external surface of the entire bone except the joint surfaces. The outer *fibrous layer* of the periosteum is dense irregular connective tissue. The inner *osteogenic layer* next to the bone surface contains *osteoprogenitor cells* (primitive stem cells that give rise to most bone cells). It also has bone-destroying cells (osteoclasts) and bone-forming cells (osteoblasts).

The periosteum is richly supplied with nerve fibers and blood vessels, which is why broken bones are painful and bleed profusely. *Perforating fibers*—bundles of collagen fibers that extend into the bone matrix—secure the periosteum to the underlying bone (Figure 6.5c). The periosteum also provides anchoring points for tendons and ligaments. At these points the perforating fibers are exceptionally dense.

A delicate connective tissue membrane called the **endosteum** (en-dos′te-um; "within the bone") covers internal bone surfaces (Figure 6.5). The endosteum covers the trabeculae of spongy bone and lines the canals that pass through the compact bone. The endosteum contains the same cell types as the inner layer of the periosteum.

Blood Vessels and Nerves   Unlike cartilage, bones are well vascularized. The main vessels serving the diaphysis are a *nutrient artery* (Figure 6.5c) and a *nutrient vein*. Together these run through a hole in the wall of the diaphysis, the **nutrient foramen** (fo-ra′men; "opening"). The nutrient artery runs inward to supply the bone marrow and the spongy bone. Branches then extend outward to supply the compact bone. Several *epiphyseal arteries* and *veins* serve each epiphysis in the same way. Nerves accompany blood vessels through the nutrient foramen into the bone.

Hematopoietic Tissue in Bones

Hematopoietic (blood-forming) tissue is also called **red marrow**. It is found in different locations in infants and adults. In infants, the medullary cavity of the diaphysis and all areas of spongy bone contain red bone marrow. In adults, much of the red marrow, particularly in long bones, has been replaced by yellow marrow. In most adult long bones, the fat-containing yellow marrow extends well into the epiphysis, and little red marrow is present in the spongy bone cavities. As a result, red marrow in adults is only found in the cavities between trabeculae of spongy bones in:

- The flat bones of the skull, as well as the sternum, ribs, clavicles, scapulae, hip bones, and vertebrae
- The heads of the femur (thigh bone) and humerus (long bone of the arm)

Compared to long bones (femur and humerus), the red marrow found in the spongy bone of flat bones (such as the sternum) and in some irregular bones (such as the hip bone) is much more active in hematopoiesis. When clinicians suspect problems with the blood-forming tissue, they obtain red marrow samples from these sites. However, yellow marrow in the medullary cavity can revert to red marrow if a person becomes very anemic (has low oxygen-carrying capacity) and needs more red blood cells.

Bone Markings

The external surface of a bone is rarely smooth and featureless. Instead, it has distinct **bone markings** that provide a wealth of information about how that bone and its attached muscles and ligaments work together. Bone markings fit into three categories: (1) *projections* that are sites of muscle and ligament attachment, (2) *surfaces* that form joints, and (3) *depressions* and *openings* for blood vessels and nerves. **Table 6.2** on p. 180 describes the most important bone markings. These will help you understand the detailed structure of the individual bones you will study in Chapter 7.

Figure 6.5 The structure of a long bone (humerus of arm). (a) Anterior view with bone sectioned frontally to show the interior at the proximal end. **(b)** Enlarged view of spongy bone and compact bone of the epiphysis of (a). **(c)** Enlargement of the diaphysis from (a). Note that the external surface of the diaphysis is covered by periosteum, but the articular surface of the epiphysis is covered with hyaline cartilage.

Table 6.2   Bone Markings

NAME OF BONE MARKING	DESCRIPTION	ILLUSTRATIONS
Projections That Are Sites of Muscle and Ligament Attachment		
Tuberosity (too″bĕ-ros′ĭ-te)	Large rounded projection; may be roughened	
Crest	Narrow ridge of bone; usually prominent	
Trochanter (tro-kan′ter)	Very large, blunt, irregularly shaped process (the only examples are on the femur)	
Line	Narrow ridge of bone; less prominent than a crest	
Tubercle (too′ber-kl)	Small rounded projection or process	
Epicondyle (ep″ĭ-kon′dĭl)	Raised area on or above a condyle	
Spine	Sharp, slender, often pointed projection	
Process	Any bony prominence	

Iliac crest
Trochanter
Intertrochanteric line
Ischial spine
Hip bone
Ischial tuberosity
Adductor tubercle
Vertebra
Femur of thigh
Medial epicondyle
Condyle
Spinous process

Surfaces That Help to Form Joints		
Head	Bony expansion carried on a narrow neck	
Facet	Smooth, nearly flat articular (joint) surface	
Condyle (kon′dĭl)	Rounded articular projection; often articulates with a corresponding fossa	

Head
Condyle
Facets
Rib
Mandible

Depressions and Openings		
For Passage of Blood Vessels and Nerves		
Groove	Furrow	
Fissure	Narrow, slitlike opening	
Foramen (fo-ra′men)	Round or oval opening through a bone	
Notch	Indentation at the edge of a structure	
Others		
Meatus (me-a′tus)	Canal-like passageway	
Sinus	Cavity within a bone, filled with air and lined with mucous membrane	
Fossa (fos′ah)	Shallow, basinlike depression in a bone, often serving as an articular surface	

Meatus
Sinus
Inferior orbital fissure
Fossa
Notch
Foramen
Groove
Skull

Microscopic Anatomy of Bone

Cells of Bone Tissue

Five major cell types populate bone tissue: osteoprogenitor cells, osteoblasts, osteocytes, bone lining cells, and osteoclasts (**Figure 6.6**). All of these except for the osteoclasts originate from embryonic connective tissue cells. The presence of bone cells is what makes bone a dynamic living tisue because these cells continuously resorb (break down) and deposit bone in a process called *remodeling*.

Osteoprogenitor Cells   **Osteoprogenitor cells**, also called **osteogenic cells**, are mitotically active stem cells found in the membranous periosteum and endosteum. In growing bones

From bone cell lineage

From white blood cell lineage

Some osteoprogenitor cells become

Some osteoblasts become

(a) Osteoprogenitor cell
- Stem cell

(b) Osteoblast
- Matrix-synthesizing cell
- Responsible for bone growth

(c) Osteocyte
- Mature bone cell
- Monitors and maintains the mineralized bone matrix

(d) Osteoclast
- Bone-resorbing cell

Figure 6.6 Types of bone cells and their derivation. Bone lining cells (not illustrated) are derived from osteoprogenitor cells and look very similar.

they are flattened or squamous cells. When stimulated, these cells differentiate into osteoblasts, while others persist as osteoprogenitor cells.

Osteoblasts   **Osteoblasts** are bone-forming cells that secrete the bone matrix. Like their close relatives, the fibroblasts and chondroblasts, they are actively mitotic. The unmineralized bone matrix they secrete includes collagen (90% of bone protein) and calcium-binding proteins that make up the initial unmineralized bone, or *osteoid*. As described later, osteoblasts also play a role in matrix calcification.

When actively depositing matrix, osteoblasts are cube shaped. When inactive, they resemble the flattened osteoprogenitor cells or may differentiate into bone lining cells. When the osteoblasts become completely surrounded by the matrix being secreted, they become osteocytes.

Osteocytes   The spidery **osteocytes** (Figure 6.6c) are mature bone cells that occupy spaces (lacunae) that conform to their shape. Osteocytes monitor and maintain the bone matrix. They also act as stress or strain "sensors" and respond to mechanical stimuli (bone loading, bone deformation, weightlessness). Osteocytes communicate this information to the cells responsible for bone remodeling (osteoblasts and osteoclasts) so that bone matrix can be made or degraded as mechanical stresses dictate. Osteocytes can also trigger bone remodeling to maintain calcium homeostasis as we will see shortly.

Bone Lining Cells   **Bone lining cells** are flat cells found on bone surfaces where bone remodeling is not going on. Like osteocytes, they are thought to help maintain the matrix.

Osteoclasts   Osteoclasts (**Figures 6.7** and 6.6d) are giant multinucleate cells located at sites of bone resorption. They are derived from the same white blood cell lineage that gives rise to macrophages. When actively resorbing (breaking down) bone, the osteoclasts lie in a shallow depression they have carved out. They have a distinctive *ruffled border* that directly contacts the

bone. The deep plasma membrane infoldings of the ruffled border tremendously increase the surface area for enzymatically degrading the bones and seal off that area from the surrounding matrix.

Microscopic Anatomy of Compact Bone

Although compact bone looks solid, a microscope reveals that it is riddled with passageways that serve as conduits for nerves and blood vessels (see Figure 6.9). (Remember the concentric "tree rings" of hard matrix that allowed you to identify bone tissue in Chapter 4, ◀ **p. 136.**)

Osteon   The structural unit of compact bone is called the **osteon** (os′te-on), or the *Haversian system* (ha-ver′zhen). Each osteon is an elongated cylinder oriented parallel to the long

Bone matrix

Osteocyte within a lacuna

Ruffled border of osteoclast

Nuclei

Figure 6.7 An osteoclast. Photomicrograph of an osteoclast destroying bone tissue (520×).

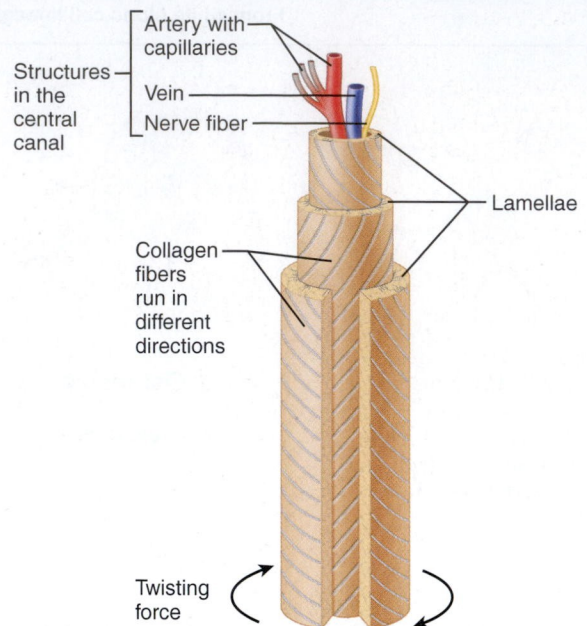

Structures in the central canal
- Artery with capillaries
- Vein
- Nerve fiber

Lamellae

Collagen fibers run in different directions

Lamellae

Twisting force

Figure 6.8 A single osteon. The osteon is drawn so that the individual lamellae are illustrated.

axis of the bone. Functionally, osteons are tiny weight-bearing pillars.

As shown in the "exploded" view in **Figure 6.8**, an osteon is a group of hollow tubes of bone matrix, one placed outside the next like the growth rings of a tree trunk. Each matrix tube is a **lamella** (lah-mel′ah; "little plate"), and for this reason compact bone is often called *lamellar bone*. Although all of the collagen fibers in a particular lamella run in a single direction, the collagen fibers in adjacent lamellae always run in different directions. This alternating pattern beautifully withstands torsional stresses—the adjacent lamellae reinforce one another to resist twisting. You can think of the osteon's design as a "twister resister."

Collagen fibers are not the only part of bone lamellae that are beautifully ordered. The tiny crystals of bone salts align between the collagen fibers and also alternate their direction in adjacent lamellae.

Canals and Canaliculi Running through the core of each osteon is the **central canal,** or *Haversian canal*, containing small blood vessels and nerve fibers that serve the osteon's cells. Canals of a second type called **perforating canals,** or *Volkmann's canals* (folk′mahnz), lie at right angles to the long axis of the bone and connect the blood and nerve supply of the medullary cavity to the central canals (**Figure 6.9a**). Unlike the central canals of osteons, the perforating canals are not surrounded by concentric lamellae, but like all other internal bone cavities, these canals are lined with endosteum.

Spider-shaped osteocytes (Figures 6.6c and 6.9b) occupy **lacunae** (*lac* = hollow; *una* = little) at the junctions of the lamellae. Hairlike canals called **canaliculi** (kan″ah-lik′u-li) radiate from the lacunae, connecting them to each other and to the central canal.

The manner in which canaliculi are formed is interesting. As bone forms, the osteoblasts maintain contact with one another and existing osteocytes by gap junctions (◄ p. 68) in their tentacle-like extensions. The osteoblasts secrete bone matrix, and are trapped within it as it hardens, becoming osteocytes. This leaves a system of tiny canals—the *canaliculi*—filled with tissue fluid and containing the osteocyte extensions. The canaliculi tie all the osteocytes in a mature osteon together, allowing them to communicate and permitting nutrients and wastes to be relayed from one osteocyte to the next throughout the osteon. It's as if the osteocytes are holding hands and passing notes to each other. Although bone matrix is hard and impermeable to nutrients, the canaliculi and gap junctions allow bone cells to be well nourished.

Interstitial and Circumferential Lamellae Not all the lamellae in compact bone are part of complete osteons. Lying between intact osteons are incomplete lamellae called **interstitial lamellae** (in″ter-stish′al) (Figure 6.9c, right). They either fill the gaps between forming osteons or are remnants of osteons that have been cut through by bone remodeling (discussed later). **Circumferential lamellae**, located just deep to the periosteum and just superficial to the endosteum, extend around the entire circumference of the diaphysis (Figure 6.9a) and effectively resist twisting of the long bone.

Microscopic Anatomy of Spongy Bone

In contrast to compact bone, spongy bone looks like a poorly organized, even haphazard, tissue (see Figures 6.4 and 6.5b). However, the trabeculae in spongy bone align precisely along lines of stress and help the bone resist stress. These tiny bone struts are as carefully positioned as the cables on a suspension bridge.

Only a few cells thick, trabeculae contain irregularly arranged lamellae and osteocytes interconnected by canaliculi. No osteons are present. Nutrients reach the osteocytes of spongy bone by diffusing through the canaliculi from capillaries in the endosteum surrounding the trabeculae.

Chemical Composition of Bone

Bone contains both organic and inorganic substances. When these two substances are present in the right proportions, bone is extremely strong and durable without being brittle.

- Its soft *organic components*—including bone cells and *osteoid*—allow it to resist tension (stretch).

- Its hard *inorganic components*—mineral salts—allow it to resist compression.

Healthy bone is half as strong as steel in resisting compression and fully as strong as steel in resisting tension.

Organic Components

The organic components of bone include its cells (osteoprogenitor cells, osteoblasts, osteocytes, bone lining cells, and osteoclasts) and **osteoid** (os′te-oid), the organic part of the matrix. Osteoid, which makes up approximately one-third of the matrix, includes ground substance (◄ p. 128) and collagen

Figure 6.9 Microscopic anatomy of compact bone. (a) Diagram of a section of long bone diaphysis. **(b)** Close-up of a portion of one osteon. **(c)** SEM (left) of cross-sectional view of an osteon (410×). Light photomicrograph (right) of a cross-sectional view of an osteon (400×).

fibers, both of which are secreted by osteoblasts. These organic substances, particularly collagen, contribute both to a bone's structure and to the flexibility and tensile strength that allow it to resist stretch and twisting.

Bone's resilience is thought to come from *sacrificial bonds* in or between collagen molecules. These bonds stretch and break easily on impact, dissipating energy to prevent the force from rising to a fracture value. In the absence of continued or additional trauma, most of the sacrificial bonds re-form.

Inorganic Components

The balance of bone tissue (65% by mass) consists of inorganic *hydroxyapatites* (hi-drok″se-ap′ah-tītz), or *mineral salts*, largely calcium phosphates present as tiny, tightly packed, needle-like crystals in and around collagen fibers in the extracellular matrix. The crystals account for the most notable characteristic of bone—its exceptional hardness, which allows it to resist compression.

Because of the mineral salts they contain, bones last long after death and provide an enduring "monument." In fact, skeletal remains many centuries old reveal the shapes and sizes of ancient peoples, the kinds of work they did, and many of the ailments they suffered, such as arthritis.

Check Your Understanding

9. In the figure below, name the parts of the bone indicated by a and b, and the membranes found at c and d. What kind of cartilage is found at e? Is this bone from an adult or a child? How can you tell?

10. **WHAT IF?** What would happen to a bone if you removed all of the inorganic components (for example, by soaking it in vinegar for several days)? Would it be able to resist compression? Tension?

11. **DRAW** Draw a cross section midway along the diaphysis of a typical long bone in an adult (omit any spongy bone). Label the medullary cavity and indicate the type of marrow found in this cavity. Label the compact bone. Draw and label the periosteum and endosteum.

12. **MAKE CONNECTIONS** Which cell has a ruffled border and acts to break down bone matrix? From your knowledge of organelles (Chapter 3), state which organelle would be the likely source of the enzymes that can digest bone matrix.

For answers, see Answers Appendix.

6.5   Bones develop either by intramembranous or endochondral ossification

Learning Outcomes

▶ **Compare and contrast intramembranous ossification and endochondral ossification.**

▶ **Describe how epiphyseal plates allow long bones to grow.**

Ossification and **osteogenesis** (os″te-o-jen′ĕ-sis) are both names for the process of bone tissue formation (*os* = bone, *genesis* = beginning). In embryos this process leads to the formation of the bony skeleton. Later another form of ossification known as *bone growth* goes on until early adulthood as the body increases in size. Bones are capable of growing thicker throughout life. However, ossification in adults serves mainly for bone remodeling and repair.

Formation of the Bony Skeleton

Before week 8, the embryonic skeleton consists only of fibrous connective tissue membranes and hyaline cartilage. Bone tissue begins to develop at about this time and eventually replaces most of the existing fibrous or cartilage structures.

- In *endochondral ossification* (*endo* = within, *chondro* = cartilage), a bone develops by replacing hyaline cartilage. The resulting bone is called an **endochondral bone**.

- In *intramembranous ossification*, a bone develops from a fibrous membrane and the bone is called a **membranous bone**.

The beauty of using flexible structures (membranes and cartilages) to fashion the embryonic skeleton is that they can accommodate mitosis. If the early skeleton was composed of calcified bone tissue from the outset, growth would be much more difficult.

Endochondral Ossification

Except for the clavicles, essentially all bones below the base of the skull form by **endochondral ossification** (en″do-kon′dral). Beginning late in the second month of development, this process uses hyaline cartilage "bones" formed earlier as models, or patterns, for bone construction. It is more complex than intramembranous ossification because the hyaline cartilage must be broken down as ossification proceeds.

Let's examine the formation of a long bone as an example of endochondral ossification (**Figure 6.10**).

① **A bone collar forms around the diaphysis of the hyaline cartilage model.** Like all cartilage, the cartilage model is initially surrounded by perichondrium. Ossification begins when the underlying *mesenchymal cells* (cells of the embryonic connective tissue, ◄ p. 129) in the deep layer of the perichondrium specialize into osteoblasts. As a result, the perichondrium becomes periosteum. Osteoblasts of the newly converted periosteum secrete osteoid against the hyaline cartilage diaphysis, encasing it in a collar of bone. As the bone collar forms, chondrocytes within the shaft hypertrophy (enlarge). This is the **primary ossification center**.

② **Cartilage calcifies in the center of the diaphysis and then develops cavities.** The hypertrophied chondrocytes calcify the surrounding cartilage matrix. Then, because calcified cartilage matrix is impermeable to diffusing nutrients, the chondrocytes die and the matrix begins to deteriorate. This deterioration opens up cavities, but the bone collar stabilizes the hyaline cartilage model. Elsewhere, the cartilage remains healthy and continues to grow briskly, causing the cartilage model to elongate.

③ **The periosteal bud invades the internal cavities and spongy bone forms.** In month 3, the forming cavities are invaded by a collection of elements called the **periosteal bud**, which contains a nutrient artery and vein, nerve fibers, red marrow elements, osteoprogenitor cells, and osteoclasts. The entering osteoclasts partially erode the calcified cartilage matrix. The osteoprogenitor cells become osteoblasts that secrete osteoid around the remaining calcified fragments of hyaline cartilage. In this way, bone-covered cartilage trabeculae, the earliest version of spongy bone, is formed.

④ **The diaphysis elongates and a medullary cavity forms.** As the primary ossification center enlarges, osteoclasts break down the newly formed spongy bone and open up a medullary cavity in the center of the diaphysis. Throughout the fetal period (week 9 until birth), the rapidly growing epiphyses consist only of cartilage, and the hyaline cartilage models continue to elongate by division of viable cartilage cells at the epiphyses. Ossification "chases" cartilage formation along the length of the shaft as cartilage calcifies,

erodes, and then is replaced by little bony spikes on the epiphyseal surfaces facing the medullary cavity.

At birth, most of our long bones have a bony diaphysis surrounding remnants of spongy bone, a widening medullary cavity, and two cartilaginous epiphyses.

Shortly before or after birth, **secondary ossification centers** develop in one or both epiphyses. Typically, the large long bones form secondary centers in both epiphyses, whereas the small long bones form only one secondary ossification center. (In short bones, only the primary ossification center is formed. Most irregular bones develop from several distinct ossification centers.)

⑤ **The epiphyses ossify.** Secondary ossification reproduces almost exactly the events of primary ossification, except that the spongy bone in the interior is retained and no medullary cavity forms in the epiphyses.

When secondary ossification is complete, hyaline cartilage remains only at two places in a long bone:

• On the epiphyseal surfaces, as the *articular cartilages*

• At the junction of the diaphysis and epiphysis, where it forms the *epiphyseal plates*

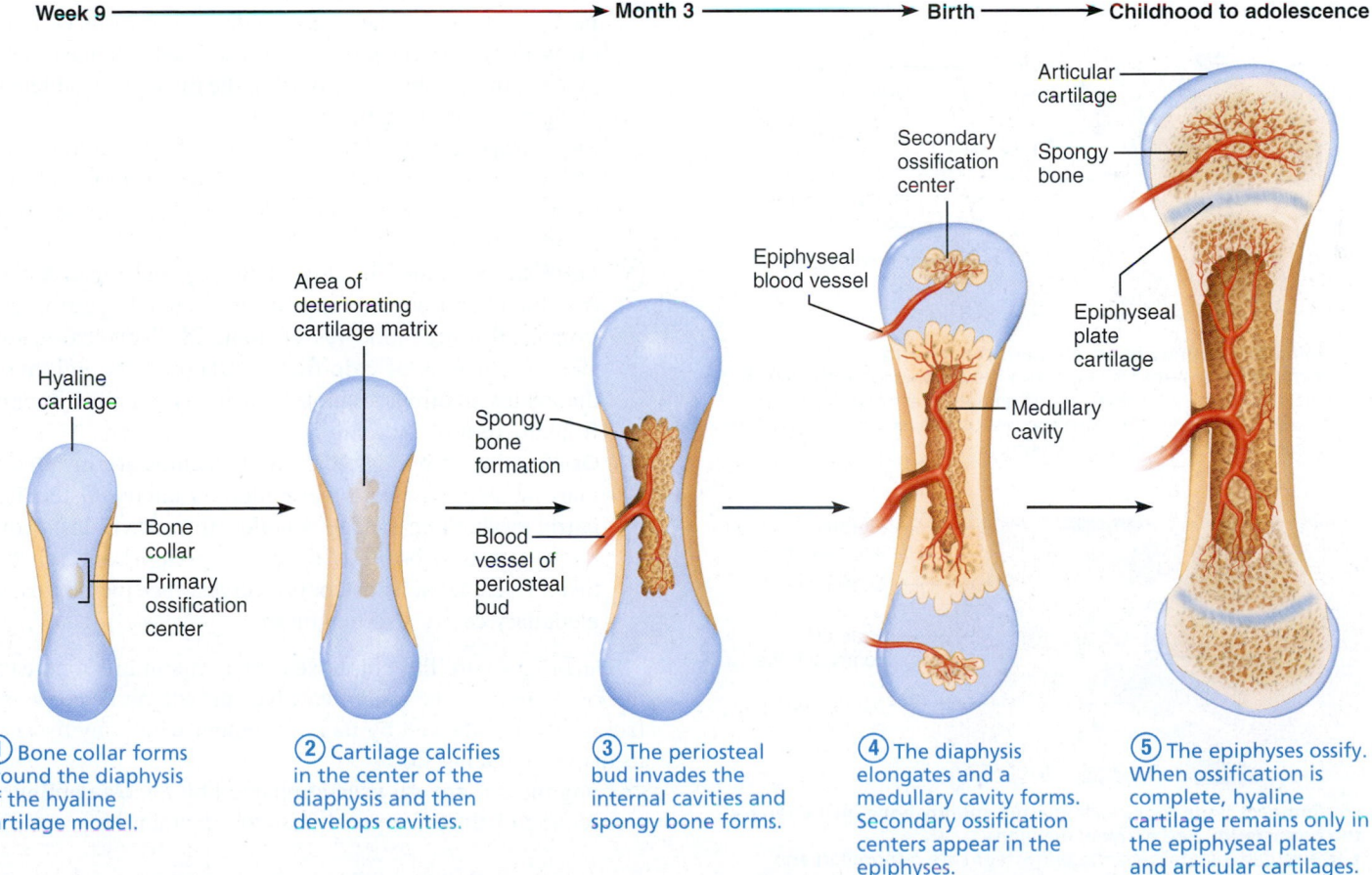

Week 9 ⟶ Month 3 ⟶ Birth ⟶ Childhood to adolescence

Hyaline cartilage

Area of deteriorating cartilage matrix

Bone collar

Primary ossification center

Spongy bone formation

Blood vessel of periosteal bud

Epiphyseal blood vessel

Secondary ossification center

Medullary cavity

Articular cartilage

Spongy bone

Epiphyseal plate cartilage

① Bone collar forms around the diaphysis of the hyaline cartilage model.

② Cartilage calcifies in the center of the diaphysis and then develops cavities.

③ The periosteal bud invades the internal cavities and spongy bone forms.

④ The diaphysis elongates and a medullary cavity forms. Secondary ossification centers appear in the epiphyses.

⑤ The epiphyses ossify. When ossification is complete, hyaline cartilage remains only in the epiphyseal plates and articular cartilages.

Figure 6.10 Endochondral ossification in a long bone.

- Mesenchymal cell
- Collagen fiber
- Ossification center
- Osteoid
- Osteoblast

① Ossification centers develop in the fibrous connective tissue membrane.
- Mesenchymal cells cluster and differentiate into osteoblasts, forming an ossification center.

- Osteoblast
- Osteoid
- Osteocyte
- Newly calcified bone matrix

② Osteoid is secreted and calcifies.
- Osteoblasts continue to secrete osteoid, which calcifies in a few days.
- Trapped osteoblasts become osteocytes.

- Mesenchyme condensing to form the periosteum
- Trabeculae of immature spongy bone
- Blood vessel

③ Immature spongy bone and periosteum form.
- Accumulating osteoid is laid down between embryonic blood vessels, forming a honeycomb of immature spongy bone.
- Vascularized mesenchyme condenses on the external face of the bone and becomes the periosteum.

- Fibrous periosteum
- Osteoblast
- Plate of compact bone
- Spongy bone cavities contain red marrow

④ Compact bone replaces immature spongy bone, just deep to the periosteum. Red marrow develops.
- Trabeculae just deep to the periosteum are remodeled and replaced with compact bone.
- The immature spongy bone in the center is remodeled into mature spongy bone that is eventually filled with red marrow.

Intramembranous Ossification

Intramembranous ossification forms the cranial bones of the skull (frontal, parietal, occipital, and temporal bones) and the clavicles. Most bones formed by this process are flat bones. At about week 8 of development, ossification begins within fibrous connective tissue membranes formed by *mesenchymal cells*. This process involves four major steps, depicted in **Figure 6.11**.

Postnatal Bone Growth

During infancy and youth, long bones lengthen entirely by interstitial growth of the epiphyseal plate cartilage and its replacement by bone, and all bones grow in thickness by appositional growth. Most bones stop growing during adolescence. However, some facial bones, such as those of the nose and lower jaw, continue to grow almost imperceptibly throughout life.

Growth in Length of Long Bones

Longitudinal bone growth mimics many of the events of endochondral ossification and depends on the presence of epiphyseal cartilage. The cartilage is relatively inactive on the side of the epiphyseal plate facing the epiphysis, a region called the *resting zone* (**Figure 6.12**). Below the resting zone, the cartilage cells form tall columns, like coins in a stack, that allow fast, efficient growth.

① **Proliferation zone:** The cells at the "top" (epiphysis-facing) side of the stack next to the resting zone comprise the *proliferation* or *growth zone*. These cells divide quickly, pushing the epiphysis away from the diaphysis and lengthening the entire long bone.

② **Hypertrophic zone:** The older chondrocytes in the stack, which are closer to the diaphysis, hypertrophy (enlarge). Their lacunae erode and enlarge, leaving large interconnecting spaces.

③ **Calcification zone:** The surrounding cartilage matrix calcifies, the chondrocytes die, and the matrix begins to deteriorate, allowing blood vessels to invade. This leaves long, slender spicules of calcified cartilage at the epiphysis-diaphysis junctions, which look like stalactites hanging from the roof of a cave.

④ **Ossification zone:** The calcified spicules are invaded by marrow elements from the medullary cavity. Osteoclasts partly erode the cartilage spicules, then osteoblasts cover them with new bone. Ultimately spongy bone replaces them. Eventually, as osteoclasts digest the spicule tips, the medullary cavity also lengthens.

During growth, the epiphyseal plate maintains a constant thickness because the rate of cartilage growth on its epiphysis-facing side is balanced by its replacement with bony tissue on its diaphysis-facing side.

Longitudinal growth is accompanied by almost continuous remodeling of the epiphyseal ends to maintain the proportion

Figure 6.11 Intramembranous ossification. Diagrams ① and ② represent much greater magnification than diagrams ③ and ④.

(a) **X-ray image of right knee, anterior view. Proximal epiphyseal plate of the tibia enlarged in part (b).**

Figure 6.12 Growth in length of a long bone occurs at the epiphyseal plate. The side of the epiphyseal plate facing the epiphysis contains resting cartilage cells. The cells of the epiphyseal plate proximal to the resting cartilage area are arranged from the earliest stage of growth ① to the region where bone is replacing the cartilage ④.

Calcified cartilage spicule

Osseous tissue

Resting zone

① **Proliferation zone**
Cartilage cells undergo mitosis.

② **Hypertrophic zone**
Older cartilage cells enlarge.

③ **Calcification zone**
Matrix becomes calcified; cartilage cells die; matrix begins deteriorating.

④ **Ossification zone**
New bone is forming.

(b) **Photomicrograph of cartilage in the epiphyseal plate (125×).**

(c) **Diagram of the zones within the epiphyseal plate.**

Practice Histology questions: MasteringA&P® > Study Area > Lab Tools > PAL

between the diaphysis and epiphyses. Bone remodeling involves both new bone formation and bone resorption (**Figure 6.13**).

As adolescence ends, the chondroblasts of the epiphyseal plates divide less often. The plates become thinner and thinner until they are entirely replaced by bone tissue. Longitudinal bone growth ends when the bone of the epiphysis and diaphysis fuses. This process, called *epiphyseal plate closure*, happens at about 18 years of age in females and 21 years of age in males. Once this has occurred, only the articular cartilage remains in bones. However, an adult bone can still widen by appositional growth if stressed by excessive muscle activity or body weight.

Growth in Width (Thickness)

Growing bones widen as they lengthen. As with cartilages, bones increase in thickness or, in the case of long bones, diameter, by appositional growth. Osteoblasts in the periosteum secrete bone matrix on the external bone surface as osteoclasts on the endosteal surface of the diaphysis remove bone (Figure 6.13). Normally there is slightly more building up than breaking down. This unequal process produces a thicker, stronger bone but prevents it from becoming too heavy.

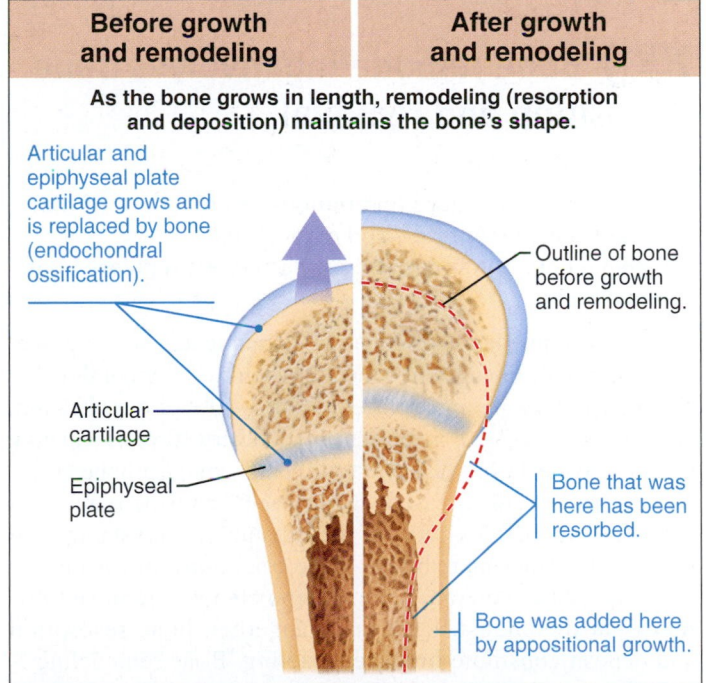

Before growth and remodeling	After growth and remodeling

As the bone grows in length, remodeling (resorption and deposition) maintains the bone's shape.

Articular and epiphyseal plate cartilage grows and is replaced by bone (endochondral ossification).

Articular cartilage

Epiphyseal plate

Outline of bone before growth and remodeling.

Bone that was here has been resorbed.

Bone was added here by appositional growth.

Figure 6.13 Long bone growth and remodeling during youth. Two points in time during growth.

Hormonal Regulation of Bone Growth

The bone growth that occurs until young adulthood is exquisitely controlled by a symphony of hormones. During infancy and childhood, the single most important stimulus of epiphyseal plate activity is *growth hormone* released by the anterior pituitary gland. Thyroid hormones modulate the activity of growth hormone, ensuring that the skeleton has proper proportions as it grows.

At puberty, sex hormones are released in increasing amounts. In both sexes, estrogens have a critical role in bone development. (In males, some circulating testosterone is converted to estrogens.) At low levels, estrogens stimulate the growth spurt typical of adolescence, but as estrogen levels increase over time, they induce epiphyseal closure, ending longitudinal bone growth. Masculinization or feminization of specific parts of the skeleton depends on the relative levels of estrogens and testosterone.

Excesses or deficits of any of these hormones can result in abnormal skeletal growth. For example, hypersecretion of growth hormone in children results in excessive height (gigantism), and deficits of growth hormone or thyroid hormone produce characteristic types of dwarfism.

Check Your Understanding

13. Bones don't begin with bone tissue. What do they begin with?
14. When describing endochondral ossification, some say "bone chases cartilage." What does that mean?
15. Where is the primary ossification center located in a long bone? Where is (are) the secondary ossification center(s) located?
16. As a long bone grows in length, what is happening in the hypertrophic zone of the epiphyseal plate?

For answers, see Answers Appendix.

6.6 Bone remodeling involves bone deposition and removal

Learning Outcomes

▶ **Compare the locations and remodeling functions of the osteoblasts, osteocytes, and osteoclasts.**
▶ **Explain how hormones and physical stress regulate bone remodeling.**

As you have just learned, bone is a dynamic and active tissue, and small-scale changes in bone architecture occur continually. Every year, remodeling replaces about 5–10% of our skeleton, and our entire skeleton is replaced about every 10 years. Spongy bone is replaced every three to four years; compact bone, every 10 years or so. This is essential because when bone remains in place for long periods, more of the calcium salts crystallize and the bone becomes more brittle—ripe conditions for fracture.

In the adult skeleton, bone is deposited and removed primarily at the endosteal surface. Together, bone resorption and deposit constitute **bone remodeling**. Bone remodeling is coordinated by cohorts of adjacent osteoblasts and osteoclasts, which respond to signals from hormones and stress-sensing osteocytes (discussed shortly).

Remodeling does not occur uniformly. For example, the distal part of the femur (thigh bone) is fully replaced every five to six months, whereas its shaft is altered much more slowly.

Bone Resorption

As noted earlier, the giant **osteoclasts** accomplish **bone resorption**. Osteoclasts move along a bone surface, digging depressions or grooves as they break down the bone matrix. The ruffled border of the osteoclast clings tightly to the bone, sealing off the area of bone destruction and secreting acid (H^+) that dissolve the bone minerals and *lysosomal enzymes* that digest the organic matrix. The digested matrix end products and dissolved minerals are then endocytosed, transported across the osteoclast (by transcytosis), and released at the opposite side. There they enter the interstitial fluid and then the blood. Osteoclasts may also phagocytize dead osteocytes and any undigested demineralized matrix. When resorption of a given area of bone is completed, the osteoclasts undergo apoptosis (cell death).

Bone Deposition

Following behind the osteoclasts, osteoblasts deposit new bone matrix. This begins as an *osteoid seam*—an unmineralized band of gauzy-looking bone matrix 10–12 micrometers (μm) wide. Between the osteoid seam and the older mineralized bone, there is an abrupt transition called the *calcification front*.

How does the osteoid become calcified? The proteins of the newly deposited osteoid bind calcium ions (Ca^{2+}), raising the local concentration of Ca^{2+}. This in turn triggers osteoblasts to release *matrix vesicles* studded with the enzyme *alkaline phosphatase*. This enzyme cuts phosphate (P_i) ions off of the osteoid proteins, raising the local concentration of P_i. When the concentrations of Ca^{2+} and P_i are high enough, tiny calcium phosphate crystals form. These crystals then act as the seed around which hydroxyapatites (the mineral salts in bones) form. The result is that calcium salts are deposited throughout the osteoid, creating calcified bone matrix.

Control of Remodeling

Remodeling goes on continuously in the skeleton, primarily regulated by two control loops that serve different purposes:

- *Maintaining Ca^{2+} homeostasis.* A hormonal negative feedback loop involving parathyroid hormone maintains Ca^{2+} homeostasis in the blood.
- *Keeping bone strong.* Mechanical and gravitational forces acting on a bone drive remodeling where it is required to strengthen that bone.

Hormonal Controls

To understand the hormonal control of bone remodeling, you need to understand calcium's importance in the body. *Maintaining extracellular fluid calcium levels within homeostatic levels is absolutely critical for maintaining the resting membrane potential of all cells.* Without normal levels of blood calcium, nerves cannot fire as needed and muscles are unable to contract.

HOMEOSTATIC IMBALANCE 6.1   CLINICAL

Minute changes from the homeostatic range for blood calcium can lead to severe neuromuscular problems. For example, *hypocalcemia* (hi″po-kal-se′me-ah; low blood Ca^{2+} levels) causes hyperexcitability. In contrast, *hypercalcemia* (hi″per-kal-se′me-ah; high blood Ca^{2+} levels) causes nonresponsiveness and inability to function. In addition, sustained high blood levels of Ca^{2+} can lead to the formation of kidney stones or undesirable deposits of calcium salts in other organs, which may hamper their function.

The human body contains 1200–1400 g of calcium. More than 99% of this calcium is contained in bone minerals, 0.9% is inside cells, and less than 0.1% is in the extracellular fluid. Hormones use the vast amount of calcium in bone as a "storage bank" from which they can make withdrawals (resorption) or deposits as needed to maintain extracellular fluid calcium levels. Hormonal control normally maintains blood Ca^{2+} within the narrow range of 9–11 mg per dl (100 ml) of blood. Calcium is absorbed from the intestine under the control of the active form of vitamin D, called *calcitriol*.

The hormonal controls primarily involve **parathyroid hormone (PTH)**, produced by the parathyroid glands. When blood levels of Ca^{2+} decline, PTH is released (**Figure 6.14**). The increased PTH level stimulates osteoclasts to resorb bone, releasing Ca^{2+} into the blood. The PTH stimulation of osteoclasts is indirect—various cells (e.g., osteoblasts) respond to PTH secretion by producing another protein (*RANK-L*) that stimulates the formation and activity of osteoclasts.

As blood concentrations of calcium rise, the stimulus for PTH release ends. The decline of PTH reverses its effects and causes blood Ca^{2+} levels to fall.

Control by PTH acts to preserve blood calcium homeostasis, not the skeleton's strength or well-being. Osteoclasts are no respecters of matrix age: When activated, they break down both old and new matrix. In fact, if blood calcium levels are low for an extended time, the bones become so demineralized that they develop large holes. We will discuss blood calcium regulation further in Chapter 16.

Another hormone, **calcitonin** (kal″sĭ-to′nin) produced by parafollicular cells of the thyroid gland, has been thought to regulate blood Ca^{2+} levels as well. However, in humans, calcitonin appears to be a hormone in search of a function. At normal levels, its effects on calcium homeostasis are negligible. When administered at pharmacological (abnormally high) doses, it does lower blood calcium levels temporarily.

Various other hormones also affect bone resorption and deposition by either stimulating or inhibiting osteoclast activity. Like PTH, glucocorticoids (hormones from the adrenal cortex) and vitamin D indirectly stimulate osteoclast activity by increasing the synthesis of RANK-L and decreasing the synthesis of its antagonist (*osteoprotegerin*). Sex hormones, on the other hand, have the opposite effect—they increase the synthesis of RANK-L's antagonist.

Response to Mechanical Stress

The second set of controls regulating bone remodeling, bone's response to mechanical stress (muscle pull and gravity), keeps the bones strong where stressors are acting.

Wolff's law holds that a bone grows or remodels in response to the demands placed on it. The first thing to understand is that a bone's anatomy reflects the common stresses it encounters. For example, a bone is loaded (stressed) whenever weight bears down on it or muscles pull on it. This loading is usually off center and tends to bend the bone. Bending compresses the

Figure 6.14 Parathyroid hormone (PTH) control of blood calcium levels.

Load here (body weight) threatens to bend this bone along this arc.

Head of femur (thigh bone)

Tension here

Compression here

Point of no stress. The tension (pull) and compression on opposite sides cancel each other. As a result, much less bone material is needed internally than superficially.

Figure 6.15 Bone anatomy and bending stress.

bone on one side and subjects it to tension (stretching) on the other (**Figure 6.15**).

In response to these mechanical stressors, the compact bone of long bones is thickest midway along the diaphysis, exactly where bending stresses are greatest (bend a stick and it will split near the middle). As Figure 6.15 shows, both compression and tension are minimal toward the center of the bone (they cancel each other out), so a bone can "hollow out" for lightness (using spongy bone instead of compact bone) without jeopardy.

Wolff's law also explains several other observations:

- Handedness (being right- or left-handed) results in the bones of one upper limb being thicker than those of the less-used limb.
- Curved bones are thickest where they are most likely to buckle.
- The trabeculae of spongy bone form trusses, or struts, along lines of compression.
- Large, bony projections occur where heavy, active muscles attach. The bones of weight lifters have enormous thickenings at the attachment sites of the most-used muscles.

Wolff's law also explains the featureless bones of the fetus and the atrophied bones of bedridden people—situations in which bones are not stressed.

How do mechanical forces communicate with the cells responsible for remodeling? Deforming bone pushes fluid containing ions through the canaliculi. This creates an electrical current. This electrical current is detected by osteocytes, which release chemical messengers that promote the formation of additional bone. This process underlies some of the electrical devices used to speed healing of fractures.

Summary of Control of Remodeling

The skeleton is continuously subjected to both hormonal influences and mechanical forces. At the risk of oversimplifying, the take-home message is:

- Hormonal controls determine *whether* and *when* remodeling occurs.
- Mechanical stress determines *where* remodeling occurs.

For example, when bone must be broken down to increase blood calcium levels, PTH is released and osteoclasts are activated. However, mechanical forces acting on osteocytes direct which osteoclasts are called into action, so that bone in the least stressed areas (which is temporarily dispensable) is broken down first.

In healthy young adults, total bone mass remains constant, an indication that the rates of bone deposit and resorption are essentially equal. Both environmental and genetic factors can tip this delicate balance in favor of bone loss or bone synthesis. Loss of this balance is the basis of several bone diseases (discussed on pp. 193–194).

Check Your Understanding

17. Which stimulus—PTH (a hormone) or mechanical forces acting on the skeleton—is more important in maintaining homeostatic blood calcium levels?

18. How do bone growth and bone remodeling differ?

19. PREDICT If osteoclasts in a long bone are more active than osteoblasts, how will bone mass change?

20. APPLY Despite daily exercise, the bones of astronauts in the International Space Station get thinner and weaker during their time in space. Why does this occur?

For answers, see Answers Appendix.

6.7 Bone repair involves hematoma and callus formation, and remodeling CLINICAL

Learning Outcome

▶ Describe the steps of fracture repair.

Despite their remarkable strength, bones are susceptible to **fractures**, or breaks. During youth, most fractures result from trauma that twists or smashes the bones (sports injuries, automobile accidents, and falls, for example). In old age, most fractures result from thinner and weaker bones. When we break bones—the most common disorder of bone homeostasis—they undergo a remarkable process of self-repair.

Fracture Classification

Fractures may be classified by:

- Position of the bone ends after fracture: In *nondisplaced fractures*, the bone ends retain their normal position. In *displaced fractures*, the bone ends are out of normal alignment.

- Completeness of the break: If the bone is broken through, the fracture is a *complete fracture*. If not, it is an *incomplete fracture*.

- Whether the bone ends penetrate the skin: If so, the fracture is an *open (compound) fracture*. If not, it is a *closed (simple) fracture*.

In addition to these three classifications, all fractures can be described in terms of the location of the fracture, its external appearance, and/or the nature of the break (**Table 6.3** on p. 192).

Fracture Treatment and Repair

Treatment begins with *reduction*, the realignment of the broken bone ends. In *closed (external) reduction*, the physician's hands coax the bone ends into position. In *open (internal) reduction*, the bone ends are secured together surgically with screws and plates.

After the broken bone is reduced, it is immobilized either by a cast or traction to allow healing. A simple fracture of small or medium-sized bones in young adults heals in six to eight weeks, but it takes much longer for large, weight-bearing bones and for bones of elderly people (because of their poorer circulation).

Repair in a simple fracture involves four major stages (**Figure 6.16**):

① **A hematoma forms.** When a bone breaks, blood vessels in the bone and periosteum (and perhaps in surrounding tissues) are torn. The hemorrhaged blood clots, forming a **hematoma** (he″mah-to′mah) at the fracture site. Soon, bone cells deprived of nutrition die, and the tissue at the site becomes swollen, painful, and inflamed.

② **Fibrocartilaginous callus forms.** Within a few days, new blood vessels grow into the clot. Fibroblasts and chondroblasts invade the fracture site from the nearby periosteum and endosteum. The fibroblasts produce collagen fibers that span the break and connect the broken bone ends. The chondroblasts secrete a cartilaginous matrix that bulges externally and later calcifies, forming a **fibrocartilaginous callus** that spans the break and connects the broken bone ends. This entire mass of repair tissue, also called the *soft callus* (kal′us; "hard skin"), splints the broken bone.

③ **Bony callus forms.** Within this mass of repair tissue, osteoblasts begin forming spongy bone. Within a week, osteoblasts begin to lay down trabeculae of new bone around and within the fibrocartilaginous callus. These trabeculae span the width of the callus and unite the two fragments of broken bone. Gradually the fibrocartilaginous callus is replaced by immature bone, converting it to a **bony (hard) callus**. Bony callus formation continues until a firm union forms about two months later. This process generally repeats the events of endochondral ossification.

④ **Bone remodeling occurs.** Beginning during bony callus formation and continuing for several months after, the bony callus is remodeled. The excess material on the diaphysis exterior and within the medullary cavity is removed. Compact bone is laid down to reconstruct the shaft walls. The repaired area resembles the original unbroken bony region because it responds to the same set of mechanical stressors.

There is evidence that daily ultrasound treatments or electrical stimulation of fracture sites may hasten repair and healing.

Check Your Understanding

21. How does an open fracture differ from a closed fracture?

For answers, see Answers Appendix.

① A hematoma forms.   ② Fibrocartilaginous callus forms.   ③ Bony callus forms.   ④ Bone remodeling occurs.

Figure 6.16 Stages in the healing of a bone fracture.

Table 6.3 Common Types of Fractures

FRACTURE TYPE	DESCRIPTION AND COMMENTS	FRACTURE TYPE	DESCRIPTION AND COMMENTS
Comminuted	Bone fragments into three or more pieces. Particularly common in the aged, whose bones are more brittle.	**Compression**	Bone is crushed. Common in porous bones (i.e., osteoporotic bones) subjected to extreme trauma, as in a fall.

Crushed vertebra

Spiral	Ragged break occurs when excessive twisting forces are applied to a bone. Common sports fracture.	**Epiphyseal**	Epiphysis separates from the diaphysis along the epiphyseal plate. Tends to occur where cartilage cells are dying and calcification of the matrix is occurring.

Depressed	Broken bone portion is pressed inward. Typical of skull fracture.	**Greenstick**	Bone breaks incompletely, much in the way a green twig breaks. Only one side of the shaft breaks; the other side bends. Common in children, whose bones have relatively more organic matrix and are more flexible than those of adults.

6.8 Bone disorders result from abnormal bone deposition and resorption

CLINICAL

Learning Outcome

▶ Contrast the disorders of bone remodeling seen in osteoporosis, osteomalacia, and Paget's disease.

Imbalances between bone deposit and bone resorption underlie nearly every disease that affects the human skeleton.

Osteomalacia and Rickets

Osteomalacia (os″te-o-mah-la′she-ah; "soft bones") includes a number of disorders in which the bones are poorly mineralized. Osteoid is produced, but calcium salts are not adequately deposited, so bones are soft and weak. The main symptom is pain when weight is put on the affected bones.

Rickets is the analogous disease in children. Because young bones are still growing rapidly, rickets is much more severe than adult osteomalacia. Bowed legs and deformities of the pelvis, skull, and rib cage are common. Because the epiphyseal plates cannot calcify, they continue to widen, and the ends of long bones become visibly enlarged and abnormally long.

Osteomalacia and rickets are caused by insufficient calcium in the diet or by a vitamin D deficiency. Increasing vitamin D intake and exposing the skin to sunlight (which spurs the body to form vitamin D) usually cure these disorders. Although rickets has been nearly eliminated in North America, it still sometimes occurs. For example, if a mother who breast-feeds her infant becomes vitamin D deficient because of sun deprivation or dreary winter weather, the infant too will be vitamin D deficient and will develop rickets.

Osteoporosis

For most of us, the phrase "bone problems of the elderly" brings to mind the stereotype of a person with osteoporosis— a hunched-over old woman shuffling behind her walker. **Osteoporosis** (os″te-o-po-ro′sis) refers to a group of diseases in which bone resorption outpaces bone deposit. The composition of the matrix remains normal but bone mass declines, and the bones become porous and light (**Figure 6.17**). The bones become so fragile that they can break from something as simple as stepping off a curb or a hearty sneeze.

Osteoporosis occurs most often in postmenopausal women. Worldwide, one in three women over 50 will experience a bone fracture due to osteoporosis. Although osteoporosis affects the entire skeleton, fractures at certain sites are most common— compression fractures of the vertebrae and fractures of the neck of the femur (a *broken hip*).

Risk Factors for Osteoporosis

Several factors can contribute to osteoporosis:

- *Decreased sex hormones.* Androgens in males and estrogens in females restrain osteoclasts and promote deposit of new bone, helping to maintain the normal density of the skeleton. Estrogen secretion decreases after menopause and this is strongly implicated in osteoporosis in older women.

(a) Normal bone

(b) Osteoporotic bone

Figure 6.17 **The contrasting architecture of normal versus osteoporotic bone.** Scanning electron micrographs, 300×.

- *Insufficient bone stress.* Low body weight or insufficient weight-bearing exercise hinders new bone deposit.
- A *diet poor in calcium, vitamin D, or protein.*
- *Smoking* (which reduces estrogen levels and calcium absorption).
- *Genetics.* Osteoporosis runs in families.
- *Hormone-related conditions* such as diabetes mellitus and hyperthyroidism.
- *Consumption of alcohol or certain medications* (such as steroids and certain cancer drugs).

Preventing and Treating Osteoporosis

Osteoporosis can be prevented, or at least delayed, by minimizing the factors listed above that you can control. In particular, have an adequate diet and get lots of load-bearing exercise. Osteoporosis is treated with weight-bearing exercise and by ensuring that calcium and vitamin D intake is adequate. A number of different categories of drugs are also available. These include bisphosphonates and the RANK-L inhibitor denosumab, both of which inhibit osteoclasts and therefore have bone-sparing properties.

Paget's Disease

Often discovered by accident when X rays are taken for some other reason, **Paget's disease** (paj′ets) is characterized by excessive and haphazard bone deposit and resorption (**Figure 6.18**). The newly formed bone, called *pagetic bone*, is hastily made and has an abnormally high ratio of spongy bone to compact bone. This, along with reduced mineralization, causes a spotty weakening of the bones. Late in the disease, osteoclast activity wanes, but osteoblasts continue to work, often forming irregular bone thickenings or filling the marrow cavity with pagetic bone.

Paget's disease may affect any part of the skeleton, but it is usually a localized condition. The spine, pelvis, femur, and skull are most often involved and become increasingly deformed and painful. It rarely occurs before age 40, and it affects about 3% of North American elderly people. Its cause is unknown, but a virus may trigger it. Drug therapies include bisphosphonates and calcitonin, which have shown success in preventing bone breakdown.

Check Your Understanding

22. Which bone disorder is characterized by excessive deposit of weak, poorly mineralized bone?
23. What are three measures that may help to maintain healthy bone density?
24. What name is given to "adult rickets"?

For answers, see Answers Appendix.

(a) Normal bone

(b) Pagetic bone

Figure 6.18 Moth-eaten appearance of bone with Paget's disease. X rays of the pelvis.

Developmental Aspects of Bones

Bones are on a precise schedule from the time they form until death. The mesoderm germ layer gives rise to embryonic mesenchymal cells, which in turn produce the membranes and cartilages that form the embryonic skeleton. These structures then ossify according to a predictable timetable that allows fetal age to be determined easily using ultrasound. Although each bone has its own developmental schedule, most long bones begin ossifying by 8 weeks after conception and have well-developed primary ossification centers by 12 weeks (**Figure 6.19**).

Birth to Young Adulthood

At birth, most long bones of the skeleton are well ossified except for their epiphyses. After birth, secondary ossification centers develop in a predictable sequence. The epiphyseal plates persist and provide for long bone growth all through childhood and the sex hormone–mediated growth spurt at adolescence. By age 25, nearly all bones are completely ossified and skeletal growth ceases.

Age-Related Changes in Bone

In children and adolescents, bone formation exceeds bone resorption. In young adults, these processes are in balance, and in old age, resorption predominates. Despite the environmental factors that influence bone density, genetics still plays the major role in determining how much a person's bone density will change over a

Figure 6.19 Fetal primary ossification centers at 12 weeks. The darker areas indicate primary ossification centers in the skeleton of a 12-week-old fetus.

lifetime. A single gene that codes for vitamin D's cellular docking site helps determine both the tendency to accumulate bone mass during early life and a person's risk of osteoporosis later in life.

Beginning in the fourth decade, bone mass decreases with age. The only exception appears to be in bones of the skull. Among young adults, skeletal mass is generally greater in males than in females. Age-related bone loss is faster in white people than in black people (who have greater bone density to begin with) and faster in females than in males.

Qualitative changes also occur: More osteons remain incompletely formed, mineralization is less complete, and the amount of nonviable bone increases, reflecting a diminished blood supply to the bones in old age. These age-related changes are also bad news because fractures heal more slowly in older adults.

RELATED CLINICAL TERMS

Achondroplasia (a-kon″dro-pla′ze-ah; *a* = without; *chondro* = cartilage; *plasi* = mold, shape) A congenital condition involving defective cartilage and endochondral bone growth so that the limbs are too short but the membrane bones are of normal size; a type of dwarfism.

Bony spur Abnormal projection from a bone due to bony overgrowth; common in aging bones.

Ostealgia (os″te-al′je-ah; *algia* = pain) Pain in a bone.

Osteitis (os″te-i′tis; *itis* = inflammation) Inflammation of bony tissue.

Osteogenesis imperfecta Also called brittle bone disease, a disorder in which the bone matrix contains inadequate collagen, putting it at risk for shattering.

Osteomyelitis (os″te-o-mi″ĕ-li′tis) Inflammation of bone and bone marrow caused by pus-forming bacteria that enter the body via a wound (e.g., compound bone fracture), or spread from an infection near the bone. Commonly affects the long bones, causing acute pain and fever. May result in joint stiffness, bone destruction, and shortening of a limb. Treatment involves antibiotics, draining any abscesses (local collections of pus), and removing dead bone fragments (which prevent healing).

Osteosarcoma (os″te-o-sar-ko′mah) A form of bone cancer typically arising in a long bone of a limb and most often in those 10–25 years of age. Grows aggressively, painfully eroding the bone; tends to metastasize to the lungs and cause secondary lung tumors. Treatment usually includes chemotherapy and surgery, and sometimes radiation. Survival rate is about 50% if detected early.

Pathologic fracture Fracture in a diseased bone involving slight (coughing or a quick turn) or no physical trauma. For example, a hip bone weakened by osteoporosis may break and cause the person to fall, rather than breaking because of the fall.

Traction ("pulling") Placing sustained tension on a body region to keep the parts of a fractured bone in proper alignment. Also prevents spasms of skeletal muscles, which would separate the fractured bone ends or crush the spinal cord in the case of vertebral column fractures.

CHAPTER SUMMARY

6.1 Hyaline, elastic, and fibrocartilage help form the skeleton (pp. 174–175)

Basic Structure, Types, and Locations (p. 174)

1. A skeletal cartilage exhibits chondrocytes housed in lacunae (cavities) within the extracellular matrix (ground substance and fibers). It contains large amounts of water (which accounts for its resilience), lacks nerve fibers, is avascular, and is surrounded by a fibrous perichondrium that resists expansion.

2. Hyaline cartilages appear glassy; the fibers are collagenous. They provide support with flexibility and resilience and are the most abundant skeletal cartilages, accounting for the articular, costal, respiratory, and nasal cartilages.

3. Elastic cartilages contain abundant elastic fibers in addition to collagen fibers, and are more flexible than hyaline cartilages. They support the outer ear and epiglottis.

4. Fibrocartilages, which contain thick collagen fibers, are the most compressible cartilages and resist stretching. They form intervertebral discs and knee joint cartilages.

Growth of Cartilage (p. 174)

5. Cartilages grow from within (interstitial growth) and by adding new cartilage tissue at the periphery (appositional growth).

6.2 Bones perform several important functions (pp. 175–176)

1. Bones give the body shape; protect and support body organs; provide levers for muscles to pull on; store calcium and other minerals; store triglyceride; and are the site of blood cell and osteocalcin production.

6.3 Bones are classified by their location and shape (p. 176)

1. Bones are classified as long, short, flat, or irregular on the basis of their shape, and as axial or appendicular based on their body location.

6.4 The gross structure of all bones consists of compact bone sandwiching spongy bone (pp. 176–184)

Gross Anatomy (pp. 177–179)

1. Flat bones consist of two thin plates of compact bone enclosing a spongy bone layer (diploë). Short and irregular bones resemble flat bones structurally.

2. A long bone is composed of a diaphysis (shaft) and epiphyses (ends). The medullary cavity of the diaphysis contains yellow marrow; the epiphyses contain spongy bone. The epiphyseal line is the remnant of the epiphyseal plate. Periosteum covers the diaphysis; endosteum lines inner bone cavities. Hyaline cartilage covers joint surfaces.

3. In adults, hematopoietic tissue (red marrow) is found within the spongy bone of flat bones and occasionally within the epiphyses of long bones. In infants, red marrow is also found in the medullary cavity.

4. Bone markings are important anatomical landmarks that reveal sites of muscle attachment, points of articulation, and sites of blood vessel and nerve passage.

6

Homeostatic Interrelationships between the Skeletal System and Other Body Systems

Endocrine System   Chapter 16
- Skeletal system provides some bony protection; stores calcium needed for second-messenger signaling mechanisms
- Hormones regulate uptake and release of calcium from bone; promote long bone growth and maturation

Cardiovascular System   Chapters 17–19
- Bone marrow cavities provide site for blood cell formation; matrix stores calcium needed for cardiac muscle activity
- Cardiovascular system delivers nutrients and oxygen to bones; carries away wastes

Lymphatic System/Immunity   Chapters 20–21
- Skeletal system provides some protection to lymphatic organs; bone marrow is site of origin for lymphocytes involved in immune response
- Lymphatic vessels pick up leaked plasma fluids and proteins from the periostea; immune cells protect against pathogens

Respiratory System   Chapter 22
- Skeletal system protects lungs by enclosure (rib cage)
- Respiratory system provides oxygen; disposes of carbon dioxide

Digestive System   Chapter 23
- Skeletal system provides some bony protection to intestines, pelvic organs, and liver
- Digestive system provides nutrients needed for bone health and growth

Urinary System   Chapters 25–26
- Skeletal system protects pelvic organs (urinary bladder, etc.)
- Kidneys activate vitamin D; dispose of nitrogenous wastes

Reproductive System   Chapter 27
- Skeletal system protects some reproductive organs by enclosure
- Gonads produce hormones that influence the form of the skeleton and epiphyseal closure

Integumentary System   Chapter 5
- Skeletal system provides support for body organs including the skin
- Skin provides vitamin D needed for proper calcium absorption and use

Muscular System   Chapters 9–10
- Skeletal system provides levers plus ionic calcium for muscle activity
- Muscle pull on bones increases bone strength and viability; helps determine bone shape

Nervous System   Chapters 11–15
- Skeletal system protects brain and spinal cord; provides depot for calcium ions needed for neural function
- Nerves innervate bone and joint capsules, providing for pain and joint sense

Microscopic Anatomy of Bone (pp. 180–182)

5. There are five types of bone cells—osteoprogenitor cells (bone stem cells), osteoblasts (matrix-synthesizing cells), osteocytes (bone matrix maintenance cells), bone lining cells (line surfaces where no bone activity is ongoing), and osteoclasts (bone destruction cells).

6. The structural unit of compact bone, the osteon, consists of a central canal surrounded by concentric lamellae of bone matrix. Osteocytes, embedded in lacunae, are connected to each other and the central canal by canaliculi.

7. Spongy bone has slender trabeculae containing irregular lamellae.

Chemical Composition of Bone (pp. 182–184)

8. Bone is composed of living cells and matrix. The extracellular matrix includes osteoid, organic substances that are secreted by osteoblasts and give the bone tensile strength. Its inorganic (mineral) components, the hydroxyapatites (calcium salts), make bone hard.

6.5 Bones develop either by intramembranous or endochondral ossification (pp. 184–188)

Formation of the Bony Skeleton (pp. 184–186)

1. Most bones are formed by endochondral ossification of a hyaline cartilage model. Osteoblasts beneath the periosteum secrete bone matrix on the cartilage model, forming the bone collar. As the cartilage model deteriorates, internal cavities open up, allowing periosteal bud entry. Bone matrix is deposited around the cartilage remnants but is later broken down.

2. Intramembranous ossification forms the clavicles and most skull bones. The ground substance of the bone matrix is deposited between collagen fibers within the fibrous membrane to form bone. Eventually, compact bone plates enclose the spongy bone.

Postnatal Bone Growth (pp. 186–188)

3. Long bones increase in length by interstitial growth of the epiphyseal plate cartilage and its replacement by bone.

4. Appositional growth increases bone diameter/thickness.

6.6 Bone remodeling involves bone deposition and removal (pp. 188–190)

1. Bone is continually deposited and resorbed in response to hormonal and mechanical stimuli. Together these processes constitute bone remodeling.

2. An unmineralized osteoid seam appears at areas of new bone deposit; calcium salts are deposited later.

3. Osteoclasts release lysosomal enzymes and acids on bone surfaces to be resorbed. The dissolved products are transcytosed to the opposite face of the osteoclast for release to the extracellular fluid.

4. The hormonal controls of bone remodeling serve blood calcium homeostasis. When blood calcium levels decline, PTH is released and stimulates osteoclasts to digest bone matrix, releasing ionic calcium. As blood calcium levels rise, PTH secretion declines.

5. Mechanical stress and gravity acting on the skeleton help maintain skeletal strength. Bones thicken, develop heavier prominences, or rearrange their trabeculae in sites where stressed.

6.7 Bone repair involves hematoma and callus formation, and remodeling (pp. 190–192)

1. Fractures are treated by open or closed reduction. The healing process involves formation of a hematoma, a fibrocartilaginous callus, a bony callus, and bone remodeling, in succession.

6.8 Bone disorders result from abnormal bone deposition and resorption (pp. 193–194)

1. Imbalances between bone formation and resorption underlie all skeletal disorders.

2. Osteomalacia and rickets occur when bones are inadequately mineralized. The bones become soft and deformed. The most frequent cause is inadequate vitamin D.

3. Osteoporosis is any condition in which bone breakdown outpaces bone formation, causing bones to become weak and porous. Postmenopausal women are particularly susceptible.

4. Paget's disease is characterized by excessive and abnormal bone remodeling.

Developmental Aspects of Bones (pp. 194–195)

1. Osteogenesis is predictable and precisely timed.

2. Longitudinal long bone growth continues until the end of adolescence. Skeletal mass increases dramatically during puberty and adolescence, when formation exceeds resorption.

3. Bone mass is fairly constant in young adulthood, but beginning in the 30s, bone resorption exceeds formation.

REVIEW QUESTIONS

To access additional practice questions using your smartphone, tablet, or computer: MasteringA&P® > Study Area > Practice Tests & Quizzes

Level 1 Remember/Understand

(Some questions have more than one correct answer. Select the best answer or answers from the choices given.)

1. Which is *not* a function of the skeletal system? (a) support, (b) hematopoietic site, (c) storage, (d) providing levers for muscle activity, (e) generating heat.

2. Sites of hematopoiesis include all but (a) red marrow of spongy bone, (b) the heads of femur and humerus in adults, (c) medullary cavities in bones of infants, (d) medullary cavities in bones of a healthy adult.

3. The organic portion of matrix is important in providing all but (a) tensile strength, (b) hardness, (c) ability to resist stretch, (d) flexibility.

4. Bone remodeling in adults is regulated and directed mainly by (a) growth hormone, (b) thyroid hormones, (c) sex hormones, (d) mechanical stress, (e) PTH.

5. Where within the epiphyseal plate are the dividing cartilage cells located? (a) between the calcification zone and the ossification zone, (b) between the hypertrophic zone and the calcification zone, (c) between the resting zone and the hypertrophic zone, (d) in the primary ossification center.

6. Wolff's law is concerned with (a) calcium homeostasis of the blood, (b) the shape of a bone being determined by mechanical stresses placed on it, (c) the electrical charge on bone surfaces.

7. Formation of the bony callus in fracture repair is followed by (a) hematoma formation, (b) fibrocartilaginous callus formation, (c) bone remodeling, (d) formation of granulation tissue.

8. The fracture type in which the bone ends are incompletely separated is **(a)** greenstick, **(b)** compound, **(c)** simple, **(d)** comminuted, **(e)** compression.
9. The disorder in which bones are porous and thin but bone composition is normal is **(a)** osteomalacia, **(b)** osteoporosis, **(c)** Paget's disease.
10. Describe in proper sequence the events of endochondral ossification in a long bone.

Level 2 Apply/Analyze

11. A bone with approximately the same width, length, and height is most likely **(a)** a long bone, **(b)** a short bone, **(c)** a flat bone, **(d)** an irregular bone.
12. Osteocytes residing in lacunae of osteons of healthy compact bone are located quite a distance from the blood vessels in the central canals, yet they are well nourished. How can this be explained?
13. As we grow, our long bones increase in diameter, but the thickness of the compact bone of the shaft remains relatively constant. Explain this phenomenon.
14. Describe the process of new bone formation in an adult bone. Use the terms osteoid seam and calcification front in your discussion.
15. **(a)** During what period of life does skeletal mass increase dramatically? Begin to decline? **(b)** Why are fractures most common in elderly individuals? **(c)** Why are greenstick fractures most common in children?
16. Yolanda is asked to review a bone slide that her professor has set up under the microscope. She sees concentric layers surrounding a central cavity. Is this bone section taken from the diaphysis or the epiphyseal plate of the specimen?
17. Why do you think wheelchair-bound people with paralyzed lower limbs have thin, weak bones of the leg and thigh?

Level 3 Evaluate/Synthesize

18. Compare bone to cartilage tissue relative to its resilience, speed of regeneration, and access to nutrients.
19. Compare and contrast controls of bone remodeling exerted by hormones and by mechanical and gravitational forces, including the purpose of each control system and changes in bone architecture that might occur.
20. Mrs. Abbruzzo brought her 4-year-old daughter to the doctor, complaining that she didn't "look right." The child's forehead was enlarged, her rib cage was knobby, and her lower limbs were bent and deformed. X rays revealed very thick epiphyseal plates. Mrs. Abbruzzo was advised to increase dietary amounts of vitamin D and milk and to get the girl outside to play in the sun. Considering the child's signs and symptoms, what disease do you think she has? Explain the doctor's instructions.
21. What would a long bone look like at the end of adolescence if bone remodeling did not occur?
22. Noah Beckenstein went to weight-lifting camp in the summer between seventh and eighth grade. He noticed that the camp trainer put tremendous pressure on him and his friends to improve their strength. After an especially vigorous workout, Noah's arm felt extremely sore and weak around the elbow. He went to the camp doctor, who took X rays and then told him that the injury was serious, for the "end of his upper arm bone was starting to twist off." What had happened? Could the same thing happen to Noah's 23-year-old sister, Karen, who was also starting a program of weight lifting? Why or why not?

CLINICAL CASE STUDY

21-Year-Old Female with Skeletal Injuries

Remember Liliana from the Clinical Case Study in Chapter 5? After we last heard about her she was admitted for further treatment. Relative to her skeletal system, the following notes have been added to her chart.

- Fracture of superior right tibia (shinbone); 3.5-cm laceration where bone protrudes. Area cleaned and the bone subjected to internal (open) reduction. Cast applied.
- Nutrient artery of tibia damaged.
- Medial meniscus (fibrocartilage disc) of right knee joint crushed; knee joint inflamed, swollen, and painful.

Relative to these notes:

1. What type of fracture does Liliana have?
2. **+ NCLEX-STYLE** When assessing Liliana's condition relative to her tibia fracture after the open reduction, the nurse will be especially alert for signs of which of these problems?
   a. Inadequate calcium for healing
   b. Infection
   c. Failure of the red bone marrow to create enough erythrocytes
   d. Bleeding
3. Which of the following statements about Liliana's condition are correct? Explain your reasoning.
   a. Her leg will not heal to its normal length if the epiphyses have been damaged.
   b. Her leg healing will involve only osteocytes, since she was fully grown before being injured.
   c. Her leg healing will involve only osteoclasts and osteoblasts.
   d. Her leg will heal by laying down cartilage in the injured area, followed by bone.
   e. Whether her leg heals will depend on whether the periosteum is damaged.
4. Given an uncomplicated recovery, approximately how long should it take before Liliana's bone has healed? What complications might be predicted by the fact that the nutrient artery is damaged?
5. **+ NCLEX-STYLE** You're talking with Liliana before her release from the hospital. Which of her statements indicates the need for further patient education?
   a. "The only supplement I should need would be calcium, so I will take a calcium supplement every morning."
   b. "My parathyroids are really important for maintaining my bones!"
   c. "Will the doctor have my vitamin D levels tested?"
   d. "At last I have an excuse for taking walks outside in the sun!"
6. How likely is it that Liliana's knee cartilage will regenerate? Why?

7 The Skeleton

In this chapter, you will learn that

The axial skeleton supports and protects, while the appendicular skeleton allows mobility

by comparing

Part 1 The Axial Skeleton ← supports → **Part 2 The Appendicular Skeleton**

and exploring

7.1 The skull **7.2 The vertebral column** **7.3 The thoracic cage**

and exploring

7.4 The pectoral girdle   and   **7.6 The pelvic girdle**

7.5 The upper limb   and   **7.7 The lower limb**

and finally, exploring

Developmental Aspects of the Skeleton

The word *skeleton* comes from the Greek word meaning "dried-up body" or "mummy," a rather unflattering description. Nonetheless, the human skeleton is a triumph of design and engineering that puts most skyscrapers to shame. It is strong, yet light, and almost perfectly adapted for the protective, locomotor, and manipulative functions it performs.

The **skeleton**, or **skeletal system**, composed of bones, cartilages, joints, and ligaments, accounts for about 20% of body mass (about 30 pounds in a 160-pound person). Bones make up most of the skeleton. Cartilages occur only in isolated areas, such as the nose, parts of the ribs, and the joints. Ligaments connect bones and reinforce joints, allowing required movements while restricting motions in other directions. Joints provide for the remarkable mobility of the skeleton. We discuss joints and ligaments separately in Chapter 8.

PART 1

THE AXIAL SKELETON

As described in Chapter 6, the skeleton is divided into *axial* and *appendicular* portions (see Figures 6.1 and 7.1). The **axial**

Figure 7.1 The human skeleton. Bones of the axial skeleton are colored green. Bones of the appendicular skeleton are gold.

skeleton is structured from 80 bones segregated into three major regions: the *skull*, *vertebral column*, and *thoracic cage* (**Figure 7.1**). The axial skeleton:

- Forms the longitudinal axis of the body
- Supports the head, neck, and trunk
- Protects the brain, spinal cord, and the organs in the thorax

As the name suggests, the bones of the appendicular skeleton, which allow us to interact with and manipulate our environment, are *appended* to the axial skeleton.

7.1 The skull consists of 8 cranial bones and 14 facial bones

Learning Outcomes

▶ Name, describe, and identify the skull bones. Identify their important markings.

▶ Compare and contrast the major functions of the cranium and the facial skeleton.

▶ Define the bony boundaries of the orbits, nasal cavity, and paranasal sinuses.

The **skull** is the body's most complex bony structure. It is formed by *cranial* and *facial bones*, 22 in all. The cranial bones, or **cranium** (kra′ne-um), enclose and protect the fragile brain and furnish attachment sites for head and neck muscles. The facial bones:

- Form the framework of the face
- Contain cavities for the special sense organs of sight, taste, and smell
- Provide openings for air and food passage
- Secure the teeth
- Anchor the facial muscles of expression, which we use to show our feelings

Most skull bones are flat bones. Except for the mandible, which is connected to the rest of the skull by freely movable joints, all bones of the adult skull are firmly united by interlocking joints called **sutures** (soo′cherz). The suture lines have a saw-toothed or serrated appearance.

The major skull sutures, the *coronal, sagittal, squamous,* and *lambdoid sutures,* connect cranial bones (Figures 7.2a, 7.4b, and 7.5a). Most other skull sutures connect facial bones and are named according to the bones they connect.

Overview of Skull Geography

It is worth surveying basic skull "geography" before describing the individual bones. With the lower jaw removed, the skull resembles a lopsided, hollow, bony sphere. The facial bones form its anterior aspect, and the cranium forms the rest of the skull (**Figure 7.2a**).

The cranium can be divided into a vault and a base.

- The *cranial vault,* also called the *calvaria* (kal-va′re-ah; "bald part of skull"), forms the superior, lateral, and posterior aspects of the skull, as well as the forehead.

- The *cranial base* forms the skull's inferior aspect. Internally, prominent bony ridges divide the base into three distinct "steps" or fossae—the *anterior, middle,* and *posterior cranial fossae* (Figure 7.2b and c). The brain sits snugly in these cranial fossae, completely enclosed by the cranial vault.

The space enclosed by the cranial vault and cranial base is called the *cranial cavity* and is occupied by the brain.

In addition to the large cranial cavity, the skull has many smaller cavities. These include the middle and internal

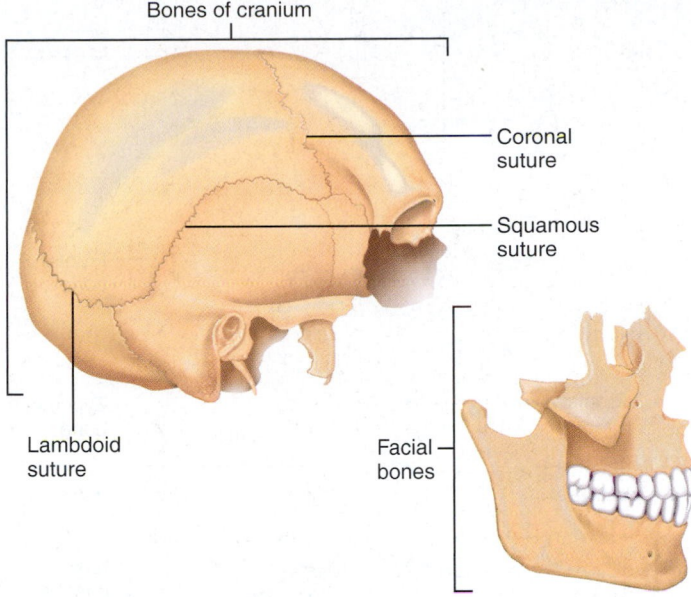

(a) **Cranial and facial divisions of the skull**

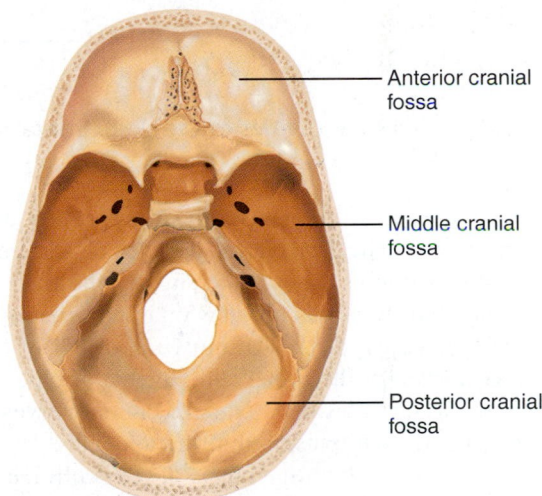

(b) **Superior view of the cranial fossae**

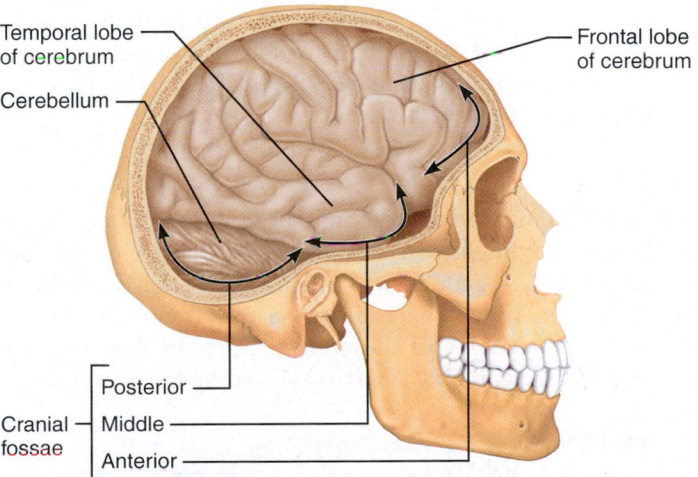

(c) **Lateral view of cranial fossae**

Figure 7.2 The skull: Cranial and facial divisions and fossae.

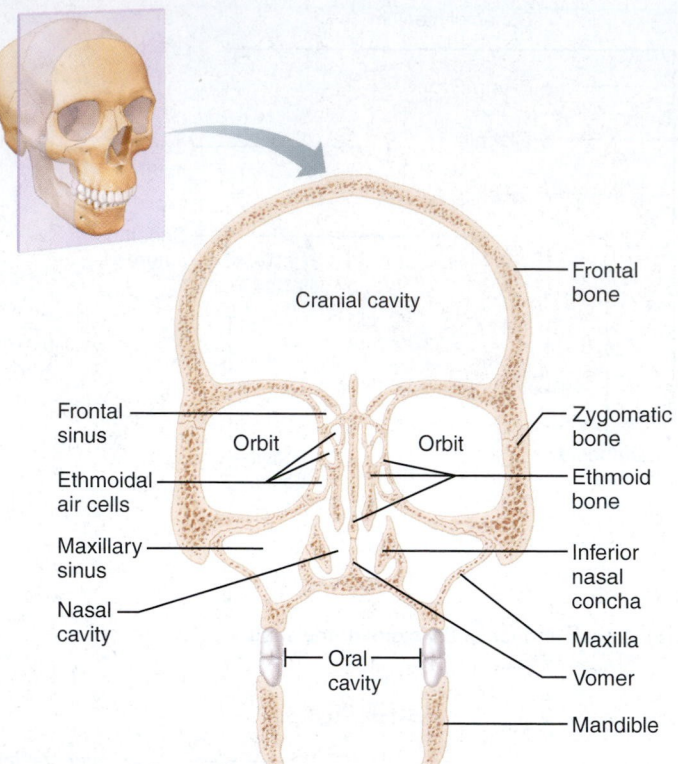

Figure 7.3   Major cavities of the skull, frontal section.

Labels in figure: Cranial cavity, Frontal bone, Frontal sinus, Orbit, Orbit, Zygomatic bone, Ethmoidal air cells, Ethmoid bone, Maxillary sinus, Inferior nasal concha, Nasal cavity, Maxilla, Oral cavity, Vomer, Mandible

ear cavities and, anteriorly, the nasal cavity and the orbits (**Figure 7.3**). The *orbits* house the eyeballs. Several skull bones contain air-filled sinuses, which lighten the skull.

The skull also has about 85 named openings (foramina, canals, fissures, etc.). The most important of these provide passageways for the spinal cord, the major blood vessels serving the brain, and the 12 pairs of cranial nerves (numbered I through XII) that transmit information to and from the brain.

As you read about the bones of the skull, locate each bone on the skull views in **Figures 7.4**, **7.5** (pp. 203–205), and **7.6** (p. 206). The skull bones and their important markings are also summarized in **Table 7.1** at the end of the skull section (pp. 216–217). To refresh your memory on bone markings, see ◄ **Table 6.2, p. 180**. Note that the color-coded boxes before a bone's name in the text and in Table 7.1 correspond to the color of that bone in the figures.

Cranium

The eight cranial bones are the paired parietal and temporal bones and the unpaired frontal, occipital, sphenoid, and ethmoid bones. Together, these construct the brain's protective bony "helmet." Because its superior aspect is curved, the cranium is self-bracing. This allows the bones to be thin, and, like an eggshell, the cranium is remarkably strong for its weight.

Frontal Bone

The shell-shaped **frontal bone** (Figures 7.4a, 7.5, and 7.7) forms the anterior cranium. It articulates (connects) posteriorly with the paired parietal bones via the prominent *coronal suture*.

The most anterior part of the frontal bone is the vertical *squamous part*, commonly called the *forehead*. The frontal squamous region ends inferiorly at the **supraorbital margins**, the thickened superior margins of the orbits that lie under the eyebrows. From here, the frontal bone extends posteriorly, forming the superior wall of the *orbits* and most of the **anterior cranial fossa** (**Figure 7.7a** and **b**, p. 207). This fossa supports the frontal lobes of the brain. Each supraorbital margin is pierced by a **supraorbital foramen (notch)**, which allows the supraorbital artery and nerve to pass to the forehead (Figure 7.4a).

The smooth portion of the frontal bone between the orbits is the **glabella** (glah-bel′ah). Just inferior to this the frontal bone meets the nasal bones at the *frontonasal suture* (Figure 7.4a). The areas lateral to the glabella contain sinuses, called the **frontal sinuses** (Figures 7.5c and 7.3).

Parietal Bones and the Major Sutures

The two large **parietal bones** are curved, rectangular bones that form most of the superior and lateral aspects of the skull; as such, they form the bulk of the cranial vault. The four largest sutures occur where the parietal bones articulate (form a joint) with other cranial bones. These sutures are:

- The **coronal suture** (kŏ-ro′nul), where the parietal bones meet the frontal bone anteriorly (Figures 7.2a and 7.5)
- The **sagittal suture**, where the parietal bones meet superiorly at the cranial midline (Figure 7.4b)
- The **lambdoid suture** (lam′doid), where the parietal bones meet the occipital bone posteriorly (Figures 7.2a, 7.4b, and 7.5)
- The **squamous suture** (one on each side), where a parietal and temporal bone meet on the lateral aspect of the skull (Figures 7.2a and 7.5)

Occipital Bone

The **occipital bone** (ok-sip′ĭ-tal) forms most of the skull's posterior wall and base. It articulates anteriorly with the paired parietal and temporal bones via the *lambdoid* and *occipitomastoid sutures*, respectively (Figure 7.5). The basilar part of the occipital bone (located nearest the base of the skull) also joins with the sphenoid bone in the cranial base (Figure 7.6a).

Internally, the occipital bone forms the walls of the **posterior cranial fossa** (Figures 7.7 and 7.2c), which supports the cerebellum of the brain. In the base of the occipital bone is the **foramen magnum** ("large hole") through which the inferior part of the brain connects with the spinal cord. The foramen magnum is flanked laterally by two occipital condyles (Figure 7.6). The curved, rockerlike **occipital condyles** articulate with the first vertebra of the spinal column in a way that permits a nodding ("yes") motion of the head. Hidden medially and superiorly to each occipital condyle is a **hypoglossal canal** (Figure 7.7a), through which a cranial nerve (XII) passes.

Just superior to the foramen magnum is a median protrusion called the **external occipital protuberance** (Figures 7.4b, 7.5c and d, and 7.6). You can feel this knoblike projection just below the most bulging part of your posterior skull. A number of inconspicuous ridges, the *external occipital crest* and the

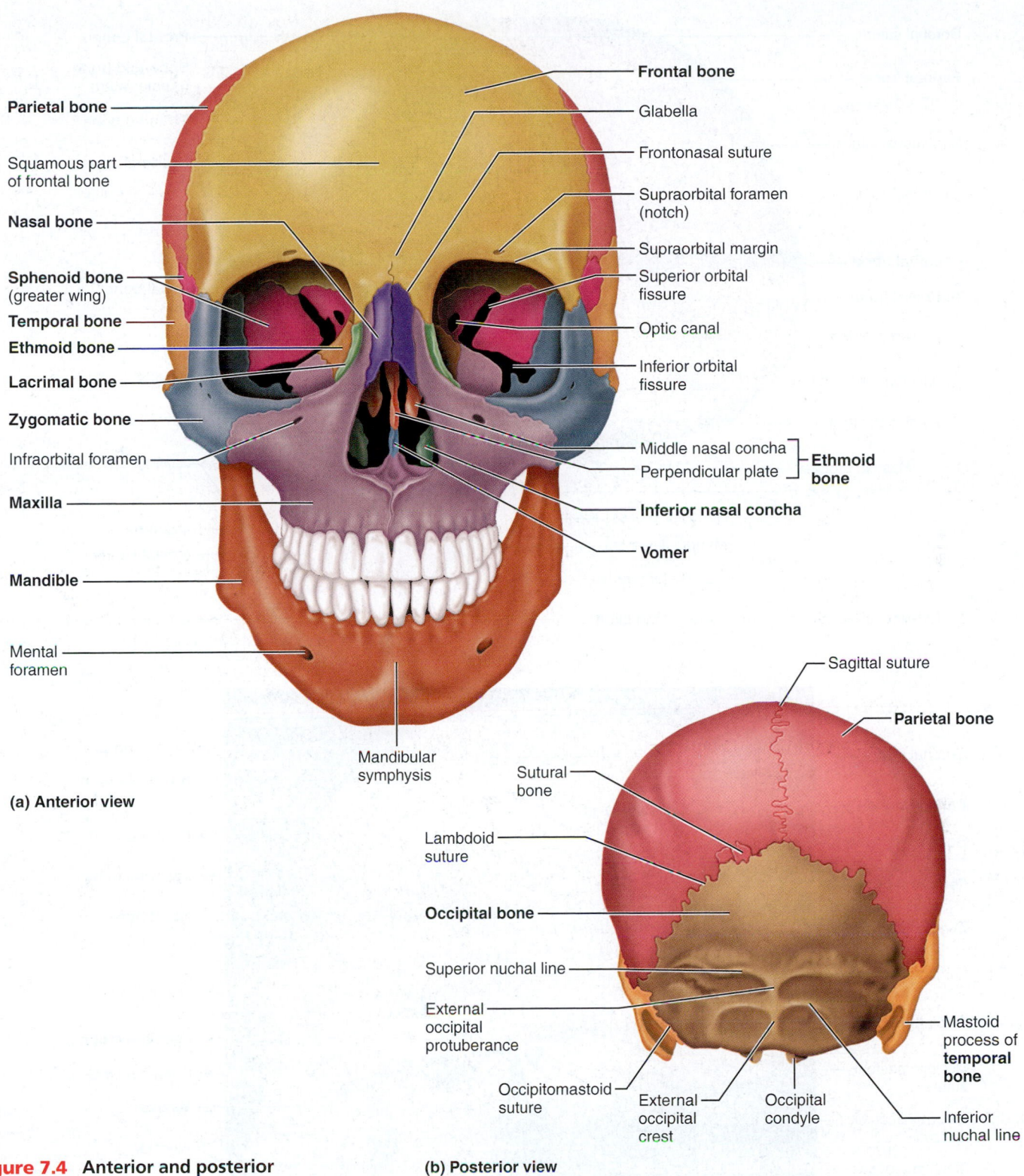

(a) Anterior view

Parietal bone

Squamous part
of frontal bone

Nasal bone

Sphenoid bone
(greater wing)

Temporal bone

Ethmoid bone

Lacrimal bone

Zygomatic bone

Infraorbital foramen

Maxilla

Mandible

Mental
foramen

Frontal bone

Glabella

Frontonasal suture

Supraorbital foramen
(notch)

Supraorbital margin

Superior orbital
fissure

Optic canal

Inferior orbital
fissure

Middle nasal concha
Perpendicular plate   **Ethmoid**
                           bone

Inferior nasal concha

Vomer

Mandibular
symphysis

Sagittal suture

Parietal bone

Sutural
bone

Lambdoid
suture

Occipital bone

Superior nuchal line

External
occipital
protuberance

Occipitomastoid
suture

External
occipital
crest

Occipital
condyle

Mastoid
process of
**temporal
bone**

Inferior
nuchal line

(b) Posterior view

Figure 7.4 **Anterior and posterior
views of the skull.**

superior and *inferior nuchal lines* (nu′kal), mark the occipital
bone near the foramen magnum. The external occipital crest
secures the *ligamentum nuchae* (lig″ah-men′tum noo′ke;
nucha = back of the neck), a sheetlike elastic ligament that connects the vertebrae of the neck to the skull. The nuchal lines,
and the bony regions between them, anchor many neck and

back muscles. The superior nuchal line marks the upper limit
of the neck.

Temporal Bones

The two **temporal bones** are best viewed on the lateral skull
surface (Figure 7.5). They lie inferior to the parietal bones and

(Text continues on p. 208.)

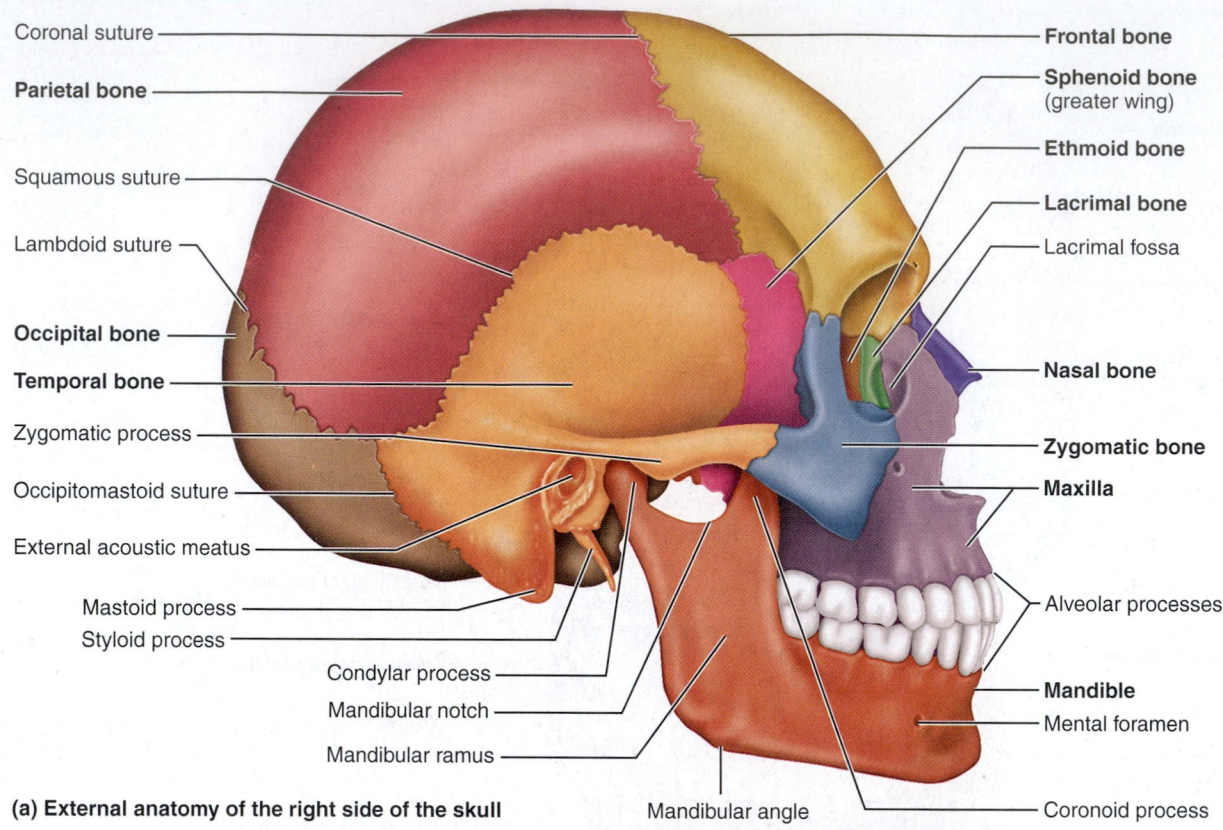

Coronal suture

Parietal bone

Squamous suture

Lambdoid suture

Occipital bone

Temporal bone

Zygomatic process

Occipitomastoid suture

External acoustic meatus

Mastoid process

Styloid process

Condylar process

Mandibular notch

Mandibular ramus

Frontal bone

Sphenoid bone
(greater wing)

Ethmoid bone

Lacrimal bone

Lacrimal fossa

Nasal bone

Zygomatic bone

Maxilla

Alveolar processes

Mandible

Mental foramen

Mandibular angle

Coronoid process

(a) External anatomy of the right side of the skull

Coronal suture

Parietal bone

Squamous suture

Temporal bone

Zygomatic process

Lambdoid suture

Occipital bone

Occipitomastoid suture

External acoustic meatus

Mastoid process

Styloid process

Condylar process

Mandibular angle

Frontal bone

Sphenoid bone
(greater wing)

Ethmoid bone

Lacrimal bone

Nasal bone

Lacrimal fossa

Zygomatic bone

Coronoid process

Maxilla

Alveolar processes

Mandible

Mental foramen

Mandibular notch

Mandibular ramus

(b) Photograph of right side of skull

Figure 7.5 Bones of the lateral aspect of the skull, external and internal views.

- **Parietal bone**
- Coronal suture
- **Frontal bone**
- Greater wing — **Sphenoid bone**
- Lesser wing
- Frontal sinus
- Crista galli
- **Nasal bone**
- Sphenoidal sinus
- **Ethmoid bone** (perpendicular plate)
- **Vomer**
- Incisive canal
- **Maxilla**
- Alveolar processes
- **Mandible**

- Squamous suture
- **Temporal bone**
- Lambdoid suture
- **Occipital bone**
- Occipitomastoid suture
- External occipital protuberance
- Internal acoustic meatus
- Sella turcica of sphenoid bone
- Pterygoid process of sphenoid bone
- Mandibular foramen
- **Palatine bone**
- Palatine process of maxilla

(c) Midsagittal section showing the internal anatomy of the left half of skull

7

- Greater wing of sphenoid bone
- Lesser wing of sphenoid bone
- Frontal sinus
- Crista galli
- **Ethmoid bone** (perpendicular plate)
- **Palatine bone**

- Petrous part of temporal bone
- External occipital protuberance
- Internal acoustic meatus
- Sella turcica and sphenoidal sinus

(d) Photo of skull cut through the midline, same view as in (c)

Figure 7.5 *(continued)*

Maxilla
(palatine process)

Hard palate

Palatine bone
(horizontal plate)

Zygomatic bone

Temporal bone
(zygomatic process)

Vomer

Mandibular fossa

Styloid process

Mastoid process

Temporal bone
(petrous part)

Basilar part of the occipital bone

Parietal bone

External occipital crest

External occipital protuberance

Incisive fossa

Median palatine suture

Infraorbital foramen

Maxilla

Sphenoid bone
(greater wing)

Foramen ovale

Foramen spinosum

Foramen lacerum

Carotid canal

External acoustic meatus

Stylomastoid foramen

Jugular foramen

Occipital condyle

Inferior nuchal line

Superior nuchal line

Occipital bone

Foramen magnum

(a) Inferior view of the skull (mandible removed)

Hard palate

Mandibular fossa

Mastoid process

Zygomatic arch

Foramen ovale

Foramen lacerum

Foramen spinosum

Carotid canal

Styloid process

Jugular foramen

Occipital condyle

Foramen magnum

Superior nuchal line

(b) Photo of inferior view of the skull

Figure 7.6 Inferior aspect of the skull, mandible removed.

View

Ethmoid bone — Cribriform plate / Crista galli

Anterior cranial fossa

Sphenoid — Lesser wing / Greater wing

Hypophyseal fossa of sella turcica

Middle cranial fossa

Temporal bone (petrous part)

Posterior cranial fossa

Parietal bone

Occipital bone

Foramen magnum

Frontal bone

Cribriform foramina

Optic canal

Foramen rotundum

Foramen ovale

Foramen spinosum

Foramen lacerum

Internal acoustic meatus

Jugular foramen

Hypoglossal canal

(a) Superior view of the skull, calvaria removed

Ethmoid bone — Crista galli / Cribriform plate

Anterior cranial fossa

Sphenoid — Lesser wing / Greater wing

Hypophyseal fossa of sella turcica

Middle cranial fossa

Temporal bone (petrous part)

Posterior cranial fossa

Parietal bone

Occipital bone

Foramen magnum

Frontal bone

Cribriform foramina

Optic canal

Foramen rotundum

Foramen ovale

Foramen spinosum

Foramen lacerum

Jugular foramen

(b) Photo of superior view of the skull, calvaria removed

Figure 7.7 **The base of the cranial cavity.**

meet them at the squamous sutures. The temporal bones form the inferolateral (inferior and lateral) aspects of the skull and parts of the cranial base. The use of the terms *temple* and *temporal*, from the Latin word *temporum*, meaning "time," came about because gray hairs, a sign of time's passing, usually appear first at the temples.

Each temporal bone has a complicated shape (**Figure 7.8**) and is described in terms of its three major parts, the *squamous*, *tympanic*, and *petrous parts*. The flaring **squamous part** ends at the squamous suture. Its barlike **zygomatic process** meets the zygomatic bone of the face anteriorly. Together, these two bony structures form the **zygomatic arch**, which you can feel as the projection of your cheek (*zygoma* = cheekbone). The small, oval **mandibular fossa** (man-dib′u-lar) on the inferior surface of the zygomatic process receives the condylar process of the mandible (lower jawbone), forming the freely movable *temporomandibular joint* (see Figure 7.11a on p. 211).

The **tympanic part** (tim-pan′ik; "eardrum") (Figure 7.8) of the temporal bone surrounds the **external acoustic meatus**, or external ear canal (*meatus* = passage). The external acoustic meatus and the eardrum at its deep end are part of the *external ear*. In a dried skull, the eardrum has been removed and part of the middle ear cavity deep to the external meatus can also be seen.

The thick **petrous part** (pet′rus) of the temporal bone houses the *middle* and *internal ear cavities*, which contain sensory receptors for hearing and balance. Extending from the occipital bone posteriorly to the sphenoid bone anteriorly, it contributes to the cranial base (Figures 7.6 and 7.7). In the floor of the cranial cavity, the petrous part of the temporal bone looks like a miniature mountain ridge (*petrous* = rocky). The posterior slope of this ridge lies in the posterior cranial fossa; the anterior slope is in the middle cranial fossa. Together, the sphenoid bone and the petrous portions of the temporal bones construct the **middle cranial fossa** (Figures 7.7 and 7.2b), which supports the temporal lobes of the brain.

Several foramina penetrate the bone of the petrous region (Figure 7.6). The large **jugular foramen** at the junction of the occipital and petrous temporal bones allows passage of the internal jugular vein and three cranial nerves (IX, X, and XI). The **carotid canal** (kah-rot′id), just anterior to the jugular foramen, transmits the internal carotid artery into the cranial cavity. The two internal carotid arteries supply blood to over 80% of the cerebral hemispheres of the brain; their closeness to the internal ear cavities explains why, during excitement or exertion, we may hear our rapid pulse as a thundering sound. The **foramen lacerum** (la′ser-um) is a jagged opening (*lacerum* = torn or lacerated) between the petrous temporal bone and the sphenoid bone. It is almost completely closed by cartilage in a living person, but it is conspicuous in a dried skull, and students usually ask its name. The **internal acoustic meatus**, positioned superolateral to the jugular foramen (Figures 7.5c and d, and 7.7), transmits cranial nerves VII and VIII.

A conspicuous feature of the petrous part of the temporal bone is the **mastoid process** (mas′toid; "breast"), which acts as an anchoring site for some neck muscles (Figures 7.5a and b, 7.6, and 7.8). This process can be felt as a lump just posterior to the ear. The needle-like **styloid process** (sti′loid; "stakelike") is an attachment point for several tongue and neck muscles and for a ligament that secures the hyoid bone of the neck to the skull (see Figure 7.12). The **stylomastoid foramen**, between the styloid and mastoid processes, allows cranial nerve VII (the facial nerve) to leave the skull (Figure 7.6).

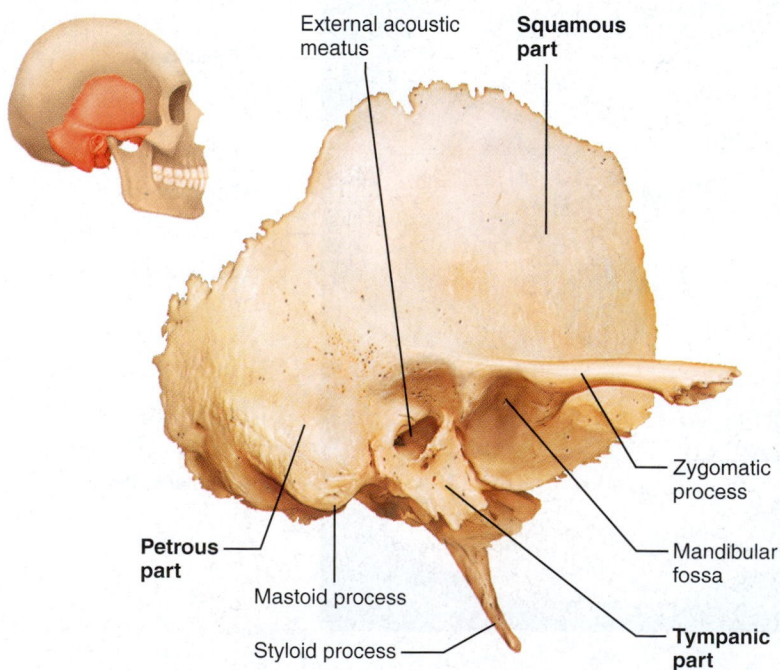

External acoustic meatus

Squamous part

Petrous part

Mastoid process

Styloid process

Zygomatic process

Mandibular fossa

Tympanic part

Figure 7.8 **The temporal bone.**

Practice Human Cadaver lab questions:
MasteringA&P® > Study Area > Lab Tools > PAL

HOMEOSTATIC IMBALANCE 7.1

CLINICAL

The mastoid process is full of air cavities (sinuses) called **mastoid air cells**. Their position adjacent to the middle ear cavity (a high-risk area for infections spreading from the throat) puts them at risk for infection themselves. A mastoid sinus infection, or *mastoiditis*, is notoriously difficult to treat. Because the mastoid air cells are separated from the brain by only a very thin bony plate, mastoid infections may spread to the brain as well.

Sphenoid Bone

The bat-shaped **sphenoid bone** (sfe'noid; *sphen* = wedge) spans the width of the middle cranial fossa (Figure 7.7). The sphenoid is the keystone bone of the cranium. (It is a *keystone* because it forms a central wedge that ties the cranial bones together. It also articulates with all of the other cranial bones.)

It is a challenging bone to study because of its complex shape. As shown in **Figure 7.9**, it consists of a central body and three pairs of processes: the greater wings, lesser wings, and pterygoid processes (ter'ĭ-goid). Within the **body** of the sphenoid are the paired **sphenoidal sinuses** (see Figures 7.5c and d, and 7.14).

The superior surface of the body bears a saddle-shaped prominence, the **sella turcica** (sel'ah ter'sĭ-kah), meaning "Turk's saddle." The seat of this saddle, called the **hypophyseal fossa**, forms a snug enclosure for the pituitary gland (hypophysis).

The **greater wings** project laterally from the sphenoid body, forming parts of (1) the middle cranial fossa (Figures 7.7 and 7.2b), (2) the posterior walls of the orbits (Figure 7.4a), and (3) the external wall of the skull, where they are seen as flag-shaped, bony areas medial to the zygomatic arch (Figure 7.5). The hornlike **lesser wings** form part of the floor of the anterior cranial fossa (Figure 7.7) and part of the medial walls of the orbits. The trough-shaped **pterygoid processes** project inferiorly from the junction of the body and greater wings (Figure 7.9b). They anchor the pterygoid muscles, which are important in chewing.

A number of openings in the sphenoid bone are visible in Figures 7.7 and 7.9. The **optic canals** lie anterior to the sella turcica; they allow the optic nerves (cranial nerves II) to pass

(a) Superior view

(b) Posterior view

Figure 7.9 The sphenoid bone.

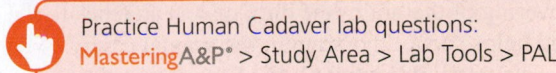
Practice Human Cadaver lab questions: MasteringA&P® > Study Area > Lab Tools > PAL

Crista galli

Cribriform plate
with cribriform
foramina

Orbital
plate

Left lateral
mass

Ethmoidal
air cells

Perpendicular
plate

Middle
nasal concha

Figure 7.10 The ethmoid bone.

7

Practice Human Cadaver lab questions:
MasteringA&P® > Study Area > Lab Tools > PAL

to the eyes. On each side of the sphenoid body is a crescent-shaped row of four openings. The anteriormost of these, the **superior orbital fissure**, is a long slit between the greater and lesser wings. It allows cranial nerves that control eye movements (III, IV, VI) to enter the orbit. This fissure is most obvious in an anterior view of the skull (Figure 7.4 and Figure 7.9b). The **foramen rotundum** and **foramen ovale** (o-va′le) provide passageways for branches of cranial nerve V to reach the face (Figure 7.7). The foramen rotundum is in the medial part of the greater wing and is usually oval, despite its name meaning "round opening." The foramen ovale, a large, oval foramen posterior to the foramen rotundum, is also visible in an inferior view of the skull (Figure 7.6). Posterolateral to the foramen ovale is the small **foramen spinosum** (Figure 7.7); it transmits the *middle meningeal artery*, which serves the internal faces of some cranial bones.

Ethmoid Bone

Like the temporal and sphenoid bones, the delicate **ethmoid bone** has a complex shape (**Figure 7.10**). Lying between the sphenoid and the nasal bones of the face, it is the most deeply situated bone of the skull. It forms most of the bony area between the nasal cavity and the orbits.

The superior surface of the ethmoid is formed by the paired horizontal **cribriform plates** (krib′rĭ-form) (see Figure 7.7), which help form the roof of the nasal cavity and the floor of the anterior cranial fossa. The cribriform plates are punctured by tiny holes (*cribr* = sieve) called *cribriform foramina* that allow the filaments of the olfactory nerves to pass from the smell receptors in the nasal cavity to the brain. Projecting superiorly between the cribriform plates is a triangular process called the **crista galli** (kris′tah gah′le; "rooster's comb"). The outermost covering of the brain (the dura mater) attaches to the crista galli and helps secure the brain in the cranial cavity.

The **perpendicular plate** of the ethmoid bone projects inferiorly in the median plane and forms the superior part of the

nasal septum, which divides the nasal cavity into right and left halves (Figure 7.5c and d). Flanking the perpendicular plate on each side is a **lateral mass** riddled with sinuses called **ethmoidal air cells** (Figures 7.10 and 7.15), for which the bone itself is named (*ethmos* = sieve). Extending medially from the lateral masses, the delicately coiled **superior** and **middle nasal conchae** (kong′ke; *concha* = shell), named after the conch shells found on warm ocean beaches, protrude into the nasal cavity (Figures 7.10 and 7.14a). The lateral surfaces of the ethmoid's lateral masses are called **orbital plates** because they contribute to the medial walls of the orbits.

Sutural Bones

Sutural bones are tiny, irregularly shaped bones or bone clusters that occur within sutures, most often in the lambdoid suture (Figure 7.4b). Not everyone has these bones and their significance is unknown.

Facial Bones

The facial skeleton is made up of 14 bones (see Figures 7.4a and 7.5a), of which only the mandible and the vomer are unpaired. The maxillae, zygomatics, nasals, lacrimals, palatines, and inferior nasal conchae are paired bones. As a rule, the facial skeleton of men is more elongated than that of women. Women's faces tend to be rounder and less angular.

Mandible

The U-shaped **mandible** (man′dĭ-bl), or lower jawbone (Figures 7.4a and 7.5, and **Figure 7.11a**), is the largest, strongest bone of the face. It has a body, which forms the chin, and two upright *rami* (*rami* = branches). Each ramus meets the body posteriorly at a **mandibular angle**. At the superior margin of each ramus are two processes separated by the **mandibular notch**. The anterior **coronoid process** (kor′o-noid;

"crown-shaped") is an insertion point for the large temporalis muscle that elevates the lower jaw during chewing. The posterior **condylar process** articulates with the mandibular fossa of the temporal bone, forming the *temporomandibular joint* on the same side.

The mandibular **body** anchors the lower teeth. Its superior border, called the **alveolar process** (al-ve´o-lar), contains the sockets (*dental alveoli*) in which the teeth are embedded. In the midline of the mandibular body is a slight ridge, the **mandibular symphysis** (sim´fih-sis), indicating where the two mandibular bones fused during infancy (Figure 7.4a).

Large **mandibular foramina**, one on the medial surface of each ramus, permit the nerves responsible for tooth sensation to pass to the teeth in the lower jaw. Dentists inject lidocaine into these foramina to prevent pain while working on the lower teeth. The **mental foramina**, openings on the lateral aspects of

the mandibular body, allow blood vessels and nerves to pass to the skin of the chin (*ment* = chin) and lower lip.

Maxillary Bones

The **maxillary bones**, or **maxillae** (mak-sil´le; "jaws") (Figures 7.4, 7.5, 7.6, and 7.11b and c), are fused medially. They form the upper jaw and the central portion of the facial skeleton. All facial bones except the mandible articulate with the maxillae. For this reason, the maxillae are considered the keystone bones of the facial skeleton.

The maxillae carry the upper teeth in their **alveolar processes**. Just inferior to the nose the maxillae meet medially, forming the pointed **anterior nasal spine** at their junction. The **palatine processes** (pă´lah-tīn) of the maxillae project posteriorly from the alveolar processes and fuse medially, forming the anterior two-thirds of the hard palate, or bony roof of the mouth (Figures 7.5c and d and 7.6). Just posterior to the teeth, a midline foramen called the **incisive fossa** leads into the **incisive canal**, a passageway for blood vessels and nerves.

The **frontal processes** extend superiorly to the frontal bone, forming part of the lateral aspects of the bridge of the nose (Figures 7.11b and 7.4a). The regions that flank the nasal cavity laterally contain the **maxillary sinuses** (see Figure 7.15), the largest of the paranasal sinuses. They extend from the orbits to the roots of the upper teeth. Laterally, the

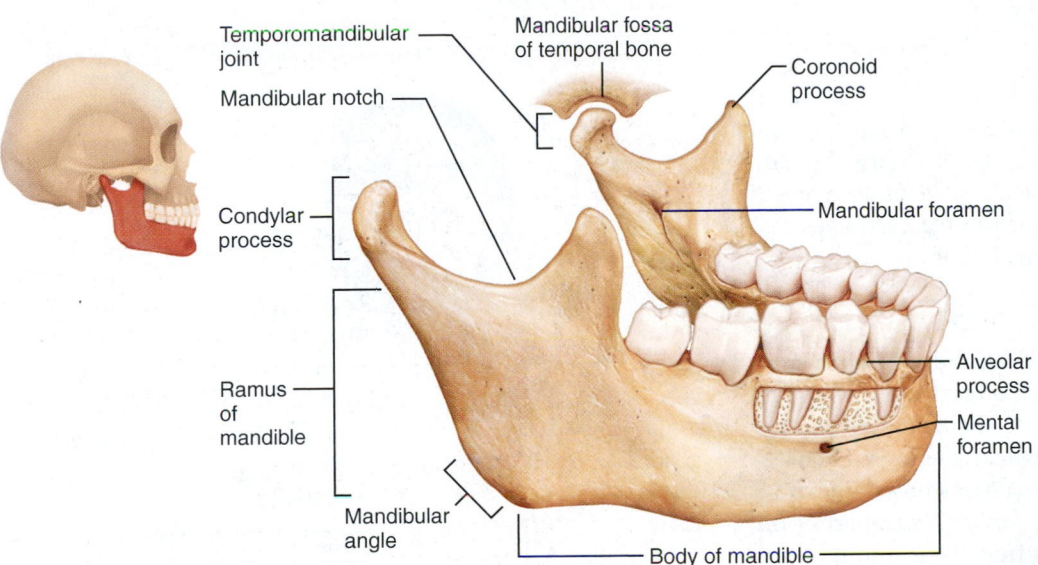

(a) Mandible, right lateral view

Temporomandibular joint
Mandibular notch
Condylar process
Ramus of mandible
Mandibular angle
Mandibular fossa of temporal bone
Coronoid process
Mandibular foramen
Alveolar process
Mental foramen
Body of mandible

(b) Maxilla, right lateral view

Orbital surface
Zygomatic process (cut)
Articulates with frontal bone
Frontal process
Infraorbital foramen
Anterior nasal spine
Alveolar process

(c) Maxilla, photo of right lateral view

Figure 7.11 Detailed anatomy of the mandible and the maxilla.

Practice Human Cadaver lab questions:
MasteringA&P® > Study Area > Lab Tools > PAL

maxillae articulate with the zygomatic bones via their **zygomatic processes**.

The **inferior orbital fissure** is located deep within the orbit (see Figure 7.13b) at the junction of the maxilla with the greater wing of the sphenoid. It permits the zygomatic nerve, the maxillary nerve (a branch of cranial nerve V), and blood vessels to pass to the face. Just below the eye socket on each side is an **infraorbital foramen** that allows the infraorbital nerve (a continuation of the maxillary nerve) and artery to reach the face.

Zygomatic Bones

The irregularly shaped **zygomatic bones** (Figures 7.4a, 7.5a, and 7.6), commonly called the cheekbones, form the prominences of the cheeks and part of the inferolateral margins of the orbits. They articulate posteriorly with the zygomatic processes of the temporal bones, superiorly with the zygomatic processes of the frontal bone, and anteriorly with the zygomatic processes of the maxillae.

Nasal Bones

The thin, basically rectangular **nasal bones** (na′zal) are fused medially, forming the bridge of the nose (Figures 7.4a and 7.5a). They articulate with the frontal bone superiorly, the maxillary bones laterally, and the perpendicular plate of the ethmoid bone posteriorly. Inferiorly they attach to the cartilages that form most of the skeleton of the external nose.

Lacrimal Bones

The delicate, fingernail-shaped **lacrimal bones** (lak′rĭ-mal) contribute to the medial walls of each orbit (Figures 7.4a and 7.5a). They articulate with the frontal bone superiorly, the ethmoid bone posteriorly, and the maxillae anteriorly. Each lacrimal bone contains a deep groove that helps form a **lacrimal fossa**. The lacrimal fossa houses the *lacrimal sac*, part of the passageway that allows tears to drain from the eye surface into the nasal cavity (*lacrima* = tears).

Palatine Bones

Each L-shaped **palatine bone** is fashioned from two bony plates, the *horizontal* and *perpendicular* (see Figures 7.14a and 7.6a). The **horizontal plates**, joined at the median palatine suture, complete the posterior portion of the hard palate. The superiorly projecting **perpendicular (vertical) plates** form part of the posterolateral walls of the nasal cavity and a small part of the orbits.

Vomer

The slender, plow-shaped **vomer** (vo′mer; "plow") lies in the nasal cavity, where it forms part of the nasal septum (see Figures 7.4a and 7.14b). It is described later in connection with the nasal cavity (p. 215).

Inferior Nasal Conchae

The paired **inferior nasal conchae** are thin, curved bones that project medially from the lateral walls of the nasal cavity, just inferior to the middle nasal conchae of the ethmoid bone (see Figures 7.4a and 7.14a). They are the largest of the three pairs of conchae and, like the others, they form part of the lateral walls of the nasal cavity.

The Hyoid Bone

Though not directly connected to the skull, the **hyoid bone** (hi′oid; "U shaped") is considered to be part of the skull. It lies in the anterior neck just inferior to the mandible, and looks like a miniature version of it (**Figure 7.12**). The hyoid bone is unique in that it is the only bone of the body that does not articulate directly with any other bone. Instead, it is anchored by the narrow *stylohyoid ligaments* to the styloid processes of the temporal bones. Horseshoe shaped, with a body and two pairs of *horns*, or *cornua*, the hyoid bone acts as a movable base for the tongue. Its body and greater horns are attachment points for neck muscles that raise and lower the larynx during swallowing and speech.

Greater horn
Lesser horn
Body

Figure 7.12 **The hyoid bone.** Anterior view.

Special Characteristics of the Orbits and Nasal Cavity

The orbits and the nasal cavity are formed from an amazing number of bones. Even though we have already described their individual bones, we give a brief summary here to pull the parts together.

The Orbits

The cone-shaped **orbits** are bony cavities in which the eyes are firmly encased and cushioned by fatty tissue. The muscles that move the eyes and the tear-producing lacrimal glands are also housed in the orbits. The walls of each orbit are formed by parts of seven bones—the frontal, sphenoid, zygomatic, maxilla, palatine, lacrimal, and ethmoid bones. Their relationships are shown in **Figure 7.13**. Also seen in the orbits are the superior and inferior orbital fissures and the optic canals, described earlier.

(a) Photograph, right orbit

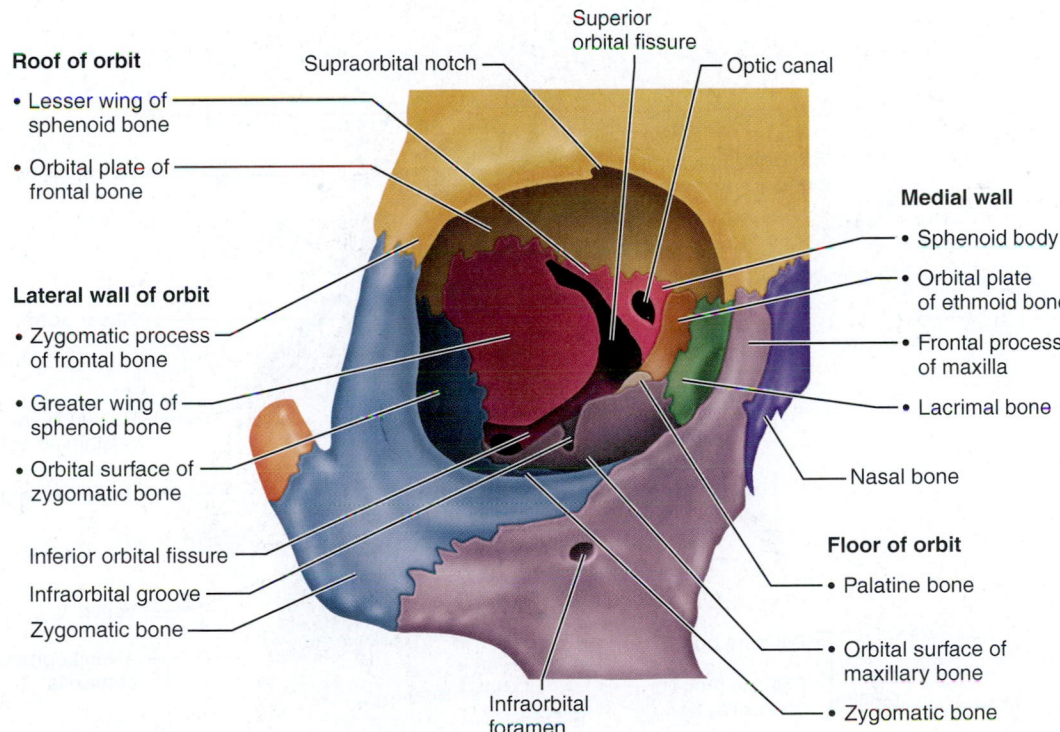

Roof of orbit

Supraorbital notch

Superior orbital fissure

Optic canal

- Lesser wing of sphenoid bone
- Orbital plate of frontal bone

Medial wall

- Sphenoid body
- Orbital plate of ethmoid bone

Lateral wall of orbit

- Zygomatic process of frontal bone
- Greater wing of sphenoid bone
- Orbital surface of zygomatic bone

- Frontal process of maxilla
- Lacrimal bone

Nasal bone

Inferior orbital fissure

Infraorbital groove

Zygomatic bone

Floor of orbit

- Palatine bone

- Orbital surface of maxillary bone

- Zygomatic bone

Infraorbital foramen

(b) Contribution of each of the seven bones forming the right orbit

Figure 7.13 Bones that form the orbits.

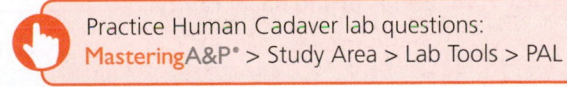

Practice Human Cadaver lab questions: MasteringA&P® > Study Area > Lab Tools > PAL

7

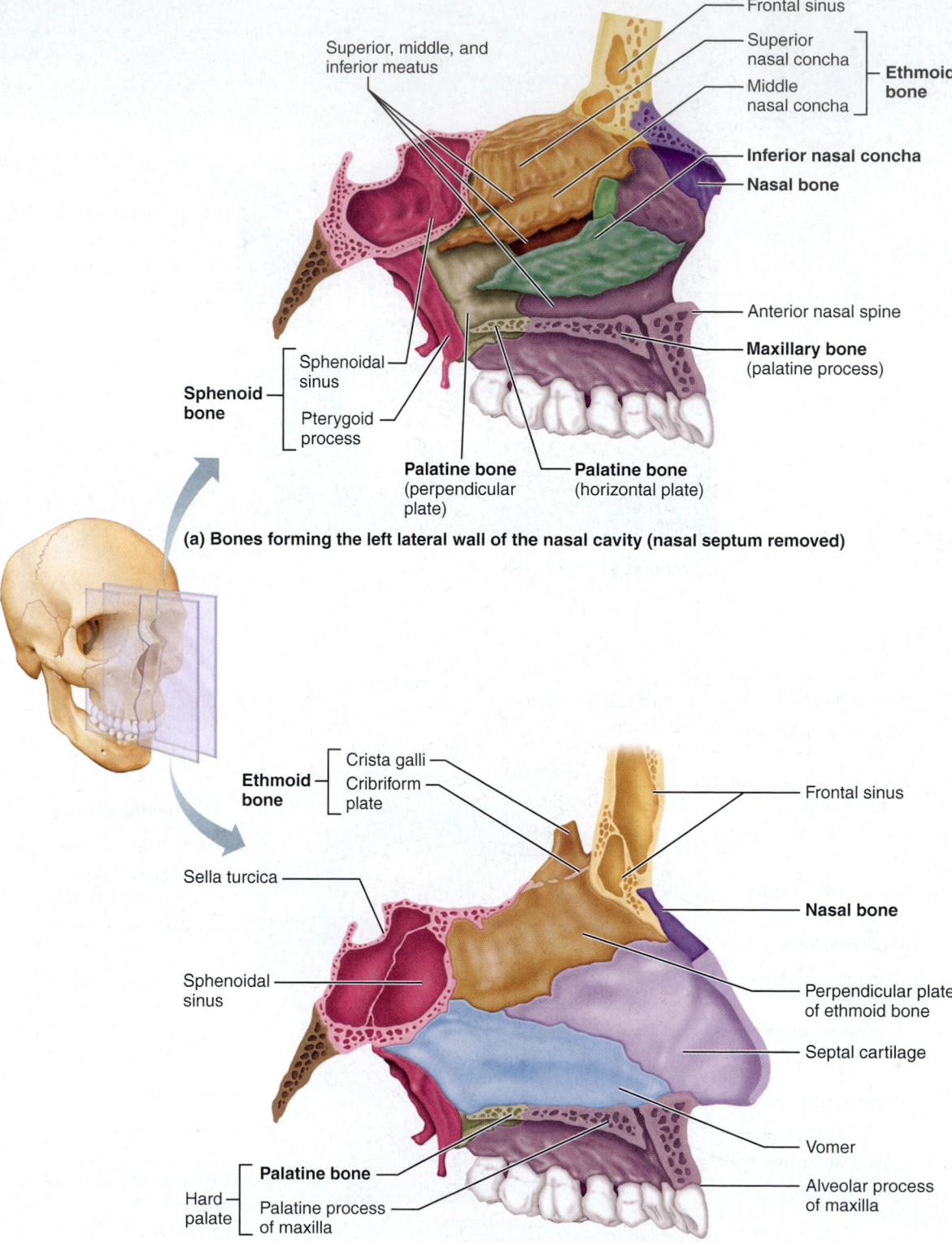

(a) Bones forming the left lateral wall of the nasal cavity (nasal septum removed)

(b) Nasal cavity with septum in place showing the contributions of the ethmoid bone, the vomer, and septal cartilage

Figure 7.14 Bones of the nasal cavity.

(a) Anterior aspect

(b) Medial aspect

Figure 7.15 Paranasal sinuses.

The Nasal Cavity

The **nasal cavity** is constructed of bone and hyaline cartilage (**Figure 7.14**). The *roof* of the nasal cavity is formed by the cribriform plates of the ethmoid. The *lateral walls* are largely shaped by the superior and middle conchae of the ethmoid bone, the perpendicular plates of the palatine bones, and the inferior nasal conchae. The depressions under cover of the conchae on the lateral walls are called *meatuses* (superior, middle, and inferior). The *floor* of the nasal cavity is formed by the palatine processes of the maxillae and the palatine bones. The nasal

cavity is divided into right and left parts by the *nasal septum*. The bony portion of the septum is formed by the vomer inferiorly and the perpendicular plate of the ethmoid bone superiorly (Figure 7.14b). A sheet of cartilage called the *septal cartilage* completes the septum anteriorly.

The nasal septum and conchae are covered with a mucus-secreting mucosa that moistens and warms the entering air and helps cleanse it of debris. The conchae increase the turbulence of air flowing through the nasal cavity. This swirling forces more of the inhaled air into contact with the warm, damp mucosa and encourages trapping of airborne particles (dust, pollen, bacteria) in the sticky mucus.

Paranasal Sinuses

Five skull bones—the frontal, sphenoid, ethmoid, and paired maxillary bones—contain mucosa-lined, air-filled sinuses called **paranasal sinuses** because they cluster around the nasal cavity (**Figure 7.15**). These sinuses give the bones a rather moth-eaten appearance on X-ray images.

Small openings connect the sinuses to the nasal cavity and allow material to flow in and out. Air enters the sinuses from the nasal cavity, and mucus formed by the sinus mucosae drains into the nasal cavity. The mucosa of the sinuses also helps to warm and humidify inhaled air. The paranasal sinuses lighten the skull and enhance the resonance of the voice.

Check Your Understanding

1. What bones are the keystone bones of the facial skeleton?
2. What bone forms the bulk of the orbit floor, and what sense organ is found in the orbit of a living person?
3. For the skull below, name sutures a and b, and bones c–f.

4. **APPLY** Nathan was vigorously exercising the only joints in the skull that are freely movable. What would you guess he was doing?
5. **WHAT IF?** Suppose you work in the coroner's office. The skeleton of a potential murder victim has arrived. It has a fractured hyoid bone. What is the likely cause of death?
6. **MAKE CONNECTIONS** You have learned about two different processes of bone formation in the embryo (see Chapter 6). Name the process that leads to formation of most of the skull bones. Name the embryonic connective tissue that is converted to bone during this process.

For answers, see Answers Appendix.

(Text continues on p. 218.)

Table 7.1   Bones of the Skull

VIEW OF SKULL	BONE WITH COMMENTS*	IMPORTANT BONE MARKINGS†

Cranial bones

Lateral view of skull (Figure 7.5)

■ **Frontal (1)**
Forms forehead, superior part of orbits, and most of the anterior cranial fossa; contains sinuses

Supraorbital foramina (notches): passageway for the supraorbital arteries and nerves

■ **Parietal (2)**
Form most of the superior and lateral aspects of the skull

■ **Occipital (1)**
Forms posterior aspect and most of the base of the skull

Foramen magnum: allows passage of the spinal cord from the brain stem to the vertebral canal

Hypoglossal canals: passageway for the hypoglossal nerve (cranial nerve XII)

Occipital condyles: articulate with the atlas (first vertebra)

External occipital protuberance and **nuchal lines:** sites of muscle attachment

External occipital crest: attachment site of ligamentum nuchae

Superior view of skull, calvaria removed (Figure 7.7)

■ **Temporal (2)**
Form inferolateral aspects of the skull and contribute to the middle cranial fossa; have squamous, tympanic, and petrous parts

Zygomatic process: contributes to the zygomatic arch, which forms the prominence of the cheek

Mandibular fossa: articular point for the condylar process of the mandible

External acoustic meatus: canal leading from the external ear to the eardrum

Styloid process: attachment site for several neck and tongue muscles and for a ligament to the hyoid bone

Mastoid process: attachment site for several neck muscles

Stylomastoid foramen: passageway for cranial nerve VII (facial nerve)

Jugular foramen: passageway for the internal jugular vein and cranial nerves IX, X, and XI

Internal acoustic meatus: passageway for cranial nerves VII and VIII

Carotid canal: passageway for the internal carotid artery

■ **Sphenoid (1)**
Keystone of the cranium; contributes to the middle cranial fossa and orbits; main parts are the body, greater wings, lesser wings, and pterygoid processes

Sella turcica: hypophyseal fossa portion is the seat of the pituitary gland

Optic canals: passageway for cranial nerve II and the ophthalmic arteries

Superior orbital fissures: passageway for cranial nerves III, IV, VI, part of V (ophthalmic division), and ophthalmic vein

Foramen rotundum (2): passageway for the maxillary division of cranial nerve V

Foramen ovale (2): passageway for the mandibular division of cranial nerve V

Foramen spinosum (2): passageway for the middle meningeal artery

■ **Ethmoid (1)**
Small contribution to the anterior cranial fossa; forms part of the nasal septum and the lateral walls and roof of the nasal cavity; contributes to the medial wall of the orbit

Crista galli: attachment point for the falx cerebri, a dural membrane fold

Cribriform plates: passageways for filaments of the olfactory nerves (cranial nerve I)

Superior and middle nasal conchae: form part of lateral walls of nasal cavity; increase turbulence of air flow

Table 7.1 (continued)

VIEW OF SKULL	BONE WITH COMMENTS*	IMPORTANT BONE MARKINGS†
	Facial bones	

Anterior view of skull (Figure 7.4)

■ **Nasal (2)**
Form the bridge of the nose

■ **Lacrimal (2)**
Form part of the medial orbit wall

Lacrimal fossa: houses the lacrimal sac, which helps to drain tears into the nasal cavity

■ **Zygomatic (2)**
Form the cheek and part of the orbit

■ **Inferior nasal concha (2)**
Form part of the lateral walls of the nasal cavity

■ **Mandible (1)**
The lower jaw

Coronoid processes: insertion points for the temporalis muscles

Condylar processes: articulate with the temporal bones to form the jaw (temporomandibular) joints

Mandibular symphysis: medial fusion point of the mandibular bones

Dental alveoli: sockets for the teeth

Mandibular foramina: passageway for the inferior alveolar nerves

Mental foramina: passageway for blood vessels and nerves to the chin and lower lip

Inferior view of skull, mandible removed (Figure 7.6)

■ **Maxilla (2)**
Keystone bones of the face; form the upper jaw and parts of the hard palate, orbits, and nasal cavity walls

Dental alveoli: sockets for teeth

Zygomatic process: helps form the zygomatic arches

Palatine process: forms the anterior hard palate; the two processes meet medially

Frontal process: forms part of lateral aspect of bridge of nose

Incisive fossa and **incisive canal:** passageway for blood vessels and nerves through anterior hard palate (fused palatine processes)

Inferior orbital fissure: passageway for maxillary branch of cranial nerve V, the zygomatic nerve, and blood vessels

Infraorbital foramen: passageway for infraorbital nerve to skin of face

■ **Palatine (2)**
Form posterior part of the hard palate and a small part of nasal cavity and orbit walls

■ **Vomer (1)**
Inferior part of the nasal septum

Auditory ossicles
(malleus, incus, and stapes) (2 each)

Found in middle ear cavity; involved in sound transmission (see Chapter 15)

*The color code beside each bone name corresponds to the bone's color in the illustrations (see Figures 7.4 to 7.14). The number in parentheses ( ) following the bone name indicates the total number of such bones in the body.

†To refresh your memory on bone markings, see ◀ **Table 6.2, p. 180,** for a description of the various kinds of bone markings.

7

7.2 The vertebral column is a flexible, curved support structure

Learning Outcomes

▶ Describe the structure of the vertebral column, list its components, and describe its curvatures.

▶ Indicate a common function of the spinal curvatures and the intervertebral discs.

▶ Discuss the structure of a typical vertebra and describe regional features of cervical, thoracic, and lumbar vertebrae.

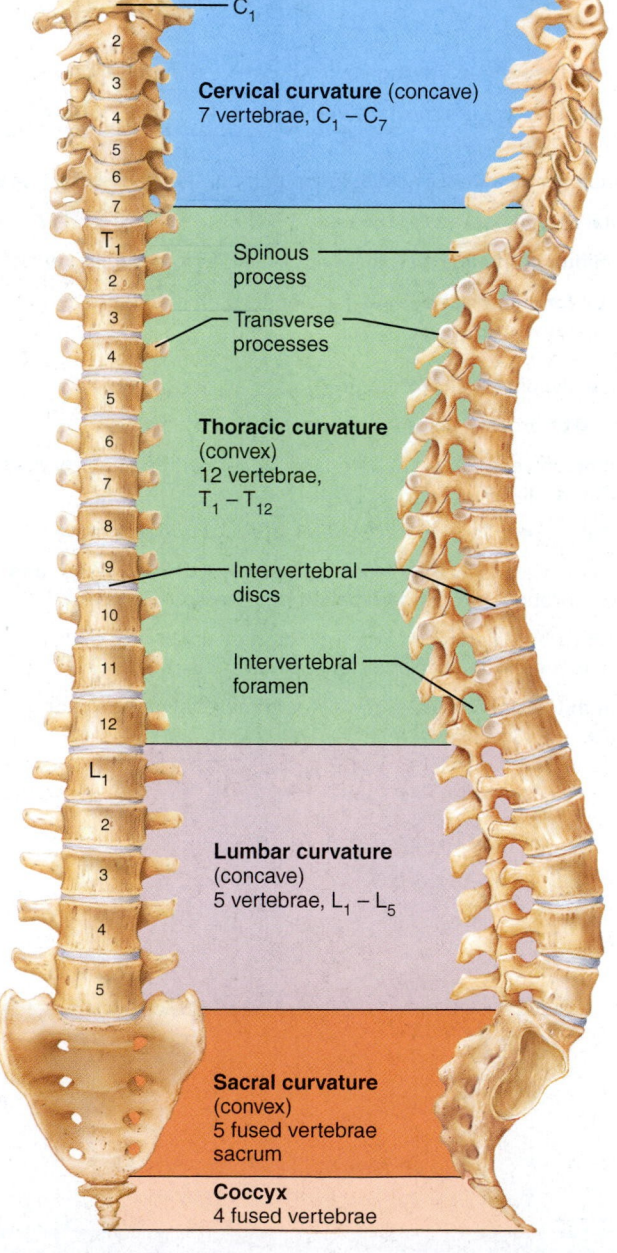

C_1
2
3
4
5
6
7

Cervical curvature (concave)
7 vertebrae, $C_1 - C_7$

T_1
2
3
4
5
6
7
8
9
10
11
12

Spinous process

Transverse processes

Thoracic curvature (convex)
12 vertebrae, $T_1 - T_{12}$

Intervertebral discs

Intervertebral foramen

L_1
2
3
4
5

Lumbar curvature (concave)
5 vertebrae, $L_1 - L_5$

Sacral curvature (convex)
5 fused vertebrae sacrum

Coccyx
4 fused vertebrae

Anterior view   Right lateral view

Figure 7.16 The vertebral column. Notice the curvatures in the lateral view. (The terms convex and concave refer to the curvature of the posterior aspect of the vertebral column.)

General Characteristics

Some people think of the **vertebral column** as a rigid supporting rod, but this is inaccurate. Also called the **spine** or **spinal column**, the vertebral column consists of 26 irregular bones connected in such a way that a flexible, curved structure results (**Figure 7.16**).

The spine extends from the skull to the pelvis, where it transmits the weight of the trunk to the lower limbs. It also surrounds and protects the delicate spinal cord and provides attachment points for the ribs and for the muscles of the back and neck.

In the fetus and infant, the vertebral column consists of 33 separate bones, or **vertebrae** (ver'tĕ-bre). Inferiorly, nine of these eventually fuse to form two composite bones, the sacrum and the tiny coccyx. The remaining 24 bones persist as individual vertebrae separated by intervertebral discs.

Regions and Curvatures

The adult vertebral column is about 70 cm (28 inches) long and has five major regions (Figure 7.16). The seven vertebrae of the neck are the **cervical vertebrae** (ser'vĭ-kal), the next 12 are the **thoracic vertebrae** (tho-ras'ik), and the five supporting the lower back are the **lumbar vertebrae** (lum'bar). Remembering common meal times—7 AM, 12 noon, and 5 PM—will help you recall the number of bones in these three regions of the spine. The vertebrae become progressively larger from the cervical to the lumbar region, as they must support greater and greater weight.

Inferior to the lumbar vertebrae is the **sacrum** (sa'krum), which articulates with the hip bones. The terminus of the vertebral column is the tiny **coccyx** (kok'siks).

All of us have the same number of cervical vertebrae. Variations in numbers of vertebrae in other regions occur in about 5% of people.

When you view the vertebral column from the side, you can see the four curvatures that give it its S, or sinusoid, shape. The **cervical** and **lumbar curvatures** are concave posteriorly; the **thoracic** and **sacral curvatures** are convex posteriorly. These curvatures increase the resilience and flexibility of the spine, allowing it to function like a spring rather than a rigid rod.

Ligaments

Like a tall, tremulous TV transmitting tower, the vertebral column cannot stand upright by itself. It must be held in place by an elaborate system of cable-like supports. In the case of the vertebral column, straplike ligaments and the trunk muscles assume this role.

The major supporting ligaments are the **anterior** and **posterior longitudinal ligaments** (**Figure 7.17**). These run as continuous bands down the front and back surfaces of the vertebrae from the neck to the sacrum. The broad anterior ligament is strongly attached to both the bony vertebrae and the discs. Along with its supporting role, it prevents hyperextension of the spine (bending too far backward). The posterior ligament, which resists hyperflexion of the spine (bending too far forward), is narrow and relatively weak. It attaches only to the discs. However, the **ligamentum flavum**, which connects adjacent vertebrae, contains elastic connective tissue and is especially strong. It stretches as we bend forward and then recoils when we resume an erect posture. Short ligaments connect each vertebra to those immediately above and below.

Intervertebral Discs

Each **intervertebral disc** is a cushionlike pad composed of two parts. The inner gelatinous **nucleus pulposus** (pul-po′sus; "pulp") acts like a rubber ball, giving the disc its elasticity and compressibility. Surrounding the nucleus pulposus is a strong collar composed of collagen fibers superficially and fibrocartilage internally, the **anulus fibrosus** (an′u-lus fi-bro′sus; "ring of fibers") (Figure 7.17a, c). The anulus fibrosus limits the expansion of the nucleus pulposus when the spine is compressed. It also acts like a woven strap to bind successive vertebrae together, withstands twisting forces, and resists tension in the spine.

Sandwiched between the bodies of neighboring vertebrae, the intervertebral discs act as shock absorbers during walking, jumping, and running. They allow the spine to flex and extend, and to a lesser extent to bend laterally. At points of compression, the discs flatten and bulge out a bit between the vertebrae. The discs are thickest in the lumbar and cervical regions, which enhances the flexibility of these regions.

Collectively the discs account for about 25% of the height of the vertebral column. They flatten somewhat during the course of the day, so we are always a few millimeters shorter at night than when we awake in the morning.

HOMEOSTATIC IMBALANCE 7.2   CLINICAL

Severe or sudden physical trauma to the spine—for example, from bending forward while lifting a heavy object—may result in herniation of one or more discs. A **herniated (prolapsed) disc** (commonly called a *slipped disc*) usually involves rupture of the anulus fibrosus followed by protrusion of the spongy nucleus pulposus through the anulus (Figure 7.17c, d). If the

Supraspinous ligament
Transverse process
Sectioned spinous process
Ligamentum flavum
Interspinous ligament
Inferior articular process

Intervertebral disc
Anterior longitudinal ligament
Intervertebral foramen
Posterior longitudinal ligament
Anulus fibrosus
Nucleus pulposus
Sectioned body of vertebra

(a) Median section of three vertebrae

Posterior longitudinal ligament
Anterior longitudinal ligament
Body of a vertebra
Intervertebral disc

(b) Anterior view of part of the spinal column

Vertebral spinous process (posterior aspect of vertebra)
Spinal nerve root
Transverse process
Herniated portion of disc
Anulus fibrosus of disc

Spinal cord
Nucleus pulposus of disc

(c) Superior view of a herniated intervertebral disc

Nucleus pulposus of intact disc
Herniated nucleus pulposus

(d) MRI of lumbar region of vertebral column in sagittal section showing herniated disc

Figure 7.17 Ligaments and fibrocartilage discs uniting the vertebrae.

protrusion presses on the spinal cord or on spinal nerves exiting from the cord, numbness or excruciating pain may result.

Herniated discs are only treated surgically if other treatments fail. Nonsurgical treatment strategies include painkillers, muscle relaxants for associated muscle spasm, ice and heat, and physiotherapy.

HOMEOSTATIC IMBALANCE 7.3   CLINICAL

There are several types of abnormal spinal curvatures (**Figure 7.18**). Some are congenital (present at birth); others result from disease, poor posture, or unequal muscle pull on the spine. *Scoliosis* (sko″le-o′sis), literally, "twisted disease," is an abnormal rotation of the spine that results in a *lateral* curvature, most often in the thoracic region. It is quite common during late childhood, particularly in girls. Other, more severe cases result from abnormal vertebral structure, lower limbs of unequal length, or muscle paralysis. If muscles on one side of the body are nonfunctional, those of the opposite side exert an unopposed pull on the spine and force it out of alignment. Scoliosis is usually treated with bracing, but more severe cases may require surgery.

Kyphosis (ki-fo′sis), or hunchback, is a *dorsally* exaggerated *thoracic* curvature. It is particularly common in elderly people because of osteoporosis, but may also reflect tuberculosis of the spine, rickets, or osteomalacia.

Lordosis, or swayback, is an accentuated *lumbar* curvature. It, too, can result from spinal tuberculosis or osteomalacia. Temporary lordosis is common in those carrying a large load up front, such as men with "potbellies" and pregnant women. In an attempt to maintain their center of gravity, these individuals automatically throw back their shoulders, accentuating their lumbar curvature.

General Structure of Vertebrae

All vertebrae have a common structural pattern (**Figure 7.19**). Each vertebra consists of a **body**, or *centrum*, anteriorly and a

(a) Scoliosis   **(b) Kyphosis**   **(c) Lordosis**

Figure 7.18 Abnormal spinal curvatures.

Figure 7.19 Typical vertebral structures. Superior view of a thoracic vertebra. Only bone features are illustrated in this and subsequent bone figures in this chapter. Articular cartilage is not depicted.

Posterior — Spinous process — Transverse process — Superior articular facet and process — Vertebral foramen — **Body** — **Vertebral arch** — • Lamina — • Pedicle — *Anterior*

vertebral arch posteriorly. The disc-shaped body is the weight-bearing region. Together, the body and vertebral arch enclose an opening called the **vertebral foramen**. Successive vertebral foramina of the articulated vertebrae form the long **vertebral canal**, through which the spinal cord passes.

The vertebral arch is a composite structure formed by two pedicles and two laminae. The **pedicles** (ped′ĭ-kelz; "little feet"), short bony pillars projecting posteriorly from the vertebral body, form the sides of the arch. The **laminae** (lam′ĭ-ne), flattened plates that fuse in the median plane, complete the arch posteriorly. The pedicles have notches on their superior and inferior borders, providing lateral openings between adjacent vertebrae called **intervertebral foramina** (see Figure 7.16). The spinal nerves issuing from the spinal cord pass through these foramina.

Seven processes project from the vertebral arch. The **spinous process** is a median posterior projection arising at the junction of the two laminae. A **transverse process** extends laterally from each side of the vertebral arch. The spinous and transverse processes are attachment sites for muscles that move the vertebral column and for ligaments that stabilize it. The paired **superior** and **inferior articular processes** protrude superiorly and inferiorly, respectively, from the pedicle-lamina junctions. The smooth joint surfaces of the articular processes, called *facets* ("little faces"), are covered with hyaline cartilage. The inferior articular processes of each vertebra form movable joints with the superior articular processes of the vertebra immediately below. Thus, successive vertebrae join both at their bodies and at their articular processes.

(a) Superior view of atlas (C₁)

(b) Inferior view of atlas (C₁)

(c) Superior view of axis (C₂)

(d) Photo of axis (C₂), superior view

Figure 7.20 The first and second cervical vertebrae.

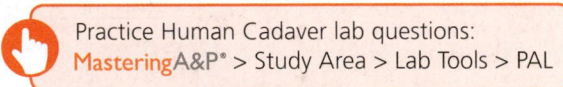

Practice Human Cadaver lab questions:
MasteringA&P® > Study Area > Lab Tools > PAL

Regional Vertebral Characteristics

Beyond their common structural features, vertebrae exhibit variations that allow different regions of the spine to perform slightly different functions and movements. In general, movements that can occur between vertebrae are (1) flexion and extension (anterior bending and posterior straightening of the spine), (2) lateral flexion (bending the *upper body* to the right or left), and (3) rotation (in which vertebrae rotate on one another in the longitudinal axis of the spine). The regional vertebral characteristics are illustrated and summarized in **Table 7.2** on p. 223.

Cervical Vertebrae

The seven cervical vertebrae, identified as C_1–C_7, are the smallest, lightest vertebrae (see Figure 7.16). The "typical" cervical vertebrae (C_3–C_7) have the following distinguishing features (see Figure 7.21 and Table 7.2):

- The body is oval—wider from side to side than in the anteroposterior dimension.
- Except in C_7, the spinous process is short, projects directly back, and is *bifid* (bi'fid), or split at its tip.
- The vertebral foramen is large and generally triangular.

- Each transverse process contains a **transverse foramen** through which the vertebral arteries pass to service the brain.

The spinous process of C_7 is not bifid and is much larger than those of the other cervical vertebrae (see Figure 7.21a). Because its spinous process is palpable through the skin, C_7 can be used as a landmark for counting the vertebrae and is called the **vertebra prominens** ("prominent vertebra").

The first two cervical vertebrae, the atlas and the axis, are somewhat more robust than the typical cervical vertebra. They have no intervertebral disc between them, and they are highly modified, reflecting their special functions. The **atlas** (C_1) has no body and no spinous process (**Figure 7.20a** and **b**). Essentially, it is a ring of bone consisting of *anterior* and *posterior arches* and a *lateral mass* on each side. Each lateral mass has articular facets on both its superior and inferior surfaces. The superior articular facets receive the occipital condyles of the skull—they "carry" the skull, just as Atlas supported the heavens in Greek mythology. These joints allow you to nod your head "yes." The inferior articular facets form joints with the axis (C_2) below.

The **axis**, which has a body and the other typical vertebral processes, is not as specialized as the atlas. In fact, its only unusual feature is the knoblike **dens** (denz; "tooth") projecting superiorly from its body. The dens is actually the "missing"

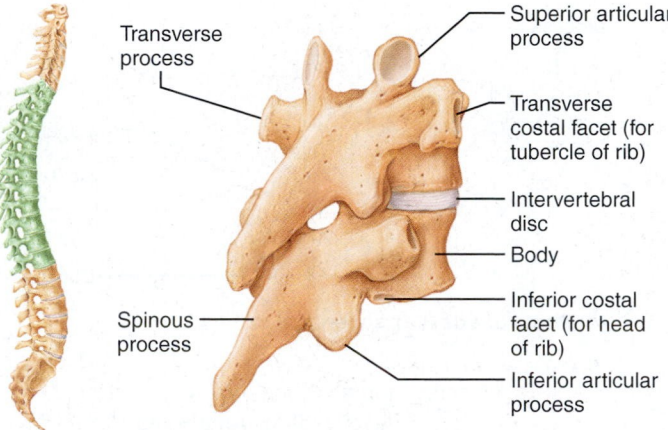

(a) Cervical vertebrae

Dens of axis
Transverse ligament of atlas
C_1 (atlas)
C_2 (axis)
C_3
Inferior articular process
Bifid spinous process
Transverse processes
C_7 (vertebra prominens)

(b) Thoracic vertebrae

Transverse process
Superior articular process
Transverse costal facet (for tubercle of rib)
Intervertebral disc
Body
Inferior costal facet (for head of rib)
Spinous process
Inferior articular process

(c) Lumbar vertebrae

Superior articular process
Transverse process
Body
Intervertebral disc
Inferior articular process
Spinous process

Figure 7.21 Posterolateral views of articulated vertebrae. Notice the bulbous tip on the spinous process of C_7, the vertebra prominens.

body of the atlas, which fuses with the axis during embryonic development. Cradled in the anterior arch of the atlas by the transverse ligament (**Figure 7.21a**), the dens acts as a pivot for the rotation of the atlas. As a result, this joint is the one that allows you to rotate your head from side to side to indicate "no."

Thoracic Vertebrae

Of the 12 thoracic vertebrae (T_1–T_{12}), the first looks much like C_7, and the last four show a progression toward lumbar vertebral structure (see Table 7.2, Figure 7.16, and Figure 7.21b). The thoracic vertebrae increase in size from the first to the last. All thoracic vertebrae articulate with the ribs. Unique characteristics of these vertebrae include:

- The body is roughly heart shaped. It typically bears two small facets, commonly called *demifacets* (half-facets), on each side, one at the superior edge (the *superior costal facet*) and the other at the inferior edge (the *inferior costal facet*). The demifacets receive the heads of the ribs. (The bodies of T_{10}–T_{12} vary from this pattern by having only a single facet to receive their respective ribs.)

- The vertebral foramen is circular.

- The spinous process is long and points sharply downward.

- With the exception of T_{11} and T_{12}, the transverse processes have facets, the *transverse costal facets*, that articulate with the tubercles (small rounded processes) of the ribs.

- The superior and inferior articular facets lie mainly in the frontal plane, a situation that limits flexion and extension, but which allows this region of the spine to rotate. Lateral flexion, though possible, is restricted by the ribs.

Lumbar Vertebrae

The lumbar region of the vertebral column, commonly referred to as the small of the back, receives the most stress. The enhanced weight-bearing function of the five lumbar vertebrae (L_1–L_5) is reflected in their sturdier structure. Their bodies are massive and kidney shaped in a superior view (see Table 7.2, Figure 7.16, and Figure 7.21c). Other characteristics are:

- The pedicles and laminae are shorter and thicker than those of other vertebrae.

- The spinous processes are short, flat, and hatchet shaped and are easily seen when a person bends forward. These processes are robust and project directly backward, adaptations for the attachment of the large back muscles.

- The vertebral foramen is triangular.

- The orientation of the facets of the articular processes of the lumbar vertebrae differs substantially from that of the other vertebra types (see Table 7.2). These modifications lock the lumbar vertebrae together and provide stability by preventing rotation of the lumbar spine. Flexion and extension are possible (as when you do sit-ups), as is lateral flexion.

Table 7.2   Regional Characteristics of Cervical, Thoracic, and Lumbar Vertebrae

CHARACTERISTIC	CERVICAL (3–7)	THORACIC	LUMBAR
Body	Small, oval, wide side to side	Larger than cervical; heart shaped; bears two costal facets	Massive; kidney shaped
Spinous process	Short; bifid (except C$_7$); projects directly posterior	Long; sharp; projects inferiorly	Short; blunt; rectangular; projects directly posteriorly
Vertebral foramen	Triangular, large	Circular	Triangular
Transverse processes	Contain foramina	Bear facets for ribs (except T$_{11}$ and T$_{12}$)	Thin and tapered
Superior and inferior articular processes	Superior facets directed superoposteriorly	Superior facets directed posteriorly	Superior facets directed posteromedially (or medially)
	Inferior facets directed inferoanteriorly	Inferior facets directed anteriorly	Inferior facets directed anterolaterally (or laterally)
Movements allowed	Flexion and extension; lateral flexion; rotation; the spine region with the greatest range of movement	Rotation; lateral flexion possible but restricted by ribs; flexion and extension limited	Flexion and extension; some lateral flexion; rotation prevented

Superior View

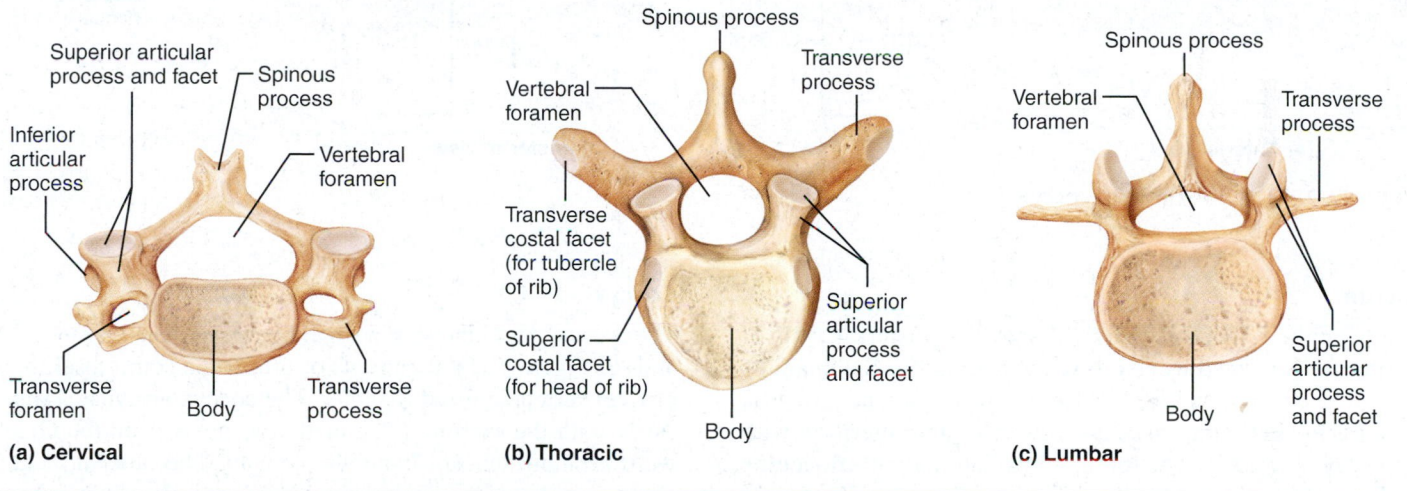

(a) Cervical

(b) Thoracic

(c) Lumbar

Right Lateral View

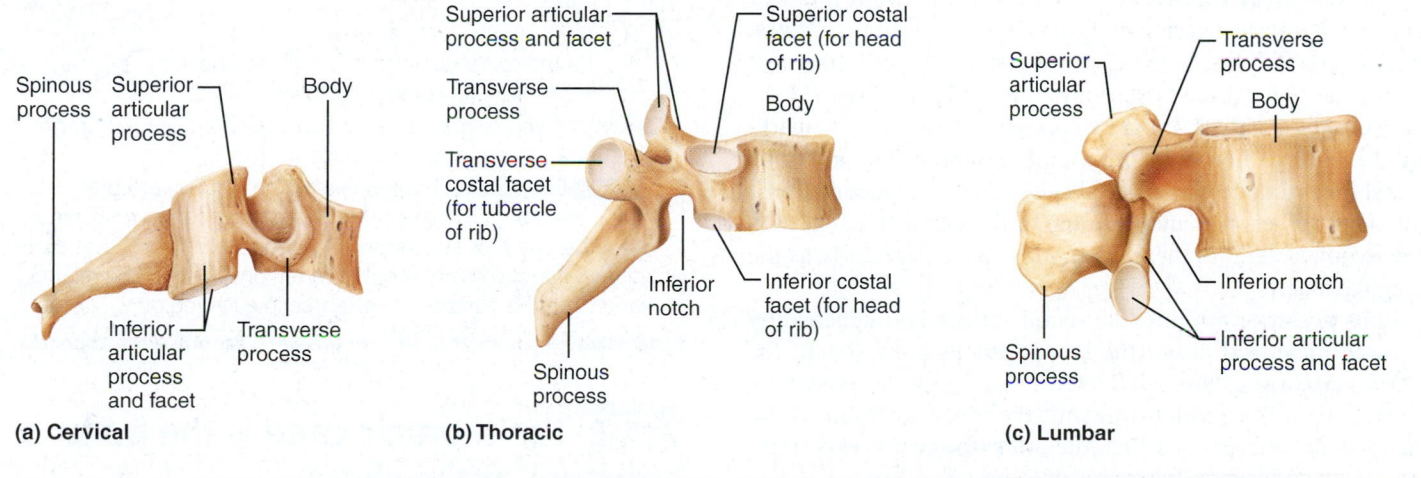

(a) Cervical

(b) Thoracic

(c) Lumbar

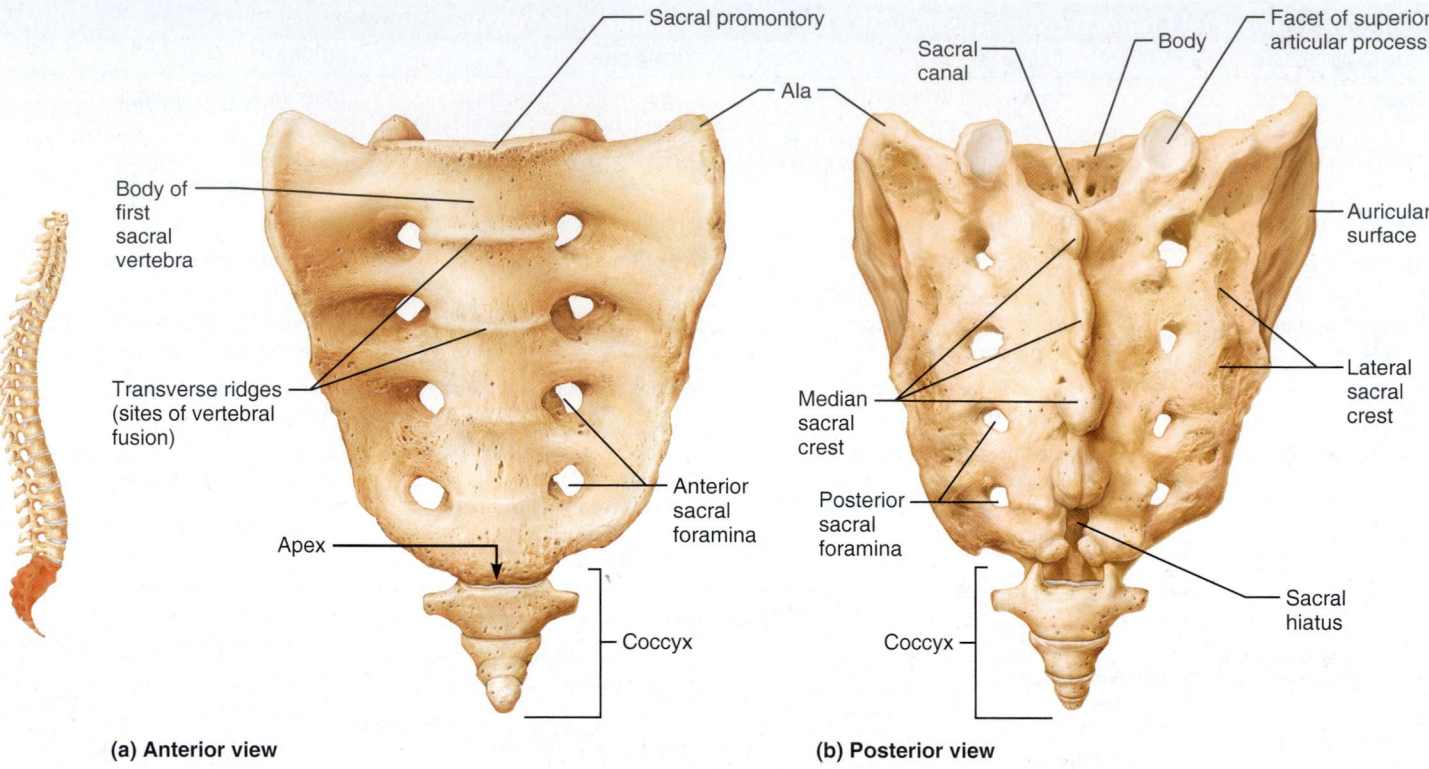

(a) Anterior view

(b) Posterior view

Figure 7.22 The sacrum and coccyx.

Sacrum

The triangular sacrum, which shapes the posterior wall of the pelvis, is formed by five fused vertebrae (S_1–S_5) in adults (**Figure 7.22**, and see Figure 7.16). It articulates superiorly (via its **superior articular processes**) with L_5 and inferiorly with the coccyx. Laterally, the sacrum articulates, via its **auricular surfaces**, with the two hip bones to form the **sacroiliac joints** (sa″kro-il′e-ak) of the pelvis.

The **sacral promontory** (prom′on-tor″e; "high point of land"), the anterosuperior margin of the first sacral vertebra, bulges anteriorly into the pelvic cavity. The body's center of gravity lies about 1 cm posterior to this landmark. Four ridges, the **transverse ridges**, cross its concave anterior aspect, marking the lines of fusion of the sacral vertebrae. The **anterior sacral foramina** lie at the lateral ends of these ridges and transmit blood vessels and anterior rami of the sacral spinal nerves. The regions lateral to these foramina expand superiorly as the winglike **alae**.

In its posterior midline, the sacral surface is roughened by the **median sacral crest** (the fused spinous processes of the sacral vertebrae). This is flanked laterally by the **posterior sacral foramina** (which transmit the posterior rami of the sacral spinal nerves) and then the **lateral sacral crests** (remnants of the transverse processes of S_1–S_5).

The vertebral canal continues inside the sacrum as the **sacral canal**. Since the laminae of the fifth (and sometimes the fourth) sacral vertebrae fail to fuse medially, an enlarged external opening called the **sacral hiatus** (hi-a′tus; "gap") is obvious at the inferior end of the sacral canal.

Coccyx

The coccyx, our tailbone, is a small triangular bone (Figure 7.22, and see Figure 7.16). It consists of four (or in some cases three or five) vertebrae fused together. The coccyx articulates superiorly with the sacrum. (The name *coccyx* is from the Greek word meaning "cuckoo" and was so named because of its fancied resemblance to a bird's beak.) Except for the slight support the coccyx affords the pelvic organs, it is a nearly useless bone.

Check Your Understanding

7. Besides the spinal curvatures, which skeletal elements help to make the vertebral column flexible?

8. How can you distinguish a lumbar vertebra from a thoracic vertebra?

9. **DRAW** Draw a line that represents the curved vertebral column. Label the two curvatures that are concave posteriorly and the two that are convex. Draw horizontal lines that divide the vertebral column into its five regions, label those regions, and state the number of vertebrae in each region.

For answers, see Answers Appendix.

7.3 The thoracic cage is the bony structure of the chest

Learning Outcomes

▶ **Name and describe the bones of the thoracic cage.**

▶ **Differentiate true from false ribs.**

Anatomically, the thorax is the chest, and its bony underpinnings are called the **thoracic cage** or *bony thorax*. Elements of the thoracic cage include the thoracic vertebrae posteriorly, the ribs laterally, and the sternum and costal cartilages anteriorly. The costal cartilages secure the ribs to the sternum (**Figure 7.23a**).

Roughly cone shaped with its broad dimension positioned inferiorly, the bony thorax forms a protective cage around the vital organs of the thoracic cavity (heart, lungs, and great blood vessels), supports the shoulder girdles and upper limbs, and provides attachment points for many muscles of the neck, back, chest, and shoulders. The *intercostal spaces* between the ribs are occupied by the intercostal muscles, which lift and then depress the thorax during breathing.

Sternum

The **sternum** (breastbone) lies in the anterior midline of the thorax. Vaguely resembling a dagger, it is a flat bone approximately 15 cm (6 inches) long, resulting from the fusion of three bones: the manubrium, the body, and the xiphoid process. The *manubrium* (mah-nu′bre-um; "knife handle") is the superior portion, which is shaped like the knot in a necktie. The manubrium articulates via its **clavicular notches** (klah-vik′u-lar) with the clavicles (collarbones) laterally, and just below this, it also articulates with the first two pairs of ribs. The *body*, or midportion, forms the bulk of the sternum. The sides of the body are notched where it articulates with the costal cartilages of the second to seventh ribs. The *xiphoid process* (zif′oid; "swordlike") forms the inferior end of the sternum. This small, variably shaped process is a plate of hyaline cartilage in youth, but it is usually ossified in adults over the age of 40. The xiphoid process articulates only with the sternal body and serves as an attachment point for some abdominal muscles.

The sternum has three important anatomical landmarks: the jugular notch, the sternal angle, and the xiphisternal joint (Figure 7.23). The easily palpated **jugular** (*suprasternal*) **notch** is the central indentation in the superior border of the manubrium. If you slide your finger down the anterior surface of your neck, it will land in the jugular notch. The jugular notch is generally in line with the disc between the second and third thoracic vertebrae and the point where the left common carotid artery branches from the aorta (Figure 7.23b).

The **sternal angle** is felt as a horizontal ridge across the front of the sternum, where the manubrium joins the sternal body. This cartilaginous joint acts like a hinge, allowing the sternal body to swing anteriorly when we inhale. The sternal angle is in line with the disc between the fourth and fifth thoracic vertebrae and at the level of the second pair of ribs. It is a handy reference point for finding the second rib. Consequently, it allows us to count the ribs during a physical examination and to listen to sounds made by specific heart valves.

Jugular notch
Clavicular notch

Sternum
• **Manubrium**
• **Sternal angle**
• **Body**
• Xiphisternal joint
• **Xiphoid process**

True ribs (1–7)

False ribs (8–12)

Floating ribs (11, 12)

L₁ Vertebra

Intercostal spaces

Costal cartilage
Costal margin

(a) Skeleton of the thoracic cage, anterior view

Jugular notch
Sternal angle

Heart

Xiphisternal joint

(b) Midsagittal section through the thorax, showing the relationship of surface anatomical landmarks of the thorax to the vertebral column

Figure 7.23 The thoracic cage.

The **xiphisternal joint** (zif″ĭ-ster′nul) is the point where the sternal body and xiphoid process fuse. It lies at the level of the ninth thoracic vertebra. The heart lies on the diaphragm just deep to this joint.

> ### HOMEOSTATIC IMBALANCE 7.4   CLINICAL
>
> In some people, the xiphoid process projects posteriorly. In such cases, **chest trauma** consisting of a blow at the level of the xiphoid process can push the process into the underlying heart or liver, causing massive hemorrhage. When doing chest compressions during cardiopulmonary resuscitation (CPR), hands are positioned midsternum to prevent breaking off the xiphoid and damaging the liver.

Ribs

Twelve pairs of **ribs** form the flaring sides of the thoracic cage (Figure 7.23a). All ribs attach posteriorly to the thoracic vertebrae (bodies and transverse processes) and curve inferiorly toward the anterior body surface. The superior seven rib pairs attach directly to the sternum by individual costal cartilages (bars of hyaline cartilage). These are **true** or **vertebrosternal ribs** (ver″tĕ-bro-ster′nal). (Notice that the anatomical name indicates the two attachment points of a rib—the posterior attachment given first.)

The remaining five pairs of ribs are called **false ribs** because they either attach indirectly to the sternum or entirely lack a sternal attachment. Rib pairs 8–10 attach to the sternum indirectly, each joining the costal cartilage immediately above it. These ribs are also called **vertebrochondral ribs** (ver″tĕ-bro-kon′dral). The inferior margin of the rib cage, or **costal margin**, is formed by the costal cartilages of ribs 7–10. Rib pairs 11 and 12 are called **vertebral ribs** or **floating ribs** because they have no anterior attachments. Instead, their costal cartilages lie embedded in the muscles of the lateral body wall.

The ribs increase in length from pair 1 to pair 7, then decrease in length from pair 8 to pair 12. Except for the first rib, which lies deep to the clavicle, the ribs are easily felt in people of normal weight.

A typical rib is a bowed flat bone (**Figure 7.24**). The bulk of a rib is simply called the *shaft*. Its superior border is smooth, but its inferior border is sharp and thin and has a *costal groove* on its inner face that lodges the intercostal nerves and blood vessels.

In addition to the shaft, each rib has a head, neck, and tubercle. The wedge-shaped *head*, the posterior end, articulates with the vertebral bodies by two facets: One joins the body of the same-numbered thoracic vertebra, the other articulates with the body of the vertebra immediately superior. The *neck* is the constricted portion of the rib just beyond the head. Lateral to this, the knoblike *tubercle* articulates with the costal facet of the transverse process of the same-numbered thoracic vertebra. Beyond the tubercle, the shaft angles sharply forward (at the angle of the rib) and then extends to attach to its costal cartilage anteriorly. The costal cartilages provide secure but flexible rib attachments to the sternum.

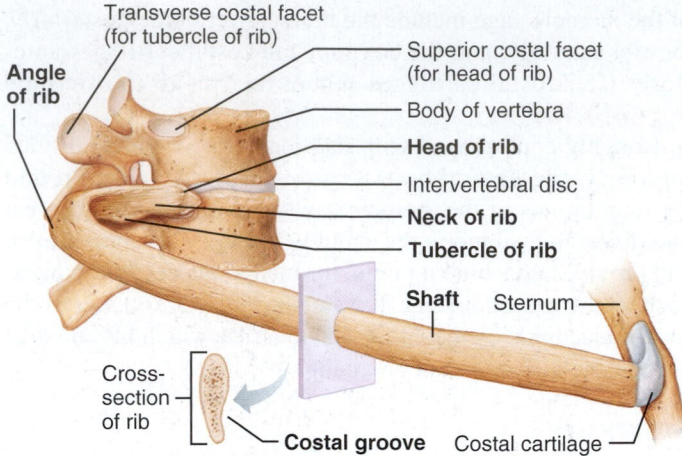

(a) Vertebral and sternal articulations of a typical true rib

(b) Superior view of the articulation between a rib and a thoracic vertebra

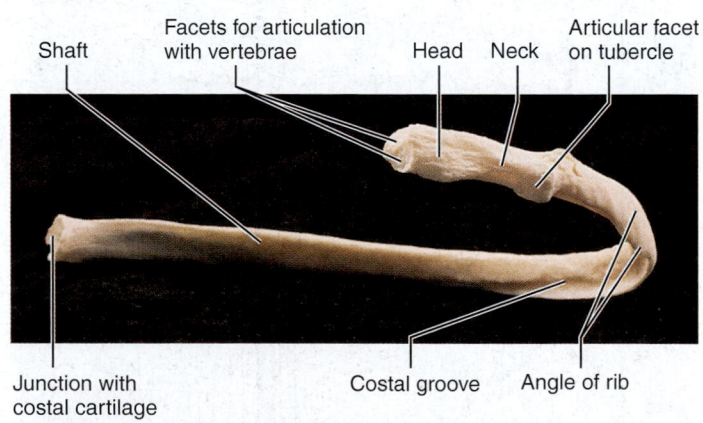

(c) A typical rib (rib 6, right), posterior view

Figure 7.24 Ribs.

> Practice Human Cadaver lab questions:
> MasteringA&P° > Study Area > Lab Tools > PAL

The first pair of ribs is atypical. They are flattened superiorly to inferiorly and are quite broad, forming a horizontal table that supports the subclavian blood vessels that serve the upper limbs. There are also other exceptions to the typical rib pattern. Rib 1 and ribs 10–12 articulate with only one vertebral body, and ribs 11 and 12 do not articulate with a vertebral transverse process.

Check Your Understanding

10. How does a true rib differ from a false rib?

11. What does the tubercle of a rib articulate with?

12. Besides the ribs and sternum, there is a third group of bones making up the thoracic cage. What is it?

13. APPLY After falling while snowboarding, Jennifer complained of pain on the right side of her chest that intensified when she took a deep breath or coughed. The emergecy medical technician suspected a broken rib. Describe the process the EMT used to determine (by palpation) the number of the rib that was injured.

For answers, see Answers Appendix.

PART 2

THE APPENDICULAR SKELETON

Bones of the limbs and their girdles are collectively called the appendicular skeleton because they are appended to the axial skeleton (see Figure 7.1). The yokelike *pectoral girdles* (pek′tor-al; "chest") attach the upper limbs to the body trunk. The more sturdy *pelvic girdle* secures the lower limbs. Although the bones of the upper and lower limbs differ in their functions and mobility, they have the same fundamental plan: Each limb is composed of three major segments connected by movable joints.

The appendicular skeleton enables us to carry out the movements typical of our freewheeling and manipulative lifestyle. Each time we take a step, throw a ball, or text a friend, we are making good use of our appendicular skeleton.

7.4 Each pectoral girdle consists of a clavicle and a scapula

Learning Outcomes

▶ **Identify bones forming the pectoral girdle and relate their structure and arrangement to the function of this girdle.**

▶ **Identify important bone markings on the pectoral girdle.**

The **pectoral girdle**, or **shoulder girdle**, consists of the *clavicle* (klav′ĭ-kl) anteriorly and the *scapula* (skap′u-lah) posteriorly (**Figure 7.25** and Table 7.3 on p. 235). The paired pectoral girdles and their associated muscles form your shoulders. Although the term *girdle* usually signifies a beltlike structure encircling the body, a single pectoral girdle, or even the pair, does not quite satisfy this description. Anteriorly, the medial end of each clavicle joins the sternum; the distal ends of the clavicles meet the scapulae laterally. However, the scapulae fail to complete the ring posteriorly, because their medial borders do not join each other or the axial skeleton. Instead, the scapulae are attached to the thorax and vertebral column only by the muscles that clothe their surfaces.

Acromio-clavicular joint

Clavicle

Scapula

Figure 7.25 The pectoral girdle with articulating bones.

The pectoral girdles attach the upper limbs to the axial skeleton and provide attachment points for many of the muscles that move the upper limbs. These girdles are very light and allow the upper limbs a degree of mobility not seen anywhere else in the body. This mobility is due to the following factors:

• Because only the clavicle attaches to the axial skeleton, the scapula can move quite freely across the thorax, allowing the arm to move with it.

• The socket of the shoulder joint (the scapula's glenoid cavity) is shallow and poorly reinforced, so it does not restrict the movement of the humerus (arm bone). Although this arrangement is good for flexibility, it is bad for stability—shoulder dislocations are fairly common.

Clavicles

The **clavicles** ("little keys"), or collarbones, are slender, S-shaped bones that can be felt along their entire course as they extend horizontally across the superior thorax (Figure 7.25). Besides anchoring many muscles, the clavicles act as braces: They hold the scapulae and arms out laterally, away from the narrower superior part of the thorax. This bracing function becomes obvious when a clavicle is fractured: The entire shoulder region collapses medially. The clavicles also transmit compression forces from the upper limbs to the axial skeleton, for example, when someone pushes a car to a gas station.

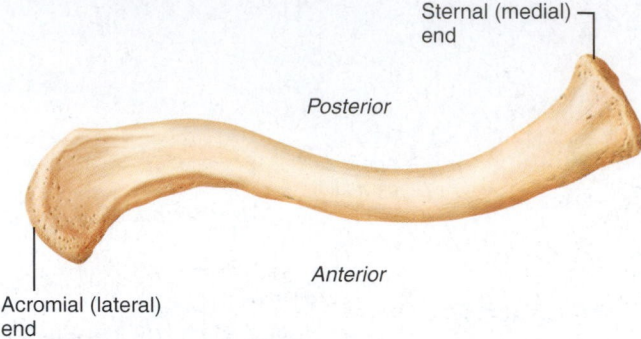

(a) Right clavicle, superior view

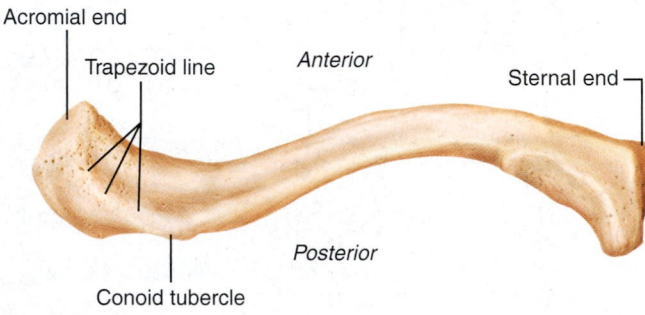

(b) Right clavicle, inferior view

Figure 7.26 **The clavicle.**

Each clavicle is cone shaped at its medial **sternal end**, which attaches to the sternal manubrium, and flattened at its lateral **acromial end** (ah-kro′me-al), which articulates with the scapula (**Figure 7.26**). The medial two-thirds of the clavicle is convex anteriorly; its lateral third is concave anteriorly. Its superior surface is fairly smooth, but the inferior surface is ridged and grooved by ligaments and by the action of the muscles that attach to it. The *trapezoid line* and the *conoid tubercle*, for example, are anchoring points for a ligament that connects the clavicle to the scapula.

The clavicles are not very strong and are likely to fracture, for example, when a person uses outstretched arms to break a fall. The curves in the clavicle ensure that it usually fractures anteriorly (outward). If it were to collapse posteriorly (inward), bone splinters would damage the subclavian artery, which passes just deep to the clavicle to serve the upper limb. The clavicles are exceptionally sensitive to muscle pull and become noticeably larger and stronger in those who perform manual labor or athletics involving the shoulder and arm muscles.

Scapulae

The **scapulae**, or *shoulder blades*, are thin, triangular flat bones (Figure 7.25 and **Figure 7.27**). Interestingly, their name derives from a word meaning "spade" or "shovel," because ancient cultures made spades from the shoulder blades of animals. The scapulae lie on the dorsal surface of the rib cage, between ribs 2 and 7.

Each scapula has three borders. The *superior border* is the shortest, sharpest border. The *medial*, or *vertebral*, *border* parallels the vertebral column. The thick *lateral*, or *axillary*, *border* is next to the armpit and ends superiorly in a small, shallow fossa, the **glenoid cavity** (gle′noid; "pit-shaped"). This cavity articulates with the humerus of the arm, forming the shoulder joint.

Like all triangles, the scapula has three corners or *angles*. The superior scapular border meets the medial border at the *superior angle* and the lateral border at the *lateral angle*. The medial and lateral borders join at the *inferior angle*. The inferior angle moves extensively as the arm is raised and lowered, and is an important landmark for studying scapular movements.

The anterior, or costal, surface of the scapula is concave and relatively featureless. Its posterior surface bears a prominent **spine** that is easily felt through the skin. The spine ends laterally in an enlarged, roughened triangular projection called the **acromion** (ah-kro′me-on; "point of the shoulder"). The acromion articulates with the acromial end of the clavicle, forming the **acromioclavicular joint**.

Projecting anteriorly from the superior scapular border is the **coracoid process** (kor′ah-coid); *corac* means "beaklike," but this process looks more like a bent little finger. The coracoid process helps anchor the biceps muscle of the arm. It is bounded by the **suprascapular notch** (a nerve passage) medially and by the glenoid cavity laterally.

Several large fossae appear on both sides of the scapula and are named according to location. The *infraspinous* and *supraspinous fossae* are inferior and superior, respectively, to the spine. The *subscapular fossa* is the shallow concavity formed by the entire anterior scapular surface. Lying within these fossae are muscles with similar names.

Check Your Understanding

14. What two bones construct each pectoral girdle?

15. Where is the single point of attachment of the pectoral girdle to the axial skeleton?

16. What is the major shortcoming of the flexibility allowed by the shoulder joint?

17. Name each of the structures a–e in the figure below. State the anatomical name for the type of rib indicated by e.

For answers, see Answers Appendix.

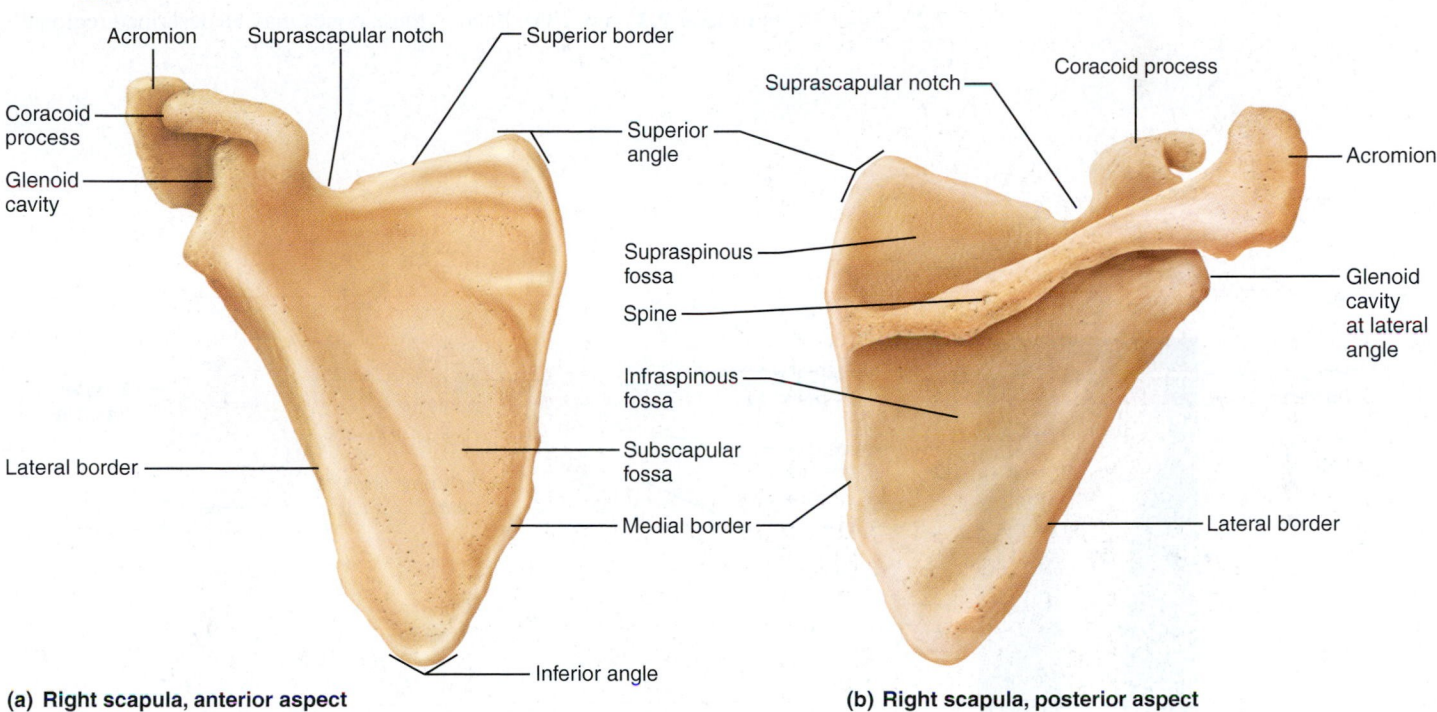

Acromion   Suprascapular notch   Superior border

Coracoid process

Coracoid process

Glenoid cavity

Suprascapular notch   Coracoid process

Superior angle

Acromion

Supraspinous fossa

Spine

Glenoid cavity at lateral angle

Infraspinous fossa

Lateral border

Subscapular fossa

Medial border

Lateral border

Inferior angle

(a) Right scapula, anterior aspect

(b) Right scapula, posterior aspect

Supraspinous fossa

Infraspinous fossa   Subscapular fossa

Posterior   Anterior

Supraspinous fossa

Acromion

Supraglenoid tubercle

Coracoid process

Spine

Glenoid cavity

Infraspinous fossa

Infraglenoid tubercle

Subscapular fossa

Inferior angle

(c) Right scapula, lateral aspect

Figure 7.27 **The scapula.** View (c) is accompanied by a schematic representation of its orientation.

7.5 The upper limb consists of the arm, forearm, and hand

Learning Outcome

▶ **Identify or name the bones of the upper limb and their important markings.**

Thirty separate bones form the bony framework of each upper limb (see Figures 7.28, 7.29, and 7.30, and **Table 7.3** on p. 235). Each of these bones may be described regionally

(a) Photo, anterior view

(b) Illustration, anterior view

(c) Illustration, posterior view

Figure 7.28 The humerus of the right arm and detailed views of articulation at the elbow.

as a bone of the arm, forearm, or hand. (Anatomically, "arm" refers only to that part of the upper limb between the shoulder and elbow.)

Arm

The **humerus** (hu'mer-us), the sole bone of the arm, is a typical long bone (**Figure 7.28**). It articulates with the scapula at the shoulder and with the radius and ulna (forearm bones) at the elbow.

At the proximal end of the humerus is its smooth, hemispherical **head**, which fits into the glenoid cavity of the scapula in a manner that allows the arm to hang freely at one's side. Immediately inferior to the head is a slight constriction, the **anatomical neck**. Just inferior to this are the lateral **greater tubercle** and the more medial **lesser tubercle**, separated by the **intertubercular sulcus**, or *bicipital groove* (bi-sip'ĭ-tal). These tubercles are sites of attachment of the rotator cuff muscles. The intertubercular sulcus guides a tendon of the biceps muscle of the arm to its attachment point at the rim of the glenoid cavity (the supraglenoid tubercle). Just distal to the tubercles is the **surgical neck**, so named because it is the most frequently fractured part of the humerus. About midway down the shaft on its lateral side is the V-shaped **deltoid tuberosity**, the roughened attachment site for the deltoid muscle of the shoulder. Nearby, the **radial groove** runs obliquely down the posterior aspect of the shaft, marking the course of the radial nerve, an important nerve of the upper limb.

At the distal end of the humerus are two condyles: a medial **trochlea** (trok'le-ah; "pulley"), which looks like an hourglass tipped on its side, and the lateral ball-like **capitulum** (kah-pit'u-lum). These condyles articulate with the ulna and the radius, respectively (Figure 7.28d and e). The condyle pair is flanked by the **medial** and **lateral epicondyles** (muscle attachment sites). Directly above these epicondyles are the **medial** and **lateral supracondylar ridges**. The ulnar nerve, which runs behind the medial epicondyle, is responsible for the painful, tingling sensation you experience when you hit your "funny bone."

Superior to the trochlea on the anterior surface is the **coronoid fossa**; on the posterior surface is the deeper **olecranon fossa** (o-lek'rah-non). These two depressions allow the corresponding processes of the ulna to move freely when the elbow is flexed and extended. A small **radial fossa**, lateral to the coronoid fossa, receives the head of the radius when the elbow is flexed.

(d) Anterior view at the elbow region

(e) Posterior view of extended elbow

Figure 7.28 *(continued)*

Forearm

Two parallel long bones, the radius and the ulna, form the skeleton of the forearm, or *antebrachium* (an″te-bra′ke-um) (**Figure 7.29**). Unless a person's forearm muscles are very bulky, these bones are easily palpated along their entire length. Their proximal ends articulate with the humerus; their distal ends form joints with bones of the wrist. The radius and ulna articulate with each other both proximally and distally at small **radio-ulnar joints** (ra″de-o-ul′nar), and they are connected along their entire length by a flat, flexible ligament, the **interosseous membrane** (in″ter-os′e-us; "between the bones").

In the anatomical position, the radius lies laterally (on the thumb side) and the ulna medially. However, when you rotate your forearm so that the palm faces posteriorly (a movement called pronation), the distal end of the radius crosses over the ulna and the two bones form an X (see Figure 8.6a, p. 261).

(a) Photo, anterior view

(b) Illustration, anterior view

(c) Illustration, posterior view

Figure 7.29 Radius and ulna of the right forearm.

Ulna

The **ulna** (ul'nah; "elbow") is slightly longer than the radius. It has the main responsibility for forming the elbow joint with the humerus. Its proximal end looks like the adjustable end of a monkey wrench: It bears two prominent processes, the **olecranon** (elbow) and the **coronoid process**, separated by a deep concavity, the **trochlear notch** (Figure 7.29d). Together, these two processes grip the trochlea of the humerus, forming a hinge joint that allows the forearm to be bent upon the arm (flexed), then straightened again (extended). When the forearm is fully extended, the olecranon "locks" into the olecranon fossa (Figure 7.28e), keeping the forearm from hyperextending (moving posteriorly beyond the anatomical position). The posterior olecranon forms the angle of the elbow when the forearm is flexed and is the bony part that rests on the table when you lean on your elbows. On the lateral side of the coronoid process is a small depression, the **radial notch**, where the ulna articulates with the head of the radius.

Distally the ulnar shaft narrows and ends in a knoblike **head** (Figure 7.29e). Medial to the head is the **ulnar styloid process**, from which a ligament runs to the wrist. The ulnar head is separated from the bones of the wrist by a disc of fibrocartilage and plays little or no role in hand movements.

Radius

The **radius** ("rod") is thin at its proximal end and wide distally—the opposite of the ulna. The **head** of the radius is shaped somewhat like the head of a nail (Figure 7.29). The superior surface of this head is concave, and it articulates with the capitulum of the humerus. Medially, the head articulates with the radial notch of the ulna (Figure 7.28d). Just inferior to the head is the rough **radial tuberosity**, which anchors the biceps muscle of the arm. Distally, where the radius is expanded, it has a medial **ulnar notch** (Figure 7.29e), which articulates with the ulna, and a lateral **radial styloid process** (an anchoring site for ligaments that run to the wrist). Between these two markings, the radius is concave where it articulates with carpal bones of the wrist.

The ulna contributes more heavily to the elbow joint, and the radius is the major forearm bone contributing to the wrist joint. When the radius moves, the hand moves with it.

HOMEOSTATIC IMBALANCE 7.5   CLINICAL

Colles' fracture is a break in the distal end of the radius. It is a common fracture when a person is falling and attempts to break the fall with outstretched hands. ● ● ● ● ● ●

(d) Proximal portion of ulna, lateral view

Olecranon
Trochlear notch
Coronoid process
Radial notch
View

(e) Distal ends of the radius and ulna at the wrist

Ulnar notch of radius
Articulation for lunate
Articulation for scaphoid
Radial styloid process
Head of ulna
Ulnar styloid process
View

Figure 7.29 *(continued)*

(a) Anterior view

(b) Posterior view

Figure 7.30 Bones of the right hand.

Hand

The skeleton of the hand (**Figure 7.30**) includes the bones of the *carpus* (wrist); the bones of the *metacarpus* (palm); and the *phalanges* (bones of the fingers). Several sesamoid bones (◄ p. 176) are also usually present.

Carpus (Wrist)

A "wrist" watch is actually worn on the distal forearm (over the lower ends of the radius and ulna), not on the wrist at all. The true wrist, or **carpus**, is the proximal part of the structure we generally call our "hand." The carpus consists of eight marble-size short bones, or **carpals** (kar′palz), closely united by ligaments. Because gliding movements occur between these bones, the carpus as a whole is quite flexible.

The carpals are arranged in two irregular rows of four bones each (Figure 7.30). In the proximal row (lateral to medial) are the **scaphoid** (skaf′oid; "boat-shaped"), **lunate** (lu′nāt; "moon-like"), **triquetrum** (tri-kwet′rum; "triangular"), and **pisiform** (pi′sĭ-form; "pea-shaped"). Of these, all but the pisiform participate in forming the wrist joint. The carpals of the distal row (lateral to medial) are the **trapezium** (trah-pe′ze-um; "little table"), **trapezoid** (tră′peh-zoid; "four-sided"), **capitate** ("head-shaped"), and **hamate** (ham′āt; "hooked").

There are numerous memory-jogging phrases to help you recall the carpals in the order given above, such as: "**S**ally **l**eft **t**he **p**arty **t**o **t**ake **C**indy **h**ome."

The arrangement of its bones is such that the carpus is concave anteriorly and a ligament roofs over this concavity, forming the notorious *carpal tunnel*. Besides the median nerve (which supplies the lateral side of the hand), several long muscle tendons crowd into this tunnel. Overuse and inflammation of the tendons cause them to swell, compressing the median nerve, which causes tingling and numbness of the areas served, and movements of the thumb weaken. Pain is greatest at night. This is called *carpal tunnel syndrome*. Risk factors include female gender, fluid retention, nerve-damaging conditions, and certain workplace activities. This condition is treated by splinting the wrist during sleep or by surgery. ●●●●●●●

Metacarpus (Palm)

Five **metacarpals** radiate from the wrist like spokes to form the **metacarpus** or palm of the hand (*meta* = beyond). These small long bones are not named, but instead are numbered I to V from thumb to little finger. The **bases** of the metacarpals articulate with the carpals proximally and each other medially and laterally (Figure 7.30). Their bulbous **heads** articulate with the proximal phalanges of the fingers. When you clench your fist, the heads of the metacarpals become prominent as your *knuckles*.

Metacarpal I, associated with the thumb, is the shortest and most mobile. It occupies a more anterior position than the other

Table 7.3   Bones of the Appendicular Skeleton, Part 1: Pectoral Girdle and Upper Limb

BODY REGION	BONES*	ILLUSTRATION	LOCATION	MARKINGS
Pectoral girdle (Figures 7.25, 7.26, 7.27)	**Clavicle** (2)		Clavicle is in superoanterior thorax; articulates medially with sternum and laterally with scapula	Acromial end; sternal end
	Scapula (2)		Scapula is in posterior thorax; forms part of the shoulder; articulates with humerus and clavicle	Glenoid cavity; spine; acromion; coracoid process; infraspinous, supraspinous, and subscapular fossae
Upper limb Arm (Figure 7.28)	**Humerus** (2)		Humerus is sole bone of arm; between scapula and elbow	Head; greater and lesser tubercles; intertubercular sulcus; radial groove; deltoid tuberosity; trochlea; capitulum; coronoid and olecranon fossae; epicondyles; radial fossa
Forearm (Figure 7.29)	**Ulna** (2)		Ulna is the medial bone of forearm between elbow and wrist; with the humerus (and radius) forms elbow joint	Coronoid process; olecranon; radial notch; trochlear notch; ulnar styloid process; head
	Radius (2)		Radius is the lateral bone of forearm; articulates with carpals to form part of the wrist joint	Head; radial tuberosity; radial styloid process; ulnar notch
Hand (Figure 7.30)	8 **Carpals** (16) Scaphoid Lunate Triquetrum Pisiform Trapezium Trapezoid Capitate Hamate		Carpals form a bony crescent at the wrist; arranged in two rows of four bones each	
	5 **Metacarpals** (10)		Metacarpals form the palm; one in line with each digit	
	14 **Phalanges** (28) Proximal Middle Distal		Phalanges form the fingers; three in digits II–V; two in digit I (the thumb)	

Anterior view of pectoral girdle and upper limb

*The number in parentheses ( ) following the bone name denotes the total number of such bones in the body.

metacarpals. Consequently, the joint between metacarpal I and the trapezium is a unique joint—it allows *opposition*, the action of touching your thumb to the tips of your other fingers.

Phalanges (Fingers)

The **fingers**, or **digits** of the upper limb, are numbered I to V beginning with the thumb, or **pollex** (pol′eks). In most people, the third finger is the longest. Each hand contains 14 miniature long bones called **phalanges** (fah-lan′jēz). Except for the thumb, each finger has three phalanges: *distal, middle,* and *proximal.* The thumb has no middle phalanx. [Phalanx (fa′langks; "a closely knit row of soldiers") is the singular term for phalanges.]

Check Your Understanding

18. Which bones play the major role in forming the elbow joint?

19. Which bones of the upper limb have a styloid process?

20. MAKE CONNECTIONS Where are carpals found and how would you classify them based on their shape? (Hint: See Chapter 6.)

For answers, see Answers Appendix.

7.6 The hip bones attach to the sacrum, forming the pelvic girdle

Learning Outcomes

▶ **Name the bones contributing to the os coxae, and relate the pelvic girdle's strength to its function.**

▶ **Describe differences in the male and female pelves and relate these to functional differences.**

The **pelvic girdle**, or **hip girdle**, is formed by the sacrum* (a part of the axial skeleton) and a pair of **hip bones**, each also called an **os coxae** (ahs kok′se), or **coxal bone** (*coxa* = hip). Each hip bone unites with its partner anteriorly and with the sacrum posteriorly (**Figure 7.31**).

The pelvic girdle attaches the lower limbs to the axial skeleton, transmits the full weight of the upper body to the lower limbs, and supports the visceral organs of the pelvis. Unlike the pectoral girdle, which is sparingly attached to the thoracic cage, the hip bones are secured to the axial skeleton by some of the strongest ligaments in the body. And unlike the shallow glenoid cavity of the scapula, the corresponding sockets of the pelvic girdle are deep and cuplike and firmly secure the head of the femur in place. Thus, even though both the shoulder and hip joints are ball-and-socket joints, very few of us can wheel or swing our legs about with the same degree of freedom as our arms. The pelvic girdle lacks the mobility of the pectoral girdle but is far more stable.

Each large, irregularly shaped hip bone consists of three separate bones during childhood: the ilium, ischium, and pubis (**Figure 7.32**). In adults, these bones are firmly fused and their boundaries are indistinguishable. Their names are retained, however, to refer to different regions of the composite hip bone.

* Here we follow the convention in *Terminologia Anatomica*, which groups the sacrum with *both* the vertebral column of the axial skeleton and the pelvic girdle of the appendicular skeleton.

At the point of fusion of the ilium, ischium, and pubis is a deep hemispherical socket called the **acetabulum** (as″ĕ-tab′u-lum; "vinegar cup") on the lateral surface of the pelvis (Figure 7.32). The acetabulum receives the head of the femur, or thigh bone, at this *hip joint.*

Ilium

The **ilium** (il′e-um; "flank") is a large flaring bone that forms the superior region of a hip bone. It consists of a **body** and a superior winglike portion called the **ala** (a′lah). When you rest your hands on your hips, you are resting them on the thickened superior margins of the alae, the **iliac crests**, to which many muscles attach. Each iliac crest ends anteriorly in the blunt **anterior superior iliac spine** and posteriorly in the sharp **posterior superior iliac spine**. (Recall that a spine is a sharp, slender, often pointed projection; ◀ Table 6.2, p. 180.)

Located below these are the less prominent *anterior* and *posterior inferior iliac spines.* All of these spines are attachment points for the muscles of the trunk, hip, and thigh. The anterior superior iliac spine is an especially important anatomical landmark. It is easily felt through the skin and is visible in thin people. The posterior superior iliac spine is difficult to palpate, but its position is revealed by a skin dimple in the sacral region.

Just inferior to the posterior inferior iliac spine, the ilium indents deeply to form the **greater sciatic notch** (si-at′ik), through which the thick cordlike sciatic nerve passes to enter the thigh. The broad posterolateral surface of the ilium, the **gluteal surface** (gloo′te-al), is crossed by three ridges, the **posterior**, **anterior**, and **inferior gluteal lines**, to which the gluteal (buttock) muscles attach.

The medial surface of the iliac ala exhibits a concavity called the **iliac fossa**. Posterior to this, the roughened **auricular surface** (aw-rik′u-lar; "ear-shaped") articulates with the same-named surface of the sacrum, forming the *sacroiliac joint* (Figure 7.31). The weight of the body is transmitted from the spine to the pelvis through the sacroiliac joints. Running inferiorly and anteriorly from the auricular surface is a robust ridge called the **arcuate line** (ar′ku-at; "bowed"). The arcuate line helps define the *pelvic brim*, the superior margin of the *true pelvis*, which we will discuss shortly. Anteriorly, the body of the ilium joins the pubis; inferiorly it joins the ischium.

Ischium

The **ischium** (is′ke-um; "hip") forms the posteroinferior part of the hip bone (Figures 7.31 and 7.32). Roughly L- or arc-shaped, it has a thicker, superior **body** adjoining the ilium and a thinner, inferior **ramus**. (A *ramus* is a branch, in this case a bar of bone.) The ramus joins the pubis anteriorly. The ischium has three important markings. Its **ischial spine** projects medially into the pelvic cavity and serves as a point of attachment of the *sacrospinous ligament* running from the sacrum. Just inferior to the ischial spine is the **lesser sciatic notch**. A number of nerves and blood vessels pass through this notch to supply the anogenital area. The inferior surface of the ischial body is rough and grossly thickened as the **ischial tuberosity**. When we sit, our weight is

Figure 7.31 Pelvis. The pelvis consists of the two hip (coxal) bones, the sacrum, and the coccyx.

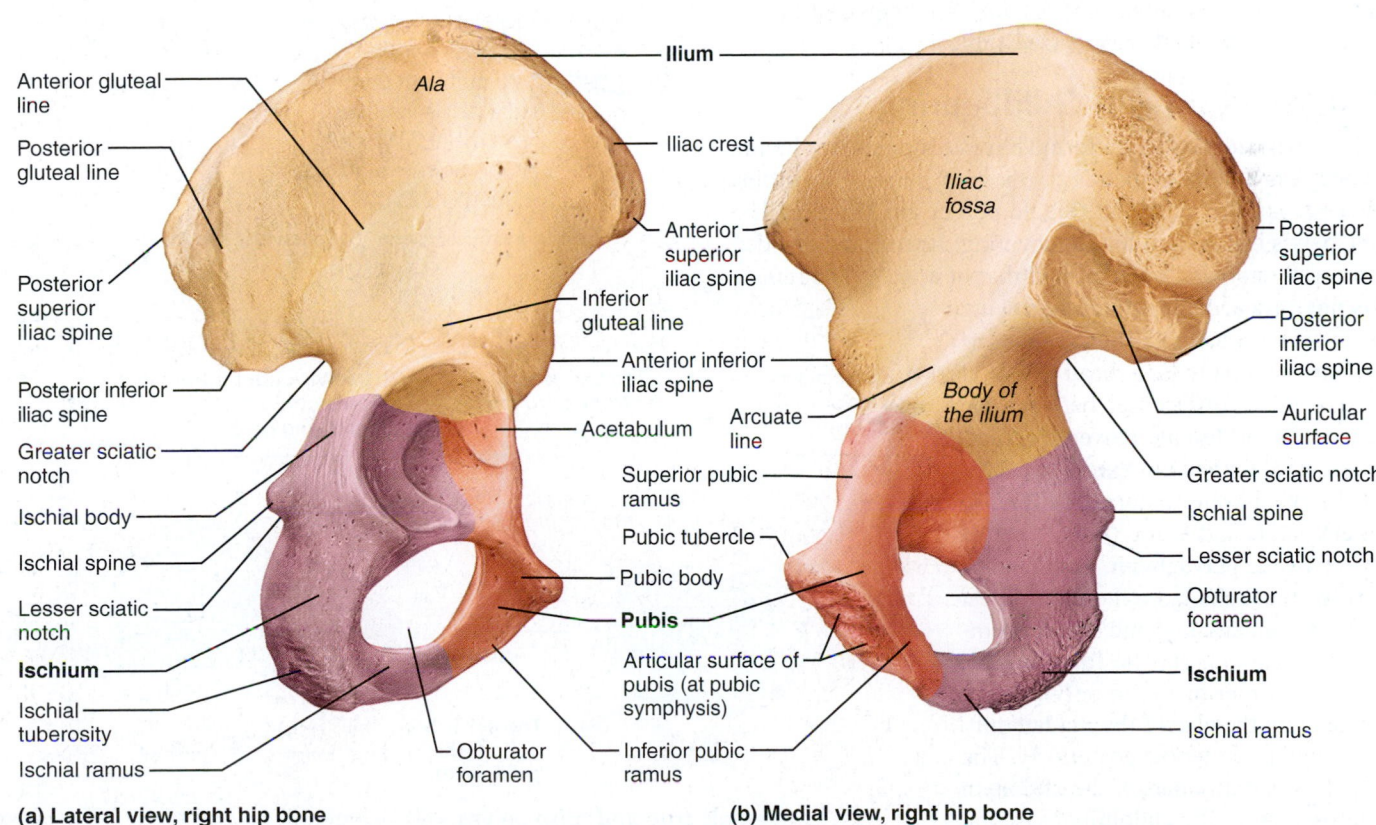

(a) Lateral view, right hip bone

(b) Medial view, right hip bone

Figure 7.32 The hip (coxal) bones. Lateral and medial views of the right hip bone. The point of fusion of the ilium (gold), ischium (violet), and pubic (red) bones at the acetabulum is indicated in the diagrams.

borne by the ischial tuberosities, which are the strongest parts of the hip bones.

A massive ligament runs from the sacrum to each ischial tuberosity. This *sacrotuberous ligament* (not illustrated) helps hold the pelvis together. The ischial tuberosity is also a site of attachment of the large hamstring muscles of the posterior thigh.

Pubis

The **pubis** (pu′bis; "sexually mature"), or *pubic bone*, forms the anterior portion of the hip bone (Figures 7.31 and 7.32). In the anatomical position, it lies nearly horizontally and the urinary bladder rests upon it. Essentially, the pubis is V shaped with **superior** and **inferior pubic rami** issuing from its flattened medial **body**. The anterior border of the pubis is thickened to form the **pubic crest**. At the lateral end of the pubic crest is the **pubic tubercle**, one of the attachments for the *inguinal ligament*. As the two rami of the pubis run laterally to join with the body and ramus of the ischium, they define a large opening in the hip bone, the **obturator foramen** (ob″tu-ra′tor), through which a few blood vessels and nerves pass. Although the obturator foramen is large, it is nearly closed by a fibrous membrane in life (*obturator* = closed up).

The bodies of the two pubic bones are joined by a fibrocartilage disc, forming the midline **pubic symphysis** joint. Inferior to this joint, the inferior pubic rami angle laterally, forming an inverted V-shaped arch called the **pubic arch** or **subpubic angle**. The acuteness of the angle of this arch helps to differentiate the male and female pelves (see Table 7.4).

Pelvic Structure and Childbearing

The deep, basinlike structure formed by the hip bones, sacrum, and coccyx is called the **pelvis** or the *bony pelvis*. The differences between the male and female pelves are striking. The female pelvis is modified for childbearing: It tends to be wider, shallower, lighter, and rounder than that of a male. The female pelvis not only accommodates a growing fetus, but it must be large enough to allow the infant's relatively large head to exit at birth. The major differences between the typical male and female pelves are summarized and illustrated in **Table 7.4**.

The pelvis is said to consist of a false (greater) pelvis and a true (lesser) pelvis separated by the **pelvic brim**, a continuous oval ridge that runs from the pubic crest through the arcuate line and sacral promontory (**Figure 7.33**). The **false pelvis**, that portion superior to the pelvic brim, is bounded by the alae of the ilia laterally and the lumbar vertebrae posteriorly. The false pelvis is really part of the abdomen and helps support the abdominal viscera. It does not restrict childbirth in any way.

The **true pelvis** is the region inferior to the pelvic brim that is almost entirely surrounded by bone. It forms a deep bowl containing the pelvic organs. Its dimensions, particularly those of its *inlet* and *outlet*, are critical to the uncomplicated delivery of a baby, and they are carefully measured by an obstetrician.

The **pelvic inlet** *is* the pelvic brim, and its widest dimension is from right to left along the frontal plane. A sacral promontory that is particularly large can impair the infant's entry into the true pelvis at the beginning of labor.

The **pelvic outlet**, illustrated in Figure 7.33 and the photos at the bottom of Table 7.4, is the inferior margin of the true pelvis. It is bounded anteriorly by the pubic arch, laterally by the ischia, and posteriorly by the sacrum and coccyx. Both the coccyx and the ischial spines protrude into the outlet opening, so a sharply angled coccyx or unusually large spines can interfere with delivery. The largest dimension of the outlet is the anteroposterior diameter.

Check Your Understanding

21. The ilium and pubis help to form the hip bone. What other bone is involved in forming the hip bone?
22. The pelvic girdle is a heavy, strong girdle. How does its structure reflect its function?
23. In the hip bone at right, name the three regions labeled a, b, and c. Name the opening indicated by d and the socket at e. What bone articulates with this socket?
24. **WHAT IF?** What if your pregnant patient had a pelvis that was shaped like a male's? Which features of this pelvis would cause difficulty during childbirth?

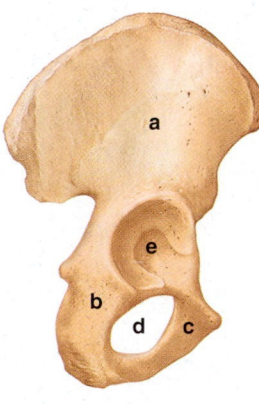

For answers, see Answers Appendix.

The **false pelvis** is superior to the pelvic brim.

The **true pelvis** is inferior to the pelvic brim.

The **pelvic outlet** is bounded by the pubic arch, ischia, sacrum, and coccyx.

Pelvic brim

The pelvic brim defines the **pelvic inlet**.

Pubic arch

Ischial spines

Coccyx

Figure 7.33 Female true and false pelves. Left: anterosuperior view. Right: inferior view.

Table 7.4   Comparison of the Male and Female Pelves

CHARACTERISTIC	FEMALE	MALE
General structure and functional modifications	Tilted forward; adapted for childbearing; true pelvis defines the extent of the birth canal; cavity of the true pelvis is broad, shallow, and has a greater capacity	Tilted less far forward; adapted for support of a male's heavier build and stronger muscles; cavity of the true pelvis is narrow and deep
Bone thickness	Less; bones lighter, thinner, and smoother	Greater; bones heavier and thicker, and markings are more prominent
Acetabula	Smaller; farther apart	Larger; closer together
Pubic arch/subpubic angle	Broader (80° to 90°); more rounded	Angle is more acute (50° to 60°)
Anterosuperior view		

Pelvic brim — Acetabulum — Pubic arch

Sacrum	Wider; shorter; less sacral curvature	Narrow; longer; more sacral curvature
Coccyx	More movable; projects inferiorly	Less movable; projects anteriorly
Greater sciatic notch	Wide and shallow	Narrow and deep
Left lateral view		

Sacrum
Coccyx
Greater sciatic notch

Pelvic inlet (brim)	Wider; oval from side to side	Narrow; basically heart shaped
Pelvic outlet	Wider; ischial spines farther apart	Narrower; ischial spines closer together and point more medially
Inferior view		

Ischial spines

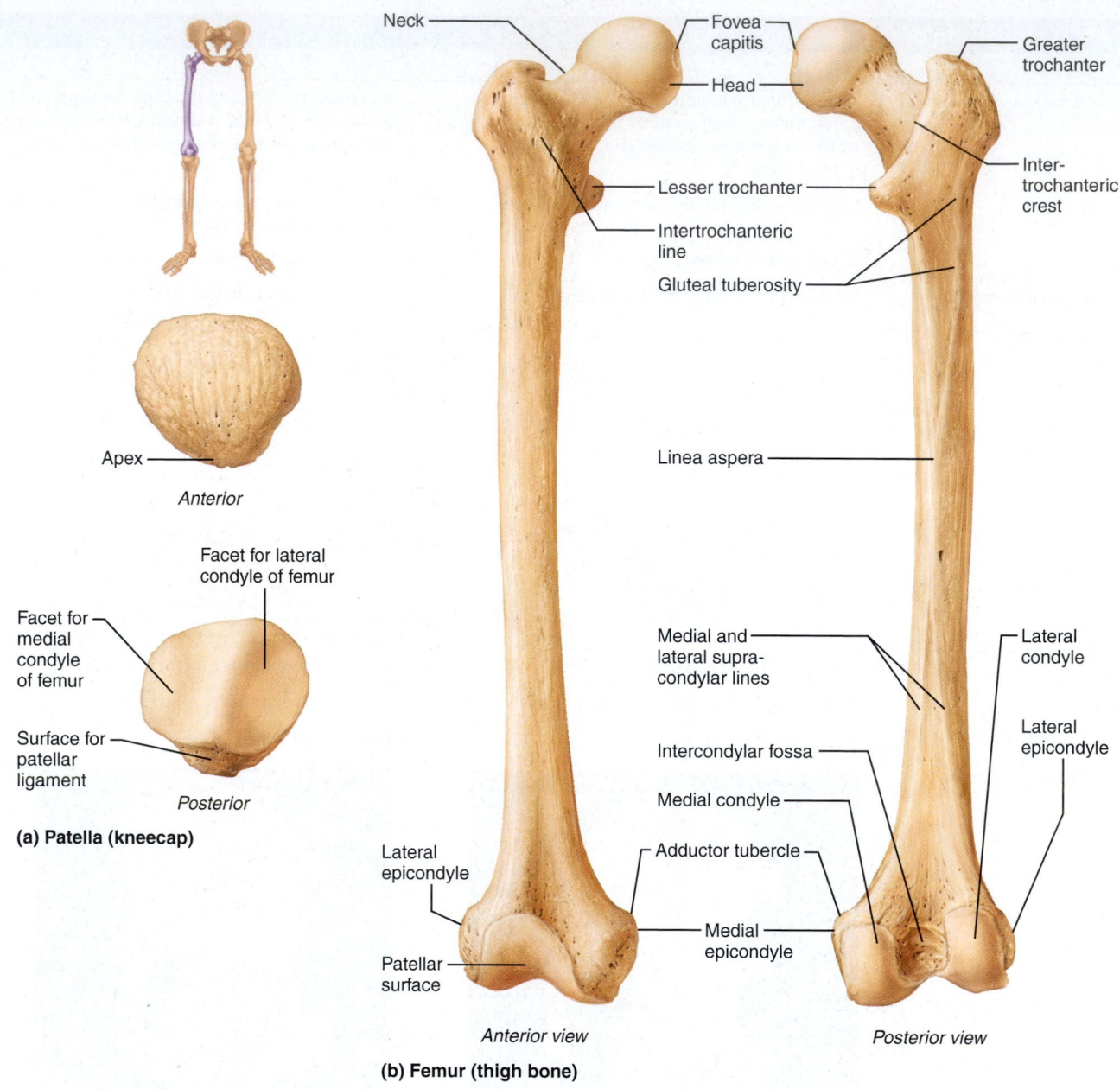

Neck
Fovea capitis
Head
Greater trochanter
Intertrochanteric crest
Lesser trochanter
Intertrochanteric line
Gluteal tuberosity
Linea aspera

Apex
Anterior

Facet for lateral condyle of femur
Facet for medial condyle of femur
Surface for patellar ligament
Posterior

(a) Patella (kneecap)

Medial and lateral supra-condylar lines
Lateral condyle
Lateral epicondyle
Intercondylar fossa
Medial condyle
Lateral epicondyle
Adductor tubercle
Patellar surface
Medial epicondyle

Anterior view   *Posterior view*

(b) Femur (thigh bone)

Figure 7.34 Bones of the right knee and thigh.

7.7 The lower limb consists of the thigh, leg, and foot

Learning Outcome

▶ **Identify the lower limb bones and their important markings.**

The lower limbs carry the entire weight of the erect body and are subjected to exceptional forces when we jump or run. Thus, it is not surprising that the bones of the lower limbs are thicker and stronger than comparable bones of the upper limbs. The three segments of each lower limb are the thigh, the leg, and the foot (see **Table 7.5**).

Thigh

The **femur** (fe′mur; "thigh"), the single bone of the thigh (**Figure 7.34**), is the largest, longest, strongest bone in the body. Its durable structure reflects the fact that the stress on the femur during vigorous jumping can reach 280 kg/cm^2 (about 2 tons per square inch)! The femur is surrounded by bulky muscles that prevent us from palpating its course down the length of the thigh. Its length is roughly one-quarter of a person's height.

Proximally, the femur articulates with the hip bone and then courses medially as it descends toward the knee. This arrangement allows the knee joints to be closer to the body's center of gravity and provides for better balance. The medial course of the two femurs is more pronounced in women because of their

Table 7.5 Bones of the Appendicular Skeleton, Part 2: Pelvic Girdle and Lower Limb

BODY REGION	BONES*	ILLUSTRATION	LOCATION	MARKINGS
Pelvic girdle (Figures 7.31, 7.32)	**Coxal** (2) (hip)		Each hip (coxal) bone is formed by the fusion of an ilium, ischium, and pubis; the hip bones articulate anteriorly at the pubic symphysis and form sacroiliac joints with the sacrum posteriorly; girdle consisting of both hip bones and the sacrum is basinlike	Iliac crest; anterior and posterior iliac spines; auricular surface; greater and lesser sciatic notches; obturator foramen; ischial tuberosity and spine; acetabulum; pubic arch; pubic crest; pubic tubercle
Lower limb Thigh (Figure 7.34)	**Femur** (2)		Femur is the sole bone of thigh; between hip joint and knee; largest bone of the body	Head; greater and lesser trochanters; neck; lateral and medial condyles and epicondyles; gluteal tuberosity; linea aspera
Kneecap (Figure 7.34)	**Patella** (2)		Patella is a sesamoid bone formed within the tendon of the quadriceps (anterior thigh) muscles	
Leg (Figure 7.35)	**Tibia** (2)		Tibia is the larger and more medial bone of leg; between knee and foot	Medial and lateral condyles; tibial tuberosity; anterior border; medial malleolus
	Fibula (2)		Fibula is the lateral bone of leg; sticklike	Head; lateral malleolus
Foot (Figure 7.36)	7 **Tarsals** (14) talus calcaneus navicular cuboid lateral cuneiform intermediate cuneiform medial cuneiform		Tarsals are seven bones forming the proximal part of the foot; the talus articulates with the leg bones at the ankle joint; the calcaneus, the largest tarsal, forms the heel	
	5 **Metatarsals** (10)		Metatarsals are five bones numbered I–V	
	14 **Phalanges** (28) distal middle proximal		Phalanges form the toes; three in digits II–V, two in digit I (the great toe)	

*The number in parentheses ( ) following the bone name denotes the total number of such bones in the body.

wider pelvis, a situation that may contribute to the greater incidence of knee problems in female athletes.

The ball-like **head** of the femur has a small central pit called the **fovea capitis** (fo′ve-ah kǎ′pĭ-tis; "pit of the head"). The short *ligament of the head of the femur* runs from this pit to the acetabulum, where it helps secure the femur. The head is carried on a *neck* that angles *laterally* to join the shaft. This arrangement reflects the fact that the femur articulates with the lateral aspect (rather than the inferior region) of the pelvis. The neck is the weakest part of the femur and is often fractured, an injury commonly called a broken hip.

At the junction of the shaft and neck are the lateral **greater trochanter** (tro-kan′ter) and posteromedial **lesser trochanter**. These projections serve as sites of attachment for thigh and buttock muscles. The two trochanters are connected by the **intertrochanteric line** anteriorly and by the prominent **intertrochanteric crest** posteriorly.

Inferior to the intertrochanteric crest on the posterior shaft is the **gluteal tuberosity**, which blends into a long vertical ridge, the **linea aspera** (lin′e-ah as′per-ah; "rough line"), inferiorly. Distally, the linea aspera diverges, forming the **medial** and **lateral supracondylar lines**. All of these markings are sites of muscle attachment. Except for the linea aspera, the femur shaft is smooth and rounded.

Distally, the femur broadens and ends in the wheel-like **lateral** and **medial condyles**, which articulate with the tibia of the leg. The **medial** and **lateral epicondyles** (sites of muscle attachment) flank the condyles superiorly. On the superior part of the medial epicondyle is a bump, the **adductor tubercle**. The smooth **patellar surface**, between the condyles on the anterior femoral surface, articulates with the *patella* (pah-tel′ah), or kneecap (see Figure 7.34 and Table 7.5). Between the condyles on the posterior aspect of the femur is the deep, U-shaped **intercondylar fossa**.

The **patella** ("small pan") is a triangular sesamoid bone enclosed in the (quadriceps) tendon that secures the anterior thigh muscles to the tibia. It protects the knee joint anteriorly and improves the leverage of the thigh muscles acting across the knee.

Leg

Two parallel bones, the tibia and fibula, form the skeleton of the leg, the region of the lower limb between the knee and the ankle (**Figure 7.35**). These two bones are connected by an *interosseous membrane* and articulate with each other both proximally and distally. Unlike the joints between the radius and ulna of the forearm, the *tibiofibular joints* (tib"e-o-fib'u-lar) of the leg allow essentially no movement. The bones of the leg

Figure 7.35 **The tibia and fibula of the right leg.**

(a) Anterior view

Intercondylar eminence
Lateral condyle
Head
Superior tibiofibular joint
Articular surface of medial condyle
Medial condyle
Tibial tuberosity
Interosseous membrane
Anterior border
Fibula
Tibia
Inferior tibiofibular joint
Lateral malleolus
Medial malleolus
Inferior articular surface

(b) Posterior view

Articular surface of lateral condyle
Head of fibula
Fibula
Lateral malleolus

(c) X ray of bimalleolar fracture

Fracture in lateral malleolus
Fracture in medial malleolus

thus form a less flexible but stronger and more stable limb than those of the forearm. The medial tibia articulates proximally with the femur to form the modified hinge joint of the knee and distally with the talus bone of the foot at the ankle. The fibula, in contrast, does not contribute to the knee joint and merely helps stabilize the ankle joint.

Tibia

The **tibia** (tib'e-ah; "shinbone") receives the weight of the body from the femur and transmits it to the foot. It is second only to the femur in size and strength. At its broad proximal end are the concave **medial** and **lateral condyles**, which look like two huge checkers lying side by side. These are separated by an irregular projection, the **intercondylar eminence**. The tibial condyles articulate with the corresponding condyles of the femur. The inferior region of the lateral tibial condyle bears a facet that indicates the site of the *superior tibiofibular joint*. Just inferior to the condyles, the tibia's anterior surface displays the rough **tibial tuberosity**, to which the patellar ligament attaches.

The tibial shaft is triangular in cross section. Neither the tibia's sharp **anterior border** nor its medial surface is covered by muscles, so they can be felt just deep to the skin along their entire length. (The anguish of a bumped shin is an experience familiar to nearly everyone.) Distally the tibia is flat where it articulates with the talus bone of the foot. Medial to that joint surface is an inferior projection, the **medial malleolus** (mah-le'o-lus; "little hammer"), which forms the medial bulge of the ankle. The **fibular notch**, on the lateral surface of the tibia, participates in the *inferior tibiofibular joint*.

Fibula

The **fibula** (fib'u-lah; "pin") is a sticklike bone with slightly expanded ends. It articulates proximally and distally with the lateral aspects of the tibia. Its proximal end is its **head**; its distal end is the **lateral malleolus**. The lateral malleolus forms the conspicuous lateral ankle bulge and articulates with the talus. The fibular shaft is heavily ridged and appears to have been twisted a quarter turn. The fibula does not bear weight, but several muscles originate from it.

HOMEOSTATIC IMBALANCE 7.7

CLINICAL

A *bimalleolar fracture* is a break in the distal ends of both the tibia and fibula. It is a common sports injury (Figure 7.35c).

Lateral condyle

Tibial tuberosity

(d) Anterior view, proximal tibia

Lateral condyle

Fibula articulates here

Line for soleus muscle

(e) Posterior view, proximal tibia

Figure 7.35 *(continued)*

Figure 7.36 Bones of the right foot.

Foot

The skeleton of the foot includes the bones of the *tarsus*, the bones of the *metatarsus*, and the *phalanges*, or toe bones (**Figure 7.36**). The foot has two important functions: It supports our body weight, and it acts as a lever to propel the body forward when we walk and run. A single bone could serve both purposes, but it would adapt poorly to uneven ground. Segmentation makes the foot flexible, avoiding this problem.

Tarsus

The **tarsus** is made up of seven bones called **tarsals** (tar′salz) that form the posterior half of the foot. It corresponds to the carpus of the hand. Body weight is carried primarily by the two largest, most posterior tarsals: the **talus** (ta′lus; "ankle"), which articulates with the tibia and fibula superiorly, and the strong **calcaneus** (kal-ka′ne-us; "heel bone"), which forms the heel of the foot and carries the talus on its superior surface. The thick *calcaneal*, or *Achilles*, *tendon* of the calf muscles attaches to the posterior surface of the calcaneus. The part of the calcaneus that touches the ground is the **calcaneal tuberosity**, and its shelf-like projection that supports part of the talus is the **sustentaculum tali** (sus″ten-tak′u-lum ta′le; "supporter of the talus") or **talar shelf**. The tibia articulates with the talus at the *trochlea* of the talus. The remaining tarsals are the lateral **cuboid**, the medial **navicular** (nah-vik′u-lar), and the anterior **medial**,

intermediate, and **lateral cuneiform bones** (ku-ne′i-form; "wedge-shaped"). The cuboid and cuneiform bones articulate with the metatarsal bones anteriorly.

Metatarsus

The **metatarsus** consists of five small, long bones called **metatarsals**. These are numbered I to V beginning on the medial (great toe) side of the foot. The first metatarsal, which plays an important role in supporting body weight, is short and thick. The arrangement of the metatarsals is more parallel than that of the metacarpals of the hands. Distally, where the metatarsals articulate with the proximal phalanges of the toes, the enlarged head of the first metatarsal forms the "ball" of the foot.

Phalanges (Toes)

The 14 phalanges of the toes are a good deal smaller than those of the fingers and so are less nimble. But their general structure and arrangement are the same. There are three phalanges in each digit except for the great toe, the **hallux**. The hallux has only two, proximal and distal.

Arches of the Foot

A segmented structure can support weight only if it is arched. The foot has three arches: two *longitudinal arches* (*medial* and *lateral*) and one *transverse arch* (**Figure 7.37**), which account for its awesome strength. These arches are maintained by the interlocking shapes of the foot bones, by strong ligaments, and by the pull of some tendons during muscle activity. The ligaments and tendons provide a certain amount of springiness. In general, the arches "give," or stretch slightly, when weight is applied to the foot and spring back when the weight is removed, which makes walking and running more economical in terms of energy use than would otherwise be the case.

If you examine your wet footprints, you will see that the medial margin from the heel to the head of the first metatarsal leaves no print. This is because the **medial longitudinal arch**

curves well above the ground. The talus is the keystone of this arch, which originates at the calcaneus, rises toward the talus, and then descends to the three medial metatarsals.

The **lateral longitudinal arch** is very low. It elevates the lateral part of the foot just enough to redistribute some of the weight to the calcaneus and the head of the fifth metatarsal (to the ends of the arch). The cuboid is the keystone bone of this arch.

The two longitudinal arches serve as pillars for the **transverse arch**, which runs obliquely from one side of the foot to the other, following the line of the joints between the tarsals and metatarsals. Together, the arches of the foot form a half-dome that distributes about half of a person's standing and walking weight to the heel bones and half to the heads of the metatarsals.

HOMEOSTATIC IMBALANCE 7.8   CLINICAL

Flatfoot, or *pes planus*, may be a congenital conditon or may be acquired later in life. Increasing age, obesity, and high-impact activities are all risk factors. Wearing proper footwear with good arch support may help prevent this problem, and orthotic shoe inserts may help manage symptoms.

Check Your Understanding

25. What lower limb bone is the second largest bone in the body?

26. The image below shows the posterior aspects of two bones. Name the bones. Are they from the right or left side of the body? Name the structures labeled a–c.

27. Which of the following sites is not a site of muscle attachment? Greater trochanter, lesser trochanter, gluteal tuberosity, lateral condyle.

28. Besides supporting our weight, what is a major function of the arches of the foot?

29. What are the two largest tarsal bones in each foot, and which one forms the heel of the foot?

For answers, see Answers Appendix.

(a) Lateral aspect of right foot

— Medial longitudinal arch
— Transverse arch
— Lateral longitudinal arch

(b) X ray, medial aspect of right foot

Figure 7.37 Arches of the foot.

Developmental Aspects of the Skeleton

The membranous bones of the skull start to ossify late in the second month of development. The rapid deposit of bone matrix at the ossification centers produces cone-shaped protrusions in the developing bones. At birth, the skull bones are still incomplete and are connected by as yet unossified remnants of fibrous membranes called **fontanelles** (fon"tah-nelz') (**Figure 7.38**). The fontanelles allow the infant's head to be compressed slightly during birth, and they accommodate brain growth in the fetus and infant. A baby's pulse can be felt surging in these "soft spots"; hence their name (*fontanelle* = little fountain). The large, diamond-shaped *anterior fontanelle* is palpable for 1½ to 2 years after birth. The others are replaced by bone by the end of the first year.

(a) Superior view

Frontal suture

Frontal bone

Anterior fontanelle

Ossification center

Parietal bone

Posterior fontanelle

Occipital bone

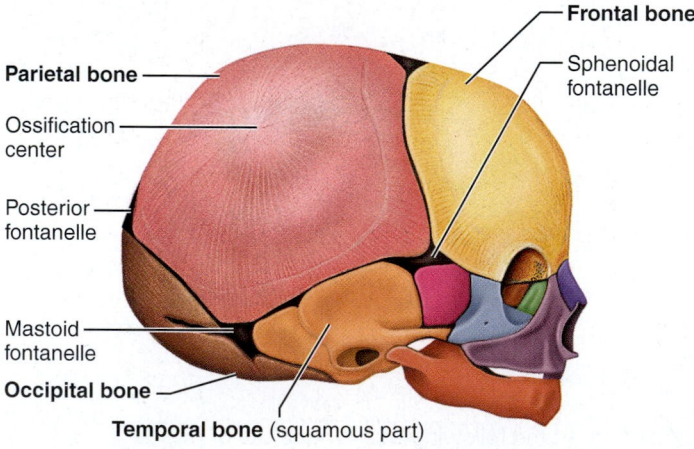

(b) Lateral view

Frontal bone

Parietal bone

Sphenoidal fontanelle

Ossification center

Posterior fontanelle

Mastoid fontanelle

Occipital bone

Temporal bone (squamous part)

Figure 7.38 Skull of a newborn. Notice that the infant's skull has more bones than that of an adult.

Figure 7.39 Cleft lip and palate. A cleft lip and palate (left) can be surgically repaired (right).

> ### HOMEOSTATIC IMBALANCE 7.9   CLINICAL
>
> Several congenital abnormalities may distort the skull. Most common is *cleft palate*, a condition in which the right and left halves of the palate fail to fuse medially (**Figure 7.39**). The persistent opening between the oral and nasal cavities interferes with sucking and can lead to aspiration (inhalation) of food into the lungs and *aspiration pneumonia*.

The skeleton changes throughout life, but the changes in childhood are most dramatic. At birth, the baby's cranium is huge relative to its face, and several bones are still unfused (e.g., the mandible and frontal bones). The maxillae and mandible are foreshortened, and the contours of the face are flat (**Figure 7.40**). By 9 months after birth, the cranium is already half of its adult size (volume) because of the rapid growth of the brain. By 8 to 9 years, the cranium has almost reached adult proportions.

Between the ages of 6 and 13, the head appears to enlarge substantially as the face literally grows out from the skull. The jaws, cheekbones, and nose become more prominent. These facial changes are correlated with the expansion of the nose and paranasal sinuses, and development of the permanent teeth. Figure 7.40 tracks how differential bone growth alters body proportions throughout life.

In the spine, only the thoracic and sacral curvatures are well developed at birth. These so-called **primary curvatures** are convex posteriorly, and an infant's spine arches, like that of a four-legged animal (**Figure 7.41**).

The **secondary curvatures**—cervical and lumbar—are convex anteriorly and are associated with a child's development. They result from reshaping of the intervertebral discs rather than from modifications of the vertebrae. The cervical curvature is present before birth but is not pronounced until the baby starts to lift its head (at about 3 months). The lumbar curvature develops when the baby begins to walk (at about 12 months). The lumbar

Human newborn   Human adult

(a)

Newborn   2 yrs   5 yrs   15 yrs   Adult

(b)

Figure 7.40 Different growth rates of body parts determine body proportions. (a) Differential growth transforms the rounded, foreshortened skull of a newborn to the sloping skull of an adult. **(b)** During growth of a human, the arms and legs grow faster than the head and trunk, as seen in this conceptualization of different-aged individuals all drawn at the same height.

curvature positions the weight of the trunk over the body's center of gravity, providing optimal balance when standing.

Vertebral problems (scoliosis or lordosis; see Figure 7.18a and c on p. 220) may appear during the early school years, when rapid growth of the limb bones stretches many muscles. During the preschool years, lordosis is often present, but this is usually rectified as the abdominal muscles become stronger and the pelvis tilts forward. The thorax grows wider, but a true "military posture" (head erect, shoulders back, abdomen in, and chest out) does not develop until adolescence.

 HOMEOSTATIC IMBALANCE 7.10　　　　　**CLINICAL**

The hips of newborns are carefully examined. Up to 2% of all newborns have abnormalities of the hip, including partial or complete dislocation of the femoral head and poorly developed acetabulum. *Developmental dysplasia of the hip* (DDH) may be suspected on physical examination, and ultrasound of the hips is

Figure 7.41 The C-shaped spine of a newborn infant.

indicated for some high-risk infants, including those presenting as breech births. Orthopedists may treat DDH with harnesses, casting, or surgery. Unrecognized DDH can lead to problems later in life, including degenerative arthritis.

During youth, growth of the skeleton not only increases overall body height but also changes body proportions (Figure 7.40). At birth, the head and trunk are approximately 1½ times as long as the lower limbs. The lower limbs grow more rapidly than the trunk from this time on, and by the age of 10, the head and trunk are approximately the same height as the lower limbs, a condition that persists thereafter. During puberty, the female pelvis broadens in preparation for childbearing, and the entire male skeleton becomes more robust. Once adult height is reached, a healthy skeleton changes very little until late middle age.

Old age affects many parts of the skeleton, especially the spine. As the discs become thinner, less hydrated, and less elastic, the risk of disc herniation increases. By age 55, a loss of several centimeters in stature is common. Further shortening can be produced by osteoporosis of the spine or by kyphosis (see Figure 7.18b on p. 220). What was done during youth may be undone in old age as the vertebral column gradually resumes its initial arc shape.

The thorax becomes more rigid with age, largely because the costal cartilages ossify. This loss of rib cage elasticity causes shallow breathing, which leads to less efficient gas exchange.

All bones, you will recall, lose mass with age. Cranial bones lose less mass than most, but changes in facial contours with age are common. As the bony tissue of the jaws declines, the jaws look small and childlike once again. If the elderly person loses his or her teeth, this loss of bone from the jaws is accelerated, because the alveolar region bone is resorbed. As bones become more porous, they are more likely to fracture, especially the vertebrae and the neck of the femur.

RELATED CLINICAL TERMS

Chiropractic (ki″ro-prak′tik) A system of treating disease by manipulating the vertebral column based on the idea that most diseases are due to pressure on nerves caused by faulty bone alignment; a specialist in this field is a chiropractor.

Clubfoot A relatively common congenital defect (one in 700 births) in which the soles of the feet face medially and the toes point inferiorly; may be genetically induced or reflect an abnormal position of the foot during fetal development.

Laminectomy Surgical removal of a vertebral lamina; most often done to relieve the symptoms of a ruptured disc.

Orthopedist (or″tho-pe′dist) or **orthopedic surgeon** A physician who specializes in restoring lost skeletal system function or repairing damage to bones and joints.

Pelvimetry Measurement of the dimensions of the inlet and outlet of the pelvis, usually to determine whether the pelvis is of adequate size to allow normal delivery of a baby.

Spina bifida (spi′nah bĭ′fĭ-dah; "cleft spine") Congenital defect of the vertebral column in which one or more of the vertebral arches are incomplete; ranges in severity from inconsequential to severe conditions that impair neural functioning and encourage nervous system infections.

Spinal fusion Surgical procedure involving insertion of bone chips (or crushed bone) to immobilize and stabilize a specific region of the vertebral column, particularly in cases of vertebral fracture and herniated discs.

CHAPTER SUMMARY

1. The axial skeleton forms the longitudinal axis of the body. Its principal subdivisions are the skull, vertebral column, and thoracic cage. It provides support and protection (by enclosure).
2. The appendicular skeleton consists of the bones of the pectoral and pelvic girdles and the limbs. It allows mobility for manipulation and locomotion.

PART 1

THE AXIAL SKELETON

7.1 The skull consists of 8 cranial bones and 14 facial bones (pp. 201–217)

1. The skull is formed by 22 bones. The cranium forms the vault and base of the skull, which protect the brain. The facial skeleton provides openings for the respiratory and digestive passages and attachment points for facial muscles.
2. Except for the temporomandibular joints, all bones of the adult skull are joined by immovable sutures.
3. **Cranium.** The eight bones of the cranium include the paired parietal and temporal bones and the single frontal, occipital, ethmoid, and sphenoid bones (see Table 7.1, pp. 216–217).
4. **Facial bones.** The 14 bones of the face include the paired maxillae, zygomatics, nasals, lacrimals, palatines, and inferior nasal conchae and the single mandible and vomer bones (Table 7.1).
5. **Hyoid bone.** The hyoid bone, supported in the neck by ligaments, serves as an attachment point for tongue and neck muscles.
6. **Orbits and nasal cavity.** Both the orbits and the nasal cavity are complicated bony regions formed of several bones.
7. **Paranasal sinuses.** Paranasal sinuses occur in the frontal, ethmoid, sphenoid, and maxillary bones.

7.2 The vertebral column is a flexible, curved support structure (pp. 218–224)

1. **General characteristics.** The vertebral column includes 24 movable vertebrae (7 cervical, 12 thoracic, and 5 lumbar) and the sacrum and coccyx.
2. The fibrocartilage intervertebral discs act as shock absorbers and provide flexibility to the vertebral column.

3. Curvatures increase spine flexibility. The thoracic and sacral curvatures are posteriorly convex; the cervical and lumbar curvatures are posteriorly concave.
4. **General structure of vertebrae.** With the exception of C_1 and C_2, all vertebrae have a body, two transverse processes, two superior and two inferior articular processes, a spinous process, and a vertebral arch.
5. **Regional vertebral characteristics.** Special features distinguish the regional vertebrae (see Table 7.2, p. 223).

7.3 The thoracic cage is the bony structure of the chest (pp. 224–227)

1. The bones of the thoracic cage include the 12 rib pairs, the sternum, and the thoracic vertebrae. The thoracic cage protects the organs of the thoracic cavity.
2. **Sternum.** The sternum consists of the fused manubrium, body, and xiphoid process.
3. **Ribs.** The first seven rib pairs are called true ribs; the rest are called false ribs. Ribs 11 and 12 are floating ribs.

PART 2

THE APPENDICULAR SKELETON

7.4 Each pectoral girdle consists of a clavicle and a scapula* (pp. 227–229)

1. Each pectoral girdle consists of one clavicle and one scapula. The pectoral girdles attach the upper limbs to the axial skeleton.
2. **Clavicles.** The clavicles hold the scapulae laterally away from the thorax. The sternoclavicular joints are the only attachment points of the pectoral girdle to the axial skeleton.
3. **Scapulae.** The scapulae articulate with the clavicles and with the humerus bones of the arms.

7.5 The upper limb consists of the arm, forearm, and hand* (pp. 230–236)

1. Each upper limb consists of 30 bones and is specialized for mobility.
2. **Arm/forearm/hand.** The skeleton of the arm is composed solely of the humerus; the skeleton of the forearm is composed of the

radius and ulna; and the skeleton of the hand consists of the carpals, metacarpals, and phalanges.

7.6 The hip bones attach to the sacrum, forming the pelvic girdle* (pp. 236–239)

1. The pelvic girdle, a heavy structure specialized for weight bearing, is composed of two hip bones and the sacrum. It secures the lower limbs to the axial skeleton.
2. Each hip bone consists of three fused bones: ilium, ischium, and pubis. The acetabulum occurs at the point of fusion.
3. **Ilium/ischium/pubis.** The ilium is the superior flaring portion of the hip bone. Each ilium forms a secure joint with the sacrum posteriorly. The ischium is a curved bar of bone; we sit on the ischial tuberosities. The V-shaped pubic bones articulate anteriorly at the pubic symphysis.
4. **Pelvic structure and childbearing.** The pelvis is the deep, basinlike structure formed by the hip bones, sacrum, and coccyx. The male pelvis is deep and narrow with larger, heavier bones than those of the female. The female pelvis is shallow and wide for childbearing.

7.7 The lower limb consists of the thigh, leg, and foot* (pp. 240–245)

1. Each lower limb consists of the thigh, leg, and foot and is specialized for weight bearing and locomotion.
2. **Thigh.** The femur is the only bone of the thigh. Its ball-shaped head articulates with the acetabulum.
3. **Leg.** The bones of the leg are the tibia, which participates in forming both the knee and ankle joints, and the fibula.

4. **Foot.** The bones of the foot include the tarsals, metatarsals, and phalanges. The most important tarsals are the calcaneus (heel bone) and the talus, which articulates with the tibia superiorly.
5. The foot is supported by three arches (lateral longitudinal, medial longitudinal, and transverse) that distribute body weight to the heel and ball of the foot.

Developmental Aspects of the Skeleton (pp. 246–247)

1. Fontanelles, which allow brain growth and ease birth passage, are present in the skull at birth. Growth of the cranium after birth is related to brain growth. Increase in size of the facial skeleton follows tooth development and enlargement of nose and paranasal sinuses.
2. The vertebral column is C shaped at birth (thoracic and sacral curvatures are present); the secondary curvatures form when the baby begins to lift its head and walk.
3. Long bones continue to grow in length until late adolescence. The head and torso, initially 1½ times the length of the lower limbs, equal their length by the age of 10.
4. Changes in the female pelvis (preparatory for childbirth) occur during puberty.
5. Once at adult height, the skeleton changes little until late middle age. With old age, the intervertebral discs thin; this, along with osteoporosis, leads to a gradual loss in height and increased risk of disc herniation. Loss of bone mass increases the risk of fractures, and thoracic cage rigidity promotes breathing difficulties.

*For associated bone markings, see the pages indicated in the module heads.

REVIEW QUESTIONS

To access additional practice questions using your smartphone, tablet, or computer: MasteringA&P® > Study Area > Practice Tests & Quizzes

Level 1 Remember/Understand

(Some questions have more than one correct answer. Select the best answer or answers from the choices given.)

1. Match the bones in column B with their description in column A. (Note that some descriptions require more than a single choice.)

Column A	Column B
___(1) connected by the coronal suture	(a) ethmoid
___(2) keystone bone of cranium	(b) frontal
___(3) keystone bone of the face	(c) mandible
___(4) form the hard palate	(d) maxillary
___(5) allows the spinal cord to pass	(e) occipital
___(6) forms the chin	(f) palatine
___(7) contain paranasal sinuses	(g) parietal
___(8) contains mastoid sinuses	(h) sphenoid
	(i) temporal

2. Match the key terms with the bone descriptions that follow.
Key:
(a) clavicle   (d) pubis   (f) scapula
(b) ilium   (e) sacrum   (g) sternum
(c) ischium

___(1) bone of the axial skeleton to which the pectoral girdle attaches
___(2) markings include glenoid cavity and acromion
___(3) features include the ala, crest, and greater sciatic notch
___(4) doubly curved; acts as a shoulder strut
___(5) hip bone that articulates with the axial skeleton
___(6) the "sit-down" bone
___(7) anteriormost bone of the pelvic girdle
___(8) part of the vertebral column

3. Use key choices to identify the bone descriptions that follow.
Key:
(a) carpals   (d) humerus   (g) tibia
(b) femur   (e) radius   (h) ulna
(c) fibula   (f) tarsals

___(1) articulates with the acetabulum and the tibia
___(2) forms the lateral aspect of the ankle
___(3) bone that "carries" the hand
___(4) the wrist bones
___(5) end shaped like a monkey wrench
___(6) articulates with the capitulum of the humerus
___(7) largest bone of this "group" is the calcaneus

4. Which vertebral curvatures are primary and which are secondary? Why are they called primary and secondary?
5. What is the function of the intervertebral discs?
6. Which ribs are floating ribs and why are they called that?
7. List three important differences between the male and female pelves.

8. Briefly describe the anatomical characteristics and impairment of function seen in cleft palate and hip dysplasia.

Level 2 Apply/Analyze

9. How do the relative proportions of the cranium and face of a fetus compare with those of an adult skull?

10. Distinguish between the anulus fibrosus and nucleus pulposus regions of a disc. Which provides durability and strength? Which provides resilience? Which part is involved in a "slipped" disc?

11. Peter Howell, a teaching assistant in the anatomy class, picked up a hip bone and pretended it was an old-fashioned telephone. He held the big hole in this bone right up to his ear and said, "Press 5 to speak to the obturator? Hello, obturator, obturator (operator, operator)." Name the structure he was helping the students to learn.

12. Justiniano worked in a poultry-packing plant where his job was cutting open chickens and stripping out their visceral organs. After work, he typed for long hours on his computer keyboard, writing a book about his work in the plant. Soon, his wrist and hand began to hurt whenever he flexed it, and he began to awaken at night with pain and tingling on the thumb half of his hand. What condition did he probably have?

13. Mary's grandmother slipped on a scatter rug and fell heavily to the floor. Her left lower limb was laterally rotated and noticeably shorter than the right, and when she attempted to get up, she winced with pain. Mary surmised that her grandmother might have "fractured her hip," which later proved to be true. What bone was probably fractured and at what site? Why is a "fractured hip" a common type of fracture in the elderly?

Level 3 Evaluate/Synthesize

14. Name the cranial and facial bones and compare and contrast the functions of the cranial and facial skeletons.

15. List at least two specific anatomical characteristics each for typical cervical, thoracic, and lumbar vertebrae that would allow anyone to identify each type correctly.

16. The major function of the shoulder girdle is flexibility. What is the major function of the pelvic girdle? Relate these functional differences to anatomical differences seen in these girdles.

17. Compare a young adult skeleton to that of an extremely aged person relative to bone mass in general and the bony structure of the skull, thorax, and vertebral column in particular.

18. Mr. Wright had polio as a boy and was partially paralyzed in one lower limb for over a year. Although no longer paralyzed, he now has a severe lateral curvature of the lumbar spine. Explain what has happened and identify his condition.

19. Mrs. Shea came up with what she considered to be a clever idea to bypass the long lines at Disney World. She had her husband rent a wheelchair and he wheeled her around from one exhibit to another for the better part of three days. As they sat on the plane, waiting to take off for Chicago, she complained to him that she had two sore spots on her buttocks. Why? What do you suppose would happen (to her buttocks) if she was wheeled around for a few more days?

CLINICAL CASE STUDY

45-Year-Old Female with Dislocated Hip

Kayla Tanner, a 45-year-old mother of four, was involved in a car accident. When emergency responders arrived on the scene and extracted her from the car, they found that her right thigh appeared shorter than her left thigh. They also noticed that even slight hip movement caused considerable pain. Suspecting a hip dislocation, they stabilized and transported her.

In the emergency department, doctors discovered a decreased ability to sense light touch in her right foot, and she was unable to move her toes or ankle. Dislocation of her right hip was confirmed by X ray. Mrs. Tanner was sedated to relax the muscles around the hip, and then doctors placed her in the supine position and performed a closed reduction ("popped" the femur back in place).

1. When the hip was pushed out of the hip socket, what specific bones and bony structures might have been injured?

2. ✚ NCLEX-STYLE After hip dislocation and reduction, the nurse would watch the patient carefully for signs of which of these problems?
   a. Sciatic nerve injury
   b. Instability of the pubic symphysis
   c. Inflammation of the sacroiliac joint
   d. Damage to the lateral epicondyle

3. Mrs. Tanner suffered an injury to the hip joint, but if you were asked to rest your hands on your hips, you would not actually touch this joint. What structure in the pelvic girdle would your hands be resting on?

4. The sedation that Mrs. Tanner was given relaxed the large gluteal muscles of the buttocks. Name the structures on the femur and the pelvic girdle where these muscles attach.

5. ✚ NCLEX-STYLE Mrs. Tanner is discussing her injury with you during recovery. Which of her remarks indicate a need for additional education about her injury?
   a. "I dislocated my shoulder twice. I guess this hip injury is just the same sort of thing—I'm surprised it hasn't happened before. I'm sort of nervous that it could just slip out any time!"
   b. "I didn't realize that so many of the hip bones were fused together."
   c. "The bottom of my spinal column is really different from the top!"
   d. "When you look at how tight those connections are, it's easy to see why I can't move my leg around at the hip the way I move my arm around at the shoulder."

8 Joints

In this chapter, you will learn that

Joints determine how bones move relative to each other

by first asking

8.1 How are joints classified?

then exploring

8.2 Fibrous joints

8.3 Cartilaginous joints

8.4 Synovial joints

looking closer at

Movement of synovial joints

focusing on

8.5 Selected synovial joints

then asking

8.6 What happens when things go wrong?

and finally, exploring

Developmental Aspects of Joints

The graceful movements of ballet dancers and the rough-and-tumble grapplings of football players demonstrate the great variety of motion allowed by **joints**, or **articulations**—the sites where two or more bones meet. Our joints have two fundamental functions: They give our skeleton mobility, and they hold it together, sometimes playing a protective role in the process.

8.1 Joints are classified into three structural and three functional categories

Learning Outcomes

▶ Define joint or articulation.
▶ Classify joints by structure and by function.

Joints are classified by structure and by function. The *structural classification* focuses on the material binding the bones together and whether or not a joint cavity is present. Structurally, there are *fibrous*, *cartilaginous*, and *synovial joints* (**Table 8.1** on p. 255). Only synovial joints have a joint cavity.

The *functional classification* is based on the amount of movement allowed at the joint. On this basis, there are **synarthroses** (sin″ar-thro′sēz; *syn* = together, *arthro* = joint), which are immovable joints; **amphiarthroses** (am″fe-ar-thro′sēz; *amphi* = on both sides), slightly movable joints; and **diarthroses** (di″ar-thro′sēz; *dia* = through, apart), or freely movable joints. Freely movable joints predominate in the appendicular skeleton (limbs). Immovable and slightly movable joints are largely restricted to the axial skeleton. This localization of functional joint types makes sense because the less movable the joint, the more stable it is likely to be.

In general, fibrous joints are immovable, and synovial joints are freely movable. However, cartilaginous joints have both rigid

(a) **Suture**	(b) **Syndesmosis**	(c) **Gomphosis**
Joint held together with very short, interconnecting fibers, and bone edges interlock. Found only in the skull.	Joint held together by a ligament. Fibrous tissue can vary in length, but is longer than in sutures.	"Peg in socket" fibrous joint. Periodontal ligament holds tooth in socket.

Suture line

Fibrous connective tissue

Fibula

Tibia

Ligament

Socket of alveolar process

Root of tooth

Periodontal ligament

Figure 8.1 Fibrous joints.

and slightly movable examples. Since the structural categories are more clear-cut, we will use the structural classification in this discussion, indicating functional properties where appropriate.

Check Your Understanding

1. What functional joint class contains the least-mobile joints?
2. How are joint mobility and stability related?

For answers, see Answers Appendix.

8.2 In fibrous joints, the bones are connected by fibrous tissue

Learning Outcome

▶ Describe the general structure of fibrous joints. Name and give an example of each of the three common types of fibrous joints.

In **fibrous joints**, the bones are joined by the collagen fibers of connective tissue. No joint cavity is present. The amount of movement allowed depends on the length of the connective tissue fibers. Most fibrous joints are immovable, although a few are slightly movable. The three types of fibrous joints are *sutures*, *syndesmoses*, and *gomphoses*.

Sutures

Sutures, literally "seams," occur only between bones of the skull (**Figure 8.1a**). The wavy articulating bone edges interlock, and

the junction is completely filled by a minimal amount of very short connective tissue fibers that are continuous with the peri-osteum (◀ p. 178). The result is nearly rigid splices that knit the bones together, yet allow the skull to expand as the brain grows during youth. During middle age, the fibrous tissue ossifies and the skull bones fuse into a single unit. At this stage, the closed sutures are more precisely called **synostoses** (sin″os-to′sēz), lit-erally, "bony junctions." Because movement of the cranial bones would damage the brain, the immovable nature of sutures is a protective adaptation.

Syndesmoses

In **syndesmoses** (sin″des-mo′sēz), the bones are connected exclusively by *ligaments* (*syndesmos* = ligament), cords or bands of fibrous tissue. The amount of movement allowed at a syndesmosis depends on the length of the connecting fibers.

Although the connecting fibers are always longer than those in sutures, they vary quite a bit in length. If the fibers are short (as in the ligament connecting the distal ends of the tibia and fibula, Figure 8.1b), little or no movement is allowed, a char-acteristic best described as "give." If the fibers are long [as in the ligament-like interosseous membrane connecting the radius and ulna (◀ **Figure 7.29, p. 232**)], a large amount of movement is possible.

Gomphoses

A **gomphosis** (gom-fo′sis) is a peg-in-socket fibrous joint (Figure 8.1c). The only example is the articulation of a tooth with its bony alveolar socket. The term *gomphosis* comes from

(a) Synchondroses

Bones united by hyaline cartilage

Epiphyseal plate (temporary hyaline cartilage joint)

Sternum (manubrium)

Joint between first rib and sternum (immovable)

(b) Symphyses

Bones united by fibrocartilage

Body of vertebra

Fibrocartilaginous intervertebral disc (sandwiched between hyaline cartilage)

Pubic symphysis

Figure 8.2 Cartilaginous joints.

the Greek *gompho*, meaning "nail" or "bolt," and refers to the way teeth are embedded in their sockets (as if hammered in). The fibrous connection in this case is the short **periodontal ligament** (Figure 23.12, p. 880).

Check Your Understanding

3. To what functional class do most fibrous joints belong?
4. **PREDICT** Babies are sometimes diagnosed with a condition called *craniosynostosis*. Using the definition you just learned, describe what you think this condition is and predict what problems it might cause.

For answers, see Answers Appendix.

8.3 In cartilaginous joints, the bones are connected by cartilage

Learning Outcome

▶ Describe the general structure of cartilaginous joints. Name and give an example of each of the two common types of cartilaginous joints.

In **cartilaginous joints** (kar″tĭ-laj′ĭ-nus), the articulating bones are united by cartilage. Like fibrous joints, they lack a joint cavity and are not highly movable. The two types of cartilaginous joints are *synchondroses* and *symphyses*.

Synchondroses

A bar or plate of *hyaline cartilage* unites the bones at a **synchondrosis** (sin″kon-dro′sis; "junction of cartilage"). Virtually all synchondroses are synarthrotic (immovable).

The most common examples of synchondroses are the epiphyseal plates in long bones of children (**Figure 8.2a**). Epiphyseal plates are temporary joints and eventually become synostoses. Another example of a synchondrosis is the immovable joint between the costal cartilage of the first rib and the manubrium of the sternum (Figure 8.2a).

Symphyses

A joint where *fibrocartilage* unites the bones is a **symphysis** (sim′fih-sis; "growing together"). Since fibrocartilage is compressible and resilient, it acts as a shock absorber and permits a limited amount of movement at the joint. Even though fibrocartilage is the main element of a symphysis, hyaline cartilage is also present in the form of articular cartilages on the bony surfaces. Symphyses are amphiarthrotic joints designed for strength with flexibility. Examples include the intervertebral joints and the pubic symphysis of the pelvis (Figure 8.2b, and see **Table 8.2** on pp. 256–257).

Check Your Understanding

5. MAKE CONNECTIONS Evan is 25 years old. Would you expect to find synchondroses at the ends of his femur? Explain. (Hint: See Chapter 6, p. 187.)

For answers, see Answers Appendix.

8.4 Synovial joints have a fluid-filled joint cavity

Learning Outcomes

▶ **Describe the structural characteristics of synovial joints.**
▶ **Compare the structures and functions of bursae and tendon sheaths.**
▶ **List three natural factors that stabilize synovial joints.**
▶ **Name and describe (or perform) the common body movements.**
▶ **Name and provide examples of the six types of synovial joints based on the movement(s) allowed.**

Synovial joints (si-no′ve-al; "joint eggs") are those in which the articulating bones are separated by a fluid-containing joint cavity. This arrangement permits substantial freedom of movement, and all synovial joints are freely movable diarthroses. Nearly all joints of the limbs—indeed, most joints of the body—fall into this class.

General Structure

Synovial joints have six distinguishing features (**Figure 8.3**):

- **Articular cartilage.** Glassy-smooth hyaline cartilage covers the opposing bone surfaces as **articular cartilage**. These thin (1 mm or less) but spongy cushions absorb compression placed on the joint and thereby keep the bone ends from being crushed.

- **Joint (articular) cavity.** The joint cavity is a feature that is unique to synovial joints. It contains a small amount of synovial fluid. The joint cavity is a *potential space* because it is normally almost nonexistent, but can expand if fluid accumulates (as happens during inflammation).

- **Articular capsule.** The joint cavity is enclosed by a two-layered **articular capsule**, or *joint capsule*. The tough external **fibrous layer** is composed of dense irregular connective tissue that is continuous with the periostea of the articulating bones. It strengthens the joint so that the bones are not pulled apart. The inner layer of the joint capsule is a **synovial membrane** composed of loose connective tissue. Besides lining the fibrous layer internally, it covers all internal joint surfaces that are not hyaline cartilage. The synovial membrane's function is to make synovial fluid.

- **Synovial fluid.** A small amount of slippery **synovial fluid** occupies all free spaces within the joint capsule. This fluid is derived largely by filtration from blood flowing through the capillaries in the synovial membrane. Synovial fluid has

Figure 8.3 General structure of a synovial joint.

Labels: Ligament; Joint cavity (contains **synovial fluid**); Articular (hyaline) cartilage; Fibrous layer; Synovial membrane (secretes synovial fluid); Articular capsule; Periosteum

a viscous, egg-white consistency (*ovum* = egg) due to hyaluronic acid secreted by cells in the synovial membrane, but it thins and becomes less viscous during joint activity.

Synovial fluid, which is also found *within* the articular cartilages, provides a slippery, weight-bearing film that reduces friction between the cartilages. Without this lubricant, rubbing would wear away joint surfaces and excessive friction could overheat and destroy the joint tissues. The synovial fluid is forced from the cartilages when a joint is compressed; then as pressure on the joint is relieved, synovial fluid seeps back into the articular cartilages like water into a sponge, ready to be squeezed out again the next time the joint is loaded (put under pressure). This process, called *weeping lubrication*, lubricates the free surfaces of the cartilages and nourishes their cells. (Remember, cartilage is avascular.) Synovial fluid also contains phagocytic cells that rid the joint cavity of microbes and cellular debris.

- **Reinforcing ligaments.** Synovial joints are reinforced and strengthened by a number of bandlike **ligaments**. Most often, these are **capsular ligaments**, which are thickened parts of the fibrous layer. In other cases, they remain distinct and are found outside the capsule (as **extracapsular ligaments**) or deep to it (as **intracapsular ligaments**). Since intracapsular ligaments are covered with synovial membrane, they do not actually lie *within* the joint cavity.

People said to be double-jointed amaze the rest of us by placing both heels behind their neck. However, they have the

Table 8.1 Summary of Joint Classes			
STRUCTURAL CLASS	**STRUCTURAL CHARACTERISTICS**	**TYPES**	**MOBILITY**
Fibrous	Adjoining bones united by collagen fibers	Suture (short fibers)	Immobile (synarthrosis)
		Syndesmosis (longer fibers)	Slightly movable (amphiarthrosis) and immobile
		Gomphosis (periodontal ligament)	Immobile
Cartilaginous	Adjoining bones united by cartilage	Synchondrosis (hyaline cartilage)	Immobile
		Symphysis (fibrocartilage)	Slightly movable
Synovial	Adjoining bones covered with articular cartilage, separated by a joint cavity, and enclosed within an articular capsule lined with synovial membrane	• Plane • Condylar • Hinge • Saddle • Pivot • Ball-and-socket	Freely movable (diarthrosis; movements depend on design of joint)

normal number of joints. It's just that their joint capsules and ligaments are more stretchy and loose than average.

- **Nerves and blood vessels.** Synovial joints are richly supplied with sensory nerve fibers that innervate the capsule. Some of these fibers detect pain, as anyone who has suffered joint injury is aware, but most monitor joint position and stretch. Monitoring joint stretch is one of several ways the nervous system senses our posture and body movements (see p. 493). Synovial joints are also richly supplied with blood vessels, most of which supply the synovial membrane. There, extensive capillary beds produce the blood filtrate that is the basis of synovial fluid.

Besides the basic components just described, certain synovial joints have other structural features. Some, such as the hip and knee joints, have cushioning **fatty pads** between the fibrous layer and the synovial membrane or bone. Others have discs or wedges of fibrocartilage separating the articular surfaces. Where present, these **articular discs**, or **menisci** (mě-nis′ki;

"crescents"), extend inward from the articular capsule and partially or completely divide the synovial cavity in two (see the menisci of the knee in Figure 8.11b, e, and f). Articular discs improve the fit between articulating bone ends, making the joint more stable and minimizing wear and tear on the joint surfaces. Besides the knees, articular discs occur in the jaw and a few other joints (see notations in the Structural Type column in Table 8.2).

Bursae and Tendon Sheaths

Bursae and tendon sheaths are not strictly part of synovial joints, but they are often found closely associated with them (**Figure 8.4**). Essentially bags of lubricant, they act as "ball bearings" to reduce friction between adjacent structures during joint activity. **Bursae** (ber′se; "purse") are flattened fibrous sacs lined with synovial membrane and containing a thin film of synovial fluid. They occur where ligaments, muscles, skin, tendons, or bones rub together.

A **tendon sheath** is essentially an elongated bursa that wraps completely around a tendon subjected to friction, like a bun around a hot dog. They are common where several tendons are crowded together within narrow canals (in the wrist, for example).

(a) Frontal section through the right shoulder joint

Acromion of scapula
Subacromial bursa
Fibrous layer of articular capsule
A **tendon sheath** is an elongated fluid-filled sac that wraps around a tendon to decrease friction.
Tendon of long head of biceps brachii muscle
Articular cartilage
Joint cavity containing synovial fluid
Synovial membrane
Fibrous layer
Humerus

(b) Enlargement of (a)

A **bursa** is a fluid-filled sac that decreases friction where a ligament (or other structure) would rub against a bone.
Bursa rolls and lessens friction.
Humerus head rolls medially as arm abducts.

Figure 8.4 Bursae and tendon sheaths.

Table 8.2 Structural and Functional Characteristics of Body Joints

ILLUSTRATION	JOINT	ARTICULATING BONES	STRUCTURAL TYPE*	FUNCTIONAL TYPE; MOVEMENTS ALLOWED
	Skull	Cranial and facial bones	Fibrous; suture	Synarthrotic; no movement
	Temporo-mandibular	Temporal bone of skull and mandible	Synovial; modified hinge† (contains articular disc)	Diarthrotic; gliding and uniaxial rotation; slight lateral movement, elevation, depression, protraction, and retraction of mandible
	Atlanto-occipital	Occipital bone of skull and atlas	Synovial; condylar	Diarthrotic; biaxial; flexion, extension, lateral flexion, circumduction of head on neck
	Atlantoaxial	Atlas (C$_1$) and axis (C$_2$)	Synovial; pivot	Diarthrotic; uniaxial; rotation of the head
	Intervertebral	Between adjacent vertebral bodies	Cartilaginous; symphysis	Amphiarthrotic; slight movement
	Intervertebral	Between articular processes	Synovial; plane	Diarthrotic; gliding
	Costovertebral	Vertebrae (transverse processes or bodies) and ribs	Synovial; plane	Diarthrotic; gliding of ribs
	Sternoclavicular	Sternum and clavicle	Synovial; shallow saddle (contains articular disc)	Diarthrotic; multiaxial (allows clavicle to move in all axes)
	Sternocostal (first)	Sternum and rib I	Cartilaginous; synchondrosis	Synarthrotic; no movement
	Sternocostal	Sternum and ribs II–VII	Synovial; double plane	Diarthrotic; gliding
	Acromio-clavicular	Acromion of scapula and clavicle	Synovial; plane (contains articular disc)	Diarthrotic; gliding and rotation of scapula on clavicle
	Shoulder (glenohumeral)	Scapula and humerus	Synovial; ball-and-socket	Diarthrotic; multiaxial; flexion, extension, abduction, adduction, circumduction, rotation of humerus
	Elbow	Ulna (and radius) with humerus	Synovial; hinge	Diarthrotic; uniaxial; flexion, extension of forearm
	Proximal radioulnar	Radius and ulna	Synovial; pivot	Diarthrotic; uniaxial; pivot (convex head of radius rotates in radial notch of ulna)
	Distal radioulnar	Radius and ulna	Synovial; pivot (contains articular disc)	Diarthrotic; uniaxial; rotation of radius around long axis of forearm to allow pronation and supination
	Wrist	Radius and proximal carpals	Synovial; condylar	Diarthrotic; biaxial; flexion, extension, abduction, adduction, circumduction of hand
	Intercarpal	Adjacent carpals	Synovial; plane	Diarthrotic; gliding
	Carpometacarpal of digit I (thumb)	Carpal (trapezium) and metacarpal I	Synovial; saddle	Diarthrotic; biaxial; flexion, extension, abduction, adduction, circumduction, opposition of metacarpal I
	Carpometacarpal of digits II–V	Carpal(s) and metacarpal(s)	Synovial; plane	Diarthrotic; gliding of metacarpals
	Metacarpo-phalangeal (knuckle)	Metacarpal and proximal phalanx	Synovial; condylar	Diarthrotic; biaxial; flexion, extension, abduction, adduction, circumduction of fingers
	Interphalangeal (finger)	Adjacent phalanges	Synovial; hinge	Diarthrotic; uniaxial; flexion, extension of fingers

Table 8.2 (continued)

ILLUSTRATION	JOINT	ARTICULATING BONES	STRUCTURAL TYPE*	FUNCTIONAL TYPE; MOVEMENTS ALLOWED
	Sacroiliac	Sacrum and coxal bone	Synovial; plane in childhood, increasingly fibrous in adult	Diarthrotic in child; amphiarthrotic in adult; (more movement during pregnancy)
	Pubic symphysis	Pubic bones	Cartilaginous; symphysis	Amphiarthrotic; slight movement (enhanced during pregnancy)
	Hip (coxal)	Hip bone and femur	Synovial; ball-and-socket	Diarthrotic; multiaxial; flexion, extension, abduction, adduction, rotation, circumduction of thigh
	Knee (tibiofemoral)	Femur and tibia	Synovial; modified hinge† (contains articular discs)	Diarthrotic; biaxial; flexion, extension of leg, some rotation allowed in flexed position
	Knee (femoropatellar)	Femur and patella	Synovial; plane	Diarthrotic; gliding of patella
	Superior tibiofibular	Tibia and fibula (proximally)	Synovial; plane	Diarthrotic; gliding of fibula
	Inferior tibiofibular	Tibia and fibula (distally)	Fibrous; syndesmosis	Synarthrotic; slight "give" during dorsiflexion
	Ankle	Tibia and fibula with talus	Synovial; hinge	Diarthrotic; uniaxial; dorsiflexion, and plantar flexion of foot
	Intertarsal	Adjacent tarsals	Synovial; plane	Diarthrotic; gliding; inversion and eversion of foot
	Tarsometatarsal	Tarsal(s) and metatarsal(s)	Synovial; plane	Diarthrotic; gliding of metatarsals
	Metatarso-phalangeal	Metatarsal and proximal phalanx	Synovial; condylar	Diarthrotic; biaxial; flexion, extension, abduction, adduction, circumduction of great toe
	Interpha-langeal (toe)	Adjacent phalanges	Synovial; hinge	Diarthrotic; uniaxial; flexion, extension of toes

*****Fibrous joints** indicated by orange circles (•); **cartilaginous joints** by blue circles (•); **synovial joints** by purple circles (•).

†These modified hinge joints are structurally bicondylar.

Factors Influencing the Stability of Synovial Joints

Joints, particularly synovial joints, are the weakest parts of the skeleton. Nonetheless, their structure resists various forces, such as crushing or tearing, that threaten to force them out of alignment. Because these joints are constantly stretched and compressed, they must be stabilized so that they do not dislocate (come out of alignment). The stability of a synovial joint depends chiefly on three factors:

- The shapes of the articular surfaces
- The number and positioning of ligaments
- Muscle tone

Articular Surfaces

The shapes of articular surfaces determine what movements are possible at a joint, but surprisingly, articular surfaces play only a minor role in joint stability. Many joints have shallow sockets or noncomplementary articulating surfaces ("misfits") that actually hinder joint stability. But when articular surfaces are large and fit snugly together, or when the socket is deep, stability is vastly improved. The ball and deep socket of the hip joint provide the best example of a joint made extremely stable by the shape of its articular surfaces.

Ligaments

The capsules and ligaments of synovial joints unite the bones and prevent excessive or undesirable motion. As a rule, the more ligaments a joint has, the stronger it is. However, when other stabilizing factors are inadequate, undue tension is placed on the ligaments and they stretch. Stretched ligaments stay stretched, like taffy, and a ligament can stretch only about 6% of its length before it snaps. Thus, when ligaments are the major means of bracing a joint, the joint is not very stable.

Muscle Tone

For most joints, the muscle tendons that cross the joint are the most important stabilizing factor. These tendons are kept under tension by the tone of their muscles. (*Muscle tone* is defined as low levels of contractile activity in relaxed muscles that keep the muscles healthy and ready to react to stimulation.) Muscle tone is extremely important in reinforcing the shoulder and knee joints and the arches of the foot.

Movements Allowed by Synovial Joints

Every skeletal muscle of the body is attached to bone or other connective tissue structures at no fewer than two points. The muscle's **origin** is attached to the immovable (or less movable) bone. Its other end, the **insertion**, is attached to the movable bone. Body movement occurs when muscles contract across joints and their insertion moves toward their origin. The movements can be described in directional terms relative to the lines, or axes, around which the body part moves and the planes of space along which the movement occurs, that is, along the transverse, frontal, or sagittal plane. (See ◀ **Chapter 1, p. 15** to review these planes.)

Range of motion allowed by synovial joints varies from **nonaxial movement** (gliding movements only) to **uniaxial movement** (movement in one plane) to **biaxial movement** (movement in two planes) to **multiaxial movement** (movement in or around all three planes of space and axes). Range of motion varies greatly. In some people, such as trained gymnasts or acrobats, range of joint movement may be extraordinary. The ranges of motion at the major joints are given in the far right column of Table 8.2.

There are three general types of movements: *gliding, angular movements,* and *rotation.* The most common body movements allowed by synovial joints are described next, illustrated in **Figure 8.5**, and summarized in **Table 8.3** on p. 260.

Gliding Movements

Gliding occurs when one flat, or nearly flat, bone surface glides or slips over another (back-and-forth and side-to-side; Figure 8.5a) without appreciable angulation or rotation. Gliding occurs at the intercarpal and intertarsal joints, and between the flat articular processes of the vertebrae (Table 8.2).

Angular Movements

Angular movements (Figure 8.5b–e) increase or decrease the angle between two bones. These movements may occur in any plane of the body and include flexion, extension, hyperextension, abduction, adduction, and circumduction.

Flexion   **Flexion** (flek'shun) is a bending movement, usually along the sagittal plane, that *decreases the angle* of the joint and brings the articulating bones closer together. Examples include bending the head forward on the chest (Figure 8.5b) and bending the body trunk or the knee from a straight to an angled position (Figure 8.5c and d). As a less obvious example, the arm

Figure 8.5 Movements allowed by synovial joints.

(a) Gliding movements at the wrist

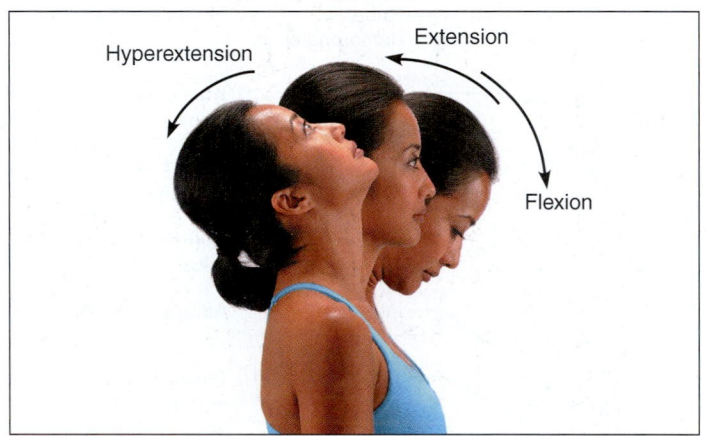

(b) Angular movements: flexion, extension, and hyperextension of the neck

(c) Angular movements: flexion, extension, and hyperextension of the vertebral column

(d) Angular movements: flexion, extension, and hyperextension at the shoulder and knee

is flexed at the shoulder when the arm is lifted in an anterior direction (Figure 8.5d).

Extension   **Extension** is the reverse of flexion and occurs at the same joints. It involves movement along the sagittal plane that *increases the angle* between the articulating bones and typically straightens a flexed limb or body part. Examples include straightening a flexed neck, body trunk, elbow, or knee (Figure 8.5b–d). Continuing such movements beyond the anatomical position is called **hyperextension** (Figure 8.5b–d).

Abduction   **Abduction** ("moving away") is movement of a limb *away* from the midline or median plane of the body, along the frontal plane. Raising the arm or thigh laterally is an example of abduction (Figure 8.5e). For the fingers or toes, abduction means spreading them apart. In this case the "midline" is the third finger or second toe. Notice, however, that lateral bending of the trunk away from the body midline in the frontal plane is called lateral flexion, not abduction.

Adduction   **Adduction** ("moving toward") is the opposite of abduction, so it is the movement of a limb *toward* the body midline or, in the case of the digits, toward the midline of the hand or foot (Figure 8.5e).

Circumduction   **Circumduction** (Figure 8.5e) is moving a limb so that it describes a cone in space (*circum* = around; *duco* = to draw). The distal end of the limb moves in a circle, while the point of the cone (the shoulder or hip joint) is more or less stationary. A pitcher winding up to throw a ball is actually circumducting his or her pitching arm. Because circumduction

(e) Angular movements: abduction, adduction, and circumduction of the upper limb at the shoulder

Figure 8.5 *(continued)*

(f) Rotation of the head, neck, and lower limb

Figure 8.5 *(continued)* **Movements allowed by synovial joints.**

Table 8.3	Movements at Synovial Joints
MOVEMENT	**DEFINITION**
Gliding	Sliding the flat surfaces of two bones across each other
Angular Movements	
Flexion	Decreasing the angle between two bones, usually in the sagittal plane
Extension	Increasing the angle between two bones, usually in the sagittal plane
Abduction	Moving a limb away from the body midline in the frontal plane
Adduction	Moving a limb toward the body midline in the frontal plane
Circumduction	Moving a limb or finger so that it describes a cone in space
Rotation	Turning a bone around its longitudinal axis
Medial rotation	Rotating toward the median plane
Lateral rotation	Rotating away from the median plane

consists of flexion, abduction, extension, and adduction performed in succession, it is the quickest way to exercise the many muscles that move the hip and shoulder ball-and-socket joints.

Rotation

Rotation is the turning of a bone around its own long axis. It is the only movement allowed between the first two cervical vertebrae and is common at the hip (Figure 8.5f) and shoulder joints. Rotation may be directed toward the midline or away from it. For example, in *medial rotation* of the thigh, the femur's anterior surface moves toward the median plane of the body; *lateral rotation* is the opposite movement.

Special Movements

Certain movements do not fit into any of the above categories and occur at only a few joints. Some of these special movements are illustrated in **Figure 8.6**.

Supination and Pronation The terms **supination** (soo″pĭ-na′shun; "turning backward") and **pronation** (pro-na′shun; "turning forward") refer to the movements of the radius around the ulna (Figure 8.6a). Rotating the forearm laterally so that the palm faces anteriorly or superiorly is supination. In the anatomical position, the hand is supinated and the radius and ulna are parallel.

In pronation, the forearm rotates medially and the palm faces posteriorly or inferiorly. Pronation moves the distal end of the radius across the ulna so that the two bones form an X. This is the forearm's position when we are standing in a relaxed manner. Pronation is a much weaker movement than supination.

A trick to help you keep these terms straight: A *pro* basketball player pronates his or her forearm to dribble the ball.

Dorsiflexion and Plantar Flexion of the Foot The up-and-down movements of the foot at the ankle are given more specific names (Figure 8.6b). Lifting the foot so that its superior surface approaches the shin is **dorsiflexion** (corresponds to wrist extension), whereas depressing the foot (pointing the toes) is **plantar flexion** (corresponds to wrist flexion).

Inversion and Eversion **Inversion** and **eversion** are special movements of the foot (Figure 8.6c). In inversion, the sole of the foot turns medially. In eversion, the sole faces laterally.

Protraction and Retraction Nonangular anterior and posterior movements in a transverse plane are called **protraction** and **retraction**, respectively (Figure 8.6d). The mandible is protracted when you jut out your jaw and retracted when you bring it back.

Elevation and Depression **Elevation** means lifting a body part superiorly (Figure 8.6e). For example, the scapulae are elevated when you shrug your shoulders. Moving the elevated part inferiorly is **depression**. During chewing, the mandible is alternately elevated and depressed.

Opposition The saddle joint between metacarpal I and the trapezium allows a movement called **opposition** of the thumb (Figure 8.6f). This movement is the action taken when you touch your thumb to the tips of the other fingers on the same hand. It is opposition that makes the human hand such a fine tool for grasping and manipulating objects.

(Text continues on p. 264.)

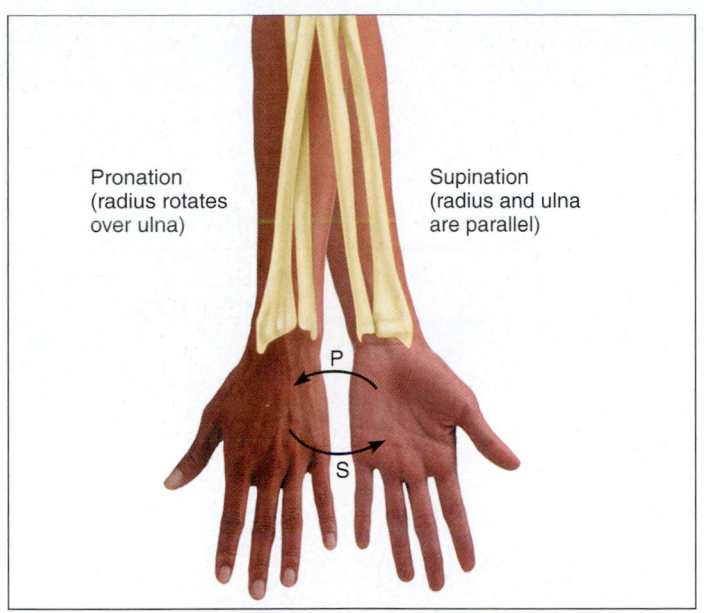

(a) Pronation (P) and supination (S)

Pronation
(radius rotates
over ulna)

Supination
(radius and ulna
are parallel)

P

S

(b) Dorsiflexion and plantar flexion

Dorsiflexion

Plantar flexion

(c) Inversion and eversion

Inversion

Eversion

(d) Protraction and retraction

Protraction
of mandible

Retraction
of mandible

(e) Elevation and depression

Elevation
of mandible

Depression
of mandible

(f) Opposition

Opposition

Figure 8.6 Special body movements.

Synovial Joints

Six types of synovial joint shapes determine
the movements that can occur at a joint.

(a) Plane joint

Nonaxial movement

Metacarpals

Carpals

Flat articular surfaces

Gliding

Examples: Intercarpal joints, intertarsal joints, joints between vertebral articular surfaces

(b) Hinge joint

Uniaxial movement

Humerus

Ulna

Cylinder

Trough

Medial/lateral axis

Flexion and extension

Examples: Elbow joints, interphalangeal joints

(c) Pivot joint

Uniaxial movement

Ulna

Radius

Sleeve (bone and ligament)

Axle (rounded bone)

Vertical axis

Rotation

Examples: Proximal radioulnar joints, atlantoaxial joint

(d) Condylar joint

Biaxial movement

Phalanges

Metacarpals

Oval articular surfaces

Medial/lateral axis

Anterior/posterior axis

Flexion and extension

Adduction and abduction

Examples: Metacarpophalangeal (knuckle) joints, wrist joints

(e) Saddle joint

Biaxial movement

Metacarpal I

Trapezium

Articular surfaces are both concave and convex

Medial/lateral axis

Anterior/posterior axis

Adduction and abduction

Flexion and extension

Example: Carpometacarpal joints of the thumbs

(f) Ball-and-socket joint

Multiaxial movement

Scapula

Humerus

Cup (socket)

Spherical head (ball)

Medial/lateral axis

Anterior/posterior axis

Vertical axis

Flexion and extension

Adduction and abduction

Rotation

Examples: Shoulder joints and hip joints

Types of Synovial Joints

Although all synovial joints have structural features in common, they do not have a common structural plan. Based on the shape of their articular surfaces, which in turn determine the movements allowed, synovial joints can be classified further into six major categories—plane, hinge, pivot, condylar (or ellipsoid), saddle, and ball-and-socket joints. The properties of these joints are summarized in *Focus on Synovial Joints* (**Focus Figure 8.1**) on pp. 262–263.

Check Your Understanding

6. How do bursae and tendon sheaths improve joint function?
7. Generally speaking, what factor is most important in stabilizing synovial joints?
8. On the basis of movement allowed, which of the following joints are uniaxial? Hinge, condylar, saddle, pivot.
9. Name the category (according to its shape and movement allowed) of each of the synovial joints labeled a–d below.

(a) (b) (c) (d)

10. **APPLY** Nick bent over to pick up a dime. What movement was occurring at his hip joint, at his knees, and between his index finger and thumb?
11. **DRAW** Sketch the general structure of a synovial joint. Label the articular cartilages, the articular capsule, and the joint cavity.

For answers, see Answers Appendix.

8.5 Five examples illustrate the diversity of synovial joints

Learning Outcome

▶ Describe the jaw, shoulder, elbow, hip, and knee joints in terms of articulating bones, anatomical characteristics of the joint, movements allowed, and joint stability.

In this section, we examine five joints in detail: temporomandibular (jaw), shoulder, elbow, hip, and knee joints. All have the six distinguishing characteristics of synovial joints, and we will not discuss these common features again. Instead, we will emphasize the unique structural features, functional abilities, and, in certain cases, functional weaknesses of each of these joints.

Temporomandibular Joint

The **temporomandibular joint** (TMJ), or jaw joint, is a modified hinge joint. It lies just anterior to the ear (**Figure 8.7**). At this joint, the condylar process of the mandible articulates with the inferior surface of the squamous part of the temporal bone (◀ p. 208). The mandible's condylar process is egg shaped, whereas the articular surface of the temporal bone has a more complex shape. Posteriorly, it forms the concave **mandibular fossa**; anteriorly it forms a dense knob called the **articular tubercle**. The lateral aspect of the loose articular capsule that encloses the joint is thickened into a **lateral ligament**. Within the capsule, an articular disc divides the synovial cavity into superior and inferior compartments (Figure 8.7a, b).

Two distinct kinds of movement occur at the TMJ. First, the concave inferior disc surface receives the condylar process of the mandible and allows the familiar hingelike movement

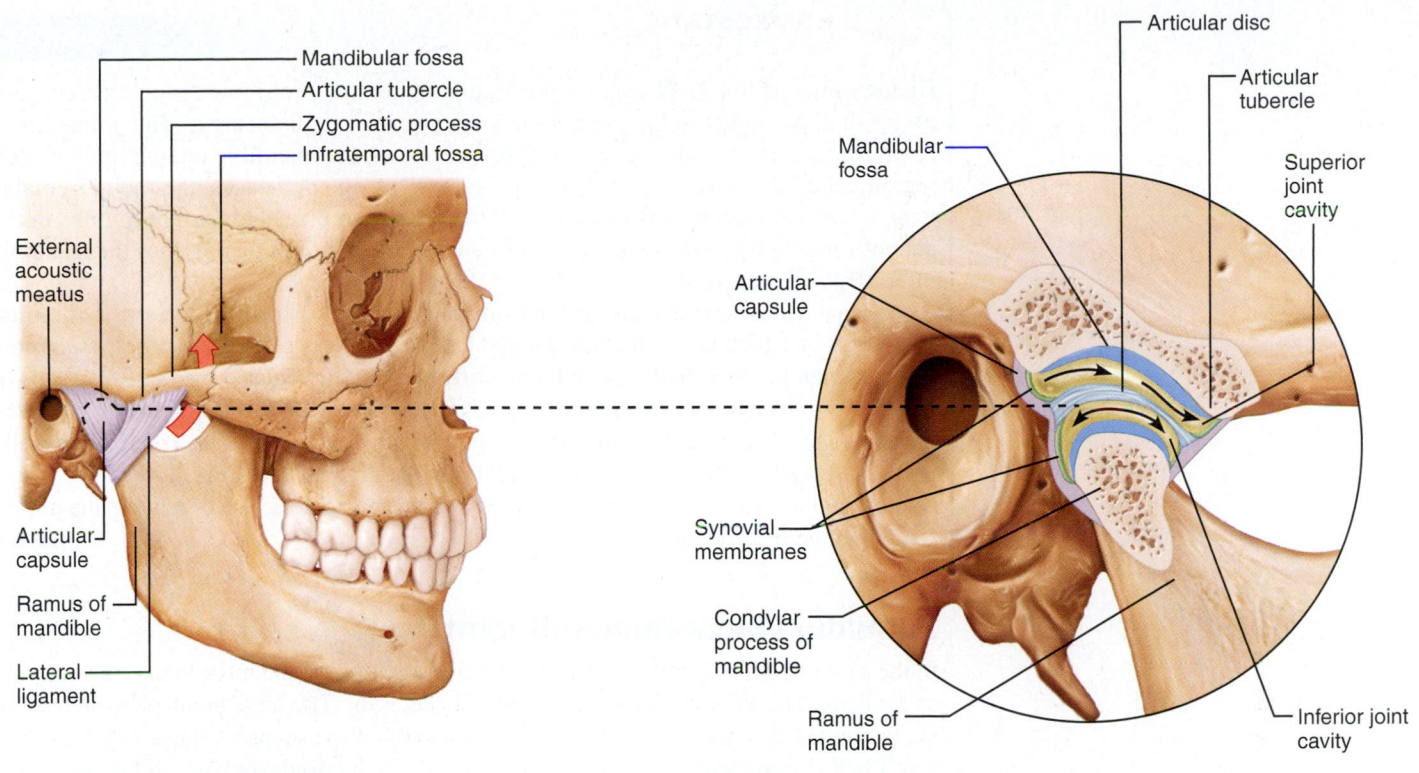

External
acoustic
meatus

Mandibular fossa
Articular tubercle
Zygomatic process
Infratemporal fossa

Articular
capsule

Ramus of
mandible

Lateral
ligament

(a) Location of the joint in the skull

Articular disc

Articular
tubercle

Superior
joint
cavity

Mandibular
fossa

Articular
capsule

Synovial
membranes

Condylar
process of
mandible

Ramus of
mandible

Inferior joint
cavity

(b) Enlargement of a sagittal section through the joint

8

Superior view

Outline of the
mandibular
fossa

(c) Lateral excursion: lateral (side-to-side) movements of the mandible

Figure 8.7 The temporomandibular (jaw) joint. Orange arrows show bone movement. In **(b)**, note that the two parts of the joint cavity allow different movements, indicated by black arrows. The inferior compartment of the joint cavity allows the condylar process of the mandible to rotate in opening and closing the mouth. The superior compartment lets the condylar process move forward to brace against the articular tubercle when the mouth opens wide, and also allows lateral excursion of this joint **(c)**.

of depressing and elevating the mandible while opening and closing the mouth (Figure 8.7a, b). Second, the superior disc surface glides anteriorly along with the condylar process when the mouth is opened wide. This anterior movement braces the condylar process against the articular tubercle, so that the mandible is not forced through the thin roof of the mandibular fossa when one bites hard foods such as nuts or hard candies. The superior compartment also allows this joint to glide from side to side. As the posterior teeth are drawn into occlusion during grinding, the mandible moves with a side-to-side movement called *lateral excursion* (Figure 8.7c). This lateral jaw movement is unique to mammals and it is readily apparent in horses and cows as they chew.

8

Dislocations of the TMJ occur more readily than any other joint dislocation because of the shallow socket in the joint. Even a deep yawn can dislocate it. This joint almost always dislocates anteriorly, the condylar process of the mandible ending up in a skull region called the *infratemporal fossa* (Figure 8.7a). In such cases, the mouth remains wide open. To realign a dislocated TMJ, the physician places his or her thumbs in the patient's mouth between the lower molars and the cheeks, and then pushes the mandible inferiorly and posteriorly.

At least 5% of Americans suffer from painful TMJ disorders, the most common symptoms of which are pain in the ear and face, tenderness of the jaw muscles, popping sounds when the mouth opens, and joint stiffness. Usually caused by painful spasms of the chewing muscles, TMJ disorders often afflict people who grind their teeth; however, it can also result from jaw trauma or from poor occlusion of the teeth. Treatment usually focuses on getting the jaw muscles to relax by using massage, muscle-relaxant drugs, heat or cold, or stress reduction techniques. For tooth grinders, use of a bite plate during sleep may be recommended.

Shoulder (Glenohumeral) Joint

In the shoulder joint, stability has been sacrificed to provide the most freely moving joint of the body. The shoulder joint is a ball-and-socket joint. The large hemispherical head of the humerus fits in the small, shallow glenoid cavity of the scapula (◀ **pp. 228–229**), like a golf ball sitting on a tee (**Figure 8.8**). Although the glenoid cavity is slightly deepened by a rim of fibrocartilage, the **glenoid labrum** (*labrum* = lip), it is only about one-third the size of the humeral head and contributes little to joint stability (Figure 8.8d).

The articular capsule enclosing the joint cavity (from the margin of the glenoid cavity to the anatomical neck of the humerus) is remarkably thin and loose, qualities that contribute to this joint's freedom of movement. The few ligaments reinforcing the shoulder joint are located primarily on its anterior aspect.

- The superiorly located **coracohumeral ligament** (kor′ah-ko-hu′mer-ul) provides the only strong thickening of the capsule and helps support the weight of the upper limb (Figure 8.8c).

- Three **glenohumeral ligaments** (glĕ″no-hu′mer-ul) strengthen the front of the capsule somewhat but are weak and may even be absent (Figure 8.8c, d).

Muscle tendons that cross the shoulder joint contribute most to this joint's stability. The "superstabilizer" is the tendon of the long head of the biceps brachii muscle of the arm (Figure 8.8c). This tendon attaches to the superior margin of the glenoid labrum, travels through the joint cavity, and then runs within the intertubercular sulcus of the humerus. It secures the head of the humerus against the glenoid cavity.

Four other tendons (and the associated muscles) make up the **rotator cuff**. This cuff encircles the shoulder joint and blends with the articular capsule. The four muscles of the rotator cuff are the subscapularis, supraspinatus, infraspinatus, and teres minor (see Figure 10.15, pp. 355–356). The rotator cuff can be severely stretched when the arm is vigorously circumducted; this is a common injury of baseball pitchers.

One price of mobility in the shoulder is that **shoulder dislocations** are common injuries. Because the structures reinforcing this joint are weakest anteriorly and inferiorly, the head of the humerus easily dislocates forward and downward. The glenoid cavity provides poor support when the humerus is rotated laterally and abducted, as when a football player uses the arm to tackle an opponent or a baseball fielder hits the ground diving for a ball. These situations may cause shoulder dislocations, as do blows to the top and back of the shoulder.

Acromion of scapula

Coracoacromial ligament

Subacromial bursa

Fibrous layer of articular capsule

Tendon sheath

Tendon of long head of biceps brachii muscle

Synovial cavity of the glenoid cavity containing synovial fluid

Articular cartilage

Synovial membrane

Fibrous layer of articular capsule

Humerus

(a) Frontal section through right shoulder joint

(b) Cadaver photo corresponding to (a)

8

Acromion

Coracoacromial ligament

Subacromial bursa

Coracohumeral ligament

Transverse humeral ligament

Tendon sheath

Tendon of long head of biceps brachii muscle

Coracoid process

Articular capsule reinforced by **glenohumeral ligaments**

Subscapular bursa

Tendon of the subscapularis muscle

Scapula

(c) Anterior view of right shoulder joint capsule

Acromion

Coracoid process

Articular capsule

Glenoid cavity

Glenoid labrum

Tendon of long head of biceps brachii muscle

Glenohumeral ligaments

Tendon of the subscapularis muscle

Scapula

Posterior   *Anterior*

(d) Lateral view of socket of right shoulder joint, humerus removed

Acromion (cut)

Glenoid cavity of scapula

Glenoid labrum

Capsule of shoulder joint (opened)

Head of humerus

Rotator cuff muscles (cut)

(e) Posterior view of an opened right shoulder joint

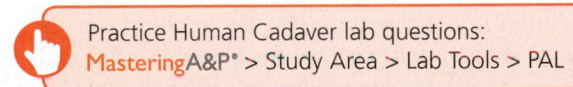
Practice Human Cadaver lab questions: MasteringA&P® > Study Area > Lab Tools > PAL

Figure 8.8 **The shoulder joint.**

(a) Median sagittal section through right elbow (lateral view)

Labels: Humerus, Fat pad, Tendon of triceps muscle, Bursa, Trochlea, Articular cartilage of the trochlear notch, Articular capsule, Synovial membrane, Synovial cavity, Articular cartilage, Coronoid process, Tendon of brachialis muscle, Ulna

(b) Lateral view of right elbow joint

Labels: Humerus, Lateral epicondyle, Articular capsule, Radial collateral ligament, Olecranon, Anular ligament, Radius, Ulna

(c) Cadaver photo of medial view of right elbow

Labels: Anular ligament, Radius, Coronoid process of ulna, Ulna, Humerus, Medial epicondyle, Articular capsule, Ulnar collateral ligament

(d) Medial view of right elbow

Labels: Articular capsule, Anular ligament, Coronoid process, Radius, Humerus, Medial epicondyle, Ulnar collateral ligament, Ulna

Figure 8.9 The elbow joint.

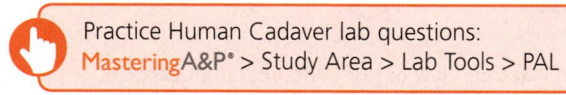

Practice Human Cadaver lab questions:
MasteringA&P® > Study Area > Lab Tools > PAL

Elbow Joint

Our upper limbs are flexible extensions that permit us to reach out and manipulate things in our environment. Besides the shoulder joint, the most prominent of the upper limb joints is the elbow. The elbow joint provides a stable and smoothly operating hinge that allows flexion and extension only (**Figure 8.9**). Within the joint, both the radius and ulna articulate with the condyles of the humerus, but it is the close gripping of the trochlea by the ulna's trochlear notch that forms the "hinge" and stabilizes this joint (Figure 8.9a). A relatively lax articular capsule extends inferiorly from the humerus to the ulna and radius, and to the **anular ligament** (an′u-lar) encircling the head of the radius (Figure 8.9b, c).

Anteriorly and posteriorly, the articular capsule is thin and allows substantial freedom for elbow flexion and extension. However, side-to-side movements are restricted by two strong capsular ligaments: the **ulnar collateral ligament** medially, and the **radial collateral ligament**, a triangular ligament on the lateral side (Figure 8.9b, c, and d). Additionally, tendons of several arm muscles, such as the biceps and triceps, cross the elbow joint and provide security.

The radius is a passive "onlooker" in the angular elbow movements. However, its head rotates within the anular ligament during supination and pronation of the forearm.

Hip Joint

The **hip** (coxal) **joint**, like the shoulder joint, is a ball-and-socket joint. It has a good range of motion, but not nearly as wide as the shoulder's range. Movements occur in all possible planes but are limited by the joint's strong ligaments and its deep socket.

The hip joint is formed by the articulation of the spherical head of the femur with the deeply cupped acetabulum of the hip bone (**Figure 8.10**) (◀ pp. 236–237). The depth of the acetabulum is enhanced by a circular rim of fibrocartilage called the

(a) Frontal section through the right hip joint

(b) Photo of the interior of the hip joint, lateral view

(c) Posterior view of right hip joint, capsule in place

(d) Anterior view of right hip joint, capsule in place

Figure 8.10 The hip joint.

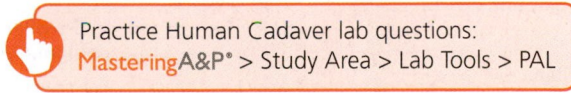

Practice Human Cadaver lab questions:
MasteringA&P® > Study Area > Lab Tools > PAL

acetabular labrum (as″ĕ-tab′u-lar) (Figure 8.10a, b). Because the diameter of the labrum is smaller than that of the head of the femur, the femur cannot easily slip out of the socket, and hip dislocations are rare.

The thick articular capsule extends from the rim of the acetabulum to the neck of the femur and completely encloses the joint. Several strong ligaments reinforce the capsule of the hip joint. These include:

- The **iliofemoral ligament** (il″e-o-fem′o-ral), a strong V-shaped ligament anteriorly (Figure 8.10c, d)
- The **pubofemoral ligament** (pu″bo-fem′o-ral), a triangular thickening of the inferior part of the capsule (Figure 8.10d)
- The **ischiofemoral ligament** (is″ke-o-fem′o-ral), a spiraling posterior ligament (Figure 8.10c).

These ligaments are arranged in such a way that they "screw" the femur head into the acetabulum when a person stands up straight, thereby providing stability.

The **ligament of the head of the femur**, also called the *ligamentum teres*, is a flat intracapsular band that runs from the femur head to the lower lip of the acetabulum (Figure 8.10a, b). This ligament is slack during most hip movements, so it is not important in stabilizing the joint. In fact, its mechanical function (if any) is unclear, but it does contain an artery that helps supply the head of the femur. Damage to this artery may lead to severe arthritis of the hip joint.

Muscle tendons that cross the joint and the bulky hip and thigh muscles that surround it contribute to its stability and strength. In this joint, however, stability comes chiefly from the deep socket that securely encloses the femoral head and the strong capsular ligaments.

(a) Sagittal section through the right knee joint

(b) Superior view of the right tibia in the knee joint, showing the menisci and cruciate ligaments

(c) Anterior view of right knee

(d) Posterior view of the joint capsule, including ligaments

Figure 8.11 The knee joint.

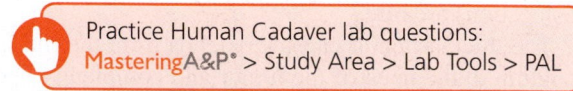
Practice Human Cadaver lab questions:
MasteringA&P® > Study Area > Lab Tools > PAL

Knee Joint

The knee joint is the largest and most complex joint in the body (**Figure 8.11**). Despite its single joint cavity, the knee consists of three joints in one: an intermediate one between the patella and the lower end of the femur (the **femoropatellar joint**), and lateral and medial joints (collectively known as the **tibiofemoral joint**) between the femoral condyles above and the C-shaped **menisci**, or *semilunar cartilages*, of the tibia below (Figure 8.11b and e). Besides deepening the shallow tibial articular surfaces, the menisci help prevent side-to-side rocking of the femur on the tibia and absorb shock transmitted to the knee joint. However, the menisci are attached only at their outer margins and are frequently torn free.

The tibiofemoral joint acts primarily as a hinge, permitting flexion and extension. However, structurally it is a bicondylar joint. Some rotation is possible when the knee is partly flexed, and when the knee is extending. But when the knee is fully extended, side-to-side movements and rotation are strongly resisted by ligaments and the menisci. The femoropatellar joint

(e) Anterior view of flexed knee, showing the cruciate ligaments (articular capsule removed, and quadriceps tendon cut and reflected distally)

(f) Photograph of an opened knee joint; view similar to (e)

Figure 8.11 *(continued)*

is a plane joint, and the patella glides across the distal end of the femur during knee flexion.

The knee joint is unique in that its joint cavity is only partially enclosed by a capsule. The relatively thin articular capsule is present only on the sides and posterior aspects of the knee, where it covers the bulk of the femoral and tibial condyles. Anteriorly, where the capsule is absent, three broad ligaments run from the patella to the tibia below. These are the **patellar ligament** flanked by the **medial** and **lateral patellar retinacula** (ret″ĭ-nak′u-lah; "retainers"), which merge imperceptibly into the articular capsule on each side (Figure 8.11c). The patellar ligament and retinacula are actually continuations of the tendon of the bulky quadriceps muscle of the anterior thigh. Physicians tap the patellar ligament to test the knee-jerk reflex.

The synovial cavity of the knee joint has a complicated shape, with several extensions that lead into "dead ends." At least a dozen bursae are associated with this joint, some of which are shown in Figure 8.11a. For example, notice the *subcutaneous prepatellar bursa*, which is often injured when the knee is bumped anteriorly.

All three types of joint ligaments (extracapsular, capsular, and intracapsular) stabilize and strengthen the capsule of the knee joint. All of the capsular and extracapsular ligaments act to prevent hyperextension of the knee and are stretched tight when the knee is extended. These include the following:

- The extracapsular **fibular** and **tibial collateral ligaments** are also critical in preventing lateral or medial rotation when the knee is extended. The broad, flat tibial collateral ligament runs from the medial epicondyle of the femur to the medial condyle of the tibial shaft below and is fused to the medial meniscus (Figure 8.11c–e).

- The **oblique popliteal ligament** (pop″lĭ-te′al) is actually part of the tendon of the semimembranosus muscle that fuses with the joint capsule and helps stabilize the posterior aspect of the knee joint (Figure 8.11d).

- The **arcuate popliteal ligament** arcs superiorly from the head of the fibula over the popliteus muscle and reinforces the joint capsule posteriorly (Figure 8.11d).

The knee's *intracapsular ligaments* are called *cruciate ligaments* (kroo′she-āt) because they cross each other, forming an X (*cruci* = cross) in the notch between the femoral condyles. They act as restraining straps to help prevent anterior-posterior displacement of the articular surfaces and to secure the articulating bones when we stand (Figure 8.11a, b, e). Although these ligaments are in the joint capsule, they are *outside* the synovial cavity, and synovial membrane nearly covers their surfaces. Note that the two cruciate ligaments both run superiorly to the femur and are named for their *tibial* attachment site.

The **anterior cruciate ligament** attaches to the *anterior* intercondylar area of the tibia (Figure 8.11b, e). From there it passes posteriorly, laterally, and upward to attach to the femur on the medial side of its lateral condyle. This ligament prevents forward sliding of the tibia on the femur and checks hyperextension of the knee. It is somewhat lax when the knee is flexed, and taut when the knee is extended.

The stronger **posterior cruciate ligament** is attached to the *posterior* intercondylar area of the tibia and passes anteriorly, medially, and superiorly to attach to the femur on the lateral side of the medial condyle (Figure 8.11a, b, e). This ligament prevents backward displacement of the tibia or forward sliding of the femur.

The knee capsule is heavily reinforced by muscle tendons. Most important are the strong tendons of the quadriceps muscles of the anterior thigh and the tendon of the semimembranosus muscle posteriorly (Figure 8.11c and d). The greater the strength and tone of these muscles, the less the chance of knee injury.

The knees have a built-in locking device that provides steady support for the body in the standing position. As we begin to stand up, the wheel-shaped femoral condyles roll like ball bearings across the tibial condyles and the flexed leg begins to extend at the knee. Because the lateral femoral condyle stops rolling before the medial condyle stops, the femur *spins* (rotates) medially on the tibia, until the cruciate and collateral ligaments of the knee are twisted and taut and the menisci are compressed. The tension in the ligaments effectively locks the joint into a rigid structure that cannot be flexed again until it is unlocked. This unlocking is accomplished by the popliteus muscle (see Figure 8.11d and Table 10.15, pp. 374–379). It rotates the femur laterally on the tibia, causing the ligaments to become untwisted and slack.

Figure 8.12 The "unhappy triad:" ruptured ACL, ruptured tibial collateral ligament, and torn meniscus. A common injury in hockey, soccer, and American football.

HOMEOSTATIC IMBALANCE 8.3 CLINICAL

Of all body joints, the knees are most susceptible to sports injuries because of their high reliance on nonarticular factors for stability and the fact that they carry the body's weight. The knee can absorb a vertical force equal to nearly seven times body weight. However, it is very vulnerable to *horizontal* blows, such as those that occur during blocking and tackling in football and in hockey.

When thinking of **common knee injuries**, remember the 3 Cs:

- Collateral ligaments
- Cruciate ligaments
- Cartilages (menisci)

Most dangerous are *lateral* blows to the extended knee. These forces tear the tibial collateral ligament and the medial meniscus attached to it, as well as the anterior cruciate ligament (ACL) (**Figure 8.12**). It is estimated that 50% of all professional football players have serious knee injuries during their careers.

Although less devastating than the injury just described, injuries that affect only the anterior cruciate ligament are becoming more common, particularly as women's sports become more vigorous and competitive. Most ACL injuries occur when a runner changes direction quickly, twisting a hyperextended knee. A torn ACL heals poorly, so repair usually requires a graft taken from either the patellar ligament, the hamstring tendon, or the calcaneal tendon.

Check Your Understanding

12. Of the five joints studied in more detail—hip, shoulder, elbow, knee, and temporomandibular—which two have menisci? Which act mainly as a uniaxial hinge? Which depend mainly on muscles and their tendons for stability?

For answers, see Answers Appendix.

8.6 Joints are easily damaged by injury, inflammation, and degeneration CLINICAL

Learning Outcomes

▶ **Name the most common joint injuries and discuss the symptoms and problems associated with each.**

▶ **Compare and contrast the common types of arthritis.**

▶ **Describe the cause and consequences of Lyme disease.**

Few of us pay attention to our joints unless something goes wrong. Although remarkably strong, joints are more likely to be injured by forces the bony skeleton can withstand. This is the price of our flexibility. Joint pain and malfunction can be caused by a number of factors besides traumatic injury, including inflammatory conditions and degenerative processes due to friction and wear.

Common Joint Injuries

For most of us, sprains and dislocations are the most common trauma-induced joint injuries. However, cartilage injuries, which are common among athletes, are the most problematic.

Cartilage Tears

Those who overdo various forms of exercise may end up feeling the snap and pop of their overstressed cartilage. Although most cartilage injuries involve tearing of the knee menisci, tears and overuse damage to the articular cartilages of other joints is becoming increasingly common in young athletes.

Cartilage tears typically occur when a meniscus is subjected to compression and shear stress at the same time. Cartilage is avascular and it rarely can obtain sufficient nourishment to repair itself, so it usually stays torn. Cartilage fragments can interfere with joint function by causing the joint to lock or bind,

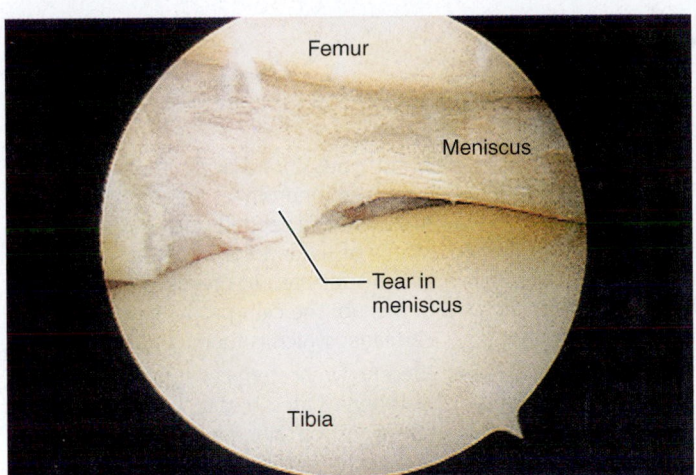

Figure 8.13 Arthroscopic photograph of a torn medial meniscus.

so most sports physicians recommend that the damaged cartilage be removed. This can be done by **arthroscopic surgery** (ar-thro-skop′ik; "looking into joints"), a procedure that enables patients to be out of the hospital the same day. The arthroscope, a small instrument bearing a tiny lens and fiber-optic light source, enables the surgeon to view the joint interior, as in **Figure 8.13**. The surgeon can then repair a ligament or remove cartilage fragments through one or more tiny slits, minimizing tissue damage and scarring. Removal of part of a meniscus does not severely impair knee joint mobility, but the joint is definitely less stable. Removal of the entire meniscus is an invitation to early onset of osteoarthritis. For younger patients a meniscal transplant may be an option to replace irreparably damaged cartilage. In the future, a tissue-engineered meniscus grown from your own stem cells may be implanted instead.

Sprains

In a **sprain**, the ligaments reinforcing a joint are stretched or torn. Common sites of sprains are the ankle, the knee, and the lumbar region of the spine. Partially torn ligaments will repair themselves, but they heal slowly because ligaments are so poorly vascularized. Sprains tend to be painful and immobilizing.

When ligaments are completely torn, there are three options:

- The torn ends of the ligament can be sewn together. This is difficult because trying to sew the hundreds of fibrous strands of a ligament together is like trying to sew two hairbrushes together.

- Certain ligaments, like the anterior cruciate ligament, are best repaired by replacing them with grafts. For example, a piece of tendon from a muscle can be attached to the articulating bones.

- For many ligaments, such as the knee's medial collateral ligament, it turns out that time and immobilization are just as effective as any surgical option.

Dislocations

A **dislocation (luxation)** occurs when bones are forced out of alignment. It is usually accompanied by sprains, inflammation,

and difficulty in moving the joint. Dislocations may result from serious falls and are common contact sports injuries. Joints of the jaw, shoulders, fingers, and thumbs are most commonly dislocated. Like fractures, dislocations must be *reduced*; that is, the bone ends must be returned to their proper positions by a physician. *Subluxation* is a partial dislocation of a joint.

Repeat dislocations of the same joint are common because the initial dislocation stretches the joint capsule and ligaments. The resulting loose capsule provides poor reinforcement for the joint.

Inflammatory and Degenerative Conditions

Inflammatory conditions that affect joints include bursitis and tendonitis, various forms of arthritis, and Lyme disease.

Bursitis and Tendonitis

Bursitis is inflammation of a bursa and is usually caused by a blow or friction. Falling on one's knee may result in a painful bursitis of the prepatellar bursa, known as *housemaid's knee* or *water on the knee*. Prolonged leaning on one's elbows may damage the bursa close to the olecranon, producing *student's elbow*, or *olecranon bursitis*. Severe cases are treated by injecting anti-inflammatory drugs into the bursa. If excessive fluid accumulates, removing some fluid by needle aspiration may relieve the pressure.

Tendonitis is inflammation of tendon sheaths, typically caused by overuse. Its symptoms (pain and swelling) and treatment (rest, ice, and anti-inflammatory drugs) mirror those of bursitis.

Arthritis

The term **arthritis** describes over 100 different types of inflammatory or degenerative diseases that damage the joints. In all its forms, arthritis is the most widespread crippling disease in North America. One in five of us suffers its ravages. To a greater or lesser degree, all forms of arthritis have the same initial symptoms: pain, stiffness, and swelling of the joint.

Acute forms of arthritis usually result from bacterial invasion and are treated with antibiotics. Chronic forms of arthritis include osteoarthritis, rheumatoid arthritis, and gouty arthritis.

Osteoarthritis Osteoarthritis (OA) is the most common chronic arthritis. A chronic degenerative condition, OA is often called "wear-and-tear arthritis." OA is most prevalent in the aged and is probably related to the normal aging process (although it is seen occasionally in younger people and some forms have a genetic basis). More women than men are affected, and nearly all of us will develop this condition by the age of 80.

Current theory holds that normal joint use prompts the release of (metalloproteinase) enzymes that break down articular cartilage, especially its collagen fibrils. In healthy individuals, this damaged cartilage is eventually replaced, but in people with OA, more cartilage is destroyed than replaced. Although its specific cause is unknown, OA may reflect the cumulative effects of years of compression and abrasion acting at joint surfaces, causing excessive amounts of the cartilage-destroying enzymes to be released. The result is softened, roughened, pitted, and eroded articular cartilages. Because this process occurs most where an uneven orientation of forces causes extensive microdamage, badly aligned or overworked joints are likely to develop OA.

Joints: From Knights in Shining Armor to Bionic Humans

The technology for fashioning joints in medieval suits of armor developed over centuries. The technology for creating the *prostheses* (artificial joints) used in medicine today developed, in relative terms, in a flash—less than 70 years. Unlike the joints in medieval armor, which was worn outside the body, today's artificial joints must function inside the body. The history of joint prostheses dates to the 1940s and 1950s, when World War II and the Korean War left large numbers of wounded who needed artificial limbs. Today, nearly 1 million Americans per year receive a total joint replacement, mostly because of the destructive effects of osteoarthritis or rheumatoid arthritis.

To produce durable, mobile joints requires substances that are strong, nontoxic, and resistant to the corrosive effects of organic acids in blood. In 1963, Sir John Charnley, an English orthopedic surgeon, revolutionized the therapy of arthritic hips with an artificial hip design that is still in use today. His device consisted of a metal ball on a stem and a cup-shaped polyethylene plastic socket anchored to the pelvis by methyl methacrylate cement. This cement proved to be exceptionally strong and relatively problem free. Hip prostheses were followed by knee prostheses, but not until 10 years later did smoothly operating total knee joint replacements become a reality. Today, the metal parts of the prostheses are strong cobalt and titanium alloys, and the number of knee replacements equals the number of hip replacements.

Replacements are now available for many other joints, including fingers, elbows, and shoulders. Most such operations are done to reduce pain and restore about 80% of original joint function. Total hip and knee replacements last about 15 to 20 years.

One common mode of failure is that the prosthesis works loose from the bone over time. A solution for this problem is to use a cementless prosthesis, which allows the bone to grow into its surface, fixing it in place. For this to happen, a precise fit in the prosthesis and the bone must be achieved. Prostheses can now be produced on 3-D printers, so they can be customized to exactly fit the patient. In addition, surgical robots can help enhance the precision of this surgery.

Joint replacement therapy has come of age. However, exciting new techniques that call on the ability of the patient's own tissues to regenerate aim to make metal prostheses obsolete:

- Bone marrow stimulation: Small holes poked through to the bone marrow allow mesenchymal stem cells from the bone marrow to migrate into the joint and produce new cartilage.
- Osteochondral grafting: Healthy bone and cartilage are removed from one part of the body and transplanted to the injured joint.
- Autologous chondrocyte implantation: Healthy chondrocytes are removed from the body and seeded onto a supporting matrix of tissue-engineered collagen. When subjected to mechanical pressure in the lab, the cells produce new cartilage, which is then implanted.
- Mesenchymal stem cell regeneration: Undifferentiated mesenchymal cells are removed from bone marrow and placed in a gel, which is packed into an area of eroded cartilage.

These techniques offer hope for a more permanent solution.

And so, through the centuries, the focus has shifted from jointed armor to artificial joints that can be put inside the body to restore lost function. Modern technology has accomplished what the armor designers of the Middle Ages never dreamed of.

A hip prosthesis.

X ray of right knee showing total knee replacement prosthesis.

As the disease progresses, the exposed bone tissue thickens and forms bony spurs (osteophytes) that enlarge the bone ends and may restrict joint movement. Patients complain of stiffness on arising that lessens somewhat with activity. The affected joints may make a crunching noise, called *crepitus* (krep′ĭ-tus), as they move and the roughened articular surfaces rub together. The joints most often affected are those of the cervical and lumbar spine and the fingers, knuckles, knees, and hips.

The course of osteoarthritis is usually slow and irreversible. In many cases, its symptoms are controllable with a mild pain reliever like aspirin or acetaminophen, along with moderate activity to keep the joints mobile. Glucosamine and chondroitin sulfate, nutritional supplements consisting of macromolecules normally present in cartilage, have been widely used by arthritis sufferers. However, several recent studies suggest that these supplements are no more effective than placebos. Osteoarthritis is rarely crippling, but it can be, particularly when the hip or knee joints are involved.

Rheumatoid Arthritis   **Rheumatoid arthritis (RA)** (roo´mah-toid) is a chronic inflammatory disorder. It usually arises between the ages of 30 and 50, but can occur at any age. It affects three times as many women as men. While not as common as osteoarthritis, rheumatoid arthritis affects millions, about 1% of all people.

In the early stages of RA, joint tenderness and stiffness are common. Many joints, particularly the small joints of the fingers, wrists, ankles, and feet, are afflicted at the same time and bilaterally. For example, if the right elbow is affected, most likely the left elbow is also affected. The course of RA is variable and marked by flare-ups (exacerbations) and remissions (*rheumat* = susceptible to change). Along with pain and swelling, its manifestations may include anemia, osteoporosis, muscle weakness, and cardiovascular problems.

RA is an *autoimmune disease*—a disorder in which the body's immune system attacks its own tissues. The initial trigger for this reaction is unknown, but various bacteria and viruses have been suspected. Perhaps these microorganisms bear molecules similar to some naturally present in the joints (possibly glycosaminoglycans, which are complex carbohydrates found in cartilage, joint fluid, and other connective tissues), and the immune system, once activated, attempts to destroy both.

RA begins with inflammation of the synovial membrane (*synovitis*) of the affected joints. Inflammatory cells (lymphocytes, macrophages, and others) migrate into the joint cavity from the blood and unleash a deluge of inflammatory chemicals that destroy body tissues when released in large amounts. Synovial fluid accumulates, causing joint swelling, and in time, the inflamed synovial membrane thickens into a **pannus** ("rag"), an abnormal tissue that clings to the articular cartilages. The pannus erodes the cartilage (and sometimes the underlying bone) and eventually scar tissue forms and connects the bone ends. Later this scar tissue ossifies and the bone ends fuse together, immobilizing the joint. This end condition, called *ankylosis* (ang˝kĭ-lo´sis; "stiff condition"), often produces bent, deformed fingers (**Figure 8.14**). Not all cases of RA progress to the severely crippling ankylosis stage, but all cases do involve restriction of joint movement and extreme pain.

The goal of current RA treatment is to go beyond simply alleviating the symptoms and instead to disrupt the relentless destruction of the joints. Steroidal and nonsteroidal anti-inflammatory drugs decrease pain and inflammation, increasing joint mobility. More powerful immune suppressants (such as methotrexate) act to slow the autoimmune reaction. Several biologic agents are available to block the action of inflammatory chemicals. An important target of some of these agents is an inflammatory chemical called *tumor necrosis factor*. Together, these drugs can dramatically slow the course of RA. As a last resort, replacing the joint with a joint prosthesis (artificial joint) may be an option to restore function (see **A Closer Look**). Indeed, some RA sufferers have over a dozen artificial joints.

Gouty Arthritis   Uric acid, a normal waste product of nucleic acid metabolism, is ordinarily excreted in urine without any problems. However, when blood levels of uric acid rise excessively (due to its excessive production or slow excretion), it may be deposited as needle-shaped urate crystals in the soft tissues of joints. An inflammatory response follows, leading to an agonizingly painful attack of **gouty arthritis** (gow´te), or **gout**. The initial attack typically affects one joint, often at the base of the great toe.

Gout is far more common in men than in women because men naturally have higher blood levels of uric acid (perhaps because estrogens increase the rate of its excretion). Because gout seems to run in families, genetic factors are definitely implicated.

Untreated gout can be very destructive; the articulating bone ends fuse and immobilize the joint. Fortunately, several drugs (colchicine, nonsteroidal anti-inflammatory drugs, glucocorticoids, and others) that terminate or prevent gout attacks are available. Patients are advised to drink plenty of water and to avoid excessive alcohol consumption (which promotes uric acid overproduction) and foods high in purine-containing nucleic acids, such as liver, kidneys, and sardines.

Lyme Disease

Lyme disease is an inflammatory disease caused by spirochete bacteria transmitted by the bite of ticks that live on mice and deer. It often results in joint pain and arthritis, especially in the knees, and is characterized by a skin rash, flu-like symptoms, and foggy thinking. If untreated, neurological disorders and irregular heartbeat may ensue.

Because symptoms vary from person to person, the disease is hard to diagnose. Antibiotic therapy is the usual treatment, and works best when initiated soon after the initial infection.

Check Your Understanding

13. What does the term "arthritis" mean?
14. What is the cause of Lyme disease?
15. **APPLY** From an arthritis patient's history and physical examination, how would you determine if he or she might have OA or RA?

For answers, see Answers Appendix.

Developmental Aspects of Joints

As bones form from mesenchyme in the embryo, the joints develop in parallel. By week 8, the synovial joints resemble adult joints in form and arrangement, and synovial fluid is being secreted. During childhood, a joint's size, shape, and flexibility are modified by use. Active joints have thicker capsules and ligaments, and larger bony supports.

Figure 8.14 A hand deformed by rheumatoid arthritis.

Injuries aside, relatively few interferences with joint function occur until late middle age. Eventually advancing years take their toll—ligaments and tendons shorten and weaken. The intervertebral discs become more likely to herniate, and osteoarthritis rears its ugly head. Many people have osteoarthritis by the time they are in their 70s. The middle years also see an increased incidence of rheumatoid arthritis.

Exercise that coaxes joints through their full range of motion, such as regular stretching and aerobics, is the key to postponing the immobilizing effects of aging on ligaments and tendons, to keeping cartilages well nourished, and to strengthening the muscles that stabilize the joints. The key word for exercising is "prudently," because excessive or abusive use of the joints guarantees early onset of osteoarthritis. The buoyancy of water relieves much of the stress on weight-bearing joints, and people who swim or exercise in a pool often retain good joint function as long as they live. As with so many medical problems, it is easier to prevent joint problems than to cure or correct them.

RELATED CLINICAL TERMS

Ankylosing spondylitis (ang′kǐ-lōz″ing spon″dǐ-li′tis; *ankyl* = crooked, bent; *spondyl* = vertebra) A variant of rheumatoid arthritis that chiefly affects males; it usually begins in the sacroiliac joints and progresses superiorly along the spine. The vertebrae become interconnected by fibrous tissue, causing the spine to become rigid.

Arthrology (ar-throl′o-je; *logos* = study) The study of joints.

Arthroplasty ("joint reforming") Replacing a diseased joint with an artificial joint.

Chondromalacia patellae (kon-dro-mal-a′sǐ-ah; "softening of cartilage by the patella") Damage and softening of the articular cartilages on the posterior patellar surface and the anterior surface of the distal femur; most often seen in adolescent athletes. Produces a sharp pain in the knee when the leg is extended (in climbing stairs, for example). May result when the quadriceps femoris, the main group of muscles on the anterior thigh, pulls unevenly on the patella, persistently rubbing it against the femur in the knee joint; often corrected by exercises that strengthen weakened parts of the quadriceps muscles.

Rheumatism A lay term referring to disease involving muscle or joint pain; may be used to apply to arthritis, bursitis, etc.

Synovitis (sin″o-vi′tis) Inflammation of the synovial membrane of a joint. Caused by injury, infection, or arthritis. Excess synovial fluid accumulates in the joint cavity, a condition called *effusion* that causes the joint to swell, limiting joint movement.

CHAPTER SUMMARY

1. Joints, or articulations, are sites where bones meet. Their functions are to hold bones together and to allow various degrees of skeletal movement.

8.1 Joints are classified into three structural and three functional categories (pp. 251–252)

1. Joints are classified structurally as fibrous, cartilaginous, or synovial. They are classed functionally as synarthrotic, amphiarthrotic, or diarthrotic. Only synovial joints have a joint cavity.

8.2 In fibrous joints, the bones are connected by fibrous tissue (pp. 252–253)

1. **Sutures/syndesmoses/gomphoses.** The major types of fibrous joints are sutures, syndesmoses, and gomphoses. Nearly all fibrous joints are synarthrotic.

8.3 In cartilaginous joints, the bones are connected by cartilage (pp. 253–254)

1. **Synchondroses/symphyses.** Cartilaginous joints include synchondroses and symphyses. Synchondroses are synarthrotic; all symphyses are amphiarthrotic.

8.4 Synovial joints have a fluid-filled joint cavity (pp. 254–264)

1. Most body joints are synovial joints, all of which are diarthrotic.

General Structure (pp. 254–255)

2. All synovial joints have: a joint cavity enclosed by a fibrous layer lined with synovial membrane and reinforced by ligaments; articulating bone ends covered with articular cartilage; and synovial fluid in the joint cavity. Some (e.g., the knee) contain fibrocartilage discs that absorb shock.

Bursae and Tendon Sheaths (p. 255)

3. Bursae are fibrous sacs lined with synovial membrane and containing synovial fluid. Tendon sheaths are similar to bursae but are cylindrical structures that surround muscle tendons. Both allow adjacent structures to move smoothly over one another.

Factors Influencing the Stability of Synovial Joints (pp. 257–258)

4. Articular surfaces providing the most stability have large surfaces and deep sockets and fit snugly together.

5. Ligaments prevent undesirable movements and reinforce the joint.

6. The tone of muscles whose tendons cross the joint is the most important stabilizing factor in many joints.

Movements Allowed by Synovial Joints (pp. 258–261)

7. When a skeletal muscle contracts, the insertion (movable attachment) moves toward the origin (immovable attachment).

8. Synovial joints differ in their range of motion. Motion may be nonaxial (gliding), uniaxial (in one plane), biaxial (in two planes), or multiaxial (in all three planes).

9. Three common types of movements can occur when muscles contract across joints: (a) gliding movements, (b) angular movements (which include flexion, extension, abduction, adduction, and circumduction), and (c) rotation.

10. Special movements include supination and pronation, inversion and eversion, protraction and retraction, elevation and depression, opposition, dorsiflexion and plantar flexion.

Types of Synovial Joints   (pp. 262–264)

11. The six major categories of synovial joints are plane joints (nonaxial movement), hinge joints (uniaxial), pivot joints (uniaxial, rotation permitted), condylar joints (biaxial with angular movements in two planes), saddle joints (biaxial, like condylar joints, but with freer movement), and ball-and-socket joints (multiaxial and rotational movement).

8.5   Five examples illustrate the diversity of synovial joints   (pp. 264–272)

1. The **temporomandibular joint** is formed by (1) the condylar process of the mandible and (2) the mandibular fossa and articular tubercle of the temporal bone. This joint allows both a hingelike opening and closing of the mouth and an anterior gliding of the mandible. It often dislocates anteriorly and exhibits a number of TMJ disorders.

2. The **shoulder joint** is a ball-and-socket joint formed by the glenoid cavity of the scapula and the humeral head. The most freely movable joint of the body, it allows all angular and rotational movements. Its articular surfaces are shallow. Its capsule is lax and poorly reinforced by ligaments. The tendons of the biceps brachii and rotator cuff muscles help to stabilize it.

3. The **elbow joint** is a hinge joint in which the ulna (and radius) articulates with the humerus, allowing flexion and extension. Its articular surfaces are highly complementary and are the most important factor contributing to joint stability.

4. The **hip joint** is a ball-and-socket joint formed by the acetabulum of the hip bone and the femoral head. It is highly adapted for weight bearing. Its articular surfaces are deep and secure. Its capsule is heavy and strongly reinforced by ligaments.

5. The **knee joint** is the largest joint in the body. It is a hinge joint formed by the articulation of the tibial and femoral condyles (and anteriorly by the patella and patellar surface of the femur). Extension, flexion, and (some) rotation are allowed. Its articular surfaces are shallow and condylar. C-shaped menisci deepen the articular surfaces. The joint cavity is enclosed by a capsule only on the sides and posterior aspect. Several ligaments help prevent displacement of the joint surfaces. Muscle tone of the quadriceps and semimembranosus muscles is important in knee stability.

8.6   Joints are easily damaged by injury, inflammation, and degeneration   (pp. 272–276)

1. Cartilage injuries, particularly of the knee, are common in contact sports and may result from excessive compression and shear stress. The avascular cartilage is unable to repair itself.

2. Sprains involve stretching or tearing of joint ligaments. Because ligaments are poorly vascularized, healing is slow.

3. Dislocations involve displacement of the articular surfaces of bones. They must be reduced.

4. Bursitis and tendonitis are inflammations of a bursa and a tendon sheath, respectively.

5. Arthritis is joint inflammation or degeneration accompanied by stiffness, pain, and swelling. Acute forms generally result from bacterial infection. Chronic forms include osteoarthritis, rheumatoid arthritis, and gouty arthritis.

6. Osteoarthritis is a degenerative condition most common in the aged. Spine, knees, hips, knuckles, and fingers are most affected.

7. Rheumatoid arthritis, the most crippling arthritis, is an autoimmune disease involving severe inflammation of the joints.

8. Gouty arthritis, or gout, is joint inflammation caused by the deposit of urate salts in soft joint tissues.

9. Lyme disease is an infectious disease caused by the bite of a tick infected with spirochete bacteria.

Developmental Aspects of Joints   (p. 276)

1. Joints form from mesenchyme and in tandem with bone development in the embryo.

2. Excluding traumatic injury, joints usually function well until late middle age, at which time symptoms of connective tissue stiffening and osteoarthritis begin to appear. Prudent exercise delays these effects, whereas excessive exercise promotes the early onset of arthritis.

REVIEW QUESTIONS

To access additional practice questions using your smartphone, tablet, or computer: MasteringA&P® > Study Area > Practice Tests & Quizzes

Level 1 Remember/Understand

(Some questions have more than one correct answer. Select the best answer or answers from the choices given.)

1. Match the key terms to the appropriate descriptions.

   Key: **(a)** fibrous joints   **(b)** cartilaginous joints
   (c) synovial joints

   _____ **(1)** exhibit a joint cavity

   _____ **(2)** types are sutures and syndesmoses

   _____ **(3)** bones connected by collagen fibers

   _____ **(4)** types include synchondroses and symphyses

   _____ **(5)** all are diarthrotic

   _____ **(6)** many are amphiarthrotic

   _____ **(7)** bones connected by a disc of hyaline cartilage or fibrocartilage

   _____ **(8)** nearly all are synarthrotic

   _____ **(9)** shoulder, hip, jaw, and elbow joints

2. Freely movable joints are **(a)** synarthroses, **(b)** diarthroses, **(c)** amphiarthroses.

3. Anatomical characteristics shared by all synovial joints include all except **(a)** articular cartilage, **(b)** a joint cavity, **(c)** an articular capsule, **(d)** presence of fibrocartilage.

4. Factors that influence the stability of a synovial joint include **(a)** shape of articular surfaces, **(b)** presence of strong reinforcing ligaments, **(c)** tone of surrounding muscles, **(d)** all of these.

5. The description "Articular surfaces deep and secure; capsule heavily reinforced by ligaments and muscle tendons; extremely stable joint" best describes **(a)** the elbow joint, **(b)** the hip joint, **(c)** the knee joint, **(d)** the shoulder joint.

6. Ankylosis means **(a)** twisting of the ankle, **(b)** tearing of ligaments, **(c)** displacement of a bone, **(d)** immobility of a joint due to fusion of its articular surfaces.

7. An autoimmune disorder in which joints are affected bilaterally and which involves pannus formation and gradual joint immobilization is **(a)** bursitis, **(b)** gout, **(c)** osteoarthritis, **(d)** rheumatoid arthritis.

8. Define joint.
9. Joint movements may be nonaxial, uniaxial, biaxial, or multiaxial. Define what each of these terms means.
10. Name two types of uniaxial, biaxial, and multiaxial joints.
11. Why are sprains and cartilage injuries a particular problem?

Level 2 Apply/Analyze

12. Compare the structure, function, and common body locations of bursae and tendon sheaths.
13. Compare and contrast the paired movements of flexion and extension with adduction and abduction.
14. How does rotation differ from circumduction?
15. What is the specific role of the menisci of the knee? Of the anterior and posterior cruciate ligaments?
16. List the functions of the following elements of a synovial joint: fibrous layer of the capsule, synovial fluid, articular cartilage.
17. Sonya worked cleaning homes for 30 years so she could send her two children to college. Several times, she had been forced to call her employers to tell them she could not come in to work because one of her kneecaps was swollen and painful. What is Sonya's condition, and what probably caused it?
18. Mrs. Bell, a 45-year-old woman, appeared at her physician's office complaining of unbearable pain in the interphalangeal joint of her right great toe. The joint was red and swollen. When asked about previous episodes, she recalled a similar attack two years earlier that disappeared as suddenly as it had come. Her diagnosis was arthritis. **(a)** What type? **(b)** What is the precipitating cause of this particular type of arthritis?
19. Tony Bowers, an exhausted biology student, was attending a lecture. After 30 minutes or so, he lost interest and began to doze. As the lecture ended, the hubbub aroused him and he let go with a tremendous yawn. To his great distress, he couldn't close his mouth—his lower jaw was "stuck" open. What do you think had happened?

Level 3 Evaluate/Synthesize

20. Discuss the relative value (to body homeostasis) of immovable, slightly movable, and freely movable joints.
21. The knee has been called "a beauty and a beast." Provide several reasons that might explain the negative (beast) part of this description.
22. As Jose was running down the road, he tripped and his left ankle twisted violently to the side. When he picked himself up, he was unable to put any weight on that ankle. The diagnosis was severe dislocation and sprains of the left ankle. The orthopedic surgeon stated that she would perform a closed reduction of the dislocation and attempt ligament repair by using arthroscopy. **(a)** Is the ankle joint normally a stable joint? **(b)** What does its stability depend on? **(c)** What is a closed reduction? **(d)** Why is ligament repair necessary? **(e)** What does arthroscopy entail? **(f)** How will the use of this procedure minimize Jose's recuperation time (and suffering)?
23. Grace heard on the evening TV news that the deer population in her state had been increasing rapidly in the past few years and it was common knowledge that deer walked the streets at night. After the program, she suddenly exclaimed, "So that's why those three boys in my son's class got Lyme disease last year." Explain what she meant by that comment.

CLINICAL CASE STUDY

45-Year-Old Female with Dislocated Hip

In the previous chapter, you met Kayla Tanner, a 45-year-old mother of four who suffered a dislocated right hip in a car accident. Prior to the closed reduction, the doctors noted that her right thigh was flexed at the hip, adducted, and medially rotated. A widened joint space in the postreduction X ray showed that the reduction was not complete, but no bone fragments were visible in the joint space. Mrs. Tanner was scheduled for immediate surgery.

The surgeons discovered that the acetabular labrum had detached from the rim of the acetabulum and was lying deep within the joint space. The detached portion of the labrum was excised, and the hip was surgically reduced. During the early healing phase (first two weeks), Mrs. Tanner was kept in traction with the thigh abducted.

1. How would you structurally and functionally classify the joint involved in the injury in this case?

2. In the previous case study, you found out that Mrs. Tanner had previously suffered two dislocations of her shoulder. Create a table to compare the following aspects of shoulder and hip joints: (a) What bony structures articulate to form the joints? (b) What is the name of the labrum? (c) Is the articular capsule loose or tight? (d) What ligaments reinforce the joint?

3. There were no bone fragments in the joint space. What is normally found in this space?

4. Surgeons had to remove part of Mrs. Tanner's acetabular labrum. What is this structure and what is its function?

5. Mrs. Tanner's thigh was flexed at the hip, adducted, and medially rotated. Describe what this means in terms of the position of her leg using layperson's terms.

6. Hip dislocations are classified as anterior or posterior depending on which direction the head of the femur is facing after it dislocates. Based on the description from question 5, which type of dislocation did Mrs. Tanner suffer?

7. **+ NCLEX-STYLE** As part of Mrs. Tanner's rehabilitation, clinicians will assess all of the movements that normally occur at the hip. This would include all EXCEPT:
   a. Flexion of the thigh
   b. Abduction of the thigh
   c. Inversion of the thigh
   d. Circumduction of the thigh

8. **+ NCLEX-STYLE** As Mrs. Tanner recovers, she will do exercises to strengthen her muscles and regain her range of motion. To reduce the risk of dislocating the hip during these exercises, the nurse will make sure she does not do which of the following?
   a. Plantar flex the foot
   b. Flex the leg at the knee
   c. Adduct her thigh until the knee moves past the midline
   d. Tighten her gluteal muscles

9 Muscles and Muscle Tissue

In this chapter, you will learn that

Muscles use actin and myosin molecules to convert the energy of ATP into force

beginning with

9.1 Overview of muscle types, special characteristics, and functions

next exploring

Skeletal muscle

and investigating

9.2 Gross anatomy

and

9.3 Microscopic anatomy and sliding filament model

then asking

9.4 How does a nerve impulse cause a muscle fiber to contract?

and

9.5 What are the properties of whole muscle contraction?

and

9.6 How do muscles generate ATP?

and

9.7 What determines the force, velocity, and duration of contraction?

and

9.8 How does skeletal muscle respond to exercise?

then exploring

Smooth muscle

and asking

9.9 How does smooth muscle differ from skeletal muscle?

and finally, exploring

Developmental Aspects of Muscles

CAREER CONNECTION

Watch a video to learn how the chapter content is used in a real health care setting. Go to **Mastering** A&P® > Study Area > Animations and Videos or use quick access URL https://goo.gl/88srfV

9

Because flexing muscles look like mice scurrying beneath the skin, some scientist long ago dubbed them *muscles*, from the Latin *mus* meaning "little mouse." Indeed, we tend to think of the rippling muscles of professional boxers or weight lifters when we hear the word *muscle*. But muscle is also the dominant tissue in the heart and in the walls of other hollow organs. In all its forms, muscle tissue makes up nearly half the body's mass.

Muscles are distinguished by their ability to transform chemical energy (ATP) into directed mechanical energy. In so doing, they become capable of exerting force.

9.1 There are three types of muscle tissue

Learning Outcomes

▶ **Compare and contrast the three basic types of muscle tissue.**

▶ **List four important functions of muscle tissue.**

Types of Muscle Tissue

Chapter 4 introduced the three types of muscle tissue—*skeletal*, *cardiac*, and *smooth* (◀ **pp. 138–140**). In this chapter, we first examine the structure and function of skeletal muscle. Then we consider smooth muscle more briefly, largely by comparing it with skeletal muscle. We describe cardiac muscle in detail in Chapter 18, but for easy comparison, Table 9.3 on pp. 312–313 summarizes the characteristics of all three muscle types.

Let's introduce some terminology before we describe each type of muscle.

- Skeletal and smooth muscle cells (but not cardiac muscle cells) are elongated, and are called **muscle fibers**.

- Whenever you see the prefixes **myo** or **mys** (both are word roots meaning "muscle") or **sarco** (flesh), the reference is to muscle. For example, the plasma membrane of muscle cells is called the *sarcolemma* (sar″ko-lem′ah), literally, "muscle" (sarco) "husk" (lemma), and muscle cell cytoplasm is called *sarcoplasm*.

Okay, let's get to it.

Skeletal Muscle

Skeletal muscle tissue is packaged into the *skeletal muscles*, organs that attach to and cover the skeleton. Skeletal muscle fibers are the longest muscle cells and have obvious stripes called *striations*. Although it is often activated by reflexes, skeletal muscle is called *voluntary muscle* because it is the only type subject to conscious control.

- When you think of skeletal muscle tissue, the key words to keep in mind are *skeletal*, *striated*, and *voluntary*.

Skeletal muscle is responsible for overall body mobility. It can contract rapidly, but it tires easily and must rest after short periods of activity. Nevertheless, it can exert tremendous power. Skeletal muscle is also remarkably adaptable. For example, your forearm muscles can exert a force of a fraction of an ounce to pick up a paper clip—or a force of about 6 pounds to pick up this book!

Cardiac Muscle

Cardiac muscle tissue occurs only in the heart, where it constitutes the bulk of the heart walls. Like skeletal muscle cells, cardiac muscle cells are striated, but cardiac muscle is not voluntary. Indeed, it can and does contract without being stimulated by the nervous system. Most of us have no conscious control over how fast our heart beats.

- Key words to remember for cardiac muscle are *cardiac*, *striated*, and *involuntary*.

Cardiac muscle usually contracts at a fairly steady rate set by the heart's pacemaker, but neural controls allow the heart to speed up for brief periods, as when you race across the tennis court to make that overhead smash.

Smooth Muscle

Smooth muscle tissue is found in the walls of hollow visceral organs, such as the stomach, urinary bladder, and respiratory passages. Its role is to force fluids and other substances through internal body channels. Smooth muscle also forms valves to regulate the passage of substances through internal body openings, dilates and constricts the pupils of your eyes, and forms the arrector pili muscles attached to hair follicles.

Like skeletal muscle, smooth muscle consists of elongated cells, but smooth muscle has no striations. Like cardiac muscle, smooth muscle is not subject to voluntary control. Its contractions are slow and sustained.

- We can describe smooth muscle tissue as *visceral*, *nonstriated*, and *involuntary*.

Characteristics of Muscle Tissue

What enables muscle tissue to perform its duties? Four special characteristics are key.

- **Excitability**, also termed **responsiveness**, is the ability of a cell to receive and respond to a stimulus by changing its membrane potential. In the case of muscle, the stimulus is usually a chemical—for example, a neurotransmitter released by a nerve cell.

- **Contractility** is the ability to shorten forcibly when adequately stimulated. This ability sets muscle apart from all other tissue types.

- **Extensibility** is the ability to extend or stretch. Muscle cells shorten when contracting, but they can be stretched, even beyond their resting length, when relaxed.

- **Elasticity** is the ability of a muscle cell to recoil and resume its resting length after stretching.

Muscle Functions

Muscles perform at least four important functions for the body:

- **Produce movement.** Skeletal muscles are responsible for all locomotion and manipulation. They enable you to respond quickly to jump out of the way of a car, direct your eyes, and smile or frown.

Blood courses through your body because of the rhythmically beating cardiac muscle of your heart and the smooth muscle in the walls of your blood vessels, which helps maintain blood pressure. Smooth muscle in organs of the digestive, urinary, and reproductive tracts propels substances (foodstuffs, urine, semen) through the organs and along the tract.

- **Maintain posture and body position.** We are rarely aware of the skeletal muscles that maintain body posture. Yet these muscles function almost continuously, making one tiny adjustment after another to counteract the never-ending downward pull of gravity.
- **Stabilize joints.** Even as they pull on bones to cause movement, they strengthen and stabilize the joints of the skeleton.
- **Generate heat.** Muscles generate heat as they contract, which plays a role in maintaining normal body temperature.

Check Your Understanding

1. When describing muscle, what does "striated" mean?
2. **MAKE CONNECTIONS** At right are photomicrographs of three types of muscle tissue (introduced in Chapter 4). For each, identify the type of muscle, state whether it is striated, whether it is voluntary, and its major location. Create a short table that summarizes this information.

For answers, see Answers Appendix.

(a)

(b)

(c)

9.2 A skeletal muscle is made up of muscle fibers, nerves, blood vessels, and connective tissues

Learning Outcome

▶ Describe the gross structure of a skeletal muscle.

Each **skeletal muscle** is a discrete organ, made up of several kinds of tissues. Skeletal muscle fibers predominate, but blood vessels, nerve fibers, and substantial amounts of connective tissue are also present. We can easily examine a skeletal muscle's shape and its attachments in the body without a microscope.

Nerve and Blood Supply

In general, one nerve, one artery, and one or more veins serve each muscle. These structures all enter or exit near the central part of the muscle and branch profusely through its connective tissue sheaths (described below). Unlike cells of cardiac and smooth muscle tissues, which can contract without nerve stimulation, every skeletal muscle fiber is supplied with a nerve ending that controls its activity.

Skeletal muscle has a rich blood supply. This is understandable because contracting muscle fibers use huge amounts of energy and require almost continuous delivery of oxygen and nutrients via the arteries. Muscle cells also give off large amounts of metabolic wastes that must be removed through veins if contraction is to remain efficient. Capillaries, the smallest of the body's blood vessels, take a long and winding path through muscle, and have numerous cross-links, features that accommodate changes in muscle length. They straighten when the muscle stretches and contort when the muscle contracts.

Connective Tissue Sheaths

In an intact muscle, there are several different connective tissue sheaths. Together these sheaths support each cell and reinforce and hold together the muscle, preventing the bulging muscles from bursting during exceptionally strong contractions.

9

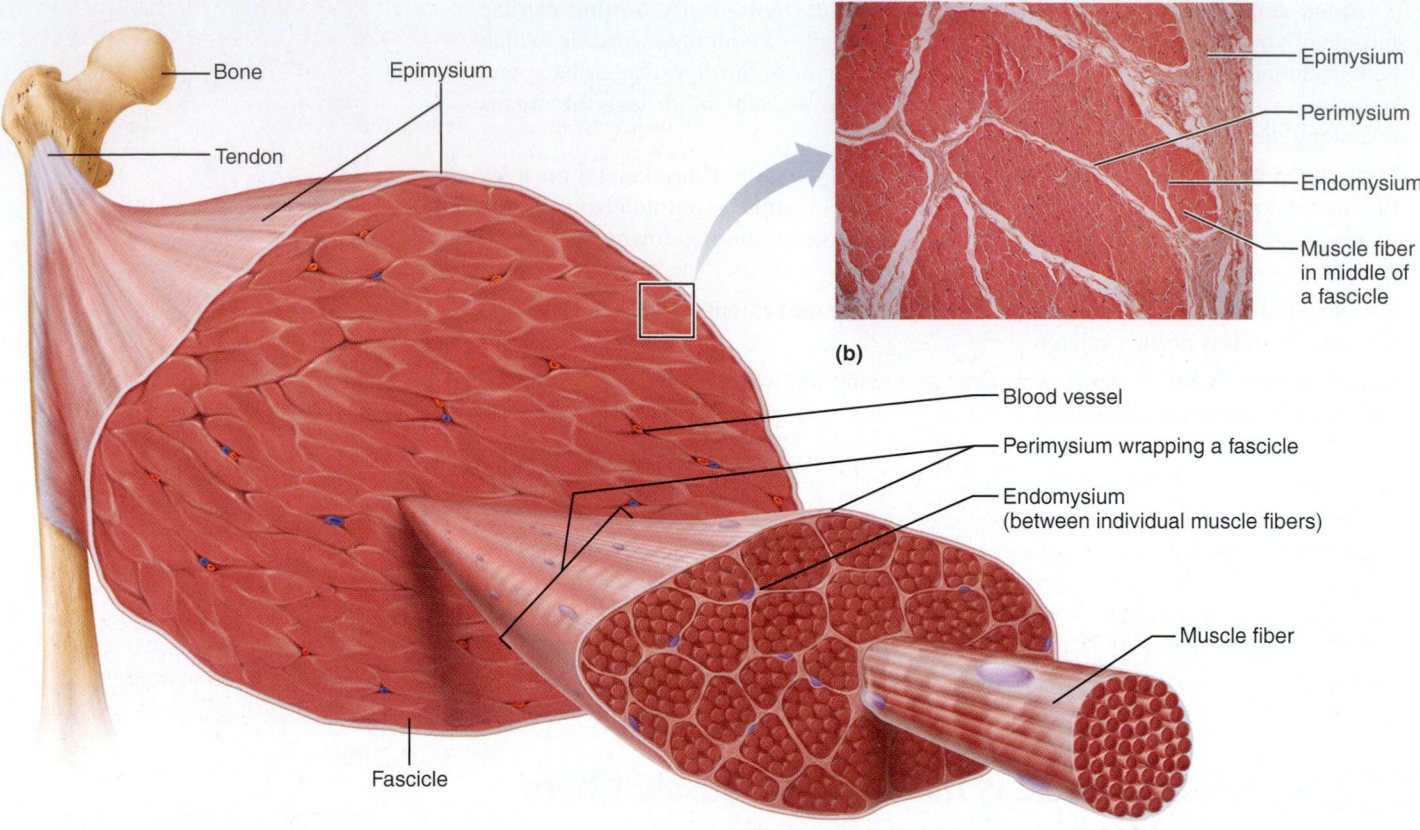

(a)

(b)

Figure 9.1 **Connective tissue sheaths of skeletal muscle: epimysium, perimysium, and endomysium. (b)** Photomicrograph of a cross section of part of a skeletal muscle (30×).

Let's consider these connective tissue sheaths from external to internal (see **Figure 9.1** and the top three rows of **Table 9.1**).

- **Epimysium.** The **epimysium** (ep″ĭ-mis′e-um; "outside the muscle") is an "overcoat" of dense irregular connective tissue that surrounds the whole muscle. Sometimes it blends with the deep fascia that lies between neighboring muscles or the superficial fascia deep to the skin (◄ p. 151).

- **Perimysium and fascicles.** Within each skeletal muscle, the muscle fibers are grouped into **fascicles** (fas′ĭ-klz; "bundles") that resemble bundles of sticks. Surrounding each fascicle is a layer of dense irregular connective tissue called **perimysium** (per″ĭ-mis′e-um; "around the muscle").

- **Endomysium.** The **endomysium** (en″do-mis′e-um; "within the muscle") is a wispy sheath of connective tissue that surrounds each individual muscle fiber. It consists of fine areolar connective tissue.

As shown in Figure 9.1, all of these connective tissue sheaths are continuous with one another as well as with the tendons that join muscles to bones. When muscle fibers contract, they pull on these sheaths, which transmit the pulling force to the bone to be moved. The sheaths contribute somewhat to the natural elasticity of muscle tissue, and also provide routes for the entry and exit of the blood vessels and nerve fibers that serve the muscle.

Attachments

Recall from Chapter 8 that most skeletal muscles span joints and attach to bones (or other structures) in at least two places. When a muscle contracts, the movable bone, the muscle's **insertion**, moves toward the immovable or less movable bone, the muscle's **origin**. In the muscles of the limbs, the origin typically lies proximal to the insertion.

Muscle attachments, whether origin or insertion, may be direct or indirect.

- In **direct**, or **fleshy**, **attachments**, the epimysium of the muscle is fused to the periosteum of a bone or perichondrium of a cartilage.

- In **indirect attachments**, the muscle's connective tissue wrappings extend beyond the muscle either as a ropelike **tendon** (Figure 9.1a) or as a sheetlike **aponeurosis** (ap″o-nu-ro′sis) (see Figure 10.12a on p. 347). The tendon or aponeurosis anchors the muscle to the connective tissue covering of a skeletal element (bone or cartilage) or to the fascia of other muscles.

Indirect attachments are much more common because of their durability and small size. Tendons are mostly tough collagen fibers, which can withstand the abrasion of rough bony projections that would tear apart the more delicate muscle tissues. Because of their relatively small size, more tendons than

Table 9.1   Structure and Organizational Levels of Skeletal Muscle

STRUCTURE AND ORGANIZATIONAL LEVEL	DESCRIPTION	CONNECTIVE TISSUE WRAPPINGS
Muscle (organ) Epimysium · Muscle · Tendon · Fascicle	A muscle consists of hundreds to thousands of muscle cells, plus connective tissue wrappings, blood vessels, and nerve fibers.	Covered externally by the epimysium
Fascicle (a portion of the muscle) Part of fascicle · Perimysium · Muscle fiber	A fascicle is a discrete bundle of muscle cells, segregated from the rest of the muscle by a connective tissue sheath.	Surrounded by perimysium
Muscle fiber (cell) Nucleus · Endomysium · Sarcolemma · Part of muscle fiber · Myofibril	A muscle fiber is an elongated multinucleate cell; it has a banded (striated) appearance.	Surrounded by endomysium
Myofibril (complex organelle composed of bundles of myofilaments) Sarcomere	Myofibrils are rodlike contractile elements that occupy most of the muscle cell volume. Composed of sarcomeres arranged end to end, they appear banded, and bands of adjacent myofibrils are aligned.	—
Sarcomere (a segment of a myofibril) Sarcomere · Thin (actin) filament · Thick (myosin) filament	A sarcomere is the contractile unit, composed of myofilaments made up of contractile proteins.	—
Myofilament, or filament (extended macromolecular structure) Thick filament · Head of myosin molecule · Thin filament · Actin molecules	Contractile myofilaments are of two types—thick and thin. Thick filaments contain bundled myosin molecules; thin filaments contain actin molecules (plus other proteins). The sliding of the thin filaments past the thick filaments produces muscle shortening. Elastic filaments (not shown here) provide elastic recoil when tension is released and help maintain myofilament organization.	—

fleshy muscles can pass over a joint—so tendons also conserve space.

Check Your Understanding

3. How does the term epimysium relate to the role and position of this connective tissue sheath?

4. MAKE CONNECTIONS What is the difference between a tendon and a ligament? (Hint: See Chapter 4, p. 133.)

For answers, see Answers Appendix.

9.3 Skeletal muscle fibers contain calcium-regulated molecular motors

Learning Outcomes

▶ **Describe the microscopic structure and functional roles of the myofibrils, sarcoplasmic reticulum, and T tubules of skeletal muscle fibers.**

▶ **Describe the sliding filament model of muscle contraction.**

Each skeletal muscle fiber is a long cylindrical cell with multiple oval nuclei just beneath its **sarcolemma** or plasma membrane (**Figure 9.2b**). Skeletal muscle fibers are huge cells. Their diameter typically ranges from 10 to 100 μm—up to ten times that of an average body cell—and their length is phenomenal, some up to 30 cm long. Their large size and multiple nuclei are not surprising once you learn that hundreds of embryonic cells fuse to produce each fiber.

Sarcoplasm, the cytoplasm of a muscle cell, is similar to the cytoplasm of other cells, but it contains unusually large amounts of **glycosomes** (granules of stored glycogen that provide glucose during muscle cell activity for ATP production) and **myoglobin**, a red pigment that stores oxygen. Myoglobin is similar to hemoglobin, the pigment that transports oxygen in blood.

In addition to the usual organelles, a muscle cell contains three specialized structures: myofibrils, sarcoplasmic reticulum, and T tubules. Let's look at these structures more closely because they play important roles in muscle contraction.

Myofibrils

A single muscle fiber contains hundreds to thousands of rodlike **myofibrils** that run parallel to its length (Figure 9.2b). The myofibrils, each 1–2 μm in diameter, are so densely packed in the fiber that mitochondria and other organelles appear to be squeezed between them. They account for about 80% of cellular volume.

Myofibrils are made up of a chain of sarcomeres linked end to end. *Sarcomeres* (shown in Figure 9.2c and decribed shortly) contain even smaller rodlike structures called *myofilaments*. Table 9.1 (bottom three rows; p. 283) summarizes these structures.

Striations

Striations, a repeating series of dark and light bands, are evident along the length of each myofibril. In an intact muscle fiber, the dark **A bands** and light **I bands** are nearly perfectly aligned, giving the cell its striated appearance.

As illustrated in Figure 9.2c:

- Each dark A band has a lighter region in its midsection called the **H zone** (*H* for *helle*; "bright").

- Each H zone is bisected vertically by a dark line called the **M line** (*M* for middle) formed by molecules of the protein myomesin.

- Each light I band also has a midline interruption, a darker area called the **Z disc** (or Z line).

Sarcomeres

The region of a myofibril between two successive Z discs is a **sarcomere** (sar′ko-měr; "muscle segment"). Averaging 2 μm long, a sarcomere is the smallest contractile unit of a muscle fiber—the *functional unit* of skeletal muscle. It contains an A band flanked by half an I band at each end. Within each myofibril, the sarcomeres align end to end like boxcars in a train.

Myofilaments

If we examine the banding pattern of a myofibril at the molecular level, we see that it arises from orderly arrangement of even smaller structures within the sarcomeres. These smaller structures, the **myofilaments** or *filaments*, are the muscle equivalents of the actin-containing microfilaments and myosin motor proteins described in Chapter 3 (◀ p. 88). As you will recall, the proteins actin and myosin play a role in motility and shape change in virtually every cell in the body. This property reaches its highest development in the contractile muscle fibers. There are two types of contractile myofilaments in a sarcomere:

- The central **thick filaments** containing myosin (red) extend the entire length of the A band (Figure 9.2c and d). They are connected in the middle of the sarcomere at the M line.

- The more lateral **thin filaments** containing actin (blue) extend across the I band and partway into the A band. The Z disc, a protein sheet, anchors the thin filaments. We describe the third type of myofilament, the *elastic filament*, in the next section.

A close look at myofibril arrangement and banding patterns reveals that:

- A hexagonal arrangement of six thin filaments surrounds each thick filament, and three thick filaments enclose each thin filament. This is shown in Figure 9.2e (far right), which shows a cross section of a sarcomere in an area where thick and thin filaments overlap.

- The H zone of the A band appears less dense because the thin filaments do not extend into this region.

- The M line in the center of the H zone is slightly darker because of the fine protein strands there that hold adjacent thick filaments together.

- The myofilaments are held in alignment at the Z discs and the M lines, and are anchored to the sarcolemma at the Z discs.

(a) Photomicrograph of **portions of two muscle fibers** (700×). Notice the striations (alternating dark and light bands).

Nuclei

Dark A band

Light I band

Fiber

(b) Diagram of **part of a muscle fiber** showing the myofibrils. One **myofibril** extends from the cut end of the fiber.

Sarcolemma

Mitochondrion

Myofibril

Dark A band   Light I band   Nucleus

(c) Small part of one **myofibril enlarged to show the myofilaments** responsible for the banding pattern. Each **sarcomere** extends from one Z disc to the next.

Thin (actin) filament

Z disc   H zone   Z disc

Thick (myosin) filament

I band   A band   I band   M line

Sarcomere

(d) Enlargement of one sarcomere (sectioned lengthwise).

Z disc   M line   Z disc

Thin (actin) filament

Elastic (titin) filaments

Thick (myosin) filament

(e) Cross sections of a sarcomere cut through in different locations.

Myosin filament

Actin filament

I band
thin filaments only

H zone
thick filaments only

M line
thick filaments linked by accessory proteins

Outer edge of A band
thick and thin filaments overlap

Figure 9.2   Microscopic anatomy of a skeletal muscle fiber.

**Longitudinal section of filaments within one
sarcomere of a myofibril**

Z disc   Z disc

In the center of the sarcomere, the thick filaments
lack myosin heads. Myosin heads are present only
in areas of myosin-actin overlap.

Thick filament	**Thin filament**
Each thick filament consists of many myosin molecules whose heads protrude at opposite ends of the filament.	A thin filament consists of two strands of actin subunits twisted into a helix plus two types of regulatory proteins (troponin and tropomyosin).
Portion of a thick filament	Portion of a thin filament

Myosin head

Tropomyosin   Troponin   Actin

Actin-binding sites

Heads   Tail

ATP-
binding
site

Flexible hinge region

Myosin molecule

Myosin-
binding sites

Actin subunits

Figure 9.3 **Composition of thick and thin filaments.**

Molecular Composition of Myofilaments

Muscle contraction depends on the myosin- and actin-containing myofilaments. As noted earlier, thick filaments are composed primarily of the protein **myosin**. Each myosin molecule consists of six polypeptide chains: two heavy (high-molecular-weight)

chains and four light chains. The heavy chains twist together to form myosin's rodlike tail, and each heavy chain ends in a globular *head* that is attached to the tail via a flexible hinge (**Figure 9.3**).

The globular heads, each associated with two light chains, are the "business end" of myosin. During contraction, they link the thick and thin filaments together, forming **cross bridges**, and swivel around their point of attachment, acting as motors to generate force. Myosin itself splits ATP (acts as an ATPase) and uses the released energy to drive movement.

Each thick filament contains about 300 myosin molecules bundled together, with their tails forming the central part of the thick filament and their heads facing outward at the end of each molecule (Figure 9.3). As a result, the central portion of a thick filament (in the H zone) is smooth, but its ends are studded with a staggered array of myosin heads.

The thin filaments are composed chiefly of the protein **actin** (blue in Figure 9.3). Actin has kidney-shaped polypeptide subunits, called *globular actin* or *G actin*. Each G actin has a *myosin-binding site* (or *active site*) to which the myosin heads attach during contraction. G actin subunits polymerize into long actin filaments called *filamentous*, or *F, actin*. Two intertwined actin filaments, resembling a twisted double strand of pearls, form the backbone of each thin filament (Figure 9.3).

Thin filaments also contain several regulatory proteins.

- Polypeptide strands of **tropomyosin** (tro″po-mi′o-sin), a rod-shaped protein, spiral about the actin core and help stiffen and stabilize it. Successive tropomyosin molecules are arranged end to end along the actin filaments, and in a relaxed muscle fiber, they block myosin-binding sites on actin so that myosin heads on the thick filaments cannot bind to the thin filaments.

- **Troponin** (tro′po-nin), the other major protein in thin filaments, is a globular protein with three polypeptide subunits (Figure 9.3). One subunit attaches troponin to actin. Another subunit binds tropomyosin and helps position it on actin. The third subunit binds calcium ions.

Both troponin and tropomyosin help control the myosin-actin interactions involved in contraction. Several other proteins help form the structure of the myofibril.

- The **elastic filament** we referred to earlier is composed of the giant protein **titin** (Figure 9.2d). Titin extends from the Z disc to the thick filament, and then runs within the thick filament (forming its core) to attach to the M line. It holds the thick filaments in place, maintaining the organization of the A band, and helps the muscle cell spring back into shape after stretching. (The part of the titin that spans the I bands is extensible, unfolding when the muscle stretches and recoiling when the tension is released.) Titin does not resist stretching in the ordinary range of extension, but it stiffens as it uncoils, helping the muscle resist excessive stretching, which might pull the sarcomeres apart.

- Another important structural protein is **dystrophin**, which links the thin filaments to the integral proteins of the sarcolemma (which in turn are anchored to the extracellular matrix).

- Other proteins that bind filaments or sarcomeres together and maintain their alignment include *nebulin*, *myomesin*, and *C proteins*. Intermediate (desmin) filaments extend from the Z disc and connect each myofibril to the next throughout the width of the muscle cell.

HOMEOSTATIC IMBALANCE 9.1 CLINICAL

The term **muscular dystrophy** refers to a group of inherited muscle-destroying diseases that generally appear during childhood. The affected muscles initially enlarge due to deposits of fat and connective tissue, but the muscle fibers atrophy and degenerate.

The most common and serious form is **Duchenne muscular dystrophy (DMD)**, which is inherited as a sex-linked recessive disease. It is expressed almost exclusively in males (one in every 3600 male births) and is diagnosed when the boy is between 2 and 7 years old (**Figure 9.4**). Active, normal-appearing boys become clumsy and fall frequently as their skeletal muscles weaken. The disease progresses relentlessly from the extremities upward, finally affecting the head and chest muscles, and cardiac muscle. The weakness continues to progress, but with supportive care, DMD patients are living into their 30s and beyond.

Figure 9.4 A boy with Duchenne muscular dystrophy (DMD). Physiotherapy can help maintain mobility.

DMD is caused by a defective gene for *dystrophin*, the cytoplasmic protein described above. Dystrophin links the cytoskeleton to the extracellular matrix and, like a girder, helps stabilize the sarcolemma. The fragile sarcolemma of DMD patients tears during contraction, allowing entry of excess Ca^{2+}, which damages the contractile fibers. Inflammatory cells (macrophages and lymphocytes) accumulate in the surrounding connective tissue. As the regenerative capacity of the muscle is lost, damaged cells undergo apoptosis, and muscle mass drops.

There is still no cure for DMD. Current treatments are aimed at preventing or reducing spine and joint deformities and helping those with DMD remain mobile as long as possible. Two newly-approved drugs may broaden the treatment options for certain patients.

9

Figure 9.5 **Relationship of the sarcoplasmic reticulum and T tubules to myofibrils of skeletal muscle.** The tubules of the SR (blue) fuse to form the saclike terminal cisterns next to the A-I junctions. The T tubules (gray) are inward invaginations of the sarcolemma that run deep into the cell between the terminal cisterns. (See detailed view in Focus Figure 9.2, pp. 294–295.) Sites of close contact of these three elements (terminal cistern, T tubule, and terminal cistern) are called triads.

Sarcoplasmic Reticulum and T Tubules

Skeletal muscle fibers contain two sets of intracellular tubules that help regulate muscle contraction: (1) the sarcoplasmic reticulum and (2) T tubules.

Sarcoplasmic Reticulum

Shown in blue in **Figure 9.5**, the **sarcoplasmic reticulum (SR)** is an elaborate smooth endoplasmic reticulum. The SR regulates intracellular levels of ionic calcium. It stores calcium and releases it on demand when the muscle fiber is stimulated to contract. As you will see, calcium provides the final "go" signal for contraction.

Interconnecting tubules of SR surround each myofibril the way the sleeve of a loosely knitted sweater surrounds your arm. Most SR tubules run longitudinally along the myofibril, communicating with each other at the H zone. Others called **terminal cisterns** ("end sacs") form larger, perpendicular cross channels at the A band–I band junctions, and they always occur in pairs. Closely associated with the SR are large numbers of mitochondria and glycogen granules, both involved in producing the energy used during contraction.

T Tubules

At each A band–I band junction, the sarcolemma of the muscle cell protrudes deep into the cell interior, forming an elongated tube called the **T tubule** (T for "transverse"), shown in gray in Figure 9.5. The *lumen* (cavity) of the T tubule is continuous with the extracellular space. As a result, T tubules tremendously increase the muscle fiber's surface area. This allows changes in the membrane potential to rapidly penetrate deep into the muscle fiber.

Along its length, each T tubule runs between the paired terminal cisterns of the SR, forming **triads** (Figure 9.5), successive groupings of the three membranous structures (terminal cistern, T tubule, and terminal cistern). As they pass from one myofibril to the next, the T tubules also encircle each sarcomere.

Muscle contraction is ultimately controlled by nerve-initiated electrical impulses that travel along the sarcolemma. Because T tubules are continuations of the sarcolemma, they conduct impulses to the deepest regions of the muscle cell and every sarcomere. These impulses trigger the release of calcium from the adjacent terminal cisterns. Think of the T tubules as a rapid communication or messaging system that ensures that every myofibril in the muscle fiber contracts at virtually the same time.

Triad Relationships

The roles of the T tubules and SR in providing signals for contraction are tightly linked. At the triads, membrane-spanning proteins from the T tubules and SR link together across the gap between the two membranes.

- The protruding integral proteins of the T tubule act as voltage sensors.

- The integral proteins of the SR form gated channels through which the terminal cisterns release Ca^{2+} (see top right of Focus Figure 9.2 on p. 295).

Sliding Filament Model of Contraction

We almost always think "shortening" when we hear the word **contraction**, but to physiologists contraction refers only to the activation of myosin's cross bridges, which are the force-generating sites. Shortening only occurs if the cross bridges generate enough tension on the thin filaments to exceed the forces that oppose shortening, such as when you lift a bowling ball. Contraction ends when the cross bridges become inactive, the tension declines, and the muscle fiber relaxes.

In a relaxed muscle fiber, the thin and thick filaments overlap only at the ends of the A band (**Figure 9.6** ①). The **sliding filament model of contraction** states that during contraction, the thin filaments slide past the thick ones so that the actin and myosin filaments overlap to a greater degree. Neither the thick nor the thin filaments change length during contraction. Here's how it works:

- When the nervous system stimulates muscle fibers, the myosin heads on the thick filaments latch onto myosin-binding sites on actin in the thin filaments, and the sliding begins.

- These cross bridge attachments form and break several times during a contraction, acting like tiny ratchets to generate tension and propel the thin filaments toward the center of the sarcomere.

- As this event occurs simultaneously in sarcomeres throughout the cell, the muscle cell shortens.

At the microscopic level, the following things occur as a muscle cell shortens:

- The I bands shorten.

- The distance between successive Z discs shortens. As the thin filaments slide centrally, the Z discs to which they attach are pulled *toward* the M line (Figure 9.6 ②).

- The H zones disappear.

- The contiguous A bands move closer together, but their length does not change.

Check Your Understanding

5. Name proteins a, b, c, and d. Which protein can act as an enzyme to hydrolyze (split) ATP? Which binds Ca^{2+}? Which must move out of the way in order for cross bridges to form?

6. Which region or organelle—cytosol, mitochondrion, or SR— contains the highest concentration of calcium ions in a resting muscle fiber? Which structure provides the ATP needed for muscle activity?

① **Fully relaxed sarcomere of a muscle fiber**

② **Fully contracted sarcomere of a muscle fiber**

Figure 9.6   Sliding filament model of contraction. At full contraction, the Z discs approach the thick filaments and the thin filaments overlap each other. The photomicrographs (top view in ① and ②) show enlargements of 33,000×.

7. DRAW Draw four thick filaments in each of two columns. Add a column of four thin filaments between the thick filaments, as they would appear in a relaxed myofibril. Then add two more columns, each with four thin filaments, on the left and right sides of the original drawing. Label an A band, an I band, and an H zone. Draw and label an M line and a Z disc.

8. MAKE CONNECTIONS Consider a phosphorus atom that is part of the membrane of the sarcoplasmic reticulum in the biceps muscle of your arm. Using the levels of structural organization (described in Chapter 1), name in order the structure that corresponds to each level of organization. Begin at the atomic level (the phosphorus atom) and end at the organ system level.

—————————— *For answers, see Answers Appendix.*

9.4 Motor neurons stimulate skeletal muscle fibers to contract

Learning Outcomes

▶ Explain how muscle fibers are stimulated to contract by describing events that occur at the neuromuscular junction.

▶ Follow the events of excitation-contraction coupling that lead to cross bridge activity.

▶ Describe the steps of a cross bridge cycle.

The sliding filament model tells us that myofilaments slide past each other as the sarcomeres contract. In this module, we will see exactly how this happens. But before we begin, let's fill in some background information that will help you understand this topic, and then look at the big picture of how muscle contraction works.

Background and Overview

Remember that skeletal muscle contractions are voluntary. For example, you *decide* when you want to contract your biceps muscle to pick up your cell phone. Making that decision involves many neurons in your brain, but the contraction of a skeletal muscle ultimately comes down to activating a few *motor neurons* in the spinal cord. Motor neurons are the way that the nervous system connects with skeletal muscles and "tells" them to contract.

Both neurons and muscles are *excitable cells*. That is, they respond to external stimuli by changing their resting membrane potential. (Remember that all cells have a resting membrane potential, which is a voltage across the plasma membrane; ◀ pp. 79–80.) These changes in membrane potential act as signals. One type of electrical signal is called an **action potential** (**AP**; sometimes called a *nerve impulse*). An AP is a large change in membrane potential that spreads rapidly over long distances within a cell. Generally, APs don't spread from cell to cell. For this reason, the signal has to be converted to a chemical signal—a chemical messenger called a *neurotransmitter* (◀ p. 81) that diffuses across the small gap between excitable cells to start the signal again. The neurotransmitter that motor neurons use to "tell" skeletal muscle to contract is **acetylcholine** (as″ĕ-til-ko′lēn), or **ACh**.

Ion Channels

Rapidly changing the membrane potential in neurons and muscle cells requires the opening and closing of membrane channel proteins that allow certain ions to pass across the membrane (◀ p. 70). The movement of ions through these ion channels changes the membrane voltage. Two classes of ion channels are important for excitation and contraction of skeletal muscle:

- **Chemically gated ion channels** are opened by chemical messengers (e.g., neurotransmitters). This class of ion channel creates small local

Chemical messenger (e.g., ACh)

Chemically gated ion channel

changes in the membrane potential (as we will see shortly). Receptors for acetylcholine are an example of this class. An **ACh receptor** is a single protein in the plasma membrane that is both a receptor and an ion channel.

- **Voltage-gated ion channels** open or close in response to changes in membrane potential. They underlie all action potentials. In skeletal muscle fibers, the initial change in membrane potential is created by chemically gated channels. In other words, chemically gated ion channels cause a small local *depolarization* (a decrease in the membrane potential) that then triggers the voltage-gated ion channels to create an action potential.

Voltage-gated ion channel

(We will describe ion channels in more detail in Chapter 11.)

Anatomy of Motor Neurons and the Neuromuscular Junction

Motor neurons that activate skeletal muscle fibers are called *somatic motor neurons*, or *motor neurons of the somatic (voluntary) nervous system* (**Figure 9.7**). These neurons reside in the spinal cord (except for those that supply the muscles of the head and neck). Each neuron has a long threadlike extension called an *axon* that extends from the *cell body* in the spinal cord to the muscle fiber it serves (◀ see Chapter 4, p. 140, to review the parts of a neuron). These axons exit the spinal cord and pass throughout the body bundled together as nerves.

The axon of each motor neuron branches profusely as it enters the muscle so that it can innervate multiple muscle fibers. When it reaches a muscle fiber, each axon divides again, giving off several short, curling branches that collectively form an oval **neuromuscular junction**, or **motor end plate**, with a single muscle fiber.

Each muscle fiber has only one neuromuscular junction, located approximately midway along its length. The end of the axon, called the **axon terminal**, and the muscle fiber are exceedingly close (50–80 nm apart), but they remain separated by a space, the **synaptic cleft** (Figure 9.7), which is filled with a gel-like extracellular substance rich in glycoproteins and collagen fibers.

Within the moundlike axon terminal are **synaptic vesicles**, small membranous sacs containing the neurotransmitter acetylcholine. The trough-like part of the muscle fiber's sarcolemma that helps form the neuromuscular junction is highly folded. These **junctional folds** provide a large surface area for the thousands of ACh receptors located there.

To summarize, the neuromuscular junction is like a sandwich: It is made up of part of a neuron (the axon terminals), part of a muscle cell (the junctional folds), and the "filler" between them (the synaptic cleft).

The Big Picture

Figure 9.7 presents an overview of skeletal muscle contraction and divides this process into four groups of steps. We will

Sequence of events leading to contraction:

A motor neuron fires an action potential (AP) down its axon.

① **Events at the neuromuscular junction** (see Focus Figure 9.1)

The motor neuron's axon terminal releases acetylcholine (ACh) into the synaptic cleft.

ACh binds receptors on the junctional folds of the sarcolemma.

ACh binding causes a local depolarization called an end plate potential (EPP).

② **Muscle fiber excitation** (see Figure 9.8)

The local depolarization (EPP) triggers an AP in the adjacent sarcolemma.

AP in sarcolemma travels down T tubules.

③ **Excitation-contraction coupling** (see Focus Figure 9.2)

Sarcoplasmic reticulum releases Ca^{2+}.

Ca^{2+} binds to troponin, which shifts tropomyosin to uncover the myosin-binding sites on actin. Myosin heads bind actin.

④ **Cross bridge cycle** (see Focus Figure 9.3)

Contraction occurs via cross bridge cycling.

Figure 9.7 **Overview of skeletal muscle contraction.**

Labels in figure:
- Brain
- Spinal cord
- **Motor neuron:**
  - Cell body
  - Axon
  - Axon terminals
- Axon of motor neuron
- The **neuromuscular junction** is the region where the motor neuron contacts the skeletal muscle. It consists of multiple **axon terminals** and the underlying junctional folds of the sarcolemma.
- Muscle fiber
- Axon of motor neuron
- Axon terminal of motor neuron
- Synaptic vesicle with ACh
- Synaptic cleft
- *Cytoplasm of skeletal muscle fiber*
- Junctional folds of the sarcolemma

explore each group of steps in more detail in the rest of this module:

① **Events at the neuromuscular junction** (see Focus Figure 9.1 on p. 292). The motor neuron releases ACh that stimulates the skeletal muscle fiber, causing a local depolarization (decrease in membrane potential) called an *end plate potential* (*EPP*).

② **Muscle fiber excitation.** The EPP triggers an action potential that travels across the entire sarcolemma (see Figure 9.8 on p. 293).

③ **Excitation-contraction coupling** (see Focus Figure 9.2 on pp. 294–295). The AP in the sarcolemma propagates along the T tubules and causes release of Ca^{2+} from the terminal cisterns of the SR. Ca^{2+} is the final trigger for contraction. It is the internal messenger that links the AP to contraction. Ca^{2+} binds to troponin and this causes the myosin-binding sites on actin to be exposed so that myosin heads can bind to actin.

④ **Cross bridge cycling** (see Focus Figure 9.3 on p. 297). The muscle contracts as a result of a repeating cycle of steps that cause myofilaments to slide relative to each other.

When a nerve impulse reaches a neuromuscular junction, acetylcholine (ACh) is released. Upon binding to sarcolemma receptors, ACh causes a change in sarcolemma permeability leading to a change in membrane potential.

Watch a 3-D animation of this process: MasteringA&P®
Study Area > Animations & Videos > A&P Flix

Action potential (AP)

Axon of motor neuron

Axon terminal of **neuromuscular junction**

Sarcolemma of the muscle fiber

① Action potential arrives at axon terminal of motor neuron.

② Voltage-gated Ca^{2+} channels open. Ca^{2+} enters the axon terminal, moving down its electrochemical gradient.

Ca^{2+}

Ca^{2+}

Synaptic vesicle containing ACh

Axon terminal of motor neuron

Synaptic cleft

③ Ca^{2+} entry causes ACh (a neurotransmitter) to be released by exocytosis.

Fusing synaptic vesicles

ACh

Junctional folds of sarcolemma

④ ACh diffuses across the synaptic cleft and binds to ACh receptors on the sarcolemma.

Sarcoplasm of muscle fiber

⑤ ACh binding opens chemically gated ion channels that allow simultaneous passage of Na^+ into the muscle fiber and K^+ out of the muscle fiber. More Na^+ ions enter than K^+ ions exit, which produces a local change in the membrane potential called the *end plate potential*.

Na^+ K^+

Postsynaptic membrane ion channel opens; ions pass.

⑥ ACh effects are terminated by its breakdown in the synaptic cleft by acetylcholinesterase and diffusion away from the junction.

ACh

Degraded ACh

Na^+

Ion channel closes; ions cannot pass.

Acetylcholinesterase

K^+

ACh-containing
synaptic vesicle

Axon terminal of
neuromuscular
junction

Ca²⁺

Ca²⁺

Synaptic
cleft

Wave of
depolarization

① **An end plate potential (EPP) is generated at the neuromuscular junction (see Focus Figure 9.1).** The EPP causes a wave of depolarization that spreads to the adjacent sarcolemma.

Figure 9.8 **Summary of events in the generation and propagation of an action potential in a skeletal muscle fiber.**

Events at the Neuromuscular Junction

How does a motor neuron stimulate a skeletal muscle fiber? *Focus on Events at the Neuromuscular Junction* (**Focus Figure 9.1**) covers this process step by step. Study this figure before continuing.

The result of the events at the neuromuscular junction is a transient change in membrane potential that causes the interior of the sarcolemma to become less negative (a depolarization). This local depolarization is called an **end plate potential** (EPP). The EPP spreads to the adjacent sarcolemma and triggers an AP there.

> Complete an interactive tutorial: **Mastering**A&P® > Study Area > Interactive Physiology

After ACh binds to the ACh receptors, its effects are quickly terminated by **acetylcholinesterase** (as″ĕ-til-ko″lin-es′ter-ās), an enzyme located in the synaptic cleft. Acetylcholinesterase breaks down ACh to its building blocks, acetic acid and choline. Removing ACh prevents continued muscle fiber contraction in the absence of additional nervous system stimulation.

HOMEOSTATIC IMBALANCE 9.2 — CLINICAL

Many toxins, drugs, and diseases interfere with events at the neuromuscular junction. For example, *myasthenia gravis* (*asthen* = weakness; *gravi* = heavy), a disease characterized by drooping upper eyelids, difficulty swallowing and talking, and generalized muscle weakness, involves a shortage of ACh receptors. Myasthenia gravis is an autoimmune disease in which the immune system destroys ACh receptors. ●●●●●●●●

Open voltage-gated Na⁺ channel   Closed voltage-gated K⁺ channel

Na⁺

K⁺

Action potential

② **Depolarization: Generating and propagating an action potential (AP).** Depolarization of the sarcolemma opens voltage-gated sodium channels. Na⁺ enters, following its electrochemical gradient. At a certain membrane voltage, an AP is generated (initiated). The AP spreads to adjacent areas of the sarcolemma and opens voltage-gated Na⁺ channels there, propagating the AP. The AP propagates along the sarcolemma in all directions, just like ripples from a pebble dropped in a pond.

Closed voltage-gated Na⁺ channel   Open voltage-gated K⁺ channel

Na⁺

K⁺

③ **Repolarization: Restoring the sarcolemma to its initial polarized state (negative inside, positive outside).** The repolarization wave is also a consequence of opening and closing ion channels—voltage-gated Na⁺ channels close and voltage-gated K⁺ channels open. The potassium ion concentration is substantially higher inside the cell than in the extracellular fluid, so K⁺ diffuses out of the muscle fiber. This restores the negatively charged conditions inside that are characteristic of a sarcolemma at rest.

Generation of an Action Potential across the Sarcolemma

Now let's consider the electrical events that trigger an action potential along the sarcolemma. An action potential is the result of a predictable sequence of electrical changes. Once initiated, an action potential sweeps along the entire surface of the sarcolemma. Three steps are involved in triggering and then propagating an action potential. These three steps—generation of an end plate potential followed by action potential depolarization and repolarization—are shown in **Figure 9.8**. A tracing of the

(Text continues on p. 296.)

Excitation-contraction (E-C) coupling is the sequence of events by which transmission of an action potential along the sarcolemma leads to the sliding of myofilaments.

Watch a 3-D animation of this process: MasteringA&P® > Study Area > Animations & Videos > A&P Flix

Setting the stage
The events at the neuromuscular junction (NMJ) set the stage for E-C coupling by providing excitation. Released acetylcholine binds to receptor proteins on the sarcolemma and triggers an action potential in a muscle fiber.

Synaptic cleft

Axon terminal of motor neuron at NMJ

ACh

Action potential is generated

Sarcolemma

T tubule

Terminal cistern of SR

Ca^{2+}

Muscle fiber

One sarcomere

One myofibril

Steps in E-C Coupling:

Sarcolemma

Voltage-sensitive tubule protein

T tubule

Ca^{2+} release channel

Terminal cistern of SR

Ca^{2+}

① The action potential (AP) propagates along the sarcolemma and down the T tubules.

② Calcium ions are released. Transmission of the AP along the T tubules of the triads causes the voltage-sensitive tubule proteins to change shape. This shape change opens the Ca^{2+} release channels in the terminal cisterns of the sarcoplasmic reticulum (SR), allowing Ca^{2+} to flow into the cytosol.

Actin

Troponin

Tropomyosin blocking myosin-binding sites

Myosin

Ca^{2+}

Myosin-binding sites exposed and ready for myosin binding

③ Calcium binds to troponin and removes the blocking action of tropomyosin. When Ca^{2+} binds, troponin changes shape, exposing myosin-binding sites on the thin filaments.

Myosin cross bridge

④ Contraction begins: Myosin binding to actin forms cross bridges and contraction (cross bridge cycling) begins. At this point, E-C coupling is over.

The aftermath

When the muscle AP ceases, the voltage-sensitive tubule proteins return to their original shape, closing the Ca^{2+} release channels of the SR. Ca^{2+} levels in the sarcoplasm fall as Ca^{2+} is continually pumped back into the SR by active transport. Without Ca^{2+}, the blocking action of tropomyosin is restored, myosin-actin interaction is inhibited, and relaxation occurs. Each time an AP arrives at the neuromuscular junction, the sequence of E-C coupling is repeated.

Figure 9.9 Recording of an action potential (AP) in a muscle fiber. An AP is a brief change in membrane potential.

9

resulting membrane potential changes is shown in **Figure 9.9**. We will describe action potentials in more detail in Chapter 11.

During repolarization, a muscle fiber is said to be in a **refractory period**, because the cell cannot be stimulated again until repolarization is complete. Note that repolarization restores only the *electrical conditions* of the resting (polarized) state. The ATP-dependent Na^+-K^+ pump restores the *ionic conditions* of the resting state, but thousands of action potentials can occur before ionic imbalances interfere with contractile activity.

Once initiated, the action potential is unstoppable. It ultimately results in contraction of the muscle fiber. Although the action potential itself lasts only a few milliseconds (ms), the contraction phase of a muscle fiber may persist for 100 ms or more and far outlasts the electrical event that triggers it.

Excitation-Contraction Coupling

Excitation-contraction (E-C) coupling is the sequence of events by which transmission of an action potential along the sarcolemma causes myofilaments to slide. The action potential is brief and ends well before any signs of contraction are obvious.

> ⟳ Complete an interactive tutorial: MasteringA&P® > Study Area>Interactive Physiology

As you will see, the electrical signal does not act directly on the myofilaments. Instead, it causes the rise in intracellular levels of calcium ions, which triggers a sequence of events that ultimately leads to sliding of the filaments.

Focus on Excitation-Contraction Coupling (**Focus Figure 9.2**) on pp. 294–295 illustrates the steps in this process. It also reveals how the integral proteins of the T tubules and terminal cisterns in the triads interact to provide the Ca^{2+} necessary for contraction to occur.

Muscle Fiber Contraction: Cross Bridge Cycling

As we have noted, cross bridge formation requires Ca^{2+}. Let's look more closely at how calcium ions promote muscle cell contraction.

When intracellular calcium levels are low, the muscle cell is relaxed because tropomyosin molecules physically block the myosin-binding sites on actin. As Ca^{2+} levels rise, the ions bind to regulatory sites on troponin. Two calcium ions must bind to a troponin, causing it to change shape and roll tropomyosin into the groove of the actin helix, away from the myosin-binding sites. In short, the tropomyosin blockade is removed when sufficient calcium is present. Once binding sites on actin are exposed, the events of the cross bridge cycle occur in rapid succession, as depicted in *Focus on the Cross Bridge Cycle* (**Focus Figure 9.3**). The cycle repeats and with each cycle, the myosin head

> ▶ Watch a 3-D animation of this process: MasteringA&P® > Study Area > Animations & Videos > A&P Flix

takes another "step" by attaching to an actin site further along the thin filament. The thin filaments continue to slide as long as calcium and adequate ATP are present.

Myosin walks along the adjacent thin filaments during muscle shortening like a centipede. The thin filaments cannot slide backward as the cycle repeats again and again because some myosin heads (the "legs") are always in contact with actin (the "ground"). Contracting muscles routinely shorten by 30–35% of their total resting length, so each myosin cross bridge attaches and detaches many times during a single contraction. It is likely that only half of the myosin heads of a thick filament are pulling at the same instant. The others are randomly seeking their next binding site.

As soon as Ca^{2+} is released from the SR, the Ca^{2+} pumps of the SR begin to reclaim it from the cytosol. As Ca^{2+} levels drop, Ca^{2+} comes off of troponin, which again changes shape and pulls tropomyosin up to block actin's myosin-binding sites. The contraction ends, and the muscle fiber relaxes. When cross bridge cycling ends, the myosin heads remain in their upright high-energy configuration (Focus Figure 9.3 ④), ready to bind actin when the muscle is stimulated to contract again.

Except for the brief period following muscle cell excitation, calcium ion concentrations in the cytosol are kept almost undetectably low. When nerve impulses arrive in quick succession, intracellular Ca^{2+} levels soar due to successive "puffs" or bursts of Ca^{2+} released from the SR. In such cases, the muscle cells do not completely relax between successive stimuli and contraction is stronger and more sustained (within limits) until nervous stimulation ceases.

HOMEOSTATIC IMBALANCE 9.3   CLINICAL

Rigor mortis (death rigor) illustrates the fact that cross bridge detachment is ATP driven. Most muscles begin to stiffen 3 to

The cross bridge cycle is the series of events during which myosin heads pull thin filaments toward the center of the sarcomere.

 Watch a 3-D animation of this process: MasteringA&P® > Study Area > Animations & Videos > A&P Flix

Actin | Ca^{2+} | Thin filament

Myosin cross bridge — ADP, P$_i$

Thick filament

Myosin

① Cross bridge formation. Energized myosin head attaches to an actin myofilament, forming a cross bridge.

ADP, P$_i$

② The power (working) stroke. ADP and P$_i$ are released and the myosin head pivots and bends, changing to its bent low-energy state. As a result it pulls the actin filament toward the M line.

In the absence of ATP, myosin heads will not detach, causing rigor mortis.

ATP

ADP, P$_i$

ATP hydrolysis

④ Cocking of the myosin head. As myosin hydrolyzes ATP to ADP and P$_i$, the myosin head returns to its prestroke high-energy, or "cocked," position.*

ATP

③ Cross bridge detachment. After ATP attaches to myosin, the link between myosin and actin weakens, and the myosin head detaches (the cross bridge "breaks").

*This cycle will continue as long as ATP is available and Ca^{2+} is bound to troponin. If ATP is not available, the cycle stops between steps ② and ③.

4 hours after death. Peak rigidity occurs at 12 hours and then gradually dissipates over the next 48 to 60 hours. Dying cells are unable to exclude calcium (which is in higher concentration in the extracellular fluid), and the calcium influx into muscle cells promotes formation of myosin cross bridges. Shortly after breathing stops, ATP synthesis ceases, but ATP continues to be consumed and cross bridge detachment is impossible once all of the ATP is gone. Actin and myosin become irreversibly cross-linked, producing the stiffness of rigor mortis, which gradually disappears as muscle proteins break down after death.

Check Your Understanding

9. From the time an action potential reaches the axon terminal of a motor neuron until cross bridge cycling begins, several sets of ion channels are activated. List these channels in the order that they are activated and state what causes each to open.

10. What is the final trigger for contraction? What is the initial trigger?

11. What prevents the filaments from sliding back to their original position each time a myosin cross bridge detaches from actin?

12. **WHAT IF?** What would happen if a muscle fiber suddenly ran out of ATP when sarcomeres had only partially contracted?

For answers, see Answers Appendix.

9.5 Temporal summation and motor unit recruitment allow smooth, graded skeletal muscle contractions

Learning Outcomes

▶ Define motor unit and muscle twitch, and describe the events occurring during the three phases of a muscle twitch.

▶ Explain how smooth, graded contractions of a skeletal muscle are produced.

▶ Differentiate between isometric and isotonic contractions.

In its relaxed state, a muscle is soft and unimpressive, not what you would expect of a prime mover of the body. However, within a few milliseconds, it can contract to become a hard elastic structure with dynamic characteristics that intrigue not only biologists but engineers and physicists as well.

Before we consider muscle contraction on the organ level, let's note two facts about muscle mechanics.

● The principles governing contraction of a single muscle fiber and of a skeletal muscle consisting of a large number of fibers are pretty much the same.

(a) **Axons of motor neurons extend from the spinal cord to the muscle.** At the muscle, each axon divides into a number of axon terminals that form neuromuscular junctions with muscle fibers scattered throughout the muscle.

(b) **Branching axon terminals form neuromuscular junctions, one per muscle fiber** (photomicrograph 330×).

Figure 9.10 A motor unit consists of one motor neuron and all the muscle fibers it innervates.

Practice Histology questions:
MasteringA&P® > Study Area > Lab Tools > PAL

- The force exerted by a contracting muscle on an object is called **muscle tension**. The opposing force exerted on the muscle by the weight of the object to be moved is called the **load**.

A skeletal muscle contracts with varying force and for different periods of time in response to our need at the time. To understand how this occurs, we must look at the nerve-muscle functional unit called a *motor unit*.

The Motor Unit

Each muscle is served by at least one *motor nerve*, and each motor nerve contains axons (fibrous extensions) of up to hundreds of motor neurons. As an axon enters a muscle, it branches into a number of endings, each of which forms a neuromuscular junction with a single muscle fiber. A **motor unit** consists of one motor neuron and all the muscle fibers it innervates, or supplies (**Figure 9.10**). When a motor neuron fires (transmits an action potential), all the muscle fibers it innervates contract.

The number of muscle fibers per motor unit may be as high as several hundred or as few as four. Muscles that exert fine control (such as those controlling the fingers and eyes) have small motor units. By contrast, large, weight-bearing muscles, whose movements are less precise (such as the hip muscles), have large motor units. The muscle fibers in a single motor unit are not clustered together but are spread throughout the muscle. As a result, stimulation of a single motor unit causes a weak but uniform contraction of the muscle.

The Muscle Twitch

Muscle contraction is investigated in the laboratory using an isolated muscle. The muscle is attached to an apparatus that produces a **myogram**, a recording of contractile activity consisting of one or more recorded lines called *tracings*. Remember that muscles can contract without shortening (an *isometric contraction*, which we will discuss further on pp. 302–303). The myograms (graphs) in Figures 9.11–9.14 show the amount of *tension* a muscle develops when its length is held constant as it contracts.

In the laboratory, a **muscle twitch** is the response of a muscle to a single stimulation. The muscle fibers contract quickly and then relax. Every twitch myogram has three distinct phases (**Figure 9.11a**).

1. **Latent period.** The **latent period** is the first few milliseconds (ms) following stimulation when excitation-contraction coupling is occurring. During this period, cross bridges begin to cycle but muscle tension is not yet measurable so the myogram does not show a response.
2. **Period of contraction.** During the period of contraction, cross bridges are active, from the onset to the peak of tension development, and the myogram tracing rises to a peak. This period lasts 10–100 ms.
3. **Period of relaxation.** This final phase, lasting 10–100 ms, is due to pumping of Ca^{2+} back into the SR. Because the number of active cross bridges is declining, contractile force is declining. Muscle tension decreases to zero and

(a) **Myogram showing the three phases of an isometric twitch**

(b) **Comparison of the relative duration of twitch responses of three muscles**

Figure 9.11 **The muscle twitch.**

the tracing returns to the baseline. Notice that a muscle contracts faster than it relaxes, as shown by the asymmetric shape of the curve in Figure 9.11a.

As you can see in Figure 9.11b, twitch contractions of some muscles are rapid and brief, as with the extraocular muscles controlling eye movements. In contrast, the fibers of fleshy calf muscles (gastrocnemius and soleus) contract more slowly and remain contracted for much longer periods. These differences between muscles reflect variations in enzymes and metabolic properties of the myofibrils.

Graded Muscle Contractions

Muscle twitches—like those single, jerky contractions provoked in a laboratory—may result from certain neuromuscular problems, but this is *not* the way our muscles normally operate. Instead, healthy muscle contractions are relatively smooth and vary in strength as different demands are placed on them. These variations, needed for proper control of skeletal movement, are referred to as **graded muscle contractions**.

In general, muscle contraction can be graded in two ways:

- An increase in the *frequency* of stimulation causes *temporal summation*. The higher the frequency, the greater the strength of contraction of a given motor unit.

- An increase in the *strength* of stimulation causes *recruitment*. The stronger the stimulation, the more motor units are activated, and the stronger the contraction.

In the laboratory, it is easy to adjust the settings used to artificially stimulate a muscle. In the body, the brain determines the strength of a muscle's contraction by changing (1) the rate of firing of action potentials along the axon of its motor neuron (frequency) and (2) the number of its motor neurons that are activated (strength). In this way, we automatically and continuously adjust the strength of our muscle contractions. We are only aware of this process when it doesn't work right. For example, if you lift what you expect to be a heavy box that is in fact empty, you will use more muscle power than you should and the box will go flying!

Muscle Response to Changes in Stimulus Frequency

The nervous system can achieve greater muscular force by increasing the firing rate of motor neurons. For example, if two identical stimuli (electrical shocks or nerve impulses) are delivered to a muscle in rapid succession, the second twitch will be stronger than the first. On a myogram the second twitch will appear to ride on the shoulders of the first (**Figure 9.12a, b**).

This phenomenon is called **temporal**, or **wave**, **summation**. It occurs because the second contraction begins before the muscle has completely relaxed. The second contraction is greater than the first because the muscle is already partially contracted and because even more calcium is squirted into the cytosol. In other words, the contractions are added together. (However, the refractory period is always honored. So if a second stimulus arrives before repolarization is complete, no wave summation occurs.)

If the muscle is stimulated at an increasingly faster rate:

- The relaxation time between twitches becomes shorter and shorter.

- The concentration of Ca^{2+} in the cytosol rises higher and higher.

- The degree of wave summation becomes greater and greater, progressing to a sustained but quivering contraction referred

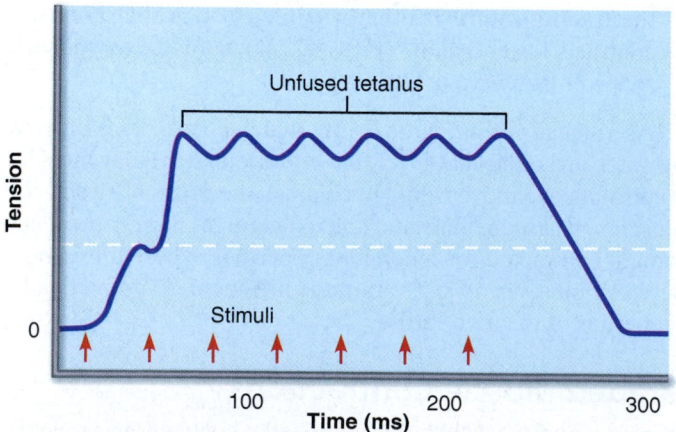

(a) Individual twitches. A second stimulus delivered after relaxation is complete does not produce summation.

(b) Temporal summation. Additional stimuli delivered before relaxation is complete produce temporal (wave) summation.

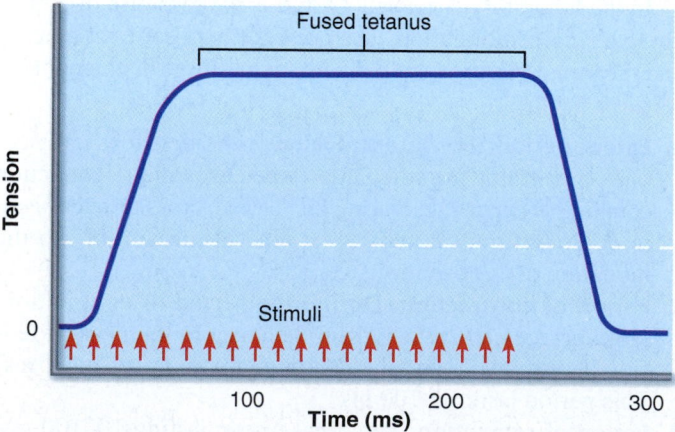

(c) Unfused tetanus. Higher stimulation frequency results in unfused tetanus.

(d) Fused tetanus. At even higher stimulus frequencies, there is no relaxation at all between stimuli. This is fused (complete) tetanus.

Figure 9.12 Temporal summation. A muscle's response to changes in stimulation frequency.

Figure 9.13 Recruitment. Relationship between stimulus intensity (graph at top) and muscle tension (myogram below). Below threshold voltage, the myogram shows no muscle response (stimuli 1 and 2). Once threshold (3) is reached, increases in voltage excite (recruit) more and more motor units until the maximal stimulus is reached (7). Further increases in stimulus voltage produce no further increase in contractile strength.

laboratory, recruitment is achieved by delivering stimuli of increasing voltage, calling more and more muscle fibers into play.

- Stimuli that produce no observable contractions are **sub-threshold stimuli**.

- The stimulus at which the first observable contraction occurs is called the **threshold stimulus** (**Figure 9.13**). Beyond this point, the muscle contracts more vigorously as the stimulus strength increases.

- The **maximal stimulus** is the strongest stimulus that increases contractile force. It represents the point at which all the muscle's motor units are recruited. In the laboratory, increasing the stimulus intensity beyond the maximal stimulus does not produce a stronger contraction.

The recruitment process is not random. Instead it is dictated by the *size principle* (**Figure 9.14**). In any muscle:

- The motor units with the smallest muscle fibers are activated first because they are controlled by the smallest, most highly excitable motor neurons.

- As motor units with larger and larger muscle fibers begin to be excited, contractile strength increases.

- The largest motor units, containing large, coarse muscle fibers, are controlled by the largest, least excitable (highest-threshold) neurons and are activated only when the most powerful contraction is necessary.

Why is the size principle important? It allows the increases in force during weak contractions (for example, those that maintain posture or slow movements) to occur in small steps, whereas gradations in muscle force are progressively greater when large amounts of force are needed for vigorous activities such as jumping or running. The size principle explains how the same hand that lightly pats your cheek can deliver a stinging slap at the volleyball during a match.

to as **unfused** or **incomplete tetanus*** (tet'ah-nus; *tetan* = rigid, tense) (Figure 9.12c).

- Finally, as the stimulation frequency continues to increase, muscle tension increases until it reaches maximal tension. At this point all evidence of muscle relaxation disappears and the contractions fuse into a smooth, sustained contraction plateau called **fused** or **complete tetanus** (Figure 9.12d). In the real world, physiological mechanisms prevent fused tetanus, so it rarely if ever occurs.

Temporal summation contributes to contractile force, but its primary function is to produce smooth, continuous muscle contractions by rapidly stimulating a specific number of muscle cells.

Muscle Response to Changes in Stimulus Strength

Recruitment, also called **multiple motor unit summation**, controls the force of contraction more precisely. In the

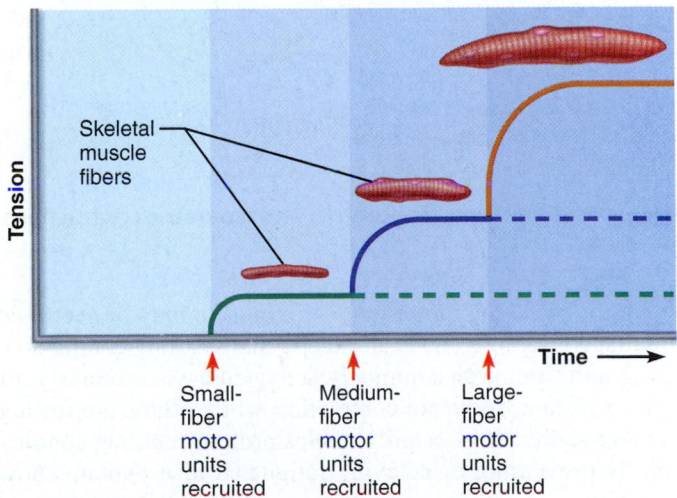

Figure 9.14 The size principle of recruitment. Motor units with the smallest fibers are recruited first, allowing fine control of contraction.

* The term *tetanus* also describes a bacterial disease (see Related Clinical Terms at the end of the chapter).

(a) Isotonic contraction (concentric)

On stimulation, muscle develops enough tension (force) to lift the load (weight). Once the resistance is overcome, the muscle shortens, and the tension remains constant for the rest of the contraction.

Tendon

Muscle contracts (isotonic contraction)

3 kg

Tendon

3 kg

Tension developed (kg)

Amount of resistance

Muscle relaxes

Peak tension developed

Muscle stimulus

Muscle length (percent of resting length)

Resting length

Time (ms)

(b) Isometric contraction

Muscle is attached to a weight that exceeds the muscle's peak tension-developing capabilities. When stimulated, the tension increases to the muscle's peak tension-developing capability, but the muscle does not shorten.

Muscle contracts (isometric contraction)

6 kg

6 kg

Tension developed (kg)

Amount of resistance

Muscle relaxes

Peak tension developed

Muscle stimulus

Muscle length (percent of resting length)

Resting length

Time (ms)

Figure 9.15 Isotonic (concentric) and isometric contractions.

Although *all* the motor units of a muscle may be recruited simultaneously to produce an exceptionally strong contraction, motor units are more commonly activated asynchronously. At a given instant, some are contracting while others are resting and recovering. This technique helps prolong a strong contraction by preventing or delaying fatigue. It also explains how weak contractions promoted by infrequent stimuli can remain smooth.

Muscle Tone

Skeletal muscles are described as voluntary, but even relaxed muscles are almost always slightly contracted, a phenomenon called **muscle tone**. Muscle tone is due to spinal reflexes that activate first one group of motor units and then another in response to activated stretch receptors in the muscles. Muscle tone does not produce active movements, but it keeps the muscles firm, healthy, and ready to respond to stimulation. Skeletal muscle tone also helps stabilize joints and maintain posture.

Isotonic and Isometric Contractions

There are two main categories of contractions—*isotonic* and *isometric*, depending on whether a muscle changes length or not. Muscles change length in isotonic, but not isometric, contractions. When studying isotonic contractions in a laboratory, the *amount of muscle shortening* is measured as shown in the bottom part of **Figure 9.15a**. More commonly, isometric contractions are studied, and what is measured is *increasing muscle tension* (Figure 9.15b, top graph).

Isotonic Contractions

If the muscle tension developed overcomes the load and muscle shortening occurs, the contraction is an **isotonic contraction** (*iso* = same; *ton* = tension), as when you lift a 5-lb bag of sugar. Once sufficient tension has developed to move the load, the tension remains relatively constant through the rest of the contractile period (Figure 9.15a).

Isotonic contractions come in two "flavors"—*concentric* and *eccentric*. **Concentric contractions** are those in which the muscle shortens and does work, such as picking up a book or kicking a ball. Concentric contractions are probably more familiar, but **eccentric contractions**, in which the muscle generates force as it lengthens, are equally important for coordination and purposeful movements.

Eccentric contractions occur in your anterior thigh muscles, for example, as you walk down a steep hill. Eccentric contractions are about 50% more forceful than concentric ones at the same load and more often cause delayed-onset muscle soreness. (Consider how your thigh muscles *feel* the day after hiking that hill.) The muscle stretching that occurs during eccentric contractions causes microtrauma in the muscles that results in soreness.

Biceps curls provide a simple example of how concentric and eccentric contractions work together in our everyday activities. When you flex your elbow to draw a weight toward your shoulder, the biceps muscle in your arm is contracting concentrically. When you straighten your arm to return the weight to the bench, the isotonic contraction of your biceps is eccentric. Basically, eccentric contractions put the body in position to contract concentrically. All jumping and throwing activities involve both types of contraction.

Isometric Contractions

A contracting muscle does not always shorten and move a load. If muscle tension develops but the load is not moved, the contraction is an **isometric contraction** (*metric* = measure). In isometric contractions, tension may build to the muscle's peak tension-producing capacity, but the muscle *neither shortens nor lengthens* (Figure 9.15b) because the load is greater than the force (tension) the muscle is able to develop—think of trying to lift a piano single-handedly. Muscles contract isometrically when they act primarily to maintain upright posture or to hold joints stationary while movements occur at other joints.

Electrochemical and mechanical events occurring within a muscle are identical in both isotonic and isometric contractions. However, the results are different. In isotonic contractions, the thin filaments slide. In isometric contractions, the cross bridges generate force but do *not* move the thin filaments, so there is no change in the banding pattern from that of the resting state. (You could say that they are "spinning their wheels" on the same actin binding sites.)

Check Your Understanding

13. What is a motor unit?

14. Identify each phase of the muscle twitch labeled a–c in the figure below. What is happening in the muscle during phase a?

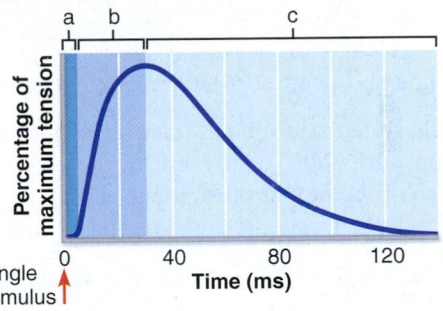

15. **APPLY** Jacob is competing in a chin-up competition. What type of muscle contractions are occurring in his biceps muscles immediately after he grabs the bar? As his body begins to move upward toward the bar? When his body begins to approach the mat?

For answers, see Answers Appendix.

9.6 ATP for muscle contraction is produced aerobically or anaerobically

Learning Outcomes

▶ **Describe three ways in which ATP is regenerated during skeletal muscle contraction.**

▶ **Define EPOC and muscle fatigue. List possible causes of muscle fatigue.**

Providing Energy for Contraction

As a muscle contracts, ATP supplies the energy to move and detach cross bridges, operate the calcium pump in the SR, and operate the Na^+-K^+ pump in the plasma membrane. Surprisingly, muscles store very limited reserves of ATP—4 to 6 seconds' worth at most, just enough to get you going. Because ATP is the *only* energy source used directly for contractile activities, it must be regenerated as fast as it is broken down if contraction is to continue.

Fortunately, after ATP is hydrolyzed to ADP and inorganic phosphate in muscle fibers, it is regenerated within a fraction of a second by one or more of the three pathways summarized in **Figure 9.16**: (a) direct phosphorylation of ADP by creatine phosphate, (b) anaerobic glycolysis, which converts glucose to lactic acid, and (c) aerobic respiration. All body cells use glycolysis and aerobic respiration to produce ATP, so we touch on them here but describe them in detail later, in Chapter 24.

Direct Phosphorylation of ADP by Creatine Phosphate (Figure 9.16a)

As we begin to exercise vigorously, the demand for ATP soars and the ATP stored in working muscles is consumed within a few twitches. Then **creatine phosphate (CP)** (kre′ah-tin), a unique high-energy molecule stored in muscles, is tapped to regenerate ATP while other metabolic pathways adjust to the sudden high demand for ATP.

Coupling CP with ADP transfers energy and a phosphate group from CP to ADP to form ATP almost instantly:

$$\text{Creatine phosphate} + \text{ADP} \xrightarrow{\text{creatine kinase}} \text{creatine} + \text{ATP}$$

Muscle cells store two to three times more CP than ATP. The CP reaction with ADP, catalyzed by the enzyme **creatine kinase**, is so efficient that the amount of ATP in muscle cells changes very little during the initial period of contraction.

Together, stored ATP and CP provide for maximum muscle power for about 15 seconds—long enough to energize a 100-meter dash (slightly longer if the activity is less vigorous). The coupled reaction is readily reversible, and to keep CP available, CP reserves are replenished during periods of rest or inactivity.

Anaerobic Pathway: Glycolysis and Lactic Acid Formation (Figure 9.16b)

As stored ATP and CP are exhausted, more ATP is generated by breaking down (catabolizing) glucose obtained from the blood or glycogen stored in the muscle. The initial phase of glucose breakdown is **glycolysis** (gli-kol′ĭ-sis; "sugar splitting"). This pathway occurs in both the presence and the absence of oxygen, but because it does not use oxygen, it is an anaerobic (an-a′er-ōb-ik; "without oxygen") pathway. During glycolysis, glucose is broken down to two *pyruvic acid* molecules, releasing enough energy to form small amounts of ATP (2 ATP per glucose).

When sufficient oxygen is present, the pyruvic acid produced during glycolysis enters the mitochondria, producing still more ATP in the oxygen-using pathway called aerobic respiration, described shortly. When blood flow and oxygen delivery are impaired during vigorous muscle contraction, most of the pyruvic acid is converted into **lactic acid**, and the overall process is referred to as **anaerobic glycolysis**. Oxygen delivery is impaired because bulging muscles compress the blood vessels within them. This happens when contractile activity reaches

(a) **Direct phosphorylation**	(b) **Anaerobic pathway**	(c) **Aerobic pathway**
Coupled reaction of creatine phosphate (CP) and ADP	Glycolysis and lactic acid formation	Aerobic respiration
Energy source: CP	**Energy source:** glucose	**Energy source:** glucose; pyruvic acid; free fatty acids from adipose tissue; amino acids from protein catabolism
Oxygen use: None **Products:** 1 ATP per CP, creatine **Duration of energy provided:** 15 seconds	**Oxygen use:** None **Products:** 2 ATP per glucose, lactic acid **Duration of energy provided:** 30–40 seconds, or slightly more	**Oxygen use:** Required **Products:** 32 ATP per glucose, CO_2, H_2O **Duration of energy provided:** Hours

Figure 9.16 Pathways for regenerating ATP during muscle activity. The fastest pathway is direct phosphorylation **(a)**, and the slowest is aerobic respiration **(c)**.

about 70% of the maximum possible (for example, when you run 600 meters with maximal effort).

Lactic acid is the end product of glucose metabolism during anaerobic glycolysis. Most of the lactic acid diffuses out of the muscles into the bloodstream. Subsequently, the liver, heart, or kidney cells pick up the lactic acid and use it as an energy source. Additionally, liver cells can reconvert it to pyruvic acid or glucose and release it back into the bloodstream for muscle use or convert it to glycogen for storage.

The anaerobic pathway is inefficient but fast. It harvests only about 5% as much ATP from each glucose molecule as the aerobic pathway, but it produces ATP about 2½ times faster. For this reason, even when large amounts of ATP are needed for moderate periods (30–40 seconds) of strenuous muscle activity, glycolysis can provide most of this ATP. Together, stored ATP and CP and the glycolysis–lactic acid pathway can support strenuous muscle activity for nearly a minute.

Although anaerobic glycolysis readily fuels spurts of vigorous exercise, it has shortcomings. Huge amounts of glucose are used to produce relatively small harvests of ATP, and the accumulating lactic acid is partially responsible for muscle soreness during intense exercise.

Aerobic Respiration (Figure 9.16c)

During rest and light to moderate exercise, even if prolonged, 95% of the ATP used for muscle activity comes from **aerobic respiration**. Aerobic respiration requires oxygen and mitochondria, and involves a sequence of chemical reactions that break the bonds of fuel molecules and release energy to make ATP.

Aerobic respiration begins with glycolysis and is followed by reactions that take place in the mitochondria. It breaks down glucose entirely to water and carbon dioxide, and generates large amounts of ATP.

$$\text{Glucose} + \text{oxygen} \rightarrow \text{carbon dioxide} + \text{water} + \text{ATP}$$

The carbon dioxide released diffuses out of the muscle tissue into the blood, to be removed from the body by the lungs.

As exercise begins, muscle glycogen provides most of the fuel. Shortly thereafter, bloodborne glucose, pyruvic acid from glycolysis, and free fatty acids are the major sources of fuels. After about 30 minutes, fatty acids become the major energy fuels. Aerobic respiration provides a high yield of ATP (about 32 ATP per glucose), but it is slow because of its many steps and it requires continuous delivery of oxygen and nutrient fuels to keep it going.

Energy Systems Used during Exercise

Which pathways predominate during exercise? As long as a muscle cell has enough oxygen, it will form ATP by the aerobic pathway. When ATP demands are within the capacity of the aerobic pathway, light to moderate muscular activity can continue for several hours in well-conditioned individuals (**Figure 9.17**). However, when exercise demands begin to exceed the ability of the muscle cells to carry out the necessary reactions quickly enough, anaerobic pathways begin to contribute more and more of the total ATP generated. The length of time a muscle can continue to contract using aerobic pathways is called **aerobic endurance**, and the point at which muscle metabolism converts to anaerobic glycolysis is called **anaerobic threshold**.

Figure 9.17 Comparison of energy sources used during short-duration exercise and prolonged-duration exercise.

Activities that require a surge of power but last only a few seconds, such as weight lifting, diving, and sprinting, rely entirely on ATP and CP stores. The slightly longer bursts of activity in tennis, soccer, and a 100-meter swim appear to be fueled almost entirely by anaerobic glycolysis (Figure 9.17). Prolonged activities such as marathon runs and bicycle touring, where endurance rather than power is the goal, depend mainly on aerobic respiration using both glucose and fatty acids as fuels. Levels of CP and ATP don't change much during prolonged exercise because ATP is generated at the same rate as it is used—a "pay as you go" system. Compared to anaerobic energy production, aerobic generation of ATP is relatively slow, but the ATP harvest is enormous.

Muscle Fatigue

Vigorous muscle activity cannot continue indefinitely. **Muscle fatigue** is a state of *physiological inability to contract* even though the muscle is still receiving stimuli. You might think that running out of ATP is the critical event that causes muscle fatigue. In fact, ATP levels inside muscle cells do drop, but muscle fatigue serves to prevent *complete* depletion of ATP in muscle, which would result in death of muscle cells and rigor mortis. (Not good!) The mechanism of muscle fatigue is complex. Although it is not fully understood, it involves alterations in excitation-contraction coupling. The following chemical changes may be involved:

- *Ionic imbalances.* Several ionic imbalances contribute to muscle fatigue. As action potentials are transmitted, potassium is lost from the muscle cells to the fluids of the T tubules and Na^+ is gained. These ionic changes disturb the membrane potential of the muscle cells. They also reduce the size of the action potential, which reduces the movement of the voltage-sensitive proteins in the T tubules and so reduces the amount of Ca^{2+} released from the SR.

- *Increased inorganic phosphate (P_i).* P_i from CP and ATP breakdown may interfere with calcium release from the SR. It may also interfere with the release of P_i from myosin and thus hamper myosin's power strokes.

- *Decreased ATP and increased magnesium (Mg^{2+}).* ATP normally binds Mg^{2+} in the cell, so as ATP levels drop, Mg^{2+} levels rise. Both low ATP and high Mg^{2+} act on the voltage-sensitive proteins in the T tubule to decrease Ca^{2+} release from the SR.

- *Decreased glycogen.* A decrease in glycogen is highly correlated with muscle fatigue.

Lactic acid has long been assumed to be a major cause of fatigue, and excessive intracellular accumulation of lactic acid raises the concentration of H^+ and alters contractile proteins. Although lactic acid and pH both contribute to the sensation of pain during intense exercise, neither seem to be directly involved in muscle fatigue.

In general, intense exercise of short duration produces fatigue rapidly, but recovery is also rapid. In contrast, the slow-developing fatigue of prolonged low-intensity exercise may require hours to days for complete recovery.

Excess Postexercise Oxygen Consumption (EPOC)

Whether or not fatigue occurs, vigorous exercise alters a muscle's chemistry dramatically. For a muscle to return to its pre-exercise state, the following must occur:

- Its oxygen reserves (stored in myoglobin) must be replenished.

- The accumulated lactic acid must be reconverted to pyruvic acid.

- Glycogen stores must be replaced.

- ATP and creatine phosphate reserves must be resynthesized.

The use of these muscle stores during anaerobic exercise simply defers when the oxygen is consumed, because replacing them requires oxygen uptake and aerobic metabolism after exercise ends. Additionally, the liver must convert any lactic acid persisting in blood to glucose or glycogen. Once exercise stops, the repayment process begins.

The extra amount of oxygen that the body must take in for these restorative processes is called the **excess postexercise oxygen consumption (EPOC)**, formerly called the oxygen debt. EPOC represents the difference between the amount of oxygen needed for totally aerobic muscle activity and the amount actually used. All anaerobic sources of ATP used during muscle activity contribute to EPOC.

Check Your Understanding

16. Chris joined the cross-country team partway through the season, and has just completed another grueling session of trying to keep up with her teammates. After the run, she is breathing heavily, her legs feel weak, and she is sweating profusely. Why is Chris breathing heavily? Which ATP-generating pathway have her working muscles been using that makes her breathless? What metabolic products might account for her muscle weakness?

For answers, see Answers Appendix.

9.7 The force, velocity, and duration of skeletal muscle contractions are determined by a variety of factors

Learning Outcomes

▶ Describe factors that influence the force, velocity, and duration of skeletal muscle contraction.
▶ Describe three types of skeletal muscle fibers and explain the relative value of each type.

Force of Muscle Contraction

The force of muscle contraction depends on the number of myosin cross bridges that are attached to actin. This in turn is affected by four factors (**Figure 9.18**), the first two of which we have already discussed:

- **Frequency of stimulation.** When a muscle is stimulated more frequently, contractions are summed (temporal summation and

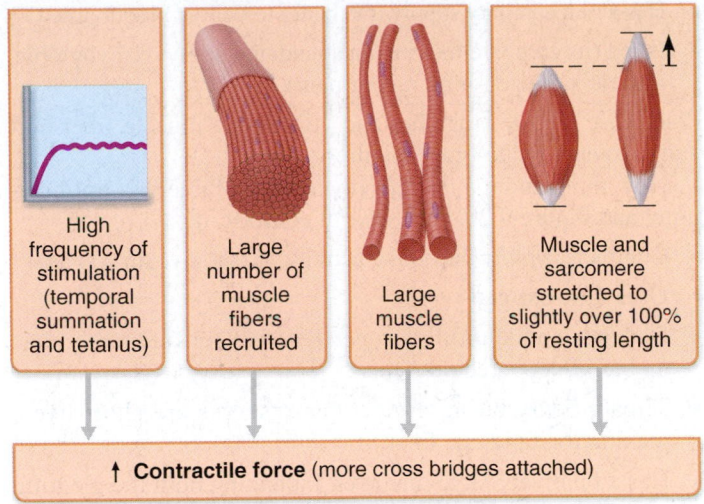

Figure 9.18 Factors that increase the force of skeletal muscle contraction.

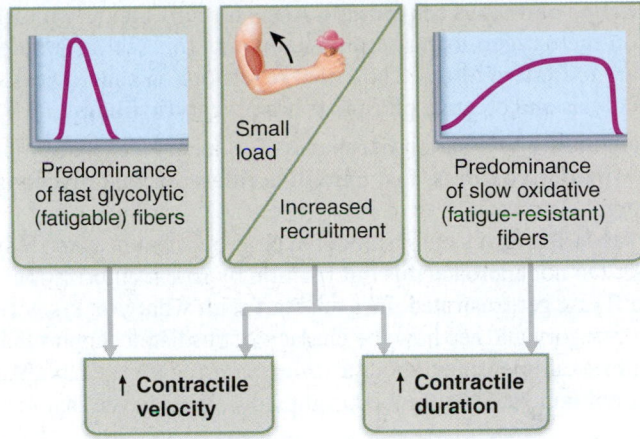

Figure 9.20 Factors influencing velocity and duration of skeletal muscle contraction.

tetany, see Figure 9.12 on p. 300). The higher the frequency of muscle stimulation, the greater the force the muscle exerts.

- **Number of muscle fibers recruited.** The more motor units recruited, the greater the force (Figure 9.13 on p. 301).

- **Size of muscle fibers.** The bulkier the muscle and the greater the cross-sectional area, the more tension it can develop. The large fibers of large motor units produce the most powerful movements. Regular resistance exercise increases muscle force by causing muscle cells to *hypertrophy* (increase in size).

- **Degree of muscle stretch.** If a muscle is stretched to various lengths and maximally stimulated, the tension the muscle can generate varies with length. The amount of tension a muscle can generate during an isometric contraction at various lengths—its **length-tension relationship**—can be shown graphically as in **Figure 9.19**. The ideal length on this curve occurs when the muscle is close to its resting length and the thin and thick filaments overlap optimally, because this permits sliding along nearly the entire length of the thin filaments. If a muscle is stretched so much that the filaments do not overlap, the myosin heads have nothing to attach to and cannot generate tension. On the other hand, if the sarcomeres are so compressed that the thin filaments interfere with one another, little or no further shortening can occur. In the body, skeletal muscles are maintained near their optimal length at rest. Our joints normally prevent bone movements that would stretch attached muscles beyond their optimal range.

Velocity and Duration of Contraction

Muscles vary in how fast they can contract and how long they can continue to contract before they fatigue. These characteristics are influenced by muscle fiber type, load, and recruitment (**Figure 9.20**).

Muscle Fiber Type

There are several ways of classifying muscle fibers, but learning about these classes will be easier if you pay attention to just two functional characteristics:

- **Speed of contraction.** On the basis of speed (velocity) of fiber shortening, there are **slow fibers** and **fast fibers**. The difference reflects how fast their myosin ATPases split ATP, and the pattern of electrical activity of their motor neurons. Contraction duration also varies with fiber type and depends on how quickly Ca^{2+} moves from the cytosol into the SR.

Figure 9.19 Length-tension relationships of sarcomeres in skeletal muscles. A muscle generates maximum force between 80 and 120% of its optimal (resting) length. Increases and decreases beyond this optimal range reduce its ability to generate tension.

- **Major pathways for forming ATP.** The cells that rely mostly on the oxygen-using aerobic pathways for ATP generation are **oxidative fibers**. Those that rely more on anaerobic glycolysis and creatine phosphate are **glycolytic fibers**.

Using these two criteria, we can classify skeletal muscle cells as: **slow oxidative fibers**, **fast oxidative fibers**, or **fast glycolytic fibers**.

Table 9.2 gives details about each group, but a word to the wise: Do not approach this information by rote memorization—you'll just get frustrated. Instead, start with what you know for any category and see how the characteristics listed support that.

For example, think about a *slow oxidative fiber* (Table 9.2, first column, and Figure 9.20, right side). We can see that it:

- Contracts *slowly* because its myosin ATPases are slow (a criterion)

- Depends on *oxygen* delivery and aerobic pathways (its major pathways for forming ATP give it *high oxidative capacity*—a criterion)

- Resists fatigue and has high endurance (typical of fibers that depend on aerobic metabolism)

- Is thin (a large amount of cytoplasm impedes diffusion of O_2 and nutrients from the blood)

- Has relatively little power (a thin cell can contain only a limited number of myofibrils)

- Has many mitochondria (actual sites of oxygen use)

- Has a rich capillary supply (the better to deliver bloodborne O_2)

- Is red (its color stems from an abundant supply of myoglobin, muscle's oxygen-binding pigment)

Add these features together and you have a muscle fiber best suited to endurance-type activities.

Now think about a *fast glycolytic fiber* (Table 9.2, third column, and Figure 9.20, left side). In contrast, it:

- Contracts *rapidly* due to the activity of fast myosin ATPases

- Uses little oxygen

- Depends on plentiful *glycogen* reserves for fuel rather than on blood-delivered nutrients

- Tires quickly because glycogen reserves are short-lived, making it a fatigable fiber

- Has a relatively large diameter, indicating both the plentiful myofilaments that allow it to contract powerfully before it "tires out" and its lack of dependence on continuous oxygen and nutrient diffusion from the blood

- Has few mitochondria, little myoglobin, and few capillaries (making it white)

For these reasons, a fast glycolytic fiber is best suited for short-term, rapid, intense movements (moving furniture across the room, for example).

Finally, consider the less common intermediate muscle fiber types, called *fast oxidative fibers* (Table 9.2, middle column).

Table 9.2 Structural and Functional Characteristics of the Three Types of Skeletal Muscle Fibers			
	SLOW OXIDATIVE FIBERS	**FAST OXIDATIVE FIBERS**	**FAST GLYCOLYTIC FIBERS**
Metabolic Characteristics			
Speed of contraction	Slow	Fast	Fast
Myosin ATPase activity	Slow	Fast	Fast
Primary pathway for ATP synthesis	Aerobic	Aerobic (some anaerobic glycolysis)	Anaerobic glycolysis
Myoglobin content	High	High	Low
Glycogen stores	Low	Intermediate	High
Recruitment order	First	Second	Third
Rate of fatigue	Slow (fatigue-resistant)	Intermediate (moderately fatigue-resistant)	Fast (fatigable)
Activities Best Suited For			
	Endurance-type activities—e.g., running a marathon; maintaining posture (antigravity muscles)	Sprinting, walking	Short-term intense or powerful movements, e.g., hitting a baseball
Structural Characteristics			
Fiber diameter	Small	Large*	Intermediate
Mitochondria	Many	Many	Few
Capillaries	Many	Many	Few
Color	Red	Red to pink	White (pale)

*In animal studies, fast glycolytic fibers were found to be the largest, but this is not true in humans.

They have many characteristics intermediate between the other two types (glycogen stores and power, for example). Like fast glycolytic fibers, they contract quickly, but like slow oxidative fibers, they are oxygen dependent and have a rich supply of myoglobin and capillaries.

Some muscles have a predominance of one fiber type, but most contain a mixture of fiber types, which gives them a range of contractile speeds and fatigue resistance. But, as might be expected, all muscle fibers in a particular *motor unit* are of the same type.

Although everyone's muscles contain mixtures of the three fiber types, some people have relatively more of one kind. These differences are genetically initiated, but can be modified by exercise and no doubt determine athletic capabilities, such as endurance versus strength, to a large extent. For example, muscles of marathon runners have a high percentage of slow oxidative fibers (about 80%), while those of sprinters contain a higher percentage (about 60%) of fast oxidative and glycolytic fibers. Interconversion between the "fast" fiber types occurs as a result of specific exercise regimes, as we'll describe below.

(a) The greater the load, the briefer the duration of muscle shortening.

(b) The greater the load, the slower the muscle shortening.

Figure 9.21 Influence of load on duration and velocity of muscle shortening.

Load and Recruitment

Because muscles are attached to bones, they are always pitted against some resistance, or load, when they contract. As you might expect, they contract fastest when there is no added load on them. A greater load results in a longer latent period, slower shortening, and a briefer duration of shortening (**Figure 9.21**).

In the same way that many hands on a project can get a job done more quickly and can keep working longer, the more motor units that are contracting, the faster and more prolonged the contraction.

Check Your Understanding

17. List two factors that influence contractile force and two that influence velocity of contraction.
18. APPLY Jordan called several friends to help him move. Would he prefer to have those with more slow oxidative muscle fibers or those with more fast glycolytic fibers as his helpers? Why?
19. APPLY There are no slow glycolytic fibers. Explain why it would not make sense to have these kinds of fibers.

For answers, see Answers Appendix.

9.8 How does skeletal muscle respond to exercise?

Learning Outcome
▶ Compare and contrast the effects of aerobic and resistance exercise on skeletal muscles.

The amount of work a muscle does is reflected in changes in the muscle itself. When used actively or strenuously, muscles may become larger or stronger, or more efficient and fatigue resistant. Exercise gains are based on the overload principle. Forcing a muscle to work hard increases its strength and endurance. As muscles adapt to greater demand, they must be overloaded to produce further gains. Inactivity, on the other hand, *always* leads to muscle weakness and atrophy.

Aerobic (Endurance) Exercise

Aerobic, or **endurance**, **exercise** such as swimming, running, fast walking, and biking results in several recognizable changes in skeletal muscles:

- The number of capillaries surrounding the muscle fibers increases.
- The number of mitochondria within the muscle fibers also increases.
- The fibers synthesize more myoglobin.

These changes occur in all fiber types, but are most dramatic in slow oxidative fibers, which depend primarily on aerobic pathways. The changes result in more efficient muscle metabolism and in greater endurance, strength, and resistance to fatigue. Regular endurance exercise may convert fast glycolytic fibers into fast oxidative fibers.

Resistance Exercise

The moderately weak but sustained muscle activity required for endurance exercise does not promote significant skeletal muscle hypertrophy, even though the exercise may go on for hours. Muscle hypertrophy—think of the bulging biceps of a professional weight lifter—results mainly from high-intensity **resistance exercise** (typically under anaerobic conditions) such as weight lifting or isometric exercise, which pits muscles against high-resistance or immovable forces. Here strength, not stamina, is important, and a few minutes every other day is sufficient to allow a proverbial weakling to put on 50% more muscle within a year.

The additional muscle bulk largely reflects the increased size of individual muscle fibers (particularly the fast glycolytic variety) rather than an increased number of muscle fibers. [However, some of the bulk may result from longitudinal splitting of the fibers and subsequent growth of these "split" cells, or from the proliferation and fusion of satellite cells (see Figure 9.27).] Vigorously stressed muscle fibers also contain more mitochondria, form more myofilaments and myofibrils, store more glycogen, and develop more connective tissue between muscle cells.

Collectively these changes promote significant increases in muscle strength and size. Resistance activities can also convert fast oxidative fibers to fast glycolytic fibers. However, if the specific exercise routine is discontinued, the converted fibers revert to their original metabolic properties.

HOMEOSTATIC IMBALANCE 9.4   CLINICAL

To remain healthy, muscles must be active. Immobilization due to enforced bed rest or loss of neural stimulation results in *disuse atrophy* (degeneration and loss of mass), which begins almost as soon as the muscles are immobilized. Under such conditions, muscle strength can decline at the rate of 5% per day!

Even at rest, muscles continually receive weak stimuli (recall that they have *muscle tone*; ◀ pp. 302–303). When totally deprived of neural stimulation, a paralyzed muscle may atrophy to one-quarter of its initial size. Fibrous connective tissue replaces the lost muscle tissue, and complete recovery usually takes longer than the period of immobilization. ●●●●●●●●

Check Your Understanding

20. How do aerobic and resistance exercise differ in their effects on muscle size and function?

For answers, see Answers Appendix.

9.9 Smooth muscle is nonstriated involuntary muscle

Learning Outcomes

▶ **Compare the gross and microscopic anatomy of smooth muscle cells to that of skeletal muscle cells.**

▶ **Compare and contrast the contractile mechanisms and the means of activation of skeletal and smooth muscles.**

▶ **Distinguish between unitary and multi unit smooth muscle structurally and functionally.**

Except for the heart, which is made of cardiac muscle, the muscle in the walls of all the body's hollow organs is almost entirely smooth muscle.

Most smooth muscle is organized into sheets of tightly packed fibers. These sheets are found in the walls of all but the smallest blood vessels and in the walls of hollow organs of the respiratory, digestive, urinary, and reproductive tracts. In most cases, there are two sheets of smooth muscle with their fibers oriented at right angles to each other, as in the intestine (**Figure 9.22**).

• In the *longitudinal layer*, the muscle fibers run parallel to the long axis of the organ. Consequently, when these fibers contract, the organ shortens.

• In the *circular layer*, the fibers run around the circumference of the organ. Contraction of this layer constricts the lumen (cavity inside) of the organ.

The alternating contraction and relaxation of these layers mixes substances in the lumen and squeezes them through the organ's internal pathway. Contraction of smooth muscle in the rectum, urinary bladder, and uterus helps those organs to expel their

(a)

(b) Cross section of the intestine showing the smooth muscle layers running at right angles to each other.

— Small intestine —

Mucosa

Longitudinal layer of smooth muscle (shows smooth muscle fibers in cross section)

Circular layer of smooth muscle (shows longitudinal views of smooth muscle fibers)

Figure 9.22 Arrangement of smooth muscle in the walls of hollow organs.

contents. Smooth muscle contraction also accounts for the constricted breathing of asthma and for stomach cramps.

The chemical and mechanical events of contraction are essentially the same in all muscle tissues, but smooth muscle is distinctive in several ways, as summarized in **Table 9.3** on pp. 312–313.

Differences between Smooth and Skeletal Muscle Fibers

There are several structural differences between skeletal and smooth muscle. In contrast to skeletal muscle:

- **Smooth muscle fibers are small spindle-shaped cells.** Each cell has one centrally located nucleus (Figure 9.22b). Typically, smooth muscle fibers have a diameter of 5–10 μm and are 30–200 μm long. Skeletal muscle fibers are up to 10 times wider and thousands of times longer.

- **Smooth muscle lacks the coarse connective tissue sheaths found in skeletal muscle.** A small amount of fine connective tissue (endomysium) is secreted by the smooth muscles themselves. It is found between smooth muscle fibers and contains blood vessels and nerves. Epimysium and perimysium are not present.

- **Smooth muscle has varicosities instead of neuromuscular junctions.** The innervating nerve fibers of the autonomic (involuntary) nervous system have numerous bulbous swellings, called **varicosities** (**Figure 9.23**). Unlike the highly structured neuromuscular junctions of skeletal muscle, varicosities form **diffuse junctions** that have wide synaptic clefts. The varicosities simply "sprinkle" neurotransmitter in the general area of the smooth muscle cells. Comparing the neural input to skeletal and smooth muscles, you could say

Figure 9.24 Sources of Ca²⁺ for smooth muscle contraction.

① Ca^{2+} enters via Ca^{2+} channels concentrated in membrane infoldings called *caveolae*.

② Ca^{2+} entry triggers Ca^{2+} release from the SR via SR Ca^{2+} channels.

that skeletal muscle gets high-speed internet while smooth muscle gets slow-speed dial-up.

- **Smooth muscle fibers have less elaborate SR and no T tubules.** The sarcoplasmic reticulum of smooth muscle fibers has no terminal cisterns and lacks a specific pattern relative to the myofilaments. Although the SR *does* release some of the Ca^{2+} that triggers contraction, most Ca^{2+} enters through calcium channels directly from the extracellular space. The sarcolemma has multiple **caveolae** (ka″ve-o′le; "little caves"), pouchlike infoldings containing large numbers of Ca^{2+} channels (**Figure 9.24**). These calcium channels are the major source of Ca^{2+} for smooth muscle contraction. This situation is quite different from what we see in skeletal muscle, which does not depend on extracellular Ca^{2+} for excitation-contraction coupling.

- **Smooth muscle fibers are usually electrically connected by gap junctions.** Gap junctions are specialized cell connections (◄ p. 68) that allow depolarization to spread from cell to cell. In contrast, skeletal muscle fibers are electrically isolated from one another.

There are no striations in smooth muscle, as its name indicates, and therefore no sarcomeres. Smooth muscle fibers do contain overlapping thick and thin filaments, but the myosin filaments are a lot shorter than the actin filaments and the type of myosin contained differs from skeletal muscle. The proportion and organization of smooth muscle myofilaments differ from skeletal muscle in the following ways:

- **Thick filaments are fewer but have myosin heads along their entire length.** The ratio of thick to thin filaments is much lower in smooth muscle than in skeletal muscle (1:13 compared to 1:2). However, thick filaments of smooth muscle contain actin-gripping myosin heads along their *entire length*, a feature that makes smooth muscle as powerful as a skeletal muscle of the same size. Also, in smooth muscle the myosin heads are oriented in one direction on one side of the filament and in the opposite direction on the other side.

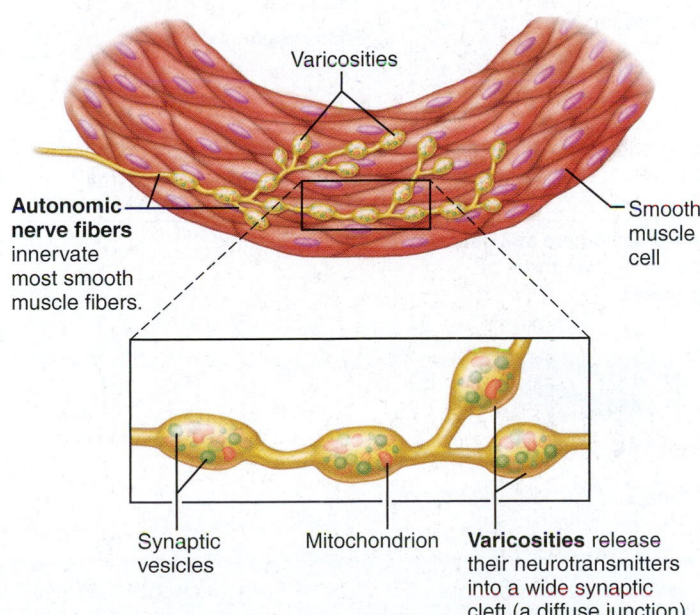

Varicosities

Autonomic nerve fibers innervate most smooth muscle fibers.

Smooth muscle cell

Synaptic vesicles   **Mitochondrion**   **Varicosities** release their neurotransmitters into a wide synaptic cleft (a diffuse junction).

Figure 9.23 Innervation of smooth muscle.

(Text continues on p. 314.)

Table 9.3   Comparison of Skeletal, Cardiac, and Smooth Muscle

CHARACTERISTIC	SKELETAL	CARDIAC	SMOOTH
Body location	Attached to bones or (some facial muscles) to skin	Walls of the heart	Unitary muscle in walls of hollow visceral organs (other than the heart); multi unit muscle in intrinsic eye muscles, airways, large arteries
Cell shape and appearance	Single, very long, cylindrical, multinucleate cells with obvious striations	Branching chains of cells; uni- or binucleate; striations	Single, spindle shaped, uninucleate; no striations
Connective tissue components	Epimysium, perimysium, and endomysium	Endomysium attached to fibrous skeleton of heart	Endomysium
Presence of myofibrils composed of sarcomeres	Yes	Yes, but myofibrils are of irregular thickness	No, but actin and myosin filaments are present throughout; dense bodies anchor actin filaments
Presence and location of T tubules	Yes; two per sarcomere at A-I junctions	Yes; one per sarcomere at Z disc; larger diameter than those of skeletal muscle	No; only caveolae

Table 9.3 (continued)			
CHARACTERISTIC	SKELETAL	CARDIAC	SMOOTH
Elaborate sarcoplasmic reticulum	Yes; large terminal cisterns	Less than skeletal muscle (1–8% of cell volume); small terminal cisterns	Equivalent to cardiac muscle (1–8% of cell volume); some SR contacts the sarcolemma
Presence of gap junctions	No	Yes; at intercalated discs	Yes; in unitary muscle
Cells exhibit individual neuromuscular junctions	Yes	No	Not in unitary muscle; yes in multi unit muscle
Regulation of contraction	Voluntary via axon terminals of the somatic nervous system	Involuntary; contraction initiated by intrinsic pacemaker cells; regulated by autonomic nervous system, hormones, and stretch	Involuntary; initiated by intrinsic pacemaker cells (in unitary muscle) or autonomic nerves (in multi unit muscle); regulated by local chemicals, hormones, and stretch
Source of Ca²⁺ for contraction	Sarcoplasmic reticulum (SR)	SR and from extracellular fluid	Extracellular fluid and from SR
Site of calcium regulation	Troponin on actin-containing thin filaments	Troponin on actin-containing thin filaments	Calmodulin in the cytosol
Presence of pacemaker(s)	No	Yes	Yes (in unitary muscle only)
Effect of nervous system stimulation	Excitation	Excitation or inhibition	Excitation or inhibition
Speed of contraction	Slow to fast	Slow	Very slow
Rhythmic contraction	No	Yes	Yes in unitary muscle
Response to stretch	Contractile strength increases with degree of stretch (to a point)	Contractile strength increases with degree of stretch	Stress-relaxation response
Metabolism	Aerobic and anaerobic	Aerobic	Mainly aerobic

(b) Contracted smooth muscle fiber

Figure 9.25 Intermediate filaments and dense bodies of smooth muscle fibers harness the pull generated by myosin cross bridges. Intermediate filaments attach to dense bodies throughout the sarcoplasm.

- **No troponin complex in thin filaments.** As in skeletal muscle, tropomyosin mechanically stabilizes the thin filaments, but smooth muscle has no calcium-binding troponin complex. Instead, a protein called *calmodulin* acts as the calcium-binding site.

- **Intermediate filament–dense body network.** Smooth muscle fibers contain a lattice-like arrangement of noncontractile *intermediate filaments* that resist tension. They attach at regular intervals to cytoplasmic structures called dense bodies (**Figure 9.25**). The **dense bodies**, which are also tethered to the sarcolemma, act as anchoring points for thin filaments and therefore correspond to Z discs of skeletal muscle.

  The intermediate filament–dense body network forms a strong, cable-like intracellular cytoskeleton that harnesses the pull generated by the sliding of the thick and thin filaments. During contraction, areas of the sarcolemma between the dense bodies bulge outward, making the cell look puffy (Figure 9.25b). Dense bodies at the sarcolemma surface also bind the muscle cell to the connective tissue fibers outside the cell and to adjacent cells. This arrangement transmits the pulling force to the surrounding connective tissue and partly accounts for the synchronous contractions of most smooth muscle.

- **Thick and thin filaments arranged diagonally.** Bundles of contractile proteins crisscross within the smooth muscle cell so they spiral down the long axis of the cell like the stripes on a candy cane). Because of this diagonal arrangement, the smooth muscle cells contract in a twisting way so that they look like tiny corkscrews (Figure 9.25b).

Contraction of Smooth Muscle

Mechanism of Contraction

In most cases, adjacent smooth muscle fibers exhibit slow, synchronized contractions, the whole sheet responding to a stimulus in unison. This synchronization reflects electrical coupling of smooth muscle cells by *gap junctions* that transmit depolarization from fiber to fiber.

Some smooth muscle fibers in the stomach and small intestine are *pacemaker cells*: Once excited, they act as "drummers" to set the pace of contraction for the entire muscle sheet. These pacemakers depolarize spontaneously in the absence of external stimuli. However, neural and chemical stimuli can modify both the rate and the intensity of smooth muscle contraction.

Contraction in smooth muscle is like contraction in skeletal muscle in the following ways:

- The final trigger for contraction is a rise in the intracellular calcium ion level.

- Actin and myosin interact by the sliding filament mechanism.

- ATP energizes the sliding process.

During excitation-contraction coupling, the SR releases Ca^{2+}, but Ca^{2+} also moves into the cell from the extracellular space via membrane channels. In all striated muscle types, calcium ions activate myosin by binding to troponin. In smooth muscle, calcium activates myosin by interacting with a regulatory molecule called **calmodulin**, a cytoplasmic calcium-binding protein. Calmodulin, in turn, interacts with a kinase enzyme called **myosin light chain kinase** or **myosin kinase**, which phosphorylates the myosin, activating it (**Figure 9.26**).

As in skeletal muscle, smooth muscle relaxes when intracellular Ca^{2+} levels drop—but getting smooth muscle to stop contracting is more complex. Events known to be involved include calcium detachment from calmodulin, active transport of Ca^{2+} into the SR and extracellular fluid, and dephosphorylation of myosin by a phosphatase enzyme, which reduces the activity of the myosin ATPases.

Energy Efficiency of Smooth Muscle Contraction

Smooth muscle takes 30 times longer to contract and relax than does skeletal muscle, but it can maintain the same contractile tension for prolonged periods at less than 1% of the energy cost. If skeletal muscle is like a speedy windup car that quickly runs down, then smooth muscle is like a steady locomotive engine that lumbers along tirelessly.

Part of the striking energy economy of smooth muscle is the sluggishness of its ATPases compared to those in skeletal muscle. Moreover, smooth muscle myofilaments may latch together during prolonged contractions. This latching also saves energy.

The smooth muscle in arterioles (small arteries) and other visceral organs routinely maintains a moderate degree of contraction, called *smooth muscle tone*, day in and day out without fatiguing. Smooth muscle has low energy requirements, and as a rule, it makes enough ATP via aerobic pathways to keep up with the demand.

Regulation of Contraction

The contraction of smooth muscle can be regulated by nerves, hormones, or local chemical changes. Let's briefly consider each of these methods.

Neural Regulation In some cases, the activation of smooth muscle by a neural stimulus is identical to that in skeletal muscle:

Extracellular fluid (ECF)

Plasma membrane

Ca²⁺

Cytoplasm

① Calcium ions (Ca²⁺) enter the cytosol from the ECF via voltage-gated or non-voltage-gated Ca²⁺ channels, or from the SR.

Sarcoplasmic reticulum (SR)

Ca²⁺

② Ca²⁺ binds to and activates calmodulin.

Ca²⁺

Inactive calmodulin

Activated calmodulin

③ Activated calmodulin activates the myosin light chain kinase enzymes.

Inactive kinase

Activated kinase

ATP

ADP

④ The activated kinase enzymes catalyze transfer of phosphate to myosin, activating the myosin ATPases.

Pᵢ

Pᵢ

Inactive myosin molecule

Activated (phosphorylated) myosin molecule

⑤ Activated myosin forms cross bridges with actin of the thin filaments just as in skeletal muscle. Shortening begins.

Thin filament

Thick filament

Figure 9.26 Sequence of events in excitation-contraction coupling of smooth muscle.

Neurotransmitter binding generates an action potential, which is coupled to a rise in calcium ions in the cytosol. However, some types of smooth muscle respond to neural stimulation with graded potentials (local electrical signals) only.

Recall that all somatic nerve endings, that is, nerve endings that excite skeletal muscle, release the neurotransmitter acetylcholine. However, different autonomic nerves serving the smooth muscle of visceral organs release different neurotransmitters, each of which may excite or inhibit a particular group of smooth muscle cells.

The effect of a specific neurotransmitter on a smooth muscle cell depends on the type of receptor molecules on the cell's sarcolemma. For example, when acetylcholine binds to ACh receptors on smooth muscle in the bronchioles (small air passageways of the lungs), the response is strong contraction that narrows the bronchioles. When norepinephrine, released by a different type of autonomic nerve fiber, binds to norepinephrine receptors on the *same* smooth muscle cells, the effect is inhibitory—the muscle relaxes, which dilates the bronchioles. However, when norepinephrine binds to smooth muscle in the walls of most blood vessels, it stimulates the smooth muscle cells to contract and constrict the vessel.

Hormones and Local Chemical Factors Some smooth muscle layers have no nerve supply at all. Instead, they depolarize spontaneously or in response to chemical stimuli that bind to G protein–linked receptors (◄ pp. 81–82). Other smooth muscle cells respond to both neural and chemical stimuli.

Several chemical factors cause smooth muscle to contract or relax without an action potential by enhancing or inhibiting Ca²⁺ entry into the sarcoplasm. They include certain hormones, histamine, excess carbon dioxide, low pH, and lack of oxygen. The direct response to these chemical stimuli alters smooth muscle activity according to local tissue needs and probably is most responsible for smooth muscle tone. For example, the hormone gastrin stimulates stomach smooth muscle to contract so it can churn foodstuffs more efficiently. We will consider activation of smooth muscle in specific organs as we discuss each organ in subsequent chapters.

Special Features of Smooth Muscle Contraction

Smooth muscle is intimately involved in the functioning of most hollow organs and has a number of unique characteristics. We have already considered some of these—smooth muscle tone, slow prolonged contractions, and low energy requirements. But smooth muscle also responds differently to stretch and can lengthen and shorten more than other muscle types. Let's take a look.

Response to Stretch Unlike skeletal muscle, smooth muscle spontaneously contracts when it is stretched. This is useful because it can move substances along an internal tract. However, the increased tension persists only briefly, and soon the muscle adapts to its new length and relaxes, while still retaining the ability to contract on demand.

This **stress-relaxation response** allows a hollow organ to fill or expand slowly to accommodate a greater volume without causing strong contractions that would expel its contents. This is an important attribute, because organs such as the stomach and intestines must store their contents long enough to digest and absorb the nutrients. Likewise, your urinary bladder must be able to store the continuously made urine until it is convenient to empty your bladder, or you would spend all your time in the bathroom.

Length and Tension Changes Smooth muscle stretches much more and generates more tension than skeletal muscles stretched to a comparable extent. The irregular, overlapping arrangement of smooth muscle filaments and the lack of sarcomeres allow them to generate considerable force, even when they are substantially stretched. The total length change that skeletal muscles can undergo and still function efficiently is about 60% (from 30% shorter to 30% longer than resting length; see Figure 9.19 on p. 307), but smooth muscle can contract when it is anywhere from half to twice its resting length—a total range of 150%. This capability allows hollow organs to tolerate tremendous changes in volume without becoming flabby when they empty.

Types of Smooth Muscle

The smooth muscle in different body organs varies substantially in its (1) fiber arrangement and organization, (2) innervation, and (3) responsiveness to various stimuli. For simplicity, however, smooth muscle is usually categorized into two major types: *unitary* and *multi unit*.

Unitary Smooth Muscle

Unitary smooth muscle, sometimes called **visceral muscle** because it is in the walls of all hollow organs except the heart, is far more common. All the smooth muscle characteristics described so far pertain to unitary smooth muscle.

For example, the cells of unitary smooth muscle:

- Are arranged in opposing (longitudinal and circular) sheets
- Are innervated by varicosities of autonomic nerve fibers and often exhibit rhythmic spontaneous action potentials
- Are electrically coupled by gap junctions and so contract as a unit (for this reason recruitment does not occur in unitary smooth muscle)
- Respond to various chemical stimuli

Multi Unit Smooth Muscle

Examples of **multi unit smooth muscle** are: the smooth muscles in the large airways to the lungs and in large arteries, the arrector pili muscles attached to hair follicles, and the internal eye muscles that adjust pupil size and allow the eye to focus.

In contrast to unitary muscle, gap junctions and spontaneous depolarizations are absent. Like skeletal muscle, multi unit smooth muscle:

- Consists of muscle fibers that are structurally independent of one another
- Is richly supplied with nerve endings, each of which forms a motor unit with a number of muscle fibers
- Responds to neural stimulation with graded contractions that involve recruitment

However, skeletal muscle is served by the somatic (voluntary) division of the nervous system. Multi unit smooth muscle, like unitary smooth muscle, is innervated by the autonomic (involuntary) division and also responds to hormones.

Check Your Understanding

21. Compare the structures of skeletal and smooth muscle fibers.
22. Calcium is the trigger for contraction of all muscle types. How does its binding site differ in skeletal and smooth muscle fibers?
23. How does the stress-relaxation response suit the role of smooth muscle in hollow organs?
24. **MAKE CONNECTIONS** Intracellular calcium performs other important roles in the body in addition to triggering muscle contraction. Give one example. (Hint: See Chapter 3.)

For answers, see Answers Appendix.

Developmental Aspects of Muscles

With rare exceptions, all three types of muscle tissue develop from cells called **myoblasts** that arise from the embryonic mesoderm (◀ Figure 4.16, p. 146). In forming skeletal muscle tissue, several myoblasts fuse to form multinuclear *myotubes* (**Figure 9.27**). Soon functional sarcomeres appear, and skeletal muscle fibers are contracting by week 7 when the embryo is only about 2.5 cm (1 inch) long.

Initially, ACh receptors "sprout" over the entire surface of the developing myoblasts. As spinal nerves invade the muscle masses, the nerve endings target individual myoblasts and release a growth factor that stimulates clustering of ACh receptors at the newly forming neuromuscular junction in each muscle fiber. Then, the nerve endings release a different chemical that eliminates the receptor sites not innervated or stabilized by the growth factor. As the somatic nervous system assumes control of muscle fibers, the number of fast and slow contractile fiber types is determined by the nerves that innervate them.

① Embryonic mesoderm cells called myoblasts undergo cell division (to increase number) and enlarge.

② Several myoblasts fuse together to form a myotube.

③ Myotube matures into skeletal muscle fiber.

Embryonic mesoderm cells

Myoblasts

Myotube (immature multinucleate muscle fiber)

Satellite cell

Mature skeletal muscle fiber

Figure 9.27 **Myoblasts fuse to form a multinucleate skeletal muscle fiber.**

Athletes Looking Good and Doing Better with Anabolic Steroids?

Society loves a winner and top athletes reap large social and monetary rewards. It is not surprising that some will grasp at anything that might increase their performance—including performance-enhancing drugs such as anabolic steroids.

These drugs are variants of the male sex hormone testosterone. They were introduced in the 1950s to treat anemia and certain muscle-wasting diseases and to prevent muscle atrophy in patients immobilized after surgery. Testosterone is responsible for the increase in muscle and bone mass and other physical changes that occur during puberty in males.

Athletes and bodybuilders were using megadoses of steroids by the early 1960s, a practice that is still going on despite drug testing programs. Investigations have stunned sports fans with revelations of steroid use by many elite athletes including Barry Bonds, Mark McGwire, Marion Jones, and Lance Armstrong.

However, steroid use is not confined to professional athletes. It is estimated that nearly one in every 10 young men has tried them, and their use has spread among young women.

It is difficult to determine the extent of anabolic steroid use because users stop doping before the event, aware that evidence of drug use is hard to find a week after they stop. "Underground" suppliers keep producing new versions of designer steroids that evade standard antidoping tests.

There is little question that many professional bodybuilders and athletes competing in events that require muscle strength are heavy users, claiming that anabolic steroids enhance muscle mass and strength, and raise oxygen-carrying capability owing to a greater volume of red blood cells.

Do the drugs do all that is claimed? Studies report increased isometric strength and body weight in steroid users. While these are results weight lifters dream about, for runners and others requiring fine muscle coordination and endurance, these changes may not translate into better performance. The jury is still out on this question.

Do the alleged advantages of steroids outweigh their risks? Absolutely not. Anabolic steroids cause: bloated faces (Cushingoid sign of steroid excess), acne and hair loss, shriveled testes and infertility, liver damage that promotes liver cancer, and changes in blood cholesterol levels that may predispose users to heart disease.

In addition, females can develop masculine characteristics such as smaller breasts, enlarged clitoris, excess body hair, and thinning scalp hair. The psychiatric hazards of anabolic steroid use may be equally threatening: One-third of users suffer serious mental problems. Depression, delusions, and manic behavior—in which users undergo Jekyll-and-Hyde personality swings and become extremely violent (termed 'roid rage)—are all common.

Some people seem willing to try almost anything to win, short of killing themselves. Are they unwittingly doing this as well?

Myoblasts producing cardiac and smooth muscle cells do not fuse but develop gap junctions at a very early embryonic stage. Cardiac muscle is pumping blood just 3 weeks after fertilization.

Regarding muscle regeneration:

- Skeletal muscles stop dividing early on. However, *satellite cells,* myoblast-like cells associated with skeletal muscle, help repair injured fibers and allow limited regeneration of dead skeletal muscle, a capability that declines with age.

- Cardiac muscle was thought to have no regenerative capability whatsoever, but recent studies suggest that cardiac cells do divide, but only at about 1% per year. As a result, injured heart muscle is repaired mostly by scar tissue.

- Smooth muscles have a good regenerative capacity, and smooth muscle cells of blood vessels divide regularly throughout life.

Both skeletal muscle and cardiac muscle retain the ability to lengthen and thicken in a growing child and to hypertrophy in response to increased load in adults.

At birth, a baby's movements are uncoordinated and largely reflexive. Muscular development reflects the level of neuromuscular coordination, which develops in a head-to-toe and proximal-to-distal direction. A baby can lift its head before it can walk, and gross movements precede fine ones.

All through childhood, our control of our skeletal muscles becomes more and more sophisticated. By midadolescence, we reach the peak of our natural neural control of muscles, but can improve it by athletic or other types of training.

A frequently asked question is whether the strength difference between women and men has a biological basis. It does. Individuals vary, but on average, women's skeletal muscles make up approximately 36% of body mass, whereas men's account for about 42%. Men's greater muscular development is due primarily to the effects of testosterone on skeletal muscle, not to the effects of exercise. Body strength per unit muscle mass, however, is the same in both sexes. Strenuous muscle exercise causes more muscle enlargement in males than in females, again because of the influence of testosterone. Some athletes take large doses of synthetic male sex hormones ("steroids") to increase their muscle mass. **A Closer Look** on p. 317 discusses this illegal and physiologically dangerous practice.

Because of its rich blood supply, skeletal muscle is amazingly resistant to infection. Given good nutrition and moderate exercise, relatively few problems afflict skeletal muscles. However, muscular dystrophy (described on p. 287) is a serious congenital condition that affects skeletal muscle.

As we age, the amount of connective tissue in our skeletal muscles increases, the number of muscle fibers decreases, and the muscles become stringier, or more sinewy. By age 30, even in healthy people, a gradual loss of muscle mass, called *sarcopenia* (sar-co-pe′ne-ah), begins. Because skeletal muscles

Homeostatic Interrelationships between the Muscular System and Other Body Systems

Endocrine System   Chapter 16
- Growth hormone and androgens influence skeletal muscle strength and mass; other hormones help regulate cardiac and smooth muscle activity

Cardiovascular System   Chapters 17–19
- Skeletal muscle activity increases efficiency of cardiovascular functioning; helps prevent atherosclerosis and causes cardiac hypertrophy
- Cardiovascular system delivers needed oxygen and nutrients to muscles

Lymphatic System/Immunity   Chapters 20–21
- Physical exercise may enhance or depress immunity depending on its intensity
- Lymphatic vessels drain leaked tissue fluids; immune system protects muscles from disease

Respiratory System   Chapter 22
- Muscular exercise increases respiratory capacity and efficiency of gas exchange
- Respiratory system provides oxygen and disposes of carbon dioxide

Digestive System   Chapter 23
- Physical activity increases gastrointestinal motility; skeletal muscle forms the voluntary sphincter of the anus
- Digestive system provides nutrients needed for muscle health; liver metabolizes lactic acid

Urinary System   Chapters 25–26
- Skeletal muscle forms the voluntary sphincter of the urethra
- Urinary system disposes of nitrogenous wastes

Reproductive System   Chapter 27
- Skeletal muscle helps support pelvic organs (e.g., uterus); assists erection of penis and clitoris
- Testicular androgen promotes increased skeletal muscle

Integumentary System   Chapter 5
- Muscular exercise enhances circulation to skin and improves skin health
- Skin protects the muscles by external enclosure; helps dissipate heat generated by the muscles

Skeletal System   Chapters 6–8
- Skeletal muscle activity maintains bone health and strength
- Bones provide levers for muscle activity

Nervous System   Chapters 11–15
- Facial muscle activity allows emotions to be expressed
- Nervous system stimulates and regulates muscle activity
- Nervous system activity maintains muscle mass

form so much of the body mass, body weight and muscle strength decline in tandem. By age 80, muscle strength usually decreases by about 50%. This "flesh wasting" condition has serious health implications for the elderly, particularly because falling becomes a common event.

Muscles can also suffer indirectly. Aging of the cardiovascular system affects nearly every organ in the body, and muscles are no exception. As atherosclerosis ("hardening of the arteries") takes its toll and begins to block distal arteries, a circulatory condition called *intermittent claudication* (klaw″dĭ-ka′shun; "limping") occurs in some individuals. This condition restricts blood delivery to the legs, leading to excruciating pains in the leg muscles during walking, which forces the person to stop and rest.

But we don't have to slow down during old age. Regular exercise helps reverse sarcopenia, and frail elders who begin to "pump iron" (lift leg and hand weights) can rebuild muscle mass and dramatically increase their strength. Performing those lifting exercises rapidly can improve our ability to carry out the "explosive" movements needed to rise from a chair. Even moderate activity, like taking a walk daily, improves neuromuscular function and enhances independent living.

Smooth muscle is remarkably trouble free. However, smooth muscle is involved in forming the plaques of atherosclerosis that can lead to strokes and heart attacks. The involvement of smooth muscle cells in this process is described in *A Closer Look* on p. 760.

RELATED CLINICAL TERMS

Fibromyalgia A group of conditions involving chronic inflammation of muscles, their connective tissue coverings and tendons, and capsules of nearby joints. Symptoms are nonspecific and involve varying degrees of tenderness associated with specific trigger points, as well as fatigue and frequent awakening from sleep.

Hernia Protrusion of an organ through its body cavity wall. May be congenital (owing to failure of muscle fusion during development), but most often is caused by heavy lifting or obesity and subsequent muscle weakening.

Myalgia (mi-al′je-ah; *algia* = pain) Muscle pain resulting from any muscle disorder.

Myofascial pain syndrome Pain caused by a tightened band of muscle fibers, which twitch when the skin over them is touched. Mostly associated with overused or strained postural muscles.

Myopathy (mi-op′ah-the; *path* = disease, suffering) Any disease of muscle.

Myotonic dystrophy A form of muscular dystrophy that is less common than DMD; in the U.S. it affects about 14 of 100,000 people. Symptoms include a gradual reduction in muscle mass and control of the skeletal muscles, abnormal heart rhythm, and diabetes mellitus. May appear at any time; not sex-linked. Underlying genetic defect is multiple repeats of a particular gene on chromosome 19. Because the number of repeats tends to increase from generation to generation, subsequent generations develop more severe symptoms. No effective treatment.

RICE Acronym for rest, ice, compression, and elevation. The standard treatment for a pulled muscle, or excessively stretched tendons or ligaments.

Spasm A sudden, involuntary twitch in smooth or skeletal muscle ranging from merely irritating to very painful; may be due to chemical imbalances. In spasms of the eyelid or facial muscles, called tics, psychological factors may be involved. Stretching and massaging the affected area may help end the spasm. A cramp is a prolonged spasm; usually occurs at night or after exercise.

Strain Commonly called a "pulled muscle," a strain is excessive stretching and possible tearing of a muscle due to muscle overuse or abuse. The injured muscle becomes painfully inflamed (myositis), and adjacent joints are usually immobilized.

Tetanus (1) A state of sustained contraction of a muscle that is a normal aspect of skeletal muscle functioning. (2) An acute infectious disease caused by the anaerobic bacterium *Clostridium tetani* and resulting in persistent painful spasms of some skeletal muscles. Progresses to fixed rigidity of the jaws (lockjaw) and spasms of trunk and limb muscles. Usually fatal due to respiratory failure.

CHAPTER SUMMARY

9.1 There are three types of muscle tissue (pp. 280–281)

Types of Muscle Tissue (p. 280)

1. Skeletal muscle is attached to the skeleton, is striated, and can be controlled voluntarily.
2. Cardiac muscle forms the heart, is striated, and is controlled involuntarily.
3. Smooth muscle, located chiefly in the walls of hollow organs, is controlled involuntarily. Its fibers are not striated.

Characteristics of Muscle Tissue (p. 280)

4. Special functional characteristics of muscle include excitability, contractility, extensibility, and elasticity.

Muscle Functions (pp. 280–281)

5. Muscles move internal and external body parts, maintain posture, stabilize joints, and generate heat.

9.2 A skeletal muscle is made up of muscle fibers, nerves, blood vessels, and connective tissues (pp. 281–284)

1. Connective tissue coverings protect and strengthen skeletal muscle fibers (cells). Superficial to deep, these are epimysium, perimysium, and endomysium.
2. Skeletal muscle attachments (origins/insertions) may be direct or indirect via tendons or aponeuroses. Indirect attachments withstand friction better.

9

9.3 Skeletal muscle fibers contain calcium-regulated molecular motors (pp. 284–289)

1. Skeletal muscle fibers are long, striated, and multinucleate.
2. Myofibrils are contractile elements that occupy most of the cell volume. Their banded appearance results from a regular alternation of dark (A) and light (I) bands. Myofibrils are chains of sarcomeres; each sarcomere contains thick (myosin) and thin (actin) myofilaments arranged in a regular array. The heads of myosin molecules form cross bridges that interact with the thin filaments.
3. The sarcoplasmic reticulum (SR) is a system of membranous tubules surrounding each myofibril. Its function is to release and then sequester calcium ions.
4. T tubules are invaginations of the sarcolemma that run between the terminal cisterns of the SR. They allow an electrical stimulus to be delivered quickly deep into the cell.
5. According to the sliding filament model, cross bridge (myosin head) activity of the thick filaments pulls the thin filaments toward the sarcomere centers.

9.4 Motor neurons stimulate skeletal muscle fibers to contract (pp. 290–298)

1. Regulation of skeletal muscle cell contraction involves (a) generating and transmitting an action potential along the sarcolemma and (b) excitation-contraction coupling.
2. An end plate potential occurs when acetylcholine released by the axon terminal of a motor neuron binds to ACh receptors on the sarcolemma, causing local changes in membrane permeability, which allow ion flows that depolarize the membrane at that site.

 Complete an interactive tutorial: MasteringA&P® > Study Area > Interactive Physiology > Muscular System: The Neuromuscular Junction

3. The flow of current from the locally depolarized area spreads to the adjacent area of the sarcolemma, opening voltage-gated Na^+ channels, which allows Na^+ influx. These events generate the action potential. Once initiated, the action potential is self-propagating and unstoppable.
4. As the action potential travels away from a region, Na^+ channels close and voltage-gated K^+ channels open, repolarizing the membrane.
5. In excitation-contraction coupling the action potential is propagated down the T tubules, causing calcium to be released from the SR into the cytosol.
6. Sliding of the filaments is triggered by a rise in intracellular calcium ion levels. Troponin binding of calcium moves tropomyosin away from myosin-binding sites on actin, allowing cross bridge binding. Myosin ATPases split ATP, which energizes the power strokes. ATP binding to the myosin head is necessary for cross bridge detachment. Cross bridge activity ends when calcium is pumped back into the SR.

 Complete an interactive tutorial: MasteringA&P® > Study Area > Interactive Physiology > Muscular System: Excitation-Contraction Coupling

9.5 Temporal summation and motor unit recruitment allow smooth, graded skeletal muscle contractions (pp. 298–303)

1. A motor unit is one motor neuron and all the muscle cells it innervates. The neuron's axon has several branches, each of which forms a neuromuscular junction with one muscle cell.
2. A motor unit's response to a single brief threshold stimulus is a twitch. A twitch has three phases: latent (preparatory events occur), contraction (the muscle tenses and may shorten), and relaxation (muscle tension declines and the muscle returns to its resting length).
3. Graded responses of muscles to rapid stimuli are temporal summation and unfused and fused tetanus. A graded response to increasingly strong stimuli is multiple motor unit summation, or recruitment. The order of motor unit recruitment follows the size principle.
4. Isotonic contractions occur when the muscle shortens (concentric contraction) or lengthens (eccentric contraction) as the load is moved. Isometric contractions occur when muscle tension produces neither shortening nor lengthening.

9.6 ATP for muscle contraction is produced aerobically or anaerobically (pp. 303–306)

1. The energy source for muscle contraction is ATP, obtained from a coupled reaction of creatine phosphate with ADP and from aerobic and anaerobic metabolism of glucose.
2. When ATP is produced by anaerobic pathways, lactic acid accumulates and ionic imbalances disturb the membrane potential. To return the muscles to their pre-exercise state, ATP must be produced aerobically and used to regenerate creatine phosphate, glycogen reserves must be restored, and accumulated lactic acid must be metabolized. Oxygen used to accomplish this repayment is called excess postexercise oxygen consumption (EPOC).

9.7 The force, velocity, and duration of skeletal muscle contractions are determined by a variety of factors (pp. 306–309)

1. The force of muscle contraction is affected by the number and size of contracting muscle cells (the more and the larger the cells, the greater the force), the frequency of stimulation, and the degree of muscle stretch.
2. When the thick and thin filaments are optimally overlapping, the muscle can generate maximum force. With excessive increase or decrease in muscle length, force declines.
3. Factors determining the velocity and duration of muscle contraction are muscle fiber types, the load (the greater the load, the slower the contraction), and how many motor units are recruited.
4. The three types of muscle fibers are: (1) fast glycolytic (fatigable) fibers, (2) slow oxidative (fatigue-resistant) fibers, and (3) fast oxidative (fatigue-resistant) fibers. Most muscles contain a mixture of fiber types. The fast muscle fiber types can interconvert with certain exercise regimens.

9.8 How does skeletal muscle respond to exercise? (pp. 309–310)

1. Regular aerobic exercise gives skeletal muscles increased efficiency, endurance, strength, and resistance to fatigue.
2. In skeletal muscle, resistance exercises cause hypertrophy and large gains in strength.
3. Immobilizing muscles leads to muscle weakness and severe atrophy.

9.9 Smooth muscle is nonstriated involuntary muscle (pp. 310–316)

1. Smooth muscle cells are most often arranged in sheets. They lack elaborate connective tissue coverings.

Differences between Smooth and Skeletal Muscle Fibers (pp. 311–314)

2. A smooth muscle fiber is spindle shaped and uninucleate, and has no striations.
3. The SR is poorly developed and T tubules are absent. Actin and myosin filaments are present, but sarcomeres are not. Intermediate filaments and dense bodies form an intracellular network that harnesses the pull generated during cross bridge activity and transfers it to the surrounding connective tissue.

Contraction of Smooth Muscle (pp. 314–316)

4. Smooth muscle fibers may be electrically coupled by gap junctions.
5. ATP energizes smooth muscle contraction, which is activated by a calcium pulse. However, calcium binds to calmodulin rather than to troponin (which is not present in smooth muscle fibers), and myosin must be phosphorylated to become active in contraction.
6. Smooth muscle contracts for extended periods at low energy cost and without fatigue.
7. Neurotransmitters of the autonomic nervous system may inhibit or stimulate smooth muscle fibers. Smooth muscle contraction may also be initiated by pacemaker cells, hormones, or local chemical factors that influence intracellular calcium levels, and by mechanical stretch.
8. Special features of smooth muscle contraction include the stress-relaxation response and the ability to generate large amounts of force when extensively stretched.

Types of Smooth Muscle (p. 316)

9. Unitary smooth muscle has electrically coupled fibers that contract synchronously and often spontaneously.
10. Multi unit smooth muscle has independent, well-innervated fibers that lack gap junctions and pacemaker cells. Stimulation occurs via autonomic nerves (or hormones). Multi unit muscle contractions are rarely synchronous.

Developmental Aspects of Muscles (pp. 316, 317, 319)

1. Muscle tissue develops from embryonic mesoderm cells called myoblasts. Several myoblasts fuse to form a skeletal muscle fiber. Smooth and cardiac muscle cells develop from single myoblasts and display gap junctions.
2. For the most part, specialized skeletal and cardiac muscle cells lose their ability to divide but retain the ability to hypertrophy. Smooth muscle regenerates well and its cells are able to divide throughout life.
3. Skeletal muscle development reflects maturation of the nervous system and occurs in head-to-toe and proximal-to-distal directions. Natural neuromuscular control reaches its peak in midadolescence.
4. Women's muscles account for about 36% of their total body weight and men's for about 42%, a difference due chiefly to the effects of male sex hormones on skeletal muscle growth.
5. Skeletal muscle is richly vascularized and quite resistant to infection, but in old age, skeletal muscles become fibrous, decline in strength, and atrophy. Regular exercise can offset some of these changes.

REVIEW QUESTIONS

 To access additional practice questions using your smartphone, tablet, or computer: MasteringA&P® > Study Area > Practice Tests & Quizzes

Level 1 Remember/Understand

(Some questions have more than one correct answer. Select the best answer or answers from the choices given.)

1. The connective tissue covering that encloses the sarcolemma of an individual muscle fiber is called the **(a)** epimysium, **(b)** perimysium, **(c)** endomysium, **(d)** periosteum.
2. A fascicle is a **(a)** muscle, **(b)** bundle of muscle fibers enclosed by a connective tissue sheath, **(c)** bundle of myofibrils, **(d)** group of myofilaments.
3. Thick and thin myofilaments have different compositions. For each descriptive phrase, indicate whether the filament is **(a)** thick or **(b)** thin.

   ___**(1)** contains actin ___**(4)** contains myosin

   ___**(2)** contains ATPases ___**(5)** contains troponin

   ___**(3)** attaches to the Z disc ___**(6)** does not lie in the I band

4. The function of the T tubules in muscle contraction is to **(a)** make and store glycogen, **(b)** release Ca^{2+} into the cell interior and then pick it up again, **(c)** transmit the action potential deep into the muscle cells, **(d)** form proteins.
5. The sites where the motor nerve impulse is transmitted from the nerve endings to the skeletal muscle cell membranes are the **(a)** neuromuscular junctions, **(b)** sarcomeres, **(c)** myofilaments, **(d)** Z discs.
6. Contraction elicited by a single brief stimulus is called **(a)** a twitch, **(b)** temporal summation, **(c)** multiple motor unit summation, **(d)** fused tetanus.

7. A smooth, sustained contraction resulting from very rapid stimulation of the muscle, in which no evidence of relaxation is seen, is called **(a)** a twitch, **(b)** temporal summation, **(c)** multiple motor unit summation, **(d)** fused tetanus.
8. Characteristics of isometric contractions include all but **(a)** shortening, **(b)** increased muscle tension throughout the contraction phase, **(c)** absence of shortening, **(d)** used in resistance training.
9. During muscle contraction, ATP is provided by **(a)** a coupled reaction of creatine phosphate with ADP, **(b)** aerobic respiration of glucose, and **(c)** anaerobic glycolysis.

   ___**(1)** Which provides ATP fastest?
   ___**(2)** Which does (do) not require that oxygen be available?
   ___**(3)** Which provides the highest yield of ATP per glucose molecule?
   ___**(4)** Which results in the formation of lactic acid?
   ___**(5)** Which has carbon dioxide and water products?
   ___**(6)** Which is most important in endurance sports?

10. The neurotransmitter released by somatic motor neurons is **(a)** acetylcholine, **(b)** acetylcholinesterase, **(c)** norepinephrine.
11. The ions that enter the skeletal muscle cell during the generation of an action potential are **(a)** calcium ions, **(b)** chloride ions, **(c)** sodium ions, **(d)** potassium ions.
12. Myoglobin has a special function in muscle tissue. It **(a)** breaks down glycogen, **(b)** is a contractile protein, **(c)** holds a reserve supply of oxygen in the muscle.

13. Aerobic exercise results in all of the following except (a) more capillaries surrounding muscle fibers, (b) more mitochondria in muscle cells, (c) increased size and strength of existing muscle cells, (d) more myoglobin.

14. The smooth muscle type found in the walls of digestive and urinary system organs and that exhibits gap junctions and pacemaker cells is (a) multi unit, (b) unitary.

Level 2 Apply/Analyze

15. Name and describe the four special functional abilities of muscle that are the basis for muscle response.

16. Distinguish between (a) direct and indirect muscle attachments and (b) a tendon and an aponeurosis.

17. What is the importance of acetylcholinesterase in muscle cell contraction?

18. Explain what is meant by the term excitation-contraction coupling.

19. Define and draw a motor unit.

20. True or false: Most muscles contain a predominance of one skeletal muscle fiber type. Explain the reasoning behind your choice.

21. Describe some cause(s) of muscle fatigue and define this term clearly.

22. When a suicide victim was found, the coroner was unable to remove the drug vial clutched in his hand. Explain the reasons for this. If the victim had been discovered three days later, would the coroner have had the same difficulty? Explain.

Level 3 Evaluate/Synthesize

23. (a) Describe the structure of a sarcomere and indicate the relationship of the sarcomere to myofilaments. (b) Explain the sliding filament model of contraction using appropriately labeled diagrams of a relaxed and a contracted sarcomere.

24. Explain how a slight (but smooth) contraction differs from a vigorous contraction of the same muscle. Use the concepts of multiple motor unit summation.

25. Smooth muscle has some unique properties, such as low energy usage, and the ability to maintain contraction over long periods. Tie these properties to the function of smooth muscle in the body.

26. Muscle-relaxing drugs are administered to a patient during major surgery. Which of the two chemicals described next would be a good skeletal muscle relaxant and why?
    - Chemical A binds to and blocks ACh receptors of muscle cells.
    - Chemical B floods the muscle cells' cytoplasm with Ca^{2+}.

CLINICAL CASE STUDY

Children with Muscular Disorders

You're doing a clinical rotation with Dr. Barr, a world-renowned specialist in children's muscle disorders. On your first day, you meet three patients with the same probable diagnosis— nemaline myopathy. Jodi, age 2, has a

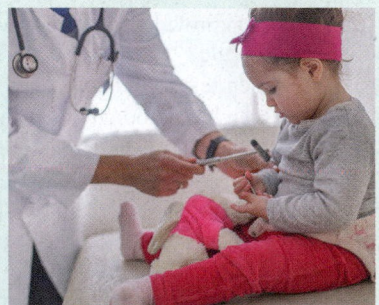

waddling walk and difficulty standing; Linda, 6 months, has pneumonia associated with problems swallowing; and Tom, 12, has progressive weakness with foot drop and difficulty walking.

"This kind of variation is not surprising," says Dr. Barr. "A mutation in one of 11 different genes related to thin filaments can cause nemaline myopathy, so you'd expect to see it present in different ways."

Dr. Barr has scheduled all three children for a muscle biopsy (surgical removal of a small piece of muscle tissue for examination). He explains to you, "We're going to use these biopsies for two things. First, we're going to look for the characteristic histology of nemaline myopathy. Second, we're going to test for malignant hyperthermia. It's not associated with nemaline myopathy but is associated with a different myopathy that has a similar presentation."

1. What does actin do? What consequences would you expect from an abnormality in actin?

2. What does troponin do?

3. All types of nemaline myopathy cause weakness, including the ones with abnormal troponin. How could abnormal troponin cause weakness?

Malignant hyperthermia is caused by a mutation in a gene related to calcium storage and release in skeletal muscle cells. In individuals with malignant hyperthermia, certain anesthetics can trigger a massive release of calcium into the cytoplasm. This leads to the generation of enough heat to substantially raise body temperature—a life-threatening reaction.

4. **+ NCLEX-STYLE** Which statement best describes the importance of calcium in skeletal muscle contraction?
    a. Extracellular calcium entering through ion channels binds to troponin to initiate muscle contraction.
    b. Extracellular calcium entering through ion channels binds to tropomyosin to initiate muscle contraction.
    c. Calcium released from the sarcoplasmic reticulum binds to tropomyosin to initiate muscle contraction.
    d. Calcium released from the sarcoplasmic reticulum binds to troponin to initiate muscle contraction.

5. What is the function of the sarcoplasmic reticulum?

6. **+ NCLEX-STYLE** Which statement best describes what might happen if a child with malignant hyperthermia is given the wrong anesthetic and the sarcoplasmic reticulum releases all its calcium?
    a. The muscles become weak and the child is unable to breathe.
    b. The blood calcium levels rise and the bones become weak.
    c. Continuous cross bridge cycling causes the muscle to contract too much.
    d. The troponin and tropomyosin stop moving off the myosin-binding sites of actin and the muscles become paralyzed.

7. Histology confirms nemaline myopathy in all three children. They are prescribed resistance training to strengthen their respiratory muscles. How will this affect their muscle cells?

9

10

The Muscular System

In this chapter, you will learn that

Skeletal muscle moves bones at joints

by asking
10.1 How do skeletal muscles act and interact?

next asking
10.2 How are skeletal muscles named?

then learning
10.5 The origin, insertion, and actions of important skeletal muscles

and exploring
10.3 Fascicle arrangements   **10.4 Lever systems**

The human body enjoys an incredibly wide range of movements. The gentle blinking of your eye, standing on tiptoe, and wielding a sledgehammer are just a sample of the actions made possible by your muscular system.

Muscle tissue includes all contractile tissues (skeletal, cardiac, and smooth muscle), but when we study the muscular system, **skeletal muscles** take center stage. This chapter focuses on these muscular "machines" that enable us to perform so many different activities. However, before describing the individual muscles in detail, we will consider:

• The manner in which muscles work with or against each other to bring about movement

• The criteria used for naming muscles

• The principles of leverage

CAREER CONNECTION

Watch a video to learn how the chapter content is used in a real health care setting. Go to Mastering A&P® > Study Area > Animations and Videos or use quick access URL https://goo.gl/88srfV

10.1 For any movement, muscles can act in one of three ways

Learning Outcomes

▶ **Describe the functions of prime movers, antagonists, and synergists.**

▶ **Explain how a muscle's position relative to a joint affects its action.**

The arrangement of body muscles permits them to work either together or in opposition to produce a wide variety of movements. As you eat, for example, you alternately raise your fork to your lips and lower it to your plate, and both sets of actions are accomplished by your arm and hand muscles. But muscles can only *pull*; they never *push*. Generally as a muscle shortens, its *insertion* (attachment on the movable bone) moves toward its *origin* (its fixed or immovable point of attachment). Whatever one muscle or muscle group can "do," another muscle or group of muscles can "undo."

Muscles can be classified into three *functional* groups:

- A muscle that has the major responsibility for producing a specific movement is a **prime mover**, or **agonist** (ag′o-nist; "leader"), of that movement. The pectoralis major muscle, which fleshes out the anterior chest (and inserts on the humerus), is a prime mover of arm flexion.

- Muscles that oppose, or reverse, a particular movement are **antagonists** (an-tag′o-nists; "against the leaders"). When a prime mover is active, the antagonist muscles may stretch or remain relaxed. Sometimes, however, antagonists help to regulate the action of a prime mover by providing some resistance. This helps to prevent the movement from going too far, or slows or stops the movement. A prime mover and its antagonist are located on opposite sides of the joint across which they act. For example, flexion of the arm by the pectoralis major muscle is antagonized by the latissimus dorsi, the prime mover for extending the arm, as shown in *Focus on Muscle Action* (**Focus Figure 10.1a, b**). Antagonists can also be prime movers in their own right: The latissimus dorsi is the prime mover of extension of the arm.

- In addition to agonists and antagonists, most movements involve the action of one or more **synergists** (sin′er-jists; *syn* = together, *erg* = work). Synergists help prime movers by adding a little extra force to the same movement or by reducing undesirable or unnecessary movements.

Let's look at synergists more closely. When a muscle crosses two or more joints, its contraction causes movement at all of the spanned joints unless other muscles act to stabilize those joints. For example, the finger flexor muscles cross both the wrist and the finger joints, but you can make a fist without bending your wrist because synergistic muscles stabilize the wrist. Additionally, as some flexors act, they may cause several other (undesirable) movements at the same joint. Synergists can prevent this, allowing the prime mover's force to be exerted in the desired direction.

When synergists immobilize a bone, or a muscle's origin so that the prime mover has a stable base on which to act, they are called **fixators** (fik′sa-terz). For example, the scapula is held to the axial skeleton only by muscles and is quite freely movable (◀ p. 227). The fixator muscles that run from the axial skeleton to the scapula can immobilize (fix) the scapula so that only the desired movements occur at the mobile shoulder joint. Additionally, muscles that help maintain upright posture are fixators.

In summary, although prime movers seem to get all the credit for causing certain movements, antagonistic and synergistic muscles are also important in producing smooth, coordinated, and precise movements. Keep in mind that a muscle may act as a prime mover in one movement, an antagonist for another movement, and as a synergist for a third movement.

Check Your Understanding

1. **APPLY** The gluteus medius muscle crosses the hip on the lateral side of the joint and abducts the thigh. What action do its antagonist muscles produce and on which side of the hip joint are they located?

For answers, see Answers Appendix.

10.2 How are skeletal muscles named?

Learning Outcome

▶ **List the criteria used in naming muscles. Provide an example to illustrate the use of each criterion.**

Skeletal muscles are named according to a number of criteria. To simplify the task of learning muscle names and actions, pay attention to the following cues.

- **Muscle location.** Some muscle names indicate the bone or body region with which the muscle is associated.

  Examples: The temporalis (tem″por-ă′lis) muscle overlies the temporal bone, and the brachialis muscle is in the arm (*brachium* = arm).

- **Muscle shape.** Some muscles are named for their distinctive shapes.

  Examples: The deltoid (del′toid) muscle is roughly triangular (*deltoid* = triangle); the right and left trapezius (trah-pe′ze-us) muscles together form a trapezoid.

- **Muscle size.** Terms such as *maximus* (largest), *minimus* (smallest), *longus* (long), and *brevis* (short) are often used in muscle names.

  Examples: The gluteus maximus and gluteus minimus are the large and small gluteus muscles, respectively.

- **Direction of muscle fibers.** The names of some muscles reveal the direction in which their fibers (and fascicles) run in reference to some imaginary line, usually the midline of the body or the longitudinal axis of a limb bone. In muscles with the term *rectus* (straight) in their names, the fibers run parallel to that imaginary line (axis). *Transversus* indicates that the muscle fibers run at right angles to that line, and *oblique* indicates that the fibers run obliquely to it.

  Examples: The rectus femoris is a straight muscle of the thigh (which contains the femur) and the transversus abdominis is the transverse muscle of the abdomen.

The action of a muscle can be inferred by the position of the muscle relative to the joint it crosses. (The examples given relate to the shoulder joint.)

 Watch a 3-D animation of this process: MasteringA&P® > Study Area > Animations & Videos > A&P Flix

(a) A muscle that crosses on the anterior side of a joint produces flexion*

Example: Pectoralis major (anterior view)

(b) A muscle that crosses on the posterior side of a joint produces extension*

Example: Latissimus dorsi (posterior view)

The latissimus dorsi is an antagonist of the pectoralis major.

(c) A muscle that crosses on the lateral side of a joint produces abduction

Example: Deltoid middle fibers (anterolateral view)

(d) A muscle that crosses on the medial side of a joint produces adduction

Example: Teres major (posterolateral view)

The teres major is an antagonist of the deltoid.

*These generalities do not apply to the knee and ankle because the lower limb is rotated during development. The muscles that cross these joints posteriorly produce flexion, and those that cross anteriorly produce extension.

- **Number of origins.** When *biceps*, *triceps*, or *quadriceps* forms part of a muscle's name, you can assume that the muscle has two, three, or four *heads*, respectively, each attached to a different origin.

    Example: The biceps brachii (bra′ke-i) muscle of the arm has two origins.

- **Location of the attachments.** Some muscles are named according to their points of origin and insertion. The origin is always named first.

    Example: The sternocleidomastoid (ster″no-kli″do-mas′-toid) muscle of the neck has a dual origin on the sternum (*sterno*) and clavicle (*cleido*), and it inserts on the mastoid process of the temporal bone.

- **Muscle action.** Muscles that are named for the movement they produce have action words such as *flexor*, *extensor*, or *adductor* as part of the muscle's name.

    Example: The adductor longus, located on the medial thigh, brings about thigh adduction. (To review the terminology for various actions, see ◀ **Chapter 8, Figures 8.5 and 8.6, pp. 258–261.**)

Often, several criteria are combined in naming a muscle. For instance, the name *extensor carpi radialis longus* tells us the muscle's action (extensor), what joint it acts on (*carpi* = wrist), and that it lies close to the radius of the forearm (radialis). It also hints at the size (longus) relative to other wrist extensor muscles. Unfortunately, not all muscle names are this descriptive.

Check Your Understanding

2. MAKE CONNECTIONS Using the criteria for naming muscles and your knowledge of bones from Chapter 7, state what the name indicates about each of the following muscles: adductor brevis, quadriceps femoris, tibialis anterior, iliacus.

────────────────────────

For answers, see Answers Appendix.

10.3 Fascicle arrangements help determine muscle shape and force

Learning Outcome

▶ Name the common patterns of muscle fascicle arrangement and relate them to power generation.

All skeletal muscles consist of fascicles (bundles of fibers), but fascicle arrangements vary, resulting in muscles with different shapes and functional capabilities. The most common patterns of fascicle arrangement are circular, convergent, parallel, and pennate (**Figure 10.1**).

Circular

The fascicular pattern is **circular** when the fascicles are arranged in concentric rings (Figure 10.1a). Muscles with this arrangement surround external body openings, which they close by contracting. A general term for such muscles is *sphincters* ("squeezers"). Examples are the orbicularis muscles surrounding the eyes and the mouth.

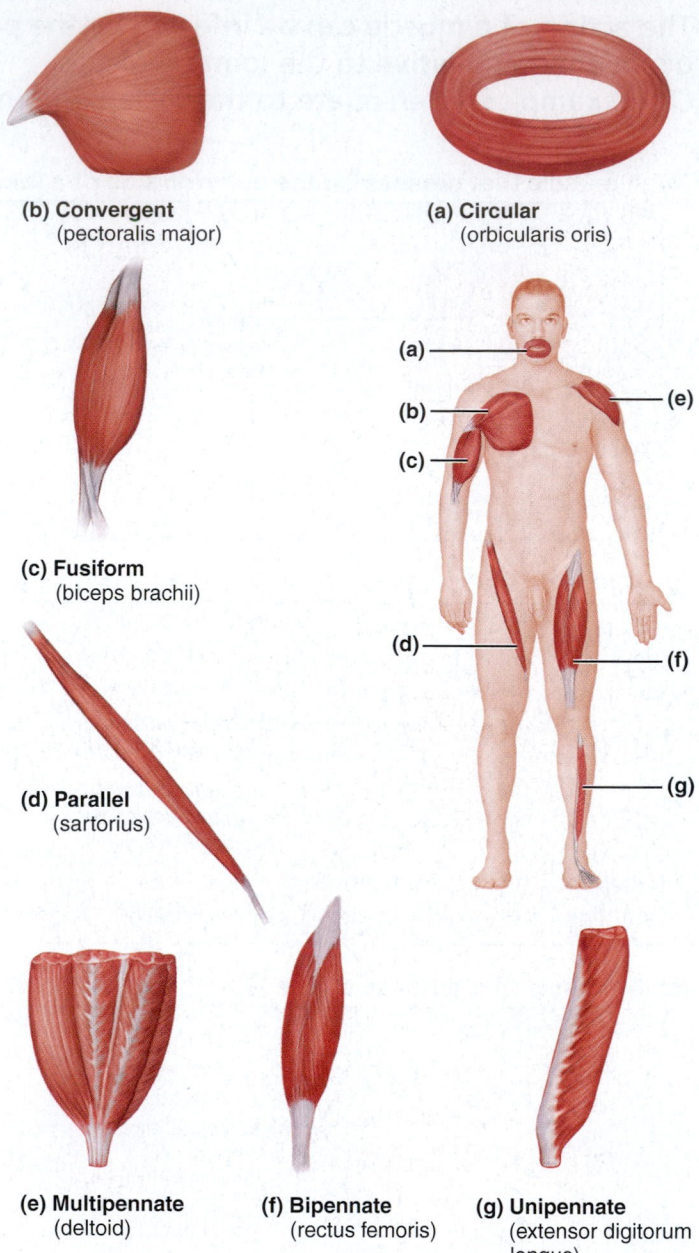

(b) Convergent (pectoralis major)

(a) Circular (orbicularis oris)

(c) Fusiform (biceps brachii)

(d) Parallel (sartorius)

(e) Multipennate (deltoid)

(f) Bipennate (rectus femoris)

(g) Unipennate (extensor digitorum longus)

Figure 10.1 Patterns of fascicle arrangement in muscles.

Convergent

A **convergent** muscle has a broad origin, and its fascicles *converge* toward a single tendon of insertion. Such a muscle is triangular or fan shaped like the pectoralis major muscle of the anterior thorax (Figure 10.1b).

Parallel

In a **parallel** arrangement, the length of the fascicles runs parallel to the long axis of the muscle. Such muscles are either *straplike* like the sartorius muscle of the thigh (Figure 10.1d), or *spindle shaped* (tapered on both ends) with an expanded *belly* (midsection), like the biceps brachii muscle of the arm (Figure 10.1c). However, some anatomists classify spindle-shaped muscles into a separate class as **fusiform muscles** (*fusi* = spindle). This is the approach we use in Figure 10.1.

Pennate

In a **pennate** (pen'āt) pattern, the fascicles (and muscle fibers) are short and they attach obliquely (*penna* = feather) to a central tendon that runs the length of the muscle. Pennate muscles come in three forms:

- *Unipennate*, in which the fascicles insert into only one side of the tendon, as in the extensor digitorum longus muscle of the leg (Figure 10.1g).
- *Bipennate*, in which the fascicles insert into the tendon from opposite sides so the muscle looks like a feather (Figure 10.1f). The rectus femoris of the thigh is bipennate.
- *Multipennate*, which looks like many feathers side by side, with all their quills inserted into one large tendon. The deltoid muscle, which forms the roundness of the shoulder, is multipennate (Figure 10.1e).

The arrangement of a muscle's fascicles determines its range of motion and its power. Because skeletal muscle fibers may shorten to about 70% of their resting length when they contract, the longer and the more nearly parallel the muscle fibers are to a muscle's long axis, the more the muscle can shorten. Muscles with parallel fascicle arrangements shorten the most, but are not usually very powerful. Muscle power depends more on the total number of muscle fibers in the muscle: The greater the number of muscle fibers, the greater the power. The stocky bipennate and multipennate muscles, which "pack in" the most fibers, shorten very little but tend to be very powerful.

Check Your Understanding

3. Of the muscles illustrated in Figure 10.1, which could shorten most? Which two would likely be most powerful? Why?

4. For each of the muscles a–d in the figure at right, identify the pattern of muscle fascicle arrangement.

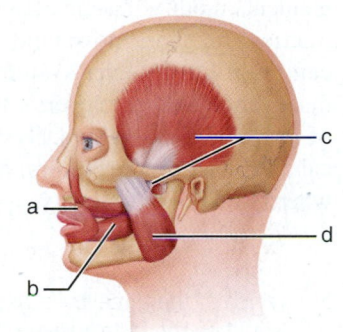

For answers, see Answers Appendix.

10.4   Muscles acting with bones form lever systems

Learning Outcomes

▶ Define lever, and explain how a lever operating at a mechanical advantage differs from one operating at a mechanical disadvantage.

▶ Name the three types of lever systems and indicate the arrangement of effort, fulcrum, and load in each. Also note the advantages of each type of lever system.

The operation of most skeletal muscles involves leverage—using a lever to move some object. A **lever** is a rigid bar that moves on a fixed point called the **fulcrum**, when a force is applied to it. The applied force, or **effort**, is used to move a resistance, or **load**. In your body, your joints are the fulcrums, and your bones act as levers. Muscle contraction provides the effort that is applied at the muscle's insertion point on a bone. The load is the bone itself, along with overlying tissues and anything else you are trying to move with that lever.

Levers: Power versus Speed

A lever allows a given effort either to (1) move a heavier load, or (2) move a load farther or faster. In the first case, if the load is close to the fulcrum and the effort is applied far from the fulcrum, a small effort can move a large load (**Figure 10.2a**). Such a lever is called a *power lever*. Power levers operate at a **mechanical advantage**. That is, the large load is moved only a small distance but the effort required is also small. For example, when using a pry bar to move a rock, each large downward push of the pry bar shifts the heavy rock only a little, and little muscular effort is required (Figure 10.2a).

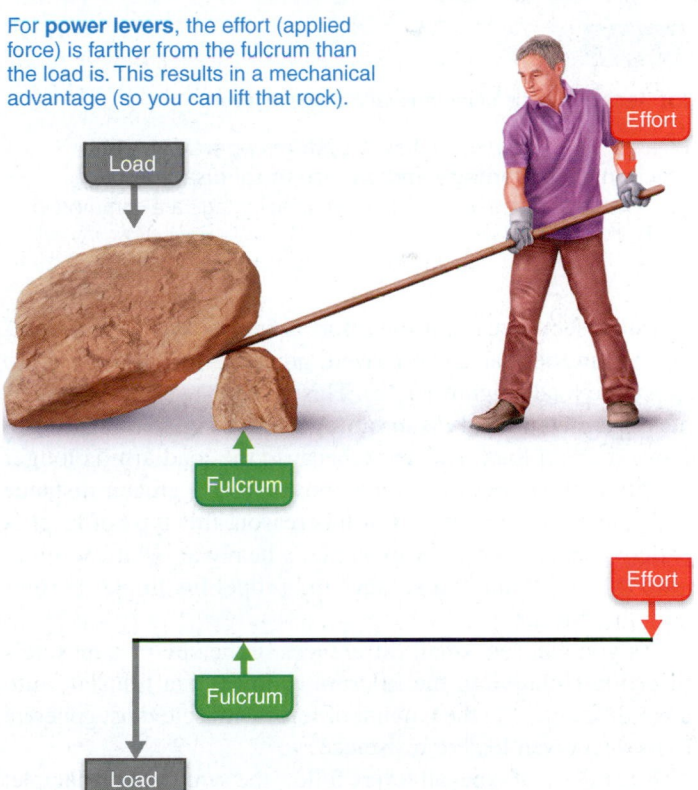

For **power levers**, the effort (applied force) is farther from the fulcrum than the load is. This results in a mechanical advantage (so you can lift that rock).

(a) Mechanical advantage with a power lever

Figure 10.2 Lever systems operating at a mechanical advantage and a mechanical disadvantage.

(Figure continues on p. 328.)

For **speed levers**, the effort (applied force) is closer to the fulcrum than the load is. This results in a mechanical disadvantage, but an increase in speed (that will send the acrobat flying).

(b) Mechanical disadvantage with a speed lever

Figure 10.2 *(continued)* **Lever systems operating at a mechanical advantage and a mechanical disadvantage.** **(b)** Levers operating at a mechanical disadvantage are common in the body.

In the second case, if the effort is applied closer to the fulcrum than the load to be moved, greater effort is required to move the load (Figure 10.2b). This type of lever system operates at a **mechanical disadvantage**, requiring a large effort to move a small load. However, because the load arm is longer than the effort arm, the load is moved over a greater distance and also at greater speed. For this reason, this type of lever is called a *speed lever*. For example, a heavy acrobat can jump onto the short end of a seesaw and propel his lighter partner high into the air.

As you can see, small differences in the site of a muscle's insertion (relative to the fulcrum or joint) can translate into large differences in the amount of force a muscle must generate to move a given load or resistance.

Regardless of type, all levers follow the same basic principle:

$$\frac{\text{Effort farther than}}{\text{load from fulcrum}} = \frac{\text{lever operates at a}}{\text{mechanical advantage}}$$

$$\frac{\text{Effort nearer than}}{\text{load to fulcrum}} = \frac{\text{lever operates at a}}{\text{mechanical disadvantage}}$$

Classes of Levers

Depending on the relative position of the three elements—effort, fulcrum, and load—a lever belongs to one of three classes.

- In a **first-class lever**, the effort is applied at one end of the lever and the load is at the other, with the fulcrum somewhere between. Seesaws and scissors are first-class levers. First-class leverage also occurs when you lift your head off your chest (**Figure 10.3a**). Some first-class levers in the body operate at a mechanical advantage (for strength), but others operate at a mechanical disadvantage (for speed and distance).

- In a **second-class lever**, the effort is applied at one end of the lever and the fulcrum is located at the other, with the load between them. A wheelbarrow demonstrates this type of lever system. Second-class levers are uncommon in the body. The best example is the act of standing on your toes (Figure 10.3b). All second-class levers in the body work at a mechanical advantage because the muscle insertion is always farther from the fulcrum than is the load. Second-class levers are levers of strength, but speed and range of motion are sacrificed for that strength.

- In a **third-class lever**, the effort is applied between the load and the fulcrum. These speedy levers always operate at a mechanical disadvantage—think of tweezers and forceps. Most skeletal muscles of the body act in third-class lever systems. An example is the activity of the biceps muscle of the arm, lifting the distal forearm and anything carried in the hand (Figure 10.3c). Third-class lever systems permit a muscle to be inserted very close to the joint across which movement occurs, which allows rapid, extensive movements (as in throwing) with relatively little shortening of the muscle. Muscles involved in third-class levers tend to be thicker and more powerful.

In conclusion, differences in the positioning of the three elements modify muscle activity with respect to speed of contraction, range of movement, and the weight of the load that can be lifted. In lever systems that operate at a mechanical disadvantage (speed levers), force is lost but speed and range of movement are gained. Systems that operate at a mechanical advantage (power levers) are slower, more stable, and used where strength is a priority.

Check Your Understanding

5. Which of the three lever systems involved in muscle mechanics would be the fastest lever—first-, second-, or third-class?

6. **APPLY** If you were using a lever to pry an old fence post out of the ground, would you want the fulcrum to be closer to you or closer to the post? Explain.

7. **DRAW** The motion produced by the deltoid muscle is shown at right. Draw two arrows on this figure (showing the locations of the load and effort), and a fulcrum. (See Figure 10.3 for an example.) What class of lever system is this? Does it operate at a mechanical advantage or disadvantage?

For answers, see Answers Appendix.

(Text continues on p. 332.)

(a) First-class lever

- Arrangement of the elements is load-fulcrum-effort
- Depending on fulcrum location, can be mechanical advantage or disadvantage

Example: scissors

In the body: A first-class lever system raises your head off your chest. The posterior neck muscles provide the effort, the atlanto-occipital joint is the fulcrum, and the weight to be lifted is the facial skeleton.

(b) Second-class lever

- Arrangement of the elements is fulcrum-load-effort
- Mechanical advantage (ideal for small, powerful movements)

Example: wheelbarrow

In the body: Second-class leverage is exerted when you stand on tip-toe. The effort is exerted by the calf muscles pulling upward on the heel; the joints of the ball of the foot are the fulcrum; and the weight of the body is the load.

(c) Third-class lever

- Arrangement of the elements is load-effort-fulcrum
- Mechanical disadvantage (ideal for fast, large movements)

Example: tweezers or forceps

In the body: Flexing the forearm by the biceps brachii muscle exemplifies third-class leverage. The effort is exerted on the proximal radius of the forearm, the fulcrum is the elbow joint, and the load is the hand and distal end of the forearm.

Figure 10.3 Lever systems.

10

10

Facial
Epicranius, frontal belly
Orbicularis oculi
Zygomaticus
Orbicularis oris

Head
Temporalis
Masseter

Neck
Platysma
Sternohyoid
Sternocleidomastoid

Shoulder
Trapezius

Thorax
Pectoralis minor
Pectoralis major
Serratus anterior
Intercostals

Deltoid

Arm
Triceps brachii
Biceps brachii
Brachialis

Forearm
Pronator teres
Brachioradialis
Flexor carpi radialis
Palmaris longus

Abdomen
Rectus abdominis
External oblique
Internal oblique
Transversus abdominis

Pelvis/thigh
Iliopsoas
Pectineus

Thigh
Tensor fascia lata
Sartorius
Adductor longus
Gracilis

Thigh
Rectus femoris
Vastus lateralis
Vastus medialis

Leg
Fibularis longus
Extensor digitorum longus
Tibialis anterior

Leg
Gastrocnemius
Soleus

Figure 10.4 Superficial muscles of the body: Anterior view. The abdominal surface has been partially dissected on the right side of the illustration to show somewhat deeper muscles.

Neck
Epicranius, occipital belly
Sternocleidomastoid

Splenius capitis
Splenius cervicis
Levator scapulae
Rhomboid minor
Rhomboid major

Trapezius

Shoulder
Deltoid
Infraspinatus
Teres major

Rhomboid major

Arm
Triceps brachii
Brachialis

Latissimus dorsi

Forearm
Brachioradialis
Extensor carpi radialis longus
Flexor carpi ulnaris
Extensor carpi ulnaris
Extensor digitorum

Hip
Gluteus medius
Gluteus maximus

Iliotibial tract

Thigh
Adductor magnus
Gracilis
Hamstrings:
Biceps femoris
Semitendinosus
Semimembranosus

Leg
Gastrocnemius
Soleus
Fibularis longus
Calcaneal (Achilles) tendon

Figure 10.5 Superficial muscles of the body: Posterior view. Trapezius removed on left side to illustrate deeper muscles.

10.5 A muscle's origin and insertion determine its action

Learning Outcome

▶ Name and identify the muscles described in Tables 10.1 to 10.17. State the typical origin, insertion, and action of each.

The overall organization of the muscular system is all the more impressive because of the sheer number of skeletal muscles in the body. There are more than 600 of them, many more than the superficial muscles shown in **Figures 10.4** and **10.5**, pp. 330–331, combined! Remembering all the names, locations, and actions of these muscles is a monumental task. Take heart; here we consider only the principal muscles (approximately 125 pairs of them). Although this number is far fewer than 600, the job of learning about all these muscles will still require a concerted effort on your part.

Memorization will be easier if you apply what you learn in a practical, or clinical, way; that is, with a *functional anatomy focus*. Once you are satisfied that you have learned the name of a muscle and can identify it on a cadaver, model, or diagram, flesh out your learning by asking yourself, "What does it do?" [It might be a good idea to review body movements (◀ pp. 258–261) before you get to this point.]

The tables that follow (**Tables 10.1–10.17**) group the muscles of the body by function and by location, roughly from head to foot. Each table is keyed to a particular figure or group of figures illustrating the muscles it describes. The legend at the beginning of each table summarizes the types of movements produced by the muscles listed and gives pointers on the way those muscles interact with one another. The table itself describes each muscle's shape, location relative to other muscles, origin and insertion, primary actions, and innervation. (Some instructors want students to defer learning the muscle innervations until they study the nervous system, so check on what is expected of you in this regard.)

As you consider each muscle, we recommend the following plan of action:

1. Be alert to the information provided by the muscle's name.
2. Read its entire description and identify the muscle on the corresponding figure. If it's a superficial muscle, also identify it on Figure 10.4 or Figure 10.5. Doing so will help you to link the description in the table to a visual image of the muscle's location in the body.
3. Relate the muscle's attachments and location to its actions. This will focus your attention on functional details that often escape student awareness. For example, both the elbow and knee joints are hinge joints that allow flexion and extension. However, the knee flexes to the dorsum of the body (the calf moves toward the posterior thigh), whereas elbow flexion carries the forearm toward the anterior aspect of the arm. In order for this to happen, leg flexors are located on the posterior thigh, while forearm flexors are found on the anterior aspect of the humerus. A visual example of this relationship is shown in *Focus on Muscle Action* (Focus Figure 10.1a and b on p. 325) with the action of the muscles acting at the shoulder. Because many muscles have several actions, we indicate the primary action of each muscle in blue type in the tables.
4. Finally, keep in mind the *best* way to learn muscle actions: Act out their movements yourself while feeling for the muscles contracting (bulging) beneath your skin.

Now you are ready to tackle the tables.

(*Text continues on p. 386.*)

MUSCLE GALLERY

Table 10.1   Muscles of the Head, Part I: Facial Expression   (Figures 10.6 and 10.7)

The muscles that promote facial expression lie in the scalp and face just deep to the skin. They are thin and variable in shape and strength, and adjacent muscles tend to be fused. They are unusual muscles in that they insert into skin (or other muscles), not bones.

In the scalp, the main muscle is the **epicranius**, which has distinct anterior and posterior parts. Muscles clothing the facial bones lift the eyebrows, flare the nostrils, open and close the eyes and mouth, and provide one of the best tools for influencing others — the smile (Figure 10.7). The tremendous importance of

facial muscles in nonverbal communication becomes especially clear when they are paralyzed, as in some stroke victims and in the expressionless "mask" of patients with Parkinson's disease.

Cranial nerve VII, the *facial nerve*, innervates all muscles listed in this table (see Table 13.2).

Chapter 15 describes the external muscles of the eyes, which direct the eyeballs, and the levator palpebrae superioris muscles, which raise the eyelids.

MUSCLE	DESCRIPTION	ORIGIN (O) AND INSERTION (I)	ACTION	NERVE SUPPLY
MUSCLES OF THE SCALP				
Epicranius (occipitofrontalis) (ep"ĭ-kra'ne-us; ok-sip"ĭ-to-fron-ta'lis) (*epi* = over; *cran* = skull)	Two-part muscle consisting of the frontal and occipital bellies connected by the epicranial aponeurosis. The alternate actions of these two muscles pull scalp forward and backward.			
• **Frontal belly** (fron'tal) (*front* = forehead)	Covers forehead and dome of skull; no bony attachments	O—epicranial aponeurosis I—skin of eyebrows and root of nose	With aponeurosis fixed, raises the eyebrows (as in surprise). Wrinkles forehead skin horizontally	Facial nerve (cranial VII)
• **Occipital belly** (ok-sip'ĭ-tal) (*occipito* = base of skull)	Overlies posterior occiput; by pulling on the epicranial aponeurosis, fixes origin of frontal belly	O—occipital and temporal bones I—epicranial aponeurosis	Fixes aponeurosis and pulls scalp posteriorly	Facial nerve
MUSCLES OF THE FACE				
Corrugator supercilii (kor'ah-ga-ter soo"per-sĭ'le-i) (*corrugo* = wrinkle; *supercilium* = eyebrow)	Small muscle; acts with orbicularis oculi	O—arch of frontal bone above nasal bone I—skin of eyebrow	Pulls eyebrows medially and inferiorly; wrinkles skin of forehead vertically (as in frowning)	Facial nerve

Corrugator supercilii

Platysma

Frontal belly

Epicranial aponeurosis

Occipital belly

Epicranius

Sternocleidomastoid

(a)

Practice Human Cadaver lab questions:
MasteringA&P® > Study Area>Lab Tools> PAL

Figure 10.6 Lateral view of muscles of the scalp, face, and neck. (a) Cadaver dissection.

MUSCLE GALLERY

Table 10.1 Muscles of the Head, Part I: Facial Expression (Figures 10.6 and 10.7) *(continued)*

MUSCLE	DESCRIPTION	ORIGIN (O) AND INSERTION (I)	ACTION	NERVE SUPPLY
Orbicularis oculi (or-bik′u-lar-is ok′u-li) (*orb* = circular; *ocul* = eye)	Thin, flat sphincter muscle of eyelid; surrounds rim of the orbit	O—frontal and maxillary bones and ligaments around orbit I—tissue of eyelid	Closes eye; produces blinking and squinting; pulls eyebrows inferiorly	Facial nerve (cranial VII)
Zygomaticus—major and minor (zi-go-mat′ĭ-kus) (*zygomatic* = cheekbone)	Muscle pair extending diagonally from cheekbone to corner of mouth	O—zygomatic bone I—skin and muscle at corner of mouth	Raises lateral corners of mouth (smiling muscle)	Facial nerve
Risorius (ri-zor′e-us) (*risor* = laughter)	Slender muscle inferior and lateral to zygomaticus	O—lateral fascia associated with masseter muscle I—skin at angle of mouth	Pulls corner of lip laterally; tenses lips; synergist of zygomaticus	Facial nerve
Levator labii superioris (lĕ-va′tor la′be-i soo-per″e-or′is) (*leva* = raise; *labi* = lip; *superior* = above, over)	Thin muscle between orbicularis oris and inferior eye margin	O—zygomatic bone and infraorbital margin of maxilla I—skin and muscle of upper lip	Opens lips; raises and furrows upper lip	Facial nerve
Depressor labii inferioris (de-pres′or la′be-i in-fer″e-or′is) (*depressor* = depresses; *infer* = below)	Small muscle running from mandible to lower lip	O—body of mandible lateral to its midline I—skin and muscle of lower lip	Pulls lower lip inferiorly (as in a pout)	Facial nerve
Depressor anguli oris (ang′gu-li or-is) (*angul* = angle, corner; *or* = mouth)	Small muscle lateral to depressor labii inferioris	O—body of mandible below incisors I—skin and muscle at angle of mouth below insertion of zygomaticus	Pulls corners of mouth down and laterally (a grimace); zygomaticus antagonist	Facial nerve
Orbicularis oris	Multilayered muscle of the lips with fibers that run in many different directions; most run circularly	O—arises indirectly from maxilla and mandible; fibers blend with fibers of other facial muscles associated with the lips I—encircles mouth; inserts into muscle and skin at angles of mouth	Closes lips; purses and protrudes lips; kissing and whistling	Facial nerve
Mentalis (men-ta′lis) (*ment* = chin)	One of the muscle pair forming a V-shaped muscle mass on chin	O—mandible below incisors I—skin of chin	Wrinkles chin; protrudes lower lip (as in a pout)	Facial nerve
Buccinator (buk′sĭ-na″tor) (*bucc* = cheek or "trumpeter")	Thin, horizontal cheek muscle; principal muscle of cheek; deep to masseter (see also Figure 10.8)	O—molar region of maxilla and mandible I—orbicularis oris	Compresses cheek (as in whistling and sucking); holds food between teeth during chewing; pulls corner of mouth laterally; well developed in nursing infants	Facial nerve
Platysma (plah-tiz′mah) (*platy* = broad, flat)	Unpaired, thin, sheetlike superficial neck muscle; not strictly a head muscle, but plays a role in facial expression	O—fascia of chest (over pectoral muscles and deltoid) I—lower margin of mandible, and skin and muscle at corner of mouth	Tenses skin of neck (as during shaving); helps depress mandible; pulls lower lip back and down, producing downward sag of mouth	Facial nerve

10

MUSCLE GALLERY

Table 10.1 *(continued)*

Epicranial aponeurosis

Frontal belly ⎤
Occipital belly ⎦ Epicranius

Corrugator supercilii

Orbicularis oculi

Levator labii superioris

Zygomaticus minor and major

Buccinator

Risorius

Orbicularis oris

Mentalis

Depressor labii inferioris

Depressor anguli oris

Platysma

Temporalis

Masseter

Sternocleidomastoid

Trapezius

Splenius capitis

(b)

Figure 10.6 *(continued)*
Lateral view of muscles of the scalp, face, and neck. (b) Superficial muscles.

Zygomaticus major (smile)

Orbicularis oris (pucker)

Mentalis (pout)

Platysma (tense neck)

Corrugator supercilii (angry eyebrows)

Orbicularis oculi (blink)

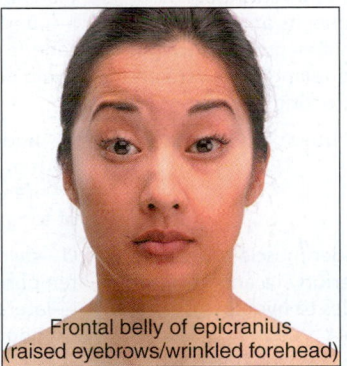

Frontal belly of epicranius (raised eyebrows/wrinkled forehead)

Figure 10.7 Muscles used in facial expressions.

MUSCLE GALLERY

Table 10.2   Muscles of the Head, Part II: Mastication and Tongue Movement   (Figure 10.8)

Four pairs of muscles are involved in mastication (chewing). All are innervated by the *mandibular branch* of *cranial nerve V*.

The prime movers of jaw closure (and biting) are the powerful **masseter** and **temporalis** muscles, which are easily palpated when the teeth are clenched (Figure 10.8a). The **pterygoid** muscles produce side-to-side grinding (Figure 10.8b). The **buccinator** muscles (see Table 10.1) also play a role in chewing. Normally, gravity is sufficient to depress the mandible, but if there is resistance to jaw opening, neck muscles such as the digastric and mylohyoid muscles (see Table 10.3) are activated.

The tongue is composed of muscle fibers that curl, squeeze, and fold the tongue during speaking and chewing. These **intrinsic tongue muscles** change the shape of the tongue and contribute to its exceptional nimbleness, but they do not really move the tongue. We discuss them in Chapter 23 with the digestive system.

In this table, we consider only the **extrinsic tongue muscles**, which anchor and move the tongue (Figure 10.8c). *Cranial nerve XII,* the *hypoglossal nerve,* innervates all extrinsic tongue muscles (see Table 13.2).

MUSCLE	DESCRIPTION	ORIGIN (O) AND INSERTION (I)	ACTION	NERVE SUPPLY
MUSCLES OF MASTICATION				
Masseter (mah-se′ter) (*maseter* = chewer)	Powerful muscle that covers lateral aspect of mandibular ramus	O—zygomatic arch and zygomatic bone I—angle and ramus of mandible	Prime mover of jaw closure; elevates mandible	Trigeminal nerve (cranial V)
Temporalis (tem″por-ă′lis) (*tempora* = time; pertaining to the temporal bone)	Fan-shaped muscle that covers parts of the temporal, frontal, and parietal bones	O—temporal fossa I—coronoid process of mandible via a tendon that passes deep to zygomatic arch	Closes jaw; elevates and retracts mandible; maintains position of the mandible at rest; deep anterior part may help protract mandible	Trigeminal nerve
Medial pterygoid (me′de-ul ter′ĭ-goid) (*medial* = toward median plane; *pterygoid* = winglike)	Deep two-headed muscle that runs along internal surface of mandible and is largely concealed by that bone	O—medial surface of lateral pterygoid plate of sphenoid bone, maxilla, and palatine bone I—medial surface of mandible near its angle	Acts with the lateral pterygoid muscle to protract (pull anteriorly) the mandible and promote side-to-side grinding movements; synergist of temporalis and masseter muscles in elevation of the mandible	Trigeminal nerve
Lateral pterygoid (*lateral* = away from median plane)	Deep two-headed muscle; lies superior to medial pterygoid muscle	O—greater wing and lateral pterygoid plate of sphenoid bone I—condylar process of mandible and capsule of temporomandibular joint	Provides side-to-side grinding movements and protracts mandible	Trigeminal nerve
Buccinator	See Table 10.1	See Table 10.1	Compresses the cheek; keeps food between grinding surfaces of teeth during chewing	Facial nerve (cranial VII)
MUSCLES PROMOTING TONGUE MOVEMENTS (EXTRINSIC TONGUE MUSCLES)				
Genioglossus (je″ne-o-glah′sus) (*geni* = chin; *glossus* = tongue)	Fan-shaped muscle; forms bulk of inferior part of tongue; its attachment to mandible prevents tongue from falling backward and obstructing breathing	O—internal surface of mandible near symphysis I—inferior aspect of the tongue and body of hyoid bone	Protracts tongue; can depress or act in synergy with other extrinsic muscles to retract tongue	Hypoglossal nerve (cranial XII)
Hyoglossus (hi′o-glos″us) (*hyo* = pertaining to hyoid bone)	Flat, quadrilateral muscle	O—body and greater horn of hyoid bone I—lateral inferior aspect of tongue	Depresses tongue and pulls its sides inferiorly	Hypoglossal nerve
Styloglossus (sti-lo-glah′sus) (*stylo* = pertaining to styloid process)	Slender muscle running superiorly to and at right angles to hyoglossus	O—styloid process of temporal bone I—lateral inferior aspect of tongue	Retracts and elevates tongue	Hypoglossal nerve

(a)

(b)

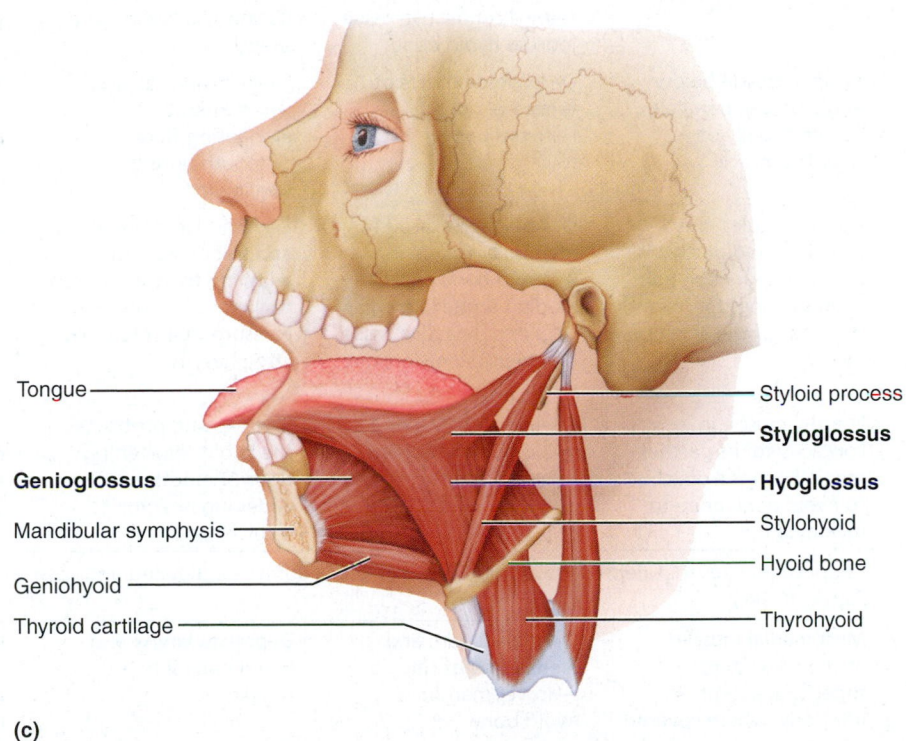

(c)

Figure 10.8 Muscles of mastication and tongue movement. Left lateral view.
(a) Superficial muscles of mastication. **(b)** Deep muscles of mastication. **(c)** Extrinsic muscles of the tongue and associated suprahyoid muscles of the throat.

MUSCLE GALLERY

Table 10.3   Muscles of the Anterior Neck and Throat: Swallowing   (Figure 10.9)

The sternocleidomastoid muscle divides the neck into two triangles (anterior and posterior; Figure 10.9a). The muscles of the *anterior* triangle are divided into **suprahyoid** and **infrahyoid muscles** (above and below the hyoid bone, respectively). Most of these muscles, with the pharyngeal constrictor muscles, participate in swallowing.

Swallowing begins when the tongue and buccinator muscles of the cheeks squeeze the food back along the roof of the mouth toward the pharynx. A rapid series of sequential muscular movements in the posterior mouth and pharynx complete the process:

1. The *suprahyoid muscles* pull the hyoid bone superiorly and anteriorly toward the mandible, which widens the pharynx

to receive the food (Figure 10.9c). The larynx is also pulled superiorly and anteriorly under the cover of the flaplike epiglottis, a maneuver that closes off the respiratory passageway (larynx) so that food is not inhaled into the lungs.

2. Small muscles that elevate the soft palate close off the nasal passages to prevent food from entering the superior nasal cavity. (These muscles, the *tensor* and *levator veli palatini*, are not described in the table but are illustrated in Figure 10.9c.)

3. The **pharyngeal constrictor muscles** in the wall of the pharynx propel food inferiorly into the esophagus.

4. The *infrahyoid muscles* pull the hyoid bone and larynx to their more inferior positions as swallowing ends.

MUSCLE	DESCRIPTION	ORIGIN (O) AND INSERTION (I)	ACTION	NERVE SUPPLY
SUPRAHYOID MUSCLES (soo"prah-hi′oid)	Muscles that help form floor of oral cavity, anchor tongue, elevate hyoid, and move larynx superiorly during swallowing; lie superior to hyoid bone			
Digastric (di-gas′trik) (*di* = two; *gaster* = belly)	Consists of two bellies united by an intermediate tendon, forming a V shape under the chin	O—lower margin of mandible (anterior belly) and mastoid process of temporal bone (posterior belly) I—by a connective tissue loop to hyoid bone	Open mouth and depress mandible; acting in synergy, the digastric muscles elevate hyoid bone and steady it during swallowing and speech	Mandibular branch of trigeminal nerve (cranial V) for anterior belly; facial nerve (cranial VII) for posterior belly
Stylohyoid (sti"lo-hi′oid) (also see Figure 10.8c)	Slender muscle below angle of jaw; parallels posterior belly of digastric muscle	O—styloid process of temporal bone I—hyoid bone	Elevates and retracts hyoid, thereby elongating floor of mouth during swallowing	Facial nerve
Mylohyoid (mi"lo-hi′oid) (*myle* = molar)	Flat, triangular muscle just deep to digastric muscle; this muscle pair makes a sling that forms the floor of the anterior mouth	O—medial surface of mandible I—hyoid bone and median raphe (a median strip of connective tissue between the mylohyoid muscles)	Elevates hyoid bone and floor of mouth, enabling tongue to exert backward and upward pressure that forces food into pharynx	Mandibular branch of trigeminal nerve
Geniohyoid (je′ne-o-hi″oid) (*geni* = chin) (also see Figure 10.8c)	Narrow muscle in contact with its partner medially; runs from chin to hyoid bone deep to mylohyoid	O—inner surface of mandibular symphysis I—hyoid bone	Elevates and protracts the hyoid, shortening floor of mouth and widening pharynx to receive food	First cervical spinal nerve via hypoglossal nerve (cranial XII)
INFRAHYOID MUSCLES	Straplike muscles that depress the hyoid bone and larynx during swallowing and speaking (see also Figure 10.10c)			
Sternohyoid (ster"no-hi′oid) (*sterno* = sternum)	Most medial muscle of the neck: thin; superficial except inferiorly, where covered by sternocleidomastoid	O—manubrium and medial end of clavicle I—lower margin of hyoid bone	Depresses larynx and hyoid bone if mandible is fixed	Cervical spinal nerves 1–3 (C_1–C_3) through ansa cervicalis (slender nerve root in cervical plexus)
Sternothyroid (ster′no-thi′roid) (*thyro* = thyroid cartilage)	Lateral and deep to sternohyoid	O—posterior surface of manubrium of sternum I—thyroid cartilage	Depresses larynx and hyoid bone	As for sternohyoid
Omohyoid (o"mo-hi′oid) (*omo* = shoulder)	Straplike muscle with two bellies united by an intermediate tendon; lateral to sternohyoid	O—superior surface of scapula I—hyoid bone, lower border	Depresses and retracts hyoid bone	As for sternohyoid
Thyrohyoid (thi″ro-hi′oid) (also see Figure 10.8c)	Appears as a superior continuation of sternothyroid muscle	O—thyroid cartilage I—hyoid bone	Depresses hyoid bone or elevates larynx if hyoid is fixed	First cervical nerve via hypoglossal nerve

10

MUSCLE GALLERY

Table 10.3 *(continued)*

MUSCLE	DESCRIPTION	ORIGIN (O) AND INSERTION (I)	ACTION	NERVE SUPPLY
PHARYNGEAL CONSTRICTOR MUSCLES (far-rin′je-al)				
Superior, middle, and inferior pharyngeal constrictor muscles	Three paired muscles whose fibers run circularly in pharynx wall; superior muscle is innermost and inferior one is outermost; substantial overlap	O—attached anteriorly to mandible and medial pterygoid plate (superior), hyoid bone (middle), and laryngeal cartilages (inferior) I—posterior median raphe of pharynx	Constrict pharynx during swallowing, which propels food to esophagus (via a wavelike squeezing action called peristalsis)	Pharyngeal plexus [branches of vagus nerve (X)]

Figure 10.9 Muscles of the anterior neck and throat used in swallowing. (a) Anterior view of the suprahyoid and infrahyoid muscles. The sternocleidomastoid muscle (not involved in swallowing) is shown on the right side of the illustration to provide an anatomical landmark. Deeper neck muscles are illustrated on the left side. Note that a raphe is a seam of fibrous tissue. **(b)** Cadaver dissection of suprahyoid and infrahyoid muscles. **(c)** Lateral view of the constrictor muscles of the pharynx shown in relationship to the buccinator and the hyoglossus muscle (which moves the tongue).

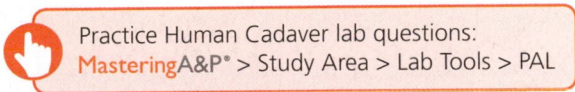

Practice Human Cadaver lab questions: MasteringA&P® > Study Area > Lab Tools > PAL

MUSCLE GALLERY

Table 10.4	Muscles of the Neck and Vertebral Column: Head Movements and Trunk Extension (Figure 10.10)

Head movements

Muscles originating from the axial skeleton move the head. The major head flexors are the **sternocleidomastoid muscles** (Figure 10.10a, c), with help from the suprahyoid and infrahyoid muscles described in Table 10.3. Lateral head movements (rotating or tilting the head) result when the muscles on only one side of the neck contract. These actions are produced by the sternocleidomastoids and a number of deeper neck muscles, considered in this table. Head extension is aided by the trapezius muscles of the back, but the main extensors of the head are the **splenius** muscles deep to the trapezius muscles (Figure 10.10b).

Trunk extension

Trunk extension is brought about by the *deep* or *intrinsic back muscles* associated with the vertebral column. These muscles also maintain the normal curvatures of the spine, acting as postural muscles. As you consider these back muscles, keep in mind that they are deep. The superficial back muscles that cover them are concerned primarily with moving the shoulder girdle and upper limbs (see Tables 10.8 and 10.9).

The deep muscles of the back form a broad, thick column extending from the sacrum to the skull. Many muscles of varying length contribute to this mass. Think of each muscle as a string that when pulled causes one or several vertebrae to extend or to rotate on the vertebrae below. The largest deep back muscle group is the **erector spinae** group (Figure 10.10d). Because the origins and insertions of the different muscle groups overlap extensively, and many of these muscles are long, large regions of the vertebral column can be moved simultaneously and smoothly. Acting together, the deep back muscles extend (or hyperextend) the spine, but muscle contraction on only one side causes the spine to bend laterally (flex). Lateral flexion is automatically accompanied by some degree of rotation of the vertebral column. During vertebral movements, the articular facets of the vertebrae glide on each other.

In addition to the long back muscles, there are a number of short muscles that extend from one vertebra to the next. These small muscles act primarily as synergists in extending and rotating the spine and as spine stabilizers. They are not described in the table but you can deduce their actions by examining their origins and insertions in Figure 10.10e.

The trunk muscles that we consider in this table are *extensors*. The more superficial muscles, which have other functions, are considered in subsequent tables.

MUSCLE	DESCRIPTION	ORIGIN (O) AND INSERTION (I)	ACTION	NERVE SUPPLY
ANTEROLATERAL NECK MUSCLES (Figure 10.10a and c)				
Sternocleidomastoid (ster"no-kli"do-mas'toid) (*sterno* = breastbone; *cleido* = clavicle; *mastoid* = mastoid process)	Two-headed muscle located deep to platysma on anterolateral surface of neck; fleshy parts on either side of neck delineate limits of anterior and posterior triangles; key muscular landmark in neck	O—manubrium of sternum and medial portion of clavicle I—mastoid process of temporal bone and superior nuchal line of occipital bone	Flexes and laterally rotates the head; simultaneous contraction of both muscles flexes neck, as when raising head when lying on back; acting alone, each muscle rotates head toward shoulder on opposite side and tilts or laterally flexes head to its own side	Accessory nerve (cranial nerve XI) and branches of cervical spinal nerves C_2 and C_3 (ventral rami)
Scalenes (ska'lēnz)— anterior, middle, and posterior (*scalene* = uneven)	Located more laterally than anteriorly on neck; deep to platysma and sternocleidomastoid	O—transverse processes of cervical vertebrae I—anterolaterally on first two ribs	Elevate first two ribs (aid in inspiration); flex and rotate neck	Cervical spinal nerves
INTRINSIC MUSCLES OF THE BACK (Figure 10.10b, d, e)				
Splenius (sple'ne-us)—capitis and cervicis portions (kă'pĭ-tis; ser-vis'us) (*splenion* = bandage; *caput* = head; *cervi* = neck) (Figures 10.10b and 10.6b)	Broad two-part superficial muscle (capitis and cervicis parts) extending from upper thoracic vertebrae to skull; capitis portion known as "bandage muscle" because it covers and holds down deeper neck muscles	O—ligamentum nuchae,* spinous processes of vertebrae C_7–T_6 I—mastoid process of temporal bone and occipital bone (capitis); transverse processes of C_2–C_4 vertebrae (cervicis)	Extend or hyperextend head; when splenius muscles on one side are activated, head rotates and bends laterally toward same side	Cervical spinal nerves (dorsal rami)

*The ligamentum nuchae (lig"ah-men'tum noo'ke) is a strong, elastic ligament extending from the occipital bone of the skull along the tips of the spinous processes of the cervical vertebrae (Figure 10.10d). It binds the cervical vertebrae together and inhibits excessive head and neck flexion, thus preventing damage to the spinal cord in the vertebral canal.

MUSCLE GALLERY

Table 10.4   *(continued)*

(a) Anterior

- 1st cervical vertebra
- Sternocleido-mastoid
- Base of occipital bone
- Mastoid process
- **Middle scalene**
- **Anterior scalene**
- **Posterior scalene**

(b) Posterior

- Mastoid process
- **Splenius capitis**
- Spinous processes of the vertebrae
- **Splenius cervicis**

(c)

- Platysma (cut)
- **Sternocleidomastoid** (cut)
- Internal jugular vein
- **Omohyoid**
- **Sternohyoid**
- **Sternothyroid**
- **Sternocleidomastoid**
- Pectoralis major

Figure 10.10   Muscles of the neck and vertebral column that move the head and trunk. (a) Muscles of the anterolateral neck; superficial platysma muscle and the deeper neck muscles removed. **(b)** Deep muscles of the posterior neck; superficial muscles removed. **(c)** Cadaver dissection of the anterior and lateral regions of the neck.

Practice Human Cadaver lab questions:
MasteringA&P° > Study Area > Lab Tools > PAL

(Figure continues on p. 343)

MUSCLE GALLERY

Table 10.4	Muscles of the Neck and Vertebral Column: Head Movements and Trunk Extension (Figure 10.10) *(continued)*			
MUSCLE	**DESCRIPTION**	**ORIGIN (O) AND INSERTION (I)**	**ACTION**	**NERVE SUPPLY**
Erector spinae (e-rek′tor spi′ne) (Figure 10.10d, left side)	Prime mover of back extension. Each side consists of three columns—the iliocostalis, longissimus, and spinalis muscles—forming the intermediate layer of intrinsic back muscles. Erector spinae provide resistance that helps control action of bending forward at the waist and act as powerful extensors to promote return to erect position. During full flexion (i.e., when touching fingertips to floor), erector spinae are relaxed and strain is borne entirely by ligaments of back. On reversing the movement, these muscles are initially inactive, and extension is initiated by hamstring muscles of thighs and gluteus maximus muscles of buttocks. As a result of this peculiarity, lifting a load or moving suddenly from a bent-over position can injure muscles and ligaments of back and intervertebral discs. Erector spinae muscles readily go into painful spasms following injury to back structures.			
• **Iliocostalis** (il″e-o-kos-tă′lis)—lumborum, thoracis, and cervicis portions (lum′bor-um; tho-ra′sis) (*ilio* = ilium; *cost* = rib; *thorac* = thorax)	Most lateral muscle group of erector spinae muscles; extend from pelvis to neck	O—iliac crests (lumborum); inferior 6 ribs (thoracis); ribs 3 to 6 (cervicis) I—angles of ribs (lumborum and thoracis); transverse processes of cervical vertebrae C_4–C_6 (cervicis)	Acting together, extend the vertebral column (maintain erect posture); acting on one side, laterally flex the vertebral column toward same side	Spinal nerves (dorsal rami)
• **Longissimus** (lon-jis′ĭ-mus)—thoracis, cervicis, and capitis parts (*longissimus* = longest)	Intermediate three-part muscle group of erector spinae; extend by many muscle slips from lumbar region to skull; mainly pass between transverse processes of vertebrae	O—transverse processes of lumbar through cervical vertebrae I—transverse processes of thoracic or cervical vertebrae and to ribs superior to origin as indicated by name; capitis inserts into mastoid process of temporal bone	Acting together, extend the vertebral column; acting on one side, laterally flex the vertebral column toward the same side; capitis extends and rotates the head toward the same side	Spinal nerves (dorsal rami)
• **Spinalis** (spi-nă′lis)—thoracis and cervicis parts (*spin* = vertebral column, spine)	Most medial muscle column of erector spinae; cervicis usually rudimentary and poorly defined	O—spinous process of upper lumbar and lower thoracic vertebrae I—spinous process of upper thoracic and cervical vertebrae	Extends vertebral column	Spinal nerves (dorsal rami)
Semispinalis (sem′e-spĭ-nă′lis)—thoracis, cervicis, and capitis regions (*semi* = half) (Figure 10.10d, right side)	Composite muscle forming part of deep layer of intrinsic back muscles; extends from thoracic region to head	O—transverse processes of C_7–T_{12} I—occipital bone (capitis) and spinous processes of cervical (cervicis) and thoracic vertebrae T_1–T_4 (thoracis)	Extends vertebral column and head and rotates them to opposite side; acts synergistically with sternocleidomastoid muscle of opposite side	Spinal nerves (dorsal rami)
Quadratus lumborum (kwod-ra′tus lum-bor′um) (*quad* = four-sided; *lumb* = lumbar region) (See also Figure 10.20a)	Fleshy muscle forming part of posterior abdominal wall	O—iliac crest and lumbar fascia I—transverse processes of lumbar vertebrae L_1–L_4 and lower margin of 12th rib	Laterally flexes vertebral column when acting separately; when pair acts jointly, lumbar spine is extended and 12th rib is fixed; maintains upright posture; assists in forced inspiration	T_{12} and upper lumbar spinal nerves (ventral rami)

Mastoid process
of temporal bone

Longissimus capitis

Iliocostalis cervicis
Longissimus cervicis

Iliocostalis thoracis

**Longissimus
thoracis**

Spinalis thoracis

**Erector
spinae** — [Iliocostalis]
[Longissimus]
[Spinalis]

**Iliocostalis
lumborum**

External oblique

Ligamentum
nuchae

**Semispinalis
capitis**

**Semispinalis
cervicis**

**Semispinalis
thoracis**

Multifidus

**Quadratus
lumborum**

(d)

O = origin
I = insertion

Intertransversarius

Rotatores

Multifidus

Interspinales

Figure 10.10 *(continued)* **Muscles of the neck and vertebral
column that move the head and trunk. (d)** Deep muscles of the
back; superficial, intermediate, and splenius muscles removed. The
erector spinae muscles are shown on the left; deeper semispinalis
muscles are shown on the right. **(e)** The deepest muscles of the back
associated with the vertebral column.

(e)

MUSCLE GALLERY

Table 10.5   Deep Muscles of the Thorax: Breathing   (Figure 10.11)

The deep muscles of the thorax promote movements necessary for breathing. Breathing consists of two phases—inspiration (inhaling) and expiration (exhaling)—produced by cyclic changes in the volume of the thoracic cavity.

The thoracic muscles are very short: Most run only from one rib to the next. They form three layers in the wall of the thorax. The **external intercostal muscles** form the more superficial layer (Figure 10.11a). They lift the rib cage, which increases the anterior to posterior and side-to-side dimensions. For this reason, the external intercostals function during inspiration. The **internal intercostal muscles** form the intermediate muscle layer. They may aid active (forced) expiration by depressing the rib cage. (However, quiet expiration is largely passive, resulting from relaxation of the external intercostals and diaphragm and elastic recoil of the lungs.) The third and deepest muscle layer of the thoracic wall attaches to the internal surfaces of the ribs. It has three discontinuous parts. These are small and their function is unclear, so they are not listed in this table.

The **diaphragm**, the most important muscle of inspiration, forms a muscular partition between the thoracic and abdominopelvic cavities (Figure 10.11b, c). In the relaxed state, the diaphragm is dome shaped. When it contracts it moves inferiorly and flattens, increasing the volume of the thoracic cavity, which pulls air into the respiratory passageways. The alternating rhythmic contraction and relaxation of the diaphragm also causes pressure changes in the abdominopelvic cavity below that facilitate the return of blood to the heart.

In addition you can contract the diaphragm voluntarily to push down on the abdominal viscera and increase the pressure in the abdominopelvic cavity to help expel the contents of pelvic organ (urine, feces, or a baby) or lift weights. When you take a deep breath to fix the diaphragm, the abdomen becomes a firm pillar that will not buckle under the weight being lifted. Needless to say, it is important to have good control of the urinary bladder and anal sphincters during such maneuvers.

With the exception of the diaphragm, which is served by the *phrenic nerves*, the muscles listed in this table are served by the *intercostal nerves*, which run between the ribs.

Forced breathing (as during exercise) calls into play a number of other muscles that insert into the ribs. For example, during forced inspiration the scalene and sternocleidomastoid muscles of the neck help lift the ribs. Forced expiration is aided by muscles that pull the ribs inferiorly and those that push the diaphragm superiorly by compressing the abdominal contents (abdominal wall muscles).

MUSCLE	DESCRIPTION	ORIGIN (O) AND INSERTION (I)	ACTION	NERVE SUPPLY
External intercostals (in″ter-kos′talz) (*external* = toward the outside; *inter* = between; *cost* = rib)	11 pairs lie between ribs; fibers run obliquely (down and forward) from each rib to rib below; in lower intercostal spaces, fibers are continuous with external oblique muscle, forming part of abdominal wall	O—inferior border of rib above I—superior border of rib below	With first ribs fixed by scalene muscles, pull ribs toward one another to elevate rib cage; aid in inspiration; synergists of diaphragm	Intercostal nerves
Internal intercostals (*internal* = toward the inside, deep)	11 pairs lie between ribs; fibers run deep to and at right angles to those of external intercostals (i.e., run downward and posteriorly); lower internal intercostal muscles are continuous with fibers of internal oblique muscle of abdominal wall	O—superior border of rib below I—inferior border (costal groove) of rib above	With 12th ribs fixed by quadratus lumborum, muscles of posterior abdominal wall, and oblique muscles of the abdominal wall, they pull ribs together and depress rib cage; aid forced expiration; antagonistic to external intercostals	Intercostal nerves
Diaphragm (di′ah-fram) (*dia* = across; *phragm* = partition)	Broad muscle pierced by the aorta, inferior vena cava, and esophagus; forms floor of thoracic cavity; dome shaped in relaxed state; fibers converge from margins of thoracic cage toward a boomerang-shaped central tendon	O—inferior, internal surface of rib cage and sternum, costal cartilages of last six ribs and lumbar vertebrae I—central tendon	Prime mover of inspiration; flattens on contraction, increasing vertical dimensions of thorax; when strongly contracted, dramatically increases intra-abdominal pressure	Phrenic nerves

MUSCLE GALLERY

Table 10.5 *(continued)*

Figure 10.11 Muscles of respiration. (a) Deep muscles of the thorax, right lateral view. The external intercostals (inspiratory muscles) are superficial to the internal intercostals (expiratory muscles). These two muscle layers run obliquely and at right angles to each other. **(b)** Inferior view of the diaphragm, the prime mover of inspiration. Notice that its muscle fibers converge toward a central tendon, an arrangement that causes the diaphragm to flatten and move inferiorly as it contracts. **(c)** Cadaver dissection of the diaphragm, superior view.

Practice Human Cadaver lab questions:
MasteringA&P® > Study Area > Lab Tools > PAL

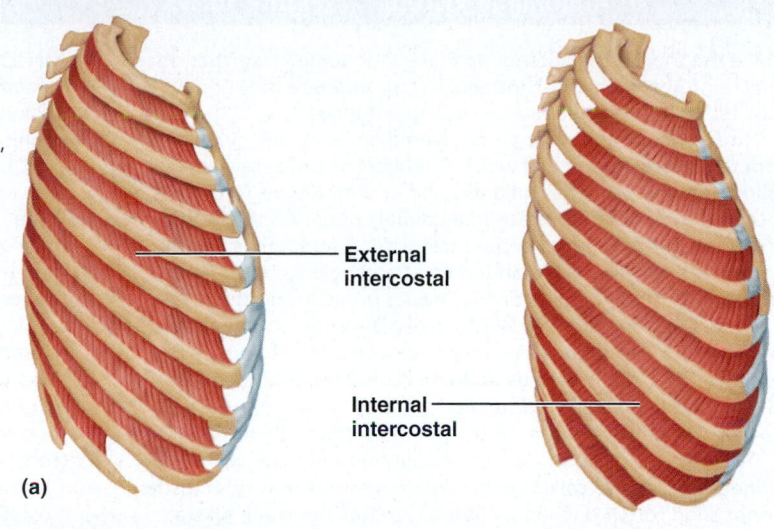

External intercostal

Internal intercostal

(a)

Xiphoid process of sternum

Foramen for inferior vena cava

Foramen for esophagus

Costal cartilage

Central tendon of diaphragm

Diaphragm

Foramen for aorta

Lumbar vertebra

12th rib

Quadratus lumborum

Psoas major

(b)

Central tendon of diaphragm

Body of thoracic vertebra

Aorta

Diaphragm (muscular part)

Inferior vena cava

Esophagus

Pericardial sac

Pericardium (cut)

(c)

MUSCLE GALLERY

Table 10.6	Muscles of the Abdominal Wall: Trunk Movements and Compression of Abdominal Viscera (Figure 10.12)

Unlike the thorax, the anterior and lateral abdominal wall has no bony reinforcements (ribs). Instead, it is a composite of four paired muscles, their associated fasciae, and aponeuroses.

Three broad flat muscle pairs, layered one over the next, form the lateral abdominal walls: The fibers of the **external oblique muscle** run inferomedially and at right angles to those of the **internal oblique**, which immediately underlies it (Figure 10.12a, b). The fibers of the deep **transversus abdominis muscle** run horizontally across the abdomen at an angle to both. This alternation of fascicle directions provides great strength.

These three muscles blend into broad insertion aponeuroses anteriorly. The aponeuroses, in turn, enclose a fourth muscle pair medially, the straplike **rectus abdominis muscles**, and then fuse, forming the **linea alba** ("white line"), a tendinous raphe (seam) that runs from the sternum to the pubic symphysis (Figure 10.12a, c). The snug enclosure of the rectus abdominis muscles within the aponeuroses, the so-called *rectus sheath*, ensures that this muscle cannot snap forward when it contracts to flex the trunk (like a bowstring on a bow). Table 10.4 describes the quadratus lumborum muscles of the *posterior* abdominal wall.

The abdominal muscles protect and support the viscera most effectively when they are well toned. When weak or severely stretched (as during pregnancy), they allow the abdomen to become pendulous (i.e., to form a potbelly). Additional functions include lateral flexion and rotation of the trunk and flexion of the trunk against resistance (as in sit-ups).

During inspiration, the abdominal muscles relax, allowing the descending diaphragm to push the abdominal viscera inferiorly. When these abdominal muscles contract, several different activities may occur. For example, when they contract simultaneously, they pull the ribs inferiorly and compress the abdominal contents. This forces the visceral organs upward on the diaphragm, aiding forced expiration. When the abdominal muscles contract with the airway closed (the Valsalva maneuver), the increased intra-abdominal pressure promotes urination, defecation, childbirth, vomiting, coughing, sneezing, burping, and nose blowing. (Next time you perform one of these activities, feel your abdominal muscles contract under your skin.)

These muscles also contract during heavy lifting— sometimes so forcefully that hernias result. Contracting the abdominal muscles as the deep back muscles contract helps prevent hyperextension of the spine and splints (stiffens) the entire body trunk.

MUSCLE	DESCRIPTION	ORIGIN (O) AND INSERTION (I)	ACTION	NERVE SUPPLY
MUSCLES OF THE ANTERIOR AND LATERAL ABDOMINAL WALL	Four paired flat muscles; important in supporting and protecting abdominal viscera; promote lateral flexion and flexion of vertebral column			
Rectus abdominis (rek′tus ab-dom′ĭ-nis) (*rectus* = straight; *abdom* = abdomen)	Medial superficial muscle pair; extend from pubis to rib cage; ensheathed by aponeuroses of lateral muscles; segmented by three tendinous intersections	O—pubic crest and symphysis I—xiphoid process and costal cartilages of ribs 5–7	Flex and rotate lumbar region of vertebral column; fix and depress ribs, stabilize pelvis during walking, increase intra-abdominal pressure; used in sit-ups, curls	Intercostal nerves (T_6 or T_7–T_{12})
External oblique (o-blek′) (*external* = toward outside; *oblique* = running at an angle)	Largest and most superficial of the three lateral muscles; fibers run downward and medially (same direction outstretched fingers take when hands are in pants pockets); aponeurosis turns under inferiorly, forming inguinal ligament	O—by fleshy strips from outer surfaces of lower eight ribs I—most fibers insert into linea alba via a broad aponeurosis; some insert into pubic crest and tubercle and iliac crest	Flex vertebral column and compress abdominal wall when pair contracts simultaneously; acting individually, aid muscles of back in trunk rotation and lateral flexion; used in oblique curls	Intercostal nerves (T_7–T_{12})
Internal oblique (*internal* = toward the inside; deep)	Most fibers run upward and medially; however, the muscle fans so its inferior fibers run downward and medially	O—lumbar fascia, iliac crest, and inguinal ligament I—linea alba, pubic crest, last three or four ribs, and costal margin	As for external oblique	Intercostal nerves (T_7–T_{12}) and L_1
Transversus abdominis (trans-ver′sus) (*transverse* = running straight across)	Deepest (innermost) muscle of abdominal wall; fibers run horizontally	O—inguinal ligament, lumbar fascia, cartilages of last six ribs; iliac crest I—linea alba, pubic crest	Compresses abdominal contents	Intercostal nerves (T_7–T_{12}) and L_1

10

Pectoralis major

Serratus anterior

Linea alba

Tendinous intersection

Transversus abdominis

Rectus abdominis

Internal oblique

External oblique

Aponeurosis of the external oblique

Inguinal ligament (formed by free inferior border of the external oblique aponeurosis)

(a)

External oblique

Iliac crest

Rectus abdominis

Internal oblique

Lumbar fascia

Pubic tubercle

Transversus abdominis

Lumbar fascia

Inguinal ligament

(b)

Transversus abdominis

Peritoneum

Linea alba

Rectus abdominis

External oblique

Internal oblique

Aponeuroses

Skin

(c)

10

Figure 10.12 Muscles of the abdominal wall. (a) Anterior view of the muscles forming the anterolateral abdominal wall; superficial muscles partially cut away to reveal the deeper muscles. **(b)** Lateral view of the trunk, illustrating the fiber direction and attachments of the abdominal muscles. Although shown in the same view, the rectus abdominis is deep to the fascia of the internal oblique muscle. **(c)** Transverse section through the anterolateral abdominal wall (midregion), showing how the aponeuroses of the lateral abdominal muscles contribute to the rectus abdominis sheath.

MUSCLE GALLERY

| Table 10.7 | Muscles of the Pelvic Floor and Perineum: Support of Abdominopelvic Organs | (Figure 10.13) |

The funnel-shaped pelvic floor, or **pelvic diaphragm**, is formed by two paired muscles, the **levator ani** and **coccygeus** (Figure 10.13a). This diaphragm (1) seals the inferior opening of the pelvis, (2) supports the pelvic organs, (3) lifts the pelvic floor superiorly to release feces, and (4) resists increased intra-abdominal pressure (which would expel contents of the urinary bladder, rectum, and uterus). The pelvic diaphragm is pierced by the rectum and urethra (urinary tube), and by the vagina in females.

The body region inferior to the pelvic diaphragm is the *perineum*. The **urogenital diaphragm** sits inferior to the muscles of the pelvic floor, and stretches between the two sides of the pubic arch in the anterior half of the perineum (Figure 10.13b).

This thin triangular sheet of muscle contains the **external urethral sphincter**, a sphincter muscle that surrounds the urethra and allows voluntary control of urination. Superficial (inferior) to the urogenital diaphragm, and covered by the skin of the perineum, is the *superficial perineal space*, which contains muscles (**ischiocavernosus** and **bulbospongiosus**) that help maintain erection of the penis and clitoris (Figure 10.13c). In the posterior half of the perineum encircling the anus is the **external anal sphincter**, which allows voluntary control of defecation. Just anterior to this sphincter is the **perineal body**, or *central tendon of the perineum*, a strong tendon into which many of the perineal muscles insert.

MUSCLE	DESCRIPTION	ORIGIN (O) AND INSERTION (I)	ACTION	NERVE SUPPLY
MUSCLES OF THE PELVIC DIAPHRAGM (Figure 10.13a)				
Levator ani (lĕ-va′tor a′ne) (*levator* = raises; *ani* = anus)	Broad, thin, three-part muscle (pubococcygeus, puborectalis, and iliococcygeus); its fibers extend inferomedially, forming a muscular "sling" around prostate (male) or vagina (female), urethra, and anorectal junction before meeting in the median plane	O—extensive linear origin inside pelvis from pubis to ischial spine I—inner surface of coccyx, levator ani of opposite side, and (in part) into the structures that penetrate it	Supports and maintains position of pelvic organs; resists downward thrusts that accompany rises in intrapelvic pressure during coughing, vomiting, and expulsive efforts of abdominal muscles; forms sphincters at anorectal junction and vagina; lifts anal canal during defecation	S$_3$, S$_4$, and inferior rectal nerve (branch of pudendal nerve)
Coccygeus (kok-sij′e-us) (*coccy* = coccyx)	Small triangular muscle lying posterior to levator ani; forms posterior part of pelvic diaphragm	O—spine of ischium I—sacrum and coccyx	Supports pelvic organs; supports coccyx and pulls it forward after it has been reflected posteriorly by defecation and childbirth	S$_4$ and S$_5$
MUSCLES OF THE UROGENITAL DIAPHRAGM (Figure 10.13b)				
Deep transverse perineal muscle (per″ĭ-ne′al) (*deep* = far from surface; *transverse* = across; *perine* = near anus)	Together the pair spans distance between ischial rami; in females, lies posterior to vagina	O—ischial rami I—midline central tendon of perineum; some fibers into vaginal wall in females	Supports pelvic organs; steadies perineal body	Pudendal nerve
External urethral sphincter (*sphin* = squeeze)	Muscle encircling urethra and vagina (female)	O—ischiopubic rami I—midline raphe	Constricts urethra; allows voluntary inhibition of urination; helps support pelvic organs	Pudendal nerve
MUSCLES OF THE SUPERFICIAL PERINEAL SPACE (Figure 10.13c)				
Ischiocavernosus (is″ke-o-kav″ern-o′sus) (*ischi* = hip; *caverna* = hollow chamber)	Runs from pelvis to base of penis or clitoris	O—ischial tuberosities I—crus of corpora cavernosa of male penis or female clitoris	Retards venous drainage and maintains erection of penis or clitoris	Pudendal nerve
Bulbospongiosus (bul″bo-spun″je-o′sus) (*bulbon* = bulb; *spongio* = sponge)	Encloses base of penis (bulb) in males and lies deep to labia in females	O—perineal body and midline raphe of penis I—anteriorly into corpora cavernosa of penis or clitoris	Empties male urethra; assists in erection of penis and of clitoris	Pudendal nerve
Superficial transverse perineal muscles (*superficial* = closer to surface)	Paired muscle bands posterior to urethral (and in females, vaginal) opening; variable; sometimes absent	O—ischial tuberosity I—perineal body	Stabilizes and strengthens perineal body	Pudendal nerve

Figure 10.13 Muscles of the pelvic floor and perineum. (a) Superior view of muscles of the pelvic diaphragm (levator ani and coccygeus) in a female pelvis. **(b)** Inferior view of muscles of the urogenital diaphragm of the perineum (external urethral sphincter and deep transverse perineal muscles). **(c)** Muscles of the superficial space of the perineum (ischiocavernosus, bulbospongiosus, and superficial transverse perineal muscle), which lie just deep to the skin of the perineum. Note that a raphe is a seam of fibrous tissue.

(a)

Anterior

Levator ani
- Pubococcygeus
- Iliococcygeus

Symphysis pubis
Urogenital diaphragm
Urethra
Vagina
Anal canal

Obturator internus
Coccyx
Piriformis

Levator ani
Coccygeus
} Pelvic diaphragm

Posterior

(b)

Inferior pubic ramus
External urethral sphincter
Deep transverse perineal muscle
Perineal body
Anus
External anal sphincter

Urethral opening
Vaginal opening

Male   Female

(c)

Penis
Midline raphe
Ischiocavernosus
Bulbospongiosus
Superficial transverse perineal muscle
Levator ani
Gluteus maximus
Anus

Clitoris
Urethral opening
Vaginal opening

Male   Female

MUSCLE GALLERY

Table 10.8 Superficial Muscles of the Anterior and Posterior Thorax: Movements of the Scapula and Arm   (Figure 10.14)

Most superficial thorax muscles are *extrinsic shoulder muscles*, which run from the ribs and vertebral column to the shoulder girdle. They can fix the scapula in place or move it to increase the range of arm movements.

The anterior muscles of this group are the **pectoralis major**, **pectoralis minor**, **serratus anterior**, and **subclavius** (Figure 10.14a). Except for the pectoralis major, which inserts into the humerus, all muscles of the anterior group insert into the pectoral girdle.

The posterior muscles are the **latissimus dorsi** and **trapezius muscles** superficially and the underlying **levator scapulae** and **rhomboids** (Figure 10.14c). The latissimus dorsi, like the pectoralis major muscles anteriorly, insert into the humerus. They are more concerned with moving the arm than the scapula, so we defer their consideration to Table 10.9 (arm-moving muscles).

The important movements of the pectoral girdle involve displacing the scapula, i.e., its elevation and depression, rotation, lateral (forward) movements, and medial (backward) movements. The clavicles rotate around their own axes to provide stability and precision to scapular movements.

Except for the serratus anterior, the anterior muscles stabilize and depress the shoulder girdle. Most scapular movements are promoted by the serratus anterior muscles anteriorly and by posterior thoracic muscles. The arrangement of muscle attachments to the scapula is such that one muscle cannot bring about a simple (linear) movement on its own. To effect scapular movements, several muscles must act in combination:

Elevation. The prime movers of shoulder (scapular) elevation are the levator scapulae and the superior fibers of the trapezius. When acting together to shrug the shoulder, their opposite rotational effects counterbalance each other.

Depression. The scapula is depressed largely by gravity (weight of the arm), but when it is depressed against resistance, the inferior part of the trapezius, the pectoralis minor, and the serratus anterior (along with the latissimus dorsi, Table 10.9) are active.

Abduction/Protraction. Anterolateral movements (abduction and protraction) of the scapula on the thorax wall, as in pushing or punching, mainly reflect serratus anterior activity.

Adduction/Retraction. Posteromedial movement (adduction and retraction) of the scapula is effected mainly by the middle part of the trapezius together with the rhomboids.

Upward rotation. Upward rotation of the scapula so that the glenoid cavity points upward (as in lifting the arm above the head) is accomplished mainly by the serratus anterior and the superior part of the trapezius.

Downward rotation. Downward rotation of the scapula so that the glenoid cavity points downward (as in paddling a canoe) is accomplished by the rhomboids and levator scapulae.

MUSCLE	DESCRIPTION	ORIGIN (O) AND INSERTION (I)	ACTION	NERVE SUPPLY
MUSCLES OF THE ANTERIOR THORAX (Figure 10.14a)				
Pectoralis minor (pek″to-ra′lis mi′nor) (*pectus* = chest, breast; *minor* = lesser)	Flat, thin muscle directly beneath and obscured by pectoralis major	O—anterior surfaces of ribs 3–5 (or 2–4) I—coracoid process of scapula	With ribs fixed, pulls scapula forward and downward; with scapula fixed, pulls rib cage superiorly	Medial and lateral pectoral nerves (C_6–C_8)
Serratus anterior (ser-a′tus) (*serratus* = saw)	Fan-shaped muscle; lies deep to scapula, deep and inferior to pectoral muscles on lateral rib cage; forms medial wall of axilla; origins have serrated (sawtooth) appearance; paralysis results in "winging" of vertebral border of scapula away from chest wall, making arm elevation impossible	O—by a series of muscle slips (branches) from ribs 1–8 (or 9) I—entire anterior surface of medial border of scapula	Rotates scapula so its inferior angle moves laterally and upward; prime mover to protract and hold scapula against chest wall; raises point of shoulder; important role in abducting and raising arm and in horizontal arm movements (pushing, punching); called "boxer's muscle"	Long thoracic nerve (C_5–C_7)
Subclavius (sub-kla′ve-us) (*sub* = under, beneath; *clav* = clavicle)	Small cylindrical muscle extending from rib 1 to clavicle	O—costal cartilage of rib 1 I—groove on inferior surface of clavicle	Helps stabilize and depress pectoral girdle	Nerve to subclavius (C_5 and C_6)

(a)

(b)

Figure 10.14 Superficial muscles of the thorax and shoulder acting on the scapula and arm. Anterior view. **(a)** The superficial muscles, which move the arm, are shown on the left side of the illustration. These muscles are removed on the right to show the muscles that stabilize or move the pectoral girdle. **(b)** Cadaver dissection.

Practice Human Cadaver lab questions:
MasteringA&P® > Study Area > Lab Tools > PAL

MUSCLE GALLERY

Table 10.8	Superficial Muscles of the Anterior and Posterior Thorax: Movements of the Scapula and Arm (Figure 10.14) *(continued)*			
MUSCLE	**DESCRIPTION**	**ORIGIN (O) AND INSERTION (I)**	**ACTION**	**NERVE SUPPLY**
MUSCLES OF THE POSTERIOR THORAX (Figure 10.14c and d)				
Trapezius (trah-pe′ze-us) (*trapezion* = irregular four-sided figure)	Most superficial muscle of posterior thorax; flat and triangular; upper fibers run inferiorly to scapula; middle fibers run horizontally to scapula; lower fibers run superiorly to scapula	O—occipital bone, ligamentum nuchae, and spinous processes of C_7 and all thoracic vertebrae I—a continuous insertion along acromion and spine of scapula and lateral third of clavicle	Stabilizes, elevates, retracts, and rotates scapula; middle fibers retract (adduct) scapula; superior fibers elevate scapula (as in shrugging the shoulders) or help extend head with scapula fixed; inferior fibers depress scapula (and shoulder)	Accessory nerve (cranial nerve XI); C_3 and C_4
Levator scapulae (skap′u-le) (*levator* = raises)	Located at back and side of neck, deep to trapezius; thick straplike muscle	O—transverse processes of C_1–C_4 I—medial border of scapula, superior to spine	Elevates and adducts scapula in synergy with superior fibers of trapezius; tilts glenoid cavity downward when scapula is fixed, flexes neck to same side	Cervical spinal nerves and dorsal scapular nerve (C_3–C_5)
Rhomboids (rom′boidz)—major and minor (*rhomboid* = diamond shaped)	Two roughly diamond-shaped muscles lying deep to trapezius and inferior to levator scapulae; rhomboid minor is the more superior muscle	O—spinous processes of C_7 and T_1 (minor) and spinous processes of T_2–T_5 (major) I—medial border of scapula	Stabilize scapula; act together (and with middle trapezius fibers) to retract (adduct) scapula, thus "squaring shoulders"; rotate scapula so that glenoid cavity is downward (as when lowering arm against resistance; e.g., paddling a canoe)	Dorsal scapular nerve (C_4 and C_5)

Figure 10.14 *(continued)* **Superficial muscles of the thorax and shoulder acting on the scapula and arm. (c)** Posterior view. The superficial muscles are shown on the left side of the illustration. Superficial muscles are removed on the right side to reveal the deeper muscles acting on the scapula, and the rotator cuff muscles that help stabilize the shoulder joint. **(d)** Cadaver dissection showing view similar to (c).

MUSCLE GALLERY

Table 10.8 *(continued)*

- Levator scapulae
- Trapezius
- Supraspinatus
- Deltoid
- Clavicle
- Rhomboid minor
- Spine of scapula
- Rhomboid major
- Infraspinatus
- Teres minor
- Teres major
- Humerus
- Latissimus dorsi

(c)

Superficial muscles:

- Trapezius
- Deltoid
- Teres major
- Latissimus dorsi

Deeper muscles:

- Levator scapulae
- Supraspinatus
- Rhomboid minor
- Spine of scapula
- Teres minor
- Infraspinatus
- Rhomboid major

(d)

10

MUSCLE GALLERY

Table 10.9   Muscles Crossing the Shoulder Joint: Movements of the Arm (Humerus)   (Figure 10.15)

Recall that the ball-and-socket shoulder joint is the most flexible joint in the body, but pays the price of instability. Several muscles cross each shoulder joint to insert on the humerus. Most muscles acting on the humerus originate from the pectoral girdle. However, two of these—the latissimus dorsi and pectoralis major—primarily originate on the axial skeleton.

As you consider the arm movements, remember that the term *arm* refers to the region that contains the humerus. Of the nine muscles covered here, only the **pectoralis major**, **latissimus dorsi**, and **deltoid muscles** are prime movers of arm movements (Figure 10.15a, b). The remaining six are synergists and fixators.

There are four *rotator cuff* muscles: **supraspinatus**, **infraspinatus**, **teres minor**, and **subscapularis** (marked with an asterisk* in Figure 10.15b, d). They originate on the scapula, and their tendons blend with the fibrous capsule of the shoulder joint en route to the humerus. Although the rotator cuff muscles act as synergists in angular and rotational movements of the arm, their main function is to reinforce the capsule of the shoulder joint to prevent dislocation of the humerus.

Two small muscles, the **teres major** and **coracobrachialis**, cross the shoulder joint but do not reinforce it.

The interactions among the nine muscles that cross the shoulder joint are complex and each contributes to several movements. Table 10.12 (Part I) summarizes their action:

Flexion. Muscles that originate *anterior* to the shoulder joint *flex* the arm (lift it anteriorly). These flexors include the pectoralis major, the anterior fibers of the deltoid, and the coracobrachialis.

Extension. Muscles that originate *posterior* to the shoulder joint extend the arm. These include the latissimus dorsi and posterior fibers of the deltoid muscles (both prime movers of arm extension) and the teres major. (Note that the pectoralis and latissimus dorsi muscles are *antagonists* of one another in flexing and extending the arm.)

Abduction. The middle region of the fleshy deltoid muscle of the shoulder is the prime mover of arm *abduction*. It extends over the superior and lateral side of the humerus to pull it laterally.

Adduction. The main arm *adductors* are the pectoralis major anteriorly and latissimus dorsi posteriorly.

Rotation. Lateral and medial rotation of the arm are primarily brought about by the rotator cuff muscles.

MUSCLE	DESCRIPTION	ORIGIN (O) AND INSERTION (I)	ACTION	NERVE SUPPLY
Pectoralis major (pek"to-ra′lis ma′jer) (*pectus* = breast, chest; *major* = larger)	Large, fan-shaped muscle covering superior portion of chest; forms anterior axillary fold; divided into clavicular and sternal parts	O—sternal end of clavicle, sternum, cartilage of ribs 1–6 (or 7), and aponeurosis of external oblique muscle I—fibers converge to insert by a short tendon into intertubercular sulcus and greater tubercle of humerus	Adducts and medially rotates arm against resistance; clavicular part assists in flexion when arm is extended and sternal part assists in extension when the arm is flexed; with scapula (and arm) fixed, pulls rib cage upward, thus can help in climbing, throwing, pushing, and forced inspiration	Lateral and medial pectoral nerves (C_5–C_8 and T_1)
Deltoid (del′toid) (*delta* = triangular)	Thick, multipennate muscle forming rounded shoulder muscle mass; a common site for intramuscular injection, particularly in males, where it tends to be quite fleshy	O—lateral third of clavicle; acromion and spine of scapula (opposite insertion of the trapezius) I—deltoid tuberosity of humerus	Prime mover of arm abduction when all its fibers contract simultaneously; antagonist of pectoralis major and latissimus dorsi, which adduct the arm; if only anterior fibers are active, can act powerfully in flexing and rotating arm medially, therefore is a synergist of pectoralis major; if only posterior fibers are active, causes extension and lateral rotation of arm; active during rhythmic arm-swinging movements while walking	Axillary nerve (C_5 and C_6)

MUSCLE GALLERY

Table 10.9   *(continued)*

MUSCLE	DESCRIPTION	ORIGIN (O) AND INSERTION (I)	ACTION	NERVE SUPPLY
Latissimus dorsi (lah-tis′ĭ-mus dor′si) (*latissimus* = widest; *dorsi* = back)	Broad, flat, triangular muscle of lower back (lumbar region); extensive superficial origins; covered by trapezius superiorly; contributes to the posterior wall of axilla	O—indirect attachment via lumbodorsal fascia into spines of lower six thoracic vertebrae, lumbar vertebrae, lower 3 to 4 ribs, and iliac crest I—spirals around teres major to insert in floor of intertubercular sulcus of humerus	Prime mover of arm extension; powerful arm adductor; medially rotates arm; plays important role in lowering arm in a power stroke, as in striking a blow, hammering, swimming, and rowing; with arms fixed overhead, it pulls the rest of the body upward and forward, as in chin-ups	Thoracodorsal nerve (C_6–C_8)

Clavicle

Deltoid

Sternum

Pectoralis major

Coracobrachialis

Triceps brachii:
  Lateral head
  Long head
  Medial head

Biceps brachii

Brachialis

Brachio-radialis

(a) Anterior view

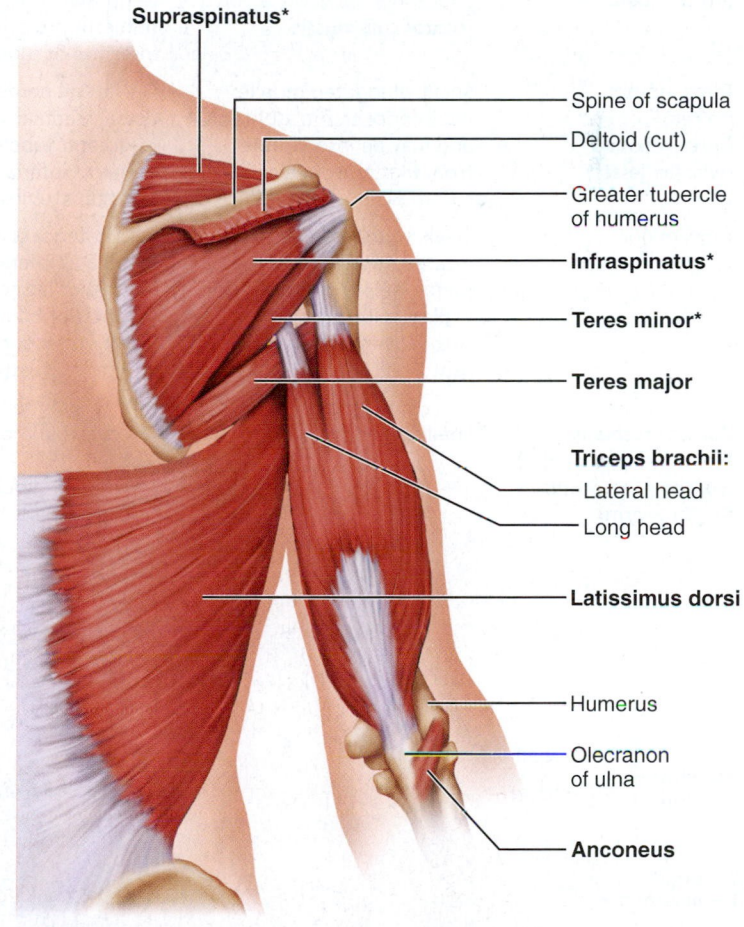

Supraspinatus*

Spine of scapula

Deltoid (cut)

Greater tubercle of humerus

Infraspinatus*

Teres minor*

Teres major

Triceps brachii:
  Lateral head
  Long head

Latissimus dorsi

Humerus

Olecranon of ulna

Anconeus

(b) Posterior view

Figure 10.15 Muscles crossing the shoulder and elbow joints acting on the arm and forearm, respectively. Right side. **(a)** Superficial muscles of the anterior thorax, shoulder, and arm, anterior view. **(b)** Muscles of the posterior thorax and arm. The deltoid muscle of the shoulder and the trapezius muscle have been removed.

Practice Human Cadaver lab questions: MasteringA&P® > Study Area > Lab Tools > PAL

*Rotator cuff muscles

MUSCLE GALLERY

Table 10.9 Muscles Crossing the Shoulder Joint: Movements of the Arm (Humerus) (Figure 10.15) (continued)

MUSCLE	DESCRIPTION	ORIGIN (O) AND INSERTION (I)	ACTION	NERVE SUPPLY
Subscapularis* (sub-scap"u-lar'is) (*sub* = under; *scapular* = scapula)	Forms part of posterior wall of axilla; tendon of insertion passes in front of shoulder joint; a rotator cuff muscle	O—subscapular fossa of scapula I—lesser tubercle of humerus	Rotates arm medially (prime mover), assisted by pectoralis major; helps hold head of humerus in glenoid cavity, stabilizing shoulder joint	Subscapular nerves (C_5–C_7)
Supraspinatus* (soo"prah-spi-nah'tus) (*supra* = above, over; *spin* = spine)	Named for its location on posterior aspect of scapula; deep to trapezius; a rotator cuff muscle	O—supraspinous fossa of scapula I—superior part of greater tubercle of humerus	Initiates abduction of arm, stabilizes shoulder joint; helps prevent downward dislocation of humerus, as when carrying a heavy suitcase	Suprascapular nerve
Infraspinatus* (in"frah-spi-nah'tus) (*infra* = below)	Partially covered by deltoid and trapezius; named for its scapular location; a rotator cuff muscle	O—infraspinous fossa of scapula I—greater tubercle of humerus posterior to insertion of supraspinatus	Rotates arm laterally; helps hold head of humerus in glenoid cavity, stabilizing the shoulder joint	Suprascapular nerve
Teres minor* (te'rēz) (*teres* = round; *minor* = lesser)	Small, elongated muscle; lies inferior to infraspinatus and may be inseparable from that muscle; a rotator cuff muscle	O—lateral border of dorsal scapular surface I—greater tubercle of humerus inferior to infraspinatus insertion	Same action(s) as infraspinatus muscle	Axillary nerve
Teres major	Thick, rounded muscle; located inferior to teres minor; helps form posterior wall of axilla (along with latissimus dorsi and subscapularis)	O—posterior surface of scapula at inferior angle I—crest of lesser tubercle on anterior humerus; insertion tendon fused with that of latissimus dorsi	Extends, medially rotates, and adducts arm; synergist of latissimus dorsi	Lower subscapular nerve (C_6 and C_7)
Coracobrachialis (kor"ah-ko-bra"ke-al'is) (*coraco* = coracoid; *brachi* = arm)	Small, cylindrical muscle	O—coracoid process of scapula I — medial surface of humerus shaft	Flexes and adducts arm; synergist of pectoralis major	Musculocutaneous nerve (C_5–C_7)

Figure 10.15 *(continued)* **Muscles crossing the shoulder and elbow joints acting on the arm and forearm, respectively. (c)** The isolated biceps brachii muscle of the anterior arm. **(d)** The brachialis muscle and the coracobrachialis and subscapularis muscles shown in isolation in the diagram on the left, and in a dissection on the right.

*Rotator cuff muscles

MUSCLE GALLERY

Table 10.10   Muscles Crossing the Elbow Joint: Flexion and Extension of the Forearm   (Figure 10.15)

This table focuses on muscles that lie in the arm but move the forearm. These muscles cross the elbow joint to insert on the forearm bones, which they flex or extend. Walls of fascia divide the arm into two muscle compartments—the *posterior extensors* and *anterior flexors.*

Extension. The main forearm extensor is the bulky **triceps brachii** muscle, which forms nearly the entire musculature of the posterior compartment (Figure 10.15a, b).

Flexion. All anterior arm muscles flex the forearm (elbow). In order of decreasing strength, these are the **brachialis**, **biceps brachii**, and **brachioradialis** (Figure 10.15a, c, d). The brachialis and biceps insert (respectively) into the ulna and radius and contract simultaneously during flexion; they are the chief forearm flexors. The biceps brachii,

a muscle that bulges when the forearm is flexed, is familiar to almost everyone. The brachialis, which lies deep to the biceps, is less known but is equally important in flexing the elbow. Because the brachioradialis arises from the distal humerus and inserts on the distal forearm, it resides mainly in the forearm. Its force is exerted far from the fulcrum (elbow), so the brachioradialis is a weak forearm flexor. The biceps muscle also supinates the forearm and is ineffective in flexing the elbow when the forearm *must* stay pronated. (This is why doing chin-ups with palms facing anteriorly is harder than with palms facing posteriorly.)

Table 10.12 (Part II) summarizes the actions of the muscles described here.

MUSCLE	DESCRIPTION	ORIGIN (O) AND INSERTION (I)	ACTION	NERVE SUPPLY
POSTERIOR MUSCLES				
Triceps brachii (tri′seps bra′ke-i) (*triceps* = three heads; *brachi* = arm)	Large fleshy muscle; the only muscle of posterior compartment of arm; three-headed origin; long and lateral heads lie superficial to medial head	O—long head: infraglenoid tubercle of scapula; lateral head: posterior shaft of humerus; medial head: posterior humeral shaft distal to radial groove I—by common tendon into olecranon of ulna	Powerful forearm extensor (prime mover, particularly medial head); antagonist of forearm flexors; long and lateral heads mainly active in extending the forearm against resistance; long head tendon may help stabilize shoulder joint and assist in arm adduction	Radial nerve (C_6–C_8)
Anconeus (an-ko′ne-us) (*ancon* = elbow) (see Figure 10.17)	Short triangular muscle; partially blended with distal end of triceps on posterior humerus	O—lateral epicondyle of humerus I—lateral aspect of olecranon of ulna	Synergist of triceps brachii in elbow extension; may control ulnar abduction during forearm pronation	Radial nerve
ANTERIOR MUSCLES				
Biceps brachii (bi′seps) (*biceps* = two heads)	Two-headed fusiform muscle; bellies unite as insertion point is approached; tendon of long head helps stabilize shoulder joint	O—short head: coracoid process; long head: supraglenoid tubercle and lip of glenoid cavity; tendon of long head runs within shoulder joint capsule and into intertubercular sulcus of humerus I—by common tendon into radial tuberosity	Flexes and supinates forearm; these actions usually occur at same time (e.g., when you open a bottle of wine, it turns the corkscrew and pulls the cork); weak flexor of arm at shoulder	Musculocutaneous nerve (C_5 and C_6)
Brachialis (bra′ke-al-is)	Strong muscle that is immediately deep to biceps brachii on distal humerus	O—anterior surface of distal humerus; embraces insertion of deltoid muscle I—coronoid process of ulna and capsule of elbow joint	A major forearm flexor (lifts ulna as biceps lifts the radius)	Musculocutaneous nerve
Brachioradialis (bra″ke-o-ra″de-al′is) (*radi* = radius, ray) (also see Figure 10.16)	Superficial muscle of lateral forearm; forms lateral boundary of cubital fossa; extends from distal humerus to distal forearm	O—lateral supracondylar ridge at distal end of humerus I—base of radial styloid process	Synergist in flexing forearm; acts to best advantage when forearm is partially flexed and semipronated; stabilizes elbow during rapid flexion *and* extension	Radial nerve (an important exception: the radial nerve typically serves extensor muscles)

10

MUSCLE GALLERY

Table 10.11   Muscles of the Forearm: Movements of the Wrist, Hand, and Fingers   (Figures 10.16 and 10.17)

The many muscles in the forearm perform several basic functions. Some move the hand at the wrist, some move the fingers and thumb, and a few help pronate and supinate the forearms. Most of these muscles are fleshy proximally and have long tendons distally, almost all of which insert in the hand. Most muscles that move the hand are located in the forearm and "operate" the fingers via their long tendons, like operating a puppet by strings. This arrangement makes the hand less bulky and enables it to perform finer movements.

At the wrist, these tendons are securely anchored by bandlike thickenings of deep fascia called **flexor** and **extensor retinacula** ("retainers") (Figure 10.16a). These "wrist bands" keep the tendons from jumping outward when tensed. Crowded together in the wrist and palm, the muscle tendons are surrounded by slippery tendon sheaths that minimize friction as they slide against one another.

Although many forearm muscles arise from the humerus (and so cross both the elbow and wrist joints), their actions on the elbow are slight. At the wrist joint, the forearm muscles bring about flexion, extension, abduction, and adduction of the hand. At the interphalangeal joints, these muscles just flex and extend the fingers.

Other finger movements are brought about by small muscles in the hand itself. (These *intrinsic muscles of the hand* are described in Table 10.13.)

Fascia subdivides the forearm muscles into two main compartments: the *anterior flexors* and *posterior extensors*.
Anterior flexor compartment. Most flexors in the anterior compartment originate from a common tendon on the humerus and are innervated largely by the median nerve. Two anterior compartment muscles are not flexors but pronators, the **pronator teres** and **pronator quadratus** (Figure 10.16a–c). Pronation is one of the most important forearm movements.
Posterior extensor compartment. Muscles of the posterior compartment extend the hand and fingers. One exception is the **supinator** muscle, which helps the biceps brachii muscle supinate the forearm (Figures 10.16b, c and 10.17b). (Also residing in the posterior compartment is the brachioradialis muscle, the weak forearm flexor considered in Table 10.10.) Most muscles of the posterior compartment arise from a common tendon on the humerus. The radial nerve supplies all posterior forearm muscles.

Table 10.12 (Parts II and III) summarizes the actions of the forearm muscles.

MUSCLE	DESCRIPTION	ORIGIN (O) AND INSERTION (I)	ACTION	NERVE SUPPLY
PART I: ANTERIOR MUSCLES (Figure 10.16)	These eight muscles of the anterior fascial compartment are listed from the lateral to the medial aspect. Most arise from a common flexor tendon attached to the medial epicondyle of the humerus and have additional origins as well. Most of their tendons of insertion are held in place at the wrist by a thickening of deep fascia called the *flexor retinaculum*.			
Superficial Muscles				
Pronator teres (pro-na′tor te′rēz) (*pronation* = turning palm posteriorly, or down; *teres* = round)	Two-headed muscle; seen in superficial view between proximal margins of brachioradialis and flexor carpi radialis; forms medial boundary of cubital fossa	O—medial epicondyle of humerus; coronoid process of ulna I—by common tendon into lateral radius, midshaft	Pronates forearm; weak flexor of forearm	Median nerve
Flexor carpi radialis (flek′sor kar′pe ra″de-al′is) (*flex* = decrease angle between two bones; *carpi* = wrist; *radi* = radius)	Runs diagonally across forearm; midway, its fleshy belly is replaced by a flat tendon that becomes cordlike at wrist	O—medial epicondyle of humerus I—base of second and third metacarpals; insertion tendon easily seen and provides guide to position of radial artery at wrist (used for taking pulse)	Powerful flexor of hand; abducts hand; weak synergist of forearm flexion	Median nerve
Palmaris longus (pahl-ma′ris lon′gus) (*palma* = palm; *longus* = long)	Small fleshy muscle with a long insertion tendon; often absent; may be used as guide to find median nerve that lies lateral to it at wrist	O—medial epicondyle of humerus I—fascia of palm (palmar aponeurosis)	Tenses skin and fascia of palm during hand movements; weak hand flexor; weak synergist for forearm flexion	Median nerve
Flexor carpi ulnaris (ul-na′ris) (*ulnar* = ulna)	Most medial muscle of this group; two-headed; ulnar nerve lies lateral to its tendon	O—medial epicondyle of humerus; olecranon and posterior surface of ulna I—pisiform and hamate bones and base of fifth metacarpal	Powerful flexor and adductor of hand in synergy with extensor carpi ulnaris (posterior muscle); stabilizes wrist during finger extension	Ulnar nerve (C_8 and T_1)

MUSCLE GALLERY

Table 10.11 *(continued)*

MUSCLE	DESCRIPTION	ORIGIN (O) AND INSERTION (I)	ACTION	NERVE SUPPLY
Flexor digitorum superficialis (dĭ″jĭ-tor′um soo″per-fish″e-al′is) (*digit* = finger, toe; *superficial* = close to surface)	Two-headed muscle; more deeply placed (therefore, actually forms an intermediate layer); overlain by muscles above but visible at distal end of forearm	O—medial epicondyle of humerus, coronoid process of ulna; shaft of radius I—by four tendons into middle phalanges of second to fifth fingers	Flexes hand and middle phalanges of second to fifth fingers; the important finger flexor for speed and flexion against resistance	Median nerve (C_7, C_8, and T_1)
Deep Muscles				
Flexor pollicis longus (pah′lĭ-sis) (*pollix* = thumb)	Partly covered by flexor digitorum superficialis; parallels flexor digitorum profundus laterally	O—anterior surface of radius and interosseous membrane I—distal phalanx of thumb	Flexes distal phalanx of thumb	Branch of median nerve (C_8, T_1)
Flexor digitorum profundus (pro-fun′dus) (*profund* = deep)	Extensive origin; overlain entirely by flexor digitorum superficialis	O—coronoid process, anteromedial surface of ulna, and interosseous membrane I—by four tendons into distal phalanges of second to fifth fingers	Flexes distal interphalangeal joints; slow-acting flexor of any or all fingers; helps flex wrist	Medial half by ulnar nerve; lateral half by median nerve

Superficial transverse ligament of palm

Palmar aponeurosis

Flexor retinaculum

Pronator quadratus

Flexor digitorum superficialis

Flexor pollicis longus

Flexor carpi ulnaris

Palmaris longus

Flexor carpi radialis

Medial epicondyle of humerus

Medial head of triceps brachii

Extensor carpi radialis longus

Brachioradialis

Pronator teres

Tendon of biceps brachii

Biceps brachii

(a)

Tendon of flexor digitorum superficialis

Tendon of brachioradialis (cut)

Tendon of flexor carpi ulnaris (cut)

Tendon of flexor carpi radialis (cut)

Flexor digitorum superficialis

Extensor carpi radialis longus

Supinator

Tendon of biceps brachii (cut)

(b)

Tendon of flexor digitorum profundus

Tendon of flexor digitorum superficialis (cut)

Lumbricals

Tendon of flexor pollicis longus

Thenar muscles of thumb

Pronator quadratus

Flexor pollicis longus

Flexor digitorum profundus

(c)

Figure 10.16 Muscles of the anterior fascial compartment of the forearm acting on the right wrist and fingers. **(a)** Superficial view. **(b)** The brachioradialis, flexors carpi radialis and ulnaris, and palmaris longus muscles have been removed to reveal the flexor digitorum superficialis. **(c)** Deep muscles of the anterior compartment. The lumbricals and thenar muscles (intrinsic hand muscles) are also illustrated.

10

MUSCLE GALLERY

Table 10.11	Muscles of the Forearm: Movements of the Wrist, Hand, and Fingers (Figures 10.16 and 10.17) *(continued)*			
MUSCLE	**DESCRIPTION**	**ORIGIN (O) AND INSERTION (I)**	**ACTION**	**NERVE SUPPLY**
Pronator quadratus (kwod-ra′tus) (*quad* = square, four-sided)	Deepest muscle of distal forearm; passes downward and laterally; only muscle that arises solely from ulna and inserts solely into radius	O—distal portion of anterior ulnar shaft I—distal surface of anterior radius	Prime mover of forearm pronation; acts with pronator teres; also helps hold ulna and radius together	Median nerve (C_7 and C_8)
PART II: POSTERIOR MUSCLES (Figure 10.17)	These muscles of the posterior fascial compartment are listed from the lateral to the medial aspect. They are all innervated by the radial nerve or its branches. More than half of the posterior compartment muscles arise from a common extensor origin tendon attached to the posterior surface of the lateral epicondyle of the humerus and adjacent fascia. The extensor tendons are held in place at the posterior aspect of the hand by the *extensor retinaculum,* which prevents "bowstringing" of these tendons when the wrist is hyperextended. The *extensor* muscles of the fingers end in a broad hood over the dorsal side of the digits, the *extensor expansion.*			
Superficial Muscles				
Brachioradialis	See Table 10.10	See Table 10.10	See Table 10.10	See Table 10.10
Extensor carpi radialis longus (ek-sten′sor) (*extend* = increase angle between two bones)	Parallels brachioradialis on lateral forearm, and may blend with it	O—lateral supracondylar ridge of humerus I—base of second metacarpal	Extends hand in conjunction with extensor carpi ulnaris and abducts hand in conjunction with flexor carpi radialis	Radial nerve (C_6 and C_7)
Extensor carpi radialis brevis (brĕ′vis) (*brevis* = short)	Shorter than extensor carpi radialis longus and lies deep to it	O—lateral epicondyle of humerus I—base of third metacarpal	Extends and abducts hand; acts synergistically with extensor carpi radialis longus to steady wrist during finger flexion	Deep branch of radial nerve
Extensor digitorum	Lies medial to extensor carpi radialis brevis; a detached portion of this muscle, called *extensor digiti minimi,* extends little finger	O—lateral epicondyle of humerus I—by four tendons into extensor expansions and distal phalanges of second to fifth fingers	Prime mover of finger extension; extends hand; can abduct (flare) fingers	Posterior interosseous nerve, a branch of radial nerve (C_7 and C_8)
Extensor carpi ulnaris	Most medial of superficial posterior muscles; long, slender muscle	O—lateral epicondyle of humerus and posterior border of ulna I—base of fifth metacarpal	Extends hand in conjunction with extensor carpi radialis and adducts hand in conjunction with flexor carpi ulnaris	Posterior interosseous nerve
Deep Muscles				
Supinator (soo″pǐ-na′tor) (*supination* = turning palm anteriorly or upward)	Deep muscle at posterior aspect of elbow; largely concealed by superficial muscles	O—lateral epicondyle of humerus; proximal ulna I—proximal end of radius	Assists biceps brachii to forcibly supinate forearm; works alone in slow supination; antagonist of pronator muscles	Posterior interosseous nerve
Abductor pollicis longus (ab-duk′tor) (*abduct* = movement away from median plane)	Lateral and parallel to extensor pollicis longus; just distal to supinator	O—posterior surface of radius and ulna; interosseous membrane I—base of first metacarpal and trapezium	Abducts and extends thumb	Posterior interosseous nerve
Extensor pollicis brevis and longus	Deep muscle pair with a common origin and action; overlain by extensor carpi ulnaris	O—dorsal shaft of radius and ulna; interosseous membrane I—base of proximal (brevis) and distal (longus) phalanx of thumb	Extends thumb	Posterior interosseous nerve

10

MUSCLE GALLERY

Table 10.11   *(continued)*

MUSCLE	DESCRIPTION	ORIGIN (O) AND INSERTION (I)	ACTION	NERVE SUPPLY
Extensor indicis (in′dĭ-sis) (*indicis* = index finger)	Tiny muscle arising close to wrist	O—posterior surface of distal ulna; interosseous membrane I—extensor expansion of index finger; joins tendon of extensor digitorum	Extends index finger (digit II) and assists in extending wrist	Posterior interosseous nerve

(a)

10

(b)

Figure 10.17 Muscles of the posterior fascial compartment of the right forearm acting on the wrist and fingers. (a) Superficial muscles, posterior view. **(b)** Deep posterior muscles, superficial muscles removed. The interossei, the deepest layer of intrinsic hand muscles, are also illustrated.

MUSCLE GALLERY

Table 10.12 Summary: Actions of Muscles Acting on the Arm, Forearm, and Hand (Figure 10.18)

PART I: Muscles Acting on the Arm (Humerus) (PM = Prime Mover)

	Actions at the Shoulder					
	Flexion	Extension	Abduction	Adduction	Medial Rotation	Lateral Rotation
Pectoralis major	×			×(PM)	×	
Latissimus dorsi		× (PM)		× (PM)	×	
Teres major		×		×	×	
Deltoid	× (PM) (anterior fibers)	× (PM) (posterior fibers)	× (PM)		× (anterior fibers)	× (posterior fibers)
Subscapularis					× (PM)	
Supraspinatus			×			
Infraspinatus						× (PM)
Teres minor				× (weak)		× (PM)
Coracobrachialis	×			×		
Biceps brachii	× (weak)					
Triceps brachii				×		

PART II: Muscles Acting on the Forearm

	Actions on the Forearm			
	Elbow Flexion	Elbow Extension	Pronation	Supination
Biceps brachii	× (PM)			×
Brachialis	× (PM)			
Triceps brachii		× (PM)		
Anconeus		×		
Pronator teres	× (weak)		×	
Pronator quadratus			× (PM)	
Supinator				×
Brachioradialis	×			

PART III: Muscles Acting on the Wrist and Fingers

	Actions at the Wrist				Actions on the Fingers	
	Flexion	Extension	Abduction	Adduction	Flexion	Extension
Anterior Compartment						
Flexor carpi radialis	× (PM)		×			
Palmaris longus	× (weak)					
Flexor carpi ulnaris	× (PM)			×		
Flexor digitorum superficialis	× (PM)				×	
Flexor pollicis longus					× (thumb)	
Flexor digitorum profundus	×				×	
Posterior Compartment						
Extensor carpi radialis longus and brevis		×	×			
Extensor digitorum		×				× (PM, and abducts)
Extensor carpi ulnaris		×		×		
Abductor pollicis longus			×		(abducts thumb)	
Extensor pollicis longus and brevis						× (thumb)
Extensor indicis		× (weak)				× (index finger)

10

Key:
- Extensors
- Flexors
- Others

(a)

(b)

Triceps brachii
- Lateral head
- Long head
- Medial head

Humerus

Brachialis

Short head — Biceps brachii
Long head

(a) Muscles of the arm

Posterior compartment of arm
(extends forearm at elbow);
innervated by radial nerve

Anterior compartment of arm
(flexes forearm at elbow); innervated
by musculocutaneous nerve

Posterior compartment of forearm
(extends hand and fingers);
innervated by radial nerve

Radius

Others
- Abductor pollicis longus
- Pronator teres
- Brachioradialis (elbow flexor)

Extensors

Ulna

Flexors

(b) Muscles of the forearm

Anterior compartment of forearm
(flexes hand and fingers);
innervated by median or ulnar nerve

10

Figure 10.18   Summary: Actions of muscles of the right arm and forearm.

MUSCLE GALLERY

Table 10.13   Intrinsic Muscles of the Hand: Fine Movements of the Fingers   (Figure 10.19)

In this table we consider the small muscles that lie entirely in the hand. All are in the palm, none on the hand's dorsal side. All move the metacarpals and fingers. Because they are small, weak muscles, they mostly control precise movements (such as threading a needle). This leaves the powerful movements of the fingers ("power grip") to the forearm muscles.

The intrinsic muscles include the main abductors and adductors of the fingers, as well as muscles that produce the movement of opposition—moving the thumb toward another digit of the same hand—that enables you to grip objects. Many palm muscles are specialized to move the thumb, and surprisingly many move the little finger.

Thumb movements are defined differently from movements of other fingers because the thumb lies at a right angle to the rest of the hand. The thumb flexes by bending medially along the palm, not by bending anteriorly, as do the other fingers (Figure 10.19a). (To

demonstrate this difference, start with your hand in the anatomical position or this will not be clear!) The thumb extends by pointing laterally (as in hitchhiking), not posteriorly, as do the other fingers. To abduct the fingers is to splay them laterally, but to abduct the thumb is to point it anteriorly. Adduction of the thumb brings it back posteriorly.

The intrinsic muscles of the palm are divided into three groups, those in:

- The *thenar eminence* (ball of the thumb)
- The *hypothenar eminence* (ball of the little finger)
- The midpalm

Thenar and hypothenar muscles are almost mirror images of each other, each containing a small flexor, an abductor, and an opponens muscle. The midpalmar muscles, called **lumbricals** and **interossei**, extend our fingers at the interphalangeal joints. The interossei are also the main finger abductors and adductors.

MUSCLE	DESCRIPTION	ORIGIN (O) AND INSERTION (I)	ACTION	NERVE SUPPLY
THENAR MUSCLES IN BALL OF THUMB (the'nar) (*thenar* = palm)				
Abductor pollicis brevis (*pollex* = thumb)	Lateral muscle of thenar group; superficial	O—flexor retinaculum and nearby carpals I—lateral base of thumb's proximal phalanx	Abducts thumb (at carpometacarpal joint)	Median nerve (C_8, T_1)
Flexor pollicis brevis	Medial and deep muscle of thenar group	O—flexor retinaculum and nearby carpals I—lateral side of base of proximal phalanx of thumb	Flexes thumb (at carpometacarpal and metacarpophalangeal joints)	Median (or occasionally ulnar) nerve (C_8, T_1)
Opponens pollicis (o-pōn'enz) (*opponens* = opposition)	Deep to abductor pollicis brevis, on metacarpal I	O—flexor retinaculum and trapezium I—whole anterior side of metacarpal I	Opposition: moves thumb to touch tip of another finger of the same hand	Median (or occasionally ulnar) nerve
Adductor pollicis	Fan-shaped with horizontal fibers; distal to other thenar muscles; oblique and transverse heads	O—capitate bone and bases of metacarpals II–IV; front of metacarpal III I—medial side of base of proximal phalanx of thumb	Adducts and helps to oppose thumb	Ulnar nerve (C_8, T_1)
HYPOTHENAR MUSCLES IN BALL OF LITTLE FINGER				
Abductor digiti minimi (dĭ'jĭ-ti min'ĭ-mi) (*digiti minimi* = little finger)	Medial muscle of hypothenar group; superficial	O—pisiform bone I—medial side of proximal phalanx of little finger	Abducts little finger at metacarpophalangeal joint	Ulnar nerve
Flexor digiti minimi brevis	Lateral deep muscle of hypothenar group	O—hamate bone and flexor retinaculum I—same as abductor digiti minimi	Flexes little finger at metacarpophalangeal joint	Ulnar nerve
Opponens digiti minimi	Deep to abductor digiti minimi	O—same as flexor digiti minimi brevis I—most of length of medial side of metacarpal V	Helps in opposition: brings metacarpal V toward thumb to cup the hand	Ulnar nerve

MUSCLE GALLERY

Table 10.13   *(continued)*

Flexion: bends medially across palm

Extension: points laterally

Abduction: points anteriorly

Adduction: moves posteriorly

Opposition: touches tip of another finger

(a) Movements of the thumb

Tendons of:
- **Flexor digitorum profundus**
- **Flexor digitorum superficialis**

Third lumbrical

Fourth lumbrical

Opponens digiti minimi

Flexor digiti minimi brevis

Abductor digiti minimi

Pisiform bone

Flexor carpi ulnaris tendon

Flexor digitorum superficialis tendons

Fibrous sheath

Second lumbrical

Dorsal interossei

First lumbrical

Adductor pollicis

Flexor pollicis brevis

Abductor pollicis brevis

Opponens pollicis

Flexor retinaculum

Abductor pollicis longus

Tendons of:
- Palmaris longus
- Flexor carpi radialis
- Flexor pollicis longus

(b) First superficial layer

Flexor digitorum profundus tendon

Flexor digitorum superficialis tendon

Dorsal interossei

Adductor pollicis

Flexor pollicis brevis

Abductor pollicis brevis

Palmar interossei

Opponens digiti minimi

Flexor digiti minimi brevis (cut)

Abductor digiti minimi (cut)

Opponens pollicis

Flexor pollicis longus tendon

(c) Second layer

Figure 10.19 **Intrinsic muscles of the hand. (a)** Movements of the thumb. **(b, c)** Hand muscles, ventral views of right hand.

MUSCLE GALLERY

| Table 10.13 | Intrinsic Muscles of the Hand: Fine Movements of the Fingers | | (Figure 10.19) *(continued)* | |

MUSCLE	DESCRIPTION	ORIGIN (O) AND INSERTION (I)	ACTION	NERVE SUPPLY
MIDPALMAR MUSCLES				
Lumbricals (lum′brĭ-klz) (*lumbric* = earthworm)	Four worm-shaped muscles in palm, one to each finger (except thumb); unusual because they originate from the tendons of another muscle	O—lateral side of each tendon of flexor digitorum profundus in palm I—lateral edge of extensor expansion on proximal phalanx of second to fifth fingers	Flex fingers at metacarpophalangeal joints but extend fingers at interphalangeal joints	Median nerve (lateral two) and ulnar nerve (medial two)
Palmar interossei (in″ter-os′e-i) (*interossei* = between bones)	Four long, cone-shaped muscles in the spaces between the metacarpals; lie ventral to the dorsal interossei	O—the side of each metacarpal that faces the midaxis of the hand (metacarpal III) but absent from metacarpal III I—extensor expansion on first phalanx of each finger (except third finger), on side facing midaxis of hand	Adduct fingers: pull fingers in toward third digit; act with lumbricals to extend fingers at interphalangeal joints and flex them at metacarpophalangeal joints	Ulnar nerve
Dorsal interossei	Four bipennate muscles filling spaces between the metacarpals; deepest palm muscles, also visible on dorsal side of hand (Figure 10.17b)	O—sides of metacarpals I—extensor expansion over proximal phalanx of second to fourth fingers on side opposite midaxis of hand (third finger), but on *both* sides of third finger	Abduct (diverge) fingers; extend fingers at interphalangeal joints and flex them at metacarpophalangeal joints	Ulnar nerve

Palmar interossei

(d) Palmar interossei (isolated)

Dorsal interossei

(e) Dorsal interossei (isolated)

Figure 10.19 *(continued)* **Intrinsic muscles of the hand. (d, e)** Hand muscles, ventral view of right hand.

MUSCLE GALLERY

Table 10.14 Muscles Crossing the Hip and Knee Joints: Movements of the Thigh and Leg (Figures 10.20 and 10.21)

The muscles fleshing out the thigh are difficult to segregate into groups on the basis of action. Some thigh muscles act only at the hip joint, others only at the knee, while still others act at both joints.

- *Most anterior* muscles of the hip and thigh flex the thigh at the hip and extend the leg at the knee, producing the foreswing phase of walking.
- *Posterior* muscles of the hip and thigh extend the thigh and flex the leg—the backswing phase of walking.
- A third group of muscles in this region, the *medial*, or *adductor*, muscles, all adduct the thigh; they have no effect on the leg.

In the thigh, the anterior, posterior, and adductor muscles are separated by walls of fascia into *anterior*, *posterior*, and *medial compartments* (see Figure 10.26a). The deep fascia of the thigh, the *fascia lata*, surrounds and encloses all three groups of muscles like a support stocking.

Movements of the thigh

Movements of the thigh (occurring at the hip joint) are accomplished largely by muscles anchored to the pelvic girdle. Like the shoulder joint, the hip joint is a ball-and-socket joint permitting flexion, extension, abduction, adduction, circumduction, and rotation. Muscles effecting these movements are among the most powerful muscles of the body.

Flexion. For the most part, the thigh *flexors* pass in front of the hip joint. The most important thigh flexors are the **iliopsoas** (the prime mover), **tensor fascia lata**, and **rectus femoris** (Figure 10.20a). They are assisted in this action by the **adductor muscles** of the medial thigh and the straplike **sartorius**.

Extension. Thigh *extension* is effected primarily by the massive **hamstring muscles** of the posterior thigh (Figure 10.21a). During forceful extension, the **gluteus maximus** of the buttocks is called into play.

Abduction. Buttock muscles that lie lateral to the hip joint (**gluteus medius** and **minimus**) *abduct* the thigh (Figure 10.21c).

Adduction. Thigh adduction is the role of the adductor muscles of the medial thigh. Abduction and adduction of the thighs are extremely important during walking to shift the trunk from side to side and balance the body's weight over the limb that is on the ground.

Rotation. Many different muscles bring about medial and lateral rotation of the thigh.

Movements of the leg

At the knee joint, flexion and extension are the main movements. The sole leg *extensor* is the **quadriceps femoris** muscle of the anterior thigh, the most powerful muscle in the body (Figure 10.20a). The quadriceps is antagonized by the hamstrings of the posterior compartment, which are the prime movers of leg flexion.

Table 10.17 (Part I) summarizes the actions of these muscles. As you study these actions, remember that "leg" refers only to the distal part of the lower limb (between the knee and the ankle; ◀ p. 242). This is in contrast to "arm," which refers to the proximal upper limb (between the shoulder and elbow; ◀ p. 231).

MUSCLE	DESCRIPTION	ORIGIN (O) AND INSERTION (I)	ACTION	NERVE SUPPLY
PART I: ANTERIOR AND MEDIAL MUSCLES (Figure 10.20)				
Origin on the Pelvis or Vertebral Column				
Iliopsoas (il″e-o-so′us)	Iliopsoas is a composite of two closely related muscles (iliacus and psoas major) whose fibers pass under the inguinal ligament (see Figure 10.12) to insert via a common tendon on the femur.			
• **Iliacus** (il-e-ak′us) (*iliac* = ilium)	Large, fan-shaped, more lateral muscle	O—iliac fossa and crest, ala of sacrum I—lesser trochanter of femur via iliopsoas tendon	Iliopsoas is the prime mover for flexing thigh, or for flexing trunk on thigh as during a bow	Femoral nerve (L_2 and L_3)
• **Psoas major** (so′us) (*psoa* = loin muscle; *major* = larger)	Longer, thicker, more medial muscle of the pair (butchers refer to this muscle as the tenderloin in animals)	O—by fleshy slips from transverse processes, bodies, and discs of lumbar vertebrae and T_{12} I—lesser trochanter of femur via iliopsoas tendon	As above; also flexes vertebral column laterally; important postural muscle	Ventral rami (L_1–L_3)
Sartorius (sar-tor′e-us) (*sartor* = tailor)	Straplike superficial muscle running obliquely across anterior surface of thigh to knee; longest muscle in body; crosses both hip and knee joints	O—anterior superior iliac spine I—winds around medial aspect of knee and inserts into medial aspect of proximal tibia	Flexes, abducts, and laterally rotates thigh; a weak knee flexor; helps produce the cross-legged position	Femoral nerve

MUSCLE GALLERY

Table 10.14 Muscles Crossing the Hip and Knee Joints: Movements of the Thigh and Leg (Figures 10.20 and 10.21) *(continued)*

10

(a)

(b)

O = origin
I = insertion

(c)

Figure 10.20 Anterior and medial muscles that move the thigh and leg. (a) Anterior view of the deep muscles of the pelvis and superficial muscles of the right thigh. **(b)** Adductor muscles of the medial compartment of the thigh, isolated. **(c)** Vastus muscles of the quadriceps group, isolated.

MUSCLE GALLERY

Table 10.14 *(continued)*				
MUSCLE	**DESCRIPTION**	**ORIGIN (O) AND INSERTION (I)**	**ACTION**	**NERVE SUPPLY**

Muscles of the Medial Compartment of the Thigh

Adductors	Large muscle mass consisting of three muscles (magnus, longus, and brevis) forming medial aspect of thigh; arise from inferior part of pelvis and insert at various levels on femur. All are used in movements that press thighs together, as when astride a horse; important in pelvic tilting movements that occur during walking and in fixing the hip when the knee is flexed and the foot is off the ground. Obturator nerve innervates entire group. Strain or stretching of this muscle group is called a "pulled groin."			
• **Adductor magnus** (mag′nus) (*adduct* = move toward midline; *magnus* = large)	Triangular muscle with a broad insertion; a composite muscle that is part adductor and part hamstring in action	O—ischial and pubic rami and ischial tuberosity I—linea aspera and adductor tubercle of femur	Anterior part adducts and medially rotates and flexes thigh; posterior part is a synergist of hamstrings in thigh extension	Obturator nerve and sciatic nerve (L₂–L₄)
• **Adductor longus** (*longus* = long)	Overlies middle aspect of adductor magnus; most anterior of adductor muscles	O—pubis near pubic symphysis I—linea aspera	Adducts, flexes, and medially rotates thigh	Obturator nerve (L₂–L₄)
• **Adductor brevis** (*brevis* = short)	In contact with obturator externus muscle; largely concealed by adductor longus and pectineus	O—body and inferior pubic ramus I—linea aspera above adductor longus	Adducts, flexes, and medially rotates thigh	Obturator nerve
Pectineus (pek-tin′e-us) (*pecten* = comb)	Short, flat muscle; overlies adductor brevis on proximal thigh; adjacent to adductor longus medially	O—pubis (and superior ramus) I—from lesser trochanter inferior to the linea aspera on posterior aspect of femur	Adducts, flexes, and medially rotates thigh	Femoral and sometimes obturator nerve
Gracilis (grah-si′lis) (*gracilis* = slender)	Long, thin, superficial muscle of medial thigh	O—inferior ramus and body of pubis and adjacent ischial ramus I—medial surface of tibia just inferior to its medial condyle	Adducts thigh, flexes and medially rotates leg, especially during walking	Obturator nerve

Muscles of the Anterior Compartment of the Thigh

Quadriceps femoris* (kwod′ri-seps fem′o-ris)	Arises from four separate heads (*quadriceps* = four heads) that form the flesh of front and sides of thigh. These heads (rectus femoris, and lateral, medial, and intermediate vasti muscles) have a common insertion tendon, the *quadriceps tendon*, which inserts into the patella and then via the *patellar ligament* into tibial tuberosity. The quadriceps is a powerful leg extensor used in climbing, jumping, running, and rising from seated position. The tone of quadriceps plays an important role in strengthening the knee joint.			
• **Rectus femoris** (rek′tus) (*rectus* = straight; *femoris* = femur)	Superficial muscle of anterior thigh; runs straight down thigh; longest head and only muscle of group to cross hip joint	O—anterior inferior iliac spine and superior margin of acetabulum I—patella and tibial tuberosity via patellar ligament	Extends leg and flexes thigh at hip	Femoral nerve (L₂–L₄)
• **Vastus lateralis** (vas′tus lat″er-a′lis) (*vastus* = large; *lateralis* = lateral)	Largest head of the group, forms lateral aspect of thigh; a common intramuscular injection site, particularly in infants (who have poorly developed buttock and arm muscles)	O—greater trochanter, intertrochanteric line, linea aspera I—as for rectus femoris	Extends leg and stabilizes knee	Femoral nerve
• **Vastus medialis** (me″de-a′lis) (*medialis* = medial)	Forms inferomedial aspect of thigh	O—linea aspera, intertrochanteric and medial supracondylar lines I—as for rectus femoris	Extends leg	Femoral nerve

*Another small muscle belonging to the quadriceps femoris group has recently been discovered. The *tensor vastus intermedius*, located between vastus lateralis and vastus intermedius, varies in shape between individuals. Will the quadriceps femoris be renamed the quinticeps femoris someday? Time will tell.

MUSCLE GALLERY

Table 10.14	Muscles Crossing the Hip and Knee Joints: Movements of the Thigh and Leg (Figures 10.20 and 10.21) *(continued)*			
MUSCLE	**DESCRIPTION**	**ORIGIN (O) AND INSERTION (I)**	**ACTION**	**NERVE SUPPLY**
• **Vastus intermedius** (in″ter-me′de-us) (*intermedius* = intermediate)	Obscured by rectus femoris; lies between vastus lateralis and vastus medialis on anterior thigh	O—anterior and lateral surfaces of proximal femur shaft I—as for rectus femoris	Extends leg	Femoral nerve
Tensor fascia lata (ten′sor fă′she-ah la′tah) (*tensor* = to make tense; *fascia* = band; *lata* = wide)	Enclosed between fascia layers of anterolateral aspect of thigh; functionally associated with medial rotators and flexors of thigh	O—anterior aspect of iliac crest and anterior superior iliac spine I—iliotibial tract*	Steadies the leg and trunk on thigh by tightening iliotibial tract; flexes and abducts thigh; rotates thigh medially	Superior gluteal nerve (L$_4$ and L$_5$)

PART II: POSTERIOR MUSCLES (Figure 10.21)

Gluteal Muscles—Origin on Pelvis

Gluteus maximus (gloo′te-us mak′sĭ-mus) (*glutos* = buttock; *maximus* = largest)	Largest and most superficial gluteus muscle; forms bulk of buttock mass; fascicles are thick and coarse; site of intramuscular injection (dorsal gluteal site); overlies large sciatic nerve; covers ischial tuberosity only when standing; when sitting, moves superiorly, leaving ischial tuberosity exposed in the subcutaneous position	O—dorsal ilium, sacrum, and coccyx I—gluteal tuberosity of femur; iliotibial tract	Major extensor of thigh; complex, powerful, and most effective when thigh is flexed and force is necessary, as in rising from a forward flexed position and in thrusting the thigh posteriorly in climbing stairs and running; generally inactive during standing and walking; laterally rotates and abducts thigh	Inferior gluteal nerve (L$_5$, S$_1$, and S$_2$)
Gluteus medius (me′de-us) (*medius* = middle)	Thick muscle largely covered by gluteus maximus; important site for intramuscular injections (ventral gluteal site); considered safer than dorsal gluteal site because less chance of injuring sciatic nerve	O—between anterior and posterior gluteal lines on lateral surface of ilium I—by short tendon into lateral aspect of greater trochanter of femur	Abducts and medially rotates thigh; steadies pelvis; its action is extremely important in walking; i.e., muscle of limb planted on ground tilts or holds pelvis in abduction so that pelvis on side of swinging limb does not sag and foot of swinging limb can clear the ground	Superior gluteal nerve (L$_4$, L$_5$, S$_1$)
Gluteus minimus (mĭ′nĭ-mus) (*minimus* = smallest)	Smallest and deepest gluteal muscle	O—between anterior and inferior gluteal lines on external surface of ilium I—anterior border of greater trochanter of femur	As for gluteus medius	Superior gluteal nerve (L$_4$, L$_5$, S$_1$)

*The iliotibial tract is a thickened lateral portion of the *fascia lata* (the fascia that ensheathes all the muscles of the thigh). It extends as a tendinous band from the iliac crest to the knee (see Figure 10.21a and b).

Gluteus medius

Gluteus maximus

Adductor magnus

Gracilis

Iliotibial tract

Biceps femoris
• Long head
• Short head

Semitendinosus

Semimembranosus

Gastrocnemius

(a)

(b)

Figure 10.21 Posterior muscles of the right hip and thigh.
(a) Superficial view showing the gluteus muscles of the buttock and hamstring muscles of the thigh. **(b)** Cadaver dissection of view similar to (a).

Practice Human Cadaver lab questions:
MasteringA&P® > Study Area > Lab Tools > PAL

10

MUSCLE GALLERY

Table 10.14	Muscles Crossing the Hip and Knee Joints: Movements of the Thigh and Leg (Figures 10.20 and 10.21) *(continued)*			
MUSCLE	**DESCRIPTION**	**ORIGIN (O) AND INSERTION (I)**	**ACTION**	**NERVE SUPPLY**
Lateral Rotators				
Piriformis (pir'ĭ-form-is) (*piri* = pear; *forma* = shape)	Pyramidal muscle located on posterior aspect of hip joint; inferior to gluteus minimus; issues from pelvis via greater sciatic notch	O—anterolateral surface of sacrum (opposite greater sciatic notch) I—superior border of greater trochanter of femur	Rotates extended thigh laterally; because inserted above head of femur, can also help abduct thigh when hip is flexed; stabilizes hip joint	S_1 and S_2, L_5
Obturator externus (ob"tu-ra'tor ek-ster'nus) (*obturator* = obturator foramen; *externus* = outside)	Flat, triangular muscle deep in superomedial aspect of thigh	O—outer surfaces of obturator membrane, pubis, and ischium, margins of obturator foramen I—by a tendon into trochanteric fossa of posterior femur	As for piriformis	Obturator nerve
Obturator internus (in-ter'nus) (*internus* = inside)	Surrounds obturator foramen within pelvis; leaves pelvis via lesser sciatic notch and turns acutely forward to insert on femur	O—inner surface of obturator membrane, greater sciatic notch, and margins of obturator foramen I—greater trochanter in front of piriformis	As for piriformis	L_5 and S_1

(c)

(d)

Figure 10.21 *(continued)* **Posterior muscles of the right hip and thigh. (c)** Deep muscles of the gluteal region, which primarily act to rotate the thigh laterally. The superficial gluteus maximus and medius have been removed. **(d)** Anterior view of the isolated obturator externus muscle.

MUSCLE GALLERY

Table 10.14 *(continued)*				

MUSCLE	DESCRIPTION	ORIGIN (O) AND INSERTION (I)	ACTION	NERVE SUPPLY
Gemellus (jĕ-mĕ′lis)—superior and inferior (*gemin* = twin, double; *superior* = above; *inferior* = below)	Two small muscles with common insertions and actions; considered extrapelvic portions of obturator internus	O—ischial spine (superior); ischial tuberosity (inferior) I—greater trochanter of femur	As for piriformis	L_5 and S_1
Quadratus femoris (*quad* = four-sided square)	Short, thick muscle; most inferior lateral rotator muscle; extends laterally from pelvis	O—ischial tuberosity I—intertrochanteric crest of femur	Rotates thigh laterally and stabilizes hip joint	L_5 and S_1

Muscles of the Posterior Compartment of the Thigh

Hamstrings	The hamstrings are fleshy muscles of the posterior thigh (biceps femoris, semitendinosus, and semimembranosus). They cross both the hip and knee joints and are prime movers of thigh extension and leg flexion. The group has a common origin site and is innervated by sciatic nerve (composed of two nerves, the tibial and common fibular nerves wrapped in a common sheath). Ability of hamstrings to act on one of the two joints spanned depends on which joint is fixed—if knee is fixed (extended), they extend hip; if hip is extended, they flex knee. However, when hamstrings are stretched, they tend to restrict full accomplishment of antagonistic movement; e.g., if knees are fully extended, it is difficult to flex the hip fully (and touch your toes), and when the thigh is fully flexed as in kicking a football, it is almost impossible to extend the knee fully at the same time (without considerable practice). Name of this muscle group comes from old butchers' practice of using their tendons to hang hams for smoking. "Pulled hamstrings" are common sports injuries in those who run very hard, e.g., sprinters.			
• **Biceps femoris** (*biceps* = two heads)	Most lateral muscle of the group; arises from two heads	O—ischial tuberosity (long head); linea aspera, lateral supracondylar line, and distal femur (short head) I—common tendon passes downward and laterally (forming lateral border of popliteal fossa) to insert into head of fibula and lateral condyle of tibia	Extends thigh and flexes leg; laterally rotates leg, especially when knee is flexed	Sciatic nerve—tibial nerve to long head, common fibular nerve to short head (L_5–S_2)
• **Semitendinosus** (sem″e-ten″dĭ-no′sus) (*semi* = half; *tendinosus* = tendon)	Lies medial to biceps femoris; although its name suggests that this muscle is largely tendinous, it is quite fleshy; its long slender tendon begins about two-thirds of the way down thigh	O—ischial tuberosity in common with long head of biceps femoris I—medial aspect of upper tibial shaft	Extends thigh and flexes leg; with semimembranosus, medially rotates leg	Sciatic nerve—tibial nerve portion (L_5–S_2)
• **Semimembranosus** (sem″e-mem″brah-no′sus) (*membranosus* = membrane)	Deep to semitendinosus	O—ischial tuberosity I—medial condyle of tibia; via oblique popliteal ligament to lateral condyle of femur	Extends thigh and flexes leg; medially rotates leg	Sciatic nerve—tibial nerve portion (L_5–S_2)

10

MUSCLE GALLERY

Table 10.15 Muscles of the Leg: Movements of the Ankle and Toes   (Figures 10.22 to 10.24)

The deep fascia of the leg is continuous with the fascia lata that ensheathes the thigh. Like a snug "knee sock" beneath the skin, the leg fascia binds the leg muscles tightly, preventing excessive muscle swelling during exercise and aiding venous return. Its inward extensions segregate the leg muscles into *anterior, lateral,* and *posterior* compartments (see Figure 10.26b), each with its own nerve and blood supply. Distally the leg fascia thickens to form the **flexor, extensor,** and **fibular** (or peroneal) **retinacula,** "ankle brackets" that hold the tendons in place where they run to the foot (Figures 10.22a, 10.23a).

The various muscles of the leg promote movements at the ankle joint (dorsiflex and plantar flex), at the intertarsal joints (invert and evert the foot), and/or at the toes (flex and extend).

Anterior extensor compartment. Muscles in the *anterior extensor compartment of the leg* are primarily toe extensors and ankle dorsiflexors. Although dorsiflexion is not a powerful movement, it is important in preventing the toes from dragging during walking.

Lateral compartment. Lateral compartment muscles are the **fibular,** formerly *peroneal* (*peron* = fibula), **muscles** that plantar flex and evert the foot.

Posterior flexor compartment. Muscles of the *posterior flexor compartment* primarily plantar flex the foot and flex the toes (Figure 10.24). Plantar flexion is the most powerful movement of the ankle (and foot) because it lifts the entire weight of our body. It is essential for standing on tiptoe and provides the forward thrust when walking and running. The **popliteus muscle,** which crosses the knee, has a unique function. It medially rotates the extended knee, "unlocking" it in preparation for flexion (Figure 10.24b, f). We consider the tiny intrinsic muscles of the sole of the foot (lumbricals, interossei, and others) in Table 10.16.

Table 10.17 (Part II) summarizes the actions of the muscles in this table.

MUSCLE	DESCRIPTION	ORIGIN (O) AND INSERTION (I)	ACTION	NERVE SUPPLY
PART I: MUSCLES OF THE ANTERIOR COMPARTMENT (Figures 10.22 and 10.23)	All muscles of the anterior compartment dorsiflex the ankle. Paralysis of the anterior muscle group causes *foot drop,* which requires that the leg be lifted unusually high during walking to prevent tripping over one's toes.			
Tibialis anterior (tib"e-a'lis) (*tibial* = tibia; *anterior* = toward the front)	Superficial muscle of anterior leg; laterally parallels sharp anterior margin of tibia	O—lateral condyle and upper 2/3 of tibial shaft; interosseous membrane I—by tendon into inferior surface of medial cuneiform and first metatarsal bone	Prime mover of dorsiflexion; inverts foot; helps support medial longitudinal arch of foot	Deep fibular nerve (L_4 and L_5)
Extensor digitorum longus (*extensor* = increases angle at a joint; *digit* = finger or toe; *longus* = long)	Unipennate muscle on anterolateral surface of leg; lateral to tibialis anterior muscle	O—lateral condyle of tibia; proximal 3/4 of fibula; interosseous membrane I—middle and distal phalanges of second to fifth toes via extensor expansion	Prime mover of toe extension (acts mainly at metatarsophalangeal joints); dorsiflexes foot	Deep fibular nerve (L_5 and S_1)
Fibularis (peroneus) tertius (fib-u-lar'ris ter'shus) (*fibular* = fibula; *tertius* = third)	Small muscle; usually continuous and fused with distal part of extensor digitorum longus; not always present	O—distal anterior surface of fibula and interosseous membrane I—tendon inserts on dorsum of fifth metatarsal	Dorsiflexes and everts foot	Deep fibular nerve (L_5 and S_1)
Extensor hallucis longus (hal'u-sis) (*hallux* = great toe)	Deep to extensor digitorum longus and tibialis anterior; narrow origin	O—anteromedial fibula shaft and interosseous membrane I—tendon inserts on distal phalanx of great toe	Extends great toe; dorsiflexes foot	Deep fibular nerve (L_5 and S_1)

10

Figure 10.22 Muscles of the anterior compartment of the right leg. (a) Superficial view of anterior leg muscles. **(b–d)** Anterior leg muscles in isolation to show their origins and insertions.

Fibularis longus

Gastrocnemius

Tibia

Tibialis anterior

Extensor digitorum longus

Soleus

Extensor hallucis longus

Fibularis tertius

Superior and inferior extensor retinacula

Extensor hallucis brevis

Extensor digitorum brevis

(a)

Extensor hallucis longus

Fibularis tertius

(c)

Tibialis anterior

(b)

Extensor digitorum longus

O = origin
I = insertion

(d)

10

MUSCLE GALLERY

Table 10.15   Muscles of the Leg: Movements of the Ankle and Toes   (Figures 10.22 to 10.24)   *(continued)*

O = origin
I = insertion

(a)

(b)

(c)

Figure 10.23   Muscles of the lateral compartment of the right leg. (a) Superficial view of lateral aspect of the leg. **(b)** View of fibularis longus in isolation; inset illustrates the insertion of the fibularis longus on the plantar surface of the foot. **(c)** View of fibularis brevis in isolation.

MUSCLE GALLERY

Table 10.15 *(continued)*

MUSCLE	DESCRIPTION	ORIGIN (O) AND INSERTION (I)	ACTION	NERVE SUPPLY
PART II: MUSCLES OF THE LATERAL COMPARTMENT (Figures 10.23 and 10.24) Besides plantar flexion and foot eversion, these muscles stabilize the lateral ankle and lateral longitudinal arch of the foot.				
Fibularis (peroneus) longus (See also Figure 10.22)	Superficial lateral muscle; overlies fibula	O—head and upper portion of lateral fibula I—by long tendon that curves under foot to first metatarsal and medial cuneiform	Plantar flexes and everts foot; may help keep foot flat on ground	Superficial fibular nerve (L$_5$–S$_2$)
Fibularis (peroneus) brevis (*brevis* = short)	Smaller muscle; deep to fibularis longus; enclosed in a common sheath	O—distal fibula shaft I—by tendon running behind lateral malleolus to insert on proximal end of fifth metatarsal	Plantar flexes and everts foot	Superficial fibular nerve
PART III: MUSCLES OF THE POSTERIOR COMPARTMENT (Figure 10.24) The muscles of the posterior compartment act together to plantar flex the foot.				
Superficial Muscles				
Triceps surae (tri″seps sur′e) (See also Figure 10.23)	Refers to muscle pair (gastrocnemius and soleus) that shapes the posterior calf and inserts via a common tendon into the calcaneus of the heel; this *calcaneal* or *Achilles tendon* is the largest tendon in the body. Prime movers of foot plantar flexion.			
• **Gastrocnemius** (gas″truk-ne′me-us) (*gaster* = belly; *kneme* = leg)	Superficial muscle of pair; two prominent bellies that form proximal curve of calf	O—by two heads from medial and lateral condyles of femur I—posterior calcaneus via calcaneal tendon	Plantar flexes foot when leg is extended; because it also crosses knee joint, it can flex leg when foot is dorsiflexed	Tibial nerve (S$_1$, S$_2$)
• **Soleus** (so′le-us) (*soleus* = fish)	Broad, flat muscle, deep to gastrocnemius on posterior surface of calf	O—extensive origin from superior tibia, fibula, and interosseous membrane I—as for gastrocnemius	Plantar flexes foot; important locomotor and postural muscle during walking, running, and dancing	Tibial nerve
Plantaris (plan-tar′is) (*planta* = sole of foot)	Generally a small, feeble muscle, but varies in size and extent; may be absent	O—posterior femur above lateral condyle I—via a long, thin tendon into calcaneus or its tendon	Helps to flex leg and plantar flex foot	Tibial nerve
Deep Muscles (Figure 10.24c–f)				
Popliteus (pop-lit′e-us) (*poplit* = back of knee)	Thin, triangular muscle at posterior knee; passes downward and medially to tibial surface	O—lateral condyle of femur and lateral meniscus of knee I—proximal tibia	Flexes and rotates leg medially to unlock extended knee when flexion begins; with tibia fixed, rotates thigh laterally	Tibial nerve (L$_4$–S$_1$)
Flexor digitorum longus (*flexor* = decreases angle at a joint)	Long, narrow muscle; runs medial to and partially overlies tibialis posterior	O—extensive origin on the posterior tibia I—tendon runs behind medial malleolus and inserts into distal phalanges of second to fifth toes	Plantar flexes and inverts foot; flexes toes; helps foot "grip" ground	Tibial nerve (L$_5$–S$_2$)
Flexor hallucis longus (See also Figure 10.23)	Bipennate muscle; lies lateral to inferior aspect of tibialis posterior	O—midshaft of fibula; interosseous membrane I—tendon runs under foot to distal phalanx of great toe	Plantar flexes and inverts foot; flexes great toe at all joints; "push off" muscle during walking	Tibial nerve (L$_5$–S$_2$)
Tibialis posterior (*posterior* = toward the back)	Thick, flat muscle deep to soleus; placed between posterior flexors	O—superior tibia and fibula and interosseous membrane I—tendon passes behind medial malleolus and under arch of foot; inserts into several tarsals and metatarsals II–IV	Prime mover of foot inversion; plantar flexes foot; stabilizes medial longitudinal arch of foot (as during ice skating)	Tibial nerve (L$_4$ and L$_5$)

Table 10.15   Muscles of the Leg: Movements of the Ankle and Toes   (Figures 10.22 to 10.24) *(continued)*

10

(a) Superficial view of the posterior leg.

(b) The gastrocnemius has been removed to show the soleus immediately deep to it.

Figure 10.24 Muscles of the posterior compartment of the right leg.

MUSCLE GALLERY

Table 10.15 *(continued)*

O = origin
I = insertion

Plantaris (cut)

Gastrocnemius
lateral head (cut)

Gastroc-
nemius
medial head
(cut)

Popliteus

Soleus (cut)

Tibialis posterior

Fibula

Fibularis
longus

Flexor
digitorum
longus

Flexor hallucis
longus

Fibularis brevis

Tendon of
tibialis posterior

Medial
malleolus

Calcaneal
tendon (cut)

Calcaneus

Tibialis
posterior

Popliteus

Flexor
hallucis
longus

Flexor
digitorum
longus

Flexor
digitorum
longus

(c) The triceps surae has been removed to show the
deep muscles of the posterior compartment.

**(d) Isolated tibialis
posterior.**

**(e) Isolated flexor
digitorum longus.**

**(f) Isolated
popliteus and
flexor hallucis
longus.**

10

Figure 10.24 *(Continued)*

MUSCLE GALLERY

The intrinsic muscles of the foot help to flex, extend, abduct, and adduct the toes. Collectively, along with the tendons of some leg muscles that enter the sole, the foot muscles help support the arches of the foot. There is a single muscle on the foot's dorsum (superior aspect), and several muscles on the plantar aspect (the sole). The plantar muscles occur in four layers, from superficial to deep. Overall, the foot muscles are remarkably similar to those in the palm of the hand.

MUSCLE	DESCRIPTION	ORIGIN (O) AND INSERTION (I)	ACTION	NERVE SUPPLY
MUSCLES ON DORSUM OF FOOT				
Extensor digitorum brevis (Figures 10.22a and 10.23a)	Small, four-part muscle on dorsum of foot; deep to the tendons of extensor digitorum longus; corresponds to the extensor indicis and extensor pollicis muscles of forearm	O—anterior part of calcaneus bone; extensor retinaculum I—base of proximal phalanx of great toe; extensor expansions on second to fourth or fifth toes	Helps extend toes at metatarsophalangeal joints	Deep fibular nerve (L_5 and S_1)
MUSCLES ON SOLE OF FOOT—FIRST LAYER (MOST SUPERFICIAL) (Figure 10.25)				
Flexor digitorum brevis	Bandlike muscle in middle of sole; corresponds to flexor digitorum superficialis of forearm and inserts into digits in the same way	O—calcaneal tuberosity I—middle phalanx of second to fourth toes	Helps flex toes	Medial plantar nerve (a branch of tibial nerve, S_1 and S_2)
Abductor hallucis (hal'u-sis) (*hallux* = great toe)	Lies medial to flexor digitorum brevis (recall the similar thumb muscle, abductor pollicis brevis)	O—calcaneal tuberosity and flexor retinaculum I—proximal phalanx of great toe, medial side, in the tendon of flexor hallucis brevis)	Abducts great toe	Medial plantar nerve
Abductor digiti minimi	Most lateral of the three superficial sole muscles (recall similar abductor muscle in palm)	O—calcaneal tuberosity I—lateral side of base of little toe's proximal phalanx	Abducts and flexes little toe	Lateral plantar nerve (a branch of tibial nerve, S_1, S_2, and S_3)
MUSCLES ON SOLE OF FOOT—SECOND LAYER				
Flexor accessorius (quadratus plantae)	Rectangular muscle just deep to flexor digitorum brevis in posterior half of sole; two heads (see also Figure 10.25c)	O—medial and lateral sides of calcaneus I—tendon of flexor digitorum longus in midsole	Straightens out the oblique pull of flexor digitorum longus	Lateral plantar nerve
Lumbricals	Four little "worms" (like lumbricals in hand)	O—from each tendon of flexor digitorum longus I—extensor expansion on proximal phalanx of second to fifth toes, medial side	By pulling on extensor expansion, flex toes at metatarsophalangeal joints and extend toes at interphalangeal joints	Medial plantar nerve (first lumbrical) and lateral plantar nerve (second to fourth lumbrical)

MUSCLE GALLERY

Table 10.16 *(continued)*

Tendon of
flexor hallucis longus
Lumbricals
Flexor hallucis
brevis
Flexor digiti
minimi brevis
Abductor hallucis
**Flexor digitorum
brevis**
Flexor accessorius
**Abductor digiti
minimi**
Calcaneal
tuberosity

(a) First layer (plantar aspect)

Lumbricals
Flexor hallucis
brevis
Flexor hallucis
longus tendon
Flexor digitorum
longus (tendon)
Flexor digiti
minimi brevis
Abductor digiti
minimi
Flexor accessorius
Fibularis longus (tendon)
Flexor digitorum
longus (tendon)
Flexor hallucis
longus (tendon)

(b) Second layer (plantar aspect)

Figure 10.25 Muscles of the right foot.

10

MUSCLE GALLERY

Table 10.16	Intrinsic Muscles of the Foot: Toe Movement and Arch Support		(Figure 10.25) *(continued)*	
MUSCLE	**DESCRIPTION**	**ORIGIN (O) AND INSERTION (I)**	**ACTION**	**NERVE SUPPLY**
MUSCLES ON SOLE OF FOOT—THIRD LAYER				
Flexor hallucis brevis	Covers metatarsal I; splits into two bellies (recall flexor pollicis brevis of thumb)	O—lateral cuneiform and cuboid bones I—via two tendons onto base of the proximal phalanx of great toe; each insertion tendon contains a sesamoid bone	Flexes great toe at metatarsophalangeal joint	Medial plantar nerve
Adductor hallucis	Oblique and transverse heads; deep to lumbricals (recall adductor pollicis in thumb)	O—from bases of metatarsals II–IV and from fibularis longus tendon sheath (oblique head); from a ligament across metatarsophalangeal joints (transverse head) I—base of proximal phalanx of great toe, lateral side	Helps maintain the transverse arch of foot; weak adductor of great toe	Lateral plantar nerve (S_2 and S_3)
Flexor digiti minimi brevis	Covers metatarsal V (recall same muscle in hand)	O—base of metatarsal V and tendon sheath of fibularis longus I—base of proximal phalanx of fifth toe	Flexes little toe at metatarsophalangeal joint	Lateral plantar nerve
MUSCLES ON SOLE OF FOOT—FOURTH LAYER (DEEPEST)				
Plantar (3) and dorsal (4) interossei	Similar to palmar and dorsal interossei of hand in locations, attachments, and actions; however, these muscles orient around the second digit, not the third	See palmar and dorsal interossei (Table 10.13)	See palmar and dorsal interossei (Table 10.13)	Lateral plantar nerve

Adductor hallucis
(transverse head)

Adductor hallucis
(oblique head)

Interosseous
muscles

**Flexor hallucis
brevis**

**Flexor digiti
minimi brevis**

Fibularis longus (tendon)

Flexor accessorius

Flexor digitorum
longus (tendon)

Flexor hallucis
longus (tendon)

(c) Third layer (plantar aspect)

Plantar
interossei

**(d) Fourth layer (plantar aspect):
plantar interossei**

Dorsal
interossei

10

**(e) Fourth layer (dorsal aspect):
dorsal interossei**

Figure 10.25 *(continued)* **Muscles of the right foot.**

MUSCLE GALLERY

Table 10.17   Summary: Actions of Muscles Acting on the Thigh, Leg, and Foot   (Figure 10.26)

PART I: Muscles Acting on the Thigh and Leg (PM = Prime Mover)	Actions at the Hip Joint						Actions at the Knee	
	Flexion	Extension	Abduction	Adduction	Medial Rotation	Lateral Rotation	Flexion	Extension
Anterior and Medial Muscles								
Iliopsoas	× (PM)							
Sartorius	×		×			×	×	
Tensor fascia lata	×		×		×			
Rectus femoris	×							× (PM)
Vastus muscles								× (PM)
Adductor magnus	×	×		×	×			
Adductor longus	×			×	×			
Adductor brevis	×			×	×			
Pectineus	×			×	×			
Gracilis				×	×		×	
Posterior Muscles								
Gluteus maximus		× (PM)	×			×		
Gluteus medius			× (PM)		×			
Gluteus minimus			×		×			
Piriformis			×			×		
Obturator internus			×			×		
Obturator externus			×			×		
Gemelli			×			×		
Quadratus femoris						×		
Biceps femoris		× (PM)					× (PM)	
Semitendinosus		×					× (PM)	
Semimembranosus		×					× (PM)	
Gastrocnemius							×	
Plantaris							×	
Popliteus							× (and rotates leg medially)	

PART II: Muscles Acting on the Ankle and Toes	Actions at the Ankle Joint				Actions at the Toes	
	Plantar Flexion	Dorsiflexion	Inversion	Eversion	Flexion	Extension
Anterior Compartment						
Tibialis anterior		× (PM)	×			
Extensor digitorum longus		×				× (PM)
Fibularis tertius		×		×		
Extensor hallucis longus		×	× (weak)			× (great toe)
Lateral Compartment						
Fibularis longus and brevis	×			×		
Posterior Compartment						
Gastrocnemius	× (PM)					
Soleus	× (PM)					
Plantaris	×					
Flexor digitorum longus	×		×		× (PM)	
Flexor hallucis longus	×		×		× (great toe)	
Tibialis posterior	×		× (PM)			

10

MUSCLE GALLERY

Table 10.17   (continued)

(a)

(b)

Adductors

Hamstrings

Vastus lateralis

Femur

Vastus intermedius

Rectus femoris

Vastus medialis

Posterior compartment of thigh (flexes leg and extends thigh); innervated by tibial nerve (portion of sciatic nerve)

Medial compartment of thigh (adducts thigh); innervated by obturator nerve

Anterior compartment of thigh (extends leg); innervated by femoral nerve

(a) Muscles of the thigh

Key:

Posterior compartment muscles

Anterior compartment muscles

Medial compartment muscles of thigh and lateral compartment muscles of leg

Triceps surae

Fibula

Fibularis muscles

Tibialis anterior

Tibia

Posterior compartment of leg (plantar flexes foot, flexes toes); innervated by tibial nerve

Lateral compartment of leg (plantar flexes and everts foot); innervated by superficial fibular nerve

Anterior compartment of leg (dorsiflexes foot, extends toes); innervated by deep fibular nerve

(b) Muscles of the leg

Figure 10.26 Summary: Actions of muscles of the thigh and leg.

10

Check Your Understanding

8. As Mario listened to Brent tell an amusing story, he raised his eyebrows and then winked at Sarah. What facial muscles was he using?

9. How can the deltoid muscles both extend and flex the arm? Aren't these antagonistic movements?

10. Which of the thenar muscles does not have an insertion on bones of the thumb?

11. APPLY What muscles would you contract to make a "sad clown's face"?

12. MAKE CONNECTIONS Table 10.14 shows that many muscles cross the knee. In addition to providing movement, these muscles perform another important function (as described in Chapter 8). What is that function?

13. APPLY The figures below show three muscles that are common sites for intramuscular injections. Name each of the muscles. Which of these is most often used in infants? Why?

(a)   (b)   (c)

For answers, see Answers Appendix.

RELATED CLINICAL TERMS

Charley horse Painful muscle spasm. Can result from stretching or tearing muscle, muscle strain, or contusion (bleeding in the muscle).

Electromyography Recording and interpretation of the electrical activity of contracting muscles. Electrodes inserted into the muscles record the action potentials that pass over muscle cell membranes during contraction. Useful for diagnosing some neuromuscular diseases.

Hernia An abnormal protrusion of an organ or tissue (such as coils of the small intestine) through the muscle or connective tissue that normally contains it. Common abdominal hernias include inguinal and umbilical hernias. In these examples, the hernia penetrates the muscle wall but not the skin, and can be made worse by increased intra-abdominal pressure.

Quadriceps and hamstring strains Also called quad and hamstring pulls, these conditions involve tearing these muscles, often at or near their tendons during strenuous activity. Muscle tightness, muscle imbalance, poor conditioning, and muscle fatigue may all predispose athletes to these injuries during forceful, high-speed movements. Treatment includes rest, ice, compression, and elevation often followed by physical therapy. Contrast these with sprains of joints described on p. 273.

Ruptured calcaneal tendon Although the calcaneal (Achilles) tendon is the largest, strongest tendon in the body, its rupture is surprisingly common, particularly during recreational sports. The rupture is accompanied by abrupt pain. A gap may be seen just above the heel, and the calf bulges as the gastrocnemius and soleus are released from their insertion. Plantar flexion is weak or impossible, but dorsiflexion is exaggerated. Usually repaired surgically.

Shin splints Also called medial tibial stress syndrome, it is a painful inflammation of the muscles, tendons, and connective tissue surounding the tibia. Most often, the medial edge of the tibia is tender. A result of exercise (usually running) without sufficient conditioning. Usually treated with rest and ice. Two other condtions may mimic shin spints: tibial stress fractures and chronic exertional compartment syndrome (pain due to pressure caused by muscle swelling within its tightly wrapping fascia).

Tennis elbow. Also called lateral epicondylitis. Tenderness due to overuse of the tendon of origin of the forearm extensor muscles at the lateral epicondyle of the humerus. A repetitive stress injury caused and aggravated when these muscles contract to extend the hand at the wrist—as in executing a tennis backhand or lifting a loaded snow shovel. Despite its name, tennis elbow does not involve the elbow joint; most cases caused by work activities (e.g., carpenters, plumbers, painters).

Torticollis (tor″tĭ-kol′is; *tort* = twisted) A condition in which the neck stays rotated to one side, keeping the head tilted in that direction. All of the various forms of this condition result from problems with a sternocleidomastoid muscle on one side of the neck only. The adult-onset form, or *spasmodic torticollis*, is caused by various factors that result in sustained muscle spasms. In infants, the most common type is *congenital muscular torticollis*. It may be due to the position of the fetus in the uterus before birth, trauma to the muscle at birth, or abnormal development of the muscle. Early identification and treatment with physical therapy are effective.

CHAPTER SUMMARY

10.1 For any movement, muscles can act in one of three ways (p. 324)

1. Skeletal muscles are arranged in opposing groups across body joints so that one group can reverse or modify the action of the other.
2. Muscles are classified as prime movers (agonists), antagonists, and synergists, which include fixators.

10.2 How are skeletal muscles named? (pp. 324–326)

1. Criteria used to name muscles include a muscle's location, shape, relative size, fiber (fascicle) direction, number of origins, attachment sites (origin/insertion), and action. Several criteria are combined to name some muscles.

10.3 Fascicle arrangements help determine muscle shape and force (pp. 326–327)

1. Common patterns of fascicle arrangement are circular, convergent, parallel, fusiform, and pennate. Muscles with fibers that run parallel to the long axis of the muscle shorten most; stocky pennate muscles shorten little but are the most powerful muscles.

10.4 Muscles acting with bones form lever systems (pp. 327–329)

1. A lever is a bar that moves on a fulcrum. When an effort is applied to the lever, a load is moved. In the body, bones are the levers, joints are the fulcrums, and skeletal muscles exert the effort at their insertions.
2. When the effort is farther from the fulcrum than is the load, the lever operates at a mechanical advantage (it's slow and strong). When the effort is exerted closer to the fulcrum than is the load, the lever operates at a mechanical disadvantage (it's fast and promotes a large degree of movement).
3. First-class levers (effort-fulcrum-load) may operate at a mechanical advantage or disadvantage. Second-class levers (fulcrum-load-effort) all operate at a mechanical advantage. Third-class levers (fulcrum-effort-load) always operate at a mechanical disadvantage. Most skeletal muscles of the body act in third-class lever systems.

10.5 A muscle's origin and insertion determine its action (pp. 332–386)

1. Muscles of the head that produce facial expression tend to be small and to insert into soft tissue (skin and other muscles) rather than into bone. These muscles open and close the eyes and mouth, compress the cheeks, and allow smiling and other types of facial language (see Table 10.1*).
2. Muscles of the head involved in mastication include the masseter and temporalis that elevate the mandible and two deep muscle pairs that promote grinding and sliding jaw movements (see Table 10.2*). Extrinsic muscles of the tongue anchor the tongue and control its movements.
3. Deep muscles of the anterior neck promote swallowing movements, including elevation/depression of the hyoid bone, closing off the respiratory passages, and peristalsis of the pharynx (see Table 10.3*).

4. Neck muscles and deep muscles of the vertebral column promote head and trunk movements (see Table 10.4*). The deep muscles of the posterior trunk can extend large regions of the vertebral column (and head) simultaneously. The anteriorly located sternocleidomastoid and scalene muscles effect head flexion and rotation.
5. The diaphragm and external intercostal muscles of the thorax promote movements of quiet breathing (see Table 10.5*). Downward movement of the diaphragm increases intra-abdominal pressure.
6. The four muscle pairs forming the abdominal wall are layered like plywood to form a natural muscular girdle that protects, supports, and compresses abdominal contents. These muscles also flex and laterally rotate the trunk (see Table 10.6*).
7. Muscles of the pelvic floor and perineum (see Table 10.7*) support the pelvic viscera, resist increases in intra-abdominal pressure, inhibit urination and defecation, and aid erection.
8. Except for the pectoralis major and the latissimus dorsi, the superficial muscles of the thorax fix or promote movements of the scapula (see Table 10.8*). Scapular movements are effected primarily by posterior thoracic muscles.
9. Nine muscles cross the shoulder joint to move the arm (see Table 10.9*). Of these, seven originate on the scapula and two arise from the axial skeleton. Four muscles contribute to the "rotator cuff" helping to stabilize the multiaxial shoulder joint. Generally speaking, muscles located anteriorly flex, rotate, and adduct the arm. Those located posteriorly extend, rotate, and adduct the arm. The deltoid muscle of the shoulder is the prime mover of shoulder abduction.
10. Muscles that move the forearm form the flesh of the arm (see Table 10.10*). Anterior arm muscles are forearm flexors; posterior muscles are forearm extensors.
11. Muscles originating on the forearm mainly move the hand and fingers (see Table 10.11*). Except for the two pronator muscles, the anterior forearm muscles are wrist and/or finger flexors; those of the posterior compartment are wrist and finger extensors.
12. The intrinsic muscles of the hand aid in precise movements of the fingers (Table 10.13*) and in opposition, which helps us grip things. These small muscles are divided into thenar, hypothenar, and midpalmar groups.
13. Muscles crossing the hip and knee joints move the thigh and leg (see Table 10.14*). Anteromedial muscles flex and/or adduct the thigh and usually extend the leg. Muscles of the posterior gluteal region extend and rotate the thigh. Posterior thigh muscles extend the hip and flex the leg.
14. Muscles in the leg act on the ankle and toes (see Table 10.15*). Anterior compartment muscles are largely ankle dorsiflexors. Lateral compartment muscles are plantar flexors and foot everters. Those of the posterior leg are plantar flexors.
15. The intrinsic muscles of the foot (Table 10.16*) support the foot arches and help move the toes. Most occur in the sole, arranged in four layers. They resemble the small muscles in the palm of the hand.

* See specific table cited for detailed description of each muscle in the group.

10

REVIEW QUESTIONS

 To access additional practice questions using your smartphone, tablet, or computer: MasteringA&P® > Study Area > Practice Tests & Quizzes

Level 1 Remember/Understand

(Some questions have more than one correct answer. Select the best answer or answers from the choices given.)

1. A muscle that assists an agonist by causing a like movement or by stabilizing a joint over which an agonist acts is a(n) **(a)** antagonist, **(b)** prime mover, **(c)** synergist, **(d)** agonist.

2. The arrangement of muscle fibers in which the fibers are arranged at an angle to a central longitudinal tendon is **(a)** circular, **(b)** longitudinal, **(c)** pennate, **(d)** parallel.

3. Match the muscle names in column B to the facial muscles described in column A.

Column A	Column B
___(1) squints the eyes	**(a)** corrugator supercilii
___(2) raises the eyebrows	**(b)** depressor anguli oris
___(3) smiling muscle	**(c)** frontal belly of epicranius
___(4) puckers the lips	**(d)** occipital belly of epicranius
___(5) pulls the scalp posteriorly	**(e)** orbicularis oculi
	(f) orbicularis oris
	(g) zygomaticus

4. The prime mover of inspiration is the **(a)** diaphragm, **(b)** internal intercostals, **(c)** external intercostals, **(d)** abdominal wall muscles.

5. The arm muscle that both flexes and supinates the forearm is the **(a)** brachialis, **(b)** brachioradialis, **(c)** biceps brachii, **(d)** triceps brachii.

6. The chewing muscles that protract the mandible and produce side-to-side grinding movements are the **(a)** buccinators, **(b)** masseters, **(c)** temporalis, **(d)** pterygoids.

7. Muscles that depress the hyoid bone and larynx include all but the **(a)** sternohyoid, **(b)** omohyoid, **(c)** geniohyoid, **(d)** sternothyroid.

8. Intrinsic muscles of the back that promote extension of the spine (or head) include all but **(a)** splenius muscles, **(b)** semispinalis muscles, **(c)** scalene muscles, **(d)** erector spinae.

9. Several muscles act to move and/or stabilize the scapula. Which of the following are small rectangular muscles that square the shoulders as they act together to retract the scapula? **(a)** levator scapulae, **(b)** rhomboids, **(c)** serratus anterior, **(d)** trapezius.

10. A prime mover of thigh flexion is the **(a)** rectus femoris, **(b)** iliopsoas, **(c)** vastus muscles, **(d)** gluteus maximus.

11. The prime mover of thigh extension *against* resistance is the **(a)** gluteus maximus, **(b)** gluteus medius, **(c)** biceps femoris, **(d)** semimembranosus.

12. Which criterion (or criteria) is/are used in naming the gluteus medius? **(a)** relative size, **(b)** muscle location, **(c)** muscle shape, **(d)** action, **(e)** number of origins.

13. Which of the following is a large, deep muscle that protracts the scapula during punching? **(a)** serratus anterior, **(b)** rhomboids, **(c)** levator scapulae, **(d)** subscapularis.

14. Which muscles act to propel food down the length of the pharynx to the esophagus?

15. Name and describe the action of muscles used to shake your head no; to nod yes.

16. **(a)** Name two forearm muscles that are powerful extensors and abductors of the hand. **(b)** Name the sole forearm muscle that can flex the distal interphalangeal joints.

Level 2 Apply/Analyze

17. Name four criteria used in naming muscles, and provide an example (other than those used in the text) that illustrates each criterion.

18. Differentiate between the arrangement of elements (load, fulcrum, and effort) in first-, second-, and third-class levers.

19. **(a)** Name the four muscle pairs that act in unison to compress the abdominal contents. **(b)** How does their arrangement (fiber direction) contribute to the strength of the abdominal wall? **(c)** Which of these muscles can effect lateral rotation of the spine? **(d)** Which can act alone to flex the spine?

20. List all six possible movements that can occur at the shoulder joint and name the prime mover(s) of each movement. Then name their antagonists.

21. Name two muscles in each of the following compartments or regions: **(a)** thenar eminence (ball of thumb), **(b)** posterior compartment of forearm, **(c)** anterior compartment of forearm—deep muscle group, **(d)** anterior muscle group in the arm, **(e)** muscles of mastication, **(f)** third muscle layer of the foot, **(g)** posterior compartment of leg, **(h)** medial compartment of thigh, **(i)** posterior compartment of thigh.

22. When Mrs. O'Brien returned to her doctor for a follow-up visit after childbirth, she complained that she was incontinent (having problems controlling her urine flow) when she sneezed. The physician asked his nurse to give Mrs. O'Brien instructions on how to perform exercises to strengthen the muscles of the pelvic floor. To which muscles was he referring?

23. As Kendra's coach watched her sail over the high jump bar, he contracted his right orbicularis oculi muscle, raised his arm, and contracted his opponens pollicis. Was he pleased or displeased with her performance? How do you know?

Level 3 Evaluate/Synthesize

24. Assume you have a 10-lb weight in your right hand. Explain why it is easier to flex the right elbow when your forearm is supinated than when it is pronated.

25. Mr. Ahmadi, an out-of-shape 45-year-old man, was advised by his physician to lose weight and to exercise on a regular basis. He followed his diet faithfully and began to jog daily. One day, while on his morning jog, he heard a snapping sound that was immediately followed by a severe pain in his right lower calf. When his leg was examined, a gap was seen between his swollen upper calf region and his heel, and he was unable to plantar flex that ankle. What do you think happened? Why was the upper part of his calf swollen?

26. What class of lever system do the following activities describe? **(a)** The soleus muscle plantar flexes the foot. **(b)** The triceps brachii is strained while doing pushups.

CLINICAL CASE STUDY

45-Year-Old Female with Motor Nerve Damage

In Chapters 7 and 8, you met Kayla Tanner, who suffered a dislocated right hip in a car accident. Six weeks later, Mrs. Tanner was still unable to walk or run without hip pain, and had weakness in flexion at the knee, excessive foot inversion and plantar flexion, and significant "foot drop" (the inability to dorsiflex the foot).

Electromyography (measurement of muscle electrical activity) and nerve conduction studies (measurement of the speed of nerve impulses) revealed sciatic nerve damage, most likely a result of the nerve being compressed when the hip was dislocated. This large nerve innervates many lower limb muscles. Since her surgery, Mrs. Tanner has been undergoing intense physical therapy and has shown significant improvement.

1. During her initial visit, the physical therapist focused on the prime movers and synergists of the hip, knee, and ankle. What are "prime movers" and "synergists"?

2. To assess the function and strength of a specific muscle, a physical therapist will often apply resistance (push against the moving limb) to mimic the action of an antagonist muscle. What is an antagonist muscle, and why would the therapist mimic its action?

3. **✚ NCLEX-STYLE** Mrs. Tanner's difficulty flexing her leg at the knee is most likely due to problems with:
   a. The gastrocnemius
   b. The quadriceps femoris
   c. The biceps femoris
   d. The biceps brachii

4. For each description below, name the muscle (or muscles) that the therapist was assessing.
   a. With Mrs. Tanner in the seated position, the therapist positioned Mrs. Tanner's knees shoulder-width apart, then asked Mrs. Tanner to bring her knees together while the therapist applied resistance to the right thigh.
   b. The therapist applied resistance to the top of Mrs. Tanner's foot and asked her to pull her forefoot up toward her shin.
   c. With Mrs. Tanner in the prone position (on her stomach), the therapist applied resistance to the leg while Mrs. Tanner tried to bring her heel up toward her buttocks (flex her leg at the knee).

5. Using descriptions similar to those listed in the question above, explain how you would assess the function of the following muscles:
   a. Extensor hallucis longus
   b. Fibularis (peroneus) longus
   c. Gastrocnemius

6. **✚ NCLEX-STYLE** A muscle that originates on the posterior surface of the ischium and inserts on the posterior surface of the tibia will:
   a. Flex the thigh at the hip
   b. Extend the thigh at the hip
   c. Flex the leg at the knee
   d. Extend the leg at the knee
   e. Flex the thigh and extend the leg
   f. Flex the leg and extend the thigh

10

11

Fundamentals of the Nervous System and Nervous Tissue

In this chapter, you will learn that

Neurons send electrical signals along their length, and communicate using chemical signals

by first asking

11.1 What does the nervous system do, and how is it organized?

and then asking

What are the cells of the nervous system?

looking closer at

11.2 Neuroglia

11.3 Neurons

How do neurons send electrical signals?

looking closer at

11.4 How is the resting membrane potential generated?

then asking

11.5 How do graded potentials act as short-distance signals?

11.6 How do action potentials act as long-distance signals?

How do neurons send and receive chemical signals?

looking closer at

11.7 The synapse

11.8 Postsynaptic potentials and synaptic integration

11.9 Neurotransmitters and their receptors

11.10 How do neurons work together?

and finally, exploring

Developmental Aspects of Neurons

CAREER CONNECTION

Watch a video to learn how the chapter content is used in a real health care setting. Go to Mastering A&P® > Study Area > Animations and Videos or use quick access URL https://goo.gl/88srfV

You are in a crowded restaurant and hear the crash of falling plates. You immediately look toward the sound. Ayesha smiles at the text message "lol" she's just received, knowing her friend enjoyed the joke. You are dozing but you awaken instantly when your infant son cries softly.

What do these three events have in common? They are all everyday examples of the functioning of your nervous system, which has your body cells humming with activity nearly all the time.

The **nervous system** is the master controlling and communicating system of the body. Every thought, action, and emotion reflects its activity. Its cells communicate by electrical and chemical signals, which are rapid and specific, and usually cause almost immediate responses.

We begin this chapter with a brief overview of the functions and organization of the nervous system. Then we focus on the functional anatomy of nervous tissue, especially the nerve cells, or *neurons*, which are the key to neural communication.

11.1 The nervous system receives, integrates, and responds to information

Learning Outcomes

▶ **List the basic functions of the nervous system.**

▶ **Explain the structural and functional divisions of the nervous system.**

The nervous system has three overlapping functions, illustrated by the example of a thirsty person seeing and then lifting a glass of water (**Figure 11.1**):

1. **Sensory input.** The nervous system uses its millions of sensory receptors to monitor changes occurring both inside and outside the body. The gathered information is called **sensory input**.

2. **Integration.** The nervous system processes and interprets sensory input and decides what should be done at each moment—a process called **integration**.

3. **Motor output.** The nervous system activates *effector organs*—the muscles and glands—to cause a *response*, called **motor output.**

Here's another example: You are driving and see a red light ahead (sensory input). Your nervous system integrates this information (red light means "stop"), and your foot hits the brake (motor output).

We have one highly integrated nervous system. For convenience, it is divided into two principal parts, *central* and *peripheral* (**Figure 11.2**).

The **central nervous system (CNS)** consists of the *brain* and *spinal cord*, which occupy the dorsal body cavity. The CNS is the integrating and control center of the nervous system. It interprets sensory input and dictates motor output based on reflexes, current conditions, and past experience.

The **peripheral nervous system (PNS)** is the part of the nervous system *outside* the CNS. The PNS consists mainly of nerves (bundles of axons) that extend from the brain and spinal cord, and *ganglia* (collections of neuron cell bodies). *Spinal nerves* carry impulses to and from the spinal cord, and *cranial nerves* carry impulses to and from the brain. These

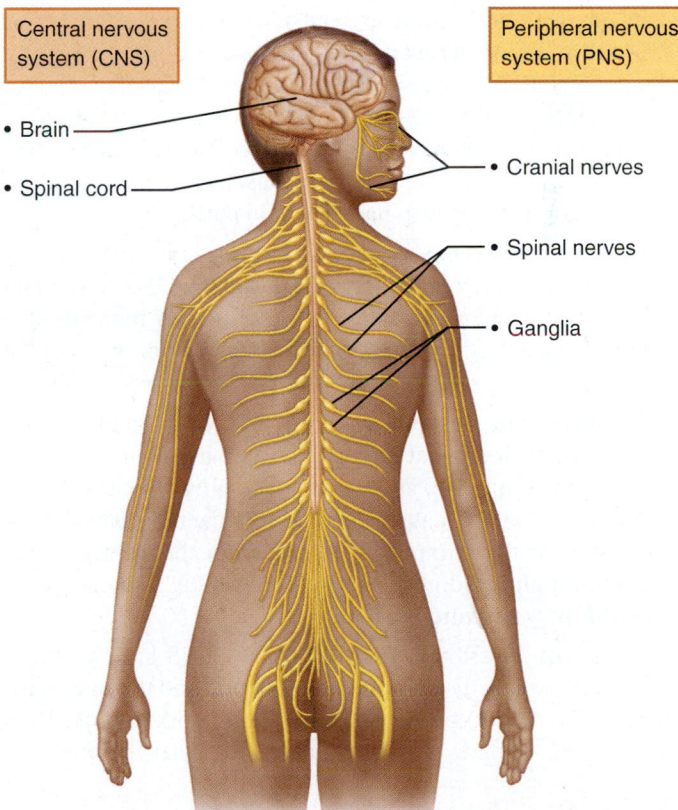

Central nervous system (CNS)

• Brain
• Spinal cord

Peripheral nervous system (PNS)

• Cranial nerves
• Spinal nerves
• Ganglia

Figure 11.2 The nervous system. The brain and spinal cord (tan) make up the central nervous system. The peripheral nervous system (dark gold) mostly consists of pairs of cranial nerves, spinal nerves, and associated ganglia.

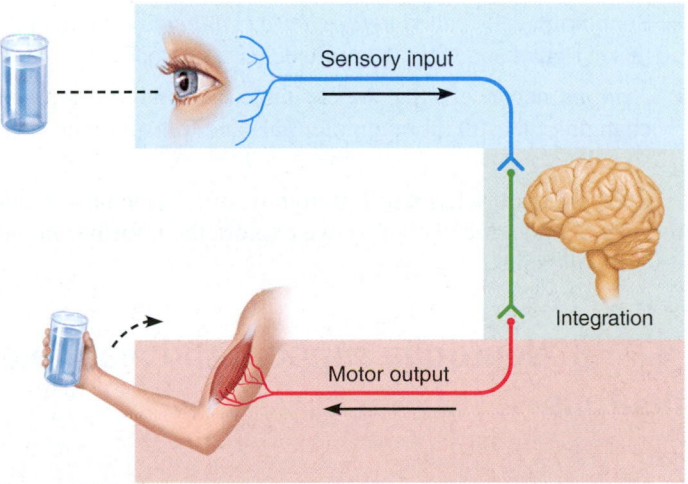

Sensory input

Integration

Motor output

Figure 11.1 The nervous system's functions.

Figure 11.3 Organization of the nervous system.

peripheral nerves serve as communication lines that link all parts of the body to the CNS.

The PNS has two functional subdivisions, as **Figure 11.3** shows. The **sensory**, or **afferent**, **division** (af′er-ent; "carrying toward") consists of nerve fibers (axons) that convey impulses *to* the central nervous system from sensory receptors located throughout the body.

- *Somatic sensory fibers* convey impulses from the skin, skeletal muscles, and joints (*soma* = body)

- *Visceral sensory fibers* transmit impulses from the visceral organs (organs within the ventral body cavity)

The sensory division keeps the CNS constantly informed of events going on both inside and outside the body.

The **motor**, or **efferent**, **division** (ef′er-ent; "carrying away") of the PNS transmits impulses *from* the CNS to effector organs, which are the muscles and glands. These impulses activate muscles to contract and glands to secrete. In other words, they *effect* (bring about) a motor response.

The motor division also has two main parts:

- The **somatic nervous system** is composed of somatic motor nerve fibers that conduct impulses from the CNS to skeletal muscles. It is often referred to as the **voluntary nervous system** because it allows us to consciously control our skeletal muscles.

- The **autonomic nervous system (ANS)** consists of visceral motor nerve fibers that regulate the activity of smooth muscles, cardiac muscle, and glands. *Autonomic* means "a law unto itself," and because we generally cannot control such activities as the pumping of our heart or the movement of food through our digestive tract, the ANS is also called the **involuntary nervous system**.

As we will describe in Chapter 14, the ANS has two functional subdivisions, the **sympathetic division** and the **parasympathetic division**. Typically these divisions work in opposition to each other—whatever one stimulates, the other inhibits.

Check Your Understanding

1. What is meant by "integration," and does it primarily occur in the CNS or the PNS?

2. **APPLY** Which subdivision of the PNS is involved in (a) relaying the feeling of a "full stomach" after a meal, (b) contracting the muscles to lift your arm, and (c) increasing your heart rate?

For answers, see Answers Appendix.

The nervous system consists mostly of nervous tissue, which is highly cellular. For example, less than 20% of the CNS is extracellular space, which means that the cells are densely packed and tightly intertwined. Although it is very complex, nervous tissue is made up of just two principal types of cells:

- Supporting cells called *neuroglia* (or *glial cells*), small cells that surround and wrap the more delicate neurons

- *Neurons*, nerve cells that are excitable (respond to stimuli by changing their membrane potential) and transmit electrical signals

Look at Figure 4.13 (◀ **p. 140**) to refresh your memory about these two kinds of cells before we explore them further in the next two modules.

11.2 Neuroglia support and maintain neurons

Learning Outcome

▶ List the types of neuroglia and cite their functions.

Neurons associate closely with much smaller cells called **neuroglia** (nu-rog′le-ah; "nerve glue"), or **glial cells** (gle′al). There are six types of neuroglia—four in the CNS and two in the PNS (**Figure 11.4**). Once considered merely the "glue" or scaffolding that supports the neurons, neuroglia are now known to have many other important and unique functions.

Neuroglia in the CNS

Neuroglia in the CNS outnumber neurons and include *astrocytes*, *microglial cells*, *ependymal cells*, and *oligodendrocytes* (Figure 11.4a–d). Like neurons, most neuroglia have branching processes (extensions) and a central cell body. They can be distinguished, however, by their much smaller size and their darker-staining nuclei.

Astrocytes

Shaped like delicate branching sea anemones, **astrocytes** (as′tro-sītz; "star cells") are the most abundant and versatile glial cells. Their numerous radiating processes cling to neurons and their synaptic endings, and cover nearby capillaries. They support and brace the neurons and anchor them to their nutrient supply lines (Figure 11.4a).

Astrocytes play a role in making exchanges between capillaries and neurons, helping determine capillary permeability. They guide the migration of young neurons and formation of synapses (junctions) between neurons. Astrocytes also control the chemical environment around neurons, where their most important job is "mopping up" leaked potassium ions and recapturing and recycling released neurotransmitters. Furthermore, astrocytes have been shown to respond to nearby nerve impulses and released neurotransmitters.

Connected by gap junctions, astrocytes signal each other with slow-paced intracellular calcium pulses (calcium waves), and by releasing extracellular chemical messengers. They also influence neuronal functioning and therefore participate in information processing in the brain.

Microglial Cells

Microglial cells (mi-kro′gle-al) are small and ovoid with relatively long "thorny" processes (Figure 11.4b). Their processes touch nearby neurons, monitoring their health, and when they sense that certain neurons are injured or in other trouble, the microglial cells migrate toward them. Where invading microorganisms or dead neurons are present, the microglial cells transform into a special type of macrophage that phagocytizes the microorganisms or neuronal debris. This protective role is important because cells of the immune system have limited access to the CNS.

Ependymal Cells

Ependymal cells (ĕ-pen′dĭ-mul; "wrapping garment") range in shape from squamous to columnar, and many are ciliated (Figure 11.4c). They line the central cavities of the brain and the spinal cord, where they form a fairly permeable barrier between the cerebrospinal fluid that fills those cavities and the tissue fluid bathing the cells of the CNS. The beating of their

(a) Astrocytes are the most abundant CNS neuroglia.

(b) Microglial cells are defensive cells in the CNS.

(c) Ependymal cells line cerebrospinal fluid–filled CNS cavities.

(d) Oligodendrocytes have processes that form myelin sheaths around CNS nerve fibers.

(e) Satellite cells and Schwann cells (which form myelin sheaths) surround neurons in the PNS.

Figure 11.4 Neuroglia. (a–d) The four types of neuroglia of the CNS. **(e)** Neuroglia of the PNS.

cilia helps to circulate the cerebrospinal fluid that cushions the brain and spinal cord.

Oligodendrocytes

Though they also branch, the **oligodendrocytes** (ol″ĭ-go-den′dro-sīts) have fewer processes (*oligo* = few; *dendr* = branch) than astrocytes. Oligodendrocytes line up along the thicker nerve fibers in the CNS and wrap their processes tightly around the fibers, producing an insulating covering called a *myelin sheath* (Figure 11.4d).

Neuroglia in the PNS

The two kinds of PNS neuroglia—*satellite cells* and *Schwann cells*—differ mainly in location.

   Satellite cells surround neuron cell bodies located in the peripheral nervous system (Figure 11.4e), and are thought to have many of the same functions in the PNS as astrocytes do in the CNS. Their name comes from a fancied resemblance to the moons (satellites) around a planet.

   Schwann cells (also called *neurolemmocytes*) surround all nerve fibers in the PNS and form myelin sheaths around the thicker nerve fibers (Figures 11.4e and 11.5a). In this way, they are functionally similar to oligodendrocytes. (We describe the formation of myelin sheaths later in this chapter.) Schwann cells are vital to regeneration of damaged peripheral nerve fibers.

Check Your Understanding

3. Which type of neuroglia controls the extracellular fluid environment around neuron cell bodies in the CNS? In the PNS?

4. Which two types of neuroglia form insulating coverings called myelin sheaths?

For answers, see Answers Appendix.

11.3 Neurons are the structural units of the nervous system

Learning Outcomes

▶ **Define neuron, describe its important structural components, and relate each to a functional role.**

▶ **Differentiate between (1) a nucleus and a ganglion, and (2) a nerve and a tract.**

▶ **Explain the importance of the myelin sheath and describe how it is formed in the central and peripheral nervous systems.**

▶ **Classify neurons by structure and by function.**

Neurons, also called *nerve cells*, are the structural units of the nervous system. There are billions of these (typically) large, highly specialized cells that conduct messages in the form of nerve impulses from one part of the body to another. Besides their excitability, they have three other special characteristics:

- Neurons have *extreme longevity*. Given good nutrition, they can function optimally for a lifetime.

- Neurons are *amitotic*. As neurons assume their roles as communicating links of the nervous system, they lose their ability to divide. We pay a high price for this feature because neurons cannot be replaced if destroyed. There *are* exceptions to this rule. For example, olfactory epithelium and some hippocampal regions of the brain contain stem cells that can produce new neurons throughout life.

- Neurons have an exceptionally *high metabolic rate* and require continuous and abundant supplies of oxygen and glucose. They cannot survive for more than a few minutes without oxygen.

   Although neurons vary in structure, they all have a *cell body* and one or more slender *processes* (**Figure 11.5**).

Neuron Cell Body

The **neuron cell body** consists of a spherical nucleus (with a conspicuous nucleolus) surrounded by cytoplasm. Also called the **perikaryon** (*peri* = around, *kary* = nucleus) or **soma**, the cell body ranges in diameter from 5 to 140 μm.

   In most neurons, the plasma membrane of the cell body acts as part of the *receptive region* that receives information from other neurons.

   The cell body is the major *biosynthetic center* and *metabolic center* of a neuron. In addition to abundant mitochondria, it contains many structures you are already familiar with, including:

- *Protein- and membrane-making machinery*. Neuron cell bodies (not axons) have the organelles needed to synthesize proteins—rough endoplasmic reticulum (ER), free ribosomes, and Golgi apparatus. The rough ER, also called the **chromatophilic substance** (*chromatophilic* = color loving) or *Nissl bodies* (nis′l), stains darkly with basic dyes.

- *Cytoskeletal elements*. Microtubules and **neurofibrils**, which are bundles of intermediate filaments (*neurofilaments*), maintain cell shape and integrity. They form a network throughout the cell body and its processes.

- *Pigment inclusions*. Pigments sometimes found inside neuron cell bodies include black melanin, a red iron-containing pigment, and a golden-brown pigment called *lipofuscin* (lip″o-fu′sin). Lipofuscin, a harmless by-product of lysosomal activity, is sometimes called the "aging pigment" because it accumulates in neurons of elderly individuals.

   Most neuron cell bodies are located in the CNS, where they are protected by the bones of the skull and vertebral column. Clusters of cell bodies in the CNS are called **nuclei**, whereas those that lie along the nerves in the PNS are called **ganglia** (gang′gle-ah; *ganglion* = "knot on a string," "swelling").

Neuron Processes

Armlike **processes** extend from the cell body of all neurons. The brain and spinal cord (CNS) contain both neuron cell bodies and their processes. The PNS consists chiefly of neuron processes (whose cell bodies are in the CNS).

   The two types of neuron processes, *dendrites* and *axons* (ak′sonz), differ in the structure and function of their plasma membranes. The convention is to describe these processes using

Figure 11.5 Structure of a motor neuron. (a) Illustration. **(b)** Digital reconstruction of a neuron (1000×).

a motor neuron as an example. We shall follow this practice, but keep in mind that many sensory neurons and some tiny CNS neurons differ from the "typical" pattern we present here.

Dendrites

Dendrites of motor neurons are short, tapering, diffusely branching extensions. Typically, motor neurons have hundreds of twiglike dendrites clustering close to the cell body. Virtually all organelles present in the cell body also occur in dendrites.

Dendrites, the main **receptive** or **input regions**, provide an enormous surface area for receiving signals from other neurons. In many brain areas, the finer dendrites are highly specialized for collecting information. They bristle with *dendritic spines*—thorny appendages with bulbous or spiky ends—which represent points of close contact (synapses) with other neurons (Figure 11.5b).

Dendrites convey incoming messages *toward* the cell body. These electrical signals are usually *not* action potentials (nerve impulses) but are short-distance signals called *graded potentials*, as we will describe shortly.

The Axon: Structure

A neuron never has more than a single **axon** (*axo* = axis, axle). The axon arises from a cone-shaped area of the cell body called the **axon hillock** ("little hill"). The **initial segment of the axon**

narrows to form a slender process that is uniform in diameter for the rest of its length (Figure 11.5a). In some neurons, the axon is very short or absent, but in others it accounts for nearly the entire length of the neuron. For example, axons of the motor neurons controlling the skeletal muscles of your big toe extend a meter or more (3–4 feet) from the lumbar region of your spine to your foot, making them among the longest cells in the body. Any long axon is also called a **nerve fiber**. Bundles of axons are called **tracts** in the CNS and **nerves** in the PNS (**Table 11.1**).

Although only one axon arises from the cell body, the axon may have occasional branches along its length. These branches, called **axon collaterals**, extend from the axon at more or less right angles. An axon usually branches profusely at its end (terminus): 10,000 or more **terminal branches** (also called *terminal arborizations*) per neuron is not unusual. The knob-like distal endings of the terminal branches are called **axon terminals**.

Table 11.1 Key CNS versus PNS Terminology

TERM	DEFINITION
Nucleus	A collection of neuron cell bodies in the CNS
Ganglion	A collection of neuron cell bodies in the PNS
Tract	A bundle of axons in the CNS
Nerve	A bundle of axons in the PNS

The Axon: Functional Characteristics

The axon is the **conducting region** of the neuron (Figure 11.5). It *generates nerve impulses* and *transmits them*, typically away from the cell body, along the plasma membrane, or **axolemma** (ak″so-lem′ah). In motor neurons, the nerve impulse is generated at the initial segment of the axon (the *trigger zone*) and conducted along the axon to the axon terminals, which are the **secretory region** of the neuron.

When the impulse reaches the axon terminals, it causes *neurotransmitters*—signaling chemicals—to be released into the extracellular space. [You are already familiar with one neurotransmitter: acetylcholine (◀ **p. 290**)]. The neurotransmitters either excite or inhibit neurons (or muscle or gland cells) with which the axon is in close contact. Each neuron receives signals from and sends signals to scores of other neurons, carrying on "conversations" with many different neurons at the same time.

An axon contains the same cytoplasmic organelles found in the dendrites and cell body with two important exceptions—it lacks rough endoplasmic reticulum and a Golgi apparatus, the structures involved with protein synthesis and packaging. Consequently, an axon depends on (1) its cell body to renew the necessary proteins and membrane components, and (2) efficient transport mechanisms to distribute them. Axons quickly decay if cut or severely damaged.

Axonal Transport

Because axons are often very long, the task of moving molecules along their length might appear difficult. However, through the cooperative efforts of motor proteins and cytoskeletal elements (mostly microtubules), substances travel continuously along the axon in both directions:

- *Anterograde movement* is movement away from the cell body. Substances moved in this direction include mitochondria, cytoskeletal elements, membrane components (vesicles) used to renew the axon plasma membrane, and enzymes needed to synthesize certain neurotransmitters. (Some neurotransmitters are synthesized in the cell body, packaged into vesicles, and then transported to the axon terminals.)

- *Retrograde movement* is movement toward the cell body. Substances moved in this direction are mostly organelles returning to the cell body to be degraded or recycled. Retrograde transport is also an important means of intracellular communication. It allows the cell body to be advised of conditions at the axon terminals. It also delivers vesicles to the cell body containing signal molecules (such as nerve growth factor, which activates certain nuclear genes promoting growth).

A single basic bidirectional transport mechanism is responsible for axonal transport. It uses different ATP-dependent "motor" proteins (kinesin or dynein), depending on the direction of transport. These proteins propel cellular components along the microtubules like trains along tracks at speeds up to 40 cm (15 inches) per day.

HOMEOSTATIC IMBALANCE 11.1   CLINICAL

Certain viruses and bacterial toxins that damage neural tissues use retrograde axonal transport to reach the cell body. This transport mechanism has been demonstrated for polio, rabies, and herpes simplex viruses and for tetanus toxin. Researchers are investigating using retrograde transport to treat genetic diseases by introducing viruses containing "corrected" genes or microRNA to suppress defective genes.

Myelin Sheath

Many nerve fibers, particularly those that are long or large in diameter, are covered with a whitish, fatty (protein-lipoid), segmented **myelin sheath** (mi′ĕ-lin). Myelin protects and electrically insulates fibers, and it increases the transmission speed of nerve impulses. **Myelinated fibers** (axons bearing a myelin sheath) conduct nerve impulses rapidly, whereas **nonmyelinated fibers** conduct impulses more slowly. Note that myelin sheaths are associated only with axons. Dendrites are *always* nonmyelinated.

Myelination in the PNS

Myelin sheaths in the PNS are formed by Schwann cells, which indent to receive an axon and then wrap themselves around it in a jelly roll fashion (**Figure 11.6**). Initially the wrapping is loose, but the Schwann cell cytoplasm is gradually squeezed from between the membrane layers.

When the wrapping process is complete, many concentric layers of Schwann cell plasma membrane enclose the axon, much like gauze wrapped around an injured finger. This tight coil of wrapped membranes is the myelin sheath, and its thickness depends on the number of spirals. The nucleus and most of the cytoplasm of the Schwann cell end up as a bulge just external to the myelin sheath. This portion is called the *outer collar of perinuclear cytoplasm* (formerly known as the *neurilemma*) (Figure 11.6b).

Plasma membranes of myelinating cells contain much less protein than those of most body cells. Channel and carrier proteins are notably absent, making myelin sheaths exceptionally good electrical insulators. Another unique characteristic of these membranes is the presence of specific protein molecules that interlock to form a sort of molecular Velcro® between adjacent myelin membranes.

Adjacent Schwann cells do not touch one another, so there are gaps in the sheath. These **myelin sheath gaps**, or **nodes of Ranvier** (ran′vē-ā″), occur at regular intervals (about 1 mm apart) along a myelinated axon. Axon collaterals can emerge at these gaps.

Sometimes Schwann cells surround peripheral nerve fibers but the coiling process does not occur. In such instances, a single Schwann cell can partially enclose 15 or more axons, each of which occupies a separate recess in the Schwann cell surface. Nerve fibers associated with Schwann cells in this manner are said to be *nonmyelinated* and are typically thin fibers.

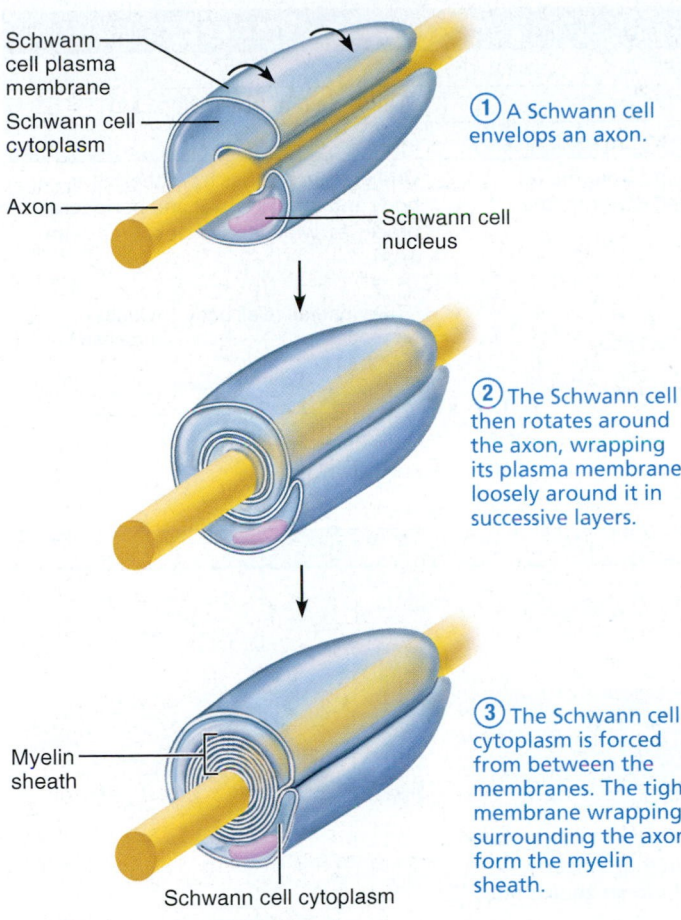

① A Schwann cell envelops an axon.

Schwann cell plasma membrane

Schwann cell cytoplasm

Axon

Schwann cell nucleus

② The Schwann cell then rotates around the axon, wrapping its plasma membrane loosely around it in successive layers.

③ The Schwann cell cytoplasm is forced from between the membranes. The tight membrane wrappings surrounding the axon form the myelin sheath.

Myelin sheath

Schwann cell cytoplasm

(a) Myelination of a nerve fiber (axon)

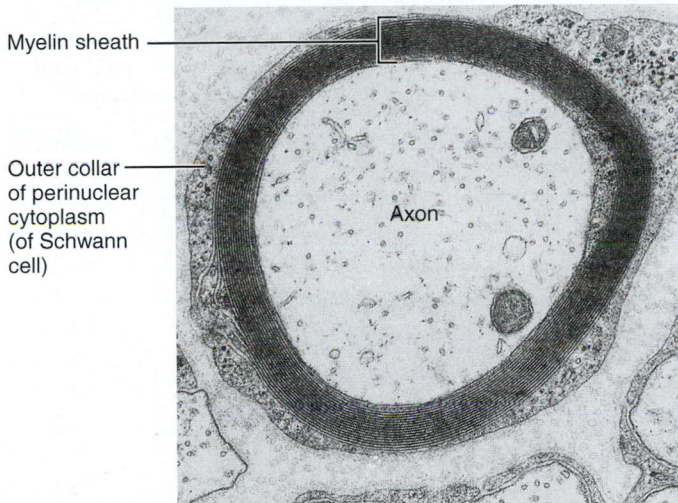

Myelin sheath

Outer collar of perinuclear cytoplasm (of Schwann cell)

Axon

(b) Cross-sectional view of a myelinated axon (electron micrograph 24,000×)

Figure 11.6 PNS nerve fiber myelination.

Myelination in the CNS

The central nervous system contains both myelinated and non-myelinated axons. However, in the CNS, it is the oligodendrocytes that form myelin sheaths (Figure 11.4d).

Unlike a Schwann cell, which forms only one segment of a myelin sheath, an oligodendrocyte has multiple flat processes that can coil around as many as 60 axons at the same time. As in the PNS, myelin sheath gaps separate adjacent sections of an axon's myelin sheath. However, CNS myelin sheaths lack an outer collar of perinuclear cytoplasm because cell extensions do the coiling and the squeezed-out cytoplasm is forced back toward the centrally located nucleus instead of peripherally.

As in the PNS, the smallest-diameter axons are nonmyelinated. These nonmyelinated axons are covered by the long extensions of adjacent glial cells.

Classification of Neurons

Neurons are classified both structurally and functionally. We describe both classifications here but use the functional classification in most discussions.

Structural Classification

Neurons are grouped structurally according to the number of processes extending from their cell body (**Table 11.2**, pp. 398–399).

- **Multipolar neurons** (*polar* = end, pole) have three or more processes—one axon and the rest dendrites. They are the most common neuron type in humans, with more than 99% of neurons in this class. Multipolar neurons are the major neuron type in the CNS.

Cell body

Dendrites

Axon

- **Bipolar neurons** have two processes—an axon and a dendrite—that extend from opposite sides of the cell body. These rare neurons are found in some of the special sense organs such as in the retina of the eye and in the olfactory mucosa.

Cell body

Dendrite

Axon

- **Unipolar neurons** have a single short process that emerges from the cell body and divides T-like into proximal and distal branches. The more distal **peripheral process** is often associated with a sensory receptor. The **central process** enters the CNS. Unipolar neurons are more accurately called **pseudounipolar neurons** (*pseudo* = false) because they originate as bipolar neurons. During early embryonic development, the two processes converge and partially fuse to form the short single process that issues from the cell body. Unipolar neurons are found chiefly in ganglia in the PNS, where they function as sensory neurons.

Peripheral process

Cell body

Central process

Receptive endings

Axon

11

Table 11.2 Comparison of Structural Classes of Neurons

	NEURON TYPE	
MULTIPOLAR	**BIPOLAR**	**UNIPOLAR (PSEUDOUNIPOLAR)**
Structural Class: Neuron Type According to the Number of Processes Extending from the Cell Body		
Many processes extend from the cell body. All are dendrites except for a single axon.	Two processes extend from the cell body. One is a fused dendrite, the other is an axon.	One process extends from the cell body and forms central and peripheral processes, which together comprise an axon.

Relationship of Anatomy to the Three Functional Regions

- ▉ Receptive region (receives stimulus).
- ▉ Conducting region (generates/transmits action potential).
- ▉ Secretory region (axon terminals release neurotransmitters).

(Many bipolar neurons do not generate action potentials. In those that do, the location of the trigger zone is not universal.)

Relative Abundance and Location in Human Body

Most abundant in body. Major neuron type in the CNS.	Rare. Found in some special sensory organs (olfactory mucosa, eye, ear).	Found mainly in the PNS. Common only in dorsal root ganglia of the spinal cord and sensory ganglia of cranial nerves.

Structural Variations

Multipolar	Bipolar	Unipolar
Purkinje cell of cerebellum Pyramidal cell	Olfactory cell Retinal cell	Dorsal root ganglion cell

Table 11.2 (continued)		
	NEURON TYPE	
MULTIPOLAR	**BIPOLAR**	**UNIPOLAR (PSEUDOUNIPOLAR)**
Functional Class: Neuron Type According to Direction of Impulse Conduction		
1. Most multipolar neurons are **interneurons** that conduct impulses within the CNS, integrating sensory input or motor output. May be one of a chain of CNS neurons, or a single neuron connecting sensory and motor neurons. 2. Some multipolar neurons are **motor neurons** that conduct impulses along the efferent pathways from the CNS to an effector (muscle/gland).	Essentially all bipolar neurons are **sensory neurons** that are located in some special sense organs. For example, bipolar cells of the retina are involved in transmitting visual inputs from the eye to the brain (via an intermediate chain of neurons).	Most unipolar neurons are **sensory neurons** that conduct impulses along afferent pathways to the CNS for interpretation. (These sensory neurons are called primary or first-order sensory neurons.)
		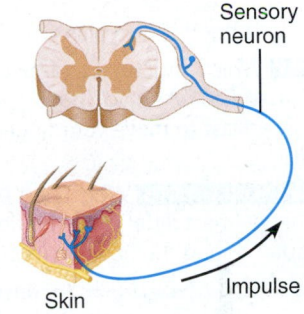

The fact that the fused peripheral and central processes of unipolar neurons are continuous and function as a single fiber might make you wonder whether they are axons or dendrites. The central process is definitely an axon because it conducts impulses away from the cell body (one definition of axon). However, the peripheral process is perplexing. Three facts favor classifying it as an axon: (1) It generates and conducts an impulse (functional definition of axon); (2) when large, it is heavily myelinated; and (3) it has a uniform diameter and is indistinguishable microscopically from an axon. But the older definition of a dendrite as a process that transmits signals *toward* the cell body conflicts with that conclusion.

So which is it? We have chosen to emphasize the newer definition of an axon as generating and transmitting an impulse. For *unipolar neurons*, we will refer to the combined length of the peripheral and central process as an axon. In place of "dendrites," unipolar neurons have *receptive endings* (sensory terminals) at the end of the peripheral process.

Functional Classification

This scheme groups neurons according to the direction in which the nerve impulse travels relative to the central nervous system. Based on this criterion, there are sensory neurons, motor neurons, and interneurons (Table 11.2, last section).

Sensory, or **afferent**, **neurons** transmit impulses from sensory receptors in the skin or internal organs *toward* or *into* the central nervous system. Except for certain neurons found in some special sense organs, virtually all sensory neurons are unipolar, and their cell bodies are located in sensory ganglia

outside the CNS. Only the most distal parts of these unipolar neurons act as the receptive region, and the peripheral processes are often very long. For example, fibers carrying sensory impulses from the skin of your big toe travel for more than a meter before they reach their cell bodies in a ganglion close to the spinal cord.

In some sensory neurons, the receptive endings function directly as sensory receptors. In other sensory neurons, receptive endings are associated with larger sensory receptors that include other cell types. We describe the various types of general sensory receptor end organs, such as those of the skin, in Chapter 13. The special sensory receptors (of the ear, eye, etc.) are the topic of Chapter 15.

Motor, or **efferent**, **neurons** carry impulses *away from* the CNS to the effector organs (muscles and glands) of the body. Motor neurons are multipolar. Except for some neurons of the autonomic nervous system, their cell bodies are located in the CNS.

Interneurons, or *association neurons*, lie between motor and sensory neurons in neural pathways and shuttle signals through CNS pathways where integration occurs. Most interneurons are confined within the CNS. They make up over 99% of the neurons of the body, including most of those in the CNS.

Almost all interneurons are multipolar, but there is considerable diversity in size and fiber-branching patterns. The Purkinje and pyramidal cells illustrated in Table 11.2 are just two examples of their variety.

11

Check Your Understanding

5. How does a nucleus within the brain differ from a nucleus within a neuron?

6. How is a myelin sheath formed in the CNS, and what is its function?

7. What is the structural classification of the neuron shown below? What is its usual functional classification? Name the parts labeled a–d.

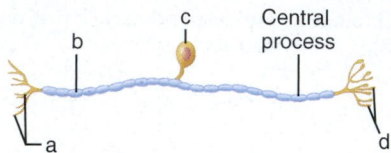

8. **APPLY** Which structural and functional type of neuron is activated first when you burn your finger? Which type is activated last to move your finger away from the source of heat?

9. **MAKE CONNECTIONS** Which part of the neuron is its fiber? How do nerve fibers differ from the fibers of connective tissue (see Chapter 4) and the fibers in muscle (see Chapter 9)?

For answers, see Answers Appendix.

11.4 The resting membrane potential depends on differences in ion concentration and permeability

Learning Outcomes

▶ **Describe the relationship between current, voltage, and resistance.**

▶ **Identify different types of membrane ion channels.**

▶ **Define resting membrane potential and describe its electrochemical basis.**

Like all cells, neurons have a *resting membrane potential.* However, unlike most other cells, neurons can rapidly change their membrane potential. This ability underlies the function of neurons throughout the nervous system. In order to understand how neurons work, let's first explore some basic priciples of electricity and revisit the resting membrane potential.

Basic Principles of Electricity

The human body is electrically neutral—it has the same number of positive and negative charges. However, there are regions where one type of charge predominates, making those regions positively or negatively charged. Because opposite charges attract, energy must be used (work must be done) to separate them. On the other hand, the coming together of opposite charges liberates energy that can be used to do work. For this reason, any situation in which there are separated electrical charges of opposite sign has the potential to do work. We call this potential energy.

Some Definitions: Voltage, Resistance, Current

Voltage, the measure of potential energy generated by separated electrical charges, is measured in either *volts* (V) or *millivolts* (1 mV = 0.001 V). Voltage is always measured between two points and is called the **potential difference** or simply the **potential** between the points. The greater the difference in charge between two points, the higher the voltage.

The flow of electrical charge from one point to another is a **current**, and it can be used to do work—for example, to power a flashlight. The amount of charge that moves between the two points depends on two factors: voltage and resistance. **Resistance** is the hindrance to charge flow provided by substances through which the current must pass. Substances with high electrical resistance are *insulators*, and those with low resistance are *conductors*.

Ohm's law gives the relationship between voltage, current, and resistance:

$$\text{Current } (I) = \frac{\text{voltage } (V)}{\text{resistance } (R)}$$

Ohm's law tells us three things:

- Current (*I*) is directly proportional to voltage: The greater the voltage (potential difference), the greater the current.

- There is no net current flow between points that have the same potential.

- Current is inversely related to resistance: The greater the resistance, the smaller the current.

In the body, electrical currents reflect the flow of ions across cellular membranes. (Unlike the electrons flowing along your house wiring, there are no free electrons "running around" in a living system.) Recall that there is a slight difference in the numbers of positive and negative ions on the two sides of cellular plasma membranes (a charge separation), so there is a potential across those membranes. The plasma membranes provide the resistance to current flow.

Role of Membrane Ion Channels

Recall that plasma membranes are peppered with a variety of membrane proteins that act as *ion channels* (◀ p. 70). Each of these channels is selective as to the type of ion (or ions) it allows to pass. For example, a potassium ion channel allows only potassium ions to pass.

Membrane channels are large proteins, often with several subunits. Some channels, **leakage** or **nongated channels**, are always open. Other channels are *gated*: Part of the protein forms a molecular "gate" that changes shape to open and close the channel in response to specific signals. There are three main types of gated channels:

- **Chemically gated channels**, also known as **ligand-gated channels**, open when the appropriate chemical (in this case a neurotransmitter) binds (**Figure 11.7a**).

- **Voltage-gated channels** open and close in response to changes in the membrane potential (Figure 11.7b).

Figure 11.7 Operation of gated channels. (a) A chemically gated channel permeable to both Na^+ and K^+, and **(b)** a voltage-gated Na^+ channel.

- **Mechanically gated channels** open in response to physical deformation of the receptor (as in sensory receptors for touch and pressure).

When gated ion channels open, ions diffuse quickly across the membrane. The direction an ion moves (into or out of the cell) is determined by the **electrochemical gradient**. The electrochemical gradient has two components:

- The *concentration gradient*. Ions move along chemical concentration gradients from an area of their higher concentration to an area of lower concentration.
- The *electrical gradient*. Ions move toward an area of opposite electrical charge.

The two gradients do not necessarily work together to drive an ion in the same direction. Often, the two gradients oppose each other, each trying to drive ions in the opposite direction. Whichever gradient is the strongest "wins" and drives the *net* flow of ions in its direction.

Flowing ions create electrical currents and voltage changes across the membrane. These voltage changes are described by the rearranged Ohm's law equation:

$$\text{Voltage } (V) = \text{current } (I) \times \text{resistance } (R)$$

Figure 11.8 Measuring membrane potential in neurons. The potential difference between an electrode inside a neuron and the ground electrode in the extracellular fluid is approximately −70 mV (inside negative).

Generating the Resting Membrane Potential

A voltmeter is used to measure the potential difference between two points. When one microelectrode of the voltmeter is inserted into a neuron and the other is in the extracellular fluid, it records a voltage across the membrane of approximately −70 mV (**Figure 11.8**). The minus sign indicates that the cytoplasmic side (inside) of the membrane is

Generating a resting membrane potential depends on (1) differences in K^+ and Na^+ concentrations inside and outside cells, and (2) differences in permeability of the plasma membrane to these ions.

 Watch a 3-D animation of this process: MasteringA&P® > Study Area > Animations & Videos > A&P Flix

The concentrations of Na^+ and K^+ on each side of the membrane are different.

The Na^+ concentration is higher outside the cell.

The K^+ concentration is higher inside the cell.

Outside cell

K^+ (5 mM)

Na^+ (140 mM)

K^+ (140 mM)

Na^+ (15 mM)

Inside cell

Na^+-K^+ pumps maintain the concentration gradients of Na^+ and K^+ across the membrane.

The permeabilities of Na^+ and K^+ across the membrane are different. In the next three panels, we will build the resting membrane potential step by step.

Suppose a cell has only K^+ channels...

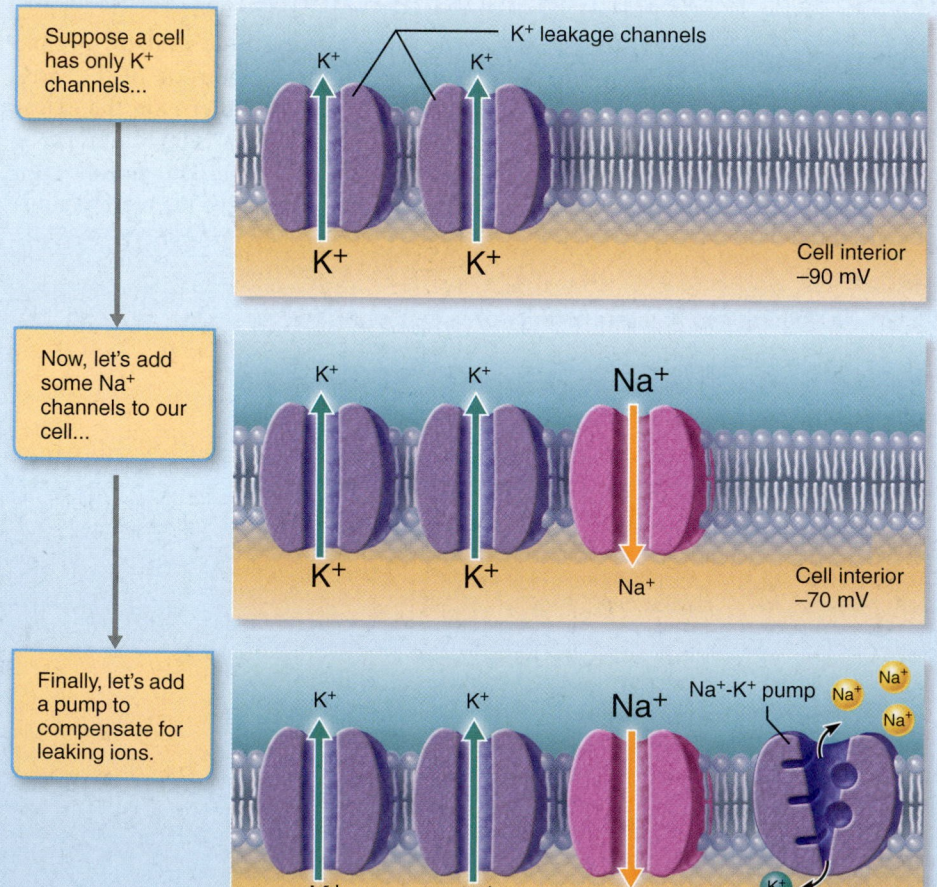

K^+ leakage channels

K^+ K^+

K^+ K^+

Cell interior −90 mV

K^+ loss through abundant leakage channels establishes a negative membrane potential.

• The membrane is highly permeable to K^+, so K^+ flows down its concentration gradient.

• As positive K^+ leaks out, a negative voltage (electrical gradient) develops on the membrane interior. This electrical gradient pulls K^+ back in.

• At −90 mV, the concentration and electrical gradients for K^+ are balanced.

Now, let's add some Na^+ channels to our cell...

K^+ K^+ Na^+

K^+ K^+

Na^+

Cell interior −70 mV

Na^+ entry through a few leakage channels reduces the negative membrane potential slightly.

• Adding Na^+ channels creates a small Na^+ permeability that brings the membrane potential to −70 mV.

Finally, let's add a pump to compensate for leaking ions.

K^+ K^+ Na^+ Na^+-K^+ pump Na^+ Na^+ Na^+

K^+ K^+ Na^+ K^+

K^+ Cell interior −70 mV

Na^+-K^+ pumps maintain the concentration gradients, resulting in the resting membrane potential.

• A cell at rest is like a leaky boat: K^+ leaks out and Na^+ leaks in through open channels.

• The "bailing pump" for this boat is the **Na^+-K^+ pump**, which transports Na^+ out and K^+ in.

negatively charged relative to the outside. This potential difference in a resting neuron (V_r) is called the **resting membrane potential**, and the membrane is said to be **polarized**. The value of the resting membrane potential varies (from −40 mV to −90 mV) in different types of neurons.

The resting potential exists only across the membrane; the solutions inside and outside the cell are electrically neutral. Two factors generate the resting membrane potential: differences in the ionic composition of the intracellular and extracellular fluids, and differences in the plasma membrane's permeability to those ions.

Differences in Ionic Composition

First, let's compare the ionic makeup of the intracellular and extracellular fluids, as shown in *Focus on Resting Membrane Potential* (**Focus Figure 11.1**). The cell cytosol contains a lower concentration of Na^+ and a higher concentration of K^+ than the extracellular fluid. Negatively charged (anionic) proteins (not shown) help to balance the positive charges of intracellular cations (primarily K^+). In the extracellular fluid, the positive charges of Na^+ and other cations are balanced chiefly by chloride ions (Cl^-). Although there are many other solutes (glucose, urea, and other ions) in both fluids, potassium (K^+) plays the most important role in generating the membrane potential.

Differences in Plasma Membrane Permeability

Next, let's consider the differential permeability of the membrane to various ions (Focus Figure 11.1, bottom). At rest the membrane is impermeable to the large anionic cytoplasmic proteins, very slightly permeable to sodium, approximately 25 times more permeable to potassium than to sodium, and quite permeable to chloride ions. These resting permeabilities reflect the properties of the leakage ion channels in the membrane. Potassium ions diffuse out of the cell along their *concentration gradient* much more easily than sodium ions can enter the cell along theirs. K^+ flowing out of the cell causes the cell to become more negative inside. Na^+ trickling into the cell makes the cell just slightly more positive than it would be if only K^+ flowed. Therefore, at resting membrane potential, the negative interior of the cell is due to a much greater ability for K^+ to diffuse out of the cell than for Na^+ to diffuse into the cell.

Because some K^+ is always leaking out of the cell and some Na^+ is always leaking in, you might think that the concentration gradients would eventually "run down," resulting in equal concentrations of Na^+ and K^+ inside and outside the cell. This does not happen because the ATP-driven sodium-potassium pump first ejects three Na^+ from the cell and then transports two K^+ back into the cell (◄ p. 80). In other words, the **sodium-potassium pump (Na^+-K^+ ATPase)** stabilizes the resting membrane potential by maintaining the concentration gradients for sodium and potassium (Focus Figure 11.1, bottom).

Changing the Resting Membrane Potential

Neurons use changes in their membrane potential as signals to receive, integrate, and send information. A change in membrane potential can be produced by (1) anything that alters ion concentrations on the two sides of the membrane, or (2) anything that changes membrane permeability to any ion. However, only permeability changes (changes in the number of open channels) are important for transferring information.

Changes in membrane potential can produce two types of signals:

- *Graded potentials*—usually incoming signals operating over short distances that have variable (graded) strength
- *Action potentials*—long-distance signals of axons that always have the same strength

The terms *depolarization* and *hyperpolarization* describe changes in membrane potential *relative to resting membrane potential*.

Depolarization is a decrease in membrane potential: The inside of the membrane becomes *less negative* (moves closer to zero) than the resting potential. For instance, a

Complete an interactive tutorial: MasteringA&P® > Study Area > Interactive Physiology

Depolarizing stimulus

(a) Depolarization: The membrane potential moves toward 0 mV, the inside becoming less negative (more positive).

Hyperpolarizing stimulus

(b) Hyperpolarization: The membrane potential increases, the inside becoming more negative.

Figure 11.9 Depolarization and hyperpolarization of the membrane. The resting membrane potential is approximately −70 mV (inside negative) in neurons.

change in resting potential from −70 mV to −65 mV is a depolarization (**Figure 11.9a**). Depolarization also includes events in which the membrane potential reverses and moves above zero to become positive.

Hyperpolarization is an increase in membrane potential: The inside of the membrane becomes *more negative* (moves farther from zero) than the resting potential. For example, a change from −70 mV to −75 mV is hyperpolarization (Figure 11.9b). As we will describe shortly, depolarization increases the probability of producing nerve impulses, whereas hyperpolarization reduces this probability.

Check Your Understanding

10. For an open channel, what factors determine in which direction ions will move through that channel?

11. For which cation are there the largest number of leakage channels in the plasma membrane?

12. PREDICT Your patient, Ms. Young, has failing kidneys that cause an increase in her extracellular fluid K^+ concentration. What would this do to the resting membrane potential of her neurons and muscle cells? Explain.

For answers, see Answers Appendix.

11.5 Graded potentials are brief, short-distance signals within a neuron

Learning Outcome

▶ **Describe graded potentials and name several examples.**

Graded potentials are short-lived, localized changes in membrane potential, usually in dendrites or the cell body. They can be either depolarizations or hyperpolarizations. These changes cause current flows that decrease in magnitude with distance. Graded potentials are called "graded" because their magnitude varies directly with stimulus strength. The stronger the stimulus, the more the voltage changes and the farther the current flows.

Graded potentials are triggered by some change (a stimulus) in the neuron's environment that opens gated ion channels. Graded potentials are given different names, depending on where they occur and the functions they perform.

- A *receptor potential* or a *generator potential* is produced when a sensory receptor is excited by its stimulus (e.g., light, pressure, chemicals). We will consider these two types of graded potentials in Chapter 13.

- A *postsynaptic potential* is produced when the stimulus is a neurotransmitter released by another neuron. Here, the neurotransmitter is released into a fluid-filled gap called a synapse and influences the neuron beyond the synapse. We will discuss postsynaptic potentials in Module 11.8.

Fluids inside and outside cells are fairly good conductors, and current, carried by ions, flows through these fluids whenever voltage changes. Suppose a stimulus depolarizes a small area of a neuron's plasma membrane (**Figure 11.10a**). Current (ions) flows on both sides of the membrane between the depolarized (active) membrane area and the adjacent polarized (resting) areas. Positive ions migrate toward more negative areas (the direction of cation movement is the direction of current flow), and negative ions simultaneously move toward more positive areas (Figure 11.10b).

For our patch of plasma membrane, positive ions (mostly K^+) inside the cell move away from the depolarized area and accumulate on the neighboring membrane areas, where they neutralize negative ions. Meanwhile, positive ions (mostly Na^+)

(a) Depolarization: A small patch of the membrane (red area) depolarizes.

(b) Depolarization spreads: Opposite charges attract each other. This creates local currents (black arrows) that depolarize adjacent membrane areas, spreading the wave of depolarization.

(c) Membrane potential decays with distance: Because current is lost through the "leaky" plasma membrane, the voltage declines with distance from the stimulus (the voltage is *decremental*). Consequently, graded potentials are short-distance signals.

Figure 11.10 The spread and decay of a graded potential.

on the outside of the membrane move toward the depolarized region, which is momentarily less positive. As these positive ions move, their "places" on the membrane become occupied by negative ions (such as Cl^- and HCO_3^-), sort of like ionic musical chairs. In this way, at regions next to the depolarized region, the inside becomes less negative and the outside becomes less positive. The depolarization spreads as the neighboring membrane patch is, in turn, depolarized.

As just explained, the flow of current to adjacent membrane areas changes the membrane potential there as well. However, the plasma membrane is permeable like a leaky garden hose, and most of the charge is quickly lost through leakage channels.

Consequently, the current dies out within a few millimeters of its origin and is said to be *decremental* (Figure 11.10c).

Because the current dissipates quickly and decays (declines) with increasing distance from the site of initial depolarization, graded potentials can act as signals only over very short distances such as in the dendrites and cell body. Nonetheless, they are essential in initiating action potentials, the long-distance signals of axons.

Check Your Understanding

13. What determines the size of a graded potential?
14. **MAKE CONNECTIONS** Recall (from Chapter 9) that, at a neuromuscular junction, the AP in the muscle fiber is also triggered by a graded potential. Name that graded potential.

For answers, see Answers Appendix.

11.6 Action potentials are brief, long-distance signals within a neuron

Learning Outcomes

▶ **Compare and contrast graded potentials and action potentials.**
▶ **Explain how action potentials are generated and propagated along neurons.**
▶ **Define absolute and relative refractory periods.**
▶ **Define saltatory conduction and explain how it differs from continuous conduction.**

The principal way neurons send signals over long distances is by generating and propagating (transmitting) action potentials. Only cells with *excitable membranes*—neurons and muscle cells—can generate action potentials.

An **action potential (AP)** is a brief reversal of membrane potential with a total amplitude (change in voltage) of about 100 mV (from −70 mV to +30 mV). Depolarization is followed by repolarization and often a short period of hyperpolarization. The whole event is over in a few milliseconds. Unlike graded potentials, action potentials do not decay with distance.

In a neuron, an AP is also called a **nerve impulse**, and is typically generated *only in axons*. A neuron generates a nerve impulse only when adequately stimulated. The stimulus changes the permeability of the neuron's membrane by opening specific voltage-gated channels on the axon. These voltage-gated channels are generally found only on axons, where they are critical for AP formation—no voltage-gated channels means no AP.

Voltage-gated channels open and close in response to changes in the membrane potential. They are initially activated by local currents (graded potentials) that spread toward the axon along the dendritic and cell body membranes.

In many neurons, the transition from local graded potential to long-distance action potential takes place at the *initial segment* of the axon. In sensory neurons, the action potential is generated by the peripheral (axonal) process just proximal to the receptor region (see Table 11.2). However, for simplicity, we will just use the term axon in our discussion. We'll look first at the generation of an action potential and then at its propagation.

(Text continues on p. 408.)

The action potential (AP) is a brief change in membrane potential in a patch of membrane that is depolarized by local currents.

Watch a 3-D animation of this process: MasteringA&P® > Study Area > Animations & Videos > A&P Flix

The big picture

What does this graph show? During the course of an action potential (below), voltage changes over time at a given point within the axon.

② **Depolarization** is caused by Na⁺ flowing into the cell.

③ **Repolarization** is caused by K⁺ flowing out of the cell.

④ **Hyperpolarization** is caused by K⁺ continuing to leave the cell.

① **Resting state.** No ions move through voltage-gated channels.

Threshold

Membrane potential (mV)
+30
0
−55
−70

Time (ms)
0   1   2   3   4

The key players

Voltage-gated Na⁺ channels have two gates and alternate between three different states.

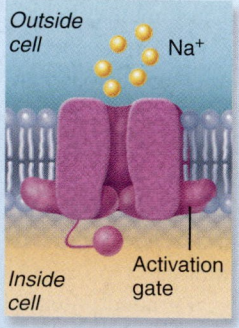

Outside cell   Na⁺

Inside cell   Activation gate

Closed at the resting state, so no Na⁺ enters the cell through them

Na⁺

Opened by depolarization, allowing Na⁺ to enter the cell

Na⁺

Inactivation gate

Inactivated—channels automatically blocked by inactivation gates soon after they open

Voltage-gated K⁺ channels have one gate and two states.

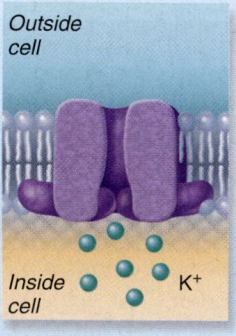

Outside cell

Inside cell   K⁺

Closed at the resting state, so no K⁺ exits the cell through them

K⁺

Opened by depolarization, after a delay, allowing K⁺ to exit the cell

The events

Each step corresponds to one part of the AP graph.

Na⁺
Sodium channel
Potassium channel
Activation gates
Inactivation gate
K⁺

① **Resting state:** All gated Na⁺ and K⁺ channels are closed.

Na⁺

② **Depolarization:** Na⁺ channels open, allowing Na⁺ entry.

Na⁺
K⁺

④ **Hyperpolarization:** Some K⁺ channels remain open, and Na⁺ channels reset.

③ **Repolarization:** Na⁺ channels are inactivating. K⁺ channels open, allowing K⁺ to exit.

Generating an Action Potential

Focus on an Action Potential (**Focus Figure 11.2**) on pp. 406–407 describes how an action potential is generated. Let's start with an axon in the resting (polarized) state.

① **Resting state: All voltage-gated Na⁺ and K⁺ channels are closed.** Only the leakage channels are open, maintaining resting membrane potential. Each Na⁺ channel has two gates: a voltage-sensitive *activation gate* that is closed at rest and responds to depolarization by opening, and an *inactivation gate* that blocks the channel once it is open. Thus, *depolarization opens and then inactivates sodium channels.*

Both gates must be open for Na⁺ to enter, but the closing of *either* gate effectively closes the channel. In contrast, each voltage-gated potassium channel has a single voltage-sensitive gate that is closed in the resting state and opens slowly in response to depolarization.

② **Depolarization: Voltage-gated Na⁺ channels open.** As local currents depolarize the axon membrane, the voltage-gated sodium channels open and Na⁺ rushes into the cell. This influx of positive charge depolarizes that local patch of membrane further, opening more Na⁺ channels so the cell interior becomes progressively less negative.

When depolarization reaches a critical level called **threshold** (often between −55 and −50 mV), depolarization becomes self-generating, urged on by positive feedback. As more Na⁺ enters, the membrane depolarizes further and opens still more channels until all Na⁺ channels are open. At this point, Na⁺ permeability is about 1000 times greater than in a resting neuron. As a result, the membrane potential becomes less and less negative and then overshoots to about +30 mV as Na⁺ rushes in along its electrochemical gradient. This rapid depolarization and polarity reversal produces the sharp upward *spike* of the action potential (Focus Figure 11.2).

Earlier, we stated that membrane potential depends on membrane permeability, but here we say that membrane permeability depends on membrane potential. Can both statements be true? Yes, because these two relationships establish a *positive feedback cycle*: Increasing Na⁺ permeability due to increased channel openings leads to greater depolarization, which increases Na⁺ permeability, and so on. This explosive positive feedback cycle is responsible for the rising (depolarizing) phase of an action potential—it puts the "action" in the action potential.

③ **Repolarization: Na⁺ channels are inactivating, and voltage-gated K⁺ channels open.** The explosively rising phase of the action potential persists for only about 1 ms. It is self-limiting because the inactivation gates of the Na⁺ channels begin to close at this point. As a result, the membrane permeability to Na⁺ declines to resting levels, and the net influx of Na⁺ stops completely. Consequently, the AP spike stops rising.

As Na⁺ entry declines, the slow voltage-gated K⁺ channels open and K⁺ rushes out of the cell, following its electrochemical gradient. This restores the internal negativity of the resting neuron, an event called **repolarization**. Both the abrupt decline in Na⁺ permeability and the increased permeability to K⁺ contribute to repolarization.

④ **Hyperpolarization: Some K⁺ channels remain open, and Na⁺ channels reset.** The period of increased K⁺ permeability typically lasts longer than needed to restore the resting state. As a result of the excessive K⁺ efflux before the potassium channels close, a hyperpolarization is seen on the AP curve as a slight dip following the spike. At the end of this phase, the Na⁺ channels have reset to their original position by changing shape to reopen their inactivation gates and close their activation gates.

Repolarization restores resting electrical conditions. While you might think that the ionic conditions change as a result of movement of massive numbers of Na⁺ and K⁺ ions during an action potential, this is not the case. Only small amounts of sodium and potassium cross the membrane. (The Na⁺ influx required to reach threshold produces only a 0.012% change in intracellular Na⁺ concentration.) These small ionic changes are quickly corrected because an axon membrane has thousands of Na⁺-K⁺ pumps.

> ⟳ Complete an interactive tutorial: MasteringA&P® > Study Area > Interactive Physiology

Threshold and the All-or-None Phenomenon

Not all local depolarization events produce APs. The depolarization must reach threshold values if an axon is to "fire." What determines the *threshold point*?

One explanation is that threshold is the membrane potential at which the outward current created by K⁺ movement is exactly equal to the inward current created by Na⁺ movement. Threshold is typically reached when the membrane has been depolarized by 15 to 20 mV from the resting value. This depolarization status represents an unstable equilibrium state. If one more Na⁺ enters, further depolarization occurs, opening more Na⁺ channels and allowing more Na⁺ to enter. If, on the other hand, one more K⁺ leaves, the membrane potential is driven away from threshold, Na⁺ channels close, and K⁺ continues to diffuse outward until the potential returns to its resting value.

Recall that local depolarizations are graded potentials and their magnitude increases when stimuli become more intense. Brief weak stimuli (*subthreshold stimuli*) produce subthreshold depolarizations that are not translated into nerve impulses. On the other hand, stronger *threshold stimuli* produce depolarizing currents that push the membrane potential toward and beyond the threshold voltage. As a result, Na⁺ permeability rises to such an extent that entering sodium ions "swamp" (exceed) the outward movement of K⁺, establishing the positive feedback cycle and generating an AP.

The critical factor here is the total amount of current that flows through the membrane during a stimulus (electrical charge × time). Strong stimuli depolarize the membrane to threshold quickly. Weaker stimuli must be applied for longer periods to provide the crucial amount of current flow. Very weak stimuli do not trigger an AP because the local current flows they produce are so slight that they dissipate long before threshold is reached.

An AP is an **all-or-none phenomenon**: It either happens completely or doesn't happen at all. We can compare the generation of an AP to lighting a match under a small dry twig. The changes occurring where the twig is heated are analogous to the change in membrane permeability that initially allows more Na^+ to enter the cell. When that part of the twig becomes hot enough (when enough Na^+ enters the cell), it reaches the flash point (threshold) and the flame consumes the entire twig, even if you blow out the match. Similarly, the AP is generated and propagated whether or not the stimulus continues. But if you blow out the match before the twig reaches the threshold temperature, ignition will not take place. Likewise, if the number of Na^+ ions entering the cell is too low to achieve threshold, no AP will occur.

Propagation of an Action Potential

If it is to serve as the neuron's signaling device, an AP must be **propagated** along the axon's entire length. As we have seen, the AP is generated by the influx of Na^+ through a given area of the membrane. This influx establishes local currents that depolarize adjacent membrane areas in the forward direction (away from the origin of the nerve impulse), which opens voltage-gated channels and triggers an action potential there (**Figure 11.11**). This process repeats down the length of the axon so that an AP is *self-propagating* and continues along the axon at a constant velocity—something like falling dominos.

Following depolarization, each segment of axon membrane repolarizes, restoring the resting membrane potential in that region. Because these electrical changes also set up local currents, the repolarization wave chases the depolarization wave down the length of the axon.

The propagation process we have just described occurs on nonmyelinated axons. On p. 411, we will describe propagation along myelinated axons.

Although the phrase *conduction of a nerve impulse* is commonly used, nerve impulses are not really conducted in the same way that an insulated wire conducts current. In fact, neurons are fairly poor conductors, and local current flows decline with distance because the charges leak through the membrane (Figure 11.10). The expression *propagation of a nerve impulse* is more accurate, because the AP is *regenerated anew* by the voltage-gated channels at each membrane patch, and every subsequent AP is identical to the one that was generated initially. Without voltage-gated channels, propagation cannot occur (see Figure 11.14a).

> ▶ Watch a 3-D animation of this process: MasteringA&P® > Study Area > Animations & Videos > A&P Flix

(a) Time = 0 ms. Action potential has not yet reached the recording electrode.

(b) Time = 2 ms. Action potential peak reaches the recording electrode.

(c) Time = 4 ms. Action potential peak has passed the recording electrode. Membrane at the recording electrode is still hyperpolarized.

Resting potential

Peak of action potential

Hyperpolarization

Figure 11.11 Propagation of an action potential (AP). Recordings at three successive times as an AP propagates along an axon (from left to right). The arrows show the direction of local current flow generated by the movement of positive ions. This current brings the resting membrane at the leading edge of the AP to threshold, propagating the AP forward.

Coding for Stimulus Intensity

Once generated, all APs are independent of stimulus strength, and all APs are alike. So how can the CNS determine whether a particular stimulus is intense or weak? For example, how can you tell the difference between a quiet hum and a loud roar?

The answer is really quite simple: Strong stimuli generate nerve impulses more *often* in a given time interval than do weak stimuli. Stimulus intensity is coded for by the number of impulses per second—that is, by the *frequency of action potentials*—rather than by increases in the strength (amplitude) of the individual APs (**Figure 11.12**).

Figure 11.13 Absolute and relative refractory periods in an AP.

Figure 11.12 Relationship between stimulus strength and action potential frequency.

Refractory Periods

When a patch of neuron membrane is generating an AP and its voltage-gated sodium channels are open, the neuron cannot respond to another stimulus, no matter how strong. Called the **absolute refractory period**, this period begins with the opening of the Na^+ channels and ends when the Na^+ channels begin to reset to their original resting state (**Figure 11.13**). The absolute refractory period:

- Ensures that each AP is a separate, *all-or-none event*
- Enforces one-way transmission of the AP. Because the area where the AP originated has just generated an AP, the Na^+ channels in that area are inactivated and no new AP is generated there. For this reason, the AP propagates away from its point of origin. In the body, APs are initiated at one end of the axon and conducted away from that point toward the other end. (If an axon is stimulated by an electrode at a location that is not at either end of the axon, the AP will move away from the point of stimulus in both directions along the axon.)

The **relative refractory period** follows the absolute refractory period. During the relative refractory period, most Na^+ channels have returned to their resting state, some K^+ channels are still open, and repolarization is occurring. The axon's threshold for AP generation is substantially elevated, so a stimulus that would normally generate an AP is no longer sufficient. An exceptionally strong stimulus can reopen the Na^+ channels that have already returned to their resting state and generate another AP. Strong stimuli trigger more frequent APs by intruding into the relative refractory period.

It might help you to remember the difference between the absolute and relative refractory periods if you know that the word "refractory" means stubborn or unmanageable. Think of a dog that is refractory (disobedient). If he is absolutely refractory, he won't come no matter how loudly you call. But if he is only relatively refractory, he may come, but only if you call loudly enough that he knows you really mean it.

Conduction Velocity

How fast do APs travel? Conduction velocities of neurons vary widely. Nerve fibers that transmit impulses most rapidly (100 m/s or more) are found in neural pathways where speed is essential, such as those that mediate postural reflexes. Axons that conduct impulses more slowly typically serve internal organs (the gut, glands, blood vessels), where slower responses are not a handicap. The rate of impulse propagation depends largely on two factors:

- **Axon diameter.** As a rule, the larger the axon's diameter, the faster it conducts impulses. Larger axons conduct more rapidly because they offer less resistance to the flow of local currents, bringing adjacent areas of the membrane to threshold more quickly.

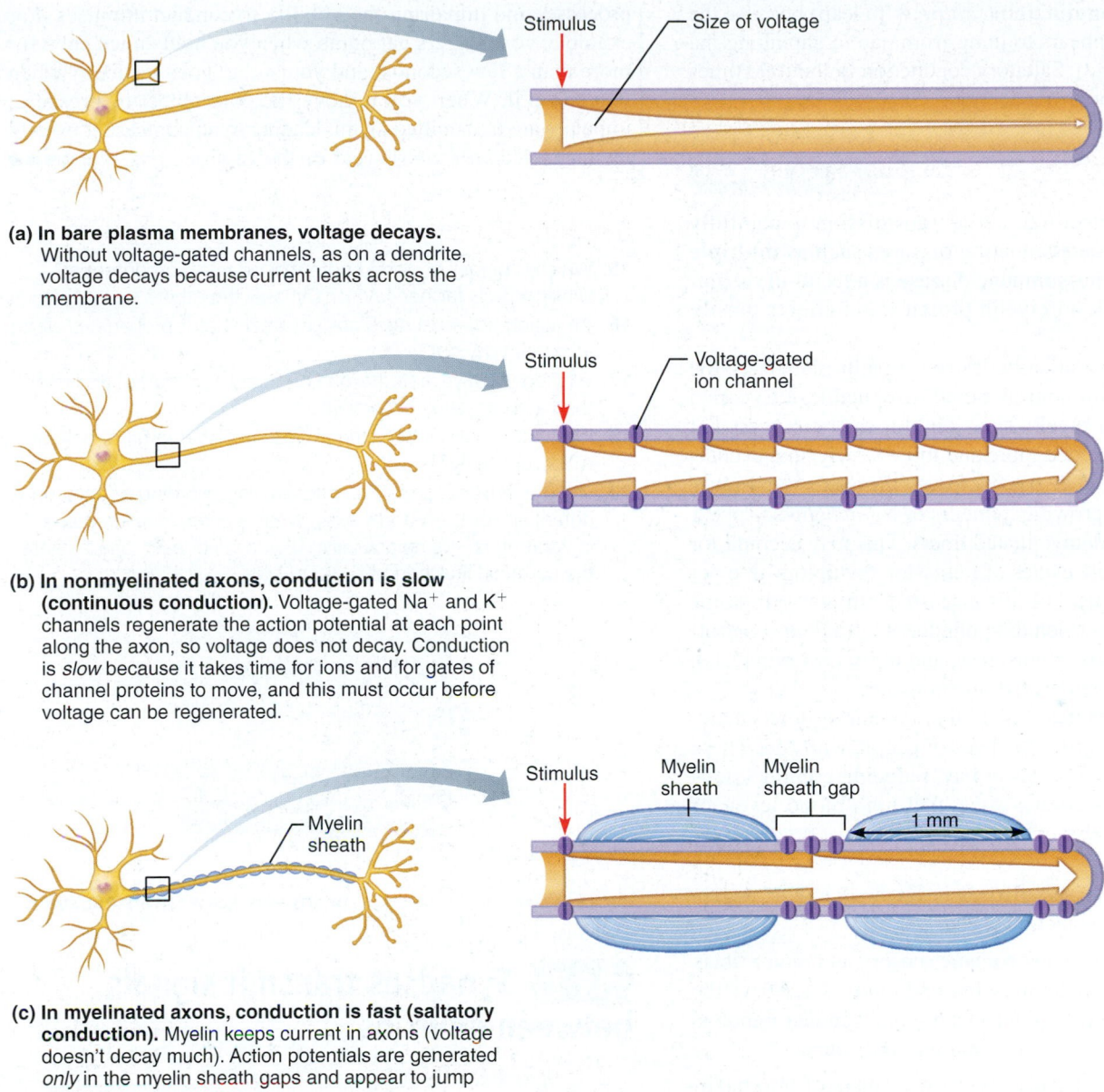

(a) In bare plasma membranes, voltage decays. Without voltage-gated channels, as on a dendrite, voltage decays because current leaks across the membrane.

(b) In nonmyelinated axons, conduction is slow (continuous conduction). Voltage-gated Na⁺ and K⁺ channels regenerate the action potential at each point along the axon, so voltage does not decay. Conduction is *slow* because it takes time for ions and for gates of channel proteins to move, and this must occur before voltage can be regenerated.

(c) In myelinated axons, conduction is fast (saltatory conduction). Myelin keeps current in axons (voltage doesn't decay much). Action potentials are generated *only* in the myelin sheath gaps and appear to jump *rapidly* from gap to gap.

Figure 11.14 Action potential propagation in nonmyelinated and myelinated axons.

- **Degree of myelination.** The presence of a myelin sheath dramatically increases the speed of propagation. The conduction velocity increases with the degree of myelination—lightly myelinated fibers conduct more slowly than heavily myelinated fibers.

  Action potentials can be propagated in one of two ways, depending on whether myelin is present or absent from the axon:

- **Continuous conduction.** Action potential propagation in nonmyelinated axons occurs by **continuous conduction** because the voltage-gated channels in the membrane are immediately adjacent to each other (**Figure 11.14b**). Continuous conduction is relatively slow.

- **Saltatory conduction.** When an AP is generated in a myelinated fiber, the local depolarizing current does not dissipate through the adjacent membrane regions, which are nonexcitable. Instead, the current is maintained and moves rapidly to the next myelin sheath gap, a distance of approximately 1 mm, where it triggers another AP. Consequently, APs are triggered only at the gaps. This type of conduction is called **saltatory conduction** (*saltare* = to leap) because the electrical signal appears to jump from gap to gap along the axon (Figure 11.14c). Saltatory conduction is about 30 times faster than continuous conduction.

HOMEOSTATIC IMBALANCE 11.2 CLINICAL

The importance of myelin to nerve transmission is painfully clear to people with demyelinating diseases such as **multiple sclerosis (MS)**. This autoimmune disease is a result of the immune system's attack on myelin proteins and affects mostly young adults.

Multiple sclerosis gradually destroys myelin sheaths in the CNS, reducing them to nonfunctional hardened lesions called *scleroses*. The loss of myelin short-circuits the current so that successive gaps are excited more and more slowly, and eventually impulse conduction ceases. However, the axons themselves are not damaged and growing numbers of Na^+ channels appear spontaneously in the demyelinated fibers. This may account for the remarkably variable cycles of remission (symptom-free periods) and relapse typical of this disease. Common symptoms are visual disturbances (including blindness), problems controlling muscles (weakness, clumsiness, and ultimately paralysis), speech disturbances, and urinary incontinence.

The advent of drugs that modify the immune system's activity will continue to improve the lives of people with MS. These drugs seem to hold symptoms at bay, reducing complications and disability. Recent studies show that high blood levels of vitamin D reduce the risk of developing MS.

Nerve fibers may be classified according to diameter, degree of myelination, and conduction speed.

- **Group A fibers** are mostly somatic sensory and motor fibers serving the skin, skeletal muscles, and joints. They have the largest diameter, thick myelin sheaths, and conduct impulses at speeds up to 150 m/s (over 300 miles per hour).

- **Group B fibers** are lightly myelinated fibers of intermediate diameter. They transmit impulses at an average rate of 15 m/s (about 30 mi/h).

- **Group C fibers** have the smallest diameter. They are nonmyelinated, so they are incapable of saltatory conduction and conduct impulses at a leisurely pace—1 m/s (2 mi/h) or less.

The B and C fiber groups include autonomic nervous system motor fibers serving the visceral organs; visceral sensory fibers; and the smaller somatic sensory fibers that transmit sensory impulses from the skin (such as pain and small touch fibers).

HOMEOSTATIC IMBALANCE 11.3 CLINICAL

Impaired impulse propagation is caused by a number of chemical and physical factors. Local anesthetics like those used by your dentist act by blocking voltage-gated Na^+ channels. As we have seen, no Na^+ entry—no AP.

Cold and continuous pressure interrupt blood circulation, hindering the delivery of oxygen and nutrients to neuron

processes and impairing their ability to conduct impulses. For example, your fingers get numb when you hold an ice cube for more than a few seconds, and your foot "goes to sleep" when you sit on it. When you remove the cold object or pressure, impulses are transmitted again, leading to an unpleasant prickly feeling called *paresthesia* (par"es-the'ze-ah).

Check Your Understanding

15. Which is bigger, a graded potential or an action potential? Which travels farther? Which initiates the other?
16. An action potential does not get smaller as it propagates along an axon. Why not?
17. Why does a myelinated axon conduct action potentials faster than a nonmyelinated axon?
18. If an axon receives two stimuli close together in time, only one AP occurs. Why?
19. The graph below shows an intracellular recording of an action potential. For the x and y axes, what is being measured and in what units? Name phases a, b, and c. For each phase, state the permeability change(s) responsible.

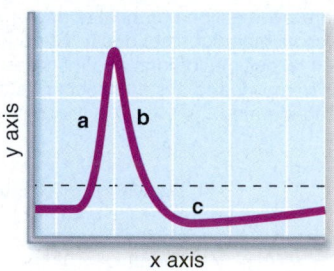

For answers, see Answers Appendix.

11.7 Synapses transmit signals between neurons

Learning Outcomes

▶ Define synapse.
▶ Distinguish between electrical and chemical synapses by structure and by the way they transmit information.

The operation of the nervous system depends on the flow of information through chains of neurons functionally connected by synapses (**Figure 11.15**). A **synapse** (sin'aps), from the Greek *syn*, "to clasp or join," is a junction that mediates information transfer from one neuron to the next or from a neuron to an effector cell—it's where the action is.

The neuron conducting impulses toward the synapse is the **presynaptic neuron**, and the neuron transmitting the electrical signal away from the synapse is the **postsynaptic neuron**. At a given synapse, the presynaptic neuron sends the information, and the postsynaptic neuron receives the information. As you might anticipate, most neurons function as both presynaptic and postsynaptic neurons. Neurons have anywhere from 1000 to 10,000 axon terminals making synapses and are stimulated by an equal number of other neurons. Outside the central nervous system, the postsynaptic cell may be either another neuron or an effector cell (a muscle cell or gland cell).

(a)

Axodendritic synapses

Dendrites

Cell body

Axoaxonal synapses

Axon

Axosomatic synapses

Axon of presynaptic neuron

Synapses

Cell body (soma) of postsynaptic neuron

(b)

Figure 11.15 Synapses. (a) Axodendritic, axosomatic, and axoaxonal synapses.
(b) Scanning electron micrograph (5300×).

Synapses between the axon endings of one neuron and the dendrites of other neurons are **axodendritic synapses**. Those between axon endings of one neuron and the cell body (soma) of another neuron are **axosomatic synapses** (Figure 11.15a). Less common (and far less understood) are synapses between axons (*axoaxonal*), between dendrites (*dendrodendritic*), or between cell bodies and dendrites (*somatodendritic*).

There are two types of synapses: *electrical* and *chemical*.

Electrical Synapses

Electrical synapses are much less common than chemical synapses. They consist of gap junctions like those found between certain other body cells (◄ p. 68). Their channel proteins (connexons) connect the cytoplasm of adjacent neurons and allow ions and small molecules to flow directly from one neuron to the next. These neurons are *electrically coupled*, and transmission across these synapses is very rapid. Depending on the nature of the synapse, communication may be unidirectional or bidirectional.

Electrical synapses between neurons provide a simple means of synchronizing the activity of all interconnected neurons. In adults, electrical synapses are found in regions of the brain responsible for certain stereotyped movements, such as the normal jerky movements of the eyes. They also occur in axoaxonal synapses in the hippocampus, a brain region involved in emotions and memory.

Electrical synapses are far more abundant in embryonic nervous tissue, where they permit exchange of guiding cues during early neuronal development so that neurons can connect properly with one another. As the nervous system develops, chemical synapses replace some electrical synapses and become the vast majority of all synapses. For this reason, we will focus on chemical synapses from now on.

Chemical Synapses

Chemical synapses are the most common type of synapse. They are specialized to allow the release and reception of chemical messengers known as *neurotransmitters*. A typical chemical synapse is made up of two parts:

- A knoblike *axon terminal* of the presynaptic neuron, which contains many tiny, membrane-bound sacs called **synaptic vesicles**, each containing thousands of neurotransmitter molecules

- A neurotransmitter *receptor region* on the postsynaptic neuron's membrane, usually located on a dendrite or the cell body

Although close to each other, presynaptic and postsynaptic membranes are separated by the **synaptic cleft**, a fluid-filled space approximately 30 to 50 nm (about one-millionth of an inch) wide.

Let's take a moment to compare the two methods of communication between neurons. An electrical synapse is like a doorway: Messages (ions) can move directly from one room (neuron) to another. The synaptic cleft of a chemical synapse, on the other hand, is like a lake that the two neurons shout across.

Because the current from the presynaptic membrane dissipates in the fluid-filled cleft, chemical synapses prevent a nerve impulse from being *directly* transmitted from one neuron to another. Instead, an impulse is transmitted via a *chemical event* that depends on the release, diffusion, and receptor binding of neurotransmitter molecules and results in *unidirectional* communication between neurons.

In short, transmission of nerve impulses along an axon and across electrical synapses is a purely electrical event. However, chemical synapses convert the electrical signals to chemical signals (neurotransmitters) that travel across the synapse to the postsynaptic cells, where they are converted back into electrical signals.

Information Transfer across Chemical Synapses

In Chapter 9 we introduced a specialized chemical synapse called a neuromuscular junction (◀ p. 290). The chain of events that occurs at the neuromuscular junction is simply one example of the general process that we will discuss next and show in *Focus on a Chemical Synapse* (**Focus Figure 11.3**):

1. **Action potential arrives at axon terminal.** Neurotransmission begins with the arrival of an AP at the presynaptic axon terminal.

2. **Voltage-gated Ca^{2+} channels open and Ca^{2+} enters the axon terminal.** Depolarization of the membrane by the action potential opens not only Na^+ channels but voltage-gated Ca^{2+} channels as well. During the brief time the Ca^{2+} channels are open, Ca^{2+} floods down its electrochemical gradient from the extracellular fluid into the terminal.

3. **Ca^{2+} entry causes synaptic vesicles to release neurotransmitter by exocytosis.** The surge of Ca^{2+} into the axon terminal acts as an intracellular messenger. A Ca^{2+}-sensing protein (*synaptotagmin*) on the vesicle binds Ca^{2+} and interacts with the SNARE proteins that control membrane fusion (◀ **Figure 3.12, p. 78**). As a result, synaptic vesicles fuse with the axon membrane and empty their contents by exocytosis into the synaptic cleft. Ca^{2+} is then quickly removed from the terminal—either taken up into the mitochondria or ejected from the neuron by an active Ca^{2+} pump.

For each nerve impulse reaching the presynaptic terminal, many vesicles (perhaps 300) empty into the synaptic cleft. The higher the impulse frequency (that is, the more intense the stimulus), the greater the number of synaptic vesicles that release their contents, and the greater the effect on the postsynaptic cell.

4. **Neurotransmitter diffuses across the synaptic cleft and binds to specific receptors on the postsynaptic membrane.**

5. **Binding of neurotransmitter opens ion channels, creating graded potentials.** When a neurotransmitter binds to the receptor protein, this receptor changes its shape. This change in turn opens ion channels and creates graded potentials. Postsynaptic membranes often contain receptor proteins and ion channels packaged together as chemically gated ion channels. Depending on the receptor protein to which the neurotransmitter binds and the type of channel the receptor controls, the postsynaptic neuron may be either excited or inhibited.

6. **Neurotransmitter effects are terminated.** Binding of a neurotransmitter to its receptor is reversible. As long as it is bound to a postsynaptic receptor, a neurotransmitter continues to affect membrane permeability and block reception of additional signals from presynaptic neurons. For this reason, some means of "wiping the postsynaptic slate clean" is necessary. The effects of neurotransmitters generally last a few milliseconds before being terminated in one of three ways, depending on the particular neurotransmitter:

- *Reuptake* by astrocytes or the presynaptic terminal, where the neurotransmitter is stored or destroyed by enzymes (as with the neurotransmitter norepinephrine)

- *Degradation* by enzymes associated with the postsynaptic membrane or present in the synaptic cleft (as with acetylcholine)

- *Diffusion* away from the synapse

Synaptic Delay

An impulse may travel at speeds of up to 150 m/s (300 mi/h) down an axon, but neural transmission across a chemical synapse is comparatively slow. It reflects the time required for neurotransmitter to be released, diffuse across the synaptic cleft, and bind to receptors. Typically, this **synaptic delay** lasts 0.3–5.0 ms, making transmission across the chemical synapse the *rate-limiting* (slowest) step of neural transmission. Synaptic delay helps explain why transmission along neural pathways involving only two or three neurons occurs rapidly, but transmission along multisynaptic pathways typical of higher mental functioning occurs much more slowly. However, in practical terms these differences are not noticeable.

Check Your Understanding

20. **DRAW** Draw a chemical synapse, showing the locations of voltage-gated and chemically gated ion channels. What causes each to open?

21. What structure joins two neurons at an electrical synapse?

For answers, see Answers Appendix.

Chemical synapses transmit signals from one neuron to another using neurotransmitters.

Presynaptic neuron

Postsynaptic neuron

Presynaptic neuron

① Action potential arrives at axon terminal.

② Voltage-gated Ca²⁺ channels open and Ca²⁺ enters the axon terminal.

Ca^{2+}

Ca^{2+}

Mitochondrion

Ca^{2+}

Ca^{2+}

③ Ca²⁺ entry causes synaptic vesicles to release neurotransmitter by exocytosis.

Axon terminal

Synaptic cleft

Synaptic vesicles

④ Neurotransmitter diffuses across the synaptic cleft and binds to specific receptors on the postsynaptic membrane.

Postsynaptic neuron

Ion movement

Graded potential

⑤ Binding of neurotransmitter opens ion channels, resulting in graded potentials.

Enzymatic degradation

Reuptake

Diffusion away from synapse

⑥ Neurotransmitter effects are terminated by reuptake through transport proteins, enzymatic degradation, or diffusion away from the synapse.

415

Table 11.3   Comparison of Graded Potentials and Action Potentials

	GRADED POTENTIAL (GP)	ACTION POTENTIAL (AP)
Location of event	Cell body and dendrites, typically	Axon (beginning at the initial segment)

Distance traveled	Short distance—typically within cell body to initial segment of axon (0.1–1.0 mm)	Long distance—from trigger zone at initial segment through entire length of axon (a few mm to over a meter)

Amplitude (size)	Various sizes (graded); decays with distance	Always the same size (all-or-none); does not decay with distance
Stimulus for opening ion channels	Chemical (neurotransmitter) or sensory stimulus (e.g., light, pressure, temperature)	Voltage (depolarization, triggered by GP reaching threshold)
Positive feedback cycle	Absent	Present
Repolarization	Voltage independent; occurs when stimulus is no longer present	Voltage regulated; occurs when Na^+ channels inactivate and K^+ channels open
Summation	Stimulus responses can summate to increase amplitude of graded potential	Does not occur; an all-or-none phenomenon

Temporal: increased frequency of stimuli   Spatial: stimuli from multiple sources

11.8 Postsynaptic potentials excite or inhibit the receiving neuron

Learning Outcomes

▶ **Distinguish between excitatory and inhibitory postsynaptic potentials.**

▶ **Describe how synaptic events are integrated and modified.**

Many receptors on postsynaptic membranes at chemical synapses are specialized to open ion channels, in this way converting chemical signals to electrical signals. Unlike the voltage-gated ion channels responsible for APs, these chemically gated channels are relatively insensitive to changes in membrane potential. Consequently, channel opening at postsynaptic membranes cannot become self-amplifying or self-generating. Instead, neurotransmitter receptors mediate graded potentials—local changes in membrane potential that are *graded* (vary in strength) based on the amount of neurotransmitter released and how long it remains in the area. **Table 11.3** compares graded potentials and action potentials.

Chemical synapses are either excitatory or inhibitory, depending on how they affect the membrane potential of the postsynaptic neuron.

Table 11.3 *(continued)*

	GRADED POTENTIAL (GP)		ACTION POTENTIAL (AP)
	POSTSYNAPTIC POTENTIAL (A TYPE OF GP)		
	EXCITATORY (EPSP)	**INHIBITORY (IPSP)**	
Function	Short-distance signaling; depolarization that spreads to axon hillock; moves membrane potential *toward* threshold for generating an AP	Short-distance signaling; hyperpolarization that spreads to axon hillock; moves membrane potential *away* from threshold for generating an AP	Long-distance signaling; constitutes the nerve impulse
Initial effect of stimulus	Opens chemically gated channels that allow simultaneous Na^+ and K^+ fluxes	Opens chemically gated K^+ or Cl^- channels	Opens voltage-gated channels; first opens Na^+ channels, then K^+ channels
Peak membrane potential	Depolarizes; moves toward 0 mV	Hyperpolarizes; moves toward −90 mV	+30 to +50 mV

Excitatory Synapses and EPSPs

At excitatory synapses, neurotransmitter binding depolarizes the postsynaptic membrane. In contrast to what happens on axon membranes, *chemically gated* ion channels open on postsynaptic membranes (those of dendrites and neuronal cell bodies). Each channel allows Na^+ and K^+ to diffuse *simultaneously* through the membrane but in opposite directions (Table 11.3).

Although this two-way cation flow may appear to be self-defeating when depolarization is the goal, remember that the electrochemical gradient for sodium is much steeper than that for potassium. As a result, Na^+ influx is greater than K^+ efflux, and *net* depolarization occurs.

If enough neurotransmitter binds, depolarization of the postsynaptic membrane can reach 0 mV—well above an axon's threshold (about −50 mV) for firing an AP. However, unlike axons which have voltage-gated channels that make an AP possible, *postsynaptic membranes generally do not generate APs.* The dramatic polarity reversal seen in axons never occurs in membranes containing *only* chemically gated channels because the opposite movements of K^+ and Na^+ prevent excessive

positive charge from accumulating inside the cell. For this reason, instead of APs, depolarizing graded potentials called **excitatory postsynaptic potentials (EPSPs)** occur at excitatory postsynaptic membranes. This is shown on the left page of **Focus Figure 11.4**, *Postsynaptic Potentials and Their Summation*, pp. 418–419.

Each EPSP lasts a few milliseconds and then the membrane returns to its resting potential. The only function of EPSPs is to help trigger an AP distally at the initial segment of the postsynaptic neuron's axon. Although currents created by individual EPSPs decline with distance, they can and often do spread all the way to the axon's initial segment. If currents reaching the initial segment are strong enough to depolarize the axon to threshold, axonal voltage-gated channels open and an AP is generated.

Inhibitory Synapses and IPSPs

As the name suggests, binding of neurotransmitters at inhibitory synapses *reduces* a postsynaptic neuron's ability to generate an AP. Most inhibitory neurotransmitters hyperpolarize the

(Text continues on p. 420.)

One postsynaptic neuron can receive input (as EPSPs and IPSPs) from thousands of other neurons. It integrates these inputs at its initial segment to determine whether it will fire an action potential (AP).

What does one neuron "say" to another neuron?

One of two things: "Fire" an action potential (AP) or "Don't fire" an action potential. They "speak" by releasing neurotransmitters that create graded potentials (small, local changes in membrane potential) in the receiving neuron's dendrites or cell body. There are two types of graded potentials at synapses:

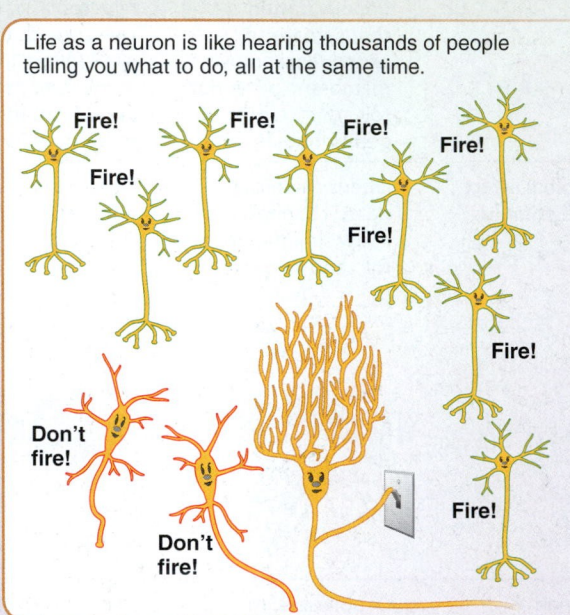

Life as a neuron is like hearing thousands of people telling you what to do, all at the same time.

Fire! Fire! Fire! Fire!
Fire! Fire!
Don't fire!
Don't fire! Fire!
Don't fire!

Excitatory postsynaptic potential (EPSP)

An **EPSP** is a local depolarization of the postsynaptic membrane (it says "**Fire!**")

EPSPs bring the neuron closer to AP threshold.

Membrane potential (mV): +30, 0, −55, −70
Threshold
Stimulus
Time (ms): 10 20 30

Neurotransmitter binding opens chemically gated ion channels, allowing Na⁺ and K⁺ to pass simultaneously. The electrochemical gradient for Na⁺ is bigger than for K⁺, so more Na⁺ enters than K⁺ exits and the result is depolarization.

Na⁺
K⁺

Inhibitory postsynaptic potential (IPSP)

An **IPSP** is a hyperpolarization of the postsynaptic membrane (it says "**Don't fire!**")

IPSPs drive the neuron away from AP threshold.

Membrane potential (mV): +30, 0, −55, −70
Threshold
Stimulus
Time (ms): 10 20 30

Neurotransmitter binding opens chemically gated ion channels, allowing either K⁺ or Cl⁻ to pass. The result is hyperpolarization.

Cl⁻
K⁺

How is a poor neuron to decide?

Neural integration—the **summation** of many **EPSPs** and **IPSPs** at the axon's initial segment—determines if a neuron reaches threshold to fire an AP down the axon. Graded potentials decay with time and distance as they spread down dendrites and across the cell body to the initial segment, where they sum together. An AP fires if the initial segment reaches threshold. Let's look at four simplified examples.

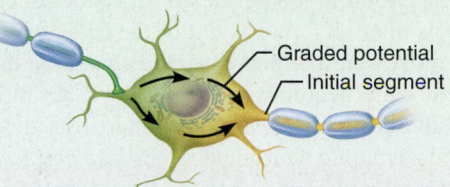

Graded potential
Initial segment

No summation

A slowly firing presynaptic neuron causes EPSPs that are far apart in time.

Two EPSPs separated in time do not add together.

Threshold

Threshold is not reached, so no AP is generated.

Fire!...

Fire!...

It's like someone casually shouting "fire!" every few minutes.

Temporal summation

A rapidly firing presynaptic neuron causes EPSPs that are **close in time**.

Two EPSPs close in time add together (temporal = time).

Summation brings the axon's initial segment to threshold and an AP fires.

Fire!
Fire!
Fire!
Fire!

It's like someone shouting "fire!" quickly over and over.

Spatial summation

If more than one presynaptic neuron fires at the same time, EPSPs are generated at **different locations** on the neuron.

Two EPSPs at the same time from different locations add together (spatial = location).

Summation brings the axon's initial segment to threshold and an AP fires.

Fire!   Fire!

Fire!

It's like many people shouting "fire!" at the same time.

Spatial summation of EPSPs and IPSPs

If a presynaptic neuron creates an **IPSP**, it can override the **EPSP** created by another neuron.

An EPSP brings the neuron closer to threshold.

An IPSP brings the neuron farther from the threshold.

Together, they (nearly) cancel each other out.

Fire!

Don't fire!

It's like one person shouting "fire!" while another shouts "don't fire!"

postsynaptic membrane by making the membrane more permeable to K^+ or Cl^-. Sodium ion permeability is not affected.

If K^+ channels open, K^+ moves out of the cell. If Cl^- channels open, Cl^- moves in. In either case, the charge on the inner face of the membrane becomes more negative. As the membrane potential increases and is driven farther from the axon's threshold, the postsynaptic neuron becomes *less and less likely* to "fire," and larger depolarizing currents are required to induce an AP. Hyperpolarizing changes in potential are called **inhibitory postsynaptic potentials (IPSPs)**.

Integration and Modification of Synaptic Events

Summation by the Postsynaptic Neuron

A single EPSP cannot induce an AP in the postsynaptic neuron. But if thousands of excitatory axon terminals fire on the same postsynaptic membrane, or if a small number of terminals deliver impulses rapidly, the probability of reaching threshold soars. Otherwise, nerve impulses would never result. EPSPs and IPSPs can add together, or **summate**, to influence the activity of a postsynaptic neuron. Focus Figure 11.4 (pp. 418–419) explains temporal and spatial summation—the two types of summation that occur in neurons. Study this figure before reading further.

Most neurons receive both excitatory and inhibitory inputs from thousands of other neurons. How is all this conflicting information sorted out?

Each neuron's initial segment keeps a running account of all the signals it receives. Not only do EPSPs summate and IPSPs summate, but also EPSPs summate with IPSPs [Focus Figure 11.4 (right)]. In other words, the axon's initial segment functions as a *neural integrator*, and the potential there at any time reflects the sum of all incoming neural information:

- If the stimulatory effects of EPSPs dominate the membrane potential enough to reach threshold, the neuron will fire.
- If summation yields only subthreshold depolarization, the neuron is **facilitated**. It does not fire an action potential, but is more easily excited by successive depolarization events because it is already near threshold.
- If summation yields hyperpolarization, the neuron cannot generate an AP.

Because EPSPs and IPSPs are graded potentials that decay the farther they spread, the most effective synapses are those closest to the axon's initial segment. Specifically, inhibitory synapses are most effective when located between the site of excitatory inputs and the site of action potential generation (the axon's initial segment). Accordingly, inhibitory synapses occur most often on the cell body and excitatory synapses occur most often on the dendrites (Focus Figure 11.4).

Synaptic Potentiation

Repeated or continuous use of a synapse (even for short periods) enhances the presynaptic neuron's ability to excite the postsynaptic neuron, producing larger-than-expected EPSPs. This phenomenon is **synaptic potentiation**. The presynaptic

terminals at such synapses contain relatively high Ca^{2+} concentrations, a condition that triggers the release of more neurotransmitter, which in turn produces larger EPSPs.

Synaptic potentiation also brings about Ca^{2+} influx into the postsynaptic neuron via dendritic spines. As Ca^{2+} floods into the cell, it activates certain kinase enzymes that promote changes resulting in more effective responses to subsequent stimuli.

Synaptic potentiation can be viewed as a learning process that increases the efficiency of neurotransmission along a particular pathway. Indeed, the hippocampus of the brain, which plays a special role in memory and learning, exhibits an important type of synaptic plasticity called *long-term potentiation* (LTP).

Presynaptic Inhibition

Events at the presynaptic membrane can also influence postsynaptic activity. **Presynaptic inhibition** occurs when the release of excitatory neurotransmitter by one neuron is inhibited by the activity of another neuron via an axoaxonal synapse. More than one mechanism is involved, but the end result is that less neurotransmitter is released and bound, forming smaller EPSPs.

In contrast to postsynaptic inhibition by IPSPs, which decreases the excitability of the postsynaptic neuron, presynaptic inhibition decreases the excitatory stimulation of the postsynaptic neuron. In this way, presynaptic inhibition temporarily turns off specific synapses.

Check Your Understanding

22. Which ions flow through chemically gated channels to produce IPSPs? EPSPs?
23. What is the difference between temporal summation and spatial summation?

For answers, see Answers Appendix.

11.9 The effect of a neurotransmitter depends on its receptor

Learning Outcomes
▶ **Define neurotransmitter and classify neurotransmitters by chemical structure and by function.**
▶ **Describe the action of neurotransmitters at channel-linked and G protein–coupled receptors.**

Neurotransmitters, along with electrical signals, are the "language" of the nervous system—the means by which neurons communicate, integrate, and send messages to the rest of the body. Sleep, thought, rage, hunger, memory, movement, and even your smile reflect the actions of these versatile molecules. Most factors that affect synaptic transmission do so by enhancing or inhibiting neurotransmitter release or destruction, or by blocking their binding to receptors. Just as speech defects may hinder interpersonal communication, anything that interferes with neurotransmitter activity may short-circuit the brain's "conversations" or internal talk (see **A Closer Look** on p. 429).

More than 50 neurotransmitters or neurotransmitter candidates have been identified. Although some neurons produce and release only one kind of neurotransmitter, most make two or more and may release any one or all of them at a given time.

Table 11.4 Neurotransmitters and Neuromodulators

NEUROTRANSMITTER	FUNCTIONAL CLASSES	SITES WHERE SECRETED	COMMENTS
Acetylcholine (ACh)			
Acetylcholine* (acts at two classes of receptor: *nicotinic* and *muscarinic*)	Excitatory or inhibitory depending on subtype of receptor (nicotinic receptors are always excitatory) Direct (at nicotinic receptors) or indirect (at muscarinic receptors)	CNS: widespread throughout cerebral cortex, hippocampus, and brain stem PNS: all neuromuscular junctions with skeletal muscle; some autonomic motor endings (all preganglionic and parasympathetic postganglionic fibers)	Effects prolonged when AChE blocked by nerve gas or organophosphate insecticides (malathion), leading to tetanic muscle spasms. Release inhibited by botulinum toxin; binding to nicotinic ACh receptors inhibited by curare (a muscle paralytic agent) and to muscarinic ACh receptors by atropine. ACh levels decrease in certain brain areas in Alzheimer's disease; nicotinic ACh receptors destroyed in myasthenia gravis.
Biogenic Amines			
Norepinephrine (NE)*	Excitatory or inhibitory depending on receptor type bound Indirect action via second messengers	CNS: brain stem, particularly in the locus coeruleus of the midbrain; limbic system; some areas of cerebral cortex PNS: main neurotransmitter of postganglionic neurons in the sympathetic nervous system	A "feel good" neurotransmitter. Release enhanced by amphetamines; removal from synapse blocked by tricyclic antidepressants and cocaine. Brain levels reduced by reserpine (an antihypertensive drug), sometimes leading to depression.
Dopamine	Excitatory or inhibitory depending on the receptor type bound Indirect action via second messengers	CNS: substantia nigra of midbrain; hypothalamus; the principal neurotransmitter of indirect motor pathways PNS: some sympathetic ganglia	A "feel good" neurotransmitter. Release enhanced by L-dopa and amphetamines; reuptake blocked by cocaine. Deficient in Parkinson's disease; dopamine neurotransmission increases in schizophrenia.
Serotonin (5-HT)	Mainly inhibitory Indirect action via second messengers; direct action at 5-HT$_3$ receptors	CNS: brain stem, especially midbrain; hypothalamus; limbic system; cerebellum; pineal gland; spinal cord	Plays a role in sleep, appetite, nausea, migraine headaches, and regulating mood. Drugs that block its uptake relieve anxiety and depression. Activity blocked by LSD and enhanced by ecstasy.
Histamine	Excitatory or inhibitory depending on receptor type bound Indirect action via second messengers	CNS: hypothalamus	Involved in wakefulness, appetite control, and learning and memory.
Amino Acids			
GABA (γ-aminobutyric acid)	Generally inhibitory Direct and indirect actions via second messengers	CNS: cerebral cortex, hypothalamus, Purkinje cells of cerebellum, spinal cord, granule cells of olfactory bulb, retina	Principal inhibitory neurotransmitter in the brain; important in presynaptic inhibition. Inhibitory effects augmented by alcohol, antianxiety drugs of the benzodiazepine class, and barbiturates. Substances that block its synthesis, release, or action induce convulsions.
Glutamate	Generally excitatory Direct action	CNS: spinal cord; widespread in brain where it represents the major excitatory neurotransmitter	Important in learning and memory. The "stroke neurotransmitter": excessive release produces excitotoxicity—neurons literally stimulated to death; most commonly caused by ischemia (oxygen deprivation, usually due to a blocked blood vessel).

* See Chapter 14 for a discussion of acetylcholine and norepinephrine receptor subtypes.

Table 11.4 *(continued)*

NEUROTRANSMITTER	FUNCTIONAL CLASSES	SITES WHERE SECRETED	COMMENTS
Glycine	Generally inhibitory Direct action	CNS: spinal cord and brain stem, retina	Principal inhibitory neurotransmitter of the spinal cord. Strychnine blocks glycine receptors, resulting in uncontrolled convulsions and respiratory arrest.
Peptides			
Endorphins, e.g., beta endorphin, dynorphin, enkephalins	Generally inhibitory Indirect action via second messengers	CNS: widely distributed in brain (hypothalamus; limbic system; pituitary) and spinal cord	Natural opiates; inhibit pain by inhibiting substance P. Effects mimicked by morphine, heroin, and methadone.
Tachykinins: substance P, neurokinin A (NKA)	Excitatory Indirect action via second messengers	CNS: basal nuclei, midbrain, hypothalamus, cerebral cortex PNS: certain sensory neurons of dorsal root ganglia (pain afferents), enteric neurons	Substance P mediates pain transmission in the PNS. In the CNS, tachykinins are involved in respiratory and cardiovascular controls and in mood.
Somatostatin	Generally inhibitory Indirect action via second messengers	CNS: widely distributed in brain (hypothalamus, basal nuclei, hippocampus, cerebral cortex) Pancreas	Often released with GABA. A gut-brain peptide. Inhibits growth hormone release.
Cholecystokinin (CCK)	Generally excitatory Indirect action via second messengers	Throughout CNS Small intestine	Involved in anxiety, pain, memory. A gut-brain peptide hormone. Inhibits appetite.
Purines			
ATP	Excitatory or inhibitory depending on receptor type bound Direct and indirect actions via second messengers	CNS: basal nuclei, induces Ca^{2+} wave propagation in astrocytes PNS: dorsal root ganglion neurons	ATP released by sensory neurons (as well as that released by injured cells) provokes pain sensation.
Adenosine	Generally inhibitory Indirect action via second messengers	Throughout CNS and PNS	Caffeine stimulates by blocking brain adenosine receptors. May be involved in sleep-wake cycle and terminating seizures. Dilates arterioles, increasing blood flow to heart and other tissues as needed.
Gases and Lipids			
Nitric oxide (NO)	Excitatory or inhibitory Indirect action via second messengers	CNS: brain, spinal cord PNS: adrenal gland; nerves to penis	Its release potentiates stroke damage. Some types of male impotence treated by enhancing NO action [e.g., with sildenafil (Viagra®)].
Carbon monoxide (CO)	Excitatory or inhibitory Indirect action via second messengers	Brain and some neuromuscular and neuroglandular synapses	
Endocannabinoids, e.g., 2-arachidonoylglycerol, anandamide	Inhibitory Indirect action via second messengers	Throughout CNS	Involved in memory (as a retrograde messenger), appetite control, nausea and vomiting, neuronal development. Receptors activated by THC, the principal active ingredient of cannabis.

It appears that in most cases, different neurotransmitters are released at different stimulation frequencies. This avoids producing a jumble of nonsense messages. However, co-release of two neurotransmitters from the same vesicles has been documented. The coexistence of more than one neurotransmitter in a single neuron makes it possible for that cell to exert several different influences.

Neurotransmitters are classified chemically and functionally. **Table 11.4** provides a detailed overview and key groups are discussed in the following sections. Many neurotransmitters can act at more than one type of receptor, and so can produce different effects depending on the type of receptor to which they bind.

Classification of Neurotransmitters by Chemical Structure

Neurotransmitters are grouped into several classes based on molecular structure.

Acetylcholine

Acetylcholine (ACh) (as″ĕ-til-ko′lēn), the first neurotransmitter identified, is still the best understood because it is released at neuromuscular junctions, which are much easier to study than synapses buried in the CNS.

ACh is released by all neurons that stimulate skeletal muscles and by many neurons of the autonomic nervous system. ACh-releasing neurons are also found in the CNS.

ACh is synthesized from acetic acid (as acetyl CoA) and choline by the enzyme *choline acetyltransferase*, then transported into synaptic vesicles for later release. Once released by the presynaptic terminal, ACh binds briefly to the postsynaptic receptors. It is then released and degraded to acetic acid and choline by the enzyme **acetylcholinesterase (AChE)**, located in the synaptic cleft and on postsynaptic membranes. Presynaptic terminals recapture the released choline and reuse it to synthesize more ACh.

Biogenic Amines

The **biogenic amines** (bi″o-jen′ik) include the **catecholamines** (kat″ĕ-kol′ah-mēnz), such as dopamine, norepinephrine (NE), and epinephrine, and the **indolamines**, which include serotonin and histamine. *Dopamine* and *NE* are synthesized from the amino acid tyrosine in a common pathway. The epinephrine-releasing cells of the brain and adrenal medulla use the same pathway. *Serotonin* is synthesized from the amino acid tryptophan. *Histamine* is synthesized from the amino acid histidine.

Biogenic amine neurotransmitters are broadly distributed in the brain, where they play a role in emotional behavior and help regulate the biological clock. Additionally, some motor neurons of the autonomic nervous system release catecholamines, particularly NE. Imbalances of these neurotransmitters are associated with mental illness. For example, overactive dopamine signaling occurs in schizophrenia. Additionally, certain psychoactive drugs (LSD and mescaline) can bind to biogenic amine receptors and induce hallucinations.

Amino Acids

It is difficult to prove a neurotransmitter role when the suspect is an amino acid, because amino acids occur in all cells of the body and are important in many biochemical reactions. The amino acids for which a neurotransmitter role is certain include **glutamate, aspartate, glycine,** and **gamma (γ)-aminobutyric acid (GABA),** and there may be others.

Peptides

The **neuropeptides**, essentially strings of amino acids, include a broad spectrum of molecules with diverse effects. For example, a neuropeptide called **substance P** is an important mediator of pain signals. By contrast, **endorphins**, which include **beta endorphin**, **dynorphin**, and **enkephalins** (en-kef′ah-linz), act as natural opiates, reducing our perception of pain under stressful conditions. Enkephalin activity increases dramatically in pregnant women in labor. Endorphin release is at least partially responsible for the "runner's high."

These painkilling neurotransmitters remained undiscovered until investigators began to ask why morphine and other opiates reduce anxiety and pain. They found that these drugs attach to the same receptors that bind natural opiates, producing similar but stronger effects.

Some neuropeptides, known as **gut-brain peptides**, are also produced by nonneural body tissues and are widespread in the gastrointestinal tract. Examples include somatostatin and cholecystokinin (CCK).

Purines

Purines are one of the classes of nitrogen-containing bases that make up DNA and RNA. Two purines, ATP and adenosine, also have well-established roles as chemical messengers:

- **Adenosine triphosphate (ATP)**, the cell's universal form of energy, is now recognized as a major neurotransmitter (perhaps the most primitive one) in both the CNS and PNS. Like the receptors for glutamate and acetylcholine, certain receptors produce fast excitatory responses when ATP binds, while other ATP receptors trigger slow, second-messenger responses. Upon binding to receptors on astrocytes, ATP mediates Ca^{2+} influx.

- **Adenosine**, a part of ATP, also acts outside of cells on adenosine receptors. Adenosine is a potent inhibitor in the brain. Caffeine's well-known stimulatory effects result from blocking these adenosine receptors.

Gases and Lipids

Not so long ago, it would have been scientific suicide to suggest that small, short-lived, toxic gas molecules might be neurotransmitters. Nonetheless, the discovery of these unlikely messengers has opened up a new chapter in the story of neurotransmission.

Gasotransmitters These gases—the so-called "gasotransmitters" nitric oxide, carbon monoxide, and hydrogen sulfide—defy all the classical descriptions of neurotransmitters. Rather than being stored in vesicles and released by exocytosis, they are synthesized on demand and diffuse out of the cells that make them. Instead of attaching to surface receptors, they zoom through the plasma membrane of nearby cells to bind with intracellular receptors.

Both **nitric oxide (NO)** and **carbon monoxide (CO)** activate *guanylate cyclase*, the enzyme that makes the second messenger *cyclic GMP*. NO and CO are found in different brain

11

regions and appear to act in different pathways, but their mode of action is similar. NO participates in a variety of processes in the brain, including the formation of new memories by increasing the strength of certain synapses.

Excessive release of NO is thought to contribute to the brain damage seen in stroke patients. In the PNS, NO causes blood vessels and intestinal smooth muscle to relax.

Less is known about **hydrogen sulfide (H_2S)**, the most recently discovered gasotransmitter. Unlike NO and CO, it appears to act directly on ion channels and other proteins to alter their function.

Endocannabinoids   Just as there are natural opiate neurotransmitters in the brain, our brains make **endocannabinoids** (en″do-kă-nă′bĭ-noids) that act at the same receptors as tetrahydrocannabinol (THC), the active ingredient in marijuana. Their receptors, the *cannabinoid receptors*, are the most common G protein–coupled receptors in the brain. Like the gasotransmitters, the endocannabinoids are lipid soluble and are synthesized on demand, rather than stored and released from vesicles. Like NO, they are thought to be involved in learning and memory. We are only beginning to understand the many other processes these neurotransmitters may be involved in, which include neuronal development, controlling appetite, and suppressing nausea.

Classification of Neurotransmitters by Function

In this text we can only sample the incredible diversity of functions that neurotransmitters mediate. We limit our discussion here to two broad ways of classifying neurotransmitters according to function, adding more details in subsequent chapters.

The important idea to keep in mind is this: The function of a neurotransmitter is determined by the receptor to which it binds.

Effects: Excitatory versus Inhibitory

Some neurotransmitters are excitatory (cause depolarization). Some are inhibitory (cause hyperpolarization). Others exert both effects, depending on the specific receptor types with which they interact.

For example, the amino acids GABA and glycine are usually inhibitory, whereas glutamate is typically excitatory (Table 11.3). On the other hand, ACh and NE each bind to at least two receptor types that cause opposite effects. For example, acetylcholine is excitatory at neuromuscular junctions in skeletal muscle and inhibitory in cardiac muscle.

Actions: Direct versus Indirect

Neurotransmitters that act *directly* are those that bind to and open ion channels. These neurotransmitters provoke rapid responses in postsynaptic cells by altering membrane potential. ACh and the amino acid neurotransmitters are typically direct-acting neurotransmitters.

Neurotransmitters that act *indirectly* promote broader, longer-lasting effects by acting through intracellular *second-messenger* molecules, typically via G protein pathways

(◄ *Focus on G Proteins*, Focus Figure 3.2, p. 82). In this way their action is similar to that of many hormones. The biogenic amines, neuropeptides, and dissolved gases are indirect neurotransmitters.

Neuromodulator is a term used to describe a chemical messenger released by a neuron that does not directly cause EPSPs or IPSPs but instead affects the strength of synaptic transmission. A neuromodulator may act presynaptically to influence the synthesis, release, degradation, or reuptake of neurotransmitters. Alternatively, it may act postsynaptically by altering the sensitivity of the postsynaptic membrane to neurotransmitters.

Receptors for neuromodulators are not necessarily found at a synapse. Instead, a neuromodulator may be released from one cell to act at many cells in its vicinity, similar to paracrines (chemical messengers that act locally and are quickly destroyed). The distinction between neurotransmitters and neuromodulators is fuzzy, but chemical messengers such as NO, adenosine, and a number of neuropeptides are often referred to as neuromodulators.

Neurotransmitter Receptors

In Chapter 3, we introduced the various receptors involved in cell signaling (◄ p. 81). Now we are ready to pick up that thread again as we examine the action of receptors that bind neurotransmitters. For the most part, neurotransmitter receptors are either channel-linked receptors, which mediate fast synaptic transmission, or G protein–coupled receptors, which oversee slow synaptic responses.

Channel-Linked Receptors

Channel-linked receptors (*ionotropic receptors*) are ligand-gated ion channels that mediate direct neurotransmitter action. They are composed of several protein subunits in a "rosette" around a central pore. As the ligand binds to one (or more) receptor subunits, the proteins change shape. This event opens the central channel and allows ions to pass (**Figure 11.16**). As a result, the membrane potential of the target cell changes.

Channel-linked receptors are always located precisely opposite sites of neurotransmitter release, and their ion channels open instantly upon ligand binding and remain open 1 ms

Figure 11.16 Channel-linked receptors cause rapid synaptic transmission. Ligand binding *directly* opens chemically gated ion channels.

or less while the ligand is bound. At excitatory receptor sites (such as certain receptor subtypes for ACh, glutamate, aspartate, and ATP), the channel-linked receptors are cation channels that allow small cations (Na^+, K^+, Ca^{2+}) to pass, but Na^+ entry contributes most to membrane depolarization. Channel-linked receptors that respond to GABA and glycine, and allow Cl^- to pass, mediate fast inhibition (hyperpolarization).

G Protein–Coupled Receptors

Unlike responses to neurotransmitter binding at channel-linked receptors, which are immediate, simple, and brief, the activity mediated by **G protein–coupled receptors** is indirect, complex, slow (hundreds of milliseconds or more), and often prolonged—ideal as a basis for some types of learning. Receptors in this class are transmembrane protein complexes. They include receptors for the biogenic amines and neuropeptides as well as the muscarinic subtype of ACh receptor. Because their effects tend to bring about widespread metabolic changes, G protein–coupled receptors are commonly called *metabotropic receptors*.

When a neurotransmitter binds to a G protein–coupled receptor, the G protein is activated (**Figure 11.17**). (To orient yourself, refer back to the simpler G protein explanation in ◀ *Focus on G Proteins*, **Focus Figure 3.3, p. 82**.) Activated G proteins typically work by controlling the production of second messengers such as **cyclic AMP**, **cyclic GMP**, **diacylglycerol**, or **Ca^{2+}**.

These second messengers, in turn, act as go-betweens to regulate the opening or closing of ion channels or activate kinase enzymes that initiate a cascade of reactions in the target cells. Some second messengers modify (activate or inactivate) other proteins, including channel proteins, by attaching phosphate groups to them. Others interact with nuclear proteins that activate genes and induce synthesis of new proteins in the target cell.

Check Your Understanding

24. APPLY ACh excites skeletal muscle and yet it inhibits heart muscle. How can this be?

25. Why is cyclic AMP called a second messenger?

For answers, see Answers Appendix.

G protein signaling mechanisms are like a molecular relay race.

Ligand (1st messenger)   Receptor   G protein   Enzyme   2nd messenger

① Neurotransmitter (1st messenger) binds and activates receptor.

Adenylate cyclase

Closed ion channel   Open ion channel

Receptor

G protein

ATP

GTP   GTP

GDP GTP

cAMP

⑤a cAMP changes membrane permeability by opening or closing ion channels.

⑤c cAMP activates specific genes.

② Receptor activates G protein.

③ G protein activates adenylate cyclase.

④ Adenylate cyclase converts ATP to cAMP (2nd messenger).

⑤b cAMP activates enzymes.

Active enzyme

Nucleus

Figure 11.17 G protein–coupled receptors cause the formation of intracellular second messengers. The neurotransmitter acts indirectly—via the second messenger cyclic AMP (cAMP) in this example—to bring about the cell's response. (For the basics of G protein signaling mechanisms, see Focus Figure 3.3 on p. 82.)

11.10 Neurons act together, making complex behaviors possible

Learning Outcomes

▶ Describe common patterns of neuronal organization and processing.

▶ Distinguish between serial and parallel processing.

Until now, we have concentrated on the activities of individual neurons. However, neurons function in groups, and each group contributes to still broader neural functions. In this way, the organization of the nervous system is hierarchical.

Any time you have a large number of *anything*—people included—there must be *integration*. In other words, the parts must be fused into a smoothly operating whole.

In this module, we move to the first level of **neural integration**: *neuronal pools* and their patterns of communicating with other parts of the nervous system. In Chapter 12 we discuss the highest levels of neural integration—how we think and remember. With this understanding of the basics and of the larger picture, in Chapter 13 we examine how sensory inputs interface with motor activity.

Organization of Neurons: Neuronal Pools

The billions of neurons in the CNS are organized into **neuronal pools**. These functional groups of neurons integrate incoming information from receptors or different neuronal pools and then forward the processed information to other destinations.

In a simple type of neuronal pool (**Figure 11.18**), one incoming presynaptic fiber branches profusely as it enters the pool and then synapses with several different neurons in the pool. When the incoming fiber is excited, it will excite some postsynaptic neurons and facilitate others by bringing them closer to threshold. Neurons most likely to generate impulses are those closely associated with the incoming fiber, because they receive the bulk of the synaptic contacts. Those neurons are in the *discharge zone* of the pool.

Neurons farther from the center are not usually excited to threshold, but they are facilitated and can easily be brought to threshold by stimuli from another source. For this reason, the periphery of the pool is the *facilitated zone*. Keep in mind, however, that our figure is a gross oversimplification. Most neuronal pools consist of thousands of neurons and include inhibitory as well as excitatory neurons.

Patterns of Neural Processing

Input processing is both *serial* and *parallel*. In serial processing, the input travels along one pathway to a specific destination. In parallel processing, the input travels along several different pathways to be integrated in different CNS regions. Each mode has unique advantages, but as an information processor, the brain derives its power from its ability to process in parallel.

Serial Processing

In **serial processing**, the whole system works in a predictable all-or-nothing manner. One neuron stimulates the next, which stimulates the next, and so on, eventually causing a specific, anticipated response. The most clear-cut examples of serial processing are spinal reflexes. Straight-through sensory pathways from receptors to the brain are also examples. Because reflexes are the functional units of the nervous system, it is important that you understand them early on.

Reflexes are rapid, automatic responses to stimuli, in which a particular stimulus always causes the same response. Reflex activity, which produces the simplest behaviors, is stereotyped

![Figure 11.18 neuronal pool diagram showing presynaptic input fiber branching to several neurons with facilitated and discharge zones]

Figure 11.18 Simple neuronal pool. Postsynaptic neurons in the discharge zone receive more synapses and are more likely to discharge (generate APs). Postsynaptic neurons in the facilitated zone receive fewer synapses and are facilitated (brought closer to threshold).

Figure 11.19 A simple reflex arc. Receptors detect a change in the internal or external environment that elicits a rapid stereotyped response. Effectors are muscles or glands.

and dependable. For example, if you touch a hot object you jerk your hand away, and an object approaching your eye triggers a blink. Reflexes occur over neural pathways called **reflex arcs** that have five essential components—receptor, sensory neuron, CNS integration center, motor neuron, and effector (**Figure 11.19**).

Parallel Processing

In **parallel processing**, inputs are segregated into many pathways, and different parts of the neural circuitry deal simultaneously with the information delivered by each pathway. For example, smelling a pickle (the input) may cause you to remember picking cucumbers on a farm; or it may remind you that you don't like pickles or that you must buy some at the market; or perhaps it will call to mind *all* these thoughts. For each person, parallel processing triggers unique pathways. The same stimulus—pickle smell, in our example—promotes many responses beyond simple awareness of the smell. Parallel processing is not repetitious because the pathways do different things with the information. Each pathway or "channel" is decoded in relation to all the others to produce a total picture.

Think, for example, about what happens when you step on a sharp object while walking barefoot. The serially processed withdrawal reflex causes you to withdraw your foot immediately. At the same time, pain and pressure impulses are speeding up to your brain along parallel pathways that allow you to decide whether to simply rub the hurt spot or seek first aid.

Parallel processing is extremely important for higher-level mental functioning—for putting the parts together to understand the whole. For example, you can recognize a dollar bill in a split second. This task takes a computer (which only processes serially) a fairly long time, but your recognition is rapid because you use parallel processing. A single neuron sends information along several pathways instead of just one, so you process a large amount of information much more quickly.

Types of Circuits

Individual neurons in a neuronal pool both send and receive information, and synaptic contacts may cause either excitation or inhibition. The patterns of synaptic connections in neuronal pools, called **circuits**, determine the pool's functional capabilities. **Figure 11.20** illustrates four basic circuit patterns and their properties: diverging, converging, reverberating, and parallel after-discharge circuits.

Check Your Understanding

26. Which types of neural circuits would give a prolonged output after a single input?

27. APPLY What pattern of neural processing occurs when your finger accidentally touches a hot grill? What is this type of pathway called?

28. APPLY What pattern of neural processing occurs when we smell freshly baked apple pie and remember Thanksgiving at our grandparents' house, the odor of freshly cooked turkey, sitting by the fire, and other such memories?

For answers, see Answers Appendix.

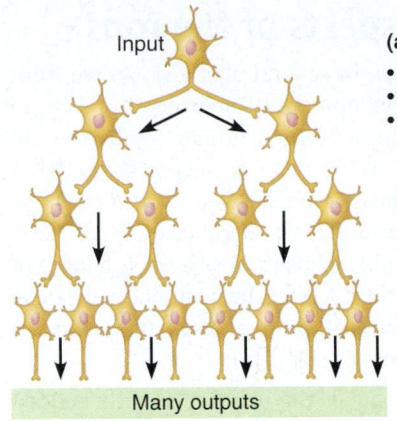

(a) Diverging circuit
- One input, many outputs
- An *amplifying* circuit
- **Example:** A single neuron in the brain can activate 100 or more motor neurons in the spinal cord and thousands of skeletal muscle fibers

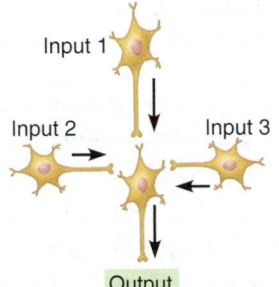

(b) Converging circuit
- Many inputs, one output
- A *concentrating* circuit
- **Example:** Different sensory stimuli can all elicit the same memory

(c) Reverberating circuit
- Signal travels through a chain of neurons, each feeding back to previous neurons
- An *oscillating* circuit
- Controls rhythmic activity
- **Example:** Involved in breathing, sleep-wake cycle, and repetitive motor activities such as walking

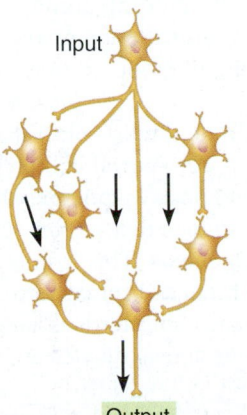

(d) Parallel after-discharge circuit
- Signal stimulates neurons arranged in parallel arrays that eventually converge on a single output cell
- Impulses reach output cell at different times, causing a burst of impulses called an *after-discharge*
- **Example:** May be involved in exacting mental processes such as mathematical calculations

Figure 11.20 Types of circuits in neuronal pools.

Developmental Aspects of Neurons

We cover the nervous system in several chapters, so we limit our attention here to the development of neurons. Let's begin with two questions, How do neurons originate? and How do they mature?

The nervous system originates from a dorsal *neural tube* and the *neural crest*, formed from surface ectoderm (see Figure 12.37 ②, ③, p. 482). The neural tube, whose walls begin as a layer of *neuroepithelial* cells, becomes the CNS. The neuroepithelial cells then begin a three-phase process of differentiation, which occurs largely in the second month of development:

1. *Proliferation*. Mitosis produces the number of cells needed for nervous system development.
2. *Migration*. The resulting **neuroblasts** (future neurons) can no longer divide and move toward their characteristic positions.
3. *Connection*. Once in position, a neuroblast sprouts an axon to connect with its functional target and in so doing becomes a neuron.

How does a neuroblast's growing axon "know" where to go—and once it gets there, where to connect? The growth of an axon toward an appropriate target requires multiple steps and is guided by multiple signals.

The growing tip of an axon, called a **growth cone**, is a prickly, fanlike structure that gives the axon the ability to interact with its environment (**Figure 11.21**). Extracellular and cell surface adhesion proteins such as laminin, integrin, and *nerve cell adhesion molecule (N-CAM)* provide anchor points for the growth cone, saying, "It's okay to grow here." *Neurotropins* are chemicals that signal to the growth cone "come this way" (netrin) or "go away" (ephrin, slit) or "stop here" (semaphorin). Throughout this growth and development, neurotrophic factors such as *nerve growth factor (NGF)* must be present to keep the neuroblast alive.

Failure of any of these guiding signals results in catastrophic developmental problems. For example, lack of N-CAM action causes developing neural tissue to fall into a tangled, spaghetti-like mass and hopelessly impairs neural function.

The growth cone gropes along like an amoeba. Oozing processes called *filopodia* detect the guiding signals in the surrounding environment. Receptors for these signals generate second messengers that cause the filopodia to move by rearranging their actin protein cores.

Once the axon reaches its target area, it must select the right site on the target cell to form a synapse. Special cell adhesion molecules couple the presynaptic and postsynaptic membranes together and generate intracellular signals that recruit vesicles containing preformed synaptic components. This results in the rapid formation of a synapse. In the brain and spinal cord, astrocytes provide physical support and the cholesterol essential for constructing synapses. Both dendrites and astrocytes are active

Figure 11.21 A neuronal growth cone. Fluorescent stains show the locations of cannabinoid receptors (green), tubulin (blue), and actin (pink) in this photomicrograph (1400×).

partners in the process of synapse formation. In the presence of thrombospondin released by astrocytes, dendrites actually reach out and grasp migrating axons, and synapses begin sprouting.

If neurons fail to make appropriate or functional synaptic contacts, they die. Besides cell death resulting from unsuccessful synapse formation, *apoptosis* (programmed cell death) is also a normal part of the developmental process. Of the neurons formed during the embryonic period, perhaps two-thirds die before we are born or shortly thereafter. Those that remain constitute most of our neural endowment for life.

During childhood and adolescence, learning reinforces certain synapses and prunes away others. This is a normal occurrence. However, recent evidence suggests that genes that promote excessive synaptic pruning may predispose an individual to schizophrenia.

The generally amitotic nature of neurons is important because their activity depends on the synapses they've formed. If neurons were to divide, their connections might be hopelessly disrupted. This aside, there *do* appear to be some special neuronal populations where stem cells are found and new neurons can be formed—notably olfactory neurons and some cells of the hippocampus, a brain region involved in learning and memory.

Pleasure Me, Pleasure Me!

Sex! Drugs! Rock 'n' roll! Eat, drink, and be merry! Why do we find these activities so compelling? Our brains are wired to reward us with pleasure when we engage in behavior that is necessary for our own and our species' survival. This reward system consists of neurons that release the neurotransmitter dopamine in areas of the brain called the *ventral tegmental area* (VTA), the *nucleus accumbens*, and the *amygdala*.

Our ability to "feel good" involves brain neurotransmitters in this reward system. For example, the ecstasy of romantic love may be just a brain bath of neurotransmitters such as glutamate and norepinephrine that act on the reward system to release dopamine.

Drugs of abuse work because they are chemical cousins of the neurotransmitters of the reward system and so can subvert this powerful system. For example, people who use "crystal meth" (methamphetamine) artificially stimulate their brains to provide a highly addictive pleasure flush. However, their pleasure is short-lived because when the brain is flooded with neurotransmitter-like chemicals from the outside, it makes less of its own (why bother?) and fewer receptors for it (see photo).

Cocaine, another reward system titillater, has been around since ancient times. Once a toy of the rich, its granular form is inhaled, or "snorted." A cheaper, more potent, smokable form, "crack," is available to the masses. For $50 or so, a novice user can experience a rush of intense pleasure. But crack is treacherous and intensely addictive. It produces not only a higher high than the inhaled form of cocaine, but also a deeper crash that leaves the user desperate for more.

How does cocaine produce its effects? It first stimulates the reward system and then "squeezes it dry." Cocaine produces its rush by hooking up to the dopamine reuptake transporter protein, blocking the reabsorption of dopamine. The neurotransmitter remains in the synapse and stimulates the postsynaptic receptor cells again and again, allowing the body to feel its effects over a prolonged period. This sensation is accompanied by increases in heart rate, blood pressure, and sexual appetite.

As repeated doses of cocaine continue to block dopamine uptake, the body releases less and less dopamine and the reward system effectively goes dry. The cocaine user becomes anxious and, in a very real sense, unable to experience pleasure without the drug. A vicious cycle of addiction begins: The addict craves cocaine, needing it to experience pleasure, but using it suppresses dopamine release even more.

These out-of-control, desperate cravings are notoriously difficult to manage. Drug abusers call it "jonesing." Unfortunately, antiaddiction drugs take so long to reduce the cravings that users commonly drop out of treatment programs.

Many people believe that certain drugs are safe or innocuous. Take, for example, ecstasy. In reality, ecstasy (MDMA) targets serotonin-releasing neurons. The "rush" of pleasure and energy that users feel is due to release of serotonin and other neurotransmitters. However, it damages and may destroy these neurons, causing the loss of verbal and spatial memory. Depression, sleeplessness, and memory problems may be permanent consequences—a steep price for a few moments of pleasure!

People who want pure, effective, and "safe" drugs of abuse don't get them on the street. They get them from doctors.

Even people who would never dream of taking illicit drugs can be caught in the addictive cycle of prescription drugs. Prescribed legitimately to relieve severe pain, OxyContin® (oxycodone) was so widely abused that it had to be reformulated to try to prevent its abuse. Fentanyl, a powerful synthetic opioid, is another drug with legitimate medical uses (for example, for severe cancer pain) that is also widely abused. Because it is powerful, inexpensive, and easy to produce, its use to dilute other, more expensive drugs on the street has become widespread. Its close cousin, the elephant tranquilizer carfentanil, is 10,000 times more potent than morphine. Despite this, drug dealers mix it in with other drugs and sell the mix to unsuspecting customers—often with fatal results. Synthetic opioids have become such a problem that their antidote, the opiate antagonist naloxone, is now carried by many first responders.

The brain is strongly driven to seek out pleasure, but its complex biochemistry always circumvents attempts to keep it in a euphoric haze. Perhaps this means that pleasure must be transient by nature, experienced only against a background of its absence.

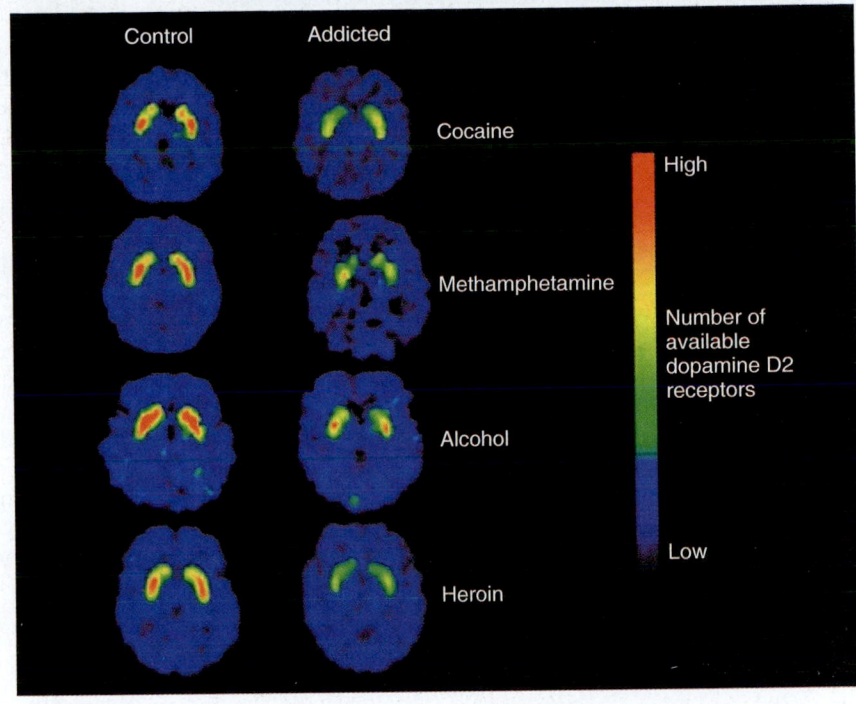

Control   Addicted

Cocaine

Methamphetamine

Alcohol

Heroin

High

Number of available dopamine D2 receptors

Low

RELATED CLINICAL TERMS

Neuroblastoma (nu″ro-blas-to′mah; *oma* = tumor) A malignant tumor in children; arises from cells that retain a neuroblast-like structure. These tumors sometimes arise in the brain, but most occur in the peripheral nervous system.

Neurologist (nu-rol′o-jist) A medical specialist in the study of the nervous system and its functions and disorders.

Neuropathy (nu-rop′ah-the) Any disease of nervous tissue, but particularly degenerative diseases of nerves.

Neuropharmacology (nu″ro-far″mah-kol′o-je) Scientific study of the effects of drugs on the nervous system.

Neurotoxin Substance that is poisonous or destructive to nervous tissue, e.g., botulinum and tetanus toxins.

Rabies (*rabies* = madness) A viral infection of the nervous system transmitted by the bite of an infected mammal (such as a dog, bat, or skunk). After entry, the virus travels via axonal transport in peripheral nerve axons to the CNS, where it causes brain inflammation, delirium, and death. A vaccine- and antibody-based treatment is effective if given before symptoms appear. Rabies in humans is very rare in the United States and Canada.

Shingles (herpes zoster) Inflammation of virally infected sensory neurons serving the skin. Caused by the varicella-zoster virus, which causes chicken pox (generally during childhood); during the initial infection the virus is transported from the skin lesions to the sensory neuron cell bodies in the sensory ganglia. Typically, the immune system holds the virus in check. It remains dormant until the immune system is weakened, often by stress. Then viral particles multiply, causing nerve pain (neuralgia), and travel back to the skin, producing characteristic scaly blisters. This rash is usually confined to one side of the body trunk. Attacks last several weeks, alternating between periods of healing and relapse. Seen mostly in those over 50 years old. Vaccination can prevent occurrence or minimize pain.

CHAPTER SUMMARY

11.1 The nervous system receives, integrates, and responds to information (pp. 391–392)

1. The nervous system bears a major responsibility for maintaining body homeostasis. It monitors, integrates, and responds to information in the environment.
2. The nervous system is divided anatomically into the central nervous system (brain and spinal cord) and the peripheral nervous system (mainly cranial and spinal nerves).
3. The major functional divisions of the PNS are the sensory (afferent) division, which conveys impulses to the CNS, and the motor (efferent) division, which conveys impulses from the CNS.
4. The efferent division includes the somatic (voluntary) system, which serves skeletal muscles, and the autonomic (involuntary) system, which innervates smooth and cardiac muscle and glands.

11.2 Neuroglia support and maintain neurons (pp. 392–394)

1. Neuroglia (supporting cells) segregate and insulate neurons and assist neurons in various other ways.
2. CNS neuroglia include astrocytes, microglial cells, ependymal cells, and oligodendrocytes. PNS neuroglia include Schwann cells (neurolemmocytes) and satellite cells.

11.3 Neurons are the structural units of the nervous system (pp. 394–400)

1. Neurons have a cell body and cytoplasmic processes called axons and dendrites.

Neuron Cell Body (p. 394)

2. The cell body is the biosynthetic (and receptive) center of the neuron. Except for those found in ganglia, cell bodies are found in the CNS.
3. A collection of cell bodies is called a nucleus in the CNS and a ganglion in the PNS.

Neuron Processes (pp. 394–397)

4. Most neurons have many dendrites, receptive processes that conduct signals from other neurons toward the nerve cell body. With few exceptions, all neurons have one axon, which generates and conducts nerve impulses away from the neuron cell body. Axon terminals release neurotransmitter.
5. A long axon is also called a nerve fiber. A bundle of nerve fibers is called a tract in the CNS and a nerve in the PNS.
6. Bidirectional transport along axons uses ATP-dependent motor proteins moving along microtubule tracks. It moves vesicles, mitochondria, and cytosolic proteins toward the axon terminals and conducts substances destined for degradation back to the cell body.
7. Large nerve fibers (axons) are myelinated. The myelin sheath is formed in the PNS by Schwann cells and in the CNS by oligodendrocytes. The myelin sheath gaps are also called nodes of Ranvier. Nonmyelinated fibers are surrounded by supporting cells, but the membrane-wrapping process does not occur.

Classification of Neurons (pp. 397–400)

8. Anatomically, neurons are classified according to the number of processes issuing from the cell body as multipolar, bipolar, or unipolar.
9. Functionally, neurons are classified according to the direction of nerve impulse conduction. Sensory neurons conduct impulses toward the CNS, motor neurons conduct away from the CNS, and interneurons (association neurons) lie between sensory and motor neurons in the neural pathways.

11.4 The resting membrane potential depends on differences in ion concentration and permeability (pp. 400–404)

Basic Principles of Electricity (pp. 400–401)

1. The measure of the potential energy of separated electrical charges is called voltage (V) or potential. Current (I) is the flow of electrical charge from one point to another. Resistance (R)

is hindrance to current flow. Ohm's law gives the relationship among these: $I = V/R$.

2. In the body, ions provide the electrical charges; cellular plasma membranes provide resistance to ion flow. The membranes contain leakage channels (nongated, always open) and gated channels.

Generating the Resting Membrane Potential (pp. 401–403)

3. A resting neuron exhibits a resting membrane potential, which is -70 mV (inside negative). It is due both to differences in sodium and potassium ion concentrations inside and outside the cell and to differences in permeability of the membrane to these ions.

4. The ionic concentration differences result from the operation of the sodium-potassium pump, which ejects 3 Na^+ from the cell for each 2 K^+ transported in.

Changing the Resting Membrane Potential (pp. 403–404)

5. Depolarization is a reduction in membrane potential (inside becomes less negative); hyperpolarization is an increase in membrane potential (inside becomes more negative).

 Complete an interactive tutorial: MasteringA&P® > Study Area > Interactive Physiology > Nervous System I: Resting Membrane Potential

11.5 Graded potentials are brief, short-distance signals within a neuron (pp. 404–405)

1. Graded potentials are small, brief, local changes in membrane potential that act as short-distance signals. The current produced dissipates with distance.

11.6 Action potentials are brief, long-distance signals within a neuron (pp. 405–412)

1. An action potential (AP), or nerve impulse, is a large, but brief, depolarization signal (and polarity reversal) that underlies long-distance neural communication. It is an all-or-none phenomenon.

2. In the AP graph, an AP begins and ends at resting membrane potential. Depolarization to approximately $+30$ mV (inside positive) is caused by Na^+ influx. Depolarization ends when Na^+ channels inactivate. Repolarization and hyperpolarization are caused by K^+ efflux.

 Complete an interactive tutorial: MasteringA&P® > Study Area > Interactive Physiology > Nervous System I: Generation of an Action Potential

3. If threshold is reached, an AP is generated. If not, depolarization remains local.

4. In nerve impulse propagation, each AP provides the depolarizing stimulus for triggering an AP in the next membrane patch. Regions that have just generated APs are refractory; for this reason, the nerve impulse propagates in one direction only.

5. APs are independent of stimulus strength: Strong stimuli cause APs to be generated more frequently but not with greater amplitude.

6. During the absolute refractory period, a neuron cannot respond to another stimulus because it is already generating an AP.

During the relative refractory period, the neuron's threshold is elevated because repolarization is ongoing.

7. In nonmyelinated fibers, APs are produced in a wave all along the axon, that is, by continuous conduction. In myelinated fibers, APs are generated only at myelin sheath gaps and are propagated more rapidly by saltatory conduction.

11.7 Synapses transmit signals between neurons (pp. 412–415)

1. A synapse is a functional junction between neurons. The information-transmitting neuron is the presynaptic neuron; the information-receiving neuron is the postsynaptic neuron.

Electrical Synapses (p. 413)

2. Electrical synapses allow ions to flow directly from one neuron to another; the cells are electrically coupled.

Chemical Synapses (pp. 413–415)

3. Chemical synapses are sites of neurotransmitter release and binding. When the impulse reaches the presynaptic axon terminals, voltage-gated Ca^{2+} channels open, and Ca^{2+} enters the cell and mediates neurotransmitter release. Neurotransmitters diffuse across the synaptic cleft and attach to postsynaptic membrane receptors, opening ion channels. After binding, the neurotransmitters are removed from the synapse by diffusion, enzymatic breakdown, or reuptake into the presynaptic terminal or astrocytes.

11.8 Postsynaptic potentials excite or inhibit the receiving neuron (pp. 416–420)

1. Binding of neurotransmitter at excitatory chemical synapses results in local graded potentials called EPSPs, caused by the opening of channels that allow simultaneous passage of Na^+ and K^+.

2. Neurotransmitter binding at inhibitory chemical synapses results in hyperpolarizations called IPSPs, caused by the opening of K^+ or Cl^- channels. IPSPs drive the membrane potential farther from threshold.

3. EPSPs and IPSPs summate temporally and spatially. The membrane of the axon's initial segment acts as a neuronal integrator.

4. Synaptic potentiation, which enhances the postsynaptic neuron's response, is produced by intense repeated stimulation. Ionic calcium appears to mediate such effects, which may be the basis of learning.

5. Presynaptic inhibition is mediated by axoaxonal synapses that reduce the amount of neurotransmitter released by the inhibited neuron.

11.9 The effect of a neurotransmitter depends on its receptor (pp. 420–425)

1. The major classes of neurotransmitters based on chemical structure are acetylcholine, biogenic amines, amino acids, peptides, purines, dissolved gases, and lipids.

2. Functionally, neurotransmitters are classified as (1) inhibitory or excitatory (or both) and (2) direct or indirect action. Direct-acting neurotransmitters bind to and open ion channels. Indirect-acting neurotransmitters act through second messengers. Neuromodulators also act indirectly presynaptically or postsynaptically to change synaptic strength.

3. Neurotransmitter receptors are either channel-linked receptors that open ion channels, leading to fast changes in membrane

potential, or G protein–coupled receptors that oversee slow synaptic responses mediated by G proteins and intracellular second messengers. Second messengers can act directly on ion channels or activate kinases, which in turn activate or inactivate other proteins, causing a variety of effects.

11.10 Neurons act together, making complex behaviors possible (pp. 426–427)

1. CNS neurons are organized into several types of neuronal pools, each with distinguishing patterns of synaptic connections called circuits.

2. In serial processing, one neuron stimulates the next in sequence, producing specific, predictable responses, as in spinal reflexes. A reflex is a rapid, involuntary motor response to a stimulus.

3. Reflexes are mediated over neural pathways called reflex arcs. The minimum number of elements in a reflex arc is five: receptor, sensory neuron, integration center, motor neuron, and effector.

4. In parallel processing, which underlies complex mental functions, impulses travel along several pathways to different integration centers.

5. The four basic circuit types are diverging, converging, reverberating, and parallel after-discharge.

Developmental Aspects of Neurons (p. 428)

1. Neuron development involves proliferation, migration, and the formation of interconnections. The formation of interconnections involves axons finding their targets and forming synapses, and the synthesis of specific neurotransmitters.

2. Axon outgrowth and synapse formation are guided by other neurons, glial cells, and chemicals (such as N-CAM and nerve growth factor). Neurons that do not make appropriate synapses die, and approximately two-thirds of neurons formed in the embryo undergo programmed cell death.

REVIEW QUESTIONS

To access additional practice questions using your smartphone, tablet, or computer: MasteringA&P® > Study Area > Practice Tests & Quizzes

Level 1 Remember/Understand

(Some questions have more than one correct answer. Select the best answer or answers from the choices given.)

1. Which of the following structures is not part of the central nervous system? **(a)** the brain, **(b)** a nerve, **(c)** the spinal cord, **(d)** a tract.

2. Match the names of the supporting cells found in column B with the appropriate descriptions in column A.

Column A	Column B
_____ **(1)** myelinates nerve fibers in the CNS	**(a)** astrocyte
_____ **(2)** lines brain cavities	**(b)** ependymal cell
_____ **(3)** myelinates nerve fibers in the PNS	**(c)** microglial cell
_____ **(4)** CNS phagocyte	**(d)** oligodendrocyte
_____ **(5)** helps regulate the ionic composition of CNS extra-cellular fluid	**(e)** satellite cell
	(f) Schwann cell

3. Assume that an EPSP is being generated on the dendritic membrane. Which will occur? **(a)** specific Na^+ channels will open, **(b)** specific K^+ channels will open, **(c)** a single type of channel will open, permitting simultaneous flow of Na^+ and K^+, **(d)** Na^+ channels will open first and then close as K^+ channels open.

4. Chemical synapses are characterized by all of the following except **(a)** the release of neurotransmitter by the presynaptic membranes, **(b)** postsynaptic membranes bearing receptors that bind neurotransmitter, **(c)** ions flowing through protein channels from the presynaptic to the postsynaptic neuron, **(d)** a fluid-filled gap separating the neurons.

5. Biogenic amine neurotransmitters include all but **(a)** norepinephrine, **(b)** acetylcholine, **(c)** dopamine, **(d)** serotonin.

6. Neuropeptides that act as natural opiates include **(a)** substance P, **(b)** somatostatin and cholecystokinin, **(c)** tachykinins, **(d)** enkephalins.

7. Inhibition of acetylcholinesterase by poisoning blocks neurotransmission at the neuromuscular junction because **(a)** ACh is no longer released by the presynaptic terminal, **(b)** ACh synthesis in the presynaptic terminal is blocked, **(c)** ACh is not degraded, so prolonged depolarization is enforced on the postsynaptic cell, **(d)** ACh is blocked from attaching to the postsynaptic ACh receptors.

8. The anatomical region of a multipolar neuron where the AP is initiated is the **(a)** soma, **(b)** dendrites, **(c)** axon's initial segment, **(d)** axon terminals.

9. An IPSP is inhibitory because **(a)** it hyperpolarizes the postsynaptic membrane, **(b)** it reduces the amount of neurotransmitter released by the presynaptic terminal, **(c)** it prevents calcium ion entry into the presynaptic terminal, **(d)** it changes the threshold of the neuron.

10. Identify the neuronal circuits described by choosing the correct response from the key.

    Key: **(a)** converging   **(c)** parallel after-discharge
    　　　　(b) diverging   **(d)** reverberating

    _____ **(1)** Impulses continue around and around the circuit until one neuron stops firing.
    _____ **(2)** One or a few inputs ultimately influence large numbers of neurons.
    _____ **(3)** Many neurons influence a few neurons.
    _____ **(4)** May be involved in exacting types of mental activity.

11. **(a)** What is myelin? **(b)** How does the myelination process differ in the CNS and PNS?

12. Since all APs generated by a given nerve fiber have the same magnitude, how does the CNS "know" whether a stimulus is strong or weak?

13. During a neurobiology lecture, a professor repeatedly refers to group A and group B fibers, absolute refractory period, and myelin sheath gaps. Define these terms.

14. Briefly describe the three stages of neuron development.

Level 2 Apply/Analyze

15. The velocity of nerve impulse conduction is greatest in **(a)** heavily myelinated, large-diameter fibers, **(b)** myelinated, small-diameter fibers, **(c)** nonmyelinated, small-diameter fibers, **(d)** nonmyelinated, large-diameter fibers.

16. **(a)** Describe the composition and function of the cell body. **(b)** How are axons and dendrites alike? In what ways (structurally and functionally) do they differ?

17. **(a)** Contrast unipolar, bipolar, and multipolar neurons structurally. **(b)** Indicate where each is most likely to be found.
18. What is the polarized membrane state? How is it maintained? (Note the relative roles of both passive and active mechanisms.)
19. Describe the events that must occur to generate an AP. Relate the sequence of changes in permeability to changes in the ion channels, and explain why the AP is an all-or-none phenomenon.
20. **(a)** Explain the difference between an EPSP and an IPSP. **(b)** What specifically determines whether an EPSP or IPSP will be generated at the postsynaptic membrane?
21. Since at any moment a neuron is likely to have thousands of neurons releasing neurotransmitters at its surface, how is neuronal activity (to fire or not to fire) determined?
22. The effects of neurotransmitter binding are very brief. Explain.
23. When admitted to the emergency room, Sean was holding his right hand, which had a deep puncture hole in its palm. He explained that he had fallen on a nail while exploring a barn. Sean was given an antitetanus shot to prevent neural complications. Tetanus bacteria fester in deep, dark wounds, but how do their toxins travel in neural tissue?

Level 3 Evaluate/Synthesize

24. Explain both the anatomical and functional divisions of the nervous system. Include the subdivisions of each.
25. Local anesthetics block voltage-gated Na^+ channels. General anesthetics are thought to activate chemically gated Cl^- channels, thereby rendering the nervous system quiescent while surgery is performed. What specific process do anesthetics impair, and how does this interfere with nerve impulse transmission?
26. Rochelle developed multiple sclerosis when she was 27. After eight years she had lost a good portion of her ability to control her skeletal muscles. How did this happen?
27. In the Netherlands a young man named Jan was admitted to the emergency room. He and his friends had been to a rave. His friends say he started twitching and having muscle spasms which progressed until he was "stiff as a board." On examination, staff found a marked increase in muscle tone and hyperreflexia involving facial and limb muscles. In his pocket, he had unmarked dark yellow tablets with dark flecks. Analysis of the tablets showed them to contain a mixture of ecstasy and strychnine. Ecstasy would not cause this clinical picture, but strychnine, which blocks glycine receptors, could. Explain how.

CLINICAL CASE STUDY

Nursing Student with Neuropathic Pain

Tamara Costa broke her right tibia and has undergone two separate surgeries to repair it. Although the bone has healed, she suffers increasing pain around the incision sites. The painful area covers the lateral surface of her right leg. She can't stand wearing anything over it or even having a sheet touch it. Her diagnosis is postsurgical neuropathic pain. Pain medication has not helped.

Tamara is a second-year nursing student and has done some reading to try to understand her problem and perhaps find a solution. She has found an article that says neuropathic pain may be caused by a decreased threshold for action potential generation in pain-detecting neurons. However, she has forgotten some of her physiology and needs some help in understanding what she is reading. She has logged into the "Ask a Nurse" chat line—and she got you!

1. **+ NCLEX-STYLE** Tamara asks about action potential thresholds, so you begin by asking her what she knows about the topic. Which statement indicates an area where she needs further instruction?
   a. "Threshold is the voltage at which the nerve actually fires an action potential."
   b. "Sodium is entering the nerve cell at threshold, but it was entering before too."
   c. "They call it threshold when the pain gets so bad the painkillers won't stop it, but I always have that!"
   d. "I know that what happens at threshold is called depolarization."

2. Tamara wants to know how a decrease in threshold in pain-detecting neurons would affect their firing. Draw a graph of an intracellular recording of a neuron showing the response to a below-threshold and an above-threshold stimulus. Then, using this graph, show how decreasing the threshold could lead to additional action potentials.

3. The paper Tamara read claims that drugs that block voltage-gated sodium channels can help with neuropathic pain. Explain how this might work.

4. **+ NCLEX-STYLE** The article suggests that it might also be possible that Tamara's pain is due to the lack of the usual inhibition of transmission of pain signals through Tamara's spinal cord. Which of these statements best describes how a postsynaptic neuron is inhibited by a presynaptic neuron?
   a. The frequency of action potentials in both the presynaptic and postsynaptic neurons decreases.
   b. The presynaptic neuron becomes more negatively charged when it fires instead of becoming more positively charged.
   c. The presynaptic neuron stops releasing its neurotransmitter.
   d. The neurotransmitter from the presynaptic neuron causes a local hyperpolarization in the postsynaptic neuron, which reduces the likelihood that the postsynaptic neuron fires an action potential.

5. You've found a copy of the same article, and see that it identifies other ways pain inhibition could decrease. Explain each of them:
   a. A decrease in the number of receptors for the inhibitory neurotransmitter GABA.
   b. A local decrease in the amount of Cl^- outside the cell. (Hint: GABA opens Cl^- channels, allowing Cl^- influx into the postsynaptic cell.)

11

12

The Central Nervous System

In this chapter, you will learn that

The central nervous system consists of the brain and spinal cord

by first exploring

The brain

then exploring

12.10 The spinal cord

by asking

12.1 How does embryonic development explain adult brain structure?

and considering

12.7 Higher mental functions

and asking

12.8 How is the brain protected?

and

12.9 What happens when things go wrong?

then examining

12.2 Cerebral hemispheres

12.3 Diencephalon

12.4 Brain stem

12.5 Cerebellum

12.6 Functional brain systems

together form

12.11 Neuronal pathways

communicate with

Peripheral nervous system Ch. 13

and finally, exploring

Developmental Aspects of the Central Nervous System

Interconnecting and directing a dizzying number of incoming and outgoing calls, the central switchboard of a telephone system historically seemed an apt comparison for the **central nervous system (CNS)**—the brain and spinal cord. Today, many people compare the CNS to a networked cloud of computers. These analogies may explain some workings of the spinal cord, but neither does justice to the fantastic complexity of the human brain—it is one of the most amazing things known.

During the course of animal evolution, **cephalization** (sĕ"fah-lĭ-za'shun) has occurred. That is, there has been an elaboration of the *rostral* ("toward the snout"), or anterior, portion of the CNS, along with an increase in the number of neurons in the head.

In this chapter, we examine the structure of the CNS and the functions associated with its various regions. We also touch on complex integrative functions, such as sleep-wake cycles and memory.

The unimpressive appearance of the human **brain** gives few hints of its remarkable abilities. It is about two good fistfuls of quivering pinkish gray tissue, wrinkled like a walnut, with a consistency somewhat like cold oatmeal. The average adult human brain has a mass of about 1500 g (3.3 lb).

12.1 Folding during development determines the complex structure of the adult brain

Learning Outcomes

▶ **Describe how space constraints affect brain development.**
▶ **Name the major regions of the adult brain.**
▶ **Name and locate the ventricles of the brain.**

We begin with an introduction to brain embryology, as the terminology used for the structural divisions of the adult brain is easier to follow when you understand brain development.

The brain and spinal cord begin as an embryonic structure called the **neural tube** (**Figure 12.1a**). As soon as the neural

tube forms, its anterior (rostral) end begins to expand and constrictions appear that mark off the three **primary brain vesicles** (Figure 12.1b):

- **Prosencephalon** (pros"en-sef'ah-lon), or **forebrain**
- **Mesencephalon** (mes"en-sef'ah-lon), or **midbrain**
- **Rhombencephalon** (romb"en-sef'ah-lon), or **hindbrain**

(*Encephalo* means "brain.") The remaining *caudal* ("toward the tail"), or posterior, portion of the neural tube becomes the spinal cord, which we will discuss later in the chapter.

The primary vesicles give rise to the **secondary brain vesicles** (Figure 12.1c). The forebrain divides into the **telencephalon** ("endbrain") and **diencephalon** ("interbrain"), and the hindbrain constricts, forming the **metencephalon** ("afterbrain") and **myelencephalon** ("spinal brain"). The midbrain remains undivided.

Each of the five secondary vesicles then develops rapidly to produce the major structures of the adult brain (Figure 12.1d). The greatest change occurs in the telencephalon, which sprouts two lateral swellings that look like Mickey Mouse's ears. These become the two *cerebral hemispheres*, referred to collectively as the **cerebrum** (ser'ĕ-brum). The diencephalon specializes to form the *hypothalamus* (hi"po-thal'ah-mus), *thalamus*, *epithalamus*, and *retina* of the eye. Less dramatic changes occur in the mesencephalon, metencephalon, and myelencephalon as these regions transform into the *midbrain*, the *pons* and *cerebellum*, and the *medulla oblongata*, respectively. All these midbrain and hindbrain structures, except the cerebellum, form the **brain stem**.

The central cavity of the neural tube remains continuous and enlarges in four areas to form the fluid-filled *ventricles* (*ventr* = little belly) of the brain (Figure 12.1e). We will describe the ventricles shortly.

(a) Neural tube (contains neural canal)	(b) Primary brain vesicles	(c) Secondary brain vesicles	(d) Adult brain structures	(e) Adult neural canal regions
Anterior (rostral)	Prosencephalon (forebrain)	Telencephalon	Cerebrum: cerebral hemispheres (cortex, white matter, basal nuclei)	Lateral ventricles
		Diencephalon	Diencephalon (thalamus, hypothalamus, epithalamus), retina	Third ventricle
	Mesencephalon (midbrain)	Mesencephalon	Brain stem: midbrain	Cerebral aqueduct
	Rhombencephalon (hindbrain)	Metencephalon	Brain stem: pons	Fourth ventricle
			Cerebellum	
Posterior (caudal)		Myelencephalon	Brain stem: medulla oblongata	
			Spinal cord	Central canal

Figure 12.1 Embryonic development of the human brain. (a) Formed by week 4, the neural tube quickly subdivides into **(b)** the primary brain vesicles, which subsequently form **(c)** the secondary brain vesicles by week 5. These five vesicles differentiate into **(d)** the adult brain structures. **(e)** The adult structures derived from the neural canal.

Because the brain grows more rapidly than the membranous skull that contains it, it folds up to occupy the available space. The *midbrain* and *cervical flexures* move the forebrain toward the brain stem (**Figure 12.2a**). The cerebral hemispheres are forced to take a horseshoe-shaped course and grow posteriorly and laterally (indicated by black arrows in Figure 12.2b). As a result, they grow back over and almost completely envelop the diencephalon and midbrain. By week 26, the continued growth

Anterior (rostral)   Posterior (caudal)

Metencephalon
Mesencephalon
Diencephalon   Midbrain
                                                                        Flexures
Telencephalon   Cervical

Myelencephalon   Spinal cord

(a) Week 5: Two major flexures form, causing the telencephalon and diencephalon to angle toward the brain stem.

Cerebral hemisphere
Outline of diencephalon
Midbrain
Cerebellum
Pons
Medulla oblongata
Spinal cord

(b) Week 13: Cerebral hemispheres develop and grow posterolaterally (black arrows) to enclose the diencephalon and the rostral brain stem.

Cerebral hemisphere

Diencephalon

Cerebellum

Brain stem
• Midbrain
• Pons
• Medulla oblongata

(c) Birth: Shows adult pattern of structures and convolutions.

Figure 12.2 Brain development. Initially, the cerebral surface is smooth. Folding begins in month 6, and convolutions become more obvious as development continues. See-through view in (b) and (c).

of the cerebral hemispheres causes their surfaces to crease and fold into *convolutions* (Figure 12.2c), which increases their surface area and allows more neurons to occupy the limited space.

Brain Regions and Organization

Some textbooks discuss brain anatomy in terms of the *embryonic scheme* (see Figure 12.1c), but in this text, we will consider the brain in terms of the medical scheme and the four adult brain regions shown in Figure 12.2c: (1) cerebral hemispheres, (2) diencephalon, (3) brain stem (midbrain, pons, and medulla oblongata), and (4) cerebellum.

A slice of brain tissue reveals patterns of darker and lighter areas called gray matter and white matter:

• **Gray matter** consists of short, nonmyelinated neurons and neuron cell bodies.

• **White matter** consists mostly of myelinated axons with some nonmyelinated axons, primarily in fiber tracts. The dense coating of fatty myelin is what gives white matter its color.

Gray and white matter have a unique distribution in the brain. The basic pattern of the CNS is a central cavity surrounded by gray matter, external to which is white matter. As shown in **Figure 12.3**:

① The spinal cord exhibits this basic pattern. This pattern changes with ascent into the brain stem.

② The brain stem has additional gray matter nuclei scattered within the white matter.

③ The cerebral hemispheres and the cerebellum have an outer layer or "bark" of gray matter called a *cortex*.

Knowledge of the basic pattern of the CNS will help you explore the brain, moving from the most rostral region (cerebrum) to the most caudal (brain stem). But first, let's explore the central hollow cavities that lie deep within the brain—the ventricles.

Ventricles

The brain **ventricles** are continuous with one another and with the central canal of the spinal cord (**Figure 12.4**). The hollow ventricular chambers are filled with cerebrospinal fluid and lined by *ependymal cells*, a type of neuroglia (◀ **Figure 11.4c, p. 393**).

The paired **lateral ventricles**, one deep within each cerebral hemisphere, are large C-shaped chambers that reflect the pattern of cerebral growth. Anteriorly, the lateral ventricles lie close together, separated only by a thin median membrane called the **septum pellucidum** (pĕ-lu′sid-um; "transparent wall"). (See Figure 12.11a, p. 447.)

Each lateral ventricle communicates with the narrow **third ventricle** in the diencephalon via a channel called an **interventricular foramen**.

The third ventricle is continuous with the **fourth ventricle** via the canal-like **cerebral aqueduct** that runs through the midbrain. The fourth ventricle lies in the hindbrain dorsal to the pons and superior medulla. It is continuous with the central canal of the spinal cord inferiorly. Three openings mark the walls of the fourth ventricle: the paired **lateral apertures** in its side walls and the **median aperture** in its roof. These apertures

Central cavity
(spinal canal)   Inner gray
matter

Central cavity
(lateral ventricle)

White matter

Gray matter
(brain nuclei)

Cortex of
gray matter

Outer white
matter

① The **spinal cord** exhibits the
basic pattern: a central cavity
surrounded by gray matter, with
outer white matter.

③ The **cerebrum** (shown here) and **cerebellum**
have islands of gray matter (nuclei) within white
matter, as well as an outer cortex of gray matter.

Central cavity (4th ventricle)

Inner gray matter

Outer white matter

Gray matter:
• Neuron cell bodies
• Short nonmyelinated neurons
White matter:
• Mostly myelinated axons (with some nonmyelinated axons)

② The **brain stem** has additional
regions of gray matter (brain nuclei)
buried within the white matter.

12

Figure 12.3 Pattern of distribution of gray and white matter in the CNS.

Lateral ventricle
• Anterior horn

• Posterior horn

• Inferior horn

Septum pellucidum

Interventricular foramen

Interventricular foramen

Third ventricle

Third ventricle

Cerebral aqueduct

Lateral aperture

Fourth ventricle

Lateral aperture

Median aperture

Central canal

(a) Left lateral view

(b) Anterior view

Figure 12.4 Ventricles of the brain. Different regions of the large lateral ventricles are
labeled anterior horn, posterior horn, and inferior horn.

connect the ventricles to the *subarachnoid space* (sub"ah-rak'noid), a fluid-filled space surrounding the brain.

Check Your Understanding

1. Which ventricle is surrounded by the diencephalon?

2. Which two areas of the adult brain have an outside layer of gray matter in addition to central gray matter and surrounding white matter?

3. What is the function of convolutions of the brain?

For answers, see Answers Appendix.

(a) Superior view

(b) Left lateral view

(c) Lobes and sulci of the cerebrum

(d) Location of the insula (or insular lobe)

Figure 12.5 Lobes, sulci, and fissures of the cerebral hemispheres.

12.2 The cerebral hemispheres consist of cortex, white matter, and the basal nuclei

Learning Outcomes

▶ List the major lobes, fissures, and functional areas of the cerebral cortex.

▶ Explain lateralization of cortical function.

▶ Differentiate between commissures, association fibers, and projection fibers.

▶ Describe the general function of the basal nuclei (basal ganglia).

The **cerebral hemispheres** form the superior part of the brain (**Figure 12.5**) and are the most conspicuous parts of an intact brain. Together they account for about 83% of total brain mass. Picture how a mushroom cap covers the top of its stalk, and you have a good idea of how the paired cerebral hemispheres cover and obscure the diencephalon and the top of the brain stem (see Figure 12.2c).

Elevated ridges of tissue called **gyri** (ji′ri; singular: *gyrus;* "twisters")—separated by shallow grooves called **sulci** (sul′ki; singular: *sulcus;* "furrows")—mark nearly the entire surface of the cerebral hemispheres. Deeper grooves, called **fissures**, separate large regions of the brain (Figure 12.5c):

- The median **longitudinal fissure** separates the cerebral hemispheres (Figure 12.5a).

- The **transverse cerebral fissure** separates the cerebral hemispheres from the cerebellum below (Figure 12.5b, c).

The more prominent gyri and sulci are also important anatomical landmarks that are similar in all humans. Several sulci divide each hemisphere into five lobes—frontal, parietal, temporal, occipital, and insula (Figure 12.5c, d). All but the last are named for the cranial bones that overlie them (◀ **Figure 7.5, pp. 204–205**).

- The **central sulcus**, which lies in the frontal plane, separates the **frontal lobe** from the **parietal lobe**. Bordering the central sulcus are the **precentral gyrus** anteriorly and the **postcentral gyrus** posteriorly.

- The **parieto-occipital sulcus** (pah-ri″ĕ-to-ok-sip′ĭ-tal), located more posteriorly on the medial surface of the hemisphere, separates the **occipital lobe** from the parietal lobe.

- The deep **lateral sulcus** outlines the flaplike **temporal lobe** and separates it from the parietal and frontal lobes. A fifth lobe of the cerebral hemisphere, the **insula** (in′su-lah; "island"), is buried deep within the lateral sulcus and forms part of its floor (Figure 12.5d). The insula is covered by portions of the temporal, parietal, and frontal lobes.

The cerebral hemispheres fit snugly in the skull. The frontal lobes lie in the anterior cranial fossa, and the anterior parts of the temporal lobes fill the middle cranial fossa (◀ **Figure 7.2b, c, p. 201**). The posterior cranial fossa, however, houses the brain stem and cerebellum. The occipital lobes are located well superior to that cranial fossa.

Each of the cerebral hemispheres has three basic regions (Figure 12.3 ③):

- A superficial *cerebral cortex* of gray matter, which looks gray in fresh brain tissue

- Internal *white matter*

- *Basal nuclei*, islands of gray matter situated deep within the white matter

We consider these regions next.

Cerebral Cortex

The **cerebral cortex** is the "executive suite" of the nervous system, where our *conscious mind* is found. It enables us to be aware of ourselves and our sensations, to communicate, remember, understand, and initiate voluntary movements.

The cerebral cortex is composed of gray matter: neuron cell bodies, dendrites, associated glia and blood vessels, but no fiber tracts. It contains billions of neurons arranged in six layers. Although only 2–4 mm (about 1/8 inch) thick, it accounts for roughly 40% of total brain mass. Its many convolutions effectively triple its surface area.

Modern imaging techniques allow us to see the brain in action—PET scans show maximal metabolic activity in the brain, and functional MRI scans reveal blood flow (**Figure 12.6**). They show that specific motor and sensory functions are localized in discrete cortical areas called *domains*. However, many higher mental functions, such as memory and language, appear to be spread over large areas of the cortex in overlapping domains.

Before we examine the functional regions of the cerebral cortex, let's consider four generalizations:

Figure 12.6 Functional neuroimaging (fMRI) of the cerebral cortex. Red and orange areas show increased blood flow in the cortical regions responsible for carrying out each activity.

Motor areas

Primary motor cortex

Premotor cortex

Frontal eye field

Broca's area (outlined by dashes)

Anterior association area (prefrontal cortex)

Working memory for spatial tasks

Executive area for task management

Working memory for object-recall tasks

Solving complex, multitask problems

Central sulcus

Sensory areas and related association areas

Primary somatosensory cortex ⎤

Somatosensory association cortex ⎦ Somatic sensation

Gustatory cortex (in insula) — Taste

Vestibular cortex (in insula) — Equilibrium

Wernicke's area (outlined by dashes within the **posterior association area**)

Primary visual cortex ⎤

Visual association area ⎦ Vision

Auditory association area ⎤

Primary auditory cortex ⎦ Hearing

(a) Lateral view, left cerebral hemisphere

Premotor cortex

Corpus callosum

Frontal eye field

Prefrontal cortex

Cingulate gyrus

Primary motor cortex

Central sulcus

Primary somatosensory cortex

Parietal lobe

Somatosensory association cortex

Parieto-occipital sulcus

Processes emotions related to personal and social interactions

Orbitofrontal cortex

Olfactory bulb

Olfactory tract

Fornix

Temporal lobe

Primary olfactory cortex

Uncus

Parahippocampal gyrus

Calcarine sulcus

Occipital lobe

Visual association area

Primary visual cortex

(b) Parasagittal view, right cerebral hemisphere

🟥 Primary motor cortex 🟧 Motor association cortex 🟪 Primary sensory cortex 🟦 Sensory association cortex 🟪 Multimodal association cortex

Figure 12.7 Functional and structural areas of the cerebral cortex.

- The cerebral cortex contains three kinds of functional areas: *motor areas*, *sensory areas*, and *association areas*. As you read about these areas, do not confuse the sensory and motor areas of the cortex with sensory and motor neurons. All neurons in the cortex are interneurons.

- Each hemisphere is chiefly concerned with the sensory and motor functions of the **contralateral** (opposite) side of the body.

- Although largely symmetrical in structure, the two hemispheres are not entirely equal in function. Instead, there is *lateralization* (specialization) of cortical functions.

- And finally, keep in mind that our approach is a gross oversimplification. No functional area of the cortex acts alone, and conscious behavior involves the entire cortex in one way or another.

Motor Areas

The following **motor areas** of the cortex, which control voluntary movement, lie in the posterior part of the frontal lobes: primary motor cortex, premotor cortex, Broca's area, and the frontal eye field (**Figure 12.7a**, dark and light red areas).

Primary Motor Cortex The **primary (somatic) motor cortex** is located in the precentral gyrus of the frontal lobe of each hemisphere (Figure 12.7, dark red area). Large neurons, called **pyramidal cells**, in these gyri allow us to consciously control the precise or skilled voluntary movements of our skeletal muscles. Their long axons, which project to the spinal cord, form the massive voluntary motor tracts called *pyramidal tracts* or *corticospinal tracts* (kor″tĭ-ko-spi′nal). All other descending motor tracts come from brain stem nuclei and consist of chains of two or more neurons.

The entire body is represented spatially in the primary motor cortex of each hemisphere. For example, the pyramidal cells that control foot movements are in one place and those that control hand movements are in another. Such mapping of the body in CNS structures is called **somatotopy** (so″mah-to-to′pe).

As illustrated in **Figure 12.8** (p. 442), the body is represented upside down—with the head at the inferolateral part of the precentral gyrus, and the toes at the superomedial end. Most of the neurons in these gyri control muscles in body areas having the most precise motor control—that is, the face, tongue, and hands. Consequently, these regions are disproportionately large in the caricature-like **motor homunculus** (ho-mung′ku-lus; "little man") in Figure 12.8. The motor innervation of the body is contralateral: In other words, the left primary motor gyrus controls muscles on the right side of the body, and vice versa.

The motor homunculus view of the primary motor cortex (Figure 12.8, left) implies a one-to-one correspondence between cortical neurons and the muscles they control, but this is somewhat misleading. In fact, a given muscle is controlled by multiple spots on the cortex, and individual cortical neurons actually send impulses to more than one muscle. In other words, individual pyramidal motor neurons control muscles that work together in a synergistic way to perform a given movement.

For example, reaching forward with one arm involves some muscles acting at the shoulder and some acting at the elbow.

Instead of the discrete map offered by the motor homunculus, the primary motor cortex map is an orderly but fuzzy map with neurons arranged in useful ways to control and coordinate sets of muscles. Neurons controlling the arm, for instance, intermingle and overlap with those controlling the hand and shoulder. However, neurons controlling unrelated movements, such as those controlling the arm and those controlling body trunk muscles, do not cooperate in motor activity.

Thus, the motor homunculus is useful to show that broad areas of the primary cortex are devoted to the leg, arm, torso, and head, but neuron organization within those broad areas is much more diffuse than initially imagined.

Premotor Cortex Just anterior to the precentral gyrus in the frontal lobe is the **premotor cortex** (Figure 12.7, light red area). The premotor cortex helps plan movements. This region selects and sequences basic motor movements into more complex tasks, such as playing a musical instrument or typing. Using highly processed sensory information received from other cortical areas, it can control voluntary actions that depend on sensory feedback, such as feeling for a light switch in a dark room.

The premotor cortex coordinates the movement of several muscle groups either simultaneously or sequentially, mainly by sending activating impulses to the primary motor cortex. However, the premotor cortex also influences motor activity more directly by supplying about 15% of pyramidal tract fibers. Think of this region as the staging area for skilled motor activities.

Broca's Area **Broca's area** (bro′kahz) lies anterior to the inferior region of the premotor area (Figure 12.7a). It has long been considered to be (1) present in one hemisphere only (usually the left) and (2) a special *motor speech area* that directs the muscles involved in speech production. However, imaging studies indicate that Broca's area also becomes active as we prepare to speak and even as we think about (plan) many voluntary motor activities other than speech.

Frontal Eye Field The **frontal eye field** is located partially in and anterior to the premotor cortex and superior to Broca's area (Figure 12.7a). This cortical region controls voluntary movement of the eyes.

HOMEOSTATIC IMBALANCE 12.1 **CLINICAL**

Damage to localized areas of the *primary motor cortex* (as from a stroke) paralyzes the body muscles controlled by those areas. If the lesion is in the right hemisphere, the left side of the body will be paralyzed. Only *voluntary* control is lost, however, as the muscles can still contract reflexively.

Destruction of the *premotor cortex*, or part of it, results in loss of the motor skill(s) programmed by that region, but does not impair muscle strength and the ability to perform the discrete individual movements. For example, if the premotor area controlling the flight of your fingers over a computer keyboard were damaged, you couldn't type with your usual speed, but you could still make the same movements with your fingers. Reprogramming the skill into another set of premotor neurons would require practice, just as the initial learning process did.

Figure 12.8 **Body maps in the primary motor cortex and somatosensory cortex of the cerebrum.** The relative amount and location of cortical tissue devoted to each function is proportional to the distorted body diagrams (homunculi).

Sensory Areas

Areas concerned with conscious awareness of sensation, the sensory areas of the cortex, occur in the parietal, insular, temporal, and occipital lobes (see Figure 12.7, dark and light blue areas).

Primary Somatosensory Cortex The **primary somatosensory cortex** resides in the postcentral gyrus of the parietal lobe, just posterior to the primary motor cortex. Neurons in this gyrus receive information from the general (somatic) sensory receptors in the skin and from proprioceptors. [**Proprioceptors** (pro″pre-o-sep′torz; *propria* = one's own), located in skeletal muscles, joints, and tendons, inform the brain of the body's position in space.] The neurons then identify the body region being stimulated, an ability called **spatial discrimination**.

As with the primary motor cortex, the body is represented spatially and upside down according to the site of stimulus input, and the right hemisphere receives input from the left side of the body. The amount of sensory cortex devoted to a particular body region is related to that region's sensitivity (that is,

to how many receptors it has), not its size. In humans, the face (especially the lips) and fingertips are the most sensitive body areas, so these regions are the largest parts of the **somatosensory homunculus** (Figure 12.8, right).

Somatosensory Association Cortex The **somatosensory association cortex** lies just posterior to the primary somatosensory cortex and has many connections with it. The major function of this area is to integrate sensory inputs (temperature, pressure, and so forth) relayed to it via the primary somatosensory cortex to produce an understanding of an object being felt: its size, texture, and the relationship of its parts.

For example, when you reach into your pocket, your somatosensory association cortex draws upon stored memories of past sensory experiences to perceive the objects you feel as coins or keys. Someone with damage to this area could not recognize these objects without looking at them.

Visual Areas The **primary visual (striate) cortex** is seen on the extreme posterior tip of the occipital lobe, but most of it is buried deep in the *calcarine sulcus* in the medial aspect of the occipital lobe (Figure 12.7b). The primary visual cortex is the largest cortical sensory area. It receives visual information that originates on the retina of the eye. The visual space on the opposite side of the body is mapped to the primary visual cortex just like the body surface is mapped onto the somatosensory cortex.

The **visual association area** surrounds the primary visual cortex and covers much of the occipital lobe. Communicating with the primary visual cortex, the visual association area uses past visual experiences to interpret visual stimuli (color, form, and movement), enabling us to recognize a flower or a person's face and to appreciate what we are seeing. We do our "seeing" with these cortical neurons. However, complex visual processing involves the entire posterior half of the cerebral hemispheres.

Auditory Areas Each **primary auditory cortex** is located in the superior margin of the temporal lobe next to the lateral sulcus. Sound energy exciting the hearing receptors of the inner ear causes impulses to be transmitted to the primary auditory cortex, where they are interpreted as pitch, loudness, and location.

The more posterior **auditory association area** then permits the perception of the sound stimulus, which we "hear" as speech, a scream, music, thunder, and so on. Memories of sounds heard in the past appear to be stored here for reference. Wernicke's area, which we describe later, includes parts of the auditory cortex.

Vestibular (Equilibrium) Cortex Imaging studies show that the part of the cortex responsible for conscious awareness of balance (the position of the head in space) is located in the posterior part of the insula and adjacent parietal cortex (see Figure 12.7a).

Olfactory Cortex The **primary olfactory (smell) cortex** lies on the medial aspect of the temporal lobe in a small region called the *piriform lobe* which is dominated by the hooklike *uncus* (Figure 12.7b). Afferent fibers from smell receptors in the superior nasal cavity send impulses along the olfactory tracts that are ultimately relayed to the olfactory cortices. The outcome is conscious awareness of different odors.

The olfactory cortex is part of the primitive **rhinencephalon** (ri″nen-sef′ah-lon; "nose brain"), which includes all parts of the cerebrum that receive olfactory signals—the orbitofrontal cortex, uncus, and associated regions located on or in the medial aspects of the temporal lobes, and the protruding olfactory tracts and bulbs that extend to the nose. During the course of evolution, most of the "old" rhinencephalon has taken on new functions concerned chiefly with emotions and memory. It has become part of the "newer" emotional brain, called the *limbic system*, which we will consider later in this chapter. This explains why certain odors can evoke strong emotions. The only portions of the human rhinencephalon still devoted to smell are the olfactory bulbs and tracts (described in Chapter 13) and the relatively small olfactory cortices.

Gustatory Cortex The **gustatory (taste) cortex** (gus′tah-tor-e), a region involved in perceiving taste stimuli, is located in the insula just deep to the temporal lobe (Figure 12.7a).

Visceral Sensory Area The cortex of the insula just posterior to the gustatory cortex is involved in conscious perception of visceral sensations. These include upset stomach, full bladder, and the feeling that your lungs will burst when you hold your breath too long.

HOMEOSTATIC IMBALANCE 12.2   CLINICAL

Damage to the *primary visual cortex* (Figure 12.7) results in functional blindness. Although unable to consciously see, individuals with this damage may still have certain visual reflexes. By contrast, individuals with a damaged visual association area can see, but they do not comprehend what they are looking at. ●●●●●●●

Multimodal Association Areas

The association areas that we have considered so far (light red or light blue in Figure 12.7) have all been tightly tied to one kind of primary motor or sensory cortex (dark red or dark blue). Most of the cortex, though, consists of complexly connected **multimodal association areas** (light violet in Figure 12.7) that receive inputs from multiple senses and send outputs to multiple areas.

In general, information flows as follows:

sensory receptors
↓
primary sensory cortex
↓
sensory association cortex
↓
multimodal association cortex

Multimodal association cortex allows us to give meaning to the information that we receive, store it in memory, tie it to previous experience and knowledge, and decide what action to take. Those decisions are relayed to the premotor cortex, which in turn communicates with the motor cortex. The multimodal association cortex seems to be where sensations, thoughts, and emotions become conscious. It is what makes us who we are.

Suppose, for example, you drop a bottle of acid in the chem lab and it splashes on you. You see the bottle shatter, hear the crash, feel your skin burning, and smell the acid fumes. These individual perceptions come together in the multimodal association areas. Along with feelings of panic, these perceptions are woven into a seamless whole, which (hopefully) recalls instructions about what to do in this situation. As a result your premotor and primary motor cortices direct your legs to propel you to the safety shower.

The multimodal association areas can be broadly divided into three parts: the anterior association, posterior association, and limbic association areas.

Anterior Association Area The **anterior association area** in the frontal lobe, also called the **prefrontal cortex**, is the most complicated cortical region of all (Figure 12.7). It is involved with intellect, complex learning abilities (called cognition), recall, and personality. It contains working memory, which is necessary for abstract ideas, judgment, reasoning, persistence, and planning. These abilities develop slowly in children, which implies that the prefrontal cortex matures slowly and depends heavily on feedback from our social environment.

Posterior Association Area   The **posterior association area** is a large region encompassing parts of the temporal, parietal, and occipital lobes. This area plays a role in recognizing patterns and faces, localizing us and our surroundings in space, and binding different sensory inputs into a coherent whole. In the spilled acid example above, your awareness of the entire scene originates from this area. Attention to an area of space or an area of one's own body is also a function of this part of the brain. Many parts of the posterior association area (including Wernicke's area, Figure 12.7a) are also involved in understanding written and spoken language.

Limbic Association Area   The **limbic association area** includes the cingulate gyrus, parahippocampal gyrus, and hippocampus (see Figures 12.7b and 12.17). Part of the limbic system (which we describe later), the limbic association area provides the emotional impact that makes a scene important to us. In our example above, it provides the sense of "danger" when the acid splashes on our legs. The hippocampus establishes memories that allow us to remember this incident. More on this later.

HOMEOSTATIC IMBALANCE 12.3   CLINICAL

Tumors or other lesions of the *anterior association area* may cause mental and personality disorders including loss of judgment, attentiveness, and inhibitions. The affected individual may be oblivious to social restraints, perhaps becoming careless about personal appearance, or rashly attacking a 7-foot opponent rather than running.

Different problems arise for individuals with lesions in the part of the *posterior association area* that provides awareness of self in space. They may refuse to wash or dress the side of their body opposite to the lesion because "that doesn't belong to me."

Lateralization of Cortical Functioning

We use both cerebral hemispheres for almost every activity, and the hemispheres appear nearly identical. Nonetheless, there is a division of labor, and each hemisphere has abilities not completely shared by its partner. This phenomenon is called **lateralization**.

Although one cerebral hemisphere or the other "dominates" each task, the term **cerebral dominance** designates the hemisphere that is *dominant for language*. In most people (about 90%), the left hemisphere has greater control over language abilities, math, and logic. This so-called dominant hemisphere is working when we compose a sentence, add numbers, and memorize a list. The other hemisphere (usually the right) is more free-spirited, more involved in visual-spatial skills, intuition, emotion, and artistic and musical skills. It is the poetic, creative, and the "Ah-ha!" (insightful) side of our nature.

Most individuals with left cerebral dominance are right-handed. In the remaining 10% of people, the roles of the hemispheres are reversed or the hemispheres share their functions equally. Typically, right-cerebral-dominant people are

left-handed. Some "lefties" who have a cerebral cortex that functions bilaterally are ambidextrous.

The two cerebral hemispheres have almost instantaneous communication with each other via connecting fiber tracts, as well as complete functional integration. And while each hemisphere is better than the other at certain functions, neither hemisphere is better at everything.

Cerebral White Matter

The second of the three basic regions of each cerebral hemisphere is the internal **cerebral white matter**. From what we have already described, you know that communication within the brain is extensive. The white matter deep to the cortical gray matter is responsible for communication between cerebral areas and between the cerebral cortex and lower CNS centers.

White matter consists largely of myelinated fibers bundled into large tracts. These fibers and tracts are classified according to the direction in which they run as *association, commissural,* or *projection* (**Figure 12.9**).

- **Association fibers** connect different parts of the same hemisphere. Short association fibers connect adjacent gyri. Long association fibers are bundled into tracts and connect different cortical lobes.

- **Commissural fibers** connect corresponding gray areas of the two hemispheres. These **commissures** (kom′ĭ-shūrz) allow the two hemispheres to function as a coordinated whole. The largest commissure is the **corpus callosum** (kah-lo′sum; "thickened body"), which lies superior to the lateral ventricles, deep within the longitudinal fissure. Less prominent examples are the **anterior** and **posterior commissures** (see Figure 12.11, pp. 447–448).

- **Projection fibers** either enter the cerebral cortex from lower brain or cord centers or descend from the cortex to lower areas. Sensory information reaches the cerebral cortex and motor output leaves it through these projection fibers. They tie the cortex to the rest of the nervous system and to the body's receptors and effectors. In contrast to commissural and association fibers, which run horizontally, projection fibers run vertically (Figure 12.9a).

At the top of the brain stem, the projection fibers on each side form a compact band, the **internal capsule**, that passes between the thalamus and some of the basal nuclei. Superior to that point, the fibers radiate fanlike through the cerebral white matter to the cortex. This distinctive arrangement is called the **corona radiata** ("radiating crown").

Basal Nuclei

Deep within the cerebral white matter is the third basic region of each hemisphere, a group of subcortical nuclei called the **basal nuclei** or **basal ganglia**.* They are primarily involved in the control of movement. Although the definition of the precise structures forming the basal nuclei is controversial, most

---

*We follow the guidelines of *Terminologia Anatomica* and use basal nuclei in this text. However, the use of basal ganglia is widespread in clinical practice.

Superior

Longitudinal fissure

Association fibers
(within hemisphere)

Commissural fibers
(between hemispheres)

Lateral
ventricle

• Corpus callosum

Projection fibers
(cerebral cortex to
or from lower area)

Basal nuclei
• Caudate
• Putamen
• Globus
pallidus

• Corona
radiata

• Internal
capsule

Thalamus

Gray matter

White matter

Third
ventricle

Pons

Medulla oblongata

Decussation
(crossover)
of pyramids

(a) Frontal section

Association fibers

Commissural fibers
• Corpus callosum

Gray
matter

Projection fibers
• Corona radiata
• Internal capsule

(b) Parasagittal section and dissection

Figure 12.9 White fiber tracts of the cerebral hemispheres. Commissural, association, and projection fibers run within the cerebrum, and between the cerebrum and lower CNS centers. In all views, notice the tight band of projection fibers, called the internal capsule, that passes between the thalamus and the basal nuclei, and then fans out as the corona radiata.

Practice Human Cadaver lab questions:
MasteringA&P® > Study Area > Lab Tools > PAL

12

anatomists agree that each hemisphere's basal nuclei include the **caudate nucleus** (kaw'dāt), **putamen** (pu-ta'men), and **globus pallidus** (glo'bis pal'ĭ-dus) (**Figure 12.10**).

The comma-shaped caudate nucleus arches superiorly over the diencephalon. Together with the putamen, it forms the **striatum** (stri-a'tum), so called because the fibers of the internal capsule passing through them create a striped appearance. Although the putamen ("pod") and globus pallidus ("pale

globe") together form a lens-shaped mass, sometimes called the *lentiform nucleus*, these two nuclei are functionally separate.

The basal nuclei are functionally associated with the *subthalamic nuclei* (located in the lateral "floor" of the diencephalon) and the *substantia nigra* of the midbrain (see Figure 12.15a).

The basal nuclei receive input from the entire cerebral cortex, as well as from other subcortical nuclei and each other. Via

Figure 12.10 Basal nuclei. (a) Three-dimensional view of the basal nuclei (basal ganglia), showing their position in the cerebrum. **(b)** Transverse section of cerebrum and diencephalon showing the relationship of the basal nuclei to the thalamus and the lateral and third ventricles.

relays through the thalamus, the output nucleus of the basal nuclei (globus pallidus) and the substantia nigra project to the premotor and prefrontal cortices and so influence muscle movements directed by the primary motor cortex. The basal nuclei have no direct access to motor pathways.

The precise role of the basal nuclei has been elusive because of their inaccessible location and because their motor functions overlap with those of the cerebellum. In addition to their motor functions, the basal nuclei play a role in cognition and emotion. In all of these cases, the basal nuclei seem to filter out incorrect or inappropriate responses, passing only the best response on to the cortex.

For example, in motor activity, the basal nuclei are particularly important in starting, stopping, and monitoring the intensity of movements executed by the cortex, especially those that are relatively slow or stereotyped, such as arm-swinging during walking. Additionally, they inhibit antagonistic or unnecessary movements. Disorders of the basal nuclei include Huntington's disease and Parkinson's disease (see p. 469).

Check Your Understanding

4. What anatomical landmark of the cerebral cortex separates primary motor areas from somatosensory areas?
5. Mike, who is left-handed, decided to wear his favorite T-shirt to his anatomy class. On his T-shirt were the words "Only left-handed people are in their right minds." What does this statement mean?
6. Which type of fiber allows the two cerebral hemispheres to "talk to each other"?
7. Name the components of the basal nuclei.

For answers, see Answers Appendix.

12.3 The diencephalon includes the thalamus, hypothalamus, and epithalamus

Learning Outcome

▶ Describe the location of the diencephalon, and name its subdivisions and functions.

Forming the central core of the forebrain and surrounded by the cerebral hemispheres, the **diencephalon** consists largely of three paired structures—the thalamus, hypothalamus, and epithalamus. These gray matter areas enclose the third ventricle (**Figure 12.11**).

Thalamus

The **thalamus** consists of bilateral egg-shaped nuclei, which form the superolateral walls of the third ventricle (Figures 12.9 and 12.11). In most people, an **interthalamic adhesion (intermediate mass)** connects the nuclei. *Thalamus* is a Greek word

(a)

Figure 12.11 Midsagittal section of the brain. (a) Illustration highlighting the diencephalon (purple) and brain stem (green).

(Figure continues on p. 448.)

Lateral ventricle

Corpus callosum

Fornix

Thalamus

Posterior commissure — **Epithalamus**

Pineal gland

Third ventricle

Anterior commissure

Corpora quadrigemina — **Midbrain**

Cerebral aqueduct

Hypothalamus

Optic chiasma

Arbor vitae

Fourth ventricle

Cerebellum

Mammillary body

Pons

Medulla oblongata

(b)

Figure 12.11 *(continued)* **Midsagittal section of the brain. (b)** Photo

12

meaning "inner room," which well describes this deep, well-hidden brain region that makes up 80% of the diencephalon.

The thalamus is *the* relay station for information coming into the cerebral cortex. Within the thalamus are a large number of nuclei, most named according to their location (**Figure 12.12a**). Each nucleus has a functional specialty, and each projects fibers to and receives fibers from a specific region of the cerebral cortex.

Afferent impulses from all senses and all parts of the body converge on the thalamus and synapse with at least one of its nuclei. For example, the ventral posterolateral nuclei receive impulses from the general somatic sensory receptors (touch, pressure, pain, etc.), and the *lateral* and *medial geniculate bodies* (jĕ-nik′u-lāt; "knee shaped") are important visual and auditory relay centers, respectively.

Within the thalamus, information is sorted out and "edited." Impulses having to do with similar functions are relayed as a group via the internal capsule to the appropriate area of the sensory cortex and to specific cortical association areas. As the afferent impulses reach the thalamus, we have a crude recognition of the sensation as pleasant or unpleasant. However, specific stimulus localization and discrimination occur in the cerebral cortex.

In addition to sensory inputs, virtually *all* other inputs ascending to the cerebral cortex funnel through thalamic nuclei. These include:

- Inputs that help regulate emotion and visceral function from the hypothalamus (via the anterior nuclei)

- Instructions that help direct the activity of the motor cortices from the cerebellum and basal nuclei (via the ventral lateral and ventral anterior nuclei, respectively)

- Inputs for memory or sensory integration that are projected to specific association cortices (via pulvinar, lateral dorsal, and lateral posterior nuclei)

In summary, the thalamus plays a key role in mediating sensation, motor activities, cortical arousal, learning, and memory. It is truly the gateway to the cerebral cortex.

Hypothalamus

Named for its position below (*hypo*) the thalamus, the **hypothalamus** caps the brain stem and forms the inferolateral walls of the third ventricle (Figure 12.11). Merging into the midbrain inferiorly, the hypothalamus extends from the optic chiasma (crossover point of the optic nerves) to the posterior margin of the mammillary bodies. Like the thalamus, the hypothalamus contains many functionally important nuclei (Figure 12.12b).

The **mammillary bodies** (mam′mil-er-e; "little breast"), paired pealike nuclei that bulge ventrally from the hypothalamus, are relay stations in the olfactory pathways. Between the optic chiasma and mammillary bodies is the **infundibulum** (in″fun-dib′u-lum), a stalk of hypothalamic tissue that connects the **pituitary gland** to the base of the hypothalamus.

Despite its small size, the hypothalamus is the main visceral control center of the body and is vitally important to overall

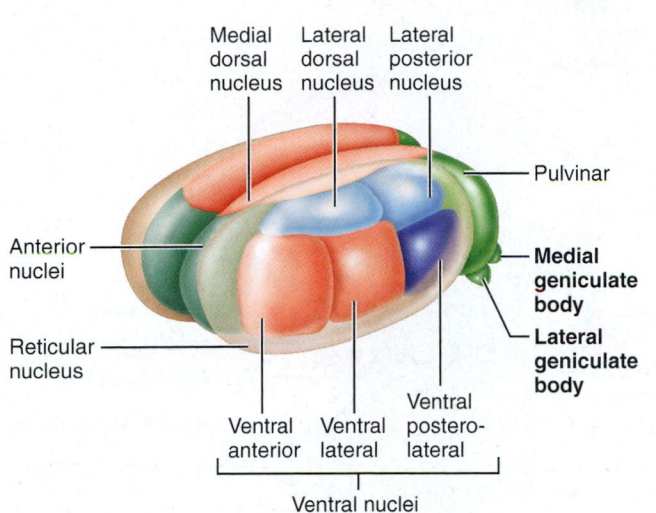

Medial dorsal nucleus   Lateral dorsal nucleus   Lateral posterior nucleus

Pulvinar

Anterior nuclei

Medial geniculate body

Lateral geniculate body

Reticular nucleus

Ventral anterior   Ventral lateral   Ventral postero-lateral

Ventral nuclei

(a) The main thalamic nuclei. (The reticular nuclei that "cap" the thalamus laterally are depicted as curving translucent structures.)

Paraventricular nucleus

Anterior commissure

Fornix

Dorsomedial nucleus

Preoptic nucleus

Anterior hypothalamic nucleus

Posterior hypothalamic nucleus

Supraoptic nucleus

Lateral hypothalamic area

Suprachiasmatic nucleus

Ventromedial nucleus

Optic chiasma

Arcuate nucleus

Mammillary body

Infundibulum (stalk of the pituitary gland)

Pituitary gland

(b) The main hypothalamic nuclei.

Figure 12.12 **Selected structures of the diencephalon.**

body homeostasis. Few tissues in the body escape its influence. Its chief homeostatic roles are to:

- **Control the autonomic nervous system.** Recall that the autonomic nervous system (ANS) is a system of peripheral nerves that regulates cardiac and smooth muscle and secretion by the glands. The hypothalamus regulates ANS activity by controlling the activity of centers in the brain stem and spinal cord. In this role, the hypothalamus influences blood pressure, rate and force of heartbeat, digestive tract motility, eye pupil size, and many other visceral activities.

- **Initiate physical responses to emotions.** The hypothalamus lies at the "heart" of the limbic system (the emotional part of the brain). It contains nuclei involved in perceiving pleasure, fear, and rage, as well as those involved in biological rhythms and drives (such as the sex drive).

    The hypothalamus acts through ANS pathways to initiate most physical expressions of emotion. For example, a fearful person has a pounding heart, high blood pressure, pallor, sweating, and a dry mouth.

- **Regulate body temperature.** The body's thermostat is in the hypothalamus. Hypothalamic neurons monitor blood temperature and receive input from other thermoreceptors in the brain and body periphery. Accordingly, the hypothalamus initiates cooling (sweating) or heat-generating actions

(shivering) as needed to maintain a relatively constant body temperature (see Figure 24.28, p. 966).

- **Regulate food intake.** In response to changing blood levels of certain nutrients (glucose and amino acids) or hormones (cholecystokinin, ghrelin, and others), the hypothalamus regulates feelings of hunger and satiety.

- **Regulate water balance and thirst.** When body fluids become too concentrated, hypothalamic neurons called *osmoreceptors* are activated. Osmoreceptors excite hypothalamic nuclei that trigger the release of antidiuretic hormone (ADH) from the posterior pituitary. ADH causes the kidneys to retain water. The same conditions also stimulate hypothalamic neurons in the *thirst center*, causing us to feel thirsty and drink more fluids.

- **Regulate sleep-wake cycles.** Acting with other brain regions, the hypothalamus helps regulate sleep. Its *suprachiasmatic nucleus* (our biological clock) sets the timing of the sleep cycle primarily in response to daylight-darkness cues received from the visual pathways.

- **Control endocrine system function.** The hypothalamus acts as the helmsman of the endocrine system in two important ways. First, its *releasing* and *inhibiting hormones* control the secretion of hormones by the anterior pituitary gland. Second, its *supraoptic* and *paraventricular nuclei* produce the hormones ADH and oxytocin.

12

Thalamus ⎤
Hypothalamus ⎦ Diencephalon

Midbrain ⎤
Pons ⎥ Brain stem
Medulla oblongata ⎦

View (a) →   ← View (c)

View (b)

Diencephalon
• Thalamus
• Hypothalamus
Mammillary body
Oculomotor nerve (III)
Trochlear nerve (IV)

Middle cerebellar peduncle
Abducens nerve (VI)
Vestibulocochlear nerve (VIII)
Pyramid
Ventral root of first cervical nerve
Decussation of pyramids

Spinal cord

Optic chiasma
Optic nerve (II)
Optic tract

Infundibulum
Pituitary gland

Crus cerebri of cerebral peduncles (midbrain)
Trigeminal nerve (V)
Pons
Facial nerve (VII)
Abducens nerve (VI)
Glossopharyngeal nerve (IX)
Hypoglossal nerve (XII)
Vagus nerve (X)
Accessory nerve (XI)

Thalamus

Superior colliculus
Inferior colliculus
Trochlear nerve (IV)
Superior cerebellar peduncle
Middle cerebellar peduncle
Inferior cerebellar peduncle
Vestibulocochlear nerve (VIII)
Olive

(a) Ventral view

(b) Left lateral view

Figure 12.13 Three views of the brain stem (green) and the diencephalon (purple).

12

HOMEOSTATIC IMBALANCE 12.4   **CLINICAL**

Hypothalamic disturbances cause a number of disorders including severe body wasting, obesity, sleep disturbances, dehydration, and emotional imbalances. The hypothalamus can be damaged by tumors, radiation, surgery, or trauma. ● ● ● ● ● ● ●

Epithalamus

The most dorsal portion of the diencephalon, the **epithalamus**, forms the roof of the third ventricle. Extending from its posterior border and visible externally is the **pineal gland** or **body** (pin′e-al; "pine cone shaped") (see Figures 12.11 and 12.13c). The pineal gland secretes the hormone *melatonin* (a sleep-inducing signal and antioxidant; ◀ p. 627) and, along with hypothalamic nuclei, helps regulate the sleep-wake cycle.

The *posterior commissure* forms the caudal border of the epithalamus.

Check Your Understanding

8. Why is the thalamus called the "gateway to the cerebral cortex"?
9. The hypothalamus oversees a branch of the peripheral nervous system. Which branch?

For answers, see Answers Appendix.

12.4 **The brain stem consists of the midbrain, pons, and medulla oblongata**

Learning Outcome

▶ Identify the three major regions of the brain stem, and note the functions of each area.

Figure 12.13 *(continued)*

From superior to inferior, the brain stem regions are midbrain, pons, and medulla oblongata (Figure 12.11, **Figure 12.13**, and **Figure 12.14** on p. 452). Each roughly an inch long, collectively they account for only 2.5% of total brain mass. The tissues of the brain stem are organized in a similar manner to that of the spinal cord—deep gray matter surrounded by white matter fiber tracts. However, the brain stem has nuclei of gray matter embedded in the white matter, a feature not found in the spinal cord.

Brain stem centers produce the rigidly programmed, automatic behaviors necessary for survival. Positioned between the cerebrum and the spinal cord, the brain stem also provides a pathway for fiber tracts running between higher and lower neural centers. Additionally, brain stem nuclei are associated with 10 of the 12 pairs of cranial nerves, so the brain stem is heavily involved with innervating the head. We discuss the cranial nerves and their functions in Chapter 13.

Midbrain

The **midbrain** is located between the diencephalon and pons (Figures 12.13 and 12.14). On its ventral aspect two bulging **cerebral peduncles** (pě-dung′klz; "little feet of the cerebrum") form vertical pillars that seem to hold up the cerebrum (Figure 12.13a, b, and **Figure 12.15a** on p. 453). The *crus cerebri* ("leg of the cerebrum") of each peduncle contains a large pyramidal (corticospinal) motor tract descending toward the spinal cord. The *superior cerebellar peduncles*, also fiber tracts, connect the midbrain to the cerebellum dorsally (Figure 12.13b, c).

Running through the midbrain is the hollow *cerebral aqueduct*, which connects the third and fourth ventricles (Figures 12.11 and

12.15a). It delineates the cerebral peduncles ventrally from the *tectum*, the midbrain's roof. Surrounding the aqueduct is the *periaqueductal gray matter*, which is involved in pain suppression and links the fear-perceiving amygdaloid body and ANS pathways that control the "fight-or-flight" response. The periaqueductal gray matter also includes nuclei that control two cranial nerves, the *oculomotor* and the *trochlear nuclei* (trok′le-ar).

Nuclei are also scattered in the surrounding white matter of the midbrain. The **corpora quadrigemina** (kor′por-ah kwod″rĭ-jem′ĭ-nah; "quadruplets"), the largest midbrain nuclei, raise four domelike protrusions on the dorsal midbrain surface (Figures 12.11 and 12.13c).

- The **superior colliculi** (kŏ-lik′ū-li), are visual reflex centers that coordinate head and eye movements when we visually follow a moving object, even if we are not consciously looking at it.

- The **inferior colliculi** are part of the auditory relay from the hearing receptors of the ear to the sensory cortex. They also act in reflexive responses to sound, such as the *startle reflex* that causes you to turn your head toward an unexpected noise.

Also embedded in each side of the midbrain white matter are two pigmented nuclei, the substantia nigra and red nucleus. The bandlike **substantia nigra** (sub-stan′she-ah ni′grah) is located deep to the cerebral peduncle (Figure 12.15a). Its dark (*nigr* = black) color reflects a high content of melanin pigment, a precursor of the neurotransmitter (dopamine) released by these neurons. The substantia nigra is functionally linked to the basal nuclei (its axons project to the putamen), and many scientists

Frontal lobe

Olfactory bulb
(synapse point of
cranial nerve I)

Optic chiasma

Optic nerve (II)

Optic tract

Mammillary body

Midbrain

Pons

Temporal
lobe

**Medulla
oblongata**

Cerebellum

Spinal cord

Figure 12.14 Inferior view of the brain, showing the three parts of the brain stem: midbrain, pons, and medulla oblongata. Only a small portion of the midbrain is visible.

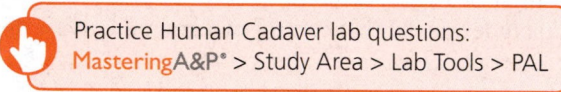

Practice Human Cadaver lab questions:
MasteringA&P® > Study Area > Lab Tools > PAL

consider it part of the basal nuclear complex. Degeneration of the dopamine-releasing neurons of the substantia nigra is the ultimate cause of Parkinson's disease.

The oval **red nucleus** lies deep to the substantia nigra (Figure 12.15a). Its reddish hue is due to its rich blood supply and to the presence of iron pigment in its neurons. The red nuclei are relay nuclei in some descending motor pathways that cause limb flexion, and they are embedded in the *reticular formation*, a system of small nuclei scattered through the core of the brain stem (see pp. 457–458).

Pons

The **pons** is the bulging brain stem region wedged between the midbrain and the medulla oblongata (see Figures 12.11, 12.13, and 12.14). Dorsally, the fourth ventricle separates it from the cerebellum.

As its name suggests (*pons* = bridge), the pons is chiefly composed of conduction tracts. They are oriented in two different directions:

- The deep projection fibers run longitudinally as part of the pathway between higher brain centers and the spinal cord.

- The more superficial ventral fibers are oriented transversely and dorsally. They form the *middle cerebellar peduncles* and connect the pons bilaterally with the two sides of the cerebellum dorsally (Figure 12.13). These fibers issue from numerous *pontine nuclei*, which relay "conversations" between the motor cortex and cerebellum.

Several cranial nerve pairs issue from pontine nuclei. They include the *trigeminal* (tri-jem′ĭ-nal), *abducens* (ab-du′senz), and *facial nerves* (Figures 12.13a, b and 12.15b).

Other important pontine nuclei are part of the reticular formation and some help the medulla oblongata maintain the normal rhythm of breathing.

Medulla Oblongata

The conical **medulla oblongata** (mĕ-dul′ah ob″long-gah′tah), or simply **medulla**, is the most inferior part of the brain stem. It blends imperceptibly into the spinal cord at the level of the foramen magnum of the skull (Figures 12.11a and 12.14; ◀ Figure 7.6, p. 206).

The central canal of the spinal cord continues upward into the medulla, where it broadens out to form the cavity of the fourth ventricle. Together, the medulla and the pons form the ventral wall of the fourth ventricle. [The dorsal ventricular wall is formed by a thin capillary-rich membrane called a choroid plexus which is next to the cerebellum dorsally (Figure 12.11a).]

Structures of the Medulla Oblongata

Flanking the midline on the medulla's ventral aspect are two longitudinal ridges called **pyramids**, formed by the large pyramidal (corticospinal) tracts descending from the motor cortex (Figure 12.15c). Just above the medulla–spinal cord junction, most of these fibers cross over to the opposite side before continuing into the spinal cord. This crossover point is called the **decussation of the pyramids** (de″kus-sa′shun; "a crossing"). As a result of this crossover, each cerebral hemisphere chiefly controls the voluntary movements of muscles on the opposite side of the body.

Several other structures are visible externally. The *inferior cerebellar peduncles* are fiber tracts that connect the medulla to the cerebellum dorsally. Situated lateral to the pyramids, the **olives** are oval swellings (which *do* resemble olives) (Figure 12.13b). These swellings are caused mainly by the wavy folds of gray matter of the underlying **inferior olivary nuclei** (Figure 12.15c). These nuclei relay sensory information on the degree of stretch in muscles and joints to the cerebellum. The rootlets of the *hypoglossal nerves* emerge from the groove between the pyramid and olive on each side of the brain stem. Other cranial nerves associated with the medulla are the *glossopharyngeal nerves* and *vagus nerves*. Additionally, the fibers

12

(a) Midbrain

Labels (clockwise): Dorsal; Superior colliculus; Tectum; Periaqueductal gray matter; Cerebral aqueduct; Oculomotor nucleus (III); Reticular formation; Medial lemniscus; Red nucleus; Substantia nigra; Fibers of pyramidal tract; Ventral; Crus cerebri of cerebral peduncle

(b) Pons

Labels: Superior cerebellar peduncle; Fourth ventricle; Trigeminal main sensory nucleus; Reticular formation; Trigeminal motor nucleus; Middle cerebellar peduncle; Trigeminal nerve (V); Pontine nuclei; Medial lemniscus; Fibers of pyramidal tract

(c) Medulla oblongata

Labels: Hypoglossal nucleus (XII); Fourth ventricle; Solitary nucleus; Dorsal motor nucleus of vagus (X); Choroid plexus; Vestibular nuclei (VIII); Inferior cerebellar peduncle; Reticular formation; Lateral nuclear group; Cochlear nuclei (VIII); Medial nuclear group; Nucleus ambiguus; Raphe nucleus; Inferior olivary nucleus; Medial lemniscus; Pyramid

Figure 12.15 Cross sections through different regions of the brain stem.

of the *vestibulocochlear nerves* (ves-tib″u-lo-kok′le-ar) synapse with the **cochlear nuclei** (auditory relays), and with numerous **vestibular nuclei** in both the pons and medulla. The vestibular nuclei mediate responses that maintain equilibrium.

Also housed in the medulla are several nuclei associated with ascending sensory tracts. The most prominent are the dorsally located **nucleus gracilis** (grah-sĭ′lis) and **nucleus cuneatus** (ku′ne-āt-us), associated with a tract called the *medial lemniscus* (Figure 12.15c). These serve as relay nuclei in a pathway

by which general somatic and proprioceptive sensory information ascends from the spinal cord to the somatosensory cortex.

Functions of the Medulla Oblongata

The small size of the medulla belies its crucial role as an autonomic reflex center involved in maintaining homeostasis. The medulla contains these important functional groups of visceral motor nuclei:

- **Cardiovascular center.** This includes the *cardiac center*, which adjusts the force and rate of heart contraction to meet the body's needs, and the *vasomotor center*, which changes blood vessel diameter to regulate blood pressure.

- **Respiratory centers.** These generate the respiratory rhythm and (together with pontine centers) control the rate and depth of breathing.

- **Various other centers.** Additional centers regulate such activities as vomiting, hiccuping, swallowing, coughing, and sneezing.

Notice that many functions listed above are also attributed to the hypothalamus (pp. 448–449). The overlap is easily explained. The hypothalamus controls many visceral functions by relaying its instructions through medullary reticular centers.

Check Your Understanding

10. What are the pyramids of the medulla? What is the result of decussation of the pyramids?

11. Which region of the brain stem is associated with the cerebral peduncles and the superior and inferior colliculi?

For answers, see Answers Appendix.

12.5 The cerebellum adjusts motor output, ensuring coordination and balance

Learning Outcome

▶ Describe the structure and function of the cerebellum.

The cauliflower-like **cerebellum** (ser″ĕ-bel′um; "small brain"), exceeded in size only by the cerebrum, accounts for about 11% of total brain mass. The cerebellum is located dorsal to the pons and medulla (and to the intervening fourth ventricle). It protrudes under the occipital lobes of the cerebral hemispheres, from which it is separated by the transverse cerebral fissure (see Figure 12.5b).

By processing inputs received from the cerebral motor cortex, various brain stem nuclei, and sensory receptors, the cerebellum provides the precise timing and appropriate patterns of skeletal muscle contraction for smooth, coordinated movements and agility needed for our daily living—driving, typing, and for some of us, playing the tuba. Cerebellar activity occurs subconsciously—we have no awareness of it.

Cerebellar Anatomy

The cerebellum is bilaterally symmetrical. The wormlike **vermis** connects its two apple-sized **cerebellar hemispheres** medially (**Figure 12.16**). Its surface is heavily convoluted, with fine, transversely oriented pleatlike gyri known as **folia** ("leaves"). Deep fissures subdivide each hemisphere into **anterior**, **posterior**, and **flocculonodular lobes** (flok″u-lo-nod′u-lar). The small propeller-shaped flocculonodular lobes, situated deep to the vermis and posterior lobe, cannot be seen in a surface view.

Like the cerebrum, the cerebellum has a thin outer cortex of gray matter, internal white matter, and small, deeply situated,

paired masses of gray matter, the most familiar of which are the *dentate nuclei*. Several types of neurons populate the cerebellar cortex, including **Purkinje cells** (◀ Table 11.2, p. 398). These large cells, with their extensively branched dendrites, are the only cortical neurons that send axons through the white matter to synapse with the central nuclei of the cerebellum. The distinctive pattern of white matter in the cerebellum resembles a branching tree, a pattern fancifully called the **arbor vitae** (ar′bor vi′te; "tree of life") (Figure 12.16a, b).

The anterior and posterior lobes of the cerebellum, which coordinate body movements, have three sensory maps of the entire body as indicated by the homunculi in Figure 12.16d. The part of the cerebellar cortex that receives sensory input from a body region influences motor output to that region.

- The medial portions influence the motor activities of the trunk and girdle muscles.

- The intermediate parts of each hemisphere influence the distal parts of the limbs and skilled movements.

- The lateralmost parts of each hemisphere integrate information from the association areas of the cerebral cortex and appear to play a role in planning movements rather than executing them.

- The flocculonodular lobes receive inputs from the equilibrium apparatus of the inner ears, and adjust posture to maintain balance.

Cerebellar Peduncles

As noted earlier, three paired fiber tracts—the cerebellar peduncles—connect the cerebellum to the brain stem (see Figures 12.13 and 12.16b). Unlike the contralateral fiber distribution to and from the cerebral cortex, virtually all fibers entering and leaving the cerebellum are **ipsilateral** (*ipsi* = same)—from and to the same side of the body.

- The **superior cerebellar peduncles** connecting cerebellum and midbrain carry instructions from neurons in the deep cerebellar nuclei to the cerebral motor cortex via thalamic relays. Like the basal nuclei, the cerebellum has no *direct* connections to the cerebral cortex.

- The **middle cerebellar peduncles** carry one-way communications from the pons to the cerebellum, advising the cerebellum of voluntary motor activities initiated by the motor cortex (via relays in the pontine nuclei).

- The **inferior cerebellar peduncles** connect medulla and cerebellum. These peduncles convey sensory information to the cerebellum from (1) muscle proprioceptors throughout the body, and (2) the vestibular nuclei of the brain stem, which are concerned with equilibrium (balance).

Cerebellar Processing

Cerebellar processing fine-tunes motor activity as follows:

1. The motor areas of the cerebral cortex, via relay nuclei in the brain stem, notify the cerebellum of their intent to initiate voluntary muscle contractions.

Figure 12.16 Cerebellum. (a) Photo of midsagittal section. **(b)** Drawing of parasagittal section. **(c)** Photo of the posterior view of the cerebellum. **(d)** Three body maps of the cerebellar cortex (in the form of homunculi).

Practice Human Cadaver lab questions:
MasteringA&P® > Study Area > Lab Tools > PAL

2. At the same time, the cerebellum receives information from proprioceptors throughout the body (regarding tension in the muscles and tendons, and joint position) and from visual and equilibrium pathways. This information enables the cerebellum to evaluate body position and momentum—where the body is and where it is going.

3. The cerebellar cortex calculates the best way to coordinate the force, direction, and extent of muscle contraction to prevent overshoot, maintain posture, and ensure smooth, coordinated movements.

4. Then, via the superior peduncles, the cerebellum dispatches to the cerebral motor cortex its "blueprint" for coordinating movement. Cerebellar fibers also send information to brain stem nuclei, which in turn influence motor neurons of the spinal cord.

Just as an automatic pilot compares a plane's instrument readings with the planned course, the cerebellum continually compares the body's performance with the higher brain's intention and sends out messages to initiate appropriate corrective measures. Cerebellar injury results in loss of muscle tone and clumsy, uncertain movements.

Cognitive Functions of the Cerebellum

Neuroanatomy, imaging studies, and observations of patients with cerebellar injuries suggest that the cerebellum also plays a role in thinking, language, and emotion. As in the motor system, the cerebellum may compare the actual output of these systems with the expected output and adjust accordingly. Much still remains to be discovered about the precise role of the cerebellum in nonmotor functions.

Check Your Understanding

12. In what ways are the cerebellum and the cerebrum similar? In what ways are they different?

13. PREDICT Five-year-old Kayla in the pediatric neurology ward has a brain tumor that is interfering with her cerebellar function. What problems would you expect her to have?

For answers, see Answers Appendix.

12.6 Functional brain systems span multiple brain structures

Learning Outcome

▶ Locate the limbic system and the reticular formation, and explain the role of each functional system.

Functional brain systems are networks of neurons that work together but span relatively large distances in the brain, so they cannot be localized to specific regions. The *limbic system* and the *reticular formation* are excellent examples. **Table 12.1** on pp. 458–459 summarizes their functions, as well as those of the cerebral hemispheres, diencephalon, brain stem, and cerebellum.

The Limbic System

The **limbic system** is a group of structures located on the medial aspect of each cerebral hemisphere and diencephalon. Its cerebral structures encircle (*limbus* = ring) the upper part of the brain stem (**Figure 12.17**). The limbic system includes the **amygdaloid body** (ah-mig′dah-loid), an almond-shaped nucleus that sits on the tail of the caudate nucleus, and other parts of the rhinencephalon (*cingulate gyrus, septal nuclei,* the C-shaped *hippocampus, dentate gyrus,* and *parahippocampal gyrus*). In the diencephalon, the main limbic structures are the *hypothalamus* and the *anterior thalamic nuclei.* The **fornix** ("arch") and other fiber tracts link these limbic system regions together.

The limbic system is our *emotional-visceral brain.* We will explore its connection to our emotions first and then see how it can act via the hypothalamus to alter autonomic (visceral) function.

The amygdaloid body and the anterior part of the **cingulate gyrus** seem especially important in emotions. The amygdaloid body is critical for responding to perceived threats (such as angry or fearful facial expressions) with fear or aggression. The cingulate gyrus plays a role in expressing our emotions through

Diencephalic structures of the limbic system
- Anterior thalamic nuclei (flanking 3rd ventricle)
- Hypothalamus
- Mammillary body

Septum pellucidum

Corpus callosum

Fiber tracts connecting limbic system structures
- Fornix
- Anterior commissure

Cerebral structures of the limbic system
- Cingulate gyrus
- Septal nuclei
- Amygdaloid body
- Hippocampus
- Dentate gyrus
- Parahippocampal gyrus

Olfactory bulb

Figure 12.17 The limbic system. Medial view of the cerebrum and diencephalon illustrating the structures of the limbic system, the emotional-visceral brain.

gestures and in resolving mental conflicts when we are frustrated.

Odors often trigger emotional reactions and memories. These responses reflect the origin of much of the limbic system in the primitive "smell brain" (rhinencephalon). Our reactions to odors are rarely neutral (a skunk smells *bad* and repulses us), and odors often recall memories of emotion-laden events.

Extensive connections between the limbic system and lower and higher brain regions allow the system to integrate and respond to a variety of environmental stimuli. Most limbic system output is relayed through the hypothalamus. Because the hypothalamus is the neural clearinghouse for both autonomic (visceral) function and emotional response, it is not surprising that some people under acute or unrelenting emotional stress fall prey to visceral illnesses, such as high blood pressure and heartburn. Disorders with physical symptoms that originate from emotional causes are known as **psychosomatic illnesses**.

③ The continuous stream of sensory stimuli keeps the cerebrum aroused and alert.

② RAS neurons relay sensory stimuli to the cerebrum through the thalamus.

① Sensory axons synapse on reticular activating system (RAS) neurons in the brain stem.

Reticular activating system (RAS)

Radiations to cerebral cortex

Reticular formation nuclei in brain stem

Visual impulses

Auditory impulses

Ascending general sensory tracts (touch, pain, temperature)

Descending projections
• From reticular formation nuclei to the spinal cord
• Help regulate skeletal and visceral muscle activity

Figure 12.18 The reticular formation. This functional brain system (purple) extends the length of the brain stem. Part of this formation, the reticular activating system (RAS), maintains alert wakefulness of the cerebral cortex.

Because the limbic system interacts with the prefrontal cortex, there is an intimate relationship between our feelings (mediated by the emotional brain) and our thoughts (mediated by the cognitive brain). As a result, we (1) react emotionally to things we consciously understand to be happening, and (2) are consciously aware of the emotional richness of our lives. Communication between the cerebral cortex and limbic system explains why emotions sometimes override logic and, conversely, why reason can stop us from expressing our emotions inappropriately. Particular limbic system structures—the **hippocampus** and amygdaloid body—also play a role in memory.

The Reticular Formation

The **reticular formation** extends through the central core of the medulla oblongata, pons, and midbrain (**Figure 12.18**). It is composed of loosely clustered neurons in what is otherwise white matter. These neurons form three broad columns along the length of the brain stem (Figure 12.15c): (1) the midline **raphe nuclei** (ra′fe; *raphe* = seam or crease), which are flanked laterally by (2) the **medial (large cell) group of nuclei** and (3) the **lateral (small cell) group of nuclei**.

The outstanding feature of the reticular neurons is their far-flung axonal connections. Individual reticular neurons project to the hypothalamus, thalamus, cerebral cortex, cerebellum, and spinal cord, making reticular neurons ideal for governing the arousal of the brain as a whole.

For example, unless inhibited by other brain areas, the neurons of the part of the reticular formation known as the **reticular activating system (RAS)** send a continuous stream of impulses to the cerebral cortex, keeping the cortex alert and conscious and enhancing its excitability. Impulses from all the great ascending sensory tracts synapse with RAS neurons, keeping them active and enhancing their arousing effect on the cerebrum. (This may explain why some students, stimulated by a bustling environment, like to study in a busy coffeehouse.)

The RAS also filters this flood of sensory inputs. Repetitive, familiar, or weak signals are filtered out, but unusual, significant, or strong impulses do reach consciousness. For example, you are probably unaware of the shoes on your feet, but would immediately notice it if one came off. Between them, the RAS and the cerebral cortex disregard perhaps 99% of all sensory stimuli as unimportant. If this filtering did not occur, the sensory overload would drive us crazy. The drug LSD interferes with these sensory dampers, promoting an often overwhelming sensory overload.

• Take a moment to become aware of all the stimuli in your environment. Notice all the colors, shapes, odors, sounds, and so on. How many of these sensory stimuli are you usually aware of?

The RAS is inhibited by sleep centers located in the hypothalamus and other neural regions, and is depressed by alcohol,

Table 12.1   Functions of Major Brain Regions

REGION	FUNCTION

Cerebral Hemispheres (pp. 439–447)

■ **Cortical gray matter:**

- Localizes and interprets sensory inputs
- Controls voluntary and skilled skeletal muscle activity
- Functions in intellectual and emotional processing

Basal nuclei (ganglia):

- Subcortical motor centers
- Help control skeletal muscle movements

Diencephalon (pp. 447–450)

■ **Thalamus:**

- Relays sensory impulses to cerebral cortex for interpretation
- Relays impulses between cerebral motor cortex and lower (subcortical) motor centers, including cerebellum
- Involved in memory processing

■ **Hypothalamus:**

- Chief integration center of autonomic (involuntary) nervous system
- Regulates body temperature, food intake, water balance, thirst, and biological rhythms and drives
- Regulates hormonal output of anterior pituitary gland
- Acts as an endocrine organ, producing posterior pituitary hormones ADH and oxytocin

■ **Limbic system (pp. 456–457)—A functional system:**

- Includes cerebral and diencephalon structures (e.g., hypothalamus and anterior thalamic nuclei)
- Mediates emotional response
- Involved in memory processing

---

sleep-inducing drugs, and tranquilizers. Severe injury to this system, as might follow a knockout punch that twists the brain stem, results in permanent unconsciousness (irreversible *coma*). Although the RAS is central to wakefulness, some of its nuclei are also involved in sleep, which we will discuss later in this chapter.

The reticular formation also has a *motor* arm. Some of its motor nuclei project to motor neurons in the spinal cord via the *reticulospinal tracts*, and help control skeletal muscles during coarse limb movements. Other reticular motor nuclei, such as the vasomotor, cardiac, and respiratory centers of the medulla, are autonomic centers that regulate visceral motor functions.

Check Your Understanding

14. The limbic system is sometimes called the emotional-visceral brain. Which part of the limbic system is responsible for the visceral connection?

15. APPLY When Taylor begins to feel drowsy while driving, she opens her window, turns up the volume of the car stereo, and sips ice-cold water. How do these actions help keep her awake?

For answers, see Answers Appendix.

12.7   ## The interconnected structures of the brain allow higher mental functions

Learning Outcomes

▶ **Identify the brain areas involved in language and memory.**
▶ **Identify factors affecting the formation of long-term memories.**
▶ **Define EEG and distinguish between alpha, beta, theta, and delta brain waves.**
▶ **Describe consciousness clinically.**
▶ **Compare and contrast the events and importance of slow-wave and REM sleep.**

During the last five decades, an exciting exploration of our "inner space," or what we commonly call the *mind*, has been going on. But researchers in the field of cognition are still struggling to understand how the mind's qualities spring from living tissue and electrical impulses. Souls and synapses are hard to reconcile!

Table 12.1 (continued)	
REGION	FUNCTION

Brain Stem (pp. 450–454)

■ **Midbrain:**

- Contains visual (superior colliculi) and auditory (inferior colliculi) reflex centers
- Contains subcortical motor centers (substantia nigra and red nuclei)
- Contains nuclei for cranial nerves III and IV
- Contains projection fibers (e.g., fibers of the pyramidal tracts)

■ **Pons:**

- Relays information from the cerebrum to the cerebellum
- Cooperates with the medullary respiratory centers to control respiratory rate and depth
- Contains nuclei of cranial nerves V–VII
- Contains projection fibers

■ **Medulla oblongata:**

- Relays ascending sensory pathway impulses from skin and proprioceptors through nuclei cuneatus and gracilis
- Contains visceral nuclei controlling heart rate, blood vessel diameter, respiratory rate, vomiting, coughing, etc.
- Relays sensory information to the cerebellum through inferior olivary nuclei
- Contains nuclei of cranial nerves VIII–X and XII
- Contains projection fibers
- Site of decussation of pyramids

■ **Reticular formation** (pp. 457–458)—A functional system:

- Maintains cerebral cortical alertness (reticular activating system)
- Filters out repetitive stimuli
- Helps regulate skeletal and visceral muscle activity

Cerebellum (pp. 454–456)

■ **Cerebellum:**

- Processes information from cerebral motor cortex, proprioceptors, and visual and equilibrium pathways
- Provides "instructions" to cerebral motor cortex and subcortical motor centers, resulting in smooth, coordinated skeletal muscle movements
- Responsible for balance and posture

Our ability to discuss "souls and synapses" is one of the things that makes us human. We will begin by looking at the mental functions of language and memory that give us this ability. Then we will explore brain waves, which reflect the underlying electrical activity of mental functions, followed by the related topics of consciousness and sleep.

Language

Language is such an important function of the brain that it involves practically all of the association cortex on the left side in one way or another. Pioneering studies of patients with *aphasias* (the loss of language abilities due to damage to specific areas of the brain) pointed to two critically important regions, Broca's area and Wernicke's area (see areas outlined by dashes in Figure 12.7a).

Patients with lesions involving **Broca's area** can understand language but have difficulty speaking (and sometimes cannot write or type or use sign language). On the other hand, patients with lesions involving **Wernicke's area** are able to speak but produce a type of nonsense often referred to as "word salad." They also have great difficulty understanding language.

In fact, this picture is clinically useful, but oversimplified. Broca's and Wernicke's areas together with the basal nuclei form a single language implementation system that analyzes incoming and produces outgoing word sounds and grammatical structures. A surrounding set of cortical areas forms a bridge between this system and the regions of cortex that hold concepts and ideas, which are distributed throughout the remainder of the association cortices.

The corresponding areas in the right or non-language-dominant hemisphere are involved in "body language"—the nonverbal emotional components of language. These areas allow the lilt or tone of our voice and our gestures to express our emotions when we speak, and permit us to comprehend the emotional content of what we hear. For example, a soft, melodious response to your question conveys quite a different meaning than a sharp reply.

Memory

Memory is the storage and retrieval of information. Memories are essential for learning and incorporating our experiences into behavior and are an integral part of our consciousness. Stored somewhere in your 3 pounds of wrinkled brain are zip codes, the face of your grandfather, and the taste of yesterday's pizza. Your memories reflect your lifetime.

There are different kinds of memory: *declarative* (*fact*) *memory* (names, faces, words, and dates), *procedural* (*skills*) *memory* (piano playing), *motor memory* (riding a bike), and *emotional memory* (your pounding heart when you hear a rattlesnake nearby). Let's focus on declarative memory.

Declarative memory storage involves two distinct stages: short-term memory and long-term memory (**Figure 12.19**).

Short-term memory (STM), also called *working memory*, is the preliminary step, as well as the power that lets you look up a telephone number, dial it, and then never think of it again. STM is limited to seven or eight chunks of information, such as the digits of a telephone number or the sequence of words in an elaborate sentence.

In contrast, **long-term memory (LTM)** seems to have a limitless capacity. Although our STM cannot recall numbers much longer than a telephone number, we can remember the lyrics to many songs by committing them to LTM. However, long-term memories can be forgotten, and so our memory bank continually changes with time. Furthermore, our ability to store and retrieve information declines with aging.

We do not remember or even consciously notice much of what is going on around us. As sensory inputs flood into our cerebral cortex, they are processed (yellow box in Figure 12.19). Some 5% of this information is selected for transfer to STM (light green box). STM serves as a temporary holding bin for data that we may or may not want to retain.

Information is then transferred from STM to LTM (dark green box). Many factors can influence this transfer, including:

- **Emotional state.** We learn best when we are alert, motivated, surprised, or aroused. For example, when we witness shocking events, transferral is almost immediate. Norepinephrine, a neurotransmitter involved in memory processing of emotionally charged events, is released when we are excited or "stressed out," which helps to explain this phenomenon.

- **Rehearsal.** Rehearsing or repeating the material enhances memory.

Figure 12.19 Memory processing.

- **Association.** Tying "new" information to "old" information already stored in LTM appears to be important in remembering facts.

- **Automatic memory.** Not all impressions that become part of LTM are consciously formed. A student concentrating on a lecturer's speech may record an automatic memory of the pattern of the lecturer's tie.

Memories transferred to LTM take time to become permanent. The process of **memory consolidation** apparently involves fitting new facts into the categories of knowledge already stored in the cerebral cortex. The hippocampus and surrounding temporal cortical areas play a major role in memory consolidation by communicating with the thalamus and the prefrontal cortex.

 HOMEOSTATIC IMBALANCE 12.5 **CLINICAL**

Damage to the hippocampus and surrounding medial temporal lobe structures on either side results in only slight memory loss, but bilateral destruction causes widespread amnesia. Consolidated memories are not lost, but new sensory inputs cannot be associated with old, and the person lives in the here and now from that point on. This condition is called *anterograde amnesia* (an′ter-o-grād″), in contrast to *retrograde amnesia*, which is the loss of memories formed in the distant past. You could carry on an animated conversation with a person with anterograde amnesia, excuse yourself, return five minutes later, and that person would not remember you.

Specific pieces of each memory are thought to be stored near regions of the brain that need them so new inputs can be quickly associated with the old. Accordingly, visual memories are stored in the occipital cortex, memories of music in the temporal cortex, and so on.

Brain Wave Patterns and the EEG

Brain waves reflect the electrical activity on which higher mental functions are based because normal brain function involves continuous electrical activity of neurons. An **electroencephalogram** (e-lek″tro-en-sef′ah-lo-gram), or **EEG**, records some aspects of this activity. EEGs are used for diagnosing epilepsy and sleep disorders, in research on brain function, and to determine brain death.

An EEG is made by placing electrodes on the scalp and connecting the electrodes to an apparatus that measures voltage differences between various cortical areas (**Figure 12.20a**). The patterns of neuronal electrical activity recorded, called **brain waves**, are generated by synaptic activity at the surface of the cortex, rather than by action potentials in the white matter.

Each of us has a brain wave pattern that is as unique as our fingerprints. For simplicity, however, we can group brain waves into the four frequency classes shown in Figure 12.20b. Each wave is a continuous train of peaks and troughs, and the wave frequency, expressed in hertz (Hz), is the number of peaks in one second. A frequency of 1 Hz means that one peak occurs each second.

The amplitude or intensity of any wave is represented by how high the wave peaks rise and how low the troughs dip. The amplitude of brain waves reflects the synchronous activity of many neurons and not the degree of electrical activity of individual neurons. Usually, brain waves are complex and low amplitude. During some stages of sleep, neurons tend to fire synchronously, producing similar, high-amplitude brain waves.

- **Alpha waves** (8–13 Hz) are relatively regular and rhythmic, low-amplitude, synchronous waves. In most cases, they indicate a brain that is "idling"—a calm, relaxed state of wakefulness.

- **Beta waves** (14–30 Hz) are also rhythmic, but less regular than alpha waves and with a higher frequency. Beta waves occur when we are mentally alert, as when concentrating on some problem or visual stimulus.

- **Theta waves** (4–7 Hz) are still more irregular. Though common in children, theta waves are uncommon in awake adults but may appear when concentrating.

- **Delta waves** (4 Hz or less) are high-amplitude waves seen during deep sleep and when the reticular activating system is suppressed, such as during anesthesia. In awake adults, they indicate brain damage.

Brain waves whose frequency is too high or too low suggest problems with cerebral cortical functions, and unconsciousness occurs at both extremes. Because spontaneous brain waves are always present, even during unconsciousness and coma, their absence—a "flat EEG"—is clinical evidence of *brain death*.

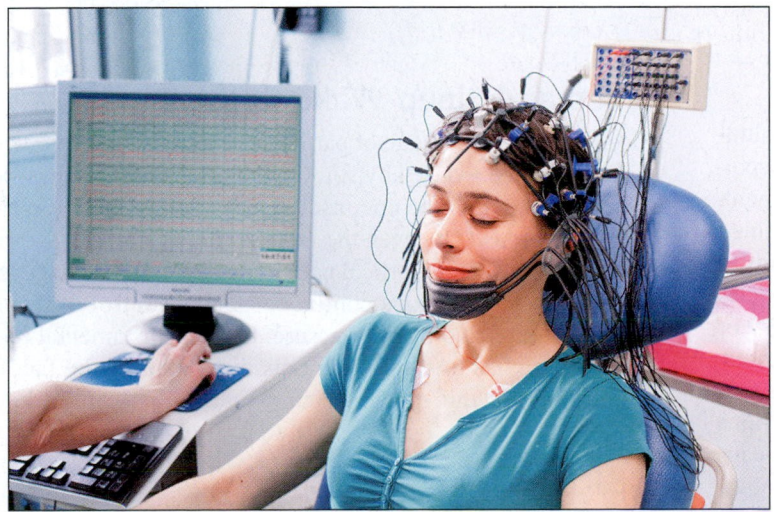

(a) Scalp electrodes are used to record brain wave activity.

(b) Brain waves shown in EEGs fall into four general classes.

Figure 12.20 Electroencephalography (EEG) and brain waves.

HOMEOSTATIC IMBALANCE 12.6 — CLINICAL

Almost without warning, a person with *epilepsy* may lose consciousness and fall stiffly to the ground, wracked by uncontrollable jerking. These **epileptic seizures** reflect a torrent of electrical discharges by groups of brain neurons, and during their uncontrolled activity no other messages can get through.

Epilepsy, manifested by one out of 100 of us, is not associated with, nor does it cause, intellectual impairment. Genetic factors induce some cases, but epilepsy can also result from brain injuries caused by blows to the head, stroke, infections, or tumors.

Epileptic seizures vary tremendously in their expression and severity.

- *Absence seizures*, formerly known as *petit mal*, are mild forms in which the expression goes blank for a few seconds as consciousness disappears. These are typically seen in young children and usually disappear by age 10.
- *Tonic-clonic seizures*, formerly called *grand mal*, are the most severe, convulsive form of epileptic seizures. The person loses consciousness, sometimes breaking bones during the intense convulsions, showing the incredible strength of these muscle contractions. Loss of bowel and bladder control and severe biting of the tongue are common. The seizure lasts for a few minutes, then the muscles relax and the person awakens but remains disoriented for several minutes.

Many seizure sufferers experience a sensory hallucination, such as a taste, smell, or flashes of light, just before the seizure begins. This phenomenon, called an **aura**, is helpful because it gives the person time to lie down and avoid falling.

Epilepsy can usually be controlled by anticonvulsive drugs. If drugs fail to control the seizures, a *vagus nerve stimulator* or *deep brain stimulator* can be implanted. These devices deliver pulses to the vagus nerve or directly to the brain at predetermined intervals to stabilize the brain's electrical activity. ●●●●●●

Consciousness

Consciousness encompasses perception of sensations, voluntary initiation and control of movement, and capabilities associated with higher mental processing (memory, logic, judgment, perseverance, and so on). Clinically, consciousness is defined on a continuum that grades behavior in response to stimuli as (1) *alertness*, (2) *drowsiness* or *lethargy* (which proceeds to sleep), (3) *stupor*, and (4) *coma*. Alertness is the highest state of consciousness and cortical activity, and coma the most depressed.

Consciousness is difficult to define. And to be frank, reducing our response to a glorious sunset to a series of interactions between dendrites, axons, and neurotransmitters does not capture what makes that event so special. A sleeping person obviously lacks something that he or she has when awake, and we call this "something" consciousness.

Current suppositions about consciousness are as follows:

- **Consciousness involves simultaneous activity of large areas of the cerebral cortex.**

- **It is superimposed on other types of neural activity.** At any time, specific neurons and neuronal pools are involved both in localized activities (such as motor control) and in cognition.
- **It is holistic and totally interconnected.** Information for "thought" can be claimed from many locations in the cerebrum simultaneously. For example, retrieval of a specific memory can be triggered by several routes—a smell, a place, a particular person, and so on.

HOMEOSTATIC IMBALANCE 12.7 — CLINICAL

Except during sleep, unconsciousness is always a signal that brain function is impaired. A brief loss of consciousness is called **fainting** or **syncope** (sing′ko-pe; "cut short"). Most often, syncope indicates inadequate cerebral blood flow due to low blood pressure, as might follow hemorrhage or sudden emotional stress.

Significant unresponsiveness to sensory stimuli for an extended period is called **coma**. Coma is *not* deep sleep. During sleep, the brain remains active and oxygen consumption resembles that of the waking state. In coma patients, oxygen use is always below normal resting levels.

Factors that can induce coma include: (1) blows to the head that cause widespread cerebral or brain stem trauma, (2) tumors or infections that invade the brain stem, (3) metabolic disturbances such as hypoglycemia (abnormally low blood sugar levels), (4) drug overdose, (5) liver or kidney failure. Strokes rarely cause coma unless they are massive and accompanied by extreme swelling of the brain, or are located in the brain stem.

When the brain has suffered irreparable damage, irreversible coma occurs. The result is **brain death**, a dead brain in an otherwise living body. Because life support can be removed only after death, physicians must determine whether a patient in an irreversible coma is legally alive or dead. ●●●●●●

Sleep and Sleep-Wake Cycles

Sleep is defined as a state of partial unconsciousness from which a person can be aroused by stimulation. This distinguishes sleep from coma, a state of unconsciousness from which a person *cannot* be aroused by even the most vigorous stimuli.

For the most part, cortical activity is depressed during sleep, but brain stem functions continue, such as control of respiration, heart rate, and blood pressure. Even environmental monitoring continues to some extent, as illustrated by the fact that strong stimuli ("things that go bump in the night") immediately arouse us.

Types of Sleep

The two major types of sleep, which alternate through most of the sleep cycle, are **non–rapid eye movement (NREM) sleep** and **rapid eye movement (REM) sleep**. During the first 30 to 45 minutes of the sleep cycle, EEGs show that we pass through the first two stages of NREM sleep and into NREM stages 3 and 4, also

Awake

REM: Skeletal muscles (except ocular muscles and diaphragm) are actively inhibited; most dreaming occurs.

NREM stage 1: Relaxation begins; EEG shows alpha waves; arousal is easy.

NREM stage 2: Irregular EEG with sleep spindles (short high-amplitude bursts); arousal is more difficult.

NREM stage 3: Sleep deepens; theta and delta waves appear; vital signs decline.

NREM stage 4: EEG is dominated by delta waves; arousal is difficult; bed-wetting, night terrors, and sleepwalking may occur.

(a) Typical EEG patterns

(b) Typical progression of an adult through one night's sleep stages

Figure 12.21 Types and stages of sleep. The four stages of non–rapid eye movement (NREM) sleep and rapid eye movement (REM) sleep are shown.

called **slow-wave sleep** (**Figure 12.21**). As we pass through these stages and slip into deeper and deeper sleep, the frequency of the EEG waves declines, but their amplitude increases. Blood pressure and heart rate also progressively decrease.

About 90 minutes after sleep begins, after reaching NREM stage 4, the EEG pattern changes abruptly. It becomes very irregular and backtracks quickly through the stages until alpha waves (more typical of the awake state) reappear, indicating the onset of REM sleep. This brain wave change is coupled with increases in heart rate, respiratory rate, and blood pressure and a decrease in gastrointestinal motility. Oxygen use by the brain is tremendous during REM—greater than during the awake state.

Although the eyes move rapidly under the lids during REM, most of the body's skeletal muscles are actively inhibited and go limp. Most dreaming occurs during REM sleep, and this temporary paralysis prevents us from acting out our dreams.

How Sleep Is Regulated

The alternating cycles of sleep and wakefulness reflect a natural *circadian*, or 24-hour, *rhythm*. The hypothalamus is responsible for the timing of the sleep cycle. Its *suprachiasmatic nucleus* (a biological clock) regulates its *preoptic nucleus* (a sleep-inducing center). By inhibiting the brain stem's reticular activating system (RAS; see Figure 12.18), the preoptic nucleus puts the cerebral cortex to sleep. However, sleep is much more than simply turning off the arousal system. RAS centers not only help maintain the awake state but also mediate some sleep stages, especially dreaming sleep.

Just before we wake, hypothalamic neurons release peptides called *orexins*, which act as "wake-up" chemicals. As a result, certain neurons of the brain stem reticular formation fire at maximal rates, arousing the sleepy cortex. A large number of chemical substances in the body cause sleepiness, but the relative importance of these various sleep-inducing substances is not known.

Importance of Sleep

Why do we sleep? We don't really know, but sleep appears to have a number of important functions. During sleep we may consolidate new memories and discard memories that are no longer accessed (in other words, we forget). Sleep is presumed to be restorative—the time when most neural activity can wind down to basal levels. When deprived of sleep, we spend more time than usual in both REM and slow-wave sleep during the next sleep episode in an attempt to catch up.

HOMEOSTATIC IMBALANCE 12.8   CLINICAL

Narcolepsy is a neurological disorder characterized by extreme daytime sleepiness and sudden attacks of sleep. A related condition, cataplexy, may cause an abrupt loss of muscle tone for seconds to minutes while a person is awake. Such episodes may be triggered by strong emotions or laughter.

Cells in the hypothalamus that secrete peptides called orexins (mentioned above as a wake-up chemical; also called hypocretins) are selectively destroyed in patients with narcolepsy, probably by the patient's own immune system. Replacing the orexins may be a key to future treatments.

Conversely, drugs that block the actions of orexin and promote sleep are used to treat **insomnia**, a chronic inability to obtain the *amount* or *quality* of sleep needed to function adequately during the day. While sleep requirements vary, most adults need 7–8 hours of sleep for optimal health and function.

A number of factors can contribute to insomnia. These include psychological disturbances such as depression or anxiety, the overuse of caffeine, and using a television, computer, or cell phone close to bedtime.

Check Your Understanding

16. Name three factors that can enhance transfer of information from short-term memory to long-term memory.
17. When would you see delta waves in an EEG?
18. Which two states of consciousness are between alertness and coma?
19. During which sleep stage are most skeletal muscles actively inhibited?

For answers, see Answers Appendix.

12.8 The brain is protected by bone, meninges, cerebrospinal fluid, and the blood brain barrier

Learning Outcomes

▶ **Describe how meninges, cerebrospinal fluid, and the blood brain barrier protect the CNS.**

▶ **Explain how cerebrospinal fluid is formed and describe its circulatory pathway.**

Nervous tissue is soft and delicate, and even slight pressure can injure neurons. However, the brain is protected by bone (the skull, discussed in Chapter 7), membranes (the meninges), and a watery cushion (cerebrospinal fluid). Furthermore, the blood brain barrier protects the brain from harmful substances in the blood.

Meninges

The **meninges** (mĕ-nin′jēz; *mening* = membrane) are three connective tissue membranes that lie just external to the CNS organs. The meninges:

- Cover and protect the CNS
- Protect blood vessels and enclose venous sinuses
- Contain cerebrospinal fluid
- Form partitions in the skull

From external to internal, the meninges (singular: **meninx**) are the dura mater, arachnoid mater, and pia mater (**Figure 12.22**).

Dura Mater

The leathery **dura mater** (du′rah ma′ter), meaning "tough mother," is the strongest meninx. Where it surrounds the brain, it is a two-layered sheet of fibrous connective tissue. The more superficial *periosteal layer* attaches to the inner surface of the skull (the periosteum). (There is no dural periosteal layer surrounding the spinal cord.) The deeper *meningeal layer* forms the true external covering of the brain and continues caudally in the vertebral canal as the spinal dura mater. The brain's two dural layers are fused together except in certain areas, where they separate to enclose **dural venous sinuses** that collect venous blood from the brain and direct it into the internal jugular veins of the neck (**Figure 12.23**).

In several places, the meningeal dura mater extends inward to form flat partitions that subdivide the cranial cavity. These **dural septa**, which limit excessive movement of the brain within the cranium, include the following (Figure 12.23a):

Figure 12.22 Meninges: dura mater, arachnoid mater, and pia mater. The meningeal dura forms the falx cerebri fold. A dural sinus, the superior sagittal sinus, is enclosed by the dural membranes superiorly. Arachnoid granulations return cerebrospinal fluid to the dural sinus. (Frontal section.)

Labels:
- Skin of scalp
- Periosteum
- Bone of skull
- **Dura mater**
  - • Periosteal layer
  - • Meningeal layer
- **Arachnoid mater**
- **Pia mater**
- Arachnoid granulation
- Blood vessel
- Falx cerebri (in longitudinal fissure only)
- Superior sagittal sinus
- Subdural space
- Subarachnoid space

(a) Midsagittal view

(b) Posterior dissection

Figure 12.23 Dural septa and dural venous sinuses. (a) Dural septa are partitioning folds of dura mater in the cranial cavity. **(b)** Dural venous sinuses (injected with blue latex) are spaces between the periosteal and meningeal dura containing venous blood.

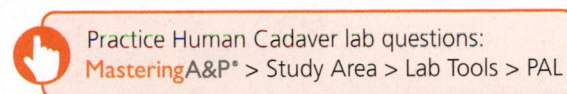

Practice Human Cadaver lab questions:
MasteringA&P® > Study Area > Lab Tools > PAL

- **Falx cerebri** (falks ser′ĕ-bri). A large sickle-shaped (*falx* = sickle) fold that dips into the longitudinal fissure between the cerebral hemispheres. Anteriorly, it attaches to the crista galli of the ethmoid bone.

- **Falx cerebelli** (ser″ĕ-bel′i). Continuing inferiorly from the posterior falx cerebri, this small midline partition runs along the vermis of the cerebellum.

- **Tentorium cerebelli** (ten-to′re-um; "tent"). Resembling a tent over the cerebellum, this nearly horizontal dural fold extends into the transverse fissure between the cerebral hemispheres (which it helps to support) and the cerebellum.

Arachnoid Mater

The middle meninx, the **arachnoid mater** (ah-rak′noid), forms a loose brain covering, never dipping into the sulci at the cerebral surface. It is separated from the dura mater by a narrow serous cavity, the **subdural space**, which contains a film of fluid. Beneath the arachnoid membrane is the wide **subarachnoid space**. Spiderweb-like extensions span this space and secure the arachnoid mater to the underlying pia mater (*arachnida* means "spider"). The subarachnoid space is filled with cerebrospinal fluid and also contains the largest blood vessels serving the brain. Because the arachnoid mater is fine and elastic, these blood vessels are poorly protected.

Knoblike projections of the arachnoid mater called **arachnoid granulations** protrude superiorly through the dura mater and into the superior sagittal sinus (Figure 12.22). These granulations absorb cerebrospinal fluid into the venous blood of the sinus.

Pia Mater

The **pia mater** (pi′ah), meaning "gentle mother," is composed of delicate connective tissue and contains many tiny blood vessels. It is the only meninx that clings tightly to the brain like plastic wrap, following its every convolution. Small arteries entering the brain tissue carry ragged sheaths of pia mater inward with them for short distances.

HOMEOSTATIC IMBALANCE 12.9 **CLINICAL**

Meningitis, inflammation of the meninges, is a serious threat to the brain because a bacterial or viral meningitis may spread to the CNS. Brain inflammation is called *encephalitis* (en-sef″ah-li′tis). Meningitis is usually diagnosed by obtaining a sample of cerebrospinal fluid via a lumbar tap (see Figure 12.29, p. 471) and examining it for signs of infection.

Cerebrospinal Fluid (CSF)

Cerebrospinal fluid, found in and around the brain and spinal cord, forms a liquid cushion that gives buoyancy to CNS structures. By floating the jellylike brain, the CSF effectively reduces brain weight by 97% and prevents the delicate brain from crushing under its own weight. CSF also protects the brain and spinal cord from blows and other trauma. Additionally, although the brain has a rich blood supply, CSF helps nourish the brain, and there is some evidence that it carries chemical signals (such as hormones and sleep- and appetite-inducing molecules) from one part of the brain to another.

CSF is a watery "broth" similar in composition to blood plasma, from which it is formed. However, it contains less protein than plasma and its ion concentrations are different. For example, CSF contains more Na^+, Cl^-, and H^+ than does blood plasma, and less Ca^{2+} and K^+.

The **choroid plexuses** that hang from the roof of each ventricle form CSF. These plexuses are frond-shaped clusters of broad, thin-walled capillaries (*plex* = interwoven) enclosed first by pia mater and then by a layer of ependymal cells lining the ventricles

12

Superior sagittal sinus

Choroid plexus

Interventricular foramen

Third ventricle

Arachnoid granulation

Subarachnoid space

Arachnoid mater

Meningeal dura mater

Periosteal dura mater

Right lateral ventricle (deep to cut)

Choroid plexus of fourth ventricle

Cerebral aqueduct

Lateral aperture

Fourth ventricle

Median aperture

Central canal of spinal cord

① The choroid plexus of each ventricle produces CSF.

② CSF flows through the ventricles and into the subarachnoid space via the median and lateral apertures.

③ CSF flows through the subarachnoid space.

④ CSF is absorbed into the dural venous sinuses via the arachnoid granulations.

(a) CSF circulation

Ependymal cells

Capillary

Connective tissue of pia mater

Section of choroid plexus

Wastes and unnecessary solutes absorbed

Cavity of ventricle

CSF forms as a filtrate containing glucose, oxygen, vitamins, and ions (Na^+, Cl^-, Mg^{2+}, etc.)

(b) CSF formation by choroid plexuses

Figure 12.24 Formation, location, and circulation of CSF.
(a) Location and circulatory pattern of cerebrospinal fluid (CSF). Arrows indicate the direction of flow. **(b)** Each choroid plexus consists of a knot of porous capillaries surrounded by a single layer of ependymal cells joined by tight junctions and bearing long cilia. Fluid leaking from porous capillaries is processed by the ependymal cells to form the CSF in the ventricles.

(**Figure 12.24b**). These capillaries are fairly permeable, and tissue fluid filters continuously from the bloodstream. However, the choroid plexus ependymal cells are joined by tight junctions, and they have ion pumps that allow them to modify this filtrate by actively transporting only certain ions across their membranes into the CSF pool. This careful regulation of CSF composition is important because CSF mixes with the extracellular fluid

bathing neurons and influences the composition of this fluid. Ion pumping also sets up ionic gradients that cause water to diffuse into the ventricles.

In adults, the total CSF volume of about 150 ml (about half a cup) is replaced every 8 hours or so. About 500 ml of CSF is formed daily. The choroid plexuses also help cleanse the CSF by removing waste products and unnecessary solutes.

Once produced, CSF moves freely through the ventricles. CSF enters the subarachnoid space via the lateral and median apertures in the walls of the fourth ventricle (Figure 12.24a). The long cilia of the ependymal cells lining the ventricles help keep the CSF in constant motion. In the subarachnoid space, CSF bathes the outer surfaces of the brain and spinal cord and then returns to the blood in the dural sinuses via the arachnoid granulations.

HOMEOSTATIC IMBALANCE 12.10
CLINICAL

When CSF is produced faster than it drains, it builds up in the ventricles. The increased pressure on the delicate brain tissues causes brain damage. In adults, the skull is rigid and the head cannot enlarge, so the increase in pressure can be fatal. In infants whose cranial sutures have not yet fused, the increase in pressure enlarges the child's head. This condition is called *hydrocephalus* ("water on the brain") (**Figure 12.25**).

Obstruction of normal CSF flow can be caused by blood clots in the ventricles (especially in premature infants), a tumor, meningitis, or trauma. Inserting a flexible tube, or shunt, into the ventricle relieves the pressure by allowing CSF to drain into the abdominal cavity or the cardiovascular system.

Figure 12.25 Baby with hydrocephalus.

Blood Brain Barrier

No other body tissue is so absolutely dependent on a constant internal environment as is the brain. In other body regions, the extracellular concentrations of hormones, amino acids, and ions are in constant flux, particularly after eating or exercise. If the brain were exposed to such chemical variations, its neurons would fire uncontrollably, because some hormones and amino acids serve as neurotransmitters and certain ions (particularly K$^+$) modify the threshold for neuronal firing.

The **blood brain barrier** is the protective mechanism that helps maintain the brain's stable environment. Exceptionally impermeable *tight junctions* between capillary endothelial cells are its major component (**Figure 12.26**). Surrounding the endothelial cells are the bulbous feet of astrocytes and smooth muscle–like cells called *pericytes*. Together these supporting cells regulate and maintain the endothelial cells. In particular, they signal the endothelial cells to form tight junctions that are very tight.

The presence of tight junctions prevents molecules from moving between the endothelial cells from the blood to the brain (◄ p. 67). Instead, everything must pass *through* the plasma membrane into an endothelial cell and out the other side.

- **Simple diffusion allows lipid-soluble subtances to pass freely.** Fats, fatty acids, oxygen, carbon dioxide, and other lipid-soluble molecules diffuse easily through the plasma membranes (◄ p. 70). This explains why drugs that affect the brain (including alcohol, nicotine, and anesthetics) are lipid soluble.

- **Specific transport mechanisms move substances important to the brain.** Nutrients such as glucose, amino acids, and specific ions move passively by facilitated diffusion. Transcytosis (◄ p. 76) moves larger substances into and out of the brain. Bloodborne metabolic wastes, proteins, certain toxins, and most drugs are denied entry. Small nonessential amino acids and potassium ions not only are prevented from entering the brain, but are also actively pumped from the brain across the capillary endothelium.

The *basement membrane* of the endothelial cells forms the last part of the blood brain barrier. It contains enzymes that destroy certain chemicals (such as epinephrine and norepinephrine) that would otherwise activate brain neurons.

The structure of the blood brain barrier is not completely uniform. In some brain areas surrounding the third and fourth ventricles, the barrier is entirely absent and the capillary endothelium is quite permeable, allowing bloodborne molecules easy access to the neural tissue. One such region is the vomiting center of the brain stem, which monitors the blood for poisonous substances. Another is in the hypothalamus, which regulates water balance, body temperature, and many other metabolic activities. Lack of a blood

Figure 12.26 The blood brain barrier.

Most capillaries	Brain capillaries
Interstitial fluid	Neuron
Endothelial cells	Astrocyte feet
Blood plasma	Astrocytes (with pericytes) induce formation of the tight junctions.
Pericytes	
Basement membrane	
Small gaps (intercellular clefts) between endothelial cells allow mixing between blood plasma and interstitial fluid.	**Tight junctions** between endothelial cells: • Are the key structures of the blood brain barrier • Exclude molecules except those that are lipid soluble from brain tissue

brain barrier here is essential to allow the hypothalamus to sample the chemical composition of the blood. The barrier is incomplete in newborn and premature infants, and potentially toxic substances can enter the CNS and cause problems not seen in adults.

Injury to the brain, whatever the cause, may result in a localized breakdown of the blood brain barrier. Most likely, this breakdown reflects some change in the capillary endothelial cells or their tight junctions.

Check Your Understanding

20. What is CSF? Where is it produced? What are its functions?
21. Identify structures a–e in the frontal section below. What is the function of structure (b)? Between which two structures would you find CSF? Name the two parts of structure (c).

For answers, see Answers Appendix.

12.9 Brain injuries and disorders have devastating consequences

CLINICAL

Learning Outcomes

▶ Describe the cause (if known) and major signs and symptoms of cerebrovascular accidents, Alzheimer's disease, Parkinson's disease, and Huntington's disease.
▶ List and explain several techniques used to diagnose brain disorders.

Brain dysfunctions are unbelievably varied and extensive. We have mentioned some of them already, but here we will focus on traumatic brain injuries, cerebrovascular accidents, and degenerative brain disorders.

Traumatic Brain Injuries

Head injuries are a leading cause of accidental death in North America. Consider, for example, what happens if you forget to fasten your seat belt and then rear-end another car. Your head is moving and then stops suddenly as it hits the windshield. Brain damage is caused not only by localized injury at the site of the blow, but also by the ricocheting brain hitting the opposite end of the skull.

A **concussion** is an alteration in brain function, usually temporary, following a blow to the head. The victim may be dizzy or lose consciousness. Although typically mild and short-lived,

even a seemingly mild concussion can be damaging, and multiple concussions over time produce cumulative damage.

More serious concussions can bruise the brain and cause permanent neurological damage, a condition called a **contusion**. In cortical contusions, the individual may remain conscious. Severe brain stem contusions always cause coma, lasting from hours to a lifetime because of injury to the reticular activating system.

Following a head injury, death may result from **subdural** or **subarachnoid hemorrhage** (bleeding from ruptured vessels into those spaces). Individuals who are initially lucid and then begin to deteriorate neurologically are, in all probability, hemorrhaging intracranially. Blood accumulating in the skull increases intracranial pressure and compresses brain tissue. If the pressure forces the brain stem inferiorly through the foramen magnum, control of blood pressure, heart rate, and respiration is lost. Intracranial hemorrhages are treated by surgically removing the hematoma (localized blood mass) and repairing the ruptured vessels.

Another consequence of traumatic head injury is **cerebral edema**, swelling of the brain. At best, cerebral edema aggravates the injury. At worst, it can be fatal in and of itself.

Cerebrovascular Accidents (CVAs)

The single most common nervous system disorder and the third leading cause of death in North America are **cerebrovascular accidents (CVAs)** (ser″ĕ-bro-vas′ku-lar), also called *strokes*. CVAs occur when blood circulation to a brain area is blocked and brain tissue dies of **ischemia** (is-ke′me-ah), a reduction of blood supply that impairs the delivery of oxygen and nutrients.

The most common cause of CVA is a blood clot that blocks a cerebral artery. A clot can originate outside the brain (from the heart, for example) or form on the roughened interior wall of a brain artery narrowed by atherosclerosis. Less frequently, strokes are caused by bleeding, which compresses brain tissue.

Many who survive a CVA are paralyzed on one side of the body (*hemiplegia*). Others commonly exhibit sensory deficits or have difficulty understanding or vocalizing speech. Even so, the picture is not hopeless. Some patients recover at least part of their lost faculties, because undamaged neurons sprout new branches that spread into the injured area and take over some lost functions. If motor function is impaired, physical therapy should start as soon as possible to prevent muscle contractures (abnormally shortened muscles due to differences in strength between opposing muscle groups).

Not all strokes are "completed." Temporary episodes of reversible cerebral ischemia, called **transient ischemic attacks (TIAs)**, are common. TIAs last from 5 to 50 minutes and are characterized by temporary numbness, paralysis, or impaired speech. These deficits are not permanent, but TIAs do constitute "red flags" that warn of impending, more serious CVAs.

A CVA is like an undersea earthquake. It's not the initial temblor that does most of the damage, it's the tsunami that floods the coast later. Similarly, the initial vascular blockage during a stroke is not usually disastrous because there are many blood vessels in the brain that can pick up the slack. Rather, it's the neuron-killing events outside the initial ischemic zone that wreak the most havoc.

Experimental evidence indicates that the main culprit is *glutamate*, an excitatory neurotransmitter. Glutamate plays a key role in learning and memory, as well as other critical brain functions. However, after brain injury, neurons totally deprived of oxygen begin to disintegrate, unleashing the cellular equivalent of a tidal wave of glutamate. Under these conditions, glutamate acts as an *excitotoxin*, literally exciting surrounding cells to death.

At present, the most successful treatment for stroke is tissue plasminogen activator (tPA), which dissolves blood clots in the brain. Alternatively, a mechanical device can drill into a blood clot and pull it from a blood vessel like a cork from a bottle.

Degenerative Brain Disorders

Alzheimer's Disease

Alzheimer's disease (AD) (altz′hi-merz), a progressive degenerative disease of the brain, ultimately results in **dementia** (mental deterioration). AD patients represent nearly half of the people living in nursing homes. Between 5 and 15% of people over 65 develop this condition, and for up to half of those over 85 it is a major contributing cause in their deaths.

AD patients exhibit memory loss (particularly for recent events), shortened attention span, disorientation, and eventual language loss. Over a period of several years, formerly good-natured people may become irritable, moody, and confused.

Examinations of brain tissue reveal senile plaques littering the brain like shrapnel between the neurons. The plaques consist of extracellular aggregations of *beta-amyloid peptide*, which is cut from a normal membrane precursor protein (APP) by enzymes. One form of Alzheimer's disease is caused by an inherited mutation in the gene for APP, which suggests that too much beta-amyloid may be toxic. Current clinical trials focus on using the immune system to clear away beta-amyloid peptide.

Another hallmark of Alzheimer's disease is the presence of *neurofibrillary tangles* inside neurons. These tangles involve a protein called tau, which functions like railroad ties to bind microtubule "tracks" together. In the brains of AD patients, tau abandons its microtubule-stabilizing role and grabs on to other tau molecules, forming spaghetti-like neurofibrillary tangles, which kill the neurons by disrupting their transport mechanisms.

Both plaques and tangles come about because the proteins that comprise them have misfolded. These misfolded proteins clump together, and also catalyze the misfolding of normally folded copies of the same proteins. Evidence suggests that this protein misfolding spreads in a predictable way from one region of the brain to another. This helps explain both the different types of dementia and their progression, as neurons in more and more brain regions die.

As the brain cells die, their functions are lost and the brain shrinks. Particularly vulnerable brain areas include the hippocampus, the basal forebrain, and association areas of the cortex, all regions involved in thinking and memory (**Figure 12.27**). Neurons that use the neurotransmitter acetylcholine are particularly vulnerable, and their loss is associated with a shortage of this neurotransmitter. Drugs that inhibit breakdown of acetylcholine slightly enhance cognitive function in AD patients.

Parkinson's Disease

Parkinson's disease results from a degeneration of the dopamine-releasing neurons of the substantia nigra. As those neurons deteriorate, the dopamine-deprived basal nuclei they target become overactive. Afflicted individuals have a persistent tremor at rest (exhibited by "pill-rolling" movements of the fingers and wrist), a forward-bent walking posture and shuffling gait, and a stiff facial expression. They are slow initiating and executing movement.

Parkinson's disease becomes more common with age. Its cause is still unknown, but multiple factors may interact to destroy dopamine-releasing neurons. Recent evidence points to abnormalities in certain mitochondrial proteins and protein degradation pathways. The drug L-dopa helps to alleviate some symptoms. It passes through the blood brain barrier and is then converted into dopamine.

However, as more and more neurons die off, L-dopa becomes ineffective. Mixing L-dopa with drugs that inhibit the breakdown of dopamine can prolong its effectiveness. Early in the disease, these drugs alone slow the neurological deterioration to some extent and delay the need to administer L-dopa.

Deep brain stimulation via implanted electrodes shuts down abnormal brain activity and can alleviate tremors. This treatment (for patients who no longer respond to drug therapy) is expensive and risky, but it works. Another possible future treatment is to use gene therapy to insert into adult brain cells the genes that would cause them to secrete the inhibitory neurotransmitter GABA. GABA then inhibits the abnormal brain activity just as the electrical stimulation does. Replacing dead or damaged cells by implanting stem cells is promising, but results to date are no better than more conventional treatments.

Huntington's Disease

Huntington's disease, a fatal hereditary disorder, strikes during middle age. Mutant *huntingtin* protein accumulates in brain cells and the tissue dies, leading to massive degeneration of the basal nuclei and later of the cerebral cortex. Its initial symptoms in many are wild, jerky, almost continuous "flapping"

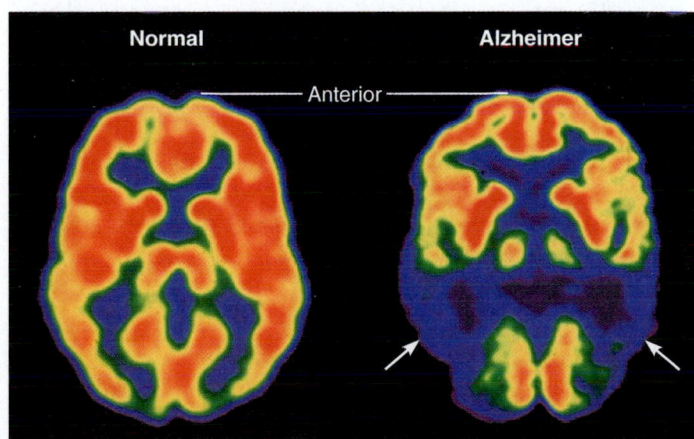

Figure 12.27 Brain activity is decreased by Alzheimer's disease. In these PET scans, high neural activity is indicated by reds and yellows. Arrows point to large decreases in parietal cortex activity.

movements called *chorea* (Greek for "dance"). Although the movements appear to be voluntary, they are not. Late in the disease, marked mental deterioration occurs. Huntington's disease is progressive and usually fatal within 15 years.

The hyperkinetic manifestations of Huntington's disease are essentially the opposite of those of Parkinson's disease (over-stimulation rather than inhibition of the motor drive). Huntington's is usually treated with drugs that block, rather than enhance, dopamine's effects.

Diagnostic Procedures for Assessing CNS Dysfunction

If you have had a routine physical examination, you are familiar with the reflex tests done to assess neural function. A tap with a reflex hammer stretches your quadriceps tendon and your anterior thigh muscles contract. This produces the knee-jerk response, which is a basic screening test for spinal cord and upper brain centers.

Abnormal responses to a reflex test may indicate such serious disorders as intracranial hemorrhage, multiple sclerosis, or hydrocephalus. They indicate that more sophisticated neurological tests are needed to identify the problem.

New imaging techniques have revolutionized the diagnosis of brain lesions (◀ *A Closer Look*, pp. 16–17). Together, various *CT* and *MRI scanning techniques* allow quick identification of most tumors, intracranial lesions, multiple sclerosis plaques, and areas of dead brain tissue (infarcts). PET scans can localize brain lesions that generate seizures (epileptic tissue), and radiotracer dyes that bind to beta-amyloid allow earlier, more reliable diagnosis of Alzheimer's disease.

Check Your Understanding

22. What is a transient ischemic attack (TIA) and how is it different from a stroke?

23. **APPLY** Mrs. Lee, a neurology patient, seldom smiles, has a shuffling, stooped gait, and often spills her coffee. What degenerative brain disorder might she have?

For answers, see Answers Appendix.

12.10 The spinal cord is a reflex center and conduction pathway

Learning Outcomes

▶ Describe the gross and microscopic structure of the spinal cord.

▶ Distinguish between flaccid and spastic paralysis, and between paralysis and paresthesia.

Gross Anatomy and Protection

The spinal cord, enclosed in the vertebral column, extends from the foramen magnum of the skull to the first or second lumbar vertebra, just inferior to the ribs (**Figure 12.28**). About 42 cm (17 inches) long and 1.8 cm (3/4 of an inch) thick, the glistening-white **spinal cord** provides a two-way conduction pathway to

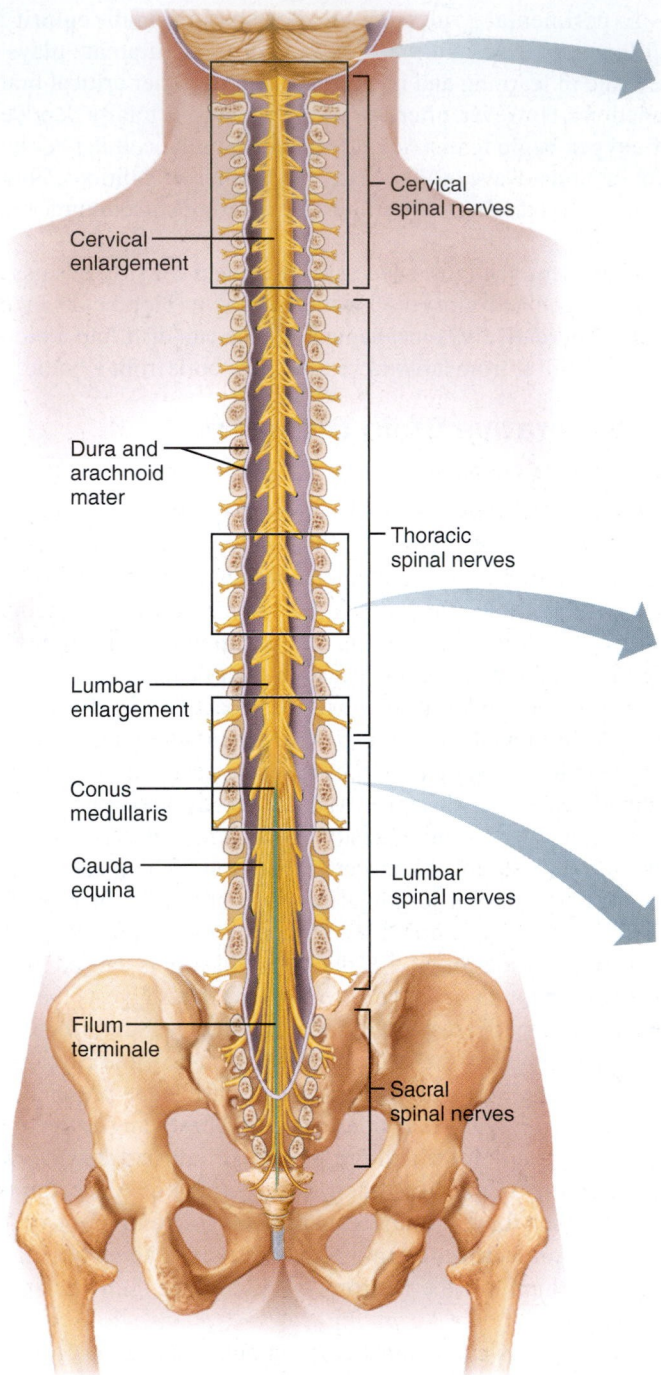

(a) The spinal cord and its nerve roots, with the bony vertebral arches removed. The dura mater and arachnoid mater are cut open and reflected laterally.

Figure 12.28 Gross structure of the spinal cord, dorsal view.

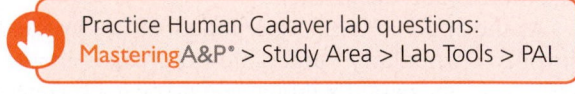

Practice Human Cadaver lab questions:
MasteringA&P® > Study Area > Lab Tools > PAL

Foramen magnum

Terminus of medulla oblongata of brain

Sectioned pedicles of cervical vertebrae

Spinal nerve rootlets

Dorsal median sulcus of spinal cord

(b) Cervical spinal cord.

Spinal cord

Vertebral arch (cut)

Cut edge of dura mater

Dorsal root ganglion

Dorsal median sulcus

Dorsal root

(c) Thoracic spinal cord.

Spinal cord

First lumbar vertebral arch (cut across)

Conus medullaris

Cauda equina

Filum terminale

(d) Inferior end of spinal cord, showing conus medullaris, cauda equina, and filum terminale.

and from the brain. It is a major reflex center: Spinal reflexes are initiated and completed at the spinal cord level. We discuss reflex functions and motor activity of the cord in subsequent chapters. In this module we focus on the anatomy of the spinal cord and its ascending and descending tracts.

Like the brain, the spinal cord is protected by bone, meninges, and cerebrospinal fluid. The single-layered **spinal dura mater** (Figure 12.28c) is not attached to the bony walls of the vertebral column. Between the bony vertebrae and the spinal dura mater is an **epidural space** filled with a soft padding of fat and a network of veins (see Figure 12.31a). Cerebrospinal fluid fills the subarachnoid space between the *arachnoid* and *pia mater* meninges.

Inferiorly, the dural and arachnoid membranes extend to the level of S_2, well beyond the end of the spinal cord. The spinal cord typically ends between L_1 and L_2 (Figure 12.28a). For this reason, the subarachnoid space within the meningeal sac inferior to that point provides an ideal spot for removing cerebrospinal fluid for testing, a procedure called a **lumbar puncture** (**Figure 12.29**). Because the spinal cord is absent there and the delicate nerve roots drift away from the point of needle insertion, there is little or no danger of damaging the cord (or spinal roots) beyond L_3.

Inferiorly, the spinal cord terminates in a tapering cone-shaped structure called the **conus medullaris** (ko′nus mĕ″dul-ar′is). The **filum terminale** (fi′lum ter″mĭ-nah′le; "terminal filament"), a fibrous extension of the conus covered by pia mater, extends inferiorly from the conus medullaris to the coccyx, where it anchors the spinal cord so it is not jostled by body movements (Figure 12.28a, d). Furthermore, saw-toothed

12

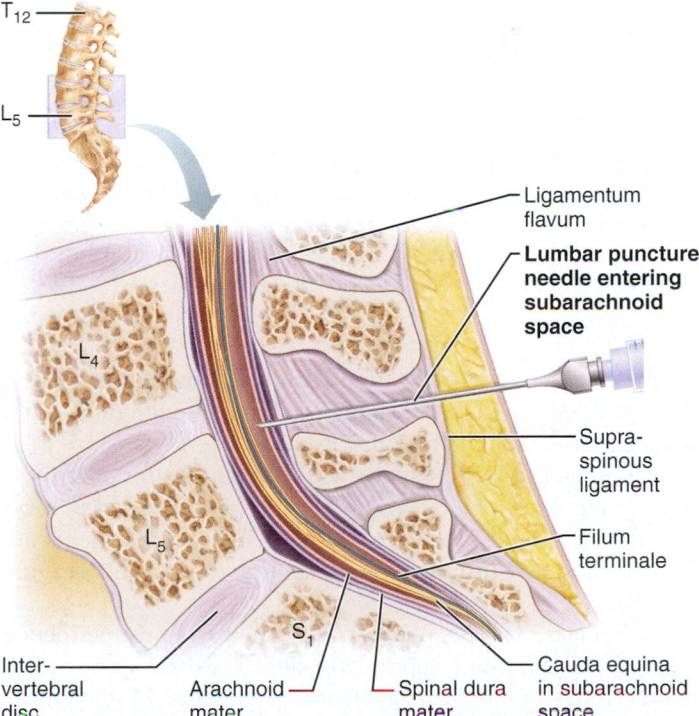

T_{12}

L_5

Ligamentum flavum

Lumbar puncture needle entering subarachnoid space

L_4

L_5

Supra-spinous ligament

Filum terminale

Inter-vertebral disc

Arachnoid mater

Spinal dura mater

S_1

Cauda equina in subarachnoid space

Figure 12.29 Diagram of a lumbar puncture. This procedure removes CSF for testing.

Dorsal (posterior)   Ventral (anterior)

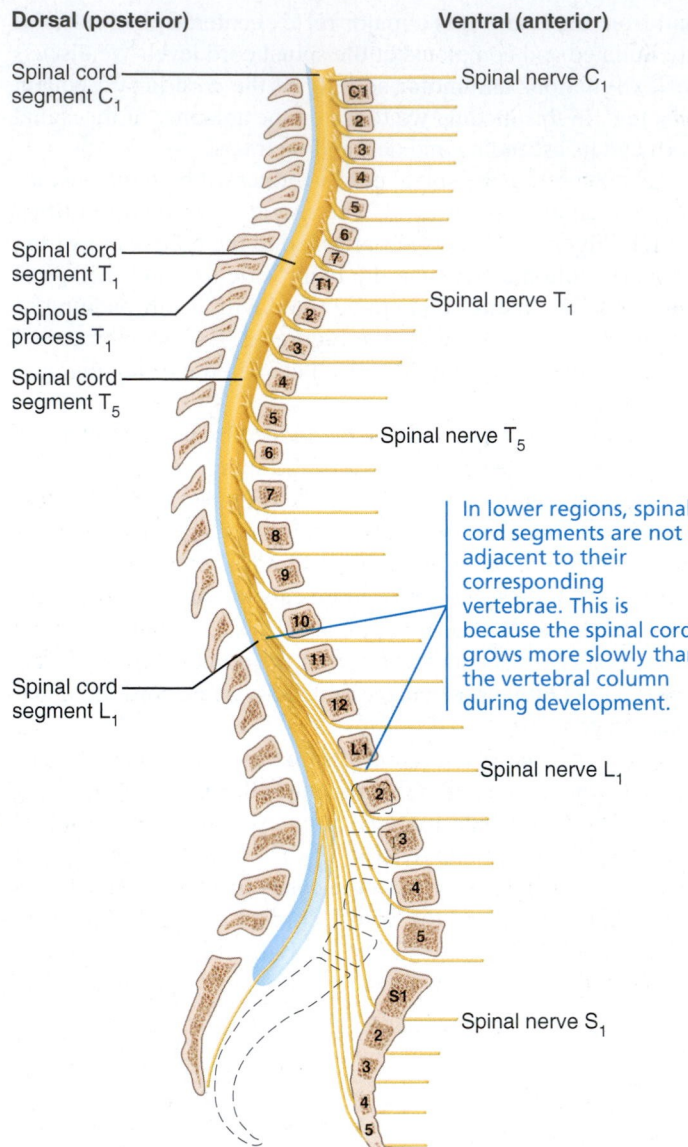

Spinal cord segment C₁ — Spinal nerve C₁

Spinal cord segment T₁ —
Spinous process T₁ —
Spinal cord segment T₅ —

Spinal nerve T₁

Spinal nerve T₅

In lower regions, spinal cord segments are not adjacent to their corresponding vertebrae. This is because the spinal cord grows more slowly than the vertebral column during development.

Spinal cord segment L₁ —

Spinal nerve L₁

Spinal nerve S₁

Figure 12.30 Spinal cord segments. Diagram of the spinal cord in sagittal view showing the position of spinal cord segments and the spinal nerves. The vertebral bodies, shown on the right, are numbered. Those in the lumbar and sacral regions are displaced anteriorly to illustrate the emergence of the spinal nerves.

shelves of pia mater called **denticulate ligaments** (den-tik′u-lāt; "toothed") secure the spinal cord to the tough dura mater meninx throughout its length.

The spinal cord is about the width of a thumb for most of its length, but it has obvious enlargements where the nerves serving the upper and lower limbs arise. These enlargements are the **cervical** and **lumbar enlargements**, respectively (Figure 12.28a).

In humans, 31 pairs of *spinal nerves*—part of the peripheral nervous system (see Chapter 13)—attach to the cord by paired **roots**. Each *spinal cord segment* is designated by the paired spinal nerves that arise from it—for example, the first thoracic cord segment (spinal cord segment T₁) is where the first thoracic

nerves (spinal nerve T₁) emerge from the spinal cord (**Figure 12.30**). While each nerve pair defines a segment of the cord, the spinal cord is, in fact, continuous throughout its length and its internal structure changes gradually. Each nerve exits from the vertebral column by passing adjacent to its corresponding vertebra via the intervertebral foramen, and travels to the body region it serves. (We will discuss the spinal nerves in more detail in Chapter 13.)

Because the cord does not reach the end of the vertebral column, the spinal cord segments are located superior to where their corresponding spinal nerves emerge through the intervertebral foramina (Figure 12.30). The lumbar and sacral spinal nerve roots angle sharply downward and travel inferiorly through the vertebral canal for some distance before reaching their intervertebral foramina. The collection of nerve roots at the inferior end of the vertebral canal is named the **cauda equina** (kaw′da e-kwi′nuh) because it resembles a horse's tail (Figure 12.28a, d). This strange arrangement reflects the fact that during fetal development, the vertebral column grows faster than the spinal cord, forcing the lower spinal nerve roots to "chase" their exit points inferiorly through the vertebral canal.

Spinal Cord Cross-Sectional Anatomy

The spinal cord is somewhat flattened from front to back and two grooves mark its surface: the wide **ventral (anterior) median fissure** and the narrower **dorsal (posterior) median sulcus** (**Figure 12.31b**). These grooves run the length of the cord and partially divide it into right and left halves. The gray matter of the cord is located in its core, the white matter outside. The cerebrospinal fluid–filled **central canal** runs the length of the spinal cord.

Gray Matter and Spinal Roots

In cross section the gray matter of the cord looks like the letter H or like a butterfly (Figure 12.31b). It consists of mirror-image lateral gray masses connected by a crossbar of gray matter, the **gray commissure**, that encloses the central canal. The two dorsal projections of the gray matter are the **dorsal (posterior) horns**, and the ventral pair are the **ventral (anterior) horns**. In 3-D, these horns form columns of gray matter that run the entire length of the spinal cord. The thoracic and superior lumbar segments of the cord have an additional pair of gray matter columns, the small **lateral horns**.

All neurons whose cell bodies are in the spinal cord gray matter are multipolar. The dorsal horns consist entirely of interneurons. The ventral horns have some interneurons but mainly house cell bodies of somatic motor neurons. These motor neurons send their axons out to the skeletal muscles (their effector organs) via the *ventral rootlets* that fuse together to become the **ventral roots** of the spinal cord (Figure 12.31b).

The amount of ventral gray matter present at a given level of the spinal cord reflects the amount of skeletal muscle innervated at that level. As a result, the ventral horns are largest in the limb-innervating cervical and lumbar regions of the cord and are responsible for the cord enlargements seen in those regions.

Epidural space (contains fat)

Subdural space

Subarachnoid space (contains CSF)

Pia mater
Arachnoid mater — Spinal meninges
Dura mater

Bone of vertebra

Dorsal root ganglion

Body of vertebra

(a) Cross section of spinal cord and vertebra (cervical region)

White columns
  Dorsal funiculus
  Ventral funiculus
  Lateral funiculus

Dorsal root ganglion

Spinal nerve

Dorsal root (fans out into dorsal rootlets)

Ventral root (derived from several ventral rootlets)

Dorsal median sulcus
Gray commissure
Dorsal horn
Ventral horn — Gray matter
Lateral horn

Central canal

Ventral median fissure

Pia mater

Arachnoid mater

Spinal dura mater

(b) The spinal cord and its meningeal coverings (thoracic region)

Figure 12.31 Anatomy of the spinal cord. (a) Cross section through the spinal cord illustrating its relationship to the surrounding vertebral column. **(b)** Anterior view of the spinal cord and its meningeal coverings.

The lateral horns consist mostly of the cell bodies of autonomic (sympathetic division) motor neurons that serve visceral organs. Their axons leave the cord via the ventral root along with those of the somatic motor neurons. Because the ventral roots contain both somatic and autonomic efferent fibers, they serve both motor divisions of the peripheral nervous system (see Figure 12.32). **Table 12.2** on p. 474 summarizes the characteristics of the three horns in the spinal cord.

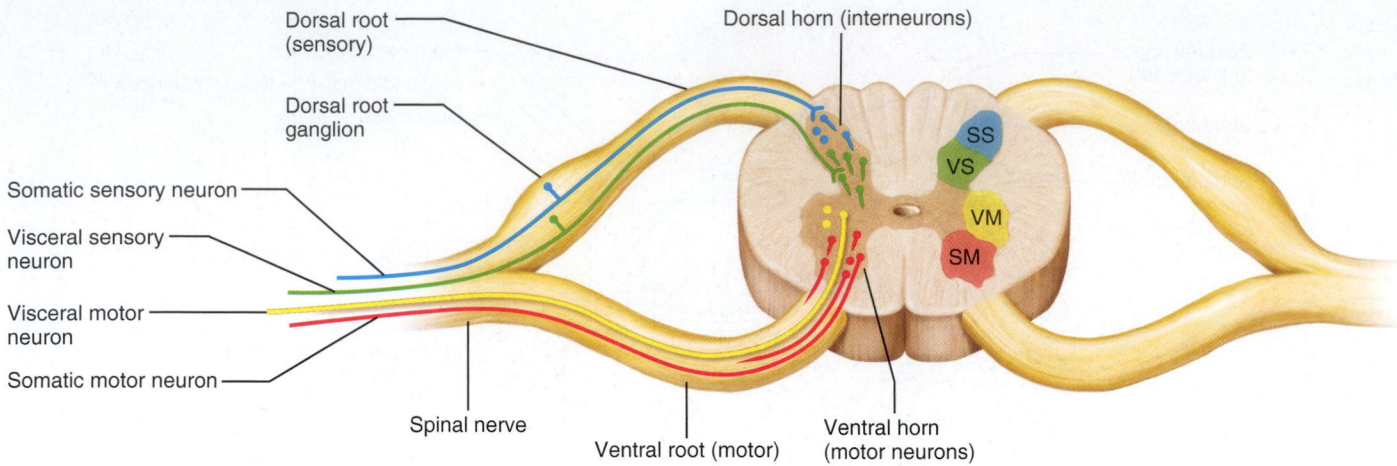

Figure 12.32 Organization of the gray matter of the spinal cord. The gray matter of the spinal cord is divided into a sensory half dorsally and a motor half ventrally. Note that the dorsal and ventral roots are part of the PNS, not of the spinal cord.

SS	Interneurons receiving input from somatic sensory neurons
VS	Interneurons receiving input from visceral sensory neurons
VM	Visceral motor (autonomic) neurons
SM	Somatic motor neurons

Afferent fibers carrying impulses from peripheral sensory receptors form the **dorsal roots** of the spinal cord that fan out as the *dorsal rootlets* before they enter the spinal cord (Figure 12.31). The cell bodies of the associated sensory neurons are found in an enlarged region of the dorsal root called the **dorsal root ganglion** or **spinal ganglion**. After entering the cord, the axons of these neurons may take a number of routes. Some enter the dorsal white matter of the cord directly and travel to synapse at higher cord or brain levels. Others synapse with interneurons in the dorsal horns of the spinal cord gray matter at their entry level. The dorsal and ventral roots are very short and fuse laterally to form the **spinal nerves** (see Chapter 13).

The spinal gray matter can be divided further according to its neurons' relative involvement in innervating the somatic and visceral regions of the body. Spinal gray matter has the following four zones (**Figure 12.32**): **somatic sensory (SS)**, **visceral sensory (VS)**, **visceral (autonomic) motor (VM)**, **somatic motor (SM)**.

White Matter

The white matter of the spinal cord is composed of myelinated and nonmyelinated nerve fibers that allow communication

between different parts of the spinal cord and between the cord and brain. These fibers run in three directions:

- *Ascending*—up to higher centers (sensory inputs)
- *Descending*—down to the cord from the brain or within the cord to lower levels (motor outputs)
- *Transverse*—across from one side of the cord to the other (commissural fibers)

Ascending and descending tracts make up most of the white matter.

The white matter on each side of the cord is divided into three **white columns**, or **funiculi** (fu-nik′u-li; "long ropes"), named according to their position as **dorsal (posterior)**, **lateral**, and **ventral (anterior) funiculi** (Figure 12.31b). Each funiculus contains several fiber tracts, and each tract is made up of axons with similar destinations and functions. With a few exceptions, the names of the spinal tracts reveal both their origin and destination. **Figure 12.33** schematically illustrates the principal ascending and descending tracts of the spinal cord in cross-sectional view.

Spinal Cord Trauma and Disorders

CLINICAL

Spinal Cord Trauma

The spinal cord is elastic, stretching with every turn of the head or bend of the trunk, but it is exquisitely sensitive to direct pressure. Any localized injury to the spinal cord or its roots leads to some functional loss. Damage to the dorsal roots or sensory tracts results in either loss of sensation or **paresthesias** (abnormal sensations). Damage to ventral roots or ventral horn cells results in **paralysis** (loss of motor function). Two types of paralysis may occur:

Table 12.2	Characteristics of Spinal Cord Horns	
HORN	**IMPORTANT NEURON CELL BODIES**	**ASSOCIATED NERVE ROOT**
Dorsal horns	Interneurons (receive input from somatic and visceral sensory neurons)	Dorsal roots
Lateral horns (when present)	Autonomic (visceral) motor neurons	Ventral roots
Ventral horns	Somatic motor neurons	Ventral roots

Ascending tracts

Dorsal white column
• Fasciculus gracilis
• Fasciculus cuneatus

Spinocerebellar tracts
• Dorsal
• Ventral

Spinothalamic tracts
• Lateral
• Ventral

Descending tracts

Corticospinal tracts (pyramidal tracts)
• Lateral
• Ventral

Rubrospinal tract

Reticulospinal tracts
• Medial
• Lateral

Vestibulospinal tract

Tectospinal tract

Figure 12.33 Major ascending (sensory) and descending (motor) tracts of the spinal cord, cross-sectional view.

• **Flaccid paralysis** (flak′sid) of the skeletal muscles occurs when the spinal cord or ventral roots are injured. Nerve impulses do not reach the affected muscles, which consequently cannot move either voluntarily or involuntarily. Without stimulation, the muscles atrophy.

• **Spastic paralysis** occurs if only the upper motor neurons of the primary motor cortex (or their axons in the spinal cord) are damaged. In this case, the spinal motor neurons remain intact and spinal reflex activity continues to stimulate the muscles irregularly. As a result, the muscles remain healthy longer, but their movements are no longer subject to voluntary control. In many cases, the muscles shorten permanently.

Transection (cross sectioning) of the spinal cord at any level results in total motor and sensory loss in body regions inferior to the site of damage.

• If the transection occurs between T_1 and L_1, both lower limbs are affected, resulting in **paraplegia** (par″ah-ple′je-ah; *para* = beside, *plegia* = a blow).

• If the injury occurs in the cervical region, all four limbs are affected and the result is **quadriplegia**.

Hemiplegia, paralysis of one side of the body, usually reflects brain injury rather than spinal cord injury.

Anyone with a spinal cord injury must be watched for symptoms of **spinal shock**, a transient period of functional loss that follows the injury. Spinal shock immediately depresses all reflex activity caudal to the lesion site. Bowel and bladder reflexes stop, blood pressure falls, and all muscles (somatic and visceral alike) below the injury are paralyzed and insensitive. Neural function usually returns within a few hours following injury. If function does not resume within 48 hours, paralysis is permanent in most cases.

Poliomyelitis

Poliomyelitis (po″le-o-mi″ĕ-li′tis; *polio* = gray matter; *myelitis* = inflammation of the spinal cord) results from the poliovirus, which typically enters the body in feces-contaminated water and destroys ventral horn motor neurons. Early symptoms include fever, headache, muscle pain and weakness, and loss of certain somatic reflexes. Later, paralysis develops and the muscles served atrophy. The patient may die from paralyzed respiratory muscles. Fortunately, vaccines have nearly eliminated this disease and a global effort is ongoing to eradicate it completely.

However, many survivors of the great polio epidemic of the late 1940s and 1950s have begun to experience extreme lethargy, sharp burning pains in their muscles, and progressive muscle weakness and atrophy. These disturbing symptoms are referred to as **postpolio syndrome**. The cause of postpolio syndrome is not known, but a likely explanation is that postpolio patients, like the rest of us, continue to lose neurons throughout life. While a healthy nervous system can recruit nearby neurons to compensate, polio survivors have already drawn on that "pool" and have few neurons left to take over. Ironically, those who worked hardest to overcome their disease are the ones most affected later in life.

Amyotrophic Lateral Sclerosis (ALS)

Amyotrophic lateral sclerosis (ALS) (a-mi″o-trof′ik), also called Lou Gehrig's disease, is a devastating neuromuscular condition that progressively destroys ventral horn motor neurons and fibers of the pyramidal tracts (a major motor pathway). As the disease progresses, the patient loses the ability to speak, swallow, and breathe. Death typically occurs within five years.

Environmental and genetic factors interact to cause ALS. In 10% of cases mutations are inherited; spontaneous mutations are probably involved in the rest. Recently, the mutations have been localized to genes that are involved in RNA processing. While the exact mechanism is not clear, the presence of excess extracellular glutamate suggests that excitotoxic cell death is involved. Two drugs, one that interferes with glutamate signaling and a new drug that scavanges free radicals, are the only treatment options so far.

Check Your Understanding

24. What is the explanation for the cervical and lumbar enlargements of the spinal cord?

25. **APPLY** Trevor was tackled while playing football. After hitting the ground, he was unable to move his lower limbs. What is a loss of motor function called? What level of his spinal cord do you think was injured (cervical, thoracic, lumbar, or sacral)? Is this a permanent injury?

26. **DRAW** On the cross section of the spinal cord below, label the central canal, and dorsal and ventral roots and horns. Draw in a sensory neuron, showing the location of its cell body and axon terminals. What is in the central canal?

For answers, see Answers Appendix.

12.11 Neuronal pathways carry sensory and motor information to and from the brain

Learning Outcomes
▶ List the key characteristics of neuronal pathways.
▶ Identify the major ascending and descending pathways.

All major spinal tracts are part of *multineuron pathways* that connect the brain to the body periphery. These great ascending and descending pathways contain not only spinal cord neurons but also parts of peripheral neurons and neurons in the brain. There are four key points in regard to spinal tracts and pathways:

- **Decussation.** Most pathways cross from one side of the CNS to the other (decussate) at some point along their journey.

- **Relay.** Most pathways consist of a chain of two or three neurons (a relay) that contribute to successive tracts of the pathway.

- **Somatotopy.** Most pathways exhibit *somatotopy*, a precise spatial relationship among the tract fibers that reflects the orderly mapping of the body. For example, fibers transmitting pain and temperature information from sensory receptors in superior body regions lie medial to those from inferior body regions within the same tract.

- **Symmetry.** All pathways and tracts are paired symmetrically (right and left), with a member of the pair present on each side of the spinal cord or brain.

Ascending Pathways to the Brain

The ascending pathways conduct sensory impulses upward, typically through chains of three successive neurons (first-, second-, and third-order neurons) to various areas of the brain. Note that both second- and third-order neurons are interneurons.

- **First-order neurons**, whose cell bodies reside in a ganglion (dorsal root or cranial), conduct impulses from the cutaneous receptors of the skin and from proprioceptors to the spinal cord or brain stem, where they synapse with second-order neurons. Impulses from the facial area are transmitted by cranial nerves, and spinal nerves conduct somatic sensory impulses from the rest of the body to the CNS. First-order neurons entering the spinal cord are shown at the bottom of **Figure 12.34**.

- The cell bodies of **second-order neurons** (Figure 12.34, middle) reside in the dorsal horn of the spinal cord or in medullary nuclei. They transmit impulses to the thalamus or to the cerebellum where they synapse.

- **Third-order neurons** have cell bodies in the thalamus (Figure 12.34, top). They relay impulses to the somatosensory cortex of the cerebrum. (There are no third-order neurons in the cerebellum.)

In general, somatosensory information travels along three main pathways on each side of the spinal cord. Two of these pathways (the *dorsal column–medial lemniscal* and *spinothalamic pathways*) transmit impulses via the thalamus to the sensory cortex for conscious interpretation. Collectively the inputs of these sister tracts provide *discriminative touch* and *conscious proprioception*. Both pathways decussate—the first in the medulla and the second in the spinal cord.

The third pathway, the *spinocerebellar pathway*, terminates in the cerebellum, and does not contribute to sensory perception. Let's examine these pathways more closely.

- **Dorsal column–medial lemniscal pathways.** The **dorsal column–medial lemniscal pathways** (lem-nis′kul; "ribbon") mediate precise, straight-through transmission of inputs from a single type (or a few related types) of sensory receptor that can be localized precisely on the body surface, such as discriminative touch and vibrations. It also transmits information from proprioceptors. These pathways are formed by the paired tracts of the **dorsal white column** of the spinal cord—**fasciculus cuneatus** and **fasciculus gracilis**—and the **medial lemniscus**.

  The medial lemniscus arises in the medulla and terminates in the thalamus (Figure 12.34a and **Table 12.3** on p. 480). From the thalamus, impulses are forwarded to specific areas of the somatosensory cortex.

- **Spinothalamic pathways.** The **spinothalamic pathways** receive input from many different types of sensory receptors and make multiple synapses in the brain stem. These pathways consist of the **lateral** and **ventral (anterior) spinothalamic tracts** (see Figure 12.34b and Table 12.3). Their fibers cross over in the spinal cord.

  The fibers in these pathways primarily transmit impulses for pain and temperature, but also for coarse touch and

Primary somatosensory cortex

Axons of third-order neurons

Cerebrum

Thalamus

Midbrain

Cerebellum

Pons

Dorsal spinocerebellar tract (axons of second-order neurons)

Medial lemniscus (tract) (axons of second-order neurons)

Nucleus gracilis

Nucleus cuneatus

Medulla oblongata

Fasciculus cuneatus (axon of first-order sensory neuron)

Lateral spinothalamic tract (axons of second-order neurons)

Joint stretch receptor (proprioceptor)

Pain receptors

Axon of first-order neuron

Muscle spindle (proprioceptor)

Cervical spinal cord

Fasciculus gracilis (axon of first-order sensory neuron)

Axons of first-order neurons

Lumbar spinal cord

Temperature receptors

Touch receptor

(a) Spinocerebellar pathway   Dorsal column–medial lemniscal pathway   **(b)** Spinothalamic pathway

Figure 12.34 Pathways of selected ascending spinal cord tracts. Cross sections up to the cerebrum, which is shown in frontal section. **(a)** The spinocerebellar pathway (left) transmits proprioceptive information only to the cerebellum, and so is subconscious. The dorsal column–medial lemniscal pathway transmits discriminative touch and proprioception signals to the cerebral cortex. **(b)** The lateral spinothalamic pathway transmits pain and temperature information. See also Table 12.3 on p. 480.

12

pressure. All are sensations that we are aware of but have difficulty localizing precisely on the body surface.

- **Spinocerebellar pathways.** The third ascending pathway consists of the **ventral** and **dorsal spinocerebellar tracts**. They convey information about muscle or tendon stretch to the cerebellum, which uses this information to coordinate skeletal muscle activity (see Figure 12.34a and Table 12.3). As noted earlier, these pathways do not contribute to conscious sensation. The fibers of the spinocerebellar pathways either do not decussate or else cross over twice (thus "undoing" the decussation).

Descending Pathways and Tracts

The descending pathways that deliver efferent impulses from the brain to the spinal cord are divided into two groups: (1) the *direct pathways,* which are the pyramidal tracts, and (2) the *indirect pathways*, essentially all others. Motor pathways involve two neurons, referred to as the upper and lower motor neurons:

- *Upper motor neurons* are the pyramidal cells of the motor cortex (◄ p. 441) and the neurons of subcortical motor nuclei.

- *Lower motor neurons* are the ventral horn motor neurons. These directly innervate the skeletal muscles (their effectors).

We give a brief overview of these pathways below. See **Table 12.4** on p. 481 for more information.

Direct (Pyramidal) Pathways

The direct pathways originate mainly with the pyramidal cells located in the precentral gyri. These neurons send impulses through the brain stem via the large **pyramidal (corticospinal) tracts** (**Figure 12.35a**). The direct pathways are so called because their axons descend without synapsing from the pyramidal cells to the spinal cord. There they synapse either with interneurons or with ventral horn motor neurons.

Stimulation of the ventral horn neurons activates the skeletal muscles with which they are associated. The direct pathway primarily regulates fast and fine (or skilled) movements such as texting or playing an instrument.

Indirect Pathways

The indirect pathways include brain stem motor nuclei and *all motor pathways except* the pyramidal pathways. These pathways were formerly lumped together as the *extrapyramidal system* because their nuclei of origin were presumed to be independent of ("extra to") the pyramidal tracts. This term is still widely used clinically. However, pyramidal tract neurons are now known to project to and influence the activity of most "extrapyramidal" nuclei, so modern anatomists refer to them as **indirect**, or **multineuronal**, **pathways**, or simply use the names of the individual motor pathways.

Indirect motor pathways are complex and multisynaptic. They are most involved in regulating:

- Axial muscles that maintain balance and posture
- Muscles controlling coarse limb movements
- Head, neck, and eye movements that follow objects in the visual field

Many of the activities controlled by subcortical motor nuclei depend heavily on reflex activity. Figure 12.35b illustrates one of these tracts, the rubrospinal tract.

Overall, the **reticulospinal** and **vestibulospinal tracts** maintain balance by varying the tone of postural muscles (Table 12.4). The **rubrospinal tracts** control flexor muscles, whereas the **tectospinal tracts** and the *superior colliculi* mediate head movements in response to visual stimuli.

• • •

(a) Pyramidal (lateral and ventral corticospinal) pathways

(b) Rubrospinal tract

Figure 12.35 Three descending pathways by which the brain influences movement.
Cross sections up to the cerebrum, which is shown in frontal section. **(a)** Pyramidal (lateral and ventral corticospinal) pathways are direct pathways that control skilled voluntary movements. **(b)** The rubrospinal tract, one of the indirect pathways, helps regulate muscle tone. See also Table 12.4 on p. 481.

12

Table 12.3 **Major Ascending (Sensory) Pathways and Spinal Cord Tracts**

SPINAL CORD TRACT	LOCATION (FUNICULUS)	ORIGIN	TERMINATION	FUNCTION
Dorsal Column–Medial Lemniscal Pathways				
Fasciculus cuneatus and fasciculus gracilis (dorsal white column)	Dorsal	Central axons of sensory (first-order) neurons enter dorsal root of the spinal cord and branch. Branches enter dorsal white column on same side without synapsing.	By synapse with second-order neurons in nucleus cuneatus and nucleus gracilis in medulla. Fibers of medullary neurons cross over and ascend in medial lemniscus to thalamus, where they synapse with third-order neurons. Thalamic neurons then transmit impulses to somatosensory cortex.	Both tracts transmit sensory impulses from general sensory receptors of skin and proprioceptors, which are interpreted as discriminative touch, pressure, and "body sense" (limb and joint position) in opposite somatosensory cortex. Cuneatus transmits afferent impulses from upper limbs, upper trunk, and neck. Gracilis carries impulses from lower limbs and inferior body trunk.
Spinothalamic Pathways				
Lateral spinothalamic	Lateral	Interneurons (second-order neurons) in dorsal horns. Fibers cross to opposite side before ascending.	By synapse with third-order neurons in thalamus. Thalamic neurons then convey impulses to somatosensory cortex.	Transmits impulses concerned with pain and temperature to opposite side of brain for interpretation by somatosensory cortex.
Ventral spinothalamic	Ventral	Interneurons (second-order neurons) in dorsal horns. Fibers cross to opposite side before ascending.	By synapse with third-order neurons in thalamus. Thalamic neurons eventually convey impulses to somatosensory cortex.	Transmits impulses concerned with crude touch and pressure to opposite side of brain for interpretation by somatosensory cortex.
Spinocerebellar Pathways				
Dorsal spinocerebellar*	Lateral (dorsal part)	Interneurons (second-order neurons) in dorsal horn on same side of cord. Fibers ascend without crossing.	By synapse in cerebellum	Transmits impulses from trunk and lower limb proprioceptors on one side of body to same side of cerebellum for subconscious proprioception.
Ventral spinocerebellar*	Lateral (ventral part)	Interneurons (second-order neurons) of dorsal horn. Contains crossed fibers that cross back to the opposite side in the pons.	By synapse in cerebellum	Transmits impulses from the trunk and lower limb on the same side of body to cerebellum for subconscious proprioception.

*These spinocerebellar tracts carry information from the lower limbs and trunk only. The corresponding tracts for the upper limb and neck (rostral spinocerebellar and others) are beyond the scope of this book.

Our discussion of ascending and descending pathways has focused on the innervation of body regions inferior to the head. In general, the pathways to and from the head are similar to those of the trunk and limbs, with two exceptions:

- The axons of the tracts servicing the head are located in cranial nerves.
- The cell bodies are located in the brain stem rather than in the spinal cord.

A relatively new imaging technique, diffusion tensor MRI, allows visualization of fiber tracts in the CNS. These images are artifically colored to show the direction of the fiber (**Figure 12.36**).

Table 12.4 Major Descending (Motor) Pathways and Spinal Cord Tracts

SPINAL CORD TRACT	LOCATION (FUNICULUS)	ORIGIN	TERMINATION	FUNCTION
Direct (Pyramidal) Pathways				
Lateral corticospinal	Lateral	Pyramidal cells of motor cortex of the cerebrum. Cross over in pyramids of medulla.	By synapse with ventral horn interneurons that influence motor neurons and occasionally with ventral horn motor neurons directly.	Transmits motor impulses from cerebrum to spinal cord motor neurons (which activate skeletal muscles on opposite side of body). A voluntary motor tract.
Ventral corticospinal	Ventral	Pyramidal cells of motor cortex. Fibers cross over at the spinal cord level.	Ventral horn (as above).	Same as lateral corticospinal tract.
Indirect Pathways				
Tectospinal	Ventral	Superior colliculus of midbrain of brain stem (fibers cross to opposite side of cord).	Ventral horn (as above).	Turns head so eyes can follow a moving object.
Vestibulospinal	Ventral	Vestibular nuclei in medulla of brain stem (fibers descend without crossing).	Ventral horn (as above).	Transmits motor impulses that maintain muscle tone and activate ipsilateral limb and trunk extensor muscles and muscles that move head. Helps maintain balance during standing and moving.
Rubrospinal	Lateral	Red nucleus of midbrain of brain stem (fibers cross to opposite side just inferior to the red nucleus).	Ventral horn (as above).	In animals, transmits motor impulses concerned with muscle tone of distal limb muscles (mostly flexors) on opposite side of body. In humans, functions are largely assumed by corticospinal tracts.
Reticulospinal (medial and lateral)	Medial and lateral	Reticular formation of brain stem (medial nuclear group of pons and medulla). Both crossed and uncrossed fibers.	Ventral horn (as above).	Transmits impulses concerned with muscle tone and many visceral motor functions. May control most unskilled movements.

Figure 12.36 Diffusion tensor MRI reveals fiber tracts in the brain and spinal cord. An image of the fiber tracts has been superimposed on an image of the brain.

Check Your Understanding

27. Where are the cell bodies of the first-, second-, and third-order sensory neurons in the spinothalamic pathway located?

28. Three-year-old Jessica proudly shows you how she can wiggle her left big toe in the sand. Where precisely are the pyramidal cells that allow her to perform this movement? (Name the side, lobe, and region of the brain.) Where are the cell bodies of the neurons that these pyramidal cells synapse with?

29. `MAKE CONNECTIONS` Figure 12.34 shows both tracts and nerves, as well as nuclei and ganglia. How do tracts and nerves differ? How do nuclei and ganglia differ? (Hint: See Chapter 11, pp. 394–395.)

For answers, see Answers Appendix.

Developmental Aspects of the Central Nervous System

Figure 12.37 shows the earliest phase of CNS development. Starting in the three-week-old embryo:

① The ectoderm (cell layer at the dorsal surface) thickens along the dorsal midline axis of the embryo to form the **neural plate**. The neural plate invaginates, forming a **neural groove** flanked by **neural folds**.

② Small groups of neural fold cells migrate laterally from between the surface ectoderm and the neural groove, forming the **neural crest**. Neural crest cells give rise (among other things) to some neurons destined to reside in ganglia.

③ As the neural groove deepens, the superior edges of the neural folds fuse, forming the **neural tube**, which soon detaches from the surface ectoderm and sinks to a deeper position.

The neural tube, formed by the fourth week of pregnancy, differentiates rapidly into the CNS. The brain forms rostrally, as we described on p. 435, and the spinal cord develops from the caudal portion of the neural tube.

By week six, each side of the developing spinal cord has two recognizable clusters of neuroblasts that have migrated outward from the original neural tube: a dorsal **alar plate** (a′lar) and a ventral **basal plate** (**Figure 12.38**). Alar plate neuroblasts become interneurons. Basal plate neuroblasts develop into motor neurons and sprout axons that grow out to the effector organs. Axons that emerge from alar plate cells (and some basal plate cells) form the white matter of the cord. As development progresses, these plates expand dorsally and ventrally to produce the H-shaped central mass of gray matter in the adult spinal cord.

Neural crest cells that come to lie alongside the cord form the *dorsal root ganglia* containing sensory neuron cell bodies. These sensory neurons send their axons into the dorsal aspect of the cord.

As the brain and spinal cord grow and mature throughout the prenatal period, gender-specific areas appear. For example, certain hypothalamic nuclei concerned with regulating typical male sexual behavior and clusters of neurons in the spinal cord that serve the external genitals are much larger in males. The key to CNS gender-specific development is whether or not the fetus is secreting testosterone. If it is, the male pattern develops. The causes of other gender differences in the adult brain (for example, in the corpus callosum and language and auditory

① The neural plate forms from surface ectoderm. It then invaginates, forming the neural groove flanked by neural folds.

② Neural fold cells migrate to form the neural crest, which will form much of the PNS and many other structures.

③ The neural groove becomes the neural tube, which will form CNS structures.

Figure 12.37 Development of the neural tube from embryonic ectoderm. Dorsal surface views (left) and transverse sections (right) of the embryo at days 17, 20, and 22.

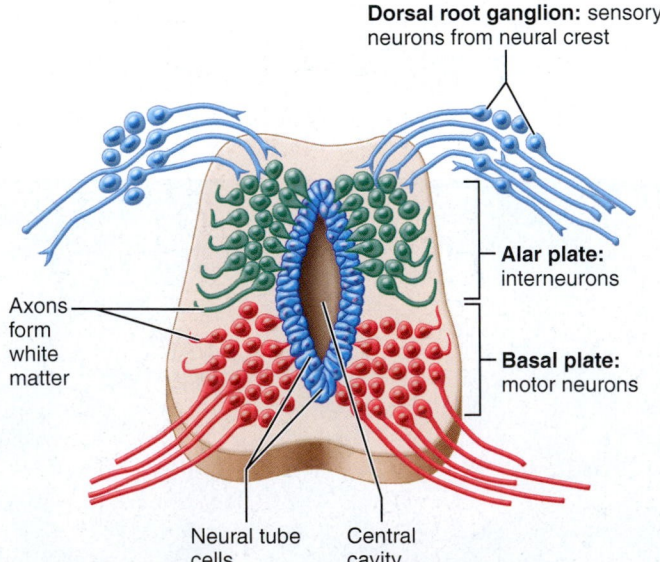

Figure 12.38 Structure of the embryonic spinal cord. At six weeks of development, aggregations of gray matter called the alar plates (future interneurons) and basal plates (future motor neurons) have formed. Dorsal root ganglia (future sensory neurons) have arisen from neural crest cells.

areas of the cerebral cortex) are not as clear, and may be due to life experiences more than hormones.

Maternal exposure to radiation, various drugs, alcohol, and infections can harm a developing infant's nervous system, particularly during the initial formative stages. For example, rubella (German measles) often leads to deafness and other types of CNS damage in the newborn. Smoking decreases the amount of oxygen in the blood, and lack of oxygen for even a few minutes destroys neurons. A mother who smokes may sentence her infant to some degree of brain damage.

HOMEOSTATIC IMBALANCE 12.11
CLINICAL

Cerebral palsy is a neuromuscular disability in which the voluntary muscles are poorly controlled or paralyzed as a result of brain damage. It may be caused by temporary lack of oxygen during a difficult delivery, or by any of the factors listed above.

In addition to spasticity, speech difficulties, and other motor impairments, about half of individuals with cerebral palsy have seizures, half are intellectually disabled, and about a third have some degree of deafness. Visual impairments are also common. Cerebral palsy does not get worse over time, but its deficits are irreversible. It is the largest single cause of physical disability in children, affecting three out of every 1000 births.

The CNS is plagued by a number of other congenital malformations triggered by genetic or environmental factors during early development. The most serious are congenital hydrocephalus (discussed previously), anencephaly, and spina bifida.

In **anencephaly** (an″en-sef′ah-le; "without brain"), the cerebrum and part of the brain stem never develop because the neural folds fail to fuse rostrally. The child is totally vegetative, unable to see, hear, or process sensory inputs. Muscles are flaccid, and no voluntary movement is possible. Mental life as we know it does not exist. Affected infants are stillborn or die within hours after birth.

Spina bifida (spi′nah bif′ĭ-dah; "forked spine") results from incomplete formation of the vertebral arches and typically involves the lumbosacral region. The technical definition is that laminae and spinous processes are missing on at least one vertebra. If the condition is severe, neural deficits occur as well.

Spina bifida occulta occurs when only one or a few vertebrae are abnormal, leaving the neural structures and coverings intact. A sacral dimple or patch of hair may hint at the underlying anomaly.

If the meninges protrude through the dorsal spine in a saclike structure called a *meningocele* (mĕ-ning′go-sēl), the situation is more serious. If the spinal cord or spinal nerve roots also protrude inside the sac, the problem is called a *myelomeningocele* (**Figure 12.39**). Infants born with these conditions undergo surgery within the first day or two of life to close the defect in the skin and prevent CNS infection. Most of the children will have significant neurological problems, including paralysis, hydrocephalus, and bowel and bladder dysfunction.

While the damage cannot be totally repaired, recent advances in fetal surgery have shown great promise in reducing disability. When myelomeningocele is diagnosed during pregnancy,

Figure 12.39 Newborn with a lumbar myelomeningocele.

surgeons open the uterus and repair the fetal defect, then return the fetus to the uterus to complete gestation.

Mothers with normal levels of folic acid are up to 70% less likely to deliver infants with neural tube defects like anencephaly or spina bifida. Mandatory supplementation of grain and cereal products have helped to ensure adequate maternal folic acid intake. In addition, the U.S. Public Health Service recommends that all women of childbearing age should take supplemental folic acid every day.

One of the last CNS areas to mature is the hypothalamus. Since the hypothalamus contains body temperature regulatory centers, premature babies have problems controlling their loss of body heat and must be kept in temperature-controlled environments. PET scans reveal that the thalamus and somatosensory cortex are active in a 5-day-old baby, but the visual cortex is not. This explains why infants of this age respond to touch but have poor vision. By 11 weeks, more of the cortex is active, and the baby can reach for a rattle. By 8 months, the cortex is very active and the child can think about what he or she sees.

Growth and maturation of the nervous system continue throughout childhood and largely reflect progressive myelination. As described in Chapter 9, neuromuscular coordination progresses in a superior-to-inferior direction and in a proximal-to-distal direction, and we know that myelination also occurs in this sequence.

The brain reaches its maximum weight in the young adult. Over the next 60 years or so, neurons are damaged and die, and brain weight and volume steadily decline. However, the number of neurons lost over the decades is normally only a small percentage of the total, and the remaining neurons can change their synaptic connections, providing for continued learning throughout life.

Although age brings some cognitive declines in spatial ability, speed of perception, decision making, reaction time, and working memory, these losses are not significant in the healthy individual until after the seventh decade. Then the brain becomes increasingly fragile, presumably due to less efficient calcium clearance in aging neurons (elevated Ca^{2+} levels are neurotoxic).

However, mathematical skills, verbal fluency, and the ability to build on experience do not decline with age, and many people continue to enjoy intellectual lives and work at mentally demanding tasks their entire life. Fewer elderly people than you might think demonstrate true dementia. Sadly, many cases of "reversible dementia" are caused by prescription drug side effects, low blood pressure, poor nutrition, hormone imbalances, depression, and/or dehydration that go undiagnosed.

Although eventual shrinking of the brain is normal and accelerates in old age, alcoholics and professional boxers hasten the process. Whether a boxer wins the match or not, the likelihood of brain damage and atrophy increases with every blow. And it is well known that alcohol profoundly affects both the mind and the body. CT scans of alcoholics reveal reduced brain size and density starting at a fairly early age. Both boxers and alcoholics exhibit signs of mental deterioration unrelated to the aging process.

RELATED CLINICAL TERMS

Autism A complex developmental neurological disorder, typically appearing in the first three years of life, characterized by difficulty in communicating, forming relationships with others, and responding appropriately to the environment. A wide variety of mutations in functionally related genes can give rise to autism and its related disorders. Occurs in about two per 1000 people, and early behavioral intervention is beneficial.

Cordotomy (kor-dot′o-me) A procedure in which a tract in the spinal cord is severed surgically; usually done to relieve unremitting pain.

Dyslexia A learning disability in 5 to 15% of the population that specifically affects the ability of otherwise intelligent people to read. This deficit in visual symbol and language processing is thought to result from errors arising in one hemisphere. Several genes that predispose children to dyslexia have been identified, but dyslexia can also be acquired by brain injury or degeneration.

Encephalopathy (en-sef″ah-lop′ah-the; *enceph* = brain; *path* = disease) Any disease or disorder of the brain.

Hypersomnia (*hyper* = excess; *somnus* = sleep) A condition in which affected individuals sleep as much as 15 hours daily.

Microcephaly (mi″kro-sef′ah-le; *micro* = small) Condition involving the formation of a small brain, as evidenced by

reduced skull size; most microcephalic children are intellectually disabled. Can be caused by Zika virus infection during gestation.

Myelitis (mi″ĕ-li′tis; *myel* = spinal cord; *itis* = inflammation) Inflammation of the spinal cord.

Myelogram (*gram* = recording) X ray of the spinal cord after injection of a contrast medium.

Myoclonus (mi″o-klo′nus; *myo* = muscle; *clon* = violent motion, tumult) Sudden contraction of a muscle or muscle part, usually involving muscles of the limbs. Myoclonal jerks can occur in normal individuals as they are falling asleep; others may be due to diseases of the reticular formation or cerebellum.

Neuroses (nu-ro′sēs) A less debilitating class of mental illness; examples include severe anxiety (panic attacks), phobias (irrational fears), and obsessive-compulsive behaviors (e.g., washing one's hands every few minutes). However, the affected individual retains contact with reality.

Psychoses (si-ko′sēs) A class of severe mental illness in which affected individuals lose touch with reality and exhibit bizarre behaviors; the legal word for psychotic behavior is insanity. Psychoses include schizophrenia (skit-so-fre′ne-ah), bipolar disorder, and some forms of depression.

CHAPTER SUMMARY

1. The brain provides for voluntary movements, interpretation and integration of sensation, consciousness, and cognitive function.

12.1 Folding during development determines the complex structure of the adult brain (pp. 435–438)

1. Early brain development yields the three primary brain vesicles: the prosencephalon (cerebral hemispheres and diencephalon), mesencephalon (midbrain), and rhombencephalon (pons, medulla oblongata, and cerebellum).
2. As a result of cephalization, the diencephalon and superior brain stem are enveloped by the cerebral hemispheres.

Brain Regions and Organization (p. 436)

3. In a widely used system, the adult brain is divided into the cerebral hemispheres, diencephalon, brain stem, and cerebellum.
4. The cerebral hemispheres and cerebellum have gray matter nuclei surrounded by white matter and an outer cortex of gray matter. The diencephalon and brain stem lack a cortex.

Ventricles (pp. 436–438)

5. The brain contains four ventricles filled with cerebrospinal fluid. The lateral ventricles are in the cerebral hemispheres; the third ventricle is in the diencephalon; the fourth ventricle is between the brain stem and the cerebellum and connects with the central canal of the spinal cord.

12.2 The cerebral hemispheres consist of cortex, white matter, and the basal nuclei (pp. 439–447)

1. The two cerebral hemispheres exhibit gyri, sulci, and fissures. The longitudinal fissure partially separates the hemispheres; other fissures or sulci subdivide each hemisphere into lobes.
2. Each cerebral hemisphere consists of the cerebral cortex, the cerebral white matter, and basal nuclei (ganglia).

Cerebral Cortex (pp. 439–444)

3. Each cerebral hemisphere receives sensory impulses from, and dispatches motor impulses to, the opposite side of the body. The

body is represented in an upside-down fashion in the sensory and motor cortices.

4. Functional areas of the cerebral cortex include (1) motor areas: primary motor and premotor cortices of the frontal lobe, the frontal eye field, and Broca's area in the frontal lobe of one hemisphere (usually the left); (2) sensory areas: primary somatosensory cortex and somatosensory association cortex in the parietal lobe; visual areas in the occipital lobe; olfactory and auditory areas in the temporal lobe; gustatory, visceral, and vestibular areas in the insula; (3) association areas: anterior association area in the frontal lobe, and posterior and limbic association areas spanning several lobes.

5. The cerebral hemispheres show lateralization of cortical function. In most people, the left hemisphere is dominant (i.e., specialized for language and mathematical skills); the right hemisphere is more concerned with visual-spatial skills and creative endeavors.

Cerebral White Matter   (p. 444)

6. Fiber tracts of the cerebral white matter include commissures, association fibers, and projection fibers.

Basal Nuclei   (pp. 444–447)

7. The paired basal nuclei (also called basal ganglia) include the globus pallidus, putamen, and caudate nucleus. The basal nuclei are subcortical nuclei that help control movements. Functionally they are closely associated with the substantia nigra of the midbrain.

12.3 The diencephalon includes the thalamus, hypothalamus, and epithalamus   (pp. 447–450)

1. The diencephalon includes the thalamus, hypothalamus, and epithalamus and encloses the third ventricle.

2. The thalamus is the major relay station for (1) sensory impulses ascending to the sensory cortex, (2) inputs from subcortical motor nuclei and the cerebellum traveling to the cerebral motor cortex, and (3) impulses traveling to association cortices from lower centers.

3. The hypothalamus is an important control center of the autonomic nervous system and a pivotal part of the limbic system. It maintains water balance and regulates thirst, eating behavior, gastrointestinal activity, body temperature, and the activity of the anterior pituitary gland.

4. The epithalamus includes the pineal gland, which secretes the hormone melatonin.

12.4 The brain stem consists of the midbrain, pons, and medulla oblongata   (pp. 450–454)

1. The midbrain contains the corpora quadrigemina (visual and auditory reflex centers), the red nucleus (subcortical motor centers), and the substantia nigra. The periaqueductal gray matter is involved in pain suppression and contains the motor nuclei of cranial nerves III and IV. The cerebral peduncles on the midbrain's ventral face house the pyramidal fiber tracts. The midbrain surrounds the cerebral aqueduct.

2. The pons is mainly a conduction area. Its nuclei contribute to regulating respiration and cranial nerves V–VII.

3. The pyramids (descending corticospinal tracts) form the ventral face of the medulla oblongata. These fibers cross over (decussation of the pyramids) before entering the spinal cord. Important nuclei in the medulla regulate respiratory rhythm, heart rate, and blood pressure and serve cranial nerves VIII–X and XII. The olivary nuclei and cough, sneezing, swallowing, and vomiting centers are also in the medulla.

12.5 The cerebellum adjusts motor output, ensuring coordination and balance   (pp. 454–456)

1. The cerebellum consists of two hemispheres, marked by convolutions and separated by the vermis. It is connected to the brain stem by superior, middle, and inferior peduncles.

2. The cerebellum processes and interprets impulses from the motor cortex and sensory pathways and coordinates motor activity so that smooth, well-timed movements occur. It also plays a poorly understood role in cognition.

12.6 Functional brain systems span multiple brain structures   (pp. 456–458)

1. The limbic system consists of numerous structures that encircle the brain stem. It is the "emotional-visceral brain." It also plays a role in memory.

2. The reticular formation is a diffuse network of neurons and nuclei spanning the length of the brain stem. It maintains the alert state of the cerebral cortex (RAS), and its motor nuclei serve both somatic and visceral motor activities.

12.7 The interconnected structures of the brain allow higher mental functions   (pp. 458–464)

Language   (pp. 459–460)

1. In most people the left hemisphere controls language. The language implementation system, which includes Broca's and Wernicke's areas and the basal nuclei, analyzes incoming and produces outgoing language. The opposite hemisphere deals with the nonverbal emotional content of language.

Memory   (pp. 460–461)

2. Memory is the storage and retrieval of information. It is essential for learning and is part of consciousness.

3. Memory storage has two stages: short-term memory (STM) and long-term memory (LTM). Transfer of information from STM to LTM takes minutes to hours, but more time is required for LTM consolidation.

Brain Wave Patterns and the EEG   (pp. 461–462)

4. Patterns of electrical activity of the brain are called brain waves; a record of this activity is an electroencephalogram (EEG). Brain wave patterns, identified by their frequencies, include alpha, beta, theta, and delta waves.

5. Epilepsy results from abnormal electrical activity of brain neurons. Involuntary muscle contractions and sensory auras are sometimes associated with epileptic seizures.

Consciousness   (p. 462)

6. Consciousness is described clinically on a continuum from alertness to drowsiness to stupor and finally to coma.

7. Human consciousness is thought to involve holistic information processing, which is (1) not localizable, (2) superimposed on other types of neural activity, and (3) totally interconnected.

8. Fainting (syncope) is a temporary loss of consciousness that usually reflects inadequate blood delivery to the brain. Coma is loss of consciousness in which the victim is unresponsive to stimuli.

Sleep and Sleep-Wake Cycles   (pp. 462–464)

9. Sleep is a state of partial unconsciousness from which a person can be aroused by stimulation. The two major types of sleep are non–rapid eye movement (NREM) sleep and rapid eye movement (REM) sleep.

12

10. During stages 1–4 of NREM sleep, brain wave frequency decreases and amplitude increases until delta wave sleep (stage 4) is achieved. REM sleep is indicated by a return to alpha waves on the EEG. During REM, the eyes move rapidly under the lids. NREM and REM sleep alternate throughout the night.

11. Sleep is restorative. Evidence suggests that it is important for memory consolidation.

12. Narcolepsy is involuntary lapses into sleep that occur without warning during waking periods. Insomnia is a chronic inability to obtain the amount or quality of sleep needed to function adequately.

12.8 The brain is protected by bone, meninges, cerebrospinal fluid, and the blood brain barrier (pp. 464–468)

1. The delicate brain is protected by bone, meninges, cerebrospinal fluid, and the blood brain barrier.

Meninges (pp. 464–465)

2. The meninges from superficial to deep are the dura mater, arachnoid mater, and pia mater. They enclose the brain and spinal cord and their blood vessels. Inward folds of the inner layer of the dura mater secure the brain to the skull.

Cerebrospinal Fluid (pp. 465–467)

3. Cerebrospinal fluid (CSF), formed by the choroid plexuses from blood plasma, circulates through the ventricles and into the subarachnoid space. It returns to the dural venous sinuses via the arachnoid granulations. CSF supports and cushions the brain and spinal cord and helps to nourish them.

Blood Brain Barrier (pp. 467–468)

4. The blood brain barrier consists mainly of the tight junctions between endothelial cells of capillaries of the brain. It allows water, respiratory gases, essential nutrients, and lipid-soluble molecules to enter the neural tissue, but blocks other, water-soluble, potentially harmful substances.

12.9 Brain injuries and disorders have devastating consequences (pp. 468–470)

1. Head trauma may cause brain injuries called concussions or, in severe cases, contusions (bruising). When the brain stem is affected, unconsciousness (temporary or permanent) occurs. Trauma-induced brain injuries may be aggravated by intracranial hemorrhage or cerebral edema, both of which compress brain tissue.

2. Cerebrovascular accidents (strokes) result when blood circulation to brain neurons is impaired and brain tissue dies. The result may be hemiplegia, sensory deficits, or speech impairment.

3. Alzheimer's disease is a degenerative brain disease in which beta-amyloid peptide deposits and neurofibrillary tangles appear. Marked by a deficit of ACh, it results in slow, progressive loss of memory and motor control and increasing dementia.

4. Parkinson's disease and Huntington's disease are neurodegenerative disorders of the basal nuclei. Both involve abnormalities of the neurotransmitter dopamine (too little or too much secreted) and are characterized by abnormal movements.

5. Diagnostic procedures used to assess neurological condition and function range from routine reflex testing to sophisticated techniques such as CT scans, MRI scans, and PET scans.

12.10 The spinal cord is a reflex center and conduction pathway (pp. 470–476)

Gross Anatomy and Protection (pp. 470–472)

1. The spinal cord, a two-way conduction pathway and a reflex center, resides within the vertebral column and is protected by meninges and cerebrospinal fluid. It extends from the foramen magnum to the end of the first or second lumbar vertebra.

2. Thirty-one pairs of spinal nerves issue from the cord. The cord is enlarged in the cervical and lumbar regions, where spinal nerves serving the limbs arise.

Spinal Cord Cross-Sectional Anatomy (pp. 472–474)

3. The central gray matter of the cord is H shaped. Ventral horns mainly contain somatic motor neurons. Lateral horns contain visceral (autonomic) motor neurons. Dorsal horns contain interneurons.

4. Axons of neurons of the lateral and ventral horns emerge in common from the cord via the ventral roots. Axons of sensory neurons (with cell bodies located in the dorsal root ganglion) form the dorsal roots and enter the dorsal aspect of the cord. The ventral and dorsal roots combine to form the spinal nerves.

5. Each side of the white matter of the cord has dorsal, lateral, and ventral columns (funiculi), and each funiculus contains a number of ascending and descending tracts. All tracts are paired and most decussate.

Spinal Cord Trauma and Disorders (pp. 474–476)

6. Injury to the ventral horn neurons or the ventral roots results in flaccid paralysis. (Injury to the upper motor neurons in the brain results in spastic paralysis.) If the dorsal roots or sensory tracts are damaged, loss of sensation or paresthesia occurs.

7. Poliomyelitis results from inflammation and destruction of the ventral horn neurons by the poliovirus. Paralysis and muscle atrophy ensue.

8. Amyotrophic lateral sclerosis (ALS) results from destruction of the ventral horn neurons and the pyramidal tracts. The patient eventually loses the ability to swallow, speak, and breathe. Death generally occurs within five years.

12.11 Neuronal pathways carry sensory and motor information to and from the brain (pp. 476–481)

1. Ascending (sensory) tracts include the fasciculi gracilis and cuneatus, spinothalamic tracts, and spinocerebellar tracts.

2. The dorsal column–medial lemniscal pathway consists of the dorsal white column (fasciculus cuneatus, fasciculus gracilis) and the medial lemniscus, which are concerned with straight-through, precise transmission of one or a few related types of sensory input. The spinothalamic pathways transmit pain, temperature, and coarse touch, and permit brain stem processing of ascending impulses. The spinocerebellar tracts, which terminate in the cerebellum, serve muscle sense, not conscious sensory perception.

3. Descending tracts include the pyramidal tracts (ventral and lateral corticospinal tracts) and a number of motor tracts originating from subcortical motor nuclei. These descending fibers issue from the brain stem motor areas [indirect system] and cortical motor areas [direct (pyramidal) system].

Developmental Aspects of the Central Nervous System (pp. 482–484)

1. The CNS develops from the embryonic neural tube—the brain from the rostral part and the spinal cord from the caudal part.
2. The gray matter of the spinal cord forms from the alar and basal plates. Fiber tracts form the outer white matter. The neural crest forms the sensory (dorsal root) ganglia.
3. Maternal and environmental factors may impair embryonic brain development, and oxygen deprivation destroys brain cells. Severe congenital brain disorders include cerebral palsy, anencephaly, hydrocephalus, and spina bifida.
4. Premature babies have trouble regulating body temperature because the hypothalamus is one of the last brain areas to mature prenatally.
5. Development of motor control indicates progressive myelination and maturation of a child's nervous system.
6. Brain growth ends in young adulthood. Neurons die throughout life and most are not replaced; brain weight and volume decline with age.
7. Healthy elders maintain nearly optimal intellectual function. Repeated brain trauma (as in boxing, for example) and alcoholism can accelerate deterioration of the aging brain.

REVIEW QUESTIONS

 To access additional practice questions using your smartphone, tablet, or computer: MasteringA&P® > Study Area > Practice Tests & Quizzes

Level 1 Remember/Understand

(Some questions have more than one correct answer. Select the best answer or answers from the choices given.)

1. The primary motor cortex, Broca's area, and the premotor cortex are located in which lobe? **(a)** frontal, **(b)** parietal, **(c)** temporal, **(d)** occipital.
2. The innermost layer of the meninges, delicate and adjacent to the brain tissue, is the **(a)** dura mater, **(b)** corpus callosum, **(c)** arachnoid mater, **(d)** pia mater.
3. Cerebrospinal fluid is formed by **(a)** arachnoid granulations, **(b)** dura mater, **(c)** choroid plexuses, **(d)** all of these.
4. Choose the correct term from the key to respond to the statements describing various brain areas.

Key:
(a) cerebellum	**(d)** striatum	**(g)** midbrain
(b) corpora quadrigemina	**(e)** hypothalamus	**(h)** pons
(c) corpus callosum	**(f)** medulla	**(i)** thalamus

____ **(1)** basal nuclei involved in fine control of motor activities
____ **(2)** region where there is a crossover of fibers of descending pyramidal tracts
____ **(3)** control of temperature, autonomic nervous system reflexes, hunger, and water balance
____ **(4)** houses the substantia nigra and cerebral aqueduct
____ **(5)** relay stations for visual and auditory stimuli input; found in midbrain
____ **(6)** houses vital centers for control of the heart, respiration, and blood pressure
____ **(7)** brain area through which all the sensory input is relayed to get to the cerebral cortex
____ **(8)** brain area most concerned with equilibrium, body posture, and coordination of motor activity

5. Destruction of the ventral horn cells of the spinal cord results in loss of **(a)** integrating impulses, **(b)** sensory impulses, **(c)** voluntary motor impulses, **(d)** all of these.
6. Fiber tracts that allow neurons within the same cerebral hemisphere to communicate are **(a)** association fibers, **(b)** commissures, **(c)** projection fibers.
7. A professor unexpectedly blew a loud horn in his anatomy and physiology class. The students looked up, startled. The reflexive movements of their eyes were mediated by the **(a)** cerebral cortex, **(b)** inferior olives, **(c)** raphe nuclei, **(d)** superior colliculi, **(e)** nucleus gracilis.
8. **(a)** What is the function of the basal nuclei? **(b)** Which basal nuclei form the striatum? **(c)** Which arches over the diencephalon?

9. **(a)** Localize the reticular formation in the brain. **(b)** What does RAS mean, and what is its function?
10. List four ways in which the CNS is protected.
11. What constitutes the blood brain barrier?

Level 2 Apply/Analyze

12. A patient has suffered a cerebral hemorrhage that has caused dysfunction of the precentral gyrus of his right cerebral cortex. As a result, **(a)** he cannot voluntarily move his left arm or leg, **(b)** he feels no sensation on the left side of his body, **(c)** he feels no sensation on his right side.
13. A number of brain structures are listed below. If an area is primarily gray matter, write **a** in the answer blank; if mostly white matter, respond with **b**.

____ **(1)** cerebral cortex
____ **(2)** corpus callosum and corona radiata
____ **(3)** red nucleus
____ **(4)** medial and lateral nuclear groups
____ **(5)** medial lemniscus
____ **(6)** cranial nerve nuclei
____ **(7)** spinothalamic tract
____ **(8)** fornix
____ **(9)** cingulate and precentral gyri

14. All of the following descriptions refer to dorsal column–medial lemniscal ascending pathways except one: **(a)** they include the fasciculus gracilis and fasciculus cuneatus; **(b)** they include a chain of three neurons; **(c)** their connections are diffuse and poorly localized; **(d)** they are concerned with precise transmission of one or a few related types of sensory input.
15. **(a)** What is the advantage of having a cerebrum that is highly convoluted? **(b)** What term is used to indicate its grooves? Its outward folds? **(c)** Which groove divides the cerebrum into two hemispheres? **(d)** What divides the parietal from the frontal lobe? The parietal from the temporal lobe?
16. **(a)** Make a rough drawing of the lateral aspect of the left cerebral hemisphere. **(b)** You may be thinking, "But I just can't draw!" So, name the hemisphere involved with most people's ability to draw. **(c)** On your drawing, locate the following areas and provide the major function of each: primary motor cortex, premotor cortex, somatosensory association cortex, primary somatosensory cortex, visual and auditory areas, prefrontal cortex, Wernicke's and Broca's areas.
17. Explain how the cerebellum is physically connected to the brain stem.

12

18. Describe the stages of sleep and outline the order in which we progress through these stages during a typical night's sleep.

19. **(a)** How is cerebrospinal fluid formed and drained? Describe its pathway within and around the brain. **(b)** What happens if CSF does not drain properly? Why is this consequence more harmful in adults?

20. Describe the functional problems that would be experienced by a person in which these fiber tracts have been cut: **(a)** lateral spinothalamic, **(b)** ventral and dorsal spinocerebellar, **(c)** tectospinal.

21. **(a)** What factors account for brain growth after birth? **(b)** List some structural brain changes observed with aging.

22. Mrs. Jones has had a progressive decline in her mental capabilities in the last five or six years. At first her family attributed her occasional memory lapses, confusion, and agitation to grief over her husband's death six years earlier. When examined, Mrs. Jones was aware of her cognitive problems and was shown to have an IQ score approximately 30 points less than would be predicted by her work history. A CT scan showed diffuse cerebral atrophy. The physician prescribed an acetylcholinesterase inhibitor and Mrs. Jones showed slight improvement. What is Mrs. Jones's problem? Why did the acetylcholinesterase inhibitor help?

23. Robert, a brilliant computer analyst, suffered a blow to his anterior skull from a falling rock while mountain climbing. Shortly thereafter, it was obvious to his coworkers that his behavior had undergone a dramatic change. Although previously a smart dresser, he was now unkempt. One morning, he was observed defecating into the wastebasket. Which region of Robert's brain was affected by the cranial blow?

Level 3 Evaluate/Synthesize

24. **(a)** What does lateralization of cortical functioning mean? **(b)** Why is the term cerebral dominance a misnomer?

25. Describe the role of the cerebellum in maintaining smooth, coordinated skeletal muscle activity.

26. **(a)** Where is the limbic system located? **(b)** Which structures make up this system? **(c)** How is the limbic system important in behavior?

27. Compare and contrast short-term memory (STM) and long-term memory (LTM) relative to storage capacity and duration of the memory.

28. Describe the spinal cord, depicting its extent, its composition of gray and white matter, and its spinal roots.

29. **(a)** Define cerebrovascular accident or CVA. **(b)** Describe its possible causes and consequences.

30. A 10-month-old infant has an enlarging head circumference and delayed overall development. She has a bulging anterior fontanelle and her CSF pressure is elevated. Based on these findings, answer the following questions: **(a)** What are the possible cause(s) of an enlarged head? **(b)** Which tests might be helpful in obtaining information about this infant's problem? **(c)** Assuming the tests conducted showed the cerebral aqueduct to be constricted, which ventricles or CSF-containing areas would you expect to be enlarged? Which would likely not be visible? Respond to the same questions based on a finding of obstructed arachnoid granulations.

31. Tara, a 16-year-old girl, was rushed to the hospital after taking a bad spill off the parallel bars. After she had a complete neurological workup, her family was told that she would be permanently paralyzed from the waist down. During rehabilitation, her neurologist outlined for Tara and her parents the importance of preventing complications in such cases. Common complications include urinary infection, bed sores, and muscular spasms. Using your knowledge of neuroanatomy, explain the underlying reasons for these complications.

CLINICAL CASE STUDY

39-Year-Old Female with Traumatic Brain Injury

Margaret Bryans, a 39-year-old female, lost control while skiing and hit her head on a tree. She regained consciousness en route to the hospital and appeared agitated and combative. Paramedics observed that she had a right hemiparesis (muscle weakness), with a nearly complete paresis of her right upper extremity and a partial paresis of the right lower extremity. A head CT scan revealed an acute subdural hematoma and an extensive subarachnoid hemorrhage. Doctors noted that she was able to follow commands from medical personnel. However, she could only speak haltingly, using simple words.

Surgery to remove large clots from the subarachnoid space was performed immediately. Two weeks after the surgery, she showed significant improvement in her speech and motor function.

1. The adult brain consists of four main anatomical regions (see Table 12.1, pp. 458–459). Based on the observed signs in this case, which of these four brain regions is involved? What evidence did you use to determine this?

2. Which side of the brain is involved in Mrs. Bryans's injury? What evidence did you use to determine this?

3. **+ NCLEX-STYLE** What specific parts of the brain region you identified in question 1 have been affected by the injury so as to cause the muscle weakness and language problems? (Hint: Use Figures 12.7 and 12.8.)
   a. The left midlateral motor cortex
   b. The superior area of her right motor cortex
   c. The superior area of her left motor cortex and Broca's area
   d. The inferolateral area of her left motor cortex and Wernicke's area

4. What are the three membranes that make up the meninges? Describe their positions relative to the brain.

5. **+ NCLEX-STYLE** Mrs. Bryan's subdural and subarachnoid bleeding most likely interfered directly with which of the following?
   a. The blood supply to the scalp
   b. The blood supply to the cerebral cortex
   c. The removal of cerebrospinal fluid
   d. The creation of cerebrospinal fluid

13

The Peripheral Nervous System and Reflex Activity

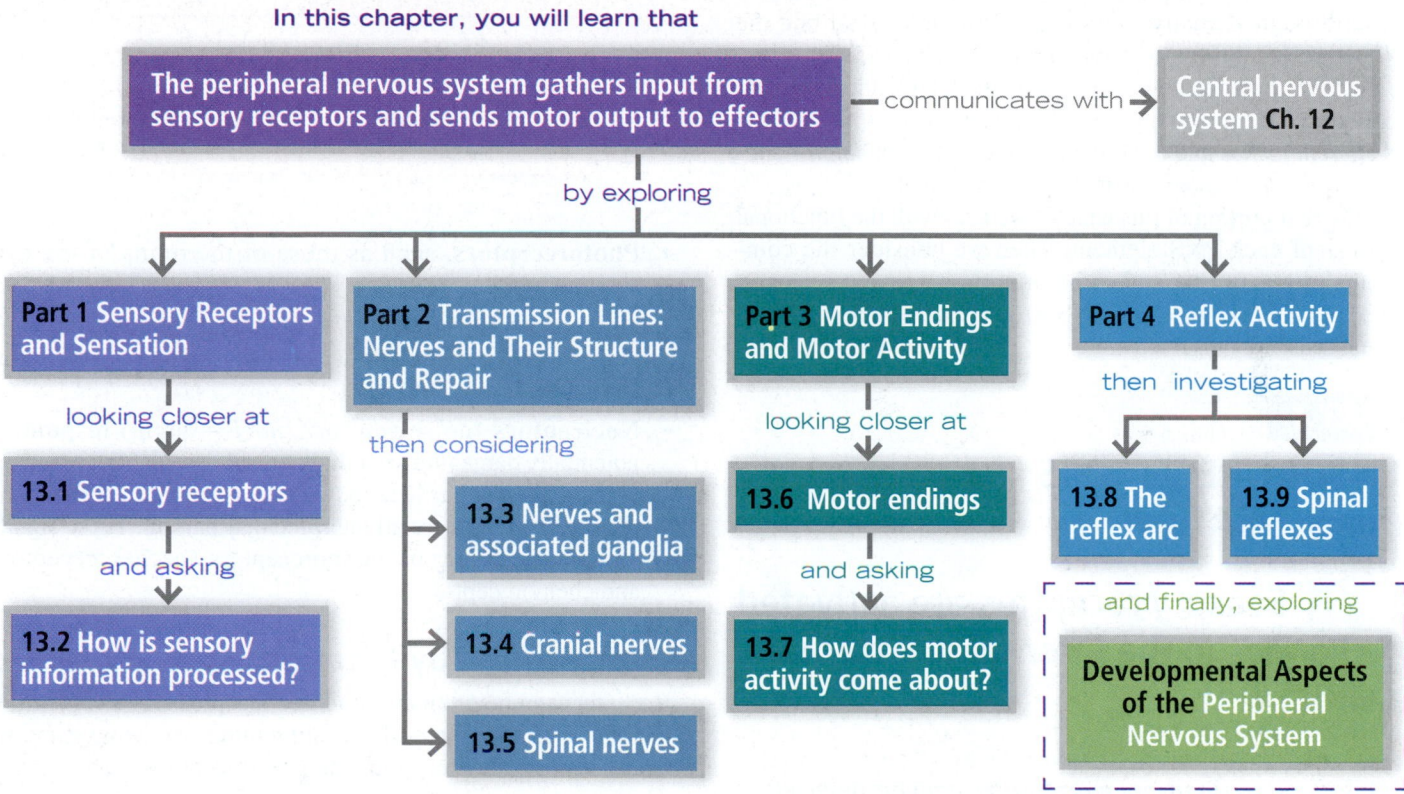

In this chapter, you will learn that

The peripheral nervous system gathers input from sensory receptors and sends motor output to effectors ── communicates with → Central nervous system **Ch. 12**

by exploring

Part 1 Sensory Receptors and Sensation

looking closer at

13.1 Sensory receptors

and asking

13.2 How is sensory information processed?

Part 2 Transmission Lines: Nerves and Their Structure and Repair

then considering

13.3 Nerves and associated ganglia

13.4 Cranial nerves

13.5 Spinal nerves

Part 3 Motor Endings and Motor Activity

looking closer at

13.6 Motor endings

and asking

13.7 How does motor activity come about?

Part 4 Reflex Activity

then investigating

13.8 The reflex arc

13.9 Spinal reflexes

and finally, exploring

Developmental Aspects of the Peripheral Nervous System

Learning Outcome

▶ Define peripheral nervous system and list its components.

The human brain, for all its sophistication, would be useless without its links to the outside world. Consider one experiment that illustrates this point. Volunteers hallucinated when they were deprived of sensory input by being blindfolded and suspended in warm water in a sensory deprivation tank. One saw charging pink and purple elephants. Another heard a chorus, still others had taste hallucinations. Our very sanity depends on a continuous flow of information from the outside.

Equally important to our well-being are the orders sent from the CNS to voluntary muscles and other effectors of the body, which allow us to move and take care of our own needs. The

Figure 13.1 Place of the PNS in the structural organization of the nervous system.

peripheral nervous system (PNS) provides these links from and to the world outside our bodies. Ghostly white nerves thread through virtually every part of the body, enabling the CNS to receive information and carry out its decisions.

The PNS includes all neural structures outside the brain and spinal cord. These structures are the *sensory receptors*, *afferent* and *efferent nerves* and their associated *ganglia*, and *motor endings*. **Figure 13.1** diagrams its basic components.

In the first portion of this chapter we deal with the functional anatomy of each PNS element. Then we consider the components of reflex arcs and some important somatic reflexes, played out almost entirely in PNS structures, that help maintain homeostasis.

PART 1

SENSORY RECEPTORS AND SENSATION

13.1 Sensory receptors are activated by changes in the internal or external environment

Learning Outcome

▶ Classify general sensory receptors by stimulus detected, body location, and structure.

Sensory receptors are specialized to respond to changes in their environment, which are called **stimuli**. Typically, activation of a sensory receptor by an adequate stimulus results in graded potentials that in turn trigger nerve impulses along the afferent PNS fibers coursing to the CNS. *Sensation* (awareness of the stimulus) and *perception* (interpretation of the meaning of the stimulus) occur in the brain. But we are getting ahead of ourselves here.

For now, let's just examine how sensory receptors are classified. Basically, there are three ways to classify sensory receptors: (1) by the type of stimulus they detect; (2) by their body location; and (3) by their structural complexity.

Classification by Stimulus Type

These categories are easy to remember because the name usually indicates the stimulus that activates the receptor.

- **Mechanoreceptors** respond to mechanical force such as touch, pressure (including blood pressure), vibration, and stretch.
- **Thermoreceptors** respond to temperature changes.

- **Photoreceptors**, such as those of the retina of the eye, respond to light.
- **Chemoreceptors** respond to chemicals in solution (molecules smelled or tasted, or changes in blood or interstitial fluid chemistry).
- **Nociceptors** (no″se-sep′torz; *noci* = harm) respond to potentially damaging stimuli that result in pain. For example, searing heat, extreme cold, excessive pressure, and inflammatory chemicals are all interpreted as painful. These signals stimulate subtypes of thermoreceptors, mechanoreceptors, and chemoreceptors.

Classification by Location

Receptors can be grouped into three receptor classes according to either their location or the location of the activating stimulus: exteroceptors, interoceptors, and proprioceptors.

- **Exteroceptors** (ek″ster-o-sep′torz) are sensitive to stimuli arising outside the body (*extero* = outside), so most exteroceptors are near or at the body surface. They include touch, pressure, pain, and temperature receptors in the skin and most receptors of the special senses (vision, hearing, equilibrium, smell, and taste).
- **Interoceptors** (in″ter-o-sep′torz), also called *visceroceptors*, respond to stimuli within the body (*intero* = inside), such as from the internal viscera and blood vessels. Interoceptors monitor a variety of stimuli, including chemical changes, tissue stretch, and temperature. Sometimes their activity causes us to feel pain, discomfort, hunger, or thirst. However, we are usually unaware of their workings.
- **Proprioceptors**, like interoceptors, respond to internal stimuli. However, their location is much more restricted. Proprioceptors occur in skeletal muscles, tendons, joints, and ligaments and in connective tissue coverings of bones and muscles. (Some authorities include the equilibrium receptors of the inner ear in this class.) Proprioceptors constantly advise the brain of our body movements by monitoring how much the organs containing these receptors are stretched.

Table 13.1 General Sensory Receptors Classified by Structure and Function

STRUCTURAL CLASS	ILLUSTRATION	FUNCTIONAL CLASSES ACCORDING TO LOCATION (L) AND STIMULUS TYPE (S)	BODY LOCATION
Nonencapsulated			
Free nerve endings of sensory neurons		L: Exteroceptors, interoceptors, and proprioceptors S: Thermoreceptors (warm and cool), chemoreceptors (itch, pH, etc.), mechanoreceptors (pressure), nociceptors (pain)	Most body tissues; most dense in connective tissues (ligaments, tendons, dermis, joint capsules, periostea) and epithelia (epidermis, cornea, mucosae, and glands)
Modified free nerve endings: Epithelial tactile complexes (Merkel cells and discs)	Tactile epithelial cell — Epithelial tactile complex	L: Exteroceptors S: Mechanoreceptors (light pressure); slowly adapting	Basal layer of epidermis
Hair follicle receptors		L: Exteroceptors S: Mechanoreceptors (hair deflection); rapidly adapting	Surrounding hair follicles

13

Classification by Receptor Structure

The overwhelming majority of sensory receptors belong to the **general senses** and are simply the modified dendritic endings of sensory neurons. They are found throughout the body and monitor most types of general sensory information.

Receptors for the **special senses** (vision, hearing, equilibrium, smell, and taste) are housed in complex **sense organs**. For example, the sense organ we know as the eye is composed not only of sensory neurons but also of nonneural cells that form its supporting wall, lens, and other associated structures.

Though the special senses are most familiar to us, the simple sensory receptors associated with the general senses are no less important, and we will concentrate on their structure and function in this chapter. The special senses are the topic of Chapter 15.

Simple Receptors of the General Senses

The widely distributed general sensory receptors are involved in tactile sensation (a mix of touch, pressure, stretch, and vibration), temperature monitoring, and pain, as well as the "muscle sense" provided by proprioceptors. As you read about these receptors, notice that there is no perfect "one-receptor–one-function" relationship. Instead, one type of receptor can respond to several different kinds of stimuli. For example, a free nerve ending can respond to both heat and chemicals. Likewise, different types of receptors can respond to similar stimuli. Anatomically, general sensory receptors are nerve endings that are either *nonencapsulated*

(*free*) or *encapsulated*. **Table 13.1** illustrates the general sensory receptors.

Nonencapsulated (Free) Nerve Endings Present nearly everywhere in the body, **nonencapsulated (free) nerve endings** of sensory neurons are particularly abundant in epithelia and connective tissues. Most of these sensory fibers are nonmyelinated, small-diameter group C fibers (◄ p. 412), and their distal endings (the sensory terminals) usually have small knoblike swellings.

Free nerve endings respond chiefly to painful stimuli and changes in temperature, but some respond to tissue movements caused by pressure as well. Nerve endings that respond to cold (10–40°C, or 50–104°F) are located in the superficial dermis. Those responding to heat (32–48°C, or 90–120°F) are deeper in the dermis.

Heat or cold outside the range of thermoreceptors activates nociceptors and is perceived as painful. Nociceptors also respond to pinch and chemicals released from damaged tissue. A key player in detecting painful stimuli is a plasma membrane protein called the *vanilloid receptor*. This protein is an ion channel that is opened by heat, low pH, and various chemicals including capsaicin, the substance found in chili peppers.

Another sensation mediated by free nerve endings is itch. Located in the dermis, the *itch receptor* escaped detection for years because of its thin diameter. A number of chemicals—notably histamine—present at inflamed sites activate these nerve endings.

Table 13.1   General Sensory Receptors Classified by Structure and Function *(continued)*

STRUCTURAL CLASS	ILLUSTRATION	FUNCTIONAL CLASSES ACCORDING TO LOCATION (L) AND STIMULUS TYPE (S)	BODY LOCATION
Encapsulated			
Tactile (Meissner's) corpuscles		L: Exteroceptors S: Mechanoreceptors (light pressure, discriminative touch, vibration of low frequency); rapidly adapting	Dermal papillae of hairless skin, particularly nipples, external genitalia, fingertips, soles of feet, eyelids
Lamellar (Pacinian) corpuscles		L: Exteroceptors, interoceptors, and some proprioceptors S: Mechanoreceptors (deep pressure, stretch, vibration of high frequency); rapidly adapting	Dermis and hypodermis; periostea, mesentery, tendons, ligaments, joint capsules; most abundant on fingers, soles of feet, external genitalia, nipples
Bulbous corpuscles (Ruffini endings)		L: Exteroceptors and proprioceptors S: Mechanoreceptors (deep pressure and stretch); slowly or nonadapting	Deep in dermis, hypodermis, and joint capsules
Muscle spindles	Intrafusal fibers	L: Proprioceptors S: Mechanoreceptors (muscle stretch, length)	Skeletal muscles, particularly in the extremities
Tendon organs		L: Proprioceptors S: Mechanoreceptors (tendon stretch, tension)	Tendons
Joint kinesthetic receptors		L: Proprioceptors S: Mechanoreceptors and nociceptors	Joint capsules of synovial joints

Other nonencapsulated nerve endings include:

- **Epithelial tactile complexes (Merkel cells** and **discs),** which lie at the junction of the epidermis and dermis, function as light pressure receptors. Certain disclike free nerve endings (Merkel discs) associate with a spiky, hemispherical epidermal cell (tactile epithelial cell, ◄ p. 153, or Merkel cell), forming a complex.

- **Hair follicle receptors,** free nerve endings that wrap basket-like around hair follicles, are light touch receptors that detect bending of hairs. The tickle of a mosquito landing on your skin is mediated by hair follicle receptors.

Encapsulated Nerve Endings All **encapsulated nerve endings** consist of one or more fiber terminals of sensory neurons enclosed in a connective tissue capsule. Virtually all encapsulated

receptors are mechanoreceptors, but they vary greatly in shape, size, and distribution in the body. They include tactile corpuscles, lamellar corpuscles, bulbous corpuscles, muscle spindles, tendon organs, and joint kinesthetic receptors (Table 13.1):

- **Tactile corpuscles** or *Meissner's corpuscles* are small receptors in which a few spiraling sensory terminals are surrounded by Schwann cells and then by a thin egg-shaped connective tissue capsule. Tactile corpuscles are found just beneath the epidermis in the dermal papillae and are especially numerous in sensitive and hairless skin areas such as the nipples, fingertips, and soles of the feet. They are receptors for discriminative touch, and apparently play the same role in sensing light touch in hairless skin that hair follicle receptors do in hairy skin.

- **Lamellar corpuscles**, also called *Pacinian corpuscles*, are scattered deep in the dermis, and in subcutaneous tissue underlying the skin. Although they are mechanoreceptors stimulated by deep pressure, they respond only when the pressure is first applied, and thus are best suited to monitoring vibration (an "on/off" pressure stimulus). They are the largest corpuscular receptors. Some are over 3 mm long and half as wide and are visible to the naked eye as white, egg-shaped bodies. In section, a lamellar corpuscle resembles a cut onion. Its single dendrite is surrounded by a capsule containing up to 60 layers of collagen fibers and flattened supporting cells.

- **Bulbous corpuscles** or *Ruffini endings*, which lie in the dermis, subcutaneous tissue, and joint capsules, contain a spray of receptor endings enclosed by a flattened capsule. They bear a striking resemblance to tendon organs (which monitor tendon stretch) and probably play a similar role in other dense connective tissues where they respond to deep and *continuous* pressure.

- **Muscle spindles** are fusiform (spindle-shaped) proprioceptors found throughout the perimysium that wraps individual fascicles of skeletal muscle. Each muscle spindle consists of a bundle of modified skeletal muscle fibers, called *intrafusal fibers* (in″trah-fu′zal), enclosed in a connective tissue capsule. Muscle spindles detect muscle stretch and initiate a reflex that resists the stretch (see pp. 520–524).

- **Tendon organs** are proprioceptors located in tendons, close to the junction between the skeletal muscle and the tendon. They consist of small bundles of tendon (collagen) fibers enclosed in a layered capsule, with sensory terminals coiling between and around the fibers. When muscle contraction stretches the tendon fibers, the resulting compression of the nerve fibers activates the tendon organs. This initiates a reflex that causes the contracting muscle to relax (see p. 524).

- **Joint kinesthetic receptors** (kin″es-thet′ik) are proprioceptors that monitor stretch in the articular capsules that enclose synovial joints. This receptor category contains at least four receptor types: lamellar corpuscles, bulbous corpuscles, free nerve endings, and receptors resembling tendon organs. Together these receptors provide information on joint position and motion (*kines* = movement), a sensation of which we are highly conscious.

Close your eyes and flex and extend your fingers—you can feel exactly which joints are moving.

Check Your Understanding

1. Your PNS mostly consists of nerves. What else belongs to your PNS?
2. You've cut your finger on a broken beaker in your A&P lab. Using stimulus type, location, and receptor structure, classify the sensory receptors that allow you to feel the pain.
3. **PREDICT** Amber, your 2-year-old patient, has no pain receptors (a condition called *congenital insensitivity to pain*). What

problems would you expect Amber to have over the course of her life?

4. **MAKE CONNECTIONS** Identify structures a–f on the section of skin below. For each structure, state the type of stimulus to which it best responds. (Note that structure b is not normally found in hairy skin.)

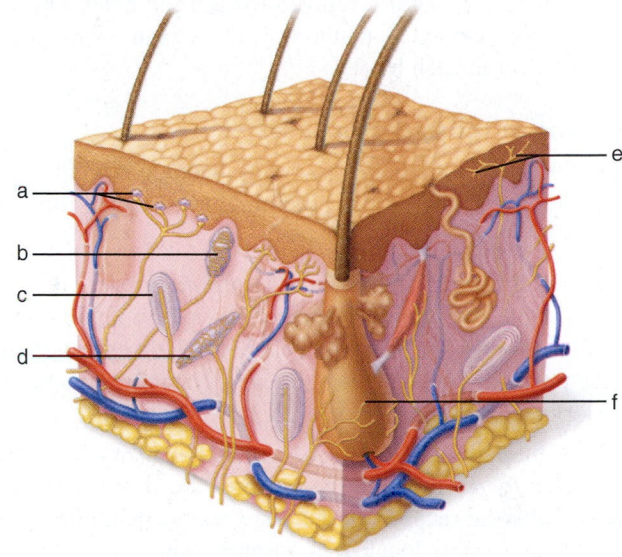

For answers, see Answers Appendix.

13.2 Receptors, ascending pathways, and cerebral cortex process sensory information

Learning Outcomes
▶ Outline the events that lead to sensation and perception.
▶ Describe receptor and generator potentials and sensory adaptation.
▶ Describe the main aspects of sensory perception.

Our survival depends not only on **sensation** (awareness of changes in the internal and external environments) but also on **perception** (conscious interpretation of those stimuli). For example, a pebble in your shoe causes the *sensation* of localized deep pressure, but your *perception* of it is an awareness of discomfort. Perception in turn determines how we will respond: In this case, you take off your shoe to get rid of the pesky pebble.

General Organization of the Somatosensory System

The **somatosensory system**—the part of the sensory system serving the body wall and limbs—receives inputs from exteroceptors, proprioceptors, and interoceptors. Consequently, it transmits information about several different sensory modalities, or types of sensation.

Three main levels of neural integration operate in the somatosensory (or any sensory) system (**Figure 13.2**):

① **Receptor level:** sensory receptors
② **Circuit level:** processing in ascending pathways
③ **Perceptual level:** processing in cortical sensory areas

Sensory input is generally relayed toward the head, but note that it is also processed along the way. Let's examine the events that must occur at each level.

Processing at the Receptor Level

Generating a Signal   For sensation to occur, a stimulus must excite a receptor and action potentials must reach the CNS (Figure 13.2 ①). For this to happen:

- The stimulus energy must match the *specificity* of the receptor. For example, a touch receptor may be sensitive to mechanical pressure, stretch, and vibration, but not to light energy (which is the province of receptors in the eye). The more complex the sensory receptor, the more specific it is.

- The stimulus must be applied within a sensory receptor's *receptive field*—the area the receptor monitors. Typically, the smaller the receptive field, the greater the ability of the brain to accurately localize the stimulus site.

- The stimulus energy must be converted into the energy of a *graded potential*, a process called **transduction**. This graded potential may be depolarizing or hyperpolarizing, similar to the EPSPs or IPSPs generated at postsynaptic membranes in response to neurotransmitter binding (◀ **p. 417**).

  Receptors can produce one of two types of graded potentials. When the receptor region is part of a sensory neuron (as with free dendrites or the encapsulated receptors of most general sense receptors), the graded potential is called a **generator potential** because it generates action potentials in a sensory neuron.

  When the receptor is a separate cell (as in most special senses), the graded potential is called a **receptor potential** because it occurs in a separate receptor cell. The receptor potential changes the amount of neurotransmitter released by the receptor cell onto the sensory neuron. The neurotransmitters then generate graded potentials in the sensory neuron.

- Graded potentials in the first-order sensory neuron must reach *threshold* so that voltage-gated sodium channels on the axon are opened and nerve impulses are generated and propagated to the CNS.

Adaptation   Information about a stimulus—its strength, duration, and pattern—is encoded in the frequency of nerve impulses: the greater the frequency, the stronger the stimulus. Many but not all sensory receptors exhibit **adaptation**, a change in sensitivity (and nerve impulse generation) in the presence of a constant stimulus. For example, when you step into bright sunlight from a darkened room, your eyes are initially dazzled, but your photoreceptors rapidly adapt, allowing you to see both bright areas and dark areas in the scene.

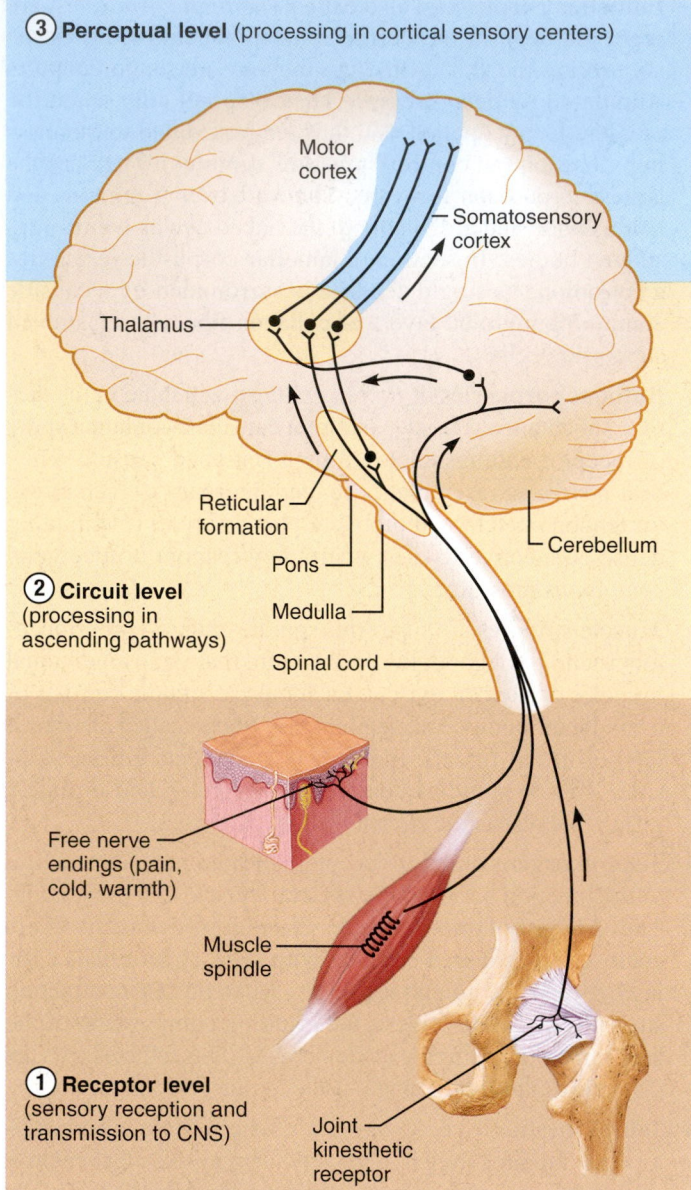

Figure 13.2   Three basic levels of neural integration in sensory systems.

Phasic receptors are *fast adapting*, often giving bursts of impulses at the beginning and end of the stimulus. Phasic receptors report *changes* in the internal or external environment. Examples are lamellar and tactile corpuscles.

Tonic receptors provide a sustained response with little or no adaptation. Nociceptors and most proprioceptors are tonic receptors because of the protective importance of their information.

Processing at the Circuit Level

At the second level of integration, the circuit level, the task is to deliver impulses to the appropriate region of the cerebral cortex for localization and perception of the stimulus (Figure 13.2 ②).

Recall from Chapter 12 that ascending sensory pathways typically consist of a chain of three neurons called first-, second-, and third-order sensory neurons (◄ p. 476). The axons of first-order sensory neurons, whose cell bodies are in the dorsal root or cranial ganglia, link the receptor and circuit levels of processing. Central processes of first-order neurons branch diffusely when they enter the spinal cord. Some branches take part in local spinal cord reflexes. Others synapse with second-order sensory neurons, which then synapse with the third-order sensory neurons that take the message to the cerebral cortex.

As we saw in Chapter 12, the different ascending pathways (spinothalamic, dorsal column–medial lemniscal, and spinocerebellar) carry various types of information to different destinations in the brain. You may wish to review this information (◄ pp. 476–478).

Processing at the Perceptual Level

Sensory input is interpreted in the cerebral cortex (Figure 13.2 ③). The ability to identify and appreciate sensations depends on the location of the target neurons in the sensory cortex, not on the nature of the message (which is, after all, just an action potential). Each sensory fiber is analogous to a "labeled line" that tells the brain "who" is calling—a taste bud or a pressure receptor—and from "where." The brain always interprets the activity of a specific sensory receptor ("who") as a specific sensation, no matter how it is activated.

For example, pressing on your eyeball activates photoreceptors, but what you "see" is light. The exact point in the cortex that is activated always refers to the same "where," regardless of how it is activated, a phenomenon called **projection**. Electrically stimulating a particular spot in the visual cortex causes you to "see" light in a particular place.

Let's examine the major features of sensory perception.

- **Perceptual detection** is the ability to detect that a stimulus has occurred. This is the simplest level of perception. As a general rule, inputs from several receptors must be summed for perceptual detection to occur.

- **Magnitude estimation** is the ability to detect how *intense* the stimulus is. Perceived intensity increases as stimulus intensity increases because of frequency coding (◄ **Figure 11.12, p. 410**).

- **Spatial discrimination** allows us to identify the site or pattern of stimulation. A common tool for studying this quality in the laboratory is the **two-point discrimination** test. The test determines how close together two points on the skin can be and still be perceived as two points rather than as one. This test provides a crude map of the density of tactile receptors in the various regions of the skin. The distance between perceived points varies from less than 5 mm on highly sensitive body areas (tip of the tongue) to more than 50 mm on less sensitive areas (the back).

- **Feature abstraction** is the mechanism by which a neuron or circuit is tuned to one feature, or property, of a stimulus in preference to others. Sensation usually involves an interplay of several stimulus features.

  For example, one touch tells us that velvet is warm, compressible, and smooth but not completely continuous, each a feature

that contributes to our perception of "velvet." Feature abstraction enables us to identify more complex aspects of a sensation.

- **Quality discrimination** is the ability to differentiate the submodalities of a particular sensation. Each sensory modality has several **qualities**, or submodalities. For example, taste is a sensory modality and its submodalities include sweet and bitter.

- **Pattern recognition** is the ability to take in the scene around us and recognize a familiar pattern, an unfamiliar one, or one that has special significance for us. For example, we can look at an image made of dots and recognize it as the portrait of a familiar face. We can listen to music and hear a melody, not just a string of notes.

Perception of Pain

Everyone has suffered pain—the cruel persistence of a headache, the smart of a bee sting or a cut finger. Although we may not appreciate it at the time, pain is invaluable because it warns us of actual or impending tissue damage and motivates us to take protective action. Managing a patient's pain can be difficult because pain is an intensely personal experience that cannot be measured objectively.

Pain receptors are activated by extremes of pressure and temperature as well as a veritable soup of chemicals released from injured tissue. Histamine, K^+, ATP, acids, and bradykinin are among the most potent pain-producing chemicals. All of these chemicals act on small-diameter fibers.

When you cut your finger, you may have noticed that you first felt a sharp pain followed some time later by burning or aching. Sharp pain is carried by the smallest of the myelinated sensory fibers, the A delta fibers, while burning pain is carried more slowly by small nonmyelinated C fibers. Both types of fibers release the neurotransmitters *glutamate* and *substance P*, which activate second-order sensory neurons. Axons from these second-order neurons ascend to the brain via the spinothalamic tract and other pathways.

If you cut your finger while fighting off an attacker, you might not notice the cut at all. How can that be? The brain has its own pain-suppressing analgesic system. The endogenous opioids such as *endorphins* and *enkephalins* (◄ p. 423) play a key role in this system. Various nuclei in the brain stem, including the periaqueductal gray matter of the midbrain, relay descending cortical and hypothalamic pain-suppressing signals. Descending fibers activate interneurons in the spinal cord, which release enkephalins. Enkephalins are inhibitory neurotransmitters that quash the pain signals generated by nociceptors.

Pain Tolerance

We all have the same *pain threshold*—that is, we begin to perceive pain at roughly the same stimulus intensity. However, our tolerance to pain varies widely. When we say that someone is "sensitive" to pain, we mean that the person has a low *pain tolerance* rather than a low pain threshold.

A number of genes help determine a person's pain tolerance and response to pain medications. The genetics of pain

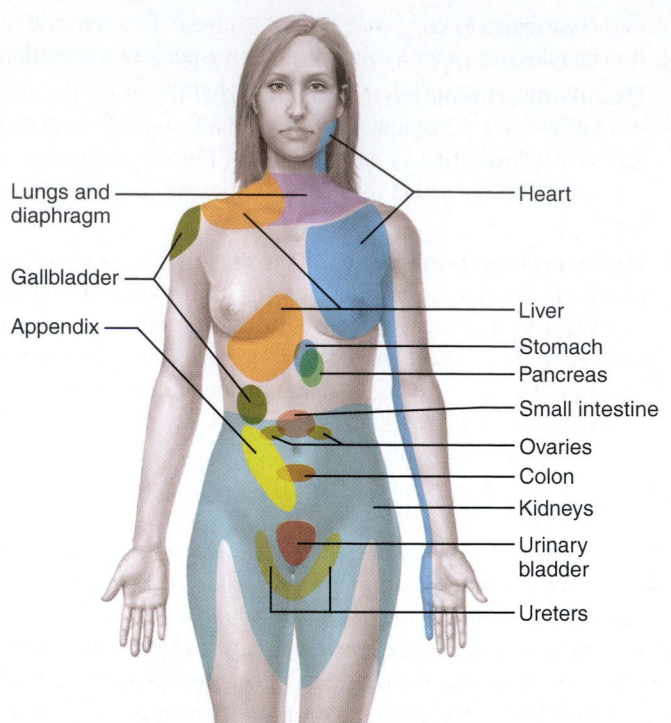

Lungs and diaphragm

Gallbladder

Appendix

Heart

Liver

Stomach

Pancreas

Small intestine

Ovaries

Colon

Kidneys

Urinary bladder

Ureters

Figure 13.3 Map of referred pain. This map shows the anterior skin areas to which pain is referred from certain visceral organs.

13

is currently an area of intense research, aimed at allowing an individual's genes to determine the best pain treatment.

HOMEOSTATIC IMBALANCE 13.1

CLINICAL

Normally the body maintains a steady state that correlates injury and pain. Long-lasting or very intense pain inputs, such as limb amputation, can disrupt this system, leading to **hyperalgesia** (pain amplification), chronic pain, and **phantom limb pain**. Intense or long-lasting pain activates *NMDA receptors*, the same receptors that strengthen neural connections during certain kinds of learning. Essentially, the spinal cord *learns* hyperalgesia. In light of this, it is crucial that health professionals effectively manage pain early to prevent chronic pain from becoming established.

Phantom limb pain (pain perceived in tissue that is no longer present) is a curious example of hyperalgesia. Until recently, surgical limb amputations were conducted under general anesthesia only and the spinal cord still experienced the pain of amputation. Epidural anesthetics block neurotransmission in the spinal cord, and using them during surgery greatly reduces the incidence of phantom limb pain.

Visceral and Referred Pain

Visceral pain results from noxious stimulation of receptors in the organs of the thorax and abdominal cavity. Like deep somatic pain, it is usually a vague sensation of dull aching, gnawing, or burning. Important stimuli for visceral pain are extreme stretching of tissue, ischemia (low blood flow), irritating chemicals, and muscle spasms.

The fact that visceral pain afferents travel along the same pathways as somatic pain fibers helps explain the phenomenon of **referred pain**, in which pain stimuli arising in one part of the body are perceived as coming from another part. For example, a person experiencing a heart attack may feel pain that radiates along the medial aspect of the left arm. Because the same spinal cord segments (T_1–T_5) innervate both the heart and arm, the brain interprets these inputs as coming from the more common somatic pathway. **Figure 13.3** shows cutaneous areas to which visceral pain is commonly referred.

Check Your Understanding

5. What are the three levels of sensory integration?

6. What is the key difference between tonic and phasic receptors? Why are pain receptors tonic?

7. Your cortex decodes incoming action potentials from sensory pathways. How does it tell the difference between hot and cold? Between cool and cold? Between ice on your finger and ice on your foot?

For answers, see Answers Appendix.

PART 2

TRANSMISSION LINES: NERVES AND THEIR STRUCTURE AND REPAIR

13.3 Nerves are cordlike bundles of axons that conduct sensory and motor impulses

Learning Outcomes

▶ Describe the general structure of a nerve.

▶ Define ganglion and indicate the general body location of ganglia.

▶ Follow the process of nerve regeneration.

Structure and Classification

A **nerve** is a cordlike organ that is part of the peripheral nervous system. Nerves are classified as *cranial* or *spinal* depending on whether they arise from the brain or spinal cord. Nerves vary in size, but every nerve consists of parallel bundles of peripheral axons (some myelinated and some not) enclosed by successive wrappings of connective tissue (**Figure 13.4**):

- Each *axon* (or *nerve fiber*) is surrounded by **endoneurium** (en″do-nu′re-um), a delicate layer of loose connective tissue that also encloses the fiber's associated Schwann cells (◀ p. 394).

- A coarser connective tissue wrapping, the **perineurium**, binds groups of axons into bundles called **fascicles** .

- A tough fibrous sheath, the **epineurium**, encloses all the fascicles to form the nerve.

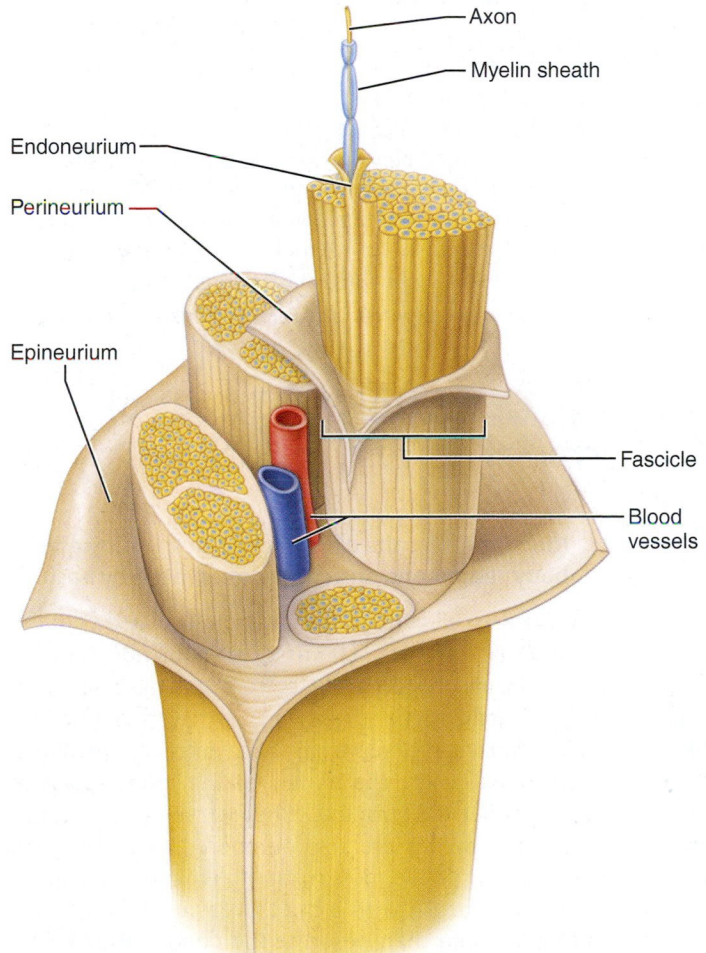

(b)

Figure 13.4 Structure of a nerve. (a) Scanning electron micrograph of a cross section of a portion of a nerve (90×). **(b)** Three-dimensional view of a portion of a nerve, showing connective tissue wrappings.

Axons constitute only a small fraction of a nerve's bulk. The balance consists chiefly of myelin, the protective connective tissue wrappings, blood vessels, and lymphatic vessels.

Recall that the PNS is divided into *sensory* (afferent) and *motor* (efferent) divisions. Nerves are also classified according to the direction in which they transmit impulses:

- **Mixed nerves** contain both sensory and motor fibers and transmit impulses both to and from the central nervous system.
- **Sensory (afferent) nerves** carry impulses only toward the CNS.
- **Motor (efferent) nerves** carry impulses only away from the CNS.

Most nerves are mixed. Pure sensory or motor nerves are rare.

Because mixed nerves often carry both somatic and autonomic (visceral) nervous system fibers, the fibers in them may be classified according to the region they innervate as *somatic afferent*, *somatic efferent*, *visceral afferent*, and *visceral efferent*.

Recall from Chapter 11 that **ganglia** are collections of neuron cell bodies associated with nerves in the PNS, whereas *nuclei* are collections of neuron cell bodies in the CNS. Ganglia associated with *afferent* nerve fibers contain cell bodies of sensory neurons. (These are the *dorsal root ganglia*, ◄ **p. 474**.) Ganglia associated with *efferent* nerve fibers mostly contain cell bodies of autonomic motor neurons (discussed in Chapter 14).

Regeneration of Nerve Axons

CLINICAL

Damage to nervous tissue is serious because, as a rule, mature neurons do not divide. If the damage is severe or close to the cell body, the entire neuron may die, and other neurons that are normally stimulated by its axon may die as well. However, if the cell body remains intact, axons of peripheral nerves can regenerate but axons in the CNS cannot. Let's take a closer look.

CNS Axons

Most CNS axons never regenerate following injury. Consequently, damage to the brain or spinal cord has been viewed as irreversible. This is why spinal cord injuries, for example, are so devastating: Permanent paralysis and sensory deficits are the usual result. The inability to regenerate seems to have less to do with the neurons themselves than with the "company they keep." As we will see shortly, Schwann cells actively help peripheral axons regenerate. Oligodendrocytes, the supporting cells of the CNS, on the other hand, actively suppress CNS axon regeneration.

Oligodendrocytes are studded with growth-inhibiting proteins. Consequently, the growing end of the damaged axon collapses and the axon fails to regrow. Moreover, astrocytes at the site of injury form scar tissue that blocks axon regrowth.

This is a big problem for the clinician treating spinal cord injury because multiple inhibitory processes need to be blocked simultaneously to promote axon regrowth. To date, a number of different experimental approaches have been tried, including neutralizing the myelin-bound growth inhibitors, blocking the receptors for the inhibitory proteins, or enzymatically destroying scar tissue components. None have yet made it past clinical trials.

13

13

① **The axon fragments.**
- The cut axon ends seal themselves off.
- Axon transport is interrupted, causing the cut ends to swell.
- Without access to the cell body, the axon (and its myelin sheath) begins to disintegrate distal to the injury.
- Degeneration of the distal end of the cut axon, called **Wallerian degeneration**, spreads down the axon.

② **Schwann cells and macrophages clean out the dead axon distal to the injury.**
- Surviving Schwann cells engulf the myelin fragments and secrete chemicals that recruit macrophages.
- Macrophages help dispose of the debris and release chemicals that stimulate Schwann cells to divide.

③ **Axon filaments grow through a regeneration tube.**
- Schwann cells release growth factors and express cell adhesion molecules (CAMs) that encourage axon growth.
- Schwann cells line up along the tube of remaining endoneurium, forming a *regeneration tube* that guides the regenerating axon "sprouts" across the gap to their original contacts.

④ **The axon regenerates and a new myelin sheath forms.**
- The Schwann cells protect and support the regenerating axon and ultimately produce a new myelin sheath.

Figure 13.5 Regeneration of a nerve fiber in a peripheral nerve.

PNS Axons

Unlike CNS axons, PNS axons that are cut or crushed can regenerate successfully. **Figure 13.5** describes what happens to a PNS axon after it has been cut.

The neuronal cell body also changes after the axon has been destroyed. Within two days, its rough endoplasmic reticulum breaks apart, and then the cell body swells as protein synthesis revs up to support regeneration of its axon.

Axons regenerate at the approximate rate of 1.5 mm a day. The greater the distance between the severed ends, the less the chance of recovery because adjacent tissues block growth by protruding into the gaps, and axon sprouts fail to find the regeneration tube. Neurosurgeons align cut nerve ends surgically to promote successful regeneration, and scaffolding devices can help guide the axons.

Whatever the measures taken, post-trauma axon regrowth never exactly matches what existed before the injury. Patients must often "retrain" the nervous system to respond appropriately so that stimulus and response are coordinated.

Check Your Understanding

8. What are ganglia?

9. What is in a nerve besides axons?

10. Will's femoral nerve was crushed while clinicians tried to control bleeding from his femoral artery. This resulted in loss of function and sensation in his leg, which gradually returned over the course of a year. Which cells were important in his recovery?

For answers, see Answers Appendix.

13.4 There are 12 pairs of cranial nerves

Learning Outcome

▶ Name the 12 pairs of cranial nerves; indicate the body region and structures innervated by each.

Twelve pairs of **cranial nerves** are associated with the brain (**Figure 13.6**). The first two pairs attach to the forebrain, and the rest are associated with the brain stem. Other than the vagus nerves, which extend into the abdomen, cranial nerves serve only head and neck structures.

In most cases, the names of the cranial nerves reveal either the structures they serve or their functions. The nerves are also numbered (using Roman numerals) from the most rostral to the most caudal.

We will begin with a brief overview of each cranial nerve and then describe the composition of all cranial nerves. After that, you will be ready to tackle **Table 13.2** (pp. 500–506), which provides detailed descriptions of the cranial nerves. Notice that it describes the pathways of the purely or mostly sensory nerves (I, II, and VIII) from the receptors to

Filaments of olfactory nerve (I)

Olfactory bulb

Olfactory tract

Optic nerve (II)

Optic chiasma

Optic tract

Oculomotor nerve (III)

Trochlear nerve (IV)

Trigeminal nerve (V)

Abducens nerve (VI)

Cerebellum

Medulla oblongata

Frontal lobe

Temporal lobe

Infundibulum

Facial nerve (VII)

Vestibulocochlear nerve (VIII)

Glossopharyngeal nerve (IX)

Vagus nerve (X)

Accessory nerve (XI)

Hypoglossal nerve (XII)

(a)

Cranial nerves I – VI	Sensory function	Motor function	PS* fibers
I Olfactory	Yes (smell)	No	No
II Optic	Yes (vision)	No	No
III Oculomotor	No	Yes	Yes
IV Trochlear	No	Yes	No
V Trigeminal	Yes (general sensation)	Yes	No
VI Abducens	No	Yes	No

Cranial nerves VII – XII	Sensory function	Motor function	PS* fibers
VII Facial	Yes (taste)	Yes	Yes
VIII Vestibulocochlear	Yes (hearing and balance)	Some	No
IX Glossopharyngeal	Yes (taste)	Yes	Yes
X Vagus	Yes (taste)	Yes	Yes
XI Accessory	No	Yes	No
XII Hypoglossal	No	Yes	No

*PS = parasympathetic

(b)

Figure 13.6 Location and function of cranial nerves. (a) Ventral view of the human brain, showing the cranial nerves. **(b)** Summary of cranial nerves by function. All cranial nerves that innervate muscles also carry afferent fibers from proprioceptors in the muscles served. Only sensory functions other than proprioception are indicated.

the brain, and the pathways of the other nerves in the opposite direction (from the brain distally). Although we mention autonomic efferents of cranial nerves, in this chapter we will focus on somatic functions. We will defer the discussion of the autonomic nervous system and its visceral functions to Chapter 14.

Overview of Cranial Nerves

I. Olfactory. These are the tiny sensory nerves (filaments) of smell, which run from the nasal mucosa to synapse with the olfactory bulbs. Note that the olfactory bulbs and tracts, shown in Figure 13.6a, are brain structures and not part of cranial nerve I. (See Table 13.2 art for the olfactory nerve filaments.)

II. Optic. Because this sensory nerve of vision develops as an outgrowth of the brain, it is really a brain tract.

III. Oculomotor. The name *oculomotor* means "eye mover." This nerve supplies four of the six extrinsic muscles that move the eyeball in the orbit.

IV. Trochlear. The term *trochlear* means "pulley" and it innervates an extrinsic eye muscle that loops through a pulley-shaped ligament in the orbit.

(Text continues on p. 506.)

Table 13.2 Cranial Nerves

I Olfactory Nerves (ol-fak′to-re)

Origin and course: Olfactory nerve fibers arise from olfactory sensory neurons located in olfactory epithelium of nasal cavity and pass through cribriform plate of ethmoid bone to synapse in olfactory bulb. Fibers of olfactory bulb neurons extend posteriorly as olfactory tract, which runs beneath frontal lobe to enter cerebral hemispheres and terminates in primary olfactory cortex. See also Figure 15.20.

Function: Purely sensory; carry afferent impulses for sense of smell.

CLINICAL TESTING: Ask subject to sniff and identify aromatic substances, such as oil of cloves and vanilla.

🔺 **HOMEOSTATIC IMBALANCE** Fracture of ethmoid bone or lesions of olfactory fibers may result in partial or total loss of smell, a condition known as *anosmia* (an-oz′me-ah). ••••••

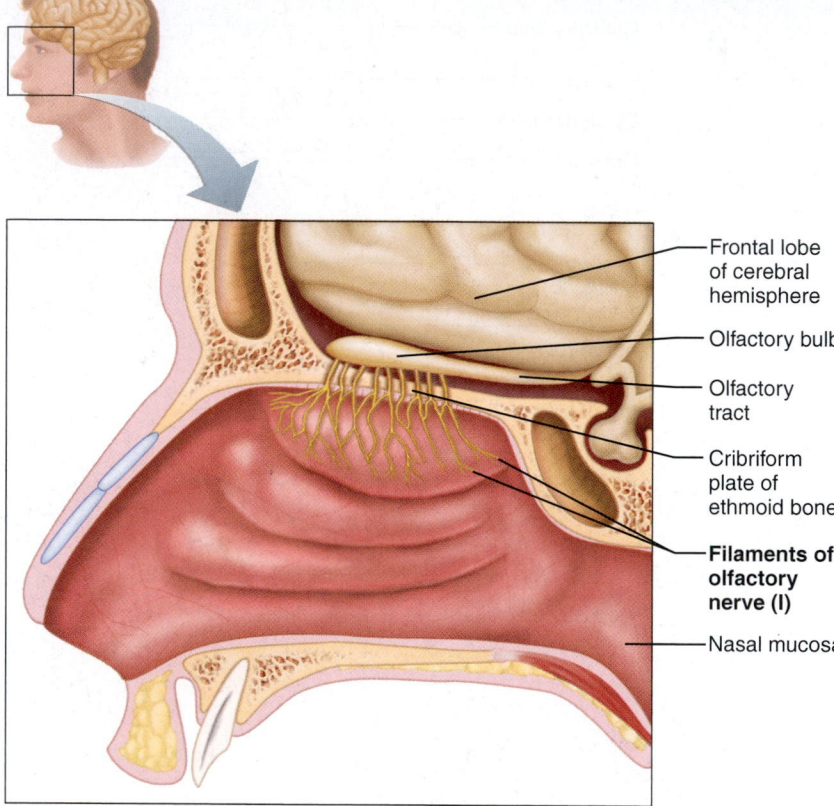

- Frontal lobe of cerebral hemisphere
- Olfactory bulb
- Olfactory tract
- Cribriform plate of ethmoid bone
- **Filaments of olfactory nerve (I)**
- Nasal mucosa

II Optic Nerves

Origin and course: Fibers arise from retina of eye to form optic nerve, which passes through optic canal of orbit. The optic nerves converge to form the optic chiasma (ki-az′mah) where fibers partially cross over, continue on as optic tracts, enter thalamus, and synapse there. Thalamic fibers run (as the optic radiation) to occipital (visual) cortex, where visual interpretation occurs. See also Figure 15.19.

Function: Purely sensory; carry afferent impulses for vision.

CLINICAL TESTING: Assess vision and visual field with eye chart and by testing the point at which the person first sees an object (finger) moving into the visual field. View fundus of eye with ophthalmoscope to detect papilledema (swelling of optic disc, the site where the optic nerve leaves the eyeball) and examine optic disc and retinal blood vessels.

🔺 **HOMEOSTATIC IMBALANCE** Damage to optic nerve results in blindness in eye served by nerve. Damage to visual pathway beyond the optic chiasma results in partial visual losses. Visual defects are called *anopsias* (ah-nop′se-ahz). ••••••

- Eyeball
- Retina
- **Optic nerve (II)**
- Optic chiasma
- Optic tract
- Lateral geniculate nucleus of thalamus
- Optic radiation
- Visual cortex

13

Table 13.2 *(continued)*

III Oculomotor Nerves (ok″u-lo-mo′tor)

Origin and course: Fibers extend from ventral midbrain (near its junction with pons) and pass through bony orbit, via superior orbital fissure, to eye.

Function: Chiefly motor nerves (*oculomotor* = motor to the eye); contain a few proprioceptive afferents. Each nerve includes the following:

- Somatic motor fibers to four of the six extrinsic eye muscles (inferior oblique and superior, inferior, and medial rectus muscles) that help direct eyeball, and to levator palpebrae superioris muscle, which raises upper eyelid.

- Parasympathetic (autonomic) motor fibers to sphincter pupillae (circular muscles of iris), which cause pupil to constrict, and to ciliary muscle, controlling lens shape for visual focusing. Some parasympathetic cell bodies are in the ciliary ganglia.

- Sensory (proprioceptor) afferents, which run from the skeletal muscles served to the midbrain.

CLINICAL TESTING: Examine pupils for size, shape, and equality. Test pupillary reflex with penlight (pupils should constrict when illuminated). Test convergence for near vision and subject's ability to follow objects with the eyes.

HOMEOSTATIC IMBALANCE In oculomotor nerve paralysis, eye cannot be moved up, down, or inward. At rest, eye rotates laterally [*external strabismus* (strah-biz′mus)] because the actions of the two extrinsic eye muscles not served by cranial nerves III are unopposed. Upper eyelid droops (*ptosis*), and the person has double vision and trouble focusing on close objects. ●●●●●●

Medial rectus muscle
Superior rectus muscle
Levator palpebrae muscle
Inferior oblique muscle
Ciliary ganglion
Inferior rectus muscle
Superior orbital fissure
Pons
Parasympathetic motor fibers
Oculomotor nerve (III)

IV Trochlear Nerves (trok′le-ar)

Origin and course: Fibers emerge from dorsal midbrain and course ventrally around midbrain to enter orbit through superior orbital fissure along with oculomotor nerves.

Function: Primarily motor nerves; supply somatic motor fibers to (and carry proprioceptor fibers from) one of the extrinsic eye muscles, the superior oblique muscle, which passes through the pulley-shaped trochlea.

CLINICAL TESTING: Test with cranial nerve III (oculomotor).

HOMEOSTATIC IMBALANCE Damage to a trochlear nerve results in double vision (*diplopia*) and impairs ability to rotate eye inferolaterally. ●●●●●●

Trochlea
Superior oblique muscle
Superior orbital fissure
Pons
Trochlear nerve (IV)

➤

Table 13.2 Cranial Nerves (continued)

V Trigeminal Nerves

Largest cranial nerves; fibers extend from pons to face, and form three divisions (*trigemina* = threefold): ophthalmic (V_1), maxillary (V_2), and mandibular (V_3) divisions. As main general sensory nerves of face, transmit afferent impulses from touch, temperature, and pain receptors. Cell bodies of sensory neurons of all three divisions are located in large *trigeminal ganglion*.

The mandibular division also contains motor fibers that innervate chewing muscles.

Dentists desensitize upper and lower jaws by injecting local anesthetic (such as procaine) into alveolar branches of maxillary and mandibular divisions, respectively. Since this blocks pain-transmitting fibers of teeth, the surrounding tissues become numb.

	Ophthalmic division (V_1)	Maxillary division (V_2)	Mandibular division (V_3)
Origin and course	Fibers run from face to pons via superior orbital fissure.	Fibers run from face to pons via foramen rotundum.	Fibers pass through skull via foramen ovale.
Function	Conveys sensory impulses from skin of anterior scalp, upper eyelid, and nose, and from nasal cavity mucosa, cornea, and lacrimal gland.	Conveys sensory impulses from nasal cavity mucosa, palate, upper teeth, skin of cheek, upper lip, lower eyelid.	Conveys sensory impulses from anterior tongue (except taste buds), lower teeth, skin of chin, temporal region of scalp. Supplies motor fibers to, and carries proprioceptor fibers from, muscles of mastication.
CLINICAL TESTING	Corneal reflex test: Touching cornea with wisp of cotton should elicit blinking.	Test sensations of pain, touch, and temperature with safety pin and hot and cold objects.	Assess motor branch by asking person to clench his teeth, open mouth against resistance, and move jaw side to side.

HOMEOSTATIC IMBALANCE *Trigeminal neuralgia* (nu-ral′je-ah), or *tic douloureux* (tik doo″loo-roo′; *tic* = twitch, *douloureux* = painful), caused by inflammation of trigeminal nerve, is widely considered to produce most excruciating pain known. The stabbing pain lasts for a few seconds to a minute, but it can be relentless, occurring a hundred times a day. Usually provoked by some sensory stimulus, such as brushing teeth or even a passing breeze hitting the face. Thought to be caused by a loop of artery or vein that compresses the trigeminal nerve near its exit from the brain stem. Several drugs are used to treat this frustrating condition. In severe cases, traditional or gamma knife surgery relieves the agony—either by moving the compressing vessel or by destroying the nerve. Nerve destruction results in loss of sensation on that side of face.

(b) Distribution of sensory fibers of each division

(a) Distribution of the trigeminal nerve

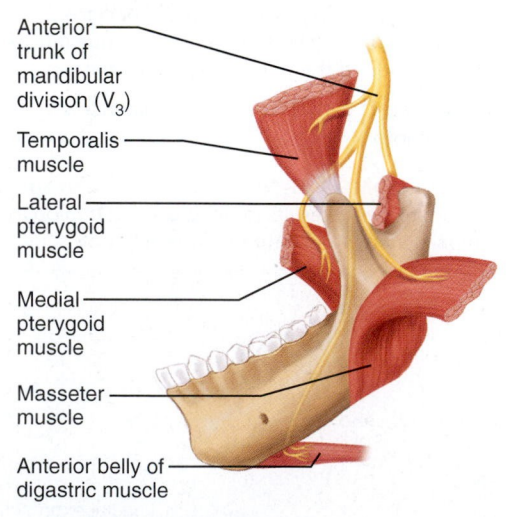

(c) Motor branches of the mandibular division (V_3)

Table 13.2 *(continued)*

VI Abducens Nerves (ab-du'senz)

Origin and course: Fibers leave inferior pons and enter orbit via superior orbital fissure to run to eye.

Function: Primarily motor; supply somatic motor fibers to lateral rectus muscle, an extrinsic muscle of the eye. Convey proprioceptor impulses from same muscle to brain.

CLINICAL TESTING: Test in common with cranial nerve III (oculomotor).

⚖ **HOMEOSTATIC IMBALANCE** In abducens nerve paralysis, eye cannot be moved laterally. At rest, eyeball rotates medially (*internal strabismus*). ●●●●●●

VII Facial Nerves

Origin and course: Fibers issue from pons, just lateral to abducens nerves (see Figure 13.6), enter temporal bone via *internal acoustic meatus*, and run within bone (and through inner ear cavity) before emerging through *stylomastoid foramen*. Nerve then courses to lateral aspect of face.

Function: Mixed nerves that are the chief motor nerves of face. Five major branches: temporal, zygomatic, buccal, mandibular, and cervical (see **c** on next page).

- Convey motor impulses to skeletal muscles of face (muscles of facial expression), except for chewing muscles served by trigeminal nerves, and transmit proprioceptor impulses from same muscles to pons (see **b** and pp. 333–335).

- Transmit parasympathetic (autonomic) motor impulses to lacrimal (tear) glands, nasal and palatine glands, and submandibular and sublingual salivary glands. Some of the cell bodies of these parasympathetic motor neurons are in

pterygopalatine (ter"eh-go-pal'ah-tīn) and *submandibular ganglia* on the trigeminal nerve (see **a**).

- Convey sensory impulses from taste buds of anterior two-thirds of tongue; cell bodies of these sensory neurons are in *geniculate ganglion* (see **a**).

CLINICAL TESTING: Test anterior two-thirds of tongue for ability to taste sweet (sugar), salty, sour (vinegar), and bitter (quinine) substances. Check symmetry of face. Ask subject to close eyes, smile, whistle, and so on. Assess tearing with ammonia fumes.

⚖ **HOMEOSTATIC IMBALANCE** *Bell's palsy* is characterized by paralysis of facial muscles on affected side and partial loss of taste sensation. May develop rapidly (often overnight). Caused by inflamed and swollen facial nerve, possibly due to herpes simplex 1 viral infection. Lower eyelid droops, corner of mouth sags (making it difficult to eat or speak normally), tears drip continuously from eye and eye cannot be completely closed (conversely, dry-eye syndrome may occur). Treated with corticosteroids. Recovery is complete in 70% of cases. ●●●●●●

(a) Parasympathetic efferents and sensory afferents

(b) Motor branches to muscles of facial expression and scalp muscles

Table 13.2   Cranial Nerves *(continued)*

VII Facial Nerves *(continued)*

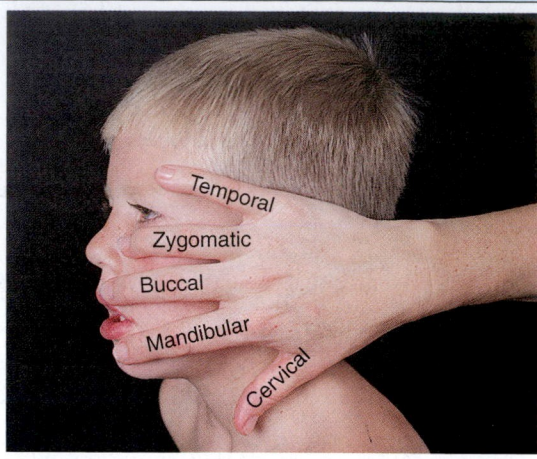

(c) A simple way to remember the courses of the five major branches of the facial nerve

VIII Vestibulocochlear Nerves (ves-tib"u-lo-kok′le-ar)

Origin and course: Fibers arise from hearing and equilibrium apparatus located within inner ear of temporal bone and pass through internal acoustic meatus to enter brain stem at pons-medulla border. Afferent fibers from hearing receptors in cochlea form the *cochlear division* (cochlear nerve); those from equilibrium receptors in semicircular canals and vestibule form the *vestibular division* (vestibular nerve). The two divisions merge to form vestibulocochlear nerve. See also Figure 15.26.

Function: Mostly sensory. Vestibular division transmits afferent impulses for sense of equilibrium, and sensory nerve cell bodies are located in *vestibular ganglia.* Cochlear division transmits afferent impulses for sense of hearing, and sensory nerve cell bodies are located in *spiral ganglion* within cochlea. Small motor component adjusts the sensitivity of sensory receptors. See also Figure 15.27.

CLINICAL TESTING: Check hearing by air and bone conduction using tuning fork.

HOMEOSTATIC IMBALANCE Lesions of cochlear nerve or cochlear receptors result in *central,* or *nerve, deafness.* Damage to vestibular division produces dizziness, rapid involuntary eye movements, loss of balance, nausea, and vomiting.

Vestibulocochlear nerve (VIII)

Table 13.2 (continued)

IX Glossopharyngeal Nerves (glos"o-fah-rin'je-al)

Origin and course: Fibers emerge from medulla and leave skull via *jugular foramen* to run to throat.

Function: Mixed nerves that innervate part of tongue and pharynx. Provide somatic motor fibers to, and carry proprioceptor fibers from, a superior pharyngeal muscle called the *stylopharyngeus*, which elevates the pharynx in swallowing. Provide parasympathetic motor fibers to parotid salivary glands (some of the nerve cell bodies of these parasympathetic motor neurons are located in *otic ganglion*).

Sensory fibers conduct taste and general sensory (touch, pressure, pain) impulses from pharynx and posterior tongue, impulses from chemoreceptors in the carotid body (which monitor O_2 and CO_2 levels in the blood and help regulate respiratory rate and depth), and impulses from baroreceptors of carotid sinus (which monitor blood pressure). Sensory neuron cell bodies are located in *superior* and *inferior ganglia*.

CLINICAL TESTING: Check position of uvula; check gag and swallowing reflexes. Ask subject to speak and cough. Test posterior third of tongue for taste.

HOMEOSTATIC IMBALANCE Injured or inflamed glossopharyngeal nerves impair swallowing and taste. ● ● ● ● ● ●

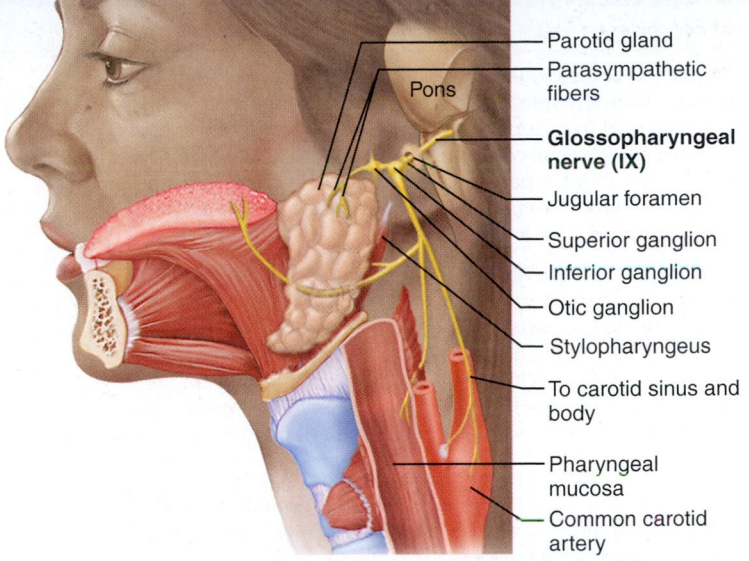

Parotid gland
Parasympathetic fibers
Pons
Glossopharyngeal nerve (IX)
Jugular foramen
Superior ganglion
Inferior ganglion
Otic ganglion
Stylopharyngeus
To carotid sinus and body
Pharyngeal mucosa
Common carotid artery

X Vagus Nerves (va'gus)

Origin and course: The only cranial nerves to extend beyond head and neck region. Fibers emerge from medulla, pass through skull via jugular foramen, and descend through neck region into thorax and abdomen. See also Figure 14.4.

Function: Mixed nerves. Nearly all motor fibers are parasympathetic efferents, except those serving skeletal muscles of pharynx and larynx (involved in swallowing). Parasympathetic motor fibers supply heart, lungs, and abdominal viscera and are involved in regulating heart rate, breathing, and digestive system activity. Transmit sensory impulses from thoracic and abdominal viscera, from the aortic arch baroreceptors (for blood pressure) and the carotid and aortic bodies (chemoreceptors for respiration), and taste buds on the epiglottis. Also carry general somatic sensory information from small area of skin on external ear. Carry proprioceptor fibers from muscles of larynx and pharynx.

CLINICAL TESTING: As for cranial nerve IX (glossopharyngeal); IX and X are tested in common, since they both innervate muscles of throat and mouth.

HOMEOSTATIC IMBALANCE Since laryngeal branches of the vagus innervate nearly all muscles of the larynx ("voice box"), vagal nerve paralysis can lead to hoarseness or loss of voice. Other symptoms are difficulty swallowing and impaired digestive system motility. These parasympathetic nerves are important for maintaining the normal state of visceral organ activity. Without their influence, the sympathetic nerves, which mobilize and accelerate vital body processes (and shut down digestion), would dominate. ● ● ● ● ● ●

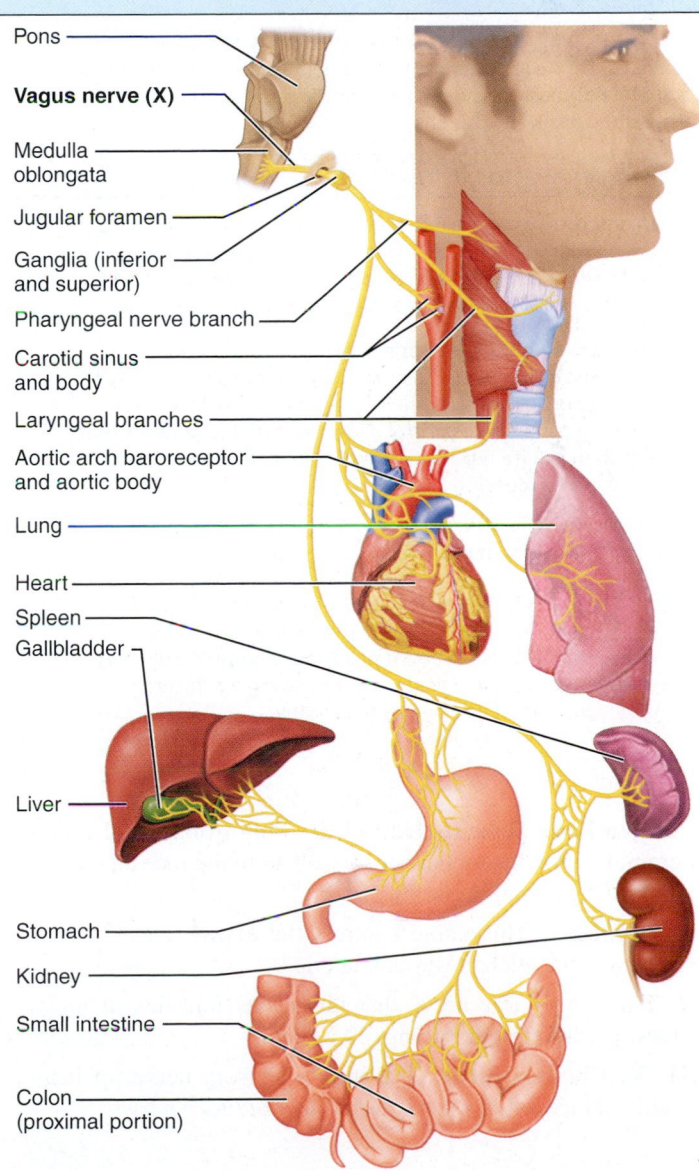

Pons
Vagus nerve (X)
Medulla oblongata
Jugular foramen
Ganglia (inferior and superior)
Pharyngeal nerve branch
Carotid sinus and body
Laryngeal branches
Aortic arch baroreceptor and aortic body
Lung
Heart
Spleen
Gallbladder
Liver
Stomach
Kidney
Small intestine
Colon (proximal portion)

13

➤

Table 13.2 Cranial Nerves (continued)

XI Accessory Nerves

Origin and course: Unique in that they form from rootlets that emerge from the spinal cord, not the brain stem. These rootlets arise laterally from superior region (C_1–C_5) of spinal cord, pass upward along spinal cord, and enter the skull as the accessory nerves via foramen magnum. The accessory nerves exit from skull through *jugular foramen* together with the vagus nerves, and supply two large neck muscles.

Until recently, these nerves were considered to have both a cranial and spinal portion, but the cranial rootlets are actually part of the vagus nerves. This raises an interesting question: Should the accessory nerves still be considered cranial nerves? Some anatomists say "no" because they don't arise from the brain. Others say "yes" because their origin is different from a typical spinal nerve and they pass through the skull. Stay tuned!

Function: Mixed nerves, but primarily motor in function. Supply motor fibers to trapezius and sternocleidomastoid muscles, which together move head and neck, and convey proprioceptor impulses from same muscles.

CLINICAL TESTING: Check strength of sternocleidomastoid and trapezius muscles by asking person to rotate head and shrug shoulders against resistance.

HOMEOSTATIC IMBALANCE Injury to one accessory nerve causes head to turn toward the injured side as a result of sternocleidomastoid muscle paralysis. Shrugging that shoulder (role of trapezius muscle) becomes difficult.

Jugular foramen
Foramen magnum
Accessory nerve (XI)
Sternocleidomastoid muscle
Rootlets of accessory nerve emerging from spinal cord (C_1–C_5)
Trapezius muscle

XII Hypoglossal Nerves (hi″po-glos′al)

Origin and course: As their name implies (*hypo* = below; *glossal* = tongue), hypoglossal nerves mainly serve the tongue. Fibers arise by a series of roots from medulla and exit from skull via *hypoglossal canal* to travel to tongue. See also Figure 13.6.

Function: Mixed nerves, but primarily motor in function. Carry somatic motor fibers to intrinsic and extrinsic muscles of tongue, and proprioceptor fibers from same muscles to brain stem. Hypoglossal nerve control allows tongue movements that mix and manipulate food during chewing, and contribute to swallowing and speech.

CLINICAL TESTING: Ask subject to protrude and retract tongue. Note any deviations in position.

HOMEOSTATIC IMBALANCE Damage to hypoglossal nerves causes difficulties in speech and swallowing. If both nerves are impaired, the person cannot protrude tongue. If only one side is affected, tongue deviates (points) toward affected side; eventually paralyzed side begins to atrophy.

Medulla oblongata
Intrinsic muscles of the tongue
Hypoglossal canal
Hypoglossal nerve (XII)
Extrinsic muscles of the tongue

V. Trigeminal. Three (*tri*) branches spring from this, the largest cranial nerve. It supplies sensory fibers to the face and motor fibers to the chewing muscles.

VI. Abducens. This nerve controls the extrinsic eye muscle that *abducts* the eyeball (turns it laterally).

VII. Facial. A large nerve that innervates muscles of *facial* expression (among other things).

VIII. Vestibulocochlear. This mostly sensory nerve for hearing and balance was formerly called the *auditory nerve*.

IX. Glossopharyngeal. The name *glossopharyngeal* means "tongue and pharynx," the structures that this nerve helps to innervate.

X. Vagus. This nerve's name means "wanderer" or "vagabond," and it is the only cranial nerve to extend beyond the head and neck to the thorax and abdomen.

XI. Accessory. Considered an *accessory* part of the vagus nerve, this nerve was formerly called the *spinal accessory nerve*.

XII. Hypoglossal. The name *hypoglossal* means under the tongue. This nerve runs inferior to the tongue and innervates the tongue muscles.

You might make up your own saying to remember the first letters of the cranial nerves in order, or use the following memory jog sent by a student: "**O**n **o**ccasion, **o**ur **t**rusty **t**ruck **a**cts **f**unny—**v**ery **g**ood **v**ehicle **a**ny**h**ow."

Composition of Cranial Nerves

In the last chapter, we described how ventral (motor) and dorsal (sensory) roots fuse to form spinal nerves. They are called mixed nerves because they carry both sensory and motor information. Cranial nerves, on the other hand, vary markedly in their composition.

Many cranial nerves are mixed nerves, as shown in Figure 13.6b. However, two nerve pairs (the olfactory and optic) associated with special sense organs are generally considered purely sensory. The cell bodies of the sensory neurons in the olfactory and optic nerves are located *within* their respective special sense organs. For other sensory neurons contributing to cranial nerves (V, VII, VIII, IX, and X), the cell bodies are located in **cranial sensory ganglia** just outside the brain. Some cranial nerves have a single sensory ganglion, others have several, and still others have none.

Several of the mixed cranial nerves contain both somatic and autonomic motor fibers because they serve both skeletal muscles and visceral organs. Except for some autonomic motor neurons located in ganglia, the cell bodies of motor neurons contributing to the cranial nerves are located in the ventral gray matter regions (nuclei) of the brain stem.

Remembering the primary functions of the 12 cranial nerves (as **s**ensory, **m**otor, or **b**oth) can be a challenge. This sentence might help: "**S**ome **s**ay **m**arry **m**oney, **b**ut **m**y **b**rother **b**elieves (it's) **b**ad **b**usiness (to) **m**arry **m**oney."

Check Your Understanding

11. APPLY Name the cranial nerve(s) most involved in each of the following: moving your eyeball; sticking out your tongue; controlling your heart rate and digestive activity; shrugging your shoulders.

For answers, see Answers Appendix.

13.5 31 pairs of spinal nerves innervate the body

Learning Outcomes

▶ **Describe the general structure of a spinal nerve and the general distribution of its rami.**

▶ **Define plexus. Name the major plexuses and describe the distribution and function of the peripheral nerves arising from each plexus.**

Thirty-one pairs of **spinal nerves**, each containing thousands of nerve fibers, arise from the spinal cord and supply all parts of the body except the head and some areas of the neck. All are mixed nerves.

Spinal nerves are named according to where they emerge from the spinal cord (**Figure 13.7**). The spinal nerves include:

- 8 pairs of cervical nerves (C_1–C_8)
- 12 pairs of thoracic nerves (T_1–T_{12})
- 5 pairs of lumbar nerves (L_1–L_5)
- 5 pairs of sacral nerves (S_1–S_5)
- 1 pair of tiny coccygeal nerves (Co_1)

Notice that there are eight pairs of cervical nerves but only seven cervical vertebrae. This "discrepancy" is easily explained. The first seven pairs exit the vertebral canal *superior* to the vertebrae for which they are named, but C_8 emerges *inferior* to the seventh cervical vertebra (between C_7 and T_1). Below the cervical level, each spinal nerve leaves the vertebral column *inferior* to the same-numbered vertebra.

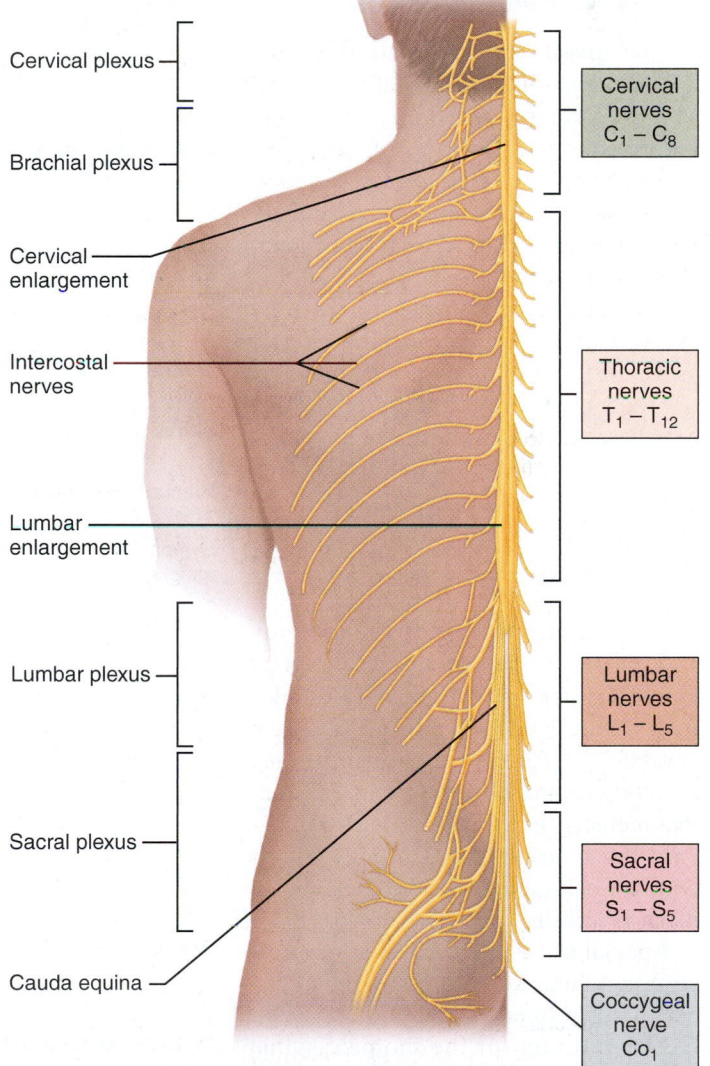

Cervical plexus

Brachial plexus

Cervical enlargement

Intercostal nerves

Lumbar enlargement

Lumbar plexus

Sacral plexus

Cauda equina

Cervical nerves C_1 – C_8

Thoracic nerves T_1 – T_{12}

Lumbar nerves L_1 – L_5

Sacral nerves S_1 – S_5

Coccygeal nerve Co_1

Figure 13.7 Spinal nerves. (Posterior view.) The spinal nerves are shown at right; their ventral rami are shown at left. Most ventral rami form nerve plexuses (cervical, brachial, lumbar, and sacral).

As described in Chapter 12, each spinal nerve connects to the spinal cord by a dorsal root and a ventral root (**Figure 13.8**). Each root forms from a series of **rootlets** that attach along the length of the corresponding spinal cord segment (Figure 13.8a).

- **Ventral roots** contain *motor* (efferent) fibers that arise from ventral horn motor neurons and extend to and innervate the skeletal muscles. (In Chapter 14, we describe autonomic nervous system efferents that are also contained in the ventral roots.)

- **Dorsal roots** contain *sensory* (afferent) fibers that arise from sensory neurons in the dorsal root ganglia and conduct impulses from peripheral receptors to the spinal cord.

The spinal roots pass laterally from the cord and unite just distal to the dorsal root ganglion to form a spinal nerve before emerging from the vertebral column via their respective intervertebral foramina. Because motor and sensory fibers mingle in a spinal nerve, it contains both efferent and afferent fibers.

The spinal roots become progressively longer from the superior to the inferior aspect of the cord. In the cervical region, the roots are short and run horizontally, but the roots of the lumbar and sacral nerves extend inferiorly for some distance through the lower vertebral canal as the *cauda equina* before exiting the vertebral column (Figures 13.7, 12.28, and 12.30).

A spinal nerve is quite short (only 1–2 cm). Almost immediately after emerging from its foramen, it divides into a small **dorsal ramus** (ra′mus; "branch"), a larger **ventral ramus**, and a tiny **meningeal branch** (mĕ-nin′je-al) that reenters the vertebral canal to innervate the meninges and blood vessels within. Each ramus, like the spinal nerve itself, is mixed.

Special rami called **rami communicantes**, which contain autonomic (visceral) nerve fibers, attach to the base of the ventral rami of the thoracic spinal nerves (Figure 13.8).

(a) Anterior view showing spinal cord, associated nerves, and vertebrae. The dorsal and ventral roots arise medially as rootlets and join laterally to form the spinal nerve.

(b) Cross section of thorax showing the main roots and branches of a spinal nerve.

Figure 13.8 Formation of spinal nerves and rami distribution. Notice the dorsal and ventral roots and rami. In the thorax, each ventral ramus continues as an intercostal nerve. (The small meningeal branch is not illustrated.)

Innervation of Specific Body Regions

The spinal nerve rami and their main branches supply the entire somatic region of the body (skeletal muscles and skin) from the neck down. The dorsal rami supply the posterior body trunk. The thicker ventral rami supply the rest of the trunk and the limbs.

Let's review the difference between roots and rami:

- Roots lie medial to and form the spinal nerves. Each root is strictly sensory or motor.
- Rami lie distal to and are lateral branches of the spinal nerves. Like spinal nerves, rami carry both sensory and motor fibers.

Before we get into the specifics of how the body is innervated, it is important for you to understand some points about the ventral rami of the spinal nerves. Except for T_2–T_{12}, all ventral rami branch and join one another lateral to the vertebral column, forming complicated interlacing nerve networks called **nerve plexuses** (see Figure 13.7). Nerve plexuses occur in the cervical, brachial, lumbar, and sacral regions and primarily serve the limbs. Notice that *only ventral rami form plexuses.*

Within a plexus, fibers from the various ventral rami crisscross one another and become redistributed so that (1) each resulting branch of the plexus contains fibers from several spinal nerves and (2) fibers from each ventral ramus travel to the body periphery via several routes. As a result, each muscle in a limb receives its nerve supply from more than one spinal nerve. An advantage of this rearrangement is that damage to one spinal segment or root cannot completely paralyze any limb muscle.

In the rest of this section, we summarize the major groups of skeletal muscles served. For more specific information on muscle innervations, see Tables 10.1–10.17.

Cervical Plexus and the Neck

The looping **cervical plexus** is buried deep in the neck under the sternocleidomastoid muscle (**Figure 13.9**). The cervical plexus is formed from the ventral rami of the first four cervical nerves (with a minor contribution from C_5). Most of its branches are **cutaneous nerves** that supply only the skin (**Table 13.3**). They transmit sensory impulses from the skin of

☐ Ventral rami

Segmental branches
Hypoglossal nerve (XII)
Lesser occipital nerve
Greater auricular nerve
Transverse cervical nerve
Ansa cervicalis
Accessory nerve (XI)
Phrenic nerve
Supraclavicular nerves

Ventral rami:
C_1
C_2
C_3
C_4
C_5

Figure 13.9 The cervical plexus. The nerves colored gray connect to the plexus but do not belong to it. See Table 13.3 for structures served. (Anterior view.)

the neck, the ear area, the back of the head, and the shoulder. Other branches innervate muscles of the anterior neck.

The single most important nerve from this plexus is the **phrenic nerve** (fren'ik), which receives fibers from C_3, C_4, and C_5. The phrenic nerve runs inferiorly through the thorax and supplies both motor and sensory fibers to the diaphragm (*phren* = diaphragm), which is the chief muscle causing breathing movements.

Table 13.3 Branches of the Cervical Plexus (See Figure 13.9)		
NERVES	**VENTRAL RAMI**	**STRUCTURES SERVED**
Cutaneous Branches (Superficial)		
Lesser occipital	C_2 (C_3)	Skin on posterolateral aspect of head and neck
Greater auricular	C_2, C_3	Skin of ear, skin over parotid gland
Transverse cervical	C_2, C_3	Skin on anterior and lateral aspect of neck
Supraclavicular (medial, intermediate, and lateral)	C_3, C_4	Skin of shoulder and clavicular region
Motor Branches (Deep)		
Ansa cervicalis (superior and inferior roots)	C_1–C_3	Infrahyoid muscles of neck (omohyoid, sternohyoid, and sternothyroid)
Segmental and other muscular branches	C_1–C_5	Deep muscles of neck (geniohyoid and thyrohyoid) and portions of scalenes, levator scapulae, trapezius, and sternocleidomastoid muscles
Phrenic	C_3–C_5	Diaphragm (sole motor nerve supply)

13

HOMEOSTATIC IMBALANCE 13.2   CLINICAL

Irritation of the phrenic nerve causes spasms of the diaphragm, or hiccups. If both phrenic nerves are severed, or if the C_3–C_5 region of the spinal cord is crushed or destroyed, the diaphragm is paralyzed and respiratory arrest occurs. (Remember this saying: "C3, 4, and 5 keep the diaphragm alive.") Mechanical ventilators keep victims alive by forcing air into their lungs—literally breathing for them.

Brachial Plexus and Upper Limb

The large, important **brachial plexus**, situated partly in the neck and partly in the axilla, gives rise to virtually all the nerves that innervate the upper limb (**Table 13.4**). It can be palpated (felt) in a living person just superior to the clavicle at the lateral border of the sternocleidomastoid muscle. Ventral rami of C_5–C_8 and most of the T_1 ramus intermix to form the brachial plexus. Additionally, it often receives fibers from C_4 or T_2 or both.

The brachial plexus is very complex (some consider it the anatomy student's nightmare). Perhaps the simplest approach is to master the terms used for its four major groups of branches (**Figure 13.10a, b**). From medial to lateral, these are the ventral rami (misleadingly called *roots*) which form *trunks*, which form *divisions*, which form *cords*. You might want to use the saying "**R**eally **t**ired? **D**rink **c**offee" to help you remember this branching sequence.

The five **roots** (ventral rami C_5–T_1) of the brachial plexus lie deep to the sternocleidomastoid muscle. At the lateral border of that muscle, these roots unite to form **upper**, **middle**, and **lower trunks**, each of which divides almost immediately into an **anterior division** and a **posterior division**.

The anterior and posterior divisions, which generally indicate which fibers serve the front or back of the limb, pass deep to the clavicle and enter the axilla. There they give rise to three large fiber bundles called the **lateral**, **medial**, and **posterior cords**. (The cords are named for their relationship to the axillary artery, which runs through the axilla; see Figure 19.25.) All along the plexus, small nerves branch off. These supply the muscles and skin of the shoulder and superior thorax.

HOMEOSTATIC IMBALANCE 13.3   CLINICAL

Injuries to the brachial plexus are common. When severe, they weaken or paralyze the entire upper limb. Such injuries may occur when the upper limb is pulled hard, stretching the plexus (as when a football tackler yanks the arm of the running back), or by blows to the top of the shoulder that force the humerus inferiorly (as when a cyclist pitches headfirst off a bicycle and his shoulder grinds into the pavement).

Table 13.4	Branches of the Brachial Plexus (See Figure 13.10)	
NERVES	**CORD AND VENTRAL RAMI**	**STRUCTURES SERVED**
Musculocutaneous	Lateral cord (C_5–C_7)	Muscular branches: flexor muscles in anterior arm (biceps brachii, brachialis, coracobrachialis) Cutaneous branches: skin on lateral forearm (extremely variable)
Median	By two branches, one from medial cord (C_8, T_1) and one from the lateral cord (C_5–C_7)	Muscular branches to flexor group of anterior forearm (palmaris longus, flexor carpi radialis, flexor digitorum superficialis, flexor pollicis longus, lateral half of flexor digitorum profundus, and pronator muscles); intrinsic muscles of lateral palm and digital branches to the fingers Cutaneous branches: skin of lateral two-thirds of hand on ventral side and dorsum of fingers 2 and 3
Ulnar	Medial cord (C_8, T_1)	Muscular branches: flexor muscles in anterior forearm (flexor carpi ulnaris and medial half of flexor digitorum profundus); most intrinsic muscles of hand Cutaneous branches: skin of medial third of hand, both anterior and posterior aspects
Radial	Posterior cord (C_5–C_8, T_1)	Muscular branches: posterior muscles of arm and forearm (triceps brachii, anconeus, supinator, brachioradialis, extensors carpi radialis longus and brevis, extensor carpi ulnaris, and several muscles that extend the fingers) Cutaneous branches: skin of posterolateral surface of entire limb (except dorsum of fingers 2 and 3)
Axillary	Posterior cord (C_5, C_6)	Muscular branches: deltoid and teres minor muscles Cutaneous branches: some skin of shoulder region
Dorsal scapular	Branches of C_5 rami	Rhomboid muscles and levator scapulae
Long thoracic	Branches of C_5–C_7 rami	Serratus anterior muscle
Subscapular	Posterior cord; branches of C_5 and C_6 rami	Teres major and subscapularis muscles
Suprascapular	Upper trunk (C_5, C_6)	Shoulder joint; supraspinatus and infraspinatus muscles
Pectoral (lateral and medial)	Branches of lateral and medial cords (C_5–T_1)	Pectoralis major and minor muscles

13

Anterior divisions | Posterior divisions | Trunks | Roots

(a) Roots (rami C₅–T₁), trunks, divisions, and cords

Dorsal scapular
Nerve to subclavius
Suprascapular
Posterior divisions

Roots (ventral rami)

C_4
C_5
C_6
C_7
C_8
T_1

Upper
Middle — **Trunks**
Lower

Cords
Lateral
Posterior
Medial

Axillary
Musculo-cutaneous
Radial
Median
Ulnar

Long thoracic
Medial pectoral
Lateral pectoral
Upper subscapular
Lower subscapular
Thoracodorsal
Medial cutaneous nerves of the arm and forearm

Axillary nerve

Humerus
Radial nerve
Musculo-cutaneous nerve
Ulna
Radius
Ulnar nerve
Median nerve
Radial nerve (superficial branch)
Dorsal branch of ulnar nerve
Superficial branch of ulnar nerve
Digital branch of ulnar nerve
Muscular branch — Median nerve
Digital branch

(c) The major nerves of the upper limb

Major terminal branches (peripheral nerves)	Cords	Divisions	Trunks	Roots (ventral rami)
Musculocutaneous	Lateral	Anterior	Upper	C_5
Median		Posterior		C_6
Ulnar	Medial	Anterior	Middle	C_7
Radial		Posterior		C_8
Axillary	Posterior	Anterior	Lower	T_1
		Posterior		

(b) Flowchart summarizing relationships within the brachial plexus

Musculocutaneous nerve
Axillary nerve
Radial nerve
Median nerve
Ulnar nerve

Lateral cord
Posterior cord
Medial cord

(d) Cadaver photo

Figure 13.10 The brachial plexus. (Anterior view.)

Practice Human Cadaver lab questions:
MasteringA&P® > Study Area > Lab Tools > PAL

13

The brachial plexus ends in the axilla, where its three cords wind along the axillary artery and give rise to the main nerves of the upper limb (Figure 13.10c, d). Five of these nerves are especially important: the axillary, musculocutaneous, median, ulnar, and radial nerves. We describe their distribution and targets briefly here. For more detail, see Table 13.4.

Axillary Nerve   The **axillary nerve** branches off the posterior cord and runs posterior to the surgical neck of the humerus. It innervates the deltoid and teres minor muscles and the skin and joint capsule of the shoulder.

Musculocutaneous Nerve   The **musculocutaneous nerve**, the major end branch of the lateral cord, courses inferiorly in the anterior arm, supplying motor fibers to the biceps brachii, brachialis, and coracobrachialis muscles. Distal to the elbow, it provides cutaneous sensation in the lateral forearm.

Median Nerve   The **median nerve** descends through the arm to the anterior forearm, where it gives off branches to the skin and to most flexor muscles. On reaching the hand, it innervates five intrinsic muscles of the lateral palm. The median nerve activates muscles that pronate the forearm, flex the hand and fingers, and oppose the thumb.

HOMEOSTATIC IMBALANCE 13.4   CLINICAL

Median nerve injury makes it difficult to use the pincer grasp (opposed thumb and index finger) to pick up small objects. Because this nerve runs down the midline of the forearm and wrist, it is a frequent casualty of accidental or intentional lacerations (cuts). In carpal tunnel syndrome (◄ p. 234), the median nerve is compressed.   ••••••••

Ulnar Nerve   The **ulnar nerve** branches off the medial cord of the plexus. It descends along the medial aspect of the arm toward the elbow, swings behind the medial epicondyle, and then follows the ulna along the medial forearm. There it supplies the flexor carpi ulnaris and the medial part of the flexor digitorum profundus (the flexors not supplied by the median nerve).

The ulnar nerve continues into the hand, where it innervates most intrinsic hand muscles and the skin of the medial aspect of the hand. It causes the hand and fingers to flex, and (with the median nerve) adducts and abducts the medial fingers.

HOMEOSTATIC IMBALANCE 13.5   CLINICAL

Where it takes a superficial course, the ulnar nerve is very vulnerable to injury. Striking the "funny bone"—the spot where this nerve rests against the medial epicondyle—makes the little finger tingle. Severe or chronic damage can lead to sensory loss, paralysis, and muscle atrophy. Affected individuals have trouble making a fist and gripping objects. As the little and ring fingers become hyperextended at the knuckles and flexed at the distal interphalangeal joints, the hand contorts into a *claw hand* (shown at right).

••••••••

Radial Nerve   The **radial nerve**, the largest branch of the brachial plexus, is a continuation of the posterior cord. This nerve wraps around the humerus (in the radial groove), and then runs anteriorly around the lateral epicondyle at the elbow. There it divides into a superficial branch that follows the lateral edge of the radius to the hand and a deep branch (not illustrated) that runs posteriorly. It supplies the posterior skin of the limb along its entire course. Its motor branches innervate essentially all the extensor muscles of the upper limb. Muscles controlled by the radial nerve extend the elbow, supinate the forearm, extend the hand and fingers, and abduct the thumb.

HOMEOSTATIC IMBALANCE 13.6   CLINICAL

Trauma to the radial nerve results in *wrist drop* (shown at right), inability to extend the hand at the wrist. Improper use of a crutch or "Saturday night paralysis," in which an intoxicated person falls asleep with an arm draped over the back of a chair or sofa edge, can compress the radial nerve and impair its blood supply.

Lumbosacral Plexus and Lower Limb

The sacral and lumbar plexuses overlap substantially. Because many fibers of the lumbar plexus contribute to the sacral plexus via the **lumbosacral trunk**, the two plexuses are often referred to as the **lumbosacral plexus**. Although the lumbosacral plexus serves mainly the lower limb, it also sends some branches to the abdomen, pelvis, and buttock.

Lumbar Plexus   The **lumbar plexus** arises from spinal nerves L_1–L_4 and lies within the psoas major muscle (**Figure 13.11**). Its proximal branches innervate parts of the abdominal wall muscles and the psoas muscle, but its major branches descend to innervate the anterior and medial thigh.

The **femoral nerve**, the largest terminal nerve of this plexus, runs deep to the inguinal ligament to enter the thigh and then divides into several large branches. The motor branches innervate anterior thigh muscles (quadriceps), which are the principal thigh flexors and leg extensors. The cutaneous branches serve the skin of the anterior thigh and the medial surface of the leg from knee to foot.

The **obturator nerve** (ob″tu-ra′tor) enters the medial thigh via the obturator foramen and innervates the adductor muscles. **Table 13.5** summarizes the branches of the lumbar plexus.

HOMEOSTATIC IMBALANCE 13.7   CLINICAL

When the spinal roots of the lumbar plexus are compressed, as by a herniated disc, gait problems occur because the femoral nerve serves the prime movers that flex the thigh and extend the leg. Other symptoms are pain or numbness of the anterior thigh and (if the obturator nerve is impaired) of the medial thigh.   •••••••

☐ Ventral rami

Iliohypogastric

Ilioinguinal

Genitofemoral

Lateral femoral cutaneous

Obturator

Femoral

Lumbosacral trunk

Ventral rami:
L_1
L_2
L_3
L_4
L_5

(a) Ventral rami and major branches of the lumbar plexus

Iliohypogastric
Ilioinguinal

Femoral

Lateral femoral cutaneous

Obturator

Anterior femoral cutaneous

Saphenous

(b) Distribution of the major nerves from the lumbar plexus to the lower limb

Figure 13.11 **The lumbar plexus.** (Anterior view.)

Table 13.5	Branches of the Lumbar Plexus (See Figure 13.11)	
NERVES	**VENTRAL RAMI**	**STRUCTURES SERVED**
Femoral	L_2–L_4	Skin of anterior and medial thigh via *anterior femoral cutaneous* branch; skin of medial leg and foot, hip and knee joints via *saphenous* branch; motor to anterior muscles (quadriceps and sartorius) of thigh and to pectineus, iliacus
Obturator	L_2–L_4	Motor to adductor magnus (part), longus, and brevis muscles, gracilis muscle of medial thigh, obturator externus; sensory for skin of medial thigh and for hip and knee joints
Lateral femoral cutaneous	L_2, L_3	Skin of lateral thigh; some sensory branches to peritoneum
Iliohypogastric	L_1	Skin on side of buttock and above pubis; muscles of anterolateral abdominal wall (internal obliques and transversus abdominis)
Ilioinguinal	L_1	Skin of external genitalia and proximal medial aspect of the thigh; inferior abdominal muscles
Genitofemoral	L_1, L_2	Skin of scrotum in males, of labia majora in females, and of anterior thigh inferior to middle portion of inguinal region; cremaster muscle in males

13

☐ Ventral rami

Ventral rami:

L₄

L₅

S₁

S₂

S₃

S₄

S₅

Co₁

Superior gluteal

Lumbosacral trunk

Inferior gluteal

Common fibular

Tibial

Posterior femoral cutaneous

Pudendal

Sciatic

(a) Ventral rami and major branches of the sacral plexus, anterior view

Superior gluteal

Inferior gluteal

Pudendal

Sciatic

Posterior femoral cutaneous

Common fibular

Tibial

Sural (cut)

Deep fibular

Superficial fibular

Plantar branches

(b) Distribution of the major nerves from the sacral plexus to the lower limb

Gluteus maximus

Piriformis

Inferior gluteal nerve

Common fibular nerve

Tibial nerve

Pudendal nerve

Posterior femoral cutaneous nerve

Sciatic nerve

(c) Dissection of the gluteal region, posterior view

Figure 13.12   The sacral plexus.

Practice Human Cadaver lab questions: MasteringA&P® > Study Area > Lab Tools > PAL

Sacral Plexus   The **sacral plexus** arises from spinal nerves L₄–S₄ and lies immediately caudal to the lumbar plexus (**Figure 13.12**). Some fibers of the lumbar plexus contribute to the sacral plexus via the *lumbosacral trunk*, as mentioned earlier.

The sacral plexus has about a dozen named branches. About half of these serve the buttock and lower limb; the others innervate pelvic structures and the perineum. Here we describe the most important branches. **Table 13.6** summarizes all but the smallest ones. (**Table 13.7** summarizes the rami and major nerves of all of the spinal nerve plexuses.)

The largest branch of the sacral plexus is the **sciatic nerve** (si-at′ik), the thickest and longest nerve in the body. It supplies the entire lower limb, except the anteromedial thigh.

The sciatic nerve is actually two nerves—the *tibial* and *common fibular*—wrapped in a common sheath. The sciatic nerve leaves the pelvis via the greater sciatic notch. It courses deep to the gluteus maximus muscle and enters the posterior thigh just medial to the hip joint (*sciatic* = of the hip). There it gives off motor branches to the hamstring muscles (all thigh extensors and leg flexors) and to the adductor magnus. Immediately above the knee, the two divisions of the sciatic nerve diverge.

The **tibial nerve** continues through the popliteal fossa (the region just posterior to the knee joint) and supplies the posterior

Table 13.6 Branches of the Sacral Plexus (See Figure 13.12)

NERVES	VENTRAL RAMI	STRUCTURES SERVED
Sciatic nerve	L_4, L_5, S_1–S_3	Composed of two nerves (tibial and common fibular) in a common sheath; they diverge just proximal to the knee
• Tibial (including sural, medial and lateral plantar, and medial calcaneal branches)	L_4–S_3	Cutaneous branches: to skin of posterior surface of leg and sole of foot Motor branches: to muscles of back of thigh, leg, and foot [hamstrings (except short head of biceps femoris), posterior part of adductor magnus, triceps surae, tibialis posterior, popliteus, flexor digitorum longus, flexor hallucis longus, and intrinsic muscles of foot]
• Common fibular (superficial and deep branches)	L_4–S_2	Cutaneous branches: to skin of anterior and lateral surface of leg and dorsum of foot Motor branches: to short head of biceps femoris of thigh, fibular muscles of lateral compartment of leg, tibialis anterior, and extensor muscles of toes (extensor hallucis longus, extensors digitorum longus and brevis)
Superior gluteal	L_4, L_5, S_1	Motor branches: to gluteus medius and minimus and tensor fascia lata
Inferior gluteal	L_5–S_2	Motor branches: to gluteus maximus
Posterior femoral cutaneous	S_1–S_3	Skin of inferior buttock, posterior thigh, and popliteal region; length varies; may also innervate part of skin of calf and heel
Pudendal	S_2–S_4	Supplies most of skin and muscles of perineum (region encompassing external genitalia and anus and including clitoris, labia, and vaginal mucosa in females, and scrotum and penis in males); external anal sphincter

compartment muscles of the leg and the skin of the posterior calf and sole of the foot.

- In the vicinity of the knee, the tibial nerve gives off the **sural nerve**, which serves the skin of the posterolateral leg.

- At the ankle the tibial nerve divides into the **medial** and **lateral plantar nerves**, which serve most of the foot.

The **common fibular nerve**, or *common peroneal nerve (perone = fibula)*, descends from its point of origin, wraps around the neck of the fibula, and then divides into superficial and deep branches. These branches innervate the knee joint, skin of the anterior and lateral leg and dorsum of the foot, and muscles of the anterolateral leg (the extensors that dorsiflex the foot).

The next largest sacral plexus branches are the **superior** and **inferior gluteal nerves**. Together, they innervate the buttock (gluteal) and tensor fascia lata muscles. The **pudendal nerve** (pu-den′dal; "shameful") innervates the muscles and skin of the perineum, and helps stimulate erection and control urination. Other branches of the sacral plexus supply the thigh rotators and muscles of the pelvic floor.

HOMEOSTATIC IMBALANCE 13.8 — CLINICAL

Injury to the proximal part of the sciatic nerve—as might follow a fall, disc herniation, or badly placed injection into the buttock—can impair the lower limbs in a variety of ways depending on the nerve roots injured.

Sciatica (si-at′ĭ-kah), characterized by stabbing pain radiating over the course of the sciatic nerve, is common. When the nerve is transected, the leg is nearly useless. The leg cannot be flexed (because the hamstrings are paralyzed), and the foot and ankle cannot move at all. The foot drops into plantar flexion (it dangles), a condition called *footdrop*. Recovery from sciatic nerve injury is usually slow and incomplete.

If the lesion occurs below the knee, thigh muscles are spared. When the tibial nerve is injured, the paralyzed calf muscles cannot plantar flex the foot and a shuffling gait develops. The common fibular nerve is susceptible to injury largely because of its superficial location at the head and neck of the fibula. Even a tight leg cast, or lying too long on your side on a firm mattress, can compress this nerve and cause footdrop. ● ● ● ● ● ●

Anterolateral Thorax and Abdominal Wall

Only in the thorax are the ventral rami arranged in a simple segmental pattern corresponding to that of the dorsal rami. The ventral rami of T_1–T_{12} mostly course anteriorly, deep to each rib, as the **intercostal nerves**. These nerves supply the intercostal muscles, the muscle and skin of the anterolateral thorax, and most of the abdominal wall. Along their course, these nerves give off *cutaneous branches* to the skin (see Figure 13.8b).

Two thoracic nerves are unusual: the tiny T_1 (most fibers enter the brachial plexus) and T_{12}, which lies inferior to the twelfth rib, making it a **subcostal nerve**.

Back

The dorsal rami innervate the posterior body trunk in a neat, segmented pattern. Via its several branches, each dorsal ramus

Table 13.7 Summary of Major Spinal Nerve Plexuses

PLEXUS	VENTRAL RAMI	MAJOR NERVES
Cervical	C_1–C_4 (some C_5)	Phrenic
Brachial	C_5–T_1	Axillary, musculocutaneous, median, radial, ulnar
Lumbar	L_1–L_4	Femoral, obturator
Sacral	L_4–S_4	Sciatic (composed of tibial and common fibular)

Figure 13.13 Map of dermatomes. (a) Anterior view. **(b)** Posterior view. Each dermatome is an area of skin innervated by cutaneous sensory fibers of a single spinal nerve. All spinal nerves except C$_1$ participate in the innervation of the dermatomes.

innervates the narrow strip of muscle (and skin) in line with where it emerges from the spinal column (see Figure 13.8b).

Innervation of Skin: Dermatomes

A **dermatome** (der′mah-tōm; "skin segment") is an area of skin innervated by the cutaneous branches of a single spinal nerve. Every spinal nerve except C$_1$ innervates dermatomes. In patients with spinal cord injuries, you can pinpoint the damaged nerves and the injured region of the spinal cord by determining which dermatomes are affected.

It's crucial that clinicians understand the general pattern of sensory nerve distribution. For example, in areas where several dermatomes overlap, two or three spinal nerves must be blocked (anesthetized) to perform local surgery.

Adjacent dermatomes on the body trunk are fairly uniform in width, almost horizontal, and in direct line with their spinal nerves (**Figure 13.13**). The dermatome arrangement in the limbs is less obvious. (It is also more variable—different clinicians have mapped a variety of areas for the same dermatomes.) The skin of the upper limbs is supplied by ventral rami of C$_5$–T$_1$ (or T$_2$). The lumbar nerves supply most of the anterior surfaces of the thighs and legs, and the sacral nerves serve most of the posterior surfaces of the lower limbs. (This distribution basically reflects the areas supplied by the lumbar and sacral plexuses, respectively.)

Adjacent dermatomes are not as cleanly separated as a typical dermatome map indicates. On the trunk, neighboring dermatomes overlap considerably (about 50%). As a result, destruction of a single spinal nerve will not cause complete numbness anywhere. In the limbs, the overlap is less complete and some skin regions are innervated by just one spinal nerve.

Innervation of Joints

The easiest way to remember which nerves serve which synovial joint is to use **Hilton's law**: *Any nerve serving a muscle that produces movement at a joint also innervates the joint and the skin over the joint.* This means that once you learn which nerves serve the various major muscles and muscle groups, no new learning is necessary.

For example, the quadriceps, gracilis, and hamstring muscles all cross the knee. The nerves to these muscles are the femoral nerve anteriorly and branches of the sciatic and obturator nerves posteriorly. Consequently, these nerves innervate the knee joint as well.

Check Your Understanding

12. Spinal nerves have both *dorsal roots* and *dorsal rami*. How are these different from each other in location and composition?

13. APPLY After his horse-riding accident, the actor Christopher Reeve was unable to breathe on his own. Which ventral rami, nerve plexus, and major nerve were involved?

For answers, see Answers Appendix.

PART 3

MOTOR ENDINGS AND MOTOR ACTIVITY

13.6 Peripheral motor endings connect nerves to their effectors

Learning Outcome

▶ **Compare and contrast the motor endings of somatic and autonomic nerve fibers.**

So far we have covered the sensory receptors that detect stimuli and the nerves containing the afferent and efferent fibers that deliver impulses to and from the CNS. We now turn to **motor endings**, the PNS elements that activate effectors by releasing neurotransmitters. Because we discussed that topic in Chapter 9 with the innervation of body muscles, all we need to do here is recap. We will follow the recap with a brief overview of motor integration.

Innervation of Skeletal Muscle

Recall that the terminals of somatic motor fibers that innervate voluntary muscles form elaborate **neuromuscular junctions** with their effector cells (◀ *Focus on Events at the Neuromuscular Junction*, **Focus Figure 9.1, p. 292**). As each axon branch reaches its target, which is a single muscle fiber, the ending splits into a cluster of *axon terminals* that branch treelike over the junctional folds of the sarcolemma of the muscle fiber. The axon terminals contain mitochondria and synaptic vesicles filled with the neurotransmitter acetylcholine (ACh).

When a nerve impulse reaches an axon terminal, ACh is released by exocytosis, diffuses across the fluid-filled synaptic cleft, and attaches to ACh receptors on the sarcolemma at the junction. ACh binding opens ligand-gated channels that allow both Na^+ and K^+ to pass. Because more Na^+ enters the cell than K^+ leaves, the muscle cell interior at that point depolarizes. The resulting graded potential is called an *end plate potential*.

The end plate potential spreads to adjacent areas of the membrane where it triggers the opening of voltage-gated sodium channels. This event causes an action potential to propagate along the sarcolemma, which stimulates the muscle fiber to contract. The synaptic cleft at neuromuscular junctions contains *acetylcholinesterase*, the enzyme that breaks down ACh.

Innervation of Visceral Muscle and Glands

The junctions between autonomic motor endings and their effectors (smooth and cardiac muscle and glands) are much simpler than the junctions formed between somatic fibers and skeletal muscle cells. The autonomic motor axons branch repeatedly, each branch forming *synapses en passant* ("synapses in passing") with its effector cells. Instead of a cluster of bulblike terminals, an axon ending serving smooth muscle or a gland (but not cardiac muscle) has a series of **varicosities**, knoblike swellings containing mitochondria and synaptic vesicles, that make it look like a string of beads (◀ **Figure 9.23, p. 311**).

The autonomic synaptic vesicles typically contain either acetylcholine or norepinephrine, both of which act indirectly on their targets via second messengers. Consequently, visceral motor responses tend to be slower than those induced by somatic motor fibers, which directly open ion channels.

Check Your Understanding

14. What are varicosities and where would you find them?

For answers, see Answers Appendix.

13.7 There are three levels of motor control

Learning Outcomes

▶ **Outline the three levels of the motor hierarchy.**
▶ **Compare the roles of the cerebellum and basal nuclei in controlling motor activity.**

How does integration in the motor system compare with integration in sensory systems? In the motor system, we have motor endings serving effectors (muscle fibers) instead of sensory receptors, descending efferent circuits instead of ascending afferent circuits, and motor behavior instead of perception. However, as in sensory systems, the basic mechanisms of motor systems operate at three levels.

The cerebral cortex is at the highest level of our conscious motor pathways, but it is *not* the ultimate planner and coordinator of complex motor behaviors. The cerebellum and basal nuclei (ganglia) play this role and are therefore at the top of the motor control hierarchy. Motor control exerted by lower levels is mediated by *reflex arcs* in some cases, but complex motor activity, such

(a) Levels of motor control and their interactions

(b) Structures involved

Figure 13.14 Hierarchy of motor control. Dashed lines in (a) indicate feedback.

as walking and swimming, depends on more complex patterns. Currently, we define three levels of motor control: the *segmental level*, *projection level*, and *precommand level* (**Figure 13.14**).

The Segmental Level

The lowest level of the motor hierarchy, the **segmental level**, consists of reflexes and spinal cord circuits that control automatic movements. A segmental circuit activates a network of ventral horn neurons in a group of cord segments, causing them to stimulate specific groups of muscles.

Circuits that control locomotion and other specific and oft-repeated motor activities are called **central pattern generators (CPGs)**. CPGs consist of networks of oscillating inhibitory and excitatory neurons, which set crude rhythms and alternating patterns of movement.

The Projection Level

The spinal cord is under the direct control of the **projection level** of motor control. The projection level consists of neurons acting through the direct and indirect motor pathways (◄ **pp. 478–479**), which use descending projection fibers (◄ **p. 444**):

- *Upper motor neurons* of the motor cortex initiate the *direct (pyramidal) pathways*. Axons of direct pathway neurons produce discrete voluntary movements of the skeletal muscles.

- Brain stem motor nuclei oversee the *indirect pathways*. Axons of these pathways help control reflex and CPG-controlled motor actions, modifying and controlling the activity of the segmental apparatus.

Projection motor pathways convey information to lower motor neurons, and send a copy of that information as *internal feedback* to higher command levels, continually informing them of what is supposed to happen.

The Precommand Level

Two other systems of brain neurons, located in the cerebellum and basal nuclei, regulate motor activity. They precisely start or stop movements, coordinate movements with posture, block unwanted movements, and monitor muscle tone. Collectively called **precommand areas**, these systems *control the outputs* of the cortex and brain stem motor centers and stand at the highest level of the motor hierarchy.

The key center for "online" sensorimotor integration and control is the **cerebellum**. The cerebellum lacks direct connections to the spinal cord. It acts on motor pathways through the projection areas of the brain stem and on the motor cortex via the thalamus to fine-tune motor activity.

The **basal nuclei** receive inputs from *all* cortical areas and send their output back mainly to premotor and prefrontal cortical areas via the thalamus. Compared to the cerebellum, the basal nuclei appear to be involved in more complex aspects of motor control. Under resting conditions, the basal nuclei inhibit various motor centers of the brain. When the motor centers are released from inhibition, coordinated motions can begin.

Cells in both the basal nuclei and the cerebellum are involved in this unconscious planning and discharge *in advance* of willed movements (which is why this is called the precommand level).

When you actually move your fingers, both the precommand areas and the primary motor cortex are active. At the risk of oversimplifying, it appears that the cortex says, "I want to do this," and then lets the precommand areas take over to provide the proper timing and patterns to execute the desired movements. The precommand areas control the motor cortex and provide its readiness to initiate a voluntary act. The conscious cortex then chooses to act or not act, but the groundwork has already been laid.

Check Your Understanding

15. Which parts of the nervous system ultimately plan and coordinate complex motor activities?

For answers, see Answers Appendix.

For answers, see Answers Appendix.

PART 4

REFLEX ACTIVITY

13.8 The reflex arc enables rapid and predictable responses

Learning Outcome

▶ Name the components of a reflex arc and distinguish between autonomic and somatic reflexes.

Many of the body's control systems are reflexes, which can be either inborn or learned.

An *inborn* (*intrinsic*) *reflex* is a rapid, predictable motor response to a stimulus. It is unlearned, unpremeditated, and involuntary, and is built into our neural anatomy. Reflexes prevent us from having to *think* about all the little details of staying upright, intact, and alive—helping us maintain posture, avoid pain, and control visceral activities.

For instance, what happens when you splash a pot of boiling water on your arm? You are likely to drop the pot instantly and involuntarily even before you feel any pain. Your response is triggered by an inborn spinal reflex without any help from the brain. In many cases we are aware of the final response of a basic reflex (you know you've dropped the pot of boiling water). In other cases, reflex activities go on without any awareness on our part. This is typical of many visceral reflexes, which are regulated by the subconscious lower regions of the CNS, specifically the brain stem and spinal cord.

The second type of reflex, a *learned* (*acquired*) *reflex*, results from practice or repetition. Take, for instance, the complex sequence of reactions that occurs when an experienced driver drives a car. The process is largely automatic, but only because substantial time and effort were expended to acquire driving skills.

In reality, the distinction between inborn and learned reflexes is not clear-cut and most inborn reflex actions can be modified by learning and conscious effort. For instance, if a 3-year-old child was standing by your side when you scalded your arm, you most likely would set the pot down (rather than just letting go) because you consciously recognized the danger to the child.

Recall the discussion in Chapter 11 about serial and parallel processing of sensory input (◀ pp. 426–427). What happens when you scald your arm is a good example of how these two processing modes work together. You drop the pot before feeling any pain, but the pain signals picked up by the interneurons of the spinal cord are quickly transmitted to the brain, so that within the next few seconds you do become aware of pain, and you also know what happened to cause it. The withdrawal reflex is serial processing mediated by the spinal cord, and pain awareness reflects simultaneous parallel processing of the sensory input.

Components of a Reflex Arc

As you learned in Chapter 11, reflexes occur over highly specific neural paths called reflex arcs. All reflex arcs have five essential components (**Figure 13.15**):

① **Receptor:** Site of the stimulus action.

② **Sensory neuron:** Transmits afferent impulses to the CNS.

③ **Integration center:** In simple reflex arcs, the integration center may be a single synapse between a sensory neuron and a motor neuron (**monosynaptic reflex**). More complex reflex arcs involve multiple synapses with chains of interneurons (**polysynaptic reflex**). The integration center for each reflex we will describe in this chapter is within the CNS.

④ **Motor neuron:** Conducts efferent impulses from the integration center to an effector organ.

⑤ **Effector:** Muscle fiber or gland cell that responds to the efferent impulses (by contracting or secreting).

Reflexes are classified functionally as **somatic reflexes** if they activate skeletal muscle, or as **autonomic (visceral) reflexes** if they activate visceral effectors (smooth or cardiac muscle or glands). Here we describe some common somatic reflexes mediated by the spinal cord. We will consider autonomic reflexes

Figure 13.15 The five basic components of all reflex arcs. The reflex arc illustrated is polysynaptic.

in later chapters along with the visceral processes they help to regulate.

Check Your Understanding

16. Which component of the reflex arc brings about the response?

17. DRAW Using the spinal cord cross section below, draw and label the five components of a reflex arc. Make sure the cell bodies are placed correctly. Name the structure that contains the cell body of the sensory neuron.

For answers, see Answers Appendix.

13.9 Spinal reflexes are somatic reflexes mediated by the spinal cord

Learning Outcomes

▶ Compare and contrast stretch, flexor, crossed-extensor, and tendon reflexes.

▶ Describe two superficial reflexes.

Many **spinal reflexes** occur without the direct involvement of higher brain centers. Generally, these reflexes are even present in animals whose brains have been destroyed as long as the spinal cord is still functional.

However, the brain is "advised" of most spinal reflex activity and can facilitate, inhibit, or adapt it, depending on the circumstances (as we described in the example of the hot water–filled pot). Moreover, continuous facilitating signals from the brain are required for normal spinal reflex activity. As we saw in Chapter 12, *spinal shock* occurs when the spinal cord is transected, immediately depressing all functions controlled by the cord.

Tests of somatic reflexes are important clinically to assess the condition of the nervous system. Exaggerated, distorted, or absent reflexes indicate degeneration or pathology of specific nervous system regions, often before other signs appear. The most commonly assessed reflexes are the stretch, flexor, and superficial reflexes.

Stretch and Tendon Reflexes

Stretch and tendon reflexes help your nervous system smoothly coordinate the activity of your skeletal muscles. What information does your nervous system need in order to do this? Two types of information about the current state of a muscle are required. The nervous system needs to know:

- The length of the muscle. This information comes from the *muscle spindles* in skeletal muscles.

- The amount of tension in the muscle and its associated tendons. *Tendon organs* provide this information.

(Just remember, muscle spindles measure length and tendon organs measure tension.)

These two types of proprioceptors play an important role in spinal reflexes and also provide essential feedback to the cerebral cortex and cerebellum, so that the brain can compare what actually happened with what was supposed to happen.

Functional Anatomy of Muscle Spindles

Each muscle spindle consists of three to ten modified skeletal muscle fibers called **intrafusal muscle fibers** (*intra* = within; *fusal* = the spindle) enclosed in a connective tissue capsule (**Figure 13.16**). These fibers are less than one-quarter the size of the effector fibers of the muscle, called **extrafusal muscle fibers**.

The central regions of the intrafusal fibers lack myofilaments and are noncontractile. These regions are the receptive surfaces

Figure 13.16 Anatomy of the muscle spindle and tendon organ. Myelin has been omitted from all nerve fibers for clarity.

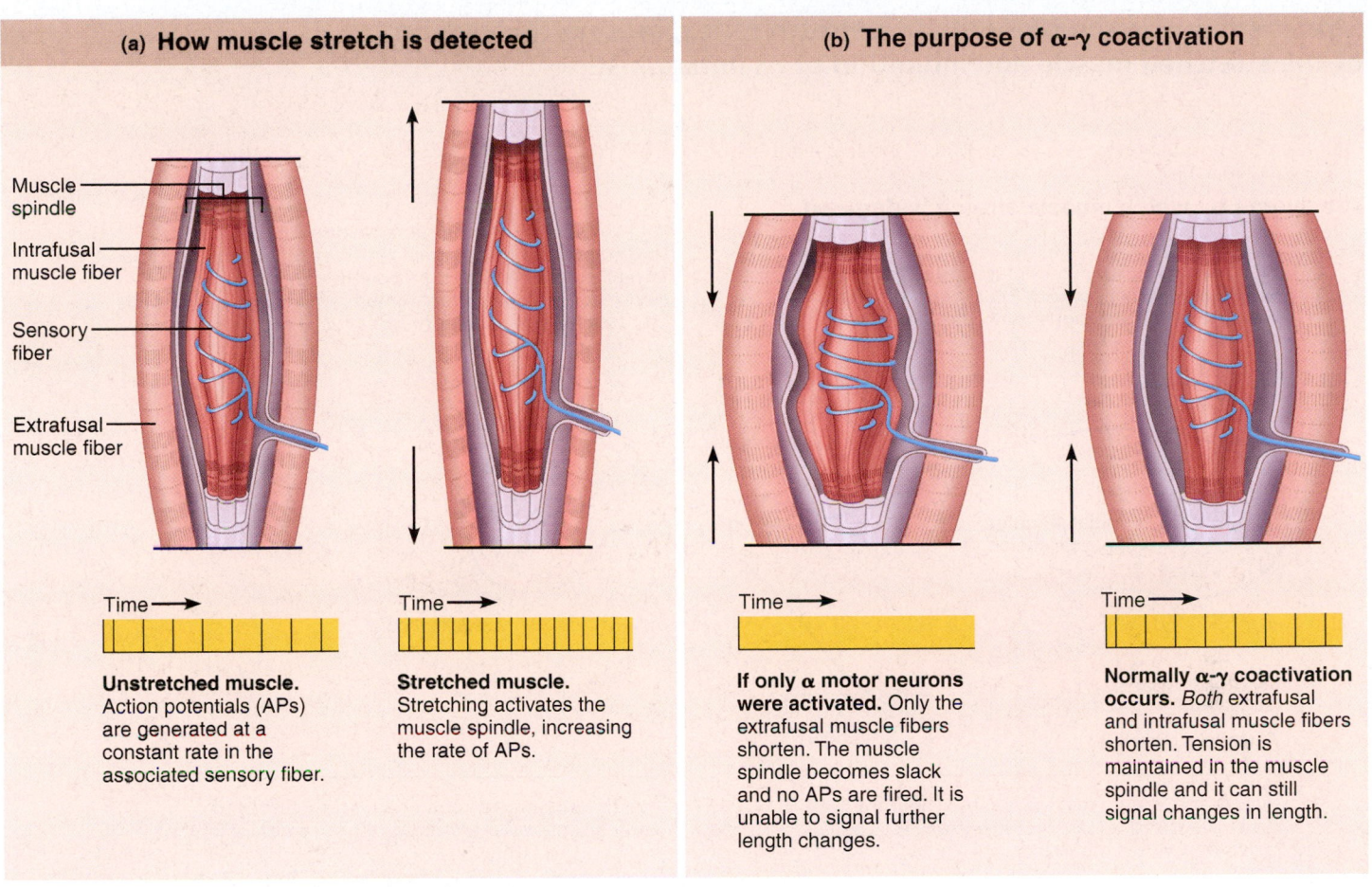

(a) How muscle stretch is detected

Muscle spindle
Intrafusal muscle fiber
Sensory fiber
Extrafusal muscle fiber

Time ➝

Unstretched muscle. Action potentials (APs) are generated at a constant rate in the associated sensory fiber.

Time ➝

Stretched muscle. Stretching activates the muscle spindle, increasing the rate of APs.

(b) The purpose of α-γ coactivation

Time ➝

If only α motor neurons were activated. Only the extrafusal muscle fibers shorten. The muscle spindle becomes slack and no APs are fired. It is unable to signal further length changes.

Time ➝

Normally α-γ coactivation occurs. *Both* extrafusal and intrafusal muscle fibers shorten. Tension is maintained in the muscle spindle and it can still signal changes in length.

Figure 13.17 Operation of the muscle spindle. The action potentials generated in the sensory fibers are shown for each case as black lines in yellow bars.

of the spindle. Two types of afferent endings send sensory inputs to the CNS:

- **Anulospiral endings** (also called *primary sensory endings*) are the endings of large axons that wrap around the spindle center. They are stimulated by both the rate and degree of stretch.

- **Flower spray endings** (also called *secondary sensory endings*) are formed by smaller axons that supply the spindle ends. They are stimulated only by degree of stretch.

The intrafusal muscle fibers have contractile regions at their ends, which are the only areas containing actin and myosin myofilaments. These regions are innervated by **gamma (γ) efferent fibers** that arise from small motor neurons in the ventral horn of the spinal cord. These γ motor fibers, which maintain spindle sensitivity (as described shortly), are distinct from the **alpha (α) efferent fibers** of the large **alpha (α) motor neurons** that stimulate the extrafusal muscle fibers to contract.

Whenever the muscle spindle is stretched, its associated sensory neurons transmit impulses at higher frequency to the spinal cord (**Figure 13.17a**). The muscle spindle is stretched (and excited) in one of two ways:

- By applying an external force that lengthens the entire muscle, such as when we carry a heavy weight or when antagonistic muscles contract (external stretch)

- By activating the γ motor neurons that stimulate the distal ends of the intrafusal fibers to contract, thereby stretching the middle of the spindle (internal stretch)

During voluntary skeletal muscle contraction, the muscle shortens. If the intrafusal muscle fibers didn't contract along with the extrafusal fibers, the muscle spindle would go slack and cease generating action potentials (Figure 13.17b). At this point it would be unable to signal further changes in muscle length, so it would be useless.

Fortunately, **α-γ coactivation** prevents this from happening. Descending fibers of motor pathways synapse with *both* α and γ motor neurons so that motor impulses are simultaneously sent to the large extrafusal fibers and to muscle spindle intrafusal fibers. Stimulating the intrafusal fibers maintains the spindle's tension (and sensitivity) during muscle contraction, so that the brain continues to be notified of changes in the muscle length (Figure 13.17b). Without this information about changes in muscle length you would, for example, find it much harder to balance on a log to cross a stream.

13

Stretched muscle spindles initiate a stretch reflex, causing contraction of the stretched muscle and inhibition of its antagonist.

The events by which muscle stretch is damped

① When stretch activates muscle spindles, the associated sensory neurons (blue) transmit afferent impulses at higher frequency to the spinal cord.

② The sensory neurons synapse directly with alpha motor neurons (red), which excite extrafusal fibers of the stretched muscle. Sensory fibers also synapse with interneurons (green) that inhibit motor neurons (purple) controlling antagonistic muscles.

Sensory neuron

Cell body of sensory neuron

Initial stimulus (muscle stretch)

Spinal cord

Muscle spindle (stretched)

Antagonist muscle

③a Efferent impulses of alpha motor neurons cause the stretched muscle to contract, which resists or reverses the stretch.

③b Efferent impulses of alpha motor neurons to antagonist muscles are reduced (reciprocal inhibition).

The patellar (knee-jerk) reflex—an example of a stretch reflex

Quadriceps (extensors)

③a ③b

①

Muscle spindle (stretched)

Patella

Spinal cord (L₂–L₄)

Hamstrings (flexors)

Patellar ligament

① Tapping the patellar ligament stretches the quadriceps and excites its muscle spindles.

② Afferent impulses (blue) travel to the spinal cord, where synapses occur with motor neurons and interneurons.

③a The motor neurons (red) send activating impulses to the quadriceps causing it to contract, extending the knee.

③b The interneurons (green) make inhibitory synapses with ventral horn neurons (purple) that prevent the antagonist muscles (hamstrings) from resisting the contraction of the quadriceps.

+ Excitatory synapse

– Inhibitory synapse

The Stretch Reflex

By sending commands to the motor neurons, the brain essentially sets a muscle's length. The **stretch reflex** makes sure that the muscle stays at that length. For example, the **patellar** (pah-tel′ar) or **knee-jerk reflex** is a stretch reflex that helps keep your knees from buckling when you are standing upright. As your knees begin to buckle and the quadriceps lengthens, the stretch reflex causes the quadriceps to contract without your having to think about it. *Focus on the Stretch Reflex* (**Focus Figure 13.1**) shows the stretch reflex and a specific example—the knee-jerk reflex.

The stretch reflex is important for maintaining muscle tone and adjusting it reflexively. It is most important in the large extensor muscles that sustain upright posture and in postural muscles of the trunk. For example, stretch reflexes initiated first on one side of the spine and then on the other regulate contractions of the postural muscles of the spine almost continuously.

Let's look at how the stretch reflex works. As we've just seen in Figure 13.17, when stretch activates sensory neurons of muscle spindles, they transmit impulses at a higher frequency to the spinal cord. There the sensory neurons synapse directly with α motor neurons, which rapidly excite the extrafusal muscle fibers of the stretched muscle (Focus Figure 13.1). The reflexive muscle contraction that follows (an example of serial processing) resists further muscle stretching.

Branches of the afferent fibers also synapse with interneurons that inhibit motor neurons controlling antagonistic muscles (parallel processing), and the resulting inhibition is called **reciprocal inhibition**. Consequently, the stretch stimulus causes the antagonists to relax so that they cannot resist the shortening of the "stretched" muscle caused by the main reflex arc. While this spinal reflex is occurring, information on muscle length and the speed of muscle shortening is being relayed (mainly via the dorsal white columns) to higher brain centers (more parallel processing).

The most familiar clinical example of a stretch reflex is the knee-jerk reflex we have just described. Stretch reflexes can be elicited in any skeletal muscle by a sudden jolt to the tendon or the muscle itself. All stretch reflexes are **monosynaptic** and **ipsilateral** (*ipsi* = same; *lateral* = side). In other words, they involve a single synapse and motor activity on the same side of the body. Stretch reflexes are the only monosynaptic reflexes in the body. However, even in these reflexes, the part of the reflex arc that inhibits the motor neurons serving the antagonistic muscles is polysynaptic.

A positive knee jerk (or a positive result for any other stretch reflex test) provides two important pieces of information. First, it proves that the sensory and motor connections between that muscle and the spinal cord are intact. Second, the vigor of the response indicates the degree of excitability of the spinal cord. When the spinal motor neurons are highly facilitated by impulses descending from higher centers, just touching the muscle tendon produces a vigorous reflex response. On the other hand, when inhibitory signals bombard the lower motor neurons, even pounding on the tendon may fail to trigger the reflex response.

HOMEOSTATIC IMBALANCE 13.9

CLINICAL

Stretch reflexes tend to be hypoactive or absent in cases of peripheral nerve damage or ventral horn injury involving the tested area. These reflexes are absent in those with chronic diabetes mellitus or neurosyphilis and during coma. However, they are hyperactive when lesions of the corticospinal tract reduce the inhibitory effect of the brain on the spinal cord (as in stroke patients). ●●●●●●

Adjusting Muscle Spindle Sensitivity

The motor supply to the muscle spindle allows the brain to voluntarily modify the stretch reflex response and the firing rate of α motor neurons. When the γ neurons are vigorously stimulated by impulses from the brain, the spindle is stretched and highly sensitive, and muscle contraction force is maintained or increased. When the γ motor neurons are inhibited, the spindle resembles a loose rubber band and is nonresponsive, and the extrafusal muscles relax.

① Quadriceps strongly contracts. Tendon organs are activated.

② Afferent fibers synapse with interneurons in the spinal cord.

Quadriceps (extensors)

Interneurons

Tendon organ

Spinal cord

Hamstrings (flexors)

+ Excitatory synapse

− Inhibitory synapse

③a Efferent impulses to muscle with stretched tendon are damped. Muscle relaxes, reducing tension.

③b Efferent impulses to antagonist muscle cause it to contract.

Figure 13.18 The tendon reflex.

The ability to modify the stretch reflex is important in many situations. As the speed and difficulty of a movement increase, the brain increases γ motor output to make the muscle spindles more sensitive. This sensitivity is highest when balance reflexes must be razor sharp, as for a gymnast on a balance beam. On the other hand, if you want to wind up to pitch a baseball, it is essential to suppress the stretch reflex so that your muscles can produce a large degree of motion (i.e., circumduct your pitching arm). Other athletes who require movements of maximum force learn to stretch muscles as much and as quickly as possible just before the movement. This advantage is demonstrated by the crouch that athletes assume just before jumping or running.

The Tendon Reflex

Stretch reflexes cause muscle contraction in response to increased muscle length (stretch). The polysynaptic **tendon reflexes**, on the other hand, produce exactly the opposite effect: Muscles relax and lengthen in response to tension.

When muscle tension increases substantially during contraction or passive stretching, high-threshold *tendon organs* may be activated. Afferent impulses are transmitted to the spinal cord, and then to the cerebellum, where the information is used to adjust muscle tension. Simultaneously, motor neurons in spinal cord circuits supplying the contracting muscle are inhibited and antagonist muscles are activated, a phenomenon called **reciprocal activation**. As a result, the contracting muscle relaxes as its antagonist is activated (**Figure 13.18**).

Tendon organs help to prevent muscles and tendons from tearing when they are subjected to potentially damaging stretching force. Tendon organs also function at normal muscle tensions, helping to ensure smooth onset and termination of muscle contraction.

The Flexor and Crossed-Extensor Reflexes

A painful stimulus initiates the **flexor**, or **withdrawal**, **reflex**, which causes automatic withdrawal of the threatened body part from the stimulus (**Figure 13.19**, left). Think of the response that occurs with a paper cut. Flexor reflexes are ipsilateral and polysynaptic, the latter a necessity when several muscles must be recruited to withdraw the injured body part.

Because flexor reflexes are protective and important to our survival, they override the spinal pathways and prevent any other reflexes from using them at the same time. However, like other spinal reflexes, descending signals from the brain can override flexor reflexes. This happens when you are expecting a painful stimulus, for example a skin prick as a lab technician prepares to draw blood from a vein.

The **crossed-extensor reflex** often accompanies the flexor reflex in weight-bearing limbs and is particularly important in maintaining balance. It is a complex spinal reflex consisting of an ipsilateral withdrawal reflex and a contralateral extensor reflex. Incoming afferent fibers synapse with interneurons that control the flexor withdrawal response on the same side of the body and with other interneurons that control the extensor muscles on the opposite side.

The crossed-extensor reflex is obvious when you step barefoot on broken glass. The ipsilateral response causes you to quickly lift your injured foot, while the contralateral response activates the extensor muscles of your opposite leg to support the weight suddenly shifted to it (Figure 13.19). The crossed-extensor reflex also occurs when someone unexpectedly grabs your arm. The grasped arm is withdrawn as the opposite arm pushes you away from the attacker.

+ Excitatory synapse
− Inhibitory synapse

Interneurons

Efferent fibers

Afferent fiber

Motor neurons to **extensors are inhibited** in the spinal cord.

Motor neurons to **flexors are excited** in the spinal cord.

Motor neurons to **extensors are excited** in the spinal cord.

Motor neurons to **flexors are inhibited** in the spinal cord.

Leg movement: *Flexes*

Leg movement: *Extends*

Site of stimulus: A noxious stimulus causes a **flexor reflex** on the same side, withdrawing that limb.

Site of reciprocal activation: At the same time, the extensor muscles on the opposite side are activated.

Figure 13.19 The crossed-extensor reflex. If you step on a piece of glass, you will reflexively withdraw that foot while the opposite leg extends to keep you upright.

Superficial Reflexes

Superficial reflexes are elicited by gentle cutaneous stimulation, such as that produced by stroking the skin with a tongue depressor. These clinically important reflexes depend both on functional upper motor pathways and on cord-level reflex arcs. The best known are the abdominal and plantar reflexes.

Abdominal Reflexes

Stroking the skin of the lateral abdomen above, to the side, or below the umbilicus induces a reflex contraction of the abdominal muscles in which the umbilicus moves toward the stimulated site. These reflexes, called **abdominal reflexes**, check the integrity of the spinal cord and ventral rami from T_8 to T_{12}.

Abdominal reflexes vary in intensity from one person to another. However, their absence indicates lesions in the corticospinal tract.

Plantar Reflex

The **plantar reflex** tests the integrity of the spinal cord from L_4 to S_2 and indirectly determines if the corticospinal tracts are functioning properly. To elicit the plantar reflex, draw a blunt object from heel to toe along the lateral aspect of the plantar surface (sole) of the foot. The normal response is for the toes to flex downward (curl).

HOMEOSTATIC IMBALANCE 13.10

CLINICAL

If the primary motor cortex or corticospinal tract is damaged, the plantar reflex is replaced by an abnormal reflex called **Babinski's sign**, in which the great toe dorsiflexes and the smaller toes fan laterally.

Infants exhibit Babinski's sign until they are about a year old because their nervous systems are incompletely myelinated. Despite its clinical significance, the physiological mechanism of Babinski's sign is not understood.

Check Your Understanding

18. What is the role of the stretch reflex? The flexor reflex?

19. Juan injured his back in a fall. When his ER physician stroked the bottom of Juan's foot, she noted that his big toe pointed up and his other toes fanned out. What is this response called and what does it indicate?

20. **MAKE CONNECTIONS** A technician is drawing blood from your arm for blood tests. As you feel the pain of the needle, you suppress your flexor reflex. Name the ascending pathway that carries pain signals and the region of the cortex that receives them (use Figure 12.34 on p. 477). Next, name the descending pathway you would use to inhibit the flexor reflex (use Figure 12.35 on p. 479).

For answers, see Answers Appendix.

Developmental Aspects of the Peripheral Nervous System

Most skeletal muscles derive from paired blocks of mesoderm (somites) distributed segmentally down the posteromedial aspect of the embryo. The spinal nerves branch from the developing spinal cord and adjacent neural crest and exit between the forming vertebrae, and each nerve becomes associated with the adjacent muscle mass. The spinal nerves supply both sensory and motor fibers to the developing muscles and help direct their maturation. Cranial nerves innervate muscles of the head in a comparable manner.

The distribution of cutaneous nerves to the skin follows a similar pattern. The trigeminal nerves innervate most of the scalp and facial skin. Spinal nerves supply cutaneous branches to specific (adjacent) dermatomes that eventually become dermal segments. As a result, the distribution and growth of the spinal nerves correlate with the segmented body plan, which is established by the fourth week of embryonic development.

Growth of the limbs and unequal growth of other body areas result in an adult pattern of dermatomes with unequal sizes and shapes and varying degrees of overlap. Because embryonic muscle cells migrate extensively, much of the early segmental pattern is lost.

Sensory receptors atrophy to some degree with age, and muscle tone decreases in the face and neck. Reflexes occur a bit more slowly during old age. This deterioration seems to reflect a general loss of neurons, fewer synapses per neuron, and a slowdown in central processing rather than any major changes in the peripheral nerve fibers. In fact, peripheral nerves remain viable and normally functional throughout life unless subjected to traumatic injury or ischemia (inadequate blood supply). The most common symptoms of ischemia are sensations of tingling or numbness in the affected region.

RELATED CLINICAL TERMS

Analgesia (an″al-je′zeah; *an* = without; *algos* = pain) Reduced ability to feel pain, but without losing consciousness. An analgesic is a pain-relieving drug.

Dysarthria (dis-ar′thre-ah) Difficulty in articulating speech due to motor pathway disorders that result in weakness, uncoordinated motion, or altered respiration or rhythm. For example, lesions of cranial nerves IX, X, and XII result in nasal, breathy speech, and lesions in upper motor pathways produce a hoarse, strained voice. Not to be confused with *dysphasia* or *aphasia*, which are disorders of language processing.

Dystonia (dis-to′ne-ah) Impaired muscle tone.

Nerve conduction studies Diagnostic tests that assess nerve integrity as indicated by their conduction velocities. The nerve is stimulated at one point, activity is recorded at a second point a known distance away, and the time required for the response to reach the recording electrode is measured. Used to assess suspected *peripheral neuropathies* (see entry in this list).

Neuralgia (nu-ral′je-ah; *neuro* = nerve) Sharp spasmlike pain along the course of one or more nerves. May be caused by inflammation or injury to the nerve(s) (for example, trigeminal neuralgia).

Paresthesia (par″es-the′ze-ah) An abnormal sensation (burning, numbness, tingling) in the absence of stimuli. Usually caused by a sensory nerve disorder.

Peripheral neuropathy (nu-rop′ah-the) Disease of the peripheral nerves characterized by muscle weakness and atrophy, pain, and numbness. Diabetes is a common cause. Other causes are genetic, other metabolic disorders, infections or inflammation, and toxins.

Tabes dorsalis (ta′bēz dor-să′lis) A slowly progressive condition caused by deteriorating dorsal tracts (gracilis and cuneatus) and associated dorsal roots. A late sign of the neurological damage caused by the syphilis bacterium. Because joint proprioceptor tracts are destroyed, affected individuals have poor muscle coordination and an unstable gait. Bacterial invasion of the sensory (dorsal) roots results in pain, which ends when dorsal roots have been completely destroyed.

CHAPTER SUMMARY

1. The peripheral nervous system consists of sensory receptors, nerves conducting impulses to and from the CNS, their associated ganglia, and motor endings.

PART 1

SENSORY RECEPTORS AND SENSATION

13.1 **Sensory receptors are activated by changes in the internal or external environment** (pp. 490–493)

1. Sensory receptors are specialized to respond to environmental changes (stimuli).
2. Sensory receptors include the receptors for the general senses—pain, touch, pressure, and temperature receptors in the skin—as well as receptors in skeletal muscles and tendons and in the visceral organs. Sense organs contain sensory receptors and other cells that serve the special senses (vision, hearing, equilibrium, smell, and taste).
3. Receptors are classified according to stimulus detected as mechanoreceptors, thermoreceptors, photoreceptors, chemoreceptors, and nociceptors, and according to location as exteroceptors, interoceptors, and proprioceptors.
4. The general sensory receptors are classified structurally as free or encapsulated nerve endings of sensory neurons. The free endings are mainly receptors for temperature and pain, with two exceptions: epithelial tactile complexes detect light pressure and hair follicle receptors respond to hair deflection. The encapsulated endings, which are mechanoreceptors, include tactile corpuscles, lamellar corpuscles, bulbous corpuscles, muscle spindles, tendon organs, and joint kinesthetic receptors.

13.2 **Receptors, ascending pathways, and cerebral cortex process sensory information** (pp. 493–496)

1. Sensation is awareness of internal and external stimuli. Perception is conscious interpretation of those stimuli.
2. The three levels of sensory integration are the receptor, circuit, and perceptual levels. These levels are functions of the sensory receptors, the ascending pathways, and the cerebral cortex, respectively.
3. Sensory receptors transduce (convert) stimulus energy via receptor or generator potentials into action potentials. Stimulus strength is frequency coded. Adaptation (decreased response to a continuous or unchanging stimulus) occurs in all general receptors except pain and proprioceptors.
4. The circuit level consists of the ascending pathways—the axons of the first-, second-, and third-order sensory neurons. These were discussed in Chapter 12.
5. Perception—the internal, conscious image of the stimulus that serves as the basis for response—is the result of cortical processing.
6. The main features of sensory perception are perceptual detection, magnitude estimation, spatial discrimination, feature abstraction, quality discrimination, and pattern recognition.

PART 2

TRANSMISSION LINES: NERVES AND THEIR STRUCTURE AND REPAIR

13.3 **Nerves are cordlike bundles of axons that conduct sensory and motor impulses** (pp. 496–498)

1. A nerve is a bundle of axons in the PNS. Each fiber is enclosed by an endoneurium, fascicles of fibers are wrapped by a perineurium, and the whole nerve is bundled by the epineurium.
2. Nerves are classified according to the direction of impulse conduction as sensory, motor, or mixed; most nerves are mixed. Efferent fibers may be somatic or autonomic.
3. Ganglia are collections of neuron cell bodies associated with nerves in the PNS. Examples are the dorsal root (sensory) ganglia and autonomic (motor) ganglia.
4. Fibers in the CNS do not normally regenerate because the oligodendrocytes inhibit axon sprouting and regrowth. Injured PNS fibers often regenerate. Macrophages enter the area, phagocytize the debris, and promote the proliferation of Schwann cells. Schwann cells then form a channel and secrete chemicals to guide axon sprouts to their original contacts.

13.4 **There are 12 pairs of cranial nerves** (pp. 498–507)

1. Twelve pairs of cranial nerves issue through the skull to innervate the head and neck. Only the vagus nerves extend into the thoracic and abdominal cavities. All but the accessory nerves originate from the brain.
2. Cranial nerves are (generally) numbered from rostral to caudal in order of emergence from the brain. Their names reflect structures served or function or both. The cranial nerves and their numbers are:
    - **I.** Olfactory nerves: purely sensory. Concerned with the sense of smell.
    - **II.** Optic nerves: purely sensory. Transmit visual impulses from the retina to the thalamus.
    - **III.** Oculomotor nerves: primarily motor. Emerge from the midbrain and serve four extrinsic eye muscles, the levator palpebrae superioris of the eyelid, and the intrinsic ciliary muscle of the eye and constrictor fibers of the iris. Also carry proprioceptive impulses from the skeletal muscles served.
    - **IV.** Trochlear nerves: primarily motor. Issue from the dorsal midbrain and carry motor and proprioceptor impulses to and from superior oblique muscles of the eyeballs.
    - **V.** Trigeminal nerves: mixed nerves. Emerge from the lateral pons as the main general sensory nerves of the face. Each has three sensory divisions: ophthalmic, maxillary, and mandibular. The mandibular branch also contains motor fibers that innervate the chewing muscles.
    - **VI.** Abducens nerves: primarily motor. Emerge from the pons and serve the motor and proprioceptive functions of the lateral rectus muscles of the eyeballs.
    - **VII.** Facial nerves: mixed nerves. Emerge from the pons as the major motor nerves of the face. Also carry sensory impulses from the taste buds of anterior two-thirds of the tongue.
    - **VIII.** Vestibulocochlear nerves: mostly sensory. Transmit impulses from the hearing and equilibrium receptors of the inner ears.
    - **IX.** Glossopharyngeal nerves: mixed nerves. Issue from the medulla. Transmit sensory impulses from the taste buds of the posterior tongue, the pharynx, and chemo- and baroreceptors of the carotid bodies and sinuses. Innervate some pharyngeal muscles and parotid glands.
    - **X.** Vagus nerves: mixed nerves. Arise from the medulla. Almost all motor fibers are autonomic parasympathetic

13

fibers; motor efferents to, and sensory fibers from, the pharynx, larynx, and visceral organs of the thoracic and abdominal cavities.

XI. Accessory nerves: primarily motor. Arise as spinal rootlets from the cervical spinal cord and enter the foramen magnum. Supply somatic efferents to the trapezius and sternocleidomastoid muscles of the neck and carry proprioceptor afferents from the same muscles.

XII. Hypoglossal nerves: primarily motor. Issue from the medulla and carry somatic motor efferents to, and proprioceptive fibers from, the tongue muscles.

13.5   31 pairs of spinal nerves innervate the body (pp. 507–517)

1. The spinal nerves (all mixed nerves) are numbered successively according to the region of the spinal cord from which they issue.
2. Dorsal and ventral roots of the spinal cord fuse to form spinal nerves. Spinal nerves are short, confined to the intervertebral foramina.
3. Branches of each spinal nerve include dorsal and ventral rami, a meningeal branch, and in the thoracic region, rami communicantes (ANS branches).
4. Ventral rami, except T_2–T_{12}, form plexuses that serve the limbs.
5. The cervical plexus (C_1–C_4) innervates the muscles and skin of the neck and shoulder. Its phrenic nerve serves the diaphragm.
6. The brachial plexus serves the shoulder, some thorax muscles, and the upper limb. It arises primarily from C_5–T_1. Proximal to distal, the brachial plexus has roots, trunks, divisions, and cords. The main nerves arising from the cords are the axillary, musculocutaneous, median, radial, and ulnar nerves.
7. The lumbar plexus (L_1–L_4) provides the motor supply to the anterior and medial thigh muscles and the cutaneous supply to the anterior thigh and part of the leg. Its chief nerves are the femoral and obturator.
8. The sacral plexus (L_4–S_4) supplies the posterior muscles and skin of the lower limb. Its principal nerve is the large sciatic nerve composed of the tibial and common fibular nerves.
9. Dorsal rami serve the muscles and skin of the posterior body trunk. T_1–T_{12} ventral rami give rise to intercostal nerves that serve the thorax wall and abdominal surface.
10. Joints are innervated by the same nerves that serve the muscles acting at the joint. All spinal nerves except C_1 innervate specific segments of the skin called dermatomes.

PART 3
MOTOR ENDINGS AND MOTOR ACTIVITY

13.6   Peripheral motor endings connect nerves to their effectors (p. 517)

1. Motor endings of somatic nerve fibers (axon terminals) form neuromuscular junctions with skeletal muscle cells. Axon terminals contain synaptic vesicles filled with acetylcholine, which (when released) signals the muscle cell to contract.
2. Autonomic motor endings, called varicosities, are functionally similar, but structurally simpler, beaded terminals that innervate smooth muscle and glands. They do not form specialized neuromuscular junctions and the motor responses elicited are generally slower.

13.7   There are three levels of motor control (pp. 517–519)

1. Motor mechanisms operate at the level of the effectors (muscle fibers), descending circuits, and control levels of motor behavior.
2. The motor control hierarchy consists of the segmental level, the projection level, and the precommand level.
3. The segmental level is the spinal cord circuitry that activates ventral horn motor neurons to stimulate the muscles. It consists of reflexes and central pattern generators (CPGs), segmental circuits controlling locomotion.
4. The projection level consists of descending fibers that project to and control the segmental level. These fibers issue from the brain stem motor areas (indirect pathway) and cortical motor areas [direct (pyramidal) pathway]. Neurons in the brain stem appear to turn CPGs on and off, or to modulate them.
5. The cerebellum and basal nuclei constitute the precommand areas that subconsciously integrate mechanisms mediated by the projection level.

PART 4
REFLEX ACTIVITY

13.8   The reflex arc enables rapid and predictable responses (pp. 519–520)

1. A reflex is a rapid, involuntary motor response to a stimulus. The reflex arc has five elements: receptor, sensory neuron, integration center, motor neuron, and effector.

13.9   Spinal reflexes are somatic reflexes mediated by the spinal cord (pp. 520–526)

1. Testing of somatic spinal reflexes provides information on the integrity of the reflex pathway and the degree of excitability of the spinal cord.
2. Somatic spinal reflexes include stretch, tendon, flexor, crossed-extensor, and superficial reflexes.
3. A stretch reflex, initiated by stretching of muscle spindles, causes contraction of the stimulated muscle and inhibits its antagonist. It is monosynaptic and ipsilateral. Stretch reflexes maintain muscle tone and body posture.
4. Tendon reflexes, initiated by stimulation of tendon organs by muscle tension, are polysynaptic reflexes. They cause relaxation of the stimulated muscle and contraction of its antagonist to prevent muscle and tendon damage.
5. Flexor reflexes are initiated by painful stimuli. They are polysynaptic, ipsilateral reflexes that are protective in nature.
6. Crossed-extensor reflexes consist of an ipsilateral flexor reflex and a contralateral extensor reflex.
7. Superficial reflexes (abdominal and plantar reflexes) are elicited by cutaneous stimulation. They require functional cord reflex arcs and corticospinal pathways.

Developmental Aspects of the Peripheral Nervous System (p. 526)

1. Each spinal nerve provides the sensory and motor supply of an adjacent muscle mass (destined to become skeletal muscles) and the cutaneous supply of a dermatome (skin segment).
2. Reflexes slow down with age; this probably reflects neuronal loss or sluggish CNS integration circuits.

REVIEW QUESTIONS

To access additional practice questions using your smartphone, tablet, or computer: MasteringA&P® > Study Area > Practice Tests & Quizzes

Level 1 Remember/Understand

(Some questions have more than one correct answer. Select the best answer or answers from the choices given.)

1. The large onion-shaped receptors that are found deep in the dermis and in subcutaneous tissue and that respond to deep pressure are **(a)** epithelial tactile complexes, **(b)** lamellar corpuscles, **(c)** free nerve endings, **(d)** muscle spindles.

2. Proprioceptors include all of the following except **(a)** muscle spindles, **(b)** tendon organs, **(c)** epithelial tactile complexes, **(d)** joint kinesthetic receptors.

3. The aspect of sensory perception by which the cerebral cortex identifies the site or pattern of stimulation is **(a)** perceptual detection, **(b)** feature abstraction, **(c)** pattern recognition, **(d)** spatial discrimination.

4. The neural machinery of the spinal cord is at the **(a)** precommand level, **(b)** projection level, **(c)** segmental level.

5. Dorsal root ganglia contain **(a)** cell bodies of somatic motor neurons, **(b)** axon terminals of somatic motor neurons, **(c)** cell bodies of autonomic motor neurons, **(d)** axon terminals of sensory neurons, **(e)** cell bodies of sensory neurons.

6. Match the receptor type in column B to the correct description in column A.

Column A	Column B
____ **(1)** pain, itch, and temperature receptors	**(a)** bulbous corpuscles
____ **(2)** contains intrafusal fibers and anulospiral and flower spray endings	**(b)** tendon organ
	(c) muscle spindle
	(d) free nerve endings
____ **(3)** discriminative touch receptor in hairless skin (fingertips)	**(e)** lamellar corpuscle
____ **(4)** contains receptor endings wrapped around thick collagen bundles	**(f)** tactile corpuscle
____ **(5)** rapidly adapting deep-pressure receptor	
____ **(6)** slowly adapting deep-pressure receptor	

7. Match the names of the cranial nerves in column B to the appropriate description in column A.

Column A	Column B
____ **(1)** causes pupillary constriction	**(a)** abducens
____ **(2)** the major sensory nerve of the face	**(b)** accessory
	(c) facial
____ **(3)** serves the sternocleido-mastoid and trapezius muscles	**(d)** glossopharyngeal
	(e) hypoglossal
____ **(4)** purely sensory (two nerves)	**(f)** oculomotor
____ **(5)** serves the tongue muscles	**(g)** olfactory
____ **(6)** allows you to chew your food	**(h)** optic
	(i) trigeminal
____ **(7)** impaired in Bell's palsy	**(j)** trochlear
____ **(8)** helps regulate heart activity	**(k)** vagus
____ **(9)** helps you hear and maintain your balance	**(l)** vestibulocochlear
____,____,____,____ **(10)** contain parasympathetic motor fibers (four nerves)	

8. For each of the following muscles or body regions, identify the plexus and the peripheral nerve(s) (or branch of one) involved. Use choices from keys A and B.

____;____ **(1)** the diaphragm
____;____ **(2)** muscles of the posterior leg
____;____ **(3)** anterior thigh muscles
____;____ **(4)** medial thigh muscles
____;____ **(5)** anterior arm muscles that flex the forearm
____;____ **(6)** muscles that flex the hand and digits (two nerves)
____;____ **(7)** muscles that extend the hand and digits
____;____ **(8)** skin and extensor muscles of the posterior arm
____;____ **(9)** fibularis muscles, tibialis anterior, and toe extensors
____;____,____,____ **(10)** elbow joint

Key A: Plexuses
(a) brachial
(b) cervical
(c) lumbar
(d) sacral

Key B: Nerves
(1) common fibular
(2) femoral
(3) median
(4) musculocutaneous
(5) obturator
(6) phrenic
(7) radial
(8) tibial
(9) ulnar

9. The connective tissue sheath that surrounds a fascicle of nerve fibers is the **(a)** epineurium, **(b)** endoneurium, **(c)** perineurium, **(d)** epimysium.

10. A reflex that causes reciprocal activation of the antagonist muscle is the **(a)** crossed-extensor, **(b)** flexor, **(c)** tendon, **(d)** muscle stretch.

11. List the structural components of the peripheral nervous system, and describe the function of each component.

12. **(a)** Define plexus. **(b)** Indicate the ventral rami of origin of the four major nerve plexuses.

13. Differentiate between ipsilateral and contralateral reflexes.

Level 2 Apply/Analyze

14. Characterize each receptor activity described below by choosing the appropriate letter and number(s) from keys A and B.

____,____ **(1)** You are enjoying an ice cream cone.
____,____ **(2)** You have just scalded yourself with hot coffee.
____,____ **(3)** The retinas of your eyes are stimulated.
____,____ **(4)** You bump (lightly) into someone.
____,____ **(5)** You are in a completely dark room and reaching toward the light switch.
____,____ **(6)** You feel uncomfortable after a large meal.

Key A:
(a) exteroceptor
(b) interoceptor
(c) proprioceptor

Key B:
(1) chemoreceptor
(2) mechanoreceptor
(3) nociceptor
(4) photoreceptor
(5) thermoreceptor

15. What is the functional relationship of the peripheral nervous system to the central nervous system?

16. Differentiate clearly between sensation and perception.

17. Central pattern generators (CPGs) are found at the segmental level of motor control. **(a)** What is the job of the CPGs? **(b)** What controls them, and where is this control localized?

18. Make a diagram of the hierarchy of motor control. Position the CPGs, motor cortex, brain stem nuclei, cerebellum, and basal nuclei in this scheme.

19. Explain why damage to peripheral nerve fibers is often reversible, whereas damage to CNS fibers rarely is.
20. **(a)** Describe the formation and composition of a spinal nerve.
    (b) Name the branches of a spinal nerve (other than the rami communicantes), and indicate their distribution.
21. Compare and contrast flexor and crossed-extensor reflexes.
22. Marcus, a football quarterback, suffered torn menisci in his right knee joint when tackled from the side. The same injury crushed his common fibular nerve against the head of the fibula. What locomotor problems did Marcus have after this?
23. As Luke fell off a ladder, he grabbed a tree branch with his right hand, but unfortunately lost his grip and fell heavily to the ground. Days later, Luke complained that his right upper limb was numb. What was damaged in his fall?
24. Mr. Frank, a former stroke victim who had made a remarkable recovery, suddenly began to have problems reading. He complained of seeing double and also had problems navigating steps. He was unable to move his left eye downward and laterally. Which cranial nerve was the site of lesion? (Right or left?)
25. Fumiko, a 19-year-old nursing student, had had a runny nose and sore throat for several days. Upon waking, her face felt "twisted." When she examined her face in the mirror, she noticed that the right side looked "droopy" and she was unable to move the facial muscles on that side. This made it difficult to speak clearly or eat. Which cranial nerve was affected and on which side? What is a common cause of this condition?
26. Mr. Jake was admitted to the hospital with excruciating pain in his left shoulder and arm. He was found to have suffered a heart attack. Explain the phenomenon of referred pain as exhibited by Mr. Jake.

Level 3 Evaluate/Synthesize

27. What is the homeostatic value of flexor reflexes?
28. What clinical information can be gained by conducting somatic reflex tests?
29. What is the structural and functional relationship between spinal nerves, skeletal muscles, and dermatomes?
30. In 1962 a boy playing in a train yard fell under a train. The train wheel cleanly cut off his right arm. Surgeons reattached the arm, sewing nerves and vessels back together. The boy was told he should eventually regain the use of his arm but that it would never be strong enough to pitch a baseball. Explain why full recovery of strength was unlikely.
31. One of a group of rabbit hunters was accidentally sprayed with buckshot in both of his gluteal prominences. When his companions saw that he would survive, they laughed and joked about where he had been shot. They were horrified and ashamed a week later when they learned their friend would be permanently paralyzed and without sensation in both legs from the knee down, as well as on the back of his thighs. What had happened?

CLINICAL CASE STUDY

44-Year-Old Male with Skull Fracture

William Hancock, 44, fell from the scaffolding at a construction site. When emergency personnel arrived, they found him unconscious but with stable vital signs. As paramedics placed him on a backboard to stabilize his head, neck, and back, they noted watery blood leaking from his right ear. In the hospital, Mr. Hancock regained consciousness and was treated for deep lacerations on his scalp and face. Head CT

scans revealed both longitudinal and transverse fractures of the right petrous temporal and sphenoid bones that extended through the foramen rotundum and foramen ovale.

The following observations were recorded on Mr. Hancock's chart on admission:

- Complete loss of hearing in the right ear.
- Paresthesia (sensation of "pins and needles") at the right corner of the mouth, extending to the lower lip and chin.
- Numbness of the right upper lip, lower eyelid, and cheek.
- Right eye turned slightly inward when looking straight ahead. Diplopia (double vision), particularly when looking to the right.

1. **+ NCLEX-STYLE** Which of the following choices best describes the fluid leaking from Mr. Hancock's ear and the cause of that leakage?
   a. Blood and mucus are leaking because the sphenoid bone is fractured.
   b. Blood and mucus are leaking because the petrous part of the temporal bone is fractured.
   c. Blood and cerebrospinal fluid are leaking because the petrous part of the temporal bone is fractured.
   d. Blood and sweat are leaking because the sphenoid bone is fractured and because his temperature was elevated.

2. Each of the four observations on Mr. Hancock's chart indicates damage to a cranial nerve. Identify each cranial nerve involved. If applicable, identify which specific branch of that nerve is involved.

Mr. Hancock was given a course of antibiotics, the head of his bed was elevated by 30°, and he was placed under close observation. After 24 hours, doctors noted that the right side of Mr. Hancock's face showed signs of drooping, with incomplete eye closure and asymmetric facial expressions. Mr. Hancock's right eye showed minimal tear production. The weakness and asymmetry on the right side of his face began to subside after a few days, and the leak of fluid from his ear stopped, but he continued to complain of paresthesia, diplopia, and an inability to hear with his right ear.

3. Why would Mr. Hancock's doctors give him antibiotics based on these observations? Why was the head of his bed elevated?

4. **+ NCLEX-STYLE** The observations after 24 hours suggest that yet another cranial nerve has been damaged. Which of the following choices best explains the new observations including the lack of tear production?
   a. The oculomotor nerve has been damaged, including its parasympathetic branch.
   b. The facial nerve has been damaged, including its parasympathetic branch.
   c. The oculomotor nerve has been damaged, including its sympathetic branch.
   d. The facial nerve has been damaged, including its sympathetic branch.

13

14

The Autonomic Nervous System

In this chapter, you will learn that

The autonomic nervous system (ANS) involuntarily controls smooth muscle, cardiac muscle, and glands to maintain homeostasis

part of → Peripheral nervous system Ch. 13

by asking

14.1 How is the ANS different from the somatic nervous system?

14.2 What are the two parts of the ANS?

14.6 What neurotransmitters and receptors does the ANS use?

14.7 How do the sympathetic and parasympathetic divisions interact?

14.8 How does the central nervous system control the ANS?

looking closer at

14.3 Parasympathetic division

14.4 Sympathetic division

then asking

14.9 What happens when things go wrong?

acting through

14.5 Visceral reflexes

and finally, exploring

Developmental Aspects of the Autonomic Nervous System

Figure 14.1 Place of the ANS in the structural organization of the nervous system.

Learning Outcome

▶ Define autonomic nervous system and explain its relationship to the peripheral nervous system.

The human body is exquisitely sensitive to changes in its internal environment, and engages in a lifelong struggle to balance competing demands for resources under ever-changing conditions. Although all body systems contribute, the stability of our internal environment depends largely on the **autonomic nervous system (ANS)**, the system of motor neurons that innervates smooth and cardiac muscle and glands (**Figure 14.1**).

At every moment, signals stream from visceral organs into the CNS, and autonomic nerves make adjustments as necessary to ensure optimal support for body activities. In response to changing conditions, the ANS shunts blood to "needy" areas, speeds or slows heart rate, adjusts blood pressure and body temperature, and increases or decreases stomach secretions.

Most of this fine-tuning occurs without our awareness or attention. Can you tell when your arteries are constricting or your pupils are dilating? Probably not—but if you've ever been stuck in a checkout line, and your full bladder was contracting as if it had a mind of its own, you've been very aware of visceral activity. The ANS controls all these functions, both those we're aware of and those we're not. Indeed, as the term *autonomic* (*auto* = self; *nom* = govern) implies, this motor subdivision of the peripheral nervous system has a certain amount of functional independence. The ANS is also called the **involuntary nervous system**, which reflects its subconscious control, or the **general visceral motor system**, which indicates the location of most of its effectors. Its two branches, the *sympathetic* and *parasympathetic divisions*, typically act in opposition to each other to control these effectors.

14.1 The ANS differs from the somatic nervous system in that it can stimulate or inhibit its effectors

Learning Outcome

▶ Compare the somatic and autonomic nervous systems relative to effectors, efferent pathways, and neurotransmitters released.

In our previous discussions of motor nerves, we have focused largely on the somatic nervous system. There are some key differences between the somatic and autonomic systems as well as areas of functional overlap. Both systems have motor fibers, but the somatic and autonomic nervous systems differ in: (1) their effectors, (2) their efferent pathways and ganglia, and (3) target organ responses to their neurotransmitters. Consult **Figure 14.2** for a summary of the differences.

Effectors

The somatic nervous system stimulates skeletal muscles, whereas the ANS innervates cardiac and smooth muscle and glands. Differences in the physiology of the effector organs account for most of the remaining differences between somatic and autonomic effects on their target organs.

Efferent Pathways and Ganglia

In the somatic nervous system, the motor neuron cell bodies are in the CNS, and their axons extend in spinal or cranial nerves all the way to the skeletal muscles they activate. Somatic motor fibers are typically thick, heavily myelinated group A fibers that conduct nerve impulses rapidly.

In contrast, the ANS uses a *two-neuron chain* to reach its effectors:

1. The cell body of the first neuron, the **preganglionic neuron**, resides in the brain stem or spinal cord. Its axon, the **preganglionic axon**, synapses with the second motor neuron.
2. The **postganglionic neuron** (sometimes called the *ganglionic neuron*) is the second motor neuron. Its cell body is in an **autonomic ganglion** outside the CNS. Its axon, the **postganglionic axon**, extends to the effector organ.

If you think about the meanings of all these terms while referring to Figure 14.2, understanding the rest of the chapter will be much easier.

Figure 14.2 Comparison of motor neurons in the somatic and autonomic nervous systems.

Preganglionic axons are thin, lightly myelinated fibers, and postganglionic axons are even thinner and nonmyelinated. Consequently, conduction through the autonomic efferent chain is slower than conduction in the somatic motor system. For most of their course, many pre- and postganglionic fibers are incorporated into spinal or cranial nerves.

Keep in mind that autonomic ganglia are *motor* ganglia, containing the cell bodies of motor neurons. Technically, they are sites of synapse and information transmission from preganglionic to postganglionic neurons. Also, remember that the somatic motor division *lacks* ganglia entirely. The dorsal root ganglia are part of the sensory, not the motor, division of the PNS.

Neurotransmitter Effects

All somatic motor neurons release **acetylcholine (ACh)** at their synapses with skeletal muscle fibers. The effect is *always excitatory*. Normally, stimulation reaches threshold and the muscle fibers contract.

Autonomic postganglionic fibers release two neurotransmitters: **norepinephrine (NE)** secreted by most sympathetic fibers, and ACh secreted by parasympathetic fibers. Depending on the type of receptors on the target organ, the effect may be excitatory or inhibitory.

Overlap of Somatic and Autonomic Function

Higher brain centers regulate and coordinate both somatic and autonomic motor activities, and most spinal nerves (and many cranial nerves) contain both somatic and autonomic fibers. Moreover, most of the body's adaptations to changing internal and external conditions involve both skeletal muscles and visceral organs. For example, when skeletal muscles are working hard, they need more oxygen and glucose, so autonomic control mechanisms speed up heart rate and dilate airways to meet these needs and maintain homeostasis.

Check Your Understanding

1. Name the three types of effectors of the autonomic nervous system.
2. Which relays instructions from the CNS to muscles more quickly, the somatic nervous system or the ANS? Explain why.
3. **MAKE CONNECTIONS** The cell bodies of autonomic postganglionic neurons are found in ANS ganglia. The cell bodies of another class of neuron are also found in ganglia (but not ANS ganglia). What are these other ganglia called? Determine the structural and functional classification of the neurons found in these other ganglia (use Table 11.2 on pp. 398–399).

For answers, see Answers Appendix.

14.2 The ANS consists of the parasympathetic and sympathetic divisions

Learning Outcome

▶ Compare and contrast the functions of the parasympathetic and sympathetic divisions.

The ANS has two arms, parasympathetic and sympathetic. The *parasympathetic division* promotes maintenance functions and conserves body energy, whereas the *sympathetic division* mobilizes the body during activity.

Both divisions generally serve the same visceral organs but cause opposite effects: While one stimulates certain smooth muscles to contract or a gland to secrete, the other inhibits that action. Through this **dual innervation**, the two divisions counterbalance each other to keep body systems running smoothly. Let's focus briefly on extreme situations in which each division exerts primary control.

Role of the Parasympathetic Division

The **parasympathetic division**, sometimes called the "rest and digest" system, keeps body energy use as low as possible, even as it directs vital "housekeeping" activities like digesting food and eliminating feces and urine. (This explains why it is a good idea to relax after a heavy meal: so sympathetic activity does not interfere with digestion.)

Parasympathetic activity is best illustrated in a person who relaxes after a meal and reads a magazine. Blood pressure and heart rate are regulated at low normal levels, and the gastrointestinal tract is actively digesting food. In the eyes, the pupils are constricted and the lenses are accommodated for close vision to improve the clarity of the close-up image.

Role of the Sympathetic Division

The activity of the **sympathetic division** (often called the "fight-or-flight" system) is evident when we are excited or find ourselves in emergency or threatening situations, such as being

Figure 14.3 Key anatomical differences between ANS divisions.

*Although sympathetic innervation to the skin is mapped to the cervical region here, all nerves to the periphery carry postganglionic sympathetic fibers.

confronted and robbed late at night. A rapidly pounding heart; deep breathing; dry mouth; cold, sweaty skin; and dilated pupils are sure signs of sympathetic nervous system mobilization.

During any type of vigorous physical activity, the sympathetic division also promotes a number of other adjustments.

- It constricts visceral (and sometimes cutaneous) blood vessels, shunting blood to active skeletal muscles and the vigorously working heart.
- It dilates the bronchioles in the lungs, increasing air flow (and thus increasing oxygen delivery to body cells).
- It stimulates the liver to release more glucose into the blood to accommodate the increased energy needs of body cells.

At the same time, the sympathetic division temporarily reduces nonessential activities, such as gastrointestinal tract motility. If you are running from a mugger, digesting lunch can wait! It is far more important to give your muscles everything they need to get you out of danger. In such active situations, the sympathetic division enables the body to cope with potential threats to homeostasis. It provides the optimal conditions for an appropriate response, whether that response is to run, see distant objects better, or think more clearly.

We have just looked at two extreme situations in which one or the other branch of the ANS dominates. Think of the parasympathetic division as the **D** division [digestion, defecation, and diuresis (urination)], and the sympathetic division as the **E** division (exercise, excitement, emergency, embarrassment). Table 14.5 (p. 546) presents a more detailed summary of how each division affects various organs.

Remember, however, that the two ANS divisions rarely work in an all-or-none fashion as described above. A dynamic antagonism exists between the divisions, and both make continuous fine adjustments to maintain homeostasis.

Key Anatomical Differences

Before we explain the anatomy of the divisions of the ANS in detail, let's look at their key anatomical differences (**Figure 14.3**). The sympathetic and parasympathetic divisions differ in:

① **Sites of origin.** Parasympathetic fibers are craniosacral—they originate in the brain (cranium) and sacral spinal cord. Sympathetic fibers are thoracolumbar—they originate in the thoracic and lumbar regions of the spinal cord.

② **Relative lengths of their fibers.** The parasympathetic division has long preganglionic and short postganglionic fibers. The sympathetic division has the opposite condition—the preganglionic fibers are short and the postganglionic fibers are long.

③ **Location of their ganglia.** Most parasympathetic ganglia are located in or near the visceral effector organs. Sympathetic ganglia lie close to the spinal cord.

Table 14.1 summarizes the major anatomical and physiological differences. We will examine these differences in detail in the next two modules.

Check Your Understanding

4. **APPLY** Which branch of the ANS would predominate if you were lying on the beach enjoying the sun and the sound of the waves? Which branch would predominate if you were on a surfboard and a shark appeared within a few feet of you?

For answers, see Answers Appendix.

Table 14.1	**Anatomical and Physiological Differences between the Parasympathetic and Sympathetic Divisions**	
CHARACTERISTIC	**PARASYMPATHETIC**	**SYMPATHETIC**
Origin	Craniosacral part: brain stem nuclei of cranial nerves III, VII, IX, and X; spinal cord segments S_2–S_4.	Thoracolumbar part: lateral horns of gray matter of spinal cord segments T_1–L_2.
Location of ganglia	Ganglia (terminal ganglia) are within the visceral organ (intramural) or close to the organ served.	Ganglia are within a few centimeters of CNS: alongside vertebral column (sympathetic trunk ganglia) and anterior to vertebral column (collateral, or prevertebral, ganglia).
Relative length of pre- and postganglionic fibers	Long preganglionic; short postganglionic.	Short preganglionic; long postganglionic.
Rami communicantes (see p. 538)	None.	Gray and white rami communicantes. White rami contain myelinated preganglionic fibers. Gray rami contain nonmyelinated postganglionic fibers.
Degree of branching of preganglionic fibers	Minimal.	Extensive.
Functional role	Maintenance functions; conserves and stores energy; "rest and digest."	Prepares body for activity; "fight or flight."
Neurotransmitters	All preganglionic and postganglionic fibers release ACh (cholinergic fibers).	All preganglionic fibers release ACh. Most postganglionic fibers release norepinephrine (adrenergic fibers). Postganglionic fibers serving sweat glands release ACh. Neurotransmitter activity is augmented by release of adrenal medullary hormones (epinephrine and norepinephrine).

14.3 Long preganglionic parasympathetic fibers originate in the craniosacral CNS

Learning Outcome

▶ For the parasympathetic division, describe the site of CNS origin, locations of ganglia, and general fiber pathways.

We begin our detailed exploration of the ANS with the anatomically simpler parasympathetic division. The parasympathetic division is also called the **craniosacral division** because its preganglionic fibers spring from opposite ends of the CNS—the brain stem and the sacral region of the spinal cord (**Figure 14.4**). The preganglionic axons extend from the CNS nearly all the way to the structures they innervate. There the axons synapse with postganglionic neurons located in **terminal ganglia** that lie close to or within the target organs. Very short postganglionic axons issue from the terminal ganglia and synapse with effector cells in their immediate area.

Cranial Part of Parasympathetic Division

Preganglionic fibers run in the oculomotor, facial, glossopharyngeal, and vagus cranial nerves. Their cell bodies lie in the brain stem in motor nuclei of the associated cranial nerves (◀ **Figure 12.13, pp. 450–451,** and ◀ **Figure 12.15, p. 453**).

Cranial nerves III, VII, and IX supply the entire parasympathetic innervation of the head, whereas the vagus nerves have a much more widespread distribution. Note that for cranial nerves III, VII, and IX, only the *preganglionic* fibers lie within these three pairs of cranial nerves—*postganglionic* fibers do not. Many of the postganglionic fibers "hitch a ride" with branches of the *trigeminal nerve* (V), taking advantage of its wide distribution.

Next, let's look at where the neurons of the cranial parasympathetics are located.

Oculomotor Nerves

The parasympathetic fibers of the **oculomotor nerves (III)** innervate smooth muscles in the eyes that cause the pupils to constrict and the lenses to bulge—actions needed to focus on close objects. The preganglionic axons found in the oculomotor nerves issue from the *accessory oculomotor* (*Edinger-Westphal*) *nuclei* in the midbrain. The cell bodies of the postganglionic neurons are in the **ciliary ganglia** within the eye orbits (◀ **Table 13.2, p. 501**).

Facial Nerves

The parasympathetic fibers of the **facial nerves (VII)** stimulate many large glands in the head.

Fibers that activate the nasal glands and the lacrimal glands of the eyes originate in the *lacrimal nuclei* of the pons. The preganglionic fibers synapse with postganglionic neurons in the **pterygopalatine ganglia** (ter″eh-go-pal′ah-tīn) just posterior to the maxillae.

The preganglionic neurons that stimulate the submandibular and sublingual salivary glands originate in the *superior*

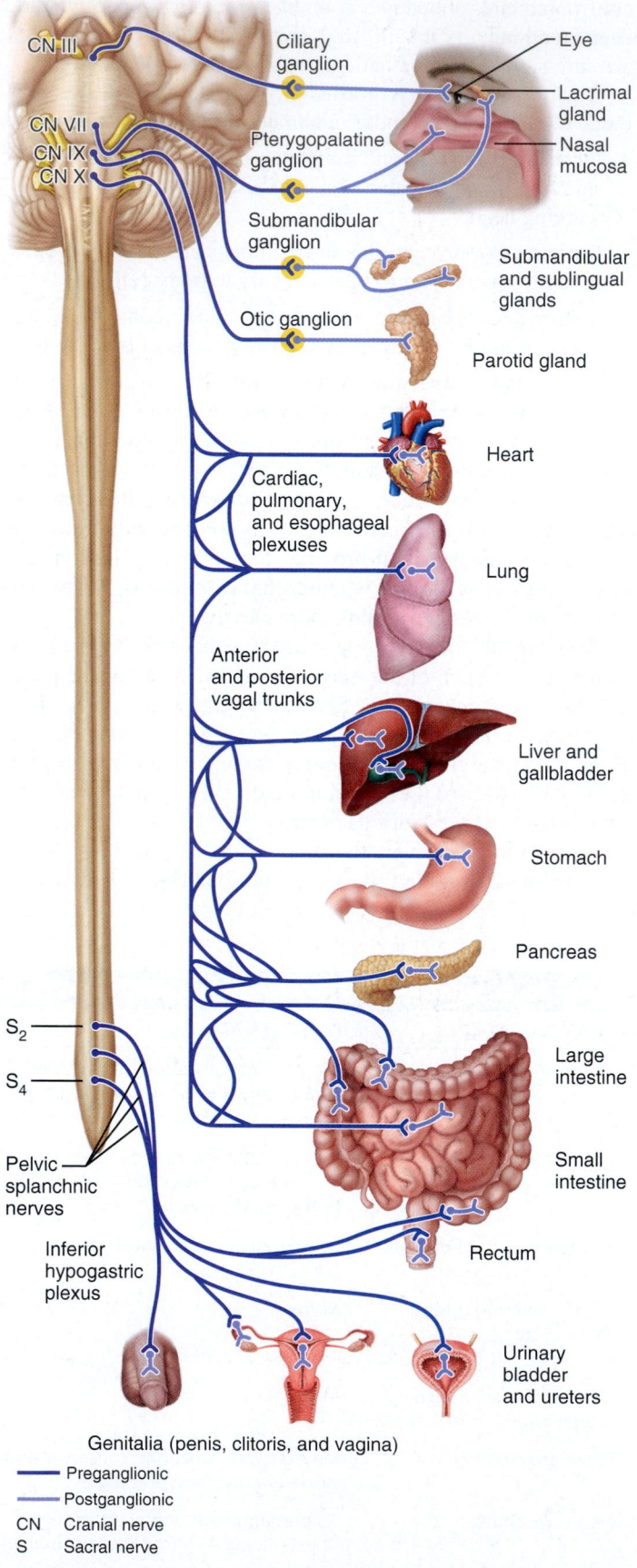

Figure 14.4 Parasympathetic division of the ANS.

salivatory nuclei of the pons. They synapse with postganglionic neurons in the **submandibular ganglia**, deep to the mandibular angles (◀ **Table 13.2, p. 503**).

Glossopharyngeal Nerves

The parasympathetics in the **glossopharyngeal nerves (IX)** originate in the *inferior salivatory nuclei* of the medulla and synapse in the **otic ganglia**, located just inferior to the foramen ovale of the skull. The postganglionic fibers course to and activate the parotid salivary glands anterior to the ears (◀ **Table 13.2, p. 505**).

Vagus Nerves

Between them, the two **vagus nerves (X)** account for about 90% of all preganglionic parasympathetic fibers in the body. They provide fibers to the neck and to nerve plexuses (interweaving networks of nerves) that serve virtually every organ in the thoracic and abdominal cavities. Note, however, that these nerve plexuses contain fibers from *both* the parasympathetic and sympathetic divisions, which travel through them to thoracic and abdominal organs.

The vagal nerve fibers (preganglionic axons) arise mostly from the *dorsal motor nuclei* of the medulla and synapse in terminal ganglia usually located in the walls of the target organ. Most of the terminal ganglia are collectively called *intramural ganglia*, literally, "ganglia within the walls."

As the vagus nerves pass into the thorax, they send branches through: (1) the *cardiac plexus*, which supplies fibers to the heart that slow heart rate; (2) the *pulmonary plexus*, which serves the airways of the lungs; and (3) the *esophageal plexus* (ĕ-sof″ah-je′al), which serves the esophagus and part of the stomach.

When the main trunks of the vagus nerves reach the esophagus, their fibers intermingle, forming the *anterior* and *posterior vagal trunks*. These vagal trunks then "ride" the esophagus down to the abdominal cavity. There they send fibers *through* the large *abdominal aortic plexus* before giving off branches to the abdominal viscera. The vagus nerves innervate the liver, gallbladder, stomach, small intestine, pancreas, and the proximal half of the large intestine.

Sacral Part of Parasympathetic Division

The sacral part of the parasympathetic division serves the pelvic organs and the distal half of the large intestine. It arises from neurons located in the lateral gray matter of spinal cord segments S_2–S_4. Axons of these neurons run in the ventral roots of the spinal nerves to the ventral rami and then branch off to form the **pelvic splanchnic nerves** (splank′nik; *splanchni* = viscera), which pass through the **inferior hypogastric (pelvic) plexus** in the pelvic floor (Figure 14.4). Some preganglionic fibers synapse with ganglia in this plexus, but most synapse in intramural ganglia in the walls of the following organs: distal half of the large intestine, urinary bladder, ureters, and reproductive organs.

Check Your Understanding

5. In general terms, where are the cell bodies of preganglionic parasympathetic neurons that innervate the head? Where are the cell bodies of postganglionic parasympathetic neurons innervated by the vagus nerve?

6. **DRAW** In the diagram of a preganglionic and a postganglionic parasympathetic neuron below, draw a circle to indicate the location of a ganglion. What is the name of this type of ganglion?

For answers, see Answers Appendix.

14.4 Short preganglionic sympathetic fibers originate in the thoracolumbar CNS

Learning Outcome

▶ For the sympathetic division, describe the site of CNS origin, locations of ganglia, and general fiber pathways.

The sympathetic division is anatomically more complex than the parasympathetic division, partly because it innervates more organs. Like the parasympathetic nervous system, the sympathetic nervous system supplies the smooth muscles, cardiac muscle, and glands in the internal body cavities. In addition, it supplies all smooth muscles and glands in the superficial (somatic) part of the body. These superficial structures are innervated *only* by the sympathetic nervous system. These structures include:

- Sweat glands (both eccrine and apocrine)
- The hair-raising arrector pili muscles of the skin
- Smooth muscle in the walls of all arteries and veins, both deep and superficial (This will be a key point for you to remember when you study the cardiovascular system.)

We will explain exactly how this works later—let's get on with the basic anatomy of the sympathetic division.

All preganglionic fibers of the sympathetic division arise from cell bodies of preganglionic neurons in spinal cord segments T_1 through L_2 (Figure 14.3). For this reason, the sympathetic division is also referred to as the **thoracolumbar division** (tho-rah″ko-lum′bar).

The numerous cell bodies of preganglionic sympathetic neurons in the gray matter of the spinal cord form the **lateral horns** (◀ **Figures 12.31b, p. 473**, and ◀ **12.32, p. 474**). The lateral horns are just posterolateral to the ventral horns that house somatic motor neurons. (Parasympathetic preganglionic neurons in the sacral cord are far less abundant than the comparable sympathetic neurons in the thoracolumbar regions, so there are no lateral horns in the sacral region of the spinal cord. This is a major anatomical difference between the two divisions.)

After leaving the cord via the ventral root, preganglionic sympathetic fibers pass through a **white ramus communicans** [plural: **rami communicantes** (kom-mu″nĭ-kan′tēz)] to enter an adjoining **sympathetic trunk ganglion** forming part of the **sympathetic trunk** (or *sympathetic chain*) (**Figure 14.5**). Looking like strands of glistening white beads, the sympathetic trunks flank each side of the vertebral column. They consist of

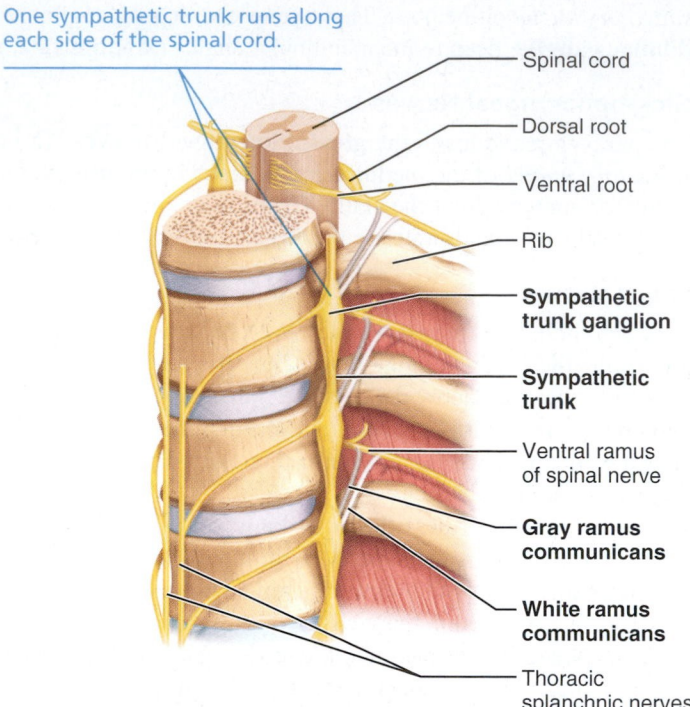

One sympathetic trunk runs along each side of the spinal cord.

- Spinal cord
- Dorsal root
- Ventral root
- Rib
- **Sympathetic trunk ganglion**
- **Sympathetic trunk**
- Ventral ramus of spinal nerve
- **Gray ramus communicans**
- **White ramus communicans**
- Thoracic splanchnic nerves

Figure 14.5 Location of the sympathetic trunk. The left sympathetic trunk in the posterior thorax.

the sympathetic ganglia and fibers running from one ganglion to another. The sympathetic trunk ganglia are also called *chain ganglia* or *paravertebral* ("near the vertebrae") *ganglia*.

Although the sympathetic *trunks* extend from neck to pelvis, sympathetic *fibers* arise only from the thoracic and lumbar cord segments, as shown in Figure 14.3. The ganglia vary in size, position, and number, but typically there are 23 in each sympathetic trunk—3 cervical, 11 thoracic, 4 lumbar, 4 sacral, and 1 coccygeal.

Once a preganglionic axon reaches a trunk ganglion, one of three things can happen (**Figure 14.6**). The preganglionic and postganglionic neurons can:

1. **Synapse at the same level.** In this case, the synapse is in the same trunk ganglion.
2. **Synapse at a higher or lower level.** The preganglionic axon ascends or descends the sympathetic trunk to another trunk ganglion.
3. **Synapse in a distant collateral ganglion.** The preganglionic axon passes through the trunk ganglion and emerges from the sympathetic trunk without synapsing. These

Table 14.2	Summary of Autonomic Ganglia	
NAME	**DIVISION**	**LOCATION**
Terminal ganglia	Parasympathetic nervous system	Within wall of organ served (*intramural ganglia*) or close to organ
Sympathetic trunk ganglia	Sympathetic nervous system	Paired, beside spinal cord
Collateral ganglia (prevertebral ganglia)	Sympathetic nervous system	Unpaired, anterior to spinal cord

① **Synapse in trunk ganglion at the same level**

Lateral horn (visceral motor zone)

Dorsal root

Dorsal root ganglion

Dorsal ramus of spinal nerve

Ventral ramus of spinal nerve

Gray ramus communicans

White ramus communicans

Ventral root

Sympathetic trunk ganglion

Sympathetic trunk

② **Synapse in trunk ganglion at a higher or lower level**

Splanchnic nerve

Collateral ganglion (such as the celiac)

③ **Pass through sympathetic trunk to synapse in a collateral ganglion anterior to the vertebral column**

Arrector pili muscles and sweat glands of skin

Smooth muscle of blood vessels

Effectors

Smooth muscle and glands of abdominal organs (e.g., intestine)

Figure 14.6 Three pathways of sympathetic innervation. Synapses between preganglionic and postganglionic sympathetic neurons can occur at three different locations.

preganglionic fibers help form several *splanchnic nerves* and synapse in **collateral**, or *prevertebral*, **ganglia** located anterior to the vertebral column. Unlike sympathetic trunk ganglia, the collateral ganglia are neither paired nor segmentally arranged. They occur only in the abdomen and pelvis.

Regardless of where the synapse occurs, all sympathetic ganglia are close to the spinal cord, and their postganglionic fibers are typically much longer than their preganglionic fibers. Recall that the opposite condition exists in the parasympathetic division, an important anatomical distinction. Autonomic ganglia are compared in **Table 14.2**.

Sympathetic Pathways with Synapses in Trunk Ganglia

When synapses are made in sympathetic trunk ganglia, the postganglionic axons enter the ventral (or dorsal) ramus of the adjoining spinal nerves by way of communicating branches called **gray rami communicantes** (Figure 14.6). From there they travel via branches of the rami to their effectors, including sweat glands and arrector pili muscles of the skin. Anywhere along their path, the postganglionic axons may transfer over to nearby blood vessels and innervate the vascular smooth muscle all the way to the final branches of the blood vessels.

Notice that the names of the rami communicantes—*white* or *gray*—reflect their appearance, revealing whether or not their fibers are myelinated. Preganglionic fibers composing the white rami are myelinated. Postganglionic axons forming the gray rami are not. (This has no relationship to the white and gray matter of the CNS.)

The white rami, which carry preganglionic axons to the sympathetic trunks, are found only in the T_1–L_2 cord

14

14

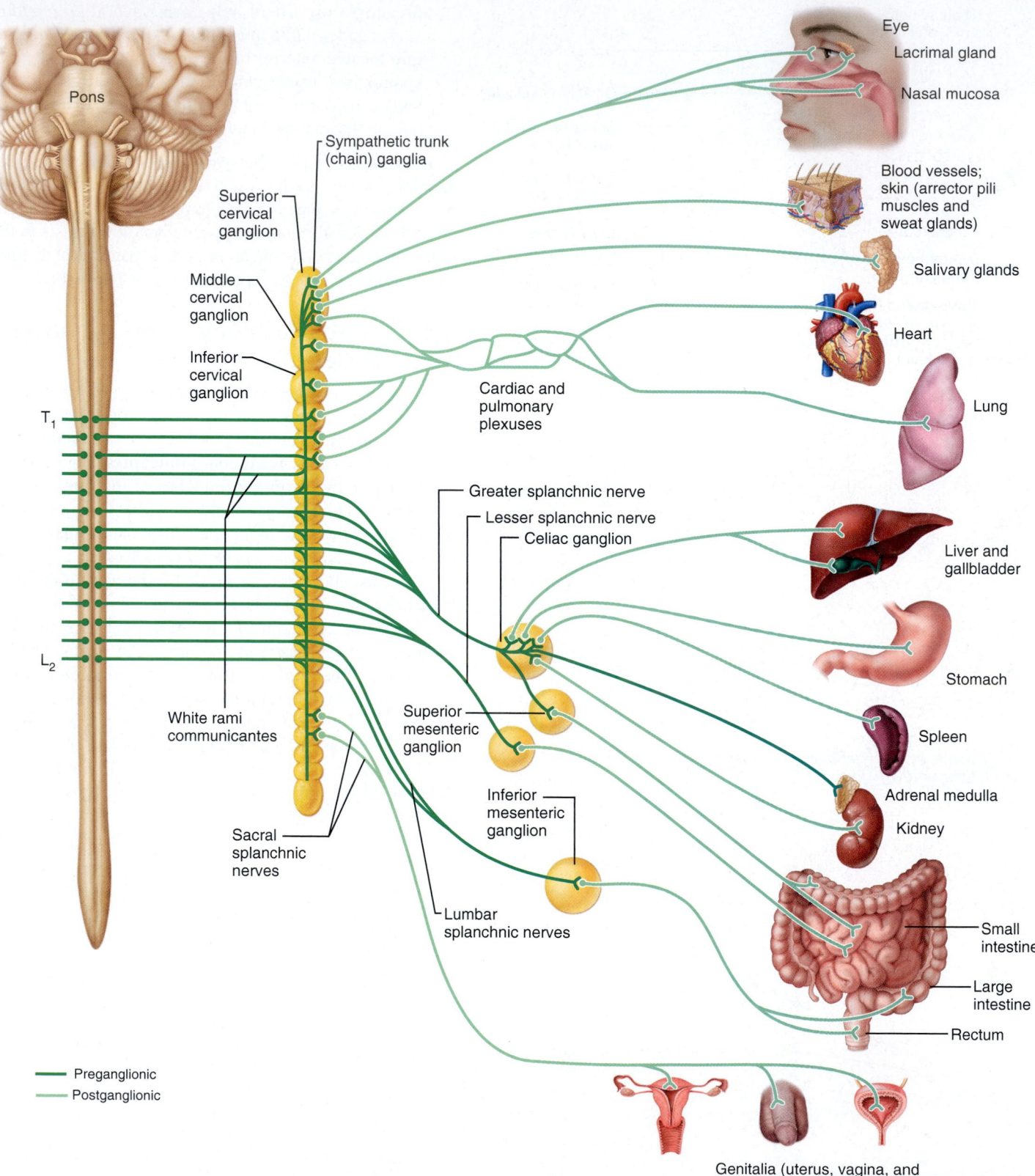

Pons

Sympathetic trunk (chain) ganglia

Superior cervical ganglion

Middle cervical ganglion

Inferior cervical ganglion

T₁

L₂

White rami communicantes

Sacral splanchnic nerves

Cardiac and pulmonary plexuses

Greater splanchnic nerve

Lesser splanchnic nerve

Celiac ganglion

Superior mesenteric ganglion

Inferior mesenteric ganglion

Lumbar splanchnic nerves

Eye

Lacrimal gland

Nasal mucosa

Blood vessels; skin (arrector pili muscles and sweat glands)

Salivary glands

Heart

Lung

Liver and gallbladder

Stomach

Spleen

Adrenal medulla

Kidney

Small intestine

Large intestine

Rectum

Genitalia (uterus, vagina, and penis) and urinary bladder

— Preganglionic
— Postganglionic

Figure 14.7 Sympathetic division of the ANS. Sympathetic innervation to peripheral structures (blood vessels, glands, and arrector pili muscles) occurs in all areas but is shown only in the cervical area.

segments, regions of sympathetic outflow. However, gray rami carrying postganglionic fibers headed for the periphery issue from every trunk ganglion from the cervical to the sacral region, allowing sympathetic output to reach all parts of the body.

Note that *rami communicantes are associated only with the sympathetic division.* They never carry parasympathetic fibers.

Pathways to the Head

Sympathetic preganglionic fibers serving the head emerge from spinal cord segments T_1–T_4 and ascend the sympathetic trunk to synapse with postganglionic neurons in the **superior cervical ganglion** (**Figure 14.7**). This ganglion contributes sympathetic fibers that run in several cranial nerves and in the upper three or four cervical spinal nerves.

Besides serving the skin and blood vessels of the head, fibers from the superior cervical ganglion stimulate the dilator muscles of the irises of the eyes, inhibit the nasal and salivary glands (the reason your mouth goes dry when you are scared), and innervate the smooth (tarsal) muscle that lifts the upper eyelid. The superior cervical ganglion also sends direct branches to the heart.

Pathways to the Thorax

Sympathetic preganglionic fibers innervating the thoracic organs originate at T_1–T_6. Some of these axons synapse in the nearest sympathetic trunk ganglion, and the postganglionic axons run directly to the organ being supplied (Figure 14.6 ①). These postganglionic axons pass through the pulmonary, esophageal, or cardiac plexuses. Fibers to the lungs and esophagus take this direct route, as do some axons to the heart. Many of the fibers to the heart, however, take a less direct path (Figure 14.6 ②). The preganglionic axons ascend in the sympathetic trunk to synapse in the **middle** and **inferior cervical ganglia** (Figure 14.7). From there, the postganglionic axons descend through the cardiac plexus and into the heart wall.

Sympathetic Pathways with Synapses in Collateral Ganglia

Most of the preganglionic fibers from T_5 down synapse in collateral ganglia, and so most of these fibers enter and leave the sympathetic trunks without synapsing. They form several nerves called **splanchnic nerves**, including the **greater**, **lesser**, and **least splanchnic nerves** (*thoracic splanchnic nerves*) and the **lumbar** and **sacral splanchnic nerves**.

The splanchnic nerves contribute to a number of interweaving nerve plexuses known collectively as the **abdominal aortic plexus**, which clings to the surface of the abdominal aorta. This complex plexus contains several ganglia that together serve the abdominopelvic viscera. From superior to inferior, the most important of these ganglia (and related subplexuses) are the **celiac**, **superior mesenteric**, and **inferior mesenteric**, named for the arteries with which they most closely associate (Figure 14.7). Postganglionic fibers issuing from these ganglia generally travel to their target organs in the company of the arteries serving these organs.

Pathways to the Abdomen

Sympathetic preganglionic fibers from T_5 to L_2 innervate the abdomen. They travel in the thoracic splanchnic nerves to synapse mainly at the celiac and superior mesenteric ganglia. Postganglionic fibers issuing from these ganglia serve the stomach, intestines (except the distal half of the large intestine), liver, spleen, and kidneys.

Pathways to the Pelvis

Preganglionic fibers innervating the pelvis originate from T_{10} to L_2 and then descend in the sympathetic trunk to the lumbar and sacral trunk ganglia. Some fibers synapse there and the postganglionic fibers run in lumbar and sacral splanchnic nerves to plexuses on the lower aorta and in the pelvis. Other preganglionic fibers pass directly to these autonomic plexuses and synapse in collateral ganglia, such as the inferior mesenteric ganglion.

Postganglionic fibers proceed from these plexuses to the pelvic organs (the urinary bladder and reproductive organs) and also the distal half of the large intestine. For the most part, sympathetic fibers *inhibit* the activity of muscles and glands in the abdominopelvic visceral organs.

Sympathetic Pathways with Synapses in the Adrenal Medulla

Some fibers traveling in the thoracic splanchnic nerves pass through the celiac ganglion without synapsing and terminate by synapsing with the hormone-producing medullary cells of the adrenal gland (**Figure 14.8**). When stimulated by preganglionic fibers, the medullary cells secrete *epinephrine* and *norepinephrine* (also called *adrenaline* and *noradrenaline*, respectively) into the blood, producing the excitatory effects we have all felt as a "surge of adrenaline."

Embryologically, sympathetic ganglia and the adrenal medulla arise from the same tissue. For this reason, the adrenal medulla is sometimes viewed as a "misplaced" sympathetic ganglion, and its hormone-releasing cells, although lacking nerve processes, are considered equivalent to postganglionic sympathetic neurons.

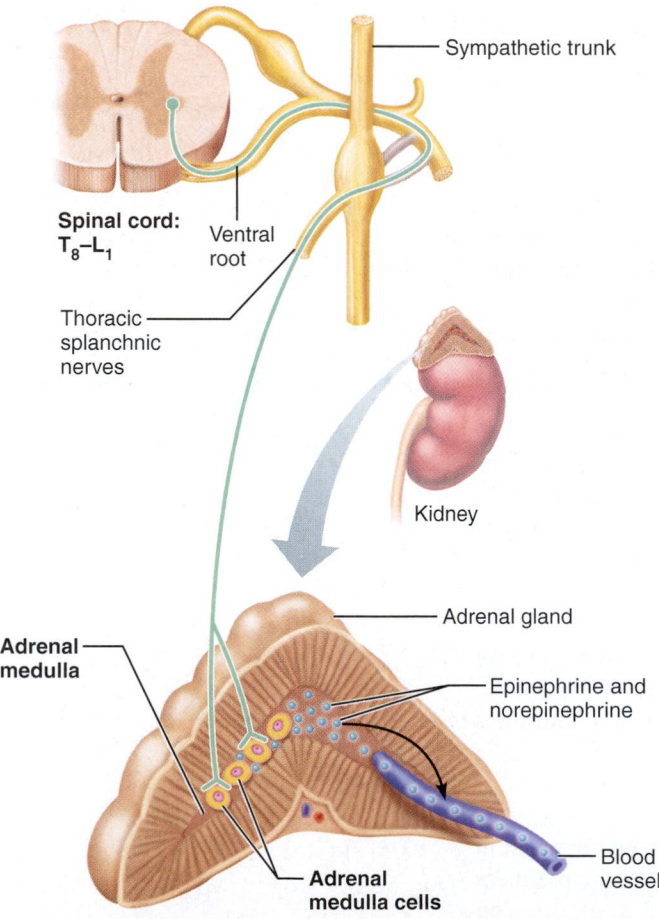

Figure 14.8 Sympathetic innervation of the adrenal medulla.

Labels: Sympathetic trunk; Spinal cord: T₈–L₁; Ventral root; Thoracic splanchnic nerves; Kidney; Adrenal gland; Adrenal medulla; Epinephrine and norepinephrine; Adrenal medulla cells; Blood vessel

Check Your Understanding

7. State whether each of the following is a characteristic of the sympathetic or parasympathetic nervous system: short preganglionic fibers; origin from thoracolumbar region of spinal cord; terminal ganglia; collateral ganglia; innervates adrenal medulla.

For answers, see Answers Appendix.

14.5 Visceral reflex arcs have the same five components as somatic reflex arcs

Learning Outcome

▶ **Compare visceral reflexes to somatic reflexes.**

Like all reflex arcs, visceral reflex arcs begin with sensory input. In this case, visceral information comes from visceral sensory neurons.

Visceral Sensory Neurons

Visceral sensory neurons send information about chemical changes, stretch, temperature, and irritation of the viscera. The brain interprets this information as hunger, fullness, pain, or nausea. Almost all receptors for these visceral senses are free (nonencapsulated) nerve endings that are widely scattered throughout the visceral organs.

Like somatic sensory neurons, the cell bodies of visceral sensory neurons are located in dorsal root ganglia and in the sensory ganglia of cranial nerves. The axons of visceral sensory neurons travel in the same nerves as the autonomic motor fibers. For example, many visceral sensory neurons accompany the parasympathetic fibers of the vagus nerve. They monitor the conditions in the various organs served by the vagus nerve. Other visceral sensory fibers accompany sympathetic fibers. (Most visceral pain fibers follow this sympathetic route to the CNS.) As they pass through sympathetic ganglia on their way to the spinal cord, branches of these visceral sensory fibers may synapse with the postganglionic neurons located in the ganglia.

Visceral sensory neurons are also involved in the phenomenon of referred pain described in Chapter 13 (◀ p. 496). This phenomenon occurs because some of the visceral sensory information converges on the same second-order sensory neurons that receive somatic sensory information arriving in the same spinal cord segment.

Visceral Reflexes

Visceral reflex arcs have essentially the same components as somatic reflex arcs—receptor, sensory neuron, integration center, motor neuron, and effector. However, there are three key differences:

- A visceral reflex arc has *two consecutive* neurons in its motor component (**Figure 14.9**; compare ◀ Figure 13.15, p. 519).
- The afferent fibers are visceral sensory neurons rather than somatic sensory neurons.

Stimulus

① Receptor in viscera

② Visceral sensory neuron

③ Integration center
• May be preganglionic neuron (as shown)
• May be a dorsal horn interneuron
• May be within walls of gastrointestinal tract

④ Motor neuron (two-neuron chain)
• Preganglionic neuron
• Postganglionic neuron

⑤ Visceral effector

Response

Dorsal root ganglion

Spinal cord

Autonomic ganglion

Figure 14.9 Visceral reflexes. Visceral reflex arcs have the same five elements as somatic reflex arcs.

• The effectors are smooth muscle, cardiac muscle, and glands rather than skeletal muscles.

Two examples of visceral reflexes are the reflexes that empty the rectum and bladder, which we discuss in Chapters 23 and 25.

Complete three-neuron reflex arcs (with sensory neurons, interneurons, and motor neurons) exist entirely within the walls of the gastrointestinal tract. Neurons composing these reflex arcs make up the **enteric nervous system**, which plays an important role in controlling gastrointestinal tract activity. We will discuss the enteric nervous system in more detail in Chapter 23.

Check Your Understanding

8. How does a visceral reflex differ from a somatic reflex?

For answers, see Answers Appendix.

14.6 Acetylcholine and norepinephrine are the major ANS neurotransmitters

Learning Outcomes

▶ Define cholinergic and adrenergic fibers, and list the different types of their receptors.

▶ Describe the clinical importance of drugs that mimic or inhibit adrenergic or cholinergic effects.

As we saw in Figure 14.2, the major neurotransmitters released by ANS neurons are *acetylcholine* (*ACh*) and *norepinephrine* (*NE*).

• **Cholinergic fibers** (ko″lin-er′jik)—those that release ACh—include (1) all ANS preganglionic axons and (2) all parasympathetic postganglionic axons at synapses with their effectors.

• **Adrenergic fibers** (ad″ren-er′jik)—those that release norepinephrine—include the majority of sympathetic postganglionic axons. An exception is sympathetic postganglionic fibers that secrete ACh onto sweat glands.

The effects of ACh and NE on their effectors are not consistently excitatory or inhibitory. Why not? Because the action of any neurotransmitter depends on the receptor to which it binds. Each autonomic neurotransmitter binds with two or more kinds of receptors, allowing it to exert different effects (activation or inhibition) at different body targets. **Table 14.3** summarizes the receptor types that we introduce next, and **Table 14.4** describes some of the many drugs that act upon them. Both are on p. 544.

Cholinergic Receptors

The two types of cholinergic (ACh-binding) receptors are named for drugs that bind to them and mimic acetylcholine's effects. **Nicotinic receptors** (nik″o-tin′ik) respond to nicotine. **Muscarinic receptors**, the other set of ACh receptors, can be activated by the mushroom poison *muscarine* (mus′kah-rin). All ACh receptors are either nicotinic or muscarinic.

Nicotinic Receptors

Nicotinic receptors are found on:

• *All* postganglionic neurons (cell bodies and dendrites), both sympathetic and parasympathetic

• The hormone-producing cells of the adrenal medulla

• The sarcolemma of skeletal muscle cells at neuromuscular junctions (which are somatic and not autonomic targets)

When ACh binds to nicotinic receptors, the effect is *always* stimulatory. Just as at the sarcolemma of skeletal muscle (examined in Chapter 9), ACh binding to any nicotinic receptor directly opens ion channels, depolarizing the postsynaptic cell.

Muscarinic Receptors

Muscarinic receptors occur on all effector cells stimulated by postganglionic cholinergic fibers—that is, all parasympathetic target organs and a few sympathetic targets, such as eccrine sweat glands. When ACh binds to muscarinic receptors, the effect can be either inhibitory or stimulatory, depending on the subclass of muscarinic receptor on the target organ. For example, ACh binding to cardiac muscle receptors slows heart activity, whereas ACh binding to receptors on smooth muscle of the gastrointestinal tract increases its motility.

Adrenergic Receptors

There are also two major classes of adrenergic (NE-binding) receptors: **alpha (α)** and **beta (β)**. These receptors are further divided into subclasses (α_1 and α_2; β_1, β_2, and β_3). Organs that respond to NE (or to epinephrine) have one or more of these receptor subtypes.

14

Table 14.3 Cholinergic and Adrenergic Receptors

NEUROTRANSMITTER	RECEPTOR TYPE	MAJOR LOCATIONS*	EFFECT OF BINDING
Acetylcholine (ACh)	**Cholinergic**		
	Nicotinic	All postganglionic neurons; adrenal medullary cells (also neuromuscular junctions of skeletal muscle)	Excitation
	Muscarinic	All parasympathetic target organs	Excitation in most cases; inhibition of cardiac muscle
		Limited sympathetic targets (e.g., eccrine sweat glands†)	Activation
Norepinephrine (NE) (and epinephrine released by adrenal medulla)	**Adrenergic**		
	β_1	Heart predominantly, but also kidneys and adipose tissue	Increases heart rate and force of contraction; stimulates kidneys to release renin
	β_2	Lungs and most other sympathetic target organs; abundant on blood vessels serving the heart, liver, and skeletal muscle	Effects mostly inhibitory; dilates blood vessels and bronchioles; relaxes smooth muscle walls of digestive and urinary visceral organs; relaxes uterus
	β_3	Adipose tissue	Stimulates lipolysis by fat cells
	α_1	Virtually all sympathetic target organs, especially blood vessels serving the skin, mucosae, abdominal viscera, kidneys, and salivary glands	Constricts blood vessels and visceral organ sphincters; dilates pupils of the eyes
	α_2	Membrane of adrenergic axon terminals; pancreas	Inhibits NE release from adrenergic terminals; inhibits insulin secretion

*Note that all of these receptor subtypes are also found in the CNS.
†Sympathetic cholinergic vasodilator fibers are found in other animals, but do not appear to be present in humans.

Table 14.4 Selected Drug Classes That Influence the Autonomic Nervous System

DRUG CLASS	RECEPTOR BOUND	EFFECTS	EXAMPLE	CLINICAL APPLICATION
Nicotinic agents	Nicotinic ACh receptors on all postganglionic neurons and in CNS	Typically stimulates sympathetic effects; blood pressure rises	Nicotine	Smoking cessation products
Parasympathomimetic agents (muscarinic agents)	Muscarinic ACh receptors	Enhance parasympathetic activity by mimicking effects of ACh	Pilocarpine	Glaucoma (opens aqueous humor drainage pores)
			Bethanechol	Difficulty urinating (increases bladder contraction)
Acetylcholinesterase inhibitors	None; bind to the enzyme (AChE) that degrades ACh	Indirect effect at all ACh receptors; prolong the effect of ACh	Neostigmine	Myasthenia gravis (increases availability of ACh)
			Sarin	Similar to widely used insecticides; used as chemical warfare agent
Sympathomimetic agents	Adrenergic receptors	Enhance sympathetic activity by binding to adrenergic receptors or increasing NE release	Albuterol (Ventolin®)	Asthma (dilates bronchioles by binding to β_2 receptors)
			Phenylephrine	Colds (nasal decongestant, binds to α_1 receptors)
Sympatholytic agents	Adrenergic receptors	Decrease sympathetic activity by blocking adrenergic receptors	Propranolol	Hypertension (drugs called *beta-blockers* block β receptors, decreasing blood pressure)

14

NE or epinephrine can be either excitatory or inhibitory depending on which subclass of receptor predominates in the target organ. For example, NE binding to the β_1 receptors of cardiac muscle prods the heart into more vigorous activity, whereas epinephrine binding to β_2 receptors in bronchiole smooth muscle causes it to relax, dilating the bronchiole.

Check Your Understanding

9. Would you find nicotinic receptors on skeletal muscle? Smooth muscle? Eccrine sweat glands? The adrenal medulla? CNS neurons?

10. **DRAW** Go to Figure 14.2 on p. 533. For each site in this figure where an ANS neurotransmitter acts, write "N" if it acts at nicotinic receptors, "M" if it acts at muscarinic receptors, and "α/β" if it acts at adrenergic receptors.

For answers, see Answers Appendix.

14.7 The parasympathetic and sympathetic divisions usually produce opposite effects

Learning Outcome

▶ State the effects of the parasympathetic and sympathetic divisions on the following organs: heart, blood vessels, gastrointestinal tract, lungs, adrenal medulla, and external genitalia.

As we mentioned earlier, most visceral organs receive *dual innervation* from both the parasympathetic and sympathetic divisions. Normally, both ANS divisions are partially active. Action potentials continually fire down both sympathetic and parasympathetic axons, producing a dynamic antagonism that precisely controls visceral activity. However, one division or the other usually predominates in given circumstances, and in a few cases, the two divisions actually cooperate with each other. **Table 14.5** (p. 546) contains an organ-by-organ summary of their effects.

Antagonistic Interactions

Antagonistic effects are most clearly seen on the activity of the heart, respiratory system, and gastrointestinal organs. In a fight-or-flight situation, the sympathetic division increases heart rate, dilates airways, and inhibits digestion and elimination. When the emergency is over, the parasympathetic division restores heart rate and airway diameter to resting levels and then attends to processes that refuel body cells and discard wastes.

Sympathetic and Parasympathetic Tone

We have described the parasympathetic division as the "rest and digest" division, but the sympathetic division is the major actor in controlling blood pressure, even at rest. With few exceptions, blood vessels are entirely innervated by sympathetic fibers that keep the blood vessels in a continual state of partial constriction called **sympathetic**, or **vasomotor**, **tone**.

When blood pressure is too low to maintain blood flow, sympathetic fibers called **vasomotor fibers** fire more rapidly. This causes blood vessels to constrict and raises blood pressure. When blood pressure becomes too high, these sympathetic vasomotor fibers fire less rapidly and the vessels dilate.

During circulatory shock (inadequate blood flow to body tissues), or when more blood is needed to meet the soaring needs of working skeletal muscles, blood vessels serving the skin and abdominal viscera strongly constrict. This blood "shunting" helps maintain circulation to vital organs and skeletal muscles.

In contrast to the sympathetic division's dominance of blood vessels, parasympathetic effects normally dominate the heart and the smooth muscle of digestive and urinary tract organs. These organs exhibit **parasympathetic tone**. The parasympathetic division slows the heart and dictates the normal activity levels of the digestive and urinary tracts. However, the sympathetic division can override these parasympathetic effects during times of stress. Drugs that block parasympathetic responses increase heart rate and cause fecal and urinary retention. Parasympathetic fibers activate most glands, except for the adrenal glands and sweat glands of the skin.

Cooperative Effects

The best example of cooperative ANS effects occurs in the external genitalia. Parasympathetic stimulation dilates blood vessels in the external genitalia, producing the erection of the male penis or female clitoris during sexual excitement. (This may explain why anxiety can impair sexual performance—the sympathetic division is in charge.) Sympathetic stimulation then causes ejaculation of semen by the penis or reflex contractions of the vagina.

HOMEOSTATIC IMBALANCE 14.1 **CLINICAL**

Autonomic neuropathy (damage to autonomic nerves) is a common complication of diabetes mellitus. One of the earliest and most troubling symptoms is sexual dysfunction. Up to 75% of male diabetics experience erectile dysfunction, and female diabetics often experience reduced vaginal lubrication. Other frequent manifestations of autonomic neuropathy include dizziness after standing suddenly (poor blood pressure control), urinary incontinence, sluggish eye pupil reactions, and impaired sweating. Maintaining tight control of blood glucose levels is the best way to prevent diabetic neuropathy. ●●●●●●●

Unique Roles of the Sympathetic Division

The adrenal medulla, sweat glands and arrector pili muscles of the skin, the kidneys, and most blood vessels receive only sympathetic fibers. It is easy to remember that the sympathetic system innervates these structures because most of us sweat under stress, our scalp "prickles" during fear, and our blood pressure skyrockets (from widespread constriction of blood vessels) when we get excited.

We have already described how sympathetic control of blood vessels regulates blood pressure and shunting of blood in the vascular system. We will now consider several other uniquely sympathetic functions.

Table 14.5 **Effects of the Parasympathetic and Sympathetic Divisions on Various Organs**

TARGET ORGAN OR SYSTEM	PARASYMPATHETIC EFFECTS	SYMPATHETIC EFFECTS
Eye (iris)	Stimulates sphincter pupillae muscles; constricts pupils	Stimulates dilator pupillae muscles; dilates pupils
Eye (ciliary muscle)	Stimulates muscle, which makes lens bulge for close vision	Weakly inhibits muscle, which flattens lens for far vision
Glands (nasal, lacrimal, gastric, pancreas)	Stimulates secretory activity	Inhibits secretory activity; constricts blood vessels supplying the glands
Salivary glands	Stimulates secretion of watery saliva	Stimulates secretion of thick, viscous saliva
Sweat glands	No effect (no innervation)	Stimulates copious sweating (cholinergic fibers)
Adrenal medulla	No effect (no innervation)	Stimulates medulla cells to secrete epinephrine and norepinephrine
Arrector pili muscles attached to hair follicles	No effect (no innervation)	Stimulates contraction (erects hairs and produces "goosebumps")
Heart (muscle)	Decreases rate (slows heart)	Increases rate and force of heartbeat
Heart (coronary blood vessels)	No effect (no innervation)	Dilates blood vessels (vasodilation)*
Urinary bladder/urethra	Contracts smooth muscle of bladder wall; relaxes urethral sphincter; promotes voiding	Relaxes smooth muscle of bladder wall; constricts urethral sphincter; inhibits voiding
Lungs	Constricts bronchioles	Dilates bronchioles*
Digestive tract organs	Increases motility (peristalsis) and amount of secretion by digestive organs; relaxes sphincters to allow foodstuffs to move through tract	Decreases activity of glands and muscles of digestive system; constricts sphincters (e.g., anal sphincter)
Liver	Increases glucose uptake from blood	Stimulates release of glucose to blood*
Gallbladder	Excites (gallbladder contracts to expel bile)	Inhibits (gallbladder is relaxed)
Kidney	No effect (no innervation)	Promotes renin release; causes vasoconstriction; decreases urine output
Penis	Causes erection (vasodilation)	Causes ejaculation
Vagina/clitoris	Causes erection (vasodilation) of clitoris; increases vaginal lubrication	Causes vagina to contract
Blood vessels	Little or no effect (except blood vessels to external genitalia)	Constricts most vessels and increases blood pressure; constricts vessels of abdominal viscera and skin to divert blood to muscles, brain, and heart when necessary; epinephrine weakly dilates vessels of skeletal muscles during exercise*
Blood coagulation	No effect (no innervation)	Increases coagulation*
Cellular metabolism	No effect (no innervation)	Increases metabolic rate*
Adipose tissue	No effect (no innervation)	Stimulates lipolysis (fat breakdown)

*Effects are mediated by epinephrine release into the bloodstream from the adrenal medulla.

- **Thermoregulatory responses to heat.** The sympathetic division mediates reflexes that regulate body temperature. For example, applying heat to the skin causes blood vessels in that area to dilate reflexively. When systemic body temperature rises, sympathetic nerves (1) dilate the skin's blood vessels, allowing heat to escape from skin flushed with warm blood, and (2) activate the sweat glands to help cool the body. When body temperature falls, skin blood vessels constrict, preventing heat loss from the skin.

- **Release of renin from the kidneys.** Sympathetic impulses stimulate the kidneys to release *renin*, an enzyme that causes the formation of potent blood pressure–increasing hormones (see Chapters 19 and 25).

- **Metabolic effects.** Through both direct neural stimulation and release of adrenal medullary hormones, the sympathetic division promotes a number of metabolic effects not reversed by parasympathetic activity. It (1) increases the metabolic rate of body cells; (2) raises blood glucose levels; and (3) mobilizes fats for use as fuels.

The medullary hormones also cause skeletal muscle to contract more strongly and quickly. As a side effect, this stimulates muscle spindles more often and, consequently, nerve impulses traveling to the muscles occur more synchronously. These neural bursts, which put muscle contractions on a "hair trigger," are great if you have to make a quick jump or run, but they can be disabling to the nervous musician or surgeon.

Localized versus Diffuse Effects

In the parasympathetic division, one preganglionic neuron synapses with one (or at most a few) postganglionic neurons. Additionally, all parasympathetic fibers release ACh, which is quickly destroyed (hydrolyzed) by acetylcholinesterase. Consequently, the parasympathetic division exerts short-lived, highly localized control over its effectors.

In contrast, in the sympathetic division, preganglionic axons branch profusely as they enter the sympathetic trunk, and they synapse with postganglionic neurons at several levels. As a result, when the sympathetic division is activated, it responds in a diffuse and highly interconnected way. Indeed, the literal translation of sympathetic (*sym* = together; *pathos* = feeling) relates to the bodywide mobilization this division provokes. Nevertheless, parts of the sympathetic nervous system can be activated individually. For example, just because your eye pupils dilate in dim light doesn't necessarily mean that your heart rate also speeds up.

Sympathetic activation produces much longer-lasting effects than parasympathetic activation. Adrenal medullary cells secrete epinephrine and NE into the blood when the sympathetic division is mobilized. These hormones reinforce and prolong the effects of the sympathetic nervous system. They have essentially the same effects as NE released by sympathetic neurons, although epinephrine is more potent at increasing heart rate and raising blood glucose levels and metabolic rate. In fact, circulating adrenal medullary hormones produce 25–50% of all the sympathetic effects acting on the body at a given time. These effects continue for several minutes until the liver destroys the hormones.

In short, sympathetic nerve impulses act only briefly, but the hormonal effects they provoke linger. The widespread and prolonged effect of sympathetic activation explains why we need time to "come down" after an extremely stressful experience.

Check Your Understanding

11. Name the division of the ANS that does each of the following: increases digestive activity; increases blood pressure; dilates bronchioles; decreases heart rate; stimulates the adrenal medulla to release its hormones; causes ejaculation.

For answers, see Answers Appendix.

14.8 The hypothalamus oversees ANS activity

Learning Outcome

▶ Describe autonomic nervous system controls.

Although the ANS is not usually considered to be under voluntary control, its activity is regulated by CNS controls in the spinal cord, brain stem, hypothalamus, and cerebral cortex (**Figure 14.10**). In general, the hypothalamus is the integrative center at the top of the ANS control hierarchy. From there, orders flow to lower and lower CNS centers for execution. Although the cerebral cortex may modify the workings of the ANS, it does so at the subconscious level and by acting through limbic system structures on hypothalamic centers.

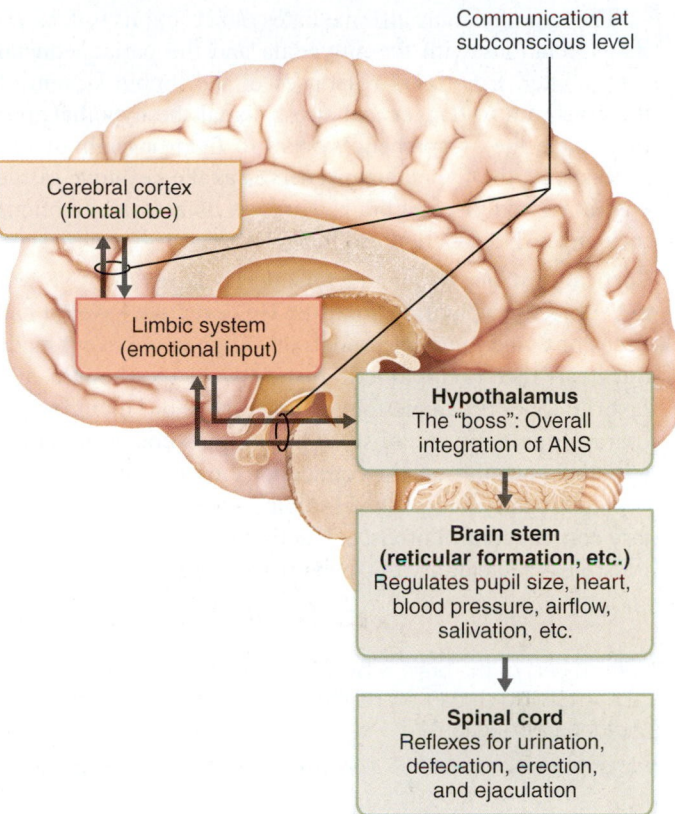

Figure 14.10 Levels of ANS control. The hypothalamus stands at the top of the control hierarchy as the integrator of ANS activity, but it is influenced by subconscious cerebral inputs via limbic system connections.

- **Brain stem and spinal cord controls.** The hypothalamus is the "boss," but the brain stem reticular formation appears to exert the most *direct* influence over autonomic functions (◀ **Figure 12.15, p. 453**). For example, certain motor centers in the ventrolateral medulla (*cardiac* and *vasomotor centers*) reflexively regulate heart rate and blood vessel diameter. Other medullary regions oversee gastrointestinal activities. Most sensory impulses involved in these autonomic reflexes reach the brain stem via vagus nerve afferents. Midbrain centers (*oculomotor nuclei*) control the muscles concerned with pupil diameter and lens focus.

  Defecation and micturition (urination) reflexes that empty the rectum and urinary bladder are integrated at the spinal cord level but are subject to conscious inhibition. We will describe all of these autonomic reflexes in later chapters in relation to the organ systems they serve.

- **Hypothalamic controls.** As we noted, the hypothalamus is the main integration center of the autonomic nervous system. In general, anterior hypothalamic regions direct parasympathetic functions, and posterior areas direct sympathetic functions. Hypothalamic centers exert their effects both directly and via relays through the *reticular formation*, which in turn influences the preganglionic motor neurons in the brain stem and spinal cord (Figure 14.10). The hypothalamus, acting through the ANS, coordinates heart activity, blood pressure, body temperature, water balance, and endocrine activity.

The hypothalamus also mediates our reactions to fear via its associations with the amygdala and the periaqueductal gray matter. Emotional responses of the limbic system of the cerebrum to danger and stress signal the hypothalamus to activate the sympathetic system to fight-or-flight status. In this way, the hypothalamus serves as the keystone of the emotional and visceral brain. Through its centers, emotions influence ANS function and behavior.

- **Cortical controls.** Originally, scientists believed the ANS was not subject to voluntary controls. However, we have all had occasions when just remembering a frightening event made our heart race (sympathetic response) or the thought of a favorite food, pecan pie for example, made our mouth water (parasympathetic response). These inputs converge on the hypothalamus through its connections to the limbic system.

  Additionally, biofeedback studies have shown that voluntary cortical control of visceral activities such as heart rate is possible—a capability untapped by most people.

Check Your Understanding

12. Which part of the brain is the main integration center of the ANS? Which part exerts the most direct influence over autonomic functions?

For answers, see Answers Appendix.

Figure 14.11 Raynaud's disease.

14.9 Most ANS disorders involve abnormalities in smooth muscle control CLINICAL

Learning Outcome

▶ Explain the relationship of some types of hypertension, Raynaud's disease, and autonomic dysreflexia to disorders of autonomic function.

The ANS is involved in nearly every important process that goes on in the body, so it is not surprising that abnormalities of autonomic function can have far-reaching effects. Most autonomic disorders reflect exaggerated or deficient controls of smooth muscle activity. The most devastating involve blood vessels and include conditions such as hypertension, Raynaud's disease, and autonomic dysreflexia.

- *Hypertension*, or high blood pressure, may result from an overactive sympathetic vasoconstrictor response promoted by continuous high levels of stress. Hypertension is always serious because it forces the heart to work harder, which may precipitate heart disease, and increases the wear and tear on artery walls. Hypertension is sometimes treated with adrenergic receptor–blocking drugs that counteract the effects of the sympathetic nervous system on the cardiovascular system. We will discuss hypertension in more detail in Chapter 19.

- *Raynaud's disease* is characterized by intermittent attacks causing the skin of the fingers and toes to become pale, then cyanotic (bluish) and painful (**Figure 14.11**). Commonly provoked by exposure to cold or emotional stress, it is an exaggerated vasoconstriction response. The severity

of Raynaud's disease ranges from merely uncomfortable to severe blood vessel constriction that causes ischemia and gangrene (tissue death).

- *Autonomic dysreflexia* is a life-threatening condition involving uncontrolled activation of autonomic neurons. It occurs in a majority of individuals with quadriplegia and in others with spinal cord injuries above the T_6 level, usually in the first year after injury. The usual trigger is a painful stimulus to the skin or an overfilled visceral organ, such as the urinary bladder. Arterial blood pressure skyrockets to life-threatening levels, which may rupture a blood vessel in the brain, precipitating stroke. Symptoms include headache, flushed face, sweating above the level of the injury, and cold, clammy skin below. The precise mechanism of autonomic dysreflexia is not yet clear.

Check Your Understanding

13. APPLY Jackson works long, stress-filled shifts as an air traffic controller at a busy airport. His doctor has prescribed a beta-blocker. Why might his doctor have done this? What does a beta-blocker do?

For answers, see Answers Appendix.

Developmental Aspects of the ANS

ANS preganglionic neurons derive from the embryonic *neural tube*, as do somatic motor neurons. ANS structures in the PNS—postganglionic neurons, the adrenal medulla, and all autonomic ganglia—derive from the **neural crest** (along with all sensory neurons) (◀ **Figure 12.37** ②, **p. 482**).

Neural crest cells reach their ultimate destinations by migrating along growing axons. Forming ganglia receive axons from preganglionic neurons in the spinal cord or brain and send their axons to synapse with their effector cells in the body periphery. This process depends on the presence of **nerve growth factor**, and is guided by a number of signaling chemicals similar to those acting in the CNS.

During youth, impairments of ANS function are usually due to injuries to the spinal cord or autonomic nerves. In old age, the efficiency of the ANS declines. At least part of the problem is due to structural changes in some preganglionic axon terminals, which become congested with neurofilaments.

Many elderly people complain of constipation (a result of reduced gastrointestinal tract motility), and of dry eyes and frequent eye infections (both a result of a diminished ability to form tears). Additionally, when they stand up they may have fainting episodes due to **orthostatic hypotension** (*ortho* = straight; *stat* = standing), low blood pressure following changes in position. Orthostatic hypotension occurs because aging pressure receptors become less responsive to changes in blood pressure, and aging cardiovascular centers fail to maintain healthy blood pressure.

These problems are distressing, but not usually life threatening, and most can be managed by lifestyle changes or products such as artificial tears for eye problems.

RELATED CLINICAL TERMS

Atonic bladder (a-ton′ik; *a* = without; *ton* = tone, tension) A condition in which the urinary bladder becomes flaccid and overfills, allowing urine to dribble through the sphincters. Results from temporary loss of the micturition reflex following spinal cord injury.

Horner's syndrome A condition due to damage to the superior sympathetic trunk on one side of the body. On the affected side, the upper eyelid droops (ptosis) and the pupil constricts, and the person does not sweat on that side of the head.

Vagotomy (va-got′o-me) Cutting or severing the vagus nerve to decrease secretion of gastric juice. Treatment for peptic ulcers that do not respond to medication.

Vasovagal syncope Vasovagal syncope, also called *neurocardiogenic syncope*, is the most common cause of fainting. Although it may be provoked by emotional stress, pain, or dehydration, it typically occurs during prolonged standing. Fainting is due to a drop in blood pressure, which decreases blood flow to the brain.

CHAPTER SUMMARY

1. The autonomic nervous system (ANS) is the motor division of the PNS that controls visceral activities, with the goal of maintaining internal homeostasis.

14.1 The ANS differs from the somatic nervous system in that it can stimulate or inhibit its effectors (pp. 532–533)

1. The somatic (voluntary) nervous system provides motor fibers to skeletal muscles. The autonomic (involuntary or visceral motor) nervous system provides motor fibers to smooth and cardiac muscles and glands.
2. In the somatic division, a single motor neuron forms the efferent pathway from the CNS to the effectors. The efferent pathway of the autonomic division consists of a two-neuron chain with the cell bodies of the preganglionic neurons in the CNS and those of the postganglionic neurons in a ganglion.
3. Acetylcholine, the neurotransmitter of somatic motor neurons, is stimulatory to skeletal muscle fibers. Neurotransmitters released by autonomic motor neurons (acetylcholine and norepinephrine) may excite or inhibit target cells.

14.2 The ANS consists of the parasympathetic and sympathetic divisions (pp. 534–535)

1. The ANS consists of two divisions, parasympathetic and sympathetic, which normally exert antagonistic effects on many of the same target organs.
2. The parasympathetic division (the rest and digest system) conserves body energy and maintains body activities at basal levels.
3. Parasympathetic effects include constricted pupils, glandular secretion, increased digestive tract motility, and smooth muscle activity leading to elimination of feces and urine.
4. The sympathetic division prepares the body for activity (the fight-or-flight system).

5. Sympathetic responses include dilated pupils, increased heart rate, increased blood pressure, dilated bronchioles of the lungs, increased blood glucose levels, and sweating. During exercise, sympathetic vasoconstriction shunts blood from the skin and digestive viscera to the heart, brain, and skeletal muscles.

14.3 Long preganglionic parasympathetic fibers originate in the craniosacral CNS (pp. 536–537)

1. Parasympathetic preganglionic neurons arise from the brain stem and from the sacral (S_2–S_4) region of the spinal cord.
2. Preganglionic fibers synapse with postganglionic neurons in terminal ganglia located in (intramural ganglia) or close to their effector organs. Preganglionic fibers are long; postganglionic fibers are short.
3. Cranial fibers arise in the brain stem nuclei of cranial nerves III, VII, IX, and X and synapse in ganglia of the head, thorax, and abdomen. The vagus nerves serve virtually all organs of the thoracic and abdominal cavities.
4. Sacral fibers (S_2–S_4) issue from the lateral region of the cord and form pelvic splanchnic nerves that innervate the pelvic viscera. The preganglionic axons do not travel within rami communicantes.

14.4 Short preganglionic sympathetic fibers originate in the thoracolumbar CNS (pp. 538–542)

1. Preganglionic sympathetic neurons arise from the lateral horns of the spinal cord from the level of T_1 through L_2.
2. Preganglionic axons leave the cord via white rami communicantes and enter the sympathetic trunk ganglia in the sympathetic trunk. An axon may synapse in a trunk ganglion at the same or at a different level, or it may pass through the sympathetic trunk without synapsing. Preganglionic fibers are short; postganglionic fibers are long.

14

Homeostatic Interrelationships between the Nervous System and Other Body Systems

Integumentary System   Chapter 5
- Sympathetic division of the ANS regulates sweat glands and blood vessels of skin (therefore heat loss/retention)
- Skin serves as heat loss surface

Skeletal System   Chapters 6–8
- Nerves innervate bones and joints, providing for pain and joint sense
- Bones serve as depot for calcium needed for neural function; skeletal system protects CNS structures

Muscular System   Chapters 9–10
- Somatic division of nervous system activates skeletal muscles; maintains muscle health
- Skeletal muscles are the effectors of the somatic division

Endocrine System   Chapter 16
- Sympathetic division of the ANS activates the adrenal medulla; hypothalamus helps regulate the activity of the anterior pituitary gland and produces the two posterior pituitary hormones
- Hormones influence neuronal metabolism

Cardiovascular System   Chapters 17–19
- ANS helps regulate heart rate and blood pressure
- Cardiovascular system provides blood containing oxygen and nutrients to the nervous system and carries away wastes

Lymphatic System/Immunity   Chapters 20–21
- Nerves innervate lymphoid organs; the brain plays a role in regulating immune function
- Lymphatic vessels carry away leaked tissue fluids from nervous system structures and surrounding tissues; immune elements protect all body organs from pathogens (CNS has additional mechanisms as well)

Respiratory System   Chapter 22
- Nervous system initiates and regulates respiratory rhythm and depth
- Respiratory system provides life-sustaining oxygen; disposes of carbon dioxide

Digestive System   Chapter 23
- ANS (particularly the parasympathetic division) regulates digestive motility and glandular activity
- Digestive system provides nutrients needed for neuronal health

Urinary System   Chapters 25–26
- ANS regulates bladder emptying and renal blood pressure
- Kidneys help dispose of metabolic wastes and maintain proper electrolyte composition and pH of blood for neural functioning

Reproductive System   Chapter 27
- ANS regulates sexual erection and ejaculation in males; erection of the clitoris in females
- Testosterone causes masculinization of the brain and underlies sex drive and aggressive behavior

3. When the synapse occurs in a trunk ganglion, the postganglionic fiber enters the spinal nerve ramus via a gray ramus communicans to travel to the body periphery. Postganglionic fibers issuing from the cervical ganglia also serve visceral organs and blood vessels of the head, neck, and thorax.

4. When synapses do not occur in the trunk ganglia, the preganglionic fibers form splanchnic nerves (thoracic, lumbar, and sacral). Most splanchnic nerve fibers synapse in collateral ganglia, and the postganglionic fibers serve abdominal viscera. Exceptions are that (1) some splanchnic nerve fibers synapse with cells of the adrenal medulla, and (2) some lumbar and sacral splanchnic nerve fibers *do* synapse in trunk ganglia.

14.5 Visceral reflex arcs have the same five components as somatic reflex arcs (pp. 542–543)

1. Visceral reflex arcs have the same components as somatic reflexes: receptor, sensory neuron, integration center, motor neurons, effector.

2. Cell bodies of visceral sensory neurons are located in dorsal root ganglia or in sensory ganglia of cranial nerves. Visceral afferents are located in spinal nerves and in virtually all autonomic nerves.

14.6 Acetylcholine and norepinephrine are the major ANS neurotransmitters (pp. 543–545)

1. Autonomic motor neurons release two major neurotransmitters, acetylcholine (ACh) and norepinephrine (NE). Based on the neurotransmitter they release, fibers are classified as cholinergic (ACh) or adrenergic (NE).

2. ACh is released by all preganglionic fibers and all parasympathetic postganglionic fibers. NE is released by almost all sympathetic postganglionic fibers. An exception is those serving the sweat glands of the skin.

3. Neurotransmitter effects depend on the receptors to which the neurotransmitter binds. Cholinergic (ACh) receptors are classified as nicotinic or muscarinic. Adrenergic (NE) receptors are classified as α_1 or α_2, or β_1, β_2, or β_3.

14.7 The parasympathetic and sympathetic divisions usually produce opposite effects (pp. 545–547)

1. Most visceral organs are innervated by both parasympathetic and sympathetic divisions. The divisions interact in various ways but usually exert a dynamic antagonism. Antagonistic interactions mainly involve the heart, respiratory system, and gastrointestinal organs. Sympathetic activity increases heart activity, dilates bronchioles, and depresses gastrointestinal activity. Parasympathetic activity reverses these effects.

2. Most blood vessels are innervated only by sympathetic fibers and exhibit vasomotor tone. Parasympathetic activity dominates the heart and muscles of the gastrointestinal tract (which normally exhibit parasympathetic tone) and glands.

3. The two ANS divisions exert cooperative effects on the external genitalia.

4. Roles unique to the sympathetic division are regulating blood pressure and body temperature, shunting blood in the vascular system, stimulating renin release, and metabolic effects.

5. Activation of the sympathetic division can cause widespread, long-lasting mobilization of the fight-or-flight response. Parasympathetic effects are highly localized and short-lived.

14.8 The hypothalamus oversees ANS activity (pp. 547–548)

1. Autonomic function is controlled at several levels: (1) The spinal cord and brain stem (particularly medullary) centers mediate reflex activity. (2) Hypothalamic integration centers interact with both higher and lower centers to orchestrate autonomic, somatic, and endocrine responses. (3) Cortical centers influence autonomic function via connections with the limbic system.

14.9 Most ANS disorders involve abnormalities in smooth muscle control (p. 548)

1. Most autonomic disorders reflect problems with smooth muscle control. Abnormalities in vascular control, such as hypertension, Raynaud's disease, and autonomic dysreflexia, are devastating.

Developmental Aspects of the ANS (pp. 548–549)

1. Preganglionic neurons develop from the neural tube; postganglionic neurons develop from the embryonic neural crest.

2. The efficiency of the autonomic nervous system declines in old age, as reflected by decreased gland secretion, reduced gastrointestinal motility, and slower sympathetic vasomotor responses to changes in position.

REVIEW QUESTIONS

To access additional practice questions using your smartphone, tablet, or computer: MasteringA&P® > Study Area > Practice Tests & Quizzes

Level 1 Remember/Understand

(Some questions have more than one correct answer. Select the best answer or answers from the choices given.)

1. All of the following characterize the ANS except (a) a two-neuron efferent chain, (b) presence of neuron cell bodies in the CNS, (c) presence of neuron cell bodies in the ganglia, (d) innervation of skeletal muscles.

2. Relate each of the following terms or phrases to either the sympathetic (**S**) or parasympathetic (**P**) division of the autonomic nervous system:
   ____ (1) short preganglionic, long postganglionic fibers
   ____ (2) intramural ganglia
   ____ (3) craniosacral part
   ____ (4) adrenergic fibers
   ____ (5) cervical ganglia
   ____ (6) otic and ciliary ganglia
   ____ (7) generally short-duration action
   ____ (8) increases heart rate and blood pressure
   ____ (9) increases gastric motility and secretion of lacrimal, salivary, and digestive juices
   ____ (10) innervates blood vessels
   ____ (11) most active when you are relaxing in a hammock
   ____ (12) active when you are running in the Boston Marathon

3. The white rami communicantes contain what kind of fibers? (a) preganglionic parasympathetic, (b) postganglionic parasympathetic, (c) preganglionic sympathetic, (d) postganglionic sympathetic.

4. Collateral sympathetic ganglia are involved with innervating (a) abdominal organs, (b) thoracic organs, (c) head, (d) arrector pili, (e) all of these.

5. Which ANS fibers release acetylcholine? Which release norepinephrine?

6. Where are the cell bodies of visceral sensory neurons?

Level 2 Apply/Analyze

7. Briefly explain why the following terms are sometimes used to refer to the autonomic nervous system: involuntary nervous system and emotional-visceral system.

8. Describe the anatomical relationship of the white and gray rami communicantes to the spinal nerve, and indicate the kind of fibers found in each ramus type.

9. Indicate the results of sympathetic activation of the following structures: sweat glands, eye pupils, adrenal medullae, heart, bronchioles of the lungs, liver, blood vessels of vigorously working skeletal muscles, blood vessels of digestive viscera, salivary glands.

10. Which of the effects listed in response to question 9 would be reversed by parasympathetic activity?

11. Describe the meaning and importance of sympathetic tone and parasympathetic tone.

12. Describe the importance of the hypothalamus in controlling the autonomic nervous system.

13. Tiffany, a 21-year-old college student, is having trouble sleeping, cries frequently, and has recurrent thoughts of suicide. An antidepressant is prescribed. Like many such drugs, this antidepressant has anticholinergic side effects. What side effects might Tiffany experience in the first week of treatment?

14. As the aroma of freshly brewed coffee drifted by dozing Henry's nose, his mouth started to water and his stomach began to rumble. Explain his reactions in terms of ANS activity.

Level 3 Evaluate/Synthesize

15. Mr. Johnson suffers from urinary retention and a hypoactive urinary bladder. Bethanechol, a drug that mimics acetylcholine's autonomic effects, is prescribed to manage his problem. First explain the rationale for prescribing bethanechol, and then predict which of the following adverse effects Mr. Johnson might experience while taking this drug (select all that apply): dizziness, low blood pressure, deficient tear formation, wheezing, increased mucus production in bronchi, deficient salivation, diarrhea, cramping, excessive sweating, undesirable erection of penis.

16. A 32-year-old woman complains of intermittent aching pains in the medial two fingers of both hands. During such episodes, the fingers become blanched and then blue. Her history is taken, and it is noted that she is a heavy smoker. The physician advises her that she must stop smoking and states that she will not prescribe any medication until the patient has discontinued smoking for a month. What is this patient's condition, and why was she told to stop smoking?

CLINICAL CASE STUDY

10-Year-Old Boy with Spinal Injury

On arrival at Holyoke Hospital, Jimmy Chin, a 10-year-old boy, is immobilized on a rigid stretcher so that he is unable to move his head or trunk. The paramedics report that he was awake and alert after his accident at the skateboard park, but crying and complaining that he couldn't get up and that he had a "wicked headache." He has severe bruises on his upper back and head, and lacerations of his back and scalp. His

blood pressure is low, body temperature is below normal, lower limbs are paralyzed, and he is insensitive to painful stimuli below the nipples. Although still alert on arrival, Jimmy soon begins to drift in and out of unconsciousness.

Jimmy is immediately scheduled for a CT scan, and an operating room is reserved.

Relative to Jimmy's condition:

1. Why were his head and torso immobilized for transport to the hospital?

2. What do his worsening neurological signs (drowsiness, incoherence, etc.) probably indicate? Relate this to the type of surgery that will be performed.

3. Assuming that Jimmy's sensory and motor deficits are due to a spinal cord injury, at what level do you expect to find a spinal cord lesion?

4. **✚ NCLEX-STYLE** Based on your understanding of the anatomy of the nervous system, what functions is Jimmy most likely to lose?
   a. Conscious control of bladder and bowel
   b. Visceral reflexes involving digestion
   c. Reflex control of heart rate
   d. Ability to have an erection

Two days after his surgery, Jimmy is alert and his MRI scan shows no residual brain injury, but does show pronounced swelling and damage to the spinal cord at T_4. On physical examination, Jimmy shows no reflex activity below the level of the spinal cord injury. His blood pressure is still low.

5. Why are there no reflexes in his lower limbs and abdomen? Why is his blood pressure low?

6. Over the next few days, his reflexes return in his lower limbs and become exaggerated. He is incontinent. Why is Jimmy hyperreflexive and incontinent?

On one occasion, Jimmy complains of a massive headache and his blood pressure is far above normal. On examination, he is sweating intensely above the nipples but has cold, clammy skin below the nipples and his heart rate is very slow.

7. **✚ NCLEX-STYLE** Which of the following names his condition and best describes what the nurse should assess for?
   a. This condition is called spinal shock, and the nurse should assess for lack of responses below the level of spinal cord injury.
   b. This condition is called Raynaud's disease, and the nurse should check for blanched fingers and toes.
   c. This condition is called autonomic neuropathy, and the nurse should check for sluggish pupil reactions and impaired sweating.
   d. This condition is called autonomic dysreflexia, and the nurse should assess for an overfilled visceral organ or a painful stimulus to his skin.

8. How does Jimmy's excessively high blood pressure put him at risk?

15

The Special Senses

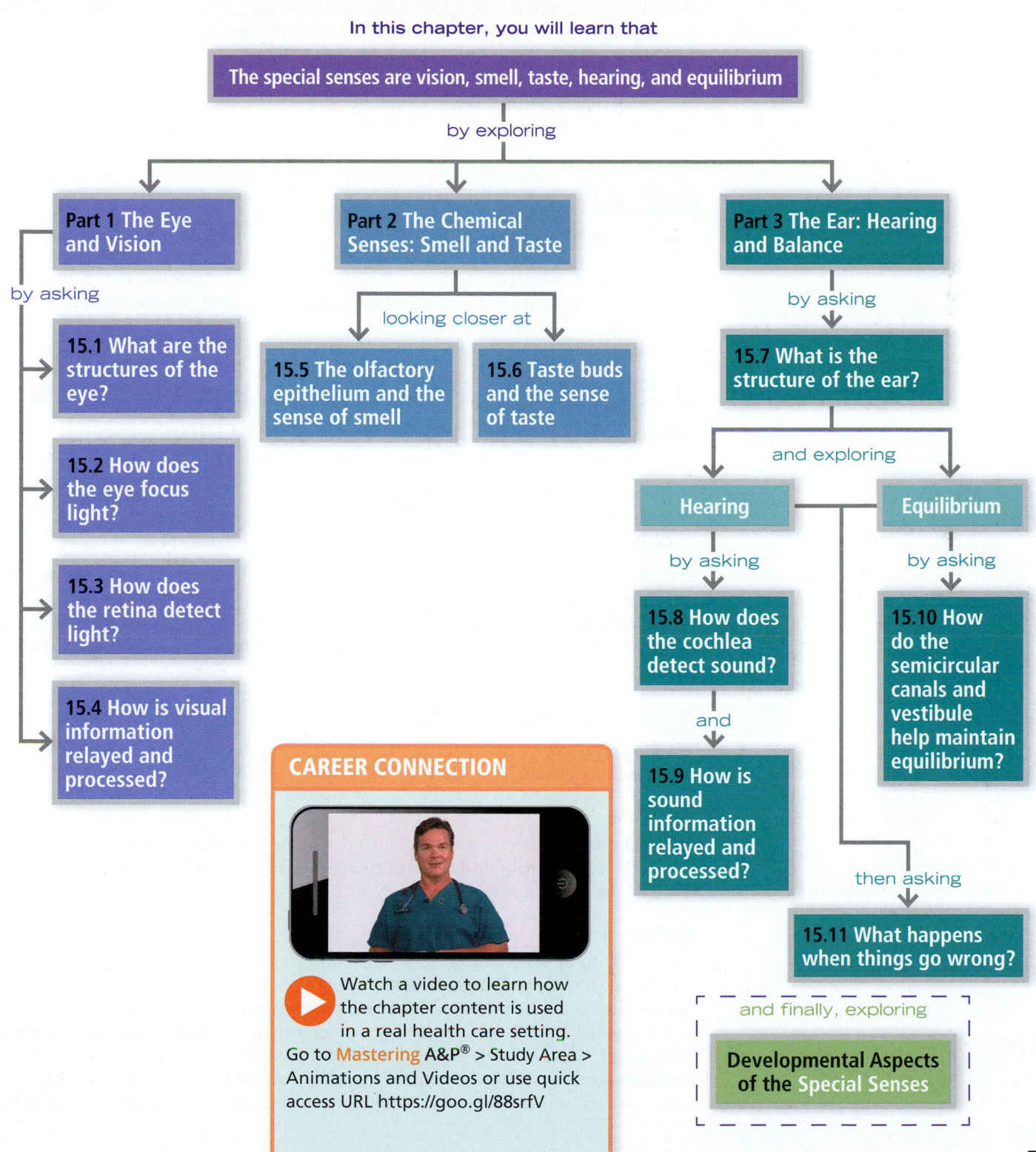

In this chapter, you will learn that

The special senses are vision, smell, taste, hearing, and equilibrium

by exploring

Part 1 The Eye and Vision

Part 2 The Chemical Senses: Smell and Taste

Part 3 The Ear: Hearing and Balance

by asking

15.1 What are the structures of the eye?

15.2 How does the eye focus light?

15.3 How does the retina detect light?

15.4 How is visual information relayed and processed?

looking closer at

15.5 The olfactory epithelium and the sense of smell

15.6 Taste buds and the sense of taste

by asking

15.7 What is the structure of the ear?

and exploring

Hearing

Equilibrium

by asking

15.8 How does the cochlea detect sound?

and

15.9 How is sound information relayed and processed?

by asking

15.10 How do the semicircular canals and vestibule help maintain equilibrium?

then asking

15.11 What happens when things go wrong?

and finally, exploring

Developmental Aspects of the Special Senses

CAREER CONNECTION

Watch a video to learn how the chapter content is used in a real health care setting. Go to Mastering A&P® > Study Area > Animations and Videos or use quick access URL https://goo.gl/88srfV

People are responsive creatures. The aroma of freshly baked bread makes our mouths water. A sudden clap of thunder makes us jump. These stimuli and many others continually greet us and are interpreted by our nervous systems.

When people are asked to list the senses, they usually come up with five: vision, taste, smell, hearing, and touch. Actually, touch reflects the combined activity of the general senses that we considered in Chapter 13. The remaining senses—*vision, taste, smell, hearing,* and *equilibrium*—are called **special senses**. Most of us tend to forget the sense of equilibrium, whose receptors are housed in the ear along with the organ of hearing.

In contrast to the widely distributed general receptors (most of which are modified nerve endings of sensory neurons), the **special sensory receptors** are distinct *receptor cells*. These receptor cells are highly localized in the head, either housed within complex sensory organs (eyes and ears) or in distinct epithelial structures (taste buds and olfactory epithelium).

While we discuss each special sense individually, keep in mind that our perceptions of sensory inputs overlap. What we finally experience—our "feel" of the world—is a blending of stimulus effects.

PART 1

THE EYE AND VISION

Vision is our dominant sense: Some 70% of all the sensory receptors in the body are in the eyes, and nearly half of the cerebral cortex is involved in some aspect of visual processing.

15.1 The eye has three layers, a lens, and humors, and is surrounded by accessory structures

Learning Outcomes

▶ Describe the structure and function of accessory eye structures, eye layers, the lens, and humors of the eye.

▶ Outline the causes and consequences of cataracts and glaucoma.

The adult **eye** is a sphere with a diameter of about 2.5 cm (1 inch). Only the anterior one-sixth of the eye's surface is visible (**Figure 15.1a**). The rest is enclosed and protected by a cushion of fat and the walls of the bony orbit. The fat pad occupies nearly all of the orbit not occupied by the eye itself. Before turning our attention to the eye itself, let's consider the accessory structures that protect it or aid its function.

Accessory Structures of the Eye

The **accessory structures** of the eye include the eyebrows, eyelids, conjunctiva, lacrimal apparatus, and extrinsic eye muscles.

Eyebrows

The **eyebrows** are short, coarse hairs that overlie the supraorbital margins of the skull (Figure 15.1). They help shade the eyes from sunlight and prevent perspiration trickling down the forehead from reaching the eyes.

Eyelids

Anteriorly, the eyes are protected by the **eyelids** or **palpebrae** (pal′pĕ-bre). The eyelids are separated by the **palpebral fissure** ("eyelid slit") and meet at the medial and lateral angles of the eye—the **medial** and **lateral commissures** (*canthi*) (Figure 15.1a).

The medial commissure sports a fleshy elevation called the **lacrimal caruncle** (kar′ung-kl; "a bit of flesh"). The caruncle contains sebaceous and sweat glands and produces the whitish, oily secretion (fancifully called the Sandman's eye-sand) that sometimes collects at the medial commissure, especially during sleep. In most Asian

Eyebrow

Eyelid

Eyelashes

Site where
conjunctiva
merges with
cornea

Palpebral
fissure

Lateral
commissure

Iris

Eyelid

Pupil   Sclera   Lacrimal   Medial
(covered by caruncle commissure
conjunctiva)

(a) Surface anatomy of the right eye

Levator palpebrae
superioris muscle

Orbicularis
oculi muscle

Eyebrow

Tarsal plate

Palpebral
conjunctiva

Tarsal glands

Cornea

Palpebral
fissure

Eyelashes

Bulbar
conjunctiva

Conjunctival
sac

Orbicularis
oculi muscle

(b) Lateral view; some structures shown in sagittal section

Figure 15.1 The eye and accessory structures.

peoples, a vertical fold of skin called the *epicanthic fold* commonly appears on both sides of the nose and sometimes covers the medial commissure.

The eyelids are thin, skin-covered folds supported internally by connective tissue sheets called **tarsal plates** (Figure 15.1b). The tarsal plates also anchor the **orbicularis oculi** and **levator palpebrae superioris** muscles that run within the eyelid. The orbicularis muscle encircles the eye, and the eye closes when it contracts. The upper eyelid is much more mobile, mainly because of the levator palpebrae superioris muscle, which raises that eyelid to open the eye.

The eyelid muscles are activated reflexively to cause blinking every 3–7 seconds and to protect the eye from foreign objects. Each time we blink, accessory structure secretions (oil, mucus, and saline solution) spread across the eyeball surface, keeping the eyes moist.

Projecting from the free margin of each eyelid are the **eyelashes**. The follicles of the eyelash hairs are richly innervated by nerve endings (hair follicle receptors), and anything that touches the eyelashes (even a puff of air) triggers reflex blinking.

Several types of glands are associated with the eyelids. The **tarsal glands** are embedded in the tarsal plates (Figure 15.1b), and their ducts open at the eyelid edge just posterior to the eyelashes. These modified sebaceous glands produce an oily secretion that lubricates the eyelid and the eye and prevents the eyelids from sticking together. A number of smaller, more typical sebaceous glands are associated with the eyelash follicles. Modified sweat glands called ciliary glands lie between the hair follicles (*cilium* = eyelash).

HOMEOSTATIC IMBALANCE 15.1   CLINICAL

Obstructed tarsal glands may cause a firm, usually painless bump called a *chalazion* (kah-la′ze-on; "swelling"). A **sty** is a painful inflammation of the sebaceous glands at the base of the eyelashes.   ●●●●●●

Conjunctiva

The **conjunctiva** (kon″junk-ti′vah; "joined together") is a transparent mucous membrane. It lines the eyelids as the **palpebral conjunctiva** and folds back over the anterior surface of the eyeball as the **bulbar conjunctiva** (Figure 15.1b). The bulbar conjunctiva covers only the white of the eye, not the cornea. The bulbar conjunctiva is very thin, and blood vessels are clearly visible beneath it. (They are even more visible in irritated "bloodshot" eyes.)

When the eye is closed, a slitlike space occurs between the conjunctiva-covered eyeball and eyelids. This so-called **conjunctival sac** is where a contact lens lies, and eye medications are often administered into its inferior recess. The major function of the conjunctiva is to produce a lubricating mucus that prevents the eyes from drying out.

HOMEOSTATIC IMBALANCE 15.2   CLINICAL

Inflammation of the conjunctiva, called *conjunctivitis*, results in reddened, irritated eyes. *Pinkeye*, a conjunctival infection caused by bacteria or viruses, is highly contagious.   ●●●●●●

Lacrimal Apparatus

The **lacrimal apparatus** (lak′rĭ-mal; "tear") consists of the lacrimal gland and the ducts that drain lacrimal secretions into the nasal cavity (**Figure 15.2**). The **lacrimal gland** lies in the orbit above the lateral end of the eye and is visible through the conjunctiva when the lid is everted. It continually releases a dilute saline solution called **lacrimal secretion**—or, more commonly, **tears**—into the superior part of the conjunctival sac through several small excretory ducts.

Blinking spreads the tears downward and across the eyeball to the medial commissure, where they enter the paired **lacrimal canaliculi** via two tiny openings called **lacrimal puncta** (literally, "prick points"), visible as tiny red dots on the medial margin of each eyelid. From the lacrimal canaliculi, the tears drain into the **lacrimal sac** and then into the **nasolacrimal duct**, which empties into the nasal cavity at the inferior nasal meatus.

Lacrimal fluid contains mucus, antibodies, and **lysozyme**, an enzyme that destroys bacteria. As a result, it cleanses and protects the eye surface as it moistens and lubricates it. When lacrimal secretion increases substantially, tears spill over the eyelids and fill the nasal cavities, causing congestion and the "sniffles." This spillover (tearing) happens when the eyes are irritated or when we are emotionally upset. In the case of eye irritation, enhanced tearing washes away or dilutes the irritating substance. The importance of emotionally induced tears is poorly understood.

① The **lacrimal gland** produces and secretes tears (lacrimal secretions).

② Tears enter the conjunctival sac via the **excretory ducts of the lacrimal gland.**

③ Tears flow down and across the eyeball.

④ Tears enter the **lacrimal canaliculi** at openings called **lacrimal puncta.**

⑤ Tears drain into the **lacrimal sac.**

⑥ From the **lacrimal sac, tears empty via the **nasolacrimal duct** into the inferior meatus of the nasal cavity.

Lacrimal punctum

Lacrimal canaliculus

Inferior meatus of nasal cavity

Nostril

Figure 15.2 The lacrimal apparatus. Arrows indicate the flow of lacrimal fluid (tears) from the lacrimal gland to the nasal cavity.

HOMEOSTATIC IMBALANCE 15.3   **CLINICAL**

Because the nasal cavity mucosa is continuous with that of the lacrimal duct system, a cold or nasal inflammation often causes the lacrimal mucosa to swell. This swelling constricts the ducts and prevents tears from draining, causing "watery" eyes. ●●●●●●●

Extrinsic Eye Muscles

How do our eyes move? Six straplike **extrinsic eye muscles** control the movement of each eyeball. These muscles originate from the walls of the orbit and insert into the outer surface of the eyeball (**Figure 15.3**). They allow the eyes to follow a moving object, help maintain the shape of the eyeball, and hold it in the orbit.

The four *rectus muscles* originate from the **common tendinous ring** (annular ring) at the back of the orbit and run straight to their insertion on the eyeball. Their locations and the movements that they promote are clearly indicated by their names: **superior, inferior, lateral,** and **medial rectus muscles.**

The actions of the two *oblique muscles* are less easy to deduce because they take rather strange paths through the orbit. They move the eye in the vertical plane when the eyeball is already turned medially by the rectus muscles. The **superior oblique muscle** originates in common with the rectus muscles, runs along the medial wall of the orbit, and then makes a right-angle turn and passes through a fibrocartilaginous loop called the **trochlea** (trok′le-ah; "pulley") suspended from the frontal bone before inserting on the superolateral aspect of the eyeball. It rotates the eye downward and somewhat laterally.

The **inferior oblique muscle** originates from the medial orbit surface and runs laterally and obliquely to insert on the inferolateral eye surface. It rotates the eye up and laterally.

The four rectus muscles would seem to provide all of the eye movements we require—medial, lateral, superior, and inferior—so why the two oblique muscles? The simplest answer is that the superior and inferior recti cannot elevate or depress the eye *without also turning it medially* because they approach the eye from a posteromedial direction. For an eye to be *directly* elevated or depressed, the lateral pull of the oblique muscles is necessary to cancel the medial pull of the superior and inferior recti.

Except for the lateral rectus and superior oblique muscles, which are innervated respectively by the *abducens* and *trochlear nerves*, the *oculomotor nerves* serve all extrinsic eye muscles.

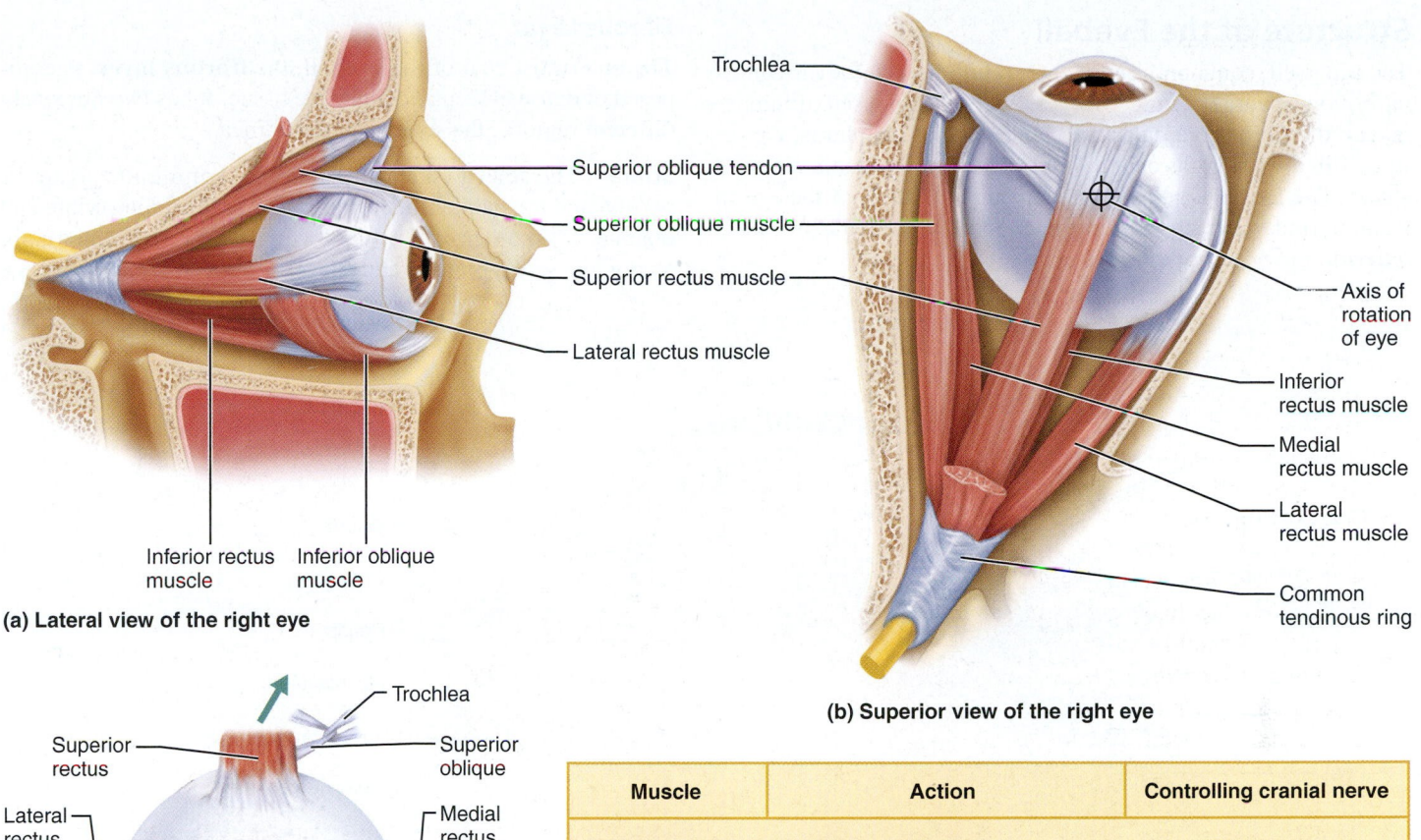

Trochlea

Superior oblique tendon

Superior oblique muscle

Superior rectus muscle

Lateral rectus muscle

Inferior rectus muscle | Inferior oblique muscle

(a) Lateral view of the right eye

Axis of rotation of eye

Inferior rectus muscle

Medial rectus muscle

Lateral rectus muscle

Common tendinous ring

(b) Superior view of the right eye

Trochlea

Superior rectus

Superior oblique

Lateral rectus

Medial rectus

Inferior oblique

Inferior rectus

(c) Anterior view of the right eye

Muscle	Action	Controlling cranial nerve
Lateral rectus	Moves eye laterally	VI (abducens)
Medial rectus	Moves eye medially	III (oculomotor)
Superior rectus	Elevates eye and turns it medially	III (oculomotor)
Inferior rectus	Depresses eye and turns it medially	III (oculomotor)
Inferior oblique	Elevates eye and turns it laterally	III (oculomotor)
Superior oblique	Depresses eye and turns it laterally	IV (trochlear)

(d) Summary of muscle actions and innervating cranial nerves

Figure 15.3 Extrinsic eye muscles.

Figure 15.3d summarizes the actions and nerve supply of these muscles. Table 13.2 (◄ pp. **501, 503**) illustrates the courses of the associated cranial nerves.

The extrinsic eye muscles are among the most precisely and rapidly controlled skeletal muscles in the entire body. This precision reflects their high axon-to-muscle-fiber ratio: The motor units of these muscles contain only 8 to 12 muscle cells and in some cases as few as two or three.

HOMEOSTATIC IMBALANCE 15.4   **CLINICAL**

When movements of the external muscles of the two eyes are not perfectly coordinated, a person cannot properly focus the images of the same area of the visual field from each eye and so sees two images instead of one. This condition is called **diplopia** (dĭ-plo′pe-ah), or *double vision*. It can result from paralysis or weakness of certain extrinsic muscles, or neurological disorders.

Congenital weakness of the external eye muscles may cause **strabismus** (strah-biz′mus; "squinting"), in which the affected eye rotates medially or laterally. To compensate, the eyes may alternate in focusing on objects. In other cases, only the controllable eye is used, and the brain begins to disregard inputs from the deviant eye, which (unless treated early) can then become functionally blind.

In addition to the extrinsic eye muscles, the eye also contains **intrinsic eye muscles**—the ciliary body and the iris—that are controlled by the autonomic nervous system. We will discuss these muscles shortly.

Structure of the Eyeball

The eye itself, commonly called the **eyeball**, is a slightly irregular hollow sphere (**Figure 15.4**). Its wall is composed of three layers: the fibrous, vascular, and inner layers. Its internal cavity is filled with fluids called *humors* that help to maintain its shape. The lens, the adjustable focusing apparatus of the eye, is supported vertically within the eyeball, and divides it into *anterior* and *posterior segments*.

Fibrous Layer

The outermost coat of the eyeball, the **fibrous layer**, is composed of dense avascular connective tissue. It has two obviously different regions: the sclera and the cornea.

Sclera The **sclera** (skler'ah; "hard"), forming the posterior portion and the bulk of the fibrous layer, is glistening white and opaque. Seen anteriorly as the "white of the eye," the tough, tendonlike sclera protects and shapes the eyeball and provides a

Ora serrata
Ciliary body
Ciliary zonule (suspensory ligament)
Cornea
Iris
Pupil
Anterior segment (contains aqueous humor)
Lens
Scleral venous sinus
Posterior segment (contains vitreous humor)

Sclera
Choroid
Retina
Macula lutea
Fovea centralis
Optic nerve
Central artery and vein of the retina
Optic disc (blind spot)

(a) Diagrammatic view. The vitreous humor is illustrated only in the bottom part of the eyeball.

Ciliary body
Ciliary processes
Iris
Margin of pupil
Anterior segment
Lens
Cornea
Ciliary zonule (suspensory ligament)

Vitreous humor in posterior segment
Retina
Choroid
Sclera
Fovea centralis
Optic disc
Optic nerve

(b) Photograph of the human eye.

Figure 15.4 Internal structure of the eye (sagittal section).

sturdy anchoring site for the extrinsic eye muscles. Posteriorly, where the sclera is pierced by the optic nerve (cranial nerve II), it is continuous with the dura mater of the brain.

Cornea   The anterior sixth of the fibrous layer is modified to form the transparent **cornea**, which bulges anteriorly from its junction with the sclera. The crystal-clear cornea forms a window that lets light enter the eye, and is a major part of the light-bending apparatus of the eye.

Epithelial sheets cover both faces of the cornea:

- The external sheet, a stratified squamous epithelium that protects the cornea from abrasion, merges with the bulbar conjunctiva at the corneoscleral junction. Epithelial cells that continually renew the cornea are located here.

- The deep *corneal endothelium*, composed of simple squamous epithelium, lines the inner face of the cornea. Its cells have active sodium pumps that maintain the clarity of the cornea by keeping its water content low.

The cornea is well supplied with nerve endings, most of which are pain receptors. (This is why some people can't tolerate contact lenses.) When the cornea is touched, blinking and increased tearing occur reflexively. Even so, the cornea is the most exposed part of the eye and is vulnerable to damage from dust, slivers, and the like. Luckily, its capacity for regeneration and repair is extraordinary.

The cornea has no blood vessels and so it is beyond the reach of the immune system. As a result, the cornea is the only tissue in the body that can be transplanted from one person to another with little risk of rejection.

Vascular Layer

The **vascular layer** forms the middle coat of the eyeball. Also called the *uvea* (u've-ah; "grape"), this pigmented layer has three regions: choroid, ciliary body, and iris (Figure 15.4).

Choroid   The **choroid** is a blood vessel–rich, dark brown membrane (*choroid* = membrane) that forms the posterior five-sixths of the vascular layer. Its blood vessels nourish all eye layers. Its brown pigment, produced by melanocytes, helps absorb light, preventing it from scattering and reflecting within the eye (which would cause visual confusion). The choroid has a posterior opening where the optic nerve leaves the eye.

Ciliary Body   Anteriorly, the choroid becomes the **ciliary body**, a thickened ring of tissue that encircles the lens and has three parts:

- **Ciliary muscles. Ciliary muscles** make up most of the ciliary body and consist of interlacing smooth muscle bundles that control lens shape.

- **Ciliary processes.** Near the lens, the posterior surface of the ciliary body has radiating folds called **ciliary processes**, which secrete the fluid that fills the cavity of the anterior segment of the eyeball.

- **Ciliary zonule.** The **ciliary zonule** (*suspensory ligament*) extends from the ciliary processes to the lens. This halo of

Parasympathetic +		Sympathetic +
Sphincter pupillae muscle contracts: Pupil constricts (size decreases).	**Iris** (two muscles) • Sphincter pupillae • Dilator pupillae	Dilator pupillae muscle contracts: Pupil dilates (size increases).

Figure 15.5 Pupil constriction and dilation.
(+ means activation.)

fine fibers encircles the circumference of the lens and helps hold it in its upright position. It also transmits tension from the ciliary muscle to the lens.

Iris   The **iris**, the colored part of the eye, is the most anterior portion of the vascular layer. Shaped like a flattened doughnut, it lies between the cornea and the lens and is continuous with the ciliary body posteriorly. Its round central opening, the **pupil**, allows light to enter the eye. The iris is made up of two smooth muscle layers with bunches of sticky elastic fibers that congeal into a random pattern before birth. Its muscle fibers allow it to act as a reflexively activated diaphragm to vary pupil size (**Figure 15.5**). In close vision and bright light, the *sphincter pupillae* (circular muscles) contracts and the pupil constricts. In distant vision and dim light, the *dilator pupillae* (radial muscles) contracts and the pupil dilates, allowing more light to enter. Sympathetic fibers control pupillary dilation, and parasympathetic fibers control constriction.

Changes in pupil size may also reflect our interests and emotional reactions. Our pupils often dilate when we see something that appeals to us, when we feel fear, and during problem solving. (Computing your taxes should make your pupils get bigger and bigger.) On the other hand, boredom or viewing something unpleasant causes pupils to constrict.

Although irises come in different colors (*iris* = rainbow), they contain only brown pigment. When they have a lot of pigment, the eyes appear brown or black. If the amount of pigment is small and restricted to the posterior surface of the iris, the unpigmented parts simply scatter the shorter wavelengths of light and the eyes appear blue, green, or gray. Most newborn babies' eyes are slate gray or blue because their iris pigment is not yet developed.

Inner Layer (Retina)

The innermost layer of the eyeball is the delicate **retina** (ret'ĭ-nah), which develops from an extension of the brain. It contains (1) millions of photoreceptors that transduce (convert) light energy, (2) other neurons involved in processing responses to light, and (3) glia. The retina consists of two layers: an outer

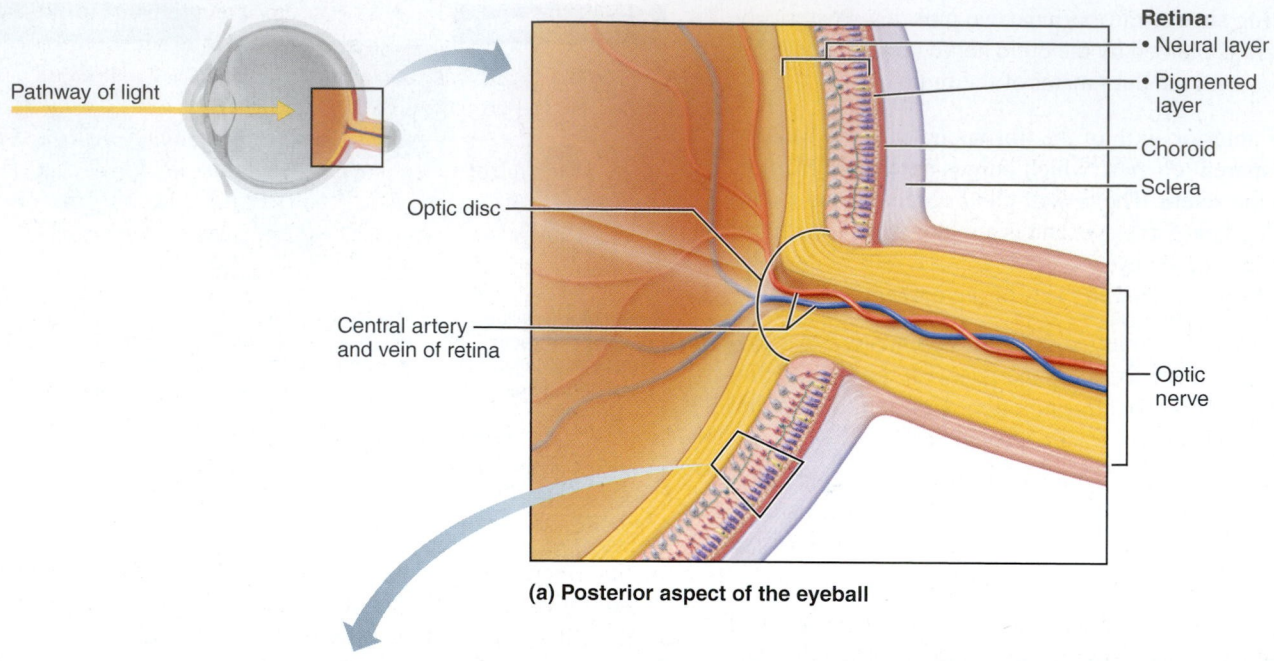

Pathway of light

Retina:
• Neural layer
• Pigmented layer

Choroid

Sclera

Optic disc

Central artery and vein of retina

Optic nerve

(a) Posterior aspect of the eyeball

Ganglion cells

Bipolar cells

Photoreceptors
• Rod
• Cone

Axons of ganglion cells

Amacrine cell

Horizontal cell

Pathway of signal output

Pathway of light

Pigmented layer of retina

(b) Cells of the neural layer of the retina

Nuclei of ganglion cells

Outer segments of rods and cones

Choroid

Axons of ganglion cells

Nuclei of bipolar cells

Nuclei of rods and cones

Pigmented layer of retina

(c) Photomicrograph of retina

Figure 15.6 Microscopic anatomy of the retina. (a) The axons of the ganglion cells form the optic nerve, which leaves the back of the eyeball at the optic disc. **(b)** Light (indicated by the yellow arrow) passes through the retina to excite the photoreceptor cells (rods and cones). Information (output signals; gray arrow) flows in the opposite direction via bipolar and ganglion cells. **(c)** Photomicrograph (145×).

Practice Histology questions: MasteringA&P® > Study Area > Lab Tools > PAL

pigmented layer and an inner *neural layer* (**Figure 15.6**). Although the pigmented and neural layers are very close together, they are not fused. Only the neural layer of the retina plays a direct role in vision.

Pigmented Layer of the Retina   The outer **pigmented layer**, a single-cell-thick lining, is next to the choroid, and extends anteriorly to cover the ciliary body and the posterior face of the iris. These pigment cells, like those of the choroid, absorb light and prevent it from scattering in the eye. They also act as phagocytes participating in photoreceptor cell renewal (described on p. 569), and store vitamin A needed by the photoreceptor cells.

Neural Layer of the Retina   The transparent inner **neural layer** extends anteriorly to the posterior margin of the ciliary body. This junction is called the **ora serrata**, literally, the sawtoothed margin (see Figure 15.4).

From posterior to anterior, the neural layer is composed of three main types of neurons: **photoreceptors**, **bipolar cells**, and **ganglion cells** (Figure 15.6). Signals are produced in response to light and spread from the photoreceptors (next to the pigmented layer) to the bipolar cells and then to the innermost ganglion cells, where action potentials are generated. The ganglion cell axons make a right-angle turn at the inner face of the retina, then leave the posterior aspect of the eye as the thick optic nerve. The retina also contains other types of neurons—horizontal cells and amacrine cells—which play a role in visual processing (see p. 575).

The **optic disc**, where the optic nerve exits the eye, is a weak spot in the **fundus** (posterior wall) of the eye because it is not reinforced by the sclera. The optic disc is also called the **blind spot** because it lacks photoreceptors, so light focused on it cannot be seen. We do not usually notice these gaps in our vision because the brain uses a sophisticated process called *filling in* to deal with absence of input.

The quarter-billion photoreceptors found in the neural layer are of two types: rods and cones.

- **Rods** are our dim-light and peripheral vision receptors. They are more numerous and far more sensitive to light than cones, but they do not provide sharp images or color vision. This is why colors disappear and the edges of objects appear fuzzy in dim light and at the edges of our visual field.

- **Cones**, in contrast, are our vision receptors for bright light and provide high-resolution color vision.

Lateral to the blind spot of each eye is an oval region called the **macula lutea** (mak′u-lah lu′te-ah; "yellow spot") with a minute (0.4 mm) pit in its center called the **fovea centralis** (see Figure 15.4). In this region, the retinal structures next to the vitreous humor are displaced to the sides. This allows light to pass almost directly to the photoreceptors rather than through several retinal layers, greatly enhancing visual acuity (the ability to resolve detail). The fovea contains only cones, the macula contains mostly cones, and from the edge of the macula toward the retina periphery, cone density declines gradually. The retina periphery contains mostly rods, which decrease in density from there to the macula.

Only the foveae have a sufficient cone density to provide detailed color vision, so anything we wish to view critically is focused on the foveae. Because each fovea is only about the size of the head of a pin, only a thousandth of the entire visual field is in *hard focus* (foveal focus) at a given moment. Consequently, for us to visually comprehend a scene that is rapidly changing (as when we drive in traffic), our eyes must flick rapidly back and forth to provide the foveae with images of different parts of the visual field.

The neural layer of the retina receives its blood supply from two sources. Vessels in the choroid supply the outer third (containing photoreceptors). The inner two-thirds is served by the **central artery** and **central vein of the retina**, which enter and leave the eye through the center of the optic nerve (see Figure 15.4a). Radiating outward from the optic disc, these vessels give rise to a rich vascular network. This is the only place where small blood vessels are visible in a living person (**Figure 15.7**). Physicians may observe these tiny vessels with an ophthalmoscope for signs of hypertension, diabetes, and other vascular diseases.

Lateral   Medial

Figure 15.7 Part of the posterior wall (fundus) of the right eye as seen with an ophthalmoscope.

Central artery and vein emerging from the optic disc

Optic disc

Macula lutea

Retina

 HOMEOSTATIC IMBALANCE 15.5   **CLINICAL**

The pattern of vascularization of the retina makes it susceptible to *retinal detachment*. This condition, in which the pigmented and neural layers separate (detach) and allow the jellylike vitreous humor to seep between them, can cause permanent blindness because it deprives the photoreceptors of nutrients.

Retinal detachment may happen when the retina is torn during a traumatic blow to the head or when the head stops moving suddenly and then jerks in the opposite direction (as in bungee jumping). The symptom that victims most often describe is "a curtain being drawn across the eye," but some people see soot-like spots or light flashes. If diagnosed early, it is often possible to reattach the retina with a laser before photoreceptors are permanently damaged.

Cornea — — Lens

Iris
Lens epithelium
Lens
Cornea
Corneal epithelium
Corneal endothelium
Aqueous humor

Anterior segment (contains aqueous humor)
⌈ Anterior chamber
⌊ Posterior chamber

Posterior segment (contains vitreous humor)

Ciliary zonule (suspensory ligament)

Scleral venous sinus
Corneoscleral junction

Ciliary processes

Ciliary body

Ciliary muscle

Bulbar conjunctiva
Sclera

① Aqueous humor forms by filtration from the capillaries in the ciliary processes.

② Aqueous humor flows from the posterior chamber through the pupil into the anterior chamber. Some also flows through the vitreous humor (not shown).

③ Aqueous humor is reabsorbed into the venous blood by the scleral venous sinus.

Figure 15.8 Circulation of aqueous humor. The arrows indicate the circulation pathway.

15

Internal Chambers and Fluids

As we noted earlier, the lens and its halolike ciliary zonule divide the eye into two segments, the anterior segment in front of the lens and the larger posterior segment behind it (**Figure 15.8** and Figure 15.4a). The **posterior segment** is filled with a clear gel called **vitreous humor** (*vitre* = glassy) that binds tremendous amounts of water. Vitreous humor:

*   Transmits light

*   Supports the posterior surface of the lens and holds the neural layer of the retina firmly against the pigmented layer

*   Contributes to intraocular pressure, helping to counteract the pulling force of the extrinsic eye muscles

Vitreous humor forms in the embryo and lasts for a lifetime.

The iris divides the **anterior segment** into the **anterior chamber** (between the cornea and the iris) and the **posterior chamber** (between the iris and the lens) (Figure 15.8). The *entire* anterior segment is filled with **aqueous humor**, a clear fluid similar in composition to blood plasma. Unlike the vitreous humor, aqueous humor forms and drains continually. In Figure 15.8, you can follow its movement from the posterior chamber to the **scleral venous sinus**, an unusual venous channel that encircles the eye in the angle at the corneoscleral junction.

Normally, aqueous humor forms and drains at the same rate, maintaining a constant intraocular pressure of about 16 mm Hg, which helps to support the eyeball internally. Aqueous humor supplies nutrients and oxygen to the lens and cornea and to some cells of the retina, and it carries away metabolic wastes.

HOMEOSTATIC IMBALANCE 15.6   CLINICAL

If the drainage of aqueous humor is blocked, fluid backs up as in a clogged sink. Pressure within the eye may increase to dangerous levels and compress the retina and optic nerve—a condition called **glaucoma** (glaw-ko'mah). The eventual result is blindness unless the condition is detected and treated early. Unfortunately, many forms of glaucoma steal sight so slowly and painlessly that people do not realize they have a problem until the damage is done. Late signs include seeing halos around lights and blurred vision.

The glaucoma examination is simple. The intraocular pressure is determined by directing a puff of air at the cornea and measuring the amount of corneal deformation it causes. This exam should be done yearly after age 40. The most common treatment is eye drops that increase the rate of aqueous humor drainage or decrease its production. Laser therapy or surgery can also help. •••••••

Lens

The **lens** is convex on both sides. It is a transparent, flexible structure that can change shape to precisely focus light on the retina. It is enclosed in a thin, elastic capsule and held in place just posterior to the iris by the ciliary zonule (Figure 15.8). Like the cornea, the lens is avascular; blood vessels interfere with transparency.

The lens has two regions: the **lens epithelium** and the lens fibers. The lens epithelium, confined to the anterior lens surface, consists of cuboidal cells that eventually differentiate into the **lens fibers** that form the bulk of the lens. The lens fibers, which are packed tightly together like the layers in an onion, contain no nuclei and few organelles. They do, however, contain transparent, precisely folded proteins called **crystallins** that form the body of the lens. Since new lens fibers are continually added, the lens enlarges throughout life, becoming denser, more

Figure 15.9 Photograph of a cataract. The lens, not the cornea, is milky and opaque.

convex, and less elastic, all of which gradually impair its ability to focus light properly.

HOMEOSTATIC IMBALANCE 15.7   CLINICAL

A **cataract** ("waterfall") is a clouding of the lens that causes the world to appear distorted, as if seen through frosted glass (**Figure 15.9**). Some cataracts are congenital, but most result from age-related hardening and thickening of the lens or are a secondary consequence of diabetes mellitus. Heavy smoking, frequent exposure to intense sunlight, and steroid use increase the risk for cataracts.

Oxidative stress and metabolic changes in the deeper lens fibers promote clumping of the crystallin proteins. Fortunately, the offending lens can be surgically removed and an artificial lens implanted to save the patient's sight. •••••••

Check Your Understanding

1. What are tears and what structure secretes them?
2. What is the blind spot and why is it blind?
3. **DRAW** On the drawing of the retina below, add an arrow showing the path that light takes to the photoreceptors. Add another arrow to show the path that information takes to the brain. Identify cells a–d and layer e. What layer is superficial to e?

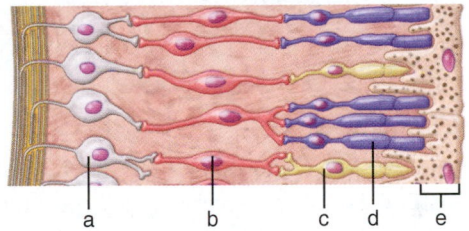

For answers, see Answers Appendix.

15.2 The cornea and lens focus light on the retina

Learning Outcomes

▶ Trace the pathway of light through the eye to the retina, and explain how light is focused for distant and close vision.

▶ Outline the causes and consequences of astigmatism, myopia, hyperopia, and presbyopia.

Overview: Light and Optics

To comprehend the function of the eye as a photoreceptor organ, we need to understand the properties of light.

Wavelength and Color

Electromagnetic radiation includes all energy waves, from long radio waves (with wavelengths measured in meters) to very short waves [gamma (γ) waves and X rays with wavelengths of 1 nm and less]. (1 nm = 10^{-9} m, or one-billionth of a meter.) Our

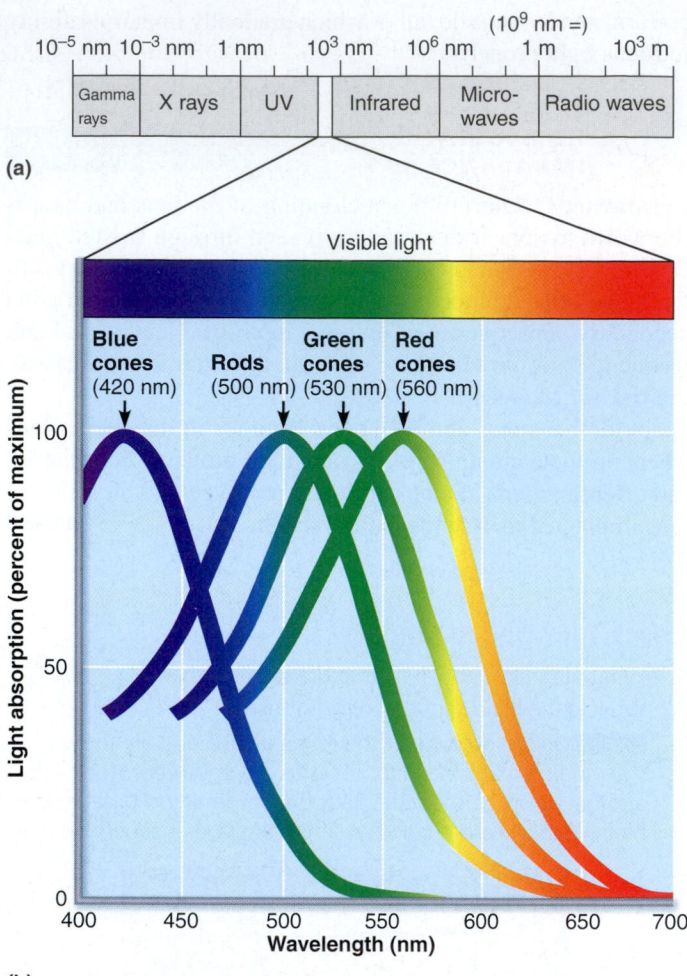

(a)

(b)

Figure 15.10 The electromagnetic spectrum and photo-receptor sensitivities. (a) The electromagnetic spectrum, of which visible light constitutes only a small portion. (nm = nanometers.) **(b)** Sensitivities of rods and the three cone types to the different wavelengths of the visible spectrum.

eyes respond to the part of the spectrum called **visible light**, which has a wavelength range of approximately 400–700 nm (**Figure 15.10a**).

Visible light travels in the form of waves, and its wavelengths can be measured very accurately. Light can also be envisioned as packets of energy, called **photons**, traveling in a wavelike fashion at very high speeds (300,000 km/s or about 186,000 mi/s). We can think of light as a vibration of pure energy ("a bright wiggle").

Visible light contains a mixture of wavelengths. When light passes through a prism, each of its component waves bends to a different degree. This disperses the beam of light into a **visible spectrum**, or band of colors (Figure 15.10b). (A rainbow that appears during a summer shower represents the collective prismatic effects of all the tiny water droplets suspended in air.) Red wavelengths are the longest and have the lowest energy, while violet wavelengths are the shortest and most energetic.

Like sound, light can reflect, or bounce, off a surface. This *reflection* of light by objects in our environment accounts for most of the light reaching our eyes. Objects have color because

they absorb some wavelengths and reflect others. A red apple reflects mostly red light, while grass reflects more of the green. Things that look white reflect all wavelengths of light, whereas black objects absorb them all.

Refraction and Lenses

Light travels in straight lines (rays) and is blocked by any non-transparent object. When light travels in a given medium, its speed is constant. But when it passes from one transparent medium into another with a different density, its speed changes. Light speeds up as it passes into a less dense medium and slows as it passes into a denser medium. Because of these changes in speed, bending or **refraction** of a light ray occurs when it meets the surface of a different medium at an oblique angle rather than at a right angle. The greater this angle, the greater the amount of bending. **Figure 15.11** shows refraction: A straw in a glass of water appears to break at the air-water interface.

A lens is a transparent object curved on one or both surfaces. Since light hits the curve at an angle, it is refracted. If the lens surface is convex, that is, thickest in the center like a camera lens, the light rays bend so that they converge or intersect to focus at a single point (**Figure 15.12a**). If the light is from a distant source, the rays are nearly parallel and the point at which they focus is called the **focal point**. In general, the more convex the lens, the more the light bends and the shorter the focal distance (distance between the lens and focal point). The image formed by a convex lens, called a **real image**, is inverted—upside down and reversed from left to right (Figure 15.12b).

Concave lenses, which are thicker at the edges than at the center, diverge the light (bend it outward) so that the light rays move away from each other. Consequently, concave lenses prevent light from focusing and extend the focal distance.

Figure 15.11 Refraction. A straw standing in a glass of water appears to be broken at the water-air interface. This occurs because light bends when it travels from a less dense medium (such as air) to a more dense medium (such as water).

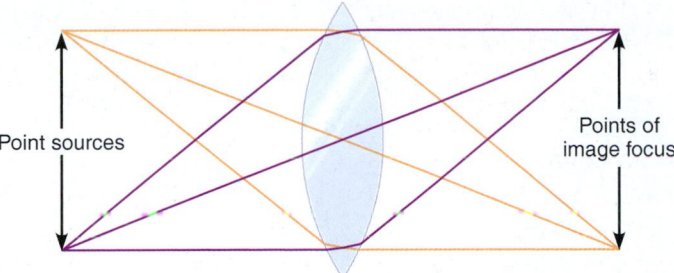

(a) Focusing of two points of light.

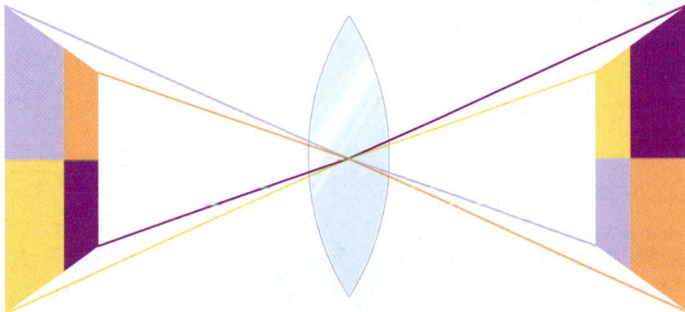

(b) The image is inverted—upside down and reversed.

Figure 15.12 Light is focused by a convex lens. The lens bends, or refracts, light rays so they converge to a point.

Focusing Light on the Retina

As light passes from air into the eye, it moves sequentially through the cornea, aqueous humor, lens, and vitreous humor, and then passes *through the entire neural layer of the retina* to excite the photoreceptors that are next to the pigmented layer (see Figures 15.4 and 15.6). During its passage, light is bent three times: (1) entering the cornea, (2) entering the lens, and (3) leaving the lens.

The cornea accounts for the majority of the refractory power of the eye. However, the refractory power of the cornea is constant. On the other hand, the lens is highly elastic, and its curvature and light-bending power can actively change to allow fine focusing.

Focusing for Distant Vision

Our eyes are best adapted ("preset to focus") for distant vision. To look at distant objects, we need only aim our eyeballs so that they are both fixated on the same spot. The **far point of vision** is that distance beyond which no change in lens shape (accommodation) is needed for focusing. For the normal or **emmetropic** (em″ĕ-tro′pik) eye, the far point is 6 m (20 feet).

Any object being viewed can be said to consist of many small points, with light radiating outward in all directions from each point. However, because distant objects appear smaller, light from an object at or beyond the far point of vision approaches the eyes as nearly parallel rays. The cornea and the at-rest lens focus the light from these distant objects precisely on the retina (**Figure 15.13b**).

During distant vision, the sphincterlike ciliary muscles are completely relaxed, and tension in the ciliary zonule stretches the lens flat. Consequently, the lens is as thin as it gets and is at its lowest refractory power when the ciliary muscle is at rest. The ciliary muscles relax when sympathetic input to them increases and parasympathetic input decreases.

Focusing for Close Vision

Light from close objects (less than 6 m away) diverges as it approaches the eyes and comes to a focal point farther from the

(a) The ciliary muscle and the ciliary zonule focus an image by changing the shape of the lens.
• They are arranged sphincterlike around the lens.
• Ciliary muscle contraction loosens the ciliary zonule fibers and relaxation tightens them.

View

Ciliary muscle
Lens
Ciliary zonule (suspensory ligament)

(b) The lens flattens for distant vision.
Sympathetic input relaxes the ciliary muscle. This tightens the ciliary zonule and flattens the lens.

Relaxed ciliary muscle
Tightened ciliary zonule
Flattened lens

Nearly parallel rays from distant object
Image

(c) The lens bulges for close vision.
Parasympathetic input contracts the ciliary muscle. This loosens the ciliary zonule and allows the lens to bulge.

Contracted ciliary muscle
Loosened ciliary zonule
Bulging lens

Divergent rays from close object
Image

Figure 15.13 Focusing for distant and close vision.

Figure 15.14 Problems of refraction. The refractive power of the cornea, which normally supplies about two-thirds of the light-bending power of the eye, is ignored here.

lens. For this reason, close vision demands that the eye make active adjustments. To restore focus, three processes must occur simultaneously: accommodation of the lenses, constriction of the pupils, and convergence of the eyeballs. The signal that induces this trio of reflex responses is blurring of the retinal image.

- **Accommodation of the lenses. Accommodation** is the process that increases the refractory power of the lens. The ciliary muscles contract, pulling the ciliary body anteriorly toward the pupil and inward, releasing tension in the ciliary zonule. No longer stretched, the elastic lens recoils and

bulges, providing the shorter focal length needed to focus the image of a close object on the retina (Figure 15.13c). Parasympathetic fibers of the oculomotor nerves control the contraction of the ciliary muscles.

The closest point on which we can focus clearly is called the **near point of vision**, and it represents the maximum bulge the lens can achieve. In young adults with emmetropic vision, the near point is 10 cm (4 inches) from the eye. However, it is closer in children and gradually recedes with age, explaining why children can hold their books very close to their faces while many elderly people must hold the newspaper at arm's length. The gradual loss of accommodation with age reflects the lens's decreasing elasticity. In many people over age 50, the lens is nonaccommodating, a condition known as **presbyopia** (pres″be-o′pe-ah), literally "old person's vision."

- **Constriction of the pupils.** The sphincter pupillae muscles of the iris enhance the effect of accommodation by reducing the size of the pupil toward 2 mm (see Figure 15.5). This **accommodation pupillary reflex**, mediated by parasympathetic fibers of the oculomotor nerves, prevents the most divergent light rays from entering the eye. Such rays would pass through the extreme edge of the lens and would not focus properly, causing blurred vision.

- **Convergence of the eyeballs.** The visual goal is always to keep the object being viewed focused on the retinal foveae. When we look at distant objects, both eyes are directed either straight ahead or to one side to the same degree, but when we fixate on a close object, our eyes converge. **Convergence**, controlled by somatic motor fibers of the oculomotor nerves, is medial rotation of the eyeballs by the medial rectus muscles so that each is directed toward the object being viewed. The closer that object, the greater the degree of convergence required. For example, when you focus on the tip of your nose, you "go cross-eyed."

Reading or viewing a smartphone screen requires almost continuous accommodation, pupillary constriction, and convergence. This is why these activities can tire the eye muscles and result in eyestrain. When your eyes tire, it helps to look up and stare into the distance occasionally to relax your intrinsic eye muscles.

⚖ **HOMEOSTATIC IMBALANCE 15.8**　　　　　　　**CLINICAL**

The vast majority of refractive problems are related to eyeball shape—either too long or too short—and not to a lens that is too strong or too weak.

Myopia (mi-o′pe-ah; "short vision"), or *nearsightedness*, occurs when distant objects focus in front of the retina, rather than on it (**Figure 15.14**, left). Myopic people see close objects without problems because they can focus them on the retina, but distant objects are blurred. Myopia typically results from an eyeball that is too long.

Correction has traditionally involved using concave lenses that diverge the light before it enters the eye. Laser procedures to flatten the cornea slightly offer other treatment options.

Hyperopia (hi″per-o′pe-ah; "far vision"), or *farsightedness*, occurs when the parallel light rays from distant objects focus *behind* the retina (Figure 15.14, right). Hyperopic individuals can see distant objects because their ciliary muscles contract almost continuously to increase the light-bending power of the lens, which moves the focal point forward onto the retina. However, diverging light rays from *nearby* objects focus so far behind the retina that the lens cannot bring the focal point onto the retina even at its full refractory power. As a result, close objects appear blurry, and convex corrective lenses are needed to converge the light more strongly for close vision. Hyperopia usually results from an eyeball that is too short.

Unequal curvatures in different parts of the cornea or lens also lead to blurry images. This refractory problem is **astigmatism** (*astigma* = not a point). Special cylindrically ground lenses or laser procedures are used to correct this problem. ●●●●●●●

Check Your Understanding

4. Arrange the following in the order that light passes through them to reach the photoreceptors (rods and cones): lens, bipolar cells, vitreous humor, cornea, aqueous humor, ganglion cells. (Hint: See Figure 15.6 if you need a reminder of where ganglion cells and bipolar cells are.)

5. Why does your near point of vision move farther away as you age?

6. **DRAW** You have been reading this book for a while now and your eyes are beginning to tire. Which intrinsic eye muscles relax as you stare thoughtfully into the distance? Label them on the image below.

7. **DRAW** Create a simple linear flowchart illustrating the sequence of steps that lead to a decrease in the lens's focal length to allow close vision. Begin with parasympathetic nerve activation .

8. **MAKE CONNECTIONS** Figures 15.5 and 15.13 show the effects sympathetic nervous system (SNS) activation has on intrinsic eye muscles. These effects make sense in situations (as discussed in Chapter 14) that cause massive SNS activation. Explain how these SNS actions on intrinsic eye muscles might help during such situations.

For answers, see Answers Appendix.

15.3 Phototransduction begins when light activates visual pigments in retinal photoreceptors

Learning Outcomes

▶ **Describe the events that convert light into a neural signal.**
▶ **Compare and contrast the roles of rods and cones in vision.**
▶ **Compare and contrast light and dark adaptation.**

Once light is focused on the retina, the photoreceptors come into play. First, we will describe the functional anatomy of the rod and cone photoreceptor cells, then the chemistry and response of their visual pigments to light, and finally, photoreceptor activation, phototransduction, and information processing in the retina.

15

Functional Anatomy of the Photoreceptors

Photoreceptors are modified neurons, but structurally they resemble tall epithelial cells turned upside down with their "tips" immersed in the pigmented layer of the retina (**Figure 15.15a**). These "tips" are the receptive regions of rods and cones and are called the **outer segments**. Moving from the pigmented layer into the neural layer, a connecting cilium joins the outer segment of a rod or cone to the **inner segment**. The inner segment then connects to the *cell body*, which is continuous with an *inner fiber* bearing *synaptic terminals*.

In rods, the outer segment is slender and rod shaped and the inner segment connects to the cell body by the *outer fiber*. In

(a) The outer segments of rods and cones are embedded in the pigmented layer of the retina.

Process of bipolar cell

Light

Synaptic terminals

Inner fibers

Rod cell body

Rod cell body

Nuclei

Cone cell body

Mitochondria

Outer fiber

Connecting cilia

Inner segment

Apical microvillus

Outer segment

Discs containing visual pigments

Discs being phagocytized

Pigmented layer

Pigment cell nucleus

Melanin granules

Basal lamina (border with choroid)

(b) Rhodopsin, the visual pigment in rods, is embedded in the membrane that forms discs in the outer segment.

Rod discs

Visual pigment consists of
• Retinal
• Opsin

Figure 15.15 Photoreceptors of the retina.

contrast, the cones have a short conical outer segment and the inner segment directly joins the cell body.

The outer segments contain an elaborate array of **visual pigments (photopigments)** that change shape as they absorb light. These pigments are embedded in areas of the plasma membrane that form discs (Figure 15.15b). Folding the plasma membrane into discs increases the surface area available for trapping light. In rods, the discs are discontinuous—stacked inside a cylinder of plasma membrane like pennies in a coin wrapper. In cones, the disc membranes are continuous with the plasma membrane, so the interiors of the cone discs are continuous with the extracellular space.

Photoreceptor cells are highly vulnerable to damage and immediately begin to degenerate if the retina becomes detached. They are also destroyed by intense light. How is it, then, that we do not all gradually go blind? The answer lies in the photoreceptors' unique system for renewing their light-trapping outer segment. Every 24 hours, new components are synthesized in the cell body and added to the base of the outer segment. As new discs are made, the discs at the tip of the outer segment fragment off and pigment cells phagocytize them.

Comparing Rod and Cone Vision

Rods and cones have different thresholds for activation. Rods, for example, are very sensitive (they respond to very dim light—a single photon), making them best suited for night vision and peripheral vision. Cones, on the other hand, need bright light for activation (have low sensitivity), but react more rapidly. Cones have one of three different pigments that furnish a vividly colored view of the world, but rods contain a single kind of visual pigment so their inputs are perceived only in gray tones.

Rods and cones also differ in their "wiring" to other retinal neurons. Rods participate in converging pathways, and as many as 100 rods may ultimately feed into each ganglion cell. As a result, rod effects are summated and considered collectively, resulting in vision that is fuzzy and indistinct. (The visual cortex has no way of knowing exactly *which* rods of the large number influencing a particular ganglion cell are actually activated.)

In contrast, each cone in the fovea (or at most a few) has a straight-through pathway via its "own personal bipolar cell" to a ganglion cell (see Figure 15.6b). Essentially, each foveal cone has its own direct line to the higher visual centers—permitting detailed, high-acuity (high-resolution) views of very small areas of the visual field.

Because rods are absent from the foveae and cones do not respond to low-intensity light, we see dimly lit objects best when we do not look at them directly, and we recognize them best when they move.

Table 15.1 summarizes the differences between rods and cones.

Visual Pigments

How do photoreceptors translate incoming light into electrical signals? The key is a light-absorbing molecule called **retinal** that combines with proteins called **opsins** to form four types of visual pigments. Depending on the type of opsin to which it is bound, retinal absorbs different wavelengths of the visible spectrum.

The cone opsins differ both from the opsin of the rods and from one another. The naming of cones reflects the colors (that is, wavelengths) of light that each cone variety absorbs best. Blue cones respond maximally to wavelengths around 420 nm, green cones to wavelengths of 530 nm, and red cones to wavelengths at or close to 560 nm (see Figure 15.10b).

How do we see other colors besides blue, green, and red? Cones' absorption spectra overlap, and our perception of intermediate hues, such as yellow, orange, and purple, results from differential activation of more than one type of cone at the same time. For example, yellow light stimulates both red and green cone receptors, but if the red cones are stimulated more than the green cones, we see orange instead of yellow. When all cones are stimulated equally, we see white.

HOMEOSTATIC IMBALANCE 15.9   **CLINICAL**

Color blindness is due to a congenital lack of one or more cone pigments. It is far more common in males than in females because it is inherited as an X-linked condition (see Chapter 29). As many as 8–10% of males have some form of color blindness.

The most common type is red-green color blindness, resulting from a deficit or absence of either red or green cone pigments. Red and green are seen as the same color—either red or green, depending on the cone pigment present. Many color-blind people are unaware of their condition because they have learned to rely on other cues—such as different intensities of the same color—to distinguish something green from something red, such as traffic signals.   ●●●●●●

Retinal is chemically related to vitamin A and is made from it. The cells of the pigmented layer of the retina absorb vitamin A from the blood and serve as the local vitamin A depot for rods and cones.

Retinal can assume a variety of three-dimensional forms, each form called an isomer. When bound to opsin, retinal has a bent shape called **11-*cis*-retinal**, as shown at the top of Figure 15.16. However, when the pigment absorbs a photon of light, retinal

Table 15.1	Comparison of Rods and Cones
RODS	**CONES**
Noncolor vision (one visual pigment)	Color vision (three visual pigments)
High sensitivity; function in dim light	Low sensitivity; function in bright light
Low acuity (many rods converge onto one ganglion cell)	High acuity (one cone per ganglion cell in fovea)
More numerous (20 rods for every cone)	Less numerous
Mostly in peripheral retina	Mostly in central retina

15

11-*cis*-retinal

① **Pigment synthesis:** 11-*cis*-retinal, derived from vitamin A, is combined with opsin to form rhodopsin.

2H$^+$

Oxidation

Vitamin A

Reduction

2H$^+$

11-*cis*-retinal

Rhodopsin

② **Pigment bleaching:** Light absorption by rhodopsin triggers a rapid series of steps in which retinal changes shape (11-*cis* to all-*trans*) and eventually releases from opsin.

③ **Pigment regeneration:** Enzymes slowly convert all-*trans*-retinal to its 11-*cis* form in cells of the pigmented layer; requires ATP.

Dark **Light**

Opsin and

All-*trans*-retinal

All-*trans*-retinal

Figure 15.16 The formation and breakdown of rhodopsin.
11-*cis*-retinal can either be regenerated from all-*trans*-retinal or made from vitamin A.

twists and snaps into a new configuration, **all-*trans*-retinal** (Figure 15.16, bottom). This change, in turn, causes opsin to change shape and assume its activated form.

The capture of light by visual pigments is the *only* light-dependent stage, and this simple photochemical event initiates a whole chain of chemical and electrical reactions in rods and cones that ultimately causes action potentials to be transmitted along the optic nerve. Let's look more closely at these events in rods and cones.

Phototransduction

Phototransduction is the process by which light energy is converted into a graded receptor potential. It begins when a visual pigment captures a photon of light.

Capturing Light

The visual pigment of rods is a deep purple pigment called **rhodopsin** (ro-dop′sin; *rhodo* = rose, *opsis* = vision). (Do you suppose the person who coined the expression "looking at the world through rose-colored glasses" knew the meaning of "rhodopsin"?) Rhodopsin molecules are arranged in a single layer in the membranes of each of the thousands of discs in the rods' outer segments. The formation and breakdown of rhodopsin follows the process shown in **Figure 15.16**.

① **Pigment synthesis:** Rhodopsin forms and accumulates in the dark. Vitamin A is oxidized (and isomerized) to the 11-*cis*-retinal form and then combined with opsin to form rhodopsin.

② **Pigment bleaching:** When rhodopsin absorbs light, retinal changes shape to its all-*trans* isomer, allowing the surrounding protein to quickly relax like an uncoiling spring into its light-activated form. Eventually, the retinal-opsin combination breaks down, allowing retinal and opsin to separate. The breakdown of rhodopsin to retinal and opsin is known as the **bleaching of the pigment**.

③ **Pigment regeneration:** Once the light-struck all-*trans*-retinal detaches from opsin, enzymes within the pigmented epithelium reconvert it to its 11-*cis* isomer. Then, retinal heads "homeward" again to the photoreceptor

cells' outer segments. Rhodopsin is regenerated when 11-*cis*-retinal is rejoined to opsin.

The breakdown and regeneration of visual pigments in cones is essentially the same as for rhodopsin. However, cones are about a hundred times less sensitive than rods, which means that it takes higher-intensity (brighter) light to activate cones.

Light Transduction Reactions

What happens when light changes opsin's shape? An enzymatic cascade occurs that ultimately results in closing cation channels that are normally kept open in the dark. **Figure 15.17** illustrates this process in detail, but in short, light-activated rhodopsin activates a G protein called **transducin**. Transducin, in turn, activates *PDE* (*phosphodiesterase*), the enzyme that breaks down **cyclic GMP (cGMP)**. In the dark, cGMP binds to cation channels in the outer segments of photoreceptor cells, holding them open. This allows Na^+ and Ca^{2+} to enter, depolarizing the cell to its *dark potential* of about −40 mV. In the light, PDE breaks down cGMP, the cation channels close, Na^+ and Ca^{2+} stop entering the cell, and the cell hyperpolarizes to about −70 mV.

This arrangement can seem bewildering, to say the least. Here we have receptors built to detect light that depolarize in the dark and hyperpolarize when exposed to light! However, all

G protein signaling mechanisms are like a molecular relay race.

Light (1st messenger) — Receptor — G protein — Enzyme — 2nd messenger

① Retinal absorbs light and changes shape. Visual pigment activates.

Visual pigment — Light — All-*trans*-retinal — Phosphodiesterase (PDE)

11-*cis*-retinal — Transducin (a G protein)

Ca^{2+} Na^+ Ca^{2+} Na^+

cGMP GMP cGMP cGMP

cGMP-gated cation channel open in dark — cGMP-gated cation channel closed in light

② Visual pigment activates transducin (G protein).

③ Transducin activates phosphodiesterase (PDE).

④ PDE converts cGMP into GMP, causing cGMP levels to fall.

⑤ As cGMP levels fall, cGMP-gated cation channels close, resulting in hyperpolarization.

Figure 15.17 Events of phototransduction. A portion of photoreceptor disc membrane is shown. The G protein conversion of GTP to GDP has been omitted for clarity. For simplicity, the channels gated by cyclic GMP (cGMP) are shown on the same membrane as the visual pigment instead of in the plasma membrane. (For the basics of G protein signaling mechanisms, see *Focus on G Proteins*, Focus Figure 3.3 on p. 82.)

15

Below, we look at a tiny column of retina. The outer segment of the rod, closest to the back of the eye and farthest from the incoming light, is at the top.

Light

In the dark

In the light

Light

1 cGMP-gated channels open, allowing cation influx. Photoreceptor depolarizes.

Na⁺
Ca²⁺

1 cGMP-gated channels close, so cation influx stops. Photoreceptor hyperpolarizes.

Photoreceptor cell (rod)

−40 mV

−70 mV

2 Voltage-gated Ca²⁺ channels open in synaptic terminals.

2 Voltage-gated Ca²⁺ channels close in synaptic terminals.

Ca²⁺

3 Neurotransmitter is released continuously.

3 No neurotransmitter is released.

4 Neurotransmitter causes IPSPs in bipolar cell. Hyperpolarization results.

4 Lack of IPSPs in bipolar cell results in depolarization.

Bipolar cell

5 Hyperpolarization closes voltage-gated Ca²⁺ channels, inhibiting neurotransmitter release.

5 Depolarization opens voltage-gated Ca²⁺ channels; neurotransmitter is released.

Ca²⁺

6 No EPSPs occur in ganglion cell.

6 EPSPs occur in ganglion cell.

Ganglion cell

7 No action potentials occur along the optic nerve.

7 Action potentials propagate along the optic nerve.

Figure 15.18 Signal transmission in the retina. (EPSP = excitatory postsynaptic potential; IPSP = inhibitory postsynaptic potential)

15

that is required is a signal and hyperpolarization is just as good a signal as depolarization.

Information Processing in the Retina

How is the hyperpolarization of the photoreceptors transmitted through the retina and on to the brain? **Figure 15.18** illustrates this process. Notice that the photoreceptors do not generate action potentials (APs), and neither do the bipolar cells that are next in line. They only generate graded potentials. Photoreceptors generate receptor potentials, and bipolar cells generate excitatory or inhibitory postsynaptic potentials (EPSPs or IPSPs).

This is not surprising if you remember that the primary function of APs is to carry information rapidly over long distances. Retinal cells are small cells that are very close together. Graded potentials can serve quite adequately as signals that directly regulate neurotransmitter release at the synapse by opening or closing voltage-gated Ca^{2+} channels.

As shown in the right panel of Figure 15.18, for example, light hyperpolarizes photoreceptors, which then stop releasing their inhibitory neurotransmitter (glutamate). No longer inhibited, bipolar cells depolarize and release neurotransmitter onto ganglion cells. Once the signal reaches the ganglion cells, it is converted into an AP. This AP is transmitted to the brain along the ganglion cell axons that make up the optic nerve.

Light and Dark Adaptation

Rhodopsin is amazingly sensitive. Even starlight bleaches some of its molecules. As long as the light is low intensity, relatively little rhodopsin bleaches and the retina continues to respond to light stimuli. However, in high-intensity light, there is wholesale bleaching of the pigment, and rhodopsin bleaches as fast as it is re-formed. At this point, the rods are nonfunctional, but cones still respond. As a result, retinal sensitivity automatically adjusts to the amount of light present.

Light Adaptation

Light adaptation occurs when we move from darkness into bright light, as when leaving a movie matinee. We are momentarily dazzled—all we see is white light—because the sensitivity of the retina is still "set" for dim light. Both rods and cones are strongly stimulated, and large amounts of the visual pigments break down almost instantaneously, producing a flood of signals that accounts for the glare.

Under such conditions, compensations occur. The rod system turns off—all of the transducins migrate to the inner segment, uncoupling rhodopsin from the rest of the transduction cascade. Without transducin in the outer segment, light hitting rhodopsin cannot produce a signal. At the same time, the less sensitive cone system and other retinal neurons rapidly adapt, and retinal sensitivity decreases dramatically. Within about 60 seconds, the cones, initially overexcited by the bright light, are sufficiently desensitized to take over. Visual acuity and color vision continue to improve over the next 5–10 minutes. Thus, during light adaptation, we lose retinal sensitivity (rod function) but gain visual acuity.

Dark Adaptation

Dark adaptation, essentially the reverse of light adaptation, occurs when we go from a well-lit area into a dark one. Initially, we see nothing but velvety blackness because (1) our cones stop functioning in low-intensity light, and (2) the bright light bleached our rod pigments, and the rods are still turned off.

But once we are in the dark, rhodopsin accumulates, transducin returns to the outer segment, and retinal sensitivity increases. Dark adaptation is much slower than light adaptation and can go on for hours. However, there is usually enough rhodopsin within 20–30 minutes to allow adequate dim-light vision.

During both light and dark adaptation, reflexive changes occur in pupil size. Bright light shining in one or both eyes constricts both pupils (elicits the *pupillary* and *consensual light reflexes*). These pupillary reflexes are mediated by the pretectal nucleus of the midbrain and by parasympathetic fibers. In dim light, the pupils dilate, allowing more light to enter the eye.

> ### HOMEOSTATIC IMBALANCE 15.10 — CLINICAL
>
> *Night blindness*, or *nyctalopia* (nic″tă-lo′pe-uh), is a condition in which rod function is seriously hampered, impairing one's ability to drive safely at night. In countries where malnutrition is common, the most common cause of night blindness is prolonged vitamin A deficiency, which leads to rod degeneration. Vitamin A supplements restore function if they are administered early.
>
> In countries where nutrition isn't a problem, *retinitis pigmentosa*—a group of degenerative retinal diseases that destroy rods—are the most common causes of night blindness. Retinitis pigmentosa results from pigment epithelial cells that are unable to recycle the tips of the rods as they get sloughed off. ●●●●●●

Check Your Understanding

9. For each of the following, indicate whether it applies to rods or cones: vision in bright light; only one type of visual pigment; most abundant in the periphery of the retina; many feed into one ganglion cell; color vision; higher sensitivity; higher acuity.

10. What does bleaching of the pigment mean and when does it happen?

For answers, see Answers Appendix.

15.4 Visual information from the retina passes through relay nuclei to the visual cortex

Learning Outcome

▶ Trace the visual pathway to the visual cortex, and briefly describe the steps in visual processing.

The Visual Pathway to the Brain

As we described earlier, the axons of the retinal ganglion cells exit the eye in the **optic nerves**. At the X-shaped **optic chiasma**

Supra-chiasmatic nucleus

Pretectal nucleus

Right eye

Left eye

Optic nerve

Optic chiasma

Optic tract

Lateral geniculate nucleus

Superior colliculus (sectioned)

Uncrossed (ipsilateral) fiber

Crossed (contralateral) fiber

Optic radiation

Lateral geniculate nucleus of thalamus

Superior colliculus

Primary visual cortex (occipital lobe)

Corpus callosum

(a) The visual fields of the two eyes overlap considerably. Note that fibers from the lateral portion of each retinal field do not cross at the optic chiasma.

(b) Photo of a human brain. The right side has been dissected to reveal internal structures.

Figure 15.19 Visual pathway to the brain and visual fields, inferior view.

Practice Human Cadaver lab questions:
MasteringA&P® > Study Area > Lab Tools > PAL

(*chiasm* = cross), fibers from the medial aspect of each eye cross over to the opposite side and then continue on via the **optic tracts** (**Figure 15.19**). As a result, each optic tract:

- Contains fibers from the lateral (temporal) aspect of the eye on the same side and fibers from the medial (nasal) aspect of the opposite eye
- Carries all the information from the same half of the visual field

Notice that, because the lens system of each eye reverses all images, the medial half of each retina receives light rays from the *temporal* (lateralmost) part of the visual field (from the far left or far right rather than from straight ahead), and the lateral

half of each retina receives an image of the nasal (central) part of the visual field. Consequently, the left optic tract carries a complete representation of the right half of the visual field, and the opposite is true for the right optic tract.

The paired optic tracts sweep posteriorly around the hypothalamus and send most of their axons to synapse with neurons in the **lateral geniculate nuclei** (contained within the lateral geniculate bodies) of the thalamus. The lateral geniculate nuclei maintain the fiber separation established at the chiasma, but they balance and combine the retinal input for delivery to the visual cortex. Axons of these thalamic neurons project through the internal capsule to form the **optic radiation** of fibers in the cerebral white matter (Figure 15.19). These fibers project to the

primary visual cortex in the occipital lobes, where conscious perception of visual images (seeing) occurs.

Some nerve fibers in the optic tracts send branches to the midbrain. One set of these fibers ends in the **superior colliculi**, visual reflex centers controlling the extrinsic muscles of the eyes. Another set comes from a small subset of ganglion cells in the retina that contain the visual pigment *melanopsin*, dubbed the circadian pigment. These ganglion cells respond directly to light stimuli and their fibers project to the **pretectal nuclei**, which mediate pupillary light reflexes, and to the **suprachiasmatic nucleus** of the hypothalamus, which functions as the "timer" to set our daily biorhythms.

Depth Perception

Each eye's visual field is about 170 degrees. The two visual fields overlap considerably, but each eye sees a slightly different view (Figure 15.19a). The visual cortex fuses the slightly different images delivered by the two eyes, providing us with **depth perception** (or **three-dimensional vision**), an accurate means of locating objects in space.

In contrast, many animals (pigeons, rabbits, and others) have *panoramic vision*. Their eyes are placed more laterally on the head, so that the visual fields overlap very little. Each visual cortex receives input principally from a single eye and a totally different visual field.

Depth perception depends on the two eyes working together. If only one eye is used, depth perception is lost, and the person must learn to judge an object's position based on learned cues (for example, nearer objects appear larger, and parallel lines converge with distance).

HOMEOSTATIC IMBALANCE 15.11   CLINICAL

Loss of an eye or destruction of one optic nerve eliminates 3-D vision, the major component of depth perception, and peripheral vision on the damaged side.

If neural destruction occurs beyond the optic chiasma—in an optic tract, the thalamus, or visual cortex—then part or all of the opposite half of the visual field is lost. For example, a stroke affecting the left visual cortex leads to blindness in the right half of the visual field. ●●●●●●

Visual Processing

How does information received by rods and cones become vision? Visual processing begins in the retina.

Retinal cells simplify and condense the information from rods and cones, splitting it into a number of different "channels," each with its own type of ganglion cell. These "channels" include information about color and brightness, but also about more complex aspects of what we see—the angle, direction and speed of movement of *edges* (sudden changes in brightness or color). Edges are detected by a kind of contrast enhancement called *lateral inhibition*, which is the job of the amacrine and horizontal cells mentioned on pp. 560–561.

The ganglion cells pass the processed information to the lateral geniculate nuclei of the thalamus. There, information from each eye is combined in preparation for depth perception and input from cones is emphasized.

The *primary visual cortex* contains an accurate topographical map of the retina, with the left visual cortex receiving input from the right visual field and vice versa. Visual processing here occurs at a relatively basic level, with the processing neurons responding to dark and bright edges (contrast information) and object orientation. Surrounding areas process form, color, and movement.

Functional neuroimaging of humans has revealed that complex visual processing extends well forward into the temporal, parietal, and frontal lobes via two parallel streams, one that identifies objects in the visual field and another that assesses the location of objects in space. Output from these regions then passes to the frontal cortex for further processing.

Check Your Understanding

11. **PREDICT** Which part of the visual field would be affected by a tumor in the right visual cortex? By a tumor compressing the right optic nerve?

For answers, see Answers Appendix.

PART 2

THE CHEMICAL SENSES: SMELL AND TASTE

Smell and taste are gritty, primitive senses that alert us to whether that "stuff" nearby (or in our mouth) is to be savored or avoided. The receptors for smell (olfaction) and taste (gustation) are **chemoreceptors** (they respond to chemicals in an aqueous solution). They complement each other and respond to different classes of chemicals. Smell receptors are excited by airborne chemicals that dissolve in fluids coating nasal membranes, and taste receptors are excited by food chemicals dissolved in saliva.

15.5 Airborne chemicals are detected by olfactory receptors in the nose

Learning Outcome

▶ Describe the location, structure, and afferent pathways of smell receptors, and explain how these receptors are activated.

Although our olfactory sense (*olfact* = to smell) is far less acute than that of many other animals, the human nose is still no slouch in picking up small differences in odors. Some people capitalize on this ability by becoming wine tasters.

Location and Structure of Olfactory Receptors

The organ of smell is a yellow-tinged patch (about 5 cm²) of pseudostratified epithelium, called the **olfactory epithelium**,

Figure 15.20 Olfactory receptors. (a) Site of olfactory epithelium in the superior nasal cavity. **(b)** An enlarged view of the olfactory epithelium showing the course of the fibers [filaments of the olfactory nerve (I)] through the ethmoid bone. These synapse in the glomeruli of the overlying olfactory bulb. The mitral cells are the output cells of the olfactory bulb.

located in the roof of the nasal cavity (**Figure 15.20a**). Air entering the nasal cavity must make a hairpin turn to stimulate olfactory receptors before entering the respiratory passageway below, so the human olfactory epithelium is in a poor position for doing its job. (This is why sniffing, which draws more air superiorly across the olfactory epithelium, intensifies the sense of smell.)

The olfactory epithelium covers the superior nasal concha on each side of the nasal septum, and contains millions of bowling pin–shaped receptor cells—the **olfactory sensory neurons**. These are surrounded and cushioned by columnar **supporting cells**, which make up the bulk of the penny-thin epithelial membrane (Figure 15.20b). The supporting cells contain a yellow-brown pigment similar to lipofuscin, which gives the olfactory epithelium its yellow hue. At the base of the epithelium lie the short **olfactory stem cells**.

The olfactory sensory neurons are unusual bipolar neurons. Each has a thin apical dendrite that terminates in a knob from which several long cilia radiate. These **olfactory cilia**, which substantially increase the receptive surface area, typically lie flat on the nasal epithelium and are covered by a coat of thin mucus produced by the supporting cells and by olfactory glands in the underlying connective tissue. This mucus is a solvent that

"captures" and dissolves airborne odorants. Unlike other cilia in the body, which beat rapidly in a coordinated manner, olfactory cilia are largely nonmotile.

The slender, nonmyelinated axons of the olfactory sensory neurons are gathered into small fascicles that collectively form the **filaments of the olfactory nerve** (cranial nerve I). They project superiorly through the openings in the cribriform plate of the ethmoid bone, where they synapse in the overlying olfactory bulbs.

Olfactory sensory neurons are also unusual because they are one of the few types of *neurons* that undergo noticeable turnover throughout adult life. Their superficial location puts them at risk for damage, and their typical life span is 30–60 days. Olfactory stem cells in the olfactory epithelium differentiate to replace them.

Specificity of Olfactory Receptors

Smell is difficult to research because any given odor (say, tobacco smoke) may be made up of hundreds of different chemicals (called *odorants*). Taste has been neatly packaged into five taste qualities as you will see, but science has yet to discover any similar means for classifying smell. How many different odors can we distinguish? Estimates range from 10,000 to 1 trillion.

① Odorant binds to its receptor.

Odorant

G protein (Golf)

Receptor

GDP

GTP

GTP

② Receptor activates G protein (Golf).

GTP

③ G protein activates adenylate cyclase.

Adenylate cyclase

ATP

④ Adenylate cyclase converts ATP to cAMP.

cAMP

cAMP

Na⁺ Ca²⁺

Open cAMP-gated cation channel

⑤ cAMP opens a cation channel, allowing Na⁺ and Ca²⁺ influx and causing depolarization.

Figure 15.21 Olfactory transduction process. A portion of olfactory cilium membrane is shown.

However, our olfactory sensory neurons are stimulated by combinations of a more limited number of olfactory qualities.

There are about 350 different odorant receptors in humans. Each receptor responds to one or more odorants and each odorant binds to several different receptor types. However, each receptor cell has only one type of receptor protein.

Olfactory neurons are exquisitely sensitive—in some cases, just a few molecules activate them. Some of what we call smell is really pain. The nasal cavities contain pain and temperature receptors that respond to irritants such as the sharpness of ammonia, the hotness of chili peppers, and the "chill" of menthol. Impulses from these receptors reach the central nervous system via afferent fibers of the trigeminal nerves.

Physiology of Smell

For us to smell a particular odorant, it must be *volatile*—that is, it must be in the gaseous state as it enters the nasal cavity. Additionally, it must dissolve in the fluid coating the olfactory epithelium.

Activation of Olfactory Sensory Neurons

Dissolved odorants stimulate olfactory sensory neurons by binding to receptor proteins in the olfactory cilium membranes, opening cation channels and generating a receptor potential. Ultimately (assuming threshold stimulation) an action potential is conducted to the first relay station in the olfactory bulb.

Smell Transduction

Transduction of odorants uses a receptor linked to a G protein. The events that follow odorant binding will be easy to remember if you compare them to what you already know: general mechanisms of receptors and G proteins (◄ Figure 11.17, p. 425) and phototransduction (see Figure 15.17).

As **Figure 15.21** shows, olfactory transduction begins when an odorant binds to a receptor. This event activates G proteins

(Golf), which activate enzymes (adenylate cyclases) that synthesize cyclic AMP (cAMP) as a second messenger. Cyclic AMP then acts directly on plasma membrane cation channels, causing them to open, allowing Na⁺ and Ca²⁺ to enter.

Na⁺ influx leads to depolarization and impulse transmission. Ca²⁺ influx causes the transduction process to adapt, decreasing its response to a sustained stimulus. This *olfactory adaptation* helps explain how a person working in a paper mill or sewage treatment plant can still enjoy lunch!

The Olfactory Pathway

As we have already noted, axons of the olfactory sensory neurons form the olfactory nerves that synapse in the overlying **olfactory bulbs**, the distal ends of the olfactory tracts (◄ Table 13.2, p. 500). There, the filaments of the olfactory nerves synapse with **mitral cells** (mi′tral), which are second-order sensory neurons, in complex structures called **glomeruli** (glo-mer′u-li; "little balls") (Figure 15.20).

Axons from neurons bearing the same kind of receptor converge on a given type of glomerulus. That is, each glomerulus represents a single aspect of an odor (like one note in a chord) but each odor activates a unique set of glomeruli (the chord itself). Different odors activate different subsets of glomeruli (making different chords which may have some of the same notes). The mitral cells refine the signal, amplify it, and then relay it. The olfactory bulbs also house *amacrine granule cells* that inhibit mitral cells, so that only highly excitatory olfactory impulses are transmitted.

When the mitral cells are activated, impulses flow from the olfactory bulbs via the **olfactory tracts** (composed mainly of mitral cell axons) to the piriform lobe of the olfactory cortex (◄ p. 443). From there, two major pathways take information to various parts of the brain:

- One pathway brings information to part of the frontal lobe just above the orbit, where smells are consciously interpreted and identified. Only some of this information passes through the thalamus.

- The other pathway flows to the hypothalamus, amygdaloid body, and other regions of the limbic system. There, emotional responses to odors are elicited. Smells associated with danger—smoke, cooking gas, or skunk scent—trigger the sympathetic fight-or-flight response. Appetizing odors stimulate salivation and the digestive tract, and unpleasant odors can trigger protective reflexes such as sneezing or choking.

HOMEOSTATIC IMBALANCE 15.12   CLINICAL

Most olfactory disorders, or *anosmias* (an-oz′me-ahz; "without smells"), result from head injuries that tear the olfactory nerves, the aftereffects of nasal cavity inflammation, and neurological disorders such as Parkinson's disease. Anosmia may also be an early sign of Alzheimer's disease. Some people have olfactory hallucinations during which they experience a particular (usually unpleasant) odor, such as rotting meat. These are usually caused by temporal lobe epilepsy involving the olfactory cortex and can occur as olfactory auras before the seizure begins or as olfactory hallucinations during the seizure.

Check Your Understanding

12. How do the cilia of olfactory sensory neurons help these cells perform their function?

For answers, see Answers Appendix.

15.6   Dissolved chemicals are detected by receptor cells in taste buds

Learning Outcome

▶ Describe the location, structure, and afferent pathways of taste receptors, and explain how these receptors are activated.

The word *taste* comes from the Latin *taxare*, meaning "to touch, estimate, or judge." When we taste things, we are in fact intimately testing or judging our environment, and the sense of taste is considered by many to be the most pleasurable of the special senses.

Location and Structure of Taste Buds

Most of our 10,000 or so **taste buds**—the sensory organs for taste—are located on the tongue. A few taste buds are scattered on the soft palate, inner surface of the cheeks, pharynx, and epiglottis of the larynx, but most are found in **papillae** (pah-pil′e), peglike projections of the tongue mucosa that make the tongue surface slightly abrasive (**Figure 15.22a,b**). Three kinds of papillae have taste buds:

- Mushroom-shaped **fungiform papillae** (fun′jĭ-form) are scattered over the entire tongue surface, but have only one to five taste buds each.

- **Vallate papillae** (val′āt) are the largest and least numerous papillae, but have many taste buds. Eight to 12 vallate papillae form an inverted V at the back of the tongue.

- **Foliate papillae**, located laterally on the tongue, contain many taste buds during childhood, but fewer with age.

Each flask-shaped taste bud consists of 50 to 100 *epithelial cells* of two major types: *gustatory epithelial cells* and *basal epithelial cells* (Figure 15.22c).

Gustatory Epithelial Cells

The **gustatory epithelial cells** are the receptor cells for taste—the *taste cells*. Long microvilli called **gustatory hairs** project from the tips of all gustatory epithelial cells and extend through a **taste pore** to the surface of the epithelium, where they are bathed by saliva. The gustatory hairs are the sensitive portions (*receptor membranes*) of the gustatory epithelial cells. Coiling intimately around the gustatory epithelial cells are sensory dendrites that represent the initial part of the gustatory pathway to the brain. Each afferent fiber receives signals from several gustatory epithelial cells within the taste bud.

There are at least three kinds of gustatory epithelial cells. One kind forms traditional synapses with the sensory dendrites and releases the neurotransmitter serotonin. The other two kinds lack synaptic vesicles, but at least one of these kinds releases ATP that acts as a neurotransmitter.

Basal Epithelial Cells

Because of their location, taste bud cells are subjected to friction and are routinely burned by hot foods. Luckily, they are replaced every seven to ten days. **Basal epithelial cells** act as stem cells, dividing and differentiating into new gustatory epithelial cells.

Basic Taste Sensations

Normally, our taste sensations are complicated mixtures of qualities. However, all taste sensations can be grouped into one of five basic modalities: sweet, sour, salty, bitter, and umami.

- *Sweet* taste is elicited by many organic substances including sugars, saccharin, alcohols, some amino acids, and some lead salts (such as those found in lead paint).

- *Sour* taste is produced by acids, specifically their hydrogen ions (H^+) in solution.

- *Salty* taste is produced by metal ions (inorganic salts); table salt (sodium chloride) tastes the "saltiest."

- *Bitter* taste is elicited by alkaloids (such as quinine, nicotine, caffeine, morphine, and strychnine) as well as a number of nonalkaloid substances, such as aspirin.

- *Umami* (oo-mam′e; "delicious"), a subtle taste discovered by the Japanese, is elicited by the amino acids glutamate and aspartate, which are responsible for the "beef taste" of steak, the characteristic tang of aging cheese, and the flavor of the food additive monosodium glutamate (MSG).

In addition, there is growing evidence for our ability to taste long-chain fatty acids from lipids. This possible sixth modality may help explain our liking for fatty foods.

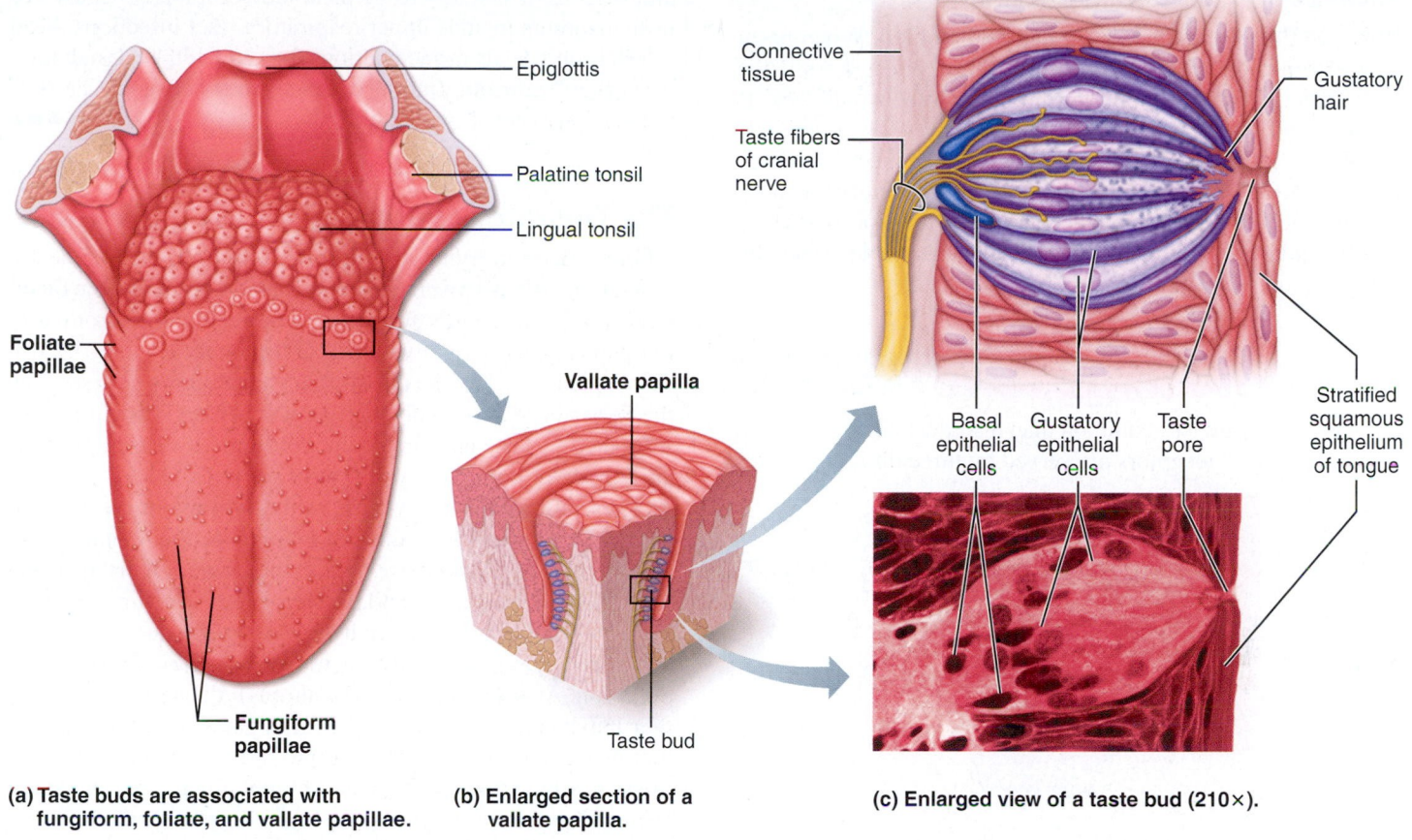

(a) **Taste buds are associated with fungiform, foliate, and vallate papillae.**

(b) **Enlarged section of a vallate papilla.**

(c) **Enlarged view of a taste bud (210×).**

Figure 15.22 Location and structure of taste buds on the tongue.

Practice Histology questions:
Mastering**A&P** > Study Area > Lab Tools > PAL

15

Keep in mind that many substances produce a mixture of these basic taste sensations, and taste buds generally respond to all five. However, it appears that a single taste cell has receptors for only one taste modality. Also, all areas of the tongue can detect all taste modalities.

Taste likes and dislikes have homeostatic value. Umami guides the intake of proteins, and a liking for sugar and salt helps satisfy the body's need for carbohydrates and minerals (as well as some amino acids). Many sour, naturally acidic foods (such as oranges, lemons, and tomatoes) are rich sources of vitamin C, an essential vitamin. On the other hand, intensely sour tastes warn us of spoilage. Likewise, many natural poisons and spoiled foods are bitter. Consequently, our dislike for intense sourness and bitterness is protective.

Physiology of Taste

For a chemical to be tasted it must dissolve in saliva, diffuse into a taste pore, and contact the gustatory hairs.

Activation of Taste Receptors

Gustatory epithelial cells contain neurotransmitters. When a food chemical, or *tastant*, binds to receptors in the gustatory epithelial cell membrane, it induces a graded depolarizing potential that causes neurotransmitter release. Binding of the

neurotransmitter to the associated sensory dendrites triggers generator potentials that elicit action potentials in these fibers.

The different gustatory epithelial cells have different thresholds for activation. In line with their protective nature, the bitter receptors detect substances present in minute amounts. The other receptors are less sensitive. Taste receptors adapt rapidly, with partial adaptation in 3–5 seconds and complete adaptation in 1–5 minutes.

Taste Transduction

The mechanisms of taste transduction are only now beginning to become clear. Three different mechanisms underlie how we taste.

- Salty taste is due to Na^+ influx through Na^+ channels, which directly depolarizes gustatory epithelial cells.

- Sour is mediated by H^+, which acts intracellularly to open channels that allow other cations to enter.

- Bitter, sweet, and umami responses share a common mechanism, but each occurs in a different cell. Each taste's unique set of receptors is coupled to a common G protein called *gustducin*. Activation leads to the release of Ca^{2+} from intracellular stores, which causes cation channels in the plasma membrane to open, thereby depolarizing the cell and releasing the neurotransmitter ATP.

Influence of Other Sensations on Taste

What we think of as taste is actually 80% smell. When nasal congestion blocks access to your olfactory receptors, food tastes bland. Without smell, our morning coffee would lack its richness and simply taste bitter.

The mouth also contains thermoreceptors, mechanoreceptors, and nociceptors, and the temperature and texture of foods can enhance or detract from their taste. "Hot" foods such as chili peppers actually bring about their pleasurable effects by exciting pain receptors in the mouth.

HOMEOSTATIC IMBALANCE 15.13   CLINICAL

Taste disorders are less common than disorders of smell, in part because the taste receptors are served by three different nerves

Figure 15.23 The gustatory pathway. Taste signals are relayed from the taste buds to the gustatory area of the cerebral cortex.

Gustatory cortex (in insula)

Thalamic nucleus (ventral posteromedial nucleus)

Pons

Facial nerve (VII)

Glossopharyngeal nerve (IX)

Solitary nucleus in medulla oblongata

Vagus nerve (X)

and thus are less likely to be disabled completely. Causes of taste disorders include upper respiratory tract infections, head injuries, chemicals or medications, or head and neck radiation for cancer treatment. Zinc supplements may help some cases of radiation-induced taste disorders.   ●●●●●●●

The Gustatory Pathway

Afferent fibers carrying taste information from the tongue are found primarily in two cranial nerve pairs. A branch of the **facial nerve** (VII), the *chorda tympani*, transmits impulses from taste receptors in the anterior two-thirds of the tongue (**Figure 15.23**). The lingual branch of the **glossopharyngeal nerve** (IX) services the posterior third and the pharynx just behind. Taste impulses from the few taste buds in the epiglottis and the lower pharynx are conducted primarily by the **vagus nerve** (X).

These afferent fibers synapse in the **solitary nucleus** of the medulla, and from there impulses stream to the thalamus and ultimately to the *gustatory cortex* in the insula. Fibers also project to the hypothalamus and limbic system structures, regions that determine our appreciation of what we are tasting.

An important role of taste is to trigger reflexes involved in digestion. As taste impulses pass through the solitary nucleus, they initiate reflexes (via synapses with parasympathetic nuclei) that increase secretion of saliva into the mouth and of gastric juice into the stomach. Saliva contains mucus that moistens food and digestive enzymes that begin digesting starch. When we ingest foul-tasting substances, the taste may initiate protective reactions such as gagging or vomiting.

Check Your Understanding

13. Name the five taste modalities. Name the three types of papillae that have taste buds.

For answers, see Answers Appendix.

PART 3

THE EAR: HEARING AND BALANCE

At first glance, the machinery for hearing and balance appears very crude—fluids must be stirred to stimulate the mechanoreceptors of the internal ear. Nevertheless, our hearing apparatus allows us to hear an extraordinary range of sound, and our equilibrium (balance) receptors continually inform the nervous system of head movements and position. Although the organs serving these two senses are structurally interconnected within the ear, their receptors respond to different stimuli and are activated independently of one another.

15.7   The ear has three major areas

Learning Outcome

▶ Describe the structure and general function of the outer, middle, and internal ears.

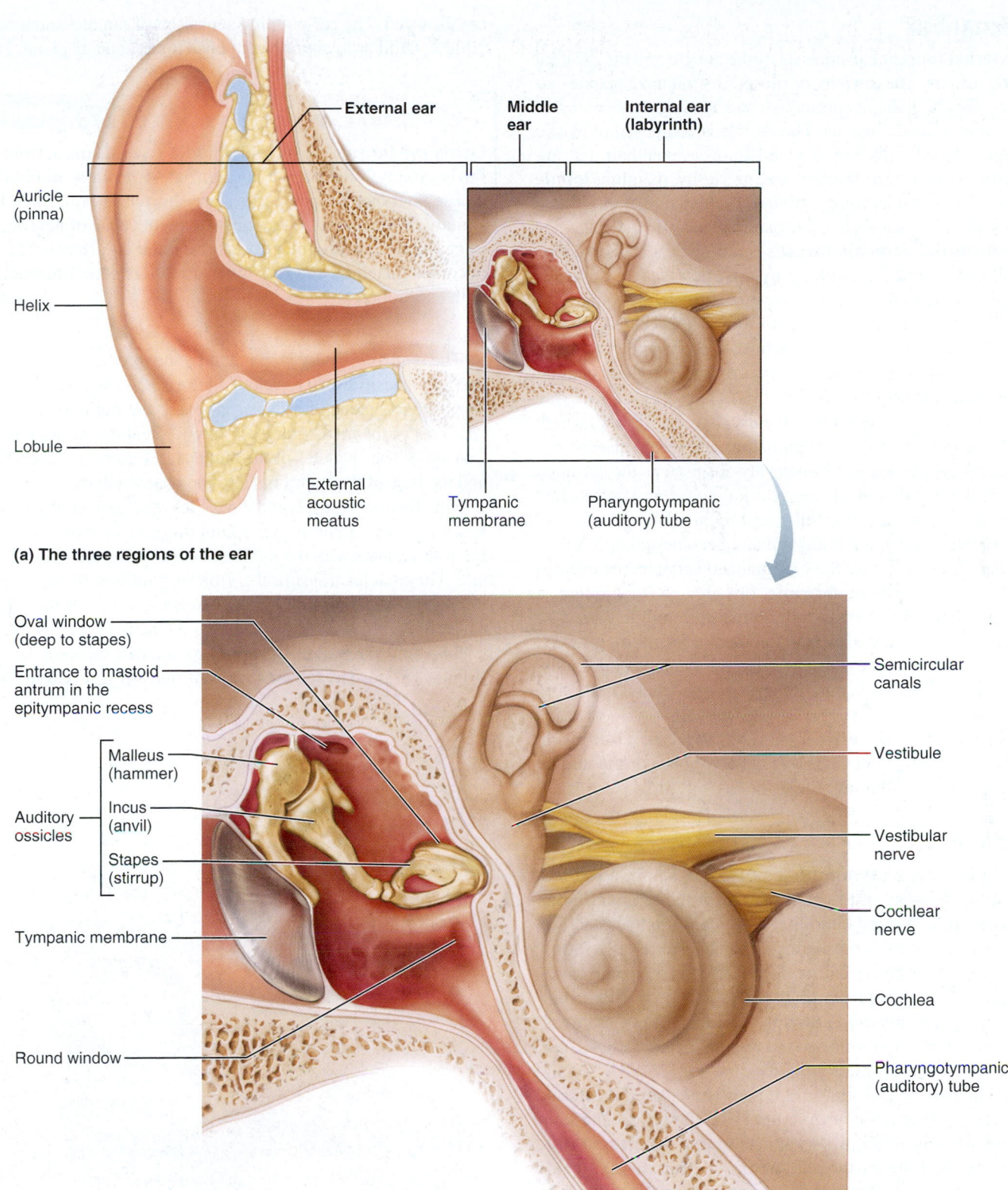

External ear

Middle ear

Internal ear (labyrinth)

Auricle (pinna)

Helix

Lobule

External acoustic meatus

Tympanic membrane

Pharyngotympanic (auditory) tube

(a) The three regions of the ear

Oval window (deep to stapes)

Entrance to mastoid antrum in the epitympanic recess

Auditory ossicles
- Malleus (hammer)
- Incus (anvil)
- Stapes (stirrup)

Tympanic membrane

Round window

Semicircular canals

Vestibule

Vestibular nerve

Cochlear nerve

Cochlea

Pharyngotympanic (auditory) tube

(b) Middle and internal ear

Figure 15.24 Structure of the ear. The inner ear structures in (b) appear as though they are veiled because they are cavities within the temporal bone.

The ear is divided into three major areas: external ear, middle ear, and internal ear (**Figure 15.24a**). The external and middle ear structures are involved with hearing only and are rather simply engineered. The internal ear functions in both equilibrium and hearing and is extremely complex.

15

External Ear

The **external (outer) ear** consists of the auricle and the external acoustic meatus. The **auricle**, or **pinna**, is what most people call the ear—the shell-shaped projection surrounding the opening of the external acoustic meatus. The auricle is composed of elastic cartilage covered with thin skin and an occasional hair. Its rim, the **helix**, is somewhat thicker, and its fleshy, dangling **lobule** ("earlobe") lacks supporting cartilage. The function of the auricle is to funnel sound waves into the external acoustic meatus.

The **external acoustic meatus** (auditory canal) is a short, curved tube (about 2.5 cm long by 0.6 cm wide) that extends from the auricle to the eardrum. Near the auricle, its framework is elastic cartilage; the remainder of the canal is carved into the temporal bone. The entire canal is lined with skin bearing hairs, sebaceous glands, and modified apocrine sweat glands called **ceruminous glands** (sĕ-roo′mĭ-nus). These glands secrete yellow-brown waxy **cerumen**, or earwax (*cere* = wax), which provides a sticky trap for foreign bodies and repels insects.

In many people, the ear is naturally cleansed as the cerumen dries and then falls out of the external acoustic meatus. Jaw movements as a person eats, talks, and so on, move the wax out. In other people, cerumen builds up and becomes compacted.

Sound waves entering the external acoustic meatus eventually hit the **tympanic membrane**, or *eardrum* (*tympanum* = drum), the boundary between the outer and middle ears. The eardrum is a thin, translucent, connective tissue membrane, covered by skin on its external face and by mucosa internally. Shaped like a flattened cone, its apex protrudes medially into the middle ear.

Sound waves make the eardrum vibrate. The eardrum, in turn, transfers the sound energy to the tiny bones of the middle ear and sets them vibrating.

Middle Ear

The **middle ear**, or **tympanic cavity**, is a small, air-filled, mucosa-lined cavity in the petrous part of the temporal bone (◀ p. 208). It is flanked laterally by the eardrum and medially by a bony wall with two openings, the superior **oval window** and the inferior **round window**. Superiorly the tympanic cavity arches upward as the **epitympanic recess**, the "roof" of the middle ear cavity. The **mastoid antrum**, a canal in the posterior wall of the tympanic cavity, allows it to communicate with *mastoid air cells* housed in the mastoid process.

The anterior wall of the middle ear is next to the internal carotid artery (the main artery supplying the brain) and contains the opening of the **pharyngotympanic (auditory) tube** (formerly called the eustachian tube). The pharyngotympanic tube runs obliquely downward to link the middle ear cavity with the nasopharynx (the superiormost part of the throat), and the mucosa of the middle ear is continuous with that lining the pharynx (throat).

Normally, the pharyngotympanic tube is flattened and closed, but swallowing or yawning opens it briefly to equalize pressure in the middle ear cavity with external air pressure. This is important because the eardrum vibrates freely only if the pressure on both of its surfaces is the same; otherwise sounds are distorted. The ear-popping sensation of the pressures equalizing is familiar to anyone who has flown in an airplane.

The tympanic cavity is spanned by the three smallest bones in the body: the **auditory ossicles** (Figure 15.24 and **Figure 15.25**). These bones, named for their shape, are the **malleus** (mal′e-us; "hammer"); the **incus** (ing′kus; "anvil"); and the **stapes** (sta′pēz; "stirrup"). The "handle" of the malleus is secured to the eardrum, and the base of the stapes fits into the oval window.

Tiny ligaments suspend the ossicles, and mini synovial joints link them into a chain that spans the middle ear cavity. The incus articulates with the malleus laterally and the stapes medially. The ossicles transmit the vibratory motion of the eardrum to the oval window, which in turn sets the fluids of the internal ear into motion, eventually exciting the hearing receptors.

Two tiny skeletal muscles are associated with the ossicles (Figure 15.25). The **tensor tympani** (ten′sor tim′pah-ni) arises

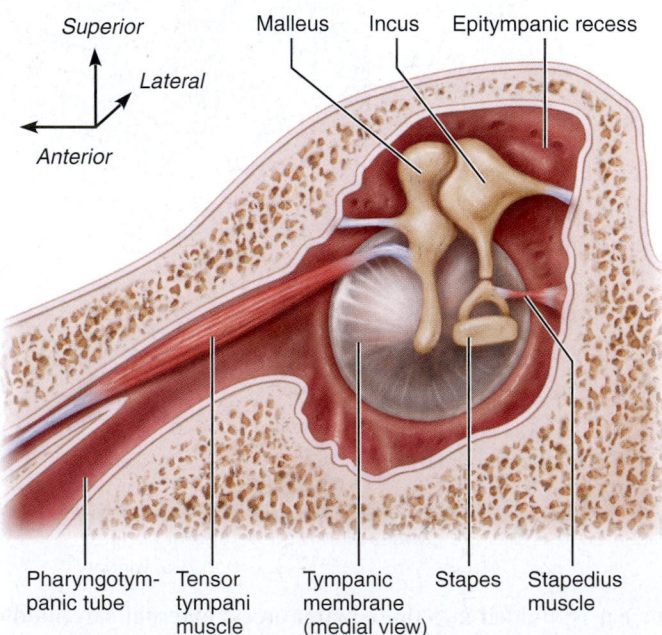

Superior

Lateral

Anterior

Malleus   Incus   Epitympanic recess

Pharyngotym-panic tube   Tensor tympani muscle   Tympanic membrane (medial view)   Stapes   Stapedius muscle

Figure 15.25 The three auditory ossicles and associated skeletal muscles. Right middle ear, medial view.

from the wall of the pharyngotympanic tube and inserts on the malleus. The **stapedius** (stah-pe′de-us) runs from the posterior wall of the middle ear to the stapes. When loud sounds assault the ears, these muscles contract reflexively to limit the ossicles' vibration and minimize damage to the hearing receptors.

Internal Ear

The **internal ear** is also called the **labyrinth** ("maze") because of its complicated shape (**Figure 15.26**). It lies deep in the temporal bone behind the eye socket and provides a secure site for all of the delicate receptor machinery.

The internal ear has two major divisions: the bony labyrinth and the membranous labyrinth.

- The **bony labyrinth** is a system of tortuous channels worming through the bone. Illustrations such as Figure 15.24 are, by necessity, somewhat misleading because we are trying to depict a *cavity* (the hollow space inside the bony labyrinth). The representation in Figure 15.24 can be compared to a plaster of paris cast of the cavity or hollow space inside the bony labyrinth.

- The **membranous labyrinth** is a continuous series of membranous sacs and ducts contained within the bony labyrinth and (more or less) following its contours (Figure 15.26).

The bony labyrinth is filled with **perilymph**, a fluid similar to cerebrospinal fluid and continuous with it. The membranous labyrinth is suspended in the surrounding perilymph, and its interior contains **endolymph**, which is chemically similar to K⁺-rich intracellular fluid. These two fluids conduct the sound vibrations involved in hearing and respond to the mechanical forces occurring during changes in body position and acceleration.

The bony labyrinth has three regions: the *vestibule*, the *semicircular canals*, and the *cochlea*.

Vestibule

The **vestibule** is the central egg-shaped cavity of the bony labyrinth. It lies posterior to the cochlea, anterior to the semicircular canals, and flanks the middle ear medially. In its lateral wall is the oval window.

Suspended in the vestibular perilymph and united by a small duct are two membranous labyrinth sacs, the **saccule** and **utricle** (u′trĭ-kl) (Figure 15.26). The smaller saccule is continuous with the membranous labyrinth extending anteriorly into the cochlea (the *cochlear duct*), and the utricle is continuous with the semicircular ducts extending into the semicircular canals posteriorly. The saccule and utricle house equilibrium receptor regions called *maculae* that respond to the pull of gravity and report on changes of head position.

Semicircular Canals

The **semicircular canals** lie posterior and lateral to the vestibule. The cavities of the bony semicircular canals project from the posterior aspect of the vestibule, each oriented in one of the three planes of space. Accordingly, there is an *anterior*, a *posterior*, and a *lateral* semicircular canal in each internal ear. The anterior and posterior canals are oriented at right angles to each other in the vertical plane, whereas the lateral canal lies horizontally (Figure 15.26).

Snaking through each semicircular canal is a corresponding membranous **semicircular duct**, which communicates with the utricle anteriorly. Each of these ducts has an enlarged swelling at one end called an **ampulla**, which houses an equilibrium

Figure 15.26 Membranous labyrinth of the internal ear. The membranous labyrinth (blue) lies within the chambers of the bony labyrinth (tan). The locations of the sensory organs for hearing (spiral organ) and equilibrium (maculae and cristae ampullares) are shown in purple.

(a)

Helicotrema at apex

Modiolus

Cochlear nerve, division of the vestibulocochlear nerve (VIII)

Spiral ganglion

Osseous spiral lamina

Vestibular membrane

Cochlear duct (scala media)

(b)

Vestibular membrane

Tectorial membrane

Cochlear duct (**scala media**; contains endolymph)

Stria vascularis

Spiral organ

Basilar membrane

Osseous spiral lamina

Spiral ganglion

Scala vestibuli (contains perilymph)

Scala tympani (contains perilymph)

(c)

Tectorial membrane

Hairs (stereocilia)

Outer hair cells

Supporting cells

Inner hair cell

Afferent nerve fibers

Fibers of cochlear nerve

Basilar membrane

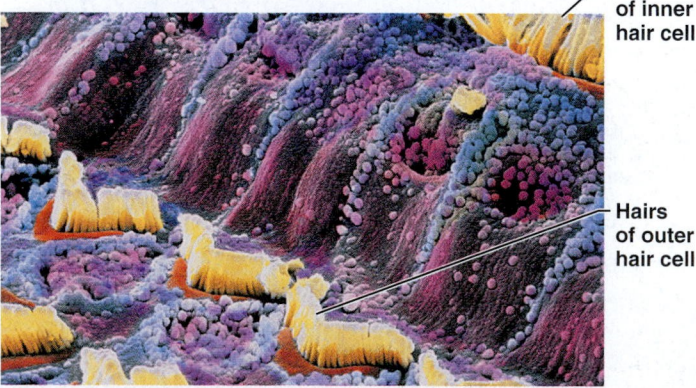

Hairs of inner hair cell

Hairs of outer hair cell

(d)

Figure 15.27 Anatomy of the cochlea. (a) Lateral view of part of the internal ear with a wedge-shaped section removed from the cochlea. **(b)** Magnified cross section of one turn of the cochlea, showing the relationship of the three scalae. This cross section has been rotated from its position in (a). **(c)** Detailed structure of the spiral organ. **(d)** Electron micrograph of cochlear hair cells (550×).

Table 15.2	**Summary of the Internal Ear**		
BONY LABYRINTH	**MEMBRANOUS LABYRINTH**	**FUNCTION**	**RECEPTOR REGION**
Semicircular canals	Semicircular ducts	Equilibrium: rotational (angular) acceleration	Crista ampullaris
Vestibule	Utricle and saccule	Equilibrium: head position relative to gravity, linear acceleration	Macula
Cochlea	Cochlear duct (scala media)	Hearing	Spiral organ

receptor region called a *crista ampullaris* (literally, crest of the ampulla). These receptors respond to rotational (angular) movements of the head.

Cochlea

The **cochlea** (kok′le-ah), from the Latin "snail," is a spiral, conical, bony chamber about the size of a split pea. It extends from the anterior part of the vestibule and coils for about 2½ turns around a bony pillar called the **modiolus** (mo-di′o-lus) (**Figure 15.27a**). Running through its center like a wedge-shaped worm is the membranous **cochlear duct**, which ends blindly at the cochlear apex (Figure 15.26). The cochlear duct houses the receptor organ of hearing, called the **spiral organ** or the *organ of Corti* (Figure 15.27b).

The cochlear duct and the **osseous spiral lamina**, a thin shelflike extension of bone that spirals up the modiolus like the thread on a screw, together divide the cavity of the bony cochlea into three separate chambers or **scalae** (*scala* = ladder).

- The **scala vestibuli** (ska′lah ves-tī′bu-li), part of the bony labyrinth, is continuous with the vestibule and begins at the oval window. It contains perilymph.

- The middle **scala media**, part of the membranous labyrinth, is the cochlear duct itself. It contains endolymph.

- The **scala tympani**, part of the bony labyrinth, terminates at the membrane-covered round window. It contains perilymph.

The perilymph-containing chambers are continuous with each other at the cochlear apex, a region called the **helicotrema** (hel″ĭ-ko-tre′mah; "the hole in the spiral") (see Figure 15.30 on p. 587).

The "roof" of the cochlear duct, separating the scala media from the scala vestibuli, is the **vestibular membrane** (Figure 15.27b). The duct's external wall, the **stria vascularis**, is composed of an unusual richly vascularized mucosa that secretes endolymph. The "floor" of the cochlear duct is composed of the osseous spiral lamina and the flexible, fibrous basilar membrane, which supports the spiral organ. The **basilar membrane**, which plays a critical role in sound reception, is narrow and thick near the oval window and gradually widens and thins as it approaches the cochlear apex.

The spiral organ, which rests atop the basilar membrane, is composed of supporting cells and hearing receptor cells called *cochlear hair cells*. The hair cells are arranged functionally— specifically, one row of **inner hair cells** and three rows of **outer hair cells**—sandwiched between the tectorial and basilar membranes (Figure 15.27c). (We will describe the tectorial membrane and hair cells further when we discuss sound

transduction.) Afferent fibers of the **cochlear nerve** [a division of the vestibulocochlear nerve (VIII)] coil about the bases of the hair cells and run from the spiral organ through the modiolus to the brain.

Table 15.2 summarizes the structures of the internal ear and their functions.

Check Your Understanding

14. Apart from the bony boundaries, which structure separates the external from the middle ear? Which two (nonbone) structures separate the middle from the inner ear?

For answers, see Answers Appendix.

15.8 Sound is a pressure wave that stimulates mechanosensitive cochlear hair cells

Learning Outcomes

▶ **Describe the sound conduction pathway to the fluids of the internal ear.**

▶ **Describe sound transduction.**

We can summarize human hearing in a single sprawling sentence: Sounds set up vibrations in air that beat against the eardrum that pushes a chain of tiny bones that press fluid in the internal ear against membranes that set up shearing forces that pull on the tiny hair cells that stimulate nearby neurons that give rise to action potentials that travel to the brain, which interprets them—and you hear. Before we unravel this intriguing sequence, let us describe sound, the stimulus for hearing.

Properties of Sound

Light can be transmitted through a vacuum (for instance, outer space), but sound depends on an *elastic* medium—one that physically transmits vibrations. Sound also travels much more slowly than light. Its speed in dry air is only about 331 m/s (0.2 mi/s), as opposed to about 300,000 km/s (186,000 mi/s) for light. A lightning flash is almost instantly visible, but the sound it creates (thunder) reaches our ears much more slowly. (For each second between the lightning bolt and the roll of thunder, the storm is 1/5 mile farther away.) The speed of sound is fastest in solids and slowest in gases, but it is constant in a given medium.

Sound is a pressure disturbance—alternating areas of high and low pressure—produced by a vibrating object and propagated by the molecules of the medium. Consider a vibrating tuning fork

15

(a) **A struck tuning fork alternately compresses and rarefies the air molecules around it.**

(b) **Sound waves radiate outward in all directions.**

Figure 15.28 Sound: Source and propagation. The graph in (a) shows that plotting the oscillating air pressures yields a sine wave.

(**Figure 15.28a**). If the tuning fork is struck on the left, its prongs will move first to the right, creating an area of high pressure by compressing the air molecules there. Then, as the prongs rebound to the left, the air on the left will be compressed, and the region on the right will be a *rarefied*, or low-pressure, area (since most of its air molecules have been pushed farther to the right).

As the fork vibrates alternately from right to left, it produces a series of compressions and rarefactions, collectively called a *sound wave*, which moves outward in all directions (Figure 15.28b). However, the individual air molecules just vibrate back and forth for short distances as they bump other molecules and rebound. Because the outward-moving molecules give up kinetic energy to the molecules they bump, energy is always transferred in the direction the sound wave is traveling. For this reason, the energy of the wave declines with time and distance (just as ripples in a pond get smaller as they move away from the pebble you threw).

We can illustrate a sound wave as an S-shaped curve, or *sine wave*, in which the compressed areas are crests and the rarefied areas are troughs (Figure 15.28a). Sound can be described in terms of two physical properties: frequency and amplitude.

Frequency

Frequency is defined as the number of waves that pass a given point in a given time. The sine wave of a pure tone has crests and troughs that repeat at specific intervals. The distance between two consecutive crests (or troughs) is the **wavelength** of the sound and is constant for a particular tone. The shorter the wavelength, the higher the frequency of the sound (Figure 15.28 and **Figure 15.29a**).

The frequency range of human hearing is from 20 to 20,000 waves per second, or *hertz (Hz)*. Our ears are most sensitive to frequencies between 1500 and 4000 Hz and, in that range, we can distinguish frequencies differing by only 2–3 Hz. We perceive different sound frequencies as differences in **pitch**: the higher the frequency, the higher the pitch.

A tuning fork produces a *tone*—a pure (but bland) sound with a single frequency—but most sounds are mixtures of several frequencies. This characteristic of sound, called **quality**, enables us to distinguish between the same musical note—say, high C—sung by a soprano or played on a piano. Sound quality also provides the richness and complexity of sounds (and music) that we hear.

Amplitude

The **amplitude**, or height, of the sine wave crests reveals a sound's intensity, which is related to its energy, or the pressure differences between its compressed and rarefied areas (Figures 15.28 and 15.29b).

Loudness refers to our subjective interpretation of sound intensity. Because we can hear such an enormous range of intensities, from the proverbial pin drop to a jet engine 10 million times more intense, sound intensity (and loudness) is measured in logarithmic units called **decibels (dB)** (des′ĭ-belz).

On a clinical audiometer, the decibel scale is arbitrarily set to begin at 0 dB, which is the threshold of hearing (barely audible

— High frequency (short wavelength) = high pitch
— Low frequency (long wavelength) = low pitch

(a) **Frequency is perceived as pitch.**

— High amplitude = loud
— Low amplitude = soft

(b) **Amplitude (size or intensity) is perceived as loudness.**

Figure 15.29 Frequency and amplitude of sound waves.

Auditory ossicles

Malleus   Incus   Stapes

Cochlear nerve

Scala vestibuli

Oval window

Helicotrema

Scala tympani

Cochlear duct

Basilar membrane

Tympanic membrane

Round window

① Sound waves vibrate the tympanic membrane.

② Auditory ossicles vibrate. Pressure is amplified.

③ Pressure waves created by the stapes pushing on the oval window move through fluid in the scala vestibuli.

④a Sounds with frequencies below the hearing range travel through the helicotrema and do not excite hair cells.

④b Sounds in the hearing range go through the cochlear duct, vibrating the basilar membrane and deflecting hairs on inner hair cells.

Figure 15.30 Pathway of sound waves. The cochlea is drawn as if uncoiled.

sound) for normal ears. Each 10-dB increase represents a ten-fold increase in sound intensity. A sound of 10 dB has 10 times more energy than one of 0 dB, and a 20-dB sound has 100 times (10 × 10) more energy. However, the same 10-dB increase represents only a twofold increase in loudness. In other words, most people would report that a 20-dB sound seems about twice as loud as a 10-dB sound. The normal range of hearing (from barely audible to the loudest sound we can process without excruciating pain) extends over a range of 120 dB. (The threshold of pain is 120 dB.)

Severe hearing loss occurs with frequent or prolonged exposure to sounds with intensities greater than 90 dB. That number becomes more meaningful when you realize that a normal conversation is in the 50-dB range, a noisy restaurant has 70-dB levels, and amplified concert music is often 120 dB or more, far above the 90-dB danger zone.

Transmission of Sound to the Internal Ear

Hearing occurs when the auditory area of the temporal lobe cortex is stimulated. However, before this can happen, sound waves must be propagated through air, membranes, bones, and fluids to stimulate receptor cells in the spiral organ (**Figure 15.30**).

① **Tympanic membrane.** Sound waves entering the external acoustic meatus strike the tympanic membrane and set it vibrating at the same frequency. The greater the intensity, the farther the membrane is displaced in its vibratory motion.

② **Auditory ossicles.** The motion of the tympanic membrane is amplified and transferred to the oval window by the ossicle lever system, which acts much like a piston to transfer the force striking the eardrum to the oval window. Because the tympanic membrane is 17–20 times larger than the oval

window, the pressure (force per unit area) actually exerted on the oval window is about 20 times that on the tympanic membrane. This increased pressure overcomes the stiffness and inertia of cochlear fluid and sets it into wave motion.

This situation can be roughly compared to the difference in pressure relayed to the floor by the broad rubber heels of a man's shoes versus a woman's tiny spike heels. The man's weight is spread over several square inches, and his heels will not make dents in a pliable vinyl floor. But spike heels concentrate the same force in an area of about 2.5 cm² (1 square inch) and *will* dent the floor.

③ **Scala vestibuli.** As the stapes rocks back and forth against the oval window, it sets the perilymph in the scala vestibuli into a similar back-and-forth motion. A pressure wave travels through the perilymph from the basal end toward the helicotrema, much as a piece of rope held horizontally can be set into wave motion by movements initiated at one end. Fluids cannot be compressed, so each time the stapes forces the fluid adjacent to the oval window medially, the membrane of the round window bulges laterally into the middle ear cavity and acts as a pressure valve.

④a **Helicotrema path.** Sounds of very low frequency (below 20 Hz) create pressure waves that take the complete route through the cochlea—up the scala vestibuli, around the helicotrema, and back toward the round window through the scala tympani (Figure 15.30). These low-frequency sounds do not activate the spiral organ and so are below the range of hearing.

④b **Basilar membrane path.** In contrast, sounds with frequencies high enough to hear create pressure waves that take a "short-cut" and are transmitted through the cochlear duct into the perilymph of the scala tympani. As a pressure wave descends through the flexible cochlear duct, it vibrates the basilar membrane. This vibration activates hair cells (receptor cells) at the site of this "shortcut" through the basilar membrane, causing action potentials to be sent to the brain.

Resonance of the Basilar Membrane

Sound waves make the basilar membrane vibrate. Maximal displacement of the membrane occurs where the fibers of the basilar membrane are "tuned" to a particular sound frequency (**Figure 15.31**). This characteristic of maximum movement at a particular frequency is called *resonance*. The fibers of the basilar membrane span its width like the strings of a harp. The fibers near the oval window (cochlear base) are short and stiff, and they resonate in response to high-frequency pressure waves (Figure 15.31). The longer, more floppy basilar membrane fibers near the cochlear apex resonate in time with lower-frequency pressure waves. As a result, the resonance of the basilar membrane mechanically processes sound signals before the signals ever reach the receptors.

Sound Transduction

Excitation of Inner Hair Cells

How does movement of the basilar membrane result in action potentials being sent to the brain? Transduction of sound stimuli occurs after localized movements of the basilar membrane "tweak" or deflect the hairs of the inner hair cells (**Figure 15.32**). The hairs of the hair cells protrude into the K^+-rich endolymph and the longest are enmeshed in the overlying stiff gel-like **tectorial membrane**. As the basilar membrane vibrates, the hair cells move with it. Their hairs pivot, opening and closing ion channels and generating an electrical signal. Let's look at how this happens in more detail.

The "hairs" protruding from the apex of a cochlear hair cell are actually microvilli called **stereocilia**. These stereocilia are longest on one side of the cell and shortest on the other. The stereocilia of the hair cells are stiffened by actin filaments so they bend *only* at their bases—that is, they pivot. The stereocilia are bound together by fine fibers called *tip links* that connect to mechanically gated ion channels. Pulling on the tip links opens the ion channels.

Pivoting the stereocilia toward the tallest stereocilium increases the distance between the tips of adjacent stereocilia and so puts tension on the tip links. Like a rope pulling on a trap door, this opens cation channels in the adjacent shorter stereocilia. K^+ and Ca^{2+} from the endolymph flow through the open channels, creating a graded depolarization (receptor potential). Pivoting the stereocilia toward the shortest stereocilium relaxes the tip links, closes the mechanically gated ion channels, and allows repolarization and even a graded hyperpolarization (Figure 15.32).

Depolarization increases intracellular Ca^{2+} and so increases the hair cells' release of neurotransmitter (glutamate). This causes the afferent cochlear fibers to transmit more impulses to the brain. Hyperpolarization produces the exact opposite effect.

The vast majority of the sensory fibers of the spiral ganglia (90–95%) service the inner hair cells, which shoulder nearly the entire responsibility for sending auditory messages to the brain.

Role of Outer Hair Cells

Most fibers coiling around the more numerous outer hair cells are *efferent* fibers that convey messages from brain to ear. So what do the outer hair cells do? The outer hair cells act on the

Let's uncoil the cochlea to see how it separates different frequencies of sound so that we can hear different pitches.

The properties of the basilar membrane change along its length.

As a result, different frequencies vibrate the basilar membrane in different places.

Low-frequency sounds can't move the short, stiff fibers at the base. They continue to the longer, floppier apex fibers.

Medium-frequency sounds vibrate the basilar membrane near its middle.

High-frequency sounds vibrate the basilar membrane near its base.

| 20,000 | 2000 | 200 | 20 |

Frequency (Hz)

Each frequency that we hear corresponds to a specific place on the basilar membrane. As a result, the brain determines the frequency of a sound wave by the *location* of the hair cells activated by the vibrating basilar membrane.

Figure 15.31 Basilar membrane function. The basilar membrane separates sound frequencies, allowing pitch discrimination.

basilar membrane itself. When outer hair cells depolarize and hyperpolarize as the basilar membrane moves, they contract and stretch in a type of cellular dance called *fast motility*. Their strange behavior changes the stiffness of the basilar membrane.

Outer hair cell motility serves two functions:

- It increases the responsiveness of the inner hair cells by amplifying the motion of the basilar membrane—a kind of cochlear tuning that increases the cochlea's ability to distinguish small differences in frequency.

- It may help protect the inner hair cells from damage. Loud sounds activate a negative feedback loop from the brain stem to the outer hair cells via the efferent fibers, which release

Figure 15.32 Pivoting of stereocilia (hairs) opens or closes mechanically gated ion channels in hair cells.

neurotransmitters that cause the outer hair cells to stiffen. This dampens the motion of the basilar membrane and spreads the sound energy over a wider area.

Check Your Understanding

15. Which structure inside the spiral organ allows us to differentiate sounds of different pitch?

16. The stapes pushes on the fluid in the scala vestibuli but cannot compress the fluid. What is the "pressure valve" that allows fluid to move within the cochlea?

17. PREDICT Sound frequencies too low to hear pass through the helicotrema without displacing the basilar membrane. Sound frequencies that we hear do displace the basilar membrane. Predict what happens to high-frequency sound waves that we cannot hear.

18. MAKE CONNECTIONS Opening cation channels causes depolarization. The influx of which ion usually causes this? (Hint: See the discussion on p. 417 of Chapter 11.) Which ions cause depolarization in hair cells? Why does K^+ move into these cells, rather than out?

For answers, see Answers Appendix.

15.9 Sound information is processed and relayed through brain stem and thalamic nuclei to the auditory cortex

Learning Outcomes

▶ Describe the pathway of impulses traveling from the cochlea to the auditory cortex.

▶ Explain how we are able to differentiate pitch and loudness, and to localize the source of sounds.

The Auditory Pathway

The ascending auditory pathway transmits auditory information primarily from the cochlear receptors (the inner hair cells) to

Figure 15.33 The auditory pathway. This simplified diagram shows only the pathway from the right ear.

the cerebral cortex. Action potentials generated in the cochlea pass through the **spiral ganglion**, where the auditory bipolar cell bodies reside, and along the afferent fibers of the cochlear nerve to the **cochlear nuclei** of the medulla (**Figure 15.33**).

From there, neurons project to the **superior olivary nucleus**, which lies at the junction of the medulla and pons. Beyond this, axons ascend in the **lateral lemniscus** (a fiber tract) to the **inferior colliculus** (auditory reflex center in the midbrain), which projects to the **medial geniculate nucleus** of the thalamus. Axons of the thalamic neurons then project to the **primary auditory cortex**, which provides conscious awareness of sound.

The auditory pathway is unusual because not all of the fibers from each ear cross over to the other side of the brain. For this reason, each auditory cortex receives impulses from both ears.

Auditory Processing

If you are at a Broadway musical, the sound of the instruments, the actors' voices, rustling of clothing, and closing of doors are all intermingled in your awareness. Yet, your auditory cortex can distinguish the separate parts of this auditory jumble. Whenever the difference between sound wavelengths is sufficient for discrimination, you hear two separate and distinct tones.

Cortical processing of sound stimuli is complex. For example, certain cortical cells depolarize at the beginning of a particular tone, and others depolarize when the tone ends. Some cortical cells depolarize continuously, and others appear to have high thresholds (low sensitivity), and so on. Here we will concentrate on the more straightforward aspects of cortical perception of pitch, loudness, and sound location.

Perception of Pitch

As we explained, sound waves of different frequencies activate hair cells in different positions along the length of the basilar membrane (Figure 15.31). Impulses from specific hair cells are interpreted as specific pitches. When a sound is composed of tones of many frequencies, it activates several populations of cochlear hair cells and cortical cells simultaneously, and we perceive multiple tones.

Detection of Loudness

Louder sounds cause larger movements of the tympanic membrane, auditory ossicles, and oval window, and pressure waves of greater amplitude in the fluids of the cochlea. These larger waves in turn cause larger movements of the basilar membrane, larger deflections of the hairs on the hair cells, and larger graded potentials in the hair cells. As a result, they release more neurotransmitter and generate more frequent action potentials. The brain interprets more frequent action potentials as greater loudness. In addition, as more neurotransmitter is released, more of the 10 or so bipolar cells connected to a given hair cell are recruited to fire action potentials.

Localization of Sound

Several brain stem nuclei (most importantly the superior olivary nuclei) help us localize a sound's source in space by means of two cues: the *relative intensity* and the *relative timing* of sound waves reaching the two ears. If the sound source is directly in front, in back, or overhead, the intensity and timing cues are the same for both ears. However, when sound comes from one side, it activates the receptors of the nearer ear slightly earlier and also more vigorously.

Check Your Understanding

19. **WHAT IF?** If the brain stem did not receive input from both ears, what would you not be able to do?

For answers, see Answers Appendix.

15.10 Hair cells in the maculae and cristae ampullares monitor head position and movement

Learning Outcome

▶ Explain how the balance organs of the semicircular canals and the vestibule help maintain equilibrium.

The equilibrium sense is not easy to describe because it does not "see," "hear," or "feel," but responds (frequently without our awareness) to various head movements. Furthermore, this sense depends not only on inputs from the internal ear but also on vision and information from stretch receptors of muscles and tendons.

The **vestibular apparatus** consists of the semicircular canals, the vestibule (utricle and saccule), and the receptors that they contain. Under normal conditions, they send signals to the brain that initiate reflexes needed to make the simplest changes in position as well as more complex movements such as pitching a baseball.

The equilibrium receptors of the vestibular apparatus can be divided into two functional arms. The receptors in the vestibule monitor linear acceleration and the position of the head with respect to gravity. Because gravity is constant, this is called our sense of *static equilibrium*. The semicircular canals monitor changes in head rotation, called our sense of *dynamic equilibrium*.

The Maculae

The **maculae** (mak'u-le; "spots"), one in each saccule wall and one in each utricle wall (**Figure 15.34**), are sensory receptor organs that monitor the position of the head in space. In so doing, they play a key role in controlling posture. They respond to *linear* acceleration, that is, changes in straight-line speed and direction, but not rotation.

Anatomy of a Macula

Each macula is a flat epithelial patch containing **hair cells**. Like the cochlear hair cells, vestibular hair cells have stereocilia (microvilli). In addition, they have a true cilium, called a **kinocilium**, next to the tallest stereocilia. **Supporting cells** surround the macula's scattered hair cells (Figure 15.34). The "hairs" of the hair cells are embedded in the overlying **otolith membrane** (o'to-lith), a jellylike mass studded with tiny stones (calcium carbonate crystals) called **otoliths** ("ear stones"). The otoliths, though small, are dense and they increase the membrane's weight and its inertia (resistance to changes in motion).

In the utricle, the macula is horizontal, and the hairs are vertically oriented when the head is upright (Figure 15.34). For this reason, the utricular maculae respond best to acceleration in the horizontal plane and tilting the head, because vertical

Figure 15.34 Structure and function of a macula. (a) Structure. **(b)** Function of a macula of the utricle in signaling head position.

(up-down) movements do not displace their horizontal otolith membrane.

In the saccule, on the other hand, the macula is nearly vertical, and the hairs protrude horizontally into the otolith membrane. The saccular maculae respond best to vertical movements, such as the sudden acceleration of an elevator.

The hair cells synapse with fibers of the **vestibular nerve**, whose endings coil around their bases. Like the cochlear nerve, the vestibular nerve is a subdivision of the vestibulocochlear nerve (VIII). The cell bodies of the sensory neurons are located in the nearby **superior** and **inferior vestibular ganglia** (see Figure 15.26).

Activating Receptors of a Macula

What happens in the maculae that leads to sensory transduction? When your head starts or stops moving in a linear direction, inertia causes the otolith membrane to slide backward or forward like a greased plate over the hair cells, bending the hairs. For example, when you start to run, the otolith membranes of the utricle maculae lag behind, bending the hairs backward. When you suddenly stop, the otolith membrane slides abruptly forward (just as you slide forward in your car when you brake),

bending the hairs forward. Likewise, when you nod your head or fall, the otoliths sag inferiorly, bending the hairs of the maculae in the saccules.

The hair cells release neurotransmitter continuously, but movement of their hairs modifies the amount they release just as it does in the cochlea (see Figure 15.32).

- When the hairs bend toward the kinocilium, the hair cells depolarize, stepping up their pace of neurotransmitter release, and more impulses travel up the vestibular nerve to the brain.

- When the hairs bend away from the kinocilium, the receptors hyperpolarize and release less neurotransmitter, generating fewer impulses.

In either case, the brain is informed of the changing position of the head in space (Figure 15.34b).

It is important to understand that the maculae respond to *changes* in the velocity of head movement (linear acceleration or deceleration). Because most hair cells adapt quickly (resuming their basal level of neurotransmitter release), they do not report on unchanging head positions. In this way, the maculae help us to maintain normal head position with respect to gravity.

15

The Cristae Ampullares

The receptor for rotational acceleration, called the **crista ampullaris** or simply *crista*, is a minute elevation in the ampulla of each semicircular duct (see Figure 15.26). Like the maculae, the cristae are excited by head movement (acceleration and deceleration), but in this case the major stimuli are rotational (angular) movements. When you twirl on the dance floor or suffer through a rough boat ride, these gyroscope-like receptors are working overtime. Since the semicircular canals are located in all three planes of space, all rotational movements of the head disturb one or another *pair* of cristae (one in each ear).

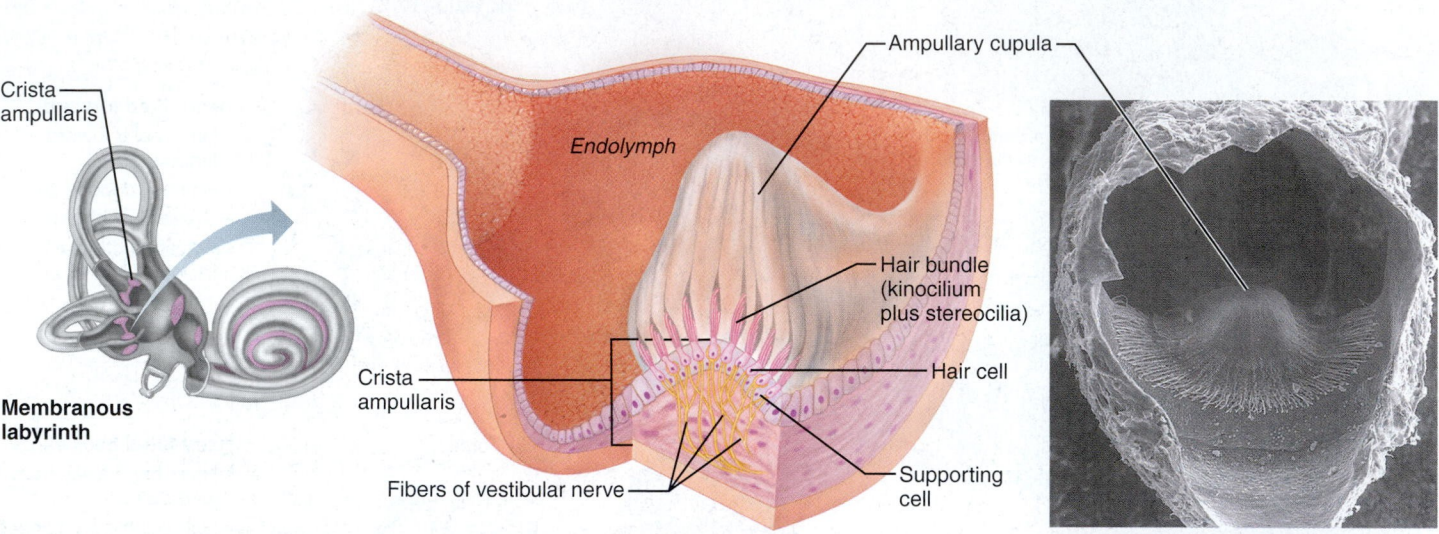

(a) Anatomy of a crista ampullaris in a semicircular duct

(b) Scanning electron micrograph of a crista ampullaris (200×)

At rest, the cupula stands upright.

During rotational acceleration, endolymph moves inside the semicircular canals in the direction opposite the rotation (it lags behind due to inertia). Endolymph flow bends the cupula and excites the hair cells.

As rotational movement slows, endolymph keeps moving in the direction of rotation. Endolymph flow bends the cupula in the opposite direction from acceleration and inhibits the hair cells.

(c) Movement of the ampullary cupula during rotational acceleration and deceleration

Figure 15.35 Location, structure, and function of a crista ampullaris in the internal ear.

Anatomy of a Crista Ampullaris

Each crista is composed of supporting cells and hair cells whose structure and function are basically the same as the hair cells of the cochlea and maculae. In this case, the gelled mass is an **ampullary cupula** (ku′pu-lah), which resembles a pointed cap (**Figure 15.35**). The cupula is a delicate, loosely organized network of gelatinous strands that radiate outward to contact the "hairs" of each hair cell. Dendrites of vestibular nerve fibers encircle the base of the hair cells.

Activating Receptors of the Crista Ampullaris

The cristae respond to *changes* in the velocity of rotational movements of the head. Because of its inertia, the endolymph in the semicircular ducts moves briefly in the direction *opposite* the body's rotation, deforming the crista in the duct. As the hairs bend, the hair cells depolarize and impulses reach the brain at a faster rate (Figure 15.35c, center). Bending the cilia in the opposite direction causes hyperpolarization and generates fewer impulses. Because the axes of the hair cells in the complementary semicircular ducts are opposite, rotation in a given direction depolarizes the receptors in one ampulla of the pair, and hyperpolarizes the receptors in the other.

If the body continues to rotate at a constant rate, the endolymph eventually comes to rest—it moves along at the same speed as the body, and the hair cells are no longer stimulated. Consequently, if we are blindfolded, we cannot tell whether we are moving at a constant speed or not moving at all after the first few seconds of rotation. However, when we suddenly stop moving, the endolymph keeps on going, in effect reversing its direction within the canal. This sudden reversal in the direction of hair bending alters membrane voltage in the receptor cells and modifies the rate of impulse transmission, which tells the brain that we have slowed or stopped (Figure 15.35c, right).

The key point to remember when considering both types of equilibrium receptors is that the rigid bony labyrinth moves with the body, while the fluids (and gels) within the membranous labyrinth are free to move at various rates, depending on the forces (gravity, acceleration, and so on) acting on them.

Vestibular Nystagmus

Information transmitted from the semicircular canals is particularly important to reflex movements of the eyes. **Vestibular nystagmus** is a complex of rather strange eye movements that occurs during and immediately after rotation.

As you rotate, your eyes slowly drift in the opposite direction, as though fixed on some object in the environment. This reaction relates to the backflow of endolymph in the semicircular ducts. Then, because of CNS compensating mechanisms, the eyes jump rapidly toward the direction of rotation to establish a new fixation point. These alternating eye movements continue until the endolymph comes to rest.

When you stop rotating, at first your eyes continue to move in the direction of the previous spin, and then they jerk rapidly in the opposite direction. This sudden change is caused by the change in the direction in which the cristae bend after you stop. Nystagmus is often accompanied by vertigo, a false sensation of movement.

The Equilibrium Pathway to the Brain

Our responses to body imbalance, such as when we stumble, must be fast and reflexive. By the time we "thought about" correcting our fall, we would already be on the ground! Accordingly, information from equilibrium receptors goes directly to reflex centers in the brain stem, rather than to the cerebral cortex as with the other special senses.

The nerve pathways connecting the vestibular apparatus with the brain are complex. The transmission sequence begins when the hair cells in the vestibular apparatus are activated. As depicted in **Figure 15.36**, impulses travel initially to one of two destinations: the **vestibular nuclei** in the brain stem or the **cerebellum**. The vestibular nuclei, the major integrative center for balance, also receive inputs from the visual and somatic receptors, particularly

Input: Information about the body's position in space comes from three main sources and is fed into two major processing areas in the central nervous system.

Vestibular receptors

Visual receptors

Somatic receptors (skin, muscle and joints)

Cerebellum

Vestibular nuclei (brain stem)

Central nervous system processing

Oculomotor control (cranial nerve nuclei III, IV, VI)

(eye movements)

Spinal motor control (cranial nerve XI nuclei and vestibulospinal tracts)

(neck, limb, and trunk movements)

Output: Responses by the central nervous system provide fast reflexive control of the muscles serving the eyes, neck, limbs, and trunk.

Figure 15.36 Neural pathways of the balance and orientation system.

from proprioceptors in neck muscles that report on the position of the head. These nuclei integrate this information and then send commands to brain stem motor centers that control the extrinsic eye muscles (cranial nerve nuclei III, IV, and VI) and reflex movements of the neck, limb, and trunk muscles (via the vestibulospinal tracts). The ensuing reflex movements of the eyes (the *vestibulo-ocular reflex*) and body allow us to remain focused on the visual field and quickly adjust our body position to maintain or regain balance. The vestibulo-ocular reflex is a normal reflex that occurs when the head rotates. Involuntary eye movements compensate for the head movement so as to keep the image steady.

The cerebellum also integrates inputs from the eyes and somatic receptors (as well as from the cerebrum). It coordinates skeletal muscle activity and regulates muscle tone to maintain head position, posture, and balance, often in the face of rapidly changing inputs. Its "specialty" is fine control of delicate postural movements and timing.

Notice that the vestibular apparatus *does not automatically compensate* for forces acting on the body. Its job is to send warning signals to the CNS, which initiates the appropriate compensations.

HOMEOSTATIC IMBALANCE 15.15 — CLINICAL

Equilibrium problems are usually obvious and unpleasant. Nausea, dizziness, and loss of balance are common and there may be nystagmus in the absence of rotational stimuli.

Motion sickness appears to be due to sensory input mismatches. For example, if you are inside a ship during a storm, visual inputs indicate that your body is fixed with reference to a stationary environment (your cabin). But as rough seas toss the ship about, your vestibular apparatus detects movement and sends impulses that disagree with the visual information. The brain thus receives conflicting signals, and its "confusion" somehow leads to motion sickness. Warning signals include excessive salivation, pallor, rapid deep breathing, and profuse sweating. Removal of the stimulus usually ends the symptoms. Over-the-counter drugs, such as meclizine, depress vestibular inputs and help alleviate the symptoms. ● ● ● ● ● ● ●

Check Your Understanding

20. For the maculae shown below, which best shows the position of the otolith membrane in the saccule of a person in an elevator that is just beginning to go up? Will this cause an increase or decrease in action potential frequency in the axons of the vestibular nerve from this saccule? How can you tell?

(a)   (b)

21. For each of the following phrases, indicate whether it applies to a macula or a crista ampullaris: inside a semicircular canal; contains otoliths; responds to linear acceleration and deceleration; has a cupula; responds to rotational acceleration and deceleration; inside a saccule.

For answers, see Answers Appendix.

15.11 Ear abnormalities can affect hearing, equilibrium, or both — CLINICAL

Learning Outcome

▶ List possible causes and symptoms of otitis media, deafness, and Ménière's syndrome.

Deafness

Any hearing loss, no matter how slight, is **deafness** of some sort. Deafness is classified as conduction or sensorineural deafness, according to its cause.

Conduction deafness occurs when something hampers sound conduction to the fluids of the internal ear. For example, compacted earwax can block the external acoustic meatus, or a *perforated* (*ruptured*) eardrum can prevent sound conduction from the eardrum to the ossicles. But the most common causes of conduction deafness are middle ear inflammations (otitis media) and **otosclerosis** (o″to-sklĕ-ro′sis) of the ossicles.

Otosclerosis ("hardening of the ear") occurs when overgrowth of bony tissue fuses the base of the stapes to the oval window or welds the ossicles to one another. In such cases, vibrations of the skull bones conduct sound to the receptors of that ear, which is far less satisfactory. Otosclerosis is treated surgically.

Sensorineural deafness results from damage to neural structures at any point from the cochlear hair cells to and including the auditory cortical cells. This type of deafness typically results from the gradual loss of hair cells throughout life. Hair cells can also be destroyed by a single explosively loud noise or prolonged exposure to high-intensity sounds, such as music or industrial noise, which literally tear off their cilia. Other causes of sensorineural deafness are degeneration of the cochlear nerve, strokes, and tumors in the auditory cortex.

Hair cells don't normally regenerate in mammals, but researchers are seeking ways to prod supporting cells into becoming hair cells. For congenital defects or age- or noise-related cochlear damage, cochlear implants (devices that convert sound energy into electrical signals) can be inserted into the temporal bone. Modern

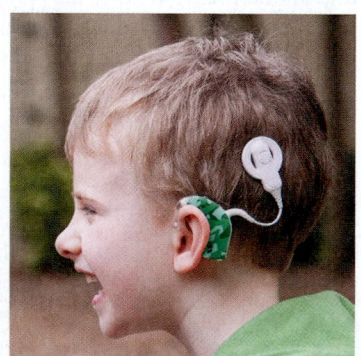

Figure 15.37 Boy with a cochlear implant. The parts visible on the boy's head are a microphone and processor that transmit signals to electrodes embedded in the temporal bone.

implants are so effective that even children born deaf can hear well enough to learn to speak well (**Figure 15.37**).

Tinnitus

Tinnitus (tĭ-ni′tus) is a ringing, buzzing, or clicking sound in the ears in the absence of auditory stimuli. It is usually a symptom rather than a disease. For example, tinnitus is one of the first symptoms of cochlear nerve degeneration. It may also signal inflammation of the middle or internal ears and is a side effect of some medications, such as aspirin.

Recent evidence suggests that tinnitus is analogous to phantom limb pain. In other words, it is "phantom cochlear noise" caused by destruction of some neurons in the auditory pathway and the subsequent ingrowth of nearby neurons whose signals are interpreted as noise by the CNS.

Ménière's Syndrome

Classic **Ménière's syndrome** (men″ē-ārz′) is a labyrinth disorder that seems to affect all three parts of the internal ear. The afflicted person has repeated attacks of vertigo, nausea, and vomiting. Balance is so disturbed that standing erect is nearly impossible. Tinnitus may be present, and hearing may be impaired or lost completely.

Mild cases can usually be managed by antimotion drugs or a low-salt diet and diuretics to decrease endolymph fluid volume. In severe cases, draining the excess endolymph from the internal ear may help. A last resort is removal of the entire malfunctioning labyrinth.

Check Your Understanding

22. Six-year-old Mohammed has a cold and says his ears feel "full" and he "can't hear well." Explain what has happened in Mohammed's ears. Which type of deafness does Mohammed have—conduction or sensorineural?

For answers, see Answers Appendix.

Developmental Aspects of the Special Senses

Taste and Smell

All the special senses are functional, to a greater or lesser degree, at birth. Smell and taste are sharp, and infants relish food that adults consider bland. Some researchers claim that smell is just as important as touch in guiding newborn infants to their mother's breast. However, very young children seem indifferent to odors.

There are few problems with the chemical senses during childhood and young adulthood. Women generally have a more acute sense of smell than men, and nonsmokers have a sharper sense of smell than smokers.

Beginning in the fourth decade of life, our ability to taste and smell declines due to the gradual loss of receptors, which are replaced more slowly as we age. More than half of people over age 65 have serious problems detecting odors, and their sense of taste is poor. This makes food taste bland and contributes to loss of appetite.

Vision

By the fourth week of development, the eyes begin forming as the **optic vesicles** that protrude from the diencephalon (◀ **Figure 12.1c, p. 435**). Soon these hollow vesicles indent to form double-layered **optic cups**, and their stalks form the optic nerves and provide a pathway for blood vessels to reach the eye interior.

Once an optic vesicle reaches the overlying surface ectoderm, it induces the ectoderm to thicken and then form a **lens vesicle** that pinches off into the cavity of the optic cup, where it becomes the lens. The lining (internal layer) of the optic cup becomes the neural layer of the retina, and the outer layer forms the pigmented layer. The rest of the eye tissues and the vitreous humor are formed by mesenchymal cells derived from the mesoderm that surrounds the optic cup.

In the darkness of the uterus, the fetus cannot see. Nonetheless, even before the light-sensitive portions of the photoreceptors develop, the central nervous system connections have been made and are functional. During infancy, synaptic connections are fine-tuned, and the typical cortical fields that allow binocular vision are established.

HOMEOSTATIC IMBALANCE 15.16   CLINICAL

Congenital problems of the eyes are relatively uncommon, but their incidence is dramatically increased by certain maternal infections, particularly rubella (German measles) during the critical first three months of pregnancy. Common rubella sequels include blindness and cataracts. •••••••

As a rule, vision is the only special sense not fully functional at birth. Because the eyeballs are foreshortened, most babies are hyperopic. The newborn sees only in gray tones, eye movements are uncoordinated, and often only one eye at a time is used. The lacrimal glands are not completely developed until about two weeks after birth, so babies are tearless for this period, even though they may cry lustily. By 5 months, infants can follow moving objects with their eyes, but visual acuity is still poor.

By the age of 3 years, depth perception is present and color vision is well developed. Because the eyeball has grown, visual acuity has improved, providing a readiness to begin reading. The eye reaches its adult size at 8–9 years of age. Presbyopia begins to set in around age 40 owing to decreasing lens elasticity.

With age, the lens loses its clarity and discolors. As a result, it begins to scatter light, causing a glare that is distressing when driving at night. The dilator pupillae muscles become less efficient, so the pupils stay partly constricted. These two changes decrease the amount of light reaching the retina, and visual acuity is dramatically lower in people over 70. In addition, the lacrimal glands are less active and the eyes tend to be dry and more susceptible to infection. Elderly persons are also at risk for certain conditions that cause blindness, such as macular degeneration, glaucoma, cataracts, atherosclerosis, and diabetes mellitus.

Hearing and Balance

The ear begins to develop in the three-week embryo. The internal ears develop first, from thickenings of the surface ectoderm called the **otic placodes** (o'tik plak'ōds), which lie lateral to the hindbrain on each side. The otic placode invaginates, forming the **otic pit** and then the **otic vesicle**, which detaches from the surface epithelium. The otic vesicle develops into the membranous labyrinth. The surrounding mesenchyme forms the bony labyrinth.

The middle ear cavity and pharyngotympanic tube of the middle ear develop from **pharyngeal pouches**, lateral outpocketings of the endoderm lining the pharynx. The auditory ossicles develop from neural crest cells.

The external acoustic meatus and external face of the tympanic membrane of the external ear differentiate from the **pharyngeal cleft** (*branchial groove*), an indentation of the surface ectoderm, and the auricle develops from swellings of the surrounding tissue.

Newborn infants can hear, but early responses to sound are mostly reflexive—for example, crying in response to a startling noise. By the fourth month, infants will turn to the voices of family members. Critical listening begins as toddlers increase their vocabulary, and good language skills are closely tied to the ability to hear well.

HOMEOSTATIC IMBALANCE 15.17   **CLINICAL**

Congenital abnormalities of the ears are fairly common. Examples include partly or completely missing pinnae and closed or absent external acoustic meatuses. Maternal rubella during the first trimester commonly results in sensorineural deafness. ●●●●●●

Except for ear inflammations, mostly due to infections, few problems affect the ears during childhood and adult life. By the 60s, however, deterioration of the spiral organ becomes noticeable. We are born with approximately 30,000 hair cells, but their number declines as they are damaged or destroyed by loud noises, disease, or drugs (for example, the antibiotic streptomycin). The hair cells *are* replaced, but at such a slow rate that there really is no functional regeneration.

The ability to hear high-pitched sounds leaves us first. This condition, called **presbycusis** (pres"bĭ-ku'sis), is a type of sensorineural deafness. Although presbycusis is considered a disability of old age, it is becoming much more common in younger people as our world grows noisier.

RELATED CLINICAL TERMS

Ageusia (ah-gu'ze-ah) Loss or impairment of the taste sense.

Age-related macular degeneration (ARMD) Progressive deterioration of the macula lutea that destroys central vision; the main cause of vision loss in those over age 65. Buildup of pigments in the macula impairs functioning of the pigmented epithelium. Continued accumulation of pigment leads to the "dry" form of ARMD, in which many pigment cells and macular photoreceptors die. The dry form is largely untreatable, although a specific cocktail of vitamins and zinc slows its progression. Less common is the "wet" form in which new blood vessels from the choroid grow into the retina. Blood and fluids leak from these vessels, scarring and detaching the retina. The cause of wet ARMD is unknown, but several treatment options can slow its progression—laser treatments that destroy some of the invading vessels and drugs that prevent blood vessel growth.

Blepharitis (blef"ah-ri'tis; *blephar* = eyelash; *itis* = inflammation) Inflammation of the margins of the eyelids.

Enucleation (e-nu"kle-a'shun) Surgical removal of an eyeball.

Exophthalmos (ek"sof-thal'mos; *ex* = out; *ophthalmo* = the eye) Anteriorly bulging eyeballs. Seen in some cases of hyperthyroidism. See Figure 16.10b on p. 620.

Labyrinthitis Inflammation of the labyrinth.

Ophthalmology (of"thal-mol'o-je) The science that studies the eye and eye diseases. An ophthalmologist is a medical doctor who specializes in treating eye disorders.

Optometrist A licensed nonphysician who measures vision and prescribes corrective lenses.

Otalgia (o-tal'je-ah; *algia* = pain) Earache.

Otitis externa Inflammation and infection of the external acoustic meatus, caused by bacteria or fungi that enter the canal from outside, especially when the canal is moist (e.g., after swimming).

Otoacoustic emissions Sounds generated by the movement of outer hair cells in the cochlea. Stimulated otoacoustic emissions are an inexpensive way to screen newborns for hearing defects.

Papilledema (pap"il-ĕ-de'mah; *papill* = nipple; *edema* = swelling) Protrusion of the optic disc into the eyeball, which can be observed by ophthalmoscopic examination; caused by conditions that increase intracranial pressure.

Scotoma (sko-to'mah; *scoto* = darkness) A blind spot other than the normal (optic disc) blind spot. Has many causes, including stroke or a brain tumor pressing on fibers of the visual pathway.

Trachoma (trah-ko'mah; *trach* = rough) A highly contagious bacterial (chlamydial) infection of the conjunctiva and cornea. Common worldwide, it blinds millions of people in poor countries of Africa and Asia. Treated with eye ointments containing antibiotic drugs.

Weber's test Hearing test during which a sounding tuning fork is held to the forehead. In those with normal hearing, the tone is heard equally in both ears. The tone will be heard best in the "good" ear if sensorineural deafness is present, and in the "bad" ear if conduction deafness is present.

CHAPTER SUMMARY

PART 1

THE EYE AND VISION

15.1 The eye has three layers, a lens, and humors, and is surrounded by accessory structures (pp. 554–563)

1. The eyes are enclosed in bony orbits and cushioned by fat.

Accessory Structures of the Eye (pp. 554–557)

2. Eyebrows help to shade and protect the eyes.
3. Eyelids protect and lubricate the eyes by reflex blinking. Within the eyelids are the orbicularis oculi and levator palpebrae superioris muscles, and modified sebaceous and sweat glands.
4. The conjunctiva is a mucosa that lines the eyelids and covers the anterior eyeball surface. Its mucus lubricates the eyeball surface.
5. The lacrimal apparatus consists of the lacrimal gland (which produces a saline solution containing mucus, lysozyme, and antibodies), the lacrimal canaliculi, the lacrimal sac, and the nasolacrimal duct.
6. The extrinsic eye muscles (superior, inferior, lateral, and medial rectus and superior and inferior oblique) move the eyeballs.

Structure of the Eyeball (pp. 558–563)

7. The wall of the eyeball is made up of three layers. The outermost fibrous layer consists of the sclera and the cornea. The sclera protects the eye and gives it shape; the cornea allows light to enter the eye.
8. The middle, pigmented vascular layer (uvea) consists of the choroid, the ciliary body, and the iris. The choroid provides nutrients to the eye and prevents light scattering within the eye. The ciliary muscles of the ciliary body control lens shape; the iris controls the size of the pupil.
9. The inner layer, or retina, consists of an outer pigmented layer and an inner neural layer. The neural layer contains photoreceptors (rods and cones), bipolar cells, and ganglion cells. Ganglion cell axons form the optic nerve, which exits via the optic disc ("blind spot").
10. The posterior segment of the eyeball, behind the lens, contains vitreous humor, which helps support the eyeball and keep the retina in place. The anterior segment, anterior to the lens, is filled with aqueous humor, formed by capillaries in the ciliary processes and drained into the scleral venous sinus. Aqueous humor is a major factor in maintaining intraocular pressure.
11. The biconvex lens is suspended within the eye by the ciliary zonule attached to the ciliary body. The lens is the only adjustable refractory structure of the eye.

15.2 The cornea and lens focus light on the retina (pp. 563–567)

1. Visible light is made up of those wavelengths of the electromagnetic spectrum that excite the photoreceptors.
2. Light is refracted (bent) when passing from one transparent medium to another of different density. Concave lenses disperse light; convex lenses converge light and bring its rays to a focal point. The greater the lens curvature, the more light bends.
3. As light passes through the eye, the cornea and lens bend and focus it on the retina. The cornea accounts for most of the refraction, but the lens allows active focusing for different distances.
4. Focusing for distant vision requires no special movements of the eye structures. Focusing for close-up vision requires accommodation (bulging of the lens), pupillary constriction, and convergence of the eyeballs (all controlled by cranial nerve III).
5. Refractory problems include presbyopia, myopia, hyperopia, and astigmatism.

15.3 Phototransduction begins when light activates visual pigments in retinal photoreceptors (pp. 567–573)

1. The outer segments of the photoreceptors contain light-absorbing visual pigment in membrane-bounded discs.
2. The light-absorbing molecule retinal is combined with various opsins to form the visual pigments. Rod visual pigment, rhodopsin, is a combination of retinal and opsin. The three types of cones all contain retinal, but each has a different type of opsin. Each cone type responds maximally to one color of light: red, blue, or green.
3. Rods respond to low-intensity light and provide night and peripheral vision. Cones are bright-light, high-discrimination receptors that provide color vision. Anything that must be viewed precisely is focused on the cone-rich fovea centralis.
4. When struck by light, retinal changes shape (11-*cis* to all-*trans*) and activates opsin. Activated opsin activates transducin (a G protein) which in turn activates PDE, an enzyme that breaks down cGMP, allowing the cation channels to close. This hyperpolarizes the receptor cells and inhibits their release of neurotransmitter.
5. Photoreceptors and bipolar cells generate graded potentials only; ganglion cells generate action potentials.
6. During light adaptation, visual pigments are bleached and rods are inactivated; then, as cones decrease their light sensitivity, high-acuity vision ensues. In dark adaptation, cones cease functioning, and visual acuity decreases; rod function begins when sufficient rhodopsin has accumulated.

15.4 Visual information from the retina passes through relay nuclei to the visual cortex (pp. 573–575)

1. The visual pathway to the brain begins with the optic nerve fibers (ganglion cell axons) from the retina. At the optic chiasma, fibers from the medial half of each retina cross over and continue in the optic tracts to the thalamus. Thalamic neurons project to the visual cortex via the optic radiation. Fibers also project from the retina to the midbrain pretectal nuclei and the superior colliculi, and to the suprachiasmatic nucleus of the hypothalamus.
2. Each eye receives a slightly different view of the visual field. The visual cortices fuse these views to provide depth perception.
3. Retinal processing involves the selective destruction of inputs so as to emphasize bright/dark or color contrasts (edges). Thalamic processing subserves high-acuity color vision and depth perception. Cortical processing involves color, form, and movement. Visual processing proceeds anteriorly in the "what" and "where" streams.

15

PART 2

THE CHEMICAL SENSES: SMELL AND TASTE

15.5 Airborne chemicals are detected by olfactory receptors in the nose (pp. 575–578)

1. The olfactory epithelium is located in the roof of the nasal cavity. The olfactory sensory neurons are ciliated bipolar neurons. Their axons are the filaments of the olfactory nerve (cranial nerve I).
2. Individual olfactory neurons have one type of receptor protein that can respond to more than one chemical. Olfactory neurons bearing the same odorant receptors synapse in the same glomerulus type.
3. Olfactory neurons are excited by volatile chemicals that bind to receptors in the olfactory cilia.
4. Action potentials of the olfactory nerve filaments are transmitted to the olfactory bulb where the filaments synapse with mitral cells. The mitral cells send impulses via the olfactory tract to the olfactory cortex. Fibers carrying impulses from the olfactory bulb also project to the limbic system.

15.6 Dissolved chemicals are detected by receptor cells in taste buds (pp. 578–580)

1. The taste buds are scattered in the oral cavity and pharynx but are most abundant on the tongue papillae.
2. Gustatory epithelial cells, the receptor cells of the taste buds, have gustatory hairs (microvilli) that serve as the receptor regions. The gustatory epithelial cells are excited by the binding of tastants (food chemicals) to receptors on their microvilli.
3. The five basic taste qualities are sweet, sour, salty, bitter, and umami. Lipid may be a sixth taste quality.
4. The taste sense is served by cranial nerves VII, IX, and X, which send impulses to the solitary nucleus of the medulla. From there, impulses are sent to the thalamus and the gustatory cortex.

PART 3

THE EAR: HEARING AND BALANCE

15.7 The ear has three major areas (pp. 580–585)

1. The auricle and external acoustic meatus compose the external ear. The tympanic membrane, the boundary between the outer and middle ears, transmits sound waves to the middle ear.
2. The middle ear is a small chamber within the temporal bone, connected by the pharyngotympanic tube to the nasopharynx. The ossicles, which help amplify sound, span the middle ear cavity and transmit sound vibrations from the tympanic membrane to the oval window.
3. The internal ear consists of the bony labyrinth, within which the membranous labyrinth is suspended. The bony labyrinth chambers contain perilymph; the membranous labyrinth ducts and sacs contain endolymph.
4. The vestibule contains the saccule and utricle. The semicircular canals extend posteriorly from the vestibule in three planes. They contain the semicircular ducts.
5. The cochlea houses the cochlear duct (scala media), containing the spiral organ, the receptor organ for hearing. Within the cochlear duct, the hair (receptor) cells rest on the basilar membrane, and their hairs project into the gelatinous tectorial membrane.

15.8 Sound is a pressure wave that stimulates mechanosensitive cochlear hair cells (pp. 585–589)

1. Sound originates from a vibrating object and travels in waves consisting of alternating areas of compression and rarefaction of the medium.

2. The distance from crest to crest on a sine wave is the sound's wavelength; the shorter the wavelength, the higher the frequency (measured in hertz). Frequency is perceived as pitch.
3. The amplitude of sound is the height of the peaks of the sine wave, which reflect the sound's intensity. Sound intensity is measured in decibels. Intensity is perceived as loudness.
4. Sound passing through the external acoustic meatus sets the tympanic membrane into vibration at the same frequency. The ossicles amplify and deliver the vibrations to the oval window.
5. Pressure waves in cochlear fluids set specific locations on the basilar membrane into resonance. At points of maximal membrane vibration, the vibratory motion alternately depolarizes and hyperpolarizes the hair cells of the spiral organ. Movements of the stereocilia toward the tallest cilium depolarize the hair cells and increase the rate of impulse generation in the auditory nerve fibers. Movements away from the tallest cilium have the opposite effect. High-frequency sounds stimulate hair cells near the oval window; low-frequency sounds stimulate hair cells near the apex. Inner hair cells send most auditory inputs to the brain. Outer hair cells amplify the inner hair cells' responsiveness.

15.9 Sound information is processed and relayed through brain stem and thalamic nuclei to the auditory cortex (pp. 589–590)

1. Impulses generated along the cochlear nerve travel to the cochlear nuclei of the medulla and from there through several brain stem nuclei to the medial geniculate nucleus of the thalamus and then the auditory cortex. Each auditory cortex receives impulses from both ears.
2. Auditory processing is analytic; each tone is perceived separately. Perception of pitch is related to the position of the excited hair cells along the basilar membrane. Intensity perception reflects the fact that as sound intensity increases, basilar membrane motion increases, and the frequency of impulse transmission to the cortex increases. Cues for sound localization include the intensity and timing of sound arriving at each ear.

15.10 Hair cells in the maculae and cristae ampullares monitor head position and movement (pp. 590–594)

1. The equilibrium receptor regions of the internal ear are called the vestibular apparatus.
2. The receptors for linear acceleration and gravity are the maculae of the saccule and utricle. A macula consists of hair cells with stereocilia and a kinocilium embedded in an overlying otolith membrane. Linear movements cause the otolith membrane to move, pulling on the hair cells and changing the rate of impulse generation in the vestibular nerve fibers.
3. The receptor within each semicircular duct, the crista ampullaris, responds to angular or rotational acceleration in one plane. It consists of a tuft of hair cells whose stereocilia are embedded in the gelatinous ampullary cupula. Rotational movements cause the endolymph to flow in the opposite direction, bending the cupula and either exciting or inhibiting the hair cells.
4. Impulses from the vestibular apparatus are sent via vestibular nerve fibers mainly to the vestibular nuclei of the brain stem and the cerebellum. These centers initiate responses that fix the eyes on objects and activate muscles to maintain balance.

15.11 Ear abnormalities can affect hearing, equilibrium, or both (pp. 594–595)

1. Conduction deafness results from interference with conduction of sound vibrations to the fluids of the internal ear. Sensorineural deafness reflects damage to neural structures.

2. Tinnitus is an early sign of sensorineural deafness; it may also result from inflammation or certain drugs.
3. Ménière's syndrome is a disorder of the membranous labyrinth. Symptoms include tinnitus, deafness, and vertigo. Excessive endolymph accumulation is the suspected cause.

Developmental Aspects of the Special Senses   (pp. 595–596)

Taste and Smell   (p. 595)

1. The chemical senses are sharpest at birth and gradually decline with age as receptor cells are replaced more slowly.

Vision   (p. 595)

2. Congenital eye problems are uncommon, but maternal rubella can cause blindness.
3. The eye starts as an optic vesicle, an outpocketing of the diencephalon that invaginates to form the optic cup, which becomes the retina. Overlying ectoderm folds to form the lens vesicle, which gives rise to the lens. The remaining eye tissues and the accessory structures are formed by mesenchyme.
4. The eye is foreshortened at birth and reaches adult size at the age of 8–9 years. Depth perception and color vision develop before age 3.

5. With age, the lens loses its elasticity and clarity, there is a decline in the ability of the iris to dilate, and visual acuity decreases. The elderly are at risk for eye problems resulting from dry eyes and disease.

Hearing and Balance   (p. 596)

6. The membranous labyrinth develops from the otic placode, an ectodermal thickening lateral to the hindbrain. Mesenchyme forms the surrounding bony structures. Pharyngeal pouch endoderm, in conjunction with mesenchyme, forms most middle ear structures; the external ear is formed largely by ectoderm.
7. Congenital ear problems are fairly common. Maternal rubella can cause deafness.
8. Response to sound in infants is reflexive. By the fourth month, an infant can locate sound. Critical listening develops in toddlers.
9. The spiral organ deteriorates throughout life as noise, disease, and drugs destroy cochlear hair cells. Age-related loss of hearing (presbycusis) occurs in the 60s and 70s.

REVIEW QUESTIONS

 To access additional practice questions using your smartphone, tablet, or computer: MasteringA&P° > Study Area > Practice Tests & Quizzes

Level 1 Remember/Understand

(Some questions have more than one correct answer. Select the best answer or answers from the choices given.)

1. Accessory glands that produce an oily secretion are the (a) conjunctiva, (b) lacrimal glands, (c) tarsal glands.
2. The portion of the fibrous layer that is white and opaque is the (a) choroid, (b) cornea, (c) retina, (d) sclera.
3. Which sequence best describes a normal route for the flow of tears from the eyes into the nasal cavity? (a) lacrimal canaliculi, lacrimal sacs, nasolacrimal ducts; (b) lacrimal ducts, lacrimal canaliculi, nasolacrimal ducts; (c) nasolacrimal ducts, lacrimal canaliculi, lacrimal sacs.
4. The phenomenon of dark adaptation is best explained by the fact that (a) rhodopsin does not function in dim light, (b) rhodopsin breakdown occurs slowly, (c) rods exposed to intense light need time to generate rhodopsin, (d) cones are stimulated to function by bright light.
5. Nearsightedness is more properly called (a) myopia, (b) hyperopia, (c) presbyopia, (d) emmetropia.
6. Of the neurons in the retina, the axons of which of these form the optic nerve? (a) bipolar cells, (b) ganglion cells, (c) cone cells, (d) horizontal cells.
7. Which reactions occur when a person looks at a distant object? (a) pupils constrict, ciliary zonule (suspensory ligament) relaxes, lenses become less convex; (b) pupils dilate, ciliary zonule becomes taut, lenses become less convex; (c) pupils dilate, ciliary zonule becomes taut, lenses become more convex; (d) pupils constrict, ciliary zonule relaxes, lenses become more convex.
8. The blind spot of the eye is (a) where more rods than cones are found, (b) where the macula lutea is located, (c) where only cones occur, (d) where the optic nerve leaves the eye.
9. Taste buds are found on the (a) anterior part of the tongue, (b) posterior part of the tongue, (c) palate, (d) all of these.

10. Olfactory nerve filaments are found (a) in the optic bulbs, (b) passing through the cribriform plate of the ethmoid bone, (c) in the optic tracts, (d) in the olfactory cortex.
11. The transmission of sound vibrations through the internal ear occurs chiefly through (a) nerve fibers, (b) air, (c) fluid, (d) bone.
12. Which of the following statements does *not* correctly describe the spiral organ? (a) Sounds of high frequency stimulate hair cells at the basal end, (b) the "hairs" of the receptor cells are embedded in the tectorial membrane, (c) the basilar membrane acts as a resonator, (d) the more numerous outer hair cells are largely responsible for our perception of sound.
13. The structure that allows pressure in the middle ear to be equalized with atmospheric pressure is the (a) pinna, (b) pharyngotympanic tube, (c) tympanic membrane, (d) oval window.
14. Otoliths (ear stones) are (a) a cause of deafness, (b) a type of hearing aid, (c) important in equilibrium, (d) the rock-hard petrous part of the temporal bones.
15. How do rods and cones differ functionally?
16. Name the cranial nerves that serve the sense of taste.

Level 2 Apply/Analyze

17. Damage to the medial recti muscles would probably affect (a) accommodation, (b) refraction, (c) convergence, (d) pupil constriction.
18. Blockage of the scleral venous sinus might result in (a) a sty, (b) glaucoma, (c) conjunctivitis, (d) a cataract.
19. Pitch is to frequency of sound as loudness is to (a) quality, (b) intensity, (c) overtones, (d) all of these.
20. Which of the following is important in maintaining the balance of the body? (a) visual cues, (b) semicircular canals, (c) the saccule, (d) proprioceptors, (e) all of these.
21. Which of the following is not a possible cause of conduction deafness? (a) impacted cerumen, (b) middle ear infection, (c) cochlear nerve degeneration, (d) otosclerosis.

22. Why do you often have to blow your nose after crying?

23. Where are the olfactory sensory neurons, and why is that site poorly suited for their job?

24. David Norris, an engineering student, has been working in a nightclub to earn money to pay for his education. After about eight months, he notices that he is having problems hearing high-pitched tones. What is the cause-and-effect relationship here?

Level 3   Evaluate/Synthesize

25. Describe the response of rhodopsin to light stimuli. What is the outcome of this cascade of events?

26. Since there are only three types of cones, how can you explain the fact that we see many more colors?

27. Describe the effect of aging on the special sense organs.

28. During an ophthalmoscopic examination, Mrs. James was found to have bilateral papilledema. Further investigation indicated that this condition resulted from a rapidly growing intracranial tumor. First, define papilledema (see Related Clinical Terms). Then explain its presence in terms of Mrs. James's diagnosis.

29. Maria, a 9-year-old girl, told the clinic physician that her "ear lump hurt" and she kept "getting dizzy and falling down." As she told her story, she pointed to her mastoid process. An otoscopic examination of the external acoustic meatus revealed a red, swollen eardrum, and her throat was inflamed. Her condition was described as mastoiditis with secondary labyrinthitis (inflammation of the labyrinth). Describe the most likely route of infection and the infected structures in Maria's case. Also explain the cause of her dizziness and falling.

30. Mrs. Orlando has been noticing flashes of light and tiny specks in her right visual field. When she begins to see a "veil" floating before her right eye, she makes an appointment to see the eye doctor. What is your diagnosis? Is the condition serious? Explain.

31. Assume that a tumor in the pituitary gland or hypothalamus is protruding inferiorly and compressing the optic chiasma. What could be the visual outcome?

32. Henri, a chef in a five-star French restaurant, has been diagnosed with leukemia. He is about to undergo chemotherapy, which will kill rapidly dividing cells in his body. He needs to continue working between bouts of chemotherapy. What consequences of chemotherapy would you predict that might affect his job as a chef?

CLINICAL CASE STUDY

42-Year-Old Male with Recurring Vertigo

Brian Rhen, 42, has experienced recurring episodes of vertigo and so was referred to a neurologist. (Vertigo is a sensation of motion or movement while the person is stationary, and can be accompanied by nausea and vomiting.)

Mr. Rhen was diagnosed with benign paroxysmal positional vertigo (BPPV). With this condition, which can be caused by head trauma, vertigo can be provoked by specific changes in head position. Mr. Rhen reported that his vertigo usually occurred when rolling over in bed, or when turning his head from side to side while sitting up, and that these movements provoked the sensation of a spinning room, which led to nausea. The neurologist confirmed the diagnosis of BPPV by using a test called the Dix-Hallpike maneuver. During this test, the neurologist looks for nystagmus (involuntary, jerking eye movements) as he makes specific rotational changes to Mr. Rhen's head position.

1. The ear is divided into three major areas (compartments). What are these three areas, and which of these areas is involved in Mr. Rhen's BPPV?

2. **+ NCLEX-STYLE** Which of the following mechanisms is a likely explanation for Mr. Rhen's condition?
   a. Damage to the stereocilia of all the hair cells in his internal ear
   b. Damage to the tectorial membrane
   c. An abnormal buildup of endolymph in the cochlear duct
   d. Dislodged otoliths pressing against a cupula

3. What are the three main sources of sensory input that the body uses in order to control balance and equilibrium?

4. Name the two functional divisions of the vestibular apparatus. Identify the sensory receptor associated with each division, and state which aspect of equilibrium each receptor senses.

5. **+ NCLEX-STYLE** As you discuss Mr. Rhen's condition with him, he makes several statements. Which of these statements indicates the need for further education about his problem?
   a. "I guess this was coming, I always have trouble clearing my ears when I go up in an airplane. This is just the same thing, getting worse."
   b. "My wife always said I had rocks in my head!"
   c. "I was reading this web site called Wobblers Anonymous, and they said my problem could give me some trouble with vision as well."
   d. "I'm afraid I'll get seasick whenever I walk."

6. Explain why nystagmus is associated with the Dix-Hallpike maneuver.

15

16

The Endocrine System

In this chapter you will learn that

16.1 The endocrine system is one of two major control systems of the body

by first examining

16.2 Hormone chemical structure

and asking

16.3 How do hormones act?

16.4 What stimuli cause hormone release?

16.5 What determines cell responses to hormones?

then looking at

Central nervous system **Ch. 12** ←*part of*— **16.6 The hypothalamus**

controls

The posterior pituitary

The anterior pituitary

then exploring

16.7 The thyroid gland

16.8 The parathyroid glands

16.9 The adrenal glands

16.10 The pineal gland

and hormones released by

16.11 Organs with other major functions

and finally, exploring

Developmental Aspects of the Endocrine System

You don't have to watch a blockbuster to experience action-packed drama. Molecules and cells inside your body have dynamic adventures on microscopic levels all the time.

For instance, when insulin molecules carried along in the blood attach to protein receptors of nearby cells, the response is dramatic: Glucose molecules disappear from the blood into the cells, and cellular activity accelerates. Such is the power of the second great control system of the body, the **endocrine system**, which interacts with the nervous system to coordinate and integrate the activity of body cells.

601

16.1 The endocrine system is one of the body's two major control systems

Learning Outcomes

▶ Indicate important differences between hormonal and neural controls of body functioning.

▶ List the major endocrine organs, and describe their body locations.

▶ Distinguish between hormones, paracrines, and autocrines.

As we have seen, the nervous system regulates the activity of muscles and glands via electrochemical impulses delivered by neurons, and those organs respond within milliseconds. The means of control and speed of the endocrine system are very different: The endocrine system influences metabolic activity by means of *hormones* (*hormone* = to excite). Hormones are chemical messengers secreted by cells into the extracellular fluids. These messengers travel through the blood and regulate the metabolic function of other cells in the body. Binding of a hormone to cellular receptors initiates responses that typically occur after a lag period of seconds or even days. But, once initiated, those responses tend to last much longer than those induced by the nervous system. **Table 16.1** summarizes the differences between the nervous sytem and the endocrine system.

Hormones ultimately target most cells of the body, producing widespread and diverse effects. The major processes that these "mighty molecules" control and integrate include:

- Reproduction
- Growth and development
- Maintenance of electrolyte, water, and nutrient balance of the blood
- Regulation of cellular metabolism and energy balance
- Mobilization of body defenses

As you can see, the endocrine system orchestrates processes that go on for relatively long periods, in some instances continuously. The scientific study of hormones and the endocrine organs is called **endocrinology**.

Compared with other organs, the endocrine system's organs are small and unimpressive, but their influence is powerful.

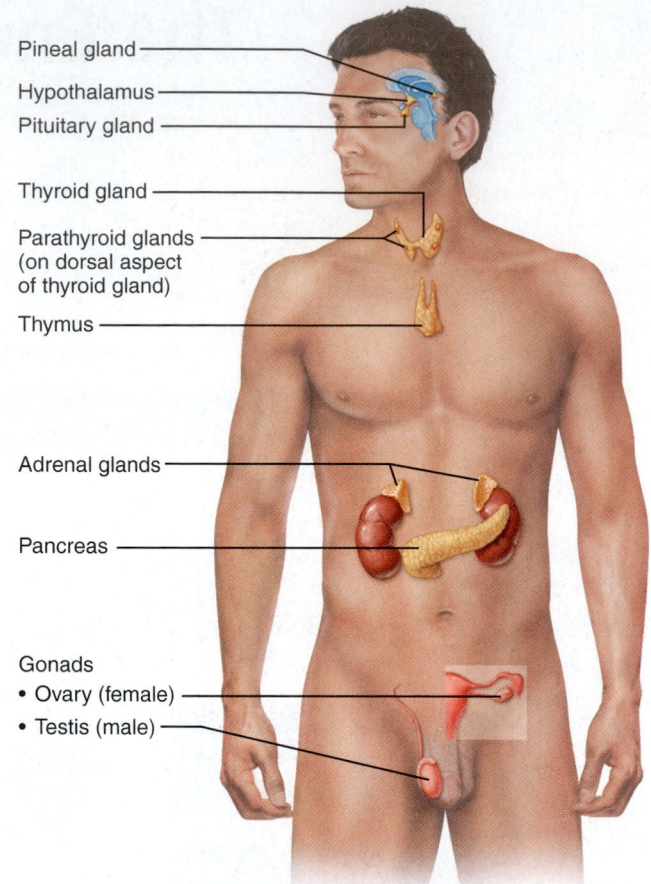

Figure 16.1 **Location of selected endocrine organs of the body.**

Pineal gland
Hypothalamus
Pituitary gland
Thyroid gland
Parathyroid glands (on dorsal aspect of thyroid gland)
Thymus
Adrenal glands
Pancreas
Gonads
• Ovary (female)
• Testis (male)

Table 16.1 Comparison of Nervous and Endocrine Systems	
NERVOUS SYSTEM	**ENDOCRINE SYSTEM**
Initiates responses **rapidly**	Initiates responses **slowly**
Short-duration responses	**Long-duration** responses
Acts via action potentials and **neurotransmitters**	Acts via **hormones** released into the blood
Acts at **specific locations** determined by axon pathways	Acts at **diffuse locations**— targets can be anywhere blood reaches
Neurotransmitters act over very **short distances**	Hormones act over **long distances**

Unlike most organ systems, the endocrine organs are not grouped together but are widely scattered about the body.

As we explained in Chapter 4 (◀ **pp. 123–124**), there are two kinds of glands:

- *Exocrine glands* produce nonhormonal substances, such as sweat and saliva, and have ducts that carry these substances to a membrane surface.

- **Endocrine glands**, also called *ductless glands*, produce hormones and lack ducts. They release their hormones into the surrounding tissue fluid (*endo* = within; *crine* = to secrete), and typically have a rich vascular and lymphatic drainage that receives their hormones. Most of the hormone-producing cells in endocrine glands are arranged in cords and branching networks, which maximizes contact between them and the surrounding capillaries.

The endocrine glands include the pituitary, thyroid, parathyroid, adrenal, and pineal glands (**Figure 16.1**). The hypothalamus, along with its neural functions, also produces and releases hormones, so we consider the hypothalamus a **neuroendocrine organ**. In addition, several organs, such as the pancreas, gonads (ovaries and testes), and placenta, contain endocrine tissue. Most other organs also contain scattered endocrine cells or small clusters of endocrine cells.

Some physiologists include local chemical messengers—autocrines and paracrines—as part of the endocrine system, but that is not the consensus.

- **Hormones** are long-distance chemical signals that travel in blood or lymph throughout the body.
- **Autocrines** are short-distance chemical signals that exert their effects on the same cells that secrete them. For example, certain prostaglandins released by smooth muscle cells cause those smooth muscle cells to contract.
- **Paracrines** are also short-distance chemical signals. They act locally (within the same tissue) but affect cell types other than those releasing the paracrine chemicals. For example, somatostatin released by one population of pancreatic cells inhibits the release of insulin by a different population of pancreatic cells.

Check Your Understanding

1. For each of the following statements, indicate whether it applies more to the endocrine system or the nervous system: rapid; discrete responses; controls growth and development; long-lasting responses.
2. Which two endocrine glands are found in the neck?
3. What is the difference between a hormone and a paracrine?
4. **MAKE CONNECTIONS** Which gland below is an endocrine gland and which is an exocrine gland? How can you tell? (Hint: See Chapter 4.)

(a)   (b)

For answers, see Answers Appendix.

16.2 The chemical structure of a hormone determines how it acts

Learning Outcome

▶ Describe how hormones are classified chemically.

A hormone's chemical structure determines one of its critical properties: its solubility in water. Its water solubility in turn affects how that hormone is transported in the blood (which is mostly water), how long it lasts before it is degraded, and what receptors it can act upon. Although a large variety of hormones are produced, nearly all of them can be classified chemically either as based on amino acids or as steroids.

- **Amino acid based:** Most hormones are based on amino acids (◀ p. 48). Molecular size varies widely in this group—from simple amino acid derivatives [which include biogenic amines (e.g., epinephrine), and thyroxine], to peptides (short chains of amino acids), to proteins (long polymers of amino

acids). These hormones are usually water soluble and cannot cross the plasma membrane.

- **Steroids:** Steroid hormones are synthesized from cholesterol (◀ pp. 47–48). Of the hormones produced by the major endocrine organs, only gonadal and adrenocortical hormones are steroids. These hormones are all lipid soluble and can cross the plasma membrane.

Some researchers add a third class, **eicosanoids** (i-ko′să-noyds), which include *leukotrienes* and *prostaglandins* (◀ p. 48). Nearly all cell membranes release these biologically active lipids (made from arachidonic acid). Leukotrienes are signaling chemicals that mediate inflammation and some allergic reactions. Prostaglandins have multiple targets and effects, ranging from raising blood pressure and increasing the expulsive uterine contractions of birth to enhancing blood clotting, pain, and inflammation.

Because the effects of eicosanoids are typically highly localized, affecting only nearby cells, they generally act as paracrines and autocrines and do not fit the definition of true hormones, which influence distant targets. For this reason, we will not consider these hormonelike chemicals here, but will discuss them in later chapters as appropriate.

Check Your Understanding

5. **MAKE CONNECTIONS** Where in the cell are steroid hormones synthesized? (Hint: Recall cell components from Chapter 3.) Where are peptide hormones synthesized? Which of these two types of hormone could be stored in vesicles and released by exocytosis?

For answers, see Answers Appendix.

16.3 Hormones act through second messengers or by activating specific genes

Learning Outcome

▶ Describe the two major mechanisms by which hormones bring about their effects on their target tissues.

All major hormones circulate to virtually all tissues, but a hormone influences the activity of only those tissue cells that have receptors for it. These cells are its **target cells**. Hormones bring about their characteristic effects by *altering* target cell activity, increasing or decreasing the rates of normal cellular processes.

The precise response depends on the target cell type. For example, when the hormone epinephrine binds to certain smooth muscle cells in blood vessel walls, it stimulates them to contract. Epinephrine binding to cells other than muscle cells does not cause those cells to contract and instead has a totally different effect.

A hormone typically produces one or more of the following changes:

- Alters plasma membrane permeability or membrane potential, or both, by opening or closing ion channels
- Stimulates synthesis of enzymes and other proteins within the cell

- Activates or deactivates enzymes
- Induces secretory activity
- Stimulates mitosis

How does a hormone communicate with its target cell? In other words, how is hormone receptor binding harnessed to the intracellular machinery needed for hormone action? The answer depends on the chemical nature of the hormone and the cellular location of the receptor. In general, hormones act at receptors in one of two ways.

- *Water-soluble hormones* (all amino acid–based hormones except thyroid hormone) act on *receptors in the plasma membrane*. These receptors are usually coupled via regulatory molecules called G proteins to one or more intracellular second messengers, which mediate the target cell's response.
- *Lipid-soluble hormones* (steroid and thyroid hormones) act on *receptors inside the cell*, which directly activate genes.

This will be easy for you to remember if you think about why the hormones must bind where they do. Receptors for water-soluble hormones must be in the plasma membrane since these hormones *cannot* diffuse across the plasma membrane, and receptors for lipid-soluble steroid and thyroid hormones are inside the cell because these hormones *can* diffuse across the plasma membrane into the cell.

Plasma Membrane Receptors and Second-Messenger Systems

With the exception of thyroid hormone, amino acid–based hormones exert their signaling effects through intracellular **second messengers** generated when a hormone binds to a receptor in the plasma membrane. You are already familiar with one of these second messengers, **cyclic AMP (cAMP)**, which is used by neurotransmitters (◀ p. 425) and olfactory receptors (◀ p. 577).

The Cyclic AMP Signaling Mechanism

As you recall, this mechanism involves the interaction of three plasma membrane components— a hormone receptor, a G protein, and an effector enzyme (adenylate cyclase)—to determine intracellular levels of cyclic AMP. **Figure 16.2** illustrates these steps:

> ▶ Watch a 3-D animation of this process: **Mastering**A&P® > Study Area > Animations & Videos > A&P Flix

① **Hormone binds receptor.** The hormone, acting as the **first messenger**, binds to its receptor in the plasma membrane.

② **Receptor activates G protein.** Hormone binding causes the receptor to change shape, allowing it to bind a nearby inactive **G protein**. This in turn allows the G protein to be activated as the guanosine diphosphate (GDP) bound to it is displaced by the high-energy compound *guanosine triphosphate* (*GTP*). The G protein behaves like a light switch: It is "off" when GDP is bound to it, and "on" when GTP is bound.

③ **G protein activates adenylate cyclase.** The activated G protein (moving along the membrane) binds to the effector enzyme **adenylate cyclase**. Some G proteins (G_s) *stimulate* adenylate cyclase (as shown in Figure 16.2), but others (G_i) *inhibit* adenylate cyclase. Eventually, the GTP bound to the G protein is hydrolyzed to GDP and the G protein becomes inactive once again. (The G protein splits the terminal phosphate group off GTP in much the same way that ATPase enzymes hydrolyze ATP.)

④ **Adenylate cyclase converts ATP to cyclic AMP.** For as long as activated G_s is bound to it, adenylate cyclase generates the *second messenger* cAMP from ATP.

⑤ **Cyclic AMP activates protein kinases.** cAMP, which is free to diffuse throughout the cell, triggers a cascade of chemical reactions by activating protein kinases. **Protein kinases** are enzymes that *phosphorylate* (add a phosphate group to) various proteins, many of which are other enzymes. Because phosphorylation activates some of these proteins and inhibits others, it may affect a variety of processes in the same target cell at the same time.

This type of intracellular enzymatic cascade has a huge amplification effect. Each activated adenylate cyclase generates large numbers of cAMP molecules, and a single kinase enzyme can catalyze hundreds of reactions. As the reaction cascades through one enzyme intermediate after another, the number of product molecules increases dramatically at each step. A single hormone molecule binding to a receptor can generate millions of final product molecules!

Cyclic AMP is a very common second messenger that has many different effects in many different cells. What determines the effect that cAMP has in a given cell? This is determined by the specific protein kinases that cell contains, and the substrates within that cell available for phosphorylation. For example, a hormone that increases cAMP in thyroid cells triggers the synthesis of the thyroid hormone thyroxine. A hormone that increases cAMP in liver cells, on the other hand, activates enzymes that break down glycogen, releasing glucose to the blood. Since some G proteins inhibit rather than activate adenylate cyclase, thereby reducing the cytoplasmic concentration of cAMP, even slight changes in levels of hormones that have opposite effects on cAMP levels can influence a target cell's activity.

The action of cAMP persists only briefly because the molecule is rapidly degraded by the intracellular enzyme **phosphodiesterase**. While at first glance this may appear to be a problem, it is quite the opposite. Because of the amplification effect, most hormones need to be present only briefly to cause results, and the quick work of phosphodiesterase means that no extracellular controls are necessary to stop the activity.

The PIP$_2$-Calcium Signaling Mechanism

Certain hormones use second messengers other than cAMP. For example, in the PIP$_2$-calcium signaling mechanism, intracellular calcium ions act as a second messenger.

Like the cAMP signaling mechanism, the PIP$_2$-calcium signaling mechanism involves a G protein (G_q) and

G protein signaling mechanisms are like a molecular relay race.

Hormone   Receptor   G protein   Enzyme   2nd
(1st messenger)   messenger

(1) Hormone (1st messenger) binds receptor.

Adenylate cyclase

Extracellular fluid

G protein (G$_s$)

Receptor

GDP   GTP

GTP

GTP

ATP

cAMP

(5) cAMP activates protein kinases.

Inactive protein kinase

Active protein kinase

(2) Receptor activates G protein (G$_s$).

(3) G protein activates adenylate cyclase.

(4) Adenylate cyclase converts ATP to cAMP (2nd messenger).

Triggers responses of target cell (activates enzymes, stimulates cellular secretion, opens ion channel, etc.)

Cytoplasm

Figure 16.2 Cyclic AMP second-messenger mechanism of water-soluble hormones. (For the basics of G protein signaling mechanisms, see Focus Figure 3.3 on p. 82.)

a membrane-bound effector, in this case an enzyme called **phospholipase C**. Phospholipase C splits a plasma membrane phospholipid called **PIP$_2$** (**p**hosphatidyl **i**nositol bis**p**hosphate) into two second messengers: **diacylglycerol (DAG)** and **inositol trisphosphate (IP$_3$)**. DAG, like cAMP, activates a protein kinase enzyme, which triggers responses within the target cell. In addition, IP$_3$ releases Ca^{2+} from intracellular storage sites.

The liberated Ca^{2+} also takes on a second-messenger role, either by directly altering the activity of specific enzymes and channels or by binding to the intracellular regulatory protein **calmodulin**. Once Ca^{2+} binds to calmodulin, it activates enzymes that amplify the cellular response.

Other Signaling Mechanisms

Other hormones that bind plasma membrane receptors act on their target cells through different signaling mechanisms. For example, cyclic guanosine monophosphate (cGMP) is a second messenger for selected hormones.

Still other hormones, such as insulin and certain growth factors, work without second messengers. The insulin receptor is a *tyrosine kinase* enzyme that is activated by adding phosphates to several of its own tyrosines when insulin binds. The activated insulin receptor provides docking sites for intracellular *relay proteins* that, in turn, initiate a series of protein phosphorylations that trigger specific cell responses.

Intracellular Receptors and Direct Gene Activation

Unlike water-soluble hormones, the lipid-soluble steroid hormones and thyroid hormone are able to diffuse through the plasma membrane into their target cells. There, they bind to and activate intracellular receptors (**Figure 16.3**). The activated receptor-hormone complex then makes its way to the nuclear chromatin and binds to a specific region of DNA. (There are exceptions to these generalizations. For example, thyroid hormone receptors are always bound to DNA even in the absence of thyroid hormone.)

When the receptor-hormone complex binds to DNA, it "turns on" a gene; that is, it prompts transcription of DNA to produce a messenger RNA (mRNA). The mRNA is then translated on the cytoplasmic ribosomes, producing specific proteins. These proteins include enzymes that promote the metabolic activities induced by that particular hormone and, in some cases, promote synthesis of either structural proteins or proteins to be exported from the target cell.

Check Your Understanding

6. Which class of hormones consists entirely of lipid-soluble hormones? Name the only hormone in the other chemical class that is lipid soluble.

7. **DRAW** Make a simple drawing of a cell. Draw and label a receptor for a lipid-soluble hormone and one for a water-soluble hormone in their correct locations in the cell. State what happens as a result of a hormone binding to each receptor.

For answers, see Answers Appendix.

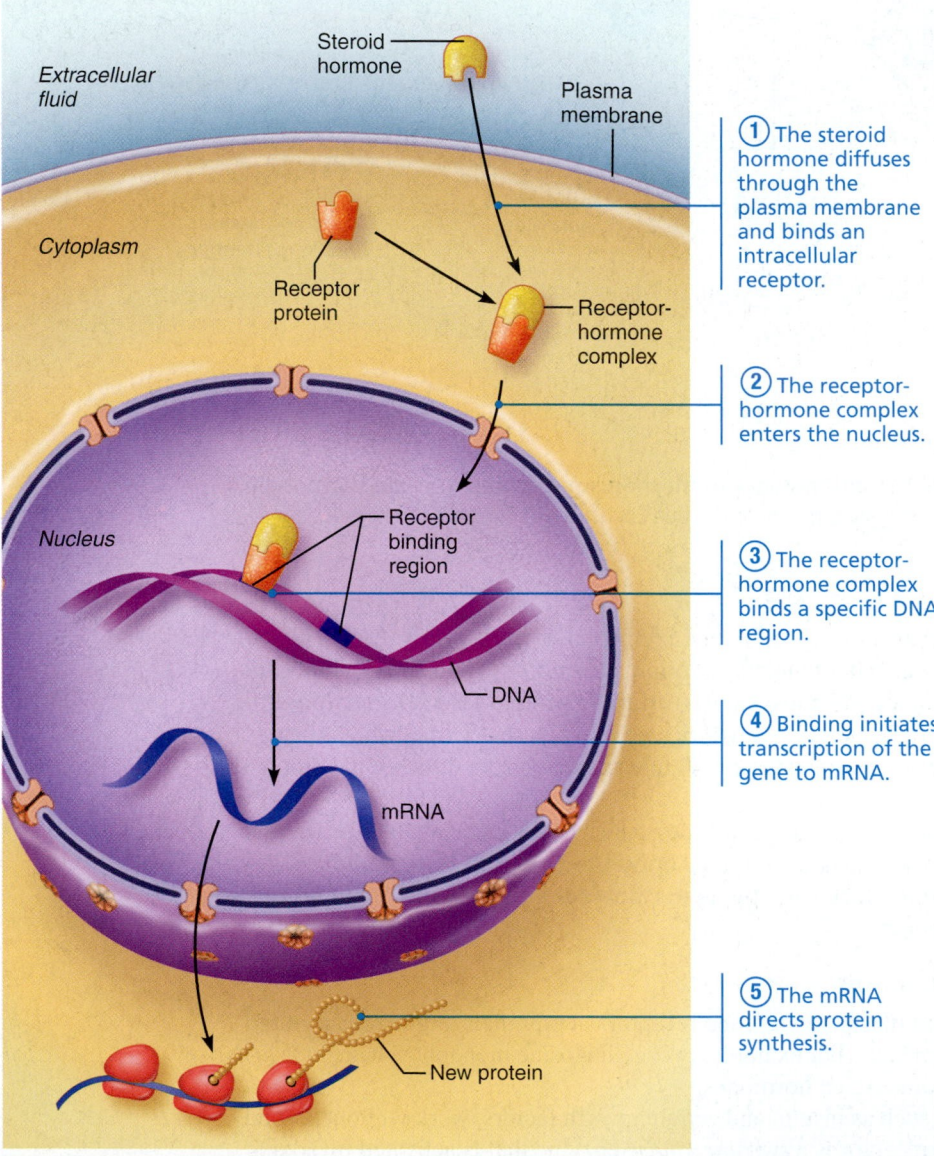

Extracellular fluid

Steroid hormone

Plasma membrane

Cytoplasm

Receptor protein

Receptor-hormone complex

Nucleus

Receptor binding region

DNA

mRNA

New protein

① The steroid hormone diffuses through the plasma membrane and binds an intracellular receptor.

② The receptor-hormone complex enters the nucleus.

③ The receptor-hormone complex binds a specific DNA region.

④ Binding initiates transcription of the gene to mRNA.

⑤ The mRNA directs protein synthesis.

Figure 16.3 Direct gene activation mechanism of lipid-soluble hormones. Receptors may be located in the nucleus, or in the cytoplasm as shown.

(a) Humoral Stimulus	(b) Neural Stimulus	(c) Hormonal Stimulus
Hormone release caused by altered levels of certain critical ions or nutrients.	Hormone release caused by neural input.	Hormone release caused by another hormone (a tropic hormone).

(a) Humoral Stimulus

Capillary (low Ca^{2+} in blood)

Thyroid gland (posterior view)

Parathyroid glands

Parathyroid glands

PTH

Stimulus: Low concentration of Ca^{2+} in capillary blood.
Response: Parathyroid glands secrete parathyroid hormone (PTH), which increases blood Ca^{2+}.

(b) Neural Stimulus

CNS (spinal cord)

Preganglionic sympathetic fibers

Medulla of adrenal gland

Capillary

Stimulus: Action potentials in preganglionic sympathetic fibers to adrenal medulla.
Response: Adrenal medulla cells secrete epinephrine and norepinephrine.

(c) Hormonal Stimulus

Hypothalamus

Anterior pituitary gland

Thyroid gland

Adrenal cortex

Gonad (Testis)

Stimulus: Hormones from hypothalamus.
Response: Anterior pituitary gland secretes hormones that stimulate other endocrine glands to secrete hormones.

Figure 16.4 Three types of endocrine gland stimuli.

16.4 Three types of stimuli cause hormone release

Learning Outcome

▶ Explain how hormone release is regulated.

The synthesis and release of most hormones are regulated by some type of **negative feedback mechanism** (◀ p. 10). In such a mechanism, some internal or external stimulus triggers hormone secretion. As levels of a hormone rise, it causes target organ effects, which then feed back to inhibit further hormone release. As a result, blood levels of many hormones vary only within a narrow range.

Endocrine Gland Stimuli

Three types of stimuli trigger endocrine glands to manufacture and release their hormones: *humoral* (hu′mer-ul), *neural*, and *hormonal stimuli*. Some endocrine organs respond to more than one type of stimulus.

Humoral Stimuli

Some endocrine glands secrete their hormones in direct response to changing blood levels of certain critical ions and nutrients. These stimuli are called *humoral stimuli* (from the Latin term *humor*, which refers to moisture or bodily fluids).

Humoral stimuli are the simplest endocrine controls. For example, cells of the parathyroid glands monitor the body's crucial blood Ca^{2+} levels and release parathyroid hormone as needed (**Figure 16.4a**). Other hormones released in response to humoral stimuli include insulin (released in response to increased blood glucose) and aldosterone (released in response to low Na^+ or high K^+ blood levels).

Neural Stimuli

In a few cases, nerve fibers stimulate hormone release. The classic example of neural stimuli is the response to stress, in which the sympathetic nervous system stimulates the adrenal medulla to release norepinephrine and epinephrine (Figure 16.4b).

Hormonal Stimuli

Many endocrine glands release their hormones in response to hormones produced by other endocrine organs. For example,

16

releasing and inhibiting hormones produced by the hypothalamus regulate the secretion of most anterior pituitary hormones, and many anterior pituitary hormones in turn stimulate other endocrine organs to release their hormones (Figure 16.4c). As blood levels of the hormones produced by the final target glands increase, they inhibit the release of anterior pituitary hormones and thus their own release.

This hypothalamic–pituitary–target endocrine organ feedback loop lies at the very core of endocrinology, and it will come up many times in this chapter. Hormonal stimuli promote rhythmic hormone release, with hormone blood levels rising and falling in a specific pattern.

Nervous System Modulation

The nervous system can modify both "turn-on" factors (hormonal, humoral, and neural stimuli) and "turn-off" factors (feedback inhibition and others) that affect the endocrine system. Without this added safeguard, endocrine system activity would be strictly mechanical, much like a household thermostat. A thermostat can maintain the temperature at or around its set value, but it cannot sense that your grandmother visiting from Florida feels cold at that temperature and reset itself accordingly. You must make that adjustment. In your body, it is the nervous system that makes certain adjustments to maintain homeostasis by overriding normal endocrine controls.

For example, the action of insulin and several other hormones normally keeps blood glucose levels in the range of 90–110 mg/100 ml of blood. However, when your body is under severe stress, blood glucose levels rise because the hypothalamus and sympathetic nervous system centers are strongly activated. In this way, the nervous system ensures that body cells have sufficient fuel in case vigorous activity is required.

Check Your Understanding

8. What are the three types of stimuli that control hormone release?

_____ *For answers, see Answers Appendix.*

16.5 Cells respond to a hormone if they have a receptor for that hormone

Learning Outcomes

▶ **Identify factors that influence activation of a target cell by a hormone.**

▶ **List three kinds of interaction of different hormones acting on the same target cell.**

In order for a target cell to respond to a hormone, the cell must have *specific* receptor proteins on its plasma membrane or in its interior to which that hormone can bind. For example, receptors for adrenocorticotropic hormone (ACTH) are normally found only on certain cells of the adrenal cortex. By contrast, thyroxine is the principal hormone stimulating cellular metabolism, and nearly all body cells have thyroxine receptors.

A hormone receptor responds to hormone binding by prompting the cell to perform, or turn on, some gene-determined "preprogrammed" function. As such, hormones

are molecular triggers rather than informational molecules. Although binding of a hormone to a receptor is *required*, the degree of target cell activation depends equally on three other factors:

- Blood levels of the hormone
- Relative numbers of receptors for that hormone on or in the target cells
- *Affinity* (strength) of the binding between the hormone and the receptor

The first two factors change rapidly in response to various stimuli and changes within the body. As a rule, for a given level of hormone in the blood, having a large number of high-affinity receptors produces a pronounced hormonal effect, and having a smaller number of low-affinity receptors reduces the target cell response or causes outright endocrine dysfunction.

Receptors are dynamic structures. For example, persistently low levels of a hormone can cause its target cells to form additional receptors for that hormone. This is called **up-regulation**. Likewise, prolonged exposure to high hormone concentrations can decrease the number of receptors for that hormone. This **down-regulation** desensitizes the target cells, so they respond less vigorously to hormonal stimulation, preventing them from overreacting to persistently high hormone levels. Receptors can also be uncoupled from their signaling mechanism, altering the sensitivity of the response.

Hormones influence not only the number of their own receptors but also the number of receptors that respond to other hormones. For example, progesterone down-regulates estrogen receptors in the uterus, thus antagonizing estrogens' actions. On the other hand, estrogens cause the same cells to produce more progesterone receptors, enhancing their ability to respond to progesterone.

Half-Life, Onset, and Duration of Hormone Activity

Hormones are potent chemicals, and they exert profound effects on their target organs even at very low concentrations. Hormones circulate in the blood in two forms—free, or bound to a protein carrier. In general, lipid-soluble hormones (steroids and thyroid hormone) are not water soluble. As a result, they must travel in blood attached to proteins of the blood plasma (*plasma proteins*). Most others circulate without carriers.

The concentration of a circulating hormone in blood at any time reflects (1) its rate of release and (2) the speed at which it is inactivated and removed from the body. Some hormones are rapidly degraded by enzymes in their target cells. However, most hormones are removed from the blood by the kidneys or liver, and the body excretes their breakdown products in urine or, to a lesser extent, in feces. As a result, the length of time for a hormone's blood level to decrease by half, referred to as its **half-life**, varies from a fraction of a minute to a week. Water-soluble hormones have the shortest half-lives because they are rapidly removed from the blood by the kidneys.

How long does it take for a hormone to have an effect? It varies. Some hormones provoke target organ responses almost immediately, while others, particularly steroid hormones,

Table 16.2 Comparison between Lipid- and Water-Soluble Hormones

	LIPID-SOLUBLE HORMONES	WATER-SOLUBLE HORMONES
Consist of	All steroid hormones and thyroid hormone	All amino acid–based hormones except thyroid hormone
Sources	Adrenal cortex, gonads, and thyroid gland	All other endocrine glands
Stored in secretory vesicles	No	Yes
Transport in blood	Bound to plasma proteins	Usually free in plasma
Half-life in blood	Long (most need to be metabolized by liver)	Short (most can be removed by kidneys)
Location of receptors	Usually inside cell	On plasma membrane
Mechanism of action at target cell	Activate genes, causing synthesis of new proteins	Usually act through second-messenger systems

require hours to days before their effects are seen. Additionally, some hormones are secreted in a relatively inactive form and must be activated in the target cells.

The duration of hormone action is limited, ranging from 10 seconds to several hours. Effects may disappear rapidly as blood levels drop, or they may persist for hours even at very low levels. Because of these many variations, hormonal blood levels must be precisely and individually controlled to meet the continuously changing needs of the body.

As you can see, many characteristics of a hormone (such as its half-life and the time it takes to have an effect) depend on its solubility in water or lipids. **Table 16.2** compares the characteristics of lipid- and water-soluble hormones.

Interaction of Hormones at Target Cells

Understanding hormonal effects is a bit more complicated than you might expect because multiple hormones may act on the same target cells at the same time. In many cases the result of such an interaction is not predictable, even when you know the effects of the individual hormones. Here we will look at three types of hormone interaction—permissiveness, synergism, and antagonism.

- **Permissiveness** is the situation in which one hormone cannot exert its full effects without another hormone being present. For example, reproductive system hormones largely regulate the development of the reproductive system, as we might expect. However, thyroid hormone is also necessary (has a permissive effect) for normal *timely* development of reproductive structures. Lack of thyroid hormone delays reproductive development.

- **Synergism** occurs when more than one hormone produces the same effects at the target cell and their combined effects are amplified. For example, both glucagon and epinephrine cause the liver to release glucose to the blood. When they act together, the amount of glucose released is about 150% of what is released if each hormone acts independently.

- **Antagonism** occurs when one hormone opposes the action of another. For example, insulin, which lowers blood glucose levels, is antagonized by glucagon, which raises blood glucose levels. How does antagonism occur? Antagonists may compete for the same receptors, act through different metabolic pathways, or even cause down-regulation of the receptors for the antagonistic hormone.

Check Your Understanding

9. Which type of hormone generally stays in the blood longer following its secretion, lipid soluble or water soluble?

For answers, see Answers Appendix.

16.6 The hypothalamus controls release of hormones from the pituitary gland in two different ways

Learning Outcomes
▶ Describe structural and functional relationships between the hypothalamus and the pituitary gland.
▶ Discuss the structure of the posterior pituitary, and describe the effects of the two hormones it releases.
▶ List and describe the chief effects of anterior pituitary hormones.

Securely seated in the sella turcica of the sphenoid bone (◀ p. 209), the tiny **pituitary gland**, or **hypophysis** (hi-pof′ĭ-sis; "to grow under"), secretes at least eight hormones. This gland is the size and shape of a pea on a stalk. Its stalk, the funnel-shaped **infundibulum**, connects the gland to the hypothalamus superiorly as shown in *Focus on Hypothalamus and Pituitary Interactions* (**Focus Figure 16.1** on pp. 610–611).

In humans, the pituitary gland has two major lobes.

- The **posterior pituitary** (lobe) is composed largely of neural tissue such as pituicytes (glia-like supporting cells) and nerve fibers. It releases **neurohormones** (hormones secreted by neurons) received ready-made from the hypothalamus. Consequently, this lobe is a hormone-storage area and not a true endocrine gland that manufactures hormones. The posterior lobe plus the infundibulum make up the region called the **neurohypophysis** (nu″ro-hi-pof′ĭ-sis).

- The **anterior pituitary** (lobe), or **adenohypophysis** (ad″ĕ-no-hi-pof′ĭ-sis), is composed of glandular tissue (*adeno* =

The hypothalamus controls release of hormones from the pituitary gland in two different ways.

Posterior Pituitary: *Action potentials* travel down the axons of hypothalamic neurons, causing hormone release from their axon terminals in the posterior pituitary.

Posterior lobe of pituitary

Paraventricular nucleus

Hypothalamus

Optic chiasma

Supraoptic nucleus

Infundibulum (connecting stalk)

Hypothalamic-hypophyseal tract

Inferior hypophyseal artery

Axon terminals

Posterior lobe of pituitary

① Hypothalamic neurons synthesize oxytocin or antidiuretic hormone (ADH).

② Oxytocin and ADH are transported down the axons of the hypothalamic-hypophyseal tract to the posterior pituitary.

③ Oxytocin and ADH are stored in axon terminals in the posterior pituitary.

④ When associated hypothalamic neurons fire, action potentials arriving at the axon terminals cause oxytocin or ADH to be released into the blood.

Oxytocin
Antidiuretic hormone (ADH)

gland). It manufactures and releases a number of hormones (**Table 16.3** on pp. 612–613).

Hypophyseal branches of the internal carotid arteries deliver arterial blood to the pituitary. The veins leaving the pituitary drain into the dural sinuses.

Pituitary-Hypothalamic Relationships

The contrasting histology of the two pituitary lobes reflects the dual origin of this tiny gland during embryonic development. The posterior lobe is actually part of the brain. It derives from hypothalamic tissue that grows downward, maintaining its neural connection with the hypothalamus via a bundle of axons called the **hypothalamic-hypophyseal tract**. This tract runs through the infundibulum (Focus Figure 16.1).

This tract arises from neurons in the **paraventricular** and **supraoptic nuclei** of the hypothalamus. These neurosecretory cells synthesize one of two neurohormones and transport them along their axons to the posterior pituitary. When these hypothalamic neurons fire, they release the stored hormones into a capillary bed in the posterior pituitary for distribution throughout the body.

The glandular anterior lobe originates from epithelial tissue as a superior outpocketing of the oral mucosa. After touching the posterior lobe, the anterior lobe adheres to the neurohypophysis and loses its connection with the oral mucosa. There is no direct neural connection between the anterior lobe and hypothalamus, but there is a vascular connection. Specifically, the **primary capillary plexus** in the infundibulum communicates inferiorly via the small **hypophyseal portal veins** with a

Anterior Pituitary: *Hypothalamic hormones* released into special blood vessels (the hypophyseal portal system) control the release of anterior pituitary hormones.

Anterior lobe of pituitary

Hypothalamus

Hypothalamic neurons synthesize releasing and inhibiting hormones (GHRH, GHIH, TRH, CRH, GnRH, PIH).

1 When appropriately stimulated, hypothalamic neurons secrete releasing or inhibiting hormones into the primary capillary plexus.

2 Hypothalamic hormones travel through portal veins to the anterior pituitary where they stimulate or inhibit release of hormones made in the anterior pituitary.

Superior hypophyseal artery

Hypophyseal portal system

- Primary capillary plexus
- Hypophyseal portal veins
- Secondary capillary plexus

A portal system is two capillary plexuses (beds) connected by veins.

3 In response to releasing hormones, the anterior pituitary secretes hormones into the secondary capillary plexus. This in turn empties into the general circulation.

Growth hormone (GH)
Thyroid-stimulating hormone (TSH)
Adrenocorticotropic hormone (ACTH)
Follicle-stimulating hormone (FSH)
Luteinizing hormone (LH)
Prolactin (PRL)

Anterior lobe of pituitary

---

secondary capillary plexus in the anterior lobe. The primary and secondary capillary plexuses and the intervening hypophyseal portal veins make up the *hypophyseal portal system* (Focus Figure 16.1). Note that a *portal system* is an unusual arrangement of blood vessels in which a capillary bed feeds into veins, which in turn feed into a second capillary bed.

Via the hypophyseal portal system, **releasing** and **inhibiting hormones** secreted by neurons in the ventral hypothalamus circulate to the anterior pituitary, where they regulate secretion of its hormones. The portal system ensures that the minute quantities of hormones released by the hypothalamus arrive rapidly at the anterior pituitary without being diluted by the systemic circulation. All these hypothalamic regulatory hormones are amino acid based, but they vary in size from a single amine to peptides to proteins.

The Posterior Pituitary and Hypothalamic Hormones

The posterior pituitary consists largely of axon terminals of hypothalamic neurons whose cell bodies are located in the supraoptic or paraventricular nuclei. The paraventricular neurons primarily make oxytocin (ok″sĭ-to′sin), and the supraoptic neurons mainly produce antidiuretic hormone (ADH). Axon terminals in the posterior pituitary release these hormones "on demand" in response to action potentials that travel down the axons of these same hypothalamic neurons.

Oxytocin and ADH, each composed of nine amino acids, are almost identical. They differ in only two amino acids, and yet they have dramatically different physiological effects, summarized in Table 16.3 and described next.

Table 16.3   Pituitary Hormones: Summary of Regulation and Effects

HORMONE (CHEMICAL STRUCTURE AND CELL TYPE)	REGULATION OF RELEASE	TARGET ORGAN AND EFFECTS	EFFECTS OF HYPOSECRETION ↓ AND HYPERSECRETION ↑
Posterior Pituitary Hormones (Made by Hypothalamic Neurons and Stored in Posterior Pituitary)			
Oxytocin (Peptide, mostly from neurons in paraventricular nucleus of hypothalamus)	**Stimulated** by impulses from hypothalamic neurons in response to stretching of the uterine cervix or suckling of infant at breast **Inhibited** by lack of appropriate neural stimuli	Uterus: stimulates uterine contractions; initiates labor Breast: initiates milk ejection	Unknown
Antidiuretic hormone (ADH) or vasopressin (Peptide, mostly from neurons in supraoptic nucleus of hypothalamus)	**Stimulated** by impulses from hypothalamic neurons in response to increased blood solute concentration or decreased blood volume; also stimulated by pain, some drugs, low blood pressure **Inhibited** by adequate hydration of the body and by alcohol	Kidneys: stimulate kidney tubule cells to reabsorb water from the forming urine back into blood	↓: Diabetes insipidus ↑: Syndrome of inappropriate ADH secretion (SIADH)
Anterior Pituitary Hormones			
Growth hormone (GH) (Protein, somatotropic cells)	**Stimulated** by GHRH* release, which is triggered by low blood levels of GH as well as by a number of secondary triggers including deep sleep, hypoglycemia, increases in blood levels of amino acids, low levels of fatty acids, exercise, and other types of stressors **Inhibited** by feedback inhibition exerted by GH and insulin-like growth factors (IGFs), and by hyperglycemia, hyperlipidemia, obesity, and emotional deprivation via either increased GHIH* (somatostatin) or decreased GHRH* release	Liver, muscle, bone, cartilage, and other tissues: anabolic hormone; stimulates somatic growth; mobilizes fats; spares glucose Growth-promoting effects mediated indirectly by IGFs	↓: Pituitary dwarfism in children ↑: Gigantism in children; acromegaly in adults

Oxytocin

A strong stimulant of uterine contraction, **oxytocin** is released in significantly higher amounts during childbirth (*oxy* = rapid; *tocia* = childbirth) and in nursing women. The number of oxytocin receptors in the uterus peaks near the end of pregnancy, and uterine smooth muscle becomes more and more sensitive to the hormone's stimulatory effects. Stretching of the cervix of the uterus as birth nears dispatches afferent impulses to the hypothalamus. The hypothalamus responds by synthesizing oxytocin and triggering its release from the posterior pituitary. Oxytocin acts via the PIP$_2$-Ca^{2+} second-messenger system to mobilize Ca^{2+}, allowing stronger contractions. As blood levels of oxytocin rise, the expulsive contractions of labor gain momentum and finally end in birth.

Oxytocin also acts as the hormonal trigger for milk ejection (the "letdown" reflex) in women whose breasts are producing milk in response to prolactin. Suckling causes a reflex-initiated release of oxytocin, which targets specialized myoepithelial cells surrounding the milk-producing glands. These cells contract and force milk from the breast into the infant's mouth. Both childbirth and milk ejection result from *positive feedback mechanisms*, which Chapter 28 describes in more detail.

Table 16.3 *(continued)*

HORMONE (CHEMICAL STRUCTURE AND CELL TYPE)	REGULATION OF RELEASE	TARGET ORGAN AND EFFECTS	EFFECTS OF HYPOSECRETION ↓ AND HYPERSECRETION ↑
Thyroid-stimulating hormone (TSH) (Glycoprotein, thyrotropic cells)	**Stimulated** by TRH* and in infants indirectly by cold temperature **Inhibited** by feedback inhibition exerted by thyroid hormones on anterior pituitary and hypothalamus and by GHIH*	Thyroid gland: stimulates thyroid gland to release thyroid hormones	↓: Hypothyroidism; may cause myxedema ↑: Hyperthyroidism; effects similar to those of Graves' disease, in which antibodies mimic TSH
Adrenocorticotropic hormone (ACTH) (Peptide, corticotropic cells)	**Stimulated** by CRH*; stimuli that increase CRH release include fever, hypoglycemia, and other stressors **Inhibited** by feedback inhibition exerted by glucocorticoids	Adrenal cortex: promotes release of glucocorticoids and androgens (mineralocorticoids to a lesser extent)	↓: Rare ↑: Cushing's disease
Follicle-stimulating hormone (FSH) (Glycoprotein, gonadotropic cells)	**Stimulated** by GnRH* **Inhibited** by feedback inhibition exerted by inhibin, and estrogens in females and testosterone in males	Ovaries and testes: in females, stimulates ovarian follicle maturation and production of estrogens; in males, stimulates sperm production	↓: Failure of sexual maturation ↑: No important effects
Luteinizing hormone (LH) (Glycoprotein, gonadotropic cells)	**Stimulated** by GnRH* **Inhibited** by feedback inhibition exerted by estrogens and progesterone in females and testosterone in males	Ovaries and testes: in females, triggers ovulation and stimulates ovarian production of estrogens and progesterone; in males, promotes testosterone production	As for FSH
Prolactin (PRL) (Protein, prolactin cells)	**Stimulated** by decreased PIH*; release enhanced by estrogens, birth control pills, breast-feeding, and dopamine-blocking drugs **Inhibited** by PIH* (dopamine)	Breast secretory tissue: promotes lactation	↓: Poor milk production in nursing women ↑: Inappropriate milk production (galactorrhea); cessation of menses in females; impotence in males

*Indicates hypothalamic releasing and inhibiting hormones: GHRH = growth hormone–releasing hormone; GHIH = growth hormone–inhibiting hormone; TRH = thyrotropin-releasing hormone; CRH = corticotropin-releasing hormone; GnRH = gonadotropin-releasing hormone; PIH = prolactin-inhibiting hormone

Both natural and synthetic oxytocic drugs are used to induce labor or to hasten labor that is progressing slowly. Less frequently, oxytocic drugs are used to stop postpartum bleeding.

Oxytocin also acts as a neurotransmitter in the brain. There, it is involved in sexual and affectionate behavior (as the "cuddle hormone"), and promotes nurturing, couple bonding, and trust.

Antidiuretic Hormone (ADH)

Diuresis (di″u-re′sis) is urine production, so an *antidiuretic* is a substance that inhibits or prevents urine formation. **Antidiuretic hormone (ADH)** prevents wide swings in water balance, helping the body avoid dehydration and water overload.

Hypothalamic neurons called *osmoreceptors* continually monitor the solute concentration (and thus the water concentration) of the blood. When solutes threaten to become too concentrated (as might follow excessive perspiration or inadequate fluid intake), the osmoreceptors transmit excitatory impulses to the hypothalamic neurons, which release ADH. ADH targets the kidney tubule cells, which respond by reabsorbing more water from the forming urine and returning it to the bloodstream. As a result, less urine is produced and the solute concentration of the blood declines. As solute levels fall, the osmoreceptors stop depolarizing, effectively ending ADH release. Other stimuli triggering ADH release include pain, low blood pressure, and such drugs as nicotine, morphine, and barbiturates.

16

16

Deep sleep (as part of day-night rhythm), ↓ blood glucose, exercise, stressors, and other stimuli act via the hypothalamus

Hypothalamus
stimulates the anterior pituitary to release GH by:
↑ Release of growth hormone–releasing hormone (GHRH)
↓ Release of growth hormone–inhibiting hormone (GHIH, or somatostatin)

Inhibits GHRH release
Stimulates GHIH release

Anterior pituitary

Inhibits GH synthesis and release

Negative feedback

Growth hormone (GH)

Direct actions
(metabolic, anti-insulin)

Indirect actions
(growth-promoting)

Liver and other tissues

Produce

Insulin-like growth factors (IGFs)

Effects

Effects

Metabolic actions:
• Increases blood glucose and has other anti-insulin effects
• Increases fat breakdown and release

Growth-promoting actions:
• Increases skeletal growth
• Increases cartilage growth
• At extraskeletal tissues, promotes protein synthesis and cell growth and proliferation

Increases, stimulates
Reduces, inhibits
Initial stimulus
Physiological response
Result

Figure 16.5 **Growth-promoting and metabolic actions of growth hormone (GH).**

Drinking alcoholic beverages inhibits ADH secretion and causes copious urine output. The dry mouth and intense thirst of a "hangover" reflect this dehydrating effect. Drinking lots of water also inhibits ADH release.

Under certain conditions, such as severe blood loss, exceptionally large amounts of ADH are released, causing vasoconstriction and raising blood pressure. This response targets different ADH receptors found on vascular smooth muscle. For this reason, ADH is also called **vasopressin**.

> ### HOMEOSTATIC IMBALANCE 16.1 CLINICAL
>
> One result of ADH deficiency is **diabetes insipidus**, a syndrome marked by intense thirst and huge urine output. The name of this condition (*diabetes* = overflow; *insipidus* = tasteless) distinguishes it from diabetes mellitus (*mel* = honey), in which insulin deficiency causes large amounts of blood glucose to be lost in the urine. At one time, a clinician would taste the patient's urine to determine the type of diabetes.
>
> Diabetes insipidus can be caused by a pituitary tumor or by a blow to the head that damages the hypothalamus or the posterior pituitary. Though inconvenient, the condition is not serious when the thirst center is operating properly and the person drinks enough water to prevent dehydration. However, it can be life threatening in unconscious or comatose patients, so accident victims with head trauma must be carefully monitored.
>
> The opposite problem, hypersecretion of ADH, can occur in children with meningitis, or in adults who have neurosurgery, hypothalamic injury, or cancer (particularly lung cancer) in which cancer cells are additional sources of ADH. It also may occur after general anesthesia or administration of certain drugs. The resulting condition, *syndrome of inappropriate ADH secretion (SIADH)*, is marked by retention of fluid, headache and disorientation due to brain edema, weight gain, and decreased solute concentration in the blood. SIADH management requires restricting fluids and carefully monitoring blood sodium levels.

Anterior Pituitary Hormones

The anterior pituitary has traditionally been called the "master endocrine gland" because many of the numerous hormones it produces regulate the activity of other endocrine glands. In recent years, however, it has been dethroned by the hypothalamus, which is now known to control the activity of the anterior pituitary.

The anterior pituitary releases six hormones, all of them peptides or proteins—growth hormone, thyroid-stimulating hormone, adrenocorticotropic hormone, follicle-stimulating hormone, luteinizing hormone, and prolactin (Table 16.3). When the anterior pituitary receives an appropriate chemical stimulus from the hypothalamus, it releases one or more of its hormones. Although many different hormones pass from the hypothalamus to the anterior lobe, each target cell distinguishes the messages directed to it and responds in kind—secreting the proper hormone in response to specific releasing hormones, and shutting off hormone release in response to specific inhibiting hormones.

Four of the six anterior pituitary hormones—thyroid-stimulating hormone, adrenocorticotropic hormone, follicle-stimulating hormone, and luteinizing hormone—are **tropic hormones** or **tropins** that regulate the secretory action of other endocrine glands (*tropi* = turn on, change). All anterior pituitary hormones except growth hormone affect their target cells via a cyclic AMP second-messenger system.

Growth Hormone (GH)

Somatotropic cells of the anterior lobe produce **growth hormone** (**GH**, also called *somatotropin*). GH is essentially an anabolic (tissue building) hormone that has both metabolic and growth-promoting actions (**Figure 16.5**).

Direct Actions on Metabolism Acting directly, GH exerts metabolic effects. It mobilizes fats from fat depots for transport to cells, increasing blood levels of fatty acids and encouraging their use for fuel. It also decreases the rate of glucose uptake and metabolism, conserving glucose. In the liver, it encourages glycogen breakdown and release of glucose to the blood. This *glucose sparing* action, which raises blood glucose levels, is called the *anti-insulin effect* of GH because its effects oppose those of insulin. In addition, GH increases amino acid uptake into cells and their incorporation into proteins.

Indirect Actions on Growth GH mediates most of its growth-enhancing effects indirectly via a family of growth-promoting proteins called **insulin-like growth factors (IGFs)**. The liver, skeletal muscle, bone, and other tissues produce IGFs in response to GH. IGFs produced by the liver act as hormones, while IGFs made in other tissues act locally within those tissues (as paracrines).

IGFs stimulate actions required for growth:

- Uptake of nutrients from the blood and their incorporation into proteins and DNA, allowing growth by cell division
- Formation of collagen and deposition of bone matrix

Although GH stimulates most body cells to enlarge and divide, its major targets are bone and skeletal muscle. Stimulation of the epiphyseal plate leads to long bone growth, and stimulation of skeletal muscles increases muscle mass.

Regulation of Secretion Secretion of GH is regulated chiefly by two hypothalamic hormones with antagonistic effects.

- **Growth hormone–releasing hormone (GHRH)** stimulates GH release. Typically, GHRH, and therefore GH, secretion has a daily cycle with the highest levels occurring during evening sleep. The total amount secreted daily peaks during adolescence and then declines with age. There are also a number of secondary triggers that cause GHRH secretion (see Table 16.3 on p. 612).

- **Growth hormone–inhibiting hormone (GHIH)**, also called **somatostatin** (so"mah-to-stat'in), inhibits GH release. GHIH release is triggered by the feedback of GH and IGFs. Rising levels of GH also feed back to inhibit its own release. GHIH is also produced in various locations in the gut, where it inhibits the release of virtually all gastrointestinal and pancreatic secretions—both endocrine and exocrine.

> **HOMEOSTATIC IMBALANCE 16.2** **CLINICAL**
>
> Both hypersecretion and hyposecretion of GH may result in structural abnormalities. Hypersecretion in children results in **gigantism** because GH targets the still-active epiphyseal (growth) plates. The person becomes abnormally tall, often reaching a height of 2.4 m (8 feet), but has relatively normal body proportions (**Figure 16.6**).

Figure 16.6 Disorders of pituitary growth hormone. An individual exhibiting gigantism (center) is flanked by a pituitary dwarf (left) and a woman of normal height (right).

If excessive GH is secreted after the epiphyseal plates have closed, **acromegaly** (ak″ro-meg′ah-le) results. Literally translated as "enlarged extremities," this condition is characterized by overgrowth of bones of the hands, feet, and face. Hypersecretion usually results from an anterior pituitary tumor that churns out excessive GH. The usual treatment is surgical removal of the tumor, but this surgery does not reverse anatomical changes that have already occurred.

Hyposecretion of GH in adults usually causes no problems, but GH deficiency in children slows long bone growth, a condition called **pituitary dwarfism** (Figure 16.6). Such individuals attain a maximum height of 1.2 m (4 feet), but usually have fairly normal body proportions. Lack of GH is often accompanied by deficiencies of other anterior pituitary hormones, and if thyroid-stimulating hormone and gonadotropins are lacking, the individual will be malproportioned and will fail to mature sexually as well. Fortunately, human GH is produced commercially by genetic engineering techniques. When pituitary dwarfism is diagnosed before puberty, growth hormone replacement therapy can promote nearly normal growth.

The availability of synthetic GH also has a downside. Athletes and the elderly have been tempted to use GH for its body-building properties, and some parents seek to give their children the hormone in an attempt to make them taller. However, while muscle mass increases, there is no objective evidence for an increase in muscle strength in either athletes or the elderly, and only minimal increases in stature occur in normal children. Moreover, taking GH can lead to fluid retention, joint and muscle pain, diabetes, and may promote cancer. ●●●●●●●●

Thyroid-Stimulating Hormone (TSH)

Thyroid-stimulating hormone (TSH), or **thyrotropin**, is a tropic hormone that stimulates normal development and secretory activity of the thyroid gland. Its release follows the hypothalamic–pituitary–target endocrine organ feedback loop described earlier and shown specifically for TSH in **Figure 16.7**.

The hypothalamic peptide **thyrotropin-releasing hormone (TRH)** triggers the release of TSH from **thyrotropic cells** of the anterior pituitary. Rising blood levels of thyroid hormones act on both the pituitary and the hypothalamus to inhibit TSH secretion. GHIH also inhibits TSH secretion.

Adrenocorticotropic Hormone (ACTH)

Adrenocorticotropic hormone (ACTH) (ah-dre″no-kor″tĭ-ko-trōp′ik), or **corticotropin**, is secreted by the **corticotropic cells** of the anterior pituitary. It is split from a *prohormone* (a large precursor molecule) with the tongue-twisting name **pro-opiomelanocortin (POMC)** (pro″o″pe-o-mah-lan″o-kor′tin). ACTH stimulates the adrenal cortex to release corticosteroid hormones, most importantly glucocorticoids that help the body resist stressors.

ACTH release, elicited by hypothalamic **corticotropin-releasing hormone (CRH)**, has a daily rhythm, with levels peaking in the morning, shortly before awakening. Rising levels of glucocorticoids feed back and block secretion of CRH and

Figure 16.7 Regulation of thyroid hormone secretion. (TRH = thyrotropin-releasing hormone, TSH = thyroid-stimulating hormone)

ACTH release. Internal and external factors that alter the normal ACTH rhythm by triggering CRH release include fever, hypoglycemia (low blood glucose levels), and stressors of all types.

Gonadotropins (FSH and LH)

Follicle-stimulating hormone (FSH) and **luteinizing hormone (LH)** (lu′te-in-īz″ing) are referred to collectively as **gonadotropins**. They regulate the function of the gonads (ovaries and testes). In both sexes, FSH stimulates production of gametes (sperm or eggs) and LH promotes production of gonadal hormones. In females, LH works with FSH to cause an egg-containing ovarian follicle to mature. LH then triggers ovulation and promotes synthesis and release of ovarian hormones. In males, LH stimulates the interstitial cells of the testes to produce the male hormone testosterone.

Gonadotropins are virtually absent from the blood of prepubertal boys and girls. During puberty, the **gonadotropic cells** of the anterior pituitary are activated and gonadotropin levels rise, causing the gonads to mature. In both sexes, **gonadotropin-releasing hormone (GnRH)** produced by the hypothalamus prompts gonadotropin release. Gonadal hormones, produced in response to the gonadotropins, feed back to suppress FSH and LH release. (Chapter 27 covers these hormones in more detail.)

Prolactin (PRL)

Prolactin (PRL) is a protein hormone structurally similar to GH. Produced by **prolactin cells**, PRL's only well-documented effect in humans is to stimulate milk production by the breasts (*pro* = for; *lact* = milk). The role of prolactin in males is not well understood.

Unlike other anterior pituitary hormones, PRL release is controlled primarily by an inhibitory hormone, **prolactin-inhibiting hormone (PIH)**, now known to be **dopamine**, which prevents prolactin secretion. Decreased PIH secretion leads to a surge in PRL release. There are a number of *prolactin-releasing factors*, including TRH, but their exact roles are not well understood.

In females, prolactin levels rise and fall in rhythm with estrogen blood levels. Estrogens stimulate prolactin release, both directly and indirectly. A brief rise in prolactin levels just before the menstrual period partially accounts for the breast swelling and tenderness some women experience at that time, but because this PRL stimulation is so brief, the breasts do not produce milk. In pregnant women, PRL blood levels rise dramatically toward the end of pregnancy, and milk production becomes possible. After birth, the infant's suckling stimulates release of prolactin-releasing factors in the mother, encouraging continued milk production.

HOMEOSTATIC IMBALANCE 16.3   CLINICAL

Hypersecretion of prolactin is more common than hyposecretion (which is not a problem in anyone except women who choose to nurse). In fact, *hyperprolactinemia* is the most frequent abnormality of anterior pituitary tumors. Clinical signs include inappropriate lactation, lack of menses, infertility in females, and impotence in males. ●●●●●●●

Check Your Understanding

10. What is the key difference between the way the hypothalamus communicates with the anterior pituitary and the way it communicates with the posterior pituitary?

11. Zoe drank too much alcohol one night and suffered from a headache and nausea the next morning. What caused these "hangover" effects?

12. List the four anterior pituitary hormones that are tropic hormones and name their target glands.

For answers, see Answers Appendix.

16.7 The thyroid gland controls metabolism

Learning Outcomes

▶ Describe the effects of the two groups of hormones produced by the thyroid gland.

▶ Follow the process of thyroxine formation and release.

Location and Structure

The butterfly-shaped **thyroid gland** is located in the anterior neck, on the trachea just inferior to the larynx (Figure 16.1 and **Figure 16.8a**). A median tissue mass called the *isthmus* (is′mus) connects its two lateral *lobes*. The thyroid gland is the largest pure endocrine gland in the body. Its prodigious blood supply (from the *superior* and *inferior thyroid arteries*) makes thyroid surgery a painstaking (and bloody) endeavor.

Internally, the gland is composed of hollow, spherical **follicles** (Figure 16.8b). The walls of each follicle are formed largely by cuboidal or squamous epithelial cells called *follicular cells*, which produce the glycoprotein **thyroglobulin** (thi″ro-glob′u-lin). The central cavity, or lumen, of the follicle stores **colloid**, an amber-colored, sticky material consisting of thyroglobulin molecules with attached iodine atoms. *Thyroid hormone* is derived from this iodinated thyroglobulin.

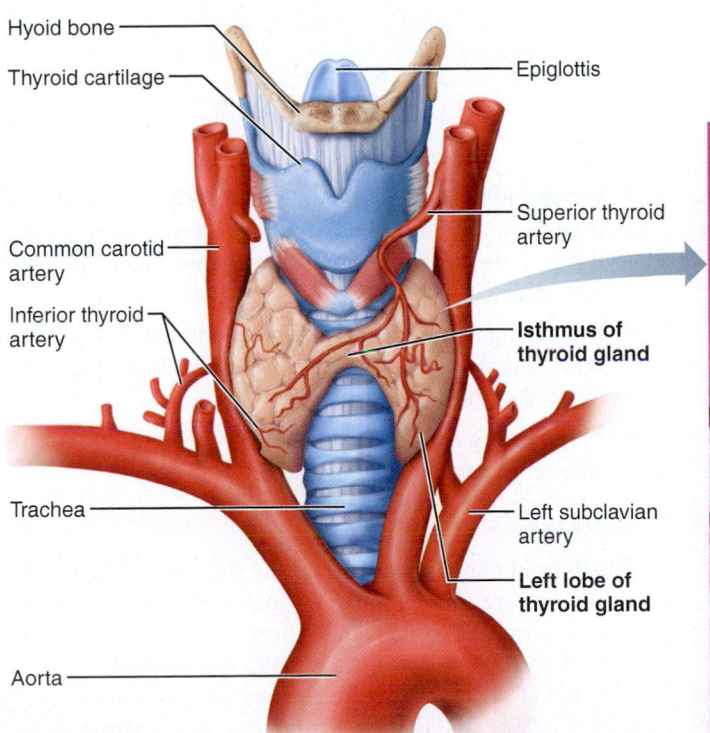

(a) Gross anatomy of the thyroid gland, anterior view

Hyoid bone
Thyroid cartilage
Epiglottis
Common carotid artery
Superior thyroid artery
Inferior thyroid artery
Isthmus of thyroid gland
Trachea
Left subclavian artery
Left lobe of thyroid gland
Aorta

(b) Photomicrograph of thyroid gland follicles (315×)

Parafollicular cells (secrete calcitonin)
Follicular cells (secrete thyroid hormone)
Colloid-filled follicles

Figure 16.8 The thyroid gland.

Practice Histology questions: MasteringA&P° > Study Area > Lab Tools > PAL

Table 16.4   Major Effects of Thyroid Hormone (T₄ and T₃) in the Body

PROCESS OR SYSTEM AFFECTED	NORMAL PHYSIOLOGICAL EFFECTS	EFFECTS OF HYPOSECRETION	EFFECTS OF HYPERSECRETION
Basal metabolic rate (BMR)/temperature regulation	Promotes normal oxygen use and BMR; calorigenesis; enhances effects of sympathetic nervous system	BMR below normal; decreased body temperature, cold intolerance; decreased appetite; weight gain; reduced sensitivity to catecholamines	BMR above normal; increased body temperature, heat intolerance; increased appetite; weight loss
Carbohydrate/lipid/protein metabolism	Promotes glucose catabolism; mobilizes fats; essential for protein synthesis; enhances liver's synthesis of cholesterol	Decreased glucose metabolism; elevated cholesterol/triglyceride levels in blood; decreased protein synthesis; edema	Enhanced catabolism of glucose, proteins, and fats; weight loss; loss of muscle mass
Nervous system	Promotes normal development of nervous system in fetus and infant; promotes normal adult nervous system function	In infant, slowed/deficient brain development, intellectual disability; in adult, mental dulling, depression, paresthesias, memory impairment, hypoactive reflexes	Irritability, restlessness, insomnia, personality changes, exophthalmos (in Graves' disease)
Cardiovascular system	Promotes normal functioning of the heart	Decreased efficiency of heart's pumping action; low heart rate and blood pressure	Increased sensitivity to catecholamines can lead to rapid heart rate, palpitations, high blood pressure, and ultimately heart failure
Muscular system	Promotes normal muscular development and function	Sluggish muscle action; muscle cramps; myalgia	Muscle atrophy and weakness
Skeletal system	Promotes normal growth and maturation of the skeleton	In child, growth retardation, skeletal stunting and retention of child's body proportions; in adult, joint pain	In child, excessive skeletal growth initially, followed by early epiphyseal closure and short stature; in adult, demineralization of skeleton
Gastrointestinal (GI) system	Promotes normal GI motility and tone; increases secretion of digestive juices	Depressed GI motility, tone, and secretory activity; constipation	Excessive GI motility; diarrhea
Reproductive system	Promotes normal female reproductive ability and lactation	Depressed ovarian function; sterility; depressed lactation	In females, depressed ovarian function; in males, impotence
Integumentary system	Promotes normal hydration and secretory activity of skin	Skin pale, thick, and dry; facial edema; hair coarse and thick	Skin flushed, thin, and moist; hair fine and soft; nails soft and thin

The *parafollicular cells*, another population of endocrine cells in the thyroid gland, produce *calcitonin*. The parafollicular cells lie in the follicular epithelium but protrude into the soft connective tissue that separates and surrounds the thyroid follicles.

Thyroid Hormone (TH)

Often referred to as the body's major metabolic hormone, **thyroid hormone (TH)** is actually two iodine-containing amine hormones, **thyroxine** (thi-rok′sin), or **T₄**, and **triiodothyronine** (tri″i-o″do-thi′ro-nēn), or **T₃**. T₄ is the major hormone secreted by the thyroid follicles. Most T₃ is formed at the target tissues by conversion of T₄ to T₃. Both T₄ and T₃ are constructed from two linked tyrosine amino acids, but T₄ has four bound iodine atoms, and T₃ has three (and so the names T₄ and T₃).

TH affects virtually every cell in the body (**Table 16.4**). Like steroids, TH enters a target cell. It binds to intracellular receptors within the cell's nucleus and initiates transcription of mRNA for protein synthesis. Effects of thyroid hormone include:

- Increasing basal metabolic rate and body heat production, by turning on transcription of genes concerned with glucose oxidation. This is the hormone's **calorigenic effect** (*calorigenic* = heat producing).

- Regulating tissue growth and development. TH is critical for normal skeletal and nervous system development and maturation and for reproductive capabilities.

- Maintaining blood pressure by increasing the number of adrenergic receptors in blood vessels.

Synthesis

The thyroid gland is unique among the endocrine glands in its ability to store its hormone extracellularly and in large quantities. A normal thyroid gland stores enough colloid to provide normal levels of hormone for two to three months.

When TSH from the anterior pituitary binds to receptors on follicular cells, their *first* response is to secrete stored thyroid hormone. Their *second* response is to begin synthesizing

Figure 16.9 Synthesis of thyroid hormone. Only a few tyrosines of the thyroglobulins in the colloid are illustrated. The unstructured yellow substance in the follicle lumen is colloid. (MIT = monoiodotyrosine; DIT = diiodotyrosine)

more colloid to "restock" the follicle lumen. As a general rule, TSH levels are lower during the day, peak just before sleep, and remain high during the night. Consequently, thyroid hormone release and synthesis follows a similar pattern.

Let's examine how follicular cells synthesize thyroid hormone (**Figure 16.9**):

1. **Thyroglobulin is synthesized and discharged into the follicle lumen.** After being synthesized on the ribosomes of the follicular cell's rough endoplasmic reticulum, thyroglobulin is transported to the Golgi apparatus, where sugar molecules are attached and the thyroglobulin is packed into transport vesicles. These vesicles move to the apex of the follicular cell, where they discharge their contents into the follicle lumen to become part of the stored colloid.

2. **Iodide is trapped.** To produce the functional iodinated hormones, the follicular cells must accumulate iodides (anions of iodine, I^-) from the blood. Iodide trapping depends on active transport. (The concentration of I^- is over 30 times

higher inside the cell than in blood.) Once trapped inside the follicular cell, iodide then moves into the follicle lumen by facilitated diffusion.

3. **Iodide is oxidized to iodine.** At the border of the follicular cell and colloid, iodides are oxidized (by removal of electrons) and converted to iodine (I_2).

4. **Iodine is attached to tyrosine.** Iodine is attached to tyrosine amino acids that are part of the thyroglobulin molecule. This iodination reaction, mediated by peroxidase enzymes, occurs at the junction of the follicular cell and the colloid. Attachment of one iodine to a tyrosine produces **monoiodotyrosine** (**MIT**), and attachment of two iodines produces **diiodotyrosine** (**DIT**).

5. **Iodinated tyrosines are linked together to form T_3 and T_4.** Enzymes in the colloid link MIT and DIT together. Two linked DITs result in T_4, and coupling of MIT and DIT produces T_3. At this point, the hormones are still part of the thyroglobulin colloid.

(6) **Thyroglobulin colloid is endocytosed.** To secrete the hormones, the follicular cells must reclaim iodinated thyroglobulin by endocytosis and combine the vesicles with lysosomes.

(7) **Lysosomal enzymes split T$_4$ and T$_3$ from thyroglobulin and the hormones diffuse from the follicular cell into the bloodstream.** The main hormonal product secreted is T$_4$. Some T$_4$ is converted to T$_3$ before secretion, but most T$_3$ is generated in the peripheral tissues.

Transport and Regulation

Most T$_4$ and T$_3$ released into the blood immediately binds to *thyroxine-binding globulins* (*TBGs*) and other transport proteins produced by the liver. Both T$_4$ and T$_3$ bind to target tissue receptors, but T$_3$ binds more tightly and is about 10 times more active. Most peripheral tissues have the enzymes needed to convert T$_4$ to T$_3$ by removing one iodine atom.

Figure 16.7 (◄ p. 616) shows the negative feedback loop that regulates blood levels of TH. Falling TH blood levels trigger release of *thyroid-stimulating hormone* (*TSH*), and ultimately of more TH. Rising TH levels feed back to inhibit the hypothalamic–anterior pituitary axis, temporarily shutting off the stimulus for TSH release.

In infants, exposure to cold stimulates the hypothalamus to secrete *thyrotropin-releasing hormone* (*TRH*), which triggers TSH release. The thyroid gland then releases larger amounts of thyroid hormones, enhancing body metabolism and heat production. Factors that inhibit TSH release include GHIH, dopamine, and rising levels of glucocorticoids. Excessively high blood iodide concentrations also inhibit TH release.

HOMEOSTATIC IMBALANCE 16.4
CLINICAL

Both overactivity and underactivity of the thyroid gland can cause severe metabolic disturbances. Hypothyroid disorders may result from some thyroid gland defects or secondarily from inadequate TSH or TRH release. They also occur when the thyroid gland is removed surgically and when dietary iodine is inadequate.

In adults, the full-blown hypothyroid syndrome is called **myxedema** (mik″sĕ-de′mah; "mucous swelling"). Symptoms include a low metabolic rate; feeling chilled; constipation; thick, dry skin and puffy eyes; edema; lethargy; and mental sluggishness. A **goiter** (an enlarged protruding thyroid gland) occurs if myxedema results from lack of iodine (**Figure 16.10a**). The follicular cells produce colloid but cannot iodinate it and make functional hormones. The pituitary gland secretes increasing amounts of TSH in a futile attempt to stimulate the thyroid to produce TH, but the only result is that the follicles accumulate more and more *unusable* colloid. Depending on the cause, iodine supplements or hormone replacement therapy can reverse myxedema.

Before iodized salt became available, the midwestern United States was called the "goiter belt." Goiters were common because this area had iodine-poor soil and no access to iodine-rich seafood. In places where goiters are especially common, these goiters are called *endemic goiters*.

Thyroid hormone is critically important for normal growth and brain development in early childhood, so newborns are routinely screened for hypothyroidism. *Congenital hypothyroidism* is usually caused by poor development of the thyroid gland. Rarely, a pituitary problem or maternal medications may affect the baby's ability to make TH. In the first weeks of life, an infant with congenital hypothyroidism may have no symptoms at all, or may present with a weak cry, poor feeding, constipation, or prolonged jaundice. Early treatment with oral thyroid hormone replacement is crucial and usually allows a normal, healthy life. Most affected children will require lifelong TH replacement.

The most common hyperthyroid pathology is **Graves' disease**. In this autoimmune condition, a person makes abnormal antibodies directed against thyroid follicular cells. Rather than marking these cells for destruction as antibodies normally do, these antibodies paradoxically mimic TSH and continuously stimulate TH release.

Typical symptoms of Graves' disease include elevated metabolic rate; sweating; rapid, irregular heartbeat; nervousness; and weight loss despite adequate food. Eyeballs may protrude

(a) An enlarged thyroid (goiter); due to iodine deficiency

(b) Bulging eyes (exophthalmos) of Graves' disease

Figure 16.10 Thyroid disorders.

(*exophthalmos*) if the tissue behind the eyes becomes edematous and fibrous (Figure 16.10b). Treatments include surgically removing the thyroid gland or ingesting radioactive iodine (^{131}I), which destroys the most active thyroid cells. ●●●●●●●

Calcitonin

Calcitonin, a polypeptide hormone released by the **parafollicular,** or **C, cells** of the thyroid gland in response to a rise in blood Ca^{2+} levels, does not have a known physiological role in humans. In fact, calcitonin does not need to be replaced in patients whose thyroid gland has been removed.

However, at pharmacological doses (doses higher than normally found in the body), calcitonin has a bone-sparing effect and is given therapeutically to patients to treat Paget's disease and sometimes osteoporosis (bone diseases; ◄ pp. 193–194). At these high concentrations, calcitonin targets the skeleton, where it (1) inhibits osteoclast activity, inhibiting bone resorption and release of Ca^{2+} from the bony matrix, and (2) stimulates Ca^{2+} uptake and incorporation into bone matrix.

Check Your Understanding

13. What is the difference between T_3 and T_4? Which one of these is referred to as thyroxine?

14. Why does TH require plasma proteins in order to move through blood? Where are the receptors for TSH located?

For answers, see Answers Appendix.

16.8 The parathyroid glands are primary regulators of blood calcium levels

Learning Outcome

▶ Indicate the general functions of parathyroid hormone.

The tiny, yellow-brown **parathyroid glands** are nearly hidden from view in the posterior aspect of the thyroid gland (**Figure 16.11a**). There are usually four of these glands, but the number varies—as many as eight have been reported in some individuals, and some may be located in other regions of the neck or even in the thorax.

The parathyroid's glandular cells are arranged in thick, branching cords containing scattered *oxyphil cells* and large numbers of smaller **parathyroid cells** (Figure 16.11b). The

Pharynx
(posterior
aspect)

Thyroid
gland

Parathyroid
glands

Esophagus

Trachea

Capillary

Parathyroid cells (secrete parathyroid hormone)

Oxyphil cells

(a)

(b)

Figure 16.11 The parathyroid glands. (a) The parathyroid glands are located on the posterior aspect of the thyroid gland and are more inconspicuous than depicted. **(b)** Photomicrograph of parathyroid gland tissue (160×).

Practice Histology questions: MasteringA&P® > Study Area > Lab Tools > PAL

parathyroid cells secrete parathyroid hormone. The function of the oxyphil cells is unclear.

Parathyroid hormone (PTH), or *parathormone*, the protein hormone of these glands, is the single most important hormone controlling calcium balance in the blood. Precise control of calcium levels is critical because Ca^{2+} homeostasis is essential for so many functions, including transmission of nerve impulses, muscle contraction, and blood clotting.

Falling blood Ca^{2+} levels trigger PTH release, and rising blood Ca^{2+} levels inhibit its release. PTH increases Ca^{2+} levels in blood by stimulating three target organs: the skeleton, the kidneys, and the intestine (**Figure 16.12**). PTH release:

- Stimulates osteoclasts (bone-resorbing cells) to digest some of the calcium-rich bony matrix and release ionic calcium and phosphates to the blood (◀ **Figure 6.14, p. 189**).

- Enhances the kidney's reabsorption of Ca^{2+} from the forming urine into the blood [and excretion of phosphate (PO_4^{3-})].

- Promotes activation of vitamin D, thereby increasing absorption of Ca^{2+} by intestinal mucosal cells. Vitamin D is required

for absorption of Ca^{2+} from food, but first the kidneys must convert it to its active vitamin D_3 form, *calcitriol* (1,25-dihydroxycholecalciferol). PTH stimulates this transformation.

Hyperparathyroidism (excess PTH) is the third most common endocrine disorder. It is often caused by a parathyroid gland tumor. Calcium leaches from the bones, which soften and deform as fibrous connective tissue replaces their mineral salts. In *osteitis fibrosa cystica*, a severe form of this disorder, the bones have a moth-eaten appearance on X rays and tend to fracture spontaneously. The resulting hypercalcemia (abnormally elevated blood Ca^{2+} level) has many outcomes, but the two most notable are (1) it depresses the nervous system, which leads to abnormal reflexes and weak skeletal muscles, and (2) excess calcium salts precipitate in the kidney tubules, forming kidney stones. Calcium deposits may also form in soft tissues throughout the body and severely impair vital organ functioning.

Hypoparathyroidism (PTH deficiency) most often follows parathyroid gland trauma or removal during thyroid surgery. The resulting hypocalcemia (low blood Ca^{2+}) makes neurons more excitable and accounts for the classic signs and symptoms of tingling sensations, *tetany* (twitching muscle), and convulsions. Untreated, the symptoms progress to respiratory paralysis and death.

Check Your Understanding

15. What is the major effect of thyroid hormone? Parathyroid hormone? Calcitonin?

16. Name the cells that release each of the three hormones listed above.

For answers, see Answers Appendix.

16.9 The adrenal glands produce hormones involved in electrolyte balance and the stress response

Learning Outcome

▶ List hormones produced by the adrenal gland, and cite their physiological effects.

The paired **adrenal glands** are pyramid-shaped organs perched atop the kidneys (*ad* = near; *renal* = kidney), where they are enclosed in a fibrous capsule and a cushion of fat (see Figure 16.1 and **Figure 16.13**). They are also called the **suprarenal glands** (*supra* = above).

Each adrenal gland is structurally and functionally two endocrine glands. The inner **adrenal medulla**, more like a knot of nervous tissue than a gland, is part of the sympathetic nervous system (◀ **p. 542**). The outer **adrenal cortex**, encapsulating the medulla and forming the bulk of the gland, is glandular tissue derived from embryonic mesoderm. Each region produces its own set of hormones summarized in **Table 16.5** (p. 626), but all adrenal hormones help us cope with stressful situations.

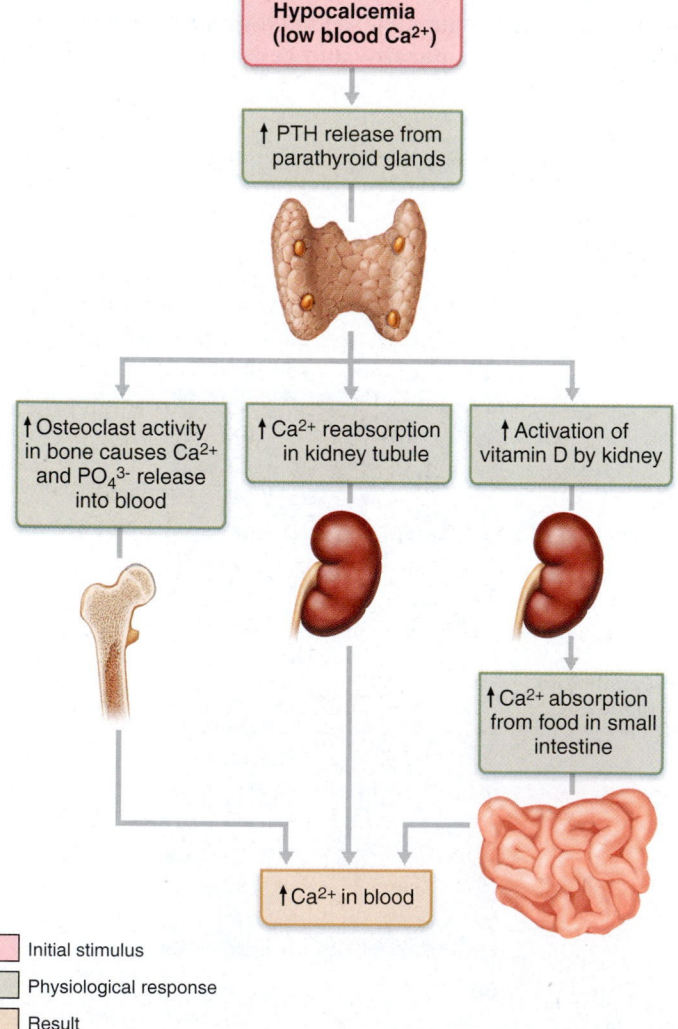

Figure with boxes:

Hypocalcemia (low blood Ca^{2+})

↑ PTH release from parathyroid glands

↑ Osteoclast activity in bone causes Ca^{2+} and PO_4^{3-} release into blood

↑ Ca^{2+} reabsorption in kidney tubule

↑ Activation of vitamin D by kidney

↑ Ca^{2+} absorption from food in small intestine

↑ Ca^{2+} in blood

☐ Initial stimulus
☐ Physiological response
☐ Result

Figure 16.12 Effects of parathyroid hormone on bone, the kidneys, and the intestine.

(a) Drawing of the histology of the adrenal cortex and a portion of the adrenal medulla

(b) Photomicrograph (115×)

Figure 16.13 Microscopic structure of the adrenal gland.

Practice Histology questions: MasteringA&P® > Study Area > Lab Tools > PAL

The Adrenal Cortex

The adrenal cortex synthesizes well over two dozen steroid hormones, collectively called **corticosteroids**. The multistep steroid synthesis pathway begins with cholesterol, and involves varying intermediates depending on the hormone being formed. Unlike the amino acid–based hormones, steroid hormones are not stored in cells. Consequently, their rate of release depends on their rate of synthesis.

The large, lipid-laden cortical cells are arranged in three layers or zones (Figure 16.13). From the outside in, they are:

- **Zona glomerulosa** (zo′nah glo-mer″u-lo′sah). The cell clusters forming this superficial layer produce *mineralocorticoids*, hormones (such as aldosterone) that help control the balance of minerals and water in the blood.

- **Zona fasciculata** (fah-sik″u-la′tah). The cells of this middle layer, arranged in more or less linear cords, mainly produce the metabolic hormones called *glucocorticoids* (such as cortisol).

- **Zona reticularis** (rĕ-tik″u-lar′is). The cells of this innermost layer, next to the adrenal medulla, have a netlike arrangement. They mainly produce small amounts of adrenal sex hormones, or *gonadocorticoids*.

Note, however, that the two innermost layers of the adrenal cortex share production of glucocorticoids and gonadocorticoids, although each layer predominantly produces one type.

Mineralocorticoids

The essential function of **mineralocorticoids** is to regulate the electrolyte (mineral salt) concentrations in extracellular fluids, particularly of Na^+ and K^+. The single most abundant cation in extracellular fluid is Na^+, and the amount of Na^+ in the body largely determines the volume of the extracellular fluid—where Na^+ goes, water follows. Changes in Na^+ concentration lead to changes in blood volume and blood pressure. Moreover, the regulation of Na^+ is coupled to the regulation of many other ions, including K^+, H^+, HCO_3^- (bicarbonate), and Cl^- (chloride).

The extracellular concentration of K^+ is also critical—it sets the resting membrane potential of all cells and determines how easily action potentials are generated in nerve and muscle. Not surprisingly, Na^+ and K^+ regulation are crucial to overall body homeostasis. Their regulation is the primary job of **aldosterone** (al-dos′ter-ōn), the most potent mineralocorticoid. Aldosterone accounts for more than 95% of the mineralocorticoids produced and is essential for life.

Aldosterone reduces excretion of Na^+ from the body. Its primary target is the kidney tubules, where it:

- Stimulates Na^+ reabsorption (causing increased blood volume and blood pressure because water follows Na^+)

- Causes K^+ secretion into the tubules for elimination from the body

In some instances, aldosterone can alter the acid-base balance of the blood (by increasing H^+ excretion). Aldosterone also enhances Na$^+$ reabsorption from perspiration, saliva, and gastric juice.

Aldosterone's regulatory effects are brief (approximately 20 minutes), allowing plasma electrolyte balance to be precisely controlled and continuously modified. The mechanism of aldosterone activity involves the synthesis and activation of proteins required for Na$^+$ transport such as Na$^+$-K$^+$ ATPase, the pump that exchanges Na$^+$ for K$^+$ across plasma membranes.

Decreasing blood volume and blood pressure, and rising blood levels of K$^+$, stimulate aldosterone secretion. The reverse conditions inhibit its secretion. Four mechanisms regulate aldosterone secretion, but two of them—the renin-angiotensin-aldosterone mechanism and plasma concentrations of potassium—are by far the most important (**Figure 16.14**).

① **The renin-angiotensin-aldosterone mechanism.** The renin-angiotensin-aldosterone mechanism (re′nin an″je-o-ten′sin) influences both blood volume and blood pressure by regulating the release of aldosterone and therefore Na$^+$ and water reabsorption by the kidneys.

1. When blood pressure (or blood volume) falls, specialized cells of the *juxtaglomerular complex* in the kidneys are excited.
2. These cells respond by releasing **renin** into the blood.
3. Renin splits off part of the plasma protein **angiotensinogen** (an″je-o-ten′sin-o-jen), triggering an enzymatic cascade that forms **angiotensin II**, which stimulates the glomerulosa cells to release aldosterone.

However, the renin-angiotensin-aldosterone mechanism does much more than trigger aldosterone release, and all of its effects ultimately raise blood pressure. We describe these additional effects in Chapters 19 and 26.

② **Plasma concentration of potassium.** Fluctuating blood levels of K$^+$ directly influence the zona glomerulosa cells in the adrenal cortex. Increased K$^+$ stimulates aldosterone release, whereas decreased K$^+$ inhibits it.

③ **ACTH.** Under normal circumstances, ACTH released by the anterior pituitary has little or no effect on aldosterone release. However, when a person is severely stressed, the hypothalamus secretes more corticotropin-releasing hormone (CRH), and the resulting rise in ACTH blood levels steps up the rate of aldosterone secretion to a small extent. The resulting

increase in blood volume and blood pressure helps deliver nutrients and respiratory gases during the stressful period.

④ **Atrial natriuretic peptide (ANP).** Atrial natriuretic peptide, a hormone secreted by the heart when blood pressure rises, fine-tunes blood pressure and sodium-water balance of the body. One of its major effects is to inhibit the renin-angiotensin-aldosterone mechanism. It blocks renin and aldosterone secretion and inhibits other angiotensin-induced mechanisms that enhance water and Na$^+$ reabsorption. Consequently, ANP's overall influence is to decrease blood pressure by allowing Na$^+$ (and water) to flow out of the body in urine (*natriuretic* = producing salty urine).

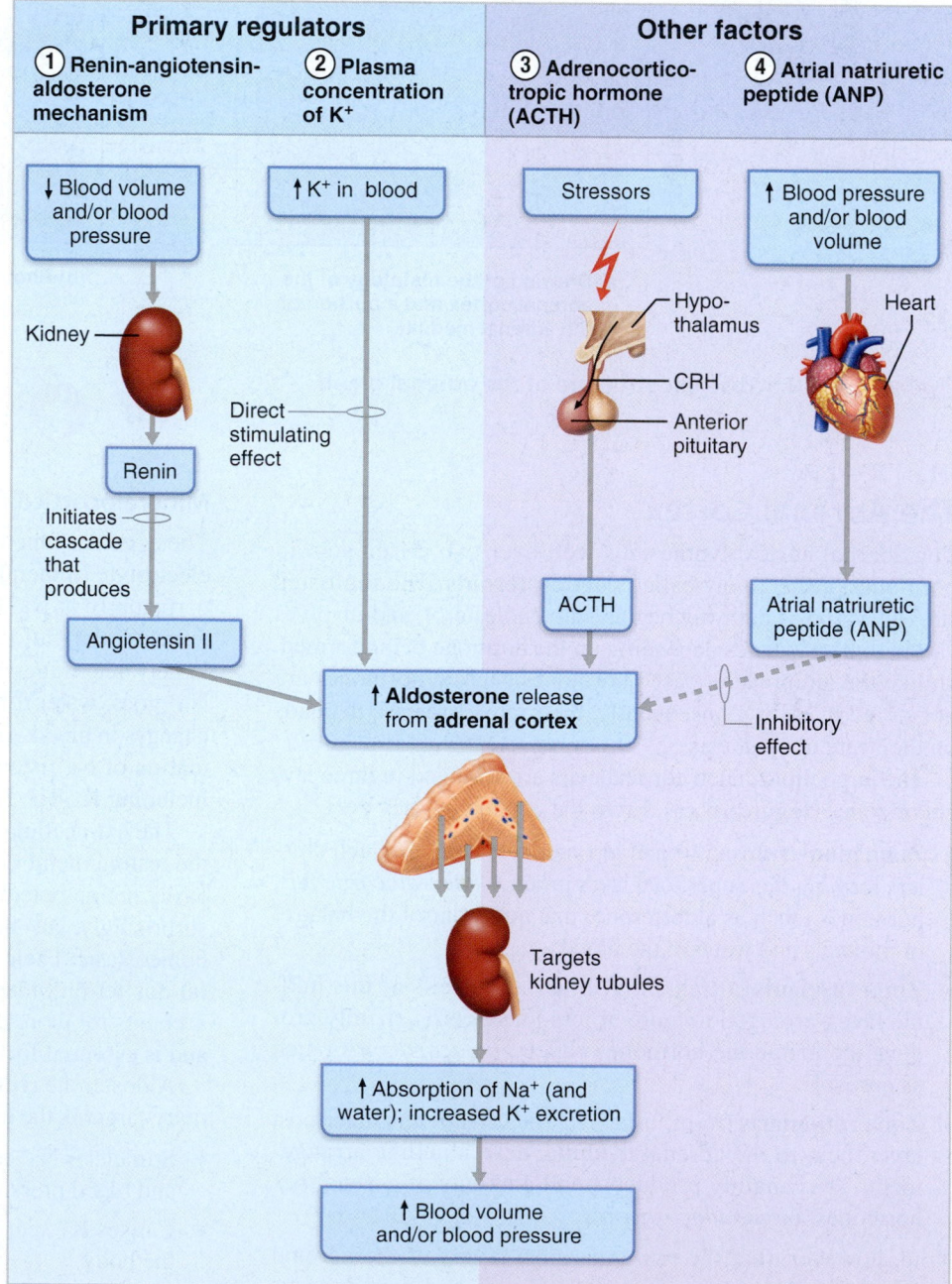

Figure 16.14 Major mechanisms controlling aldosterone release.

HOMEOSTATIC
IMBALANCE 16.6   CLINICAL

Hypersecretion of aldosterone, a condition called aldosteronism, typically results from adrenal tumors. Two major sets of problems result: (1) hypertension and edema due to excessive Na⁺ and water retention, and (2) accelerated excretion of potassium ions. If K^+ loss is extreme, neurons become nonresponsive, leading to muscle weakness and eventually paralysis. •••••••

Glucocorticoids

The **glucocorticoids** influence the energy metabolism of most body cells and help us resist stressors. Under normal circumstances, they help the body adapt to intermittent food intake by keeping blood glucose levels fairly constant, and help maintain blood pressure.

Glucocorticoid hormones include **cortisol (hydrocortisone)**, **cortisone**, and **corticosterone**, but only cortisol is secreted in significant amounts in humans. As with all steroid hormones, glucocorticoids act on target cells by modifying gene activity.

Regulation of Secretion   Negative feedback regulates glucocorticoid secretion. Cortisol release is promoted by ACTH. ACTH release is triggered in turn by the hypothalamic releasing hormone CRH. Rising cortisol levels feed back to act on both the hypothalamus and the anterior pituitary, preventing CRH release and shutting off ACTH and cortisol secretion.

Cortisol secretory bursts, driven by patterns of eating and activity, occur in a definite pattern throughout the day and night. Cortisol blood levels peak shortly before we rise in the morning. The lowest levels occur in the evening just before and shortly after we fall asleep.

Various stressors (for example, hemorrhage, infection, or physical or emotional trauma) interrupt the normal cortisol rhythm. Higher CNS centers override the inhibitory effects of elevated cortisol levels and trigger CRH release. The resulting increase in ACTH blood levels causes an outpouring of cortisol from the adrenal cortex.

Actions   The dramatically higher output of glucocorticoids during stress is essential for negotiating the crisis. Cortisol provokes a marked rise in blood levels of glucose, fatty acids, and amino acids. Cortisol's prime metabolic effect is to provoke *gluconeogenesis*, that is, the formation of glucose from fats and proteins. In order to "save" glucose for the brain, cortisol mobilizes fatty acids from adipose tissue and encourages their increased use for energy. Under cortisol's influence, proteins are broken down to provide building blocks for repair or to make enzymes for metabolic processes. Cortisol's second critical function is to enhance the sympathetic nervous system's vasoconstrictive effects, helping to maintain blood pressure.

Note that *ideal amounts of glucocorticoids promote normal function*, but too much cortisol exerts significant anti-inflammatory and anti-immune effects. Excessive levels of glucocorticoids:

- Depress cartilage and bone formation
- Inhibit inflammation by decreasing the release of inflammatory chemicals

- Depress the immune system
- Disrupt normal cardiovascular, neural, and gastrointestinal function

Glucocorticoid drugs can control symptoms of many chronic inflammatory disorders, such as rheumatoid arthritis and allergic responses. However, these potent drugs are a double-edged sword because they also cause the undesirable effects of excessive levels of these hormones.

HOMEOSTATIC
IMBALANCE 16.7   CLINICAL

The pathology of glucocorticoid excess, **Cushing's syndrome**, may be caused by an ACTH-releasing pituitary tumor (in which case it is called **Cushing's disease**); by an ACTH-releasing malignancy of the lungs, pancreas, or kidneys; or by a tumor of the adrenal cortex. However, it most often results from the clinical administration of glucocorticoid drugs.

The syndrome is characterized by persistent elevated blood glucose levels (*steroid diabetes*), dramatic losses in muscle and bone protein, and water and salt retention, leading to hypertension and edema. The so-called *cushingoid signs* (**Figure 16.15**) include a swollen "moon" face, redistribution of fat to the abdomen and the posterior neck (causing a "buffalo hump"), easy bruising, and poor wound healing. Because of enhanced anti-inflammatory effects, infections may become overwhelmingly severe before producing recognizable symptoms. Eventually, muscles weaken and spontaneous fractures force the person to become bedridden. The only treatment is to remove the cause—be it surgically removing the tumor or discontinuing the drug.

Addison's disease, the major hyposecretory disorder of the adrenal cortex, usually involves deficits in both glucocorticoids and mineralocorticoids. Affected individuals tend to lose

(a) Patient before onset

(b) Same patient with Cushing's syndrome. The white arrow shows the characteristic "buffalo hump" of fat on the upper back.

Figure 16.15 The effects of excess glucocorticoid.

Table 16.5 Adrenal Gland Hormones: Summary of Regulation and Effects

HORMONE	REGULATION OF RELEASE	TARGET ORGAN AND EFFECTS	EFFECTS OF HYPERSECRETION ↑ AND HYPOSECRETION ↓
Adrenocortical Hormones			
Mineralocorticoids (chiefly aldosterone)	**Stimulated** by renin-angiotensin-aldosterone mechanism (activated by decreasing blood volume or blood pressure), elevated blood K+ levels, and ACTH (minor influence) **Inhibited** by increased blood volume and pressure, and decreased blood K+ levels	Kidneys: increase blood levels of Na+ and decrease blood levels of K+; since water reabsorption usually accompanies sodium retention, blood volume and blood pressure rise	↑: Aldosteronism ↓: Addison's disease
Glucocorticoids (chiefly cortisol)	**Stimulated** by ACTH **Inhibited** by feedback inhibition exerted by cortisol	Body cells: promote gluconeogenesis and hyperglycemia; mobilize fats for energy metabolism; stimulate protein catabolism; assist body to resist stressors; depress inflammatory and immune responses	↑: Cushing's syndrome ↓: Addison's disease
Gonadocorticoids (chiefly androgens, converted to testosterone or estrogens after release)	**Stimulated** by ACTH; mechanism of inhibition incompletely understood, but feedback inhibition not seen	Insignificant effects in males; contributes to female libido; development of pubic and axillary hair in females; source of estrogens after menopause	↑: Masculinization of females (adrenogenital syndrome) ↓: No effects known
Adrenal Medullary Hormones			
Catecholamines (epinephrine and norepinephrine)	**Stimulated** by preganglionic fibers of the sympathetic nervous system	Sympathetic nervous system target organs: effects mimic sympathetic nervous system activation; increase heart rate and metabolic rate; increase blood pressure by promoting vasoconstriction	↑: Prolonged fight-or-flight response; hypertension ↓: Unimportant

weight; plasma glucose and sodium levels drop, and potassium levels rise. Severe dehydration and hypotension are common. Corticosteroid replacement therapy is the usual treatment.

A common early sign of Addison's disease is a characteristic bronzing of the skin. This occurs because the lack of negative feedback by corticosteroids increases ACTH release by the anterior pituitary. At high levels, ACTH triggers melanin production in melanocytes.

Gonadocorticoids (Adrenal Sex Hormones)

Most **gonadocorticoids** secreted by the adrenal cortex are weak **androgens**, or male sex hormones, such as *androstenedione* and *dehydroepiandrosterone* (*DHEA*). Most are converted in tissue cells to more potent male hormones, such as *testosterone*, and some are converted to estrogens. The amount of gonadocorticoids produced by the adrenal cortex is insignificant compared with the amounts made by the gonads during late puberty and adulthood.

The exact role of the adrenal sex hormones is still in question, but we know that they contribute to axillary and pubic hair development. In adult women, adrenal androgens are thought to contribute to the sex drive, and they largely account for the estrogens produced after menopause when ovarian estrogens are no longer produced. Control of gonadocorticoid secretion is not completely understood. ACTH stimulates their release, but the gonadocorticoids do not appear to exert feedback inhibition on ACTH release.

HOMEOSTATIC IMBALANCE 16.8   CLINICAL

Since androgens predominate, hypersecretion of gonadocorticoids causes *adrenogenital syndrome* (masculinization). In adult males, elevated gonadocorticoid levels may not be noticeable since testicular testosterone has already produced masculinization, but in prepubertal males and in females, the results can be dramatic. In boys, the reproductive organs mature and secondary sex characteristics appear early, and the sex drive emerges with a vengeance. Females develop a beard and a masculine distribution of body hair, and the clitoris grows to resemble a small penis.

The Adrenal Medulla

We discussed the adrenal medulla in Chapter 14 as part of the autonomic nervous system (◄ p. 542), so our coverage here is brief. The spherical **medullary chromaffin cells** (kro′mah-fin), which crowd around porous blood-filled capillaries, are modified postganglionic sympathetic neurons. The cells synthesize the *catecholamines* **epinephrine** and **norepinephrine (NE)** via a molecular sequence from tyrosine to dopamine to NE to epinephrine.

When a short-term stressor activates the body to fight-or-flight status, the sympathetic nervous system is mobilized. Blood vessels constrict and the heart beats faster (together raising the blood pressure), and blood is diverted from temporarily non-essential organs to the heart and skeletal muscles. Blood glucose levels rise, and preganglionic sympathetic nerve endings weaving through the adrenal medulla signal for release of catecholamines, which reinforce and prolong the fight-or-flight response.

Unequal amounts of the two hormones are stored and released. Approximately 80% is epinephrine and 20% norepinephrine. With a few exceptions, the two hormones exert the same effects (◄ Table 14.3, p. 544). Epinephrine is the more potent stimulator of metabolic activities and dilator of small airways (bronchioles), but norepinephrine has a greater influence on peripheral vasoconstriction and blood pressure. Epinephrine is used clinically as a heart stimulant and to dilate the bronchioles during acute asthmatic attacks and severe allergic reactions.

Unlike hormones from the adrenal cortex, which promote long-lasting body responses to stressors, catecholamines cause fairly brief responses. *Focus on Stress and the Adrenal Gland* (**Focus Figure 16.2** on pp. 628–629) illustrates how our bodies respond to stressors in the short and long term.

HOMEOSTATIC IMBALANCE 16.9   CLINICAL

A deficiency in adrenal medulla hormones is not a problem because these hormones merely intensify activities set into motion by the sympathetic nervous system neurons. Unlike glucocorticoids and mineralocorticoids, adrenal catecholamines are not essential for life.

On the other hand, hypersecretion of catecholamines, sometimes arising from a medullary chromaffin cell tumor called a *pheochromocytoma* (fe-o-kro″mo-si-to′mah), produces symptoms of uncontrolled sympathetic nervous system activity—increased metabolic rate, rapid heartbeat and palpitations, hypertension, intense nervousness, and sweating. ●●●●●●●

Check Your Understanding

17. List the three classes of hormones released from the adrenal cortex and briefly state the major effect(s) of each.
18. What are the two major stimuli for aldosterone release? What is the major stimulus for cortisol release?
19. Which part of the gland(s) below releases steroid hormones? Amino acid–based hormones? Name the hormone released from

(c) that acts on (a). Histologically, what do (b) and (d) have in common that differentiates them from other glands?

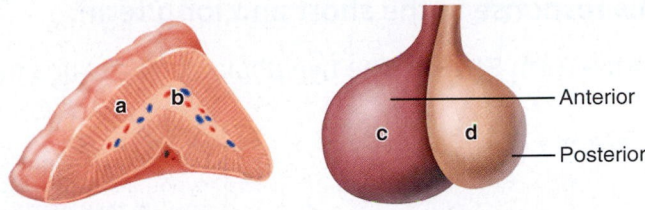

- Anterior
- Posterior

For answers, see Answers Appendix.

16.10 The pineal gland secretes melatonin

Learning Outcome

▶ Briefly describe the importance of melatonin.

The tiny, pinecone-shaped **pineal gland** hangs from the roof of the third ventricle in the diencephalon (see Figure 16.1 and ◄ Figure 12.11, p. 447). Its secretory cells, called **pinealocytes**, are arranged in compact cords and clusters. Lying between pinealocytes in adults are dense particles containing calcium salts. These salts are radiopaque, making the pineal gland a handy landmark for determining brain orientation in X rays.

Although many peptides and amines have been isolated from this minute gland, its only major secretory product is **melatonin** (mel″ah-to′nin), an amine hormone derived from serotonin. Melatonin concentrations in the blood rise and fall in a diurnal (daily) cycle. Peak levels occur during the night and make us drowsy, and lowest levels occur around noon. Recent evidence suggests that melatonin also controls the production of protective antioxidant and detoxification molecules within cells.

The pineal gland indirectly receives input from the visual pathways (retina → suprachiasmatic nucleus of hypothalamus → superior cervical ganglion → pineal gland) concerning the intensity and duration of daylight. In some animals, mating behavior and gonadal size vary with relative lengths of light and dark periods, and melatonin mediates these effects.

In children, melatonin may have an antigonadotropic effect. In other words, it may affect the timing of puberty and inhibit precocious (too early) sexual maturation.

The *suprachiasmatic nucleus* of the hypothalamus, an area referred to as our "biological clock," is richly supplied with melatonin receptors, and exposure to bright light (known to suppress melatonin secretion) can reset the clock timing. As a result, changing melatonin levels may influence rhythmic variations in physiological processes such as body temperature, sleep, and appetite.

Check Your Understanding

20. **APPLY** Synthetic melatonin supplements are available, although their efficacy and long-term safety have not been proved. What do you think they might be used for?

For answers, see Answers Appendix.

16

Stress is a bodywide nonspecific response to stressors—actual or potential threats to homeostasis. The hypothalamus coordinates this response in the short and long term.

Short-Term Stress: The hypothalamus activates the *adrenal medulla* via *neural stimuli.*

Stressors

Hypothalamus

① Action potentials (APs) triggered by the hypothalamus in response to stressors activate the sympathetic nervous system.

Spinal cord

② APs travel along preganglionic sympathetic axons to the adrenal medulla.

Adrenal medulla

③ The adrenal medulla secretes amino acid-based hormones: epinephrine (80%) and some norepinephrine (20%).

Blood vessel

④ Epinephrine and norepinephrine reinforce other sympathetic responses to ready the body for exertion ("fight or flight").

Cardiovascular effects

↑ Heart rate

↑ Blood pressure

Redirect blood flow away from nonessential organs (digestive, urinary)

Respiratory effects

Bronchiole

Bronchioles dilate to facilitate airflow in and out of lungs and help oxygenate blood.

Metabolic effects

Glucose

↑ Blood glucose (liver converts glycogen to glucose), making more energy available to essential organs

↑ Metabolic rate

The short-term stress response helps the body successfully respond to a short-term emergency (such as running away from a fire).

Long-Term Stress: The hypothalamus activates the *adrenal cortex* via *hormonal stimuli.* This greatly reinforces the continuing short-term stress response.

Hypothalamus

CRH

Anterior pituitary

ACTH

Adrenal gland

Glucocorticoids (some mineralocorticoids)

Target cells

Stressors

1 Stressors cause hypothalamic neurons to release corticotropin-releasing hormone (CRH).

2 CRH travels via the portal system of blood vessels to the anterior pituitary.

3 Anterior pituitary cells (corticotrophs) release adrenocorticotropic hormone (ACTH).

4 ACTH travels in blood to the adrenal cortex.

ACTH

5 The adrenal cortex synthesizes and releases glucocorticoids, e.g., cortisol (and some mineralocorticoids, e.g., aldosterone).

Adrenal cortex

Glucocorticoids

Mineralocorticoids

Metabolic effects

Renal effects

• Protein breakdown → ↑Blood amino acids (for new protein synthesis if required for emergency)

↑Blood glucose (fuel for brain)

• Fat breakdown → ↑Blood fatty acids (fuel for muscles)

Also:
• Supresses immune system
• Maintains blood pressure

• Kidneys retain sodium (and water)
• Blood volume and blood pressure rise

The long-term stress response helps the body successfully deal with some long-term stressors (such as starvation). However, if the stress response is prolonged, it can become detrimental. It may lead to high blood pressure, muscle loss, altered immune function, and other problems.

16.11 The pancreas, gonads, and most other organs secrete hormones

Learning Outcomes

▶ **Compare and contrast the effects of the two major pancreatic hormones.**

▶ **Describe the functional roles of hormones of the testes, ovaries, and placenta.**

▶ **State the location of enteroendocrine cells.**

▶ **Briefly explain the hormonal functions of the heart, kidney, skin, adipose tissue, bone, and thymus.**

So far, we've examined the endocrine role of the hypothalamus and of glands dedicated solely to endocrine function. We will now consider a set of organs that contain endocrine tissue but also have other major functions. These include the pancreas, gonads, and placenta.

The Pancreas

Located partially behind the stomach in the abdomen, the soft, tadpole-shaped **pancreas** is a mixed gland composed of both endocrine and exocrine gland cells (see Figure 16.1). Along with the thyroid and parathyroids, it develops as an outpocketing of the epithelial lining of the gastrointestinal tract. *Acinar cells*, forming the bulk of the gland, produce an enzyme-rich juice that is carried by ducts to the small intestine during digestion.

Scattered among the acinar cells are approximately a million **pancreatic islets** (also called *islets of Langerhans*), tiny cell clusters that produce pancreatic hormones (**Figure 16.16**). The islets contain two major populations of hormone-producing cells:

Pancreatic islet (endocrine)

• α (Glucagon-producing) cells

• β (Insulin-producing) cells

Pancreatic acinar cells (exocrine)

Figure 16.16 Photomicrograph of differentially stained pancreatic tissue. A pancreatic islet is surrounded by acinar cells, which produce the exocrine product (enzyme-rich pancreatic juice) (190×).

Practice Histology questions: MasteringA&P° > Study Area > Lab Tools > PAL

• **Alpha (α) cells** release glucagon and are fewer in number

• **Beta (β) cells** release insulin and are more numerous.

These cells act as tiny fuel sensors, secreting glucagon and insulin appropriately during the fasting and fed states.

Insulin and glucagon are intimately but independently involved in regulating blood glucose levels. Their effects are antagonistic: Glucagon is a *hyperglycemic* hormone (increases blood glucose), whereas insulin is a *hypoglycemic* hormone (decreases blood glucose) (**Figure 16.17**). Some islet cells also synthesize other peptides in small amounts, including *somatostatin*, *pancreatic polypeptide* (*PP*), and others. However, here we will focus on glucagon and insulin.

Glucagon

Glucagon (gloo′kah-gon), a 29-amino-acid polypeptide, is an extremely potent hyperglycemic agent: One molecule can cause the release of 100 million glucose molecules into the blood! The major target of glucagon is the liver, where it promotes the following actions:

• Breakdown of glycogen to glucose (*glycogenolysis*) (Figure 16.17)

• Synthesis of glucose from lactic acid and from noncarbohydrate molecules (*gluconeogenesis*)

• Release of glucose to the blood by liver cells, causing blood glucose levels to rise

A secondary effect is to lower blood levels of amino acids as the liver cells sequester these molecules to make new glucose molecules.

Humoral stimuli, mainly falling blood glucose levels, prompt the alpha cells to secrete glucagon. However, sympathetic nervous system stimulation and rising amino acid levels (as might follow a protein-rich meal) are also stimulatory. Glucagon release is suppressed by rising blood glucose levels, insulin, and somatostatin.

Insulin

Insulin is a small (51-amino-acid) protein consisting of two amino acid chains linked by disulfide (–S–S–) bonds. It is synthesized as part of a larger polypeptide chain called **proinsulin**. Enzymes then excise the middle portion of this chain, releasing functional insulin. This "clipping" process occurs in the secretory vesicles just before the beta cell releases insulin.

Insulin's effects are most obvious when we have just eaten. Its main effect is to lower blood glucose levels (Figure 16.17), but it also promotes protein synthesis and fat storage. Circulating insulin lowers blood glucose levels in three ways.

• It enhances membrane transport of glucose (and other simple sugars) into most body cells, especially muscle and fat cells.

• It inhibits the breakdown of glycogen to glucose.

• It inhibits the conversion of amino acids or fats to glucose. These inhibiting effects counter any metabolic activity that would increase plasma levels of glucose.

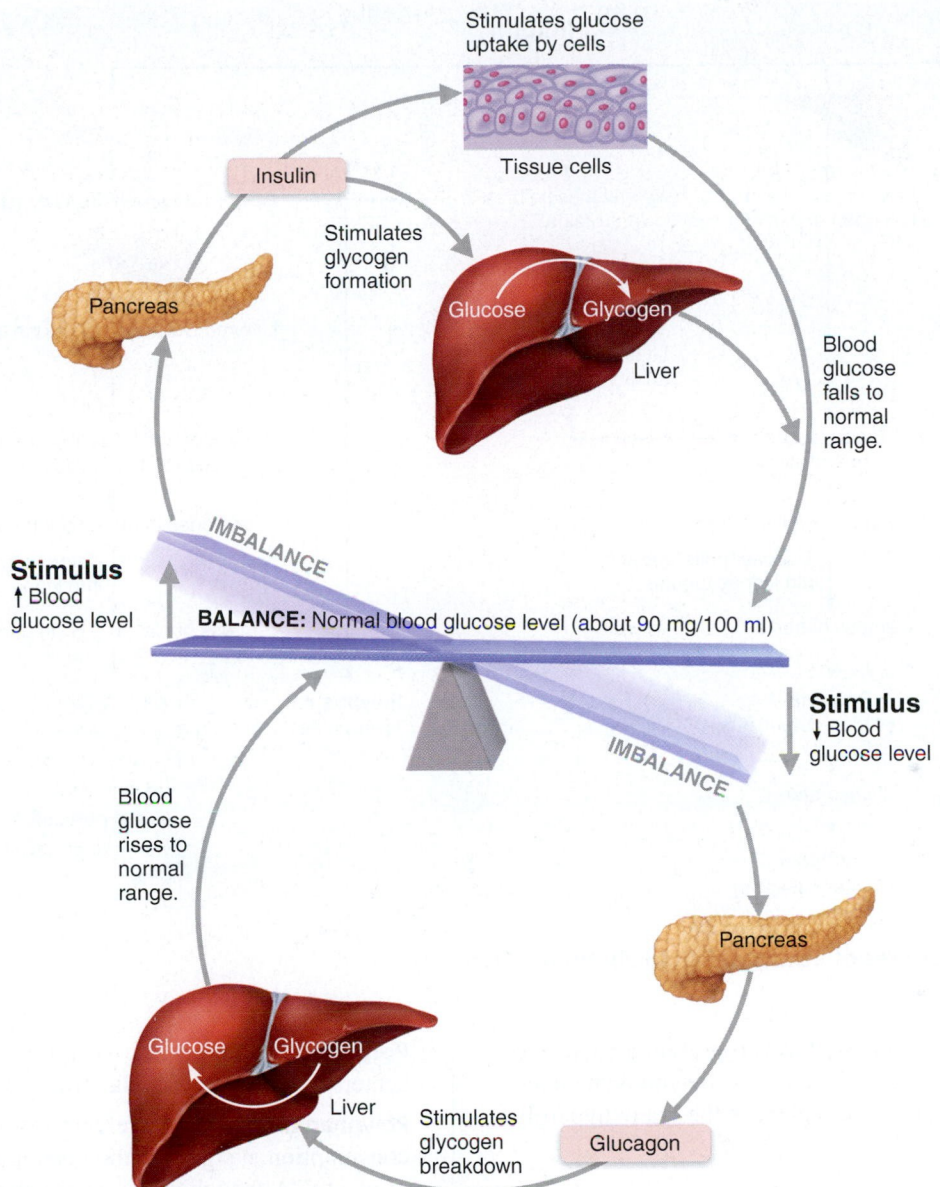

Figure 16.17 Insulin and glucagon from the pancreas regulate blood glucose levels.

Insulin is *not* needed for glucose entry into liver, kidney, and brain tissue, all of which have easy access to blood glucose regardless of insulin levels. However, insulin does have important roles in the brain—it participates in neuronal development, feeding behavior, and learning and memory.

Insulin activates its receptor (a tyrosine kinase enzyme), which phosphorylates specific proteins, beginning the cascade that promotes glucose uptake and insulin's other effects. After glucose enters a target cell, insulin binding triggers enzymatic activities that:

- Catalyze the oxidation of glucose for ATP production
- Join glucose molecules together to form glycogen
- Convert glucose to fat (particularly in adipose tissue)

As a rule, energy needs are met first, followed by glycogen formation. Finally, if excess glucose is still available, it is converted

to fat. Insulin also stimulates amino acid uptake and protein synthesis in muscle tissue.

Factors That Influence Insulin Release Pancreatic beta cells secrete insulin when stimulated by:

- Elevated blood glucose levels. This is the chief controlling factor.
- Rising blood levels of amino acids and fatty acids.
- Acetylcholine released by parasympathetic nerve fibers.
- Hyperglycemic hormones (such as glucagon, epinephrine, growth hormone, thyroxine, or glucocorticoids). This effect is indirect and occurs because all of these hormones increase blood glucose levels.

Somatostatin and sympathetic nervous system activation depress insulin release.

↓ **Insulin**

All tissues	Liver breaks down glycogen to glucose	Skeletal muscle breaks down proteins

Adipocytes break down fat (lipolysis)

↓ Glucose uptake (and usage)

Liver converts amino acids to glucose

Liver converts fats to ketone bodies

Blood
↑ Blood glucose (hyperglycemia)

↓ Blood pH due to ketone bodies (ketoacidosis)

Urine
Glucose in urine (glycosuria)

Ketones in urine (ketonuria)

Glucose "pulls" water into kidney tubules

Ketones "pull" cations into kidney tubules

Osmotic diuresis

Loss of Na⁺, K⁺, H⁺ in urine

Signs and symptoms

Polyuria
(↑ Urine output)

Polyphagia
(↑ Appetite)

- Heart rhythm abnormalities
- Nausea, vomiting, abdominal pain
- Central nervous system depression, coma

- Acetone breath
- ↑ Rate and depth of breathing

Dehydration

Polydipsia
(↑ Water intake)

Figure 16.18 Consequences of insulin deficit (diabetes mellitus).

As you can see, blood glucose levels represent a balance of humoral, neural, and hormonal influences. Insulin is the major hypoglycemic factor that counterbalances the many hyperglycemic hormones.

HOMEOSTATIC IMBALANCE 16.10 — CLINICAL

Diabetes mellitus (DM) results from either hyposecretion or hypoactivity of insulin. When insulin is absent, the result is *type 1 diabetes mellitus*. If insulin is present, but its effects are deficient, the result is *type 2 diabetes mellitus*. In either case, blood glucose levels remain high after a meal because glucose is unable to enter most tissue cells.

Diabetes is so common that you will encounter many patients with this disease in your career. It is described in more detail in **A Closer Look**. The three cardinal signs of diabetes mellitus (outlined in **Figure 16.18**) are:

- **Polyuria.** Excessive glucose in the blood leads to excessive glucose in the kidney filtrate where it acts as an osmotic diuretic (that is, it inhibits water reabsorption by the kidney tubules). The end result is **polyuria**, a huge urine output that decreases blood volume and causes dehydration.

- **Polydipsia.** Dehydration stimulates hypothalamic thirst centers, causing **polydipsia**, or excessive thirst.

- **Polyphagia. Polyphagia** refers to excessive hunger and food consumption, a sign that the person is "starving in the land of plenty." Although plenty of glucose is available, the body cannot use it. Instead, the body breaks down protein and fat to supply energy, and this is thought to stimulate appetite.

When sugars cannot be used as cellular fuel, more fats are mobilized, resulting in high fatty acid levels in the blood, a condition called lipidemia. In severe cases of diabetes mellitus, blood levels of fatty acids and their metabolites (acetoacetic acid, acetone, and others) rise dramatically. The fatty acid metabolites, collectively called **ketones** (ke′tōnz) or **ketone bodies**, are organic acids. When they accumulate in the blood, the blood pH drops, resulting in **ketoacidosis**, and ketone bodies begin to spill into the urine (*ketonuria*).

Severe ketoacidosis is life threatening. The nervous system responds by initiating rapid deep breathing (hyperpnea) to blow off carbon dioxide from the blood and increase blood pH. (We will explain the physiological basis of this mechanism in Chapter 22.) Serious electrolyte losses also occur as the body rids itself of excess ketone bodies. Ketone bodies are negatively charged and carry positive ions out with them, so sodium and

Sweet Revenge: Taming the Diabetes Monster?

Few medical breakthroughs have been as electrifying as the discovery of insulin in 1921, an event that changed diabetes mellitus (DM) from a death sentence to a survivable disease. Nonetheless, DM is still a huge health problem: Determining blood glucose levels accurately and maintaining desirable levels sorely challenge our present biotechnology. Let's take a closer look at the characteristics and challenges of type 1 and type 2, the major forms of diabetes mellitus.

More than 1 million Americans have **type 1 diabetes mellitus**, formerly called *insulin-dependent diabetes mellitus*. Symptoms appear suddenly, usually before age 15, following a long asymptomatic period during which the immune system destroys the pancreas's beta cells. Consequently, type 1 diabetics effectively lack insulin.

Type 1 diabetes susceptibility genes have been localized on several chromosomes, indicating that type 1 diabetes is an example of a multigene autoimmune response. However, some investigators believe that *molecular mimicry* is at least part of the problem: Some foreign substance (for example, a virus) has entered the body and is so similar to certain self (beta cell) proteins that the immune system attacks the beta cells as well as the invader. Sadly, once beta cells are destroyed, they're gone for good.

The goal of several lines of current research is to halt the destruction of beta cells. While broad-spectrum immunosuppressants have been used, these have serious side effects, and more targeted approaches are being pursued. Unfortunately, beta cell destruction is rapid and virtually complete by the time of diagnosis.

Type 1 diabetes patients typically develop long-term vascular and neural problems. The lipidemia and high blood cholesterol levels typical of the disease can lead to severe vascular complications including atherosclerosis, strokes, heart attacks, kidney shutdown, gangrene, and blindness. Nerve damage leads to loss of sensation, impaired bladder function, and impotence. Female type 1 diabetics also tend to have lumpy breasts and to undergo premature menopause, which increases their risk for cardiac problems.

Hyperglycemia is the culprit behind these complications, and the closer to normal blood glucose levels are held, the less likely complications are. Continuous glucose monitors make this much easier than relying on finger pricks. Currently, frequent insulin injections (up to four times daily, or better yet, by a continuous infusion pump) are recommended to reduce vascular and renal complications. Nevertheless, coupling glucose sensors to an insulin pump to make a true artificial pancreas has been surprisingly difficult. It was only in 2016 that the first such device was approved in the United States.

Pancreatic islet cell transplants have become increasingly successful in helping type 1 diabetics. Still, only about 50% of patients need no injected insulin after two years. The need for long-term immunosuppression limits this treatment to only those diabetics who cannot control their blood glucose by any other means.

Over 90% of DM cases are **type 2 diabetes mellitus**, formerly called *non-insulin-dependent diabetes mellitus*. Type 2 diabetics produce insulin, but their insulin receptors are unable to respond to it, a phenomenon called **insulin resistance**. Type 2 DM grows increasingly common with age and with the increasing size of our waistlines. About 12 million people in the U.S. have been diagnosed with type 2 diabetes, and roughly half as many are believed to be undiagnosed. Type 2 diabetics

are at risk for the same complications as type 1 diabetics—heart disease, amputations, kidney failure, and blindness.

A hereditary predisposition is particularly striking in this diabetic group. Mutations in any one of several genes could lead to insulin resistance. About 25–30% of Americans carry a gene that predisposes them to type 2 diabetes, with nonwhites affected to a much greater extent. If an identical twin has type 2 diabetes mellitus, the probability that the other twin will have the disease is virtually 100%.

Lifestyle factors also play a key role: Type 2 diabetics are almost always overweight and sedentary. Adipose tissue of obese people overproduces a number of signaling chemicals including *tumor necrosis factor alpha* and *resistin*, which may alter the enzymatic cascade triggered by insulin binding. It is now well established that weight loss and regular exercise can lower the risk of type 2 diabetes dramatically, even for people at high risk.

In many cases type 2 diabetes can be managed solely by exercise, weight loss, and a healthy diet. Some type 2 diabetics also benefit from oral medications that lower blood glucose or reduce insulin resistance. However, many type 2 diabetics must eventually inject insulin.

While we cannot yet cure diabetes, biotechnology promises to continue to improve control of blood glucose levels and thereby tame the monster that is diabetes.

potassium ions are also lost. The electrolyte imbalance leads to abdominal pain and possibly vomiting. If untreated, ketoacidosis disrupts heart activity and oxygen transport, and severe depression of the nervous system leads to coma and death.

The opposite condition to diabetes mellitus is *hyperinsulinism*, or excessive insulin secretion. It causes low blood glucose levels, or **hypoglycemia**. This condition triggers the release of hyperglycemic hormones, which cause anxiety, nervousness, tremors, and weakness. Insufficient glucose delivery to the brain causes disorientation, progressing to convulsions, unconsciousness, and even death. In rare cases, hyperinsulinism results from an islet cell tumor. More commonly, it is caused by an overdose of insulin and is easily treated by ingesting sugar.

The Gonads and Placenta

The male and female **gonads** (testes and ovaries) produce steroid sex hormones identical to those produced by adrenal cortical cells (see Figure 16.1). The major distinction is the source and relative amounts produced. As described earlier, gonadotropins regulate the release of gonadal hormones.

The paired *ovaries* are small, oval organs located in the female's abdominopelvic cavity. Besides producing ova, or eggs, the ovaries produce several hormones, most importantly **estrogens** and **progesterone** (pro-jes′tĕ-rōn). Estrogens are responsible for maturation of the reproductive organs and the appearance of the secondary sex characteristics of females at puberty. Acting with progesterone, estrogens promote breast development and cyclic changes in the uterine mucosa (the menstrual cycle).

The male *testes*, located in an extra-abdominal skin pouch called the scrotum, produce sperm and male sex hormones, primarily **testosterone** (tes-tos′tĕ-rōn). During puberty, testosterone initiates the maturation of the male reproductive organs and the appearance of secondary sex characteristics and sex drive. In addition, testosterone is necessary for normal sperm production and maintains the reproductive organs in their mature functional state in adult males.

The *placenta* is a temporary endocrine organ. Besides sustaining the fetus during pregnancy, it secretes several steroid and protein hormones that influence the course of pregnancy. Placental hormones include estrogens, progesterone, and human chorionic gonadotropin (hCG).

We will discuss the roles of the gonadal, placental, and gonadotropic hormones in Chapters 27 and 28, where we consider the reproductive system and pregnancy.

Hormone Secretion by Other Organs

Other hormone-producing cells occur in various organs including the heart, gastrointestinal tract, kidneys, skin, adipose tissue, skeleton, and thymus (**Table 16.6**).

Adipose Tissue

Adipose cells release **leptin**, which serves to tell your body how much stored energy (as fat) you have. The more fat you have, the more leptin there will be in your blood. As we describe in Chapter 24, leptin binds to CNS neurons concerned with appetite control, producing a sensation of satiety. It also appears to stimulate increased energy expenditure.

Two other hormones released by adipose cells affect the sensitivity of cells to insulin. *Resistin* is an insulin antagonist, while *adiponectin* enhances sensitivity to insulin.

Gastrointestinal Tract

Enteroendocrine cells are hormone-secreting cells sprinkled in the mucosa of the gastrointestinal tract. These scattered cells release several peptide hormones that help regulate a wide variety of digestive functions, as we will see in Chapter 23. Enteroendocrine cells also release amines such as serotonin, which act as paracrines, diffusing to and influencing nearby target cells without first entering the bloodstream. Enteroendocrine cells have been referred to as *paraneurons* because they are similar in certain ways to neurons and many of their hormones and paracrines are chemically identical to neurotransmitters.

Heart

Chambers of the heart called *atria* contain specialized cardiac muscle cells that secrete **atrial natriuretic peptide (ANP)**. As noted on p. 624, ANP decreases the amount of sodium in the extracellular fluid, thereby reducing blood volume and blood pressure.

Kidneys

Interstitial cells in the kidneys secrete **erythropoietin** (ĕ-rith″ro-poi′ĕ-tin; "red-maker"), a glycoprotein hormone that signals the bone marrow to increase production of red blood cells. The kidneys also release **renin**, which acts as an enzyme to initiate the renin-angiotensin-aldosterone mechanism of aldosterone release described earlier.

Skeleton

Osteoblasts in bone secrete *osteocalcin*, a hormone that prods pancreatic beta cells to divide and secrete more insulin. It also restricts fat storage by adipocytes, and triggers the release of adiponectin. This improves glucose handling and reduces body fat.

Interestingly, insulin promotes the conversion of inactive osteocalcin to active osteocalcin in bone, forming a two-way conversation between bone and the pancreas. Osteocalcin levels are low in type 2 diabetes, and increasing its level may offer a new treatment approach.

Skin

The skin produces **cholecalciferol**, an inactive form of vitamin D_3, when modified cholesterol molecules in epidermal cells are exposed to ultraviolet radiation. This compound then enters the blood via the dermal capillaries, is modified in the liver, and becomes fully activated in the kidneys. The active form of vitamin D_3, **calcitriol**, is an essential regulator of the carrier system that intestinal cells use to absorb Ca^{2+} from food. Without

Table 16.6 Selected Examples of Hormones Produced by Organs Other than the Major Endocrine Organs

SOURCE	HORMONE	CHEMICAL COMPOSITION	TRIGGER	TARGET ORGAN AND EFFECTS
Adipose tissue	Leptin	Peptide	Secretion proportional to fat stores; increased by nutrient uptake	Brain: suppresses appetite; increases energy expenditure
Adipose tissue	Resistin, adiponectin	Peptides	Secretion proportional to fat stores for resistin, inversely proportional for adiponectin	Fat, muscle, liver: resistin antagonizes insulin's action and adiponectin enhances it
Gastrointestinal (GI) tract mucosa				
• Stomach	Gastrin, ghrelin	Peptides	See Table 23.1 on p. 888; Figure 24.25 on p. 961	
• Duodenum (of small intestine)	Secretin, cholecystokinin (CCK), glucose-dependent insulinotropic peptide (GIP)	Peptides	See Table 23.1 on p. 888	
Heart (atria)	Atrial natriuretic peptide (ANP)	Peptide	Secreted in response to stretching of atria (by rising blood pressure)	Kidney: inhibits sodium ion reabsorption and renin release
				Adrenal cortex: inhibits secretion of aldosterone; decreases blood pressure
Kidney	Erythropoietin (EPO)	Glycoprotein	Secreted in response to hypoxia	Red bone marrow: stimulates production of red blood cells
Skeleton	Osteocalcin	Peptide	Unknown; insulin promotes its activation	Increases insulin production and insulin sensitivity
Skin (epidermal cells)	Cholecalciferol (provitamin D_3)	Steroid	Activated by the kidneys to active vitamin D_3 (calcitriol) in response to parathyroid hormone	Intestine: stimulates active transport of dietary calcium across cell membranes of small intestine
Thymus	Thymulin, thymopoietins, thymosins	Peptides	Unknown	Mostly act locally as paracrines; involved in T lymphocyte development and in immune responses

this vitamin, bones become soft and weak. In addition, most cells throughout the body have vitamin D receptors. Vitamin D modulates immune functions, decreases inflammation, and may act as an anticancer agent.

Thymus

Located deep to the sternum in the thorax is the **thymus** (see Figure 16.1). Large and conspicuous in infants and children, the thymus shrinks throughout adulthood. By old age, it is composed largely of adipose and fibrous connective tissues.

Thymic epithelial cells secrete several different families of peptide hormones, including *thymosins* (thi′mo-sinz), *thymulin*, and *thymopoietins*. These hormones are thought to be involved in the normal development of *T lymphocytes* and the immune response, but their roles are not well understood. Although called hormones, they mainly act locally as paracrines. We

describe the thymus in Chapter 20 in our discussion of lymphoid organs and tissues.

Check Your Understanding

21. Which hormone does the heart produce and what is its function?

22. What is the main function of the hormone produced by the skin?

23. **MAKE CONNECTIONS** Diabetes mellitus and diabetes insipidus are both due to lack of a hormone. Which hormone causes which? What symptom do they have in common? What would you find in the urine of a patient with one but not the other?

24. **MAKE CONNECTIONS** Which of the two chemical classes of hormones introduced at the beginning of this chapter do the gonadal hormones belong to? Which major endocrine gland secretes hormones of this same chemical class?

For answers, see Answers Appendix.

Developmental Aspects of the Endocrine System

Hormone-producing glands arise from all three embryonic germ layers. Endocrine glands derived from mesoderm produce steroid hormones. All others produce amines, peptides, or protein hormones.

Effects of Environmental Pollutants

Exposure to environmental pollutants—many pesticides, industrial chemicals, arsenic, dioxin, and other soil and water pollutants—can disrupt endocrine function. So far, sex hormones, thyroid hormone, and glucocorticoids have proved vulnerable to the effects of such pollutants. Interference with glucocorticoids, which turn on many genes that may suppress cancer, may help to explain high cancer rates in populations exposed to various types of toxins.

Endocrine Function throughout Life

Barring exposure to environmental pollutants, and hypersecretory and hyposecretory disorders, most endocrine organs operate smoothly until old age. Aging may alter the rates of hormone secretion, breakdown, and excretion, or the sensitivity of target cell receptors.

Structural changes in the anterior pituitary occur with age. The amount of connective tissue increases, vascularization decreases, and the number of hormone-secreting cells declines. These changes may or may not affect hormone production. In women, for example, blood levels and the release rhythm of ACTH remain constant, but levels of gonadotropins increase with age. GH levels decline in both sexes, which partially explains muscle atrophy in old age.

The adrenal glands also show structural changes with age, but normal controls of cortisol persist as long as a person is healthy and not stressed. Chronic stress, on the other hand, drives up blood levels of cortisol and appears to contribute to hippocampal (and memory) deterioration. Plasma levels of aldosterone fall by half in old age, but this change may reflect a decline in renin release by the kidneys, which become less responsive to renin-evoking stimuli. No age-related differences have been found in the release of catecholamines by the adrenal medulla.

The gonads, particularly the ovaries, undergo significant changes with age. In late middle age, the ovaries become smaller and unresponsive to gonadotropins. As female hormone production declines dramatically, the ability to bear children ends, and problems associated with estrogen deficiency appear, such as atherosclerosis and osteoporosis. Testosterone production by the testes also wanes with age, but this effect usually is not seen until very old age.

Glucose tolerance (the ability to dispose of a glucose load effectively) begins to deteriorate as early as the fourth decade of life. Blood glucose levels rise higher and return to resting levels more slowly in the elderly than in young adults. The fact that the islet cells continue to secrete near-normal amounts of insulin leads researchers to conclude that decreasing glucose tolerance with age may reflect declining receptor sensitivity to insulin (pre–type 2 diabetes).

Thyroid hormone synthesis and release diminish somewhat with age. Typically, the follicles are loaded with colloid in the elderly, and the gland becomes fibrosed. Basal metabolic rate declines with age. Mild hypothyroidism is only one cause of this decline. The increase in body fat relative to muscle is equally important, because muscle tissue is more active metabolically than fat.

The parathyroid glands change little with age, and PTH levels remain fairly normal throughout life. Estrogens protect women against the demineralizing effects of PTH, but estrogen production wanes after menopause, leaving older women vulnerable to the bone-demineralizing effects of PTH and to osteoporosis.

RELATED CLINICAL TERMS

Galactorrhea Inappropriate lactation (discharge of milk from the nipple). May occur in women who are not breastfeeding, and in men. Usually due to excess prolactin.

Hirsutism (her′soot-izm; *hirsut* = hairy, rough) Excessive hair growth; usually refers to this phenomenon in women and reflects excessive androgen production.

Hypophysectomy (hi-pof″ĭ-sek′to-me) Surgical removal of the pituitary gland.

Prolactinoma (pro-lak″tĭ-no′mah; *oma* = tumor) The most common type (30–40% or more) of pituitary gland tumor; evidenced by hypersecretion of prolactin and menstrual disturbances in women.

Thyroid storm (thyroid crisis) A sudden and dangerous increase in all of the manifestations of hyperthyroidism due to excessive amounts of circulating TH. Signs include fever, rapid heart rate, high blood pressure, dehydration, nervousness, and tremors. Precipitating factors include severe infection, excessive intake of TH supplements, or trauma.

Homeostatic Interrelationships between the Endocrine System and Other Body Systems

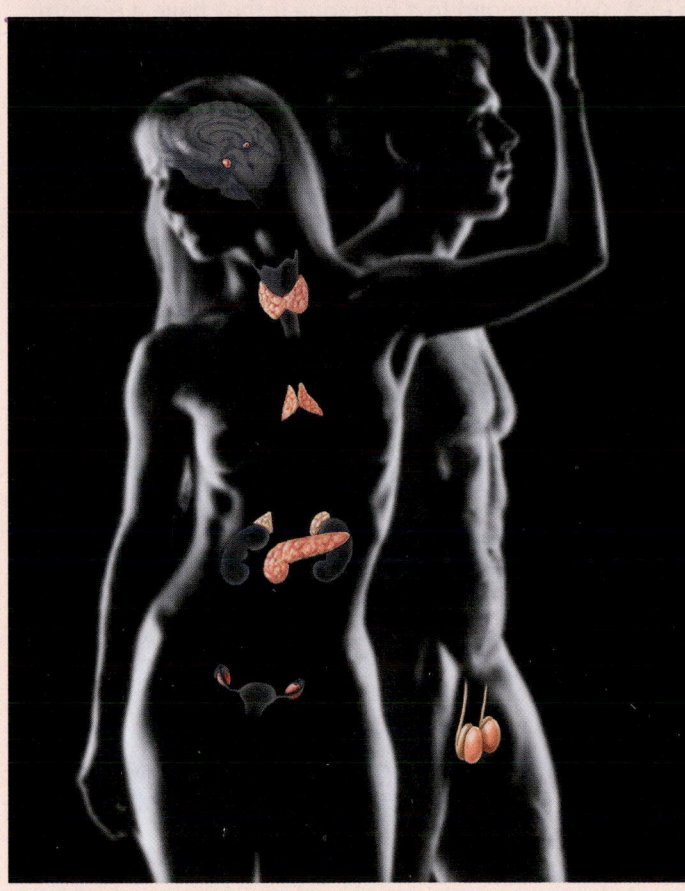

Integumentary System   Chapter 5
- Androgens cause activation of sebaceous glands; estrogens increase skin hydration
- Skin produces cholecalciferol (provitamin D)

Skeletal System   Chapters 6–8
- PTH regulates calcium blood levels; growth hormone, T_3, T_4, and sex hormones are necessary for normal skeletal development
- The skeleton provides some protection to endocrine organs, especially to those in the brain, chest, and pelvis

Muscular System   Chapters 9–10
- Growth hormone is essential for normal muscular development; other hormones (thyroxine and catecholamines) influence muscle metabolism
- Muscular system mechanically protects some endocrine glands; muscular activity elicits catecholamine release

Nervous System   Chapters 11–15
- Many hormones (growth hormone, thyroxine, sex hormones) influence normal maturation and function of the nervous system
- Hypothalamus controls anterior pituitary function and produces two hormones released from the posterior pituitary

Cardiovascular System   Chapters 17–19
- Several hormones influence blood volume, blood pressure, and heart contractility; erythropoietin stimulates red blood cell production
- Blood is the main transport medium of hormones; heart produces atrial natriuretic peptide

Lymphatic System/Immunity   Chapters 20–21
- Lymphocytes "programmed" by thymic hormones seed the lymph nodes; glucocorticoids depress the immune response and inflammation
- Chemical messengers of the immune system stimulate the release of cortisol and ACTH; lymph provides a route for transport of hormones

Respiratory System   Chapter 22
- Epinephrine influences ventilation (dilates bronchioles)
- Respiratory system provides oxygen; disposes of carbon dioxide; converting enzyme in lungs converts angiotensin I to angiotensin II

Digestive System   Chapter 23
- GI hormones and paracrines influence GI function; activated vitamin D necessary for absorption of calcium from diet; catecholamines influence digestive motility and secretory activity
- Digestive system provides nutrients to endocrine organs

Urinary System   Chapters 25–26
- Aldosterone and ADH influence renal function
- Kidneys activate vitamin D (considered a hormone); erythropoietin released by kidneys influences red blood cell formation

Reproductive System   Chapters 27–28
- Hypothalamic, anterior pituitary, and gonadal hormones direct reproductive system development and function; oxytocin and prolactin involved in birth and breast-feeding
- Gonadal hormones feed back to influence endocrine system function

16

CHAPTER SUMMARY

16.1 The endocrine system is one of the body's two major control systems (pp. 602–603)

1. The nervous and endocrine systems are the major controlling systems of the body. The nervous system exerts rapid controls via nerve impulses; the endocrine system exerts more prolonged effects via hormones.
2. Hormonally regulated processes include reproduction; growth and development; maintaining electrolyte, water, and nutrient balance; regulating cellular metabolism and energy balance; and mobilizing body defenses.
3. Endocrine organs are ductless, well-vascularized glands that release hormones directly into the blood or lymph. They are small and widely separated in the body.
4. The purely endocrine organs are the pituitary, thyroid, parathyroid, adrenal, and pineal glands. The hypothalamus is a neuroendocrine organ. The pancreas, gonads, and placenta also have endocrine tissue.
5. Local chemical messengers, not generally considered part of the endocrine system, include autocrines, which act on the cells that secrete them, and paracrines, which act on a different cell type nearby.

16.2 The chemical structure of a hormone determines how it acts (p. 603)

1. Most hormones are steroids or amino acid based. Steroids are lipid soluble. All amino acid–based hormones are water soluble, except for thyroid hormone.

16.3 Hormones act through second messengers or by activating specific genes (pp. 603–606)

1. Hormones alter cell activity by stimulating or inhibiting characteristic cellular processes of their target cells.
2. Cell responses to hormone stimulation may involve changes in membrane permeability; enzyme synthesis, activation, or inhibition; secretory activity; and mitosis.
3. Second-messenger mechanisms employing G proteins and intracellular second messengers are a common means by which amino acid–based hormones interact with their target cells. In the cyclic AMP system, the hormone binds to a plasma membrane receptor that couples to a G protein. When the G protein is activated, it couples to adenylate cyclase, which catalyzes the synthesis of cyclic AMP from ATP. Cyclic AMP initiates reactions that activate protein kinases and other enzymes, leading to cellular responses. The PIP_2-calcium signaling mechanism is another important second-messenger system. Other second messengers are cyclic GMP and calcium.
4. Steroid hormones (and thyroid hormone) enter their target cells and effect responses by activating DNA, which initiates messenger RNA formation leading to protein synthesis.

16.4 Three types of stimuli cause hormone release (pp. 607–608)

1. Humoral, neural, or hormonal stimuli activate endocrine organs to release their hormones. Negative feedback is important in regulating hormone levels in the blood.
2. The nervous system, acting through hypothalamic controls, can in certain cases override or modulate hormonal effects.

16.5 Cells respond to a hormone if they have a receptor for that hormone (pp. 608–609)

1. The ability of a target cell to respond to a hormone depends on the presence of receptors, on its plasma membrane or within the cell, to which the hormone can bind.
2. Hormone receptors are dynamic structures. High or low levels of stimulating hormones can change the number and/or sensitivity of hormone receptors.

Half-Life, Onset, and Duration of Hormone Activity (pp. 608–609)

3. Blood levels of hormones reflect a balance between secretion and degradation/excretion. The liver and kidneys are the major organs that degrade hormones; breakdown products are excreted in urine and feces.
4. Hormone half-life and duration of activity are limited and vary from hormone to hormone.

Interaction of Hormones at Target Cells (p. 609)

5. Permissiveness is the situation in which one hormone must be present in order for another hormone to exert its full effects.
6. Synergism occurs when two or more hormones produce the same effects in a target cell and their results together are amplified.
7. Antagonism occurs when a hormone opposes or reverses the effect of another hormone.

16.6 The hypothalamus controls release of hormones from the pituitary gland in two different ways (pp. 609–617)

Pituitary-Hypothalamic Relationships (pp. 610–611)

1. The pituitary gland hangs from the base of the brain and is enclosed by bone. It consists of a hormone-producing glandular portion (anterior pituitary or adenohypophysis) and a neural portion (posterior pituitary or neurohypophysis), which is an extension of the hypothalamus. The neurohypophysis includes the infundibulum (stalk) and the posterior pituitary.
2. The hypothalamus (a) synthesizes two hormones that it exports to the posterior pituitary for storage and later release and (b) regulates the hormonal output of the anterior pituitary via releasing and inhibiting hormones.

The Posterior Pituitary and Hypothalamic Hormones (pp. 611–614)

3. The posterior pituitary stores and releases two hypothalamic hormones, oxytocin and antidiuretic hormone (ADH).
4. Oxytocin stimulates powerful uterine contractions, which trigger labor and delivery of an infant, and milk ejection in nursing women. Its release is mediated reflexively by the hypothalamus and represents a positive feedback mechanism.
5. Antidiuretic hormone stimulates the kidney tubules to reabsorb and conserve water, resulting in small volumes of highly concentrated urine and decreased plasma solute concentration. ADH is released in response to high solute concentrations in the blood and inhibited by low solute concentrations in the blood. Hyposecretion results in diabetes insipidus.

Anterior Pituitary Hormones (pp. 614–617)

6. Four of the six anterior pituitary hormones are tropic hormones that regulate the function of other endocrine organs. Most

anterior pituitary hormones exhibit a diurnal rhythm of release, which is subject to modification by stimuli influencing the hypothalamus.

7. Growth hormone (GH) is an anabolic hormone that stimulates growth of all body tissues but especially skeletal muscle and bone. It may act directly, or indirectly, via insulin-like growth factors (IGFs). GH mobilizes fats, stimulates protein synthesis, and inhibits glucose uptake and metabolism. Its secretion is regulated by growth hormone–releasing hormone (GHRH) and growth hormone–inhibiting hormone (GHIH), or somatostatin. GH hypersecretion causes gigantism in children and acromegaly in adults; hyposecretion in children causes pituitary dwarfism.

8. Thyroid-stimulating hormone (TSH) promotes normal development and activity of the thyroid gland. Thyrotropin-releasing hormone (TRH) stimulates release of TSH; negative feedback of thyroid hormone inhibits it.

9. Adrenocorticotropic hormone (ACTH) stimulates the adrenal cortex to release corticosteroids. Corticotropin-releasing hormone (CRH) triggers ACTH release; rising glucocorticoid levels inhibit it.

10. The gonadotropins—follicle-stimulating hormone (FSH) and luteinizing hormone (LH)—regulate the functions of the gonads in both sexes. FSH stimulates sex cell production; LH stimulates gonadal hormone production. Gonadotropin levels rise in response to gonadotropin-releasing hormone (GnRH). Negative feedback of gonadal hormones inhibits gonadotropin release.

11. Prolactin (PRL) promotes milk production in humans. Its secretion is inhibited by prolactin-inhibiting hormone (PIH).

16.7 The thyroid gland controls metabolism (pp. 617–621)

1. The thyroid gland is located in the anterior neck. Thyroid follicles store colloid containing thyroglobulin, a glycoprotein from which thyroid hormone is derived.

2. Thyroid hormone (TH) includes thyroxine (T_4) and triiodothyronine (T_3), which increase the rate of cellular metabolism. Consequently, oxygen use and heat production rise.

3. Secretion of thyroid hormone, prompted by TSH, requires the follicular cells to take up the stored colloid and split the hormones from the colloid for release. Rising levels of thyroid hormone feed back to inhibit the anterior pituitary and hypothalamus.

4. Most T_4 is converted to T_3 (the more active form) in the target tissues. These hormones act by turning on gene transcription and protein synthesis.

5. Graves' disease is the most common cause of hyperthyroidism. Hyposecretion can be congenital in infants and can cause myxedema in adults.

6. The parafollicular (C) cells of the thyroid gland produce calcitonin. It is not normally important in calcium homeostasis. At pharmacological levels, it inhibits bone matrix resorption and enhances calcium deposit in bone.

16.8 The parathyroid glands are primary regulators of blood calcium levels (pp. 621–622)

1. The parathyroid glands, located on the dorsal aspect of the thyroid gland, secrete parathyroid hormone (PTH), which increases blood calcium levels. It targets bone, the kidneys, and the small intestine (indirectly via vitamin D activation). PTH is the key hormone for calcium homeostasis.

2. Falling blood calcium levels trigger PTH release; rising blood calcium levels inhibit its release.

3. Hyperparathyroidism results in hypercalcemia and extreme bone wasting. Hypoparathyroidism leads to hypocalcemia, evidenced by tetany and respiratory paralysis.

16.9 The adrenal glands produce hormones involved in electrolyte balance and the stress response (pp. 622–627)

1. The paired adrenal (suprarenal) glands sit atop the kidneys. Each adrenal gland has two functional portions, the cortex and the medulla.

The Adrenal Cortex (pp. 623–626)

2. The cortex produces three groups of steroid hormones from cholesterol.

3. Mineralocorticoids (primarily aldosterone) regulate sodium ion reabsorption and potassium ion excretion by the kidneys. Sodium ion reabsorption usually leads to water reabsorption, and raises blood volume and blood pressure. Release of aldosterone is stimulated by the renin-angiotensin-aldosterone mechanism, rising potassium ion levels in the blood, and ACTH. Atrial natriuretic peptide inhibits aldosterone release.

4. Glucocorticoids (primarily cortisol) are important metabolic hormones that help the body resist stressors by increasing blood glucose, fatty acid and amino acid levels, and blood pressure. High levels of glucocorticoids depress the immune system and the inflammatory response. ACTH is the major stimulus for glucocorticoid release.

5. Gonadocorticoids (mainly androgens) are produced in small amounts throughout life.

6. Hypoactivity of the adrenal cortex results in Addison's disease. Hypersecretion can result in aldosteronism, Cushing's syndrome, and adrenogenital syndrome.

The Adrenal Medulla (p. 627)

7. The adrenal medulla produces catecholamines (epinephrine and norepinephrine) in response to sympathetic nervous system stimulation. Catecholamines enhance and prolong the fight-or-flight response to short-term stressors. Hypersecretion leads to symptoms typical of sympathetic nervous system overactivity.

16.10 The pineal gland secretes melatonin (p. 627)

1. The pineal gland is located in the diencephalon. Its primary hormone is melatonin, which influences daily rhythms in body temperature, sleep, and appetite and may have an antigonadotropic effect in children.

16.11 The pancreas, gonads, and most other organs secrete hormones (pp. 630–635)

The Pancreas (pp. 630–634)

1. The pancreas, located in the abdomen close to the stomach, is both an exocrine and an endocrine gland. The endocrine portion (pancreatic islets) releases insulin and glucagon and smaller amounts of other hormones to the blood.

2. Glucagon, released by alpha (α) cells when blood levels of glucose are low, stimulates the liver to release glucose to the blood.

3. Insulin is released by beta (β) cells when blood levels of glucose (and amino acids) are rising. It increases the rate of glucose uptake and metabolism by most body cells. Hyposecretion or hypoactivity of insulin results in diabetes mellitus; cardinal signs are polyuria, polydipsia, and polyphagia.

The Gonads and Placenta (p. 634)

4. The ovaries of the female, located in the pelvic cavity, release two main hormones. The ovarian follicles begin to secrete estrogens at puberty under the influence of FSH. Estrogens stimulate maturation of the female reproductive system and development

of the secondary sex characteristics. Progesterone is released in response to high blood levels of LH. It works with estrogens in establishing the menstrual cycle.

5. The testes of the male begin to produce testosterone at puberty in response to LH. Testosterone promotes maturation of the male reproductive organs, development of secondary sex characteristics, and production of sperm by the testes.

6. The placenta produces hormones of pregnancy—estrogens, progesterone, and others.

Hormone Secretion by Other Organs (pp. 634–635)

7. Many body organs not normally considered endocrine organs contain cells that secrete hormones. Examples include the heart (atrial natriuretic peptide); gastrointestinal tract organs (gastrin, secretin, and others); the kidneys (erythropoietin); skin (cholecalciferol); adipose tissue (leptin, resistin, and adiponectin); bone (osteocalcin); and thymus (thymic hormones).

8. The thymus, located in the upper thorax, declines in size and function with age. It releases the paracrines thymulin, thymosins, and thymopoietins, which are important to the normal development of the immune response.

Developmental Aspects of the Endocrine System (p. 636)

1. Endocrine glands derive from all three germ layers. Those derived from mesoderm produce steroid hormones; the others produce the amino acid–based hormones.

2. The natural decrease in function of the female's ovaries during late middle age results in menopause.

3. All endocrine glands gradually become less efficient as aging occurs. This change leads to a generalized increase in the incidence of diabetes mellitus and a lower metabolic rate.

REVIEW QUESTIONS

To access additional practice questions using your smartphone, tablet, or computer: MasteringA&P® > Study Area > Practice Tests & Quizzes

Level 1 Remember/Understand

(Some questions have more than one correct answer. Select the best answer or answers from the choices given.)

1. The major stimulus for release of parathyroid hormone is (a) hormonal, (b) humoral, (c) neural.

2. The anterior pituitary secretes all but (a) antidiuretic hormone, (b) growth hormone, (c) gonadotropins, (d) TSH.

3. A hormone not involved in glucose metabolism is (a) glucagon, (b) cortisone, (c) aldosterone, (d) insulin.

4. Parathyroid hormone (a) increases bone formation and lowers blood calcium levels, (b) increases calcium excretion from the body, (c) decreases calcium absorption from the gut, (d) demineralizes bone and raises blood calcium levels.

5. Choose from the following key to identify the hormones described.

Key: (a) aldosterone (e) oxytocin
 (b) antidiuretic hormone (f) prolactin
 (c) growth hormone (g) T_4 and T_3
 (d) luteinizing hormone (h) TSH

_____ (1) important anabolic hormone; many of its effects mediated by IGFs

_____ (2) cause the kidneys to conserve water and/or salt (two choices)

_____ (3) stimulates milk production

_____ (4) tropic hormone that stimulates the gonads to secrete sex hormones

_____ (5) increases uterine contractions during birth

_____ (6) major metabolic hormone(s) of the body

_____ (7) causes reabsorption of sodium ions by the kidneys

_____ (8) tropic hormone that stimulates the thyroid gland to secrete thyroid hormone

_____ (9) secreted by the posterior pituitary (two choices)

_____ (10) the only steroid hormone in the list

6. A hypodermic injection of epinephrine would (a) increase heart rate, increase blood pressure, dilate the bronchioles of the lungs, and increase digestive activity, (b) decrease heart rate, decrease blood pressure, constrict the bronchioles, and increase digestive activity, (c) decrease heart rate, increase blood pressure, constrict the bronchioles, and decrease digestive activituy, (d) increase heart rate, increase blood pressure, dilate the bronchioles, and decrease digestive activity.

7. Testosterone is to the male as which hormone is to the female? (a) luteinizing hormone, (b) progesterone, (c) estrogen, (d) prolactin.

8. If anterior pituitary secretion is deficient in a growing child, the child will (a) develop acromegaly, (b) become a dwarf but have fairly normal body proportions, (c) mature sexually at an earlier than normal age, (d) be in constant danger of becoming dehydrated.

9. If there is adequate carbohydrate intake, secretion of insulin results in (a) lower blood glucose levels, (b) increased cell utilization of glucose, (c) storage of glycogen, (d) all of these.

10. Hormones (a) are produced by exocrine glands, (b) are carried to all parts of the body in blood, (c) remain at constant concentration in the blood, (d) affect only non-hormone-producing organs.

11. Some hormones act by (a) increasing the synthesis of enzymes, (b) converting an inactive enzyme into an active enzyme, (c) affecting only specific target organs, (d) all of these.

12. Absence of thyroid hormone would result in (a) increased heart rate and increased force of heart contraction, (b) depression of the CNS and lethargy, (c) exophthalmos, (d) high metabolic rate.

13. Define hormone.

14. Name a hormone secreted by a muscle cell and two hormones secreted by neurons.

Level 2 Apply/Analyze

15. (a) Describe the body location of each of the following endocrine organs: anterior pituitary, pineal gland, pancreas, ovaries, testes, and adrenal glands. (b) List the hormones produced by each organ.

16. The anterior pituitary is often referred to as the master endocrine organ, but it, too, has a "master." What controls the release of anterior pituitary hormones?

17. The posterior pituitary is not really an endocrine gland. Why not? What is it?

18. Endemic goiter is not really the result of a malfunctioning thyroid gland. What does cause it?

19. Roger Proulx has severe arthritis and has been taking prednisone (a glucocorticoid) for two months. He isn't feeling

well, complains of repeated "colds," and is extremely "puffy" (edematous). Explain the reason for these symptoms.

Level 3 Evaluate/Synthesize

20. Which type of hormone receptor—plasma membrane bound or intracellular—would be expected to provide the most long-lived response to hormone binding and why?

21. Name two endocrine glands (or regions) that are important in the stress response, and explain why they are important.

22. How are the hyperglycemia and lipidemia of insulin deficiency linked?

23. List some problems that elderly people might have as a result of decreasing hormone production.

24. Mary Morgan has just been brought into the emergency room of City General Hospital. She is perspiring profusely and is breathing rapidly and irregularly. Her breath smells like acetone (sweet and fruity), and her blood glucose tests out at 650 mg/100 ml of blood. She is in acidosis. Which hormone drug should be administered, and why?

25. Kyle, a 5-year-old boy, has been growing by leaps and bounds; his height is 100% above normal for his age. He has been complaining of headaches and vision problems. A CT scan reveals a large pituitary tumor. **(a)** Which hormone is being secreted in excess? **(b)** What condition will Kyle exhibit if corrective measures are not taken? **(c)** What is the probable cause of his headaches and visual problems?

26. Aaron, a 42-year-old single father, goes to his physician complaining of nausea and chronic fatigue. He reports having felt fatigued and listless for about half a year, but he had attributed this to stress. He has lost considerable weight and, strangely, his skin looks tanned, even though he spends long hours at work and rarely ventures outside. His doctor finds very low blood pressure and a rapid, weak pulse. Blood tests show that Aaron does not have anemia, but his plasma glucose, cortisol, and Na^+ are low, and his plasma K^+ is high. His doctor orders an ACTH stimulation test, in which Aaron's secretion of cortisol is measured after he is given a synthetic form of ACTH. **(a)** What would account for Aaron's low plasma Na^+ and high plasma K^+? **(b)** What is the reason for doing an ACTH stimulation test? **(c)** Which gland is primarily affected if ACTH does not cause a normal elevation of cortisol secretion? What is this abnormality called? **(d)** Which gland is primarily affected if ACTH does cause an elevation of cortisol secretion?

70-Year-Old Male with Polyuria

Mr. Gutteman, a 70-year-old male, was brought into the ER. He had been sick several days with the flu, and was found confused and barely conscious by his daughter.

Mr. Gutteman is breathing rapidly and has a fever of 39°C (102°F). His skin is dry and flaccid, his mucous membranes are dry, and his eyes are sunken. The physician ordered:

- IV (intravenous) fluid and electrolyte replacement
- Blood and urine tests for presence of glucose and ketones
- Strict I&O [careful measurement of fluid intake (e.g., IV, drinking) and output (e.g., urine)]

1. **+ NCLEX-STYLE** You would expect high levels of blood glucose and the presence of glucose and ketones in Mr. Gutteman's urine if:
   a. His pancreas is secreting too much insulin
   b. His liver is secreting too little insulin
   c. His pancreas is secreting too little insulin
   d. His liver is secreting too much glucagon

   Mr. Gutteman's blood and urine tests are negative for glucose and ketones. However, Mr. Gutteman is losing large amounts of water in urine, and the volume lost is being replaced (via IV).

2. **+ NCLEX-STYLE** Given this new information, Mr. Gutteman's excessive fluid loss in his urine most likely indicates:
   a. An increased blood level of aldosterone
   b. A decreased blood level of antidiuretic hormone
   c. An increased blood level of antidiuretic hormone
   d. A decreased blood level of insulin

3. Mr. Gutteman's diagnosis is diabetes insipidus. His daughter is concerned about this, because she knows diabetes is a dangerous disease. She asks whether he will have to follow a diabetic diet. What's your answer?

4. Is diabetes insipidus life threatening? Explain your answer.

17

Blood

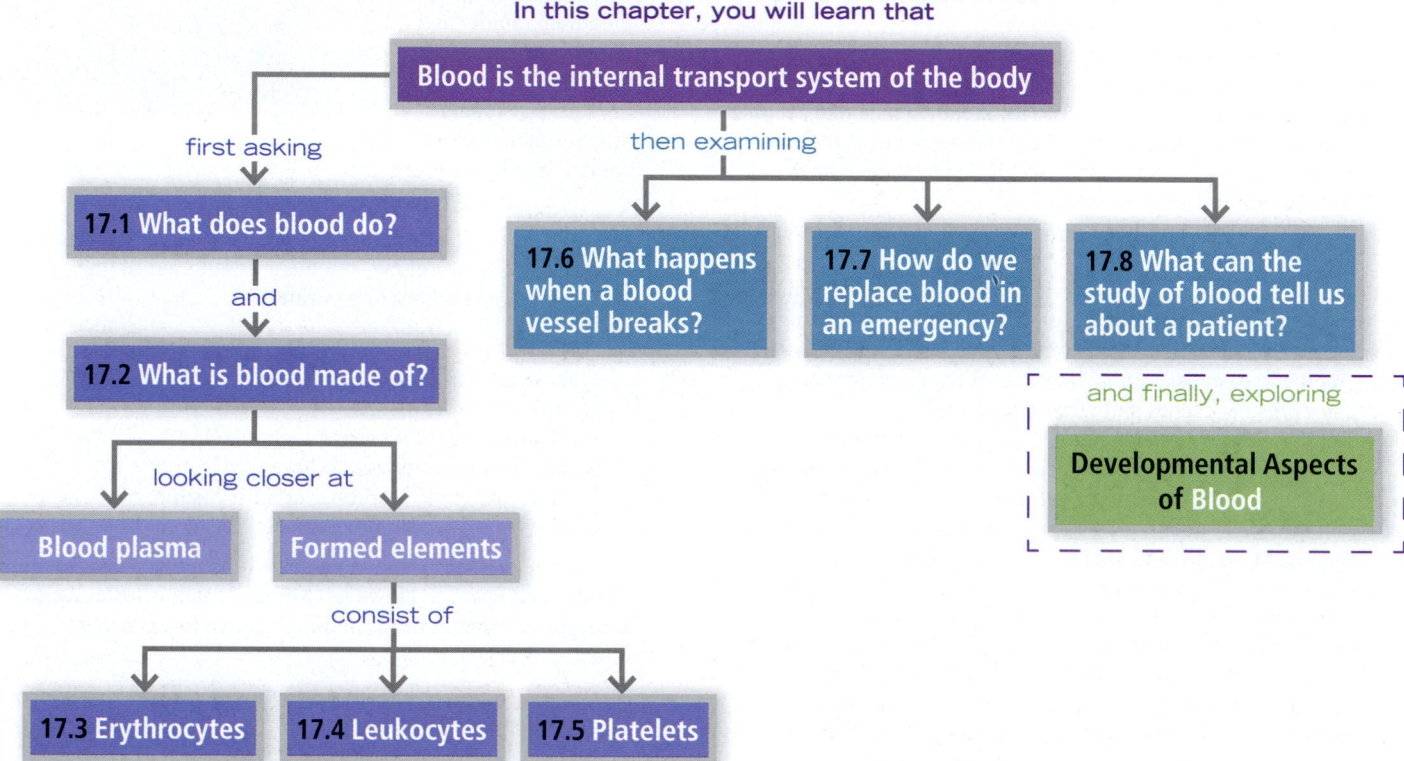

In this chapter, you will learn that

Blood is the internal transport system of the body

first asking

17.1 What does blood do?

and

17.2 What is blood made of?

looking closer at

Blood plasma   **Formed elements**

consist of

17.3 Erythrocytes   **17.4 Leukocytes**   **17.5 Platelets**

then examining

17.6 What happens when a blood vessel breaks?

17.7 How do we replace blood in an emergency?

17.8 What can the study of blood tell us about a patient?

and finally, exploring

Developmental Aspects of Blood

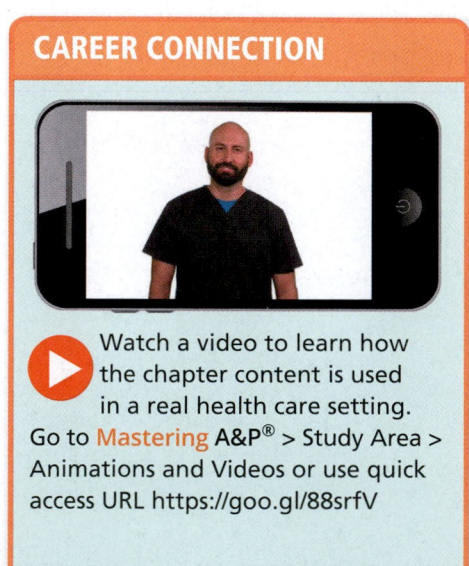
Blood is the river of life that surges within us, transporting nearly everything that must be carried from one place to another. Long before modern medicine, blood was viewed as magical—an elixir that held the mystical force of life—because when it drained from the body, life departed as well. Today, blood still has enormous importance in the practice of medicine. Clinicians examine it more often than any other tissue when trying to determine the cause of disease in their patients.

In this chapter, we describe the composition and functions of this life-sustaining fluid that serves as a transport "vehicle" for the organs of the cardiovascular system (*cardio* = heart, *vasc* = blood vessels). To get started, we need a brief overview of blood circulation, which is initiated by the pumping action of the heart. Blood exits the *heart* via *arteries*, which branch repeatedly until they become tiny *capillaries* that service all of the tissues in the body. Oxygen and nutrients leave the blood, crossing capillary walls into the body's tissues; carbon dioxide and wastes move from the tissues to the capillaries. Oxygen-deficient blood flows from the capillaries into larger *veins* that carry blood back to the heart. The returning blood then flows from the heart to the lungs, where it picks up oxygen and then returns to the heart to be pumped throughout the body once again.

17.1 The functions of blood are transport, regulation, and protection

Learning Outcome

▶ **List the functions of blood.**

Blood performs a number of functions, all concerned in one way or another with transporting substances, regulating blood levels of particular substances, or protecting the body.

Transport

Transport functions of blood include:

- Delivering oxygen from the lungs and nutrients from the digestive tract to all body cells.
- Transporting metabolic waste products from cells to elimination sites (to the lungs to eliminate carbon dioxide, and to the kidneys to dispose of nitrogenous wastes in urine).
- Transporting hormones from the endocrine organs to their target organs.

Regulation

Regulatory functions of blood include:

- Maintaining appropriate body temperature by absorbing and distributing heat throughout the body and to the skin surface to encourage heat loss.
- Maintaining normal pH in body tissues. Many blood proteins and other bloodborne solutes act as buffers to prevent excessive or abrupt changes in blood pH that could jeopardize normal cell activities. Blood also acts as the reservoir for the body's "alkaline reserve" of bicarbonate ions.
- Maintaining adequate fluid volume in the circulatory system. As we will see, blood proteins prevent excessive fluid loss from the bloodstream into the tissue spaces. As a result, the fluid volume in the blood vessels remains ample to support efficient blood circulation to all parts of the body.

Protection

Protective functions of blood include:

- Preventing blood loss. When a blood vessel is damaged, platelets and plasma proteins initiate clot formation, halting blood loss.
- Preventing infection. Drifting along in blood are antibodies, complement proteins, and leukocytes (white blood cells), all of which help defend the body against foreign invaders such as bacteria and viruses.

Check Your Understanding

1. List two protective functions of blood.

For answers, see Answers Appendix.

17.2 Blood consists of plasma and formed elements

Learning Outcomes

▶ **Describe the composition and physical characteristics of whole blood. Explain why it is classified as a connective tissue.**

▶ **Discuss the composition and functions of plasma.**

Blood is the only fluid tissue in the body. It appears to be a thick, homogeneous liquid, but the microscope reveals that it has both cellular and liquid components. Blood is a specialized connective tissue in which living blood cells, called the *formed elements*, are suspended in a nonliving fluid matrix called *plasma* (plaz′mah). Blood lacks the collagen and elastic fibers typical of other connective tissues, but dissolved fibrous proteins become visible as fibrin strands during blood clotting.

If we spin a sample of blood in a centrifuge, centrifugal force packs down the heavier formed elements and the less dense plasma remains at the top (**Figure 17.1**). Most of the reddish mass at the bottom of the tube is *erythrocytes* (ĕ-rith′ro-sīts; *erythro* = red), the red blood cells that transport oxygen. A thin, whitish layer called the **buffy coat** is present at the erythrocyte-plasma junction. This layer contains *leukocytes* (*leuko* = white),

① Withdraw blood and place in tube. ② Centrifuge the blood sample.

Plasma
- 55% of whole blood
- Least dense component

Buffy coat
- Leukocytes and platelets
- <1% of whole blood

Erythrocytes
- 45% of whole blood (hematocrit)
- Most dense component

Formed elements

Figure 17.1 The major components of whole blood.

the white blood cells that act in various ways to protect the body, and *platelets*, cell fragments that help stop bleeding.

Erythrocytes normally constitute about 45% of the total volume of a blood sample, a percentage known as the **hematocrit** (he-mat′o-krit; "blood fraction"). Normal hematocrit values vary. In healthy males the norm is 47% ± 5%; in females it is 42% ± 5%. Leukocytes and platelets contribute less than 1% of blood volume. Plasma makes up most of the remaining 55% of whole blood.

Physical Characteristics and Volume

Blood is a sticky, opaque fluid with a characteristic metallic taste. As children, we discover its saltiness the first time we stick a cut finger into our mouth. Depending on the amount of oxygen it is carrying, the color of blood varies from scarlet (oxygen rich) to dark red (oxygen poor). It is slightly alkaline, with a pH between 7.35 and 7.45. Blood is more dense than water and about five times more viscous, largely because of its formed elements.

Erythrocytes are the major factor contributing to blood viscosity. Women typically have a lower red blood cell count than men [4.2–5.4 million cells per microliter (1 μl = 1 mm^3) of blood versus 4.7–6.1 million cells/μl respectively]. When the number of red blood cells increases beyond the normal range, blood becomes more viscous and flows more slowly. Similarly, as the number of red blood cells drops below the lower end of the range, the blood thins and flows more rapidly.

Blood accounts for approximately 8% of body weight. Its average volume in healthy adult males is 5–6 L (about 1.5 gallons), somewhat greater than in healthy adult females (4–5 L).

Blood Plasma

Blood **plasma** is a straw-colored, sticky fluid (Figure 17.1). Although it is mostly water (about 90%), plasma contains over 100 different dissolved solutes, including nutrients, gases, hormones, wastes and products of cell activity, proteins, and inorganic ions (electrolytes). Electrolytes (Na$^+$, Cl$^-$, etc.) vastly outnumber the other solutes. **Table 17.1** summarizes the major plasma components.

Although outnumbered by the lighter electrolytes, the heavier **plasma proteins** are the most abundant plasma solutes by weight, accounting for about 8% of plasma weight. Except for hormones and gamma globulins (antibodies), most plasma proteins are produced by the liver. Plasma proteins serve a variety of functions, but they are *not* normally taken up by cells to be used as fuels or metabolic nutrients as are most other organic solutes, such as glucose, fatty acids, and amino acids.

Albumin (al-bu′min) accounts for some 60% of plasma protein. It acts as a carrier to shuttle certain molecules through the circulation, and is the major blood protein contributing to the plasma osmotic pressure (the pressure that helps to keep water in the bloodstream). We will learn more about this critical role of plasma proteins in Chapter 19.

The composition of plasma varies continuously as cells remove or add substances to the blood. However, assuming a healthy diet, plasma composition is kept relatively constant by various homeostatic mechanisms. For example, when blood

protein levels drop undesirably, the liver makes more proteins. When the blood starts to become too acidic (acidosis), both the lungs and the kidneys are called into action to restore plasma's normal, slightly alkaline pH. Body organs make dozens of adjustments, day in and day out, to maintain the many plasma solutes at life-sustaining levels.

Formed Elements

The **formed elements** of blood—*erythrocytes, leukocytes,* and *platelets*—have some unusual features.

Table 17.1	Composition of Plasma
CONSTITUENT	**DESCRIPTION AND IMPORTANCE**
Water	90% of plasma volume; dissolving and suspending medium for solutes of blood; absorbs heat
Solutes	
Electrolytes	Most abundant solutes by number; cations include sodium, potassium, calcium, magnesium; anions include chloride, phosphate, sulfate, and bicarbonate; help to maintain plasma osmotic pressure and normal blood pH
Plasma proteins	8% (by weight) of plasma; all contribute to osmotic pressure and maintain water balance in blood and tissues; all have other functions (transport, enzymatic, etc.) as well
• Albumin	60% of plasma proteins; produced by liver; main contributor to osmotic pressure
• Globulins	36% of plasma proteins
alpha, beta	Produced by liver; most are transport proteins that bind to lipids, metal ions, and fat-soluble vitamins
gamma	Antibodies released by plasma cells during immune response
• Fibrinogen	4% of plasma proteins; produced by liver; forms fibrin threads of blood clot
Nonprotein nitrogenous substances	By-products of cellular metabolism, such as urea, uric acid, creatinine, and ammonium salts
Nutrients (organic)	Materials absorbed from digestive tract and transported for use throughout body; include glucose and other simple carbohydrates, amino acids (protein digestion products), fatty acids, glycerol and triglycerides (fat digestion products), cholesterol, and vitamins
Respiratory gases	Oxygen and carbon dioxide; oxygen mostly bound to hemoglobin inside RBCs; carbon dioxide transported dissolved as bicarbonate ion or CO_2, or bound to hemoglobin in RBCs
Hormones	Steroid and thyroid hormones carried by plasma proteins

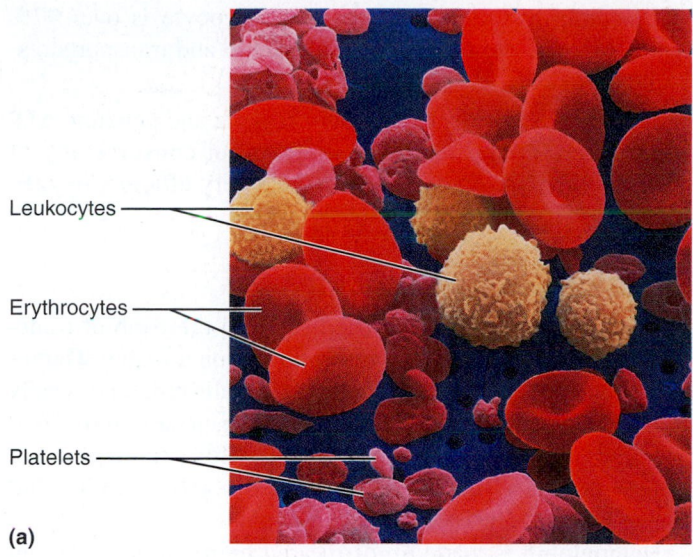

Leukocytes

Erythrocytes

Platelets

(a)

Erythrocytes

Platelets

Monocyte   Lymphocyte   Neutrophil   Eosinophil

Leukocytes (white blood cells)

(b)

Figure 17.2 Blood cells. (a) SEM (1800×, artificially colored).
(b) Photomicrograph (610×, Wright's stain).

> Practice Histology questions: MasteringA&P® >
> Study Area > Lab Tools > PAL

- Two of the three are not even true cells: Erythrocytes have no nuclei or organelles, and platelets are cell fragments. Only leukocytes are complete cells.
- Most types of formed elements survive in the bloodstream for only a few days.
- Most blood cells do not divide. Instead, stem cells divide continuously in red bone marrow to replace them.

If you examine a stained smear of human blood under the light microscope, you will see disc-shaped red blood cells, a variety of gaudily stained spherical leukocytes, and some scattered platelets that look like debris (**Figure 17.2**). Erythrocytes vastly outnumber the other types of formed elements. Table 17.2 on p. 656 summarizes the important characteristics of the formed elements.

Check Your Understanding

2. What is the hematocrit? What is its normal value?

3. **WHAT IF?** Suppose you are caring for a patient in liver failure due to chronic alcoholism. What problems would you expect the reduced synthesis of plasma proteins to cause in this patient?

For answers, see Answers Appendix.

17.3 Erythrocytes play a crucial role in oxygen and carbon dioxide transport

Learning Outcomes

▶ **Describe the structure, function, and production of erythrocytes.**

▶ **Describe the chemical composition of hemoglobin.**

▶ **Give examples of disorders caused by abnormalities of erythrocytes. Explain what goes wrong in each disorder.**

Structural Characteristics

Erythrocytes or **red blood cells (RBCs)** are small cells, about 7.5 μm in diameter. They are shaped like biconcave discs—flattened discs with depressed centers (**Figure 17.3**). Under the microscope, they appear lighter in color at their thin

17

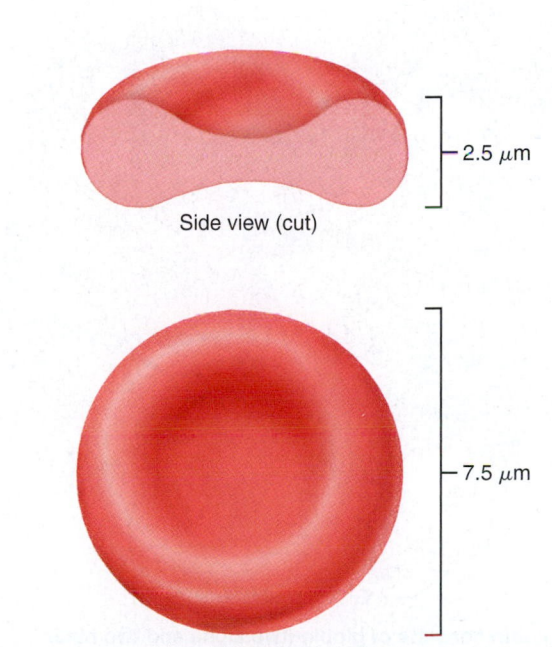

2.5 μm

Side view (cut)

7.5 μm

Top view

Figure 17.3 Structure of erythrocytes (red blood cells). Notice the distinctive biconcave shape.

centers than at their edges, making them look like miniature doughnuts (Figure 17.2).

Mature erythrocytes are bound by a plasma membrane, but lack a nucleus (are *anucleate*) and have essentially no organelles. In fact, they are little more than "bags" of *hemoglobin* (*Hb*), the RBC protein that functions in gas transport. Other proteins are present, such as antioxidant enzymes that rid the body of harmful oxygen radicals, but most function as structural proteins, allowing the RBC to deform yet spring back into shape.

For example, a network of proteins, especially one called *spectrin*, attached to the cytoplasmic face of RBC plasma membranes maintains the biconcave shape of an erythrocyte. The spectrin net is deformable, allowing erythrocytes to change shape as necessary—to twist, turn, and become cup shaped as they are carried passively through capillaries with diameters smaller than themselves—and then to resume their biconcave shape.

The erythrocyte is a superb example of complementarity of structure and function. It picks up oxygen in the capillaries of the lungs and releases it to tissue cells across other capillaries throughout the body. It also transports some 20% of the carbon dioxide released by tissue cells back to the lungs. Three structural characteristics contribute to erythrocyte gas transport functions:

- Its small size and shape provide a huge surface area relative to volume (about 30% more surface area than comparable spherical cells). The disc shape is ideally suited for gas exchange because no point within the cytoplasm is far from the surface.

- Discounting water content, an erythrocyte is over 97% hemoglobin, the molecule that binds to and transports respiratory gases.

- Because erythrocytes lack mitochondria and generate ATP by anaerobic mechanisms, they do not consume any of the oxygen they carry, making them very efficient oxygen transporters.

Functions of Erythrocytes

Erythrocytes are completely dedicated to their job of transporting respiratory gases (oxygen and carbon dioxide). **Hemoglobin**, the protein that makes red blood cells red, binds easily and reversibly with oxygen, and most oxygen carried in blood is bound to hemoglobin. Normal values for hemoglobin are 13–18 grams per 100 milliliters of blood (g/100 ml) in adult males, and 12–16 g/100 ml in adult females.

Hemoglobin is made up of the red **heme** pigment bound to the protein **globin**. Globin consists of four polypeptide chains—two alpha (α) and two beta (β)—each binding a red ringlike heme group (**Figure 17.4a**). Each heme group bears an iron ion (Fe^{2+}) set like a jewel in its center (Figure 17.4b). A hemoglobin molecule can transport four molecules of oxygen because each iron atom can combine reversibly with one molecule of oxygen. A single red blood cell contains about 250 million hemoglobin molecules, so each of these tiny cells can scoop up about 1 billion molecules of oxygen!

The fact that hemoglobin is contained in erythrocytes, rather than existing free in plasma, prevents it from leaking out of the

(a) Hemoglobin consists of globin (two alpha and two beta polypeptide chains) and four heme groups.

(b) Iron-containing heme pigment.

Figure 17.4 **Structure of hemoglobin.** Hemoglobin's structure makes it a highly efficient oxygen carrier.

bloodstream (through porous capillary walls) and clogging up the kidneys. Large amounts of free hemoglobin in the blood would also make blood more viscous and would draw water from the tissues into the bloodstream by osmosis.

Oxygen loading occurs in the lungs, and the direction of transport is from lungs to tissue cells. As oxygen-deficient blood moves through the lungs, oxygen diffuses from the air sacs of the lungs into the blood and then into the erythrocytes, where it binds to the iron in hemoglobin. The iron is partially oxidized by the oxygen (◀ p. 37). As a result, the protein, now called **oxyhemoglobin**, assumes a new three-dimensional shape and becomes ruby red.

In body tissues, the process is reversed. Oxygen detaches from iron, hemoglobin resumes its former shape, and the resulting **deoxyhemoglobin**, or *reduced hemoglobin*, becomes dark red. The released oxygen diffuses from the blood into the tissue fluid and then into tissue cells.

About 20% of the carbon dioxide transported in the blood combines with hemoglobin, but it binds to globin's amino acids rather than to the heme group. This formation of **carbamino-hemoglobin** (kar-bam″ĭ-no-he″muh″glo′bin) occurs more readily when hemoglobin is in the reduced state (dissociated from oxygen). Carbon dioxide loading occurs in the tissues, and the direction of transport is from tissues to lungs, where carbon dioxide is eliminated from the body (see Chapter 22).

Production of Erythrocytes

Blood cell formation is referred to as **hematopoiesis** (hem″ah-to-poy-e′sis; *hemato* = blood; *poiesis* = to make). Hematopoiesis occurs in the **red bone marrow**, which is composed largely of a soft network of reticular connective tissue (◀ p. 133) bordering on wide blood capillaries called *blood sinusoids*. Within this network are immature blood cells, macrophages, fat cells, and *reticular cells* (which secrete the connective tissue fibers). In adults, red marrow is found chiefly in the bones of the axial skeleton and girdles, and in the proximal epiphyses of the humerus and femur.

The production of each type of blood cell varies in response to changing body needs and regulatory factors. As blood cells mature, they migrate through the porous walls of the sinusoids to enter the bloodstream. On average, the marrow turns out an ounce of new blood containing 100 billion new cells every day.

The various formed elements have different functions, but there are similarities in their life histories. All arise from the **hematopoietic stem cell**, sometimes called a *hemocytoblast* (*cyte* = cell, *blast* = bud). These undifferentiated precursor cells reside in the red bone marrow. However, the maturation pathways of the various formed elements differ, and once a cell is *committed* to a specific blood cell pathway, it cannot change. This commitment is signaled by the appearance of membrane surface receptors that respond to specific hormones or growth factors, which in turn "push" the cell toward further specialization.

Stages of Erythropoiesis

Erythrocyte production, or **erythropoiesis** (ĕ-rith″ro-poy-e′sis), begins when a hematopoietic stem cell descendant called a **myeloid stem cell** transforms into a **proerythroblast** (**Figure 17.5**). Proerythroblasts, in turn, give rise to **basophilic erythroblasts** that produce huge numbers of ribosomes. During these first two phases, the cells divide many times. Hemoglobin is synthesized and iron accumulates as the basophilic erythroblast transforms into a **polychromatic erythroblast** and then an **orthochromatic erythroblast**. The "color" of the cell cytoplasm changes as the blue-staining ribosomes become masked by the pink color of hemoglobin. When an orthochromatic erythroblast has accumulated almost all of its hemoglobin, it ejects most of its organelles. Additionally, its nucleus degenerates and is pinched off, allowing the cell to collapse inward and eventually assume the biconcave shape. The result is the **reticulocyte** (essentially a young erythrocyte), so named because it still contains a scant *reticulum* (network) of clumped ribosomes.

The entire process from hematopoietic stem cell to reticulocyte takes about 15 days. The reticulocytes, filled almost to bursting with hemoglobin, enter the bloodstream to begin their

Figure 17.5 Erythropoiesis: formation of red blood cells. Reticulocytes are released into the bloodstream. The myeloid stem cell, the phase intermediate between the hematopoietic stem cell and the proerythroblast, is not illustrated.

task of oxygen transport. Usually they become fully mature erythrocytes within two days of release as their ribosomes are degraded by intracellular enzymes.

Reticulocytes account for 1–2% of all erythrocytes in the blood of healthy people. **Reticulocyte counts** provide a rough index of the *rate* of RBC formation—reticulocyte counts below or above this range indicate abnormal rates of erythrocyte formation.

Regulation and Requirements for Erythropoiesis

The number of circulating erythrocytes in a given individual is remarkably constant and reflects a balance between red blood cell production and destruction. This balance is important because having too few erythrocytes leads to tissue hypoxia (oxygen deprivation), whereas having too many makes the blood undesirably viscous.

To ensure that the number of erythrocytes in blood remains within the homeostatic range, new cells are produced at the incredibly rapid rate of more than 2 million per second in healthy people. This process is controlled hormonally and depends on adequate supplies of iron, amino acids, and certain B vitamins.

Hormonal Controls

Erythropoietin (EPO), a glycoprotein hormone, stimulates the formation of erythrocytes (**Figure 17.6**). Normally, a small amount of EPO circulates in the blood at all times and sustains red blood cell production at a basal rate. The kidneys play the major role in EPO production, although the liver also produces some. When certain kidney cells become *hypoxic* (oxygen deficient), oxygen-sensitive enzymes are unable to carry out their normal functions of degrading an intracellular

signaling molecule called hypoxia-inducible factor (HIF). As HIF accumulates, it accelerates the synthesis and release of erythropoietin.

The drop in normal blood oxygen levels that triggers EPO formation can result from:

- Reduced numbers of red blood cells due to hemorrhage (bleeding) or excessive RBC destruction
- Insufficient hemoglobin per RBC (as in iron deficiency)
- Reduced availability of oxygen, as might occur at high altitudes or during pneumonia

Conversely, too many erythrocytes or excessive oxygen in the bloodstream depresses erythropoietin production. Note that it is not the number of erythrocytes in blood that controls the rate of erythropoiesis. Instead, control is based on their ability to transport enough oxygen to meet tissue demands.

Bloodborne erythropoietin stimulates red marrow cells that *are already committed* to becoming erythrocytes, causing them to mature more rapidly. Two to three days after erythropoietin levels rise in the blood, the rate of reticulocyte release and the reticulocyte count rise markedly. Notice that hypoxia does not activate the bone marrow directly. Instead it stimulates the kidneys, which in turn provide the hormonal stimulus that activates the bone marrow.

HOMEOSTATIC IMBALANCE 17.1   CLINICAL

Some athletes abuse recombinant EPO—particularly professional bike racers and marathon runners seeking increased stamina and performance. However, the consequences can be deadly. By injecting EPO, healthy athletes increase their normal hematocrit from 45% to as much as 65%. Then, with the dehydration that occurs in a long race, the blood concentrates even

Figure 17.6 Erythropoietin (EPO) mechanism for regulating erythropoiesis.

1. Low O_2 levels in blood stimulate kidneys to produce erythropoietin.

2. Erythropoietin levels rise in blood.

3. Erythropoietin and necessary raw materials in blood promote erythropoiesis in red bone marrow.

4. New erythrocytes enter bloodstream; function about 120 days.

5. Aged and damaged red blood cells are engulfed by macrophages of spleen, liver, and bone marrow; the hemoglobin is broken down.

Hemoglobin

Heme   Globin

Bilirubin is picked up by the liver.

Iron is stored as ferritin or hemosiderin.

Amino acids

Iron is bound to transferrin and released to blood from liver as needed for erythropoiesis.

Bilirubin is secreted into intestine in bile where it is metabolized to stercobilin by bacteria.

Circulation

6. Raw materials are made available in blood for erythrocyte synthesis.

Stercobilin is excreted in feces.

Food nutrients (amino acids, Fe, B_{12}, and folic acid) are absorbed from intestine and enter blood.

Figure 17.7 Life cycle of red blood cells.

further, becoming a thick, sticky "sludge" that can cause clotting, stroke, or heart failure.

The male sex hormone *testosterone* also enhances the kidneys' production of EPO. Because female sex hormones do not have similar stimulatory effects, testosterone may be at least partially responsible for the higher RBC counts and hemoglobin levels seen in males.

Dietary Requirements

The raw materials required for erythropoiesis include:

- *Nutrients.* Amino acids, lipids, and carbohydrates are essential for the synthesis of all cells.
- *Two B-complex vitamins.* Vitamin B_{12} and folic acid are necessary for normal DNA synthesis. Even slight deficits jeopardize rapidly dividing cell populations, such as developing erythrocytes.
- *Iron.* Iron is essential for hemoglobin synthesis.

Iron is available from the diet, and intestinal cells precisely control its absorption into the bloodstream in response to changing body stores of iron. Approximately 65% of the body's iron supply (about 4000 mg) is in hemoglobin. Most of the remainder is stored in the liver, spleen, and (to a much lesser extent) bone marrow. Free iron ions (Fe^{2+}, Fe^{3+}) are toxic, so iron is stored inside cells as protein-iron complexes such as **ferritin** (fer´ĭ-tin) and **hemosiderin** (he″mo-sid´er-in). In blood, iron is transported loosely bound to a transport protein called **transferrin**, and developing erythrocytes take up iron as needed to form hemoglobin (**Figure 17.7**).

Small amounts of iron are lost each day in feces, urine, and perspiration. The average daily loss of iron is 1.7 mg in women and 0.9 mg in men. In women, the menstrual flow accounts for the additional losses.

Fate and Destruction of Erythrocytes

Red blood cells have a useful life span of 100 to 120 days. Their anucleate condition carries with it some important limitations. Red blood cells are unable to synthesize new proteins, grow, or divide. Erythrocytes become "old" as they lose their flexibility, become increasingly rigid and fragile, and their hemoglobin begins to degenerate. They become trapped and fragment in smaller circulatory channels, particularly in those of the spleen. For this reason, the spleen is sometimes called the "red blood cell graveyard."

Macrophages engulf and destroy dying erythrocytes. The heme of their hemoglobin is split off from globin (Figure 17.7). Its core of iron is salvaged, bound to protein (as ferritin or hemosiderin), and stored for reuse. The balance of the heme group is degraded to **bilirubin** (bil″ĭ-roo′bin), a yellow pigment that is released to the blood and binds to albumin for transport. Liver cells pick up bilirubin and in turn secrete it (in bile) into the intestine, where it is metabolized to *urobilinogen*. Most of this degraded pigment leaves the body in feces, as a brown pigment called *stercobilin*. The protein (globin) part of hemoglobin is metabolized or broken down to amino acids, which are released to the circulation.

17

Erythrocyte Disorders

CLINICAL

Most erythrocyte disorders can be classified as anemia or polycythemia.

Anemia

Anemia (ah-ne′me-ah; "lacking blood") is a condition in which the blood's oxygen-carrying capacity is too low to support normal metabolism. It is a *sign* of some disorder rather than a disease in itself. Its hallmark is blood oxygen levels that are inadequate to support normal metabolism. Anemic individuals are fatigued, often pale, short of breath, and chilled.

The causes of anemia can be divided into three groups: blood loss, not enough red blood cells produced, or too many of them destroyed.

Blood Loss *Hemorrhagic anemia* (hem″o-raj′ik) is caused by blood loss. In acute hemorrhagic anemia, blood loss is rapid (as might follow a severe stab wound); it is treated by replacing the lost blood. Slight but persistent blood loss (due to hemorrhoids or an undiagnosed bleeding ulcer, for example) causes chronic hemorrhagic anemia. Once the primary problem is resolved, normal erythropoietic mechanisms replace the lost blood cells.

Not Enough Red Blood Cells Produced A number of problems can decrease erythrocyte production. These problems range from lack of essential raw materials (such as iron) to complete failure of the red bone marrow.

- *Iron-deficiency anemia* is generally a secondary result of hemorrhagic anemia, but it also results from inadequate intake of iron-containing foods or impaired iron absorption. The erythrocytes produced under these conditions, called **microcytes**, are small and pale because they cannot synthesize their normal complement of hemoglobin. The treatment is to increase iron intake in diet or through iron supplements.

- *Pernicious anemia* is an autoimmune disease that most often affects the elderly. The immune system of these individuals destroys cells of their own stomach mucosa. These cells normally produce a substance called **intrinsic factor** that must be present for vitamin B_{12} to be absorbed by intestinal cells. Without vitamin B_{12}, the developing erythrocytes grow but cannot divide, and large, pale cells called **macrocytes** result. Treatment involves regular intramuscular injections of vitamin B_{12} or application of a B_{12}-containing gel to the nasal lining once a week.

  As you might expect, lack of vitamin B_{12} in the diet also leads to anemia. However, this is usually a problem only in strict vegetarians because meats, poultry, and fish provide ample vitamin B_{12}.

- *Renal anemia* is caused by the lack of EPO, the hormone that controls red blood cell production. Renal anemia frequently accompanies renal disease because damaged or diseased kidneys cannot produce enough EPO. Fortunately, it can be treated by administering EPO.

- *Aplastic anemia* may result from destruction or inhibition of the red marrow by certain drugs and chemicals,

ionizing radiation, or viruses. In most cases, though, the cause is unknown. Because marrow destruction impairs formation of *all* formed elements, anemia is just one of its signs. Defects in blood clotting and immunity are also present. Blood transfusions provide a stopgap treatment until stem cells harvested from a donor's blood, bone marrow, or umbilical cord blood can be transplanted.

Too Many Red Blood Cells Destroyed In *hemolytic anemias* (he″mo-lit′ik), erythrocytes rupture, or lyse, prematurely. Hemoglobin abnormalities, transfusion of mismatched blood, and certain bacterial and parasitic infections are possible causes. Here we focus on the hemoglobin abnormalities.

Production of abnormal hemoglobin usually has a genetic basis. Two such examples, thalassemia and sickle-cell anemia, can be serious, incurable, and sometimes fatal diseases. In both diseases the globin part of hemoglobin is abnormal and the erythrocytes produced are fragile and rupture prematurely.

Thalassemias (thal″ah-se′me-ahs; "sea blood") typically occur in people of Mediterranean ancestry. One of the globin chains is absent or faulty, and the erythrocytes are thin, delicate, and deficient in hemoglobin. There are many subtypes of thalassemia that range in severity from mild to so severe that monthly blood transfusions are required.

In **sickle-cell anemia**, the havoc caused by the abnormal hemoglobin, *hemoglobin S* (*HbS*), results from a change in just one of the 146 amino acids in a beta chain of the globin molecule! (See **Figure 17.8**.) This alteration causes the beta chains to link together under low-oxygen conditions, forming stiff rods so that hemoglobin S becomes spiky and sharp. This, in turn, causes the red blood cells to become crescent shaped when they unload oxygen molecules or when the oxygen content of the blood is lower than normal, as during vigorous exercise and other activities that increase metabolic rate.

The stiff, deformed erythrocytes rupture easily and tend to jam up in small blood vessels. These events interfere with oxygen delivery, leaving the victims gasping for air and in extreme pain. Bone and chest pain are particularly severe, and infection and stroke often follow. Blood transfusion is still the standard treatment for an acute sickle-cell crisis, but preliminary results using inhaled nitric oxide to dilate blood vessels are promising.

Sickle-cell anemia occurs chiefly in black people who live in the malaria belt of Africa and among their descendants. It strikes nearly one of every 500 African-American newborns.

Why would such a dangerous genetic trait persist in a population? Globally, about 250 million people are infected with malaria and about a million die each year. While individuals with two copies of the sickle-cell gene have sickle-cell anemia, individuals with only one copy of the gene (sickle-cell trait) have a better chance of surviving malaria. Their cells only sickle under abnormal circumstances, most importantly when they are infected with malaria. Sickling reduces the malaria parasites' ability to survive and enhances macrophages' ability to destroy infected RBCs and the parasites they contain.

Several treatment approaches for sickle-cell anemia focus on preventing RBCs from sickling. Fetal hemoglobin (HbF) does

17

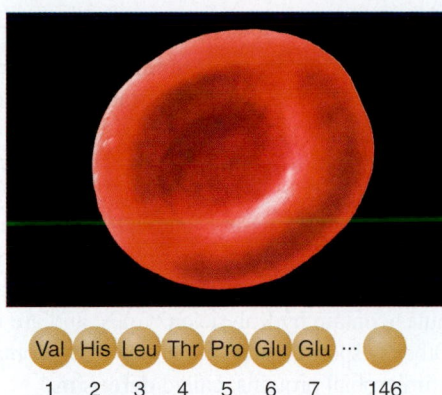

(a) Normal erythrocyte has normal hemoglobin amino acid sequence in the beta chain.

(b) Sickled erythrocyte results from a single amino acid change in the beta chain of hemoglobin.

Figure 17.8 **Sickle-cell anemia.** Scanning electron micrographs (4950×).

not "sickle," even in those destined to have sickle-cell anemia. *Hydroxyurea*, a drug used to treat chronic leukemia, switches the fetal hemoglobin gene back on. This drug dramatically reduces the excruciating pain and overall severity and complications of sickle-cell anemia (by 50%). In children who are severely affected, bone marrow stem cell transplants offer a complete cure, but carry high risks.

Polycythemia

Polycythemia (pol″e-si-the′me-ah; "many blood cells") is an abnormal excess of erythrocytes that increases blood viscosity, causing it to flow sluggishly. *Polycythemia vera*, a bone marrow cancer, is characterized by dizziness and an exceptionally high RBC count (8–11 million cells/μl). The hematocrit may be as high as 80% and blood volume may double, causing the vascular system to become engorged with blood and severely impairing circulation. Severe polycythemia is treated by removing some blood (a procedure called a therapeutic phlebotomy).

Secondary polycythemias result when less oxygen is available or EPO production increases. The secondary polycythemia that appears in individuals living at high altitudes is a normal physiological response to the reduced atmospheric pressure and lower oxygen content of the air in such areas. RBC counts of 6–8 million/μl are common in such people.

Blood doping, practiced by some athletes competing in aerobic events, is artificially induced polycythemia. Some of the athlete's red blood cells are drawn off and stored. The body quickly replaces these erythrocytes because removing blood triggers the erythropoietin mechanism. Then, when the stored blood is reinfused a few days before the athletic event, a temporary polycythemia results.

Since red blood cells carry oxygen, the additional infusion should translate into increased oxygen-carrying capacity due to a higher hematocrit, and greater endurance and speed should result. Other than the risk of stroke and heart failure due to high hematocrit and high blood viscosity, blood doping seems to work. However, the practice is considered unethical and has been banned from competitive sports.

Check Your Understanding

4. How many molecules of oxygen can each hemoglobin molecule transport? What part of the hemoglobin binds the oxygen?

5. **APPLY** Patients with advanced kidney disease often have anemia. Explain the connection.

For answers, see Answers Appendix.

17.4 Leukocytes defend the body

Learning Outcomes

▶ **List the classes, structural characteristics, and functions of leukocytes.**

▶ **Describe how leukocytes are produced.**

▶ **Give examples of leukocyte disorders, and explain what goes wrong in each disorder.**

General Structural and Functional Characteristics

Leukocytes (*leuko* = white), or **white blood cells (WBCs)**, are the only formed elements that are complete cells, with nuclei and the usual organelles. Accounting for less than 1% of total blood volume, leukocytes are far less numerous than red blood cells. On average, there are 4800–10,800 WBCs/μl of blood.

Leukocytes are crucial to our defense against disease. They form a mobile army that helps protect the body from damage by bacteria, viruses, parasites, toxins, and tumor cells. As such, they have special functional characteristics. Red blood cells are confined to the bloodstream, and they carry out their functions in the blood. However, white blood cells are able to slip out of the capillary blood vessels—a process called **diapedesis** (di″ah-pĕ-de′sis; "leaping across"). The circulatory system is simply their means of transport to areas of the body (mostly loose connective tissues or lymphoid tissues) where they mount inflammatory or immune responses.

As we explain in more detail in Chapter 21, the signals that prompt WBCs to leave the bloodstream at specific locations are cell adhesion molecules displayed by endothelial cells

17

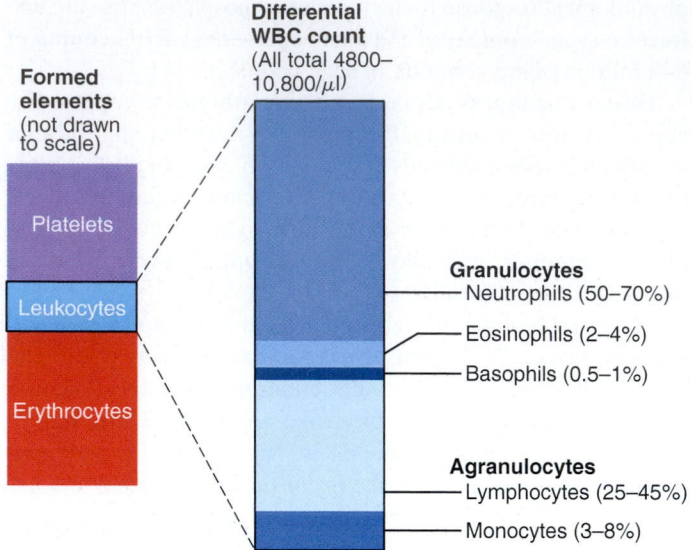

Figure 17.9 Types and relative percentages of leukocytes in normal blood. Erythrocytes comprise nearly 98% of the formed elements, and leukocytes and platelets together account for the remaining 2+%.

forming the capillary walls at sites of inflammation. Once out of the bloodstream, leukocytes move through the tissue spaces by **amoeboid motion** (they form flowing cytoplasmic extensions that move them along). By following the chemical trail of molecules released by damaged cells or other leukocytes, a phenomenon called **positive chemotaxis**, they pinpoint areas of tissue damage and infection and gather there in large numbers to destroy foreign substances and dead cells.

Whenever white blood cells are mobilized for action, the body speeds up their production and their numbers may double within a few hours. A *white blood cell count* of over 11,000 cells/μl is **leukocytosis**. This condition is a normal response to an infection in the body.

Leukocytes are grouped into two major categories on the basis of structural and chemical characteristics (**Figure 17.9**, and **Table 17.2** on p. 656):

- *Granulocytes* contain obvious membrane-bound cytoplasmic granules.

- *Agranulocytes* lack obvious granules.

Students are often asked to list the leukocytes in order from most abundant to least abundant. The following phrase may help you with this task: **N**ever **l**et **m**onkeys **e**at **b**ananas (neutrophils, lymphocytes, monocytes, eosinophils, basophils).

Granulocytes

Granulocytes (gran′u-lo-sīts), which include neutrophils, eosinophils, and basophils, are all roughly spherical in shape. They are larger and much shorter-lived (in most cases) than erythrocytes. They characteristically have lobed nuclei (rounded nuclear masses connected by thinner strands of nuclear material), and their membrane-bound cytoplasmic granules stain quite specifically with Wright's stain.

Neutrophils

Neutrophils (nu′tro-filz), the most numerous white blood cells, account for 50–70% of the WBC population. Neutrophils are about twice as large as erythrocytes.

The neutrophil cytoplasm contains very fine granules (of two varieties) that are difficult to see (Table 17.2 and **Figure 17.10a**). Neutrophils get their name (literally, "neutral-loving") because their granules take up both *basic* (blue) and *acidic* (red) *dyes*. Together, the two types of granules give the cytoplasm a lilac color. Some of these granules contain hydrolytic enzymes, and are regarded as lysosomes. Others, especially the smaller granules, contain a potent "brew" of antimicrobial proteins, called **defensins**.

Neutrophil nuclei typically have three to six lobes. Because of this nuclear variability, they are often called **polymorphonuclear leukocytes** (*PMNs*) or simply *polys* (*polymorphonuclear* = many shapes of the nucleus).

Neutrophils are our body's bacteria slayers, and their numbers increase explosively during acute bacterial infections such as meningitis and appendicitis. Neutrophils are chemically attracted to sites of inflammation and are active phagocytes. They are especially partial to ingesting bacteria and some fungi, which they enclose in a vesicle called a phagosome (◀ **p. 77**). One way that neutrophils kill bacteria is a process called a **respiratory burst**. During the respiratory burst, the cells metabolize oxygen to produce potent germ-killer oxidizing substances such as bleach and hydrogen peroxide. In addition, defensin-containing granules merge with the microbe-containing phagosome. As a result, defensins form peptide "spears" that pierce holes in the membrane of the ingested "foe" and the bacterium lyses.

Eosinophils

Eosinophils (e″o-sin′o-filz) account for 2–4% of all leukocytes and are approximately the size of neutrophils. Their nucleus usually has two lobes connected by a broad band of nuclear material (Table 17.2 and Figure 17.10b) and so resembles earmuffs.

Large, coarse granules that stain from brick red to crimson with acid (eosin) dyes pack the cytoplasm. These granules are lysosome-like and filled with a unique variety of digestive enzymes. However, unlike typical lysosomes, they lack enzymes that specifically digest bacteria.

The most important role of eosinophils is to lead the counterattack against parasitic worms, such as flatworms (tapeworms and flukes) and roundworms (pinworms and hookworms) that are too large to be phagocytized. These worms are ingested in food (especially raw fish) or invade the body via the skin and then typically burrow into the intestinal or respiratory mucosae. Eosinophils reside in the loose connective tissues at the same body sites, and when they encounter a parasitic worm "prey," they gather around and release the enzymes from their cytoplasmic granules onto the parasite's surface, digesting it away.

Eosinophils have complex roles in many other diseases including allergies and asthma. While they contribute to the tissue damage that occurs in many immune processes, we are also beginning to recognize them as important modulators of the immune response.

Granulocytes   Agranulocytes

(a) Neutrophils (2 shown): Multilobed nucleus, pale red and blue cytoplasmic granules

(b) Eosinophil: Bilobed nucleus, red cytoplasmic granules

(c) Basophil: Bilobed nucleus, purplish-black cytoplasmic granules

(d) Lymphocyte (small): Large spherical nucleus, thin rim of pale blue cytoplasm

(e) Monocyte: Kidney-shaped nucleus, abundant pale blue cytoplasm

Figure 17.10 Leukocytes. In each case the leukocytes are surrounded by erythrocytes. (All 2100×, Wright's stain.)

 Practice Histology questions: MasteringA&P® > Study Area > Lab Tools > PAL

Basophils

Basophils are the rarest white blood cells, accounting for only 0.5–1% of the leukocyte population. Their cytoplasm contains large, coarse, histamine-containing granules that have an affinity for the basic dyes (*basophil* = base loving) and stain purplish-black (Figure 17.10c). *Histamine* is an inflammatory chemical that acts as a vasodilator (makes blood vessels dilate) and attracts other white blood cells to the inflamed site; drugs called antihistamines counter this effect. The deep purple nucleus is generally U or S shaped with one or two constrictions.

Granulated cells similar to basophils, called *mast cells*, are found in connective tissues. Although mast cell nuclei tend to be more oval than lobed, the cells are similar microscopically, and both cell types bind to a particular antibody (immunoglobulin E) that causes the cells to release histamine. However, mast cells and basophils arise from different cell lines.

Agranulocytes

The **agranulocytes** include lymphocytes and monocytes, WBCs that lack *visible* cytoplasmic granules. Although similar to each other structurally, lymphocytes and monocytes are functionally distinct and unrelated cell types. Their nuclei are typically spherical or kidney shaped.

Lymphocytes

Lymphocytes, accounting for 25% or more of the WBC population, are the second most numerous leukocytes in the blood. When stained, a typical lymphocyte has a large, dark-purple nucleus that occupies most of the cell volume. The nucleus is usually spherical but may be slightly indented, and it is surrounded by a thin rim of pale-blue cytoplasm (Table 17.2 and Figure 17.10d). Lymphocytes are often classified by size

(diameter) as small (5–8 μm), medium (10–12 μm), and large (14–17 μm).

Large numbers of lymphocytes exist in the body, but relatively few (mostly the small lymphocytes) are found in the bloodstream. In fact, lymphocytes are so called because most are closely associated with lymphoid tissues (lymph nodes, spleen, etc.), where they play a crucial role in immunity.

- **T lymphocytes (T cells)** function in the immune response by acting directly against virus-infected cells and tumor cells.

- **B lymphocytes (B cells)** give rise to *plasma cells*, which produce **antibodies** (immunoglobulins) that are released to the blood. (We describe B and T lymphocyte functions in Chapter 21.)

Monocytes

Monocytes account for 3–8% of WBCs. With an average diameter of 18 μm, they are the largest leukocytes. They have abundant pale-blue cytoplasm and a darkly staining purple nucleus, which is often U or kidney shaped (Table 17.2 and Figure 17.10e).

When circulating monocytes leave the bloodstream and enter the tissues, they differentiate into highly mobile **macrophages** with prodigious appetites. Macrophages are actively phagocytic, and they are crucial in the body's defense against viruses, certain intracellular bacterial parasites, and *chronic* infections such as tuberculosis. As we explain in Chapter 21, macrophages can also activate lymphocytes to mount an immune response.

Production and Life Span of Leukocytes

Like erythropoiesis, **leukopoiesis**, or the production of white blood cells, is stimulated by chemical messengers. These messengers, which can act either as paracrines or hormones, are

17

glycoproteins that fall into two families of hematopoietic factors, **interleukins** and **colony-stimulating factors**, or **CSFs**. The interleukins are numbered (e.g., IL-3, IL-5), but most CSFs are named for the leukocyte population they stimulate—for example, *granulocyte-CSF* (*G-CSF*) stimulates production of granulocytes. Hematopoietic factors, released by supporting cells of the red bone marrow and mature WBCs, not only prompt the white blood cell precursors to divide and mature, but also enhance the protective potency of mature leukocytes.

HOMEOSTATIC IMBALANCE 17.2 · CLINICAL

Many of the hematopoietic hormones (EPO and several of the CSFs) are used clinically. These hormones stimulate the bone marrow of cancer patients who are receiving chemotherapy (which suppresses the marrow) and of those who have received stem cell transplants. They are also used to beef up the protective responses of AIDS patients. ●●●●●●●

Figure 17.11 shows the pathways of leukocyte differentiation, starting with the hematopoietic stem cell that gives rise to all of the formed elements in the blood. An early branching of the pathway divides the **lymphoid stem cells**, which produce lymphocytes, from the **myeloid stem cells**, which give rise to all other formed elements. In each granulocyte line, the committed cells, called **myeloblasts** (mi′ĕ-lo-blasts″), accumulate lysosomes, becoming **promyelocytes**. The distinctive granules of each granulocyte type appear next in the **myelocyte** stage and then cell division stops. Next, the nuclei become U shaped, producing the **band cell** stage. Just before granulocytes leave the marrow and enter the circulation, their nuclei constrict, beginning the process of nuclear segmentation.

The bone marrow stores mature granulocytes and usually contains about ten times more granulocytes than are found in the blood. The normal ratio of granulocytes to erythrocytes produced is about 3:1, which reflects granulocytes' much shorter life span (0.25 to 9.0 days). Most die fighting invading microorganisms.

Despite their similar appearance, the two types of agranulocytes have very different lineages.

- Monocytes are derived from myeloid stem cells, and share a common precursor with neutrophils that is not shared with the other granulocytes.* Cells following the monocyte line pass through the **monoblast** and **promonocyte** stages before leaving the bone marrow and becoming monocytes (Figure 17.11d).

- T and B lymphocytes are derived from **T** and **B lymphocyte precursors**, which arise from the lymphoid stem cell. The T lymphocyte precursors leave the bone marrow and travel to the thymus, where their further differentiation occurs (as we describe in Chapter 21). B lymphocyte precursors remain and mature in the bone marrow.

*Recent research suggests that monocytes may share a common precursor with lymphocytes rather than granulocytes.

Monocytes may live for several months, whereas the life span of lymphocytes varies from a few hours to decades.

Leukocyte Disorders · CLINICAL

Overproduction of abnormal leukocytes occurs in leukemia and infectious mononucleosis. At the opposite pole, **leukopenia** (loo″ko-pe′ne-ah) is an abnormally low white blood cell count (*penia* = poverty), commonly induced by drugs, particularly glucocorticoids and anticancer agents.

Leukemias

The term *leukemia*, literally "white blood," refers to a group of cancerous conditions involving overproduction of abnormal white blood cells. As a rule, the renegade leukocytes are members of a single *clone* (descendants of a single cell) that remain unspecialized and proliferate out of control, impairing normal red bone marrow function. The leukemias are named according to the rate at which they progress and the type of cell primarily involved. For example:

- *Acute leukemia* is quickly advancing because it derives from stem cells.
- *Chronic leukemia* is slowly advancing because it involves proliferation of later cell stages.
- *Myeloid leukemia* involves myeloblast descendants.
- *Lymphocytic leukemia* involves lymphocytes.

The more serious acute forms primarily affect children. Chronic leukemia occurs more often in elderly people. Without therapy, all leukemias are fatal, and only the time course differs.

In all leukemias, cancerous leukocytes fill the red bone marrow and immature WBCs flood into the bloodstream. The other blood cell lines are crowded out, so severe anemia and bleeding problems result. Other symptoms include fever, weight loss, and bone pain. Although tremendous numbers of leukocytes are produced, they are nonfunctional and cannot defend the body in the usual way. The most common causes of death are internal hemorrhage and overwhelming infections.

Irradiation and antileukemic drugs can destroy the rapidly dividing cells and induce remissions (symptom-free periods) lasting from months to years. Stem cell transplants are used in selected patients when compatible donors are available.

Infectious Mononucleosis

Sometimes called the "kissing disease," *infectious mononucleosis* is a highly contagious viral disease most often seen in young adults. Caused by the Epstein-Barr virus, the disease has a hallmark of excessive numbers of lymphocytes. Many of these lymphocytes are so large and atypical that they were originally misidentified as monocytes, and the disease was mistakenly named mononucleosis. The affected individual complains of being tired and achy, and has a chronic sore throat and a low-grade fever. There is no cure, but with rest the condition typically runs its course to recovery in four to six weeks.

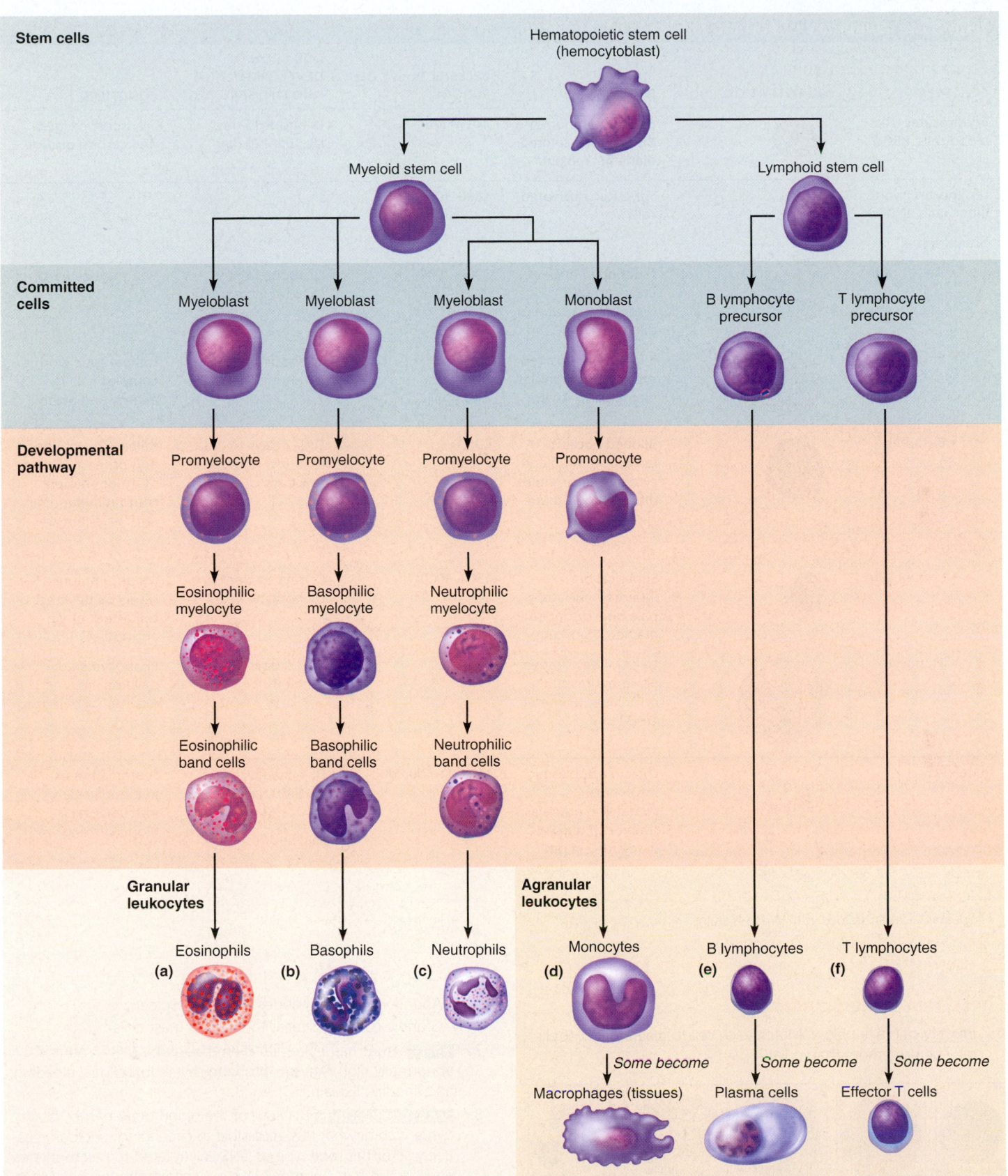

Stem cells

Hematopoietic stem cell
(hemocytoblast)

Myeloid stem cell

Lymphoid stem cell

Committed cells

Myeloblast   Myeloblast   Myeloblast   Monoblast   B lymphocyte precursor   T lymphocyte precursor

Developmental pathway

Promyelocyte   Promyelocyte   Promyelocyte   Promonocyte

Eosinophilic myelocyte   Basophilic myelocyte   Neutrophilic myelocyte

Eosinophilic band cells   Basophilic band cells   Neutrophilic band cells

Granular leukocytes

Agranular leukocytes

Eosinophils   Basophils   Neutrophils   Monocytes   B lymphocytes   T lymphocytes

(a)   **(b)**   **(c)**   **(d)**   **(e)**   **(f)**

Some become   *Some become*   *Some become*

Macrophages (tissues)   Plasma cells   Effector T cells

Figure 17.11 Leukocyte formation.
Leukocytes arise from ancestral stem cells called hematopoietic stem cells. **(a–c)** Granular leukocytes develop via a sequence involving myeloblasts. **(d)** Monocytes, like granular leukocytes, are progeny of the myeloid stem cell and share a common precursor with neutrophils (not shown). **(e, f)** Only lymphocytes arise via the lymphoid stem cell line.

17

Table 17.2 Summary of Formed Elements of the Blood

CELL TYPE	ILLUSTRATION	DESCRIPTION*	CELLS/µL (mm³) OF BLOOD	DURATION OF DEVELOPMENT (D) AND LIFE SPAN (LS)	FUNCTION
Erythrocytes (red blood cells, RBCs)		Biconcave, anucleate disc; salmon-colored; diameter 7–8 µm	4–6 million	D: about 15 days LS: 100–120 days	Transport oxygen and carbon dioxide
Leukocytes (white blood cells, WBCs)		Spherical, nucleated cells	4800–10,800		
Granulocytes					
• Neutrophil		Multilobed nucleus; inconspicuous cytoplasmic granules; diameter 10–12 µm	3000–7000	D: about 14 days LS: 6 hours to a few days	Phagocytize bacteria
• Eosinophil		Bilobed nucleus; red cytoplasmic granules; diameter 10–14 µm	100–400	D: about 14 days LS: about 5 days	Kill parasitic worms; complex role in allergy and asthma
• Basophil		Bilobed nucleus; large purplish-black cytoplasmic granules; diameter 10–14 µm	20–50	D: 1–7 days LS: a few hours to a few days	Release histamine and other mediators of inflammation; contain heparin, an anticoagulant
Agranulocytes					
• Lymphocyte		Spherical or indented nucleus; pale blue cytoplasm; diameter 5–17 µm	1500–3000	D: days to weeks LS: hours to years	Mount immune response by direct cell attack or via antibodies
• Monocyte		U- or kidney-shaped nucleus; gray-blue cytoplasm; diameter 14–24 µm	100–700	D: 2–3 days LS: months	Phagocytosis; develop into macrophages in the tissues
Platelets		Discoid cytoplasmic fragments containing granules; stain deep purple; diameter 2–4 µm	150,000–400,000	D: 4–5 days LS: 5–10 days	Seal small tears in blood vessels; instrumental in blood clotting

*Appearance when stained with Wright's stain.

Check Your Understanding

6. Identify cells a–e below. Match each cell to a statement. (Cells may be used more than once.)

(a)   (b)   (c)   (d)   (e)

__, __ 1. Are or become important phagocytes.
__, __, __ 2. Granulocytes.

__ 3. Abundant in lymph nodes and in the spleen.
__ 4. Granule contents similar to those of mast cells.

7. **APPLY** Amos has leukemia. Even though his WBC count is abnormally high, Amos is prone to severe infections, bleeding, and anemia. Explain.

8. **MAKE CONNECTIONS** Because of the blood brain barrier, brain tissue is somewhat less accessible to circulating macrophages. Because of this, one type of CNS cell (related to macrophages and discussed in Chapter 11), can become phagocytic. Which type of cell is this?

For answers, see Answers Appendix.

Stem cell   Developmental pathway

Hematopoietic stem   Megakaryoblast   Megakaryocyte   Megakaryocyte   Platelets
cell (hemocytoblast)   (stage I megakaryocyte) (stage II/III)   (stage IV)

Figure 17.12 Formation of platelets. The hematopoietic stem cell gives rise to cells that undergo several mitotic divisions unaccompanied by cytoplasmic division to produce megakaryocytes. The plasma membrane of the megakaryocyte fragments, liberating the platelets. (Intermediate stages between the hematopoietic stem cell and megakaryoblast are not illustrated.)

17.5 Platelets are cell fragments that help stop bleeding

Learning Outcome

▶ **Describe the structure and function of platelets.**

Platelets are not cells in the strict sense. Rather, they are fragments of extraordinarily large cells (up to 60 μm in diameter) called **megakaryocytes** (meg″ah-kar′e-o-sītz). Platelets are about one-fourth the diameter of a lymphocyte. In blood smears, each platelet exhibits a blue-staining outer region and an inner area containing granules that stain purple. The granules contain an impressive array of chemicals that act in the clotting process, including serotonin, Ca^{2+}, a variety of enzymes, ADP, and platelet-derived growth factor (PDGF).

Platelets are essential for the clotting process that occurs in plasma when blood vessels are ruptured or their lining is injured. By sticking to the damaged site, platelets form a temporary plug that helps seal the break. (We explain this process shortly.) Because they are anucleate, platelets age quickly and degenerate in about 10 days if they are not involved in clotting. In the meantime, they circulate freely, kept mobile but inactive by molecules (nitric oxide, prostacyclin) secreted by endothelial cells lining the blood vessels.

A hormone called **thrombopoietin** regulates the formation of platelets. Their immediate ancestral cells, the megakaryocytes, are progeny of the hematopoietic stem cell and the myeloid stem cell, but their formation is quite unusual (**Figure 17.12**). In this line, repeated mitoses of the **megakaryoblast**

Endothelial cells line blood vessels.

Platelet

Blood vessel

Normal, undamaged endothelial cells secrete molecules (such as nitric oxide and prostacyclin) that keep platelets from sticking.

(also called a stage I megakaryocyte) occur, but cytokinesis does not. The final result is the mature (stage IV) megakaryocyte (literally "big nucleus cell"), a bizarre cell with a huge, multilobed nucleus and a large cytoplasmic mass.

After it forms, the megakaryocyte presses against a sinusoid (the specialized type of capillary in the red marrow) and sends cytoplasmic extensions through the sinusoid wall into the bloodstream. These extensions rupture, releasing platelets like leaves blowing off a tree, seeding the blood with platelets. The plasma membranes associated with each fragment quickly seal around the cytoplasm to form the grainy, roughly disc-shaped platelets (see Table 17.2), each with a diameter of 2–4 μm. Each microliter of blood contains 150,000 to 400,000 tiny platelets.

Check Your Understanding

9. What is a megakaryocyte? What does its name mean?

For answers, see Answers Appendix.

17.6 Hemostasis prevents blood loss

Learning Outcomes

▶ **Describe the process of hemostasis. List factors that limit clot formation and prevent undesirable clotting.**

▶ **Give examples of hemostatic disorders. Indicate the cause of each condition.**

Normally, blood flows smoothly past the intact blood vessel lining (endothelium). But if a blood vessel wall breaks, a whole series of reactions is set in motion to accomplish **hemostasis** (he″mo-sta′sis), which stops the bleeding (*stasis* = halting). Without this plug-the-hole defensive reaction, we would quickly bleed out our entire blood volume from even the smallest cuts.

The hemostasis response is fast, localized, and carefully controlled. It involves many *clotting factors* normally present in plasma as well as several substances that are released by platelets and injured tissue cells. During hemostasis, three steps occur in

17

① **Vascular spasm**
• Smooth muscle contracts, causing vasoconstriction.

② **Platelet plug formation**
• Injury to endothelium lining the vessel exposes underlying collagen fibers; platelets adhere.

Collagen fibers

Endothelium

• Platelets release chemicals that make nearby platelets sticky; platelet plug forms.

Platelets

③ **Coagulation**
• Fibrin proteins form a mesh that traps red blood cells and platelets, forming the clot.

Fibrin

Figure 17.13 Events of hemostasis.

rapid sequence (**Figure 17.13**): ① vascular spasm, ② platelet plug formation, and ③ coagulation (blood clotting). Following hemostasis, the clot retracts. It then dissolves as it is replaced by fibrous tissue that permanently prevents blood loss.

Step 1: Vascular Spasm

In the first step, the damaged blood vessels respond to injury by constricting (vasoconstriction) (Figure 17.13 ①). Factors that trigger this **vascular spasm** include direct injury to vascular smooth muscle, chemicals released by damaged endothelial cells and activated platelets (discussed next), and reflexes initiated by local pain receptors. The spasm mechanism becomes more and more efficient as the amount of tissue damage increases, and is most effective in the smaller blood vessels. The spasm response is valuable because a strongly constricted artery can significantly reduce blood

loss for 20–30 minutes, allowing time for the next two steps to occur.

Step 2: Platelet Plug Formation

In the second step, platelets play a key role in hemostasis by aggregating (sticking together), forming a plug that temporarily seals the break in the vessel wall (Figure 17.13 ②). They also help orchestrate subsequent events that form a blood clot.

As a rule, platelets do not stick to each other or to the smooth endothelium that lines blood vessels. (Recall that endothelium is a type of epithelium; ◄ **p. 121**.) Intact endothelial cells release nitric oxide and a prostaglandin called **prostacyclin** (or PGI_2). Both chemicals prevent platelet aggregation (clumping) in undamaged tissue and restrict aggregation to the site of injury.

However, when the endothelium is damaged and the collagen fibers of the underlying connective tissue (◄ **p. 128**) are exposed, platelets adhere tenaciously to the collagen fibers. A large plasma protein called *von Willebrand factor* stabilizes bound platelets by forming a bridge between collagen and platelets. Platelets become *activated*: They swell, form spiked processes, and become stickier. In addition, they release chemical messengers including the following:

• **Adenosine diphosphate (ADP)**—a potent aggregating agent that causes more platelets to stick to the area and release their contents

• **Serotonin** and **thromboxane A$_2$** (throm-boks'ān; a short-lived prostaglandin derivative)—messengers that enhance vascular spasm and platelet aggregation

As more platelets aggregate, they release more chemicals, aggregating more platelets, and so on, in a positive feedback cycle (◄ **Figure 1.7, p. 12**). Within one minute, a platelet plug is built up, further reducing blood loss. Platelets alone are sufficient for sealing the thousands of minute rips and holes that occur unnoticed as part of the daily wear and tear in your smallest blood vessels. Larger breaks, however, need additional reinforcement because platelet plugs are loosely knit.

Step 3: Coagulation

The third step, **coagulation** or **blood clotting**, reinforces the platelet plug with fibrin threads that act as a "molecular glue" for the aggregated platelets (Figure 17.13 ③). The resulting blood clot (fibrin mesh) is quite effective in sealing larger breaks in a blood vessel. Blood is transformed from a liquid to a gel when dissolved blood proteins assemble into fibrin threads. This multistep process involves a series of substances called **clotting factors**, or **procoagulants** (**Table 17.3** on p. 661).

Most clotting factors are plasma proteins synthesized by the liver. They are numbered I to XIII according to the order of their discovery; as a result, the numerical order does not reflect their reaction sequence. All (except tissue factor) normally circulate in blood in inactive form until mobilized. Although vitamin K is not directly involved in coagulation, this fat-soluble vitamin is required for synthesizing four of the clotting factors (Table 17.3).

The coagulation sequence shows how clotting factors work together to form a clot (**Figure 17.14**). This sequence will look intimidating at first glance, but don't panic—two things will help you cope with its complexity. First, realize that in most cases, *activation turns clotting factors into enzymes* by clipping off a piece of the protein, causing it to change shape. Once one clotting factor is activated, it activates the next in sequence, and so on, in a cascade. (Two important exceptions to this generalization are fibrinogen and Ca^{2+}.)

The second strategy that will help you cope is to recognize that coagulation occurs in three phases. Each phase has a specific end point, as you will see next.

Phase 1: Two Pathways to Prothrombin Activator

Coagulation may be initiated by either the **intrinsic** or the **extrinsic pathway** (Figure 17.14). In the body, the same tissue-damaging events usually trigger both pathways. Before we examine how these pathways are different, let's see what they have in common.

Crucial components in both pathways are negatively charged membranes. In particular, activated platelets display negatively charged phosphatidylserine, once known as *PF3* (platelet factor 3). Many intermediates of both pathways can be activated only when on such phospholipid surfaces. The intermediate steps of each pathway *cascade* toward a common intermediate, factor X (Figure 17.14). Once factor X has been activated, it complexes with calcium ions and factor V on a phospholipid surface to form **prothrombin activator**. This is usually the slowest step of the blood clotting process, but once prothrombin activator is present, the clot forms in 10 to 15 seconds.

The intrinsic and extrinsic pathways usually work together and are interconnected in many ways, but there are significant differences between them. The *intrinsic pathway* is:

- Called *intrinsic* because the factors needed for clotting are present *within* (intrinsic to) the blood.

- Triggered by negatively charged surfaces such as activated platelets, collagen, or glass. (The negative charges on glass explain why this is the pathway that initiates clotting outside the body in test tubes.)

- Slower because it has many intermediate steps.

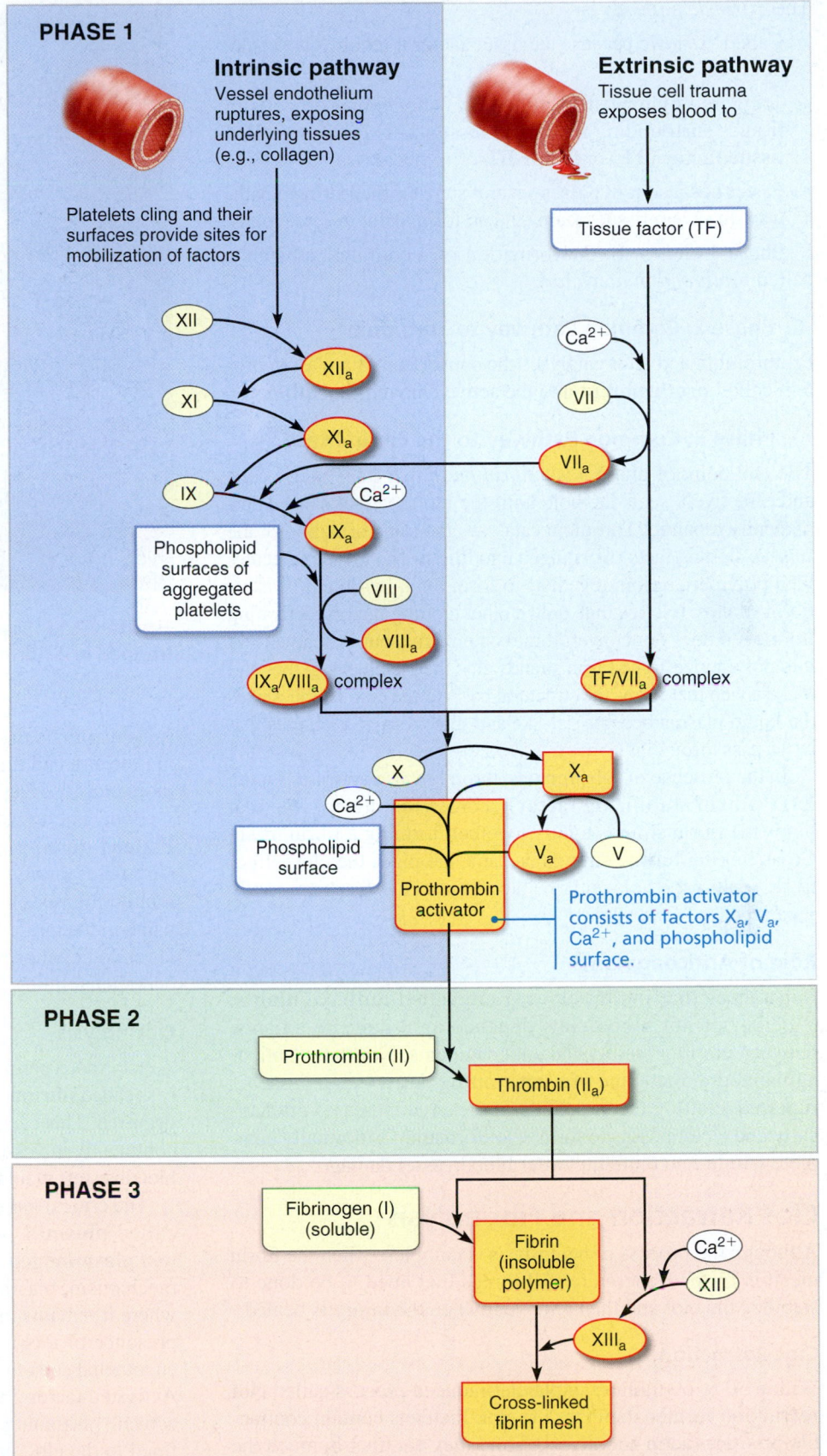

Figure 17.14 The intrinsic and extrinsic pathways of blood clotting (coagulation). The subscript "a" indicates the activated clotting factor (procoagulant).

The *extrinsic pathway* is:

- Called *extrinsic* because the tissue factor it requires is *outside* of blood.

- Triggered by exposing blood to a factor found on cells in tissues surrounding the blood vessel. This factor is called **tissue factor (TF)** or **factor III**.

- Faster because it bypasses several steps of the intrinsic pathway. In severe tissue trauma, it can form a clot in 15 seconds.

Phase 1 ends with the formation of a complex substance called *prothrombin activator*.

Phase 2: Common Pathway to Thrombin

Prothrombin activator catalyzes the conversion of a plasma protein called **prothrombin** into the active enzyme **thrombin**.

Phase 3: Common Pathway to the Fibrin Mesh

The end point of phase 3 is a *fibrin mesh* that traps blood cells and effectively seals the hole until the blood vessel can be permanently repaired. Thrombin catalyzes the transformation of the *soluble* clotting factor **fibrinogen** into **fibrin**. The fibrin molecules then polymerize (join together) to form long, hairlike, *insoluble* fibrin strands. (Notice that, unlike other clotting factors, activating fibrinogen does not convert it into an enzyme, but instead allows it to polymerize.) The fibrin strands glue the platelets together and make a web that forms the structural basis of the clot. Fibrin makes the liquid plasma become gel-like and traps formed elements that try to pass through it (**Figure 17.15**).

In the presence of calcium ions, thrombin also activates **factor XIII (fibrin stabilizing factor)**, a cross-linking enzyme that binds the fibrin strands tightly together, forming a fibrin mesh. Cross-linking further strengthens and stabilizes the clot, effectively sealing the hole until the blood vessel can be permanently repaired.

Role of Anticoagulants

Substances that inhibit clotting are called **anticoagulants**. Whether or not blood clots depends on a delicate balance between clotting factors and anticoagulants. Normally, anticoagulants dominate and prevent clotting, but when a vessel is ruptured, clotting factor activity in that area increases dramatically and a clot begins to form. Clot formation is normally complete within 3 to 6 minutes after blood vessel damage.

Clot Retraction and Fibrinolysis

Although the process of hemostasis is complete when the fibrin mesh is formed, there are still things that need to be done to stabilize the clot and then remove it when the injury is healed.

Clot Retraction

Within 30 to 60 minutes, a platelet-induced process called **clot retraction** further stabilizes the clot. Platelets contain contractile proteins (actin and myosin), and they contract in much the same manner as smooth muscle cells. As the platelets contract, they pull on the surrounding fibrin strands, squeezing **serum**

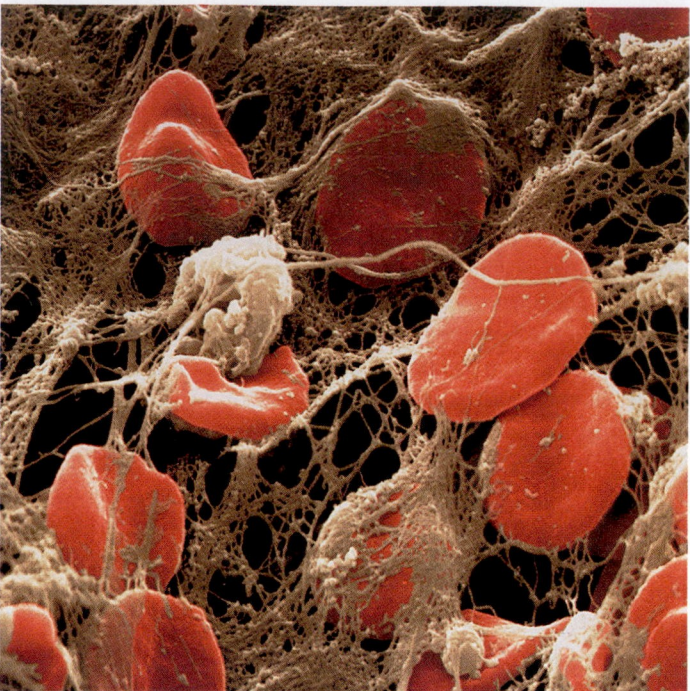

Figure 17.15 Scanning electron micrograph of erythrocytes trapped in a fibrin mesh. (2700×).

(plasma minus the clotting proteins) from the mass, compacting the clot and drawing the ruptured edges of the blood vessel more closely together.

Even as clot retraction is occurring, the vessel is healing. **Platelet-derived growth factor (PDGF)** released by platelets stimulates smooth muscle cells and fibroblasts to divide and rebuild the vessel wall. As fibroblasts form a connective tissue patch in the injured area, endothelial cells, stimulated by vascular endothelial growth factor (VEGF), multiply and restore the endothelial lining.

Fibrinolysis

A clot is not a permanent solution to blood vessel injury, and a process called **fibrinolysis** removes unneeded clots when healing has occurred. This cleanup detail is crucial because small clots form continually in vessels throughout the body. Without fibrinolysis, blood vessels would gradually become completely blocked.

The critical natural "clot buster" is a fibrin-digesting enzyme called **plasmin**, which is produced when the plasma protein **plasminogen** is activated. Like a built-in self-destruction mechanism, plasminogen is incorporated into a forming clot, where it remains inactive until appropriate signals reach it. The presence of a clot in and around the blood vessel causes the endothelial cells to secrete **tissue plasminogen activator (tPA).** Activated factor XII and thrombin released during clotting also activate plasminogen. As a result, most plasmin activity is confined to the clot, and circulating enzymes quickly destroy any plasmin that strays into the plasma. Fibrinolysis begins within two days and continues slowly over several days until the clot finally dissolves.

Table 17.3   Blood Clotting Factors (Procoagulants)

FACTOR NUMBER	FACTOR NAME	NATURE	SOURCE	PATHWAY; FUNCTION
I	Fibrinogen	Plasma protein	Liver	Common pathway; converted to fibrin (insoluble weblike substance of clot)
II	Prothrombin	Plasma protein	Liver*	Common pathway; converted to thrombin (converts fibrinogen to fibrin)
III	Tissue factor (TF)	Plasma membrane glycoprotein	Tissue cells	Activates extrinsic pathway
IV	Calcium ions (Ca^{2+})	Inorganic ion	Plasma	Needed for essentially all stages of coagulation process; always present
V	Proaccelerin	Plasma protein	Liver, platelets	Common pathway
VI[†]				
VII	Proconvertin	Plasma protein	Liver*	Both extrinsic and intrinsic pathways
VIII	Antihemophilic factor (AHF)	Plasma protein	Liver, lung capillaries	Intrinsic pathway; deficiency results in hemophilia A
IX	Plasma thromboplastin component (PTC)	Plasma protein	Liver*	Intrinsic pathway; deficiency results in hemophilia B
X	Stuart factor	Plasma protein	Liver*	Common pathway
XI	Plasma thromboplastin antecedent (PTA)	Plasma protein	Liver	Intrinsic pathway; deficiency results in hemophilia C
XII	Hageman factor	Plasma protein; activated by negatively charged surfaces (e.g., glass)	Liver	Intrinsic pathway; activates plasmin; initiates clotting *in vitro*; activation initiates inflammation
XIII	Fibrin stabilizing factor (FSF)	Plasma protein	Liver, bone marrow	Cross-links fibrin, forming a strong, stable clot

*Synthesis requires vitamin K

[†]Number no longer used; substance now believed to be same as factor V

Factors Limiting Clot Growth or Formation

Factors Limiting Normal Clot Growth

Once the clotting cascade has begun, it continues until a clot forms. Normally, two homeostatic mechanisms prevent clots from becoming unnecessarily large: (1) swift removal of clotting factors, and (2) inhibition of activated clotting factors. For clotting to occur, the concentration of activated clotting factors must reach certain critical levels. Clots do not usually form in rapidly moving blood because the activated clotting factors are washed away and diluted. (This is why vasoconstriction is so important to assist clot formation and also why you apply pressure to a wound to slow bleeding.) For the same reasons, a clot stops growing when it contacts blood flowing normally.

Other mechanisms block the final step in which fibrinogen is polymerized into fibrin. They work by restricting thrombin to the clot or by inactivating it if it escapes into the general circulation. As a clot forms, almost all of the thrombin produced is bound onto the fibrin threads. This is an important safeguard because thrombin also exerts positive feedback effects on the coagulation process prior to the common pathway. Not only does it speed up the production of prothrombin activator by acting indirectly through factor V, but it also accelerates the earliest steps of the intrinsic pathway by activating platelets. By binding thrombin, fibrin effectively acts as an anticoagulant, preventing the clot from enlarging and thrombin from acting elsewhere.

Antithrombin III, a protein present in plasma, quickly inactivates any thrombin not bound to fibrin. Antithrombin III and **protein C**, another protein produced in the liver, also inhibit the activity of other intrinsic pathway clotting factors.

Heparin, the natural anticoagulant contained in basophil and mast cell granules, is also found on the surface of endothelial cells. It inhibits thrombin by enhancing the activity of antithrombin III. Like most other clotting inhibitors, heparin also inhibits the intrinsic pathway.

Factors Preventing Undesirable Clotting

As long as the endothelium is smooth and intact, platelets are prevented from clinging and piling up. Also, antithrombic substances—nitric oxide, prostacyclin, and others—secreted by the endothelial cells normally prevent platelet adhesion.

Disorders of Hemostasis   CLINICAL

Blood clotting is one of nature's most elegant creations, but it sometimes goes awry. The two major disorders of hemostasis are at opposite poles. **Thromboembolic disorders** result from

17

conditions that cause undesirable clot formation. **Bleeding disorders** arise from abnormalities that prevent normal clot formation. **Disseminated intravascular coagulation (DIC)**, which has characteristics of both types of disorder, involves both widespread clotting and severe bleeding.

Thromboembolic Disorders

Despite the body's many safeguards, undesirable intravascular clotting sometimes occurs.

Thrombi and Emboli A clot that develops and persists in an *unbroken* blood vessel is called a **thrombus**. If the thrombus is large enough, it may block circulation to the cells beyond the blockage and lead to death of those tissues. For example, if the blockage occurs in the coronary circulation of the heart (coronary thrombosis), the consequences may be death of heart muscle and a fatal heart attack.

If the thrombus breaks away from the vessel wall and floats freely in the bloodstream, it becomes an **embolus** (plural: *emboli*). An embolus ("wedge") is usually no problem until it encounters a blood vessel too narrow for it to pass through. Then it becomes an **embolism**, obstructing the vessel. For example, emboli that become trapped in the lungs (pulmonary embolisms) dangerously impair the body's ability to obtain oxygen. A cerebral embolism may cause a stroke.

Conditions that roughen the vessel endothelium, such as atherosclerosis or inflammation, cause thromboembolic disease by allowing platelets to gain a foothold. Slowly flowing blood or blood stasis is another risk factor, particularly in bedridden patients and those taking a long flight without moving around. In this case, clotting factors are not washed away as usual and accumulate, allowing clots to form.

Anticoagulant Drugs A number of drugs—most importantly aspirin, heparin, and warfarin—are used clinically to prevent undesirable clotting.

- *Aspirin.* **Aspirin** is an antiprostaglandin drug that inhibits thromboxane A_2 formation (blocking platelet aggregation and platelet plug formation). Clinical studies of men taking low-dose aspirin over several years demonstrated a 50% reduction in incidence of heart attack.

- *Heparin.* Heparin, administered in injectable form, is the anticoagulant most used in the hospital (for heart patients undergoing various procedures and to prevent or treat venous thrombosis).

- *Warfarin and direct oral anticoagulants.* **Warfarin** and *direct oral anticoagulants* are mainstays of outpatient treatment to reduce the risk of stroke in those prone to atrial fibrillation, a condition in which blood pools in the heart. Warfarin interferes with the action of vitamin K in the production of some clotting factors (see Impaired Liver Function below). Because treatment with warfarin is difficult to manage, the introduction of direct oral anticoagulants using other mechanisms has been welcomed.

A Closer Look in Chapter 19 (p. 760) describes another drug that dissolves blood clots (tPA; see also p. 660) and innovative medical techniques for treating clots.

Bleeding Disorders

Anything that interferes with the clotting mechanism can result in abnormal bleeding. The most common causes are platelet deficiency (thrombocytopenia) and deficits of some clotting factors, which can result from impaired liver function or genetic conditions such as hemophilia.

Thrombocytopenia **Thrombocytopenia** (throm″bo-si″to-pe′ne-ah) is characterized by a low number of circulating platelets (less than 150,000/µl of blood). In severe cases, it causes spontaneous bleeding from small blood vessels all over the body. Even normal movement leads to widespread hemorrhage, evidenced by many small purplish spots, called *petechiae* (pe-te′ke-e), on the skin (**Figure 17.16**).

Thrombocytopenia can arise from any condition that suppresses or destroys the red bone marrow, such as bone marrow malignancy, exposure to ionizing radiation, or certain drugs. Transfusions of concentrated platelets provide temporary relief from bleeding.

Impaired Liver Function When the liver is unable to synthesize its usual supply of clotting factors, abnormal and often severe bleeding occurs. The causes can range from an easily resolved vitamin K deficiency (common in newborns) to nearly total impairment of liver function (as in hepatitis or cirrhosis).

Liver cells require vitamin K to produce clotting factors. Although intestinal bacteria make some vitamin K, we obtain most of it from vegetables in our diet and dietary deficiencies are rarely a problem. However, vitamin K deficiency can occur if fat absorption is impaired, because vitamin K is a fat-soluble vitamin that is absorbed into the blood along with fats. In liver disease, the nonfunctional liver cells fail to produce not only the clotting factors, but also bile that is required to absorb fat and vitamin K.

Hemophilias The term **hemophilia** refers to several hereditary bleeding disorders that have similar signs and symptoms.

- *Hemophilia A* results from a deficiency of **factor VIII (antihemophilic factor)**. It accounts for 77% of cases.

Figure 17.16 Petechiae. Thrombocytopenia causes small, round dots called petechiae to appear. These are the result of everyday capillary damage that would ordinarily be plugged by platelets.

- *Hemophilia B* results from a deficiency of factor IX.
- *Hemophilia C*, a less severe form seen in both sexes, is due to a lack of factor XI.

Hemophilia A and B occur primarily in males (they are X-linked genetic conditions, discussed in Chapter 29). The relative mildness of hemophilia C reflects the fact that the clotting factor (factor IX) that the missing factor XI activates can also be activated by factor VII (see Figure 17.14).

Symptoms of hemophilia begin early in life. Even minor tissue trauma causes prolonged and potentially life-threatening bleeding into tissues. Commonly, the person's joints become seriously disabled and painful because of repeated bleeding into the joint cavities after exercise or trauma. Hemophilias are managed clinically by transfusions of fresh plasma or injections of the appropriate purified clotting factor. These therapies provide relief for several days but are expensive and inconvenient.

In addition, dependence on transfusions or injections has caused other problems. In the past, many hemophilia patients became infected by the hepatitis virus and, beginning in the early 1980s, by HIV, a blood-transmitted virus that depresses the immune system and causes AIDS (see Chapter 21). New infections are now avoided due to the availability of genetically engineered clotting factors, hepatitis vaccines, and new testing methods for HIV.

Disseminated Intravascular Coagulation (DIC)

DIC involves both widespread clotting and severe bleeding. Clotting occurs in intact blood vessels and the residual blood becomes unable to clot. Blockage of blood flow accompanied by severe bleeding follows. DIC most commonly happens as a complication of pregnancy or a result of septicemia or incompatible blood transfusions.

Check Your Understanding

10. What are the three steps of hemostasis?

11. Name the enzyme that catalyzes each of the steps in the final common pathway shown in (a) and (b) below. What is the key difference between prothrombin and thrombin? Between fibrinogen and fibrin? Between most factors before and after they are activated?

(a)   (b)

12. Which bleeding disorder results from not having enough platelets? From absence of clotting factor VIII?

For answers, see Answers Appendix.

17.7 Transfusion can replace lost blood   CLINICAL

Learning Outcomes

▶ Describe the ABO and Rh blood groups. Explain the basis of transfusion reactions.

▶ Describe fluids used to replace blood volume and the circumstances for their use.

The human cardiovascular system minimizes the effects of blood loss by (1) reducing the volume of the affected blood vessels, and (2) stepping up production of red blood cells. However, the body can compensate for only so much blood loss. Losing 15–30% causes pallor and weakness. Losing more than 30% of blood volume results in severe shock, which can be fatal.

Restoring Blood Volume

When a patient's blood volume is so low that death from shock is imminent, there may not be time to type blood (see below), or appropriate whole blood may be unavailable. Such emergencies demand that blood *volume* be replaced immediately to restore adequate circulation.

Fundamentally, blood consists of proteins and cells suspended in a salt solution. Replacing lost blood volume essentially consists of replacing that isotonic salt solution. *Normal saline* or a *multiple electrolyte solution* that mimics the electrolyte composition of plasma (for example, *Ringer's solution*) are the preferred choices.

Volume replacement restores adequate circulation but cannot, of course, replace the oxygen-carrying capacity of the lost red blood cells. To date, only transfusion of red blood cells can do that (see below), but research on ways to replace that capability by using artificial blood substitutes is ongoing.

Transfusing Red Blood Cells

Whole blood transfusions are rarely used—only when blood loss is rapid and substantial. In all other cases, infusions of **packed red blood cells** (**PRBCs**; whole blood from which most of the plasma and leukocytes have been removed) are preferred for restoring oxygen-carrying capacity. The usual blood bank procedure involves collecting blood from a donor and mixing it with an anticoagulant that prevents clotting by binding calcium ions. The shelf life of collected blood is about 35 days. Because blood is such a valuable commodity, it is most often separated into its component parts so that each component can be used when and where it is needed.

Human Blood Groups

People have different blood types, and transfusion of incompatible blood can be fatal. RBC plasma membranes, like those of all body cells, bear highly specific molecular markers at their external surfaces, which identify each of us as unique from all others. These glycoprotein and glycolipid markers are called *antigens.* An antigen is anything the body perceives as foreign and that generates an immune response. Examples are toxins and molecules on the surfaces of bacteria, viruses, and cancer cells—and mismatched RBCs.

One person's RBC antigens may be recognized as foreign if transfused into someone with a different red blood cell type, and the transfused cells may be agglutinated (clumped together) and destroyed. Since these RBC antigens promote agglutination, they are more specifically called **agglutinogens** (ag″loo-tin′o-jenz).

17

Table 17.4 ABO Blood Groups

BLOOD GROUP	RBC ANTIGENS (AGGLUTINOGENS)	PLASMA ANTIBODIES (AGGLUTININS)		BLOOD THAT CAN BE RECEIVED	FREQUENCY (% OF U.S. POPULATION)				
					WHITE	BLACK	ASIAN	HISPANIC	NATIVE AMERICAN
AB	A B	None		A, B, AB, O "Universal recipient"	4	4	7	2	<1
B	B	Anti-A (a)		B, O	11	19	25	10	4
A	A	Anti-B (b)		A, O	40	26	28	31	16
O	None	Anti-A (a) Anti-B (b)		O "Universal donor"	45	51	40	57	79

At least 30 groups of naturally occurring RBC antigens (blood groups) are found in humans, and many variants occur in individual families ("private antigens") rather than in the general population. The presence or absence of various antigens allows a person's blood cells to be classified into each of these different blood groups. Antigens determining the ABO and Rh blood groups cause vigorous transfusion reactions (in which the foreign erythrocytes are destroyed) when they are improperly transfused. For this reason, blood typing for these antigens is always done before blood is transfused.

Other antigens (such as those in the MNS, Duffy, Kell, and Lewis groups) are mainly of legal or academic importance. Because these factors rarely cause transfusion reactions, blood is not specifically typed for them unless the person is expected to need several transfusions, in which case reactions are more likely to occur. Here we describe only the ABO and Rh blood groups.

ABO Blood Groups The **ABO blood groups** are based on the presence or absence of two agglutinogens, type A and type B (**Table 17.4**). Depending on which of these a person inherits, his or her ABO blood group will be one of the following: A, B, AB, or O. The O blood group, which has neither agglutinogen, is the most common ABO group in North America. AB, with both antigens, is least prevalent. The presence of either the A or the B agglutinogen results in group A or B, respectively.

Unique to the ABO blood groups is the presence in the plasma of *preformed antibodies* called **agglutinins**. The agglutinins act against RBCs carrying ABO antigens that are *not* present on a person's own red blood cells. A newborn lacks these antibodies,

but they begin to appear in the plasma within two months and reach adult levels between 8 and 10 years of age. As indicated in Table 17.4, a person with neither the A nor the B antigen (group O) possesses both anti-A and anti-B antibodies, also called *a* and *b agglutinins* respectively. Those with group A blood have anti-B antibodies, while those with group B have anti-A antibodies. AB individuals have neither antibody.

Rh Blood Groups There are 52 named Rh agglutinogens, each of which is called an **Rh factor**. Only five of these, the C, D, E, c, and e antigens, are fairly common. The Rh blood typing system is so named because one Rh antigen (agglutinogen D) was originally identified in *rhesus* macaques. Later, the same antigen was discovered in humans.

About 85% of Americans are Rh^+ (Rh positive), meaning that their RBCs carry the D antigen. The remaining 15% of individuals do not carry the D antigen and so are Rh^- (Rh negative). As a rule, a person's ABO and Rh blood groups are reported together, for example, O^+, A^-, and so on.

Unlike the ABO system antibodies, anti-Rh antibodies do not spontaneously form in the blood of Rh^- individuals. However, if an Rh^- person receives Rh^+ blood, the immune system becomes sensitized and begins producing anti-Rh antibodies against the foreign antigen soon after the transfusion. Hemolysis does not occur after the first such transfusion because it takes time for the body to react and start making antibodies. But the second time, and every time thereafter, a typical transfusion reaction occurs in which the recipient's antibodies attack and rupture the donor RBCs.

17

HOMEOSTATIC IMBALANCE 17.3   CLINICAL

An important problem related to the Rh factor occurs in pregnant Rh⁻ women who are carrying Rh⁺ babies. The first such pregnancy usually results in the delivery of a healthy baby. But during this pregnancy and particularly during delivery, the mother may be sensitized by her baby's Rh⁺ antigens that pass into her bloodstream. If so, she will form anti-Rh antibodies unless treated with RhoGAM during pregnancy and around the time of delivery. (The same precautions are taken in women who have miscarried or aborted the fetus.) RhoGAM is a serum containing anti-Rh antibodies. By agglutinating the Rh factor, it blocks the mother's immune response and prevents her sensitization.

If the mother is not treated and becomes pregnant again with an Rh⁺ baby, her antibodies will cross through the placenta and destroy the baby's RBCs, producing a condition known as **hemolytic disease of the newborn**, or **erythroblastosis fetalis**. The baby becomes anemic and hypoxic. In severe cases, brain damage and even death may result unless transfusions are done *before* birth to provide the fetus with more erythrocytes for oxygen transport. Additionally, one or two *exchange transfusions* (see Related Clinical Terms, p. 666) are done after birth. The baby's Rh⁺ blood is removed, and Rh⁻ blood is infused. Within six weeks, the transfused Rh⁻ erythrocytes have been broken down and replaced with the baby's own Rh⁺ cells. ●●●●●●

Transfusion Reactions

When mismatched blood is infused, a **transfusion reaction** occurs in which the recipient's plasma antibodies attack the donor's red blood cells. (Note that the donor's plasma antibodies may also agglutinate the recipient's RBCs, but these antibodies are so diluted in the recipient's circulation that this does not usually present a problem.)

The initial event, agglutination of the foreign red blood cells, clogs small blood vessels throughout the body. During the next few hours, the clumped red blood cells begin to rupture or are destroyed by phagocytes, and their hemoglobin is released into the bloodstream. When the transfusion reaction is exceptionally severe, the RBCs are lysed almost immediately.

These events lead to two easily recognized problems: (1) The transfused blood cells cannot transport oxygen, and (2) the clumped red blood cells in small vessels hinder blood flow to tissues beyond those points. Less apparent, but more devastating, is the consequence of hemoglobin that escapes into the bloodstream. Circulating hemoglobin passes freely into the kidney tubules, causing cell death and kidney shutdown. If shutdown is complete (acute renal failure), the recipient may die.

Transfusion reactions can also cause fever, chills, low blood pressure, rapid heartbeat, nausea, vomiting, and general toxicity, but in the absence of kidney shutdown, these reactions are rarely lethal. Treatment of transfusion reactions focuses on preventing kidney damage by administering fluid and diuretics to increase urine output, diluting and washing out the hemoglobin.

As indicated in Table 17.4, group O red blood cells bear neither the A nor the B antigen, so theoretically group O is

the **universal donor**. Indeed, some laboratories are developing methods to enzymatically clip off the sugar molecules that specifiy A or B blood type, converting the blood to type O. Since group AB plasma is devoid of antibodies to both A and B antigens, group AB people are theoretically **universal recipients** and can receive blood transfusions from any of the ABO groups. However, these classifications are misleading, because they do not take into account the other agglutinogens in blood that can trigger transfusion reactions.

The risk of transfusion reactions and transmission of life-threatening infections (particularly with HIV) from pooled blood transfusions has increased public interest in **autologous transfusions** (*auto* = self). In autologous transfusions, the patient *predonates* his or her own blood, and it is stored and immediately available if needed during surgery.

Blood Typing

It is crucial to determine the blood group of both the donor and the recipient *before* blood is transfused. **Figure 17.17** briefly outlines the general procedure for determining ABO blood type. Because it is so critical that blood groups be compatible, cross matching is also done. *Cross matching* tests whether the recipient's serum will agglutinate the donor's RBCs or the donor's serum will agglutinate the recipient's

Blood being tested   **Serum**

Type AB (contains agglutinogens A and B; agglutinates with both sera)

Anti-A   Anti-B

RBCs

Type A (contains agglutinogen A; agglutinates with anti-A)

Type B (contains agglutinogen B; agglutinates with anti-B)

Type O (contains no agglutinogens; does not agglutinate with either serum)

Figure 17.17 Blood typing of ABO blood types. When serum containing anti-A or anti-B agglutinins is added to a blood sample diluted with saline, agglutination will occur between the agglutinin and the corresponding agglutinogen (A or B).

RBCs. Typing for Rh factors is done in the same manner as ABO blood typing.

Check Your Understanding

13. Nigel is told he has type A blood. Which ABO antibodies does he have in his plasma? Which agglutinogens are on his RBCs? Could he donate blood to an AB recipient? Could he receive blood from an AB donor? Explain.

14. DRAW Draw a red blood cell. Show and label the location of the antigens (agglutinogens) that must be present in a person with type B⁺ blood. Also show the type of antibodies (agglutinins) that would be present in the plasma.

For answers, see Answers Appendix.

17.8 Blood tests give insights into a patient's health

Learning Outcome

▶ **Explain the diagnostic importance of blood testing.**

A laboratory examination of blood yields information that can be used to evaluate a person's health. For example, in some anemias, the blood is pale and has a low hematocrit. A high fat content (*lipidemia*) gives blood plasma a yellowish hue and forecasts problems in those with heart disease. Blood glucose tests indicate how well a diabetic is controlling diet and blood sugar levels. Leukocytosis signals infections; severe infections yield larger-than-normal buffy coats in the hematocrit.

Microscopic studies of blood can reveal variations in the size and shape of erythrocytes that indicate iron deficiency or pernicious anemia. A **differential white blood cell count**, which determines the relative proportions of individual leukocyte types, is a valuable diagnostic tool. For example, a high eosinophil count may indicate a parasitic infection or an allergic response somewhere in the body.

A number of tests provide information on the status of the hemostasis system. For example, clinicians determine the **prothrombin time** to assess the ability of blood to clot, or do a **platelet count** when thrombocytopenia is suspected.

Two batteries of tests—a **CMP (comprehensive metabolic panel)** and a **complete blood count (CBC)**—are routinely ordered during physical examinations and before hospital admissions. CMP is a blood *chemistry* profile that measures various electrolytes,

glucose, and markers of liver and kidney disorders. The CBC includes counts of the different types of formed elements, the hematocrit, measurements of hemoglobin content, and size of RBCs. Together these tests provide a comprehensive picture of a person's general health in relation to normal blood values.

Appendix F lists normal values for selected blood tests.

Check Your Understanding

15. PREDICT Emily Wong, 17, is brought to the ER with a fever, headache, and stiff neck. You suspect bacterial meningitis. Would you expect to see an elevated neutrophil count in a differential WBC count? Explain.

For answers, see Answers Appendix.

Developmental Aspects of Blood

Early in fetal development, blood cells form at many sites—the fetal yolk sac, liver, and spleen, among others—but by the seventh month, the red marrow has become the primary hematopoietic area and remains so (barring serious illness) throughout life. If there is a severe need for blood cell production, however, the liver and spleen may resume their fetal blood-forming roles. Additionally, inactive yellow bone marrow regions (essentially fatty tissue) may revert to active red marrow.

Blood cells develop from collections of mesenchymal cells, called *blood islands*, derived from the mesoderm germ layer. The fetus forms a unique hemoglobin, **hemoglobin F**, that has a higher affinity for oxygen than does adult hemoglobin (hemoglobin A) so that the fetus can effectively transfer oxygen from the mother to itself. It contains two alpha and two gamma (γ) polypeptide chains per globin molecule, instead of the paired alpha and beta chains typical of hemoglobin A. After birth, the liver rapidly destroys fetal erythrocytes carrying hemoglobin F, and the baby's erythroblasts begin producing hemoglobin A.

The most common blood diseases that appear during aging are chronic leukemias, anemias, and clotting disorders. However, these and most other age-related blood disorders are usually precipitated by disorders of the heart, blood vessels, or immune system. For example, the increased incidence of leukemias in old age is believed to result from the waning efficiency of the immune system, and abnormal thrombus and embolus formation reflects atherosclerosis, which roughens the blood vessel walls.

RELATED CLINICAL TERMS

Blood chemistry tests Chemical analysis of substances in the blood, e.g., glucose, iron, calcium, protein, bilirubin, and pH.

Blood fraction Any one of the components of whole blood that has been separated out from the other blood components, such as platelets or clotting factors.

Bone marrow biopsy A sample of red bone marrow is obtained by needle aspiration (typically from the anterior or posterior iliac crest) and examined to diagnose disorders of blood-cell formation, leukemia, various marrow infections, and anemias resulting from damage to or failure of the marrow.

Exchange transfusion A technique of removing the patient's blood and infusing donor blood until a large fraction of the patient's blood has been replaced; used to treat fetal blood incompatibilities and poisoning victims.

Hematology (hem″ah-tol′o-je) Study of blood.

Hematoma (hem″ah-to′mah) Accumulated, clotted blood in the tissues usually resulting from injury; visible as "black and blue" marks or bruises; eventually absorbed naturally unless infections develop.

17

Hemochromatosis (he″mo-kro″mah-to′sis) An inherited disorder of iron overload in which the intestine absorbs too much iron from the diet. The iron builds up in body tissues, where it oxidizes to form compounds that poison those organs (especially joints, liver, and pancreas).

Myeloproliferative disorder All-inclusive term for a group of proliferative disorders (disorders in which normal cell division controls are lost) including leukoerythroblastic anemia involving fibrosis of the bone marrow, polycythemia vera, and leukemia.

Plasmapheresis (plaz″mah-fĕ-re′sis) A process in which blood is removed, its plasma is separated from formed elements, and the formed elements are returned to the patient or donor. The most important application is removal of antibodies or immune complexes from the blood of individuals with autoimmune disorders (multiple sclerosis, myasthenia gravis, and others). Also used by blood banks to collect plasma for burn victims and to obtain plasma components for therapeutic use.

Septicemia (sep″tĭ-se′me-ah; *septos* = rotten) Excessive and harmful levels of bacteria or their toxins in the blood. Also called blood poisoning.

CHAPTER SUMMARY

17.1 The functions of blood are transport, regulation, and protection (p. 643)

1. Transport functions include delivering oxygen and nutrients to body tissues, removing metabolic wastes, and transporting hormones.
2. Regulation functions include maintaining body temperature, constant blood pH, and adequate fluid volume.
3. Protective functions include hemostasis and prevention of infection.

17.2 Blood consists of plasma and formed elements (pp. 643–645)

1. Blood is composed of formed elements and plasma. The hematocrit is a measure of one formed element, erythrocytes, as a percentage of total blood volume.
2. Blood is a viscous, slightly alkaline fluid representing about 8% of total body weight. Blood volume of a normal adult is about 5 L.
3. Plasma is a straw-colored, viscous fluid and is 90% water. The remaining 10% is solutes, such as nutrients, respiratory gases, electrolytes, hormones, and proteins. Plasma makes up 55% of whole blood.
4. Plasma proteins, most made by the liver, include albumin, globulins, and fibrinogen. Albumin is the most important contributor to the osmotic pressure of blood and carries certain molecules through the blood.
5. Formed elements, accounting for 45% of whole blood, are erythrocytes, leukocytes, and platelets. All formed elements arise from hematopoietic stem cells in red bone marrow.

17.3 Erythrocytes play a crucial role in oxygen and carbon dioxide transport (pp. 645–651)

1. Erythrocytes (red blood cells, RBCs) are small, biconcave cells containing large amounts of hemoglobin. They have no nucleus and few organelles. Spectrin allows the cells to change shape as they pass through tiny capillaries.
2. Oxygen transport is the major function of erythrocytes. In the lungs, oxygen binds to iron atoms in hemoglobin molecules, producing oxyhemoglobin. In body tissues, oxygen dissociates from iron, producing deoxyhemoglobin.
3. Red blood cells begin as hematopoietic stem cells and, through erythropoiesis, proceed from the proerythroblast (committed cell) stage to the basophilic, polychromatic and orthochromatic erythroblast, and reticulocyte stages. During this process, hemoglobin accumulates and the organelles and nucleus are extruded. Differentiation of reticulocytes is completed in the bloodstream.
4. Erythropoietin and testosterone enhance erythropoiesis.
5. Iron, vitamin B_{12}, and folic acid are essential for production of hemoglobin.
6. Red blood cells have a life span of approximately 120 days. Macrophages of the spleen and liver remove old and damaged erythrocytes from the circulation. Released iron from hemoglobin is stored as ferritin or hemosiderin to be reused. The balance of the heme group is degraded to bilirubin and secreted in bile. Amino acids of globin are metabolized or recycled.
7. Erythrocyte disorders include anemia and polycythemia.

17.4 Leukocytes defend the body (pp. 651–656)

1. Leukocytes are white blood cells (WBCs). All are nucleated, and all have crucial roles in defending against disease. Two main categories exist: granulocytes and agranulocytes.
2. Granulocytes include neutrophils, eosinophils, and basophils. Neutrophils are active phagocytes. Eosinophils attack parasitic worms, and their numbers increase during allergic reactions. Basophils contain histamine, which promotes vasodilation and enhances migration of leukocytes to inflammatory sites.
3. Agranulocytes have crucial roles in immunity. They include lymphocytes—the "immune cells"—and monocytes, which differentiate into macrophages.
4. Leukopoiesis is directed by colony-stimulating factors and interleukins released by supporting cells of the red bone marrow and mature WBCs.
5. Leukocyte disorders include leukemias and infectious mononucleosis.

17.5 Platelets are cell fragments that help stop bleeding (p. 657)

1. Platelets are fragments of large megakaryocytes formed in red marrow. When a blood vessel is damaged, platelets form a plug to help prevent blood loss and play a central role in the clotting cascade.

17.6 Hemostasis prevents blood loss (pp. 657–663)

1. Hemostasis is prevention of blood loss. The three major steps of hemostasis are vascular spasm, platelet plug formation, and blood coagulation.
2. Spasms of smooth muscle in blood vessel walls and accumulation of platelets (platelet plug) at the site of vessel injury stop or slow down blood loss temporarily until coagulation occurs.
3. Coagulation of blood may be initiated by either the intrinsic or the extrinsic pathway. Platelet phospholipid membranes are crucial to both pathways. Tissue factor (factor III) exposed by tissue injury allows the extrinsic pathway to bypass many steps

17

of the intrinsic pathway. A series of activated clotting factors oversees the intermediate steps of each cascade. The pathways converge as prothrombin is converted to thrombin.

4. After a clot is formed, clot retraction occurs. Serum is squeezed out and the ruptured vessel edges are drawn together. Smooth muscle, connective tissue, and endothelial cell proliferation and migration repair the injured blood vessel.

5. When healing is complete, clot digestion (fibrinolysis) occurs.

6. Abnormal expansion of clots is prevented by removal of coagulation factors in contact with rapidly flowing blood and by inhibition of activated blood factors. Prostacyclin (PGI$_2$) and nitric oxide secreted by the endothelial cells help prevent undesirable (unnecessary) clotting.

7. Thromboembolic disorders involve undesirable clot formation, which can block vessels.

8. Thrombocytopenia, a deficit of platelets, causes spontaneous bleeding from small blood vessels. Hemophilia is caused by a genetic deficiency of certain coagulation factors. Liver disease can also cause bleeding disorders because many coagulation proteins are formed by the liver.

9. Disseminated intravascular coagulation (DIC) is a condition of bodywide clotting in undamaged blood vessels and subsequent hemorrhages.

17.7 Transfusion can replace lost blood (pp. 663–666)

1. Plasma volume can be replaced with balanced electrolyte solutions.

2. Whole blood transfusions are sometimes given to replace severe and rapid blood loss. Packed RBCs are given to replace lost O$_2$-carrying capacity.

3. Blood group is based on agglutinogens (antigens) present on red blood cell membranes.

4. When mismatched blood is transfused, the recipient's agglutinins (plasma antibodies) clump the foreign RBCs. The clumped RBCs may block blood vessels temporarily and then are lysed. Released hemoglobin may cause kidney shutdown.

5. Before whole blood can be transfused, it must be typed and cross matched to prevent transfusion reactions. The most important blood groups for which blood must be typed are the ABO and Rh groups.

17.8 Blood tests give insights into a patient's health (p. 666)

1. Diagnostic blood tests can provide valuable information about the current status of the blood and of the body as a whole.

Developmental Aspects of Blood (p. 666)

1. Fetal hematopoietic sites include the yolk sac, liver, and spleen. By the seventh month of development, the red bone marrow is the primary blood-forming site.

2. Blood cells develop from blood islands derived from mesoderm. Fetal blood contains hemoglobin F. After birth, hemoglobin A is formed.

3. The major blood-related problems associated with aging are leukemia, anemia, and thromboembolic disease.

REVIEW QUESTIONS

To access additional practice questions using your smartphone, tablet, or computer: **Mastering**A&P® > Study Area > Practice Tests & Quizzes

Level 1 Remember/Understand

(Some questions have more than one correct answer. Select the best answer or answers from the choices given.)

1. The blood volume in an adult averages approximately **(a)** 1 L, **(b)** 3 L, **(c)** 5 L, **(d)** 7 L.

2. The hormonal stimulus that prompts red blood cell formation is **(a)** serotonin, **(b)** heparin, **(c)** erythropoietin, **(d)** thrombopoietin.

3. All of the following are true of RBCs except **(a)** biconcave disc shape, **(b)** life span of approximately 120 days, **(c)** contain hemoglobin, **(d)** contain nuclei.

4. The most numerous WBC is the **(a)** eosinophil, **(b)** neutrophil, **(c)** monocyte, **(d)** lymphocyte.

5. Blood proteins play an important part in **(a)** blood clotting, **(b)** immunity, **(c)** maintenance of blood volume, **(d)** all of the above.

6. The white blood cell that releases histamine and other inflammatory chemicals is the **(a)** basophil, **(b)** neutrophil, **(c)** monocyte, **(d)** eosinophil.

7. The blood cell that can become an antibody-secreting cell is the **(a)** lymphocyte, **(b)** megakaryocyte, **(c)** neutrophil, **(d)** basophil.

8. Which of the following does not promote multiple steps in the clotting pathway? **(a)** platelet phospholipids, **(b)** factor XI, **(c)** thrombin, **(d)** Ca^{2+}.

9. The normal pH of the blood is about **(a)** 8.4, **(b)** 7.8, **(c)** 7.4, **(d)** 4.7.

10. Suppose your blood is AB positive. This means that **(a)** agglutinogens A and B are present on your red blood cells, **(b)** there are no anti-A or anti-B antibodies in your plasma, **(c)** your blood is Rh$^+$, **(d)** all of the above.

11. **(a)** List the three major categories of formed elements. **(b)** Which is least numerous? **(c)** Which comprise(s) the buffy coat in a hematocrit tube?

12. **(a)** What is a transfusion reaction and why does it happen? **(b)** What are its possible consequences?

Level 2 Apply/Analyze

13. If you had a high hematocrit, would you expect your hemoglobin determination to be low or high? Why?

14. **(a)** Describe the process of erythropoiesis. **(b)** What name is given to the immature cell type released to the circulation? **(c)** How does it differ from a mature erythrocyte?

15. Besides the ability to move by amoeboid motion, what other physiological attributes contribute to the function of white blood cells in the body?

16. **(a)** If you had a severe infection, would you expect your WBC count to be closest to 5000, 10,000, or 15,000/μl? **(b)** What is this condition called?

17. **(a)** Describe the appearance of platelets and state their major function. **(b)** Why should platelets not be called "cells"?

18. **(a)** Define fibrinolysis. **(b)** What is the importance of this process?

19. **(a)** How is clot overgrowth usually prevented? **(b)** List two conditions that may lead to unnecessary (and undesirable) clot formation.

20. How can poor nutrition lead to anemia?

21. Alan Forsythe, a middle-aged college professor from Boston, is in the Swiss Alps studying astronomy during his sabbatical leave. He has been there for two days and plans to stay the entire year. However, he notices that he is short of breath when he walks up steps and tires easily with any physical activity. His symptoms gradually disappear, and he feels fine after less than a month. Upon returning to the United States, he has a complete physical exam and is told that his erythrocyte count is higher than normal.

(a) Attempt to explain this finding. (b) Will his RBC count remain at this higher-than-normal level? Why or why not?

Level 3 Evaluate/Synthesize

22. Discuss hemoglobin relative to its chemical structure, its function, and the color changes it undergoes during loading and unloading of oxygen.

23. (a) Define hemostasis. (b) List the three major phases of coagulation. Explain what initiates each phase and what the phase accomplishes. (c) In what general way do the intrinsic and extrinsic mechanisms of clotting differ? (d) Which ion is essential to virtually all stages of coagulation?

24. Cancer patients being treated with chemotherapeutic drugs designed to destroy rapidly dividing cells are monitored closely for changes in their red and white blood counts. Why so?

25. Amanda Healy, a young woman with severe vaginal bleeding, is admitted to the emergency room. She is three months pregnant, and the physician is concerned about the volume of blood she is losing. (a) What type of transfusion will probably be given to this patient? (b) Which blood tests will be performed before starting the transfusion?

26. A reticulocyte count indicated that 5% of Tyler's red blood cells were reticulocytes. His blood test also indicated he had polycythemia and a hematocrit of 65%. Explain the connection between these three facts.

27. Mr. Chu has been scheduled for surgery to have his arthritic hip replaced. His surgeon tells him he must switch from aspirin to acetaminophen for pain control before his surgery. Why?

CLINICAL CASE STUDY

20-Year-Old-Male with Abdominal Trauma

Richard Watkins, 20, fell off his roof while replacing some shingles. Upon arrival at the scene, paramedics make the following observations:

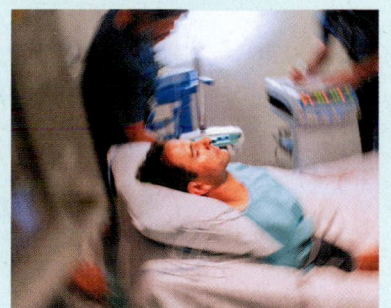

- Abdominal pain, right upper quadrant
- Cyanotic, cool & clammy skin
- Falling blood pressure, 100/60 mm Hg; rapid pulse, 100 beats/min

Paramedics start an IV to rapidly infuse a 0.9% sodium chloride solution (normal saline). They transport him to a small rural hospital where Mr. Watkins's blood pressure continues to fall and his cyanosis worsens. The local physician begins infusing O negative packed red blood cells (PRBCs) and arranges transport by helicopter to a trauma center. She sends additional PRBC units in the helicopter for transfusion en route. After arrival at the trauma center, the following notes were added to Mr. Watkins's chart:

- Abdomen firm and distended
- Blood drawn for typing and cross matching; packed A positive blood cells infused
- Emergency FAST (focused assessment with sonography for trauma) ultrasound is positive for intraperitoneal fluid

A positive FAST scan indicates intra-abdominal bleeding. Mr. Watkins's condition continues to deteriorate, so he is prepared for surgery. The operation reveals a lacerated liver, which is repaired. Mr. Watkins's vital signs stabilize.

1. Mr. Watkins was going into shock because of blood loss, so paramedics infused a saline solution. Why would this help?

2. **+ NCLEX-STYLE** Why was Mr. Watkins given PRBCs in addition to normal saline solution? What problem does the infusion of PRBCs address that the saline solution could not?
   a. PRBCs help restore blood ion composition to normal.
   b. PRBCs increase blood osmotic pressure, making the infused fluid stay in the vessels rather than move into the cells.
   c. PRBCs enhance clotting, sealing off the injured vessels.
   d. PRBCs contain hemoglobin, so they are able to carry oxygen to the tissues.

3. Why was the physician able to use O negative blood before the results of the blood type tests were obtained?

4. **+ NCLEX-STYLE** Mr. Watkins's blood type was determined to be A positive. Which of the following types of blood can he receive?
   a. A positive, A negative, O positive, and O negative
   b. A positive, O positive, and AB positive
   c. A positive but not A negative
   d. Only A positive and O negative

5. What would happen if doctors had infused type B PRBCs into Mr. Watkins's circulation?

17

18 The Cardiovascular System: The Heart

In this chapter, you will learn that

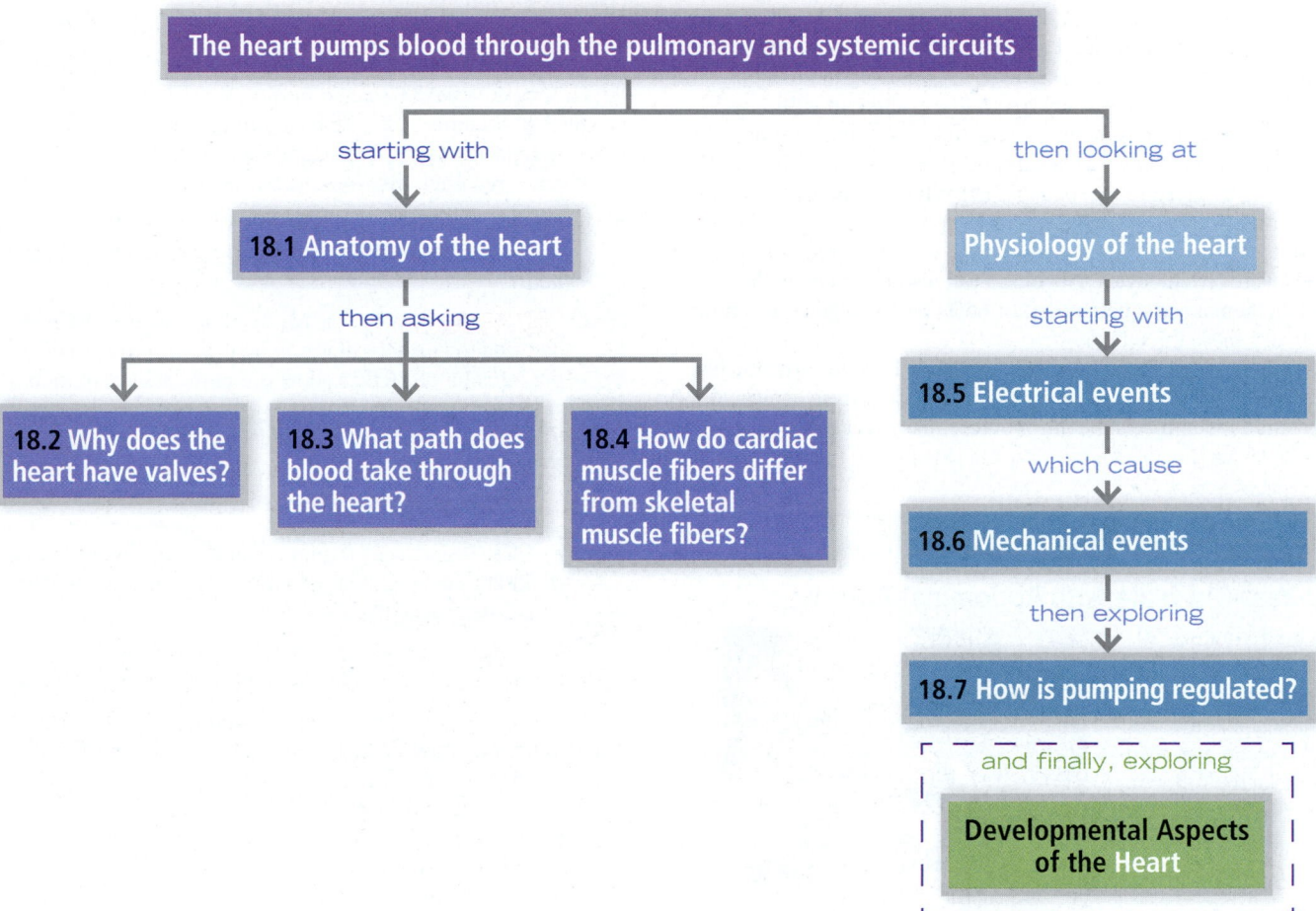

The heart pumps blood through the pulmonary and systemic circuits

starting with

18.1 Anatomy of the heart

then asking

18.2 Why does the heart have valves?

18.3 What path does blood take through the heart?

18.4 How do cardiac muscle fibers differ from skeletal muscle fibers?

then looking at

Physiology of the heart

starting with

18.5 Electrical events

which cause

18.6 Mechanical events

then exploring

18.7 How is pumping regulated?

and finally, exploring

Developmental Aspects of the Heart

Our ceaselessly beating heart has intrigued people for centuries. The ancient Greeks believed the heart was the seat of intelligence. Others thought it was the source of emotions. While these ideas have proved false, we do know that emotions affect heart rate. When your heart pounds or skips a beat, you become acutely aware of how much you depend on this dynamic organ for your very life.

Despite its vital importance, the heart does not work alone. Indeed, it is only part of the cardiovascular system, which includes the miles of blood vessels that run through your body. Day and night, tissue cells take in nutrients and oxygen and excrete wastes. Cells can make such exchanges only with their immediate environment, so some means of changing and renewing that environment is necessary to ensure a continual supply of nutrients and prevent a buildup of wastes. The cardiovascular system provides the transport system "hardware" that keeps blood continuously circulating to fulfill this critical homeostatic need.

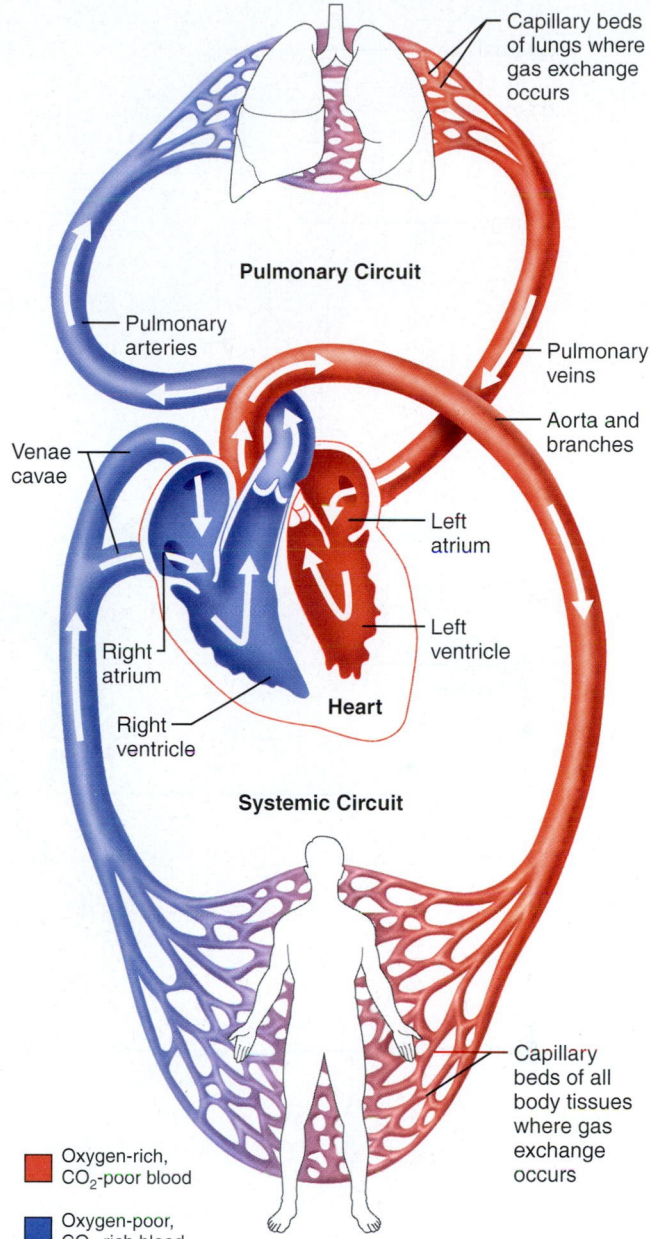

Figure 18.1 **The systemic and pulmonary circuits.** The right side of the heart pumps blood through the pulmonary circuit and the left side of the heart pumps blood through the systemic circuit. The arrows indicate the direction of blood flow. For simplicity, the actual number of two pulmonary arteries and four pulmonary veins has been reduced to one each.

18.1 The heart has four chambers and pumps blood through the pulmonary and systemic circuits

Learning Outcomes

▶ Describe the size, shape, location, and orientation of the heart in the thorax.

▶ Name the coverings of the heart.

▶ Describe the structure and function of each of the three layers of the heart wall.

▶ Describe the structure and functions of the four heart chambers. Name each chamber and provide the name and general route of its associated great vessel(s).

The Pulmonary and Systemic Circuits

Stripped of its romantic cloak, the **heart** is no more than the transport system pump, and the blood vessels are the delivery routes. In fact, the heart is actually two pumps side by side (**Figure 18.1**).

- The *right side* of the heart receives oxygen-poor blood from body tissues and then pumps this blood to the lungs to pick up oxygen and dispel carbon dioxide. The blood vessels that carry blood to and from the lungs form the **pulmonary circuit** (*pulmo* = lung).

- The *left side* of the heart receives the oxygenated blood returning from the lungs and pumps this blood throughout the body to supply oxygen and nutrients to body tissues. The blood vessels that carry blood to and from all body tissues form the **systemic circuit**.

The heart has two receiving chambers:

- The *right atrium* receives blood returning from the systemic circuit.

- The *left atrium* receives blood returning from the pulmonary circuit.

The heart also has two main pumping chambers:

- The *right ventricle* pumps blood into the pulmonary circuit.

- The *left ventricle* pumps blood into the systemic circuit.

(a)

(b)

Posterior

(c)

Figure 18.2 Location of the heart in the mediastinum. (a) Relationship of the heart to the sternum, ribs, and diaphragm in a person who is lying down (the heart is slightly inferior to this position in a standing person). **(b)** Inferior view of a cross section showing the heart's relative position in the thorax. **(c)** Relationship of the heart and great vessels to the lungs.

Size, Location, and Orientation of the Heart

The modest size and weight of the heart belie its incredible strength and endurance. About the size of a fist, the hollow, cone-shaped heart has a mass of 250 to 350 grams—less than a pound (**Figure 18.2**).

Snugly enclosed within the **mediastinum** (me″de-ah-sti′num), the medial cavity of the thorax, the heart extends obliquely for 12 to 14 cm (about 5 inches) from the second rib to the fifth intercostal space (Figure 18.2a). As it rests on the superior surface of the diaphragm, the heart lies anterior to the vertebral column and posterior to the sternum. Approximately two-thirds of its mass lies to the left of the midsternal line; the balance projects to the right. The lungs flank the heart laterally and partially obscure it (Figure 18.2b, c).

The broad, flat **base**, or posterior surface, of the heart is about 9 cm (3.5 in) wide and directed toward the right shoulder. Its **apex** points inferiorly toward the left hip. If you press your fingers between the fifth and sixth ribs just below the left nipple, you can easily feel the **apical impulse** caused by your beating heart's apex where it touches the chest wall.

18

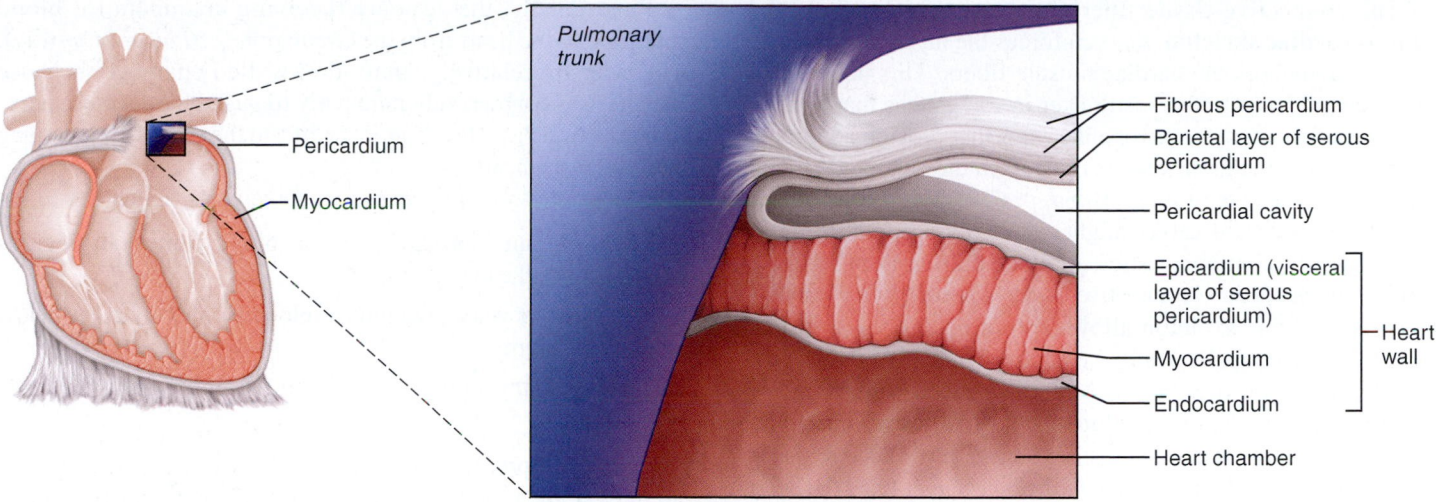

Figure 18.3 The layers of the pericardium and of the heart wall.

Coverings of the Heart

The heart is enclosed in a double-walled sac called the **pericardium** (per″ĭ-kar′de-um; *peri* = around, *cardi* = heart) (**Figure 18.3**). The loosely fitting superficial part of this sac is the **fibrous pericardium**. This tough, dense connective tissue layer (1) protects the heart, (2) anchors it to surrounding structures, and (3) prevents overfilling of the heart with blood.

Deep to the fibrous pericardium is the **serous pericardium**, a thin, slippery, two-layer serous membrane that forms a closed sac around the heart (◄ **Figure 1.11, p. 19**). Its **parietal layer** lines the internal surface of the fibrous pericardium. At the superior margin of the heart, the parietal layer attaches to the large arteries exiting the heart, and then turns inferiorly and continues over the external heart surface as the **visceral layer**, also called the *epicardium* ("upon the heart"), which is an integral part of the heart wall.

Between the parietal and visceral layers is the slitlike **pericardial cavity**, which contains a film of serous fluid. The serous membranes, lubricated by the fluid, glide smoothly past one another, allowing the mobile heart to work in a relatively friction-free environment.

HOMEOSTATIC IMBALANCE 18.1

CLINICAL

Pericarditis, inflammation of the pericardium, roughens the serous membrane surfaces. Consequently, as the beating heart rubs against its pericardial sac, it creates a creaking sound (*pericardial friction rub*) that can be heard with a stethoscope. Pericarditis is characterized by pain deep to the sternum. Over time, it may lead to adhesions in which the visceral and parietal pericardia stick together and impede heart activity.

In severe cases, large amounts of inflammatory fluid seep into the pericardial cavity. This excess fluid compresses the heart and limits its ability to pump blood, a condition called *cardiac tamponade* (tam″pŏ-nād′), literally, "heart plug." Physicians treat cardiac tamponade by inserting a syringe into the pericardial cavity and draining off the excess fluid. ●●●●●●●●

Layers of the Heart Wall

The heart wall, richly supplied with blood vessels, is composed of three layers: the epicardium, myocardium, and endocardium (Figure 18.3).

As we have noted, the superficial **epicardium** is the visceral layer of the serous pericardium. It is often infiltrated with fat, especially in older people.

The middle layer, the **myocardium** ("muscle heart"), is composed mainly of cardiac muscle and forms the bulk of the heart. This is the layer that contracts. In the myocardium, the branching cardiac muscle cells are tethered to one another by crisscrossing connective tissue fibers and arranged in spiral or circular *bundles* (**Figure 18.4**). These interlacing bundles effectively link all parts of the heart together.

Figure 18.4 The circular and spiral arrangement of cardiac muscle bundles in the myocardium of the heart.

18

The connective tissue fibers form a dense network, the fibrous **cardiac skeleton**, that reinforces the myocardium internally and anchors the cardiac muscle fibers. This network of collagen and elastic fibers is thicker in some areas than others. For example, it constructs ropelike rings that provide additional support where the great vessels issue from the heart and around the heart valves (see Figure 18.6a, p. 678). Without this support, the vessels and valves might eventually become stretched because of the continuous stress of blood pulsing through them. Additionally, because connective tissue is not electrically excitable, the cardiac skeleton allows action potentials to spread only via specific pathways in the heart.

The third layer of the heart wall, the **endocardium** ("inside the heart"), is a glistening white sheet of endothelium (squamous epithelium; ◀ p. 121) resting on a thin connective tissue layer. Located on the inner myocardial surface, it lines the heart chambers and covers the fibrous skeleton of the valves. The endocardium is continuous with the endothelial linings of the blood vessels leaving and entering the heart.

Chambers and Associated Great Vessels

The heart has four chambers (**Figure 18.5e** on p. 677)—two superior **atria** (a′tre-ah) and two inferior **ventricles** (ven′trĭ-klz). The internal partition that divides the heart longitudinally is called the **interatrial septum** where it separates the atria, and the **interventricular septum** where it separates the ventricles. The right ventricle forms most of the anterior surface of the heart. The left ventricle dominates the inferoposterior aspect of the heart and forms the heart apex.

Two grooves visible on the heart surface indicate the boundaries of its four chambers and carry the blood vessels supplying the myocardium. The **coronary sulcus** (Figure 18.5b, d), or *atrioventricular groove*, encircles the junction of the atria and ventricles like a crown (*corona* = crown). The **anterior interventricular sulcus**, cradling the anterior interventricular artery, marks the anterior position of the septum separating the right and left ventricles. It continues as the **posterior interventricular sulcus**, which provides a similar landmark on the heart's posteroinferior surface.

Atria: The Receiving Chambers

Except for small, wrinkled, protruding appendages called **auricles** (or′ĭ-klz; *auricle* = little ear), which increase the atrial volume somewhat, the right and left atria are remarkably free of distinguishing surface features. Internally, the right atrium has two basic parts (Figure 18.5c): a smooth-walled posterior part and an anterior portion in which bundles of muscle tissue form ridges in the walls. These muscle bundles are called **pectinate muscles** because they look like the teeth of a comb (*pectin* = comb). The posterior and anterior regions of the right atrium are separated by a C-shaped ridge called the *crista terminalis* ("terminal crest").

In contrast, the left atrium is mostly smooth and pectinate muscles are found only in the auricle. The interatrial septum bears a shallow depression, the **fossa ovalis** (o-vă′lis) (Figure 18.5c, e), that marks the spot where an opening, the *foramen ovale*, existed in the fetal heart (see p. 700).

Functionally, the atria are receiving chambers for blood returning to the heart from the circulation (*atrium* = entryway). The atria are relatively small, thin-walled chambers because they need to contract only minimally to push blood "downstairs" into the ventricles. They contribute little to the propulsive pumping activity of the heart.

Blood enters the *right atrium* via three veins (Figure 18.5c–e):

- The **superior vena cava** returns blood from body regions superior to the diaphragm.
- The **inferior vena cava** returns blood from body areas below the diaphragm.
- The **coronary sinus** collects blood draining from the myocardium.

Four **pulmonary veins** enter the *left atrium*, which makes up most of the heart's base. These veins, which transport blood from the lungs back to the heart, are best seen in a posterior view (Figure 18.5d).

Ventricles: The Discharging Chambers

Together the ventricles (*ventr* = underside) make up most of the volume of the heart. As already mentioned, the right ventricle forms most of the heart's anterior surface and the left ventricle dominates its posteroinferior surface. Irregular ridges of muscle called **trabeculae carneae** (trah-bek′u-le kar′ne-e; "crossbars of flesh") mark the internal walls of the ventricular chambers. Other muscle bundles, the **papillary muscles**, which play a role in valve function, project into the ventricular cavity (Figure 18.5e).

The ventricles are the discharging chambers, the actual pumps of the heart. Their walls are much more massive than the atrial walls, reflecting the difference in function between the atria and ventricles (Figure 18.5e and f). When the ventricles contract, they propel blood out of the heart into the circulation. The right ventricle pumps blood into the **pulmonary trunk**, which routes the blood to the lungs where gas exchange occurs. The left ventricle ejects blood into the **aorta** (a-or′tah), the largest artery in the body.

Check Your Understanding

1. The heart is in the mediastinum. Just what is the mediastinum?
2. From inside to outside, list the layers of the heart wall and the coverings of the heart.
3. What is the purpose of the serous fluid inside the pericardial cavity?
4. In the frontal section of the heart at right, name the structures indicated by a–g. Which heart chamber receives oxygen-poor blood from veins? Which chamber pumps blood into the systemic circulation? Which blood vessel transports blood toward the lungs?

For answers, see Answers Appendix.

(Text continues on p. 679.)

Left subclavian artery

Left common carotid artery

Brachiocephalic trunk

Ascending aorta

Right atrium

Right coronary artery
(in coronary sulcus)

Right ventricle

Aortic arch

Ligamentum arteriosum

Pulmonary trunk

Auricle of left atrium

Anterior interventricular artery
(in anterior interventricular
sulcus)

Additional branch off left
coronary artery, normal
variation

Left ventricle

Apex of heart
(left ventricle)

(a) Anterior view (pericardium removed)

Brachiocephalic trunk

Superior vena cava

Right pulmonary artery

Ascending aorta

Pulmonary trunk

Right pulmonary veins

Right atrium

Right coronary artery
(in coronary sulcus)

Anterior cardiac vein

Right ventricle

Right marginal artery

Small cardiac vein

Inferior vena cava

Left common carotid
artery

Left subclavian artery

Aortic arch

Ligamentum arteriosum

Left pulmonary artery

Left pulmonary veins

Auricle of
left atrium

Circumflex artery

Left coronary artery
(in coronary sulcus)

Left ventricle

Great cardiac vein

Anterior interventricular
artery (in anterior
interventricular sulcus)

Apex

(b) Anterior view

Figure 18.5 Gross anatomy of the heart. In diagrammatic views, vessels transporting
oxygen-rich blood are red; those transporting oxygen-poor blood are blue.

18

Superior vena cava

Pectinate muscles

Crista terminalis

Interatrial septum

Fossa ovalis

Opening of inferior vena cava (cut)

Opening of coronary sinus

Interventricular sulcus

Ascending aorta

Myocardium of left ventricle

(c) Internal aspect of the right atrium, anterior view

Aorta

Left pulmonary artery

Left pulmonary veins

Auricle of left atrium

Left atrium

Great cardiac vein

Posterior vein of left ventricle

Left ventricle

Apex

Superior vena cava

Right pulmonary artery

Right pulmonary veins

Right atrium

Inferior vena cava

Coronary sinus

Right coronary artery (in coronary sulcus)

Posterior interventricular artery (in posterior interventricular sulcus)

Middle cardiac vein

Right ventricle

(d) Posterior view

Figure 18.5 (continued) **Gross anatomy of the heart.** In (c), the anterior wall of the atrium has been opened and folded superiorly.

18

Aorta

Left pulmonary artery

Superior vena cava

Right pulmonary artery

Left atrium

Pulmonary trunk

Left pulmonary veins

Right atrium

Mitral (bicuspid) valve

Right pulmonary veins

Fossa ovalis

Aortic valve

Pectinate muscles

Pulmonary valve

Tricuspid valve

Left ventricle

Right ventricle

Papillary muscle

Chordae tendineae

Interventricular septum

Trabeculae carneae

Epicardium

Inferior vena cava

Myocardium

Endocardium

(e) Frontal section

Left subclavian artery

Left common carotid artery

Brachiocephalic trunk

Aortic arch

Ligamentum arteriosum

Superior vena cava

Pulmonary trunk

Pulmonary valve

Right auricle

Pulmonary vein

Left auricle

Tricuspid valve

Chordae tendineae
of mitral valve

Chordae tendineae

Papillary muscle

Papillary muscle

Myocardium of left ventricle

Myocardium of right ventricle

Trabeculae carneae

Interventricular septum

(f) Internal aspect of ventricles; dissection of view similar to (e)

Figure 18.5 *(continued)*

18

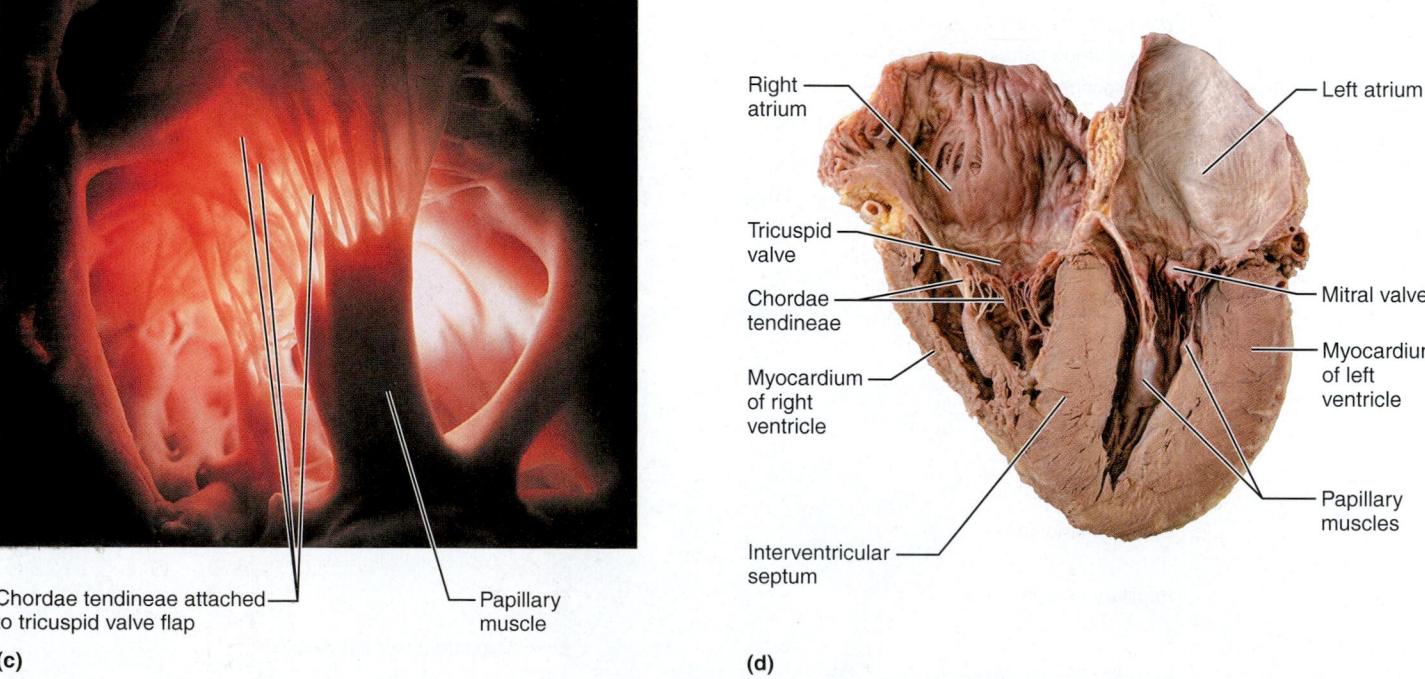

Figure 18.6 **Heart valves. (a)** Superior view of the two sets of heart valves (atria removed). The paired atrioventricular valves are located between atria and ventricles; the two semilunar valves are located at the junction of the ventricles and the arteries issuing from them. **(b)** Photograph of the heart valves, superior view. **(c)** Photograph of the tricuspid valve. This bottom-to-top view shows the valve as seen from the right ventricle. **(d)** Frontal section of the heart.

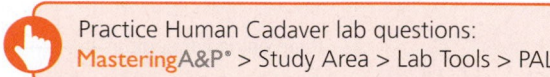

Practice Human Cadaver lab questions:
MasteringA&P® > Study Area > Lab Tools > PAL

18.2 Heart valves make blood flow in one direction

Learning Outcome

▶ Name the heart valves and describe their location, function, and mechanism of operation.

Blood flows through the heart in one direction: from atria to ventricles and out the great arteries leaving the superior aspect of the heart. Four valves enforce this one-way traffic (Figure 18.5e and **Figure 18.6**). They open and close in response to differences in blood pressure on their two sides.

Atrioventricular (AV) Valves

The two **atrioventricular (AV) valves**, one located at each atrial-ventricular junction, prevent backflow into the atria when the ventricles contract.

- The right AV valve, the **tricuspid valve** (tri-kus′pid), has three flexible cusps (flaps of endocardium reinforced by connective tissue cores).

- The left AV valve, with two cusps, is called the **mitral valve** (mi′tral) because it resembles the two-sided bishop's miter (tall, pointed hat). It is sometimes called the *bicuspid valve*.

Attached to each AV valve flap are tiny white collagen cords called **chordae tendineae** (kor′de ten″dĭ′ne-e; "tendinous cords"), "heart strings" which anchor the cusps to the papillary muscles protruding from the ventricular walls (Figure 18.6c, d).

When the heart is completely relaxed, the AV valve flaps hang limply into the ventricular chambers below. During this time, blood flows into the atria and then through the open AV valves into the ventricles (**Figure 18.7a**). When the ventricles contract, compressing the blood in their chambers, the intraventricular pressure rises, forcing the blood superiorly against the valve flaps. As a result, the flap edges meet, closing the valve (Figure 18.7b).

The chordae tendineae and the papillary muscles act as tethers that anchor the valve flaps in their *closed* position. If the cusps were not anchored, they would be blown upward (everted) into the atria, in the same way an umbrella is blown inside out by a gusty wind. The papillary muscles contract with the other ventricular musculature so that they take up the slack on the chordae tendineae as the full force of ventricular contraction hurls the blood against the AV valve flaps.

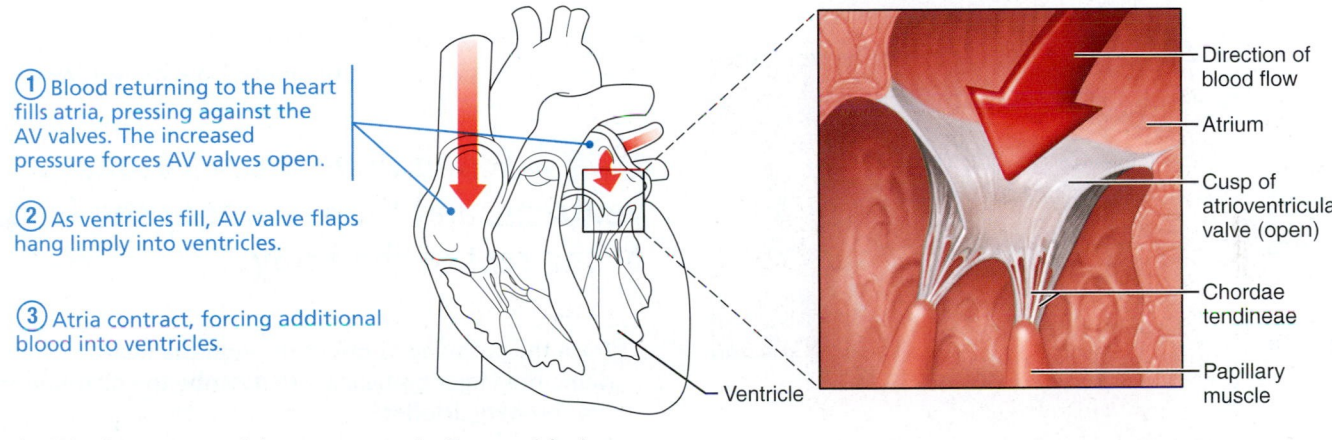

① Blood returning to the heart fills atria, pressing against the AV valves. The increased pressure forces AV valves open.

② As ventricles fill, AV valve flaps hang limply into ventricles.

③ Atria contract, forcing additional blood into ventricles.

Direction of blood flow

Atrium

Cusp of atrioventricular valve (open)

Chordae tendineae

Papillary muscle

Ventricle

(a) AV valves open; atrial pressure greater than ventricular pressure

① Ventricles contract, forcing blood against AV valve cusps.

② AV valves close.

③ Papillary muscles contract and chordae tendineae tighten, preventing valve flaps from everting into atria.

Atrium

Cusps of atrioventricular valve (closed)

Blood in ventricle

(b) AV valves closed; atrial pressure less than ventricular pressure

Figure 18.7 The function of the atrioventricular (AV) valves.

Semilunar (SL) Valves

The **aortic** and **pulmonary (semilunar, SL) valves** guard the bases of the large arteries issuing from the ventricles (aorta and pulmonary trunk, respectively) and prevent backflow into the associated ventricles. Each SL valve is fashioned from three pocketlike cusps, each shaped roughly like a crescent moon (*semilunar* = half-moon).

Like the AV valves, the SL valves open and close in response to differences in pressure. When the ventricles contract and intraventricular pressure rises above the pressure in the aorta and pulmonary trunk, the SL valves are forced open and their cusps flatten against the arterial walls as blood rushes past them (**Figure 18.8a**). When the ventricles relax, and the blood flows backward toward the heart, it fills the cusps and closes the valves (Figure 18.8b).

We complete the valve story by noting what seems to be an important omission—there are no valves guarding the entrances of the venae cavae and pulmonary veins into the right and left atria, respectively. Small amounts of blood *do* spurt back into these vessels during atrial contraction, but backflow is minimal because of the inertia of the blood and because as it contracts, the atrial myocardium compresses (and collapses) these venous entry points.

As ventricles contract and intraventricular pressure rises, blood is pushed up against semilunar valves, forcing them open.

(a) Semilunar valves open

As ventricles relax and intraventricular pressure falls, blood flows back from arteries, filling the cusps of semilunar valves and forcing them to close.

(b) Semilunar valves closed

Figure 18.8 The function of the semilunar (SL) valves.

Heart valves are simple devices, and the heart—like any mechanical pump—can function with "leaky" valves as long as the impairment is not too great. However, severe valve deformities can seriously hamper cardiac function.

An *incompetent*, or *insufficient*, *valve* forces the heart to repump the same blood over and over because the valve does not close properly and blood backflows. In valvular *stenosis* ("narrowing"), the valve flaps become stiff and constrict the opening. [Stenosis is typically due to calcium salt deposits or scar tissue that forms following endocarditis (see Related Clinical Terms, p. 702).] The constricted opening compels the heart to contract more forcibly than normal. Both types of valve problems increase the heart's workload and may weaken the heart severely over time. The mitral and aortic valves are most often affected.

The faulty valve can be replaced with a mechanical valve, a pig or cow heart valve chemically treated to prevent rejection, or cryopreserved valves from human cadavers. Heart valves tissue-engineered from a patient's own cells grown on a biodegradable scaffold are being developed. ●●●●●●●

Check Your Understanding

5. What is the function of the papillary muscles and chordae tendineae?
6. Name the valve that has just two cusps.

For answers, see Answers Appendix.

18.3 Blood flows from atrium to ventricle, and then to either the lungs or the rest of the body

Learning Outcomes
▶ Trace the pathway of blood through the heart.
▶ Name the major branches and describe the distribution of the coronary arteries.

Having covered the basic anatomy of the heart, we can now follow the path that blood takes through the heart and its associated circuits. *Focus on Blood Flow through the Heart* (**Focus Figure 18.1**) follows a single "spurt" of blood as it passes through all four chambers of the heart and both blood circuits in its ever-repeating journey.

As you work your way through this figure, keep in mind that the left side of the heart is the *systemic circuit pump* and the right side is the *pulmonary circuit pump*. Notice how unique the pulmonary circuit is. Elsewhere in the body, veins carry relatively oxygen-poor blood to the heart, and arteries transport oxygen-rich blood from the heart. The opposite oxygenation conditions exist in veins and arteries of the pulmonary circuit.

Equal volumes of blood are pumped to the pulmonary and systemic circuits at any moment, but the two ventricles have very unequal workloads. The pulmonary circuit, served by the right ventricle, is a short, low-pressure circulation. In contrast, the systemic circuit, associated with the left ventricle, takes a

The heart is a double pump, each side supplying its own circuit.

- ■ Oxygen-poor blood
- ■ Oxygen-rich blood

Superior vena cava (SVC) / Inferior vena cava (IVC) / Coronary sinus → Right atrium → (Tricuspid valve) → Right ventricle → (Pulmonary semilunar valve) → Pulmonary trunk

SVC

Coronary sinus

Right atrium

IVC

Tricuspid valve

Right ventricle

Pulmonary arteries

Pulmonary trunk

Pulmonary semilunar valve

To heart

Oxygen-poor blood returns from the body tissues back to the heart.

Oxygen-poor blood is carried in two pulmonary arteries to the lungs (*pulmonary circuit*) to be oxygenated.

To lungs

Systemic capillaries

Pulmonary capillaries

To body

Oxygen-rich blood is delivered to the body tissues (*systemic circuit*).

Oxygen-rich blood returns to the heart via the four pulmonary veins.

To heart

Aorta

Aortic semilunar valve

Mitral valve

Left ventricle

Pulmonary veins

Left atrium

Aorta ← (Aortic semilunar valve) ← Left ventricle ← (Mitral valve) ← Left atrium ← Four pulmonary veins

Right ventricle:
- Thinner wall than left ventricle
- Crescent shape
- Wraps around left ventricle

Left ventricle:
- Thicker wall than right ventricle
- Round shape

Interventricular septum

Figure 18.9 Anatomical differences between the right and left ventricles.

long pathway through the entire body and encounters about five times as much friction, or resistance to blood flow.

This functional difference is revealed in the anatomy of the two ventricles (Figure 18.5e and **Figure 18.9**):

- The walls of the left ventricle are three times thicker than those of the right ventricle, and its cavity is nearly circular.

- The right ventricular cavity is flattened into a crescent shape that partially encloses the left ventricle, much the way a hand might loosely grasp a clenched fist.

Consequently, the left ventricle can generate much more pressure than the right and is a far more powerful pump.

Coronary Circulation

Although the heart is continuously filled with various amounts of blood, this blood provides little nourishment to heart tissue. (The myocardium is too thick to make diffusion a practical means of delivering nutrients.) How, then, does the heart get nourishment? It does so through the **coronary circulation**, the functional blood supply of the heart and the shortest circulation in the body.

Coronary Arteries

The *left* and *right coronary arteries* both arise from the base of the aorta and encircle the heart in the coronary sulcus. They provide the arterial supply of the coronary circulation (**Figure 18.10a**).

The **left coronary artery** runs toward the left side of the heart and then divides into two major branches:

- The **anterior interventricular artery** (also known clinically as the *left anterior descending artery*) follows the anterior

interventricular sulcus and supplies blood to the interventricular septum and anterior walls of both ventricles.

- The **circumflex artery** supplies the left atrium and the posterior walls of the left ventricle.

The **right coronary artery** courses to the right side of the heart, where it also gives rise to two branches:

- The **right marginal artery** serves the myocardium of the lateral right side of the heart.

- The **posterior interventricular artery** runs to the heart apex and supplies the posterior ventricular walls. Near the apex of the heart, this artery merges (anastomoses) with the anterior interventricular artery.

Together the branches of the right coronary artery supply the right atrium and nearly all the right ventricle.

The arterial supply of the heart varies considerably. For example, in 15% of people, the left coronary artery gives rise to *both* interventricular arteries. In about 4% of people, a single coronary artery supplies the whole heart. Additionally, there may be both right and left marginal arteries. There are many anastomoses (junctions) among the coronary arterial branches. These fusing networks provide additional (*collateral*) routes for blood delivery to the heart muscle, but are not robust enough to supply adequate nutrition when a coronary artery is suddenly occluded (blocked). Complete blockage leads to tissue death and heart attack.

The coronary arteries provide an intermittent, pulsating blood flow to the myocardium. These vessels and their main branches lie in the epicardium and send branches inward to nourish the myocardium. They deliver blood when the heart is relaxed, but are fairly ineffective when the ventricles are contracting because they are compressed by the contracting

18

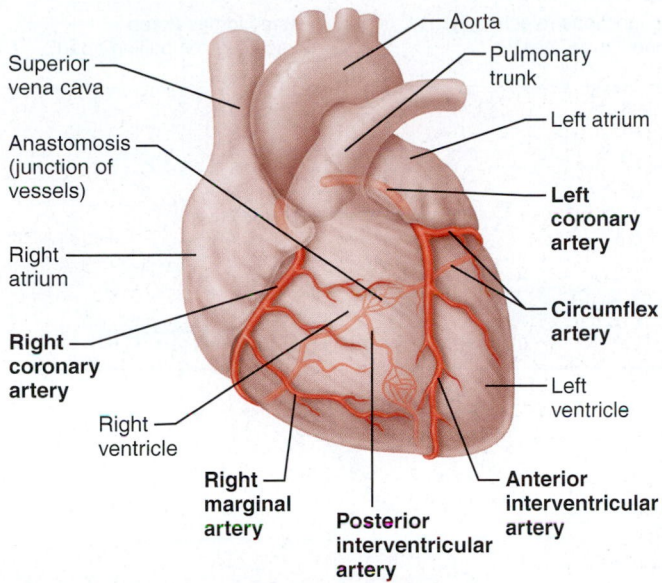

(a) The major coronary arteries.

Labels: Superior vena cava; Anastomosis (junction of vessels); Right atrium; **Right coronary artery**; Right ventricle; **Right marginal artery**; **Posterior interventricular artery**; Aorta; Pulmonary trunk; Left atrium; **Left coronary artery**; **Circumflex artery**; Left ventricle; **Anterior interventricular artery**

(b) The major cardiac veins.

Labels: Superior vena cava; **Anterior cardiac veins**; Small cardiac vein; Middle cardiac vein; Great cardiac vein; Coronary sinus

Figure 18.10 Coronary circulation. In both drawings, lighter-tinted vessels are more posterior in the heart.

myocardium. Although the heart represents only about 1/200 of the body's weight, it requires about 1/20 of the body's blood supply. As might be expected, the left ventricle receives the most plentiful blood supply.

Coronary Veins

After passing through the capillary beds of the myocardium, the venous blood is collected by the **cardiac veins**, whose paths roughly follow those of the coronary arteries. These veins join

to form an enlarged vessel called the **coronary sinus**, which empties the blood into the right atrium. The coronary sinus is obvious on the posterior aspect of the heart (Figure 18.10b).

The sinus has three large tributaries: the **great cardiac vein** in the anterior interventricular sulcus; the **middle cardiac vein** in the posterior interventricular sulcus; and the **small cardiac vein**, running along the heart's right inferior margin. Additionally, several **anterior cardiac veins** empty directly into the right atrium anteriorly.

HOMEOSTATIC IMBALANCE 18.3   **CLINICAL**

Blockage of the coronary arterial circulation can be serious and sometimes fatal. **Angina pectoris** (an-ji′nah pek′tor-is; "choked chest") is thoracic pain caused by a fleeting deficiency in blood delivery to the myocardium. It may result from stress-induced spasms of the coronary arteries or from increased physical demands on the heart. The myocardial cells are weakened by the temporary lack of oxygen but do not die.

Prolonged coronary blockage is far more serious because it can lead to a **myocardial infarction (MI)**, commonly called a **heart attack**, in which cells *do* die. Since adult cardiac muscle is essentially amitotic, most of the dead tissue is replaced with noncontractile scar tissue. Whether or not a person survives a myocardial infarction depends on the extent and location of the damage. Damage to the left ventricle—the systemic pump—is most serious.

Check Your Understanding

7. Which side of the heart acts as the pulmonary pump? The systemic pump?

8. Which of the following statements are true? (a) The left ventricle wall is thicker than the right ventricle wall. (b) The left ventricle pumps blood at a higher pressure than the right ventricle. (c) The left ventricle pumps more blood with each beat than the right ventricle. Explain.

9. Name the two main branches of the right coronary artery.

For answers, see Answers Appendix.

18.4 Intercalated discs connect cardiac muscle fibers into a functional syncytium

Learning Outcomes

▶ Describe the structural and functional properties of cardiac muscle, and explain how it differs from skeletal muscle.

▶ Briefly describe the events of excitation-contraction coupling in cardiac muscle cells.

Although similar to skeletal muscle, cardiac muscle displays some special anatomical features that reflect its unique blood-pumping role.

18

(a)

(b)

Cardiac muscle cell   Nucleus   Intercalated discs

Gap junctions (electrically connect myocytes)

Desmosomes (keep myocytes from pulling apart)

Cardiac muscle cell
Intercalated disc

Mitochondrion   Nucleus

Mitochondrion
T tubule
Sarcoplasmic reticulum
Z disc
Nucleus
Sarcolemma

I band   A band   I band

Figure 18.11 Microscopic anatomy of cardiac muscle. (a) Photomicrograph of cardiac muscle (290×). Notice that the cardiac muscle cells are short, branched, and striated. The dark-staining areas are intercalated discs, or junctions, between adjacent cells. **(b)** Components of intercalated discs and cardiac muscle fibers.

Microscopic Anatomy

Like skeletal muscle, **cardiac muscle** is striated and contracts by the sliding filament mechanism. However, in contrast to the long, cylindrical, multinucleate skeletal muscle fibers, cardiac cells are short, fat, branched, and interconnected. Each *cardiac myocyte* (*myo* = muscle, *cyte* = cell) contains one or at most two large, pale, *centrally* located nuclei (**Figure 18.11a**). The intercellular spaces are filled with a loose connective tissue matrix (the *endomysium*) containing numerous capillaries. This delicate matrix is connected to the fibrous cardiac skeleton, which acts both as a tendon and as an insertion, giving the cardiac cells something to pull or exert their force against.

Skeletal muscle fibers are independent of one another both structurally and functionally. In contrast, the plasma membranes of adjacent cardiac cells interlock like the ribs of two sheets of corrugated cardboard at dark-staining junctions called **intercalated discs** (in-ter′kah-la″ted; *intercala* = insert) (Figure 18.11). Intercalated discs contain anchoring *desmosomes* and *gap junctions* (cell junctions discussed in Chapter 3, ◄ pp. 67–68). The desmosomes prevent adjacent cells from separating during contraction, and the gap junctions allow ions to pass from cell to cell, transmitting current across the entire heart. Because gap junctions electrically couple cardiac cells, the myocardium *behaves* as a single coordinated unit, or **functional syncytium** (sin-sit′e-um; *syn* = together, *cyt* = cell).

Large mitochondria account for 25–35% of the volume of cardiac cells (compared with only 2% in skeletal muscle), a characteristic that makes cardiac cells highly resistant to fatigue. Most of the remaining volume is occupied by myofibrils composed of fairly typical sarcomeres. The sarcomeres have Z discs, A bands, and I bands that reflect the arrangement of the thick (myosin) and thin (actin) filaments composing them. However, in contrast to skeletal muscle, the myofibrils of cardiac muscle cells vary greatly in diameter and branch extensively, accommodating the abundant mitochondria between them. This difference produces a banding pattern less dramatic than that seen in skeletal muscle.

The system for delivering Ca^{2+} is less elaborate in cardiac muscle cells. The cardiac sarcoplasmic reticulum is simpler and lacks the large terminal cisterns seen in skeletal muscle. As a result, cardiac muscle fibers do not have *triads*. In addition, the T tubules are wider and fewer than in skeletal muscle. They enter the cells once per sarcomere at the Z discs. (Recall that T tubules are invaginations of the sarcolemma. In skeletal muscle, the T tubules invaginate twice per sarcomere, at the A band–I band junctions; ◀ p. 288.)

How Does the Physiology of Skeletal and Cardiac Muscle Differ?

Both skeletal muscle and cardiac muscle are contractile tissues, and in both types of muscle the contraction is preceded by a depolarization in the form of an action potential (AP). Transmission of the depolarization wave down the T tubules (ultimately) causes the sarcoplasmic reticulum (SR) to release Ca^{2+} into the sarcoplasm. Excitation-contraction coupling occurs as Ca^{2+} provides the signal (via troponin binding) for cross bridge activation. This sequence of events couples the depolarization wave to the sliding of the myofilaments in both skeletal and cardiac muscle cells. However, cardiac muscle fibers differ from skeletal muscle fibers as summarized in **Table 18.1** and ◀ **Table 9.3 (pp. 312–313)**, and as described below.

- **Some cardiac muscle cells are self-excitable.** The heart contains two kinds of myocytes. Almost all of the myocytes are *contractile cardiac muscle cells*, responsible for the heart's pumping activity. However, certain locations in the heart contain special noncontractile cells, called *pacemaker cells*, that spontaneously depolarize. This property is called **automaticity** or **autorhythmicity**. Because heart cells are electrically joined together by gap junctions, these cells can initiate not only their own depolarization, but also that of the rest of the heart. No neural input is required. As demonstrated by transplanted hearts, you can cut all of the nerves to the heart and it still beats. In contrast, each skeletal muscle fiber must be stimulated by a nerve ending to contract, and cutting the nerves results in paralysis.

- **The heart contracts as a unit.** As we just learned, gap junctions tie cardiac muscle cells together to form a functional syncytium. This allows the wave of depolarization to travel from cell to cell across the heart. As a result, either all fibers in the heart contract as a unit or the heart doesn't contract at all. In skeletal muscle, on the other hand, impulses do not spread from cell to cell. Only skeletal muscle fibers that are individually stimulated by nerve fibers contract, and the strength of the contraction increases as more motor units are recruited. Such recruitment cannot happen in the heart because it acts as a single huge motor unit. Contraction of all of the cardiac myocytes ensures effective pumping by the heart—a half-hearted contraction would just not do.

- **The influx of Ca^{2+} from extracellular fluid triggers Ca^{2+} release from the SR.** Recall that in skeletal muscle, the wave of depolarization directly causes release from the SR of all the Ca^{2+} required for contraction. In cardiac muscle, depolarization opens special Ca^{2+} channels in the plasma membrane. These *slow Ca^{2+} channels* allow entry of 10–20% of the Ca^{2+} needed for contraction. Once inside, this influx of Ca^{2+} triggers Ca^{2+}-sensitive channels in the SR to release bursts of Ca^{2+} ("calcium sparks") that account for the other 80–90% of the Ca^{2+} needed for contraction.

- **Tetanic contractions cannot occur in cardiac muscles.** The absolute refractory period is the period during an action potential when another action potential cannot be triggered. In skeletal muscle, the absolute refractory period is much shorter than the contraction, allowing multiple contractions to summate (tetanic contractions). If the heart were to contract tetanically, it would be unable to relax and fill, and so would be useless as a pump. The absolute refractory period in the heart is nearly as long as the contraction itself, which prevents tetanic contractions. We will examine the mechanism

Table 18.1	Key Differences between Skeletal and Cardiac Muscle	
	SKELETAL MUSCLE	**CARDIAC MUSCLE**
Structure	Striated, long, cylindrical, multinucleate	Striated, short, branched, one or two nuclei per cell
Gap junctions between cells	No	Yes
Contracts as a unit	No, motor units must be stimulated individually	Yes, gap junctions create a functional syncytium
T tubules	Abundant	Fewer, wider
Sarcoplasmic reticulum	Elaborate; has terminal cisterns	Less elaborate; no terminal cisterns
Source of Ca^{2+} for contraction	Sarcoplasmic reticulum only	Sarcoplasmic reticulum and extracellular fluid
Ca^{2+} binds to troponin	Yes	Yes
Pacemaker cells present	No	Yes
Tetanus possible	Yes	No
Supply of ATP	Aerobic and anaerobic (fewer mitochondria)	Aerobic only (more mitochondria)

underlying the long refractory period of contractile cardiac myocytes shortly.

- **The heart relies almost exclusively on aerobic respiration.** Cardiac muscle has more mitochondria than skeletal muscle does, reflecting its greater dependence on oxygen for its energy metabolism. The heart relies almost exclusively on aerobic respiration, so cardiac muscle cannot operate effectively for long without oxygen. This is in contrast to skeletal muscle, which can contract for prolonged periods by carrying out anaerobic respiration, and then restore its reserves of oxygen and fuel using excess postexercise oxygen consumption (EPOC).

  Both types of muscle tissue use multiple fuel molecules, including glucose and fatty acids. But cardiac muscle is much more adaptable and readily switches metabolic pathways to use whatever nutrients are available, including lactic acid generated by skeletal muscle activity. Consequently, the real danger of an inadequate blood supply to the myocardium is not lack of nutrients, but lack of oxygen.

Check Your Understanding

10. For each of the following, state whether it applies to skeletal muscle, cardiac muscle, or both: (a) refractory period is almost as long as the contraction; (b) source of Ca^{2+} for contraction is only SR; (c) has troponin; (d) has triads.

11. **MAKE CONNECTIONS** Identify the type of muscle represented by structures a and b below. How can you tell?

(a) (b)

For answers, see Answers Appendix.

18.5 Pacemaker cells trigger action potentials throughout the heart

Learning Outcomes

▶ Describe and compare action potentials in cardiac pacemaker and contractile cells.

▶ Name the components of the conduction system of the heart, and trace the conduction pathway.

▶ Draw a diagram of a normal electrocardiogram tracing. Name the individual waves and intervals, and indicate what each represents. Name some abnormalities that can be detected on an ECG tracing.

Although the ability of the heart to depolarize and contract is intrinsic (no nerves required), the healthy heart *is* supplied with autonomic nerve fibers that alter its basic rhythm. In this module, we examine how the basic rhythm is generated and modified.

Setting the Basic Rhythm: The Intrinsic Conduction System

The independent, but coordinated, activity of the heart is a function of (1) the presence of gap junctions, and (2) the activity of the heart's "in-house" conduction system. The **intrinsic**

cardiac conduction system consists of noncontractile cardiac cells specialized to initiate and distribute impulses throughout the heart, so that it depolarizes and contracts in an orderly, sequential manner. Let's look at how this system works.

Action Potential Initiation by Pacemaker Cells

Unstimulated contractile cells of the heart (and neurons and skeletal muscle fibers) maintain a stable resting membrane potential. However, about 1% of cardiac fibers are autorhythmic ("self-rhythmic") **cardiac pacemaker cells**, having the special ability to depolarize spontaneously and so pace the heart. Pacemaker cells are a part of the intrinsic conduction system. They have an *unstable resting potential* that continuously depolarizes, drifting slowly toward threshold. These spontaneously changing membrane potentials, called **pacemaker potentials** or **prepotentials**, initiate the action potentials that spread throughout the heart to trigger its rhythmic contractions. Let's look at the three parts of an action potential in typical pacemaker cells as shown in **Figure 18.12**.

① **Pacemaker potential.** The pacemaker potential is due to the special properties of the ion channels in the sarcolemma. In these cells, hyperpolarization at the end of an action potential both closes K^+ channels and opens slow Na^+ channels. The Na^+ influx alters the balance between K^+ loss and Na^+ entry, and the membrane interior becomes less and less negative (more positive).

② **Depolarization.** Ultimately, at threshold (approximately -40 mV), **Ca^{2+} channels** open, allowing explosive entry of Ca^{2+} from the extracellular space. As a result, in pacemaker cells, it is the influx of Ca^{2+} (rather than Na^+) that produces the rising phase of the action potential and reverses the membrane potential.

③ **Repolarization.** Ca^{2+} channels inactivate. As in other excitable cells, the falling phase of the action potential and repolarization reflect opening of K^+ channels and K^+ efflux from the cell.

Once repolarization is complete, K^+ channels close, K^+ efflux declines, and the slow depolarization to threshold begins again.

Sequence of Excitation

Typical cardiac pacemaker cells are found in the *sinoatrial* (si″no-a′tre-al) and *atrioventricular nodes* (**Figure 18.13**). In addition, cells of the atrioventricular bundle, right and left bundle branches, and subendocardial conducting network (Purkinje fibers) can sometimes act as pacemakers. Impulses pass across the heart in order from ① to ⑤ following the yellow pathway in Figure 18.13a.

> ⊙ Complete an interactive tutorial: MasteringA&P® > Study Area > Interactive Physiology

① **Sinoatrial (SA) node.** The crescent-shaped **sinoatrial node** is located in the right atrial wall, just inferior to the entrance of the superior vena cava. A minute cell mass with a mammoth job, the SA node typically generates impulses about 75 times every minute. The SA node sets the pace for the

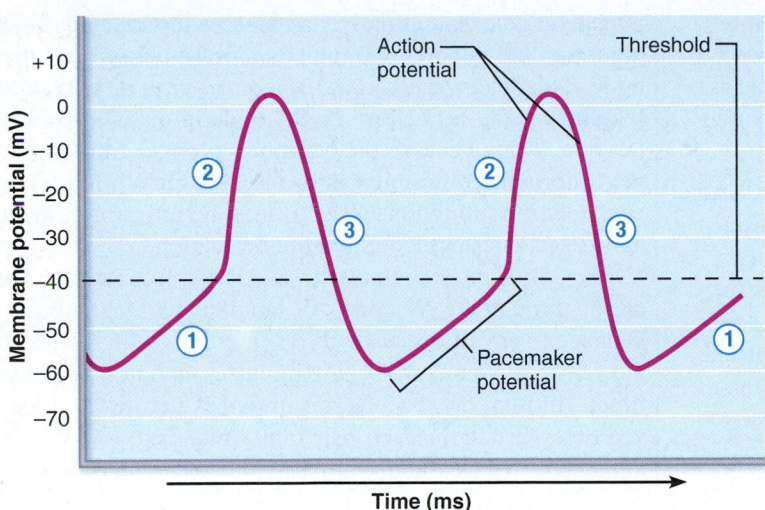

① **Pacemaker potential** This slow depolarization is due to both opening of Na⁺ channels and closing of K⁺ channels. Notice that the membrane potential is never a flat line.

② **Depolarization** The action potential begins when the pacemaker potential reaches threshold. Depolarization is due to Ca²⁺ influx through Ca²⁺ channels.

③ **Repolarization** is due to Ca²⁺ channels inactivating and K⁺ channels opening. This allows K⁺ efflux, which brings the membrane potential back to its most negative voltage.

Figure 18.12 Pacemaker and action potentials of typical cardiac pacemaker cells.

heart as a whole because no other region of the conduction system or the myocardium has a faster depolarization rate. For this reason, it is the heart's **pacemaker**, and its characteristic rhythm, called **sinus rhythm**, determines heart rate.

② **Atrioventricular (AV) node.** From the SA node, the depolarization wave spreads via gap junctions throughout the atria and via the *internodal pathway* to the **atrioventricular node**, located in the inferior portion of the interatrial septum immediately above the tricuspid valve. At the AV node, the impulse is delayed for about 0.1 second, allowing the atria to respond and complete their contraction before the ventricles contract. This delay reflects the smaller diameter of the fibers here and the fact that they have fewer gap junctions for current flow. Consequently, the AV node conducts impulses more slowly than other parts of the system, just as traffic slows when cars are forced to merge from four lanes into two. Once through the AV node, the signaling impulse passes rapidly through the rest of the system.

③ **Atrioventricular (AV) bundle.** From the AV node, the impulse sweeps to the **atrioventricular bundle** (also called the *bundle of His*) in the superior part of the interventricular septum. Although the atria and ventricles are adjacent to each other, they are *not* connected by gap junctions. The AV bundle is the *only* electrical connection between them. The fibrous cardiac skeleton is nonconducting and insulates the rest of the AV junction.

④ **Right and left bundle branches.** The short AV bundle splits into two pathways—the **right** and **left bundle branches**, which

course along the interventricular septum toward the heart apex.

⑤ **Subendocardial conducting network.** Essentially long strands of barrel-shaped cells with few myofibrils, the **subendocardial conducting network**, also called *Purkinje fibers* (pur-kin′je), completes the pathway through the interventricular septum, penetrates into the heart apex, and then turns superiorly into the ventricular walls. The bundle branches excite the cells of the interventricular septum, but the bulk of ventricular depolarization depends on the large

① The **sinoatrial (SA) node** (pacemaker) generates impulses.

② The impulses pause (0.1 s) at the **atrioventricular (AV) node.**

③ The **atrioventricular (AV) bundle** connects the atria to the ventricles.

④ The **bundle branches** conduct the impulses through the interventricular septum.

⑤ The **subendocardial conducting network** depolarizes the contractile cells of both ventricles.

Figure 18.13 The intrinsic cardiac conduction system.

fibers of the conducting network and, ultimately, on cell-to-cell transmission of the impulse via gap junctions between the ventricular muscle cells. Because the left ventricle is much larger than the right, the subendocardial conducting network is more elaborate in that side of the heart.

The total time between initiation of an impulse by the SA node and depolarization of the last of the ventricular muscle cells is approximately 0.22 s (220 ms) in a healthy human heart.

Ventricular contraction almost immediately follows the ventricular depolarization wave. The wringing motion of contraction begins at the heart apex and moves toward the atria, following the direction of the excitation wave through the ventricle walls. This contraction ejects some of the contained blood *superiorly* into the large arteries leaving the ventricles.

The various cardiac pacemaker cells have different rates of spontaneous depolarization. The SA node normally drives the heart at a rate of 75 beats per minute. Without SA node input, the AV node would depolarize only about 50 times per minute. Without input from the AV node, the atypical pacemakers of the AV bundle and the subendocardial conducting network would depolarize only about 30 times per minute. Note that these slower pacemakers cannot dominate the heart unless faster pacemakers stop functioning.

The cardiac conduction system coordinates and synchronizes heart activity. Without it, impulses would travel much more slowly. This slower rate would allow some muscle fibers to contract long before others, reducing pump effectiveness.

HOMEOSTATIC IMBALANCE 18.4   CLINICAL

Defects in the intrinsic conduction system can cause irregular heart rhythms, or **arrhythmias** (ah-rith′me-ahz). They may also cause uncoordinated atrial and ventricular contractions, or even **fibrillation**, a condition of rapid and irregular or out-of-sync contractions in which control of heart rhythm by the SA node is disrupted by rapid activity in other heart regions. The heart in fibrillation has been compared with a squirming bag of worms. Fibrillating ventricles are useless as pumps; and unless the heart is defibrillated quickly, circulation stops and brain death occurs.

Defibrillation is accomplished by electrically shocking the heart, which interrupts its chaotic twitching by depolarizing the entire myocardium. The hope is that "with the slate wiped clean" the SA node will begin to function normally and sinus rhythm will be reestablished. Implantable cardioverter defibrillators (ICDs) can continually monitor heart rhythms and slow an abnormally fast heart rate or emit an electrical shock if the heart begins to fibrillate.

A defective SA node may have several consequences. An **ectopic focus** (ek-top′ik) (an abnormal pacemaker) may appear and take over the pacing of heart rate, or the AV node may become the pacemaker. The pace set by the AV node (**junctional rhythm**) is 40 to 60 beats per minute, slower than sinus rhythm but still adequate to maintain circulation.

Occasionally, ectopic pacemakers appear even when the SA node is operating normally. A small region of the heart becomes hyperexcitable, sometimes as a result of too much caffeine or nicotine, and generates impulses more quickly than the SA node. This leads to a *premature contraction* or **extrasystole** (ek″strah-sis′to-le) before the SA node initiates the next contraction. Then, because the heart has a longer time to fill, the next (normal) contraction is felt as a thud. While premature *ventricular* contractions (PVCs) may be unsettling, most are harmless. Persistent PVCs require evaluation.

The only route for impulse transmission from atria to ventricles is through the AV node, AV bundle, and bundle branches. Damage to any of these structures interferes with the ability of the ventricles to receive pacing impulses, and may cause **heart block**. In total heart block, no impulses get through and the ventricles beat at their intrinsic rate, which is too slow to maintain adequate circulation. In partial heart block, only some of

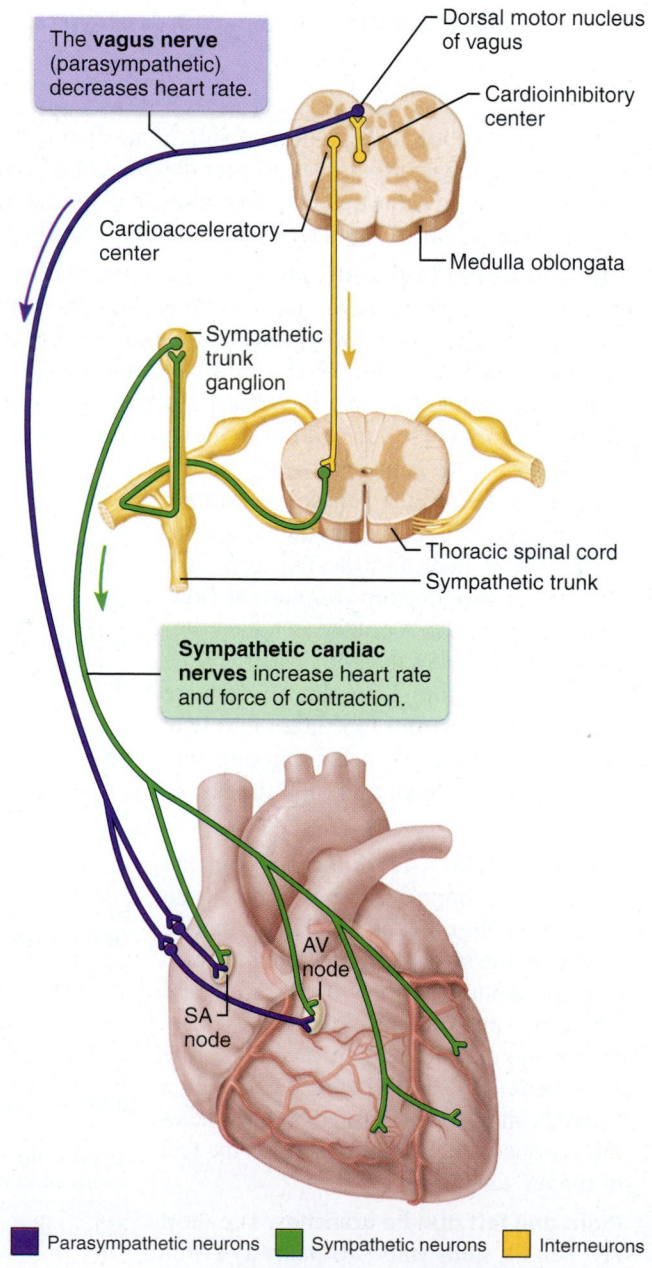

Figure 18.14 Autonomic innervation of the heart.

Labels in figure:
The **vagus nerve** (parasympathetic) decreases heart rate.
Dorsal motor nucleus of vagus
Cardioinhibitory center
Cardioacceleratory center
Medulla oblongata
Sympathetic trunk ganglion
Thoracic spinal cord
Sympathetic trunk
Sympathetic cardiac nerves increase heart rate and force of contraction.
SA node
AV node

■ Parasympathetic neurons   ■ Sympathetic neurons   ■ Interneurons the atrial impulses reach the ventricles. In both cases, artificial pacemakers are implanted to recouple the atria to the ventricles as necessary. These programmable devices speed up in response to increased physical activity just as a normal heart would, and many can send diagnostic information to the patient's doctor. ●●●●●●

Modifying the Basic Rhythm: Extrinsic Innervation of the Heart

Although the intrinsic conduction system sets the basic heart rate, fibers of the autonomic nervous system modify the march-like beat and introduce a subtle variability from one beat to the next. The sympathetic nervous system (the "accelerator") increases both the rate and the force of the heartbeat. Parasympathetic activation (the "brakes") slows the heart. We explain these neural controls later—here we discuss the anatomy of the nerve supply to the heart.

The cardiac centers are located in the medulla oblongata (◄ p. 454). The **cardioacceleratory center** projects to sympathetic neurons in the T_1–T_5 level of the spinal cord. These preganglionic neurons, in turn, synapse with postganglionic neurons in the cervical and upper thoracic sympathetic trunk (**Figure 18.14**). From there, postganglionic fibers run through the cardiac plexus to the heart where they innervate the SA and AV nodes, heart muscle, and coronary arteries.

The **cardioinhibitory center** sends impulses to the parasympathetic dorsal vagus nucleus in the medulla, which in turn sends inhibitory impulses to the heart via branches of the vagus nerves. Most parasympathetic postganglionic motor neurons lie in ganglia in the heart wall and their fibers project most heavily to the SA and AV nodes.

Action Potentials of Contractile Cardiac Muscle Cells

The bulk of heart muscle is composed of *contractile muscle fibers* responsible for the heart's pumping activity. As we have seen, the sequence of events leading to contraction of these cells is similar to that in skeletal muscle fibers. However, the action potential has a characteristic "hump" or *plateau* as shown in **Figure 18.15**.

(1) Depolarization opens a few **fast voltage-gated Na⁺ channels** in the sarcolemma, allowing extracellular Na⁺ to enter. This influx initiates a positive feedback cycle that causes the rising phase of the action potential (and reversal of the membrane potential from −90 mV to nearly +30 mV). The period of Na⁺ influx is very brief, because the sodium channels quickly inactivate and the Na⁺ influx stops.

(2) When Na⁺-dependent membrane depolarization occurs, the voltage change also opens channels that allow Ca²⁺ to enter from the extracellular fluid. These channels are called **slow Ca²⁺ channels** because their opening is delayed a bit. The Ca²⁺ surge across the sarcolemma prolongs the depolarization, producing a **plateau** in the action potential tracing. Not many voltage-gated K⁺ channels are open yet, so the plateau is prolonged. As long as Ca²⁺ is entering, the cells continue to contract. Notice in Figure 18.15 that muscle tension develops during the plateau, and peaks just after the plateau ends.

(3) After about 200 ms, the slope of the action potential tracing falls rapidly. This repolarization results from inactivation of Ca²⁺ channels and opening of voltage-gated K⁺ channels. The rapid loss of potassium from the cell through K⁺

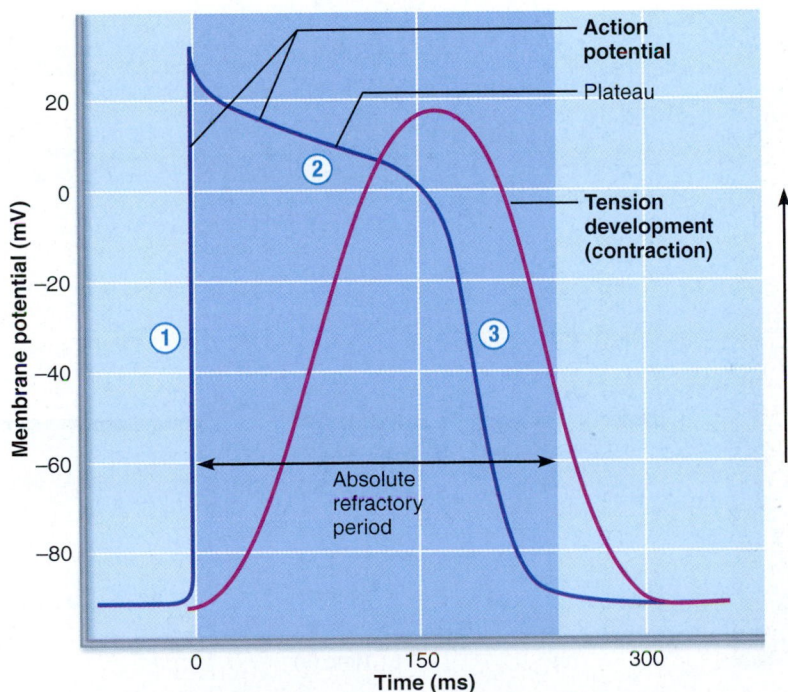

① **Depolarization** is due to Na⁺ influx through fast voltage-gated Na⁺ channels. A positive feedback cycle rapidly opens many Na⁺ channels, reversing the membrane potential. Channel inactivation ends this phase.

② **Plateau phase** is due to Ca²⁺ influx through slow Ca²⁺ channels. This keeps the cell depolarized because most K⁺ channels are closed.

③ **Repolarization** is due to Ca²⁺ channels inactivating and K⁺ channels opening. This allows K⁺ efflux, which brings the membrane potential back to its resting voltage.

Figure 18.15 The action potential of contractile cardiac muscle cells. Relationship between the action potential, period of contraction, and absolute refractory period in a single ventricular cell.

channels restores the resting membrane potential. During repolarization, Ca²⁺ is pumped back into the SR and the extracellular space.

Notice that the action potential and contractile phase lasts much longer in cardiac muscle than in skeletal muscle. In skeletal muscle, the action potential typically lasts 1–2 ms and the contraction (for a single stimulus) 15–100 ms (◀ **p. 299**). In cardiac muscle, the action potential lasts 200 ms or more (because of the plateau), and tension development persists for 200 ms or more. This long plateau in cardiac muscle has two consequences:

- It ensures that the contraction is sustained so that blood is ejected efficiently from the heart.
- It ensures that there is a long absolute refractory period, so that tetanic contractions cannot occur and the heart can fill again for the next beat.

Electrocardiography

The electrical currents generated in and transmitted through the heart spread throughout the body and can be detected with a device called an **electrocardiograph**. An **electrocardiogram (ECG)** is a graphic record of heart activity. An ECG is a composite of all the action potentials generated by nodal and contractile cells at a given time (**Figure 18.16b**)—*not*, as sometimes assumed, a tracing of a single action potential.

To record an ECG, recording electrodes are placed at various sites on the body surface. In a typical 12-lead ECG, three electrodes form bipolar leads that measure the voltage

(a) **Skin electrodes are used to record an ECG.** (The round antenna near the patient's left shoulder reads his pacemaker data.)

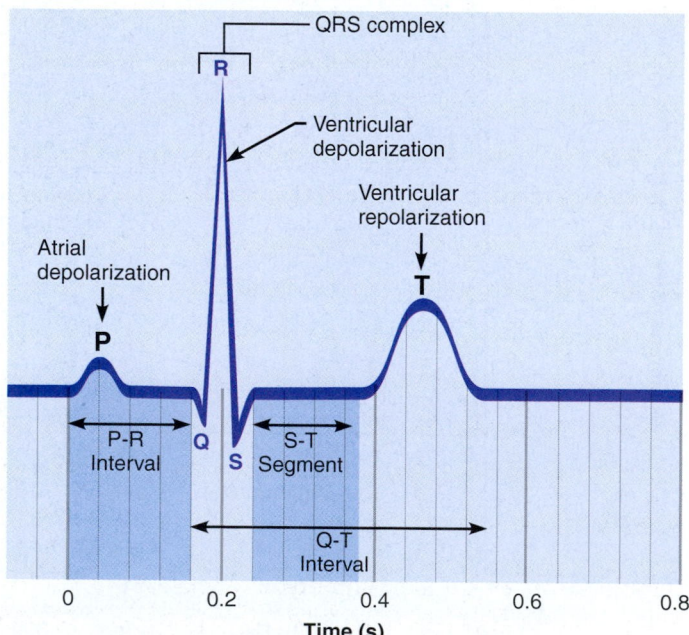

(b) **An ECG tracing.** The labels identify the three normally recognizable waves and the important intervals.

Figure 18.16 The electrocardiogram (ECG).

difference either between the arms or between an arm and a leg, and nine form unipolar leads (Figure 18.16a). Together the 12 leads provide a comprehensive picture of the heart's electrical activity.

A typical ECG has three almost immediately distinguishable waves or *deflections*: the P wave, the QRS complex, and the T wave (Figure 18.16b).

The P Wave

The first, the small **P wave**, lasts about 0.08 s and results from movement of the depolarization wave from the SA node through the atria. Approximately 0.1 s after the P wave begins, the atria contract.

The QRS Complex

The large **QRS complex** results from ventricular depolarization and precedes ventricular contraction. It has a complicated shape because the paths of the depolarization waves through the ventricular walls change continuously, producing corresponding changes in current direction. Additionally, the time required for each ventricle to depolarize depends on its size relative to the other ventricle. Average duration of the QRS complex is 0.08 s.

The T Wave

The **T wave**, caused by ventricular repolarization, typically lasts about 0.16 s. Repolarization is slower than depolarization, so the T wave is more spread out and has a lower amplitude (height) than the QRS complex. Because atrial repolarization takes place during the period of ventricular excitation, the wave representing atrial repolarization is normally obscured by the large QRS complex being recorded at the same time.

The P-R Interval

The **P-R interval** is the time (about 0.16 s) from the beginning of atrial excitation to the beginning of ventricular excitation. If the Q wave is visible (which is often not the case), it marks the beginning of ventricular excitation, and for this reason this interval is sometimes called the **P-Q interval**. The P-R interval includes atrial depolarization (and contraction) as well as the passage of the depolarization wave through the rest of the conduction system.

The S-T Segment

During the **S-T segment** of the ECG, when the action potentials of the ventricular myocytes are in their plateau phases, the entire ventricular myocardium is depolarized. The **Q-T interval**, lasting about 0.38 s, is the period from the beginning of ventricular depolarization through ventricular repolarization.

Depolarization and Repolarization in the Heart

Figure 18.17 relates the parts of an ECG to the sequence of depolarization and repolarization in the heart. Note that these electrical events always precede the mechanical events of contraction and relaxation that they trigger.

1. Atrial depolarization, initiated by the SA node, causes the P wave.

2. With atrial depolarization complete, the impulse is delayed at the AV node.

3. Ventricular depolarization begins at apex, causing the QRS complex. Atrial repolarization occurs.

4. Ventricular depolarization is complete.

5. Ventricular repolarization begins at apex, causing the T wave.

6. Ventricular repolarization is complete.

☐ Depolarization   ☐ Repolarization

Figure 18.17 The sequence of depolarization and repolarization of the heart related to the ECG waves.

Figure 18.18 Normal and abnormal ECG tracings.

HOMEOSTATIC IMBALANCE 18.5   **CLINICAL**

Clinicians look for variations in the pattern or timing of the ECG, which tend to be consistent in a healthy heart. Changes alert them to damage or disease, as well as problems with the conduction system (**Figure 18.18**). For example, an enlarged R wave hints of enlarged ventricles, an S-T segment that is elevated or depressed indicates cardiac ischemia (insufficient blood flow), and a prolonged Q-T interval reveals a repolarization abnormality that increases the risk of ventricular arrhythmias. •

Check Your Understanding

12. Cardiac muscle cannot go into tetany. Why?

13. Describe the electrical event in the heart that occurs during each of the following: (a) the QRS complex of the ECG; (b) the T wave of the ECG; (c) the P-R interval of the ECG.

14. MAKE CONNECTIONS In skeletal muscle, every muscle fiber must be innervated so it can contract. In cardiac muscle, the intrinsic conduction system does not connect to every cardiac myocyte. Why is this not necessary? Which part of the intrinsic conduction system directly excites ventricular myocardial cells? In which direction does the depolarization wave travel across the ventricles?

15. MAKE CONNECTIONS Refer to graphs a–c of action potentials in the next column. Two of these occur in the heart, and one occurs in skeletal muscle (as you learned in Chapter 9).

Which one comes from a contractile cardiac muscle cell? A skeletal muscle cell? A cardiac pacemaker cell? For each one, state which ion is responsible for the depolarization phase and which ion is responsible for the repolarization phase.

For answers, see Answers Appendix.

18.6 The cardiac cycle describes the mechanical events associated with blood flow through the heart

Learning Outcomes

▶ Describe the timing and events of the cardiac cycle.
▶ Describe normal heart sounds, and explain how heart murmurs differ.

The heart undergoes some dramatic writhing movements as it alternately contracts, forcing blood out of its chambers, and then relaxes, allowing its chambers to refill with blood. The term **systole** (sis'to-le) refers to these periods of contraction, and **diastole** (di-as'to-le) refers to those of relaxation. The **cardiac cycle** includes *all* events associated with the blood flow through the heart during one complete heartbeat—atrial systole and diastole followed by ventricular systole and diastole.

The graphical representation of the cardiac cycle is quite complex and can be daunting to students who see it for the first time. *Focus on the Cardiac Cycle* (**Focus Figure 18.2**, pp. 694–695) will lead you step by step through the cardiac cycle so that you can understand the logical relationships between each of the parts of this cycle. After you work your way through this figure, come back and read the summary of the events that occur during each phase of the cardiac cycle (below).

⟳ Complete an interactive tutorial: MasteringA&P® > Study Area > Interactive Physiology

Phases of the Cardiac Cycle

The cardiac cycle is marked by a succession of pressure and blood volume changes in the heart. Because blood circulates continuously, we must choose an arbitrary starting point for one turn of the cardiac cycle. We begin with the heart in total relaxation: Atria and ventricles are quiet, and it is mid-to-late diastole (bottom right of Focus Figure 18.2 on p. 695).

① **Ventricular filling: mid-to-late diastole.** Pressure in the heart is low, blood returning from the circulation is flowing passively through the atria and the open AV valves into the ventricles, and the aortic and pulmonary valves are closed. More than 80% of ventricular filling occurs during this period. (The remaining 20% of the blood is delivered to the ventricles when the atria contract toward the end of this phase.)

Now the stage is set for atrial systole. Following depolarization (P wave of ECG), the atria contract, compressing the blood in their chambers. This causes a sudden slight rise in atrial pressure, which propels residual blood out of the atria into the ventricles. At this point the ventricles are in the last part of their diastole and have the maximum volume of blood they will contain in the cycle, an amount called the *end diastolic volume* (*EDV*). Then the atria relax and the ventricles depolarize (QRS complex). Atrial diastole persists through the rest of the cycle.

② **Isovolumetric contraction.** As the atria relax, the ventricles begin contracting. Their walls close in on the blood in their chambers, and ventricular pressure rises rapidly and sharply, closing the AV valves. The **isovolumetric contraction phase** (i″so-vol″u-met″rik) is the split-second period when the ventricles are completely closed chambers and the blood volume in the chambers remains constant as the ventricles contract.

As ventricular pressure continues to rise, it finally exceeds the pressure in the large arteries issuing from the ventricles. The isovolumetric stage ends as the SL valves are forced open.

③ **Ventricular ejection.** During the ventricular ejection phase, blood rushes from the ventricles into the aorta and pulmonary trunk. The pressure in the aorta normally reaches about 120 mm Hg.

④ **Isovolumetric relaxation: early diastole.** Following the peak of the T wave, the ventricles relax. Because the blood remaining in their chambers, referred to as the *end systolic volume* (*ESV*), is no longer compressed, ventricular pressure drops rapidly and blood in the aorta and pulmonary trunk flows back toward the heart, closing the SL valves. Once again, the ventricles are completely closed chambers, and so this event marks the beginning of the **isovolumetric relaxation phase**. Closure of the aortic valve raises aortic pressure briefly as backflowing blood rebounds off the closed valve cusps, an event beginning at the **dicrotic notch** shown on the pressure graph (p. 694).

All during ventricular systole, the atria have been in diastole. They have been filling with blood and the intra-atrial pressure has been rising. When blood pressure on the atrial side of the AV valves exceeds that in the ventricles, the AV valves are forced open and ventricular filling, phase ①, begins again. Atrial pressure drops to its lowest point and ventricular pressure begins to rise, completing the cycle.

Assuming the average heart beats 75 times each minute, the cardiac cycle lasts about 0.8 s, with atrial systole accounting for 0.1 s and ventricular systole 0.3 s. The remaining 0.4 s is a period of total heart relaxation, the **quiescent period**.

Notice two important points: (1) Blood flow through the heart is controlled entirely by pressure changes, and (2) blood flows down a pressure gradient through any available opening. The pressure changes, in turn, reflect the alternating contraction and relaxation of the myocardium and cause the heart valves to open, which keeps blood flowing in the forward direction.

The situation in the right side of the heart is essentially the same as in the left side *except* for pressure. The pulmonary circulation is a low-pressure circulation as evidenced by the much thinner myocardium of its right ventricle. So, typical systolic and diastolic pressures for the pulmonary artery are 24 and 10 mm Hg, compared to systolic and diastolic pressures of 120 and 80 mm Hg, respectively, for the aorta. However, the two sides of the heart eject the same blood volume with each heartbeat.

Heart Sounds

Auscultating (listening to) the thorax with a stethoscope will reveal two sounds during each heartbeat. These **heart sounds**, often described as lub-dup, are associated with the heart valves closing. (The timing of heart sounds in the cardiac cycle is shown in Focus Figure 18.2 on p. 694.)

The basic rhythm of the heart sounds is lub-dup, pause, lub-dup, pause, and so on, with the pause indicating the period when the heart is relaxing. The first sound occurs as the AV valves close. It signifies the point when ventricular pressure rises above atrial pressure (the beginning of ventricular systole). The first sound tends to be louder, longer, and more resonant than the second. The second sound occurs as the SL valves snap shut at the beginning of ventricular relaxation (diastole), resulting in a short, sharp sound.

(Text continues on p. 696.)

18

The cardiac cycle consists of the events during one heartbeat. By following pressure changes in an atrium, a ventricle, and a major artery, you can understand when cardiac valves open and close, allowing blood to flow.

Let's measure pressures in the left atrium, left ventricle, and aorta. Pressures from the right side would look the same, but would be lower.

Pressures and Valves
Recall that blood flows from high to low pressure, but only if valves are open. When lines cross on the pressure curve, valves either open or close. If you understand this, the other parts of the cardiac cycle will fall into place.

① **Atrioventricular (AV) valves close** when the ventricular pressure exceeds the atrial pressure.

② **Semilunar (SL) valves open** when the ventricular pressure exceeds the aortic pressure.

Dicrotic notch

Aorta

120

80

Left ventricle

40

Left atrium

0

Pressure (mm Hg)

Time ⟶

③ **SL valves close** when the ventricular pressure drops below the aortic pressure. (Blood in the aorta rebounds against the closed valve, causing the pressure to rise briefly at the dicrotic notch.)

④ **AV valves open** when the ventricular pressure drops below the atrial pressure.

Isovolumetric phases
There are two periods when all four valves are closed and volumes cannot change. These periods are isovolumetric (iso = same; metric = measure).

During the **isovolumetric contraction phase**, the ventricles are contracting and building up pressure.

During the **isovolumetric relaxation phase**, the ventricles are relaxing and pressures fall.

Heart sound
As valves close, the resulting turbulent blood flow creates a sound.

The **first heart sound** is caused by the AV valves closing.

The **second heart sound** is caused by the SL valves closing.

Heart sounds

Systole and Diastole

Diastole

Systole (ventricular contraction) occurs between the first and second heart sounds.

Diastole (ventricular relaxation) occurs between the second heart sound and the first heart sound of the next cycle.

Now that you understand the pressure changes, let's look at how they correspond to the electrical events and volume changes in the heart.

The P wave (atrial depolarization) precedes atrial contraction.

The QRS complex (ventricular depolarization) precedes ventricular contraction.

The T wave (ventricular repolarization) precedes ventricular relaxation.

ECG
Recall that the **electrocardiogram** (ECG) tells us about the electrical activity of the heart and that the electrical events precede the mechanical events they cause.

Aorta

Left ventricle

Left atrium

The rise in pressure is caused by atrial contraction.

Pressure (mm Hg)

120

80

40

0

Volumes
Pressure changes cause volume changes (if valves are open). Let's look at blood volume in a ventricle (it's the same for both ventricles).

Ventricular volume (ml)

120

50

The volume of blood in the ventricle is greatest at the end of diastole. This is the **end diastolic volume (EDV).**

The amount of blood ejected from each ventricle is the **stroke volume (SV).**

The volume of blood in the ventricle is smallest at the end of systole. This is the **end systolic volume (ESV).**

Blood flow
Let's follow blood through the four phases of the cardiac cycle.

Left atrium

Right atrium

Left ventricle

Right ventricle

Passive   Atrial contraction

① Ventricular filling phase

② Isovolumetric contraction phase

③ Ventricular ejection phase

④ Isovolumetric relaxation phase

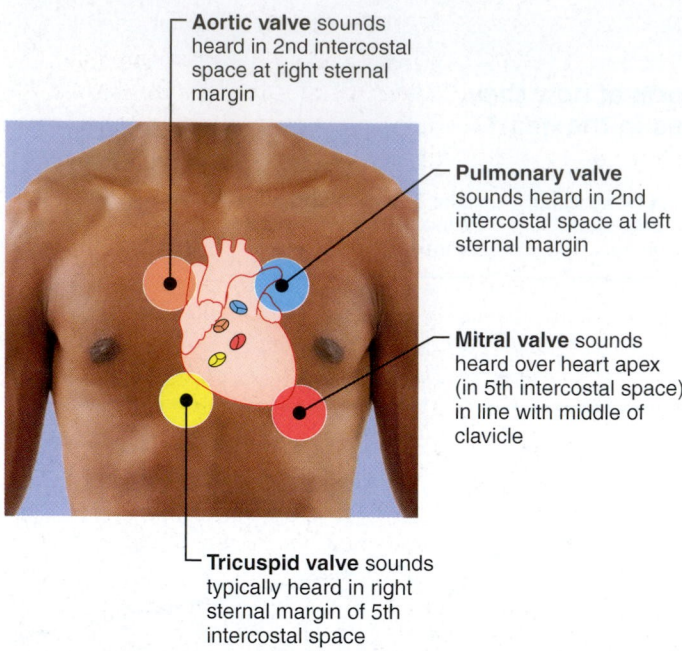

Aortic valve sounds heard in 2nd intercostal space at right sternal margin

Pulmonary valve sounds heard in 2nd intercostal space at left sternal margin

Mitral valve sounds heard over heart apex (in 5th intercostal space) in line with middle of clavicle

Tricuspid valve sounds typically heard in right sternal margin of 5th intercostal space

Figure 18.19 Areas of the thoracic surface where the sounds of individual valves are heard most clearly.

Because the mitral valve closes slightly before the tricuspid valve does, and the aortic SL valve generally snaps shut just before the pulmonary valve, it is possible to distinguish the individual valve sounds by auscultating four specific regions of the thorax (**Figure 18.19**). Notice that these four points, while not directly superficial to the valves (because the sounds take oblique paths to reach the chest wall), handily define the four corners of the normal heart. Knowing normal heart size and location is essential for recognizing an enlarged (and often diseased) heart.

HOMEOSTATIC IMBALANCE 18.6   **CLINICAL**

Blood flows silently as long as its flow is smooth and uninterrupted. If blood strikes obstructions, however, its flow becomes turbulent and generates abnormal heart sounds, called **heart murmurs**, that can be heard with a stethoscope. Heart murmurs are quite common in young children (and some elderly people) with perfectly healthy hearts, probably because their heart walls are relatively thin and vibrate with rushing blood.

Sometimes, especially in adults, murmurs indicate valve problems. An *incompetent* or *insufficient* valve fails to close completely. There is a swishing sound as blood backflows or regurgitates through the partially open valve *after* the valve has (supposedly) closed.

A *stenotic* valve fails to open completely and its narrow opening restricts blood flow *through* the valve. In a stenotic aortic valve, for instance, a high-pitched sound or click can be detected when the valve should be wide open during ventricular contraction, but is not.

⬤⬤⬤⬤⬤⬤⬤

Check Your Understanding

16. The second heart sound is associated with the closing of which valve(s)?

---

17. **PREDICT** If the mitral valve were insufficient, would you expect to hear the murmur (of blood flowing through the valve that should be closed) during ventricular systole or diastole?

18. **DRAW** The lines in the graph at right represent pressures in the left heart and aorta. Label each of the three lines. Label the points on the graph where AV valves open or close and where SL valves open or close. Label two periods during which all four valves are closed, using the appropriate terms.

Pressure

Time

For answers, see Answers Appendix.

18.7   **Stroke volume and heart rate are regulated to alter cardiac output**

Learning Outcomes

▶ **Name and explain the effects of various factors regulating stroke volume and heart rate.**

▶ **Explain the role of the autonomic nervous system in regulating cardiac output.**

Cardiac output (CO) is the amount of blood pumped out by *each* ventricle in 1 minute. It is the product of heart rate (HR) and stroke volume (SV). **Stroke volume** is defined as the volume of blood pumped out by one ventricle with each beat. In general, stroke volume correlates with the force of ventricular contraction.

> ⟳ Complete an interactive tutorial: MasteringA&P° > Study Area > Interactive Physiology

Using normal resting values for heart rate (75 beats/min) and stroke volume (70 ml/beat), the average adult cardiac output can be computed:

$$CO = HR \times SV = \frac{75 \text{ beats}}{\text{min}} \times \frac{70 \text{ ml}}{\text{beat}}$$
$$= \frac{5250 \text{ ml}}{\text{min}} = \frac{5.25 \text{ L}}{\text{min}}$$

The normal adult blood volume is about 5 L (a little more than 1 gallon). As you can see, the entire blood supply passes through each side of the heart once each minute.

Notice that cardiac output varies directly with SV and HR. This means that CO increases when the stroke volume increases or the heart beats faster or both, and it decreases when either or both of these factors decrease.

Cardiac output is highly variable and increases markedly in response to special demands, such as running to catch a bus. **Cardiac reserve** is the difference between resting and maximal CO. In nonathletic people, cardiac reserve is typically four to five times resting CO (20–25 L/min), but CO in trained athletes during competition may reach 35 L/min (seven times resting CO).

How does the heart accomplish such tremendous increases in output? To understand this feat, let's look at how stroke volume and heart rate are regulated. See **Figure 18.20** for an overview of the factors that affect stroke volume and heart rate, and consequently, cardiac output. Work your way up from the bottom of the figure as we add more details in the following discussion.

Regulation of Stroke Volume

Mathematically, stroke volume (SV) represents the difference between **end diastolic volume (EDV)**, the amount of blood that collects in a ventricle during diastole, and **end systolic volume (ESV)**, the volume of blood remaining in a ventricle *after* it has contracted. The EDV, determined by how long ventricular diastole lasts and by venous pressure, is normally about 120 ml. (An increase in either factor *raises* EDV.) The ESV, determined by arterial blood pressure and the force of ventricular contraction, is approximately 50 ml. (The higher the arterial blood pressure, the higher the ESV.) To figure normal stroke volume, simply plug these values into this equation:

$$SV = EDV - ESV = \frac{120\ ml}{beat} - \frac{50\ ml}{beat} = \frac{70\ ml}{beat}$$

As you can see, each ventricle pumps out about 70 ml of blood with each beat, which is about 60% of the blood in its chambers.

So what is important here—how do we make sense out of this alphabet soup (SV, ESV, EDV)? Although many factors affect SV by altering EDV or ESV, the three most important are *preload, contractility*, and *afterload*. As we describe in detail next, preload affects EDV, whereas contractility and afterload affect the ESV.

Preload: Degree of Stretch of Heart Muscle

The degree to which cardiac muscle cells are stretched just before they contract, called the **preload**, controls stroke volume. In a normal heart, the higher the preload, the higher the stroke volume. This relationship between preload and stroke volume is called the **Frank-Starling law of the heart**.

Recall that at an *optimal length* of muscle fibers (and sarcomeres) (1) the maximum number of active cross bridge attachments is possible between actin and myosin, and (2) the force of contraction is maximal (◄ **Figure 9.19, p. 307**). Cardiac muscle, like skeletal muscle, exhibits a *length-tension relationship*.

Resting skeletal muscle fibers are kept near optimal length for developing maximal tension. In contrast, resting cardiac cells are normally *shorter* than optimal length. As a result, stretching cardiac cells can produce dramatic increases in contractile force. The most important factor stretching cardiac muscle is **venous return**, the amount of blood returning to the heart and distending its ventricles.

Anything that increases venous return increases EDV and, consequently, SV and contraction force (Figure 18.21). Basically:

$$\uparrow \text{Venous} \rightarrow \uparrow \text{EDV} \rightarrow \uparrow \text{SV} \rightarrow \uparrow \text{Cardiac}$$
$$\text{return} \quad\quad (\text{preload}) \quad\quad\quad \text{output}$$

Frank-Starling law

Both exercise and increased filling time increase EDV. Exercise increases venous return because both increased sympathetic nervous system activity and the squeezing action of the skeletal muscles compress the veins, decreasing the volume of blood they contain and returning more blood to the heart. During vigorous exercise, SV may double as a result of increased venous

Figure 18.20 Factors involved in determining cardiac output. (EDV = end diastolic volume, ESV = end systolic volume)

Figure 18.21 Norepinephrine increases heart contractility via a cyclic AMP second-messenger system. Cyclic AMP activates protein kinases that phosphorylate proteins that determine cytoplasmic Ca^{2+} concentrations.

return. Conversely, low venous return might result from severe blood loss or an extremely rapid heart rate that does not allow enough time for ventricular filling. Low venous return decreases EDV, causing the heart to beat less forcefully and lowering SV.

Because the systemic and pulmonary circulations are in series, the intrinsic mechanism we just described ensures equal outputs of the two ventricles and proper distribution of blood volume between the two circuits. If one side of the heart suddenly begins to pump more blood than the other, the increased venous return to the opposite ventricle forces that ventricle—through increased cardiac muscle stretch—to pump out an equal volume, preventing backup or accumulation of blood in the circulation.

Contractility

EDV is the major *intrinsic factor* influencing SV, but *extrinsic factors* that increase heart muscle contractility can also enhance SV. **Contractility** is defined as the contractile strength achieved at a given muscle length. Note in Figure 18.20 that contractility is *independent* of muscle stretch and EDV. Contractility rises when more Ca^{2+} enters the cytoplasm from the extracellular fluid and the SR. Enhanced contractility means more blood is

ejected from the heart (greater SV), and so ESV is lower.

Increased sympathetic stimulation increases contractility. As noted on p. 689, sympathetic fibers serve not only the intrinsic conduction system but the entire heart. One effect of norepinephrine or epinephrine binding is to initiate a cyclic AMP second-messenger system that increases Ca^{2+} entry, which in turn promotes more cross bridge binding and enhances ventricular contractility (**Figure 18.21**).

A battery of other chemicals also influence contractility. Substances that increase contractility are called *positive inotropic agents* (*ino* = muscle, fiber). The hormones epinephrine, thyroxine, and glucagon; the drug digitalis; and high levels of extracellular Ca^{2+} are all positive inotropic agents. *Negative inotropic agents,* which impair or decrease contractility, include acidosis (excess H^+), rising extracellular K^+ levels, and drugs called calcium channel blockers.

Afterload: Back Pressure Exerted by Arterial Blood

Afterload is the pressure that the ventricles must overcome to eject blood. It is essentially the back pressure that arterial blood exerts on the aortic and pulmonary valves (about 80 mm Hg in the aorta and 10 mm Hg in the pulmonary trunk as we will see in Chapter 19).

In healthy individuals, afterload is not a major determinant of stroke volume because it is relatively constant. However, in people with hypertension (high blood pressure), afterload is important because it reduces the ability of the ventricles to eject blood. Consequently, more blood remains in the heart after systole, increasing ESV and reducing stroke volume.

Regulation of Heart Rate

Given a healthy cardiovascular system, SV tends to be relatively constant. However, when blood volume drops sharply or the heart is seriously weakened, SV declines and CO is maintained by increasing HR and contractility. Temporary stressors can also influence HR—and consequently CO—by acting through homeostatic mechanisms induced neurally, chemically, and physically. Factors that increase HR are called *positive chronotropic factors* (*chrono* = time), and those that decrease HR are *negative chronotropic factors.*

Autonomic Nervous System Regulation of Heart Rate

The autonomic nervous system exerts the most important extrinsic controls affecting heart rate, as shown on the right side of

Figure 18.20. When emotional or physical stressors (such as fright, anxiety, or exercise) activate the sympathetic nervous system, sympathetic nerve fibers release norepinephrine at their cardiac synapses. Norepinephrine binds to β_1-adrenergic receptors in the heart, causing threshold to be reached more quickly. As a result, the SA node fires more rapidly and the heart responds by beating faster.

Sympathetic stimulation also enhances contractility and speeds relaxation. It does this by enhancing Ca^{2+} movements in the contractile cells as we described above and in Figure 18.21. Enhanced contractility lowers ESV, so SV does not decline as it would if only heart *rate* increased. (Remember, when the heart beats faster, there is less time for ventricular filling and so a lower EDV.)

The parasympathetic division opposes sympathetic effects and effectively reduces heart rate when a stressful situation has passed. Parasympathetic-initiated cardiac responses are mediated by acetylcholine, which hyperpolarizes the membranes of its effector cells by *opening* K^+ channels. Because innervation of the ventricles by the vagus nerve is sparse, parasympathetic activity has little effect on cardiac contractility.

Under resting conditions, both autonomic divisions continuously send impulses to the SA node, but the *dominant* influence is inhibitory. For this reason, the heart is said to exhibit **vagal tone**, and heart rate is generally slower than it would be if the vagus nerves were not innervating it. Cutting the vagus nerves results in an almost immediate increase in heart rate of about 25 beats/min, reflecting the inherent rate (100 beats/min) of the pacemaking SA node.

When sensory input from various parts of the cardiovascular system activates either division of the autonomic nervous system more strongly, the other division is temporarily inhibited. Most such sensory input is generated by *baroreceptors* which respond to changes in systemic blood pressure, as we will discuss in Chapter 19. Another example, the **atrial (Bainbridge) reflex**, is an autonomic reflex initiated by increased venous return and increased atrial filling. Stretching the atrial walls increases heart rate by stimulating both the SA node and the atrial stretch receptors. Stretch receptor activation triggers reflexive adjustments of autonomic output to the SA node, increasing heart rate.

Increased or decreased CO results in corresponding changes to systemic blood pressure, so blood pressure regulation often involves reflexive controls of heart rate. In Chapter 19 we describe in more detail neural mechanisms that regulate blood pressure.

Chemical Regulation of Heart Rate

Chemicals normally present in the blood and other body fluids may influence heart rate.

- **Hormones.** *Epinephrine*, liberated by the adrenal medulla during sympathetic nervous system activation, produces the same cardiac effects as norepinephrine released by the sympathetic nerves: It enhances heart rate and contractility.

  Thyroxine is a thyroid gland hormone that increases metabolic rate and production of body heat. When released in large quantities, it causes a sustained increase in heart rate. Thyroxine acts directly on the heart but also *enhances* the effects of epinephrine and norepinephrine.

- **Ions.** Normal heart function depends on having normal levels of intracellular and extracellular ions. Plasma electrolyte imbalances pose real dangers to the heart.

HOMEOSTATIC IMBALANCE 18.7 — CLINICAL

Reduced blood Ca^{2+} levels (*hypocalcemia*) depress the heart. Conversely, above-normal levels (*hypercalcemia*) increase heart rate and contractility—up to a point. Very high Ca^{2+} levels disrupt heart function and can cause life-threatening arrhythmias.

High or low blood K^+ levels are particularly dangerous and arise in a number of clinical conditions. Excessive K^+ (*hyperkalemia*) alters electrical activity in the heart by depolarizing the resting membrane potential, and may lead to heart block and cardiac arrest. *Hypokalemia* is also life threatening, in that the heart beats feebly and arrhythmically. ●●●●●●●●

Other Factors That Regulate Heart Rate

Age, gender, exercise, and body temperature also influence HR, although they are less important than neural factors. Resting heart rate is fastest in the fetus (140–160 beats/min) and gradually declines throughout life. Average heart rate is faster in females (72–80 beats/min) than in males (64–72 beats/min).

Exercise raises HR by acting through the sympathetic nervous system (Figure 18.20). Exercise also increases systemic blood pressure and routes more blood to the working muscles. However, resting HR in the physically fit tends to be substantially lower than in those who are out of condition, and in trained athletes it may be as slow as 40 beats/min. We explain this apparent paradox below.

Heat increases HR by enhancing the metabolic rate of cardiac cells. This explains the rapid, pounding heartbeat you feel when you have a high fever and also accounts, in part, for the effect of exercise on HR (remember, working muscles generate heat). Cold directly decreases heart rate.

HOMEOSTATIC IMBALANCE 18.8 — CLINICAL

HR varies with changes in activity, but marked and persistent rate changes usually signal cardiovascular disease.

Tachycardia (tak″e-kar′de-ah; "heart hurry") is an abnormally fast resting heart rate (more than 100 beats/min) that may result from elevated body temperature, stress, certain drugs, or heart disease. Persistent tachycardia is considered pathological because tachycardia occasionally promotes fibrillation.

Bradycardia (brad″e-kar′de-ah; *brady* = slow) is a resting heart rate slower than 60 beats/min. It may result from low body temperature, certain drugs, or parasympathetic nervous system activation. It is a known, and desirable, consequence of endurance training. With physical and cardiovascular conditioning, the heart hypertrophies and SV increases, allowing a lower resting heart rate while still providing the same cardiac output. However, in poorly conditioned people persistent bradycardia may result in grossly inadequate blood circulation to body tissues. Bradycardia is often a warning of brain edema after head trauma. ●●●●●●●●

18

Homeostatic Imbalance of Cardiac Output CLINICAL

The heart's pumping action ordinarily maintains a balance between cardiac output and venous return. Were this not so, a dangerous damming up of blood (blood congestion) would occur in the veins returning blood to the heart.

In **congestive heart failure (CHF)**, the heart is such an inefficient pump that blood circulation is inadequate to meet tissue needs. This progressively worsening disorder reflects weakening of the myocardium by various conditions that damage it in different ways. The most common causes include:

- **Coronary atherosclerosis.** Coronary atherosclerosis, essentially fatty buildup that clogs the coronary arteries, impairs blood and oxygen delivery to cardiac cells. The heart becomes increasingly hypoxic and begins to contract ineffectively.

- **Persistent high blood pressure.** Normally, pressure in the aorta during diastole is 80 mm Hg, and the left ventricle exerts only slightly over that amount of force to eject blood from its chamber. When aortic diastolic blood pressure rises to 90 mm Hg or more, the myocardium must exert more force to open the aortic valve and pump out the same amount of blood. If afterload is chronically elevated, ESV rises and the myocardium hypertrophies. Eventually, the stress takes its toll and the myocardium becomes progressively weaker.

- **Multiple myocardial infarctions.** A succession of MIs (heart attacks) depresses pumping efficiency because noncontractile fibrous (scar) tissue replaces the dead heart cells.

- **Dilated cardiomyopathy (DCM).** In this condition, the ventricles stretch and become flabby and the myocardium deteriorates, often for unknown reasons. Drug toxicity or chronic inflammation may be involved.

Because the heart is a double pump, each side can initially fail independently of the other. If the left side fails, **pulmonary congestion** occurs. The right side continues to propel blood to the lungs, but the left side does not adequately eject the returning blood into the systemic circulation. Blood vessels in the lungs become engorged with blood, the pressure in them increases, and fluid leaks from the blood vessels into the lung tissue, causing pulmonary edema. If the congestion is untreated, the person suffocates.

If the right side of the heart fails, **peripheral congestion** occurs. Blood stagnates in body organs, and pooled fluids in the tissue spaces impair the ability of body cells to obtain adequate nutrients and oxygen and rid themselves of wastes. The resulting edema is most noticeable in the extremities (feet, ankles, and fingers).

Failure of one side of the heart puts a greater strain on the other side, and ultimately the whole heart fails. A seriously weakened, or *decompensated*, heart is irreparable. Treatment is directed primarily toward (1) removing the excess leaked fluid with *diuretics* (drugs that increase the kidneys' excretion of Na$^+$ and water), (2) reducing afterload with drugs that drive down blood pressure, and (3) increasing contractility with digitalis derivatives. Heart transplants and other surgical or mechanical remedies to replace damaged heart muscle provide additional hope for some cardiac patients.

Check Your Understanding

19. **APPLY** After running to catch a bus, Josh noticed that his heart was beating faster than normal and was pounding forcefully in his chest. How did his increased HR and SV come about?

20. What problem of cardiac output might ensue if the heart beats far too rapidly for an extended period, that is, if tachycardia occurs? Why?

For answers, see Answers Appendix.

Developmental Aspects of the Heart

The human heart, derived from mesoderm and guided by powerful signaling molecules, begins as two simple endothelial tubes. They soon fuse to form a single chamber or heart tube that is busily pumping blood by the 22nd day of gestation (**Figure 18.22**).

The tube develops four slightly bulged areas that represent the earliest heart chambers. From tail to head, following the direction of blood flow, the four primitive chambers are the following (Figure 18.22b):

1. **Sinus venosus** (ven-o'sus). This chamber initially receives all the venous blood of the embryo. It will become the smooth-walled part of the right atrium and the coronary sinus. It also gives rise to the sinoatrial node, which takes control and sets heart rate early in embryonic development.

2. **Atrium.** This embryonic chamber eventually becomes the pectinate muscle–ridged parts of the atria.

3. **Ventricle.** This is the strongest pumping chamber of the early heart and becomes the *left* ventricle.

4. **Bulbus cordis.** This chamber plus its cranial extension, the *truncus arteriosus* (labeled 4a in Figure 18.22b), give rise to the pulmonary trunk, the first part of the aorta, and most of the *right* ventricle.

During the next three weeks, the heart "tube" contorts and major structural changes convert it into a four-chambered organ capable of acting as a double pump—all without missing a beat! After the second month, few changes other than growth occur until birth.

The interatrial septum of the fetal heart is incomplete. The **foramen ovale** (literally, "oval door") connects the two atria and allows blood entering the right heart to bypass the pulmonary circuit and the collapsed, nonfunctional fetal lungs (Figure 18.22e). Another lung bypass, the **ductus arteriosus**, exists between the pulmonary trunk and the aorta. At or shortly after birth, these shunts close, completing the separation between the right and left sides of the heart. We give a more complete description of the fetal and newborn circulation in Chapter 28 (see Focus Figure 28.2, pp. 1108–1109).

HOMEOSTATIC IMBALANCE 18.9 CLINICAL

Building a perfect heart is difficult. Each year about 40,000 infants are born in the U.S. with one or more of 30 different **congenital heart defects**, making them the most common of

18

(a) Day 20: Endothelial tubes begin to fuse.

(b) Day 22: Heart starts pumping.

(c) Day 24: Heart continues to bend and elongate.

(d) Day 28: Bending continues as ventricle moves caudally and atrium moves cranially.

(e) Day 35: Bending is complete.

Figure 18.22 Development of the human heart. Ventral views, with the cranial direction toward the top of the figures. Arrows within the heart show the direction of blood flow. Days are approximate. **(b)** 1 is the sinus venosus; 2, the atrium; 3, the ventricle; 4, the bulbus cordis; and 4a, the truncus arteriosus.

all birth defects. Some congenital heart problems are traceable to environmental influences, such as maternal infection or drug intake during month 2 when the major events of heart formation occur.

The most prevalent abnormalities produce two basic kinds of disorders in the newborn. They either (1) lead to mixing of oxygen-poor blood with oxygenated blood (so that inadequately oxygenated blood reaches the body tissues) or (2) involve narrowed valves or vessels that greatly increase the workload on the heart.

Examples of the first type of defect are *septal defects* (**Figure 18.23a**) and *patent ductus arteriosus*, in which the connection between the aorta and pulmonary trunk remains open. *Coarctation of the aorta* (Figure 18.23b) is an example of the second type of problem. *Tetralogy of Fallot* (te-tral′o-je ov fal-o′), a serious condition in which the baby becomes cyanotic (turns blue) within minutes of birth, encompasses both types of disorders (Figure 18.23c). Surgery can usually correct these congenital defects.

(a) Ventricular septal defect. The superior part of the interventricular septum fails to form, allowing blood to mix between the two ventricles. More blood is shunted from left to right because the left ventricle is stronger.

Occurs in about 1 in every 500 births

Narrowed aorta

(b) Coarctation of the aorta. A part of the aorta is narrowed, increasing the workload of the left ventricle.

Occurs in about 1 in every 1500 births

(c) Tetralogy of Fallot. Multiple defects (*tetra* = four): (1) Pulmonary trunk too narrow and pulmonary valve stenosed, resulting in (2) hypertrophied right ventricle; (3) ventricular septal defect; (4) aorta receives blood from both ventricles.

Occurs in about 1 in every 2000 births

Figure 18.23 Three examples of congenital heart defects. Tan areas indicate the locations of the defects.

18

Heart Function throughout Life

The heart functions admirably throughout a long lifetime for most people. Homeostatic mechanisms are normally so efficient that people rarely notice when the heart is working harder.

In people who exercise regularly and vigorously, the heart gradually adapts to the increased demand by enlarging and becoming more efficient and more powerful. Aerobic exercise also helps clear fatty deposits from blood vessel walls throughout the body, retarding atherosclerosis and coronary heart disease. This beneficial cardiac response to exercise persists into old age.

Because of the incredible amount of work the heart does over the course of a lifetime, certain structural changes are inevitable. Age-related changes include the following:

- **Valve flaps thicken and become sclerotic (stiff).** This change occurs particularly where the stress of blood flow is greatest (mitral valve). For this reason, heart murmurs are more common in elderly people.

- **Cardiac reserve declines.** Although the passing years seem to cause little change in resting heart rate, the aged heart is less able to respond to both sudden and prolonged stressors that demand increased output. In addition, the maximum HR declines as sympathetic control of the heart becomes less efficient. These changes are less of a problem in physically active seniors.

- **Cardiac muscle becomes fibrosed (scarred).** As a person ages, more and more cardiac cells die and are replaced with fibrous tissue. As a result, the heart stiffens and fills less efficiently, reducing stroke volume. The nodes of the heart's conduction system may also become fibrosed, which increases the incidence of arrhythmias and other conduction problems.

- **Atherosclerosis.** The insidious progress of atherosclerosis begins in childhood, but inactivity, smoking, and stress accelerate it. The most serious consequences to the heart are hypertensive heart disease and coronary artery occlusion, both of which increase the risk of heart attack and stroke. Although the aging process itself leads to changes in blood vessel walls that promote atherosclerosis, many investigators feel that diet, not aging, is the single most important contributor to cardiovascular disease. We can lower our risk by consuming less animal fat, sugar, and salt.

RELATED CLINICAL TERMS

Asystole (a-sis′to-le) Situation in which the heart fails to contract.

Cardiac catheterization Diagnostic procedure that involves passing a fine catheter (tubing) through a blood vessel into the heart. Oxygen content of blood, blood flow, and pressures within the heart can be measured. Findings help to detect valve problems, heart deformities, and other heart malfunctions.

Commotio cordis ("concussion of the heart") Situation in which a relatively mild blow to the chest causes heart failure and sudden death because it occurs during a vulnerable interval (2 ms) when the heart is repolarizing. Explains those rare instances when youngsters drop dead on the playing field after being hit in the chest by a ball.

Cor pulmonale (kor pul-mun-nă′le; *cor* = heart, *pulmo* = lung) A condition of right-sided heart failure resulting from elevated blood pressure in the pulmonary circuit (pulmonary hypertension). Acute cases may develop suddenly due to a pulmonary embolism; chronic cases are usually associated with chronic lung disorders such as emphysema.

Endocarditis (en″do-kar-di′tis) Inflammation of the endocardium, usually confined to the endocardium of the heart valves. Endocarditis often results from infection by bacteria that have entered the bloodstream but may result from fungal infection or an autoimmune response. Drug addicts may develop endocarditis by injecting themselves with contaminated needles.

Heart palpitation A heartbeat that is unusually strong, fast, or irregular so that the person becomes aware of it; may be caused by certain drugs, emotional stress ("nervous heart"), or heart disorders.

Hypertrophic cardiomyopathy (HCM) The leading cause of sudden death in young athletes, this condition, which is usually inherited, causes the cardiac muscle cells to enlarge, thickening the heart wall. The heart pumps strongly but doesn't relax well during diastole when the heart is filling.

Mitral valve prolapse Valve disorder affecting up to 1% of the population; most often seen in young women. It appears to have a genetic basis resulting in abnormal chordae tendineae or a malfunction of the papillary muscles. One or more of the mitral valve flaps become incompetent and billow into the left atrium during ventricular systole, allowing blood regurgitation. Occasionally requires valve replacement surgery.

Myocarditis (mi″o-kar-di′tis; *myo* = muscle, *card* = heart, *itis* = inflammation) Inflammation of the cardiac muscle layer (myocardium) of the heart; sometimes follows an untreated streptococcal infection in children. May weaken the heart and impair its ability to pump effectively.

Paroxysmal atrial tachycardia (PAT) Bursts of atrial contractions with little pause between them.

Ventricular tachycardia (VT or V-tac) Rapid ventricular contractions that are not coordinated with atrial activity.

18

CHAPTER SUMMARY

18.1 The heart has four chambers and pumps blood through the pulmonary and systemic circuits (pp. 671–678)

1. The right side of the heart is the pulmonary circuit pump. It pumps blood through the lungs, where the blood picks up oxygen and dumps carbon dioxide. The left side of the heart is the systemic circuit pump. It pumps blood through the body's tissues, supplying them with oxygen and nutrients and removing carbon dioxide.

2. The human heart, about the size of a clenched fist, is located obliquely within the mediastinum of the thorax.

3. The heart is enclosed within a double sac made up of the outer fibrous pericardium and the inner serous pericardium (parietal and visceral layers). The pericardial cavity between the serous layers contains lubricating serous fluid.

4. Layers of the heart wall, from the interior out, are the endocardium, myocardium (reinforced by a fibrous cardiac skeleton), and epicardium (visceral layer of the serous pericardium).

5. The heart has two superior atria and two inferior ventricles. Functionally, the heart is a double pump.

6. Entering the right atrium are the superior vena cava, inferior vena cava, and coronary sinus. Four pulmonary veins enter the left atrium.

7. The right ventricle discharges blood into the pulmonary trunk; the left ventricle pumps blood into the aorta.

18.2 Heart valves make blood flow in one direction (pp. 679–680)

1. The atrioventricular (AV) valves (tricuspid and mitral) prevent backflow into the atria when the ventricles are contracting; the semilunar (SL) valves (pulmonary and aortic) prevent backflow into the ventricles when the ventricles are relaxing.

18.3 Blood flows from atrium to ventricle, and then to either the lungs or the rest of the body (pp. 680–683)

1. Oxygen-poor systemic blood enters the right atrium, passes into the right ventricle, through the pulmonary trunk to the lungs, and back to the left atrium via the pulmonary veins. Oxygen-laden blood entering the left atrium from the lungs flows into the left ventricle and then into the aorta, which provides the functional supply of all body organs. Systemic veins return the oxygen-depleted blood to the right atrium.

2. The right and left coronary arteries branch from the aorta and supply the heart itself via their main branches (anterior and posterior interventricular, right marginal, and circumflex arteries). Venous blood, collected by the cardiac veins (great, middle, and small), empties into the coronary sinus.

3. Blood delivery to the myocardium occurs during heart relaxation.

18.4 Intercalated discs connect cardiac muscle fibers into a functional syncytium (pp. 683–686)

1. Cardiac muscle cells are branching, striated, generally uninucleate cells. They contain myofibrils consisting of typical sarcomeres.

2. Intercalated discs containing desmosomes and gap junctions connect adjacent cardiac cells. The myocardium behaves as a functional syncytium because of electrical coupling provided by gap junctions.

3. Ca^{2+} released by the SR and entering from the extracellular space couples the action potential to sliding of the myofilaments. Compared to skeletal muscle, cardiac muscle has a prolonged refractory period that prevents tetany.

4. Cardiac muscle has abundant mitochondria and depends almost entirely on aerobic respiration to form ATP.

18.5 Pacemaker cells trigger action potentials throughout the heart (pp. 686–692)

1. Certain noncontractile cardiac muscle cells exhibit automaticity and rhythmicity and can independently initiate action potentials. Such cells have an unstable resting potential called a pacemaker potential that gradually depolarizes, drifting toward threshold for firing. These cells compose the intrinsic conduction system of the heart.

2. The conduction system of the heart consists of the SA and AV nodes, the AV bundle and bundle branches, and the subendocardial conducting network. This system coordinates the depolarization of the heart and ensures that the heart beats as a unit. The SA node has the fastest rate of spontaneous depolarization and acts as the heart's pacemaker; it sets the sinus rhythm.

3. Defects in the intrinsic conduction system can cause arrhythmias, fibrillation, and heart block.

4. The autonomic nervous system innervates the heart. Cardiac centers in the medulla include the cardioacceleratory center, which projects to the T_1–T_5 region of the spinal cord, which in turn projects to the cervical and upper thoracic sympathetic trunk. Postganglionic fibers innervate the SA and AV nodes and the cardiac muscle fibers. The cardioinhibitory center exerts its influence via the parasympathetic vagus nerves (X), which project to the heart wall. Most parasympathetic fibers serve the SA and AV nodes.

5. The membrane depolarization of contractile myocytes causes opening of sodium channels and allows sodium to enter, which is responsible for the rising phase of the action potential curve. Depolarization also opens slow Ca^{2+} channels; Ca^{2+} entry prolongs the period of depolarization (creates the plateau).

6. An electrocardiogram (ECG) is a graphic representation of the cardiac conduction cycle. The P wave reflects atrial depolarization. The QRS complex indicates ventricular depolarization; the T wave represents ventricular repolarization.

Complete an interactive tutorial: MasteringA&P° > Study Area > Interactive Physiology > Cardiovascular System: Electrical Activity of the Heart

18.6 The cardiac cycle describes the mechanical events associated with blood flow through the heart (pp. 692–696)

1. A cardiac cycle consists of the events occurring during one heartbeat. During mid-to-late diastole, the ventricles fill and the atria contract. Ventricular systole consists of the isovolumetric contraction phase and the ventricular ejection phase. During early diastole, the ventricles are relaxed and are closed chambers until the atrial pressure exceeds the ventricular pressure, forcing the AV valves open. Then the cycle begins again. At a normal heart rate of 75 beats/min, a cardiac cycle lasts 0.8 s.

2. Pressure changes promote blood flow and valve opening and closing.

 Complete an interactive tutorial: MasteringA&P® > Study Area > Interactive Physiology > Cardiovascular System: Cardiac Cycle

3. Normal heart sounds arise chiefly from turbulent blood flow during the closing of heart valves. Abnormal heart sounds, called murmurs, usually reflect valve problems.

18.7 Stroke volume and heart rate are regulated to alter cardiac output (pp. 696–700)

1. Cardiac output, typically 5 L/min, is the amount of blood pumped out by each ventricle in 1 minute. Stroke volume is the amount of blood pumped out by a ventricle with each contraction. Cardiac output = heart rate × stroke volume.

2. Stroke volume depends to a large extent on the degree to which venous return stretches cardiac muscle. Approximately 70 ml, it is the difference between end diastolic volume (EDV) and end systolic volume (ESV). Anything that influences heart rate or blood volume influences venous return, hence stroke volume.

3. Activation of the sympathetic nervous system increases heart rate and contractility; parasympathetic activation decreases heart rate but has little effect on contractility. Ordinarily, the heart exhibits vagal tone.

4. Chemical regulation of the heart is effected by hormones (epinephrine and thyroxine) and ions (particularly potassium and calcium). Ion imbalances severely impair heart activity.

5. Other factors influencing heart rate are age, sex, exercise, and body temperature.

6. Congestive heart failure occurs when the pumping ability of the heart cannot provide adequate circulation to meet body needs. Right heart failure leads to systemic edema; left heart failure results in pulmonary edema.

 Complete an interactive tutorial: MasteringA&P® > Study Area > Interactive Physiology > Cardiovascular System: Cardiac Output

Developmental Aspects of the Heart (pp. 700–702)

1. The heart begins as a simple (mesodermal) tube that is pumping blood by the fourth week of gestation. The fetal heart has two lung bypasses: the foramen ovale and the ductus arteriosus.

2. Congenital heart defects are the most common of all birth defects. The most common of these disorders lead to inadequate oxygenation of blood or increase the workload of the heart.

3. Age-related changes include sclerosis and thickening of the valve flaps, declines in cardiac reserve, fibrosis of cardiac muscle, and atherosclerosis.

4. Risk factors for cardiac disease include dietary factors, excessive stress, cigarette smoking, and lack of exercise.

REVIEW QUESTIONS

 To access additional practice questions using your smartphone, tablet, or computer: MasteringA&P® > Study Area > Practice Tests & Quizzes

Level 1 Remember/Understand

(Some questions have more than one correct answer. Select the best answer or answers from the choices given.)

1. When the semilunar valves are open, which of the following are occurring?
   (1) coronary arteries fill
   (2) AV valves are closed
   (3) ventricles are in systole
   (4) ventricles are in diastole
   (5) blood enters aorta
   (6) blood enters pulmonary arteries
   (7) atria contract
   (a) 2, 3, 5, 6, (b) 1, 2, 3, 7, (c) 1, 3, 5, 6, (d) 2, 4, 5, 7.

2. The portion of the intrinsic conduction system located in the superior interventricular septum is the (a) AV node, (b) SA node, (c) AV bundle, (d) subendocardial conducting network.

3. An ECG provides information about (a) cardiac output, (b) movement of the excitation wave across the heart, (c) coronary circulation, (d) valve impairment.

4. The sequence of contraction of the heart chambers is (a) random, (b) left chambers followed by right chambers, (c) both atria followed by both ventricles, (d) right atrium, right ventricle, left atrium, left ventricle.

5. The fact that the left ventricular wall is thicker than the right reveals that it (a) pumps a greater volume of blood, (b) pumps blood against greater resistance, (c) expands the thoracic cage, (d) pumps blood through a smaller valve.

6. The chordae tendineae (a) close the atrioventricular valves, (b) prevent the AV valve flaps from everting, (c) contract the papillary muscles, (d) open the semilunar valves.

7. In the heart, which of the following apply? (1) Action potentials are conducted from cell to cell across the myocardium via gap junctions, (2) the SA node sets the pace for the heart as a whole, (3) spontaneous depolarization of cardiac cells can occur in the absence of nerve stimulation, (4) cardiac muscle can continue to contract for long periods in the absence of oxygen. (a) all of the above, (b) 1, 3, 4, (c) 1, 2, 3, (d) 2, 3.

8. Freshly oxygenated blood is first received by the (a) right atrium, (b) left atrium, (c) right ventricle, (d) left ventricle.

9. Describe the location and position of the heart in the body.

10. The refractory period of cardiac muscle is much longer than that of skeletal muscle. Why is this a desirable functional property?

11. Draw a normal ECG pattern. Label and explain the significance of its deflection waves.

12. Define cardiac cycle, and follow the events of one cycle.

13. What is cardiac output, and how is it calculated?

14. As Cara worked at her dissection, she became frustrated that several of the structures she had to learn about had more than one common name. Provide another name for each of these structures: (a) atrioventricular groove, (b) tricuspid valve, (c) bicuspid valve (give two synonyms), and (d) atrioventricular bundle.

Level 2 Apply/Analyze

15. The activity of the heart depends on intrinsic properties of cardiac muscle and on neural factors. Thus, (a) vagus nerve stimulation of the heart reduces heart rate, (b) sympathetic nerve stimulation of the heart decreases time available for ventricular filling, (c) sympathetic stimulation of the heart increases its force of contraction, (d) all of the above.

16. Describe the pericardium and distinguish between the fibrous and the serous pericardia relative to histological structure and location.

18

17. Trace one drop of blood from the time it enters the right atrium until it enters the left atrium. What is this circuit called?

18. **(a)** Describe how heart contraction and relaxation influence coronary blood flow. **(b)** Name the major branches of the coronary arteries, and note the heart regions served by each.

19. **(a)** Name the elements of the intrinsic conduction system of the heart in order, beginning with the pacemaker. **(b)** What is the important function of this conduction system?

20. Discuss how the Frank-Starling law of the heart helps to explain the influence of venous return on stroke volume.

21. **(a)** Describe the common function of the foramen ovale and the ductus arteriosus in a fetus. **(b)** What problems result if these shunts remain patent (open) after birth?

22. You have been called upon to demonstrate the technique for listening to valve sounds. **(a)** Explain where you would position your stethoscope to auscultate (1) the aortic valve of a patient with severe aortic valve insufficiency and (2) a stenotic mitral valve. **(b)** During which period(s) would you hear these abnormal valve sounds most clearly? (During atrial diastole, ventricular systole, ventricular diastole, or atrial systole?) **(c)** What cues would you use to differentiate between an insufficient and a stenotic valve?

23. Gabriel, a heroin addict, feels tired, is weak and feverish, and has vague aches and pains. Terrified that he has AIDS, he goes to a doctor and is informed that he is suffering not from AIDS, but from a heart murmur accompanied by endocarditis. What is the most likely way that Gabriel contracted endocarditis? (Hint: See Related Clinical Terms.)

Level 3 Evaluate/Synthesize

24. Florita Santos, a middle-aged woman, is admitted to the coronary care unit with a diagnosis of left ventricular failure resulting from a myocardial infarction. Her history indicates that she awoke in the middle of the night with severe chest pain. Her skin is pale and cold, and moist sounds are heard over the lower regions of both lungs. Explain how failure of the left ventricle can cause these signs and symptoms.

25. Hannah, a newborn baby, needs surgery because she was born with an aorta that arises from the right ventricle and a pulmonary trunk that issues from the left ventricle, a condition called transposition of the great vessels. What are the physiological consequences of this defect?

CLINICAL CASE STUDY

54-Year-Old Female with a Defective Heart Valve

Dr. El-Sheikh is a 54-year-old Egyptian female who complains of weakness, fatigue, shortness of breath, chest pain, and dizziness. Her heart rate is 96 beats/min, and her blood pressure is 148/54 mm Hg. You notice that she nods her head with every heartbeat, and her radial pulse is very strong and abrupt. Her heart sounds are abnormal, with a long diastolic murmur at the right sternal margin in the second intercostal space.

Her doctor orders an echocardiogram, which shows that Dr. El-Sheikh has severe aortic regurgitation (backflow of blood through an incompetent valve) and an enlarged left ventricle.

1. Why does an incompetent aortic valve cause a diastolic murmur?

2. How will blood leaking through the incompetent aortic valve during diastole affect the preload and stroke volume of the left ventricle?

3. Why is Dr. El-Sheikh's systolic blood pressure abnormally high and her diastolic blood pressure abnormally low?

4. **+ NCLEX-STYLE** In a patient with an incompetent aortic valve, the nurse expects the patient to show signs and symptoms of which of the following?
   a. Peripheral congestion
   b. Pulmonary congestion
   c. Cerebral congestion
   d. Hepatic congestion

5. The additional muscle in Dr. El-Sheikh's enlarged left ventricle means that it requires more oxygen. Which arteries are most important for supplying this ventricle?

6. **+ NCLEX-STYLE** Although Dr. El-Sheikh has a high systolic blood pressure, her heart rate is also high. This is most likely due to:
   a. The Bainbridge reflex, caused by increased atrial filling
   b. Parasympathetic activation
   c. Excessive vagus nerve firing
   d. Sympathetic stimulation

18

19

The Cardiovascular System: Blood Vessels

In this chapter, you will learn that

Blood vessels are dynamic structures that control the delivery of blood to body tissues

by exploring

Part 1 Blood Vessel Structure and Function

by examining

19.1 Structure of blood vessel walls

forms

- **19.2 Arteries**
- **19.3 Capillaries**
- **19.4 Veins**

some form

19.5 Anastomoses

Part 2 Physiology of Circulation

by asking

19.6 How are flow, pressure, and resistance related?

and

19.7 What is blood pressure and how does it differ in arteries, capillaries, and veins?

19.9 How is blood flow through tissues controlled?

looking closer at

19.10 Capillary exchange

and

19.8 How is blood pressure regulated?

Neural controls Ch. 14

Hormonal controls Ch. 16

Short-term control

Long-term control

Renal mechanisms Ch. 26

Part 3 Circulatory Pathways

by exploring

19.11 Principal vessels of the systemic circulation

Arteries

Veins

and finally, exploring

Developmental Aspects of Blood Vessels

CAREER CONNECTION

Watch a video to learn how the chapter content is used in a real health care setting. Go to **Mastering** A&P® > Study Area > Animations and Videos or use quick access URL https://goo.gl/88srfV

Blood vessels are sometimes compared to a system of pipes with blood circulating in them, but this analogy is only a starting point. Unlike rigid pipes, blood vessels are dynamic structures that pulsate, constrict, relax, and even proliferate (multiply). In this chapter we examine the structure and function of these important circulatory passageways.

The **blood vessels** of the body form a closed delivery system that begins and ends at the heart. The idea that blood circulates in the body dates back to the 1620s with the inspired experiments of William Harvey, an English physician. Prior to that time, people thought, as proposed by the ancient Greek physician Galen, that blood moved through the body like an ocean tide, first moving out from the heart and then ebbing back in the same vessels.

- In the pulmonary circulation, the arteries, still defined as the vessels leading away from the heart, carry oxygen-poor blood to the lungs, and the veins carry oxygen-rich blood from the lungs to the heart.

- In the special umbilical vessels of a fetus, the roles of veins and arteries also differ with respect to oxygenation.

Of all the blood vessels, only the capillaries have intimate contact with tissue cells and directly serve cellular needs. Exchanges between the blood and tissue cells occur primarily through the gossamer-thin capillary walls.

Figure 19.1 summarizes how these vascular channels relate to one another and to vessels of the lymphatic system. The lymphatic system recovers fluids that leak from the blood vessels and is described in Chapter 20.

PART 1

BLOOD VESSEL STRUCTURE AND FUNCTION

The three major types of blood vessels are *arteries*, *capillaries*, and *veins*. As the heart contracts, it forces blood into the large arteries leaving the ventricles. The blood then moves into successively smaller arteries, finally reaching their smallest branches, the *arterioles* (ar-te′re-ōlz; "little arteries"), which feed into the capillary beds of body organs and tissues. Blood drains from the capillaries into *venules* (ven′ūlz), the smallest veins, and then on into larger and larger veins that merge to form the large veins that ultimately empty into the heart. Altogether, the blood vessels in the adult human stretch for about 100,000 km (60,000 miles) through the internal body landscape!

Arteries carry blood *away from* the heart, so they are said to "branch," "diverge," or "fork" as they form smaller and smaller divisions. **Veins**, by contrast, carry blood *toward* the heart and so are said to "join," "merge," and "converge" into the successively larger vessels approaching the heart. In the systemic circulation, arteries always carry oxygenated blood and veins always carry oxygen-poor blood. The opposite is true in two special locations:

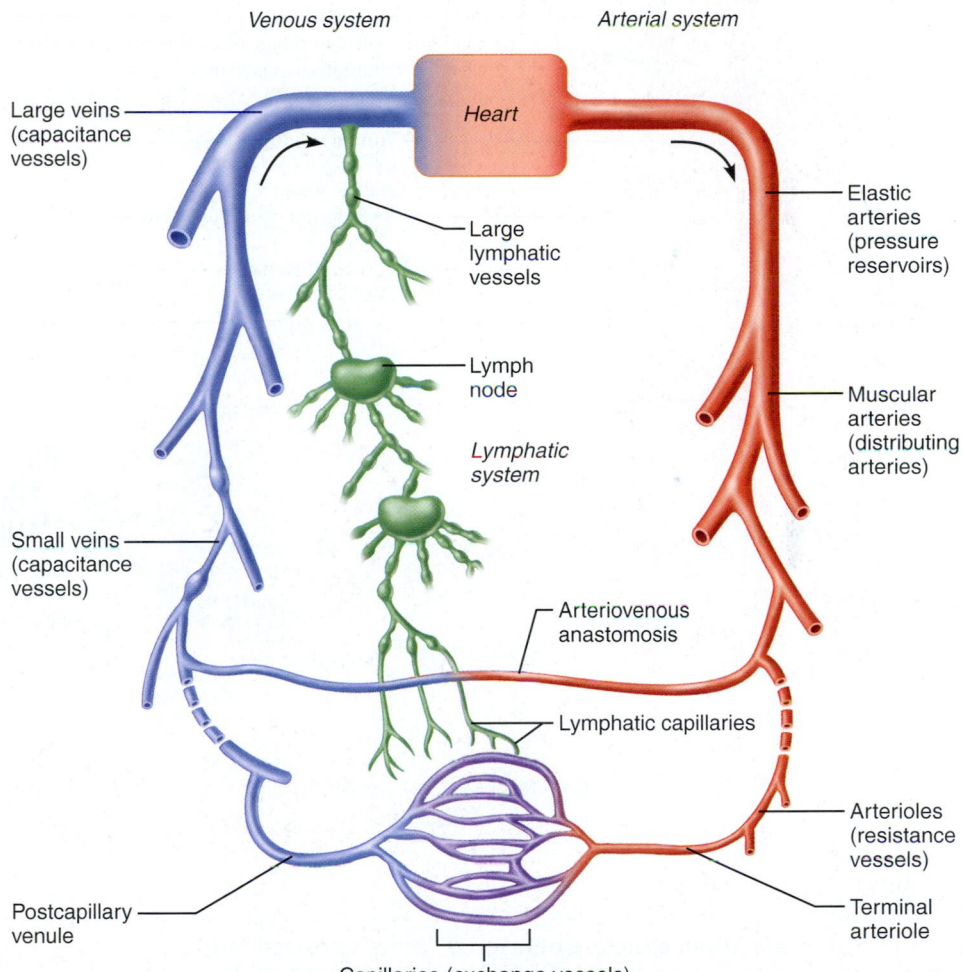

Figure 19.1 The relationship of blood vessels to each other and to lymphatic vessels. Lymphatic vessels recover excess tissue fluid and return it to the blood.

Labels in figure:
Venous system
Arterial system
Large veins (capacitance vessels)
Heart
Large lymphatic vessels
Elastic arteries (pressure reservoirs)
Lymph node
Muscular arteries (distributing arteries)
Lymphatic system
Small veins (capacitance vessels)
Arteriovenous anastomosis
Lymphatic capillaries
Postcapillary venule
Arterioles (resistance vessels)
Terminal arteriole
Capillaries (exchange vessels)

19

(a)

Artery   Vein

Artery   **Vein**

Tunica intima
• Endothelium
• Subendothelial layer
• Internal elastic membrane

Tunica media
(smooth muscle and
elastic fibers)
• External elastic membrane

Tunica externa
(collagen fibers)
• Vasa vasorum

Valve

Capillary network

Lumen   Lumen

Basement membrane

Capillary   Endothelial cells

(b)

Figure 19.2 Generalized structure of arteries, veins, and capillaries.
(a) Light photomicrograph of a muscular artery and the corresponding vein in cross
section (6×). **(b)** Comparison of wall structure of arteries, veins, and capillaries.
Note that the tunica media is thicker than the tunica externa in arteries and that the
opposite is true in veins.

Practice Histology questions: MasteringA&P® >
Study Area > Lab Tools > PAL

19

19.1 Most blood vessel walls have three layers

Learning Outcomes

▶ Describe the three layers that typically form the wall of a blood vessel, and state the function of each.

▶ Define vasoconstriction and vasodilation.

The walls of all blood vessels, except the very smallest, have three distinct layers, or *tunics* ("coverings"), that surround a central blood-containing space, the vessel **lumen** (**Figure 19.2**).

The innermost tunic is the **tunica intima** (in'tĭ-mah). The name is easy to remember once you know that this tunic is in *intimate* contact with the blood in the lumen. The tunica intima contains the **endothelium**, the simple squamous epithelium that lines the lumen of all vessels (◀ **p. 121**). The endothelium is continuous with the endocardial lining of the heart, and its flat cells fit closely together, forming a slick surface that minimizes friction as blood moves through the lumen. In vessels larger than 1 mm in diameter, a *subendothelial layer*, consisting of a basement membrane and loose connective tissue, supports the endothelium.

The middle tunic, the **tunica media** (me'de-ah), is mostly circularly arranged smooth muscle cells and sheets of elastin. The activity of the smooth muscle is regulated by sympathetic *vasomotor nerve fibers* of the autonomic nervous system and a whole battery of chemicals. Depending on the body's needs at any given moment, regulation causes either **vasoconstriction** (lumen diameter decreases as the smooth muscle contracts) or **vasodilation** (lumen diameter increases as the smooth muscle relaxes). The activities of the tunica media are critical in regulating circulatory dynamics because small changes in vessel diameter greatly influence blood flow and blood pressure. Generally, the tunica media is the bulkiest layer in arteries, which bear the chief responsibility for maintaining blood pressure and circulation.

The outermost layer of a blood vessel wall, the **tunica externa** (also called the *tunica adventitia*; ad"ven-tish'e-ah; "coming from outside"), is composed largely of loosely woven collagen fibers that protect and reinforce the vessel, and anchor it to surrounding structures. The tunica externa is infiltrated with nerve fibers, lymphatic vessels, and, in larger veins, a network of elastic fibers. In larger vessels, the tunica externa contains a system of tiny blood vessels, the **vasa vasorum** (va'sah va-sor'um)—literally, "vessels of the vessels"—that nourish the more external tissues of the blood vessel wall. The innermost (luminal) portion of the vessel obtains nutrients directly from blood in the lumen.

The three vessel types vary in length, diameter, wall thickness, and tissue makeup (see **Table 19.1**).

Check Your Understanding

1. Which branch of the autonomic nervous system innervates blood vessels? Which layer of the blood vessel wall do these nerves innervate? What are the effectors (cells that carry out the response)?

2. When vascular smooth muscle contracts, what happens to the diameter of the blood vessel? What is this called?

For answers, see Answers Appendix.

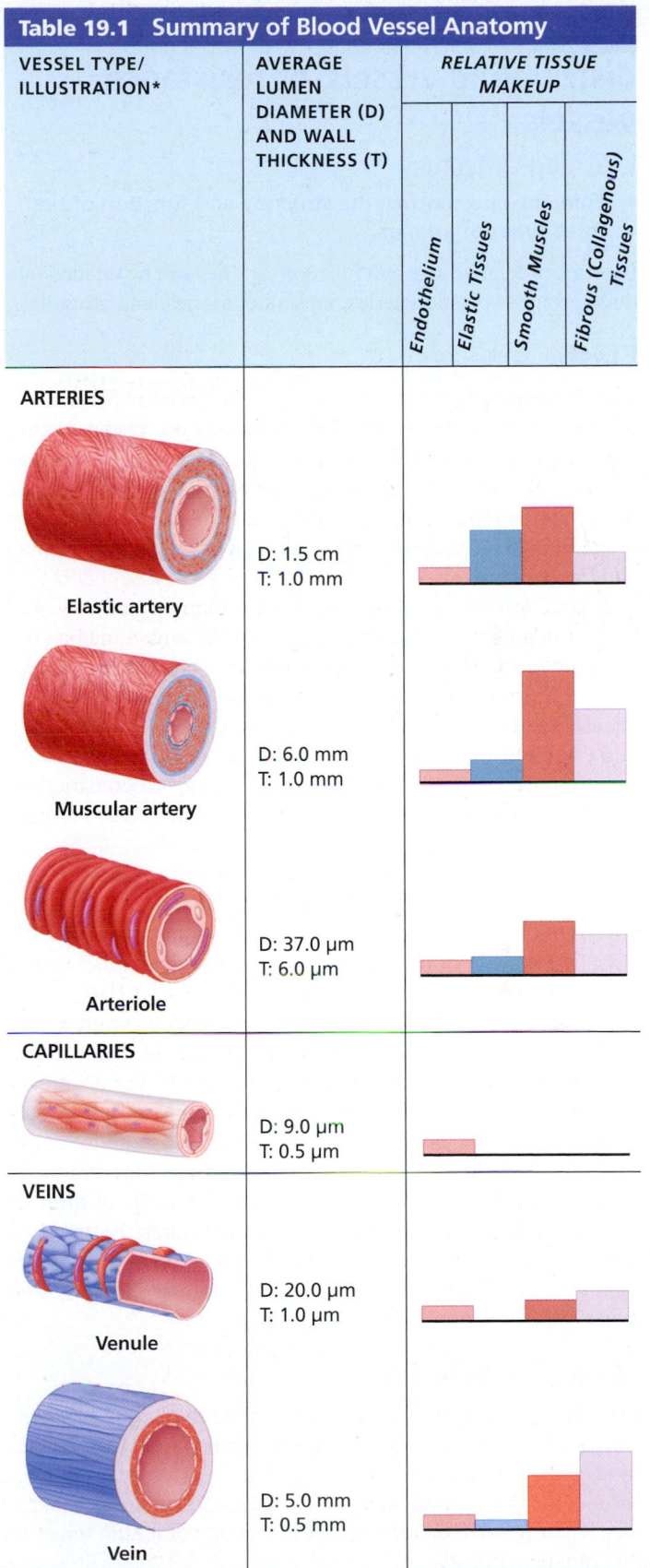

Table 19.1   Summary of Blood Vessel Anatomy

VESSEL TYPE/ ILLUSTRATION*	AVERAGE LUMEN DIAMETER (D) AND WALL THICKNESS (T)	RELATIVE TISSUE MAKEUP			
		Endothelium	Elastic Tissues	Smooth Muscles	Fibrous (Collagenous) Tissues
ARTERIES					
Elastic artery	D: 1.5 cm T: 1.0 mm				
Muscular artery	D: 6.0 mm T: 1.0 mm				
Arteriole	D: 37.0 μm T: 6.0 μm				
CAPILLARIES	D: 9.0 μm T: 0.5 μm				
VEINS					
Venule	D: 20.0 μm T: 1.0 μm				
Vein	D: 5.0 mm T: 0.5 mm				

*Size relationships are not proportional. Smaller vessels are drawn relatively larger so detail can be seen. See column 2 for actual dimensions.

19

19.2 Arteries are pressure reservoirs, distributing vessels, or resistance vessels

Learning Outcome

▶ Compare and contrast the structure and function of the three types of arteries.

In terms of relative size and function, arteries can be divided into three groups—elastic arteries, muscular arteries, and arterioles.

Elastic Arteries

Elastic arteries are the thick-walled arteries near the heart—the aorta and its major branches (Figure 19.1). These arteries are the largest in diameter, ranging from 2.5 cm to 1 cm, and the most elastic (Table 19.1). Because their large lumens make them low-resistance pathways that conduct blood from the heart to medium-sized arteries, elastic arteries are sometimes called *conducting arteries*.

Elastic arteries contain more elastin than any other vessel type. It is present in all three tunics, but the tunica media contains the most. There the elastin constructs concentric "holey" sheets of elastic connective tissue that look like slices of Swiss cheese sandwiched between layers of smooth muscle cells. Although elastic arteries also contain substantial amounts of smooth muscle, they are relatively inactive in vasoconstriction. Thus, in terms of function, they can be visualized as simple elastic tubes.

Elastic arteries are *pressure reservoirs*, expanding and recoiling as the heart ejects blood. Consequently, blood flows fairly continuously rather than starting and stopping with the pulsating rhythm of the heartbeat. If the blood vessels become hard and unyielding, as in atherosclerosis, blood flows more intermittently, similar to the way water flows through a hard rubber garden hose attached to a faucet. When the faucet is on, the high pressure makes the water gush out of the hose. But when the faucet is shut off, the water flow abruptly becomes a trickle and then stops, because the hose walls cannot recoil to keep the water under pressure. Also, without the pressure-smoothing effect of the elastic arteries, the walls of arteries throughout the body experience higher pressures. Battered by high pressures, the arteries eventually weaken and may balloon out (as an *aneurysm*) or even burst (see **A Closer Look** on p. 760).

Muscular Arteries

Distally the elastic arteries give way to the **muscular arteries**, which deliver blood to specific body organs (and so are sometimes called *distributing arteries*). Muscular arteries account for most of the named arteries studied in the anatomy laboratory. Their internal diameter ranges from that of a little finger to that of a pencil lead.

Proportionately, muscular arteries have the thickest tunica media of all vessels. Their tunica media contains relatively more smooth muscle and less elastic tissue than do elastic arteries

(Table 19.1). For this reason, they are more active in vasoconstriction and less capable of stretching. In muscular arteries, however, there *is* an *elastic membrane* on each face of the tunica media.

Arterioles

The smallest of the arteries, **arterioles** have a lumen diameter ranging from 0.3 mm down to 10 μm. Larger arterioles have all three tunics, but their tunica media is chiefly smooth muscle with a few scattered elastic fibers. Smaller arterioles, which lead into the capillary beds, are little more than a single layer of smooth muscle cells spiraling around the endothelial lining.

Minute-to-minute blood flow into the capillary beds is determined by arteriolar diameter, which varies in response to changing neural, hormonal, and local chemical influences. Changing diameter changes resistance to blood flow, and so arterioles are called *resistance vessels*. When arterioles constrict, the tissues served are largely bypassed. When arterioles dilate, blood flow into the local capillaries increases dramatically.

Check Your Understanding

3. Name the type of artery that matches each description: major role in dampening the pulsatile pressure of heart contractions; vasodilation or constriction determines blood flow to individual capillary beds; have the thickest tunica media relative to their lumen size.

For answers, see Answers Appendix.

19.3 Capillaries are exchange vessels

Learning Outcome

▶ Describe the structure and function of a capillary bed.

The microscopic **capillaries** are the smallest blood vessels (Figure 19.2b). Their exceedingly thin walls consist of just a thin tunica intima surrounded by a basement membrane (◀ p. 118). In some cases, a single endothelial cell forms the entire circumference of the capillary wall. At strategic locations along the outer surface of some capillaries are spider-shaped **pericytes**, contractile stem cells that can generate new vessels or scar tissue, stabilize the capillary wall, and help control capillary permeability (**Figure 19.3a**).

Average capillary length is 1 mm and average lumen diameter is 8–10 μm, just large enough for red blood cells to slip through in single file. Most tissues have a rich capillary supply, but there are exceptions. Tendons and ligaments are poorly vascularized (and so heal poorly). Cartilage and epithelia lack capillaries but receive nutrients from blood vessels in nearby connective tissues, and the avascular cornea and lens of the eye receive nutrients from the aqueous humor.

If we compare arteries and arterioles to expressways and roads, capillaries are the back alleys and driveways that provide direct access to nearly every cell in the body. Given their location and thin walls, capillaries are ideally suited for their role—exchange of materials (gases, nutrients, hormones, and so on) between the blood and the interstitial fluid. We describe

(a) Continuous capillary

Continuous capillaries are the least permeable and most common.

- Abundant in skin, muscles, lungs, and CNS.
- Often have associated pericytes.
- Pinocytotic vesicles ferry fluid across the endothelial cell.
- Most continuous capillaries have intercellular clefts between endothelial cells. However, brain capillary endothelial cells lack intercellular clefts and have tight junctions around their entire perimeter. (This is the structural basis of the *blood brain barrier* described in Chapter 12.)

(b) Fenestrated capillary

Fenestrated capillaries have large fenestrations (pores) that increase permeability.

- Occur in areas of active filtration (e.g., kidney) or absorption (e.g., small intestine), and areas of endocrine hormone secretion.
- Fenestrations are Swiss cheese–like holes that tunnel *through* endothelial cells.
- Fenestrations are usually covered by a very thin layer of condensed extracellular glycoproteins. This layer has little effect on solute and fluid movement.
- In some digestive tract organs, the number of fenestrations in capillaries increases during active absorption of nutrients.

(c) Sinusoid capillary

Sinusoid capillaries are the most permeable and occur in limited locations.

- Occur in liver, bone marrow, spleen, and adrenal medulla.
- Have large intercellular clefts as well as fenestrations; few tight junctions.
- Have incomplete basement membranes.
- Are irregularly shaped and have larger lumens than other capillaries.
- Allow large molecules and even cells to pass across their walls.
- Blood flows *slowly* through their tortuous channels.
- Macrophages may extend processes through the clefts to catch "prey" or, in liver, form part of the sinusoid wall.

Figure 19.3 Capillary structure.

19

these exchanges later in this chapter. Here, we focus on capillary structure.

Types of Capillaries

Structurally, there are three types of capillaries—*continuous, fenestrated,* and *sinusoid.* As you study their properties in Figure 19.3, notice that all three types have *tight junctions* that join their endothelial cells together. However, these junctions are usually incomplete and leave gaps of unjoined membrane called **intercellular clefts**, which allow limited passage of fluids and small solutes. Leakier capillaries have specialized passageways that increase fluid movement.

Capillary Beds

Capillaries do not function independently. Instead they form interweaving networks called **capillary beds**. The flow of blood from an arteriole to a venule—that is, through a capillary bed—is called the **microcirculation**. In most body regions, a **terminal arteriole** branches into 10 to 20 **capillaries** (*exchange vessels*) that form the capillary bed. These then drain into a **postcapillary venule** (**Figure 19.4**). Blood flow through the capillary bed is controlled by the diameter of the terminal arteriole as well as by all of the arterioles upstream from it. As blood flows through the capillaries, it takes part in exchanges of gases, nutrients, and wastes with the surrounding tissue cells.

Local chemical conditions and arteriolar vasomotor nerve fibers regulate the amount of blood entering a capillary bed. A bed may be flooded with blood or almost completely bypassed, depending on conditions in the body or in that specific organ. For example, suppose you have just eaten and are sitting relaxed, listening to your favorite musical group. Food is being digested, and blood is circulating freely through the capillaries of your gastrointestinal organs to receive the breakdown products of digestion. Between meals, however, most of these same capillary pathways are closed.

To take another example, when you exercise vigorously, blood is rerouted from your digestive organs (food or no food) to the

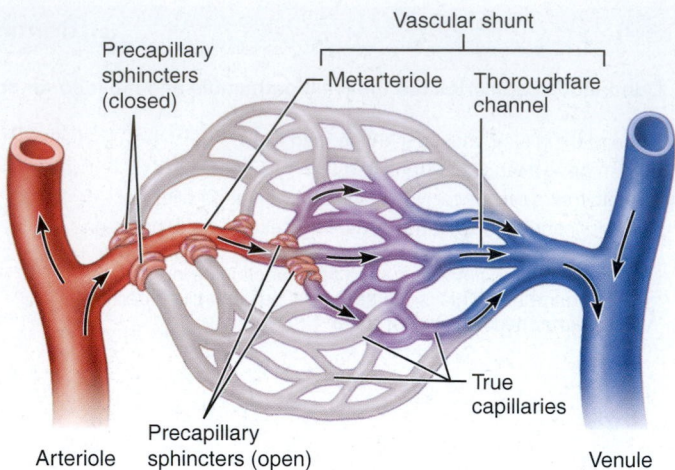

Figure 19.5 Anatomy of a special (mesenteric) capillary bed.

capillary beds of your skeletal muscles where it is more immediately needed. This rerouting helps explain why vigorous exercise right after a meal can cause indigestion or abdominal cramps.

In the serous membranes that hold the intestines in place (mesenteries), there are two additional features that form a special arrangement of capillaries (**Figure 19.5**):

- There is a *vascular shunt* that directly connects the terminal arteriole to the postcapillary venule, so that blood can bypass the *true capillaries*. This shunt consists of a *metarteriole* and a *thoroughfare channel*.

- As it branches from the metarteriole, each true capillary is surrounded by a cuff of smooth muscle called a **precapillary sphincter**. The precapillary sphincter acts as a valve to regulate blood flow into the capillary. The precapillary sphincters are not innervated, and so are controlled only by local chemical conditions.

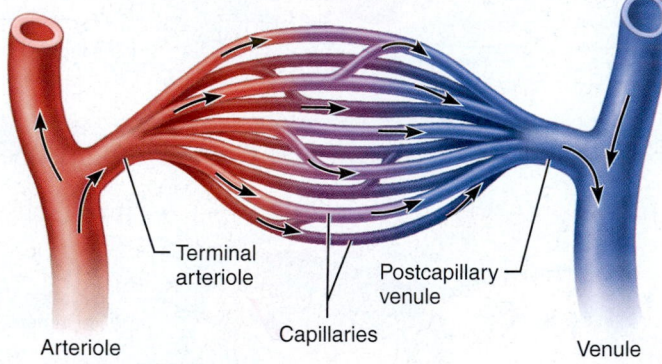

(a) Arterioles dilated—blood flows through capillaries.

Check Your Understanding

4. **APPLY** Look at Figure 19.4 and assume that the capillary bed depicted is in your calf muscle. Which condition—(a) or (b)—would the bed be in if you were doing calf raises at the gym?

For answers, see Answers Appendix.

(b) Arterioles constricted—no blood flows through capillaries.

Figure 19.4 Anatomy of a typical capillary bed.

19.4 Veins are blood reservoirs that return blood toward the heart

Learning Outcome

▶ Describe the structure and function of veins, and explain how veins differ from arteries.

Veins carry blood from the capillary beds toward the heart. Along the route, the diameter of successive venous vessels increases, and their walls gradually thicken as they progress from venules to larger and larger veins.

19

Venules

Capillaries unite to form **venules**, which range from 8 to 100 μm in diameter. The smallest venules, the *postcapillary venules*, consist entirely of endothelium around which pericytes congregate. Postcapillary venules are extremely porous (more like capillaries than veins in this way), and fluid and white blood cells move easily from the bloodstream through their walls. Indeed, a well-recognized sign of inflammation is adhesion of white blood cells to the postcapillary venule endothelium, followed by their migration through the wall into the inflamed tissue.

Larger venules have a thin tunica media (one or two layers of smooth muscle cells) and a thin tunica externa as well.

Veins

Venules join to form *veins*. Veins usually have three distinct tunics, but their walls are always thinner and their lumens larger than those of corresponding arteries (see Figure 19.2 and Table 19.1). Consequently, in histological preparations, veins are usually collapsed and their lumens appear slitlike.

There is relatively little smooth muscle or elastin in the tunica media, which is poorly developed and tends to be thin even in the largest veins. The tunica externa is the heaviest wall layer. Consisting of thick longitudinal bundles of collagen fibers and elastic networks, it is often several times thicker than the tunica media. In the largest veins—the venae cavae, which return blood directly to the heart—longitudinal bands of smooth muscle make the tunica externa even thicker.

With their large lumens and thin walls, veins can accommodate a fairly large blood volume. Veins are called *capacitance vessels* and *blood reservoirs* because they can hold up to 65% of the body's blood supply at any time (**Figure 19.6**). Even so, these distensible vessels are usually not filled to capacity.

The walls of veins can be much thinner than arterial walls without danger of bursting because the blood pressure in veins is low. However, the low-pressure condition demands several structural adaptations to ensure that veins return blood to the heart at the same rate it was pumped into the circulation. One such adaptation is their large-diameter lumens, which offer relatively little resistance to blood flow.

Venous Valves

Venous valves prevent blood from flowing backward in veins just as valves do in the heart, and represent another adaptation to compensate for low venous pressure. They are formed from folds of the tunica intima and resemble the semilunar valves of the heart (◄ **p. 680**) (see Figure 19.2). Venous valves are most abundant in the veins of the limbs, where gravity opposes the upward flow of blood. They are usually absent in veins of the thoracic and abdominal body cavities.

The effectiveness of venous valves is demonstrated by this simple experiment: Hang one hand by your side until the blood vessels on its dorsal aspect distend with blood. Next place two fingertips against one of the distended veins, and pressing firmly, move the superior finger proximally along the vein and then release that finger. The vein will remain collapsed (flat) despite the pull of gravity. Finally, remove your distal fingertip and watch the vein refill with blood.

> **HOMEOSTATIC IMBALANCE 19.1**   **CLINICAL**
>
> **Varicose veins** are veins that are tortuous and dilated because of incompetent (leaky) valves. More than 15% of adults suffer from varicose veins, usually in the lower limbs.
>
> Several factors contribute, including heredity and conditions that hinder venous return, such as prolonged standing in one position, obesity, or pregnancy. Both the "potbelly" of an overweight person and the enlarged uterus of a pregnant woman exert downward pressure on vessels of the groin, restricting return of blood to the heart. Consequently, blood pools in the lower limbs, and with time, the valves weaken and the venous walls stretch. Superficial veins, which receive little support from surrounding tissues, are especially susceptible.
>
> Elevated venous pressure can also cause varicose veins. For example, straining to deliver a baby or have a bowel movement raises intra-abdominal pressure, preventing blood from draining from anal veins. The resulting varicosities in the anal veins are called *hemorrhoids* (hem′ŏ-roidz). ●●●●●●

Venous Sinuses

Venous sinuses are highly specialized, flattened veins with extremely thin walls composed only of endothelium. They are supported by the tissues that surround them, rather than by any additional tunics. (Do not confuse venous sinuses with the

Pulmonary blood vessels 12%

Systemic arteries and arterioles 15%

Pulmonary blood vessels supply the lungs.

Heart 8%

Capillaries 5%

Systemic veins and venules 60%

Systemic veins:
- Supply all of the body except the lungs
- Are distensible
- Contain a large proportion of the blood volume and so are called **capacitance vessels** or **blood reservoirs**

Figure 19.6 Relative proportion of blood volume throughout the cardiovascular system.

19

air-filled spaces in bones, the *paranasal sinuses*.) Examples of venous sinuses include the coronary sinus of the heart and the dural venous sinuses of the brain. The dural venous sinuses, which receive cerebrospinal fluid and blood draining from the brain, are reinforced by the tough dura mater that covers the brain surface (◀ p. 464).

Check Your Understanding

5. What is the function of venous valves? What forms the valves?

6. In the systemic circuit, which contains more blood—arteries or veins—or is it the same?

For answers, see Answers Appendix.

19.5   Anastomoses are special interconnections between blood vessels

Learning Outcome

▶ Explain the importance of vascular anastomoses.

Blood vessels form special interconnections called **vascular anastomoses** (ah-nas″to-mo′sēz; "coming together"). Most organs receive blood from more than one arterial branch, and arteries supplying the same territory often merge, forming **arterial anastomoses**. These anastomoses provide alternate pathways, called **collateral channels**, for blood to reach a given body region. If one branch is cut or blocked by a clot, the collateral channel can often provide sufficient blood to the area.

Arterial anastomoses occur around joints, where active movement may hinder blood flow through one channel. They are also common in abdominal organs, the heart, and the brain (for example, the *cerebral arterial circle* in Figure 19.24c on p. 741). Arteries that supply the retina, kidneys, and spleen either do not anastomose or have a poorly developed collateral circulation. If their blood flow is interrupted, cells supplied by such vessels die.

The metarteriole–thoroughfare channel shunts of some capillary beds that connect arterioles and venules are examples of **arteriovenous anastomoses**. Veins interconnect much more freely than arteries, and **venous anastomoses** are common. (You may be able to see venous anastomoses through the skin on the dorsum of your hand.) Because venous anastomoses are abundant, an occluded vein rarely blocks blood flow or leads to tissue death.

Check Your Understanding

7. Which have more anastomoses, arteries or veins?

For answers, see Answers Appendix.

PART 2

PHYSIOLOGY OF CIRCULATION

Have you ever climbed a mountain? Well, get ready to climb a metaphorical mountain as you learn about circulatory dynamics.

Like scaling a mountain, tackling blood pressure regulation and other topics of cardiovascular physiology is challenging while you're doing it, and exhilarating when you succeed. Let's begin the climb.

To sustain life, blood must be kept circulating. By now, you are aware that the heart is the pump, the arteries are pressure reservoirs and conduits, the arterioles are resistance vessels that control distribution, the capillaries are exchange sites, and the veins are conduits and blood reservoirs. Now for the dynamics of this system.

19.6   Blood flows from high to low pressure against resistance

Learning Outcome

▶ Define blood flow, blood pressure, and resistance, and explain the relationships between these factors.

First we need to define three physiologically important terms—blood flow, blood pressure, and resistance—and examine how these factors relate to the physiology of blood circulation.

Definition of Terms

Blood Flow

Blood flow is the volume of blood flowing through a vessel, an organ, or the entire circulation in a given period (ml/min). If we consider the entire vascular system, blood flow is equivalent to cardiac output (CO), and under resting conditions, it is relatively constant. At any given moment, however, blood flow through *individual* body organs may vary widely according to their immediate needs.

Blood Pressure (BP)

Blood pressure (BP), the force per unit area exerted on a vessel wall by the contained blood, is expressed in millimeters of mercury (mm Hg). For example, a blood pressure of 120 mm Hg is equal to the pressure exerted by a column of mercury 120 mm high.

Unless stated otherwise, the term *blood pressure* means systemic arterial blood pressure in the largest arteries near the heart. The hydrostatic pressure gradient—the *differences* in blood pressure within the vascular system—provides the driving force that keeps blood moving, always from an area of higher pressure to an area of lower pressure, through the body.

Resistance

Resistance is opposition to flow and is a measure of the amount of friction blood encounters as it passes through the vessels. Because most friction is encountered in the peripheral (systemic) circulation, well away from the heart, we generally use the term **total peripheral resistance (TPR)**.

There are three important sources of resistance: blood viscosity, vessel length, and vessel diameter. You already know more about these sources of resistance than you think you do. If you wanted to drink this thick (viscous) milkshake, would you pick the short, wide straw or the long, narrow straw? We bet you picked the short, wide straw. Let's see why that makes sense.

Blood Viscosity The internal resistance to flow that exists in all fluids is *viscosity* (vis-kos′ĭ-te) and is related to the thickness or "stickiness" of a fluid. The greater the viscosity, the less easily molecules slide past one another and the more difficult it is to get and keep the fluid moving. Like our milkshake, blood is much more viscous than water. Because it contains formed elements and plasma proteins, it flows more slowly under the same conditions.

Blood viscosity is fairly constant. However, conditions such as polycythemia (excessive numbers of red blood cells) can increase blood viscosity and so resistance increases. On the other hand, if the red blood cell count is low, as in some anemias, blood is less viscous and total peripheral resistance declines.

Total Blood Vessel Length The relationship between total blood vessel length and resistance is straightforward: the longer the vessel, the greater the resistance. (That's why a shorter straw is easier to drink from.) For example, an infant's blood vessels lengthen as he or she grows to adulthood, and so both total peripheral resistance and blood pressure increase.

Blood Vessel Diameter The relationship between blood vessel diameter and resistance is also straightforward: the smaller the diameter, the greater the resistance. (That's why a wider straw is easier to drink from.)

Blood viscosity and vessel length are normally unchanging in the short term, and so the influence of these factors can be considered constant. However, blood vessel diameter changes frequently and significantly alters total peripheral resistance. How so? The answer lies in principles of fluid flow. Fluid close to the wall of a tube or channel is slowed by friction as it passes along the wall, whereas fluid in the center of the tube flows more freely and faster. You can verify this by watching the flow of water in a river. Water close to the bank hardly seems to move, while that in the middle of the river flows quite rapidly.

In a tube of a given size, the relative speed and position of fluid in the different regions of the tube's cross section remain constant, a phenomenon called *laminar flow* or *streamlining*. The smaller the tube, the greater the friction, because relatively more of the fluid contacts the tube wall, where its movement is impeded.

Resistance varies *inversely* with the *fourth power* of the vessel radius (one-half the diameter). This means, for example, that if the radius of a vessel doubles, the resistance drops to one-sixteenth of its original value ($r^4 = 2 \times 2 \times 2 \times 2 = 16$ and $1/r^4 = 1/16$). For this reason, the large arteries close to the heart, which do not change dramatically in diameter, contribute little to total peripheral resistance. Instead, the small-diameter arterioles, which can enlarge or constrict in response to neural and chemical controls, are the major determinants of total peripheral resistance.

When blood encounters either an abrupt change in vessel diameter or rough or protruding areas of the tube wall (such as the fatty plaques of atherosclerosis), the smooth laminar blood flow is replaced by *turbulent flow*, that is, irregular fluid motion where blood from the different laminae (different layers of the tube's cross section) mixes. Turbulence dramatically increases resistance.

Relationship between Flow, Pressure, and Resistance

Now that we have defined these terms, let's summarize the relationships between them.

- Blood flow (*F*) is *directly* proportional to the difference in blood pressure (ΔP) between two points in the circulation, that is, the blood pressure, or hydrostatic pressure, gradient. Consequently, when ΔP increases, blood flow increases, and when ΔP decreases, blood flow declines.

- Blood flow is *inversely* proportional to the total peripheral resistance (TPR) in the systemic circulation; if TPR increases, blood flow decreases.

We can express these relationships by the formula

$$F = \frac{\Delta P}{\text{TPR}}$$

Of these two factors influencing blood flow, TPR is far more important than ΔP in influencing local blood flow because TPR can easily be changed by altering blood vessel diameter. For example, when the arterioles serving a particular tissue dilate (decreasing the resistance), blood flow to that tissue increases, even though the systemic pressure is unchanged or may actually be falling.

Check Your Understanding

8. List three factors that determine resistance in a vessel. Which of these factors is physiologically most important?

9. APPLY Suppose vasoconstriction decreases the diameter of a vessel to one-third its size. What happens to the rate of flow through that vessel? Calculate the expected size of the change.

10. APPLY Here are three blood vessels. Assuming that the difference in pressure along the vessel length is the same for each, which would have the greatest blood flow? The least blood flow? Explain.

(a)

(b)

(c)

For answers, see Answers Appendix.

19.7 Blood pressure decreases as blood flows from arteries through capillaries and into veins

Learning Outcome

▶ Describe how blood pressure differs in the arteries, capillaries, and veins.

Any fluid driven by a pump through a circuit of closed channels operates under pressure, and the nearer the fluid is to the pump, the greater the pressure exerted on the fluid. Blood flow in blood vessels is no exception, and blood flows through the blood vessels along a pressure gradient, always moving from higher- to lower-pressure areas. Fundamentally, *the pumping action of the heart generates blood flow. Pressure results when flow is opposed by resistance.*

As illustrated in **Figure 19.7**, systemic blood pressure is highest in the aorta and declines throughout the pathway to finally reach 0 mm Hg in the right atrium. The steepest drop in blood pressure occurs in the arterioles, which offer the greatest resistance to blood flow. However, as long as a pressure gradient exists, no matter how small, blood continues to flow until it completes the circuit back to the heart.

Arterial Blood Pressure

Arterial blood pressure reflects two factors: (1) how much the elastic arteries close to the heart can stretch (their *compliance* or *distensibility*) and (2) the volume of blood forced into them at any time. If the amounts of blood entering and leaving the elastic arteries in a given period were equal, arterial pressure would be constant. Instead, as Figure 19.7 reveals, blood pressure is *pulsatile*—it rises and falls in a regular fashion—in the elastic arteries near the heart.

As the left ventricle contracts and expels blood into the aorta, it imparts kinetic energy to the blood, which stretches the elastic

aorta as aortic pressure reaches its peak. Indeed, if the aorta were opened during this period, blood would spurt upward 5 or 6 feet! This pressure peak generated by ventricular contraction is called the **systolic pressure** (sis-tah′lik) and averages 120 mm Hg in healthy adults. Blood moves forward into the arterial bed because the pressure in the aorta is higher than the pressure in the more distal vessels.

During diastole, the aortic valve closes, preventing blood from flowing back into the heart. The walls of the aorta (and other elastic arteries) recoil, maintaining sufficient pressure to keep the blood flowing forward into the smaller vessels. During this time, aortic pressure drops to its lowest level (approximately 70 to 80 mm Hg in healthy adults). This is called the **diastolic pressure** (di-as-tah′lik). You can picture the elastic arteries as pressure reservoirs that operate as auxiliary pumps. They keep blood circulating throughout the period of diastole, when the heart is relaxing. Essentially, the volume and energy of blood stored in the elastic arteries during systole are given back during diastole.

The difference between the systolic and diastolic pressures is called the **pulse pressure**. It is felt as a throbbing pulsation in an artery (a **pulse**) during systole as ventricular contraction forces blood into the elastic arteries and expands them. Increased stroke volume and faster blood ejection from the heart (a result of increased contractility) raise pulse pressure *temporarily*. Atherosclerosis chronically increases pulse pressure because the elastic arteries become less stretchy.

Because aortic pressure fluctuates up and down with each heartbeat, the important pressure to consider is the **mean arterial pressure (MAP)**—the pressure that propels the blood to the tissues. Diastole usually lasts longer than systole, so MAP is not simply the value halfway between systolic and diastolic pressures. Instead, it is roughly equal to the diastolic pressure plus one-third of the pulse pressure.

$$\text{MAP} = \text{diastolic pressure} + \frac{\text{pulse pressure}}{3}$$

For a person with a systolic blood pressure of 120 mm Hg and a diastolic pressure of 80 mm Hg:

$$\text{MAP} = 80 \text{ mm Hg} + \frac{40 \text{ mm Hg}}{3} = 93 \text{ mm Hg}$$

MAP and pulse pressure both decline with increasing distance from the heart. The MAP loses ground to the never-ending friction between the blood and the vessel walls, and the pulse pressure is gradually phased out in the less elastic muscular arteries, where elastic rebound of the vessels ceases to occur. At the end of the arterial tree, blood flow is steady and the pulse pressure has disappeared.

Clinical Monitoring of Circulatory Efficiency

Clinicians can assess the efficiency of a person's circulation by measuring pulse and blood pressure. These values, along with measurements of respiratory rate and body temperature, are referred to collectively as the body's **vital signs**. Let's examine how vital signs are determined or measured.

Figure 19.7 Blood pressure in various blood vessels of the systemic circulation.

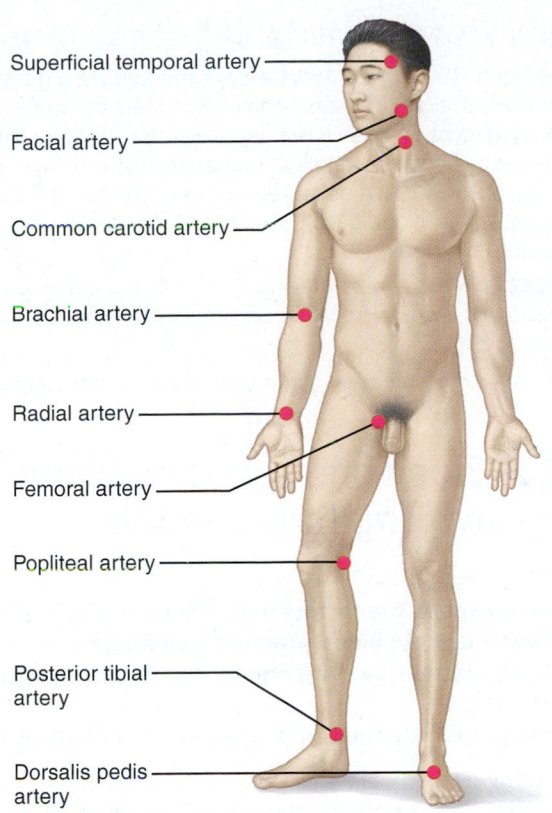

Superficial temporal artery

Facial artery

Common carotid artery

Brachial artery

Radial artery

Femoral artery

Popliteal artery

Posterior tibial artery

Dorsalis pedis artery

Figure 19.8 Body sites where the pulse is most easily palpated. (We discuss the specific arteries indicated on pp. 738–749.)

Taking a Pulse You can feel a pulse in any artery that lies close to the body surface by compressing the artery against firm tissue, and this provides an easy way to count heart rate. Because it is so accessible, the point where the radial artery surfaces at the wrist, the *radial pulse*, is routinely used to take a pulse measurement, but there are several other clinically important arterial pulse points (**Figure 19.8**).

These pulse points are also called **pressure points** because they are compressed to stop blood flow into distal tissues during hemorrhage. For example, if you seriously lacerate your hand, you can slow or stop the bleeding by compressing your radial or brachial artery.

Monitoring pulse rate is an easy way to assess the effects of activity, postural changes, and emotions on heart rate. For example, the pulse of a healthy man may be around 66 beats per minute when he is lying down, 70 when he sits up, and 80 if he suddenly stands. During vigorous exercise or emotional upset, pulse rates between 140 and 180 are not unusual because of sympathetic nervous system effects on the heart.

Measuring Blood Pressure Most often, you measure systemic arterial blood pressure indirectly in the brachial artery of the arm by the **auscultatory method** (aw-skul'tah-to"re). The steps of this procedure are:

1. Wrap the *blood pressure cuff*, or *sphygmomanometer* (sfig"mo-mah-nom'ĕ-ter; *sphygmo* = pulse), snugly around the person's arm just superior to the elbow.
2. Inflate the cuff until the cuff pressure exceeds systolic pressure. At this point, blood flow into the arm stops and a brachial pulse cannot be felt or heard.
3. Reduce the cuff pressure gradually and listen (auscultate) with a stethoscope for sounds in the brachial artery.

The pressure read when the first soft tapping sounds are heard (the first point at which a small amount of blood is spurting through the constricted artery) is systolic pressure. As the cuff pressure is reduced further, these sounds, called the *sounds of Korotkoff*, become louder and more distinct. However, when the artery is no longer constricted and blood flows freely, the sounds can no longer be heard. The pressure at which the sounds disappear is the diastolic pressure.

Capillary Blood Pressure

As Figure 19.7 shows, by the time blood reaches the capillaries, blood pressure has dropped to approximately 35 mm Hg and by the end of the capillary beds is only around 17 mm Hg. Such low capillary pressures are desirable because:

- Capillaries are fragile and high pressures would rupture them.
- Most capillaries are extremely permeable and so even the low capillary pressure can force solute-containing fluids (filtrate) out of the bloodstream into the interstitial space.

As we describe later in this chapter, these fluid flows are important for continuously refreshing the interstitial fluid.

Venous Blood Pressure

Unlike arterial pressure, which pulsates with each contraction of the left ventricle, venous blood pressure is steady and changes very little during the cardiac cycle. The pressure gradient in the veins, from venules to the termini of the venae cavae, is only about 15 mm Hg (that from the aorta to the ends of the arterioles is about 60 mm Hg).

The difference in pressure between an artery and a vein becomes very clear when the vessels are cut. If a vein is cut, the blood flows evenly from the wound, but a lacerated artery spurts blood. The very low pressure in the venous system reflects the cumulative effects of total peripheral resistance, which dissipates most of the energy of blood pressure (as heat) during each circuit.

Despite the structural modifications of veins (large lumens and valves), venous pressure is normally too low to promote adequate venous return. For this reason, three functional adaptations are critically important to venous return:

- **The muscular pump.** The **muscular pump** consists of skeletal muscle activity. As the skeletal muscles surrounding the deep veins contract and relax, they squeeze or "milk" blood

19

Venous valve
(open)

Contracted
skeletal
muscle

Venous valve
(closed)

Vein

Direction of
blood flow

Figure 19.9 The muscular pump. When contracting skeletal muscles press against a vein, they force open the valves proximal to the area of contraction and blood is propelled toward the heart. Backflowing blood closes the valves distal to the area of contraction.

toward the heart, and once blood passes each successive valve, it cannot flow back (**Figure 19.9**). People who earn their living in "standing professions," such as hairdressers, often have swollen ankles because blood pools in their feet and legs. Indeed, standing for prolonged periods may cause fainting because skeletal muscle inactivity reduces venous return.

- **The respiratory pump.** The **respiratory pump** moves blood up toward the heart as pressure changes in the ventral body cavity during breathing. As we inhale, abdominal pressure increases, squeezing local veins and forcing blood toward the heart. At the same time, the pressure in the chest decreases, allowing thoracic veins to expand and speeding blood entry into the right atrium.

- **Sympathetic venoconstriction.** Sympathetic venoconstriction reduces the volume of blood in the veins—the capacitance vessels. As the layer of smooth muscle around the veins constricts under sympathetic control, venous volume is reduced and blood is pushed toward the heart.

All three of these functional adaptations increase venous return, which increases stroke volume (by the Frank-Starling mechanism, ◄ **p. 697**) and therefore increases cardiac output.

Check Your Understanding

11. Use Figure 19.7 on p. 716 to answer the following questions: **(a)** In which type of vessel is the largest pressure drop? **(b)** Which type of vessel is the first in which there is no difference in pressure between systole and diastole? Why is this a good thing? **(c)** How much pressure remains at the start of the venules to get blood back to the heart? What else helps blood return to the heart?

12. APPLY Cole has a systolic pressure of 140 and a diastolic pressure of 80 mm Hg. What is his mean arterial pressure? His pulse pressure?

For answers, see Answers Appendix.

19.8 Blood pressure is regulated by short- and long-term controls

Learning Outcomes

▶ **List and explain the factors that influence blood pressure, and describe how blood pressure is regulated.**

▶ **Define hypertension. Describe its manifestations and consequences.**

▶ **Define circulatory shock. List several possible causes.**

Maintaining a steady flow of blood from the heart to the toes is vital for organs to function properly. In fact, making sure a person jumping out of bed in the morning does not keel over from inadequate blood flow to the brain requires the finely tuned cooperation of the heart, blood vessels, and kidneys—all supervised by the brain.

Maintaining blood pressure is critical for cardiovascular system homeostasis. Its regulation involves three key variables:

- Cardiac output
- Total peripheral resistance
- Blood volume

To see why these are the central variables, we use the formula about blood flow presented on p. 715. In the cardiovascular system, flow (F) is cardiac output (CO)—the blood flow of the entire circulation. P is blood pressure (MAP), and TPR is the total peripheral resistance (resistance of the blood vessels in the systemic circulation). If we rearrange the formula for blood flow, we can see how CO and TPR relate to blood pressure:

$$F = \Delta P / \text{TPR} \quad \text{or} \quad \text{CO} = \Delta P / \text{TPR} \quad \text{or} \quad \Delta P = \text{CO} \times \text{TPR}$$

As you can see, blood pressure varies *directly* with CO and TPR. Anything that increases cardiac output or total peripheral resistance increases blood pressure. Blood pressure also varies directly with blood volume because CO depends on blood volume (the heart can't pump out what doesn't enter its chambers).

From Chapter 18 (◄ **Figure 18.20, p. 697**), you know that CO is equal to *stroke volume* (ml/beat) times *heart rate* (beats/min), so anything that increases these two variables will also increase blood pressure. During stress, for example, the cardio-acceleratory center activates the sympathetic nervous system, which increases both heart rate (by acting on the SA node) and

Figure 19.10 Major factors that increase MAP. In addition, cardiac output increases as blood volume increases (not shown).

stroke volume (by enhancing cardiac muscle contractility). The resulting increase in CO increases MAP.

We also know that total peripheral resistance is determined by three variables, the most important of which is blood vessel diameter (◀ pp. 714–715). **Figure 19.10** summarizes the relationships between the factors controlling CO and resistance. Keep these relationships in mind as you read through the sections that follow, because each blood pressure regulation mechanism acts on one or more of these variables.

Also be aware that things aren't quite that simple in real life. A change in any variable that threatens blood pressure homeostasis is usually compensated for by changes in the other variables so that a constant blood pressure is maintained.

We will now explore two classes of mechanisms that regulate blood pressure. *Short-term regulation* by the nervous system and bloodborne hormones alters blood pressure by changing total peripheral resistance and CO. *Long-term regulation* alters blood volume via the kidneys. Figure 19.13 (p. 723) summarizes the influence of nearly all of the important factors.

Short-Term Regulation: Neural Controls

Neural controls alter both cardiac output and total peripheral resistance. We discussed neural control of cardiac output in Chapter 18, so we will focus on total peripheral resistance here. Neural controls of total peripheral resistance are directed at two main goals:

- Maintaining adequate MAP by altering blood vessel diameter on a moment-to-moment basis. (Remember, very small changes in blood vessel diameter cause substantial changes in total peripheral resistance, and so in systemic blood pressure.) Under conditions of low blood volume, all vessels except those supplying the heart and brain are constricted to allow as much blood as possible to flow to those two vital organs.

- Altering blood distribution to respond to specific demands of various organs. For example, during exercise blood is shunted temporarily from the digestive organs to the skeletal muscles.

Most neural controls operate via reflex arcs involving *baroreceptors* (pressure-sensitive mechanoreceptors that respond to changes in arterial pressure and stretch) and associated afferent fibers. These reflexes are integrated in the cardiovascular center of the medulla, and their output travels via autonomic fibers to the heart and vascular smooth muscle. Occasionally, inputs from *chemoreceptors* (receptors that respond to changes in blood levels of carbon dioxide, H^+, and oxygen) and higher brain centers also influence the neural control mechanism.

Role of the Cardiovascular Center

Several clusters of neurons in the medulla oblongata act together to integrate blood pressure control by altering cardiac output and blood vessel diameter. This **cardiovascular center** consists of the *cardiac centers* (the cardioacceleratory and cardioinhibitory centers discussed in Chapter 18, ◀ p. 689) and the **vasomotor center** that controls the diameter of blood vessels.

The vasomotor center transmits impulses at a fairly steady rate along sympathetic efferents called **vasomotor fibers**. These fibers exit from the T_1 through L_2 levels of the spinal cord and innervate the smooth muscle of blood vessels, mainly arterioles. As a result, the arterioles are almost always in a state of moderate constriction, called **vasomotor tone** (◀ p. 545).

The degree of vasomotor tone varies from organ to organ. Generally, arterioles of the skin and digestive viscera receive vasomotor impulses more frequently and tend to be more strongly constricted than those of skeletal muscles. Any increase in sympathetic activity produces generalized vasoconstriction and raises blood pressure. Decreased sympathetic activity allows the vascular smooth muscle to relax somewhat and lowers blood pressure to basal levels.

Cardiovascular center activity is modified by inputs from baroreceptors, chemoreceptors, and higher brain centers. Let's take a look.

Baroreceptor Reflexes

When arterial blood pressure rises, it activates **baroreceptors**. These stretch receptors are located in the *carotid sinuses* (dilations in the internal carotid arteries, which provide the major blood supply to the brain), in the *aortic arch*, and in the walls of nearly every large artery of the neck and thorax. When stretched, baroreceptors send a rapid stream of impulses to the

19

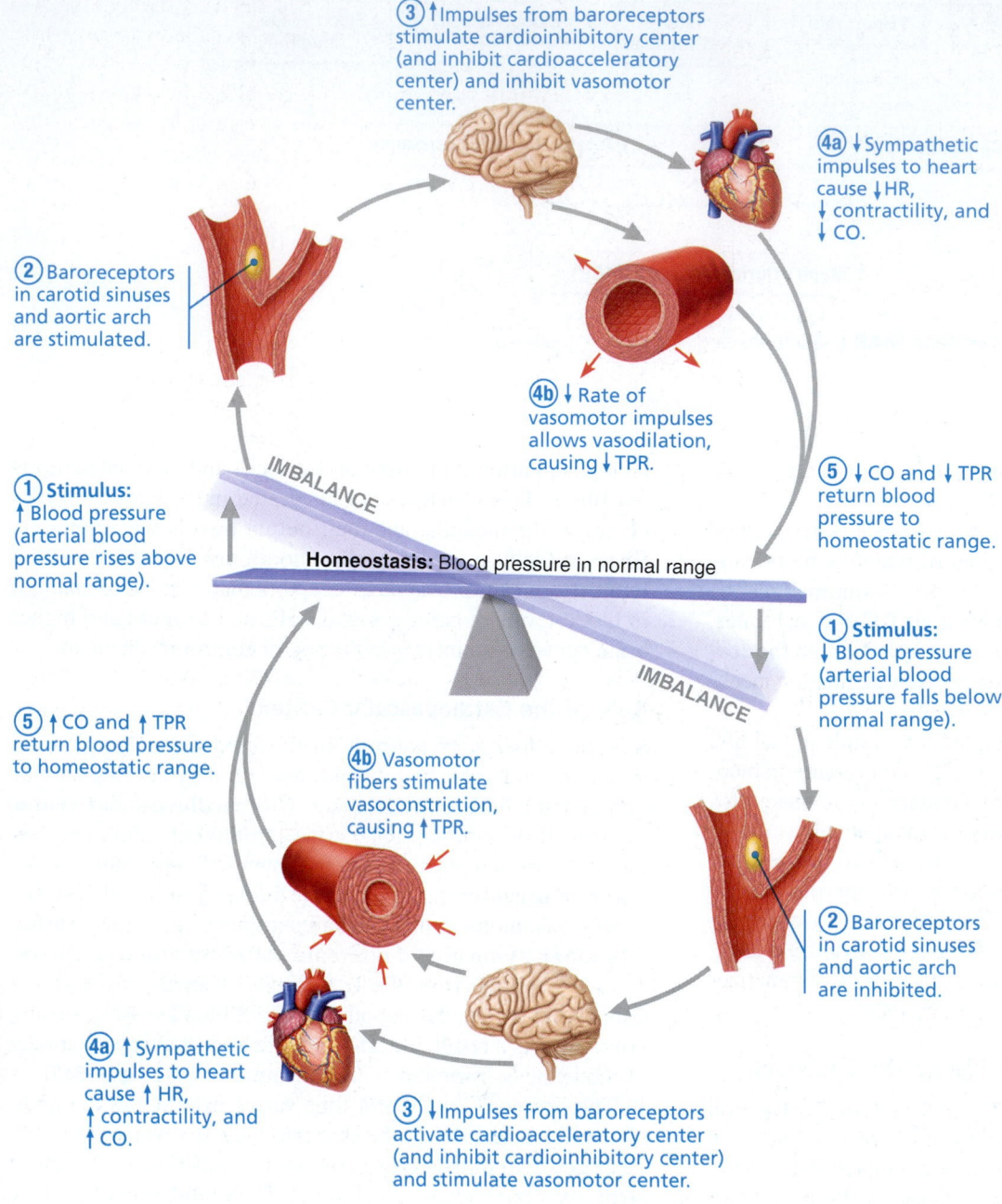

(3) ↑ Impulses from baroreceptors stimulate cardioinhibitory center (and inhibit cardioacceleratory center) and inhibit vasomotor center.

(2) Baroreceptors in carotid sinuses and aortic arch are stimulated.

(4a) ↓ Sympathetic impulses to heart cause ↓HR, ↓ contractility, and ↓ CO.

(1) Stimulus: ↑ Blood pressure (arterial blood pressure rises above normal range).

(4b) ↓ Rate of vasomotor impulses allows vasodilation, causing ↓TPR.

IMBALANCE

(5) ↓ CO and ↓ TPR return blood pressure to homeostatic range.

Homeostasis: Blood pressure in normal range

IMBALANCE

(1) Stimulus: ↓ Blood pressure (arterial blood pressure falls below normal range).

(5) ↑ CO and ↑ TPR return blood pressure to homeostatic range.

(4b) Vasomotor fibers stimulate vasoconstriction, causing ↑TPR.

(2) Baroreceptors in carotid sinuses and aortic arch are inhibited.

(4a) ↑ Sympathetic impulses to heart cause ↑ HR, ↑ contractility, and ↑ CO.

(3) ↓Impulses from baroreceptors activate cardioacceleratory center (and inhibit cardioinhibitory center) and stimulate vasomotor center.

Figure 19.11 Baroreceptor reflexes that help maintain blood pressure homeostasis. (CO = cardiac output; TPR = total peripheral resistance; HR = heart rate; BP = blood pressure)

cardiovascular center, inhibiting the vasomotor and cardio-acceleratory centers and stimulating the cardioinhibitory center. The result is a decrease in blood pressure (**Figure 19.11**).

Two mechanisms bring this about:

- **Vasodilation.** Decreased output from the vasomotor center allows arterioles and veins to dilate. **Arteriolar vasodilation** reduces total peripheral resistance, so MAP falls. **Venodilation** shifts blood to the venous reservoirs, which decreases venous return and cardiac output.

- **Decreased cardiac output.** Impulses to the cardiac centers inhibit sympathetic activity and stimulate parasympathetic

activity, reducing heart rate and contractile force. As CO falls, so does MAP.

In the opposite situation, a decline in MAP initiates reflex vasoconstriction and increases cardiac output, bringing blood pressure back up. In this way, total peripheral resistance and cardiac output are regulated in tandem to minimize changes in blood pressure.

Rapidly responding baroreceptors protect the circulation against short-term (acute) changes in blood pressure. For example, blood pressure falls (particularly in the head) when you stand up after reclining. Baroreceptors taking part in the **carotid sinus reflex** protect the blood supply to your brain, whereas those activated in the **aortic reflex** help maintain adequate blood pressure in your systemic circuit as a whole. Failure of the baroreceptor reflex results in orthostatic hypotension (◄ **p. 549**).

Baroreceptors are relatively *ineffective* in protecting us against sustained pressure changes, as evidenced by the fact that many people develop chronic hypertension. In such cases, the baroreceptors are "reprogrammed" (adapt) to monitor pressure changes at a higher set point.

Chemoreceptor Reflexes

When the carbon dioxide levels rise, or the pH falls, or oxygen content of the blood drops sharply, **chemoreceptors** in the aortic arch and large arteries of the neck transmit impulses to the cardioacceleratory center, which then increases cardiac output. Chemoreceptors also activate the vasomotor center, which causes reflex vasoconstriction. The rise in blood pressure that follows speeds the return of blood to the heart and lungs.

The most prominent chemoreceptors are the *carotid* and *aortic bodies* located close by the baroreceptors in the carotid sinuses and aortic arch. Chemoreceptors play a larger role in regulating respiratory rate than blood pressure, so we consider their function in Chapter 22.

Influence of Higher Brain Centers

Reflexes that regulate blood pressure are integrated in the medulla oblongata of the brain stem. Although the cerebral cortex and hypothalamus are not involved in routine controls of blood pressure, these higher brain centers can modify arterial pressure via relays to the medullary centers.

For example, the fight-or-flight response mediated by the hypothalamus has profound effects on blood pressure. (Even the simple act of speaking can make your blood pressure jump if the person you are talking to makes you anxious.) The hypothalamus also mediates the redistribution of blood flow and other cardiovascular responses that occur during exercise and changes in body temperature.

Short-Term Regulation: Hormonal Controls

Hormones also help regulate blood pressure, both in the short term via changes in total peripheral resistance and in the long term via changes in blood volume (**Table 19.2**). Paracrines (local chemicals), on the other hand, primarily serve to match blood flow to the metabolic need of a particular tissue. In rare instances, massive release of paracrines can affect blood pressure. We will discuss these paracrines later—here we will examine the short-term effects of hormones.

- **Adrenal medulla hormones.** During periods of stress, the adrenal gland releases **epinephrine** and **norepinephrine** (**NE**) to the blood. Both hormones enhance the sympathetic response by increasing cardiac output and promoting generalized vasoconstriction.

- **Angiotensin II.** When blood pressure or blood volume are low, the kidneys release renin. Renin acts as an enzyme, ultimately generating **angiotensin II** (an″je-o-ten′sin), which stimulates intense vasoconstriction, promoting a rapid rise in systemic blood pressure. It also stimulates release of aldosterone and ADH, which act in long-term regulation of blood pressure as described on pp. 722–723.

- **Atrial natriuretic peptide (ANP).** The atria of the heart produce the hormone **atrial natriuretic peptide (ANP)**, which leads to a reduction in blood volume and blood pressure. As noted in Chapter 16 (◀ p. 624), ANP antagonizes aldosterone and prods the kidneys to excrete more sodium and water from the body, reducing blood volume. It also causes generalized vasodilation.

- **Antidiuretic hormone (ADH).** Produced by the hypothalamus, **antidiuretic hormone** (**ADH**, also called **vasopressin**) stimulates the kidneys to conserve water (◀ p. 613). It is not usually important in short-term blood pressure regulation. However, when blood pressure falls to dangerously low levels (as during severe hemorrhage), much more ADH is released and helps restore arterial pressure by causing intense vasoconstriction.

Long-Term Regulation: Renal Mechanisms

Unlike short-term controls of blood pressure that alter total peripheral resistance and cardiac output, long-term controls alter blood volume. Renal mechanisms mediate long-term regulation by the kidneys.

Although baroreceptors respond to short-term changes in blood pressure, they quickly adapt to prolonged or chronic episodes of high or low pressure. This is where the kidneys step in to restore and maintain blood pressure homeostasis by regulating blood volume. Although blood volume varies with age, body size, and sex, renal mechanisms usually keep it close to 5 L.

As we noted earlier, blood volume is a major determinant of cardiac output [via its influence on venous return, end diastolic volume (EDV), and stroke volume]. An increase in blood volume is followed by a rise in blood pressure. Anything that increases blood volume—such as excessive salt intake, which promotes water retention—raises MAP because of the greater fluid load in the vascular tree.

By the same token, decreased blood volume translates to a fall in blood pressure. Dehydration that occurs during vigorous exercise and blood loss are common causes of reduced blood volume. A sudden drop in blood pressure often signals internal bleeding and blood volume too low to support normal circulation.

However, these assertions—increased blood volume increases BP and decreased blood volume decreases BP—do not tell the whole story because we are dealing with a dynamic system. Increases in blood volume that raise blood pressure also stimulate the kidneys to eliminate water, which reduces blood volume and consequently blood pressure. Likewise, falling blood volume triggers renal mechanisms that increase blood volume and blood pressure. As you can see, blood pressure can be stabilized or maintained within normal limits only when blood volume is stable.

Table 19.2	Effects of Selected Hormones on Blood Pressure			
HORMONE	**EFFECT ON BP**	**VARIABLE AFFECTED**		**SITE OF ACTION**
Epinephrine and norepinephrine (NE)	↑	↑ CO (HR and contractility)		Heart (β₁ receptors)
		↑ Total peripheral resistance (vasoconstriction)		Arterioles (α receptors)
Angiotensin II	↑	↑ Total peripheral resistance (vasoconstriction)		Arterioles
Antidiuretic hormone (ADH)	↑	↑ Total peripheral resistance (vasoconstriction)		Arterioles
		↑ Blood volume (↓ water loss)		Kidney tubule cells
Aldosterone	↑	↑ Blood volume (↓ salt and water loss)		Kidney tubule cells
Atrial natriuretic peptide (ANP)	↓	↓ Blood volume (↑ salt and water loss)		Kidney tubule cells
		↓ Total peripheral resistance (vasodilation)		Arterioles

19

Figure 19.12 Direct and indirect (hormonal) mechanisms for renal control of blood pressure. Low blood pressure also triggers other actions not shown here that increase BP: additional mechanisms of renin release (described in Chapter 25) and short-term actions of the sympathetic nervous system.

The kidneys act both directly and indirectly to regulate arterial pressure and provide the major long-term mechanisms of blood pressure control.

Direct Renal Mechanism

The *direct renal mechanism* alters blood volume independently of hormones. When either blood volume or blood pressure rises, the rate at which fluid filters from the bloodstream into the kidney tubules speeds up. In such situations, the kidneys cannot reabsorb the filtrate rapidly enough, and more of it leaves the body in urine. As a result, blood volume and blood pressure fall.

When blood pressure or blood volume is low, water is conserved and returned to the bloodstream, and blood pressure rises (**Figure 19.12**). As blood volume goes, so goes the arterial blood pressure.

Indirect Renal Mechanism

The kidneys can also regulate blood pressure *indirectly* via the **renin-angiotensin-aldosterone mechanism**. When arterial

Figure 19.13 Factors that increase MAP.

blood pressure declines, certain cells in the kidneys release the enzyme **renin** into the blood. Renin enzymatically splits **angiotensinogen**, a plasma protein made by the liver, converting it to **angiotensin I**. In turn, **angiotensin converting enzyme (ACE)** converts angiotensin I to **angiotensin II**. ACE is found in the capillary endothelium in various body tissues, particularly the lungs.

Angiotensin II acts in four ways to stabilize arterial blood pressure and extracellular fluid volume (Figure 19.12).

- It stimulates the adrenal cortex to secrete **aldosterone**, a hormone that enhances renal reabsorption of sodium from the forming urine. As sodium moves back into the bloodstream, water follows, which conserves blood volume. In addition, angiotensin II directly stimulates sodium reabsorption by the kidneys.

- It prods the posterior pituitary to release ADH, which promotes more water reabsorption by the kidneys.

- It triggers the sensation of thirst by activating the hypothalamic thirst center (see Chapter 26). This encourages water consumption, ultimately restoring blood volume and so blood pressure.

- It is a potent vasoconstrictor, increasing blood pressure by increasing total peripheral resistance.

Summary of Blood Pressure Regulation

How do each of the different mechanisms that we have just explored act together to control blood pressure? **Figure 19.13** provides a summary of how mean arterial pressure is controlled in concert by short- and long-term mechanisms. Notice that the left part of the figure (the factors that control cardiac output) builds upon what you learned in Chapter 18 (◄ **Figure 18.20, p. 697**).

The goal of blood pressure regulation is to keep blood pressure high enough to provide adequate tissue perfusion (blood flow), but not so high that blood vessels are damaged. Consider the brain. If pressure is too low, then perfusion is inadequate and you lose consciousness. If pressure is too high, your fragile brain capillaries might rupture and you would have a stroke. Malfunction of blood pressure control is our next topic.

Complete an interactive tutorial: MasteringA&P® > Study Area > Interactive Physiology

19

Homeostatic Imbalances in Blood Pressure

Normal blood pressure for resting adults is a systolic pressure of less than 120 mm Hg and a diastolic pressure of less than 80 mm Hg. Transient elevations in blood pressure occur as normal adaptations during changes in posture, physical exertion, emotional upset, and fever. Age, sex, weight, and race also affect blood pressure.

Hypertension

Chronically elevated blood pressure is called **hypertension** and is characterized by a sustained increase in either systolic pressure (above 140 mm Hg) or diastolic pressure (above 90 mm Hg). The American Heart Association considers individuals to have *prehypertension* if their blood pressure values are elevated but not yet in the hypertension range. These individuals are at higher risk for developing full-blown hypertension and are often advised to change their lifestyles to reduce their risk of developing full-blown hypertension.

Chronic hypertension is a common and dangerous disease. An estimated 30% of people over age 50 are hypertensive. Although this "silent killer" is usually asymptomatic for the first 10 to 20 years, it slowly but surely strains the heart and damages the arteries. Prolonged hypertension is the major cause of heart failure, vascular disease, renal failure, and stroke. The higher the pressure, the greater the risk for these serious problems.

Because the heart is forced to pump against greater resistance, it must work harder, and over time the myocardium enlarges. When finally strained beyond its capacity, the heart weakens and its walls become flabby. Hypertension also ravages the blood vessels, accelerating the progress of atherosclerosis (see *A Closer Look* on p. 760). As the vessels become increasingly blocked, blood flow to the tissues becomes inadequate and vascular complications appear in the brain, heart, kidneys, and retinas of the eyes.

Primary Hypertension　Although hypertension and atherosclerosis are often linked, it is often difficult to blame hypertension on any distinct anatomical pathology. Indeed, about 90% of hypertensive people have **primary**, or **essential**, **hypertension**, for which no underlying cause has been identified. This is because primary hypertension is due to a rich interplay between your genes and a variety of environmental factors:

- **Heredity.** Children of hypertensive parents are twice as likely to develop hypertension as are children of normotensive parents, and more blacks than whites are hypertensive. Many of the factors listed here require a genetic predisposition, and the course of the disease varies in different population groups.
- **Diet.** Dietary factors that contribute to hypertension include high intakes of salt (NaCl), saturated fat, and cholesterol, and deficiencies in certain metal ions (K^+, Ca^{2+}, and Mg^{2+}).
- **Obesity.** Obesity causes hypertension in a number of ways that are not yet well understood. For example, adipocytes release hormones that appear to increase sympathetic tone

and interfere with the ability of endothelial cells to induce vasodilation.

- **Age.** Hypertension usually appears after age 40.
- **Diabetes mellitus.**
- **Stress.** Particularly at risk are "hot reactors," people whose blood pressure zooms upward during every stressful event.
- **Smoking.** Nicotine causes intense vasoconstriction not only by directly stimulating postganglionic sympathetic neurons but also by prompting release of large amounts of epinephrine and NE. Chemicals in cigarette smoke also damage the tunica intima, interfering with its ability to chemically regulate arteriolar diameter.

Primary hypertension cannot be cured, but most cases can be controlled. Improving diet, increasing exercise and losing weight, stopping smoking, managing stress, and taking antihypertensive drugs can all help. Drugs commonly used are diuretics, beta-blockers, calcium channel blockers, angiotensin converting enzyme (ACE) inhibitors, and angiotensin II receptor blockers. Inhibiting ACE or blocking receptors for angiotensin II suppresses the renin-angiotensin-aldosterone mechanism.

Secondary Hypertension　**Secondary hypertension** accounts for 10% of cases. It is due to identifiable conditions, for example obstructed renal arteries, kidney disease, and endocrine disorders such as hyperthyroidism and Cushing's syndrome. Treatment for secondary hypertension focuses on correcting the problem that caused it.

Hypotension

In many cases, **hypotension**, or low blood pressure (below 90/60 mm Hg), simply reflects individual variations and is no cause for concern. In fact, low blood pressure is often associated with long life and an old age free of cardiovascular disease.

Orthostatic hypotension is a temporary drop in blood pressure resulting in dizziness (due to inadequate oxygen delivery to the brain) when a person rises suddenly from a reclining or sitting position. Elderly people are prone to orthostatic hypotension because the aging sympathetic nervous system does not respond as quickly as it once did to postural changes. Blood pools briefly in the lower limbs, reducing blood pressure and consequently blood delivery to the brain. Changing position slowly gives the nervous system time to adjust and usually prevents this problem.

Occasionally, *chronic hypotension* is a sign of a serious underlying condition. Addison's disease (inadequate adrenal cortex function), hypothyroidism, or severe malnutrition can cause chronic hypotension.

Hypotension is usually a concern only if it leads to inadequate blood flow to tissues. *Acute hypotension* is one of the most important signs of circulatory shock.

Circulatory Shock

Circulatory shock is any condition in which blood vessels are inadequately filled and blood cannot circulate normally. Blood flow is inadequate to meet tissue needs. If circulatory shock persists, cells die and organ damage follows.

Hypovolemic Shock   The most common form of circulatory shock is **hypovolemic shock** (hi″po-vo-le′mik; *hypo* = low, deficient; *volemia* = blood volume), which results from large-scale blood or fluid loss, as might follow acute hemorrhage, severe vomiting or diarrhea, or extensive burns. If blood volume drops rapidly, heart rate increases in an attempt to correct the problem. A weak, "thready" (barely detectable) pulse is often the first sign of hypovolemic shock. Intense vasoconstriction also occurs, which shifts blood from the various blood reservoirs into the major circulatory channels and enhances venous return.

Blood pressure is stable at first, but eventually drops if blood loss continues. A sharp drop in blood pressure is a serious, and late, sign of hypovolemic shock. The key to managing hypovolemic shock is to replace fluid volume as quickly as possible.

Vascular Shock   In **vascular shock**, blood volume is normal, but circulation is poor as a result of extreme vasodilation. A huge drop in total peripheral resistance follows, as revealed by rapidly falling blood pressure.

There are three common types of vascular shock:

- *Anaphylactic shock* is a loss of vasomotor tone due to *anaphylaxis*, a systemic allergic reaction in which the massive release of the chemical messenger histamine triggers body-wide vasodilation.

- *Neurogenic shock* results from failure of autonomic nervous system regulation.

- *Septic shock* (septicemia) is due to severe systemic bacterial infection (bacterial toxins are notorious vasodilators).

Cardiogenic Shock   **Cardiogenic shock**, or pump failure, occurs when the heart is so inefficient that it cannot sustain adequate circulation. Its usual cause is myocardial damage, as might follow numerous myocardial infarctions (heart attacks).

Check Your Understanding

13. Describe the baroreceptor reflex changes that occur to maintain blood pressure when you rise from a lying-down to a standing position.

14. The kidneys play an important role in maintaining MAP by influencing which variable? Explain how renal artery obstruction could cause secondary hypertension.

15. APPLY Your neighbor, Bob, calls you because he thinks he is having an allergic reaction to a medication. You find Bob on the verge of losing consciousness and having trouble breathing. When paramedics arrive, they note his blood pressure is 63/38 and he has a rapid, thready pulse. Explain Bob's low blood pressure and rapid heart rate.

16. MAKE CONNECTIONS You have just learned that hypertension can be treated with a variety of different drugs including diuretics, beta-blockers, and calcium channel blockers. Using your knowledge of the autonomic nervous system (Chapter 14), smooth muscle (Chapter 9), and cardiac muscle (Chapter 18), explain how these drugs work to decrease blood pressure.

For answers, see Answers Appendix.

19.9 Intrinsic and extrinsic controls determine blood flow through tissues

Learning Outcome

▶ **Explain how blood flow through tissues is regulated in general and in specific organs.**

Blood flow through body tissues, or **tissue perfusion**, is involved in:

- Delivering oxygen and nutrients to tissue cells, and removing carbon dioxide and wastes

- Exchanging gases in the lungs

- Absorbing nutrients from the digestive tract

- Forming urine in the kidneys

The rate of blood flow to each tissue and organ is almost exactly the right amount to provide for proper function—no more, no less. This is achieved by **intrinsic controls** (*autoregulation*) acting automatically on the smooth muscle of arterioles that feed any given tissue. We will examine these intrinsic mechanisms in the next section.

First, let's step back and look at the big picture. What do you think would happen if all of the arterioles in your body dilated at once? Because there is only a finite amount of blood, blood pressure would fall. Critical tissues, such as the brain, would be deprived of the oxygen and nutrients they need and would stop functioning. **Extrinsic controls** keep this from happening by acting on arteriolar smooth muscle to maintain blood pressure. The extrinsic controls act via the nerves (sympathetic nervous system) and hormones of the nervous and endocrine systems, the two major control systems of the body. They reduce blood flow to regions that need it the least, maintaining a constant MAP and allowing intrinsic mechanisms to direct blood flow to where it is most needed.

A number of physiological processes are under both intrinsic and extrinsic control (**Figure 19.14**). (Note that these are control mechanisms, and not related in any way to the intrinsic and extrinsic pathways of blood clotting.)

Intrinsic controls	Extrinsic controls
• Control is entirely from within the tissue or organ • Uses paracrines or properties of muscle tissue • Also known as *autoregulation* or *local control*	• Control is from outside of the tissue or organ • Uses nerves or hormones

Examples where both control mechanisms occur:
- Stroke volume in heart (◀ pp. 697–698)
- Arteriolar diameter (see Figure 19.16)
- Glomerular filtration in kidneys (see Chapter 25)

Figure 19.14 A quick summary of intrinsic versus extrinsic control mechanisms.

19

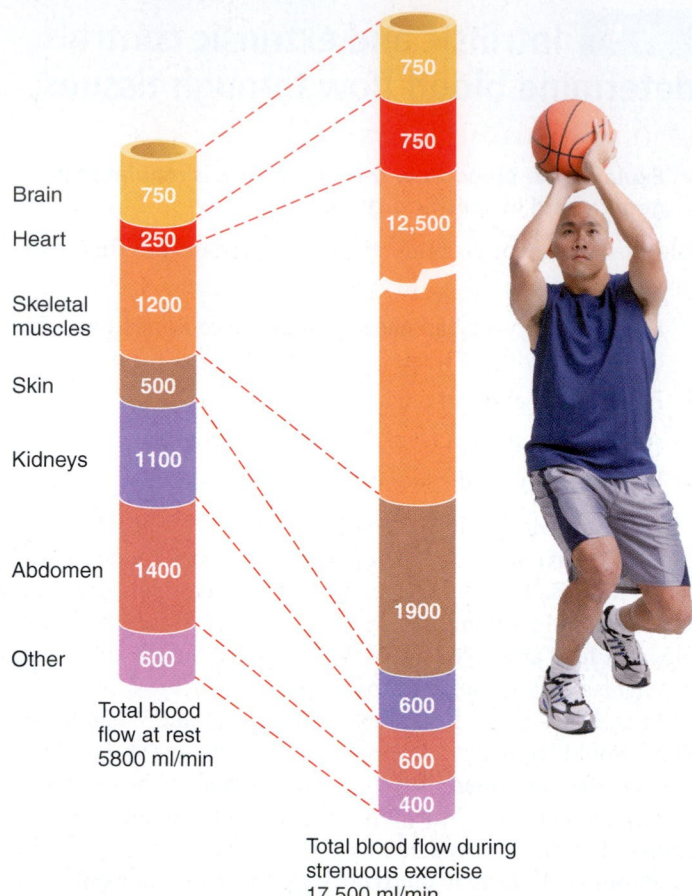

Brain

Heart

Skeletal muscles

Skin

Kidneys

Abdomen

Other

750

250

1200

500

1100

1400

600

Total blood flow at rest 5800 ml/min

750

750

12,500

1900

600

600

400

Total blood flow during strenuous exercise 17,500 ml/min

Figure 19.15 Distribution of blood flow at rest and during strenuous exercise.

The redistribution of blood during exercise provides an example of how intrinsic and extrinsic controls work together (**Figure 19.15**). When the body is at rest, the brain receives about 13% of total blood flow, the heart 4%, kidneys 20%, and abdominal organs 24%. Skeletal muscles, which make up almost half of body mass, normally receive about 20% of total blood flow. During exercise, however, nearly all of the increased cardiac output flushes into the skeletal muscles as intrinsic autoregulatory controls dilate skeletal muscle arterioles. To maintain blood pressure in spite of the widespread dilation of arterioles in skeletal muscle, the extrinsic controls act to decrease blood flow to the kidneys and digestive organs.

Autoregulation: Intrinsic (Local) Regulation of Blood Flow

As our activities change throughout the day, how does each organ or tissue manage to get the blood flow it needs? The answer is **autoregulation**. Local conditions regulate blood flow independent of control by nerves or hormones. Changes in blood flow through individual organs are controlled *intrinsically* by modifying the diameter of local arterioles feeding the capillaries.

You can compare blood flow autoregulation to water use in your home. Whether you have several taps open or none, the pressure in the main water pipe in the street remains relatively constant, as it does in the even larger water lines closer to the pumping station. Similarly, local conditions in the arterioles feeding the capillary beds of an organ have little effect on pressure in the muscular artery feeding that organ, or in the large elastic arteries. The pumping station is, of course, the heart. As long as the water company (circulatory feedback mechanisms) maintains a relatively constant water pressure (MAP), local demand regulates the amount of fluid (blood) delivered to various areas.

Organs regulate their own blood flows by varying the resistance of their arterioles. These intrinsic control mechanisms may be classed as *metabolic* (chemical) or *myogenic* (physical). Generally, both metabolic and myogenic factors determine the final autoregulatory response of a tissue. For example, **reactive hyperemia** (hi″per-e′me-ah) refers to the dramatically increased blood flow into a tissue that occurs after the blood supply to the area has been temporarily blocked. It results both from the myogenic response and from the metabolic wastes that accumulated during occlusion. **Figure 19.16** summarizes the various intrinsic (local) and extrinsic controls of arteriolar diameter.

Metabolic Controls

When blood flow is too low to meet a tissue's metabolic needs, oxygen levels decline and metabolic products (which act as paracrines) accumulate. These changes serve as stimuli that lead to automatic increases in tissue blood flow.

The metabolic factors that regulate blood flow are low oxygen levels, and increases in H^+ (from CO_2 and lactic acid), K^+, adenosine, and prostaglandins. The relative importance of these factors is not clear. Many of them act directly to relax vascular smooth muscle, but some may act by causing vascular endothelial cells to release nitric oxide.

Nitric oxide (NO) is a powerful vasodilator, but its effects are very brief because it is quickly destroyed. Even so, NO is the major player in controlling local vasodilation, often overriding sympathetic vasoconstriction when tissues need more blood flow.

The endothelium also releases potent vasoconstrictors, including the family of peptides called **endothelins**, which are among the most potent vasoconstrictors known. Normally, NO and endothelin released from endothelial cells are in a dynamic balance, but this balance tips in favor of NO when blood flow is too low for metabolic needs.

The net result of metabolically controlled autoregulation is immediate vasodilation of the arterioles serving the capillary beds of the "needy" tissues. Blood flow to the area rises temporarily, allowing blood to surge through the capillaries and become available to the tissue cells.

Inflammatory chemicals released in injury, infection, or allergic reactions also cause local vasodilation. Inflammatory vasodilation helps the defense mechanisms clear microorganisms and toxins from the area, and promotes healing.

19

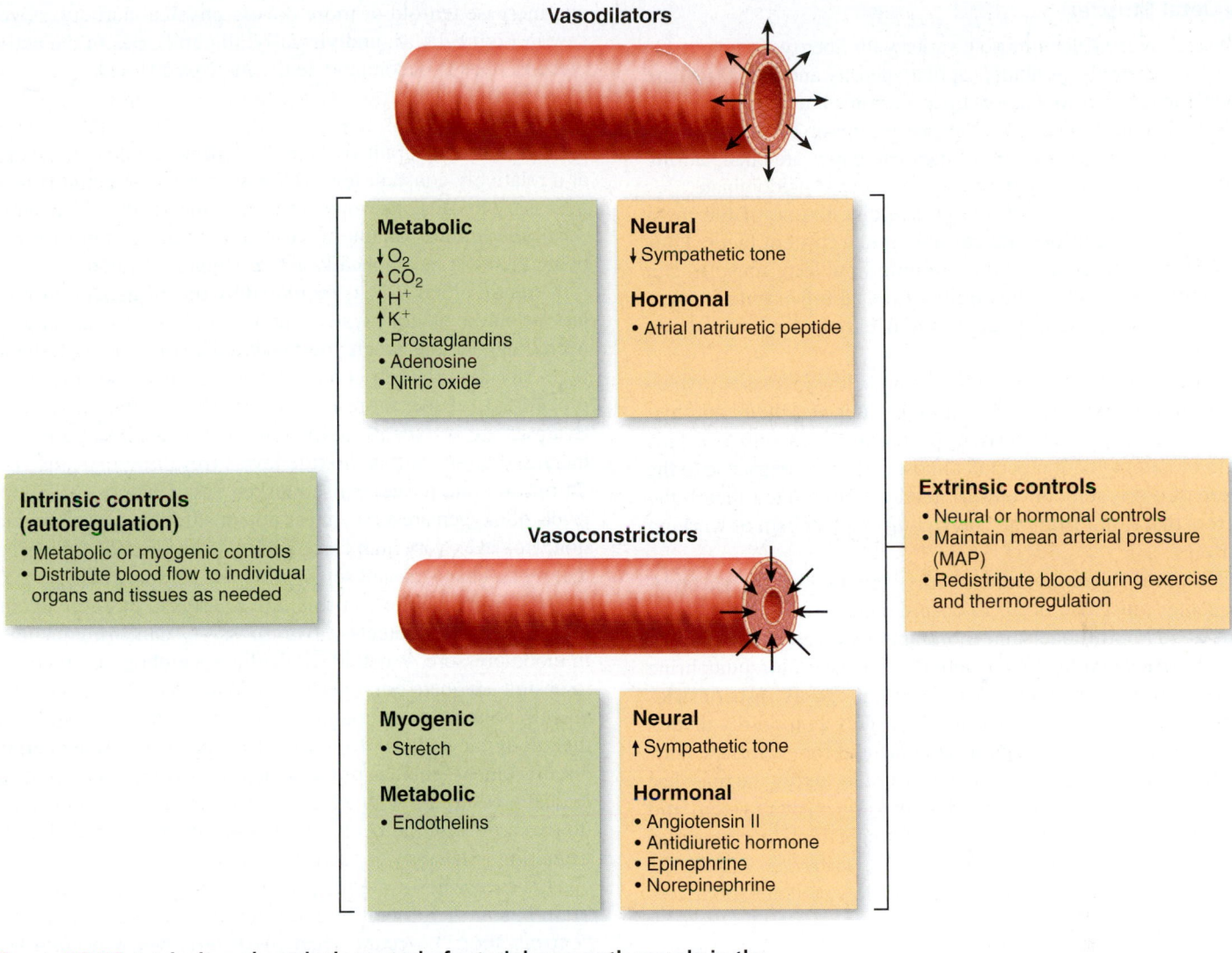

Figure 19.16 Intrinsic and extrinsic control of arteriolar smooth muscle in the systemic circulation. Controls are listed in the boxes below the arterioles. Epinephrine and norepinephrine constrict arteriolar smooth muscle by acting at α-adrenergic receptors. β-adrenergic receptors (causing vasodilation) are present in arterioles supplying skeletal and heart muscle, but their physiological relevance is minimal.

Myogenic Controls

Fluctuations in systemic blood pressure would cause problems for individual organs were it not for the **myogenic responses** (*myo* = muscle, *gen* = origin) of vascular smooth muscle. Inadequate blood perfusion through an organ is quickly followed by a decline in the organ's metabolic rate and, if prolonged, organ death. Likewise, excessively high arterial pressure can be dangerous because it may rupture more fragile blood vessels.

Fortunately, vascular smooth muscle prevents these problems by responding directly to passive stretch (caused by increased intravascular pressure) with increased tone, which resists the stretch and causes vasoconstriction. In contrast, reduced stretch (due to decreased intravascular pressure) results in vasodilation and increases blood flow into the tissue. These myogenic responses keep tissue perfusion fairly constant despite most variations in systemic pressure.

Long-Term Autoregulation

If a tissue needs more nutrients than short-term autoregulatory mechanisms can supply, a long-term autoregulatory mechanism may develop over weeks or months to enrich local blood flow still more. The number of blood vessels in the region increases, and existing vessels enlarge. This phenomenon, called *angiogenesis*, is particularly common in the heart when a coronary vessel is partially occluded. It occurs throughout the body in people who live in high-altitude areas, where the air contains less oxygen.

Blood Flow in Special Areas

Each organ has special requirements and functions that are revealed in its pattern of autoregulation. Autoregulation in the brain, heart, and kidneys is extraordinarily efficient, maintaining adequate perfusion even when MAP fluctuates.

19

Skeletal Muscles

Blood flow in skeletal muscle varies with fiber type and muscle activity. Generally speaking, capillary density and blood flow are greater in red (slow oxidative) fibers than in white (fast glycolytic) fibers. Resting skeletal muscles receive about 1 L of blood per minute, and only about 25% of their capillaries are open. During rest, myogenic and general neural mechanisms predominate.

Without question, strenuous exercise is one of the most demanding conditions the cardiovascular system faces. Ultimately, the major factor determining how long muscles can contract vigorously is the ability of the cardiovascular system to deliver adequate oxygen and nutrients and remove waste products.

When muscles become active, blood flow increases (*hyperemia*) in direct proportion to their greater *metabolic* activity, a phenomenon called **active hyperemia (Figure 19.17)**. This form of autoregulation occurs almost entirely in response to the decreased oxygen concentration and accumulated metabolic factors that result from the "revved-up" metabolism of working muscles.

However, systemic adjustments mediated by the vasomotor center must also occur to ensure that blood delivery to the muscles is both faster and more abundant. During exercise, sympathetic nervous system activity increases. Norepinephrine released from sympathetic nerve endings causes vasoconstriction of arterioles throughout the body. This temporarily diverts blood away from most regions not essential for exercise, ensuring that there is sufficient blood pressure to supply the muscles.

In skeletal muscles, the sympathetic nervous system and local metabolic controls have opposing effects on arteriolar diameter. During exercise, local controls *override* sympathetic vasoconstriction. Consequently, blood flow to skeletal muscles

Figure 19.17 Active hyperemia. Blood flow in exercising skeletal muscle is largely controlled by metabolic autoregulation.

can increase tenfold or more during physical activity, as you saw in Figure 19.15, and virtually all capillaries in the active muscles open to accommodate the increased flow.*

The Brain

Blood flow to the brain averages 750 ml/min and is maintained at a relatively constant level. Constant cerebral blood flow is necessary because neurons are totally intolerant of ischemia. Also, the brain is unable to store essential nutrients despite being the most metabolically active organ in the body.

Cerebral blood flow is regulated by one of the body's most precise autoregulatory systems and is tailored to local neuronal need. For example, when you make a fist with your right hand, the neurons in the left cerebral motor cortex controlling that movement receive more blood than the adjoining neurons. Brain tissue is exceptionally sensitive to declining pH, and increased blood carbon dioxide levels (resulting in acidic conditions in brain tissue) cause marked vasodilation. Low blood levels of oxygen are a much less potent stimulus for autoregulation. However, very high carbon dioxide levels abolish autoregulatory mechanisms and severely depress brain activity.

Besides metabolic controls, the brain also has a myogenic mechanism that protects it from possibly damaging changes in blood pressure. When MAP declines, cerebral vessels dilate to ensure adequate brain perfusion. When MAP rises, cerebral vessels constrict, protecting the small, more fragile vessels farther along the pathway from excessive pressure. Under certain circumstances, such as brain ischemia caused by rising intracranial pressure (as with a brain tumor), the brain (via the medullary cardiovascular centers) regulates its own blood flow by triggering a rise in systemic blood pressure.

However, when systemic pressure changes are extreme, the brain becomes vulnerable. Fainting, or *syncope* (sin'cuh-pe; "cutting short"), occurs when MAP falls below 60 mm Hg. Cerebral edema is the usual result of pressures over 160 mm Hg, which dramatically increase brain capillary permeability.

The Skin

Blood flow through the skin:

- Supplies nutrients to cells
- Helps regulate body temperature
- Provides a blood reservoir

Autoregulation serves the first function in response to the need for oxygen, but the other two require neural intervention. The primary function of the cutaneous circulation is to help maintain body temperature.

Below the skin surface are extensive venous plexuses (networks of intertwining vessels). The blood flow through these plexuses can change from 50 ml/min to as much as 2500 ml/min, depending on body temperature. This capability reflects neural adjustments of blood flow through arterioles and through

*Epinephrine acting at beta (β) adrenergic receptors and acetylcholine acting at cholinergic receptors were once thought to contribute to arteriolar dilation during exercise. However, these appear to have little physiological importance in controlling human skeletal muscle blood flow.

unique coiled arteriovenous anastomoses. These tiny arteriovenous shunts are located mainly in the fingertips, palms of the hands, toes, soles of the feet, ears, nose, and lips. Richly supplied with sympathetic nerve endings (unlike the shunts of some other capillary beds), they are controlled by reflexes initiated by temperature receptors or signals from higher CNS centers. The arterioles, in addition, respond to metabolic autoregulatory stimuli.

When the skin is exposed to heat, or body temperature rises for other reasons (such as vigorous exercise), the hypothalamic "thermostat" signals for reduced vasomotor stimulation of the skin vessels. As a result, warm blood flushes into the capillary beds and heat radiates from the skin surface. (This is why Figure 19.15 shows an increase in blood flow to the skin during exercise.)

When the ambient temperature is cold and body temperature drops, superficial skin vessels strongly constrict. As a result, blood almost entirely bypasses the capillaries associated with the arteriovenous anastomoses, diverting the warm blood to the deeper, more vital organs. Paradoxically, the skin may stay quite rosy because some blood gets "trapped" in the superficial capillary loops as the shunts swing into operation. The trapped blood remains red because the chilled skin cells take up less O_2.

The Lungs

Blood flow through the pulmonary circuit to and from the lungs is unusual in many ways. The pathway is relatively short, and pulmonary arteries and arterioles are structurally like veins and venules. That is, they have thin walls and large lumens. Because resistance to blood flow is low in the pulmonary arterial system, less pressure is needed to propel blood through those vessels. Consequently, arterial pressure in the pulmonary circulation is much lower than in the systemic circulation (24/10 versus 120/80 mm Hg).

In the pulmonary circulation, the autoregulatory mechanism is the *opposite* of what is seen in most tissues: Low pulmonary oxygen levels cause local vasoconstriction, and high levels promote vasodilation. While this may seem odd, it is perfectly consistent with the gas exchange role of this circulation. When the air sacs of the lungs are flooded with oxygen-rich air, the pulmonary capillaries become flushed with blood and ready to receive the oxygen load. In contrast, if the air sacs are collapsed or blocked with mucus, the oxygen content in those areas is low, and blood largely bypasses those nonfunctional areas.

The Heart

Aortic pressure and the pumping activity of the ventricles influence the movement of blood through the smaller vessels of the coronary circulation. When the ventricles contract and compress the coronary blood vessels, blood flow through the myocardium stops. As the heart relaxes, the high aortic pressure forces blood through the coronary circulation.

Under normal circumstances, the myoglobin in cardiac cells stores sufficient oxygen to satisfy the cells' oxygen needs during systole. However, an abnormally rapid heartbeat seriously reduces the ability of the myocardium to receive adequate oxygen and nutrients during diastole.

Under resting conditions, blood flow through the heart is about 250 ml/min and is controlled by a myogenic mechanism. Consequently, blood flow remains fairly constant despite wide variations (50 to 140 mm Hg) in coronary perfusion pressure. During strenuous exercise, the coronary vessels dilate in response to local accumulation of vasodilators (particularly adenosine), and blood flow may increase three to four times (see Figure 19.15). Additionally, any event that decreases the oxygen content of the blood releases vasodilators that adjust the O_2 supply to the O_2 demand.

This enhanced blood flow during increased heart activity is important because under resting conditions, cardiac cells use as much as 65% of the oxygen carried to them in blood. (Most other tissue cells use about 25% of the delivered oxygen.) Consequently, increasing the blood flow is the only way to provide more oxygen to a vigorously working heart.

Check Your Understanding

17. Suppose you are in a bicycle race. What happens to the smooth muscle in the arterioles supplying your leg muscles? What is the key mechanism in this case?

18. If many arterioles in your body dilated at once, you would expect MAP to plummet. What prevents MAP from decreasing during your bicycle race?

For answers, see Answers Appendix.

19.10 Slow blood flow through capillaries promotes diffusion of nutrients and gases, and bulk flow of fluids

Learning Outcome

▶ Outline factors involved in capillary exchange and bulk flow, and explain the significance of each.

Velocity of Blood Flow

Have you ever watched a swift river emptying into a large lake? The water's speed decreases as it enters the lake until its flow becomes almost imperceptible. This is because the total cross-sectional area of the lake is much larger than that of the river.

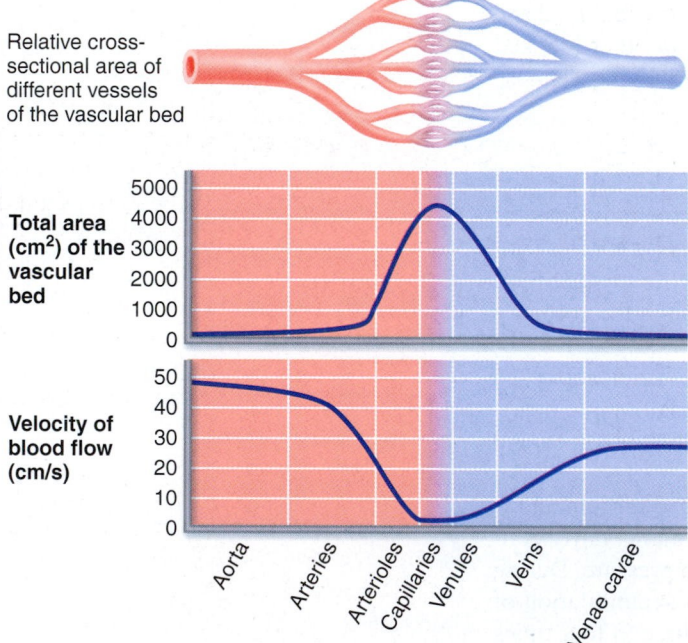

Relative cross-sectional area of different vessels of the vascular bed

Figure 19.18 Blood flow velocity and total cross-sectional area of vessels. Various blood vessels of the systemic circulation differ in their total cross-sectional area (e.g., the cross section of all systemic capillaries combined versus the cross section of all systemic arteries combined), which affects the velocity of blood flow through them.

Velocity in this case is *inversely* related to cross-sectional area. The same thing happens with blood flow inside our blood vessels.

As shown in **Figure 19.18**, the speed or velocity of blood flow changes as blood travels through the systemic circulation. It is fastest in the aorta and other large arteries (the river), slowest in the capillaries (whose large total cross-sectional area makes them analogous to the lake), and then picks up speed again in the veins (the river again).

Just as in our analogy of the river and lake, blood flows fastest where the total cross-sectional area is least. As the arterial system branches, the total cross-sectional area of the vascular bed increases, and the velocity of blood flow declines proportionately. Even though the individual branches have smaller lumens, their *combined* cross-sectional areas and thus the volume of blood they can hold are much greater than that of the aorta.

For example, the cross-sectional area of the aorta is 2.5 cm^2, but the combined cross-sectional area of all the capillaries is 4500 cm^2. This difference results in fast blood flow in the aorta (40–50 cm/s) and slow blood flow in the capillaries (about 0.03 cm/s). Slow capillary flow is beneficial because it allows adequate time for exchanges between the blood and tissue cells.

Vasomotion

Blood flow through capillaries is not only slow, it is also intermittent. The intermittent flow of blood through a capillary bed is due to **vasomotion**, the on/off constriction/dilation of arterioles, mostly in response to local chemical conditions (intrinsic control). Precapillary sphincters also respond to the same local autoregulatory signals that affect arteriolar diameter.

Capillary Exchange of Respiratory Gases and Nutrients

Oxygen, carbon dioxide, most nutrients, and metabolic wastes pass between the blood and interstitial fluid by diffusion. Recall that in **diffusion**, net movement always occurs along a concentration gradient—each substance moving from an area of its higher concentration to an area of its lower concentration. As a result, oxygen and nutrients pass from the blood, where their concentration is fairly high, through the interstitial fluid to the tissue cells. Carbon dioxide and metabolic wastes leave the cells, where their content is higher, and diffuse into the capillary blood.

There are four different routes across capillaries for different types of molecules, as **Figure 19.19** shows. ① Lipid-soluble molecules, such as respiratory gases, diffuse through the lipid bilayer of the endothelial cell plasma membranes. Recall that the plasma membrane is a major barrier to diffusion of solutes that are not lipid soluble (◀ p. 70). Small water-soluble solutes, such as amino acids and sugars, pass through ② fluid-filled intercellular capillary clefts or ③ fenestrations. ④ Some larger molecules, such as proteins, are actively transported in pinocytotic vesicles.

As we mentioned earlier, capillaries differ in their "leakiness," or permeability. Liver capillaries, for instance, are sinusoids that allow even proteins to pass freely, whereas brain capillaries are impermeable to most substances.

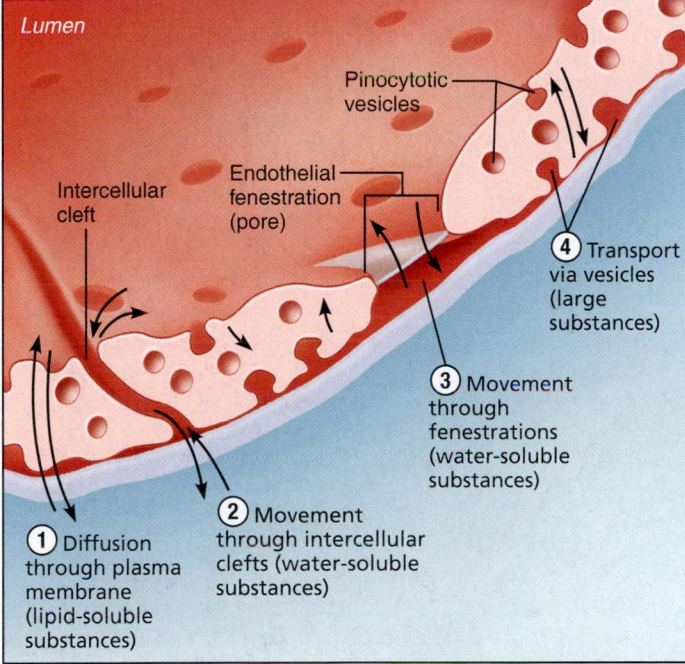

Figure 19.19 Capillary transport mechanisms. The four possible pathways or routes of transport across the endothelial cell wall of a fenestrated capillary.

Fluid Movements: Bulk Flow

While nutrient and gas exchanges are occurring across the capillary walls by diffusion, bulk fluid flows are also going on. Fluid is forced out of the capillaries through the clefts at the arterial end of the bed, but most of it returns to the bloodstream at the venous end. Though relatively unimportant to capillary exchange of nutrients and wastes, bulk flow is extremely important in determining the relative fluid volumes in the bloodstream and the interstitial space. (Approximately 20 L of fluid filter out of the capillaries each day before being returned to the blood—almost seven times the total plasma volume!)

As we describe next and show in *Focus on Bulk Flow across Capillary Walls* (**Focus Figure 19.1** on pp. 732–733), the

direction and amount of flow across capillary walls reflect the balance between two dynamic and opposing forces—hydrostatic and colloid osmotic pressures.

Hydrostatic Pressures

Hydrostatic pressure (HP) is the force exerted by a fluid pressing against a wall. In capillaries, hydrostatic pressure is the same as *capillary blood pressure*—the pressure exerted by blood on capillary walls. **Capillary hydrostatic pressure (HP$_c$)** tends to force fluids through capillary walls (a process called *filtration*), leaving behind cells and most proteins. Blood pressure drops as blood flows along a capillary bed, so HP$_c$ is higher at the arterial end of the bed (35 mm Hg) than at the venous end (17 mm Hg).

In theory, blood pressure—which forces fluid out of the capillaries—is opposed by the **interstitial fluid hydrostatic pressure (HP$_{if}$)** acting outside the capillaries and pushing fluid in. However, there is usually very little fluid in the interstitial space because the lymphatic vessels constantly withdraw it. HP$_{if}$ may vary from slightly negative to slightly positive, but traditionally it is assumed to be zero.

Colloid Osmotic Pressures

Colloid osmotic pressure (OP), the force opposing hydrostatic pressure, is created by large nondiffusible molecules, such as plasma proteins, that are unable to cross the capillary wall (◄ p. 644). Such molecules draw water toward themselves. In other words, they encourage osmosis. This is because water moves to make the solute more dilute. A quick and dirty way to remember this is "hydrostatic pressure pushes and osmotic pressure sucks."

The abundant plasma proteins in capillary blood (primarily albumin molecules) develop a **capillary colloid osmotic pressure (OP$_c$)**, also called *oncotic pressure*, of approximately 26 mm Hg. The **interstitial fluid colloid osmotic pressure (OP$_{if}$)** is substantially lower—from 0.1 to 5 mm Hg—because interstitial fluid contains few proteins. Unlike HP, OP does not vary significantly from one end of the capillary bed to the other.

Hydrostatic-Osmotic Pressure Interactions

We are now ready to calculate the **net filtration pressure (NFP)**, which considers all the forces acting at the capillary bed. As you work your way through the right-hand page of Focus Figure 19.1, notice that while net *filtration* is occurring at the arteriolar end of the capillary, a negative value for NFP at the venous end of the capillary indicates that fluid is moving *into* the capillary bed (a process called *reabsorption*). As a result, net fluid flow is *out* of the circulation at the arterial ends of capillary beds and *into* the circulation at the venous ends.

However, more fluid enters the tissue spaces than returns to the blood, resulting in a net loss of fluid from the circulation of about 1.5 ml/min. Lymphatic vessels pick up this fluid and any leaked proteins and return it to the vascular system, which accounts for the relatively low levels of both fluid and proteins in the interstitial space. Were this not so, this "insignificant" fluid loss would empty your blood vessels of plasma in about 24 hours!

(Text continues on p. 734.)

19

Bulk fluid flow across capillary walls causes continuous mixing of fluid between the plasma and the interstitial fluid compartments, and maintains the interstitial environment.

The big picture
Each day, 20 L of fluid filters from capillaries at their arteriolar end and flows through the interstitial space. Most (17 L) is reabsorbed at the venous end.

Recall from Chapter 3 (◄ pp. 71–72) that two kinds of pressure drive fluid movement:

Hydrostatic pressure (HP)
- Due to fluid pressing against a boundary (e.g., capillary wall)
- HP "pushes" fluid across the boundary
- In blood vessels, is due to blood pressure

Piston

Boundary

"Pushes"

Osmotic pressure (OP)
- Due to nondiffusible solutes that cannot cross the boundary
- OP "pulls" fluid across the boundary
- In blood vessels, is due to plasma proteins

Solute molecules (proteins)

Boundary

"Sucks"

Arteriole

Fluid moves through the interstitial space.

For all capillary beds, 20 L of fluid is filtered out per day—almost 7 times the total plasma volume!

17 L of fluid per day is reabsorbed into the capillaries at the venous end.

About 3 L per day of fluid (and any leaked proteins) are removed by the lymphatic system (see Chapter 20).

Venule

Lymphatic capillary

How do the pressures drive fluid flow across a capillary?

Net filtration occurs at the arteriolar end of a capillary.

Capillary lumen — Boundary (capillary wall) — *Interstitial fluid*

Hydrostatic pressure in capillary (HP_c) "pushes" fluid *out* of capillary.

$HP_c = 35$ mm Hg

Osmotic pressure in capillary (OP_c) "pulls" fluid *into* capillary.

$OP_c = 26$ mm Hg

$HP_{if} = 0$ mm Hg — Hydrostatic pressure (HP_{if}) in interstitial fluid "pushes" fluid *into* capillary.

$OP_{if} = 1$ mm Hg — Osmotic pressure (OP_{if}) in interstitial fluid "pulls" fluid *out* of capillary.

$NFP = 10$ mm Hg

Let's use what we know about pressures to determine the **net filtration pressure (NFP)** at any point. (NFP is the pressure driving fluid out of the capillary.) To do this we calculate the outward pressures (HP_c and OP_{if}) minus the inward pressures (HP_{if} and OP_c). So,

$$NFP = (HP_c + OP_{if}) - (HP_{if} + OP_c)$$
$$= (35 + 1) - (0 + 26)$$
$$= 10 \text{ mm Hg (net outward pressure)}$$

As a result, **fluid moves from the capillary into the interstitial space.**

Net reabsorption occurs at the venous end of a capillary.

Capillary lumen — Boundary (capillary wall) — *Interstitial fluid*

Hydrostatic pressure in capillary "pushes" fluid *out* of capillary. The pressure has dropped because of resistance encountered along the capillaries.

$HP_c = 17$ mm Hg

Osmotic pressure in capillary "pulls" fluid *into* capillary.

$OP_c = 26$ mm Hg

$HP_{if} = 0$ mm Hg — Hydrostatic pressure in interstitial fluid "pushes" fluid *into* capillary.

$OP_{if} = 1$ mm Hg — Osmotic pressure in interstitial fluid "pulls" fluid *out* of capillary.

$NFP = -8$ mm Hg

Again, we calculate the NFP:

$$NFP = (HP_c + OP_{if}) - (HP_{if} + OP_c)$$
$$= (17 + 1) - (0 + 26)$$
$$= -8 \text{ mm Hg (net inward pressure)}$$

Notice that the NFP at the venous end is a negative number. This means that *reabsorption*, not filtration, is occurring and so **fluid moves from the interstitial space into the capillary.**

When bulk flow goes wrong, *edema* can result (see Homeostatic Imbalance 19.2, p. 734).

HOMEOSTATIC IMBALANCE 19.2

CLINICAL

Edema is an abnormal increase in the amount of interstitial fluid. You will encounter it frequently in the clinic because it occurs in diverse clinical scenarios. However, it will be easy for you to discern the underlying cause of edema in any given situation if you think of it in terms of the pressures that drive bulk flow. Either an increase in outward pressure (driving fluid out of the capillaries) or a decrease in inward pressure could be the cause.

- An increase in *capillary hydrostatic pressure* accelerates fluid loss from the blood. This could result from incompetent venous valves, localized blood vessel blockage, congestive heart failure, or high blood volume. It could also result from the enlarged uterus of a pregnant woman pressing on veins that return blood to the heart. Whatever the cause, the abnormally high capillary hydrostatic pressure intensifies filtration.

- Increased *interstitial fluid osmotic pressure* can result from an inflammatory response. Inflammation increases capillary permeability, allowing plasma proteins to leak into the interstitial fluid. Together, the more porous capillaries and the increased osmolality of the interstitial fluid draw large amounts of fluid out of the capillaries, accounting for the localized swelling seen in inflammation. In an anaphylactic response (see p. 813), edema results from the massive release of the inflammatory chemical histamine.

- Decreased *capillary colloid osmotic pressure* hinders fluid return to the blood. Since plasma proteins are largely responsible for OP_c, **hypoproteinemia** (hi″po-pro″te-ĭ-ne′me-ah), a condition of unusually low levels of plasma proteins, results in tissue edema. Fluids are forced out of the capillary beds at the arteriolar ends by blood pressure as usual, but fail to return to the blood at the venous ends. As a result, the interstitial spaces become congested with fluid. Hypoproteinemia may result from liver disease, protein malnutrition, or glomerulonephritis (in which plasma proteins pass through "leaky" renal filtration membranes and are lost in urine).

- Theoretically, a decrease in *interstitial fluid hydrostatic pressure* should also be a potential cause of edema. However, this does not occur because HP_{if} is too low to decrease to any extent.

- A fourth cause of edema is decreased drainage of interstitial fluid through *lymphatic vessels* that have been blocked (e.g., by parasitic worms; see elephantiasis in the Chapter 20 Related Clinical Terms on p. 777) or surgically removed (for example, during cancer surgery).

Edema can occur anywhere in the body but is most easily visible in the skin. Excess interstitial fluid in the subcutaneous tissues generally causes *pitting edema* (**Figure 19.20**). Gravity determines where edematous fluid accumulates, so involvement of the legs and feet is common.

Edema can impair tissue function because excess fluid in the interstitial space increases the distance nutrients and oxygen must diffuse between the blood and the cells. Usually edema develops slowly, and so the fluid losses from the blood

Figure 19.20 Pitting edema. Applying pressure with a thumb leaves an indentation that remains for some time.

are compensated for by renal mechanisms that maintain blood volume and pressure. However, rapid onset of edema such as that in anaphylaxis may have serious effects on the efficiency of the circulation due to a decrease in blood volume and blood pressure.

Check Your Understanding

19. **DRAW** At a given point in a capillary, suppose that capillary hydrostatic pressure is 32 mm Hg, interstitial fluid hydrostatic pressure is 1 mm Hg, capillary colloid osmotic pressure is 25 mm Hg, and interstitial fluid osmotic pressure is 2 mm Hg. Draw a line representing the capillary wall and label the compartments on either side as "capillary lumen" and "interstitial fluid." For each of the four pressures, draw an arrow across the capillary wall pointing in the correct direction, and label it with the appropriate name and value. Calculate the net filtration pressure. Would you expect to find this point at the venous or arterial end of the capillary?

20. **PREDICT** Suppose OP_{if} rises dramatically—say because of a severe bacterial infection in the surrounding tissue. (a) Predict how fluid flow will change in this situation. (b) Now calculate the NFP at the venous end of the capillary in Focus Figure 19.1 if OP_{if} increases to 10 mm Hg. (c) In which direction does fluid flow at the venous end of the capillary now—in or out?

21. **MAKE CONNECTIONS** Your patient in right heart failure is experiencing peripheral edema. Which of the four pressures that drive bulk fluid flow at capillaries has been changed as a result of the heart failure and in which direction?

For answers, see Answers Appendix.

PART 3

CIRCULATORY PATHWAYS: BLOOD VESSELS OF THE BODY

Learning Outcomes

▶ **Trace the pathway of blood through the pulmonary circuit, and state the importance of this special circulation.**

▶ **Describe the general functions of the systemic circuit.**

The term **vascular system** is often used to describe the body's complex network of blood vessels. However, as we saw in Chapter 18, the heart is actually a double pump that serves two distinct circulations, each with its own set of arteries, capillaries, and veins. The *pulmonary circulation* is the short loop that runs from the heart to the lungs and back to the heart. The *systemic circulation* routes blood through a long loop to all parts of the body before returning it to the heart. **Table 19.3** on pp. 736–737 shows both circuits schematically.

The heart pumps all of its blood into a single systemic artery—the aorta. In contrast, blood returning to the heart is delivered largely by two terminal systemic veins, the superior and inferior venae cavae. The single exception to this is the blood draining from the myocardium of the heart, which is collected by the cardiac veins and reenters the right atrium via the coronary sinus.

In addition to these differences between arteries and veins connecting to the heart, there are three important differences between systemic arteries and veins:

- **Arteries run deep while veins are both deep and superficial.** Deep veins parallel the course of the systemic arteries and both are protected by body tissues along most of their course. With a few exceptions, these veins are named identically to their companion arteries. Superficial veins run just beneath the skin and are readily seen, especially in the limbs, face, and neck. Because there are no superficial arteries, the names of the superficial veins do not correspond to the names of any of the arteries.

- **Venous pathways are more interconnected.** Unlike the fairly distinct arterial pathways, venous pathways tend to have numerous interconnections, and many veins are represented by not one but two similarly named vessels. As a result, venous pathways are more difficult to follow.

- **The brain and digestive systems have unique venous drainage systems.** Most body regions have a similar pattern for their arterial supply and venous drainage. However, the venous drainage patterns in at least two important body areas are unique. First, venous blood draining from the brain enters large *dural venous sinuses* (◀ **pp. 464–465**) rather than typical veins. Second, blood draining from the diges-

tive organs enters a special subcirculation, the *hepatic portal system*, and perfuses through the liver before it reenters the general systemic circulation (see Table 19.12).

<table><tr><td>**19.11**</td><td></td></tr></table>

19.11 The vessels of the systemic circulation transport blood to all body tissues

Learning Outcomes

▶ **Name and give the location of the major arteries and veins in the systemic circulation.**
▶ **Describe the structure and special function of the hepatic portal system.**

The principal arteries and veins of the systemic circulation are described in **Tables 19.4** through **19.13**. (The fetal circulation is described later in the Developmental Aspects section of this chapter and in Chapter 28.)

Notice that by convention, oxygen-rich blood is shown red, while blood that is relatively oxygen-poor is depicted blue, regardless of vessel type. The schematic flowcharts (pipe diagrams) that accompany each table show the vessels that would be closer to the viewer in brighter, more intense colors than vessels deeper or farther from the viewer. For example, darker blue veins would be closer to the viewer than lighter blue veins in the body region shown.

As you examine the tables that follow and locate the various systemic arteries and veins in the illustrations, be aware of cues that make your memorization task easier. Also notice that:

- In many cases, the name of a vessel reflects the body region traversed (axillary, brachial, femoral, etc.), the organ served (renal, hepatic, gonadal), or the bone followed (vertebral, radial, tibial).

- Arteries and veins tend to run side by side and, in many places, they also run with nerves.

- The systemic vessels do not always match on the right and left sides of the body. Thus, while almost all vessels in the head and limbs are bilaterally symmetrical, some of the large, deep vessels of the trunk region are asymmetrical or unpaired.

(Text continues on p. 747.)

19

Table 19.3 **Pulmonary and Systemic Circulations**

Pulmonary Circulation

The pulmonary circulation (**Figure 19.21a**) functions only to bring blood into close contact with the alveoli (air sacs) of the lungs so that gases can be exchanged. It does not directly serve the metabolic needs of body tissues.

Oxygen-poor, dark red blood enters the pulmonary circulation as it is pumped from the right ventricle into the large **pulmonary trunk** (Figure 19.21b), which runs diagonally upward for about 8 cm and then divides abruptly to form the **right** and **left pulmonary arteries**. In the lungs, the pulmonary arteries subdivide into the **lobar arteries** (lo′bar) (three in the right lung and two in the left lung), each of which serves one lung lobe. The lobar arteries accompany the main bronchi into the lungs and then branch profusely, forming first arterioles and then the dense networks of **pulmonary capillaries** that surround and cling to the delicate air sacs. It is here that oxygen moves from the alveolar air to the blood and carbon dioxide moves from the blood to the alveolar air. As gases are exchanged and the oxygen content of the blood rises, the blood becomes bright red. The pulmonary capillary beds drain into venules, which join to form the two **pulmonary veins** exiting from each lung. The four pulmonary veins complete the circuit by unloading their precious cargo into the left atrium of the heart.

(a) Schematic flowchart.

(b) Illustration. The pulmonary arterial system is shown in blue to indicate that the blood it carries is oxygen-poor. The pulmonary venous drainage is shown in red to indicate that the blood it transports is oxygen-rich.

Figure 19.21 **Pulmonary circulation.** (RA = right atrium, RV = right ventricle, LA = left atrium, LV = left ventricle)

Table 19.3 *(continued)*

Note that any vessel with the term *pulmonary* or *lobar* in its name is part of the pulmonary circulation. All others are part of the systemic circulation.

Pulmonary arteries carry oxygen-poor, carbon dioxide–rich blood, and pulmonary veins carry oxygen-rich blood.[*] This is opposite to the systemic circulation, where arteries carry oxygen-rich blood and veins carry carbon dioxide–rich, relatively oxygen-poor blood.

Systemic Circulation

The systemic circulation provides the *functional blood supply* to all body tissues; that is, it delivers oxygen, nutrients, and other needed substances while carrying away carbon dioxide and other metabolic wastes. Freshly oxygenated blood[*] returning from the pulmonary circuit is pumped out of the left ventricle into the aorta (**Figure 19.22**).

From the aorta, blood can take various routes, because essentially all systemic arteries branch from this single great vessel. The aorta arches upward from the heart and then curves and runs downward along the body midline to its terminus in the pelvis, where it splits to form the two large arteries serving the lower extremities. The branches of the aorta continue to subdivide to produce the arterioles and, finally, the capillaries that spread throughout the organs. Venous blood draining from organs inferior to the diaphragm ultimately enters the inferior vena cava.[†] Except for some coronary and thoracic venous drainage (which enters the azygos system of veins), the superior vena cava drains body regions above the diaphragm. The venae cavae empty the carbon dioxide–laden blood into the right atrium of the heart.

Two important points concerning the two major circulations:

- Blood passes from systemic veins to systemic arteries only after first moving through the pulmonary circuit (Figure 19.21a).

- Although the entire cardiac output of the right ventricle passes through the pulmonary circulation, only a small fraction of the output of the left ventricle flows through any single organ (Figure 19.22).

The systemic circulation can be viewed as multiple circulatory channels functioning in parallel to distribute blood to all body organs.

[*]By convention, oxygen-rich blood is shown red and oxygen-poor blood is shown blue.
[†]Venous blood from the digestive viscera passes through the hepatic portal circulation (liver and associated veins) before entering the inferior vena cava.

Figure 19.22 Schematic flowchart showing an overview of the systemic circulation. The pulmonary circulation is shown in gray for comparison. (RA = right atrium, RV = right ventricle, LA = left atrium, LV = left ventricle)

Table 19.4   **The Aorta and Major Arteries of the Systemic Circulation**

Figure 19.23a diagrams the distribution of the aorta and major arteries of the systemic circulation in flowchart form, and Figure 19.23b illustrates them. See Tables 19.5 through 19.8 for fine points about the vessels arising from the aorta.

The **aorta** (a-or′tah) is the largest artery in the body. In adults, the aorta is approximately the size of a garden hose where it issues from the left ventricle of the heart. Its internal diameter is 2.5 cm, and its wall is about 2 mm thick. It decreases in size slightly as it runs to its terminus. The aortic valve guards the base of the aorta and prevents backflow of blood during diastole. Opposite each aortic valve cusp is an *aortic sinus*, which contains baroreceptors important in reflex regulation of blood pressure.

Different portions of the aorta are named according to shape or location. The first portion, the **ascending aorta**, runs posteriorly and to the right of the pulmonary trunk. It persists for only about 5 cm before curving to the left as the aortic arch. The only branches of the ascending aorta are the **right** and **left coronary**

arteries, which supply the myocardium. The **aortic arch**, deep to the sternum, begins and ends at the sternal angle (T_4 level). Its three major branches (R to L) are: (1) the **brachiocephalic trunk** (bra′ke-o-sĕ-fal″ik; "armhead"), which passes superiorly under the right sternoclavicular joint and branches into the **right common carotid artery** (kah-rot′id) and the **right subclavian artery**, (2) the **left common carotid artery**, and (3) the **left subclavian artery**. These three vessels provide the arterial supply of the head, neck, upper limbs, and part of the thorax wall.

The **descending aorta** runs along the anterior spine. Called the **thoracic aorta** from T_5 to T_{12}, it sends off numerous small arteries to the thorax wall and viscera before piercing the diaphragm. As it enters the abdominal cavity, it becomes the **abdominal aorta**. This portion supplies the abdominal walls and viscera and ends at the L_4 level, where it splits into the **right** and **left common iliac arteries**, which supply the pelvis and lower limbs.

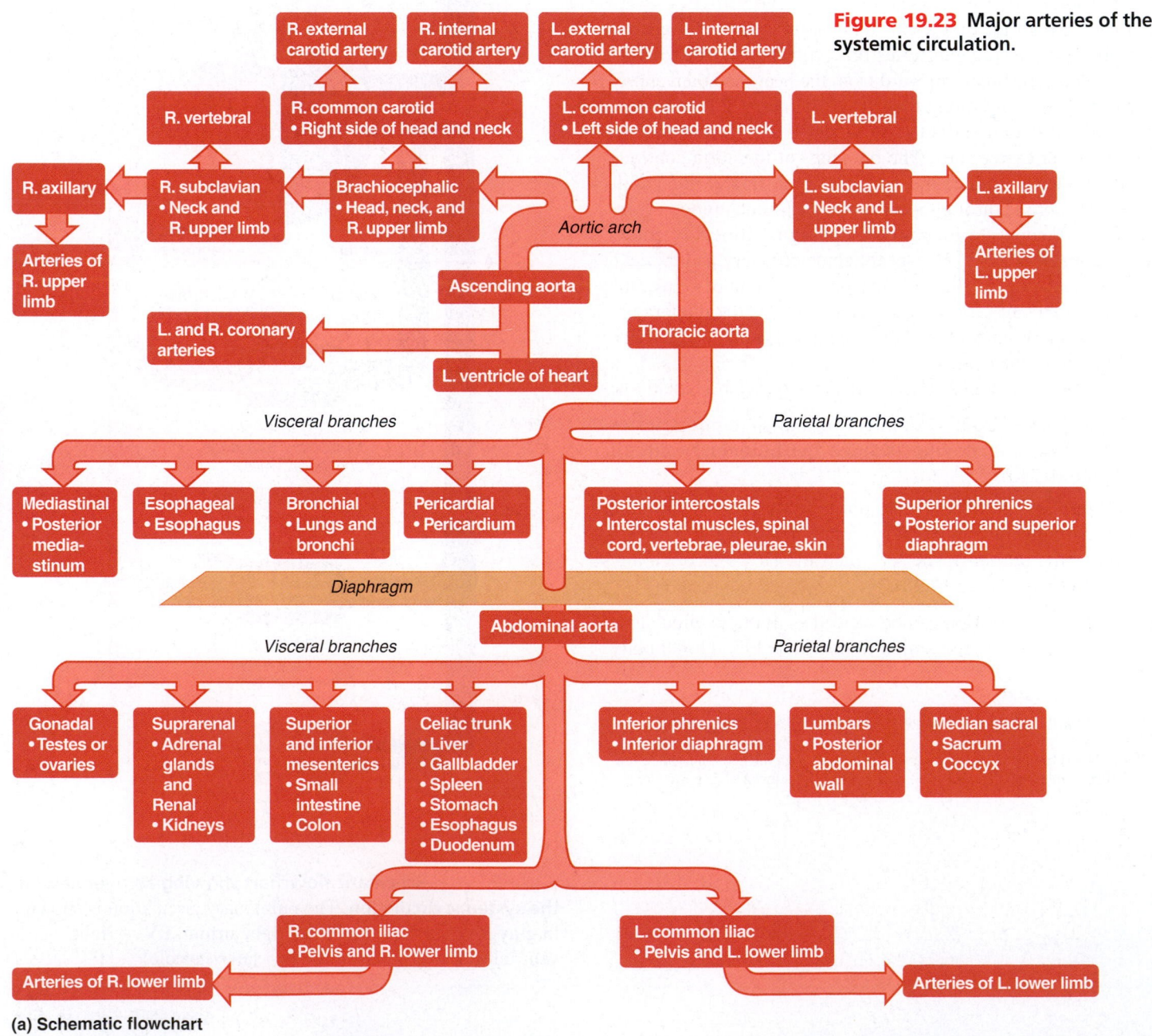

Figure 19.23 Major arteries of the systemic circulation.

(a) Schematic flowchart

19

Table 19.4 (continued)

Arteries of the head and trunk

Internal carotid artery

External carotid artery

Common carotid arteries

Vertebral artery

Subclavian artery

Brachiocephalic trunk

Aortic arch

Ascending aorta

Coronary artery

Celiac trunk

Abdominal aorta

Superior mesenteric artery

Renal artery

Gonadal artery

Inferior mesenteric artery

Common iliac artery

Internal iliac artery

Arteries that supply the upper limb

Subclavian artery

Axillary artery

Brachial artery

Radial artery

Ulnar artery

Deep palmar arch

Superficial palmar arch

Digital arteries

Arteries that supply the lower limb

External iliac artery

Femoral artery

Popliteal artery

Anterior tibial artery

Posterior tibial artery

Arcuate artery

(b) Illustration, anterior view

Figure 19.23 (continued)

19

Table 19.5   Arteries of the Head and Neck

Four paired arteries supply the head and neck. These are the common carotid arteries, plus three branches from each subclavian artery: the vertebral arteries, the thyrocervical trunks, and the costocervical trunks (**Figure 19.24b**). Of these, the common carotid arteries have the broadest distribution (Figure 19.24a).

(a) Schematic flowchart

Figure 19.24 Arteries of the head, neck, and brain.

Each common carotid divides into two major branches (the internal and external carotid arteries). At the division point, each internal carotid artery has a slight dilation, the **carotid sinus**, that contains baroreceptors that assist in reflex blood pressure control. The **carotid bodies**, chemoreceptors involved in controlling respiratory rate, are located close by. Pressing on the neck in the area of the carotid sinuses can cause unconsciousness (*carot* = stupor) because the pressure created mimics high blood pressure, eliciting vasodilation, which interferes with blood delivery to the brain.

Description and Distribution

Common carotid arteries.   The origins of these two arteries differ: The right common carotid artery arises from the brachiocephalic trunk; the left is the second branch of the aortic arch. The common carotid arteries ascend through the lateral neck, and at the superior border of the larynx (the level of the "Adam's apple"), each divides into its two major branches, the *external* and *internal carotid arteries.*

The **external carotid arteries** supply most tissues of the head except for the brain and orbit. As each artery runs superiorly, it sends branches to the thyroid gland and larynx (**superior thyroid artery**), the tongue (**lingual artery**), the skin and muscles of the anterior face (**facial artery**), and the posterior scalp (**occipital artery**). Each external carotid artery terminates by splitting into a **superficial temporal artery**, which supplies the parotid salivary gland and most of the scalp, and a **maxillary artery**, which supplies the upper and lower jaws and chewing muscles, the teeth, and the nasal cavity. A clinically important branch of the maxillary artery is the *middle meningeal artery* (not illustrated). It enters the skull through the foramen spinosum and supplies the inner surface of the parietal bone, squamous part of the temporal bone, and the underlying dura mater.

The larger **internal carotid arteries** supply the orbits and more than 80% of the cerebrum. They assume a deep course and enter the skull through the carotid canals of the temporal bones (Figure 7.6a and b, ◄ **p. 206**). Once inside the cranium, each artery gives off one main branch, the ophthalmic artery, and then divides into the anterior and middle cerebral arteries. The **ophthalmic arteries** (of-thal′mik) supply the eyes, orbits, forehead, and nose. Each **anterior cerebral artery** supplies the medial surface of the frontal and parietal lobes of the cerebral hemisphere on its side and also anastomoses with its partner on the opposite side via a short arterial shunt called the **anterior communicating artery** (Figure 19.24c). The **middle cerebral arteries** run in the lateral sulci of their respective cerebral hemispheres and supply the lateral parts of the temporal, parietal, and frontal lobes.

Vertebral arteries.   The vertebral arteries spring from the subclavian arteries at the root of the neck and ascend through foramina in the transverse processes of the cervical vertebrae to enter the skull through the foramen magnum. En route, they

Table 19.5 *(continued)*

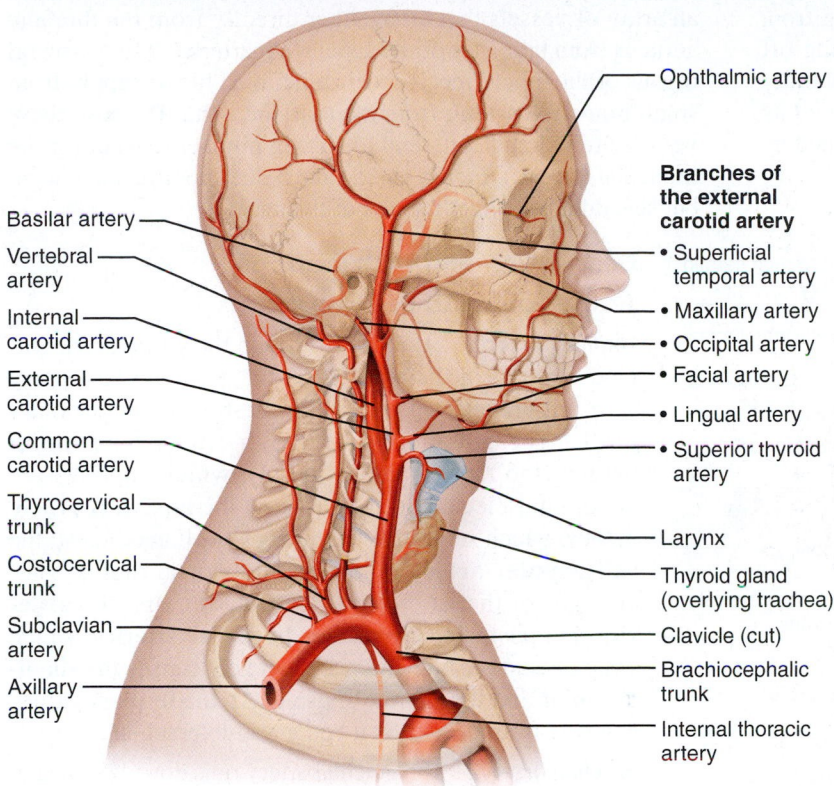

Ophthalmic artery

Basilar artery

Vertebral artery

Internal carotid artery

External carotid artery

Common carotid artery

Thyrocervical trunk

Costocervical trunk

Subclavian artery

Axillary artery

Branches of the external carotid artery
- Superficial temporal artery
- Maxillary artery
- Occipital artery
- Facial artery
- Lingual artery
- Superior thyroid artery

Larynx

Thyroid gland (overlying trachea)

Clavicle (cut)

Brachiocephalic trunk

Internal thoracic artery

(b) Arteries of the head and neck, right aspect

send branches to the vertebrae and cervical spinal cord and to some deep structures of the neck. Within the cranium, the right and left vertebral arteries join to form the **basilar artery** (bas′ĭ-lar), which ascends along the anterior aspect of the brain stem, giving off branches to the cerebellum, pons, and inner ear (Figure 19.24b and c). At the pons-midbrain border, the basilar artery divides into a pair of **posterior cerebral arteries**, which supply the occipital lobes and the inferior parts of the temporal lobes.

Arterial shunts called **posterior communicating arteries** connect the posterior cerebral arteries to the middle cerebral arteries anteriorly. The two posterior arteries and a single **anterior communicating artery** complete the formation of an arterial anastomosis called the **cerebral arterial circle** (*circle of Willis*). This structure encircles the pituitary gland and optic chiasma and unites the brain's anterior and posterior blood supplies. It also equalizes blood pressure in the two brain areas and provides alternate routes for blood to reach the brain tissue if a carotid or vertebral artery becomes occluded.

Thyrocervical and costocervical trunks. These short vessels arise from the subclavian artery just lateral to the vertebral arteries on each side (Figure 19.24b and Figure 19.25). The thyrocervical trunk mainly supplies the thyroid gland, portions of the cervical vertebrae and spinal cord, and some scapular muscles. The costocervical trunk serves deep neck and superior intercostal muscles.

Anterior

Frontal lobe

Optic chiasma

Middle cerebral artery

Internal carotid artery

Mammillary body

Temporal lobe

Pons

Occipital lobe

Posterior

Cerebral arterial circle (*circle of Willis*)
- Anterior communicating artery
- Anterior cerebral artery
- Posterior communicating artery
- Posterior cerebral artery

Basilar artery

Vertebral artery

Cerebellum

(c) Major arteries serving the brain (inferior view, right side of cerebellum and part of right temporal lobe removed)

Figure 19.24 *(continued)*

19

Table 19.6   Arteries of the Upper Limbs and Thorax

The upper limbs are supplied entirely by arteries arising from the **subclavian arteries** (**Figure 19.25a**). After giving off branches to the neck, each subclavian artery courses laterally between the clavicle and first rib to enter the axilla, where its name changes to axillary artery. The thorax wall is supplied by an array of vessels that arise either directly from the thoracic aorta or from branches of the subclavian arteries. Most visceral organs of the thorax receive their functional blood supply from small branches issuing from the thoracic aorta. Because these vessels are so small and tend to vary in number (except for the bronchial arteries), Figure 19.25a and b does not illustrate them, but several are listed at the end of this table.

(a) Schematic flowchart

Figure 19.25 Arteries of the right upper limb and thorax.

Description and Distribution

Arteries of the Upper Limb

Axillary artery.   As it runs through the axilla accompanied by cords of the brachial plexus, each axillary artery gives off branches to the axilla, chest wall, and shoulder girdle. These branches include the **thoracoacromial artery** (tho″rah-ko-ah-kro′me-al), which supplies the deltoid muscle and pectoral region; the **lateral thoracic artery**, which serves the lateral chest wall and breast; the **subscapular artery** to the scapula, dorsal thorax wall, and part of the latissimus dorsi muscle; and the **anterior** and **posterior circumflex humeral arteries**, which wrap around the humeral neck and help supply the shoulder joint and the deltoid muscle. As the axillary artery emerges from the axilla, it becomes the brachial artery.

Brachial artery.   The brachial artery runs down the medial aspect of the humerus and supplies the anterior flexor muscles of the arm. One major branch, the **deep artery of the arm**, serves the posterior triceps brachii muscle. As it nears the elbow, the brachial artery gives off several small branches that contribute to an anastomosis serving the elbow joint and that connect to the arteries of the forearm. As the brachial artery crosses the anterior midline aspect of the elbow, it provides an easily palpated pulse point (brachial pulse) (see Figure 19.8). Immediately beyond the elbow, the brachial artery splits to form the radial and ulnar arteries, which more or less follow the course of similarly named bones down the anterior forearm.

Radial artery.   The radial artery runs from the median line of the cubital fossa to the styloid process of the radius. It supplies the lateral muscles of the forearm, the wrist, and the thumb and index finger. Proximal to the carpal bones, the radial artery provides a convenient site for taking the radial pulse.

Ulnar artery.   The ulnar artery supplies the medial aspect of the forearm, fingers 3–5, and the medial aspect of the index finger. Proximally, the ulnar artery gives off a short branch, the **common interosseous artery** (in″ter-os′e-us), which runs between the radius and ulna to serve the deep flexors and extensors of the forearm.

Palmar arches.   In the palm, branches of the radial and ulnar arteries anastomose to form the **superficial** and **deep palmar arches**. The **metacarpal arteries** and **digital arteries** that supply the fingers arise from these palmar arches.

Table 19.6   *(continued)*

Vertebral artery

Thyrocervical trunk

Costocervical trunk

Suprascapular artery

Thoracoacromial artery

Axillary artery

Subscapular artery

Posterior circumflex humeral artery

Anterior circumflex humeral artery

Brachial artery

Deep artery of arm

Common interosseous artery

Radial artery

Ulnar artery

Common carotid arteries

Right subclavian artery

Left subclavian artery

Brachiocephalic trunk

Posterior intercostal arteries

Anterior intercostal artery

Internal thoracic artery

Lateral thoracic artery

Descending aorta

Deep palmar arch

Superficial palmar arch

Digital arteries

(b) Illustration, anterior view

Figure 19.25 *(continued)*

Arteries of the Thorax Wall

Internal thoracic arteries. The internal thoracic arteries (formerly called the internal mammary arteries) arise from the subclavian arteries and supply blood to most of the anterior thorax wall. Each of these arteries descends lateral to the sternum and gives off **anterior intercostal arteries**, which supply the intercostal spaces anteriorly. The internal thoracic artery also sends superficial branches to the skin and mammary glands and terminates in twiglike branches to the anterior abdominal wall and diaphragm.

Posterior intercostal arteries. The superior two pairs of posterior intercostal arteries are derived from the **costocervical trunk**. The next nine pairs issue from the thoracic aorta and course around the rib cage to anastomose anteriorly with the anterior intercostal arteries. Inferior to the 12th rib, a pair of *subcostal arteries* emerges from the thoracic aorta (not illustrated). The posterior intercostal arteries supply the posterior intercostal spaces, deep muscles of the back, vertebrae, and spinal cord. Together, the posterior and anterior intercostal arteries supply the intercostal muscles.

Superior phrenic arteries. One or more paired superior phrenic arteries serve the posterior superior aspect of the diaphragm surface.

Arteries of the Thoracic Viscera

Pericardial arteries. Several tiny branches supply the posterior pericardium.

Bronchial arteries. Two left and one right bronchial arteries supply systemic (oxygen-rich) blood to the lungs, bronchi, and pleurae.

Esophageal arteries. Four to five esophageal arteries supply the esophagus.

Mediastinal arteries. Many small mediastinal arteries serve the posterior mediastinum.

19

Table 19.7   Arteries of the Abdomen

The arterial supply to the abdominal organs arises from the abdominal aorta (**Figure 19.26a**). Under resting conditions, about half of the entire arterial flow moves through these vessels. Except for the celiac trunk, the superior and inferior mesenteric arteries, and the median sacral artery, all are paired vessels. These arteries supply the abdominal wall, diaphragm, and visceral organs of the abdominopelvic cavity. We discuss the branches in the order of their issue.

(a) Schematic flowchart.

Figure 19.26   Arteries of the abdomen.

Table 19.7 *(continued)*

Description and Distribution

Inferior phrenic arteries. The inferior phrenics emerge from the aorta at T_{12}, just inferior to the diaphragm (Figure 19.26c). They serve the inferior diaphragm surface.

Celiac trunk. This very large unpaired branch of the abdominal aorta divides almost immediately into three branches (Figure 19.26b):

- **Common hepatic artery.** The **common hepatic artery** (hĕ-pat′ik) gives off branches to the stomach, duodenum, and pancreas. Where the **gastroduodenal artery** branches off, the common hepatic becomes the **hepatic artery proper**, which splits into right and left branches that serve the liver.

- **Splenic artery.** As the **splenic artery** (splen′ik) passes deep to the stomach, it sends branches to the pancreas and stomach and terminates in branches to the spleen.

- **Left gastric artery.** The **left gastric artery** (*gaster* = stomach) supplies part of the stomach and the inferior esophagus.

The **right** and **left gastroepiploic arteries** (gas″tro-ep″ĭ-plo′ik)—branches of the gastroduodenal and splenic arteries, respectively—serve the greater curvature of the stomach. A **right gastric artery**, which supplies the stomach's lesser curvature, may arise from the common hepatic artery or from the hepatic artery proper.

(b) The celiac trunk and its major branches. The left half of the liver has been removed.

Figure 19.26 *(continued)*

Table 19.7 **Arteries of the Abdomen** (continued)

Superior mesenteric artery (mes-en-ter′ik). This large, unpaired artery arises from the abdominal aorta at the L_1 level immediately below the celiac trunk (Figure 19.26d). It runs deep to the pancreas and then enters the mesentery (a drapelike membrane that supports the small intestine), where its numerous anastomosing branches serve virtually all of the small intestine via the **intestinal arteries**, most of the large intestine—the appendix, cecum, ascending colon (via the **ileocolic** and **right colic arteries**)— and part of the transverse colon (via the **middle colic artery**).

Suprarenal arteries (soo″prah-re′nal). The **middle suprarenal arteries** flank the origin of the superior mesenteric artery as they emerge from the abdominal aorta (Figure 19.26c). They supply blood to the adrenal (suprarenal) glands overlying the kidneys. The adrenal glands also receive two sets of branches not illustrated: *superior suprarenal* branches from the nearby inferior phrenic arteries, and *inferior suprarenal* branches from the nearby renal arteries.

Renal arteries. The short but wide renal arteries, right and left, issue from the lateral surfaces of the aorta slightly below the superior mesenteric artery (between L_1 and L_2). Each serves the kidney on its side.

Gonadal arteries (go-nă′dul). The paired gonadal arteries are called **ovarian arteries** in females and **testicular arteries** in males. The ovarian arteries extend into the pelvis to serve the ovaries and part of the uterine tubes. The much longer testicular arteries descend through the pelvis and inguinal canals to enter the scrotum, where they serve the testes.

Inferior mesenteric artery. This final major branch of the abdominal aorta is unpaired and arises from the anterior aortic surface at the L_3 level. It serves the distal part of the large intestine—from the midpart of the transverse colon to the midrectum—via its **left colic**, **sigmoidal**, and **superior rectal branches** (Figure 19.26d). Looping anastomoses between the superior and inferior mesenteric arteries help ensure that blood will continue to reach the digestive viscera in cases of trauma to one of these abdominal arteries.

Lumbar arteries. Four pairs of lumbar arteries arise from the posterolateral surface of the aorta in the lumbar region (Figure 19.26c). These segmental arteries supply the posterior abdominal wall.

Median sacral artery. The unpaired median sacral artery issues from the posterior surface of the abdominal aorta at its terminus. This tiny artery supplies the sacrum and coccyx.

Common iliac arteries. At the L_4 level, the aorta splits into the right and left common iliac arteries, which supply blood to the lower abdominal wall, pelvic organs, and lower limbs.

Hiatus (opening) for inferior vena cava

Hiatus (opening) for esophagus

Adrenal (suprarenal) gland

Celiac trunk

Kidney

Abdominal aorta

Lumbar arteries

Ureter

Median sacral artery

Diaphragm

Inferior phrenic artery

Middle suprarenal artery

Renal artery

Superior mesenteric artery

Gonadal (testicular or ovarian) artery

Inferior mesenteric artery

Common iliac artery

(c) Major branches of the abdominal aorta.

Figure 19.26 (continued) **Arteries of the abdomen.**

Table 19.7 (continued)

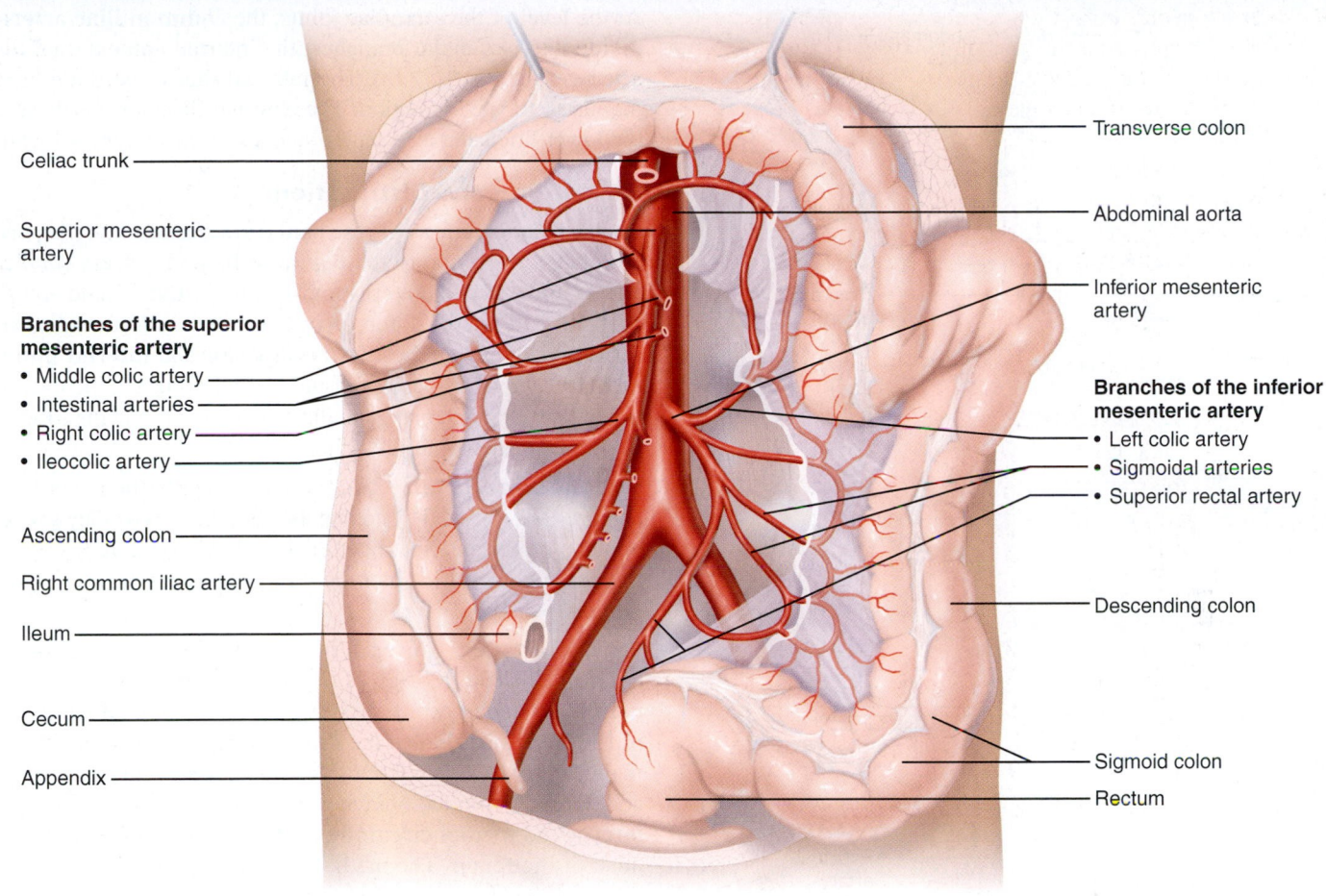

Celiac trunk

Superior mesenteric artery

Branches of the superior mesenteric artery
• Middle colic artery
• Intestinal arteries
• Right colic artery
• Ileocolic artery

Ascending colon

Right common iliac artery

Ileum

Cecum

Appendix

Transverse colon

Abdominal aorta

Inferior mesenteric artery

Branches of the inferior mesenteric artery
• Left colic artery
• Sigmoidal arteries
• Superior rectal artery

Descending colon

Sigmoid colon

Rectum

(d) Distribution of the superior and inferior mesenteric arteries. The transverse colon has been pulled superiorly.

Figure 19.26 (continued)

Check Your Understanding

22. Which paired artery supplies most of the tissues of the head except for the brain and orbits?

23. Name the arterial anastomosis at the base of the cerebrum.

24. Name the four unpaired arteries that emerge from the abdominal aorta.

For answers, see Answers Appendix.

(*Text continues on p. 749.*)

Table 19.8   Arteries of the Pelvis and Lower Limbs

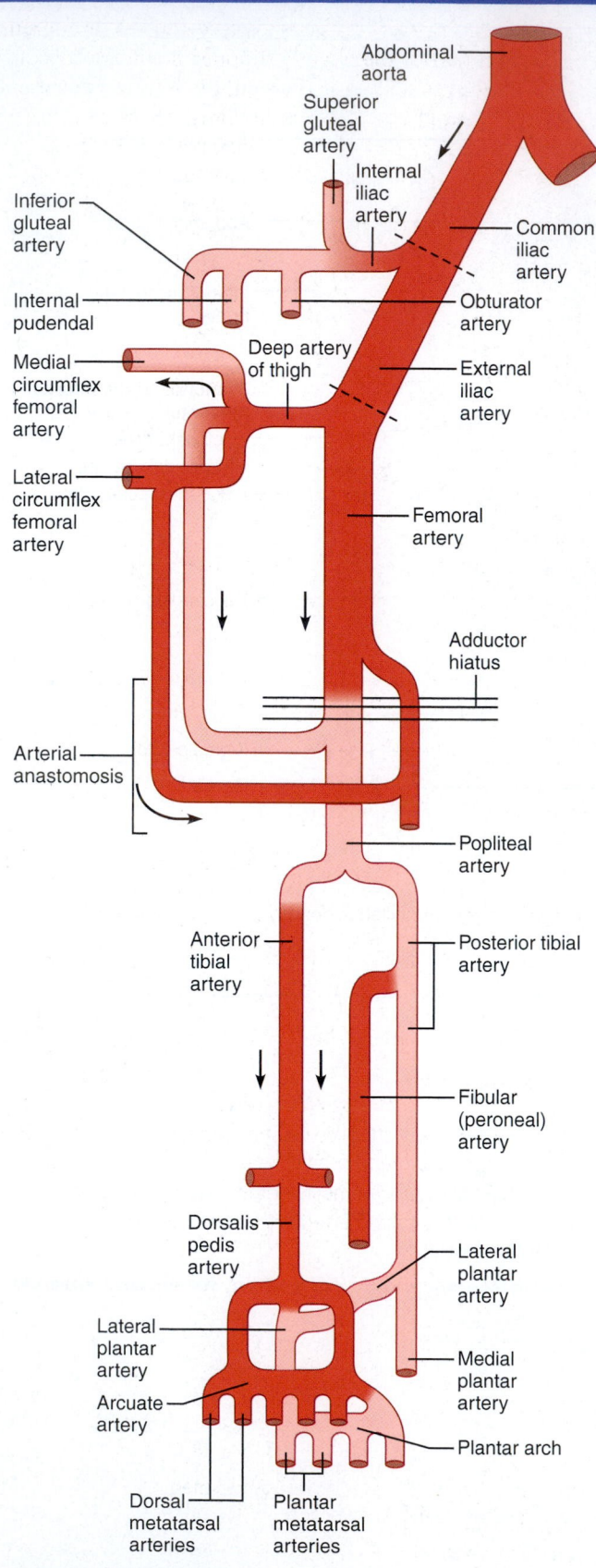

(a) Schematic flowchart

At the level of the sacroiliac joints, the **common iliac arteries** divide into two major branches, the internal and external iliac arteries (**Figure 19.27a**). The internal iliacs distribute blood mainly to the pelvic region. The external iliacs primarily serve the lower limbs but also send branches to the abdominal wall.

Description and Distribution

Internal iliac arteries.   These paired arteries run into the pelvis and distribute blood to the pelvic walls and viscera (bladder and rectum, plus the uterus and vagina in the female and the prostate and ductus deferens in the male). Additionally they serve the gluteal muscles via the **superior** and **inferior gluteal arteries**, adductor muscles of the medial thigh via the **obturator artery**, and external genitalia and perineum via the **internal pudendal artery** (not illustrated).

External iliac arteries.   These arteries supply the lower limbs (Figure 19.27b). As they course through the pelvis, they give off branches to the anterior abdominal wall. After passing under the inguinal ligaments to enter the thigh, they become the femoral arteries.

Femoral arteries.   As each of these arteries passes down the anteromedial thigh, it gives off several branches to the thigh muscles. The largest of the deep branches is the **deep artery of the thigh** (also called the *deep femoral artery*), which is the main supply to the thigh muscles (hamstrings, quadriceps, and adductors). Proximal branches of the deep femoral artery, the **lateral** and **medial circumflex femoral arteries**, encircle the neck of the femur. The medial circumflex artery is the major vessel to the head of the femur. If it is torn in a hip fracture, the bone tissue of the head of the femur dies. A long descending branch of the lateral circumflex artery supplies the vastus lateralis muscle. Near the knee the femoral artery passes posteriorly and through a gap in the adductor magnus muscle, the *adductor hiatus*, to enter the popliteal fossa, where its name changes to popliteal artery.

Popliteal artery.   This posterior vessel contributes to an arterial anastomosis that supplies the knee region and then splits into the anterior and posterior tibial arteries of the leg.

Anterior tibial artery.   The anterior tibial artery runs through the anterior compartment of the leg, supplying the extensor muscles along the way. At the ankle, it becomes the **dorsalis pedis artery**, which supplies the ankle and dorsum of the foot, and gives off a branch, the **arcuate artery**, which issues the **dorsal metatarsal arteries** to the metatarsus of the foot. The superficial dorsalis pedis ends by penetrating into the sole where it forms the medial part of the **plantar arch**. The dorsalis pedis artery provides a clinically important pulse point, the pedal pulse. If the pedal pulse is easily felt, it is fairly certain that the blood supply to the leg is good.

Figure 19.27 Arteries of the right pelvis and lower limb.

Table 19.8 *(continued)*

- Common iliac artery
- Internal iliac artery
- Superior gluteal artery
- External iliac artery
- Deep artery of thigh
- Lateral circumflex femoral artery
- Medial circumflex femoral artery
- Obturator artery
- Femoral artery
- Adductor hiatus
- Popliteal artery
- Anterior tibial artery
- Posterior tibial artery
- Fibular artery
- Dorsalis pedis artery
- Arcuate artery
- Dorsal metatarsal arteries

(b) Anterior view

Posterior tibial artery. This large artery courses through the posteromedial part of the leg and supplies the flexor muscles. Proximally, it gives off a large branch, the **fibular (peroneal) artery**, which supplies the lateral fibularis muscles of the leg. On the medial side of the foot, the posterior tibial artery divides into **lateral** and **medial plantar arteries** that serve the plantar surface of the foot. The lateral plantar artery forms the lateral end of the plantar arch. **Plantar metatarsal arteries** and digital arteries to the toes arise from the plantar arch.

- Popliteal artery
- Anterior tibial artery
- Posterior tibial artery
- Fibular artery
- Lateral plantar artery
- Dorsalis pedis artery (from top of foot)
- Medial plantar artery
- Plantar arch

(c) Posterior view

Figure 19.27 *(continued)*

Check Your Understanding

25. You are assessing the circulation in the leg of a diabetic patient at the clinic. Name the artery you palpate in each of these three locations: behind the knee, behind the medial malleolus of the tibia, on the dorsum of the foot.

For answers, see Answers Appendix.

(Text continues on p. 753.)

In our survey of the systemic veins, the major tributaries (branches) of the venae cavae are noted first in **Figure 19.28**, followed by a description in Tables 19.10 through 19.13 of the venous pattern of the various body regions. Because veins run toward the heart, the most distal veins are named first and those closest to the heart last. Deep veins generally drain the same areas served by their companion arteries, so they are not described in detail.

Description and Areas Drained

Superior vena cava. This great vein receives systemic blood draining from all areas superior to the diaphragm, except the heart wall. It is formed by the union of the **right** and **left brachiocephalic veins** and empties into the right atrium (Figure 19.28b). Notice that there are two brachiocephalic veins, but only one brachiocephalic artery (trunk). Each brachiocephalic vein is formed by the joining of the **internal jugular** and **subclavian veins** on its side. In most of the flowcharts that follow, only the vessels draining blood from the right side of the body are shown (except for the azygos circulation of the thorax).

Inferior vena cava. The widest blood vessel in the body, this vein returns blood to the heart from all body regions below the diaphragm. The abdominal aorta lies directly to its left. The paired **common iliac veins** join at L_5 to form the distal end of the inferior vena cava. From this point, it courses superiorly along the anterior aspect of the spine, receiving venous blood from the abdominal walls, gonads, kidneys, adrenal glands, and liver. Immediately above the diaphragm, the inferior vena cava ends as it enters the inferior aspect of the right atrium.

(a) Schematic flowchart

Figure 19.28 Major veins of the systemic circulation.

19

Table 19.9 *(continued)*

Veins of the head and trunk
Dural venous sinuses

External jugular vein
Vertebral vein
Internal jugular vein
Right and left
brachiocephalic veins
Superior vena cava
Great cardiac vein

Hepatic veins
Splenic vein
Hepatic portal vein
Renal vein
Superior mesenteric vein
Inferior mesenteric vein
Inferior vena cava
Common iliac vein
Internal iliac vein

**Veins that drain
the upper limb**
Subclavian vein
Axillary vein
Cephalic vein
Brachial vein
Basilic vein

Median cubital vein

Ulnar vein
Radial vein

Digital veins

Veins that drain the lower limb
External iliac vein
Femoral vein
Great saphenous vein
Popliteal vein

Posterior tibial vein
Anterior tibial vein
Small saphenous vein

Dorsal venous arch
Dorsal metatarsal veins

(b) Illustration, anterior view

Figure 19.28 *(continued)*

19

Table 19.10 **Veins of the Head and Neck**

Three pairs of veins collect most of the blood draining from the head and neck (**Figure 19.29a**):

- The external jugular veins, which empty into the subclavians
- The internal jugular veins, which join with the subclavians
- The vertebral veins, which drain into the brachiocephalic veins

Although most extracranial veins have the same names as the extracranial arteries, their courses and interconnections differ substantially.

Most veins of the brain drain into the **dural venous sinuses**, an interconnected series of enlarged chambers located between the dura mater layers (◀ p. 464). The **superior** and **inferior**

sagittal sinuses are in the falx cerebri, which dips down between the cerebral hemispheres. The inferior sagittal sinus drains into the **straight sinus** posteriorly (Figure 19.29a and c). The superior sagittal and straight sinuses then empty into the **transverse sinuses**, which run in shallow grooves on the internal surface of the occipital bone. These drain into the S-shaped **sigmoid sinuses**, which become the *internal jugular veins* as they leave the skull through the jugular foramen. The **cavernous sinuses**, which flank the sphenoid body, receive venous blood from the **ophthalmic veins** of the orbits and the facial veins, which drain the nose and upper lip area. The internal carotid artery and cranial nerves III, IV, VI, and part of V all run *through* the cavernous sinus on their way to the orbit and face.

Description and Area Drained

External jugular veins. The right and left external jugular veins drain superficial scalp and face structures served by the external carotid arteries. However, their tributaries anastomose frequently, and some of the superficial drainage from these regions enters the internal jugular veins as well. As the external jugular veins descend through the lateral neck, they pass obliquely over the sternocleidomastoid muscles and then empty into the subclavian veins.

Vertebral veins. Unlike the vertebral arteries, the vertebral veins do not serve much of the brain. Instead they drain the cervical vertebrae, the spinal cord, and some small neck muscles. They run inferiorly through the transverse foramina of the cervical vertebrae and join the brachiocephalic veins at the root of the neck.

Internal jugular veins. The paired internal jugular veins, which receive the bulk of blood draining from the brain, are the largest of the paired veins draining the head and neck. They arise from the dural venous sinuses, exit the skull via the *jugular foramina* (◀ **Figure 7.6a and b, p. 206**), and then descend through the neck alongside the internal carotid arteries. As they move inferiorly, they receive blood from some of the deep veins of the face and neck—branches of the **facial** and **superficial temporal veins** (Figure 19.29b). At the base of the neck, each internal jugular vein joins the subclavian vein on its own side to form a brachiocephalic vein. As already noted, the two brachiocephalic veins unite to form the **superior vena cava**.

(a) Schematic flowchart

Figure 19.29 Venous drainage of the head, neck, and brain.

Table 19.10 *(continued)*

- Ophthalmic vein
- Superficial temporal vein
- Facial vein
- Occipital vein
- Posterior auricular vein
- External jugular vein
- Vertebral vein
- Internal jugular vein
- Superior and middle thyroid veins
- Brachiocephalic vein
- Subclavian vein
- Superior vena cava

(b) Veins of the head and neck, right superficial aspect

- Superior sagittal sinus
- Falx cerebri
- Inferior sagittal sinus
- Straight sinus
- Cavernous sinus
- Transverse sinuses
- Sigmoid sinus
- Jugular foramen
- Right internal jugular vein

(c) Dural venous sinuses of the brain

Figure 19.29 *(continued)*

Check Your Understanding

26. In what important way does the area drained by the vertebral veins differ from the area served by the vertebral arteries?

27. Which veins drain the dural venous sinuses and where do these veins terminate?

For answers, see Answers Appendix.

(Text continues on p. 759.)

19

Table 19.11   Veins of the Upper Limbs and Thorax

(a) Schematic flowchart

Figure 19.30 Veins of the thorax and right upper limb. For clarity, the abundant branching and anastomoses of the superficial veins are not shown.

The deep veins of the upper limbs follow the paths of their companion arteries and have the same names (**Figure 19.30a**). However, except for the largest, most are paired veins that flank their artery. The superficial veins of the upper limbs are larger than the deep veins and are easily seen just beneath the skin. The median cubital vein, crossing the anterior aspect of the elbow, is commonly used to obtain blood samples or administer intravenous medications.

Blood draining from the mammary glands and the first two to three intercostal spaces enters the **brachiocephalic veins**. However, the vast majority of thoracic tissues and the thorax wall are drained by a complex network of veins called the **azygos system** (a-zi′gos). The branching nature of the azygos system provides a collateral circulation for draining the abdominal wall and other areas served by the inferior vena cava, and there are numerous anastomoses between the azygos system and the inferior vena cava.

Description and Areas Drained

Deep Veins of the Upper Limbs

The most distal deep veins of the upper limb are the radial and ulnar veins. The **deep** and **superficial venous palmar arches** of the hand empty into the **radial** and **ulnar veins** of the forearm, which then unite to form the **brachial vein** of the arm. As the brachial vein enters the axilla, it becomes the **axillary vein**, which becomes the **subclavian vein** at the level of the first rib.

Superficial Veins of the Upper Limbs

The superficial venous system begins with the *dorsal venous network* (not illustrated), a plexus of superficial veins in the dorsum of the hand. In the distal forearm, this plexus drains into two major superficial veins—the cephalic and basilic veins—which anastomose frequently as they course upward (Figure 19.30b). The **cephalic vein** bends around the radius as it travels superiorly and then continues up the lateral superficial aspect of the arm to the shoulder, where it runs in the groove between the deltoid and pectoralis muscles to join the axillary vein. The **basilic vein** courses along the posteromedial aspect of the forearm, crosses the elbow, and then joins the brachial vein in the axilla, forming the axillary vein. At the anterior aspect of the elbow, the **median cubital vein** connects the basilic and cephalic veins. The **median antebrachial vein** lies between the radial and ulnar veins in the forearm and terminates (variably) at the elbow by entering either the basilic or the cephalic vein.

The Azygos System

The azygos system consists of the following vessels, which flank the vertebral column laterally.

Azygos vein.   Located against the right side of the vertebral column, the **azygos vein** (*azygos* = unpaired) originates in the abdomen, from the **right ascending lumbar vein** that drains most of the right abdominal cavity wall and from the **right posterior intercostal veins** (except the first) that drain the chest muscles. At the T_4 level, it arches over the great vessels that run to the right lung and empties into the superior vena cava.

Table 19.11   *(continued)*

Hemiazygos vein (hĕ″me-a-zi′gos; "half the azygos").   This vessel ascends on the left side of the vertebral column. Its origin, from the **left ascending lumbar vein** and the lower (9th–11th) **posterior intercostal veins**, mirrors that of the inferior portion of the azygos vein on the right. About midthorax, the hemiazygos vein passes in front of the vertebral column and joins the azygos vein.

Accessory hemiazygos vein.   The accessory hemiazygos completes the venous drainage of the left (middle) thorax and can be thought of as a superior continuation of the hemiazygos vein. It receives blood from the 4th–8th posterior intercostal veins and then crosses to the right to empty into the azygos vein. Like the azygos, it receives oxygen-poor systemic blood from the bronchi of the lungs (*bronchial veins*).

Brachiocephalic veins

Right subclavian vein

Axillary vein

Brachial vein

Cephalic vein

Basilic vein

Median cubital vein

Median antebrachial vein

Cephalic vein

Radial vein

Internal jugular vein

External jugular vein

Left subclavian vein

Superior vena cava

Azygos vein

Accessory hemiazygos vein

Hemiazygos vein

Posterior intercostals

Inferior vena cava

Ascending lumbar vein

Basilic vein

Ulnar vein

Deep venous palmar arch

Superficial venous palmar arch

Digital veins

(b) Anterior view

Figure 19.30 *(continued)*

19

Table 19.12 Veins of the Abdomen

The **inferior vena cava** returns blood from the abdominopelvic viscera and abdominal walls to the heart (**Figure 19.31a**). Most of its venous tributaries have names that correspond to the arteries serving the abdominal organs.

Veins draining the digestive viscera empty into a common vessel, the *hepatic portal vein*, which transports this venous blood into the liver before it is allowed to enter the major systemic circulation via the hepatic veins (Figure 19.31c). Such a venous system—veins connecting two capillary beds

together—is called a *portal system* and always serves very specific needs. The **hepatic portal system** carries nutrient-rich blood (which may also contain toxins and microorganisms) from the digestive organs to the liver, where it can be "treated" before it reaches the rest of the body. As the blood percolates slowly through the liver sinusoid capillaries, hepatocytes process nutrients and toxins, and phagocytic cells rid the blood of bacteria and other foreign matter.

Description and Areas Drained

The veins of the abdomen are listed in inferior to superior order.

Lumbar veins.   Several pairs of lumbar veins drain the posterior abdominal wall. They empty both directly into the inferior vena cava and into the ascending lumbar veins of the azygos system of the thorax.

Gonadal (testicular or ovarian) veins.   The right gonadal vein drains the ovary or testis on the right side of the body and empties into the inferior vena cava. The left gonadal vein drains into the left renal vein superiorly.

Renal veins.   The right and left renal veins drain the kidneys.

Suprarenal veins.   The right suprarenal vein drains the adrenal gland on the right and empties into the inferior vena cava. The left suprarenal vein drains into the left renal vein.

Hepatic portal system.   Like all portal systems, the hepatic portal system is a series of vessels in which two separate capillary beds lie between the arterial supply and the final venous drainage. In this case, the first capillary beds are in the stomach and intestines and drain into tributaries of the hepatic portal vein, which brings them to the second capillary bed in the liver. The short **hepatic portal vein** begins at the L_2 level. Numerous tributaries from the stomach and pancreas contribute to the hepatic portal system (Figure 19.31c), but the major vessels are:

- **Superior mesenteric vein:** Drains the entire small intestine, part of the large intestine (ascending and transverse regions), and stomach.
- **Splenic vein:** Collects blood from the spleen, parts of the stomach and pancreas, and then joins the superior mesenteric vein to form the hepatic portal vein.
- **Inferior mesenteric vein:** Drains the distal portions of the large intestine and rectum and joins the splenic vein just before that vessel unites with the superior mesenteric vein to form the hepatic portal vein.

Hepatic veins.   The right, left, and middle hepatic veins carry venous blood from the liver to the inferior vena cava.

Cystic veins.   The cystic veins drain the gallbladder and join the portal veins in the liver.

Inferior phrenic veins.   The inferior phrenic veins drain the inferior surface of the diaphragm.

(a) Schematic flowchart.

Figure 19.31 Veins of the abdomen.

Table 19.12   *(continued)*

Hepatic veins

Inferior vena cava

Right suprarenal vein

Right gonadal vein

External iliac vein

Inferior phrenic vein

Left suprarenal vein

Renal veins

Left ascending lumbar vein

Lumbar veins

Left gonadal vein

Common iliac vein

Internal iliac vein

(b) Tributaries of the inferior vena cava. Venous drainage of abdominal organs not drained by the hepatic portal vein.

Hepatic veins

Liver

Hepatic portal vein

Small intestine

Rectum

Gastric veins

Spleen

Inferior vena cava

Splenic vein

Right gastroepiploic vein

Inferior mesenteric vein

Superior mesenteric vein

Large intestine

(c) The hepatic portal circulation.

Figure 19.31   *(continued)*

19

Table 19.13   Veins of the Pelvis and Lower Limbs

As in the upper limbs, most deep veins of the lower limbs have the same names as the arteries they accompany and many are double. Poorly supported by surrounding tissues, the two superficial saphenous veins (great and small) are common sites of varicosities. The great saphenous (*saphenous* = obvious) vein is frequently excised and used as a coronary bypass vessel.

Description and Areas Drained

Deep veins.   After being formed by the union of the **medial** and **lateral plantar veins**, the **posterior tibial vein** ascends deep in the calf muscle and receives the **fibular (peroneal) vein** (**Figure 19.32**). The **anterior tibial vein**, which is the superior continuation of the **dorsalis pedis vein** of the foot, unites at the knee with the posterior tibial vein to form the **popliteal vein**, which crosses the back of the knee. As the popliteal vein emerges from the knee, it becomes the **femoral vein**, which drains the deep structures of the thigh. The femoral vein becomes the **external iliac vein** as it enters the pelvis. In the pelvis, the external iliac vein unites with the **internal iliac vein** to form the **common iliac vein**. The distribution of the internal iliac veins parallels that of the internal iliac arteries.

Superficial veins.   The **great** and **small saphenous veins** (sah-fe′nus) issue from the **dorsal venous arch** of the foot (Figure 19.32b and c). These veins anastomose frequently with each other and with the deep veins along their course. The great saphenous vein is the longest vein in the body. It travels superiorly along the medial aspect of the leg to the thigh, where it empties into the femoral vein just distal to the inguinal ligament. The small saphenous vein runs along the lateral aspect of the foot and then through the deep fascia of the calf muscles, which it drains. At the knee, it empties into the popliteal vein.

(a) Schematic flowchart of the anterior and posterior veins

(b) Anterior view

(c) Posterior view

Figure 19.32 Veins of the right lower limb.

Check Your Understanding

28. Below is a schematic drawing of the hepatic portal system showing two capillary beds separated by a vein. State the locations of capillary beds a and c. What type of capillaries are found in bed c? Name the major veins labeled b, d, and e.

29. Name the leg veins that often become varicosed.

30. MAKE CONNECTIONS You learned about another portal system in Chapter 16. Name that portal system.

For answers, see Answers Appendix.

Developmental Aspects of Blood Vessels

The endothelial lining of blood vessels is formed by mesodermal cells, which collect in little masses called **blood islands** throughout the microscopic embryo. These blood islands form fragile sprouting extensions that reach toward one another and toward the forming heart to lay down the rudimentary vascular tubes. Meanwhile, adjacent mesenchymal cells, stimulated by platelet-derived growth factor, surround the endothelial tubes, forming the stabilizing muscular and fibrous coats of the vessel walls.

How do blood vessels "know" where to grow? Many blood vessels simply follow the same guidance cues that nerves follow, which is why forming vessels often snuggle closely to nerves. Whether a vessel becomes an artery or a vein depends upon the local concentration of a differentiation factor called *vascular endothelial growth factor*. As noted in Chapter 18, the heart pumps blood through the rudimentary vascular system by the fourth week of development.

In addition to the fetal shunts that bypass the nonfunctional lungs (the *foramen ovale* and *ductus arteriosus*), other vascular modifications are found in the fetus. A special vessel, the *ductus venosus*, largely bypasses the liver. Also important are the *umbilical vein* and *arteries*, large vessels that circulate blood between the fetal circulation and the placenta where gas and nutrient exchanges occur with the mother's blood (see Chapter 28). Once the fetal circulatory pattern is laid down, few vascular changes occur until birth, when the umbilical vessels and shunts are occluded (blocked).

Unlike congenital heart diseases, congenital vascular problems are rare, and blood vessels are remarkably trouble-free during youth. Vessels form as needed to support body growth and wound healing, and to rebuild vessels lost each month during a woman's menstrual cycle. As we age, signs of vascular disease begin to appear. In some, the venous valves weaken, and purple, snakelike varicose veins appear. In others, more insidious signs of inefficient circulation appear: tingling fingers and toes and cramping muscles.

Although the degenerative process of atherosclerosis begins in youth, its consequences are rarely apparent until middle to old age, when it may precipitate a myocardial infarction (heart attack) or stroke. Until puberty, the blood vessels of boys and girls look alike, but from puberty to about age 45, women have strikingly less atherosclerosis than men because of the protective effects of estrogens. Estrogens reduce resistance to blood flow and increase the production of HDL ("good" lipoprotein), which reduces the risk of atherosclerosis.

Between the ages of 45 and 65, when estrogen production wanes in women, this "gap" between the sexes closes, and males and females above age 65 are equally at risk for cardiovascular disease. You might expect that giving postmenopausal women supplementary estrogens would maintain this protective effect. Surprisingly, clinical trials have shown that this is not the case.

Blood pressure changes with age. In a newborn baby, arterial pressure is about 90/55. Blood pressure rises steadily during childhood to finally reach the adult value (120/80). After age 40, the incidence of hypertension increases dramatically, as do associated illnesses such as heart attacks, strokes, vascular disease, and renal failure.

At least some vascular disease is a product of our modern technological culture. "Blessed" with high-protein and lipid-rich diets, empty-calorie snacks, energy-saving devices, and high-stress jobs, many of us are struck down prematurely. Lifestyle modifications—a healthy diet, regular aerobic exercise, and eliminating cigarette smoking—can help prevent cardiovascular disease.

RELATED CLINICAL TERMS

Aneurysm (an'u-rizm; *aneurysm* = a widening) A balloonlike outpocketing of an artery wall that places the artery at risk for rupture; most often reflects gradual weakening of the artery by chronic hypertension or atherosclerosis. The most common sites of aneurysms are the abdominal aorta and arteries feeding the brain and kidneys.

Angiogram (an'je-o-gram"; *angio* = a vessel; *gram* = writing) Diagnostic technique involving the infusion of a radiopaque substance into the circulation for X-ray examination of specific blood vessels (◀ p. 16). The major technique for diagnosing coronary artery occlusion and risk of a heart attack.

Deep venous thrombosis Clot formation in a deep vein. An ever-present danger is that the clot may detach and form a life-threatening pulmonary embolus.

(Text continues on p. 762.)

Atherosclerosis? Get Out the Cardiovascular Drāno®

When pipes get clogged, it is usually because we've dumped something down the drain that shouldn't be there, or something is growing inside (tree roots, for example), trapping the normal sludge coming through (see top photo). In **atherosclerosis**, small patchy thickenings called *atheromas* form that make the walls of our arteries thicker and stiffer, resulting in hypertension. Atheromas can intrude into the vessel lumen, making it easy for arterial spasms or a blood clot to close the vessel completely.

Onset and Stages

Atherosclerosis indirectly causes half of the deaths in the Western world. How does this scourge of blood vessels come about? The development of a full-blown atheroma is believed to occur in several stages.

1. **The endothelium is injured.**
   The initial event is damage to the endothelium caused by turbulent blood flow, bloodborne chemicals (such as cholesterol), hypertension, components of cigarette smoke, or viral or bacterial infections. Damaged endothelial cells become dysfunctional, and this sets the stage for atherosclerosis.

2. **Lipids accumulate and oxidize in the tunica intima.** Injured endothelial cells set up a local inflammatory response, attracting white blood cells. Endothelial cells then begin to transport and modify low-density lipoproteins (LDLs) that deliver cholesterol to tissue cells from the blood. The accumulated LDL oxidizes in the inflammatory environment. This not only damages neighboring cells, but also attracts more macrophages. Some of these macrophages become so engorged with LDLs that they are transformed into lipid-laden *foam cells*.

3. **Smooth muscle cells proliferate and a fibrous cap forms.** Smooth muscle cells from the tunica media deposit collagen and elastic fibers. The thickened intima, called a **fibrous** or **atherosclerotic plaque**, has a core of dead and dying foam cells. At first the vessel walls accommodate the growing plaque by expanding outward, but eventually these fatty mounds begin to protrude into the vessel lumen, producing full-blown atherosclerosis (see bottom photo).

4. **The plaque becomes unstable.** As the plaque enlarges, the cells at its center die, calcium is deposited, and collagen fiber production declines. Now called a

Top **A pipe clogged by accumulated deposits.** *Bottom* **Atherosclerotic plaques nearly close a human artery.**

complicated plaque, it is unstable and prone to rupture.

Consequences

Plaques stiffen artery walls, causing *hypertension*. The increased pressure stresses the plaques, making them even more unstable. Plaques also constrict the vessel and cause the arterial walls to fray and ulcerate, conditions that encourage thrombus (clot) formation.

Two other factors also promote thrombus formation: (1) Injured endothelial cells make less anti-thrombotic chemicals. (2) *Lipoprotein (a)*, an altered form of LDL found in some individuals, inhibits fibrinolysis.

Plaque formation increases the risk of heart attacks and strokes, and is responsible for the pain (angina) that occurs when heart muscle is ischemic. Atherosclerosis, particularly of the aorta, can also cause *aneurysms* (ballooning of the arterial wall) that may burst.

Risk Factors

Why are some of us so troubled by atherosclerosis while others are seemingly

immune to its ravages? A large number of interacting risk factors, such as increasing age, being male, family history, high blood cholesterol, hypertension, cigarette smoking, lack of exercise, diabetes, obesity, stress, and intake of trans fats are involved.

A growing body of evidence links systemic inflammation with the formation and subsequent rupture of atherosclerotic plaques. *C-reactive protein* is a marker of systemic inflammation that is measured to predict the likelihood of future heart attacks and strokes.

Prevention and Treatment

Some risk factors are under our control. We can avoid smoking, lose weight, exercise regularly to increase blood levels of high-density lipoprotein (HDL, the "good" lipoprotein that removes cholesterol from vessel walls and carries it to the liver), and eat a healthy diet low in saturated and trans fats.

But for many of us, these measures are not enough. Cholesterol-lowering, anti-inflammatory statins (and even the humble aspirin) can help decrease the risk of heart attacks and strokes.

Plaques that partially block arteries are treated in much the same way we would treat a blocked sewer pipe—dig it up and replace it or call a plumber to drill through the obstruction. In *coronary bypass surgery*, veins or small arteries removed from other areas of the body are implanted in the heart to restore its circulation. In *balloon angioplasty*, a balloon is passed through the vessels to the obstruction and then inflated to compress the fatty mass against the vessel wall.

Angioplasty temporarily clears the path, but *restenoses* (new blockages) often occur. *Stents*, short metal-mesh tubes, are sometimes used to help hold the vessel open.

When an atheroma ruptures and induces clot formation, *thrombolytic (clot-dissolving) agents* can help. A genetically engineered form of the naturally occurring *tissue plasminogen activator* (tPA) is injected directly into the blocked vessel. tPA restores blood flow quickly and puts an early end to many heart attacks and strokes in progress.

Of course, it's best to prevent atherosclerosis from progressing in the first place by changing our lifestyles. But how many of us will give up our burgers, butter, and sedentary lifestyles for the sake of preventing atherosclerosis?

Homeostatic Interrelationships between the
Cardiovascular System and Other Body Systems

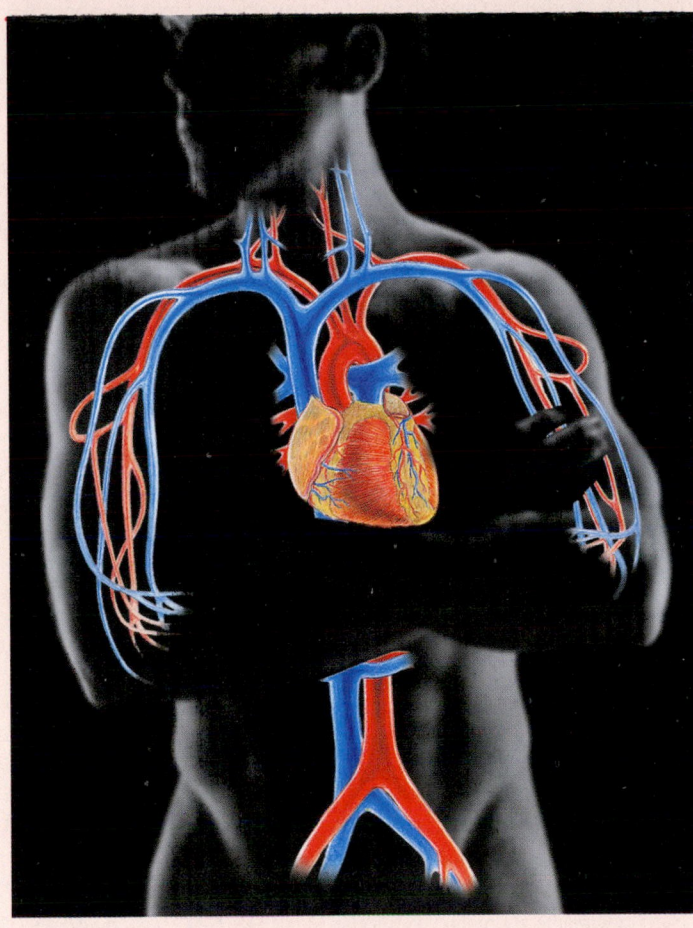

Integumentary System   Chapter 5
- The cardiovascular system delivers oxygen and nutrients; carries away wastes
- The skin vasculature is an important blood reservoir and provides a site for heat loss from the body

Skeletal System   Chapters 6–8
- The cardiovascular system delivers oxygen and nutrients; carries away wastes
- Bones are the sites of hematopoiesis; protect cardiovascular organs by enclosure; and provide a calcium depot

Muscular System   Chapters 9–10
- The cardiovascular system delivers oxygen and nutrients; carries away wastes
- The muscular "pump" aids venous return; aerobic exercise enhances cardiovascular efficiency and helps prevent atherosclerosis

Nervous System   Chapters 11–15
- The cardiovascular system delivers oxygen and nutrients; carries away wastes
- The ANS regulates cardiac rate and force; sympathetic division maintains blood pressure and controls blood flow to skin for thermoregulation

Endocrine System   Chapter 16
- The cardiovascular system delivers oxygen and nutrients; carries away wastes; blood serves as a transport vehicle for hormones
- Various hormones influence blood pressure (epinephrine, ANP, angiotensin II, thyroxine, ADH); estrogens maintain vascular health in premenopausal women

Lymphatic System/Immunity   Chapters 20–21
- The cardiovascular system delivers oxygen and nutrients to lymphatic organs, which house immune cells; provides transport medium for lymphocytes and antibodies; carries away wastes
- The lymphatic system picks up leaked fluid and plasma proteins and returns them to the cardiovascular system; immune cells protect cardiovascular organs from specific pathogens

Respiratory System   Chapter 22
- The cardiovascular system delivers oxygen and nutrients; carries away wastes
- The respiratory system carries out gas exchange: loads oxygen and unloads carbon dioxide from the blood; respiratory "pump" aids venous return

Digestive System   Chapter 23
- The cardiovascular system delivers oxygen and nutrients; carries away wastes
- The digestive system provides nutrients to the blood including iron and B vitamins essential for RBC (and hemoglobin) formation

Urinary System   Chapters 25–26
- The cardiovascular system delivers oxygen and nutrients; carries away wastes; blood pressure drives filtration in the kidneys
- The urinary system helps regulate blood volume and pressure by altering urine volume and releasing renin

Reproductive System   Chapter 27
- The cardiovascular system delivers oxygen and nutrients; carries away wastes
- Estrogens maintain vascular and bone health in women

761

Diuretic (*diure* = urinate) A chemical that promotes urine formation, thus reducing blood volume. Diuretic drugs are frequently prescribed to manage hypertension.

Phlebitis (flĕ-bi′tis; *phleb* = vein; *itis* = inflammation) Inflammation of a vein accompanied by painful throbbing and redness of the skin over the inflamed vessel. It is most often caused by bacterial infection or local physical trauma.

Phlebotomy (flĕ-bot′o-me; *tomy* = cut) A venous incision or puncture made for the purpose of withdrawing blood or bloodletting.

Sclerotherapy Procedure for removing varicose or spider veins. Tiny needles are used to inject scarring agents into the abnormal vein. The vein scars, closes down, and is absorbed by the body.

Superficial thrombophlebitis Inflammation and clot formation in superficial veins, usually in the leg.

CHAPTER SUMMARY

PART 1

BLOOD VESSEL STRUCTURE AND FUNCTION

1. Blood is transported throughout the body via a continuous system of blood vessels. Arteries transport blood away from the heart; veins carry blood back to the heart. Capillaries carry blood to tissue cells and are exchange sites.

19.1 Most blood vessel walls have three layers (p. 709)

1. All blood vessels except capillaries have three layers: tunica intima, tunica media, and tunica externa. Capillary walls are composed of the tunica intima only.

19.2 Arteries are pressure reservoirs, distributing vessels, or resistance vessels (p. 710)

1. Elastic (conducting) arteries are the large arteries close to the heart that expand during systole, acting as pressure reservoirs, and then recoil during diastole to keep blood moving. Muscular (distributing) arteries carry blood to specific organs; they are less stretchy and more active in vasoconstriction. Arterioles regulate blood flow into capillary beds.
2. Atherosclerosis is a degenerative vascular disease that decreases the elasticity of arteries.

19.3 Capillaries are exchange vessels (pp. 710–712)

1. Capillaries are microscopic vessels with very thin walls. Most exhibit intercellular clefts, which aid in the exchange between blood and interstitial fluid.
2. The most permeable capillaries are sinusoid capillaries (wide, tortuous channels). Fenestrated capillaries with pores are next most permeable. Least permeable are continuous capillaries, which lack pores.
3. Most capillary beds consist of a terminal arteriole leading into capillaries drained by a postcapillary venule. The diameter of the terminal arteriole and upstream arterioles determines the amount of blood flowing through the capillaries. In select capillary beds (e.g., mesenteric capillary beds), vascular shunts (metarteriole–thoroughfare channels) connect the terminal arteriole and postcapillary venule at opposite ends of a capillary bed. In this case, the amount of blood flowing into the true capillaries is regulated by precapillary sphincters.

19.4 Veins are blood reservoirs that return blood toward the heart (pp. 712–714)

1. Veins have comparatively larger lumens than arteries, and a system of valves prevents backflow of blood.
2. Normally most veins are not filled to capacity; for this reason, they can serve as blood reservoirs.

19.5 Anastomoses are special interconnections between blood vessels (p. 714)

1. The joining together of blood vessels to provide alternate channels in the same organ is called an anastomosis. Vascular anastomoses form between arteries, between veins, and between arterioles and venules.

PART 2

PHYSIOLOGY OF CIRCULATION

19.6 Blood flows from high to low pressure against resistance (pp. 714–715)

1. Blood flow is the amount of blood flowing through a vessel, an organ, or the entire circulation in a given period of time. Blood pressure (BP) is the force per unit area exerted on a vessel wall by the contained blood. Resistance is opposition to blood flow; blood viscosity and blood vessel length and diameter contribute to resistance.
2. Blood flow is directly proportional to blood pressure and inversely proportional to resistance.

 Complete an interactive tutorial: MasteringA&P® > Study Area > Interactive Physiology > Cardiovascular System: Factors Affecting Blood Pressure.

19.7 Blood pressure decreases as blood flows from arteries through capillaries and into veins (pp. 716–718)

1. Systemic blood pressure is highest in the aorta and lowest in the venae cavae. The steepest drop in BP occurs in the arterioles, where resistance is greatest.
2. Arterial BP depends on compliance of the elastic arteries and on how much blood is forced into them. Arterial blood pressure is pulsatile, and peaks during systole; this is measured as systolic pressure. During diastole, as blood is forced distally in the circulation by the rebound of elastic arteries, arterial BP drops to its lowest value, called the diastolic pressure.
3. Pulse pressure is systolic pressure minus diastolic pressure. The mean arterial pressure (MAP) = diastolic pressure plus one-third of pulse pressure and is the pressure that keeps blood moving throughout the cardiac cycle.
4. Pulse and blood pressure measurements are used to assess cardiovascular efficiency.
5. The pulse is the alternating expansion and recoil of arterial walls with each heartbeat. Pulse points are also pressure points.
6. Blood pressure is routinely measured by the auscultatory method. Normal BP in adults is 120/80 mm Hg (systolic/diastolic).
7. Low capillary pressure (35 to 17 mm Hg) protects the delicate capillaries from rupture while still allowing adequate exchange across the capillary walls.

8. Venous pressure is nonpulsatile and low (declining to zero) because of the cumulative effects of resistance. Venous valves, large lumens, functional adaptations (muscular and respiratory pumps), and sympathetic nervous system activity promote venous return.

19.8 Blood pressure is regulated by short- and long-term controls   (pp. 718–725)

1. Blood pressure varies directly with cardiac output, total peripheral resistance (TPR), and blood volume. Vessel diameter is the major factor determining resistance, and small changes in the diameter of vessels (chiefly arterioles) significantly affect blood pressure.

2. BP is regulated by autonomic neural reflexes involving baroreceptors or chemoreceptors, the cardiovascular center (a medullary center that includes the cardiac and vasomotor centers), and autonomic fibers to the heart and vascular smooth muscle.

3. Activation of the receptors by falling BP (and to a lesser extent by a rise in blood CO_2, or falling blood pH or O_2 levels) stimulates the vasomotor center to increase vasoconstriction and the cardioacceleratory center to increase heart rate and contractility. Rising BP inhibits the vasomotor center (permitting vasodilation) and activates the cardioinhibitory center.

4. Higher brain centers (cerebrum and hypothalamus) may modify neural controls of BP via medullary centers.

5. Hormones that increase BP by promoting vasoconstriction include epinephrine and NE (these also increase heart rate and contractility), ADH, and angiotensin II (generated in response to renin release by kidney cells). Atrial natriuretic peptide is a hormone that promotes vasodilation and also causes a decline in blood volume.

6. The kidneys regulate blood pressure by regulating blood volume. Rising BP directly enhances filtrate formation and fluid losses in urine; falling BP causes the kidneys to retain more water, increasing blood volume.

7. Indirect renal regulation of blood volume involves the renin-angiotensin-aldosterone mechanism, a hormonal mechanism. When BP falls, the kidneys release renin, which triggers the formation of angiotensin II. Angiotensin II causes (1) release of aldosterone, stimulating salt and water retention, (2) vasoconstriction, (3) release of ADH, and (4) thirst.

8. Chronic hypertension (high blood pressure) is persistent BP readings of 140/90 or higher. It indicates increased total peripheral resistance, which strains the heart and promotes vascular complications of other organs, particularly the eyes and kidneys. It is a major cause of myocardial infarction, stroke, and renal disease. Risk factors are high-fat, high-salt diet, obesity, diabetes mellitus, advanced age, smoking, stress, and being a member of the black race or a family with a history of hypertension.

9. Hypotension, or low blood pressure (below 90/60 mm Hg), is rarely a problem except in circulatory shock.

10. Circulatory shock occurs when blood perfusion of body tissues is inadequate. Most cases of shock reflect low blood volume (hypovolemic shock), abnormal vasodilation (vascular shock), or pump failure (cardiogenic shock).

19.9 Intrinsic and extrinsic controls determine blood flow through tissues   (pp. 725–730)

1. Intrinsic controls (autoregulation) involve local adjustment of blood flow to individual organs based on their immediate requirements. Extrinsic controls (nerves and hormones) maintain MAP and redistribute blood during exercise and thermoregulation.

2. Autoregulation involves myogenic controls that maintain flow despite changes in blood pressure, and local chemical factors. Vasodilators include increased CO_2, H^+, and nitric oxide. Decreased O_2 concentrations also cause vasodilation. Other factors, including endothelins, decrease blood flow.

3. In most instances, autoregulation is controlled by the accumulation of local metabolites and the lack of oxygen. However, autoregulation in the brain is controlled primarily by a drop in pH and by myogenic mechanisms; and pulmonary circuit vessels dilate in response to high levels of oxygen.

19.10 Slow blood flow through capillaries promotes diffusion of nutrients and gases, and bulk flow of fluids   (pp. 730–734)

1. Blood flows fastest where the cross-sectional area of the vascular bed is least (aorta), and slowest where the total cross-sectional area is greatest (capillaries). The slow flow in capillaries allows time for nutrient-waste exchanges.

2. Nutrients, gases, and other solutes smaller than plasma proteins cross the capillary wall by diffusion; larger molecules are actively transported via pinocytotic vesicles. Water-soluble substances move through the clefts or fenestrations; fat-soluble substances pass through the lipid portion of the endothelial cell membrane.

3. Bulk flow of fluids at capillary beds determines the distribution of fluids between the bloodstream and the interstitial space. It reflects the relative effects of hydrostatic and osmotic pressures acting at the capillary (outward minus inward pressures). In general, fluid flows out of the capillary bed at the arteriolar end and reenters the capillary blood at the venule end.

4. Lymphatic vessels collect the small net loss of fluid and protein into the interstitial space and return it to the cardiovascular system.

5. Edema is an abnormal accumulation of fluid in the interstitial space as a result of imbalances in pressures that drive bulk flow or a block of lymphatic drainage.

PART 3
CIRCULATORY PATHWAYS: BLOOD VESSELS OF THE BODY

1. The pulmonary circulation transports O_2-poor, CO_2-laden blood to the lungs for oxygenation and carbon dioxide unloading. Blood returning to the right atrium of the heart is pumped by the right ventricle to the lungs via the pulmonary trunk. Blood issuing from the lungs is returned to the left atrium by the pulmonary veins. (See Table 19.3 and Figure 19.21.)

2. The systemic circulation transports oxygenated blood from the left ventricle to all body tissues via the aorta and its branches. Venous blood returning from the systemic circuit is delivered to the right atrium via the venae cavae.

3. All arteries are deep, while veins are both deep and superficial. Superficial veins tend to have numerous interconnections. Dural venous sinuses and the hepatic portal circulation are unique venous drainage patterns.

19.8 The vessels of the systemic circulation transport blood to all body tissues   (pp. 735–759)

1. Tables 19.3 to 19.13 and Figures 19.22 to 19.32 illustrate and describe vessels of the systemic circulation.

19

Developmental Aspects of Blood Vessels (p. 759)

1. The fetal vasculature develops from embryonic blood islands and mesenchyme and functions in blood delivery by the fourth week.
2. Fetal circulation differs from circulation after birth. The pulmonary and hepatic shunts and special umbilical vessels are normally occluded shortly after birth.
3. Blood pressure is low in infants and rises to adult values. Age-related vascular problems include varicose veins, hypertension, and atherosclerosis. Hypertension and associated atherosclerosis are the most important causes of cardiovascular disease in the aged.

REVIEW QUESTIONS

To access additional practice questions using your smartphone, tablet, or computer: MasteringA&P® > Study Area > Practice Tests & Quizzes

Level 1 Remember/Understand

(Some questions have more than one correct answer. Select the best answer or answers from the choices given.)

1. Which statement does not accurately describe veins? (a) have less elastic tissue and smooth muscle than arteries, (b) contain more fibrous tissue than arteries, (c) most veins in the extremities have valves, (d) always carry deoxygenated blood.
2. Total peripheral resistance (a) is inversely proportional to the length of the vascular bed, (b) increases in anemia, (c) decreases in polycythemia, (d) is inversely related to the diameter of the arterioles.
3. Which of the following can lead to decreased venous return of blood to the heart? (a) an increase in blood volume, (b) an increase in venous pressure, (c) damage to the venous valves, (d) increased muscular activity.
4. Arterial blood pressure increases in response to (a) increasing stroke volume, (b) increasing heart rate, (c) atherosclerosis, (d) rising blood volume, (e) all of these.
5. Which of the following would *not* result in the dilation of the terminal arterioles and upstream arterioles in systemic capillary beds? (a) a decrease in local tissue O_2 content, (b) an increase in local tissue CO_2, (c) a local increase in histamine, (d) a local increase in pH.
6. The structure of a capillary wall differs from that of a vein or an artery because (a) it has two tunics instead of three, (b) there is less smooth muscle, (c) it has a single tunic—only the tunica intima, (d) none of these.
7. The baroreceptors in the carotid sinus and aortic arch are sensitive to (a) a decrease in CO_2, (b) changes in arterial pressure, (c) a decrease in O_2, (d) all of these.
8. The myocardium receives its blood supply directly from the (a) aorta, (b) coronary arteries, (c) coronary sinus, (d) pulmonary arteries.
9. Blood flow in the capillaries is steady despite the rhythmic pumping of the heart because of the (a) elasticity of the large arteries, (b) small diameter of capillaries, (c) thin walls of the veins, (d) venous valves.
10. Using the letters from column B, match the artery descriptions in column A. (Note that some require more than a single choice.)

Column A	Column B
____ (1) unpaired branch of abdominal aorta	(a) right common carotid
____ (2) second branch of aortic arch	(b) superior mesenteric
____ (3) branch of internal carotid	(c) left common carotid
____ (4) branch of external carotid	(d) external iliac
____ (5) origin of femoral arteries	(e) inferior mesenteric
	(f) superficial temporal
	(g) celiac trunk
	(h) facial
	(i) ophthalmic
	(j) internal iliac

11. Which of the following do not drain directly into the inferior vena cava? (a) inferior phrenic veins, (b) hepatic veins, (c) inferior mesenteric vein, (d) renal veins.

Level 2 Apply/Analyze

12. Tracing the blood from the heart to the right hand, we find that blood leaves the heart and passes through the aorta, the right subclavian artery, the axillary and brachial arteries, and through either the radial or ulnar artery to arrive at the hand. Which artery is missing from this sequence? (a) coronary, (b) brachiocephalic, (c) cephalic, (d) right common carotid.
13. How is the anatomy of capillaries and capillary beds well suited to their function?
14. (a) Define blood pressure. Differentiate between systolic and diastolic blood pressure. (b) What is the normal blood pressure value for an adult?
15. Describe the neural mechanisms responsible for controlling blood pressure.
16. How are nutrients, wastes, and respiratory gases transported to and from the blood and tissue spaces?
17. (a) What blood vessels contribute to the formation of the hepatic portal circulation? (b) Why is a portal circulation a "strange" circulation?
18. Physiologists often consider capillaries and postcapillary venules together. (a) What functions do these vessels share? (b) Structurally, how do they differ?

Level 3 Evaluate/Synthesize

19. Distinguish between elastic arteries, muscular arteries, and arterioles relative to location, histology, and functional adaptations.
20. Write an equation showing the relationship between total peripheral resistance, blood flow, and blood pressure.
21. Explain the reasons for the observed changes in blood flow velocity in the different regions of the circulation.
22. How does the control of blood flow to the skin for the purpose of regulating body temperature differ from the control of nutrient blood flow to skin cells?
23. Describe neural and chemical (both systemic and local) effects exerted on the blood vessels when you are fleeing from a mugger. (Be careful, this is more involved than it appears at first glance.)
24. A 60-year-old man is unable to walk more than 100 yards without experiencing severe pain in his left leg; the pain is relieved by resting for 5–10 minutes. He is told that the arteries of his leg are becoming occluded with fatty material and is advised to have the sympathetic nerves serving that body region severed. Explain how such surgery might help to relieve this man's problem.
25. Your friend Jillian, who knows little about science, is reading a magazine article about a patient who had an "aneurysm at the base of his brain that suddenly grew much larger." The surgeons' first goal was to "keep it from rupturing," and the second goal was to

"relieve the pressure on the brain stem and cranial nerves." The surgeons were able to "replace the aneurysm with a section of plastic tubing," so the patient recovered. Jillian asks you what all this means. Explain. (Hint: Check this chapter's Related Clinical Terms, p. 759.)

26. The Agawam High School band is playing some lively marches while the coaches are giving pep talks to their respective football squads. Although it is September, it is unseasonably hot (88°F/31°C) and the band uniforms are wool. Suddenly Ryan, the tuba player, becomes light-headed and faints. Explain his fainting in terms of vascular events.

27. When we are cold or the external temperature is low, most venous blood returning from the distal part of the arm travels in the deep veins where it picks up heat (by countercurrent exchange) from the nearby brachial artery en route. However, when we are hot, and especially during exercise, venous return from the distal arm travels in the superficial veins and those veins tend to bulge superficially in a person who is working out. Explain why venous return takes a different route in the second situation.

28. Edema is a common clinical problem. On your first day of a clinical rotation, you encounter four patients who have edema for different reasons. Your challenge is to explain the edema in terms of either an increase or a decrease in one of the four pressures that causes bulk flow (see Focus Figure 19.1 on pp. 732–733).

(1) First you encounter Mrs. Taylor in the medical unit awaiting a liver transplant. What is the connection between liver failure and her edema?

(2) Next in the obstetric ward, Mrs. So is experiencing premature labor and has edema in her legs. Which bulk flow pressures might be altered here?

(3) In emergency, Mr. Herrera is in anaphylactic shock. His capillaries have become leaky, allowing plasma proteins that are normally kept inside the blood vessels to escape into the interstitial fluid. Which of the bulk flow pressures is altered in this case and in what direction is the change?

(4) Finally, in oncology Mrs. O'Leary is recovering from breast cancer surgery. Her right breast and all of her axillary lymph nodes were removed. Unfortunately, this severed most of the lymphatic vessels draining her right arm. You notice that this arm is quite edematous. Why? Mrs. O'Leary is given a compression sleeve to wear on this arm to help relieve the edema. Which of the bulk flow pressures will be altered by the compression sleeve?

CLINICAL CASE STUDY

62-Year-Old Male Astronaut with Hypotension

You're working in the ER when a crowd of security guards and photographers arrive! They're all here with Captain Fallowell, an astronaut who has spent the last eight months aboard the International Space Station. He passed out in the middle of his first speaking engagement.

Captain Fallowell is a fit-looking 62-year-old male, alert and oriented. His heart rate is 85 beats/min, but he admits it was racing during his speech. His pulse is weak and his BP is 116/66 mm Hg while lying down, but his heart rate is 112 beats/min and BP is 94/52 mm Hg when standing. Reflexes are normal and there is no sign of head trauma. He admits to occasional dizziness and blackouts upon standing.

1. Calculate Captain Fallowell's pulse pressure and MAP when he is lying down and when he is standing.

2. **+ NCLEX-STYLE** When the patient stood up, his heart rate increased. This was most likely caused by which of the following?
   a. Parasympathetic activation due to the change in posture
   b. The baroreceptor reflex activated by blood pooling in his lower body due to gravity
   c. A decrease in sympathetic tone due to the stress of movement
   d. An increase in the amount of blood the heart has to pump, because the movements of his leg muscles forced venous blood up into the right atrium

3. As Captain Fallowell stands up, he complains of dizziness and you notice that he turns pale. What causes this, and why does it happen?

4. Two of the nurses are discussing the case. One says "He must be fit. They make them exercise two hours a day up there!" The second nurse says, "Yes, but I heard the blood and tissue fluids all shift up into their head and torso and they get skinny 'chicken legs.'" What effect would increased circulating blood volume in the head and torso have on the sympathetic and parasympathetic nervous systems and on the heart rate?

5. **+ NCLEX-STYLE** Captain Fallowell's drop in blood pressure when standing up is most likely due to:
   a. Decreased total peripheral resistance due to lowered blood viscosity
   b. Increased total peripheral resistance due to vasodilation
   c. Decreased cardiac output due to low heart rate
   d. Decreased cardiac output due to low stroke volume

6. When Captain Fallowell was in the space station, he had more blood circulating through his torso and to his internal organs. As a result, what happened to each of the following measures for him?
   a. Renin levels
   b. Aldosterone levels
   c. ADH levels
   d. Thirst
   e. Atrial natriuretic peptide
   f. Urine volume
   g. Blood volume

19

The Lymphatic System and Lymphoid Organs and Tissues

In this chapter, you will learn that

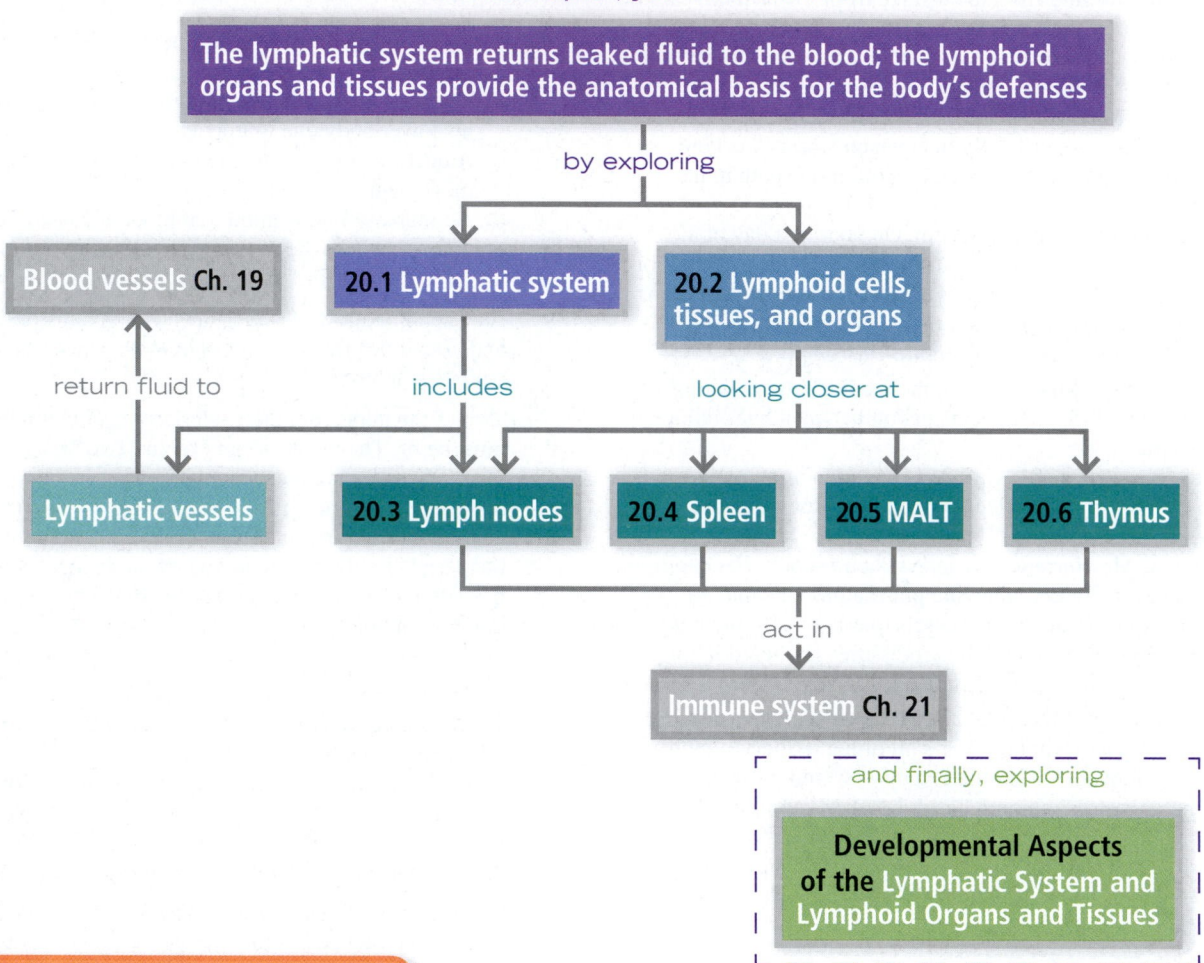

The lymphatic system returns leaked fluid to the blood; the lymphoid organs and tissues provide the anatomical basis for the body's defenses

by exploring

| Blood vessels Ch. 19 | 20.1 Lymphatic system | 20.2 Lymphoid cells, tissues, and organs |

return fluid to · includes · looking closer at

| Lymphatic vessels | 20.3 Lymph nodes | 20.4 Spleen | 20.5 MALT | 20.6 Thymus |

act in

Immune system Ch. 21

and finally, exploring

Developmental Aspects of the Lymphatic System and Lymphoid Organs and Tissues

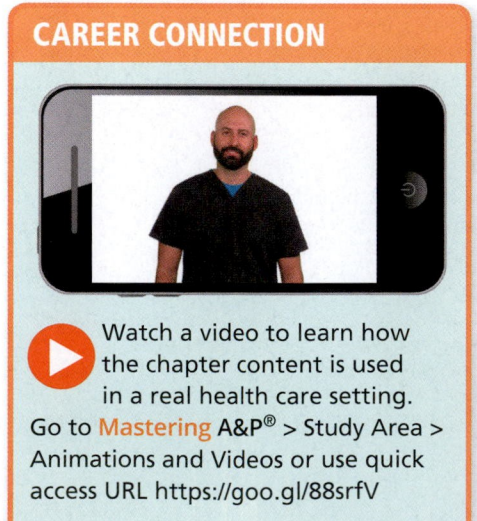
They can't all be superstars! When we mentally tick off the names of the body's organ systems, the lymphatic (lim-fat′ik) system and the lymphoid organs and tissues are probably not the first to come to mind. Yet if they failed their quiet background work, our cardiovascular system would stop working and our immune system would be hopelessly impaired.

In this chapter, we will explore two functionally different but structurally overlapping systems, the *lymphatic system* and the *lymphoid organs and tissues*. The **lymphatic system** returns fluids that have leaked from the vascular system back to the blood. It consists of three parts:

- A meandering network of *lymphatic vessels*
- *Lymph*, the fluid contained in those vessels
- *Lymph nodes* that cleanse the lymph as it passes through them

The **lymphoid organs and tissues** provide the structural basis of the immune system. These organs and tissues play essential roles in the body's defense mechanisms and its resistance to disease. These structures include the spleen, thymus, tonsils, and other lymphoid tissues scattered throughout the body. The lymph nodes are also part of this system, and like a keystone, they have important roles to play in both the lymphoid organs and tissues and the lymphatic system.

Let's begin by looking at the lymphatic system.

20.1 ## The lymphatic system includes lymphatic vessels, lymph, and lymph nodes

Learning Outcomes

▶ **List the functions of the lymphatic vessels.**

▶ **Describe the structure and distribution of lymphatic vessels.**

▶ **Describe the source of lymph and mechanism(s) of lymph transport.**

As blood circulates through the body, nutrients, wastes, and gases are exchanged between the blood and the interstitial fluid. As we explained in *Focus on Bulk Flow across Capillary Walls* (◀ **Focus Figure 19.1, pp. 732–733**), the hydrostatic and colloid osmotic pressures operating at capillary beds force fluid out of the blood at the arterial ends of the beds ("upstream") and cause most of it to be reabsorbed at the venous ends ("downstream"). The fluid that remains behind in the tissue spaces, as much as 3 L daily, becomes part of the interstitial fluid.

This leaked fluid, plus any plasma proteins that escape from the bloodstream, must somehow be returned to the blood to ensure that the cardiovascular system has sufficient blood volume. This problem of circulatory dynamics is resolved by the **lymphatic vessels**, or **lymphatics**, elaborate networks of drainage vessels that collect the excess protein-containing interstitial fluid and return it to the bloodstream. Once interstitial fluid enters the lymphatic vessels, it is called **lymph** (*lymph* = clear water).

Distribution and Structure of Lymphatic Vessels

The lymphatic vessels form a one-way system in which lymph flows only toward the heart.

Lymphatic Capillaries

The transport of lymph begins in microscopic blind-ended **lymphatic capillaries** (**Figure 20.1a**). These capillaries weave between the tissue cells and blood capillaries in the loose connective tissues of the body. Lymphatic capillaries are widespread, but they are absent from bones (including bone marrow) and

(a) Structural relationship between a capillary bed of the blood vascular system and lymphatic capillaries.

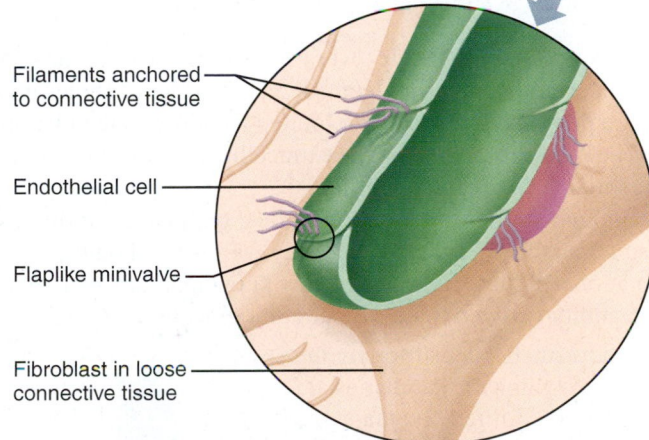

(b) Lymphatic capillaries are blind-ended tubes in which adjacent endothelial cells overlap each other, forming flaplike minivalves.

Figure 20.1 Distribution and special features of lymphatic capillaries. Arrows in (a) indicate direction of fluid movement.

The right lymphatic duct:
- Drains lymph from the right upper limb, and the right side of the head and thorax
- Empties into the blood at the junction of the right internal jugular and subclavian veins

Internal jugular veins

Right jugular trunk

Right subclavian trunk

Right subclavian vein

Right broncho-mediastinal trunk

Brachiocephalic veins

Superior vena cava

Azygos vein

Cisterna chyli

Right lumbar trunk

Inferior vena cava

Esophagus

Trachea

Left jugular trunk

Left subclavian trunk

Left subclavian vein

Left broncho-mediastinal trunk

The thoracic duct:
- Drains lymph from the rest of the body
- Empties into the blood at the junction of the left internal jugular and subclavian veins

Ribs

Thoracic duct

Hemiazygos vein

Left lumbar trunk

Intestinal trunk

Figure 20.2 Major lymphatic trunks and ducts. Anterior view of thoracic and abdominal wall. See Figure 20.3 for regions drained by each duct.

20

teeth. Lymphatic capillaries were long thought to be absent from the central nervous system, but recent evidence shows that they are present but restricted to limited locations in the meninges. These vessels help drain the extracellular fluids in the brain (interstitial and cerebrospinal fluids).

Although similar to blood capillaries, lymphatic capillaries are so remarkably permeable that they were once thought to be open at one end like a straw. We now know that they owe their permeability to two unique structural modifications:

- The endothelial cells forming the walls of lymphatic capillaries are not tightly joined. Instead, the edges of adjacent cells overlap each other loosely, forming easily opened, flaplike *minivalves* (shown at right).

- Collagen filaments anchor the endothelial cells to surrounding structures so that any increase in interstitial fluid volume

Anchoring filaments

Endothelial cell

Flaplike minivalve

opens the minivalves, rather than causing the lymphatic capillaries to collapse.

So, what we have is a system analogous to one-way swinging doors in the lymphatic capillary wall. When fluid pressure in the interstitial space is greater than the pressure in the lymphatic capillary, the minivalve flaps gape open, allowing fluid to enter the lymphatic capillary. However, when the pressure is greater *inside* the lymphatic capillary, it forces the endothelial minivalve flaps shut, preventing lymph from leaking back out as the pressure moves it along the vessel.

Proteins in the interstitial space are unable to enter blood capillaries, but they enter lymphatic capillaries easily. In addition, when tissues become inflamed, lymphatic capillaries develop openings that permit uptake of even larger particles such as cell debris, pathogens (disease-causing microorganisms such as bacteria and viruses), and cancer cells. These particles travel with the lymph to the lymph nodes, where they are removed by cells of the immune system. The process of pathogens meeting up with immune cells in lymph nodes is critical for activating an immune response. In some cases, unfortunately, pathogens or cancer cells manage to pass through the lymph nodes and can then travel throughout the body.

A special set of lymphatic capillaries called **lacteals** (lak′te-alz) transports absorbed fat from the small intestine to the bloodstream (see Figure 23.31b, p. 902). Lacteals are so called because of the milky white lymph that drains through them (*lact* = milk). This fatty lymph, called **chyle** ("juice"), drains from the fingerlike villi of the intestinal mucosa.

Larger Lymphatic Vessels

From the lymphatic capillaries, lymph flows through successively larger and thicker-walled channels:

Lymphatic capillaries → Collecting lymphatic vessels → Lymphatic trunks → Lymphatic ducts

The **collecting lymphatic vessels** have the same three tunics as veins, but the collecting vessels have thinner walls and more internal valves, and they anastomose more. In general, lymphatics in the skin travel along with superficial *veins*, while the deep lymphatic vessels of the trunk and digestive viscera travel with the deep *arteries*. The exact anatomical distribution of lymphatic vessels varies greatly between individuals, even more than it does for veins.

The largest collecting vessels unite to form **lymphatic trunks**, which drain fairly large areas of the body. The major trunks, named mostly for the regions from which they drain lymph, are the paired **lumbar**, **bronchomediastinal**, **subclavian**, and **jugular trunks**, and the single **intestinal trunk** (**Figure 20.2**).

Lymph is eventually delivered to one of two large *ducts* in the thoracic region. The **right lymphatic duct** drains lymph from the right upper limb and the right side of the head and thorax (**Figure 20.3**). The much larger **thoracic duct** receives lymph from the rest of the body. It collects lymph from the two large lumbar trunks that drain the lower limbs and from the intestinal trunk that drains the digestive organs. In about half of individuals, the thoracic duct begins as an enlarged sac, the **cisterna chyli** (sis-ter′nah ki′li), located in the region between the last thoracic and second lumbar vertebrae. As the thoracic duct runs superiorly, it receives lymphatic drainage from the left side of the thorax, left upper limb, and the left side of the head. Each terminal duct empties its lymph into the venous circulation at the junction of the internal jugular vein and subclavian vein on its own side of the body (Figure 20.2).

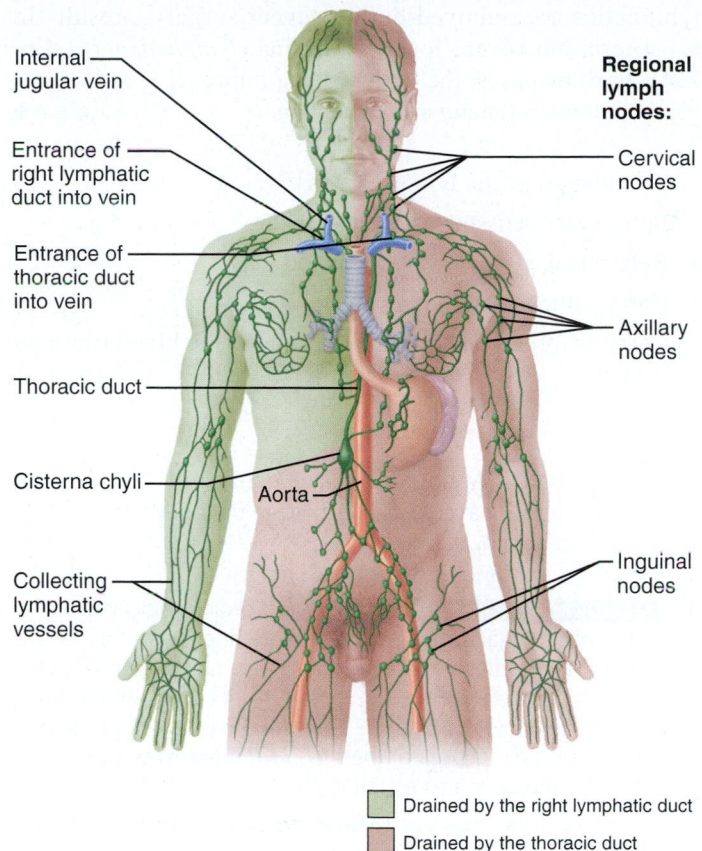

Figure labels:
- Internal jugular vein
- Entrance of right lymphatic duct into vein
- Entrance of thoracic duct into vein
- Thoracic duct
- Cisterna chyli
- Aorta
- Collecting lymphatic vessels
- Regional lymph nodes:
- Cervical nodes
- Axillary nodes
- Inguinal nodes
- Drained by the right lymphatic duct
- Drained by the thoracic duct

Figure 20.3 The lymphatic system. General distribution of collecting lymphatic vessels and regional lymph nodes.

Lymph Transport

The lymphatic system lacks an organ that acts as a pump. Under normal conditions, lymphatic vessels are low-pressure conduits, and the same mechanisms that promote venous return in blood vessels act here as well—the milking action of active skeletal muscles, pressure changes in the thorax during breathing, and valves to prevent backflow. Lymphatic vessels are usually bundled together in connective tissue sheaths along with blood vessels, and pulsations of nearby arteries also promote lymph flow. In addition to these mechanisms, smooth muscle in the walls of all but the smallest lymphatic vessels contracts rhythmically, helping to pump the lymph along.

Even so, lymph transport is sporadic and slow. Movement of adjacent tissues is extremely important in propelling lymph through the lymphatics. When physical activity or passive movements increase, lymph flows much more rapidly (balancing the greater rate of fluid loss from the blood in such situations). For this reason, it is a good idea to immobilize a badly infected body part to hinder flow of inflammatory material from that region.

HOMEOSTATIC IMBALANCE 20.1   CLINICAL

Like the larger blood vessels, the larger lymphatics receive their nutrient blood supply from a branching vasa vasorum. When lymphatic vessels are severely inflamed, the related vessels of the vasa vasorum become congested with blood. As a result, the pathway of the associated superficial lymphatics becomes visible through the skin as red streaks that are tender to the touch. This unpleasant condition is called *lymphangitis* (lim″fan-ji′tis; *angi* = vessel).

HOMEOSTATIC IMBALANCE 20.2   CLINICAL

Anything that prevents the normal return of lymph to the blood—such as when tumors block the lymphatics or lymphatics are removed during cancer surgery—results in short-term but severe localized edema (*lymphedema*) (◀ p. 734). In some cases, the lymphedema improves if some lymphatic pathways remain and can enlarge. ●●●●●●●●

To summarize, the lymphatic vessels:

- Return excess tissue fluid to the blood
- Return leaked proteins to the blood
- Carry pathogens to lymph nodes
- Carry absorbed fat from the intestine to the blood (through lacteals)

Check Your Understanding

1. What is lymph? Where does it come from?
2. Name two lymphatic ducts and indicate the body regions usually drained by each.
3. What is the driving force for lymph movement?
4. **MAKE CONNECTIONS** A tumor in the left groin is blocking lymphatic drainage from Mr. Thomas's left leg, causing obvious edema. Which two of the functions of lymphatic vessels listed above no longer work in that leg? Think about the pressures driving bulk flow (Chapter 19) and state which two pressures are affected by the loss of these two lymphatic vessel functions and how the pressures are affected.

For answers, see Answers Appendix.

20.2 Lymphoid cells and tissues are found in lymphoid organs and in connective tissue of other organs

Learning Outcome

▶ Describe the basic structure and cellular population of lymphoid tissue. Differentiate between diffuse and follicular lymphoid tissues.

To understand the role of the lymphoid organs in the body, we need to investigate their components—lymphoid cells and lymphoid tissues—before considering the organs themselves.

Lymphoid Cells

The lymphoid cells consist of immune system cells found in lymphoid tissues together with the supporting cells that form the "scaffolding" of those tissues.

Lymphocytes are the main warriors of the immune system. There are two main varieties of lymphocytes—**T cells (T lymphocytes)** and **B cells (B lymphocytes)**—that protect the body against antigens. (*Antigens* are anything that provokes an immune response, such as bacteria and their toxins, viruses, mismatched RBCs, or cancer cells.) Activated T cells manage the immune response, and some of them directly attack and destroy infected cells. Activated B cells protect the body by producing **plasma cells**, daughter cells that secrete antibodies into the blood (or other body fluids). Antibodies mark antigens for destruction. Chapter 21 explores the roles of the lymphocytes in immunity.

Macrophages play a crucial role in body protection and the immune response by phagocytizing foreign substances and by helping to activate T cells. So, too, do the spiny-looking **dendritic cells** that capture antigens and bring them back to the lymph nodes.

Last but not least are the **reticular cells**, fibroblast-like cells that produce the reticular fiber **stroma** (stroʹmah), which is the network that supports the other cell types in lymphoid organs and tissues (**Figure 20.4**).

Lymphoid Tissue

Lymphoid tissue is an important component of the immune system, mainly because it:

- Houses lymphocytes and provides a site where they can be activated and proliferate (multiply)
- Furnishes an ideal surveillance vantage point for lymphocytes and macrophages

Lymphoid tissue, largely composed of loose connective tissue called **reticular connective tissue** (◀ p. 133), dominates all the lymphoid organs except the thymus. Macrophages live on the fibers of the reticular connective tissue network. Huge numbers of lymphocytes squeeze through the walls of postcapillary venules (◀ p. 712) coursing through this network. The lymphocytes temporarily occupy the spaces in the network before returning to the bloodstream via the lymphatics so they can patrol the body again (Figure 20.4). The cycling of lymphocytes between the circulatory vessels, lymphoid tissues, and loose connective tissues of the body ensures that lymphocytes reach infected or damaged sites quickly.

Lymphoid tissue comes in various "packages":

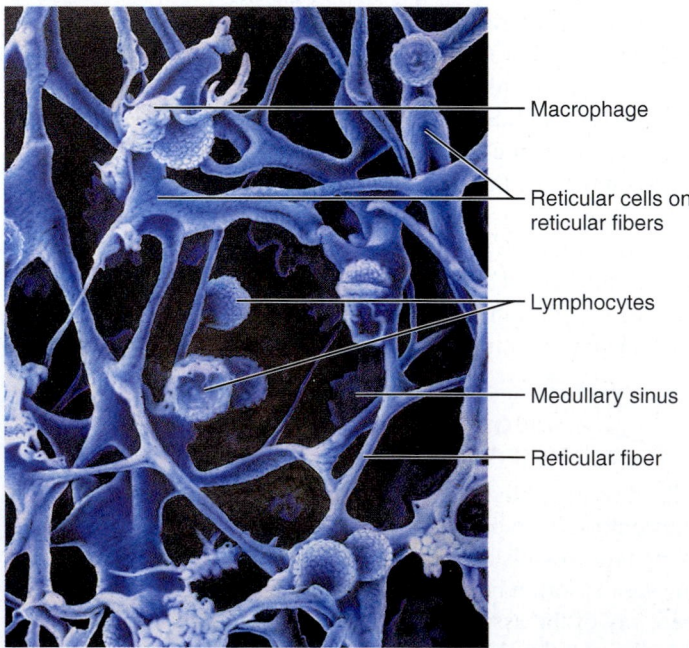

Figure 20.4 Reticular connective tissue in a human lymph node. Scanning electron micrograph (690×).

Macrophage

Reticular cells on reticular fibers

Lymphocytes

Medullary sinus

Reticular fiber

- **Diffuse lymphoid tissue**—a loose arrangement of lymphoid cells and some reticular fibers—is found in virtually every body organ. Larger collections appear in the lamina propria (loose connective tissue, ◄ p. 141) of mucous membranes such as those lining the digestive tract.

- **Lymphoid follicles (lymphoid nodules)** are solid, spherical bodies consisting of tightly packed lymphoid cells and reticular fibers. Follicles often have lighter-staining **germinal centers** where proliferating (rapidly dividing) B cells predominate (see Figure 20.6). These centers enlarge dramatically when the B cells are dividing rapidly and producing plasma cells. In many cases, the follicles form part of larger lymphoid organs, such as lymph nodes. However, isolated aggregations of lymphoid follicles occur in the intestinal wall as Peyer's patches (aggregated lymphoid nodules) and in the appendix (see p. 775).

Lymphoid Organs

The **lymphoid organs** (**Figure 20.5**) are grouped into two functional categories.

- The **primary lymphoid organs** are where B and T cells mature—the *red bone marrow* and the *thymus*. While both B and T cells originate in the red bone marrow, B cells mature in the red bone marrow and T cells mature in the thymus. We described red marrow and its role in blood cell formation earlier (◄ p. 647) and we will describe the thymus shortly.

- The **secondary lymphoid organs** are where mature lymphocytes first encounter their antigens and are activated. They include the lymph nodes, the spleen, and the collections of mucosa-associated lymphoid tissue (MALT) that form the tonsils, Peyer's patches (aggregated lymphoid nodules) in the small intestine, and the appendix. Lymphocytes also encounter antigens and are activated in the diffuse lymphoid tissues.

Although all lymphoid organs help protect the body, only the lymph nodes cleanse the lymph. The other secondary lymphoid organs typically have efferent lymphatics draining them, but lack afferent lymphatics.

Check Your Understanding

5. What are the primary lymphoid organs and what makes them special?

6. MAKE CONNECTIONS Both B and T cells originate in red bone marrow. Where do you find red marrow in adults? (Hint: See Chapter 6.)

For answers, see Answers Appendix.

20.3 Lymph nodes cleanse lymph and house lymphocytes

Learning Outcome

▶ Describe the general location, histological structure, and functions of lymph nodes.

The most important of the secondary lymphoid organs in the body are the **lymph nodes**, which cluster along the lymphatic

Primary lymphoid organs

- Thymus
- Red bone marrow

Secondary lymphoid organs

- Lymph nodes (see Figure 20.3)
- Tonsils
- Spleen
- Peyer's patches (aggregated lymphoid nodules in small intestine)
- Appendix

Figure 20.5 Lymphoid organs.

vessels of the body. There are hundreds of these small organs, but because they are usually embedded in connective tissue, they are not ordinarily visible. Large clusters of lymph nodes occur near the body surface in the inguinal, axillary, and cervical regions, places where the collecting lymphatic vessels converge to form trunks (see Figure 20.3).

Lymph nodes have two basic protective functions:

- **Cleansing the lymph.** As lymph is transported back to the bloodstream, the lymph nodes act as lymph "filters." Macrophages in the nodes remove and destroy microorganisms and other debris that enter the lymph from the loose connective tissues, preventing them from being delivered to the blood and spreading to other parts of the body.

- **Immune system activation.** Lymph nodes and other lymphoid organs are strategically located sites where lymphocytes encounter antigens and are activated to mount an attack against them. Many antigens are brought to lymph nodes from their point of entry into the body by dendritic cells. Dendritic cells use these antigens to activate T cells in the lymph nodes.

Structure of a Lymph Node

The structure of a lymph node supports its defensive functions. Lymph nodes vary in shape and size, but most are bean shaped and less than 2.5 cm (1 inch) in length. Each node is surrounded by a dense fibrous **capsule** from which connective tissue strands called **trabeculae** extend inward to divide the node into

20

Afferent lymphatic vessels

Cortex
- Lymphoid follicle
- Germinal center
- Subcapsular sinus

Efferent lymphatic vessels

Hilum

Medulla
- Medullary cord
- Medullary sinus

Trabeculae

Capsule

(a) Longitudinal view of the internal structure of a lymph node and associated lymphatics

Lymphoid follicles

Trabecula

Subcapsular sinus

Capsule

Medullary cords

Medullary sinuses

(b) Photomicrograph of part of a lymph node (72×)

Figure 20.6 Lymph node. In (a), notice that several afferent lymphatics converge on its convex side, whereas fewer efferent lymphatics exit at its hilum.

a number of compartments (**Figure 20.6**). The node's internal framework, or stroma, of reticular fibers physically supports its ever-changing population of lymphocytes.

A lymph node has two histologically distinct regions:

- **Cortex.** The superficial part of the **cortex** contains densely packed follicles, many with germinal centers heavy with dividing B cells. The deeper part of the cortex primarily houses T cells in transit. T cells circulate continuously between the blood, lymph nodes, and lymph, performing their surveillance role. Dendritic cells are abundant in the cortex and intimately associated with both B and T cells.

- **Medulla.** The **medulla** consists of **medullary cords**, thin inward extensions from the cortical lymphoid tissue. Both B and T lymphocytes are found here.

Throughout the node are **lymph sinuses** (e.g., the subcapsular and medullary sinuses described below). These sinuses are large lymphatic capillaries spanned by crisscrossing reticular fibers. Numerous macrophages reside on these reticular fibers and phagocytize foreign matter in the lymph as it flows by in the sinuses. Additionally, some of the lymph-borne antigens in the percolating lymph leak into the surrounding lymphoid tissue, where they activate lymphocytes to mount an immune attack against them.

Circulation in the Lymph Nodes

Lymph enters the convex side of a lymph node through a number of **afferent lymphatic vessels**. It then moves through a large, baglike sinus, the **subcapsular sinus**, into a number of smaller sinuses that cut through the cortex and enter the medulla. The lymph meanders through these **medullary sinuses** and finally exits the node at its **hilum** (hi′lum), the indented region on the concave side, via **efferent lymphatic vessels**.

There are fewer efferent vessels draining the node than afferent vessels feeding it, so the flow of lymph through the node stagnates somewhat, allowing time for the lymphocytes and macrophages to carry out their protective functions. Lymph passes through several nodes before it is completely cleansed.

HOMEOSTATIC IMBALANCE 20.3   **CLINICAL**

Sometimes lymph nodes are overwhelmed by the agents they are trying to destroy. For example, when large numbers of bacteria are trapped in the nodes, the nodes become inflamed, swollen, and tender to the touch, a condition often referred to (erroneously) as swollen "glands" but correctly called *lymphadenopathy*. Such infected lymph nodes (often pus-filled) are called *buboes* (bu′bōz). (The bubonic plague was named for these buboes.)

Lymph nodes can also become secondary cancer sites, particularly when metastasizing cancer cells enter lymphatic vessels and become trapped there. Cancer-infiltrated lymph nodes are swollen but usually not painful, a fact that helps distinguish cancerous nodes from those infected by microorganisms.

Check Your Understanding

7. What is a lymphoid follicle? What type of lymphocyte predominates in follicles, especially in their germinal centers?
8. What is the benefit of having fewer efferent than afferent lymphatics in lymph nodes?
9. **DRAW** Sketch a lymph node including its afferent and efferent lymphatic vessels. Label the hilum, cortex, and medulla. Draw and label some lymphoid follicles and their germinal centers.

For answers, see Answers Appendix.

20.4 The spleen removes bloodborne pathogens and aged red blood cells

Learning Outcome

▶ Compare and contrast the structure and function of the spleen and lymph nodes.

The soft, blood-rich **spleen** is about the size of a fist and is the largest lymphoid organ. Located in the left side of the abdominal cavity just beneath the diaphragm, it curls around the anterior aspect of the stomach (Figure 20.5 and **Figure 20.7**). It is served by the large *splenic artery* and *vein*, which enter and exit the *hilum* on its slightly concave anterior surface.

The spleen provides a site for lymphocyte proliferation and immune surveillance and response. But perhaps even more important are its blood-cleansing functions. Besides extracting aged and defective blood cells and platelets from the blood, its macrophages remove debris and foreign matter. The spleen also performs three additional, and related, functions. The spleen:

(a) Diagram of the spleen, anterior view.

Splenic artery
Splenic vein
Hilum

Capsule
Trabecula
Splenic cords
Splenic sinusoids
Arterioles and capillaries
Red pulp
White pulp
Central artery
Splenic artery
Splenic vein

(b) Diagram of spleen histology.

Diaphragm
Spleen
Adrenal gland
Left kidney
Splenic artery
Pancreas

(c) Photograph of the spleen in its normal position in the abdominal cavity, anterior view.

Capsule
White pulp
Red pulp

(d) **Photomicrograph of spleen tissue (30×).** The white pulp, a lymphoid tissue with many lymphocytes, is surrounded by red pulp containing abundant erythrocytes.

Figure 20.7 The spleen.

Practice Histology questions: MasteringA&P° >
Study Area > Lab Tools > PAL

20

- Recycles the breakdown products of red blood cells for later reuse. It releases the breakdown products to the blood for processing by the liver and stores some of the iron salvaged from hemoglobin.

- Stores blood platelets and monocytes for release into the blood when needed.

- May be a site of erythrocyte production in the fetus.

Like lymph nodes, the spleen is surrounded by a fibrous capsule and has trabeculae that extend inward. Histologically, the spleen consists of two components: white pulp and red pulp.

- **White pulp** is where immune functions take place, so it is composed mostly of lymphocytes suspended on reticular fibers. The white pulp clusters or forms "cuffs" around central arteries (small branches of the splenic artery). These clusters of white pulp look like islands in a sea of red pulp.

- **Red pulp** is where worn-out red blood cells and blood-borne pathogens are destroyed, so it contains huge numbers of erythrocytes and the macrophages that engulf them. It is essentially all splenic tissue that is not white pulp. It consists of **splenic cords**, regions of reticular connective tissue, that separate the blood-filled **splenic sinusoids** (venous sinuses).

The names of the pulp regions reflect their appearance in fresh spleen tissue rather than their staining properties. Indeed, as you can see in Figure 20.7d, the white pulp sometimes appears darker than the red pulp due to the darkly staining nuclei of the densely packed lymphocytes.

HOMEOSTATIC IMBALANCE 20.4 — CLINICAL

Because the spleen's capsule is relatively thin, a direct blow or severe infection may cause it to rupture, spilling blood into the peritoneal cavity. Once, *splenectomy* (surgical removal of the ruptured spleen) was the standard treatment and thought necessary to prevent life-threatening hemorrhage and shock. However, surgeons have discovered that, if left alone, the spleen can often repair itself. As a result, the frequency of emergency splenectomies has decreased dramatically. If the spleen must be removed, the liver and bone marrow take over most of its functions. In children younger than 12, the spleen will regenerate if a small part of it is left in the body.

Check Your Understanding

10. List several functions of the spleen.

11. Identify structures a and b in this diagram of the spleen. With respect to regions c and d: Which region is the site of red blood cell destruction? Which region is composed mostly of lymphocytes? Which region has splenic cords and splenic sinusoids?

12. **MAKE CONNECTIONS** From which blood vessel does the splenic artery arise? The splenic vein empties into which blood vessel? (Hint: See Chapter 19.)

For answers, see Answers Appendix.

20.5 MALT guards the body's entryways against pathogens

Learning Outcome

▶ Define MALT and list its major components.

Mucosa-associated lymphoid tissues (MALT) are a set of distributed lymphoid tissues strategically located in mucous membranes throughout the body (◀ **Figure 4.14b, p. 141**, introduced mucous membranes). MALT helps protect us from the never-ending onslaught of pathogens that seek to enter our bodies. Here

Pharyngeal tonsil
Palatine tonsil
Lingual tonsil

Germinal centers in lymphoid follicles

Tonsillar crypt

Figure 20.8 Histology of the palatine tonsil. The exterior surface of the tonsil is covered by stratified squamous epithelium, which invaginates deeply to form tonsillar crypts (10×).

Practice Histology questions: MasteringA&P® > Study Area > Lab Tools > PAL

we will consider the largest collections of MALT—the tonsils, Peyer's patches, and appendix. In addition to these large named collections, MALT also occurs in the mucosa of the respiratory and genitourinary organs as well as the rest of the digestive tract.

Tonsils

The **tonsils** form a ring of lymphoid tissue around the entrance to the pharynx (throat), where they appear as swellings of the mucosa (**Figure 20.8** and Figure 22.5 on p. 823). The tonsils are named according to location.

- The paired **palatine tonsils** are located on either side at the posterior end of the oral cavity. These are the largest tonsils and the ones most often infected.
- The **lingual tonsil** is the collective term for a lumpy collection of lymphoid follicles at the base of the tongue.
- The **pharyngeal tonsil** (referred to as the *adenoids* if enlarged) is in the posterior wall of the nasopharynx.
- The tiny **tubal tonsils** surround the openings of the auditory tubes into the pharynx.

The tonsils gather and remove many of the pathogens entering the pharynx in food or in inhaled air.

The lymphoid tissue of the tonsils contains follicles with obvious germinal centers surrounded by diffusely scattered lymphocytes. The tonsils are not fully encapsulated, and the epithelium overlying them invaginates deep into their interior, forming blind-ended **tonsillar crypts** (Figure 20.8). The crypts trap bacteria and particulate matter, and the bacteria work their way through the mucosal epithelium into the lymphoid tissue, where most are destroyed. It seems a bit dangerous to "invite" infection this way, but this strategy produces a wide variety of immune cells that have a "memory" for the trapped pathogens. In other words, the body takes a calculated risk early on (during childhood) for the benefits of heightened immunity and better health later.

Peyer's Patches

Peyer's patches (pi′erz), or **aggregated lymphoid nodules**, are large clusters of lymphoid follicles, structurally similar to the tonsils. They are located in the wall of the distal portion of the small intestine (Figure 20.5 and **Figure 20.9**).

Appendix

The **appendix** is a tubular offshoot of the first part of the large intestine and contains a high concentration of lymphoid follicles. Like Peyer's patches, the appendix is in an ideal position (1) to prevent bacteria (present in large numbers in the intestine) from breaching the intestinal wall, and (2) to generate many "memory" lymphocytes for long-term immunity.

Check Your Understanding

13. What is MALT? List several components of MALT.

_____ *For answers, see Answers Appendix.*

Figure 20.9 Peyer's patch (aggregated lymphoid nodules). Cross section of wall of the ileum of the small intestine (20×).

Small intestine

Mucous membrane lining small intestine

Follicles of a Peyer's patch (aggregated lymphoid nodules)

Smooth muscle in the intestinal wall

20

Practice Histology questions: MasteringA&P® > Study Area > Lab Tools > PAL

Capsule

Cortex

Medulla

Thymic corpuscles

Figure 20.10 **The thymus.** The photomicrograph of a portion of the thymus shows part of a lobule with cortical and medullary regions (85×).

Practice Histology questions: MasteringA&P° > Study Area > Lab Tools > PAL

20.6 T lymphocytes mature in the thymus

Learning Outcome

▶ Describe the structure and function of the thymus.

The bilobed **thymus** (thi'mus) has important functions primarily during the early years of life. It is found in the inferior neck and extends into the superior thorax, where it partially overlies the heart deep to the sternum (see Figure 20.5 and **Figure 20.10**). In the thymus, T lymphocyte precursors mature to become immunocompetent lymphocytes. In other words, the thymus is where T lymphocytes become able to defend us against specific pathogens in the immune response.

Prominent in newborns, the thymus continues to increase in size during the first year, when it is highly active. After puberty, it gradually atrophies and by old age it has been replaced almost entirely by fibrous and fatty tissue and is difficult to distinguish from surrounding connective tissue. Even though it atrophies, the thymus continues to produce immunocompetent cells as we age, although at a declining rate.

To understand thymic histology, it helps to compare the thymus to a cauliflower head—the flowerets represent *thymic lobules*, each containing an outer cortex and an inner medulla (Figure 20.10). Most thymic cells are lymphocytes. In the cortical regions the rapidly dividing lymphocytes are densely packed, with a few macrophages scattered among them.

The lighter-staining medullary areas contain fewer lymphocytes plus some bizarre structures called **thymic corpuscles**. Consisting of concentric whorls of keratinized epithelial cells, they were thought to be sites of T cell destruction. Thymic corpuscles are involved in the development of *regulatory T cells*, a class of T lymphocytes that are important for preventing autoimmune responses.

The thymus is a primary lymphoid organ and differs from secondary lymphoid organs in three important ways:

• The thymus has no follicles because it lacks B cells.

• The thymus does not *directly* fight antigens. Instead, the thymus functions strictly as a maturation site for T lymphocyte precursors. These precursors must be kept isolated from foreign antigens to prevent their premature activation. In fact, there is a *blood thymus barrier* that keeps bloodborne antigens out of the thymus.

• The stroma of the thymus consists of epithelial cells rather than reticular fibers. These epithelial cells provide the physical and chemical environment in which T lymphocytes mature.

Table 20.1 summarizes the properties of the thymus and compares it with other lymphoid organs and tissues.

Check Your Understanding

14. **MAKE CONNECTIONS** T cells mature in the thymus. Where do B cells mature?

For answers, see Answers Appendix.

Developmental Aspects of the Lymphatic System and Lymphoid Organs and Tissues

By the fifth week of embryonic development, the beginnings of the lymphatic vessels and the main lymph node clusters are apparent. These usually arise as **lymph sacs** that bud from developing veins and form a branching system of lymphatic vessels throughout the thorax, upper extremities, and head. The two main connections to the venous system are retained and become the right lymphatic duct and, on the left, the superior part of the thoracic duct. Caudal lymphatics bud largely from the primitive inferior vena cava.

Except for the thymus, which is an endodermal derivative, the lymphoid organs develop from mesodermal mesenchymal cells that migrate to particular body sites and develop into reticular tissue. The thymus, the first lymphoid organ to appear, is an outgrowth of the primitive pharynx lining. It detaches and migrates caudally to the thorax where it is infiltrated with lymphocyte precursors. Shortly after birth, the other lymphoid organs become heavily populated by lymphocytes.

20

	MAJOR FUNCTIONS	CAPSULE	CORTEX AND MEDULLA	LYMPHOID FOLLICLES	STROMA	SPECIAL FEATURES
Lymph nodes	• Cleanse lymph • Site for lymphocyte activation and proliferation	Yes	Yes	Yes (in cortex)	Reticular connective tissue	Both afferent and efferent lymphatic vessels
Spleen	• Cleanses blood and removes aged or defective red blood cells • Site for lymphocyte activation and proliferation • Stores platelets, moncytes, and iron	Yes	No	Yes (in white pulp)	Reticular connective tissue	Red and white pulp
MALT†	• Prevent pathogens from penetrating mucous membrane • Site for lymphocyte activation and proliferation	No	No	Yes	Reticular connective tissue	Diffuse lymphatic tissue in addition to follicles
Thymus	• Site of T cell maturation	Yes	Yes	No	Epithelial tissue	Thymic corpuscles

*Red bone marrow, a primary lymphoid organ, is not included. †MALT = mucosa-associated lymphoid tissues.

RELATED CLINICAL TERMS

Elephantiasis (el″lĕ-fan-ti′ah-sis) Typically a tropical disease in which the lymphatics (particularly those of the lower limbs and scrotum) become clogged with parasitic roundworms, an infectious condition called filariasis. Swelling (due to edema) reaches enormous proportions.

Hodgkin's lymphoma A malignancy of lymphoid tissue; symptoms include swollen, nonpainful lymph nodes, fatigue, and often intermittent fever and night sweats. Characterized by presence of giant malignantly transformed B cells called Reed-Sternberg cells. Infection with Epstein-Barr virus (see mononucleosis below) and genetic susceptibility appear to be predisposing factors. Treated with chemotherapy and radiation; high cure rate.

Lymphadenopathy (lim-fad″ĕ-nop′ah-the; *adeno* = a gland; *pathy* = disease) Any disease of the lymph nodes.

Lymphangiography (lim-fan″je-og′rah-fe) Diagnostic procedure in which the lymphatic vessels are injected with radiopaque dye and then visualized with X rays.

Lymphoma Any neoplasm (tumor) of the lymphoid tissue, whether benign or malignant.

Mononucleosis ◀ Chapter 17, p. 654.

Non-Hodgkin's lymphoma Includes all cancers of lymphoid tissues except Hodgkin's lymphoma. Involves uncontrolled multiplication and metastasis of undifferentiated lymphocytes, with swelling of the lymph nodes, spleen, and Peyer's patches; other organs may eventually become involved. The seventh most common cancer. A rapidly progressing type, which primarily affects young people, grows quickly but responds to chemotherapy; up to an 88% 5-year survival rate. A slowly progressing type, which affects the elderly, resists chemotherapy and so is often fatal.

Sentinel node The first node that receives lymph drainage from a body area suspected of being cancerous. When examined for presence of cancer cells, this node gives the best indication of whether metastasis through the lymphatic vessels has occurred.

Splenomegaly (sple″no-meg′ah-le; *mega* = big) Enlargement of the spleen. May be due to accumulation of infectious microorganisms; typically caused by septicemia, mononucleosis, malaria, or leukemia.

Tonsillitis (ton″sĭ-li′tis; *itis* = inflammation) Inflammation of the tonsils, typically due to bacterial infection. Tonsils become red, swollen, and sore.

20

Homeostatic Interrelationships between the Lymphatic System/Immunity and Other Body Systems

Nervous System   Chapters 11–15
- Lymphatic vessels pick up leaked plasma fluid and proteins; immune cells protect nervous system structures from specific pathogens
- The nervous system innervates larger lymphatics; opiate neuropeptides influence immune functions; the brain helps regulate immune response

Endocrine System   Chapter 16
- Lymphatic vessels pick up leaked fluids and proteins; lymph helps distribute hormones; immune cells protect endocrine organs
- Stress hormones depress immune activity; the thymus produces hormones and paracrines that are involved in the maturation of T cells

Cardiovascular System   Chapters 17–19
- Lymphatic vessels pick up leaked plasma and proteins; spleen removes and destroys aged RBCs and debris, and stores iron, platelets, and monocytes; immune cells protect cardiovascular organs from specific pathogens
- Blood is the source of lymph; most lymphatics develop from veins; blood circulates immune elements

Respiratory System   Chapter 22
- Lymphatic vessels pick up leaked fluids and proteins from respiratory organs; immune cells protect respiratory organs from specific pathogens; the tonsils and plasma cells in the respiratory mucosa (which secrete the antibody IgA) prevent pathogen invasion
- The lungs provide O_2 needed by lymphoid/immune cells and eliminate CO_2; the pharynx houses the tonsils; respiratory "pump" aids lymph flow

Digestive System   Chapter 23
- Lymphatic vessels pick up leaked fluids and proteins from digestive organs; lymph transports some products of fat digestion to the blood; lymphoid follicles in the intestinal wall prevent invasion of pathogens
- The digestive system digests and absorbs nutrients needed by cells of lymphoid organs; gastric acidity inhibits pathogens' entry into blood

Urinary System   Chapters 25–26
- Lymphatic vessels pick up leaked fluid and proteins from urinary organs; immune cells protect urinary organs from specific pathogens
- Urinary system eliminates wastes and maintains homeostatic balances of water, pH, and electrolytes in the blood for lymphoid/immune cell functioning; urine flushes some pathogens out of the urinary tract

Reproductive System   Chapter 27
- Lymphatic vessels pick up leaked fluid and proteins; immune cells protect against pathogens
- Reproductive organs' hormones may influence immune functioning; acidity of vaginal secretions is bacteriostatic

Integumentary System   Chapter 5
- Lymphatic vessels pick up leaked plasma fluid and proteins from the dermis; lymphocytes in lymphoid organs and tissues enhance the skin's protective role by defending against specific pathogens via the immune response
- The skin's keratinized epithelium provides a mechanical barrier to pathogens; epithelial dendritic cells and dermal macrophages act as antigen presenters in the immune response; acid pH of skin secretions inhibits growth of bacteria on the skin

Skeletal System   Chapters 6–8
- Lymphatic vessels pick up leaked plasma fluid and proteins from the periostea; immune cells protect bones from pathogens
- The bones house hematopoietic tissue, which produces the lymphocytes (and macrophages) that populate lymphoid organs and provide immunity

Muscular System   Chapters 9–10
- Lymphatic vessels pick up leaked fluids and proteins; immune cells protect muscles from pathogens
- The skeletal muscle "pump" aids the flow of lymph; shivering produces heat at the beginning of a fever

CHAPTER SUMMARY

1. Lymphatic vessels, lymph nodes, and lymph make up the lymphatic system. Lymphatic vessels return fluids that have leaked from the blood vascular system back to the blood. Lymphoid organs and tissues protect the body by removing foreign material from the lymph and bloodstream, and provide a site for immune surveillance.

20.1 The lymphatic system includes lymphatic vessels, lymph, and lymph nodes (pp. 767–770)

1. Lymphatic vessels form a one-way network—lymphatic capillaries, collecting vessels, trunks, and ducts—in which fluid flows only toward the heart. The right lymphatic duct drains lymph from the right arm and right side of the upper body; the thoracic duct receives lymph from the rest of the body. These ducts empty into the blood vascular system at the junction of the internal jugular and subclavian veins in the neck.
2. Lymphatic capillaries are exceptionally permeable, admitting proteins and particulate matter from the interstitial space.
3. Lymphatic vessels carry pathogens to lymph nodes where they can be phagocytized and may activate lymphocytes.
4. Pathogens and cancer cells may spread through the body via the lymphatic stream.
5. The flow of lymphatic fluid is slow; it is maintained by skeletal muscle contraction, pressure changes in the thorax, and contractions of the lymphatic vessels. Valves prevent backflow.

20.2 Lymphoid cells and tissues are found in lymphoid organs and in connective tissue of other organs (pp. 770–771)

1. The cells in lymphoid tissues include lymphocytes (cells called T cells or B cells), plasma cells (antibody-producing offspring of B cells), macrophages and dendritic cells (cells that capture antigens and initiate an immune response), and reticular cells that form the lymphoid tissue stroma.
2. Lymphoid tissue is reticular connective tissue. It houses macrophages and a continuously changing population of lymphocytes. It is an important element of the immune system.
3. Lymphoid tissue may be diffuse or packaged into dense follicles. Follicles often display germinal centers (areas where B cells are proliferating).
4. Primary lymphoid organs—the red bone marrow and thymus—are sites where lymphocytes develop and mature. Secondary lymphoid organs include lymph nodes, the spleen, and MALT. They are sites where lymphocytes encounter their antigens and are activated.

20.3 Lymph nodes cleanse lymph and house lymphocytes (pp. 771–773)

1. Lymph nodes, the principal lymphoid organs, are clustered along lymphatic vessels. Lymph nodes "filter" lymph and help activate the immune system.
2. Each lymph node has a fibrous capsule, a cortex, and a medulla. The cortex contains mostly lymphocytes, which act in immune responses; the medulla contains macrophages, which engulf and destroy viruses, bacteria, and other foreign debris, as well as lymphocytes.
3. Lymph enters the lymph nodes via afferent lymphatic vessels and exits via efferent vessels. There are fewer efferent vessels; therefore, lymph flow stagnates within the lymph node, allowing time for its cleansing.

20.4 The spleen removes bloodborne pathogens and aged red blood cells (pp. 773–774)

1. The spleen provides a site for lymphocyte proliferation and immune function, and destroys aged or defective red blood cells and bloodborne pathogens. It also recycles the breakdown products of hemoglobin, stores platelets and monocytes, and may be a hematopoietic site in the fetus.

20.5 MALT guards the body's entryways against pathogens (pp. 774–775)

1. Peyer's patches of the intestinal wall, lymphoid follicles of the appendix, tonsils of the pharynx and oral cavity, and follicles in the genitourinary and respiratory tract mucosae are known as MALT (mucosa-associated lymphoid tissue). They prevent pathogens in these passages from penetrating mucous membrane lining.

20.6 T lymphocytes mature in the thymus (pp. 776–777)

1. The thymus provides the environment in which T lymphocytes mature and become immunocompetent. The thymus is most functional during youth.

Developmental Aspects of the Lymphatic System and Lymphoid Organs and Tissues (p. 777)

1. Lymphatics usually develop as outpocketings of developing veins. The thymus develops from endoderm; the other lymphoid organs derive from mesenchymal cells of mesoderm.
2. The thymus is the first lymphoid organ to develop.
3. Lymphoid organs are populated by lymphocytes, which arise from hematopoietic tissue.

20

REVIEW QUESTIONS

To access additional practice questions using your smartphone, tablet, or computer: MasteringA&P® > Study Area > Practice Tests & Quizzes

Level 1 Remember/Understand

(Some questions have more than one correct answer. Select the best answer or answers from the choices given.)

1. Lymphatic vessels (a) serve as sites for immune surveillance, (b) filter lymph, (c) transport leaked plasma proteins and fluids to the cardiovascular system, (d) are represented by vessels that resemble arteries, capillaries, and veins.

2. The sac that often forms the initial portion of the thoracic duct is the (a) lacteal, (b) right lymphatic duct, (c) cisterna chyli, (d) lymph sac.

3. Entry of lymph into the lymphatic capillaries is promoted by which of the following? (a) one-way minivalves formed by overlapping endothelial cells, (b) the respiratory pump, (c) the skeletal muscle pump, (d) greater fluid pressure in the interstitial space.

4. The structural framework of lymphoid organs is (**a**) areolar connective tissue, (**b**) hematopoietic tissue, (**c**) reticular tissue, (**d**) adipose tissue.

5. Lymph nodes are densely clustered in all of the following body areas *except* (**a**) the brain, (**b**) the axillae, (**c**) the groin, (**d**) the cervical region.

6. The germinal centers in lymph nodes are largely sites of (**a**) macrophages, (**b**) proliferating B lymphocytes, (**c**) T lymphocytes, (**d**) all of these.

7. The red pulp areas of the spleen are sites of (**a**) splenic sinusoids, macrophages, and red blood cells, (**b**) clustered lymphocytes, (**c**) connective tissue septa.

8. The lymphoid organ that functions primarily during youth and then begins to atrophy is the (**a**) spleen, (**b**) thymus, (**c**) palatine tonsils, (**d**) bone marrow.

9. Collections of lymphoid tissue (MALT) that guard mucosal surfaces include all of the following except (**a**) appendix follicles, (**b**) the tonsils, (**c**) Peyer's patches, (**d**) the thymus.

Level 2 Apply/Analyze

10. Compare and contrast blood, interstitial fluid, and lymph.

11. Compare the structure and functions of a lymph node to those of the spleen.

12. A friend tells you that she has tender, swollen "glands" along the left side of the front of her neck. You notice that she has a bandage on her left cheek that is not fully hiding a large infected cut there. Exactly what are her swollen "glands," and how did they become swollen?

Level 3 Evaluate/Synthesize

13. (**a**) Which anatomical characteristic ensures that the flow of lymph through a lymph node is slow? (**b**) Why is this desirable?

14. There are no lymphatic arteries. Why isn't this a problem?

15. Mrs. Jackson, a 59-year-old woman, has undergone a left radical mastectomy (removal of the left breast and left axillary lymph nodes and vessels). Her left arm is severely swollen and painful, and she is unable to raise it to more than shoulder height. (**a**) Explain her signs and symptoms. (**b**) Can she expect to have relief from these symptoms in time? How so?

16. Once almost a rite of childhood, tonsillectomy (surgical removal of the tonsils) is now rarely performed. Similarly, while ruptured spleens were once routinely removed, they are now conserved whenever possible. Why should these lymphoid organs be preserved when possible?

20

27-Year-Old Male with a Crushed Hand

Andrew Hutchinson, 27, had surgery to repair his right hand after it was crushed by his garage door. The routine complete blood count (CBC) performed on admission revealed both a dangerously low total leukocyte count and a low proportion of lymphocytes.

One day postsurgery, he complains of pain in his injured right ring finger. When examined, the affected finger and the dorsum of the right hand are red, warm, and edematous, and red streaks radiate proximally on his right forearm. High doses of antibiotics are prescribed, and a sling is applied to the affected arm. Nurses are instructed to wear gloves and gown for Andrew's protection (called reverse isolation) when tending to him.

Relative to these observations:

1. How might the low lymphocyte count, orders for antibiotic therapy, and reverse isolation be related? Why is it important that Andrew not move the affected arm excessively (i.e., why was the sling ordered)?

2. **+ NCLEX-STYLE** Which of the statements best explains the presence of red streaks on Andrew's forearm?
   a. The lymphatic vessels of his forearm are inflamed.
   b. The major arteries of his forearm are inflamed.
   c. The lymphatic vessels of his forearm are dilated.
   d. The major arteries of his forearm are dilated.

3. **+ NCLEX-STYLE** The day after Andrew's surgery, the lymph nodes in his right armpit become enlarged and tender. This was most likely caused by which of the following?
   a. This is due to an allergic reaction to his antibiotics.
   b. His low lymphocyte count has triggered lymphocyte proliferation in his right armpit lymph nodes.
   c. This is due to an infiltration of his lymph nodes by cancer cells.
   d. This is due to infection of his lymph nodes by bacteria.

4. Do you predict that Andrew's recovery will be uneventful or problematic? Why?

21

The Immune System: Innate and Adaptive Body Defenses

In this chapter, you will learn that

The immune system defends the body from disease-causing organisms and cancerous cells

by comparing

Part 1 Innate Defenses

looking closer at

21.1 Surface barriers: Skin and mucosae

21.2 Innate internal defenses: Cells and chemicals

Part 2 Adaptive Defenses

beginning with

21.3 Antigens

identified by

21.4 Lymphocytes and antigen-presenting cells

trigger

21.5 Humoral immune response

21.6 Cellular immune response

then asking

21.7 What happens when things go wrong?

and finally, exploring

Developmental Aspects of the Immune System

CAREER CONNECTION

Watch a video to learn how the chapter content is used in a real health care setting. Go to Mastering A&P® > Study Area > Animations and Videos or use quick access URL https://goo.gl/88srfV

Every second of every day, armies of hostile bacteria, fungi, and viruses swarm on our skin and yet we stay amazingly healthy most of the time. The body has evolved a single-minded approach to such foes—if you're not with us, you're against us! To implement that stance, it relies heavily on two intrinsic defense systems that act both independently and cooperatively to provide resistance to disease, or **immunity** (*immun* = free).

1. The **innate (nonspecific) defense system**, like a lowly foot soldier, is always prepared, responding within minutes to protect the body from foreign substances. This system has two "barricades." The *first line of defense* consists of surface barriers—intact skin and mucosae. The *second line of defense*, called into action whenever the first line has been penetrated, relies on internal defenses such as antimicrobial proteins, phagocytes, and other cells to inhibit the invaders' spread throughout the body. The hallmark of the second line of defense is *inflammation*.

2. The **adaptive (specific) defense system** functions like an elite fighting force equipped with high-tech weapons to attack *identified* foreign substances. The adaptive defense response, which provides the body's *third line of defense*, takes considerably longer to mount than the innate defense response.

Although we consider them separately, the innate and adaptive systems always work hand in hand. An overview of these two systems is shown in **Figure 21.1**. Small portions of this diagram will reappear in subsequent figures to let you know which part of the immune system we're dealing with.

Although certain organs of the body (notably lymphoid organs) are intimately involved in the immune response, the **immune system** is a *functional system* rather than an organ system in an anatomical sense. Its "structures" are a diverse array of molecules plus trillions of immune cells (especially lymphocytes) that inhabit lymphoid tissues and circulate in body fluids.

Once, the term *immune system* was equated with the adaptive defense system only. However, we now know that the innate and adaptive defenses are deeply intertwined. Specifically:

- The innate and adaptive systems release and recognize (bind to) many of the same defensive molecules.
- The innate responses are not as nonspecific as once thought. Indeed, they have specific pathways to target certain foreign substances.
- Proteins released during innate responses alert cells of the adaptive system to the presence of specific foreign molecules in the body.

When the immune system is operating effectively, it protects the body from most infectious microorganisms, cancer cells, and (unfortunately) transplanted organs and grafts. It does this both directly, by cell attack, and indirectly, by releasing mobilizing chemicals and protective antibody molecules.

Figure 21.1 Simplified overview of innate and adaptive defenses. Humoral immunity (primarily involving B lymphocytes) and cellular immunity (involving T lymphocytes) are distinct but overlapping areas of adaptive immunity.

PART 1

INNATE DEFENSES

You could say we come fully equipped with innate defenses. The mechanical barriers that cover body surfaces and the cells and chemicals that act on the initial internal battlefronts are in place at birth, ready to ward off invading **pathogens** (harmful or disease-causing microorganisms).

Many times, our innate defenses alone ward off infection. In other cases, the adaptive immune system is called into action to reinforce and enhance the innate defenses. Either way, the innate defenses reduce the workload of the adaptive system by preventing the entry and spread of microorganisms in the body.

21.1 Surface barriers act as the first line of defense to keep invaders out of the body

Learning Outcome

▶ Describe surface membrane barriers and their protective functions.

The body's first line of defense—the *skin* and the *mucous membranes* (◀ p. 141), along with the secretions these membranes produce—is highly effective. As long as the epidermis is unbroken, this heavily keratinized epithelial membrane is a formidable physical barrier to most microorganisms. Keratin is also resistant to most weak acids and bases and to bacterial enzymes and toxins. Intact mucosae provide similar mechanical barriers within the body. Recall that mucous membranes line all body cavities that open to the exterior: the digestive, respiratory, urinary, and reproductive tracts.

Table 21.1   The First Line of Defense: Surface Membrane Barriers

CATEGORY/ASSOCIATED ELEMENTS	PROTECTIVE MECHANISM
Intact skin epidermis	Forms mechanical barrier that prevents entry of pathogens and other harmful substances into body
• Acid mantle of skin	Skin secretions (sweat and sebum) make epidermal surface acidic, which inhibits bacterial growth; also contain various bactericidal chemicals
• Keratin	Provides resistance against acids, alkalis, and bacterial enzymes
Intact mucous membranes	Form mechanical barrier that prevents entry of pathogens
• Mucus	Traps microorganisms in respiratory and digestive tracts
• Nasal hairs	Filter and trap microorganisms in nasal passages
• Cilia	Propel debris-laden mucus away from nasal cavity and lower respiratory passages
• Gastric juice	Contains concentrated hydrochloric acid and protein-digesting enzymes that destroy pathogens in stomach
• Acid mantle of vagina	Inhibits growth of most bacteria and fungi in female reproductive tract
• Lacrimal secretion (tears); saliva	Continuously lubricate and cleanse eyes (tears) and oral cavity (saliva); contain lysozyme, an enzyme that destroys microorganisms
• Urine	Normally acid pH inhibits bacterial growth; cleanses the lower urinary tract as it flushes from the body

Besides serving as physical barriers, skin and mucous membranes produce a variety of protective chemicals:

- **Acid.** The acidity of skin, vaginal, and stomach secretions—the *acid mantle*—inhibits bacterial growth.

- **Enzymes.** *Lysozyme*—found in saliva, respiratory mucus, and lacrimal fluid of the eye—destroys bacteria. Protein-digesting enzymes in the stomach kill many different microorganisms.

- **Mucin.** *Mucin* dissolved in water forms thick, sticky mucus that lines the digestive and respiratory passageways. This mucus traps many microorganisms. Likewise, the mucin in watery saliva traps microorganisms and washes them out of the mouth into the stomach where they are digested.

- **Defensins.** Mucous membranes and skin secrete small amounts of broad-spectrum antimicrobial peptides called *defensins*. Defensin output increases dramatically in response to inflammation when surface barriers are breached. Using various mechanisms, such as disruption of microbial membranes, defensins help to control bacterial and fungal colonization in the exposed areas.

- **Other chemicals.** In the skin, some lipids in sebum and *dermcidin* in eccrine sweat are toxic to bacteria.

The respiratory tract mucosae also have structural modifications that counteract potential invaders. Tiny mucus-coated hairs inside the nose trap inhaled particles, and cilia on the mucosa of the upper respiratory tract sweep dust- and bacteria-laden mucus toward the mouth, preventing it from entering the lower respiratory passages.

Although these surface barriers (summarized in **Table 21.1**) are quite effective, they are breached by everyday nicks and cuts, for example, when you brush your teeth or shave. When this happens and microorganisms invade deeper tissues, your *internal* innate defenses—the second line of defense—come into play.

Check Your Understanding

1. What distinguishes the innate defense system from the adaptive defense system?
2. What is the first line of defense against disease?

For answers, see Answers Appendix.

21.2 Innate internal defenses are cells and chemicals that act as the second line of defense

Learning Outcomes

▶ Explain the importance of phagocytosis, natural killer cells, and fever in innate body defense.

▶ Describe the inflammatory process. Identify several inflammatory chemicals and indicate their specific roles.

▶ Name the body's antimicrobial substances and describe their functions.

The body uses an enormous number of nonspecific cellular and chemical means to protect itself, including phagocytes, natural killer cells, antimicrobial proteins, and fever. The inflammatory response enlists macrophages, mast cells (◀ p. 129), all types of white blood cells, and dozens of chemicals that kill pathogens and help repair tissue. These protective tactics identify potentially harmful substances by recognizing (binding to) molecules with specific shapes that are part of infectious organisms (bacteria, viruses, fungi, and various parasites) but not normal human cells. The receptors that do this are called **pattern recognition receptors**.

One class of these receptors, called **Toll-like receptors (TLRs)**, plays a central role in triggering immune responses. There are 11 types of human TLRs, each recognizing a particular class of attacking microbe. For example, one type responds to a glycolipid in cell walls of the tuberculosis bacterium and

21

Innate defenses ⟶ Internal defenses

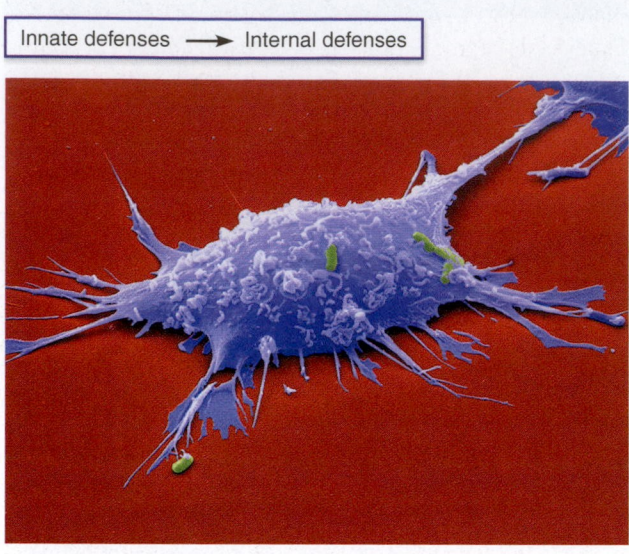

(a) A macrophage (purple) about to engulf rod-shaped bacteria (green). Scanning electron micrograph (2000×).

① Phagocyte adheres to pathogens or debris (using receptors).

② Phagocyte forms pseudopods that eventually engulf the particles, forming a phagosome.

③ Lysosome fuses with the phagocytic vesicle, forming a phagolysosome.

④ Toxic compounds and lysosomal enzymes destroy pathogens.

⑤ Sometimes exocytosis of the vesicle removes indigestible and residual material.

Phagosome (phagocytic vesicle)

Lysosome

Acid hydrolase enzymes

(b) Events of phagocytosis.

Figure 21.2 Phagocytosis.

another to a component of gram-negative bacteria such as *Salmonella*. A variety of cells have TLRs, including macrophages and cells of certain boundary tissues such as the epithelial cells lining the gastrointestinal and respiratory tracts. These pattern recognition receptors allow the cells to recognize invaders and sound a chemical "alarm" that initiates inflammation (which we will discuss shortly).

Phagocytes

Pathogens that get through the skin or mucosae into the underlying connective tissue are confronted by *phagocytes* (*phago* = eat). **Neutrophils**, the most abundant type of white blood cell, become phagocytic on encountering infectious material in the tissues. However, the most voracious phagocytes are **macrophages** ("big eaters"). Macrophages derive from white blood cells called **monocytes** that leave the bloodstream, enter the tissues, and develop into macrophages.

- *Free macrophages* wander throughout the tissue spaces in search of cellular debris or "foreign invaders."

- *Fixed macrophages*, such as *stellate macrophages* in the liver, are permanent residents of particular organs.

Phagocytosis

A phagocyte engulfs particulate matter much the way an amoeba ingests a food particle. Receptors on its flowing cytoplasmic extensions bind to the particle. The particle is then pulled inside, enclosed within a membrane-lined vesicle (**Figure 21.2a**). The resulting **phagosome** then fuses with a *lysosome* to form a **phagolysosome** (steps ①–③ in Figure 21.2b).

Neutrophils and macrophages generally kill ingested prey by acidifying the phagolysosome and digesting its contents with lysosomal enzymes. However, some pathogens such as the tuberculosis bacillus and certain parasites are resistant to lysosomal enzymes and can even multiply within the phagolysosome. In this case, other immune cells called helper T cells release chemicals that stimulate the macrophage. This activates additional enzymes that produce a lethal **respiratory burst**. The respiratory burst promotes killing of pathogens by:

- Liberating a deluge of highly destructive free radicals (◀ **p. 86**)

- Producing oxidizing chemicals (hydrogen peroxide and a substance identical to household bleach)

- Increasing the phagolysosome's pH and osmolarity, which activates other protein-digesting enzymes that digest the invader

Neutrophils also pierce the pathogen's membrane by using *defensins*, the antimicrobial peptides we mentioned earlier.

Phagocytic attempts are not always successful. In order for a phagocyte to ingest a pathogen, the phagocyte must first *adhere* to that pathogen, a feat made possible by recognizing the pathogen's carbohydrate "signature." Many bacteria have external capsules that conceal their carbohydrate signatures, allowing them to elude capture because phagocytes cannot bind to them.

Our immune systems get around this problem by coating pathogens with **opsonins**. Opsonins are complement proteins (discussed shortly) or antibodies. Both provide "handles" to which phagocyte receptors can bind. Any pathogen can be coated with opsonins, a process called **opsonization** ("to make tasty"), which greatly accelerates phagocytosis of that pathogen.

21

When phagocytes are unable to ingest their targets (because of size, for example), they can release their toxic chemicals into the extracellular fluid. Whether killing ingested or extracellular targets, neutrophils rapidly destroy themselves in the process. In contrast, macrophages are more robust and can survive to kill another day.

Natural Killer (NK) Cells

Natural killer (NK) cells, which "police" the body in blood and lymph, are a unique group of defensive cells that can kill cancer cells and virus-infected body cells before the adaptive immune system is activated. NK cells are part of a small group of *large granular lymphocytes*.

Unlike lymphocytes of the adaptive immune system, which only recognize and react against *specific* infected or tumor cells, NK cells are far less picky. They can eliminate a variety of infected or cancerous cells by detecting general abnormalities such as the lack of "self" cell-surface proteins called MHC, described on pp. 791–792. The name "natural" killer cells reflects their nonspecificity.

NK cells are not phagocytic. They kill by directly contacting the target cell, inducing it to undergo apoptosis (programmed cell death). This is the same method used by cytotoxic T cells (described on p. 807). NK cells also secrete potent chemicals that enhance the inflammatory response.

Inflammation: Tissue Response to Injury

Inflammation is a nonspecific response to any tissue injury. Causes include physical trauma, intense heat, irritating chemicals, or infection. Inflammation has several beneficial effects:

- It prevents the spread of damaging agents to nearby tissues.
- It disposes of cell debris and pathogens.
- It alerts the adaptive immune system.
- It sets the stage for repair.

How can you tell if inflammation is present? There are four *cardinal signs* (distinguishing indicators) of acute (short term) inflammation: *redness, heat* (*inflam* = set on fire), *swelling*, and *pain*. Some authorities consider *impaired function* to be a fifth cardinal sign. For instance, movement in an inflamed joint may be hampered temporarily, forcing it to rest, which aids healing. Inflammations are so common that many have specific names. Each name begins with the organ or tissue affected and ends with *-itis* (such as tonsillitis, appendicitis, and tendonitis).

Inflammatory Chemical Release

The inflammatory process begins with a chemical "alarm"—a flood of inflammatory chemicals released into the extracellular fluid. Inflammatory chemicals are released by injured or stressed tissue cells, and immune cells. For example, **mast cells** (◀ p. 129), a key component of the inflammatory response, release the potent inflammatory chemical **histamine** (his'tah-mēn). Inflammatory chemicals can also be formed from chemicals circulating in the blood (**Table 21.2**).

Other inflammatory chemicals include **kinins** (ki'ninz), **prostaglandins** (pros"tah-glan'dinz), and *cytokines* (a group of chemical mediators of immune responses that we will discuss later in the chapter). If pathogens provoked the inflammation, a group of plasma proteins known as **complement** is activated to form potent inflammatory chemicals.

All inflammatory chemicals dilate local arterioles and make local capillaries leakier. Many also attract phagocytes to the injured area. (We will see the importance of these effects next.) In addition, some have individual inflammatory roles, such as mobilizing lymphocytes and other elements of adaptive immunity.

Vasodilation and Increased Vascular Permeability

Vasodilation of local arterioles causes local **hyperemia** (increased blood flow). Hyperemia brings more cells and chemicals of the immune system to the injured area. It accounts for two of the cardinal signs of inflammation: the *redness* and *heat* of an inflamed region.

Inflammatory chemicals also increase the permeability of local capillaries and venules. Consequently, **exudate**—fluid containing clotting factors and antibodies—seeps from the

21

Table 21.2	Inflammatory Chemicals	
CHEMICAL	**SOURCE**	**PHYSIOLOGICAL EFFECTS**
Histamine	Granules of mast cells and basophils (◀ p. 653). Released in response to mechanical injury, presence of certain microorganisms, and chemicals released by neutrophils.	Promotes vasodilation of local arterioles, increasing blood flow to injured area. Increases permeability of local capillaries, promoting formation of exudate.
Kinins (bradykinin and others)	A plasma protein, kininogen, is split by the enzyme kallikrein found in plasma, urine, saliva, and in lysosomes of neutrophils and other types of cells. Splitting releases active kinin peptides.	Same as for histamine. Also induce chemotaxis of leukocytes and prompt neutrophils to release lysosomal enzymes, thereby enhancing generation of more kinins. Induce pain.
Prostaglandins (◀ p. 48)	Fatty acid molecules produced from arachidonic acid found in all cell membranes; generated by enzymes of neutrophils, basophils, mast cells, and others.	Same as for histamine. Also induce neutrophil chemotaxis. Induce pain. (Some prostaglandins are anti-inflammatory.)
Complement	See Table 21.3 (p. 790).	
Cytokines	See Table 21.7 (p. 805).	

blood into the tissue spaces. This influx of protein-rich fluids into the tissue spaces sweeps foreign material into the lymphatic vessels so it can be processed in the lymph nodes. It also delivers important proteins such as complement and clotting factors to the interstitial fluid. The clotting factors form a gel-like fibrin mesh (a clot; ◄ p. 658) that acts as a scaffold for permanent repair. The mesh also isolates the injured area and prevents bacteria and other harmful agents from spreading. Walling off the injured area is such an important defense strategy that some bacteria (such as *Streptococcus*) have evolved enzymes that break down the clot, allowing them to invade surrounding tissues.

The increased tissue fluid causes the local *swelling* (edema) that presses on adjacent nerve endings, contributing to a sensation of *pain*. Pain also results from the release of bacterial toxins, and the sensitizing effects of released prostaglandins and kinins. (Aspirin and some other anti-inflammatory drugs reduce pain by inhibiting prostaglandin synthesis.)

Phagocyte Mobilization

Soon after inflammation begins, phagocytes flood the damaged area. Neutrophils lead, followed by macrophages. **Figure 21.3** illustrates the four steps in which phagocytes are mobilized to infiltrate the injured site.

① **Leukocytosis.** Injured cells release chemicals called **leukocytosis-inducing factors**. In response, neutrophils enter blood from red bone marrow and within a few hours, the number of neutrophils in blood increases four- to five-fold. This **leukocytosis**, the increase in white blood cells (WBCs), is characteristic of inflammation.

② **Margination. Margination** refers to the phenomenon of phagocytes clinging to the inner walls (margins) of the capillaries and postcapillary venules. Inflamed endothelial cells sprout cell adhesion molecules (CAMs, ◄ p. 81) that signal "this is the place." As neutrophils encounter these CAMs, they bind briefly. This causes them to slow and roll along the inner surface of the blood vessel so they can be activated by inflammatory chemicals. Once activated, neutrophils sprout additional CAMs on their own plasma membranes. This allows the neutrophils to bind endothelial cells tightly, creating an initial foothold.

③ **Diapedesis.** Continued chemical signaling prompts the neutrophils to flatten and squeeze between the endothelial cells of the capillary walls—a process called **diapedesis**.

④ **Chemotaxis.** Inflammatory chemicals act as homing devices, or more precisely **chemotactic agents**. Neutrophils and other WBCs migrate up the gradient of chemotactic agents

Innate defenses ⟶ Internal defenses

Inflammatory chemicals diffusing from the inflamed site act as chemotactic agents.

④ **Chemotaxis.** Neutrophils follow chemical trail.

Capillary wall
Basement membrane
Endothelium

① **Leukocytosis.** Neutrophils enter blood from bone marrow.

② **Margination.** Neutrophils cling to capillary wall.

③ **Diapedesis.** Neutrophils flatten and squeeze out of capillaries.

Figure 21.3 Phagocyte mobilization.

to the site of injury. Within an hour after the inflammatory response has begun, neutrophils have collected at the site and are devouring any foreign material present.

As the body's counterattack continues, monocytes follow neutrophils into the injured area. Monocytes are fairly poor phagocytes, but within 12 hours of leaving the blood and entering the tissues, they swell and develop large numbers of lysosomes, becoming macrophages with insatiable appetites. These late-arriving macrophages replace the neutrophils on the battlefield.

Macrophages are the central actors in the final disposal of cell debris as acute inflammation subsides, and they predominate at sites of chronic inflammation. The ultimate goal of an inflammatory response is to clear the injured area of pathogens, dead tissue cells, and any other debris so that tissue can be repaired. Once this is accomplished, healing usually occurs quickly.

The inflammatory response is summarized in **Figure 21.4**.

HOMEOSTATIC IMBALANCE 21.1   **CLINICAL**

In severely infected areas, the battle takes a considerable toll on both sides, and creamy yellow **pus** (a mixture of dead or dying neutrophils, broken-down tissue cells, and living and dead pathogens) may accumulate in the wound. If the inflammatory mechanism fails to clear the area of debris, collagen fibers may

Innate defenses ⟶ Internal defenses

Initial stimulus

Physiological response

Signs of inflammation

Result

Tissue injury

Release of inflammatory chemicals (histamine, complement, kinins, prostaglandins, etc.)

Release of leukocytosis-inducing factors

Leukocytosis (increased numbers of white blood cells in bloodstream)

Arterioles dilate

Increased capillary permeability

Attract neutrophils, monocytes, and lymphocytes to area (chemotaxis)

Leukocytes migrate to injured area

Local hyperemia (increased blood flow to area)

Capillaries leak fluid (exudate formation)

Margination (leukocytes cling to capillary walls)

Leaked protein-rich fluid in tissue spaces

Leaked clotting proteins form interstitial clots that wall off area to prevent movement of pathogen to surrounding tissue

Diapedesis (leukocytes pass through capillary walls)

Heat   **Redness**   **Pain**   **Swelling**

Phagocytosis of pathogens and dead tissue cells

Locally increased temperature increases metabolic rate of cells

Possible temporary impairment of function

Temporary fibrin patch forms scaffolding for repair

Pus may form

Area cleared of debris

21

Healing

Figure 21.4 Events of acute inflammation. The four cardinal signs are shown in red boxes, as is impaired function, which in some cases constitutes a fifth cardinal sign.

be laid down, which walls off the sac of pus, forming an *abscess*. The abscess may need to be surgically drained before healing can occur.

Some bacteria, such as tuberculosis bacilli, resist digestion by the macrophages that engulf them. They escape the effects of prescription antibiotics by remaining snugly enclosed within their macrophage hosts. In such cases, *granulomas* form. These tumorlike growths contain a central region of infected macrophages surrounded by uninfected macrophages and an outer fibrous capsule.

A person may harbor pathogens walled off in granulomas for years without displaying any symptoms. However, if the person's resistance to infection is ever compromised, the bacteria may be activated and break free, leading to clinical disease symptoms.

Antimicrobial Proteins

A variety of **antimicrobial proteins** enhance our innate defenses by attacking microorganisms directly or by hindering their ability to reproduce. The most important antimicrobial proteins are interferons and complement proteins (**Table 21.3** on p. 790).

Interferons

Viruses—essentially nucleic acids surrounded by a protein envelope—lack the cellular machinery to generate ATP or synthesize proteins. They do their "dirty work" in the body by invading tissue cells and taking over the cellular metabolic machinery needed to reproduce themselves.

Infected cells can do little to save themselves, but some can secrete small proteins called **interferons (IFNs)** (in″ter-fēr′onz) to help protect cells that have not yet been infected. The IFNs diffuse to nearby cells, which they stimulate to synthesize proteins that block further protein synthesis and degrade viral RNA. In this way, they "interfere" with viral replication (**Figure 21.5**). Because IFN protection is *not* virus-specific, IFNs produced against a particular virus protect against other viruses, too.

The IFNs are a family of immune modulating proteins produced by a variety of body cells, each having a slightly different physiological effect. IFN alpha (α) and beta (β) have the antiviral effects that we've just described and also activate NK cells. Another interferon, IFN gamma (γ), or immune interferon, is secreted by lymphocytes and has widespread immune mobilizing effects, such as activating macrophages. Because both macrophages and NK cells can also act directly against cancerous cells, the interferons have an indirect role in fighting cancer. IFNs are used to treat several disorders including hepatitis C, genital warts, and multiple sclerosis.

Complement

The term **complement system**, or simply **complement**, refers to a group of at least 20 plasma proteins that normally circulate in the blood in an inactive state. These proteins include those numbered C1 through C9 plus several others that act as regulatory proteins and other factors.

Complement provides a major mechanism for destroying foreign substances in the body. Its activation unleashes inflammatory chemicals that amplify virtually all aspects of the inflammatory process. Activated complement also lyses and kills certain bacteria and other cell types. (Luckily our own cells are equipped with proteins that normally inhibit complement activation.) Although complement is a nonspecific defensive mechanism, it "complements" (enhances) the effectiveness of *both* innate and adaptive defenses.

Figure 21.6 outlines the three pathways by which complement can be activated.

- The **classical pathway** involves *antibodies*, water-soluble protein molecules that the adaptive immune system produces to fight off foreign invaders. When antibodies bind to pathogens, they can also bind complement components. This double binding is the first step in this complement activation pathway. (We describe this in more detail on pp. 800–801.)

- The **lectin pathway** involves *lectins*, water-soluble protein molecules that the innate immune system produces to recognize foreign invaders. When lectins bind specific sugars on the surface of microorganisms, they can then bind and activate complement.

- The **alternative pathway** is triggered when spontaneously activated C3 and other complement factors interact on the surface of microorganisms. These microorganisms lack the complement activation inhibitors our own cells have.

Like the blood clotting cascade, complement activation by any of these pathways involves a cascade in which proteins are activated in an orderly sequence—each step catalyzing the next. The three pathways converge at C3, which is split into C3a and C3b. Splitting C3 initiates a common terminal pathway that enhances inflammation, promotes phagocytosis, and can cause cell lysis.

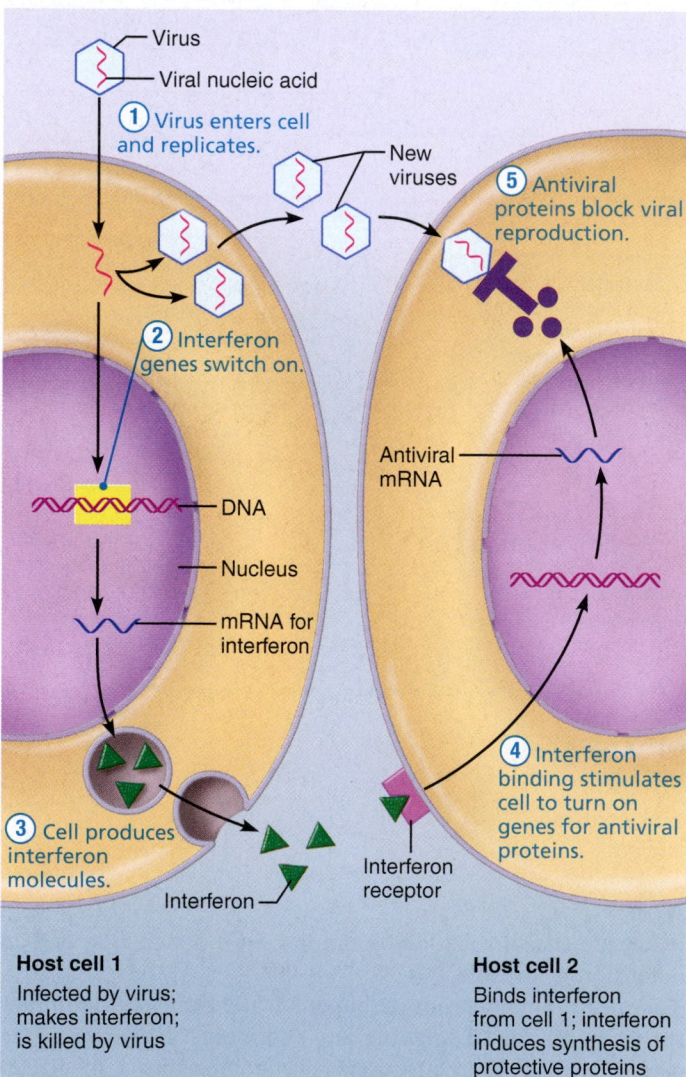

Innate defenses ⟶ Internal defenses

- Virus
- Viral nucleic acid
- ① Virus enters cell and replicates.
- New viruses
- ⑤ Antiviral proteins block viral reproduction.
- ② Interferon genes switch on.
- Antiviral mRNA
- DNA
- Nucleus
- mRNA for interferon
- ④ Interferon binding stimulates cell to turn on genes for antiviral proteins.
- ③ Cell produces interferon molecules.
- Interferon receptor
- Interferon

Host cell 1
Infected by virus; makes interferon; is killed by virus

Host cell 2
Binds interferon from cell 1; interferon induces synthesis of protective proteins

Figure 21.5 The interferon mechanism against viruses.

21

Classical pathway	Lectin pathway	Alternative pathway
Activated by **antibodies** coating target cell	Activated by **lectins** binding to specific sugars on microorganism's surface	Activated spontaneously. Lack of inhibitors on microorganism's surface allows process to proceed

All three activation pathways converge at C3 to initiate the final activation of complement (together with other complement proteins and factors). C3 is split into two active pieces: C3a and C3b.

C3

C3b   C3a

C3b

Opsonization:
Coats pathogen surfaces, which enhances phagocytosis

C5b   C5a
C6
C7
C8
C9

MAC

Enhances inflammation:
Stimulates histamine release, increases blood vessel permeability, attracts phagocytes by chemotaxis, etc.

Membrane attack complexes (MACs) form when C3b activates other complement components (C5b and C6–C9). This occurs only on the surfaces of certain target cells (usually bacteria). MACs insert into the target cell membrane, creating pores that lyse the target cell by allowing water to enter and cell contents to leave.

Pore

Complement proteins (C5b–C9)

Membrane of target cell

Figure 21.6 Complement activation.

Cell lysis begins when C3b binds to the target cell's surface and triggers the insertion of a group of complement proteins called **MAC (membrane attack complex)** into the cell's membrane. MAC forms and stabilizes a hole in the membrane that allows a massive influx of water, lysing the target cell.

The C3b molecules also act as *opsonins*. As previously described, opsonins coat the microorganism, providing "handles" that receptors on macrophages and neutrophils can adhere to. This allows them to engulf the particle more rapidly. C3a and other molecules formed during complement activation amplify the inflammatory response by stimulating mast cells and basophils to release histamine and by attracting neutrophils and other inflammatory cells to the area.

Fever

Inflammation is a localized response to infection, but sometimes the body's response to the invasion of microorganisms is more widespread. **Fever**, an abnormally high body temperature, is a systemic response to invading microorganisms.

When leukocytes and macrophages are exposed to foreign substances in the body, they release chemicals called **pyrogens** (*pyro* = fire). These pyrogens act on the body's thermostat—a cluster of neurons in the hypothalamus—raising the body's temperature above normal [37°C (98.6°F)].

Fever seems to benefit the body, but exactly how it does so is unclear. Fever causes the liver and spleen to sequester iron and zinc, which may make them less available to support bacterial growth. Additionally, fever increases the metabolic rate of tissue cells, and may speed up repair processes.

Check Your Understanding

3. What is opsonization and how does it help phagocytes? Give an example of a molecule that acts as an opsonin.

4. Under what circumstances might NK cells kill our own cells?

21

Table 21.3   The Second Line of Defense: Innate Cellular and Chemical Defenses

CATEGORY/ASSOCIATED ELEMENTS	PROTECTIVE MECHANISM
Phagocytes	Engulf and destroy pathogens that breach surface membrane barriers; macrophages also contribute to adaptive immune responses
Natural killer (NK) cells	Promote apoptosis (cell suicide) by directly attacking virus-infected or cancerous body cells; recognize general abnormalities rather than specific antigens
Inflammatory response	Prevents injurious agents from spreading to adjacent tissues, disposes of pathogens and dead tissue cells, and promotes tissue repair; released inflammatory chemicals attract phagocytes (and other immune cells) to the area
Antimicrobial proteins	
• Interferons (α, β, γ)	Proteins released by virus-infected cells and certain lymphocytes; act as chemical messengers to protect uninfected tissue cells from viral takeover; mobilize immune system
• Complement	A group of bloodborne proteins that, when activated, lyse microorganisms, enhance phagocytosis by opsonization, and intensify inflammatory and other immune responses
Fever	Systemic response initiated by pyrogens; high body temperature inhibits microbes from multiplying and enhances body repair processes

5. What are the cardinal signs of inflammation and what causes them?

6. WHAT IF? Julian has a mutation in the gene that codes for interferon. What type of pathogen is Julian particularly susceptible to?

For Answers, See Answers Appendix.

21

PART 2

ADAPTIVE DEFENSES

Most of us would find it wonderfully convenient if we could walk into a single clothing store and buy a complete wardrobe—hat to shoes—that fits perfectly regardless of any special figure problems. We know that such a service would be next to impossible to find. And yet, we take for granted our **adaptive immune system**, the body's built-in *specific defensive system* that stalks and eliminates with nearly equal precision almost any type of pathogen that intrudes into the body. This is the body's third line of defense.

When it operates effectively, the adaptive immune system protects us from a wide variety of infectious agents, as well as from abnormal body cells. When it fails, or is disabled, devastating diseases such as cancer and AIDS result. The activity of the adaptive immune system tremendously amplifies the inflammatory response and is responsible for most complement activation.

At first glance, the adaptive system seems to have a major shortcoming. Unlike the innate system, which is always ready and able to react, the adaptive system must "meet" or be primed by an initial exposure to a specific foreign substance (antigen). Only then can it protect the body against that substance, and this priming takes precious time.

Experiments in the late 1800s revealed the basis of this specific immunity. Researchers demonstrated that animals surviving a serious bacterial infection have protective factors (the proteins we now call *antibodies*) in their blood that defend against future attacks by the same pathogen. Furthermore, researchers found that if antibody-containing serum from the surviving animals was injected into animals that had not been exposed to the pathogen, the injected animals would also be protected.

At first antibodies were thought to be the sole artillery of the adaptive immune system. Then, in the mid-1900s researchers discovered that injecting antibody-containing serum did *not* always protect the recipient from diseases the serum donor had survived. In such cases, however, injecting the donor's lymphocytes *did* provide immunity. As the pieces fell into place, researchers recognized two separate but overlapping arms of adaptive immunity, each using a variety of attack mechanisms that vary with the intruder:

- **Humoral immunity** (hu′mor-ul), also called **antibody-mediated immunity**, is provided by antibodies present in the body's "humors," or fluids (blood, lymph, etc.). Though they are produced by lymphocytes, antibodies circulate freely in the blood and lymph, where they bind primarily to *extracellular* targets—bacteria, bacterial toxins, and free viruses—inactivating them temporarily and marking them for destruction by phagocytes or complement.

- When lymphocytes rather than antibodies defend the body, the immunity is called **cellular** or **cell-mediated immunity** because living cells (the lymphocytes themselves) provide the protection. Cellular immunity also has *cellular* targets—cells infected with pathogens (such as viruses, some bacteria, and parasites), cancer cells, and cells of foreign grafts. The lymphocytes act against such targets either *directly*, by killing the infected cells, or *indirectly*, by releasing chemicals. These chemicals enhance the inflammatory response, activate macrophages, or even direct the entire adaptive immune response. Indeed, many lymphocytes that belong to the cellular immunity branch act to coordinate both humoral and cellular adaptive immunity.

In summary, the adaptive immune response has four key characteristics that set it apart from the innate defenses:

- **It involves lymphocytes called B and T lymphocytes.** These lymphocytes are not involved in innate immunity.
- **It is specific.** It recognizes and targets *identified* pathogens or foreign substances that initiate the immune response.
- **It is systemic.** Adaptive immunity is not restricted to the initial infection site.
- **It has "memory."** After an initial exposure, it recognizes and mounts even stronger attacks on previously encountered pathogens.

Before describing the humoral and cellular responses, we will first consider the central role of *antigens*.

21.3 Antigens are substances that trigger the body's adaptive defenses

Learning Outcomes
▶ Define antigen and describe how antigens affect the adaptive defenses.
▶ Define complete antigen, hapten, and antigenic determinant.

Antigens (an′tĭ-jenz) are substances that can mobilize the adaptive defenses. They are the ultimate targets of all adaptive immune responses. (*Antigen* is a contraction of "*anti*body *gen*erating.") Most antigens are large, complex molecules (natural or synthetic) that are *foreign* (not normally present in the body). Consequently, as far as our immune system is concerned, they are intruders, or **nonself**.

Complete Antigens and Haptens

Antigens can be *complete* or *incomplete*. **Complete antigens** have two important functional properties:

- **Immunogenicity**, which is the ability to stimulate specific lymphocytes to proliferate (multiply).
- **Reactivity**, which is the ability to react with the activated lymphocytes and the antibodies released by immunogenic reactions.

An almost limitless variety of foreign molecules can act as complete antigens, including virtually all foreign proteins, many large polysaccharides, and some lipids and nucleic acids. Of these, proteins are the strongest antigens. Pollen grains and microorganisms—such as bacteria, fungi, and virus particles—are all immunogenic because their surfaces bear many different foreign macromolecules.

As a rule, small molecules—such as peptides, nucleotides, and many hormones—are not immunogenic. But if they link up with the body's own proteins, the adaptive immune system may recognize the *combination* as foreign and mount an attack that is harmful rather than protective. (We describe these reactions, called *hypersensitivities*, later in the chapter.) In such cases, the troublesome small molecule is called a **hapten** (hap′ten;

Figure 21.7 Most antigens have several different antigenic determinants. Antibodies (and related receptors on lymphocytes) bind to small areas on the antigen surface called antigenic determinants. In this example, three different types of antibodies react with different antigenic determinants on the same antigen molecule.

haptein = grasp) or **incomplete antigen**. Unless attached to protein carriers, haptens have reactivity but not immunogenicity. Besides certain drugs (particularly penicillin), chemicals that act as haptens are found in poison ivy, animal dander, detergents, cosmetics, and a number of common household and industrial products.

Antigenic Determinants

The ability of a molecule to act as an antigen depends on both its size and its complexity. Only certain parts of the antigen, called **antigenic determinants**, are immunogenic. Antibodies or lymphocyte receptors bind to these antigenic determinants in much the same manner that an enzyme binds to a substrate.

Most naturally occurring antigens have a variety of antigenic determinants on their surfaces, some more potent than others in provoking an immune response (**Figure 21.7**). Different lymphocytes "recognize" different antigenic determinants, so a single antigen may mobilize several lymphocyte populations and stimulate formation of many kinds of antibodies.

Large proteins have hundreds of chemically different antigenic determinants, which accounts for their high immunogenicity and reactivity. However, large simple molecules such as plastics, which have many identical, regularly repeating units, have little or no immunogenicity. Such substances are used to make artificial implants because the substances are not seen as foreign and rejected by the body.

Self-Antigens: MHC Proteins

A huge variety of protein molecules dot the external surfaces of all our cells. Assuming your immune system has been properly "programmed," your **self-antigens** are not foreign or antigenic to you, but they are strongly antigenic to other individuals. (This is the basis of transfusion reactions and graft rejection.)

Among the cell surface proteins that identify a cell as *self* is a group of glycoproteins called **MHC proteins**. Genes of the major histocompatibility complex (MHC) code for these proteins. Because millions of combinations of these genes are possible, it is unlikely that any two people except identical twins have the same MHC proteins. Each MHC protein has a deep groove that holds a peptide, either a self-antigen or a foreign antigen. As we will describe shortly, T lymphocytes can only bind antigens that are presented (displayed to them) on MHC proteins.

Check Your Understanding

7. Name four key characteristics of adaptive immunity.

8. What is the difference between a complete antigen and a hapten?

9. What marks a cell as "self" as opposed to "nonself"?

For answers, see Answers Appendix.

21.4 B and T lymphocytes and antigen-presenting cells are cells of the adaptive immune response

Learning Outcomes

▶ Compare and contrast the origin, maturation process, and general function of B and T lymphocytes.

▶ Define immunocompetence and self-tolerance, and describe their development in B and T lymphocytes.

▶ Name several antigen-presenting cells and describe their roles in adaptive defenses.

The adaptive immune system involves three crucial types of cells: two distinct populations of lymphocytes, plus *antigen-presenting cells (APCs)*.

- **B lymphocytes (B cells)** oversee humoral immunity.

- **T lymphocytes (T cells)** are non-antibody-producing lymphocytes that constitute the cellular arm of adaptive immunity.

- APCs do not respond to specific antigens as lymphocytes do. Instead, they play essential auxiliary roles. As we will see, T cells cannot recognize their antigens without APCs.

Lymphocytes

Despite their differences, B and T lymphocytes share a common pattern of development and common steps in their life cycles. Let's take a look.

Lymphocyte Development, Maturation, and Activation

The development, maturation, and activation of B and T cells share the five general steps shown in **Figure 21.8**.

① **Origin.** Like all blood cells, lymphocytes originate in red bone marrow from hematopoietic stem cells.

② **Maturation.** Lymphocytes are "educated" (go through a rigorous selection process) as they mature. The aim of this education is twofold:

- **Immunocompetence.** Each lymphocyte must become able (competent) to recognize its one specific antigen by binding to it. This ability is called **immunocompetence**. When B or T cells become immunocompetent, they display a unique type of receptor on their surface. These receptors (some 10^5 per cell) enable the lymphocyte to recognize and bind a specific antigen. Once these receptors appear, the lymphocyte is committed to react to one (and only one) distinct antigenic determinant because *all* of its antigen receptors are the same. The receptors on B cells are in fact membrane-bound antibodies. The receptors on T cells are not antibodies but are products of the same gene superfamily and have similar functions.

- **Self-tolerance.** Each lymphocyte must be relatively unresponsive to self-antigens so that it does not attack the body's own cells. This is called **self-tolerance**.

Maturation is a two- to three-day process that occurs in the red bone marrow for B cells and in the thymus for T cells. Recall from Chapter 20 that the lymphoid organs where lymphocytes mature—thymus and red bone marrow—are called **primary lymphoid organs**. All other lymphoid organs are referred to as **secondary lymphoid organs**. We will examine the selection process that underlies lymphocyte maturation on p. 794.

③ **Seeding secondary lymphoid organs and circulation.** Immunocompetent B and T cells that have not yet been exposed to antigen are called **naive**. Naive B cells and T cells are exported from the primary lymphoid organs to seed (colonize) the secondary lymphoid organs—lymph nodes, spleen, and so on—where they are likely to encounter antigens.

Lymphocytes, especially the T cells (which account for 65–85% of bloodborne lymphocytes), circulate continuously throughout the body. Circulating greatly increases a lymphocyte's chance of coming into contact with antigens located in different parts of the body, as well as with other lymphocytes and antigen-presenting cells. For example, a T cell circulates through lymph and blood and back to a lymph node about once a day. Although lymphocyte circulation appears to be random, lymphocyte movement into the tissues is highly specific, regulated by homing signals (CAMs) displayed on vascular endothelial cells.

④ **Antigen encounter and activation.** The first encounter between an immunocompetent but naive lymphocyte and an invading antigen usually takes place in a lymph node or the spleen, but it may happen in any secondary lymphoid organ. Immune cells in lymph nodes are in a strategic position to encounter a large variety of antigens because lymphatic capillaries pick up proteins and pathogens from nearly all body tissues.

When an antigen binds to the particular lymphocyte that has a receptor for it, the antigen selects that lymphocyte for further development. This is called **clonal selection**. If the proper signals (which we describe later) are present, the selected lymphocyte will activate to complete its differentiation.

Adaptive defenses → Humoral immunity
→ Cellular immunity

▢ **Primary lymphoid organs**
(red bone marrow and thymus)

▢ **Secondary lymphoid organs**
(lymph nodes, spleen, etc.)

Red bone marrow

Lymphocyte precursors

① **Origin**
• Both B and T lymphocyte precursors originate in red bone marrow.

Thymus

Red bone marrow

② **Maturation**
• Lymphocyte precursors destined to become T cells migrate (in blood) to the thymus and mature there.
• B cells mature in the bone marrow.
• During maturation lymphocytes develop immunocompetence and self-tolerance.

③ **Seeding secondary lymphoid organs and circulation**
• Immunocompetent but still naive lymphocytes leave the thymus and bone marrow.
• They "seed" the secondary lymphoid organs and circulate through blood and lymph.

Antigen

Lymph node

④ **Antigen encounter and activation**
• When a lymphocyte's antigen receptors bind its antigen, that lymphocyte can be activated.

⑤ **Proliferation and differentiation**
• Activated lymphocytes proliferate (multiply) and then differentiate into effector cells and memory cells.
• Memory cells and effector T cells circulate continuously in the blood and lymph and throughout the secondary lymphoid organs.

Figure 21.8 Lymphocyte development, maturation, and activation.

⑤ **Proliferation and differentiation.** Once activated, the lymphocyte rapidly proliferates (multiplies) to form an army of cells exactly like itself and having the same antigen specificity. This army of identical cells, all descended from the *same* ancestor cell, is called a **clone**.

Most members of the clone become **effector cells,** the cells that actually do the work of fighting infection. A few members of the clone become **memory cells** that are able to respond quickly after any subsequent encounter with the same antigen. B and T memory cells and effector T cells circulate throughout the body on continuous patrol.

How Does Antigen Receptor Diversity in Lymphocytes Come About?

We know that lymphocytes become immunocompetent *before* meeting the antigens they may later attack. *Our genes, not antigens we encounter, determine which specific foreign substances our immune system will be able to recognize and resist.* In other words, the immune cell receptors represent our genetically acquired knowledge of the microbes that are likely to be in our environment. An antigen simply determines which existing T or B cells will proliferate and mount the attack against it.

Only some of the antigens our lymphocytes are programmed to resist will ever invade our bodies. Consequently, only some

21

members of our army of immunocompetent cells are mobilized in our lifetime. The others are forever idle.

Our lymphocytes make up to a billion different types of antigen receptors. These receptors, like all other proteins, are specified by genes, so you might think that an individual must have billions of genes. Not so; each body cell only contains about 20,000 genes that code for all the proteins the cell must make.

How can a limited number of genes generate a seemingly limitless number of different antigen receptors? Molecular genetic studies have shown that the genes that dictate the structure of each antigen receptor are not present as such in lymphocyte stem cells. Instead of a complete set of "antigen receptor genes," stem cells contain a few hundred genetic bits and pieces that can be thought of as a "Lego® set" for antigen receptor genes. As each lymphocyte becomes immunocompetent, these gene segments are shuffled and combined in different ways, a process called *somatic recombination*. The information of the newly assembled genes is then expressed as the surface receptors of B and T cells and as the antibodies later released by the B cell's "offspring."

How Are Lymphocytes Educated during Maturation?

The selection process (education) that lymphocytes undergo during maturation is best understood in T cells. T cell education consists of positive and negative selection in the thymus (**Figure 21.9**).

1. **Positive selection** is the first of two tests a developing T lymphocyte must pass. It ensures that *only* T cells with receptors that are able to recognize self-MHC proteins survive. Remember that T cells cannot bind antigens unless the antigens are presented on self-MHC proteins. T cells that are unable to recognize self-MHC are eliminated by apoptosis.

2. **Negative selection**, the second test, ensures that T cells do not recognize self-antigens displayed on self-MHC. If they do, they are eliminated by apoptosis. Negative selection is the basis for immunological self-tolerance, making sure that T cells don't attack the body's own cells, which would cause autoimmune disorders. Because the self-reactive lymphocyte and all of its potential progeny are eliminated, this is called *clonal deletion*.

This education of T cells is expensive indeed—only about 2% of T cells survive it and continue to become successful immunocompetent, self-tolerant T cells.

Less is known about the factors that control B cell maturation. Like T cells, only immunocompetent B cells are allowed to mature. But selection is not perfect for either B or T cells—some self-reactive lymphocytes survive.

Antigen-Presenting Cells (APCs)

Antigen-presenting cells (APCs) engulf antigens and then present fragments of them, like signal flags, on their own surfaces where T cells can recognize them. Naive T cells can only be activated by antigens that are presented to them on MHC proteins by APCs. In other words, APCs *present antigens* to the cells that will deal with the antigens. The major types of cells acting as APCs are *dendritic cells*, *macrophages*, and *B lymphocytes*.

Adaptive defenses → Cellular immunity

1. Positive Selection

T cells **must** recognize self major histocompatibility proteins (self-MHC)

Antigen-presenting thymic cell · Developing T cell

Self-MHC · Self-antigen · T cell receptor

Failure to recognize self-MHC results in **apoptosis** (death by cell suicide).

Recognizing self-MHC results in survival. Survivors proceed to negative selection.

2. Negative Selection

T cells **must not** recognize self-antigens

Recognizing self-antigen results in **apoptosis**. This eliminates self-reactive T cells that could cause autoimmune diseases.

Failure to recognize (bind tightly to) self-antigen results in survival and continued maturation.

Figure 21.9 T cell education in the thymus.

Dendritic Cells

Dendritic cells are found at the body's frontiers (skin, for example) where they act as mobile sentinels. With their long, wispy extensions, dendritic cells are very efficient antigen catchers (**Figure 21.10**). Once they have internalized antigens by phagocytosis, they enter nearby lymphatics to get to a lymph node where they will present the antigens to T cells.

Migration of dendritic cells to secondary lymphoid organs is now recognized as the most important way of ensuring that lymphocytes encounter invading antigens. Indeed, dendritic cells are the most effective antigen presenter known—it's their only job. This early alert spares the body a good deal of tissue damage that might otherwise occur.

Figure 21.10 Dendritic cell. Scanning electron micrograph (1050×).

Dendritic cells are one of the key links between innate and adaptive immunity. They initiate adaptive immune responses tailored to the type of pathogen they have encountered.

Macrophages

Macrophages are widely distributed throughout the lymphoid organs and connective tissues. Although macrophages, like dendritic cells, can activate naive T cells, macrophages often present antigens to T cells for other reasons—to maintain T cell activation or to be activated themselves. Certain effector T cells release chemicals that prod macrophages to become *activated macrophages*. Activated macrophages are true

"killers"—insatiable phagocytes that also trigger powerful inflammatory responses and recruit additional defenses.

B Lymphocytes

Unlike dendritic cells and macrophages, B cells do not activate naive T cells. Instead, they present antigens to a certain kind of T cell (a helper T cell) in order to obtain "help" in their own activation. We will discuss this more on p. 806.

Check Your Understanding

10. What event (or observation) signals that a B or T cell has achieved immunocompetence?

11. Name three different APCs. Which is most important for T lymphocyte activation?

12. In clonal selection, "who" does the selecting? What is being selected?

13. PREDICT Which of the following T cells would survive education in the thymus? (a) one that recognizes neither MHC nor self-antigen, (b) one that recognizes both MHC and self-antigen, (c) one that recognizes MHC but not self-antigen, (d) one that recognizes self-antigen but not MHC.

For answers, see Answers Appendix.

In summary, the two-fisted adaptive immune system uses lymphocytes, APCs, and specific molecules to identify and destroy all substances—both living and nonliving—that are in the body but not recognized as self. The system's response to such threats depends on the ability of its cells to (1) recognize antigens by binding to them and (2) communicate with one another so that the whole system mounts a response specific to those antigens. As you will see, interactions between various lymphocytes, and between lymphocytes and APCs, underlie virtually all phases of the adaptive immune response.

Table 21.4 outlines major similarities and differences between B and T lymphocytes. This will help orient you as you delve into the details of humoral and cellular immunity in the next two modules.

Table 21.4	Overview of B and T Lymphocytes	
	B LYMPHOCYTES	**T LYMPHOCYTES**
Type of immune response	Humoral	Cellular
Antibody secretion	Yes	No
Primary targets	Extracellular pathogens (e.g., bacteria, fungi, parasites, some viruses in extracellular fluid)	Intracellular pathogens (e.g., virus-infected cells) and cancer cells*
Site of origin	Red bone marrow	Red bone marrow
Site of maturation	Red bone marrow	Thymus
Effector cells	Plasma cells	Cytotoxic T (T_C) cells
		Helper T (T_H) cells
		Regulatory T (T_{Reg}) cells
Memory cell formation	Yes	Yes

*In addition, helper T cells coordinate both B and T lymphocyte responses.

21.5 In humoral immunity, antibodies are produced that target extracellular antigens

Learning Outcomes

▶ Define humoral immunity.

▶ Describe the process of clonal selection of a B cell and recount the roles of plasma cells and memory cells in humoral immunity.

▶ Compare and contrast active and passive humoral immunity.

▶ Describe the structure and functions of antibodies and name the five antibody classes.

Now that you understand the common steps in lymphocyte maturation and activation, let's examine how this basic pattern applies to B lymphocytes. When a B cell encounters its antigen, that antigen provokes the *humoral immune response*, in which antibodies specific for that antigen are made.

Activation and Differentiation of B Cells

An immunocompetent but naive B lymphocyte is *activated* when matching antigens bind to its surface receptors and cross-link adjacent receptors together. Antigen binding is quickly followed by receptor-mediated endocytosis of the cross-linked antigen-receptor complexes. As we described previously, this is called *clonal selection* and is followed by proliferation and differentiation into effector cells (**Figure 21.11**). (As we will see shortly, interactions with T cells are usually required to help B cells achieve full activation.)

Most cells of the clone differentiate into **plasma cells**, the antibody-secreting *effector cells* of the humoral response.

Adaptive defenses → Humoral immunity

Primary response
(initial encounter with antigen)

Antigen

Antigen binds to a receptor on a specific B lymphocyte (clonal selection). (B lymphocytes with noncomplementary receptors remain inactive.)

Activated B cells

Proliferation to form a clone

Plasma cells (effector B cells)

Memory B cell— primed to respond to same antigen

Secreted antibody molecules

Secondary response
(can be years later)

Clone of cells identical to ancestral cells

Subsequent challenge by the same antigen results in a larger, more rapid response.

Plasma cells

Secreted antibody molecules

Memory B cells

Figure 21.11 Clonal selection of a B cell.

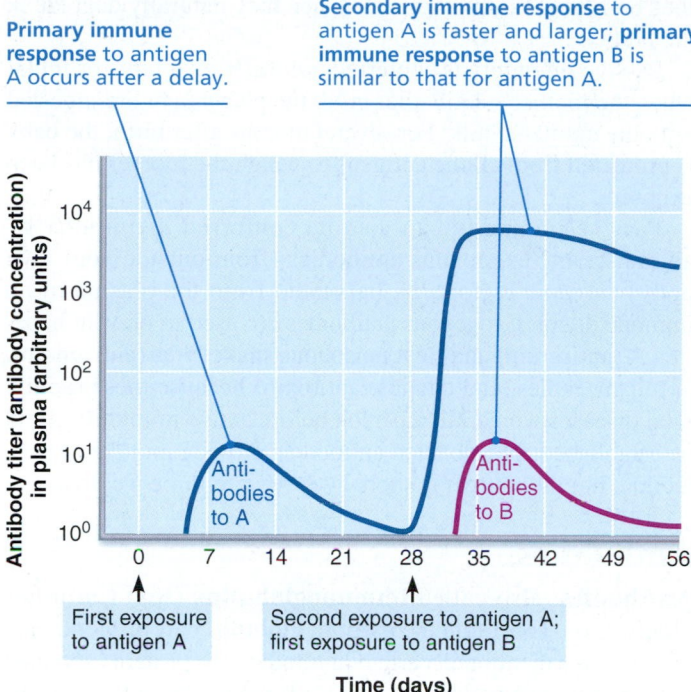

Primary immune response to antigen A occurs after a delay.

Secondary immune response to antigen A is faster and larger; **primary immune response** to antigen B is similar to that for antigen A.

↑ First exposure to antigen A

↑ Second exposure to antigen A; first exposure to antigen B

Time (days)

Figure 21.12 Primary and secondary humoral responses. The primary response to antigen A generates memory cells that give rise to the enhanced secondary response to antigen A. The response to antigen B is independent of the response to antigen A.

Plasma cells develop the elaborate internal machinery (largely rough endoplasmic reticulum) needed to secrete antibodies at the unbelievable rate of about 2000 molecules per second. Each plasma cell functions at this breakneck pace for 4 to 5 days and then dies. The secreted antibodies, each with the same antigen-binding properties as the receptor molecules on the surface of the parent B cell, circulate in the blood or lymph. There they bind to free antigens and mark them for destruction by other innate or adaptive mechanisms.

Clone cells that do not become plasma cells become long-lived **memory cells**. They can mount an almost immediate humoral response if they encounter the same antigen again in the future (Figure 21.11, bottom).

Immunological Memory

The cellular proliferation and differentiation we have just described constitute the **primary immune response**, which occurs on first exposure to a particular antigen. The primary response typically has a lag period of 3 to 6 days after the antigen encounter. This lag period mirrors the time required for the few B cells specific for that antigen to proliferate (about 12 generations) and for their offspring to differentiate into plasma cells. After the mobilization period, plasma antibody levels rise, reach peak levels in about 10 days, and then decline (**Figure 21.12**).

If (and when) someone is reexposed to the same antigen, whether it's the second or twenty-second time, a **secondary immune response** occurs. Secondary immune responses are faster, more prolonged, and more effective, because the immune system has already been primed to the antigen, and sensitized memory cells are already "on alert." These memory cells provide **immunological memory**.

Within hours after recognizing the "old enemy" antigen, a new army of plasma cells is being generated. Within 2 to 3 days the antibody concentration in the plasma, called the *antibody titer*, rises steeply to reach much higher levels than in the primary response. Secondary response antibodies not only bind with greater affinity (more tightly), but their blood levels remain high for weeks to months. (When the appropriate chemical signals are present, plasma cells can keep functioning for much longer than the 4 to 5 days seen in primary responses.) Memory cells persist for long periods and many retain their capacity to produce powerful secondary humoral responses for life.

The same general events occur in the cellular immune response: A primary response sets up a pool of effector cells (in this case, T cells) and generates memory cells that can then mount secondary responses.

Active and Passive Humoral Immunity

When your B cells encounter antigens and produce antibodies against them, you are exhibiting **active humoral immunity**. Active immunity is acquired in two ways (**Figure 21.13**). It is (1) *naturally acquired* when you get a bacterial or viral infection, during which time you may develop symptoms of the disease and suffer a little (or a lot), and (2) *artificially acquired* when you receive a **vaccine**. Indeed, once researchers realized that secondary responses are so much more vigorous than primary responses, the race was on to develop vaccines to "prime" the immune response by providing a first encounter with the antigen.

Most vaccines contain pathogens that are dead or *attenuated* (living, but extremely weakened), or their components. Vaccines provide two benefits:

Figure 21.13 Active and passive humoral immunity. Active immunity establishes immunological memory; passive immunity never does.

- Their weakened antigens provide functional antigenic determinants that are both immunogenic and reactive.

- They spare us most of the symptoms and discomfort of the disease that would otherwise occur during the primary response.

Vaccine *booster shots* are used in some cases to intensify the immune response at later encounters with the same antigen.

Vaccines have wiped out smallpox and have substantially lessened the illness caused by such former childhood killers as whooping cough, polio, and measles. Although vaccines have dramatically reduced hepatitis B, tetanus, influenza, and pneumonia in adults, immunization of adults in the U.S. has a much lower priority than that of children. As a result more than 65,000 Americans die each year from preventable infections.

Conventional vaccines have shortcomings. The biggest shortcoming is that they are not always as effective or long-lasting as we would like. In some individuals, contaminating proteins (for example, egg albumin) cause allergic responses to the vaccine.

Passive humoral immunity differs from active immunity, both in the antibody source and in the degree of protection it provides (Figure 21.13). Instead of being made by your plasma cells, ready-made antibodies are introduced into your body. As a result, your B cells are not challenged by antigens, immunological memory does not occur, and the protection provided by the "borrowed" antibodies ends when they naturally degrade in the body.

Passive immunity is conferred *naturally* on a fetus or infant when the mother's antibodies cross the placenta or are ingested with the mother's milk. For several months after birth, the baby is protected from all the antigens to which the mother has been exposed.

Passive immunity can also be conferred *artificially* by administering exogenous antibodies (from outside your own body) as *gamma globulin*, harvested from the plasma of an immune donor. Exogenous antibodies are used to prevent hepatitis A (antiserum) and treat poisonous snake bites (antivenom), botulism, rabies, and tetanus (antitoxin) because these rapidly fatal diseases would kill a person before active immunity could be established. The donated antibodies provide immediate protection, but their effect is short-lived (two to three weeks).

Antibodies

Antibodies, also called **immunoglobulins (Igs)** (im″u-no-glob′u-linz), constitute the **gamma globulin** part of blood proteins. As we mentioned earlier, antibodies are proteins secreted in response to an antigen by effector B cells called plasma cells, and the antibodies bind specifically with that antigen. They are formed in response to an incredible number of different antigens. Despite their variety, all antibodies can be grouped into

(a)

Heavy chain variable region
Heavy chain constant region
Light chain constant region
Light chain variable region
Disulfide bond

(b)

Figure 21.14 Antibody structure.
(a) Schematic antibody structure (based on IgG) consists of four polypeptides—two short *light chains* and two long *heavy chains* joined together by covalent bonds between sulfur atoms called disulfide bonds (S–S). Each chain has a variable (V) region (which differs in antibodies made by different cells) and a constant (C) region (essentially identical in different antibodies of the same class). Together, the variable regions form the antigen-binding sites—two per antibody monomer. **(b)** Computer-generated image of antibody structure.

one of five Ig classes. Before seeing how these Ig classes differ from one another, let's look at how all antibodies are alike.

Basic Antibody Structure

Regardless of its class, each antibody consists of four looping polypeptide chains linked together by disulfide (sulfur-to-sulfur) bonds. The four chains combined form a molecule, called an **antibody monomer** (mon'o-mer), with two identical halves. The molecule as a whole is T or Y shaped (**Figure 21.14**).

Two of the chains, called the **heavy (H) chains**, are identical to each other (blue chains in Figure 21.14a). The other two chains, called the **light (L) chains** (pink), are also identical to each other, but they are only about half as long as each H chain. The heavy chains have a flexible *hinge* region at their approximate "middles." The "loops" on each chain are created by disulfide bonds that cause the intervening parts of the polypeptide chains to loop out.

Each chain forming an antibody has a **variable (V) region** at one end and a **constant (C) region** at the other end. Antibodies responding to different antigens have very different V regions, but their C regions are the same (or nearly so) in all antibodies of a given class. In each arm of the monomer, the V regions of the heavy and light chains combine to form an **antigen-binding site** shaped to "fit" a specific antigenic determinant. Consequently, each antibody monomer has two such antigen-binding regions.

The C regions that form the *stem* of the antibody monomer determine the antibody class and serve common functions in all antibodies: These are the *effector regions* of the antibody that dictate (1) the cells and chemicals of the body the antibody can bind to, and (2) how the antibody class functions to eliminate antigens. For example, some antibodies activate complement, some circulate in blood while others are found primarily in body secretions, some cross the placental barrier, and so on.

Antibody Classes

The five major immunoglobulin classes are designated IgM, IgA, IgD, IgG, and IgE, on the basis of the C regions in their heavy chains. (Remember the name MADGE to recall the five Ig types.)

The antibodies of each class have different characteristics, biological roles, and locations in the body, as shown in **Table 21.5**. IgM in plasma is huge compared to the other antibodies. It is constructed from five Y-shaped units, or *monomers*, linked together to form a *pentamer* (*penta* = five). IgA occurs in both monomer and *dimer* (two linked monomers) forms. IgD, IgG, and IgE are monomers and have the same basic Y-shaped structure.

B cells usually switch from making one class of antibody to another as they become plasma cells. As a result, two or more different antibody classes having the same antigen specificity are produced. For example, the first antibody released in the primary response is IgM, and then later plasma cells begin to secrete IgG. During secondary responses, almost all of the Ig protein is IgG. Switching from IgM to IgA or IgE also occurs.

Table 21.5 Immunoglobulin Classes*	
IgM (pentamer)	• The first immunoglobulin class secreted by plasma cells during the primary response. (This fact is diagnostically useful because presence of IgM in plasma usually indicates current infection by the pathogen eliciting IgM's formation.) • Readily activates complement. • Exists in monomer and pentamer (five united monomers) forms. • The monomer serves as an antigen receptor on the B cell surface. • The pentamer circulates in blood plasma. • Numerous antigen-binding sites make it a potent agglutinating agent.
IgA (dimer)	• The dimer, referred to as **secretory IgA**, is found in body secretions such as saliva, sweat, intestinal juice, and milk. • Secretory IgA helps stop pathogens from attaching to epithelial cell surfaces (including mucous membranes and the epidermis). • The monomer exists in limited amounts in plasma.
IgD (monomer)	• Found on the B cell surface. • Functions as a B cell antigen receptor (as does IgM).
IgG (monomer)	• The most abundant antibody in plasma, accounting for 75–85% of circulating antibodies. • The main antibody of both secondary and late primary responses. • Readily activates complement. • Protects against bacteria, viruses, and toxins circulating in blood and lymph. • Crosses the placenta and confers passive immunity from the mother to the fetus.
IgE (monomer)	• Stem end binds to mast cells or basophils. Antigen binding to its receptor end triggers these cells to release histamine and other chemicals that mediate inflammation and an allergic reaction. • Secreted by plasma cells in skin, mucosae of the gastrointestinal and respiratory tracts, and tonsils. • Only traces of IgE are found in plasma. • Levels rise during severe allergic attacks or chronic parasitic infections of the gastrointestinal tract.

*Key characteristics are listed in blue type.

21

Antibody Targets and Functions

Though antibodies themselves cannot destroy antigens, they can inactivate antigens and tag them for destruction (**Figure 21.15**). The common event in all antibody-antigen interactions is formation of **antigen-antibody** (or **immune**) **complexes**. Defensive mechanisms used by antibodies include neutralization, agglutination, precipitation, and complement activation, with the first two most important.

Neutralization Neutralization, the simplest defensive mechanism, occurs when antibodies block specific sites on viruses or bacterial exotoxins (toxic chemicals secreted by bacteria). As a result, the virus or exotoxin cannot bind to receptors on tissue cells. Phagocytes eventually destroy the antigen-antibody complexes.

Agglutination Because antibodies have more than one antigen-binding site, they can bind to the same determinant on more than one antigen at a time. Consequently,

antigen-antibody complexes can be cross-linked into large lattices. When cell-bound antigens are cross-linked, the process causes clumping, or **agglutination**, of the foreign cells. IgM, with 10 antigen-binding sites, is an especially potent agglutinating agent (see Table 21.5). Recall from Chapter 17 that agglutination occurs when mismatched blood is transfused (the foreign red blood cells clump) and is the basis of tests used for blood typing.

Precipitation In **precipitation**, soluble molecules (instead of cells) are cross-linked into large complexes that settle out of solution. Like agglutinated bacteria, precipitated antigen molecules are much easier for phagocytes to capture and engulf than are freely moving antigens.

Complement Activation Complement activation is the chief antibody defense used against cellular antigens, such as bacteria or mismatched red blood cells. When several antibodies bind close together on the same cell, the complement-binding sites

Figure 21.15 Mechanisms of antibody action. Antibodies act against free viruses, red blood cell antigens, bacterial toxins, intact bacteria, fungi, and parasitic worms.

on their stem regions align and complement is activated (◀ **p. 788**). Membrane attack complexes may insert into the antigenic cell's surface, triggering cell lysis.

Additionally, as we described earlier, molecules released during complement activation tremendously amplify the inflammatory response and promote phagocytosis by opsonizing pathogens. This sets into motion a positive feedback cycle that enlists more and more defensive elements.

A quick and dirty way to remember how antibodies work is to remember they have a PLAN of action—**p**recipitation, **l**ysis (by complement), **a**gglutination, and **n**eutralization.

HOMEOSTATIC IMBALANCE 21.2
CLINICAL

Around the world, billions of people are infected by parasitic worms such as *Ascaris* and *Schistosoma*. These large pathogens are difficult for our immune systems to deal with and "PLAN" is insufficient.

Nevertheless, antibodies still play a critical role in the worm's destruction. IgE antibodies coat the surface of parasitic worms, marking them for destruction by eosinophils (◀ **p. 652**). When eosinophils encounter antibody-coated worms, they bind to the exposed stems of the IgE. This triggers the eosinophils to release the toxic contents of their large cytoplasmic granules all over their prey.

Monoclonal Antibodies as Clinical and Research Tools

Monoclonal antibodies are pure antibody preparations specific for a single antigenic determinant. They are produced by descendants of a single cell. Commercially prepared monoclonal antibodies are essential in research, clinical testing, and treatment.

Monoclonal antibodies are used to diagnose pregnancy, certain sexually transmitted infections, some cancers, hepatitis, and rabies. These monoclonal antibody tests are more specific, sensitive, and rapid than other tests. Monoclonal antibodies are also used to treat leukemia and lymphomas, cancers that are present in the circulation and so are easily accessible to injected antibodies. They also serve as "guided missiles" to deliver anticancer drugs only to cancerous tissue, and to treat certain autoimmune diseases (as we will discuss later).

Summary of Antibody Actions

At the most basic level, the race between antibody production and pathogen multiplication determines whether or not you become sick. Remember, however, that forming antigen-antibody complexes does *not* destroy the antigens. Instead, it prepares them for destruction by innate defenses.

Antibodies produced by plasma cells are in many ways the simplest, most versatile ammunition of the immune response. Nevertheless, they provide only partial immunity. Their prey are *extracellular pathogens*—intact bacteria, free viruses, and soluble foreign molecules—in other words, pathogens that are free in body secretions and tissue fluid and circulating in blood and lymph. Antibodies never invade solid tissues unless inflammation is present.

Until recently, the accepted dogma was that antibodies *only* act extracellularly. Remarkably, we now know that antibodies can act intracellularly as well. Antibodies that are attached to a virus before that virus infects a cell can "hang on" to the virus as it slips inside the cell. There, the antibodies activate intracellular mechanisms that destroy the virus. Even so, antibodies are not very effective against pathogens like viruses and tuberculosis bacilli that quickly slip inside body cells to multiply there. For these intracellular pathogens, the cellular arm of adaptive immunity comes into play.

Check Your Understanding

14. Why is the secondary response to an antigen so much faster than the primary response?
15. Which class of antibody is shown at right? Which is most abundant in blood? Which is secreted first in a primary immune response? Which is most abundant in secretions?
16. List four ways in which antibodies can bring about destruction of a pathogen.
17. **APPLY** How do vaccines protect against common childhood illnesses such as chicken pox, measles, and mumps?
18. **DRAW** Draw an antibody. Label the antigen-binding sites, and the hinge and stem regions. Which region determines the antibody's class?
19. **MAKE CONNECTIONS** What is the function of the abundant endoplasmic reticulum in plasma cells? What other organelle(s) (described in Chapter 3) would be especially abundant in plasma cells? Why?

For answers, see Answers Appendix.

21.6 Cellular immunity consists of T lymphocytes that direct adaptive immunity or attack cellular targets

Learning Outcomes
▶ Define cellular immunity and describe the process of activation and clonal selection of T cells.
▶ Describe the roles of different types of T cells.
▶ Describe T cell functions in the body.

T cells are best suited for cell-to-cell interactions. When antigens are presented to a T lymphocyte, they provoke a cellular immune response. Some activated T cells directly kill body cells that are:

- Infected by viruses or bacteria
- Cancerous or abnormal
- Foreign cells (e.g., transplanted cells)

Other T cells release chemicals that regulate the immune response.

T cells are a diverse lot, much more complex than B cells in both classification and function. There are two major populations of T cells based on which of two structurally related *cell differentiation glycoproteins*—CD4 or CD8—a mature T cell displays. The CD4 and CD8 glycoproteins are surface receptors but

Adaptive defenses → Cellular immunity

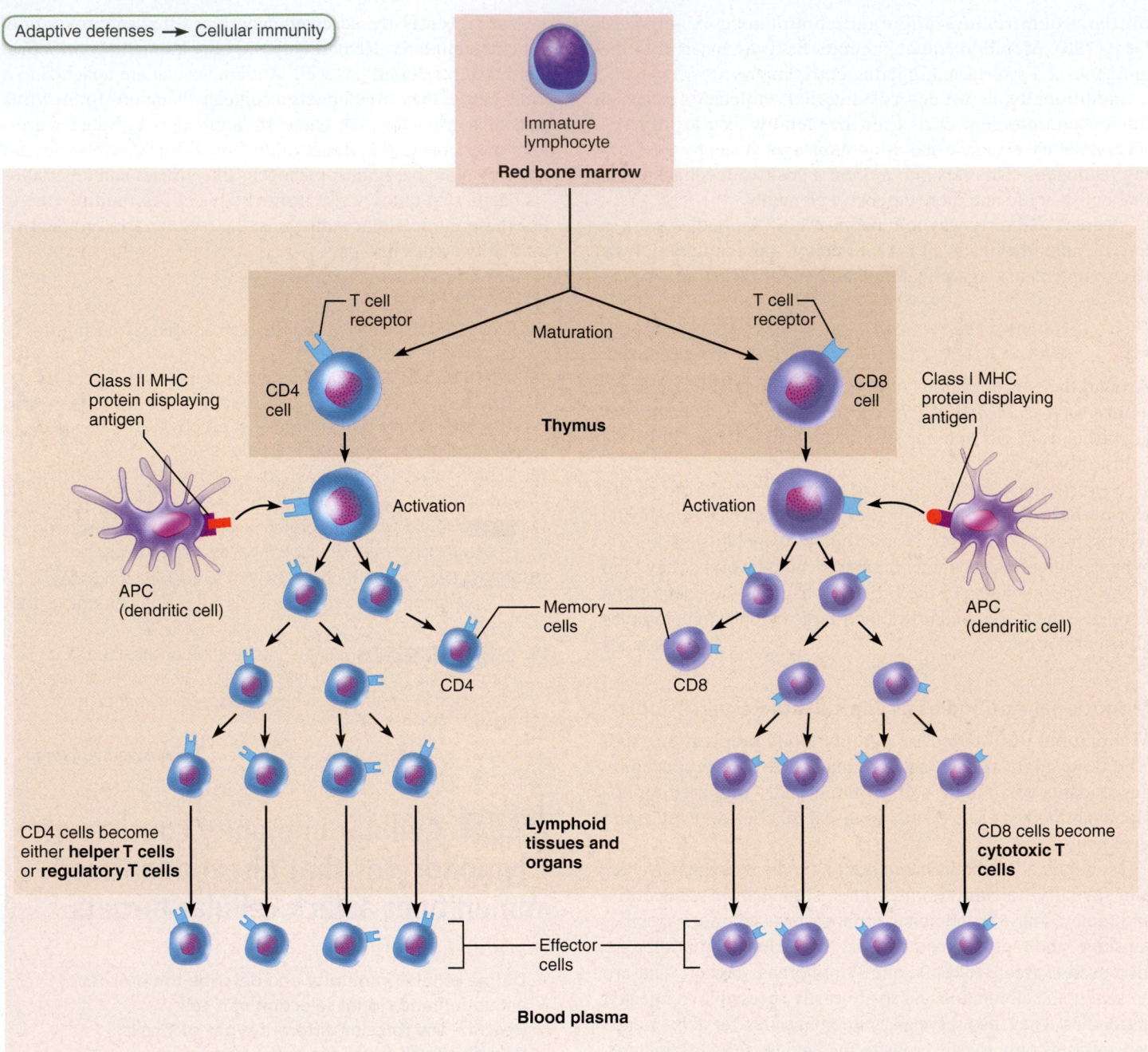

Figure 21.16 Major types of T cells.

are distinct from the T cell antigen receptors. They play a role in interactions between T cells and other cells.

When activated, CD4 and CD8 cells differentiate into the three major kinds of *effector cells* of cellular immunity (**Figure 21.16**).

- **CD4 cells** usually become *helper T (T_H) cells* that help activate B cells, other T cells, and macrophages, and direct the adaptive immune response.
- **CD8 cells** become *cytotoxic T (T_C) cells* that destroy cells in the body that harbor anything foreign.
- Some CD4 cells become *regulatory T (T_{Reg}) cells*, which moderate the immune response.

Activated CD4 and CD8 cells can also become memory T cells. Note that the names of the effector cells (helper, cytotoxic, regulatory) are reserved for *activated* T cells, while naive T cells are simply called CD4 or CD8 cells.

MHC Proteins and Antigen Presentation

Unlike B cells and antibodies, T cells cannot "see" either free antigens or antigens that are in their natural state. T cells can recognize and respond only to *processed* fragments of protein antigens displayed on surfaces of body cells (APCs and others).

Antigen presentation is necessary for both activation of naive T cells and the normal functioning of effector T cells.

21

As previously noted, the cell surface proteins on which antigens are presented to T cells are the major histocompatibility complex (MHC) proteins. It is important to understand how these critical players in antigen presentation work before we can understand how T cells work. There are two classes of MHC proteins—class I and class II—summarized in **Table 21.6**.

Class I MHC Proteins

Class I MHC proteins are found on the surface of virtually *all* body cells except red blood cells. Each class I MHC protein has a groove that holds an antigen—in this case, a protein fragment (of 8 or 9 amino acids).

Where do these protein fragments come from? All antigens displayed on class I MHC proteins are **endogenous antigens**—fragments of proteins synthesized *inside the cell*. In a healthy cell, endogenous antigens are all self-antigens, generally bits of digested cellular proteins. But in an infected cell, endogenous antigens may also include fragments of foreign antigens that are synthesized within the infected cell but "belong to" the pathogen. In a cancerous cell, endogenous antigens can include altered (cancer) proteins.

How do these endogenous antigens end up attached to class I MHCs on the outside of our cells? As proteases (protein-digesting enzymes) degrade cytoplasmic proteins as part of their natural recycling, a random sample of the resulting protein fragments is transported into the endoplasmic reticulum. Inside the ER, these peptides bind to newly synthesized class I MHC proteins. Transport vesicles then export the "loaded" class I MHC proteins to the cell surface.

Class I MHC proteins are crucial for both activating naive CD8 cells and "informing" cytotoxic T cells that infectious microorganisms are hiding in body cells. Without them, viruses and certain bacteria that thrive in cells could multiply unnoticed and unbothered.

When class I MHC proteins display fragments of our own proteins (self-antigens), cytotoxic T cells passing by get the signal "Leave this cell alone, it's ours!" and ignore them. But when class I MHC proteins display foreign antigens, they "sound a molecular alarm" that signals invasion. In this signaling, the class I MHC proteins both (1) act as antigen holders and (2) form the self part of the self-nonself complexes that cytotoxic T cells must recognize in order to kill.

Class II MHC Proteins

This second type of MHC protein is less widespread. **Class II MHC proteins** are typically found *only on the surfaces of cells that present antigens to CD4 cells*: dendritic cells, macrophages, and B cells. Like their class I MHC counterparts, class II MHC proteins are synthesized at the ER and bind to peptide fragments. However, the peptides they bind are longer (14–17 amino acids) and come from **exogenous antigens**—antigens from *outside* the cell that have been engulfed by the cell that displays them.

The engulfed exogenous antigens are broken down by proteases inside a phagolysosome. Vesicles from the ER containing class II MHC proteins fuse with the phagolysosome, and the antigen fragments bind to the groove of the MHC proteins. The vesicle is then exported to the cell surface, where the class II MHC protein displays its prize for CD4 cells to recognize.

Class II MHC proteins loaded with foreign antigens generally signal CD4 cells that help is required. It's as if they are showing the CD4 cells what the foreign invader "looks like." Most

Table 21.6	Role of MHC Proteins in Cellular Immunity	
	CLASS I MHC PROTEINS	**CLASS II MHC PROTEINS**
Displayed by	All nucleated cells	APCs (dendritic cells, macrophages, B cells)
Recognized by	Naive CD8 cells and cytotoxic T cells	Naive CD4 cells and helper T cells
Foreign antigens on MHC are	Endogenous (intracellular pathogens or proteins made by cancerous cells)*	Exogenous (phagocytized extracellular pathogens)
Cells displaying foreign antigens on MHC send this message	If the cell is an APC: "I belong to self, but have captured a foreign invader. This is what it looks like. Kill any cell that displays it." If the cell is not an APC: "I belong to self, but have been invaded or become cancerous. Kill me!"	"I belong to self, but have captured a foreign invader. This is what it looks like. Help me mount a defense against it."

* Dendritic cells are an exception because they can present *another cell's* endogenous antigens on their class I MHC proteins to activate CD8 cells.

often, the result is that the CD4 cell will help mount a defense against the antigen.

MHC Restriction

CD4 and CD8 cells have different requirements for the class of MHC protein that presents antigens to them. This constraint, acquired during the education process in the thymus, is called *MHC restriction*.

- CD4 cells (which usually become helper T cells) are restricted to binding antigens only on class II MHC proteins, which are typically displayed on antigen-presenting cell (APC) surfaces.

- CD8 cells (which become cytotoxic T cells) are activated by antigen fragments on class I MHC proteins, also found on the surface of APCs. Once activated, cytotoxic T cells look for this same antigen presented on class I MHC proteins located on any cell in the body.

The MHC restriction for CD8 cells presents a problem for APCs. As a rule, class I MHC proteins display endogenous antigens—those that originate inside that cell. How do APCs obtain endogenous antigens from another cell and display them on class I MHCs to activate CD8 cells? Dendritic cells have the special ability to do this. They obtain other cells' endogenous antigens either by engulfing dying virus-infected or tumor cells, or by importing antigens through temporary gap junctions with infected cells. Dendritic cells then display these antigens on *both* class I and class II MHCs.

Activation and Differentiation of T Cells

T cells can *only* be activated by APCs and this activation is actually a two-step process involving antigen binding and co-stimulation. Both steps usually occur on the surface of the same APC, and both steps are required for *clonal selection*.

Antigen Binding

This first step mostly entails what we have already described: T cell antigen receptors (**TCRs**) bind to an antigen-MHC complex on the surface of an APC. As you will recall from our discussion of T cell education, TCRs must perform *double recognition*: They must recognize both MHC (self-antigen) and the foreign antigen it displays.

The TCR that recognizes (by binding) the nonself-self complex triggers multiple intracellular signaling pathways that lead to T cell activation. Other T cell surface proteins also play a role in T cell activation. For example, the CD4 and CD8 proteins that distinguish the two major T cell groups are adhesion molecules that help bind cells together during antigen recognition.

Co-stimulation

The story isn't over yet because step 2 comes next. Once antigen binding has occurred, the T cell is stimulated but is still "idling," like a car that has been started but not put into gear. Before an "idling" T cell can be "put into gear" and proliferate to form a clone, it must also bind one or more **co-stimulatory**

signals (**Figure 21.17** ②ⓑ). These signals are yet other molecules that appear on the surfaces of APCs in tissues that are damaged or invaded by pathogens. For example, dendritic cells and macrophages sprout co-stimulatory molecules on their surfaces when the innate defenses are being mobilized. The binding of these molecules to specific receptors on a T cell is a crucial co-stimulatory signal.

What happens if a T cell binds to antigen without receiving the co-stimulatory signal? In this case, the T cell becomes tolerant to that antigen and is unable to respond. This state of unresponsiveness to antigen is called **anergy**.

The two-signal sequence acts as a kind of "double-handshake," a safeguard to prevent the immune system from destroying healthy cells. Without this safeguard, class I MHC proteins, which occur on all body cells and which display peptides from within the cell, could activate cytotoxic T cells, leading to widespread damage of healthy cells.

The important thing to understand is that along with antigen binding, co-stimulation is crucial for T cell activation. To

Adaptive defenses → Cellular immunity

Bacterial antigen

Class II MHC protein displaying processed bacterial antigen

Dendritic cell

Co-stimulatory molecule

① **Antigen presentation** Dendritic cell engulfs an exogenous antigen, processes it, and displays its fragments on class II MHC protein.

CD4 protein

T cell receptor (TCR)

Co-stimulatory molecule receptor

CD4 T cell

② **Double recognition**

②ⓐ CD4 T cell recognizes antigen-MHC complex. Both TCR and CD4 proteins bind to antigen-MHC complex.

②ⓑ Co-stimulatory molecules bind their receptors.

Clone formation

③ **Clone formation** Activated CD4 T cells proliferate (clone), and become memory and effector cells.

Memory CD4 T cell

Helper T cells

Figure 21.17 Clonal selection of T cells involves simultaneous recognition of self and nonself. Activation of CD4 cells is shown here, but activation of CD8 cells is similar.

Table 21.7	Selected Cytokines
CYTOKINE	**FUNCTION IN IMMUNE RESPONSE**
Interferons (IFNs)	
• Alpha (α) and beta (β)	Secreted by many cells. Have antiviral effects; activate NK cells.
• Gamma (γ)	Secreted by lymphocytes. Activates macrophages; stimulates synthesis and expression of more class I and II MHC proteins; promotes differentiation of T_H cells into T_H1.
Interleukins (ILs)	
• IL-1	Secreted by activated macrophages. Promotes inflammation and T cell activation; causes fever (acts as a pyrogen that resets the thermostat of the hypothalamus).
• IL-2	Secreted by T_H cells. Stimulates T and B cell proliferation, T_{Reg} cell development, and NK cell activation.
• IL-4	Secreted by some T_H cells. Promotes differentiation to T_H2; promotes B cell activation; switches antibody production to IgE.
• IL-5	Secreted by some T_H cells and mast cells. Attracts and activates eosinophils; causes plasma cells to secrete IgA antibodies.
• IL-10	Secreted by macrophages, T_H, and T_{Reg} cells. Inhibits macrophages and dendritic cells; turns down cellular and innate immune response.
• IL-12	Secreted by dendritic cells and macrophages. Stimulates T_C and NK cell activity; promotes T_H1 differentiation.
• IL-17	Secreted by T_H17 cells. Important in innate and adaptive immunity and recruiting neutrophils. Involved in inflammation in many autoimmune diseases.
Suppressor factors	A generic term for a number of cytokines that suppress the immune system, for example TGF-β and IL-10.
Transforming growth factor beta (TGF-β)	A suppressor factor similar to IL-10; stimulates T_{Reg} and T_H17 cell development.
Tumor necrosis factors (TNFs)	Produced by lymphocytes and in large amounts by macrophages. Promote inflammation; enhance phagocyte chemotaxis and nonspecific killing; slow tumor growth by selectively damaging tumor blood vessels; promote cell death by apoptosis.

go back to our idling car analogy, the car will not go anywhere unless it has been both (1) started and (2) put into gear.

Proliferation and Differentiation

Once activated by antigen binding and co-stimulation, a T cell enlarges and proliferates. Cytokines (which we will discuss next) released by APCs or T cells themselves promote this process. Cells of the resulting clone differentiate to perform functions according to their T cell class. This primary response peaks within a week of exposure to the triggering antigen. A period of apoptosis then occurs between days 7 and 30, during which time the activated T cells die off and effector activity wanes as the amount of antigen declines.

This wholesale disposal of T cells has a critical protective role because activated T cells that are no longer needed are potential hazards. They produce huge amounts of inflammatory cytokines, which contribute to increased cell division, and may promote cancer in chronically inflamed tissue. Thousands of clone members become memory T cells, persisting perhaps for life, and providing a reservoir of T cells that can later mediate secondary responses to the same antigen.

Cytokines

The chemical messengers involved in cellular immunity belong to a group of molecules called **cytokines**, a general term for mediators that influence cell development, differentiation, and responses in the immune system. Cytokines include interferons and interleukins. In **Table 21.7**, you can see some of the large variety of these molecules and their myriad effects on target cells.

Cytokines include hormone-like or paracrine-like glycoproteins released by a variety of cells. Some cytokines promote T cell proliferation. For example, **interleukin 1 (IL-1)**, released by macrophages, stimulates T cells to liberate **interleukin 2 (IL-2)** and to synthesize more IL-2 receptors. IL-2 is a key growth factor. Acting on the cells that release it (as well as other T cells), it sets up a positive feedback cycle that encourages activated T cells to divide even more rapidly.

Additionally, all activated T cells secrete one or more other cytokines that help amplify and regulate a variety of adaptive and innate immune responses. For example, some are inflammatory chemicals and others (such as *gamma interferon*) enhance the killing power of macrophages.

Roles of Specific Effector T Cells

Activated T cells (like B cells) can become either effector cells or memory cells. Next, we will focus on the three major groups of effector T cells: helper, cytotoxic, and regulatory T cells.

21

Figure 21.18 The central role of helper T cells in mobilizing both humoral and cellular immunity. (a) T$_H$ and B cells usually must interact directly for full B cell activation.

(b) In response to T$_H$ cell binding, dendritic cells express co-stimulatory molecules required to activate CD8 T cells. (Some types of antigens induce these co-stimulatory molecules themselves, in which case T$_H$ cell help may not be needed.) The T$_H$ cell also produces interleukin 2, which causes the CD8 cell to proliferate and differentiate into a T$_C$ cell.

Helper T Cells

Helper T (T$_H$) cells play a central role in adaptive immunity, mobilizing both its humoral and cellular arms (**Figure 21.18**). Once activated by APC presentation of antigen, T$_H$ cells help activate B and T cells, and induce B and T cells to proliferate. In fact, without the help of "director" T$_H$ cells, there is *no* adaptive immune response. Their cytokines furnish the chemical help needed to recruit other immune cells. The crucial role of T$_H$ cells in immunity is painfully evident when they are destroyed, as in AIDS (see p. 811).

Activation of B Cells   Helper T cells interact directly with B cells displaying antigen fragments bound to class II MHC receptors (Figure 21.18a). Whenever a T$_H$ cell binds to a B cell, the T$_H$ cell releases cytokines that prod the B cells into dividing more rapidly. Then, like the boss of an assembly line, the T$_H$ cell signals for antibody formation to begin. B cells continue to divide as long as T$_H$ stimulation continues. In this way, helper T cells activate B cells to help unleash their protective potential.

Some B cells may be activated solely by binding to certain antigens called **T cell–independent antigens**. However, T cell–independent antigen responses tend to be weak and short-lived. Most antigens are **T cell–dependent antigens** that require T cell help to activate the B cells to which they bind.

Activation of CD8 Cells   Like B cells, CD8 cells usually require help from T$_H$ cells to activate into destructive cytotoxic T cells. As shown in Figure 21.18b, T$_H$ cells cause dendritic

cells to express on their surfaces the co-stimulatory molecules required to activate CD8 cells.

Amplification of Innate Defenses   T$_H$ cells also amplify the responses of the innate immune system. For example, they activate macrophages to become more potent killers. The cytokines released by T$_H$ cells not only mobilize lymphocytes and macrophages but also attract other types of white blood cells into the area. As the released chemicals summon more and more cells into the battle, the immune response gains momentum, and the sheer number of immune elements overwhelms the antigens.

Subsets of T$_H$ Cells   Various subsets of helper T cells exist. Which subset develops during T$_H$ cell differentiation depends on the type of antigen and the site at which it is encountered as well as the cytokine exposure of the differentiating T$_H$ cell.

- *T$_H$1 cells* stimulate inflammation, activate macrophages, and promote differentiation of cytotoxic T cells. They mediate most aspects of cellular immunity.

- *T$_H$2 cells* mainly defend against parasitic worms. They mobilize eosinophils and activate immune responses that depend on B cells and antibody formation. They are also the cells that promote allergies.

- *T$_H$17 cells* link together adaptive and innate immunity by releasing IL-17, which promotes inflammatory responses against extracellular microbes and may underlie most autoimmune diseases.

Adaptive defenses → Cellular immunity

Cytotoxic T cell (T$_C$)

① T$_C$ identifies foreign antigens on MHC I proteins and binds tightly to target cell.

② T$_C$ releases **perforin** and **granzyme** molecules from its granules by exocytosis.

③ Perforin molecules insert into the target cell membrane, polymerize, and form transmembrane pores (cylindrical holes) similar to those produced by complement activation.

Granule

Perforin

T$_C$ cell membrane

Target cell membrane

Target cell

Perforin pore

Granzymes

⑤ The T$_C$ detaches and searches for another prey.

④ Granzymes enter the target cell via the pores. Once inside, granzymes activate enzymes that trigger apoptosis.

Cytotoxic T cell

Cancer cell

(a) A mechanism of target cell killing by T$_C$ cells.

(b) Scanning electron micrograph of a T$_C$ cell killing a cancer cell (2100×).

Figure 21.19 Cytotoxic T cells attack infected and cancerous cells. MHC proteins and T cell receptors not shown.

Cytotoxic T Cells

Cytotoxic T (T$_C$) cells (activated CD8 cells) are T cells that can directly attack and kill other cells. T$_C$ cells roam the body, circulating in and out of the blood and lymph and through lymphoid organs in search of body cells displaying antigens that the T$_C$ cells recognize. Their main targets are virus-infected cells, but they also attack tissue cells infected by certain intracellular bacteria or parasites, cancer cells, and foreign cells introduced into the body by blood transfusions or organ transplants.

Before the onslaught can begin, the cytotoxic T cell must "dock" on the target cell by binding to a self-nonself complex. Remember, all body cells display class I MHC antigens, so T$_C$ cells can destroy all infected or abnormal body cells. The attack on foreign human cells, such as those of a graft, is more difficult to explain because here *all* of the antigens are nonself. However, apparently the T$_C$ cells sometimes "see" the foreign class I MHC antigens as a combination of self class I MHC protein bound to foreign antigen.

Once cytotoxic T cells recognize their targets, how do they deliver a *lethal hit*? There are two major mechanisms. One involves **perforins** and **granzymes** (**Figure 21.19**). The other involves binding to a specific membrane receptor on the target cell that stimulates the target cell to undergo apoptosis.

NK cells, introduced earlier, use the same key mechanisms to kill their target cells. NK cells, however, do not look for foreign antigen displayed on class I MHC proteins. Instead they search for other signs of abnormality, including the *lack* of class I MHC or the presence of antibodies coating the target cell. Stressed cells also often express different surface markers, which can activate NK cells. In short, NK cells stalk abnormal or foreign cells in the body that T$_C$ cells can't "see."

NK cells and T$_C$ lymphocytes roam the body, adhering to and crawling over the surfaces of other cells, examining them for markers they might recognize, a process called **immune surveillance**. NK cells check to make sure each cell has "identity flags" (class I MHC proteins *inhibit* NK cell attack), whereas T$_C$ cells check the "identity flags" to see if they look the way they are supposed to (foreign antigens *stimulate* T$_C$ cell attack).

Regulatory T Cells

While T$_H$ cells help activate adaptive immune responses, related T cells called **regulatory T (T$_{Reg}$) cells** dampen the immune response. They act either by direct contact or by releasing inhibitory cytokines.

T$_{Reg}$ cells are important in preventing autoimmune reactions because they suppress self-reactive lymphocytes in the

21

The immune system has three lines of defense against foreign invaders.

Let's look at how the body defends against its first exposure to influenza virus.

First Line of Defense: Surface Barriers

The mucous membrane lining the airways helps prevent the virus from entering the body.

Mucus traps viruses.

Cilia sweep contaminated mucus toward the pharynx, where it is swallowed and digested.

Cilia

Mucus

Virus

Bronchiole

Second Line of Defense: Innate Internal Defenses

Virus-infected cells release **interferon** to warn nearby uninfected cells.

Phagocytes (e.g., resident macrophages) engulf viruses and "sound the alarm" by releasing inflammatory chemicals.

Inflammation brings more immune cells and plasma proteins to the area by dilating arterioles and increasing capillary permeability.

Complement activation enhances inflammation and causes opsonization of virus particles.

NK cells recognize and kill virus-infected cells.

Inflammatory chemicals

↑Capillary permeability

Arteriole dilates

Virus

C3b

Perforins

Third Line of Defense: Adaptive Defenses

① Dendritic cells (APCs) engulf viruses and dead virus-infected cells.

② Dendritic cells migrate to a lymph node where they activate T lymphocytes.

Lymph node

③ Dendritic cell activates CD4 cell, which forms a clone of helper T (T_H) cells and memory cells.

CD4 cell

Dendritic cell

Memory CD4 T cell

T_H cell

T_H cells

T_H cell

B cell

CD8 cell

④a Dendritic cell and T_H cell together activate CD8 cell, which forms a clone of cytotoxic T (T_C) cells and memory cells.

④b T_H cell activates a B cell, which forms a clone of plasma cells and memory B cells.

Memory CD8 T cell

Memory B cell

Cellular immunity (via T_C cells) targets intracellular viruses by killing infected cells.

T_C cells

Plasma cells

⑤a T_C cells migrate to site of infection. They attack and kill the infected cells.

Perforins

Antibodies

Humoral immunity targets extracellular (free) viruses.

⑤b Plasma cells release antibodies, which travel (via blood) to site of infection.

Antibodies **neutralize** viruses by covering the proteins that allow the virus to bind to and enter our cells. No entry = no infection.

Agglutination and **complement activation** enhance phagocytosis by neutrophils and macrophages.

809

Table 21.8 Cells and Molecules of the Adaptive Immune Response

ELEMENT	FUNCTION IN IMMUNE RESPONSE
Cells	
B cell	Lymphocyte that matures in bone marrow. Its progeny (clone members) form plasma cells and memory cells.
Plasma cell	Antibody-producing "machine"; produces huge numbers of antibodies. An effector B cell.
T cells	
• Helper T (T_H) cell	An effector CD4 T cell central to both humoral and cellular immunity. It stimulates production of cytotoxic T cells and plasma cells, activates macrophages, and acts both directly and indirectly by releasing cytokines.
• Cytotoxic T (T_C) cell	An effector CD8 T cell that kills invaded body cells and cancer cells.
• Regulatory T (T_{Reg}) cell	An effector CD4 T cell that slows or stops activity of immune system. Important in controlling autoimmune diseases; several different kinds.
Memory cell	Descendant of activated B cell or any class of activated T cell; generated during initial immune response. May exist in body for years, enabling it to respond quickly and efficiently to subsequent encounters with same antigen.
Antigen-presenting cell (APC)	Any of several cell types (dendritic cell, macrophage, B cell) that engulfs and digests antigens that it encounters, then presents parts of them on its plasma membrane (bound to an MHC protein) for recognition by T cells bearing receptors for the same antigen. This function, antigen presentation, is essential for activation of T cells.
Molecules	
Antigen	Substance capable of provoking an immune response. Typically a large, complex molecule (e.g., protein or modified protein) not normally present in the body.
Antibody (immunoglobulin or Ig)	Protein produced by B cell or by plasma cell. Antibodies produced by plasma cells are released into body fluids (blood, lymph, saliva, mucus, etc.), where they attach to antigens. This causes complement activation, neutralization, precipitation, or agglutination, which "marks" the antigens for destruction by phagocytes or complement.
Perforins, granzymes	Released by T_C cells. Perforins create large pores in the target cell's membrane, allowing entry of apoptosis-inducing granzymes.
Complement	Group of bloodborne proteins activated after binding to antibody-covered antigens or certain molecules on the surface of microorganisms; enhances inflammatory response and lyses some microorganisms.
Cytokines	Small proteins that act as chemical messengers between various parts of the immune system. See Table 21.7.

Labels in table figures: B cell receptor (membrane-bound antibody); CD4 or CD8 protein; T cell receptor; Dendritic cell; MHC I or MHC II protein; Perforin pore

periphery—that is, outside the lymphoid organs. T_{Reg} cells and their subpopulations are currently hot research topics. For example, researchers hope to use them to induce tolerance to transplanted tissue and to lessen the severity of autoimmune diseases.

Table 21.8 summarizes the cells and molecules of the adaptive immune response. *Focus on An Example of a Primary Immune Response* (**Focus Figure 21.1** on pp. 808–809) gives an overview of the entire primary immune response, both innate and adaptive.

Organ Transplants and Prevention of Rejection

CLINICAL

The goal of organ transplantion is to provide patients with a functional organ from a living or deceased donor. This is often the only viable option left in end-stage cardiac or renal disease. Immune rejection presents a particular problem and so transplant success depends in part on the similarity of the donor and recipient tissues. NK cells, macrophages, antibodies, and especially T cells act vigorously to destroy any tissue they recognize as foreign.

The most common type of transplant is the *allograft*. Allografts are grafts transplanted from different individuals of the same species.* Before an allograft is attempted, the ABO and other blood group antigens of donor and recipient must be determined because these antigens are also present on most body cells and mismatches lead to immediate rejection. Matching donor and recipient MHCs helps minimize long-term rejection.

Following surgery the patient is treated with *immunosuppressive therapy*. Many of the drugs used to suppress rejection kill rapidly dividing cells (such as activated lymphocytes), and all of them have severe side effects. The major problem with immunosuppressive therapy is that the patient's suppressed immune system cannot protect the body against other foreign agents. As a result, overwhelming bacterial and viral infection remains the most frequent cause of death in transplant patients. Even under the best conditions, by 10 years after receiving a transplant, roughly 50% of patients have rejected the donor organ.

Check Your Understanding

20. Class II MHC proteins display what kind of antigens? What class of T cell recognizes antigens bound to class II MHC? What types of cells display these proteins?

21. Which type of T cell is the most important in both cellular and humoral immunity? Why?

22. Two killing mechanisms that we have seen in this chapter create pores in their target. Name the proteins that form this structure if they are (a) secreted by an immune cell that binds to the target cell or (b) derived from plasma proteins. Describe how these pores kill.

23. Once activated, the lymphocyte below will become one of two types of effector cells. Name them.

CD4 protein

For answers, see Answers Appendix.

21.7 **CLINICAL**

Insufficient or overactive immune responses create problems

Learning Outcomes
▶ Give examples of immunodeficiency diseases and of hypersensitivity states.
▶ Cite factors involved in autoimmune disease.

Certain circumstances can cause the immune system to weaken, fail, or act in a way that damages the body. Most such problems can be classified as immunodeficiencies, autoimmune diseases, or hypersensitivities.

---

*Autografts, isografts, and xenografts are described in Related Clinical Terms on p. 814.

Immunodeficiencies

An **immunodeficiency** is a congenital or acquired condition that impairs the production or function of immune cells or certain molecules, such as complement or antibodies.

The most devastating *congenital immunodeficiencies* are a group of related disorders called **severe combined immunodeficiency (SCID) syndromes**, which result from various genetic defects that produce a marked deficit of B and T cells.

Children afflicted with SCID have little or no protection against disease-causing organisms. Interventions must begin in the first months of life because minor infections easily shrugged off by most children cause those children with SCID to become deathly ill. Untreated, this condition is fatal, but successful transplants of matched donor hematopoietic stem cells dramatically improve survival rates. Genetic engineering techniques using viruses to carry corrected genes into the patient's own hematopoietic stem cells can also help.

There are various *acquired immunodeficiencies*. For example, *Hodgkin's lymphoma*, a cancer of the B cells, can lead to immunodeficiency by depressing lymph node cells. Immunosuppressive drugs used in transplantation and certain drugs used to treat cancer also suppress the immune system.

The most devastating of the acquired immunodeficiencies is **acquired immune deficiency syndrome (AIDS)**, which cripples the immune system by interfering with the activity of helper T cells. AIDS is caused by the **human immunodeficiency virus (HIV)**, a virus transmitted in body secretions—especially blood, semen, and vaginal secretions. HIV commonly enters the body via blood-contaminated needles, or during sexual intercourse. An infected mother can also transmit the virus to her fetus.

HIV targets CD4 cells via their CD4 surface proteins. Once inside a CD4 cell, HIV "sets up housekeeping," using the viral enzyme *reverse transcriptase* to produce DNA from its (viral) RNA. This DNA copy, now called a *provirus*, inserts itself into the target cell's DNA and directs the cell to crank out new copies of viral RNA and proteins so that the virus can multiply and infect other cells. In the process, HIV destroys T_H cells, depressing both branches of adaptive immunity and turning the whole immune system topsy-turvy.

The HIV reverse transcriptase enzyme is not very accurate and produces errors rather frequently, causing HIV's relatively high mutation rate and its ability to develop drug resistance.

No cure for AIDS has yet been found. Fortunately, a number of classes of antiviral drugs are available, and these are often used in combination. Sadly, even combination drug therapies often fail as the virus becomes resistant.

Autoimmune Diseases

Occasionally the immune system loses its ability to distinguish friend (self) from foe (foreign antigens). When this happens, the artillery of the immune system, like friendly fire, turns against itself. The body produces antibodies (*autoantibodies*) and cytotoxic T cells that destroy its own tissues. This puzzling phenomenon is called **autoimmunity**. If a disease state results, it is referred to as **autoimmune disease**.

Some 5% of adults in North America—two-thirds of them women—are afflicted with an autoimmune disease. You have encountered examples of autoimmune diseases in earlier chapters, and you will encounter more later on. Some important autoimmune diseases are:

- *Rheumatoid arthritis*, which systematically destroys joints (◀ p. 275)
- *Myasthenia gravis*, which impairs communication between nerves and skeletal muscles (◀ p. 293)
- *Multiple sclerosis*, which destroys the myelin of the white matter of the brain and spinal cord (◀ p. 412)
- *Graves' disease*, which prompts the thyroid gland to produce excessive amounts of thyroxine (◀ pp. 620–621)
- *Type 1 (insulin-dependent) diabetes mellitus*, which destroys pancreatic beta cells, resulting in a deficit of insulin and inability to use carbohydrates (◀ pp. 632–634)
- *Systemic lupus erythematosus (SLE)*, a systemic disease that particularly affects the kidneys, heart, lungs, and skin (see Related Clinical Terms, p. 814)
- *Glomerulonephritis*, which damages the kidney's filtration membrane and severely impairs renal function (see p. 1008)

Treatment of Autoimmune Diseases

The most widely used treatments for autoimmune conditions suppress the entire immune system—for example, anti-inflammatory drugs such as corticosteroids. Newer treatments seek to target specific aspects of the immune response. Fortunately, the immune system offers many potential targets because it is so complex. Two widely used therapeutic approaches involve:

- Blocking the actions of various cytokines using antibodies against them or their receptors
- Blocking the co-stimulatory molecules required to activate effector cells

Approaches still being developed include reestablishing self-tolerance by:

- Activating regulatory T cells
- Inducing self-tolerance using vaccines
- Destroying self-reactive immune cells by directing antibodies against them

The biggest challenge is selectively blocking autoimmune responses without blocking responses necessary to combat infection.

Failure of Self-Tolerance

Every autoimmune disease represents a failure of self-tolerance. How does this happen? As you recall, maturing lymphocytes undergo an extensive education (negative selection) that weeds out self-reactive cells. This weeding is thorough, but not too thorough, since there are pathogens that look somewhat like self. Weakly self-reactive lymphocytes that can detect these pathogens are allowed to survive negative selection and so they may cause autoimmune disease if they become activated.

Recall that activation of a T cell requires a co-stimulatory signal on an APC (see p. 804), and these co-stimulatory signals are only present if the APC has received "danger" signals alerting it to damage or invasion. This important safety check and the presence of regulatory T cells generally keeps both humoral and cellular immunity under control.

Normally, these mechanisms are sufficient, but sometimes self-reactive lymphocytes slip out of control. One of the following conditions or events may trigger this:

- **Foreign antigens resemble self-antigens.** If the determinants of a foreign antigen resemble those on a self-antigen, antibodies made against the foreign antigen can cross-react with the self-antigen. In the disease *rheumatic fever*, for instance, antibodies produced during a streptococcal infection react with heart antigens, causing lasting damage to the heart muscle and valves, as well as to joints and kidneys.
- **New self-antigens appear.** Self-proteins not previously exposed to the immune system may appear in the circulation. They may be generated by (1) gene mutations that cause new proteins to appear at the external cell surface, (2) changes in the structure of self-antigens as a result of hapten attachment or infectious damage, or (3) novel self-antigens, normally hidden behind barriers such as the blood brain barrier, that are released by trauma. These newly generated proteins then become immune system targets.

Hypersensitivities

Hypersensitivities result when the immune system damages tissue as it fights off a perceived threat (such as pollen or animal dander) that would otherwise be harmless to the body. People rarely die of hypersensitivities. They just make you miserable.

The different types of hypersensitivity reactions are distinguished by (1) their time course, and (2) whether they involve antibodies or T cells. Antibody-associated reactions cause *immediate* and *subacute hypersensitivities*. T cells cause *delayed hypersensitivities*.

Immediate Hypersensitivities

Immediate hypersensitivities, also called **acute** or **type I hypersensitivities**, are simply what most of us would call **allergies** (*allo* = altered; *erg* = reaction). An **allergen** is an antigen that causes an allergic reaction. Allergic reactions begin within seconds after contact with the allergen and last about half an hour.

The initial meeting with an allergen produces no symptoms but it sensitizes a susceptible person, causing IgE antibodies to be secreted that attach to the surfaces of the body's *mast cells* and *basophils* (**Figure 21.20**). Later encounters with the same allergen trigger an allergic reaction, in which the allergen promptly binds and cross-links the IgE antibodies on the surfaces of the mast cells and basophils. This event causes mast cells and basophils to release a flood of *histamine* and other inflammatory chemicals that together induce the inflammatory response typical of allergy.

Allergic reactions may be local or systemic (bodywide). Mast cells are abundant in connective tissues of the skin and beneath

Adaptive defenses ⟶ Humoral immunity

Sensitization stage

① Antigen (allergen) invades body.

Antigen

② Plasma cells produce large amounts of class IgE antibodies against allergen.

③ IgE antibodies attach to mast cells in body tissues (and to circulating basophils).

Mast cell with fixed IgE antibodies

IgE

Granules containing histamine

Subsequent (secondary) responses

④ More of same antigen invades body.

⑤ Antigen combines with IgE attached to mast cells (and basophils), which triggers degranulation and release of histamine (and other chemicals).

Mast cell granules release contents after antigen binds with IgE antibodies

Histamine

⑥ Histamine causes blood vessels to dilate and become leaky, which promotes edema; stimulates secretion of large amounts of mucus; and causes smooth muscles to contract. (If respiratory system is site of antigen entry, asthma may ensue.)

Outpouring of fluid from capillaries   Release of mucus   Constriction of small respiratory passages (bronchioles)

Figure 21.20 Mechanism of an acute allergic (immediate hypersensitivity) response.

the mucosa of respiratory passages and the gastrointestinal tract, and these areas are common sites of local allergic reactions.

Histamine causes blood vessels to dilate and become leaky, and is largely to blame for the best-recognized symptoms of allergy: runny nose, itching reddened skin (hives), and watery eyes. When the allergen is inhaled, symptoms of *asthma* appear because smooth muscle in the walls of bronchioles (small airways) contracts, constricting those small passages and restricting air flow. When the allergen is ingested in food or via drugs, gastrointestinal discomfort occurs. Over-the-counter antiallergy drugs contain antihistamines that counteract these effects.

The systemic response known as **anaphylactic shock** is fairly rare. It typically occurs when the allergen directly enters the blood and circulates rapidly through the body—for example, a bee sting, spider bite, or injection of a foreign substance (such as penicillin or other drugs that act as haptens).

The mechanism of anaphylactic shock is essentially the same as that of local responses, but when mast cells and basophils are enlisted throughout the body, the outcome is life threatening. The bronchioles constrict (and the tongue may swell), making it difficult to breathe, and the sudden vasodilation and fluid loss from the bloodstream may cause circulatory collapse (hypotensive shock) and death within minutes. Epinephrine is the drug of choice to reverse these histamine-mediated effects.

Subacute Hypersensitivities

Like the immediate types, **subacute hypersensitivities** are caused by antibodies (IgG and IgM rather than IgE). However, their onset is slower (1–3 hours after antigen exposure) and the reaction lasts longer (10–15 hours).

Cytotoxic (type II) reactions occur when antibodies bind to antigens on specific body cells and stimulate phagocytosis and complement-mediated lysis of the cellular antigens. Cytotoxic hypersensitivity may occur after a patient has received a transfusion of mismatched blood and complement lyses the foreign red blood cells.

Immune-complex (type III) hypersensitivities result when antigens are widely distributed through the body or blood and the huge number of insoluble antigen-antibody complexes formed cannot be cleared from a particular area. (This may reflect a persistent infection or an autoimmune disease.) An intense inflammatory reaction occurs, complete with complement-mediated cell lysis and cell killing by neutrophils that severely damages local tissues. An example of type III hypersensitivity is the glomerulonephritis of systemic lupus erythematosus (see Related Clinical Terms, p. 814).

Delayed Hypersensitivities

Delayed (type IV) hypersensitivity reactions are caused by T cells and take longer to appear (1–3 days) than antibody-mediated hypersensitivity reactions. Inflammation and tissue damage result from the action of cytokine-activated macrophages, and sometimes cytotoxic T cells.

The most familiar examples of delayed hypersensitivity reactions are those classified as **allergic contact dermatitis** which follow skin contact with poison ivy, some metals (nickel in jewelry),

21

and certain cosmetic and deodorant chemicals. These agents diffuse through the skin and attach to self-proteins as *haptens*.

Skin tests for tuberculosis depend on a delayed hypersensitivity reaction. When the tubercle antigens are introduced just under the skin, a small hard lesion forms that persists for days if the person has been sensitized to the antigen.

Check Your Understanding

24. What makes HIV particularly hard for the immune system to defeat?

25. What event triggers the release of histamine from mast cells in an allergic response?

For answers, see Answers Appendix.

Developmental Aspects of the Immune System

Stem cells of the immune system originate in the liver and spleen during weeks 1–9 of embryonic development. Later the red bone marrow becomes the predominant source of stem cells, and it persists in this role into adult life. The lymphocyte precursors become immunocompetent in the thymus and bone marrow, and then populate other lymphoid tissues.

The newborn's immune system depends primarily on antibodies and T_H2 lymphocytes. The T_H1 system is educated and gets stronger as a result of encounters with microbes—both harmful and harmless. If such exposure and education does not occur, immune balance is upset and the T_H2 system flourishes, causing the immune system to teeter toward allergies. Unhappily, the desire to keep our children squeaky clean with antibiotics that kill off both harmful and harmless bacteria may derail normal immune development.

A wide variety of factors outside the immune system influence immune function. For example:

- *Nervous system.* The study of psychoneuroimmunology—a brain-twisting term coined to describe links between the brain and the immune system—helps explain how various stressors such as depression, emotional stress, sleep deprivation, and grief can impair the immune response.

- *Diet.* Researchers have discovered that vitamin D is required for CD8 cells to become T_C cells. On the other hand, vitamin D deficiency has been linked with autoimmune diseases such as multiple sclerosis.

The immune system normally serves us well until late in life. Then its efficiency begins to wane, and its ability to fight infection declines. Old age is also accompanied by greater susceptibility to both immunodeficiency and autoimmune diseases. The greater incidence of cancer in the elderly may also reflect the progressive failure of the immune system. We do not know why the immune system begins to fail, but we do know that the thymus begins to atrophy after puberty. Memory cells accumulate, but the production of naive T and B cells declines with age, possibly because progenitor cells reach the limits of their ability to divide. This reduces the body's capacity to repond to new antigens.

In many elderly people, the immune system enters a state of sustained, low-grade inflammation due to the increased production of inflammatory chemicals. Such chronic inflammation is thought to cause or promote many diseases associated with aging, such as atherosclerosis and Alzheimer's disease.

RELATED CLINICAL TERMS

Autograft A tissue graft transplanted from one location to another in the same person.

Congenital thymic aplasia An immune deficiency disease in which the thymus fails to develop. Affected individuals have no T cells, and so have little or no immune protection; fetal thymic and bone marrow transplants have been helpful in some cases.

Eczema (ek′zĕ-mah) A clinical term for several conditions that cause "weeping" skin lesions and intense itching. One common cause, *atopic dermatitis*, has a strong familial predisposition, has features of immediate hypersensitivity, and usually begins in the first five years of life. Recent research suggests the underlying defect may be increased leakiness of the skin.

Hashimoto's thyroiditis An autoimmune disease in which both B and T lymphocytes attack the thyroid gland. It is the most common cause of hypothyroidism, affecting mostly middle-aged and elderly women. Genetic factors associated with this autoimmune disease (certain MHC variants) make individuals susceptible to environmental triggers (possibly iodine, irradiation, or trauma).

Immunization The process of rendering a subject immune (by vaccination or injecting antiserum).

Immunology The study of immunity.

Immunopathology Disease related to the immune system.

Isograft A graft transplanted from a genetically identical individual. The only example is identical twins.

Septic shock (sepsis) A dangerous condition in which the inflammatory response goes out of control. It results from especially severe bacterial infections or more ordinary infections that grow rapidly worse in patients with weakened defenses, such as the hospitalized elderly recovering from surgery. In an inflammatory response, neutrophils and other white blood cells secrete cytokines that increase capillary permeability. In sepsis, continued cytokine release makes capillaries so leaky that the bloodstream is depleted of fluid. Blood pressure falls and the body organs shut down, causing death in 50% of cases. Sepsis has proven difficult to control and its incidence remains high.

Systemic lupus erythematosus (SLE) (er″ĭ-the″mah-to′sus) A systemic autoimmune disorder that occurs mainly in young females. Diagnosis is helped by finding antinuclear (anti-DNA) antibodies in the patient's blood. DNA–anti-DNA complexes (typical of type III hypersensitivity) localize in the kidneys, blood vessels, brain, and synovial membranes. This results in glomerulonephritis (see Related Clinical Terms, p. 1008), vascular problems, loss of memory and mental sharpness, and painful arthritis. Reddened skin lesions, particularly a "butterfly rash" (the sign of the wolf, or lupus) on the face, are common.

Xenograft A tissue or organ graft taken from a different species. Generally not viable except for acellular grafts such as heart valves.

CHAPTER SUMMARY

PART 1
INNATE DEFENSES

21.1 Surface barriers act as the first line of defense to keep invaders out of the body (pp. 782–783)

1. Skin and mucous membranes constitute the first line of defense. Protective membranes line all body cavities and organs exposed to the exterior.
2. Surface membranes provide mechanical barriers to pathogens. Some have structural modifications and produce secretions that enhance their defensive effects: The skin's acidity, lysozyme, mucus, keratin, and ciliated cells are examples.

21.2 Innate internal defenses are cells and chemicals that act as the second line of defense (pp. 783–790)

Phagocytes (pp. 784–785)

1. Phagocytes (macrophages and neutrophils) engulf and destroy pathogens that breach epithelial barriers. This process is facilitated when opsonins (antibodies or complement to which the phagocyte's receptors can bind) attach to the pathogen's surface. The respiratory burst enhances cell killing.

Natural Killer (NK) Cells (p. 785)

2. Natural killer (NK) cells are large granular lymphocytes that act nonspecifically to kill virus-infected and cancerous cells.

Inflammation: Tissue Response to Injury (pp. 785–787)

3. The inflammatory response prevents the spread of harmful agents, disposes of pathogens and dead tissue cells, and promotes healing. Exudate forms; protective leukocytes enter the area; fibrin walls off the area; and tissue repair occurs.
4. The cardinal signs of inflammation are swelling, redness, heat, and pain. These result from inflammatory chemicals that induce vasodilation and make capillaries and venules more permeable. Impaired function may be a fifth cardinal sign.

Antimicrobial Proteins (pp. 788–789)

5. Interferons are a group of related proteins synthesized by virus-infected cells and certain immune cells that prevent viruses from multiplying in other body cells.
6. Activation of complement (a group of plasma proteins) on the membrane of a foreign cell promotes phagocytosis of that cell, enhances inflammation, and sometimes causes lysis of the target cell.

Fever (p. 789)

7. Fever enhances the body's fight against pathogens.

PART 2
ADAPTIVE DEFENSES

1. The adaptive immune system recognizes something as foreign and acts to immobilize, neutralize, or remove it. The adaptive immune response involves B and T lymphocytes, is antigen-specific, systemic, and has memory. It provides the body's third line of defense.

21.3 Antigens are substances that trigger the body's adaptive defenses (pp. 791–792)

1. Complete antigens have both immunogenicity and reactivity. Incomplete antigens, or haptens, must combine with a body protein before becoming immunogenic.
2. Antigenic determinants are the portions of antigen molecules that are recognized as foreign. Most antigens have many such sites.
3. Major histocompatibility complex (MHC) proteins are membrane-bound glycoproteins that mark our cells as "self."

21.4 B and T lymphocytes and antigen-presenting cells are cells of the adaptive immune response (pp. 792–795)

Lymphocytes (pp. 792–794)

1. Lymphocytes arise from hematopoietic stem cells of the bone marrow and are educated to develop immunocompetence and self-tolerance. T cells are educated in the thymus and provide cellular immunity. B cells are educated in the bone marrow and provide humoral immunity. Immunocompetence is signaled by the appearance of antigen-specific receptors on the surface of the lymphocyte. Immunocompetent lymphocytes seed the secondary lymphoid organs, where the antigen encounter occurs, and circulate between the blood, lymph, and lymphoid organs. When naive lymphocytes encounter their antigen, clonal selection, proliferation, and differentiation occur. Most of the clone members become effector cells, but some become memory cells.
2. In both B and T lymphocytes, antigen receptor diversity is generated by shuffling gene fragments.

Antigen-Presenting Cells (APCs) (pp. 794–795)

3. Antigen-presenting cells (APCs) include dendritic cells, macrophages, and B lymphocytes. They internalize antigens and present antigenic determinants on their surfaces for recognition by T cells.

21.5 In humoral immunity, antibodies are produced that target extracellular antigens (pp. 796–801)

Activation and Differentiation of B Cells (pp. 796–797)

1. When B cells are activated, most of the clone members become effector cells called plasma cells, which secrete antibodies. This is the primary adaptive immune response.

Immunological Memory (p. 797)

2. Other clone members become memory B cells, capable of mounting a rapid attack against the same antigen in subsequent encounters (secondary immune responses). The memory B cells provide humoral immunological memory.

Active and Passive Humoral Immunity (pp. 797–798)

3. Active humoral immunity is acquired during an infection or via vaccination and provides immunological memory. Passive immunity is conferred when a donor's antibodies are injected into the bloodstream, or when the mother's antibodies cross the placenta. Its protection is short-lived; immunological memory is not established.

Antibodies (pp. 798–801)

4. The antibody monomer consists of four polypeptide chains, two heavy and two light. Each chain has both a constant and a

21

variable region. Constant regions determine antibody function and class. Variable regions enable the antibody to recognize its specific antigen.

5. Five classes of antibodies exist: IgM, IgA, IgD, IgG, and IgE. They differ structurally and functionally.

6. Antibody functions include complement activation and antigen neutralization, precipitation, and agglutination.

7. Monoclonal antibodies are pure preparations of a single antibody type useful in diagnostic tests and treating some types of cancer.

21.6 Cellular immunity consists of T lymphocytes that direct adaptive immunity or attack cellular targets (pp. 801–811)

MHC Proteins and Antigen Presentation (pp. 802–804)

1. MHC proteins present antigens to T cells. Class I MHC proteins are found on all nucleated cells, but class II MHC proteins are found only on APCs.

Activation and Differentiation of T Cells (pp. 804–805)

2. Immunocompetent CD4 and CD8 T cells are activated by binding to an antigen-MHC complex on the surface of an APC. A co-stimulatory signal is also essential. The resulting clone members differentiate into either the appropriate effector T cells that mount the primary immune response (e.g., helper or cytotoxic T cells) or memory T cells.

3. The immune response is enhanced by cytokines such as interleukin 1 released by macrophages, and interleukin 2, gamma interferon, and others released by activated T cells.

Roles of Specific Effector T Cells (pp. 805–810)

4. Helper T cells are required for full activation of most B and T cells, activate macrophages, and release essential cytokines. Cytotoxic T cells directly attack and kill infected cells and cancer cells. Together with NK cells, they conduct immune surveillance. Regulatory T (T_{Reg}) cells help to maintain tolerance.

Organ Transplants and Prevention of Rejection (p. 810–811)

5. Cell-mediated responses reject grafts and foreign organ transplants unless the recipient is immunosuppressed. Infections are major complications in such patients.

21.7 Insufficient or overactive immune responses create problems (pp. 811–814)

1. Immunodeficiency diseases include acquired immune deficiency syndrome (AIDS), caused by HIV, and severe combined immunodeficiency (SCID) syndromes. Minor infections become fatally overwhelming because the immune system is unable to combat them.

2. Autoimmune disease occurs when the body regards its own tissues as foreign and mounts an immune attack against them. Examples include rheumatoid arthritis and multiple sclerosis.

3. Hypersensitivity is an abnormally intense reaction to an otherwise harmless antigen. Immediate hypersensitivities (allergies) are mounted by IgE antibodies. Subacute hypersensitivities, involving both antibodies and complement, include antibody-mediated cytotoxic and immune-complex hypersensitivities. Cell-mediated hypersensitivity is called delayed hypersensitivity.

Developmental Aspects of the Immune System (p. 814)

1. The liver and spleen are the initial sources of immune stem cells before the red bone marrow takes over.

2. The nervous system plays an important role in regulating immune responses. Various stressors can impair immune function.

3. With aging, the immune system becomes less responsive. The elderly more often suffer from immunodeficiency, autoimmune diseases, and cancer.

REVIEW QUESTIONS

 To access additional practice questions using your smartphone, tablet, or computer: MasteringA&P® > Study Area > Practice Tests & Quizzes

Level 1 Remember/Understand

(Some questions have more than one correct answer. Select the best answer or answers from the choices given.)

1. All of the following are considered innate body defenses *except* (a) complement, (b) phagocytosis, (c) antibodies, (d) lysozyme, (e) inflammation.

2. The process by which neutrophils squeeze through capillary walls in response to inflammatory signals is called (a) diapedesis, (b) chemotaxis, (c) margination, (d) opsonization.

3. Antibodies released by plasma cells are involved in (a) humoral immunity, (b) immediate hypersensitivity reactions, (c) autoimmune disorders, (d) all of the above.

4. Which antibody class is abundant in body secretions? (a) IgA, (b) IgD, (c) IgE, (d) IgG, (e) IgM.

5. Small molecules that must combine with large proteins to become immunogenic are called (a) complete antigens, (b) kinins, (c) antigenic determinants, (d) haptens.

6. Lymphocytes that develop immunocompetence in the bone marrow are (a) T lymphocytes, (b) B lymphocytes, (c) NK cells, (d) B and T lymphocytes.

7. Cells that can directly attack target cells include all of the following *except* (a) macrophages, (b) cytotoxic T cells, (c) helper T cells, (d) natural killer cells.

8. Which of the following is not involved in the activation of a B cell? (a) antigen, (b) helper T cell, (c) cytokine, (d) cytotoxic T cell.

9. Complement activation promotes all of the following except (a) cell lysis, (b) inflammation, (c) opsonization, (d) interferon release, (e) chemotaxis of neutrophils and other cells.

10. Using the letters from column B, match the cell description in column A. (Note that all require more than a single choice.)

Column A	Column B
____ (1) phagocytic "cleanup crew"	(a) natural killer cell
____ (2) releases histamine	(b) neutrophil
____ (3) releases perforins	(c) dendritic cell
____ (4) lymphocyte	(d) mast cell
____ (5) effector cells of adaptive immunity	(e) cytotoxic T cell
____ (6) antigen-presenting cell	(f) B cell
	(g) macrophage
	(h) helper T cell
	(i) basophil

11. The cell type most often invaded by HIV is a(n) (**a**) eosinophil, (**b**) cytotoxic T cell, (**c**) natural killer cell, (**d**) helper T cell, (**e**) B cell.

12. Name the five antibody classes and describe where each is most likely to be found in the body.

Level 2 Apply/Analyze

13. Besides acting as mechanical barriers, the skin epidermis and mucosae of the body have other attributes that contribute to their protective roles. Cite the common body locations and the importance of mucus, lysozyme, keratin, acid pH, and cilia.

14. What is complement? How does it cause bacterial lysis? What are some of the other roles of complement?

15. Interferons are referred to as antiviral proteins. What stimulates their production, and how do they protect uninfected cells? Which cells of the body secrete interferons?

16. Define immunocompetence and self-tolerance. How is self-tolerance achieved?

17. Differentiate between a primary and a secondary immune response. Which is more rapid and why?

18. Define antibody. Using an appropriately labeled diagram, describe the structure of an antibody monomer. Indicate and label variable and constant regions, heavy and light chains.

19. How do antibodies help defend the body?

20. Describe the specific roles of helper, regulatory, and cytotoxic T cells in normal cellular immunity.

21. Name several cytokines and describe their role in the immune response.

22. Define hypersensitivity. List three types of hypersensitivity reactions. For each, note whether antibodies or T cells are involved and provide examples.

23. What accounts for the declining efficiency of the immune system with age?

Level 3 Evaluate/Synthesize

24. Differentiate between humoral and cellular adaptive immunity.

25. Although the adaptive immune system has two arms, it has been said, "no T cells, no immunity." Explain.

26. Isabella, a 6-year-old child who has been raised in a germ-free environment from birth, has one of the most severe examples of an abnormal immune system. Isabella also suffers from cancer caused by the Epstein-Barr virus. Relative to this case: (**a**) Why is Isabella's brother chosen as the hematopoietic stem cell donor? (**b**) Why is her physician planning to use umbilical cord blood as a source of stem cells for transplant if her brother's stem cells fail (what are the hoped-for results)? (**c**) Attempt to explain Isabella's cancer. (**d**) What similarities and dissimilarities exist between Isabella's illness and AIDS?

27. Capillary permeability increases and plasma proteins leak into the interstitial fluid as part of the inflammatory process. Why is this desirable?

28. Caroline, a pregnant 29-year-old, has been HIV-positive for at least 10 years, dating back to when she was homeless and injecting heroin. While she currently has no symptoms of AIDS, she is taking several medications and is worried about the possibility that her baby might be infected. How do you think the HIV virus might be transferred from a mother to her offspring? Which of Caroline's cells are infected by the virus and why is the viral attack on these cells so devastating? Why is Caroline taking medication even though she has no symptoms?

CLINICAL CASE STUDY

52-Year-Old Male with Multiple Sclerosis

Mr. Ayers was diagnosed with an aggressive form of multiple sclerosis 18 years ago. His illness has not responded to conventional treatments so he's been referred to your clinic for a clinical trial. The doctor explains that his treatment will involve giving Mr. Ayers a drug that will release stem cells from his bone marrow so they can be collected from his blood. Then Mr. Ayers will be given (1) drugs that will destroy the blood-forming cells in his bone marrow, and (2), antibodies against his T lymphocytes. Finally, the stem cells that were collected from him will be used to repopulate his bone marrow.

The doctor says, "This treatment avoids problems, such as transplant rejection, that are associated with other kinds of transplants."

1. In organ transplantation, the transplanted organ is called a graft. What type of graft would it be if Mr. Ayers received someone else's bone marrow? What type would it be if he receives his own bone marrow?

2. One reason for using Mr. Ayers's own tissue is that it can be very difficult to find a donor whose bone marrow cells match the patient's class I and II MHC proteins. What is an MHC protein? What is the difference between class I and class II MHC proteins?

3. **+ NCLEX-STYLE** The nurse explains to Mr. Ayers that a bone marrow transplant from another donor could result in the transplanted immune cells attacking his own cells. The best explanation for this is:
   a. Transplants from another person are never perfect matches. Any organ transplant could cause this problem.
   b. A donor transplant would be a problem because Mr. Ayers's own immune system will be destroyed, so he wouldn't be able to fight off mismatched cells from a donor transplant.
   c. Bone marrow is where immune cells are made. If Mr. Ayers received mismatched immune cells, these cells could attack his body.
   d. Because Mr. Ayers has an autoimmune disorder, his body will attack anything transplanted into it.

4. **+ NCLEX-STYLE** You're talking with Mr. Ayers about his course of treatment. Which of his statements indicates a need for further instruction?
   a. "It's ironic that they're giving me antibodies against my T cells, when I already attack my own tissues."
   b. "I'm gonna ask the doc to not give me the stem cells after they destroy my bone marrow. After all, the immune cells are what's hurting me!"
   c. "My T cells are just brutal. Gotta get those bad boys under control!"
   d. "I'm guessing the real trouble was something that happened to my T cells."

21

22

The Respiratory System

In this chapter, you will learn that

The respiratory system supplies cells with oxygen and eliminates carbon dioxide

by first examining

then exploring

Part 1 Functional Anatomy

Part 2 Respiratory Physiology

looking closer at

by asking

and then exploring

22.1 The upper respiratory system

22.4 What causes air to move in and out of the lungs?

22.8 Control of respiration

and

22.2 The lower respiratory system

22.5 How do we assess ventilation?

22.9 Exercise and high altitude

includes

22.3 The lungs and pleurae

22.6 How do gases move between the lungs, blood, and tissues?

22.10 What happens when things go wrong?

and finally, exploring

Developmental Aspects of the Respiratory System

and

The cardiovascular system Ch. 17–19

transported through

22.7 How does blood carry oxygen and carbon dioxide?

Far from self-sustaining, our bodies depend on the external environment, both as a source of substances we need to survive and as a catch basin for wastes. We require a continuous supply of oxygen so that our trillions of cells can carry out their vital functions. We can live without food or water for days, but we cannot do without oxygen for even a little while.

As cells use oxygen, they give off carbon dioxide, a waste product the body must get rid of. They also generate dangerous free radicals, the inescapable by-products of living in a world full of oxygen.

The major function of the **respiratory system** is to supply the body with oxygen and dispose of carbon dioxide. To accomplish this function, at least four processes, collectively called **respiration**, must happen. These processes are shown in **Figure 22.1**.

The respiratory system is responsible for only the first two processes, but it cannot accomplish its primary goal of obtaining oxygen and eliminating carbon dioxide unless the third and fourth processes also occur. As you can see, the respiratory and circulatory systems are closely coupled, and if either system fails, the body's cells begin to die from oxygen starvation.

The actual use of oxygen and production of carbon dioxide by tissue cells, known as *cellular respiration*, is the cornerstone of all energy-producing chemical reactions in the body. We discuss cellular respiration, which is not a function of the respiratory system, in the metabolism section of Chapter 24.

Because it moves air, the respiratory system is also involved with the sense of smell and with speech.

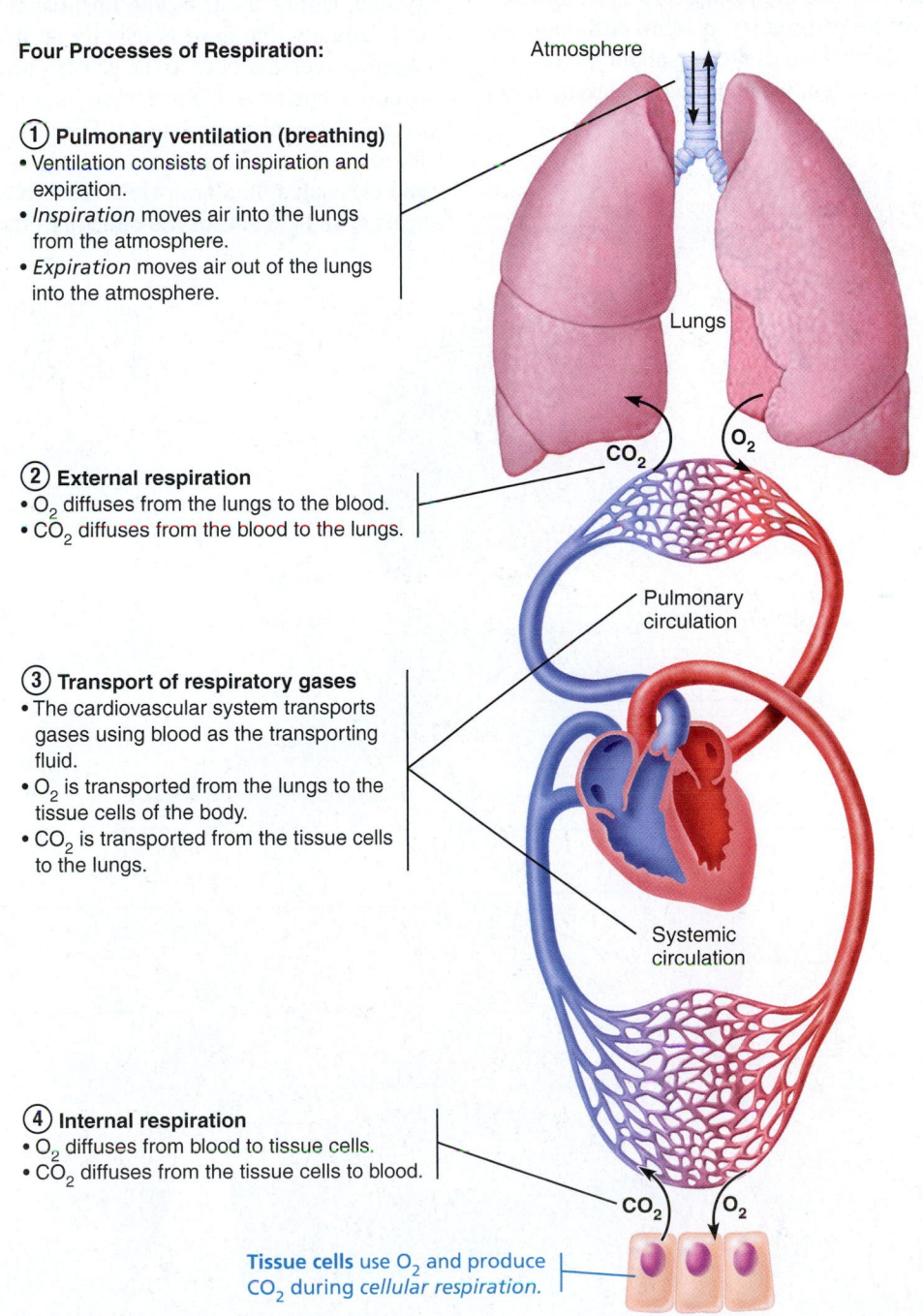

Four Processes of Respiration:

(1) Pulmonary ventilation (breathing)
• Ventilation consists of inspiration and expiration.
• *Inspiration* moves air into the lungs from the atmosphere.
• *Expiration* moves air out of the lungs into the atmosphere.

(2) External respiration
• O_2 diffuses from the lungs to the blood.
• CO_2 diffuses from the blood to the lungs.

(3) Transport of respiratory gases
• The cardiovascular system transports gases using blood as the transporting fluid.
• O_2 is transported from the lungs to the tissue cells of the body.
• CO_2 is transported from the tissue cells to the lungs.

(4) Internal respiration
• O_2 diffuses from blood to tissue cells.
• CO_2 diffuses from the tissue cells to blood.

Atmosphere

Lungs

CO_2 O_2

Pulmonary circulation

Systemic circulation

CO_2 O_2

Tissue cells use O_2 and produce CO_2 during *cellular respiration.*

Figure 22.1 Respiration consists of four processes.

22

FUNCTIONAL ANATOMY

The respiratory system (**Figure 22.2**) includes the *nose* and *paranasal sinuses*; the *pharynx*; the *larynx*; the *trachea*; the *bronchi* and their smaller branches; and the *lungs*, which contain tiny air sacs called *alveoli*. The *upper respiratory system* (summarized in **Table 22.1**, p. 824) consists of all the structures from the nose to the larynx. The *lower respiratory system* (**Table 22.2**, p. 831) consists of the larynx and all the structures below it.

Some authorities also include the respiratory muscles (diaphragm, etc.) as part of the respiratory system. Although we will consider how these skeletal muscles bring about the volume changes that promote ventilation, we continue to classify them as part of the *muscular system*.

22.1 The upper respiratory system warms, humidifies, and filters air

Learning Outcomes

▶ Describe the location, structure, and function of each of the following: nose, paranasal sinuses, and pharynx.
▶ List and describe several protective mechanisms of the respiratory system.

The Nose and Paranasal Sinuses

The nose is the only externally visible part of the respiratory system. Unlike the eyes and lips, facial features often referred to poetically, the nose is usually an irreverent target. We are urged to keep our nose to the grindstone and keep it out of other people's business. Considering its important functions, however, it deserves more esteem. The nose (1) provides an airway for respiration, (2) moistens and warms entering air, (3) filters and cleans inspired air, (4) serves as a resonating chamber for speech, and (5) houses the olfactory (smell) receptors.

22

Figure 22.2 The major respiratory organs in relation to surrounding structures.

(a) Surface anatomy

- Epicranius, frontal belly
- Root and bridge of nose
- Dorsum nasi
- Ala of nose
- Apex of nose
- Naris (nostril)

(b) External skeletal framework

- Frontal bone
- Nasal bone
- Septal cartilage
- Maxillary bone (frontal process)
- Lateral process of septal cartilage
- Minor alar cartilages
- Dense fibrous connective tissue
- Major alar cartilages

Figure 22.3 The external nose.

The structures of the nose are divided into the *external nose* and the internal *nasal cavity* for ease of consideration.

External Nose

The surface features of the external nose include the *root* (area between the eyebrows), *bridge*, and *dorsum nasi* (anterior margin), the latter terminating in the *apex* (tip of the nose) (**Figure 22.3a**). The external openings of the nose, the **nostrils** or **nares** (na′rēz), are bounded laterally by the flared *alae*.

The skeletal framework of the external nose is fashioned by the nasal and frontal bones superiorly (forming the bridge and root, respectively), the maxillary bones laterally, and flexible plates of hyaline cartilage (the alar and septal cartilages, and the lateral processes of the septal cartilage) inferiorly (Figure 22.3b). Noses vary a great deal in size and shape, largely because of differences in nasal cartilages. The skin covering the nose's anterior and lateral aspects is thin and contains many sebaceous glands.

Nasal Cavity

The internal **nasal cavity** lies in and posterior to the external nose. During breathing, air enters the cavity by passing through the *nostrils* or *nares* (Figure 22.3a and **Figure 22.4**). The nasal cavity is divided by a midline **nasal septum**, formed anteriorly by the septal cartilage and posteriorly by the vomer bone and perpendicular plate of the ethmoid bone (◀ **Figure 7.14b, p. 214**). The nasal cavity is continuous posteriorly with the nasal

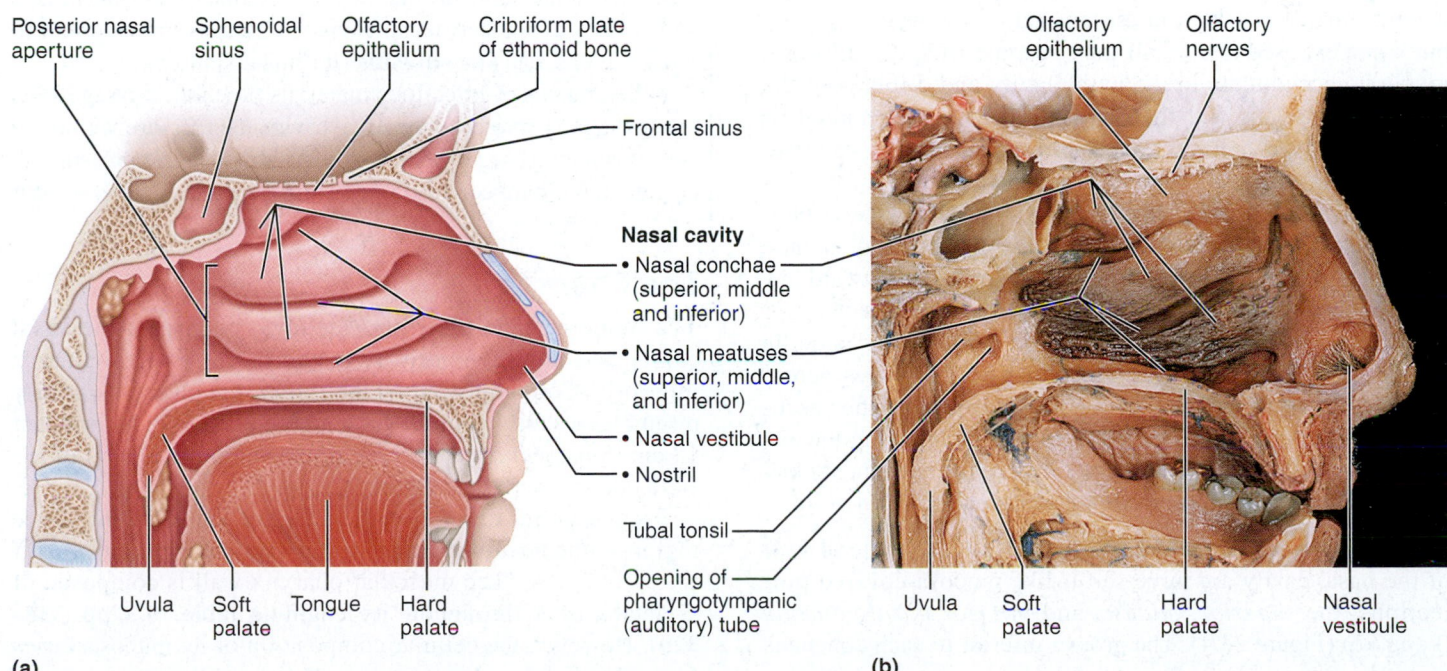

(a)

- Posterior nasal aperture
- Sphenoidal sinus
- Olfactory epithelium
- Cribriform plate of ethmoid bone
- Frontal sinus
- **Nasal cavity**
  - • Nasal conchae (superior, middle and inferior)
  - • Nasal meatuses (superior, middle, and inferior)
  - • Nasal vestibule
  - • Nostril
- Tubal tonsil
- Opening of pharyngotympanic (auditory) tube
- Uvula
- Soft palate
- Tongue
- Hard palate

(b)

- Olfactory epithelium
- Olfactory nerves
- Uvula
- Soft palate
- Hard palate
- Nasal vestibule

Figure 22.4 The nasal cavity. Midsagittal section of the head. **(a)** Illustration. **(b)** Photo.

portion of the pharynx through the **posterior nasal apertures**, also called the *choanae* (ko-a′ne; "funnels").

The roof of the nasal cavity is formed by the ethmoid and sphenoid bones of the skull. The floor is formed by the *palate*, which separates the nasal cavity from the oral cavity below. Anteriorly, where the palate is supported by the palatine bones and processes of the maxillary bones, it is called the **hard palate**. The unsupported posterior portion is the muscular **soft palate**.

The part of the nasal cavity just superior to each nostril, called the **nasal vestibule**, is lined with skin containing sebaceous and sweat glands and numerous hair follicles. The hairs, or **vibrissae** (vi-bris′e; *vibro* = to quiver), filter coarse particles (dust, pollen) from inspired air. The rest of the nasal cavity is lined with two types of mucous membrane.

- The small patch of **olfactory mucosa** lines the slitlike superior region of the nasal cavity and contains smell receptors in its **olfactory epithelium** (◀ p. 576).

- The **respiratory mucosa** lines most of the nasal cavity. The respiratory mucosa is a pseudostratified ciliated columnar epithelium, containing scattered *goblet cells*, that rests on a lamina propria richly supplied with seromucous *nasal glands*.

Seromucous nasal glands contain mucus-secreting mucous cells and serous cells that secrete a watery fluid containing enzymes. Each day, these glands secrete about a quart (a liter) of mucus containing *lysozyme*, an antibacterial enzyme. The sticky mucus traps inspired dust, bacteria, and other debris, while lysozyme attacks and chemically destroys bacteria. The epithelial cells of the respiratory mucosa also secrete *defensins*, natural antibiotics that help kill invading microbes. Additionally, the high water content of the mucus film humidifies incoming air.

The ciliated cells of the respiratory mucosa create a gentle current that moves the sheet of contaminated mucus posteriorly toward the throat, where it is swallowed and digested. We are usually unaware of this important action of our nasal cilia, but when exposed to cold air they become sluggish, allowing mucus to accumulate in the nasal cavity and dribble out the nostrils. This, along with the fact that water vapor in expired air tends to condense at lower temperatures, explains why you have a runny nose on a crisp wintry day.

The nasal mucosa is richly supplied with sensory nerve endings, and contact with irritating particles (dust, pollen, and the like) triggers a sneeze reflex. The sneeze forces air outward in a violent burst—a crude but effective way to expel irritants.

Rich plexuses of capillaries and thin-walled veins underlie the nasal epithelium and warm incoming air as it flows across the mucosal surface. When the inspired air is cold, the vascular plexus becomes engorged with blood, intensifying the air-heating process. Because these blood vessels are abundant and located superficially, nosebleeds are common and often profuse.

Nasal Conchae Protruding medially from each lateral wall of the nasal cavity are three scroll-like mucosa-covered projections, the *superior*, *middle*, and *inferior nasal conchae* (kong′ke) (Figure 22.4). The groove inferior to each concha is a *nasal meatus* (me-a′tus).

The curved conchae greatly increase the mucosal surface area exposed to air and enhance air turbulence in the cavity. The gases in inhaled air swirl through the twists and turns, deflecting heavier, nongaseous particles onto the mucus-coated surfaces, where they become trapped. As a result, few particles larger than 6 μm make it past the nasal cavity.

The conchae and nasal mucosa not only function during inhalation to filter, heat, and moisten the air, but also act during exhalation to reclaim this heat and moisture. In other words, inspired air cools the conchae, then during exhalation these cooled conchae precipitate moisture and extract heat from the humid air flowing over them. This reclamation process minimizes the amount of moisture and heat lost from the body through breathing, helping us to survive in dry and cold climates.

Paranasal Sinuses

The nasal cavity is surrounded by a ring of **paranasal sinuses** (Figure 22.4a). They are located in the frontal, sphenoid, ethmoid, and maxillary bones (◀ **Figure 7.15, p. 215**). The sinuses lighten the skull, and they may help warm and moisten the air. The mucus they produce ultimately flows into the nasal cavity, and the suctioning effect created by nose blowing helps drain the sinuses.

HOMEOSTATIC IMBALANCE 22.1   **CLINICAL**

Cold viruses, streptococcal bacteria, and various allergens can cause *rhinitis* (ri-ni′tis), inflammation of the nasal mucosa accompanied by excessive mucus production, nasal congestion, and postnasal drip. The nasal mucosa is continuous with the mucosa of the other respiratory passageways, explaining the typical nose to throat to chest progression of colds. Because the mucosa extends tentacle-like into the nasolacrimal (tear) ducts and paranasal sinuses, nasal cavity infections often spread to those regions, causing **sinusitis** (inflamed sinuses).

When mucus or infectious materials block the passageways connecting the sinuses to the nasal cavity, the pressure within the sinuses may no longer be the same as atmospheric pressure. A change in pressure can result in a *sinus headache*. ●●●●●●

The Pharynx

The funnel-shaped **pharynx** (far′ingks) connects the nasal cavity and mouth superiorly to the larynx and esophagus inferiorly. Commonly called the *throat*, the pharynx vaguely resembles a short length of garden hose as it extends for about 13 cm (5 inches) from the base of the skull to the level of the sixth cervical vertebra (Figure 22.2).

From superior to inferior, the pharynx is divided into three regions—the *nasopharynx, oropharynx*, and *laryngopharynx* (**Figure 22.5a**). The muscular pharynx wall is composed of skeletal muscle throughout its length (◀ **Table 10.3, pp. 338–339**). However, the cellular composition of its mucosa varies from one pharyngeal region to another.

The Nasopharynx

The **nasopharynx** is posterior to the nasal cavity, inferior to the sphenoid bone, and superior to the level of the soft palate. Because it lies above the point where food enters the body, it serves *only* as an air passageway. During swallowing, the soft palate and its pendulous *uvula* (u′vu-lah; "little grape") move superiorly, an action that closes off the nasopharynx and prevents food from entering the nasal cavity. (When we giggle, this sealing action fails and fluids being swallowed can end up spraying out the nose.)

The nasopharynx is continuous with the nasal cavity through the posterior nasal apertures (Figure 22.5b). Its pseudostratified ciliated epithelium takes over the job of propelling mucus where the nasal mucosa leaves off. High on its posterior wall is the **pharyngeal tonsil** (far-rin′je-al) (or *adenoids*), which traps and destroys pathogens entering the nasopharynx in air.

> #### HOMEOSTATIC IMBALANCE 22.2   CLINICAL
>
> Infected and swollen adenoids block air passage in the nasopharynx, making it necessary to breathe through the mouth. As a result, the air is not properly moistened, warmed, or filtered before reaching the lungs. When the adenoids are chronically enlarged, both speech and sleep may be disturbed. ● ● ● ● ● ●

The *pharyngotympanic (auditory) tubes*, which drain the middle ear cavities and allow middle ear pressure to equalize with atmospheric pressure, open into the lateral walls of the nasopharynx (Figures 22.4b and 22.5b). A ridge of pharyngeal mucosa posterior to each of these openings marks the site of the *tubal tonsil* (Figure 22.5b). Its strategic location helps protect the middle ear against infections likely to spread from the nasopharynx.

The Oropharynx

The **oropharynx** lies posterior to the oral cavity and is continuous with it through an archway called the **isthmus of the fauces** (faw′sēz; "throat") (Figure 22.5b). Because the oropharynx extends inferiorly from the level of the soft palate to the epiglottis, both swallowed food and inhaled air pass through it.

As the nasopharynx blends into the oropharynx, the epithelium changes from pseudostratified columnar to a more protective stratified squamous epithelium. This structural adaptation accommodates the increased friction and chemical trauma (characteristic of hot and spicy foods) accompanying food passage.

The paired **palatine tonsils** lie embedded in the lateral walls of the oropharyngeal mucosa just posterior to the oral cavity. The **lingual tonsil** covers the posterior surface of the tongue.

The Laryngopharynx

Like the oropharynx above it, the **laryngopharynx** (lah-ring″go-far′ingks) serves as a passageway for food and air and is lined with a stratified squamous epithelium. It lies

Nasopharynx

Oropharynx

Laryngopharynx

(a) Regions of the pharynx

Posterior nasal aperture

Nasopharynx
• Pharyngeal tonsil
• Tubal tonsil
• Opening of pharyngotympanic tube

Oropharynx
• Palatine tonsil
• Isthmus of the fauces

Laryngopharynx

Esophagus

Trachea

Hard palate

Soft palate

Tongue

Lingual tonsil

Hyoid bone

Larynx
• Epiglottis
• Vestibular fold
• Thyroid cartilage
• Vocal fold
• Cricoid cartilage

Thyroid gland

(b) Structures of the pharynx and larynx

Figure 22.5 The pharynx, larynx, and upper trachea. Midsagittal section of the head and neck.

Table 22.1 The Upper Respiratory System

STRUCTURE	DESCRIPTION, GENERAL AND DISTINCTIVE FEATURES	FUNCTION
Nose (external nose and nasal cavity)	Jutting external portion is supported by bone and cartilage. Internal nasal cavity is divided by midline nasal septum and lined with mucosa.	Produces mucus; filters, warms, and moistens incoming air; resonance chamber for speech
	Roof of nasal cavity contains olfactory epithelium.	Receptors for sense of smell
Paranasal sinuses	Mucosa-lined, air-filled cavities in cranial bones surrounding nasal cavity (◀ p. 215).	Lighten skull; may also warm, moisten, and filter incoming air
Pharynx	Passageway connecting nasal cavity to larynx and oral cavity to esophagus. Three subdivisions: nasopharynx, oropharynx, and laryngopharynx.	Passageway for air and food
	Houses tonsils (lymphoid tissue masses involved in protection against pathogens).	Facilitates exposure of immune system to inhaled antigens

directly posterior to the larynx, where the respiratory and digestive pathways diverge, and extends to the inferior edge of the cricoid cartilage. The laryngopharynx is continuous with the esophagus posteriorly.

The esophagus conducts food and fluids to the stomach; air enters the larynx anteriorly. During swallowing, food has the "right of way," and air passage temporarily stops.

Check Your Understanding

1. Air moving from the nose to the larynx passes by a number of structures. List (in order) as many of these structures as you can.
2. Name the two types of mucous membrane found in the nasal cavity.
3. **DRAW** Create a summary table to help you study the pharynx by comparing and contrasting its three parts. For each part, identify what it conducts (air, food, or both), the type of epithelium found there, and the associated tonsils.

For answers, see Answers Appendix.

22.2 The lower respiratory system consists of conducting and respiratory zone structures

Learning Outcomes

▶ Distinguish between conducting and respiratory zone structures.
▶ Describe the structure, function, and location of the larynx, trachea, and bronchi.
▶ Describe the makeup of the respiratory membrane, and relate structure to function.
▶ Identify the organs forming the respiratory passageway(s) in descending order until you reach the alveoli.

Anatomically, the lower respiratory system consists of the *larynx*, *trachea*, *bronchi*, and *lungs*. Functionally, the respiratory system as a whole consists of two zones:

- The **respiratory zone**, the actual site of gas exchange, is composed of the respiratory bronchioles, alveolar ducts, and alveoli, all microscopic structures.

- The **conducting zone** consists of all of the respiratory passageways from the nose to the respiratory bronchioles. These provide fairly rigid conduits for air to reach the gas exchange sites. The conducting zone organs also cleanse, humidify, and warm incoming air. As a result, air reaching the lungs has fewer irritants (dust, bacteria, etc.) than when it entered the body, and it is warm and damp, like the air of the tropics.

The Larynx

Basic Anatomy

The **larynx** (lar'ingks), or voice box, extends for about 5 cm (2 inches) from the level of the third to the sixth cervical vertebra. Superiorly it attaches to the hyoid bone and opens into the laryngopharynx. Inferiorly it is continuous with the trachea (Figure 22.5b).

The larynx has three functions:

- Provide a *patent* (open) airway
- Act as a switching mechanism to route air and food into the proper channels
- Voice production [because it houses the vocal folds (vocal cords)]

The framework of the larynx is an intricate arrangement of nine cartilages connected by membranes and ligaments (**Figure 22.6**). Except for the epiglottis, all laryngeal cartilages are hyaline cartilages.

The large, shield-shaped **thyroid cartilage** is formed by the fusion of two cartilage plates at the midline. It resembles an upright open book, with the book's "spine" lying in the anterior midline of the neck. This "book spine" is the **laryngeal prominence** (lah-rin'je-al), which can be seen externally as the *Adam's apple* (Figure 22.6a). The thyroid cartilage is typically larger in males than in females because male sex hormones stimulate its growth during puberty. Inferior to the thyroid cartilage is the ring-shaped **cricoid cartilage** (kri'koid), perched atop and anchored to the trachea inferiorly.

Three pairs of small cartilages—**arytenoid** (ar″ĭ-te'noid), **cuneiform** (ku-ne'ĭ-form), and **corniculate cartilages**—form part of the lateral and posterior walls of the larynx. The most important of these are the pyramid-shaped arytenoid cartilages, which anchor the vocal folds.

Epiglottis The ninth cartilage, the flexible, spoon-shaped **epiglottis** (ep″ĭ-glot′is; "above the glottis"), is composed of elastic cartilage and is almost entirely covered by a taste bud–containing mucosa. The epiglottis extends from the posterior aspect of the tongue to its anchoring point on the anterior rim of the thyroid cartilage (Figure 22.6c and d).

When only air is flowing into the larynx, the inlet to the larynx is wide open and the free edge of the epiglottis projects upward. During swallowing, the larynx is pulled superiorly and the epiglottis tips to cover the laryngeal inlet. Because this action keeps food out of the lower respiratory passages, the epiglottis has been called the guardian of the airways. Anything other than air entering the larynx initiates the cough reflex to expel the substance. This protective reflex does not work when we are unconscious, so it is never a good idea to administer liquids when attempting to revive an unconscious person.

Vocal Folds Lying under the laryngeal mucosa on each side are the **vocal ligaments**, which attach the arytenoid cartilages to the thyroid cartilage. These ligaments, composed largely of elastic fibers, form the core of mucosal folds called the

(a) Surface view

(b) Anterior view

(c) Photograph of cartilaginous framework of the larynx, posterior view

(d) Sagittal section (anterior on the right)

Figure 22.6 The larynx.

vocal folds, or **true vocal cords**, which appear pearly white because they lack blood vessels (**Figure 22.7**).

The vocal folds vibrate, producing sounds as air rushes up from the lungs. The vocal folds and the medial opening between them through which air passes are called the **glottis**. Superior to the vocal folds is a similar pair of mucosal folds called the **vestibular folds**, or **false vocal cords**. These play no direct part in sound production but help to close the glottis when we swallow.

Epithelium of Larynx   Stratified squamous epithelium lines the superior portion of the larynx, an area subject to food contact. Below the vocal folds the epithelium is a pseudostratified ciliated columnar type that filters dust. The power stroke of its cilia is directed upward toward the pharynx to continually move mucus *away* from the lungs. We help move mucus up and out of the larynx when we "clear our throat."

Voice Production

Speech involves the intermittent release of expired air as the glottis opens and closes. The length of the vocal folds and the size of the glottis change with the action of the intrinsic laryngeal muscles that clothe the cartilages. Most of these muscles move the arytenoid cartilages. As the length and tension of the vocal folds change, the pitch of the sound varies. Generally, the tenser the vocal folds, the faster they vibrate and the higher the pitch.

As a boy's larynx enlarges during puberty, his vocal folds become longer and thicker. Because this causes them to vibrate more slowly, his voice becomes deeper. Until the young man learns to control his newly enlarged vocal folds, his voice "cracks."

Loudness of the voice depends on the force with which the airstream rushes across the vocal folds. The greater the force, the stronger the vibration and the louder the sound. The vocal folds do not move at all when we whisper, but they vibrate vigorously when we yell. The muscles of the chest, abdomen, and back provide the power for the airstream.

The vocal folds actually produce buzzing sounds. The perceived quality of the voice depends on the coordinated activity of many structures above the glottis. For example, the entire length of the pharynx acts as a resonating chamber, to amplify and enhance the sound quality. The oral, nasal, and sinus cavities also contribute to vocal resonance. In addition, good enunciation depends on muscles in the pharynx, tongue, soft palate, and lips that "shape" sound into recognizable consonants and vowels.

HOMEOSTATIC IMBALANCE 22.3   CLINICAL

Inflammation of the vocal folds, or **laryngitis**, causes the vocal folds to swell, interfering with their vibration. This changes the vocal tone, causing hoarseness, or in severe cases limiting us to a whisper. Laryngitis is most often caused by viral infections, but may also be due to overusing the voice, very dry air, bacterial infections, tumors on the vocal folds, or inhalation of irritating chemicals. • • • • • • • •

Sphincter Functions of the Larynx

Under certain conditions, the vocal folds act as a sphincter that prevents air passage. During abdominal straining associated with defecation, the glottis closes to prevent exhalation and the abdominal muscles contract, causing the intra-abdominal pressure to rise. These events, collectively known as the **Valsalva maneuver**, help empty the rectum and can also splint (stabilize) the body trunk when lifting a heavy load.

The Trachea

The **trachea** (tra′ke-ah), or *windpipe*, descends from the larynx through the neck and into the mediastinum. It ends by dividing into the two main bronchi at midthorax (see Figure 22.2 on

(a) Vocal folds in closed position; closed glottis

Epiglottis
Vestibular fold (false vocal cord)
Vocal fold (true vocal cord)
Glottis
Inner lining of trachea
Cuneiform cartilage
Corniculate cartilage

(b) Vocal folds in open position; open glottis

Figure 22.7 Movements of the vocal folds. Superior view of the larynx and vocal folds, as seen through a laryngoscope.

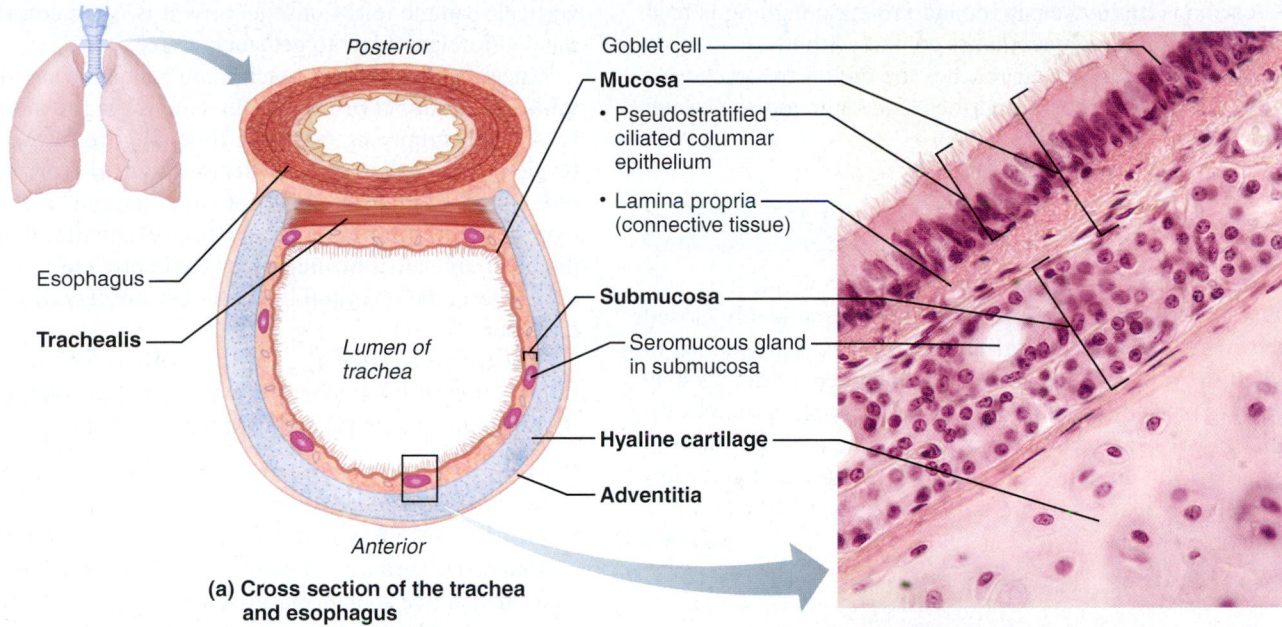

(a) Cross section of the trachea and esophagus

Goblet cell
Mucosa
• Pseudostratified ciliated columnar epithelium
• Lamina propria (connective tissue)
Esophagus
Trachealis
Lumen of trachea
Submucosa
Seromucous gland in submucosa
Hyaline cartilage
Adventitia
Posterior
Anterior

(b) Photomicrograph of the tracheal wall (320×)

Figure 22.8 Tissue composition of the tracheal wall. In the scanning electron micrograph in (c), the cilia appear as yellow, grass-like projections. Mucus-secreting goblet cells (orange) with short microvilli are interspersed between the ciliated cells.

Practice Histology questions:
MasteringA&P® > Study Area > Lab Tools > PAL

p. 820). In humans, it is 10–12 cm (about 4 inches) long and 2 cm (3/4 inch) in diameter, and very flexible and mobile.

The tracheal wall consists of several layers that are common to many tubular body organs—the *mucosa, submucosa,* and *adventitia*—plus a layer of hyaline cartilage (**Figure 22.8**). The **mucosa** has the same goblet cell–containing pseudostratified epithelium that occurs throughout most of the respiratory tract. Its cilia continually propel debris-laden mucus toward the pharynx. This epithelium rests on a fairly thick lamina propria that has a rich supply of elastic fibers.

(c) Scanning electron micrograph of cilia in the trachea (2500×)

HOMEOSTATIC IMBALANCE 22.4   CLINICAL

Smoking inhibits and ultimately destroys cilia. Without ciliary activity, coughing is the only way to prevent mucus from accumulating in the lungs. When someone stops smoking, ciliary function usually recovers within a few weeks, and the morning "smoker's cough" subsides.

The **submucosa**, a connective tissue layer deep to the mucosa, contains seromucous glands that help produce the mucus "sheets" within the trachea. The submucosa is supported by 16 to 20 C-shaped rings of hyaline cartilage encased by the **adventitia**, the outermost layer of connective tissue (Figure 22.8).

The trachea's elastic elements make it flexible enough to stretch and move inferiorly during inspiration and recoil during expiration, but the cartilage rings prevent it from collapsing and keep the airway patent despite the pressure changes that occur during breathing. The open posterior parts of the cartilage rings, which lie next to the esophagus (Figure 22.8a), are connected by smooth muscle fibers of the **trachealis** and by soft connective tissue. Because this portion of the tracheal wall is flexible, the esophagus can expand anteriorly as swallowed food passes through it.

Contraction of the trachealis decreases the trachea's diameter, causing expired air to rush upward from the lungs with greater force. This action helps expel mucus from the trachea when we cough by accelerating the exhaled air to speeds of 100 mph!

The last tracheal cartilage is expanded, and a spar of cartilage, called the **carina** (kar-ri′nah; "keel"), projects posteriorly from its inner face, marking the point where the trachea branches into the two *main bronchi* (Figure 22.2). The mucosa

22

of the carina is highly sensitive, and violent coughing is triggered when a foreign object makes contact with it.

By the time incoming air reaches the end of the trachea, it is warm, cleansed of most impurities, and saturated with water vapor.

HOMEOSTATIC IMBALANCE 22.5   CLINICAL

Tracheal obstruction is life threatening. Many people have suffocated after choking on a piece of food that suddenly closed off their trachea. The **Heimlich maneuver**, a procedure in which air in the victim's lungs is used to "pop out," or expel, an obstructing piece of food, has saved many people from becoming victims of "café coronaries." The maneuver is simple to learn and easy to do. However, it is best learned by demonstration because cracked ribs are a distinct possibility when it is done incorrectly.

The Bronchi and Subdivisions

The air passageways in the lungs branch and branch again, about 23 times overall, in a pattern often called the **bronchial tree** (**Figure 22.9**). At the tips of the bronchial tree, conducting zone structures give way to respiratory zone structures.

Conducting Zone Structures

The trachea divides to form the **right** and **left main (primary) bronchi** (brong′ki) approximately at the level of T_7 in an erect (standing) person. Each bronchus runs obliquely in the mediastinum before plunging into the medial depression (hilum) of its lung. The right main bronchus is wider, shorter, and more vertical than the left. Consequently, it is more common for an inhaled foreign object to get stuck there.

Once inside the lungs, each main bronchus subdivides into **lobar (secondary) bronchi**—three on the right and two on the left—each supplying one lung lobe (Figure 22.9). The lobar bronchi branch into third-order **segmental (tertiary) bronchi**, which divide repeatedly into smaller and smaller bronchi (fourth-order, fifth-order, etc.). Passages smaller than 1 mm in diameter are called **bronchioles** ("little bronchi"), and the tiniest of these, the **terminal bronchioles**, are less than 0.5 mm in diameter.

The tissue composition of the walls of the main bronchi mimics that of the trachea. However, as the conducting tubes become smaller, the following structural changes occur:

- **Support structures change.** Irregular patches, or *plates*, of cartilage replace the cartilage rings, and by the time the bronchioles are reached, the tube walls no longer contain supportive cartilage. However, the tube walls throughout the bronchial tree contain elastic fibers.

- **Epithelium type changes.** The mucosal epithelium thins as it changes from pseudostratified columnar to simple columnar and then to simple cuboidal in the terminal bronchioles. Mucus-producing cells and cilia are sparse in the bronchioles. For this reason, most airborne debris found at or below the level of the bronchioles must be removed by macrophages in the alveoli.

- **Amount of smooth muscle increases.** The relative amount of smooth muscle in the tube walls increases as the passageways become smaller. A complete layer of circular smooth muscle in the bronchioles and the lack of supporting cartilage (which would hinder constriction) allows the bronchioles to

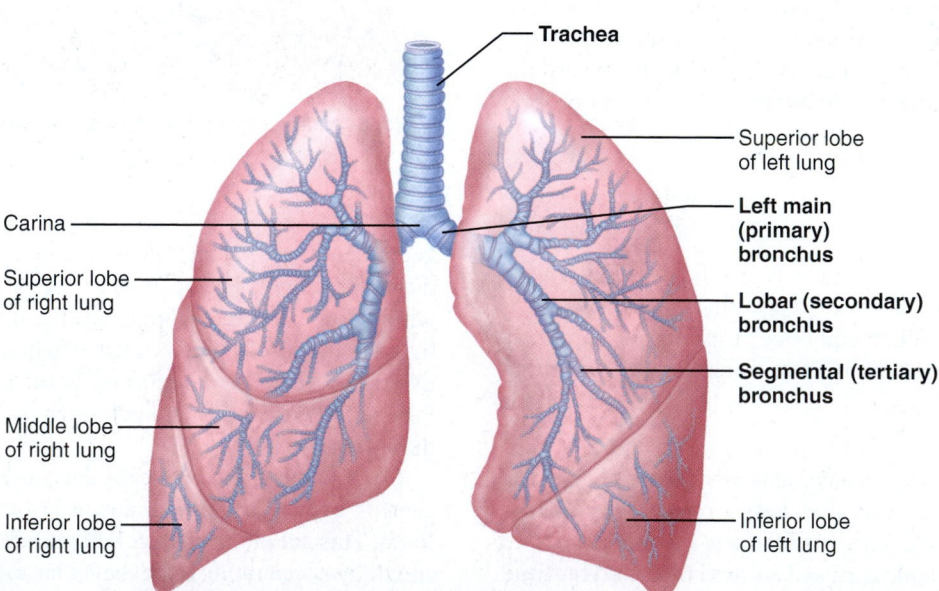

Figure 22.9 Conducting zone passages. The air pathway inferior to the larynx consists of the trachea and the main, lobar, and segmental bronchi, which branch into the smaller bronchi and bronchioles until reaching the terminal bronchioles of the lungs.

(a)

Alveoli

Alveolar duct

Respiratory bronchioles

Alveolar duct

Terminal bronchiole

Alveolar sac

(b)

Respiratory bronchiole

Alveolar duct

Alveolar pores

Alveoli

Alveolar sac

Figure 22.10 Respiratory zone structures. (a) Diagram of respiratory bronchioles, alveolar ducts, alveolar sacs, and alveoli. **(b)** Photomicrograph of a section of the lung (80×). Notice the thinness of the alveolar walls.

Practice Histology questions:
MasteringA&P° > Study Area > Lab Tools > PAL

22

provide substantial resistance to air passage under certain conditions (as we will describe later).

Respiratory Zone Structures

Defined by the presence of thin-walled air sacs called **alveoli** (al-ve′o-li; *alveol* = small cavity), the respiratory zone begins as the terminal bronchioles feed into **respiratory bronchioles** within the lung (**Figure 22.10**). Protruding from these smallest bronchioles are scattered alveoli. The respiratory bronchioles lead into winding **alveolar ducts**, whose walls consist of diffusely arranged rings of smooth muscle cells, connective tissue fibers, and outpocketing alveoli. The alveolar ducts lead into terminal clusters of alveoli called **alveolar sacs** or **alveolar saccules**.

Many people mistakenly equate alveoli, the site of gas exchange, with alveolar sacs, but they are not the same thing. The alveolar sac is analogous to a bunch of grapes, and the alveoli are the individual grapes. The 300 million or so gas-filled alveoli in the lungs account for most of our lung volume and provide a tremendous surface area for gas exchange.

The Respiratory Membrane The walls of the alveoli are composed primarily of a single layer of squamous epithelial cells, called **type I alveolar cells**, surrounded by a flimsy basement membrane. The thinness of their walls is hard to imagine, but a sheet of tissue paper is 15 times thicker!

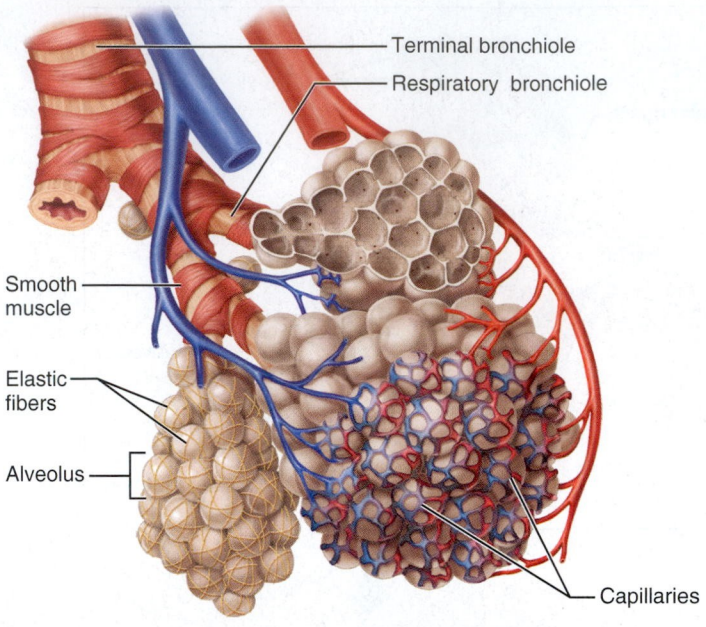

Terminal bronchiole
Respiratory bronchiole
Smooth muscle
Elastic fibers
Alveolus
Capillaries

(a) Diagrammatic view of capillary-alveoli relationships

(b) Scanning electron micrograph of pulmonary capillary casts (300×)

Nucleus of type I alveolar cell
Alveolar pores
Capillary
Macrophage
Endothelial cell nucleus
Respiratory membrane
Alveolus
Red blood cell
O_2
CO_2
Capillary
Alveolus
Alveolar epithelium
Fused basement membranes of alveolar epithelium and capillary endothelium
Capillary endothelium

Alveoli (gas-filled air spaces)
Red blood cell in capillary
Type II alveolar cell (secretes surfactant)
Type I alveolar cell

(c) Detailed anatomy of the respiratory membrane

Figure 22.11 Alveoli and the respiratory membrane. Elastic fibers and capillaries surround all alveoli, but for clarity they are shown only on some alveoli in (a). In (b), the tissue forming the alveoli has been removed. Only the capillary network remains.

The external surfaces of the alveoli are densely covered with a "cobweb" of pulmonary capillaries (**Figure 22.11**). Together, the capillary and alveolar walls and their fused basement membranes form the **respiratory membrane**, a 0.5-μm-thick *blood air barrier* that has blood flowing past on one side and gas on the other (Figure 22.11c). Gas exchanges occur readily by simple diffusion across the respiratory membrane—O_2 passes from the alveolus into the blood, and CO_2 leaves the blood to enter the gas-filled alveolus.

22

Table 22.2 The Lower Respiratory System

STRUCTURE		DESCRIPTION, GENERAL AND DISTINCTIVE FEATURES	FUNCTION
Larynx		Connects pharynx to trachea. Has framework of cartilage and dense connective tissue. Opening (glottis) can be closed by epiglottis or vocal folds.	Air passageway; prevents food from entering lower respiratory tract
		Houses vocal folds (true vocal cords).	Voice production
Trachea		Flexible tube running from larynx and dividing inferiorly into two main bronchi. Walls contain C-shaped cartilages that are incomplete posteriorly where connected by trachealis.	Air passageway; cleans, warms, and moistens incoming air
Bronchial tree		Consists of right and left main bronchi, which subdivide within the lungs to form lobar and segmental bronchi and bronchioles. Bronchiolar walls lack cartilage but contain complete layer of smooth muscle. Constriction of this muscle impedes expiration.	Air passageways connecting trachea with alveoli; cleans, warms, and moistens incoming air
Alveoli		Microscopic chambers at termini of bronchial tree. Walls of simple squamous epithelium overlie thin basement membrane. External surfaces are intimately associated with pulmonary capillaries.	Main sites of gas exchange
		Special alveolar cells produce surfactant.	Surfactant reduces surface tension; helps prevent lung collapse
Lungs		Paired composite organs that flank mediastinum in thorax. Composed primarily of alveoli and respiratory passageways. Stroma is elastic connective tissue, allowing lungs to recoil passively during expiration.	House respiratory passages smaller than the main bronchi
Pleurae		Serous membranes. Parietal pleura lines thoracic cavity; visceral pleura covers external lung surfaces.	Produce lubricating fluid and compartmentalize lungs

Alveoli   Three major types of cells are found in alveoli:

- **Type I alveolar cells** (described above) are squamous epithelial cells that form the major part of the alveolar walls.
- **Type II alveolar cells** are cuboidal epithelial cells that are scattered among the type I cells (Figure 22.11c). They secrete a fluid containing a detergent-like substance called *surfactant* that coats the gas-exposed alveolar surfaces. (We describe surfactant's role in reducing the surface tension of the alveolar fluid later in this chapter.) Type II alveolar cells also secrete a number of antimicrobial proteins that are important elements of innate immunity.
- **Alveolar macrophages** crawl freely along the internal alveolar surfaces consuming bacteria, dust, and other debris (Figure 22.11c).

The alveoli have two other significant features: (1) They are surrounded by fine elastic fibers of the same type that surround the entire bronchial tree. (2) Open **alveolar pores** connecting adjacent alveoli allow air pressure throughout the lung to be equalized and provide alternate air routes to any alveoli whose bronchi have collapsed due to disease.

Although huge numbers of infectious microorganisms are continuously carried into the alveoli, alveolar surfaces are usually sterile. Because the alveoli are "dead ends," aged and dead macrophages must be prevented from accumulating in them. Most macrophages simply get swept up by the ciliary current of superior regions and carried to the pharynx. In this manner, we clear and swallow over 2 million alveolar macrophages per hour!

Check Your Understanding

4. Which structure seals the larynx when we swallow?

5. Which structural features of the trachea allow it to expand and contract, yet keep it from collapsing?

6. What features of the alveoli and their respiratory membranes suit them to their function of exchanging gases by diffusion?

7. APPLY A 3-year-old boy is brought to the emergency department after aspirating (inhaling) a peanut. Bronchoscopy confirms the suspicion that the peanut is lodged in a bronchus and then it is successfully extracted. Which main bronchus was the peanut most likely to be in? Why?

For answers, see Answers Appendix.

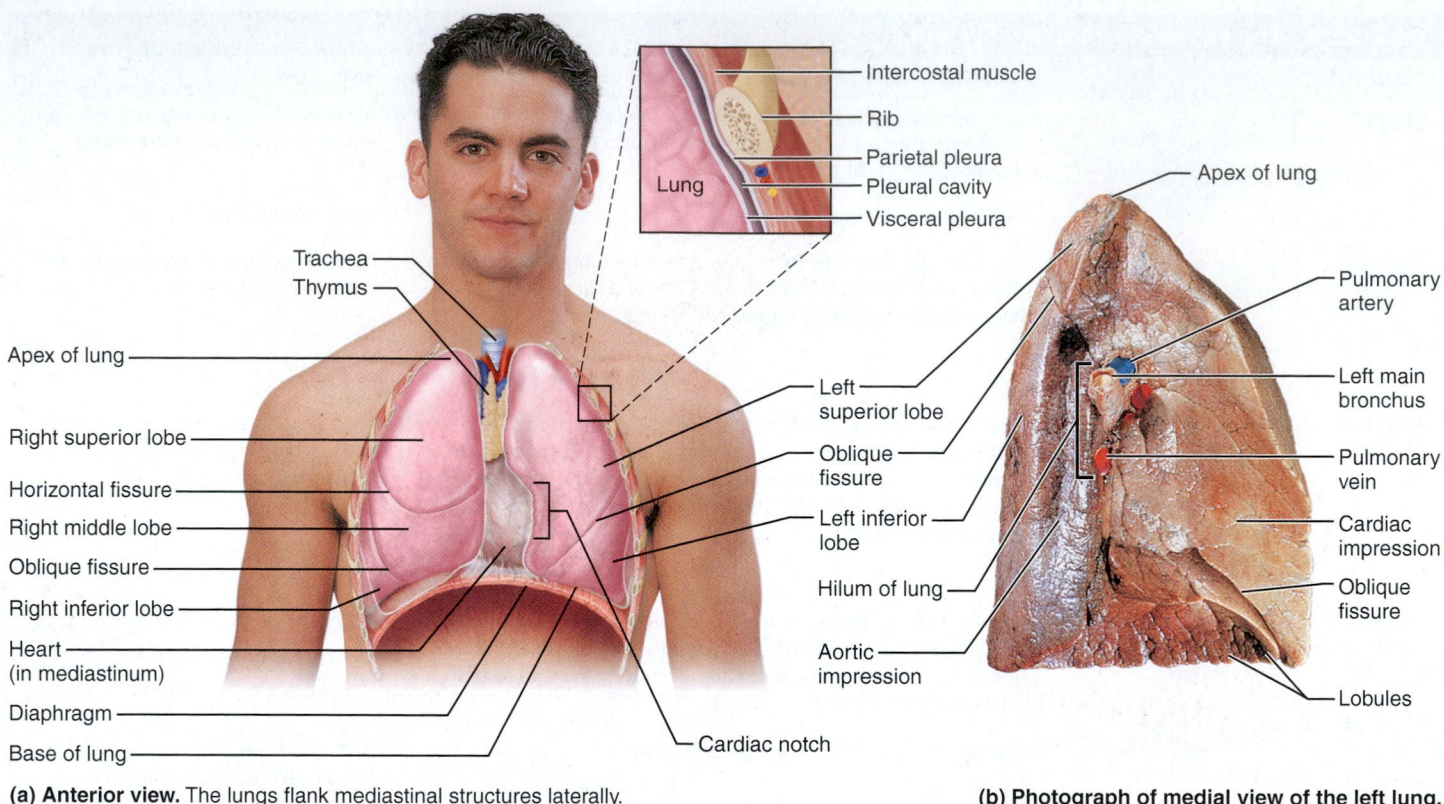

Intercostal muscle
Rib
Parietal pleura
Pleural cavity
Visceral pleura
Lung

Trachea
Thymus
Apex of lung
Right superior lobe
Horizontal fissure
Right middle lobe
Oblique fissure
Right inferior lobe
Heart (in mediastinum)
Diaphragm
Base of lung

Apex of lung
Left superior lobe
Oblique fissure
Left inferior lobe
Hilum of lung
Aortic impression
Cardiac notch

Apex of lung
Pulmonary artery
Left main bronchus
Pulmonary vein
Cardiac impression
Oblique fissure
Lobules

(a) Anterior view. The lungs flank mediastinal structures laterally.

(b) Photograph of medial view of the left lung.

Posterior

Vertebra
Right lung
Parietal pleura
Visceral pleura
Pleural cavity
Pericardial membranes
Sternum

Esophagus (in mediastinum)
Root of lung at hilum
• Left main bronchus
• Left pulmonary artery
• Left pulmonary vein
Left lung
Thoracic wall
Pulmonary trunk
Heart (in mediastinum)
Anterior mediastinum

Anterior

(c) Transverse section through the thorax, viewed from above. Lungs, pleural membranes, and major organs in the mediastinum are shown.

Figure 22.12 **Anatomical relationships of organs in the thoracic cavity.**
In (c), the size of the pleural cavity is exaggerated for clarity.

22

22.3 Each multilobed lung occupies its own pleural cavity

Learning Outcome

▶ Describe the gross structure of the lungs and pleurae.

The paired **lungs** occupy all of the thoracic cavity except the mediastinum, which houses the heart, great blood vessels, bronchi, esophagus, and other organs (**Figure 22.12**).

Gross Anatomy of the Lungs

Each cone-shaped lung is surrounded by pleurae and connected to the mediastinum by vascular and bronchial attachments, collectively called the lung **root**. The anterior, lateral, and posterior lung surfaces lie in close contact with the ribs and form the continuously curving **costal surface**. Just deep to the clavicle is the **apex**, the narrow superior tip of the lung. The concave, inferior surface that rests on the diaphragm is the **base**.

On the mediastinal surface of each lung is an indentation, the **hilum**, through which pulmonary and systemic blood vessels, bronchi, lymphatic vessels, and nerves enter and leave the lungs. Each main bronchus plunges into the hilum on its own side and begins to branch almost immediately. All conducting and respiratory passageways distal to the main bronchi are found in the lungs.

The two lungs differ slightly in shape and size because the apex of the heart is slightly to the left of the median plane. The left lung is smaller than the right, and the **cardiac notch**—a concavity in its medial aspect—is molded to and accommodates the heart (Figure 22.12a). The left lung is subdivided into superior and inferior **lobes** by the *oblique fissure*, whereas the right lung is partitioned into superior, middle, and inferior lobes by the *oblique* and *horizontal fissures*.

Each lobe contains a number of pyramid-shaped **bronchopulmonary segments** separated from one another by connective tissue septa. The right lung has 10 bronchopulmonary segments, but the left lung is more variable and consists of 8 to 10 segments (**Figure 22.13**). Each segment is served by its own artery and vein and receives air from an individual segmental (tertiary) bronchus.

The bronchopulmonary segments are clinically important because pulmonary disease is often confined to one or a few segments. Their connective tissue partitions allow diseased segments to be surgically removed without damaging neighboring segments or impairing their blood supply.

The smallest subdivisions of the lung visible with the naked eye are the **lobules**, which appear at the lung surface as hexagons ranging from the size of a pencil eraser to the size of a penny (Figure 22.12b). A large bronchiole and its branches serve each lobule. In most city dwellers and in smokers, the connective tissue that separates the individual lobules is blackened with carbon.

As we mentioned earlier, the lungs consist largely of air spaces. The balance of lung tissue, or its **stroma** ("mattress" or "bed"), is mostly elastic connective tissue. As a result, the lungs are soft, spongy, elastic organs that together weigh just over 1 kg (2.2 lb). The elasticity of healthy lungs reduces the work of breathing, as we will describe shortly.

Right lung — Left lung

Right superior lobe (3 segments)

Left superior lobe (4 segments)

Right middle lobe (2 segments)

Right inferior lobe (5 segments)

Left inferior lobe (5 segments)

Figure 22.13 A cast of the bronchial tree. The individual bronchopulmonary segments have been painted different colors.

22

Blood Supply and Innervation of the Lungs

The lungs are perfused by two circulations, the pulmonary and the bronchial, which differ in size, origin, and function.

Pulmonary Circulation of the Lungs

Systemic venous blood that is to be oxygenated in the lungs is delivered by the **pulmonary arteries**, which lie anterior to the main bronchi (Figure 22.12c). In the lungs, the pulmonary arteries branch profusely along with the bronchi and finally feed into the **pulmonary capillary networks** surrounding the alveoli (see Figure 22.11a).

The **pulmonary veins** convey the freshly oxygenated blood from the respiratory zone of the lungs to the heart. Their tributaries course back to the hilum both with the corresponding bronchi and in the connective tissue septa separating the bronchopulmonary segments.

The pulmonary circuit is a low-pressure, high-volume circulation. Because *all* of the body's blood passes through the lungs about once each minute, the lung capillary endothelium is an ideal location for enzymes that act on materials in the blood. Examples include *angiotensin converting enzyme*, which activates an important blood pressure hormone, and enzymes that inactivate certain prostaglandins.

Bronchial Circulation of the Lungs

In contrast to the pulmonary circulation, the **bronchial arteries** provide oxygenated systemic blood to lung tissue. The bronchial arteries arise from the aorta, enter the lungs at the hilum, and then run along the branching bronchi. They provide a high-pressure, low-volume supply of oxygenated blood to all lung tissues except the alveoli. The tiny bronchial veins drain some systemic venous blood from the lungs, but there are multiple anastomoses between the two circulations, and most venous blood returns to the heart via the pulmonary veins.

Innervation of the Lungs

The lungs are innervated by parasympathetic and sympathetic motor fibers, and visceral sensory fibers. These nerve fibers enter each lung through the **pulmonary plexus** on the lung root and run along the bronchial tubes and blood vessels in the lungs. Parasympathetic fibers cause the air tubes to constrict, whereas sympathetic fibers dilate them.

The Pleurae

The **pleurae** (ploo're; "sides") form a thin, double-layered serosa (◄ pp. 18–19). The layer called the **parietal pleura** covers the thoracic wall and superior face of the diaphragm (Figure 22.12a, c). It continues around the heart and between the lungs, forming the lateral walls of the mediastinal enclosure and snugly enclosing the lung root. From here, the pleura extends as the layer called the **visceral pleura** to cover the external lung surface, dipping into and lining its fissures.

The pleurae produce **pleural fluid**, which fills the slitlike **pleural cavity** between them. This lubricating secretion allows the lungs to glide easily over the thorax wall during our breathing movements. Although the pleurae slide easily across each other, the surface tension of the pleural fluid strongly resists their separation. Consequently, the lungs cling tightly to the thorax wall and expand and recoil passively as the volume of the thoracic cavity alternately increases and decreases during breathing.

The pleurae also help divide the thoracic cavity into three chambers—the central mediastinum and the two lateral pleural compartments, each containing a lung. This compartmentalization helps prevent one mobile organ (for example, the lung or heart) from interfering with another. It also limits the spread of local infections.

HOMEOSTATIC IMBALANCE 22.6 — CLINICAL

Pleurisy (ploo'rĭ-se), inflammation of the pleurae, often results from pneumonia. Inflamed pleurae become rough, resulting in friction and stabbing pain with each breath. As the disease progresses, the pleurae may produce excessive amounts of fluid. This increased fluid relieves the pain caused by pleural surfaces rubbing together, but may exert pressure on the lungs and hinder breathing movements.

Other fluids that may accumulate in the pleural cavity include blood (leaked from damaged blood vessels) and blood filtrate (the watery fluid that oozes from the lung capillaries when left-sided heart failure occurs). The term for fluid accumulation in the pleural cavity is *pleural effusion*. ●●●●●●

Check Your Understanding

8. The lungs are perfused by two different circulations. Name these circulations and indicate their roles in the lungs.

9. **MAKE CONNECTIONS** Where is angiotensin converting enzyme found and why is this a good location for this enzyme? Name the blood pressure–increasing hormone cascade (described in Chapter 19) to which this enzyme belongs.

For answers, see Answers Appendix.

PART 2

RESPIRATORY PHYSIOLOGY

22.4 Volume changes cause pressure changes, which cause air to move

Learning Outcomes

▶ Explain the functional importance of the partial vacuum that exists in the intrapleural space.
▶ Relate Boyle's law to events of inspiration and expiration.
▶ Explain the relative roles of the respiratory muscles and lung elasticity in producing the volume changes that cause air to flow into and out of the lungs.
▶ List several physical factors that influence pulmonary ventilation.

Breathing, or **pulmonary ventilation**, consists of two phases: **inspiration**, the period when air flows into the lungs, and **expiration**, the period when gases exit the lungs.

Pressure Relationships in the Thoracic Cavity

Before we discuss the breathing process, it is important to understand that *respiratory pressures are always described relative to* **atmospheric pressure** (P_{atm}), which is the pressure exerted by the air (gases) surrounding the body. At sea level, atmospheric pressure is 760 mm Hg (the pressure exerted by a column of mercury 760 mm high). This pressure can also be expressed in atmosphere units: atmospheric pressure = 760 mm Hg = 1 atm.

- A negative respiratory pressure in any respiratory area indicates that the pressure in that region is lower than atmospheric pressure. For instance, a respiratory pressure of –4 mm Hg indicates a pressure that is lower than atmospheric pressure by 4 mm Hg (760 – 4 = 756 mm Hg). In this case, 756 mm Hg is the absolute pressure in that region.

- A positive respiratory pressure is higher than atmospheric pressure.

- A zero respiratory pressure is equal to atmospheric pressure.

Now we are ready to examine the pressure relationships that normally exist in the thoracic cavity.

Intrapulmonary Pressure (P_{pul})

The **intrapulmonary** (intra-alveolar) **pressure** (P_{pul}) is the pressure in the alveoli. Intrapulmonary pressure rises and falls with the phases of breathing, but it *always* equalizes with the atmospheric pressure eventually (**Figure 22.14**).

Intrapleural Pressure (P_{ip})

The pressure in the pleural cavity, the **intrapleural pressure** (P_{ip}), also fluctuates with breathing phases, but is always about 4 mm Hg less than P_{pul}. That is, P_{ip} is *always* negative relative to P_{pul}.

What causes this negative intrapleural pressure? To answer this question, let's examine the forces that exist in the thorax. First of all, we know there are opposing forces. Two forces act to pull the lungs (visceral pleura) away from the thorax wall (parietal pleura) and cause the lungs to collapse:

- **The lungs' natural tendency to recoil.** Because of their elasticity, lungs always assume the smallest size possible.

- **The surface tension of the alveolar fluid.** The molecules of the fluid lining the alveoli attract each other. This produces *surface tension* (◄ p. 35) that constantly acts to draw the alveoli to their smallest possible dimension.

However, these lung-collapsing forces are opposed by the natural elasticity of the chest wall, a force that tends to pull the thorax outward and enlarge the lungs. So which force wins? In a healthy person, the answer is neither, because of the strong

Figure 22.14 Intrapulmonary and intrapleural pressure relationships. Values shown are pressures relative to atmospheric pressure (760 mm Hg at sea level). Absolute pressures are given in parentheses. These pressures are at the end of a normal expiration. For illustration, the size of the pleural cavity has been greatly exaggerated.

adhesive force between the parietal and visceral pleurae. Pleural fluid secures the pleurae together in the same way a drop of water holds two glass slides together. The pleurae slide from side to side easily, but they remain stuck together, and separating them requires extreme force. The net result of the dynamic interplay between these forces is a negative P_{ip}.

The amount of pleural fluid in the pleural cavity must remain minimal to maintain a negative P_{ip}. The pleural fluid is actively pumped out of the pleural cavity into the lymphatics continuously. If it wasn't, fluid would accumulate in the intrapleural space (remember, fluids move from high to low pressure), producing a positive pressure in the pleural cavity.

Transpulmonary Pressure

The **transpulmonary pressure** is the difference between the intrapulmonary and intrapleural pressures ($P_{pul} - P_{ip}$). It is this pressure that keeps the air spaces of the lungs open or, phrased another way, keeps the lungs from collapsing. Moreover, *the size of the transpulmonary pressure determines the size of the lungs* at any time—the greater the transpulmonary pressure, the larger the lungs. We cannot overemphasize the importance of negative pressure in the intrapleural space and the tight coupling of the lungs to the thorax wall. Any condition that equalizes P_{ip} with

Rupturing either the parietal or the visceral pleura can cause pneumothorax:

Punctured parietal pleura
(e.g., knife wound)

Ruptured visceral pleura
(often spontaneous)

Parietal pleura

Visceral pleura

Pleural cavity
(Intrapleural pressure
= −4 mm Hg)

Intrapulmonary
pressure (0 mm Hg)

Atmospheric pressure
0 mm Hg (760 mm Hg)

Pneumothorax (air
in pleural cavity): causes
intrapleural pressure to
become equal to
intrapulmonary pressure.
As a result, the lungs
collapse.

Intrapleural
pressure
(−4 mm Hg)

Intrapulmonary
pressure (0 mm Hg)

Collapsed lung
(atelectasis)

Figure 22.15 Pneumothorax. One lung can collapse without affecting the other because the lungs are in separate cavities.

the intrapulmonary (or atmospheric) pressure causes immediate *lung collapse* (**Figure 22.15**).

HOMEOSTATIC IMBALANCE 22.7

CLINICAL

Atelectasis (at″ĕ-lek′tah-sis), or lung collapse, occurs when a bronchiole becomes plugged (as may follow pneumonia). Its associated alveoli then absorb all of their air and collapse. Atelectasis can also occur when air enters the pleural cavity either through a chest wound or a rupture of the visceral pleura, which allows air from the respiratory tract to enter the pleural cavity.

The presence of air in the pleural cavity is referred to as a **pneumothorax** (nu″mo-tho′raks; "air thorax"), and is reversed by drawing air out of the intrapleural space with chest tubes. This procedure allows the pleurae to heal and the lung to reinflate and resume normal function.　●●●●●●●

Pulmonary Ventilation

Pulmonary ventilation, consisting of inspiration and expiration, is a mechanical process that depends on volume changes in the thoracic cavity. A rule to keep in mind is that *volume changes*

lead to *pressure changes*, and pressure changes lead to the *flow of gases* to equalize the pressure.

Boyle's law gives the relationship between the pressure and volume of a gas: At constant temperature, the pressure of a gas varies inversely with its volume. That is,

$$P_1V_1 = P_2V_2$$

where P is the pressure of the gas, V is its volume, and subscripts 1 and 2 represent the initial and resulting conditions respectively.

Gases always *fill* their container. Consequently, in a large container, the molecules in a given amount of gas will be far apart and the pressure will be low. But if the volume of the container is reduced, the gas molecules will be forced closer together and the pressure will rise.

A good example is an inflated automobile tire. The tire is hard and strong enough to bear the weight of a car because air is compressed to about one-third of its atmospheric volume inside the tire, providing high pressure.

Now let's see how this relates to inspiration and expiration.

Inspiration

Visualize the thoracic cavity as a gas-filled box with a single entrance at the top, the tubelike trachea. The volume of this box can be increased by enlarging all of its dimensions, thereby decreasing the gas pressure inside it. This drop in pressure causes air to rush into the box from the atmosphere, because gases always flow down their pressure gradients.

The same thing happens during normal quiet (resting) inspiration, when the **inspiratory muscles**—the diaphragm and external intercostal muscles—are activated. Here's how quiet inspiration works:

- **Action of the diaphragm.** When the dome-shaped diaphragm contracts, it moves inferiorly and flattens out (**Figure 22.16**, top). As a result, the superior-inferior dimension (height) of the thoracic cavity increases.

- **Action of the intercostal muscles.** When the external intercostal muscles contract, they lift the rib cage and pull the sternum superiorly (Figure 22.16, top). Because the ribs curve downward as well as forward around the chest wall, the broadest lateral and anteroposterior dimensions of the rib cage are normally directed obliquely downward. But when the ribs are raised and drawn together, they swing outward, expanding the diameter of the thorax both laterally and in the anteroposterior plane. This is much like the

Sequence of events	Changes in anterior-posterior and superior-inferior dimensions	Changes in lateral dimensions (superior view)
Inspiration ① Inspiratory muscles contract (diaphragm descends; rib cage rises). ↓ ② Thoracic cavity volume increases. ↓ ③ Lungs are stretched; intrapulmonary volume increases. ↓ ④ Intrapulmonary pressure drops (to −1 mm Hg). ↓ ⑤ Air (gases) flows into lungs down its pressure gradient until intrapulmonary pressure is 0 (equal to atmospheric pressure).	Ribs are elevated and sternum flares as external intercostals contract. Diaphragm moves inferiorly during contraction.	External intercostals contract.
Expiration ① Inspiratory muscles relax (diaphragm rises; rib cage descends due to recoil of costal cartilages). ↓ ② Thoracic cavity volume decreases. ↓ ③ Elastic lungs recoil passively; intrapulmonary volume decreases. ↓ ④ Intrapulmonary pressure rises (to +1 mm Hg). ↓ ⑤ Air (gases) flows out of lungs down its pressure gradient until intrapulmonary pressure is 0.	Ribs and sternum are depressed as external intercostals relax. Diaphragm moves superiorly as it relaxes.	External intercostals relax.

Figure 22.16 Changes in thoracic volume and sequence of events during inspiration and expiration. The sequence of events in the left column includes volume changes during inspiration (top) and expiration (bottom). The lateral views in the middle column show changes in the superior-inferior dimension (as the diaphragm alternately contracts and relaxes, see black arrows) and in the anterior-posterior dimension (as the external intercostal muscles alternately contract and relax). The superior views of transverse thoracic sections in the right column show lateral dimension changes resulting from alternate contraction and relaxation of the external intercostal muscles.

action that occurs when a curved bucket handle is raised—it moves outward as it moves upward, as shown at right.

Although these actions expand the thoracic dimensions by only a few millimeters

along each plane, this is enough to increase thoracic volume by almost 500 ml—the usual volume of air that enters the lungs during a normal quiet inspiration. Of the two types of inspiratory muscles, the diaphragm is far more important in producing these volume changes that lead to normal quiet inspiration.

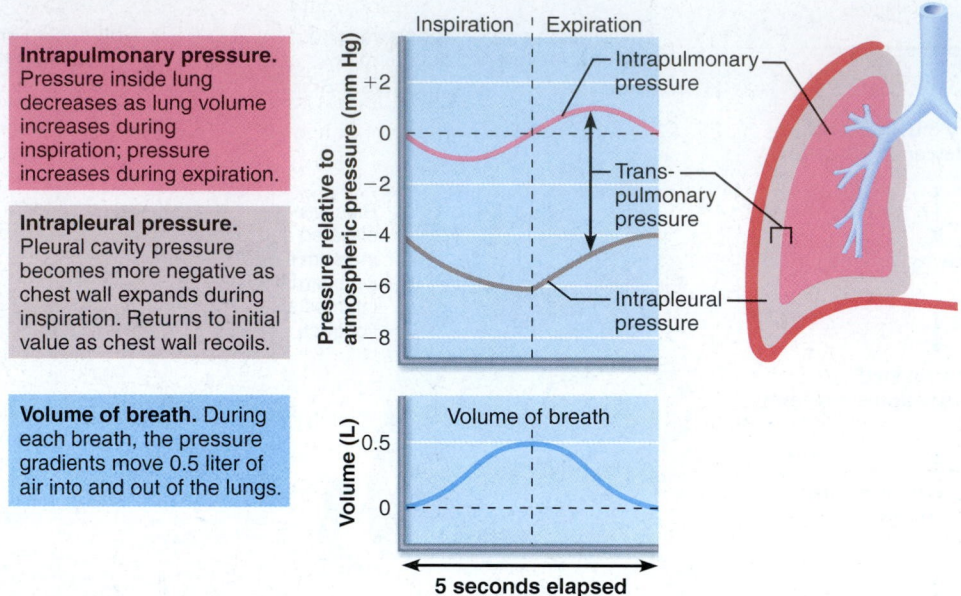

Intrapulmonary pressure. Pressure inside lung decreases as lung volume increases during inspiration; pressure increases during expiration.

Intrapleural pressure. Pleural cavity pressure becomes more negative as chest wall expands during inspiration. Returns to initial value as chest wall recoils.

Volume of breath. During each breath, the pressure gradients move 0.5 liter of air into and out of the lungs.

Figure 22.17 Changes in intrapulmonary and intrapleural pressures during inspiration and expiration. Notice that normal atmospheric pressure (760 mm Hg) is given a value of 0 on the scale.

As the thoracic dimensions increase during inspiration, the lungs are stretched and the intrapulmonary volume increases. As a result, P_{pul} drops about 1 mm Hg relative to P_{atm}. Anytime the intrapulmonary pressure is less than the atmospheric pressure ($P_{pul} < P_{atm}$), air rushes into the lungs along the pressure gradient. Inspiration ends when $P_{pul} = P_{atm}$. During the same period, P_{ip} declines to about −6 mm Hg relative to P_{atm} (**Figure 22.17**).

During the *deep* or *forced inspirations* that occur during vigorous exercise and in some chronic obstructive pulmonary diseases, accessory muscles further increase thoracic volume. Several muscles, including the scalenes and sternocleido-mastoid muscles of the neck and the pectoralis minor of the chest, raise the ribs even more than during quiet inspiration. Additionally, the back extends as the erector spinae muscles straighten the thoracic curvature.

Expiration

In healthy individuals, quiet expiration is a passive process that depends more on lung elasticity than on muscle contraction. As the inspiratory muscles relax and resume their resting length, the rib cage descends and the lungs recoil (Figure 22.16, bottom). As a result, both the thoracic and intrapulmonary volumes decrease. This volume decrease compresses the alveoli, and P_{pul} rises to about 1 mm Hg above atmospheric pressure (Figure 22.17). When $P_{pul} > P_{atm}$, the pressure gradient forces gases to flow out of the lungs.

Forced expiration is an active process produced by contracting abdominal wall muscles, primarily the oblique and transversus muscles. These contractions (1) increase the intra-abdominal pressure, which forces the abdominal organs superiorly against the diaphragm, and (2) depress the rib cage. The internal intercostal muscles also help depress the rib cage and decrease thoracic volume.

To precisely regulate air flow from the lungs, it is necessary to control the accessory muscles of expiration. For instance, the ability of a trained vocalist to hold a musical note depends on the coordinated activity of several muscles normally used in forced expiration.

Nonrespiratory Air Movements

Many processes other than breathing move air into or out of the lungs, altering the normal respiratory rhythm. These **nonrespiratory air movements** occur whenever you cough, sneeze, cry, laugh, hiccup, or yawn. Some can be produced voluntarily, but some (such as sneezing and hiccups) are reflexive.

Physical Factors Influencing Pulmonary Ventilation

As we have seen, the lungs are stretched during inspiration and recoil passively during expiration. The inspiratory muscles consume energy to enlarge the thorax. Energy is also used to overcome various factors that hinder air passage and pulmonary ventilation. We examine these factors next.

Airway Resistance

The major *nonelastic* source of resistance to gas flow is friction, or drag, encountered in the respiratory passageways. The following equation gives the relationship between gas flow (F), pressure (P), and resistance (R):

$$F = \frac{\Delta P}{R}$$

Notice that the factors determining gas flow in the respiratory passages and blood flow in the blood vessels are equivalent.

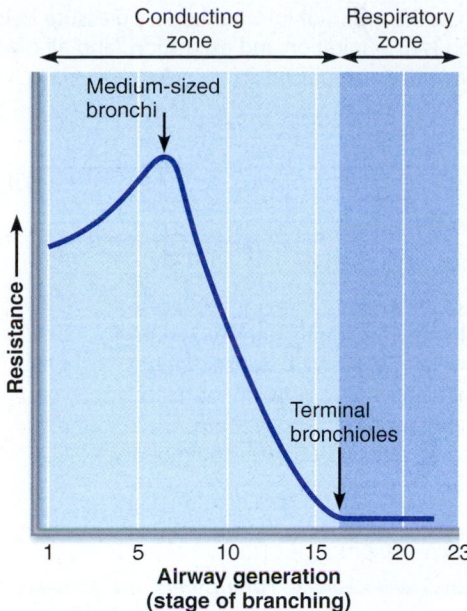

Conducting zone   Respiratory zone

Medium-sized bronchi

Terminal bronchioles

Airway generation (stage of branching)

Figure 22.18 Resistance in respiratory passageways. Airway resistance peaks in the medium-sized bronchi and then declines sharply as the total cross-sectional area of the airways increases rapidly.

The amount of gas flowing into and out of the alveoli is directly proportional to ΔP, the *difference* in pressure, or pressure gradient, between the external atmosphere and the alveoli.

Normally, very small differences in pressure produce large changes in gas flow. The average pressure gradient during normal quiet breathing is 2 mm Hg or less, and yet it is sufficient to move 500 ml of air in and out of the lungs with each breath.

But, as the equation also indicates, gas flow changes *inversely* with resistance. In other words, gas flow decreases as resistance increases. As in the cardiovascular system, resistance in the respiratory tree is determined mostly by the diameters of the conducting tubes. However, as a rule, airway resistance is insignificant for two reasons:

- Airway diameters in the first part of the conducting zone are huge, relative to the low viscosity of air.

- As the airways get progressively smaller, there are progressively more branches. As a result, although individual bronchioles are tiny, there are an enormous number of them in parallel, so the total cross-sectional area is huge.

Consequently, the greatest resistance to gas flow occurs in the medium-sized bronchi (**Figure 22.18**). At the terminal bronchioles, gas flow stops and diffusion takes over as the main force driving gas movement, so resistance is no longer an issue.

HOMEOSTATIC IMBALANCE 22.8   CLINICAL

Smooth muscle of the bronchiolar walls is exquisitely sensitive to neural controls and certain chemicals. For example, inhaled irritants activate a reflex of the parasympathetic division of the nervous system that causes vigorous constriction of the

bronchioles and dramatically reduces air passage. During an acute *asthma attack*, histamine and other inflammatory chemicals can cause such strong bronchoconstriction that pulmonary ventilation almost completely stops, regardless of the pressure gradient. Conversely, epinephrine released during sympathetic nervous system activation or administered as a drug dilates bronchioles and reduces airway resistance. Local accumulations of mucus, infectious material, or solid tumors in the passageways are important sources of airway resistance in those with respiratory disease.

Whenever airway resistance rises, breathing movements become more strenuous, but such compensation has its limits. When the bronchioles are severely constricted or obstructed, even the most extreme respiratory efforts cannot restore ventilation to life-sustaining levels.   ●●●●●●●

Alveolar Surface Tension

At any gas-liquid boundary, the molecules of the liquid are more strongly attracted to each other than to the gas molecules. This unequal attraction produces a state of tension at the liquid surface, called **surface tension**, that (1) draws the liquid molecules closer together and reduces their contact with the dissimilar gas molecules, and (2) resists any force that tends to increase the surface area of the liquid.

Water is composed of highly polar molecules and has a very high surface tension (◀ p. 35). As the major component of the liquid film that coats the alveolar walls, water is always acting to reduce the alveoli to their smallest possible size.

If the film were pure water, the alveoli would collapse between breaths. But the alveolar film also contains **surfactant** (ser-fak′tant), a detergent-like complex of lipids and proteins produced by the type II alveolar cells (◀ p. 831). Surfactant decreases the cohesiveness of water molecules, much the way laundry detergent reduces the attraction of water for water, allowing water to interact with and pass through fabric. As a result, the surface tension of alveolar fluid is reduced, and less energy is needed to overcome those forces to expand the lungs and discourage alveolar collapse. Breaths that are deeper than normal stimulate type II cells to secrete more surfactant.

HOMEOSTATIC IMBALANCE 22.9   CLINICAL

When too little surfactant is present, surface tension can collapse the alveoli. Once this happens, the alveoli must be completely reinflated during each inspiration, an effort that uses tremendous amounts of energy. This is the problem faced by newborns with **infant respiratory distress syndrome (IRDS)**, a condition common in premature babies. Since fetal lungs do not produce adequate amounts of surfactant until the last two months of development, babies born prematurely often are unable to keep their alveoli inflated between breaths.

IRDS is treated by spraying natural or synthetic surfactant into the newborn's respiratory passageways. In addition, devices that maintain positive airway pressure throughout the respiratory cycle can keep the alveoli open between breaths. Severe cases require mechanical ventilators.

Many IRDS survivors suffer from *bronchopulmonary dysplasia*, a chronic lung disease, during childhood and beyond. This condition seems to be caused by a combination of factors including prolonged therapy with oxygen and mechanical ventilation.

Lung Compliance

Healthy lungs are unbelievably stretchy, and this distensibility is called **lung compliance**. Specifically, lung compliance (C_L) is a measure of the change in lung volume (ΔV_L) that occurs with a given change in transpulmonary pressure [$\Delta(P_{pul} - P_{ip})$]. This relationship is stated as

$$C_L = \frac{\Delta V_L}{\Delta(P_{pul} - P_{ip})}$$

The more a lung expands for a given rise in transpulmonary pressure, the greater its compliance. Said another way, the higher the lung compliance, the easier it is to expand the lungs at any given transpulmonary pressure.

Lung compliance is determined largely by two factors:

- Distensibility of the lung tissue
- Alveolar surface tension

Because lung distensibility is generally high and surfactant keeps alveolar surface tension low, healthy lungs tend to have high compliance, which favors efficient ventilation.

HOMEOSTATIC IMBALANCE 22.10   **CLINICAL**

Any decrease in the natural resilience of the lungs diminishes lung compliance. Chronic inflammation, or infections such as tuberculosis, can cause nonelastic scar tissue to replace normal lung tissue (*fibrosis*). Decreased production of surfactant can also impair lung compliance. The lower the lung compliance, the more energy is needed just to breathe.

Since the lungs are contained within the thoracic cavity, we also need to consider the compliance (distensibility) of the thoracic wall. Factors that reduce the compliance of the thoracic wall hinder lung expansion. The total compliance of the respiratory system is comprised of lung compliance and thoracic wall compliance.

HOMEOSTATIC IMBALANCE 22.11   **CLINICAL**

Deformities of the thorax, ossified costal cartilages (common during old age), and paralyzed intercostal muscles all hinder thoracic expansion, reducing total respiratory compliance.

Check Your Understanding

10. What is the driving force for pulmonary ventilation?

11. What causes the partial vacuum (negative pressure) inside the pleural cavity? What happens to a lung if air enters the pleural cavity? What is the clinical name for this condition?

12. Premature infants often lack adequate surfactant. How does this affect their ability to breathe?

13. DRAW On the graph of intrapulmonary pressure below, label the periods of inspiration and expiration, and all of the points where air flow is zero. Explain.

14. DRAW On the diagram at right, indicate where the following pressures would be measured and what their values would be at the middle of inspiration: atmospheric pressure, intrapulmonary pressure, intrapleural pressure, and transpulmonary pressure.

For answers, see Answers Appendix.

22.5 Measuring respiratory volumes, capacities, and flow rates helps us assess ventilation

Learning Outcomes

▶ Explain and compare the various lung volumes and capacities.

▶ Define dead space.

▶ Indicate types of information that can be gained from pulmonary function tests.

The amount of air flushed in and out of the lungs depends on the conditions of inspiration and expiration. Consequently, several respiratory volumes can be described. Specific combinations of these respiratory volumes, called *respiratory capacities*, are measured to gain information about a person's respiratory status.

Lung volumes and capacities are often abnormal in people with pulmonary disorders. The original clinical measuring tool, a **spirometer** (spi-rom′ĕ-ter), was a cumbersome instrument utilizing a hollow bell inverted over water. Now patients simply blow into a small electronic measuring device.

Respiratory Volumes

The four **respiratory volumes** of interest are tidal, inspiratory reserve, expiratory reserve, and residual. The values recorded in **Figure 22.19a** represent normal values. Figure 22.19b provides average values for males and females.

During normal quiet breathing, about 500 ml of air moves into and out of the lungs with each breath. This respiratory volume is the **tidal volume (TV)**. The amount of air that can be inspired forcibly beyond the tidal volume (2100 to 3200 ml) is the **inspiratory reserve volume (IRV)**.

The **expiratory reserve volume (ERV)** is the amount of air—normally 1000 to 1200 ml—that can be expelled from the lungs after a normal tidal volume expiration. Even after the most strenuous expiration, about 1200 ml of air remains

(a) Spirographic record for a male

Measurement	Adult male average value	Adult female average value	Description
Respiratory volumes			
Tidal volume (TV)	500 ml	500 ml	Amount of air inhaled or exhaled with each breath under resting conditions
Inspiratory reserve volume (IRV)	3100 ml	1900 ml	Amount of air that can be forcefully inhaled after a normal tidal volume inspiration
Expiratory reserve volume (ERV)	1200 ml	700 ml	Amount of air that can be forcefully exhaled after a normal tidal volume expiration
Residual volume (RV)	1200 ml	1100 ml	Amount of air remaining in the lungs after a forced expiration
Respiratory capacities			
Total lung capacity (TLC)	6000 ml	4200 ml	Maximum amount of air contained in lungs after a maximum inspiratory effort: TLC = TV + IRV + ERV + RV
Vital capacity (VC)	4800 ml	3100 ml	Maximum amount of air that can be expired after a maximum inspiratory effort: VC = TV + IRV + ERV
Inspiratory capacity (IC)	3600 ml	2400 ml	Maximum amount of air that can be inspired after a normal tidal volume expiration: IC = TV + IRV
Functional residual capacity (FRC)	2400 ml	1800 ml	Volume of air remaining in the lungs after a normal tidal volume expiration: FRC = ERV + RV

(b) Summary of respiratory volumes and capacities for males and females

Figure 22.19 Respiratory volumes and capacities. Idealized spirographic record of respiratory volumes in (a) is for a healthy adult male (70 kg, age 20).

in the lungs; this is the **residual volume (RV)**, which helps to keep the alveoli open and prevent lung collapse.

Respiratory Capacities

The **respiratory capacities** include inspiratory, functional residual, vital, and total lung capacities (Figure 22.19). The respiratory capacities always consist of two or more lung volumes.

- **Inspiratory capacity (IC)** is the total amount of air that can be inspired after a normal tidal volume expiration, so it is the sum of TV and IRV.
- **Functional residual capacity (FRC)** represents the amount of air remaining in the lungs after a normal tidal volume expiration and is the combined RV and ERV.

- **Vital capacity (VC)** is the total amount of exchangeable air. It is the sum of TV, IRV, and ERV.
- **Total lung capacity (TLC)** is the sum of all lung volumes.

As indicated in Figure 22.19b, lung volumes and capacities (with the possible exception of TV) tend to be smaller in women than in men because of women's smaller size.

Dead Space

Some of the inspired air fills the conducting respiratory passageways and never contributes to gas exchange in the alveoli. The volume of these conducting zone conduits, which make up the **anatomical dead space**, typically amounts to about 150 ml. (The rule of thumb is that the anatomical dead space volume in

22

Table 22.3 Effects of Breathing Rate and Depth on Alveolar Ventilation of Three Hypothetical Patients

BREATHING PATTERN OF HYPOTHETICAL PATIENT	DEAD SPACE VOLUME (DSV)	TIDAL VOLUME (TV)	RESPIRATORY RATE*	MINUTE VENTILATION (MVR)	ALVEOLAR VENTILATION (AVR)	% EFFECTIVE VENTILATION (AVR/MVR)
I—Normal rate and depth	150 ml	500 ml	20/min	10,000 ml/min	7000 ml/min	70%
II—Slow, deep breathing	150 ml	1000 ml	10/min	10,000 ml/min	8500 ml/min	85%
III—Rapid, shallow breathing	150 ml	250 ml	40/min	10,000 ml/min	4000 ml/min	40%

*Respiratory rate values are artificially adjusted to provide equivalent minute ventilation as a baseline for comparing alveolar ventilation.

a healthy young adult is equal to 1 ml per pound of ideal body weight.) This means that if TV is 500 ml, only 350 ml of it is involved in alveolar ventilation. The remaining 150 ml of the tidal breath is in the anatomical dead space.

If some alveoli cease to act in gas exchange (due to alveolar collapse or obstruction by mucus, for example), the **alveolar dead space** is added to the anatomical dead space. The sum of the nonuseful volumes is the **total dead space**.

Pulmonary Function Tests

Spirometry is most useful for evaluating losses in respiratory function and for following the course of certain respiratory diseases. It cannot provide a specific diagnosis, but it can distinguish between *obstructive pulmonary diseases* involving increased airway resistance (such as chronic bronchitis) and *restrictive diseases* involving reduced total lung capacity. (These changes might be due to diseases such as tuberculosis, or to fibrosis due to exposure to certain environmental agents such as asbestos). In obstructive diseases, TLC, FRC, and RV may increase because the lungs hyperinflate, whereas in restrictive diseases, VC, TLC, FRC, and RV decline because lung expansion is limited.

We can obtain more information by assessing the *rate* at which gas moves into and out of the lungs.

- **Forced vital capacity (FVC)** measures the amount of gas expelled when a subject takes a deep breath and then forcefully exhales maximally and as rapidly as possible.

- **Forced expiratory volume (FEV)** determines the amount of air expelled during specific time intervals of the FVC test. For example, the volume exhaled during the first second is FEV_1. Those with healthy lungs can exhale about 80% of the FVC within 1 second. Those with obstructive pulmonary disease exhale considerably less than 80% of the FVC within 1 second, while those with restrictive disease can exhale 80% or more of FVC in 1 second even though their FVC is reduced.

Alveolar Ventilation

The **minute ventilation** is the total amount of gas that flows into or out of the respiratory tract in 1 minute. During normal quiet breathing, the minute ventilation in healthy people is about 6 L/min (500 ml per breath multiplied by 12 breaths per minute). During vigorous exercise, the minute ventilation may reach 200 L/min.

Minute ventilation values provide a rough yardstick for assessing respiratory efficiency, but the **alveolar ventilation rate (AVR)** is a better index of effective ventilation. The AVR takes into account the volume of air wasted in the dead space and measures the flow of fresh gases in and out of the alveoli during a particular time interval. We can compute AVR using this equation:

$$\underset{\text{(ml/min)}}{\text{AVR}} = \underset{\text{(breaths/min)}}{\text{frequency}} \times \underset{\text{(ml/breath)}}{(\text{TV} - \text{dead space})}$$

In healthy people, AVR is usually about 12 breaths per minute times the difference of 500 − 150 ml per breath, or 4200 ml/min.

Because anatomical dead space is constant in a particular individual, increasing the volume of each inspiration (breathing depth) enhances AVR and gas exchange more than raising the respiratory rate. AVR drops dramatically during rapid shallow breathing because most of the inspired air never reaches the exchange sites. Furthermore, as tidal volume approaches the dead space value, effective ventilation approaches zero, regardless of how fast a person is breathing. **Table 22.3** summarizes the effects of breathing rate and breathing depth on alveolar ventilation for three hypothetical patients.

Check Your Understanding

15. Explain why slow, deep breaths ventilate the alveoli more effectively than do rapid, shallow breaths.

16. What is the difference between respiratory volumes and respiratory capacities?

For answers, see Answers Appendix.

22.6 Gases exchange by diffusion between the blood, lungs, and tissues

Learning Outcomes

▶ State Dalton's law of partial pressures and Henry's law.
▶ Describe how atmospheric and alveolar air differ in composition, and explain these differences.
▶ Relate Dalton's and Henry's laws to events of external and internal respiration.

As you've discovered, during *external respiration* oxygen enters and carbon dioxide leaves the blood in the lungs by diffusion.

At the body tissues, where the process is called *internal respiration*, the same gases move in opposite directions, also by diffusion. To understand these processes, let's examine the physical properties of gases and consider the composition of alveolar gas.

Basic Properties of Gases

Beyond Boyle's law, two more gas laws provide most of the information we need—*Dalton's law of partial pressures* reveals how a gas behaves when it is part of a mixture of gases, and *Henry's law* helps us understand how gases move into and out of solution.

Dalton's Law of Partial Pressures

Dalton's law of partial pressures states that the total pressure exerted by a mixture of gases is the sum of the pressures exerted independently by each gas in the mixture. Further, the pressure exerted by each gas—its **partial pressure**—is directly proportional to the percentage of that gas in the gas mixture.

As indicated in **Table 22.4**, nitrogen makes up about 79% of air, and the partial pressure of nitrogen P_{N_2} is 78.6% × 760 mm Hg, or 597 mm Hg. Oxygen, which accounts for nearly 21% of air, has a partial pressure P_{O_2} of 159 mm Hg (20.9% × 760 mm Hg). Together nitrogen and oxygen contribute about 99% of the total atmospheric pressure. Air also contains 0.04% carbon dioxide, up to 0.5% water vapor, and insignificant amounts of inert gases (such as argon and helium).

At high altitudes, partial pressures decline in direct proportion to the decrease in atmospheric pressure. For example, at 10,000 feet above sea level where the atmospheric pressure is 523 mm Hg, P_{O_2} is 110 mm Hg.

Moving in the opposite direction, atmospheric pressure increases by 1 atm (760 mm Hg) for each 33 feet of descent (in water) below sea level. At 99 feet below sea level, the total pressure exerted on the body is equivalent to 4 atm, or 3040 mm Hg, and the partial pressure exerted by each component gas is also quadrupled.

Henry's Law

Henry's law states that when a gas is in contact with a liquid, the gas will dissolve in the liquid in proportion to its partial pressure. Accordingly, the greater the concentration of a particular gas in the gas phase, the more and the faster that gas will go into solution in the liquid.

At equilibrium, the partial pressures in the gas and liquid phases are the same. If, however, the partial pressure of the gas later becomes greater in the liquid than in the adjacent gas phase, some of the dissolved gas molecules will reenter the gaseous phase. So the direction and amount of movement of a gas are determined by its partial pressure in the two phases. This flexible situation is exactly what occurs when gases are exchanged in the lungs and tissues. For example, when P_{CO_2} in the pulmonary capillaries is higher than in the lungs, CO_2 diffuses out of the blood and enters the air in the alveoli.

How much of a gas will dissolve in a liquid at any given partial pressure also depends on the *solubility* of the gas in the liquid and the *temperature* of the liquid. The gases in air have very different solubilities in water (and in blood plasma). Carbon dioxide is most soluble. Oxygen is only 1/20 as soluble as CO_2, and N_2 is only half as soluble as O_2. For this reason, at a given partial pressure, much more CO_2 than O_2 dissolves in water, and practically no N_2 goes into solution.

When a liquid's temperature rises, gas solubility decreases. Think of club soda, which is produced by forcing CO_2 gas to dissolve in water under high pressure. If you take the cap off a bottle of club soda and leave it in the fridge, it will slowly go flat. But if you leave it at room temperature, it will very quickly go flat. In both cases, you end up with plain water—all the CO_2 gas has escaped from solution.

In summary, the amount of gas that will dissolve in a liquid depends on:

- The partial pressure of the gas in contact with the liquid
- The solubility of the gas in the liquid
- The temperature of the liquid

Hyperbaric oxygen chambers provide clinical applications of Henry's law. These chambers contain O_2 gas at pressures higher than 1 atm and are used to force greater-than-normal amounts of O_2 into the blood of patients suffering from carbon monoxide poisoning (see p. 852). (They are also used to treat gas gangrene and tissue damage following radiation therapy.)

Scuba diving provides another illustration of Henry's law. If divers rise rapidly from the depths, dissolved nitrogen forms bubbles in their blood, causing "the bends."

Table 22.4	Comparison of Gas Partial Pressures and Approximate Percentages in the Atmosphere and in the Alveoli				
	ATMOSPHERE (SEA LEVEL)			*ALVEOLI*	
GAS	APPROXIMATE PERCENTAGE	PARTIAL PRESSURE (mm Hg)		APPROXIMATE PERCENTAGE	PARTIAL PRESSURE (mm Hg)
N_2	78.6	597		74.9	569
O_2	20.9	159		13.7	104
CO_2	0.04	0.3		5.2	40
H_2O	0.46	3.7		6.2	47
	100.0%	760		100.0%	760

Composition of Alveolar Gas

As shown in Table 22.4, the gaseous makeup of the atmosphere is quite different from that in the alveoli. The atmosphere is almost entirely O_2 and N_2; the alveoli contain more CO_2 and water vapor and much less O_2. These differences reflect the effects of:

- Gas exchanges occurring in the lungs (O_2 diffuses from the alveoli into the pulmonary blood and CO_2 diffuses in the opposite direction).

- Humidification of air by conducting passages.

- The mixing of alveolar gas that occurs with each breath. Because only 500 ml of air enter with each tidal inspiration, gas in the alveoli is actually a mixture of newly inspired gases and gases remaining in the respiratory passageways between breaths.

The alveolar partial pressures of O_2 and CO_2 are easily changed by increasing breathing depth and rate. A high AVR brings more O_2 into the alveoli, increasing alveolar P_{O_2} and rapidly eliminating CO_2 from the lungs.

External Respiration

During external respiration (pulmonary gas exchange), dark red blood flowing through the pulmonary circuit is transformed into the scarlet river that is returned to the heart for distribution by systemic arteries to all body tissues. This color change is due to O_2 uptake and binding to hemoglobin in red blood cells (RBCs), but CO_2 exchange (unloading) is occurring equally fast.

The following three factors influence external respiration:

- Partial pressure gradients and gas solubilities

- Thickness and surface area of the respiratory membrane

- Ventilation-perfusion coupling (matching alveolar ventilation with pulmonary blood perfusion)

Let's look at these factors one by one.

Partial Pressure Gradients and Gas Solubilities

Partial pressure gradients of O_2 and CO_2 drive the diffusion of these gases across the respiratory membrane. A steep gradient for the partial pressure of oxygen exists across the respiratory membrane because the P_{O_2} of deoxygenated blood in the pulmonary arteries is only 40 mm Hg, as opposed to a P_{O_2} of approximately 104 mm Hg in the alveoli. As a result, O_2 diffuses rapidly from the alveoli into the pulmonary capillary blood (**Figure 22.20**).

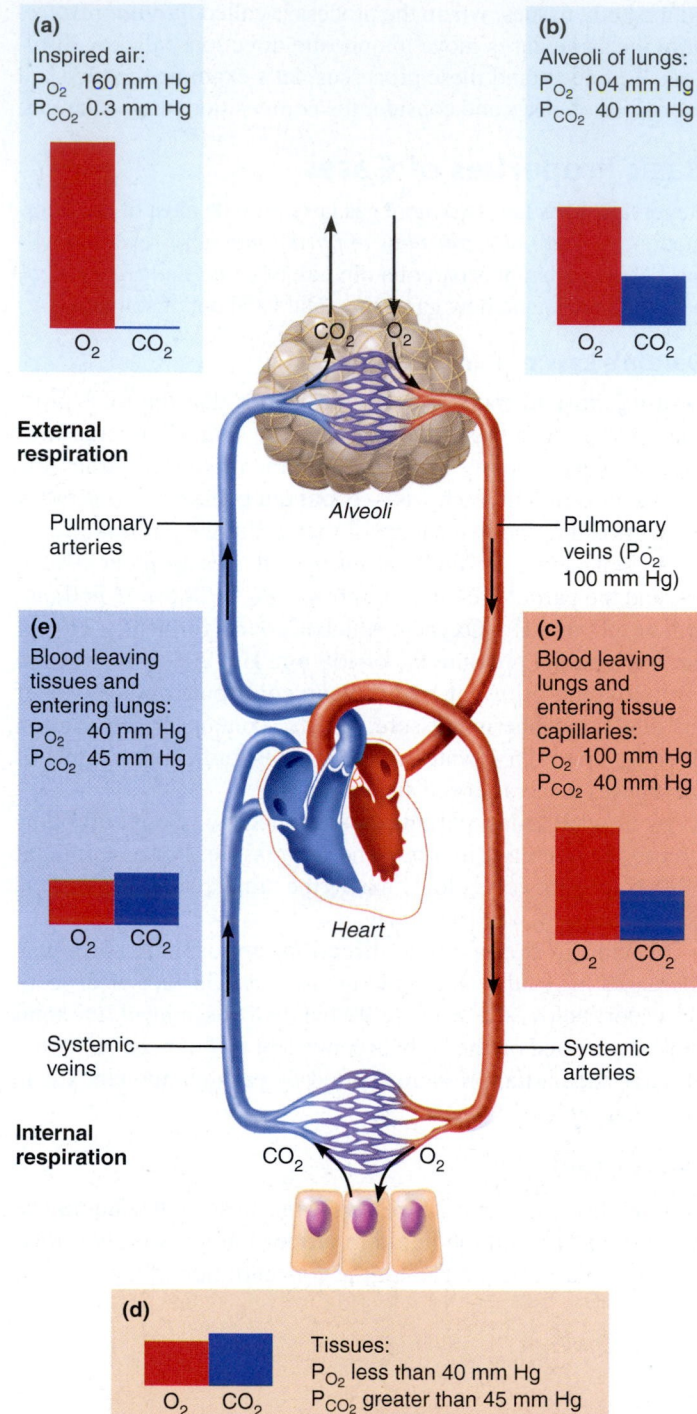

(a) Inspired air:
P_{O_2} 160 mm Hg
P_{CO_2} 0.3 mm Hg

O_2 CO_2

(b) Alveoli of lungs:
P_{O_2} 104 mm Hg
P_{CO_2} 40 mm Hg

O_2 CO_2

External respiration

Pulmonary arteries

Alveoli

Pulmonary veins (P_{O_2} 100 mm Hg)

(e) Blood leaving tissues and entering lungs:
P_{O_2} 40 mm Hg
P_{CO_2} 45 mm Hg

O_2 CO_2

(c) Blood leaving lungs and entering tissue capillaries:
P_{O_2} 100 mm Hg
P_{CO_2} 40 mm Hg

O_2 CO_2

Heart

Systemic veins

Systemic arteries

Internal respiration

CO_2 O_2

(d)

O_2 CO_2

Tissues:
P_{O_2} less than 40 mm Hg
P_{CO_2} greater than 45 mm Hg

Figure 22.20 Partial pressure gradients promoting gas movements in the body. Gradients promoting O_2 and CO_2 exchange across the respiratory membrane in the lungs [**(b)**↔**(e)**] and across systemic capillary membranes in body tissues [**(c)**↔**(d)**]. (The small decrease in P_{O_2} in blood leaving lungs is due to partial dilution of pulmonary capillary blood with less oxygenated blood.)

Equilibrium—that is, a P_{O_2} of 104 mm Hg on both sides of the respiratory membrane—usually occurs in 0.25 second, which is about one-third of the time a red blood cell spends in a pulmonary capillary (**Figure 22.21**). The lesson here is that

Figure 22.21 Oxygenation of blood in the pulmonary capillaries at rest. Oxygen loading only takes about one-third of the time a red blood cell spends in the pulmonary capillary.

blood can flow through the pulmonary capillaries three times as quickly and still be adequately oxygenated.

Carbon dioxide diffuses in the opposite direction along a much gentler partial pressure gradient of about 5 mm Hg (45 mm Hg to 40 mm Hg) until equilibrium occurs at 40 mm Hg. Expiration then gradually expels carbon dioxide from the alveoli.

Even though the O_2 pressure gradient for oxygen diffusion is much steeper than the CO_2 gradient, equal amounts of these gases are exchanged. Why? The reason is that CO_2 is 20 times more soluble in plasma and alveolar fluid than O_2.

Thickness and Surface Area of the Respiratory Membrane

In healthy lungs, the respiratory membrane is only 0.5 to 1 μm thick, and gas exchange is usually very efficient.

HOMEOSTATIC IMBALANCE 22.13   **CLINICAL**

The effective thickness of the respiratory membrane increases dramatically if the lungs become waterlogged and edematous, as in pneumonia or left heart failure (◄ **p. 700**). Under such conditions, even the 0.75 s that red blood cells spend in transit through the pulmonary capillaries may not be enough for adequate gas exchange, and body tissues suffer from oxygen deprivation. ● ● ● ● ● ● ● ●

The greater the surface area of the respiratory membrane, the more gas can diffuse across it in a given time period. In healthy lungs, the alveolar surface area is enormous. Spread flat, the total gas exchange surface of these tiny sacs in an adult male's lungs is about 90 m² —approximately 40 times greater than the surface area of his skin!

HOMEOSTATIC IMBALANCE 22.14   **CLINICAL**

Certain pulmonary diseases drastically reduce the alveolar surface area. For instance, in *emphysema* the walls of adjacent alveoli break down and the alveolar chambers enlarge (**Figure 22.22**). Tumors, mucus, or inflammatory material also reduce surface area by blocking gas flow into the alveoli. ● ● ● ● ● ● ●

Ventilation-Perfusion Coupling

For optimal gas exchange, there must be a close match, or coupling, between *ventilation* (the amount of gas reaching the alveoli) and *perfusion* (the blood flow in pulmonary capillaries). Both are controlled by local autoregulatory mechanisms that continuously respond to local conditions. For the most part:

- P_{O_2} controls perfusion by changing *arteriolar* diameter.
- P_{CO_2} controls ventilation by changing *bronchiolar* diameter.

(a) Normal lung (150×)

(b) Lung with emphysema (150×)

Figure 22.22 Tissue changes in emphysema. Scanning electron micrographs. Notice the enlargement of the alveoli due to the destruction of alveolar walls.

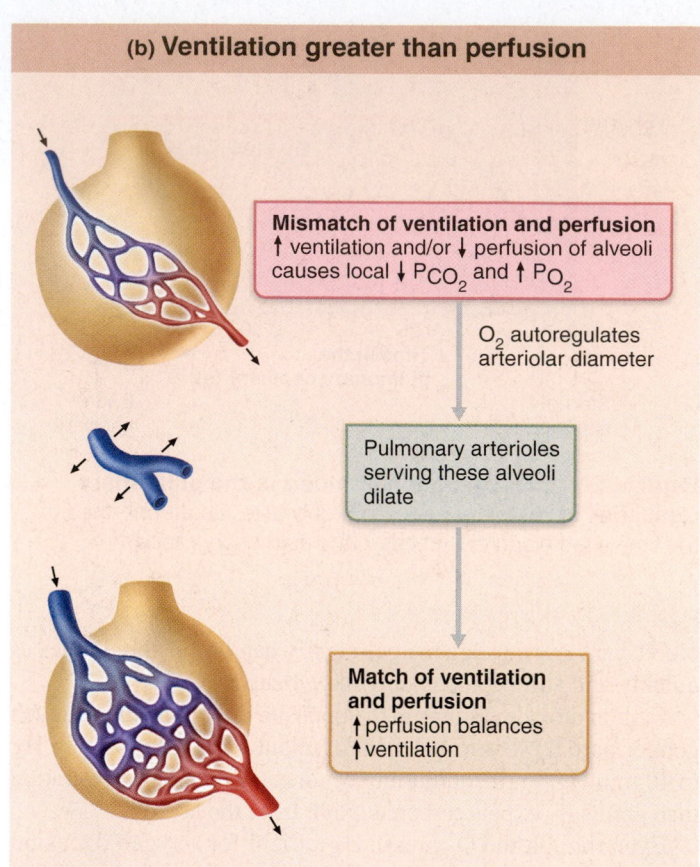

Figure 22.23 Ventilation-perfusion coupling. Autoregulatory events result in local matching of blood flow (perfusion) through the pulmonary capillaries with the amount of alveolar ventilation.

Influence of Local P_{O_2} on Perfusion We begin with perfusion because we introduced its autoregulatory control in Chapter 19 (◄ p. 729). If alveolar ventilation is inadequate, local P_{O_2} is low because blood takes O_2 away more quickly than ventilation can replenish it (**Figure 22.23a**). As a result, the terminal arterioles constrict, redirecting blood to respiratory areas where P_{O_2} is high and oxygen pickup is more efficient.

In alveoli where ventilation is maximal, high P_{O_2} dilates pulmonary arterioles and blood flow into the associated pulmonary capillaries increases (Figure 22.23b). Notice that the autoregulatory mechanism controlling pulmonary vascular muscle is the opposite of the mechanism controlling arterioles in the systemic circulation.

Influence of Local P_{CO_2} on Ventilation Bronchioles servicing areas where alveolar CO_2 levels are high dilate, allowing CO_2 to be eliminated from the body more rapidly. Bronchioles serving areas where P_{CO_2} is low constrict.

Balancing Ventilation and Perfusion The changing diameter of local bronchioles and arterioles synchronizes alveolar ventilation and pulmonary perfusion. Poor alveolar ventilation results in low oxygen and high carbon dioxide levels in the alveoli. Consequently, pulmonary arterioles constrict and airways dilate, bringing blood flow and air flow into closer physiological match. High P_{O_2} and low P_{CO_2} in the alveoli cause bronchioles serving the alveoli to constrict, and promote flushing of blood into the pulmonary capillaries.

Although these homeostatic mechanisms provide appropriate conditions for efficient gas exchange, they never completely balance ventilation and perfusion in every alveolus due to other factors. In particular, (1) gravity causes regional variations in blood and air flow in the lungs, and (2) the occasional alveolar duct plugged with mucus creates unventilated areas. These factors, together with blood shunted from the bronchial veins, account for the slight drop in P_{O_2} from alveolar air (104 mm Hg) to pulmonary venous blood (100 mm Hg), as shown in Figure 22.20.

Internal Respiration

Internal respiration involves capillary gas exchange in body tissues. In internal respiration, the partial pressure and diffusion gradients are reversed from the situation we have just described for external respiration and pulmonary gas exchange. However, the factors promoting gas exchanges between systemic capillaries and tissue cells are essentially identical to those acting in the lungs (see Figure 22.20).

Tissue cells continuously use O_2 for their metabolic activities and produce CO_2. Because P_{O_2} is always lower in tissues than it is in systemic arterial blood (40 mm Hg versus 100 mm Hg), O_2 moves rapidly from blood into tissues until equilibrium is reached. At the same time, CO_2 moves quickly along its pressure gradient into blood. As a result, venous blood draining the tissue capillary beds and returning to the heart has a P_{O_2} of 40 mm Hg and a P_{CO_2} of 45 mm Hg.

In summary, the gas exchanges that occur between blood and alveoli and between blood and tissue cells take place by simple diffusion. They are driven by the partial pressure gradients of O_2 and CO_2 that exist on the opposite sides of the exchange membranes.

Check Your Understanding

17. P_{O_2} in the alveoli is about 56 mm Hg lower than in the inspired air. Explain this difference.

18. Suppose a patient is receiving oxygen by mask. Are the arterioles leading into the O_2-enriched alveoli dilated or constricted? What is the advantage of this response?

19. WHAT IF? Suppose you are given a sealed container of water and air. The P_{CO_2} and P_{O_2} in the air are both 100 mm Hg. What are the P_{CO_2} and P_{O_2} in the water? Which gas has more molecules dissolved in the water? Why?

For answers, see Answers Appendix.

22.7 Oxygen is transported by hemoglobin, and carbon dioxide is transported in three different ways

Learning Outcomes

▶ Describe how oxygen is transported in blood.

▶ Explain how temperature, pH, BPG, and P_{CO_2} affect oxygen loading and unloading.

▶ Describe carbon dioxide transport in the blood.

We have considered external and internal respiration consecutively to emphasize their similarities, but keep in mind that it is blood that transports O_2 and CO_2 between these two exchange sites.

Oxygen Transport

Molecular oxygen is carried in blood in two ways: bound to hemoglobin within red blood cells and dissolved in plasma. Oxygen is poorly soluble in water, so only about 1.5% of the oxygen transported is carried in the dissolved form. Indeed, if this were the *only* means of oxygen transport, a P_{O_2} of 3 atm or a cardiac output of 15 times normal would be required to provide the oxygen levels needed by body tissues! Hemoglobin, of course, solves this problem—98.5% of the oxygen is carried from lungs to tissues in a loose chemical combination with hemoglobin.

Association of Oxygen and Hemoglobin

As we described in Chapter 17, hemoglobin (Hb) is composed of four polypeptide chains, each bound to an iron-containing heme group (◀ **Figure 17.4, p. 646**). Because the iron atoms bind oxygen, each hemoglobin molecule can combine with four molecules of O_2, and oxygen loading is rapid and reversible.

The hemoglobin-oxygen combination, called **oxyhemoglobin** (ok″sĭ-he″mo-glo′bin), is written **HbO$_2$**. Hemoglobin that has released oxygen is called **reduced hemoglobin**, or **deoxyhemoglobin**, and is written **HHb**. A single reversible equation describes the loading and unloading of O_2:

$$HHb + O_2 \xrightleftharpoons[\text{Tissues}]{\text{Lungs}} HbO_2 + H^+$$

After the first O_2 molecule binds to iron, the Hb molecule changes shape. As a result, it more readily takes up two more O_2 molecules, and uptake of the fourth is even more facilitated. When one, two, or three oxygen molecules are bound, a hemoglobin molecule is *partially saturated*. When all four of its heme groups are bound to O_2, the hemoglobin is *fully saturated*.

By the same token, unloading of one oxygen molecule enhances the unloading of the next, and so on. In this way, the *affinity* (binding strength) of hemoglobin for oxygen changes with the extent of oxygen saturation, and both loading and unloading of oxygen are very efficient.

The rate at which Hb reversibly binds or releases O_2 is regulated by P_{O_2}, temperature, blood pH, P_{CO_2}, and blood concentration of an organic chemical called BPG. These factors interact to ensure that adequate O_2 is delivered to tissue cells.

Influence of P_{O_2} on Hemoglobin Saturation The **oxygen-hemoglobin dissociation curve** shows how local P_{O_2} controls oxygen loading and unloading from hemoglobin. *Focus on the Oxygen-Hemoglobin Dissociation Curve* (**Focus Figure 22.1** on pp. 848–849) walks you through this graph step by step, explaining how hemoglobin ensures adequate oxygen delivery under a variety of conditions.

Under normal resting conditions (P_{O_2} = 100 mm Hg), arterial blood hemoglobin is 98% saturated, and 100 ml of systemic arterial blood contains about 20 ml of O_2. This *oxygen content* of arterial blood is written as 20 vol % (volume percent). As arterial blood flows through systemic capillaries, it releases about 5 ml of O_2 per 100 ml of blood, yielding an Hb saturation of 75% and an O_2 content of 15 vol % in venous blood. This means that substantial amounts of O_2 are normally still available in venous blood (the *venous reserve*), which can be used if needed.

The nearly complete saturation of Hb in arterial blood explains why breathing deeply increases both the alveolar and arterial blood P_{O_2} but causes very little increase in the O_2 saturation of hemoglobin. Remember, P_{O_2} measurements indicate only the amount of O_2 dissolved in plasma, not the amount bound to hemoglobin. However, P_{O_2} values are a good index of lung function, and when arterial P_{O_2} is significantly less than alveolar P_{O_2} some degree of respiratory impairment exists.

Influence of Other Factors on Hemoglobin Saturation Temperature, blood pH, P_{CO_2}, and the amount of BPG in the blood all influence hemoglobin saturation at a given P_{O_2}. Red blood cells (RBCs) produce BPG (2,3-bisphosphoglycerate) as they

(Text continues on p. 850.)

The Oxygen-Hemoglobin Dissociation Curve

The amount of oxygen carried by hemoglobin depends on the P_{O_2} (the amount of oxygen) available locally. This relationship ensures optimal oxygen pickup and delivery.

The oxygen-hemoglobin dissociation curve will help you understand how the properties of hemoglobin (Hb) affect oxygen binding in the lungs and oxygen release in the tissues.

This axis tells you how much O_2 is bound to Hb. At 100%, each Hb molecule has 4 bound oxygen molecules.

Hemoglobin

Oxygen

In the lungs, where P_{O_2} is high (100 mm Hg), Hb is almost fully saturated (98%) with O_2.

If more O_2 is present, more O_2 is bound. However, because of Hb's properties (O_2 binding strength changes with saturation), this is an S-shaped curve.

Percent O_2 saturation of hemoglobin (y-axis: 0, 20, 40, 60, 80, 100)

P_{O_2} (mm Hg) (x-axis: 0, 20, 40, 60, 80, 100)

This axis tells you the relative amount (partial pressure) of O_2 dissolved in the fluid surrounding the Hb.

In the tissues of other organs, where P_{O_2} is low (40 mm Hg), Hb is less saturated (75%) with O_2.

In the lungs

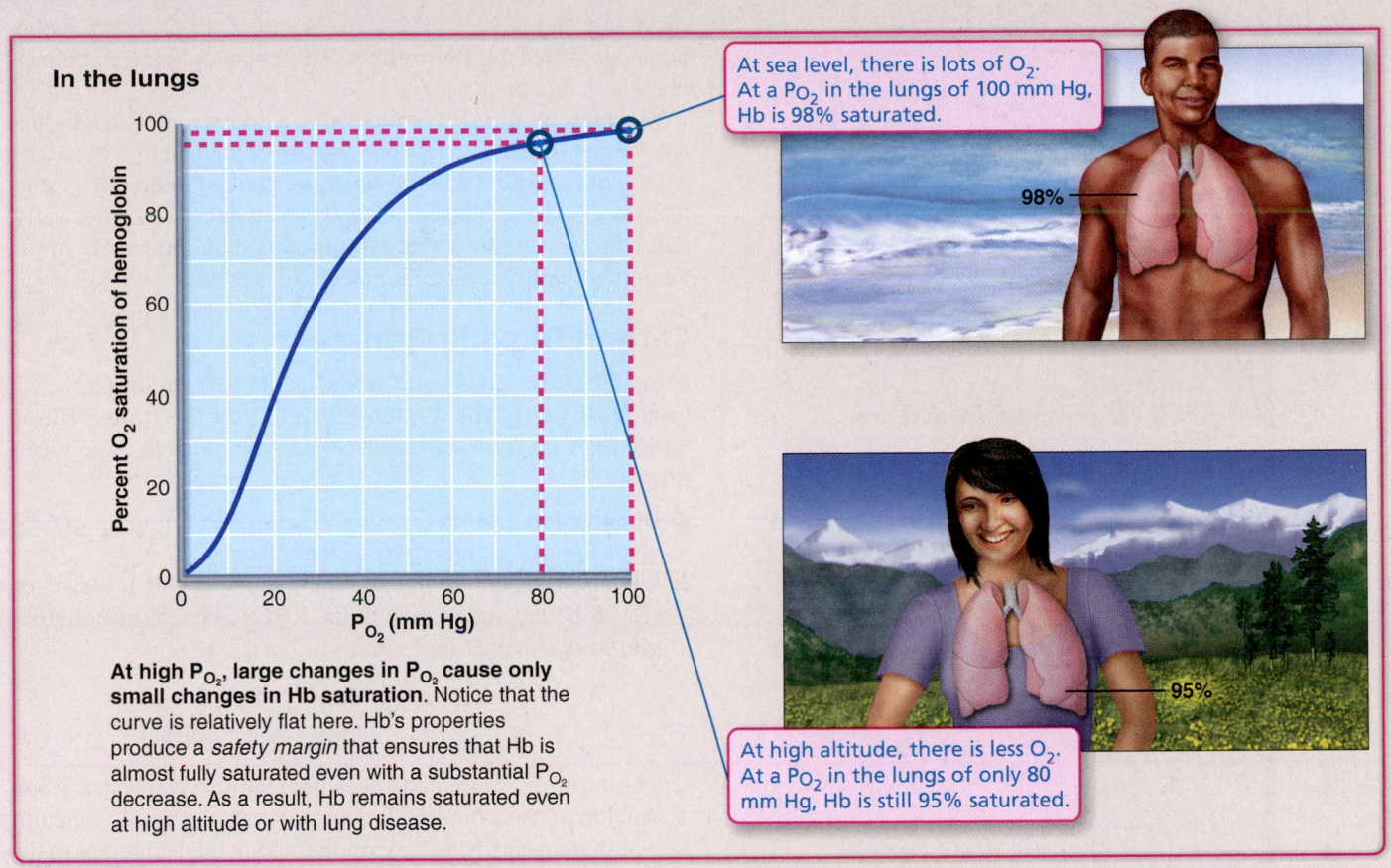

Percent O_2 saturation of hemoglobin vs P_{O_2} (mm Hg)

At sea level, there is lots of O_2. At a P_{O_2} in the lungs of 100 mm Hg, Hb is 98% saturated.

98%

At high altitude, there is less O_2. At a P_{O_2} in the lungs of only 80 mm Hg, Hb is still 95% saturated.

95%

At high P_{O_2}, large changes in P_{O_2} cause only small changes in Hb saturation. Notice that the curve is relatively flat here. Hb's properties produce a *safety margin* that ensures that Hb is almost fully saturated even with a substantial P_{O_2} decrease. As a result, Hb remains saturated even at high altitude or with lung disease.

In the tissues

Percent O_2 saturation of hemoglobin vs P_{O_2} (mm Hg)

75%

In resting tissues, at a P_{O_2} of 40 mm Hg, Hb is 75% saturated—only 23% of O_2 carried by Hb is released.

40%

In metabolically active tissues (e.g., exercising muscle), the P_{O_2} is even lower. At a P_{O_2} of 20 mm Hg, Hb is only 40% saturated—an additional 35% of O_2 has been unloaded for tissue use.

At low P_{O_2}, large changes in P_{O_2} cause large changes in Hb saturation. Tissues other than lungs have a low P_{O_2} because they consume O_2. Notice that the curve is relatively steep at low P_{O_2}. Hb's properties ensure that oxygen is delivered where it is most needed—when tissues need more, they get more.

(a)

(b)

Figure 22.24 Effect of temperature, P_{CO_2}, and blood pH on the oxygen-hemoglobin dissociation curve. Oxygen unloading is enhanced by **(a)** increased temperature, **(b)** increased P_{CO_2}, and/or hydrogen ion concentration (decreased pH), causing the dissociation curve to shift to the right. This response is called the Bohr effect.

metabolize glucose. BPG binds reversibly with hemoglobin, and its levels rise when oxygen levels are chronically low.

All of these factors influence Hb saturation by modifying hemoglobin's three-dimensional structure, thereby changing its affinity for O_2. An *increase* in temperature, P_{CO_2}, H^+, or BPG levels in blood lowers Hb's affinity for O_2, enhancing oxygen unloading from the blood. This is shown by the rightward shift of the oxygen-hemoglobin dissociation curve in **Figure 22.24**.

Conversely, a *decrease* in any of these factors increases hemoglobin's affinity for oxygen, decreasing oxygen unloading. This change shifts the dissociation curve to the left.

If you give a little thought to how these factors are related, you'll realize that they all tend to be highest in the systemic capillaries, where oxygen unloading is the goal. As cells metabolize glucose and use O_2, they release CO_2, which increases the P_{CO_2} and H^+ levels in capillary blood. Both declining blood pH

(acidosis) and increasing P_{CO_2} weaken the Hb-O_2 bond, a phenomenon called the **Bohr effect**. This enhances oxygen unloading where it is most needed.

Heat is a by-product of metabolic activity, and active tissues are warmer than less active ones. A rise in temperature decreases hemoglobin's affinity for O_2 both directly and indirectly (via its influence on RBC metabolism and BPG synthesis). Collectively, these factors see to it that Hb unloads much more O_2 in the vicinity of hard-working tissue cells.

Carbon Dioxide Transport

Normally active body cells produce about 200 ml of CO_2 each minute—exactly the amount excreted by the lungs. Blood transports CO_2 from the tissue cells to the lungs in three forms (**Figure 22.25**):

1. **Dissolved in plasma** (7–10%). The smallest amount of CO_2 is transported simply dissolved in plasma.
2. **Chemically bound to hemoglobin** (just over 20%). Dissolved CO_2 is bound and carried in the RBCs as **carbaminohemoglobin** (kar-bam″ĭ-no-he″mo-glo′bin):

$$CO_2 + Hb \rightleftharpoons HbCO_2$$
carbaminohemoglobin

This reaction is rapid and does not require a catalyst. Carbon dioxide transport in RBCs does not compete with oxyhemoglobin transport because carbon dioxide binds directly to the amino acids of globin (not to the heme).

CO_2 loading and unloading are directly influenced by the P_{CO_2} and the degree of Hb oxygenation. Carbon dioxide rapidly dissociates from hemoglobin in the lungs, where the P_{CO_2} of alveolar air is lower than that in blood. Carbon dioxide readily binds with hemoglobin in the tissues, where the P_{CO_2} is higher than that in blood. Deoxygenated hemoglobin combines more readily with carbon dioxide than does oxygenated hemoglobin, as we will see in the discussion of the Haldane effect below.

3. **As bicarbonate ions in plasma** (about 70%). Most carbon dioxide molecules entering the plasma quickly enter RBCs. The reactions that convert carbon dioxide to **bicarbonate ions (HCO_3^-)** for transport mostly occur inside RBCs. As illustrated in Figure 22.25a, when dissolved CO_2 diffuses into RBCs, it combines with water, forming carbonic acid (H_2CO_3). H_2CO_3 is unstable and dissociates into hydrogen ions and bicarbonate ions:

$$CO_2 + H_2O \rightleftharpoons H_2CO_3 \rightleftharpoons H^+ + HCO_3^-$$
carbon water carbonic hydrogen bicarbonate
dioxide acid ion ion

Although this reaction also occurs in plasma, it is thousands of times faster in RBCs because they (and not plasma) contain **carbonic anhydrase** (kar-bon′ik an-hi′drās), an enzyme that reversibly catalyzes the conversion of carbon dioxide and water to carbonic acid. Hydrogen ions released during the reaction (as well as CO_2 itself) bind to Hb, triggering the Bohr effect. In this way CO_2 loading enhances O_2 release. Because of the buffering

22

(a) Oxygen release and carbon dioxide pickup at the tissues

(b) Oxygen pickup and carbon dioxide release in the lungs

Figure 22.25 Transport and exchange of CO₂ and O₂. Relative sizes of the transport arrows indicate the proportionate amounts of O_2 and CO_2 moved by each method. (H_2CO_3 = carbonic acid, HCO_3^- = bicarbonate, HHb = reduced hemoglobin, HbO_2 = oxyhemoglobin)

effect of Hb, the liberated H⁺ causes little change in pH under resting conditions. As a result, blood becomes only slightly more acidic (the pH declines from 7.4 to 7.34) as it passes through the tissues.

Once generated, HCO_3^- moves quickly from the RBCs into the plasma, where it is carried to the lungs. To counterbalance the rapid outrush of these anions from the RBCs, chloride ions (Cl⁻) move from the plasma into the RBCs. This ion exchange process, called the **chloride**

shift, occurs via facilitated diffusion through an RBC membrane protein.

In the lungs, the process is reversed (Figure 22.25b). As blood moves through the pulmonary capillaries, its P_{CO_2} declines from 45 mm Hg to 40 mm Hg. For this to occur, CO_2 must first be freed from its "bicarbonate housing." HCO_3^- reenters the RBCs (and Cl^- moves into the plasma) and binds with H^+ to form carbonic acid. Carbonic anhydrase then splits carbonic acid to release CO_2 and water. This CO_2, along with that released from hemoglobin and from solution in plasma, then diffuses along its partial pressure gradient from the blood into the alveoli.

The Haldane Effect

The amount of carbon dioxide transported in blood is markedly affected by the degree to which blood is oxygenated. The lower the P_{O_2} and the lower the Hb saturation with oxygen, the more CO_2 that blood can carry. This phenomenon, called the **Haldane effect**, reflects the greater ability of reduced hemoglobin to form carbaminohemoglobin and to buffer H^+ by combining with it. As CO_2 enters the systemic bloodstream, it causes more oxygen to dissociate from Hb (Bohr effect). The dissociation of O_2 allows more CO_2 to combine with Hb (Haldane effect).

The Haldane effect encourages CO_2 exchange in both the tissues and lungs. In the pulmonary circulation, the situation that we just described is reversed—uptake of O_2 facilitates release of CO_2 (Figure 22.25b). As Hb becomes saturated with O_2, the H^+ released combines with HCO_3^-, helping to unload CO_2 from the pulmonary blood.

Influence of CO_2 on Blood pH

Typically, the H^+ released during carbonic acid dissociation is buffered by Hb or other proteins within the RBCs or in plasma. The HCO_3^- generated in the red blood cells diffuses into the plasma, where it acts as the *alkaline reserve* (weak base) part of the blood's bicarbonate buffer system (◀ p. 41).

The **bicarbonate buffer system** is very important in resisting shifts in blood pH, as shown in the equation in point 3 concerning CO_2 transport (see p. 850). For example, if the hydrogen ion concentration in blood begins to rise, excess H^+ is removed by combining with HCO_3^- to form carbonic acid (a weak acid). If H^+ concentration in blood drops below desirable levels, carbonic acid dissociates, releasing hydrogen ions and lowering the pH again.

Changes in respiratory rate or depth can alter blood pH dramatically by altering the amount of carbonic acid in blood. Slow, shallow breathing allows CO_2 to accumulate in blood. As a result, carbonic acid levels increase and blood pH drops. Conversely, rapid, deep breathing quickly flushes CO_2 out of blood, reducing carbonic acid levels and increasing blood pH.

In this way, respiratory ventilation provides a fast-acting system to adjust blood pH (and P_{CO_2}) when it is disturbed by metabolic factors. Respiratory adjustments play a major role in the acid-base balance of the blood, as we will discuss in Chapter 26.

22

Hypoxia

Inadequate oxygen delivery to body tissues is called **hypoxia** (hi-pok′se-ah). Hypoxia is more visible in fair-skinned people because their skin and mucosae take on a bluish cast (become *cyanotic*) when Hb saturation falls below 75%. The mucosae and nail beds are often the easiest places to observe this color change.

Hypoxia is classified based on cause:

- **Anemic hypoxia** reflects poor O_2 delivery resulting from too few RBCs or from RBCs that contain abnormal or too little Hb.

- **Ischemic (stagnant) hypoxia** results from impaired or blocked blood circulation.

- **Histotoxic hypoxia** occurs when body cells are unable to use O_2 even though adequate amounts are delivered. Metabolic poisons, such as cyanide, can cause histotoxic hypoxia.

- **Hypoxemic hypoxia** is indicated by reduced arterial P_{O_2}. Possible causes include disordered or abnormal ventilation-perfusion coupling, pulmonary diseases that impair ventilation, and breathing air containing scant amounts of O_2.

- **Carbon monoxide poisoning** is a unique type of hypoxemic hypoxia and a leading cause of death from fire. Carbon monoxide (CO) is an odorless, colorless gas. Because Hb's affinity for CO is more than 200 times greater than its affinity for oxygen, CO outcompetes O_2 for the heme binding sites. Even at minuscule partial pressures, carbon monoxide can drastically reduce hemoglobin's ability to carry O_2.

  CO poisoning is particularly dangerous because it does not produce the characteristic signs of hypoxia—cyanosis and respiratory distress. Instead the victim is confused and has a throbbing headache. In rare cases, fair skin becomes cherry red. Patients with CO poisoning are given hyperbaric therapy (if available) or 100% O_2 until the CO has been cleared from the body.

Check Your Understanding

20. List the three ways CO_2 is transported in blood.

21. What is the relationship between CO_2 and pH in the blood? Explain.

22. The dotted lines in the two graphs below represent a shift in the oxygen-hemoglobin dissociation curve. Which shift would allow more oxygen delivery to the tissues? [Hint: Ask yourself which curve allows Hb to "let go" of more O_2 (hold *less* O_2).] Name three conditions in the tissues that would cause the curve to shift this way.

(a)

(b)

For answers, see Answers Appendix.

22.8   Respiratory centers in the brain stem control breathing with input from chemoreceptors and higher brain centers

Learning Outcomes

▶ Describe the neural controls of respiration.

▶ Compare and contrast the influences of arterial pH, arterial partial pressures of oxygen and carbon dioxide, lung reflexes, volition, and emotions on respiratory rate and depth.

Although our tidelike breathing seems so beautifully simple, its control is more complex than you might think. Higher brain centers, chemoreceptors, and other reflexes all modify the basic respiratory rhythms generated in the brain stem.

Neural Mechanisms

Control of respiration primarily involves neurons in the reticular formation of the medulla and pons. Because the medulla sets the respiratory rhythm, we will begin there.

Medullary Respiratory Centers

Clustered neurons in two areas of the medulla oblongata appear to be critically important in respiration (**Figure 22.26**). These are:

• The **ventral respiratory group (VRG)**, a network of neurons that extends in the ventral brain stem from the spinal cord to the pons-medulla junction

• The **dorsal respiratory group (DRG)**, located dorsally near the root of cranial nerve IX

Ventral Respiratory Group (VRG)   The VRG appears to be a rhythm-generating and integrative center. It contains groups of neurons that fire during inspiration and others that fire during expiration in a dance of mutual inhibition. (Recent evidence suggests that forced expiration and passive expiration may involve separate groups of neurons.)

When its inspiratory neurons fire, a burst of impulses travels along the **phrenic** and **intercostal nerves** to excite the diaphragm and external intercostal muscles, respectively (Figure 22.26). As a result, the thorax expands and air rushes into the lungs. When the VRG's expiratory neurons fire, the output stops, and expiration occurs as the inspiratory muscles relax and the lungs recoil.

This cyclic on/off activity of the inspiratory and expiratory neurons repeats continuously and produces a respiratory rate of 12–16 breaths per minute, with inspiratory phases lasting about 2 seconds followed by expiratory phases lasting about 3 seconds. This normal respiratory rate and rhythm is called **eupnea** (ūp-ne′ah; *eu* = good, *pne* = breath).

During severe hypoxia, VRG networks generate gasping (perhaps in a last-ditch effort to restore O_2 to the brain). Respiration stops when a certain cluster of VRG neurons is completely suppressed, as by an overdose of morphine or alcohol.

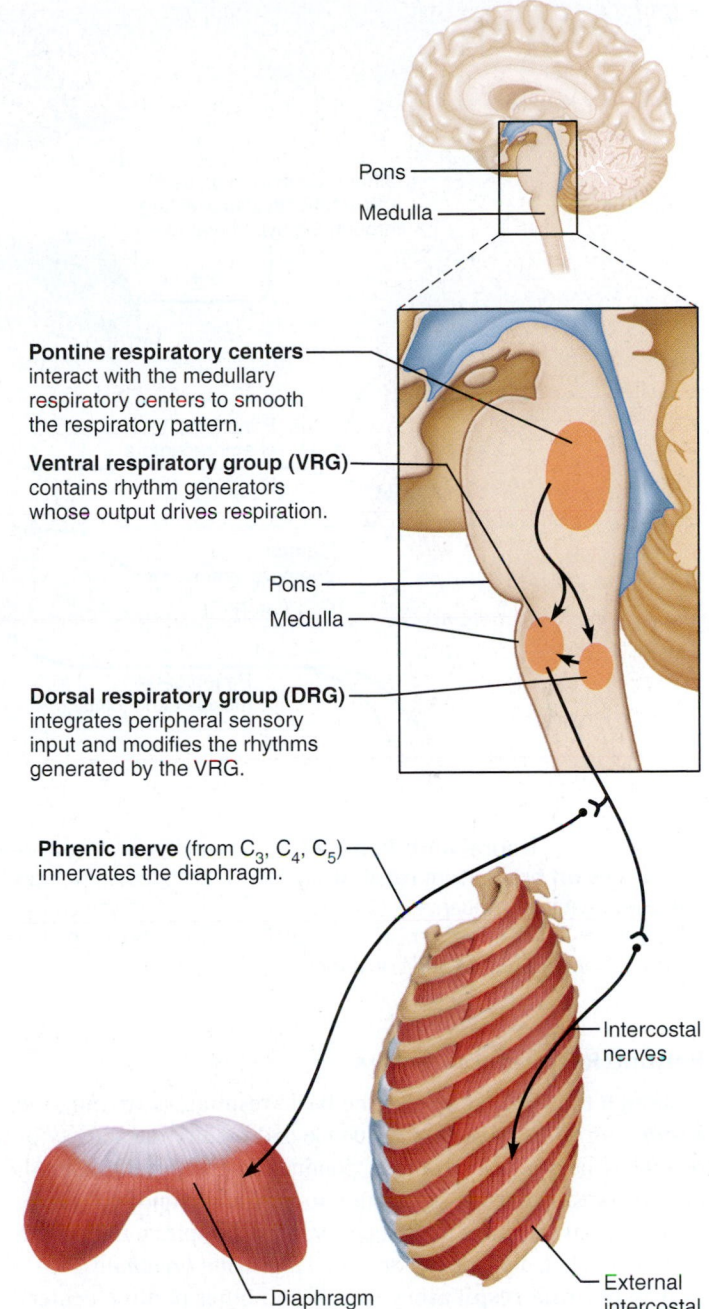

Pons
Medulla

Pontine respiratory centers interact with the medullary respiratory centers to smooth the respiratory pattern.

Ventral respiratory group (VRG) contains rhythm generators whose output drives respiration.

Pons
Medulla

Dorsal respiratory group (DRG) integrates peripheral sensory input and modifies the rhythms generated by the VRG.

Phrenic nerve (from C_3, C_4, C_5) innervates the diaphragm.

Intercostal nerves

Diaphragm

External intercostal muscles

Figure 22.26   Respiratory centers in the brain stem. The synapses in the spinal cord with somatic motor neurons to the inspiratory muscles are shown schematically.

Dorsal Respiratory Group (DRG)   Until recently, it was thought that the DRG acts as an inspiratory center, performing many of the tasks now known to be performed by the VRG. In almost all mammals, including humans, the DRG integrates input from peripheral stretch and chemoreceptors (which we will describe shortly) and communicates this information to the VRG. It may seem surprising, but many of the details of this system so essential to life are still being worked out.

Figure 22.27 Neural and chemical influences on brain stem respiratory centers. Excitatory influences (+) increase the frequency of impulses sent to the muscles of respiration and recruit additional motor units, resulting in deeper, faster breathing. Inhibitory influences (–) have the reverse effect. In some cases, the influences may be excitatory or inhibitory (±), depending on which receptors or brain regions are activated. The cerebral cortex also directly innervates respiratory muscle motor neurons (not shown).

Pontine Respiratory Centers

Although the VRG generates the basic respiratory rhythm, the *pontine respiratory centers* influence and modify the activity of medullary neurons. For example, pontine centers appear to smooth out the transitions from inspiration to expiration, and vice versa. When lesions are made in its superior region, inspirations become very prolonged, a phenomenon called *apneustic breathing*.

The **pontine respiratory group** and other pontine centers transmit impulses to the VRG of the medulla (Figure 22.26). This input modifies and fine-tunes the breathing rhythms generated by the VRG during certain activities such as vocalization, sleep, and exercise. As you would expect from these functions, the pontine respiratory centers, like the DRG, receive input from higher brain centers and from various sensory receptors in the periphery.

Generation of the Respiratory Rhythm

There is little question that breathing is rhythmic, but we still cannot fully explain the origin of its rhythm. One hypothesis is that there are *pacemaker neurons,* which have intrinsic (automatic) rhythmicity like the pacemaker cells found in the heart. Pacemaker-like activity has been demonstrated in certain VRG neurons, but suppressing their activity does not abolish breathing.

This leads us to the second (and more widely accepted) hypothesis: Normal respiratory rhythm results from reciprocal inhibition of interconnected neuronal networks in the medulla. Rather than a single set of pacemaker neurons, there are two (or possibly three) sets that inhibit each other and cycle their activity to generate the rhythm.

Factors Influencing Breathing Rate and Depth

Inspiratory depth is determined by how actively the respiratory centers stimulate the motor neurons serving the respiratory muscles. The greater the stimulation, the greater the number of motor units excited and the greater the force of respiratory muscle contractions. Respiratory rate is determined by how long the inspiratory center is active or how quickly it is switched off.

Changing body demands can modify depth and rate of breathing. The respiratory centers in the medulla and pons are sensitive to both excitatory and inhibitory stimuli, as summarized in **Figure 22.27**.

Chemical Factors

Among the factors that influence breathing rate and depth, the most important are changing levels of CO_2, O_2, and H^+ in

Figure legend:
- Initial stimulus
- Physiological response
- Result

Figure 22.28 Changes in P$_{CO_2}$ regulate ventilation by a negative feedback mechanism.

Flowchart text:
- ↑ Arterial P$_{CO_2}$
- ↑ P$_{CO_2}$ decreases pH in brain extracellular fluid (ECF)
- Central chemoreceptors in brain stem respond to ↑H$^+$ in brain ECF (mediate 70% of the CO$_2$ response)
- Peripheral chemoreceptors in carotid and aortic bodies (mediate 30% of the CO$_2$ response)
- Afferent impulses
- Medullary respiratory centers
- Efferent impulses
- Respiratory muscles
- ↑ Ventilation (more CO$_2$ exhaled)
- Arterial P$_{CO_2}$ and pH return to normal

As P$_{CO_2}$ levels rise in the blood, a condition referred to as **hypercapnia** (hi″per-kap′ne-ah), CO$_2$ accumulates in the brain. As CO$_2$ accumulates, it is hydrated to form carbonic acid. The acid dissociates, H$^+$ is liberated, and the pH drops. This is the same reaction that occurs when CO$_2$ enters RBCs (see p. 850).

The increase in H$^+$ excites the central chemoreceptors, which make abundant synapses with the respiratory regulatory centers. As a result, the depth and rate of breathing increase. This enhanced alveolar ventilation quickly flushes CO$_2$ out of the blood, raising blood pH.

An elevation of only 5 mm Hg in arterial P$_{CO_2}$ doubles alveolar ventilation, even when arterial O$_2$ levels and pH haven't changed. When P$_{O_2}$ and pH are below normal, the response to elevated P$_{CO_2}$ is even greater. Increased ventilation is normally self-limiting, ending when homeostatic blood P$_{CO_2}$ levels are restored.

Notice that while rising blood CO$_2$ levels act as the initial stimulus, it is rising levels of H$^+$ generated within the brain that prod the central chemoreceptors into increased activity. (CO$_2$ readily diffuses across the blood brain barrier between the brain and the blood, but H$^+$ does not.) In the final analysis, control of breathing during rest is aimed primarily at *regulating the H$^+$ concentration in the brain.*

HOMEOSTATIC IMBALANCE 22.15   CLINICAL

Hyperventilation is an increase in the rate and depth of breathing that exceeds the body's need to remove CO$_2$. A person experiencing an anxiety attack may hyperventilate involuntarily. As the person blows off CO$_2$, the low CO$_2$ levels in the blood (**hypocapnia**) constrict cerebral blood vessels. This reduces brain perfusion, producing cerebral ischemia that causes dizziness or fainting. Earlier symptoms of hyperventilation are tingling and involuntary muscle spasms (tetany) in the hands and face caused by blood Ca^{2+} levels falling as pH rises.

The symptoms of hyperventilation may be averted by breathing into a paper bag. The air being inspired from the bag is expired air, rich in carbon dioxide, which causes carbon dioxide to be retained in the blood. ●●●●●●●

arterial blood. Sensors responding to such chemical fluctuations, called **chemoreceptors**, are found in two major body locations:

- **Central chemoreceptors** are located throughout the brain stem, including the ventrolateral medulla.
- **Peripheral chemoreceptors** are found in the aortic arch and carotid arteries.

Influence of P$_{CO_2}$ Of all the chemicals influencing respiration, CO$_2$ is the most potent and the most closely controlled. Normally, arterial P$_{CO_2}$ is 40 mm Hg and is maintained within ±3 mm Hg of this level by an exquisitely sensitive homeostatic mechanism that is mediated mainly by the effect of rising CO$_2$ levels on the central chemoreceptors of the brain stem (**Figure 22.28**).

When P$_{CO_2}$ is abnormally low, respiration is inhibited and becomes slow and shallow. In fact, periods of **apnea** (breathing cessation) may occur until arterial P$_{CO_2}$ rises and again stimulates respiration.

Sometimes swimmers voluntarily hyperventilate so they can hold their breath longer during swim meets. This is dangerous. Blood O$_2$ content rarely drops much below 60% of normal during regular breath-holding, because as P$_{O_2}$ drops, P$_{CO_2}$ rises enough to make breathing unavoidable. However, strenuous hyperventilation can lower P$_{CO_2}$ so much that a lag period occurs before P$_{CO_2}$ rebounds enough to stimulate respiration again. This lag may allow oxygen levels to fall well below 50 mm Hg, causing the swimmer to black out (and perhaps drown) before he or she has the urge to breathe.

Influence of P$_{O_2}$ The peripheral chemoreceptors—found in the **aortic bodies** of the aortic arch and in the **carotid bodies**

22

- Brain
- Sensory nerve fiber in cranial nerve IX (pharyngeal branch of glossopharyngeal)
- External carotid artery
- Internal carotid artery
- **Carotid body**
- Common carotid artery
- Cranial nerve X (vagus nerve)
- Sensory nerve fiber in cranial nerve X
- **Aortic bodies** in aortic arch
- Aorta
- Heart

Figure 22.29 Location and innervation of the peripheral chemoreceptors in the carotid and aortic bodies.

at the bifurcation of the common carotid arteries—contain cells sensitive to arterial O_2 levels (**Figure 22.29**). The main oxygen sensors are in the carotid bodies.

Under normal conditions, declining P_{O_2} has only a slight effect on ventilation, mostly limited to enhancing the sensitivity of peripheral receptors to increased P_{CO_2}. Arterial P_{O_2} must drop *substantially*, to at least 60 mm Hg, before O_2 levels become a major stimulus for increased ventilation.

This is not as strange as it may appear. Remember, there is a huge reservoir of O_2 bound to Hb, and Hb remains almost entirely saturated unless or until the P_{O_2} of alveolar gas and arterial blood falls below 60 mm Hg. The brain stem centers then begin to suffer from O_2 starvation, and their activity is depressed. At the same time, the peripheral chemoreceptors become excited and stimulate the respiratory centers to increase

ventilation, even if P_{CO_2} is normal. In this way, the peripheral chemoreceptor system can maintain ventilation even though the brain stem centers are depressed by hypoxia.

Influence of Arterial pH Changes in arterial pH can modify respiratory rate and rhythm even when CO_2 and O_2 levels are normal. Because H^+ does not cross the blood brain barrier, the increased ventilation that occurs in response to falling arterial pH is mediated through the peripheral chemoreceptors.

Although changes in P_{CO_2} and H^+ concentration are interrelated, they are distinct stimuli. A drop in blood pH may reflect CO_2 retention, but it may also result from metabolic causes, such as accumulation of lactic acid during exercise or of fatty acid metabolites (ketone bodies) in patients with poorly controlled diabetes mellitus. Regardless of cause, as arterial pH declines, respiratory system controls attempt to compensate and raise the pH. They do this by increasing respiratory rate and depth to eliminate CO_2 (and carbonic acid) from the blood.

Summary of Interactions of P_{CO_2}, P_{O_2}, and Arterial pH The body's need to rid itself of CO_2 is the most important stimulus for breathing in a healthy person. However, CO_2 does not act in isolation, and various chemical factors enforce or inhibit one another's effects. These interactions are summarized here:

- *Rising CO_2 levels are the most powerful respiratory stimulant.* As CO_2 is hydrated in brain tissue, liberated H^+ acts directly on the central chemoreceptors, causing a reflexive increase in breathing rate and depth. Low P_{CO_2} levels depress respiration.

- *Under normal conditions, blood P_{O_2} affects breathing only indirectly.* It does so by influencing peripheral chemoreceptor sensitivity to changes in P_{CO_2}. Low P_{O_2} augments P_{CO_2} effects, and high P_{O_2} levels diminish the effectiveness of CO_2 stimulation.

- *When arterial P_{O_2} falls below 60 mm Hg, it becomes the major stimulus for respiration.* The peripheral chemoreceptors respond, reflexively increasing ventilation. This may increase O_2 loading into the blood, but it also causes hypocapnia (low P_{CO_2} blood levels) and an increase in blood pH, both of which inhibit respiration.

- *Changes in arterial pH due to CO_2 retention or metabolic factors act through the peripheral chemoreceptors.* The resulting changes in ventilation in turn modify arterial P_{CO_2} and pH. Arterial pH does not influence the central chemoreceptors directly.

Influence of Higher Brain Centers

Hypothalamic Controls Acting through the hypothalamus and the rest of the limbic system, strong emotions and pain send signals to the respiratory centers, modifying respiratory rate and depth. For example, have you ever touched something cold and clammy and gasped? That response was mediated through the hypothalamus. So too is the breath-holding that occurs when we are angry and the increased respiratory rate that occurs when we are excited. A rise in body temperature raises the respiratory rate, while a drop in body temperature produces the opposite effect. Sudden chilling (a dip in the North Atlantic Ocean in late

October) can stop your breathing (apnea)—or at the very least, leave you gasping.

Cortical Controls Although the brain stem respiratory centers normally regulate breathing involuntarily, we can also exert conscious (voluntary) control over the rate and depth of our breathing. We can choose to hold our breath or take an extra-deep breath, for example. During voluntary control, the cerebral motor cortex sends signals to the motor neurons that stimulate the respiratory muscles, bypassing the medullary centers.

Our ability to voluntarily hold our breath is limited, however, because the brain stem respiratory centers automatically reinitiate breathing when the blood concentration of CO_2 reaches critical levels. That explains why drowning victims typically have water in their lungs.

Pulmonary Irritant Reflexes

The lungs contain receptors that respond to an enormous variety of irritants. When activated, these receptors communicate with the respiratory centers via vagal nerve afferents. Accumulated mucus, inhaled debris such as dust, or noxious fumes stimulate receptors in the bronchioles that promote reflex constriction of those air passages. In the trachea or bronchi, the same irritants stimulate a cough. Stimulation in the nasal cavity triggers a sneeze.

The Inflation Reflex

The visceral pleurae and conducting passages in the lungs contain numerous stretch receptors that are vigorously stimulated when the lungs are inflated. These receptors signal the medullary respiratory centers via afferent fibers of the vagus nerves, sending inhibitory impulses that end inspiration and allow expiration to occur.

As the lungs recoil, the stretch receptors become quiet, and inspiration is initiated once again. This reflex, called the **inflation reflex**, or **Hering-Breuer reflex** (her'ing broy'er), is thought to be more a protective response (to prevent the lungs from being stretched excessively) than a normal regulatory mechanism.

Check Your Understanding

23. Which brain stem respiratory area is thought to generate the respiratory rhythm?

24. Which chemical factor in blood normally provides the most powerful stimulus to breathe? Which chemoreceptors are most important for this response?

For answers, see Answers Appendix.

22.9 Exercise and high altitude bring about respiratory adjustments

Learning Outcomes

▶ **Compare and contrast the hyperpnea of exercise with hyperventilation.**

▶ **Describe the process and effects of acclimatization to high altitude.**

Exercise

Respiratory adjustments during exercise are geared to both the intensity and duration of the exercise. Working muscles consume tremendous amounts of O_2 and produce large amounts of CO_2, so ventilation can increase 10- to 20-fold during vigorous exercise. Increased ventilation in response to metabolic needs is called **hyperpnea** (hi"perp-ne'ah).

How does hyperpnea differ from hyperventilation? The respiratory changes in hyperpnea do not alter blood O_2 and CO_2 levels significantly. In contrast, hyperventilation is excessive ventilation, and is characterized by low P_{CO_2} and alkalosis.

Exercise-enhanced ventilation does *not* appear to be prompted by rising P_{CO_2} and declining P_{O_2} and pH in the blood for two reasons.

- Ventilation increases abruptly as exercise begins, followed by a gradual increase, and then reaches a steady state. When exercise stops, there is a small but abrupt decline in ventilation rate, followed by a gradual decrease to the pre-exercise value.

- Although venous levels change, arterial P_{CO_2} and P_{O_2} levels remain surprisingly constant during exercise. In fact, P_{CO_2} may fall below normal and P_{O_2} may rise slightly because the respiratory adjustments are so efficient.

The most widely accepted explanation for the abrupt increase in ventilation that occurs as exercise begins reflects interaction of three neural factors:

- Psychological stimuli (our conscious anticipation of exercise)
- Simultaneous cortical motor activation of skeletal muscles and respiratory centers
- Excitatory input reaching respiratory centers from proprioceptors in moving muscles, tendons, and joints

The subsequent gradual increase and then plateauing of respiration probably reflect the rate of CO_2 delivery to the lungs (the "CO_2 flow").

The rise in lactic acid levels during exercise results from anaerobic respiration. However, it is *not* a result of inadequate respiratory function, because alveolar ventilation and pulmonary perfusion are as well matched during exercise as during rest (hemoglobin remains fully saturated). Rather, it reflects cardiac output limitations or inability of the skeletal muscles to further increase their oxygen consumption.

In light of this fact, the practice of inhaling pure O_2 by mask, used by some football players to replenish their "oxygen-starved" bodies as quickly as possible, is useless. The panting athlete *does* need more oxygen, but inspiring extra oxygen will not help, because the shortage is in the muscles—not the lungs.

High Altitude

Most people live between sea level and an altitude of approximately 2400 m (8000 feet). In this range, differences in atmospheric pressure are not great enough to cause healthy people any problems when they spend brief periods in higher-altitude areas.

However, if you travel quickly from sea level to elevations above 8000 ft, where atmospheric pressure and P_{O_2} are lower,

22

your body responds with symptoms of *acute mountain sickness* (*AMS*)—headaches, shortness of breath, nausea, and dizziness. AMS is sometimes seen in travelers to ski resorts such as Vail, Colorado (8120 ft). In severe cases of AMS, lethal pulmonary and cerebral edema may occur.

When you move on a *long-term* basis from sea level to the mountains, your body makes respiratory and hematopoietic adjustments via an adaptive response called **acclimatization**. As we have already explained, decreases in arterial P_{O_2} cause the peripheral chemoreceptors to become more responsive to increases in P_{CO_2}, and a substantial decline in P_{O_2} directly stimulates them. As a result, ventilation increases as the brain attempts to restore gas exchange. Increased ventilation also reduces arterial CO_2 levels, so the P_{CO_2} of individuals living at high altitudes is typically below 40 mm Hg (its value at sea level).

High-altitude conditions result in lower-than-normal hemoglobin saturation levels because less O_2 is available to be loaded. For example, at about 19,000 ft above sea level, O_2 saturation of arterial blood is only 67% (compared to nearly 98% at sea level). But Hb unloads only 20–25% of its oxygen at sea level, which means that even at the reduced saturations at high altitudes, the O_2 needs of the tissues are still met under resting conditions.

Additionally, at high altitudes hemoglobin's affinity for O_2 is reduced because BPG concentrations increase. This releases more O_2 to the tissues during each circulatory round.

When blood O_2 levels decline, the kidneys produce more erythropoietin, which stimulates bone marrow production of RBCs (◀ p. 648). This phase of acclimatization, which occurs slowly, provides long-term compensation for living at high altitudes.

Check Your Understanding

25. What long-term adjustments does the body make when living at high altitude?

26. **APPLY** An injured soccer player arrives by ambulance in the emergency room. She is in obvious distress, breathing rapidly. Her blood P_{CO_2} is 26 mm Hg and pH is 7.5. Is she suffering from hyperventilation or hyperpnea? Explain.

For answers, see Answers Appendix.

22.10 Respiratory diseases are major causes of disability and death   CLINICAL

Learning Outcome

▶ Compare the causes and consequences of chronic bronchitis, emphysema, asthma, tuberculosis, and lung cancer.

The respiratory system is particularly vulnerable to infectious diseases because it is wide open to airborne pathogens. However, some of the most disabling respiratory disorders are *chronic obstructive pulmonary disease* (*COPD*), *asthma, tuberculosis,* and *lung cancer*. COPD and lung cancer are living proof of the devastating effects of tobacco smoke on the body. Long known to promote cardiovascular disease, smoking is perhaps even more effective at destroying the lungs.

Chronic Obstructive Pulmonary Disease (COPD)

The **chronic obstructive pulmonary diseases (COPD)**, exemplified best by emphysema and chronic bronchitis, are a major cause of disability and death in North America. The key physiological feature of these diseases is an irreversible decrease in the ability to force air out of the lungs. Other features they share in common are (**Figure 22.30**):

- More than 80% of patients have a history of smoking.
- **Dyspnea** (disp-ne′ah), difficult or labored breathing often referred to as "air hunger," gets progressively worse.
- Coughing and frequent pulmonary infections are common.
- Most COPD patients develop respiratory failure manifested as **hypoventilation** (insufficient ventilation in relation to metabolic needs, causing them to retain CO_2), respiratory acidosis, and hypoxemia.

Emphysema

Emphysema is distinguished by permanent enlargement of the alveoli, accompanied by destruction of the alveolar walls (see Figure 22.22 on p. 845). Invariably the lungs lose their elasticity. This has three important consequences:

- Accessory muscles must be enlisted to breathe, and victims are perpetually exhausted because breathing requires 15–20% of their total body energy supply (as opposed to 5% in healthy individuals).

Figure 22.30 The pathogenesis of COPD.

- For complex reasons, the bronchioles open during inspiration but collapse during expiration, trapping huge volumes of air in the alveoli. This hyperinflation leads to development of a permanently expanded "barrel chest" and flattens the diaphragm, thus reducing ventilation efficiency.

- Damage to the pulmonary capillaries as the alveolar walls disintegrate increases resistance in the pulmonary circuit, forcing the right ventricle to overwork and consequently become enlarged.

Emphysema is usually caused by smoking, but hereditary factors (e.g., alpha-1 antitrypsin deficiency) cause emphysema in some patients.

Chronic Bronchitis

In **chronic bronchitis**, inhaled irritants lead to chronic production of excessive mucus. The mucosae of the lower respiratory passageways become inflamed and fibrosed. These responses obstruct the airways, severely impairing lung ventilation and gas exchange. Pulmonary infections are frequent because bacteria thrive in the stagnant pools of mucus. Smoking is a major risk factor. Environmental pollution also promotes chronic bronchitis.

COPD: Symptoms and Treatments

In the clinical setting you might see two very different patterns that represent the extremes of patients with COPD. One pattern has traditionally been called the "pink puffer": These patients work so hard to maintain adequate ventilation that they lose weight, becoming thin but still having nearly normal blood gases. In contrast, "blue bloaters," commonly of stocky build, become sufficiently hypoxic that they are obviously cyanotic. The hypoxia causes constriction of pulmonary blood vessels, leading to pulmonary hypertension and right-sided heart failure.

Traditionally, "pink puffers" were associated with emphysema while "blue bloaters" were associated with chronic bronchitis. As usual, things are not that clear-cut. It turns out that patients with the same underlying disease can display either of these clinical patterns, and this may depend on a third factor—the strength of their innate respiratory drive. Most COPD patients fall between these two clinical extremes.

COPD is routinely treated with inhaled bronchodilators and corticosteroids. Severe dyspnea and hypoxia mandate oxygen use, but oxygen must be administered with care. In some of these patients, giving pure oxygen can increase the blood P_{CO_2} (and lower blood pH) to life-threatening levels. The solution is to use the minimum concentration of oxygen that relieves the patient's hypoxia.

Asthma

Asthma is characterized by episodes of coughing, dyspnea, wheezing, and chest tightness—alone or in combination. A sense of panic accompanies most acute attacks. Although sometimes classed with COPD because it is an obstructive disorder, asthma is marked by acute episodes followed by symptom-free periods—that is, the obstruction is *reversible*.

The cause of asthma has been hard to pin down. Initially it was viewed as a consequence of bronchospasms triggered by various factors such as cold air, exercise, or allergens. However, researchers have found that in allergic asthma (the most common kind), active inflammation of the airways comes first. The inflammation is an immune response controlled by a subset of T lymphocytes that stimulate the production of IgE antibodies and recruit inflammatory cells to the site.

Once someone has allergic asthma, the inflammation persists even during symptom-free periods and makes the airways hypersensitive. (The most common triggers are in the home—the allergens from dust mites, cockroaches, cats, dogs, and fungi.) Once the airway walls are thickened with inflammatory exudate, the effect of bronchospasm is vastly magnified and can dramatically reduce air flow.

About one in ten people in North America suffer from asthma—children more than adults. Over the past 20 years, the number of cases rose dramatically, plateaued, and may now be declining.

While asthma remains a major health problem, better treatment options have reduced the number of asthma-related deaths. Instead of merely treating the symptoms with fast-acting bronchodilators, we now treat the underlying inflammation using inhaled corticosteroids and a variety of other drugs.

Tuberculosis (TB)

Tuberculosis (TB), the infectious disease caused by the bacterium *Mycobacterium tuberculosis*, is spread by coughing and primarily enters the body in inhaled air. TB mostly affects the lungs but can spread through the lymphatics to other organs.

One-third of the world's population is infected, but most people never develop active TB because a massive inflammatory and immune response usually contains the primary infection in fibrous, or calcified, nodules (tubercles) in the lungs. However, the bacteria survive in the nodules and when the person's immunity is weakened, they may break out and cause symptomatic TB. Symptoms include fever, night sweats, weight loss, racking cough, and coughing up blood.

Deadly strains of drug-resistant (even multidrug-resistant) TB can develop when treatment is incomplete or inadequate. Resistant strains are found elsewhere in the world and have appeared in North America.

Lung Cancer

Lung cancer is the leading cause of cancer death for both men and women in North America, killing more people every year than breast, prostate, and colorectal cancer combined. This is tragic, because lung cancer is largely preventable—nearly 90% of cases result from smoking.

The cure rate for lung cancer is notoriously low, with most people dying within one year of diagnosis. The five-year survival rate is about 16%. Because lung cancer is aggressive and metastasizes rapidly and widely, most cases are not diagnosed until they are well advanced.

Lung cancer appears to follow closely the oncogene-activating steps outlined in *A Closer Look* in Chapter 4 (◄ **p. 144**). Ordinarily, nasal hairs, sticky mucus, and cilia do a fine job of protecting the lungs from chemical and biological irritants, but when a person smokes, these defenses are overwhelmed and eventually stop functioning. In particular, smoking paralyzes the cilia that clear mucus from the airways, allowing irritants and pathogens to accumulate. The "cocktail" of free radicals and other carcinogens in tobacco smoke eventually translates into lung cancer.

The most common types of lung cancer are:

- **Adenocarcinoma** (about 40% of cases), which originates in peripheral lung areas as solitary nodules that develop from bronchial glands and alveolar cells.

- **Squamous cell carcinoma** (about 20%), which arises in the epithelium of the bronchi or their larger subdivisions and tends to form masses that may cavitate (hollow out) and bleed.

- **Small cell carcinoma** (about 15%), round lymphocyte-sized cells that originate in the main bronchi and grow aggressively in small grapelike clusters within the mediastinum. Metastasis from the mediastinum is especially rapid. Some small cell carcinomas cause additional problems because they produce certain hormones. For example, some secrete antidiuretic hormone (ADH), resulting in the syndrome of inappropriate ADH secretion (◄ **p. 614**).

Because lung cancers metastasize aggressively and early, the key to survival is early detection. If the cancer has not metastasized before it is discovered, complete removal of the diseased lung has the greatest potential for prolonging life and providing a cure. With metastatic lung cancer, radiation therapy and chemotherapy are the only options, but these have low success rates.

Sleep Apnea

Sleep apnea is a common disorder that is characterized by the temporary cessation of breathing during sleep. Although they are usually not aware of it, people with sleep apnea wake themselves up many times each night (in severe cases, more than 30 times per hour). As a result, people with sleep apnea may feel as tired when they get up as when they went to sleep, leaving them with excessive daytime sleepiness. Loud snoring, particularly when respiration resumes, is often reported by others sleeping in the same room because it disturbs their sleep as well.

Obstructive sleep apnea, the most common kind, is caused by collapse of the upper airway. This occurs because the muscles of the pharynx relax during sleep, which allows the soft tissues of the pharynx to sag and obstruct the airway. Obstructive sleep apnea is more common in men, and is made worse by obesity.

With what you have learned about the regulation of respiration, it will not surprise you to learn that there is another kind of sleep apnea—**central sleep apnea**. Central sleep apnea is caused by a reduced drive from the respiratory centers of the brain stem during sleep rather than by obstruction of the airway.

Consequences of disrupted sleep are an increased susceptibility to accidents and an increase in a variety of chronic illnesses such as hypertension, heart disease, stroke, and diabetes.

Obstructive sleep apnea is often treated using a continuous positive airway pressure (CPAP) device. By gently blowing air into the airway, it provides a positive air pressure that holds the airway open and prevents its collapse.

Check Your Understanding

27. What distinguishes the obstruction in asthma from that in chronic bronchitis?

28. MAKE CONNECTIONS Your patient, Mr. Singh, has COPD. As a result, his pulmonary alveoli are poorly ventilated. Would you expect his pulmonary arterial blood pressure to be higher or lower than normal? Explain. Over the long term, this change in blood pressure may cause Mr. Singh to develop heart failure (Chapter 18). Which side of the heart will fail? Edema is one sign of heart failure. Where would you expect to see edema in Mr. Singh?

For answers, see Answers Appendix.

Developmental Aspects of the Respiratory System

Because embryos develop in a cephalocaudal (head-to-tail) direction, the upper respiratory structures appear first. By the fourth week of development, two thickened plates of ectoderm, the **olfactory placodes** (plak′ōds), are present on the anterior aspect of the head (**Figure 22.31**). These quickly invaginate to form **olfactory pits** that form the nasal cavity. The olfactory pits then extend posteriorly to connect with the developing pharynx, which forms at the same time from the endodermal germ layer.

The epithelium of the lower respiratory organs develops as an outpocketing of the foregut endoderm, which becomes the pharyngeal mucosa. This protrusion, called the **laryngo-tracheal bud**, is present by the fifth week of development. The proximal part of the bud forms the tracheal lining, and its distal end splits and forms the mucosae of the bronchi and all their subdivisions, including (eventually) the lung alveoli. Mesoderm covers these endoderm-derived linings and forms the walls of the respiratory passageways and the stroma of the lungs.

By 28 weeks, the respiratory system has developed sufficiently to allow most babies born prematurely to breathe on their own. As we noted earlier, infants born at or before this time often exhibit infant respiratory distress syndrome resulting from inadequate surfactant production.

During fetal life, the lungs are filled with fluid and the placenta makes all respiratory exchanges. Vascular shunts cause circulating blood to largely bypass the lungs (see Chapter 28). At birth, the respiratory passageways fill with air. As the P_{CO_2} in the baby's blood rises, the respiratory centers are excited, causing the baby to take its first breath. The alveoli inflate and begin to function in gas exchange, but it is nearly two weeks before the lungs are fully inflated.

— Frontonasal elevation

Olfactory placode

— Stomodeum
(future mouth)

(a) 4 weeks: anterior superficial view
of the embryo's head

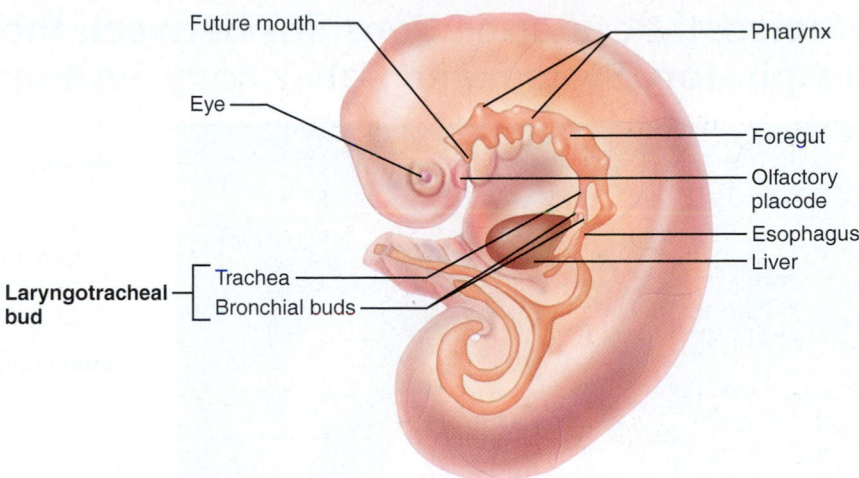

Future mouth ─

Eye ─

Pharynx

Foregut

Olfactory
placode

Esophagus

Liver

Laryngotracheal
bud

[Trachea ─
[Bronchial buds ─

(b) 5 weeks: left lateral view of the developing lower
respiratory passageway mucosae

Figure 22.31 Embryonic development of the respiratory system.

HOMEOSTATIC IMBALANCE 22.16   CLINICAL

Important birth defects of the respiratory system include *cleft palate* (◀ p. 246) and cystic fibrosis. **Cystic fibrosis (CF)**, the most common lethal genetic disease in North America, strikes in one out of every 3000 births. In CF, abnormally viscous mucus clogs the respiratory passages, providing a breeding ground for airborne bacteria and predisposing the child to respiratory infections.

At the root of CF is a faulty gene that codes for the *CFTR* (cystic fibrosis transmembrane conductance regulator) *protein*. The normal CFTR protein works as a membrane channel to control Cl^- flow in and out of cells. In one common mutation, CFTR lacks a critical amino acid and so does not fold correctly. As a result, it gets "stuck" in the endoplasmic reticulum, is marked for degradation, and never reaches the plasma membrane to perform its normal role. Consequently, less Cl^- is secreted and less water follows, resulting in the thick mucus typical of CF.

This thick mucus forms a perfect harbor for bacterial infection. By early adulthood, 80% of patients are colonized by *Pseudomonas aeruginosa*, which causes chronic inflammation and triggers the disabled cells to churn out a thick sludge of abnormal mucus. Repeated cycles of infection and inflammation eventually result in extensive tissue damage that can be treated only by a lung transplant.

Defects in the CFTR protein can affect other organ systems, too. The ducts of the pancreas become clogged with secretions, impairing food digestion. Obstructed reproductive ducts render 97% of males with CF infertile. Characteristically, sweat glands of CF patients produce extremely salty perspiration.

Conventional therapy for CF has included mucus-dissolving drugs, "clapping" the chest to loosen the thick mucus, and antibiotics to prevent infection. The goal of CF research is to restore normal salt and water movements. This can be done using a number of approaches:

- Introducing normal CFTR genes into respiratory tract mucosa cells
- Prodding another channel protein to take over the duties of transporting Cl^-
- Developing techniques to free the CFTR protein from the ER
- Using hypertonic saline to osmotically move water. This is a novel and surprisingly simple approach that involves inhaling hypertonic saline droplets. This draws water into the mucus, making it more liquid.

Alone or in combination, these therapies provide new hope to patients with CF.   ● ● ● ● ● ●

The respiratory rate is highest in newborn infants (40–80 breaths per minute). At five years of age it is around 25 per minute, and in adults it is between 12 and 16 per minute. In old age, the rate often increases again.

At birth, only about one-sixth of the final number of alveoli are present. The lungs continue to mature and form more alveoli until young adulthood. However, if a person begins smoking in the early teens, the lungs never completely mature, and those additional alveoli are lost forever.

In infants, the ribs take a nearly horizontal course. For this reason, infants rely almost entirely on descent of the diaphragm to increase thoracic volume for inspiration. By the second year, the ribs are more obliquely positioned, and the adult form of breathing is established.

The maximum amount of oxygen we can use during aerobic metabolism, $\dot{V}_{O_2max}$, declines about 9% per decade in inactive people beginning in their mid-20s. In those who remain active, $\dot{V}_{O_2max}$ still declines but much less. As we age, the thoracic wall becomes more rigid and the lungs gradually lose their elasticity, decreasing the ability to ventilate the lungs. Vital capacity declines by about one-third by age 70. Blood O_2 levels decline slightly, and many elderly people tend to become hypoxic during sleep. Sleep apnea becomes more common.

22

Homeostatic Interrelationships between the Respiratory System and Other Body Systems

Integumentary System   Chapter 5

- Respiratory system provides oxygen and disposes of carbon dioxide
- Skin protects respiratory system organs by forming surface barriers

Skeletal System   Chapters 6–8

- Respiratory system provides oxygen and disposes of carbon dioxide
- Bones protect lungs and bronchi by enclosure

Muscular System   Chapters 9–10

- Respiratory system provides oxygen needed for muscle activity and disposes of carbon dioxide
- Activity of the diaphragm and intercostal muscles essential for producing volume changes that lead to pulmonary ventilation; regular exercise increases respiratory efficiency

Nervous System   Chapters 11–15

- Respiratory system provides oxygen needed for normal neuronal activity and disposes of carbon dioxide
- Medullary and pontine centers regulate respiratory rate and depth; stretch receptors in lungs and chemoreceptors provide feedback

Endocrine System   Chapter 16

- Respiratory system provides oxygen and disposes of carbon dioxide; angiotensin converting enzyme in lungs converts angiotensin I to angiotensin II
- Epinephrine dilates the bronchioles; testosterone promotes laryngeal enlargement in pubertal males; glucocorticoids promote surfactant production

Cardiovascular System   Chapters 17–19

- Respiratory system provides oxygen and disposes of carbon dioxide; carbon dioxide present in blood as HCO_3^- and H_2CO_3 contributes to blood buffering
- Blood is the transport medium for respiratory gases

Lymphatic System/Immunity   Chapters 20–21

- Respiratory system provides oxygen and disposes of carbon dioxide; tonsils in pharynx house immune cells
- Lymphatic system helps to maintain blood volume required for respiratory gas transport; immune system protects respiratory organs from bacteria, bacterial toxins, viruses, protozoa, fungi, and cancer

Digestive System   Chapter 23

- Respiratory system provides oxygen and disposes of carbon dioxide
- Digestive system provides nutrients needed by respiratory system organs

Urinary System   Chapters 25–26

- Respiratory system provides oxygen and disposes of carbon dioxide to provide short-term pH homeostasis
- Kidneys dispose of metabolic wastes of respiratory system organs (other than carbon dioxide) and maintain long-term pH homeostasis

Reproductive System   Chapter 27

- Respiratory system provides oxygen and disposes of carbon dioxide

The number of glands in the nasal mucosa decreases with age, as does blood flow to this mucosa. For this reason, the nose dries and produces a thick mucus that makes us want to clear our throat. Additionally, many of the respiratory system's protective mechanisms become less effective. Mucosal cilia are less active, and the macrophages in the lungs become sluggish. The net result is that the elderly are more at risk for respiratory tract infections, particularly pneumonia and influenza.

RELATED CLINICAL TERMS

Adenoidectomy (adenotonsillectomy) Surgical removal of an infected pharyngeal tonsil (adenoids).

Adult respiratory distress syndrome (ARDS) A dangerous lung condition that can develop after severe illness or injury to the body. Neutrophils leave the body's capillaries in large numbers and then secrete chemicals that increase capillary permeability. The capillary-rich lungs are heavily affected. As the lungs fill with fluid, the patient suffocates. Even with mechanical ventilation, ARDS is hard to control and often lethal.

Aspiration (as″pĭ-ra′shun) (1) Inhaling or drawing something into the lungs or respiratory passages. (2) Withdrawing fluid by suction (use of an aspirator); done during surgery to keep an area free of blood or other body fluids. Mucus is aspirated from the trachea of tracheotomy patients.

Bronchoscopy (*scopy* = viewing) Use of a viewing tube inserted through the nose or mouth to examine the internal surface of the main bronchi in the lung. Forceps attached to the tip of the tube can remove trapped objects or take samples of mucus for examination.

Cheyne-Stokes breathing (chān′stōks) Abnormal breathing pattern sometimes seen just before death (the "death rattle") and in people with combined neurological and cardiac disorders. It consists of bursts of tidal volume breaths (increasing and then decreasing in depth) alternating with periods of apnea.

Deviated septum Condition in which the nasal septum takes a more lateral course than usual and may obstruct breathing.

Endotracheal tube A thin plastic tube threaded into the trachea through the nose or mouth; used to deliver oxygen to patients who are breathing inadequately, in a coma, or under anesthesia.

Epistaxis (ep″ĭ-stak′sis; *epistazo* = to bleed at the nose) Nosebleed; commonly follows trauma to the nose or excessive nose blowing. Most nasal bleeding is from the highly vascularized anterior septum and can be stopped by pinching the nostrils closed or packing them with cotton.

Nasal polyps Mushroomlike benign growths of the nasal mucosa; sometimes caused by infections, but most often cause is unknown. May block air flow.

Orthopnea (or″thop-ne′ah; *ortho* = straight, upright) Inability to breathe in the horizontal (lying down) position.

Otorhinolaryngology (o″to-ri″no-lar″in-gol′o-je; *oto* = ear; *rhino* = nose) Branch of medicine that deals with diagnosis and treatment of diseases of the ears, nose, and throat.

Pneumonia Infectious inflammation of the lungs, in which fluid accumulates in the alveoli; the eighth most common cause of death in the United States. Most of the more than 50 varieties of pneumonia are viral or bacterial.

Pulmonary embolism Obstruction of the pulmonary artery or one of its branches by an embolus (most often a blood clot that has been carried from the lower limbs and through the right side of the heart into the pulmonary circulation). Symptoms are chest pain, productive bloody cough, tachycardia, and rapid, shallow breathing. Can cause sudden death unless treated quickly.

Sudden infant death syndrome (SIDS) Unexpected death of an apparently healthy infant during sleep. Commonly called crib death, SIDS is one of the most frequent causes of death in infants under 1 year old. Believed to be a problem of immaturity of the respiratory control centers. Most cases occur in infants placed in a prone position (on their abdomen) to sleep—a position that may result in hypoxia and hypercapnia due to rebreathing exhaled (CO_2-rich) air.

Tracheotomy (tra″ke-ot′o-me) Surgical opening of the trachea; done to provide an alternate route for air to reach the lungs when more superior respiratory passageways are obstructed (as by food or a crushed larynx).

CHAPTER SUMMARY

1. Respiration involves four processes: pulmonary ventilation, external respiration, transport of respiratory gases in the blood, and internal respiration. Both the respiratory system and the cardiovascular system are involved in respiration.

PART 1

FUNCTIONAL ANATOMY

22.1 The upper respiratory system warms, humidifies, and filters air (pp. 820–824)

The Nose and Paranasal Sinuses (pp. 820–822)

1. The nose provides an airway for respiration; warms, moistens, and cleanses incoming air; and houses the olfactory receptors.
2. Bone and cartilage plates shape the external nose. The nasal cavity, which opens to the exterior, is divided by the nasal septum. Paranasal sinuses and nasolacrimal ducts drain into the nasal cavities.

The Pharynx (pp. 822–824)

3. The pharynx extends from the base of the skull to the level of C_6. The nasopharynx is an air conduit; the oropharynx and laryngopharynx are common passageways for food and air. Tonsils are found in the oropharynx and nasopharynx.

22.2 The lower respiratory system consists of conducting and respiratory zone structures (pp. 824–831)

1. Respiratory system organs are divided functionally into conducting zone structures (nose to terminal bronchioles), which filter, warm, and moisten incoming air; and respiratory zone structures (respiratory bronchioles to alveoli), where gas exchanges occur.

The Larynx (pp. 824–826)

2. The larynx, or voice box, contains the vocal folds (cords). It also provides a patent airway and serves as a switching mechanism to route food and air into the proper channels.

3. The epiglottis prevents food or liquids from entering the respiratory channels during swallowing.

The Trachea (pp. 826–828)

4. The trachea extends from the larynx to the main bronchi. C-shaped cartilage rings reinforce the trachea and keep the trachea patent. Its mucosa is ciliated.

The Bronchi and Subdivisions (pp. 828–831)

5. The right and left main bronchi run into their respective lungs, within which they subdivide into smaller and smaller passageways.

6. The terminal bronchioles lead into respiratory zone structures: respiratory bronchioles, alveolar ducts, alveolar sacs, and finally alveoli. Gas exchange occurs in the alveoli, across the respiratory membrane.

7. As the respiratory conduits become smaller, the amount of cartilage decreases and is finally lost; the mucosa thins, and smooth muscle in the walls increases.

22.3 Each multilobed lung occupies its own pleural cavity (pp. 833–834)

1. The lungs, the paired organs of gas exchange, flank the mediastinum in the thoracic cavity. Each is suspended in pleurae via its root and has a base, an apex, and medial and costal surfaces. The right lung has three lobes; the left has two.

2. The lungs are primarily air passageways/chambers, supported by an elastic connective tissue stroma.

3. The pulmonary arteries carry deoxygenated blood returned from the systemic circulation to the lungs, where gas exchange occurs. The pulmonary veins return newly oxygenated (and most bronchial venous) blood back to the heart to be distributed throughout the body. The bronchial arteries provide the nutrient blood supply of the lungs.

4. The parietal pleura lines the thoracic wall and mediastinum; the visceral pleura covers external lung surfaces. Pleural fluid reduces friction during breathing movements.

PART 2

RESPIRATORY PHYSIOLOGY

22.4 Volume changes cause pressure changes, which cause air to move (pp. 834–840)

Pressure Relationships in the Thoracic Cavity (pp. 835–836)

1. Intrapulmonary pressure is the pressure within the alveoli. Intrapleural pressure is the pressure within the pleural cavity; normally it is negative relative to intrapulmonary pressures.

Pulmonary Ventilation (pp. 836–838)

2. Gases travel from an area of higher pressure to an area of lower pressure.

3. Inspiration occurs when the diaphragm and external intercostal muscles contract, increasing the dimensions (and volume) of the thorax. As the intrapulmonary pressure drops, air rushes into the lungs until the intrapulmonary and atmospheric pressures are equalized.

4. Expiration is largely passive, occurring as the inspiratory muscles relax and the lungs recoil. When intrapulmonary pressure exceeds atmospheric pressure, gases flow from the lungs.

Physical Factors Influencing Pulmonary Ventilation (pp. 838–840)

5. Friction in the air passageways causes resistance, which decreases air passage and causes breathing movements to become more strenuous. The greatest resistance to air flow occurs in the midsize bronchi.

6. Surface tension of alveolar fluid acts to reduce alveolar size and collapse the alveoli. Surfactant reduces this tendency.

7. Premature infants have problems keeping their lungs inflated owing to the lack of surfactant in their alveoli, resulting in infant respiratory distress syndrome (IRDS). Surfactant formation begins late in fetal development.

8. Total respiratory compliance depends on elasticity of lung tissue and flexibility of the bony thorax. When compliance is impaired, inspiration becomes more difficult.

22.5 Measuring respiratory volumes, capacities, and flow rates helps us assess ventilation (pp. 840–842)

1. The four respiratory volumes are tidal, inspiratory reserve, expiratory reserve, and residual. The four respiratory capacities are vital, functional residual, inspiratory, and total lung. Spirometry measures respiratory volumes and capacities.

2. Anatomical dead space is the air-filled volume (about 150 ml) of the conducting passageways. If alveoli become nonfunctional in gas exchange, their volume is added to the anatomical dead space, and the sum is the total dead space.

3. The FVC and FEV tests, which determine the rate at which VC air can be expelled, are particularly valuable in distinguishing between obstructive and restrictive disease.

4. Alveolar ventilation is the best index of ventilation efficiency because it accounts for anatomical dead space.

$$AVR = (TV - dead\ space) \times respiratory\ rate$$

22.6 Gases exchange by diffusion between the blood, lungs, and tissues (pp. 842–847)

Basic Properties of Gases (pp. 843–844)

1. Dalton's law states that each gas in a mixture of gases exerts pressure in proportion to its percentage in the total mixture.

2. Henry's law states that the amount of gas that will dissolve in a liquid is proportional to the partial pressure of that gas. Other important factors are the solubility of the gas in the liquid and the temperature of the liquid.

Composition of Alveolar Gas (p. 844)

3. Alveolar gas contains more carbon dioxide and water vapor and considerably less oxygen than atmospheric air.

External Respiration (pp. 844–846)

4. External respiration is the process of gas exchange that occurs in the lungs. Oxygen enters the pulmonary capillaries; carbon dioxide leaves the blood and enters the alveoli. Factors influencing this process include the partial pressure gradients, the thickness of the respiratory membrane, surface area available, and ventilation-perfusion coupling (matching alveolar ventilation with pulmonary perfusion).

22

Internal Respiration (pp. 846–847)

5. Internal respiration is the gas exchange that occurs between the systemic capillaries and the tissues. Carbon dioxide enters the blood, and oxygen leaves the blood and enters the tissues.

22.7 Oxygen is transported by hemoglobin, and carbon dioxide is transported in three different ways (pp. 847–852)

Oxygen Transport (pp. 847–850)

1. Molecular oxygen is carried bound to hemoglobin in the red blood cells. The amount of oxygen bound to hemoglobin depends on the P_{O_2} and P_{CO_2} of blood, blood pH, the presence of BPG, and temperature. A small amount of oxygen gas is transported dissolved in plasma.

Carbon Dioxide Transport (pp. 850–852)

2. CO_2 is transported in the blood dissolved in plasma, chemically bound to hemoglobin, and (primarily) as bicarbonate ions in plasma. Loading and unloading of O_2 and CO_2 are mutually beneficial.

3. Accumulating CO_2 lowers blood pH; depletion of CO_2 from blood raises blood pH.

Hypoxia (p. 852)

4. Hypoxia occurs when inadequate amounts of oxygen are delivered to body tissues. When this occurs, the skin and mucosae may become cyanotic.

22.8 Respiratory centers in the brain stem control breathing with input from chemoreceptors and higher brain centers (pp. 853–857)

Neural Mechanisms (pp. 853–854)

1. Medullary respiratory centers are the ventral and dorsal respiratory groups. The ventral respiratory group is likely responsible for the rhythmicity of breathing.

2. The pontine respiratory centers influence the activity of the medullary respiratory centers.

Factors Influencing Breathing Rate and Depth (pp. 854–857)

3. Important chemical factors modifying baseline respiratory rate and depth are arterial levels of CO_2, H^+, and O_2.

4. An increasing arterial P_{CO_2} level (hypercapnia) is the most powerful respiratory stimulant. It acts (via formation of H^+ in brain tissue) on central chemoreceptors to cause a reflexive increase in the rate and depth of breathing.

5. Hypocapnia depresses respiration and results in decreased ventilation and, possibly, apnea.

6. Arterial P_{O_2} levels below 60 mm Hg strongly stimulate peripheral chemoreceptors.

7. Decreased pH and a decline in blood P_{O_2} act on peripheral chemoreceptors and enhance the response to CO_2.

8. Emotions, pain, body temperature changes, and other stressors can alter respiration by acting through hypothalamic centers. Respiration can also be controlled voluntarily for short periods.

9. Dust, mucus, fumes, and pollutants initiate pulmonary irritant reflexes.

10. The inflation (Hering-Breuer) reflex is a protective reflex initiated by extreme overinflation of the lungs; it acts to terminate inspiration.

22.9 Exercise and high altitude bring about respiratory adjustments (pp. 857–858)

Exercise (p. 857)

1. As exercise begins, there is an abrupt increase in ventilation (hyperpnea) followed by a more gradual increase. When exercise stops, there is an abrupt decrease in ventilation followed by a gradual decline to baseline values.

2. P_{O_2}, P_{CO_2}, and blood pH remain quite constant during exercise and hence do not appear to account for changes in ventilation. Inputs from higher centers and proprioceptors may contribute.

High Altitude (pp. 857–858)

3. At high altitudes, arterial P_{O_2} and hemoglobin saturation levels fall because of the decrease in atmospheric pressure compared to sea level. Increased ventilation helps restore P_{O_2} to physiological levels.

4. Long-term acclimatization involves increased erythropoiesis.

22.10 Respiratory diseases are major causes of disability and death (pp. 858–860)

1. Two major respiratory disorders are COPD (emphysema and chronic bronchitis) and lung cancer; smoking is a significant cause. A third major disorder is asthma. Multidrug-resistant tuberculosis may become a major public health problem.

Chronic Obstructive Pulmonary Disease (COPD) (pp. 858–859)

2. COPD is characterized by an irreversible decrease in the ability to force air out of the lungs.

3. In emphysema, alveoli enlarge permanently and disintegrate. The lungs lose their elasticity, and expiration becomes an active process.

4. Chronic bronchitis is characterized by excessive mucus production in the lower respiratory passageways, which severely impairs ventilation and gas exchange.

Asthma (p. 859)

5. Asthma is a reversible obstructive condition caused by an immune response that causes its victims to wheeze and gasp for air as their inflamed respiratory passages constrict. It is marked by acute episodes and symptom-free periods.

Tuberculosis (TB) (p. 859)

6. Tuberculosis, an infectious disease caused by an airborne bacterium, mainly affects the lungs. Although most infected individuals remain asymptomatic by walling off the bacteria in nodules (tubercles), symptoms appear when immunity is depressed. Some patients' failure to complete drug therapy has produced multidrug-resistant TB strains.

Lung Cancer (pp. 859–860)

7. Lung cancer, promoted by free radicals and other carcinogens in tobacco smoke, is extremely aggressive and metastasizes rapidly.

Sleep Apnea (p. 860)

8. Sleep apnea is usually caused by obstruction of the pharynx during sleep.

22

Developmental Aspects of the Respiratory System
(pp. 860–863)

1. The mucosa of the nasal cavity develops from the invagination of the ectodermal olfactory placodes. The mucosa of the pharynx and lower respiratory passageways develops from an outpocketing of the endodermal foregut lining. Mesoderm forms the walls of the respiratory conduits and the lung stroma.

2. Cystic fibrosis (CF), the most common fatal hereditary disease in North America, results from an abnormal CFTR protein that fails to form a chloride channel. The result is thick mucus, which clogs respiratory passages and invites infection.

3. With age, the thorax becomes more rigid, the lungs become less elastic, and vital capacity declines. In addition, sleep apnea becomes more common, and respiratory system protective mechanisms are less effective.

REVIEW QUESTIONS

 To access additional practice questions using your smartphone, tablet, or computer: MasteringA&P® > Study Area > Practice Tests & Quizzes

Level 1 Remember/Understand
(Some questions have more than one correct answer. Select the best answer or answers from the choices given.)

1. Cutting the phrenic nerves will result in (a) air entering the pleural cavity, (b) paralysis of the diaphragm, (c) stimulation of the diaphragmatic reflex, (d) paralysis of the epiglottis.

2. Which of the following laryngeal cartilages is/are not paired? (a) epiglottis, (b) arytenoid, (c) cricoid, (d) cuneiform, (e) corniculate.

3. Under ordinary circumstances, the inflation reflex is initiated by (a) noxious chemicals, (b) the ventral respiratory group, (c) overinflation of the alveoli and bronchioles, (d) the pontine respiratory centers.

4. The detergent-like substance that keeps the alveoli from collapsing between breaths because it reduces the surface tension of the water film in the alveoli is called (a) lecithin, (b) bile, (c) surfactant, (d) reluctant.

5. Which of the following determines the direction of gas movement? (a) solubility in water, (b) partial pressure gradient, (c) temperature, (d) molecular weight and size of the gas molecule.

6. When the inspiratory muscles contract, (a) the size of the thoracic cavity increases in diameter, (b) the size of the thoracic cavity increases in length, (c) the volume of the thoracic cavity decreases, (d) the size of the thoracic cavity increases in both length and diameter.

7. The nutrient blood supply of the lungs is provided by (a) the pulmonary arteries, (b) the aorta, (c) the pulmonary veins, (d) the bronchial arteries.

8. Oxygen and carbon dioxide are exchanged in the lungs and through all cell membranes by (a) active transport, (b) diffusion, (c) filtration, (d) osmosis.

9. Most oxygen carried in the blood is (a) in solution in the plasma, (b) combined with plasma proteins, (c) chemically combined with the heme in red blood cells, (d) in solution in the red blood cells.

10. Which of the following has the greatest stimulating effect on the respiratory centers in the brain? (a) oxygen, (b) carbon dioxide, (c) calcium, (d) willpower.

Level 2 Apply/Analyze

11. In mouth-to-mouth artificial respiration, the rescuer blows air from his or her own respiratory system into that of the victim. Which of the following statements are correct?

(1) Expansion of the victim's lungs is brought about by blowing air in at higher than atmospheric pressure (positive-pressure breathing).

(2) During inflation of the lungs, the intrapleural pressure increases.

(3) This technique will not work if the victim has a hole in the chest wall, even if the lungs are intact.

(4) Expiration during this procedure depends on the elasticity of the alveolar and thoracic walls.

(a) all of these, (b) 1, 2, 4, (c) 1, 2, 3, (d) 1, 4.

12. A baby holding its breath will (a) have brain cells damaged because of low blood oxygen levels, (b) automatically start to breathe again when the carbon dioxide levels in the blood reach a high enough value, (c) suffer heart damage because of increased pressure in the carotid sinus and aortic arch areas, (d) be called a "blue baby."

13. Damage to which of the following would most likely result in cessation of breathing? (a) the pontine respiratory group, (b) the ventral respiratory group of the medulla, (c) the stretch receptors in the lungs, (d) the dorsal respiratory group of the medulla.

14. The bulk of carbon dioxide is carried (a) chemically combined with the amino acids of hemoglobin as carbaminohemoglobin in the red blood cells, (b) as the ion HCO_3^- in the plasma after first entering the red blood cell, (c) as carbonic acid in the plasma, (d) chemically combined with the heme portion of Hb.

15. Briefly explain the anatomical "reason" why most men have deeper voices than boys or women.

16. (a) Differentiate clearly between minute ventilation and alveolar ventilation rate. (b) Which provides a more accurate measure of ventilatory efficiency, and why?

17. (a) Define hyperventilation. (b) If you hyperventilate, do you retain or expel more carbon dioxide? (c) What effect does hyperventilation have on blood pH?

18. Describe age-related changes in respiratory function.

19. A member of the "Blues" gang was rushed into an emergency room after receiving a knife wound in the left side of his thorax. The diagnosis was pneumothorax and a collapsed lung. Explain exactly (a) why the lung collapsed, and (b) why only one lung (not both) collapsed.

Level 3 Evaluate/Synthesize

20. Trace the route of air from the nares to an alveolus. Name subdivisions of organs where applicable, and differentiate between conducting and respiratory zone structures.

21. **(a)** Why is it important that the trachea is reinforced with cartilage rings? **(b)** Why is it advantageous that the rings are incomplete posteriorly?

22. Describe the functional relationships between volume changes and gas flow into and out of the lungs.

23. Daniel, the swimmer with the fastest time on the Springfield College swim team, routinely hyperventilates before a meet, as he says, "to sock some more oxygen into my lungs so I can swim longer without having to breathe." First of all, what basic fact about oxygen loading has Daniel forgotten (a lapse leading to false thinking)? Second, how is Daniel jeopardizing not only his time but his life?

24. A surgeon removed three adjacent bronchopulmonary segments from the left lung of a patient with TB. Almost half of the lung was removed, yet there was no severe bleeding, and relatively few blood vessels had to be cauterized (closed off). Why was the surgery so easy to perform?

25. After a week of scuba diving in the Bahamas, Mary Ann boards an airplane. During her flight home, she develops aching joints, nausea, and dyspnea, which resolve upon landing. During the flight, the cabin pressure was equivalent to an altitude of 8000 feet. Explain her problems.

CLINICAL CASE STUDY

35-Year-Old Female with Chest Trauma

Barbara Joley was in a major car accident. When EMS managed to free her from the wreckage, she was cyanotic and apneic (not breathing). Her heart was still beating, but her pulse was fast and thready. The emergency medical technician reported that when Barbara was found, her head was cocked at a peculiar angle and it looked like she had a fracture at the level of the C_2 vertebra. The following questions refer to these observations.

1. How might the "peculiar" head position explain Barbara's cessation of breathing?

2. **+ NCLEX-STYLE** Barbara's cyanosis can best be explained by which of the following?
   a. Trauma has caused bleeding under the skin, and the accumulated blood is turning black and blue.
   b. Her lungs must be damaged, and as a result, the right side of her heart is sending blood to the body instead of the lungs.
   c. The level of CO_2 in her blood has increased because her breathing has stopped.
   d. The level of O_2 in her blood has decreased because her breathing has stopped.

Barbara is artificially ventilated and survives transport to the hospital. The admission notes include the following observations.
- Right thorax compressed; ribs 7 to 9 fractured
- Right lung pneumothorax; trachea deviation (shift) to the left

Relative to these notes:

3. What is pneumothorax and why would it cause the right lung to collapse? Why hasn't the left lung collapsed?

4. **+ NCLEX-STYLE** How do Barbara's recorded injuries relate to the pneumothorax?
   a. The pain from her broken ribs is causing her to decrease breathing, so the right lung is not inflating.
   b. Barbara's fractured ribs probably punctured her lung tissue and allowed air within the lung to enter the pleural cavity.
   c. When the level of O_2 in the blood decreases, a negative feedback loop involving the central chemoreceptors causes lung inflation to decrease as well.
   d. Her traumatic injury has caused her peripheral chemoreceptors to stop working, so they are not causing her to increase ventilation.

5. The physician inserts a chest tube into the right intrapleural cavity and hooks it up to suction. How might this alleviate the pneumothorax?

22

23

The Digestive System

In this chapter, you will learn that

The digestive system breaks down food and absorbs the components for use by body cells

by exploring

Part 1 Overview of the Digestive System

by asking

23.1 What major processes occur during digestive system activity?

23.2 How is the digestive system organized?

23.3 How is the digestive system controlled?

Part 2 Functional Anatomy of the Digestive System

looking closer at

23.4 The mouth and associated organs

23.5 The pharynx and esophagus

23.6 The stomach

23.7 The liver, gallbladder, and pancreas

23.8 The small intestine

23.9 The large intestine

Part 3 Physiology of Digestion and Absorption

by first asking

23.10 What are the basic mechanisms of digestion and absorption?

then asking

23.11 How is each type of nutrient processed?

and finally, exploring

Developmental Aspects of the Digestive System

Children are fascinated by the workings of the digestive system. They relish crunching a potato chip, delight in making "mustaches" with milk, and giggle when their stomach "growls." As adults, we know that a healthy digestive system is essential to life, because it converts foods into the raw materials that build and fuel our body's cells. Specifically, the **digestive system** takes in food, breaks it down into nutrient molecules, absorbs these molecules into the bloodstream, and then rids the body of the indigestible remains.

CAREER CONNECTION

Watch a video to learn how the chapter content is used in a real health care setting. Go to Mastering A&P® > Study Area > Animations and Videos or use quick access URL https://goo.gl/88srfV

Mouth (oral cavity)

Tongue*

Parotid gland

Sublingual gland

Submandibular gland

Salivary glands*

Esophagus

Pharynx

Stomach

Pancreas*

(Spleen)

Liver*

Gallbladder*

Transverse colon

Descending colon

Duodenum

Small intestine — Jejunum

Ileum

Ascending colon

Cecum

Large intestine

Sigmoid colon

Rectum

Appendix

Anus

Anal canal

Figure 23.1 Alimentary canal and related accessory digestive organs. Organs with asterisks are accessory organs. Those without asterisks are alimentary canal organs (except the spleen, a lymphoid organ).

PART 1

OVERVIEW OF THE DIGESTIVE SYSTEM

Learning Outcome

▶ **Describe the functions of the digestive system, and differentiate between organs of the alimentary canal and accessory digestive organs.**

The organs of the digestive system fall into two main groups: (1) those of the *alimentary canal* (al″ĭ-men′tar-e; *aliment* = nourish) and (2) *accessory digestive organs* (**Figure 23.1**).

The **alimentary canal**, also called the **gastrointestinal (GI) tract** or gut, is the continuous muscular tube that winds through the body from the mouth to the anus. It **digests** food—breaks it down into smaller fragments (*digest* = dissolve)—and **absorbs** the digested fragments through its lining into the blood.

The organs of the alimentary canal are the *mouth, pharynx, esophagus, stomach, small intestine,* and *large intestine.* The large intestine leads to the terminal opening, or *anus.* In a cadaver, the alimentary canal is approximately 9 m (about 30 ft) long, but in a living person, it is considerably shorter because of its muscle tone. Food material in this tube is technically outside the body because the canal is open to the external environment at both ends.

23

The **accessory digestive organs** are the *teeth*, *tongue*, *gallbladder*, and a number of large digestive glands—the *salivary glands*, *liver*, and *pancreas*. The teeth and tongue are in the mouth, or oral cavity, while the digestive glands and gallbladder lie outside the GI tract and connect to it by ducts. The accessory digestive glands produce a variety of secretions that help break down foodstuffs.

23.1 What major processes occur during digestive system activity?

Learning Outcome

▶ List and define the major processes occurring during digestive system activity.

We can view the digestive tract as a "disassembly line" in which food becomes less complex at each step of processing and its nutrients become available to the body. The processing of food by the digestive system involves six essential activities (**Figure 23.2**):

■ **Ingestion** is taking food into the digestive tract (eating).

■ **Propulsion**, which moves food through the alimentary canal, includes *swallowing*, which is initiated voluntarily, and *peristalsis* (per″ĭ-stal′sis), an involuntary process. **Peristalsis** (*peri* = around; *stalsis* = constriction), the major means of propulsion, involves alternating waves of contraction and relaxation of muscles in the organ walls (**Figure 23.3a**). Its main effect is to squeeze food along the tract, but some mixing occurs as well. In fact, peristaltic waves are so powerful that, once swallowed, food and fluids will reach your stomach even if you stand on your head.

Figure 23.2 Gastrointestinal tract activities. Note that sites of digestion produce enzymes or receive enzymes or other secretions made by accessory organs outside the alimentary canal.

Figure 23.3 Peristalsis and segmentation.

■ **Mechanical breakdown** increases the surface area of ingested food, physically preparing it for digestion by enzymes. Mechanical processes include chewing, mixing food with saliva by the tongue, churning food in the stomach, and **segmentation** (rhythmic local constrictions of the small intestine, Figure 23.3b). Segmentation mixes food with digestive juices and makes absorption more efficient by repeatedly moving different parts of the food mass over the intestinal wall.

■ **Digestion** involves a series of steps in which enzymes secreted into the lumen (cavity) of the alimentary canal break down complex food molecules to their chemical building blocks. That is, digestion is a catabolic process (◄ p. 36).

■ **Absorption** is the passage of digested end products (plus vitamins, minerals, and water) from the lumen of the GI tract through the mucosal cells by active or passive transport into the blood or lymph.

■ **Defecation** eliminates indigestible substances from the body via the anus in the form of feces.

Some of these processes are the job of a single organ. For example, only the mouth ingests and only the large intestine defecates. But most digestive system activities require the cooperation of several organs and occur bit by bit as food moves along the tract as described later.

Check Your Understanding

1. Name one organ of the alimentary canal found in the thorax. Name three organs located in the abdominal cavity.
2. What is the usual site of ingestion?
3. Which digestive system activity actually moves nutrients from the outside to the inside of the body?

For answers, see Answers Appendix.

23.2 The GI tract has four layers and is usually surrounded by peritoneum

Learning Outcomes

▶ **Describe the location and function of the peritoneum.**
▶ **Define retroperitoneal and name the retroperitoneal organs of the digestive system.**
▶ **Define splanchnic circulation and indicate the importance of the hepatic portal system.**
▶ **Describe the tissue composition and general function of each of the four layers of the alimentary canal.**

Relationship of the Digestive Organs to the Peritoneum

Most digestive system organs reside in the abdominopelvic cavity. Recall from Chapter 1 that all ventral body cavities contain slippery *serous membranes* (◀ pp. 18–19). The **peritoneum** of the abdominopelvic cavity is the most extensive of these membranes (**Figure 23.4**). The **visceral peritoneum** covers the external surfaces of most digestive organs and is continuous with the **parietal peritoneum** that lines the body wall (see Figure 23.34d, p. 908). Between the two peritoneums is the **peritoneal cavity**, a slitlike potential space containing a slippery fluid secreted by the serous membranes. The serous fluid allows the mobile digestive organs to glide easily across one another and along the body wall as they carry out their activities.

A **mesentery** (mes′en-ter″e) is a double layer of peritoneum—a sheet of two serous membranes fused back to back—that extends to the digestive organs from the body wall. Mesenteries have the following functions:

- They provide routes for blood vessels, lymphatics, and nerves to reach the digestive viscera.
- They hold organs in place.
- They store fat.

In most places the mesentery is *dorsal* and attaches to the posterior abdominal wall, but there are *ventral* mesenteries too, such as the one that extends from the liver to the anterior abdominal wall (Figure 23.4, middle). Some digestive organ mesenteries have specific names (such as the *omenta*), or are called "ligaments" (even though these peritoneal folds are nothing like the fibrous ligaments that connect bones).

Not all alimentary canal organs are suspended by a mesentery. For example, during development, some regions of the small intestine adhere to the dorsal abdominal wall (Figure 23.4, bottom). In so doing, they lose their mesentery and come to lie posterior to the peritoneum. These organs, which include most of the pancreas and duodenum (the first part of the small intestine) and parts of the large intestine, are called **retroperitoneal organs** (*retro* = behind). By contrast, digestive organs (like the stomach) that keep their mesentery and remain in the peritoneal cavity are called **intraperitoneal** or **peritoneal organs**.

HOMEOSTATIC IMBALANCE 23.1 — CLINICAL

Peritonitis is inflammation of the peritoneum. It can arise from a piercing abdominal wound, a perforating ulcer that leaks stomach juices into the peritoneal cavity, or poor sterile technique during abdominal surgery. However, most commonly it results from a burst appendix that sprays bacteria-containing feces all over the peritoneum. In peritonitis, the peritoneal coverings tend to stick together around the infection site. This localizes the infection, providing time for macrophages to prevent the inflammation from spreading.

If peritonitis becomes widespread within the peritoneal cavity, it is dangerous and often lethal. Treatment includes removing as much infectious debris as possible and administering megadoses of antibiotics. ●●●●●●●●

Histology of the Alimentary Canal

Each digestive organ has only a share of the work of digestion. Consequently, it helps to consider structural characteristics that promote similar functions in all parts of the alimentary canal before we consider the functional anatomy of each digestive organ.

From the esophagus to the anal canal, the walls of the alimentary canal have the same four basic layers, or *tunics*—*mucosa, submucosa, muscularis externa*, and *serosa* (**Figure 23.5**). Each layer contains a predominant tissue type that plays a specific role in food breakdown.

The Mucosa

The innermost layer is the **mucosa**, or **mucous membrane**, a moist epithelial membrane that lines the alimentary canal lumen from mouth to anus. Its major functions are to:

- *Secrete* mucus, digestive enzymes, and hormones
- *Absorb* the end products of digestion into the blood
- *Protect* against infectious disease

The mucosa in a particular region of the GI tract may perform one or all three of these functions.

Abdominopelvic cavity

Parietal peritoneum
Visceral peritoneum
Peritoneal cavity
Alimentary canal organ

Most digestive organs are intraperitoneal and are suspended from the body wall by a **dorsal mesentery.**

Dorsal mesentery
Liver

Some intraperitoneal digestive organs are also suspended from the body wall by **ventral mesenteries.**

Site of lost (resorbed) mesentery

Some digestive organs are **retroperitoneal** because they have lost their mesentery during development.

Figure 23.4 The peritoneum and mesenteries. Note that the peritoneal cavity is much smaller than depicted here.

Intrinsic nerve plexuses
• Myenteric nerve plexus
• Submucosal nerve plexus

Glands in submucosa

Mucosa
• Epithelium
• Lamina propria
• Muscularis mucosae

Submucosa

Muscularis externa
• Circular layer
• Longitudinal layer

Serosa
• Connective tissue
• Epithelium (mesothelium)

Lumen

Mucosa-associated
lymphoid tissue

Nerve
Artery
Vein
Lymphatic vessel
Mesentery

Gland in mucosa

Duct of gland outside
alimentary canal

Figure 23.5 Basic structure of the alimentary canal. Its four basic layers are the mucosa, submucosa, muscularis externa, and serosa.

More complex than most other mucosae in the body, the typical digestive mucosa consists of three sublayers: (1) a lining epithelium, (2) a lamina propria, and (3) a muscularis mucosae. Except for that of the mouth, esophagus, and anus where it is stratified squamous, the **epithelium** of the mucosa is a *simple columnar epithelium* rich in mucus-secreting cells (◄ p. 120). The slippery mucus it produces protects certain digestive organs from being digested by enzymes working within their cavities and eases food passage along the tract. In the stomach and small intestine, the mucosa also contains both enzyme-synthesizing and hormone-secreting cells. In such sites, the mucosa is a diffuse endocrine organ as well as part of the digestive organ.

The **lamina propria** (*proprius* = one's own), which underlies the epithelium, is loose areolar connective tissue (◄ p. 130). Its capillaries nourish the epithelium and absorb digested nutrients. Its isolated lymphoid follicles, part of **MALT** (the mucosa-associated lymphoid tissue; ◄ p. 774), help defend us against bacteria and other pathogens, which have rather free access to our digestive tract. Particularly large collections of lymphoid follicles occur within the pharynx (as the tonsils) and in the appendix.

External to the lamina propria is the **muscularis mucosae**, a scant layer of smooth muscle cells that produces local movements of the mucosa that can enhance absorption and secretion.

The Submucosa

The **submucosa**, just external to the mucosa, is areolar connective tissue containing a rich supply of blood and lymphatic vessels, lymphoid follicles, and nerve fibers that supply the surrounding tissues of the GI tract wall. Its abundant elastic fibers enable the stomach, for example, to regain its normal shape after temporarily storing a large meal.

The Muscularis Externa

Surrounding the submucosa is the **muscularis externa**, also simply called the **muscularis**. This layer is responsible for segmentation and peristalsis. It typically has an inner *circular layer* and an outer *longitudinal layer* of smooth muscle cells (◄ **Figure 9.22, p. 310**, and Figure 23.5). In several places along the tract, the circular layer thickens, forming *sphincters* that act as valves to control food passage from one organ to the next and prevent backflow.

The Serosa

The **serosa**, the outermost layer of the intraperitoneal organs, is the *visceral peritoneum*. In most alimentary canal organs, it is formed of areolar connective tissue covered with *mesothelium*, a single layer of squamous epithelial cells (◄ **Figure 4.11a** and ◄ **Figure 4.4a**, respectively).

23

In the esophagus, which is located in the thoracic instead of the abdominopelvic cavity, the serosa is replaced by an **adventitia** (ad"ven-tish'e-ah), ordinary dense connective tissue that binds the esophagus to surrounding structures. Retroperitoneal organs have *both* an adventitia (on the side facing the dorsal body wall) and a serosa (on the side facing the peritoneal cavity).

Blood Supply: The Splanchnic Circulation

The **splanchnic circulation** includes those arteries that branch off the abdominal aorta to serve the digestive organs and the *hepatic portal circulation*. The arterial supply—the branches of the celiac trunk that serve the spleen, liver, and stomach, and the mesenteric arteries that serve the small and large intestines (◀ **pp. 745–747**)—normally receives one-quarter of the cardiac output. This percentage increases after a meal. The hepatic portal circulation (◀ **pp. 756–757**) collects nutrient-rich venous blood draining from the digestive viscera and delivers it to the liver.

Check Your Understanding

4. How does the location of the visceral peritoneum differ from that of the parietal peritoneum?
5. Of the following organs, which is/are retroperitoneal? Stomach, pancreas, liver.
6. What name is given to the venous portion of the splanchnic circulation?
7. **DRAW** Sketch a cross section of the alimentary canal and label the four major tissue layers and the three parts of the innermost layer.
8. **MAKE CONNECTIONS** The two types of smooth muscle are unitary and multi unit (see Chapter 9). Which type would you expect to find in the muscularis externa, and what characteristics make it well suited for this location?

For answers, see Answers Appendix.

23.3 The GI tract has its own nervous system called the enteric nervous system

Learning Outcome

▶ Describe stimuli and controls of digestive activity.

A theme we have stressed in this book is the body's efforts to maintain a constant internal environment. Most organ systems respond to changes in that environment either by attempting to restore some plasma variable to its former levels or by changing their own function.

The digestive system, however, must create the optimal environment for its functioning within the lumen of the GI tract, an area that is actually *outside* the body. Essentially all digestive tract regulatory mechanisms control luminal conditions so that food breakdown and absorption can occur there as effectively as possible.

In order to accomplish this, the GI tract has its own *enteric nervous system* (sometimes also called the *gut brain*), which consists of over 100 million neurons. You could truly say that

the gut, with more neurons than the entire spinal cord, has a mind of its own! We will describe the enteric nervous system next, and then outline key concepts that govern regulation of digestive activity.

Enteric Nervous System

The **enteric nervous system** (*enter* = gut) is the in-house nerve supply of the alimentary canal. It is staffed by **enteric neurons** that communicate widely with one another to regulate digestive system activity. These semiautonomous enteric neurons constitute the bulk of the two major *intrinsic nerve plexuses* (ganglia interconnected by unmyelinated fiber tracts) found in the walls of the alimentary canal: the submucosal and myenteric nerve plexuses (Figure 23.5). These plexuses interconnect like chicken wire all along the GI tract and regulate digestive activity throughout its length (**Figure 23.6**).

The **submucosal nerve plexus** occupies the submucosa, and the large **myenteric nerve plexus** (mi-en-ter'ik; "intestinal muscle") lies between the circular and longitudinal muscle layers of the muscularis externa. Enteric neurons of these plexuses provide the major nerve supply to the GI tract wall and control GI tract motility (motion).

The enteric nervous system participates in both short and long reflex arcs (**Figure 23.7**).

- **Short reflexes** are mediated entirely by enteric nervous system plexuses in response to stimuli within the GI tract.

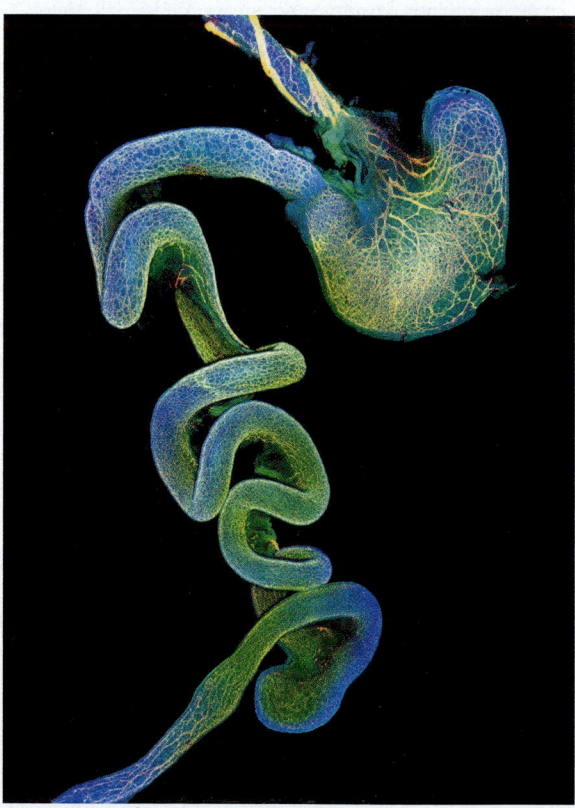

Figure 23.6 The enteric nervous system. The nerves of the enteric nervous sytem (yellow) extend throughout the alimentary canal. (Photo of immature mouse gut.)

Figure 23.7 Neural reflex pathways initiated by stimuli inside or outside the gastrointestinal tract.

Control of the patterns of segmentation and peristalsis is largely automatic, involving pacemaker cells and reflex arcs between enteric neurons in the same or different organs.

- **Long reflexes** involve CNS integration centers and extrinsic autonomic nerves. The enteric nervous system sends information to the central nervous system via visceral sensory fibers. It receives sympathetic and parasympathetic motor fibers from the autonomic nervous system. These enter the intestinal wall to synapse with neurons in the intrinsic plexuses. Long reflexes can be initiated by stimuli arising inside or outside of the GI tract. In these reflexes, the enteric nervous system acts as a way station for the autonomic nervous system, allowing extrinsic controls to influence digestive activity (Figure 23.7). Generally speaking, parasympathetic inputs enhance digestive activity and sympathetic inpulses inhibit them.

Basic Concepts of Regulating Digestive Activity

Three key concepts govern regulation of digestive activity:

- **Digestive activity is provoked by a range of mechanical and chemical stimuli.** Receptors involved in controlling GI tract activity are located in the walls of the tract's organs. These receptors respond to several stimuli, most importantly stretching of the organ by food in the lumen, changes in osmolarity (solute concentration) and pH of the contents, and the presence of substrates and end products of digestion.

- **Effectors of digestive activity are smooth muscle and glands.** When stimulated, receptors in the GI tract initiate reflexes that stimulate smooth muscle of the GI tract walls to mix lumen contents and move them along the tract. Reflexes can also activate or inhibit glands that secrete digestive juices into the lumen or hormones into the blood.

- **Neurons (intrinsic and extrinsic) and hormones control digestive activity.** The nervous system controls digestive activity via both *intrinsic controls* (involving short reflexes entirely within the enteric nervous system as described above) and *extrinsic controls* (involving long reflexes).

  The stomach and small intestine also contain hormone-producing cells. When stimulated, these cells release their products to the interstitial fluid in the extracellular space. Blood and interstitial fluid distribute these hormones to their target cells in the same or different digestive tract organs, where they affect secretion or contraction.

Check Your Understanding

9. When sensors in the GI tract are stimulated, they trigger reflexes. What types of digestive activity may be put into motion via those reflexes?
10. The term "gut brain" does not really mean there is a brain in the digestive system. What does it refer to?
11. Jerry has been given a drug that inhibits parasympathetic stimulation of his digestive tract. Should he "eat hearty" or temporarily refrain from eating, and why?

For answers, see Answers Appendix.

PART 2

FUNCTIONAL ANATOMY OF THE DIGESTIVE SYSTEM

Now that we have summarized some points that unify the digestive system organs, let's take a tour down the alimentary canal and examine the special structural and functional capabilities of each organ of this system. Figure 23.1 shows most of these organs in their normal body positions, so you may find it helpful to refer back to that illustration from time to time as you read the following modules.

As we tour the digestive system organs, we will examine how each participates in the six basic digestive processes. (This information was introduced on pp. 870–871 and is summarized in Table 23.2 on p. 904.) Since we cover digestion and absorption in a special physiology section later in the chapter, we will focus on the other digestive processes on our tour.

23.4 Ingestion occurs only at the mouth

Learning Outcomes

▶ Describe the gross and microscopic anatomy and the basic functions of the mouth and its associated organs.
▶ Describe the composition and functions of saliva, and explain how salivation is regulated.
▶ Explain the dental formula and differentiate clearly between deciduous and permanent teeth.

Aside from ingestion, the digestive functions associated with the mouth mostly reflect the activity of the related accessory organs, such as teeth, salivary glands, and tongue. In the mouth we chew food and mix it with saliva containing enzymes that begin the process of digestion. The mouth also begins the propulsive process of swallowing, which carries food through the pharynx and esophagus to the stomach.

The Mouth

The **mouth** is also called the **oral cavity**, or *buccal cavity* (buk′al). Its boundaries are the lips anteriorly, cheeks laterally, palate superiorly, and tongue inferiorly (**Figure 23.8**). Its anterior opening is the **oral orifice**. Posteriorly, the oral cavity is continuous with the *oropharynx*.

The walls of the mouth are lined with a thick stratified squamous epithelium (◀ **Figure 4.4e, p. 122**), which withstands considerable friction. The epithelium on the gums, hard palate, and dorsum of the tongue is slightly keratinized for extra protection against abrasion during eating.

The Lips and Cheeks

The **lips (labia)** and the **cheeks**, which help keep food between the teeth when we chew, are composed of a core of skeletal muscle covered externally by skin. The *orbicularis oris muscle* forms the fleshy lips; the cheeks are formed largely by the *buccinators*. The recess bounded externally by the lips and cheeks and internally by the gums and teeth is the **oral vestibule** ("porch"). The area that lies within the teeth and gums is the **oral cavity proper**. The **labial frenulum** (fren′u-lum) is a median fold that joins the internal aspect of each lip to the gum (Figure 23.8b).

The Palate

The **palate**, forming the roof of the mouth, has two distinct parts: the hard palate anteriorly and the soft palate posteriorly (Figure 23.8). The **hard palate** is underlain by the palatine bones

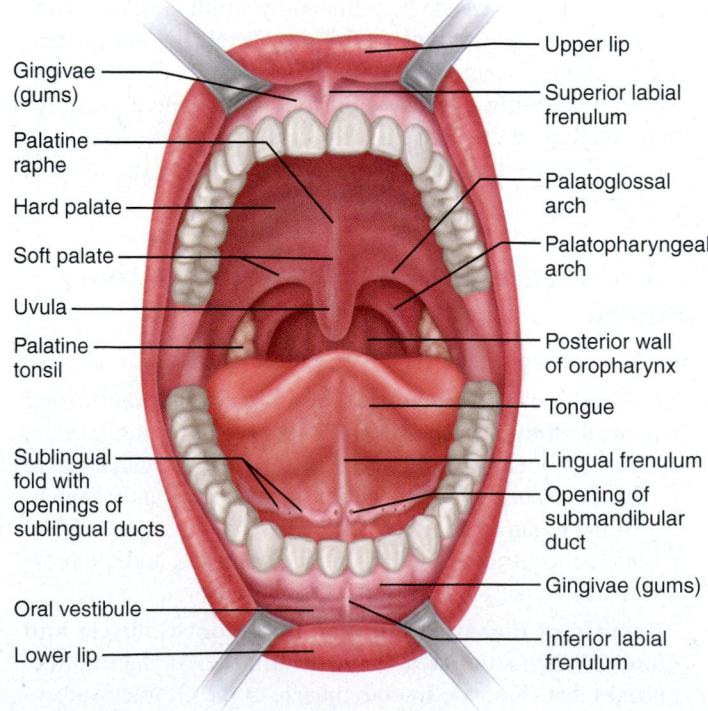

(a) Sagittal section of the oral cavity and pharynx

Soft palate
Palatoglossal arch
Uvula
Hard palate
Oral cavity
Palatine tonsil
Tongue
Oropharynx
Lingual tonsil
Epiglottis
Hyoid bone
Laryngopharynx
Esophagus
Trachea

(b) Anterior view

Gingivae (gums)
Palatine raphe
Hard palate
Soft palate
Uvula
Palatine tonsil
Sublingual fold with openings of sublingual ducts
Oral vestibule
Lower lip
Upper lip
Superior labial frenulum
Palatoglossal arch
Palatopharyngeal arch
Posterior wall of oropharynx
Tongue
Lingual frenulum
Opening of submandibular duct
Gingivae (gums)
Inferior labial frenulum

Figure 23.8 Anatomy of the oral cavity (mouth).

and the palatine processes of the maxillae, and it forms a rigid surface against which the tongue forces food during chewing. The mucosa on either side of its *raphe* (ra'fe), a midline ridge, is slightly corrugated, which helps create friction.

The **soft palate** is a mobile fold formed mostly of skeletal muscle that rises reflexively to close off the nasopharynx when we swallow.

- To demonstrate this action, try to breathe and swallow at the same time.

Laterally, the soft palate is anchored to the tongue by the **palatoglossal arches** and to the wall of the oropharynx by the more posterior **palatopharyngeal arches**. These two paired folds form the boundaries of the **fauces** (faw'sēz; *fauc* = throat), the arched area of the oropharynx that contains the palatine tonsils. Projecting downward from the free edge of the soft palate is the fingerlike **uvula** (u'vu-lah).

The Tongue

The **tongue** occupies the floor of the mouth (Figure 23.8). It is composed of interlacing bundles of skeletal muscle fibers. During chewing, the tongue grips the food and constantly repositions it between the teeth. The tongue also mixes food with saliva, forming it into a compact mass called a **bolus** (bo'lus; "a lump"), and then initiates swallowing by pushing the bolus posteriorly into the pharynx. The versatile tongue also helps us form consonants (k, d, t, and so on) when we speak.

The tongue has both intrinsic and extrinsic skeletal muscle fibers. The **intrinsic muscles** are confined in the tongue and are not attached to bone. Their muscle fibers, which run in several different planes, allow the tongue to change its shape (but not its position), becoming thicker, thinner, longer, or shorter as needed for speech and swallowing.

The **extrinsic muscles** extend to the tongue from their points of origin on bones of the skull or the soft palate, as described in Chapter 10 (◀ **Table 10.2** and **Figure 10.8**). The extrinsic muscles alter the tongue's position. They protrude it, retract it, and move it from side to side. The tongue has a median septum of connective tissue, and each half contains identical muscle groups. A fold of mucosa called the **lingual frenulum** secures the tongue to the floor of the mouth and limits its posterior movements.

HOMEOSTATIC IMBALANCE 23.2   CLINICAL

Children born with an extremely short lingual frenulum are sometimes referred to as "tongue-tied." Restricted tongue movement may interfere with breast-feeding. This congenital condition, called *ankyloglossia* ("fused tongue"), is sometimes corrected surgically by snipping the frenulum. ●●●●●●

The superior tongue surface bears papillae, peglike projections of the underlying mucosa (**Figure 23.9**).

- The conical **filiform papillae** roughen the tongue surface, helping us lick semisolid foods (such as ice cream) and providing friction for manipulating foods. These papillae, the

smallest and most numerous type, align in parallel rows on the tongue dorsum. They contain keratin, which stiffens them and gives the tongue its whitish appearance.

- The mushroom-shaped **fungiform papillae** are scattered widely over the tongue surface. Each has a vascular core that gives it a reddish hue.

- Eight to twelve large **vallate papillae** are located in a V-shaped row at the back of the tongue. They resemble the fungiform papillae but have an additional surrounding furrow.

- Pleatlike **foliate papillae** are located on the lateral aspects of the posterior tongue.

The fungiform, vallate, and foliate papillae house taste buds (◀ **pp. 578–579**).

Immediately posterior to the vallate papillae is the **terminal sulcus**, a groove that distinguishes the portion of the tongue that lies in the oral cavity (its body) from its posterior portion in the oropharynx (its root). The mucosa covering the root of the tongue lacks papillae, but it is still bumpy because of the nodular *lingual tonsil*, which lies just deep to its mucosa (Figure 23.9).

The Salivary Glands

A number of glands associated with the oral cavity secrete **saliva**. Saliva:

- Cleanses the mouth
- Dissolves food chemicals so they can be tasted
- Moistens food and helps compact it into a bolus
- Contains the enzyme **amylase** that begins the digestion of starchy foods

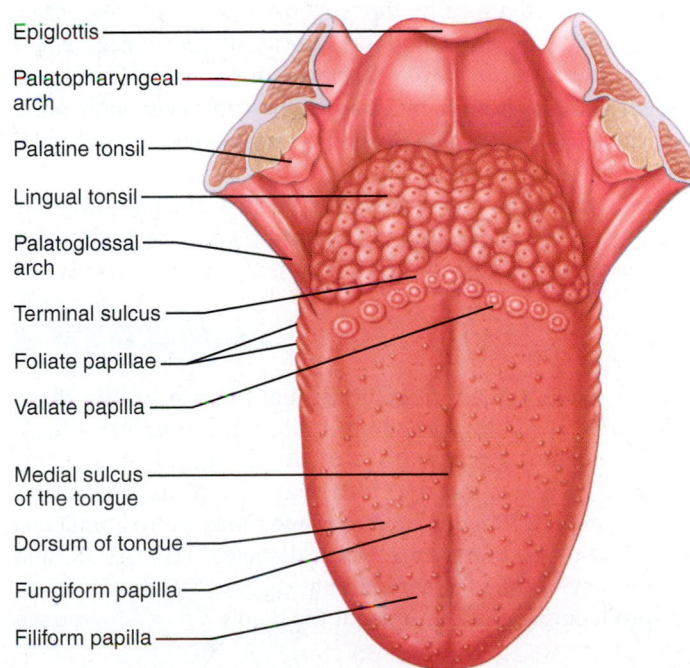

Epiglottis

Palatopharyngeal arch

Palatine tonsil

Lingual tonsil

Palatoglossal arch

Terminal sulcus

Foliate papillae

Vallate papilla

Medial sulcus of the tongue

Dorsum of tongue

Fungiform papilla

Filiform papilla

Figure 23.9 Dorsal surface of the tongue, and the tonsils.

(a)

Tongue

Teeth

Ducts of sublingual gland

Frenulum of tongue

Sublingual gland

Mylohyoid muscle (cut)

Anterior belly of digastric muscle

Parotid gland

Parotid duct

Masseter muscle

Body of mandible (cut)

Posterior belly of digastric muscle

Submandibular duct

Submandibular gland

(b)

Mucous cells   Serous cells

Figure 23.10 The salivary glands. (a) The parotid, submandibular, and sublingual glands and their ducts. **(b)** Photomicrograph of the sublingual gland (150×), a mixed salivary gland containing mostly mucous cells (light blue) with a few serous cells (purple).

Practice Histology questions:
MasteringA&P° > Study Area > Lab Tools > PAL

23

Most saliva is produced by the **major** or **extrinsic salivary glands** that lie outside the oral cavity and empty their secretions into it. **Minor** or **intrinsic salivary glands** (buccal glands and others) scattered throughout the oral cavity mucosa augment the output slightly.

The major salivary glands are paired compound alveolar or tubuloalveolar glands that develop from the oral mucosa and remain connected to it by ducts (**Figure 23.10a**). The large, roughly triangular **parotid gland** (pah-rot'id; *par* = near, *oto* = the ear) lies anterior to the ear between the masseter muscle and the skin. Its prominent duct parallels the zygomatic arch, pierces the buccinator muscle, and opens into the vestibule next to the second upper molar.

Branches of the facial nerve run through the parotid gland on their way to the muscles of facial expression (◀ **p. 503**). For this reason, surgery on this gland can result in facial paralysis.

HOMEOSTATIC IMBALANCE 23.3   **CLINICAL**

Normal salivary gland function is vital for oral health. If too little saliva is made, an uncomfortably dry mouth results, a condition called *xerostomia* (ze-ro-sto′me-ah; *zer* = dry, *stom* = mouth). This lack of moisture may lead to difficulty with chewing and swallowing, and oral infections. Xerostomia can be caused by certain medications, diabetes, HIV/AIDS, and Sjögren's syndrome (an autoimmune disease that affects moisture-producing glands throughout the body).

About the size of a walnut, the **submandibular gland** lies along the medial aspect of the mandibular body. Its duct runs

beneath the mucosa of the oral cavity floor and opens at the base of the lingual frenulum (see Figure 23.8b). The small, almond-shaped **sublingual gland** lies anterior to the submandibular gland under the tongue and opens via 10–20 ducts into the floor of the mouth (Figure 23.10a).

The salivary glands are composed of two types of secretory cells: serous and mucous (Figure 23.10b). **Serous cells** produce a watery secretion containing enzymes, ions, and a tiny bit of mucin, whereas **mucous cells** produce **mucus**, a stringy, viscous solution. The parotid and submandibular glands contain mostly serous cells. Buccal glands have approximately equal numbers of serous and mucous cells. The sublingual glands contain mostly mucous cells.

Composition of Saliva

Saliva is largely water—97 to 99.5%—and therefore is hypo-osmotic. Its osmolarity depends on the specific glands that are active and the stimulus for salivation. As a rule, saliva is slightly acidic (pH 6.75 to 7.00), but its pH may vary. Its solutes include:

- Electrolytes (Na^+, K^+, Cl^-, PO_4^{3-}, and HCO_3^-)
- The digestive enzymes salivary amylase and lingual lipase (lingual lipase makes only a minor contribution to overall fat digestion)
- The proteins mucin, lysozyme, and IgA
- Metabolic wastes (urea and uric acid)

When dissolved in water, the glycoprotein *mucin* forms thick mucus that lubricates the oral cavity and hydrates foodstuffs (◀ **p. 783**).

Saliva protects against microorganisms because it contains:

- *IgA antibodies* (◄ **Table 21.5, p. 799**).

- *Lysozyme*, a bactericidal enzyme that inhibits bacterial growth in the mouth and may help prevent tooth decay.

- *Defensins* (◄ **p. 783**). Besides acting as a local antibiotic, defensins function as cytokines to call defensive cells (lymphocytes, neutrophils, etc.) into the mouth for battle.

Control of Salivation

The minor salivary glands secrete saliva continuously in amounts just sufficient to keep the mouth moist. But when food enters the mouth, the major glands are activated and large amounts of saliva pour out. The average output of saliva is about 1500 ml per day, but can be much higher when salivary glands are appropriately stimulated.

Salivation is controlled primarily by the parasympathetic division of the autonomic nervous system. When we ingest food, chemoreceptors and mechanoreceptors in the mouth send signals to the **salivatory nuclei** in the brain stem (pons and medulla). As a result, parasympathetic nervous system activity increases. Impulses sent via motor fibers in the *facial* (VII) and *glossopharyngeal* (IX) *nerves* dramatically increase the output of watery (serous), enzyme-rich saliva.

The chemoreceptors are activated most strongly by acidic substances. The mechanoreceptors are activated by virtually any mechanical stimulus in the mouth—even chewing rubber bands.

Sometimes just the sight or smell of food is enough to get the juices flowing. The mere thought of hot fudge sauce on peppermint ice cream will make many a mouth water! Irritation of the lower GI tract by bacterial toxins, spicy foods, or hyperacidity also increases salivation. This response may help wash away or neutralize the irritants.

In contrast to parasympathetic controls, the sympathetic division (specifically fibers in T_1–T_3) causes release of a thick, mucin-rich saliva. Strong activation of the sympathetic division constricts blood vessels serving the salivary glands and almost completely inhibits saliva release, causing a dry mouth. Dehydration also inhibits salivation because low blood volume reduces filtration pressure at capillary beds.

The Teeth

The **teeth** lie in sockets (alveoli) in the gum-covered margins of the mandible and maxilla. We *masticate*, or chew, by opening and closing our jaws and moving them from side to side while using our tongue to move the food between our teeth. In the process, the teeth tear and grind the food, physically breaking it down into smaller fragments.

Dentition and the Dental Formula

Ordinarily by age 21, two sets of teeth, the **primary** and **permanent dentitions**, have formed (**Figure 23.11**). The primary dentition consists of the **deciduous teeth** (de-sid′u-us; *decid* = falling off), also called **milk** or **baby teeth**. The first teeth to appear, at about age 6 months, are the lower central incisors.

Additional pairs of teeth erupt at one- to two-month intervals until about 24 months, when all 20 milk teeth have emerged.

HOMEOSTATIC IMBALANCE 23.4 CLINICAL

Although primary teeth will eventually be replaced by permanent teeth, these first teeth nevertheless deserve careful attention. *Decaying primary teeth* can be painful and may lead to serious infection. Primary teeth also serve as important "place holders" for developing permanent teeth. Primary teeth can be kept healthy by brushing and by limiting exposure to sugary liquids, especially from prolonged bottle feeding. ●●●●●●

As the deep-lying **permanent teeth** enlarge and develop, the roots of the milk teeth are resorbed from below, causing them to loosen and fall out between ages 6 and 12. Generally, all the permanent teeth but the third molars have erupted by the end of adolescence. The third molars, also called *wisdom teeth*, emerge

Incisors
Central (6–8 mo)
Lateral (8–10 mo)
Canine (eyetooth)
(16–20 mo)
Molars
First molar
(10–15 mo)
Second molar
(about 2 yr)

Deciduous (milk) teeth

Incisors
Central (7 yr)
Lateral (8 yr)
Canine (eyetooth)
(11 yr)
Premolars (bicuspids)
First premolar
(11 yr)
Second premolar
(12–13 yr)
Molars
First molar (6–7 yr)
Second molar
(12–13 yr)
Third molar
(wisdom tooth)
(17–25 yr)

Permanent teeth

Figure 23.11 Human dentition. Teeth of the lower jaw: the deciduous and permanent sets. Approximate age at which tooth erupts is shown in parentheses. The shapes of individual teeth are shown on the right.

between ages 17 and 25. There are usually 32 permanent teeth in a full set, but sometimes the wisdom teeth never erupt or are completely absent.

> ### HOMEOSTATIC IMBALANCE 23.5 — CLINICAL
>
> When a tooth remains trapped in the jawbone, it is said to be *impacted*. Impacted teeth can cause a good deal of pressure and pain and must be removed surgically. Wisdom teeth are most commonly involved.

Teeth are classified according to their shape and function as incisors, canines, premolars, and molars (Figure 23.11). The chisel-shaped **incisors** are adapted for cutting or nipping off pieces of food. The conical or fanglike **canines** (cuspids or eye-teeth) tear and pierce. The **premolars** (bicuspids) and **molars** have broad crowns with rounded cusps (tips) best suited for grinding or crushing. The molars, with four or five cusps, are the best grinders. During chewing, the upper and lower molars repeatedly lock together, an action that generates tremendous crushing forces.

The **dental formula** is a shorthand way of indicating the numbers and relative positions of the different types of teeth. This formula is written as a ratio, uppers over lowers, for *one-half* of the mouth. Since the other side is a mirror image, we obtain total dentition by multiplying the dental formula by 2.

The primary dentition consists of two incisors (I), one canine (C), and two molars (M) on each side of each jaw, and its dental formula is written as

$$\frac{2I, 1C, 2M \text{ (upper jaw)}}{2I, 1C, 2M \text{ (lower jaw)}} \times 2 \quad (20 \text{ teeth})$$

Similarly, the permanent dentition [two incisors, one canine, two premolars (PM), and three molars] is

$$\frac{2I, 1C, 2PM, 3M}{2I, 1C, 2PM, 3M} \times 2 \quad (32 \text{ teeth})$$

Tooth Structure

Each tooth has two major regions: the crown and the root (**Figure 23.12**). The enamel-covered **crown** is the exposed part of the tooth above the **gingiva** (jin′jĭ-vah), or **gum**, which surrounds the tooth like a tight collar. **Enamel**, a brittle ceramic-like material thick as a dime, directly bears the force of chewing. Enamel is heavily mineralized with calcium salts, and its densely packed hydroxyapatite (mineral) crystals are oriented in force-resisting columns perpendicular to the tooth's surface. This makes it the hardest substance in the body. The cells that produce enamel degenerate when the tooth erupts; consequently, decayed or cracked areas of enamel will not heal and must be artificially filled.

The **root** is the portion of the tooth embedded in the jawbone. Canine teeth, incisors, and premolars have one root, although the first upper premolars commonly have two. The first two upper molars have three roots, while the corresponding lower molars have two. The root pattern of the third molar varies, but a fused single root is most common.

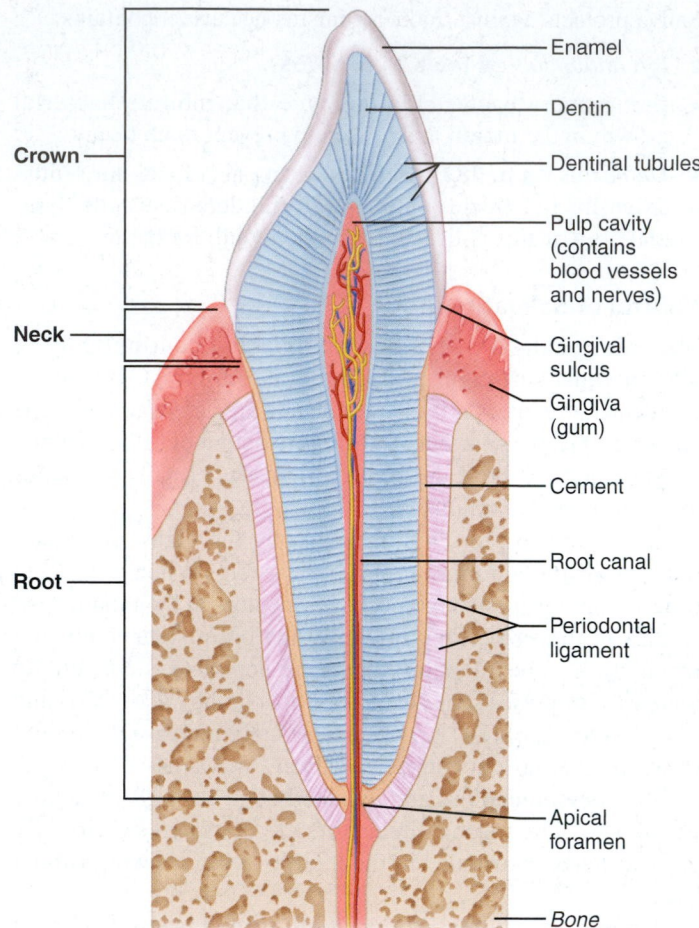

Figure 23.12 Longitudinal section of a canine tooth within its bony socket (alveolus).

Labels: Enamel, Dentin, Dentinal tubules, Pulp cavity (contains blood vessels and nerves), Gingival sulcus, Gingiva (gum), Cement, Root canal, Periodontal ligament, Apical foramen, Bone, Crown, Neck, Root

A constricted tooth region called the **neck** connects the crown and root. **Cement**, a calcified connective tissue, covers the outer surface of the root and attaches the tooth to the thin **periodontal ligament** (per″e-o-don′tal; "around the tooth"). This ligament anchors the tooth in the bony socket (alveolus) of the jaw, forming a fibrous joint called a *gomphosis*. Where the gingiva borders on a tooth, it dips downward to form a shallow groove called the *gingival sulcus*.

Dentin, a protein-rich bonelike material, underlies the enamel cap and forms the bulk of a tooth. More resilient (flexible) than enamel, dentin acts as a shock absorber during biting and chewing. Dentin surrounds a central **pulp cavity** containing a number of soft tissue structures (connective tissue, blood vessels, and nerve fibers) collectively called **pulp**. Pulp supplies nutrients to the tooth tissues and provides tooth sensation. Where the pulp cavity extends into the root, it becomes the **root canal**. At the proximal end of each root canal is an **apical foramen** that allows blood vessels, nerves, and other structures to enter the pulp cavity.

The teeth are served by the superior and inferior alveolar nerves, branches of the trigeminal nerve (◀ Table 13.2, p. 502). The superior and inferior alveolar arteries, branches of the maxillary artery (◀ Figure 19.24b, p. 741), supply blood.

Dentin contains unique radial striations called *dentinal tubules* (Figure 23.12). Each tubule contains an elongated process of an **odontoblast** (o-don'to-blast; "tooth former"), the cell type that secretes and maintains the dentin. The odontoblasts line the pulp cavity just deep to the dentin. Dentin forms throughout adult life and gradually encroaches on the pulp cavity. New dentin can also be laid down fairly rapidly to compensate for tooth damage or decay.

Enamel, dentin, and cement are all calcified and resemble bone (to differing extents), but they differ from bone because they are avascular. Enamel differs from cement and dentin because it lacks collagen and is almost entirely mineral.

Tooth and Gum Disease CLINICAL

Dental caries (kār'ēz; "rottenness"), or **cavities**, result from bacterial action that gradually demineralizes enamel and underlying dentin. Decay begins when **dental plaque** (a film of sugar, bacteria, and other mouth debris) adheres to the teeth. Bacterial metabolism of the trapped sugars produces acids, which dissolve the calcium salts of the teeth. Once the salts are leached out, enzymes released by the bacteria readily digest the remaining organic matrix of the tooth. Frequent brushing helps prevent caries by removing plaque.

More serious than tooth decay is the effect of unremoved plaque on the gums. As dental plaque accumulates, it calcifies, forming **calculus** (kal'ku-lus; "stone") or tartar. These stony-hard deposits disrupt the seal between gingivae and teeth, deepening the sulcus and putting the gums at risk for infection by pathogenic anaerobic bacteria. In the early stages of such an infection, called **gingivitis** (jin"jĭ-vi'tis), the gums are red, sore, swollen, and may bleed.

Gingivitis is reversible if the calculus is removed, but if it is neglected the bacteria eventually form pockets of infection which become inflamed. Neutrophils and other immune cells attack not only the intruders but also body tissues, carving deep pockets around the teeth, destroying the periodontal ligament, and activating osteoclasts which dissolve the bone. This serious condition, **periodontal disease** or **periodontitis**, affects up to 95% of all people over age 35 and accounts for 80–90% of tooth loss in adults.

Tooth loss from periodontitis is not inevitable. Various treatments can alleviate the bacterial infestations and encourage the surrounding tissues to reattach to the teeth and bone.

Periodontal disease may jeopardize more than just teeth. Some contend that it increases the risk of heart disease and stroke in at least two ways: (1) the chronic inflammation promotes atherosclerotic plaque, and (2) bacteria entering the blood from infected gums stimulate the formation of clots that clog coronary and cerebral arteries. Risk factors for periodontal disease include smoking, diabetes mellitus, and oral (tongue or lip) piercing.

Digestive Processes of the Mouth

The mouth and its accessory digestive organs are involved in four of the six digestive processes described earlier. The mouth (1) ingests, (2) begins mechanical breakdown by chewing, (3) initiates propulsion by swallowing, and (4) starts the digestion of polysaccharides. Absorption does not occur in the mouth except for a few drugs that are absorbed through the oral mucosa (for example, nitroglycerine used to alleviate the pain of angina, ◀ **p. 683**).

Chewing and swallowing are the processes that promote mechanical breakdown and propulsion, respectively. We describe chewing next, but because the mouth participates in only the first phase of swallowing, we will postpone its discussion until the end of the next module.

Mastication (Chewing)

As food enters the mouth, its mechanical breakdown begins with **mastication**, or chewing. The cheeks and closed lips hold food between the teeth, the tongue mixes food with saliva to soften it, and the teeth cut and grind solid foods into smaller morsels.

Mastication is partly voluntary and partly reflexive. We voluntarily put food into our mouths and contract the muscles that close our jaws. The pattern and rhythm of continued jaw movements are controlled mainly by stretch reflexes and in response to pressure inputs from receptors in the cheeks, gums, and tongue, but they can also be voluntary if desired.

Check Your Understanding

12. Which structure forms the roof of the mouth?

13. Besides preparing food for swallowing, the tongue has another role. What is it?

14. Name three antimicrobial substances found in saliva.

15. Which tooth substance is harder than bone? Which tooth region includes nervous tissue and blood vessels?

16. Identify each of the glands a–c. Which has a duct that opens at the base of the lingual frenulum? Which has a duct that pierces the buccinator muscle?

For answers, see Answers Appendix.

23.5 The pharynx and esophagus move food from the mouth to the stomach

Learning Outcomes

▶ **Describe the anatomy and basic functions of the pharynx and esophagus.**

▶ **Describe the mechanism of swallowing.**

The Pharynx

From the mouth, food passes posteriorly into the **oropharynx** and then the **laryngopharynx** (◄ **Figure 22.5** and Figure 23.8a), both common passageways for food, fluids, and air. (The nasopharynx has no digestive role.)

The histology of the pharyngeal wall resembles that of the oral cavity. The mucosa contains a friction-resistant stratified squamous epithelium well supplied with mucus-producing glands. The external muscle layer consists of two *skeletal muscle* layers. The cells of the inner layer run longitudinally. Those of the outer layer, the *pharyngeal constrictor* muscles, encircle the wall like three stacked fists (◄ **Figure 10.9c, p. 339**). Contractions of these muscles propel food into the esophagus below.

The Esophagus

The **esophagus** (ĕ-sof′ah-gus; "carry food") is a muscular tube about 25 cm (10 inches) long and is collapsed when not involved in food propulsion (**Figure 23.13**). As food moves through the laryngopharynx, the epiglottis closes off the larynx and incoming food is routed posteriorly into the esophagus.

As shown in Figure 23.1, the esophagus takes a fairly straight course through the mediastinum of the thorax. It pierces the diaphragm at the **esophageal hiatus** (hi-a′tus; "gap") to enter the abdomen. It joins the stomach at the **cardial orifice** within the abdominal cavity. The cardial orifice is surrounded by the **gastroesophageal** or **cardiac sphincter** (gas″tro-ĕ-sof″ah-je′al), which is a *physiological* sphincter (see Figure 23.14). That is, it acts as a sphincter, but the only structural evidence of this sphincter is a slight thickening of the circular smooth muscle at that point. The muscular diaphragm, which surrounds this

- Lumen
- **Mucosa** (stratified squamous epithelium)
- **Submucosa** (areolar connective tissue)
- **Muscularis externa**
  - Circular layer
  - Longitudinal layer
- **Adventitia** (fibrous connective tissue)

Esophagus-stomach junction

Simple columnar epithelium of stomach

(a)

(b)

Figure 23.13 Microscopic structure of the esophagus. (a) Cross-sectional view of the esophagus taken from the region close to the stomach junction (10×). Here, the muscularis is composed of smooth muscle. **(b)** Longitudinal section through the esophagus-stomach junction (130×). Notice the abrupt transition from the stratified squamous epithelium of the esophagus (top) to the simple columnar epithelium of the stomach (bottom).

Practice Histology questions:
MasteringA&P® > Study Area > Lab Tools > PAL

sphincter, helps keep it closed when food is not being swallowed. Mucous cells on both sides of the sphincter help protect the esophagus from reflux of stomach acid.

HOMEOSTATIC IMBALANCE 23.6   **CLINICAL**

Heartburn, the first symptom of *gastroesophageal reflux disease* (*GERD*), is the burning, radiating substernal pain that occurs when stomach acid regurgitates into the esophagus. Symptoms are so similar to those of a heart attack that many first-time sufferers of heartburn are rushed to the emergency room. Heartburn is most likely when a person has eaten or drunk to excess, and in conditions that force abdominal contents superiorly, such as extreme obesity, pregnancy, and running, which splashes stomach contents upward with each step.

Heartburn is also common in those with a **hiatal hernia**, a structural abnormality (most often due to abnormal relaxation or weakening of the gastroesophageal sphincter) in which the superior part of the stomach protrudes slightly above the diaphragm. Since the diaphragm no longer reinforces the sphincter, gastric juice may enter the esophagus, particularly when lying down. If the episodes are frequent and prolonged, *esophagitis* (inflammation of the esophagus) and *esophageal ulcers* may result. An even more threatening sequel is esophageal cancer. Treatment varies, but GERD is usually addressed with lifestyle and dietary modifications, along with antacids and certain prescription drugs.   ●●●●●●●

Unlike the mouth and pharynx, the esophagus wall has all four of the basic alimentary canal layers described earlier. Some features of interest:

- The esophageal mucosa contains a *nonkeratinized* stratified squamous epithelium. At the esophagus-stomach junction, that abrasion-resistant epithelium changes abruptly to the simple columnar epithelium of the stomach, which is specialized for secretion (Figure 23.13b).

- The submucosa contains mucus-secreting *esophageal glands*. As a bolus moves through the esophagus, it compresses these glands, causing them to secrete mucus that "greases" the esophageal walls and aids food passage.

- The muscularis externa is skeletal muscle in its superior third, a mixture of skeletal and smooth muscle in its middle third, and entirely smooth muscle in its inferior third.

- Instead of a serosa, the esophagus has a fibrous adventitia composed entirely of connective tissue, which blends with surrounding structures along its route.

Digestive Processes: Swallowing

The pharynx and esophagus merely serve as conduits to pass food from the mouth to the stomach. Their single digestive system function is food propulsion, accomplished by **deglutition** (deg"loo-tish'un), or swallowing.

To send food on its way from the mouth, it is first compacted by the tongue into a bolus and is then swallowed. This complicated process involves the coordinated activity of over 22 separate muscle groups. Before we examine the steps of this process in detail, let's outline the two major phases involved in deglutition.

- The **buccal phase** occurs in the mouth and is voluntary. It ends when a food bolus or a "bit of saliva" leaves the mouth and stimulates tactile receptors in the posterior pharynx, initiating the next phase.

- The **pharyngeal-esophageal phase** is involuntary and is controlled by the swallowing center in the brain stem (medulla and lower pons). Various cranial nerves, most importantly the vagus nerves, transmit motor impulses from the swallowing center to the muscles of the pharynx and esophagus. Once food enters the pharynx, respiration is momentarily inhibited and all routes except the desired one into the digestive tract are blocked off. Solid foods pass from the oropharynx to the stomach in about 8 seconds, and fluids, aided by gravity, pass in 1 to 2 seconds.

① **Buccal phase:**
- The upper esophageal sphincter is contracted (closed).
- The tongue presses against the hard palate, forcing the food bolus into the oropharynx.

② **Pharyngeal-esophageal phase begins:**
- The tongue blocks the mouth.
- The soft palate and its uvula rise, closing off the nasopharynx.
- The larynx rises so that the epiglottis blocks the trachea.
- The upper esophageal sphincter relaxes; food enters the esophagus.

③ **Pharyngeal-esophageal phase continues (steps ③–⑤):**
- The constrictor muscles of the pharynx contract, forcing food into the esophagus inferiorly.
- The upper esophageal sphincter contracts after food enters.

④ Peristalsis moves food through the esophagus to the stomach.

⑤ The gastroesophageal sphincter surrounding the cardial orifice opens. After food enters the stomach, the sphincter closes, preventing regurgitation.

Figure 23.14 **Deglutition (swallowing).** The process of swallowing consists of a buccal (voluntary) phase (step ①) and a pharyngeal-esophageal (involuntary) phase (steps ②–⑤).

Figure 23.14 illustrates each phase of deglutition and the protective mechanisms that prevent food from being inhaled. If we talk or inhale while swallowing, these protective mechanisms may be short-circuited and food may enter the respiratory passageways instead. This typically triggers the cough reflex.

Check Your Understanding

17. To which two organ systems does the pharynx belong?
18. How is the muscularis externa of the esophagus unique in the body?
19. What is the functional significance of the epithelial change seen at the esophagus-stomach junction?
20. How are the respiratory passages blocked during swallowing?

For answers, see Answers Appendix.

23.6 The stomach temporarily stores food and begins protein digestion

Learning Outcomes

▶ Describe stomach structure and indicate changes in the basic alimentary canal structure that aid its digestive function.

▶ Name the cell types responsible for secreting the various components of gastric juice and indicate the importance of each component in stomach activity.

▶ Explain how gastric secretion and stomach motility are regulated.

▶ Define and account for the alkaline tide.

23

Below the esophagus, the GI tract expands to form the **stomach** (see Figure 23.1), a temporary "storage tank." Besides serving as a holding area for ingested food, the stomach continues the demolition job begun in the oral cavity by further degrading food both physically and chemically. As a result, food is converted into a slurry called **chyme** (kīm; "juice"). It then delivers chyme into the small intestine.

Gross Anatomy of the Stomach

The stomach lies in the upper left quadrant of the peritoneal cavity, nearly hidden by the liver and diaphragm. The adult stomach varies from 15 to 25 cm (6 to 10 inches) long, but its diameter and volume depend on how much food it contains. An empty stomach has a volume of about 50 ml and a cross-sectional diameter only slightly larger than the large intestine, but when it is really distended it can hold about 4 L (1 gallon) of food and may extend nearly to the pelvis! When empty, the stomach collapses inward, throwing its mucosa (and submucosa) into large, longitudinal folds called **rugae** (roo′ge; *ruga* = wrinkle, fold).

Figure 23.15a shows the major regions of the stomach. The small **cardial part**, or **cardia** ("near the heart"), surrounds the cardial orifice through which food enters the stomach from the esophagus. The **fundus** is the stomach's dome-shaped part, tucked beneath the diaphragm, that bulges superolaterally to the cardia. The **body**, or the midportion of the stomach, is continuous inferiorly with the funnel-shaped **pyloric part**. The wider and more superior area of the pyloric part, the **pyloric antrum** (*antrum* = cave), narrows to form the **pyloric canal**, which terminates at the **pylorus**. The pylorus is continuous with the duodenum through the **pyloric sphincter** or **valve**, which controls stomach emptying (*pylorus* = gatekeeper).

The convex lateral surface of the stomach is its **greater curvature**, and its concave medial surface is the **lesser curvature**. Extending from these curvatures are two mesenteries, called *omenta* (o-men′tah), that help tether the stomach to other digestive organs and the body wall (see Figure 23.34, p. 908). The **lesser omentum** runs from the liver to the lesser curvature of the stomach, where it becomes continuous with the visceral peritoneum covering the stomach. The **greater omentum** drapes inferiorly from the greater curvature of the stomach to cover the coils of the small intestine. It then runs dorsally and superiorly, wrapping the spleen and the transverse portion of the large intestine before blending with the *mesocolon*, a dorsal mesentery that secures the large intestine to the parietal peritoneum of the posterior abdominal wall.

(a)

(b)

Figure 23.15 Anatomy of the stomach. (a) Gross internal anatomy (frontal section). **(b)** Photograph of external aspect of stomach.

The greater omentum is riddled with fat deposits (*oment* = fatty skin) that give it the appearance of a lacy apron. It also contains large collections of lymph nodes. The immune cells and macrophages in these nodes "police" the peritoneal cavity and intraperitoneal organs.

The stomach is served by the autonomic nervous system. Sympathetic fibers from thoracic splanchnic nerves are relayed through the celiac ganglion (◄ pp. 540–541). Parasympathetic fibers are supplied by the vagus nerve. The arterial supply of the stomach is provided by branches (gastric and splenic) of the celiac trunk (◄ Figure 19.26, p. 745). The

(a) Layers of the stomach wall

- Surface epithelium
- **Mucosa**
- Lamina propria
- Muscularis mucosae
- **Submucosa** (contains submucosal plexus)
- Oblique layer
- **Muscularis externa** (contains myenteric plexus)
- Circular layer
- Longitudinal layer
- **Serosa**

Stomach wall

(b) Enlarged view of gastric pits and gastric glands

- Gastric pits
- Surface epithelium (mucous cells)
- Mucous neck cells
- Parietal cell
- Chief cell
- Gastric pit
- Gastric gland
- Enteroendocrine cell

(c) Part of a gastric gland

Pepsinogen → Pepsin
HCl

- **Parietal cell** (secretes HCl and intrinsic factor)
- **Chief cell** (secretes pepsinogen)
- **Enteroendocrine cell** (secretes hormones and paracrines)

Figure 23.16 Microscopic anatomy of the stomach.

corresponding veins are part of the hepatic portal system and ultimately drain into the hepatic portal vein (◄ **Figure 19.31c, p. 757**).

Microscopic Anatomy of the Stomach

The stomach wall contains the four layers typical of most of the alimentary canal, but its muscularis and mucosa are modified for the special roles of the stomach. Besides the usual circular and longitudinal layers of smooth muscle, the muscularis externa has an incomplete innermost layer of smooth muscle fibrils that runs *obliquely* (Figure 23.15a and **Figure 23.16a**). Together, the smooth muscles of the stomach allow the stomach not only to mix, churn, and move food along the tract, but also

to pummel the food, physically breaking it down into smaller fragments and ramming it into the small intestine.

The lining epithelium of the stomach mucosa is a simple columnar epithelium composed entirely of mucous cells. They produce a cloudy, protective two-layer coat of alkaline mucus in which the surface layer consists of viscous, insoluble mucus that traps a layer of bicarbonate-rich fluid beneath it. This otherwise smooth lining is dotted with millions of deep **gastric pits**, which lead into tubular **gastric glands** that produce the stomach secretion called **gastric juice** (Figure 23.16).

The cells forming the walls of the gastric pits are primarily mucous cells, but those composing the gastric glands vary in different stomach regions. For example, the cells in the glands of the cardia and pylorus primarily secrete mucus, whereas cells of the pyloric antrum produce mucus and several hormones including most of the stimulatory hormone called gastrin.

Types of Gland Cells

Glands of the stomach fundus and body, where most digestion occurs, are substantially larger and produce the majority of the stomach secretions. The glands in these regions contain a variety of secretory cells, including mucous neck, parietal, chief, and enteroendocrine cells.

Mucous Neck Cells **Mucous neck cells** are found primarily in the "neck," but are also scattered deeper within the glands. They produce a thin, soluble mucus quite different from that secreted by the mucous cells of the surface epithelium (Figure 23.16b). It is not yet understood what special function this *acidic* mucus performs.

Parietal Cells **Parietal cells** are scattered among the chief cells (described next), mainly in the more apical (closer to the lumen) region of the glands. Parietal cells simultaneously secrete *hydrochloric acid* (*HCl*) and *intrinsic factor* (Figure 23.16b, c). Although they appear oval when viewed with a light microscope, parietal cells actually have three prongs that bear dense microvilli (they look like fuzzy pitchforks!). This structure provides a huge surface area for secreting H+ and Cl− into the stomach lumen.

HCl makes the stomach contents extremely acidic (pH 1.5–3.5), a condition necessary for activation and optimal activity of the protein-digesting enzyme *pepsin*. The acidity also helps digest food by denaturing proteins and breaking down cell walls of plant foods, and is harsh enough to kill many of the bacteria ingested with foods. Intrinsic factor is a glycoprotein required for vitamin B_{12} absorption in the small intestine.

Chief Cells **Chief cells** occur mainly in the basal regions of the gastric glands. The cuboidal chief cells produce *pepsinogen* (pep-sin′o-jen), the inactive form of the **pepsin**. When these cells are stimulated, the first pepsinogen molecules they release are activated by HCl encountered in the apical region of the gland (Figure 23.16c). But once pepsin is present, it also catalyzes the conversion of pepsinogen to pepsin. The activation process involves removing a small peptide fragment from pepsinogen, causing it to change shape and expose its active site. This positive feedback process is limited only by the amount of pepsinogen present.

Chief cells also secrete lipases (fat-digesting enzymes) that account for about 15% of overall GI lipolysis.

Enteroendocrine Cells **Enteroendocrine cells** (en″ter-o-en′do-krin; "gut endocrine"), typically located deep in the gastric glands (Figure 23.16b, c), release a variety of chemical messengers directly into the interstitial fluid of the lamina propria. Some of these, for example **histamine** and **serotonin**, act locally as paracrines. Others, such as **somatostatin**, act both as paracrines locally and as hormones that diffuse into the blood capillaries to influence several digestive system target organs (**Table 23.1**, p. 888). **Gastrin**, a hormone, plays essential roles in regulating stomach secretion and motility, as we will describe shortly.

The Mucosal Barrier

The stomach mucosa is exposed to some of the harshest conditions in the entire digestive tract. Gastric juice is corrosively acidic (the H+ concentration in the stomach can be 100,000 times that found in blood), and its protein-digesting enzymes can digest the stomach itself.

However, the stomach protects itself by producing the **mucosal barrier**. Three factors create this barrier:

- *A thick coating of bicarbonate-rich mucus* builds up on the stomach wall.
- *The epithelial cells of the mucosa are joined together by tight junctions* (◀ p. 67) that prevent gastric juice from leaking into underlying tissue layers.
- *Damaged epithelial mucosal cells are shed and quickly replaced* by division of undifferentiated *stem cells* that reside where the gastric pits join the gastric glands. The stomach surface epithelium of mucous cells is completely renewed every three to six days, but the more sheltered glandular cells deep within the gastric glands have a much longer life span.

⚖ HOMEOSTATIC IMBALANCE 23.7 — CLINICAL

Anything that breaches the gel-like mucosal barrier causes inflammation of the stomach wall, a condition called *gastritis*. Persistent damage to the underlying tissues can promote **peptic ulcers**, specifically called **gastric ulcers** when they are erosions of the stomach wall (**Figure 23.17**). The most distressing symptom of gastric ulcers is gnawing epigastric pain that seems to bore through to your back. The pain typically occurs 1–3 hours after eating and is often relieved by eating again. The danger posed by ulcers is perforation of the stomach wall, leading to peritonitis and, perhaps, massive hemorrhage.

For years, ulcers were blamed on factors that increased HCl production or reduced mucus secretion, including aspirin and non-steroidal anti-inflammatory drugs (NSAIDs such as ibuprofen), smoking, spicy food, alcohol, coffee, and stress. Although acidic conditions *are* necessary for ulcers to form, acidity in itself is not sufficient to cause them. Ninety percent of recurrent ulcers are the

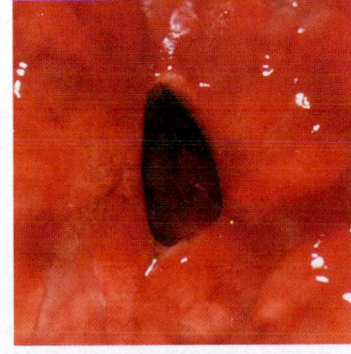

Figure 23.17 A gastric ulcer.

Table 23.1   Hormones and Paracrines That Act in Digestion*

HORMONE	SITE OF PRODUCTION	STIMULUS FOR PRODUCTION	TARGET ORGAN	ACTIVITY
Cholecystokinin (CCK)[†]	Duodenal mucosa	Fatty chyme (also partially digested proteins)	Stomach	• Inhibits stomach's secretory activity
			Liver/pancreas	• Potentiates secretin's actions on these organs
			Pancreas	• Increases output of enzyme-rich pancreatic juice
			Gallbladder	• Stimulates organ to contract and expel stored bile
			Hepatopancreatic sphincter	• Relaxes sphincter to allow entry of bile and pancreatic juice into duodenum
Glucose-dependent insulinotropic peptide (GIP) (or gastric inhibitory peptide)[†]	Duodenal mucosa	Fatty chyme	Stomach	• Inhibits HCl production (minor effect)
			Pancreas (beta cells)	• Stimulates insulin release
Gastrin	Stomach mucosa (G cells)	Food (particularly partially digested proteins) in stomach (chemical stimulation); acetylcholine released by nerve fibers	Stomach (parietal cells)	• Increases HCl secretion
				• Stimulates gastric emptying (minor effect)
			Small intestine	• Stimulates contraction of intestinal muscle
			Ileocecal valve	• Relaxes ileocecal valve
			Large intestine	• Stimulates mass movements
Histamine	Stomach mucosa	Food in stomach	Stomach	• Activates parietal cells to release HCl
Intestinal gastrin	Duodenal mucosa	Acidic and partially digested foods in duodenum	Stomach	• Stimulates gastric glands and motility
Motilin	Duodenal mucosa	Fasting; periodic release every 1½–2 hours by neural stimuli	Proximal duodenum	• Stimulates migrating motor complex
Secretin[†]	Duodenal mucosa	Acidic chyme (also partially digested proteins and fats)	Stomach	• Inhibits gastric gland secretion and gastric motility
			Pancreas	• Increases output of pancreatic juice rich in bicarbonate ions; potentiates CCK's action
			Liver	• Increases bile output
Serotonin	Stomach mucosa	Food in stomach	Stomach	• Causes contraction of stomach muscle
Somatostatin	Stomach mucosa; duodenal mucosa	Food in stomach; stimulation by sympathetic nerve fibers	Stomach	• Inhibits gastric secretion of all products
			Pancreas	• Inhibits secretion
			Small intestine	• Inhibits GI blood flow; thus inhibits intestinal absorption
			Gallbladder and liver	• Inhibits contraction and bile release
Vasoactive intestinal peptide (VIP)	Enteric neurons	Chyme containing partially digested foods	Small intestine	• Stimulates buffer secretion
				• Increases blood flow through intestinal capillaries
				• Relaxes intestinal smooth muscle
			Pancreas	• Increases secretion
			Stomach	• Inhibits acid secretion

*Except for somatostatin, all of these polypeptides also stimulate the growth (particularly of the mucosa) of the organs they affect.
[†]Secretin, cholecystokinin, and GIP are enterogastrones (p. 891). VIP may also play this role.

23

Bacteria

Mucosa layer of stomach

Figure 23.18 Photomicrograph of *H. pylori*, the bacteria that most commonly cause gastric ulcers.

work of a strain of acid-resistant, corkscrew-shaped *Helicobacter pylori* bacteria (**Figure 23.18**), which burrow like a drill bit through the mucus and destroy the protective mucosal layer. Even more troubling are studies that link this bacterium to some stomach cancers.

More than half of the population harbors *H. pylori*, but these pathological effects occur in only 10–20% of infected individuals. The antimicrobial activity of gastric mucin appears to protect most of us from *H. pylori*'s invasive attacks.

A breath test can easily detect the presence of *H. pylori*. A two-week-long course of antibiotics kills the bacteria, promotes healing of the ulcers, and prevents recurrence. For active ulcers, a blocker for H_2 (histamine) receptors may also help because it inhibits HCl secretion by blocking histamine's effects.

The relatively few peptic ulcers not caused by *H. pylori* generally result from long-term use of NSAIDs. In such noninfectious cases, blocking HCl secretion either directly (with pump inhibitors) or indirectly [with H_2 (histamine) receptor blockers] is the therapy of choice.

Digestive Processes in the Stomach

Except for ingestion and defecation, the stomach is involved in the whole "menu" of digestive activities. We will briefly outline the digestive activities here, and these are summarized in **Table 23.2** (p. 904).

- **Propulsion.** Like the esophagus, the stomach exhibits peristalsis.
- **Mechanical breakdown.** The churning action provided by the stomach's smooth muscle during peristalsis causes mechanical breakdown of stomach contents. We discuss stomach contractions further on p. 892.
- **Digestion.** Protein digestion begins in the stomach and is the main type of enzymatic breakdown that occurs there. HCl produced by stomach glands denatures dietary proteins in

preparation for enzymatic digestion. (The unfolded amino acid chain is more accessible to the enzymes.) The most important protein-digesting enzyme produced by the gastric mucosa is pepsin. In infants, however, the stomach glands also secrete *rennin*, an enzyme that acts on milk protein (casein), converting it to a curdy substance that looks like soured milk.

Fat digestion occurs primarily in the small intestine, but gastric and lingual lipases acting in the acidic pH of the stomach also contribute.

- **Absorption.** Not much is absorbed in the stomach, but two common lipid-soluble substances—alcohol and aspirin—pass easily through the stomach mucosa into the blood.

Despite the obvious benefits of preparing food to enter the intestine, the only stomach function essential to life is secretion of intrinsic factor. **Intrinsic factor** is required for intestinal absorption of vitamin B_{12}, needed to produce mature erythrocytes. In its absence, *pernicious anemia* results. However, if vitamin B_{12} is administered by injection, individuals can survive with minimal digestive problems even after total gastrectomy (stomach removal).

In the sections that follow, we will examine (1) the regulation of secretion by gastric glands, (2) the mechanism of HCl secretion, and (3) stomach motility and emptying.

Regulation of Gastric Secretion

Under normal conditions the gastric mucosa pours out as much as 3 L of gastric juice—an acid brew so potent it can dissolve nails—every day. Both neural and hormonal mechanisms control gastric secretion.

- Neural controls consist of both long (vagus nerve–mediated) and short (local enteric) nerve reflexes (see Figure 23.7, p. 875). In each case, acetylcholine (ACh) is released, stimulating the output of gastric juice. When the stomach is stimulated by the vagus nerves (which are parasympathetic, ◄ p. 537), secretory activity of virtually all of its glands increases. (In contrast, activation of sympathetic nerves depresses secretory activity.)

- Hormonal control of gastric secretion is largely the province of *gastrin*. It stimulates secretion of HCl by the stomach, and of hormones (mostly gastrin antagonists) by the small intestine.

Control of HCl-secreting parietal cells is multifaceted. It is stimulated by three chemicals: ACh, gastrin, and histamine. When only one of the three chemicals binds to parietal cell receptors, HCl secretion is scanty, but when all three bind, HCl pours forth. As we noted earlier, antihistamines that bind to and block the H_2 (histamine) receptors of parietal cells are used to treat gastric ulcers caused by hyperacidity. So, as you might guess, histamine (acting as a paracrine) is a key player here.

Stimuli acting at three distinct sites—the head, stomach, and small intestine—provoke or inhibit gastric secretions. Accordingly, the three phases of gastric secretion are called the *cephalic, gastric,* and *intestinal phases.* One or more phases may occur at the same time.

23

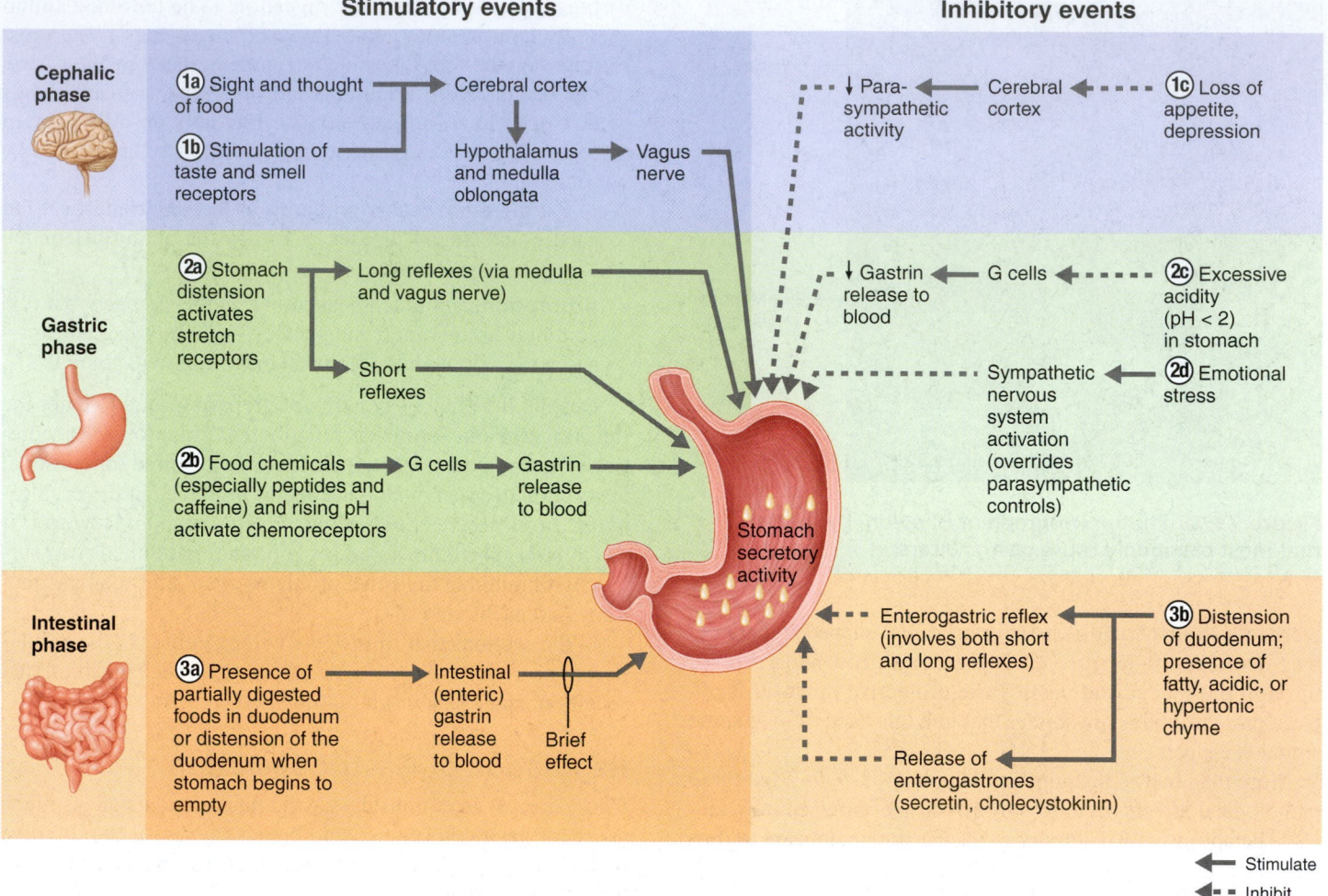

Figure 23.19 Neural and hormonal mechanisms that regulate release of gastric juice.
Stimulatory factors are shown on the left; inhibitory factors are shown on the right.

Cephalic (Reflex) Phase

The **cephalic**, or **reflex**, **phase** of gastric secretion occurs *before* food enters the stomach (**Figure 23.19**). Only a few minutes long, this phase is triggered by the aroma, taste, sight, or thought of food. These triggers act via the vagus nerve to stimulate gastric glands, getting the stomach ready for its digestive chore.

Gastric Phase

Once food reaches the stomach, local neural and hormonal mechanisms initiate the **gastric phase**. This phase lasts three to four hours and provides about two-thirds of the gastric juice released.

Stimulation   The most important secretory stimuli are distension, peptides, and low acidity. As shown in Figure 23.19:

2a Stomach distension activates stretch receptors and initiates both short and long reflexes. In the long reflexes, impulses travel to the medulla and then back to the stomach via vagal fibers.

2b Chemical stimuli provided by partially digested proteins, caffeine, and rising pH directly activate gastrin-secreting enteroendocrine cells called **G cells** in the stomach antrum.

During this phase, gastrin plays a major role in stimulating parietal cells to secrete HCl. It prods parietal cells to spew out HCl (1) by acting directly on receptors on these cells, and (2) by stimulating enteroendocrine cells to release histamine.

When protein foods are in the stomach, the pH of the gastric contents generally rises because proteins act as buffers to tie up H^+. The rise in pH stimulates gastrin secretion and subsequently HCl release, which in turn provides the acidic conditions needed to digest proteins. The more protein in the meal, the greater the amount of gastrin and HCl released. As proteins are digested, the gastric contents gradually become more acidic, which again inhibits the gastrin-secreting cells. This negative feedback mechanism helps maintain optimal pH and working conditions for gastric enzymes.

Inhibition   As shown in Figure 23.19 **2c**, highly acidic (pH below 2) gastric contents *inhibit* gastrin secretion—a situation that commonly occurs between meals. Stress, fear, anxiety, or anything that triggers the fight-or-flight response inhibits gastric secretion because the sympathetic division overrides parasympathetic (vagal) controls of digestion (Figure 23.19 **2d**).

Intestinal Phase

The **intestinal phase** of gastric secretion begins with a brief stimulatory component followed by inhibition (Figure 23.19 ③).

Stimulation   The initial stimulatory part of the intestinal phase is set into motion as partially digested food fills the first part (duodenum) of the small intestine. This stimulates intestinal mucosal cells to release **intestinal (enteric) gastrin**, a hormone that encourages the gastric glands to continue their secretory activity. This stimulatory effect is brief because it is overridden by inhibitory stimuli as the intestine fills.

Inhibition   Four main factors in the duodenum cause it to put the "brakes" on gastric secretion. *Distension* of the duodenum or the presence of *acidic, fatty,* or *hypertonic chyme* all trigger both neuronal and hormonal signals to tell the stomach "whoa, enough already!" As we will see on p. 893, these same four factors also decrease gastric emptying. These brakes on gastric activity protect the small intestine from excessive acidity. They also prevent a massive influx of chyme from overwhelming the digestive and absorptive capacities of the duodenum by matching the amount of entering chyme to the processing abilities of the small intestine. Inhibition is achieved in two ways:

- **Enterogastric reflex:** The duodenum inhibits acid secretion in the stomach by short reflexes through the enteric nervous system and by long reflexes involving sympathetic and vagus nerves.

- **Enterogastrones:** The enterogastrone hormones are released by a scattering of enteroendocrine cells in the duodenal mucosal epithelium. The two most important enterogastrones are **secretin** (se-kre′tin) and **cholecystokinin (CCK)** (ko″le-sis″to-ki′nin). The enterogastrones inhibit gastric secretion and also play other roles (see Table 23.1, p. 888).

Mechanism of HCl Secretion

The process of HCl formation within the parietal cells is shown in **Figure 23.20**. When parietal cells are appropriately stimulated, H^+ is actively pumped into the stomach lumen by H^+-K^+ ATPases (*proton pumps*). As acid is pumped into the stomach, base (HCO_3^-) is exported into the blood. This flow of base is called the **alkaline tide**.

① H^+ and HCO_3^- (bicarbonate ions) are generated from the dissociation of carbonic acid (H_2CO_3) produced from CO_2 and H_2O by carbonic anhydrase.

② H^+-K^+ ATPase pumps H^+ into the lumen and K^+ into the cell. K^+ returns to the lumen through membrane channels.

③ HCO_3^- leaves the cell (**alkaline tide**) in exchange for interstitial fluid Cl^-.

④ Cl^- diffuses through membrane channels into the lumen.

Figure 23.20 Mechanism of HCl secretion by parietal cells.

Regulation of Gastric Motility and Emptying

Stomach contractions not only accommodate its filling and cause its emptying, but they also compress, knead, and mix the food with gastric juice to produce chyme. The processes of mechanical breakdown and propulsion are inseparable in the stomach due to a unique type of peristalsis.

Response of the Stomach to Filling

The stomach stretches to accommodate incoming food, but internal stomach pressure remains constant until about 1.5 L of food have been ingested. Thereafter, the pressure rises. The relatively unchanging pressure in a filling stomach is due to two factors:

- **Receptive relaxation** of smooth muscle in the stomach fundus and body occurs both in anticipation of and in response to food moving through the esophagus and into the stomach. The swallowing center of the brain stem coordinates this process, which is mediated by the vagus nerves.

- **Gastric accommodation** is the intrinsic ability of visceral smooth muscle to exhibit the *stress-relaxation response*. In other words, the stomach can stretch without greatly increasing its tension and contracting expulsively. As we described in Chapter 9 (◄ pp. 315–316), this capability is very important in hollow organs, like the stomach, that must serve as temporary reservoirs.

Gastric Contractile Activity

Like the esophagus, the stomach exhibits peristalsis. After a meal, peristalsis begins near the gastroesophageal sphincter, where it produces gentle rippling movements of the thin stomach wall. But as the contractions approach the pylorus, where

the stomach musculature is thicker, they become much more powerful. Consequently, the contents of the fundus and body (food storage area) remain relatively undisturbed, while foodstuffs in and around the pyloric antrum receive a lively pummeling and mixing.

The pyloric part of the stomach, which holds about 30 ml of chyme, acts as a dynamic filter that allows only liquids and small particles to pass through the barely open pyloric valve. Normally, each peristaltic wave reaching the pyloric muscle squirts 3 ml or less of chyme into the small intestine. Because the contraction also *closes* the valve, the rest (about 27 ml) is propelled backward into the stomach, where it is mixed further **(Figure 23.21)**. This back-and-forth pumping action (retropulsion) effectively breaks up solids.

Although the strength of the stomach's peristaltic waves can be modified, their rate is constant—always around three per minute. This contractile rhythm is set by *enteric pacemaker cells*, muscle-like noncontractile cells formerly called *interstitial cells of Cajal* (kă-hal′). Located in the myenteric plexus (◄ p. 874), the pacemaker cells depolarize and repolarize spontaneously three times each minute, establishing the so-called *cyclic slow waves* of the stomach, or its **basic electrical rhythm (BER)**. Since gap junctions (◄ p. 68) couple the pacemakers electrically to the rest of the smooth muscle sheet, their "beat" is transmitted efficiently and quickly to the entire muscularis.

The pacemakers set the maximum frequency of contraction, but they do not initiate the contractions or regulate their force. Instead, they generate subthreshold depolarization waves, which are then "ignited" (enhanced by further depolarization and brought to threshold) by neural and hormonal factors.

The same factors that increase gastric secretions also enhance the strength of stomach contractions. Distension of the stomach wall by food activates stretch receptors and gastrin-secreting cells, both of which ultimately stimulate gastric smooth muscle

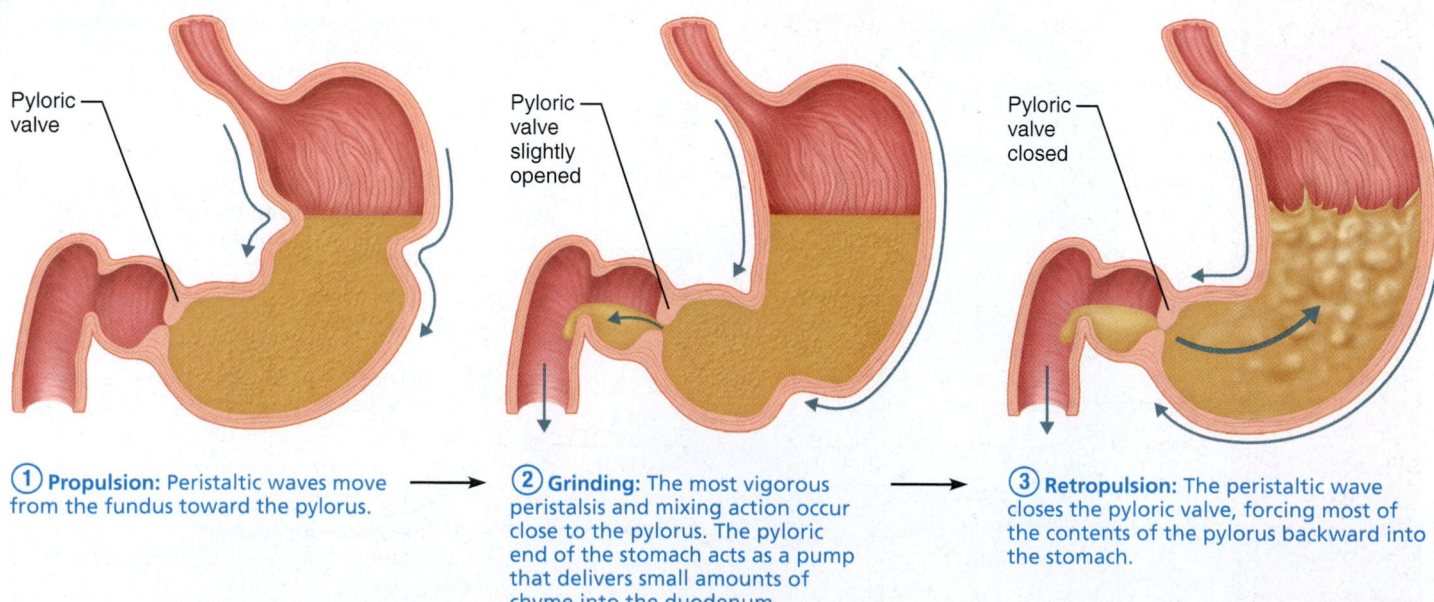

① **Propulsion:** Peristaltic waves move from the fundus toward the pylorus.

② **Grinding:** The most vigorous peristalsis and mixing action occur close to the pylorus. The pyloric end of the stomach acts as a pump that delivers small amounts of chyme into the duodenum.

③ **Retropulsion:** The peristaltic wave closes the pyloric valve, forcing most of the contents of the pylorus backward into the stomach.

Pyloric valve

Pyloric valve slightly opened

Pyloric valve closed

Figure 23.21 Peristaltic waves in the stomach.

and increase gastric motility. For this reason, the more food there is in the stomach, the more vigorous the stomach mixing and emptying movements will be—within certain limits—as we describe next.

Regulation of Gastric Emptying

The stomach usually empties completely within four hours after a meal. However, the larger the meal (the greater the stomach distension) and the more liquid its contents, the faster the stomach empties. Fluids pass quickly through the stomach. Solids linger, remaining until they are well mixed with gastric juice and converted to the liquid state.

The rate of gastric emptying also depends as much—and perhaps more—on the contents of the duodenum as on what is happening in the stomach. The stomach and duodenum act in tandem. As chyme enters the duodenum, receptors in its wall respond to chemical signals and to stretch. This initiates the enterogastric reflex and the hormonal (enterogastrone) mechanisms that inhibit gastric secretion as we described earlier. These mechanisms also prevent further duodenal filling by reducing the force of pyloric contractions (**Figure 23.22**).

A carbohydrate-rich meal moves through the duodenum rapidly, but fats form an oily layer at the top of the chyme and are digested more slowly by enzymes acting in the intestine. For this reason, when chyme entering the duodenum is fatty, reflexes slow stomach emptying, and food may remain in the stomach six hours or more.

> ### HOMEOSTATIC IMBALANCE 23.8
> **CLINICAL**
>
> **Vomiting**, or **emesis**, is an unpleasant experience that empties the stomach by a different route. Many factors signal the stomach to "launch lunch," but the most common are extreme stretching of the stomach or intestine or irritants such as bacterial toxins, excessive alcohol, spicy foods, and certain drugs.
>
> Bloodborne molecules and sensory impulses stream from the irritated sites to the **emetic center** (e-met′ik) of the medulla where they initiate a number of motor responses. Before vomiting, an individual typically feels nauseated, is pale, and salivates excessively. A deep inspiration directly precedes vomiting. The diaphragm and abdominal wall muscles contract, increasing intra-abdominal pressure. The gastroesophageal sphincter relaxes, and the soft palate rises to close off the nasal passages. As a result, the stomach (and perhaps duodenal) contents are forced upward through the esophagus and pharynx and out the mouth.
>
> Excessive vomiting can cause dehydration and severely disrupt the body's electrolyte and acid-base balance. Because large amounts of HCl are lost in vomitus, the blood becomes alkaline as the stomach attempts to replace its lost acid. ●●●●●●

Check Your Understanding

21. Describe the process by which solids are broken apart into smaller pieces in the stomach.

22. Two substances secreted by cells of the gastric glands are needed to produce the active protein-digesting enzyme pepsin. What are these substances and which cells secrete them?

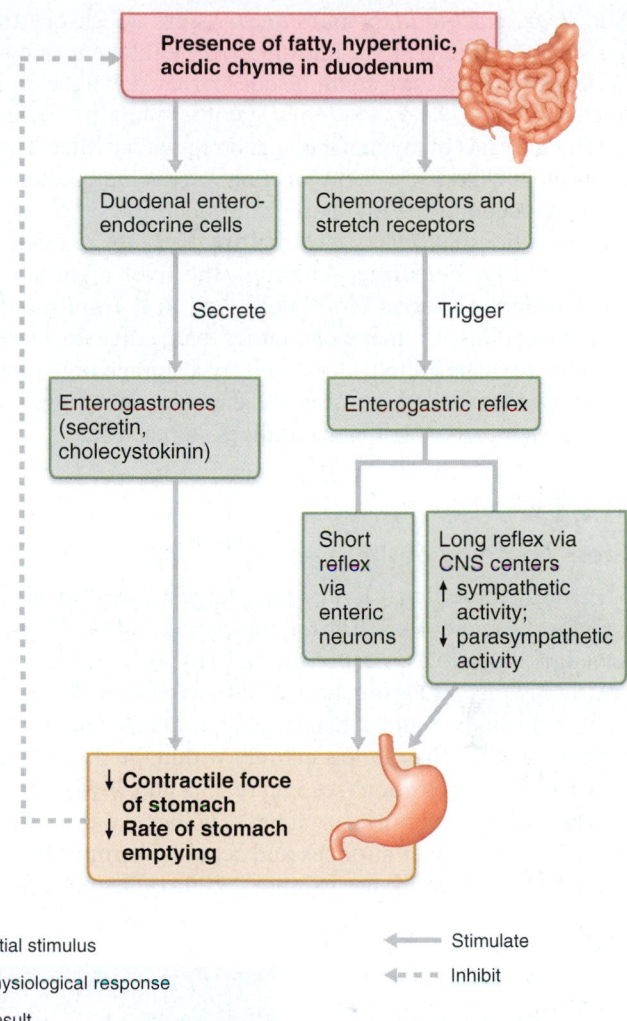

Figure 23.22 Neural and hormonal factors that inhibit gastric emptying. These controls ensure that the food is well liquefied in the stomach and prevent the small intestine from being overwhelmed.

23. Name the three phases of gastric secretion.

24. How does the presence of food in the small intestine inhibit gastric secretion and motility?

For answers, see Answers Appendix.

23.7 The liver secretes bile; the pancreas secretes digestive enzymes

Learning Outcomes

▶ **Describe the histologic anatomy of the liver and pancreas.**

▶ **State the roles of bile and pancreatic juice in digestion.**

▶ **Describe the role of the gallbladder.**

▶ **Describe how bile and pancreatic juice secretion into the small intestine are regulated.**

The *liver*, *gallbladder*, and *pancreas* are accessory organs associated with the small intestine. We will take a side trip to these organs before we continue our journey down the digestive tract.

The liver has many metabolic and regulatory roles. Here we focus on its *digestive* system function: producing bile for export to the duodenum (the first part of the small intestine). Bile is a fat emulsifier that breaks fats into tiny particles to make them more readily digestible. Although the liver also processes nutrient-laden venous blood delivered to it from the digestive organs, this is a metabolic rather than a digestive role (see Chapter 24). The gallbladder is chiefly a storage organ for bile. The pancreas supplies most of the enzymes that digest chyme as well as bicarbonate that neutralizes stomach acid.

The Liver

Gross Anatomy of the Liver

The ruddy, blood-rich **liver** is the largest gland in the body, weighing about 1.4 kg (3 lb) in the average adult. Shaped like a wedge, it occupies most of the right hypochondriac and epigastric regions (◀ **Figure 1.13, p. 20**), extending farther to the right of the body midline than to the left. Located under the diaphragm, the liver lies almost entirely within the rib cage, which provides some protection (see Figure 23.1 and **Figure 23.23**).

The liver has four primary lobes. The largest, the *right lobe*, is visible on all liver surfaces and separated from the smaller *left lobe* by a deep fissure (Figure 23.23a). The posteriormost

caudate lobe and the *quadrate lobe*, which lies inferior to the left lobe, are visible in an inferior view of the liver (Figure 23.23b, c).

A mesentery, the **falciform ligament**, separates the right and left lobes anteriorly and suspends the liver from the diaphragm and anterior abdominal wall. Running along the inferior edge of the falciform ligament is the **round ligament**, or **ligamentum teres** (te′rēz; "round"), a fibrous remnant of the fetal umbilical vein. Except for the superiormost liver area (the *bare area*), which touches the diaphragm, the entire liver is enclosed by the visceral peritoneum.

The lesser omentum anchors the liver to the lesser curvature of the stomach (see Figure 23.34b). The **hepatic artery proper** and the **hepatic portal vein**, which enter the liver at the **porta hepatis** ("gateway to the liver"), and the common hepatic duct, which runs inferiorly from the liver, all travel through the lesser omentum to reach their destinations. The gallbladder rests in a recess on the inferior surface of the right liver lobe (Figure 23.23b, c).

The traditional scheme of defining liver lobes is based on superficial features of the liver. Hepatic surgeons use a different system that divides the liver into eight segments based on its internal anatomy relative to its vascular and biliary supply. This system delineates sections that can be removed while encountering the fewest major vascular structures and the lowest risk.

Bile leaves the liver lobes through the *right* and *left hepatic ducts*. These fuse to form the large **common hepatic duct**, which travels downward toward the duodenum. Along its course, that duct fuses with the **cystic duct** draining the gallbladder to form the **bile duct** (Figure 23.23c).

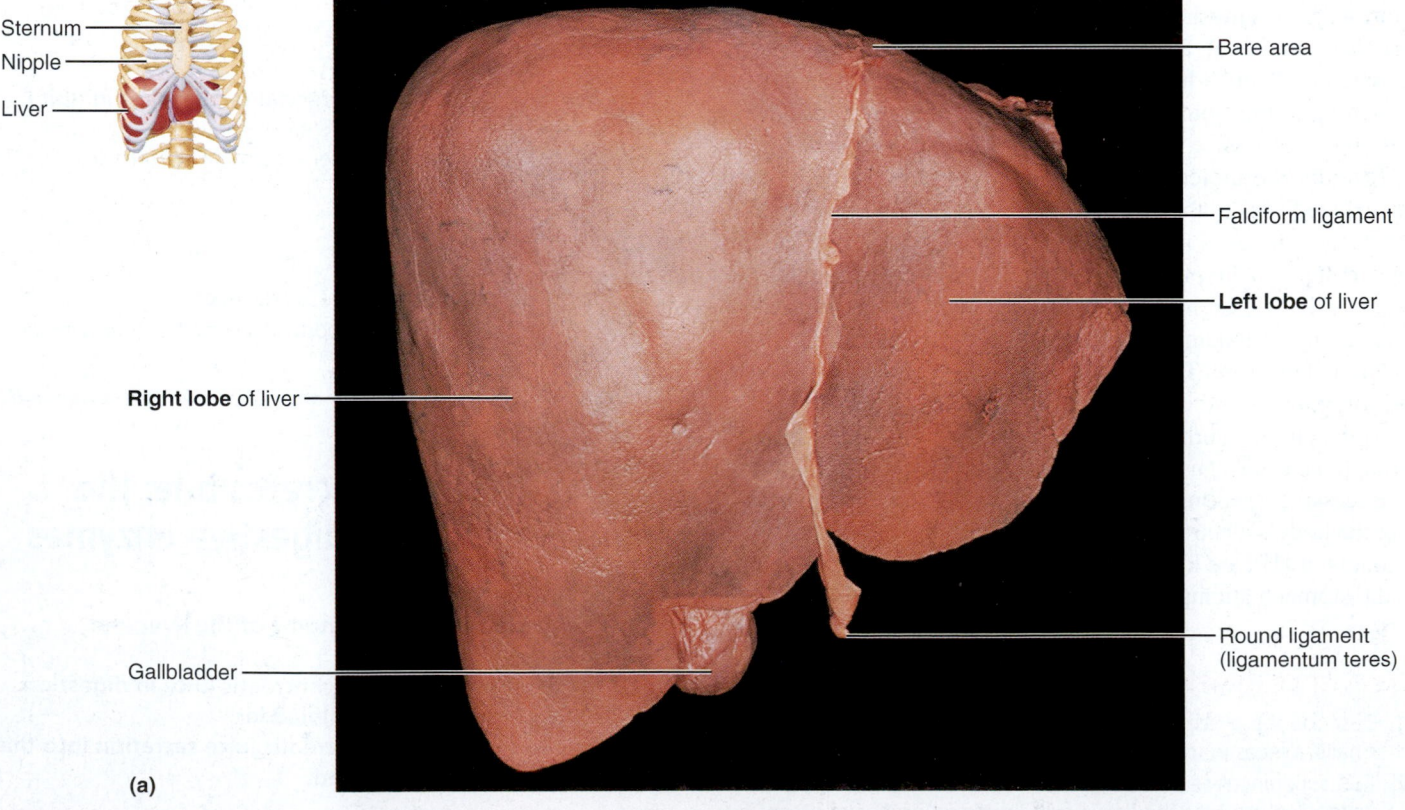

Sternum
Nipple
Liver

Bare area

Falciform ligament

Left lobe of liver

Right lobe of liver

Round ligament
(ligamentum teres)

Gallbladder

(a)

Figure 23.23 Gross anatomy of the human liver. (a) Anterior view of the liver.

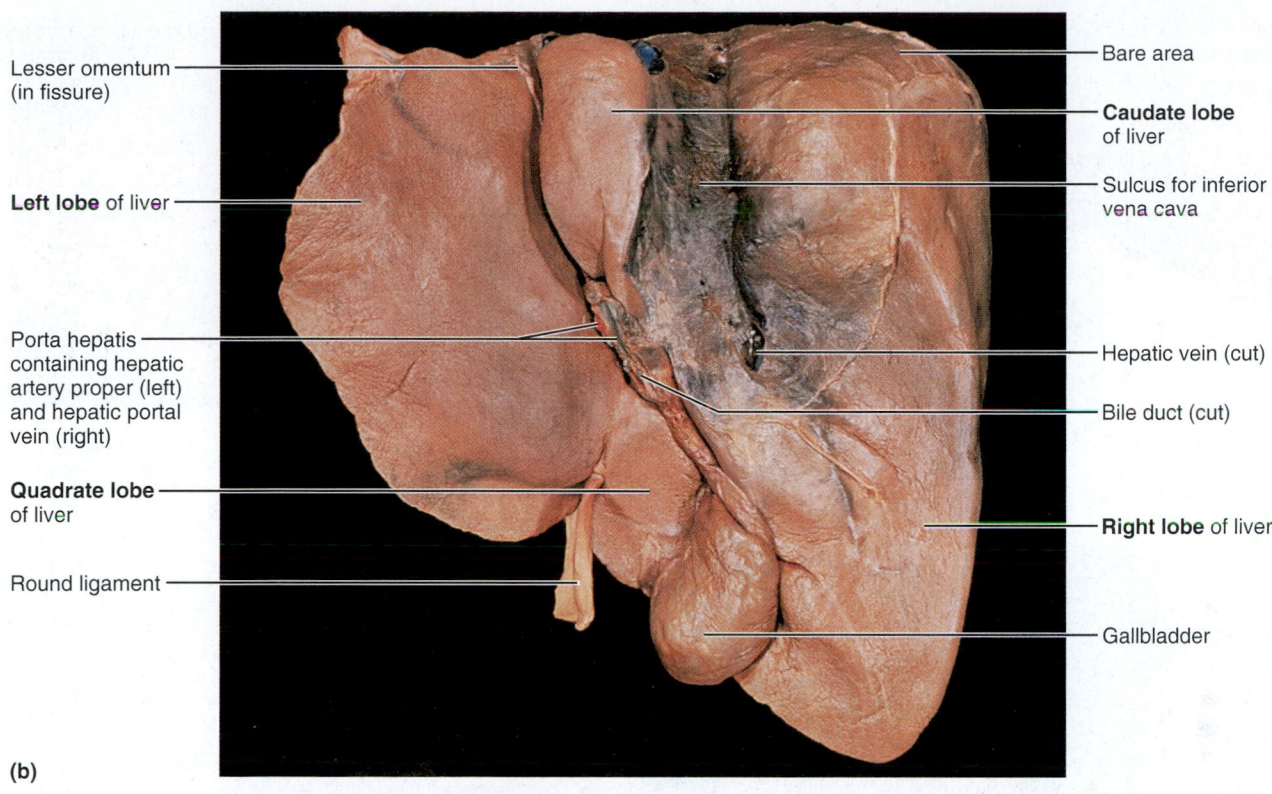

Lesser omentum
(in fissure)

Bare area

Caudate lobe
of liver

Left lobe of liver

Sulcus for inferior
vena cava

Porta hepatis
containing hepatic
artery proper (left)
and hepatic portal
vein (right)

Hepatic vein (cut)

Bile duct (cut)

Quadrate lobe
of liver

Right lobe of liver

Round ligament

Gallbladder

(b)

Hepatic veins

Inferior vena cava

Caudate lobe

Bare area

Left lobe

Porta hepatis
allows passage of
• Hepatic portal vein
• Hepatic artery proper
• Common hepatic duct

Cystic duct

Bile duct

Falciform ligament

Gallbladder

Round ligament

Right lobe

(c)

Quadrate lobe

Figure 23.23 *(continued)* **(b)** Photograph of posteroinferior view of the liver.
(c) Illustration of posteroinferior view of the liver. In these views a group of fissures
separate the four liver lobes. The porta hepatis is a deep fissure that contains the
hepatic portal vein, hepatic artery proper, common hepatic duct, and lymphatics.

Practice Human Cadaver lab questions:
MasteringA&P® > Study Area > Lab Tools > PAL

23

(a) Lobule

(b) Central vein   Connective tissue septum

Interlobular veins (to hepatic vein)

Central vein

Sinusoids

Plates of hepatocytes

Bile canaliculi

Bile duct (receives bile from bile canaliculi)

Fenestrated lining (endothelial cells) of sinusoids

Stellate macrophages in sinusoid walls

Bile duct

Branch of portal vein

Branch of hepatic artery

Portal triad

Portal vein

(c)

23

Figure 23.24 Microscopic anatomy of the liver. (a) Classic lobular pattern of a pig liver. **(b)** Enlarged view of one liver lobule. **(c)** Three-dimensional representation of a small portion of one liver lobule, showing the structure of sinusoids. Arrows indicate the direction of blood flow.

Practice Histology questions: MasteringA&P® > Study Area > Lab Tools > PAL

Microscopic Anatomy of the Liver

The liver is composed of sesame seed–sized structural and functional units called **liver lobules**. Each lobule is a roughly hexagonal (six-sided) structure consisting of plates of *liver cells*, or **hepatocytes** (hep-ah′to-sīts), organized like bricks in a garden wall (**Figure 23.24**). The hepatocyte plates radiate outward from a **central vein** running in the longitudinal axis of the lobule. To make a rough model of a liver lobule, open a thick paperback book until its two covers meet: The pages represent the plates of hepatocytes, and the hollow cylinder formed by the rolled spine represents the central vein.

If you keep in mind that the liver's main function is to process the nutrient-rich blood delivered to it, its histology makes a lot of sense. At each of the six corners of a lobule is a **portal triad** (*portal tract* region), so named because it contains three basic structures (Figure 23.24c):

- A branch of the *hepatic artery* (supplying oxygen-rich arterial blood to the liver)
- A branch of the *hepatic portal vein* (carrying venous blood laden with nutrients from the digestive viscera)
- A *bile duct*

Between the hepatocyte plates are enlarged, heavily fenestrated **liver sinusoids**. Blood from both the hepatic portal vein and the hepatic artery proper percolates from the triad regions through these sinusoids and empties into the central vein. From the central veins blood eventually enters the hepatic veins, which drain the liver, and empties into the inferior vena cava. Forming part of the sinusoid walls are star-shaped **stellate macrophages**, also called **hepatic macrophages** (Figure 23.24c). They remove debris such as bacteria and worn-out blood cells from the blood as it flows past.

The versatile hepatocytes have large amounts of both rough and smooth ER, Golgi apparatus, peroxisomes, and mitochondria. Equipped in this way, the hepatocytes can:

- Secrete some 900 ml of bile daily
- Process bloodborne nutrients in various ways (e.g., they store glucose as glycogen and use amino acids to make plasma proteins)
- Store fat-soluble vitamins
- Play important roles in detoxification, such as ridding the blood of ammonia by converting it to urea (Chapter 24)

Secreted bile flows through tiny canals, called **bile canaliculi** (kan″ah-lik′u-li;), that run between adjacent hepatocytes toward the bile duct branches in the portal triads (Figure 23.24c). Although most illustrations show the canaliculi as discrete tubular structures (shown here in green), their walls are actually formed by the apical membranes of adjoining hepatocytes.

Notice that blood and bile flow in opposite directions in the liver lobule. Bile entering the bile ducts eventually leaves the liver via the common hepatic duct to travel toward the duodenum.

Bile: Composition and Enterohepatic Circulation

Bile is a yellow-green, alkaline solution containing bile salts, bile pigments, cholesterol, triglycerides, phospholipids (lecithin and others), and a variety of electrolytes. Of these, *only* bile salts and phospholipids aid the digestive process.

Bile salts, primarily salts of cholic and chenodeoxycholic acids, are cholesterol derivatives. They play a crucial role in both the digestion and absorption of fats, as we will see later.

Many substances secreted in bile leave the body in feces, but bile salts are not among them. Instead, a recycling mechanism called the *enterohepatic circulation*, shown in **Figure 23.25**, conserves bile salts. This helps minimize the amount of new bile salts that must be synthesized, which is important because they are recirculated 4 to 12 times per day.

The chief *bile pigment* is **bilirubin** (bil″ĭ-roo′bin), a yellow waste product of the heme of hemoglobin formed during the breakdown of worn-out erythrocytes (◄ p. 649). The globin and iron parts of hemoglobin are saved and recycled. Bilirubin, however, is absorbed from the blood by liver cells, excreted into bile, and metabolized in the small intestine by resident bacteria. One of its breakdown products, *stercobilin* (ster′ko-bi″lin), gives feces a brown color. In the absence of bile, feces are gray-white and have fatty streaks because essentially no fats are digested or absorbed.

Homeostatic Imbalances of the Liver

Hepatitis (hep″ah-ti′tis), or inflammation of the liver, has many causes including the toxic effects of alcohol, drugs, and wild

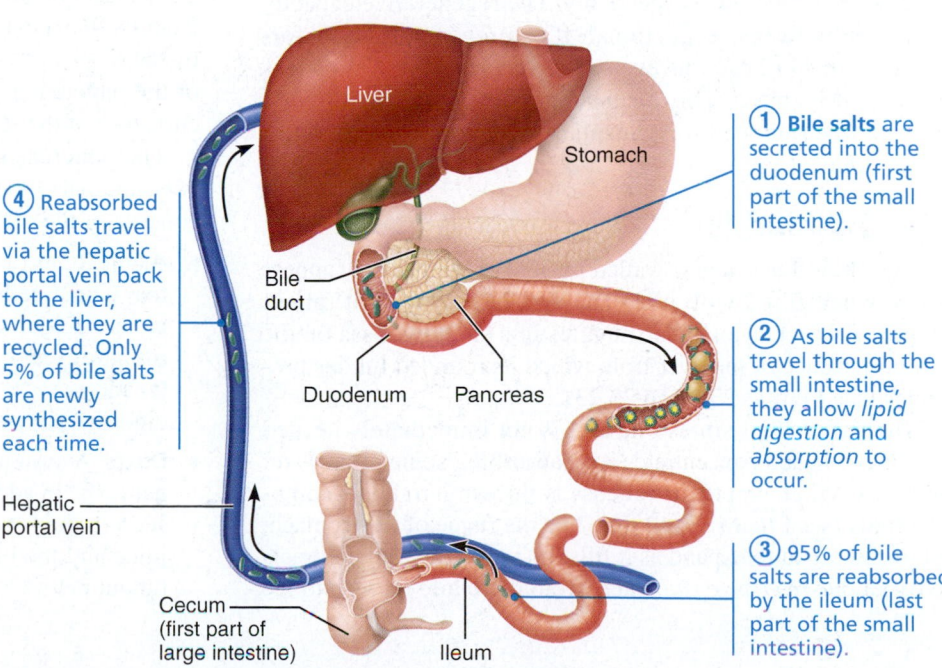

④ Reabsorbed bile salts travel via the hepatic portal vein back to the liver, where they are recycled. Only 5% of bile salts are newly synthesized each time.

Hepatic portal vein

Cecum (first part of large intestine)

Bile duct

Liver

Stomach

Duodenum   Pancreas

Ileum

① **Bile salts** are secreted into the duodenum (first part of the small intestine).

② As bile salts travel through the small intestine, they allow *lipid digestion* and *absorption* to occur.

③ 95% of bile salts are reabsorbed by the ileum (last part of the small intestine).

Figure 23.25 The enterohepatic circulation.

mushrooms. Six viruses that cause hepatitis have been identified and named HVA to HVF. Two of these (HVA and HVE) are transmitted enterically (acquired through eating contaminated food), and the infections they cause tend to be self-limiting. Those transmitted via blood—most importantly HVB and HVC—are linked to chronic hepatitis, liver cirrhosis, and cancer.

Hepatitis C has emerged as the most important infectious liver disease in the United States because it produces persistent or chronic liver infections. More than 4 million Americans are infected and over 10,000 die annually due to sequels of HVC infection. However, the life-threatening C form of hepatitis can now be successfully (and very expensively) treated by a 12-week combination drug therapy.

Outpacing even hepatitis C and alcohol-associated liver damage, **non-alcoholic fatty liver disease (NAFLD)** has become the most common liver disease in North America. It affects about 30% of the general population, but 70% of the obese. Obesity and increased insulin resistance are associated with abnormal lipid metabolism and liver inflammation, which cause NAFLD. There are usually no symptoms associated with NAFLD, but it predisposes the patient to develop full-blown cirrhosis or even liver cancer.

Cirrhosis (sĭr-ro′sis; "orange colored") is the last stage of progressive chronic inflammation of the liver. It typically results from severe chronic hepatitis due to chronic alcoholism, NAFLD, or viral hepatitis. While damaged hepatocytes can regenerate, the liver's connective (scar) tissue regenerates faster. Liver activity is depressed and the liver becomes fibrous with scar tissue. The scar tissue obstructs blood flow throughout the hepatic portal system, causing **portal hypertension**.

Liver transplants are the only clinically proven effective treatment for patients with end-stage liver disease. The one- and five-year survival rate of such transplants is approximately 90% and 75%, respectively. The regenerative capacity of a healthy liver is exceptional. It can regenerate to its former size in 6–12 months even after surgical removal or loss of 80% of its mass. This means that part of a living donor's liver can be removed for transplant without long-term harm to the donor.

The Gallbladder

The **gallbladder** is a thin-walled muscular sac that may appear green when filled with bile. The size of a kiwi fruit (about 10 cm or 4 inches long), it snuggles in a shallow fossa on the inferior surface of the liver from which its rounded fundus protrudes (see Figures 23.1 and 23.23).

The gallbladder stores bile that is not immediately needed for digestion and concentrates it by absorbing some of its water and ions. When empty, its mucosa is thrown into honeycomb-like folds (see Figure 23.28) that, like the rugae of the stomach, allow the organ to expand as it fills. Its muscular wall contracts to expel bile into the *cystic duct*. From there bile flows into the bile duct. The gallbladder, like most of the liver, is covered by visceral peritoneum.

HOMEOSTATIC IMBALANCE 23.9   **CLINICAL**

Bile is the major vehicle for excreting cholesterol from the body, and bile salts keep the cholesterol dissolved within bile. Too much cholesterol or too few bile salts allows the cholesterol to crystallize, forming **gallstones**. Gallstones can obstruct the flow of bile from the gallbladder, a condition called *cholelithiasis* (ko-lĕ-lĭ-thi′ah-sis). When the gallbladder or its duct contracts, the sharp crystals cause agonizing pain that radiates to the right thoracic region.

Gallstones are easy to diagnose because they show up well with ultrasound imaging. Treatments for gallstones include dissolving the crystals with drugs, pulverizing them with ultrasound vibrations (lithotripsy), vaporizing them with lasers, and the classical treatment, surgically removing the gallbladder (cholecystectomy). When the gallbladder is removed, the bile duct enlarges to assume the bile-storing role.

Bile duct blockage prevents both bile salts and bile pigments from entering the intestine. As a result, yellow bile pigments accumulate in blood and eventually are deposited in the skin, causing it to become yellow, or *jaundiced*. Jaundice caused by blocked ducts is called *obstructive jaundice*, but jaundice may also reflect liver disease (in which the liver is unable to carry out its normal metabolic duties). ●●●●●●●

The Pancreas

The **pancreas** (pan′kre-as; *pan* = all, *creas* = flesh, meat) is important to the digestive process because it produces enzymes that break down all categories of foodstuffs. The pancreas is a soft, tadpole-shaped gland that extends across the abdomen from its *tail* (next to the spleen) to its *head*, which is encircled by the C-shaped duodenum (see Figures 23.1 and 23.28). Most of the pancreas is retroperitoneal and lies deep to the greater curvature of the stomach.

The pancreas contains exocrine and endocrine parts. The exocrine part of the pancreas produces **pancreatic juice** and consists of the following (**Figure 23.26**):

- **Acini.** Acini (as′ĭ-ni; singular: acinus) are clusters of secretory acinar cells that produce the enzyme-rich component of pancreatic juice. Acinar cells are full of rough endoplasmic reticulum and exhibit deeply staining **zymogen granules** (zi′mo-jen; "fermenting"). These granules contain inactive digestive enzymes (proenzymes).

- **Ducts.** A system of ducts transports the secretions of the acinar cells. In addition, the epithelial cells of the smallest ducts secrete the water that makes up the bulk of the pancreatic juice and the bicarbonate that makes this secretion alkaline (about pH 8).

(a)

(b)

Figure 23.26 Structure of the enzyme-producing tissue of the pancreas. (a) Schematic view of one acinus (a secretory unit). The acinar cells contain abundant zymogen (enzyme-containing) granules and dark-staining rough ER (typical of gland cells producing large amounts of protein for export). **(b)** Photomicrograph of pancreatic acinar tissue (155×).

The endocrine part of the pancreas is a scattering of mini-endocrine glands called *pancreatic islets*. As we saw in Chapter 16, these islets release insulin and glucagon, hormones that play an important role in carbohydrate metabolism.

Composition of Pancreatic Juice

Approximately 1200 to 1500 ml of clear pancreatic juice is produced daily. It consists mainly of water, and contains enzymes and electrolytes (primarily bicarbonate ions). The high pH of pancreatic fluid helps neutralize acidic chyme entering the duodenum and provides the optimal environment for intestinal and pancreatic enzymes. The pancreatic enzymes include:

- **Proteases** (for proteins)
- **Amylase** (for starch)
- **Lipases** (for fats)
- **Nucleases** (for nucleic acids)

Like pepsin of the stomach, pancreatic proteases are produced and released in inactive forms that are activated in the duodenum, where they do their work. This protects the pancreas from digesting itself.

For example, within the duodenum, **enteropeptidase** (formerly called *enterokinase*), an enzyme bound to the plasma membrane of duodenal epithelial cells, activates *trypsinogen* to **trypsin**. Trypsin, in turn, activates more trypsinogen and two other pancreatic proteases (*procarboxypeptidase* and *chymotrypsinogen*) to their active forms, **carboxypeptidase** (kar-bok″se-pep′tĭ-dās) and **chymotrypsin** (ky″mo-trip′sin), respectively (**Figure 23.27**).

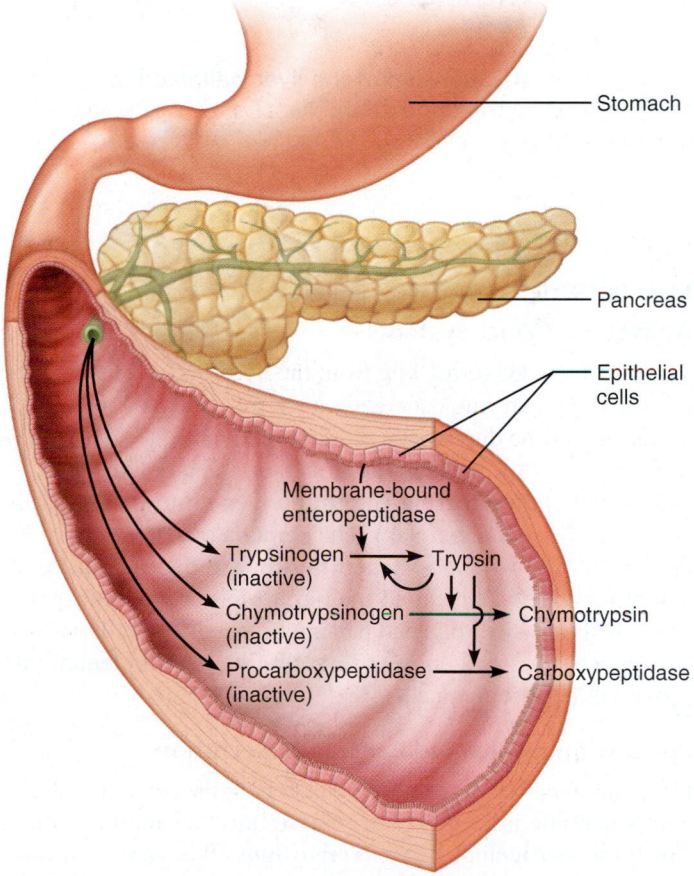

Figure 23.27 Activation of pancreatic proteases in the small intestine. Pancreatic proteases are secreted in an inactive form and are activated in the duodenum.

23

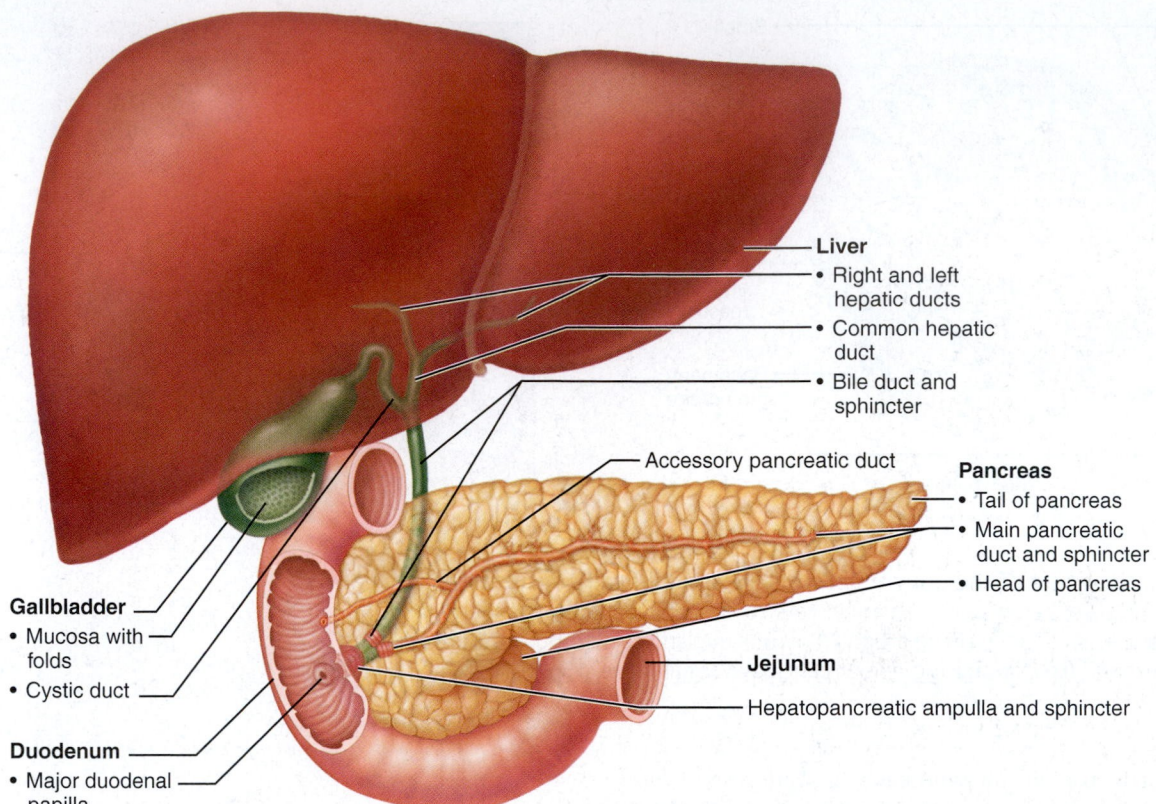

Figure 23.28 **Relationship of the liver, gallbladder, and pancreas to the duodenum.** Ducts from the pancreas, gallbladder, and liver empty into the duodenum.

Bile and Pancreatic Secretion into the Small Intestine

Anatomy of Duct Systems

The bile duct, delivering bile from the liver, and the **main pancreatic duct**, carrying pancreatic juice from the pancreas, unite in the wall of the duodenum, the first section of the small intestine (**Figure 23.28**). They fuse together at a bulblike structure called the **hepatopancreatic ampulla** (hep″ah-to-pan″kre-at′ik am-pul′ah; *ampulla* = flask). The ampulla opens into the duodenum via the volcano-shaped **major duodenal papilla**. A smooth muscle valve called the **hepatopancreatic sphincter** controls the entry of bile and pancreatic juice. A smaller *accessory pancreatic duct* empties directly into the duodenum just proximal to the main duct.

Regulation of Bile and Pancreatic Secretion

Hormones and neural stimuli regulate both the secretion of bile and pancreatic juice and their release into the small intestine. The hormones include two *enterogastrones* that you are already familiar with—*cholecystokinin* and *secretin* (◀ p. 891). **Figure 23.29** summarizes the hormonal and neural mechanisms that control the secretion and release of bile and pancreatic juice.

Bile salts themselves are the major stimulus for enhanced bile secretion (Figure 23.29). After a fatty meal, when the

enterohepatic circulation is returning large amounts of bile salts to the liver, its output of bile rises dramatically. Secretin also stimulates liver cells to secrete bile.

When no digestion is occurring, the hepatopancreatic sphincter is closed and the released bile backs up the cystic duct into the gallbladder, where it is stored until needed. Although the liver makes bile continuously, bile does not usually enter the small intestine until the gallbladder contracts.

Check Your Understanding

25. What is the importance of the enterohepatic circulation?
26. What is the functional difference between pancreatic acini and islets?
27. What is the makeup of the fluid in the pancreatic duct? In the cystic duct? In the bile duct?
28. What stimulates CCK release and what are its effects on the digestive process?
29. **DRAW** Sketch a liver lobule. Add the central vein and portal triads. Label the three parts of a portal triad. Draw and label arrows to show the directions of blood and bile flow.

For answers, see Answers Appendix.

23.8 The small intestine is the major site for digestion and absorption

Learning Outcomes

▶ Identify and describe structural modifications of the wall of the small intestine that enhance the digestive process.
▶ Differentiate between the roles of the various cell types of the intestinal mucosa.
▶ Describe the functions of intestinal hormones and paracrines.

① **CCK and secretin are secreted by duodenal enteroendocrine cells.**
• Cholecystokinin (CCK) release is stimulated by proteins and fats in chyme.
• Secretin release is stimulated by acidic chyme.
• CCK and secretin enter the circulation and cause the following four events (②–⑤).

② **Pancreatic secretion:**
• CCK induces secretion by acinar cells of enzyme-rich pancreatic juice.
• Secretin causes secretion by duct cells of HCO_3^--rich pancreatic juice.
• Vagus nerve weakly stimulates during cephalic and gastric phases.

③ **Bile secretion by liver:**
• Bile salts returning from enterohepatic circulation are the most powerful stimulus for bile secretion.
• Secretin is a minor stimulus.

④ **Gallbladder contraction:**
• CCK causes gallbladder contraction.
• Vagus nerve stimulates weak gallbladder contraction during cephalic and gastric phases.

⑤ **Hepatopancreatic sphincter relaxation:**
• CCK causes hepatopancreatic sphincter to relax. Bile and pancreatic juice enter duodenum.

•• CCK
°° Secretin

Figure 23.29 Mechanisms promoting secretion and release of bile and pancreatic juice.

The **small intestine** is the body's major digestive organ. Within its twisted passageway, digestion is completed (with the help of bile and pancreatic enzymes) and virtually all absorption occurs.

Gross Anatomy

The small intestine is a convoluted tube extending from the pyloric sphincter to the **ileocecal valve (sphincter)** (il″e-o-se′kal) where it joins the large intestine (**Figure 23.30**). It is the longest part of the alimentary canal, but is only about half the diameter of the large intestine, ranging from 2.5 to 4 cm (1–1.6 inches). Although 6–7 m long (approximately 20 ft) in a cadaver, the small intestine is only 2–4 m (7–13 ft) long during life because of muscle tone.

The small intestine has three subdivisions: the duodenum, which is mostly retroperitoneal, and the jejunum and ileum, both intraperitoneal organs (Figure 23.30). The relatively immovable **duodenum** (du″o-de′num; "twelve finger widths long"), which curves around the head of the pancreas, is about 25 cm (10 inches) long (Figure 23.28). Although it is the shortest intestinal subdivision, the duodenum has the most features of interest, including the *major duodenal papilla* mentioned earlier.

The **jejunum** (jĕ -joo′num; "empty"), about 2.5 m (8 ft) long, extends from the duodenum to the ileum. The **ileum** (il′e-um;

"twisted"), approximately 3.6 m (12 ft) in length, joins the large intestine at the ileocecal valve. The jejunum and ileum hang in sausagelike coils in the central and lower part of the abdominal cavity, suspended from the posterior abdominal wall by a fan-shaped *mesentery* (see Figure 23.34). The large intestine encircles these more distal parts of the small intestine.

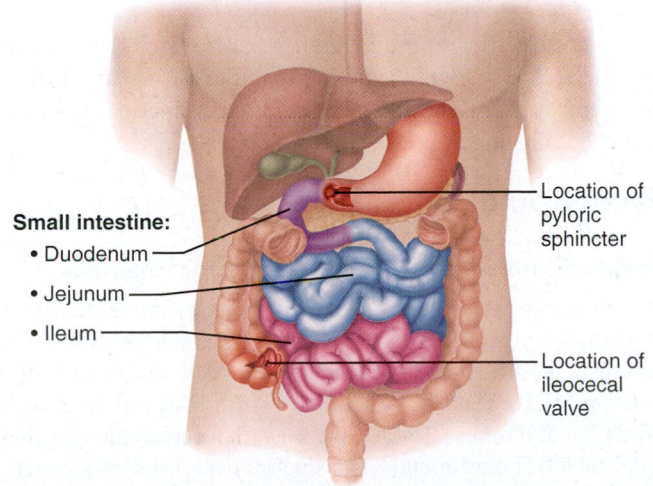

Small intestine:
• Duodenum
• Jejunum
• Ileum

Location of pyloric sphincter

Location of ileocecal valve

Figure 23.30 The small intestine.

Vein carrying blood to hepatic portal vessel

Muscle layers

Circular folds

Lumen

Villi

(a)

Microvilli (brush border)

Enterocytes (absorptive cells)

Lacteal

Goblet cell

Blood capillaries

Mucosa-associated lymphoid tissue

Intestinal crypt

Muscularis mucosae

Duodenal gland

Villus

Entero-endocrine cells

Paneth cells

Venule

Lymphatic vessel

Submucosa

(b)

Practice Histology questions:
MasteringA&P® > Study Area > Lab Tools > PAL

Figure 23.31 Structural modifications of the small intestine that increase its surface area for digestion and absorption. **(a)** Enlargement of a few circular folds, showing associated finger-like villi (muscularis and serosa layers not indicated). **(b)** Structure of a villus. Enlargement shows absorptive enterocytes that exhibit microvilli on their free (apical) surface. **(c)** Photomicrograph of the mucosa, showing villi (250×).

The arterial supply of the small intestine is primarily from the superior mesenteric artery (◀ pp. 746–747). The veins parallel the arteries and typically drain into the superior mesenteric vein. From there, the nutrient-rich venous blood from the small intestine drains into the hepatic portal vein, which carries it to the liver.

Nerve fibers serving the small intestine include parasympathetics from the vagus and sympathetics from the thoracic splanchnic nerves, both relayed through the superior mesenteric (and celiac) plexus.

Microscopic Anatomy

Modifications of the Small Intestine for Absorption

The small intestine is highly adapted for absorbing nutrients. Its length alone provides a huge surface area, and its wall has three structural modifications—circular folds, villi, and microvilli—that amplify its absorptive surface enormously (by a factor of more than 600 times). In fact, the intestinal surface area is about equal to 200 square meters, the size of a singles tennis court!

- The **circular folds** are deep, permanent folds of the mucosa and submucosa (**Figure 23.31a**). Nearly 1 cm tall, these folds

Enterocytes (absorptive cells)

Goblet cells

Villi

(c) Intestinal crypt

force chyme to spiral through the lumen, slowing its movement and allowing time for full nutrient absorption.

- **Villi** (vil'i; "tufts of hair") are fingerlike projections of the mucosa, over 1 mm high, that give it a velvety texture, much

Mucus granules

Microvilli forming the brush border

Apical membrane

Enterocyte (absorptive cell)

Figure 23.32 Microvilli of the small intestine. Colored electron micrograph. Microvilli (28,000×) appear as red projections from the surface of the absorptive cell.

- *Goblet cells* are mucus-secreting cells found in the epithelia of the villi and crypts.

- *Enteroendocrine cells* are the source of the enterogastrones discussed earlier (p. 891)—secretin and cholecystokinin to name two. They are mostly found scattered in the crypts but some are also found in the villi.

- *Paneth cells*, found deep in the crypts, are specialized secretory cells that fortify the small intestine's defenses by releasing antimicrobial agents such as defensins and lysozyme. These secretions destroy certain bacteria and help to determine which bacteria colonize the intestinal lumen.

- *Stem cells* continuously divide in the depths of the crypts. Their daughter cells differentiate to become all of the other cell types. Most of these daughter cells (except Paneth cells) differentiate as they gradually migrate up the villi. Paneth cells, on the other hand, migrate to the very bottom of the crypts. Enterocytes at the tips of the villi undergo apoptosis and are shed, renewing the villus epithelium every three to five days.

HOMEOSTATIC IMBALANCE 23.10   **CLINICAL**

Treatments for cancer, such as radiation therapy and chemotherapy, preferentially target rapidly dividing cells. They kill cancer cells, but also nearly obliterate the rapidly dividing GI tract epithelium. Many patients suffer nausea, vomiting, and diarrhea after each treatment. ●●●●●●●

like the soft nap of a towel (Figure 23.31). The villi are large and leaflike in the duodenum (the intestinal site of most active absorption) and gradually narrow and shorten along the length of the small intestine. In the core of each villus is a dense capillary bed and a wide lymphatic capillary called a **lacteal** (◄ p. 769). Digested foodstuffs are absorbed through the enterocytes into both the capillary blood and the lacteal.

- **Microvilli** are long, densely packed cytoplasmic extensions of the absorptive cells of the mucosa that give the mucosal surface a fuzzy appearance called the **brush border** (Figure 23.31b enlargement and **Figure 23.32**). The plasma membranes of the microvilli bear enzymes referred to as **brush border enzymes**, which complete the digestion of carbohydrates and proteins in the small intestine.

Histology of the Small Intestine Wall

The four layers typical of the GI tract are also seen in the small intestine, but the mucosa and submucosa are modified to reflect the intestine's functions in the digestive pathway.

Between the villi, which are specialized for absorption, the small intestine mucosa is studded with tubular glands called **intestinal crypts** (see Figure 23.31b, c). The crypts decrease in number along the length of the small intestine.

Five major types of cells are found in the mucosal epithelium of the villi and crypts:

- *Enterocytes* form the bulk of the epithelium. They are simple columnar absorptive cells bound by tight junctions and richly endowed with microvilli. These cells bear the primary responsibility for absorbing nutrients and electrolytes in the villi. In the crypts, enterocytes are primarily secretory cells that secrete *intestinal juice*, a watery mixture that contains mucus and serves as a carrier fluid for absorbing nutrients from chyme.

Mucosa-associated lymphoid tissue (MALT) includes both individual lymphoid follicles and *aggregated lymphoid nodules*, the latter called **Peyer's patches** (pi′erz; ◄ p. 775). Peyer's patches are primarily located in the lamina propria but occasionally protrude into the submucosa below. Their increasing abundance toward the distal end of the small intestine reflects the fact that this region contains huge numbers of bacteria that must be prevented from entering the bloodstream. The lamina propria of the mucosa contains large numbers of immunoglobulin A (IgA)-secreting plasma cells that help protect against intestinal pathogens (◄ p. 799).

The submucosa is typical areolar connective tissue. Elaborate mucus-secreting **duodenal glands** in the submucosa of the duodenum produce an alkaline (bicarbonate-rich) mucus that helps neutralize acidic chyme moving in from the stomach. When this protective mucus barrier is inadequate, the intestinal wall erodes and *duodenal ulcers* result.

The muscularis is typical and bilayered. Except for the bulk of the duodenum, which is retroperitoneal and has an adventitia, visceral peritoneum (serosa) covers the external intestinal surface.

Intestinal Juice

The intestinal glands normally secrete 1 to 2 L of intestinal juice daily. The major stimulus for its production comes from hypertonic or acidic chyme. Normally, intestinal juice is slightly alkaline (7.4–7.8) and isotonic with blood plasma.

23

Table 23.2 Overview of the Functions of the Gastrointestinal Organs

ORGAN	MAJOR FUNCTIONS*	COMMENTS/ADDITIONAL FUNCTIONS
Mouth and associated accessory organs	■ Ingestion: food is voluntarily placed into oral cavity ■ Propulsion: voluntary (buccal) phase of deglutition (swallowing) initiated by tongue; propels food into pharynx ■ Mechanical breakdown: mastication (chewing) by teeth and mixing movements by tongue ■ Digestion: salivary amylase in saliva, produced by salivary glands, begins digestion of starch	• Mouth serves as a receptacle; most functions performed by associated accessory organs. • Mucus in saliva helps dissolve food so it can be tasted and moistens food so that tongue can compact it into a bolus that can be swallowed. Saliva cleanses and lubricates oral cavity and teeth.
Pharynx and esophagus	■ Propulsion: peristaltic waves move food bolus to stomach, thus accomplishing the pharyngeal-esophageal (involuntary) phase of deglutition	• Primarily food chutes. • Mucus produced helps to lubricate food passageways.
Stomach	■ Mechanical breakdown and ■ propulsion: peristaltic waves mix food with gastric juice and propel it into the duodenum ■ Digestion: pepsin begins the digestion of proteins ■ Absorption: absorbs a few fat-soluble substances (aspirin, alcohol, some drugs)	• Also stores food until it can be moved into the duodenum. • Hydrochloric acid produced is a bacteriostatic agent and activates protein-digesting enzymes. • Mucus produced helps lubricate and protect stomach from self-digestion. • Intrinsic factor produced is required for intestinal absorption of vitamin B_{12}.
Small intestine and associated accessory organs (liver, gallbladder, pancreas)	■ Mechanical breakdown and ■ propulsion: segmentation by smooth muscle of the small intestine continually mixes contents with digestive juices and, along with short-distance peristaltic waves, moves food along tract, allowing sufficient time for digestion and absorption ■ Digestion: digestive enzymes delivered from pancreas and brush border enzymes attached to microvilli membranes complete digestion of all classes of foods ■ Absorption: breakdown products of carbohydrate, protein, fat, and nucleic acid digestion, plus vitamins, electrolytes, and water, are absorbed by active and passive mechanisms	• Small intestine is highly modified for digestion and absorption (circular folds, villi, and microvilli). • Alkaline mucus produced by intestinal glands and bicarbonate-rich juice ducted in from pancreas help neutralize acidic chyme and provide proper environment for enzymatic activity. • Bile produced by liver emulsifies fats and enhances (1) fat digestion and (2) absorption of fatty acids, monoglycerides, cholesterol, phospholipids, and fat-soluble vitamins. • Gallbladder stores and concentrates bile, releasing it to small intestine in response to hormonal signals.
Large intestine	■ Digestion: some remaining food residues are digested by gut bacteria (which also produce some vitamin K and some B vitamins) ■ Absorption: absorbs most remaining water, electrolytes (largely NaCl), and vitamins produced by bacteria ■ Propulsion: propels feces toward rectum by mass movements ■ Defecation: reflex triggered by rectal distension; eliminates feces from body	• Temporarily stores and concentrates residues until defecation can occur. • Copious mucus produced by goblet cells eases passage of feces through colon.

*The colored boxes beside the functions correspond to the color coding of digestive functions (gastrointestinal tract activities) illustrated in Figure 23.2.

Intestinal juice is largely water but also contains some mucus, which is secreted both by the duodenal glands and by goblet cells of the mucosa.

Digestive Processes in the Small Intestine

By the time food reaches the small intestine it is unrecognizable, but still far from being digested. Carbohydrates and proteins are partially degraded, but fat digestion has only begun. The process of digestion accelerates during the chyme's tortuous three- to six-hour journey through the small intestine, and it is here that most of the water and virtually all nutrients are absorbed.

Since we cover the actual chemistry of digestion and absorption in detail later, here we will examine the source of enzymes for digestion and how intestinal motility mixes and moves chyme along the length of the small intestine.

Sources of Enzymes for Digestion

Most of the substances required for digestion—bile, digestive enzymes (except for the brush border enzymes), and bicarbonate ions (to provide the proper pH for enzymatic catalysis)—are *imported* from the liver and pancreas. For this reason, anything that impairs liver or pancreatic function or delivery of their juices to the small intestine severely hinders our ability to digest food and absorb nutrients.

Brush border enzymes perform the final digestion of food into the simple components that can be absorbed by intestinal cells. Note that the brush border enzymes are *not* secreted. Instead, they remain bound to the plasma membranes of the enterocytes.

Regulating Chyme Entry

Chyme entering the duodenum is usually hypertonic. For this reason, if large amounts of chyme rushed into the small intestine, the osmotic water loss from the blood into the intestinal lumen would result in dangerously low blood volume. Additionally, the low pH of entering chyme must be adjusted upward and the chyme must be well mixed with bile and pancreatic juice for digestion to continue. These modifications take time. Feedback via the enterogastric reflex and enterogastrones to

the stomach pylorus carefully controls food movement into the small intestine to prevent the duodenum from being overwhelmed (see Figure 23.22).

Notice that the feedback mechanisms regulating chyme entry (the enterogastric reflex and enterogastrones) are the same as those that decrease gastric secretion (Figure 23.19).

Motility of the Small Intestine

There are two motility patterns in the small intestine. After a meal, *segmentation* is the principal form of motility. This motor pattern ensures that chyme is thoroughly mixed with bile and pancreatic and intestinal juices. It also ensures that the absorbable products of digestion come in contact with the mucosa for absorption. Between meals, the primary motor pattern is a form of peristalsis called the *migrating motor complex*. This peristalsis is largely a housekeeping function that sweeps debris toward the large intestine.

After a Meal If we examine the small intestine with X-ray fluoroscopy after it is "loaded" with a meal, it looks like the intestinal contents are being massaged—alternately contracting and relaxing rings of smooth muscle simply move the chyme backward and forward a few centimeters at a time (see Figure 23.3b). As with stomach peristalsis, intrinsic pacemaker cells (in the myenteric plexus) initiate these segmenting movements. The intensity of segmentation is altered by long and short reflexes, which parasympathetic activity enhances and sympathetic activity decreases, and by hormones (**Table 23.3**). The more intense the contractions, the greater the mixing.

We usually think of segmentation as purely a mixing movement. How, then, does chyme move along the small intestine? First, some very weak, short-distance peristalsis does occur even in the full small intestine. Second, the pacemakers in the duodenum depolarize more frequently than those in the ileum. As a result, segmentation also moves intestinal contents slowly and steadily toward the ileocecal valve at a rate that allows ample time to complete digestion and absorption. Unlike the strength of contraction, the basic contractile rhythm of each intestinal region is not affected by reflexes or hormones.

Between Meals True peristalsis occurs only after most nutrients have been absorbed. At this point, segmenting movements

Table 23.3 Control of Small Intestinal Motility		
STIMULUS	**EFFECT ON SMALL INTESTINE MOTILITY**	**MECHANISM**
↑ Gastric motility and emptying	• ↑ Motility in ileum • Relaxation of ileocecal valve	Gastroileal reflex (long neural reflex) and gastrin release (both cause same effects)
Distension of small intestine	↑ Strength of segmentation	Long and short neural reflexes
Reduced intestinal volume; fasting	Initiates migrating motor complex (peristalsis); repeats until next meal	Motilin (initiates long and short neural reflexes)

wane and the duodenal mucosa begins to release the hormone **motilin**.

As motilin blood levels rise, peristaltic waves are initiated in the proximal duodenum every 90 to 120 minutes and sweep slowly along the intestine, moving 50–70 cm (about 2 ft) before dying out. Each successive wave begins a bit more distally, a pattern of peristaltic activity called the **migrating motor complex (MMC)**. A complete "trip" from duodenum to ileum takes about two hours. The process then repeats itself, sweeping the last remnants of the meal plus bacteria, sloughed-off mucosal cells, and other debris into the large intestine.

This "housekeeping" function prevents bacteria in the large intestine from entering the small intestine. As food again enters the stomach with the next meal, segmentation replaces peristalsis.

Ileocecal Valve Control   Most of the time, the ileocecal valve is closed. However, two mechanisms cause it to relax and allow food residues to enter the cecum when ileal motility increases:

- The **gastroileal reflex** (gas″tro-il′e-ul), a long neural reflex triggered by stomach activity, increases the force of segmentation in the ileum and relaxes the ileocecal valve.
- Gastrin, a hormone released by the stomach, increases the motility of the ileum and relaxes the ileocecal valve.

Once the chyme has passed through, it exerts backward pressure that closes the valve's flaps, preventing regurgitation into the ileum. This reflex sweeps the contents of the previous meal completely out of the stomach and small intestine as the next meal is eaten.

Check Your Understanding

30. What common advantage do circular folds, villi, and microvilli provide to the digestive process?

31. What are brush border enzymes?

32. Distension of the stomach and duodenal walls have different effects on stomach secretory activity. What are these effects?

33. After a meal, what is the small intestine's most common motility pattern? Why?

34. What is the MMC and why is it important?

For answers, see Answers Appendix.

23.9 The large intestine absorbs water and eliminates feces

Learning Outcomes

▶ List the major functions of the large intestine.
▶ Describe the regulation of defecation.

The **large intestine** frames the small intestine on three sides and extends from the ileocecal valve to the anus (see Figure 23.1). Its diameter, at about 7 cm, is greater than that of the small intestine, but it is much shorter (1.5 m versus 6 m). Its major digestive functions are to absorb most of the remaining water from indigestible food residues, store the residues temporarily, and then eliminate them from the body as semisolid **feces**

(fe′sēz), also called *stool*. It also absorbs metabolites produced by resident bacteria as they ferment carbohydrates not absorbed in the small intestine.

Gross Anatomy

The large intestine exhibits three features not seen elsewhere— teniae coli, haustra, and epiploic appendages. Except for its terminal end, the longitudinal muscle layer of its muscularis is mostly reduced to three bands of smooth muscle called **teniae coli** (ten′ne-e ko′li; "ribbons of the colon"). Their tone puckers the wall of the large intestine into pocketlike sacs called **haustra** (haw′strah; singular: *haustrum*; "to draw up"). Another obvious feature of the large intestine is its **epiploic appendages** (ep″ĭ-plo′ik; "membrane covered"), which are small fat-filled pouches of visceral peritoneum that hang from the surface of the large intestine (**Figure 23.33a**). Their significance is not known.

Subdivisions of the Large Intestine

The large intestine has the following subdivisions: cecum, appendix, colon, rectum, and anal canal. The saclike **cecum** (se′kum; "blind pouch"), which lies below the ileocecal valve in the right iliac fossa, is the first part of the large intestine (Figure 23.33a).

Attached to the posteromedial surface of the cecum is the blind, wormlike **appendix**. The appendix contains masses of lymphoid tissue, and as part of MALT (◀ p. 774) it plays an important role in body immunity. Additionally, it serves as a storehouse of bacteria and recolonizes the gut when needed. However, the appendix has an important structural shortcoming— its twisted structure makes it susceptible to blockage.

HOMEOSTATIC IMBALANCE 23.11   **CLINICAL**

Acute inflammation of the appendix, or **appendicitis**, results from a blockage (often by feces) that traps infectious bacteria in its lumen. Unable to empty its contents, the appendix swells, squeezing off venous drainage, which may lead to ischemia and necrosis (low blood flow and tissue death) of the appendix. If the appendix ruptures, feces containing bacteria spray over the abdominal contents, causing *peritonitis*.

The symptoms of appendicitis vary, but the first symptom is usually pain in the umbilical region. Loss of appetite, nausea and vomiting, and pain relocalization to the lower right abdominal quadrant follow. Appendicitis is treated with surgical removal of the appendix (appendectomy), or, in select cases, with antibiotics. Appendicitis is most common during adolescence, when the entrance to the appendix is at its widest.   ●●●●●●

The **colon** has several distinct regions. Proximally, as the **ascending colon**, it travels up the right side of the abdominal cavity to the level of the right kidney. Here it makes a right-angle turn—the **right colic (hepatic) flexure**—and travels across the abdominal cavity as the **transverse colon**. Directly anterior to the spleen, it bends acutely at the **left colic (splenic) flexure** and descends down the left side of the posterior abdominal wall as

Right colic
(hepatic) flexure

Transverse colon

Superior
mesenteric
artery

Haustrum

Ascending colon

Ileum

Ileocecal valve

Cecum

Appendix

(a)

Left colic
(splenic) flexure

Transverse mesocolon

Epiploic appendages

Descending colon

Cut edge of mesentery

Tenia coli

Sigmoid colon

Rectum

Anal canal

External anal sphincter

Rectal valve

Rectum

Hemorrhoidal
veins

Levator ani muscle

Anal canal

External anal
sphincter

Internal anal
sphincter

Anal columns

Anal sinuses

Anus

(b)

Figure 23.33 Gross anatomy of the large intestine.
(a) Large intestine. **(b)** Structure of the anal canal.

the **descending colon**. Inferiorly, it enters the pelvis,
where it becomes the S-shaped **sigmoid colon**.

In the pelvis, at the level of the third sacral vertebra,
the sigmoid colon joins the **rectum**, which runs pos-
teroinferiorly just in front of the sacrum.

Despite its name (*rectum* = straight), the rectum
has three lateral curves or bends, represented internally
as three transverse folds called **rectal valves** (Figure
23.33b). These valves stop feces from being passed
along with gas (flatus).

The **anal canal**, the last segment of the large intestine, lies in the perineum, entirely external to the
abdominopelvic cavity. About 3 cm long, it begins
where the rectum penetrates the levator ani muscle
of the pelvic floor and opens to the body exterior at
the **anus**. The anal canal has two sphincters, an involuntary **internal anal sphincter** composed of smooth
muscle (part of the muscularis), and a voluntary **external anal sphincter** composed of skeletal muscle.
The sphincters, which act rather like purse strings to

23

(a)

Falciform ligament
Liver
Gallbladder
Lesser omentum
Spleen
Stomach
Duodenum
Round ligament
Transverse colon
Greater omentum
Small intestine
Cecum
Urinary bladder

(b)

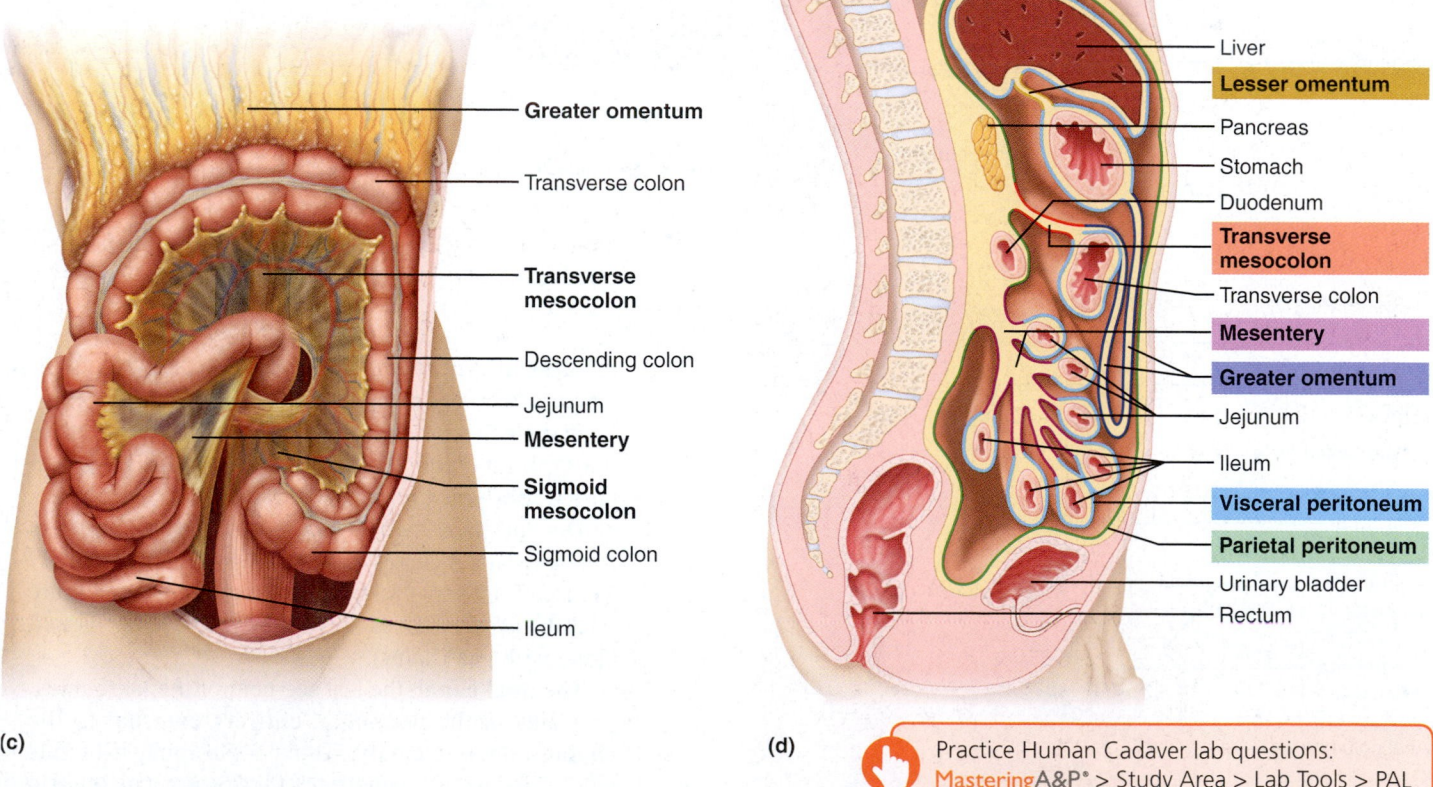

(c)

Greater omentum
Transverse colon
Transverse mesocolon
Descending colon
Jejunum
Mesentery
Sigmoid mesocolon
Sigmoid colon
Ileum

(d)

Liver
Lesser omentum
Pancreas
Stomach
Duodenum
Transverse mesocolon
Transverse colon
Mesentery
Greater omentum
Jejunum
Ileum
Visceral peritoneum
Parietal peritoneum
Urinary bladder
Rectum

23

Practice Human Cadaver lab questions:
MasteringA&P® > Study Area > Lab Tools > PAL

Figure 23.34 Mesenteries of the abdominal digestive organs. (a) The greater omentum, a dorsal mesentery, is shown in its normal position covering the abdominal viscera. **(b)** The liver and gallbladder have been reflected superiorly and the greater omentum removed to reveal the lesser omentum, a ventral mesentery attaching the liver to the lesser curvature of the stomach. **(c)** The greater omentum has been reflected superiorly to reveal the mesentery attachments of the small and large intestine. **(d)** Sagittal section of the abdominopelvic cavity of a male. Mesentery labels appear in colored boxes.

open and close the anus, are ordinarily closed except during defecation.

The rectum and anal canal lack teniae coli and haustra. However, the rectum's muscularis muscle layers are complete and well developed, consistent with its role in generating strong contractions to expel feces.

Relationship of the Large Intestine to the Peritoneum

The colon is retroperitoneal, except for its transverse and sigmoid parts. These parts are intraperitoneal and anchored to the posterior abdominal wall by mesentery sheets called **meso-colons** (**Figure 23.34c, d**). The rectum (and sometimes the cecum) is also retroperitoneal.

Microscopic Anatomy

The wall of the large intestine differs in several ways from that of the small intestine:

- There are no circular folds, villi, or brush border because most food is absorbed before reaching the large intestine.
- Its mucosa is thicker, its abundant crypts are deeper, and the crypts contain tremendous numbers of goblet cells. Mucus produced by goblet cells eases the passage of feces and protects the intestinal wall from irritation by acids and gases released by resident bacteria.

Like the small intestine, the large intestine *mucosa* is simple columnar epithelium except in the anal canal.

The mucosa of the anal canal, a stratified squamous epithelium, merges with the true skin surrounding the anus and is quite different from the mucosa in the rest of the colon, reflecting the greater abrasion that this region receives. Superiorly, it hangs in long ridges or folds called **anal columns**. **Anal sinuses**, recesses between the anal columns, exude mucus when compressed by feces, which aids in emptying the anal canal (Figure 23.33b).

Together, the inferior margins of the anal sinuses form a wavy line. Superior to this line, visceral sensory fibers innervate the mucosa, and so it is relatively insensitive to pain. The area inferior to the line, however, is very sensitive to pain because it is innervated by somatic sensory fibers.

Two superficial venous plexuses are associated with the anal canal, one with the anal columns and the other with the anus itself. If these (hemorrhoidal) veins become dilated and inflamed, itchy varicosities called *hemorrhoids* result.

Bacterial Microbiota

The **bacterial microbiota** (also called the **bacterial flora**) of the large intestine consists of over a thousand different types of bacteria and accounts for a couple of pounds of our body weight. Some of these bacteria colonize the colon via the anus, but others enter from the small intestine still "alive and kicking" after running the gauntlet of antimicrobial defenses (lysozyme, defensins, HCl, and protein-digesting enzymes). We provide a home for these bacteria, but what's in it for us? We are just beginning to understand that we depend upon these bacteria just as much as they depend upon us.

Metabolic Functions

Our gut bacteria help us by recovering energy from otherwise indigestible foods and synthesizing some vitamins.

- **Fermentation.** Gut bacteria ferment some indigestible carbohydrates and the mucin in gut mucus. The resulting short-chain fatty acids can be absorbed and used for fuel by the body's cells. Unfortunately, fermentation also produces a mixture of gases (including dimethyl sulfide, H_2, N_2, CH_4, and CO_2). Some of these gases, such as dimethyl sulfide, are quite odorous (smelly). About 500 ml of gas (flatus) is produced each day, much more when we eat foods (such as beans) rich in indigestible carbohydrates.

- **Vitamin synthesis.** B complex vitamins and some of the vitamin K the liver needs in order to produce several clotting proteins are synthesized by gut bacteria.

Keeping Pathogenic Bacteria in Check

The immune system and the gut bacteria live in a dynamic equilibrium. The immune system destroys any bacteria that threaten to breach the mucosal barrier. The gut bacteria, on the other hand, instruct the immune system not to overreact to their presence in the lumen.

Potentially harmful bacteria in our large intestine are kept in check in two ways. First, beneficial bacteria outcompete and actively suppress harmful bacteria, and as a result normally vastly outnumber them. Second, our immune system prevents bacteria from entering the body through the gut epithelium. An elegant system keeps the bacteria from breaching the mucosal barrier. Dendritic cells sample the microbial antigens in the lumen. They then migrate to the nearby lymphoid follicles within the gut mucosa (MALT) and trigger an IgA antibody–mediated response restricted to the gut lumen. This prevents the bacteria from straying into tissues deep to the mucosa where they might elicit a much more widespread systemic response.

> **HOMEOSTATIC IMBALANCE 23.12**   CLINICAL   23
>
> *Clostridium difficile*, an anaerobic bacterium, is the most common cause of *antibiotic-associated diarrhea*, accounting for 14,000 deaths per year in the U.S. Where does it come from? For some people, *C. difficile* is a normal, but small, fraction of the gut's bacteria. Other people acquire *C. difficile* through the fecal-oral route (poor hand washing), particularly in hospital or long-term care settings. In either case, when other bacteria are wiped out by antibiotics, *C. difficile* flourishes in the gut and may cause pseudomembranous colitis (inflammation of the colon) that leads to bowel perforation and sepsis.
>
> Because *C. difficile* infections are resistant to many antibiotics, they are notoriously difficult to treat and often recur. Instead of using ever more powerful antibiotics, a new treatment strategy seeks to restore competitive bacteria to the gut's ecosystem by performing a *fecal transplant*. Transferring fecal bacteria from an uninfected donor to the patient (e.g., by enema or during colonoscopy) cures *C. difficile* infections 90–100% of the time. In spite of the considerable "yuck factor," this treatment

is becoming more and more mainstream. It is likely that optimized mixtures of cultured bacteria will become available for transplantation to treat *C. difficile* in the future. ●●●●●●●

While the immune system keeps the gut bacteria in check, the gut bacteria also profoundly shape our immune system responses. For example, the type of bacteria present influences the balance between subtypes of T cells, which in turn affects the balance between pro- and anti-inflammatory responses.

The coexistence of the bacterial microbiota with our immune system does sometimes fail. When that happens, the painful and debilitating condition known as inflammatory bowel disease (see Related Clinical Terms, p. 921) may result.

Gut Bacteria in Health and Disease

We are only just beginning to learn how gut bacteria affect the body. There is mounting evidence that the kinds and proportions of the bacteria in our gut can influence our body weight, our susceptibility to various diseases (including diabetes, atherosclerosis, fatty liver disease), and even our mood. Manipulating our various gut bacteria may become a routine health-care strategy in the near future.

Digestive Processes in the Large Intestine

What is finally delivered to the large intestine contains few nutrients, but it still has 12 to 24 hours more to spend there. Except for a small amount of digestion of that residue by the gut bacteria, no further food breakdown occurs in the large intestine.

The large intestine harvests vitamins made by the gut bacteria and reclaims most of the remaining water and some of the electrolytes (particularly sodium and chloride). However, nutrient absorption is not its *major* function. As mentioned, the primary concerns of the large intestine are propulsive activities that force fecal material toward the anus and eliminate it from the body (defecation).

HOMEOSTATIC IMBALANCE 23.13 **CLINICAL**

The large intestine is important for our comfort, but it is not essential for life. If the colon is removed, the terminal ileum can be brought out to the abdominal wall in a procedure called an *ileostomy* (il″e-os′to-me). From there food residues are eliminated into a sac attached to the abdominal wall. ●●●●●●●

Motility of the Large Intestine

When food residue enters the colon through the ileocecal valve, the colon becomes motile, but its contractions are sluggish or short-lived compared to those of the small intestine. The movements most seen in the colon are **haustral contractions**, slow segmenting movements that last about one minute and occur every 30 minutes or so.

These contractions, which occur mainly in the ascending and transverse colon, reflect local controls of smooth muscle within the walls of the individual haustra. As a haustrum fills

with food residue, the distension stimulates its muscle to contract. These movements mix the residue, which aids in water absorption.

Mass movements (mass peristalsis) are long, slow-moving, but powerful contractile waves that move over large areas of the colon three or four times daily and force the contents toward the rectum. Typically, they occur during or just after eating. The presence of food in the stomach activates the gastroileal reflex in the small intestine and the propulsive **gastrocolic reflex** in the colon.

Haustral contractions in the descending and sigmoid colon promote the final drying out of the feces. This part of the colon also stores feces until mass movements propel the feces into the rectum. Fiber in the diet strengthens colon contractions and softens the feces, allowing the colon to act like a well-oiled machine.

HOMEOSTATIC IMBALANCE 23.14 **CLINICAL**

When the diet lacks fiber and the volume of residues in the colon is small, the colon narrows and its contractions become more powerful, increasing the pressure on its walls. This promotes formation of **diverticula** (di″ver-tik′u-lah), small herniations of the mucosa through the colon walls.

This condition, called **diverticulosis**, most commonly occurs in the sigmoid colon, and affects over half of people over age 70. In 4–10% of cases, diverticulosis progresses to **diverticulitis**, in which the diverticula become inflamed and may rupture, leaking into the peritoneal cavity, which can be life threatening.

Irritable bowel syndrome (*IBS*) is a functional GI disorder not explained by anatomical or biochemical abnormalities. Affected individuals have recurring (or persistent) abdominal pain that is relieved by defecation. Additionally, they may have changes in the consistency and frequency of their stools, and varying complaints of bloating, flatulence, nausea, and depression. Stress is a common precipitating factor, and stress management is an important aspect of treatment. ●●●●●●●

The semisolid feces delivered to the rectum contain undigested food residues, mucus, sloughed-off epithelial cells, millions of bacteria, and just enough water to allow their smooth passage. Of the 500 ml or so of food residue entering the cecum daily, approximately 150 ml becomes feces.

Defecation

The rectum is usually empty, but when mass movements force feces into it, stretching of the rectal wall initiates the **defecation reflex**. This parasympathetic spinal reflex causes the sigmoid colon and the rectum to contract, and the internal anal sphincter to relax (**Figure 23.35** ① and ②). As feces are forced into the anal canal, messages reach the brain allowing us to decide whether the external (voluntary) anal sphincter should open or remain constricted to stop passage of feces temporarily (Figure 23.35 ③).

If defecation is delayed, the reflex contractions end within a few seconds and the rectal walls relax. The next mass movement

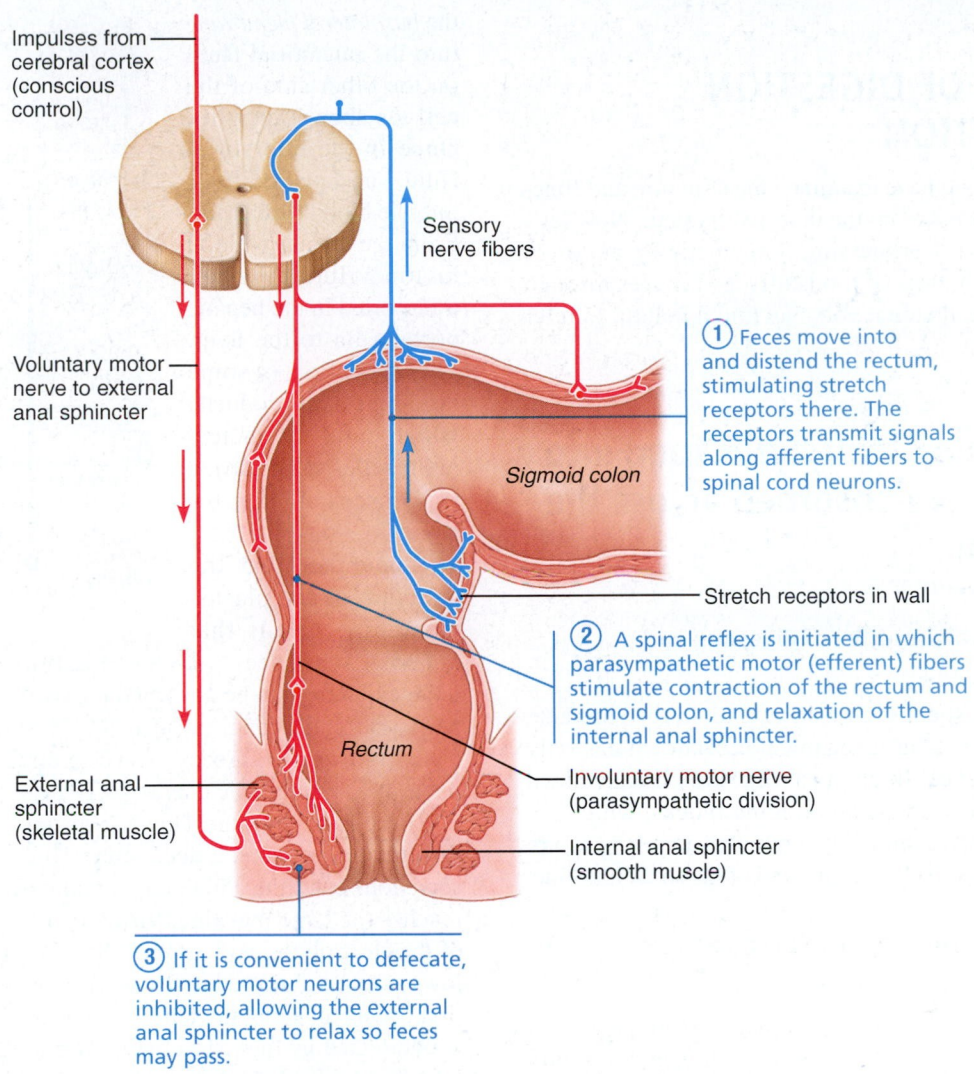

Impulses from cerebral cortex (conscious control)

Sensory nerve fibers

Voluntary motor nerve to external anal sphincter

Sigmoid colon

1 Feces move into and distend the rectum, stimulating stretch receptors there. The receptors transmit signals along afferent fibers to spinal cord neurons.

Stretch receptors in wall

2 A spinal reflex is initiated in which parasympathetic motor (efferent) fibers stimulate contraction of the rectum and sigmoid colon, and relaxation of the internal anal sphincter.

Rectum

External anal sphincter (skeletal muscle)

Involuntary motor nerve (parasympathetic division)

Internal anal sphincter (smooth muscle)

3 If it is convenient to defecate, voluntary motor neurons are inhibited, allowing the external anal sphincter to relax so feces may pass.

Figure 23.35 Defecation reflex.

initiates the defecation reflex again—and so on, until the person chooses to defecate or the urge becomes irresistible.

During defecation, the muscles of the rectum contract to expel the feces. We aid this process voluntarily by closing the glottis and contracting our diaphragm and abdominal wall muscles to increase the intra-abdominal pressure (a procedure called the *Valsalva maneuver*). We also contract the levator ani muscle (◄ **pp. 348–349**), which lifts the anal canal superiorly. This lifting action leaves the feces below the anus—and outside the body. Involuntary or automatic defecation (fecal incontinence) occurs in infants because they have not yet gained control of their external anal sphincter. It also occurs in those with spinal cord transections.

HOMEOSTATIC IMBALANCE 23.15   **CLINICAL**

Watery stools, or **diarrhea**, result from any condition that rushes food residue through the large intestine before that organ has had sufficient time to absorb the remaining water. Causes

include irritation of the colon by bacteria or, less commonly, prolonged physical jostling of the digestive viscera (occurs in marathon runners). Prolonged diarrhea may result in dehydration and electrolyte imbalance (acidosis and loss of potassium).

Conversely, when food remains in the colon for extended periods, too much water is absorbed and the stool becomes hard and difficult to pass. This condition, called **constipation**, may result from insufficient fiber or fluid in the diet, improper bowel habits (failing to heed the "call"), lack of exercise, or laxative abuse.

Check Your Understanding

35. Name and briefly describe the types of motility that occur in the large intestine.

36. What is the result of stimulation of stretch receptors in the rectal walls?

37. In what ways are gut bacteria important to our nutrition?

For answers, see Answers Appendix.

PART 3

PHYSIOLOGY OF DIGESTION AND ABSORPTION

So far in this chapter, we have examined the structure and function of the organs that make up the digestive system. Now let's investigate the chemical processing (enzymatic breakdown) and absorption of each class of foodstuffs as it moves through the GI tract. As you read along, you may find it helpful to refer to the summary in **Figure 23.36**.

23.10 Digestion hydrolyzes food into nutrients that are absorbed across the gut epithelium

Learning Outcome

▶ Describe the general processes of digestion and absorption.

After foodstuffs have spent even a short time in the stomach, they are unrecognizable, but mechanical breakdown has only changed their appearance. In contrast, digestion breaks down ingested foods into their chemical building blocks, which are very different molecules chemically. Only these molecules are small enough to be absorbed across the wall of the small intestine.

Mechanism of Digestion: Enzymatic Hydrolysis

Digestion is a catabolic process that breaks down large food molecules to *monomers* (chemical building blocks). Digestion is accomplished by enzymes secreted into the lumen of the alimentary canal by intrinsic and accessory glands. Recall from Chapter 2 that enzymatic breakdown of any food molecule is **hydrolysis** (hi-drol′ĭ-sis) because it involves adding a water molecule to each molecular bond to be broken (lysed).

Most digestion is done in the small intestine. Pancreatic enzymes break large chemicals (usually polymers) into smaller pieces that are, in turn, broken down into individual components by the intestinal (brush border) enzymes. Alkaline pancreatic juice neutralizes the acidic chyme that enters the small intestine from the stomach. This provides the proper environment for operation of the enzymes. Both pancreatic juice (the main source of lipases) and bile are necessary for fat breakdown.

Mechanisms of Absorption

Absorption is the process of moving substances from the lumen of the gut into the body. Recall from Chapter 4 (◀ p. 117) that epithelia are polarized and have an apical and basal side. Because tight junctions join the epithelial cells (enterocytes) of the intestinal mucosa at their apical surfaces, substances usually cannot move *between* cells. Instead, materials must pass *through* the enterocytes. Materials enter an enterocyte through its *apical membrane* from the lumen of the gut, and exit through

the *basolateral membrane* into the interstitial fluid on the other side of the cell (as shown at right). Once in the interstitial fluid, substances diffuse into the blood capillaries. From the capillary blood in the villus they are transported in the hepatic portal vein to the liver. The exception is some lipid digestion products, which enter the lacteal in the villus to be carried via lymphatic fluid to the blood.

Remember that the structure of the plasma membrane means that nonpolar substances, which can dissolve in the lipid core of the membrane, can be absorbed passively. All other substances need a carrier mechanism (◀ p. 70). Most nutrients are absorbed by *active transport* processes driven directly or indirectly (secondarily) by metabolic energy (ATP; ◀ p. 75).

There is much more flowing through the alimentary tube than food monomers. Indeed, up to 10 L of food, drink, and GI secretions enter the alimentary canal daily, but only 1 L or less reaches the large intestine. Virtually all of the foodstuffs, 80% of the electrolytes, and most of the water (remember water follows salt) are absorbed in the small intestine. Although absorption occurs all along the length of the small intestine, most of it is completed by the time chyme reaches the ileum. The major absorptive role of the ileum is to reclaim bile salts to be recycled back to the liver for resecretion. The absorptive capacity of the small intestine is truly remarkable and it is virtually impossible to exceed.

Check Your Understanding

38. In order to be absorbed, nutrients must pass through two plasma membranes. Name these membranes.

39. **MAKE CONNECTIONS** Name the layer and sublayer of the alimentary canal wall that houses the capillaries into which nutrients are absorbed.

For answers, see Answers Appendix.

23.11 How is each type of nutrient processed?

Learning Outcomes

▶ List the enzymes involved in digestion; name the foodstuffs on which they act.

▶ List the end products of protein, fat, carbohydrate, and nucleic acid digestion.

▶ Describe the processes by which breakdown products of foodstuffs are absorbed in the small intestine.

Foodstuff	Enzyme(s) and source	Site of action	Path of absorption

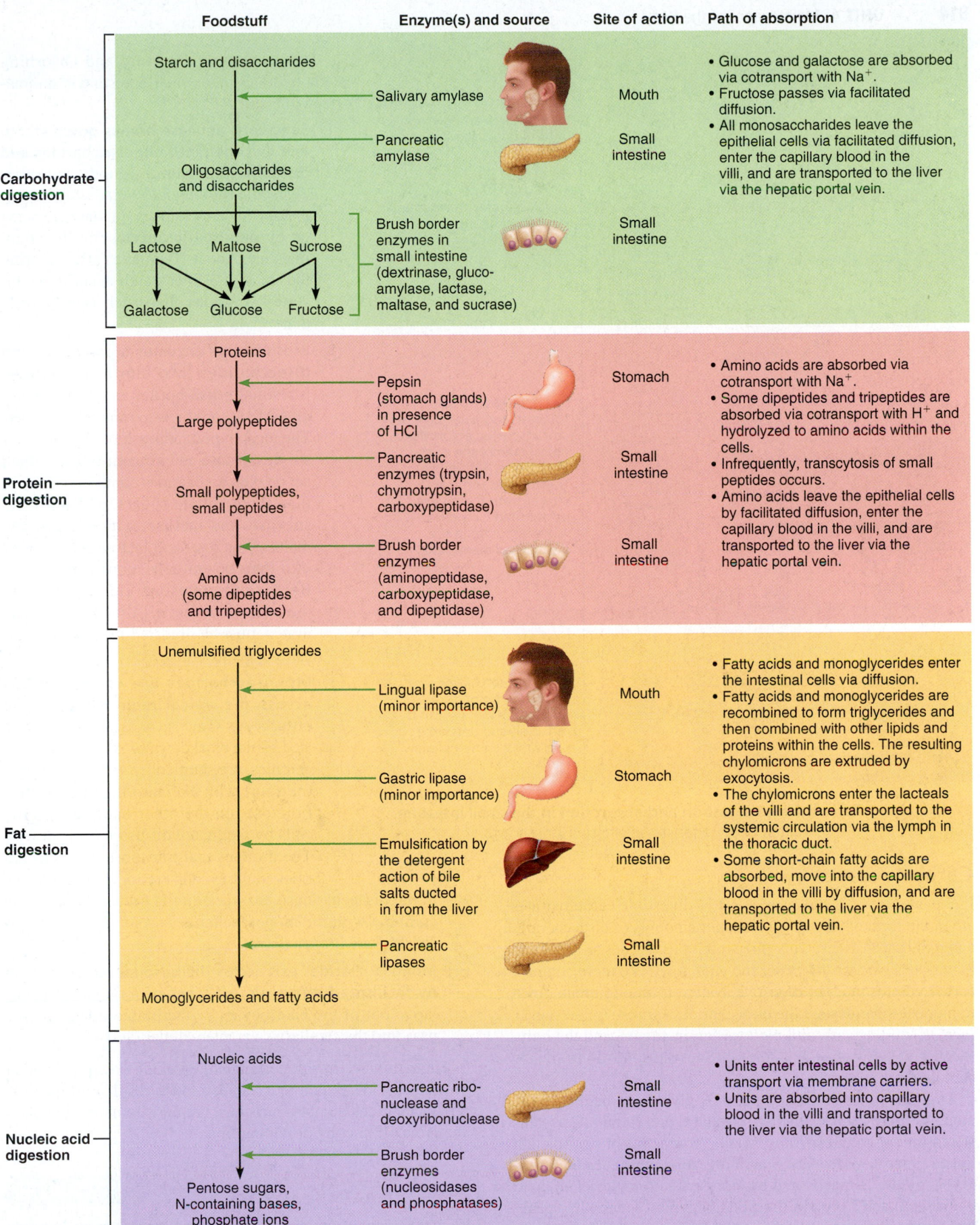

Carbohydrate digestion

Starch and disaccharides
→ Salivary amylase — Mouth
→ Pancreatic amylase — Small intestine

Oligosaccharides and disaccharides

Lactose   Maltose   Sucrose

Brush border enzymes in small intestine (dextrinase, gluco-amylase, lactase, maltase, and sucrase) — Small intestine

Galactose   Glucose   Fructose

- Glucose and galactose are absorbed via cotransport with Na$^+$.
- Fructose passes via facilitated diffusion.
- All monosaccharides leave the epithelial cells via facilitated diffusion, enter the capillary blood in the villi, and are transported to the liver via the hepatic portal vein.

Protein digestion

Proteins
→ Pepsin (stomach glands) in presence of HCl — Stomach

Large polypeptides
→ Pancreatic enzymes (trypsin, chymotrypsin, carboxypeptidase) — Small intestine

Small polypeptides, small peptides
→ Brush border enzymes (aminopeptidase, carboxypeptidase, and dipeptidase) — Small intestine

Amino acids (some dipeptides and tripeptides)

- Amino acids are absorbed via cotransport with Na$^+$.
- Some dipeptides and tripeptides are absorbed via cotransport with H$^+$ and hydrolyzed to amino acids within the cells.
- Infrequently, transcytosis of small peptides occurs.
- Amino acids leave the epithelial cells by facilitated diffusion, enter the capillary blood in the villi, and are transported to the liver via the hepatic portal vein.

Fat digestion

Unemulsified triglycerides
→ Lingual lipase (minor importance) — Mouth
→ Gastric lipase (minor importance) — Stomach
→ Emulsification by the detergent action of bile salts ducted in from the liver — Small intestine
→ Pancreatic lipases — Small intestine

Monoglycerides and fatty acids

- Fatty acids and monoglycerides enter the intestinal cells via diffusion.
- Fatty acids and monoglycerides are recombined to form triglycerides and then combined with other lipids and proteins within the cells. The resulting chylomicrons are extruded by exocytosis.
- The chylomicrons enter the lacteals of the villi and are transported to the systemic circulation via the lymph in the thoracic duct.
- Some short-chain fatty acids are absorbed, move into the capillary blood in the villi by diffusion, and are transported to the liver via the hepatic portal vein.

Nucleic acid digestion

Nucleic acids
→ Pancreatic ribo-nuclease and deoxyribonuclease — Small intestine
→ Brush border enzymes (nucleosidases and phosphatases) — Small intestine

Pentose sugars, N-containing bases, phosphate ions

- Units enter intestinal cells by active transport via membrane carriers.
- Units are absorbed into capillary blood in the villi and transported to the liver via the hepatic portal vein.

Figure 23.36 Flowchart of digestion and absorption of foodstuffs.

23

Figure 23.37 Carbohydrate digestion and absorption in the small intestine. Note that fructose enters enterocytes via facilitated diffusion (not shown).

1. Pancreatic amylase breaks down starch and glycogen into oligosaccharides and disaccharides.

2. Brush border enzymes break oligosaccharides and disaccharides into monosaccharides.

3. Monosaccharides (glucose and galactose) are cotransported across the apical membrane of the enterocyte. This active transport uses the Na$^+$ concentration gradient established by the Na$^+$-K$^+$ ATPase (pump) in the basolateral membrane.

4. Monosaccharides exit across the basolateral membrane by facilitated diffusion and enter the capillary via intercellular clefts.

Carbohydrates

In the average diet, most (up to 60%) digestible carbohydrates are in the form of starch, with smaller amounts of disaccharides and monosaccharides. Only three monosaccharides are common in our diet: *glucose, fructose,* and *galactose.* The more complex carbohydrates that our digestive system is able to break down to monosaccharides are the disaccharides *sucrose* (table sugar), *lactose* (milk sugar), and *maltose* (grain sugar), and the poly-saccharides *glycogen* and *starch.* (See ◄ pp. 43–44 to review carbohydrate structures.)

Digestion of starch (and perhaps glycogen) begins in the mouth (Figure 23.36). **Salivary amylase**, present in saliva, splits starch into *oligosaccharides*, smaller fragments of two to eight linked glucose molecules. Starch digestion continues until salivary amylase is inactivated by stomach acid and broken apart by the stomach's protein-digesting enzymes. Generally speaking, the larger the meal, the longer salivary amylase continues to work in the stomach because foodstuffs in its relatively immobile fundus are poorly mixed with gastric juices.

The process of digesting and absorbing carbohydrates in the small intestine is summarized in **Figure 23.37**.

1. **Pancreatic amylase breaks down starch and glycogen into oligosaccharides and disaccharides.** Starchy foods and other digestible carbohydrates that escape being broken down by salivary amylase are acted on by **pancreatic amylase** in the small intestine. About 10 minutes after entering the small intestine, starch is entirely converted to various oligosaccharides, mostly maltose.

2. **Brush border enzymes break oligo- and disaccharides into monosaccharides.** Intestinal brush border enzymes further digest these products to monosaccharides. The most important brush border enzymes are **dextrinase** and **glucoamylase**, which act on oligosaccharides composed of more than three simple sugars, and **maltase**, **sucrase**, and **lactase**, which hydrolyze maltose, sucrose, and lactose respectively into their constituent monosaccharides. Because the intestine can absorb only monosaccharides, all dietary carbohydrates must be digested to monosaccharides to be absorbed.

3. **Monosaccharides are cotransported across the apical membrane of the enterocyte.** Glucose and galactose, liberated by the breakdown of starch and disaccharides, are shuttled by secondary active transport with Na$^+$ into the enterocytes. Fructose, on the other hand, enters the cells by facilitated diffusion (not shown). The proteins that transport monosaccharides into the cells are located very close to the disaccharidase enzymes on the brush border. They combine with the monosaccharides as soon as the disaccharides are broken down.

4. **Monosaccharides exit across the basolateral membrane by facilitated diffusion.** All types of monosaccharides move out of the enterocytes by facilitated diffusion and pass into the capillaries via intercellular clefts.

Humans lack enzymes capable of breaking down most other polysaccharides, such as cellulose. As a result, indigestible polysaccharides do not nourish us, but they do help move food along the GI tract by providing fiber.

HOMEOSTATIC IMBALANCE 23.16 — CLINICAL

In some people, intestinal lactase is present at birth but then becomes deficient due to genetic factors. When people with *lactose intolerance* consume lactose, the undigested disaccharides create osmotic gradients that prevent water from being absorbed in the intestines and also pull water from the interstitial space into the intestines. The result is diarrhea. Bacterial metabolism of the undigested solutes produces large amounts of gas that result in bloating, flatulence, and cramping pain. For the most part, the solution to this problem is simple—add lactase enzyme "drops" to your milk or take a lactase tablet before consuming milk products.

Proteins

Proteins digested in the GI tract include:

- Dietary proteins (typically about 125 g per day)
- Enzyme proteins secreted into the GI tract by its various glands (15–25 g)
- Protein derived from sloughed and disintegrating mucosal cells

Healthy individuals digest much of this protein all the way to its **amino acid** monomers.

Protein digestion begins in the stomach when pepsinogen secreted by the chief cells is activated to **pepsin** (Figure 23.36). Pepsin functions optimally in the acidic pH range found in the stomach: 1.5–2.5. It preferentially cleaves bonds involving the amino acids tyrosine and phenylalanine, breaking the proteins down into polypeptides and free amino acids. Pepsin, which hydrolyzes 10–15% of ingested protein, is inactivated by the high pH in the duodenum, so its proteolytic activity is restricted to the stomach.

The process of digesting and absorbing proteins in the small intestine is summarized in **Figure 23.38**.

① **Pancreatic proteases break down proteins and protein fragments into smaller pieces and some individual amino acids.** Protein fragments entering the small intestine are greeted by a host of proteolytic enzymes. **Trypsin** and **chymotrypsin** cleave the proteins into smaller peptides. **Carboxypeptidases** split off one amino acid at a time from the end of the polypeptide chain that bears the carboxyl group.

② **Brush border enzymes break oligo- and dipeptides into amino acids.** A variety of brush border peptidases liberate individual amino acids from either end of a peptide chain (*carboxypeptidases* and *aminopeptidases*), while *dipeptidases* break pairs of amino acids apart. Carboxypeptidases and aminopeptidases can independently dismantle a protein, but the teamwork between these

enzymes and between trypsin and chymotrypsin, which attack the more internal parts of the protein, speeds up the process tremendously.

③ **Amino acids are cotransported across the apical membrane of the enterocyte.** Several types of carriers transport the different amino acids resulting from protein digestion. Most of these carriers, like those for glucose and galactose, are coupled to the active transport of sodium. Short chains of two or three amino acids (dipeptides and tripeptides, respectively) are also actively absorbed using H^+-dependent cotransport. They are digested to their amino acids within the enterocytes.

④ **Amino acids exit across the basolateral membrane via facilitated diffusion.** They then enter capillaries via intercellular clefts.

Figure 23.38 Protein digestion and absorption in the small intestine.

Lumen of intestine

Pancreatic proteases

Amino acids of protein fragments

Brush border enzymes

Na^+

Apical membrane (microvilli)

Na^+

Amino acid carrier

Basolateral membrane

Na^+ ATP

Na^+-K^+ ATPase

Na^+ K^+

Capillary

① Pancreatic proteases break down proteins and protein fragments into smaller pieces and some individual amino acids.

② Brush border enzymes break protein fragments into amino acids.

③ Amino acids are cotransported across the apical membrane of the enterocyte. This active transport uses the Na^+ concentration gradient established by the Na^+-K^+ ATPase (pump) in the basolateral membrane.

④ Amino acids exit across the basolateral membrane via facilitated diffusion and enter the capillary via intercellular clefts.

23

Lipids

Triglycerides are the most abundant fats in the diet (◀ p. 45). The small intestine is the primary site of lipid digestion because the pancreas is the major source of fat-digesting enzymes, or **lipases** (see Figure 23.36).

The steps of lipid digestion and absorption in the small intestine are shown in **Figure 23.39**.

① **Emulsification.** Because triglycerides and their breakdown products are insoluble in water, fats need special "pretreatment" with bile salts to be digested in the watery environment of the small intestine. In aqueous solutions, triglycerides aggregate to form large fat globules, and only the triglyceride molecules at the surfaces of such fatty masses are accessible to the water-soluble lipase enzymes. However, bile salts vastly increase the surface area exposed to pancreatic lipases by breaking large fat globules into many smaller droplets. Without bile, lipids could not be completely digested during the time food spends in the small intestine.

Bile salts have both nonpolar and polar regions. Their nonpolar (hydrophobic) parts cling to the fat molecules, and their polar (ionized hydrophilic) parts allow them to repel each other and interact with water. As a result, fatty droplets are pulled off the large fat globules, forming a stable *emulsion*—an aqueous suspension of fatty droplets, each about 1 μm in diameter. Emulsification does *not* break chemical bonds. It just reduces the attraction between fat molecules so they can be more widely dispersed, just as dish detergent breaks up a pool of fat drippings.

② **Digestion.** Pancreatic lipases catalyze the breakdown of triglycerides by splitting off two of the fatty acid chains, yielding free **fatty acids** and **monoglycerides** (glycerol with one fatty acid chain attached).

③ **Micelle formation.** Just as bile salts accelerate lipid digestion, they are also essential for the *absorption* of its end products. As the water-insoluble products of fat digestion—the monoglycerides and free fatty acids—are liberated by lipase activity, they quickly become associated with bile salts and *lecithin* (a phospholipid found in bile) to form micelles. **Micelles** (mi-selz′) are collections of fatty elements clustered together with bile salts in such a way that the polar (hydrophilic) ends of the molecules face the water and the nonpolar portions form the core. Also nestled in the hydrophobic core are cholesterol molecules and fat-soluble vitamins. Although micelles are similar to emulsion droplets, they are about 500 times smaller and easily diffuse between microvilli to come into close contact with the apical cell surface. Without micelles, the lipids would simply float on the surface of the chyme (like oil on water), inaccessible to the absorptive surfaces of the enterocytes.

④ **Diffusion.** Upon reaching the enterocytes, the various lipid substances leave the micelles and move through the apical plasma membrane by simple diffusion.

⑤ **Chylomicron formation.** Once the free fatty acids and monoglycerides enter the enterocytes, the smooth ER

Fat globule (in small intestine)

Bile salts

Fat droplets coated with bile salts

Enterocytes of small intestine

Lacteal

① **Emulsification.** Bile salts in the duodenum break large fat globules into smaller fat droplets, increasing the surface area available to lipase enzymes.

② **Digestion.** Pancreatic lipases hydrolyze triglycerides, yielding monoglycerides and free fatty acids.

③ **Micelle formation.** Free fatty acids and monoglycerides assemble with bile salts, forming micelles. Micelles ferry their contents to enterocytes.

④ **Diffusion.** Fatty acids and monoglycerides diffuse from micelles into enterocytes.

⑤ **Chylomicron formation.** Fatty acids and monoglycerides are recombined and packaged with other fatty substances and proteins to form chylomicrons.

⑥ **Chylomicron transport.** Chylomicrons are extruded from enterocytes by exocytosis, enter lacteals, and are carried away from the intestine in lymph.

Figure 23.39 Emulsification, digestion, and absorption of fats.

converts them back into triglycerides. The triglycerides are then combined with lecithin and other phospholipids and cholesterol, and coated with a "skin" of proteins to form water-soluble lipoprotein droplets called **chylomicrons** (ki″-lo-mi′kronz). This series of events is quite different from the absorption of amino acids and simple sugars, which pass through the enterocytes unchanged.

⑥ **Chylomicron transport.** The milky-white chylomicrons are too large to pass through either the plasma membrane of the enterocyte or the basement membrane of a blood capillary. Instead, the chylomicron-containing vesicles migrate to the basolateral membrane and are extruded by exocytosis. They then enter the more permeable lacteals (◀ p. 769).

Thus, most fat enters the lymphatic stream for distribution in the lymph. Eventually the chylomicrons are emptied into the venous blood via the thoracic duct, which drains the lymphatics of the digestive viscera.

While in the bloodstream, the triglycerides of the chylomicrons are hydrolyzed to free fatty acids and glycerol by **lipoprotein lipase**, an enzyme associated with capillary endothelium. The fatty acids and glycerol can then pass through the capillary walls to be used by tissue cells for energy or stored as fats in adipose tissue. Liver cells then endocytose and process the residual chylomicron material.

Passage of short-chain fatty acids is quite different from what we have just described. These fat breakdown products do not depend on the presence of bile salts or micelles and are not recombined to form triglycerides within the intestinal cells. They simply diffuse into the portal blood for distribution.

Generally, fat absorption is completed in the ileum, but in the absence of bile (as might occur when a gallstone blocks the cystic duct), it happens so slowly that most of the fat passes into the large intestine and is lost in feces.

Nucleic Acids

The nuclei of the cells of ingested foods contain DNA and RNA. **Pancreatic nucleases** in pancreatic juice hydrolyze the nucleic acids to their **nucleotide** monomers. Intestinal brush border enzymes (**nucleosidases** and **phosphatases**) then break the nucleotides apart to release their nitrogenous bases, pentose sugars, and phosphate ions (see Figure 23.36).

Special carriers in the epithelium of the villi actively transport the breakdown products of nucleic acid digestion across the epithelium. These then enter the blood.

Absorption of Vitamins, Electrolytes, and Water

Vitamin Absorption

The small intestine absorbs dietary vitamins, and the large intestine absorbs some of the K and B vitamins made by its gut bacterial "guests." As we already noted, fat-soluble vitamins (A, D, E, and K) dissolve in dietary fats, become incorporated into the micelles, and move across the villus epithelium passively (by diffusion). It follows that gulping pills containing fat-soluble vitamins without simultaneously eating some fat-containing food results in little or no absorption of these vitamins.

Most water-soluble vitamins (B vitamins and vitamin C) are absorbed via specific active or passive transporters. The exception is vitamin B_{12}, which is a very large, charged molecule. *Intrinsic factor*, produced by the stomach (◄ **p. 889**), binds to vitamin B_{12}. The vitamin B_{12}–intrinsic factor complex then binds to specific mucosal receptor sites in the terminal ileum, which trigger its active uptake by endocytosis.

Electrolyte Absorption

Absorbed electrolytes come from both ingested foods and gastrointestinal secretions. Most ions are actively absorbed along the entire length of the small intestine. But absorption of iron and calcium is largely limited to the duodenum.

For most nutrients, the amount *reaching* the intestine is the amount absorbed, regardless of the nutritional state of the body. In contrast, absorption of iron and calcium is intimately related to the body's need for them at the time.

Sodium, Chloride, and Bicarbonate   As we mentioned earlier, absorption of sodium ions in the small intestine is coupled to active absorption of glucose and amino acids. For the most part, anions passively follow the electrical potential established by sodium transport. In other words, Na^+ is actively pumped out of the enterocytes by a Na^+-K^+ pump after entering those cells. Usually, chloride ions passively follow Na^+. In the terminus of the small intestine, HCO_3^- is actively secreted into the lumen in exchange for Cl^-.

Potassium   Potassium ions move across the intestinal mucosa passively by facilitated diffusion (or leaky tight junctions). As water is absorbed from the lumen, rising potassium levels in chyme create a concentration gradient for its absorption. Anything that interferes with water absorption (resulting in diarrhea) not only reduces potassium absorption but also "pulls" K^+ from the interstitial space into the intestinal lumen.

Iron   Ionic iron, essential for hemoglobin production, is actively transported into the mucosal cells, where it binds to the protein **ferritin** (fer′ĭ-tin). The intracellular iron-ferritin complexes then serve as local storehouses for iron. When body reserves of iron are adequate, only 10–20% is allowed to pass into the portal blood, and most of the stored iron is lost as the enterocytes later slough off. However, when iron reserves are depleted (as during acute or chronic hemorrhage), iron uptake from the intestine and its release to the blood accelerate. In the blood, iron binds to **transferrin**, a plasma protein that transports it in the circulation.

Menstrual bleeding is a major route of iron loss in females, and premenopausal women require about 50% more iron in their diets. The intestinal enterocytes of women have about four times as many iron transport proteins as do those of men, and little iron is lost from the body other than that lost in menses.

Calcium   Calcium absorption is closely related to blood levels of ionic calcium. The active form of **vitamin D** promotes active calcium absorption. Decreased blood levels of ionic calcium prompt *parathyroid hormone (PTH)* release from the parathyroid glands (◄ **p. 622**). Besides facilitating the release of calcium ions from bone matrix and enhancing the reabsorption of calcium by the kidneys, PTH stimulates activation of vitamin D to calcitriol by the kidneys, which in turn accelerates calcium ion absorption in the small intestine.

Water Absorption

Approximately 9 L of water, mostly derived from GI tract secretions, enter the small intestine daily. Water is the most abundant substance in chyme, and 95% of it is absorbed in the small intestine by osmosis. Most of the rest is absorbed in the large intestine, leaving only about 0.1 L to soften the feces.

The normal rate of water absorption is 300 to 400 ml per hour. Water moves freely in both directions across the intestinal mucosa, but *net osmosis* occurs whenever a concentration gradient is established by the active transport of solutes (particularly Na⁺) into the mucosal cells. In this way, water uptake is effectively coupled to solute uptake and, in turn, affects the absorption of substances that normally pass by diffusion. As water moves into mucosal cells, these substances follow along their concentration gradients.

HOMEOSTATIC IMBALANCE 23.17

CLINICAL

Malabsorption, or impaired nutrient absorption, has many and varied causes. It can result from anything that interferes with the delivery of bile or pancreatic juice to the small intestine. Factors that damage the intestinal mucosa (severe bacterial infections and some antibiotics) or reduce its absorptive surface area are also common causes.

A common malabsorption syndrome is *gluten-sensitive enteropathy* or *celiac disease*, which affects one in 100 people. This chronic genetic condition is caused by an immune reaction to gluten, a protein plentiful in all grains but corn and rice. Breakdown products of gluten interact with molecules of the immune system in the GI tract, forming complexes. These complexes activate T cells, which then attack the intestinal lining, damaging intestinal villi and reducing the surface area of the brush border. Bloating, diarrhea, pain, and malnutrition result.

The usual treatment is to eliminate gluten-containing grains from the diet. In recent years, many gluten-free products have become available.

●●●●●●●●

Check Your Understanding

40. Fill in the blank: Amylase is to starch as ___ is to fats.

41. **WHAT IF?** Suppose you had a patient that was unable to make bile salts. What digestive processes would be affected and what might be some of the consequences?

For answers, see Answers Appendix.

Developmental Aspects of the Digestive System

As previously described, the very young embryo is flat and consists of three germ layers. From top to bottom, they are ectoderm, mesoderm, and endoderm. This flattened cell mass soon folds to form a cylindrical body, and its internal cavity becomes the cavity of the alimentary canal, which is initially closed at both ends.

The epithelial lining of the developing alimentary canal, or **primitive gut**, forms from endoderm (**Figure 23.40a**). The rest of the wall arises from mesoderm. The anteriormost endoderm (that of the foregut) touches a depressed area of the surface ectoderm called the **stomodeum** (sto″mo-de′um). The two membranes fuse, forming the **oral membrane**, which soon breaks through to form the opening of the mouth. Similarly, the end of the hindgut fuses with an ectodermal depression, the **proctodeum** (*procto* = anus), to form the **cloacal membrane** (*cloaca* = sewer), which then breaks through to form the anus.

By week 5, the alimentary canal is a continuous "tube" extending from mouth to anus and opens to the external environment at each end. Shortly after, the glandular organs (salivary glands,

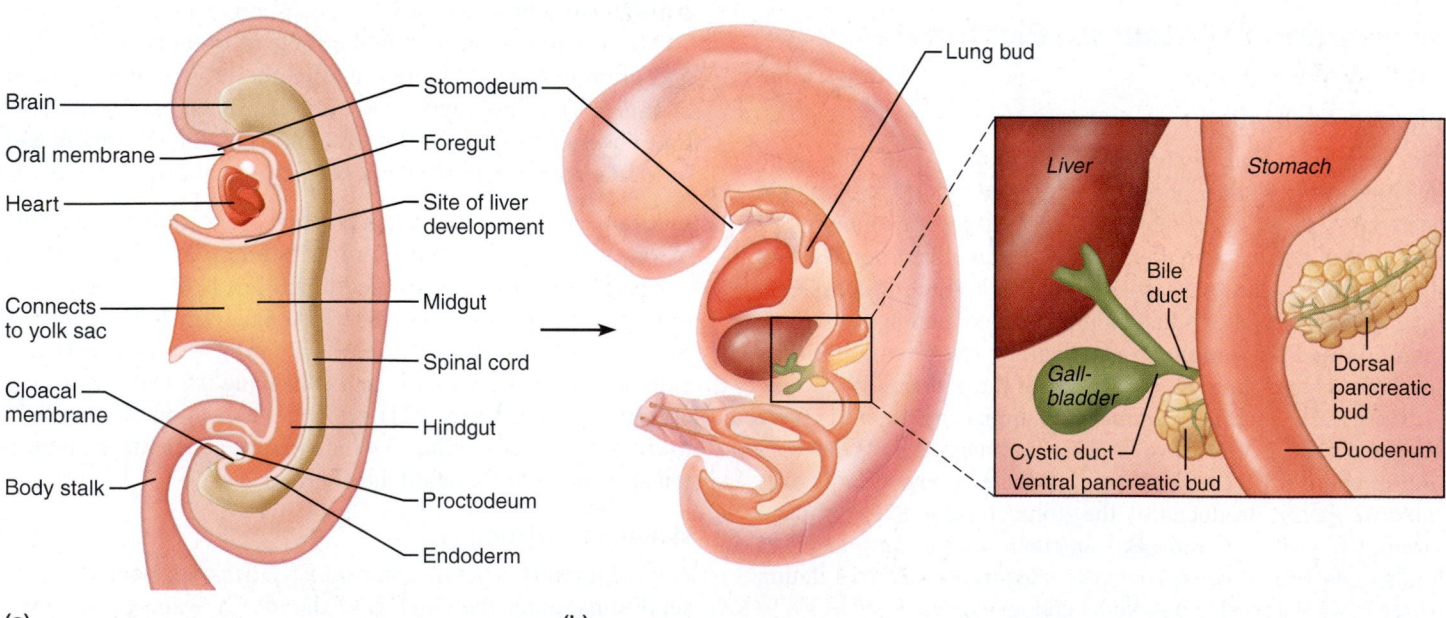

(a)   (b)

Figure 23.40 Embryonic development of the digestive system. (a) Three-week-old embryo. The endoderm has folded, and the foregut and hindgut have formed. (The midgut is still open and continuous with the yolk sac.) **(b)** By five weeks of development, the accessory organs are budding out from the endodermal layer, as shown in the enlargement.

liver with gallbladder, and pancreas) bud out from the mucosa at various points (Figure 23.40b). These glands retain their connections, which become ducts leading into the digestive tract.

HOMEOSTATIC IMBALANCE 23.18   **CLINICAL**

The digestive system is susceptible to many congenital defects that interfere with feeding. The most common are **cleft palate**, in which the palatine bones or palatine processes of the maxillae (or both) fail to fuse, and **cleft lip**, which often occur together (◀ **Figure 7.39, p. 246**). Of the two, cleft palate is far more serious because the child is unable to suck properly.

Another common defect is *tracheoesophageal fistula*, in which there is an opening between the esophagus and the trachea, and the esophagus often lacks a connection to the stomach. The baby chokes and becomes cyanotic during feedings because food enters the respiratory passageways. Surgery can usually correct these defects.

Cystic fibrosis (described in more detail in Chapter 22, ◀ **p. 861**) primarily affects the lungs, but it also impairs the activity of the pancreas. In this genetic disease, the mucous glands produce abnormally thick mucus, which blocks ducts and passageways of organs. Blockage of the pancreatic duct prevents pancreatic juice from reaching the small intestine. This impairs digestion, and most fats and fat-soluble vitamins are not digested or absorbed. Consequently the stools are bulky and fat laden. The pancreatic problems can be handled by administering pancreatic enzymes with meals. •••••••

During fetal life, the developing infant receives all of its nutrients through the placenta. Nonetheless, the fetal GI tract is "trained" in utero for future food digestion as the fetus naturally swallows some of the surrounding amniotic fluid. This fluid contains several chemicals that stimulate GI maturation, including gastrin and epidermal growth factor.

Digestive System after Birth

Feeding is a newborn baby's most important activity, and several reflexes enhance the infant's ability to obtain food: The *rooting reflex* helps the infant find the nipple, and the *sucking reflex* helps the baby hold onto the nipple and swallow.

Newborn babies tend to double their birth weight within six months, and their caloric intake and food processing ability are extraordinary. For example, a 6-week-old infant weighing about 4 kg (less than 9 lb) drinks about 600 ml of milk daily. A 65-kg adult (143 lb) would have to drink 10 L of milk to ingest a corresponding volume of fluid! However, the stomach of a newborn infant is very small, so feeding must be frequent (every 3–4 hours). Peristalsis is inefficient, and vomiting is not unusual. As the teeth break through the gums, the infant progresses to solid foods and is usually eating an adult diet by the age of 2 years.

As a rule, the digestive system operates throughout childhood and adulthood with relatively few problems. However, contaminated food or extremely spicy or irritating foods sometimes cause **gastroenteritis** (gas″tro-en″tĕ-ri′tis), inflammation of the GI tract. Ulcers and gallbladder problems—inflammation or **cholecystitis** (ko″le-sis-ti′tis), and gallstones—are problems of middle age.

Aging and the Digestive System

During old age, GI tract activity declines. Fewer digestive juices are produced, absorption is less efficient, and peristalsis slows. The result is less frequent bowel movements and, often, constipation. Taste and smell are less acute, and periodontal disease often develops. Many elderly people live alone or on a reduced income. These factors, along with increasing physical disability, tend to make eating less appealing, and many of our elderly citizens are poorly nourished.

Diverticulosis, fecal incontinence, and cancer of the GI tract are fairly common problems of the aged. Stomach and colon cancers rarely have early signs, and often metastasize before a person seeks medical attention. (*A Closer Look*, ◀ **p. 144**, describes the development of colorectal cancer.) Should metastasis occur, secondary cancer of the liver is almost guaranteed because of the "detour" the splanchnic venous blood takes through the liver via the hepatic portal circulation.

However, when detected early, most GI tract cancers are treatable. The best advice is to have regular dental and medical checkups. Most oral cancers are detected during routine dental examinations, 50% of all rectal cancers can be felt digitally, and nearly 80% of colon cancers can be seen and removed during a colonoscopy. Screening for colon cancer is recommended after age 50.

23

RELATED CLINICAL TERMS

Achalasia (a-kah-la′ze-ah) A disorder in which swallowing is hindered or prevented. Botox injections can relax the esophageal sphincter.

Ascites (ah-si′tēz; *asci* = bag, bladder) Abnormal accumulation of fluid within the peritoneal cavity; if excessive, causes visible bloating of the abdomen. May result from portal hypertension caused by liver cirrhosis or by heart or kidney disease.

Barrett's esophagus A pathological change in the epithelium of the lower esophagus from stratified squamous to a columnar epithelium. A possible sequel to untreated chronic gastroesophageal reflux due to hiatal hernia, it predisposes the individual to aggressive esophageal cancer (adenocarcinoma).

Bruxism (bruck′sizm) Grinding or clenching of teeth, usually at night during sleep in response to stress. Can wear down and crack the teeth.

Bulimia (bu-lim′e-ah; *bous* = ox; *limos* = hunger) Binge-purge behavior—episodes of overeating followed by purging (self-induced vomiting, taking laxatives or diuretics, or excessive exercise). Most common in women of high school or college age. Often associated with stress and depression. Consequences include eroded tooth enamel, stomach trauma or rupture (from vomiting), and severe electrolyte disturbances, which impair heart activity. Therapy includes hospitalization to control behavior, and nutritional counseling.

Homeostatic Interrelationships between the Digestive System and Other Body Systems

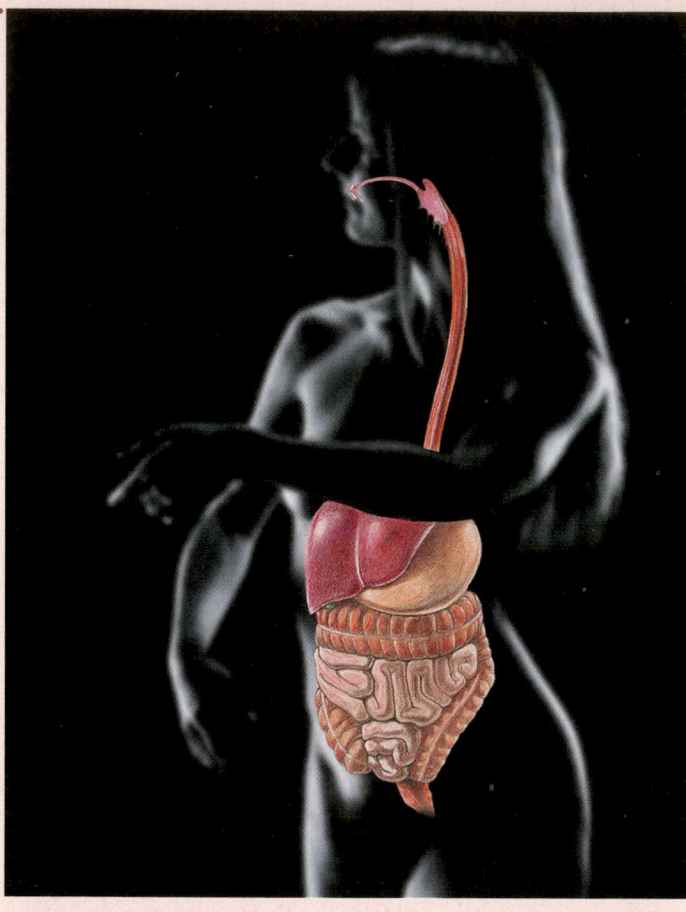

- Nervous system controls digestive function; in general, parasympathetic fibers accelerate and sympathetic fibers inhibit digestive activity; reflex and voluntary controls of defecation

Endocrine System   Chapter 16

- Liver removes hormones from blood, ending their activity; digestive system provides nutrients needed for energy, growth, and repair; pancreas, stomach, and small intestine have hormone-producing cells
- Hormones and paracrines help regulate digestive function

Cardiovascular System   Chapters 17–19

- Digestive system provides nutrients to heart and blood vessels; absorbs iron needed for hemoglobin synthesis; absorbs water necessary for normal plasma volume; liver secretes bilirubin resulting from hemoglobin breakdown in bile and stores iron for reuse
- Cardiovascular system transports nutrients absorbed by alimentary canal to all tissues of body; distributes hormones of the digestive tract

Lymphatic System/Immunity   Chapters 20–21

- Digestive system provides nutrients for normal functioning; lysozyme of saliva and HCl of stomach provide nonspecific protection against bacteria
- Lacteals drain fatty lymph from digestive tract organs and convey it to blood; Peyer's patches and lymphoid tissue in mesentery house macrophages and immune cells that protect digestive tract organs against infection; plasma cells provide IgA in digestive tract secretions

Integumentary System   Chapter 5

- Digestive system provides nutrients needed for energy, growth, and repair; supplies fats that provide insulation in the dermis and subcutaneous tissue
- The skin synthesizes vitamin D needed for calcium absorption from the intestine; protects by enclosure

Skeletal System   Chapters 6–8

- Digestive system provides nutrients needed for energy, growth, and repair; absorbs calcium needed for bone salts
- Skeletal system protects some digestive organs by bone; bones and bone cavities store some nutrients (e.g., calcium, fats)

Muscular System   Chapters 9–10

- Digestive system provides nutrients needed for energy, growth, and repair; liver removes lactic acid, resulting from muscle activity, from the blood
- Skeletal muscle activity increases motility of GI tract

Nervous System   Chapters 11–15

- Digestive system provides nutrients needed for normal neural functioning; nutrient signals influence neural regulation of satiety

Respiratory System   Chapter 22

- Digestive system provides nutrients needed for energy, growth, and repair
- Respiratory system provides oxygen to and disposes of carbon dioxide produced by digestive system organs

Urinary System   Chapters 25–26

- Digestive system provides nutrients for energy, growth, and repair
- Kidneys transform vitamin D to its active form, which is needed for calcium absorption

Reproductive System   Chapter 27

- Digestive system provides nutrients for energy, growth, and repair and extra nutrition needed to support fetal growth

Dysphagia (dis-fa′je-ah; *dys* = difficult, abnormal; *phag* = eat) Difficulty swallowing; usually due to obstruction or physical trauma to the esophagus.

Endoscopy (en-dos′ko-pe; *endo* = within, inside; *scopy* = viewing) Visual examination of a ventral body cavity or the interior of a visceral organ with a flexible tubelike device called an endoscope, which contains a light source and a lens. A general term for a colonoscopy (viewing the colon), sigmoidoscopy (viewing the sigmoid colon), etc.

Enteritis (*enteron* = intestine) Inflammation of the intestine, especially the small intestine.

Hemochromatosis (he″mo-kro″mah-to′sis; *hemo* = blood; *chroma* = color; *osis* = condition of) A disorder in iron metabolism due to excessive/prolonged iron intake or a breakdown of the mucosal iron barrier; excess iron is deposited in the tissues, increasing skin pigmentation and the risk of hepatic cancer and liver cirrhosis. Also called bronze diabetes and iron storage disease.

Ileus (il′e-us) A condition in which all GI tract movement stops and the gut appears to be paralyzed. Can result from electrolyte imbalances and blockade of parasympathetic impulses by drugs (such as those commonly used during abdominal surgery); usually reversed when these interferences end. The reappearance of intestinal sounds (gurgling, etc.) indicates restoration of motility.

Inflammatory bowel disease (IBD) A noncontagious, periodic inflammation of the intestinal wall now understood to be an abnormal immune and inflammatory response to bacterial antigens that normally occur in the intestine. Linked to T_H17 cells, certain cytokines, and the loss of a normal epithelial barrier. Afflicts up to two of every 1000 people. Symptoms include cramping, diarrhea, weight loss, and intestinal bleeding. Two subtypes occur: (1) Crohn's disease is a syndrome characterized by relapsing and remitting periods. Deep ulcers and fissures can develop anywhere along the GI tract, but mostly occur in the terminal ileum. (2) Ulcerative colitis is characterized by inflammation of the large-intestinal mucosa, mainly in the rectum. Both types are treated with anti-inflammatory and immunosuppressant drugs and sometimes probiotic agents and antibiotics. Extremely severe cases of ulcerative colitis are treated by colectomy (removal of a portion of the colon).

Laparoscopy (lap″ah-ros′ko-pe; *lapar* = the flank; *scopy* = observation) Examination of the peritoneal cavity and its organs with an endoscope inserted through the anterior abdominal wall. Often used to assess the condition of the digestive organs and the pelvic reproductive organs of females.

Mumps An inflammation of the parotid glands caused by the mumps virus (myxovirus). A once-common childhood illness now prevented by vaccination with the MMR (measles, mumps, rubella) vaccine. In 25% of mumps cases in adult males, the testes are also infected, which can lead to sterility.

Orthodontics (or″tho-don′tiks; *ortho* = straight) Branch of dentistry that prevents and corrects misaligned teeth.

Pancreatitis (pan″kre-ah-ti′tis) A rare but extremely serious inflammation of the pancreas. May result from excessively high levels of fat in the blood or excessive alcohol ingestion, but most acute cases arise from gallstones that block the bile duct. Pancreatic enzymes are activated in the pancreatic duct, causing the pancreatic tissue and duct to be digested (literally, eaten from within). This painful condition can lead to nutritional deficiencies because pancreatic enzymes are essential to food digestion in the small intestine.

Peptic ulcers Term referring to gastric and duodenal ulcers.

Proctology (prok-tol′o-je; *procto* = rectum, anus; *logy* = study of) Branch of medicine dealing with treatment of diseases of the colon, rectum, and anus.

Pyloric stenosis (pi-lor′ik stĕ-no′sis; *stenosis* = narrowing, constriction) Congenital abnormality in which the pyloric sphincter is abnormally constricted. There is usually no problem until the baby begins to take solid food, and then projectile vomiting begins. Corrected surgically.

Vagotomy (va-got′o-me) Cutting or severing of the vagus nerve to decrease secretion of gastric juice in those with peptic ulcers that do not respond to medication.

CHAPTER SUMMARY

PART 1
OVERVIEW OF THE DIGESTIVE SYSTEM

1. The digestive system includes organs of the alimentary canal (mouth, pharynx, esophagus, stomach, small and large intestines) and accessory digestive organs (teeth, tongue, salivary glands, liver, gallbladder, and pancreas).

23.1 What major processes occur during digestive system activity? (pp. 870–871)

1. Digestive system activities include six processes: ingestion (food intake); propulsion (movement of food through the tract); mechanical breakdown (processes that physically mix or break foods down into smaller fragments); digestion (food breakdown by enzymatic action); absorption (transport of products of digestion through the intestinal mucosa into the blood); and defecation (elimination of the undigested residues [feces] from the body).

23.2 The GI tract has four layers and is usually surrounded by peritoneum (pp. 871–874)

1. The parietal and visceral layers of the peritoneum are continuous with one another via several extensions (mesenteries, falciform ligament, lesser and greater omenta), and are separated by a potential space containing serous fluid, which decreases friction during organ activity.

2. All organs of the GI tract have the same basic pattern of tissue layers in their walls; all have a mucosa, submucosa, muscularis externa, and serosa (or adventitia). Intrinsic nerve plexuses (enteric nervous system) are found within the wall.

3. The digestive viscera are served by the splanchnic circulation, consisting of arterial branches of the celiac trunk and aorta and the hepatic portal circulation.

23.3 The GI tract has its own nervous system called the enteric nervous system (pp. 874–875)

1. The digestive system controls the environment within its lumen to ensure optimal conditions for digestion and absorption of foodstuffs.
2. Receptors and hormone-secreting cells in the alimentary canal wall respond to stretch and chemical signals that result in stimulation or inhibition of GI secretory activity or motility. The alimentary canal has an intrinsic nerve supply.

PART 2

FUNCTIONAL ANATOMY OF THE DIGESTIVE SYSTEM

23.4 Ingestion occurs only at the mouth (pp. 876–881)

1. Food enters the GI tract via the mouth, which is continuous with the oropharynx posteriorly. The boundaries of the mouth are the lips and cheeks, palate, and tongue.
2. The oral mucosa is stratified squamous epithelium, an adaptation seen where abrasion occurs.
3. The tongue is mucosa-covered skeletal muscle. Its intrinsic muscles allow it to change shape; its extrinsic muscles allow it to change position.
4. Saliva is produced by many minor salivary glands and three pairs of major salivary glands—parotid, submandibular, and sublingual—that secrete their product into the mouth via ducts. Largely water, saliva also contains ions, proteins, metabolic wastes, lysozyme, defensins, IgA, salivary amylase, and mucin.
5. Saliva moistens and cleanses the mouth; moistens foods, aiding their compaction; dissolves food chemicals to allow for taste; and begins digestion of starch (salivary amylase). Saliva output is increased by parasympathetic reflexes initiated by activation of chemical and pressure receptors in the mouth and by the sight or smell of food. The sympathetic nervous system depresses salivation.
6. The 20 deciduous teeth begin to be shed at the age of 6 and are gradually replaced during childhood and adolescence by the 32 permanent teeth.
7. Teeth are classed as incisors, canines, premolars, and molars. Each tooth has an enamel-covered crown and a cement-covered root. The bulk of the tooth is dentin, which surrounds the central pulp cavity. A periodontal ligament secures the tooth to the bony alveolus.
8. The mouth and associated accessory organs accomplish food ingestion and mechanical breakdown (chewing and mixing), initiate the digestion of starch (salivary amylase), and propel food into the pharynx (buccal phase of swallowing).
9. Teeth masticate food. Chewing is initiated voluntarily and then controlled reflexively.

23.5 The pharynx and esophagus move food from the mouth to the stomach (pp. 881–884)

1. Food propelled from the mouth passes through the oropharynx and laryngopharynx. The mucosa of the pharynx is stratified squamous epithelium; skeletal muscles in its wall (constrictor muscles) move food toward the esophagus.
2. The esophagus extends from the laryngopharynx and joins the stomach at the cardial orifice, which is surrounded by the gastroesophageal sphincter.
3. The esophageal mucosa is stratified squamous epithelium. Its muscularis is skeletal muscle superiorly and changes to smooth muscle inferiorly. It has an adventitia rather than a serosa.
4. The pharynx and esophagus are primarily food conduits that conduct food to the stomach by peristalsis.

5. Swallowing is initiated by the mouth (the buccal phase) after the tongue has compacted food and saliva into a bolus. The pharyngeal-esophageal phase is controlled reflexively by the swallowing center in the medulla and pons. When the peristaltic wave approaches the gastroesophageal sphincter, the sphincter relaxes to allow food to enter the stomach.

23.6 The stomach temporarily stores food and begins protein digestion (pp. 884–893)

1. The J-shaped stomach lies in the upper left quadrant of the abdomen. Its major regions are the cardia, fundus, body, and pyloric part. When empty, its internal surface exhibits rugae.
2. The stomach muscularis contains a third (oblique) layer of smooth muscle.
3. The stomach mucosa is simple columnar epithelium dotted with gastric pits that lead into gastric glands. Secretory cells in the gastric glands include pepsinogen-producing chief cells; parietal cells, which secrete hydrochloric acid and intrinsic factor; mucous neck cells, which produce mucus; and enteroendocrine cells, which secrete hormones and paracrines.
4. The mucosal barrier, which protects the stomach from self-digestion and HCl, reflects the fact that the mucosal cells are connected by tight junctions, secrete a thick mucus, and are quickly replaced when damaged.
5. Protein digestion is initiated in the stomach by activated pepsin and requires acidic conditions (provided by HCl). Few substances are absorbed in the stomach.
6. Both nervous and hormonal factors control gastric secretory activity. The three phases of gastric secretion are cephalic, gastric, and intestinal. Most food-related stimuli acting on the head and stomach (cephalic and gastric, respectively) stimulate gastric secretion. Most stimuli acting on the small intestine trigger the enterogastric reflex and release of secretin and CCK, both of which inhibit gastric secretory activity. Sympathetic activity also inhibits gastric secretion.
7. Mechanical breakdown in the stomach is triggered by stomach distension and coupled to food propulsion and stomach emptying. Food movement into the duodenum is controlled by the pylorus and feedback signals from the small intestine. Pacemaker cells in the smooth muscle sheet set the rate of peristalsis.

23.7 The liver secretes bile; the pancreas secretes digestive enzymes (pp. 893–900)

1. The liver is a lobed organ overlying the stomach. Its digestive role is to produce bile, which it secretes into the common hepatic duct.
2. The structural and functional units of the liver are the liver lobules. Blood flowing to the liver via the hepatic artery proper and hepatic portal vein flows into its sinusoids, from which stellate macrophages remove debris and hepatocytes remove nutrients. Hepatocytes store glucose as glycogen, use amino acids to make plasma proteins, and detoxify metabolic wastes and drugs.
3. Bile is made continuously by the hepatocytes. Bile salts and secretin stimulate bile production.
4. The gallbladder, a muscular sac that lies beneath the right liver lobe, stores and concentrates bile.
5. Bile contains electrolytes, a variety of fatty substances, bile salts, and bile pigments in an aqueous medium. Bile salts are emulsifying agents; they disperse fats and form water-soluble micelles, which solubilize the products of fat digestion.
6. The pancreas is retroperitoneal between the spleen and small intestine. Its exocrine product, pancreatic juice, is carried to the duodenum via the pancreatic duct.

7. The bile duct and pancreatic duct join to form the hepatopancreatic ampulla and empty their secretions into the duodenum through the hepatopancreatic sphincter.

8. Pancreatic juice is a HCO_3^--rich fluid containing enzymes that digest all categories of foods. Intestinal hormones and the vagus nerves control secretion of pancreatic juice.

9. Cholecystokinin released by the small intestine stimulates the gallbladder to contract and the hepatopancreatic sphincter to relax, allowing bile (and pancreatic juice) to enter the duodenum.

23.8 The small intestine is the major site for digestion and absorption (pp. 900–906)

1. The small intestine extends from the pyloric sphincter to the ileocecal valve. Its three subdivisions are the duodenum, jejunum, and ileum.

2. Circular folds, villi, and microvilli increase the intestinal surface area for digestion and absorption.

3. The duodenal submucosa contains elaborate mucus-secreting duodenal glands. The mucosa of the ileum contains Peyer's patches (aggregated lymphoid nodules). The duodenum is covered not with a serosa but with an adventitia.

4. Intestinal juice is largely water with mucus. The major stimuli for its release are hypertonic and acidic chyme.

5. Segmentation is the principal motility pattern of the small intestine after a meal. It promotes mechanical breakdown and mixing of chyme with digestive juices and bile. Together with short-distance peristalsis, it also causes slow propulsion. Pacemaker cells set the rate of segmentation. Ileocecal valve opening is controlled by the gastroileal reflex and gastrin.

6. Between meals, the basic motor pattern is the migrating motor complex (a type of peristalsis) that sweeps any remaining contents of the small intestine toward the large intestine.

23.9 The large intestine absorbs water and eliminates feces (pp. 906–911)

1. The subdivisions of the large intestine are the cecum (and appendix), colon (ascending, transverse, descending, and sigmoid portions), rectum, and anal canal. It opens to the body exterior at the anus.

2. The longitudinal muscle in the muscularis is reduced to three bands (teniae coli), which pucker its wall, producing haustra. The mucosa of most of the large intestine is simple columnar epithelium containing abundant goblet cells.

3. The major functions of the large intestine are absorption of water, some electrolytes, and vitamins made by gut bacteria, and defecation (evacuation of food residues from the body).

4. The defecation reflex is triggered when feces enter the rectum. It involves parasympathetic reflexes leading to contraction of the rectal walls and is aided by the Valsalva maneuver.

PART 3
PHYSIOLOGY OF DIGESTION AND ABSORPTION

23.10 Digestion hydrolyzes food into nutrients that are absorbed across the gut epithelium (p. 912)

1. Digestion is accomplished by hydrolysis catalyzed by enzymes.

2. Virtually all of the foodstuffs and most of the water and electrolytes are absorbed in the small intestine. Most nutrients are absorbed by active transport processes.

23.11 How is each type of nutrient processed? (pp. 912–918)

1. Carbohydrates and proteins are digested to monomers, taken up into enterocytes of the small intestine by cotransport with Na^+, and passively transported across the basolateral membane into the capillaries.

2. Fat breakdown products are emulsified for digestion. The products of digestion are solubilized by bile salts (in micelles), resynthesized to triglycerides in the intestinal mucosal cells, and combined with other lipids and protein as chylomicrons that enter the lacteals.

3. Nucleic acids are digested into their components and actively transported into enterocytes.

4. Fat-soluble vitamins are absorbed by diffusion. Water-soluble vitamins are absorbed by active or passive transport.

5. Absorbed substances other than fat enter the villus blood capillaries and are transported to the liver via the hepatic portal vein.

Developmental Aspects of the Digestive System (pp. 918–919)

1. The mucosa of the alimentary canal develops from the endoderm, which folds to form a tube. The remaining three layers of the alimentary canal wall are formed by mesoderm. The glandular accessory organs (salivary glands, liver, pancreas, and gallbladder) form from outpocketings of the mucosa.

2. Important congenital abnormalities of the digestive tract include cleft palate/lip, tracheoesophageal fistula, and cystic fibrosis. All interfere with normal nutrition.

3. Various inflammations plague the digestive system throughout life. Appendicitis is common in adolescents, gastroenteritis and food poisoning can occur at any time (given the proper irritating factors), and ulcers and gallbladder problems increase in middle age.

4. The efficiency of all digestive system processes declines in the elderly, and periodontal disease is common. Diverticulosis, fecal incontinence, and GI tract cancers such as stomach and colon cancer appear with increasing frequency in an aging population.

23

REVIEW QUESTIONS

 To access additional practice questions using your smartphone, tablet, or computer: MasteringA&P® > Study Area > Practice Tests & Quizzes

Level 1 Remember/Understand

(Some questions have more than one correct answer. Select the best answer or answers from the choices given.)

1. The peritoneal cavity (a) is the same thing as the abdominopelvic cavity, (b) is filled with air, (c) like the pleural and pericardial cavities is a potential space containing serous fluid, (d) contains the pancreas and all of the duodenum.

2. Obstruction of the hepatopancreatic sphincter impairs digestion by reducing the availability of (a) bile and HCl, (b) HCl and intestinal juice, (c) pancreatic juice and intestinal juice, (d) pancreatic juice and bile.

3. The lamina propria forms part of the (a) muscularis externa, (b) submucosa, (c) serosa, (d) mucosa.

4. Carbohydrates are acted on by (a) peptidases, trypsin, and chymotrypsin, (b) amylase, maltase, and sucrase, (c) lipases, (d) peptidases, lipases, and galactase.

5. The parasympathetic nervous system influences digestion by (a) relaxing smooth muscle, (b) stimulating peristalsis and secretory activity, (c) constricting sphincters, (d) none of these.

6. The digestive juice product containing enzymes capable of digesting all four major foodstuff categories is (a) pancreatic, (b) gastric, (c) salivary, (d) biliary.

7. The site of production of cholecystokinin is (a) the stomach, (b) the small intestine, (c) the pancreas, (d) the large intestine.

8. Which of the following is not characteristic of the colon? (a) It is divided into ascending, transverse, and descending portions; (b) it contains abundant bacteria, some of which synthesize certain vitamins; (c) it is the main absorptive site; (d) it absorbs much of the water and salts remaining in the wastes.

9. The gallbladder (a) produces bile, (b) is attached to the pancreas, (c) stores and concentrates bile, (d) produces secretin.

10. The sphincter between the stomach and duodenum is (a) the pyloric sphincter, (b) the gastroesophageal sphincter, (c) the hepatopancreatic sphincter, (d) the ileocecal valve.

11. Name the six functional activities of the digestive system.

12. (a) What is the normal number of permanent teeth? Of deciduous teeth? (b) What substance covers the tooth crown? Its root? (c) What substance makes up the bulk of a tooth? (d) What and where is pulp?

Level 2 Apply/Analyze

13. Someone has eaten a meal of buttered toast, cream, and eggs. Which of the following would you expect to happen? (a) Compared to the period shortly after the meal, gastric motility and secretion of HCl decrease when the food reaches the duodenum; (b) gastric motility increases even as the person is chewing the food (before swallowing); (c) fat will be emulsified in the duodenum by the action of bile; (d) all of these.

In items 14–17, trace the path of a single protein molecule that has been ingested.

14. The protein molecule will be digested by enzymes made by (a) the mouth, stomach, and colon, (b) the stomach, liver, and small intestine, (c) the small intestine, mouth, and liver, (d) the pancreas, stomach, and small intestine.

15. The protein molecule must be digested before it can be transported to and utilized by the cells because (a) protein is only useful directly, (b) protein has a low pH, (c) proteins in the circulating blood produce an adverse osmotic pressure, (d) the protein is too large to be readily absorbed.

16. The products of protein digestion enter the bloodstream largely through cells lining (a) the stomach, (b) the small intestine, (c) the large intestine, (d) the bile duct.

17. Before the blood carrying the products of protein digestion reaches the heart, it first passes through capillary networks in (a) the spleen, (b) the lungs, (c) the liver, (d) the brain.

18. Make a simple line drawing of the organs of the alimentary canal and label each organ. Then add three labels to your drawing—salivary glands, liver, and pancreas—and use arrows to show where each of these organs empties its secretions into the alimentary canal.

19. Lara was on a diet but she could not eat less and kept claiming her stomach had a mind of its own. She was joking, but indeed, there is a "gut brain" called the enteric nervous system. Is it part of the parasympathetic and sympathetic nervous system? Explain.

20. (a) Describe the boundaries of the oral cavity. (b) Why do you suppose its mucosa is stratified squamous epithelium rather than the more typical simple columnar epithelium?

21. Describe the two phases of swallowing, noting the organs involved and the activities that occur.

22. Describe the role of these cells found in gastric glands: parietal, chief, mucous neck, and enteroendocrine.

23. Describe the regulation of the cephalic, gastric, and intestinal phases of gastric secretion.

24. (a) What is the relationship between the cystic, common hepatic, bile, and pancreatic ducts? (b) What is the point of fusion of the bile and pancreatic ducts called?

25. Explain why fatty stools result from the absence of bile or pancreatic juice.

26. Indicate the function of the stellate macrophages and the hepatocytes of the liver.

27. Explain why activation of pancreatic enzymes is delayed until they reach the small intestine.

Level 3 Evaluate/Synthesize

28. After a heavy meal rich with fried foods, Debby Collins, an overweight 45-year-old woman, was rushed to the emergency room with severe spasmodic pains in her epigastric region that radiated to the right side of her rib cage. She indicated that the attack came on suddenly, and her abdomen was found to be tender to the touch and somewhat rigid. What do you think is this patient's problem and why is her pain discontinuous (colicky)? What are the treatment options and what might happen if the problem is not resolved?

29. A baby is admitted to the hospital with a history of diarrhea and watery feces occurring over the last three days. The baby has sunken fontanelles, indicating extreme dehydration. Tests indicate that the baby has a bacterium-induced colitis, and antibiotics are prescribed. Because of the baby's loss of intestinal juices, do you think that his blood pH would indicate acidosis or alkalosis? Explain your reasoning.

30. Troy Francis, a middle-aged salesman, complains of a burning pain in the "pit of his stomach," usually beginning about two hours after eating and abating after drinking a glass of milk. When asked to indicate the site, he points to his epigastric region. The GI tract is examined by X-ray fluoroscopy. A gastric ulcer is visualized, and drug therapy using a proton pump inhibitor and antibiotics is recommended. (a) Why is this treatment suggested? (b) What are the possible consequences of nontreatment?

31. Dr. Dolan used an endoscope to view Mr. Habib's colon. He noted the presence of several polyps and removed them during the same procedure. What is an endoscope? Why did Dr. Dolan opt to remove the polyps immediately? (Hint: See *A Closer Look* on pp. 144–145.)

32. Mr. Holden has had severe diarrhea all day and is severely weakened. Explain why his nurse is concerned about his present condition.

23

48-Year-Old Female with Gallstones

Ms. Burton, 48, presents at the clinic complaining of abdominal pain. She is a middle-aged woman, overweight, pale and sweating, with jaundice (yellowish skin and sclerae). Her heart rate is 102 beats/min and BP is 145/98 mm Hg. She describes the pain as "severe" and "steady" over the past night, accompanied by nausea and vomiting. She sits leaning forward and indicates that the pain is in the epigastric region radiating up to the tip of the right shoulder.

Ms. Burton's medical history shows that she has had several episodes of biliary colic (pain caused by gallstones going through the bile duct). She says her current pain is "sort of like that, but it's lasted longer." She says her last meal was a seven-cheese pizza the previous night, and that the pain began 3–4 hours after eating. She denies drinking or smoking.

1. Do any of Ms. Burton's signs and symptoms suggest that her current problem is related to the biliary system? Why or why not?

2. Why would pain related to gallstones begin after a high-fat meal?

3. When Ms. Burton previously had gallstones, they passed through her biliary ducts into her duodenum and the pain resolved. What structures (in order) did the stones pass through?

4. **+ NCLEX-STYLE** If Ms. Burton's current problem is related to obstruction of the biliary ducts, which other signs or symptoms would you expect her to develop?
   a. Blood in her stool
   b. Dizziness and decreased consciousness
   c. Fatty and gray stools
   d. Low RBC count

Ms. Burton's doctor orders a complete blood count, tests for liver and pancreatic enzymes in her blood, and bedside ultrasonography. The ultrasonography reveals dilation of her bile and pancreatic ducts and a gallstone lodged in the hepatopancreatic ampulla. Blood tests come back positive for pancreatic amylase and pancreatic lipase.

5. **+ NCLEX-STYLE** Which of the following correctly interprets these results?
   a. The location of the gallstone is good news because it shows that the stone has almost passed into the duodenum.
   b. The location of the gallstone is a problem because it blocks bile from entering the duodenum and forces bile up into the pancreas.
   c. The presence of pancreatic enzymes in the blood is good news because it indicates that the pancreas is still functioning.
   d. The dilation of the ducts is good news because that will make it easier for gallstones to pass.

Ms. Burton was diagnosed with acute pancreatitis secondary to a gallstone and taken to surgery where the gallstone and her gallbladder were removed. After surgery, her abdominal pain resolved and her blood values returned to normal.

23

24

Nutrition, Metabolism, and Energy Balance

In this chapter, you will learn that

Nutrients are metabolized (used for building blocks or energy), and energy intake and output of the body are regulated

by investigating

Part 1 Nutrients

by asking

24.1 What are carbohydrates, lipids, and proteins used for?

24.2 What are vitamins and minerals used for?

Part 2 Metabolism

by asking

24.3 What is metabolism?

looking closer at

24.4 Carbohydrate metabolism

24.5 Lipid metabolism

24.6 Protein metabolism

24.7 Metabolic states of the body

24.8 The metabolic role of the liver

Part 3 Energy Balance

by asking

24.9 How is food intake regulated?

24.10 What factors affect metabolic rate?

24.11 How is body temperature regulated?

and finally, exploring

Developmental Aspects of Nutrition and Metabolism

CAREER CONNECTION

Watch a video to learn how the chapter content is used in a real health care setting. Go to Mastering A&P® > Study Area > Animations and Videos or use quick access URL https://goo.gl/88srfV

Are you a food lover? We are too. In fact, most people fall into one of two groups—those who live to eat and those who eat to live. The saying "you are what you eat" is true in that part of the food we eat is converted to our living flesh. In other words, our bodies use some nutrients to build cell structures, replace worn-out parts, and synthesize functional molecules. However, most nutrients we ingest are used as metabolic fuel. That is, they are oxidized and transformed to **ATP**, the chemical energy form used by cells (◄ **pp. 55–56**).

In Chapter 23, we talked about how foods are digested and absorbed, but what happens to these foods once they enter the blood? We will answer this question as we examine both the nature of nutrients and their metabolic roles.

PART 1
NUTRIENTS

Learning Outcomes
▶ **Define nutrient, essential nutrient, and calorie.**
▶ **List the five major nutrient categories. Note important sources and main cellular uses.**

A **nutrient** is a substance in food the body uses to promote normal growth, maintenance, and repair. The nutrients needed for health divide into five categories. Three of these—carbohydrates, lipids, and proteins—are **macronutrients** that make up most of what we eat. The fourth and fifth categories, vitamins and minerals, though equally crucial for health, are **micronutrients** required in only minute amounts.

Water accounts for about 60% by volume of the food we eat. We described its importance in the body in Chapter 2, so here we consider only the five nutrient categories listed above.

About 40 molecules, called **essential nutrients**, cannot be made fast enough to meet the body's needs, so our diet must provide them. As long as we ingest all the essential nutrients, the body can synthesize the hundreds of additional molecules required for life and good health. The ability of cells, especially liver cells, to convert one type of molecule to another is truly remarkable. These interconversions allow the body to use a wide range of foods and to adjust to varying food intakes. While "essential" is a standard way to describe the chemicals that must be obtained from outside sources, both essential and nonessential nutrients are equally vital for normal functioning.

Most foods offer a combination of nutrients. A balanced diet of foods from each of the different food groups normally guarantees adequate amounts of all the needed nutrients and adequate energy.

The energy value of foods is measured in **kilocalories (kcal)**. One kilocalorie is the amount of heat energy needed to raise the temperature of 1 kilogram of water 1°C (1.8°F). This unit is the "calorie" (C) that dieters count so conscientiously. Not all foods have the same amount of energy. Carbohydrates and proteins each have 4 kcal per gram, but lipids have almost 9 kcal per gram.

Various organizations release dietary recommendations. Most of these emphasize eating more vegetables, whole grains, and fruits.

Figure 24.1 USDA's MyPlate food guide.

For example, the U.S. Department of Agriculture (USDA) guidelines are represented as portions of a dinner plate (**Figure 24.1**). This image suggests how consumers might plan their meals relative to amounts and variety of foods from each food group. The MyPlate website provides details on healthy choices in each food group as well as personalized information according to age, sex, and activity level (www.choosemyplate.gov).

Nutrition advice is constantly in flux and often mired in the self-interest of food companies. Nonetheless, basic dietary principles have not changed in years and are not in dispute: Eat only what you need; eat plenty of fruits, vegetables, and whole grains; avoid junk food.

24.1 Carbohydrates, lipids, and proteins supply energy and are used as building blocks

Learning Outcomes
▶ **Distinguish between simple and complex carbohydrate sources.**
▶ **Distinguish between saturated, unsaturated, and trans fatty acid sources.**
▶ **Distinguish between nutritionally complete and incomplete proteins.**
▶ **Define nitrogen balance and indicate possible causes of positive and negative nitrogen balance.**
▶ **Indicate the major uses of carbohydrates, lipids, and proteins in the body.**

Carbohydrates

Dietary Sources
Except for milk sugar (lactose) and negligible amounts of glycogen in meats, all the carbohydrates we ingest are derived from plants. Sugars (monosaccharides and disaccharides;

24

◄ **pp. 43–44**) come from fruits, sugar cane, sugar beets, honey, and milk. The polysaccharide starch is found in grains and vegetables.

Two varieties of polysaccharides provide fiber (◄ **p. 44**). Cellulose, plentiful in most vegetables, is not digested by humans but provides roughage, or *insoluble fiber*, which increases the bulk of the stool and facilitates defecation. *Soluble fiber*, such as pectin found in apples and citrus fruits, reduces blood cholesterol levels.

Uses in the Body

The monosaccharide **glucose** is *the* carbohydrate molecule ultimately used as fuel by body cells to produce ATP. Carbohydrate digestion also yields fructose and galactose, but the liver converts these monosaccharides to glucose before they enter the general circulation.

Many body cells also use fats as energy sources, but neurons and red blood cells rely almost entirely on glucose for their energy needs. Because even a temporary shortage of blood glucose can severely depress brain function and lead to neuron death, the body carefully monitors and regulates blood glucose levels. Any glucose in excess of what is needed for ATP synthesis is converted to glycogen or fat and stored for later use.

There are few other uses for carbohydrates. Small amounts of pentose sugars are used to synthesize nucleic acids, and short chains of sugars are attached to externally facing plasma membrane proteins and lipids to form the glycocalyx (◄ **pp. 66–67**).

Dietary Requirements

The low-carbohydrate diet of the Inuit (Eskimos) and the high-carbohydrate diet of peoples in the Far East indicate that humans can be healthy even with wide variations in carbohydrate intake. The recommended intake to maintain health is 45–65% of total calorie intake, with the emphasis on *complex* carbohydrates (whole grains and vegetables) rather than simple carbohydrates (monosaccharides and disaccharides).

American adults typically consume about 46% of dietary food energy in the form of carbohydrates. Because starchy foods (rice, pasta, breads) cost less than meat and other high-protein foods, carbohydrates often make up a greater percentage of the diet in low-income groups. Highly processed carbohydrate foods such as candy and soft drinks only provide concentrated energy sources—so-called empty calories. Eating refined, sugary foods instead of more complex carbohydrates may cause nutritional deficiencies as well as obesity. **Table 24.1** (p. 930) lists other possible consequences of excessive intake of carbohydrates.

Lipids

Dietary Sources

The most abundant dietary lipids are triglycerides (◄ **p. 45**). We eat saturated fats (◄ **p. 46**) in animal products such as meat and dairy foods, in a few tropical plant products such as coconut, and in hydrogenated oils (trans fats) such as margarine and solid shortenings used in baking. Unsaturated fats (◄ **p. 46**) are present in seeds, nuts, olive oil, and most vegetable oils.

Major sources of cholesterol (◄ **p. 48**) are egg yolk, meats and organ meats, shellfish, and milk products. However, the liver produces about 85% of blood cholesterol regardless of dietary intake.

The liver is also adept at converting one fatty acid to another, but there are two *essential fatty acids* that it cannot synthesize:

- *Linoleic acid* (lin″o-le′ik), an omega-6 fatty acid that is a component of *lecithin* (les′ĭ-thin)
- *Linolenic acid*, an omega-3 fatty acid

Fortunately, most vegetable oils contain both linoleic and linolenic acids.

Uses in the Body

Fats make foods tender, flaky, or creamy, and make us feel full and satisfied. Fats in the body are necessary for several reasons:

- Fatty deposits in adipose tissue provide (1) a protective cushion around body organs, (2) an insulating layer beneath the skin, and (3) an easy-to-store concentrated source of energy.
- Phospholipids (◄ **p. 47**) are an integral component of myelin sheaths and cellular membranes.
- Cholesterol is a stabilizing component of plasma membranes and is the precursor from which bile salts, steroid hormones, and other important molecules are formed. Unlike triglycerides, cholesterol is not used for energy.
- *Prostaglandins* (pros″tah-glan′dinz; ◄ **p. 48**), regulatory molecules formed from linoleic acid via arachidonic acid (ah″rah-kĭ-don′ik), play a role in smooth muscle contraction, control of blood pressure, and inflammation.
- Triglycerides are the major energy fuel of skeletal muscle and hepatocytes.
- Fats help the body absorb fat-soluble vitamins.

Dietary Requirements

Fats represent over 40% of the calories in the typical American diet. Once viewed as detrimental, fat intake in the diet is being reconsidered. Most current guidelines suggest:

- Fats should represent 20–35% of total caloric intake.
- Saturated fats should be limited to 10% or less of total caloric intake.

Because the body can synthesize as much cholesterol as it needs, cholesterol is not required in the diet. Many guidelines still recommend keeping cholesterol intake as low as possible. This is particularly important for individuals who already have high total blood cholesterol levels, which have been associated with cardiovascular disease. Table 24.1 summarizes sources of the various lipid classes and consequences of deficiency or excessive intake.

Proteins

Dietary Sources

Proteins in eggs, milk, fish, and most meats are **complete proteins** that meet all the body's amino acid (◄ **p. 48**) requirements for tissue maintenance and growth (Table 24.1). Legumes (beans and peas), nuts, and cereals are protein-rich, but their proteins are nutritionally incomplete because they are low in one or more of the essential amino acids. The exception to this generalization is soybeans, which provide plant-derived complete proteins. Animal products contain the highest-quality proteins, in other words, those with the greatest amount and best ratios of *essential amino acids* (**Figure 24.2**).

Strict vegetarians must carefully plan their diets to obtain all the essential amino acids and prevent protein malnutrition. When ingested together, cereal grains and legumes provide all the essential amino acids (Figure 24.2b). Some combination of these foods is found in the diets of all cultures (for instance, the rice and beans seen on nearly every plate in a Mexican restaurant). For nonvegetarians, grains and legumes are useful as partial substitutes for more expensive animal proteins.

Uses in the Body

Proteins are important structural materials of the body, including, for example, keratin in skin, collagen and elastin in connective tissues, and muscle proteins. In addition, functional proteins such as enzymes and some hormones regulate an incredible variety of body functions. Whether amino acids are used to synthesize new proteins or burned for energy depends on a number of factors:

- **The all-or-none rule.** All amino acids needed to make a particular protein must be present in a cell at the same time and in sufficient amounts. If one is missing, the protein cannot be made. Because essential amino acids cannot be stored,

those not used immediately to build proteins are oxidized for energy or converted to carbohydrates or fats.

- **Adequacy of caloric intake.** For optimal protein synthesis, the diet must supply sufficient carbohydrate or fat calories for ATP production. When it doesn't, dietary and tissue proteins are used for energy.

- **Hormonal controls.** Certain hormones, called *anabolic hormones*, accelerate protein synthesis and growth. The effects of these hormones vary continually throughout life. For example, pituitary growth hormone stimulates tissue growth during childhood and conserves protein in adults, and estrogens trigger the growth spurt of adolescence (◄ **p. 188**). Other hormones, such as the adrenal glucocorticoids released during stress, enhance protein breakdown and conversion of amino acids to glucose.

In healthy adults the rate of protein synthesis equals the rate of protein breakdown and loss, a homeostatic state called **nitrogen balance**. The body is in nitrogen balance when the amount of nitrogen ingested in proteins equals the amount excreted in urine and feces.

The body is in *positive nitrogen balance* when the amount of protein incorporated into tissue is greater than the amount being broken down and used for energy—the normal situation in growing children and pregnant women. A positive balance also occurs when tissues are being repaired following illness or injury.

In *negative nitrogen balance*, protein breakdown for energy exceeds the amount of protein being incorporated into tissues.

(a) Essential amino acids

(b) Corn and beans together can provide all eight essential amino acids

Figure 24.2 Essential amino acids. (a) The essential amino acids represent only a small percentage of the total recommended protein intake. Histidine and arginine are essential in infants but not in adults. **(b)** Incomplete proteins from different sources can be combined in one meal to provide all essential amino acids.

Table 24.1   Summary of Carbohydrate, Lipid, and Protein Nutrients

FOOD SOURCES	RECOMMENDED DAILY ALLOWANCE (RDA) FOR ADULTS	PROBLEMS	
		EXCESSES	DEFICITS
Carbohydrates			
Total Digestible	130 g 45–65% of total caloric intake	Obesity; diabetes mellitus; nutritional deficits; dental caries; gastrointestinal irritation; elevated triglycerides in plasma	Tissue wasting (in extreme deprivation); metabolic acidosis resulting from accelerated fat use for energy
• **Complex carbohydrates** (starches): bread, cereal, crackers, flour, pasta, rice, potatoes			
• **Simple carbohydrates (sugars):** carbonated drinks, candy, fruit, ice cream, pudding, young (immature) vegetables			
Total Fiber (indigestible)	25–30 g		
Lipids			
Total: Animal sources (such as meat and dairy products) and plant sources (such as oils from nuts and seeds)	65 g 20–35% of total caloric intake	Obesity; may increase risk of cardiovascular disease (particularly with trans fat)	Weight loss; fat stores and tissue proteins catabolized to provide metabolic energy; problems controlling heat loss (due to depletion of subcutaneous fat)
• **Linoleic acid (an omega-6 fatty acid):** nuts, seeds, and vegetable oils (e.g., corn, soy, safflower)	11–17 g		
• **Linolenic acid (an omega-3 fatty acid):** fish oil, vegetable oils (e.g., canola, soy, flax), walnuts	1.1–1.6 g	Excess dietary intake of omega-3 fatty acid may increase risk of atherosclerosis	Poor growth; skin lesions (eczema-like); depression
• **Cholesterol:** organ meats (liver, kidneys, brains), egg yolks, fish roe; smaller concentrations in milk products and meat	As low as possible	Levels of blood cholesterol and low-density lipoproteins may increase; correlated with increased risk of cardiovascular disease	Possible increased risk of stroke (CVA) in susceptible individuals
Proteins			
• **Complete proteins:** eggs, milk, milk products, meat (fish, poultry, pork, beef, lamb), soybeans	0.8 g/kg body weight 12–20% of total caloric intake	Obesity; enhanced calcium excretion and bone loss; high cholesterol levels in blood; kidney stones	Profound weight loss and tissue wasting; retarded growth in children; anemia; edema (due to deficits of plasma proteins)
• **Incomplete proteins:** legumes (lima beans, kidney beans, lentils); nuts and seeds; grains and cereals; vegetables			During pregnancy: miscarriage or premature birth

24

This occurs during physical and emotional stress (for example, infection, injury, or burns), when the quality or quantity of dietary protein is poor, or during starvation.

Dietary Requirements

Besides supplying essential amino acids, dietary proteins furnish the raw materials for making nonessential amino acids and various nonprotein nitrogen-containing substances. The amount of protein a person needs to ingest reflects his or her age, size, metabolic rate, and the need to build new proteins (whether the body is in positive nitrogen balance). As a rule of thumb, nutritionists recommend a daily intake of 0.8 g per kilogram of body weight.

Check Your Understanding

1. What are the five major nutrient categories?
2. Why is it important to include cellulose in a healthy diet even though we do not digest it?
3. How does the body use triglycerides? Cholesterol?
4. **APPLY** Jared eats nothing but baked bean sandwiches. Is he getting all the essential amino acids he needs in this restricted diet?

For answers, see Answers Appendix.

24.2 Most vitamins act as coenzymes; minerals have many roles in the body

Learning Outcomes

▶ **Distinguish between fat- and water-soluble vitamins, and list the vitamins in each group.**

▶ **For each vitamin, list important sources, body functions, and important consequences of its deficit or excess.**

▶ **List minerals essential for health.**

▶ **Indicate important dietary sources of minerals and describe how each is used.**

Vitamins

Vitamins (*vita* = life) are organic compounds needed in minute amounts for growth and good health. Unlike other organic nutrients, vitamins do not serve as an energy source nor as building blocks, but they are crucial in helping the body use those nutrients that do. Without vitamins, all the carbohydrates, proteins, and fats we eat would be useless.

Most vitamins function as **coenzymes** (or parts of coenzymes), which act with an enzyme to accomplish a particular chemical task. For example, the B vitamins act as coenzymes when glucose is oxidized for energy.

Most vitamins are not made in the body, so we must ingest them in foods or vitamin supplements. The exceptions are vitamin D made in the skin, and small amounts of B vitamins and vitamin K synthesized by intestinal bacteria. In addition, the body can convert *beta-carotene* (kar′o-tēn), the orange pigment in carrots and other foods, to vitamin A. (For this reason, beta-carotene and substances like it are called *provitamins*.)

Vitamins are found in all major food groups, but no one food contains all the required vitamins. A balanced diet is the best way to ensure a full vitamin complement.

Initially vitamins were given letter designations that indicated the order of their discovery. Although more chemically descriptive names have been assigned to them, this earlier terminology is still commonly used.

Vitamins are either water soluble or fat soluble. **Water-soluble** vitamins—the B-complex vitamins and vitamin C—are absorbed along with water from the gastrointestinal tract. (The exception is vitamin B_{12}: To be absorbed, it must bind to *intrinsic factor*, a stomach secretion; ◀ **p. 917**.) The body's lean tissue stores insignificant amounts of water-soluble vitamins, and any ingested amounts not taken up by cells within an hour or so are excreted in urine. Consequently, health problems resulting from excessive levels of these vitamins are rare.

Fat-soluble vitamins (A, D, E, and K) are absorbed together with other lipids in the gut. Anything that interferes with lipid absorption also interferes with the uptake of fat-soluble vitamins. Except for vitamin K, fat-soluble vitamins are stored in the body, and pathologies due to fat-soluble vitamin toxicity, particularly excess vitamin A, are well documented.

Metabolism uses oxygen, and sometimes these reactions generate potentially tissue-damaging free radicals (molecules with an unpaired electron; ◀ **p. 86**). Vitamins A, C, E, and the mineral selenium participate in *antioxidant reactions* that neutralize free radicals. The whole story of how antioxidants interact in the body is still murky. Chemists propose that antioxidants pass the dangerous free electron from one molecule to the next. Eventually, a chemical such as glutathione absorbs and neutralizes the unpaired electron. Broccoli, cabbage, cauliflower, and brussels sprouts are all good sources of vitamins A and C.

The notion that megadoses of vitamin supplements are the road to eternal youth and glowing health is useless at best—and at worst, may cause serious health problems, particularly in the case of fat-soluble vitamins. **Table 24.2** (p. 932) contains an overview of the roles of vitamins in the body.

Minerals

The body requires moderate amounts of seven **minerals** (calcium, phosphorus, potassium, sulfur, sodium, chlorine, magnesium) and trace amounts of about a dozen others (**Table 24.3** on p. 933). Minerals make up about 4% of the body by weight, with calcium and phosphorus (as bone salts) accounting for about three-quarters of this amount.

Minerals, like vitamins, are not used for fuel but work with other nutrients to ensure a smoothly functioning body. Incorporating minerals into structures makes them stronger. For example, calcium, phosphorus, and magnesium salts harden the teeth and strengthen the skeleton.

Most minerals are ionized in body fluids or bound to organic compounds to form phospholipids, hormones, and various proteins. For example, iron is essential to the oxygen-binding heme of hemoglobin, and sodium and chloride ions are the major electrolytes in blood. The amount of a particular mineral in the body gives very few clues to its importance in body function.

Table 24.2   Vitamins

VITAMIN	MAJOR DIETARY SOURCES	MAJOR FUNCTIONS IN THE BODY	SYMPTOMS OF DEFICIENCY OR EXTREME EXCESS
Water-Soluble Vitamins			
Vitamin B$_1$ (thiamine)	Pork, legumes, peanuts, whole grains	Coenzyme used in removing CO_2 from organic compounds	Beriberi (nerve disorder—tingling, poor coordination, reduced heart function)
Vitamin B$_2$ (riboflavin)	Dairy products, meats, enriched grains, vegetables	Component of coenzymes FAD and FMN	Skin lesions such as cracks at corners of mouth
Vitamin B$_3$ (niacin)	Nuts, meats, grains	Component of coenzymes NAD^+ and $NADP^+$	Skin and gastrointestinal lesions, nervous disorders Liver damage
Vitamin B$_5$ (pantothenic acid)	Most foods: meats, dairy products, whole grains, etc.	Component of coenzyme A	Fatigue, numbness, tingling of hands and feet
Vitamin B$_6$ (pyridoxine)	Meats, vegetables, whole grains	Coenzyme used in amino acid metabolism	Irritability, convulsions, muscular twitching, anemia Unstable gait, numb feet, poor coordination
Vitamin B$_7$ (biotin)	Legumes, other vegetables, meats	Coenzyme in synthesis of fat, glycogen, and amino acids	Scaly skin inflammation, neuromuscular disorders
Vitamin B$_9$ (folic acid)	Green vegetables, oranges, nuts, legumes, whole grains	Coenzyme in nucleic acid and amino acid metabolism	Anemia, birth defects May mask deficiency of vitamin B$_{12}$
Vitamin B$_{12}$	Meats, eggs, dairy products	Coenzyme in nucleic acid metabolism; maturation of red blood cells	Anemia, nervous system disorders (numbness, loss of balance)
Vitamin C (ascorbic acid)	Fruits and vegetables, especially citrus fruits, broccoli, tomatoes	Used in collagen synthesis (such as for bone, cartilage, gums); antioxidant	Scurvy (degeneration of skin, teeth, blood vessels), weakness, delayed wound healing Gastrointestinal upset
Fat-Soluble Vitamins			
Vitamin A (retinol)	Provitamin A (beta-carotene) in deep green and orange vegetables and fruits; retinol in dairy products	Component of visual pigments; maintenance of epithelial tissues; antioxidant	Blindness, skin disorders, impaired immunity Headache, irritability, vomiting, hair loss, blurred vision, liver and bone damage
Vitamin D	Dairy products, egg yolk; also made in human skin in presence of sunlight	Aids in absorption and use of calcium and phosphorus	Rickets (bone deformities) in children, bone softening in adults Brain, cardiovascular, and kidney damage
Vitamin E (tocopherol)	Vegetable oils, nuts, seeds	Antioxidant; helps prevent damage to cell membranes	Degeneration of the nervous system
Vitamin K (phylloquinone)	Green vegetables, tea; also made by colon bacteria	Important in blood clotting	Defective blood clotting Liver damage and anemia

Modified from: Urry et al., CAMPBELL BIOLOGY, 11th Edition, © 2017. Reprinted by permission of Pearson Education, Inc., Upper Saddle River, N.J.

For example, just a few milligrams of iodine (required for thyroid hormone synthesis) can make a critical difference to health.

A fine balance between uptake and excretion is crucial for retaining needed amounts of minerals while preventing toxic overload. Sodium present in virtually all natural and minimally processed foods poses little or no health risk. However, the large amounts added to processed foods and sprinkled on food prior to eating may contribute to fluid retention and high blood pressure.

Fats and sugars are practically devoid of minerals, and highly refined cereals and grains are poor sources. The most mineral-rich foods are legumes and other vegetables, milk, and some meats.

Check Your Understanding

5. Vitamins are not used for energy fuels. What are they used for?
6. Which mineral is essential for thyroxine synthesis? For making bones hard? For hemoglobin synthesis?
7. **MAKE CONNECTIONS** Which B vitamin requires the help of a product made in the stomach to be absorbed? What is that gastric product and which cells in the gastric mucosa secrete it? What part of the small intestine ultimately absorbs this B vitamin? (Hint: See Chapter 23.) Lack of this B vitamin causes what kind of anemia? (Hint: See Chapter 17.)

For answers, seeAnswers Appendix.

Table 24.3   Minerals in the Body

MINERAL	MAJOR DIETARY SOURCES	MAJOR FUNCTIONS IN THE BODY	SYMPTOMS OF DEFICIENCY*
Greater than 200 mg per Day Required			
Calcium (Ca)	Dairy products, dark green vegetables, legumes	Bone and tooth formation, blood clotting, nerve and muscle function	Impaired growth, possibly loss of bone mass
Phosphorus (P)	Dairy products, meats, grains	Bone and tooth formation, acid-base balance, nucleotide synthesis	Weakness, loss of minerals from bone, calcium loss
Sulfur (S)	Proteins from many sources	Component of certain amino acids	Symptoms of protein deficiency
Potassium (K)	Meats, dairy products, many fruits and vegetables, grains	Nerve function, acid-base balance	Muscular weakness, paralysis, nausea, heart failure
Chlorine (Cl)	Table salt	Acid-base balance, formation of gastric juice, nerve function, osmotic balance	Muscle cramps, reduced appetite
Sodium (Na)	Table salt	Water balance, blood pressure, nerve function	Muscle cramps, reduced appetite
Magnesium (Mg)	Whole grains, green leafy vegetables	Cofactor; ATP bioenergetics	Nervous system disturbances
Trace Amounts Required			
Iron (Fe)	Meats, eggs, legumes, whole grains, green leafy vegetables	Component of hemoglobin and of electron carriers in energy metabolism; enzyme cofactor	Iron-deficiency anemia, weakness, impaired immunity
Fluorine (F)	Drinking water, tea, seafood	Maintenance of tooth (and probably bone) structure	Higher frequency of tooth decay
Zinc (Zn)	Meats, seafood, grains	Component of certain digestive enzymes and other proteins	Growth failure, skin abnormalities, reproductive failure, impaired immunity
Copper (Cu)	Seafood, nuts, legumes, organ meats	Enzyme cofactor in iron metabolism, melanin synthesis, electron transport	Anemia, cardiovascular abnormalities
Manganese (Mn)	Nuts, grains, vegetables, fruits, tea	Enzyme cofactor	Abnormal bone and cartilage
Iodine (I)	Seafood, iodized salt	Component of thyroid hormones	Goiter (enlarged thyroid)
Cobalt (Co)	Meats and dairy products	Component of vitamin B_{12}	None, except as B_{12} deficiency
Selenium (Se)	Seafood, meats, whole grains	Enzyme cofactor for antioxidant enzymes	Muscle pain, possibly heart muscle deterioration
Chromium (Cr)	Brewer's yeast, liver, seafood, meats, some vegetables	Involved in glucose and energy metabolism	Impaired glucose metabolism
Molybdenum (Mo)	Legumes, grains, some vegetables	Enzyme cofactor	Disorder in excretion of nitrogen-containing compounds

*All of these minerals are also harmful when consumed in excess.

Modified from: Urry et al., CAMPBELL BIOLOGY, 11th Edition, © 2017. Reprinted by permission of Pearson Education, Inc., Upper Saddle River, N.J.

24

PART 2

METABOLISM

Once inside body cells, nutrients become involved in an incredible variety of biochemical reactions known collectively as **metabolism** (*metabol* = change). During metabolism, substances are constantly built up and torn down. Cells use energy to extract more energy from foods, and then use some of this extracted energy to drive their activities. Even at rest, the body uses energy on a grand scale.

24.3 Metabolism is the sum of all biochemical reactions in the body

Learning Outcomes

▶ Define metabolism. Explain how catabolism and anabolism differ.

▶ Define oxidation and reduction and indicate the importance of these reactions in metabolism.

▶ Indicate the role of coenzymes used in cellular oxidation reactions.

▶ Explain the difference between substrate-level phosphorylation and oxidative phosphorylation.

Anabolism and Catabolism

Metabolic processes are either *anabolic* (synthetic, building up) or *catabolic* (degradative, tearing down). **Anabolism** (ah-nab′o-lizm) is the general term for all reactions that build larger molecules or structures from smaller ones, such as the bonding together of amino acids to build proteins. **Catabolism** (kah-tab′o-lizm) refers to all processes that break down complex structures to simpler ones—for example, the hydrolysis of foods in the digestive tract.

Three major stages are involved in processing energy-containing nutrients in the body (**Figure 24.3**).

- *Stage 1* is digestion and absorption in the gastrointestinal tract (◀ p. 912). The absorbed nutrients are then transported in blood to the tissue cells.

- *Stage 2* occurs in the cytoplasm of tissue cells. Newly delivered nutrients are either built into lipids, proteins, and glycogen by anabolic pathways or broken down by catabolic pathways to smaller fragments. Many molecules end up as *pyruvic acid* (pi-roo′vik), an important metabolic intermediate.

- *Stage 3*, which occurs in the mitochondria, is almost entirely catabolic. It requires oxygen and completes the breakdown of stage 2 products, most of which are first converted to *acetyl CoA* (as′ĕ-til ko-a′). This process produces carbon dioxide and water and harvests large amounts of ATP.

The group of catabolic reactions collectively called **cellular respiration** consists of glycolysis, the citric acid cycle, and oxidative phosphorylation. As we will see shortly, these reactions take food fuels, particularly glucose, and break them down. Some of the energy released is captured to form ATP, the cells' energy currency that links energy-releasing catabolic reactions to cellular work.

Recall from Chapter 2 (◀ p. 55) that reactions driven by ATP are coupled. As ATP is hydrolyzed, enzymes shift its high-energy phosphate groups to other molecules, which are then said to be **phosphorylated** (fos″for′ĭ-la-ted). Phosphorylation primes a molecule, changing it in a way that increases its activity, produces motion, or does work. For example, phosphorylation activates many regulatory enzymes that catalyze key steps in metabolic pathways.

The body can also store energy in fuels, such as glycogen and fats, and mobilize these stores later to produce ATP for cellular use.

You do not need to memorize Figure 24.3, but you may want to refer to it often as a cohesive summary of nutrient processing and metabolism in the body.

Oxidation-Reduction Reactions and the Role of Coenzymes

Many of the reactions that take place within cells are **oxidation reactions**. *Oxidation* was originally defined as the combination of oxygen with other elements, seen in the rusting of iron (the slow formation of iron oxide) and the burning of wood. In burning, oxygen combines rapidly with carbon, releasing carbon dioxide, water, and an enormous amount of energy as heat and light.

However, oxidation also occurs when hydrogen atoms are *removed* from compounds. So *oxidation is the gain of oxygen or the loss of hydrogen*. As explained in Chapter 2 (◀ p. 37), whichever way oxidation occurs, the oxidized substance always *loses* (or nearly loses) electrons as they move to (or toward) a substance that more strongly attracts them.

To explain this loss of electrons, let's review the consequences of different electron-attracting abilities of atoms (◀ pp. 31–35). Consider a molecule made up of a hydrogen atom plus some other kinds of atoms. Hydrogen is very electropositive, so its lone electron usually spends more time orbiting the other atoms of the molecule. But when a hydrogen *atom* is removed, its electron goes with it, and the molecule as a whole loses that electron. Conversely, oxygen is very electron-hungry (electronegative), so when oxygen binds with other atoms the shared electrons spend more time in oxygen's vicinity. Again, the rest of the molecule loses electrons.

Essentially all oxidation of food fuels involves the step-by-step removal of pairs of hydrogen atoms (with their electrons) from the substrate molecules, eventually leaving only carbon dioxide (CO_2). Molecular oxygen (O_2) is the final electron acceptor. It combines with the removed hydrogen atoms at the very end of the process, to form water (H_2O).

Whenever one substance loses electrons (is oxidized), another substance gains them (is reduced). For this reason, oxidation and reduction are coupled reactions and we speak of **oxidation-reduction (redox) reactions**. The key understanding about redox reactions is that "oxidized" substances *lose* energy and "reduced" substances *gain* energy as energy-rich electrons are transferred from one substance to the next. Consequently, as food fuels are oxidized, their energy is transferred to a series of other molecules and ultimately to ADP to form energy-rich ATP.

Like all other chemical reactions in the body, redox reactions are catalyzed by enzymes. Those that catalyze redox reactions in which hydrogen atoms are removed are called **dehydrogenases** (de-hi′dro-jen-ās″ez), while enzymes catalyzing the transfer of oxygen are **oxidases**.

Most of these enzymes require the help of a specific coenzyme, typically derived from one of the B vitamins. Although

24

Stage 1: GI Tract

Nutrients are:
• Digested into absorbable units.
• Absorbed into the blood and transported to tissue cells.

PROTEINS   CARBOHYDRATES   FATS

Amino acids   Glucose and other sugars   Glycerol   Fatty acids

Stage 2: Tissue Cells

Anabolism or catabolism:
• In anabolism, nutrients are built into macromolecules.
• In catabolism, nutrients are partially broken down.
• Glycolysis is the major catabolic pathway and ends with pyruvate.

Proteins

Glucose

Glycogen

Triglycerides

Glycolysis

NH_3

Pyruvic acid

CO_2

Stage 3: Mitochondria

Oxidative breakdown of stage 2 products:
• CO_2 is released.
• The H atoms removed are ultimately delivered to molecular oxygen, forming water.
• Some of the energy released is used to form ATP.
• The citric acid cycle and oxidative phoshorylation are the major pathways.

Acetyl CoA

Citric acid cycle

Infrequent

CO_2

O_2

H → **Oxidative phosphorylation** (in electron transport chain) → H_2O

→ Catabolic reactions
→ Anabolic reactions

ATP ATP ATP

Figure 24.3 Three stages of metabolism of energy-containing nutrients.

24

the enzymes catalyze the removal of hydrogen atoms to oxidize a substance, they cannot *accept* the hydrogen (hold on or bond to it). Their *coenzymes*, however, can act as hydrogen (or electron) acceptors, becoming reduced each time a substrate is oxidized.

Two very important coenzymes of the oxidative pathways are **nicotinamide adenine dinucleotide (NAD$^+$)** (nik″o-tin′ah-mīd), based on *niacin*, and **flavin adenine dinucleotide (FAD)**, derived from *riboflavin*. The oxidation of succinic acid to fumaric acid and the simultaneous reduction of FAD to FADH$_2$, an example of a coupled redox reaction, is shown on the right.

FAD (oxidized)   FADH$_2$ (reduced)

COOH
|
H — C — H
|
H — C — H   oxidation (−2H)
|
COOH

Succinic acid (reduced)

2 H atoms are transferred from succinic acid to the coenzyme FAD, reducing it to FADH$_2$.

COOH
|
C — H
‖
C — H
|
COOH

Fumaric acid (oxidized)

ATP Synthesis

How do our cells capture some of the energy liberated during cellular respiration to make ATP molecules? There are two mechanisms—substrate-level phosphorylation and oxidative phosphorylation.

Substrate-level phosphorylation occurs when high-energy phosphate groups are transferred directly from phosphorylated substrates (metabolic intermediates such as glyceraldehyde 3-phosphate) to ADP (**Figure 24.4a**). Essentially, this process occurs because the high-energy bonds attaching the phosphate groups to the substrates are even more unstable than those in ATP. ATP is synthesized by this route twice during glycolysis, and once during each turn of the citric acid cycle. The enzymes catalyzing substrate-level phosphorylations are located in both the cytosol (where glycolysis occurs) and in the watery matrix inside the mitochondria (where the citric acid cycle takes place) (**Figure 24.5**).

Oxidative phosphorylation is more complicated, but it releases most of the energy that is eventually captured in ATP bonds during cellular respiration. Oxidative phosphorylation, which is carried out by electron transport proteins embedded in the inner mitochondrial membranes, is an example of a chemiosmotic process. **Chemiosmotic processes** couple the movement of substances across membranes to chemical reactions.

In this case, some of the energy released during the oxidation of food fuels (the "chemi" part of chemiosmotic) is used to pump (*osmo* = push) protons (H^+) across the inner mitochondrial membrane into the intermembrane space (Figure 24.4b).

This creates a steep concentration gradient for protons across the membrane. Then, when H^+ flows back across the membrane (through a membrane channel protein called *ATP synthase*), some of this gradient energy is captured and used to attach phosphate groups to ADP.

Check Your Understanding

8. What is a redox reaction?
9. How are anabolism and catabolism linked by ATP?
10. What is the energy source for the proton pumps of oxidative phosphorylation?
11. Briefly, how do substrate-level and oxidative phosphorylation differ?

For answers, see Answers Appendix.

24.4 Carbohydrate metabolism is the central player in ATP production

Learning Outcomes

▶ Summarize important events and products of glycolysis, the citric acid cycle, and electron transport.
▶ Define glycogenesis, glycogenolysis, and gluconeogenesis.

The story of carbohydrate metabolism is really a tale of glucose metabolism because all food carbohydrates are eventually transformed to glucose. Glucose enters tissue cells by facilitated

(a) Substrate-level phosphorylation

- A high-energy phosphate group is transferred directly from a substrate to ADP to form ATP.
- Occurs in the cytosol and mitochondrial matrix.

P

ADP   Substrate

Enzyme

Catalysis

ATP

Product

Enzyme

(b) Oxidative phosphorylation

- Electron transport proteins "pump" protons, creating a proton gradient.
- ATP synthase uses the energy of the proton gradient to bind phosphate groups to ADP.
- Occurs only in the mitochondrial matrix.

High H^+ concentration in intermembrane space

Inner mitochondrial membrane

H^+ H^+

Proton pumps (electron transport chain)

ATP synthase

H^+

Energy from food

ADP + P_i ATP

Low H^+ concentration in mitochondrial matrix

Figure 24.4 Mechanisms of phosphorylation.

① Glycolysis, in the cytosol, breaks down each glucose molecule into two molecules of pyruvic acid.

② The pyruvic acid then enters the mitochondrial matrix, where the citric acid cycle oxidizes it to CO_2. During glycolysis and the citric acid cycle, substrate-level phosphorylation forms small amounts of ATP.

③ Energy-rich electrons picked up by coenzymes are transferred to the electron transport chain, built into the inner mitochondrial membrane. The electron transport chain carries out oxidative phosphorylation, which generates most of the ATP in cellular respiration.

Figure 24.5 During cellular respiration, ATP is formed in the cytosol and in the mitochondria.

diffusion, a process that is greatly enhanced by insulin (◀ p. 630). Immediately after entering a cell, glucose is phosphorylated to *glucose-6-phosphate* by transfer of a phosphate group to its sixth carbon during a coupled reaction with ATP:

$$\text{Glucose} + \text{ATP} \rightarrow \text{glucose-6-PO}_4 + \text{ADP}$$

Phosphorylation effectively traps glucose inside the cell (it can't diffuse out) and most body cells lack the enzymes needed to reverse this reaction. Because glucose-6-phosphate is a *different* molecule from simple glucose, the reaction also keeps intracellular glucose levels low, maintaining a concentration gradient for glucose entry. Only intestinal epithelial cells, kidney tubule cells, and liver cells have the enzymes needed to reverse this phosphorylation reaction, which reflects their central roles in glucose uptake *and* release. The catabolic and anabolic pathways for carbohydrates all begin with glucose-6-phosphate.

Oxidation of Glucose

Glucose is the pivotal fuel molecule in the oxidative (ATP-producing) pathways. Glucose is catabolized via the reaction

$$C_6H_{12}O_6 + 6O_2 \rightarrow 6H_2O + 6CO_2 + 32 \text{ ATP} + \text{heat}$$

glucose   oxygen   water   carbon
                                    dioxide

This equation gives few hints that glucose breakdown is complex and involves three of the pathways featured in Figures 24.3 and 24.5:

1. Glycolysis (color-coded orange throughout the chapter)
2. The citric acid cycle (color-coded green)
3. The electron transport chain and oxidative phosphorylation (color-coded lavender)

These metabolic pathways occur sequentially.

■ Glycolysis

Also called the *glycolytic pathway*, **glycolysis** (gli-kol'ĭ-sis; "sugar splitting") occurs in the cytosol of cells. This pathway, a series of ten chemical steps, converts glucose to two *pyruvic acid* molecules. All steps are fully reversible except the first, during which glucose entering the cell is phosphorylated to glucose-6-phosphate.

Glycolysis is an *anaerobic process*. Although this term is sometimes mistakenly interpreted to mean the pathway occurs only in the absence of oxygen, it actually means that glycolysis *does not use oxygen and occurs whether or not oxygen is present.*

24

Figure 24.6 The three major phases of glycolysis. The fate of pyruvic acid depends on whether or not molecular O_2 is available.

- **Phase 1. Sugar activation.** In phase 1, glucose is phosphorylated twice and converted to fructose-1,6-bisphosphate. These reactions use two ATP molecules (to be recouped later), which provide the *activation energy* needed to prime the later stages of the pathway. Phase 1 can be thought of as the *energy investment phase*.

- **Phase 2. Sugar cleavage.** During phase 2, fructose-1,6-bisphosphate is split into two 3-carbon fragments that exist (interconvertibly) as one of two isomers: glyceraldehyde (glis″er-al′dĕ-hīd) 3-phosphate or dihydroxyacetone (di″hi-drok″se-as′ĕ-tōn) phosphate.

- **Phase 3. Sugar oxidation and ATP formation.** In phase 3, two major events happen. First, the two 3-carbon fragments are oxidized by the removal of hydrogen, which NAD^+ picks up. In this way, some of glucose's energy is transferred to NAD^+. Second, inorganic phosphate groups (P_i) are attached to each oxidized fragment by high-energy bonds. Later, when these terminal phosphates are split off, enough energy is captured to form four ATP molecules. As we noted earlier, formation of ATP this way is called *substrate-level phosphorylation*.

The final products of glycolysis are two molecules of **pyruvic acid**, two molecules of reduced NAD^+ (NADH), and two H^+. There is a net gain of two ATP molecules per glucose molecule. Four ATPs are produced, but remember that two are consumed in phase 1 to activate the reactions. Each pyruvic acid molecule has the formula $C_3H_4O_3$, and glucose is $C_6H_{12}O_6$. Between them the two pyruvic acid molecules have lost four hydrogen atoms, whose electrons are now bound to two molecules of NAD^+. NAD carries a positive charge (NAD^+), so when it accepts a hydrogen pair, $NADH + H^+$ is the resulting reduced product. Although a small amount of ATP has been harvested, the other two products of glucose oxidation (H_2O and CO_2) have yet to appear.

The fate of pyruvic acid, which still contains most of glucose's chemical energy, depends on the availability of oxygen at the time the pyruvic acid is produced. Because the supply of NAD^+ is limited, glycolysis can continue only if the reduced coenzymes (NADH, together with its H^+) formed during glycolysis are relieved of their extra hydrogens. Only then can they continue to act as hydrogen acceptors.

When oxygen is readily available, this is no problem. NADH + H^+ delivers its burden of hydrogen atoms to the enzymes of the electron transport chain in the mitochondria, which deliver them to O_2, forming water. However, when oxygen is not present in sufficient amounts, as might occur during strenuous exercise, NADH + H^+ unloads its hydrogen atoms *back onto pyruvic acid*, reducing it. This addition of two hydrogen atoms to pyruvic acid yields **lactic acid** (see bottom right of Figure 24.6). Some of this lactic acid diffuses out of the cells and is transported to the liver for processing.

When oxygen is again available, lactic acid is oxidized back to pyruvic acid and enters the **aerobic pathways** (the oxygen-requiring citric acid cycle and electron transport chain within the mitochondria), and is completely oxidized to water and carbon dioxide. The liver may also convert lactic acid all the way back to glucose-6-phosphate (reverse glycolysis). Glucose-6-phosphate

The three major phases of glycolysis shown in **Figure 24.6** are described next. Appendix D shows the complete glycolytic pathway.

can be either stored as glycogen or freed of its phosphate and released to the blood if blood sugar levels are low.

Except for red blood cells (which typically carry out *only* glycolysis), prolonged anaerobic metabolism ultimately results in acid-base problems. Consequently, *totally* anaerobic conditions resulting in lactic acid formation provide only a temporary route for rapid ATP production. Totally anaerobic conditions can go on without tissue damage for the longest periods in skeletal muscle, for much shorter periods in cardiac muscle, and almost not at all in the brain. Although glycolysis generates ATP rapidly, each glucose molecule yields only 2 ATP as compared to the 30 to 32 ATP when a glucose molecule is completely oxidized.

Citric Acid Cycle

The **citric acid cycle** (or Krebs cycle) is the next stage of glucose oxidation and is named for its first substrate. The citric acid cycle occurs in the mitochondrial matrix and is fueled largely by pyruvic acid produced during glycolysis and by fatty acids resulting from fat breakdown.

Because pyruvic acid is a charged molecule, it must enter the mitochondrion by active transport with the help of a transport protein. Once in the mitochondrion, the first order of business is a **transitional phase** that converts pyruvic acid to acetyl CoA. This occurs via a three-step process (**Figure 24.7**, top):

Figure 24.7 Simplified version of the citric acid (Krebs) cycle. During each turn of the cycle, two carbon atoms are removed from the substrates as CO_2 (decarboxylation reactions); four oxidations by removal of hydrogen atoms occur, producing four molecules of reduced coenzymes (3 NADH + 3 H^+ and 1 $FADH_2$); and one ATP is synthesized by substrate-level phosphorylation. An additional decarboxylation and an oxidation reaction occur in the transitional phase (at top) that converts pyruvic acid, the product of glycolysis, to acetyl CoA, the molecule that enters the citric acid cycle pathway.

Carbon atom

P_i Inorganic phosphate

CoA   Coenzyme A

Cytosol

Mitochondrion (matrix)

Transitional phase

Pyruvic acid from glycolysis

CO_2 NAD$^+$

CoA   NADH + H^+

Acetyl CoA

Oxaloacetic acid (pickup molecule)

CoA

Citric acid (initial reactant)

NADH + H^+

NAD$^+$

Malic acid

Isocitric acid

NAD$^+$

Citric acid cycle

CO_2 NADH + H^+

Fumaric acid

α-Ketoglutaric acid

CoA

$FADH_2$

CO_2 NAD$^+$

FAD

Succinic acid

Succinyl-CoA

NADH + H^+

CoA

GTP   GDP + P_i

ADP   ATP

Glycolysis

Citric acid cycle

Electron transport chain and oxidative phosphorylation

ATP   ATP   ATP

24

1. **Decarboxylation.** In this step, one of pyruvic acid's carbons is removed and released as carbon dioxide gas, a process called **decarboxylation**. CO_2 diffuses out of the cells into the blood to be expelled by the lungs. This is the first time that CO_2 is released during cellular respiration.

2. **Oxidation.** The remaining 2-carbon fragment is oxidized to acetic acid by removing hydrogen atoms, which are picked up by NAD^+.

3. **Formation of acetyl CoA.** Acetic acid is combined with *coenzyme A* to produce the reactive final product, **acetyl coenzyme A (acetyl CoA)**. Coenzyme A is a sulfur-containing coenzyme derived from vitamin B_5.

Acetyl CoA is now ready to enter the citric acid cycle and be broken down completely by mitochondrial enzymes. Coenzyme A shuttles the 2-carbon acetic acid to an enzyme that joins it to a 4-carbon acid called **oxaloacetic acid** (ok″sah-lo″ah-sēt′ik) to produce the 6-carbon **citric acid**.

As the cycle moves through its eight successive steps, the atoms of citric acid are rearranged to produce different intermediate molecules, most called **keto acids** (Figure 24.7). The acetic acid that enters the cycle is broken apart carbon by carbon (decarboxylated) and oxidized, generating $NADH + H^+$ and $FADH_2$. At the end of the cycle, acetic acid has been totally disposed of and oxaloacetic acid, the *pickup molecule*, is regenerated.

For each turn of the cycle, we get:

- Two CO_2 molecules that come from two *decarboxylations*.

- Four molecules of reduced coenzymes (3 $NADH + 3 H^+$ and 1 $FADH_2$). The addition of water at certain steps accounts for some of the released hydrogen.

- One molecule of ATP (via substrate-level phosphorylation).

The detailed events of each of the eight steps of the citric acid cycle are described in Appendix D.

Now let's back up and account for the pyruvic acid molecules entering the mitochondria. We need to consider the products of both the transitional phase and the citric acid cycle itself. Altogether, each pyruvic acid yields three CO_2 molecules and five molecules of reduced coenzymes—1 $FADH_2$ and 4 NADH + 4 H^+ (equal to removing 10 hydrogen atoms). The products of glucose oxidation in the citric acid cycle are twice that (remember 1 glucose = 2 pyruvic acids): six CO_2, ten molecules of reduced coenzymes, and two ATP molecules.

Notice that it is these citric acid cycle reactions that produce the CO_2 released during glucose oxidation. The reduced coenzymes, which carry their extra electrons in high-energy linkages, must now be oxidized if the citric acid cycle and glycolysis are to continue.

Although glycolysis is exclusive to carbohydrate oxidation, breakdown products of carbohydrates, fats, and proteins can feed into the citric acid cycle to be oxidized for energy. On the other hand, some citric acid cycle intermediates can be siphoned off to make fatty acids and nonessential amino acids. In this way, the citric acid cycle is a source of building materials for anabolic reactions, as well as the final common pathway for oxidizing food fuels.

Electron Transport Chain and Oxidative Phosphorylation

Like glycolysis, none of the reactions of the citric acid cycle use oxygen directly. This is the exclusive function of the **electron transport chain**, which carries out the final catabolic reactions that occur on the inner mitochondrial membrane. However, because the reduced coenzymes produced in the citric acid cycle are the substrates for the electron transport chain, these two pathways are coupled, and both are *aerobic*, meaning they require oxygen.

In the electron transport chain, the hydrogens removed during the oxidation of food fuels are combined with O_2 to form water, and the energy released during those reactions is harnessed to attach P_i groups to ADP, forming ATP. As we noted earlier, this type of phosphorylation process is called *oxidative phosphorylation*. Let's peek under the hood of a cell's power plant and see how this rather complicated process works.

Just as in a car, what we see is a complicated structure with many parts. Why are there so many parts? In a car engine, we burn fuel for energy. If released all at once, that energy would result in a big, hot explosion. All of the complex engine parts make it possible to capture some of this energy to do useful work. In the same way, the electron transport chain harvests the energy of food fuels a bit at a time in a step-by-step way so we can ultimately make ATP.

Most components of the electron transport chain are proteins that bind metal atoms (known as *cofactors*). These proteins vary in composition and form multiprotein complexes that are firmly embedded in the inner mitochondrial membrane as shown in *Focus on Oxidative Phosphorylation* (**Focus Figure 24.1**). For example, some of the proteins, the **flavins**, contain flavin mononucleotide (FMN) derived from the vitamin riboflavin, and others contain both sulfur (S) and iron (Fe). Most of these proteins, however, are brightly colored iron-containing pigments called **cytochromes** (si′to-krōmz; *cyto* = cell, *chrom* = color), including complexes III and IV depicted in Focus Figure 24.1. Neighboring carriers are clustered together to form four **respiratory enzyme complexes** that are alternately reduced and oxidized as they pick up electrons and pass them on to the next complex in the sequence.

As Focus Figure 24.1 shows, the first such complex accepts hydrogen atoms from $NADH + H^+$, oxidizing it to NAD^+. $FADH_2$ transfers its hydrogen atoms slightly farther along the chain to the small complex II. The hydrogen atoms that the reduced coenzymes deliver to the electron transport chain are quickly split into protons (H^+) plus electrons (e^-). The electrons are shuttled along the inner mitochondrial membrane from one complex to the next, losing energy with each transfer. The protons escape into the watery matrix, only to be picked up and "pumped" across the inner mitochondrial membrane into the intermembrane space by one of the three major respiratory enzyme complexes (I, III, or IV).

Ultimately two electrons are delivered to half a molecule of O_2 (in other words, to an oxygen atom), creating oxygen ions (O^-) that strongly attract H^+ and form water, as indicated by the reaction

$$2H^+ + 2e^- + \tfrac{1}{2}O_2 \rightarrow H_2O$$

Oxidative phosphorylation has two phases:
Phase 1: The electron transport chain creates a proton (H$^+$) gradient across the inner mitochondrial membrane using high-energy electrons (e$^-$) removed from food fuels.
Phase 2: Chemiosmosis uses the energy of the proton gradient to synthesize ATP.

Phase 1:
Electron transport creates the proton gradient.

Phase 2:
Chemiosmosis uses the proton gradient to synthesize ATP.

Outer mitochondrial membrane

Intermembrane space

Inner mitochondrial membrane (crista)

Cyt c

2 H$^+$ + ½ O$_2$ H$_2$O

NADH + H$^+$

NAD$^+$

FADH$_2$

FAD

ATP synthase   V

ADP + P$_i$ ATP

Mitochondrial matrix

① Reduced coenzymes (NADH + H$^+$ and FADH$_2$) deliver electrons to respiratory enzyme complexes I and II.

② The electrons are transferred from one complex to another in the membrane.
• Each complex is reduced and then oxidized.
• The energy released pumps H$^+$ into the intermembrane space, creating an electrochemical gradient between the matrix and the intermembrane space.
• Coenzyme Q (ubiquinone) and cytochrome c shuttle electrons between the larger complexes.

③ At respiratory enzyme complex IV, two electrons combine with two protons (H$^+$) and a half molecule of O$_2$, forming water.

④ ATP synthase (complex V) harnesses the energy of the proton gradient to synthesize ATP. As H$^+$ flows back across the membrane through ATP synthase, the synthase rotor spins, causing P$_i$ to attach to ADP, forming ATP.

Glycolysis

Citric acid cycle

Electron transport chain and oxidative phosphorylation

ATP   ATP   ATP

Virtually all the water resulting from glucose oxidation is formed during oxidative phosphorylation. Because NADH + H⁺ and FADH₂ are oxidized as they release their cargo of hydrogen atoms, the net reaction for the electron transport chain is

$$\text{Coenzyme-2H} + \tfrac{1}{2}O_2 \rightarrow \text{coenzyme} + H_2O$$

reduced coenzyme   oxidized coenzyme

Each successive carrier has a greater affinity (binding strength, or "pull") for electrons than those preceding it. For this reason, the electrons cascade "downhill" from NADH + H⁺ to progressively lower energy levels until they are finally delivered to oxygen, which has the greatest affinity of all for electrons. You could say that oxygen "pulls" the electrons down the chain (**Figure 24.8**).

The electron transport chain functions as an energy converter by using the stepwise release of electronic energy to pump protons from the matrix into the intermembrane space. Because the inner mitochondrial membrane is nearly impermeable to H⁺, this chemiosmotic process creates an electrochemical **proton (H⁺) gradient** across that membrane, a gradient that has potential energy and the capacity to do work.

Figure 24.8 Energy is harvested at each step in the electron transport chain. The electron transport chain takes the large overall energy change between NADH + H⁺ or FADH₂ (from food fuels) and oxygen and harvests smaller, more manageable amounts of energy at each step.

The proton gradient:

- Creates a pH gradient, with the H⁺ concentration in the matrix much lower than that in the intermembrane space

- Generates a voltage across the membrane that is negative on the matrix side and positive between the mitochondrial membranes

Both conditions strongly attract H⁺ back into the matrix. But how can they get there?

The only areas of the inner mitochondrial membrane freely permeable to H⁺ are large enzyme-protein complexes (complex V) called **ATP synthases** (**Figure 24.9**). These complexes lay claim to being nature's smallest rotary motors. Each of these motors drives a molecular mill that fuses ADP and P_i together into ATP. How this molecular complex works is shown in **Figure 24.10**.

Notice something here: The ATP synthase works like an ion pump running in reverse. Recall from Chapter 3 (◄ p. 75) that ion pumps use ATP as their energy source to transport ions against an electrochemical gradient. Here we have ATP synthases using the energy of a proton gradient to power ATP synthesis.

The proton gradient also supplies energy to pump needed metabolites (ADP, pyruvic acid, inorganic phosphate) and calcium ions across the relatively impermeable inner mitochondrial membrane. The supply of energy from oxidation is not limitless, however, so when more of the gradient energy is used to drive these transport processes, less is available to make ATP.

HOMEOSTATIC IMBALANCE 24.1   CLINICAL

Studies of metabolic poisons support the chemiosmotic model of oxidative phosphorylation. For example, cyanide (the gas used in gas chambers) disrupts oxidative phosphorylation by binding to cytochrome oxidase and blocking electron flow from complex IV to oxygen (see Figure 24.8). Poisons called "uncouplers" abolish the proton gradient by making the inner mitochondrial membrane permeable to H⁺. Consequently, although the electron transport chain continues to deliver electrons to oxygen at a furious pace and oxygen consumption rises, no ATP is made. ● ● ● ● ● ●

Summary of ATP Production

The average person at rest uses energy at the rate of roughly 100 kcal/hour, which is equal to 116 watts, or slightly more than an old-fashioned lightbulb. This may seem a tiny amount, but from a biochemical standpoint it places a staggering power demand on our mitochondria. Luckily, they are up to the task.

When O_2 is present, cellular respiration is remarkably efficient. Of the 686 kcal of energy present in 1 mole of glucose, as much as 262 kcal can be captured in ATP bonds. (The rest is liberated as heat.) This corresponds to an energy capture of about 38%, making cells far more efficient than any human-made machines, which capture only 10–30% of the energy available to them.

During cellular respiration, most energy flows in this sequence:

$$\text{Glucose} \rightarrow \text{NADH} + \text{H}^+ \rightarrow \text{electron transport chain} \rightarrow$$
$$\text{proton gradient energy} \rightarrow \text{ATP}$$

Figure 24.9 Atomic force microscopy reveals the structure of energy-converting ATP synthase rotor rings.

Intermembrane space

① Protons (H⁺) flow down their electrochemical gradient through the stationary **stator**.

② Each H⁺ binds to a subunit of the **rotor**, causing it to change shape and making the rotor spin.

③ The spinning rotor turns the connecting **rod**.

④ As the rod spins, it activates catalytic sites in the **knob** that join P_i to ADP to make ATP.

ADP + P_i

ATP

Mitochondrial matrix

Figure 24.10 Structure and function of ATP synthase.

24

Figure 24.11 Energy yield during cellular respiration.

Let's do a little bookkeeping to summarize the net energy gain from one glucose molecule (**Figure 24.11**).

1. **Substrate-level phosphorylation.** Substrate-level phosphorylation gives us a net gain of 4 ATP (2 during glycolysis and 2 during the citric acid cycle).

2. **Oxidative phosphorylation.** NADH + H$^+$ and FADH$_2$ feeding into oxidative phosphorylation give us about 28 ATP.

   For each NADH + H$^+$, the proton gradient generates about 2½ ATP molecules. The 2 NADH + H$^+$ generated during glycolysis yield 5 ATP. The 8 NADH + H$^+$ produced during the transition reaction and citric acid cycle generate 20 ATP.

   The oxidation of FADH$_2$ is less efficient because it doesn't donate electrons to the "top" of the electron transport chain as does NADH + H$^+$, but to a lower energy level (at complex II). Each FADH$_2$ generates only about 1½ ATP, and so the 2 FADH$_2$ from the citric acid cycle are "worth" a total of 3 ATP.

Overall, complete oxidation of 1 glucose molecule to CO_2 and H_2O by both substrate-level phosphorylation and oxidative phosphorylation yields a maximum of 32 molecules of ATP. However, there is uncertainty about how much ATP is generated from the NADH + H$^+$ that comes from glycolysis that occurs *outside* the mitochondria. The inner mitochondrial membrane is not permeable to reduced NADH, so NADH + H$^+$ formed during glycolysis uses a *shuttle molecule* to deliver its electron pair to the electron transport chain. Some shuttles (for example, the glycerol phosphate shuttle) have an energy cost of 1 ATP per NADH, whereas other shuttles seem to provide a "free ride."

So, if we deduct 2 ATP to cover the average "fare" for the shuttle, our bookkeeping comes up with a grand total of 30 ATP per glucose as the typical energy yield. (Actually our figures are probably still too high because the proton gradient energy is also used to do other work and so the electron transport chain cannot recover all of the energy in a glucose molecule.)

Glycogenesis, Glycogenolysis, and Gluconeogenesis

Although most glucose is used to generate ATP molecules, unlimited amounts of glucose do *not* result in unlimited ATP synthesis, because cells cannot store large amounts of ATP. Aside from cellular respiration, the goal of carbohydrate metabolism is to make sure that just the right amount of glucose is present in the blood. Three processes with similar sounding names are required. **Figure 24.12** will help you keep them straight.

Glycolysis: Converts glucose to pyruvic acid
Glycogenesis: Polymerizes glucose to form glycogen
Glycogenolysis: Hydrolyzes glycogen to glucose monomers
Gluconeogenesis: Forms glucose from noncarbohydrate precursors

Figure 24.12 Quick summary of carbohydrate reactions.

Glycogenesis
- Synthesizes glycogen from glucose
- Occurs when glucose supplies exceed demand for ATP

Glycogenolysis
- Breaks down glycogen to release glucose
- Stimulated by low blood glucose

Figure 24.13 Glycogenesis and glycogenolysis.

When more glucose is available than can immediately be oxidized, rising intracellular ATP concentrations eventually inhibit glucose catabolism and cause glucose to be stored as glycogen or fat. Because the body can store much more fat than glycogen, fats account for 80–85% of stored energy.

Glycogenesis

When high ATP levels begin to "turn off" glycolysis, glucose molecules are combined in long chains to form glycogen, the animal carbohydrate storage product. This process is called **glycogenesis** (*glyco* = sugar; *genesis* = origin) (**Figure 24.13**, left side).

Glycogenesis begins as glucose entering cells is phosphorylated to glucose-6-phosphate and then converted to its isomer, *glucose-1-phosphate*. The terminal (last) phosphate group is split off as the enzyme *glycogen synthase* catalyzes the attachment of glucose to the growing glycogen chain. Liver and skeletal muscle cells are most active in glycogen synthesis and storage.

Glycogenolysis

On the other hand, when blood glucose levels drop, glycogen lysis (splitting) occurs. This process is known as **glycogenolysis** (gli″ko-jĕ-nol′ĭ-sis) (Figure 24.13, right side). The enzyme *glycogen phosphorylase* oversees phosphorylation and splitting of glycogen to release glucose-1-phosphate, which is then

converted to glucose-6-phosphate, a form that can enter the glycolysis pathway to be oxidized for energy.

In muscle cells and most other cells, the glucose-6-phosphate resulting from glycogenolysis is trapped because it cannot cross the cell membrane. However, hepatocytes (and some kidney and intestinal cells) contain *glucose-6-phosphatase*, an enzyme that removes the terminal phosphate, producing free glucose. Because glucose can then readily diffuse from the cell into the blood, the liver can use its glycogen stores to provide blood sugar for other organs when blood glucose levels drop. Liver glycogen is also an important energy source for skeletal muscles that have depleted their own glycogen reserves.

Gluconeogenesis

When too little glucose is available to stoke the "metabolic furnace," glycerol and amino acids are converted to glucose. **Gluconeogenesis**, the process of forming new (*neo*) glucose from *noncarbohydrate* molecules, occurs in the liver.

Gluconeogenesis takes place when dietary sources and glucose reserves have been used up and blood glucose levels are beginning to drop. Gluconeogenesis protects the body, especially the nervous system, from the damaging effects of low blood sugar (*hypoglycemia*) by ensuring that ATP synthesis can continue.

Check Your Understanding

12. What happens in glycolysis if oxygen and pyruvic acid are absent and NADH + H$^+$ cannot transfer its cargo of hydrogen to pyruvic acid?

13. What two major kinds of chemical reactions occur in the citric acid cycle, and how are these reactions indicated symbolically?

14. What name is given to the chemical reaction in which glycogen is broken down to its glucose subunits?

15. **DRAW** In the drawing of a cell and mitochondrion below, label the inner mitochondrial membrane, the outer mitochondrial membrane, the intermembrane space, and the matrix of the mitochondrion. Point to a location where each of the following processes takes place: glycolysis, citric acid cycle, oxidative phosphorylation. Identify the region where you would find the lowest pH.

For answers, see Answers Appendix.

24.5 Lipid metabolism is key for long-term energy storage and release

Learning Outcomes

▶ Describe the process by which fatty acids are oxidized for energy.

▶ Define ketone bodies, and indicate the stimulus for their formation.

Fats are the body's most concentrated source of energy. They contain very little water, and the energy yield from fat catabolism is more than twice that from either glucose or protein catabolism—9 kcal per gram of fat versus 4 kcal per gram of carbohydrate or protein. Most products of fat digestion are transported in lymph in the form of fatty-protein droplets called *chylomicrons* (see Chapter 23). Eventually, enzymes on capillary endothelium hydrolyze the lipids in the chylomicrons, and the resulting fatty acids and glycerol are taken up by body cells and processed in various ways. **Figure 24.14** summarizes the key metabolic reactions for lipids.

Oxidation of Glycerol and Fatty Acids

Of the various lipids, only triglycerides (◀ p. 45) are routinely oxidized for energy. Their catabolism involves the separate oxidation of their two different building blocks: glycerol and fatty acid chains (**Figure 24.15**).

Most body cells easily convert glycerol to glyceraldehyde 3-phosphate (a glycolysis intermediate) and eventually to acetyl CoA, which enters the citric acid cycle. Glyceraldehyde is equal to half a glucose molecule, and so the ATP energy harvest from its complete oxidation is approximately half that of glucose (15 ATP/glycerol).

Beta oxidation, the initial phase of fatty acid oxidation, occurs in the mitochondria. The net result is that the fatty acid chains are broken apart into two-carbon *acetic acid* fragments, and coenzymes (FAD and NAD+) are reduced (Figure 24.15, right side). Each acetic acid molecule is fused to coenzyme A, forming acetyl CoA. The term "beta oxidation" reflects the fact that the carbon in the beta (third) position is oxidized each time a two-carbon fragment is broken off. Acetyl CoA then enters the citric acid cycle where it is oxidized to CO_2 and H_2O.

Beta oxidation: Converts fatty acids to acetyl CoA
Lipolysis: Breaks down lipids to fatty acids and glycerol
Lipogenesis: Forms lipids from acetyl CoA and glyceraldehyde 3-phosphate

Figure 24.14 Quick summary of lipid reactions.

Figure 24.15 Lipid oxidation.

Lipogenesis

There is a continuous turnover of triglycerides in adipose tissue. New fats are stored for later use, and stored fats are broken down and released to the blood. That bulge of fatty tissue you see today does *not* contain the same fat molecules it did a month ago.

Glycerol and fatty acids from dietary fats not immediately needed for energy are recombined into triglycerides and stored. About 50% ends up in subcutaneous tissue, and the balance is stockpiled in other fat depots of the body.

Triglyceride synthesis, or **lipogenesis**, occurs when cellular ATP and glucose levels are high (**Figure 24.16**, magenta arrows). Excess ATP leads to an accumulation of acetyl CoA and glyceraldehyde 3-phosphate, two intermediates of glucose metabolism that would otherwise feed into the citric acid cycle. When these two metabolites are present in excess, they are channeled into triglyceride synthesis pathways.

Acetyl CoA molecules are joined together, forming fatty acid chains that grow two carbons at a time. (This accounts for

Figure 24.16 Lipid metabolism. When needed for energy, fats enter catabolic pathways (lipolysis). Excessive amounts of carbohydrates and amino acids are converted to triglycerides (lipogenesis) and stored.

the fact that almost all fatty acids in the body contain an even number of carbon atoms.) Because acetyl CoA, an intermediate in glucose catabolism, is also the *starting point* for fatty acid synthesis, glucose is easily converted to fat. Glyceraldehyde 3-phosphate is converted to glycerol, which is joined with fatty acids to form triglycerides. Consequently, even with a low-fat diet, carbohydrate intake can provide *all the raw materials* needed to make triglycerides. When blood sugar is high, lipogenesis is the major activity in adipose tissues and is also an important liver function.

Lipolysis

Lipolysis (lĭ-pol′ĭ-sis; "fat splitting"), the breakdown of stored fats into glycerol and fatty acids, is essentially lipogenesis in reverse (Figure 24.16, blue arrows). The fatty acids and glycerol are released to the blood, helping to ensure that body organs have continuous access to fat fuels for aerobic respiration. (The liver, cardiac muscle, and resting skeletal muscles actually prefer fatty acids as an energy fuel.)

The adage "Fats burn in the flame of carbohydrates" becomes clear when carbohydrate intake is inadequate. Under such conditions, lipolysis accelerates as the body attempts to fill the fuel gap with fats. However, the ability of acetyl CoA to enter the citric acid cycle depends on having sufficient carbohydrate intermediate molecules (see Figure 24.7). When carbohydrates are deficient, these intermediates are converted to glucose (to fuel the brain). Without them, fat oxidation is incomplete, and acetyl CoA accumulates.

Via a process called **ketogenesis**, the liver converts acetyl CoA molecules to **ketone bodies**, or **ketones**, which are released into the blood. Ketone bodies include acetoacetic acid, β-hydroxybutyric acid, and acetone. (The *keto acids* cycling through the citric acid cycle and the *ketone bodies* resulting from fat metabolism are quite different and should not be confused.)

HOMEOSTATIC IMBALANCE 24.2 **CLINICAL**

When ketone bodies accumulate in the blood, *ketosis* results and large amounts of ketone bodies are excreted in the urine. Ketosis is a common consequence of starvation and diabetes mellitus, and also occurs during carbohydrate restriction.

Because most ketone bodies are organic acids, ketosis leads to *metabolic acidosis*. The body's buffer systems cannot tie up the acids (ketones) fast enough, and blood pH drops to dangerously low levels. The person's breath smells fruity as acetone vaporizes from the lungs, and breathing becomes more rapid as the respiratory system tries to reduce blood carbonic acid by blowing off CO_2 to force the blood pH up. In severe untreated cases, the person may become comatose or even die as the acid pH depresses the nervous system. ●●●●●●

Synthesis of Structural Materials

All body cells use phospholipids and cholesterol to build their membranes. Phospholipids are important components of the myelin sheaths of neurons, and the ovaries, testes, and adrenal cortex use cholesterol to synthesize their steroid hormones. In addition, the liver:

- Synthesizes lipoproteins to transport cholesterol, fats, and other substances in the blood (see pp. 956–957)
- Synthesizes cholesterol from acetyl CoA
- Uses cholesterol to form bile salts (◄ p. 897)

Check Your Understanding

16. Which part of triglyceride molecules enters glycolysis?

17. What is the central molecule in fat metabolism?

18. What are the products of beta oxidation?

▬▬▬▬▬▬▬▬▬▬▬▬▬▬▬   *For answers, see Answers Appendix.*

24.6   Amino acids are used to build proteins or for energy

Learning Outcomes

▶ **Describe how amino acids are metabolized for energy.**

▶ **Describe the need for protein synthesis in body cells.**

Like all other biological molecules, proteins have a limited life span and must be broken down and replaced before they deteriorate. As proteins are broken down, their amino acids are recycled and used to build new proteins or modified to form a different N-containing compound. Cells actively take up newly ingested amino acids from the blood and use them to *replace* tissue proteins at the rate of about 100 grams each day.

Although popular opinion has it that excess protein can be stored by the body, nothing is farther from the truth. When more protein is available than is needed for anabolic purposes,

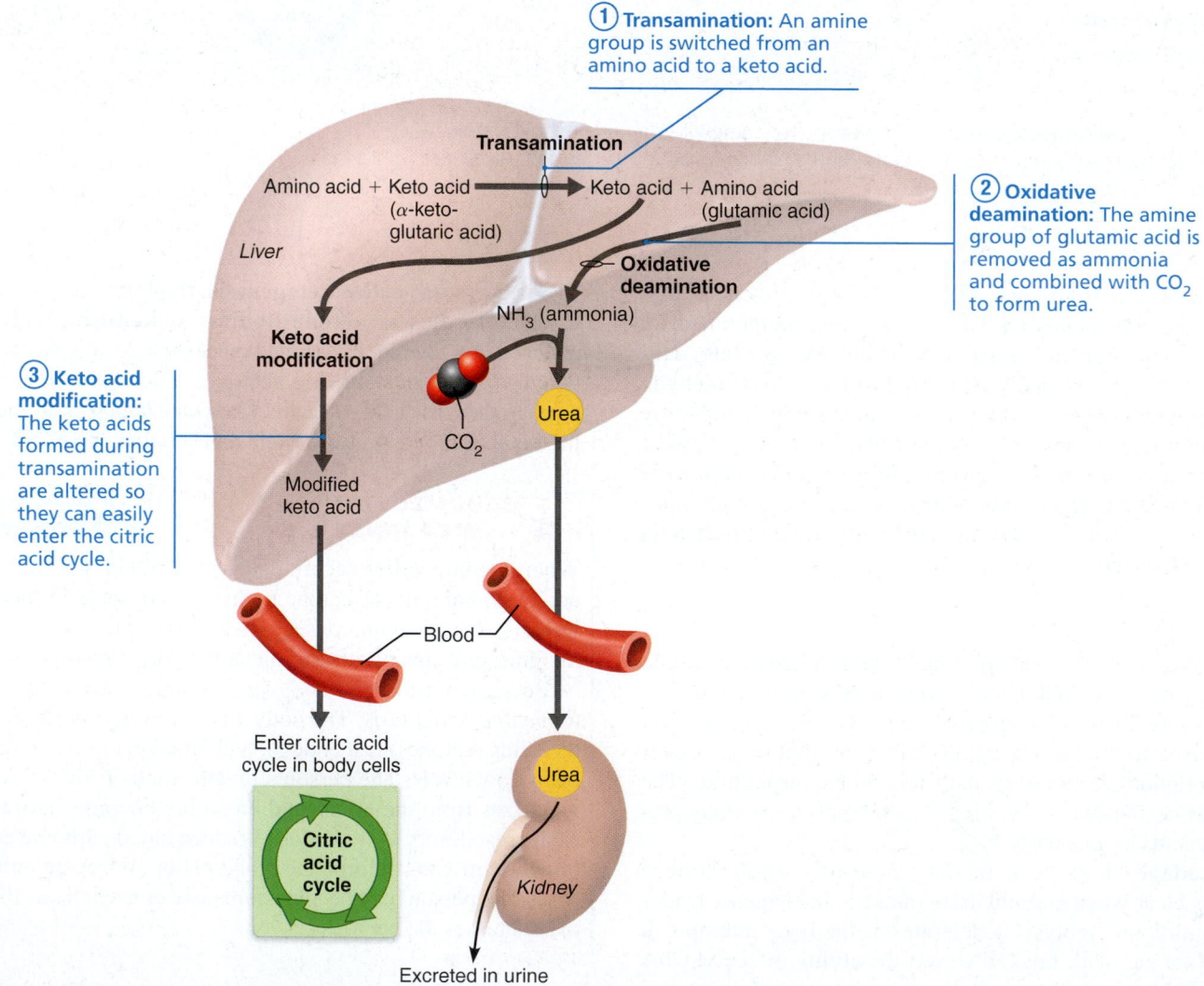

Figure 24.17 Processes that occur when amino acids are utilized for energy.

① **Transamination:** An amine group is switched from an amino acid to a keto acid.

Transamination

Amino acid + Keto acid ⟶ Keto acid + Amino acid
(α-keto-glutaric acid)   (glutamic acid)

Liver

② **Oxidative deamination:** The amine group of glutamic acid is removed as ammonia and combined with CO_2 to form urea.

Oxidative deamination

NH_3 (ammonia)

Keto acid modification

③ **Keto acid modification:** The keto acids formed during transamination are altered so they can easily enter the citric acid cycle.

CO_2

Urea

Modified keto acid

Blood

Enter citric acid cycle in body cells

Citric acid cycle

Urea

Kidney

Excreted in urine

amino acids are oxidized for energy or converted to fat for future energy needs.

Degradation of Amino Acids

The goal of amino acid degradation is to produce molecules that can be oxidized for energy in the citric acid cycle or converted to glucose. Before amino acids can be oxidized or converted, they must be *deaminated*, that is, their amine group (NH_2) must be removed. (In sulfur-containing amino acids, such as methionine and cysteine, sulfur must also be removed.) The resulting molecule is then converted to pyruvic acid or to one of the keto acid intermediates in the citric acid cycle.

The key molecule in these conversions is the amino acid *glutamic acid* (gloo-tam′ik). As **Figure 24.17** shows, the following events occur:

① **Transamination** (trans″am-ĭ-na′shun). A number of amino acids can transfer their amine group to α-ketoglutaric acid (a citric acid cycle keto acid), thereby transforming α-ketoglutaric acid to glutamic acid. In the process, the original amino acid becomes a keto acid (that is, it has an oxygen atom where the amine group formerly was). This reaction is fully reversible.

② **Oxidative deamination.** In the liver, the amine group of glutamic acid is removed as **ammonia (NH_3)**, and α-ketoglutaric acid is regenerated. The liberated NH_3 molecules are combined with CO_2, yielding **urea** and water. The urea is released to the blood and excreted from the body in urine. Because ammonia is toxic to body cells, the ease with which glutamic acid funnels amine groups into the **urea cycle** is extremely important. This cycle rids the body not only of NH_3 produced during oxidative deamination, but also of bloodborne NH_3 produced by intestinal bacteria.

③ **Keto acid modification.** Keto acids resulting from transamination are altered as necessary to produce metabolites that can enter the citric acid cycle. The most important of these metabolites are pyruvic acid, acetyl CoA, α-ketoglutaric acid, and oxaloacetic acid (see Figure 24.7). Because the reactions of glycolysis are reversible, deaminated amino acids that are converted to pyruvic acid can be reconverted to glucose and contribute to gluconeogenesis.

A quick summary of the key reactions is shown in **Figure 24.18**.

Transamination: Transfers an amine group from an amino acid to α-ketoglutaric acid, generating glutamic acid

Oxidative deamination: Removes an amine group (as ammonia) from glutamic acid and regenerates α-ketoglutaric acid

Figure 24.18 Quick summary of amino acid catabolism.

Protein Synthesis

Amino acids are the most important anabolic nutrients. Not only do they form all structural proteins, but they form most of the body's functional molecules as well. As we described in Chapter 3 (◄ pp. 104–107), protein synthesis occurs on ribosomes, where ribosomal enzymes oversee the formation of peptide bonds linking the amino acids together into protein polymers. Hormones (growth hormone, thyroxine, sex hormones, and others) precisely control the amount and type of protein synthesized. The rate of protein anabolism changes as we age and our balance of hormones changes.

During your lifetime, your cells will synthesize 225–450 kg (about 500–1000 lb) of proteins, depending on your size. However, you do not need to consume anywhere near that amount of protein because the body easily forms nonessential amino acids by siphoning keto acids from the citric acid cycle and transferring amine groups to them. Most of these transformations occur in the liver, which provides nearly all the nonessential amino acids needed to produce the relatively small amount of protein that the body synthesizes each day.

However, a complete set of amino acids must be present for protein synthesis to take place, so the diet must provide all essential amino acids. If some are lacking, the rest are oxidized for energy even though they may be needed for anabolism. In such cases, body protein is broken down to supply the essential amino acids needed, and negative nitrogen balance results.

Check Your Understanding

19. What does the liver use as its substrates when it synthesizes nonessential amino acids?

20. What happens to the ammonia removed from amino acids when the amino acids are used for energy fuel?

For answers, see Answers Appendix.

24.7 Energy is stored in the absorptive state and released in the postabsorptive state

Learning Outcomes

▶ **Explain the concept of amino acid or carbohydrate-fat pools, and describe pathways by which substances in these pools can be interconverted.**

▶ **Summarize important events of the absorptive and postabsorptive states, and explain how these events are regulated.**

Catabolic-Anabolic Balance of the Body

Now that we have examined metabolism at cellular levels, let's step back to look at the body as a system that deploys metabolic processes to provide fuel. Your body exists in a *dynamic catabolic-anabolic balance*, that is, its organic molecules are continuously broken down and rebuilt—frequently at a head-spinning rate.

The body draws on its **nutrient pools**—the current stocks of amino acids, carbohydrates, and fats—to meet its varying needs

(**Figure 24.19**). These pools are interconvertible because their pathways are linked by key intermediates. The liver, adipose tissue, and skeletal muscles are the primary effector organs or tissues determining the amounts and direction of the conversions shown in the figure.

The **amino acid pool** is the body's total supply of free amino acids. Our bodies lose small amounts of amino acids and proteins daily in urine and in sloughed hairs and skin cells. Typically, we replace these lost molecules via our diet. Otherwise, amino acids arising from tissue breakdown return to the pool. This pool is the source of amino acids used to synthesize proteins and form amino acid derivatives. In addition, as we described above, gluconeogenesis can convert deaminated amino acids to glucose. Not all events of amino acid metabolism occur in all cells. For example, *only* the liver forms urea. Nonetheless, the concept of a common amino acid pool is valid because all cells are connected by the blood.

Because carbohydrates are easily and frequently converted to fats, the **carbohydrate** and **fat pools** are usually considered together. There are two major differences between the carbohydrate-fat pool and the amino acid pool:

- Fats and carbohydrates are oxidized directly to produce cellular energy, whereas amino acids can be used to supply energy *only after being converted to a citric acid cycle intermediate* (a keto acid).

- Excess carbohydrate and fat can be stored as such, whereas excess amino acids are *not* stored as protein. Instead, they are oxidized for energy or converted to fat or glycogen for storage.

Metabolic controls act to equalize blood concentrations of energy sources between two nutritional states—the *absorptive state* and the *postabsorptive state*.

Absorptive State

The absorptive state, or *fed state,* lasts about four hours after eating begins, when nutrients are flushing into the blood from the GI tract. During the absorptive state, anabolism exceeds catabolism and nutrients are stored (**Figure 24.20**). Glucose is the major energy fuel. Dietary amino acids and fats are used to remake degraded body protein or fat, and small amounts are oxidized to provide ATP. Excess metabolites, regardless of source, are transformed to fat if not used for anabolism. Let's look at the fate and hormonal control of each nutrient group during this phase.

Carbohydrates

Absorbed monosaccharides are delivered directly to the liver, where fructose and galactose are converted to glucose. Glucose,

Figure 24.19 Interconversion of carbohydrates, fats, and proteins. These interconversions primarily occur in the liver, adipose tissue, and skeletal muscle.

in turn, is released to the blood or converted to glycogen and fat. Glycogen formed in the liver is stored there, but most fat synthesized there is packaged with proteins as *very low-density lipoproteins* (*VLDLs*; see p. 956) and released to the blood to be picked up by adipose tissues for storage.

Bloodborne glucose not sequestered by the liver enters body cells to be metabolized for energy. Excess glucose is stored in skeletal muscle cells as glycogen or in adipose cells as fat.

Triglycerides

Nearly all products of fat digestion enter the lymph in the form of chylomicrons (◄ p. 916). Triglycerides are hydrolyzed to fatty acids and glycerol, which can then pass through the capillary walls. *Lipoprotein lipase*, the enzyme that catalyzes fat hydrolysis, is particularly active in the capillaries of muscle and fat tissues (◄ p. 917).

Adipose cells, skeletal and cardiac muscle cells, and liver cells use triglycerides as their primary energy source, but other cells also begin to oxidize more fat for energy when dietary carbohydrates are limited. Although some fatty acids and glycerol are used for anabolic purposes by tissue cells, most enter adipose tissue to be reconverted to triglycerides and stored.

Amino Acids

Absorbed amino acids are delivered to the liver, which deaminates some of them to keto acids. The keto acids may flow into the citric acid cycle to be used for ATP synthesis, or they may be converted to liver fat stores. The liver also uses some

(a) Major events of the absorptive state

(b) Principal pathways of the absorptive state

Figure 24.20 Major events and principal metabolic pathways of the absorptive state.
Although not indicated in (b), amino acids are also taken up by tissue cells and used for protein
synthesis, and fats (triglycerides) are the primary energy fuel of muscle, liver cells, and adipose tissue.

of the amino acids to synthesize plasma proteins, including albumin, clotting proteins,
and transport proteins. However, most amino acids flushing through the liver sinusoids
remain in the blood for uptake by other body cells, where they are used to synthesize
proteins.

Hormonal Control of the Absorptive State

Insulin directs essentially all events of the absorptive state (**Figure 24.21**). After a meal, rising blood glucose and amino acid levels stimulate the beta cells of the pancreatic islets to secrete more insulin (◄ **Figure 16.17, p. 631**). The GI tract hormone *glucose-dependent insulinotropic peptide* (*GIP*) and parasympathetic stimulation also promote the release of insulin.

Insulin binds to membrane receptors of its target cells. This stimulates the shuttling of glucose transporters to the plasma membrane, which enhances the carrier-mediated facilitated diffusion of glucose into those cells. Within minutes, the rate of glucose entry into tissue cells (particularly muscle and adipose cells) increases about 20-fold. The exception is brain and liver cells, which take up glucose whether or not insulin is present.

Once glucose enters tissue cells, insulin enhances glucose oxidation for energy and stimulates its conversion to glycogen and, in adipose tissue, to triglycerides. Insulin also "revs up" the active transport of amino acids into cells, promotes protein synthesis, and inhibits liver export of glucose and virtually all liver enzymes that promote gluconeogenesis.

As you can see, insulin is a **hypoglycemic hormone** (hi″po-gli-se′mik). It sweeps glucose out of the blood into tissue cells, lowering blood glucose levels. It also enhances glucose oxidation or storage while inhibiting any process that might raise blood glucose levels.

HOMEOSTATIC IMBALANCE 24.3 CLINICAL

Diabetes mellitus is a disorder of inadequate insulin production or abnormal insulin receptors. Without insulin or receptors that "recognize" it, glucose is unavailable to most body cells. Blood glucose levels remain high, and large amounts of glucose are excreted in urine. Metabolic acidosis, protein wasting, and weight loss occur as large amounts of fats and tissue proteins are used for energy. (Diabetes mellitus is described in more detail in Chapter 16, ◄ **pp. 632–634**.) ●●●●●●●

Postabsorptive State

The postabsorptive state, or *fasting state,* is the period when the GI tract is empty and body reserves are broken down to supply energy. Net synthesis of fat, glycogen, and proteins ends, and catabolism of these substances begins (**Figure 24.22a**).

The primary goal during the postabsorptive state is to maintain blood glucose levels within the homeostatic range (70–110 mg of glucose per 100 ml). Remember that constant blood glucose is important because the brain almost always uses glucose as its energy source. Most events of the postabsorptive state either (1) make glucose available to the blood or (2) make certain organs (such as skeletal muscle) switch over to using fats instead of glucose to spare glucose for organs that can't use fats.

Sources of Blood Glucose

So where does blood glucose come from in the postabsorptive state? Sources include stored glycogen in the liver and skeletal muscles, tissue proteins, and, in limited amounts, fats (Figure 24.22b).

① **Glycogenolysis in the liver.** The liver's glycogen stores (about 100 g) are the first line of glucose reserves. They are mobilized quickly and can maintain blood sugar levels for about four hours during the postabsorptive state.

② **Glycogenolysis in skeletal muscle.** Glycogen stores in skeletal muscle are approximately equal to those of the liver. Before liver glycogen is exhausted, glycogenolysis begins in skeletal muscle (and to a lesser extent in other tissues). However, the glucose-6-phosphate produced is not released to the blood because, unlike the liver, skeletal muscle lacks the enzymes needed to dephosphorylate glucose. Instead, glucose is partly oxidized to pyruvic acid (or, during anaerobic conditions, lactic acid), which enters the blood, is reconverted to glucose by the liver, and is released to the blood again.

Figure 24.21 Insulin directs nearly all events of the absorptive state. (Note: Not all effects shown occur in all cells.)

(a) Major events of the postabsorptive state

(b) Principal pathways of the postabsorptive state

Figure 24.22 Major events and principal metabolic pathways of the postabsorptive state.

③ **Lipolysis in adipose tissues and the liver.** Adipose and liver cells produce glycerol by lipolysis, and the liver converts the glycerol to glucose (gluconeogenesis) and releases it to the blood. Because acetyl CoA, a product of the beta oxidation of fatty acids, is produced beyond the *reversible* steps of glycolysis, fatty acids *cannot* be used to bolster blood glucose levels.

④ **Catabolism of cellular protein.** Tissue proteins become the major source of blood glucose during prolonged fasting when glycogen and fat stores are nearly exhausted. Cellular amino acids are deaminated and converted to glucose in the liver. During fasts lasting several weeks, the kidneys also carry out gluconeogenesis and contribute as much glucose to the blood as the liver.

In these situations, the body sets priorities. Muscle proteins are the first to be catabolized. Movement is not nearly as important as maintaining wound healing and the immune response. Of course there are limits to the amount of tissue protein that can be catabolized before the body stops functioning. The heart is almost entirely muscle protein, and when it is severely catabolized, the result is death. In general, the amount of fat the body contains determines the time a person can survive without food.

Glucose Sparing

During prolonged periods without food, the body can adapt to burn more fats and proteins, which enter the citric acid cycle along with glucose breakdown products. The increased use of noncarbohydrate fuel molecules (especially triglycerides) to conserve glucose is called **glucose sparing**.

As the body progresses from the absorptive to the postabsorptive state, the brain continues to take its share of blood glucose, but virtually every other organ switches to fatty acids as its major energy source, sparing glucose for the brain. During this transition phase, lipolysis begins in adipose tissues. Tissue cells pick up released fatty acids and oxidize them for energy. In addition, the liver oxidizes fats to ketone bodies and releases them to the blood for use by tissue cells.

If fasting continues for longer than four or five days, the brain too begins to use large quantities of ketone bodies as well as glucose as its energy fuel. The brain's ability to use an alternative fuel source has survival value—much less tissue protein has to be ravaged to form glucose.

Hormonal and Neural Controls of the Postabsorptive State

Several hormones and the sympathetic nervous system together control events of the postabsorptive state. Consequently, regulation of this state is much more complex than that of the absorptive state when a single hormone, insulin, holds sway.

An important trigger for initiating postabsorptive events is reduced insulin release, which occurs as blood glucose levels drop. Falling insulin levels inhibit all insulin-induced cellular responses.

Glucagon   Declining glucose levels also stimulate the alpha cells of the pancreatic islets to release the insulin antagonist **glucagon**. Like other hormones acting during the postabsorptive state, glucagon is a **hyperglycemic hormone**—it raises blood glucose levels.

Glucagon targets the liver and adipose tissue (**Figure 24.23**).

- Hepatocytes respond by accelerating glycogenolysis and gluconeogenesis.
- Adipose cells mobilize their fatty stores (lipolysis) and release fatty acids and glycerol to the blood. In this way, glucagon "refurbishes" blood energy sources by enhancing *both* glucose and fatty acid levels. Under certain hormonal conditions and persistent low glucose levels or prolonged fasting, most of the fat mobilized is converted to ketone bodies.

Glucagon release is inhibited after the next meal or whenever blood glucose levels rise and insulin secretion begins again.

So far, the picture is pretty straightforward. Increasing blood glucose levels trigger insulin release, which "pushes" glucose out of blood and into cells. This drop in blood glucose stimulates secretion of glucagon, which "pulls" glucose from the cells into blood. However, there is more than a push-pull mechanism here because rising amino acid levels in the blood stimulate the release of *both* insulin and glucagon.

This effect is insignificant when we eat a balanced meal, but it has an important adaptive role when we eat a high-protein, low-carbohydrate meal. In this instance, the stimulus for insulin release is strong, and if it were not counterbalanced, the brain might be damaged by the abrupt onset of hypoglycemia as glucose rushes out of the blood. Simultaneous release of glucagon modulates the effects of insulin and helps stabilize blood glucose levels.

Sympathetic Nervous System   The sympathetic nervous system also plays a crucial role in supplying fuel quickly when

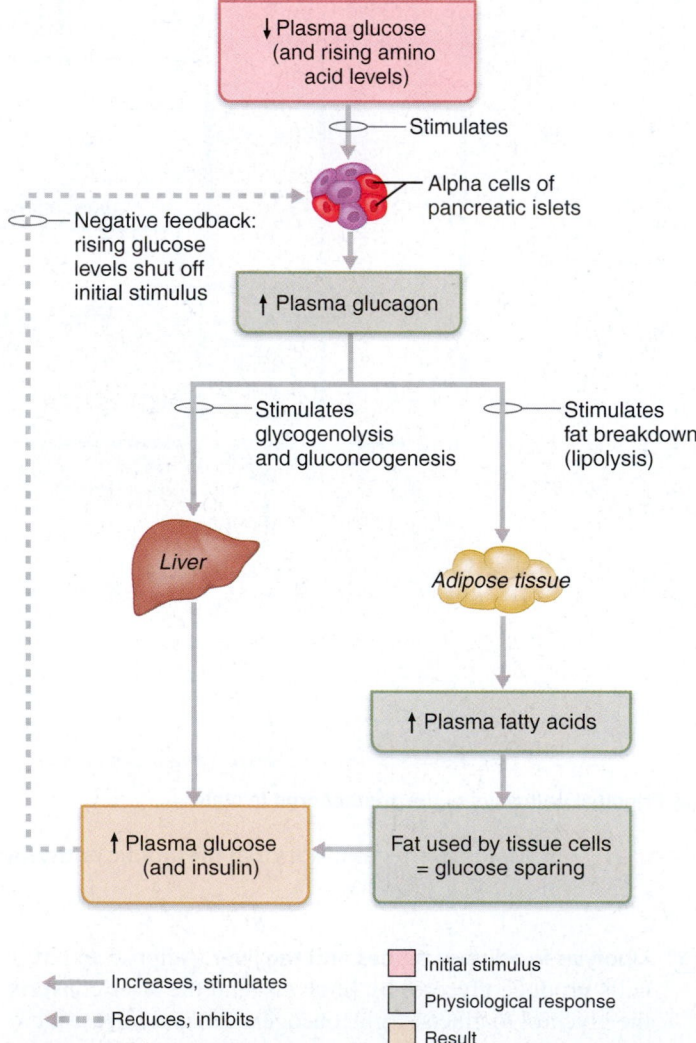

Figure 24.23 Glucagon is a hyperglycemic hormone that stimulates a rise in blood glucose levels.

Table 24.4 Summary of Normal Hormonal Influences on Metabolism

HORMONE'S EFFECTS	INSULIN	GLUCAGON	EPINEPHRINE	GROWTH HORMONE	THYROXINE	CORTISOL	TESTOSTERONE
Stimulates glucose uptake by cells	✓				✓		
Stimulates amino acid uptake by cells	✓			✓			
Stimulates glucose catabolism for energy	✓				✓		
Stimulates glycogenesis	✓						
Stimulates lipogenesis and fat storage	✓						
Inhibits gluconeogenesis	✓						
Stimulates protein synthesis (anabolic)	✓			✓	✓		✓
Stimulates glycogenolysis		✓	✓				
Stimulates lipolysis and fat mobilization		✓	✓	✓	✓	✓	
Stimulates gluconeogenesis		✓	✓	✓		✓	
Stimulates protein breakdown (catabolic)						✓	

blood glucose levels drop suddenly. Adipose tissue is well supplied with sympathetic fibers, and epinephrine released by the adrenal medulla in response to sympathetic activation acts on the liver, skeletal muscle, and adipose tissues. Together, these stimuli mobilize fat and promote glycogenolysis—essentially the same effects prompted by glucagon.

Injury, anxiety, or any other stressor that mobilizes the fight-or-flight response will trigger this control pathway, as does exercise. During exercise, large amounts of fuels must be made available for muscles, and the metabolic profile is essentially the same as that of a fasting person when glucagon and the sympathetic nervous system are in control except that the facilitated diffusion of glucose into muscle is enhanced.

Other Hormones A number of other hormones—including growth hormone, thyroxine, sex hormones, and corticosteroids—influence metabolism and nutrient flow. Prolonged fasting or rapid declines in blood glucose levels enhance growth hormone secretion, which exerts important anti-insulin effects (◄ Figure 16.5, p. 614).

However, the release and activity of most of these hormones are not specifically related to absorptive or postabsorptive metabolic events. **Table 24.4** summarizes typical metabolic effects of various hormones.

Check Your Understanding

21. Which three organs or tissues are the primary effector organs determining the amounts and directions of interconversions in the nutrient pools?
22. Generally speaking, what kinds of reactions and events characterize the absorptive state? The postabsorptive state?
23. Which hormone is glucagon's main antagonist?
24. Which event increases both glucagon and insulin release?

For answers, see Answers Appendix.

24.8 The liver metabolizes, stores, and detoxifies

Learning Outcomes

▶ **Describe several metabolic functions of the liver.**
▶ **Differentiate between LDLs and HDLs relative to their structures and major roles in the body.**

The liver is one of the most biochemically complex organs in the body. It processes nearly every class of nutrients and plays a major role in regulating plasma cholesterol levels. In a pinch, mechanical contraptions can stand in for a failed heart, lungs, or kidney, but the only thing that can do the versatile liver's work is a hepatocyte.

The hepatocytes carry out some 500 or more intricate metabolic functions. A description of all of these functions is well beyond the scope of this text, but **Table 24.5** on p. 956 provides a brief summary.

Cholesterol Metabolism and Regulation of Blood Cholesterol Levels

Cholesterol has received little attention in this discussion so far, primarily because it is not used as an energy source. It serves instead as the structural basis of bile salts, steroid hormones, and vitamin D and is a major component of plasma membranes.

About 15% of blood cholesterol comes from the diet. The liver—and to a lesser extent other body cells, particularly intestinal cells—make the other 85% from acetyl CoA. Cholesterol

Table 24.5   Summary of Metabolic Functions of the Liver

METABOLIC PROCESSES TARGETED	FUNCTIONS
Carbohydrate Metabolism	
Particularly important in maintaining blood glucose homeostasis	• Converts galactose and fructose to glucose • Stores glucose as glycogen when blood glucose levels are high; in response to hormonal controls, performs glycogenolysis and releases glucose to blood • Gluconeogenesis: converts amino acids and glycerol to glucose when glycogen stores are exhausted and blood glucose levels are falling • Converts glucose to fats for storage
Fat Metabolism	
Although most cells are capable of some fat metabolism, liver bears the major responsibility	• Primary body site of beta oxidation (breakdown of fatty acids to acetyl CoA) • Converts excess acetyl CoA to ketone bodies for release to tissue cells • Stores fats • Synthesizes lipoproteins for transport of fatty acids, fats, and cholesterol in blood • Synthesizes cholesterol from acetyl CoA; catabolizes cholesterol to bile salts, which are secreted in bile (◄ p. 897)
Protein Metabolism	
Protein metabolism by the liver is essential for life (e.g., clotting protein synthesis, ammonia disposal)	• Synthesizes urea to remove ammonia from body; inability to perform this function (e.g., in cirrhosis or hepatitis) allows ammonia to accumulate in the blood • Synthesizes most plasma proteins (exceptions are antibodies and some hormones and enzymes); inability to perform this function (e.g., in cirrhosis) causes edema (◄ p. 734) • Deaminates amino acids for their conversion to glucose or for ATP synthesis (amount of deamination that occurs outside the liver is unimportant) • Transamination: interconversion of nonessential amino acids (amount that occurs outside liver is unimportant)
Vitamin/Mineral Storage	
	• Stores vitamin A (1–2 years' supply) • Stores sizable amounts of vitamins D and B_{12} (1–4 months' supply) • Stores iron; other than iron bound to hemoglobin, most of body's supply is stored in liver as ferritin
Biotransformation Functions	
	• Metabolizes alcohol, drugs, and other toxins by inactivating them for excretion by the kidneys, and performs reactions that may result in products which are more active, changed in activity, or less active • Processes bilirubin resulting from RBC breakdown and excretes bile pigments in bile (◄ p. 649) • Metabolizes bloodborne hormones to forms that can be excreted in urine

is lost from the body when it is catabolized and secreted in bile salts, which are eventually excreted in feces (◄ p. 897).

Cholesterol Transport

Because triglycerides and cholesterol are insoluble in water, they do not circulate free in the blood. Instead, they are transported to and from tissue cells bound to small lipid-protein complexes called **lipoproteins**. These complexes solubilize the hydrophobic lipids, and contain signals that regulate lipid entry and exit at specific target cells.

Lipoproteins vary in their relative fat-protein composition, but they all contain triglycerides, phospholipids, and cholesterol in addition to protein (**Figure 24.24**). In general, the higher the percentage of lipid in the lipoprotein, the lower its density; and the greater the proportion of protein, the higher its density. On this basis, there are **very low-density lipoproteins (VLDLs), low-density lipoproteins (LDLs),** and **high-density lipoproteins (HDLs)**. *Chylomicrons*, which transport absorbed lipids from the GI tract, have the lowest density of all.

The liver is the primary source of VLDLs, which transport triglycerides from the liver to peripheral (nonliver) tissues,

Protein

Phospholipid

Cholesterol

Cholesterol bound to fatty acid

Triglyceride

| **Chylomicron** carries absorbed lipids from intestine to liver. 80–95% triglyceride 3–6% phospholipid 2–7% cholesterol 1–2% protein 1000-nm diameter | **Very low-density lipoprotein (VLDL)** carries triglycerides from liver, mostly to adipose tissue. 55–65% triglyceride 15–20% phospholipid 10–15% cholesterol 5–10% protein 70-nm diameter | **Low-density lipoprotein (LDL)** carries cholesterol from liver to tissues. 10% triglyceride 20% phospholipid 45% cholesterol 25% protein 20-nm diameter | **High-density lipoprotein (HDL)** carries cholesterol from tissues to liver. 5% triglyceride 30% phospholipid 20% cholesterol 45–50% protein 10-nm diameter |

Figure 24.24 Composition and function of lipoproteins. The VLDL has been cut open to show its contents.

mostly to *adipose tissues*. Once the triglycerides are unloaded, the residues are called LDLs, which are cholesterol-rich. The job of the LDLs is to transport cholesterol *to peripheral tissues*, making it available to tissue cells to synthesize membranes or hormones, and to store it for later use.

Most cells other than liver and intestinal cells obtain the bulk of the cholesterol they need for membrane synthesis from the blood. When a cell needs cholesterol, it makes membrane receptor proteins for LDL. LDL binds to the receptors, is engulfed by endocytosis, and the endocytotic vesicles fuse with lysosomes, where the cholesterol is freed for use. When excessive cholesterol accumulates in a cell, it inhibits both the cell's own cholesterol synthesis and its synthesis of LDL receptors.

The major function of HDLs, which are particularly rich in phospholipids and proteins, is to scoop up and transport excess cholesterol *from peripheral tissues to the liver*, where it is broken down and becomes part of bile. The liver makes the protein envelopes of the HDL particles and then ejects them into the bloodstream in collapsed form, rather like deflated beach balls. Once in the blood, these still-incomplete HDL particles fill with cholesterol picked up from tissue cells and "pulled" from artery walls. Surprisingly, HDL also provides the steroid-producing organs, like the ovaries and adrenal glands, with their raw material (cholesterol).

Blood Levels of Total Cholesterol, HDL, and LDL

For adults, a total cholesterol level below 200 mg/dl of blood is generally considered desirable. Blood cholesterol levels above 200 mg/dl have been linked to atherosclerosis, which clogs

the arteries and causes strokes and heart attacks (◄ p. 760). However, it is not enough to simply measure total cholesterol. How cholesterol is packaged for transport in the blood is more important clinically.

As a rule, high levels of HDLs are considered *good* because the transported cholesterol is destined for degradation (think H for healthy). HDL levels above 60 mg/dl are thought to protect against heart disease, and levels below 40 are considered undesirable. In the United States, HDL levels average 40–50 in males and 50–60 in women.

High LDL levels (160 mg/dl or above) are generally considered *bad* (think L for lousy) because when LDLs are excessive, potentially lethal cholesterol deposits are laid down in the artery walls. A good rule of thumb is that HDL levels can't be too high and LDL levels can't be too low.

Factors Regulating Blood Cholesterol Levels

A negative feedback loop partially adjusts the amount of cholesterol produced by the liver according to the amount of cholesterol in the diet. A high cholesterol intake inhibits its synthesis by the liver, but it is not a one-to-one relationship because the liver produces a basal amount of cholesterol even when dietary intake is high. Conversely, severely restricting dietary cholesterol, although helpful, does not markedly reduce blood cholesterol levels.

However, the relative amounts of saturated and unsaturated fatty acids in the diet do have an important effect on blood cholesterol levels. Saturated fatty acids *stimulate liver synthesis* of cholesterol and *inhibit its excretion* from the body. In contrast,

24

Obesity: Magical Solution Wanted

Fat—unwanted, unloved, and yet often overabundant. Besides the physical toll, the social stigma and economic disadvantages of obesity are legendary. A fat person pays higher insurance premiums, is discriminated against in the job market, has fewer clothing choices, and endures frequent humiliation throughout life.

What Causes Obesity?

It's a fair bet that few people choose to become obese, and yet too many of us are. So what causes it? Fundamentally, obesity is caused by consuming more calories than you use. With easy access to large amounts of calorie-dense foods and less physical activity, our waistlines have expanded. However, it is much more complicated than that. Some of the other factors involved are:

1. **Obese people are more fuel efficient and more effective "fat storers."** Although it is often assumed that people who are obese eat more, this is not necessarily true—many actually eat less than people of normal weight. When yo-yo dieters lose weight, their metabolic rate falls sharply. But when they subsequently gain weight, their metabolic rate temporarily increases like a furnace being stoked. Each successive weight loss occurs more slowly, but they regain weight faster.

   To make matters worse, "fat" fat cells of overweight people:

   - Sprout more alpha-adrenergic receptors, which favor fat accumulation.
   - Send different molecular messages than "thin" fat cells. They spew out

   inflammatory cytokines that can promote insulin resistance and they release less adiponectin, a hormone that improves the action of insulin in glucose uptake and storage.

   - Have exceptionally efficient lipoprotein lipases, which unload fat from the blood (usually to fat cells).

2. **Genetic predisposition.** A true genetic predisposition for "fatness" appears to account for only about 5% of obese Americans. For these people, extra calories are always laid down as fat, as opposed to people who lay down more muscle with some of the excess.

3. **Gut bacteria.** New evidence suggests that the kinds and proportions of the various bacteria that live in the gut can have a profound influence on the development and maintenance of obesity. The gut bacteria of obese individuals are less diverse, leading to higher levels of inflammation. In addition, evidence suggests that the gut bacteria of obese individuals generate fewer short-chain fatty acids. Since short-chain fatty acids inhibit fat deposition and enhance satiety hormones, obese individuals lay down more fat. The good news is that you may be able to change your gut bacteria through changes in your diet.

4. **Environmental chemicals.** A number of environmental contaminants such as pesticides, fungicides, and chemicals found in plastics interact with hormonal pathways. In animal models, these chemicals cause obesity. While it has been much more difficult to establish a link in humans, environmental chemicals may well play a role in the increased prevalence of obesity.

Treatments— The Bad and the Good

Some so-called treatments for obesity are almost more dangerous than the disease itself. Unfortunate strategies include the following.

1. **"Water pills."** Diuretics prompt the kidneys to excrete more water and may cause a few pounds of weight loss for a few hours. They can also cause serious dehydration and electrolyte imbalance.

2. **Liposuction.** Liposuction reshapes the body by suctioning off fat deposits, but it is not a good choice for losing weight. It carries all the risks of surgery, and

unsaturated fatty acids (found in olive and most other vegetable oils) *enhance excretion* of cholesterol and its catabolism to bile salts, thereby reducing total cholesterol levels.

Trans fats are made from "healthy" oils that have been hydrogenated. This makes them more solid, such as some margarines. Trans fats have a worse effect on blood cholesterol levels than saturated fats do. The trans fatty acids spark a greater increase in LDLs and a greater reduction in HDLs, producing the unhealthiest combination.

The unsaturated omega-3 fatty acids found in especially large amounts in some cold-water fish (such as salmon) lower the proportions of both saturated fats and cholesterol. The omega-3

fatty acids also make blood platelets less sticky, thus helping prevent spontaneous clotting that can block blood vessels. In addition, they appear to lower blood pressure.

Factors other than diet also influence blood cholesterol levels. For example, cigarette smoking and stress lower HDL levels, whereas regular aerobic exercise and estrogen lower LDL levels and increase HDL levels. Interestingly, body shape provides clues to risky blood levels of cholesterol and fats. "Apples" (people with upper body and abdominal fat, seen more often in men) tend to have higher levels of cholesterol and LDLs than "pears" (whose fat is localized in the hips and thighs, a pattern more common in women).

unless eating habits change, fat deposits elsewhere in the body overfill.

3. **Diet drugs and weight-loss supplements.** Older weight-loss drugs (e.g., phentermine) are generally amphetamine-like stimulants that decrease hunger but also increase blood pressure and heart rate.

Another drug, orlistat, interferes with pancreatic lipase so that part of the fat eaten is not digested or absorbed. This also interferes with the absorption of fat-soluble vitamins. It is effective as a weight-loss agent, but its side effects (diarrhea and anal leakage) are unpleasant to say the least. It also carries the risk of severe liver injury.

Several over-the-counter weight-loss supplements that claim to increase metabolism and burn calories have proved to be very dangerous. For example:

- Capsules containing usnic acid damage hepatocytes and have led to liver failure in a few cases.
- Ephedra-containing supplements are notorious—over 100 deaths and 16,000 cases of problems including strokes, seizures, and headaches have been reported.

When it comes to supplements, the burden of proof is on the U.S. FDA to show that the product is unsafe. For this reason, the true extent of problems caused by weight-loss supplements is not known—while the worst cases attract attention, less serious ones go unreported or undiagnosed.

Unfortunately, there is no quick and easy solution, no "magic pill." For most of us who want to lose weight, diet is where we begin. But which diet?

There is a long-standing duel between those promoting low-carbohydrate (high-protein and -fat) regimens such as the Atkins Diet™ and South Beach Diet®, and those espousing the traditional low-fat (high–complex carbohydrates) diet.

Clinical studies show that people on the low-carbohydrate diets lose weight more quickly at first, but plateau at 6 months. When dieting continued for a year, those on the low-fat diets lost just as much weight as did those on the low-carbohydrate diets. Although there was concern that the low-carbohydrate diets would promote undesirable plasma cholesterol and lipid values, for the most part this has not been the case.

The oldie-but-goodie Weight Watchers® diet, which has dieters counting points, still works and allows virtually any food choice as long as the allowed point count isn't exceeded.

For some of us, more extreme measures may be required.

1. **New drugs.** Several new diet drugs have recently been approved by the FDA. These drugs are not a panacea, however. Their side effects limit their use to only the very obese, and weight loss (even with diet and exercise) tends to be moderate.

2. **Surgery.** For the severely obese, there are surgical solutions: reducing stomach volume by banding, and gastric bypass surgery.

The most common type of gastric bypass surgery (Roux-en-Y) rearranges the digestive tract. The surgeon forms a small pouch from the upper part of the stomach and attaches the jejunum. The duodenum is attached farther down the jejunum, allowing pancreatic enzymes to act on chyme. However, bypassing chyme around the duodenum reduces nutrient absorption.

These surgical procedures have been remarkably effective at promoting weight loss and restoring health by reducing the effects of the *metabolic syndrome* (see p. 968). Blood pressure returns to normal in many who were originally hypertensive, and sleep apnea is reduced. Additionally, up to 86% of patients with long-standing type 2 diabetes mellitus find they are suddenly diabetes free or demonstrate a dramatic improvement in their regulation of the disease within weeks after the procedure. This result may prove to be the single greatest benefit.

New treatments are always on the horizon. For example, animal studies have shown that activating brown adipose tissue counteracts weight gain and type 2 diabetes. (Brown adipose tissue is a special kind of adipose tissue that uses energy to generate heat.) The possibility that activated brown adipose tissue could deter or reverse obesity is definitely worth exploring. Can we find a way to activate this pathway in humans?

At present, the only realistic way to lose weight is to take in fewer calories and increase physical activity. Fidgeting helps and so does resistance exercise, which increases muscle mass. Physical exercise suppresses appetite and increases metabolic rate not only during activity but also for some time after. The only way to keep the weight off is to make dietary and exercise changes lifelong habits.

HOMEOSTATIC IMBALANCE 24.4   **CLINICAL**

Previously, high cholesterol and LDL:HDL ratios were considered the most valid predictors of risk for atherosclerosis, cardiovascular disease, and heart attack. However, almost half of those who get heart disease have normal cholesterol levels, while others with poor lipid profiles remain free of heart problems. Presently, LDL levels and assessments of other cardiovascular disease risk factors are believed to be more accurate indicators of whether treatment is needed, and many physicians recommend dietary changes regardless of total cholesterol or HDL levels.

Cholesterol-lowering drugs such as *statins* are routinely prescribed for people with elevated LDL levels. It is estimated that more than 10 million Americans are now taking statins. ● ● ● ● ● ●

Check Your Understanding

25. If you had your choice, would you prefer to have high blood levels of HDLs or LDLs? Explain your answer.

26. What are trans fats and how do they affect LDL and HDL levels?

For answers, see Answers Appendix.

PART 3

ENERGY BALANCE

Learning Outcome

▶ **Explain what is meant by body energy balance.**

When any fuel is burned, it consumes oxygen and liberates heat. The "burning" of food fuels by our cells is no exception. As we described in Chapter 2 (◀ **pp. 24–25**), energy can be neither created nor destroyed—only converted from one form to another. If we apply this principle (actually the *first law of thermodynamics*) to cell metabolism, it means that bond energy released when foods are catabolized (energy input) must be precisely balanced by the total energy output of the body. A dynamic balance exists between the body's energy intake and energy output:

Energy intake = energy output
(heat + work + energy storage)

Energy intake is the energy liberated during food oxidation. **Energy output** includes energy (1) immediately lost as heat (about 60% of the total), (2) used to do work (driven by ATP), and (3) stored as fat or glycogen. Because losses of organic molecules in urine, feces, and sweat are very small in healthy people, they are usually ignored in calculating energy output.

Nearly all the energy derived from foodstuffs is eventually converted to heat. Heat is lost during every cellular activity—when ATP bonds are formed and when they are broken to do work, as muscles contract, and through friction as blood flows through blood vessels. Though cells cannot use this energy to do work, the heat warms the tissues and blood and helps maintain the homeostatic body temperature that allows metabolic reactions to occur efficiently. Energy storage is an important part of the equation only during periods of growth and net fat deposit.

24.9 Neural and hormonal factors regulate food intake

Learning Outcome

▶ **Describe several theories of food intake regulation.**

When energy intake and energy output are balanced, body weight remains stable. When they are not, weight is either gained or lost. Unhappily for many people, the body's weight-controlling systems appear to be designed more to protect us against weight loss than weight gain.

Obesity

How fat is too fat? What distinguishes a person who is obese from one who is merely overweight? Let's take a look.

The bathroom scale is an inaccurate guide because body weight tells little of body composition. Dense bones and well-developed muscles can make a fit, healthy person technically overweight. A competitive bodybuilder, for example, may tip the scales at a hefty 250 lb.

Body mass index (BMI) is a formula for determining obesity based on a person's weight relative to height. To estimate BMI, multiply weight in pounds by 705 and then divide by your height in inches squared:

$$\text{BMI} = \text{wt(lb)} \times 705/\text{ht(inches)}^2$$

Overweight is defined by a BMI between 25 and 30 and carries some health risk. Obesity is a BMI greater than 30 and has a markedly increased health risk.

A body fat content of 18–20% of body weight (males and females respectively) is deemed normal for adults.

However it's defined, obesity is perplexing and poorly understood, and the economic toll of obesity-related disease is staggering. Chronic low-grade systemic inflammation accompanies obesity and contributes to insulin resistance and type 2 diabetes mellitus. People who are obese also have a higher incidence of atherosclerosis, hypertension, heart disease, and osteoarthritis.

The U.S. is big and getting bigger, at least around its middle. Two out of three adults are overweight, one out of three is obese, and one in ten has diabetes. U.S. kids are getting fatter too: 40 years ago, 5% were obese; today over 17% are. **A Closer Look** on pp. 958–959 discusses the obesity epidemic.

Regulation of Food Intake

Control of food intake poses difficult questions for researchers. For example, what type of receptor could sense the body's total calorie content and alert us to start eating or put down that fork? Despite heroic research efforts, no such single receptor type has been found.

Current theories of how feeding behavior and hunger are regulated focus mainly on (1) neural signals from the digestive tract, (2) hormones, and (3) blood levels of nutrients. To a smaller degree, body temperature and psychological factors also play a role. All these factors appear to operate through feedback signals to the feeding centers of the brain.

It has been known for some time that the hypothalamus, particularly its *arcuate nucleus* (*ARC*) and two other areas—the *lateral hypothalamic area* (*LHA*) and the *ventromedial nucleus* (*VMN*)—release several peptides that influence feeding behavior (◀ **p. 449**). Most importantly, this influence ultimately reflects the activity of two opposing sets of neurons:

- **Hunger-promoting neurons.** The NPY/AgRP neurons of ARC release neuropeptide Y (NPY) and agouti-related peptide (AgRP), which enhance appetite by stimulating the neurons of the LHA that release orexins (**Figure 24.25**, right side).

- **Satiety-promoting neurons.** The other neuron group in ARC consists of the POMC/CART neurons, which suppress appetite by releasing the peptides α-melanocyte stimulating hormone [α-MSH, derived from pro-opiomelanocortin (POMC)] and cocaine- and amphetamine-regulated transcript (CART). α-MSH and CART act on the ventromedial nucleus, causing its neurons to release CRH (corticotropin-releasing hormone), an important appetite suppressor.

The hypothalamic nuclei play an essential role in regulating hunger and satiety, but brain stem areas are also involved. Sensors in peripheral locations have also been suggested, with the liver and gut itself the prime candidates. Controls of food intake come in two varieties—short term and long term.

Short-Term Regulation of Food Intake

Short-term regulation of appetite and feeding behavior involves neural signals from the GI tract, blood levels of nutrients, and GI tract hormones. For the most part, the short-term signals target hypothalamic centers via the solitary tract (and nucleus) of the brain stem (Figure 24.25).

Neural Signals from the Digestive Tract   One way the brain evaluates the contents of the gut depends on vagal nerve fibers that carry on a two-way conversation between gut and brain. For example, clinical tests show that ingesting protein produces a 30–40% larger and longer response in vagal afferents than ingesting the same amount of glucose. Furthermore, activating stretch receptors ultimately inhibits appetite, because GI tract distension sends signals along vagus nerve afferents that suppress the appetite-enhancing or hunger center. Using these signals, together with others it receives, the brain can decode what is eaten and how much.

Nutrient Signals Related to Energy Stores   Blood levels of glucose, amino acids, and fatty acids provide information to the brain that may help adjust energy intake to energy output. Nutrient signals that indicate fullness or satiety include:

- *Rising blood glucose levels.* When we eat, rising blood glucose levels ultimately depress eating (Figure 24.25). During fasting and hypoglycemia, this signal is absent, resulting in hunger and "turning on" food-seeking behavior.

  The brain shows other responses to glucose and high-caloric food as well. When we ingest sugar, the brain's reward (pleasure) system "gives off fireworks" in the form of rising dopamine levels to a greater or lesser degree. Perhaps this response is the genesis of overeating (hedonistic) behavior.

- *Elevated blood levels of amino acids* depress eating, but the precise mechanism is unknown.

- *Blood concentrations of fatty acids* influence hunger. The more fatty acids in the blood, the more eating behavior is inhibited.

Hormones   Gut hormones, including insulin and cholecystokinin (CCK) released during food absorption, act as satiety signals to depress hunger. Cholecystokinin (◀ p. 891), which blocks the appetite-inducing effect of NPY, is most important.

In contrast, glucagon and epinephrine levels rise during fasting and stimulate hunger. Ghrelin (Ghr), produced by the stomach, is a powerful appetite stimulant. In fact, ghrelin appears to be the "dinner bell" or trigger for initiating a meal. Its level peaks just before mealtime, signaling the brain that it is time to eat, and then the level drops after the meal.

Figure 24.25 Model for hypothalamic command of appetite and food intake. The arcuate nucleus (ARC) of the hypothalamus contains two sets of neurons with opposing effects. Activating the NPY and orexin neurons enhances appetite, whereas activating POMC/CART neurons has the opposite effect. These centers connect with other brain centers, and signals are transmitted through the brain stem to the body. Many appetite-regulating hormones act through the ARC, though they may also exert direct effects on other brain centers. (AgRP = agouti-related peptide, CART = cocaine- and amphetamine-regulated transcript, CCK = cholecystokinin, CRH = corticotropin-releasing hormone, LHA = lateral hypothalamic area, α-MSH = α-melanocyte-stimulating hormone, NPY = neuropeptide Y, POMC = pro-opiomelanocortin, VMN = ventromedial nucleus)

24

Long-Term Regulation of Food Intake

A key component of the long-term controls of feeding behavior is the hormone **leptin** ("thin"). Secreted exclusively by adipose cells in response to an increase in body fat mass, leptin indicates the body's total energy stores in fat tissue. (Thus adipose tissue acts as a "Fat-o-Stat" that sends chemical messages to the brain in the form of leptin.)

When levels of leptin rise in the blood, leptin binds to receptors in ARC that specifically (1) suppress the release of NPY and (2) stimulate the expression of CART. NPY is the most potent appetite stimulant known. By blocking its release, leptin prevents the release of the appetite-enhancing orexins from the LHA. This decreases appetite and food intake, eventually promoting weight loss.

When fat stores shrink, leptin blood levels drop, an event that exerts opposite effects on the two sets of ARC neurons. Consequently, appetite and food intake increase, and (eventually) weight gain occurs.

Initially it seemed that leptin was the magic bullet that obesity researchers were looking for, but their hopes were soon dashed. Rising leptin levels do promote weight loss, but only to a certain point. Furthermore, individuals who are obese have higher-than-normal leptin blood levels, but for some unknown reason, they are resistant to its action. Now the consensus is that leptin's main role is to protect against weight loss in times of nutritional deprivation.

Although leptin has received the most attention as a long-term appetite and metabolism regulator, there are several other players. Insulin, like leptin, inhibits NPY release in non-insulin-resistant individuals, but its effect is less potent.

Additional Regulatory Factors

There is no easy answer to explain how body weight is regulated, but theories abound. Rising ambient temperature discourages food seeking, whereas cold temperature activates the hunger center. Depending on the individual, stress may increase or decrease food-seeking behavior, but chronic stress in combination with a junk food (high in sugar and fat) diet sharply increases release of NPY.

Psychological factors are thought to be important in people who are obese, but even when psychological factors contribute to obesity, individuals do not continue to gain weight endlessly. Controls still operate but at a higher set-point weight. The composition of gut bacteria, sleep deprivation, and even certain adenovirus infections are additional factors that may affect a person's fat mass and weight regulation.

Whew! How does all this information fit together? So far we only have bits and pieces of the story, but Figure 24.25 shows the best current model.

Check Your Understanding

27. What three groups of stimuli influence short-term regulation of feeding behavior?

28. What is the most important long-term regulator of feeding behavior and appetite?

29. **PREDICT** Suppose your patient has had a stroke that damaged the ventromedial nucleus (VMN). Would you expect this patient's weight to increase, decrease, or stay the same? Why? (Hint: See Figure 24.25 on p. 961.)

For answers, see Answers Appendix.

24.10 Thyroxine is the major hormone that controls basal metabolic rate

Learning Outcome

▶ Define basal metabolic rate and total metabolic rate. Name factors that influence each.

The body's rate of energy output, the **metabolic rate**, is the total heat produced by all the chemical reactions and mechanical work of the body. We can measure metabolic rate directly or indirectly. In the *direct method*, a person enters a chamber called a **calorimeter** and water circulating around the chamber absorbs heat liberated by the body. The rise in water temperature is directly related to the heat produced by the person's body. The *indirect method* uses a **respirometer** to measure oxygen consumption, which is directly proportional to heat production. For each liter of oxygen used, the body produces about 4.8 kcal of heat.

Basal Metabolic Rate (BMR)

Because many factors influence metabolic rate, it is usually measured under standardized conditions. The person is in a postabsorptive state (has not eaten for at least 12 hours), is reclining, and is mentally and physically relaxed. The temperature of the room is a comfortable 20–25°C. The measurement obtained under these circumstances, the **basal metabolic rate (BMR)**, reflects the energy the body needs to perform only its most essential activities, such as breathing and maintaining resting levels of organ function. Although named the *basal* metabolic rate, this measurement is *not* the lowest metabolic state of the body. That situation occurs during sleep, when skeletal muscles are completely relaxed.

The BMR, often referred to as the "energy cost of living," is reported in kilocalories per square meter of body surface per hour (kcal/m^2/h). Why do we take surface area into account in this formula? It is because the rate at which energy is used depends on the body surface area. As the ratio of body surface area to body volume increases, heat loss to the environment increases and the metabolic rate must be higher to replace the lost heat. For this reason, if two people weigh the same, the taller or thinner person will use more energy.

A 70-kg adult male of average height has a BMR of approximately 66 kcal/h. You can approximate your BMR by multiplying weight in kilograms (2.2 lb = 1 kg) by 1 if you are male and by 0.9 if you are female.

Several factors influence BMR, including age, gender, body temperature, stress, and the hormone thyroxine.

Age and Gender

In general, the younger the person, the higher the BMR. Children and adolescents require large amounts of energy for growth. In old age, BMR declines dramatically as skeletal muscles begin to atrophy.

Gender also plays a role. Metabolic rate is higher in males than in females, mainly because males typically have more muscle, which is very active metabolically even during rest. Fatty tissue, present in greater relative amounts in females, is metabolically more sluggish than muscle.

Body Temperature

Body temperature and BMR tend to rise and fall together. Fever markedly increases metabolic rate above basal levels.

Stress

Whether physical or emotional, stress increases BMR by mobilizing the sympathetic nervous system. As epinephrine and norepinephrine flood into the blood from the adrenal medulla, they raise BMR primarily by stimulating fat catabolism.

Thyroxine

The amount of **thyroxine** produced by the thyroid gland (◀ p. 618) is probably the most important hormonal factor in determining BMR. For this reason, thyroxine has been dubbed the "metabolic hormone." Its direct effect on most body cells (except brain cells) is to increase O_2 consumption and heat production, in part by accelerating the use of ATP to operate the sodium-potassium pump. As ATP reserves decline, cellular respiration accelerates. Thus, the more thyroxine produced, the higher the BMR.

HOMEOSTATIC IMBALANCE 24.5   **CLINICAL**

Hyperthyroidism causes a host of problems resulting from the excessive BMR it produces. The body catabolizes stored fats and tissue proteins and, despite increased hunger and food intake, the person often loses weight. Bones weaken and muscles, including the heart, begin to atrophy.

In contrast, *hypothyroidism* results in slowed metabolism, obesity, and diminished thought processes (◀ Table 16.4, p. 618).

Total Metabolic Rate (TMR)

The **total metabolic rate (TMR)** is the rate of kilocalorie consumption needed to fuel *all* ongoing activities—involuntary and voluntary. BMR accounts for a surprisingly large part of TMR. For example, a woman whose total energy needs per day are about 2000 kcal may spend 1400 kcal or so supporting vital body activities.

Skeletal muscles make up nearly half of body mass, so skeletal muscle activity causes the most dramatic short-term changes in TMR. Even slight increases in muscular work can cause remarkable leaps in TMR and heat production. When a well-trained athlete exercises vigorously for several minutes, TMR may increase 20-fold and remain elevated for several hours after exercise stops.

Food ingestion also induces a rapid increase in TMR. This effect, called **food-induced thermogenesis**, is greatest when proteins and alcohol are ingested. The heightened metabolic activity of the liver during such periods probably accounts for the bulk of additional energy use. In contrast, fasting or very low caloric intake depresses TMR and results in a slower breakdown of body reserves.

Check Your Understanding

30. Which of the following contributes to a person's BMR? Kidney function, breathing, jogging, eating, fever.

31. MAKE CONNECTIONS Samantha tells her doctor that she has been feeling jittery and has had trouble falling asleep. She complains that her roommates always set the thermostat too high. Blood tests reveal antibodies that suggest an autoimmune disease. Explain Samantha's symptoms. What organ are the antibodies binding to and how does this cause her symptoms? (Hint: See Chapter 16.)

For answers, see Answers Appendix.

24.11 The hypothalamus acts as the body's thermostat

Learning Outcomes

▶ **Distinguish between core and shell body temperature.**

▶ **Describe how body temperature is regulated, and indicate the common mechanisms regulating heat production/retention and heat loss from the body.**

As shown in **Figure 24.26**, body temperature represents the balance between heat production and heat loss. All body tissues produce heat, but those most active metabolically produce the

Figure 24.26 Body temperature remains constant as long as heat production and heat loss are balanced.

greatest amounts. When the body is at rest, most heat is generated by the liver, heart, brain, kidneys, and endocrine organs, with the inactive skeletal muscles accounting for only 20–30%.

This situation changes dramatically with even slight changes in muscle tone. When we are cold, shivering helps warm us up, and during vigorous exercise, skeletal muscles can produce 30 to 40 times more heat than the rest of the body. A change in muscle activity is one of the most important means of modifying body temperature.

Body temperature averages 37°C ± 0.5°C (98.6°F) and is usually maintained within the range 35.8–38.2°C (96–101°F), despite considerable change in external (air) temperature. A healthy individual's body temperature fluctuates approximately 1°C (1.8°F) in 24 hours, lowest in early morning and highest in late afternoon or early evening.

The adaptive value of temperature homeostasis becomes apparent when we consider how temperature affects enzymatic activity. At normal body temperature, conditions are optimal for enzymatic activity. Rising body temperature accelerates enzymatic catalysis: With each rise of 1°C, the rate of chemical reactions increases about 10%. If temperature rises above the homeostatic range, neurons are depressed and proteins begin to denature. Children below the age of 5 may go into convulsions when body temperature reaches 41°C (106°F), and 43°C (about 109°F) appears to be the absolute limit for life.

In contrast, most body tissues can withstand marked reductions in temperature if other conditions are carefully controlled. This fact underlies the use of body cooling during open heart surgery when the heart must be stopped. Low body temperature reduces metabolic rate (and consequently oxygen and nutrient requirements of body tissues and the heart), allowing more time for surgery without incurring tissue damage.

Core and Shell Temperatures

Different body regions have different resting temperatures. The body's **core** (organs within the skull and the thoracic and abdominal cavities) has the highest temperature and its **shell** (essentially the skin) has the lowest temperature in most circumstances. Of the two body sites used most often to obtain body temperature clinically, the rectum typically has a temperature about 0.4°C (0.7°F) higher than the mouth and is a better indicator of core temperature.

It is core temperature that is precisely regulated. Blood serves as the major *agent of heat exchange* between the core and shell. Whenever the shell is warmer than the external environment, the body loses heat as warm blood is allowed to flush into skin capillaries. On the other hand, when heat must be conserved, blood largely bypasses the skin. This reduces heat loss and allows the shell temperature to fall toward that of the environment.

For this reason, core temperature stays relatively constant, but the temperature of the shell may fluctuate substantially. For example, it can vary between 20°C (68°F) and 40°C (104°F), as it adapts to changes in body activity and external temperature. (You really *can* have cold hands and a warm heart.)

Mechanisms of Heat Exchange

Heat exchange between our skin and the external environment works in the same way as heat exchange between inanimate objects. It helps to think of an object's temperature—whether that object is a radiator or your skin—as a guide to its heat content (think "heat concentration"). Then just remember that heat always flows down its concentration gradient from a warmer region to a cooler region. The body uses four mechanisms of heat transfer—radiation, conduction, convection, and evaporation (**Figure 24.27**).

Radiation

Radiation is the loss of heat in the form of infrared waves (thermal energy). Any object that is warmer than objects in its environment—for example, a radiator and (usually) the

Figure 24.27 Mechanisms of heat exchange. A woman enjoying a hot tub illustrates several mechanisms of heat exchange between the body and the environment—conduction, radiation, convection, and (possibly) evaporation.

body—will transfer heat to those objects. Under normal conditions, close to half of your body heat loss occurs by radiation.

Because radiant energy flows from warmer to cooler, radiation explains why a cold room warms up shortly after it fills with people. Your body also gains heat by radiation, as demonstrated by the way your skin warms when you sunbathe.

Conduction

Conduction transfers heat from a warmer object to a cooler one when the two are in direct contact with each other. For example, when you step into a hot tub, some of the heat of the water transfers to your skin, and warm buttocks transfer heat to the seat of a chair by conduction.

Convection

Convection is the transfer of heat that occurs because warm air expands and rises and cool air, being denser, falls. Consequently, the warmed air enveloping the body is continually replaced by cooler air molecules. Convection substantially enhances heat transfer from the body surface to the air because the cooler air absorbs heat by conduction more rapidly than the already-warmed air.

Together, conduction and convection account for 15–20% of heat loss to the environment. These processes are enhanced by anything that moves air more rapidly across the body surface, such as wind or a fan, in other words, by *forced convection*.

Evaporation

The fourth mechanism by which the body loses heat is **evaporation**. Water evaporates because its molecules absorb heat from the environment and become energetic enough—in other words, vibrate fast enough—to escape as a gas, which we know as water vapor. The heat absorbed by water during evaporation is called **heat of vaporization** (◀ p. 38). The evaporation of

water from body surfaces removes large amounts of body heat. Every gram of water that evaporates removes about 0.58 kcal of heat from the body.

There is a basal level of body heat loss due to the continuous evaporation of water from the lungs and oral mucosa, and through the skin. The unnoticeable water loss occurring via these routes is called **insensible water loss**, and the accompanying heat loss is **insensible heat loss**. Insensible heat loss dissipates about 10% of the basal heat production of the body and is a constant not subject to body temperature controls.

Evaporative heat loss becomes an active or *sensible* process when body temperature rises and sweating produces increased amounts of water for vaporization. Extreme emotional states activate the sympathetic nervous system, causing body temperature to rise by one degree or so, and vigorous exercise can raise body temperature as much as 2–3°C (3.6–5.4°F). Vigorous muscular activity can produce and evaporate 1–2 L/h of sweat, removing 600–1200 kcal of heat from the body each hour. This is more than 30 times the amount of heat lost via insensible heat loss!

Role of the Hypothalamus

Although other brain regions contribute, the hypothalamus (◀ p. 449), particularly its anterior region, is the main integrating center for thermoregulation. Together the **heat-loss center** (located more anteriorly) and the **heat-promoting center** make up the brain's **thermoregulatory centers**.

The hypothalamus receives afferent input from (1) **peripheral thermoreceptors** located in the shell (the skin), and (2) **central thermoreceptors** sensitive to blood temperature and located in the body core including the anterior portion of the hypothalamus. Much like a thermostat, the hypothalamus responds to this input by reflexively initiating appropriate heat-promoting or heat-loss activities.

24

The central thermoreceptors have more influence than the peripheral ones, but varying inputs from the shell probably alert the hypothalamus to the need to prevent temperature changes in the core. In other words, they allow the hypothalamus to anticipate changes to be made.

Heat-Promoting Mechanisms

When the external temperature is low or blood temperature falls for any reason, the heat-promoting center is activated. It triggers one or more of the following mechanisms to maintain or increase core body temperature (**Figure 24.28**, bottom).

- **Constriction of cutaneous blood vessels.** Activation of the sympathetic vasoconstrictor fibers serving the blood vessels of the skin causes strong vasoconstriction. This restricts blood to deep body areas and largely bypasses the skin. Because a layer of insulating subcutaneous (fatty) tissue separates the skin from deeper organs, heat loss from the core through the shell is dramatically lowered. As a result, the shell temperature is closer to that of the external environment.

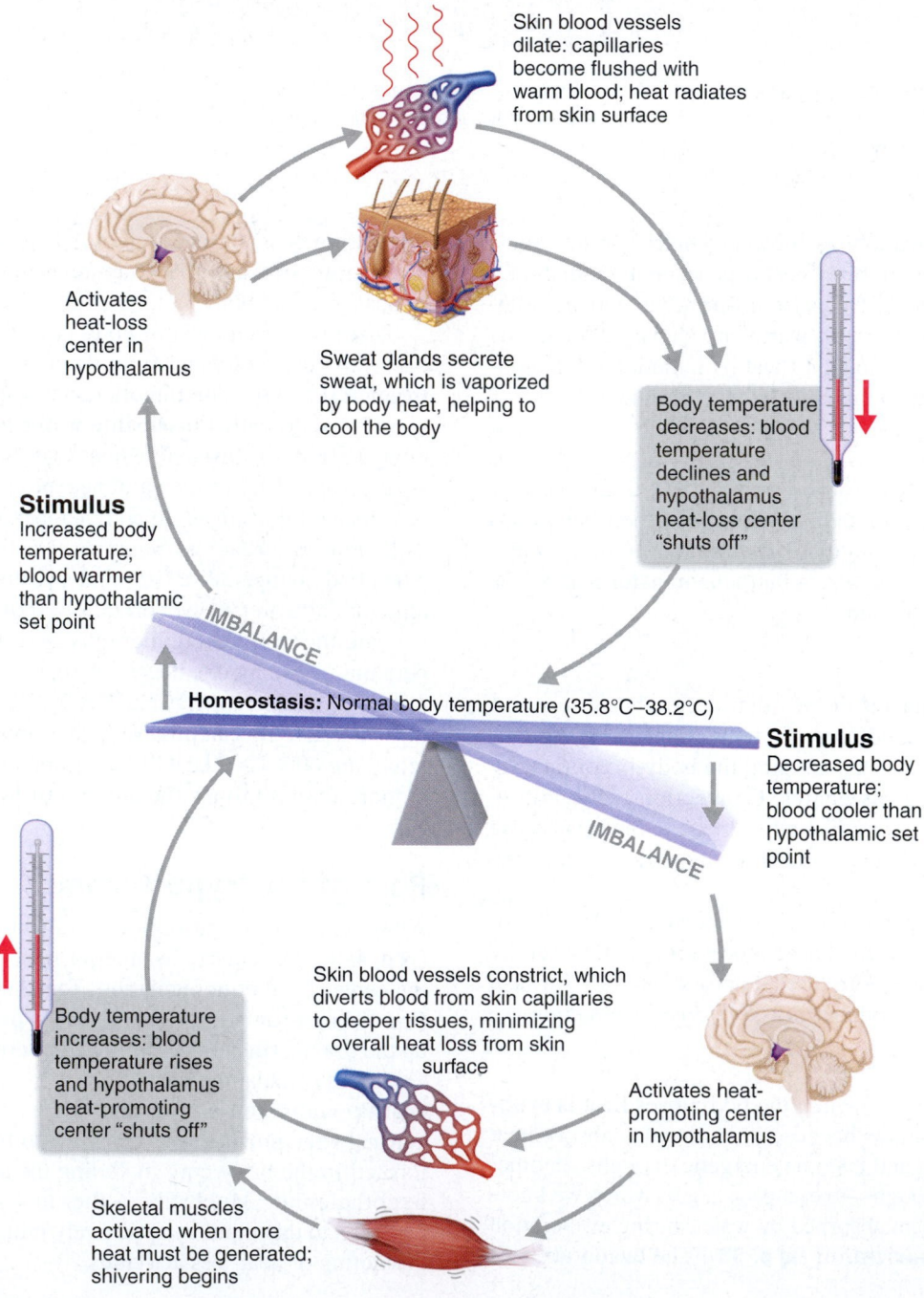

Skin blood vessels dilate: capillaries become flushed with warm blood; heat radiates from skin surface

Activates heat-loss center in hypothalamus

Sweat glands secrete sweat, which is vaporized by body heat, helping to cool the body

Body temperature decreases: blood temperature declines and hypothalamus heat-loss center "shuts off"

Stimulus
Increased body temperature; blood warmer than hypothalamic set point

IMBALANCE

Homeostasis: Normal body temperature (35.8°C–38.2°C)

Stimulus
Decreased body temperature; blood cooler than hypothalamic set point

IMBALANCE

Body temperature increases: blood temperature rises and hypothalamus heat-promoting center "shuts off"

Skin blood vessels constrict, which diverts blood from skin capillaries to deeper tissues, minimizing overall heat loss from skin surface

Activates heat-promoting center in hypothalamus

Skeletal muscles activated when more heat must be generated; shivering begins

Figure 24.28 Mechanisms of body temperature regulation.

HOMEOSTATIC IMBALANCE 24.6 — CLINICAL

Restricting blood flow to the skin is not a problem for a brief period. If it is prolonged (as during exposure to very cold weather), the lack of heat transfer can lead to the formation of ice crystals in and around cells, causing them to die. This extremely serious condition is *frostbite*.

- **Shivering.** Shivering—involuntary shuddering contractions—is triggered when the hypothalamus activates brain centers that cause an increase in muscle tone. When muscle tone reaches sufficient levels, stretch receptors are alternately stimulated in antagonistic muscles. Shivering raises body temperature because skeletal muscle activity produces large amounts of heat.

- **Increase in metabolic rate.** Cold increases sympathetic stimulation of the adrenal medulla, releasing epinephrine and norepinephrine (◄ p. 627). These hormones elevate the metabolic rate and enhance heat production. This mechanism, called **chemical (nonshivering) thermogenesis**, occurs primarily in infants. Small deposits of *brown adipose tissue*, a special kind of adipose tissue that dissipates energy and produces heat by this mechanism, are also found in adults.

- **Enhanced release of thyroxine.** When environmental temperature decreases gradually, as in the transition from summer to winter, the hypothalamus of infants releases *thyrotropin-releasing hormone* (◄ p. 616). This hormone activates the anterior pituitary to release *thyroid-stimulating hormone*, which induces the thyroid to liberate more thyroid hormone to the blood. Because thyroid hormone raises metabolic rate, body heat production rises. Adults do not show a similar TSH response to cold exposure.

Besides these involuntary adjustments, we humans make a number of *behavioral modifications* to prevent overcooling of our body core:

- Putting on more or warmer clothing to restrict heat loss (hat, gloves, and insulated outer garments)
- Drinking hot fluids
- Changing posture to reduce exposed body surface area (hunching over or clasping the arms across the chest)
- Increasing physical activity to generate more heat (jumping up and down, clapping the hands)

Heat-Loss Mechanisms

How do heat-loss mechanisms protect the body from excessively high temperatures? Whenever core body temperature rises above normal, it inhibits the hypothalamic heat-promoting center. At the same time, it activates the heat-loss center and triggers one or both of the following (Figure 24.28, top):

- **Dilation of cutaneous blood vessels.** Inhibiting the vasomotor fibers serving blood vessels of the skin allows the vessels to dilate. As the blood vessels swell with warm blood, the shell loses heat by radiation, conduction, and convection.

- **Enhanced sweating.** If the body is extremely overheated or if the environment is so hot—over 33°C (about 92°F)—that heat cannot be lost by other means, evaporation becomes necessary. Sympathetic fibers activate the sweat glands to spew out large amounts of sweat.

Evaporation of sweat is an efficient means of ridding the body of surplus heat as long as the air is dry. However, when the relative humidity is high, evaporation occurs much more slowly. In such cases, the heat-liberating mechanisms cannot work well, and we feel miserable and irritable. Behavioral or voluntary measures commonly taken to reduce body heat in such circumstances include:

- Reducing activity ("laying low")
- Seeking a cooler environment (a shady spot) or using a device to increase convection (a fan) or cooling (an air conditioner)
- Wearing light-colored, loose clothing that reflects radiant energy. (This is actually cooler than being nude because bare skin absorbs most of the radiant energy striking it.)

HOMEOSTATIC IMBALANCE 24.7 — CLINICAL

Overexposure to a hot and humid environment makes normal heat-loss processes ineffective. The resulting **hyperthermia** (elevated body temperature) depresses the hypothalamus. At a core temperature of around 41°C (105°F), heat-control mechanisms are suspended, creating a vicious *positive feedback cycle*. Increasing temperatures increase the metabolic rate, which increases heat production. The skin becomes hot and dry and, as the temperature continues to spiral upward, multiple organ damage becomes a distinct possibility, including brain damage. This condition, called **heat stroke**, can be fatal unless corrective measures are initiated immediately (immersing the body in cool water and administering fluids).

The terms *heat exhaustion* and *exertion-induced heat exhaustion* are often used to describe the heat-associated extreme sweating and collapse of an individual during or following vigorous physical activity. This condition, evidenced by elevated body temperature and mental confusion and/or fainting, is due to dehydration and consequent low blood pressure. As heat-loss mechanisms struggling to function in heat exhaustion further give way, heat exhaustion merges into heat stroke. Death results if the body is not cooled and rehydrated promptly.

Hypothermia (hi″po-ther′me-ah) is low body temperature resulting from prolonged uncontrolled exposure to cold. Vital signs (respiratory rate, blood pressure, and heart rate) decrease as cellular enzymes become sluggish. Drowsiness sets in and, oddly, the person becomes comfortable even though previously he or she felt extremely cold. Shivering stops at a core temperature of 30–32°C (87–90°F) when the body has exhausted its heat-generating capabilities. Uncorrected, hypothermia progresses to coma and finally death (by cardiac arrest), when body temperatures approach 21°C (70°F).

Fever

Fever is *controlled hyperthermia*. Most often, it results from infection somewhere in the body, but it may be caused by cancer, allergic reactions, or CNS injuries.

Whatever the cause, macrophages and other cells release cytokines that act as *pyrogens* (literally, "fire starters"). These chemicals act on the hypothalamus, causing release of prostaglandins which reset the hypothalamic thermostat to a higher-than-normal temperature, so that heat-promoting mechanisms kick in. As a result of vasoconstriction, heat loss from the body surface declines, the skin cools, and shivering begins to generate heat. These "chills" are a sure sign body temperature is rising.

The temperature rises until it reaches the new setting, and then is maintained at that setting until natural body defenses or medications reverse the disease process. Then, heat-loss mechanisms swing into action. Sweating begins and the skin becomes flushed and warm. Physicians have long recognized these signs as signals that body temperature is falling (the fever has broken). As we explained in Chapter 21, fever speeds healing by increasing the metabolic rate, and it also appears to inhibit bacterial growth.

Check Your Understanding

32. What is the body's core?

33. Andrea is flushed and her teeth are chattering even though her bedroom temperature is 72°F. Why do you think this is happening?

34. How does convection differ from conduction in causing heat loss?

For answers, see Answers Appendix.

Developmental Aspects of Nutrition and Metabolism

Good nutrition is essential *in utero*, as well as throughout life. If the mother is ill nourished, the development of her infant is affected. Most serious is the lack of adequate calories, proteins, and vitamins needed for fetal tissue growth, especially brain growth. Additionally, inadequate nutrients during the first three years after birth will lead to mental deficits or learning disorders because brain growth continues during this time. Proteins are needed for muscle and bone growth, and calcium is required for strong bones. Although anabolic processes are less critical after growth is completed, sufficient nutrients are still essential to maintain normal tissue replacement and metabolism.

HOMEOSTATIC IMBALANCE 24.8 **CLINICAL**

There are many inborn errors of metabolism (or genetic disorders), but one of the most common is *phenylketonuria* (*PKU*) (fen″il-ke″to-nu′re-ah).

In PKU, tissue cells are unable to use the amino acid phenylalanine (fen″il-al′ah-nēn), which is present in all protein foods. The enzyme that converts phenylalanine to tyrosine is defective.

Phenylalanine cannot be metabolized, so it and its deaminated products accumulate in the blood and act as neurotoxins that cause brain damage. These consequences are uncommon today because most states require a urine or blood test to identify affected newborns, and these children are put on a diet low in phenylalanine. People with PKU must restrict their intake of phenylalanine throughout life, especially during pregnancy.

There are a number of other enzyme defects, classified according to the impaired process as carbohydrate, lipid, or mineral metabolic disorders. The carbohydrate deficit *galactosemia* results from an abnormality in or lack of the liver enzymes needed to transform galactose to glucose. Galactose accumulates in the blood and leads to mental deficits.

In *glycogen storage disease*, glycogen synthesis is normal, but one of the enzymes needed to convert it back to glucose is missing. As excessive amounts of glycogen are stored, its storage organs (liver and skeletal muscles) become glutted with glycogen and enlarge tremendously. ●●●●●●●

With the exception of *type 1 diabetes mellitus*, children free of genetic disorders rarely exhibit metabolic problems. However, by middle age and particularly old age, *type 2 diabetes mellitus* becomes a major problem, particularly in people who are obese.

HOMEOSTATIC IMBALANCE 24.9 **CLINICAL**

Metabolic syndrome is defined as a cluster of five risk factors (**Figure 24.29**). Together, they double the chance of getting heart disease and stroke, and increase the chance of developing type 2 diabetes by five times. About 32% of the U.S. population has the dubious distinction of qualifying for this syndrome by having three of the five risk factors.

Metabolic syndrome reflects abnormal adipose tissue function, and visceral fat in the abdominal cavity is particularly susceptible. Increased visceral fat (as in the "apple" body shape) is associated with increased inflammatory cytokines and insulin resistance, which contribute to elevated blood glucose and

Metabolic syndrome:

- ↑ Waist circumference
- ↑ Blood pressure
- ↑ Blood glucose
- ↑ Blood triglycerides
- ↓ Blood HDL cholesterol

Figure 24.29 Metabolic syndrome. Increased abdominal fat deposition (the "apple" body shape) is associated with metabolic syndrome.

triglycerides, and the development of atherosclerosis. On the other hand, subcutaneous fat (as in the "pear" body shape where fat is deposited in the hips) is less susceptible to dysfunction.

What can those of us with metabolic syndrome do? The key is to reduce abdominal fat. As always, changes to diet and increased exercise are the first steps. In addition, drugs are usually used to manage the individual components of metabolic syndrome (hypertension, insulin resistance, and abnormal blood lipid profile).

Metabolic rate declines throughout the life span. In old age, muscle and bone wasting and declining efficiency of the endocrine system take their toll. Because many elderly are also less active, their metabolic rate is sometimes so low that it becomes nearly impossible to obtain adequate nutrition without gaining weight. The elderly also use more medications than any other group, at a time of life when the liver has become less efficient in its detoxifying duties.

Many drugs and popular remedies influence nutrition. For example:

- Some diuretics prescribed for congestive heart failure or hypertension (to flush fluids out of the body) can cause severe hypokalemia by promoting excessive loss of potassium.

- Some antibiotics—for example, sulfa drugs, tetracycline, and penicillin—interfere with food digestion and absorption. They may also cause diarrhea, further decreasing absorption.

- Although physicians discourage its use because it interferes with absorption of fat-soluble vitamins, mineral oil is still a popular laxative with the elderly.

- About half the elderly in the U.S. consume alcohol. When it is substituted for food, nutrient stores can be depleted. Excessive alcohol intake leads to absorption problems, certain vitamin and mineral deficiencies, deranged metabolism, and damage to the liver and pancreas.

Although malnutrition and a waning metabolic rate present problems to some elderly, certain nutrients—notably glucose—appear to contribute to the aging process in all of us. Nonenzymatic reactions (the so-called *browning reactions*) between glucose and proteins, long known to discolor and toughen foods, may have the same effects on body proteins. When enzymes attach sugars to proteins, they do so at specific sites and the glycoproteins produced play well-defined roles in the body. In contrast, nonenzymatic binding of glucose to proteins (a process that increases with age) is haphazard and eventually causes cross-links between proteins. This type of binding probably contributes to lens clouding, and the general tissue stiffening and loss of elasticity so common in the aged.

RELATED CLINICAL TERMS

Appetite A desire for food. A psychological phenomenon dependent on memory and associations, as opposed to hunger, which is a physiological need to eat.

Familial hypercholesterolemia (hi″per-ko-les″ter-ol-e′me-ah) An inherited condition in which the LDL receptors are absent or abnormal, the uptake of cholesterol by tissue cells is blocked, and the total concentration of cholesterol (and LDLs) in the blood is enormously elevated. Atherosclerosis develops at an early age, heart attacks begin in the third or fourth decade, and most die by age 60 from coronary artery disease. Treatment entails dietary modifications, exercise, and cholesterol-reducing drugs.

Pica (pi′kah) Craving and eating nonfood substances such as clay or dirt.

Protein energy malnutrition Severe deficiency of calories or calories and protein. **Marasmus** (mah-raz′mus), caused by inadequate overall calorie intake, results in weight loss and depletion of fat and muscle. **Kwashiorkor** (kwash″e-or′kor) is caused by proportionately lower protein intake than calories in children. It is characterized by a bloated abdomen (edema) because the amount of plasma proteins is inadequate to keep fluid in the bloodstream (◄ p. 734). Skin lesions and infections are likely. In children, both conditions result in failure to grow.

Skin-fold test Clinical test of body fatness. A skin fold in the back of the arm or below the scapula is measured with a caliper. A fold over 1 inch in thickness indicates excess fat.

24

CHAPTER SUMMARY

PART 1
NUTRIENTS

1. Nutrients include carbohydrates, lipids, proteins, vitamins, and minerals. The bulk of the organic nutrients is used as fuel to produce cellular energy (ATP). The energy value of foods is measured in kilocalories (kcal).
2. Essential nutrients are those that are inadequately synthesized by body cells and must be ingested in the diet.

24.1 Carbohydrates, lipids, and proteins supply energy and are used as building blocks (pp. 927–931)

Carbohydrates (pp. 927–928)

1. Carbohydrates are obtained primarily from plant products. Absorbed monosaccharides other than glucose are converted to glucose by the liver.

2. Monosaccharides are used primarily for cellular fuel. Small amounts are used for nucleic acid synthesis and to add sugar residues to plasma membrane proteins and lipids.

3. Recommended carbohydrate intake for adults is 45–65% of daily caloric intake.

Lipids (p. 928)

4. Most dietary lipids are triglycerides. The primary sources of saturated fats are animal products, tropical oils, and hydrogenated oils; unsaturated fats are present in plant products, nuts, and cold-water fish. The major sources of cholesterol are egg yolk, meats, and milk products.

5. Linoleic and linolenic acids are essential fatty acids.

6. Triglycerides provide reserve energy, cushion body organs, and insulate the body. Phospholipids are used to synthesize plasma membranes and myelin. Cholesterol is used in plasma membranes and is the structural basis of vitamin D, steroid hormones, and bile salts.

7. Guidelines regarding optimum fat intake are in flux, but generally recommend that fat intake be limited to 20–35% of caloric intake, and that saturated and trans fats should be replaced by mono- and polyunsaturated fats as much as possible.

Proteins (pp. 928–931)

8. Animal products provide high-quality complete protein containing all essential amino acids. Most plant products lack one or more of the essential amino acids.

9. Amino acids are the structural building blocks of the body and of important regulatory molecules.

10. Protein synthesis can and will occur if all essential amino acids are present and sufficient carbohydrate (or fat) calories are available to produce ATP. Otherwise, amino acids will be burned for energy.

11. Nitrogen balance occurs when protein synthesis equals protein loss.

12. A dietary intake of 0.8 g of protein per kg of body weight is recommended for most healthy adults.

24.2 Most vitamins act as coenzymes; minerals have many roles in the body (pp. 931–932)

Vitamins (p. 931)

1. Vitamins are organic compounds needed in minute amounts. Most act as coenzymes. The richest sources are whole grains, vegetables, legumes, and fruit.

2. Except for vitamin D and the K and B vitamins made by enteric bacteria, vitamins are not made in the body.

3. Water-soluble vitamins (B and C) are not stored to excess in the body. Fat-soluble vitamins include vitamins A, D, E, and K; all but vitamin K are stored and can accumulate to toxic amounts.

Minerals (pp. 931–932)

4. The body requires calcium, phosphorus, potassium, sulfur, sodium, chloride, and magnesium, as well as trace amounts of at least a dozen other minerals.

5. Minerals are not used for energy. Some are used to mineralize bone; others are bound to organic compounds or exist as ions in body fluids, where they play various roles in cell processes and metabolism.

6. Mineral uptake and excretion are carefully regulated to prevent mineral toxicity. The richest sources of minerals are some meats, vegetables, nuts, and legumes.

PART 2
METABOLISM

24.3 Metabolism is the sum of all biochemical reactions in the body (pp. 934–936)

1. Metabolism encompasses all chemical reactions necessary to maintain life. Metabolic processes are either anabolic or catabolic.

2. Cellular respiration refers to catabolic processes during which energy is released and some is captured in ATP bonds.

3. Energy is released when organic compounds are oxidized. Cellular oxidation is accomplished primarily by removing hydrogen (or electrons). When molecules are oxidized, others are simultaneously reduced by accepting hydrogen (or electrons).

4. Most enzymes catalyzing oxidation-reduction reactions require coenzymes as hydrogen acceptors. Two important coenzymes in these reactions are NAD^+ and FAD.

5. In animal cells, the two mechanisms of ATP synthesis are substrate-level phosphorylation and oxidative phosphorylation.

24.4 Carbohydrate metabolism is the central player in ATP production (pp. 936–945)

Oxidation of Glucose (pp. 937–944)

1. Carbohydrate metabolism is essentially glucose metabolism.

2. Phosphorylation of glucose on entry into cells effectively traps it in most tissue cells.

3. Glucose is oxidized to carbon dioxide and water via three successive pathways: glycolysis, citric acid cycle, and electron transport chain. Some ATP is harvested in each pathway, but the bulk is captured in the electron transport chain.

4. Glycolysis is a reversible pathway in which glucose is converted to two pyruvic acid molecules; two molecules of reduced NAD^+ are formed, and there is a net gain of 2 ATP. Under aerobic conditions, pyruvic acid enters the citric acid cycle; under anaerobic conditions, it is reduced to lactic acid.

5. The citric acid cycle is fueled by pyruvic acid (and fatty acids). To enter the cycle, pyruvic acid is converted to acetyl CoA. The acetyl CoA is then oxidized and decarboxylated. Complete oxidation of two pyruvic acid molecules yields $6 CO_2$, $8 NADH + H^+$, $2 FADH_2$, and a net gain of 2 ATP. Much of the energy originally present in the bonds of pyruvic acid is now present in the reduced coenzymes.

6. In the electron transport chain, (a) reduced coenzymes are oxidized by delivering hydrogen to a series of oxidation-reduction acceptors; (b) hydrogen is split into hydrogen ions and electrons (as electrons run downhill from acceptor to acceptor, the energy released is used to pump H^+ into the mitochondrial intermembrane space, which creates an electrochemical proton gradient); (c) the energy stored in the electrochemical proton gradient drives H^+ back through ATP synthase, which uses the energy to form ATP; (d) H^+ and electrons are combined with oxygen to form water.

7. For each glucose molecule oxidized to carbon dioxide and water, there is a net gain of 32 ATP: 4 ATP from substrate-level phosphorylation and 28 ATP from oxidative phosphorylation. The shuttle for reduced NAD^+ produced in the cytosol may use 2 ATP of that amount.

Glycogenesis, Glycogenolysis, and Gluconeogenesis (pp. 944–945)

8. When cellular ATP reserves are high, glucose catabolism is inhibited and glucose is converted to glycogen (glycogenesis) or to fat (lipogenesis). Much more fat than glycogen is stored.

9. When blood glucose levels begin to fall, glycogenolysis occurs, in which glycogen stores are converted to glucose. The liver can also perform gluconeogenesis, the formation of glucose from noncarbohydrate (fat or protein) molecules.

24.5 Lipid metabolism is key for long-term energy storage and release (pp. 946–948)

1. End products of lipid digestion (and cholesterol) are transported in blood in the form of chylomicrons.

2. Glycerol is converted to glyceraldehyde 3-phosphate and enters the citric acid cycle or is converted to glucose.

3. Fatty acids are oxidized by beta oxidation into acetic acid fragments. These are bound to coenzyme A and enter the citric acid cycle as acetyl CoA. Dietary fats not needed for energy or structural materials are stored in adipose tissue.

4. There is a continual turnover of fats in fat depots. Breakdown of fats to fatty acids and glycerol is called lipolysis.

5. When excessive amounts of fats are used, the liver converts acetyl CoA to ketone bodies and releases them to the blood. Excessive levels of ketone bodies (ketosis) lead to metabolic acidosis.

6. All cells use phospholipids and cholesterol to build their plasma membranes. The liver forms many functional molecules from lipids.

24.6 Amino acids are used to build proteins or for energy (pp. 948–949)

1. To be oxidized for energy, amino acids are converted to keto acids that can enter the citric acid cycle. This involves transamination, oxidative deamination, and keto acid modification.

2. Amine groups removed during deamination (as ammonia) are combined with carbon dioxide by the liver to form urea. Urea is excreted in urine.

3. Deaminated amino acids may also be converted to fatty acids and glucose.

4. Amino acids are the body's most important building blocks. Nonessential amino acids are made in the liver by transamination.

5. In adults, most protein synthesis serves to replace tissue proteins and to maintain nitrogen balance.

6. Protein synthesis requires the presence of all essential amino acids. If any are lacking, amino acids are used as energy fuels.

24.7 Energy is stored in the absorptive state and released in the postabsorptive state (pp. 949–955)

Catabolic-Anabolic Balance of the Body (pp. 949–950)

1. The amino acid pool provides amino acids for synthesis of proteins and amino acid derivatives, ATP synthesis, and energy storage.

2. The carbohydrate-fat pool primarily provides fuels for ATP synthesis and other molecules that can be stored as energy reserves.

3. The nutrient pools are connected by the bloodstream; fats, carbohydrates, and proteins may be interconverted via common intermediates.

Absorptive State (pp. 950–952)

4. During the absorptive state (during and shortly after a meal), glucose is the major energy source; needed structural and functional molecules are made; excess carbohydrates, fats, and amino acids are stored as glycogen and fat.

5. Events of the absorptive state are controlled by insulin, which enhances the entry of glucose (and amino acids) into cells and accelerates its use for ATP synthesis or storage as glycogen or fat.

Postabsorptive State (pp. 952–955)

6. The postabsorptive state is the period when bloodborne fuels are provided by breakdown of energy reserves. Glucose is made available to the blood by glycogenolysis, lipolysis, and gluconeogenesis. Glucose sparing begins. During prolonged fasting, the brain also begins to metabolize ketone bodies.

7. Events of the postabsorptive state are controlled largely by glucagon and the sympathetic nervous system, which mobilize glycogen and fat reserves and trigger gluconeogenesis.

24.8 The liver metabolizes, stores, and detoxifies (pp. 955–960)

1. The liver is the body's main metabolic organ and it plays a crucial role in processing (or storing) virtually every nutrient group. It helps maintain blood energy sources, metabolizes hormones, and detoxifies drugs and other substances.

2. The liver synthesizes cholesterol, catabolizes cholesterol and secretes it in the form of bile salts, and makes lipoproteins. The liver makes a basal amount of cholesterol (85%) even when dietary cholesterol intake is excessive.

3. LDLs transport cholesterol from the liver to the tissues, whereas HDLs transport cholesterol from the tissues to the liver (for catabolism and elimination).

4. Excessively high LDL levels are implicated in atherosclerosis, cardiovascular disease, and stroke.

PART 3

ENERGY BALANCE

1. Body energy intake (derived from food oxidation) is precisely balanced by energy output (heat, work, and energy storage). Eventually, all of the energy intake is converted to heat.

24.9 Neural and hormonal factors regulate food intake (pp. 960–962)

1. When energy balance is maintained, weight remains stable. When excess amounts of energy are stored, the result is obesity (condition of excessive fat storage resulting in a BMI above 30).

2. The hypothalamus (particularly its arcuate nucleus) and other brain centers are involved in regulating eating behavior.

3. Factors thought to be involved in regulating food intake include (a) neural signals from the gut to the brain; (b) nutrient signals related to total energy storage; (c) plasma concentrations of hormones that control events of the absorptive and postabsorptive states, and hormones that provide feedback signals to brain feeding centers (leptin appears to exert the main long-term controls of appetite and energy metabolism); (d) body temperature, psychological factors, and others.

24.10 Thyroxine is the major hormone that controls basal metabolic rate (pp. 962–963)

1. Energy used by the body per hour is the metabolic rate.

2. Basal metabolic rate (BMR), reported in kcal/m^2/h, is the measurement obtained when the person is at comfortable room temperature, supine, relaxed, and in the postabsorptive state. BMR indicates energy needed to drive only the resting body processes.

24

3. Factors influencing metabolic rate include body surface area, age and gender, body temperature, stress, and thyroxine; also the specific dynamic action of foods, and muscular activity.

24.11 The hypothalamus acts as the body's thermostat
(pp. 963–968)

1. Body temperature reflects the balance between heat production and heat loss and is normally 37°C (± 0.5°C), which is optimal for physiological activities.
2. At rest, most body heat is produced by the liver, heart, brain, kidneys, and endocrine organs. Activation of skeletal muscles causes dramatic increases in body heat production.
3. The body core (organs within the skull and the ventral body cavity) generally has the highest temperature. The shell (the skin) is the heat-exchange surface, and is usually coolest.
4. Blood serves as the major heat-exchange agent between the core and the shell. When skin capillaries are flushed with blood and the skin is warmer than the environment, the body loses heat. When blood is withdrawn to deep organs, heat loss from the shell is inhibited.
5. Heat-exchange mechanisms include radiation, conduction, convection, and evaporation. Evaporation, the conversion of water to water vapor, requires the absorption of heat. For each gram of water vaporized, about 0.6 kcal of heat is absorbed.
6. The hypothalamus acts as the body's thermostat. Its heat-promotion and heat-loss centers receive inputs from peripheral and central thermoreceptors, integrate these inputs, and initiate responses leading to heat loss or heat promotion.
7. Heat-promoting mechanisms include constriction of skin vasculature, and shivering.
8. When heat must be removed from the body, dermal blood vessels dilate, allowing heat loss through radiation, conduction, and convection. When greater heat loss is mandated (or the environmental temperature is so high that radiation and conduction are ineffective), sweating is initiated. Evaporation of sweat is an efficient means of heat loss as long as the humidity is low.
9. When the body cannot rid itself of surplus heat, body temperature rises to the point where all thermoregulatory mechanisms become ineffective—a potentially lethal condition called heat stroke. Physical activity in hot temperatures can lead to heat exhaustion, indicated by a rise in temperature, a drop in blood pressure, and collapse.
10. Fever is controlled hyperthermia, which follows thermostat resetting to higher levels by prostaglandins and initiation of heat-promotion mechanisms, evidenced by the chills. When the disease process is reversed, heat-loss mechanisms are initiated.

Developmental Aspects of Nutrition and Metabolism   (pp. 968–969)

1. Good nutrition is essential for normal fetal development and normal growth during childhood.
2. Inborn errors of metabolism include PKU, glycogen storage disease, galactosemia, and many others. Hormonal disorders, such as lack of insulin or thyroid hormone, may also lead to metabolic abnormalities. Diabetes mellitus and metabolic syndrome are the most significant metabolic disorders in middle-aged and older adults.
3. In old age, metabolic rate declines as enzyme and endocrine systems become less efficient and skeletal muscles atrophy. Reduced caloric needs make it difficult to obtain adequate nutrition without becoming overweight.
4. Elderly individuals ingest more medications than any other age group, and many of these drugs negatively affect their nutrition.

REVIEW QUESTIONS

To access additional practice questions using your smartphone, tablet, or computer: MasteringA&P® > Study Area > Practice Tests & Quizzes

Level 1 Remember/Understand

(Some questions have more than one correct answer. Select the best answer or answers from the choices given.)

1. Which of the following reactions would liberate the most energy? (a) complete oxidation of a molecule of sucrose to CO_2 and water, (b) conversion of a molecule of ADP to ATP, (c) respiration of a molecule of glucose to lactic acid, (d) conversion of a molecule of glucose to carbon dioxide and water.
2. The formation of glucose from glycogen is (a) gluconeogenesis, (b) glycogenesis, (c) glycogenolysis, (d) glycolysis.
3. The net gain of ATP from the complete metabolism (aerobic) of glucose is closest to (a) 2, (b) 30, (c) 3, (d) 4.
4. Which of the following best defines cellular respiration? (a) intake of carbon dioxide and output of oxygen by cells, (b) excretion of waste products, (c) inhalation of oxygen and exhalation of carbon dioxide, (d) oxidation of substances by which energy is released in usable form to the cells.
5. What is formed during aerobic respiration when electrons are passed down the electron transport chain? (a) oxygen, (b) water, (c) glucose, (d) NADH + H⁺.
6. Metabolic rate is relatively low in (a) youth, (b) physical exercise, (c) old age, (d) fever.

7. In a temperate climate under ordinary conditions, the greatest loss of body heat occurs through (a) radiation, (b) conduction, (c) evaporation, (d) none of the above.
8. Which of the following is not a function of the liver? (a) glycogenolysis and gluconeogenesis, (b) synthesis of cholesterol, (c) detoxification of alcohol and drugs, (d) synthesis of glucagon, (e) deamination of amino acids.
9. Amino acids are essential (and important) to the body for all the following except (a) production of some hormones, (b) production of antibodies, (c) formation of most structural materials, (d) as a source of quick energy.
10. Transamination is a chemical process by which (a) protein is synthesized, (b) an amine group is transferred from an amino acid to a keto acid, (c) an amine group is split from the amino acid, (d) amino acids are broken down for energy.
11. Body temperature regulation is (a) influenced by temperature receptors in the skin, (b) influenced by the temperature of the blood perfusing the heat regulation centers of the brain, (c) subject to both neural and hormonal control, (d) all of the above.
12. Which of the following yields the greatest caloric value per gram? (a) fats, (b) proteins, (c) carbohydrates, (d) all are equal in caloric value.

Level 2 Apply/Analyze

13. A person has been on a hunger strike for seven days. Compared to normal, he has **(a)** increased release of fatty acids from adipose tissue, and ketosis, **(b)** elevated glucose concentration in the blood, **(c)** increased plasma insulin concentration, **(d)** increased glycogen synthase (enzyme) activity in the liver.

14. What is cellular respiration? What is the common role of FAD and NAD^+ in cellular respiration?

15. Describe the site, major events, and outcomes of glycolysis.

16. Define glycogenesis, glycogenolysis, gluconeogenesis, and lipogenesis. Which is (are) likely to be occurring **(a)** shortly after a carbohydrate-rich meal, **(b)** just before waking up in the morning?

17. What is the harmful result when excessive amounts of fats are burned for energy? Name two conditions that might lead to this result.

18. Distinguish between the role of HDLs and that of LDLs.

19. List some factors that influence plasma cholesterol levels. Also list the sources and fates of cholesterol in the body.

20. What is meant by "body energy balance," and what happens if the balance is not precise?

21. Calculate the number of ATP molecules that can be harvested during the complete oxidation of an 18-carbon fatty acid. (Take a deep breath and think about it—you can do it.)

Level 3 Evaluate/Synthesize

22. Make a flowchart that indicates the pivotal intermediates through which glucose can be converted to fat.

23. Explain the effect of the following on metabolic rate: thyroxine levels, eating, body surface area, muscular exercise, emotional stress, starvation.

24. Compare and contrast mechanisms of heat loss with mechanisms of heat promotion, and explain how these mechanisms determine body temperature.

25. Every year dozens of elderly people are found dead in their unheated apartments and listed as victims of hypothermia. What is hypothermia, and how does it kill? Why are the elderly more susceptible to hypothermia than the young?

26. While attempting to sail solo from Los Angeles to Tahiti, Seth encountered a storm that marooned him on an uninhabited island. He was able, using his ingenuity and a pocket knife, to obtain plenty of fish to eat, and roots were plentiful. However, the island was barren of fruits and soon his gums began to bleed and he started to develop several infections. Analyze his problem.

27. Gregor, a large, beefy man, came home from the doctor's office and complained to his wife that his blood tests "were bad." He told her that the doctor said he would have to give up some of his steaks and butter. He went on to mourn the fact that he would have to start eating more cottage cheese and olive oil instead. What kind of problem was revealed by his "bad" blood tests? What do you think of his choice of food substitutes and why? What would you suggest?

CLINICAL CASE STUDY

32-Year-Old Male with Altered Mental Status

First responders retrieve Kyle Boulard, 32, from a local mall. Bystanders reported that he was behaving erratically, appeared disoriented, and staggered as he walked. When he arrives in your emergency department, you note that his breath has a fruity acetone odor. Tests are ordered and the following results obtained:

GENERAL	BLOOD	URINE
BP: 95/58	pH: 7.1	pH: 4.3
HR: 110	Glucose: 345 mg/dl	Glucose: strongly positive
	Ketone bodies: 22 mg/dl	Ketone bodies: strongly positive; odor "fruity acetone"

1. Urine glucose and ketone bodies are usually negative. Using the information in Appendix F, look up normal values for blood pH, blood glucose, blood ketone bodies, and urine pH, and identify whether each test result is normal or abnormal.

2. A "fruity acetone" smell was detected in Kyle's breath, and in his urine. Which substance is producing the "fruity acetone" smell?

3. **+ NCLEX-STYLE** Given the lab results above, what is the most likely explanation for Kyle's condition?
   a. The presence of ketone bodies in Kyle's blood suggests that Kyle is starving.
   b. The presence of glucose in Kyle's urine suggests that he has uncontrolled diabetes insipidus.
   c. The presence of glucose in Kyle's urine suggests that he has uncontrolled diabetes mellitus.
   d. The low pH in Kyle's blood and urine suggest that Kyle has ingested a poison.

4. **+ NCLEX-STYLE** Kyle's blood contains ketone bodies. What is the most likely explanation for this?
   a. The major source for ketones is dietary, so he must have been eating or drinking them.
   b. Kyle's blood contains ketone bodies because his kidney is not excreting them at the normal rate.
   c. He was eating excess carbohydrates, and his liver turned these into ketones for storage.
   d. His body cells are not able to use carbohydrates for fuel, so the liver is converting fats into ketone bodies.

5. Explain how the pH of Kyle's blood and urine is related to the ketone bodies measured in each of these fluids.

24

25 The Urinary System

In this chapter, you will learn that

The kidneys maintain the composition of the body's extracellular fluids

starting with

25.1 Gross anatomy of kidneys

next examining

25.2 Nephrons

and asking

25.3 How do the kidneys make urine?

exploring

25.4 Step 1: Glomerular filtration

25.5 Step 2: Tubular reabsorption

25.6 Step 3: Tubular secretion

by acting on

and asking

25.7 How do the kidneys regulate urine concentration and volume?

measured by

25.8 Clinical evaluation of kidney function

and then asking

25.9 How does the body transport, store, and eliminate urine?

and finally, exploring

Developmental Aspects of the Urinary System

Every day the kidneys filter nearly 200 liters of fluid from our bloodstream, allowing toxins, metabolic wastes, and excess ions to leave the body in urine while returning needed substances to the blood. Much like a water purification plant that keeps a city's water drinkable and disposes of its wastes, the kidneys are usually unappreciated until they malfunction and body fluids become contaminated.

The kidneys perform a chemical balancing act that would be tricky even for the best chemical engineer. They maintain the body's internal environment by:

- Regulating the total volume of water in the body and the total concentration of solutes in that water (osmolality).
- Regulating the concentrations of the various ions in the extracellular fluids. (Even relatively small changes in some ion concentrations such as K^+ can be fatal.)
- Ensuring long-term acid-base balance.
- Excreting metabolic wastes and foreign substances such as drugs or toxins.

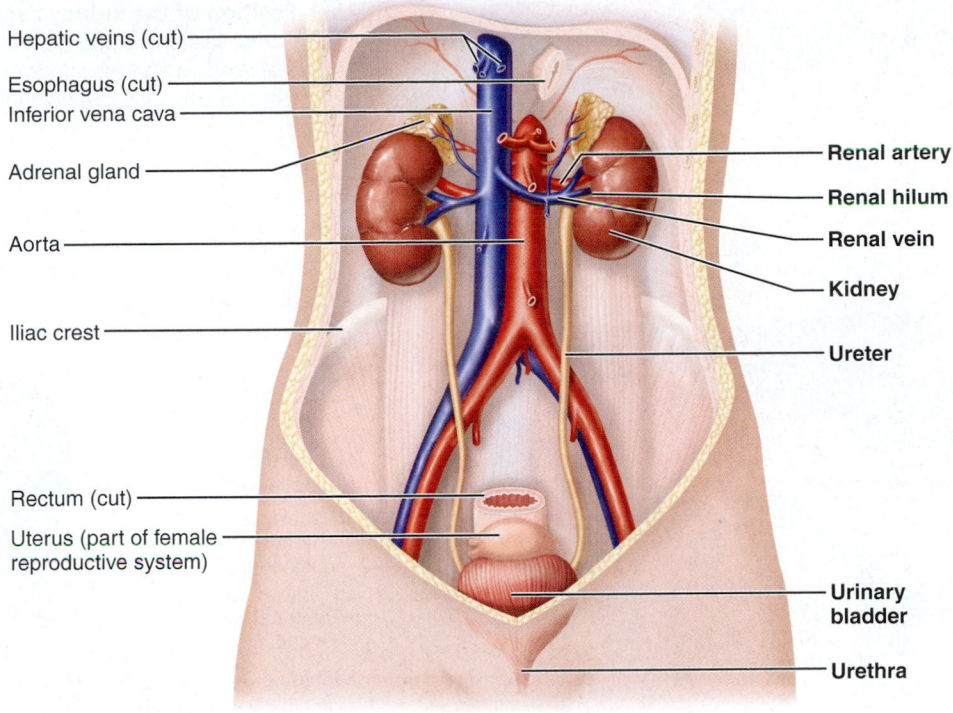

Hepatic veins (cut)
Esophagus (cut)
Inferior vena cava
Adrenal gland
Aorta
Iliac crest
Rectum (cut)
Uterus (part of female reproductive system)

Renal artery
Renal hilum
Renal vein
Kidney
Ureter
Urinary bladder
Urethra

Figure 25.1 The urinary system. Anterior view of the urinary organs in a female. (Most unrelated abdominal organs have been omitted.)

- Producing *erythropoietin* (◀ **p. 648**) and *renin* (re′nin; *ren =* kidney) (◀ **pp. 722–723**), important molecules for regulating red blood cell production and blood pressure, respectively.
- Converting vitamin D to its active form (◀ **p. 917**).
- Carrying out gluconeogenesis during prolonged fasting (◀ **p. 945**).

The urine-forming kidneys are crucial components of the **urinary system** (**Figures 25.1** and **25.2**). The urinary system also includes:

- *Ureters*—paired tubes that transport urine from the kidneys to the urinary bladder
- *Urinary bladder*—a temporary storage reservoir for urine
- *Urethra*—a tube that carries urine from the bladder to the body exterior

25.1 The kidneys have three distinct regions and a rich blood supply

Learning Outcomes
▶ Describe the gross anatomy of the kidney and its coverings.
▶ Trace the blood supply through the kidney.

Location and External Anatomy

The bean-shaped kidneys lie in a retroperitoneal position (between the dorsal body wall and the parietal peritoneum) in

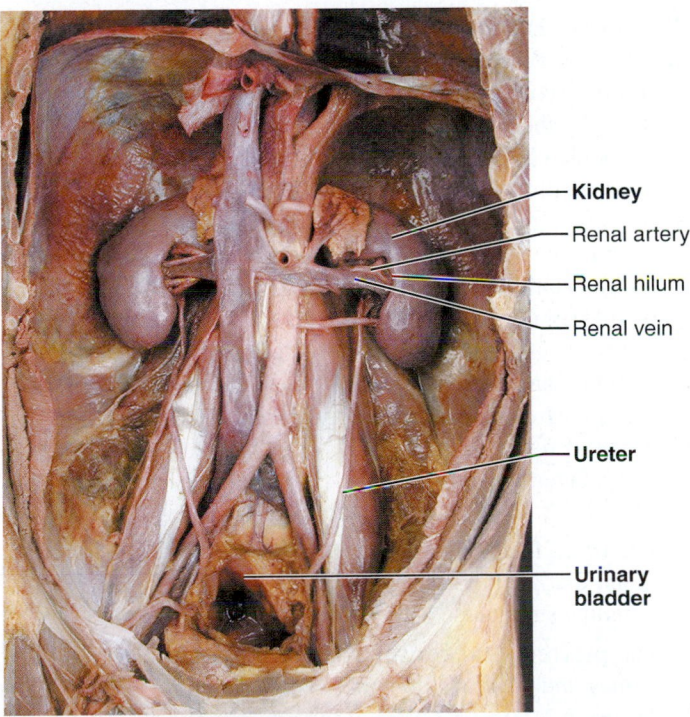

Kidney
Renal artery
Renal hilum
Renal vein
Ureter
Urinary bladder

Figure 25.2 Dissection of urinary system organs (male).

Figure 25.3 Position of the kidneys against the posterior body wall. (a) Cross section viewed from inferior direction. Note the retroperitoneal position and the supportive tissue layers of the kidney. **(b)** Posterior *in situ* view showing relationship of the kidneys to the 12th ribs.

(a)

(b)

the *superior* lumbar region (**Figure 25.3**). Extending approximately from T_{12} to L_3, the kidneys receive some protection from the lower part of the rib cage (Figure 25.3b). The right kidney is crowded by the liver and lies slightly lower than the left.

An adult's kidney has a mass of about 150 g (5 ounces) and its average dimensions are 11 cm long, 6 cm wide, and 3 cm thick—about the size of a large bar of soap. The lateral surface is convex. The medial surface is concave and has a vertical cleft called the **renal hilum** (see Figures 25.1, 25.2, and 25.4) that leads into an internal space called the *renal sinus*. The ureter, renal blood vessels, lymphatics, and nerves all join each kidney at the hilum and occupy the sinus. Atop each kidney is an *adrenal* (or *suprarenal*) *gland*, an endocrine gland that is functionally unrelated to the kidney.

Three layers of supportive tissue surround each kidney (Figure 25.3a). From superficial to deep, these are:

- The **renal fascia**, an outer layer of dense fibrous connective tissue that anchors the kidney and the adrenal gland to surrounding structures

- The **perirenal fat capsule**, a fatty mass that surrounds the kidney and cushions it against blows

- The **fibrous capsule**, a transparent capsule that prevents infections in surrounding regions from spreading to the kidney

HOMEOSTATIC IMBALANCE 25.1   CLINICAL

The upper parts of both kidneys are protected by the thoracic cage, and the perirenal fat provides cushioning. However, the lower parts of the kidneys, especially the right kidney, are susceptible to blunt trauma such as from falls, motor vehicle accidents, or contact sports injuries. The renal artery is especially vulnerable to injury from rapid deceleration during car crashes, and may be lacerated (torn) or develop a thrombosis (blood clot). The presence of *hematuria* (blood in the urine) is an important sign of such renal trauma. Occasionally, surgical treatment is required.

Internal Gross Anatomy

A frontal section through a kidney reveals three distinct regions: *cortex*, *medulla*, and *pelvis* (**Figure 25.4**). The most superficial region, the **renal cortex**, is light-colored and has a granular appearance. Deep to the cortex is the darker, reddish-brown

Renal
hilum

Renal cortex

Renal medulla

Major calyx

Papilla of
pyramid

Renal pelvis

Minor calyx

Ureter

Renal pyramid in
renal medulla

Renal column

Fibrous capsule

(a) Photograph of right kidney, frontal section

(b) Diagrammatic view

Figure 25.4 Internal anatomy of the kidney.

renal medulla, which exhibits cone-shaped tissue masses called **medullary**, or **renal, pyramids**. The broad *base* of each pyramid faces toward the cortex, and its apex, or *papilla* ("nipple"), points internally. The pyramids appear striped because they are formed almost entirely of parallel bundles of microscopic urine-collecting tubules and capillaries. The **renal columns**, inward extensions of cortical tissue, separate the pyramids. Each pyramid and its surrounding cortical tissue constitutes one of approximately eight **lobes** of a kidney.

The **renal pelvis**, a funnel-shaped tube, is continuous with the ureter leaving the hilum. Branching extensions of the pelvis form two or three **major calyces** (ka′lih-sēz; singular: calyx). Each major calyx subdivides to form several **minor calyces**, cup-shaped areas that enclose the papillae.

The calyces collect urine, which drains continuously from the papillae, and empty it into the renal pelvis. The urine then flows through the renal pelvis and into the ureter, which moves it to the bladder to be stored. The walls of the calyces, pelvis, and ureter contain smooth muscle that contracts rhythmically to propel urine by peristalsis.

Infections or inflammations that affect the kidney are called *pyelonephritis* (pi″ĕ-lo-nĕ-fri′tis). Kidney infections in females are usually caused by fecal bacteria that spread from the anal region to the urinary tract. Less often they result from blood-borne bacteria (traveling from other infected sites) that lodge and multiply in a kidney.

In severe cases of pyelonephritis, the kidney swells, abscesses form, and the pelvis fills with pus. Untreated, the kidney may be severely damaged, but antibiotic therapy can usually treat the infection successfully.

Blood and Nerve Supply

The kidneys continuously cleanse the blood and adjust its composition, so it is not surprising that they have a rich blood supply. Under normal resting conditions, the large **renal arteries** deliver one-fourth of the total cardiac output to the kidneys—about 1200 ml each minute.

25

(a) Frontal section illustrating major blood vessels

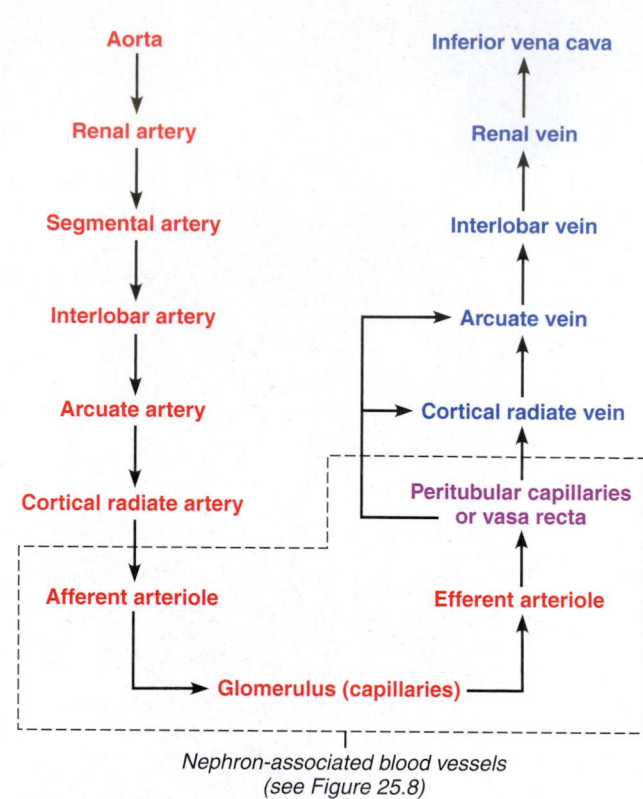

(b) Path of blood flow through renal blood vessels

Figure 25.5 Blood vessels of the kidney.

The renal arteries exit at right angles from the abdominal aorta, and the right renal artery is longer than the left because the aorta lies to the left of the midline. As each renal artery approaches a kidney, it divides into five **segmental arteries** (**Figure 25.5**). Within the renal sinus, each segmental artery branches further to form several **interlobar arteries**.

At the cortex-medulla junction, the interlobar arteries branch into the **arcuate arteries** (ar'ku-āt) that arch over the bases of the medullary pyramids. Small **cortical radiate arteries** (also called *interlobular arteries*) radiate outward from the arcuate arteries to supply the cortical tissue. More than 90% of the blood entering the kidney perfuses the renal cortex.

Afferent arterioles branching from the cortical radiate arteries begin a complex arrangement of microscopic blood vessels. These vessels are key elements of kidney function, and we will examine them in the next module when we describe the nephron.

Veins mostly trace the pathway of the arterial supply in reverse (Figure 25.5). Blood leaving the renal cortex drains sequentially into the **cortical radiate**, **arcuate**, **interlobar**, and finally **renal veins**. (There are no segmental veins.) The renal veins exit from the kidneys and empty into the inferior vena cava. Because the inferior vena cava lies to the right of the vertebral column, the left renal vein is about twice as long as the right.

The **renal plexus**, a variable network of autonomic nerve fibers and ganglia, provides the nerve supply of the kidney and its ureter. An offshoot of the celiac plexus, the renal plexus is largely supplied by sympathetic fibers from the most inferior thoracic and first lumbar splanchnic nerves, which course along with the renal artery to reach the kidney. These sympathetic vasomotor fibers regulate renal blood flow by adjusting the diameter of renal arterioles and also influence the formation of urine by the nephron.

Check Your Understanding

1. Zach is hit in the lower back by an errant baseball. What protects his kidneys from this mechanical trauma?
2. From inside to outside, list the three layers of supportive tissue that surround each kidney. Where is the parietal peritoneum in relation to these layers?
3. The lumen of the ureter is continuous with a space inside the kidney. This space has branching extensions. What are the names of this space and its extensions?

For answers, see Answers Appendix.

25.2 Nephrons are the functional units of the kidney

Learning Outcome

▶ Describe the anatomy of a nephron.

Nephrons (nef'ronz) are the structural and functional units of the kidneys. (A functional unit is the smallest physical structure capable of carrying out the function of an organ.) If you understand how a single nephron works, then you will understand

Figure 25.6 Location and structure of nephrons. Schematic view of a nephron and collecting duct depicting the structural characteristics of epithelial cells forming various regions.

how the kidneys work. Each kidney contains about 1 million nephrons acting in parallel to make urine (**Figure 25.6**). Each nephron creates a cell- and protein-free filtrate from blood. From that filtrate, it recovers chemicals the body needs while also secreting into that filtrate selected chemicals that the body

needs to get rid of. Nearby nephrons empty their processed filtrate into one of the thousands of *collecting ducts*, which convey this fluid, now called urine, into the renal pelvis.

Each nephron consists of a *renal corpuscle* and a *renal tubule*. All of the renal corpuscles are located in the renal cortex, while the renal tubules begin in the cortex and then pass into the medulla before returning to the cortex.

Renal Corpuscle

Each **renal corpuscle** consists of a tuft of capillaries called a **glomerulus** (glo-mer'u-lus; *glom* = ball of yarn) and a cup-shaped hollow structure called the **glomerular capsule** (or **Bowman's capsule**) (Figures 25.6 and 25.8). The glomerular capsule is continuous with its renal tubule and completely surrounds the glomerulus, much as a well-worn baseball glove encloses a ball.

Glomerulus

The endothelium of the glomerular capillaries is *fenestrated* (penetrated by many pores; ◀ p. 711), which makes these capillaries exceptionally porous. This property allows large amounts of solute-rich but virtually protein-free fluid to pass from the blood into the glomerular capsule. This plasma-derived fluid or **filtrate** is the raw material that the renal tubules process to form urine.

Glomerular Capsule

The glomerular capsule has an external parietal layer and a visceral layer that clings to the glomerular capillaries.

- The *parietal layer* is simple squamous epithelium (Figures 25.6, 25.7, 25.10, and 25.12a). This layer contributes to the capsule structure but plays no part in forming filtrate.

- The *visceral layer*, which clings to the glomerular capillaries, consists of highly modified, branching epithelial cells called **podocytes** (pod'o-sīts; "foot cells") (see Figure 25.12a–c). The octopus-like podocytes have **foot processes** that cling to the basement membrane of the glomerulus. The clefts or openings between the foot processes are called **filtration slits**. Through these slits, filtrate enters the **capsular space** inside the glomerular capsule.

We describe the *filtration membrane*, the filter that lies between the blood in the glomerulus and the filtrate in the capsular space, on pp. 984–986.

Renal Tubule and Collecting Duct

The **renal tubule** is about 3 cm (1.2 inches) long and has three major parts. It leaves the glomerular capsule as the elaborately coiled *proximal convoluted tubule*, drops into a hairpin loop called the *nephron loop*, and then winds and twists again as the *distal convoluted tubule* before emptying into a collecting duct. The terms *proximal* and *distal* indicate the relationship of the convoluted tubules to the renal corpuscle—filtrate from the renal corpuscle passes through the proximal convoluted tubule first and then the distal convoluted tubule, which is thus "further away" from the renal corpuscle. The meandering nature of the renal tubule increases its length and enhances its filtrate processing capabilities.

Throughout their length, the renal tubule and collecting duct consist of a single layer of epithelial cells on a basement membrane. However, each region has a unique histology that reflects its role in processing filtrate.

Proximal Convoluted Tubule (PCT)

The walls of the **proximal convoluted tubule** (**PCT**) are formed by cuboidal epithelial cells with large mitochondria. Their apical (luminal) surfaces bear dense microvilli (Figure 25.6 and **Figure 25.7**). Just as in the intestine, this *brush border* dramatically increases the surface area and capacity for reabsorbing water and solutes from the filtrate and secreting substances into it.

Nephron Loop

The U-shaped **nephron loop** (formerly called the *loop of Henle*) has **descending** and **ascending limbs**. The proximal part of the descending limb is continuous with the proximal tubule and its cells are similar. The rest of the descending limb, called the *descending thin limb*, consists of a simple squamous epithelium. The epithelium becomes cuboidal or even low columnar in the ascending part of the nephron loop, which is therefore called the *thick ascending limb*. In most nephrons, the entire ascending limb is thick, but in some nephrons, the thin segment extends around the bend as the *ascending thin limb*. The thick and thin parts of the nephron loop are also referred to as thick and thin segments.

Distal Convoluted Tubule (DCT)

The epithelial cells of the **distal convoluted tubule** (**DCT**), like those of the PCT, are cuboidal and confined to the cortex,

Practice Histology questions:
MasteringA&P® > Study Area > Lab Tools > PAL

Renal corpuscle
- Squamous epithelium of parietal layer of glomerular capsule
- Glomerular capsular space
- Glomerulus

Proximal convoluted tubule (fuzzy lumen due to long microvilli)

Distal convoluted tubule (clear lumen)

Figure 25.7 Renal cortical tissue. Photomicrograph (415×).

Cortical nephron
- Short nephron loop
- Glomerulus further from the cortex-medulla junction
- Efferent arteriole supplies peritubular capillaries

Juxtamedullary nephron
- Long nephron loop
- Glomerulus closer to the cortex-medulla junction
- Efferent arteriole supplies vasa recta

Renal corpuscle
Glomerulus (capillaries)
Glomerular capsule
Efferent arteriole
Cortical radiate vein
Cortical radiate artery
Afferent arteriole
Collecting duct
Distal convoluted tubule
Afferent arteriole
Efferent arteriole
Proximal convoluted tubule
Peritubular capillaries
Kidney
Arcuate vein
Arcuate artery
Nephron loop
- Ascending limb
- Descending limb
Cortex-medulla junction
Vasa recta

Figure 25.8 Cortical and juxtamedullary nephrons, and their blood vessels. Arrows indicate direction of blood flow. Capillary beds from adjacent nephrons overlap (not shown).

but they are thinner and almost entirely lack microvilli (Figure 25.6).

Collecting Duct

Each **collecting duct** contains two cell types. The more numerous *principal cells* have sparse, short microvilli and are responsible for maintaining the body's water and Na⁺ balance. The *intercalated cells* are cuboidal cells with abundant microvilli. There are two varieties of intercalated cells (types A and B), and each plays a role in maintaining the acid-base balance of the blood.

Each collecting duct receives filtrate from many nephrons. The collecting ducts run side by side through the medullary pyramids, giving them their striped appearance. As the collecting ducts approach the renal pelvis, they fuse together and deliver urine into the minor calyces via papillae of the pyramids.

Classes of Nephrons

Nephrons are generally divided into two major groups, cortical and juxtamedullary (**Figure 25.8**).

25

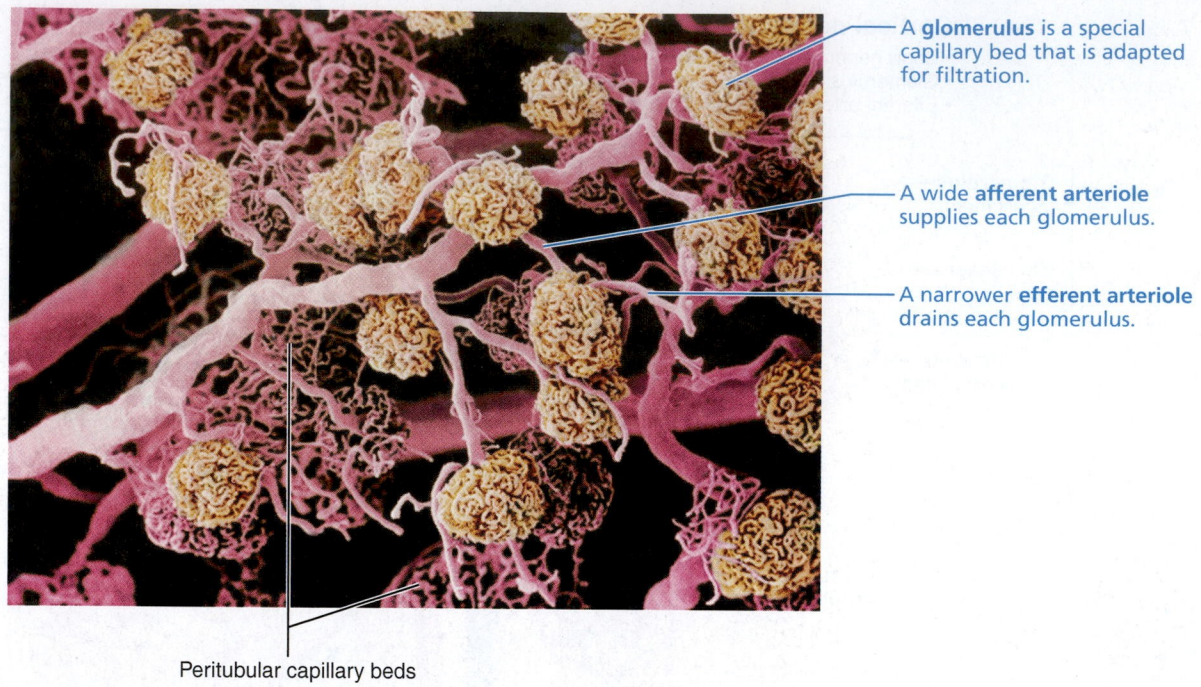

A **glomerulus** is a special capillary bed that is adapted for filtration.

A wide **afferent arteriole** supplies each glomerulus.

A narrower **efferent arteriole** drains each glomerulus.

Peritubular capillary beds

Figure 25.9 **Blood vessels of the renal cortex.** Scanning electron micrograph of a cast of blood vessels associated with nephrons (125×). View looking down onto the cortex.

- **Cortical nephrons** account for 85% of the nephrons in the kidneys. Except for small parts of their nephron loops that dip into the outer medulla, they are located entirely in the cortex.

- **Juxtamedullary nephrons** (juks″tah-mĕ′dul-ah-re) originate close to (*juxta* = near to) the cortex-medulla junction, and they play an important role in the kidneys' ability to produce urine that is concentrated (which conserves water). They have long nephron loops that deeply invade the medulla, and their ascending limbs have both thin and thick segments.

Nephron Capillary Beds

The renal tubule of every nephron is closely associated with two capillary beds. The first capillary bed (the *glomerulus*) produces the filtrate. The second (a combination of *peritubular capillaries* and *vasa recta*) reclaims most of that filtrate.

Glomerulus

The glomerulus, in which the capillaries run in parallel, is specialized for filtration. It differs from all other capillary beds in the body in that it is both fed and drained by arterioles—the **afferent arteriole** and **efferent arteriole**, respectively (Figure 25.8 and **Figure 25.9**). This arrangement maintains the high pressure in the glomerulus that is needed for filtration, a process we discuss in Module 25.4. Filtration produces a large amount of fluid, most (99%) of which is reabsorbed by the renal tubule cells and returned to the blood in the peritubular capillary beds.

The afferent arterioles arise from the *cortical radiate arteries* that run through the renal cortex. The efferent arterioles feed into either the peritubular capillaries or the vasa recta.

Peritubular Capillaries

The **peritubular capillaries** cling closely to adjacent renal tubules and empty into nearby venules. Because they arise from the efferent arterioles (which have high resistance), they only experience low pressure. As a result, these low-pressure, porous capillaries readily absorb solutes and water from the tubule cells as these substances are reclaimed from the filtrate. Renal tubules are closely packed together, so the peritubular capillaries of each nephron absorb substances from several adjacent nephrons.

Vasa Recta

Notice in Figure 25.8 that the efferent arterioles serving the juxtamedullary nephrons tend *not* to break up into meandering peritubular capillaries. Instead they form bundles of long straight vessels called **vasa recta** (va′sah rek′tah; "straight vessels") that extend deep into the medulla paralleling the longest nephron loops. Like all blood vessels, the vasa recta supply oxygen and nutrients to the tissue through which they pass (the renal medulla). However, the thin-walled vasa recta also play an important role in forming concentrated urine, as we will describe shortly.

Juxtaglomerular Complex (JGC)

Each nephron has a **juxtaglomerular complex (JGC)** (juks″-tah-glo-mer′u-lar), a region where the most distal portion of the ascending limb of the nephron loop lies against the afferent arteriole feeding the glomerulus (and sometimes the efferent arteriole) (**Figure 25.10**). Both the ascending limb and the afferent arteriole are modified at the point of contact.

Glomerular capsule — Glomerulus
Efferent arteriole
Afferent arteriole

Parietal layer of glomerular capsule
Capsular space
Efferent arteriole
Foot processes of podocytes
Podocyte cell body (visceral layer of glomerular capsule)
Red blood cell
Proximal tubule cell

Juxtaglomerular complex
• **Macula densa cells** of the ascending limb of nephron loop
• **Extraglomerular mesangial cells**
• **Granular cells**

Afferent arteriole

Lumens of glomerular capillaries
Endothelial cell of glomerular capillary
Glomerular mesangial cells

Juxtaglomerular complex   Renal corpuscle

Figure 25.10 Juxtaglomerular complex (JGC) of a nephron. Mesangial cells that surround the glomerular capillaries (glomerular mesangial cells) are not part of the JGC.

The JGC includes three populations of cells that help regulate the rate of filtrate formation and systemic blood pressure.

- The **macula densa** (mak'u-lah den'sah; "dense spot") is a group of tall, closely packed cells in the ascending limb of the nephron loop that lies adjacent to the granular cells (Figure 25.10). The macula densa cells are chemoreceptors that monitor the NaCl content of the filtrate entering the distal convoluted tubule.

- **Granular cells** [also called *juxtaglomerular (JG) cells*] are in the arteriolar walls. They are enlarged smooth muscle cells with prominent secretory granules containing the enzyme *renin* (see p. 989). Granular cells act as mechanoreceptors that sense the blood pressure in the afferent arteriole.

- *Extraglomerular mesangial cells* lie between the arteriole and tubule cells, and are interconnected by gap junctions. These cells may pass regulatory signals between macula densa and granular cells.

We discuss the physiological role of the JGC on pp. 988–989.

Check Your Understanding

4. Name the tubular components of a nephron in the order that filtrate passes through them.

5. What are the structural differences between juxtamedullary and cortical nephrons, and between their capillary beds?

6. What type of capillaries are the glomerular capillaries? What is their function?

7. For the juxtamedullary nephron and collecting duct illustrated below, name each of the structures labeled a–f.

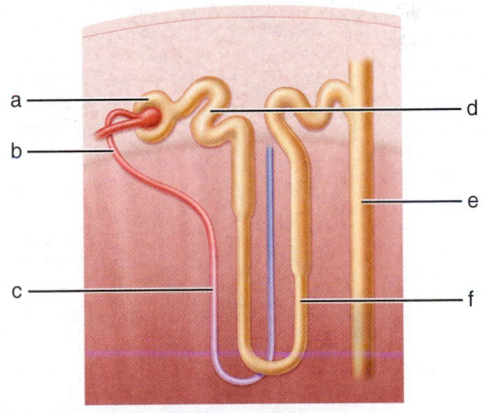

a
b
d
e
c
f

For answers, see Answers Appendix.

25.3 Overview: Filtration, absorption, and secretion are the key processes of urine formation

Learning Outcome

▶ List and define the three major renal processes.

If you had to design a system to chemically balance and cleanse the blood, how would you do it? Conceptually, it's really very simple. The body solves this problem in the following way. First, it "dumps" cell- and protein-free filtered blood into a separate "waste container." From this container, it reclaims everything the body needs to keep (which is almost everything in the container). Finally, the kidney selectively adds specific things to the container, fine-tuning the body's chemical balance. Anything left in the container becomes urine. This is basically how nephrons work.

Urine formation and the adjustment of blood composition involve three processes (**Figure 25.11**):

1 **Glomerular filtration.** *Glomerular filtration* ("dumping filtrate into the waste container") takes place in the renal corpuscle and produces a cell- and protein-free filtrate.

2 **Tubular reabsorption.** *Tubular reabsorption* ("reclaiming what the body needs to keep") is the process of selectively moving substances from the filtrate back into the blood. It takes place in the renal tubules and collecting ducts. Tubular reabsorption reclaims almost everything filtered—all of the glucose and amino acids, and some 99% of the water, salt, and other components. Anything that is *not* reabsorbed becomes urine.

3 **Tubular secretion.** *Tubular secretion* ("selectively adding to the waste container") is the process of selectively moving substances from the blood into the filtrate. Like tubular reabsorption, it occurs along the length of the tubule and collecting duct.

The kidneys process an enormous volume of blood each day. Of the approximately 1200 ml of blood that passes through the glomeruli each minute, some 650 ml is plasma, and about one-fifth of this (120–125 ml) is forced into the glomerular capsules as filtrate. This is equivalent to filtering your entire plasma volume more than 60 times each day! Considering the magnitude of their task, it is not surprising that the kidneys (which account for only 1% of body weight) consume 20–25% of all oxygen used by the body at rest.

Filtrate and urine are quite different. Filtrate contains everything found in blood plasma except proteins. **Urine** contains unneeded substances such as excess salts and metabolic wastes. The kidneys process about 180 L (47 gallons!) of blood-derived fluid daily. Of this amount, less than 1% (1.5 L) typically leaves the body as urine; the rest returns to the circulation.

Check Your Understanding

8.   **MAKE CONNECTIONS** In the kidneys, tubular secretion of a substance usually results in its excretion as well. Explain the difference between excretion (defined in Chapter 1) and tubular secretion.

For answers, see Answers Appendix.

25.4 Urine formation, step 1: The glomeruli make filtrate

Learning Outcomes

▶ **Describe the forces (pressures) that promote or counteract glomerular filtration.**

▶ **Compare the intrinsic and extrinsic controls of the glomerular filtration rate.**

Glomerular filtration is a passive process in which hydrostatic pressure forces fluids and solutes through a membrane. The glomeruli can be viewed as simple mechanical filters because filtrate formation does not directly consume metabolic energy. Let's first look at the structure of the filtration membrane and then see how it works.

The Filtration Membrane

The **filtration membrane** lies between the blood and the interior of the glomerular capsule. It is a porous membrane that allows free passage of water and solutes smaller than plasma proteins. As **Figure 25.12d** shows, its three layers are:

• **Fenestrated endothelium of the glomerular capillaries.** The fenestrations (capillary pores) allow all blood components except blood cells to pass through.

• **Basement membrane.** The basement membrane lies between the other two layers and is composed of their fused basal laminae (◀ p. 118). It forms a physical barrier that blocks all

Three major renal processes:

1 ➡ Glomerular filtration

2 ➡ Tubular reabsorption

3 ➡ Tubular secretion

Figure 25.11 The three major renal processes. A single nephron is shown schematically, as if uncoiled. Each kidney actually has more than a million nephrons acting in parallel.

Labels in figure: Afferent arteriole; Glomerular capillaries; Efferent arteriole; Cortical radiate artery; Glomerular capsule; Renal tubule and collecting duct containing filtrate; Peritubular capillary; To cortical radiate vein; Urine

Efferent arteriole

Glomerular capsular space

Afferent arteriole

Glomerular capillary covered by podocytes that form the visceral layer of glomerular capsule

Parietal layer of glomerular capsule

Proximal convoluted tubule

(a) Renal corpuscle

Cytoplasmic extensions of podocytes

Filtration slits

Podocyte cell body

Fenestrations (pores)

Glomerular capillary endothelium (podocyte covering and basement membrane removed)

Foot processes of podocyte

(b) Glomerular capillary surrounded by podocytes

Filtration slits

Podocyte cell body

Foot processes

(c) Filtration slits between the podocyte foot processes

The **filtration membrane** separates the blood in the glomerular capillaries from the filtrate in the glomerular capsule. It has three layers:

- Capillary endothelium
- Basement membrane
- Foot processes of podocyte of glomerular capsule

Filtration slit

Slit diaphragm

Blood in glomerular capillary

Filtrate in capsular space

Fenestration (pore)

Foot processes of podocyte

(d) Filtration membrane

Figure 25.12 The filtration membrane. (a) The renal corpuscle consists of a glomerulus surrounded by a glomerular capsule. **(b)** Enlargement of a glomerular capillary covered by the visceral (inner) layer of the glomerular capsule consisting of podocytes. Some podocytes and the basement membrane have been removed to show the fenestrations (pores) in the underlying capillary wall. **(c)** Scanning electron micrograph of the visceral layer (6000×). **(d)** Diagram of a section through the filtration membrane showing its three layers.

25

but the smallest proteins while still permitting most other solutes to pass. The glycoproteins of the gel-like basement membrane give it a negative charge. As a result, the basement membrane electrically repels many negatively charged macromolecular anions such as plasma proteins, reinforcing the blockade based on molecular size.

- **Foot processes of podocytes of the glomerular capsule.** The visceral layer of the glomerular capsule is made of podocytes that have filtration slits between their foot processes. If any macromolecules manage to make it through the basement membrane, *slit diaphragms*—thin membranes that extend across the filtration slits—prevent almost all of them from traveling farther.

Macromolecules that get "hung up" in the filtration membrane are engulfed by specialized pericytes (◀ **p. 710**) called *glomerular mesangial cells* (Figure 25.10).

Molecules smaller than 3 nm in diameter—such as water, glucose, amino acids, and nitrogenous wastes—pass freely from the blood into the glomerular capsule. As a result, these substances usually have similar concentrations in the blood and the glomerular filtrate. Larger molecules pass with greater difficulty, and those larger than 5 nm are generally barred from entering the tubule. Keeping the plasma proteins *in* the capillaries maintains the colloid osmotic (oncotic) pressure (◀ **p. 731**) of the glomerular blood, preventing the loss of all its water to the capsular space. The presence of proteins or blood cells in the urine usually indicates a problem with the filtration membrane.

Pressures That Affect Filtration

The principles that govern filtration from the glomerulus are the same as those that govern filtration from any capillary bed. *Focus on Bulk Flow across Capillary Walls* (◀ **Focus Figure 19.1, pp. 732–733**) shows filtration in normal capillary beds. **Figure 25.13** applies these principles to the glomerular capillaries of the nephron.

Outward Pressures

Outward pressures promote filtrate formation.

- The **hydrostatic pressure in glomerular capillaries (HP$_{gc}$)** is essentially glomerular blood pressure. It is the chief force pushing water and solutes out of the blood and across the filtration membrane. The blood pressure in the glomerulus is extraordinarily high (approximately 55 mm Hg compared to an average of 26 mm Hg or so in other capillary beds) and it remains high across the entire capillary bed. This is because the glomerular capillaries are drained by a high-resistance efferent arteriole whose diameter is smaller than the afferent arteriole that feeds them. As a result, filtration occurs along the entire length of each glomerular capillary and reabsorption does not occur as it would in other capillary beds.

- Theoretically, the *colloid osmotic pressure in the capsular space* of the glomerular capsule would "pull" filtrate into the tubule. However, this pressure is essentially zero because virtually no proteins enter the capsule, so we will not consider it further.

NFP = Net filtration pressure
= outward pressures − inward pressures
= (HP$_{gc}$) − (HP$_{cs}$ + OP$_{gc}$)
= (55) − (15 + 30)
= 10 mm Hg

Figure 25.13 Forces determining net filtration pressure (NFP). The pressure values cited in the diagram are approximate. (HP$_{gc}$ = hydrostatic pressure in glomerular capillaries, OP$_{gc}$ = osmotic pressure in glomerular capillaries, HP$_{cs}$ = hydrostatic pressure in capsular space)

Inward Pressures

Two inward forces inhibit filtrate formation by opposing HP$_{gc}$.

- The **hydrostatic pressure in the capsular space (HP$_{cs}$)** is the pressure exerted by filtrate in the glomerular capsule. HP$_{cs}$ is much higher than hydrostatic pressure surrounding most capillaries because filtrate is confined in a small space with a narrow outlet.

- The **colloid osmotic pressure in glomerular capillaries (OP$_{gc}$)** is the pressure exerted by the proteins in the blood that "sucks" water into the capillary.

As shown in Figure 25.13, the above pressures determine the **net filtration pressure (NFP)**. NFP largely determines the glomerular filtration rate, which we consider next.

Glomerular Filtration Rate (GFR)

The **glomerular filtration rate** is the volume of filtrate formed each minute by the combined activity of all 2 million glomeruli of the kidneys. GFR is directly proportional to each of the following factors:

- **Net filtration pressure.** NFP is the main controllable factor. Of the pressures determining NFP, the most important is hydrostatic pressure in the glomerulus. This pressure can be controlled by changing the diameter of the afferent (and sometimes the efferent) arterioles, as we will see shortly.

- **Total surface area available for filtration.** Glomerular capillaries have a huge surface area (collectively equal to the surface area of the skin). Glomerular mesangial cells surrounding these capillaries can fine-tune GFR by contracting to adjust the total surface area available for filtration.

- **Filtration membrane permeability.** Glomerular capillaries are thousands of times more permeable than other capillaries because of their fenestrations.

The huge surface area and high permeability of the filtration membrane explain how the relatively modest 10 mm Hg NFP can produce huge amounts of filtrate. Furthermore, the NFP in the glomerulus favors filtration over the entire length of the capillary, unlike other capillary beds where filtration occurs only at the arteriolar end and reabsorption occurs at the venous end. As a result, the adult kidneys produce about 180 L of filtrate daily, in contrast to the 2 to 4 L formed daily by all other capillary beds combined. This 180 L of filtrate per day translates to the normal GFR of 120–125 ml/min.

Regulation of Glomerular Filtration

GFR is tightly regulated to serve two crucial and sometimes opposing needs. The kidneys need a relatively constant GFR to make filtrate and do their job of maintaining extracellular homeostasis. On the other hand, the body as a whole needs a constant blood pressure, and this is closely tied to GFR in the following way: Assuming nothing else changes, an increase in GFR increases urine output, which reduces blood volume and blood pressure. The opposite holds true for a decrease in GFR.

Two types of controls serve these two different needs. *Intrinsic controls (renal autoregulation)* act locally within the kidney to maintain GFR, while *extrinsic controls* by the nervous and endocrine systems maintain blood pressure. In extreme changes of blood pressure (mean arterial pressure less than 80 or greater than 180 mm Hg), extrinsic controls take precedence over intrinsic controls in an effort to prevent damage to the brain and other crucial organs. These controls are summarized in **Table 25.1** and explained in detail on the next two pages.

GFR can be controlled by changing a single variable—glomerular hydrostatic pressure. All major control mechanisms act primarily to change this one variable. If the glomerular hydrostatic pressure rises, NFP rises and so does GFR. If the glomerular hydrostatic pressure falls by as little as 18%, GFR drops to zero. Clearly hydrostatic pressure in the glomerulus must be tightly controlled. Let's see how the intrinsic and extrinsic mechanisms accomplish this feat.

Intrinsic Controls: Renal Autoregulation

By adjusting its own resistance to blood flow, a process called **renal autoregulation**, the kidney can maintain a nearly constant GFR despite fluctuations in systemic arterial blood pressure. Renal autoregulation uses two different mechanisms: (1) a *myogenic*

Complete an interactive tutorial: MasteringA&P® > Study Area > Interactive Physiology

25

Table 25.1	Summary of Regulation of GFR	
	INTRINSIC CONTROL (RENAL AUTOREGULATION)	**EXTRINSIC CONTROL**
Purpose	Maintain GFR in spite of changes in blood pressure (so kidneys can "do their job")	Maintain systemic blood pressure
Mechanisms	• Myogenic • Tubuloglomerular feedback	• Hormonal (renin-angiotensin-aldosterone) • Neural (baroreceptor reflex via sympathetic nervous system)
Operating conditions	Mean arterial pressure between 80 and 180 mm Hg	Mean arterial pressure far outside of the normal range (<80 or >180 mm Hg)

↓ **SYSTEMIC BLOOD PRESSURE**

| Myogenic mechanism of autoregulation | Tubuloglomerular mechanism of autoregulation | Hormonal (renin-angiotensin-aldosterone) mechanism | Neural controls |

Intrinsic mechanisms directly regulate GFR despite moderate changes in blood pressure (between 80 and 180 mm Hg mean arterial pressure).

Extrinsic mechanisms indirectly regulate GFR by maintaining systemic blood pressure, which drives filtration in the kidneys.

Figure 25.14 Regulation of glomerular filtration rate (GFR) in the kidneys. (Note that while the extrinsic controls are aimed at maintaining blood pressure, they also maintain GFR since bringing blood pressure back up allows the kidneys to maintain GFR.)

mechanism and (2) a *tubuloglomerular feedback mechanism* (**Figure 25.14**, left side).

Myogenic Mechanism The **myogenic mechanism** (mi″o-jen′ik) reflects a property of vascular smooth muscle—it contracts when stretched and relaxes when not stretched (◄ p. 727). Rising systemic blood pressure stretches vascular smooth muscle in the arteriolar walls, causing the afferent arterioles to constrict. This constriction restricts blood flow into the glomerulus and keeps the GFR at a level ideal for kidney function. Declining systemic blood pressure causes dilation of afferent arterioles and raises glomerular hydrostatic pressure. Both responses help maintain normal NFP and GFR.

Tubuloglomerular Feedback Mechanism Autoregulation by the flow-dependent **tubuloglomerular feedback mechanism** is "directed" by the *macula densa cells* of the *juxtaglomerular complex* (see Figure 25.10). These cells, located in the walls of the ascending limb of the nephron loop, respond to filtrate NaCl concentration (which varies directly with filtrate flow rate). When GFR increases, there is not enough time for reabsorption and the concentration of NaCl in the filtrate remains high. The macula densa cells respond to high levels of NaCl in filtrate by releasing vasoconstrictor chemicals (ATP and others) that cause intense constriction of the afferent arteriole, reducing blood flow into the glomerulus. This drop in blood flow decreases the

NFP and GFR, slowing the flow of filtrate and allowing more time for filtrate processing (NaCl reabsorption).

In contrast, the low NaCl concentration of slowly flowing filtrate inhibits ATP release from macula densa cells, causing vasodilation of the afferent arterioles (Figure 25.14). This allows more blood to flow into the glomerulus, thus increasing NFP and GFR.

Autoregulatory mechanisms maintain a relatively constant GFR over an arterial pressure range from about 80 to 180 mm Hg. Consequently, normal day-to-day changes in our blood pressure (such as during exercise, sleep, or changes in posture) do not cause large changes in water and solute excretion. However, the intrinsic controls cannot handle extremely low systemic blood pressure, such as might result from serious hemorrhage (*hypovolemic shock*). Once the mean arterial pressure drops below 80 mm Hg, autoregulation ceases and extrinsic controls take over.

Extrinsic Controls: Neural and Hormonal Mechanisms

The purpose of the extrinsic controls regulating the GFR is to maintain systemic blood pressure (Figure 25.14, right side).

Sympathetic Nervous System Controls   Neural renal controls serve the needs of the body as a whole—sometimes to the detriment of the kidneys. When the volume of the extracellular fluid is normal and the sympathetic nervous system is at rest, the renal blood vessels are dilated and renal autoregulation mechanisms prevail. However, when the extracellular fluid volume is extremely low (as in hypovolemic shock during severe hemorrhage), it is necessary to shunt blood to vital organs, and neural controls may override autoregulatory mechanisms. This could reduce renal blood flow to the point of damaging the kidneys.

When blood pressure falls, norepinephrine released by sympathetic nerve fibers (and epinephrine released by the adrenal medulla) causes vascular smooth muscle to constrict, increasing peripheral resistance and bringing blood pressure back up toward normal. [This is the *baroreceptor reflex* that we discussed previously (◄ p. 719).] As part of this reflex, the afferent arterioles also constrict. Constriction of the afferent arterioles decreases GFR and so helps restore blood volume and blood pressure to normal.

Renin-Angiotensin-Aldosterone Mechanism   As we discussed in Chapter 19 (◄ pp. 722–723), the **renin-angiotensin-aldosterone mechanism** is the body's main mechanism for increasing blood pressure. Without adequate blood pressure (as might be due to hemorrhage, dehydration, etc.), glomerular filtration is not possible, so this mechanism regulates GFR indirectly.

Low blood pressure causes the granular cells of the juxtaglomerular complex to release **renin**. There are three pathways that stimulate granular cells:

- *Sympathetic nervous system.* As part of the baroreceptor reflex, renal sympathetic nerves activate β_1-adrenergic receptors that cause the granular cells to release renin.

- *Activated macula densa cells.* Low blood pressure or vasoconstriction of the afferent arterioles by the sympathetic nervous system reduces GFR, slowing down the flow of filtrate

through the renal tubules. When macula densa cells sense the low NaCl concentration of this sluggishly flowing filtrate, they signal the granular cells to release renin. They may signal by releasing less ATP (also thought to be the tubuloglomerular feedback messenger), by releasing *more* prostaglandin PGE_2, or both.

- *Reduced stretch.* Granular cells act as mechanoreceptors. A drop in mean arterial blood pressure reduces the tension in the granular cells' plasma membranes and stimulates them to release more renin.

Other Factors Affecting GFR

Renal cells produce a battery of chemicals, many of which act as paracrines (local signaling molecules) affecting renal arterioles. These include *adenosine* and *prostaglandin E_2* (*PGE_2*). In addition, the kidney makes its own locally acting *angiotensin II* that reinforces the effects of hormonal angiotensin II described in Chapter 19 (◄ pp. 721–723).

> **HOMEOSTATIC IMBALANCE 25.3**   **CLINICAL**
>
> Abnormally low urinary output (less than 50 ml/day), called *anuria* (ah-nu′re-ah), may indicate that glomerular blood pressure is too low to cause filtration. Renal failure and anuria can also result from situations in which the nephrons stop functioning, including acute nephritis, transfusion reactions, and crush injuries. ● ● ● ● ● ● ●

Check Your Understanding

9. Extrinsic and intrinsic controls of GFR serve two different purposes. What are they?
10. Which of the pressures that determine NFP is regulated by both intrinsic and extrinsic controls of GFR?
11. **APPLY** Calculate net filtration pressure given the following values: glomerular hydrostatic pressure = 50 mm Hg, blood colloid osmotic pressure = 25 mm Hg, capsular hydrostatic pressure = 20 Hg.

For answers, see Answers Appendix.

25.5 Urine formation, step 2: Most of the filtrate is reabsorbed into the blood

Learning Outcomes

▶ **Describe the mechanisms underlying water and solute reabsorption from the renal tubules into the peritubular capillaries.**

▶ **Describe how sodium and water reabsorption are regulated in the distal tubule and collecting duct.**

Our total plasma volume filters into the renal tubules about every 22 minutes, so all our plasma would drain away as urine in less than 30 minutes were it not for **tubular reabsorption**, which quickly reclaims most of the tubule contents and returns them to the blood.

25

The transcellular route involves:

① Transport across the apical membrane.

② Diffusion through the cytosol.

③ Transport across the basolateral membrane. (Often involves the lateral intercellular spaces because membrane transporters transport ions into these spaces.)

④ Movement through the interstitial fluid and into the capillary.

The paracellular route involves:

ⓐ Movement through leaky tight junctions, particularly in the proximal convoluted tubule.

ⓑ Movement through the interstitial fluid and into the capillary.

Figure 25.15 Tubular reabsorption occurs by transcellular and paracellular routes. Generally, water and solutes move into the peritubular capillaries through intercellular clefts. For simplicity, transporters, ion channels, intercellular clefts, and aquaporins are not depicted.

Tubular reabsorption (shown schematically at right) is a selective *transepithelial process* that begins as soon as the filtrate enters the proximal tubules.

To reach the blood, reabsorbed substances follow either the *transcellular* or *paracellular route* (**Figure 25.15**). The transcellular route is just like the route we described for absorption in the gut (◀ p. 912). Transported substances move through the *apical membrane*, the cytosol, and the *basolateral membrane* of the tubule cell and then the endothelium of the peritubular capillaries. Movement of substances in the paracellular route—*between* the tubule cells—is limited by the tight junctions connecting these cells. In the proximal nephron, however, these tight junctions are "leaky" and allow water and some important ions (Ca^{2+}, Mg^{2+}, K^+, and some Na^+) to pass through the paracellular route.

Given healthy kidneys, virtually all organic nutrients such as glucose and amino acids are completely reabsorbed to maintain or restore normal plasma concentrations. In contrast, the reabsorption of water and many ions is continuously regulated and adjusted in response to hormonal signals. Depending on the substances transported, the reabsorption process may be *active* or *passive*.

Tubular reabsorption

- **Active tubular reabsorption** requires ATP either directly (primary active transport; ◀ p. 75) or indirectly (secondary active transport; ◀ pp. 75–76) for at least one of its steps.
- **Passive tubular reabsorption** encompasses diffusion, facilitated diffusion, and osmosis (◀ pp. 70–73)—processes

in which substances move down their electrochemical gradients.

(You may wish to review these membrane transport processes before you proceed.)

Tubular Reabsorption of Sodium

Sodium ions are the single most abundant cation in the filtrate, and about 80% of the energy used for active transport is devoted to reabsorbing them. Sodium reabsorption is almost always active and via the transcellular route. Let's begin with the ATP-driven step.

Sodium Transport across the Basolateral Membrane

Na^+ is actively transported out of the tubule cell by *primary active transport*—a Na^+-K^+ ATPase pump in the basolateral membrane (**Figure 25.16** ①). From there, the bulk flow of water sweeps Na^+ into adjacent peritubular capillaries. This bulk flow of water and solutes into the peritubular capillaries is rapid because the blood there has low hydrostatic pressure and high osmotic pressure (remember, most proteins remain in the blood instead of filtering out into the tubule).

Sodium Transport across the Apical Membrane

Active pumping of Na^+ from the tubule cells results in a strong electrochemical gradient that favors its entry at the apical face via *secondary active transport (cotransport)* carriers (Figure 25.16 ②, ③) or via facilitated diffusion through channels (not illustrated). This occurs because (1) the pump maintains the intracellular Na^+ concentration at low levels, and (2) the K^+ pumped into the tubule cells almost immediately diffuses out into the interstitial fluid via leakage channels, leaving the interior of the tubule cell with a net negative charge.

Because each tubule segment plays a slightly different role in reabsorption, the precise mechanism by which Na$^+$ is reabsorbed at the apical membrane varies.

Tubular Reabsorption of Nutrients, Water, and Ions

The reabsorption of Na$^+$ by primary active transport provides the energy and the means for reabsorbing almost every other substance, including water.

> ⟳ Complete an interactive tutorial: MasteringA&P®
> > Study Area > Interactive Physiology

Secondary Active Transport

Substances reabsorbed by *secondary active transport* (the "push" comes from the gradient created by Na$^+$-K$^+$ pumping at the basolateral membrane) include glucose, amino acids, some ions, and vitamins. In nearly all these cases, an apical carrier moves Na$^+$ down its concentration gradient as it cotransports

another solute (Figure 25.16 ③). Cotransported solutes move across the basolateral membrane by facilitated diffusion via other transport proteins (not shown) before moving into the peritubular capillaries.

Passive Tubular Reabsorption of Water

The movement of Na$^+$ and other solutes establishes a strong osmotic gradient, and water moves by osmosis into the peritubular capillaries. Transmembrane proteins called **aquaporins** (◀ p. 71) aid this process by acting as water channels across plasma membranes (Figure 25.16 ④).

In continuously water-permeable regions of the renal tubules, such as the PCT, aquaporins are always present in the tubule cell membranes. Their presence "obliges" the body to absorb water in the proximal nephron regardless of its state of over- or underhydration. This water flow is referred to as **obligatory water reabsorption**.

Aquaporins are virtually absent in the apical membranes of the collecting duct unless antidiuretic hormone (ADH) is present. Water reabsorption that depends on ADH is called **facultative water reabsorption**.

Passive Tubular Reabsorption of Solutes

As water leaves the tubules, the concentration of solutes in the filtrate increases and, if able, they too follow their concentration gradients into the peritubular capillaries. This

PCT

① At the basolateral membrane, Na$^+$ is pumped into the interstitial space by the Na$^+$-K$^+$ ATPase. Active Na$^+$ transport creates concentration gradients that drive:

② "Downhill" Na$^+$ entry at the apical membrane.

③ Reabsorption of organic nutrients and certain ions by cotransport at the apical membrane.

④ Reabsorption of water by osmosis through aquaporins. Water reabsorption increases the concentration of the solutes that are left behind. These solutes can then be reabsorbed as they move down their gradients:

⑤ Lipid-soluble substances diffuse by the transcellular route.

⑥ Various ions (e.g., Cl$^-$, Ca^{2+}, K$^+$) and urea diffuse by the paracellular route.

Figure 25.16 Tubular reabsorption of water and nutrients uses active and passive transport. PCT cells (shown here) reabsorb 80% of the filtrate. Microvilli have been omitted for simplicity.

phenomenon—solutes following solvent—explains the passive reabsorption of a number of solutes present in the filtrate, such as lipid-soluble substances, certain ions, and some urea (Figure 25.16 ⑤, ⑥). It also explains in part why lipid-soluble drugs and environmental pollutants are difficult to excrete: Since lipid-soluble compounds can generally pass through membranes, they will follow their concentration gradients and be reabsorbed, even if this is not "desirable."

As Na^+ ions move through the tubule cells into the peritubular capillary blood, they also establish an electrical gradient that favors passive reabsorption of anions (primarily Cl^-) to restore electrical neutrality in the filtrate and plasma.

Transport Maximum

The transcellular transport systems for the various solutes are quite specific and *limited*. There is a **transport maximum (T_m)** for nearly every substance that is reabsorbed using a transport protein in the membrane. The T_m (reported in mg/min) reflects the number of transport proteins in the renal tubules available to ferry a particular substance. In general, there are plenty of transporters for substances such as glucose that need to be retained, and few or no transporters for substances of no use to the body.

When the transporters are saturated—that is, all bound to the substance they transport—the excess is excreted in urine. This is what happens in individuals who become hyperglycemic because of uncontrolled diabetes mellitus. As plasma levels of glucose approach and exceed 180 mg/dl, the glucose T_m is exceeded and large amounts of glucose may be lost in the urine even though the renal tubules are still functioning normally.

Reabsorptive Capabilities of the Renal Tubules and Collecting Ducts

Table 25.2 compares the reabsorptive abilities of various regions of the renal tubules and collecting ducts.

Proximal Convoluted Tubule

The entire renal tubule is involved in reabsorption to some degree, but the PCT cells are by far the most active reabsorbers and the events just described occur mainly in this tubular segment. Normally, the PCT reabsorbs *all* of the glucose and amino acids in the filtrate and 65% of the Na^+ and water. The bulk of the electrolytes are reabsorbed by the time the filtrate reaches the nephron loop. Nearly all of the uric acid and about half of the urea are reabsorbed in the proximal tubule, but both are later secreted back into the filtrate.

Nephron Loop

Beyond the PCT, the permeability of the tubule epithelium changes dramatically. Here, for the first time, water reabsorption is not coupled to solute reabsorption. Water can leave the descending limb of the nephron loop but *not* the ascending limb, where aquaporins are scarce or absent in the tubule cell membranes. These permeability differences play a vital role in the kidneys' ability to form dilute or concentrated urine.

The rule for water is that it leaves the descending (but not the ascending) limb of the nephron loop. The opposite is true for solutes. Virtually no solute reabsorption occurs in the descending limb, but solutes are reabsorbed both actively and passively in the ascending limb.

In the thin segment of the ascending limb, Na^+ moves passively down the concentration gradient created by water reabsorption. In the thick ascending limb, a Na^+-K^+-$2Cl^-$ symporter (◀ p. 76) is the main means of Na^+ entry at the apical surface. A Na^+-K^+ ATPase operates at the basolateral membrane to create

Cortex

65% of filtrate volume reabsorbed
• H_2O
• Na^+, HCO_3^-, and many other ions
• Glucose, amino acids, and other nutrients

Regulated reabsorption
• Na^+ (by aldosterone; Cl^- follows)
• Ca^{2+} (by parathyroid hormone)

• H^+ and NH_4^+
• Some drugs

Regulated secretion
• K^+ (by aldosterone)

• H_2O

• Na^+, K^+, Cl^-

Outer medulla

Regulated reabsorption
• H_2O (by ADH)
• Na^+ (by aldosterone; Cl^- follows)
• Urea (increased by ADH)

• Urea

Regulated secretion
• K^+ (by aldosterone)

Inner medulla

• Reabsorption or secretion to maintain blood pH described in Chapter 26; involves H^+, HCO_3^-, and NH_4^+

→ Reabsorption
→ Secretion

Figure 25.17 Summary of tubular reabsorption and secretion. The various regions of the renal tubule carry out reabsorption and secretion and maintain a gradient of osmolality within the medullary interstitial fluid. Color gradients represent varying osmolality at different points in the interstitial fluid.

Table 25.2   Reabsorption Capabilities of Different Segments of the Renal Tubules and Collecting Ducts

TUBULE SEGMENT	SUBSTANCE REABSORBED	MECHANISM
Proximal Convoluted Tubule (PCT)		
	Sodium ions (Na^+)	Primary active transport via basolateral Na^+-K^+ pump; crosses apical membrane through channels, symporters, or antiporters
	Virtually all nutrients (glucose, amino acids, vitamins, some ions)	Secondary active transport with Na^+
	Cl^-, K^+, Mg^{2+}, Ca^{2+}, and other ions	Passive paracellular diffusion driven by electrochemical gradient
	HCO_3^-	Secondary active transport linked to H^+ secretion and Na^+ reabsorption (see Chapter 26)
	Water	Osmosis; driven by solute reabsorption (obligatory water reabsorption)
	Lipid-soluble solutes	Passive diffusion driven by the concentration gradient created by reabsorption of water
	Urea	Primarily passive paracellular diffusion driven by chemical gradient
Nephron Loop		
Descending limb	Water	Osmosis
Ascending limb	Na^+, Cl^-, K^+	Secondary active transport of Cl^-, Na^+, and K^+ via Na^+-K^+-$2Cl^-$ cotransporter in thick portion; paracellular diffusion; Na^+-H^+ antiport
	Ca^{2+}, Mg^{2+}	Passive paracellular diffusion driven by electrochemical gradient
Distal Convoluted Tubule (DCT)		
	Na^+, Cl^-	Primary active Na^+ transport at basolateral membrane; secondary active transport at apical membrane via Na^+-Cl^- symporter and channels; aldosterone-regulated at distal portion
	Ca^{2+}	Passive uptake via PTH-modulated channels in apical membrane; primary and secondary active transport (antiport with Na^+) in basolateral membrane
Collecting Duct		
	Na^+, K^+, HCO_3^-, Cl^-	Primary active transport of Na^+ (requires aldosterone); passive paracellular diffusion of some Cl^-; cotransport of Cl^- and HCO_3^-; K^+ is both reabsorbed and secreted (aldosterone dependent), usually resulting in net K^+ secretion
	Water	Osmosis; controlled (facultative) water reabsorption; ADH required to insert aquaporins
	Urea	Facilitated diffusion in response to concentration gradient in the deep medulla region; recycles and contributes to medullary osmotic gradient

the ionic gradient that drives the symporter. The thick ascending limb also has Na^+-H^+ antiporters. In addition, some 50% of Na^+ passes via the paracellular route in this region.

Distal Convoluted Tubule and Collecting Duct

While reabsorption in the PCT and nephron loop does not vary with the body's needs, hormones fine-tune reabsorption in the DCT and collecting duct (**Figure 25.17**). Because most of the filtered water and solutes have been reabsorbed by the time the DCT is reached, only a small amount of the filtered load is subject to this fine tuning (e.g., about 10% of the originally filtered NaCl and 25% of the water). We introduce hormones that act at the DCT and collecting duct here but discuss them in more detail in Chapter 26.

- **Antidiuretic hormone (ADH).** As its name reveals, ADH inhibits *diuresis* (di"u-re′sis), or urine output. ADH makes the principal cells of the collecting ducts more permeable to water by causing aquaporins (◄ p. 71) to be inserted into their apical membranes. The amount of ADH determines the number of aquaporins, and so regulates the amount of water that is reabsorbed there. When the body is overhydrated, extracellular fluid osmolality decreases, decreasing ADH secretion by the posterior pituitary (◄ p. 613) and making the collecting ducts relatively impermeable to water. ADH also increases urea reabsorption by the collecting ducts, as we will describe later.

- **Aldosterone.** Aldosterone fine-tunes reabsorption of the remaining Na^+. Decreased blood volume or blood pressure, or high extracellular K^+ concentration (hyperkalemia), can cause the adrenal cortex to release aldosterone to the blood (◄ pp. 623–624). Except for hyperkalemia (which *directly* stimulates the adrenal cortex to secrete aldosterone), these conditions promote the renin-angiotensin-aldosterone mechanism (see Figure 25.14).

  Aldosterone targets the principal cells of the collecting ducts and cells of the distal portion of the DCT (prodding them to synthesize and retain more apical Na^+ and K^+ channels, and more basolateral Na^+-K^+ ATPases). As a result, little or no Na^+ leaves the body in urine. In the absence of aldosterone, these segments reabsorb much less Na^+ and about 2% of Na^+ filtered daily can be lost—an amount incompatible with life.

  Physiologically, aldosterone's role is to increase blood volume, and therefore blood pressure, by enhancing Na^+ reabsorption. In general, water follows Na^+ if aquaporins are present. Aldosterone also reduces blood K^+ concentrations because aldosterone-induced reabsorption of Na^+ is coupled to K^+ secretion in the principal cells of the collecting duct. That is, as Na^+ enters the cell, K^+ moves into the lumen.

- **Atrial natriuretic peptide (ANP).** In contrast to aldosterone, which acts to conserve Na^+, ANP reduces blood Na^+, thereby decreasing blood volume and blood pressure (◄ p. 624). Released by cardiac atrial cells when blood volume or blood pressure is elevated, ANP exerts several effects that lower blood Na^+ content, including direct inhibition of Na^+ reabsorption at the collecting ducts.

- **Parathyroid hormone (PTH).** Acting primarily at the DCT, PTH increases the reabsorption of Ca^{2+} (◄ p. 622).

Check Your Understanding

12. In which part of the nephron does most reabsorption occur?

13. How does the movement of Na^+ drive the reabsorption of water and solutes?

14. **DRAW** In the figure at right above, draw arrows to represent the pathway for transcellular reabsorption of Na^+. If transport proteins are required for this process, draw and label them in the appropriate locations. In each case, state whether the transport of Na^+ is primary active, secondary active, or passive.

15. **MAKE CONNECTIONS** Primary and secondary active transport processes are shown in Figure 25.16 (and were introduced in Chapter 3). How do they differ?

For answers, see Answers Appendix.

25.6 Urine formation, step 3: Certain substances are secreted into the filtrate

Learning Outcome

▶ Describe the importance of tubular secretion and list several substances that are secreted.

The most important way to clear plasma of unwanted substances is to simply not reabsorb them from the filtrate. Another way is **tubular secretion**—essentially, reabsorption in reverse (shown schematically at right). Tubular secretion moves *selected* substances [such as H^+, K^+, NH_4^+ (ammonium ion), creatinine, and certain organic acids and bases] from the peritubular capillaries through the tubule cells into the filtrate. Also, some substances (such as HCO_3^-) that are synthesized in the tubule cells are secreted.

Tubular secretion

The urine eventually excreted contains *both filtered and secreted substances*. With one major exception (K^+), the PCT is the main site of secretion, but the collecting ducts are also active (Figure 25.17).

Tubular secretion is important for:

- *Disposing of substances, such as certain drugs and metabolites, that are tightly bound to plasma proteins.* Because plasma proteins are generally not filtered, the substances they bind are not filtered and so must be secreted.

- *Eliminating undesirable substances or end products that have been reabsorbed by passive processes.* Urea and uric acid, two nitrogenous wastes, are both handled in this way. Urea handling in the nephron is complicated and will be discussed on p. 1000, but the net effect is that 40–50% of the urea in the filtrate is excreted.

- *Ridding the body of excess K^+.* Because virtually all K^+ present in the filtrate is reabsorbed in the PCT and ascending nephron loop, nearly all K^+ in urine comes from aldosterone-driven

25

active tubular secretion into the late DCT and collecting ducts.

- *Controlling blood pH.* When blood pH drops toward the acidic end of its homeostatic range, the renal tubule cells actively secrete more H^+ into the filtrate and retain and generate more HCO_3^- (a base). As a result, blood pH rises and the urine drains off the excess H^+. Conversely, when blood pH approaches the alkaline end of its range, Cl^- is reabsorbed instead of HCO_3^-, which is allowed to leave the body in urine. We will discuss the kidneys' role in pH homeostasis in more detail in Chapter 26.

Check Your Understanding

16. List several substances that are secreted into the kidney tubules.

For answers, see Answers Appendix.

25.7 The kidneys create and use an osmotic gradient to regulate urine concentration and volume

Learning Outcomes

▶ **Describe the mechanisms responsible for the medullary osmotic gradient.**

▶ **Explain how dilute and concentrated urine are formed.**

From day to day, and even hour to hour, our intake and loss of fluids can vary dramatically. For example, when you run on a hot summer day, you dehydrate as you rapidly lose fluid as sweat. On the other hand, if you drink a pitcher of lemonade while sitting on the porch, you may overhydrate. In response, the kidneys make adjustments to keep the solute concentration of body fluids constant at about 300 mOsm, the normal osmotic concentration of blood plasma. Maintaining constant osmolality of extracellular fluids is crucial for preventing cells, particularly in the brain, from shrinking or swelling from the osmotic movement of water.

Recall from Chapter 3 (pp. 71–73) that a solution's osmolality is the concentration of solute particles per kilogram of water. Because 1 osmol (equivalent to 1 mole of particles) is a fairly large unit, the milliosmol (mOsm) (mil″e-oz′mōl), equal to 0.001 osmol, is generally used. In the discussion that follows, we use mOsm to indicate mOsm/kg.

The kidneys keep the solute load of body fluids constant by regulating urine concentration and volume.

- When you dehydrate, your kidneys produce a small volume of concentrated urine.

- When you overhydrate, your kidneys produce a large volume of dilute urine.

The kidneys accomplish this feat using countercurrent mechanisms. In the kidneys, the term *countercurrent* means that fluid flows in opposite directions through adjacent segments of the same tube connected by a hairpin turn* (see *Focus on the Medullary Osmotic Gradient*, Focus Figure 25.1, pp. 996–997). This arrangement makes it possible to exchange materials between the two segments.

Two types of countercurrent mechanisms determine urine concentration and volume:

- The **countercurrent multiplier** is the interaction between the flow of filtrate through the ascending and descending limbs of the long nephron loops of juxtamedullary nephrons.

- The **countercurrent exchanger** is the flow of blood through the ascending and descending portions of the vasa recta.

These countercurrent mechanisms establish and maintain an osmotic gradient extending from the cortex through the depths of the medulla (**Figure 25.18**). This gradient—the **medullary osmotic gradient**—allows the kidneys to vary urine concentration dramatically.

How do the kidneys form the osmotic gradient? *Focus on the Medullary Osmotic Gradient* (**Focus Figure 25.1** on pp. 996–997) explores the answer to this question.

This greatly enlarged nephron and collecting duct show you the arrangement of structures that pass through the osmotic gradient in the renal medulla. If you keep this anatomy in mind, it will help you understand the process of making either very dilute or very concentrated urine.

The osmolality of the interstitial fluid in the **renal cortex** is isotonic at 300 mOsm.

The osmolality of the interstitial fluid of the **renal medulla** increases progressively from 300 mOsm at the cortex-medulla junction to 1200 mOsm at the medulla-pelvis junction.

Figure 25.18 Osmotic gradient in the renal medulla.

*The term "countercurrent" is commonly misunderstood to mean that the direction of fluid flow in the nephron loops is opposite that of the blood in the vasa recta. In fact, there is no one-to-one relationship between individual nephron loops and capillaries of the vasa recta as might be suggested by two-dimensional diagrams such as Focus Figure 25.1. Instead, there are many tubules and capillaries packed together like a bundle of straws. Each tubule is surrounded by many blood vessels, whose flow is not necessarily counter to flow in that tubule (see Figure 25.8).

(Text continues on p. 998.)

Juxtamedullary nephrons create an osmotic gradient within the renal medulla that allows the kidney to produce urine of varying concentration.

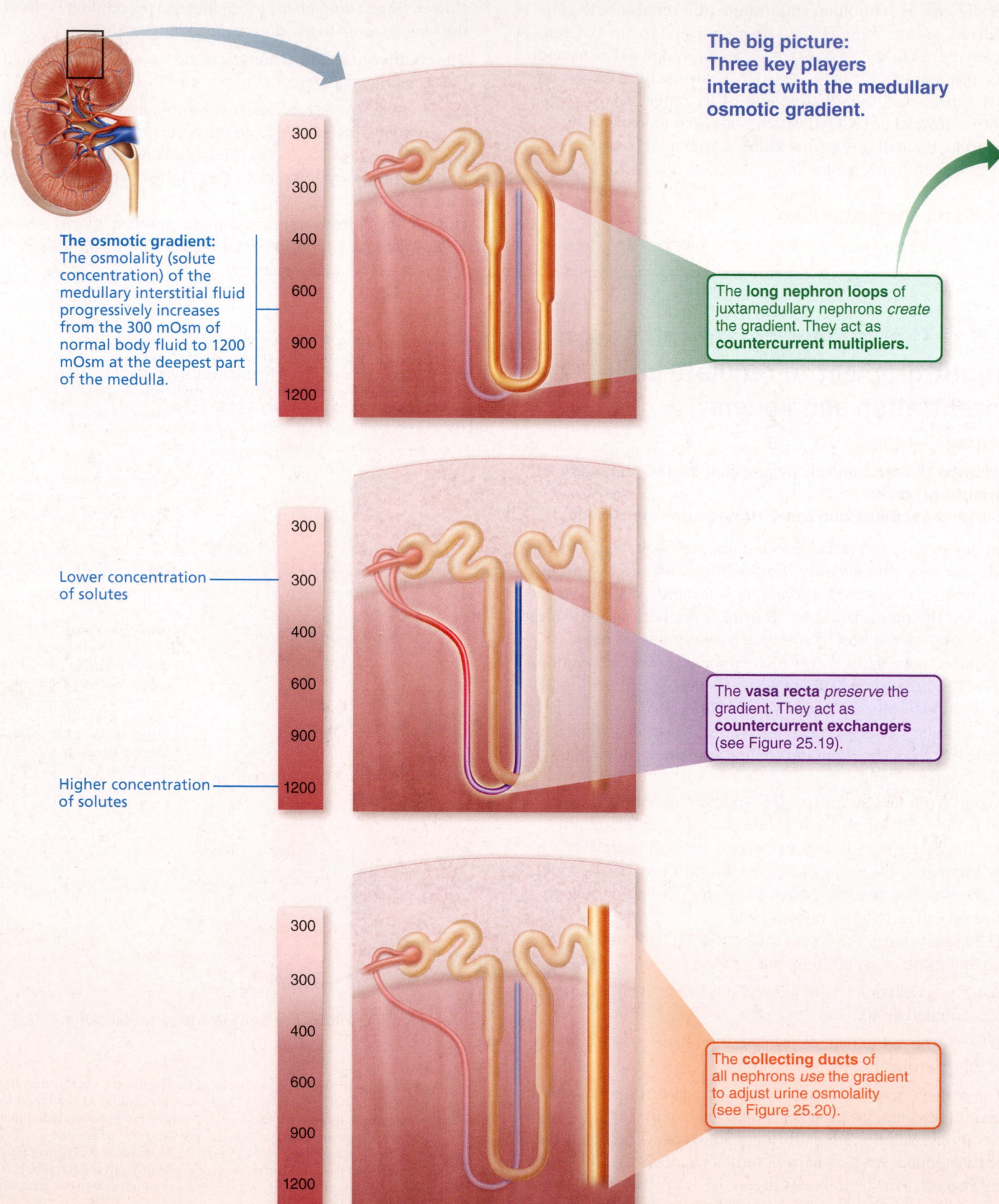

The big picture:
Three key players interact with the medullary osmotic gradient.

The osmotic gradient:
The osmolality (solute concentration) of the medullary interstitial fluid progressively increases from the 300 mOsm of normal body fluid to 1200 mOsm at the deepest part of the medulla.

300
300
400
600
900
1200

The **long nephron loops** of juxtamedullary nephrons *create* the gradient. They act as **countercurrent multipliers.**

300
300
400
600
900
1200

Lower concentration of solutes

Higher concentration of solutes

The **vasa recta** *preserve* the gradient. They act as **countercurrent exchangers** (see Figure 25.19).

300
300
400
600
900
1200

The **collecting ducts** of all nephrons *use* the gradient to adjust urine osmolality (see Figure 25.20).

Long nephron loops of juxtamedullary nephrons create the gradient.

The countercurrent multiplier depends on three properties of the nephron loop to establish the osmotic gradient.

Filtrate flows in the opposite direction (countercurrent) through two adjacent parallel sections of a nephron loop.

H₂O ← (Passive transport) NaCl → (Active transport)

The **descending limb** is permeable to water, but not to salt.

The **ascending limb** is impermeable to water, and pumps out salt.

→ Active transport
→ Passive transport
▬ Water impermeable

These properties establish a positive feedback cycle that uses the flow of fluid to multiply the power of the salt pumps.

Water leaves the descending limb

↑ Interstitial fluid osmolality

H₂O ← NaCl →

Start here

↑ Osmolality of filtrate in descending limb

Salt is pumped out of the ascending limb

↑ Osmolality of filtrate entering the ascending limb

As water and solutes are reabsorbed, the loop first concentrates the filtrate, then dilutes it.

Osmolality of interstitial fluid (mOsm)

Cortex — 300

Outer medulla — 400 / 600

Inner medulla — 900 / 1200

① Filtrate entering the nephron loop is isosmotic to both blood plasma and cortical interstitial fluid.

300 300 300 100 100 100

H₂O ← NaCl ←
H₂O ← NaCl ←
H₂O ← NaCl ←   400 200
H₂O ← NaCl ←   600 400
H₂O ←
H₂O ← NaCl ←   900 700
H₂O ←
1200

Nephron loop

② Water moves out of the filtrate in the descending limb down its osmotic gradient. This concentrates the filtrate.

③ Filtrate reaches its highest concentration at the bend of the loop.

④ Na⁺ and Cl⁻ are pumped out of the filtrate. This increases the interstitial fluid osmolality.

⑤ Filtrate is at its most dilute as it leaves the nephron loop. At 100 mOsm, it is hypo-osmotic to the interstitial fluid.

The Countercurrent Multiplier

Take some time to study the mechanism of the countercurrent multiplier in Focus Figure 25.1 (pp. 996–997). The countercurrent multiplier depends on actively transporting solutes out of the ascending limb ("Start" of the positive feedback cycle).

Although the two limbs of the nephron loop are not in direct contact with each other, they are close enough to influence each other's exchanges with the interstitial fluid they share. The more NaCl the ascending limb extrudes, the more water diffuses out of the descending limb and the saltier the filtrate in the descending limb becomes. The ascending limb then uses the increasingly "salty" filtrate left behind in the descending limb to raise the osmolality of the medullary interstitial fluid even further. This establishes a positive feedback cycle that produces the high osmolality of the fluids in the descending limb and interstitial fluid.

Notice at the bottom of the right page of Focus Figure 25.1 that there is a constant difference in filtrate concentration (200 mOsm) between the two limbs of the nephron loop, and between the ascending limb and the interstitial fluid. This

difference reflects the power of the ascending limb's NaCl pumps, which are just powerful enough to create a 200 mOsm difference between the inside and outside of the ascending limb. A 200 mOsm gradient by itself would not be enough to allow excretion of very concentrated urine. The beauty of this system lies in the fact that, because of countercurrent flow, the nephron loop is able to "multiply" these small changes in solute concentration into a gradient change along the vertical length of the loop (both inside and outside) that is closer to 900 mOsm (1200 mOsm – 300 mOsm).

Notice also that while much of the Na$^+$ and Cl$^-$ reabsorption in the ascending limb is active (via Na$^+$-K$^+$-2Cl$^-$ cotransporters in the thick ascending limb), some is passive (mostly in the thin portion of the ascending limb).

The Countercurrent Exchanger

The vasa recta act as countercurrent exchangers (**Figure 25.19**). Countercurrent exchange does not create the medullary gradient, but preserves it by (1) preventing rapid removal of salt from the medullary interstitial space, and (2) removing reabsorbed water. As a result, blood leaving and reentering the cortex via the vasa recta has nearly the same solute concentration.

The water picked up by the ascending vasa recta includes not only water lost from the descending vasa recta, but also water reabsorbed from the nephron loop and collecting duct. As a result, the volume of blood at the end of the vasa recta is greater than at the beginning.

Formation of Dilute or Concentrated Urine

As we have just seen, the kidneys go to a great deal of trouble to create the medullary osmotic gradient. But for what purpose? Without this gradient, you would not be able to raise the concentration of urine above 300 mOsm—the osmolality of interstitial fluid. As a result, you would not be able to conserve water when you are dehydrated.

Figure 25.20 shows the body's response to either overhydration or dehydration and ADH's role in controlling the production of dilute or concentrated urine. When we are overhydrated, ADH production decreases (◄ p. 613) and the osmolality of urine falls as low as 100 mOsm. If aldosterone (not shown) is present, the DCT and collecting duct cells can remove Na$^+$ and selected other ions from the filtrate, making the urine that enters the renal pelvis even more dilute. The osmolality of urine can plunge as low as 50 mOsm, about one-sixth the concentration of glomerular filtrate or blood plasma.

When we are dehydrated, the posterior pituitary releases large amounts of ADH and the solute concentration of urine may rise as high as 1200 mOsm, the concentration of interstitial fluid in the deepest part of the medulla. With maximal ADH secretion, up to 99% of the water in the filtrate is reabsorbed and returned to the blood, and only half a liter per day of highly concentrated urine is excreted. The ability of our kidneys to produce such concentrated urine is critically tied to our ability to survive for a limited time without water.

Vasa recta preserve the gradient.

- **The vasa recta are highly permeable to water and solutes.**
- **Countercurrent exchanges occur between each section of the vasa recta and its surrounding fluid. As a result:**
  - The blood within the vasa recta remains nearly isosmotic to the surrounding fluid.
  - The vasa recta are able to reabsorb water and solutes into the general circulation without undoing the osmotic gradient created by the countercurrent multiplier.

The countercurrent flow of fluid moves through two adjacent parallel sections of vasa recta.

Blood from efferent arteriole

To vein

Vasa recta

Figure 25.19 Countercurrent exchange.

Collecting ducts use the gradient.

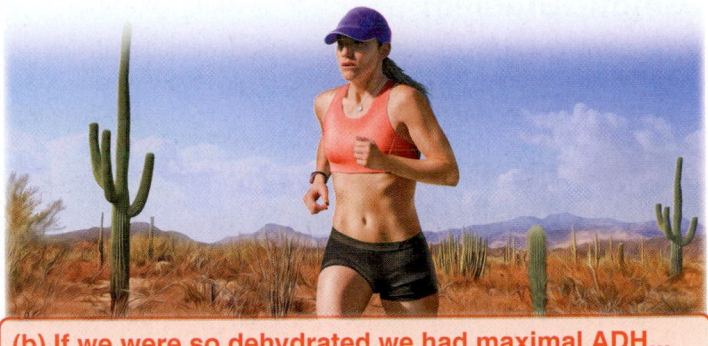

(a) If we were so overhydrated we had no ADH...

(b) If we were so dehydrated we had maximal ADH...

↓ Osmolality of extracellular fluids

↓ ADH release from posterior pituitary

↓ Number of aquaporins (H_2O channels) in collecting duct

↓ H_2O reabsorption from collecting duct

Large volume of dilute urine

↑ Osmolality of extracellular fluids

↑ ADH release from posterior pituitary

↑ Number of aquaporins (H_2O channels) in collecting duct

↑ H_2O reabsorption from collecting duct

Small volume of concentrated urine

Figure 25.20 Mechanism for forming dilute or concentrated urine. Under the control of antidiuretic hormone (ADH), the collecting ducts fine-tune urine concentration using the medullary osmotic gradient.

Urea Recycling and the Medullary Osmotic Gradient

We're not quite done yet. There's one last piece of the puzzle left—urea. We usually think of urea as simply a metabolic waste product, but conserving water is so important that the kidneys actually use urea to help form the medullary gradient (Figure 25.20).

1. Urea enters the filtrate by facilitated diffusion in the ascending thin limb of the nephron loop.
2. As the filtrate moves on, the cortical collecting duct usually reabsorbs water, leaving urea behind.
3. When filtrate reaches the portion of the collecting duct in the deep medullary region, the now highly concentrated urea moves by facilitated diffusion out of the collecting duct into the interstitial fluid of the medulla. These movements form a pool of urea that recycles back into the ascending thin limb of the nephron loop. In this way, urea contributes substantially to the high osmolality in the medulla.

Antidiuretic hormone enhances urea transport out of the medullary collecting duct. When ADH is present, it increases urea recycling and strengthens the medullary osmotic gradient, allowing more concentrated urine to be formed.

Diuretics

There are several types of **diuretics**, chemicals that enhance urinary output. Alcohol encourages diuresis by inhibiting release of ADH. Other diuretics increase urine flow by inhibiting Na^+ reabsorption and the obligatory water reabsorption that normally follows. Examples include many drugs prescribed for hypertension (◀ p. 724) or the edema of congestive heart failure (◀ p. 700). Most diuretics inhibit Na^+-associated symporters. "Loop diuretics" (like furosemide) are powerful because they inhibit formation of the medullary gradient by acting at the ascending limb of the nephron loop. Thiazides are less potent and act at the DCT. An *osmotic diuretic* is a substance that is not reabsorbed and that carries water out with it (for example, the high blood glucose of a diabetes mellitus patient).

Check Your Understanding

17. Describe the special characteristics of the descending and ascending limbs of the nephron loop that cause the formation of the medullary osmotic gradient.
18. Under what conditions is ADH released from the posterior pituitary? What effect does ADH have on the collecting ducts?

For answers, see Answers Appendix.

25.8 Renal function is evaluated by analyzing blood and urine

Learning Outcomes

▶ Define renal clearance and explain how this value summarizes the way a substance is handled by the kidney.

▶ Describe the normal physical and chemical properties of urine.

▶ List several abnormal urine components, and name the condition characterized by the presence of detectable amounts of each.

Since the time of the ancient Greek Hippocrates, physicians have examined their patients' urine for signs of disease. **Urinalysis**, the analysis of urine, can aid in the diagnosis of diseases or detect illegal substances. However, to measure how well the kidneys are functioning, we need to analyze *both blood and urine*. Renal function can quickly be assessed by measuring levels of *nitrogenous wastes* in blood. But to precisely measure kidney function, we need to determine the renal clearance, which requires that both blood and urine be tested. Normal blood and urine values for various substances are found in Appendix F.

Renal Clearance

Renal clearance refers to the volume of plasma from which the kidneys clear (completely remove) a particular substance in a given time, usually 1 minute. Renal clearance tests are done to determine the GFR, which allows us to detect glomerular damage and follow the progress of renal disease.

The renal clearance rate (*C*) of any substance, in ml/min, is calculated from the equation

$$C = UV/P$$

where

U = concentration of the substance in urine (mg/ml)

V = flow rate of urine formation (ml/min)

P = concentration of the substance in plasma (mg/ml)

Because it is freely filtered and neither reabsorbed nor secreted by the kidneys, *inulin* (in'u-lin) is the substance used to determine the GFR. Inulin is a plant polysaccharide that has a renal clearance value equal to the GFR. When inulin is infused such that its plasma concentration is 1 mg/ml ($P = 1$ mg/ml), then generally $U = 125$ mg/ml, and $V = 1$ ml/min. Therefore, its renal clearance is $C = (125 \times 1)/1 = 125$ ml/min, meaning that in 1 minute the kidneys have cleared all the inulin present in 125 ml of plasma.

The clearance value tells us about the net handling of a substance by the kidneys. There are three possible cases:

- A clearance value less than that of inulin means that the substance is reabsorbed. For example, *C* is normally 70 ml/min for urea, meaning that of the 125 ml of glomerular filtrate formed each minute, approximately 70 ml is completely cleared of urea, while the urea in the remaining 55 ml is recovered and returned to the plasma. If *C* is zero (such as for glucose in healthy individuals), reabsorption is complete or the substance is not filtered.

- If *C* is equal to that of inulin, there is no net reabsorption or secretion.

- If *C* is greater than that of inulin, the tubule cells are secreting the substance into the filtrate. This is the case with most drug metabolites. Knowing a drug's renal clearance value is essential because if it is high, the drug dosage must also be high and administered frequently to maintain a therapeutic level.

Creatinine, which has a *C* value of 140 ml/min, is freely filtered but also secreted in small amounts. Nevertheless, its measured blood value is often used to give an estimate of GFR. This is convenient because creatinine does not need to be intravenously infused into the patient as does inulin.

retains or adds needed substances, while removing wastes and excess ions.

Urine

Chemical Composition

Water accounts for about 95% of urine volume; the remaining 5% consists of solutes. The largest component of urine by weight, apart from water, is **urea**, which is derived from the normal breakdown of amino acids. Other **nitrogenous wastes** in urine include **uric acid** (an end product of nucleic acid metabolism) and **creatinine** (a metabolite of creatine phosphate, which is found in large amounts in skeletal muscle tissue where it stores energy to regenerate ATP; ◄ p. 304).

Normal solute constituents of urine, in order of decreasing concentration, are urea, Na^+, K^+, PO_4^{3-}, SO_4^{2-}, creatinine, and uric acid. Much smaller but highly variable amounts of Ca^{2+}, Mg^{2+}, and HCO_3^- are also present.

Unusually high concentrations of any solute, or the presence of abnormal substances such as plasma proteins, WBCs (pus), or bile pigments, may indicate pathology (**Table 25.3**). (Normal urine values are listed in Appendix F.)

Physical Characteristics

Color and Transparency Freshly voided urine is clear and pale to deep yellow. Its yellow color is due to **urochrome** (u′ro-krōm), a pigment that results when the body destroys hemoglobin. The more concentrated the urine, the deeper the color.

An abnormal color (such as pink, brown, or a smoky tinge) may result from eating certain foods (beets, rhubarb) or from the presence of bile pigments or blood in the urine. Additionally, some commonly prescribed drugs and vitamin supplements alter the color of urine. Cloudy urine may indicate a urinary tract infection.

Odor Fresh urine is slightly aromatic, but if allowed to stand, it develops an ammonia odor as bacteria metabolize its urea solutes. Some drugs and vegetables alter the odor of urine, as do some diseases. For example, in uncontrolled diabetes mellitus the urine smells fruity because of its acetone content.

HOMEOSTATIC IMBALANCE 25.4 | CLINICAL

Chronic renal disease, defined as a GFR of less than 60 ml/min for at least three months, often develops silently over many years. Filtrate formation decreases gradually, nitrogenous wastes accumulate in the blood, and blood pH drifts toward the acidic range. The leading cause of chronic renal disease is diabetes mellitus (44% of new cases), with hypertension a close second (28% of new cases). Other causes include repeated kidney infections, physical trauma, and heavy metal poisoning.

In **renal failure** (GFR < 15 ml/min), filtrate formation decreases or stops completely. The clinical syndrome associated with renal failure is called **uremia** (literally "urine in the blood") and includes fatigue, anorexia, nausea, mental changes, and muscle cramps.

While uremia was once attributed to accumulation of nitrogenous wastes (particularly urea), we now know that urea is not especially toxic. Rather, this multiorgan failure is caused by the interplay of multiple factors. These include ionic and hormonal imbalances (including anemia due to lack of erythropoietin; ◄ p. 648), as well as metabolic abnormalities and accumulation of various toxic molecules that interfere with normal metabolism.

Current treatment options are hemodialysis or a kidney transplant. **Hemodialysis** uses an "artificial kidney" apparatus, passing the patient's blood through a membrane tubing that is permeable only to selected substances. The tubing is immersed in a solution that differs slightly from normal cleansed plasma. As blood circulates through the tubing, substances such as nitrogenous wastes and K^+ present in the blood (but not in the bath) diffuse out of the blood into the surrounding solution. Meanwhile, substances to be added to the blood, mainly buffers for H^+ (and glucose for malnourished patients), move from the bathing solution into the blood. In this way, hemodialysis

---

Table 25.3   Abnormal Urinary Constituents

SUBSTANCE	NAME OF CONDITION	POSSIBLE CAUSES
Glucose	Glycosuria	Diabetes mellitus
Proteins	Proteinuria, albuminuria	Nonpathological: excessive physical exertion, pregnancy Pathological (over 150 mg/day): glomerulonephritis (see Related Clinical Terms on p. 1008), severe hypertension, heart failure, often an initial sign of renal disease
Ketone bodies	Ketonuria	Excessive formation and accumulation of ketone bodies, as in starvation and untreated diabetes mellitus
Hemoglobin	Hemoglobinuria	Various: transfusion reaction, hemolytic anemia, severe burns, etc.
Bile pigments	Bilirubinuria	Liver disease (hepatitis, cirrhosis) or obstruction of bile ducts from liver or gallbladder
Erythrocytes	Hematuria	Bleeding urinary tract (due to trauma, kidney stones, infection, or cancer)
Leukocytes (pus)	Pyuria	Urinary tract infection

pH Urine is usually slightly acidic (around pH 6), but changes in body metabolism or diet may cause the pH to vary from about 4.5 to 8.0. A predominantly *acidic* diet that contains large amounts of protein and whole wheat products produces acidic urine. A vegetarian (*alkaline*) diet, bacterial infection of the urinary tract, or prolonged drainage of the stomach via a nasogastric tube all cause the urine to become alkaline.

Specific Gravity The ratio of the mass of a substance to the mass of an equal volume of distilled water is its **specific gravity**. Because urine is water plus solutes, a given volume has a greater mass than the same volume of distilled water. The specific gravity of distilled water is 1.0 and that of urine ranges from 1.001 to 1.035, depending on its solute concentration.

Check Your Understanding

19. What are the three major nitrogenous wastes excreted in the urine?

20. PREDICT What would you expect the normal clearance value for amino acids to be? Explain.

For answers, see Answers Appendix.

25.9 The ureters, bladder, and urethra transport, store, and eliminate urine

Learning Outcomes

▶ **Describe the general location, structure, and function of the ureters, urinary bladder, and urethra.**

▶ **Compare the course, length, and functions of the male urethra with those of the female.**

▶ **Define micturition and describe its neural control.**

The kidneys form urine continuously and the ureters transport it to the bladder. It is usually stored in the bladder until its release through the urethra in a process called *micturition*.

Ureters

The **ureters** are slender tubes that convey urine from the kidneys to the bladder (Figures 25.1, 25.2, and 25.23). Each ureter begins at the level of L_2 as a continuation of the renal pelvis. From there, it descends behind the peritoneum and runs obliquely through the posterior bladder wall. This arrangement prevents backflow of urine because any increase in bladder pressure compresses and closes the distal ends of the ureters.

Histologically, the ureter wall has three layers (**Figure 25.21**). From the inside out:

• The *mucosa* contains a transitional epithelium (◀ p. 123) that is continuous with the mucosae of the kidney pelvis superiorly and the bladder medially.

• The *muscularis* is composed chiefly of two smooth muscle sheets: the internal longitudinal layer and the external circular layer. An additional smooth muscle layer, the external longitudinal layer, appears in the lower third of the ureter.

• The *adventitia* covering the ureter's external surface is typical fibrous connective tissue.

Lumen

Mucosa
• Transitional epithelium
• Lamina propria

Muscularis
• Longitudinal layer
• Circular layer

Adventitia

Figure 25.21 Cross section of the ureter wall (10×). The prominent mucosal folds seen in an empty ureter stretch and flatten to accommodate large pulses of urine.

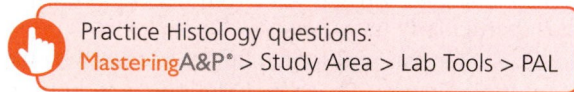

Practice Histology questions:
MasteringA&P® > Study Area > Lab Tools > PAL

The ureter plays an active role in transporting urine. Incoming urine distends the ureter and stimulates its muscularis to contract, propelling urine into the bladder. (Urine does *not* reach the bladder through gravity alone.) The strength and frequency of the peristaltic waves are adjusted to the rate of urine formation. Both sympathetic and parasympathetic fibers innervate each ureter, but neural control of peristalsis appears to be insignificant compared to the way ureteral smooth muscle responds to stretch.

HOMEOSTATIC IMBALANCE 25.5 CLINICAL

When calcium or magnesium salts or uric acid are concentrated in urine, they may crystallize and precipitate as "kidney stones," or **renal calculi** (kal'ku-li; *calculus* = little stone). If stones obstruct urine flow, they cause excruciating pain radiating from the flank to the abdomen. *Nephrolithiasis* (*litho* = stone; *iasis* = condition) refers to calculi in the kidney, whereas *ureterolithiasis* refers to calculi lodged in the ureters.

While most small calculi pass without intervention, larger calculi or those that lodge in the ureter can be removed endoscopically or surgically. Many calculi, especially those in the kidney, can be treated by *lithotripsy*, a noninvasive procedure that uses acoustic wave energy to break stones into smaller pieces that may be passed more easily.

Obesity and elevated blood calcium are risk factors for the formation of renal calculi. The best way to prevent recurrence is to maintain dilute urine by adequate hydration. ●●●●●●●

Peritoneum
Ureter
Rugae
Detrusor
Adventitia
Ureteric orifices
Trigone of bladder
Bladder neck
Internal urethral sphincter
Prostate
Prostatic urethra
Intermediate part of the urethra
External urethral sphincter
Urogenital diaphragm

Spongy urethra
Erectile tissue of penis
External urethral orifice

(a) Male. The long male urethra has three regions: prostatic, intermediate, and spongy.

Trigone
Urethra
External urethral orifice

(b) Female.

Figure 25.22 Structure of the urinary bladder and urethra. The anterior wall of the bladder has been cut away to reveal the position of the trigone.

Urinary Bladder

The **urinary bladder** is a smooth, collapsible, muscular sac that stores urine temporarily.

Urinary Bladder Anatomy

The bladder is located retroperitoneally on the pelvic floor just posterior to the pubic symphysis. The prostate (part of the male reproductive system) lies inferior to the bladder neck, which empties into the urethra. In females, the bladder is anterior to the vagina and uterus (see Figure 27.15 on p. 1061).

The interior of the bladder has openings for both ureters and the urethra (**Figure 25.22**). The smooth, triangular region of the bladder base outlined by these three openings is the **trigone** (tri′gōn; *trigon* = triangle), important clinically because infections tend to persist in this region.

The bladder wall has three layers: a mucosa containing transitional epithelium, a thick muscular layer, and a fibrous adventitia (except on its superior surface, where it is covered by the peritoneum). The muscular layer, called the **detrusor** (de-tru′sor; "to thrust out"), consists of intermingled smooth muscle fibers arranged in inner and outer longitudinal layers and a middle circular layer.

Urine Storage Capacity

The bladder is very distensible. When empty, the bladder collapses into its basic pyramidal shape. Its walls are thick and the mucosa is thrown into folds (*rugae*). As urine accumulates, the bladder expands, becomes pear shaped, and rises superiorly in the abdominal cavity. The transitional epithelium and muscular wall stretch and thin, and rugae disappear. These changes allow the bladder to store more urine without a significant rise in internal pressure.

Usually, a moderately full bladder holds approximately 500 ml (1 pint) of urine. However, if urine flow has been partially obstructed for a long time (as by an enlarged prostate), then the bladder can become enlarged to hold more than twice that amount. When tense with urine, it can be palpated well above the pubic symphysis.

25

Kidney

Renal pelvis

Ureter

Urinary bladder

Figure 25.23 Pyelogram. This X-ray image was obtained using a contrast medium to show the ureters, kidneys, and urinary bladder. (Artificially colored.)

The urinary bladder and ureters can be seen in a special X ray called a pyelogram (**Figure 25.23**).

Urethra

The **urethra** is a thin-walled muscular tube that drains urine from the bladder and conveys it out of the body. The epithelium of its mucosal lining is mostly pseudostratified columnar epithelium. However, near the bladder it becomes transitional epithelium, and near the external opening it changes to a protective stratified squamous epithelium.

At the bladder-urethra junction, the detrusor smooth muscle thickens to form the **internal urethral sphincter** (Figure 25.22). This involuntary sphincter, controlled by the autonomic nervous system, keeps the urethra closed when urine is not being passed and prevents leaking between voiding.

The **external urethral sphincter** surrounds the urethra as it passes through the *urogenital diaphragm*. This sphincter is formed of skeletal muscle and is voluntarily controlled. The *levator ani* muscle of the pelvic floor also serves as a voluntary constrictor of the urethra (◀ **Table 10.7, pp. 348–349**).

The length and functions of the urethra differ in the two sexes. In females the urethra is only 3–4 cm (1.5 inches) long and fibrous connective tissue binds it tightly to the anterior vaginal wall. Its external opening, the **external urethral orifice**, lies anterior to the vaginal opening and posterior to the clitoris.

In males the urethra is approximately 20 cm (8 inches) long and has three regions.

- The **prostatic urethra**, about 2.5 cm (1 inch) long, runs within the prostate.
- The **intermediate part of the urethra** (or *membranous urethra*), which runs through the urogenital diaphragm, extends about 2 cm from the prostate to the beginning of the penis.
- The **spongy urethra**, about 15 cm long, passes through the penis and opens at its tip via the **external urethral orifice**.

The male urethra has a double function: It carries semen as well as urine out of the body. We discuss its reproductive function in Chapter 27.

HOMEOSTATIC IMBALANCE 25.6   CLINICAL

Urinary tract infection (UTI) is a common problem, with up to 50% of women experiencing a UTI during their lives. The location of the female urethra near the anus predisposes it to colonization by fecal bacteria. Sexual activity and the use of spermicides increase the risk. In both men and women, catheterization or any condition that obstructs the normal flow of urine also predisposes an individual to acquiring a UTI.

Infection of the urinary bladder, or *cystitis*, causes urinary *frequency* (increased voiding) and *dysuria* (painful urination). The urine may be blood tinged or foul smelling. If the infection ascends to the kidney, the more serious *pyelonephritis* (see p. 977) results, accompanied by fever, back pain, and other symptoms of systemic illness. After the offending bacteria are identified by culturing the urine, antibiotics are used to treat the problem. ●●●●●●●●

Micturition

Micturition (mik″tu-rish′un; *mictur* = urinate), also called **urination** or *voiding*, is the act of emptying the urinary bladder. For micturition to occur, three things must happen simultaneously: (1) the detrusor must contract, (2) the internal urethral sphincter must open, and (3) the external urethral sphincter must open.

The detrusor and its internal urethral sphincter are composed of smooth muscle and are innervated by both the parasympathetic and sympathetic nervous systems, which have opposing actions. The external urethral sphincter, in contrast, is skeletal muscle, and therefore is innervated by the somatic nervous system.

How are the three events required for micturition coordinated? Micturition is most easily understood in infants where a spinal reflex coordinates the process. As urine accumulates, distension of the bladder activates stretch receptors in its walls. Impulses from the activated receptors travel via visceral afferent fibers to the sacral region of the spinal cord. Visceral afferent impulses, relayed by sets of interneurons, excite parasympathetic neurons and inhibit sympathetic neurons (**Figure 25.24**). As a result, the detrusor contracts and the internal sphincter opens. Visceral afferent impulses also decrease the firing rate of somatic efferents that normally keep the external urethral sphincter closed. This allows the sphincter to relax so urine can flow.

25

Figure 25.24 Control of micturition.

Between ages 2 and 3, descending circuits from the brain have matured enough to begin to override reflexive urination. The pons (◀ p. 452) has two centers that participate in control of micturition. The *pontine storage center* inhibits micturition, whereas the *pontine micturition center* promotes this reflex. Afferent impulses from bladder stretch receptors are relayed to the pons, as well as to higher brain centers that provide the conscious awareness of bladder fullness.

Lower bladder volumes primarily activate the pontine storage center, which inhibits urination by suppressing parasympathetic and enhancing sympathetic output to the bladder. When a person chooses not to void, reflex bladder contractions subside within a minute or so and urine continues to accumulate. Because the external sphincter is voluntarily controlled, we can choose to keep it closed and postpone bladder emptying temporarily. After additional urine has collected, the micturition reflex occurs again and, if urination is delayed again, is damped once more.

The urge to void gradually becomes greater and greater, and micturition usually occurs before urine volume exceeds 400 ml. After normal micturition, only about 10 ml of urine remains in the bladder.

HOMEOSTATIC IMBALANCE 25.7   CLINICAL

In adults, **urinary incontinence** (the inability to control urination) is usually a result of weakened pelvic muscles following childbirth or surgery, physical pressure during pregnancy, or nervous system problems. In *stress incontinence*, a sudden increase in intra-abdominal pressure (during laughing and coughing) forces urine through the external sphincter. This condition is common during pregnancy when the heavy uterus stretches the muscles of the pelvic floor and the urogenital diaphragm (◀ p. 348) that support the external sphincter. In *overflow incontinence*, urine dribbles from the urethra whenever the bladder overfills.

In **urinary retention**, the bladder is unable to expel its contained urine. Urinary retention is common after general anesthesia (it takes a little time for the detrusor to regain its activity). Urinary retention in men often reflects hypertrophy of the prostate, which narrows the urethra, making it difficult to void. When urinary retention is prolonged, a slender drainage tube called a **catheter** (kath'ĕ-ter) must be inserted through the urethra to drain the urine and prevent bladder trauma from excessive stretching.

Check Your Understanding

21. A kidney stone blocking a ureter would interfere with urine flow to which organ? Why would the pain occur in waves?

22. What is the trigone of the bladder, and which landmarks define its borders?

23. Name the three regions of the male urethra.

24. How does the detrusor respond to increased firing of the parasympathetic fibers that innervate it?

25. MAKE CONNECTIONS Compare the structure and regulation of the sphincters that control micturition to those that control defecation (Chapter 23).

For answers, see Answers Appendix.

Developmental Aspects of the Urinary System

In an embryo, three different sets of kidneys develop from the *urogenital ridges*, paired elevations of the intermediate mesoderm that give rise to both the urinary organs and reproductive organs (**Figure 25.25**). Only the last set persists to become adult kidneys.

During the fourth week of development, the first tubule system, the **pronephros** (pro-nef′ros; "prekidney"), forms and then quickly degenerates as a second, lower set appears. Although the pronephros never functions and is gone by the sixth week, the **pronephric duct** that connects it to the cloaca persists and is used by the later-developing kidneys. (The cloaca is the terminal part of the gut that opens to the body exterior.)

As the second renal system, the **mesonephros** (mez″o-nef′-ros; "middle kidney"), claims the pronephric duct, it comes to be called the **mesonephric duct** (Figure 25.25a, b). The mesonephric kidneys degenerate (with remnants incorporated into the male reproductive system) once the third set, the **metanephros** (met″ah-nef′ros; "after kidney"), makes its appearance (Figure 25.25b, c).

The metanephros starts to develop at about five weeks as hollow **ureteric buds** that push superiorly from the mesonephric duct into the urogenital ridge, inducing the mesoderm there to form nephrons. The distal ends of the ureteric buds form the renal pelves, calyces, and collecting ducts, and their unexpanded proximal parts, now called the **ureteric ducts**, become the ureters (Figure 25.25d).

Because the kidneys develop in the pelvis and then ascend to their final position, they receive their blood supply from

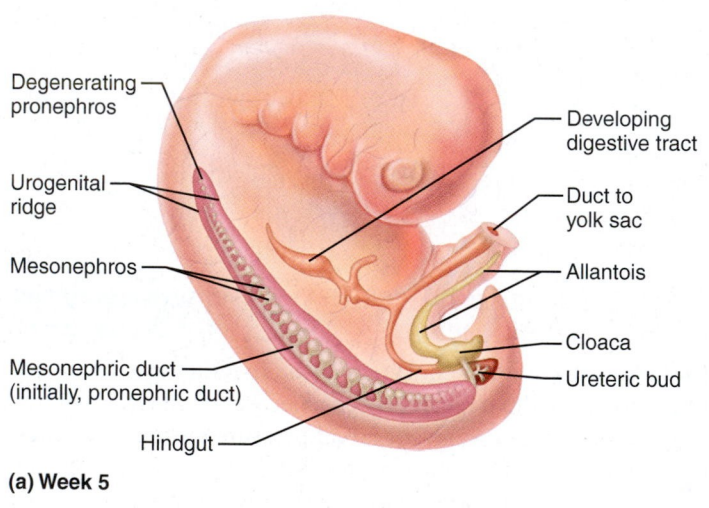

(a) Week 5

Degenerating pronephros
Urogenital ridge
Mesonephros
Mesonephric duct (initially, pronephric duct)
Hindgut
Developing digestive tract
Duct to yolk sac
Allantois
Cloaca
Ureteric bud

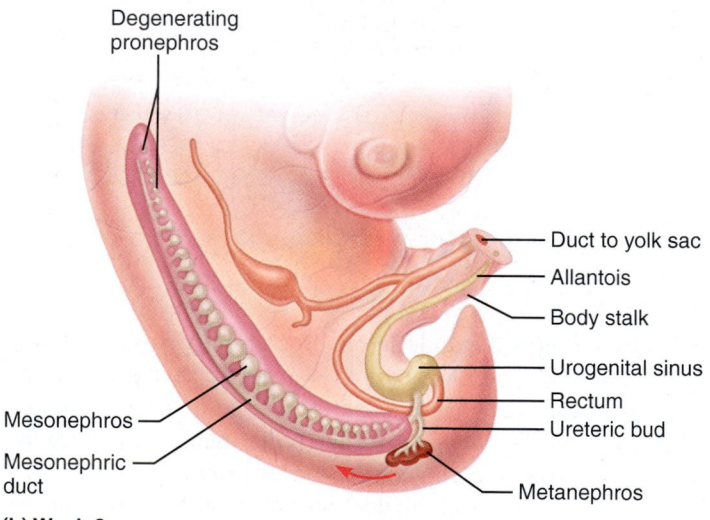

(b) Week 6

Degenerating pronephros
Mesonephros
Mesonephric duct
Duct to yolk sac
Allantois
Body stalk
Urogenital sinus
Rectum
Ureteric bud
Metanephros

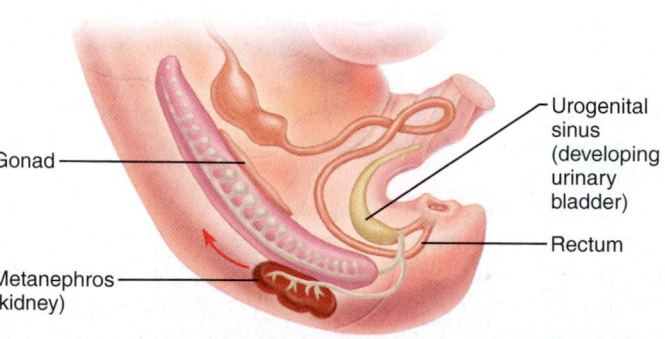

(c) Week 7

Gonad
Metanephros (kidney)
Urogenital sinus (developing urinary bladder)
Rectum

(d) Week 8

Gonad
Kidney
Ureter
Urinary bladder
Urethra
Anus
Rectum

Figure 25.25 Development of the urinary system in the embryo. Red arrows indicate the direction of metanephros migration as it develops.

successively higher sources. Although the lower blood vessels usually degenerate, they sometimes persist, and for this reason multiple renal arteries are common. The metanephric kidneys excrete urine by the third month of fetal life, and most of the amniotic fluid that surrounds a developing fetus is fetal urine. Nonetheless, the fetal kidneys do not work nearly as hard as they will after birth because exchange through the placenta allows the mother's urinary system to clear most of the undesirable substances from the fetal blood.

As the metanephros is developing, the cloaca subdivides to form the future rectum and anal canal and the **urogenital sinus**, into which the urinary and genital ducts empty. The urinary bladder and the urethra then develop from the urogenital sinus (Figure 25.25b–d).

HOMEOSTATIC IMBALANCE 25.8   CLINICAL

Three of the most common congenital abnormalities of the urinary system are horseshoe kidney, hypospadias, and polycystic kidney.

When ascending from the pelvis the kidneys are very close together, and in 1 out of 600 people they fuse across the midline, forming a single, U-shaped *horseshoe kidney*. This condition is usually asymptomatic, but it may be associated with other kidney abnormalities, such as obstructed drainage, that increase the risk of frequent kidney infections.

Hypospadias (hi″po-spa′de-as), found in male infants only, is the most common congenital abnormality of the urethra. It occurs when the urethral orifice is located on the ventral surface of the penis. This problem may be corrected surgically when the child is around 12 months old.

Polycystic kidney disease (PKD) is a group of disorders characterized by the presence of many fluid-filled cysts in the kidneys. These interfere with renal function, ultimately leading to renal failure. These disorders can be grouped into two general forms based on their pattern of inheritance:

- *Autosomal dominant PKD*, the less severe form, is much more common, affecting 1 in 500 people. The cysts develop so gradually that they produce no symptoms until about 40 years of age.

- *Autosomal recessive PKD* is more severe and much less common, affecting 1 in 20,000 people. Almost half of newborns with recessive PKD die just after birth, and survivors generally develop renal failure in early childhood. ●●●●●●●

Because its bladder is very small and its kidneys are less able to concentrate urine for the first two months, a newborn baby voids 5 to 40 times daily, depending on fluid intake. By 2 months, infants void approximately 400 ml/day, and the amount steadily increases until adolescence, when adult urine output (about 1500 ml/day) is achieved.

Incontinence is normal in infants because their nervous systems have not matured enough to control the external urethral sphincter. Reflex voiding occurs each time a baby's bladder fills enough to activate the stretch receptors. Control of the voluntary urethral sphincter goes hand in hand with nervous system development. By 15 months, most toddlers know when they have voided. By 24 months, some children are ready to begin toilet training. Daytime control usually is achieved first. It is unrealistic to expect complete nighttime control before age 4.

From childhood through late middle age, most urinary system problems are infectious conditions. *Escherichia coli* bacteria are normal residents of the digestive tract and generally cause no problems there, but these bacteria account for 80% of all urinary tract infections. *Sexually transmitted infections* can also inflame the urinary tract and clog some of its ducts. Childhood streptococcal infections such as strep throat and scarlet fever, if not treated promptly, may cause long-term inflammatory renal damage.

Only about 3% of elderly people have histologically normal kidneys, and kidney function declines with advancing age. The kidneys shrink as the nephrons decrease in size and number, and the tubule cells become less efficient. By age 80, the GFR of healthy seniors is only about 70% that of young adults. Diabetics are particularly at risk for renal disease, accounting for almost half of new cases.

Loss of bladder tone with age can cause an annoying increase in frequency of micturition. *Nocturia* (nok-tu′re-ah), the need to get up during the night to urinate, plagues almost two-thirds of this population. Many people eventually experience incontinence, which can usually be treated with exercise, medications, or surgery.

RELATED CLINICAL TERMS

Bladder cancer Bladder cancer, three times more common in men than in women, accounts for about 2% of all cancer deaths. It usually involves neoplasms of the bladder's lining epithelium and may be induced by carcinogens from the environment or the workplace that end up in urine. Smoking, exposure to industrial chemicals, and arsenic in drinking water have also been linked to bladder cancer. Blood in the urine is a common warning sign.

Cystocele (sis′to-sēl; *cyst* = a sac, the bladder; *cele* = hernia, rupture) Herniation of the urinary bladder into the vagina; a common result of tearing of the pelvic floor muscles during childbirth.

Cystoscopy (sis-tos′ko-pe; *cyst* = bladder; *scopy* = observation) Procedure in which a thin viewing tube is threaded into the bladder through the urethra to examine the bladder's mucosal surface.

Diabetes insipidus (in-sĭ′pĭ-dus; *insipid* = tasteless, bland) See Chapter 16, p. 614. Nephrogenic diabetes insipidus is due to lack of ADH receptors in the collecting duct.

25

Glomerulonephritis (GN) (glo-mer″u-lo-nef-ri′tis) Inflammation of the glomeruli, leading to increased permeability of the filtration membrane. In some cases, circulating immune complexes (antibodies bound to foreign substances, such as streptococcal bacteria) become trapped in the glomerular basement membranes. In other cases, immune responses are mounted against one's own kidney tissues, leading to glomerular damage. In either case, the inflammatory response that follows damages the filtration membrane, allowing blood proteins and even blood cells to pass into the renal tubules and into the urine. As the osmotic pressure of blood drops, fluid seeps from the bloodstream into the tissue spaces, causing bodywide edema. Renal shutdown requiring dialysis may occur temporarily, but normal renal function usually returns within a few months. If permanent glomerular damage occurs, acute GN becomes chronic GN and ultimately renal failure results.

Intravenous pyelogram (IVP) (pi′ĕ-lo-gram; *pyelo* = kidney pelvis; *gram* = written) An X ray of the kidneys and ureters obtained after intravenous injection of a contrast medium (as in Figure 25.23).

Nephrotoxin A substance (heavy metal, organic solvent, or bacterial toxin) that is toxic to the kidneys.

Nocturnal enuresis (NE) (en″u-re′sis) An inability to control urination at night during sleep; bed-wetting. In children over 6, called primary NE if control has never been achieved and secondary NE if control was achieved and then lost. Secondary NE often has psychological causes. Primary NE is more common and results from a combination of inadequate nocturnal ADH production, unusually sound sleep, or a small bladder capacity. Synthetic ADH often corrects the problem.

Renal infarct Area of dead, or necrotic, renal tissue due to blockage of the vascular supply to the kidney or hemorrhage. A common cause of localized renal infarct is an obstructed interlobar artery. Because interlobar arteries do not anastomose, their obstruction leads to ischemic necrosis of the portions of the kidney they supply.

Urologist (u-rol′o-jist) Physician who specializes in diseases of urinary structures in both sexes and in diseases of the reproductive tract of males.

CHAPTER SUMMARY

25.1 The kidneys have three distinct regions and a rich blood supply (pp. 975–978)

Location and External Anatomy (pp. 975–976)

1. The paired kidneys are retroperitoneal in the superior lumbar region.
2. A fibrous capsule, a perirenal fat capsule, and renal fascia surround each kidney. The perirenal fat capsule helps hold the kidneys in position.

Internal Gross Anatomy (pp. 976–977)

3. A kidney has a superficial cortex, a deeper medulla consisting mainly of medullary pyramids, and a medial pelvis. Extensions of the pelvis (calyces) surround and collect urine draining from the apices of the medullary pyramids.

Blood and Nerve Supply (pp. 977–978)

4. The kidneys receive 25% of the total cardiac output per minute.
5. The vascular pathway through a kidney is as follows: renal artery → segmental arteries → interlobar arteries → arcuate arteries → cortical radiate arteries → afferent arterioles → glomeruli → efferent arterioles → peritubular capillary beds → cortical radiate veins → arcuate veins → interlobar veins → renal vein.
6. The nerve supply of the kidneys is derived from the renal plexus.

25.2 Nephrons are the functional units of the kidney (pp. 978–983)

1. Nephrons are the structural and functional units of the kidneys.
2. Each nephron consists of a glomerulus (a high-pressure capillary bed), a glomerular capsule, and a renal tubule that is continuous with the capsule. Subdivisions of the renal tubule (from the glomerular capsule) are the proximal convoluted tubule, nephron loop, and distal convoluted tubule. A second capillary bed, the low-pressure peritubular capillary bed, is closely associated with the renal tubule of each nephron.
3. The more numerous cortical nephrons are located almost entirely in the cortex; only a small part of their nephron loop penetrates into the medulla. Glomeruli of juxtamedullary nephrons are located near the cortex-medulla junction, and their nephron loops dip deeply into the medulla. Instead of directly forming peritubular capillaries, the efferent arterioles of many juxtamedullary nephrons form unique bundles of straight vessels, called vasa recta, that serve tubule segments in the medulla. Juxtamedullary nephrons and the vasa recta play an important role in establishing the medullary osmotic gradient.
4. Collecting ducts receive urine from many nephrons and help concentrate urine. They form the medullary pyramids.
5. The juxtaglomerular complex is at the point of contact between the afferent arteriole and the most distal part of the ascending limb of the nephron loop. It consists of the granular cells, the macula densa, and extraglomerular mesangial cells.

25.3 Overview: Filtration, absorption, and secretion are the key processes of urine formation (pp. 983–984)

1. Functions of the nephrons include glomerular filtration, tubular reabsorption, and tubular secretion. Via these functional processes, the kidneys regulate the volume, composition, and pH of the blood, and eliminate nitrogenous metabolic wastes.

25.4 Urine formation, step 1: The glomeruli make filtrate (pp. 984–989)

1. The filtration membrane consists of the fenestrated glomerular endothelium, the intervening basement membrane, and the podocyte-containing visceral layer of the glomerular capsule. It permits free passage of substances smaller than (most) plasma proteins.
2. The glomeruli function as filters. High glomerular blood pressure (55 mm Hg) occurs because the glomeruli are fed and drained by arterioles, and the afferent arterioles are larger in diameter than the efferent arterioles.
3. About one-fifth of the plasma flowing through the kidneys is filtered from the glomeruli into the glomerular capsule.
4. Usually about 10 mm Hg, the net filtration pressure (NFP) is determined by the relationship between forces favoring filtration (glomerular hydrostatic pressure) and forces that oppose it (capsular hydrostatic pressure and blood colloid osmotic pressure).

5. The glomerular filtration rate (GFR) is directly proportional to the net filtration pressure and is about 125 ml/min (180 L/day).

6. Intrinsic renal control, or renal autoregulation, enables the kidneys to maintain a relatively constant renal blood flow and glomerular filtration rate. Intrinsic control involves a myogenic mechanism and a tubuloglomerular feedback mechanism mediated by the macula densa.

7. Extrinsic control of GFR, via nerves and hormones, maintains blood pressure. Strong sympathetic nervous system activation causes constriction of the afferent arterioles, which decreases filtrate formation.

8. The renin-angiotensin-aldosterone mechanism raises systemic blood pressure by generating angiotensin II. Renin is released from granular cells in response to (1) direct sympathetic nervous system stimulation, (2) paracrines released by the macula densa, and (3) reduced stretch of granular cell membranes.

 Complete an interactive tutorial: MasteringA&P® > Study Area > Interactive Physiology > Urinary System > Glomerular Filtration

25.5 Urine formation, step 2: Most of the filtrate is reabsorbed into the blood (pp. 989–994)

1. During tubular reabsorption, needed substances are removed from the filtrate by the tubule cells and returned to the peritubular capillary blood. The primary active transport of Na^+ by a Na^+-K^+ ATPase pump at the basolateral membrane accounts for Na^+ reabsorption and establishes the electrochemical gradient that drives the reabsorption of most other solutes and H_2O. Na^+ enters at the apical surface of the tubule cell via facilitated diffusion through channels or as part of a cotransport mechanism.

2. Passive tubular reabsorption is driven by electrochemical gradients established by active reabsorption of sodium ions. Water, many ions, and various other substances (for example, urea) are reabsorbed passively by diffusion via transcellular or paracellular pathways.

3. Secondary active tubular reabsorption occurs by cotransport with Na^+ via transport proteins. Transport of such substances is limited by the number of carriers available. Actively reabsorbed substances include glucose, amino acids, and some ions.

4. The proximal tubule cells are most active in reabsorption. Most of the nutrients, 65% of the water and sodium ions, and the bulk of actively transported ions are reabsorbed in the proximal convoluted tubules.

5. Reabsorption of additional sodium ions and water occurs in the distal tubules and collecting ducts and is hormonally controlled. Aldosterone increases the reabsorption of sodium; antidiuretic hormone (ADH) enhances water reabsorption by the collecting ducts.

 Complete an interactive tutorial: MasteringA&P® > Study Area > Interactive Physiology > Urinary System > Tubular Reabsorption and Secretion

25.6 Urine formation, step 3: Certain substances are secreted into the filtrate (pp. 994–995)

1. Tubular secretion adds substances to the filtrate (from the blood or tubule cells). It is an active process that is important in eliminating drugs, certain wastes, and excess ions and in maintaining the acid-base balance of the blood.

25.7 The kidneys create and use an osmotic gradient to regulate urine concentration and volume (pp. 995–1000)

1. The graduated hyperosmolality of the medullary fluids (largely due to the cycling of NaCl and urea) ensures that the filtrate reaching the distal convoluted tubule is dilute (hypo-osmolar). This allows urine with osmolalities ranging from 50 to 1200 mOsm to be formed.
   - The descending limb of the nephron loop is permeable to water, which leaves the filtrate and enters the medullary interstitial space. The filtrate and medullary fluid at the bend of the nephron loop are hyperosmolar.
   - The ascending limb is impermeable to water. Na^+ and Cl^- move out of the filtrate into the interstitial space, passively in the thin portion and actively in the thick portion. The filtrate becomes more dilute.
   - As filtrate flows through the collecting ducts in the inner medulla, urea diffuses into the interstitial space. From here, urea reenters the ascending thin limb and is recycled.
   - The blood flow in the vasa recta is sluggish, and the contained blood equilibrates with the medullary interstitial fluid. Hence, blood exiting the medulla in the vasa recta is nearly isotonic to blood plasma and the high solute concentration of the medulla is maintained.

2. In the absence of antidiuretic hormone, dilute urine is formed because the dilute filtrate reaching the collecting duct is simply allowed to pass from the kidneys.

3. When extracellular fluid osmolality rises, blood levels of antidiuretic hormone rise, and the collecting ducts become more permeable to water. Water moves out of the filtrate as it flows through the hyperosmotic medullary areas. Consequently, more concentrated urine is produced, and in smaller amounts.

25.8 Renal function is evaluated by analyzing blood and urine (pp. 1000–1002)

Renal Clearance (pp. 1000–1001)

1. Renal clearance is the volume of plasma that is completely cleared of a particular substance per minute. Studies of renal clearance provide information about renal function or the course of renal disease.

2. Renal failure has serious consequences: The kidneys are unable to concentrate urine, acid-base and electrolyte imbalances occur, and nitrogenous wastes accumulate in the blood.

Urine (pp. 1001–1002)

3. Urine is 95% water; solutes include nitrogenous wastes (urea, uric acid, and creatinine) and various ions (always sodium, potassium, sulfate, and phosphate).

4. Substances not normally found in urine include glucose, proteins, erythrocytes, leukocytes, hemoglobin, and bile pigments.

5. Urine is typically clear, yellow, aromatic, and slightly acidic. Its specific gravity ranges from 1.001 to 1.035.

25.9 The ureters, bladder, and urethra transport, store, and eliminate urine (pp. 1002–1006)

Ureters (p. 1002)

1. The ureters are slender tubes running retroperitoneally from each kidney to the bladder. They conduct urine by peristalsis from the renal pelvis to the urinary bladder.

25

Urinary Bladder (pp. 1003–1004)

2. The urinary bladder, which functions to store urine, is a distensible muscular sac that lies posterior to the pubic symphysis. It has two inlets (ureters) and one outlet (urethra) that outline the trigone. In males, the prostate surrounds the bladder outlet.

3. The bladder wall consists of a mucosa containing transitional epithelium; a three-layered detrusor; and an adventitia.

Urethra (p. 1004)

4. The urethra is a muscular tube that conveys urine from the bladder to the body exterior.

5. Where the urethra leaves the bladder, it is surrounded by an internal urethral sphincter, an involuntary smooth muscle sphincter. Where it passes through the urogenital diaphragm, the voluntary external urethral sphincter is formed by skeletal muscle.

6. In females the urethra is 3–4 cm long and conducts only urine. In males it is 20 cm long and conducts both urine and semen.

Micturition (pp. 1004–1005)

7. Micturition is emptying of the bladder.

8. Accumulating urine stretches the bladder wall, which initiates the micturition reflex. In infants, this is a simple spinal reflex: Parasympathetic fibers are excited (and sympathetic fibers

inhibited), causing the detrusor to contract and the internal urethral sphincter to open. In adults, pontine storage and micturition centers can override this simple reflex.

9. Because the external sphincter is voluntarily controlled, micturition can usually be delayed temporarily.

Developmental Aspects of the Urinary System (pp. 1006–1007)

1. Three sets of kidneys (pronephric, mesonephric, and metanephric) develop from the intermediate mesoderm. The metanephros excretes urine by the third month of development.

2. Common congenital abnormalities are horseshoe kidney, hypospadias, and polycystic kidney disease (PKD).

3. The kidneys of newborns are less able to concentrate urine; their bladder is small and voiding is frequent. Neuromuscular maturation generally allows toilet training for micturition to begin by 24 months of age.

4. The most common urinary system problems in children and young to middle-aged adults are bacterial infections.

5. With age, nephrons are lost, the filtration rate decreases, and tubule cells become less efficient at concentrating urine.

6. Bladder capacity and tone decrease with age, leading to frequent micturition and (often) incontinence. Urinary retention is a common problem of elderly men.

REVIEW QUESTIONS

To access additional practice questions using your smartphone, tablet, or computer: MasteringA&P® > Study Area > Practice Tests & Quizzes

Level 1 Remember/Understand

(Some questions have more than one correct answer. Select the best answer or answers from the choices given.)

1. The lowest blood concentration of nitrogenous waste occurs in the (a) hepatic vein, (b) inferior vena cava, (c) renal artery, (d) renal vein.

2. The glomerular capillaries differ from other capillary networks in the body because they (a) have a larger area of anastomosis, (b) are derived from and drain into arterioles, (c) are not made of endothelium, (d) are sites of filtrate formation.

3. Damage to the renal medulla would interfere *first* with the functioning of the (a) glomerular capsules, (b) distal convoluted tubules, (c) collecting ducts, (d) proximal convoluted tubules.

4. Which is reabsorbed by the proximal convoluted tubule cells? (a) Na^+, (b) K^+, (c) amino acids, (d) all of the above.

5. Glucose is not normally found in the urine because it (a) does not pass through the walls of the glomerulus, (b) is kept in the blood by colloid osmotic pressure, (c) is reabsorbed by the tubule cells, (d) is removed by the body cells before the blood reaches the kidney.

6. Filtration at the glomerulus is inversely related to (a) water reabsorption, (b) capsular hydrostatic pressure, (c) arterial blood pressure, (d) acidity of the urine.

7. Tubular reabsorption (a) of glucose and many other substances is a T_m-limited active transport process, (b) of chloride is always linked to the passive transport of Na^+, (c) is the movement of substances from the blood into the nephron, (d) of sodium occurs only in the proximal tubule.

8. If a freshly voided urine sample contains excessive amounts of urochrome, it has (a) an ammonia-like odor, (b) a pH below normal, (c) a dark yellow color, (d) a pH above normal.

9. Which of the following is/are true about ADH? (a) It promotes obligatory water reabsorption, (b) it is secreted in response to an increase in extracellular fluid osmolality, (c) it causes insertion of aquaporins in the PCT, (d) it promotes Na^+ reabsorption.

10. What is the importance of the perirenal fat capsule that surrounds the kidney?

11. Define micturition and describe the micturition reflex.

Level 2 Apply/Analyze

12. Describe the mechanisms that contribute to renal autoregulation.

13. Describe the mechanisms of extrinsic regulation of GFR, and their physiological role.

14. Describe what is involved in active and passive tubular reabsorption.

15. Explain how the peritubular capillaries are adapted for receiving reabsorbed substances.

16. Explain the process and purpose of tubular secretion.

17. How does urinary bladder anatomy support its storage function?

18. Describe the changes that occur in kidney and bladder anatomy and physiology in old age.

Level 3 Evaluate/Synthesize

19. Trace the pathway a creatinine molecule takes from a glomerulus to the urethra. Name every microscopic or gross structure it passes through on its journey.

20. Explain the important differences between blood plasma and glomerular filtrate, and relate the differences to the structure of the filtration membrane.

21. How does aldosterone modify the chemical composition of urine?

22. Explain why the filtrate becomes hypotonic as it flows through the ascending limb of the nephron loop. Also explain why the filtrate at the bend of the nephron loop (and the interstitial fluid of the deep portions of the medulla) is hypertonic.

23. Mrs. Bigda, a 60-year-old woman, was brought to the hospital by the police after falling to the pavement. She is found to have alcoholic hepatitis. She is put on a salt- and protein-restricted diet and diuretics are prescribed to manage her ascites (accumulated fluid in the peritoneal cavity). How will diuretics reduce this excess fluid? Name and describe the mechanisms of action of three types of diuretics. Why is her diet salt-restricted?

24. While repairing a frayed utility wire, Kevin, an experienced lineman, slips and falls to the ground. Medical examination reveals a fracture of his lower spine and transection of the lumbar region of the spinal cord. How will Kevin's micturition be controlled from this point on? Will he ever again feel the need to void? Will there be dribbling of urine between voidings? Explain the reasoning behind all your responses.

25. Patty, aged 55, is awakened by excruciating pain that radiates from her right abdomen to the loin and groin regions on the same side. The pain is not continuous but recurs at intervals of 3 to 4 minutes. Diagnose her problem, and cite factors that might favor its occurrence.

26. Why are renal failure patients undergoing dialysis at risk for anemia and osteoporosis? What medications or supplements could you give them to prevent these problems?

CLINICAL CASE STUDY

32-Year-Old Diabetic Male on a Diuretic

Let's return to Kyle Boulard, whom we met in the previous chapter. Mr. Boulard has recovered from his acute diabetic crisis. The last update on his chart before he is discharged includes the following:

- BP 150/95, HR 75, temperature 37.2°C
- Urine: pH 6.9, negative for glucose and ketones; 24-hour urine collection reveals 170 mg albumin in urine per day

Mr. Boulard is prescribed a thiazide diuretic and an angiotensin converting enzyme (ACE) inhibitor. He is counseled on the importance of taking his medications regularly and keeping his outpatient follow-up appointments.

1. What is albumin? Is it normally found in the urine? What does its presence suggest?

2. **+ NCLEX-STYLE** Mr. Boulard's blood pressure is quite high. How would this affect his nephron function? Choose the statement below that is most accurate.
   a. The kidneys normally increase their rate of filtration when blood pressure increases.
   b. The kidneys normally increase their rate of reabsorption when blood pressure increases.
   c. The kidneys normally increase both their rate of filtration and reabsorption when blood pressure increases.
   d. The kidneys release more renin when blood pressure increases, activating the renin-angiotensin-aldosterone pathway.

3. Where and how do thiazide diuretics act in the kidneys and how does this reduce blood pressure?

At his two-week appointment at the outpatient clinic, Mr. Boulard complains of fatigue, weakness, muscle cramps, and irregular heartbeats. A physical examination and lab tests produce the following observations:

- BP 133/90, HR 75, temperature 37.3°C
- Blood K^+ 2.9 mEq/L (normal 3.5–5.5 mEq/L); blood Na^+ 135 mEq/L (normal 135–145 mEq/L)
- Urine K^+ 55 mEq/L (normal <40 mEq/L); urine Na^+ 21 mEq/L (normal >20 mEq/L)

4. What is Mr. Boulard's main problem at this point? Explain how the thiazide diuretic might have caused this problem.

When asked about his medications, Mr. Boulard admits that he did not fill his ACE inhibitor prescription because it was too expensive. He could only afford the thiazide medications along with his insulin.

5. How do ACE inhibitors reduce blood pressure?

6. **+ NCLEX-STYLE** You've counseled Mr. Boulard about how to prevent recurrence of his hypokalemia. Which of the statements he makes indicates a need for further instruction?
   a. "There's lots of potassium in fruits and vegetables, so I'll eat more bananas."
   b. "There's lots of potassium in regular table salt, so I'll just use more."
   c. "I guess I need to come in regularly and have potassium levels checked."
   d. "Maybe I should take a potassium supplement."

25

26

Fluid, Electrolyte, and Acid-Base Balance

In this chapter, you will learn that

The composition and volume of body fluids is tightly regulated

by first examining

26.1 Body fluid compartments

and

Body fluid composition

looking closer at

26.2 Water balance and ECF osmolality ← → **26.3 Electrolyte balance**

including

26.4 Acid-base balance

controlled by   and asking

Chemical buffer systems   **Respiratory regulation**   **26.5 Renal regulation**   **26.6 What happens when things go wrong?**

using   using

Respiratory system Ch. 22   **Urinary system Ch. 25**

and finally, exploring

Developmental Aspects of Fluid, Electrolyte, and Acid-Base Balance

Have you ever wondered why on some days you don't urinate for hours at a time, while on others it seems like you void every few minutes? Or why on occasion you cannot seem to quench your thirst? These situations reflect one of the body's most important functions: maintaining fluid, electrolyte, and acid-base balance.

Cell function depends not only on a continuous supply of nutrients and removal of metabolic wastes, but also on the physical and chemical homeostasis of the surrounding fluids. The French physiologist Claude Bernard recognized this truth with style in 1857 when he said, "It is the fixity of the internal environment which is the condition of free and independent life."

In this chapter, we first examine the composition and distribution of fluids in the internal environment and then consider the roles of various body organs and functions in establishing, regulating, and altering this balance.

26.1 Body fluids consist of water and solutes in three main compartments

Learning Outcomes

▶ List the factors that determine body water content and describe the effect of each factor.

▶ Indicate the relative fluid volume and solute composition of the fluid compartments of the body.

▶ Contrast the overall osmotic effects of electrolytes and nonelectrolytes.

▶ Describe factors that determine fluid shifts in the body.

Body Water Content

Not all bodies contain the same amount of water. Total body water is a function not only of age and body mass but also of sex and the relative amount of body fat. Infants, with their low body fat and low bone mass, are 73% or more water. (This high level of hydration accounts for their "dewy" skin, like that of a freshly picked peach.) After infancy total body water declines throughout life, accounting for only about 45% of body mass in old age.

A healthy young man is about 60% water, and a healthy young woman about 50%. This difference between the sexes reflects the fact that females have relatively more body fat and less skeletal muscle than males. Of all body tissues, adipose tissue is *least* hydrated (less than 20% water)—even bone contains more water than does fat. In contrast, skeletal muscle is about 75% water, so people with greater muscle mass have proportionately more body water.

Fluid Compartments

Water occupies two main **fluid compartments** within the body (**Figure 26.1**). Almost two-thirds by volume is in the **intracellular fluid (ICF) compartment**, which actually consists of trillions of tiny individual "compartments": the cells. In an adult male of average size (70 kg, or 154 lb), ICF accounts for about 25 L of the 40 L of body water.

The remaining one-third or so of body water is outside cells, in the **extracellular fluid (ECF) compartment**. The ECF

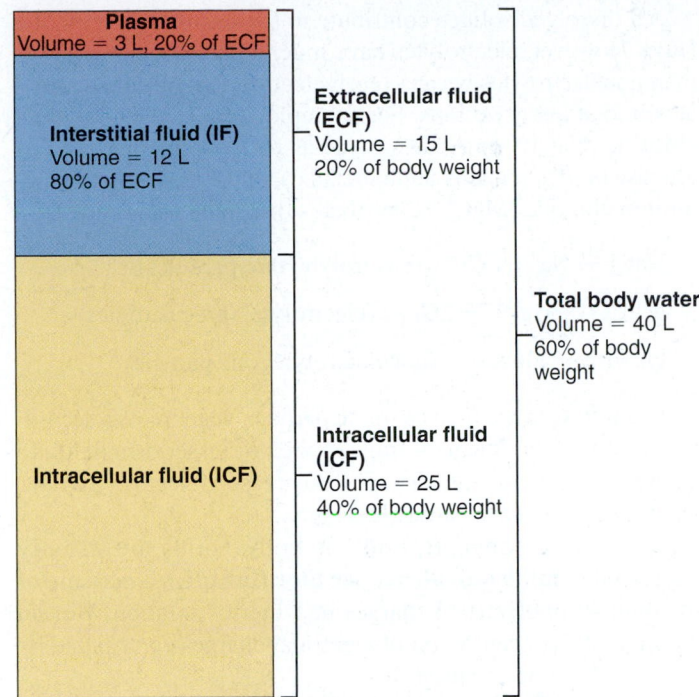

Figure 26.1 The major fluid compartments of the body. [Values are for a 70-kg (154-lb) male.]

constitutes the body's "internal environment" referred to by Claude Bernard and is the external environment of each cell. As Figure 26.1 shows, the ECF compartment is divisible into two subcompartments: (1) **plasma**, the fluid portion of blood, and (2) **interstitial fluid (IF)**, the fluid in the microscopic spaces between tissue cells. There are numerous other examples of ECF that are distinct from both plasma and interstitial fluid—lymph, cerebrospinal fluid, humors of the eye, synovial fluid, serous fluid, gastrointestinal secretions—but most of these are similar to IF and are usually considered part of it.

Composition of Body Fluids

Water serves as the *universal solvent* in which a variety of solutes are dissolved. Solutes may be classified broadly as *electrolytes* and *nonelectrolytes*.

Electrolytes and Nonelectrolytes

Electrolytes are chemical compounds that dissociate into *ions* in water. (See Chapter 2, ◀ p. 39, if you need to review these concepts of chemistry.) Because ions are charged particles, they can conduct an electrical current—and so have the name *electrolyte*. Typically, electrolytes include inorganic salts, both inorganic and organic acids and bases, and some proteins.

In contrast, **nonelectrolytes** have bonds (usually covalent bonds) that prevent them from dissociating in solution. For this reason, no electrically charged species are created when nonelectrolytes dissolve in water. Most nonelectrolytes are organic molecules—glucose, lipids, creatinine, and urea, for example.

26

All dissolved solutes contribute to the osmotic activity of a fluid. However, electrolytes have much greater osmotic power than nonelectrolytes because each electrolyte molecule dissociates into at least two ions. For example, a molecule of sodium chloride (NaCl) contributes twice as many solute particles as glucose (which remains undissociated), and a molecule of magnesium chloride ($MgCl_2$) contributes three times as many:

$$NaCl \rightarrow Na^+ + Cl^-$$ (electrolyte; two particles)

$$MgCl_2 \rightarrow Mg^{2+} + 2Cl^-$$ (electrolyte; three particles)

$$glucose \rightarrow glucose$$ (nonelectrolyte; one particle)

Regardless of the type of solute particle, water moves according to osmotic gradients—from an area of lesser osmolality to an area of greater osmolality. For this reason, electrolytes have the greatest ability to cause fluid shifts.

Electrolyte concentrations of body fluids are usually expressed in **milliequivalents per liter (mEq/L)**, a measure of the number of electrical charges in 1 liter of solution. We can compute the concentration of electrical charges contributed by any ion using the equation

$$mEq/L = \frac{\text{ion concentration (mg/L)}}{\text{atomic weight of ion (mg/mmol)}} \times \begin{array}{l} \text{no. of} \\ \text{electrical} \\ \text{charges on} \\ \text{one ion} \end{array}$$

(Recall from p. 30 that 1 millimole (mmol) = 0.001 mole.)

For example, to calculate the mEq/L of sodium or calcium ions in solution in plasma, we would determine the normal concentration of these ions in plasma, look up their atomic weights in the periodic table (see Appendix E), and plug these values into the equation:

$$Na^+: \quad \frac{3300 \text{ mg/L}}{23 \text{ mg/mmol}} \times 1 = 143 \text{ mEq/L}$$

$$Ca^{2+}: \quad \frac{100 \text{ mg/L}}{40 \text{ mg/mmol}} \times 2 = 5 \text{ mEq/L}$$

Notice that for ions with a single charge, 1 mEq is equal to 1 mmol, which, when dissolved in 1 kg of water, produces 1 mOsm (see p. 995). On the other hand, 1 mEq of ions with a double charge (like calcium) is equal to 1/2 mOsm. In either case, 1 mEq provides the same amount of charge.

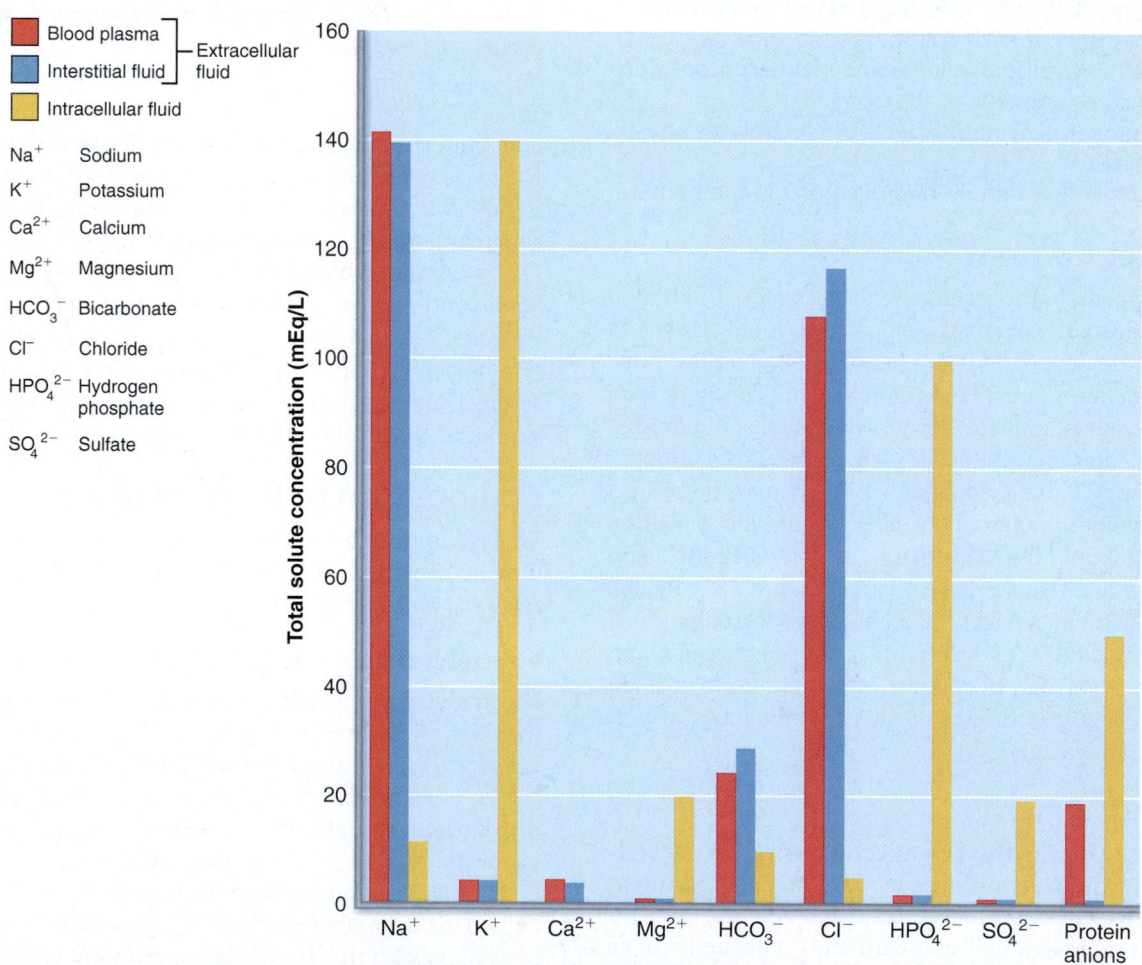

Figure 26.2 Electrolyte composition of blood plasma, interstitial fluid, and intracellular fluid. The very low intracellular Ca^{2+} concentration (10^{-7} *M*) does not include Ca^{2+} stores sequestered inside organelles. The high concentration of intracellular HPO_4^{2-} includes large amounts bound to intermediate metabolites, proteins, and lipids.

Comparison of Extracellular and Intracellular Fluids

A quick glance at the bar graphs in **Figure 26.2** reveals that each fluid compartment has a distinctive pattern of electrolytes. Except for the relatively high protein content in plasma (◄ p. 644), however, the extracellular fluids are very similar. Their chief cation is sodium, and their major anion is chloride. However, plasma contains somewhat fewer chloride ions than interstitial fluid because the nonpenetrating plasma proteins are normally anions and plasma is electrically neutral.

In contrast to extracellular fluids, the ICF contains only small amounts of Na^+ and Cl^-. Its most abundant cation is potassium, and its major anion is HPO_4^{2-}. Cells also contain substantial quantities of soluble proteins (about three times the amount found in plasma).

Notice that sodium and potassium ion concentrations in ECF and ICF are nearly opposite (Figure 26.2). The characteristic distribution of these ions on the two sides of cellular membranes reflects the activity of cellular ATP-dependent sodium-potassium pumps, which keep intracellular Na^+ concentrations low and K^+ concentrations high (◄ p. 75).

Electrolytes are the most abundant solutes in body fluids and determine most of their chemical and physical reactions, but they do not constitute the *bulk* of dissolved solutes in these fluids. Instead, the bulk of these fluids consists of large molecules such as proteins and some nonelectrolytes (phospholipids, cholesterol, and triglycerides). These account for about 90% of the mass of dissolved solutes in plasma, 60% in the IF, and 97% in the ICF.

Fluid Movement among Compartments

Osmotic and hydrostatic pressures regulate the continuous exchange and mixing of body fluids. Although water moves freely among the compartments along osmotic gradients, solutes are unequally distributed because of their size, electrical charge, or dependence on transport proteins. *Anything that changes the solute concentration in any compartment leads to net water flows.*

Figure 26.3 summarizes the exchanges of gases, solutes, and water across the body's borders and between the three fluid compartments within the body. In general, substances must pass through both the plasma and IF to reach the ICF. In the lungs, gastrointestinal tract, and kidneys, exchanges between the "outside world" and the plasma occur almost continuously. These exchanges alter plasma composition and volume, with plasma serving as the "highway" for delivering substances throughout the body. Compensating adjustments between the plasma and the other two fluid compartments follow quickly so that balance is restored.

Recall that:

- **Exchanges between plasma and IF occur across capillary walls.** We described the pressures driving these fluid movements in *Focus on Bulk Flow across Capillary Walls* (◄ **Focus Figure 19.1, pp. 732–733**). To recap, the hydrostatic pressure of blood forces nearly protein-free plasma out of the blood into the interstitial space. This filtered fluid is then almost completely reabsorbed into the bloodstream in response to the colloid osmotic pressure of plasma proteins. Under

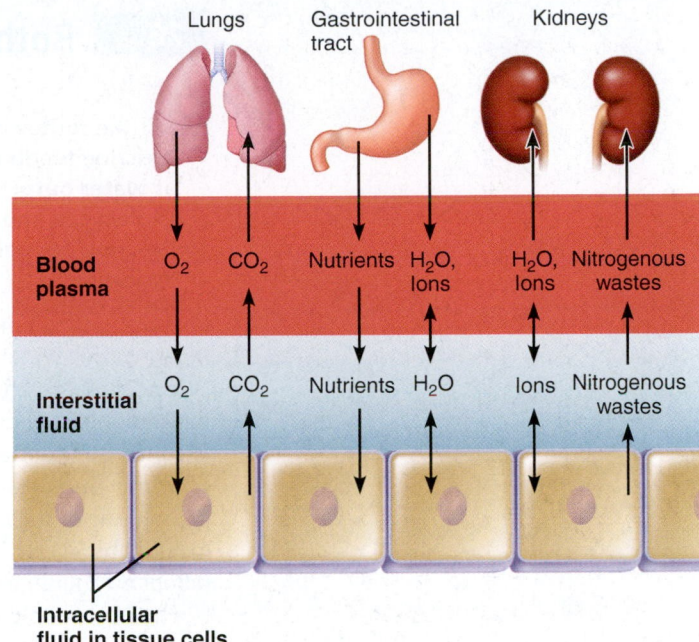

Figure 26.3 Exchange of gases, nutrients, water, and wastes between the three fluid compartments of the body.

normal circumstances, lymphatic vessels pick up the small net leakage that remains behind in the interstitial space and return it to the blood.

- **Exchanges between the IF and ICF occur across plasma membranes.** As described in Chapter 3 (◄ **pp. 68–78**), exchanges across the plasma membrane depend on its permeability properties. As a general rule, two-way osmotic flow of water is substantial. But ion fluxes are restricted and, in most cases, ions move selectively by active transport or through channels. Movements of nutrients, respiratory gases, and wastes are typically unidirectional. For example, glucose and oxygen move into the cells and metabolic wastes move out.

Many factors can change ECF and ICF volumes. Because water moves freely between compartments, however, the osmolalities of all body fluids are equal (except during the first few minutes after a change occurs in one of the fluids). Increasing the ECF solute content (mainly the NaCl concentration) causes osmotic and volume changes in the ICF—namely, a shift of water out of the cells. Conversely, decreasing ECF osmolality causes water to move into the cells. Thus, ECF solute concentration determines ICF volume.

It is important for you to understand the concepts above, because they underlie all events that control fluid balance in the body.

Check Your Understanding

1. Which do you have more of, extracellular or intracellular fluid? Plasma or interstitial fluid?
2. What is the major cation in the ECF? In ICF? What are the intracellular anion counterparts of ECF's chloride ions?
3. **PREDICT** You eat salty pretzels without drinking. What happens to the volume of your extracellular fluid? Explain.

For answers, see Answers Appendix.

26.2 Both intake and output of water are regulated

Learning Outcomes

▶ List the routes by which water enters and leaves the body.
▶ Describe feedback mechanisms that regulate water intake and hormonal controls of water output in urine.
▶ Explain the importance of obligatory water losses.
▶ Describe possible causes and consequences of dehydration and of hypotonic hydration.

For the body to remain properly hydrated, water intake must equal water output. *Water intake* varies widely from person to person and is strongly influenced by habit, but is typically about 2500 ml a day in adults (**Figure 26.4**). Most water enters the body through ingested liquids and solid foods. Body water produced by cellular metabolism is called **metabolic water** or **water of oxidation**.

Water output occurs by several routes. Water that vaporizes out of the lungs in expired air or diffuses directly through the skin is called **insensible water loss**. Any measurable water loss is called **sensible water loss**. This includes water lost as urine (about 60%), obvious sweat, and in feces.

Healthy people have a remarkable ability to maintain the osmolality of their body fluids within very narrow limits (280–300 mOsm). A rise in plasma osmolality triggers (1) thirst, which prompts us to drink water, and (2) release of antidiuretic hormone (ADH; ◀ p. 613), which causes the kidneys to conserve water and excrete concentrated urine. On the other hand, a decline in osmolality inhibits both thirst and ADH release, and the latter prompts the kidneys to excrete large volumes of dilute urine.

In the body, water and Na^+ are closely tied together. In fact, Na^+ acts as a powerful "water magnet." However, the ADH and thirst mechanisms controlling osmolality regulate water *independently* of sodium. Later, we will describe mechanisms that regulate Na^+ and water *together* for the purpose of maintaining blood volume and pressure.

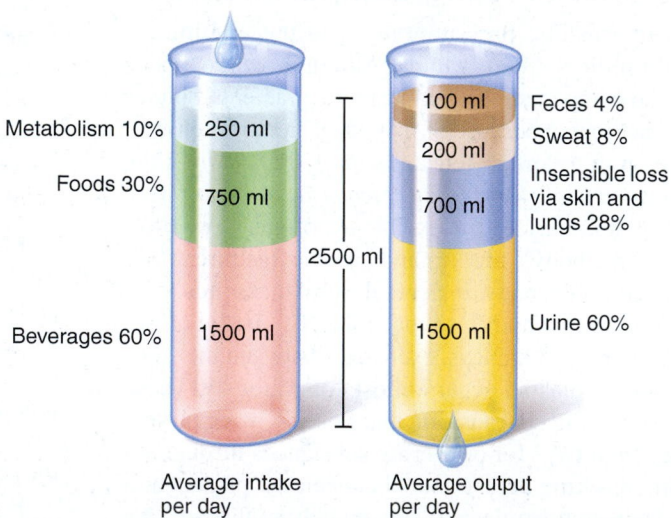

Figure 26.4 Major sources of water intake and output. When intake and output are in balance, the body is adequately hydrated.

Regulation of Water Intake

The **thirst mechanism** is the driving force for water intake. It is governed by the hypothalamic *thirst center* (◀ p. 449), which is activated by various stimuli (**Figure 26.5**):

• *Osmoreceptors*. Hypothalamic *osmoreceptors* detect ECF osmolality through changes in plasma membrane stretch that result from gaining or losing water. An increase in osmolality of only 1–2% activates these osmoreceptors.

- *Dry mouth.* When the blood's osmotic pressure increases, the salivary glands produce less saliva because the osmotic gradient drawing water from the blood into the salivary ducts is reduced.

- *A decrease in blood volume (or pressure).* A substantial decrease in blood volume or pressure (5–10%, as in hemorrhage) also triggers the thirst mechanism. These changes in volume or pressure are signaled by baroreceptors that directly activate the thirst center, and by angiotensin II as we described in Chapter 19 (◀ **p. 723**).

Collectively, these events cause a subjective sensation of thirst, which motivates us to get a drink. This mechanism explains why some bars provide free *salty* snacks to their patrons.

Curiously, thirst is quenched almost as soon as we begin drinking water, even though the water has yet to be absorbed into the blood. How does this happen? Incoming liquid moistens the mucosa of the mouth and throat and activates osmoreceptors and stretch receptors in the stomach and small intestine, providing feedback signals that inhibit the thirst center. Premature quenching of thirst prevents us from drinking more than we need and overdiluting our body fluids, and allows time for the osmotic changes to come into play as regulatory factors.

As effective as thirst is, it is not always a reliable indicator of need. This is particularly true during athletic events, when thirst can be satisfied long before sufficient liquids have been drunk to maintain the body in top form. Additionally, elderly or confused people may not recognize or heed thirst signals. In contrast, fluid-overloaded renal or cardiac patients may feel thirsty despite their condition.

Regulation of Water Output

Output of certain amounts of water is unavoidable. Such **obligatory water losses** help to explain why we cannot survive for long without drinking. Even the most heroic conservation efforts by the kidneys cannot compensate for zero water intake. Obligatory water loss includes the *insensible water losses* described above, as well as *sensible water loss* of at least 500 ml per day in urine plus the water that accompanies undigested food residues in feces.

Obligatory water loss in urine reflects the fact that human kidneys must normally flush 600 mmol per day of urine solutes (end products of metabolism and so forth) out of the body in water. The maximum concentration of urine is about 1200 mOsm, so at least 500 ml of water must be excreted.

Beyond obligatory water loss, the solute concentration and volume of urine excreted depend on fluid intake, diet, and water loss via other avenues. For example, if you sweat profusely on a hot day, you have to excrete much less urine than usual to maintain water balance. Normally, the kidneys begin to eliminate excess water about 30 minutes after it is ingested. This delay reflects the time required to inhibit ADH release. Diuresis reaches a peak about 1 hour after drinking and declines to its lowest level after 3 hours.

Influence of Antidiuretic Hormone (ADH)

The amount of water reabsorbed in the renal collecting ducts is proportional to ADH release (◀ **Figure 25.20, p. 999**). When ADH levels are low, most of the water reaching the collecting ducts is not reabsorbed because the lack of aquaporins in the apical membranes of the principal cells prevents the movement of water. Instead, the water simply flows on by. The result is dilute urine and a reduced volume of body fluids. When ADH

Figure 26.5 The thirst mechanism for regulating water intake. Only large volume changes (5–10% of plasma volume) activate the thirst mechanism. Osmolality changes are the normal stimuli.

Figure 26.6 Mechanisms and consequences of ADH release. (Vasoconstrictor effects of ADH are not shown.)

levels are high, aquaporins are inserted in the principal cell apical membranes, nearly all of the filtered water is reabsorbed, and a small volume of concentrated urine is excreted (**Figure 26.6**).

Osmoreceptors of the hypothalamus sense the ECF solute concentration and trigger or inhibit ADH release from the posterior pituitary accordingly (◄ **p. 613**). An increase in ECF osmolality prompts ADH release by stimulating the hypothalamic osmoreceptors. In contrast, a decrease in ECF osmolality inhibits ADH release and allows more water to be excreted in urine, restoring normal blood osmolality.

Large changes in blood volume or blood pressure also influence ADH secretion. A decrease in blood pressure increases ADH release both directly via baroreceptors in the atria and various blood vessels, and indirectly via the renin-angiotensin-aldosterone mechanism (◄ **pp. 722–723**).

The key word here is "large" because changes in ECF osmolality are much more important as day-to-day stimulatory or inhibitory factors. Factors that greatly reduce blood volume and so trigger ADH release include excessive sweating, vomiting, or diarrhea; severe blood loss; traumatic burns; and prolonged fever. Under these conditions, high concentrations of ADH also act to constrict arterioles, directly increasing blood pressure—this accounts for its other name: *vasopressin*. For a summary of how renal mechanisms involving ADH, aldosterone, and angiotensin II tie into overall controls of blood volume and blood pressure, see Figure 26.10 (p. 1024). However, remember that the main thrust of ADH is to maintain ECF osmolality.

Disorders of Water Balance CLINICAL

Few people really appreciate the importance of water in keeping the body's "machinery" working at peak efficiency. The principal abnormalities of water balance are dehydration, hypotonic hydration, and edema—each presenting a special set of problems.

Dehydration

Clinically, **dehydration** is defined as fluid loss, either the loss of water or the loss of water and solutes together. Dehydration is a common sequel to hemorrhage, severe burns, prolonged vomiting or diarrhea, profuse sweating, water deprivation, and diuretic abuse. Dehydration may also be caused by endocrine disturbances, such as diabetes mellitus (◄ **p. 632**) or diabetes insipidus (◄ **p. 614**).

Early signs and symptoms of dehydration include a "cottony" or sticky oral mucosa, thirst, dry flushed skin, and decreased urine output (*oliguria*). Prolonged dehydration may lead to weight loss, fever, and mental confusion. Another serious consequence of water loss from plasma is inadequate blood volume to maintain normal circulation and ensuing *hypovolemic shock*.

Water and solutes can be lost together (as in hemorrhage), or more water than solutes can be lost (as in profuse sweating). If the body loses more water than solutes, water moves osmotically from the cells into the ECF (**Figure 26.7a**). This water movement equalizes the osmolality of the extracellular and intracellular fluids even though the total fluid volume has been reduced.

① Excessive loss of H_2O from ECF → ② ECF osmotic pressure rises → ③ Cells lose H_2O to ECF by osmosis; cells shrink

(a) Consequences of dehydration. If more water than solutes is lost, cells shrink.

① Excessive H_2O enters the ECF → ② ECF osmotic pressure falls → ③ H_2O moves into cells by osmosis; cells swell

(b) Consequences of hypotonic hydration (water gain). If more water than solutes is gained, cells swell.

Figure 26.7 Disturbances in water balance.

intake and output, edema is an increase in volume of *only* the IF. The extra fluid in the interstitial space can impair tissue function by increasing the distance across which oxygen and nutrients must diffuse between the blood and the cells. Edema may be caused by any event that steps up the flow of fluid out of the blood or hinders its return (see *Focus on Bulk Flow across Capillary Walls*, ◀ **Focus Figure 19.1, pp. 732–733**). Edema and its various causes are discussed in Chapter 19 (◀ **p. 734**).

Check Your Understanding

4. What change in plasma is most important for triggering thirst? Where is that change sensed?

5. ADH, by itself, cannot reduce an increase in osmolality in body fluids. Why not? What other mechanism is required?

6. **MAKE CONNECTIONS** For each of the following, state whether it might result in dehydration, hypotonic hydration, or edema (discussed in Chapter 19): (a) decreased synthesis of plasma proteins due to liver failure; (b) copious sweating; (c) using ecstasy (MDMA), which promotes ADH secretion.

For answers, see Answers Appendix.

Hypotonic Hydration

Declining ECF osmolality sets several compensatory mechanisms into motion. ADH release is inhibited, and as a result, less water is reabsorbed and excess water is quickly flushed from the body in urine. But, when there is renal insufficiency or we drink an extraordinary amount of water very quickly, a type of cellular *overhydration* called **hypotonic hydration** may occur. In either case, the ECF is diluted—its sodium *content* is normal, but excess water is present so the sodium *concentration* is low. For this reason, the hallmark of hypotonic hydration is **hyponatremia** (low ECF Na^+ concentration), which promotes net osmosis into tissue cells, causing them to swell as they become abnormally hydrated (Figure 26.7b).

Hypotonic hydration leads to severe metabolic disturbances evidenced by nausea, vomiting, muscular cramping, and cerebral edema. It is particularly damaging to neurons. Uncorrected cerebral edema quickly leads to disorientation, convulsions, coma, and death. Indeed, several marathon runners have died of overhydration after drinking too much water. Sudden and severe hyponatremia is treated by administering intravenous hypertonic saline to reverse the osmotic gradient and "pull" water out of the cells.

Edema

Edema (ĕ-de′mah; "a swelling") is an atypical accumulation of fluid in the interstitial space, leading to tissue (but not cell) swelling. Unlike hypotonic hydration, which increases the amount of fluid in all compartments due to an imbalance between water

26.3 ## Sodium, potassium, calcium, and phosphate levels are tightly regulated

Learning Outcomes

▶ Indicate routes of electrolyte entry and loss from the body.

▶ Describe the importance of sodium in the body's fluid and electrolyte balance.

▶ Describe mechanisms involved in regulating sodium balance, blood volume, and blood pressure.

▶ Explain how potassium, calcium, and anion balances in plasma are regulated.

Electrolytes include salts, acids, bases, and some proteins, but the term **electrolyte balance** usually refers to salt balance in the body. Salts are important in controlling fluid movements and provide minerals essential for excitability, secretory activity, and membrane permeability. Although many electrolytes are crucial for cellular activity, here we will specifically examine the regulation of sodium, potassium, calcium, and phosphate.

Salts enter the body in foods and fluids, and small amounts are generated during metabolic activity. For example, phosphates are liberated during catabolism of nucleic acids and bone matrix. Obtaining enough electrolytes is usually not a problem. Indeed, most of us have a far greater taste than need for salt. We shake table salt (NaCl) on our food even though natural foods contain ample amounts and processed foods contain exorbitant quantities. The

26

Table 26.1   Causes and Consequences of Electrolyte Imbalances

ION	ABNORMALITY (SERUM VALUE)	POSSIBLE CAUSES	CONSEQUENCES
Sodium	**Hypernatremia** (Na^+ excess: >145 mEq/L)	Dehydration; uncommon in healthy individuals; may occur in infants, the elderly, or any individual unable to indicate thirst; or may result from excessive intravenous NaCl administration	Thirst. CNS dehydration leads to confusion and lethargy progressing to coma; increased neuromuscular irritability evidenced by twitching and convulsions.
	Hyponatremia (Na^+ deficit: <135 mEq/L)	Solute loss, water retention, or both (e.g., excessive Na^+ loss through vomiting, diarrhea, burned skin, gastric suction, or excessive use of diuretics); deficiency of aldosterone (Addison's disease, ◀ pp. 625–626); renal disease; excess ADH release; excess H_2O ingestion	Most common signs are those of neurologic dysfunction due to brain swelling. If body sodium content is normal but water is excessive, the symptoms are the same as those of water excess: mental confusion; giddiness; coma if development occurs slowly; muscular twitching, irritability, and convulsions if the condition develops rapidly. In hyponatremia accompanied by water loss, the main signs are decreased blood volume and blood pressure (circulatory shock).
Potassium	**Hyperkalemia** (K^+ excess: >5.5 mEq/L)	Renal failure; deficit of aldosterone; rapid intravenous infusion of KCl; burns or severe tissue injuries that cause K^+ to leave cells	Nausea, vomiting, diarrhea; bradycardia; cardiac arrhythmias and arrest; skeletal muscle weakness; flaccid paralysis.
	Hypokalemia (K^+ deficit: <3.5 mEq/L)	Gastrointestinal tract disturbances (vomiting, diarrhea), gastric suction; Cushing's syndrome; inadequate dietary intake (starvation); hyperaldosteronism; diuretic therapy	Cardiac arrhythmias, flattened T wave on ECG; muscular weakness; metabolic alkalosis; mental confusion; nausea; vomiting.
Phosphate	**Hyperphosphatemia** (HPO_4^{2-} excess: >2.9 mEq/L)	Decreased urinary loss due to renal failure; hypoparathyroidism; major tissue trauma; increased intestinal absorption	Clinical symptoms arise because of reciprocal changes in Ca^{2+} levels rather than directly from changes in plasma phosphate concentrations.
	Hypophosphatemia (HPO_4^{2-} deficit: <1.6 mEq/L)	Decreased intestinal absorption; increased urinary output; hyperparathyroidism	
Chloride	**Hyperchloremia** (Cl^- excess: >105 mEq/L)	Dehydration; increased retention or intake; metabolic acidosis; hyperparathyroidism	No direct clinical symptoms; symptoms generally associated with the underlying cause, which is often related to pH abnormalities.
	Hypochloremia (Cl^- deficit: <95 mEq/L)	Metabolic alkalosis (e.g., due to vomiting or excessive ingestion of alkaline substances); aldosterone deficiency	
Calcium	**Hypercalcemia** (Ca^{2+} excess: >5.2 mEq/L or 10.5 mg%)*	Hyperparathyroidism; excessive vitamin D; prolonged immobilization; renal disease (decreased excretion); malignancy	Decreased neuromuscular excitability leading to cardiac arrhythmias and arrest, skeletal muscle weakness, confusion, stupor, and coma; kidney stones; nausea and vomiting.
	Hypocalcemia (Ca^{2+} deficit: <4.5 mEq/L or 9 mg%)*	Burns (calcium trapped in damaged tissues); hypoparathyroidism; vitamin D deficiency; renal tubular disease; renal failure; hyperphosphatemia; diarrhea; alkalosis	Increased neuromuscular excitability leading to tingling fingers, tremors, skeletal muscle cramps, tetany, convulsions; depressed excitability of the heart; osteomalacia; fractures.
Magnesium	**Hypermagnesemia** (Mg^{2+} excess: >2.2 mEq/L)	Rare; occurs in renal failure when Mg^{2+} is not excreted normally; excessive ingestion of Mg^{2+}-containing antacids	Lethargy; impaired CNS functioning, coma, respiratory depression; cardiac arrest.
	Hypomagnesemia (Mg^{2+} deficit: <1.4 mEq/L)	Alcoholism; chronic diarrhea, severe malnutrition; diuretic therapy	Tremors, increased neuromuscular excitability, tetany, convulsions.

*1 mg% = 1 mg/100 ml

26

taste for very salty foods is learned, but some liking for salt may be innate to ensure adequate intake of these two vital ions.

We lose salts from the body in sweat, feces, urine, and vomit. Even though sweat is normally hypotonic, large amounts of salt can be lost on a hot day simply because more sweat is produced. Gastrointestinal disorders can also lead to large salt losses in feces or vomitus. Consequently, the flexibility of renal mechanisms that regulate the electrolyte balance of the blood is a critical asset. **Table 26.1** summarizes several causes and consequences of electrolyte imbalance.

HOMEOSTATIC IMBALANCE 26.1   **CLINICAL**

Severe electrolyte deficiencies may cause unusual cravings. For example, in *Addison's disease* (◄ pp. 625–626), the adrenal cortex produces too little aldosterone, and so Na^+ is lost. People with this disorder often crave salty foods.

When minerals such as iron are deficient, a person may even eat substances not usually considered foods, like chalk, clay, starch, and burnt match tips. This appetite for abnormal substances is called *pica*.

The Central Role of Sodium in Fluid and Electrolyte Balance

Sodium holds a central position in fluid and electrolyte balance and overall body homeostasis. Indeed, regulating the balance between sodium input and output is one of the most important renal functions. The salts $NaHCO_3$ and $NaCl$ account for 90–95% of all solutes in the ECF, and they contribute about 280 mOsm of the total ECF solute concentration (300 mOsm).

At its normal plasma concentration of about 142 mEq/L, Na^+ is the single most abundant cation in the ECF and the only one exerting *significant* osmotic pressure. Additionally, cellular plasma membranes are relatively impermeable to Na^+ (some does manage to diffuse in and must be pumped out against its electrochemical gradient). These two qualities give sodium the primary role in controlling ECF volume and water distribution in the body.

Remember: *Water follows salt.* Because all body fluids are in osmotic equilibrium, a change in plasma Na^+ levels affects not only plasma volume and blood pressure, but also ICF and IF volumes. In addition, sodium ions continuously move back and forth between the ECF and body secretions. For example, about 8 L of Na^+-containing secretions (gastric, intestinal, and pancreatic juice, saliva, bile) are spewed into the digestive tract daily, only to be almost completely reabsorbed. Finally, renal acid-base control mechanisms (which we will discuss shortly) are coupled to Na^+ transport.

Sodium Concentration versus Sodium Content

For ions other than Na^+, their concentration in body fluids is the only important variable. For Na^+, however, both the concentration and the total body content (the total amount contained in the body) are important.

- **Concentration of Na^+.** The concentration of Na^+ in the ECF largely determines the osmolality of ECF fluids and influences electrical excitability of neurons and muscles. The concentration of Na^+ normally remains relatively stable because water immediately moves by osmosis into or out of the ICF, counteracting the change in Na^+ concentration. In the long term, the ADH and thirst mechanisms control the ECF Na^+ concentration by controlling water loss or gain.

- **Content of Na^+.** The total amount of Na^+ in the body determines the ECF volume and therefore blood pressure. Na^+ content is regulated by the renin-angiotensin-aldosterone and atrial natriuretic peptide (ANP) hormone mechanisms that control Na^+ reabsorption and excretion (Na^+ balance). We will review these mechanisms in the next section.

It helps to think about these aspects of Na^+ balance as separate and distinct because they have different roles in homeostasis (**Table 26.2**). Nevertheless, there is some overlap between them and their mechanisms are intertwined. For example, as ECF Na^+ content increases, ECF osmolality rises as well, which triggers the ADH and thirst mechanisms. This increases water retention and intake, simultaneously reducing the Na^+ concentration and increasing the ECF volume.

Regulation of Sodium Balance

Despite the crucial importance of sodium, receptors that specifically monitor the concentration or content of Na^+ in body fluids have yet to be found.

- Total body Na^+ content is indirectly monitored by baroreceptors because Na^+ content determines fluid volume and fluid volume determines blood pressure.

- ECF Na^+ concentration, in contrast, is primarily monitored by hypothalamic osmoreceptors.

Changes in these two variables trigger a variety of hormonal and neural controls that regulate Na^+ balance.

Influence of Aldosterone and Angiotensin II

The hormone **aldosterone** "has the most to say" about renal regulation of sodium ions. It is an integral part of the

Table 26.2	Sodium Concentration and Sodium Content	
	ECF Na⁺ CONCENTRATION	**BODY Na⁺ CONTENT**
Homeostatic Importance	ECF osmolality	Blood volume and blood pressure
Sensors	Osmoreceptors	Baroreceptors
Primary Regulators	ADH and thirst mechanisms	Renin-angiotensin-aldosterone and ANP hormone mechanisms*

*ADH and thirst are also required to maintain blood volume and for long-term control of blood pressure.

Figure 26.8 Mechanisms and consequences of aldosterone release.

renin-angiotensin-aldosterone mechanism—the body's main mechanism for increasing blood volume and blood pressure (◀ pp. 722–723).

Whether aldosterone is present or not, some 65% of the Na$^+$ in the renal filtrate is reabsorbed in the proximal tubules of the kidneys and another 25% is reclaimed in the nephron loops (◀ p. 992).

- **When aldosterone concentrations are high.** Essentially *all* the remaining filtered Na$^+$ is actively reabsorbed in the distal convoluted tubules and collecting ducts. Water always follows Na$^+$. In this case, the water that follows Na$^+$ comes from either the intracellular fluid or, if ADH is present, from the filtrate in the collecting ducts. One way or another, aldosterone increases ECF volume.

- **When aldosterone concentrations are low.** Virtually no Na$^+$ reabsorption occurs beyond the distal convoluted tubule.

Urinary excretion of large amounts of Na$^+$ *always* results in the excretion of large amounts of water as well, but the reverse

is *not* true. Substantial amounts of nearly sodium-free urine can be eliminated as needed to achieve water balance.

The most important trigger for aldosterone release from the adrenal cortex is the renin-angiotensin-aldosterone mechanism mediated by the juxtaglomerular complex (JGC) of nephrons (◀ p. 989). In addition, elevated K$^+$ concentration in the ECF can directly stimulate adrenal cortical cells to release aldosterone (**Figure 26.8**). The result of aldosterone release is increased reabsorption of Na$^+$ and increased secretion of K$^+$. Aldosterone brings about its effects slowly, over a period of hours.

Low blood volume and blood pressure (reflecting decreased body Na$^+$ content) trigger *renin* release from the granular cells of the JGC in three ways shown in Figure 26.10:

- Sympathetic stimulation
- Decreased filtrate NaCl concentration
- Decreased stretch of the granular cells of the afferent arterioles

Renin catalyzes the initial step in the reactions that produce angiotensin II. Angiotensin II prods the adrenal cortex to release aldosterone, and also directly increases Na$^+$ reabsorption by kidney tubules. In addition, it has a number of other actions described in Chapter 19 (◀ pp. 722–723) that are all aimed at raising blood volume and blood pressure.

HOMEOSTATIC IMBALANCE 26.2 **CLINICAL**

A deficiency of aldosterone (as in *Addison's disease*; ◀ pp. 625–626) causes enormous losses of water and NaCl in the urine. People with this condition require careful monitoring to avoid critically low blood volume (hypovolemia), especially during illness or other stresses.

Influence of Atrial Natriuretic Peptide (ANP)

We can summarize the influence of **atrial natriuretic peptide (ANP)** in one sentence: It reduces blood pressure and blood volume by inhibiting nearly all events that promote vasoconstriction and Na$^+$ and water retention (◀ p. 624) (**Figure 26.9**). A hormone that is released by certain cells of the heart atria when they are stretched by the effects of elevated blood pressure, ANP has diuretic and natriuretic (salt-excreting) effects. It promotes excretion of Na$^+$ and water by the kidneys by inhibiting the ability of the collecting ducts to reabsorb Na$^+$ and by suppressing the release of ADH, renin, and aldosterone. Additionally, ANP acts both directly and indirectly (by inhibiting renin-induced generation of angiotensin II) to relax vascular smooth muscle, and in this way it causes vasodilation. Collectively, these effects reduce body Na$^+$ content and blood pressure.

Influence of Other Hormones

Female Sex Hormones The **estrogens** are chemically similar to aldosterone and, like aldosterone, enhance NaCl reabsorption by the renal tubules. Because water follows, many women retain fluid as their estrogen levels rise during the menstrual cycle. Estrogens are also largely responsible for the edema experienced by many pregnant women. **Progesterone**, in contrast, has a mild

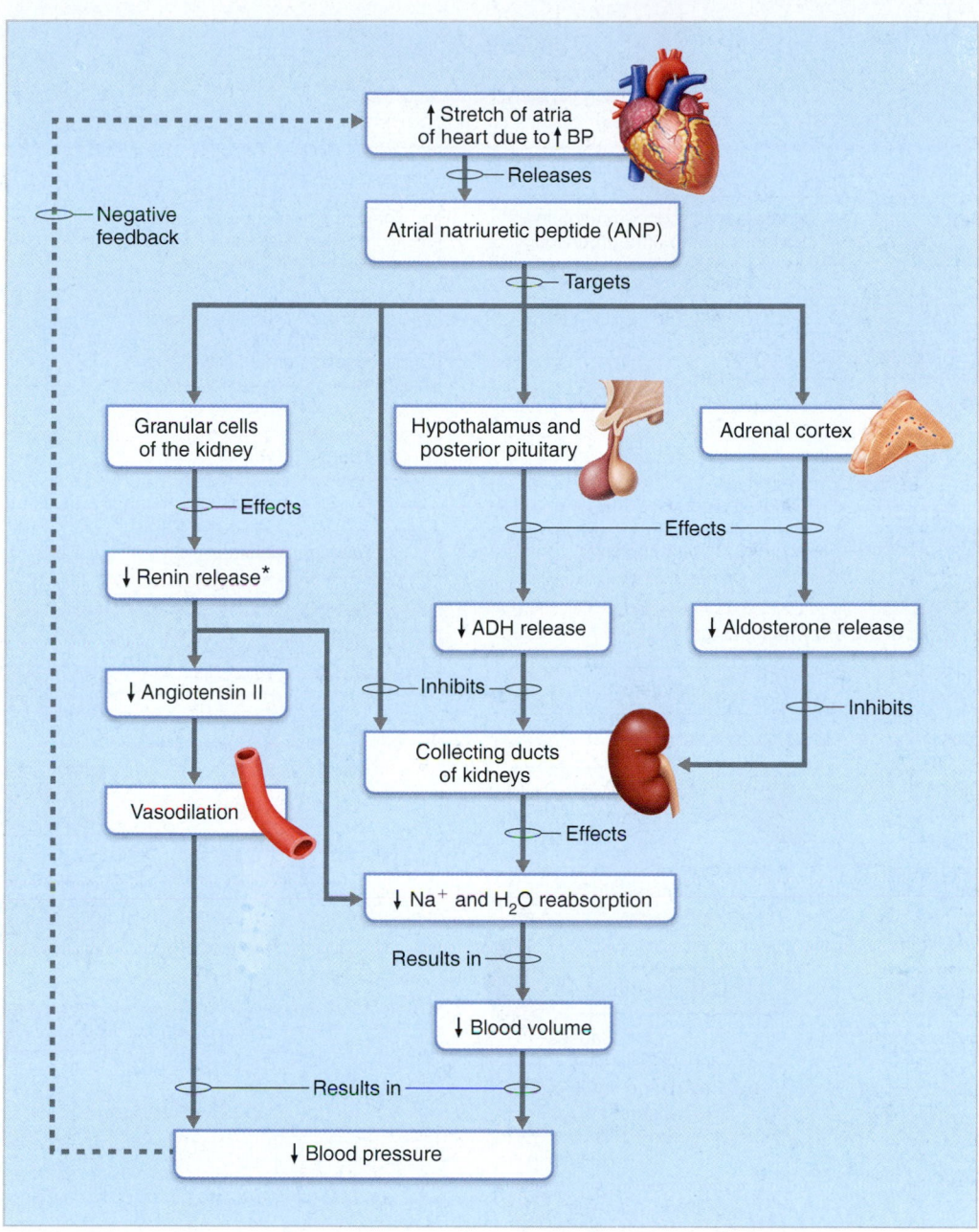

Figure 26.9 Mechanisms and consequences of ANP release.
*↓ Renin release also inhibits ADH and aldosterone release and hence the effects of those hormones.

diuretic effect which is thought to be due to its blocking aldosterone receptors.

Glucocorticoids **Glucocorticoids**, such as cortisol and hydrocortisol, enhance tubular reabsorption of Na^+, but they also promote an increased glomerular filtration rate that may mask their effects on the tubules. However, when their plasma levels are high, the glucocorticoids exhibit potent aldosterone-like effects and promote edema.

Cardiovascular Baroreceptors

Blood volume (and indirectly body Na^+ content) is carefully monitored and regulated to maintain blood pressure and cardiovascular function. As blood volume (and with it, pressure) rises, baroreceptors in the heart and the large vessels of the neck and thorax (carotid arteries and aorta) alert the cardiovascular center in the brain stem. Shortly after, sympathetic nervous system impulses to the kidneys decline, allowing the afferent arterioles to dilate. As the glomerular filtration rate rises, Na^+ output and water output increase. This phenomenon, part of the baroreceptor reflex described in Chapter 19 (◀ **pp. 719–720**), reduces blood volume and blood pressure.

Drops in systemic blood pressure lead to reflex constriction of systemic arterioles including the afferent arterioles, which reduces filtrate formation and urinary output and increases

Figure 26.10 Mechanisms regulating sodium and water balance help maintain blood pressure homeostasis. The thirst mechanism, activated by angiotensin II and other factors, is not shown (see Figure 26.5).

systemic blood pressure (**Figure 26.10**). The baroreceptors provide information on the "fullness" or volume of the circulation that is critical for maintaining cardiovascular homeostasis.

Regulation of Potassium Balance

Potassium, the chief intracellular cation, is required for normal neuromuscular functioning as well as for several essential metabolic activities. Even slight changes in K$^+$ concentration in the ECF have profound and potentially life-threatening effects on neurons and muscle fibers because the relative ICF-ECF potassium concentration directly affects the resting membrane potential of these cells (◀ **pp. 401–403**). K⁺ excess in the ECF decreases their membrane potential, causing depolarization, often followed by reduced excitability. Too little K⁺ in the ECF causes hyperpolarization and nonresponsiveness.

The heart is particularly sensitive to K⁺ levels. Both too much and too little K⁺ (hyperkalemia and hypokalemia, respectively) can disrupt electrical conduction in the heart, leading to sudden death (Table 26.1).

Potassium is also part of the body's buffer system, which resists changes in the pH of body fluids. Shifts of hydrogen ions (H⁺) into and out of cells induce corresponding shifts of K⁺ in the opposite direction to maintain cation balance. Consequently, ECF potassium levels rise with acidosis, as K⁺ leaves and H⁺ enters the cells, and fall with alkalosis, as K⁺ enters the cells and H⁺ leaves them to enter the ECF. Although these pH-driven shifts do not change the total amount of K⁺ in the body, they can seriously interfere with the activity of excitable cells.

Regulatory Site: The DCT and Collecting Duct

Like Na⁺ balance, K⁺ balance is maintained chiefly by renal mechanisms. However, there are important differences in the way this balance is achieved.

- *K⁺ handling.* The proximal tubules reabsorb about 65% of the filtered K⁺, and the thick ascending limb of the nephron loop absorbs another 25% or so regardless of need, leaving about 10% at the end of the nephron loop. The responsibility for K⁺ balance falls chiefly on the distal convoluted tubule (DCT) and collecting ducts. They achieve this balance mainly by changing the amount of K⁺ *secreted* into the filtrate. Notice that nephrons first reabsorb more K⁺ than the body needs, and then use secretion of K⁺ to fine-tune its balance.

- *Na⁺ handling.* In contrast, recall that the amount of Na⁺ reabsorbed in the tubules is precisely tailored to need, and Na⁺ is *never* secreted into the filtrate.

As a rule, K⁺ levels in the ECF are sufficiently high that K⁺ needs to be excreted, and the DCT and cortical collecting ducts secrete K⁺ into the filtrate. (At times, the amount of K⁺ excreted may actually exceed the amount filtered.) The cells responsible for K⁺ secretion are the principal cells we described in Chapter 25 (◀ **p. 981**). Note that these principal cells are the same cells that mediate aldosterone-induced reabsorption of Na⁺ and ADH-stimulated reabsorption of water.

When ECF potassium concentrations are abnormally low, the renal principal cells conserve K⁺ by reducing its secretion (and therefore excretion) to a minimum. Additionally, *type A intercalated cells*, a unique population of collecting duct cells, can reabsorb some of the K⁺ left in the filtrate (in conjunction with active secretion of H⁺), thereby helping to reestablish K⁺ (and pH) balance.

However, keep in mind that the main thrust of renal regulation of K⁺ is to *excrete* it. Because the kidneys have a limited ability to retain K⁺, it may be lost in urine even in the face of a deficiency. Consequently, people who don't eat potassium-rich foods can eventually develop a severe deficiency.

Influence of Plasma Potassium Concentration

The single most important factor influencing K⁺ secretion is the K⁺ concentration in the ECF. A high-potassium diet increases the K⁺ content of the ECF. This favors entry of K⁺ into the principal cells and prompts them to secrete K⁺ into the filtrate so that more of it is excreted. Conversely, a low-potassium diet or accelerated K⁺ loss depresses its secretion (and promotes its limited reabsorption).

Influence of Aldosterone

The second factor influencing K⁺ secretion into the filtrate is aldosterone. As it stimulates the principal cells to reabsorb Na⁺, aldosterone simultaneously enhances K⁺ secretion (see Figure 26.8). Adrenal cortical cells are *directly* sensitive to the K⁺ content of the ECF bathing them (◀ **p. 624**). When it increases even slightly, the adrenal cortex is strongly stimulated to release aldosterone, which increases K⁺ secretion. The result is that K⁺ controls its own concentrations in the ECF via feedback regulation of aldosterone release.

Aldosterone is also secreted in response to the renin-angiotensin-aldosterone mechanism previously described. Given the opposing effects of aldosterone on plasma Na⁺ and K⁺, you might expect that Na⁺- and volume-driven changes in aldosterone would disrupt K⁺ balance. This generally does not occur because other compensatory mechanisms in the kidneys maintain plasma K⁺.

HOMEOSTATIC IMBALANCE 26.3 CLINICAL

To reduce their NaCl intake, many people have turned to salt substitutes, which are high in potassium. However, heavy consumption of these substitutes is safe only when aldosterone release in the body is normal. In the absence of aldosterone, hyperkalemia is swift and lethal regardless of K⁺ intake (Table 26.1). Conversely, when a person has an adrenocortical tumor that pumps out tremendous amounts of aldosterone, ECF potassium levels fall so low that neurons all over the body hyperpolarize and paralysis occurs. ●●●●●●

Regulation of Calcium and Phosphate Balance

About 99% of the body's calcium is found in bones in the form of calcium phosphate salts, which make the skeleton rigid and strong. The bony skeleton provides a dynamic reservoir from which calcium and phosphate can be withdrawn or deposited to maintain the balance of these electrolytes in the ECF (◀ **p. 189**).

HOMEOSTATIC IMBALANCE 26.4 CLINICAL

Ionic calcium in the ECF is important for normal blood clotting, cell membrane permeability, and secretory behavior, but its most important effect by far is on neuromuscular excitability. *Hypocalcemia* increases excitability and causes muscle tetany. *Hypercalcemia* is equally dangerous because it inhibits neurons and muscle cells and may cause life-threatening cardiac arrhythmias (Table 26.1).

ECF calcium ion levels are closely regulated by **parathyroid hormone (PTH)** and rarely deviate from normal limits. [The hormone calcitonin, produced by the thyroid, is often thought of as a calcium-lowering hormone, but, as we discussed in Chapter 6 (◄ **p. 189**) its effects on blood calcium levels in humans are negligible.] Parathyroid hormone is released by the tiny parathyroid glands located on the posterior aspect of the thyroid gland in the neck. Declining plasma levels of Ca^{2+} directly stimulate the parathyroid glands to release PTH, which promotes an increase in calcium levels by targeting the following organs (see also Figure 16.12, ◄ **p. 622**):

- **Bones.** PTH activates bone-digesting osteoclasts, which break down the bone matrix, releasing Ca^{2+} and HPO_4^{2-} to the blood.

- **Kidneys.** PTH increases Ca^{2+} reabsorption by the renal tubules while decreasing phosphate ion reabsorption. In this way, calcium conservation and phosphate excretion go hand in hand. The *mathematical product* of Ca^{2+} and HPO_4^{2-} concentrations ($[Ca^{2+}] \times [HPO_4^{2-}]$) in the ECF remains constant, preventing calcium-salt deposits in bones or soft body tissues.

- **Small intestine.** PTH enhances intestinal absorption of Ca^{2+} indirectly by stimulating the kidneys to transform vitamin D to its active form, which is necessary for the small intestine to absorb Ca^{2+}.

Most Ca^{2+} is reabsorbed passively in the proximal convoluted tubule (PCT) via diffusion through the paracellular route (◄ **p. 990**) (a process driven by its electrochemical gradient). However, as with other ions, Ca^{2+} reabsorption is fine-tuned in the distal nephron. PTH-regulated Ca^{2+} channels control Ca^{2+} entry into DCT cells at the apical membrane, while Ca^{2+} pumps and antiporters export it at the basolateral membrane. Under normal circumstances about 98% of the filtered Ca^{2+} is reabsorbed owing to the action of PTH.

As a rule, 75% of the filtered phosphate ions (including $H_2PO_4^-$, HPO_4^{2-}, and PO_4^{3-}) are reabsorbed in the PCT by secondary active transport. Phosphate reabsorption is set by its transport maximum. Amounts over that maximum simply flow out in urine. PTH inhibits active transport of phosphate by decreasing its transport maximum.

When ECF calcium levels are within normal limits (9–11 mg/100 ml of blood) or higher, PTH secretion is inhibited. Consequently, release of Ca^{2+} from bone is inhibited, more Ca^{2+} is lost in feces and urine, and more phosphate is retained.

Regulation of Anions

Chloride is the major anion accompanying Na^+ in the ECF and, like sodium, Cl^- helps maintain the osmotic pressure of the blood. When blood pH is within normal limits or slightly alkaline, about 99% of filtered Cl^- is reabsorbed. In the PCT,

it moves passively and simply follows sodium ions out of the filtrate and into the peritubular capillary blood. In most other tubule segments, Na^+ and Cl^- transport are coupled.

When acidosis occurs, less Cl^- accompanies Na^+ because HCO_3^- reabsorption is stepped up to restore blood pH to its normal range. Thus, the choice between Cl^- and HCO_3^- serves acid-base regulation.

Most other anions, such as sulfates and nitrates, have transport maximums, and when their concentrations in the filtrate exceed the amount that can be reabsorbed, the excess spills into urine.

Check Your Understanding

7. In the drawing below, which arrow shows where (and in what direction) the largest fraction of Ca^{2+} moves? Which arrow represents hormonally regulated movement of Na^+? Which arrow represents hormonally regulated movement of K^+?

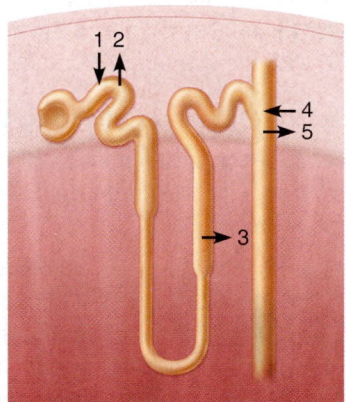

8. Which hormone is the major regulator of Ca^{2+} in the blood? What are the effects of hypercalcemia? Hypocalcemia?

9. **APPLY** Nathan has Addison's disease (insufficient aldosterone release). How does this affect his plasma Na^+ and K^+ levels? How does this affect his blood pressure? Explain.

For answers, see Answers Appendix.

26.4 Chemical buffers and respiratory regulation rapidly minimize pH changes

Learning Outcomes

▶ List important sources of acids in the body.
▶ List the three major chemical buffer systems of the body and describe how they resist pH changes.
▶ Describe the influence of the respiratory system on acid-base balance.

Because of their abundant hydrogen bonds, all proteins are influenced by H^+ concentration, and because many proteins are enzymes, nearly all biochemical reactions are influenced by the pH of their fluid environment. Consequently, the **acid-base balance** of body fluids is closely regulated (see ◄ **pp. 39–41** for a review of the basic principles).

Optimal pH varies from one body fluid to another, but not by much. The normal pH of arterial blood is 7.4, that of venous blood and IF is 7.35, and that of ICF averages 7.0. The lower pH in cells and venous blood reflects their greater amounts of acidic metabolites and of carbon dioxide, which combines with water to form carbonic acid, H_2CO_3.

Whenever the pH of arterial blood rises above 7.45, a person is said to have **alkalosis** (al″kah-lo′sis) or **alkalemia**. A drop in arterial pH below 7.35 results in **acidosis** (as″ĭ-do′sis) or **acidemia**. Because pH 7.0 is neutral, chemically speaking 7.35 is not acidic. However, it is a higher-than-optimal H^+ concentration for most cells, so any arterial pH between 7.0 and 7.35 is called *physiological acidosis*.

Although small amounts of acidic substances enter the body via ingested foods, most hydrogen ions originate as metabolic by-products or end products. For example:

- Breakdown of phosphorus-containing proteins releases *phosphoric acid* into the ECF.
- Anaerobic respiration of glucose produces *lactic acid*.
- Fat metabolism yields other organic acids, such as fatty acids and *ketone bodies*.
- Loading and transport of carbon dioxide in the blood as HCO_3^- liberates hydrogen ions.

The H^+ concentration in blood is regulated sequentially by:

1. *Chemical buffers.* Chemical buffers are the first line of defense and act within a fraction of a second to resist pH changes.
2. *Brain stem respiratory centers.* Changes in respiratory rate and depth occur to compensate for acidosis or alkalosis within 1–3 minutes.
3. *Renal mechanisms.* The kidneys are the body's most potent acid-base regulatory system, but ordinarily require hours to a day or more to alter blood pH.

The respiratory and renal systems together form the *physiological buffering systems* that control pH by regulating the amount of acid or base in the body. Although physiological buffer systems act more slowly than chemical buffer systems, they have many times the buffering power of all the body's chemical buffers combined.

Chemical Buffer Systems

Recall that acids are *proton donors*, and that the acidity of a solution reflects only the *free* hydrogen ions, not those bound to anions. *Strong acids* dissociate completely and liberate all their H^+ in water (**Figure 26.11a**). They can dramatically change a solution's pH. By contrast, *weak acids* dissociate only partially (Figure 26.11b). Accordingly, they have a much smaller effect on pH. However, weak acids are efficient at preventing pH changes, and this feature allows them to play important roles in chemical buffer systems.

Bases are *proton acceptors*. Strong bases are those that dissociate easily in water and quickly tie up H^+. Conversely, weak bases are less likely to accept protons.

A **chemical buffer** is a system of one or more compounds that resists changes in pH when a strong acid or base is added. They do this by binding H^+ ions whenever the pH drops and releasing them when pH rises.

(a) A strong acid such as HCl dissociates completely into its ions.

(b) A weak acid such as H_2CO_3 does *not* dissociate completely.

Figure 26.11 Dissociation of strong and weak acids in water. HCl = hydrochloric acid; H_2CO_3 = carbonic acid.

The three major chemical buffer systems in the body are the *bicarbonate*, *phosphate*, and *protein buffer systems*. Anything that causes a shift in H^+ concentration in one fluid compartment simultaneously causes a change in the others. As a result, the buffer systems actually buffer one another, so the *entire* buffer system resists any drifts in pH.

Bicarbonate Buffer System

The **bicarbonate buffer system** is a mixture of carbonic acid (H_2CO_3) and its salt, sodium bicarbonate ($NaHCO_3$, a weak base), in the same solution. Although it also buffers the ICF, it is the *only* important ECF buffer.

Carbonic acid, a weak acid, does not dissociate to any great extent in neutral or acidic solutions. When a strong acid such as HCl is added to this buffer system, the existing carbonic acid remains intact. However, the bicarbonate ions of the salt act as weak bases to tie up the H^+ released by the stronger acid (HCl), forming *more* carbonic acid:

$$HCl + NaHCO_3 \rightarrow H_2CO_3 + NaCl$$

strong acid   weak base   weak acid   salt

Because it is converted to the weak acid H_2CO_3, HCl lowers the pH of the solution only slightly.

When a strong base such as sodium hydroxide (NaOH) is added to the same buffer solution, a weak base such as sodium bicarbonate ($NaHCO_3$) does not dissociate further under the alkaline conditions. However, the added base forces the carbonic acid to dissociate further, donating more H^+ to tie up the OH^- released by the strong base:

$$NaOH + H_2CO_3 \rightarrow NaHCO_3 + H_2O$$

strong base   weak acid   weak base   water

The net result is that a weak base ($NaHCO_3$) replaces a strong base (NaOH), so the pH of the solution rises very little.

Although the bicarbonate salt in the example is sodium bicarbonate, other bicarbonate salts function in the same way because HCO_3^- is the important ion, not the cation it is paired with. In cells, where little Na^+ is present, potassium and magnesium bicarbonates are part of the bicarbonate buffer system.

The buffering power of this type of system is directly related to the concentrations of the buffering substances. If enough acid enters the blood so that all the available HCO_3^- ions, often referred to as the **alkaline reserve**, are tied up, the buffer system becomes ineffective and blood pH changes.

The bicarbonate ion concentration in the ECF is normally around 25 mEq/L and is closely regulated by the kidneys. The concentration of H_2CO_3 is just over 1 mEq/L but the supply of H_2CO_3 (which comes from the CO_2 released during cellular respiration) is almost limitless, so obtaining that member of the buffer pair is usually not a problem. The H_2CO_3 content of the blood is subject to respiratory controls.

Phosphate Buffer System

The operation of the **phosphate buffer system** is nearly identical to that of the bicarbonate buffer. The components of the phosphate system are the salts of dihydrogen phosphate ($H_2PO_4^-$) and monohydrogen phosphate (HPO_4^{2-}). For example NaH_2PO_4 acts as a weak acid. Na_2HPO_4, with one less hydrogen atom, acts as a weak base.

Again, H^+ released by strong acids is tied up in weak acids:

$$\underset{\text{strong acid}}{HCl} + \underset{\text{weak base}}{Na_2HPO_4} \rightarrow \underset{\text{weak acid}}{NaH_2PO_4} + \underset{\text{salt}}{NaCl}$$

and strong bases are converted to weak bases:

$$\underset{\text{strong base}}{NaOH} + \underset{\text{weak acid}}{NaH_2PO_4} \rightarrow \underset{\text{weak base}}{Na_2HPO_4} + \underset{\text{water}}{H_2O}$$

Because the phosphate buffer system is present in low concentrations in the ECF (approximately one-sixth that of the bicarbonate buffer system), it is relatively unimportant for buffering blood plasma. However, it is a very effective buffer in urine and in ICF, where phosphate concentrations are usually higher.

Protein Buffer System

Proteins in plasma and in cells are the body's **protein buffer system**. In fact, at least three-quarters of all the buffering power of body fluids resides in cells, and most of this reflects the powerful buffering activity of intracellular proteins.

As described in Chapter 2, proteins are polymers of amino acids. Some of the linked amino acids have exposed side chains of atoms called *carboxyl groups* (—COOH), organic acid groups that release H^+ when the pH begins to rise:

$$R-COOH \rightarrow R-COO^- + H^+$$

(Note that R indicates the rest of the organic molecule, which contains many atoms.)

Other amino acids have exposed groups that can act as bases and accept H^+. For example, an exposed —NH_2 group can bind with a hydrogen ion, becoming —NH_3^+:

$$R-NH_2 + H^+ \rightarrow R-NH_3^+$$

Bicarbonate Buffer System
- Mixture of H_2CO_3 (weak acid) and salts of HCO_3^- (weak base)
- Main ECF buffer; also operates in ICF

Phosphate Buffer System
- Salts of $H_2PO_4^-$ (weak acid) and HPO_4^{2-} (weak base)
- Important buffer in urine and ICF

Protein Buffer System
- Some amino acid side chains can act as weak acids (—COOH) or weak bases (e.g.,—NH_2)
- Most important buffer in ICF; also in blood plasma

Figure 26.12 Quick summary of chemical buffer systems.

Because this binding removes free hydrogen ions from the solution, it prevents the solution from becoming too acidic. Consequently, a single protein molecule can function reversibly as either an acid or a base depending on the pH of its environment. Molecules with this ability are called **amphoteric molecules** (am″fo-ter′ik).

Hemoglobin in red blood cells is an excellent example of an intracellular protein buffer. As we explained in Chapter 22 (◄ **pp. 850–851**), CO_2 released from the tissues combines with water to form carbonic acid (H_2CO_3), which dissociates to liberate H^+ and HCO_3^- in the blood. Meanwhile, hemoglobin is unloading oxygen, becoming reduced hemoglobin, which carries a negative charge. Because H^+ rapidly binds to the hemoglobin anions, pH changes are minimized. In this case, carbonic acid, a weak acid, is buffered by an even weaker acid, hemoglobin.

The three main chemical buffers are summarized in **Figure 26.12**.

Respiratory Regulation of H^+

As we described in Chapter 22, the respiratory system eliminates CO_2, an acid, from the blood while replenishing its supply of O_2. Carbon dioxide generated by cellular respiration enters erythrocytes in the circulation and is converted to bicarbonate ions for transport in the plasma (◄ **p. 850**):

$$CO_2 + H_2O \underset{}{\overset{\text{carbonic anhydrase}}{\rightleftharpoons}} \underset{\text{carbonic acid}}{H_2CO_3} \rightleftharpoons H^+ + \underset{\text{bicarbonate ion}}{HCO_3^-}$$

The first set of double arrows indicates a reversible equilibrium between dissolved carbon dioxide and water on the left and carbonic acid on the right. The second set indicates a reversible equilibrium between carbonic acid on the left and hydrogen and bicarbonate ions on the right. Because of these equilibria, an increase in any of these chemical species pushes the reaction in the opposite direction. Notice also that the right side of the equation is equivalent to the bicarbonate buffer system.

Healthy individuals expel CO_2 from the lungs at the same rate it is formed in the tissues. During carbon dioxide unloading, the reaction shifts to the left, and H^+ generated from carbonic acid is reincorporated into water. Because of the protein buffer system, H^+ produced by CO_2 transport is not allowed to accumulate and has little or no effect on blood pH.

If P_{CO_2} (the partial pressure, or level, of CO_2 in the blood) rises, it activates medullary chemoreceptors (via cerebral acidosis promoted by excessive accumulation of CO_2). As a result, respiratory rate and depth increase (◀ **Figure 22.28, p. 855**). Additionally, a rising plasma H^+ concentration resulting from any metabolic process excites the respiratory centers indirectly (via peripheral chemoreceptors) to stimulate deeper, more rapid respiration. As ventilation increases, more CO_2 is removed from the blood, pushing the reaction to the left and reducing the H^+ concentration.

When blood pH rises, the respiratory centers are depressed. As respiratory rate drops and respiration becomes shallower, CO_2 accumulates, pushing the equilibrium to the right and causing the H^+ concentration to increase. Again blood pH is restored to the normal range. These respiratory system–mediated corrections of blood pH are accomplished within a minute or so.

Changes in alveolar ventilation can produce dramatic changes in blood pH—far more than is needed. For example, doubling alveolar ventilation can raise blood pH by about 0.2 pH unit. Likewise, cutting alveolar ventilation in half can lower blood pH by the same amount. Because normal arterial pH is 7.4, a change of 0.2 pH unit yields a blood pH of 7.6 or 7.2—both well beyond the normal limits. Respiratory controls of blood pH have a tremendous reserve capacity because alveolar ventilation can rise about 15-fold or fall (briefly) to zero.

Anything that impairs respiratory system functioning causes acid-base imbalances. For example, net carbon dioxide retention (hypoventilation) leads to acidosis. On the other hand, hyperventilation, which causes net elimination of CO_2, causes alkalosis. When respiratory system problems cause the pH imbalance, the resulting condition is either *respiratory acidosis* or *respiratory alkalosis* (see Table 26.3 on p. 1033).

Check Your Understanding

10. Define acidemia and alkalemia.

11. What are the body's three major chemical buffer systems? What is the most important buffer inside cells?

12. APPLY Joanne, a diabetic patient, is at the emergency department with acidosis due to the production of ketone bodies (◀ **p. 632**). Would you expect her ventilation to be increased or decreased? Why?

For answers, see Answers Appendix.

26.5 Renal regulation is a long-term mechanism for controlling acid-base balance

Learning Outcome

▶ Describe how the kidneys regulate hydrogen and bicarbonate ion concentrations in the blood.

The ultimate acid-base regulatory organs are the kidneys, which act slowly but surely to compensate for acid-base imbalances resulting from variations in diet or metabolism, or disease. Chemical buffers can tie up excess acids or bases temporarily, but they cannot eliminate them from the body. And while the lungs can dispose of the **volatile acid** carbonic acid by eliminating CO_2, only the kidneys can rid the body of other acids generated by cellular metabolism: phosphoric, uric, and lactic acids, and ketone bodies. These acids are referred to as **nonvolatile (fixed) acids**. Additionally, only the kidneys can regulate blood levels of alkaline substances and renew chemical buffers that are used up in regulating H^+ levels in the ECF.

The kidneys regulate acid-base balance by adjusting the amount of bicarbonate in the blood. Depending on the blood's pH, the kidneys do one of the following:

- Conserve (reabsorb) or generate new HCO_3^-
- Excrete HCO_3^-

Figure legend:
- → Primary active transport
- ⇢ Secondary active transport
- → Simple diffusion
- Transport protein
- CA Carbonic anhydrase

① CO_2 combines with water within the tubule cell, forming H_2CO_3.

② H_2CO_3 is quickly split, forming H^+ and bicarbonate ion (HCO_3^-).

③a H^+ is secreted into the filtrate.

③b For each H^+ secreted, a HCO_3^- enters the peritubular capillary blood either via symport with Na^+ or via antiport with Cl^-.

④ Secreted H^+ combines with HCO_3^- in the filtrate, forming carbonic acid (H_2CO_3). HCO_3^- disappears from the filtrate at the same rate that HCO_3^- (formed within the tubule cell) enters the peritubular capillary blood.

⑤ The H_2CO_3 formed in the filtrate dissociates to release CO_2 and H_2O.

⑥ CO_2 diffuses into the tubule cell, where it triggers further H^+ secretion.

Figure 26.13 Reabsorption of filtered HCO_3^- is coupled to H^+ secretion.

*The breakdown of H_2CO_3 to CO_2 and H_2O in the tubule lumen is catalyzed by carbonic anhydrase only in the proximal convoluted tubule (PCT).

If we look back at the equation on p. 1028 for the CO_2–carbonic acid–bicarbonate buffer system, we can see that losing a HCO_3^- from the body produces the same net effect as gaining a H^+, because it pushes the equation to the right, increasing the H^+ level. By the same token, generating or reabsorbing HCO_3^- is the same as losing H^+ because it pushes the equation to the left, decreasing the H^+ level. For this reason, to reabsorb bicarbonate, the kidney has to secrete H^+, and when it excretes excess HCO_3^-, H^+ is retained (not secreted).

Because the mechanisms for regulating acid-base balance depend on H^+ being secreted into the filtrate, we consider that process first. Secretion of H^+ occurs mainly in the PCT and in type A intercalated cells of the collecting duct. The H^+ secreted comes from the dissociation of carbonic acid, which is created when CO_2 and water combine within the tubule cells in a reaction catalyzed by *carbonic anhydrase* (**Figure 26.13** ①, ②). As H^+ is secreted into the lumen of the PCT, Na^+ is reabsorbed from the filtrate, maintaining the electrical balance (Figure 26.13 ③a).

The rate of H^+ secretion rises and falls with CO_2 levels in the ECF. The more CO_2 in the peritubular capillary blood, the faster the rate of H^+ secretion. Because blood CO_2 levels directly relate to blood pH, this system can respond to both rising and falling H^+ concentrations. Notice that secreted H^+ can combine with HCO_3^- in the filtrate, generating CO_2 and water (Figure 26.13 ④, ⑤). In this case, H^+ is bound in water. The rising concentration of CO_2 in the filtrate creates a steep diffusion gradient for its entry into the tubule cell (Figure 26.13 ⑥), where it promotes still more H^+ secretion (Figure 26.13 ③a).

Conserving Filtered Bicarbonate Ions: Bicarbonate Reabsorption

Bicarbonate (HCO_3^-; a base) is an important part of the bicarbonate buffer system, a critical blood buffer. If this reservoir of base, the *alkaline reserve*, is to be maintained, the kidneys must do more than just eliminate enough hydrogen ions to counter rising blood H^+ levels. Depleted stores of HCO_3^- have to be replenished. This task is more complex than it seems because the apical membranes of the tubule cells have no transporters for HCO_3^- and so HCO_3^- cannot cross these membranes. As a result, tubule cells cannot reabsorb HCO_3^- directly.

However, the kidneys can conserve filtered HCO_3^- in a rather roundabout way. As you can see, dissociation of carbonic acid liberates HCO_3^- as well as H^+ (Figure 26.13 ②). Although the tubule cells cannot reclaim HCO_3^- directly from the filtrate, they can and do shunt HCO_3^- generated within them (as a result of splitting H_2CO_3) into the peritubular capillary blood. HCO_3^- leaves the tubule cell either accompanied by Na^+ or in exchange for Cl^- (Figure 26.13 ③b). H^+ is actively secreted, mostly by a Na^+-H^+ antiporter, but also by a H^+ ATPase (Figure 26.13 ③a). In the filtrate, H^+ combines with filtered HCO_3^- (Figure 26.13 ④, ⑤). For this reason, reabsorption of HCO_3^- depends on the active secretion of H^+.

In short, for each filtered HCO_3^- that "disappears" from the filtrate, a HCO_3^- generated within the tubule cells enters the blood—a one-for-one exchange. When large amounts of H^+ are secreted, correspondingly large amounts of HCO_3^- enter the

Filtrate in tubule lumen

Tight junction

Peritubular capillary

$H_2O + CO_2$

(1) CA

H_2CO_3

(2)

HPO_4^{2-}

(3a)   (3b)

H^+ $H^+ + HCO_3^-$ HCO_3^- (new)

ATPase

Cl^- Cl^-

(4)   Type A intercalated cell of collecting duct Cl^-

$H_2PO_4^-$

(5)

out in urine

(1) CO_2 combines with water within the type A intercalated cell, forming H_2CO_3.

(2) H_2CO_3 is quickly split, forming H^+ and bicarbonate ion (HCO_3^-).

(3a) H^+ is secreted into the filtrate by a H^+ ATPase (pump).

(3b) For each H^+ secreted, a HCO_3^- enters the peritubular capillary blood via an antiport carrier in a HCO_3^--Cl^- exchange process.

(4) Secreted H^+ combines with HPO_4^{2-} in the tubular filtrate, forming $H_2PO_4^-$.

(5) The $H_2PO_4^-$ is excreted in the urine.

→ Primary active transport
┈→ Secondary active transport
→ Simple diffusion
┈→ Facilitated diffusion

⬭ Transport protein
⬓ Ion channel
CA Carbonic anhydrase

Figure 26.14 New HCO_3^- is generated via buffering of secreted H^+ by HPO_4^{2-} (monohydrogen phosphate).

peritubular blood. The net effect is to remove HCO_3^- almost completely from the filtrate.

Generating New Bicarbonate Ions

Two renal mechanisms commonly carried out by cells of the PCT and collecting ducts generate *new* (as opposed to filtered) HCO_3^- that can be added to plasma. Both mechanisms involve renal excretion of acid, *via secretion and excretion* of either H^+ or ammonium ions in urine. Let's examine how these mechanisms differ.

As long as *filtered bicarbonate* is reclaimed, as we saw in Figure 26.13, the secreted H^+ is *not excreted or lost* from the body in urine. Instead, the H^+ is buffered by HCO_3^- in the filtrate and ultimately becomes part of water molecules (most of which are reabsorbed) (Figure 26.13 (4), (5)).

However, once the filtered HCO_3^- is "used up" (usually by the time the filtrate reaches the collecting ducts), any additional H^+ secreted is excreted in urine. More often than not, this is the case.

Reclaiming filtered HCO_3^- simply restores the bicarbonate concentration of plasma that exists at the time. However, metabolism of food normally releases new H^+ into the body. This additional H^+ uses up HCO_3^- and so must be balanced by generating *new* HCO_3^- that moves into the blood to counteract acidosis. This process of alkalinizing the blood is the way the kidneys compensate for acidosis.

Via Excretion of Buffered H^+

Binding H^+ to buffers in the filtrate minimizes the H^+ concentration gradient, allowing the proton pumps of the type A intercalated cells to secrete the large numbers of H^+ that the body must get rid of to prevent acidosis. (H^+ secretion ceases when urine pH falls to 4.5 because the proton pumps cannot pump against this large gradient.) The most important urine buffer is the *phosphate buffer system*, specifically its weak base *monohydrogen phosphate* (HPO_4^{2-}).

The components of the phosphate buffer system filter freely into the tubules, and about 75% of the filtered phosphate is reabsorbed. However, their reabsorption is inhibited during acidosis. As a result, the buffer pair becomes more and more concentrated as the filtrate moves through the renal tubules.

As shown in **Figure 26.14** (3a), the type A intercalated cells secrete H^+ actively via a H^+ ATPase pump and via a K^+-H^+ antiporter (not illustrated). The secreted H^+ combines with HPO_4^{2-}, forming $H_2PO_4^-$ which then flows out in urine (Figure 26.14 (4) and (5)).

Bicarbonate ions generated in the cells during the same reaction move into the interstitial space via a HCO_3^--Cl^- antiport process and then move passively into the peritubular capillary blood (Figure 26.14 (3b)). Notice again that when H^+ is being excreted, "brand new" bicarbonate ions are added to the blood—over and above those reclaimed from the filtrate. As you can see, in response to acidosis, the kidneys generate new HCO_3^- and add it to the blood (alkalinizing the blood) while adding an equal amount of H^+ to the filtrate (acidifying the urine).

① PCT cells metabolize glutamine to NH_4^+ and HCO_3^-.

②a This weak acid NH_4^+ (ammonium) is secreted into the filtrate, taking the place of H^+ on a Na^+-H^+ antiport carrier.

②b For each NH_4^+ secreted, a bicarbonate ion (HCO_3^-) enters the peritubular capillary blood via a symport carrier.

③ The NH_4^+ is excreted in the urine.

→ Primary active transport
▪▪▸ Secondary active transport
→ Simple diffusion
● Transport protein

Figure 26.15 New HCO_3^- is generated via glutamine metabolism and NH_4^+ secretion.

Via NH_4^+ Excretion

The second and more important mechanism for excreting acid uses the ammonium ion (NH_4^+) produced by glutamine metabolism in PCT cells. Ammonium ions are weak acids that donate few H^+ at physiological pH.

As **Figure 26.15** step ① shows, for each glutamine metabolized (deaminated, oxidized, and acidified by combining with H^+), two NH_4^+ and two HCO_3^- result. The HCO_3^- moves through the basolateral membrane into the blood (Figure 26.15 ②b). The NH_4^+, in turn, is excreted and lost in urine (Figure 26.15 ②a, ③). As with the phosphate buffer system, this buffering mechanism replenishes the alkaline reserve of the blood, because the newly made HCO_3^- enters the blood as NH_4^+ is secreted.

Bicarbonate Ion Secretion

When the body is in alkalosis, another population of intercalated cells (type B) in the collecting ducts exhibit net HCO_3^- *secretion* while reclaiming H^+ to acidify the blood. Overall we can think of the type B cells as "flipped" type A cells, and we can visualize the HCO_3^- secretion process as the exact opposite of the HCO_3^- reabsorption process illustrated in Figure 26.13. However, the predominant process in the nephrons and collecting ducts is HCO_3^- reabsorption, and even during alkalosis, much more HCO_3^- is conserved than excreted.

Check Your Understanding

13. What is the most important urinary buffer of H^+?

14. List the two mechanisms by which tubule and collecting duct cells generate new HCO_3^-.

15. **DRAW** Using arrows and the appropriate equations, show how bicarbonate is reabsorbed from the tubular lumen. You do not need to include Na^+, K^+, Cl^-, or transport proteins. Remember that there are no transport proteins for moving bicarbonate into the tubule cell.

HCO_3^-

16. **MAKE CONNECTIONS** Renal tubule cells acidify the urine and parietal cells acidify the stomach contents (see Chapter 23, ◀ **p. 891**). In each case (a) which intracellular enzyme is key, and (b) blood concentration of which ion increases?

For answers, see Answers Appendix.

26.6 Abnormalities of acid-base balance are classified as metabolic or respiratory

CLINICAL

Learning Outcomes

▶ Distinguish between acidosis and alkalosis resulting from respiratory and metabolic factors.

▶ Describe the importance of respiratory and renal compensations in maintaining acid-base balance.

All cases of acidosis and alkalosis can be classed according to cause as *respiratory* or *metabolic* (**Table 26.3**). **A Closer Look** on p. 1034 discusses methods for determining the cause of an acid-base disturbance and whether it is being compensated (whether the lungs or kidneys are taking steps to correct the imbalance).

Respiratory Acidosis and Alkalosis

Respiratory pH imbalances result from some failure of the respiratory system to perform its normal pH-balancing role. The partial pressure of carbon dioxide (P_{CO_2}) in the arteries is the single most important indicator of the adequacy of respiratory function. When respiratory function is normal, the P_{CO_2} fluctuates between 35 and 45 mm Hg. Generally speaking, values above 45 mm Hg indicate respiratory acidosis, and values below 35 mm Hg signal respiratory alkalosis.

Respiratory acidosis is a common cause of acid-base imbalance. It most often occurs when a person breathes shallowly or

Table 26.3	Causes and Consequences of Acid-Base Imbalances
CONDITION AND HALLMARK	**POSSIBLE CAUSES; COMMENTS**
Respiratory Acidosis (Hypoventilation)	
If uncompensated (uncorrected): P_{CO_2} >45 mm Hg; pH <7.35	**Impaired lung function** (e.g., chronic bronchitis, cystic fibrosis, emphysema): impaired gas exchange or alveolar ventilation
	Impaired ventilatory movement: paralyzed respiratory muscles, chest injury, extreme obesity
	Narcotic or barbiturate overdose or injury to brain stem: depression of respiratory centers, resulting in hypoventilation and respiratory arrest
Respiratory Alkalosis (Hyperventilation)	
If uncompensated: P_{CO_2} <35 mm Hg; pH >7.45	**Strong emotions:** pain, anxiety, fear, panic attack
	Hypoxemia: asthma, pneumonia, high altitude; represents effort to raise P_{O_2} at the expense of excessive CO_2 excretion
	Brain tumor or injury: abnormal respiratory controls
Metabolic Acidosis	
If uncompensated: HCO_3^- <22 mEq/L; pH <7.35	**Severe diarrhea:** bicarbonate-rich intestinal (and pancreatic) secretions rushed through digestive tract before their solutes can be reabsorbed
	Renal disease: failure of kidneys to rid body of acids formed by normal metabolic processes
	Untreated diabetes mellitus: lack of insulin or inability of tissue cells to respond to insulin, resulting in inability to use glucose; fats are used as primary energy fuel, and ketoacidosis occurs
	Starvation: lack of dietary nutrients for cellular fuels; body proteins and fat reserves are used for energy—both yield acidic metabolites as they are broken down for energy
	Excess alcohol ingestion: results in excess acids in blood
Metabolic Alkalosis	
If uncompensated: HCO_3^- >26 mEq/L; pH >7.45	**Vomiting or gastric suctioning:** loss of stomach HCl requires that H^+ be withdrawn from blood to replace stomach acid; thus H^+ decreases and HCO_3^- increases proportionately
	Selected diuretics: cause K^+ depletion and H_2O loss. Low K^+ directly stimulates tubule cells to secrete H^+. Reduced blood volume elicits the renin-angiotensin-aldosterone mechanism, which stimulates Na^+ reabsorption and H^+ secretion.
	Ingestion of excessive sodium bicarbonate (antacid): bicarbonate moves easily into ECF, where it enhances natural alkaline reserve
	Excess aldosterone (e.g., adrenal tumors): promotes excessive reabsorption of Na^+, which pulls increased amount of H^+ into urine. Hypovolemia promotes the same relative effect because aldosterone secretion is increased to enhance Na^+ (and H_2O) reabsorption.

26

Sleuthing: Using Blood Values to Determine the Cause of Acidosis or Alkalosis

Students, particularly nursing students, are often given blood values and asked to determine (1) whether the patient is in acidosis or alkalosis, (2) the cause of the condition (respiratory or metabolic), and (3) whether the condition is being compensated. If you approach these questions systematically, they are not nearly as difficult as they may appear.

To analyze a person's acid-base balance, scrutinize the blood values in the following order:

1. *Note the pH.* This tells you whether the person is in acidosis (pH below 7.35) or alkalosis (pH above 7.45), but it does not tell you the cause.

2. *Check the P_{CO_2}* to see if this is causing the acid-base imbalance. Because the respiratory system is a fast-acting system, an excessively high or low P_{CO_2} may indicate that the respiratory system is either causing the condition or compensating for it. For example, if the pH indicates acidosis and

   a. The P_{CO_2} is over 45 mm Hg, then the respiratory system is the cause of the problem and the condition is respiratory acidosis

   b. The P_{CO_2} is below normal limits (below 35 mm Hg), then the respiratory system is *not the cause but is compensating*

   c. The P_{CO_2} is within normal limits, then the condition is *neither caused nor compensated* by the respiratory system

3. *Check the bicarbonate level.* If step 2 proves that the respiratory system is not responsible for the imbalance, then the condition is metabolic and should be reflected in increased or decreased bicarbonate levels. HCO_3^- values below 22 mEq/L indicate metabolic acidosis, and values over 26 mEq/L indicate metabolic alkalosis. Notice that whereas P_{CO_2} levels vary inversely with blood pH (P_{CO_2} rises as blood pH falls), HCO_3^- levels vary directly with blood pH (increased HCO_3^- results in increased pH).

Beyond this bare-bones approach there is something else to consider when you are assessing acid-base problems. If an imbalance is fully compensated, the pH may be normal even while the patient is in trouble. So, when the pH is normal, carefully scrutinize the P_{CO_2} or HCO_3^- values for clues to what imbalance may be occurring.

Consider the following two examples of the three-step approach.

Problem 1

Blood values: pH 7.6; P_{CO_2} 24 mm Hg; HCO_3^- 23 mEq/L

Analysis:

1. The pH is elevated: alkalosis.
2. The P_{CO_2} is very low: the cause of the alkalosis.
3. The HCO_3^- value is within normal limits.

Conclusion: This is respiratory alkalosis not compensated by renal mechanisms, as might occur during short-term hyperventilation.

Problem 2

Blood values: pH 7.48; P_{CO_2} 46 mm Hg; HCO_3^- 33 mEq/L

Analysis:

1. The pH is elevated: alkalosis.
2. The P_{CO_2} is elevated: the cause of acidosis, not alkalosis. Thus, the respiratory system is compensating and is not the cause.
3. The HCO_3^- is elevated: the cause of the alkalosis.

Conclusion: This is metabolic alkalosis being compensated by respiratory acidosis (retention of CO_2 to restore blood pH to the normal range).

Use the accompanying simple chart to help you in your future sleuthing.

	Normal Range in Plasma		
Acid-Base Disturbance	pH 7.35–7.45	P_{CO_2} 35–45 mm Hg	HCO_3^- 22–26 mEq/L
Respiratory acidosis	↓	↑	↑ if compensating
Respiratory alkalosis	↑	↓	↓ if compensating
Metabolic acidosis	↓	↓ if compensating	↓
Metabolic alkalosis	↑	↑ if compensating	↑

when gas exchange is hampered by diseases such as pneumonia, cystic fibrosis, or emphysema. Under such conditions, CO_2 accumulates in the blood. Thus, respiratory acidosis is characterized by falling blood pH and rising P_{CO_2}.

Respiratory alkalosis results when carbon dioxide is eliminated from the body faster than it is produced. This is called **hyperventilation** (deeper and faster breathing than needed to remove CO_2; ◄ p. 855), and results in the blood becoming more alkaline. While respiratory acidosis is frequently associated with respiratory system pathology, respiratory alkalosis is often due to stress or pain.

Metabolic Acidosis and Alkalosis

Metabolic pH imbalances include all abnormalities of acid-base imbalance *except* those caused by too much or too little carbon dioxide in the blood. Bicarbonate ion levels below or above the normal range of 22–26 mEq/L indicate a metabolic acid-base imbalance.

Another common cause of acid-base imbalance, **metabolic acidosis**, is recognized by low blood pH and HCO_3^- levels. Typical causes are ingesting too much alcohol (which is metabolized to acetic acid) and excessive loss of HCO_3^-, as might result from persistent diarrhea. Other causes are accumulation of lactic acid during exercise or shock, the ketosis that occurs in diabetic crisis or starvation, and, infrequently, kidney failure.

Metabolic alkalosis, indicated by rising blood pH and HCO_3^- levels, is much less common than metabolic acidosis. Typical causes are vomiting the acidic contents of the stomach (or loss of those secretions through gastric suctioning) and intake of excess base (too many antacids, for example).

Effects of Acidosis and Alkalosis

The absolute blood pH limits for life are a low of 6.8 and a high of 7.8. When blood pH falls below 6.8, the central nervous system is so depressed that the person goes into coma and death soon follows. When blood pH rises above 7.8, the nervous system is overexcited, leading to muscle tetany, extreme nervousness, and convulsions. Death often results from respiratory arrest.

Respiratory and Renal Compensations

If one of the physiological buffer systems (lungs or kidneys) malfunctions and disrupts acid-base balance, the other system tries to compensate. The respiratory system attempts to compensate for metabolic acid-base imbalances, and the kidneys (although much slower) work to correct imbalances caused by respiratory disease. We can recognize these **respiratory** and **renal compensations** by the resulting changes in plasma P_{CO_2} and bicarbonate ion concentrations (see *A Closer Look*). Because the compensations act to restore normal blood pH, a patient may have a normal pH despite a significant medical problem.

Respiratory Compensations

As a rule, changes in respiratory rate and depth are evident when the respiratory system is attempting to compensate for metabolic acid-base imbalances. In metabolic acidosis, respiratory rate and depth are usually elevated—an indication that high H^+ levels are stimulating the respiratory centers. Blood pH is low (below 7.35) and the HCO_3^- level is below 22 mEq/L. As the respiratory system "blows off" CO_2 to rid the blood of excess acid, the P_{CO_2} falls below 35 mm Hg. In contrast, in respiratory acidosis, the respiratory rate is often depressed and *is the immediate cause of the acidosis* (with some exceptions such as pneumonia or emphysema where gas exchange is impaired).

Respiratory compensation for metabolic alkalosis involves slow, shallow breathing, which allows CO_2 to accumulate in the blood. Evidence of metabolic alkalosis being compensated by respiratory mechanisms includes a pH over 7.45 (at least initially), elevated bicarbonate levels (over 26 mEq/L), and a P_{CO_2} above 45 mm Hg.

Renal Compensations

When an acid-base imbalance is of respiratory origin, renal mechanisms are stepped up to compensate for the imbalance.

For example, a hypoventilating individual will exhibit acidosis. When renal compensation is occurring, both the P_{CO_2} and the HCO_3^- levels are high. The high P_{CO_2} causes the acidosis, and the rising HCO_3^- level indicates that the kidneys are retaining bicarbonate to offset the acidosis.

Conversely, a person with renal-compensated respiratory alkalosis will have a high blood pH and a low P_{CO_2}. Bicarbonate ion levels begin to fall as the kidneys eliminate more HCO_3^- from the body by failing to reclaim it or by actively secreting it. Note that the kidneys cannot compensate for alkalosis or acidosis if that condition reflects a *renal* problem.

Check Your Understanding

17. Which two abnormalities in plasma are key features of an uncompensated metabolic alkalosis? An uncompensated respiratory acidosis?

18. How do the kidneys compensate for respiratory acidosis?

For answers, see Answers Appendix.

Developmental Aspects of Fluid, Electrolyte, and Acid-Base Balance

An embryo and a very young fetus are more than 90% water, but solids accumulate as fetal development continues, and at birth an infant is "only" 70–80% water. (The average value for adults is 50–60%.) Infants have proportionately more ECF than adults and, consequently, a much higher NaCl content in relation to K^+, Mg^{2+}, and PO_4^{3-} salts.

Distribution of body water begins to change about two months after birth and achieves the adult pattern by age 2. Plasma electrolyte concentrations are similar in infants and adults, but K^+ and Ca^{2+} values are higher and Mg^{2+}, HCO_3^-, and total protein levels are lower in the first few days of life than at any other time. At puberty, sex differences in body water content become obvious as males develop relatively greater amounts of skeletal muscle.

Problems with fluid, electrolyte, and particularly acid-base balance are most common in infancy, reflecting the following conditions:

- *The very low residual volume of infant lungs* (approximately half that of adults relative to body weight). When respiration is altered, P_{CO_2} can shift rapidly and dramatically.

- *The high rate of fluid intake and output in infants* (about seven times higher than in adults). Infants may exchange fully half their ECF daily. Though infants have proportionately much more body water than adults, this does not protect them from excessive fluid shifts. Even slight alterations in fluid balance can cause serious problems. Further, although adults can live without water for about ten days, infants can survive for only three to four days.

- *The relatively high infant metabolic rate* (about twice that of adults). The higher metabolic rate yields much larger amounts of metabolic wastes and acids that the kidneys must excrete. This, along with buffer systems that are not yet fully effective, results in a tendency toward acidosis.

26

Homeostatic Interrelationships between the Urinary System and Other Body Systems

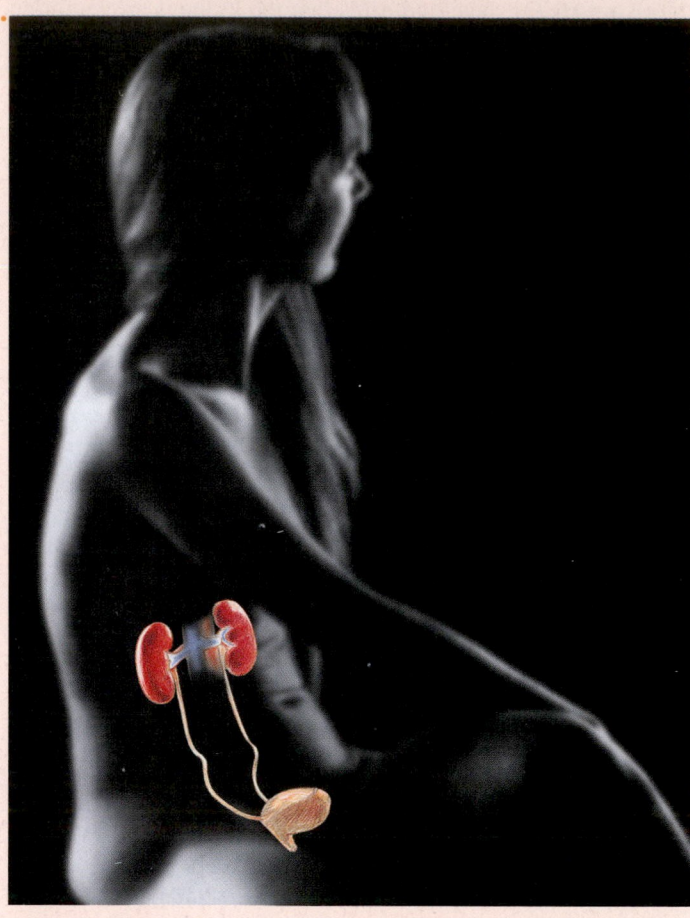

Nervous System   Chapters 11–15
- Kidneys maintain fluid, electrolyte, and acid-base balance of blood; dispose of nitrogenous wastes; renal control of K^+, Ca^{2+}, and Na^+ in ECF essential for normal neural function
- Neural controls involved in micturition; sympathetic nervous system activity triggers the renin-angiotensin-aldosterone mechanism

Endocrine System   Chapter 16
- Kidneys maintain fluid, electrolyte, and acid-base balance of blood; dispose of nitrogenous wastes; produce erythropoietin; renal regulation of Na^+ and water balance essential for blood pressure homeostasis and hormone transport in blood
- ADH, aldosterone, ANP, and other hormones help regulate renal reabsorption of water and electrolytes

Cardiovascular System   Chapters 17–19
- Kidneys maintain fluid, electrolyte, and acid-base balance of blood; dispose of nitrogenous wastes; renal regulation of Na^+ and water balance essential for blood pressure homeostasis. K^+, Ca^{2+}, and Na^+ regulation maintains cardiac excitability
- Systemic arterial blood pressure is the driving force for glomerular filtration; heart secretes atrial natriuretic peptide; blood transports nutrients, oxygen, etc. to urinary organs

Lymphatic System/Immunity   Chapters 20–21
- Kidneys maintain fluid, electrolyte, and acid-base balance of blood; dispose of nitrogenous wastes
- By returning leaked plasma fluid to cardiovascular system, lymphatic vessels help ensure normal systemic arterial pressure required for kidney function; immune cells protect urinary organs from infection and cancer

Respiratory System   Chapter 22
- Kidneys maintain fluid, electrolyte, and long-term acid-base balance of blood; dispose of nitrogenous wastes
- Respiratory system provides oxygen required by kidney cells for their high metabolic activity; disposes of carbon dioxide; rapid acid-base balance of blood; lung capillary endothelial cells convert angiotensin I to angiotensin II

Digestive System   Chapter 23
- Kidneys maintain fluid, electrolyte, and acid-base balance of blood; dispose of nitrogenous wastes; metabolize vitamin D to the active form needed for calcium absorption
- Digestive organs provide nutrients needed for kidney cell health; liver synthesizes urea and glutamine to transport waste nitrogen to the kidneys for excretion

Reproductive System   Chapter 27
- Kidneys maintain fluid, electrolyte, and acid-base balance of blood; dispose of nitrogenous wastes

Integumentary System   Chapter 5
- Kidneys maintain fluid, electrolyte, and acid-base balance of blood; dispose of nitrogenous wastes
- Skin provides external protective barrier; site of water loss (via perspiration); vitamin D synthesis site

Skeletal System   Chapters 6–8
- Kidneys maintain fluid, electrolyte, and acid-base balance of blood; dispose of nitrogenous wastes
- Bones of rib cage provide some protection to kidneys; form major store of calcium and phosphate ions

Muscular System   Chapters 9–10
- Kidneys maintain fluid, electrolyte, and acid-base balance of blood; dispose of nitrogenous wastes; renal regulation of K^+, Ca^{2+}, and Na^+ content in ECF crucial for muscle excitability and contractility
- Muscles of pelvic diaphragm and external urethral sphincter function in voluntary control of micturition; creatinine is a nitrogenous waste product of muscle metabolism that must be excreted by the kidneys

- *The high rate of insensible water loss in infants because of their larger surface area relative to body volume* (about three times that of adults). Infants lose substantial amounts of water through their skin.
- *The inefficiency of infant kidneys.* At birth, the kidneys are immature, only about half as proficient as adult kidneys at concentrating urine, and notably inefficient at ridding the body of acids.

All these factors put newborns at risk for dehydration and acidosis, at least until the end of the first month when the kidneys achieve reasonable efficiency. Bouts of vomiting or diarrhea greatly amplify the risk.

In old age, total body water often decreases (the loss is largely from the intracellular compartment) because muscle mass progressively declines and body fat rises. Few changes occur in the solute concentrations of body fluids, but the speed with which homeostasis is restored after being disrupted declines with age.

Elders may be unresponsive to thirst cues and thus are at risk for dehydration. Additionally, they are the most frequent prey of diseases that lead to severe fluid, electrolyte, or acid-base problems, such as congestive heart failure (and its attendant edema) and diabetes mellitus. Because most fluid, electrolyte, and acid-base imbalances occur when body water content is highest or lowest, the very young and the very old are most frequently affected.

RELATED CLINICAL TERMS

Antacid An agent that counteracts acidity, such as sodium bicarbonate, aluminum hydroxide gel, and magnesium trisilicate. Commonly used to manage heartburn.

Hyperaldosteronism (Conn's disease) A condition of hypersecretion of aldosterone by adrenal cortical cells accompanied by excessive loss of potassium and generalized muscular weakness, hypernatremia, and hypertension. Usual cause is adrenal tumor; usual treatment is adrenal-suppressing drugs prior to tumor removal.

Renal tubular acidosis A metabolic acidosis resulting from impaired renal reabsorption of bicarbonate; the urine is alkaline.

Syndrome of inappropriate ADH secretion (SIADH) A group of disorders associated with excessive ADH secretion in the absence of appropriate (osmotic or nonosmotic) stimuli. Characterized by hypotonic hydration (hyponatremia, fluid retention, and weight gain) and concentrated urine. Usual causes are ectopic secretion of ADH by cancer cells (e.g., small cell lung cancers) and brain disorders or trauma affecting the ADH-secreting hypothalamic neurons. Temporary management involves restricting water intake.

CHAPTER SUMMARY

26.1 Body fluids consist of water and solutes in three main compartments (pp. 1013–1015)

Body Water Content (p. 1013)
1. Water accounts for 45–75% of body weight, depending on age, sex, and amount of body fat.

Fluid Compartments (p. 1013)
2. About two-thirds (25 L) of body water is found within cells in the intracellular fluid (ICF) compartment; the balance (15 L) is in the extracellular fluid (ECF) compartment. The ECF includes plasma and interstitial fluid.

Composition of Body Fluids (pp. 1013–1015)
3. Solutes dissolved in body fluids include electrolytes and nonelectrolytes. Electrolyte concentration is expressed in mEq/L.
4. Plasma contains more proteins than does interstitial fluid; otherwise, extracellular fluids are similar. The most abundant ECF electrolytes are sodium, chloride, and bicarbonate ions.
5. Intracellular fluids contain large amounts of protein anions and potassium, phosphate, and magnesium ions.

Fluid Movement among Compartments (p. 1015)
6. Substances usually pass through the plasma and interstitial fluid to reach the intracellular fluid.
7. Osmotic and hydrostatic pressure regulate fluid exchanges between plasma and interstitial fluid compartments: (a) Hydrostatic pressure forces filtrate out of the capillaries and

colloid osmotic pressure pulls filtrate back in. (b) Water moves freely between the ECF and the ICF by osmosis, but solute movements are restricted by size, charge, and dependence on transport proteins. (c) Water flows always follow changes in ECF osmolality.

26.2 Both intake and output of water are regulated (pp. 1016–1019)

1. Sources of body water are ingested foods and fluids and metabolic water.
2. Water leaves the body primarily via the lungs, skin, gastrointestinal tract, and kidneys.

Regulation of Water Intake (pp. 1016–1017)
3. Increased plasma osmolality triggers the thirst mechanism, mediated by hypothalamic osmoreceptors. Thirst, inhibited by distension of the gastrointestinal tract by ingested water and then by reduced osmolality, may be damped before body needs for water are met.

Regulation of Water Output (p. 1017)
4. Obligatory water loss is unavoidable and includes insensible water losses from the lungs, the skin, and in feces, and about 500 ml of urine output daily.
5. Beyond obligatory water loss, the volume of urinary output depends on water intake and loss via other routes and reflects the influence of antidiuretic hormone on the renal collecting ducts.

Influence of Antidiuretic Hormone (ADH) **(pp. 1017–1018)**

6. Antidiuretic hormone causes aquaporins (water channels) to be inserted in the cell membranes of the collecting ducts, so that most filtered water is reabsorbed. ADH release is triggered if ECF osmolality is high, or if a large drop in blood volume or pressure occurs.

Disorders of Water Balance **(pp. 1018–1019)**

7. Dehydration occurs when water loss exceeds water intake over time. It is evidenced by thirst, dry skin, and decreased urine output. A serious consequence is hypovolemic shock.
8. Hypotonic hydration occurs when body fluids are excessively diluted and entering water makes cells swell. The most serious consequence is cerebral edema.
9. Edema is an abnormal accumulation of fluid in the interstitial space (increased IF), which may impair blood circulation.

26.3 **Sodium, potassium, calcium, and phosphate levels are tightly regulated** **(pp. 1019–1026)**

1. Most electrolytes are obtained from ingested foods and fluids. Salts, particularly NaCl, are often ingested in excess of need.
2. Electrolytes are lost in sweat, feces, and urine (and sometimes in vomit). The kidneys are most important in regulating electrolyte balance.

The Central Role of Sodium in Fluid and Electrolyte Balance **(p. 1021)**

3. Sodium salts are the most abundant solutes in ECF. They exert the bulk of ECF osmotic pressure and control water volume and distribution in the body.

Regulation of Sodium Balance **(pp. 1021–1024)**

4. Total body sodium ion content is linked to ECF volume and blood pressure regulation. Blood volume and blood pressure are monitored by baroreceptors and regulated by both neural and hormonal controls.
5. Na^+ concentration in ECF is monitored by osmoreceptors and regulated primarily by thirst and ADH mechanisms.
6. Aldosterone promotes Na^+ reabsorption to maintain blood volume and blood pressure.
7. Declining blood pressure and falling filtrate NaCl concentration stimulate the granular cells to release renin. Renin, via angiotensin II, enhances systemic blood pressure, Na^+ reabsorption, and aldosterone release.
8. Atrial natriuretic peptide, released by certain atrial cells in response to rising blood pressure (or blood volume), causes systemic vasodilation and inhibits renin, aldosterone, and ADH release. As a result, it enhances Na^+ and water excretion, reducing blood volume and blood pressure.
9. Estrogens and glucocorticoids increase renal retention of sodium. Progesterone promotes enhanced sodium and water excretion in urine.
10. Cardiovascular system baroreceptors sense changing arterial blood pressure, prompting changes in sympathetic vasomotor activity. Rising arterial pressure leads to vasodilation and enhanced Na^+ and water loss in urine. Falling arterial pressure promotes vasoconstriction and conserves Na^+ and water.

Regulation of Potassium Balance **(pp. 1024–1025)**

11. The more proximal regions of the nephrons reabsorb about 90% of filtered potassium.

12. The main thrust of renal regulation of K^+ is to excrete it. Aldosterone and increased plasma K^+ content enhance potassium ion secretion distal to the nephron loop. In the collecting duct, principal cells secrete K^+ whereas type A intercalated cells reabsorb small amounts of K^+ during K^+ deficit.

Regulation of Calcium and Phosphate Balance **(pp. 1025–1026)**

13. Calcium balance is regulated primarily by parathyroid hormone (PTH), which enhances blood Ca^{2+} levels by targeting the bones, kidneys, and intestine. PTH-regulated reabsorption occurs primarily in the distal convoluted tubule.
14. PTH decreases renal reabsorption of phosphate ions.

Regulation of Anions **(p. 1026)**

15. When blood pH is normal or slightly high, chloride is the major anion accompanying sodium reabsorption. In acidosis, bicarbonate replaces chloride.
16. Reabsorption of most other anions is regulated by their transport maximums (T_m).

26.4 **Chemical buffers and respiratory regulation rapidly minimize pH changes** **(pp. 1026–1029)**

1. The homeostatic pH range of arterial blood is 7.35 to 7.45. A higher pH represents alkalosis; a lower pH reflects acidosis.
2. Some acids enter the body in foods, but most are generated by breakdown of phosphorus-containing proteins, incomplete oxidation of fats or glucose, and the loading and transport of carbon dioxide in the blood.
3. Acid-base balance is achieved by chemical buffers, respiratory regulation, and in the long term by renal regulation of bicarbonate ion (and therefore, hydrogen ion) concentration of body fluids.

Chemical Buffer Systems **(pp. 1027–1028)**

4. Acids are proton (H^+) donors; bases are proton acceptors. Acids that dissociate completely in solution are strong acids; those that dissociate incompletely are weak acids. Strong bases are more effective proton acceptors than are weak bases.
5. Chemical buffers are single or paired sets (a weak acid and its salt) of molecules that act rapidly to resist excessive shifts in pH by releasing or binding H^+.
6. Chemical buffers of the body include the bicarbonate, phosphate, and protein buffer systems.

Respiratory Regulation of H^+ **(pp. 1028–1029)**

7. Respiratory regulation of acid-base balance of the blood utilizes the bicarbonate buffer system and the fact that CO_2 and H_2O are in reversible equilibrium with H_2CO_3.
8. Acidosis activates the respiratory centers to increase respiratory rate and depth, which eliminates more CO_2 and causes blood pH to rise. Alkalosis depresses the respiratory centers, resulting in CO_2 retention and a fall in blood pH.

26.5 **Renal regulation is a long-term mechanism for controlling acid-base balance** **(pp. 1029–1032)**

1. The kidneys provide the major long-term mechanism for controlling acid-base balance by maintaining stable HCO_3^- levels in the ECF. Nonvolatile acids (organic acids other than carbonic acid) can be eliminated from the body only by the kidneys.
2. Secreted hydrogen ions come from the dissociation of carbonic acid generated within the tubule cells.

3. Tubule cells are impermeable to bicarbonate in the filtrate, but they can conserve filtered bicarbonate ions indirectly by absorbing HCO_3^- generated within them (by dissociation of carbonic acid to HCO_3^- and H^+). For each HCO_3^- reabsorbed, one H^+ is secreted into the filtrate, where it combines with HCO_3^-.

4. To generate and add new HCO_3^- to plasma to counteract acidosis, either of two mechanisms may be used:
   - Secreted H^+, buffered by bases other than HCO_3^-, is excreted from the body in urine (the major urine buffer is the phosphate buffer system).
   - NH_4^+ (derived from glutamine catabolism) is excreted in urine.

5. To counteract alkalosis, bicarbonate ion is secreted into the filtrate and H^+ is reabsorbed.

26.6 Abnormalities of acid-base balance are classified as metabolic or respiratory (pp. 1033–1035)

1. Respiratory acidosis results from carbon dioxide retention. Respiratory alkalosis occurs when carbon dioxide is eliminated faster than it is produced.

2. Metabolic acidosis occurs when nonvolatile acids (lactic acid, ketone bodies, and others) accumulate in the blood or when bicarbonate is lost from the body. Metabolic alkalosis occurs when bicarbonate levels are excessive.

3. Extremes of pH for life are 6.8 and 7.8.

4. Compensations occur when the respiratory system or kidneys counteract acid-base imbalances resulting from abnormal or inadequate functioning of the alternate system. Respiratory compensations involve changes in respiratory rate and depth. Renal compensations modify blood levels of HCO_3^-.

Developmental Aspects of Fluid, Electrolyte, and Acid-Base Balance (pp. 1035–1037)

1. Infants have a higher risk of dehydration and acidosis because of their low lung residual volume, high rate of fluid intake and output, high metabolic rate, relatively large body surface area, and functionally immature kidneys at birth.

2. The elderly are at risk for dehydration because of their low percentage of body water and insensitivity to thirst cues. Diseases that promote fluid and acid-base imbalances (cardiovascular disease, diabetes mellitus, and others) are most common in the aged.

REVIEW QUESTIONS

 To access additional practice questions using your smartphone, tablet, or computer: MasteringA&P® > Study Area > Practice Tests & Quizzes

Level 1 Remember/Understand

(Some questions have more than one correct answer. Select the best answer or answers from the choices given.)

1. Body water content is greatest in (a) infants, (b) young adults, (c) elderly adults.

2. Potassium, magnesium, and phosphate ions are the predominant electrolytes in (a) plasma, (b) interstitial fluid, (c) intracellular fluid.

3. Sodium balance is regulated primarily by control of amount(s) (a) ingested, (b) excreted in urine, (c) lost in perspiration, (d) lost in feces.

4. Water balance is regulated by control of amount(s) (use choices in question 3).

Answer questions 5 through 10 by choosing responses from the following:

(a) ammonium ions	(f) magnesium
(b) bicarbonate	(g) phosphate
(c) calcium	(h) potassium
(d) chloride	(i) sodium
(e) hydrogen ions	(j) water

5. Two main substances regulated by the influence of aldosterone on the kidney tubules.

6. Two substances regulated by parathyroid hormone.

7. Two substances secreted into the proximal convoluted tubules in exchange for sodium ions.

8. Part of an important chemical buffer system in plasma.

9. Two ions produced during catabolism of glutamine.

10. Substance regulated by ADH's effects on the renal tubules.

11. Which of the following factors will enhance ADH release? (a) increase in ECF volume, (b) decrease in ECF volume, (c) decrease in ECF osmolality, (d) increase in ECF osmolality.

12. In an individual with metabolic acidosis, a clue that the respiratory system is compensating is provided by (a) high blood bicarbonate levels, (b) low blood bicarbonate levels, (c) rapid, deep breathing, (d) slow, shallow breathing.

13. Name the body fluid compartments, noting their locations and the approximate fluid volume in each.

Level 2 Apply/Analyze

14. Describe the thirst mechanism, indicating how it is triggered and terminated.

15. Explain why and how ECF osmolality is maintained.

16. Explain why and how total body sodium content, ECF volume, and blood pressure are jointly regulated.

17. Describe the role of the respiratory system in controlling acid-base balance.

18. Explain how the chemical buffer systems resist changes in pH.

19. Explain the relationship of the following to renal secretion and excretion of hydrogen ions: (a) plasma carbon dioxide levels, (b) phosphate, and (c) sodium bicarbonate reabsorption.

Level 3 Evaluate/Synthesize

20. Mr. Jessup, a 55-year-old man, is operated on for a cerebral tumor. About a month later, he appears at his physician's office complaining of excessive thirst. He claims to have been drinking about 20 liters of water daily for the past week and says he has been voiding nearly continuously. A urine sample is collected and its specific gravity is reported as 1.001. What is your diagnosis of Mr. Jessup's problem? What connection might exist between his previous surgery and his present problem?

21. For each of the following sets of blood values, name the acid-base imbalance (acidosis or alkalosis), determine its cause (metabolic or respiratory), decide whether the condition is being compensated, and cite at least one possible cause of the imbalance.
    Problem 1: pH 7.63; P_{CO_2} 19 mm Hg; HCO_3^- 19.5 mEq/L
    Problem 2: pH 7.22; P_{CO_2} 30 mm Hg; HCO_3^- 12.0 mEq/L

22. Explain how emphysema and congestive heart failure can lead to acid-base imbalance.

23. Mrs. Bush, a 70-year-old woman, is admitted to the hospital. Her history states that she has been suffering from diarrhea for three weeks. On admission, she complains of severe fatigue and

26

muscle weakness. A blood chemistry study yields the following information: Na^+ 142 mEq/L; K^+ 1.5 mEq/L; Cl^- 92 mEq/L; P_{CO_2} 32 mm Hg. Which electrolytes are within normal limits? Which are so abnormal that the patient has a medical emergency? Which of the following represents the greatest danger to Mrs. Bush? **(a)** a fall due to her muscular weakness, **(b)** edema, **(c)** cardiac arrhythmia and cardiac arrest.

24. During a routine medical checkup, Shelby, a 26-year-old physiotherapy student, is surprised to hear that her blood pressure is 180/110. She also has a rumbling systolic and diastolic abdominal bruit (murmur) that is loudest at the mid-epigastric area. Her physician suspects renal artery stenosis (narrowing). She orders an abdominal ultrasound and renal artery arteriography, which confirm that Shelby has a small right kidney and the distal part of her right renal artery is narrowed by more than 70%. Her physician prescribes diuretics and calcium channel blockers as temporary measures, and refers Shelby to a cardiovascular surgeon. Explain the connection between Shelby's renal artery stenosis and her hypertension. Why is her right kidney smaller than her left? What would you expect Shelby's blood levels of K^+, Na^+, aldosterone, angiotensin II, and renin to be?

CLINICAL CASE STUDY

72-Year-Old Male with Abdominopelvic Trauma

Mr. Heyden, 72, is brought to the emergency room after an accident at his farm. The paramedics report that his left side was pinned beneath his tractor, and that when he was freed, his left lower quadrant appeared to be compressed. His blood pressure is 90/50 mm Hg and falling, and his heart rate is 116 beats/min. His pulse is thready. Mr. Heyden complains of pain in his left side and then loses consciousness.

1. Given the values above, what would you conclude is Mr. Heyden's immediate problem?

2. **+ NCLEX-STYLE** Mr. Heyden's low blood pressure will trigger certain compensatory mechanisms. Which statement below best reflects the changes in hormone levels that will occur?
   a. Mr. Heyden's ADH, aldosterone, and renin will increase.
   b. Mr. Heyden's ADH will decrease, his aldosterone will increase, and his renin will be unchanged.
   c. Mr. Heyden's ADH will increase, and his aldosterone and renin will decrease.
   d. Mr. Heyden's ADH and renin will decrease, and his aldosterone will increase.

A CT scan reveals a ruptured spleen and a large hematoma in the upper left abdominal quadrant. Splenic repair surgery is scheduled but unsuccessful; the spleen is removed. Post-surgery urinalysis reveals some granular casts (particulate cell debris) and that the urine is brownish-red in color; other values are normal, but urine output is very low. An order is given to rapidly infuse intravenous saline solution.

3. **(a)** What might account for the low volume of urine output? (Name at least two possibilities.) **(b)** What might explain the casts and abnormal color of his urine? Can you see any possible relationship between his crush injury and these findings?

The next day, Mr. Heyden is awake and alert. His urine output is still low and his blood work indicates the beginnings of acute renal failure.

4. **+ NCLEX-STYLE** Mr. Heyden's blood K^+ levels are high. Which of the following best explains his high K^+ levels?
   a. Mr. Heyden's increased secretion of aldosterone will cause his K^+ levels to increase.
   b. Mr. Heyden's damaged kidneys can no longer adequately remove K^+ from his blood.
   c. Mr. Heyden's increased secretion of ADH will cause his K^+ levels to increase.
   d. The loss of Mr. Heyden's spleen means that his aged and fragile erythrocytes are not being removed, and so are rupturing in his blood vessels and releasing K^+.

5. Mr. Heyden's blood pH is 7.33, P_{CO_2} is 34 mm Hg, and his HCO_3^- is 20 mEq/L. Explain these findings.

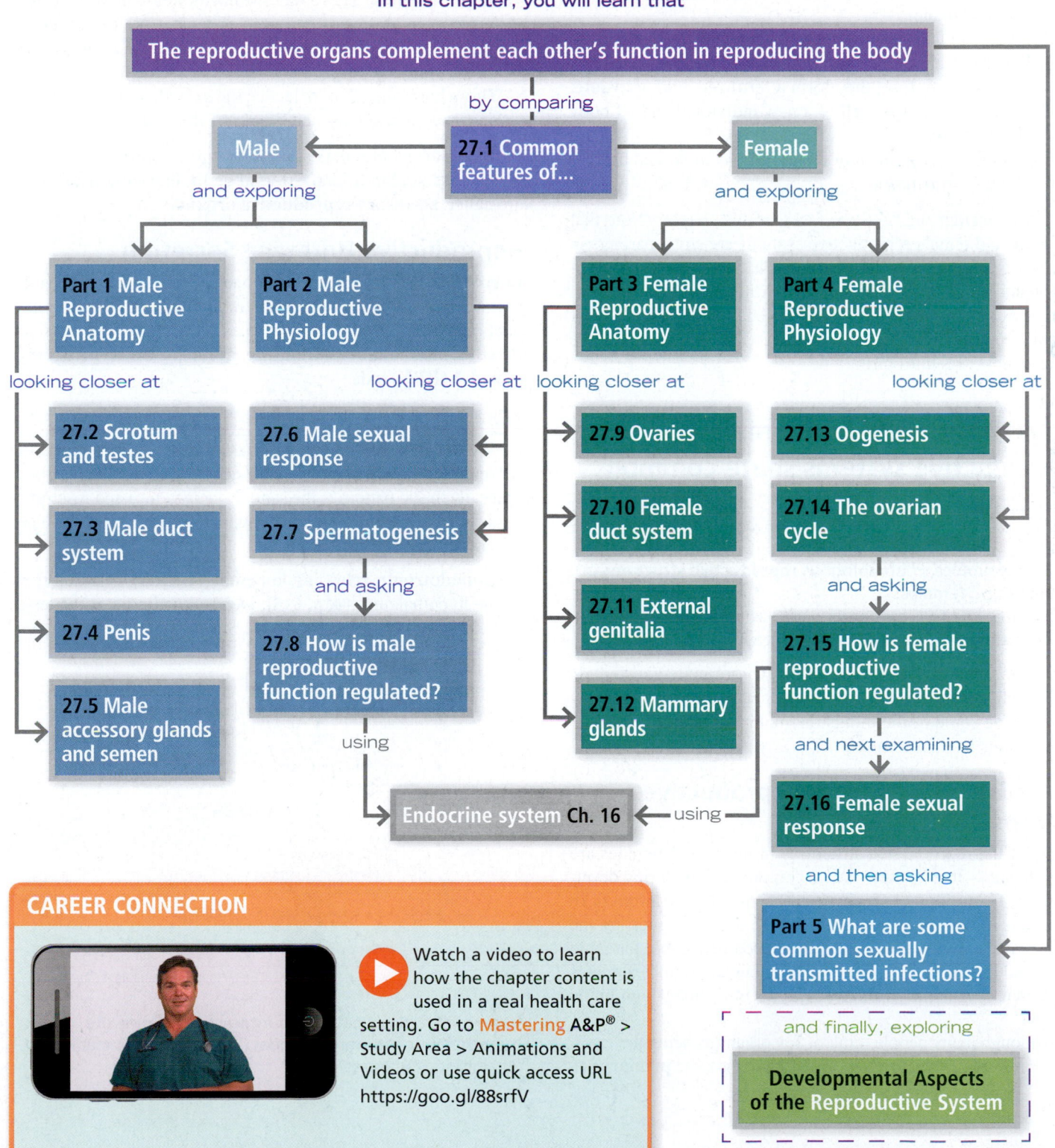

In this chapter, you will learn that

The reproductive organs complement each other's function in reproducing the body

by comparing

Male ← **27.1 Common features of...** → Female

and exploring · and exploring

Part 1 Male Reproductive Anatomy

Part 2 Male Reproductive Physiology

Part 3 Female Reproductive Anatomy

Part 4 Female Reproductive Physiology

looking closer at · looking closer at · looking closer at · looking closer at

27.2 Scrotum and testes

27.6 Male sexual response

27.9 Ovaries

27.13 Oogenesis

27.3 Male duct system

27.7 Spermatogenesis

27.10 Female duct system

27.14 The ovarian cycle

27.4 Penis

and asking

27.11 External genitalia

and asking

27.5 Male accessory glands and semen

27.8 How is male reproductive function regulated?

27.12 Mammary glands

27.15 How is female reproductive function regulated?

using

and next examining

Endocrine system Ch. 16 ← using

27.16 Female sexual response

and then asking

Part 5 What are some common sexually transmitted infections?

and finally, exploring

Developmental Aspects of the Reproductive System

CAREER CONNECTION

Watch a video to learn how the chapter content is used in a real health care setting. Go to Mastering A&P® > Study Area > Animations and Videos or use quick access URL https://goo.gl/88srfV

Most organ systems function almost continuously to maintain the well-being of the individual. The **reproductive system**, however, appears to "slumber" until puberty, after which it plays an important role in our adult lives. Although male and female reproductive organs are quite different, their common purpose is to produce offspring. To do this successfully, the reproductive system must carry out four tasks:

1. Form specialized cells for sexual reproduction called *gametes*—*sperm* in males and *ova* (eggs) in females.
2. Bring the gametes from the male and female together. The process of *sexual intercourse* (*copulation*) delivers male gametes to the female.
3. Combine the genetic information contained within the male and female gametes. This process is called *fertilization*. When a sperm and egg fuse, they form a fertilized egg, or **zygote**. The zygote is the first cell of a new individual and all body cells arise from it.
4. Support the development of the fetus (*gestation*) and birth of the baby (*parturition*).

We will consider the first two tasks in this chapter. Note that the male and female reproductive systems are equal partners in the events leading up to fertilization. The last two tasks, fertilization and pregnancy, occur entirely within the female and will be described in Chapter 28. Once fertilization has occurred, the female's uterus provides the protective environment where the embryo develops until birth.

27.1 The male and female reproductive systems share common features

Learning Outcomes

▶ Give examples of homologous reproductive structures in males and females.

▶ Describe the HPG axis and its activation at puberty.

▶ Define meiosis. Compare and contrast it to mitosis.

Although most of this chapter will explore male and female reproductive anatomy and physiology separately, let's begin by considering some of the things the male and female reproductive systems have in common.

Gonads and Accessory Reproductive Organs

Many of the reproductive structures in males and females are *homologues*—that is, they share a common origin during development. **Homologous structures** (ho-mol′ŏ-gus) often have functions in common. Some examples of homologues are the male testes and female ovaries (described next), and the male penis and female clitoris. We will point out additional examples (and their overlapping functions) as we encounter them in the chapter.

Although many people think of sex organs as being the external reproductive structures (external genitalia), the **primary sex**

organs are actually the **gonads** (go′nadz; "seeds")—the *testes* in males and the *ovaries* in females. The type of gonad present during embryonic development determines the fate (male or female pattern of development) of all of the other developing reproductive structures. The gonads of both sexes produce two essential products:

- **Gametes** (gam′ēts; "spouses"). Males produce *sperm* and females produce *ova* (*eggs*). Gamete formation depends on a special type of cell division called *meiosis* (described below).

- **Sex hormones.** These steroid hormones—mainly testosterone in males, estrogens and progesterone in females—play vital roles both in the development and function of the reproductive organs and in sexual behavior and drives. These hormones also influence the growth and development of many other organs and tissues of the body (◀ p. 634).

In both males and females, the remaining reproductive structures—ducts, glands, and external genitalia (jen-ĭ-ta′le-ah)—are called **accessory reproductive organs**.

Reproductive Hormone Secretion

In both males and females, hormonal interactions between the hypothalamus, anterior pituitary, and gonads regulate the production of gametes and sex hormones. The relationship between these components is called the **hypothalamic-pituitary-gonadal (HPG) axis**.

The HPG Axis

Between each level of the HPG axis, hormones regulate the secretion of other hormones (◀ **Figure 16.4c, p. 607**). In both sexes, the basic pattern is the same (**Figure 27.1**) and resembles patterns you are already familiar with (◀ **Figure 16.7, p. 616**). The hormones involved are:

- **Gonadotropin-releasing hormone (GnRH)**, released from the hypothalamus (◀ **p. 616**), indirectly stimulates the testes

Figure 27.1 Basic pattern of interactions along the hypothalamic-pituitary-gonadal (HPG) axis. Note that inhibin only inhibits FSH release.

or ovaries via its effect on FSH and LH release. GnRH reaches the anterior pituitary cells via the hypophyseal portal system.

- **Follicle-stimulating hormone (FSH)** and **luteinizing hormone (LH)** are gonadotropins released from the anterior pituitary (◄ p. 616). Notice that these hormones are named for their actions in the female, but act in both sexes. FSH and LH act on specific cells in the gonads where they stimulate gamete production and sex hormone secretion.

- *Sex hormones* (primarily testosterone in males and estrogens and progesterone in females) act at target tissues in the body. They also exert negative feedback on the hypothalamus and anterior pituitary.

An additional hormone, called **inhibin**, is released from the gonads of both males and females. As its name implies, this peptide hormone also exerts negative feedback, but in this case, only on FSH release from the anterior pituitary.

Understanding the basic pattern of interactions along the HPG axis will help you when you encounter variations in this pattern later in the chapter (particularly in the female).

Activation of the HPG Axis at Puberty

Puberty is the period of life when the reproductive organs grow to adult size and become functional. It represents the earliest time that reproduction is possible.

Before puberty, the circulating levels of all steroidal sex hormones (testosterone, estrogens, and progesterone) are low. Even these very low levels are sufficient to suppress secretion of GnRH by the hypothalamus. As puberty approaches, the hypothalamus becomes less sensitive to inhibition by these sex hormones. As a result, it begins to release GnRH in a rhythmic, pulselike manner. In response to the elevated GnRH levels, the anterior pituitary releases FSH and LH. These gonadotropins, in turn, stimulate the release of more sex hormones from the gonads. As the threshold for inhibition of GnRH release continues to rise, more and more sex hormones are released until the adult pattern of hormone interaction is achieved. These high levels of sex hormones bring about the bodywide changes of puberty that we describe later in this chapter.

Meiosis

Gamete formation in both sexes involves **meiosis** (mi-o′sis; "a lessening"), a unique kind of nuclear division that occurs only in the gonads. Although the type and number of gametes produced by males and females differ, the underlying process—meiosis—is the same in both sexes.

Without meiosis, sexual reproduction cannot take place. This is because reproduction normally involves fusing together one cell from each parent. If those cells were somatic cells (normal body cells), the resulting zygote would then have twice as many chromosomes as the parental cells. One main function of meiosis is to prevent this from happening by reducing the number of chromosomes in gametes by half. As we will see, its second function is to greatly increase the genetic variability of gametes.

Terminology of Meiosis

Before we describe meiosis, let's define some terms.

- **Sister chromatids.** As in mitosis, the cell's DNA is replicated during interphase (◄ pp. 96–97) before meiosis begins. As a result, each chromosome has two **sister chromatids**. These sister chromatids are identical, and joined together by a centromere (◄ p. 100). The DNA then coils and condenses, which makes the chromosome visible (as shown at right) when viewed through a light microscope.

- **Homologous pairs of chromosomes.** In humans, chromosomes come in pairs called **homologous chromosomes** (shown at right). One member of each pair is from the male parent (the *paternal chromosome*), and the other is from the female parent (the *maternal chromosome*). Generally speaking, the two homologues of each chromosome pair look alike and carry genes that code for the same traits, though not necessarily for identical expression of those traits. In Chapter 29, we will consider how our mom's and dad's genes interact to produce our visible traits.

Maternal chromosome

Paternal chromosome

Two pairs of replicated homologous chromosomes

- **Nonsister chromatids.** Within a homologous pair of chromosomes, the chromatids of one chromosome are **nonsister chromatids** to the chromatids of the other chromosome in the pair (shown at right). Nonsister chromatids are not identical to each other because one is paternal and the other is maternal. One of the special features of meiosis is that nonsister chromatids exchange DNA with each other in a process called *crossing over*. This process greatly increases genetic variability because it combines maternal and paternal genes on the same chromatid. We will discuss this in more detail on the next page.

Nonsister chromatids

- **Diploid.** The **diploid chromosomal number** (dip′loid) is the normal number of chromosomes in most body cells. It consists of two sets of chromosomes, one from each parent, and is symbolized as **2n**. In humans, the diploid chromosomal number is 46, and *diploid* cells contain 23 pairs of homologous chromosomes.

- **Haploid.** The **haploid chromosomal number** (hap′loid), or **n**, is the number of chromosomes in a gamete. In humans, the haploid chromosomal number is 23. Gametes are *haploid* and contain only one member of each homologous pair, either maternal or paternal. When sperm and egg fuse, they form a zygote that reestablishes the typical diploid chromosomal number of human cells (2n).

27

Sequence of Events in Meiosis

Meiosis consists of two consecutive nuclear divisions (*meiosis I* and *meiosis II*) that follow a single round of DNA replication. Its product is four haploid daughter cells. Each of these nuclear divisions consists of four steps, each having the same name as a corresponding step of mitosis: prophase, metaphase, anaphase, and telophase. Before we examine the details, let's step back and look at the big picture of what happens before and then during meiosis I and meiosis II (**Figure 27.2**).

- During the interphase preceding meiosis, all of the chromosomes are replicated.

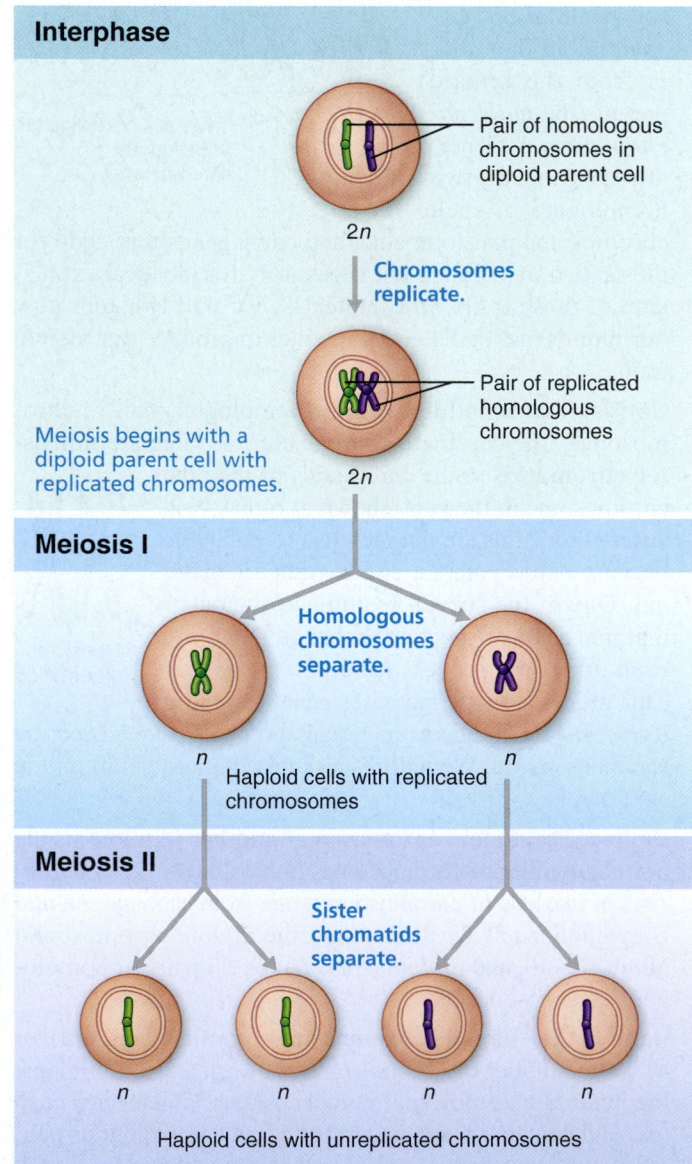

Let's follow one pair of homologous chromosomes through interphase and meiosis.

Interphase

Pair of homologous chromosomes in diploid parent cell

2n

Chromosomes replicate.

Meiosis begins with a diploid parent cell with replicated chromosomes.

Pair of replicated homologous chromosomes

2n

Meiosis I

Homologous chromosomes separate.

n   n

Haploid cells with replicated chromosomes

Meiosis II

Sister chromatids separate.

n   n   n   n

Haploid cells with unreplicated chromosomes

Figure 27.2 Overview of meiosis. For simplicity, all chromosomes are shown as condensed.

- During meiosis I, homologous pairs of chromosomes separate from each other, reducing the number of chromosomes from 2n to n. For this reason, it is sometimes called the **reduction division of meiosis**.

- During meiosis II, the sister chromatids of each chromosome separate. Each daughter cell gets an equal share of chromosomes, so this is sometimes called the **equational division of meiosis**.

Now you are ready to study **Figure 27.3**, which describes each of the steps of meiosis in detail. Meiosis is complicated, so take some time to work your way through this figure while reading the text below.

Meiosis I   As we have seen, meiosis I reduces the chromosome number from *2n* to *n*. As in mitosis, during prophase I of meiosis the chromosomes coil and condense, the nuclear envelope and nucleolus break down and disappear, and a spindle forms.

In addition, prophase I of meiosis includes an event never seen in mitosis (nor in meiosis II for that matter): The replicated chromosomes pair up with their homologous partners. (In the complicated dance of the chromosomes that is meiosis, finding your partner is the first step.) The alignment between homologous pairs of chromosomes takes place at discrete spots along their entire length—more like buttoning together than zipping. This process, called **synapsis**, forms little groups of four chromatids called **tetrads** (Figure 27.3).

During synapsis, a second unique event, called **crossing over**, occurs. These *crossovers* are called **chiasmata** (singular: chiasma), and form within each tetrad as the free ends of nonsister chromatids wrap around each other at one or more points. Crossing over allows the paired maternal and paternal chromosomes to exchange genetic material (see Figure 29.3, p. 1127). By the end of prophase I, the tetrads are attached to the spindle and are moving toward the spindle equator.

During metaphase I, the pairs of homologous chromosomes line up *randomly* at the spindle equator, so that either the paternal or maternal homologous chromosome can be on a given side of the equator (see Figure 29.2, p. 1126). (In our dance of the chromosomes, this is like partners lining up at the start of the dance with males and females randomly placed on either side of the line.)

During anaphase I, the sister chromatids representing each homologue behave as a unit—almost as if replication had not occurred. The homologous chromosomes, each still composed of two joined sister chromatids (a dyad), are distributed to opposite ends of the cell.

As a result, when meiosis I ends, each daughter cell has:

- *Two* copies of one member of each homologous pair (either the paternal or maternal) and none of the other

- A *haploid* chromosomal number (because the still-united sister chromatids are considered a single chromosome), but twice the amount of DNA in each chromosome

Meiosis II   The second meiotic division, meiosis II, mirrors mitosis in every way (◀ **pp. 100–101**), except that:

- The chromosomes are *not* replicated before it begins.

- The cells that undergo meiosis II are haploid, not diploid.

Interphase cell

Centriole pairs —

Nuclear envelope —

Chromatin —

$2n = 4$

Interphase events
As in mitosis, meiosis is preceded by DNA replication and other preparations for cell division.

MEIOSIS I

Chiasma (crossover)

Centromere

Spindle

Sister chromatids

Nuclear envelope fragments late in prophase I

Prophase I
Prophase events occur, as in mitosis. Additionally, synapsis occurs: Homologous chromosomes come together along their length to form tetrads. During synapsis, the "arms" of homologous chromatids wrap around each other, forming several chiasmata. The nonsister chromatids trade segments during crossing over. The fate of this exchanged genetic material is followed in the diagrams below.

Tetrad —

Exchanged genetic material —

Metaphase I
The tetrads align randomly on the spindle equator in preparation for anaphase.

Dyad —

Anaphase I
Unlike anaphase of mitosis, the centromeres do not separate during anaphase I of meiosis, so the sister chromatids (dyads) remain firmly attached. However, the homologous chromosomes do separate from each other and the dyads move toward opposite poles of the cell.

Chromosomes uncoil —

Nuclear envelopes re-form —

Cleavage furrow —

Telophase I
The nuclear membranes re-form around the chromosomal masses, the spindle breaks down, and the chromatin reappears as telophase and cytokinesis end. The 2 daughter cells (now haploid) enter a second interphase-like period, called interkinesis, before meiosis II occurs. There is no second replication of DNA before meiosis II.

MEIOSIS II

Prophase II

Meiosis II begins with the products of meiosis I (2 haploid daughter cells) and is a mitosis-like nuclear division process.

Metaphase II

Anaphase II

Telophase II and cytokinesis

After progressing through the phases of meiosis and cytokinesis, the product is 4 haploid cells, each genetically different from the original parent cell.

Products of meiosis: haploid daughter cells

Figure 27.3 Meiosis. The series of events in meiotic cell division for an animal cell with a diploid number (2n) of 4. The behavior of the chromosomes is emphasized.

27

MITOSIS	MEIOSIS
Prophase	**MEIOSIS I** **Prophase I**

Replicated chromosome

Parent cell (before chromosome replication)

Chromosome replication

2n = 4

Chromosome replication

Tetrad (pair of replicated homologous chromosomes) formed by synapsis

Metaphase

Individual chromosomes align at the metaphase plate.

Metaphase I

Homologous chromosomes (tetrads) align randomly at the metaphase plate.

Anaphase Telophase

Homologous chromosomes separate but sister chromatids remain together during anaphase I.

Daughter cells of meiosis I

Anaphase I Telophase I

Daughter cells of mitosis

Sister chromatids separate during anaphase.

2n   2n

Sister chromatids separate, becoming individual chromosomes during anaphase II.

MEIOSIS II

n   n   n   n

Daughter cells of meiosis II (usually gametes)

PROPERTY	MITOSIS	MEIOSIS
DNA replication	Occurs during interphase, before mitosis begins.	Occurs during interphase before meiosis I but not meiosis II.
Number of divisions	One, consisting of prophase, metaphase, anaphase, and telophase.	Two, each consisting of prophase, metaphase, anaphase, and telophase.
Synapsis of homologous chromosomes	Does not occur.	Occurs during prophase I; tetrads form, allowing crossing over to occur.
Number of daughter cells and genetic composition	Two. Each diploid (2n) cell is identical to the parent cell.	Four. Each haploid (n) cell contains half as many chromosomes as the parent cell and is genetically different from the parent cell.
Roles in the body	• Development of multicellular adult from zygote • Tissue repair and renewal • Ensures constant genetic makeup of all body cells	• Produces cells for reproduction (gametes) • Introduces genetic variability in gametes • Reduces chromosomal number by half so that fertilization can restore the normal diploid chromosome number (2n = 46).

Figure 27.4 Comparison of mitosis and meiosis in a parent cell with a diploid number (2n) of 4. (Not all the phases of mitosis and meiosis are shown, and only one possible alignment of the tetrads at metaphase I is illustrated.)

27

In meiosis II, the sister chromatids in the two daughter cells of meiosis I are simply parceled out equally among four resulting daughter cells.

Meiosis Compared with Mitosis

Now that you have followed meiosis to its conclusion, take a moment to compare mitosis and meiosis using **Figure 27.4**. Make sure you understand the different functions of these processes in the body. Mitosis ensures that all cells in the body have the same complement of DNA so that they are capable of normal growth and repair. In contrast, meiosis accomplishes two important tasks:

- It reduces the chromosomal number by half.
- It introduces genetic variability.

The random alignment of the homologous pairs during meiosis I (Figure 27.3) provides tremendous variability in the resulting gametes by scrambling genetic characteristics of the two parents in different combinations. Variability is also increased by crossing over during late prophase I. (We will revisit this process in Chapter 29.) As a result, it is likely that no two gametes are exactly alike and all are different from the original parent cells, which is the whole point of sexual reproduction.

Check Your Understanding

1. What is the HPG axis?
2. What is the final outcome of meiosis?

3. **DRAW** Consider a cell with a diploid number of six. Draw this cell at metaphase I of meiosis and its daughter cells at anaphase II. Show the correct number and placement of chromosomes.

For answers, see Answers Appendix.

PART 1

ANATOMY OF THE MALE REPRODUCTIVE SYSTEM

The sperm-producing *male gonads* are the **testes** (tes′tez) that lie within the *scrotum*. From the testes, the sperm are delivered out of the body through a system of ducts:

1. The *epididymis*
2. The *ductus deferens*
3. The *ejaculatory duct*
4. The *urethra*, which opens to the outside at the tip of the *penis*

The accessory sex glands, which empty their secretions into the ducts during ejaculation, are:

- The *seminal glands*
- *Prostate*
- The *bulbo-urethral glands*

Take a moment to trace the duct system in **Figure 27.5**, and identify the testis and accessory glands before continuing.

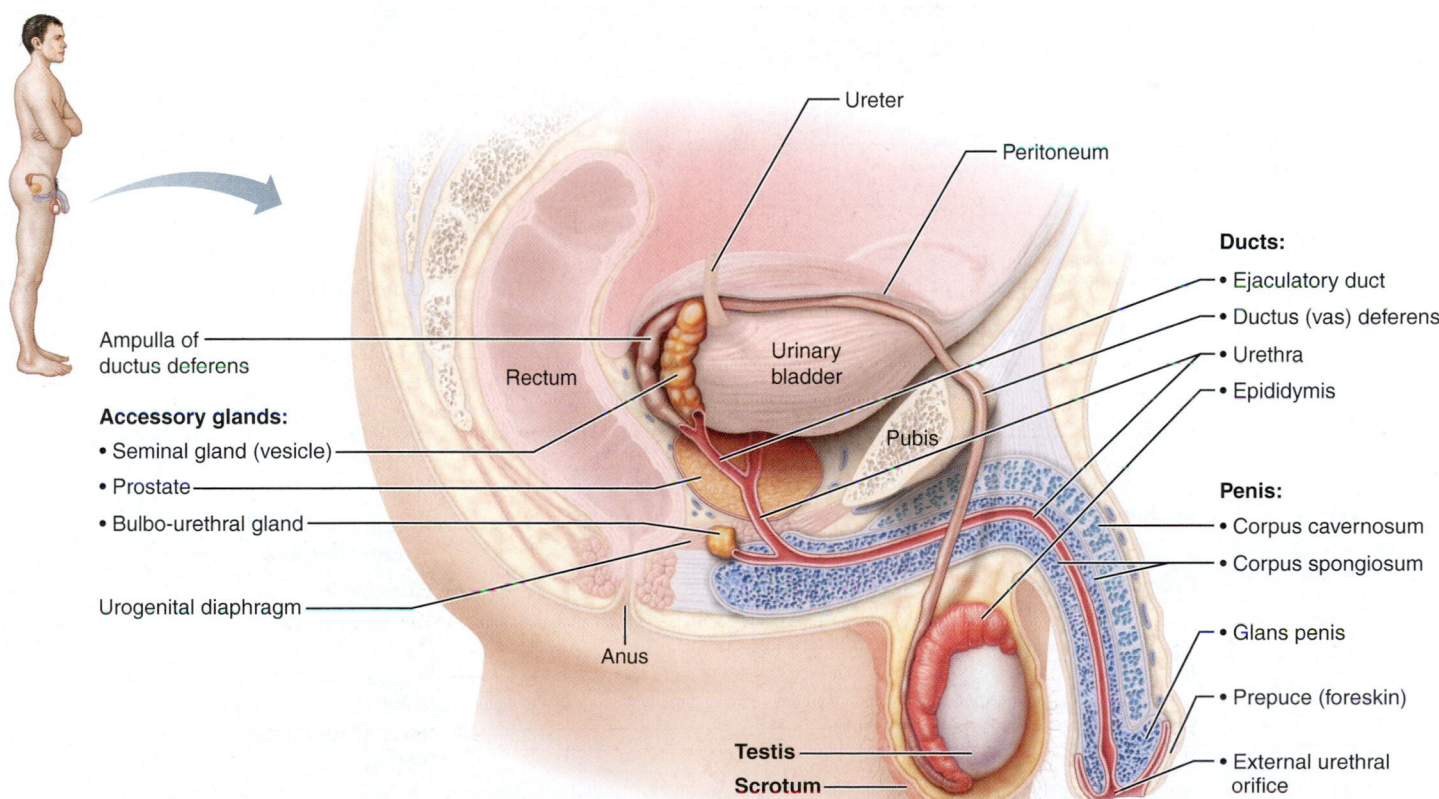

Ureter
Peritoneum

Ducts:
- Ejaculatory duct
- Ductus (vas) deferens
- Urethra
- Epididymis

Ampulla of ductus deferens

Rectum

Urinary bladder

Pubis

Accessory glands:
- Seminal gland (vesicle)
- Prostate
- Bulbo-urethral gland

Urogenital diaphragm

Anus

Penis:
- Corpus cavernosum
- Corpus spongiosum
- Glans penis
- Prepuce (foreskin)
- External urethral orifice

Testis
Scrotum

Figure 27.5 Reproductive organs of the male, sagittal view. A portion of the pubis is shown to illustrate the position of the ductus deferens as it enters the pelvic cavity.

27

27.2 The testes are enclosed and protected by the scrotum

Learning Outcome

▶ Describe the structure and function of the testes, and explain the importance of their location in the scrotum.

The Scrotum

The **scrotum** (skro'tum; "pouch") is a sac of skin and superficial fascia that hangs outside the abdominopelvic cavity at the root of the penis (Figure 27.5 and **Figure 27.6**). It is covered with sparse hairs, and contains paired oval testes. A midline *septum* divides the scrotum, providing a compartment for each testis.

This seems a rather vulnerable location for a man's testes, which contain his entire ability to father offspring. However, because viable sperm cannot be produced in abundance at core body temperature (37°C), the superficial location of the scrotum, which provides a temperature about 3°C lower, is an essential adaptation.

Furthermore, the scrotum is affected by temperature changes. When it is cold, the testes are pulled closer to the warm body wall. The scrotal skin wrinkles, decreasing its surface area and increasing its thickness to reduce heat loss. Two sets of muscles act to help maintain a fairly constant intrascrotal temperature:

- The **dartos muscle** (dar'tos; "skinned"), a layer of smooth muscle in the superficial fascia, wrinkles the scrotal skin.
- The **cremaster muscles** (kre-mas'ter; "a suspender"), bands of skeletal muscle that arise from the internal oblique muscles of the trunk, elevate the testes.

The Testes

Each testis is approximately 4 cm (1.6 inches) long by 2.5 cm (1 inch) wide and is surrounded by two tunics. The outer two-layered **tunica vaginalis** (vaj"ĭ-nal'is) is derived from an out-pocketing of the peritoneum (Figure 27.6 and **Figure 27.7a**). Deep to this serous layer is the **tunica albuginea** (al"bu-jin'e-ah; "white coat"), the fibrous capsule of the testis.

Septa extending inward from the tunica albuginea divide the testis into about 250 wedge-shaped *lobules*. Each contains one to four tightly coiled **seminiferous tubules** (sem"ĭ-nif'er-us; "sperm-carrying"), the actual "sperm factories."

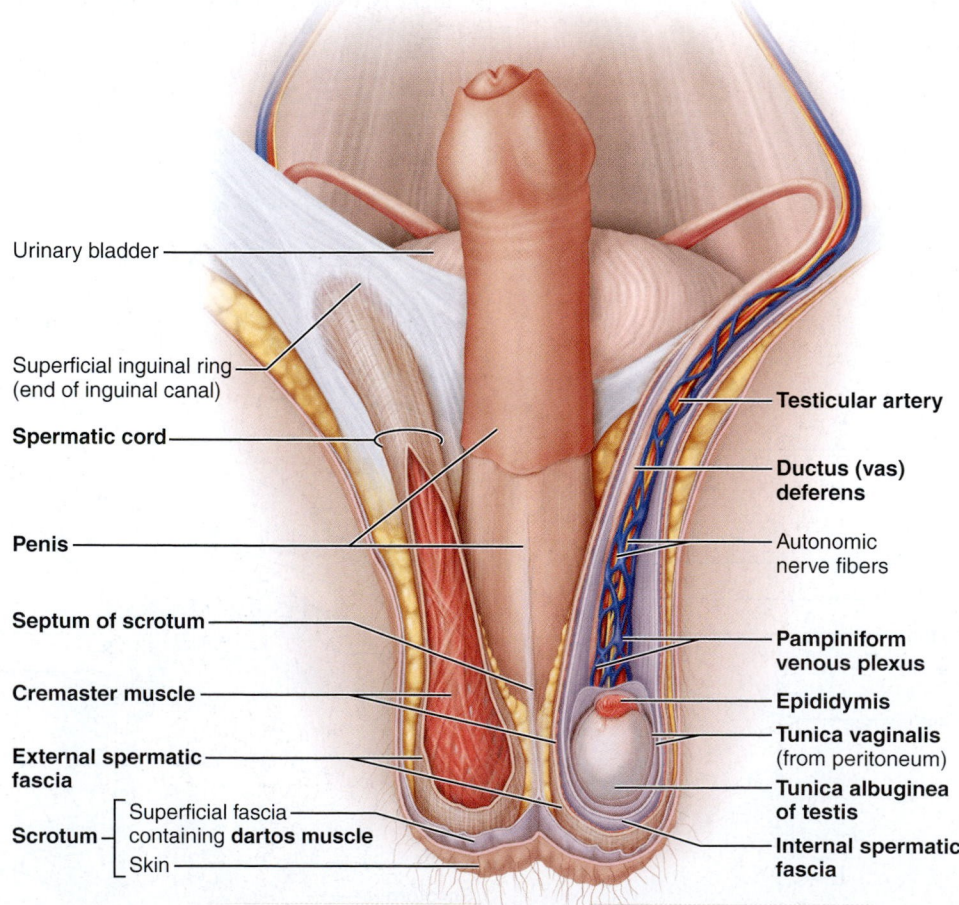

Urinary bladder
Superficial inguinal ring (end of inguinal canal)
Spermatic cord
Penis
Septum of scrotum
Cremaster muscle
External spermatic fascia
Scrotum — Superficial fascia containing **dartos muscle** — Skin

Testicular artery
Ductus (vas) deferens
Autonomic nerve fibers
Pampiniform venous plexus
Epididymis
Tunica vaginalis (from peritoneum)
Tunica albuginea of testis
Internal spermatic fascia

Figure 27.6 Relationships of the testis to the scrotum and spermatic cord. The scrotum has been opened and its anterior portion removed.

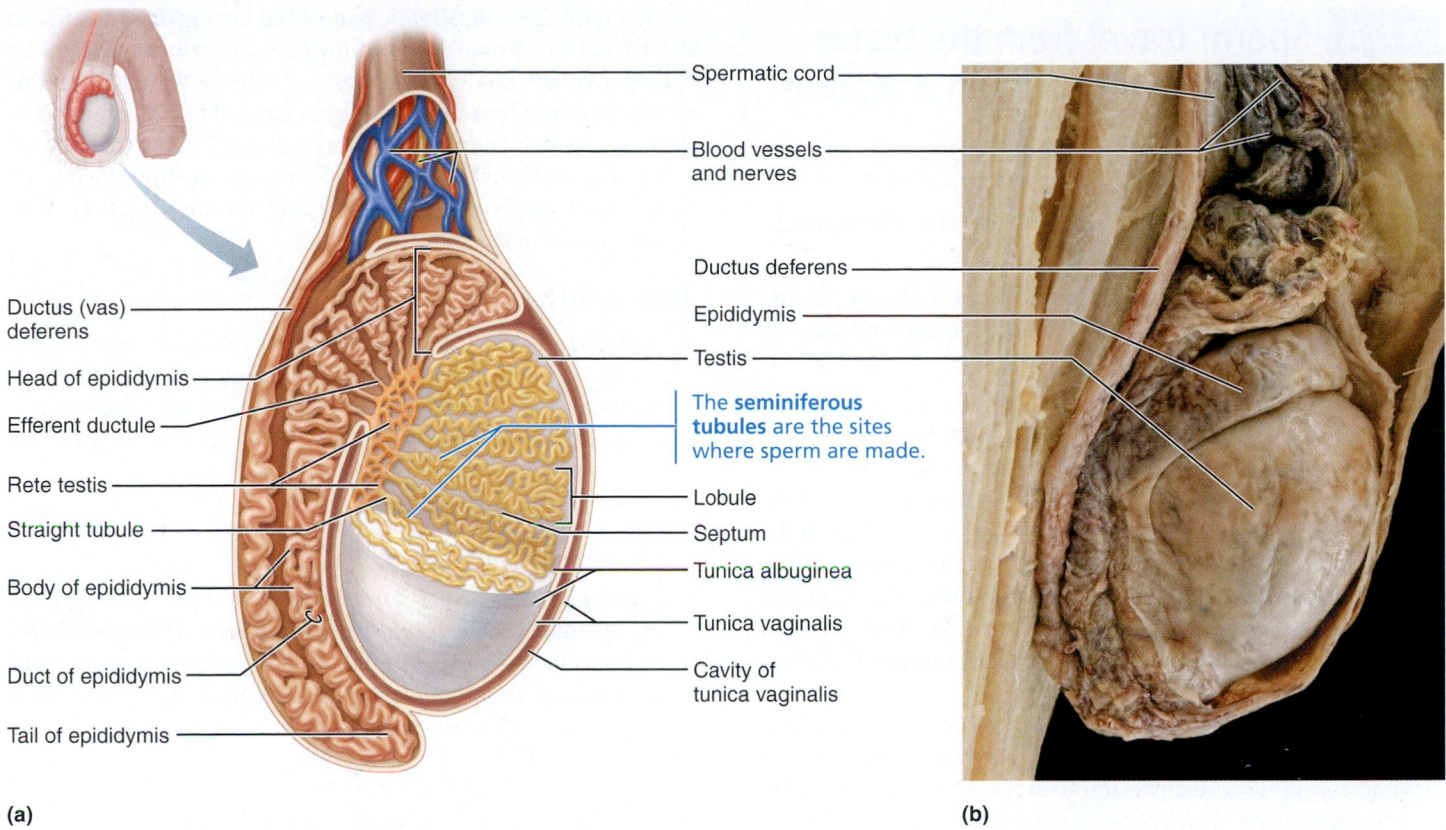

Spermatic cord

Blood vessels and nerves

Ductus deferens

Epididymis

Testis

Ductus (vas) deferens

Head of epididymis

Efferent ductule

The **seminiferous tubules** are the sites where sperm are made.

Rete testis

Lobule

Straight tubule

Septum

Body of epididymis

Tunica albuginea

Tunica vaginalis

Duct of epididymis

Cavity of tunica vaginalis

Tail of epididymis

(a)

(b)

Figure 27.7 Structure of the testis. (a) Partial sagittal section through the testis and epididymis. The anterior aspect is to the right. **(b)** External view of a testis from a cadaver; same orientation as in (a).

The seminiferous tubules of each lobule converge to form a **straight tubule** that conveys sperm into the **rete testis** (re′te), a tubular network on the posterior side of the testis. From the rete testis, sperm leave the testis through the *efferent ductules* and enter the *epididymis* (ep″ĭ-did′ĭ-mis), which hugs the external testis surface posteriorly. The immature sperm pass through the head, the body, and then move into the tail of the epididymis, where they are stored until ejaculation.

The long **testicular arteries**, which branch from the abdominal aorta superior to the pelvis (see *gonadal arteries*, ◄ **Figure 19.26c, p. 746**), supply the testes. The **testicular veins** draining the testes arise from a network called the **pampiniform venous plexus** (pam-pin′ĭ-form; "tendril-shaped") that surrounds the portion of each testicular artery within the scrotum like a climbing vine (Figure 27.6). The cooler venous blood in each pampiniform plexus absorbs heat from the arterial blood, cooling it before it enters the testes. In this way, these plexuses help to keep the testes at their cool homeostatic temperature.

Both divisions of the autonomic nervous system serve the testes, and when the testes are hit forcefully, associated sensory nerves transmit impulses that result in agonizing pain and nausea. A connective tissue sheath encloses nerve fibers, blood vessels, and lymphatics. Together with the ductus deferens, these structures make up the **spermatic cord**, which passes through the inguinal canal (Figure 27.6).

HOMEOSTATIC IMBALANCE 27.1   CLINICAL

Although *testicular cancer* is relatively rare (affecting one of every 40,000 males each year), it is the most common cancer in young men ages 15 to 35. The most important risk factor for this cancer is *cryptorchidism* (nondescent of the testes, see p. 1084).

The most common sign of testicular cancer is a painless solid mass. Testicular cancer has an impressive cure rate. Over 96% of cases are cured by surgical removal of the cancerous testis (*orchiectomy*) alone or in combination with radiation therapy or chemotherapy.

Check Your Understanding

4. Name the tubular structures in the testes that act as "sperm factories."
5. Muscle activity and the pampiniform venous plexus help to keep the temperature of the testes at homeostatic levels. How do they do that?
6. **APPLY** Carlos and Jennifer have been trying to start a family. Carlos spends his daily two-hour commute working with his laptop on his lap. How might this affect his fertility?

For answers, see Answers Appendix.

27

27.3 Sperm travel from the testes to the body exterior through a system of ducts

Learning Outcome

▶ Compare and contrast the roles of each part of the male reproductive duct system.

In order (proximal to distal), the **accessory ducts** are the epididymis, ductus deferens, ejaculatory duct, and urethra.

The Epididymis

Sperm are not yet fully functional when they leave the testes. The **epididymis** (*epi* = beside; *didym* = the testes) is where sperm mature. It is a comma-shaped organ that arches over the posterior and lateral side of the testis (Figures 27.5 and 27.7). Its *head* contains efferent ductules that empty into the **duct of the epididymis**, a highly coiled duct that completes the head and forms all of the *body* and *tail* of this organ. With an uncoiled length of over 6 m (20 feet), the duct of the epididymis is longer than the entire intestine!

Some pseudostratified epithelial cells of the duct mucosa exhibit long, nonmotile microvilli (*stereocilia*). The huge surface area of these stereocilia allows them to absorb excess testicular fluid and to pass nutrients to the many sperm stored temporarily in the lumen.

The immature, nearly nonmotile sperm that leave the testis are moved slowly along the duct of the epididymis through fluid that contains several kinds of antimicrobial proteins, including defensins. As they move along its tortuous course (a trip that takes about 20 days), the sperm gain the ability to swim.

Sperm can be stored in the epididymis for several months. If they are held longer, epithelial cells of the epididymis eventually phagocytize them.

Sperm are ejaculated from the epididymis, not the testes. When a male ejaculates, the smooth muscle in the ducts of the epididymis contracts, expelling sperm into the next segment of the duct system, the *ductus deferens*.

The Ductus Deferens and Ejaculatory Duct

The **ductus deferens** (duk′tus def′er-ens; "carrying away"), or *vas deferens*, transports sperm during ejaculation. From the epididymis, it runs upward (as part of the spermatic cord) through the inguinal canal into the pelvic cavity, a distance of about 45 cm (18 inches) (Figure 27.5). Easily palpated as it passes anterior to the pubic bone, it then loops medially over the ureter and descends along the posterior bladder wall (**Figure 27.8**). Its end expands to form the **ampulla** of the ductus deferens and then joins with the duct of the seminal gland to form the short **ejaculatory duct**. After entering the prostate, each ejaculatory duct empties into the urethra.

Like that of the epididymis, the mucosa of the ductus deferens is pseudostratified epithelium. However, its muscular layer is extremely thick and the duct feels like a hard wire when squeezed between the fingertips. During ejaculation, the smooth muscle in its walls creates strong peristaltic waves that squeeze the sperm forward along the tract and into the urethra.

As Figure 27.7 illustrates, part of the ductus deferens lies in the scrotal sac. Some men opt for permanent birth control by having a **vasectomy** (vah-sek′to-me; "cutting the vas"). In this relatively minor operation, the physician makes a small incision into the scrotum and then cuts through and ligates (ties off) or cauterizes each ductus deferens. Sperm are still produced, but they can no longer reach the body exterior. Instead, they deteriorate and are phagocytized.

The Urethra

The **urethra** is the terminal portion of the male duct system (Figures 27.5 and 27.8). It transports urine and semen (at different times), so it serves both the urinary and reproductive systems. As we saw in Chapter 25 (◀ p. 1004), its three regions are:

- *Prostatic urethra*, the portion surrounded by the prostate.
- *Intermediate part of the urethra* (or membranous urethra) in the urogenital diaphragm.
- *Spongy urethra*, which runs through the penis and opens to the outside at the *external urethral orifice*. The spongy urethra accounts for 75% of urethral length. Its mucosa contains scattered *urethral glands* that secrete lubricating mucus into the lumen just before ejaculation.

Check Your Understanding

7. Name the organs of the male duct system in order, from the epididymis to the body exterior. Which of these organs runs from the scrotum into the pelvic cavity?

8. What are two functions of the stereocilia on the epididymal epithelium?

For answers, see Answers Appendix.

27.4 The penis is the copulatory organ of the male

Learning Outcome

▶ Describe the location, structure, and function of the penis.

The **penis** is a copulatory organ that delivers sperm into the female reproductive tract (Figures 27.5 and 27.8). The penis and scrotum, which hang suspended from the perineum, make up the **external genitalia** of the male.

The penis consists of an attached root and a free *body* that ends in an enlarged tip, the **glans penis**. The skin covering the penis is loose, and extends distally around the glans to form a cuff called the **prepuce** (pre′pūs), or **foreskin**. Sometimes, the foreskin is surgically removed shortly after birth, a procedure called *circumcision* ("cutting around"). While circumcision is not medically necessary, it has been shown to decrease the risk of acquiring HIV. It also significantly reduces the risk of other reproductive system infections. About 60% of newborn boys in the United States are circumcised.

To understand penile anatomy, it is important to know that its dorsal and ventral surfaces are named in reference to the erect penis. Internally, the penis contains the spongy urethra and three long cylindrical bodies (*corpora*) of erectile tissue,

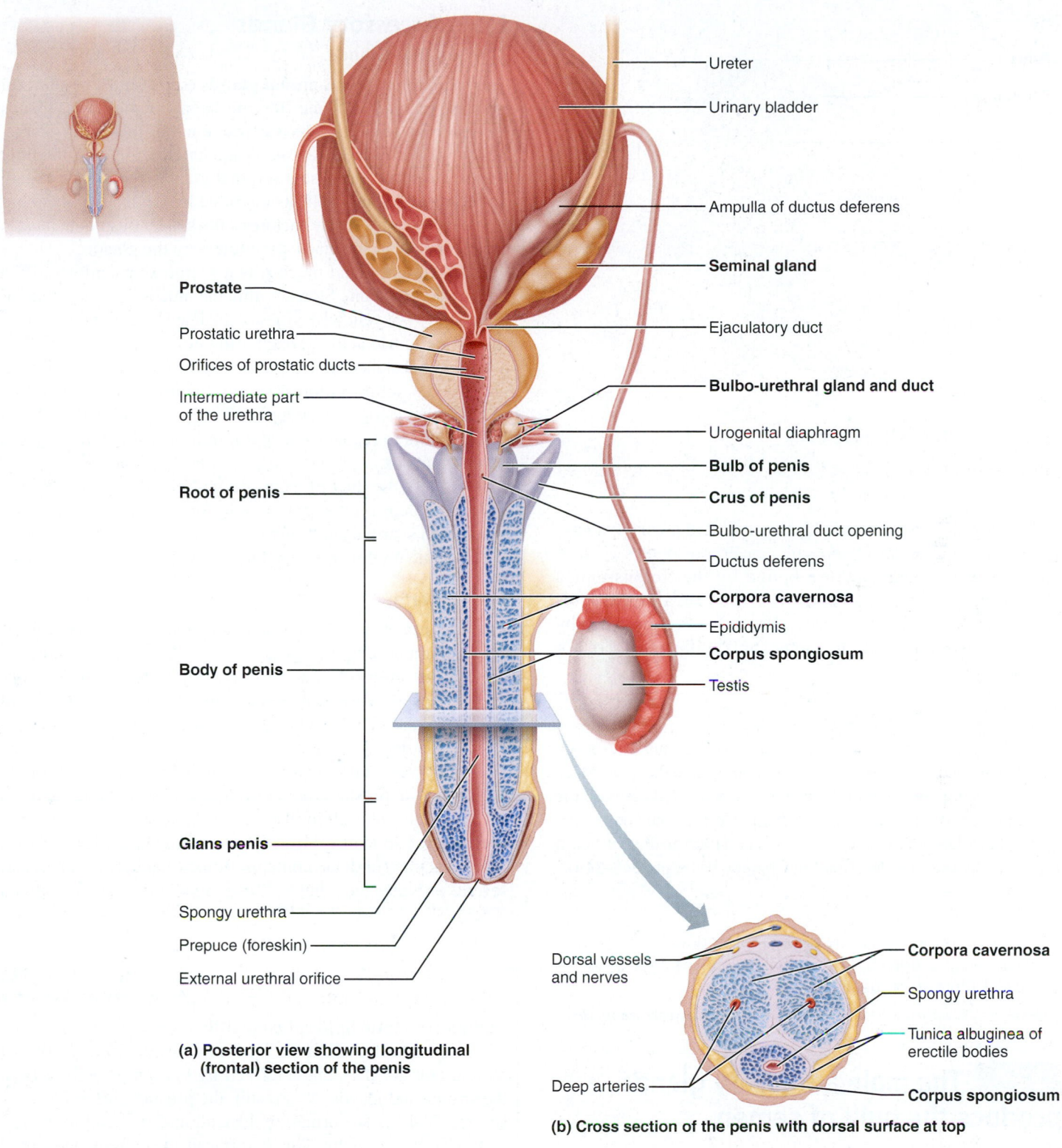

(a) Posterior view showing longitudinal (frontal) section of the penis

(b) Cross section of the penis with dorsal surface at top

Figure 27.8 Male reproductive structures.

each covered by a sheath of dense fibrous connective tissue. This *erectile tissue* is a spongy network of connective tissue and smooth muscle riddled with vascular spaces. During sexual excitement, the vascular spaces fill with blood, causing the penis to enlarge and become rigid. This condition, called *erection*, enables the penis to serve as the organ of copulation.

The midventral erectile body, the **corpus spongiosum** (spon"je-o'sum; "spongy body"), surrounds the urethra. It expands distally to form the glans and proximally to form the part of the root called the **bulb of the penis**. The sheetlike bulbospongiosus muscle covers the bulb externally and secures it to the urogenital diaphragm.

27

Penis

Scrotum

Pubic symphysis

Ischial tuberosity

Anus

Coccyx

Figure 27.9 The male perineum.

The paired dorsal erectile bodies, called the **corpora cavernosa** (kor'por-ah kă-ver-no'sah; "cavernous bodies"), make up most of the penis and are bound by the fibrous tunica albuginea. Their proximal ends form the **crura of the penis** (kroo'rah; "legs"; singular: crus). Each crus is surrounded by an ischiocavernosus muscle and anchors to the pubic arch of the bony pelvis.

The Male Perineum

The male **perineum** (per″ĭ-ne'um; "around the anus") suspends the scrotum and contains the root of the penis, and the anus. More specifically, it is the diamond-shaped region located between the pubic symphysis anteriorly, the coccyx posteriorly, and the ischial tuberosities laterally (**Figure 27.9**). The floor of the perineum is formed by muscles described in Chapter 10 (◀ pp. 348–349).

Check Your Understanding

9. What is the function of the corpora cavernosa?

10. What name is given to the external region where the penis and scrotum are suspended?

For answers, see Answers Appendix.

27.5 **The male accessory glands produce the bulk of semen**

Learning Outcomes
▶ **Compare the roles of the seminal glands and the prostate.**
▶ **Discuss the sources and functions of semen.**

The **accessory glands** include the paired seminal glands and bulbo-urethral glands and the single prostate (Figures 27.5 and 27.8).

Male Accessory Glands

The Seminal Glands

The secretions of the **seminal glands** (sem'ĭ-nul), or **seminal vesicles**, account for some 70% of semen volume. These glands lie on the posterior bladder surface. Each of these fairly large, hollow glands is about the shape and length (5–8 cm) of a little finger. However, because a seminal gland is pouched, coiled, and folded back on itself, its uncoiled length is actually about 15 cm. Its fibrous capsule encloses a thick layer of smooth muscle that contracts during ejaculation to empty the gland.

The seminal gland mucosa is a secretory pseudostratified columnar epithelium. Stored within the mucosa's honeycomb of blind alleys is a yellowish viscous alkaline fluid containing fructose sugar, citric acid, a coagulating enzyme, prostaglandins, and other substances that increase sperm motility or fertilizing ability. The yellow color of seminal fluid is due to a yellow pigment that fluoresces under UV light, a capability that allows investigators to recognize a sperm trail or residue in instances of sexual attack.

As noted, the duct of each seminal gland joins that of the ductus deferens on the same side to form the ejaculatory duct. Sperm and seminal fluid mix in the ejaculatory duct and enter the prostatic urethra together during ejaculation.

The Prostate

The **prostate** (pros'tāt) is a single doughnut-shaped gland about the size of a peach pit (Figures 27.5 and 27.8). It encircles the urethra just inferior to the bladder. Enclosed by a thick connective tissue capsule, it is made up of 20 to 30 compound tubuloalveolar glands embedded in a mass (stroma) of smooth muscle and dense connective tissue.

During ejaculation, prostatic smooth muscle contracts, squeezing the prostatic secretion into the prostatic urethra via several ducts. This fluid plays a role in activating sperm and accounts for up to one-third of the semen volume. It is a milky, slightly acidic fluid that contains citrate (a nutrient source) and several enzymes including one called prostate-specific antigen (PSA).

HOMEOSTATIC IMBALANCE 27.2 | CLINICAL

Benign prostatic hyperplasia (BPH) is a common prostatic disease, affecting about 50% of men at age 50 and 80% of men at age 70. It is characterized by benign growth of both the stroma and glands. As a result, the prostatic urethra is constricted. This makes urination difficult and can lead to incomplete bladder emptying. This in turn can cause frequent urinary tract infections and kidney damage due to the backup of urine. Treatment usually begins with drugs that relax the smooth muscle in the stroma and drugs that inhibit testosterone's stimulation of prostate growth. *Transurethral resection of the prostate* is a surgical procedure used to widen the prostatic urethra when drug treatment is insufficient.

Prostatitis, a common reason for a man to consult a urologist, is described in Related Clinical Terms on p. 1086.

Prostate cancer is the third most common cause of cancer death in men, and 1 in 6 men in the U.S. will develop it. Despite these seemingly alarming statistics, most men with prostate cancer die *with* it rather than *from* it. Prostate cancers range from extremely slow-growing tumors to highly aggressive fatal tumors. This cancer is usually symptomless in the early stages, but can be detected by a digital rectal exam. A blood test to detect the levels of prostate-specific antigen (PSA) was once used routinely for screening. However, its use for routine screening is no longer generally recommended because of its high rate of falsely positive results (elevated PSA, but no cancer).

Because many prostate cancers are slow growing and develop late in life, an acceptable treatment option for some early stage prostate cancers is simply to watch and wait. For more advanced cancers, surgery alone or in conjunction with radiation therapy is used. In a few men, prostate cancer spreads beyond the prostate, often into bone (◀ **p. 16**, PET scan). These prostate cancer metastases are often dependent on testosterone, so blocking testosterone synthesis or action can slow the progression of metastatic prostate cancer.

The Bulbo-Urethral Glands

The paired **bulbo-urethral glands** (bul″bo-u-re′thral) are pea sized. They are located inferior to the prostate in the urogenital diaphragm (Figures 27.5 and 27.8). They produce a thick, clear mucus, some of which drains into the spongy urethra where it neutralizes traces of acidic urine. In addition, this mucus lubricates the glans penis when a man becomes sexually excited.

Semen

Semen (se′men) is a milky white, somewhat sticky mixture of accessory gland secretions and sperm. The accessory gland secretions provide a transport medium and nutrients. They also contain chemicals that protect and activate the sperm and facilitate their movement. Mature sperm cells are streamlined cellular "torpedos" containing little cytoplasm or stored nutrients. Catabolism of the fructose in seminal gland secretions provides nearly all the fuel needed for sperm ATP synthesis.

Semen contains several substances that play many roles:

- Prostaglandins (◀ **pp. 47–48**) cause the viscosity of mucus guarding the entry (cervix) of the uterus to decrease. They also stimulate reverse peristalsis in the uterus, facilitating sperm movement through the female reproductive tract.
- The hormone relaxin and certain enzymes in semen enhance sperm motility.
- ATP provides energy.
- Certain ingredients suppress the immune response in the female's reproductive tract.
- Antibiotic chemicals destroy some bacteria.
- Clotting factors in semen coagulate it just after it is ejaculated. Coagulation causes the sperm to stick to the walls of the vagina and prevents the initially immobile sperm from draining out of the vagina. Soon after semen coagulates, enzymes liquefy the sticky mass and the sperm swim out and begin their journey through the female duct system.

Semen, as a whole, is alkaline (pH 7.2–8.0), which helps neutralize the acid environment of the male's urethra and the female's vagina, protecting the delicate sperm and enhancing their motility. Sperm are sluggish under acidic conditions (below pH 6).

The amount of semen propelled out of the male duct system during ejaculation is relatively small, only 2–5 ml and only 10% sperm, but there are between 20 and 150 million sperm per milliliter.

Check Your Understanding

11. For each of the following questions, name the glandular accessory organ and identify its location in the figure below: Which gland produces primarily mucus? Which gland produces the largest fraction of semen volume? Which gland is located at the junction of the ejaculatory duct and urethra?

12. What is semen?

13. **PREDICT** Arthur, age 68, has trouble urinating. What is his most probable condition?

For answers, see Answers Appendix.

PART 2

PHYSIOLOGY OF THE MALE REPRODUCTIVE SYSTEM

27.6 The male sexual response includes erection and ejaculation

Learning Outcome

▶ Describe the phases of the male sexual response.

Although there is more to it, the chief phases of the male sexual response are:

1. *Erection* of the penis, which allows it to penetrate the female vagina
2. *Ejaculation*, which expels semen into the vagina

These events are an example of cooperation between the two branches of the autonomic nervous system (◀ **p. 545**).

27

- Erection is under parasympathetic control.
- Ejaculation is under sympathetic control.

Erection

Erection, enlargement and stiffening of the penis, results from engorgement of the erectile bodies with blood. When a man is not sexually aroused, arterioles supplying the erectile tissue are constricted and the penis is flaccid. However, sexual excitement triggers a parasympathetic reflex that promotes release of nitric oxide (NO) (◀ pp. 423–424) locally. NO relaxes smooth muscle in the penile blood vessel walls, dilating these arterioles,* and the erectile bodies fill with blood. Expansion of the corpora cavernosa of the penis compresses their drainage veins, retarding blood outflow and maintaining engorgement. The corpus spongiosum expands as well, but not nearly as much as the cavernosa. Its main job is to keep the urethra open during ejaculation. Parasympathetic input also stimulates secretion from the bulbourethral glands.

Various sexual stimuli can initiate erection—tactile stimulation, and erotic sights, sounds, and smells. The CNS responds by activating parasympathetic neurons that innervate the internal pudendal arteries serving the penis. Sometimes erection is induced solely by emotional or higher mental activity (the thought of a sexual encounter). Emotions and thoughts can also inhibit erection, causing vasoconstriction and returning the penis to its flaccid state.

Ejaculation

Ejaculation (*ejac* = to shoot forth) is the propulsion of semen from the male duct system. When impulses provoking erection reach a critical level, a spinal reflex is initiated, and a massive discharge of nerve impulses occurs over the sympathetic nerves serving the genital organs (largely at the level of L_1 and L_2). As a result:

1. The ductus deferens, prostate, and seminal glands contract, emptying their contents into the prostatic urethra. At the same time, the bladder internal sphincter muscle constricts, preventing expulsion of urine or reflux of semen into the bladder.

2. Semen in the urethra triggers a spinal reflex through somatic motor neurons. The bulbospongiosus muscles of the penis undergo a rapid series of contractions, propelling semen from the urethra at a speed of about 500 cm/s (11 mi/h). These rhythmic contractions are accompanied by intense pleasure (**orgasm**, or **climax**) and many systemic changes, such as generalized muscle contraction, rapid heartbeat, and elevated blood pressure.

Ejaculation is quickly followed by **resolution**, a period of muscular and psychological relaxation. Activity of sympathetic nerve fibers constricts the *internal pudendal arteries* (and penile arterioles), reducing blood flow into the penis, and activates small muscles that squeeze the cavernous bodies,

forcing blood from the penis into the general circulation. Although the erection may persist a little longer, the penis soon becomes flaccid again. After ejaculation, there is a *latent*, or *refractory, period* that ranges from minutes to hours, during which a man is unable to achieve another orgasm. The latent period lengthens with age.

> ### ⚖ HOMEOSTATIC IMBALANCE 27.3   CLINICAL
>
> Approximately 50% of American men over age 40 have some degree of **erectile dysfunction (ED)**, the inability to attain an erection. Temporary ED can be caused by psychological factors, alcohol, or certain drugs (antihypertensives, antidepressants, and others). Chronic ED is largely the result of problems with blood vessels (atherosclerosis, varicose veins) or the nervous system (stroke, penile nerve damage, multiple sclerosis). Diabetes mellitus, which damages both blood vessels and nerves, is often an underlying cause.
>
> As we've just described, attaining erection depends on parasympathetic stimulation triggering NO release. Sildenafil (Viagra®) and similar drugs enhance the effect of existing NO. Taken orally, these drugs can enhance blood flow to the penis, allowing erection to occur. ●●●●●●●

Check Your Understanding

14. What is erection and which division of the ANS regulates it?

15. What occurs during resolution and what brings it about?

For answers, see Answers Appendix.

27.7 Spermatogenesis is the sequence of events that leads to formation of sperm

Learning Outcomes

▶ Describe the histology of the seminiferous tubules of the testes and the roles of sustentocytes.

▶ Outline the events of spermatogenesis.

Spermatogenesis (sper″mah-to-jen′ĕ-sis; "sperm formation") is the process of forming male gametes. It occurs in the seminiferous tubules of the testes and is the hallmark of sexual maturity. Spermatogenesis begins at puberty, around the age of 14 years, and continues throughout life. Every day, a healthy adult male makes about 90 million sperm. It seems that nature has made sure that the human species will not be endangered for lack of sperm!

Histology of the Seminiferous Tubules

Recall from p. 1048 that the seminiferous tubules are the "sperm factories" of the testes. Each seminiferous tubule consists of a thick stratified epithelium surrounding a central fluid-containing lumen (**Figure 27.10**). A cross section through a tubule reveals four important types of cells:

---

*Erection of the penis is one of the rare examples of parasympathetic control of arterioles.

Seminiferous
tubules inside
testis

Sperm

Sustentocytes:
• Surround, support,
and nourish
developing sperm

Spermatogenic cells:
• In tubule epithelium
• Sperm-forming cells;
surrounded by sustentocytes

Interstitial endocrine cells:
• In connective tissue
between tubules
• Secrete testosterone

Myoid cells:
• Smooth muscle–like
cells that surround
seminiferous tubules

Seminiferous tubule

(a)

(b)

Figure 27.10 Histology of the seminiferous tubule. Cross sections.
(a) Photomicrograph (270×). **(b)** Scanning electron micrograph (215×).

Practice Histology questions: MasteringA&P®
Study Area > Lab Tools > PAL

• **Sustentocytes** are large, columnar cells. They act as supporting cells and play several roles in sperm formation, as described shortly.

• **Spermatogenic cells** (*spermatogenic* = sperm forming) are cells that give rise to sperm. They are found enveloped by the sustentocytes and can be seen at many different stages of cell division. We will describe them shortly as well.

• **Myoid cells** are smooth muscle–like cells that surround each seminiferous tubule in three to five layers. By contracting rhythmically, myoid cells may help to squeeze sperm and testicular fluids through the tubules and out of the testes.

• **Interstitial endocrine cells**, also called *Leydig cells*, lie in the soft connective tissue surrounding the seminiferous tubules. These cells produce androgens (most importantly *testosterone*) and small amounts of estrogens, which they secrete into the surrounding interstitial fluid. Notice that completely different cell populations carry out the sperm-producing and hormone-producing functions of the testis.

27

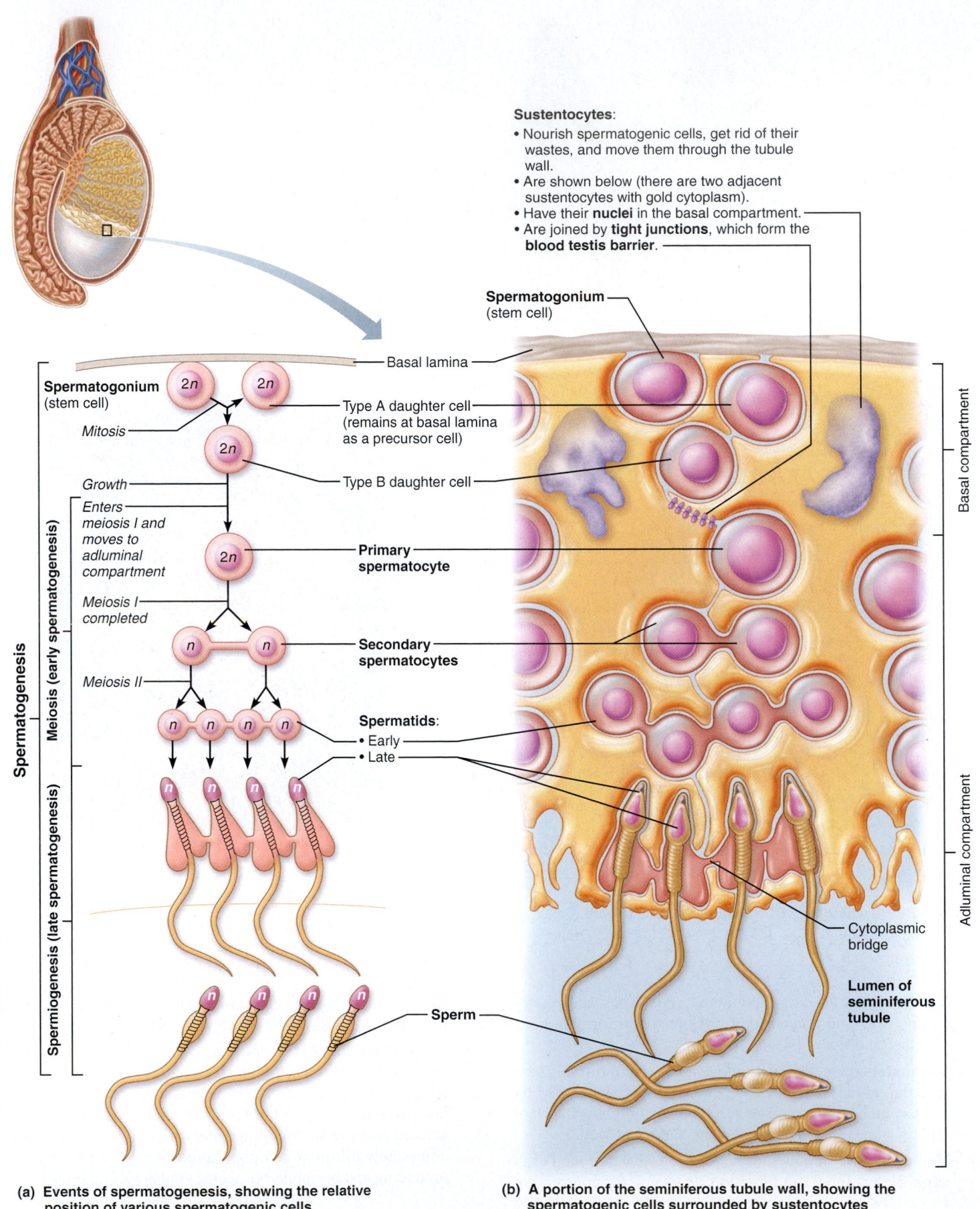

Sustentocytes:
- Nourish spermatogenic cells, get rid of their wastes, and move them through the tubule wall.
- Are shown below (there are two adjacent sustentocytes with gold cytoplasm).
- Have their **nuclei** in the basal compartment.
- Are joined by **tight junctions**, which form the **blood testis barrier**.

Spermatogonium
(stem cell)

Basal lamina

Spermatogonium
(stem cell)

Type A daughter cell
(remains at basal lamina
as a precursor cell)

Mitosis

Type B daughter cell

Growth

Enters meiosis I and moves to adluminal compartment

Primary spermatocyte

Meiosis I completed

Secondary spermatocytes

Meiosis II

Spermatids:
- Early
- Late

Cytoplasmic bridge

Sperm

Lumen of seminiferous tubule

Basal compartment

Adluminal compartment

Spermatogenesis

Meiosis (early spermatogenesis)

Spermiogenesis (late spermatogenesis)

(a) Events of spermatogenesis, showing the relative position of various spermatogenic cells

(b) A portion of the seminiferous tubule wall, showing the spermatogenic cells surrounded by sustentocytes

Figure 27.11 Spermatogenesis.

27

Spermatogenesis: Summary of Events in the Seminiferous Tubules

Spermatogenic cells give rise to sperm via the series of divisions and cellular transformations described next (**Figure 27.11**).

Mitosis of Spermatogonia: Forming Spermatocytes

The outermost tubule cells, which are in direct contact with the epithelial basal lamina, are stem cells called **spermatogonia** (sper″mah-to-go′ne-ah; "sperm seed"). The spermatogonia divide more or less continuously by *mitosis* and, until puberty, all their daughter cells become spermatogonia.

Spermatogenesis begins during puberty, and from then on, each mitotic division of a spermatogonium results in two distinctive daughter cells:

- The **type A daughter cell** remains at the basal lamina to maintain the pool of dividing germ cells.
- The **type B daughter cell** gets pushed toward the lumen, where it becomes a **primary spermatocyte** destined to produce four sperm.

(To keep these cell types straight, remember that just as the letter A is always at the beginning of our alphabet, a type A cell is always at the tubule basal lamina ready to begin a new generation of gametes.)

Meiosis: Spermatocytes to Spermatids

Each primary spermatocyte undergoes meiosis I, forming two smaller haploid cells called **secondary spermatocytes** (Figure 27.11). The secondary spermatocytes continue on rapidly into meiosis II. Their daughter cells, called **spermatids** (sper′mah-tidz), are small round cells, with large spherical nuclei, seen closer to the lumen of the tubule. Midway through spermatogenesis, the developing sperm "turn off" nearly all their genes and compact their DNA into dense pellets.

Spermiogenesis: Spermatids to Sperm

Each spermatid has the correct chromosomal number for fertilization (*n*), but is nonmotile. It still must undergo a streamlining process called **spermiogenesis**, during which it elongates, sheds its excess cytoplasmic baggage, and forms a tail. Follow the details of this process in **Figure 27.12a** (1)–(7).

Each resulting **sperm** has a head, midpiece, and tail, which correspond roughly to *genetic*, *metabolic*, and *locomotor regions*,

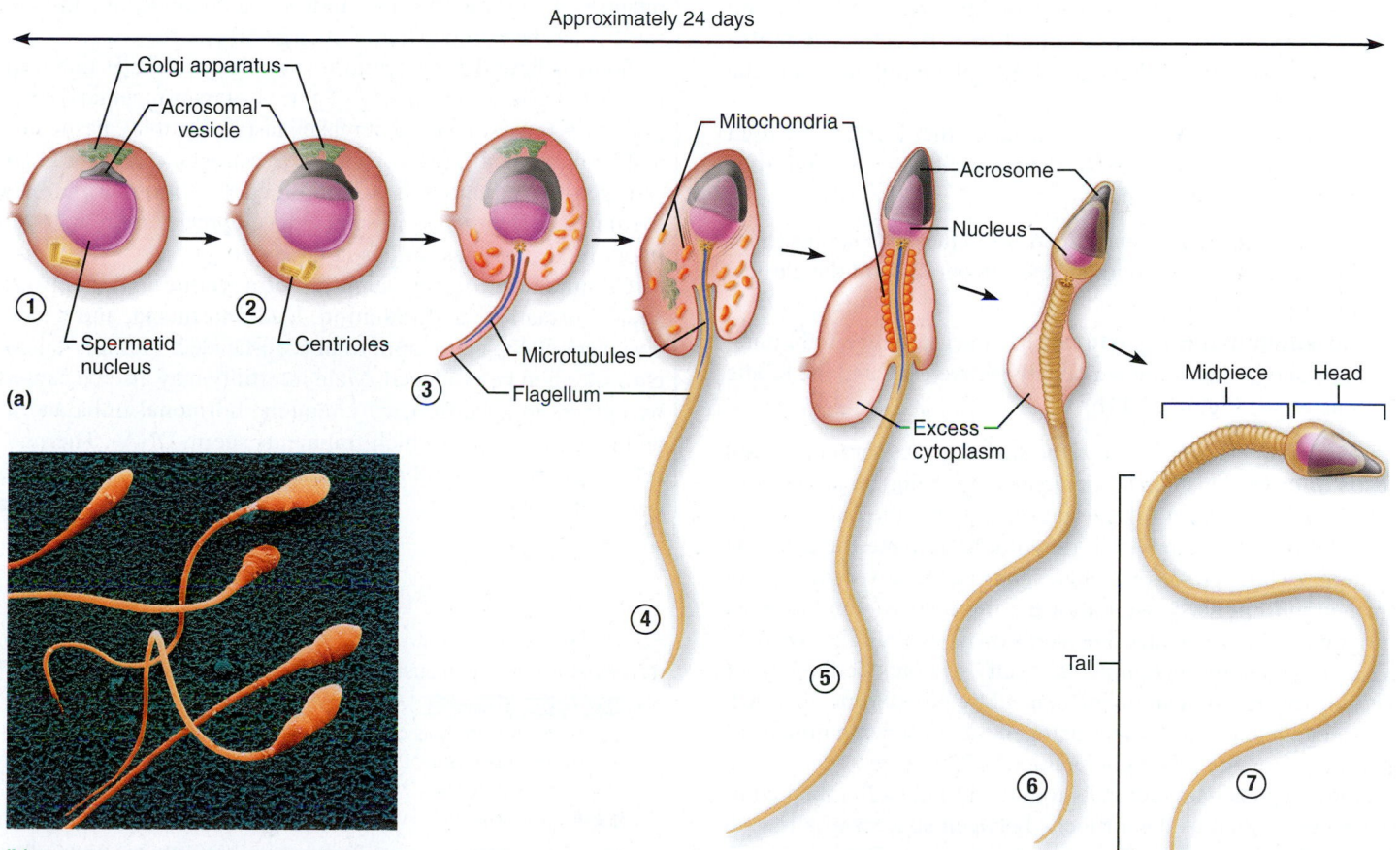

Approximately 24 days

Golgi apparatus
Acrosomal vesicle
Spermatid nucleus
Centrioles
Mitochondria
Microtubules
Flagellum
Acrosome
Nucleus
Excess cytoplasm
Midpiece   Head
Tail

(1) (2) (3) (4) (5) (6) (7)

(a)

(b)

Figure 27.12 Spermiogenesis: transformation of a spermatid into a functional sperm. (a) During the process of spermiogenesis: (1) The Golgi apparatus packages the acrosomal enzymes, (2) the acrosome forms at the anterior end of the nucleus and the centrioles gather at the opposite end, (3) microtubules form the flagellum, (4) mitochondria multiply and cluster around the proximal portion of the flagellum, and (5) excess cytoplasm sloughs off. (6) An immature sperm that has just been released from a sustentocyte. (7) Structure of a fully mature sperm. **(b)** Scanning electron micrograph of mature sperm (1400×).

respectively (Figure 27.12a ⑦). Sperm "pack" lightly—they only carry what they absolutely need.

The **head** of a sperm consists almost entirely of its flattened nucleus, which contains the compacted DNA. A helmetlike **acrosome** (ak′ro-sōm; "tip piece") adheres to the top of the nucleus (Figure 27.12a ⑤ and ⑥). Like a lysosome, the acrosome contains hydrolytic enzymes that, in this case, enable the sperm to penetrate and enter an egg.

The sperm **midpiece** contains mitochondria spiraled tightly around the microtubules of the tail. The long **tail** is a typical flagellum (◄ p. 90). The mitochondria provide the metabolic energy (ATP) needed for the whiplike movements of the tail that will propel the sperm along its way in the female reproductive tract.

Spermatogenesis—from formation of a primary spermatocyte to release of immature sperm into the lumen—takes 64 to 72 days. Sperm in the lumen are unable to "swim" and are incapable of fertilizing an egg. The pressure of the testicular fluid pushes them through the tubular system of the testes into the epididymis, where they gain increased motility and fertilizing power.

Role of the Sustentocytes

Throughout spermatogenesis, descendants of the same spermatogonium remain closely attached to one another by cytoplasmic bridges (see Figure 27.11). They are also surrounded by and connected to nonreplicating supporting cells called **sustentocytes** (or *Sertoli cells*) that extend from the basal lamina to the tubule lumen.

The sustentocytes are bound to each other laterally by tight junctions (◄ p. 67). In this way, they divide the seminiferous tubule into two compartments:

- The **basal compartment** extends from the basal lamina to the tight junctions and contains spermatogonia and the earliest primary spermatocytes.

- The **adluminal compartment** lies internal to the tight junctions and includes the meiotically active cells and the tubule lumen (see Figure 27.11b).

The tight junctions between the sustentocytes form the **blood testis barrier**. This barrier prevents the membrane antigens of differentiating sperm from escaping through the basal lamina into the bloodstream where they could activate the immune system. (This activation could occur because sperm are not formed until puberty, well after the immune system has been programmed to recognize a person's own tissues.) The spermatogonia, which are recognized as "self," are outside the barrier and for this reason can be influenced by bloodborne chemical messengers that prompt spermatogenesis. Following mitosis of the spermatogonia, the tight junctions of the sustentocytes open to allow type B daughter cells to pass into the adluminal compartment—much as the air locks between spacecraft allow an astronaut to pass.

In addition to forming the blood testis barrier, sustentocytes:

- Provide nutrients and essential signals to the dividing cells, even telling them to live or die.

- Move spermatocytes and spermatids along to the lumen.

- Secrete **testicular fluid** (rich in androgens and metabolic acids) that provides the transport medium for sperm in the lumen.

- Phagocytize faulty spermatogenic cells and excess cytoplasm sloughed off as the spermatids transform into sperm.

- Produce two proteins that help regulate spermatogenesis. **Androgen-binding protein (ABP)** keeps the concentration of testosterone near the spermatogenic cells high, which stimulates spermatogenesis. The hormone *inhibin* (◄ p. 1043) inhibits FSH release from the anterior pituitary, and so inhibits spermatogenesis. Levels of inhibin are proportional to sperm counts in males—the higher the sperm count, the higher the level of inhibin.

HOMEOSTATIC IMBALANCE 27.4 — CLINICAL

Roughly one in seven American couples seek treatment for *infertility*, often because of problems with sperm quality or quantity. According to some studies, a gradual decline in male fertility has been occurring in the past 50 years.

Some believe the main culprit is foreign molecules that have invaded our lives in a variety of forms—environmental toxins, phthalates (oily solvents that make plastics flexible), pesticides, and herbicides, all of which can have estrogenic effects. These estrogen-like compounds, which block the action of male sex hormones as they program sexual development, are now found in our food supply as well as in the air.

Common antibiotics such as tetracycline may suppress sperm formation, and radiation, lead, marijuana, and excessive alcohol can cause abnormal (two-headed, multiple-tailed, etc.) sperm to be produced. Male infertility may also be caused by defects in specific Ca^{2+} channels, hormonal imbalances, and oxidative stress (which fragments sperm DNA). Thermal-related events that inhibit sperm maturation include fever and overusing hot tubs. ●●●●●●●

Check Your Understanding

16. Describe the major structural and functional regions of a sperm.
17. What is the role of sustentocytes? Of interstitial endocrine cells?
18. **MAKE CONNECTIONS** Caleb and Sarah have been unable to conceive. An analysis of Caleb's semen reveals that his sperm are nonmotile (cannot propel themselves). Further investigation uncovers a defect in the motor protein dynein (◄ p. 90). In addition to infertility, what problems might Caleb have?

For answers, see Answers Appendix.

27.8 Male reproductive function is regulated by hypothalamic, anterior pituitary, and testicular hormones

Learning Outcome

▶ Discuss hormonal regulation of testicular function and the physiological effects of testosterone on male reproductive anatomy.

As we discussed previously (◀ p. 1042), hormonal interactions along the HPG axis regulate the production of gametes and sex hormones. You are now ready to delve into the details of the regulatory axis in males, shown schematically in **Figure 27.13**. Study this figure before you go on.

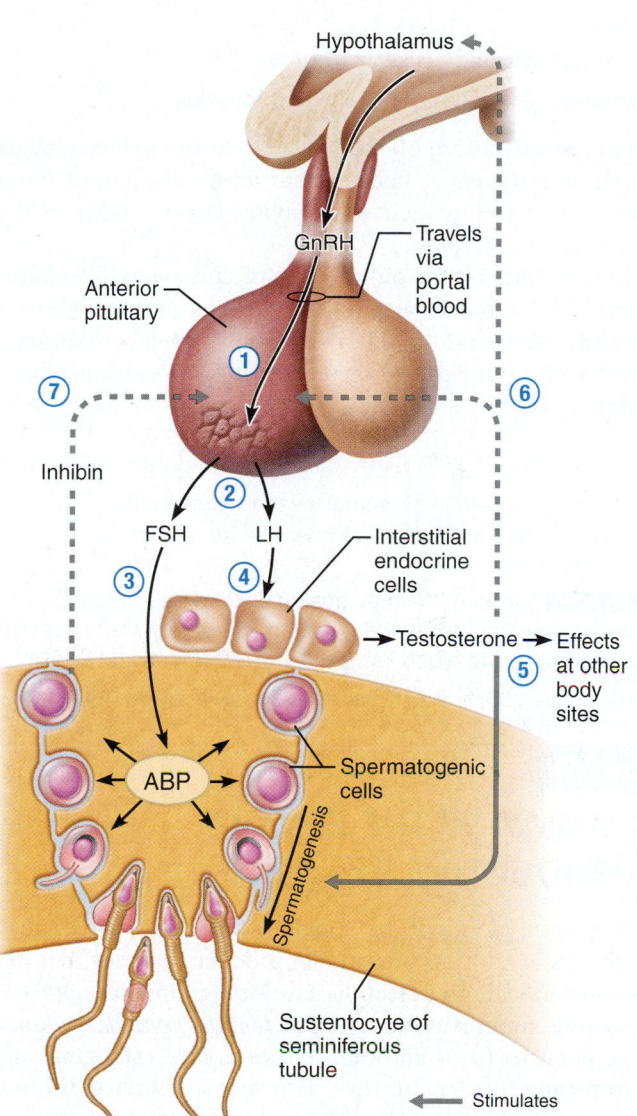

① The hypothalamus releases GnRH, which reaches the anterior pituitary via the hypophyseal portal veins.

② GnRH causes anterior pituitary gonadotropic cells to release FSH and LH.

③ FSH indirectly stimulates spermatogenesis by causing sustentocytes to release ABP, which keeps the local concentration of testosterone high.

④ LH stimulates interstitial endocrine cells to secrete testosterone, which is essential for spermatogenesis.

⑤ Testosterone acts at other body sites [e.g., to stimulate maturation of sex organs, development and maintenance of secondary sex characteristics, and libido (sex drive)].

⑥ Negative feedback by testosterone inhibits FSH and LH release from the anterior pituitary and GnRH release from the hypothalamus.

⑦ Inhibin released by sustentocytes feeds back on the anterior pituitary, decreasing FSH release.

Figure 27.13 Hormonal regulation of testicular function by the hypothalamic-pituitary-gonadal (HPG) axis. GnRH = gonadotropin-releasing hormone; FSH = follicle-stimulating hormone; LH = luteinizing hormone; ABP = androgen-binding protein.

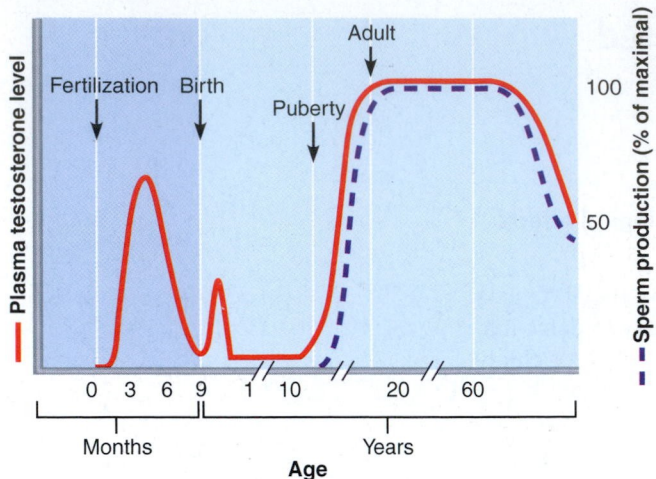

Figure 27.14 Plasma testosterone and sperm production levels versus age in males.

Because input from other brain areas also influences the hypothalamus, the whole HPG axis is under CNS control. In the absence of GnRH and gonadotropins, the testes atrophy, and sperm and testosterone production cease.

Development of male reproductive structures depends on prenatal secretion of male hormones, and for a few months before birth, a male infant has plasma gonadotropin and testosterone levels nearly two-thirds those of an adult male (**Figure 27.14**). After a brief rise in early infancy, blood levels of these hormones recede and they remain low throughout childhood. As puberty nears, higher levels of testosterone are required to suppress hypothalamic release of GnRH and the adult pattern of hormone interaction is established as we described on p. 1043.

The amount of testosterone and semen produced by the adult testes reflects a balance among the interacting hormones of the HPG axis. Once this balance is established during puberty (a process that takes about three years), the amount of testosterone and sperm produced remains fairly stable until late in life.

Mechanism and Effects of Testosterone Activity

Like all steroid hormones, testosterone is synthesized from cholesterol (◄ pp. 47–48, 603). It exerts its effects by activating specific genes, which results in enhanced synthesis of certain proteins in the target cells. (See Chapter 16, ◄ p. 606, for mechanisms of steroid hormone action.)

In some target cells, testosterone must be transformed into another steroid to exert its effects. In the prostate, testosterone is converted to *dihydrotestosterone* (*DHT*) before it can bind in the cell nucleus. In certain parts of the body (e.g., brain, bone, fat), testosterone is converted to *estradiol* (es″trah-di′ol), a female sex hormone, to bring about its effects.

As puberty ensues, testosterone not only prompts spermatogenesis but also has multiple anabolic effects throughout the body (see Table 27.2, p. 1079). It targets accessory reproductive organs—ducts, glands, and the penis—causing them to grow to

adult size and function. In adult males, normal plasma levels of testosterone maintain these organs. When the hormone is deficient or absent, all accessory organs atrophy, semen volume declines, and erection and ejaculation are impaired. Testosterone replacement therapy can remedy this situation.

Male secondary sex characteristics—that is, features induced in *nonreproductive* organs by the male sex hormones (mainly testosterone)—develop at puberty. These include:

- The appearance of pubic, axillary, and facial hair
- Enhanced hair growth on the chest or other body areas in some men
- Larynx enlargement, which causes deepening of the voice
- Thicker and oilier skin (which predisposes young men to acne)
- Increased size and density of bones
- Increased size and mass of skeletal muscles

The last two effects are often referred to as the *somatic effects* of testosterone (*soma* = body). Testosterone also boosts basal metabolic rate and influences behavior. It is the basis of the male libido (sex drive).

The testes are not the only source of androgens. The adrenal glands of both sexes also release them as gonadocorticoids (◄ p. 626). However, the relatively small amounts of adrenal androgens are insufficient to support normal testosterone-mediated functions.

Check Your Understanding

19. How does FSH indirectly stimulate spermatogenesis?

20. What are three secondary sex characteristics promoted by testosterone?

21. WHAT IF? Suppose the interstitial endocrine cells became dysregulated and secreted large amounts of testosterone. What would happen to blood levels of GnRH, FSH, and LH? Explain.

For answers, see Answers Appendix.

PART 3

ANATOMY OF THE FEMALE REPRODUCTIVE SYSTEM

The reproductive role of the female is far more complex than that of a male. Not only must she produce gametes, but her body must prepare to nurture a developing fetus for approximately nine months. **Ovaries**, the *female gonads*, produce female gametes (ova) and the sex hormones, **estrogens** and **progesterone** (pro-ges′tĕ-rōn). Estrogens include *estradiol*, *estrone*, and *estriol*, but estradiol is the most abundant and is most responsible for estrogenic effects in humans.

As illustrated in **Figure 27.15**, a female's **internal genitalia**—her ovaries and duct system—are mostly located in the pelvic cavity. The female's accessory ducts, from the vicinity of the ovary to the body exterior, are the *uterine tubes*, the *uterus*, and the *vagina*. They transport or otherwise serve the needs of the reproductive cells and a developing fetus. The external sex organs of females are referred to as **external genitalia**.

27

Figure 27.15 Organs of the female reproductive system, midsagittal section.

27.9 Immature eggs develop in follicles in the ovaries

Learning Outcome

▶ **Describe the location, structure, and function of the ovaries.**

The paired ovaries flank the uterus (**Figure 27.16**). Shaped like an almond and about twice as large, each ovary is held in place by several ligaments in the peritoneal cavity. The **ovarian ligament** anchors the ovary medially to the uterus; the **suspensory ligament** anchors it laterally to the pelvic wall; and the **mesovarium** (mez″o-va′re-um) suspends it in between (see Figures 27.15 and 27.17). The mesovarium is part of the **broad ligament**, a peritoneal fold that "tents" over the uterus and supports the uterine tubes, uterus, and vagina. The broad ligament encloses the ovarian ligaments.

The ovaries are served by the **ovarian arteries**, branches of the abdominal aorta (◀ **Figure 19.26c, p. 746**), and by the *ovarian branch of the uterine arteries*. The

Practice Histology questions:
MasteringA&P® Study Area > Lab
Tools > PAL

Figure 27.16 Photomicrograph of a mammalian ovary. (3×).

Ovarian follicles at various stages of development contain immature eggs.

Surface epithelium
Tunica albuginea
Cortex
Medulla

ovarian blood vessels reach the ovaries by traveling through the suspensory ligaments and mesovaria (**Figure 27.17a**).

Like each testis, each ovary is surrounded externally by a fibrous **tunica albuginea** (Figure 27.16). This, in turn, is covered externally by a layer of cuboidal epithelial cells called the surface epithelium (previously called the *germinal epithelium*) that is actually a continuation of the peritoneum. The ovary has an outer *cortex*, which houses the forming gametes, and an inner *medulla* containing the largest blood vessels and nerves.

Embedded in the highly vascular connective tissue of the ovary cortex are many tiny saclike structures called **ovarian follicles**. Each follicle consists of an immature egg, called an **oocyte** (o'o-sīt; *oo* = egg), encased by one or more layers of very different cells that we will describe later (pp. 1070, 1073).

As a rule, follicles of various sizes and stages of development can be seen in an ovary of a woman of child-bearing age (Figure 27.16). Each month, one of the ripening follicles ejects its oocyte from the ovary, an event called *ovulation* (see Figure 27.23 ⑤). In older women, the surfaces of the ovaries are scarred and pitted, revealing that many oocytes have been released.

Check Your Understanding

22. Briefly, what are the internal genitalia of a woman?

23. What are the two main things that the ovaries do?

For answers, see Answers Appendix.

27.10 The female duct system includes the uterine tubes, uterus, and vagina

Learning Outcome

▶ Describe the location, structure, and function of each of the organs of the female reproductive duct system.

Unlike the male duct system, which is continuous with the tubules of the testes, the female duct system has little or no actual contact with the ovaries. An ovulated oocyte is released into the peritoneal cavity, and some oocytes may be lost there.

The Uterine Tubes

The **uterine tubes** (u'ter-in), also called *fallopian tubes* or *oviducts*, form the initial part of the female duct system (Figures 27.15 and 27.17). They receive the ovulated oocyte and are the site where fertilization generally occurs.

Each uterine tube begins laterally near an ovary and ends medially where it empties into the superior part of the uterus. It has three regions:

- Laterally, the **infundibulum** (in"fun-dib'u-lum), a funnel-shaped region, is the opening of the uterine tube into the peritoneal cavity. The margin of the infundibulum is surrounded by ciliated, fingerlike projections called *fimbriae* (fim'bre-e; "fringe") that drape over the ovary.

- Next to the infundibulum is the wide **ampulla**. It forms half of the length of the uterine tube, and is the site where fertilization usually occurs.

- The **isthmus** (is'mus) forms the narrow medial third of the uterine tube. It empties into the superolateral region of the uterus.

Around the time of ovulation, the uterine tube performs a complex "dance" to capture oocytes. It bends to drape over the ovary while the fimbriae stiffen and sweep the ovarian surface. The beating cilia on the fimbriae create currents in the peritoneal fluid that carry an oocyte into the uterine tube, where it begins its journey toward the uterus.

The uterine tube contains sheets of smooth muscle, and its thick, highly folded mucosa contains both ciliated and nonciliated cells. The oocyte is carried toward the uterus by a combination of muscular peristalsis and the beating cilia. Nonciliated cells of the mucosa

Suspensory
ligament of ovary

Uterine (fallopian) tube

Ovarian
blood vessels

Fundus
of uterus

Lumen (cavity)
of uterus

Broad ligament
• Mesosalpinx
• Mesovarium
• Mesometrium

Ovary

Uterine tube
• Ampulla
• Isthmus
• Infundibulum
• Fimbriae

Ovarian
ligament

Round ligament of uterus

Wall of uterus
• Endometrium
• Myometrium
• Perimetrium

Body of uterus

Ureter

Uterine blood vessels

Isthmus

Uterosacral ligament

Internal os

Cervical canal

External os

Cardinal (transverse cervical)
ligament

Lateral fornix

Cervix

Vagina

(a) Posterior view

Figure 27.17 Internal female reproductive organs. (a) The posterior walls of the vagina,
uterus, and uterine tubes, and the broad ligament (a peritoneal fold) have been removed on the
right side to reveal the shape of the lumen of these organs.

have dense microvilli and produce a secretion that keeps the oocyte (and sperm, if present) moist and nourished.

Externally, the uterine tubes are covered by peritoneum and supported along their length by a short mesentery (part of the broad ligament) called the **mesosalpinx** (mez″o-sal′pinks; "mesentery of the trumpet"; *salpin* = trumpet), a reference to the trumpet-shaped uterine tube it supports (Figure 27.17).

HOMEOSTATIC
IMBALANCE 27.5   CLINICAL

The fact that the uterine tubes are not continuous with the ovaries places women at risk for *ectopic pregnancy*, in which an oocyte fertilized in the peritoneal cavity or distal portion of the uterine tube begins developing there. Because the tube lacks adequate mass and vascularization to support a full-term pregnancy, ectopic pregnancies tend to naturally abort, often with substantial bleeding. Because this situation is dangerous, it is important to seek medical attention.

Another potential problem is infection spreading into the peritoneal cavity from other parts of the reproductive tract. Sexually transmitted microorganisms, including chlamydia and

gonorrhea bacteria, sometimes infect the peritoneal cavity in this way, causing an extremely severe inflammation called **pelvic inflammatory disease (PID)**. Unless treated promptly with broad-spectrum antibiotics, PID can cause scarring of the narrow uterine tubes, resulting in sterility, and can even be fatal. PID is one of the major causes of female infertility. ● ● ● ● ● ●

The Uterus

The **uterus** (womb) lies in the pelvic cavity, anterior to the rectum and posterosuperior to the bladder (Figures 27.15 and 27.17). It is a hollow, thick-walled, muscular organ that receives, retains, and nourishes a fertilized ovum. In women who have never been pregnant, the uterus is about the size and shape of an inverted pear, but it is usually larger in women who have borne children. Normally, the uterus tilts anteriorly where it joins the vagina (see Figure 27.15), causing the uterus as a whole to be inclined forward, or *anteverted*. However, the organ is sometimes tilted backward, or *retroverted*.

The major portion of the uterus is referred to as the **body** (Figure 27.17). The rounded region superior to the entrance of the uterine tubes is the **fundus**, and the slightly narrowed region between the body and the cervix is the *isthmus*. The **cervix** of

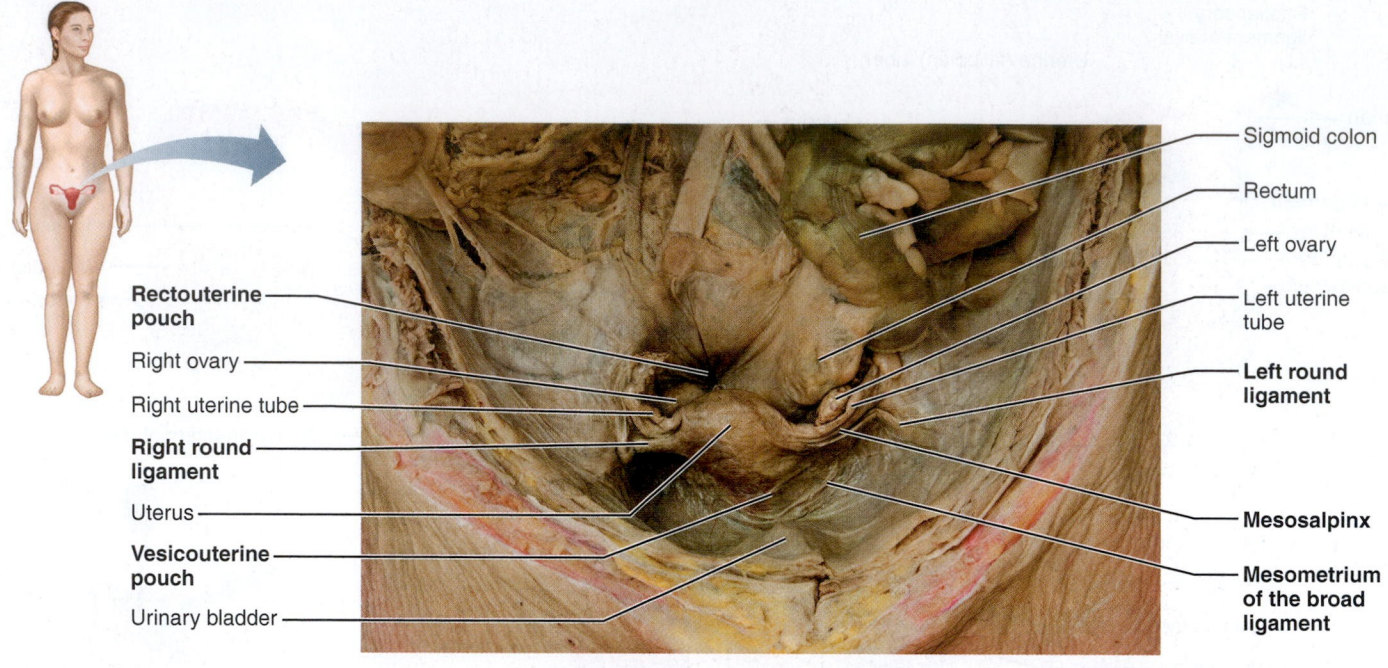

Sigmoid colon

Rectum

Left ovary

Left uterine tube

Left round ligament

Mesosalpinx

Mesometrium of the broad ligament

Rectouterine pouch

Right ovary

Right uterine tube

Right round ligament

Uterus

Vesicouterine pouch

Urinary bladder

(b) Superoanterior view

Figure 27.17 *(continued)* **(b)** Photo of female pelvic organs *in situ*.

the uterus is its narrow neck, or outlet, which projects into the vagina inferiorly.

The cavity of the cervix, called the **cervical canal**, empties into the vagina via the *external os (os = mouth)* and connects with the cavity of the uterine body via the *internal os*. The mucosa of the cervical canal contains *cervical glands* that secrete a mucus that fills the cervical canal and covers the external os. This helps block the spread of bacteria from the vagina into the uterus. Cervical mucus also blocks sperm entry except at midcycle, when it becomes less viscous and allows sperm to pass through.

HOMEOSTATIC IMBALANCE 27.6
CLINICAL

Cancer of the cervix strikes about 450,000 women worldwide each year, killing about half. It is most common among women between the ages of 30 and 50. Risk factors include frequent cervical inflammations, sexually transmitted infections [particularly human papillomavirus (HPV) that causes genital warts], and multiple pregnancies. The cancer cells arise from the epithelium covering the cervical tip.

In a *Papanicolaou (Pap) smear*, or cervical smear test, some of these cells are scraped away and examined for abnormalities. A Pap smear is the most effective way to detect this slow-growing cancer. Regular Pap smears are recommended for women between ages 21 and 65.

Vaccination can protect against HPV-induced cervical cancer. It is recommended for all 11- and 12-year-old girls. In girls who have not been exposed to HPV, the vaccine specifically blocks two cancer-causing kinds of HPV as well as two additional types not associated with cervical cancer. HPV vaccination is also recommended for boys.

Supports of the Uterus

The uterus is supported laterally by the **mesometrium** ("mesentery of the uterus") portion of the broad ligament (Figure 27.17). More inferiorly, the **cardinal (transverse cervical) ligaments** extend from the cervix and superior vagina to the lateral walls of the pelvis, and the paired **uterosacral ligaments** secure the uterus to the sacrum posteriorly (Figure 27.17a). The uterus is bound to the anterior body wall by the **round ligaments**, which run through the inguinal canals to anchor in the subcutaneous tissue of the labia majora (Figure 27.15). These flexible ligaments allow the uterus to move as the underlying bladder and rectum fill and empty.

HOMEOSTATIC IMBALANCE 27.7
CLINICAL

Despite its many anchoring ligaments, the uterus is principally supported by the muscles of the pelvic floor, namely those of the urogenital and pelvic diaphragms (◀ Table 10.7, pp. 348–349). These muscles stretch and sometimes tear during childbirth. Subsequently, the unsupported uterus may sink inferiorly, sometimes until the tip of the cervix protrudes through the external vaginal opening. This condition is called **prolapse of the uterus**.

The undulating course of the peritoneum produces several blind-ended peritoneal pouches. The most important of these are the *vesicouterine pouch* (ves″i-ko-u′ter-in) between the bladder and uterus, and the *rectouterine pouch* between the rectum and uterus (see Figure 27.15).

Lumen of uterus

Epithelium

Uterine glands

Venous sinuoids

Functional layer of the endometrium

Lamina propria (connective tissue)

Spiral arteries serve functional layer and repeatedly degenerate and regenerate.

Straight arteries serve basal layer and do not degenerate.

Basal layer of the endometrium

Endometrial vein

Radial artery

Portion of the myometrium

Smooth muscle fibers

(a)

Arcuate artery

Uterine artery

(b)

Figure 27.18 The endometrium and its blood supply. (a) Photomicrograph of the endometrium showing its functional and basal layers (28×). **(b)** Diagrammatic view. (While many spiral and straight arteries serve the endometrium, we only label one of each here.)

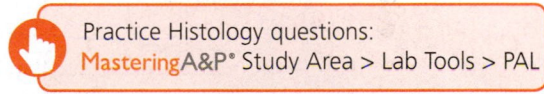

Practice Histology questions:
MasteringA&P® Study Area > Lab Tools > PAL

The Uterine Wall

The wall of the uterus is composed of three layers (Figure 27.17a):

- The **perimetrium** is the incomplete outermost serous layer.

- The **myometrium** (mi″o-me′tre-um; "muscle of the uterus") is the bulky middle layer, composed of interlacing bundles of smooth muscle. The myometrium contracts rhythmically during childbirth to expel the baby from the mother's body.

- The **endometrium** is the mucosa that lines the uterine cavity (**Figure 27.18**). This mucosa consists of a simple columnar epithelium and a thick lamina propria. If fertilization occurs, the young embryo burrows into the endometrium (implants) and resides there for the rest of its development.

The endometrium has two chief layers. The **functional layer**, or *stratum functionalis* (fungk-shun-a′lis), undergoes cyclic changes in response to blood levels of ovarian hormones and is shed during menstruation (about every 28 days). Stem cells found in the thinner, deeper **basal layer**,

or *stratum basalis* (ba-să′lis), form a new functional layer after menstruation ends. The endometrium has numerous *uterine glands* that change in length as endometrial thickness changes during the menstrual cycle (see Figure 27.25).

The vascular supply of the uterus is key to understanding the cyclic changes of the uterine endometrium. The **uterine arteries** arise from the *internal iliacs* in the pelvis, ascend along the sides of the uterus, and send branches into the uterine wall (Figures 27.17a and 27.18b). These branches break up into several **arcuate arteries** (ar′ku-āt) within the myometrium. The arcuate arteries send **radial arteries** into the endometrium, where they give off **straight arteries** to the basal layer and **spiral arteries** (or *coiled arteries*) to the functional layer. The spiral arteries repeatedly degenerate and regenerate in response to changing hormone levels. It is their spasms that actually cause the functional layer to be shed during menstruation. Veins in the endometrium are thin walled and form an extensive network with occasional sinusoidal enlargements.

27

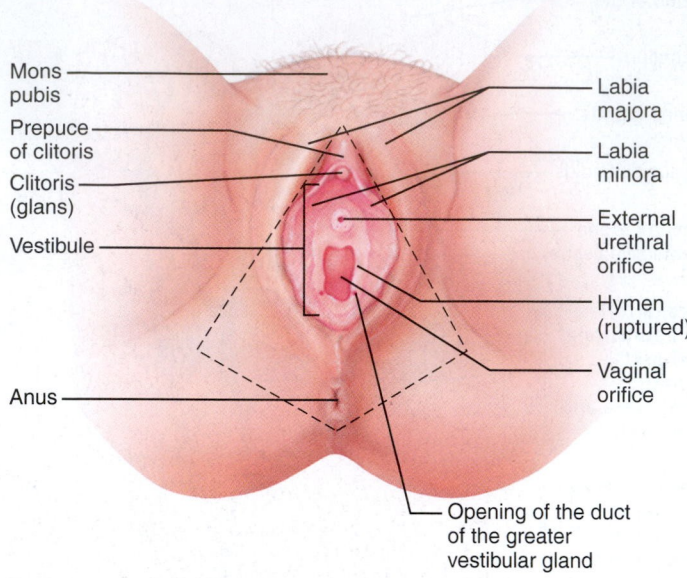

Mons pubis
Prepuce of clitoris
Clitoris (glans)
Vestibule
Anus
Labia majora
Labia minora
External urethral orifice
Hymen (ruptured)
Vaginal orifice
Opening of the duct of the greater vestibular gland

(a)

Clitoris
Labia minora
Labia majora
Anus

Pubic symphysis
Inferior pubic ramus

Clitoris:
• Body of clitoris (contains corpora cavernosa)
• Glans of clitoris
• Crus of clitoris

External urethral orifice
Vaginal orifice
Greater vestibular gland
Bulb of vestibule
Fourchette

(b)

 27

Figure 27.19 The external genitalia (vulva) of the female.
(a) Superficial structures. The region enclosed by dashed lines is the perineum. **(b)** Deep structures. The labia and associated skin have been removed to show the underlying erectile bodies. See also the associated superficial muscles (◀ **Figure 10.13, p. 349**).

The Vagina

The **vagina** ("sheath") is a thin-walled tube that lies between the bladder and the rectum. It extends from the cervix to the body exterior (see Figure 27.15). The urethra parallels its course anteriorly. The vagina provides a passageway for delivery of an infant and for menstrual flow. Because it receives the penis (and semen) during sexual intercourse, it is the *female organ of copulation*.

The distensible wall of the vagina consists of three coats: an outer fibroelastic *adventitia*, a smooth muscle *muscularis*, and an inner *mucosa* marked by transverse ridges or rugae, which stimulate the penis during intercourse. The mucosa is a stratified squamous epithelium adapted to stand up to friction.

The vaginal mucosa has no glands. Instead, it is lubricated by the cervical mucous glands and the mucosal fluid that seeps from the vaginal walls. Glycogen from epithelial cells shed into the lumen is fermented by beneficial resident bacteria into lactic acid. Consequently, the pH of a woman's vagina is normally quite *acidic*. This acidity helps keep the vagina healthy and free of infection, but it is also hostile to sperm. Although vaginal fluid of adult females is acidic, it tends to be alkaline in adolescents, predisposing sexually active teenagers to sexually transmitted infections.

In women who have never participated in sexual intercourse, the mucosa near the distal **vaginal orifice** forms an incomplete partition called the **hymen** (hi′men) (**Figure 27.19a**). The hymen may bleed when it stretches or ruptures during the first sexual intercourse. However, its durability varies. In some females, it is ruptured by sports, inserting tampons, or pelvic examinations. Occasionally, it is so tough that it must be breached surgically if intercourse is to occur.

The upper end of the vaginal canal loosely surrounds the cervix of the uterus, producing a vaginal recess called the **vaginal fornix**. The posterior part of this recess, the *posterior fornix*, is much deeper than the *lateral* and *anterior fornices* (see Figures 27.15 and 27.17a). Generally, the lumen of the vagina is quite small and, except where the cervix holds it open, its posterior and anterior walls touch each other. Although the vagina stretches considerably during copulation and childbirth, its lateral distension is limited by the ischial spines and the sacrospinous ligaments.

HOMEOSTATIC IMBALANCE 27.8   **CLINICAL**

The uterus tilts away from the vagina. For this reason, attempts by untrained persons to induce an abortion by entering the uterus with a surgical instrument may puncture the posterior wall of the vagina. This causes hemorrhage and—if the instrument is unsterile—peritonitis.

Check Your Understanding

24. Why are women more at risk for PID than men?
25. Oocytes are ovulated into the peritoneal cavity and yet women do get pregnant. What action of the uterine tubes helps to direct the oocytes into the woman's duct system?

26. In the figure at right, identify structures a–g. Which structure is the usual site of fertilization? Which has spiral arteries?

For answers, see Answers Appendix.

27.11 The external genitalia of the female include those structures that lie external to the vagina

Learning Outcome

▶ Describe the anatomy of the female external genitalia.

The female reproductive structures that lie external to the vagina are the *external genitalia* (Figure 27.19), also called the **vulva** (vul′vah; "covering") or **pudendum** ("shameful"). These structures include the mons pubis, labia, clitoris, and structures associated with the vestibule.

The **mons pubis** (mons pu′bis; "mountain on the pubis") is a fatty, rounded area overlying the pubic symphysis. After puberty, pubic hair covers the skin of this area. Running posteriorly from the mons pubis are two elongated, hair-covered fatty skin folds, the **labia majora** (la′be-ah mah-jor′ah; "larger lips"). These are the homologues of the male scrotum (recall from p. 1042 that this means that they derive from the same embryonic tissue). The labia majora enclose the **labia minora** (mi-nor′ah; "smaller"), two thin, hair-free skin folds, homologous to the spongy urethra of the penis.

The labia minora enclose a recess called the **vestibule** ("entrance hall"), which contains the external openings of the urethra and the vagina. Flanking the vaginal opening are the pea-sized **greater vestibular glands** (Figure 27.19b), homologous to the bulbourethral glands of males. These glands release mucus into the vestibule and help to keep it moist and lubricated, facilitating intercourse. At the extreme posterior end of the vestibule the labia minora come together to form a ridge called the **fourchette**.

Just anterior to the vestibule is the **clitoris** (klit′o-ris; "hill"), a small, protruding structure composed largely of erectile tissue, which is homologous to the penis of the male. Its exposed portion is called the **glans of the clitoris**. It is hooded by a skin fold called the **prepuce of the clitoris**, formed by the junction of the labia minora folds.

The clitoris is richly innervated with nerve endings sensitive to touch. It becomes swollen with blood and erect during tactile stimulation, contributing to a female's sexual arousal. Like the penis, the **body of the clitoris** has dorsal erectile columns (corpora cavernosa) attached proximally by crura, but it lacks a corpus spongiosum that conveys a urethra.

In males the urethra carries both urine and semen and runs through the penis, but the female urinary and reproductive tracts are completely separate. Instead, the **bulbs of the vestibule** (Figure 27.19b), which lie along each side of the vaginal orifice and deep to the bulbospongiosus muscles, are the homologues of the single penile bulb and corpus spongiosum of the male. During sexual stimulation the bulbs of the vestibule engorge with blood. This may help the vagina grip the penis and also squeezes the urethral orifice shut, which prevents semen (and bacteria) from traveling superiorly into the bladder during intercourse.

The Female Perineum

The female **perineum** is a diamond-shaped region located between the pubic arch anteriorly, the coccyx posteriorly, and the ischial tuberosities laterally (Figure 27.19a). The soft tissues of the perineum overlie the muscles of the pelvic outlet, and the posterior ends of the labia majora overlie the *central tendon*, into which most muscles supporting the pelvic floor insert (◀ **Table 10.7, pp. 348–349**).

Check Your Understanding

27. What is the female homologue of the bulbo-urethral glands of males?
28. Cite similarities and differences between the penis and clitoris.

For answers, see Answers Appendix.

27

27.12 The mammary glands produce milk

Learning Outcome

▶ **Describe the structure and function of the mammary glands.**

Mammary glands are modified sweat glands that are present in both sexes. They normally function only in lactating females (**Figure 27.20**). At that time, they produce milk in response to hormonal stimulation to nourish an infant.

The Mammary Glands

Each mammary gland is contained within the subcutaneous tissue of a rounded, skin-covered breast, anterior to the pectoral muscles of the thorax. Slightly below the center of each breast is a ring of pigmented skin, the **areola** (ah-re′o-lah), which surrounds a central protruding **nipple**. Large sebaceous glands in the areola make it slightly bumpy and produce sebum that reduces chapping and cracking of the skin of the nipple. Autonomic nervous system controls of smooth muscle fibers in the areola and nipple cause the nipple to become erect when stimulated by tactile or sexual stimuli, or cold temperatures.

Internally, each mammary gland consists of 15 to 25 **lobes** that radiate around and open at the nipple. The lobes are padded and separated from each other by fibrous connective tissue and fat. The interlobar connective tissue forms **suspensory ligaments** that attach the breast to the underlying muscle fascia and the overlying dermis. As suggested by their name, the suspensory ligaments provide built-in support for the breasts.

Within the lobes are smaller units called **lobules**, which contain glandular **alveoli** that produce milk when a woman is lactating. These compound alveolar glands pass the milk into the **lactiferous ducts** (lak-tif′er-us), which open to the outside at the nipple. Just deep to the areola, each duct has a dilated region called a **lactiferous sinus** where milk accumulates during nursing. We describe the process of lactation in Chapter 28.

The description of mammary glands that we have just given applies only to nursing women or women in the last trimester of pregnancy. Glandular alveoli begin to form from the smallest ducts about halfway through pregnancy. In nonpregnant women, the glandular structure of the breast is largely undeveloped and the duct system is rudimentary. For this reason, breast size is largely due to the amount of fat deposits.

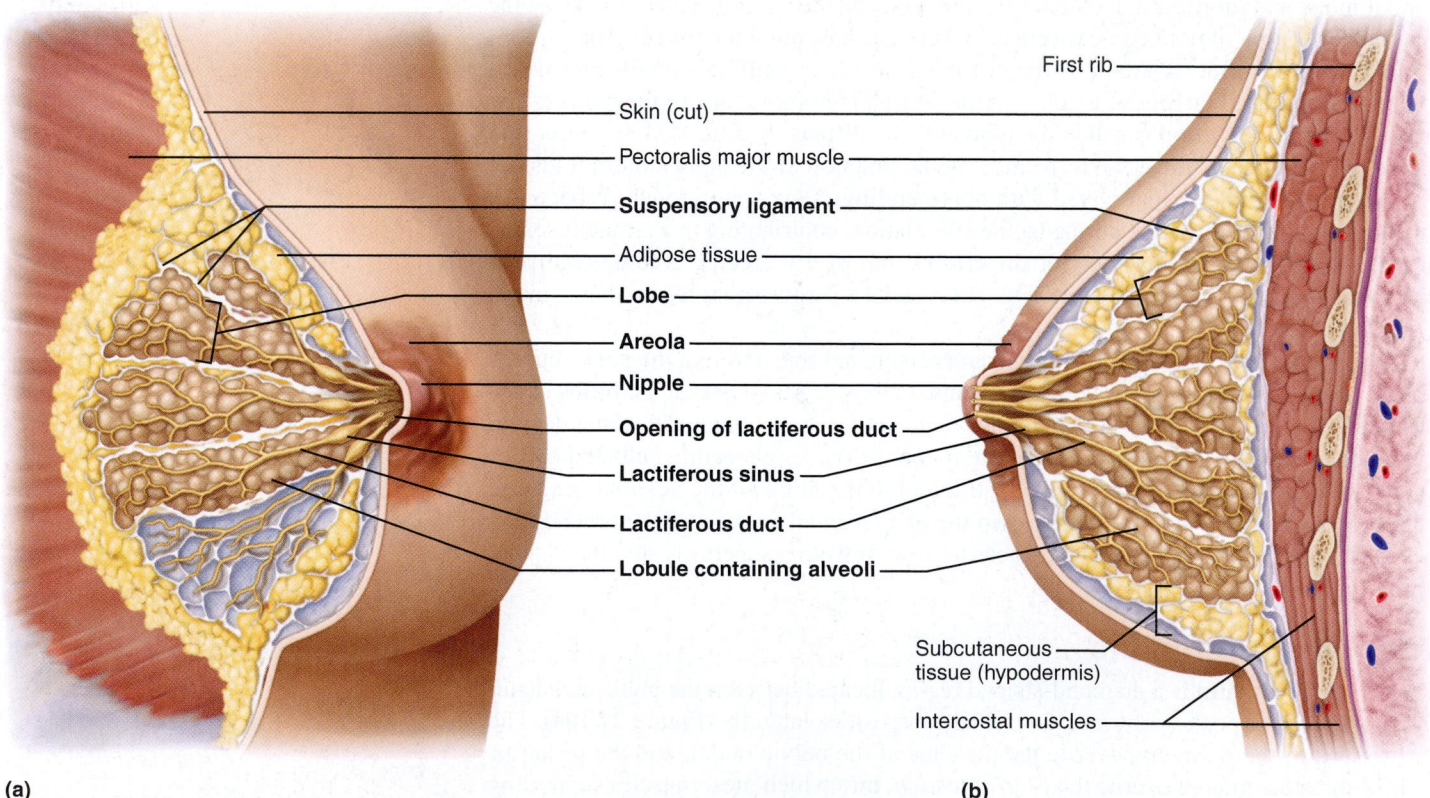

First rib

Skin (cut)

Pectoralis major muscle

Suspensory ligament

Adipose tissue

Lobe

Areola

Nipple

Opening of lactiferous duct

Lactiferous sinus

Lactiferous duct

Lobule containing alveoli

Subcutaneous tissue (hypodermis)

Intercostal muscles

(a)   (b)

Figure 27.20 Structure of lactating (milk-secreting) mammary glands. (a) Anterior view of a partially dissected breast. **(b)** Sagittal section of a breast.

27

(a) Mammogram procedure

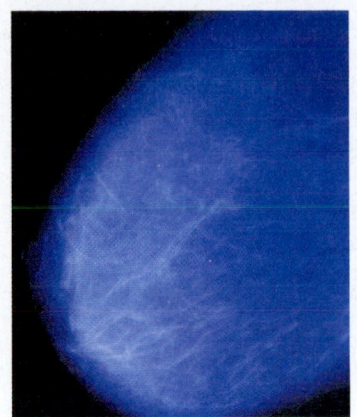

(b) Film of normal breast

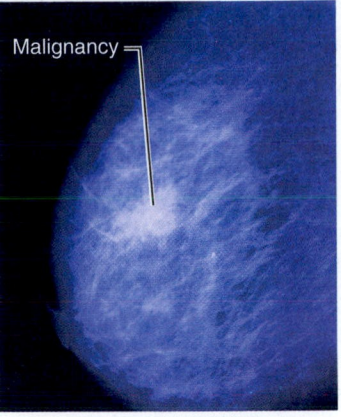

Malignancy

(c) Film of breast with tumor

Figure 27.21 Mammograms.

Breast Cancer `CLINICAL`

Breast cancer is the second most common cancer in U.S. women—about one in eight women will develop breast cancer in their lifetime. Breast cancer usually arises from the epithelial cells of the smallest ducts, eventually becoming a lump in the breast from which cells often metastasize.

The risk of breast cancer is proportional to a woman's lifetime exposure to estrogens. Early puberty, late menopause, no or delayed pregnancies, and hormone replacement therapy after menopause all increase the risk. In addition, a family history of breast cancer or genetic mutations in one of two genes (*BRCA1* and *BRCA2*) predispose women to this cancer. Nevertheless, more than 70% of women who develop breast cancer have no known risk factors for the disease.

Most breast cancers are discovered by a routine X-ray examination called a **mammogram** (**Figure 27.21**). Since most breast cancers occur later in life, the U.S. Preventive Services Task Force on Breast Cancer Screening recommends mammograms every two years for women between ages 50 and 74.

Breast cancer treatment often begins with a surgical procedure called a *lumpectomy* and may include radiation therapy, chemotherapy, or both. If the cancer cells are stimulated to divide by estrogens, drugs that block estrogen receptors or their effects can be used.

New therapies that target specific molecular characteristics of an individual's cancer cells are now in use and more are being developed. These targeted therapies are more specific than traditional chemotherapy and cause fewer side effects. Over the last 20 years, breast cancer survival rates have been steadily increasing. The 5-year survival rate is now greater than 90%.

Check Your Understanding

29. Developmentally, mammary glands are modifications of certain skin glands. Which type?

30. When does the glandular portion of the breast develop in females?

For answers, see Answers Appendix.

PART 4

PHYSIOLOGY OF THE FEMALE REPRODUCTIVE SYSTEM

27.13 Oogenesis is the sequence of events that leads to the formation of ova

Learning Outcomes

▶ **Describe the process of oogenesis and compare it to spermatogenesis.**

▶ **Describe the stages of follicle development.**

We are now ready to explore the details of **oogenesis** (o″o-gen′ĕ-sis; "the beginning of an egg"), the process by which female gametes are produced in the ovary. The process of oogenesis parallels that of spermatogenesis because the underlying processes—mitosis and then meiosis—are the same for both:

Oogonia (diploid stem cells)
↓ Divide by mitosis to produce
Primary oocytes
↓ Undergo meiosis I to produce
Secondary oocytes
↓ Undergo meiosis II to produce
Ova

Oogonia are the diploid stem cells in females, and just like spermatogonia in males, they multiply rapidly by mitosis. Then the oogonia transform into **primary oocytes** that begin the first meiotic division, just like primary spermatocytes. However, there are three key differences in the time course of events between oogenesis and spermatogenesis:

- The production of primary oocytes occurs only in the fetus.
- In primary oocytes, meiosis is arrested (stops) in late prophase I and resumes only years later (if at all).
- In secondary oocytes, meiosis is arrested in metaphase II and is only completed if fertilization occurs.

27

At birth, a female is presumed to have her lifetime supply of primary oocytes. Of the original 7 million fetal oocytes, approximately 2 million escape programmed cell death and remain in place in the cortical region of the ovary at birth. By puberty, an endowment of about 400,000 oocytes remains.

The Relationship between Oocytes and Follicles

Recall that in spermatogenesis, multiple developing sperm are surrounded by one sustentocyte. In oogenesis, a single oocyte is surrounded by many cells, and all of these cells are enclosed in a basal lamina to form an **ovarian follicle**. The surrounding cells are called **pre-granulosa cells** if a single layer of cells is present, and **granulosa cells** when more than one layer is present.

Follicles are the functional units of the ovary. They provide an environment where the developing oocyte and its surrounding cells can hold private "conversations" with each other (using paracrines and gap junctions). Follicles at different stages of maturation are distinguished by their size and by their structure (**Figure 27.22**, left side):

- **Primordial follicles** have a single layer of *squamous* pre-granulosa cells surrounding a primary oocyte.

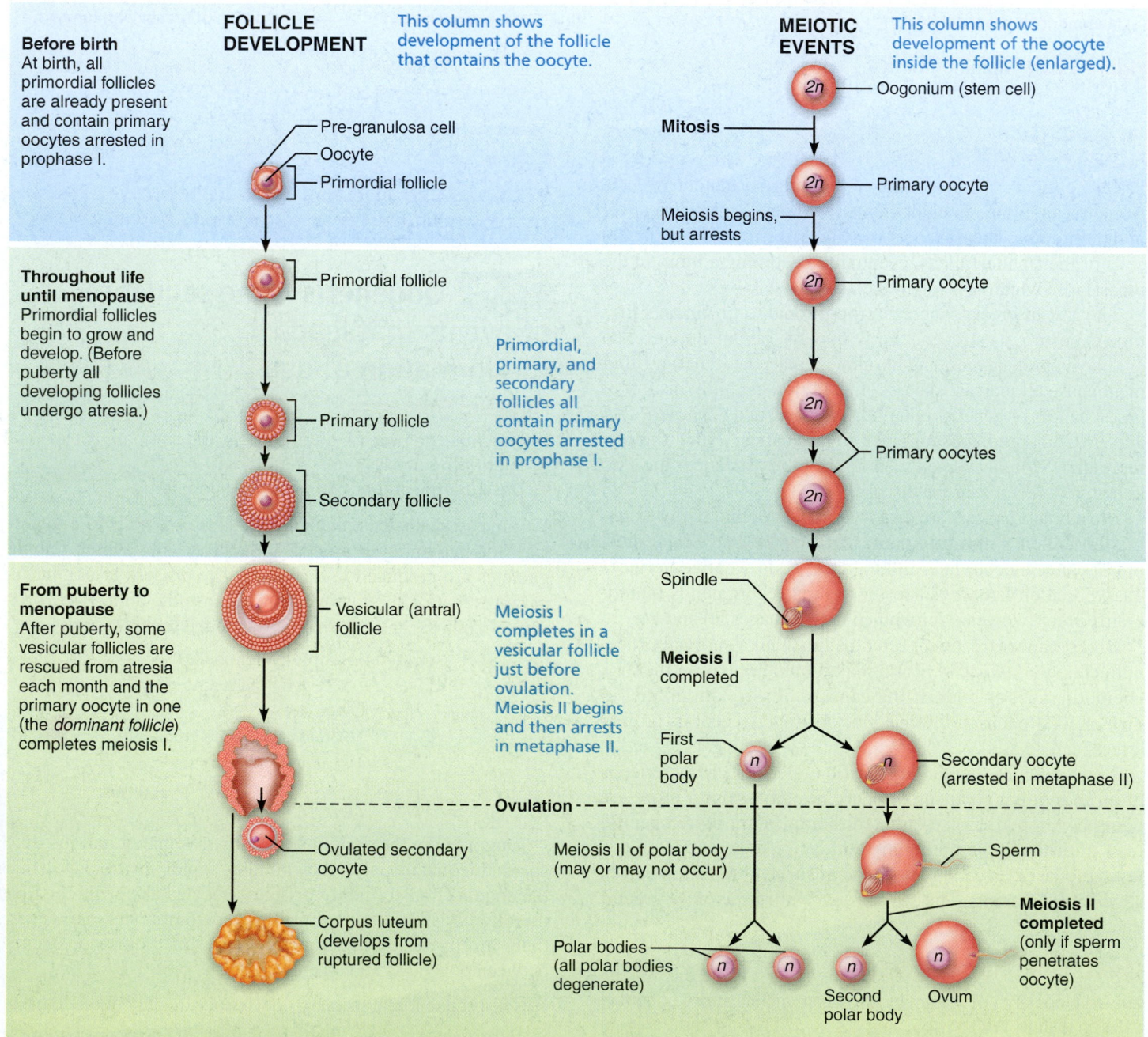

Figure 27.22 Events of oogenesis. At various stages, unsuccessful follicles undergo atresia (not shown).

- **Primary follicles** have a single layer of *cuboidal* pregranulosa cells surrounding a primary oocyte.
- **Secondary follicles** have multiple layers of granulosa cells surrounding a primary oocyte.
- **Vesicular (antral) follicles** (sometimes called *tertiary follicles*) have a fluid-filled cavity called an **antrum** ("cave"). Just before ovulation, the primary oocyte inside the vesicular follicle resumes meiosis and becomes a secondary oocyte.

The process of oogenesis begins during the fetal period and takes years to complete (Figure 27.22). The first follicles to develop in the fetus are primordial follicles. At any given time, the large majority of follicles in a woman's ovaries are dormant (functionally inactive) primordial follicles whose primary ooctyes are arrested in prophase I. Beginning in fetal life, a small fraction of the available dormant primordial follicles are *recruited* into a pool of growing follicles. This recruitment process begins during fetal life and continues throughout life until the supply of primordial follicles is effectively depleted, a time called menopause (see p. 1084). A follicle has one of two fates:

- **Atresia.** Atresia is apoptosis (programmed cell death) of the oocyte and its surrounding cells. This is the fate of over 99.9% of all follicles. Some primordial follicles undergo atresia without ever being recruited, and recruited follicles can undergo atresia at any stage in their development. Before puberty, *all* recruited follicles undergo atresia. This explains why only about 400,000 primordial follicles remain at puberty.
- **Ovulation.** After puberty, sufficiently high levels of hormones (particularly FSH) rescue a cohort (group) of fewer than 20 recruited follicles each month from atresia. Of these, typically only one follicle, the **dominant follicle**, is selected. It continues to escape atresia and achieves ovulation. As a result, only a small minority of recruited follicles are ever ovulated. Since the reproductive life of a woman is about 40 years and only one ovulation typically occurs each cycle, fewer than 500 oocytes are generally ovulated during a woman's lifetime.

Products of Meiosis in Females

As we mentioned previously, meiosis in the primary oocyte arrests in prophase I and this pause in meiosis can last for decades. Only the oocyte in the dominant follicle ever resumes meiosis I. This occurs just before ovulation and produces two haploid cells (each with 23 replicated chromosomes) that are quite dissimilar in size. The smaller cell, called the **first polar body,** may continue its development and undergo meiosis II, producing two even smaller polar bodies. The larger cell, which contains nearly all the cytoplasm of the primary oocyte, is the **secondary oocyte**.

The events of this first meiotic division ensure that the polar body receives almost no cytoplasm or organelles. Notice in Figure 27.22 (right) that a spindle forms at the very edge of the oocyte. A little "nipple" also appears at that edge, and the polar body chromosomes are cast into it.

The secondary oocyte arrests in metaphase II, and it is this cell (not a functional ovum) that is ovulated. If a sperm does not penetrate an ovulated secondary oocyte, the oocyte deteriorates. But if sperm penetration does occur, the oocyte quickly completes meiosis II, yielding one large **ovum** and a tiny **second polar body** (Figure 27.22). The union of the egg and sperm nuclei that constitutes fertilization is described in Chapter 28.

The unequal cytoplasmic divisions that occur during oogenesis ensure that a fertilized egg has ample nutrients for its six- to seven-day journey to the uterus. Lacking nutrient-containing cytoplasm, the polar bodies degenerate and die.

Comparison of Oogenesis and Spermatogenesis

There are many differences between spermatogenesis and oogenesis, which are summarized in **Table 27.1** and discussed below.

The potential end products of oogenesis are two or three tiny polar bodies, nearly devoid of cytoplasm, and one large ovum. All of these cells are haploid, but only the ovum is a *functional gamete*. This is quite different from spermatogenesis, where the product is four viable gametes—sperm.

Gamete production in males begins at puberty and continues throughout life. As a result, the number of gametes produced by a male during his lifetime is huge. In contrast, a female's total supply of eggs is already determined by the time she is born, and the time span during which she releases them extends only from puberty to menopause. (This dogma that the number of oocytes is fixed before birth has been challenged by the recent discovery in humans of ovarian stem cells that have the potential to form new oocytes under experimental conditions. However, their physiological role, if any, has yet to be established.)

Perhaps the most striking difference between male and female meiosis is the error rate. As many as 20% of oocytes (but only 5% of sperm) have the wrong number of chromosomes, a situation that often results from failure of the homologues to

Table 27.1 Spermatogenesis versus Oogenesis

	MALE	FEMALE
Total Time to Produce One Gamete	74 days	13–50 years
Occurrence during Lifetime	Puberty to old age	Begins in fetal life, ends in menopause
Number of Gametes per Meiotic Division	4 equal-size sperm	1 large ovum and 2–3 small polar bodies
Number of Gametes per Lifetime	>1 trillion	<500
Error Rate (wrong number of chromosomes)	5%	20%
Cells Surrounding Developing Gametes	One sustentocyte sustains many spermatocytes	Many granulosa cells sustain one oocyte

① **Primordial follicles**

② **Primary follicle**

Zona pellucida

③ **Secondary follicle**

④a **Early vesicular follicle**

Forming antrum

Theca folliculi

Theca folliculi

③

④a

② Corpus albicans

⑦

Primary oocyte

Zona pellucida

Antrum

Secondary oocyte

④b

⑥

Corpus luteum

⑥

⑤

Secondary oocyte

Corona radiata

Zona pellucida

Antrum

Ovary

Oocyte

⑥ **Corpus luteum (forms from ruptured follicle)**

⑤ **Follicle ruptures; secondary oocyte ovulated**

④b **Mature vesicular follicle; oocyte completes meiosis I; ready to be ovulated**

Figure 27.23 Development and fate of ovarian follicles. The numbers on the diagram indicate the sequence of follicle development, not the movement of a developing follicle within the ovary. Primordial follicles ① not shown in central diagram.

Practice Histology questions:
MasteringA&P® Study Area > Lab Tools > PAL

27

separate during meiosis I. It appears that faced with meiotic disruption, meiosis in males grinds to a halt and the sperm dies, but in females it marches on, making a defective oocyte.

Stages of Follicle Development

Oocyte and follicle development go hand in hand. Once recruited, a follicle and its oocyte require nearly a year to mature to the point that the oocyte is ready to be ovulated. As a result, follicles at many different stages of development can be observed at any given time within an ovary (◄ Figure 27.16, p. 1061). The maturation process involves two phases:

- Phase 1 is the gonadotropin-independent *preantral phase*. During this phase, intrafollicular paracrines (such as cytokines and growth factors) control oocyte and follicle development.

- Phase 2 is the *antral phase*. During this phase, the follicles are stimulated by FSH and LH (◄ p. 616) and grow tremendously, the dominant follicle is selected, and the primary oocyte in the dominant follicle resumes meiosis I.

We are now ready to look more closely at the development of ovarian follicles as shown in Figure 27.22 and **Figure 27.23**.

- **A primordial follicle becomes a primary follicle.** When a primordial follicle ① is recruited, the squamous cells surrounding the primary oocyte grow and become cuboidal cells. The oocyte enlarges. The follicle is now called a *primary follicle* ②. At the same time, the oocyte secretes a glycoprotein-rich substance that forms a thick transparent extracellular layer called the **zona pellucida** (pĕ-lu′sid-ah). The zona pellucida encapsulates the oocyte ②, ③.

- **A primary follicle becomes a secondary follicle.** The pre-granulosa cells proliferate, forming a stratified cuboidal epithelium around the oocyte. As soon as more than one cell layer is present, the follicle is called a *secondary follicle* ③. The *granulosa cells* (as they are now called) are connected to the developing oocyte by gap junctions, through which ions, metabolites, and signaling molecules can pass. From this point on, bidirectional "conversations" occur between the oocyte and granulosa cells, so they guide one another's development. One of the signals passing from the granulosa cells tells the oocyte to grow. Other signals dictate polarity in the future egg. The oocyte grows tremendously during this stage.

    As the follicle grows, a layer of connective tissue and epithelial cells condenses around the follicle, forming the **theca folliculi** (the′kah fah-lik′u-li; "box around the follicle") ④ₐ. Some of these **thecal cells** secrete steroid hormone in response to LH (see Figure 27.24).

- **A secondary follicle becomes a vesicular (antral) follicle.** The secondary follicle stage ends when a clear liquid begins to accumulate between the granulosa cells, producing the early *vesicular follicle* ④ₐ. When six to seven layers of granulosa cells are present, the fluid between the granulosa cells coalesces to form an *antrum*, a large fluid-filled cavity ④ᵦ. The presence of the antrum distinguishes vesicular follicles from

all prior follicles (*preantral follicles*). At this point, the preantral phase ends and the antral phase begins. The further maturation of the follicle depends on the presence of gonadotropic hormones.

The antrum continues to expand with fluid until it almost completely surrounds the oocyte and its surrounding capsule of granulosa cells, which are now called a **corona radiata** ("radiating crown"). The oocyte remains attached to the rest of the follicle only by a column of cells on one side of the follicle. When a follicle is full size (about 2.5 cm, or 1 inch, in diameter), it bulges from the external ovarian surface like a pimple. This usually occurs just before ovulation.

After the oocyte with its corona radiata is ejected from the follicle during ovulation, the ruptured follicle transforms into a very different-looking glandular structure called the *corpus luteum* (lu′te-um; "yellow body"; plural: corpora lutea) ⑥, discussed on the next page.

Check Your Understanding

31. How do the haploid cells arising from oogenesis differ structurally and functionally from those arising from spermatogenesis?

32. What name is given to the fluid-filled cavity of a vesicular follicle?

33. Look at Figure 27.16 on p. 1061. Identify the stage of each of the follicles indicated in the figure.

For answers, see Answers Appendix.

27.14 The ovarian cycle consists of the follicular phase and the luteal phase

Learning Outcome

▶ **Describe ovarian cycle phases, and relate them to events of oogenesis.**

The monthly series of events associated with the maturation of an egg is called the **ovarian cycle**. The ovarian cycle is best described in terms of two consecutive phases:

- The **follicular phase** is the period when a cohort of vesicular follicles begins to secrete increasing amounts of estrogens and one of these follicles is selected as the dominant follicle. This phase generally lasts from the first to the fourteenth day of the ovarian cycle, at which point ovulation typically occurs.

- The **luteal phase**, days 14–28, is the period during which the corpus luteum (discussed on the next page) is active.

The so-called typical ovarian cycle repeats at intervals of 28 days, with *ovulation* occurring midcycle. However, only 10–15% of women naturally have 28-day cycles, and cycles as long as 40 days or as short as 21 days are fairly common. In such cases, the length of the follicular phase and timing of ovulation vary, but the luteal phase remains constant: It is always 14 days from the time of ovulation to the end of the cycle.

27

Follicular Phase of the Ovarian Cycle

The first phase is called the follicular phase because many large vesicular follicles are growing in the ovary during this phase. This cohort of vesicular follicles was rescued from atresia by rising levels of FSH near the end of the previous cycle. FSH is a survival factor for vesicular follicles, allowing them to continue to survive and grow. One of these growing vesicular follicles becomes more sensitive to FSH than the others. As FSH levels begin to drop in the middle of the follicular phase, that follicle outcompetes the other follicles for FSH, and so continues to escape from atresia. It becomes the *dominant follicle*. Multiple factors, including increased sensitivity to LH, allow it to continue to grow so that it is at its peak stage of maturation when the hormonal (LH) stimulus for ovulation is given. (The other stimulated follicles eventually undergo atresia.)

As one of the final events of follicle maturation, the primary oocyte of the dominant follicle completes meiosis I to form the secondary oocyte and first polar body (see Figure 27.22). Once this has occurred, the stage is set for ovulation. At this point, the granulosa cells send another important signal to the oocyte that says, in effect, "Wait, do not complete meiosis yet!"

Ovulation

Ovulation occurs when the ballooning ovary wall ruptures and expels the secondary oocyte, still surrounded by its corona radiata, into the peritoneal cavity (Figure 27.23 ⑤). Some women experience a twinge of pain in the lower abdomen when ovulation occurs. The precise cause of this pain, called *mittelschmerz* (mit′el-shmārts; German for "middle pain"), is not known, but possible reasons include intense stretching of the ovarian wall during ovulation and irritation of the peritoneum by blood or fluid released from the ruptured follicle.

In 1–2% of all ovulations, more than one oocyte is ovulated. This phenomenon, which increases with age, can result in multiple births. Since, in such cases, different oocytes are fertilized by different sperm, the siblings are *fraternal*, or nonidentical, *twins*. *Identical twins* result from the fertilization of a single oocyte by a single sperm, followed by separation of the fertilized egg's daughter cells during early development. Additionally, it now appears that in some women oocytes may be released at times unrelated to hormone levels. This timing may help to explain why a rhythm method of contraception sometimes fails and why some fraternal twins have different conception dates.

Luteal Phase of the Ovarian Cycle

After ovulation, the ruptured follicle collapses and the antrum fills with clotted blood. This *corpus hemorrhagicum* is eventually absorbed. The remaining granulosa cells enlarge, and along with the internal theca cells they form a new, quite different endocrine structure, the **corpus luteum** (Figure 27.23 ⑥). The corpus luteum supplies the high levels of progesterone and estrogens that characterize the luteal phase.

If pregnancy does not occur, the corpus luteum starts degenerating after about 10 days and its hormonal output ends. In this case, all that ultimately remains is a scar called the *corpus albicans* (al′bĭ-kans; "white body") as shown in Figure 27.23 ⑦. The last two or three days of the luteal phase, when the endometrium is just beginning to erode, is sometimes called the *luteolytic* or *ischemic phase*.

On the other hand, if the oocyte is fertilized and pregnancy ensues, the corpus luteum persists until the placenta is ready to take over its hormone-producing duties in about three months.

Check Your Understanding

34. How do identical twins differ developmentally from fraternal twins?

35. What occurs in the luteal phase of the ovarian cycle?

For answers, see Answers Appendix.

27.15 Female reproductive function is regulated by hypothalamic, anterior pituitary, and ovarian hormones

Learning Outcomes

▶ Describe the regulation of the ovarian and uterine cycles.
▶ Discuss the physiological effects of estrogens and progesterone.

Female reproductive events are much more complicated than those occurring in males, but the hormonal controls involving the hypothalamic-pituitary-gonadal axis set into motion at puberty are similar in both sexes. Gonadotropin-releasing hormone (GnRH), pituitary gonadotropins, and, in this case, ovarian estrogens and progesterone interact to produce the cyclic events occurring in the ovaries and uterus.

However, in females another hormone plays an important role in allowing the hypothalamus to release GnRH. The onset of puberty in females is linked to the amount of adipose tissue, and the messenger from fatty tissue to the hypothalamus is *leptin* (better known for its role in appetite and energy storage, ◀ p. 634). If body levels of fat and leptin are low, puberty is delayed.

Hormonal Regulation of the Ovarian Cycle

Establishing the Ovarian Cycle

During childhood, the ovaries grow and continuously secrete small amounts of estrogens, which inhibit hypothalamic release of GnRH. Provided that leptin levels are adequate, the hypothalamus becomes less sensitive to estrogens as puberty nears and, as we described on p. 1043, this triggers the rising levels of hormones seen at puberty.

Gonadotropin (FSH and LH) levels increase for about four years. During this time, pubertal girls are still not ovulating and cannot become pregnant. Eventually, the adult cyclic pattern is achieved, and hormonal interactions stabilize. These events are heralded by the young woman's first menstrual period, referred to as **menarche** (mĕ-nar′ke; *men* = month, *arche* = first). Usually, it is not until the third year postmenarche that the cycles become regular and all are ovulatory.

Early and midfollicular phases

Hypothalamus

GnRH

Travels via portal blood

① Anterior pituitary

③ Slightly elevated estrogen and rising inhibin levels inhibit FSH secretion.

FSH   **LH**

② ②

Thecal cells

②

Androgens

Inhibin   Granulosa cells (convert androgens to estrogens)

②

↑ **Estrogens**

Late follicular and luteal phases

Hypothalamus

④ Positive feedback exerted by large ↑ in estrogen output by maturing follicle.

④   GnRH   ⑥

④

↑↑↑ **Estrogens**   **LH surge**   ↑ Progesterone ↑ Estrogens ↑ Inhibin

Ruptured follicle   ⑤   ⑥

⑤

Mature vesicular follicle   Ovulated secondary oocyte   Corpus luteum

②

——— Stimulates   - - - - Inhibits

Figure 27.24 Regulation of the ovarian cycle. Events that follow step ⑥ are not depicted, but involve a gradual deterioration of the corpus luteum and, therefore, a decline in ovarian hormone production. Ovarian hormones reach their lowest blood levels at the end of the luteal phase. Note that inhibin only inhibits FSH release.

Hormonal Interactions during the Ovarian Cycle

Next, let's look at how the cyclic pattern of hormone release in females varies from the basic pattern (◀ p. 1042). In females, levels of anterior pituitary gonadotropins (FSH and LH) and ovarian hormones wax and wane, and there is both negative and positive feedback during each cycle. **Figure 27.24** shows these events.

① **GnRH stimulates FSH and LH secretion.**

② **FSH and LH stimulate follicles to grow, mature, and secrete sex hormones.** FSH exerts its main effects on the granulosa cells of vesicular follicles, causing them to release estrogens, whereas LH (at least initially) prods the thecal cells to release androgens, which the granulosa cells convert to estrogens. Only tiny amounts of ovarian androgens enter

the blood, because they are almost completely converted to estrogens within the ovaries.

③ **Negative feedback inhibits gonadotropin release.** As estrogen levels in the plasma rise, they exert *negative feedback* on the hypothalamus and anterior pituitary, inhibiting release of FSH. *Inhibin,* released by the granulosa cells, also exerts negative feedback controls on FSH release during this period. Only the dominant follicle survives the resulting dip in FSH—the other developing follicles fail to develop further.

④ **Positive feedback stimulates gonadotropin release.** Although the initial small rise in bloodborne estrogens inhibits the hypothalamic-pituitary-gonadal axis, once the dominant follicle becomes a mature vesicular follicle just

27

before ovulation, the follicle produces high levels of estrogens. This has the opposite effect. When estrogens reach a critical blood concentration, they briefly exert *positive feedback* on the hypothalamus and anterior pituitary.

⑤ **LH surge triggers ovulation and formation of the corpus luteum.** The high estrogen level sets a cascade of events into motion. There is a sudden burst of LH (and, to a lesser extent, FSH) released by the anterior pituitary about mid-cycle (also see Figure 27.25a).

This **LH surge** rouses the primary oocyte of the mature vesicular follicle from its resting state and it completes its first meiotic division, forming a secondary oocyte that continues on to metaphase II. Around day 14, LH—aided by intraovarian paracrines—stimulates many events that lead to ovulation: It increases local vascular permeability and triggers an inflammatory response that promotes release of metalloproteinase enzymes that weaken the ovary wall. As a result, blood stops flowing through the protruding part of the follicle wall. Within minutes, that region of the follicle wall thins, bulges, and ruptures, forming a hole. The oocyte, still surrounded by its corona radiata, exits, accomplishing ovulation.

Shortly after ovulation, estrogen levels decline. This probably reflects the damage to the dominant estrogen-secreting follicle during ovulation.

The LH surge also transforms the ruptured follicle into a corpus luteum (which gives LH its name, "luteinizing" hormone). LH stimulates this newly formed endocrine structure to produce large amounts of progesterone and some estrogens almost immediately after it is formed (also see Figure 27.25c). Progesterone helps maintain the functional layer of the endometrium and is essential for maintaining pregnancy should conception occur.

⑥ **Negative feedback inhibits LH and FSH release.** Rising progesterone and estrogen blood levels exert a powerful negative feedback effect on the hypothalamus and the anterior pituitary release of LH and FSH. Inhibin is also released by the corpus luteum and enhances this inhibitory effect. Declining gonadotropin levels inhibit the maturation of new vesicular follicles and prevent additional LH surges that might cause additional oocytes to be ovulated.

When pregnancy does not occur, the stimulus for luteal activity ends when LH blood levels fall and the corpus luteum degenerates. Without the corpus luteum supplying ovarian hormones, blood estrogen and progesterone levels drop sharply. The marked decline in ovarian hormones at the end of the cycle (days 26–28) ends their blockade of FSH and LH secretion, and the cycle starts anew.

We have just described ovarian events through a 28-day cycle. Be aware, however, that the cohort of vesicular follicles that we followed from the beginning of this cycle were actually recruited to grow from primordial follicles about 290 days before one of them is ultimately ovulated. This is far longer than the 14 days you might expect from a casual look at the graph.

The Uterine (Menstrual) Cycle

Although the uterus is where an embryo implants and develops, it is receptive to implantation for only a short period each month. Not surprisingly, this brief interval is exactly the time when a developing embryo would normally begin implanting, six to seven days after ovulation. The **uterine** or **menstrual cycle** (men′stroo-al) is a series of cyclic changes that the uterine endometrium goes through each month as it responds to the waxing and waning of ovarian hormones in the blood. These endometrial changes are coordinated with the phases of the ovarian cycle, which are dictated by gonadotropins released by the anterior pituitary.

Changes in ovarian steroid hormone levels drive the events of the uterine cycle (**Figure 27.25d**) as follows:

1. **Days 0–4: Menstrual phase.** In this phase, **menstruation** (men″stroo-a′shun) or **menses**, the uterus sheds all but the deepest part of its endometrium. (Note in Figure 27.25c that at the beginning of this stage, the ovarian hormones are at their lowest normal levels.) The thick, hormone-dependent functional layer of the endometrium detaches from the uterine wall, a process accompanied by bleeding for 3–5 days. The detached tissue and blood pass out through the vagina as the menstrual flow. By day 4, the growing ovarian follicles start to produce more estrogens (Figure 27.25c).

2. **Days 5–14: Proliferative (preovulatory) phase.** In this phase, the endometrium rebuilds itself: Under the influence of rising blood levels of estrogens, the basal layer of the endometrium generates a new functional layer. As this new layer thickens, its glands enlarge and its spiral arteries increase in number (also see Figure 27.18). Consequently, the endometrium once again becomes velvety, thick, and well vascularized. During this phase, estrogens also induce the endometrial cells to synthesize progesterone receptors, readying them for interaction with progesterone.

   Normally, cervical mucus is thick and sticky, but rising estrogen levels cause it to thin and form channels that facilitate sperm passage into the uterus. *Ovulation* occurs at the end of the proliferative stage (day 14) in response to the LH surge from the anterior pituitary.

3. **Days 15–28: Secretory (postovulatory) phase.** This 14-day phase is the most constant timewise. During the secretory phase the endometrium prepares for an embryo to implant. Rising levels of progesterone from the corpus luteum act on the estrogen-primed endometrium, causing the spiral arteries to elaborate and converting the functional layer to a secretory mucosa. The endometrial glands enlarge, coil, and begin secreting nutrients into the uterine cavity. These nutrients will sustain the embryo until it has implanted in the blood-rich endometrial lining.

   As progesterone levels rise, the cervical mucus becomes viscous again, forming the *cervical plug*, which helps to block entry of sperm, pathogens, and other foreign materials. This protective role of the cervical plug is particularly important for sealing off the uterus in the event an embryo has begun to implant. Rising progesterone (and estrogen) levels inhibit LH release by the anterior pituitary.

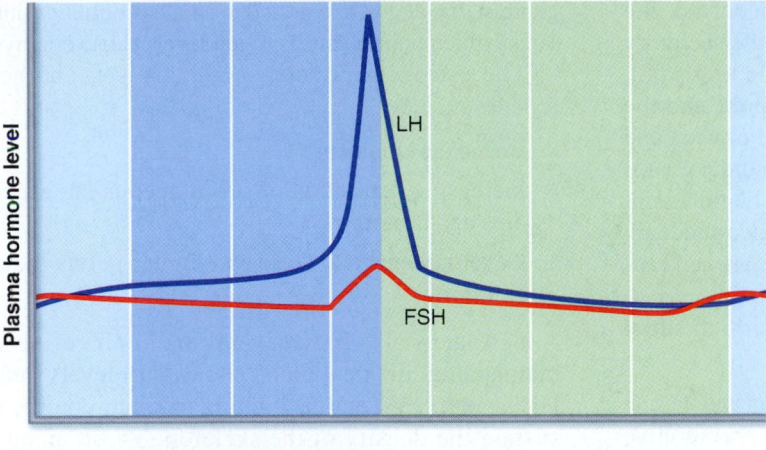

(a) Fluctuation of gonadotropin levels: Fluctuating levels of pituitary gonadotropins (follicle-stimulating hormone and luteinizing hormone) in the blood regulate the events of the ovarian cycle.

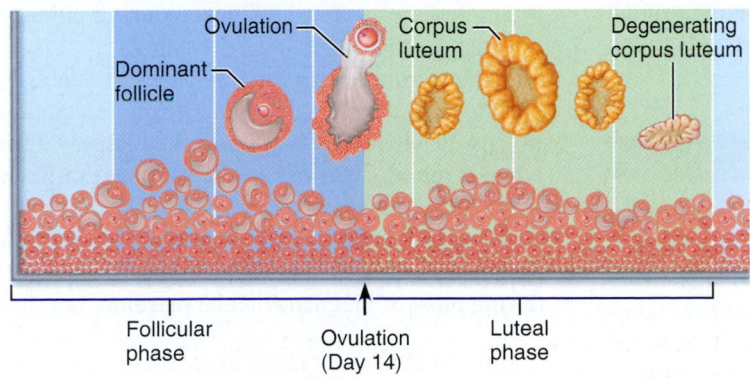

Ovarian phases:

Follicular phase   Ovulation (Day 14)   Luteal phase

(b) Ovarian cycle: Structural changes in vesicular ovarian follicles and the corpus luteum are correlated with changes in the endometrium of the uterus during the uterine cycle (d). Recall that only vesicular follicles (in their antral phase) are hormone dependent. Primary and secondary follicles are not hormone dependent and develop in waves.

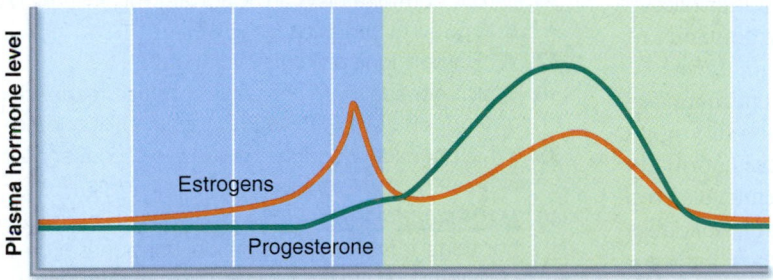

(c) Fluctuation of ovarian hormone levels: Fluctuating levels of ovarian hormones (estrogens and progesterone) cause the endometrial changes of the uterine cycle. The high estrogen levels are also responsible for the LH/FSH surge in (a).

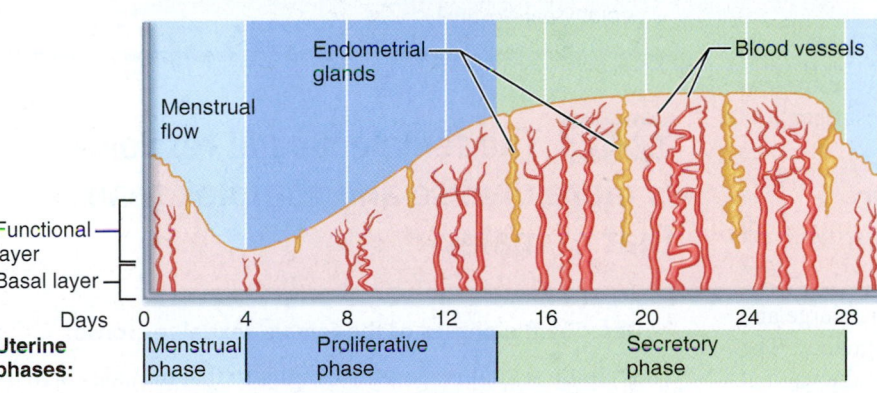

Uterine phases:

Menstrual phase | Proliferative phase | Secretory phase

(d) The three phases of the uterine cycle:
- Menstrual: The functional layer of the endometrium is shed.
- Proliferative: The functional layer of the endometrium is rebuilt.
- Secretory: Begins immediately after ovulation. Enrichment of the blood supply and glandular secretion of nutrients prepare the endometrium to receive an embryo.

Both the menstrual and proliferative phases occur before ovulation, and together they correspond to the follicular phase of the ovarian cycle. The secretory phase corresponds in time to the luteal phase of the ovarian cycle.

Figure 27.25 Correlation of anterior pituitary and ovarian hormones with structural changes of the ovary and uterus. Note that the end of day 28 of the current cycle marks the beginning (day 0) of the first day of the next cycle.

27

As noted earlier, if fertilization has not occurred, the corpus luteum degenerates toward the end of the secretory phase as LH blood levels decline. Progesterone levels fall, depriving the endometrium of hormonal support, and the spiral arteries kink and go into spasms. Denied oxygen and nutrients, the ischemic endometrial cells die and the glands regress, setting the stage for menstruation to begin again. The spiral arteries constrict one final time and then suddenly relax and open wide. As blood floods into the weakened capillary beds, they fragment, causing the functional layer to slough off. The uterine cycle starts over again on this first day of menstrual flow.

Figure 27.25b and d also illustrate how the ovarian and uterine cycles fit together. Notice that the menstrual and proliferative phases overlap the follicular phase and ovulation in the ovarian cycle, and that the uterine secretory phase corresponds to the ovarian luteal phase.

HOMEOSTATIC IMBALANCE 27.9   CLINICAL

Extremely strenuous physical activity can delay menarche in girls. Very low levels of body fat in female athletes can also cause *amenorrhea* (a-men″o-re′ah), cessation of menstruation. This is because adipose cells help convert adrenal androgens to estrogens and are the source of leptin which, as noted above, plays a critical permissive role in the onset of puberty in females. Leptin informs the hypothalamus whether energy stores are sufficient to support the high energy demands of reproduction. If not, the reproductive cycles are shut down.

Amenorrhea is usually reversible if the woman increases her body fat, but it has a worrisome consequence in young, healthy adult women: dramatic losses in bone mass normally seen only in osteoporosis of old age. Once estrogen levels drop and the menstrual cycle stops (regardless of cause), bone loss begins.

Effects of Estrogens and Progesterone

With a name meaning "generators of sexual activity," estrogens are analogous to testosterone, the male steroid. As estrogen levels rise during puberty, they:

- Promote oogenesis and follicle growth in the ovary
- Exert anabolic effects on the female reproductive tract (**Table 27.2**)

Consequently, the uterine tubes, uterus, and vagina enlarge and become functional—more ready to support a pregnancy. The uterine tubes and uterus exhibit enhanced motility; the vaginal mucosa thickens; and the external genitalia mature.

Estrogens also support the growth spurt at puberty that makes girls grow much more quickly than boys between the ages of 11 and 12. But this growth is short-lived because rising estrogen levels also cause the epiphyses of long bones to close (◄ p. 188), and so females reach their full height between the ages of 13 and 15. In contrast, the aggressive growth of males continues until ages 15 to 19, at which point rising estrogen levels cause epiphyseal closure.

The estrogen-induced secondary sex characteristics of females include:

- Breast development
- Increasing deposition of subcutaneous fat, especially in the hips and breasts
- Development of a wider and lighter pelvis (adaptations for childbirth)

Estrogens also have several metabolic effects, including maintaining low total blood cholesterol levels (and high HDL levels, ◄ p. 956) and facilitating calcium uptake, which helps sustain the density of the skeleton. These metabolic effects begin under estrogens' influence during puberty, but they are not true secondary sex characteristics.

Progesterone works with estrogens to establish and then help regulate the uterine cycle, and promotes changes in cervical mucus. Its other effects are exhibited largely during pregnancy, when it inhibits uterine motility and takes up where estrogens leave off in preparing the breasts for lactation. Indeed, progesterone is named for these important roles (*pro* = for, *gestation* = pregnancy). However, the source of progesterone and estrogens during most of pregnancy is the placenta, not the ovaries.

Check Your Understanding

36. Which hormone plays an important role in signaling the brain that puberty may occur in girls?

37. Which hormone prompts ovulation?

38. Which gonadal hormones exert positive feedback on the anterior pituitary that results in a burstlike release of LH?

39. Which gonadal hormones cause the secondary sex characteristics to appear in a young woman?

40. MAKE CONNECTIONS You've studied both bone growth and hormonal effects. Which gonadal hormones promote epiphyseal closure in both males and females, and what is the effect of epiphyseal closure?

For answers, see Answers Appendix.

27.16 The female sexual response is more diverse and complex than that of males

Learning Outcome

▶ Describe the phases of the female sexual response.

The **female sexual response** is similar to that of males in many respects. During sexual excitement, the clitoris, vaginal mucosa, bulbs of the vestibule, and breasts engorge with blood and the nipples become erect. Increased activity of the greater vestibular glands and seepage from the vaginal walls lubricate the vestibule to facilitate entry of the penis. These events, though more widespread, are analogous to the *erection* phase in men. Touch

Table 27.2   Summary of Hormonal Effects of Gonadal Estrogens, Progesterone, and Testosterone

SOURCE, STIMULUS, EFFECTS	ESTROGENS (MOSTLY ESTRADIOL)	PROGESTERONE	TESTOSTERONE
Major source	Ovary: developing follicles and corpus luteum.	Ovary: mainly the corpus luteum.	Testes: interstitial endocrine cells.
Stimulus for release	FSH (and LH).	LH.	LH.
Feedback effects exerted	Both negative and positive feedback on gonadotropin release by the anterior pituitary and GnRH release by the hypothalamus.	Negative feedback on gonadotropin release by the anterior pituitary and GnRH release by the hypothalamus.	Negative feedback on gonadotropin release by the anterior pituitary and GnRH release by the hypothalamus.
Effects on reproductive organs	Stimulate growth and maturation of reproductive organs and breasts at puberty and maintain their adult size and function. Promote the proliferative phase of the uterine cycle. Stimulate production of watery cervical mucus and activity of fimbriae and uterine tube cilia. Promote oogenesis and ovulation. During pregnancy, stimulate growth of the uterus and enlargement of the external genitalia and mammary glands.	Cooperates with estrogens in stimulating growth of breasts. Promotes the secretory phase of the uterine cycle. Stimulates production of viscous cervical mucus. Progesterone surge after ovulation enhances beating of cilia in the uterine tube, promoting meeting of sperm and oocyte. During pregnancy, quiets the myometrium and acts with estrogens to cause mammary glands to achieve their mature milk-producing state.	Stimulates formation of male reproductive ducts, glands, and external genitalia. Promotes descent of the testes. Stimulates growth and maturation of the internal and external genitalia at puberty; maintains their adult size and function. Required for normal spermatogenesis; ABP keeps its concentration high near spermatogenic cells.
Promotion of secondary sex characteristics and somatic effects	Promote long bone growth followed by epiphyseal closure, and feminization of the skeleton (particularly the pelvis); inhibit bone reabsorption. Promote female pattern of fat deposit.		Promotes (after conversion to estrogens) long bone growth followed by epiphyseal closure; promotes increased skeletal and muscle mass during adolescence. Promotes growth of the larynx and vocal cords and deepening of the voice. Enhances sebum secretion and hair growth, especially on the face, axillae, genital region, and chest.
Metabolic effects	Generally anabolic. Stimulate Na^+ reabsorption by the renal tubules, which inhibits diuresis. Enhance HDL (and reduce LDL) blood levels (cardiovascular sparing effect).	Promotes diuresis (antiestrogenic effect). Increases body temperature.	Generally anabolic. Stimulates hematopoiesis. Enhances the basal metabolic rate.
Neural effects	Along with DHEA (an androgen produced by the adrenal cortex) are partially responsible for female libido (sex drive).		Responsible for libido in males; promotes aggressiveness.

and psychological stimuli promote sexual excitement, which is mediated along the same autonomic nerve pathways as in males.

The final phase of the female sexual response, *orgasm*, is not accompanied by ejaculation. However, muscle tension increases throughout the body, pulse rate and blood pressure rise, and the uterus contracts rhythmically. As in males, orgasm is accompanied by a sensation of intense pleasure followed by relaxation. But unlike what is seen in males, a refractory period does not follow orgasm, so females may experience multiple orgasms during a single sexual experience. Female orgasm is not required for

conception. Indeed, some women never experience orgasm, yet are perfectly able to conceive.

It is unclear which hormones are most closely associated with sexual desire in females. Female libido seems to be driven by both estrogens and dehydroepiandrosterone (DHEA), an androgen produced by the adrenal cortex (see gonadocorticoids, ◀ p. 626).

Check Your Understanding

41. Which glands help to lubricate the vestibule?

For answers, see Answers Appendix.

SEXUALLY TRANSMITTED INFECTIONS

27.17 Sexually transmitted infections cause reproductive and other disorders

Learning Outcome

▶ Indicate the infectious agents and modes of transmission of chlamydia, trichomoniasis, gonorrhea, syphilis, genital warts, and genital herpes.

Sexually transmitted infections (STIs), also called *sexually transmitted diseases (STDs)*, are infectious diseases spread through sexual contact. The United States has the highest rates of infection among developed countries. About 20 million people in the U.S. each year get STIs, which are the single most important cause of reproductive disorders. The first STI you might think of is HIV, which we discussed in Chapter 21 (◀ p. 811), but there are many more common ones. The last few years have seen a dramatic rise in STIs, particularly among 15- to 24-year-olds. Transmission of all STIs can be reduced through the consistent use of condoms.

Bacterial and Parasitic Sexually Transmitted Infections

Chlamydia

Chlamydia (klah-mid′e-ah; *chlamys* = cloak) is the most common bacterial STI in the U.S. *Chlamydia trachomatis* is a bacterium with a viruslike dependence on host cells. Its incubation period may be up to several weeks.

Most infections are silent (asymptomatic)—only about 20% of women and 10% of men have symptoms. These can include urethritis (painful, frequent urination); penile or vaginal discharge; abdominal, rectal, or testicular pain; painful intercourse; and irregular menses. In women, chlamydia is responsible for 25–50% of all cases of pelvic inflammatory disease and is a major cause of sterility. Newborns infected during birth tend to develop conjunctivitis, a painful eye infection that leads to corneal scarring if untreated, and respiratory tract inflammations including pneumonia. Chlamydia can easily be cured with antibiotics.

Trichomoniasis

Trichomoniasis is a parasitic infection that is more common in women than men. It is easily and inexpensively treated. While 70% of women have no symptoms, a characteristic yellow-green vaginal discharge with a strong odor is sometimes present.

Gonorrhea

The causative agent of **gonorrhea** (gon″o-re′ah) is *Neisseria gonorrhoeae*, which invades the mucosae of the reproductive and urinary tracts. Like chlamydia, most infections are asymptomatic.

In men, the most common symptom of gonorrhea is *urethritis*, inflammation of the urethra accompanied by painful urination and discharge of pus from the penis (penile "drip"). Symptoms vary in women, but can include abdominal discomfort, vaginal discharge, abnormal uterine bleeding, and occasionally, urethral symptoms similar to those seen in males.

In men, untreated gonorrhea can cause urethral constriction and inflammation of the entire duct system. In women, it causes pelvic inflammatory disease and sterility. Gonorrhea is becoming increasingly resistant to antibiotics and now requires treatment with at least two antibiotics simultaneously.

Syphilis

Syphilis (sif′ĭ-lis), caused by the corkscrew-shaped bacterium *Treponema pallidum*, is usually transmitted sexually, but it can be contracted congenitally from an infected mother. Fetuses infected with syphilis are usually stillborn or die shortly after birth.

The bacterium easily penetrates intact mucosae and abraded skin. Within a few hours of exposure, an asymptomatic bodywide infection is in progress. After an incubation period of about three weeks, a red, painless primary lesion called a *chancre* (shang′ker) appears at the site of bacterial invasion. In males, this is typically the penis, but in females the lesion often goes undetected within the vagina or on the cervix. The chancre ulcerates and becomes crusty, and then heals spontaneously and disappears within a few weeks.

Secondary signs of untreated syphilis appear several weeks later. A pink skin rash all over the body is one of the first symptoms. Fever and joint pain are common. These signs and symptoms disappear spontaneously in three to twelve weeks. Then the disease enters the *latent period* and is detectable only by a blood test. The latent stage may last a person's lifetime (or the immune system may kill the bacteria), or it may be followed by *tertiary syphilis*.

Tertiary syphilis is characterized by destructive lesions of the CNS, blood vessels, bones, and skin. Penicillin is still the treatment of choice for all stages of syphilis.

Viral Sexually Transmitted Infections

While a number of viruses (such as hepatitis, Ebola, and Zika) can be transmitted sexually, we limit our discussion here to those viruses that are *primarily* transmitted sexually.

Human Papillomavirus (HPV)

Infection by the **human papillomavirus (HPV)**—actually a group of about 40 viruses—is the most common STI in the United States. More than half of adults are infected during their lifetime. Although HPV causes **genital warts**, the long-term concern is that HPV causes cancers in infected body regions. The virus is linked not only to 80% of all cases of invasive cervical cancer (see Homeostatic Imbalance 27.6, p. 1064) but also to penile, anal, and oropharyngeal cancers.

Vaccination can guard against infection by the most common cancer-causing strains of HPV, and is recommended for both girls and boys before they become sexually active.

Genital Herpes

The cause of **genital herpes** is the *herpes simplex virus*, and these viruses are among the most difficult human pathogens to control. In part, this is because most people who have genital herpes do not know they have it and can shed the virus while symptom free. Only about 15% of infected individuals display signs of infection.

Characteristic blisterlike lesions may flare up repeatedly over time. The painful lesions that appear on the reproductive organs of infected adults are usually more of a nuisance than a threat. However, congenital herpes infections can cause severe malformations of a fetus. The virus is transmitted via infectious secretions or direct skin-to-skin contact.

Antiviral drugs can reduce the duration and intensity of the flare-ups. However, once contracted, genital herpes never leaves. It just goes into periodic remissions.

Check Your Understanding

42. Which pathogen is most associated with cervical cancer?

43. What is the most common bacterial STI in the United States?

For answers, see Answers Appendix.

Developmental Aspects of the Reproductive System

So far, we have described the reproductive organs as they exist and operate in adults. Now we are ready to look at events that cause us to become reproductive individuals. These events begin long before birth and end, at least in women, in late middle age.

Embryological and Fetal Events

Determination of Genetic Sex

Aristotle believed that the "heat" of lovemaking determined maleness of the offspring. Not so! Genetic sex is determined at the instant the genes of a sperm combine with those of an ovum, and the determining factor is the **sex chromosomes** each gamete contains. Of the 46 chromosomes in the fertilized egg, two (one pair) are sex chromosomes. The other 44 are called **autosomes**.

Two types of sex chromosomes, quite different in size, exist in humans: the large **X chromosome** and the much smaller **Y chromosome**. The body cells of females have two X chromosomes and are designated XX, and the ovum resulting from normal meiosis in a female always contains an X chromosome (**Figure 27.26**). Males have one X chromosome and one Y in each body cell (XY).

If the fertilizing sperm delivers an X chromosome, the fertilized egg and its daughter cells will contain the female (XX) composition, and the embryo will develop ovaries. If the sperm bears a Y chromosome, the offspring will be male (XY) and will develop testes. The presence of a single gene on the Y chromosome—the **SRY** (for **s**ex-determining **r**egion of the **Y** chromosome) **gene**—determines maleness by initiating testes development. In this way, the father's gamete determines the genetic sex of the offspring. All subsequent events of sexual differentiation depend on which gonads form during embryonic life.

Figure 27.26 Genetic determination of sex. The male gamete determines the sex of the baby.

> #### HOMEOSTATIC IMBALANCE 27.10   CLINICAL
>
> When meiosis distributes the sex chromosomes to the gametes improperly (an event called **nondisjunction**, see Related Clinical Terms, p. 1135), abnormal combinations of sex chromosomes in the zygote (fertilized egg) result. If the sex chromosomes are involved, striking **disorders of sex development (DSD)** follow. These are named using the total number of chromosomes and the complement of X and Y chromosomes.
>
> For example, females with a single X chromosome (XO), a condition called *Turner's syndrome* (45,XO DSD), never develop ovaries. Women with Turner's syndrome usually have short stature and are at risk for some heart defects and specific learning disabilities.
>
> *Klinefelter's syndrome* (47,XXY DSD), which affects one out of 500 live male births, is the most common sex chromosome abnormality. Affected individuals usually have a single Y chromosome, two or more X chromosomes, and are sterile males. Although XXY males are normal (or only slightly below normal) intellectually, the incidence of intellectual disability increases as the number of X chromosomes rises.

Sexual Differentiation of the Reproductive System

Differentiation of accessory structures and the external genitalia into male or female structures depends on whether testosterone is present. When testes form in response to *SRY*, they release testosterone, which causes the development of male accessory ducts and external genitalia. In the absence of testosterone, the female ducts and external genitalia develop.

27

Figure 27.27 Development of the internal reproductive organs. (*SRY* = sex-determining region of the Y chromosome)

The gonads begin to develop during week 5 of gestation as masses of mesoderm called the **gonadal ridges** (**Figure 27.27**). These ridges bulge from the dorsal abdominal wall just medial to the mesonephros (a transient kidney system, ◄ **p. 1006**). The **paramesonephric**, or *Müllerian*, **ducts** (future female ducts) develop lateral to the **mesonephric** (*Wolffian*) **ducts** (future male ducts). Both sets of ducts empty into a common chamber called the *cloaca*. At this stage of development, the embryo is said to be in the **sexually indifferent stage**, because the gonadal ridge tissue can develop into either male or female gonads and both duct systems are present.

Shortly after the gonadal ridges appear, **primordial germ cells** migrate to them from the yolk sac (described in Chapter 28). There they seed the developing gonads with stem cells destined to become spermatogonia or oogonia. Once these cells are in

residence, the gonadal ridges form testes or ovaries, depending on the genetic makeup of the embryo.

In male embryos, the process of forming the testes begins in week 7. (1) Seminiferous tubules form in the internal part of the gonadal ridges and join the mesonephric duct via the efferent ductules. (2) Further development of the mesonephric duct produces the duct system of the male. (3) The paramesonephric ducts degenerate.

In female embryos the process of gonad formation begins about a week later. (1) The cortical part of each immature ovary forms follicles. (2) The paramesonephric ducts differentiate into the structures of the female duct system. (3) The mesonephric ducts degenerate.

Like the gonads, the external genitalia arise from the same structures in both sexes (**Figure 27.28**). During the sexually

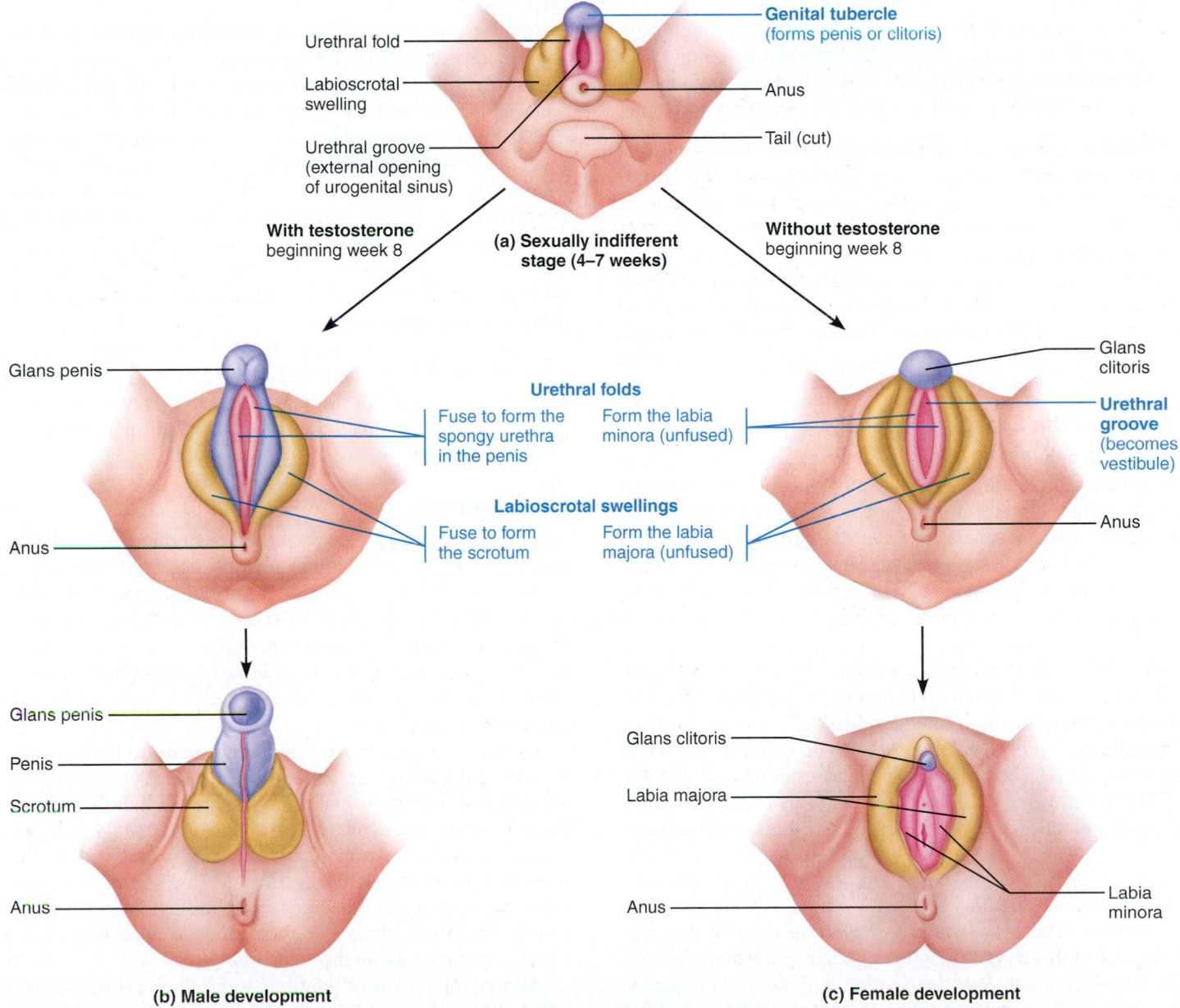

Figure 27.28 Development of homologous structures of the external genitalia in both sexes. The two pictures at the bottom of the figure show the fully developed perineum.

indifferent stage, all embryos exhibit a small external projection called the **genital tubercle**. The urogenital sinus (future bladder and urethra), which develops by subdivision of the cloaca, lies deep to the tubercle. The **urethral groove**, the external opening of the urogenital sinus, is on the tubercle's inferior surface and is flanked laterally by the **urethral folds** and then the **labio-scrotal swellings**. Figure 27.28 shows the sequence of events in males and females.

HOMEOSTATIC IMBALANCE 27.11   CLINICAL

Any interference with the normal pattern of sex hormone production in the embryo results in abnormalities. For example:

- If the embryonic testes do not produce testosterone, a genetic male develops female accessory structures and external genitalia.

- If a genetic female is exposed to testosterone (as might happen if the mother has an androgen-producing tumor of her adrenal gland), the embryo has ovaries but develops the male ducts and glands, as well as a penis and an empty scrotum.

The female pattern of reproductive structures has an intrinsic ability to develop. In the absence of testosterone it proceeds to do so, regardless of the embryo's genetic makeup. In many cases, however, the external genitalia are ambiguous (neither fully male nor female).

Individuals with external genitalia that do not "match" their gonads are *intersex*. Intersex individuals sometimes have surgery to match their outer selves (external genitalia) with their inner selves (gonads).

Descent of the Gonads

About two months before birth, the testes begin their descent toward the scrotum, dragging their supplying blood vessels and nerves along behind them. They finally exit from the pelvic cavity via the inguinal canals and enter the scrotum. Testosterone made by the fetus's testes stimulates this migration. The migration is mechanically guided by a strong fibrous cord called the **gubernaculum** ("governor"), which extends from the testis to the floor of the scrotal sac.

Initially the gubernaculum is a column of soft connective tissue, but it becomes increasingly fibrous as it continues to grow. By the seventh month of fetal development, it stops growing. The gubernaculum's cessation of growth, coupled with rapid growth of the fetal body, helps to pull the testes into the scrotum.

The *tunica vaginalis* covering of the testis is derived from a fingerlike outpocketing of the parietal peritoneum, the *vaginal process*. The accompanying blood vessels, nerves, and fascial layers form part of the *spermatic cord*, which helps suspend the testis within the scrotum.

Like the testes, the ovaries descend during fetal development, but in this case only to the pelvic brim, where the tent-like broad ligament stops their progress. Each ovary is guided in its descent by a gubernaculum that later divides, becoming the ovarian and round ligaments that help support the internal genitalia in the pelvis.

HOMEOSTATIC IMBALANCE 27.12   CLINICAL

Failure of the testes to make their normal descent leads to *cryptorchidism* (*crypt* = hidden, concealed; *orchi* = testicle). Because cryptorchidism causes sterility and increases the risk of testicular cancer, surgery is usually performed during early childhood to rectify this problem.

Puberty

The events of puberty occur in the same sequence in all individuals, but the age at which they occur varies widely. Puberty usually occurs between the ages of 10 and 15 as a consequence of the maturation of the hypothalamic-pituitary-gonadal axis (◄ p. 1043).

In males, secretion of adrenal androgens, particularly dehydroepiandrosterone (DHEA; ◄ p. 626), begins to rise several years before the testosterone surge of puberty and initiates facial, pubic, and axillary hair growth and other pubertal events. The major event that signals puberty's onset in males is enlargement of the testes and scrotum between the ages of 8 and 14. The penis grows over the next two years, and sexual maturation is evidenced by the presence of mature sperm in the semen (see Figure 27.14). In the meantime, the young man has unexpected erections and occasional nocturnal emissions ("wet dreams") as his hormones surge and the hormonal control axis struggles to achieve a normal balance.

In females, the first sign of puberty is budding breasts between the ages of 8 and 13, followed by the appearance of axillary and pubic hair. Menarche usually occurs about two years later, but dependable ovulation and fertility takes nearly two more years.

Menopause

Most women reach the peak of their reproductive abilities in their late 20s. At age 30, there are still some 100,000 oocytes in the ovaries but their quality (fertility) has begun to decline. By age 50, there are about 1000 eggs left. As the number of follicles decreases, the total amount of estrogen produced also decreases until at last the feedback loop with the anterior pituitary and hypothalamus fails. As a result, cyclical hormone changes no longer occur.

As estrogen production declines, many ovarian cycles become anovulatory, while in others 2 to 4 oocytes per month are ovulated. This sign of declining control explains why twins and triplets are more common in older women. In the perimenopausal period, menstrual periods become erratic and shorter. Eventually, ovulation and menstruation cease entirely. This normally occurs between the ages of 46 and 54, an event called *menopause*. **Menopause** is considered to have occurred when a year has passed without menstruation.

Although the ovaries continue to produce estrogens for a while after menopause, they finally stop producing hormones. Without sufficient estrogens the reproductive organs and breasts begin to atrophy, the vagina becomes dry, and vaginal

Homeostatic Interrelationships between the Reproductive System and Other Body Systems

Muscular System   Chapters 9–10
- Androgens promote increased muscle mass
- Abdominal muscles active during childbirth; muscles of the pelvic floor support reproductive organs and aid erection of penis/clitoris

Nervous System   Chapters 11–15
- Sex hormones masculinize or feminize the brain and influence sex drive
- Hypothalamus regulates timing of puberty; neural reflexes regulate events of sexual response

Endocrine System   Chapter 16
- Gonadal hormones exert feedback effects on hypothalamic-pituitary axis
- Gonadotropins (and GnRH) help regulate function of gonads; leptin signals the hypothalamus about the body's state of energy availability (in fat stores)

Cardiovascular System   Chapters 17–19
- Estrogens lower blood cholesterol levels and promote cardiovascular health in premenopausal women; pregnancy increases workload of the cardiovascular system
- Cardiovascular system transports needed substances to organs of reproductive system; local vasodilation involved in erection; blood transports sex hormones

Lymphatic System/Immunity   Chapters 20–21
- Developing embryo/fetus escapes immune surveillance (not rejected)
- Lymphatic vessels drain leaked tissue fluids; transport sex hormones; immune cells protect reproductive organs from disease; IgA is present in breast milk

Respiratory System   Chapter 22
- Pregnancy impairs descent of the diaphragm, promotes dyspnea
- Respiratory system provides oxygen; disposes of carbon dioxide; tidal volume increases during pregnancy while residual volume declines

Digestive System   Chapter 23
- Developing fetus crowds digestive organs; heartburn, constipation common during pregnancy
- Digestive system provides nutrients needed for health

Urinary System   Chapters 25–26
- Hypertrophy of the prostate inhibits urination; compression of bladder during pregnancy leads to urinary frequency and urgency
- Kidneys maintain acid-base balance of blood of mother and fetus and dispose of nitrogenous wastes; semen discharged through the urethra of the male

Integumentary System   Chapter 5
- Androgens activate oil glands that lubricate skin and hair; gonadal hormones stimulate characteristic fat distribution and appearance of pubic and axillary hair; estrogens increase skin hydration, and during pregnancy they enhance facial skin pigmentation
- Skin protects all body organs by external enclosure; mammary gland secretions (milk) nourish the infant

Skeletal System   Chapters 6–8
- Androgens masculinize the skeleton and increase bone density; estrogens feminize the skeleton and maintain bone mass in females
- The pelvis encloses some reproductive organs; if narrow, the pelvis may hinder vaginal delivery of an infant

infections become increasingly common. Other sequels due to lack of estrogens include:

- Intense vasodilation of the skin's blood vessels, which causes uncomfortable, sweat-drenching "hot flashes"
- Gradually thinning skin
- Loss of bone mass

Slowly rising total blood cholesterol levels and falling HDL levels place postmenopausal women at risk for cardiovascular disorders. Irritability and depression occurs in some women.

Estrogen-containing hormone replacement therapy (HRT) is sometimes used to help women with severe menopausal symptoms through this period. Clinical trials have shown that HRT can increase the quality of life, decrease loss of bone mass, and decrease the risk for colon cancer. Surprisingly, despite its beneficial effects on blood lipids, HRT increases the risk for heart disease. HRT also increases the risk of breast cancer because many breast cancers are sensitive to estrogens.

Nonetheless, for women who do not have existing breast cancer, the risks may be outweighed by the benefits of taking the smallest dose of HRT needed to reduce symptoms for the shortest period of time.

There is no equivalent of menopause in males, and healthy men are able to father offspring well into their eighth decade of life. However, aging men do exhibit a steady decline in testosterone secretion and experience a longer latent period after orgasm, a condition sometimes called *andropause*. Additionally, there is a noticeable difference in sperm quality and motility with aging.

RELATED CLINICAL TERMS

Dysmenorrhea (dis″men-ŏ-re′ah; *dys* = bad, *meno* = menses, a month) Painful menstruation; may reflect abnormally high prostaglandin activity during menses.

Endometrial cancer Cancer that arises from the uterine endometrium (usually from uterine glands). Most important sign is vaginal bleeding, which allows early detection. Risk factors include obesity and hormone replacement therapy.

Endometriosis (en″do-me″tre-o′sis) An inflammatory condition in which endometrial tissue occurs and grows atypically in the pelvic cavity. Characterized by abnormal uterine or rectal bleeding, dysmenorrhea, and pelvic pain. May cause sterility.

Gynecology (gi″nĕ-kol′o-je; *gyneco* = woman, *ology* = study of) Specialized branch of medicine that deals with the diagnosis and treatment of female reproductive system disorders.

Gynecomastia (gi″nĕ-ko-mas′te-ah) Development of breast tissue in the male; a consequence of adrenal cortex hypersecretion of estrogens, or certain drugs (cimetidine, spironolactone, and some chemotherapeutic agents).

Hysterectomy (his″tĕ-rek′to-me; *hyster* = uterus, *ectomy* = cut out) Surgical removal of the uterus.

Inguinal hernia Protrusion of part of the intestine into the scrotum or through a separation in the abdominal muscles in the groin region. Since the inguinal canals represent weak points in the abdominal wall, inguinal hernia may be caused by heavy lifting or other activities that increase intra-abdominal pressure.

Laparoscopy (lap″ah-ros′ko-pe; *lapar* = the flank, *scopy* = observation) Examination of the abdominopelvic cavity with a laparoscope, a viewing device at the end of a thin tube inserted through the anterior abdominal wall. Laparoscopy is often used to assess the condition of a woman's pelvic reproductive organs.

Oophorectomy (o″of-o-rek′to-me; *oophor* = ovary) Surgical removal of the ovary.

Orchitis (or-ki′tis; *orcho* = testis) Inflammation of the testes, sometimes caused by the mumps virus.

Ovarian cancer Malignancy that typically arises from the cells in the epithelial covering of the ovary. The fifth most common reproductive system cancer, its incidence increases with age. Early symptoms are nondescript and easily mistaken for other disorders (back pain, abdominal discomfort, nausea, bloating, and flatulence). Diagnosis may involve palpating a mass during a physical exam, visualizing it with ultrasound imaging, or conducting blood tests for a protein marker for ovarian cancer (CA-125). However, medical assessment is often delayed until after metastasis has occurred.

Ovarian cysts Ovarian cysts are nonovulatory, fluid-filled sacs larger than normal follicles in the ovaries. They are extremely common, and the vast majority of these cysts are benign and asymptomatic. They are usually discovered by routine palpation during a pelvic exam or by ultrasound. Some cause abdominal pain or discomfort due to their size. Twisting or rupture may lead to more severe pain. For persistent follicles of more than 5–10 cm, surgery may be required.

Polycystic ovary syndrome The most common endocrinopathy in women and the most common cause of anovulatory infertility. Affects 5–10% of women; characterized by signs of androgen excess, increased cardiovascular risk (evidenced by high blood pressure, decreased HDL cholesterol levels, and high triglycerides); and linked to obesity and some degree of insulin resistance. Treated with insulin-sensitizing drugs.

Prostatitis An acute or chronic inflammatory condition of the prostate that may (or may not) be associated with infection. Acute or chronic bacterial prostatitis is usually caused by *E. coli*.

Salpingitis (sal″pin-ji′tis; *salpingo* = uterine tube) Inflammation of the uterine tubes.

27

CHAPTER SUMMARY

1. The function of the reproductive system is to produce offspring. The gonads produce gametes (sperm or ova) and sex hormones. All other reproductive organs are accessory organs.

27.1 The male and female reproductive systems share common features (pp. 1042–1047)

1. The gonads (testes or ovaries) are the primary sex organs. They produce gametes (sperm or eggs) and sex hormones. The gonads (and many other reproductive structures) are homologous structures.
2. In both males and females, secretion of reproductive hormones is regulated by interactions along the hypothalamic-pituitary-gonadal (HPG) axis. GnRH, produced by the hypothalamus, stimulates the anterior pituitary gland to release FSH and LH. These, in turn, stimulate the secretion of steroid hormones from the gonads. Beginning at puberty, the HPG axis becomes established.
3. Meiosis, the basis of gamete production, consists of two consecutive nuclear divisions without DNA replication in between. Meiosis reduces the chromosomal number by half and introduces genetic variability. Events unique to meiosis include synapsis and crossing over of homologous chromosomes.

PART 1
ANATOMY OF THE MALE REPRODUCTIVE SYSTEM

27.2 The testes are enclosed and protected by the scrotum (pp. 1048–1049)

1. The scrotum contains the testes. It provides a temperature slightly lower than body temperature, as required to produce viable sperm.
2. Each testis is covered externally by a tunica albuginea that extends internally to divide the testis into many lobules. Each lobule contains sperm-producing seminiferous tubules.

27.3 Sperm travel from the testes to the body exterior through a system of ducts (p. 1050)

1. The epididymis hugs the external surface of the testis and serves as a site for sperm maturation and storage.
2. The ductus (vas) deferens, extending from the epididymis to the ejaculatory duct, propels sperm into the urethra by peristalsis during ejaculation. Its terminus fuses with the duct of the seminal gland, forming the ejaculatory duct, which empties into the urethra within the prostate.
3. The urethra extends from the urinary bladder to the tip of the penis. It conducts semen and urine to the body exterior.

27.4 The penis is the copulatory organ of the male (pp. 1050–1052)

1. The penis is largely erectile tissue (corpus spongiosum and corpora cavernosa). Engorgement of the erectile tissue with blood causes the penis to become rigid, an event called erection.
2. The male perineum, which suspends the scrotum and penis, is the region encompassed by the pubic symphysis, ischial tuberosities, and coccyx.

27.5 The male accessory glands produce the bulk of semen (pp. 1052–1053)

1. The accessory glands include the seminal glands, the prostate, and the bulbo-urethral glands. Semen contains fructose from the seminal glands, an activating fluid from the prostate, and mucus from the bulbo-urethral glands.
2. Semen is an alkaline fluid that dilutes and transports sperm. Important chemicals in semen are nutrients, prostaglandins, and antibiotic chemicals.

PART 2
PHYSIOLOGY OF THE MALE REPRODUCTIVE SYSTEM

27.6 The male sexual response includes erection and ejaculation (pp. 1053–1054)

1. Parasympathetic reflexes control erection.
2. Ejaculation is expulsion of semen from the male duct system, promoted by the sympathetic nervous system. Ejaculation is part of male orgasm, which also includes pleasurable sensations and increased pulse and blood pressure.

27.7 Spermatogenesis is the sequence of events that leads to formation of sperm (pp. 1054–1058)

1. Spermatogenesis, the production of male gametes in the seminiferous tubules, begins at puberty.
2. Interstitial endocrine cells, located in connective tissue between seminiferous tubules, secrete testosterone.
3. Spermatogonia divide by mitosis to maintain the germ cell line. Some of their progeny become primary spermatocytes, which undergo meiosis I to produce secondary spermatocytes. Secondary spermatocytes undergo meiosis II, each producing two haploid (n) spermatids.
4. Spermiogenesis converts spermatids to functional sperm, stripping away superfluous cytoplasm and producing an acrosome and a flagellum (tail).
5. Sustentocytes form the blood testis barrier, nourish spermatogenic cells, move them toward the lumen of the tubules, and secrete fluid for sperm transport.

27.8 Male reproductive function is regulated by hypothalamic, anterior pituitary, and testicular hormones (pp. 1059–1060)

1. FSH causes sustentocytes to produce androgen-binding protein (ABP). LH stimulates interstitial endocrine cells to release testosterone, which binds to ABP, stimulating spermatogenesis. Testosterone feeds back to inhibit the hypothalamus and anterior pituitary. Inhibin, released by sustentocytes, inhibits FSH release.
2. Maturation of hormonal controls occurs during puberty and takes about three years.
3. Testosterone stimulates maturation of the male reproductive organs and triggers the development of the secondary sex characteristics of the male. It exerts anabolic effects on the skeleton and skeletal muscles, stimulates spermatogenesis, and is responsible for male sex drive.

PART 3
ANATOMY OF THE FEMALE REPRODUCTIVE SYSTEM

1. The female reproductive system produces gametes and sex hormones and houses a developing infant until birth.

27

27.9 Immature eggs develop in follicles in the ovaries (pp. 1061–1062)

1. The ovaries flank the uterus laterally and are held in position by the ovarian and suspensory ligaments and mesovaria.
2. Within each ovary are oocyte-containing follicles at different stages of development.

27.10 The female duct system includes the uterine tubes, uterus, and vagina (pp. 1062–1067)

1. The uterine tube, supported by the mesosalpinx, extends from near the ovary to the uterus. Movements of its ciliated distal projections, called fimbriae, along with peristalsis, create currents that help move an ovulated oocyte into the uterine tube.
2. The uterus has fundus, body, and cervical regions. It is supported by the broad, cardinal, uterosacral, and round ligaments.
3. The uterine wall is composed of the outer perimetrium, the myometrium, and the inner endometrium. The endometrium consists of a functional layer (stratum functionalis), which sloughs off periodically unless an embryo has implanted, and an underlying basal layer (stratum basalis), which rebuilds the functional layer.
4. The vagina extends from the uterus to the exterior. It is the copulatory organ and allows passage of the menstrual flow or a baby.

27.11 The external genitalia of the female include those structures that lie external to the vagina (p. 1067)

1. The female external genitalia (vulva) include the mons pubis, labia majora and minora, clitoris, and the urethral and vaginal orifices. The mucus-secreting greater vestibular glands flank the vaginal opening.

27.12 The mammary glands produce milk (pp. 1068–1069)

1. The mammary glands lie over the pectoral muscles of the chest and are surrounded by adipose and fibrous connective tissue. Each mammary gland consists of many lobules, which contain milk-producing alveoli.

PART 4

PHYSIOLOGY OF THE FEMALE REPRODUCTIVE SYSTEM

27.13 Oogenesis is the sequence of events that leads to the formation of ova (pp. 1069–1073)

1. Oogenesis, the production of eggs, begins in the fetus. Oogonia, diploid stem cells that give rise to female gametes, are converted to primary oocytes before birth. The infant female's ovaries contain about 2 million primary oocytes arrested in prophase of meiosis I.
2. At puberty, meiosis resumes. Each month, one primary oocyte completes meiosis I, producing a large secondary oocyte and a tiny first polar body. Meiosis II of the secondary oocyte produces a functional ovum and a second polar body, but does not occur unless a sperm penetrates the secondary oocyte.
3. The ovum contains most of the primary oocyte's cytoplasm. The polar bodies are nonfunctional and degenerate.

27.14 The ovarian cycle consists of the follicular phase and the luteal phase (pp. 1073–1074)

1. During the follicular phase (days 1–14), several vesicular follicles are rescued from atresia and continue to grow. Generally, only one follicle per month is selected to become the

dominant follicle and completes the maturation process. Late in this phase, the oocyte in the dominant follicle completes meiosis I. Ovulation occurs about day 14, releasing the secondary oocyte into the peritoneal cavity. Other developing follicles deteriorate.
2. In the luteal phase (days 15–28), the ruptured follicle is converted to a corpus luteum, which produces progesterone and estrogens for the remainder of the cycle. If fertilization does not occur, the corpus luteum degenerates after about 10 days.

27.15 Female reproductive function is regulated by hypothalamic, anterior pituitary, and ovarian hormones (pp. 1074–1078)

Hormonal Regulation of the Ovarian Cycle (pp. 1074–1076)

1. Establishment of the mature cyclic pattern of hormone secretion, indicated by menarche, takes about four years. Leptin serves a permissive role in puberty's onset, stimulating the hypothalamus when adipose tissue is sufficient for the energy requirements of reproduction.
2. The hormonal events of each ovarian cycle are as follows: (1) GnRH stimulates the anterior pituitary to release FSH and LH, which stimulate follicle maturation and estrogen production. (2) When blood estrogens reach a certain level, positive feedback exerted on the hypothalamic-pituitary-gonadal axis causes a sudden release of LH that stimulates the primary oocyte to continue meiosis and triggers ovulation. LH then causes conversion of the ruptured follicle to a corpus luteum and stimulates its secretory activity. (3) Rising levels of progesterone and estrogens inhibit the hypothalamic-pituitary-gonadal (HPG) axis, the corpus luteum deteriorates, ovarian hormones drop to their lowest levels, and the cycle begins anew.
3. While the ovarian cycle is about 28 days, maturation of a primordial follicle from its recruitment until it is ready to be ovulated takes about 290 days.

The Uterine (Menstrual) Cycle (pp. 1076–1078)

4. Varying levels of ovarian hormones in the blood trigger events of the uterine cycle.
5. During the menstrual phase of the uterine cycle (days 0–4), the functional layer sloughs off in menses. During the proliferative phase (days 5–14), rising estrogen levels stimulate its regeneration. Under the influence of both estrogens and progesterone during the secretory phase (days 15–28), the uterine glands secrete nutrients, and endometrial vascularity increases further. This makes the uterus receptive to implantation about one week after ovulation.
6. Falling levels of ovarian hormones during the last few days of the ovarian cycle cause the spiral arteries to become spastic and cut off the blood supply of the functional layer. The uterine cycle begins again with menstruation.

Effects of Estrogens and Progesterone (p. 1078)

7. Estrogens promote oogenesis. At puberty, they stimulate the growth of the reproductive organs and the growth spurt and promote the appearance of the secondary sex characteristics.
8. Progesterone cooperates with estrogens in breast maturation and regulation of the uterine cycle.

27.16 The female sexual response is more diverse and complex than that of males (pp. 1078–1079)

1. The female sexual response is similar to that of males. Orgasm in females is not accompanied by ejaculation.

SEXUALLY TRANSMITTED INFECTIONS

27.17 Sexually transmitted infections cause reproductive and other disorders (pp. 1080–1081)

1. Sexually transmitted infections (STIs) are infectious diseases spread via sexual contact. Gonorrhea, syphilis, and chlamydia are bacterial diseases. Syphilis has broader consequences than most other sexually transmitted bacterial diseases since it can infect organs throughout the body. Trichomoniasis is a parasitic infection. Genital herpes and genital warts are viral infections; genital warts, caused by the HPV virus, are implicated in cervical cancer.

Developmental Aspects of the Reproductive System (pp. 1081–1086)

Embryological and Fetal Events (pp. 1081–1084)

1. Genetic sex is determined by the sex chromosomes: an X from the mother, an X or a Y from the father. If the fertilized egg contains XX, it is a female and develops ovaries; if it contains XY, it is a male and develops testes.

2. Gonads of both sexes arise from the mesodermal gonadal ridges. The mesonephric ducts produce the male accessory ducts and glands. The paramesonephric ducts produce the female duct system.

3. The external genitalia arise from the genital tubercle and associated structures. The development of male accessory structures and external genitalia depends on the presence of testosterone produced by the embryonic testes. In its absence, female structures develop.

4. The testes descend into the scrotum from the abdominal cavity.

Puberty (p. 1084)

5. Puberty is the interval when reproductive organs mature and become functional. It begins in males with penile and scrotal growth and in females with breast development.

Menopause (pp. 1084, 1086)

6. During menopause, ovulation and menstruation cease. Hot flashes and mood changes may occur. Postmenopausal events include atrophy of the reproductive organs, bone mass loss, and increasing risk for cardiovascular disease. Men experience andropause with milder signs and symptoms of testosterone deficit.

REVIEW QUESTIONS

To access additional practice questions using your smartphone, tablet, or computer: MasteringA&P® > Study Area > Practice Tests & Quizzes

Level 1 Remember/Understand

(Some questions have more than one correct answer. Select the best answer or answers from the choices given.)

1. The normal diploid number of human chromosomes is (a) 48, (b) 47, (c) 46, (d) 23, (e) 24.

2. Relative to differences between mitosis and meiosis, choose the statements that apply *only* to events of meiosis. (a) tetrads present, (b) produces two daughter cells, (c) produces four daughter cells, (d) occurs throughout life, (e) reduces the chromosomal number by half, (f) synapsis and crossing over of homologues occur.

3. The structures that draw an ovulated oocyte into the female duct system are (a) cilia, (b) fimbriae, (c) microvilli, (d) stereocilia.

4. The usual site of embryo implantation is (a) the uterine tube, (b) the peritoneal cavity, (c) the vagina, (d) the uterus.

5. Which of the following is correct relative to female anatomy? (a) the vaginal orifice is the most dorsal of the three openings in the perineum, (b) the urethra is between the vaginal orifice and the anus, (c) the anus is between the vaginal orifice and the urethra, (d) the urethra is the more ventral of the two orifices in the vulva.

6. Secondary sex characteristics are (a) present in the embryo, (b) a result of male or female sex hormones increasing in amount at puberty, (c) the testis in the male and the ovary in the female, (d) development of male and female external genitalia.

7. Which of the following produces the male sex hormones? (a) seminal glands, (b) corpus luteum, (c) developing follicles in the testes, (d) interstitial endocrine cells.

8. Which will occur as a result of nondescent of the testes? (a) male sex hormones will not be circulated in the body, (b) sperm will have no means of exit from the body, (c) inadequate blood supply will retard the development of the testes, (d) viable sperm will not be produced.

9. Match the key choices with the descriptive phrases below.
   Key:
   (a) androgen-binding protein   (e) inhibin
   (b) estrogens   (f) LH
   (c) FSH   (g) progesterone
   (d) GnRH   (h) testosterone

   ____ (1) Hormones that directly regulate the ovarian cycle
   ____ (2) Chemicals in males that inhibit the hypothalamic-pituitary-testicular axis
   ____ (3) Hormone that makes the cervical mucus viscous
   ____ (4) Potentiates the activity of testosterone on spermatogenic cells
   ____ (5) Chemicals in females that inhibit the hypothalamic-pituitary-ovarian axis
   ____ (6) Stimulates the secretion of testosterone

10. Which of the following does not add a secretion that makes a major contribution to semen? (a) prostate, (b) bulbo-urethral glands, (c) testes, (d) ductus deferens.

11. The corpus luteum is formed at the site of (a) fertilization, (b) ovulation, (c) menstruation, (d) implantation.

Level 2 Apply/Analyze

12. Sperm are to seminiferous tubules as oocytes are to (a) fimbriae, (b) corpus albicans, (c) ovarian follicles, (d) corpora lutea.

13. FSH is to estrogens as high levels of estrogens are to (a) progesterone, (b) LH, (c) FSH, (d) testosterone.

14. Why is the term *urogenital system* more applicable to males than to females?

15. Describe the major structural (and functional) regions of a sperm.

16. Describe the events and possible consequences of menopause.

17. Define menarche. What does it indicate?

18. Trace the pathway of a sperm from the male testes to the uterine tube of a female.

19. In menstruation, the functional layer is shed from the endometrium. Explain the hormonal and physical factors responsible for this shedding. (Hint: See Figure 27.25.)

20. Both the epithelium of the vagina and the cervical glands of the uterus help prevent the invasion and spread of vaginal pathogens. Explain how each of these mechanisms works.

Level 3 Evaluate/Synthesize

21. Oogenesis in the female results in one functional gamete—the egg, or ovum. What other cells are produced? What is the significance of this rather wasteful type of gamete production—that is, production of a single functional gamete instead of four, as seen in males?

22. Gina Marciano, a 44-year-old mother of eight children, visited her physician complaining of a "bearing down" sensation in her pelvis, low backache, and urinary incontinence. A vaginal examination showed that the external os of her cervix was just inside the vaginal orifice and her perineum exhibited large keloids (masses of scar tissue). What do you think Gina's problem is and what caused it? (Be anatomically specific.)

23. Grant, a sexually active adolescent, visited his doctor complaining of a red, painless swelling on his penis. He has no penile discharge or pain on urination. An account of his recent sexual behavior was requested and recorded. **(a)** What do you think Grant's problem is? **(b)** What is the causative agent of this disorder? **(c)** How is the condition treated, and what may happen if it isn't treated?

24. A 36-year-old mother of four is considering tubal ligation to ensure that her family gets no larger. She asks the physician if she will become "menopausal" after the surgery. **(a)** How would you answer her question and explain away her concerns? **(b)** Explain what a tubal ligation is.

25. Mr. Scanlon, a 76-year-old man, is interested in a much younger woman. Concerned because of his age, he asks his urologist if he will be able to father a child. What questions would a physician ask of this man, and what diagnostic tests might be ordered?

26. Erin had both her left ovary and her right uterine tube removed surgically at age 17 because of a cyst and a tumor in these organs. Now, at age 32, she remains healthy and is expecting her second child. How could Erin conceive a child with just one ovary and one uterine tube, widely separated on opposite sides of the pelvis like this?

CLINICAL CASE STUDY

23-Year-Old Female with Menstrual Irregularity

Jasmine Herrera, 23, has been referred to the endocrinology clinic. She is very overweight, with obvious acne and facial hair. She complains of irregular menstrual cycles. "My husband and I want to use the rhythm method," she says, "but I don't have any rhythm!"

1. A change in the levels of which hormone triggers ovulation? What triggers the onset of menstruation?

2. **+ NCLEX-STYLE** Jasmine's doctor has ordered measurement of her androgen levels. Which of the following statements is most accurate about this test?
   a. It should come back negative, because women do not produce any androgens.
   b. It depends on the phase of her cycle; androgens are only synthesized during the luteal phase of the ovarian cycle.
   c. It depends on her muscle mass; androgens are anabolic steroids and cause muscle growth, so more muscle means more androgens.
   d. Levels should be low but not absent, because the ovaries produce androgens but convert most of them into estrogens.

3. Transvaginal ultrasound reveals more than 30 enlarged vesicular follicles in Jasmine's left ovary. What are vesicular follicles?

4. Jasmine thinks that having so many follicles means she is extremely fertile. Is her assumption correct?

5. Jasmine's blood tests reveal very high levels of androgens. She is diagnosed with polycystic ovary syndrome (PCOS). In this syndrome, the woman develops high levels of androgens and many vesicular follicles, but has reduced fertility because few follicles are ovulated and the timing of their release is unpredictable. Part of the profile of PCOS is that LH levels are high. How would this lead to androgen excess, and what clinical evidence of androgen excess does Jasmine have?

6. **+ NCLEX-STYLE** Jasmine is started on oral contraceptives containing a combination of an estrogen and progesterone. These will regulate her menstrual cycles and decrease her androgen level. They will also act to prevent pregnancy by preventing ovulation. What phase of the uterine cycle is caused by sustained elevation of both estrogen and progesterone levels?
   a. Proliferative phase
   b. Follicular phase
   c. Secretory phase
   d. Menstrual phase

28 Pregnancy and Human Development

In this chapter, you will learn that

Pregnancy and development are an intricate series of events that allow a single cell to become a fully formed human being

by asking

28.1 How does fertilization occur?

and

28.2 What happens between fertilization and implantation?

and

28.3 How does the embryo implant into the uterus wall and trigger development of the placenta?

and

28.4 How does the embryo become a fetus?

28.5 How does pregnancy affect the mother?

and

28.6 How is a baby born?

and

28.7 How does the infant adjust to extrauterine life?

then examining

28.8 Lactation

28.9 How can infertile couples be helped?

CAREER CONNECTION

Watch a video to learn how the chapter content is used in a real health care setting. Go to Mastering A&P® > Study Area > Animations and Videos or use quick access URL https://goo.gl/88srfV

The birth of a baby is such a familiar event that we tend to lose sight of the wonder of this accomplishment: How does a single cell, the fertilized egg, grow to become a complex human being consisting of trillions of cells? The details of this process can fill a good-sized book. Our intention here is simply to outline the important events of gestation and to consider briefly the events occurring immediately after birth.

Let's get started by defining some terms.

- **Conceptus** (kon-sep′tus; "that which is conceived"): The pregnant woman's developing offspring (embryo or fetus).

- **Pregnancy**: The state of carrying a developing conceptus.

- **Gestation period** (*gestare* = to carry): The time during which the development of the conceptus occurs. By convention, this extends from the last menstrual period (a date the woman is likely to remember) until birth, approximately 280 days. So, at the moment of fertilization, the mother is officially (but illogically) two weeks pregnant!

From fertilization through week 8, the *embryonic period*, the conceptus is called an **embryo**, and from week 9 through birth, the *fetal period*, the conceptus is called a **fetus** ("the young in the womb"). At birth, it is an infant. **Figure 28.1** shows the changing size and shape of the conceptus as it progresses from fertilization to the early fetal stage.

28.1 Fertilization combines the sperm and egg chromosomes, forming a zygote

Learning Outcomes
▶ Describe the importance of sperm capacitation.
▶ Explain the mechanisms behind the blocks to polyspermy.
▶ Define fertilization.

Before fertilization can occur, sperm must reach the ovulated secondary oocyte. The oocyte is viable for 12 to 24 hours after it is ejected from the ovary. The chance of pregnancy drops to almost zero the next day. Although sperm may survive in the female reproductive tract for up to a week, fertilization is most likely to occur within 24 to 48 hours after ejaculation. Consequently, for successful fertilization to occur, intercourse must occur no more than two days before ovulation and no later than 24 hours after, at which point the oocyte has traveled approximately one-third of the way down the uterine tube.

Fertilization   1-week conceptus   3-week embryo (3 mm)   — Embryo   5-week embryo (10 mm)   8-week embryo (30 mm)   12-week fetus (90 mm)

Embryonic period (first 8 weeks postfertilization)   **Fetal period** [weeks 9–38 (or birth)]

Figure 28.1 Size of the conceptus up to the early fetal stage. Embryos and fetus are shown approximately life size. (Measurements are crown-to-rump length.)

Fertilization occurs when a sperm's chromosomes combine with those of an egg to form a fertilized egg, or **zygote** (zi′gōt; "yoked together"), the first cell of the new individual. Let's look at the events leading to fertilization.

Sperm Transport and Capacitation

During ejaculation, a man expels millions of sperm into his partner's vagina. Despite these vast numbers, most sperm don't reach the oocyte, even though it is only about 12 cm (5 inches) away. Millions of sperm leak from the vagina almost immediately. Of those remaining, millions more are destroyed by the vagina's acidic environment. Only a small fraction of sperm make it through the cervix and gain access to the uterus.

Sperm that do reach the uterus, propelled by their whiplike tail movements, are subjected to forceful uterine contractions that tumble them throughout the uterine cavity, where thousands more are destroyed by resident phagocytes. Only a few thousand sperm, out of the millions in the male ejaculate, are conducted by reverse peristalsis into the uterine tube, where the oocyte may be moving leisurely toward the uterus.

These difficulties aside, there is still another hurdle to overcome. Sperm freshly deposited in the vagina are incapable of penetrating an oocyte. They must first undergo **capacitation**, a biochemically delicate process that can take 2 to 10 hours. Specifically, their motility must be enhanced and their membranes must become fragile so that the hydrolytic enzymes in their acrosomes (◄ p. 1058) can be released. How does this happen? As sperm move through the cervical mucus, uterus, and uterine tubes, secretions of the female tract remove some of their protective membrane proteins, as well as the cholesterol that keeps their acrosomal membranes "tough" and stable. Even if the sperm reach the oocyte within a few minutes, they must "wait around" (so to speak) for capacitation to occur.

This elaborate mechanism prevents the spilling of acrosomal enzymes. Consider the alternative: Fragile acrosomal membranes could rupture prematurely in the male reproductive tract, causing some degree of autolysis (self-digestion) of the male reproductive organs.

How do sperm find their way to a released oocyte in the uterine tube? Sperm have receptors that sense the direction of fluid flow, temperature, and certain chemicals. These cues guide sperm "upstream" through the uterine tubes. For example, they swim toward warmer temperatures at the far end of the tube, and, like a dog following a scent trail, sperm use olfactory receptor proteins to "sniff" chemicals released by the oocyte and its surrounding cells.

Acrosomal Reaction and Sperm Penetration

Recall from Chapter 27 that the ovulated oocyte is encapsulated by the corona radiata and by the deeper zona pellucida (◄ p. 1073). Both must be breached before the oocyte can be penetrated by a sperm from the cloud of sperm that surrounds it (**Figure 28.2**). Central to this process is the sperm's **acrosomal reaction**, the release of acrosomal enzymes. *Focus on Sperm*

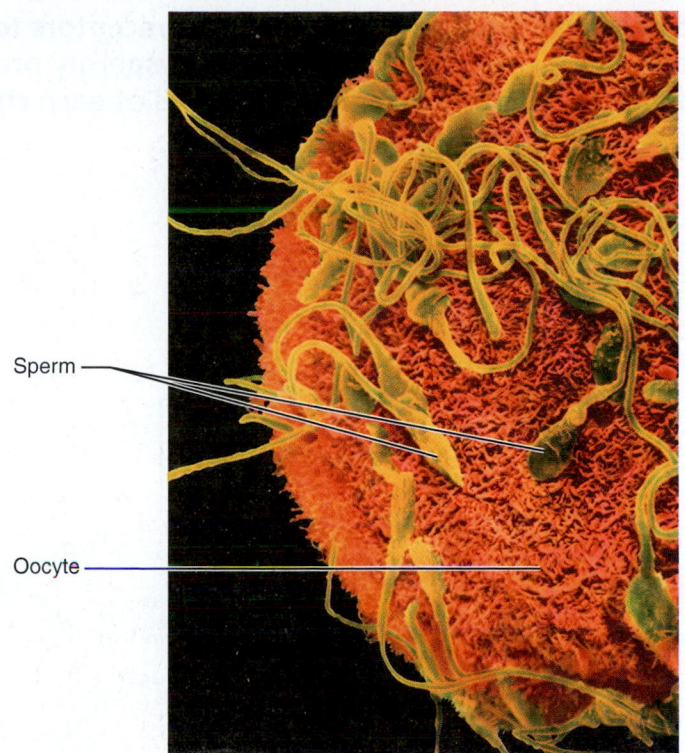

Sperm

Oocyte

Figure 28.2 Sperm surrounding an oocyte. (SEM, 2240×)

Penetration and the Blocks to Polyspermy (**Focus Figure 28.1**, pp. 1094–1095) describes the acrosomal reaction and sperm penetration.

Blocks to Polyspermy

Polyspermy (entry of several sperm into an egg) occurs in some animals, but in humans only one sperm is allowed to penetrate the oocyte, ensuring **monospermy**, the one-sperm-per-oocyte condition. In the rare cases of polyspermy that do occur, the embryos contain too much genetic material and die because they cannot successfully complete mitosis. At least two mechanisms help ensure monospermy: the *oocyte membrane block* and the *cortical reaction*.

When a sperm binds to the oocyte's sperm-binding membrane receptors, it triggers the **oocyte membrane block**. The oocyte sheds the remaining sperm-binding receptors on vesicles made from the plasma membrane. In the absence of sperm receptors, additional sperm are unable to bind to the oocyte. The released vesicles, studded with sperm-binding receptors, may act as "decoys" that attract and bind any other sperm that make it through the zona pellucida.

Once the plasma membranes of the sperm and oocyte have fused, waves of Ca^{2+} are released from the oocyte's endoplasmic reticulum into its cytoplasm. These calcium surges trigger the **cortical reaction**[*] (Focus Figure 28.1 ⑤)—exocytosis of

---

[*] The oocyte membrane block is analogous to the fast block to polyspermy seen in invertebrates. The cortical and zona reactions together are equivalent to the slow block to polyspermy of invertebrates.

(Text continues on p. 1096.)

28

Sperm use acrosomal enzymes and receptors to approach, bind, and enter the oocyte. Blocks to polyspermy prevent further sperm entry, ensuring that only two copies of each chromosome are present in the fertilized ovum.

Sperm, delivered to the vagina and capacitated in the female reproductive tract, stream toward a secondary oocyte.

① **Approach.** Aided by enzymes on its surface, a sperm cell weaves its way past granulosa cells of the corona radiata.

Extracellular space

Sperm

Sperm

Zona pellucida (glycoprotein-rich extracellular matrix)

Granulosa cells of corona radiata

Polar body

Oocyte nucleus arrested in meiotic metaphase II

Zona pellucida

Extracellular space

Oocyte plasma membrane

② **Acrosomal reaction.**
• Binding of the sperm to receptors in the zona pellucida causes Ca^{2+} levels within the sperm to rise, triggering the release of acrosomal enzymes.
• Acrosomal enzymes from many sperm digest holes through the zona pellucida, clearing a path to the oocyte plasma membrane.

Zona pellucida sperm-binding receptors

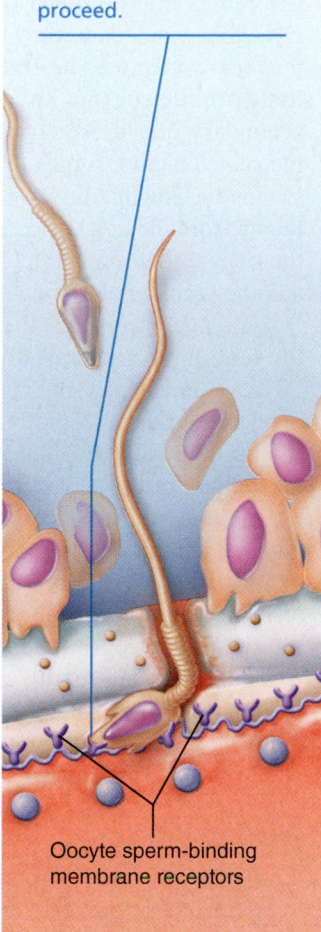

③ **Binding.** The sperm's membrane binds to the oocyte's sperm-binding receptors. This allows the next step (fusion) to proceed.

Oocyte sperm-binding membrane receptors

④ **Fusion.** Sperm and oocyte plasma membranes fuse. Sperm contents enter the oocyte.

Cortical granules

⑤ **Blocks to polyspermy.**
• **Oocyte membrane block:** Oocyte sperm-binding membrane receptors are shed.
• **Cortical reaction:** Increased Ca^{2+} levels trigger exocytosis of cortical granules. This hardens the zona pellucida and destroys its sperm-binding receptors (*zona reaction*).

Microtubules from sperm flagellum

Mitochondria

Sperm nucleus

Sperm nucleus

Extracellular space

Corona radiata

Zona pellucida

Second meiotic division of oocyte

Second meiotic division of first polar body

① After the sperm penetrates the secondary oocyte, the oocyte completes meiosis II, forming the ovum and second polar body.

Male pro-nucleus

Female pro-nucleus (swollen ovum nucleus)

Polar bodies

② Sperm and ovum nuclei swell, forming pronuclei.

Male pronucleus

Mitotic spindle

Centriole

Female pronucleus

③ The DNA in each pronucleus replicates. The pronuclei approach each other and a mitotic spindle forms between them.

Zygote

④ Chromosomes of the pronuclei intermix. Fertilization is accomplished and the cell, now called a zygote, is ready for the first cleavage division.

(a)

Male and female pronuclei

Polar bodies

(b)

Figure 28.3 Fertilization. (a) Events from sperm penetration to zygote formation. **(b)** Micrograph of an oocyte in the time between steps ③ and ④ of (a).

granules located just inside the plasma membrane. These granules contain enzymes that destroy the zona pellucida's sperm-binding receptors and harden the zona pellucida. These changes in the zona pellucida are called the **zona reaction** and they effectively clear sperm out of the zona pellucida.

Completion of Meiosis II and Fertilization

As the sperm's cytoplasmic contents enter the oocyte, the sperm loses its plasma membrane. The centrosome from its midpiece assembles microtubules which the sperm uses to propel its DNA-rich nucleus toward the oocyte nucleus. On the way, its nucleus swells to about five times its previous size to form the **male pronucleus** (pro-nu'kle-us; *pro* = before). Meanwhile the secondary oocyte is activated from its semidormant state by the same calcium surges that trigger the cortical reaction. Following the fluctuations in calcium levels, a shower of zinc ions bursts from the egg (zinc "sparks"). This efflux of zinc allows the oocyte to complete meiosis II, forming the ovum nucleus and the second polar body (**Figure 28.3a** ① and ②).

This accomplished, the ovum nucleus swells, becoming the **female pronucleus**, and the two pronuclei replicate their DNA as they approach each other. As a mitotic spindle develops between them (Figure 28.3a ③), the pronuclei membranes rupture, releasing their chromosomes together into the immediate vicinity of the newly formed spindle.

The true moment of fertilization occurs as the maternal and paternal chromosomes combine and produce the diploid *zygote*, or fertilized egg (Figure 28.3a ④). Some sources define the term fertilization simply as the act of oocyte penetration by the sperm. However, unless the chromosomes in the male and female pronuclei actually combine, the zygote is never formed in humans. The zygote, the first cell of a new individual, is now ready to undergo the first mitotic division of the conceptus.

Check Your Understanding

1. What is the cortical reaction and what does it accomplish?
2. **WHAT IF?** Suppose a sperm has penetrated a polar body (◄ p. 1071), and their nuclei fuse. Would it be likely that the resulting cell would develop into a healthy embryo? Explain.
3. **MAKE CONNECTIONS** In the diagram below, what is region **a** called and what does it contain? (Hint: See Chapter 27.) What triggers release of the contents of **a**? What process must the sperm complete before this can happen?

a

For answers, see Answers Appendix.

28.2 Embryonic development begins as the zygote undergoes cleavage and forms a blastocyst en route to the uterus

Learning Outcome

▶ Describe the process and product of cleavage.

Early embryonic development begins with fertilization and continues as the embryo travels through the uterine tube, floats free in the cavity of the uterus, and finally implants in the uterine wall. Significant events of this early embryonic period are *cleavage*, which produces a structure called a blastocyst, and *implantation* of the blastocyst.

Cleavage

Cleavage is a period of fairly rapid mitotic divisions of the zygote without intervening growth (**Figure 28.4**). The goal of this first phase of development is to produce small cells with a high surface-to-volume ratio, which enhances their uptake of nutrients and oxygen and the disposal of wastes. It also provides a large number of cells to serve as building blocks for constructing the embryo. Consider the difficulty of trying to construct a building from one huge block of granite. If you now consider how much easier it would be if instead you could use hundreds of bricks, you will quickly grasp the importance of cleavage.

Some 36 hours after fertilization, the first cleavage division of the zygote has produced two identical cells (called *blastomeres*). These divide to produce four cells (Figure 28.4b), then eight, and so on. By 72 hours after fertilization, a loose collection of cells that form a berry-shaped cluster of 16 or more cells called the **morula** (mor′u-lah; "little mulberry") has been formed (Figure 28.4c). All the while, transport of the embryo (still contained within the zona pellucida) toward the uterus continues.

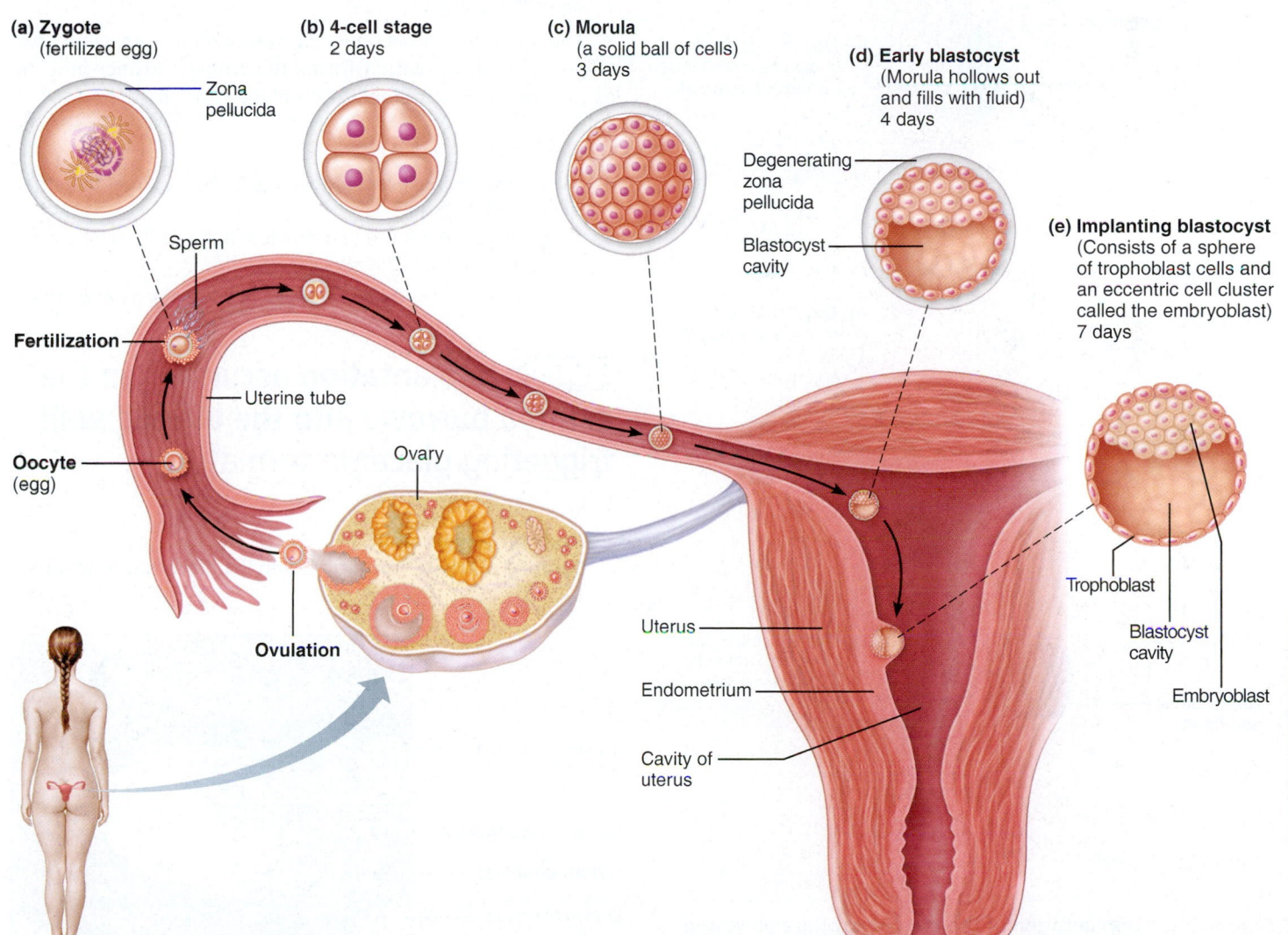

Figure 28.4 Cleavage: From zygote to blastocyst.

(a) **Zygote** (fertilized egg) — Zona pellucida

(b) **4-cell stage** 2 days

(c) **Morula** (a solid ball of cells) 3 days

(d) **Early blastocyst** (Morula hollows out and fills with fluid) 4 days — Degenerating zona pellucida, Blastocyst cavity

(e) **Implanting blastocyst** (Consists of a sphere of trophoblast cells and an eccentric cell cluster called the embryoblast) 7 days

Sperm

Fertilization

Uterine tube

Oocyte (egg)

Ovary

Ovulation

Uterus

Endometrium

Cavity of uterus

Trophoblast

Blastocyst cavity

Embryoblast

28

Endometrium   *Lumen of uterus*

Uterine
endometrial
epithelium

Blastocyst
cavity

Degenerating
zona pellucida

(a) Day 4: Early blastocyst floats in uterine cavity and "hatches" from zona pellucida.

Trophoblast

Embryoblast

(b) Day 6: Blastocyst adheres to uterine wall.

Trophoblast:
• Cytotrophoblast
• Syncytiotrophoblast

Embryoblast

(c) Day 7: Implantation begins as trophoblast invades uterine wall.

Embryonic disc:
• Epiblast
• Hypoblast

Cytotrophoblast

Syncytiotrophoblast

(d) Day 9: Implantation continues; embryoblast has become a bilayered embryonic disc.

Embryonic disc:
• Epiblast
• Hypoblast

Amniotic sac cavity

Yolk sac

Cytotrophoblast

Syncytiotrophoblast

(e) Day 11: Implantation complete; amniotic sac and yolk sac form.

Figure 28.5 Implantation of the blastocyst. The embryos are shown in section (cut in half).

Blastocyst Formation

By day 4 or 5 after fertilization, the embryo consists of about 100 cells. It floats free in the uterus and has begun accumulating fluid within an internal cavity (Figure 28.4d and **Figure 28.5a**). The zona pellucida starts to break down and the inner structure, now called a blastocyst, "hatches" from it. The **blastocyst** (blas′to-sist) is a fluid-filled hollow sphere composed of a single layer of large, flattened cells called **trophoblast cells** (tro′fo-blast) and a small cluster of 20 to 30 rounded cells, called the **embryoblast** (or *inner cell mass*), clinging to the inside (Figure 28.5b, c).

The trophoblast has several key roles during development:

● As we will see in the next section, trophoblast cells play an important role in implanting the embryo in the wall of the uterus.

● As suggested by the literal translation of trophoblast ("nourishment generator"), it makes a major contribution to the *chorion*, the embryo's portion of the placenta.

● It protects the conceptus from attack by the mother's immune cells by producing several factors with immunosuppressive effects.

The embryoblast becomes the two-layered *embryonic disc* (Figure 28.5d, e), which forms the embryo proper and the extraembryonic membranes (except the chorion).

Check Your Understanding

4. Why is the multicellular blastocyst only slightly larger than the single-celled zygote?

5. Trophoblasts contribute to a structure that lasts throughout pregnancy. What is this structure called?

For answers, see Answers Appendix.

28.3 Implantation occurs when the embryo burrows into the uterine wall, triggering placenta formation

Learning Outcomes

▶ **Describe implantation.**
▶ **Describe placenta formation, and list placental functions.**

Endometrial stroma with blood vessels and glands

Syncytiotrophoblast

Cytotrophoblast

Lumen of uterus

(f) Day 12: Photomicrograph (150×)

Implantation

While the blastocyst floats in the uterine cavity for two to three days, it is nourished by the glycoprotein-rich uterine secretions. Then, six to seven days after ovulation, **implantation** begins. The receptivity of the endometrium to implantation—the so-called *window of implantation*—is opened by the surging levels of ovarian hormones (estrogens and progesterone) in the blood. If the mucosa is receptive, cell adhesion molecules (◀ p. 81) bind the trophoblast to the endometrial epithelium, and the blastocyst implants high in the uterus. If the endometrium is not yet optimally mature, the blastocyst fails to attach and drifts to a lower level, implanting when it finds a site with the proper receptors and chemical signals.

The trophoblast cells overlying the embryoblast adhere to the endometrium (Figure 28.5b) and secrete digestive enzymes and growth factors onto the endometrial surface. The endometrium quickly thickens at the point of contact and takes on characteristics of an acute inflammatory response—the uterine blood vessels become more permeable and leaky, and inflammatory cells including lymphocytes, natural killer cells, and macrophages invade the area.

The trophoblast then proliferates and forms two distinct layers (Figure 28.5c, d):

- The **cytotrophoblast** (si″to-tro′fo-blast; "cellular part of the trophoblast") forms the innermost layer of cells.

- The **syncytiotrophoblast** (sin-sit″e-o-tro′fo-blast; *syn* = together, *cyt* = cell) forms the outermost layer. Here the cells have fused together to form a single multinuclear cytoplasmic mass, or *syncytium*. This syncytium sends out long protrusions that invade the endometrium and rapidly digest the uterine cells it contacts.

As the endometrium is eroded, the blastocyst burrows into this thick, velvety lining and is surrounded by a pool of blood leaked from degraded endometrial blood vessels. Proliferation of the endometrial cells then covers and seals off the implanted blastocyst from the uterine cavity (Figure 28.5e).

In cases where implantation fails to occur, a receptive uterus becomes nonreceptive once again. At least two-thirds of all zygotes formed fail to implant by the end of the first week or spontaneously abort. Moreover, about 30% of implanted embryos later miscarry due to genetic defects of the embryo, uterine malformation, or other problems.

Successful implantation takes about five days. It is usually completed by day 12, which is day 26 of a woman's menstrual cycle—just before the endometrium normally begins to slough off. Menstruation would flush away the newly implanted embryo, so the embryo takes steps to prevent this by rescuing the corpus luteum. [Remember that menstruation occurs because the corpus luteum degenerates and stops releasing progesterone and estrogens required to maintain the uterine lining (◀ p. 1078).]

How does the embryo rescue the corpus luteum? Its syncytiotrophoblast secretes **human chorionic gonadotropin (hCG)** (ko″re-on′ik go-nad″o-trōp′in), which acts like LH but bypasses the usual hypothalamus-pituitary-ovarian controls. Just like LH, hCG prompts the corpus luteum to continue secreting

Figure 28.6 Hormonal changes during pregnancy. The relative changes in maternal blood levels of three hormones that maintain pregnancy are depicted, rather than actual blood concentrations.

progesterone and estrogens. The syncytiotrophoblast, which becomes part of the *chorion* (a major part of the embryo's contribution to the placenta) continues to make hCG throughout pregnancy. Besides rescuing the corpus luteum, hCG also helps prevent the mother's immune system from rejecting the implanted embryo.

All pregnancy tests used today are antibody tests that detect hCG in a woman's blood or urine. hCG is usually detectable in the mother's blood two days after implantation. Blood levels of hCG continue to rise until the end of the second month. Then, blood levels decline sharply to reach a low value by 4 months, a situation that persists for the remainder of gestation (**Figure 28.6**). Between the second and third month, the placenta assumes the role of progesterone and estrogen production for the remainder of the pregnancy. The corpus luteum then degenerates and the ovaries remain inactive until after birth.

Initially, the implanted embryo acts like a parasite that obtains its nutrition by digesting the endometrial cells. By the second month, the placenta is providing nutrients and oxygen to the embryo and carrying away embryonic metabolic wastes. Since placenta formation is a continuation of the events of implantation, we will consider it next.

Formation of the Placenta

Formation of the **placenta** ("flat cake") is called **placentation** (plas″en-ta′shun). The placenta is a pancake-shaped temporary organ that originates from both embryonic and maternal (endometrial) tissues. Cells from the original embryoblast give rise to a layer of extraembryonic mesoderm that lines the inner surface

28

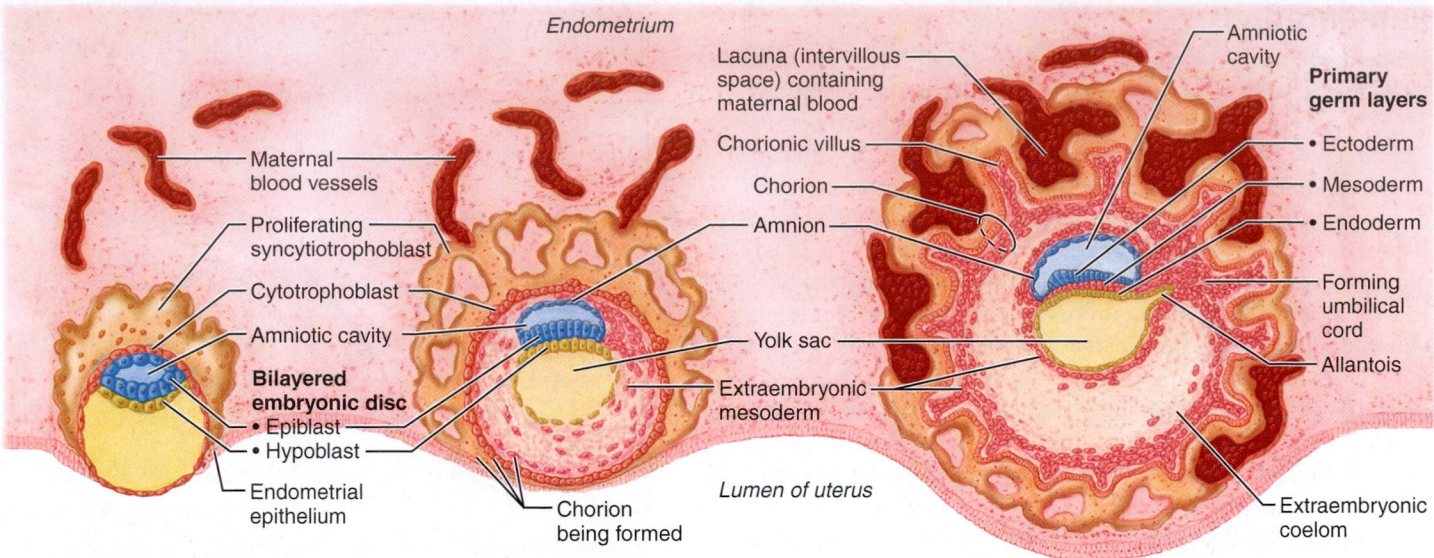

Endometrium

Lacuna (intervillous space) containing maternal blood

Chorionic villus

Chorion

Amnion

Amniotic cavity

Primary germ layers

• Ectoderm

• Mesoderm

• Endoderm

Forming umbilical cord

Allantois

Maternal blood vessels

Proliferating syncytiotrophoblast

Cytotrophoblast

Amniotic cavity

Bilayered embryonic disc

• Epiblast

• Hypoblast

Endometrial epithelium

Chorion being formed

Yolk sac

Extraembryonic mesoderm

Lumen of uterus

Extraembryonic coelom

(a) Implanting 8-day blastocyst.
• The syncytiotrophoblast erodes the endometrium.
• The embryonic disc is now separated from the amnion by a fluid-filled space.

(b) 12-day blastocyst.
• Implantation is complete.
• Extraembryonic mesoderm is forming a discrete layer beneath the cytotrophoblast.

(c) 16-day embryo.
• Trophoblast and associated mesoderm have become the chorion.
• Chorionic villi are forming.
• The embryo exhibits all three germ layers, a yolk sac, and an allantois.

Decidua basalis

Maternal blood

Chorionic villus

Umbilical blood vessels in umbilical cord

Amnion

Amniotic cavity

Yolk sac

Extraembryonic coelom

Chorion

Decidua capsularis

Lumen of uterus

(d) 4½-week embryo.
• The decidua capsularis, decidua basalis, amnion, and yolk sac are well formed.
• The chorionic villi lie in blood-filled intervillous spaces within the endometrium.
• The embryo is nourished via the umbilical vessels that connect it (through the umbilical cord) to the placenta.

Placenta

Decidua basalis

Chorionic villi

Yolk sac

Amnion

Amniotic cavity

Umbilical cord

Uterus

Lumen of uterus

Decidua capsularis

Extraembryonic coelom

(e) 13-week fetus.

Figure 28.7 Early embryonic development and formation of the placenta and extraembryonic membranes.

of the trophoblast (**Figure 28.7b**). Together these become the **chorion**. The chorion develops fingerlike **chorionic villi**, which become especially elaborate where they are in contact with maternal blood (Figure 28.7c).

Soon the mesodermal cores of the chorionic villi are invaded by newly forming blood vessels, which extend to the embryo as the umbilical arteries and vein. The continuing erosion of the endometrium produces large, blood-filled **lacunae**, or

Figure 28.8 Detailed anatomy of the vascular relationships in the mature decidua basalis. This state of development has been accomplished by the end of the third month of development.

intervillous spaces, in the functional layer of the endometrium (◀ **Figure 27.18, p. 1065**). The villi come to lie in these spaces totally immersed in maternal blood (Figure 28.7d). The part of the endometrium that lies beneath the embryo becomes the **decidua basalis** (de-sid′u-ah), and that surrounding the uterine cavity face of the implanted embryo forms the **decidua capsularis** (Figure 28.7d and e). Together, the chorionic villi and the decidua basalis form the disc-shaped placenta.

The placenta detaches and sloughs off after the infant is born, so the name of the maternal portion—*decidua* ("that which falls off")—is appropriate. During development, the decidua capsularis expands to accommodate the fetus, which eventually fills and stretches the uterine cavity. As the developing fetus grows, the villi in the decidua capsularis are compressed and

degenerate, while those in the decidua basalis increase in number and branch even more profusely.

The placenta is usually fully functional as a nutritive, respiratory, excretory, and endocrine organ by the end of the third month of pregnancy. However, well before this time, oxygen and nutrients are diffusing from maternal to embryonic blood, and embryonic metabolic wastes are passing in the opposite direction. The barriers to free passage of substances between the two blood supplies are embryonic barriers—the membranes of the chorionic villi and the endothelium of embryonic capillaries. Although the maternal and embryonic blood supplies are very close, they normally do not intermix (**Figure 28.8**).

The placenta secretes hCG from the beginning, but the ability of its syncytiotrophoblast cells (the "hormone manufacturers")

28

to produce the estrogens and progesterone of pregnancy matures much more slowly. If, for some reason, placental hormones are inadequate when hCG levels wane, the endometrium degenerates and the pregnancy spontaneously aborts. Throughout pregnancy, blood levels of estrogens and progesterone continue to increase (Figure 28.6). They encourage growth and further differentiation of the mammary glands and ready them for lactation. The placenta also produces other hormones, such as *human placental lactogen* and *relaxin*. We will describe the effects of these hormones on the mother shortly.

Check Your Understanding

6. Which portion of the trophoblast accomplishes implantation?

7. What is the composition of the chorion?

8. What endometrial decidua cooperates with the chorionic villi to form the placenta?

9. Generally speaking, when does the placenta become fully functional?

10. **APPLY** Amber, wondering if she is pregnant, buys an over-the-counter pregnancy test to assess this possibility. How will the blastocyst, if present, "make itself known"?

For answers, see Answers Appendix.

28.4 ## Embryonic events include gastrula formation and tissue differentiation, which are followed by rapid growth of the fetus

Learning Outcomes

▶ Name and describe the formation, location, and function of the extraembryonic membranes.

▶ Describe gastrulation and its consequences.

▶ Define organogenesis and indicate the important roles of the three primary germ layers in this process.

▶ Describe unique features of the fetal circulation and the changes that occur in this circulation after birth.

▶ Indicate the duration of the fetal period, and note the major events of fetal development.

Having followed placental development into the fetal stage, we will now backtrack and consider development of the embryo during and after implantation, referring to Figures 28.7 and 28.8. Even while implantation is occurring, the blastocyst is being converted to a **gastrula** (gas′troo-lah), in which the three primary germ layers form, and the extraembryonic membranes develop. The embryoblast first subdivides into two layers—the upper *epiblast* and the lower *hypoblast* (Figure 28.7a, b). The subdivided embryoblast is now called the **embryonic disc**.

Extraembryonic Membranes

The **extraembryonic membranes** that form during the first two to three weeks of development include the amnion, yolk sac, allantois, and chorion (Figures 28.7c and 28.8). The **amnion** (am′ne-on) develops when cells of the epiblast fashion themselves into a transparent membranous sac. This sac surrounds the *amniotic cavity*, which is filled with **amniotic fluid**. Eventually, the amnion extends all the way around the embryo, broken only by the umbilical cord (Figures 28.7d and 28.8).

Sometimes called the "bag of waters," the amnion provides a buoyant environment that protects the developing embryo against physical trauma, and helps maintain a constant temperature. The fluid also prevents the rapidly growing embryonic parts from adhering and fusing together and allows the embryo considerable freedom of movement. Initially, the fluid is derived from the maternal blood, but as the fetal kidneys become functional later in development, fetal urine contributes to amniotic fluid volume.

The **yolk sac** forms from cells of the primitive gut, which arrange themselves into a sac that hangs from the ventral surface of the embryo (Figure 28.7b–e). The amnion and yolk sac resemble two balloons touching one another with the embryonic disc at the point of contact. In many species, the yolk sac is the main source of nutrition for the embryo, but in humans it contains very little yolk and has a nutritive function only during the second and third weeks. Nevertheless, the yolk sac is important in humans because it:

- Forms part of the gut (digestive tube)
- Is the source of the earliest blood cells
- Is the location from which primordial germ cells (◀ p. 1083) migrate to the gonads to form sperm and ova

The **allantois** (ah-lan′to-is) forms as a small outpocketing of embryonic tissue at the caudal end of the yolk sac (Figure 28.7c). In animals that develop in shelled eggs, the allantois is a disposal site for solid metabolic wastes (excreta). In humans, the allantois is the structural base for the **umbilical cord** that links the embryo to the placenta, and ultimately it becomes part of the urinary bladder. The fully formed umbilical cord contains a core of embryonic connective tissue, the umbilical arteries and vein, and is covered externally by amniotic membrane.

We have already described the *chorion*, which helps to form the placenta (Figures 28.7c and 28.8). As the outermost membrane, the chorion encloses the embryonic body and all other membranes.

Gastrulation: Germ Layer Formation

During week 3, the two-layered embryonic disc transforms into a three-layered *embryo* in which the **primary germ layers**—*ectoderm, mesoderm,* and *endoderm*—are present (Figure 28.7c). This process, called **gastrulation** (gas″troo-la′shun), involves cellular rearrangements and migrations.

Figure 28.9 focuses on the changes that take place in the days between Figure 28.7b (12-day embryo) and Figure 28.7c (16-day embryo). Gastrulation begins when a groove with raised edges called the **primitive streak** appears on the dorsal surface of the embryonic disc. This groove establishes the longitudinal axis of the embryo (Figure 28.9e). Surface (epiblast) cells of the embryonic disc then migrate medially and enter the primitive streak. The first cells to enter the groove displace the hypoblast cells of the yolk sac and form the most inferior germ layer, the **endoderm** (Figure 28.9f). Those that follow push laterally between the cells at the upper and lower surfaces, forming the

Amnion

Bilayered embryonic disc

Yolk sac

(a)

(b) Frontal section

(c) 3-D view

(d) Section view in (e)

Head end of bilayered embryonic disc

② The first epiblast cells that migrate into the primitive streak become **endoderm**.

Primitive streak

Epiblast

Hypoblast

(f) 14–15 days

Head end

Cut edge of amnion

Yolk sac (cut edge)

Left

Right

Tail end

① Gastrulation begins with the formation of the **primitive streak**.

(e) Bilayered embryonic disc, superior view

③ The next epiblast cells that migrate into the primitive streak become **mesoderm**.

④ The epiblast surface is now called the **ectoderm**.

(g) 16 days   Mesoderm   Endoderm

Figure 28.9 Formation of the three primary germ layers. (a–d) Orienting diagrams. **(e)** Surface view of an embryonic disc, amnion and yolk sac removed. **(f, g)** Sections through the embryonic disc at the location shown in (e).

mesoderm (Figure 28.9g). As soon as the mesoderm is formed, the mesodermal cells immediately beneath the early primitive streak aggregate, forming a rod of mesodermal cells called the **notochord** (no′to-kord), the first axial support of the embryo (see Figure 28.12a). The cells that remain on the embryo's dorsal surface are the **ectoderm**. At this point, the embryo is about 2 mm long.

The three primary germ layers serve as the *primitive tissues* from which all body organs derive.

- Ectoderm ("outer skin") fashions structures of the nervous system and the skin epidermis.
- Endoderm ("inner skin") forms the epithelial linings of the digestive, respiratory, and urogenital systems, and associated glands.
- Mesoderm ("middle skin") forms virtually everything else.

Both ectoderm and endoderm consist mostly of cells that are securely joined to each other and are *epithelia*. Mesoderm, by

contrast, is a *mesenchyme* [literally, "poured into the middle (of the embryo)"], an embryonic tissue with star-shaped cells that are free to migrate widely within the embryo. Figure 28.13 lists the germ layer derivatives. Some of the details of the differentiation processes are described next.

Organogenesis: Differentiation of the Germ Layers

Gastrulation lays down the basic structural framework of the embryo. It also sets the stage for the rearrangements that occur during **organogenesis** (or″gah-no-jen′ĕ-sis), formation of body organs and organ systems. By the end of the embryonic period at 8 weeks, when the embryo is about 30 mm (slightly more than 1 inch) long from head to buttocks (referred to as the *crown-rump measurement*), all the adult organ systems are recognizable. It is truly amazing how much organogenesis occurs in such a short time in such a small amount of living matter.

28

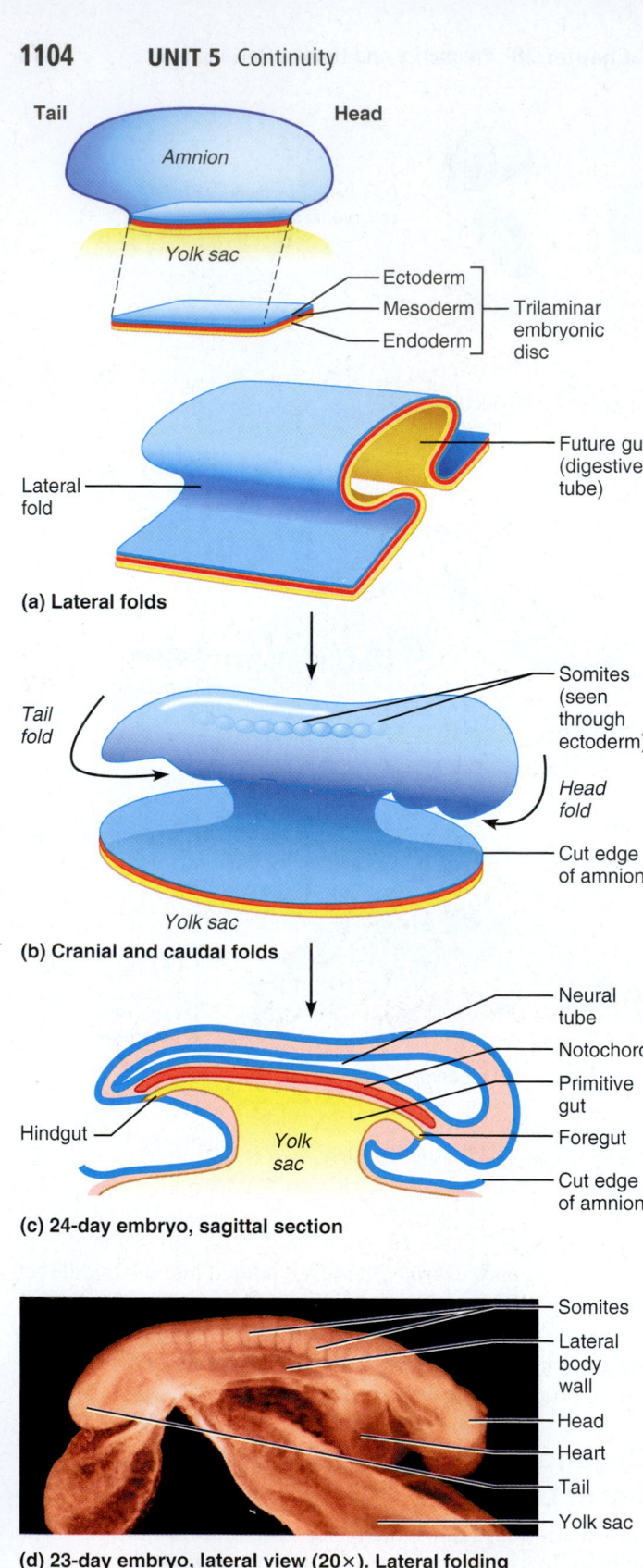

(a) Lateral folds

- Tail
- Head
- Amnion
- Yolk sac
- Ectoderm
- Mesoderm — Trilaminar embryonic disc
- Endoderm
- Lateral fold
- Future gut (digestive tube)

(b) Cranial and caudal folds

- Somites (seen through ectoderm)
- Tail fold
- Head fold
- Cut edge of amnion
- Yolk sac

(c) 24-day embryo, sagittal section

- Neural tube
- Notochord
- Primitive gut
- Foregut
- Cut edge of amnion
- Hindgut
- Yolk sac

(d) 23-day embryo, lateral view (20×). Lateral folding nearing completion.

- Somites
- Lateral body wall
- Head
- Heart
- Tail
- Yolk sac

Figure 28.10 Folding of the embryonic body. Lateral views.

As **Figure 28.10** shows, the embryo starts off as a flat plate, but as it grows, it folds to achieve a cylindrical body shape which lifts off the yolk sac and protrudes into the amniotic cavity. In the simplest sense, this process resembles three stacked sheets of paper folding laterally into a tube (Figure 28.10a). At the same time, the folding occurs from both ends (the head and tail regions) and progresses toward the central part of the embryonic body, where the yolk sac and umbilical vessels protrude. As the endoderm undercuts and its edges come together and fuse, it encloses part of the yolk sac (Figure 28.10c).

Specialization of the Endoderm

The tube of endoderm formed by the folding process described above is called the **primitive gut**. It becomes the epithelial lining of the gastrointestinal tract (**Figure 28.11**). The organs of the GI tract (pharynx, esophagus, etc.) quickly become apparent, and then the oral and anal ends of the gut open. The mucosal lining of the respiratory tract forms as an outpocketing of the *foregut*, and glands arise as endodermal outpocketings at various points farther along the tract. For example, the epithelium of the thyroid, parathyroids, and thymus forms from the pharyngeal endoderm.

Specialization of the Ectoderm

The first major event in organogenesis is **neurulation**, the differentiation of ectoderm that produces the brain and spinal cord (**Figure 28.12**). This process is *induced* (stimulated to happen) by chemical signals from the *notochord*, the rod of mesoderm (mentioned earlier) that defines the body axis. The ectoderm overlying the notochord thickens, forming the **neural plate** (Figure 28.12a). Then the ectoderm folds, forming the **neural groove** as well as **neural folds** on either side (Figure 28.12b). By day 22, the superior margins of the neural folds fuse, forming a **neural tube**, which soon pinches off from the ectodermal layer and becomes covered by surface ectoderm (Figure 28.12c).

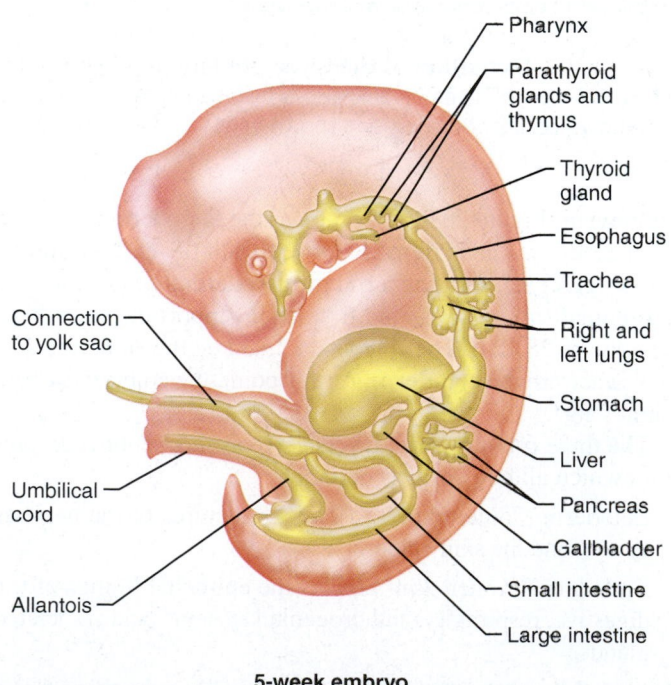

- Pharynx
- Parathyroid glands and thymus
- Thyroid gland
- Esophagus
- Trachea
- Right and left lungs
- Stomach
- Liver
- Pancreas
- Gallbladder
- Small intestine
- Large intestine
- Connection to yolk sac
- Umbilical cord
- Allantois

5-week embryo

Figure 28.11 Endodermal differentiation. Endoderm forms the epithelial linings of the digestive and respiratory tracts and associated glands.

28

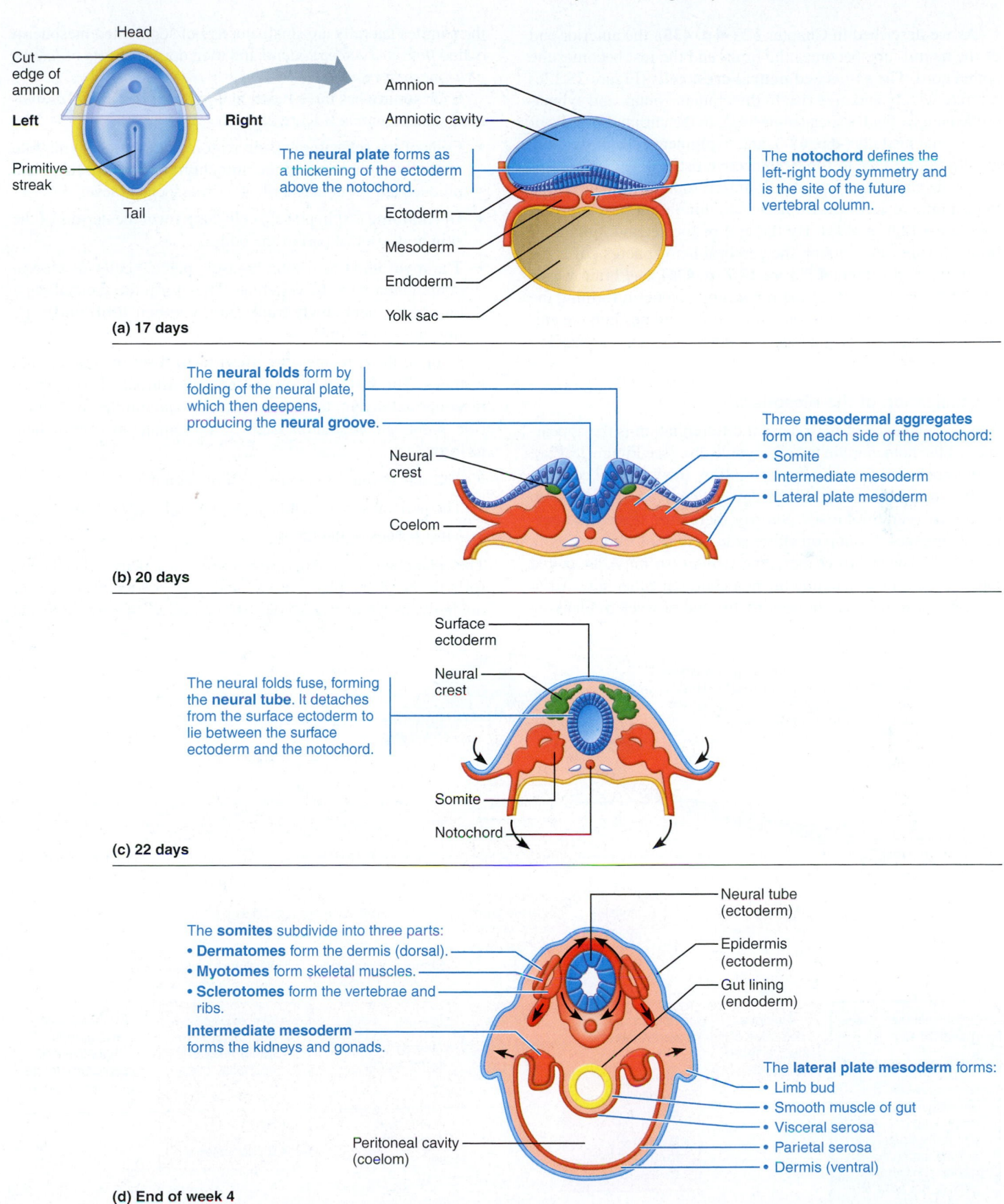

(a) 17 days

Head

Cut edge of amnion

Left **Right**

Primitive streak

Tail

Amnion

Amniotic cavity

The **neural plate** forms as a thickening of the ectoderm above the notochord.

Ectoderm

Mesoderm

Endoderm

Yolk sac

The **notochord** defines the left-right body symmetry and is the site of the future vertebral column.

(b) 20 days

The **neural folds** form by folding of the neural plate, which then deepens, producing the **neural groove**.

Neural crest

Coelom

Three **mesodermal aggregates** form on each side of the notochord:
• Somite
• Intermediate mesoderm
• Lateral plate mesoderm

(c) 22 days

Surface ectoderm

Neural crest

The neural folds fuse, forming the **neural tube**. It detaches from the surface ectoderm to lie between the surface ectoderm and the notochord.

Somite

Notochord

(d) End of week 4

The **somites** subdivide into three parts:
• **Dermatomes** form the dermis (dorsal).
• **Myotomes** form skeletal muscles.
• **Sclerotomes** form the vertebrae and ribs.

Intermediate mesoderm forms the kidneys and gonads.

Peritoneal cavity (coelom)

Neural tube (ectoderm)

Epidermis (ectoderm)

Gut lining (endoderm)

The **lateral plate mesoderm** forms:
• Limb bud
• Smooth muscle of gut
• Visceral serosa
• Parietal serosa
• Dermis (ventral)

Figure 28.12 Neurulation and early mesodermal differentiation. The embryonic body begins to undercut in (**c**) and undercutting is complete in (**d**).

28

As we described in Chapter 12 (◄ p. 435), the anterior end of the neural tube becomes the brain and the rest becomes the spinal cord. The associated **neural crest cells** (Figure 28.12c) migrate widely and give rise to the cranial, spinal, and sympathetic ganglia (and associated nerves), to the chromaffin cells of the adrenal medulla (◄ p. 627), and to pigment cells of the skin, and they contribute to some connective tissues.

By the end of the first month of development, the three primary brain vesicles (fore-, mid-, and hindbrain) are apparent (◄ Figure 12.1, p. 435). By the end of the second month, all brain flexures are evident, the cerebral hemispheres cover the top of the brain stem (◄ Figure 12.2, p. 436), and brain waves can be recorded. Most of the remaining ectoderm forming the surface layer of the embryonic body differentiates into the epidermis of the skin. The ectodermal derivatives are summarized in **Figure 28.13**.

Specialization of the Mesoderm

The first evidence of mesodermal differentiation is the appearance of the notochord in the embryonic disc (see Figure 28.12a). The notochord is eventually replaced by the vertebral column, but its remnants persist in the springy *nucleus pulposus* (◄ p. 219) of the intervertebral discs. Shortly thereafter, three mesodermal aggregates develop on either side of the notochord (Figure 28.12b, c). The largest of these, the *somites* (so′mīts), are paired mesodermal blocks that hug the notochord on either side. All of the pairs of somites are present by the end of week 5. Flanking the somites laterally are small clusters of segmented mesoderm called *intermediate mesoderm* and then double sheets of *lateral plate mesoderm*.

Each **somite** has three functional parts—*sclerotome, dermatome,* and *myotome* (Figure 28.12d).

- Cells of the **sclerotome** (skle′ro-tōm; "hard piece") migrate medially, gather around the notochord and neural tube, and produce the vertebra and rib at each associated level.

- **Dermatome** ("skin piece") cells help form the dermis of the skin in the dorsal part of the body.

- The **myotome** (mi′o-tōm; "muscle piece") cells develop in conjunction with the vertebrae. They form the skeletal muscles of the neck, body trunk, and, via their **limb buds**, the muscles of the limbs.

Cells of the **intermediate mesoderm** form the gonads and kidneys. The **lateral plate mesoderm** splits to form paired mesodermal layers: the *somatic mesoderm* and the more inferior *splanchnic mesoderm*. Cells of the somatic mesoderm help to form:

- The skin dermis of the ventral body region

- The parietal serosa that lines the ventral body cavity

- Most tissues of the limbs

The splanchnic mesoderm provides the mesenchymal cells that form the heart and blood vessels and most connective tissues of the body, as well as nearly the entire wall of the digestive and

The three **primary germ layers** (shown at right in a cross section of a 16-day embryo) specialize to form all of the body's tissues and organs.

Figure 28.13 Flowchart showing major derivatives of the embryonic germ layers.

respiratory organs. The two layers of the lateral plate mesoderm cooperate to form the serosae of the **coelom** (se′lom), or ventral body cavity. The mesodermal derivatives are summarized in Figure 28.13.

The Fetal Circulation

Embryonic development of the cardiovascular system lays the groundwork for the fetal circulatory pattern, which is converted to the adult pattern at birth. The first blood cells arise in the yolk sac. Before week 3 of development, tiny spaces appear in the splanchnic mesoderm. These are quickly lined by endothelial cells, covered with mesenchyme, and linked together into rapidly spreading vascular networks, destined to form the heart, blood vessels, and lymphatics. By the end of week 3, the embryo has a system of paired blood vessels, and the two vessels forming the heart have fused (◄ **Figure 18.22b, p. 701**). By 22 days, the miniature heart is pumping blood for an embryo less than 2 mm (0.1 inch) long.

Unique cardiovascular modifications seen only during prenatal development include the **umbilical arteries** and **vein** and three *vascular shunts*. All of these structures are occluded (blocked off) at birth. They are described in *Focus on Fetal and Newborn Circulation* (**Focus Figure 28.2** on pp. 1108–1109) and summarized in **Table 28.1**.

As you read about these structures, keep in mind that the blood is flowing from and to the fetal heart but that O_2, CO_2, nutrition, and waste exchanges occur *at the placenta*. The large umbilical vein carries freshly oxygenated blood returning from the placenta into the embryonic body, where it is conveyed to the liver. There, some of the returning blood percolates through the liver sinusoids and out the hepatic veins. Most of the blood coursing through the umbilical vein, however, enters the **ductus venosus** (duk′tus ve-no′sus), a venous shunt that bypasses the liver sinusoids. Both the hepatic veins and the ductus venosus empty into the inferior vena cava where the placental blood mixes with deoxygenated blood returning from the lower parts of the fetus's body. The vena cava in turn conveys this mixed blood directly to the right atrium of the heart.

After birth, the liver plays an important role in nutrient processing, but during embryonic life the mother's liver performs these functions. Consequently, blood flow through the fetal liver during development is important only to ensure that the liver cells remain healthy.

Blood entering and leaving the heart encounters two more shunt systems, each serving to bypass the nonfunctional lungs (Focus Figure 28.2, left side). (The lungs *do* receive adequate blood to maintain their growth.)

Occlusion of Vascular Shunts After Birth

After birth the special umbilical blood vessels and fetal shunts are no longer necessary (Focus Figure 28.2, right side). The umbilical arteries and vein constrict and become fibrosed. The proximal parts of the umbilical arteries persist as the *superior vesical arteries* that supply the urinary bladder, and their distal parts become the **median umbilical ligaments**. The remnant of the umbilical vein becomes the **round ligament** of the liver, or **ligamentum teres** (◄ **p. 894**), which attaches the umbilicus to

Table 28.1	Special Fetal Circulation Structures	
FETAL STRUCTURE	**POSTNATAL STRUCTURE**	**FUNCTION IN FETUS**
Foramen ovale	Fossa ovalis	Diverts blood from the right atrium to the left atrium
Ductus arteriosus	Ligamentum arteriosum	Diverts blood from the pulmonary trunk to the aorta
Ductus venosus	Ligamentum venosum	Carries blood from the umbilical vein through the liver (bypassing sinusoids) into the inferior vena cava
Umbilical arteries	Median umbilical ligaments	Paired vessels carry blood from the fetus to the placenta
Umbilical vein	Round ligament of the liver (ligamentum teres)	Single vessel carries oxygenated blood from the placenta to the fetus

the liver. The ductus venosus collapses as blood stops flowing through the umbilical vein and is eventually converted to the **ligamentum venosum** on the liver's undersurface.

As the pulmonary circulation becomes functional, pressure in the left side of the heart increases and that in the right side of the heart decreases, causing the pulmonary shunts to close. The flap of the **foramen ovale** is pushed to the shut position, and its edges fuse to the septal wall. Ultimately, only a slight depression, the **fossa ovalis** (◄ **p. 674**), marks its position. The **ductus arteriosus** constricts in response to rising blood oxygen levels and is converted to the cordlike **ligamentum arteriosum**, connecting the aorta and pulmonary trunk.

Except for the foramen ovale, all of the special circulatory adaptations of the fetus are functionally occluded within 30 minutes after birth. Closure of the foramen ovale is usually complete within the year. Failure of the ductus arteriosus or foramen ovale to close are common congenital heart defects.

Events of Fetal Development

By the end of the embryonic period (8 weeks), the bones have begun to ossify and the skeletal muscles are well formed and contracting spontaneously. Metanephric kidneys are developing (◄ **p. 1006**), gonads are formed, and the lungs and digestive organs are attaining their final shape and body position. Blood delivery to and from the placenta via the umbilical vessels is constant and efficient. The heart and the liver are competing for space and form a conspicuous bulge on the ventral surface of the embryo's body. All this by the end of eight weeks in an embryo about 3 cm (slightly more than 1 inch) long from crown to rump and a weight of approximately 2 g (0.06 ounce)!

During the fetal period, the developing fetus grows to about 360 mm (14 inches) and 3.2 kg (7 lb) or more. (Total body length at birth is about 550 mm, or 22 inches.) As you might expect with such tremendous growth, the changes in fetal appearance

(Text continues on p. 1110.)

Before birth, fetal blood bypasses the nonfunctional lungs and liver because their functions are provided by the mother via the placenta.

There are four essential adaptations for life *in utero*:

Placenta
The placenta is the site where O_2, nutrients, CO_2, and wastes are exchanged with the mother's circulation.

- Two **umbilical arteries** carry deoxygenated blood from the fetus to the placenta.
- One **umbilical vein** carries oxygenated blood from the placenta to the fetus.

Liver

Ductus venosus
- Carries blood from the umbilical vein to the inferior vena cava.
- Allows most blood to bypass the liver capillaries.

Umbilical vein

Abdominal aorta

Inferior vena cava

Before the lungs are inflated at birth, resistance to blood flow through them is high. This means that blood pressure is higher in the right side of the heart than in the left side, and blood will flow from right to left through any available opening (shunt). Normally there are two shunts: the foramen ovale and the ductus arteriosus.

Left atrium

Right atrium

Aortic arch

Pulmonary trunk

Foramen ovale
- A flap-covered hole between the right and left atria.
- Shunts most blood from the right atrium to the left atrium, bypassing the lungs.

Ductus arteriosus
- Carries blood from the pulmonary trunk to the aorta.
- Allows most blood to bypass the lungs.

Below is a simplified pathway of blood through the fetus:

The cardiovascular system changes at birth.
The umbilical vessels as well as the liver and lung bypasses are occluded.

Fetus

Newborn

Superior vena cava

Aortic arch

Ductus arteriosus
↓
Ligamentum arteriosum

Pulmonary artery

Pulmonary veins

Right lung

Left lung

Heart

Foramen ovale
↓
Fossa ovalis

Liver

Ductus venosus
↓
Ligamentum venosum

Umbilical vein
↓
Ligamentum teres

Hepatic portal vein

Umbilicus

Inferior vena cava

Abdominal aorta

Common iliac artery

Umbilical arteries
↓
Median umbilical ligaments

Urinary bladder

Umbilical cord

Placenta

■ High oxygenation
■ Moderate oxygenation
■ Low oxygenation
■ Very low oxygenation

are quite dramatic (**Figure 28.14**). The fetal period is a time of rapid growth of the body structures that were established in the embryo. During the first half of this period, cells are still differentiating into specific cell types to form the body's distinctive tissues and are completing the fine details of body structure. The main events of the fetal period—weeks 9 through 38—are listed chronologically in **Table 28.2**. The greatest amount of growth occurs in the first 8 weeks of life, when the embryo grows from one cell to a fetus of about 1 inch.

Check Your Understanding

11. The early embryo is flat like a three-layered pancake. What event must occur before organogenesis can get going in earnest?

Amniotic sac   Umbilical cord   Umbilical vein

Chorionic villi

Yolk sac

Cut edge of chorion

(a) Embryo at week 7, about 18 mm long.

(b) Fetus in month 3, about 6 cm long.

12. What germ layer gives rise to essentially all body tissues except nervous tissue, the epidermis, and mucosae?

13. What happens to the special fetal circulatory modifications after birth?

14. DRAW What two modifications of the fetal circulation allow most blood to bypass parts of the heart? Draw in and label these two modifications in the schematic heart and great vessels below.

15. PREDICT The amnion is crucial for development. What problems would arise if it wasn't there?

For answers, see Answers Appendix.

Figure 28.14 Photographs of a developing fetus. By birth, the fetus is typically 36 cm long from crown to rump.

(c) Fetus late in month 5, about 19 cm long.

Table 28.2   Developmental Events of the Fetal Period

TIME AFTER FERTILIZATION		CHANGES AND ACCOMPLISHMENTS
8 weeks (end of embryonic period)	 **8 weeks**	Head nearly as large as body; all major brain regions present; first brain waves in brain stem Liver disproportionately large; begins to form blood cells Limbs present; digits are initially webbed, but fingers and toes are free by the end of week 8 Ossification begun; weak, spontaneous muscle contractions occur Cardiovascular system fully functional (heart has been pumping blood since week 4) All body systems present in at least rudimentary form Approximate crown-to-rump length: 30 mm (1.2 inches); weight: 2 grams (0.06 ounce)
9–12 weeks (month 3)	 **12 weeks**	Head still dominant, but body elongating; brain continues to enlarge; cervical and lumbar enlargements apparent in spinal cord; retina of eye is present Skin epidermis and dermis obvious; facial features present in crude form Liver prominent, bile being secreted; palate is fusing Heartbeat can be detected externally; blood cell formation begins in bone marrow Lungs begin to develop; fetus inhales and exhales amniotic fluid Sex of the fetus can be determined from the genitals Approximate crown-to-rump length at end of interval: 90 mm (3.5 inches)
13–16 weeks (month 4)	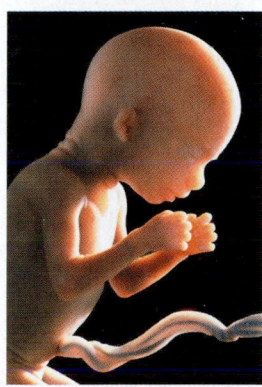	Cerebellum prominent; general sensory organs differentiated, blinking of eyes and sucking motions of lips occur Growth of the body beginning to outpace that of the head Glands developed in GI tract; intestines collecting meconium (dark green fetal intestinal contents) Kidneys attain typical structure Most bones are now distinct and joint cavities are apparent Approximate crown-to-rump length at end of interval: 140 mm (5.5 inches)
17–20 weeks (month 5)	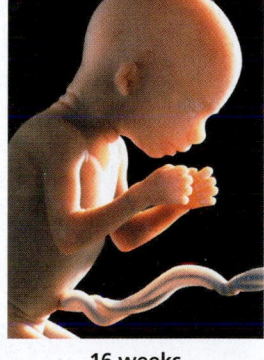 **16 weeks**	Vernix caseosa (fatty secretions of sebaceous glands) covers body; lanugo (silklike hair) covers skin Fetal position (body flexed anteriorly) assumed because of space restrictions Limbs reach near-final proportions Mother can feel fetal movements (quickening) Approximate crown-to-rump length at end of interval: 190 mm (7.5 inches)
21–30 weeks (months 6 and 7)	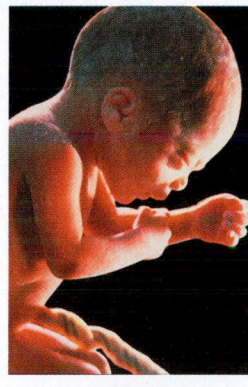	Period of substantial increase in weight (may survive if born prematurely at 27–28 weeks, but hypothalamic temperature regulation and lung production of surfactant are still inadequate) Myelination of spinal cord begins Distal limb bones are beginning to ossify Skin is wrinkled; fingernails and toenails are complete Body is lean and well proportioned Bone marrow becomes sole site of blood cell formation Testes descend to scrotum in month 7 (males) Approximate crown-to-rump length at end of interval: 280 mm (11 inches)
30–38 weeks (term) (months 8 and 9)	**At birth**	Fat accumulates in hypodermis; skin thickens Approximate crown-to-rump length at end of interval: 360 mm (14 inches); weight: 3.2 kg (7 lb)

28

28.5 During pregnancy, the mother undergoes anatomical, physiological, and metabolic changes

Learning Outcomes

▶ Describe functional changes in maternal reproductive organs and in the cardiovascular, respiratory, and urinary systems during pregnancy.

▶ Indicate the effects of pregnancy on maternal metabolism and posture.

The mean duration of pregnancy is 38 weeks from the time of ovulation to birth. Pregnancy can be a difficult time for the mother because profound adaptations occur in several body systems. Not only are there anatomical changes, but striking changes in her metabolism and physiology occur to support the pregnancy and prepare her body for delivery and lactation.

Anatomical Changes

As pregnancy progresses, the female reproductive organs become increasingly vascular and engorged with blood, and the vagina develops a purplish hue (*Chadwick's sign*). Prodded by rising levels of estrogens and progesterone, the breasts enlarge and engorge with blood, and their areolae darken. Some women develop increased pigmentation of facial skin of the nose and cheeks, a condition called *chloasma* (klo-az′mah; "to be green") or the "mask of pregnancy."

The degree of uterine enlargement during pregnancy is remarkable. Starting as a fist-sized organ, the uterus fills most of the pelvic cavity by 16 weeks (**Figure 28.15a, b**). Though the fetus is only about 140 mm long (crown-to-rump) at this time,

the placenta is fully formed, uterine muscle is hypertrophied, and amniotic fluid volume is increasing.

As pregnancy continues, the uterus pushes higher into the abdominal cavity, exerting pressure on abdominal and pelvic organs (Figure 28.15c). As birth nears, the uterus reaches the level of the sternum's xiphoid process and occupies most of the abdominal cavity (Figure 28.15d). The crowded abdominal organs press superiorly against the diaphragm, which intrudes on the thoracic cavity. As a result, the ribs flare, causing the thorax to widen.

The increasing bulkiness of the anterior abdomen changes the woman's center of gravity, and many women develop *lordosis* (accentuated lumbar curvature) and backaches during the last few months of pregnancy. **Relaxin**, produced by the corpus luteum and placenta, causes pelvic ligaments and the pubic symphysis to relax, widen, and become more flexible. This increased flexibility eases passage of the baby during birth, but it may result in a waddling gait in the meantime.

Considerable weight gain occurs during a normal pregnancy. Because some women are over- or underweight before pregnancy begins, it is almost impossible to state the ideal or desirable weight gain. However, summing up the weight increases resulting from fetal and placental growth, increased size of the maternal reproductive organs and breasts, and greater blood volume during pregnancy, a weight gain of approximately 13 kg (about 28 lb) is fairly typical.

Good nutrition is necessary all through pregnancy if the developing fetus is to have all the building materials (especially proteins, calcium, and iron) needed to form its tissues. The B vitamin folic acid reduces the risk of having a baby with neural tube problems such as spina bifida and anencephaly (◀ p. 483). A pregnant woman needs only 300 additional calories daily to

(a) Before conception
Uterus is the size of a fist and resides in the pelvis. Mammary gland structure is largely rudimentary.

(b) 4 months
Fundus of the uterus is halfway between the pubic symphysis and the umbilicus.

(c) 7 months
Fundus is well above the umbilicus.

(d) 9 months
Fundus reaches the xiphoid process. Glandular tissue of mammary glands is fully developed.

Figure 28.15 Relative size of the uterus before conception and during pregnancy.

sustain proper fetal growth. The emphasis should be on eating high-quality food, not just more food.

Metabolic Changes

As the placenta enlarges, it secretes various hormones that influence the mother's metabolism.

- **Placental growth hormone** suppresses and replaces the mother's growth hormone. It enhances nutrient availability to the fetus by stimulating lipolysis and glucose production (gluconeogenesis).

- **Human placental lactogen (hPL)** (also called *human chorionic somatomammotropin*) is secreted in amounts proportional to the placenta's size. In cooperation with estrogens and progesterone, it stimulates maturation of the breasts for lactation (Figure 28.15), promotes growth of the fetus, and exerts a glucose-sparing effect in the mother. This means that maternal cells metabolize more fatty acids and less glucose than usual, sparing glucose for use by the fetus. It also reduces insulin sensitivity. Consequently, gestational diabetes mellitus occurs in about 3–10% of pregnancies. Over half of these women go on to develop type 2 diabetes later in life.

- **Corticotropin-releasing hormone (CRH)** (◀ p. 616) levels rise dramatically toward the end of pregnancy. This rise is thought to be related to the timing of birth. As CRH levels rise, the mother's ACTH and cortisol (◀ p. 625) levels rise, too. This may protect her in times of stress, such as giving birth. Rising cortisol levels in the fetus promote maturation, particularly of the lungs.

Plasma levels of activated vitamin D rise two- to threefold during pregnancy, which increases absorption of dietary calcium. This rise in vitamin D occurs in spite of levels of parathyroid hormone that are normal or even low when intake of calcium is adequate. This ensures that the developing fetus will have adequate calcium to mineralize its bones.

Physiological Changes

Physiological changes take place in many systems during pregnancy. A few of these changes are described next.

Gastrointestinal System

The nausea and vomiting (commonly called morning sickness) suffered by many women during the first few months of pregnancy are believed to be related to elevated levels of hCG, estrogens, and progesterone. *Heartburn*, due to reflux of stomach acid into the esophagus, is common because the esophagus is displaced and the stomach is crowded by the growing uterus. *Constipation* occurs as the motility of the digestive tract decreases.

HOMEOSTATIC IMBALANCE 28.1 — CLINICAL

Because many potentially harmful substances can cross placental barriers and enter the fetal blood, a pregnant woman should be aware of what she is taking into her body, particularly during the embryonic period when the body's foundations are laid down. **Teratogens** (ter′ah-to-jenz; *terato* = monster), factors that may cause severe congenital abnormalities or even fetal death, include alcohol, nicotine, many drugs (anticoagulants, sedatives, antihypertensives, and others), and maternal infections, particularly rubella. For example, when a woman drinks alcohol, her fetus becomes inebriated as well. However, the fetal consequences may be much more lasting and result in the *fetal alcohol syndrome* (*FAS*) typified by microcephaly (small head), intellectual disability, and abnormal growth. Smoking hinders oxygen delivery to the fetus, impairing normal growth and development. ●●●●●●

Urinary System

The kidneys produce more urine during pregnancy because of the mother's increased metabolic rate, greater blood volume, and the additional burden of disposing of fetal metabolic wastes. As the growing uterus compresses the bladder, urination becomes more frequent, more urgent, and sometimes uncontrollable (*stress incontinence*).

Respiratory System

The nasal mucosa responds to estrogens by becoming edematous and congested. Thus, nasal stuffiness and occasional nosebleeds may occur. Tidal volume (◀ p. 840) increases markedly during pregnancy, while respiratory rate is relatively unchanged and residual volume (◀ p. 841) decreases. The increase in tidal volume is due to (1) the increased production of CO_2 and (2) enhancement of the sensitivity of the medullary respiratory center to CO_2 caused by progesterone. Many women exhibit *dyspnea* (disp-ne′ah), or difficult breathing, during the later stages of pregnancy when the diaphragm is pushed superiorly by the increasing size of the uterus.

Cardiovascular System

The most dramatic physiological changes occur in the cardiovascular system. Total body water rises, and blood volume may increase as much as 40% by the 32nd week to accommodate the additional needs of the fetus. The rise in blood volume also safeguards against blood loss during birth. Mean blood pressure typically decreases during midpregnancy, but then rises to normal levels during the third trimester. Cardiac output increases by 35–40% at various stages of pregnancy. This helps propel the greater circulatory volume around the body. The uterus presses on the pelvic blood vessels, which may impair venous return from the lower limbs, resulting in *varicose veins* and leg edema (◀ p. 734).

Interestingly, some fetal cells do make their way into the maternal circulation. Because these cells are undifferentiated, they can take up residence in various maternal tissues and differentiate into the types of cells that surround them. It is not clear what consequences this may have—they may predispose a woman to autoimmune diseases or they may protect her against certain diseases. Time and research will tell.

28

HOMEOSTATIC IMBALANCE 28.2   **CLINICAL**

Preeclampsia, a dangerous complication affecting about 5% of pregnancies after 20 weeks, is characterized by hypertension and proteinuria. Proteinuria often leads to edema (◄ p. 734), and severe hypertension can lead to numerous complications. Even worse, preeclampsia can lead to **eclampsia**, generalized seizures that can be fatal. The only cure is delivery of the baby.

Check Your Understanding

16. What causes the difficult breathing that some women experience during pregnancy? What causes the waddling gait seen in some?

17. What is the role of the hormone hPL?

18. MAKE CONNECTIONS Pregnant women often experience edema of their lower limbs. Explain this edema in terms of an increase or decrease in one of the pressures that drives bulk flow at the capillaries. (Hint: See Focus Figure 19.1 on pp. 732–733.)

For answers, see Answers Appendix.

28.6 The three stages of labor are the dilation, expulsion, and placental stages

Learning Outcome

▶ Explain how labor is initiated, and describe the three stages of labor.

Parturition (par″tu-rish′un; "bringing forth young") is the culmination of pregnancy—giving birth to the baby. It usually occurs within 15 days of the calculated due date (280 days from the last menstrual period). The series of events that expel the infant from the uterus are collectively called **labor**.

Initiation of Labor

The fetus largely determines its own birth date, and several events and hormones interlock to trigger labor. As the fetal hypothalamus and adrenal glands mature during the last few weeks of pregnancy, increased CRH from the fetal hypothalamus leads to increased fetal cortisol release. Fetal cortisol then begins a positive feedback loop with *placental* CRH—cortisol increases placental CRH release, which increases cortisol release even more—leading to a rapid rise in fetal cortisol levels. Fetal cortisol triggers the final maturation of the fetal lungs, which increases the production and release of *surfactant protein A (SP-A)*. SP-A, in turn, acts as an additional signal that the fetus is ready to be born. The rapid increase in cortisol levels also causes a rapid increase in estrogen production by the placenta.

Remember that the placenta produces both estrogens and progesterone, and that levels of both of these hormones steadily increase throughout pregnancy. A major function of progesterone during pregnancy is to suppress contraction of uterine smooth muscle as it is stretched by the growing fetus. This quieting action of progesterone is important for preventing preterm labor. As estrogen levels rise at the end of pregnancy, they switch the progesterone receptor to a less effective type. The increase in estrogens (together with the decrease in progesterone's action) has the following effects:

- **It prepares the cervix for dilation.** Estrogens increase the synthesis of both **prostaglandins** (pros″tah-glan′dinz) (◄ p. 48) and their receptors. Prostaglandins play a major role in the thinning and softening of the cervix just before and during labor.

- **It prepares the myometrium of the uterus for contraction.** Additional contractile machinery (including actin, myosin, and myosin light chain kinase) is synthesized so that the uterus can contract more forcefully. Like the heart, the uterus needs to contract as a unit. As the onset of labor nears, gap junctions are formed between myometrial cells to tie them together electrically. As a result, the myometrium becomes increasingly irritable. Weak, irregular uterine contractions begin to occur. These contractions, called *Braxton Hicks contractions*, have caused many women to go to the hospital, only to be told that they were in **false labor** and sent home.

- **It increases the sensitivity of the myometrium to chemical messengers that cause it to contract.** Under the influence of rising estrogen action and decreasing progesterone action, the myometrial cells make abundant receptors for **oxytocin** (ok″sĭ-to′sin) (◄ p. 612) and prostaglandin (**Figure 28.16**). These are the two chemical messengers that will drive myometrial contraction during true labor.

As birth nears, the now powerful uterine muscle is highly sensitive to prostaglandins and oxytocin. Under the influence

Figure 28.16 Hormonal induction of labor. Note that three different positive feedback loops are involved in labor.

of estrogens, the placenta releases prostaglandins that trigger the rhythmic expulsive contractions of true labor. Pressure from the baby's head dilates the cervix. Increasing cervical distension activates the mother's hypothalamus, which signals for oxytocin release by the posterior pituitary.

Several *positive feedback mechanisms* (◀ p. 11) involving prostaglandins and oxytocin are propelled into action—contractions and greater distension of the cervix cause the release of more prostaglandins and oxytocin, which cause greater contractile force, and so on (Figure 28.16). Expulsion of the baby is aided by a chemical change in *fetal fibronectin*, the natural "stickum" (adhesive protein) that bound the fetal and maternal tissues of the placenta together up to this point in pregnancy.

As mentioned earlier, prostaglandins are essential for initiating labor in humans, and interfering with their production will hinder onset of labor. For example, antiprostaglandin drugs such as ibuprofen can inhibit the early stages of labor and such drugs are used occasionally to prevent preterm births.

Stages of Labor

Labor includes the dilation, expulsion, and placental stages illustrated in **Figure 28.17**.

① **Dilation stage.** The **dilation stage** is the time from labor's onset until the cervix is fully dilated by the baby's head (about 10 cm in diameter). As labor starts, weak but regular contractions begin in the upper part of the uterus and move toward the vagina. At first, only the superior uterine muscle is active; contractions are 15–30 minutes apart, and last for 10–30 seconds. As labor progresses, the contractions become more vigorous and rapid, and the lower part of the uterus gets involved. As the infant's head is forced against the cervix with each contraction, the cervix softens, *effaces* (thins), and dilates. Eventually the amnion ruptures, releasing the amniotic fluid, an event commonly referred to as the mother's "water breaking."

The dilation stage is the longest part of labor, lasting 6–12 hours or more. Several events happen during this phase. *Engagement* occurs when the infant's head enters the true pelvis (◀ p. 238). As descent continues through the vagina (sometimes referred to as the birth canal), the baby's head rotates so that its greatest dimension is in the anteroposterior line, which allows it to navigate the narrow dimensions of the pelvic outlet (Figure 28.17 ①b).

② **Expulsion stage.** The **expulsion stage** lasts from full dilation to delivery of the infant, or actual childbirth. By the time the cervix is fully dilated, strong contractions occur every 2–3 minutes and last about 1 minute. In this stage, a mother undergoing labor without local anesthesia has an increasing urge to push or bear down with the abdominal muscles. Although this phase may last 2 hours, it is typically 50 minutes in a first birth and around 20 minutes in subsequent births.

Crowning occurs when the largest dimension of the baby's head distends the vulva. At this point, an *episiotomy* (e-piz″e-ot′o-me) is sometimes done to reduce tissue tearing. An episiotomy is an incision made to widen the vaginal orifice. The baby's neck extends as the head exits from the

①a Early dilation. Baby's head engaged; its widest dimension lies in the left-right line of the pelvic outlet.

- Umbilical cord
- Placenta
- Uterus
- Cervix
- Vagina

①b Late dilation. Baby's head rotates so widest dimension is in anteroposterior line (of pelvic outlet). Dilation nearly complete.

- Pubic symphysis
- Sacrum

② Expulsion. Baby's head extends as it is delivered.

- Perineum

③ Placental stage. After baby is delivered, the placenta detaches and is removed.

- Uterus
- Placenta (detaching)
- Umbilical cord

Figure 28.17 Stages of labor.

28

perineum, and once the head has been delivered, the rest of the baby's body is delivered much more easily. After birth, the umbilical cord is clamped and cut.

When the infant is in the usual *vertex*, or head-first, *presentation*, the skull (its largest diameter) acts as a wedge to dilate the cervix. The head-first presentation also allows the baby to be suctioned free of mucus and to breathe even before it has completely exited from the birth canal. In *breech* (buttock-first) and other nonvertex presentations, these advantages are lost and delivery is much more difficult, often requiring the use of forceps, or a C-section (see below).

③ **Placental stage.** The **placental stage**, or the delivery of the placenta and its attached fetal membranes, collectively called the **afterbirth**, is usually accomplished within 30 minutes after birth of the infant. The strong uterine contractions that continue after birth compress uterine blood vessels, limit bleeding, and shear the placenta off the uterine wall (cause placental detachment). The blood vessels in the umbilical cord are counted after the placenta is delivered to verify the presence of two umbilical arteries and one umbilical vein because the absence of one umbilical artery is often associated with cardiovascular disorders in the newborn.

HOMEOSTATIC IMBALANCE 28.3   **CLINICAL**

If a woman has a deformed or narrow malelike pelvis, labor may be prolonged and difficult. This condition is called *dystocia* (dis-to'se-ah; *dys* = difficult, *toc* = birth). Besides extreme maternal fatigue, another possible consequence of dystocia is fetal brain damage, resulting in cerebral palsy or epilepsy. To prevent these outcomes, a *cesarean (C-) section* (se-sa're-an) is performed in many such cases. A C-section is delivery of the infant through an incision made through the abdominal and uterine walls.   ●●●●●●●

Check Your Understanding

19. What is a breech presentation?

20. What chemical is most responsible for triggering true labor?

21. Why does a baby's head rotate as it descends through the true pelvis?

For answers, see Answers Appendix.

28.7 ## An infant's extrauterine adjustments include taking the first breath and closure of vascular shunts

Learning Outcome

▶ Outline the events leading to the first breath of a newborn.

The **neonatal period** is the four-week period following birth. Here we will be concerned with the events of just the first few hours after birth in a normal infant. As you might suspect, birth represents quite a shock to the infant. Exposed to physical trauma during the birth process, it is suddenly cast out of its watery, warm environment and its placental life supports are severed. Now it must do for itself all that the mother had been

doing for it—breathe, obtain nutrients, excrete, and maintain its body temperature.

Two critical adjustments must occur.

● **Taking the first breath.** The crucial first requirement is to breathe. Vasoconstriction of the umbilical arteries, initiated when they are stretched during birth, leads to loss of placental support. Once carbon dioxide is no longer removed by the placenta, it accumulates in the baby's blood, causing central acidosis. This excites respiratory control centers in the baby's brain and triggers the first inspiration. The first breath requires a tremendous effort—the airways are tiny, and the lungs are collapsed. However, once the lungs have been inflated in full-term babies, surfactant (◀ p. 839) in alveolar fluid reduces surface tension in the alveoli, and breathing is easier. The rate of respiration is rapid (about 45 respirations/min) during the first two weeks and then gradually decreases.

Keeping the lungs inflated is much more difficult for premature infants (those weighing less than 2500 g, or about 5.5 lb, at birth) because surfactant production occurs during the last months of prenatal life. Consequently, preemies are usually put on a ventilator until their lungs are mature enough to function on their own.

● **Occlusion of special fetal blood vessels and vascular shunts.** After birth the special umbilical blood vessels and fetal shunts are occluded as we described previously (see p. 1107 and Focus Figure 28.2).

At 1 and 5 minutes after birth, the infant's physical status is assessed based on five signs: heart rate, respiration, color, muscle tone, and reflexes. Each observation is given a score of 0 to 2, and the total is called the **Apgar score**. An Apgar score of 8 to 10 indicates a healthy baby. Lower scores reveal problems in one or more of the physiological functions assessed.

For 6–8 hours after birth, infants pass through an unstable **transitional period** marked by alternating periods of increased activity and sleep. During the activity periods, vital signs are irregular and the baby gags frequently as it regurgitates mucus and debris. After this, the infant stabilizes, with waking periods (dictated by hunger) occurring every 3–4 hours.

Check Your Understanding

22. What is an Apgar score?

23. MAKE CONNECTIONS Surfactant decreases surface tension in the alveoli of a newborn. What property of the lung does surfactant affect, making breathing easier? (Hint: See Chapter 22.)

For answers, see Answers Appendix.

28.8 ## Lactation is milk secretion by the mammary glands in response to prolactin

Learning Outcome

▶ Explain how the breasts are prepared for lactation.

Lactation is production of milk by the hormone-prepared mammary glands. Rising levels of (placental) estrogens, progesterone, and human placental lactogen toward the end of pregnancy stimulate the hypothalamus to release prolactin-releasing factors (PRFs). The anterior pituitary gland responds by secreting **prolactin** (◀ **p. 616**). (This mechanism is described below.) After a delay of two to three days following birth, true milk production begins.

During the initial delay (and during late gestation), the mammary glands secrete a yellowish fluid called **colostrum** (ko-los′trum). It has less lactose than milk and almost no fat, but it contains more protein, vitamin A, and minerals than true milk. Like milk, colostrum is rich in IgA antibodies (◀ **p. 799**). Since these antibodies are resistant to digestion in the stomach, they may help to protect the infant's digestive tract against bacterial infection.

After birth, prolactin release gradually wanes, and continued milk production depends on mechanical stimulation of the nipples, normally provided by the suckling infant. Mechano-receptors in the nipple send action potentials to the hypothalamus, stimulating secretion of PRFs. This results in a burstlike release of prolactin, which stimulates milk production for the next feeding.

The same action potentials also prompt hypothalamic release of oxytocin from the posterior pituitary via a *positive feedback mechanism*. Oxytocin causes the **let-down reflex**, the actual ejection of milk from the alveoli of the mammary glands (**Figure 28.18**). Let-down occurs when oxytocin binds to myoepithelial cells surrounding the glands, after which milk is ejected from *both* breasts, not just the suckled one. During nursing,

oxytocin also stimulates the recently emptied uterus to contract, helping it to return to (nearly) its prepregnant size.

Breast milk has advantages for the infant:

- *Its fats and iron are better absorbed and its amino acids are metabolized more efficiently* than those of cow's milk.

- *It has a host of beneficial chemicals*, including IgA, complement, lysozyme, interferon, and lactoperoxidase, that protect infants from life-threatening infections. Mother's milk also contains interleukins and prostaglandins that prevent over-zealous inflammatory responses, and a glycoprotein that deters the ulcer-causing bacterium (*Helicobacter pylori*) from attaching to the stomach mucosa.

- *It helps beneficial bacteria to colonize the infant's gut* by providing certain oligosaccharides (sugars) they need to grow.

- *Its natural laxative effect helps to cleanse the bowels of* **meconium** (mĕ-ko′ne-um), a tarry green-black paste containing sloughed-off epithelial cells, bile, and other substances. Since meconium, and later feces, provides the route for eliminating bilirubin from the body, clearing meconium as quickly as possible helps to prevent *physiological jaundice* (see Related Clinical Terms, p. 1120).

When nursing is discontinued, the stimulus for prolactin release and milk production ends, and the mammary glands stop producing milk. Women who nurse their infants for six months or more lose a significant amount of calcium from their bones, but those on sound diets usually replace lost bone calcium after weaning the infant.

Figure 28.18 **Milk production and the positive feedback mechanism of the milk let-down reflex.**

Contraception: To Be or Not To Be

In a world such as ours, **contraception** (*contra* = against, *cept* = taking), commonly called birth control, is often seen as a necessity. The key to birth control is dependability. As the red arrows in the accompanying flowchart show, the birth control techniques currently available have many sites of action for blocking reproduction. Generally speaking, birth control methods can be categorized into three groups based on how each prevents pregnancy—behavioral methods, barrier methods, or hormonal methods. Let's examine a few of them more closely.

Behavioral methods (coitus interruptus, the rhythm method, and abstinence) consist of altering behavior to prevent pregnancy. *Coitus interruptus*, or withdrawal of the penis just before ejaculation, is unreliable because control of ejaculation is never ensured.

Rhythm or *temporary abstinence methods* involve avoiding intercourse during periods of ovulation or fertility. Ovulation can be detected by recording daily basal body temperatures (body temperature drops slightly immediately prior to ovulation and then rises slightly after ovulation). This technique requires accurate record keeping for several cycles before it can be used with confidence, and can have a high success rate for those willing and able to take the time necessary. However, the average failure rate is 24% per year.

Barrier methods include diaphragms, cervical caps, male and female condoms, and spermicides. These techniques prevent the sperm and egg from meeting and implanting. They are quite effective, especially when used in combination—for example, condoms and spermicides.

Hormonal methods include a broad variety of hormone-containing contraceptive products (the pill, IUD, patches, vaginal ring, and implanted or injected agents). These methods are highly effective when used consistently, with a failure rate of less than 1% for oral contraceptives, and have a low risk of adverse effects in most women.

The most-used contraceptive product in North America is the *oral contraceptive pill* (OCP) or simply "the pill." The pill is available in various formulations containing estrogens and progestins (progesterone-like hormones). Some are taken daily, others according to a different schedule.

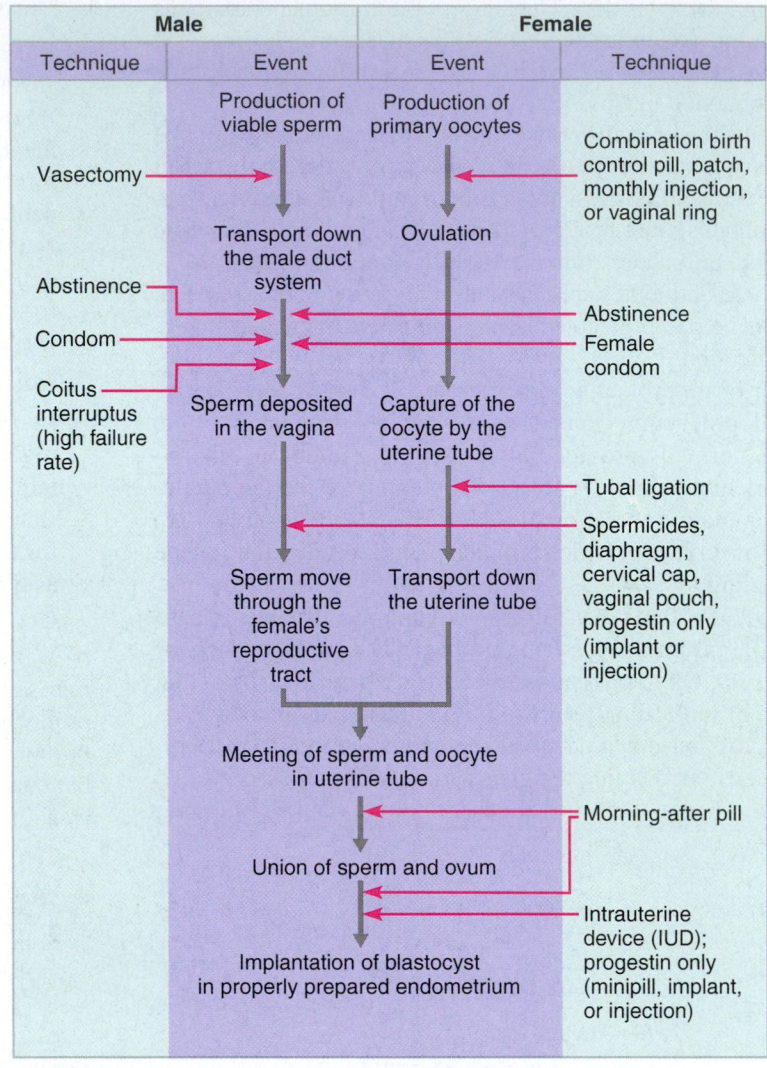

Mechanisms of contraception. Techniques or products that interfere with events from production of gametes to implantation are indicated by red arrows at the site of interference. They act to prevent the next step from occurring.

The pill tricks the hypothalamic-pituitary-gonadal axis by making it appear that the woman is pregnant (estrogens and progesterone are produced throughout pregnancy). Mature ovarian follicles do not develop, ovulation ceases, and menstrual flow is much reduced. However, since hormonal balance is one of the most precisely controlled body functions, some women cannot tolerate these changes—they become nauseated and/or hypertensive.

The pill has adverse cardiovascular effects in a small number of users, and there is still debate about whether it increases the risk of breast cancer. On the other hand, it decreases the risk of ovarian and endometrial cancers. Presently, well over 10 million U.S. women use the pill.

Other delivery methods using the combination hormone approach include two slow-release products—a flexible ring that is inserted into the vagina, and a transdermal (skin) patch. Failure rates and side effects of the vaginal ring and skin patch are comparable to those of the pill.

Other hormonal approaches to contraception use progestin-only products that thicken the cervical mucus enough to block sperm entry into the uterus, decrease the frequency of ovulation, and make the endometrium inhospitable to implantation. These include match-sized silicone rods implanted just under the skin that release progestin over a five-year period (e.g., Nexplanon®), and an injectable form that lasts for three months (Depo-Provera®). New intrauterine devices (IUDs) that are inserted into the uterus provide sustained local delivery of synthetic progesterone to the endometrium. The failure rates of progestin treatments are even less than that of the pill.

Morning-after pills, or *emergency contraceptive pills*, are based on progesterone-like hormones and used for postcoital contraception. They should be taken as soon as possible within 3–5 days after unprotected intercourse. They interfere with normal hormonal signals enough to prevent either fertilization or implantation. They have a failure rate of less than 25%.

Sterilization techniques permanently prevent gamete release. *Tubal ligation* or *vasectomy* (cutting or cauterizing the uterine tubes or ductus deferentes, respectively) are nearly foolproof and are the choice of approximately 33% of couples of childbearing age in the United States. However, these techniques are usually permanent, making them unpopular with individuals who still plan to have children but want to select the time.

This summary doesn't even begin to touch on the experimental birth control drugs now awaiting clinical trials and male contraceptives that are currently being evaluated. Other methods are sure to be developed in the near future. In the final analysis, however, the only 100% effective means of birth control is the age-old one— total abstinence.

While prolactin levels are high, the normal hypothalamic-pituitary controls of the ovarian cycle are damped, probably because stimulation of the hypothalamus by suckling causes it to release beta endorphin, a peptide hormone that inhibits hypothalamic release of GnRH and, consequently, the release of gonadotropins by the pituitary. Because of this inhibition of ovarian function, nursing has been called natural birth control. However, there is a good deal of "slippage" in these controls. Most women begin to ovulate even while continuing to nurse their infants and so *can* become pregnant again.

Check Your Understanding

24. What hormone causes the let-down reflex?

For answers, see Answers Appendix.

28.9 Assisted reproductive technology may help an infertile couple have offspring

Learning Outcome

▶ **Describe some techniques of ART including IVF, ZIFT, and GIFT.**

So far we have been describing how babies are made. But if a couple lacks that capability for some reason, what recourse do they have? Hormone therapy may increase sperm or egg production in cases where that is the problem, and surgery can open blocked uterine tubes. Perhaps the simplest intervention is *artificial insemination*. This can help by concentrating sperm (to make up for low sperm counts) and by giving the sperm a "head start" past the cervical mucus.

Beyond that are *assisted reproductive technology (ART)* procedures that entail surgically removing oocytes from a woman's ovaries following hormone stimulation, fertilizing the oocytes, and then returning them to the woman's body. These procedures, now performed worldwide in major medical centers, have produced thousands of infants. However, they are expensive, emotionally draining, and painful for the oocyte donor. Unused oocytes, sperm, and embryos can be frozen for later attempts at accomplishing pregnancy.

In the most common ART process, *in vitro fertilization (IVF)*, harvested oocytes are incubated with sperm in culture dishes (*in vitro*) for several days to allow fertilization to occur. In cases where the quality or number of sperm is low, the oocytes are injected with sperm. Embryos reaching the two-cell or blastocyst stage are then carefully transferred into the woman's uterus in the hope that implantation will occur.

In *zygote intrafallopian transfer (ZIFT)*, oocytes fertilized *in vitro* are immediately transferred to the woman's uterine (fallopian) tubes. The goal is to have development to the blastocyst stage occur followed by normal implantation in the uterus.

In *gamete intrafallopian transfer (GIFT)*, no *in vitro* procedures are used. Instead, sperm and harvested oocytes are transferred together into the woman's uterine tubes in the hope that fertilization will take place there.

Check Your Understanding

25. What is the most common ART technique?

For answers, see Answers Appendix.

In this chapter, we have focused on changes that occur during human development *in utero*. But having a baby is not always the goal, and we humans have devised a variety of techniques for preventing this outcome (see **A Closer Look** above).

28

RELATED CLINICAL TERMS

Abortion (*abort* = born prematurely) Termination of a pregnancy that is in progress; may be spontaneous or induced.

Abruptio placenta (ah-brup'she-o; *abrupt* = broken away from) Premature separation of the placenta from the uterine wall; if this occurs before labor, it can result in fetal death due to anoxia.

Ectopic pregnancy (ek-top'ik; *ecto* = outside) A pregnancy in which the embryo implants in any site other than the uterus; most often the site is a uterine tube (tubal pregnancy). Since the uterine tube (as well as most other ectopic sites) is unable to establish a placenta or accommodate growth, the uterine tube ruptures unless the condition is diagnosed early, or the pregnancy spontaneously aborts.

Hydatid (hydatidiform) mole (hi'dah-tid; *hydat* = watery) Developmental abnormality of the placenta; the conceptus degenerates and the chorionic villi convert into a mass of vesicles that resemble tapioca. Signs include vaginal bleeding, which contains some of the grapelike vesicles.

Physiological jaundice (jawn'dis) Jaundice sometimes occurring in normal newborns within three to four days after birth. Fetal erythrocytes are short-lived, and they break down rapidly after birth; the infant's liver may be unable to process the bilirubin (breakdown product of hemoglobin pigment) (◄ **p. 649**) fast enough to prevent its accumulation in blood and subsequent deposit in body tissues.

Placenta previa (pre've-ah) Placental formation adjacent to or across the internal os of the uterus. This is a problem because as the uterus and cervix stretch, separation of the placenta may occur. Additionally, the placenta precedes the infant during labor.

Ultrasonography (ul"trah-son-og'rah-fe) Noninvasive technique that uses sound waves to visualize the position and size of the fetus and placenta (◄ **A Closer Look, pp. 16–17**).

CHAPTER SUMMARY

1. The gestation period of approximately 280 days extends from the woman's last menstrual period to birth. The conceptus undergoes embryonic development for 8 weeks after fertilization, and fetal development from week 9 to birth.

28.1 Fertilization combines the sperm and egg chromosomes, forming a zygote (pp. 1092–1096)

1. An oocyte is fertilizable for up to 24 hours; sperm are best able to fertilize an oocyte within the first one to two days after intercourse.
2. Sperm must survive the hostile environment of the vagina and become capacitated (capable of reaching and fertilizing the oocyte).
3. Hundreds of sperm must release their acrosomal enzymes to break down the egg's corona radiata and zona pellucida.
4. When one sperm binds to receptors on the egg, it triggers two events that block polyspermy: the oocyte membrane block and the cortical and zona reactions.
5. Following sperm penetration, the secondary oocyte completes meiosis II. Then the ovum and sperm pronuclei fuse (fertilization), forming a zygote.

28.2 Embryonic development begins as the zygote undergoes cleavage and forms a blastocyst en route to the uterus (pp. 1097–1098)

1. Early development consists of cleavage, a rapid series of mitotic divisions without intervening growth, that begins with the zygote and ends with a blastocyst. The blastocyst consists of the trophoblast and an embryoblast. Cleavage produces a large number of cells with a favorable surface-to-volume ratio.

28.3 Implantation occurs when the embryo burrows into the uterine wall, triggering placenta formation (pp. 1098–1102)

1. The trophoblast adheres to, digests, and implants in the endometrium. Implantation is completed when the blastocyst is entirely surrounded by endometrial tissue, about 12 days after ovulation.

2. hCG released by the trophoblast (and later the chorion) maintains hormone production by the corpus luteum, preventing menses. hCG levels decline after four months.
3. The placenta acts as the respiratory, nutritive, and excretory organ of the fetus and produces the hormones of pregnancy. It is formed from embryonic (chorionic villi) and maternal (endometrial decidua) tissues. The chorion develops when the trophoblast becomes associated with extraembryonic mesoderm. Typically, the placenta is functional as an endocrine organ by the third month.

28.4 Embryonic events include gastrula formation and tissue differentiation, which are followed by rapid growth of the fetus (pp. 1102–1111)

Extraembryonic Membranes (p. 1102)

1. The fluid-filled amnion forms from cells of the superior surface (epiblast) of the embryonic disc. It protects the embryo from physical trauma and adhesion formation, provides a constant temperature, and allows fetal movements.
2. The yolk sac forms from the hypoblast of the embryonic disc; it is the source of early blood cells.
3. The allantois, a caudal outpocketing adjacent to the yolk sac, forms the structural basis of the umbilical cord.
4. The chorion is the outermost membrane and takes part in placentation.

Gastrulation: Germ Layer Formation (pp. 1102–1103)

5. Gastrulation involves cellular migrations that ultimately transform the embryoblast into a three-layered embryo (gastrula) containing ectoderm, mesoderm, and endoderm. Cells that move through the midline primitive streak become endoderm if they form the most inferior layer of the embryonic disc and mesoderm if they ultimately occupy the middle layer. Cells remaining on the superior surface become ectoderm.

Organogenesis: Differentiation of the Germ Layers (pp. 1103–1107)

6. Endoderm forms the epithelial lining of the digestive and respiratory systems, and the epithelial cells of all associated glands (thyroid, parathyroids, thymus, liver, pancreas). It becomes a continuous tube when the embryonic body undercuts and fuses ventrally.

7. Ectoderm forms the nervous system and the epidermis of the skin and its derivatives. The first event of organogenesis is neurulation, which produces the brain and spinal cord. By the eighth week, all major brain regions are formed.

8. Mesoderm forms all other organ systems and tissues. It segregates early into (1) a dorsal superior notochord, (2) paired somites that form the vertebrae, skeletal trunk muscles, and part of the dermis, and (3) paired masses of intermediate and lateral plate mesoderm. The intermediate mesoderm forms the kidneys and gonads. The lateral plate mesoderm forms the dermis of skin; parietal serosa; bones and muscles of the limbs; the cardiovascular system; and the visceral serosae.

The Fetal Circulation (p. 1107)

9. The fetal cardiovascular system is formed in the embryonic period. The umbilical vein delivers nutrient- and oxygen-rich blood to the embryo; the paired umbilical arteries return oxygen-poor, waste-laden blood to the placenta. The ductus venosus allows most of the blood to bypass the liver; the foramen ovale and ductus arteriosus are pulmonary shunts.

10. Inflation of the lungs causes pressure changes in the circulation; as a result, the umbilical arteries and vein, ductus venosus, and ductus arteriosus collapse, and the foramen ovale closes. The occluded blood vessels are converted to fibrous cords; the site of the foramen ovale becomes the fossa ovalis.

Events of Fetal Development (pp. 1107–1111)

11. All organ systems are laid down during the embryonic period; growth and tissue/organ specialization are the major events of the fetal period.

12. During the fetal period, fetal length increases from about 30 mm to 360 mm, and weight increases from less than an ounce to 7 pounds or more.

28.5 During pregnancy, the mother undergoes anatomical, physiological, and metabolic changes (pp. 1112–1114)

Anatomical Changes (pp. 1112–1113)

1. Maternal reproductive organs and breasts become increasingly vascularized during pregnancy, and the breasts enlarge.

2. The uterus eventually occupies nearly the entire abdominopelvic cavity. Abdominal organs are pushed superiorly and encroach on the thoracic cavity, causing the ribs to flare.

3. The increased abdominal mass changes the woman's center of gravity; lordosis and backache are common. A waddling gait occurs as pelvic ligaments and joints are loosened by placental relaxin.

4. A typical weight gain during pregnancy in a woman of normal weight is 28 pounds.

Metabolic Changes (p. 1113)

5. The placenta releases hormones that affect the mother's metabolism: placental growth hormone, CRH, and human placental lactogen. Human placental lactogen has anabolic effects and promotes glucose sparing in the mother.

Physiological Changes (pp. 1113–1114)

6. Many women suffer morning sickness, heartburn, and constipation during pregnancy.

7. The kidneys produce more urine, and pressure on the bladder may cause frequency, urgency, and stress incontinence.

8. Respiratory tidal volume increases, but residual volume decreases. The respiratory rate remains relatively unchanged. Dyspnea is common.

9. Total body water and blood volume increase dramatically and cardiac output rises. Blood pressure declines in midpregnancy and then rises to normal levels during the third trimester.

28.6 The three stages of labor are the dilation, expulsion, and placental stages (pp. 1114–1116)

1. Parturition encompasses a series of events called labor.

Initiation of Labor (pp. 1114–1115)

2. Rising fetal CRH levels signal that the fetus is ready for birth. CRH increases fetal cortisol secretion, which increases estrogen release from the placenta and decreases the action of progesterone. Together, the increased estrogen levels and decreased progesterone action ready the myometrium for delivery by increasing myometrial oxytocin and prostaglandin receptors, and causing the formation of gap junctions between myometrial cells.

3. Prostaglandins and oxytocin stimulate contraction of uterine muscle. Increasing distension of the cervix activates the hypothalamus, causing oxytocin release from the mother's posterior pituitary; this sets up a positive feedback loop resulting in true labor.

Stages of Labor (pp. 1115–1116)

4. The dilation stage is from the onset of rhythmic, strong contractions until the cervix is fully dilated. The head of the fetus rotates as it descends through the pelvic outlet.

5. The expulsion stage extends from full cervical dilation until birth of the infant.

6. The placental stage is the delivery of the afterbirth (the placenta and attached fetal membranes).

28.7 An infant's extrauterine adjustments include taking the first breath and closure of vascular shunts (p. 1116)

1. The infant's Apgar score is recorded immediately after birth.

2. After the umbilical cord is clamped, carbon dioxide accumulates in the infant's blood, causing respiratory centers in the brain to trigger the first inspiration.

3. Once the lungs are inflated, breathing is eased by the presence of surfactant, which decreases the surface tension of the alveolar fluid.

4. During transition, the first 8 hours after birth, the infant is physiologically unstable and adjusting. After stabilizing, the infant wakes approximately every 3–4 hours in response to hunger.

28.8 Lactation is milk secretion by the mammary glands in response to prolactin (pp. 1116–1119)

1. The breasts are prepared for lactation during pregnancy by high blood levels of estrogens, progesterone, and human placental lactogen.

2. Colostrum, a premilk fluid, is a fat-poor fluid that contains more protein, vitamin A, and minerals than does true milk. It is produced toward the end of pregnancy and for the first two to three days after birth.

3. True milk is produced around day 3 in response to suckling, which stimulates the hypothalamus to prompt anterior pituitary release of prolactin and posterior pituitary release of oxytocin. Prolactin stimulates milk production. Oxytocin triggers milk let-down. Continued breast-feeding is required for continued milk production.

4. At first, ovulation and menses are absent or irregular during nursing, but in most women the ovarian cycle is eventually reestablished while still nursing.

28.9 Assisted reproductive technology may help an infertile couple have offspring (p. 1119)

1. Assisted reproductive technology (ART) procedures assist infertile couples to bear children. Among the most-used techniques are IVF, ZIFT, and GIFT. IVF and ZIFT attempt to fertilize harvested oocytes *in vitro* and then return the embryo or zygote to the woman's body. GIFT utilizes *in vivo* methods—sperm and oocytes are transferred together to the woman's uterine tubes.

REVIEW QUESTIONS

To access additional practice questions using your smartphone, tablet, or computer: MasteringA&P® > Study Area > Practice Tests & Quizzes

Level 1 Remember/Understand

(Some questions have more than one correct answer. Select the best answer or answers from the choices given.)

1. Indicate whether each of the following statements is describing **(a)** cleavage or **(b)** gastrulation.
   ____ **(1)** period during which a morula forms
   ____ **(2)** period when vast amounts of cell migration occur
   ____ **(3)** period when the three embryonic germ layers appear
   ____ **(4)** period during which the blastocyst is formed

2. Most systems are operational in the fetus by four to six months. Which system is the exception to this generalization, affecting premature infants? **(a)** the circulatory system, **(b)** the respiratory system, **(c)** the urinary system, **(d)** the digestive system.

3. The outer layer of the blastocyst, which later attaches to the uterus, is the **(a)** decidua, **(b)** trophoblast, **(c)** amnion, **(d)** embryoblast.

4. Match each adult structure in column B with the embryonic structure it derives from in column A.

Column A	Column B
____ **(1)** notochord	**(a)** kidney
____ **(2)** ectoderm (not neural tube)	**(b)** peritoneal cavity
____ **(3)** intermediate mesoderm	**(c)** pancreas, liver
____ **(4)** lateral plate mesoderm	**(d)** parietal serosa, most tissues of limbs
____ **(5)** sclerotome	**(e)** nucleus pulposus
____ **(6)** coelom	**(f)** hair and epidermis
____ **(7)** neural tube	**(g)** brain
____ **(8)** endoderm	**(h)** ribs and vertebrae

5. In the fetus, the ductus arteriosus carries blood from **(a)** the pulmonary artery to the pulmonary vein, **(b)** the liver to the inferior vena cava, **(c)** the right ventricle to the left ventricle, **(d)** the pulmonary trunk to the aorta.

6. Which of the following changes occur in the baby's cardiovascular system after birth? **(a)** umbilical arteries and vein become fibrosed, **(b)** pulmonary circulation begins to function, and pressure in the left side of the heart increases, **(c)** the ductus venosus becomes obliterated, as does the ductus arteriosus, **(d)** all of these.

7. The umbilical vein carries **(a)** waste products to the placenta, **(b)** oxygen and food to the fetus, **(c)** oxygen and food to the placenta, **(d)** oxygen and waste products to the fetus.

8. The germ layer from which the epidermis and brain are derived is the **(a)** ectoderm, **(b)** endoderm, **(c)** mesoderm.

9. The most important hormone in initiating and maintaining lactation after birth is **(a)** estrogen, **(b)** FSH, **(c)** prolactin, **(d)** oxytocin.

10. The initial stage of labor, during which the neck of the uterus is stretched, is the **(a)** dilation stage, **(b)** expulsion stage, **(c)** placental stage.

11. Cite two problems with a breech presentation.

12. What factors are believed to bring about uterine contractions at the termination of pregnancy?

13. During Mrs. Li's labor, the obstetrician decided that it was necessary to perform an episiotomy. What is an episiotomy, and why is it done?

Level 2 Apply/Analyze

14. What is the function of hCG and why is it not important after the first trimester of pregnancy?

15. Cleavage is an embryonic event that mainly involves mitotic divisions. How does cleavage differ from mitosis occurring during life after birth, and what are its important functions?

16. The life span of the ovarian corpus luteum is extended for nearly three months after implantation, but otherwise it deteriorates. **(a)** Explain why this is so. **(b)** Explain why it is important that the corpus luteum remain functional following implantation.

17. Why is it that only one sperm out of the hundreds (or thousands) available enters the oocyte?

18. What is the function of gastrulation?

19. Explain how the flat embryonic disc takes on the cylindrical shape of a tadpole.

20. Jenna, a freshman in your dormitory, tells you she just discovered that she is three months pregnant. She recently bragged that since she came to college she has been drinking alcohol heavily and experimenting with every kind of recreational drug she could find. From the following, select the advice you would give her, and explain why it is the best choice. **(a)** She must stop taking drugs, but they could not have affected her fetus during these first few months of her pregnancy. **(b)** Harmful substances usually cannot pass from mother to embryo, so she can keep using drugs. **(c)** There could be defects in the fetus, so she should stop using drugs and visit a doctor as soon as possible. **(d)** If she has not taken any drugs in the last week, she is okay.

21. While Mark was cramming for his anatomy test, he read that some parts of the mesoderm become segmented. He suddenly realized that he could not remember what segmentation is. Define segmentation, and give two examples of segmented structures in the embryo.

Level 3 Evaluate/Synthesize

22. Fertilization involves much more than a mere restoration of the diploid chromosome number. **(a)** What does the process of fertilization entail on the part of both the egg and sperm? **(b)** What are the effects of fertilization?

23. The placenta is a marvelous, but temporary, organ. Starting with a description of its formation, show how it is an intimate part of both fetal and maternal anatomy and physiology during the gestation period.

24. A woman in substantial pain called her doctor and explained (between sobs) that she was about to have her baby "right here." The doctor calmed her and asked how she had come to that conclusion. She said that her water had broken and that her husband could see the baby's head. **(a)** Was she right? If so, what stage of labor was she in? **(b)** Do you think that she had time to make it to the hospital 60 miles away? Why or why not?

25. Claire is a heavy smoker and has ignored a friend's advice to stop smoking during her pregnancy. On the basis of what you know about the effect of smoking on physiology, describe how Claire's smoking might affect her fetus.

CLINICAL CASE STUDY

38-Year-Old Female with Preeclampsia

Ms. Johansen, 38, is 26 weeks pregnant with her first child. She was diagnosed with preeclampsia (see p. 1114) three weeks ago. Antihypertensive drugs (◄ p. 724) and daily home care monitoring were initiated. However, her blood pressure is continuing to rise and is now 162/108 mm Hg. She complains of a constant headache, exhibits edema, and her urinalysis is positive for protein.

Her doctor has her admitted and orders further blood tests to assess Ms. Johansen's platelet level and liver function. "We may have to deliver because that's the only cure for preeclampsia," she says to you during your clinical rotation in obstetrics. "But if we do, then the baby will have problems. It's a tough decision because we have to balance both the mother's and the baby's health."

1. Although blood volume increases during pregnancy, blood pressure normally decreases. You know that blood pressure depends on cardiac output (CO) and total peripheral resistance (TPR) (◄ **Figure 19.10, p. 719**). How does increased blood volume affect CO, and what factor(s) must change so that blood pressure decreases?

2. One of your fellow nurses says that preeclampsia is characterized by vasoconstriction. Does this make sense? Why or why not?

3. **+ NCLEX-STYLE** Her doctor wants to put delivery off as long as possible because premature babies have multiple problems. One possible problem is that the ductus arteriosus does not close. Why would this be a problem?
   a. If the ductus arteriosus doesn't close, the liver will not get an adequate blood supply.
   b. If the ductus arteriosus doesn't close, blood will continue to perfuse the umbilical cord with a risk of fatal hemorrhage.
   c. If the ductus arteriosus doesn't close, the right atrium will continually be overloaded with blood.
   d. If the ductus arteriosus doesn't close, the lungs will continually receive too much blood.

Ms. Johansen's tests indicate liver and kidney damage. Her doctor schedules delivery for tomorrow morning and orders an injection of cortisol for Ms. Johansen.

4. What roles does cortisol play in a developing fetus and how would an injection of cortisol help a premature infant?

5. **+ NCLEX-STYLE** Cortisol prompts the fetal lungs to begin synthesizing surfactant. Why is this important?
   a. Surfactant reduces surface tension in the alveoli, allowing the alveoli to inflate.
   b. Surfactant causes vasodilation, allowing blood to flow through the lungs.
   c. Surfactant is necessary to protect the cells lining the respiratory tract from drying out.
   d. Surfactant attracts white blood cells into the lungs to protect them against pathogens.

28

29 Heredity

In this chapter, you will learn that

Genes direct human development and characteristics

by asking	next looking at	then learning about	finally asking
29.1 What terms are used to describe genetics?	**29.2 Sexual sources of genetic variation**	**29.3 Patterns of inheritance**	**29.6 How to detect genetic disorders**

and asking

29.4 How do environmental factors affect gene expression?

then exploring

29.5 Other inheritance mechanisms

The wondrous growth and development of a new individual is guided by the gene-bearing chromosomes it receives from its parents in egg and sperm. As we described in Chapter 3, individual *genes*, or DNA segments, contain the "recipes" or genetic blueprints for proteins. Many of these proteins are enzymes that dictate the synthesis of virtually all the body's molecules. Consequently, gene expression leads to observable traits such as your hair color, sex, blood type, and so on.

Genes, however, do not act as free agents. As you will see, a gene's ability to prompt the development of a specific trait can be enhanced or inhibited by interactions with other genes, as well as by environmental factors.

The science of genetics, which studies the mechanism of heredity (*genes* = origin), is relatively young, but our understanding of how human genes act and interact has advanced considerably since Gregor Mendel first proposed the basic principles of heredity in the mid-1800s. Mendel studied characteristics that vary in an either-or fashion, which is easier to understand than the more-or-less fashion in which most human traits vary.

The urge to understand human inheritance is powerful. Our ability to manipulate DNA enables us to engineer genes to examine their expression. Research in this area has tremendous promise for more sophisticated genetic screening, and for developing personalized drugs to treat or cure disease. Although we touch upon some of these newer topics, we will concentrate on the principles of heredity discovered by Mendel more than a century ago.

29.1 Genes are the vocabulary of genetics

Learning Outcomes

▶ Define allele.
▶ Differentiate between genotype and phenotype.

The nuclei of all human cells except gametes contain the diploid number of chromosomes (46), consisting of 23 pairs of homologous chromosomes. Recall that *homologous chromosomes* are pairs of chromosomes—one from the father (sperm) and one from the mother (egg)—that look similar and carry genes for the same traits, but do not necessarily bring about the same expressions of those traits. Two of the 46 chromosomes are **sex chromosomes** (X and Y), which determine genetic sex

(male = XY; female = XX). The other 44 are the 22 pairs of **autosomes** that guide the expression of most other traits.

The complete human **karyotype** (kar′e-o-tīp), or diploid chromosomal complement displayed in homologous pairs, is illustrated in **Figure 29.1** ③. The diploid **genome** (je′nōm) is our genetic (DNA) makeup, and represents two sets of genetic instructions—one from the egg and the other from the sperm.

Gene Pairs (Alleles)

Because chromosomes are paired, it follows that the genes in them are paired as well. Consequently, each of us receives *two* genes, one from each parent (for the most part), that interact to dictate each trait. Different versions of the same genes are called **alleles** (ah-lēlz′). They occur at the same *locus* (location) on homologous chromosomes. Alleles may code for the same or for alternative forms of a given trait. Suppose, for example, that a single gene codes for the tightness of thumb ligaments. One allele might code for tight thumb ligaments and the other for loose ligaments (the double-jointed thumb condition). When the two alleles controlling a trait are the same, a person is said to be **homozygous** (ho-mo-zi′gus) for that gene. When the two alleles are different, the individual is **heterozygous** (het″er-o-zi′gus) for the gene.

  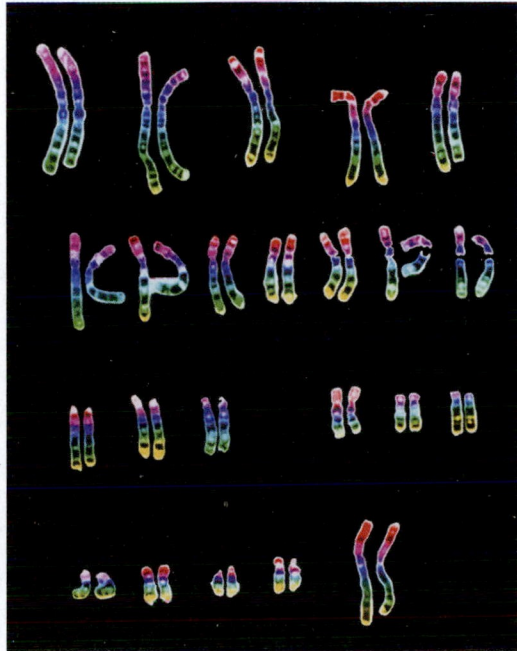

① **Chromosomes are photographed.**
• Division of lymphocytes is arrested in metaphase by a drug (colchicine).
• Chromosomes are dispersed and stained for imaging.

② **Chromosomes are analyzed and manipulated.**
• The chromosomes are electronically arranged into homologous pairs according to size and structure.
• The pattern of stained bands helps identify specific chromosomes and parts of chromosomes.

③ **The resulting display is a karyotype.**
• A karyotype is an organized display of condensed chromosomes.
• The karyotype can be screened for defective chromosomes or abnormal chromosome numbers associated with certain diseases such as Down syndrome.

Figure 29.1 Preparing a karyotype.

Sometimes, one allele masks or suppresses the expression of its partner. Such an allele is said to be **dominant**, whereas the allele that is masked is said to be **recessive**. By convention, a dominant allele is represented by a capital letter (for example, *J*), and a recessive allele by the lowercase form of the same letter (*j*). Dominant alleles are expressed, or make themselves "known," when they are present in either single or double dose. For recessive alleles to be expressed, they must be present in double dose, that is, the homozygous condition. Returning to our thumb example, a person whose genetic makeup includes either the gene pair *JJ* (the homozygous dominant condition) or the gene pair *Jj* (the heterozygous condition) would have double-jointed thumbs. The combination *jj* (the homozygous recessive condition) would be needed to produce tight thumb ligaments.

Keep in mind that dominant and recessive do not refer to how frequently an allele occurs in the population. A dominant allele that is lethal, for example, will be very rare. Even dominant alleles for relatively harmless conditions are often not widespread.

Genotype and Phenotype

A person's genetic makeup is referred to as his or her **genotype** (jēn′o-tīp). The expression of that genotype in the body is called one's **phenotype** (fe′no-tīp). In our example, the double-jointed condition is the phenotype produced by a genotype of *JJ* or *Jj*.

Check Your Understanding

1. What term refers to chromosomes other than our sex chromosomes?
2. Is an allele represented by a capital letter presumed to be dominant or recessive?

3. Adam is homozygous for the dominant alleles *HH*, *CC*, and *LL*, and heterozygous for *Bb* and *Kk*. He is blond and blue-eyed and has a very hairy chest. Which of these descriptions refer to his phenotype?

4. **MAKE CONNECTIONS** The drawings below show cells in various stages of the cell cycle (not in order). Identify the stage that would be used to make a karyotype. Why are chromosomes in this stage used? (Hint: See Chapter 3.)

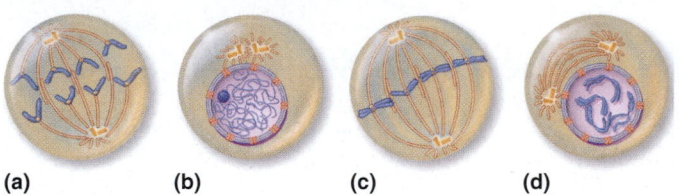

(a)   (b)   (c)   (d)

For answers, see Answers Appendix.

29.2 Genetic variation results from independent assortment, crossing over, and random fertilization

Learning Outcome

▶ Describe events that lead to genetic variability of gametes.

Before we examine how genes interact, let us consider why (with the possible exception of identical twins) each of us is one of a kind, with a unique genotype and phenotype. This variability reflects three events:

- Independent assortment of chromosomes
- Crossing over of homologous chromosomes
- Random fertilization of eggs by sperm

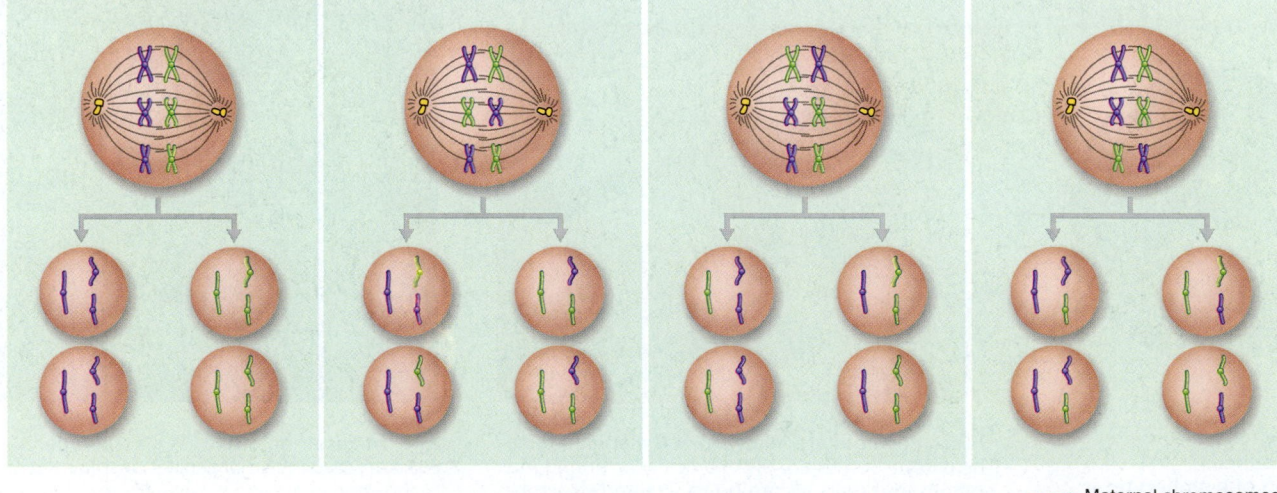

--- Maternal chromosomes
--- Paternal chromosomes

Figure 29.2 Gamete variability resulting from independent assortment. During metaphase of meiosis I, the tetrads of homologous chromosomes align independently of other tetrads. The large circles depict the possible alignments in a parent cell having a diploid number of 6. The small circles show the gametes arising from each alignment. Some gametes contain all maternal or all paternal chromosomes; others have various combinations of maternal and paternal chromosomes.

Chromosome Segregation and Independent Assortment

Recall that during meiosis, each pair of replicated homologous chromosomes synapses during meiosis I, forming a tetrad (◀ **p. 1044**). This happens during both spermatogenesis and oogenesis. Because chance determines how the tetrads align (line up) on the meiosis I metaphase spindle, maternal and paternal chromosomes are randomly distributed to daughter nuclei.

As illustrated in **Figure 29.2**, this simple event leads to an amazing amount of variation in gametes. The cell in our example has a diploid number of 6, and so three tetrads form. As you can see, the possible combinations of alignments of the three tetrads result in eight gamete possibilities. Because the way each tetrad aligns is random, and because many cells are undergoing meiosis simultaneously, each alignment and each type of gamete occurs with the same frequency as all others.

Two important points about metaphase of meiosis I:

- The two alleles determining each trait are **segregated**, which means that they are distributed to different gametes. Errors in this important process have been linked to cancer progression and Down syndrome.

- Alleles on different pairs of homologous chromosomes are distributed independently of each other. The net result is that each gamete has a single allele for each trait, and that allele represents only one of the two possible parental alleles.

The number of different gamete types resulting from this **independent assortment** of homologues during meiosis I can be calculated for any genome from the formula 2^n, where n is the number of homologous pairs. In our example, $2^n = 2^3$ ($2 \times 2 \times 2$), for a total of 8 different gamete types.

Note that the number of gamete types increases dramatically as the chromosome number increases. A cell with six pairs of homologues would produce 2^6, or 64, kinds of gametes. In a man's testes, the number of gamete types that can be produced on the basis of independent assortment alone is 2^{23}, or about 8.5 million—an incredible variety. The number of different gamete types produced in a woman's ovaries is substantially less because her ovaries complete at most 500 reduction divisions in her lifetime. Still, each ovulated oocyte will most likely be novel genetically because of independent assortment.

Crossing Over of Homologues and Gene Recombination

Another aspect of gamete variation results from the **crossing over** and exchange of chromosomal parts during meiosis I. Genes are arranged linearly along a chromosome's length, and genes on the same chromosome are said to be **linked** because they are transmitted as a unit to daughter cells during mitosis. However, during meiosis, chromosomes can break and precisely exchange gene segments with their homologous counterparts (◀ **p. 1044**). This exchange gives rise to **recombinant chromosomes** that have mixed contributions from each parent.

In the hypothetical example shown in **Figure 29.3**, the genes for hair and eye color are linked. The paternal chromosome contains alleles coding for blond hair and blue eyes, and

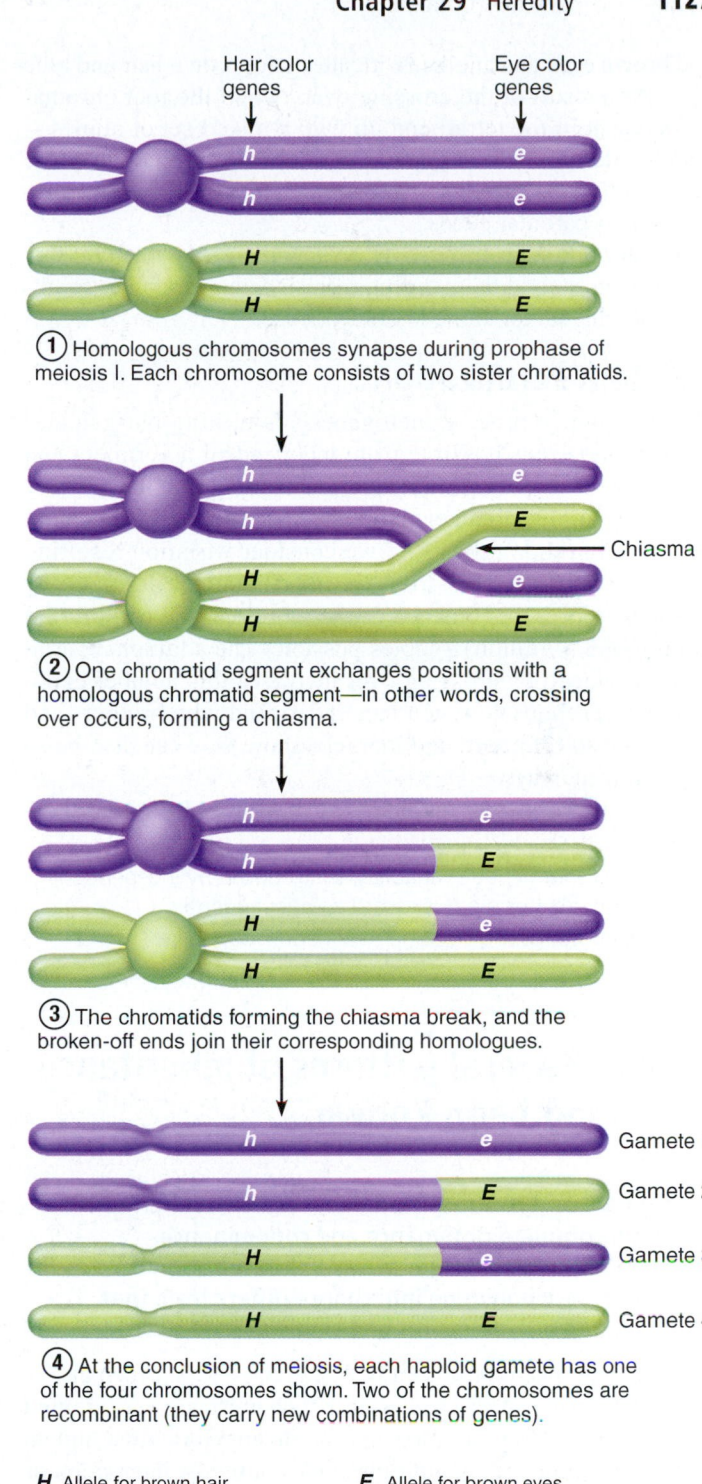

Hair color genes   Eye color genes

① Homologous chromosomes synapse during prophase of meiosis I. Each chromosome consists of two sister chromatids.

Chiasma

② One chromatid segment exchanges positions with a homologous chromatid segment—in other words, crossing over occurs, forming a chiasma.

③ The chromatids forming the chiasma break, and the broken-off ends join their corresponding homologues.

Gamete 1
Gamete 2
Gamete 3
Gamete 4

④ At the conclusion of meiosis, each haploid gamete has one of the four chromosomes shown. Two of the chromosomes are recombinant (they carry new combinations of genes).

H Allele for brown hair   **E** Allele for brown eyes
h Allele for blond hair   **e** Allele for blue eyes

Paternal chromosome ⎤
               ⎬ Homologous pair
Maternal chromosome ⎦

Figure 29.3 Crossing over and genetic recombination. These meiosis I events increase genetic variability in the gametes. For simplicity, only two chromatids are shown taking part in crossing over. Multiple chiasmata result in more complex patterns.

the maternal alleles code for brown hair and brown eyes. In the **chiasma**, or *crossover*, shown, the break occurs between these linked genes, resulting in one gamete with alleles for blond hair and brown eyes and another with alleles for brown hair and blue eyes. As a result of the crossing over, two of the four chromatids present in the tetrad end up with a mixed set of alleles—some maternal and some paternal. This means that when the chromatids segregate, each gamete will receive a unique combination of parental genes.

Because humans have 23 tetrads, with at least one chiasma going on in each of them during meiosis I, the variability resulting from this factor alone is tremendous.

Random Fertilization

At any point in time, gametogenesis is turning out gametes with all variations possible from independent assortment and random crossing over. Fertilization compounds this variety because a single human egg will be fertilized by a single sperm on a totally random basis. If we consider variation resulting only from independent assortment and random fertilization, any offspring represents one out of the more than 72 trillion (8.5 million × 8.5 million) zygotes possible. The additional variation introduced by crossing over increases this number exponentially. Perhaps now you can understand why brothers and sisters are so different, and marvel at how they can also be so alike in many ways.

Check Your Understanding

5. We said that genetic variability is introduced by independent assortment. Just what is assorting independently?

6. How does crossing over increase genetic variability?

For answers, see Answers Appendix.

29.3 Several patterns of inheritance have long been known

Learning Outcomes

▶ **Compare and contrast dominant-recessive inheritance with incomplete dominance and codominance.**

▶ **Describe the mechanism of sex-linked inheritance.**

▶ **Explain how polygene inheritance differs from that resulting from the action of a single pair of alleles.**

A few human phenotypes can be traced to a single gene pair (as we will describe shortly), but most such traits are very limited in nature, or reflect variation in a single enzyme. Most human traits are determined by multiple alleles or by the interaction of several gene pairs.

Dominant-Recessive Inheritance

Dominant-recessive inheritance reflects the interaction of dominant and recessive alleles. A simple diagram, called the **Punnett square**, is used to figure out, for a single trait, the possible gene combinations that would result from the mating of parents of known genotypes (**Figure 29.4**). In the example shown, both parents have normal pigmentation but carry the recessive allele for albinism (*a*), the lack of the pigment melanin (◀ p. 169). This means each parent is heterozygous for this trait and has the genotype *Aa*. The alleles of one parent are

Figure 29.4 Genotype and phenotype probabilities resulting from a mating of two heterozygous parents. The Punnett square shows all possible combinations of a set of alleles in the zygote. In this example, the *A* allele is dominant and determines pigmentation (green background); the *a* allele is recessive. The photo shows a boy with albinism and his normally pigmented sister.

written along one side of the Punnett square, and the alleles for the other parent are shown along an adjacent side. The alleles are then combined down and across to determine the possible gene combinations (genotypes) and their expected frequency in the offspring of these two parents.

As the completed Punnett square shows, the probability of these parents producing a homozygous dominant child (*AA*) is 25% (1 out of 4); of producing a heterozygous child (*Aa*), 50% (2 out of 4); and of producing a homozygous recessive child, 25% (1 out of 4). The *AA* and *Aa* offspring will have normal pigmentation. Only the *aa* offspring will have albinism.

The Punnett square predicts only the *probability* of a particular genotype (and phenotype). The larger the number of offspring, the greater the likelihood that the ratios will conform to the predicted values—just as the chances of getting heads half the time and tails half the time increase with the number of tosses of a coin. If we toss only twice, we may well get heads both times. Likewise, if the couple in our example had only two children, it would not be surprising if both children had the genotype *Aa*.

Table 29.1 Traits Determined by Simple Dominant-Recessive Inheritance

PHENOTYPE DUE TO EXPRESSION OF:	
DOMINANT GENES (ZZ OR Zz)	**RECESSIVE GENES (zz)**
Syndactyly (webbed digits)	Normal digits
Achondroplasia (heterozygous: dwarfism; homozygous: lethal)	Normal endochondral ossification
Huntington's disease	Absence of Huntington's disease
Normal skin pigmentation	Albinism
Absence of Tay-Sachs disease	Tay-Sachs disease
Absence of cystic fibrosis	Cystic fibrosis

What are the chances of having two children of the same genotype? To determine the probability of two events happening in succession, we must multiply the probabilities of the separate events happening. The probability of getting heads in one coin toss is 1/2, so the probability of getting two heads in a row is $1/2 \times 1/2 = 1/4$. Remember that the production of each child, like each coin toss in a series, is an *independent event* that is not influenced by the production of any other child by the same couple. If you get heads on the first toss, the chance of getting heads the second time is still 1/2. Likewise, if our couple's first child is an *aa*, they still have a 1/4 chance of getting an *aa* the next time.

Dominant Traits

Most easily observable traits in humans (such as eye color and hair color) are controlled by multiple genes and influenced by the environment. Simple dominant-recessive patterns of inheritance are rare. Indeed, many human traits once thought to be determined by single-gene dominant-recessive inheritance (such as freckles, dimples, a widow's peak, double-jointed thumbs, the ability to tongue roll or taste phenylthiocarbamide) don't actually follow this pattern. This pattern of inheritance is most clearly seen in disorders such as those listed in **Table 29.1**.

Disorders caused by dominant genes are uncommon because *lethal dominant genes* are almost always expressed and result in the death of the embryo, fetus, or child. However, in some dominant disorders the person is less impaired or at least survives long enough to reproduce. One example is *Huntington's disease*, a fatal nervous system disease involving degeneration of the basal nuclei (◄ p. 444). It involves a *delayed-action gene* that is expressed when the affected individual is about 40. Offspring of a parent with Huntington's disease have a 50% chance of inheriting the lethal gene. (The parent is heterozygous, because the dominant homozygous condition is lethal to the fetus.) Many children of such parents are, as adults, opting not to become parents themselves.

Recessive Traits

Some examples of recessive inheritance are the more desirable genetic condition. For example, normal endochondral ossification (◄ p. 184) is dictated by recessive alleles, whereas

achondroplasia (abnormal endochondral ossification; ◄ p. 195) is prescribed by dominant alleles. However, many if not most genetic disorders are inherited as simple recessive traits. These include conditions as different as *albinism*; *cystic fibrosis*, a condition of excessively thick mucus production that impairs lung and pancreatic functioning (◄ p. 861); and *Tay-Sachs disease*, a disorder of brain lipid metabolism caused by an enzyme deficit that shows itself a few months after birth (◄ p. 87).

Recessive genetic disorders are more frequent than disorders caused by dominant alleles because those who carry a *single* recessive allele for a recessive genetic disorder do not themselves express the disease. However, they can pass the gene on to offspring and so are called *carriers* of the disorder.

Incomplete Dominance

In dominant-recessive inheritance, one allele variant completely masks the other. Some traits, however, exhibit **incomplete dominance**. In such instances, the heterozygote has a phenotype intermediate between those of homozygous dominant and homozygous recessive individuals.

Perhaps the best human example is inheritance of the *sickling gene (s)*, which causes a substitution of one amino acid in the beta chain of hemoglobin. Hemoglobin molecules containing the abnormal beta chains crystallize when blood oxygen levels are low, causing the erythrocytes to assume a sickle shape (◄ Figure 17.8b, p. 651). Individuals with a double dose of the sickling allele (*ss*) have **sickle-cell anemia** (◄ p. 650), and any condition that lowers their blood oxygen level, such as respiratory difficulty or excessive exercise, can precipitate a *sickle-cell crisis*. The deformed erythrocytes jam up and fragment in small capillaries, causing intense pain.

Individuals heterozygous for the sickling gene (*Ss*) have **sickle-cell trait**. They make both normal and sickling hemoglobin, and as a rule, these individuals are healthy. However, they can suffer a crisis if there is prolonged reduction in blood oxygen levels, as might happen when traveling in high-altitude areas, and they can transmit the sickling gene to their offspring.

Multiple-Allele Inheritance

Although we inherit only two alleles for each gene, some genes exhibit more than two allele forms, leading to a phenomenon called **multiple-allele inheritance**. An example of multiple-allele inheritance in humans is ABO blood types (◄ Table 17.4, p. 664). Three alleles determine the ABO blood types in humans: I^A, I^B, and *i*. Each of us receives two of these. The I^A and I^B alleles are **codominant**, and both are expressed when present, resulting in the AB blood type. The *i* allele is recessive to the other two. Genotypes determining the four possible ABO blood groups are shown in **Table 29.2**.

Table 29.2 ABO Blood Groups

BLOOD GROUP (PHENOTYPE)	GENOTYPE
O	*ii*
A	$I^A I^A$ or $I^A i$
B	$I^B I^B$ or $I^B i$
AB	$I^A I^B$

Sex-Linked Inheritance

Inherited traits determined by genes on the sex chromosomes are said to be **sex-linked**. The X and Y sex chromosomes are not homologous in the true sense. The Y, which contains the gene that determines maleness, is much smaller than the X chromosome (**Figure 29.5**). The X bears over 1400 genes, and a disproportionately large number of them code for proteins important to brain function.

Figure 29.5 Human sex chromosomes. (9100×)

Because the Y carries only about 200 genes, it lacks many of the genes present on the X. For example, genes coding for certain clotting factors, cone pigments, and even testosterone receptors are present on X but not on Y. A gene found only on the X chromosome is said to be **X-linked**.

Only relatively short regions at either end of the Y chromosome (corresponding to about 5% of the Y's DNA) code for nonsexual characteristics corresponding to those on the X.

Those homologous regions are the only areas that can participate in crossing over with the X.

Two examples of X-linked conditions are hemophilia (◄ p. 662) and red-green color blindness (◄ p. 569). When a male inherits an X-linked recessive allele for one of these conditions, its expression is never masked or damped, because there is no corresponding allele on his Y chromosome. Consequently, the recessive gene is always expressed, even though it is present only in single dose. In contrast, females must have two X-linked recessive alleles to express such a disorder, and as a result, very few females exhibit any X-linked conditions.

X-linked traits are typically passed from mother to son, never from father to son, because males receive no X chromosome from their father. Of course, the mother can also pass the recessive allele to her daughter, but unless the daughter receives another such allele from her father, she will not express the trait (**Figure 29.6**).

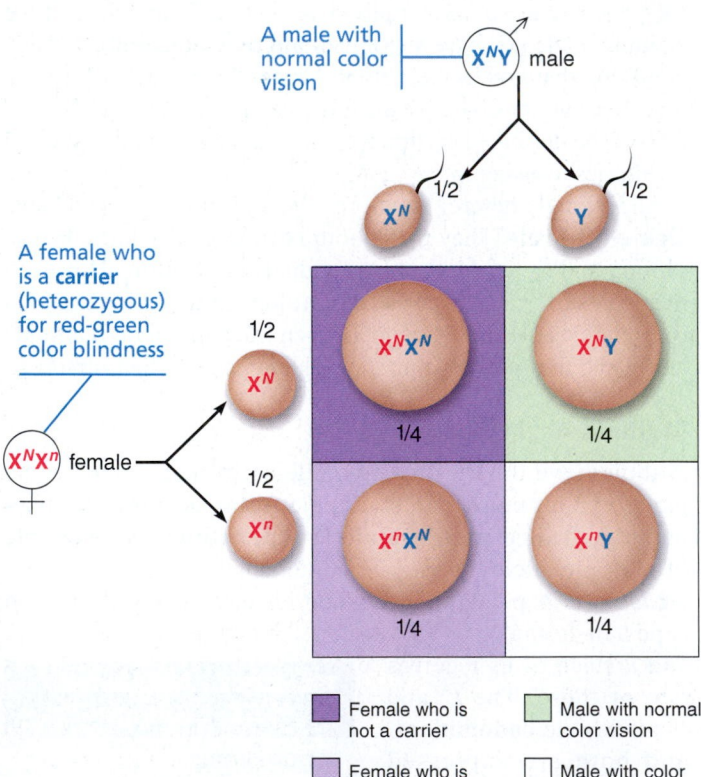

Female who is not a carrier

Male with normal color vision

Female who is a carrier

Male with color blindness

Figure 29.6 Sex-linked inheritance. If a female heterozygous for red-green color blindness and a male with normal color vision have offspring, there is a 50% chance that their sons will be color blind and a 50% chance that their daughters will be carriers. The *N* allele determines normal color vision and the *n* allele is recessive.

Figure 29.7 Simplified model for polygene inheritance of skin color based on three gene pairs. Alleles for dark skin are incompletely dominant over those for light skin. Each dominant gene (*A*, *B*, *C*) contributes 1 "unit" of darkness (indicated by a dark dot) to the phenotypes.

Here the parents are homozygotes at opposite ends of the phenotype range. Each child inherits 3 "units" of darkness (is a heterozygote) and has intermediate pigmentation. When they grow up, if they mate with people with the same alleles, their offspring (second generation) may have a wide variation in pigmentation, as shown by the plotted distribution of skin colors.

Polygene Inheritance

So far, we have considered only traits inherited by mechanisms of classical Mendelian genetics, which are fairly easy to understand, and such traits typically have two, or perhaps three, alternate forms. However, most phenotypes depend on several gene pairs at different locations acting in tandem. Such **polygene inheritance** results in *continuous*, or *quantitative*, phenotypic variation between two extremes and explains many human characteristics. Examples of polygene traits in humans include skin color, height, metabolic rate, and intelligence.

Skin color, for instance, is controlled by at least three separately inherited genes, each existing in two allelic forms: *A, a*; *B, b*; *C, c*. The *A, B*, and *C* alleles confer dark skin pigment, and their effects are additive. The *a, b*, and *c* alleles confer pale skin tone. An individual with an *AABBCC* genotype would be about as dark-skinned as a human can get, while an *aabbcc* person would be very fair. However, when individuals heterozygous for at least one of these gene pairs mate, a broad range of pigmentation is possible in their offspring. Such polygene inheritance results in a distribution of genotypes and phenotypes that, when plotted, yields a bell-shaped curve (**Figure 29.7**).

Check Your Understanding

7. How does incomplete dominance differ from codominance?
8. How can you explain why parents of average height can produce very tall or very short offspring?
9. PREDICT Which lethal allele would you expect to be more widespread in a population—one that is dominant or one that is recessive? Why?
10. DRAW Draw a Punnett square showing the potential offspring of a man with hemophilia and a woman who is a carrier for the same type of X-linked hemophilia. Summarize the results.

For answers, see Answers Appendix.

29.4 Environmental factors may influence or override gene expression

Learning Outcome

▶ Provide examples illustrating how gene expression may be modified by environmental factors.

In many situations environmental factors override or at least influence gene expression. Our genotype (discounting mutations) is as unchanging as a rock, but our phenotype is more like clay. If this were not the case, we would never get a tan, women bodybuilders would never be able to develop bulging muscles, and there would be no hope for treating genetic disorders.

Sometimes, maternal factors such as drugs or pathogens alter normal gene expression during embryonic development. Take, for example, the case of babies in the 1960s whose mothers took the sedative thalidomide to alleviate morning sickness. This drug caused the embryos to develop phenotypes (flipperlike appendages) not directed by their genes. (The same phenotype can also be caused by a genetic mutation.) Such environmentally produced phenotypes that mimic conditions that may be caused by genetic mutations (permanent transmissible changes in the DNA) are called **phenocopies**.

Equally significant are environmental factors that may influence genetic expression after birth, such as the effect of poor infant nutrition on brain growth, general body development, and height. In this way, a person with "tall genes" can be stunted by insufficient nutrition. Furthermore, part of a gene's environment consists of the influence of other genes. For example, hormonal deficits during childhood can lead to abnormal skeletal growth and proportions, as in congenital hypothyroidism, a type of dwarfism resulting from insufficient thyroid hormone secretion.

Check Your Understanding

11. Which of the following factors may alter gene expression? Other genes, measles in a pregnant woman, lack of key nutrients in the diet.

For answers, see Answers Appendix.

29.5 Factors other than nuclear DNA sequence can determine inheritance

Learning Outcomes

▶ Describe how RNA-only genes and epigenetic marks affect gene expression.
▶ Describe the basis of extranuclear (mitochondria-based) genetic disorders.

Mendel's writings underlie mainstream thinking about heredity, but some genetic outcomes do not fit his rules. Among these nontraditional types of inheritance are influences due to RNA-only genes, to chemical groups attached to DNA or histone proteins, and to *extranuclear inheritance* conferred by mitochondrial DNA.

Beyond DNA: Regulation of Gene Expression

Our genome is a biochemical system of awesome complexity, with three basic levels of controls. The protein-coding genes that we have been describing to this point only make up the first level and account for less than 2% of the DNA of a human cell. This is the part of the genome traditionally considered to be a "blueprint" for protein structure.

The second and third layers of information important in directing development lie elsewhere—in the noncoding DNA and even totally outside the DNA sequences. It takes relatively few genes to build a human being. Our complexity is the result of the mechanisms that control the expression of genes, especially during growth and differentiation. So what are these other regulatory systems?

Small Noncoding RNAs

The second layer of control consists of the abundant "RNA-only genes." About 80% of the genome, which was once believed to be "junk DNA," forms a parallel regulatory system. This system generates a variety of *small noncoding RNAs* (*sncRNAs*): microRNAs, small interfering RNAs (◄ p. 108), and piwi-interacting RNAs. These small RNA molecules are mobile "control freaks" that act in similar ways that generally reduce the expression of particular genes. All of their mechanisms depend on combining with protein enzymes and using complementary RNA-RNA base pairing to identify their targets. Binding of the small RNA–enzyme complex has three possible results:

- Destruction of the target mRNA by dicing it up into smaller fragments
- Inhibition of translation of the target mRNA
- Silencing the gene that codes for the target mRNA by modifying the histone proteins that pack up the DNA coding for that gene (◄ p. 93)

Small RNAs are particularly important during development and in helping us fight off viral infections. They can also protect us from elements of our own DNA that can act as aggressive (jumping) genes called *transposons*, which tend to copy themselves and then insert the copies into distant DNA sites, disabling or hyperactivating genes at those sites.

At present, small RNAs are widely used as research tools to experimentally interfere with the expression of certain genes. This allows researchers to find out what those genes do. Mutations in genes coding for small RNAs have already been linked to conditions as diverse as cancers and schizophrenia. Small RNAs are also on the cutting edge of drug research. Clinical trials are under way using small RNAs to treat a large variety of diseases including macular degeneration, cancers, and various liver diseases.

Epigenetic Marks

Epigenetic marks (*epi* = "over, above") form the third layer of gene controls. Chemical tags such as methyl and acetyl groups bound to DNA segments and to histone proteins (◄ p. 93) determine (mark) whether the DNA is available for transcription (acetylation) or silenced (methylation). This information, stored in the proteins and chemical groups that bind to the DNA, and in the way chromatin is packaged in the cell, is continually changing. Epigenetic marks also account for the inactivation (by methylation) of one of the female's X chromosomes in the early embryo.

Epigenetic marks or lack of them may predispose a cell for transformation from normal to cancerous, and even slight deviations in the epigenetic marks on specific chromosomes can result in devastating human illness.

For most genes, both the maternal and paternal alleles are expressed at the same time. For about 300 genes, this isn't the case and only one of the two parental alleles is expressed in every body cell. The addition of a methyl (—CH$_3$) group during gamete formation silences one of the alleles and marks it as coming from the mother or the father. This process is called **genomic imprinting**. During gametogenesis, the "old" imprinting tags are stripped away and a new set of tags are added to each gene, marking it as maternal (in ova) or paternal (in sperm).

Mutations of imprinted genes may lead to pathology and the mutated allele can have different effects depending on which parent it comes from. For example, people with Prader-Willi syndrome have mild to moderate intellectual disability, and are short and often obese. People with Angelman syndrome have severe intellectual disability, are unable to speak coherently, laugh uncontrollably, and exhibit jerky, lurching movements as if tied to a puppeteer's strings. The manifestations of these two disorders are very different, but the genetic cause is the same—deletion of a particular region of chromosome 15. If the defective chromosome comes from the father, the result is Prader-Willi syndrome; the mother's defective chromosome confers Angelman syndrome.

Unlike the part of the genome that acts as a blueprint for protein structure, epigenetic marks are more like marks on an Etch a Sketch®—a good "shake" during gametogenesis wipes away the message. However, sometimes that "shake" is insufficient to make the message disappear and the current epigenetic marks are passed on to the next generation (inherited) and cause epigenetic changes to occur there.

29

In short, protein-coding genes are not the only instructions to which cells refer. RNA matters, and so do the tiny chemical tags that attach to the chromatin.

Extranuclear (Mitochondrial) Inheritance

Although we have focused on the chromosomal basis of inheritance, remember that not all genes are in the cell's nucleus: 37 genes (referred to as mtDNA) are in mitochondria (◀ **pp. 83–84**). Mitochondrial genes are transmitted to the offspring almost exclusively by the mother because the ovum donates essentially all the cytoplasm in the fertilized egg. Additionally, sperm mitochondrial DNA is selectively destroyed in both the sperm and egg.

A growing list of disorders, all rare, is now being linked to errors (mutations) in mitochondrial genes. Most involve problems with oxidative phosphorylation within the mitochondria, but a few lead to unusual degenerative muscle disorders or neurological problems. Some researchers suggest that Alzheimer's and Parkinson's disease may be among them.

Check Your Understanding

12. What process labels genes as paternal or maternal?

13. What is the source of the genes that confer extranuclear inheritance?

For answers, see Answers Appendix.

29.6 Genetic screening is used to detect genetic disorders `CLINICAL`

Learning Outcomes

▶ **List and explain several techniques used to detect or predict genetic diseases.**

▶ **Describe briefly some approaches of gene therapy.**

Certain congenital abnormalities occur sporadically throughout the population, but can be treated successfully if they are caught at birth. For this reason, all newborns are generally examined and tested for:

- Anatomical disorders [such as developmental dysplasia of the hip (◀ p. 247) and imperforate anus]
- Metabolic diseases [such as phenylketonuria (PKU) (◀ p. 968)]

Genetic screening can be used to detect certain genetic conditions before birth. For example, a woman pregnant after age 35 may wish to know if her baby has trisomy 21, or Down syndrome (see Related Clinical Terms, p. 1135), a chromosome abnormality with a high incidence in children of older mothers. Genetic screening and *genetic counseling* provide information and options for prospective parents not even dreamed of 100 years ago.

Genetic screening and genetic counseling may also be important for adults. For example, adults with a parent who had Huntington's disease may wish to know if they will develop this condition themselves or if they might pass it on to their potential offspring.

Depending on the condition being investigated, screening can occur before conception by carrier recognition, or during fetal testing.

Carrier Recognition

There are two major avenues for identifying carriers of detrimental genes: pedigrees and blood tests. A **pedigree** traces a genetic trait through several generations and helps predict the future (**Figure 29.8**). For prospective parents, a genetic counselor collects phenotype information on as many family members as possible and uses it to construct the pedigree (often called the family tree). By working backward from current individuals and applying the rules of dominant-recessive inheritance, a counselor can deduce the genotypes of the parents and figure out the genotypes of the other individuals in their parents' generation.

Simple blood tests are used to screen for the sickling gene in heterozygotes, and sophisticated *blood chemistry tests* and *DNA probes* can detect the presence of other unexpressed recessive genes. At present, carriers of the Tay-Sachs and cystic fibrosis genes can be identified with such tests.

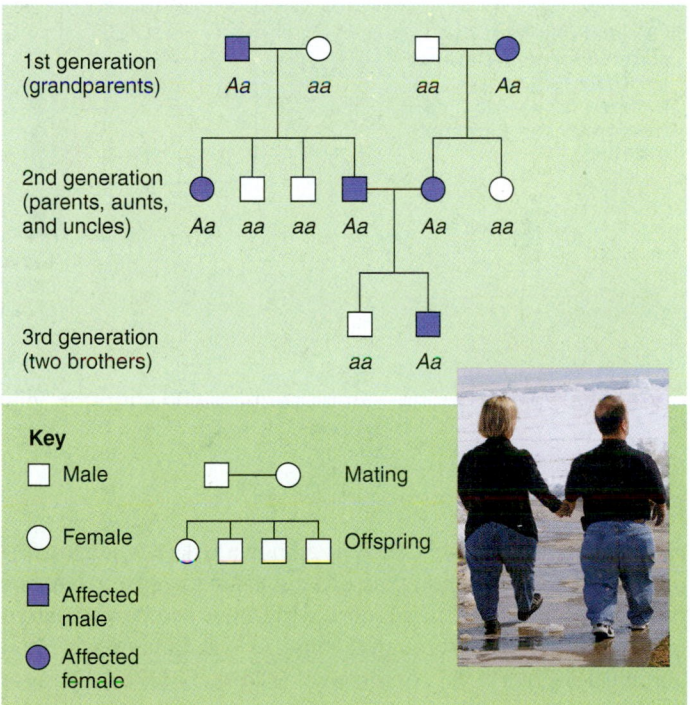

Figure 29.8 Pedigree analysis. This pedigree traces the pattern of inheritance of a dominant allele through three generations. The trait is achondroplasia, a form of dwarfism. All individuals with the dominant allele (*A*) show this trait. Notice that if both parents have achondroplasia (second generation; see photo) their children may or may not be affected. This pattern of inheritance only occurs for dominant alleles. (For this gene, the homozygous condition is lethal.)

Figure 29.9 Fetal testing—amniocentesis and chorionic villus sampling.

Fetal Testing

Fetal testing is used when there is a known risk of a genetic disorder. The most common type of fetal testing is **amniocentesis** (am″ne-o-sen-te′sis), in which a wide-bore needle is inserted into the amniotic sac through the mother's abdominal wall, and about 10 ml of fluid is withdrawn (**Figure 29.9a**). Because there is a chance of injuring the fetus before ample amniotic fluid is present, this procedure is not normally done before the 14th week of pregnancy.

The fluid is checked for enzymes and other chemicals that serve as markers for specific diseases, but most tests are done on the sloughed-off fetal cells in the fluid. These cells are cultured in laboratory dishes over a period of several weeks. Then the cells are examined for DNA markers of genetic disease and karyotyped to check for chromosomal abnormalities (see Figure 29.1).

Chorionic villus sampling (CVS) suctions off bits of the chorionic villi from the placenta for examination (Figure 29.9b). A small tube is inserted through the vagina and cervical canal and guided by ultrasound to an area where a piece of placental tissue can be removed. CVS allows testing at 8 weeks but waiting until after the 10th week is usually recommended. Karyotyping can be done almost immediately on the rapidly dividing chorionic cells, much earlier than in amniocentesis.

Both of these procedures are invasive, and they carry with them an inherent risk to both fetus and mother. (For example, increased fetal risk of finger and toe defects is linked to CVS.) These tests are routinely ordered for pregnant women over 35 (because of the higher risk of Down syndrome), but they are performed on younger women when the probability of finding a severe fetal disorder is greater than the probability of doing harm during the procedure.

Advances in DNA sequencing are about to make both of these invasive procedures obsolete. An easily obtained maternal blood sample contains free-floating maternal and fetal DNA released from the placenta. Maternal blood samples are currently being used in the clinic to test for fetal chromosomal abnormalities such as Down syndrome. The same techniques can provide detailed information about the fetus's entire genome, including mutations that increase the likelihood of certain diseases. The technology already exists and is in increasingly wide use.

Human Gene Therapy

As indicated earlier, the ability to sequence and alter DNA has opened an exciting world to scientists investigating gene expression. Advances in diagnosing human genetic diseases have been followed quickly by new applications of gene therapy to alleviate or even cure the disorders, particularly in cases traced to a single defective gene or protein. Advances in genetic engineering using the new CRISPR-Cas9 genome editing technique have greatly facilitated making specific changes to genes. However, delivering the corrected genes to where they are needed is still a daunting task.

One approach (used successfully in treatment of severe combined immunodeficiency syndromes; ◀ p. 811) is to use a modified virus to deliver a functional copy of the affected gene into cells. Another is to inject the "corrected" DNA directly into the patient's cells. Such therapies have had mixed results in treating cystic fibrosis and muscular dystrophy. In addition, genetic engineering processes raise some thorny ethical, religious, and societal questions: Who will pay? Who determines who will be treated with the new therapies? Are we playing God?

Check Your Understanding

14. Which fetal testing procedure depends on analyzing amniotic fluid?

15. APPLY Mr. and Mrs. Lehman have sought genetic counseling. Mrs. Lehman is concerned because she is unexpectedly pregnant and her husband's brother died of Tay-Sachs disease. She can recall no incidence of Tay-Sachs disease in her own family. Do you think biochemical testing should be recommended to detect the deleterious gene in Mrs. Lehman? Explain your answer.

16. MAKE CONNECTIONS What noninvasive imaging procedure is used to determine some aspects of fetal development? (You may want to check *A Closer Look* in Chapter 1, pp. 16–17, for this.)

For answers, see Answers Appendix.

RELATED CLINICAL TERMS

Deletion Chromosomal aberration in which part of a chromosome is lost.

Down syndrome This condition usually reflects the presence of an extra autosome (trisomy of chromosome 21). The child has slightly slanted eyes, flattened facial features, a large tongue, and a tendency toward short stature and stubby fingers. Some, but not all, affected people are intellectually disabled. Down syndrome individuals have mitochondrial defects, which have been associated with neurodegeneration in other disorders. A distinctive feature of Down syndrome is early onset of Alzheimer's disease. The most important risk factor appears to be advanced maternal (or paternal) age.

Mutation (*mutare* = change) A permanent structural change in a gene. The mutation may or may not affect function, depending on the precise site and nature of the alteration.

Nondisjunction Abnormal segregation of chromosomes during meiosis, resulting in gametes receiving two or no copies of a particular parental chromosome. It is more common in female meiosis than in male meiosis. If the abnormal gamete participates in fertilization, the resulting zygote will have an abnormal chromosomal complement (monosomy or trisomy) for that particular chromosome (as in Down syndrome).

CHAPTER SUMMARY

1. Genetics is the study of heredity and mechanisms of gene transmission.

29.1 Genes are the vocabulary of genetics (pp. 1125–1126)

1. A karyotype is a representation of a complete set of chromosomes; the complete genetic complement is the genome. A person's genome consists of two sets of instructions, one from each parent.

2. Genes coding for the same trait and found at the same locus on homologous chromosomes are called alleles.

3. Alleles may be the same or different in expression. When the allele pair is identical, the person is homozygous for that trait; when the alleles differ, the person is heterozygous.

4. The actual genetic makeup of cells is the genotype; phenotype is the manner in which those genes are expressed.

29.2 Genetic variation results from independent assortment, crossing over, and random fertilization (pp. 1126–1128)

1. During meiosis I of gametogenesis, tetrads align randomly on the metaphase plate, and chromatids are randomly distributed to the daughter cells. This is called independent assortment of the homologues. Each gamete receives only one allele of each gene pair.

2. Each different metaphase I alignment produces a different assortment of parental chromosomes in the gametes, and all combinations of maternal and paternal chromosomes are equally possible.

29

3. During meiosis I, nonsister chromatids of each tetrad cross over at one or more points and exchange corresponding gene segments. The recombinant chromosomes contain new gene combinations, adding to the variability arising from independent assortment.

4. The third source of genetic variation is random fertilization of eggs by sperm.

29.3 Several patterns of inheritance have long been known (pp. 1128–1131)

1. Dominant genes are expressed when present in single or double dose; recessive genes must be present in double dose to be expressed.

2. For traits following the dominant-recessive pattern, the laws of probability predict the outcome of a large number of matings.

3. Genetic disorders more often reflect the homozygous recessive condition than the homozygous dominant or heterozygous condition because dominant genes are expressed and, if they are lethal genes, the pregnancy usually results in miscarriage. Genetic disorders caused by dominant alleles include achondroplasia and Huntington's disease; recessive disorders include cystic fibrosis and Tay-Sachs disease.

4. Carriers are heterozygotes who carry a deleterious recessive gene (but do not express the trait) and have the potential of passing it on to offspring.

5. In incomplete dominance, the heterozygote exhibits a phenotype intermediate between those of the homozygous dominant and recessive individuals. Inheritance of sickle-cell trait is an example of incomplete dominance.

6. Multiple-allele inheritance involves genes that exist in more than two allelic forms in a population. Only two of the alleles are inherited, but on a random basis. Inheritance of ABO blood types is an example of multiple-allele inheritance in which the I^A and I^B alleles are codominant.

7. Traits determined by genes on the X and Y chromosomes are said to be sex-linked. The small Y chromosome lacks most genes present on the X chromosome. Recessive genes located only on the X chromosome are expressed in single dose in males. Examples of X-linked conditions, passed from mother to son, include hemophilia and red-green color blindness.

8. Polygene inheritance occurs when several gene pairs interact to produce phenotypes that vary quantitatively over a broad range. Height, intelligence, and skin pigmentation are examples.

29.4 Environmental factors may influence or override gene expression (p. 1131)

1. Environmental factors may influence the expression of genotype.

2. Maternal factors that cross the placenta may alter expression of fetal genes. Environmentally provoked phenotypes that mimic genetically determined ones are called phenocopies. Nutritional deficits or hormonal imbalances may alter anticipated growth and development during childhood.

29.5 Factors other than nuclear DNA sequence can determine inheritance (pp. 1131–1133)

1. Control of gene expression occurs at three levels: protein-coding genes, small noncoding RNAs, and epigenetic marks. The latter two account for many cases of inherited disease that fail to follow traditional inheritance patterns.

2. Small noncoding RNAs silence genes or prevent their expression, particularly during development or infection by viruses.

3. Epigenetic marks involve attachment of small chemical groups (methyl or acetyl) to DNA or histone proteins. In general, methylation prevents DNA transcription whereas acetylation allows DNA transcription. Genomic imprinting, which involves methylation of certain genes during gametogenesis, confers different effects and phenotypes on maternal and paternal genes. It is reversible, and typically occurs anew each generation.

4. Mitochondrial genes pass to offspring via the ovum and help to determine certain characteristics. Deletions or mutations in mitochondrial genes are responsible for some problems with oxidative phosphorylation and some rare genetic diseases.

29.6 Genetic screening is used to detect genetic disorders (pp. 1133–1135)

1. The likelihood of an individual carrying a deleterious recessive gene may be assessed by constructing a pedigree. Some of these genes can be detected by various blood tests and DNA probes.

2. Amniocentesis is fetal testing based on needle-aspirated samples of amniotic fluid. Fetal cells in the fluid are cultured for several weeks, then examined for chromosomal defects (karyotyped) or for DNA markers of genetic disease. Amniocentesis cannot be performed until the 14th week of pregnancy.

3. Chorionic villus sampling is fetal testing based on a sample of the chorion. This tissue is rapidly mitotic, so karyotyping can be done almost immediately. Samples may be obtained by the 8th week.

4. The ability to sequence and edit DNA has allowed research on diagnosis of genetic disease and on its treatment to surge forward. So far, gene therapy has been particularly useful for correcting single-gene disorders. The most common approach involves transferring a corrected gene via a virus to the affected cells to restore normal function.

REVIEW QUESTIONS

To access additional practice questions using your smartphone, tablet, or computer: MasteringA&P® > Study Area > Practice Tests & Quizzes

Level 1 Remember/Understand

(Some questions have more than one correct answer. Select the best answer or answers from the choices given.)

1. Match each of the following key terms (a–i) with its description below:

   Key: **(a)** alleles   **(f)** homozygote
        (b) autosomes   **(g)** phenotype
        (c) dominant allele   **(h)** recessive allele
        (d) genotype   **(i)** sex chromosomes
        (e) heterozygote

   ____ **(1)** genetic makeup
   ____ **(2)** how genetic makeup is expressed
   ____ **(3)** chromosomes that dictate most body characteristics
   ____ **(4)** alternate forms of the same gene
   ____ **(5)** an individual bearing two alleles that are the same for a particular trait
   ____ **(6)** an allele that is expressed whether in single or double dose
   ____ **(7)** an individual bearing two alleles that differ for a particular trait
   ____ **(8)** an allele that must be present in double dose to be expressed

2. Match the following types of inheritance (key terms a–f) with the descriptions below:

Key: **(a)** dominant-recessive   **(d)** polygene
      (b) incomplete dominance   **(e)** sex-linked
      (c) multiple-allele   **(f)** extranuclear

_____ **(1)** only sons show the trait
_____ **(2)** homozygotes and heterozygotes have the same phenotype
_____ **(3)** heterozygotes exhibit a phenotype intermediate between those of the homozygotes
_____ **(4)** phenotypes of offspring may be more varied than those of the parents
_____ **(5)** inheritance of ABO blood types
_____ **(6)** inheritance of stature
_____ **(7)** reflects activity of mitochondrial DNA

Level 2   Apply/Analyze

3. Describe the important mechanisms that lead to genetic variations in gametes.

4. Suppose that the ability to taste PTC (phenylthiocarbamide) depends on the presence of a dominant gene T; nontasters are homozygous for the recessive gene t. This would be a case of classical dominant-recessive inheritance. **(a)** Consider a mating between heterozygous parents producing three offspring. What proportion of the offspring are likely to be tasters? What is the chance that all three offspring will be tasters? Nontasters? What is the chance that two will be tasters and one will be a nontaster? **(b)** Consider a mating between Tt and tt parents. What is the anticipated percentage of tasters? Nontasters? What proportion can be expected to be homozygous recessive? Heterozygous? Homozygous dominant?

5. Most children with cystic fibrosis are born to normal parents. Children with cystic fibrosis are homozygous for the recessive gene (cc). What can you conclude about the genotypes of the normal parents?

6. A woman with blood type A has two children. One has type O blood and the other has type B blood. What is the genotype of the mother? What are the genotype and phenotype of the father? What is the genotype of each child?

7. In skin color inheritance, what will be the relative range of pigmentation in offspring arising from the following parental matches? **(a)** $AABBCC \times aabbcc$, **(b)** $AABBCC \times AaBbCc$, **(c)** $Aabbcc \times aabbcc$.

8. Compare and contrast amniocentesis and chorionic villus sampling as to the time at which they can be performed and the techniques used to obtain information on the fetus's genetic status.

9. The Browns are both carriers of the recessive allele that causes the metabolic disorder called phenylketonuria. What is the probability of each of the following occurring? **(a)** All three children will have the disorder. **(b)** None of their three children will have the disorder. **(c)** One or more of their children will have the disorder. **(d)** At least one of their children will be phenotypically normal.

Level 3   Evaluate/Synthesize

10. A color-blind man marries a woman with normal vision. The woman's father was also color-blind. **(a)** What is the chance that their first child will be a color-blind son? A color-blind daughter? **(b)** If they have four children, what is the chance that two will be color-blind sons? (Be careful on this one.)

11. A genetic counselor has been consulted by a couple that is concerned about a dominant form of syndactyly (fused fingers or toes) in the male partner, Paul. Paul has syndactyly, as do his three brothers. His mother, his maternal grandparents, and his paternal grandfather do not have syndactyly, but his father and paternal grandmother do. The counselor constructs a pedigree spanning three generations for Paul's family. Draw this pedigree and show the phenotype and genotype for each person.

CLINICAL CASE STUDY

27-Year-Old with Normal First Pregnancy

Ms. Rodriguez has had a normal pregnancy so far. During one of her regular visits with her doctor, Ms. Rodriguez asks what blood type her baby will have, given that her husband's blood type is type A and hers is type O. Her doctor explains to Ms. Rodriguez that she would need to know more about the blood type of Mr. Rodriguez's parents.

1. Describe the difference between genotype and phenotype. What are the phenotypes and genotypes of Mr. and Ms. Rodriguez?

2. Describe the type of inheritance that is responsible for blood type.

3. **+ NCLEX-STYLE** Mr. Rodriguez's mother has type O blood and his father has type A blood. What does this tell you about Mr. Rodriguez's genotype?
   a. He might be either $I^A I^A$ or $I^A i$, since his parents had both alleles.
   b. He must be $I^A i$, because his mother was ii.
   c. He must be $I^A I^A$, because his father was $I^A I^A$.
   d. The man who Mr. Rodriguez thinks is his biological father cannot be—the son of type A and type O parents cannot be type A.

4. What is the probability that Ms. Rodriguez's baby will have type O blood? A? B? AB?

5. **+ NCLEX-STYLE** You're talking with Ms. Rodriguez about her baby. Which of her statements indicate a need for further instruction about genetics?
   a. "My mother had Huntington's disease, but I've been tested and don't have the bad gene, and no one in my husband's family has had the disease, so luckily my baby won't get it either."
   b. "We should talk to all of our relatives and find out if any other diseases run in the family, so we know what to watch out for."
   c. "He's sure to be type A if he's a boy, because boys always inherit their father's traits."
   d. "Girls are protected from some of those inherited conditions because of their chromosomes."

29

Answers to Check Your Understanding and Level 1 Review Questions

Chapter 1

Check Your Understanding 1. The operation or function of a structure is promoted or prevented by its anatomy. For example, oxygen and carbon dioxide are exchanged across the very thin membranes of the lungs but not across the much thicker skin. **2.** Muscle shortening is a topic of physiology. The body location of the lungs is an anatomy topic. **3.** gastritis: inflammation of the stomach; leukocyte: white cell; nephropathy: kidney disease. **4.** Cytologists study the cellular level of organization. **5.** The order in the structural hierarchy is cell, tissue, organ, and organism. **6.** Living organisms can maintain their boundaries, move, respond to environmental changes, digest nutrients, carry out metabolism, dispose of wastes, reproduce, and grow. While inanimate objects may exhibit some of these properties, they do not exhibit all of them. **7.** Metabolism is the term that encompasses all the chemical reactions that occur in body cells. **8.** Boundary "a" is the plasma membrane. Fluid compartment "b" is intracellular fluid. Fluid compartment "c" is interstitial fluid, which is part of the extracellular fluid. **9.** Negative feedback mechanisms allow us to adjust to conditions outside the normal temperature range by causing heat to be lost from the body (in hot conditions) and retained or generated by the body (in cold conditions). **10.** This is a positive feedback mechanism because it enhances the change (formation of a platelet plug) set into motion by the stimulus (damage to the blood vessel). The response ends when the platelet plug has plugged the hole in the blood vessel. **11.** Thirst is part of a negative feedback control system because it prods us to drink, which ends the thirst stimulus and returns body fluid volume to the normal range. **12.** The position in which a person is standing erect with feet slightly separated and palms facing anteriorly. Knowing the anatomical position is important because directional terms refer to the body as if it is in this position. **13.** Axillary region is the armpit. Acromial area is the tip of the shoulder. **14.** A frontal (coronal) section would separate the brain into anterior and posterior parts. **15.** ❯ **16.** Of these organs, only the spinal cord is in the dorsal body cavity. **17.** As mobile organs (heart, lungs, digestive organs) work,

Midsagittal   Coronal   Transverse

friction is greatly reduced by the presence of serous fluid. Serous fluid allows the surrounding serous membranes to glide easily over one another. **18.** He may have appendicitis if the pain is in the lower right quadrant of his abdomen.

Review Questions 1. c; **2.** a; **3.** e; **4.** a, d; **5.** (a) wrist, (b) hip bone, (c) nose, (d) toes, (e) scalp; **6.** b; **7.** c; **8.** Function (physiology) reflects structure (anatomy); structure will determine and/or influence function. **9.** See Figure 1.4, which provides a summary of all the organ systems of the body. **10.** Five survival factors are: (1) nutrients—the chemical substances used for energy and cell building; (2) oxygen—used in the reactions that produce cellular energy; (3) water—the liquid environment necessary for all chemical reactions; (4) body temperature—to maintain the proper temperature for chemical reactions to proceed; and (5) atmospheric pressure—to allow gas exchange to occur. **11.** Homeostasis is the ability to maintain internal conditions within a narrow set of limits, even in the face of continuous change in the outside world. **12.** A plane refers to an imaginary flat surface, and a section refers to a cut along that imaginary surface. **13.** (a) brachial, (b) femoral, (c) thoracic, (d) digits, (e) patellar. **14.** See Figures 1.12 and 1.13. The figures illustrate the regions and quadrants and list several organs for each.

Chapter 2

Check Your Understanding 1. Foods contain chemical energy. **2.** Electrical energy is the energy used by nerve cells to transmit messages in the body. **3.** Potential energy (PE) is available when we are still. When we exercise, PE is converted to kinetic (working) energy, specifically to mechanical energy. **4.** Besides hydrogen and nitrogen, carbon and oxygen help to make up the bulk of living matter. **5.** ❯ **6.** An iron deficiency would decrease the ability of red

blood cells to transport oxygen on hemoglobin molecules. This would result in less oxygen in the blood. **7.** A molecule is 2 or more atoms held together by chemical bonds. **8.** A compound is formed when two or more different kinds of atoms chemically bond together, as in NaCl. Oxygen gas is 2 oxygen atoms (the same kind of atom) bonded together. **9.** Blood is a mixture because its components are not changed by their combination and they can be separated by physical means. **10.** Hydrogen bonds (linking H of one water molecule to O of another) form between water molecules. **11.** Argon's valence shell is full:)2e)8e)8e. For this reason it is nonreactive. **12.** Electrons would spend more time in the vicinity of the more electronegative atom in XY, whereas electrons in XX would orbit both X atoms to an equal extent. **13.** Fats are digested in the small intestine by decomposition reactions. **14.** Biochemical reactions in the body tend to be irreversible because (a) one or more of the products is removed from the reaction site or (b) reactions usually release large amounts of energy that would have to be restored for the reaction to run in reverse. **15.** Decomposition reactions in which foods are broken down for energy are oxidation-reduction (redox) reactions. **16.** Electrolytes are substances that will conduct an electrical current in aqueous solution. **17.** H^+ is responsible for acidity. **18.** You will get a large change in pH if you add a strong acid to a strong base. You will get a smaller change in pH if you add a strong acid to a weak base because the weak base will act as a buffer. **19.** Water is an excellent solvent because of its polarity. As a dipole, it can orient itself to the end of other molecules, causing them to dissociate or go into solution. **20.** Hydrolysis reactions are decomposition reactions in which molecules are broken down to simpler substances by addition of a water molecule to each bond. **21.** Monomers of carbohydrates are called monosaccharides or simple sugars. Glucose is blood sugar. **22.** The animal form of stored carbohydrate is glycogen. **23.** Triglycerides, the major source of stored energy in the body, are composed of three fatty acid chains and a glycerol molecule and are found in fat tissue. Phospholipids consist of two fatty acid chains, glycerol, a charged P-containing group, and a nitrogen-containing group. They are found in all cell membranes and form the basis of those membranes. **24.** An "amino acid" has an amine group (NH_2) and a COOH group that has acidic properties. **25.** The primary structure of proteins is the stringlike chain of amino acids. **26.** The secondary structures of proteins are the alpha helix and the beta-pleated sheet. **27.** ❯ **28.** DNA contains deoxyribose

sugar and the bases A, T, G, C. RNA contains ribose sugar and the bases A, U, G, C. **29.** The base pair C-G is less likely to come apart than the base pair A-T because it has three hydrogen bonds holding it together, whereas A-T only has two hydrogen bonds between its bases. (Hydrogen bonds are shown as dotted lines in Figure 2.25b.) **30.** ATP stores energy in smaller packets that are more readily released and transferred (during ATP hydrolysis) than the energy stored in glucose. The use of ATP as an energy source keeps energy waste to a minimum. **31.** When ATP releases energy, it loses a phosphate group and becomes ADP (also energy-rich).

Review Questions 1. d; **2.** d; **3.** a; **4.** b; **5.** d; **6.** a; **7.** c; **8.** d; **9.** d; **10.** a; **11.** b; **12.** Ca, C, H, Fe, N, O, K, Na.

Chapter 3

Check Your Understanding 1. The cell is the smallest unit of life. All organisms are made of one or more cells. Cells only arise from other cells. **2.** The three main parts of a human cell are the plasma membrane, the cytoplasm, and the nucleus. **3.** All cellular membranes consist of a double layer of phospholipids in which proteins are embedded. **4.** Structure (a) is a carbohydrate. Carbohydrates act as identity molecules, helping cells recognize each other. Structure (b) is the steroid cholesterol. Cholesterol stiffens the membrane. Structure (c) is a protein. Proteins are responsible for most specialized membrane functions such as acting as receptors, mediating transport, acting as enzymes, and attaching cells to each other and the extracellular matrix. **5.** The heart has desmosomes (anchoring junctions) that secure cardiac cells together as the heart pumps, and gap junctions (communicating junctions) that allow ions to flow from cardiac cell to cardiac cell. **6.** Unsaturated phospholipids would make the membrane more fluid. The double bonds cause the fatty acid chains to kink so that they cannot be packed closely, and this makes the membrane more fluid. **7.** Diffusion is driven by kinetic energy of the molecules. **8.** In channel-mediated diffusion, the diffusing substance moves through a channel protein. In carrier-mediated diffusion, the diffusing substance attaches to a carrier protein that moves it across the membrane. **9.** If a cell suddenly became permeable to both Na^+ and Cl^-, then these solutes have become penetrating solutes. As result, the nonpenetrating solutes inside the cell would no longer be balanced by nonpenetrating solutes outside the cell. The cell would swell (and possibly lyse) as Na^+, Cl^-, and water all move into the cell. **10.** Graph (a) represents simple diffusion. In graph (a), there is no maximum rate of diffusion. Notice that the higher the difference in concentration of a substance across the membrane, the higher will be its rate of diffusion. Graph (b) represents facilitated diffusion. There is a maximum rate of diffusion (that is shown by the rate of transport leveling off at higher solute concentrations). ❯ The graphs illustrate the transport property of saturability.

11. The Na^+-K^+ ATPase pumps K^+ into the cell. It expels three Na^+ for every two K^+ it brings in. **12.** The plasma membrane expands as a result of exocytosis. **13.** Phagocytic cells engulf debris, and a smoker's lungs would be laden with carbon particles and other debris from smoke inhalation. **14.** Cholesterol is taken in by receptor-mediated endocytosis. **15.** Diffusion of ions, mainly the diffusion of K^+ from the cell through leakage channels, establishes the resting membrane potential. **16.** In a polarized membrane of a resting cell, the inside is negative relative to its outside. **17.** Signaling chemicals that bind to membrane receptors are called ligands. G protein–coupled receptors direct intracellular events by promoting formation of second messengers. **18.** The lysosomal enzymes digest foreign substances engulfed by the cell, nonuseful or deteriorating organelles, or even the cell itself to prevent the buildup of cellular debris. The enzymes in peroxisomes detoxify harmful chemicals and neutralize free radicals. **19.** Both microfilaments and microtubules are involved in organelle movements within the cell and/or movements of the cell as a whole. **20.** The organelles are: (a) smooth endoplasmic reticulum, (b) microtubule, (c) mitochondria, (d) Golgi apparatus, and (e) rough endoplasmic reticulum. (1)b; (2)c; (3)a,d,e; (4)c; (5)a; (6)d. **21.** The protein is synthesized by a ribosome attached to the rough endoplasmic reticulum. Carbohydrates are attached to the protein in the cisterns of the rough ER, making it a glycoprotein. The glycoprotein is incorporated into a vesicle that migrates from the ER to the Golgi apparatus. Its carbohydrate may be modified or added to within the Golgi. It then exits from the *trans* face of the Golgi as part of a vesicle that migrates to and fuses with the plasma membrane. This membrane fusion results in the carbohydrate of the glycoprotein facing the outer surface of the plasma membrane. **22.** The major function of microvilli is to increase the cell's surface area for absorption of substances. **23.** If a cell loses its nucleus, it is doomed to die because it will be unable to make proteins, which include the enzymes needed for all metabolic reactions. **24.** Nucleoli are the site of synthesis of ribosomal subunits. **25.** Histone proteins provide the means to pack DNA in a compact, orderly way, and they play a role in gene regulation. **26.** DNA is synthesized during the S phase. **27.** Nuclear envelope breaks up, spindle forms, nucleoli disappear, and the chromosomes coil and condense. **28.** The base sequence of the corresponding strand will be GCTTAC. **29.** The A site is the entry site for tRNA at the ribosome. The P site is where the tRNA carrying the growing polypeptide is located. The E site is the tRNA exit site from the ribosome. **30.** DNA provides the coded instructions (is the template) for protein synthesis via the mRNA synthesized on it. **31.** The anticodon CAC binds the codon GUG. This codes for the amino acid valine, as shown in the diagram. ❯ **32.** Ubiquitin attaches to misfolded, damaged, or unneeded proteins, tagging them for destruction by proteasomes. **33.** Apoptosis is a process of programmed cell death which rids the body of cells that are stressed, damaged, old, or no longer needed.

Review Questions 1. d; **2.** a, c; **3.** b; **4.** e; **5.** a; **6.** a; **7.** b; **8.** c; **9.** d; **10.** a; **11.** The extensions found on the cells lining the trachea are cilia. Cilia are extensions of the plasma membrane made by microtubules that extend out from basal bodies. They are used to move mucus with trapped dust particles and bacteria up and away from the lungs.

Chapter 4

Check Your Understanding 1. Fixing tissue preserves it and prevents it from deteriorating. **2.** Heavy metal salts are used to stain tissues viewed by electron microscopes. **3.** Epithelial tissue lines body cavities and covers the body's external surface; thus polarity with one free (apical) surface is a requirement. **4.** Simple epithelia are "built" to provide for efficient absorption and filtration across their thin epithelial barriers. **5.** Pseudostratified epithelia appear to be stratified because their cells' nuclei lie at different distances from the basement membrane. However, all cells rest on the basement membrane. **6.** This gland is a holocrine gland. Holocrine glands have the highest rate of cell division. The secretory cells rupture when they secrete, and so they must be continuously replaced. **7.** ❯ **8.** Connective tissue functions to bind, support, protect, and insulate body organs. In addition, blood acts to transport substances throughout the body. **9.** Reticular, collagen, and

elastic fibers are found in the various connective tissues. **10.** Areolar connective tissue, because of its loose weblike nature, is capable of serving as a fluid reservoir. **11.** Dense regular connective tissue is damaged when you lacerate a tendon. **12.** Oxygen and nutrient needs are not being met because the calcified cartilage matrix is too hard to allow them to reach the cells by diffusion. Remember, cartilage is avascular. **13.** Cardiac muscle cells have striations and are branching cells. **14.** Skeletal muscle tissue is voluntary and is the muscle tissue injured when you "pull a muscle" while exercising. **15.** With extended processes, a neuron can conduct electrical signals a great distance within the body. **16.** A mucous membrane consists of both connective tissue and epithelium. It lines body cavities open to the exterior. **17.** The serous membranes called pleurae line the thorax walls and cover the lungs. **18.** These membranes differ from each other in their size, composition, and location. Plasma membranes surround cells and are tiny, made of two layers of phospholipid molecules with embedded proteins. The basement membrane is larger than the plasma membrane and is a layer of extracellular material that lies just outside the basal surface of epithelial cells. Serous and mucous membranes are much larger still. They consist of many cells and, because they consist of two types of tissues, they are considered organs. **19.** Water's high surface tension (due to its hydrogen bonds) makes the layers of serous membrane stick together. **20.** The three main steps of tissue repair are inflammation, organization, and permanent repair via regeneration and fibrosis. **21.** More severe injuries damage and destroy more tissue, requiring greater replacement with scar tissue. **22.** The liver is a gland that is derived from epithelial tissue. Epithelial tissue regenerates well. Nervous tissue in the brain and cardiac muscle exhibit virtually no functional regeneration.

Review Questions 1. (1)a, (2)c, (3)d, (4)b; **2.** c, e; **3.** b; **4.** c; **5.** b; **6.** fibrocyte, chondrocyte, osteocyte; **7.** The extracellular matrix consists of fibers (collagen, elastic, and reticular) and ground substance. Ground substance consists of interstitial fluid, cell adhesion proteins, and proteoglycans. **8.** (a) areolar connective tissue, (b) elastic cartilage, (c) elastic connective tissue, (d) mesenchyme, (e) fibrocartilage, (f) hyaline cartilage, (g) adipose tissue. **9.** Macrophages are cells of the immune system that are involved in overall body defenses. They are phagocytes that engulf and destroy bacteria, dead cells, and other debris.

Chapter 5

Check Your Understanding 1. The dermis is connective tissue, which is vascular, so its cells would be better nourished than those of the epidermis, which is avascular epithelium. **2.** Since the sole of the foot has thick skin, the layers from most superficial to deepest would be the stratum corneum, stratum lucidum, stratum granulosum, stratum spinosum, and stratum basale. **3.** The stratum basale undergoes almost continuous mitosis to replace cells lost by abrasion. **4.** The skin is subjected to a lot of abrasion and physical trauma. The desmosomes, which are connecting junctions, help to hold the cells together during such stress. The tight junctions help to reduce water movement though the skin. **5.** Some possible answers to this question are: (1) You would either be unable to grow, or there would be big cracks in the stratum corneum as you grew. (2) The thickness of the skin would impede joint movement so it would be hard to move. (3) You would not be able to lose heat through your skin as effectively as you do now. (4) Your skin would be very heavy to carry around after a while. **6.** The papillary dermis consists of areolar connective tissue, and the reticular dermis consists of dense irregular connective tissue. **7.** Because there is no bleeding, the cut has penetrated into the avascular epidermis only. **8.** Because Stefan's wound runs along a cleavage line, it is likely to heal much more cleanly than it would if it ran perpendicular to the cleavage lines. **9.** ❯ **10.** The third pig-

ment that contributes to skin color is hemoglobin, the pig-ment contained in red blood cells found in blood vessels of the dermis. **11.** Jaundice, a yellow cast to the skin due to the deposit of yellow bile pigments in body tissues, may indicate a liver disorder. **12.** There are no nerves in a hair, so cutting hair is painless. **13.** The parts of the hair labeled a and c are the hair shaft and hair root, respectively. Structures b and d are the arrector pili muscle and the hair papilla, respectively. Arrector pili muscles pull the hair (normally slanted) to the upright position (when cold or scared). Hair papillae contain a knot of capillaries that sup-plies nutrients to cells of the hair bulb.

14. ❯ **15.** The lunule of the nail is white because the thick nail matrix that under-lies it blocks the rosy color of the dermal blood supply from showing through. **16.** Nails are hard because the keratin they contain is the hard keratin variety. **17.** Sebaceous glands and apocrine glands are associated with the hair follicles. **18.** His autonomic nervous system activated his eccrine sweat glands and caused heat-induced sweating in order to cool the body. **19.** Heat-induced sweating occurs all over the body when we are overheated. A cold sweat is emotionally induced sweating that begins on the palms, soles, and armpits and then spreads to other body areas. Both types of sweating are primar-ily produced by the eccrine sweat glands. **20.** The palms of the hands and soles of the feet are thick skin areas. It would be dangerous to have oily soles, and oily palms would decrease the ability of the hands to hang onto things. **21.** The low pH of skin secretions (acid mantle) inhibits division of bacteria, and many bacte-ria are killed by dermcidin in sweat, bactericidal substances in sebum, or natural antibiotics called defensins produced by skin cells. Damaged skin secretes cathel-icidins that are effective against a certain strain of strep bacteria. **22.** The epider-mal dendritic cells play a role in body immunity. **23.** Sunlight causes the skin to produce a precursor of vitamin D from cholesterol. Vitamin D is essential for absorption of calcium from the diet, and calcium is a major component of bone. **24.** The effector in this case is smooth muscle. **25.** Basal cell carcinoma develops from the youngest epidermal cells. **26.** The ABCD rule helps one to recognize signs of melanoma. **27.** First- and second-degree burns can heal uneventfully by regeneration of epidermal cells as long as infection does not occur. Third-degree burns destroy the entire depth of skin and regeneration is usually not possible. Infection and loss of body fluid and proteins make recovery problematic. **28.** Burns to the face are serious because damage to the respiratory passageways can occur in such burns.

Review Questions 1. a; **2.** c; **3.** d; **4.** d; **5.** b; **6.** b; **7.** c; **8.** b; **9.** a; **10.** d; **11.** b; **12.** First-degree burns damage only the epidermis, second-degree burns damage the epidermis and the upper region of the dermis, and third-degree burns damage the entire thickness of the skin (the entire epidermis and dermis). **13.** Wrinkling is due to the loss of the elasticity of the skin, along with the loss of the subcutane-ous tissue, and is hastened by exposure to the sun.

Chapter 6

Check Your Understanding 1. Hyaline cartilage is the most plentiful in the adult body. **2.** The epiglottis and external ear cartilages are flexible elastic cartilage. **3.** Interstitial growth is growth from within. **4.** Skeletal muscles use bones as

levers to cause movement of the body and its parts. **5.** Bone matrix stores minerals (calcium and phosphate). Bone marrow cavities serve as sites for triglyceride (fat) storage. **6.** The components of the axial skeleton are the skull, the vertebral column, and the rib cage. **7.** The major function of the axial skeleton is to establish the long axis of the body and to protect structures that it encloses. The general function of the appendicular skeleton is to allow us mobility for propulsion and manipula-tion of our environment. **8.** The ribs and skull bones are flat bones. **9.** The parts of the bone are the epiphysis (a) and the diaphysis (b). The membranes at c and d are the endosteum and periosteum, respectively. The cartilage at e is articular cartilage, which is hyaline cartilage. This bone is from an adult. You can tell because it does not have an epiphyseal plate anymore (it just has an epiphyseal line). **10.** Because acidic vinegar soaks away a bone's mineral salts, the bone becomes flexible and bends too easily to support weight, so it would not be good at resisting compression. However, its organic components, particularly collagen, would allow it to resist tension.

11. ❯ **12.** The osteoclast fits this description. The lysosomes would contain the matrix-digesting enzymes. **13.** Bones begin as fibrous membranes or hyaline cartilages. **14.** The cartilage model grows, then breaks down

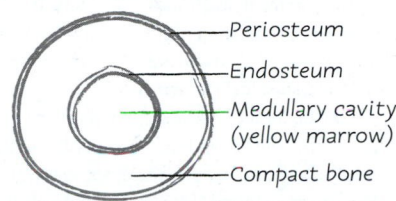

and is replaced by bone. **15.** The primary ossification center in a long bone is in the center of the shaft. The secondary ossification centers are in the epiphyses (bone ends). **16.** The chondrocytes are enlarging and their lacunae are break-ing down and leaving holes in the cartilage matrix. **17.** Parathyroid hormone maintains blood calcium levels. **18.** Bone growth increases bone mass, as during childhood or when exceptional stress is placed on the bones. Bone remodeling follows bone growth to maintain the proper proportions of the bone considering stresses placed upon it. **19.** If bone-destroying cells (osteoclasts) are more active than bone-forming cells (osteoblasts), bone mass will decrease. **20.** The bones of astronauts get thinner and weaker because they are deprived of the normal loading stresses that are placed on bones by gravity. Exercise that puts more stress on bones during space flight helps to some extent but cannot substitute for gravity. **21.** In an open fracture, the bone ends are exposed to the external environment. In a closed fracture, the bone ends do not penetrate the external boundary of the skin. **22.** Paget's disease is characterized by excessive deposit of weak, poorly mineralized bone. **23.** Sufficient vitamin D, calcium, and weight-bearing exercise all help to maintain healthy bone density. **24.** Adult rickets is called osteomalacia.

Review Questions 1. e; **2.** d; **3.** b; **4.** d, e; **5.** c; **6.** b; **7.** c; **8.** a; **9.** b; **10.** The steps in endochondral ossification in a long bone are (a) a bone collar forms around the diaphysis of the hyaline cartilage model; (b) cartilage in the center of the diaphysis calcifies and then develops cavities; (c) the periosteal bud invades the internal cavities and spongy bone forms; (d) the diaphysis elongates and a med-ullary cavity forms. Secondary ossification centers appear in the epiphyses; and (e) the epiphyses ossify.

Chapter 7

Check Your Understanding 1. The maxillae are the keystone bones of the facial skeleton. **2.** The maxillae form the bulk of the orbit floor. The eye is housed in the orbit. **3.** Sutures a and b are the coronal and squamous sutures, respectively. The bones are: (c) sphenoid bone, (d) parietal bone, (e) zygomatic bone, and (f) maxilla. **4.** Eating or talking, because the only freely movable joints of the skull are the temporomandibular joints of the jaw. **5.** Because the hyoid bone is in the neck, a fractured hyoid bone suggests death by strangulation. Alternatively, a forceful blow to the anterior neck could cause a fracture and the person could choke to death. **6.** Most of the skull bones are formed by intra-membranous ossification in which bone is formed from embryonic tissue called *mesenchyme*, which contains mesenchymal cells (see also Chapter 4, p. 129). **7.** The fibrocartilage discs contribute to the flexibility of the vertebral column. **8.** A lumbar vertebra is heavier and its massive body is kidney shaped. Its spinous processes are short and project directly back. A thoracic vertebral body is gener-ally heart shaped, its spinous process is long, sharp, and points downward, and its transverse processes have facets for articulating with the ribs. **9.** See the diagram. ❯ The number of vertebrae in each region is stated in parentheses. **10.** A true rib connects to the sternum by its own costal cartilage. A false rib connects to the sternum via costal cartilages of other ribs or not at all. **11.** The tubercle of a rib articulates with the costal facet of the transverse process of the thoracic vertebra that has the same number as the rib. **12.** The thoracic vertebrae also contribute to the thoracic cage. **13.** To determine the number of the rib that was broken, the EMT would first find the sternal

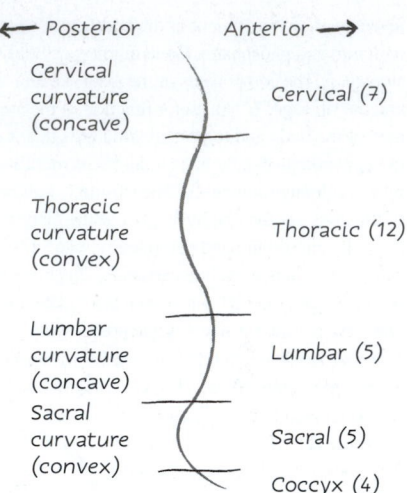

Posterior ← Anterior →

Cervical curvature (concave) — Cervical (7)

Thoracic curvature (convex) — Thoracic (12)

Lumbar curvature (concave) — Lumbar (5)

Sacral curvature (convex) — Sacral (5)

Coccyx (4)

angle on the front of the sternum (where the manubrium joins the sternal body). Because it is aligned with the second rib, the sternal angle would help the EMT locate the second rib. From there, she can count ribs to determine which rib is painful when palpated. **14.** Each pectoral girdle is formed by a scapula and a clavicle. **15.** The pectoral girdle attaches to the sternal manubrium of the axial skeleton via the medial end of its clavicle. **16.** A consequence of its flexibility is that it is easily dislocated. **17.** The structures are (a) acromioclavicular joint, (b) coracoid process, (c) manubrium of sternum, (d) body of sternum, and (e) costal cartilage. The rib indicated by (e) is a called a true rib or a vertebrosternal rib. **18.** Together the ulna and humerus form the elbow joint. **19.** The ulna and the radius each have a styloid process distally. **20.** Carpals are found in the proximal region of the palm. They are classified as short bones. **21.** The third bone that forms the hip bone is the ischium. **22.** The pelvic girdle receives the weight of the upper body (trunk, head, and upper limbs) and transmits that weight to the lower limbs. **23.** The regions of the hip bone are (a) ilium, (b) ischium, and (c) pubis. The opening at (d) is the obturator foramen and the socket at (e) is the acetabulum. The head of the femur (thigh bone) articulates with the acetabulum. **24.** A pelvis shaped like a male's would cause problems during childbirth because of its narrower overall shape. A smaller pelvic inlet would impede the infant's entry into the true pelvis at the beginning of labor. The smaller pelvic outlet would impede passage of the baby at birth. In particular, ischial spines that are closer together and point more medially and a coccyx that projects anteriorly and is less movable could cause a problem during the delivery. **25.** The tibia is the second largest bone in the body. **26.** The two bones shown are the tibia and fibula. They are from the right side of the body. The labeled structures are (a) medial condyle, (b) medial malleolus, and (c) head of fibula. **27.** The lateral condyles are not sites of muscle attachment; they are articular surfaces. **28.** Because of their springiness, the foot arches save energy during locomotion. **29.** The two largest tarsals are the talus and the calcaneus, which forms the heel.

Review Questions 1. (1)b, g; (2)h; (3)d; (4)d, f; (5)e; (6)c; (7)a, b, d, h; (8)i; **2.** (1)g, (2)f, (3)b, (4)a, (5)b, (6)c, (7)d, (8)e; **3.** (1)b, (2)c, (3)e, (4)a, (5)h, (6)e, (7)f; **4.** The thoracic and sacral curvatures are called primary curvatures because they are present at birth. The cervical and lumbar curvatures are called secondary curvatures because they develop after birth when the baby learns to lift its head (cervical curvature) and when it learns to walk (lumbar curvature). **5.** The intervertebral discs act as shock absorbers and provide flexibility to the spine, allowing it to flex and extend. **6.** Rib pairs 11 and 12 are called floating ribs (or vertebral ribs) because they are not attached to the sternum anteriorly. **7.** Differences between the male and female pelves include that (1) the female pelvic inlet and outlet are wider, (2) the female pelvis is wider, shallower, and lighter, and (3) and the female ischial spines are farther apart. **8.** Both cleft palate and hip dysplasia are congenital abnormalities affecting skeletal formation. Cleft palate is a condition in which the left and right halves of the palate fail to fuse medially. The resulting opening between the nasal and oral cavities may cause difficulty sucking. Aspiration of food and aspiration pneumonia may result. In developmental dysplasia of the hip, the acetabulum is poorly formed, and the head of

the femur may be partially or completely dislocated. Hip dysplasia can lead to degenerative arthritis of the hip joint.

Chapter 8

Check Your Understanding 1. The synarthroses are the least mobile of the joint types. **2.** In general, the more stable a joint, the less mobile it is. **3.** Most fibrous joints are synarthroses (immovable). **4.** *Cranio-* refers to the skull bones (see Chapter 7, p. 201) and *synostosis* refers to ossification of skull bone sutures. Craniosynostosis is a disorder in which the sutures in the skull close prematurely (in babies). Premature closing of sutures prevents the brain from growing normally. **5.** Evan would not have synchondroses at the ends of his femur. By age 25, his epiphyseal plates have fused and become synostoses. **6.** Bursae and tendon sheaths help to reduce friction during joint movement. **7.** The muscle tendons that cross the joint are typically the most important factor in stabilizing synovial joints. **8.** The hinge and pivot joints are uniaxial joints. **9.** The synovial joint categories are (a) ball-and-socket, (b) hinge, (c) pivot, and (d) condylar. **10.** Nick's hip joint was flexed and his knees extended and his thumb was in opposition (to his index finger). **11.** ❯ **12.** The knee and temporomandibular joints have menisci. The elbow and knee each act mainly as a uniaxial hinge. The shoulder depends largely on muscle tendons

Articular cartilages — Articular capsule — Joint cavity

for stability. **13.** Arthritis means inflammation of the joint. **14.** Lyme disease is caused by spirochete bacteria and transmitted by a tick bite. **15.** RA typically produces pain, swelling, and joint deformations that tend to be bilateral and crippling. OA patients tend to have pain, particularly on arising, which is relieved by gentle exercise, and enlarged bone ends (due to spurs) in affected joints. Affected joints may exhibit crepitus.

Review Questions 1. (1)c, (2)a, (3)a, (4)b, (5)c, (6)b, (7)b, (8)a, (9)c; **2.** b; **3.** d; **4.** d; **5.** b; **6.** d; **7.** d; **8.** Joints are defined as sites where two or more bones meet. **9.** Nonaxial movements are gliding or slipping movements only; uniaxial movements are in one plane; biaxial movements are in two planes; and multiaxial movements are in or around all three planes or axes. **10.** Uniaxial joints include hinge (elbow) and pivot (atlantoaxial and radioulnar) joints. Biaxial joints include condylar (knuckle) and saddle (thumb) joints. Multiaxial joints are ball-and-socket joints (hip and shoulder). **11.** Sprains and cartilage injuries are particularly problematic because cartilage and ligaments are poorly vascularized and tend to heal very slowly.

Chapter 9

Check Your Understanding
1. Striated means "with stripes."
2. Photomicrograph (a) is smooth muscle, (b) is skeletal muscle, and (c) is cardiac muscle. ❯ **3.** "Epimysium" literally translates to "outside

	(a) Smooth muscle	(b) Skeletal muscle	(c) Cardiac muscle
	Not striated	Striated	Striated
	Involuntary	Voluntary	Involuntary
	Walls of hollow visceral organs	Attached to skeleton	Heart

the muscle" and this connective tissue sheath is the outermost muscle sheath, enclosing the entire muscle. **4.** A tendon attaches muscles to bones, whereas a ligament connects bones at joints. **5.** The proteins are (a) troponin, (b) tropomyosin, (c) actin, and (d) myosin. Myosin hydrolyzes ATP, troponin binds Ca^{2+}, and tropomyosin must move out of the way so that cross bridges can form. **6.** In a resting muscle fiber, the SR would have the highest concentration of calcium ions. The mitochondria provide the ATP needed for muscle activity. **7.** ❯ **8.** The levels of organization are: (1) atom: phosphorus atom, (2) molecule: phospholipid

A Band   I Band   A Band

M line   Z disc   H zone

molecule, (3) organelle: sarcoplasmic reticulum, (4) cell: skeletal muscle fiber, (5) tissue: muscle tissue, (6) organ: the biceps muscle, and (7) organ system: the muscular system. **9.** There are at least four sets of ion channels involved: (1) Voltage-gated Ca^{2+} channels in the axon terminal of the motor neuron are opened by the action potential running down the axon. (2) Chemically gated channels permeable to both Na^+ and K^+ are opened by acetylcholine binding to ACh receptors on the junctional folds of the sarcolemma. (3) Voltage-gated Na^+ and K^+ channels are required for generating and propagating the action potential along the sarcolemma. The end plate potential initiates the action potential by opening the first set of voltage-gated Na^+ channels, and then the action potential itself opens more voltage-gated channels as it moves along the sarcolemma. (4) Ca^{2+} release channels in the terminal cisterns of the SR open to release Ca^{2+} into the cytosol. These channels open because an AP running along the T tubules causes the voltage-sensitive tubule proteins to change shape, and this shape change causes the Ca^{2+} release channels of the SR to open. **10.** The final trigger for contraction is a certain concentration of calcium ions in the cytosol. The initial trigger is depolarization of the sarcolemma. **11.** There are always some myosin cross bridges bound to the actin myofilament during the contraction phase. This prevents backward sliding of the actin filaments. **12.** Without ATP, rigor would occur because the myosin heads could not detach. **13.** A motor unit is an axon of a motor neuron and all the muscle fibers it innervates. **14.** The phases are (a) the latent period, (b) the period of contraction, and (c) the period of relaxation. During the latent period, events of excitation-contraction coupling are occurring. **15.** Immediately after Jacob grabs the bar, his biceps muscles are contracting isometrically. As his body moves upward toward the bar, they are contracting isotonically and concentrically. As he lowers his body toward the mat, the biceps are contracting isotonically and eccentrically. **16.** Chris is breathing heavily because it takes some time for her heart rate and overall metabolism to return to the resting state after exercise. Moreover, she has likely incurred an oxygen debt that requires her to take in extra oxygen, called EPOC, for the restorative processes. Although running can be an aerobic exercise, there is generally some anaerobic respiration that occurs as well—the amount depends on exercise intensity. The weakness Chris feels is likely in part due to muscle fatigue. Muscle fatigue is likely due to a combination of factors in her muscles including increased inorganic phosphate and magnesium, decreased ATP and glycogen, and other ionic imbalances. **17.** Factors that influence muscle contractile force include the frequency of stimulation of muscle fibers, muscle fiber size, the number of muscle fibers stimulated, and the degree of muscle stretch. Factors that influence velocity of contraction include muscle fiber type, load, and the number of motor units contracting. **18.** Fast glycolytic fibers would provide for short periods of intense strength needed to lift and move furniture. **19.** It makes sense that there are no slow glycolytic fibers because glycolysis is a fast but inefficient pathway for generating ATP. Why use such an inefficient pathway if the contraction is just going to be slow anyway (and so use up ATP slowly)? **20.** Aerobic exercise increases muscle endurance, whereas anaerobic exercise (such as high-intensity resistance exercise) increases muscle size and strength. **21.** Both skeletal and smooth muscle fibers are elongated cells, but unlike smooth muscle cells, which are spindle shaped (tapered at both ends), uninucleate, and nonstriated, skeletal muscle cells are very large cylindrical, multinucleate, striated cells. **22.** Calcium binds to troponin on the thin filaments in skeletal muscle cells. In smooth muscle cells, it binds to a cytoplasmic protein called calmodulin. **23.** Hollow organs that have smooth muscle cells helping to form their walls often must temporarily store the organ's contents (urine, food residues, etc.), an ability facilitated by the stress-relaxation response. **24.** Intracellular calcium acts as a second messenger (see Chapter 3, p. 81). In addition, calcium is involved in exocytosis in many different cells just as it is in the motor neuron axon terminal.

Review Questions 1. c; **2.** b; **3.** (1)b, (2)a, (3)b, (4)a, (5)b, (6)a; **4.** c; **5.** a; **6.** a; **7.** d; **8.** a; **9.** (1)a, (2)a, c, (3)b, (4)c, (5)b, (6)b; **10.** a; **11.** c; **12.** c; **13.** c; **14.** b

Chapter 10

Check Your Understanding 1. The antagonists of gluteus medius (a hip abductor) are hip adductors. Hip adductors are located on the medial side of the hip joint. **2.** The adductor brevis is a small muscle that causes adduction (of the

thigh); the quadriceps femoris muscle has four heads and follows the course of the femur; the tibialis anterior lies anterior to the tibial bone; and the iliacus overlies the iliac bone. **3.** Of the muscles illustrated in Figure 10.1, the one with the parallel arrangement (sartorius) could shorten to the greatest degree. The stocky bipennate (rectus femoris) and multipennate (deltoid) muscles would be most powerful because they pack in the most fibers. **4.** The patterns of muscle fascicle arrangement are (a) circular, (b) fusiform or parallel (fusiform is a subtype of parallel), (c) convergent, and (d) parallel. **5.** Third-class levers are the fastest levers. **6.** You would want the fulcrum to be closer to the post because this would give you a mechanical advantage. A lever that operates at a mechanical advantage allows the muscle to exert less force than the load being moved. **7.** ❯ The deltoid is part of a third-class lever system, which operates at a mechanical disadvantage. **8.** Mario was using the frontal belly of his epicranius to raise his eyebrows and an orbicularis oculi muscle to wink at Sarah. **9.** The deltoid has a broad origin. When only its anterior fibers contract, it

Fulcrum
Effort
Load

flexes and medially rotates the arm. When only its posterior fibers contract, it extends and laterally rotates the arm. **10.** The opponens pollicis does not have an insertion on the bones of the thumb. **11.** To make a sad clown's face you would contract your platysma, depressor anguli oris, and depressor labii inferioris muscles. **12.** The muscles that cross the knee also reinforce the joint by providing increased stability. **13.** The muscles for intramuscular injections are (a) deltoid, (b) gluteus medius, and (c) vastus lateralis. The vastus lateralis is a common injection site in infants because they have poorly developed buttock and arm muscles.

Review Questions 1. c; **2.** c; **3.** (1)e, (2)c, (3)g, (4)f, (5)d; **4.** a; **5.** c; **6.** d; **7.** c; **8.** c; **9.** b; **10.** b; **11.** a; **12.** a, b; **13.** a; **14.** The pharyngeal constrictor muscles propel food through the pharynx to the esophagus. **15.** To nod your head "yes," the sternocleidomastoid muscles contract simultaneously. Acting on one side only, the sternocleidomastoid and splenius muscles allow you to shake your head "no." **16.** (a) The extensor carpi radialis longus and brevis are strong hand extensors and abductors. (b) The forearm muscle that can flex the distal interphalangeal joints is the flexor digitorum profundus.

Chapter 11

Check Your Understanding 1. Integration involves processing and interpreting sensory information, and making a decision about motor output. Integration occurs primarily in the CNS. **2.** (a) This "full stomach" feeling would be relayed by the sensory (afferent) division of the PNS (via its visceral afferent fibers). (b) The somatic nervous system, which is part of the motor (efferent) division of the PNS, controls movement of skeletal muscle. (c) The autonomic nervous system, which is part of the motor (efferent) division of the PNS, controls the heart rate. **3.** Astrocytes control the extracellular environment around neuron cell bodies in the CNS, whereas satellite cells perform this function in the PNS. **4.** Oligodendrocytes and Schwann cells form myelin sheaths in the CNS and PNS, respectively. **5.** A nucleus within the brain is a cluster of neuron cell bodies, whereas the nucleus within each neuron is a large organelle that acts as the control center of the cell. **6.** In the CNS, a myelin sheath is formed by oligodendrocytes that wrap their plasma membranes around the axon. The myelin sheath protects and electrically insulates axons and increases the speed of transmission of nerve impulses. **7.** The neuron is a unipolar (or pseudounipolar) neuron that acts as a sensory neuron. The parts of this neuron are (a) receptive endings, (b) peripheral process (part of the axon), (c) cell body, and (d) axon terminals. **8.** Burning a finger will first activate unipolar (pseudounipolar) neurons that are sensory (afferent) neurons. The impulse to move your finger away from the heat will be carried by multipolar neurons that are motor (efferent) neurons. **9.** A nerve fiber is a long axon, an extension of the cell. In connective tissue, fibers are extracellular proteins that provide support. In muscle tissue, a muscle fiber is a muscle cell. **10.** The concentration gradient and the electrical gradient—together called the electrochemical gradient—determine the direction in which ions flow through an open membrane channel. **11.** There are more leakage channels for K^+ than for any other cation. **12.** The resting membrane potential depends on both the concentration gradients of K^+ and Na^+ across the membrane and the permeability of the membrane to K^+ and Na^+. An increase in the extracellular K^+ concentration will decrease the concentration gradient for K^+ and this will decrease the resting membrane potential of neurons and muscle cells. In other words, the inside of these cells will become less negative (be depolarized). **13.** The size of a graded potential is determined by the strength of a stimulus. **14.** The graded potential at the junctional folds of the sarcolemma is called an end plate potential (EPP). It triggers an action potential in the adjacent sarcolemma. **15.** Action potentials are larger than graded potentials and travel farther. Graded potentials generally initiate action potentials. **16.** An action potential is regenerated anew at each membrane patch. **17.** Conduction of action potentials is faster in myelinated axons because myelin allows the axon membrane between myelin sheath gaps to change its voltage rapidly, and allows current to flow only at the widely spaced gaps. **18.** If a second stimulus occurs before the end of the absolute refractory period, no AP can occur because sodium channels are still inactivated. **19.** The x axis indicates time measured in milliseconds (ms). The y axis indicates membrane potential (which is the difference in voltage between the inside and outside of the cell) in millivolts (mV). Phase (a) is depolarization, which is due to an increase in permeability to Na^+. Phase (b) is repolarization, which is due to both a decrease in permeability to Na^+ and an increase in permeability to K^+. Phase (c) is hyperpolarization, which is due to delayed closure of K^+ channels, which means that the membrane is still more permeable to K^+ than it is at rest. **20.** ❯ As shown in the diagram, voltage-gated ion (Ca^{2+}) channels are found in the presynaptic axon terminal.

Voltage-gated Ca^{2+} channel

Axon terminal of presynaptic neuron

Cell body (or dendrite) of postsynaptic neuron

Chemically gated channel

They open when an action potential reaches the axon terminal. Chemically gated ion channels are found in the postsynaptic membrane and open when neurotransmitter binds to the receptor protein. **21.** At an electrical synapse, neurons are joined by gap junctions. **22.** IPSPs result from the flow of either K^+ or Cl^- through chemically gated channels. EPSPs result from the flow of both Na^+ and K^+ through chemically gated channels. **23.** Temporal summation is summation in time of graded potentials occurring in quick succession at the postsynaptic membrane. It can result from EPSPs arising from just one synapse. Spatial summation is summation in space—a postsynaptic neuron is stimulated by a large number of terminals at the same time. **24.** ACh interacts with more than one specific receptor type, and this explains how it can excite at some synapses and inhibit at others. **25.** Cyclic AMP (cAMP) is called a second messenger because it relays the message between the first messenger (the original chemical messenger) outside of the cell and effector molecules that will ultimately bring about the desired response within the cell. **26.** Reverberating circuits and parallel after-discharge circuits both result in prolonged output. **27.** The pattern of neural processing is serial processing. The response is a reflex arc. **28.** The pattern of neural processing is parallel processing.

Review Questions 1. b; **2.** (1)d, (2)b, (3)f, (4)c, (5)a; **3.** c; **4.** c; **5.** b; **6.** d; **7.** c; **8.** c; **9.** a; **10.** (1)d, (2)b, (3)a, (4)c; **11.** (a) Myelin is a whitish, fatty, phospholipid-containing insulating material (essentially the wrapped plasma membranes of oligodendrocytes or Schwann cells). (b) CNS myelin sheaths are formed by extensions of oligodendrocytes and lack an outer collar of perinuclear cytoplasm. Each oligodendrocyte can myelinate several (up to 60) axons. PNS myelin sheaths are formed by Schwann cells. One Schwann cell is dedicated to wrapping one internode region and the myelin sheaths have an outer collar of perinuclear cytoplasm. **12.** The CNS interprets a stimulus as strong when the frequency, or rate, of action potentials arriving at the CNS is high. Conversely, a low frequency of action potentials indicates a weaker stimulus. **13.** When the professor discusses group A fibers, she is referring to fibers that have a large diameter, thick myelin sheaths, and rapid conduction. Group B fibers are lightly myelinated, have intermediate diameters, and have slower conduction velocity. The absolute refractory period is the period of time during which a neuron

cannot generate an action potential in response to another stimulus (no matter how large) because the voltage-gated sodium channels are either still open or are inactivated. When the professor refers to myelin sheath gaps, she is referring to part of an axon that is not covered by a myelin sheath. These gaps occur between Schwann cells or adjacent processes of oligodendrocytes and are about 1 mm apart. **14.** During the first developmental stage, neurons proliferate; during the second stage, neurons migrate to proper position; during the third stage, neurons connect to other neurons.

Chapter 12

Check Your Understanding 1. The third ventricle is surrounded by the diencephalon. **2.** The cerebral hemispheres and the cerebellum have an outside layer of gray matter in addition to central gray matter and its surrounding white matter. **3.** Convolutions increase the surface area of the cortex, which allows more neurons to occupy the limited space within the skull. **4.** The central sulcus separates primary motor areas from somatosensory areas. **5.** Motor functions on the left side of the body are controlled by the right hemisphere of the brain because motor tracts from the right hemisphere cross over (in the medulla oblongata) to the left side of the spinal cord to go to the left side of the body. **6.** Commissural fibers (which form commissures) allow the cerebral hemispheres to "talk to each other." **7.** The caudate nucleus, putamen, and globus pallidus together form the basal nuclei. **8.** Virtually all inputs ascending to the cerebral cortex synapse in the thalamus en route. **9.** The hypothalamus oversees the autonomic nervous system. **10.** The pyramids of the medulla are the corticospinal (pyramidal) tracts, the large voluntary motor tracts descending from the motor cortex. The result of decussation (crossing over) is that each side of the motor cortex controls the opposite side of the body. **11.** The cerebral peduncles and the colliculi are associated with the midbrain. **12.** There are many possible answers to this question—here are a few: Structurally, the cerebellum and cerebrum are similar in that they both have a thin outer cortex of gray matter, internal white matter, and deep gray matter nuclei. Also, both have body maps (homunculi) and large fiber tracts connecting them to the brain stem. Both receive sensory input and influence motor output. A major difference is that the cerebellum is almost entirely concerned with motor output, whereas the cerebrum has much broader responsibilities. Also, while a cerebral hemisphere controls the opposite side of the body, a cerebellar hemisphere controls the same side of the body. **13.** Kayla would likely have problems with executing smooth, coordinated movements. This would affect her balance so she might fall over when walking or running, and might have trouble keeping up with her peers. Her motor tone would also likely be decreased so her limbs might seem "floppy." **14.** The hypothalamus is part of the limbic system and also an autonomic (visceral) control center. **15.** Taylor is increasing the amount of sensory stimuli she receives, which will be relayed to the reticular activating system, which, in turn, will increase activation of the cerebral cortex. **16.** Transfer of memory from STM to LTM is enhanced by (1) rehearsal, (2) association (tying "new" information to "old" information), and (3) a heightened emotional state (for example, alert, motivated, surprised, or aroused). **17.** Delta waves are typically seen in deep sleep in normal adults. **18.** Drowsiness (or lethargy) and stupor are stages of consciousness between alertness and coma. **19.** Most skeletal muscles are actively inhibited during REM sleep. **20.** CSF, formed by the choroid plexuses as a filtrate of blood plasma, is a watery "broth" similar in composition to plasma. It protects the brain and spinal cord from blows and other trauma, helps nourish the brain, and carries chemical signals from one part of the brain to another. **21.** The structures are (a) dural venous sinus (if you said superior sagittal sinus, that's also correct), (b) arachnoid granulation, (c) dura mater, (d) arachnoid mater, and (e) pia mater. Arachnoid granulations absorb CSF into venous blood. You would find CSF in the subarachnoid space between structures (d) and (e). The two parts of the dura mater are the superficial periosteal layer next to the bone and the deeper meningeal layer. **22.** A TIA is a temporary loss of blood supply to brain tissue, and it differs from a stroke in that the resulting impairment is fully reversible. **23.** Mrs. Lee might have Parkinson's disease. **24.** The nerves serving the limbs arise in the cervical and lumbar enlargements of the spinal cord. **25.** A loss of motor function is called paralysis. Lower limb paralysis could be caused by a spinal cord injury in the thoracic region (between T_1 and L_1). If the spinal cord is transected, the result is paraplegia. If the cord is only bruised, he may regain function in the limbs. **26.** See the labeled diagram. ❯ Make sure that the sensory neuron you drew has a unipolar shape (there should be a clear "T-shaped junction" of the axon attached to the cell body with the central and peripheral axons; see also p. 397). The central canal contains CSF. **27.** In the spinothalamic pathway, the cell bodies of first-order sensory neurons are outside the spinal cord in a ganglion, cell bodies of second-order sensory neurons are in the dorsal horn of the spinal cord, and cell bodies of third-order sensory neurons are in the thalamus. (See also Figure 12.34b.) **28.** The pyramidal cells controlling left big toe movement are in the right primary motor cortex in the frontal lobe. They synapse with the cell bodies of ventral horn neurons in the spinal cord. **29.** A nerve is a bundle of axons in the PNS, whereas a tract is a bundle of axons in the CNS. A nucleus is a collection of neuron cell bodies in the CNS, whereas a ganglion is a collection of neuron cell bodies in the PNS.

Labels: Dorsal root; Dorsal horn; Axon terminal of sensory neuron; Cell body of sensory neuron; Ventral root; Central canal; Ventral horn

Review Questions 1. a; **2.** d; **3.** c; **4.** (1)d, (2)f, (3)e, (4)g, (5)b, (6)f, (7)i, (8)a; **5.** c; **6.** a; **7.** d; **8.** (a) The basal nuclei help to coordinate and control motor activity. They initiate slow and sustained movement. (b) The striatum is formed by the caudate nucleus and the putamen. (c) The caudate nucleus arches over the diencephalon. **9.** (a) The reticular formation extends through the central core of the medulla, pons, and midbrain. (b) RAS means reticular activating system, which is our cortical arousal mechanism. It helps to keep the cerebral cortex alert while filtering out unimportant inputs. **10.** The CNS is protected by the bony cranium, the meninges, the cerebrospinal fluid, and the blood brain barrier. **11.** The blood brain barrier is primarily the endothelial cells of the capillaries, which are joined by particularly impermeable tight junctions. This characteristic makes them highly selective, ensuring that only certain substances can gain access to neural tissue.

Chapter 13

Check Your Understanding 1. In addition to nerves, the PNS also consists of sensory receptors, motor endings, and ganglia. **2.** Nociceptors respond to painful stimuli. They are exteroceptors that are nonencapsulated (free nerve endings). **3.** Amber is likely to suffer from numerous undetected injuries. These could lead to infection (and possibly amputation). **4.** The sensory structures are as follows: (a) epithelial tactile complexes, which detect light pressure; (b) tactile corpuscle, which also detects light pressure (usually on hairless skin); (c) lamellar corpuscle, which detects deep pressure, stretch, and vibration; (d) bulbous corpuscle, which detects deep pressure and stretch; (e) free nerve endings, which detect pain and temperature; and (f) hair follicle receptor, which detects hair movement. **5.** The three levels of sensory integration are receptor level, circuit level, and perceptual level. **6.** Phasic receptors adapt, whereas tonic receptors exhibit little or no adaptation. Pain receptors are tonic so that we are reminded to protect the injured body part. **7.** Hot and cold are conveyed by different sensory receptors that are parts of separate "labeled lines." Cool and cold are two different intensities of the same stimulus, detected by frequency coding—the frequency of APs would be higher for a cold stimulus than for a cool one. Action potentials arising in the fingers and foot arrive at different locations in the somatosensory cortex via their own "labeled lines" and in this way the cortex can determine their origin. **8.** Ganglia are collections of neuron cell bodies in the PNS. **9.** Nerves also contain connective tissue, blood vessels, lymphatic vessels, and the myelin surrounding the axons. **10.** Schwann cells, macrophages, and the neurons themselves were all important in healing the nerve. **11.** The oculomotor (III), trochlear (IV), and abducens (VI) nerves control eye movements. Sticking out your tongue involves the hypoglossal nerve (XII). The vagus nerve (X) influences heart rate and digestive activity. The accessory nerve (XI) innervates the trapezius muscle, which is involved in shoulder shrugging. **12.** Roots lie medial to spinal nerves, whereas rami lie lateral to spinal nerves. Dorsal roots are purely sensory, whereas dorsal rami carry both motor and sensory fibers. **13.** The ventral rami were C_3–C_5, the nerve plexus was the cervical plexus, and the major

nerve was the phrenic nerve. The phrenic nerve is the sole motor nerve supply to the diaphragm, the primary muscle for respiration. **14.** Varicosities are the series of knoblike swellings that are the axon endings of autonomic motor neurons. You would find them on axon endings serving smooth muscle or glands. **15.** The cerebellum and basal nuclei, which form the precommand level of motor control, plan and coordinate complex motor activities. **16.** The effector (muscle or gland) brings about the response. **17.** See the labeled diagram. ❯ Make sure that in your drawing (as in the diagram), the cell body of the sensory neuron is in the dorsal root ganglion, the cell body of the interneuron is in the dorsal horn, and the cell body of the motor neuron is in the ventral horn.

Labels: (1) Receptor; (2) Sensory neuron; (3) Integration center; (5) Effector; (4) Interneuron; Motor neuron

18. The stretch reflex is important for maintaining muscle tone and adjusting it reflexively by causing muscle contraction in response to increased muscle length (stretch). It maintains posture. The flexor or withdrawal reflex is initiated by a painful stimulus and causes automatic withdrawal of the painful body part from the stimulus. It is protective. **19.** This response is called Babinski's sign and it indicates damage to the corticospinal tract or primary motor cortex. **20.** The spinothalamic pathway carries pain signals to the somatosensory cortex in the parietal lobe. The pyramidal pathways carry voluntary motor information and would be involved in inhibiting the flexor reflex.

Review Questions 1. b; **2.** c; **3.** d; **4.** c; **5.** e; **6.** (1)d, (2)c, (3)f, (4)b, (5)e, (6)a; **7.** (1)f, (2)i, (3)b, (4)g, h, (5)e, (6)i, (7)c, (8)k, (9)l, (10)c, d, f, k; **8.** (1)b 6; (2)d 8; (3)c 2; (4)c 5; (5)a 4; (6)a 3, 9; (7)a 7; (8)a 7; (9)d 1; (10)a 3, 4, 7, 9; **9.** c; **10.** c; **11.** The PNS includes all nervous tissue outside the CNS, and consists of the sensory receptors that detect specific stimuli, the peripheral nerves (cranial and spinal) that conduct impulses to and from the CNS, the ganglia that contain synapses or cell bodies outside the CNS, and motor nerve endings that innervate effector organs. **12.** (a) A plexus is a network. A *nerve* plexus is a branching nerve network formed by ventral rami from several spinal nerves. (b) The cervical plexus originates from the ventral rami of C_1 to C_4 (and some C_5) spinal nerves; the brachial plexus originates from the ventral rami of C_5 to T_1; the lumbar plexus originates from the ventral rami of L_1 to L_4; and the sacral plexus originates from L_4 to S_4. **13.** Ipsilateral reflexes involve a reflex initiated on and affecting the same side of the body; contralateral reflexes involve a reflex that is initiated on one side of the body and affects the other side.

Chapter 14

Check Your Understanding 1. The effectors of the autonomic nervous system are cardiac muscle, smooth muscle, and glands. **2.** The somatic motor system relays instructions to muscles more quickly because it involves only one motor neuron, whereas the ANS uses a two-neuron chain. Moreover, axons of somatic motor neurons are typically heavily myelinated, whereas preganglionic autonomic axons are lightly myelinated and postganglionic axons are nonmyelinated. **3.** Dorsal root ganglia also contain neuron cell bodies. These cell bodies belong to neurons that are structurally classified as unipolar (pseudounipolar) and functionally classified as sensory neurons (first-order sensory neurons in this case). **4.** While you relax in the sun on the beach, the parasympathetic branch of the ANS would probably predominate. When you perceive danger (as in a shark), the sympathetic branch of the ANS predominates. **5.** Cell bodies of preganglionic parasympathetic neurons that innervate the head are in the brain stem. Cell bodies of postganglionic parasympathetic neurons innervated by the vagus nerve are in ganglia found mostly in the walls of visceral organs. **6.** ❯ Ganglia of

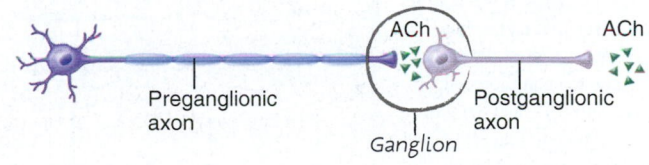

Labels: Preganglionic axon; ACh; Ganglion; Postganglionic axon; ACh

parasympathetic neurons are terminal ganglia. If you said "intramural ganglion" you are partly right—most but not all terminal ganglia are intramural ganglia (depending on whether they are in the wall of the target organ). **7.** "Short preganglionic fibers," "origin from thoracolumbar region of spinal cord," "collateral ganglia," and "innervates adrenal medulla" are all characteristic of the sympathetic nervous system. Terminal ganglia are found in the parasympathetic nervous system. **8.** The major differences are (1) the ANS has visceral afferents rather than somatic afferents, (2) the ANS has a two-neuron efferent chain, whereas the somatic nervous system has one, and (3) the effectors of the ANS are smooth muscles, cardiac muscle, and glands, whereas the effectors of the somatic nervous system are skeletal muscles. **9.** You would find nicotinic receptors on skeletal muscle and the hormone-producing cells of the adrenal medulla, but not on smooth muscle or glands. Virtually all types of receptors (including nicotinic receptors) are also found in the CNS (see Table 11.4 on pp. 421–422). **10.** ❯ As shown in the diagram, nicotinic receptors are located on skeletal muscle

(in the junctional folds of the sarcolemma), on postganglionic neurons (cell bodies and dendrites), and on the secretory cells of the adrenal medulla. Muscarinic receptors are located on the plasma membrane of target organs that respond to acetylcholine, whereas α and β receptors are found on the plasma membrane of target organs that respond to norepinephrine (and epinephrine). **11.** The parasympathetic nervous system increases digestive activity and decreases heart rate. The sympathetic nervous system increases blood pressure, dilates bronchioles, stimulates the adrenal medulla to release its hormones, and causes ejaculation. **12.** The main integration center of the ANS is the hypothalamus, although the most direct influence is through the brain stem reticular formation and the reflex centers in the pons and medulla oblongata. **13.** Jackson's doctor may have prescribed a beta-blocker because Jackson has hypertension. (Chronic stress is a factor in causing hypertension.) The beta-blocker will decrease blood pressure by blocking beta-adrenergic receptors in the heart, thereby decreasing heart rate and force of contraction, and by decreasing renin release from the kidneys.

Review Questions 1. d; **2.** (1)S, (2)P, (3)P, (4)S, (5)S, (6)P, (7)P, (8)S, (9)P, (10)S, (11)P, (12)S; **3.** c; **4.** a; **5.** All preganglionic fibers of the ANS secrete ACh. In addition, all postganglionic fibers of the parasympathetic nervous system secrete ACh, as do some postganglionic sympathetic fibers (e.g., those serving sweat glands). Only postganglionic sympathetic fibers secrete norepinephrine. **6.** The cell bodies of visceral sensory neurons are in the dorsal root ganglia and in the sensory ganglia of cranial nerves.

Chapter 15

Check Your Understanding 1. Tears (lacrimal fluid) are a dilute saline secretion that contains mucus, antibodies, and lysozyme. They are secreted by the lacrimal glands. **2.** The blind spot of the eye is the optic disc. It is the part of the retina where the optic nerve exits the eye, and it is "blind" because it is a region of the retina that lacks photoreceptors. **3.** ❯ The cells and layers are: (a) ganglion cell; (b) bipolar cell; (c) cone; (d) rod; and (e) pigmented layer. The choroid is superficial to the

Pathway of light

Pathway of information to brain

a   b   c   d   e

pigmented layer. **4.** Light passes through the cornea, aqueous humor, lens, vitreous humor, ganglion cells, and bipolar cells before it reaches the photoreceptors. **5.** The near point moves farther away as you age because the lens becomes less flexible (presbyopia), so that it is unable to assume the more rounded shape required for near vision. **6.** The ciliary muscles and sphincter pupillae of the iris relax for distant vision. (If you said the medial rectus muscles also relax, this is true, but remember that the rectus muscles are extrinsic eye muscles, not intrinsic.) To see whether you labeled the ciliary muscle and the iris correctly on the diagram, refer to Figure 15.8. **7.** ❯ **8.** Massive activation of the sympathetic

Parasympathetic activation → Ciliary muscle contracts → Ciliary zonule relaxes → Lens bulges to more convex shape → ↓ in lens focal length allows close vision

nervous system occurs in extreme situations eliciting a "fight-or-flight" response. During these extreme situations, relaxation of the ciliary muscle adjusts your eye for distant vision and stimulation of the dilator pupillae dilates your pupil, allowing more light to enter. These actions allow you to survey the entire threatening scene (helping you decide where to run, for example). **9.** The following are characteristics of cones: "vision in bright light," "color vision," and "higher acuity." The following are characteristics of rods: "only one type of visual pigment," "most abundant in the periphery of the retina," "many feed into one ganglion cell," and "higher sensitivity." **10.** Breakdown of the retinal-opsin combination is called bleaching of the pigment. It occurs with exposure to light. **11.** A tumor in the right visual cortex would affect the left visual field. A tumor compressing the right optic nerve would affect both the left and right visual fields from the right eye only. **12.** The cilia of these receptor cells greatly increase the surface area for sensory receptors. **13.** The five taste modalities are sweet, sour, bitter, salty, and umami. The fungiform, vallate, and foliate papillae contain taste buds. **14.** The tympanic membrane separates the external from the middle ear. The oval and round windows separate the middle from the inner ear. **15.** The basilar membrane allows us to differentiate sounds of different pitch. **16.** The round window acts as the "pressure valve" that allows fluid to move in the cochlea. **17.** High-frequency sounds do not enter the cochlea. You can think of this in two ways: mechanically and logically. Mechanically, the ossicles and tympanic membrane cannot move rapidly enough to transmit these high-frequency sounds, so they reflect off the tympanic membrane. The other way to think about it is that the highest-frequency sounds we can hear map to the part of the basilar membrane that is nearest to the oval window. Logically, even higher-frequency sounds would map to the other (lateral) side of the oval window where there are no sensory cells and so these sounds cannot be heard. **18.** Influx of Na^+ ions is responsible for depolarization in most cells. K^+ and Ca^{2+} ions cause depolarization in hair cells. K^+ moves into hair cells (rather than out) because the extracellular fluid is endolymph, which is rich in K^+. As a result, the electrochemical gradient drives K^+ into these cells. **19.** You would not be able to locate the origin of a sound if the brain stem did not receive input from both ears. **20.** Drawing (a) shows the position of the otolith membrane in the saccule of a person in an elevator that just began to rise. This will cause an increase in action potential frequency in the nerve fibers from this saccule because the hair cells are bending toward the kinocilium. **21.** The following apply to a macula: "contains otoliths," "responds to linear acceleration and deceleration," and "inside a saccule." The following apply to a crista ampullaris: "inside a semicircular canal," "has a cupula," "responds to rotational acceleration and deceleration." **22.** The "fullness" in Mohammed's ears is likely due to an accumulation of fluid in the middle ear as a result of his upper respiratory infection ("cold") spreading into his ear. He has a form of conduction deafness.

Review Questions 1. c; **2.** d; **3.** a; **4.** c; **5.** a; **6.** b; **7.** b; **8.** d; **9.** d; **10.** b; **11.** c; **12.** d; **13.** b; **14.** c; **15.** Rods operate mostly in dim light and provide noncolor vision with relatively low acuity. Cones provide high-acuity color vision in bright light. For other differences between rods and cones, see Table 15.1 on p. 569. **16.** Taste is served by cranial nerves VII (facial), IX (glossopharyngeal), and X (vagus).

Chapter 16

Check Your Understanding 1. The endocrine system is more closely associated with growth and development, and its responses tend to be long-lasting, whereas nervous system responses tend to be rapid and discrete. **2.** The thyroid and parathyroid glands are found in the neck. **3.** Hormones are released into the blood and transported throughout the body, whereas paracrines act locally, generally within the same tissue. **4.** Gland (a) is an exocrine gland and gland (b) is an endocrine gland. In gland (a), you can see the duct that is characteristic of exocrine glands. The duct conveys glandular secretions either out of the body onto the skin or into a hollow organ. Endocrine glands, on the other hand, do not have ducts and instead secrete hormones into the blood. Notice the prominent blood vessels around gland (b). See also Figure 4.5 on p. 123. **5.** Steroid hormones are synthesized on the membrane of the smooth endoplasmic reticulum. Peptide hormones are synthesized on rough endoplasmic reticulum. Peptide hormones can be stored in vesicles and released by exocytosis. **6.** Steroids are all lipid soluble. Thyroid hormones are the only amino acid–based hormones that are lipid soluble.

7. ❭ As shown in the drawing, receptors for water-soluble hormones are in the plasma membrane, whereas receptors for lipid-soluble hormones are inside the cell. They can be in the cytoplasm (as shown here)

or inside the nucleus. Plasma membrane receptors for water-soluble hormones are coupled most often via regulatory molecules called G proteins to intracellular second messengers. Intracellular receptors for lipid-soluble hormones directly activate genes and stimulate synthesis of specific proteins. **8.** Hormone release can be triggered by humoral, neural, or hormonal stimuli. **9.** Lipid-soluble hormones have longer half-lives, meaning that they stay in the blood longer. (They are not as readily excreted by the kidneys because they are bound to plasma proteins, and most need to be metabolized by the liver before they can be excreted.) **10.** The hypothalamus communicates with the anterior pituitary via hormones released into a special portal system of blood vessels. In contrast, it communicates with the posterior pituitary via action potentials traveling down axons that connect the hypothalamus to the posterior pituitary. **11.** Drinking alcoholic beverages inhibits ADH secretion from the posterior pituitary and causes copious urine output and dehydration. The dehydration causes the hangover effects. **12.** LH and FSH are tropic hormones that act on the gonads, TSH is a tropic hormone that acts on the thyroid, and ACTH is a tropic hormone that acts on the adrenal cortex. (If you said growth hormone, that's also a good answer, as GH causing the liver to release IGFs might also be considered a tropic effect.) **13.** T_4 has four bound iodine atoms, and T_3 has three. T_4 is the major hormone secreted, but T_3 is more potent. T_4 is referred to as thyroxine. **14.** Thyroid hormone must be carried through blood by plasma proteins because it is lipid soluble and so does not readily dissolve in blood, which is mostly water. The receptors for TSH (thyroid-stimulating hormone) are located on the plasma membrane of the thyroid follicular cells (on the sides facing away from the colloid). (Recall that TSH is a water-soluble hormone from the anterior pituitary.) **15.** Thyroid hormone increases basal metabolic rate (and heat production) in the body. Parathyroid hormone increases blood Ca^{2+} levels in a variety of ways. Calcitonin at high (pharmacological) levels has a Ca^{2+}-lowering, bone-sparing effect. (At normal blood levels its effects in humans are negligible.) **16.** Thyroid follicular cells release thyroid hormone, parathyroid cells in the parathyroid gland release parathyroid hormone, and parafollicular (C) cells in the thyroid gland release calcitonin. **17.** Glucocorticoids are stress hormones that, among many effects, increase blood glucose. Mineralocorticoids increase blood Na^+ (and blood pressure) and decrease blood K^+. Gonadocorticoids are male and female sex hormones that are thought to have a variety of effects (for example, contribute to onset of puberty, sex drive in women, pubic and axillary hair development in women). **18.** The two major stimuli for aldosterone release are (1) a decrease in blood volume or blood pressure, which triggers the renin-angiotensin-aldosterone mechanism, and (2) an increase in plasma K^+ levels. The major stimulus for cortisol secretion is ACTH from the anterior pituitary. ACTH release, in turn, is controlled by CRH from the hypothalamus. CRH release follows a diurnal (day-night) rhythm and is increased in response to various stressors, as well as by hypoglycemia and fever. **19.** The adrenal cortex (a) releases steroid hormones. The adrenal medulla (b) and anterior pituitary (c) and posterior pituitary (d) all release amino acid–based hormones. ACTH from the anterior pituitary acts to cause release of steroid hormones from the adrenal cortex. Both the adrenal medulla (b) and

posterior pituitary (d) are composed of nervous tissue rather than the glandular tissue (a type of epithelial tissue) that other glands are made of. **20.** Melatonin is used by some individuals as a sleep aid, particularly to counter jet lag. **21.** The heart produces atrial natriuretic peptide (ANP). ANP decreases blood volume and blood pressure by increasing the kidneys' production of salty urine. **22.** The major function of vitamin D_3, produced in inactive form by the skin, is to increase intestinal absorption of calcium. **23.** Diabetes mellitus is due to a lack of insulin production or action, whereas diabetes insipidus is due to a lack of ADH. Both conditions are characterized by production of copious amounts of urine. You would find glucose in the urine of a patient with diabetes mellitus, but not in the urine of a patient with diabetes insipidus. **24.** The gonadal hormones are steroid hormones. A major endocrine gland that also secretes steroid hormones is the adrenal cortex.

Review Questions 1. b; **2.** a; **3.** c; **4.** d; **5.** (1)c, (2)a and b, (3)f, (4)d, (5)e, (6)g, (7)a, (8)h, (9)b and e, (10)a; **6.** d; **7.** c; **8.** b; **9.** d; **10.** b; **11.** d; **12.** b; **13.** Hormones are chemical messengers secreted by cells into the extracellular fluid. They travel through the blood and regulate the metabolic function of other cells in the body. **14.** Atrial natriuretic peptide is a hormone released by cardiac muscle cells. Antidiuretic hormone, oxytocin, epinephrine, and norepinephrine are all hormones that are released by neurons.

Chapter 17

Check Your Understanding 1. Blood can prevent blood loss by forming clots when a blood vessel is damaged. Blood can prevent infection because it contains antimicrobial proteins and white blood cells. **2.** The hematocrit is the percentage of blood that is occupied by erythrocytes. It is normally about 45%. **3.** Plasma proteins perform many key functions in the blood (see Table 17.1). One key function is to maintain the correct osmotic pressure in the blood. Without the osmotic pressure of plasma proteins "pulling" water molecules into the bloodstream, more water enters the tissues. Your patient may have tissue swelling (called edema) due to water leaking out of the blood. In addition, some of the plasma proteins synthesized by the liver (e.g., fibrinogen) are clotting proteins, so your patient may bleed more readily in response to only minor injuries. While you might expect to see problems due to the loss of the carrier function of many plasma proteins, this doesn't generally lead to easily observable signs and symptoms. **4.** Each hemoglobin molecule can transport four O_2. The heme portion of the hemoglobin binds the O_2. **5.** The kidneys' synthesis of erythropoietin is compromised in advanced kidney disease, so RBC production decreases, causing anemia. **6.** The cells are (a) lymphocyte, (b) eosinophil, (c) monocyte, (d) basophil, (e) neutrophil. (1) Neutrophils (e) are and monocytes (c) become important phagocytes. (2) Eosinophils (b), basophils (d), and neutrophils (e) are granulocytes. (3) Lymphocytes (a) are abundant in lymph nodes and in the spleen. (4) The granule contents of basophils (d) are similar to those of mast cells. **7.** Amos's red bone marrow is spewing out many abnormal white blood cells, which are crowding out the production of normal bone marrow elements. The lack of normal white blood cells allows the infections, the low number of platelets fails to stop bleeding, and the lack of erythrocytes is anemia. **8.** Microglial cells can become phagocytes in the brain. **9.** A megakaryocyte is a cell that produces platelets. Its name means "big nucleus cell." **10.** The three steps of hemostasis are vascular spasm, platelet plug formation, and coagulation. **11.** Prothrombin activator catalyzes the conversion of prothrombin to thrombin. Thrombin catalyzes the conversion of fibrinogen to fibrin. Prothrombin is an inactive precursor, whereas thrombin acts as an enzyme. Fibrinogen is water soluble, whereas fibrin is not. Most factors are inactive in blood before activation and become enzymes upon activation. (There are exceptions, such as fibrinogen and calcium.) **12.** Thrombocytopenia (platelet deficiency) results in failure to plug the countless small tears in blood vessels, and so manifests as small purple spots (petechiae). Hemophilia A results from the absence of clotting factor VIII. **13.** Nigel has anti-B antibodies in his blood and type A agglutinogens (antigens) on his RBCs. He can donate blood to an AB recipient, but he should not receive blood from an AB donor because his anti-B antibodies will cause a transfusion reaction. **14.** See the drawing. ❭ As shown, the antigens (agglutinogens) are in the plasma membrane and the antibody (agglu-tinin) is in the plasma (outside of the RBC). The presence of agglutinogen D is Rh$^+$. Note that the shapes of the antigens and antibody are only representative. You would be correct to use any shape to represent them (as long as the anti-A antibody does not match the shape of antigen B). **15.** If Emily has a bacterial men-ingitis, a differential WBC count would likely reveal an increase in neutrophils because neutrophils are a major body defense against bacteria.

Review Questions 1. c; **2.** c; **3.** d; **4.** b; **5.** d; **6.** a; **7.** a; **8.** b; **9.** c; **10.** d; **11.** (a) The three major categories of formed elements are leukocytes, erythrocytes, and platelets. (b) The least numerous of the formed elements are the leukocytes. (c) The buffy coat in a hematocrit tube consists of the leukocytes and platelets. **12.** (a) A transfusion reaction includes agglutination of donor RBCs, blockage of small blood vessels, and lysis of RBCs. It occurs after transfusion of mismatched blood as a result of the recipient's plasma antibodies attacking the donor's RBCs. (b) Consequences of a transfusion reaction include loss of oxygen-carrying capacity, fever, chills, low blood pressure, rapid heartbeat, nausea, vomiting, and kidney failure.

Chapter 18

Check Your Understanding 1. The mediastinum is the medial cavity of the thorax within which the heart, great vessels, thymus, and parts of the trachea, bronchi, and esophagus are found. **2.** The layers of the heart wall are the endo-cardium, the myocardium, and the epicardium. The epicardium is also called the visceral layer of the serous pericardium. This is surrounded by the parietal layer of the serous pericardium and the fibrous pericardium. **3.** The serous fluid decreases friction caused by movement of the layers against one another. **4.** The structures are (a) aorta, (b) pulmonary trunk, (c) left atrium, (d) right atrium, (e) left ventricle, (f) right ventricle, (g) inferior vena cava. The right atrium (d) receives oxygen-poor blood. The left ventricle (e) pumps blood into the systemic circulation. The pulmonary trunk (b) transports blood toward the lungs. **5.** The papillary muscles and chordae tendineae keep the AV valve flaps from evert-ing into the atria as the ventricles contract. **6.** The mitral (left atrioventricular) valve has two cusps. **7.** The right side of the heart acts as the pulmonary pump, whereas the left acts as the systemic pump. **8.** (a) True. The left ventricle wall is thicker than the right. (b) True. The left ventricle pumps blood at much higher pressure than the right ventricle because the left ventricle supplies the whole body, whereas the right ventricle supplies only the lungs. (c) False. Each ventri-cle pumps the same amount of blood with each beat. If this were not true, blood would back up in either the systemic or pulmonary circulation (because the two ventricles are in series). **9.** The branches of the right coronary artery are the right marginal artery and the posterior interventricular artery. **10.** (a) The refractory period is almost as long as the contraction in cardiac muscle. (b) The source of Ca^{2+} for the contraction is only SR in skeletal muscle. (c) Both skeletal muscle and cardiac muscle have troponin. (d) Only skeletal muscle has triads. **11.** Structure (a) shows cardiac muscle and structure (b) shows skeletal muscle. You can tell that both of these structures are parts of striated (and not smooth) muscle fibers because they have sarcomeres. You can tell that (b) is skeletal muscle because it has terminal cisterns next to each of the T tubules, whereas (a) does not. Also, the T tubules in cardiac muscle (a) are aligned with the Z discs, whereas in skeletal muscle (b), they are at the A-I junctions. **12.** Cardiac muscle cannot go into tetany because the absolute refractory period is almost as long as the contraction. **13.** (a) The QRS wave occurs during ventricular depolarization. (b) The T wave of the ECG occurs during ventricular repolarization. (c) The P-R interval of the ECG occurs during atrial depolarization and the conduction of the action potential through the rest of the intrinsic conduction system. **14.** It is not necessary for the intrinsic conduction system to contact each individual cardiac myocyte. This is because cardiac myocytes are connected to each other by gap junctions and so are depolarized when an adjacent myocyte depolarizes. The subendocardial conducting network excites ventricular muscle fibers. The depo-larization wave travels upward from the apex toward the atria. **15.** (b) Represents an action potential (AP) in a contractile cardiac muscle cell, (a) represents an AP in a skeletal muscle cell, and (c) represents an AP in cardiac pacemaker cells. The depolarization phase is due to Na$^+$ influx in skeletal muscle and contractile

cardiac muscle cells, and it is due to Ca^{2+} entry in cardiac pacemaker cells. K$^+$ efflux is responsible for the repolarization in all action potentials. **16.** The second heart sound is associated with the closing of the semilunar valves. **17.** The murmur of mitral insufficiency occurs during ventricular systole (because this is when the valve should be closed, and the murmur is due to blood leaking through the incompletely closed valve into the atrium). **18.** ❯ **19.** Exercise acti-vates the sympathetic nervous system. Sym-pathetic nervous system activity increases heart rate. It also directly increases ventricular con-tractility, thereby increas-ing Josh's stroke volume. **20.** If the heart is beating very rapidly, the amount of time for ventricular filling between contractions is decreased. This decreases the end diastolic volume, decreases the stroke volume, and therefore decreases the cardiac output.

Review Questions 1. a; **2.** c; **3.** b; **4.** c; **5.** b; **6.** b; **7.** c; **8.** b; **9.** The heart is located within the mediastinum. It lies anterior to the vertebral column and pos-terior to the sternum. About 2/3 of the heart lies to the left of the midsternal line. **10.** The longer refractory period of cardiac muscle is desirable because it prevents the heart from going into prolonged or tetanic contractions, which would stop its pumping action. **11.** See Figure 18.16. The P wave results from depolarization of the atria. The QRS complex results from ventricular depolarization and precedes ventricular contraction. The T wave is caused by ventricular repolarization and precedes ventricular relaxation. **12.** The cardiac cycle includes all events associ-ated with the flow of blood through the heart during one complete heartbeat. One cycle includes a period of (1) ventricular filling (in mid-to-late diastole), (2) isovolumetric contraction, (3) ventricular ejection, and (4) isovolumetric relaxation (in early diastole). (For a more detailed description of what happens in each of these periods, see p. 693.) **13.** Cardiac output is the amount of blood pumped out of each ventricle in 1 minute. It can be calculated using the follow-ing equation: cardiac output = heart rate × stroke volume. **14.** The synonyms are: (a) atrioventricular groove = coronary sulcus, (b) tricuspid valve = right AV valve, (c) bicuspid valve = left AV valve or mitral valve, (d) atrioventricular bundle = bundle of His.

Chapter 19

Check Your Understanding 1. The sympathetic nervous system innervates blood vessels. The sympathetic nerves innervate the tunica media. The effector cells in the tunica media are smooth muscle cells. **2.** When vascular smooth muscle contracts, the diameter of the blood vessel becomes smaller. This is called vasoconstriction. **3.** Elastic arteries play a major role in dampening the pulsatile pressure of heart contractions. Dilation or constriction of arterioles determines blood flow to individual capillary beds. Muscular arteries have the thickest tunica media relative to their lumen size. **4.** If you were doing calf raises, your capillary bed would be in the condition depicted in part (a). The capillar-ies would be flushed with blood to ensure that the working calf muscles could receive the needed nutrients and dispose of their metabolic wastes. **5.** Valves prevent blood from flowing backwards in veins. They are formed from folds of the tunica intima. **6.** In the systemic circuit, veins contain more blood than arter-ies (see Figure 19.6). **7.** Veins have more anastomoses than arteries. **8.** The three factors that determine resistance are blood viscosity, vessel length, and vessel diameter. Vessel diameter is physiologically most important. **9.** The rate of flow will decrease 81-fold from its original flow (3 × 3 × 3 × 3 = 81). **10.** Vessel (c) would have the greatest blood flow and vessel (a) would have the least blood flow. This is because flow increases as resistance decreases ($F = \Delta P/\text{TPR}$), and resistance varies directly with length and inversely with diameter. Vessel (c) has the largest diameter and so would have a lower resistance than vessel (b) (which has the same length, but a smaller diameter). Vessel (a) is the longest and so would have a higher resistance than (b) (which has the same diameter, but a shorter length). **11.** (a) The largest pressure drop occurs in the arterioles (because they have the largest resistance). (b) The capillaries are the first vessels that do not have a difference in pressure between systole and diastole. This is good because pressure fluctuations might rupture fragile capillaries. In addition, smooth flow is ideal for exchange at capillaries. (c) About 15–17 mm Hg is left at the start of the venules to get blood back to the heart. Factors that help blood return to the heart include venous valves, the muscular pump, the respiratory pump, and sympathetic venoconstriction. **12.** Cole's pulse pressure is 60 mm Hg. His mean arterial pressure is $80 + 60/3 = 100$ mm Hg. **13.** When you first stand up, mean arterial pressure (MAP) temporarily decreases and this is sensed by aortic and carotid baroreceptors. Medullary cardiac and vasomotor center reflexes increase sympathetic and decrease parasympathetic outflow to the heart. Heart rate and contractility increase, increasing cardiac output and therefore MAP. Further, sympathetic constriction of arterioles increases total peripheral resistance, also increasing MAP. (In addition, increased constriction of veins increases venous return, which increases end diastolic volume, increasing stroke volume, and therefore cardiac output and MAP.) See also Figure 19.11 (bottom). **14.** The kidneys help maintain MAP by influencing blood volume. In renal artery obstruction, the blood pressure in the kidney is lower than in the rest of the body (because it is downstream from the obstruction). Low renal blood pressure triggers both direct and indirect renal mechanisms to increase blood pressure by increasing blood volume. This can cause hypertension (called "secondary hypertension" because it is secondary to a defined cause—in this case the renal artery obstruction). **15.** Bob is in vascular shock due to anaphylaxis, a systemic allergic reaction to his medication. His blood pressure is low because of widespread vasodilation triggered by the massive release of histamine. Bob's rapid heart rate is a result of the baroreceptor reflex triggered by his low blood pressure. This activates the sympathetic nervous system, increasing heart rate, in an attempt to restore blood pressure. **16.** Diuretics cause a decrease in blood volume (because more fluid is lost in urine), decreasing cardiac output, which decreases blood pressure. Beta-blockers block beta-adrenergic receptors. Their antihypertensive effects are primarily due to their action in the heart, where they decrease contractility (and therefore stroke volume) and heart rate, resulting in lower cardiac output. Ca^{2+} channel blockers decrease Ca^{2+} entry into arteriolar smooth muscle, decreasing its contraction (causing vasodilation) and resulting in decreased total peripheral resistance. By acting on Ca^{2+} channels in the heart, these drugs may also decrease contractility and heart rate. **17.** In a bicycle race, autoregulation by intrinsic metabolic controls causes arteriolar smooth muscle in your legs to relax, dilating the vessels and supplying more O_2 and nutrients to the exercising muscles. **18.** Extrinsic mechanisms, primarily the sympathetic nervous system, prevent blood pressure from plummeting by constricting arterioles elsewhere (such as the gut and kidneys). In addition, cardiac output increases, which also helps maintain MAP. **19.** ❯ The net filtration pressure (NFP) at this point in the capillary is 8 mm Hg (34 mm Hg – 26 mm Hg). This is a net outward pressure, which tells us that this point in the capillary is near

Capillary lumen	Interstitial fluid
HP_c = 32 mm Hg	
	HP_{if} = 1 mm Hg
OP_c = 25 mm Hg	
	OP_{if} = 2 mm Hg

the arterial end of the capillary. **20.** (a) An increase in interstitial fluid osmotic pressure (OP_{if}) would tend to pull more fluid out of capillaries (causing localized swelling, or edema). (b) An increase of OP_{if} to 10 mm Hg would increase the outward pressure on both the arteriolar and venous ends of the capillary. The NFP at the venous end would become 1 mm Hg (27 mm Hg – 26 mm Hg). (c) Fluid would flow out of the venous end of the capillary rather than in. **21.** In heart failure, the heart is no longer able to pump blood adequately (see p. 700). As a result, blood backs up into the veins and capillary beds and the hydrostatic pressure in the capillaries increases, causing edema. **22.** The external carotid

arteries supply most of the tissues of the head except for the brain and orbits. **23.** The cerebral arterial circle (circle of Willis) is the arterial anastomosis at the base of the cerebrum. **24.** The four unpaired arteries that emerge from the abdominal aorta are the celiac trunk, the superior and inferior mesenteric arteries, and the median sacral artery. **25.** You would palpate the popliteal artery behind the knee, the posterior tibial artery behind the medial malleolus of the tibia, and the dorsalis pedis artery on the foot. (See also Figure 19.8.) **26.** The vertebral arteries help supply the brain, but the vertebral veins do not drain much blood from the brain. **27.** The internal jugular veins drain the dural venous sinuses. Each internal jugular vein joins a subclavian vein to form a brachiocephalic vein. **28.** Capillary bed (a) is in the digestive organs and capillary bed (c) is in the liver. The liver capillaries are sinusoid capillaries. The blood vessel names are hepatic portal vein (b), hepatic vein (d), and inferior vena cava (e). **29.** The leg veins that often become varicosed are the great and small saphenous veins. **30.** In Chapter 16 (The Endocrine System), you learned about the hypophyseal portal system (see Focus Figure 16.1 on p. 611).

Review Questions 1. d; **2.** d; **3.** c; **4.** e; **5.** d; **6.** c; **7.** b; **8.** b; **9.** a; **10.** (1)b, e, g; (2)c; (3)i; (4)f, h; (5)d; **11.** c

Chapter 20

Check Your Understanding 1. Lymph is the fluid inside lymphatic vessels. It enters lymphatic vessels from interstitial fluid. Interstitial fluid, in turn, is a filtrate of blood plasma. **2.** The right lymphatic duct receives lymph from the right upper arm and the right side of the head and thorax. The thoracic duct drains lymph from the rest of the body. **3.** Lymph movement is driven by the contraction of adjacent skeletal muscles, pressure changes in the thorax during breathing, the pulsations of nearby arteries, and contraction of smooth muscle in the lymphatic vessel walls. (Valves in lymphatic vessels prevent backflow of lymph.) **4.** The blocked lymphatic vessels in Mr. Thomas's left leg no longer return excess tissue fluid or proteins to the bloodstream. The lack of fluid return from the tissues into the blood will result in a higher interstitial fluid hydrostatic pressure. This is an inward pressure, and will tend to help limit the edema. However, the lack of proteins being returned from the interstitial fluid to the blood will result in a higher interstitial fluid osmotic pressure, which will tend to pull more fluid into the tissue. **5.** The primary lymphoid organs are the red bone marrow and thymus. Primary lymphoid organs are special because they are the organs where lymphocytes originate and mature. **6.** Red bone marrow in adults is found in the cavities between trabeculae of spongy bones in the flat bones of the skull, the sternum, ribs, clavicles, scapulae, hip bones, and vertebrae. In addition it is found in the heads of the femur and humerus. **7.** Lymphoid follicles are solid, spherical bodies consisting of tightly packed reticular fibers and lymphoid cells, often with a lighter-staining central region. They are regions where B cells predominate. **8.** Having fewer efferents causes lymph to accumulate in lymph nodes, allowing more time for its cleansing. **9.** ❯ **10.** The spleen cleanses the blood, recycles breakdown products of RBCs, stores iron, stores platelets and monocytes, and is thought to be a site of erythrocyte production in the fetus. **11.** The structures are: (a) capsule and (b) trabecula. Region (d) is red pulp and so it is the site of red blood cell destruction and also the site

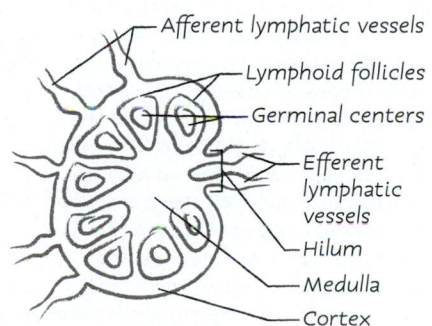

that has splenic cords and splenic sinusoids. Region (c) is white pulp and so it is the region composed mostly of lymphocytes. **12.** The splenic artery arises from the celiac trunk. The splenic vein empties into the hepatic portal vein. **13.** MALT (mucosa-associated lymphoid tissue) is lymphoid tissue found in the mucosa of the digestive, respiratory, and genitourinary tracts. It includes tonsils, Peyer's patches, and the appendix. **14.** B cells mature in the bone marrow.

Review Questions 1. c; **2.** c; **3.** a, d; **4.** c; **5.** a; **6.** b; **7.** a; **8.** b; **9.** d

Chapter 21

Check Your Understanding 1. The innate defense system is always ready to respond immediately, whereas it takes considerable time to mount the adaptive defense system. The innate defenses consist of surface barriers and internal defenses, whereas the adaptive defenses consist of humoral and cellular immunity, which rely on B and T lymphocytes. **2.** Surface barriers (the skin and mucous membranes) constitute the first line of defense. **3.** Opsonization is the process of making pathogens more susceptible to phagocytosis by decorating their surface with molecules that phagocytes can bind. Antibodies and complement proteins are examples of molecules that act as opsonins. **4.** Our own cells are killed by NK cells when they have been infected by viruses or when they have become cancerous. **5.** Redness, heat, swelling, and pain are the cardinal signs of inflammation. Redness and local heat are both caused by vasodilation of arterioles, which increases the flow of blood (warmed by the body core) to the affected area. The swelling (edema) is due to the release of histamine and other chemical mediators of inflammation, which increase capillary permeability. This increased permeability allows proteins to leak into the interstitial fluid (IF), increasing the IF osmotic pressure and drawing more fluid out of blood vessels and into the tissues, thereby causing swelling. The pain is due to two things: (1) the actions of certain chemical mediators (kinins and prostaglandins) on nerve endings, and (2) the swelling, which can compress free nerve endings. **6.** A mutation in the gene that codes for interferon would make Julian particularly susceptible to viruses. **7.** Four key characteristics of adaptive immunity are that it involves B and T lymphocytes, it is specific, it is systemic, and it has memory. **8.** A complete antigen has both immunogenicity and reactivity, whereas a hapten has reactivity but not immunogenicity. **9.** Self-antigens, particularly MHC proteins, mark a cell as self. **10.** Development of immunocompetence of a B or T cell is signaled by the appearance on its surface of specific and unique receptors for an antigen. In the case of a B cell, this receptor is a membrane-bound antibody. (In T cells, it is simply called the T cell receptor.) **11.** Dendritic cells, macrophages, and B cells can all act as APCs. Dendritic cells are most important for T cell activation. **12.** In clonal selection, the antigen does the selecting. What is being selected is a particular clone of B or T cells that has antigen receptors corresponding to that antigen. **13.** The T cell that would survive is (c), one that recognizes MHC but not self-antigen. **14.** The secondary response to an antigen is faster than the primary response because the immune system has already been "primed" and has memory cells that are specific for that particular antigen. **15.** An IgM antibody is shown at right. (You can tell because it consists of five Y-shaped parts—that is, it is a pentamer.) IgG antibody is most abundant in blood. IgM is secreted first in a primary immune response. IgA is most abundant in secretions. **16.** Antibodies can bring about destruction of pathogen via "PLAN"—**p**hagocytosis, **l**ysis (via complement), **a**gglutination, or **n**eutralization. **17.** Vaccinations protect by providing the initial encounter to an antigen—the primary response to that antigen. As a result, when the pathogen for that illness is encountered again, the pathogen elicits the much faster, more powerful secondary response, which is generally effective enough to prevent clinical illness. **18.** ❯ The stem region determines the class of antibody. **19.** Plasma cells make large amounts of antibodies—proteins that are exported from the cell. Rough endoplasmic reticulum is the site where proteins that are exported are synthesized. Ribosomes, the Golgi apparatus, and secretory vesicles are also required for protein synthesis and export, and so would also be abundant in these cells. **20.** Class II MHC proteins display exogenous antigens. Class II MHC proteins are recognized by CD4 T cells. APCs display class II MHC proteins. **21.** Helper T cells are central to both humoral and cellular immunity because they are required for activation of both cytotoxic T cells and most B cells. **22.** (a) Immune cells that secrete proteins that form a pore are cytotoxic T cells and NK cells. In this case, perforins are the proteins that form the pore. (b) Complement proteins are plasma proteins that can form a pore [which is called

Antigen-binding sites

Hinge region

Stem region

a membrane attack complex (MAC)]. Specifically, in complement-mediated cell lysis, MAC complexes cause a target cell (usually a bacterium) to become leaky. This allows water to enter and cytoplasmic contents to exit the cell, killing the cell. In contrast, cytotoxic T cells and NK cells kill their targets (usually virus-infected cells) by releasing perforins and granzymes onto the identified target cell. Perforins form a pore in the target cell membrane, and granzymes enter through this pore, activating enzymes that trigger apoptosis (cell suicide). **23.** The cell being activated is a CD4 cell and so could become either a helper T cell or a regulatory T cell. **24.** HIV is particularly hard for the immune system to defeat because (1) it destroys helper T cells, which are key players in adaptive immunity and (2) it has a high mutation rate and so it can avoid detection by the immune system by changing its surface antigens. **25.** Binding of an allergen onto specific IgE antibodies attached to mast cells triggers the mast cells to release histamine.

Review Questions 1. c; **2.** a; **3.** d; **4.** a; **5.** d; **6.** b; **7.** c; **8.** d; **9.** d; **10.** (1)b, g; (2)d, i; (3)a, e; (4)a, e, f, h; (5)e, h; (6)c, f, g; **11.** d; **12.** The antibody classes and their likely locations in the body are: IgM—attached to B cell plasma membrane as a monomer and in plasma as a pentamer; IgA—dimers found in secretions such as saliva, tears, intestinal juice, and milk; IgD—attached to B cell plasma membrane as a monomer; IgG—in plasma; IgE—attached to mast cells in skin, mucosae of the gastrointestinal and respiratory tracts, and tonsils.

Chapter 22

Check Your Understanding 1. The structures that air passes by are the nasal cavity (nares, nasal vestibule, nasal conchae), nasopharynx, oropharynx, and laryngopharynx. **2.** The nasal cavity contains both respiratory mucosa and olfactory mucosa. **3.** ❯ **4.** The epiglottis seals the larynx when we swallow. **5.** The incomplete, C-shaped cartilage rings of the trachea allow it to expand and contract and yet keep it from collapsing. **6.** The many tiny alveoli together have a

	Conducts	Epithelium	Tonsils
Nasopharynx	Air	Pseudostratified ciliated columnar	Pharyngeal Tubal
Oropharynx	Air and food	Stratified squamous	Palatine Lingual
Laryngopharynx	Air and food	Stratified squamous	— (none)

large surface area. This and the thinness of their respiratory membranes make them ideal for gas exchange. **7.** The peanut was most likely in the right main bronchus because it is wider and more vertical than the left. **8.** The two circulations of the lungs are the pulmonary circulation, which delivers deoxygenated blood to the lungs for oxygenation and returns oxygenated blood to the heart, and the bronchial circulation, which provides systemic (oxygenated) blood to lung tissue. **9.** Angiotensin converting enzyme is found in the plasma membrane of lung capillary endothelial cells. This is a good location for it because all of the blood in the body passes through the lung capillaries about once every minute. Angiotensin converting enzyme is part of the renin-angiotensin-aldosterone hormone cascade, which increases blood pressure. **10.** The driving force for pulmonary ventilation is a pressure gradient created by changes in the thoracic volume. **11.** The partial vacuum (negative pressure) inside the pleural cavity is caused by the opposing forces acting on the visceral and parietal pleurae. The visceral pleurae are pulled inward by the lungs' natural tendency to recoil and the surface tension of the alveolar fluid. The parietal pleurae are pulled outward by the elasticity of the chest wall. If air enters the pleural cavity, the lung on that side will collapse. This condition is called pneumothorax. **12.** A lack of surfactant increases surface tension in the alveoli and causes them to collapse between breaths. (In other words, it markedly decreases lung compliance.) **13.** ❯ Intrapulmonary pressure increases during expiration and decreases during inspiration. When intrapulmonary pressure is zero relative to atmospheric pressure, air flow is zero. This occurs at the three points labeled in the graph, which are between inspiration and expiration. After air flow stops, it changes direction.

14. ❯ **15.** Slow, deep breaths ventilate the alveoli more effectively because a smaller fraction of the tidal volume of each breath is spent moving air into and out of the dead space. **16.** Respiratory capacities are combinations of two or more respiratory volumes.

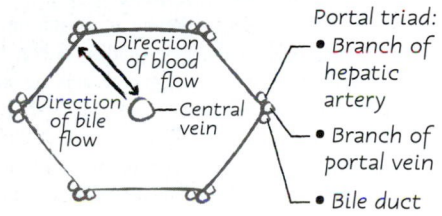

— Atmospheric pressure
0 mm Hg (760 mm Hg)

— Intrapulmonary pressure
-1 mm Hg (759 mm Hg)

— Intrapleural pressure
-5.5 mm Hg (754.5 mm Hg)

— Transpulmonary pressure
4.5 mm Hg

17. The difference in P_{O_2} between inspired air and alveolar air can be explained by (1) the gas exchange occurring in the lungs (O_2 continuously diffuses out of the alveoli into the blood), (2) the humidification of inspired air (which adds water molecules that dilute the O_2 molecules), and (3) the mixing of newly inspired air with gases already present in the alveoli. **18.** The arterioles leading into the O_2-enriched alveoli would be dilated. This response allows matching of blood flow to availability of oxygen. **19.** In a sealed container, the air and water would be at equilibrium. Therefore, the partial pressures of CO_2 and O_2 (P_{CO_2} and P_{O_2}) will be the same in the water as in the air: 100 mm Hg each. More CO_2 than O_2 molecules will be dissolved in the water (even though they are at the same partial pressure) because CO_2 is much more soluble than O_2 in water. **20.** About 70% of CO_2 is transported as bicarbonate ion (HCO_3^-) in plasma. Just over 20% is transported bound to hemoglobin in the RBCs, and 7–10% is dissolved in plasma. **21.** As blood CO_2 increases, blood pH decreases. This is because CO_2 combines with water to form carbonic acid. (However, the change in pH in blood for a given increase in CO_2 is minimized by other buffer systems.) **22.** The shift shown in graph (b) would allow more oxygen delivery to the tissues. Conditions that would cause the curve to shift this way are increased temperature, increased P_{CO_2}, decreased pH, or an increase in BPG levels. **23.** The ventral respiratory group of the medulla (VRG) is thought to be the rhythm-generating area. **24.** CO_2 in blood normally provides the most powerful stimulus to breathe. Central chemoreceptors are most important in this response (see Figure 22.28). **25.** Long-term adjustments to altitude include an increase in erythropoiesis, resulting in a higher hematocrit; an increase in BPG, which decreases Hb affinity for oxygen; and an increase in minute respiratory volume. **26.** The injured soccer player's P_{CO_2} is low. (Recall that normal P_{CO_2} = 40 mm Hg.) The low P_{CO_2} reveals that this is hyperventilation and not hyperpnea (which is not accompanied by changes in blood CO_2 levels). **27.** The obstruction in asthma is reversible, and acute exacerbations are typically followed by symptom-free periods. In contrast, the obstruction in chronic bronchitis is generally not reversible. **28.** Because Mr. Singh's alveoli are poorly ventilated, the local P_{O_2} around those alveoli would be low. This low P_{O_2} would cause the pulmonary arterioles to constrict (due to ventilation-perfusion coupling), increasing resistance and thereby increasing pulmonary blood pressure. In the long term, pulmonary hypertension can cause right heart failure (see *cor pulmonale* in Chapter 18 *Related Clinical Terms* on p. 702). When one side of the heart fails, pressure increases upstream of the failing pump—in this case in the systemic capillaries. This causes peripheral edema, which would be noticeable in the extremities such as the feet and ankles (see p. 700).

Review Questions 1. b; **2.** a and c; **3.** c; **4.** c; **5.** b; **6.** d; **7.** d; **8.** b; **9.** c; **10.** b

Chapter 23

Check Your Understanding 1. The esophagus is found in the thorax. Three alimentary canal organs found in the abdominal cavity include the stomach, small intestine, and large intestine. **2.** The usual site of ingestion is the mouth. **3.** The process of absorption moves nutrients into the body. **4.** The visceral peritoneum is the outermost layer of the digestive organ; the parietal peritoneum is the serous membrane covering the wall of the abdominal cavity. **5.** The pancreas is retroperitoneal. **6.** The hepatic portal circulation is the venous portion of the splanchnic circulation. **7.** ❯ **8.** The muscularis externa is unitary smooth muscle. Characteristics of unitary smooth muscle that make it well suited for this location are that

Mucosa
• Epithelium
• Lamina propria
• Muscularis mucosae
Submucosa
Muscularis externa
Serosa

it is electrically coupled by gap junctions and so contracts as a unit, is arranged in longitudinal and circular sheets, and exhibits rhythmic spontaneous action potentials. **9.** Reflexes associated with the GI tract promote muscle contraction and secretion of digestive juices or hormones. **10.** The term "gut brain" refers to the enteric nervous system, the web of neurons closely associated with the digestive organs. **11.** He should temporarily refrain from eating because the parasympathetic nervous system stimulates digestive activities. **12.** The palate forms the roof of the mouth. The hard palate supported by bone is anterior to the soft palate (no bony support). **13.** The tongue is important for taste and for speech, particularly for uttering consonants. **14.** Antimicrobial substances found in saliva include lysozyme, defensins, and IgA antibodies. **15.** Enamel is harder than bone. Pulp consists of nervous tissue and blood vessels. **16.** The glands are (a) parotid gland, (b) sublingual gland, and (c) submandibular gland. The duct of the submandibular gland opens at the base of the lingual frenulum. The duct of the parotid gland pierces the buccinator muscle. **17.** The pharynx is part of the digestive and respiratory systems. **18.** The esophageal muscularis externa undergoes a transformation along its length from skeletal muscle superiorly to smooth muscle near the stomach. **19.** The esophagus is merely a chute for food passage and is subjected to a good deal of abrasion, which a stratified squamous epithelium can withstand. The stomach mucosa is a secretory mucosa served well by a simple columnar epithelium. **20.** During swallowing the larynx rises and the epiglottis covers its lumen so that foodstuffs are diverted into the esophagus posteriorly. **21.** Peristaltic contractions become more and more powerful as they approach the pylorus. Only a small squirt of liquid chyme is allowed through the pyloric sphincter before this sphincter closes as part of the peristaltic wave. Once the pyloric sphincter is closed, solids in the stomach are propelled back into the stomach (a movement called retropulsion) while being pummeled and ground together. **22.** The chief cells produce pepsinogen, which is the inactive enzyme pepsin, and the parietal cells secrete HCl needed to activate pepsinogen. **23.** The three phases of gastric secretion are the cephalic, gastric, and intestinal phases. **24.** The presence of food in the duodenum inhibits gastric activity by triggering the enterogastric reflex and the secretion of enterogastrones (hormones). **25.** The enterohepatic circulation is an important recycling mechanism for retaining bile salts needed for fat absorption. **26.** Pancreatic acini produce the exocrine products of the pancreas (digestive enzymes and bicarbonate-rich juice). The islets produce pancreatic hormones, most importantly insulin and glucagon. **27.** Fluid in the pancreatic duct is bicarbonate-rich, enzyme-rich pancreatic juice. Fluid in the cystic and bile ducts is bile. **28.** CCK is secreted in response to the entry into the duodenum of chyme rich in protein and fat. It causes the pancreatic acini to secrete digestive enzymes, stimulates the gallbladder to contract, and relaxes the hepatopancreatic sphincter. **29.** ❯ **30.** All of these modifications increase the surface area of the small intestine. **31.** Brush border enzymes are enzymes associated with the microvilli

Direction of blood flow

Direction of bile flow

Central vein

Portal triad:
• Branch of hepatic artery
• Branch of portal vein
• Bile duct

of the enterocytes of the small intestine mucosa. **32.** Distension of stomach walls enhances stomach secretory activity. Distension of the walls of the small intestine reduces stomach secretory activity (to give the small intestine time to carry out its digestive and absorptive activities). **33.** Segmentation is the most common motility pattern after a meal. Segmentation mixes chyme with digestive enzymes and exposes the products of this digestion to the absorptive epithelium where brush border enzymes complete digestion and where absorption occurs. **34.** MMC is the migrating motor complex, a pattern of peristalsis seen in the small intestine that moves the last remnants of a meal plus bacteria and other debris into the large intestine. MMC is important to prevent the overgrowth of bacteria in the small intestine. **35.** Mass movements and haustral contractions occur in the large intestine. Mass movements are long, slow, powerful contractions that move over large areas of the colon three or four times a day, forcing the contents toward the rectum. Haustral contractions are a special type of segmentation. **36.** Activation of stretch receptors in the rectal walls initiates the defecation reflex. **37.** Enteric bacteria synthesize B vitamins and some of the vitamin K the liver needs to synthesize clotting proteins. **38.** To be absorbed, nutrients pass through the apical and then the basolateral membranes of enterocytes. **39.** Capillaries that receive absorbed nutrients are in the lamina propria of the mucosa of the small intestine. **40.** Amylase is to starch as lipase is to fats. **41.** Lack of bile salts would dramatically decrease both digestion and absorption of fats and fat-soluble vitamins. Without bile salts, you would expect the patient to have fatty stools. The patient might exhibit signs of weight loss and malnutrition. (In children this would be seen as lack of growth and failure to thrive.) The patient may also exhibit signs of deficiency of fat-soluble vitamins (see Table 24.2).

Review Questions 1. c; **2.** d; **3.** d; **4.** b; **5.** b; **6.** a; **7.** b; **8.** c; **9.** c; **10.** a; **11.** The six functional activities of the digestive system are ingestion, propulsion, mechanical breakdown, digestion, absorption, and defecation. **12.** (a) The normal number of permanent teeth is 32. The normal number of deciduous teeth is 20. (b) Enamel covers a tooth crown, whereas cement covers its root. (c) Dentin forms the bulk of a tooth. (d) Pulp is found in the central cavity of a tooth. It consists of connective tissue containing blood vessels and nerve fibers.

Chapter 24

Check Your Understanding 1. The five major nutrients are carbohydrates, proteins, fats, minerals, and vitamins. **2.** Cellulose provides fiber, which helps in elimination. **3.** Triglycerides are used for ATP synthesis, body insulation and protective padding, and to help the body absorb fat-soluble vitamins. Cholesterol is the basis of our steroid hormones and bile salts, and stabilizes cellular membranes. **4.** Beans (legumes) and grains (wheat) are good sources of protein, but neither is a complete one. However, together they provide all the essential amino acids. **5.** Vitamins serve as the basis for coenzymes, which work with enzymes to accomplish metabolic reactions. **6.** Iodine is essential for thyroxine synthesis. Calcium in the form of bone salts is needed to make bones hard. Iron is needed to make functional hemoglobin. **7.** Vitamin B_{12} needs intrinsic factor synthesized by parietal cells to be absorbed by the intestine. Vitamin B_{12} is absorbed in the ileum, and its lack causes pernicious anemia. **8.** A redox reaction is a combination of an oxidation and a reduction reaction. As one substance is oxidized, another is reduced. **9.** Some of the energy released during catabolism is captured in the bonds of ATP, which provides the energy needed to carry out the constructive activities of anabolism. **10.** The energy released during the oxidation of food fuels is used to pump protons across the inner mitochondrial membrane. **11.** In substrate-level phosphorylation, high-energy phosphate groups are transferred directly from phosphorylated intermediates to ADP to form ATP. In oxidative phosphorylation, electron transport proteins forming part of the inner mitochondrial membrane use energy released during oxidation of glucose to create a steep gradient for protons across this membrane. Then as protons flow back through the membrane, gradient energy is captured to attach phosphate to ADP. **12.** If oxygen and pyruvic acid are not available, glycolysis will stop because the supply of NAD^+ is limited and glycolysis can continue only if the reduced coenzymes ($NADH + H^+$) formed during glycolysis are relieved of their extra hydrogen. **13.** Oxidation (via removal of H) is common in the citric acid cycle; it is indicated by the reduction of a coenzyme (either NAD^+ or FAD). Decarboxylations are also common, and are indicated by the removal of CO_2 from the cycle.

14. Glycogenolysis is the reaction in which glycogen is broken down to its glucose monomers. **15.** ❯ Note that the area with the lowest pH is the area with the most protons (H^+). **16.** Glycerol, a breakdown product of fat metabolism, enters glycolysis. **17.** Acetyl CoA is the central molecule of fat metabolism. **18.** The products of beta

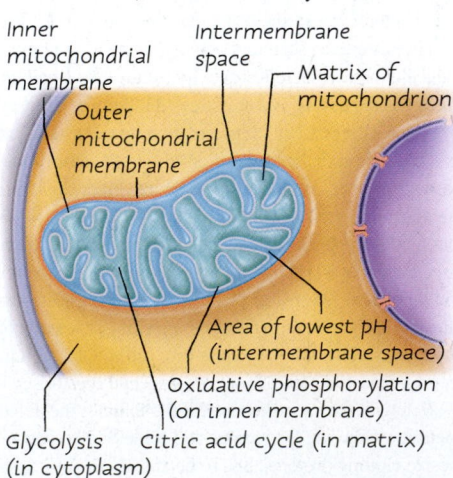

Inner mitochondrial membrane

Intermembrane space

Matrix of mitochondrion

Outer mitochondrial membrane

Area of lowest pH (intermembrane space)

Oxidative phosphorylation (on inner membrane)

Glycolysis (in cytoplasm)

Citric acid cycle (in matrix)

oxidation are acetyl CoA (acetic acid + coenzyme A), $NADH^+ + H^+$, and $FADH_2$. **19.** The liver uses keto acids drained off the citric acid cycle and amino groups (from other nonessential amino acids) as substrates to make the nonessential amino acids that the body needs. **20.** The ammonia removed from amino acids is combined with carbon dioxide to form urea, which is then eliminated by the kidneys. **21.** The three organs or tissues that regulate the directions of interconversions in the nutrient pools are the liver, skeletal muscles, and adipose tissues. **22.** Anabolic reactions and energy storage typify the absorptive state. Catabolic reactions (to increase blood sugar levels) such as lipolysis and glycogenolysis, and glucose sparing occur in the postabsorptive state. **23.** The main antagonist of glucagon is insulin. **24.** A rise in amino acid levels in blood increases both insulin and glucagon release. **25.** High HDL levels would be preferable because the cholesterol these particles transport is destined for the liver and elimination from the body. **26.** Trans fats are oils that have been hydrogenated (with H atoms). They are unhealthy because they cause LDLs to increase and HDLs to decrease—exactly the opposite of what is desirable. **27.** Among short-term stimuli influencing feeding behavior are neural signals from the digestive tract, nutrient signals related to energy stores, and GI tract hormones (CCK, insulin, glucagon, and ghrelin). **28.** Leptin is the most important long-term regulator of feeding behavior. **29.** As you can see in Figure 24.25, the ventromedial nucleus (VMN) is part of the satiety (appetite suppression) pathway. If the VMN is destroyed, the balance between this satiety pathway and the hunger pathway would be lost and the patient would never feel full and would very likely gain weight. **30.** Of the factors listed, breathing and kidney function contribute to BMR. **31.** Samantha is suffering from hypersecretion of thyroid hormone, which makes her feel restless and warm (because it raises her BMR) and gives her insomnia. The presence of autoantibodies suggests that she has Graves' disease (see p. 620), a condition in which antibodies bind to receptors for TSH (mimicking TSH) and stimulate continuous thyroid hormone release. **32.** The body's core is the organs within the skull and the thoracic and abdominal cavities. **33.** Andrea's body temperature is rising as heat-promoting mechanisms (shivering, chills) are activated. Something (an infection?) has caused the hypothalamic thermostat to be set to a higher level (fever) temporarily. **34.** In conduction, heat is transferred directly from one object to another (a hot surface to your palm). In convection, air warmed by body heat is continually removed (warm air rises) and replaced by cooler air (cool air falls), which in turn will absorb heat radiating from the body.

Review Questions 1. a; **2.** c; **3.** b; **4.** d; **5.** b; **6.** c; **7.** a; **8.** d; **9.** d; **10.** b; **11.** d; **12.** a

Chapter 25

Check Your Understanding 1. The lower part of his rib cage and the perirenal fat capsule protect his kidneys from blows. **2.** The layers of supportive tissue around each kidney are the fibrous capsule, the perirenal fat capsule, and the renal fascia. The parietal peritoneum overlies the anterior renal fascia. **3.** The renal pelvis, which has extensions called calyces, is continuous with the ureter. **4.** Filtrate is formed in the glomerular capsule and then passes through the proximal convoluted tubule (PCT), the descending and ascending limbs of the nephron loop, and the distal convoluted tubule (DCT). **5.** The structural differences are (1) juxtamedullary nephrons have long nephron loops (with long thin segments) and renal corpuscles that are near the cortex-medulla junction, whereas cortical nephrons have short nephron loops and renal corpuscles that lie more superficially in the cortex; (2) efferent arterioles of juxtamedullary nephrons supply vasa recta, while efferent arterioles of cortical nephrons supply peritubular capillaries. **6.** The glomerular capillaries are fenestrated capillaries. (See Figure 19.3 on p. 711 to refresh your memory of capillary types.) Their function is to filter large amounts of plasma into the glomerular capsule. **7.** The structures shown are (a) glomerular capsule, (b) efferent arteriole, (c) vasa recta, (d) proximal convoluted tubule (PCT), (e) collecting duct, and (f) nephron loop (ascending thin limb). **8.** Excretion is a means of removing wastes from the body, whereas tubular secretion is the process of selectively moving substances from the blood into the filtrate. While secretion often leads to excretion, this is not always the case because something could be secreted in the proximal tubules and yet reabsorbed distally. **9.** Intrinsic controls serve to maintain a nearly constant GFR in spite of changes in systemic blood pressure. Extrinsic controls serve to maintain systemic blood pressure. **10.** Hydrostatic pressure in the glomerular capillaries (HP_{gc}) is regulated by intrinsic and extrinsic controls of GFR. **11.** Net filtration pressure is 5 mm Hg [50 mm Hg – (25 mm Hg + 20 mm Hg)]. **12.** The majority of reabsorption occurs in the proximal convoluted tubule. **13.** The reabsorption of Na^+ by primary active transport drives reabsorption of amino acids and glucose by secondary active transport. It also drives passive reabsorption of chloride, and reabsorption of water by osmosis. The reabsorption of water leaves behind other solutes, which become more concentrated and can therefore be reabsorbed by diffusion. **14.** ❯ **15.** In primary active transport, the energy for the process is provided directly by ATP. In secondary active transport, the energy for the process is provided by the Na^+ concentration gradient established by active pumping of Na^+ occurring elsewhere in the cell. As Na^+ moves down its own concentration gradient, it drives the movement of another

Transport protein (either secondary active transport or passive through ion channels)

Na^+

Transport protein (primary active—the Na⁺-K⁺ ATPase)

substance (e.g., glucose) against that substance's concentration gradient. **16.** H^+, K^+, NH_4^+, creatinine, urea, and uric acid are all substances that are secreted into the kidney tubules. **17.** The descending limb of the nephron loop is permeable to water and impermeable to NaCl. The ascending limb is impermeable to water and permeable to NaCl. **18.** ADH is released from the posterior pituitary in response to hyperosmotic extracellular fluid (as sensed by hypothalamic osmoreceptors). ADH causes insertion of aquaporins into the apical membranes of the principal cells of the collecting ducts. **19.** The three major nitrogenous wastes excreted in urine are urea, creatinine, and uric acid. **20.** The normal renal clearance value for amino acids is zero. You would expect this because amino acids are valuable as nutrients and as the building blocks for protein synthesis, so it would not be good to lose them in the urine. **21.** A kidney stone blocking the ureter would interfere with urine flow to the bladder. The pain would occur in waves that coincide with the peristaltic contractions of the smooth muscle of the ureter. **22.** The trigone is a smooth triangular region at the base of the bladder. Its borders are defined by the openings for the ureters and the urethra. **23.** The prostatic urethra, the intermediate part of the urethra, and the spongy urethra are the three regions of the male urethra. **24.** The detrusor contracts in response to increased firing of parasympathetic nerves. **25.** In both cases (micturition and defecation) the external sphincters are skeletal muscle under voluntary control, and the internal sphincters are smooth muscle controlled by the autonomic nervous system.

Review Questions 1. d; **2.** b; **3.** c; **4.** d; **5.** c; **6.** b; **7.** a; **8.** c; **9.** b; **10.** A perirenal fat capsule surrounds each kidney and cushions it against blows. **11.** Micturition is the act of emptying the bladder. The micturition reflex is activated when stretch receptors sense stretch of the bladder wall. This causes firing of afferent neurons, which excite parasympathetic (and inhibit sympathetic) efferent neurons in the sacral spinal cord. Parasympathetic activation causes the detrusor muscle to contract. In addition, the afferent impulses from stretch receptors inhibit somatic motor neurons, which allows the external urethral sphincter to open. Together, these responses allow micturition.

Chapter 26

Check Your Understanding 1. You have more intracellular than extracellular fluid and more interstitial fluid than plasma. **2.** Na^+ is the major cation in the ECF and K^+ is the major cation in the ICF. The intracellular counterparts to extracellular Cl^- are HPO_4^{2-} and protein anions. **3.** If you eat salty pretzels, your extracellular fluid volume will expand even if you don't ingest fluids. This is because water will flow by osmosis from the intracellular fluid to the extracellular fluid. **4.** An increase in osmolality of the plasma is most important for triggering thirst. This change is sensed by osmoreceptors in the hypothalamus. **5.** ADH cannot add water—it can only conserve what is already there. In order to reduce an increase in osmolality of body fluids, the thirst mechanism is required.

6. (a) A decrease in plasma proteins causes edema. (b) Copious sweating causes dehydration. (c) Using ecstasy (together with drinking lots of fluids) could cause hypotonic hydration because ecstasy promotes ADH secretion, which interferes with the body's ability to get rid of extra water. **7.** Arrow 2 in the proximal convoluted tubule shows the site where most Ca^{2+} is reabsorbed. Arrow 5 corresponds to Na^+ reabsorption in the collecting duct, which is regulated by aldosterone. Arrow 4 corresponds to K^+ secretion in the collecting duct, which is regulated by aldosterone. **8.** The major regulator of calcium in the blood is parathyroid hormone. Hypercalcemia decreases excitability of neurons and muscle cells and may cause life-threatening cardiac arrhythmias. Hypocalcemia increases excitability and causes muscle tetany. **9.** Insufficient aldosterone would cause Nathan's plasma Na^+ to be decreased and his plasma K^+ to be elevated. The decrease in plasma Na^+ would cause a decrease in blood pressure, because plasma Na^+ is directly related to blood volume, which is a major determinant of blood pressure. **10.** Acidemia is an arterial pH below 7.35 and alkalemia is a pH above 7.45. **11.** The three major chemical buffer systems of the body are the bicarbonate buffer system, the phosphate buffer system, and the protein buffer system. The most important intracellular buffer is the protein buffer system. **12.** Joanne's ventilation would be increased. The acidosis caused by the accumulated ketone bodies will stimulate the peripheral chemoreceptors, and this will cause more CO_2 to be "blown off" in an attempt to restore pH to normal. **13.** The most important urinary buffer of H^+ is the phosphate buffer system.

14. The tubule and collecting duct cells generate new HCO_3^- either by excreting ammonium ions (NH_4^+) or by excreting buffered H^+ ions. **15.** ❯ **16.** In both renal tubule cells and parietal cells, (a) the enzyme carbonic

$HCO_3^- + H^+$ $H^+ + HCO_3^-$

H_2CO_3 H_2CO_3

$H_2O + CO_2$ $CO_2 + H_2O$

anhydrase is key, and (b) blood concentration of HCO_3^- increases. **17.** Key features of an uncompensated metabolic alkalosis are an increase in blood pH and an increase in blood HCO_3^-. Key features of an uncompensated respiratory acidosis are a decrease in blood pH and an increase in blood P_{CO_2}. **18.** The kidneys compensate for respiratory acidosis by excreting more H^+ and generating new HCO_3^- to buffer the acidosis.

Review Questions 1. a; **2.** c; **3.** b; **4.** a and b; **5.** h, i; **6.** c, g; **7.** a, e; **8.** b; **9.** a, b; **10.** j; **11.** b, d; **12.** c; **13.** The body fluid compartments are: (1) intracellular fluid (ICF), which is the fluid inside cells, about 25 liters; (2) extracellular fluid (ECF), which is all of the fluid outside cells, about 15 liters. The ECF can be further divided into (a) plasma, which is the fluid part of blood, about 3 liters, and (b) interstitial fluid (IF), which is the fluid that surrounds and bathes the cells in tissues and is about 12 liters.

Chapter 27

Check Your Understanding 1. The HPG axis is the hormonal interrelationship between the hypothalamus, anterior pituitary, and gonads that regulates the production of gametes and sex hormones. **2.** Meiosis reduces the chromosomal count from $2n$ to n and introduces variability. **3.** ❯ **4.** The seminiferous tubules act as the sperm factories. **5.** When the ambient temperature is cold, the associated muscles contract, bringing the testes close to the warm body wall. When temperature is high, the associated muscles relax, allowing the testes to hang away from the body wall. The pampiniform venous plexus absorbs heat from the arterial blood, cooling the blood before it enters the testes. **6.** Balancing his laptop on his lap for two hours a day might affect Carlos's fertility by increasing the temperature inside his scrotum. Increased testicular

Metaphase I

Anaphase II

temperatures decrease fertility. **7.** The organs of the male duct system in order from the epididymis to the body exterior are the ductus deferens, ejaculatory duct, prostatic urethra, intermediate part of the urethra, and spongy urethra. The ductus deferens runs from the scrotum into the pelvic cavity. **8.** These stereocilia pass nutrients to the sperm and absorb excess testicular fluid. **9.** The corpora cavernosa allow the penis to become stiff so that it can enter the female vagina to deliver sperm. **10.** The male perineum is the external region where the penis and scrotum are suspended. **11.** The bulbo-urethral glands (c) secrete primarily mucus. The seminal glands (a) produce the largest fraction of semen volume. The prostate (b) is at the junction of the ejaculatory duct and urethra. **12.** Semen is sperm plus the secretions of the male accessory glands. **13.** Arthur probably has benign prostatic hyperplasia (BPH). (Recall that that prostate cancer is usually symptomless, whereas BPH is characterized by difficult urination.) **14.** Erection is the stiffening of the penis that occurs when blood floods the cavernous tissue and is also prevented from leaving the penis. It is caused by the parasympathetic division of the autonomic nervous system. **15.** Resolution is a period of muscular and psychological relaxation that follows orgasm. It results as the sympathetic nervous system causes constriction of the internal pudendal arteries, reducing blood flow to the penis, and activates small muscles that force blood out of the penis. **16.** The sperm head is the compacted DNA-containing nucleus. The acrosome that caps the head is a lysosome-like sac of enzymes. The midpiece contains the energy-producing mitochondria. The tail, a flagellum, is the propulsive structure. **17.** Sustentocytes provide nutrients and essential developmental signals to the developing sperm and form the blood testis barrier that prevents sperm antigens from escaping into the blood. Interstitial endocrine cells secrete testosterone. **18.** Both cilia and flagella use the motor protein dynein for movement (see p. 90). Cilia line both the upper and lower respiratory tract, and so Caleb might have recurrent episodes of sinusitis, bronchitis, or pneumonia (as a result of the inability of his cilia to efficiently move the mucus that clears infectious organisms from his respiratory tract). (Cilia also line the middle ear and so patients with defective dynein often have chronic otitis media and may present with hearing problems.) **19.** Follicle-stimulating hormone indirectly stimulates spermatogenesis by prompting the sustentocytes to secrete androgen-binding protein. Androgen-binding protein keeps the concentration of testosterone high in the vicinity of the spermatogenic cells, which directly stimulates spermatogenesis. **20.** Secondary sex characteristics of males include appearance of pubic, axillary, and facial hair, deepening of the voice, increased oiliness of the skin, and increased size (length and mass) of the bones and skeletal muscles. **21.** Testosterone has negative feedback effects on GnRH release from the hypothalamus and on gonadotropin release from the anterior pituitary. So if testosterone blood levels are high, blood levels of GnRH, FSH, and LH would be low. **22.** The female's internal genitalia include the ovaries and duct system (uterine tubes, uterus, and vagina). **23.** The ovaries produce the female gametes and secrete female sex hormones (estrogens and progesterone). **24.** Women are more at risk for PID than men because the duct system of women is incomplete—there is no physical connection between the ovary and the uterine tubes, which are open to the pelvic cavity. In men, the duct system is continuous from the testes to the body exterior. **25.** The waving action of the fimbriae and currents created by the beating cilia help to direct the ovulated oocytes into the uterine tube. **26.** The labeled structures are (a) ampulla of uterine tube, (b) fimbriae of the uterine tube, (c) ovary, (d) ovarian ligament, (e) endometrium of uterus, (f) cervix, and (g) vagina. The usual site of fertilization is the uterine tube (a). The endometrium of the uterus (e) has spiral arteries. **27.** The greater vestibular glands are the female homologue of the male bulbo-urethral glands. **28.** Both the penis and clitoris are hooded by a skin fold and are largely erectile tissue. However, the clitoris lacks a corpus spongiosum containing a urethra, so the urinary and reproductive systems are completely separate in females. **29.** Developmentally, the mammary glands are modified sweat glands. **30.** The glandular portion of the breast usually develops about midway through a pregnancy. **31.** The products of meiosis in females are 2 or 3 nonfunctional polar bodies (tiny haploid cells with essentially no cytoplasm) and 1 haploid ovum (functional gamete). Meiosis in males yields 4 functional gametes, the haploid sperm. **32.** The antrum is the fluid-filled cavity of a vesicular follicle. **33.** From left to right, the three follicles indicated in Figure 27.16 on p. 1061 are (1) a vesicular (antral) follicle, (2) a secondary follicle, and (3) a primary follicle. (Fixing and staining makes the fluid

inside the first, vesicular follicle appear cloudy. The primary and primordial follicles are quite tiny in this photomicrograph, so don't worry if you weren't sure whether the last one was a primary or primordial follicle.) **34.** Identical twins develop from separation of a very young embryo (the result of fertilization of a single oocyte by a single sperm) into two parts. Fraternal twins develop when different oocytes are fertilized by different sperm. **35.** In the luteal phase, the follicle from which an oocyte has been ovulated develops into a corpus luteum, which then secretes progesterone (and some estrogens). **36.** Leptin is important in advising the brain of the girl's readiness (relative to energy stores) for puberty. **37.** LH prompts ovulation. **38.** Estrogens exert positive feedback on the anterior pituitary that leads to a burstlike release of LH. **39.** Estrogens are responsible for the secondary sex characteristics of females. **40.** Estrogens promote epiphyseal closure in both males and females. Long bones stop growing in length when the epiphyses are ossified. **41.** The vestibular glands help to lubricate the vestibule. **42.** The human papillomavirus (HPV) is most associated with cervical cancer. **43.** Chlamydia is the most common bacterial sexually transmitted infection in the U.S.

Review Questions 1. c; **2.** a, c, e, f; **3.** a and b; **4.** d; **5.** d; **6.** b; **7.** d; **8.** d; **9.** (1)c, f; (2)e, h; (3)g; (4)a; (5)b, e, and g; (6)f; **10.** d; **11.** b

Chapter 28

Check Your Understanding 1. The cortical reaction involves the release of enzymes from cortical granules to the oocyte exterior, which accomplishes destruction of the zona pellucida sperm-binding receptors and also the detachment of any sperm still bound to these receptors. These events help ensure monospermy. **2.** A cell derived from the fusion of a sperm and a polar body would not likely survive and develop into an embryo because it would lack the cytoplasm containing the nutrients required for its journey to the uterus (see p. 1071). **3.** Region (a) is the acrosome, which contains hydrolytic enzymes that the sperm uses to help it penetrate and enter the egg. The acrosomal reaction (release of the contents of the acrosome) is triggered by sperm binding to sperm-binding receptors in the zona pellucida. Before the acrosomal reaction can occur, sperm must undergo capacitation. **4.** The blastocyst is only slightly larger than the zygote because, although cell division has been going on (cleavage divisions), there is essentially no time for growth between divisions, so the resulting cells get smaller and smaller. The result is a large number of cells with a high surface-to-volume ratio. **5.** The trophoblast cells contribute to the formation of the placenta. **6.** The syncytiotrophoblast actually accomplishes implantation. **7.** The chorion develops from the trophoblast and a layer of extraembryonic mesoderm. **8.** The decidua basalis cooperates with the chorionic villi to form the placenta. **9.** The placenta is usually fully functional by the end of the third month of pregnancy. **10.** The blastocyst secretes the hormone human chorionic gonadotropin, which is detectable in the urine. **11.** Before organogenesis can occur in earnest, the embryonic body must fold and undercut to form a tubular embryo. **12.** The mesoderm gives rise to essentially all body tissues except neural and epidermal tissue and mucosae. **13.** For the most part, the special fetal circulatory modifications are occluded at birth or shortly thereafter. **14.** ❭ The foramen ovale and ductus arteriosus allow most of the blood to bypass the heart.

15. Without an amnion, the temperature of the developing fetus would fluctuate more. The fetus would also be less protected from physical trauma, and its body parts might adhere to each other during development. **16.** Difficult breathing during late pregnancy is due to the uterus pressing against and crowding the diaphragm (and hence the lungs). The waddling gait is due to the relaxing effect of relaxin on the pelvic ligaments and the pubic symphysis. **17.** The hormone hPL works with estrogens and progesterone to stimulate breast maturation in the pregnant woman, promotes fetal growth, and exerts a glucose-sparing effect on the mother. **18.** The enlarged uterus of a pregnant woman can press on the veins that return blood to the heart. Because the veins are downstream of the capillaries, the partial obstruction of venous flow would cause capillary hydrostatic pressure to rise (just like stepping on the downstream part of a hose would cause pressure inside the hose to go up). The edema of preeclampsia, on the other hand, is due to a decrease in capillary colloid osmotic pressure. (See also p. 734.) **19.** Breech presentation is a buttock-first presentation of the baby during labor. **20.** Prostaglandins are most responsible for triggering true labor. **21.** The descent of the baby's head (the largest part of its body) follows the widest dimensions of the bony pelvis. **22.** An Apgar score is a quick assessment of the baby's overall physical status that is determined at 1 and 5 minutes after birth. **23.** Surfactant increases compliance (distensibility) by decreasing the alveolar surface tension. **24.** Oxytocin causes the let-down reflex. **25.** *In vitro* fertilization is the most common ART technique.

Review Questions 1. (1)a, (2)b, (3)b, (4)a; **2.** b; **3.** b; **4.** (1)e, (2)f, (3)a, (4)d, (5)h, (6)b, (7)g, (8)c; **5.** d; **6.** d; **7.** b; **8.** a; **9.** c; **10.** a; **11.** In breech (buttocks-first) presentation, delivery may be more difficult because the widest diameter of the baby, which is the skull, is not delivered first. When the pelvis or hips of a baby are delivered first, there is a possibility that the head may become stuck in the mother's pelvis. Also, the baby will not be able to breathe until it has completely exited the birth canal. **12.** The factors that bring about uterine contractions include high levels of estrogen (brought about by high levels of fetal cortisol). These high estrogen levels oppose the quieting actions of progesterone and cause insertion of myometrial oxytocin and prostaglandin receptors and formation of gap junctions between myometrial cells. Later, high estrogen levels trigger prostaglandin secretion by the placenta, This stimulates the rhythmic contractions of true labor and thinning of the cervix. Cervical distension activates the maternal hypothalamus, triggering the positive feedback release of oxytocin by the posterior pituitary. **13.** An episiotomy is an incision made to widen the vaginal orifice in order to reduce tissue tearing as the baby's head exits from the perineum.

Chapter 29

Check Your Understanding 1. Chromosomes other than sex chromosomes are called autosomes. **2.** An allele represented by a capital letter is presumed to be a dominant allele. **3.** Descriptions of his phenotype are blond, blue-eyed, and hairy chest. **4.** Stage (c) shows metaphase, which is the phase of the cell cycle from which chromosomes are used to prepare a karyotype. During interphase (b) the DNA-containing material is in the form of dispersed strands of chromatin. As mitosis begins, the chromatin coils and condenses, becoming visible as chromosomes in prophase (d). Chromosomes continue to condense throughout prophase and are most visible during metaphase when we can arrest cell division with a drug. **5.** The alleles on different pairs of homologous chromosomes are segregated independently of each other to different gametes. **6.** Crossing over

causes separations of linked genes on the same chromosome, producing gametes with more varied genomes. **7.** In incomplete dominance, the heterozygote has a phenotype intermediate between that of the dominant and recessive alleles [for example, for the sickling gene, the dominant homozygote (SS) has no evidence of sickling; the heterozygote (Ss) has sickle-cell trait; and the homozygote for the recessive gene (ss) has sickle-cell anemia]. In codominance, both dominant alleles are expressed (as in ABO blood types). **8.** Height is an example of a trait conferred by polygene inheritance in which several genes on different chromosomes contribute to the trait. Such traits show a distribution of phenotypes that yields a bell-shaped curve. **9.** A lethal recessive allele would likely be more widespread in the population. This is because dominant alleles are expressed. So if the dominant allele is lethal, the person who has that allele would probably not live very long or would die during development, and as a result the allele would not be transferred to the next generation. In contrast, a lethal recessive allele would kill only 1/4 of the possible offspring of two carriers, and would be passed to the next generation by half of the potential offspring. **10.** ❯ As the Punnett square shows, half of the daughters of this union would be affected by hemophilia (genotype X^hX^h) and half would be carriers (X^HX^h). Also, half of the sons would be affected (X^hY) and half would have

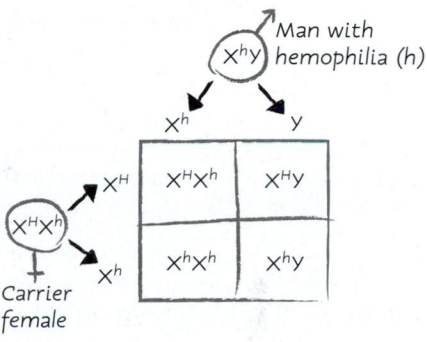

normal blood clotting (X^HY). **11.** Other genes, measles in a pregnant woman, and lack of key dietary nutrients all may alter gene expression. **12.** Genomic imprinting labels genes as maternal or paternal. **13.** Maternal mitochondrial DNA confers extranuclear inheritance. **14.** Amniocentesis analyzes chemicals and cells in amniotic fluid. **15.** Mrs. Lehman should be tested because Tay-Sachs disease is a recessive disorder. In order to be affected, the baby would need to receive one affected allele from each parent. Even though there is no incidence of the disease in her family, the recessive gene might be present and masked by the dominant gene. (Her husband should also be tested to see whether he received an affected allele from either of his parents.) **16.** Ultrasound imaging is used to determine some aspects of fetal development (fetal age for example) and is noninvasive.

Review Questions 1. (1)d, (2)g, (3)b, (4)a, (5)f, (6)c, (7)e, (8)h; **2.** (1)e, (2)a, (3)b, (4)d, (5)c, (6)d, (7)f

The Metric System

MEASUREMENT	UNIT AND ABBREVIATION	METRIC EQUIVALENT	METRIC TO ENGLISH CONVERSION FACTOR	ENGLISH TO METRIC CONVERSION FACTOR
Length	1 kilometer (km)	= 1000 (10^3) meters	1 km = 0.62 mile	1 mile = 1.61 km
	1 meter (m)	= 100 (10^2) centimeters = 1000 millimeters	1 m = 1.09 yards 1 m = 3.28 feet 1 m = 39.37 inches	1 yard = 0.914 m 1 foot = 0.305 m
	1 centimeter (cm)	= 0.01 (10^{-2}) meter	1 cm = 0.394 inch	1 foot = 30.5 cm 1 inch = 2.54 cm
	1 millimeter (mm)	= 0.001 (10^{-3}) meter	1 mm = 0.039 inch	
	1 micrometer (μm)	= 0.000001 (10^{-6}) meter		
	1 nanometer (nm)	= 0.000000001 (10^{-9}) meter		
	1 angstrom (Å)	= 0.0000000001 (10^{-10}) meter		
Area	1 square meter (m²)	= 10,000 square centimeters	1 m² = 1.1960 square yards 1 m² = 10.764 square feet	1 square yard = 0.8361 m² 1 square foot = 0.0929 m²
	1 square centimeter (cm²)	= 100 square millimeters	1 cm² = 0.155 square inch	1 square inch = 6.4516 cm²
Mass	1 metric ton (t)	= 1000 kilograms	1 t = 1.103 ton	1 ton = 0.907 t
	1 kilogram (kg)	= 1000 grams	1 kg = 2.205 pounds	1 pound = 0.4536 kg
	1 gram (g)	= 1000 milligrams	1 g = 0.0353 ounce 1 g = 15.432 grains	1 ounce = 28.35 g
	1 milligram (mg)	= 0.001 gram	1 mg = approx. 0.015 grain	
	1 microgram (μg)	= 0.000001 gram		
Volume (solids)	1 cubic meter (m³)	= 1,000,000 cubic centimeters	1 m³ = 1.3080 cubic yards 1 m³ = 35.315 cubic feet	1 cubic yard = 0.7646 m³ 1 cubic foot = 0.0283 m³
	1 cubic centimeter (cm³ or cc)	= 0.000001 cubic meter = 1 milliliter	1 cm³ = 0.0610 cubic inch	1 cubic inch = 16.387 cm³
	1 cubic millimeter (mm³)	= 0.000000001 cubic meter		
Volume (liquids and gases)	1 kiloliter (kl or kL)	= 1000 liters	1 kL = 264.17 gallons	1 gallon = 3.785 L
	1 liter (l or L)	= 1000 milliliters	1 L = 0.264 gallon 1 L = 1.057 quarts	1 quart = 0.946 L
	1 milliliter (ml or mL)	= 0.001 liter = 1 cubic centimeter	1 ml = 0.034 fluid ounce 1 ml = approx. $\frac{1}{5}$ teaspoon 1 ml = approx. 15–16 drops (gtt)	1 quart = 946 ml 1 pint = 473 ml 1 fluid ounce = 29.57 ml 1 teaspoon = approx. 5 ml
	1 microliter (μl or μL)	= 0.000001 liter		
Time	1 second (s)	= $\frac{1}{60}$ minute		
	1 millisecond (ms)	= 0.001 second		
Temperature	Degrees Celsius (°C)		°F = $\frac{9}{5}$(°C) + 32	°C = $\frac{5}{9}$(°F − 32)

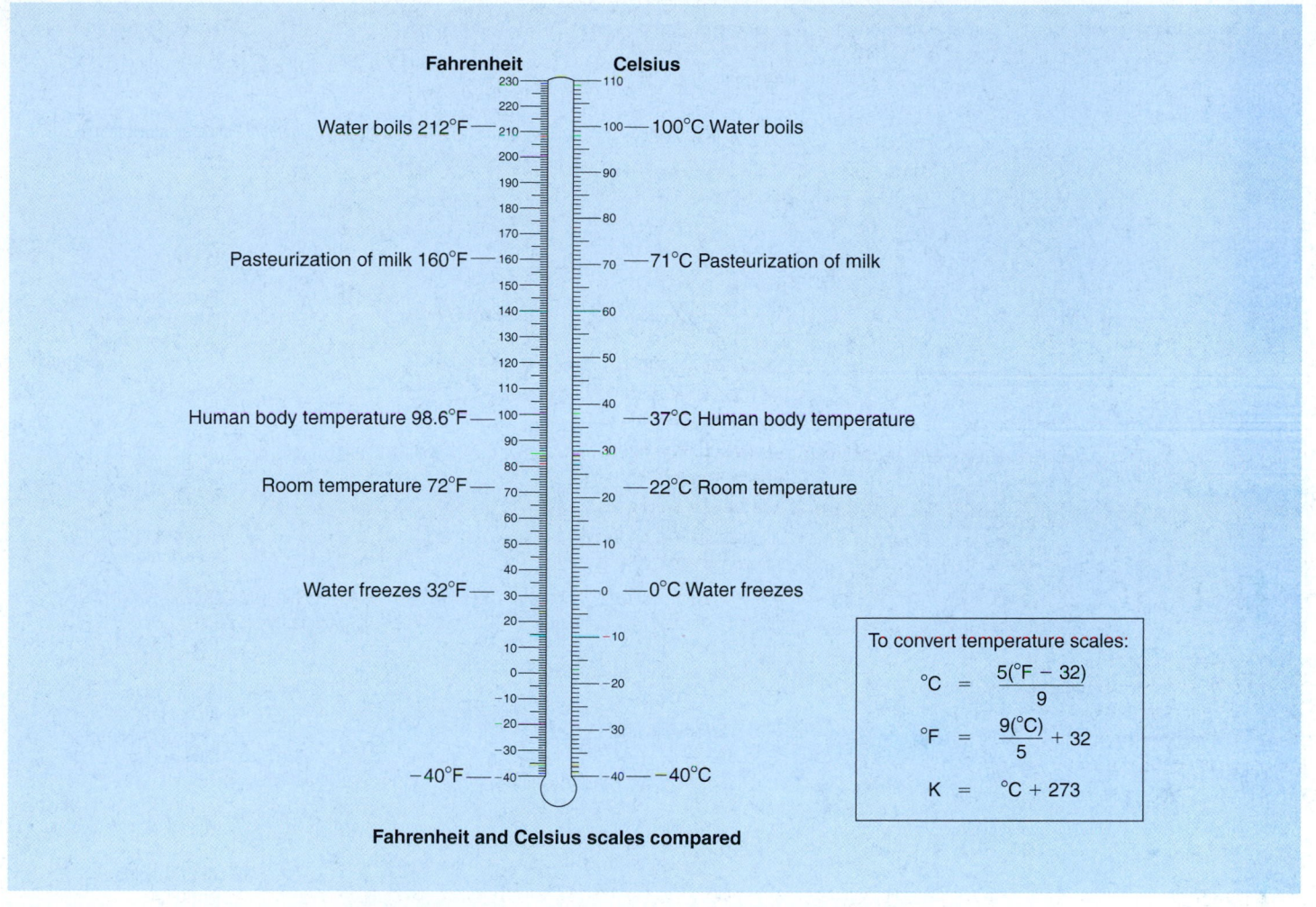

Fahrenheit and Celsius scales compared

To convert temperature scales:

$$°C = \frac{5(°F - 32)}{9}$$

$$°F = \frac{9(°C)}{5} + 32$$

$$K = °C + 273$$

Functional Groups in Organic Molecules

Functional group	General formula	Name of compounds	Example	Where else found
Hydroxyl —OH (or HO—)	—O—H	Alcohols	Ethanol	Sugars; water-soluble vitamins
Carbonyl \CO	—C=O (with H)	Aldehydes	Propanal	Some sugars; formaldehyde (a preservative)
	—C— (with O)	Ketones	Acetone	Some sugars; "ketone bodies" in urine (from fat breakdown)
Carboxyl —COOH	—C=O with OH	Carboxylic acids	Acetic acid	Amino acids; proteins; some vitamins; fatty acids
Amino —NH₂ (or H₂N—)	—N with H, H	Amines	Methylamine	Amino acids; proteins; urea in urine (from protein breakdown)

Nonpolar

Glycine (Gly)   Alanine (Ala)   Valine (Val)   Leucine (Leu)   Isoleucine (Ile)

Methionine (Met)   Phenylalanine (Phe)   Tryptophan (Trp)   Proline (Pro)

Polar

Serine (Ser)   Threonine (Thr)   Cysteine (Cys)   Tyrosine (Tyr)   Asparagine (Asn)   Glutamine (Gln)

Electrically charged

Acidic   Basic

Aspartic acid (Asp)   Glutamic acid (Glu)   Lysine (Lys)   Arginine (Arg)   Histidine (His)

Two Important Metabolic Pathways

① Glucose enters the cell and is phosphorylated by the enzyme hexokinase, which catalyzes the transfer of a phosphate group, indicated as ⓟ, from ATP to the number six carbon of the sugar, producing glucose-6-phosphate. The electrical charge of the phosphate group traps the sugar in the cell because the plasma membrane is impermeable to ions. Phosphorylation of glucose also makes the molecule more chemically reactive. Although glycolysis is supposed to *produce* ATP, ATP is actually consumed in step 1—an energy investment that will be repaid with dividends later in glycolysis.

② Glucose-6-phosphate is rearranged and converted to its isomer, fructose-6-phosphate. Isomers, remember, have the same number and types of atoms but in different structural arrangements.

③ In this step, still another molecule of ATP is used to add a second phosphate group to the sugar, producing fructose-1,6-bisphosphate. So far, the ATP ledger shows a debit of −2. With phosphate groups on its opposite ends, the sugar is now ready to be split in half.

④ This is the reaction from which glycolysis gets its name. An enzyme cleaves the sugar molecule into two different 3-carbon sugars: glyceraldehyde 3-phosphate and dihydroxyacetone phosphate. These two sugars are isomers of one another.

⑤ An isomerase enzyme interconverts the 3-carbon sugars, and if left alone in a test tube, the reaction reaches equilibrium. This does not happen in the cell, however, because the next enzyme in glycolysis uses only glyceraldehyde 3-phosphate as its substrate and not dihydroxyacetone phosphate. This pulls the equilibrium between the two 3-carbon sugars in the direction of glyceraldehyde 3-phosphate, which is removed as fast as it forms. Thus, the net result of steps 4 and 5 is cleavage of a 6-carbon sugar into two molecules of glyceraldehyde 3-phosphate; each will progress through the remaining steps of glycolysis.

⑥ An enzyme now catalyzes two sequential reactions while it holds glyceraldehyde 3-phosphate in its active site. First, the sugar is oxidized by the transfer of H from the number one carbon of the sugar to NAD^+, forming $NADH + H^+$. Here we see in metabolic context the oxidation-reduction reaction described in Chapter 24. This reaction releases substantial amounts of energy, and the enzyme capitalizes on this by coupling the reaction to the creation of a high-energy phosphate bond at the number one carbon of the oxidized substrate. The source of the phosphate is inorganic phosphate (P_i) always present in the cytosol. The enzyme releases $NADH + H^+$ and 1,3-bisphosphoglyceric acid as products. Notice in the figure that the new phosphate bond is symbolized with a squiggle (~), which indicates that the bond is at least as energetic as the high-energy phosphate bonds of ATP.

THE TEN STEPS OF GLYCOLYSIS   Each of the ten steps of glycolysis is catalyzed by a specific enzyme found dissolved in the cytoplasm. All steps are reversible. An abbreviated version of the three major phases of glycolysis appears in the lower right-hand corner of the next page.

⑦ Finally, glycolysis produces ATP. The phosphate group, with its high-energy bond, is transferred from 1,3-bisphosphoglyceric acid to ADP. For each glucose molecule that began glycolysis, step 7 produces two molecules of ATP, because every product after the sugar-splitting step (step 4) is doubled. Of course, two ATPs were invested to get sugar ready for splitting. The ATP ledger now stands at zero. By the end of step 7, glucose has been converted to two molecules of 3-phosphoglyceric acid. This compound is not a sugar. The sugar was oxidized to an organic acid back in step 6, and now the energy made available by that oxidation has been used to make ATP.

⑧ Next, an enzyme relocates the remaining phosphate group of 3-phosphoglyceric acid to form 2-phosphoglyceric acid. This prepares the substrate for the next reaction.

⑨ An enzyme forms a double bond in the substrate by removing a water molecule from 2-phosphoglyceric acid to form phosphoenolpyruvic acid, or PEP. This results in the electrons of the substrate being rearranged in such a way that the remaining phosphate bond becomes very unstable; it has been upgraded to high-energy status.

⑩ The last reaction of glycolysis produces another molecule of ATP by transferring the phosphate group from PEP to ADP. Because this step occurs twice for each glucose molecule, the ATP ledger now shows a net gain of two ATPs. Steps 7 and 10 each produce two ATPs for a total credit of four, but a debt of two ATPs was incurred from steps 1 and 3. Glycolysis has repaid the ATP investment with 100% interest. In the meantime, glucose has been broken down and oxidized to two molecules of pyruvic acid, the compound produced from PEP in step 10.

Summary

Phase 1 Sugar activates by phosphorylation

Phase 2 Sugar cleavage

Phase 3 Sugar oxidation and formation of ATP

① Two-carbon acetyl CoA is combined with oxaloacetic acid, a 4-carbon compound. The unstable bond between the acetyl group and CoA is broken as oxaloacetic acid binds and CoA is freed to prime another 2-carbon fragment derived from pyruvic acid. The product is the 6-carbon citric acid, for which the cycle is named.

② A molecule of water is removed, and another is added back. The net result is the conversion of citric acid to its isomer, isocitric acid.

③ The substrate loses a CO_2 molecule, and the remaining 5-carbon compound is oxidized, forming an α-ketoglutaric acid and reducing NAD^+.

④ This step is catalyzed by a multienzyme complex very similar to the one that converts pyruvic acid to acetyl CoA. CO_2 is lost; the remaining 4-carbon compound is oxidized by the transfer of electrons to NAD^+ to form $NADH+H^+$ and is then attached to CoA by an unstable bond. The product is succinyl CoA.

⑤ Substrate-level phosphorylation occurs in this step. CoA is displaced by a phosphate group, which is then transferred to GDP to form guanosine triphosphate (GTP). GTP is similar to ATP, which is formed when GTP donates a phosphate group to ADP. The products of this step are succinic acid and ATP.

⑥ In another oxidative step, two hydrogens are removed from succinic acid (forming fumaric acid) and transferred to FAD to form $FADH_2$. The function of this coenzyme is similar to that of $NADH+H^+$, but $FADH_2$ stores less energy. The enzyme that catalyzes this oxidation-reduction reaction is the only enzyme of the cycle that is embedded in the mitochondrial membrane. All other enzymes of the citric acid cycle are dissolved in the mitochondrial matrix.

⑦ Bonds in the substrate are rearranged in this step by the addition of a water molecule. The product is malic acid.

⑧ The last oxidative step reduces another NAD^+ and regenerates oxaloacetic acid, which accepts a 2-carbon fragment from acetyl CoA for another turn of the cycle.

CITRIC ACID CYCLE (KREBS CYCLE) All but one of the steps (step 6) occur in the mitochondrial matrix. The preparation of pyruvic acid (by oxidation, decarboxylation, and reaction with coenzyme A) to enter the cycle as acetyl CoA is shown above the cycle. Acetyl CoA is picked up by oxaloacetic acid to form citric acid; and as it passes through the cycle, it is oxidized four more times [forming three molecules of reduced NAD (NADH + H^+) and one of reduced FAD ($FADH_2$)] and decarboxylated twice (releasing 2 CO_2). Energy is captured in the bonds of GTP, which then acts in a coupled reaction with ADP to generate one molecule of ATP by substrate-level phosphorylation.

Periodic Table of the Elements

Representative (main group) elements

Representative (main group) elements

Transition metals

IA 1 H 1.0079	IIA											IIIA	IVA	VA	VIA	VIIA	VIIIA 2 He 4.003
3 Li 6.941	4 Be 9.012											5 B 10.811	6 C 12.011	7 N 14.007	8 O 15.999	9 F 18.998	10 Ne 20.180
11 Na 22.990	12 Mg 24.305	IIIB	IVB	VB	VIB	VIIB	VIIIB			IB	IIB	13 Al 26.982	14 Si 28.086	15 P 30.974	16 S 32.065	17 Cl 35.453	18 Ar 39.948
19 K 39.098	20 Ca 40.078	21 Sc 44.956	22 Ti 47.867	23 V 50.942	24 Cr 51.996	25 Mn 54.938	26 Fe 55.845	27 Co 58.933	28 Ni 58.69	29 Cu 63.546	30 Zn 65.38	31 Ga 69.723	32 Ge 72.64	33 As 74.922	34 Se 78.96	35 Br 79.904	36 Kr 83.8
37 Rb 85.468	38 Sr 87.62	39 Y 88.906	40 Zr 91.224	41 Nb 92.906	42 Mo 95.96	43 Tc 98	44 Ru 101.07	45 Rh 102.906	46 Pd 106.42	47 Ag 107.868	48 Cd 112.411	49 In 114.82	50 Sn 118.71	51 Sb 121.76	52 Te 127.60	53 I 126.905	54 Xe 131.29
55 Cs 132.905	56 Ba 137.327	57 La 138.906	72 Hf 178.49	73 Ta 180.948	74 W 183.84	75 Re 186.207	76 Os 190.23	77 Ir 192.22	78 Pt 195.08	79 Au 196.967	80 Hg 200.59	81 Tl 204.383	82 Pb 207.2	83 Bi 208.980	84 Po 209	85 At 210	86 Rn 222
87 Fr 223	88 Ra 226	89 Ac 227	104 Rf 267	105 Db 268	106 Sg 271	107 Bh 272	108 Hs 270	109 Mt 276	110 Ds 281	111 Rg 280	112 Cn 285	113 Nh 284	114 Fl 289	115 Mc 288	116 Lv 293	117 Ts 294	118 Og 294

Rare earth elements

Lanthanides	58 Ce 140.116	59 Pr 140.908	60 Nd 144.24	61 Pm 145	62 Sm 150.36	63 Eu 151.964	64 Gd 157.25	65 Tb 158.925	66 Dy 162.5	67 Ho 164.93	68 Er 167.26	69 Tm 168.934	70 Yb 173.054	71 Lu 174.967
Actinides	90 Th 232.038	91 Pa 231.036	92 U 238.029	93 Np 237.048	94 Pu 244	95 Am 243	96 Cm 247	97 Bk 247	98 Cf 251	99 Es 252	100 Fm 257	101 Md 258	102 No 259	103 Lr 262

The periodic table arranges elements according to atomic number and atomic weight into horizontal rows called periods and 18 vertical columns called groups or families. The elements in the groups are classified as being in either A or B classes.

Elements of each group of the A series have similar chemical and physical properties. This reflects the fact that members of a particular group have the same number of valence shell electrons, which is indicated by the roman numeral of the group. For example, group IA elements have one valence shell electron, group IIA elements have two, and group VA elements have five. In contrast, as you progress across a period from left to right, the properties of the elements change in discrete steps, varying gradually from the very metallic properties of groups IA and IIA elements to the nonmetallic properties seen in group VIIA (chlorine and others), and finally to the inert elements (noble gases) in group VIIIA. This change reflects the continual increase in the number of valence shell electrons seen in elements (from left to right) within a period.

Class B elements are referred to as transition elements. All transition elements are metals, and in most cases they have one or two valence shell electrons. (In these elements, some electrons occupy more distant electron shells before the deeper shells are filled.)

In this periodic table, the colors are used to convey information about the phase (solid, liquid, or gas) in which a pure element exists under standard conditions (25 degrees Celsius and 1 atmosphere of pressure). If the element's symbol is solid black, then the element exists as a solid. If its symbol is red, then it exists as a gas. If its symbol is dark blue, then it is a liquid. If the element's symbol is green, the element does not exist in nature and must be created by some type of nuclear reaction.

* Atomic weights of the elements per IUPAC Commission on Isotopic Abundances and Atomic Weights, 2007.

Reference Values for Selected Blood and Urine Studies

The reference values listed for the selected blood and urine studies are common ranges for adults, but specific "normals" are established by the laboratory performing the analysis. The values may be affected by a wide range of circumstances, including testing methods and equipment used, client age, body mass, sex, diet, activity level, medications, and extent of disease processes.

Reference values are identified in both standard or conventional units and in the system of international (SI) units. SI units (given in parentheses) are measurements of amount per volume and are used in most countries and scientific journals. SI units are often given as moles or millimoles per liter. Most clinical laboratories and textbooks in the United States use conventional or standard units, which measure mass per volume. These values are given as grams, milligrams, or milliequivalents per deciliter or liter. It is anticipated that the United States will eventually use SI units exclusively.

For enzymes, 1 international unit (IU) represents an arbitrary but defined amount of activity, whereas 1 katal (kat) is the amount of enzyme required to consume 1 mol of substrate per second.

Sample types in column 1 are serum (S), plasma (P), arterial whole blood (A), and whole blood (WB).

TEST (SAMPLE)	REFERENCE VALUES: CONVENTIONAL (SI)	PHYSIOLOGICAL INDICATION AND CLINICAL IMPLICATIONS
Blood Chemistry Studies		
Ammonia (P)	15–120 µg/dl (9–70 µmol/L)	Liver and renal function. Increased values in liver disease, renal failure, newborn hemolytic disease, heart failure, cor pulmonale. Decreased values in hypertension.
Amylase (S)	56–190 IU/L (0.4–2.1 µkat/L)	Pancreatic function. Increased values in pancreatitis, mumps, obstruction of pancreatic duct, ketoacidosis. Decreased values in kidney disease, pancreatic damage or cancer, toxemia of pregnancy.
Aspartate aminotransferase (AST, or SGOT) (S)	≤40 IU/L (≤0.7 µkat/L)	Cellular damage. Increased after myocardial infarction, acute liver disease, drug toxicity, muscle trauma. Decreased in pyridoxine (vitamin B_6) deficiency.
Bilirubin (S)	Total: 0.1–1.0 mg/dl (1.7–17.1 µmol/L) Direct: <0.4 mg/dl (<6.8 µmol/L) Indirect: 0.1–1.0 mg/dl (1.7–17.1 µmol/L) Newborn: <13.0 mg/dl (<222 µmol/L)	Liver function and red cell breakdown. Increased levels of direct in liver disease and biliary obstruction. Increased levels of indirect in hemolysis of red blood cells.
Blood urea nitrogen (S)	7–26 mg/dl (2.5–9.3 mmol/L)	Kidney function. Increased values in renal disease, dehydration, urinary obstruction, congestive heart failure, myocardial infarction, burns. Decreased values in liver failure, overhydration, impaired protein absorption, pregnancy.
Cholesterol	<200 mg/dl (<5.2 mmol/L)	Metabolism—fat utilization. Increased values in diabetes mellitus, pregnancy, use of oral contraceptives or anabolic steroids.
High-density lipoprotein (HDL) cholesterol (S)	20–30% of total >40 mg/dl (>1.0 mmol/L)	Increased levels in liver disease, aerobic exercise. Decreased levels in atherosclerotic heart disease, malnutrition.
Low-density lipoprotein (LDL) cholesterol (S)	60–70% of total <130 mg/dl (<3.4 mmol/L)	Increased values in hyperlipidemia, atherosclerotic heart disease. Decreased values in fat malabsorption and malnutrition.
Very low density lipoprotein cholesterol (VLDL) (S)	10–15% of total	Same as LDL.
Creatine kinase (CK) (S)	Female: ≤190 U/L (≤3.2 µkat/L) Male: ≤235 U/L (≤3.9 µkat/L)	Cellular damage. Increased values in myocardial infarction, muscular dystrophy, hypothyroidism, pulmonary infarction, cerebrovascular accident (CVA), shock, tissue damage, and trauma.
Creatinine (S)	0.5–1.2 mg/dl (44–106 µmol/L)	Renal function. Increased values in renal disease and acromegaly. Decreased in muscular dystrophy.

TEST (SAMPLE)	REFERENCE VALUES: CONVENTIONAL (SI)	PHYSIOLOGICAL INDICATION AND CLINICAL IMPLICATIONS
Blood Chemistry Studies (continued)		
Gases (A)		
Bicarbonate	22–26 mEq/L (22–26 mmol/L)	Acid-base balance. Increased values in metabolic alkalosis and respiratory acidosis. Decreased values in metabolic acidosis and respiratory alkalosis.
Carbon dioxide content	Arterial: 19–24 mEq/L (19–24 mmol/L) Venous: 22–30 mEq/L (22–30 mmol/L)	
Carbon dioxide partial pressure (P_{CO_2})	Arterial: 35–45 mm Hg Venous: 45 mm Hg	
Oxygen (O_2) saturation	95–98% (same)	Values increased slightly in hyperventilation. Decreased values (hypoxia) in pulmonary disease, hypoventilation, high altitude.
Oxygen partial pressure (P_{O_2})	80–105 mm Hg	
pH	7.35–7.45 (same)	Increased values in metabolic and respiratory alkalosis. Decreased values in metabolic and respiratory acidosis.
Glucose (S)	70–120 mg/dl (3.9–6.7 mmol/L)	Metabolic function. Increased values in diabetes mellitus, Cushing's syndrome, liver disease, acute stress, and acromegaly. Decreased levels in Addison's disease, insulinomas.
Immunoglobulins (S)		
IgG	560–1800 mg/dl (5.6–18 g/L)	Immune response. Increased levels in chronic infections, rheumatic fever, liver disease, rheumatoid arthritis. Decreased levels in amyloidosis, leukemia, and preeclampsia.
IgE	<43.2 µg/dl (<432 µg/L)	Allergic responses. Increased values in allergic responses. Decreased values in agammaglobulinemia.
IgA	85–563 mg/dl (0.85–5.6 g/L)	Immune integrity. Increased values in liver disease, rheumatic fever, chronic infection, inflammatory bowel disease. Decreased values in immunodeficiency disorders and immunosuppression.
IgM	55–375 mg/dl (0.5–3.8 g/L)	Immune integrity. Increased in autoimmune disease (e.g., rheumatoid arthritis), acute infections. Decreased in amyloidosis and leukemia.
IgD	0.5–14 mg/dl (5–140 mg/L)	Immune integrity. Increased values in myelomas.
Ketone bodies (S or P)	Negative Toxic level ≥20 mg/dl (0.2 g/L)	Fatty acid catabolism. Increased values (ketosis, ketoacidosis) in starvation, low-carbohydrate diet, uncontrolled diabetes mellitus, aspirin overdose.
Lactate dehydrogenase (LDH) (S)	105–333 U/L (1.7–5.6 µkat/L)	Tissue damage of organs or striated muscle. Increased in myocardial infarction, pulmonary infarction, liver disease, cerebrovascular accident, infectious mononucleosis, muscular dystrophy, fractures.
Lactic acid (lactate) (P)	9–16 mg/dl (1.0–1.8 mmol/L)	Anaerobic tissue metabolism. Increased values in congestive heart failure, shock, hemorrhage, strenuous exercise.
Osmolality (S)	280–300 mOsm/kg H_2O (280–300 mmol/kg H_2O)	Fluid and electrolyte balance. Increased levels in hypernatremia, dehydration, kidney disease, alcohol ingestion. Decreased levels in hyponatremia, overhydration, and syndrome of inappropriate ADH secretion (SIADH).
Phosphate (S) (phosphorus)	2.5–4.5 mg/dl (0.8–1.5 mmol/L)	Parathyroid function; bone disease. Increased levels in hypoparathyroidism, renal failure, bone metastasis, hypocalcemia. Decreased values in hyperparathyroidism, hypercalcemia, alcoholism, vitamin D deficiency, ketoacidosis, osteomalacia.
Potassium (S)	3.5–5.5 mEq/L (3.5–5.5 mmol/L)	Fluid and electrolyte balance. Increased levels in renal disease, Addison's disease, ketoacidosis, burns, and crush injuries. Decreased levels in vomiting, diarrhea, Cushing's syndrome, alkalosis, diuretics.

TEST (SAMPLE)	REFERENCE VALUES: CONVENTIONAL (SI)	PHYSIOLOGICAL INDICATION AND CLINICAL IMPLICATIONS
Blood Chemistry Studies (continued)		
Protein (S)		
Total	6.0–8.5 g/dl (60–85 g/L)	Osmotic pressure; immune system integrity. Increased values in multiple myeloma, dehydration, myxedema. Decreased values in protein malnutrition, burns, diarrhea, renal failure, liver failure.
Albumin	3.2–5.0 g/dl (32–50 g/L)	Osmotic pressure. Increased levels in dehydration. Decreased levels in liver disease, malnutrition, Crohn's disease, nephrotic syndrome, systemic lupus erythematosus.
Sodium (S)	135–145 mEq/L (135–145 mmol/L)	Fluid and electrolyte balance. Increased values in dehydration, diabetes insipidus, Cushing's syndrome. Decreased values in vomiting, diarrhea, burns, Addison's disease, myxedema, congestive heart failure, overhydration, syndrome of inappropriate ADH secretion (SIADH).
Triglycerides	10–150 mg/dl (0.1–1.5 g/L)	Increased values in diabetes mellitus, liver disease, nephrotic syndrome, pregnancy.
Uric acid (S)	Female: 2.0–7.3 mg/dl (119–434 µmol/L) Male: 2.1–8.5 mg/dl (125–506 µmol/L)	Renal function. Increased in lead poisoning, impaired renal function, gout, alcoholism, hematologic cancers. Decreased in Wilson's disease.
Hematology Studies		
Hemoglobin (S)	Female: 12–16 g/dl (120–160 g/L) Male: 13–18 g/dl (130–180 g/L)	Oxygenation status. Increased values in dehydration, polycythemia, congestive heart failure, chronic obstructive pulmonary disease, high altitudes. Decreased levels in anemia, hemorrhage, bone marrow cancer, renal disease, systemic lupus erythematosus, nutritional deficiency.
Hematocrit (WB)	Female: 37–47% (same) Male: 42–52% (same)	Oxygenation status. Increased levels in polycythemia, dehydration, congestive heart failure, shock, surgery. Decreased levels in anemia, hemorrhage, bone marrow disease, malnutrition, cirrhosis, rheumatoid arthritis.
Partial thromboplastin time (activated) (PTT or aPTT)	20–36 s (same)	Clotting mechanisms. Increased values in clotting factor deficiencies, cirrhosis, vitamin K deficiency, disseminated intravascular coagulation (DIC). Decreased values in early DIC, extensive cancer.
Platelet count (WB)	150,000–400,000/µl (150–400 × 10^9/L)	Clotting mechanisms. Increased values in polycythemia, cancers, rheumatoid arthritis, trauma. Decreased values in liver disease, hemolytic uremic syndrome, disseminated intravascular coagulation (DIC), idiopathic thrombocytopenic purpura (ITP), systemic lupus erythematosus (SLE).
Prothrombin time (PT) (WB)	11–12.5 s (same) Reported as INR, a ratio to a standard time	Clotting mechanisms. Increased values in liver disease, vitamin K deficiency, salicylate intoxication. Decreased values in disseminated intravascular coagulation (DIC).
Red blood cell count (RBC) (WB)	Female: 4.2–5.4 million/µl (4.2–5.4 × 10^{12}/L) Male: 4.7–6.1 million/µl (4.7–6.1 × 10^{12}/L)	Oxygenation status. Increased values in high altitudes, polycythemia, hemoconcentration, cor pulmonale. Decreased values in hemorrhage, hemolysis, anemias, chronic illness, nutritional deficiencies, leukemia, overhydration.
Reticulocyte count (WB)	0.5–2.0% (same)	Bone marrow function. Increased values in hemolytic anemia, sickle-cell anemia, leukemia, pregnancy. Decreased values in pernicious anemia, folic acid deficiency, cirrhosis, chronic infection, bone marrow depression or failure.
White blood cell count (WBC) (WB)		
Total	4800–10,800/µl (4.8–10.8 × 10^9/L)	Immune system integrity. Increased values in infection, trauma, stress, tissue necrosis, inflammation. Decreased values in bone marrow depression or failure, drug toxicity, overwhelming infection, malnutrition.

TEST (SAMPLE)	REFERENCE VALUES: CONVENTIONAL (SI)	PHYSIOLOGICAL INDICATION AND CLINICAL IMPLICATIONS
Hematology Studies (continued)		
White blood cell count, differential (WB)		
Neutrophils	50–70% (same)	Immune system integrity. Increased values in acute bacterial infections, stress, Cushing's syndrome, inflammatory disorders, ketoacidosis, gout. Decreased levels in aplastic anemia, bone marrow suppression, and overwhelming bacterial infections.
Lymphocytes	25–40% (same)	Immune system integrity. Increased values in viral infections (e.g., mumps, rubella, infectious mononucleosis, hepatitis), lymphocytic leukemia, certain bacterial infections. Decreased levels in leukemia, immunodeficiency, lupus erythematosus, bone marrow depressive drugs.
Eosinophils	1–4% (same)	Immune system integrity. Increased levels in allergic reactions, parasitic infections, leukemia. Decreased levels in excess adrenocorticosteroid production.
Monocytes	3–8% (same)	Immune system integrity. Increased levels in certain infections (e.g., tuberculosis, malaria), inflammatory disorders.
Basophils	0.5–1.0% (same)	Immune system integrity. Increased levels in myeloproliferative disorders, leukemia. Decreased levels in allergic reactions, hyperthyroidism, stress.
Urine Tests		
Amylase (24 h)	<6000 Somogyi units/24 h 0–500 U/24 h (0–8.3 μkat/24 h)	Pancreatic function. Increased values in pancreatic disease or obstruction, inflammation of the salivary glands, and cholecystitis.
Bilirubin (random)	Negative (same)	Liver function. Increased values in liver disease, extrahepatic obstruction (gallstones, tumor, inflammation).
Blood (hemoglobin) (random)	Negative (same)	Urinary system function. Increased values in cystitis, renal disease, hemolytic anemia, transfusion reaction, prostatitis, burns.
Osmolality (random or fasting)	Random: 50–1200 mOsm/kg H_2O (50–1200 mmol/kg H_2O) Fluid restriction: ≥850 mOsm/kg H_2O (≥850 mmol/kg H_2O)	Fluid and electrolyte balance, renal function, and endocrine function. Increased levels in hypernatremia, syndrome of inappropriate ADH secretion (SIADH), congestive heart failure, metabolic acidosis. Decreased levels in diabetes insipidus, water intoxication, pyelonephritis, renal tubular necrosis, aldosteronism.
Phosphate (24 h)	0.9–1.3g/24 hr (same)	Parathyroid function. Increased levels in hyperparathyroidism, osteomalacia, certain renal diseases, vitamin D deficiency. Decreased levels in hypoparathyroidism.
Potassium (24 h)	25–125 mEq/24 h (25–125 mmol/24 h)	Fluid and electrolyte balance. Increased values in renal tubular necrosis, metabolic acidosis, dehydration, aldosteronism, Cushing's syndrome. Decreased values in Addison's disease, malabsorption, acute renal failure.
Protein (random)	<8 mg/dl (<80 mg/L)	Renal function. Increased levels in nephrotic syndrome, renal trauma, hyperthyroidism, diabetic nephropathy, systemic lupus erythematosus.
Sodium (24 h)	40–220 mEq/24 h (40–220 mmol/24 h)	Fluid and electrolyte balance. Increased values in dehydration, ketoacidosis, syndrome of inappropriate ADH secretion (SIADH), adrenocortical insufficiency. Decreased levels in congestive heart failure, renal failure, diarrhea, aldosteronism.
Uric acid (24 h)	250–750 mg/24 h (1.5–4.5 mmol/24 h)	Renal function and metabolism. Increased in gout, leukemia, liver disease, ulcerative colitis. Decreased in renal disease, alcoholism, lead toxicity, folic acid deficiency.

TEST (SAMPLE)	REFERENCE VALUES: CONVENTIONAL (SI)	PHYSIOLOGICAL INDICATION AND CLINICAL IMPLICATIONS
Urine Tests *(continued)*		
Urinalysis (random)		
Color	Straw, yellow, amber	Fluid balance and renal function. Darker in dehydration. Lighter in overhydration, diabetes insipidus. Color varies with disease states, diet, and medications.
Odor	Aromatic	Metabolic function, infection. Abnormal odors in infection, ketonuria, rectal fistula, hepatic failure, phenylketonuria.
Specific gravity	1.001–1.035	An indirect measurement of urine concentration (osmolality). Same physiological indications and clinical implications as osmolality.
pH	4.5–8.0	A crude indicator of acid-base balance. Decreased by acidic diet (proteins). Increased by a vegetarian diet, prolonged vomiting, and bacterial infection of the urinary tract.
Urobilinogen (24 h)	0.2–1.0 mg/dl (2–10 mg/L)	Liver function. Increased in hemolytic anemias, hepatitis, cirrhosis, biliary disease. Decreased in common bile duct obstruction.
Volume (24 h)	800–2000 ml/24 h (0.8–2.0 L/24 h)	Fluid and electrolyte balance, renal function. Increased values in diabetes insipidus, diabetes mellitus, renal disease. Decreased values in dehydration, syndrome of inappropriate ADH secretion (SIADH), renal disease.

Glossary

Pronunciation Key

′ = Primary accent

″ = Secondary accent

Pronounce:

		as in					as in		
a, fa, āt			fate		o, no, ōt			note
ă, hă, at			hat		ŏ, frŏ, og			frog
ah			father		oo			soon
ar			tar		or			for
e, stre, ēt			street		ow			plow
ě, hě, en			hen		oy			boy
er			her		sh			she
ew			new		u, mu, ūt			mute
g			go		ŭ, sŭ, un			sun
i, bi, īt			bite		z			zebra
ĭ, hĭ, im			him		zh			measure
ng			ring					

Abduct (ab-dukt′) To move away from the midline of the body.

Absolute refractory period Period following stimulation during which no additional action potential can be evoked.

Absorption Process by which the products of digestion pass through the alimentary canal mucosa into the blood or lymph.

Accessory digestive organs Organs that contribute to the digestive process but are not part of the alimentary canal; include the tongue, teeth, salivary glands, pancreas, liver.

Accommodation The process of increasing the refractive power of the lens of the eye to focus on near objects.

Acetabulum (as″ĕ-tab′u-lum) Cuplike cavity on lateral surface of the hip bone that receives the femur.

Acetylcholine (ACh) (as″ĕ-til-ko′lēn) Chemical transmitter substance released by some nerve endings.

Acetylcholinesterase (AChE) (as″ĕ-til-ko″lin-es′ter-ās) Enzyme that degrades acetylcholine and terminates its action at the neuromuscular junction and synapses.

Achilles tendon *See* Calcaneal tendon.

Acid A substance that releases hydrogen ions when in solution (compare with Base); a proton donor.

Acid-base balance Situation in which the pH of the blood is maintained between 7.35 and 7.45.

Acidosis (as″ĭ-do′sis) State of abnormally high hydrogen ion concentration in the extracellular fluid.

Actin (ak′tin) A cytoskeletal element; a contractile protein of muscle.

Action potential (AP) A large transient depolarization event, including polarity reversal, that is propagated along the membrane of a muscle cell or an axon of a neuron.

Activation energy The amount of energy required to push a reactant to the level necessary for action.

Active immunity Immunity produced by an encounter with an antigen; provides immunological memory.

Active site Region on the surface of a functional (globular) protein where it binds and interacts chemically with other molecules of complementary shape and charge.

Active (transport) processes (1) Membrane transport processes for which ATP is directly or indirectly required, e.g., solute pumping and endocytosis. (2) "Active transport" also refers specifically to solute pumping.

Adaptation (1) Any change in structure or response to suit a new environment; (2) decline in the frequency of action potentials in a sensory nerve when a receptor is stimulated continuously and at constant stimulus strength.

Adduct (a-dukt′) To move toward the midline of the body.

Adenine (A) (ad′ĕ-nēn) One of the two major purines found in both RNA and DNA; also found in various free nucleotides of importance to the body, such as ATP.

Adenohypophysis (ad″ĕ-no-hi-pof′ĭ-sis) *See* Anterior pituitary.

Adenoids (ad′en-noids) Pharyngeal tonsil.

Adenosine triphosphate (ATP) (ah-den′o-sēn tri″fos′fāt) Organic molecule that stores and releases chemical energy for use in body cells.

Adenylate cyclase Enzyme that makes the second messenger cyclic AMP (cAMP) from ATP.

Adipocyte (ad′ĭ-po-sīt) An adipose, or fat, cell.

Adipose tissue (ad′ĭ-pōs) Areolar connective tissue modified to store fat; a connective tissue consisting chiefly of adipocytes.

Adrenal glands (ah-drē′nul) Hormone-producing glands located superior to the kidneys; each consists of medulla and cortex areas.

Adrenergic fibers (ad″ren-er′jik) Axons with axon terminals that, upon stimulation, release norepinephrine.

Adrenocorticotropic hormone (ACTH) (ah-dre″no-kor″tĭ-ko-trop′ik) Anterior pituitary hormone that influences the activity of the adrenal cortex.

Adventitia (ad″ven-tish′e-ah) Outermost layer or covering of some organs.

Aerobic (a′er-ōb″ik) Oxygen-requiring.

Aerobic endurance The length of time a muscle can continue to contract using aerobic pathways.

Aerobic respiration Respiration in which oxygen is consumed and glucose is broken down entirely; water, carbon dioxide, and large amounts of ATP are the final products.

Afferent (af′er-ent) Carrying to or toward a center.

Afferent (sensory) nerve Nerve that contains axons of sensory neurons and carries action potentials to the central nervous system.

Agglutination (ah-gloo″tĭ-na′shun) Clumping of (foreign) cells; induced by cross-linking of antigen-antibody complexes.

Agonist (ag′o-nist) Muscle that bears the major responsibility for effecting a particular movement; a prime mover.

Albumin (al-bu′min) The most abundant plasma protein.

Aldosterone (al-dos′ter-ōn) Hormone produced by the adrenal cortex that regulates Na^+ reabsorption and K^+ secretion by the kidneys.

Alimentary canal (al″ĭ-men′tar-e) The continuous hollow tube extending from the mouth to the anus; its walls are constructed by the oral cavity, pharynx, esophagus, stomach, and small and large intestines.

Alkalosis (al″kah-lo′sis) State of abnormally low hydrogen ion concentration in the extracellular fluid.

Allantois (ah″lan′to-is) Embryonic membrane; its blood vessels develop into blood vessels of the umbilical cord.

Alleles Genes coding for the same trait and found at the same locus on homologous chromosomes.

Allergy A type of hypersensitivity (overzealous immune response to an otherwise harmless antigen) that involves IgE antibodies and histamine release.

Alopecia (al″o-pe′she-ah) Baldness.

Alpha (α)-helix The most common type of secondary structure of the amino acid chain in proteins; resembles a coiled spring.

Alveolar (acinar) gland (al-ve′o-lar) A gland whose secretory cells form small, flasklike sacs.

Alveolar ventilation rate (AVR) An index of respiratory efficiency; measures volume of fresh air that flows in and out of alveoli.

Alveolus (al-ve′o-lus) (1) One of the microscopic air sacs of the lungs; (2) tiny milk-producing glandular sac in the breast; (3) tooth socket.

Alzheimer's disease (AD) (altz′hi-merz) Degenerative brain disease resulting in progressive loss of memory and motor control, and increasing dementia.

Amino acid (ah-me′no) Organic compound containing nitrogen, carbon, hydrogen, and oxygen; building block of protein.

Ammonia (NH_3) Common waste product of protein breakdown in the body; a colorless volatile gas, very soluble in water and capable of acting as a weak base; a proton acceptor.

Amnion (am′ne-on) Fetal membrane that forms a fluid-filled sac around the embryo.

Amoeboid motion (ah-me′boyd) The flowing movement of the cytoplasm of a phagocyte as it moves across a surface.

Amphiarthrosis (am″fe-ar-thro′sis) A slightly movable joint.

Ampulla (am-pul′lah) A localized dilation of a canal or duct.

Amylase Digestive system enzyme that breaks down starchy foods.

Anabolism (ah-nab′o-lizm) Energy-requiring building phase of metabolism in which simpler substances are combined to form more complex substances.

Anaerobic (an-a′er-ōb-ik) Not requiring oxygen.

Anaerobic threshold The point at which muscle metabolism converts to anaerobic glycolysis.

Anaphase Third stage of mitosis, meiosis I, and meiosis II in which chromosomes move toward each pole of a cell.

Anastomosis (ah-nas″to-mo′sis) A union or joining of nerves, blood vessels, or lymphatic vessels.

Anatomy Study of the structure of living organisms.

Androgen (an′dro-jen) A hormone such as testosterone that controls male secondary sex characteristics.

Anemia (ah-ne′me-ah) Reduced oxygen-carrying ability of blood resulting from too few erythrocytes or abnormal hemoglobin.

Aneurysm (an′u-rizm) Blood-filled sac in an artery wall caused by dilation or weakening of the wall.

Angina pectoris (an′jĭ-nah pek′tor-is) Severe suffocating chest pain caused by brief lack of oxygen supply to heart muscle.

Angiotensin II (an″je-o-ten′sin) Triggers release of aldosterone; also a potent vasoconstrictor, and stimulates thirst and antidiuretic hormone release. Part of renin-angiotensin-aldosterone mechanism.

Anion (an′i-on) An ion with a negative charge.

Anoxia (ah-nok′se-ah) Deficiency of oxygen.

Antagonist (an-tag′o-nist) (1) Muscle that reverses, or opposes, the action of another muscle. (2) Hormone that opposes the action of another hormone.

Anterior pituitary The glandular part of the pituitary gland. Also called adenohypophysis.

Antibody A protein molecule that is released by a plasma cell (a daughter cell of an activated B lymphocyte) and that binds specifically to an antigen; an immunoglobulin.

Anticodon (an″ti-ko′don) The three-base sequence complementary to the messenger RNA (mRNA) codon.

Antidiuretic hormone (ADH, also called vasopressin) (an″ti-di″yer-eh′tik) Hormone produced by the hypothalamus and released by the posterior pituitary; stimulates the kidneys to reabsorb more water, reducing urine volume.

Antigen (an′tĭ-jen) A substance or part of a substance (living or nonliving) that is recognized as foreign by the immune system, activates the immune system, and reacts with immune cells or their products.

Antigen-presenting cell (APC) A specialized cell (dendritic cell, macrophage, or B cell) that captures, processes, and presents antigens on its surface to T lymphocytes.

Anucleate cell (a-nu′kle-āt) A cell without a nucleus.

Anus (a′nus) Distal end of digestive tract; outlet of rectum.

Aorta (a-or′tah) Major systemic artery; arises from the left ventricle of the heart.

Aortic body Receptor in the aortic arch sensitive to changing oxygen, carbon dioxide, and pH levels of the blood.

Apnea Breathing cessation.

Apocrine sweat gland (ap′o-krin) The less numerous type of sweat gland; produces a secretion containing water, salts, proteins, and fatty acids.

Apoenzyme (ap′ō-en-zīm) The protein portion of an enzyme.

Aponeurosis (ap″o-nu-ro′sis) Fibrous or membranous sheet connecting a muscle and the part it moves.

Apoptosis A process of controlled cellular suicide; eliminates cells that are unneeded, stressed, or aged.

Appendicitis (ă-pen′dĭ-sī′tis) Inflammation of the appendix (wormlike sac attached to the cecum of the large intestine).

Appendicular Relating to the limbs; one of the two major divisions of the skeleton.

Appositional growth Growth accomplished by the addition of new layers onto those previously formed.

Aquaporins (ă″kwă-por′ins) Transmembrane proteins that form water channels.

Aqueous humor (a′kwe-us) Watery fluid in the anterior segment of the eye.

Arachnoid (ah-rak′noid) Weblike; specifically, the weblike arachnoid mater, the middle layer of the three meninges.

Areola (ah-re′o-lah) (1) Circular, pigmented area surrounding the nipple; (2) any small space in a tissue.

Areolar connective tissue A type of loose connective tissue.

Arrector pili (ah-rek′tor pi′li) Tiny, smooth muscles attached to hair follicles; contraction causes the hair to stand upright.

Arrhythmia (a-rith′me-ah) Irregular heart rhythm, often caused by defects in the intrinsic conduction system.

Arteries Blood vessels that conduct blood away from the heart and into the circulation.

Arteriole (ar-tēr′e-ōl) A minute artery.

Arteriosclerosis (ar-tēr′e-o-skler-o′sis) Any of a number of proliferative and degenerative changes in the arteries leading to their decreased elasticity.

Arthritis Inflammation of the joints.

Arthroscopic surgery (ar-thro-skop′ik) Procedure enabling a surgeon to repair the interior of a joint through a small incision.

Articular capsule Double-layered capsule composed of an outer fibrous layer lined by synovial membrane; encloses the joint cavity of a synovial joint.

Articular cartilage Hyaline cartilage covering bone ends at movable joints.

Articulation (joint) The junction of two or more bones.

Association areas Functional areas of the cerebral cortex that act mainly to integrate diverse information for purposeful action.

Astigmatism (ah-stig′mah-tiz-em) A condition in which unequal curvatures in different parts of the cornea or lens of the eye lead to blurred vision.

Astrocyte (as′tro-sīt) A type of CNS supporting cell; assists in exchanges between blood capillaries and neurons.

Atelectasis (at″ĕ-lek′tah-sis) Lung collapse.

Atherosclerosis (a″ther-o″skler-o′sis) Changes in the walls of large arteries consisting of lipid deposits on the artery walls; one form of arteriosclerosis.

Atmospheric pressure Force that air exerts on the surface of the body (760 mm Hg at sea level).

Atom Smallest particle of an elemental substance that exhibits the properties of that element; composed of protons, neutrons, and electrons.

Atomic number The number of protons in an atom.

Atomic symbol The one- or two-letter symbol used to indicate an element; usually the first letter(s) of the element's name.

Atomic weight The average of the mass numbers of all the isotopes of an element.

ATP *See* Adenosine triphosphate.

Atria (a′tre-ah) The two superior receiving chambers of the heart.

Atrial natriuretic peptide (ANP) (a′tre-al na″tre-u-ret′ik) A hormone released by certain cells of the heart atria that reduces blood pressure and blood volume by inhibiting nearly all events that promote Na^+ and water retention and vasoconstriction.

Atrioventricular (AV) bundle (a″tre-o-ven-trĭ′kyoo-ler) Bundle of specialized fibers that conduct impulses from the AV node to the right and left ventricles; also called bundle of His.

Atrioventricular (AV) node Specialized mass of conducting cells located at the atrioventricular junction in the heart.

Atrioventricular (AV) valve Valve that prevents backflow into the atrium when the connected ventricle is contracting.

Atrophy (at′ro-fe) Reduction in size or wasting away of an organ or cell resulting from disease or lack of use.

Auditory ossicles (ah′sĭ-kulz) The three tiny bones serving as transmitters of vibrations and located within the middle ear: the malleus, incus, and stapes.

Auditory tube *See* Pharyngotympanic tube.

Autoimmunity Production of antibodies or effector T cells that attack a person's own tissue.

Autolysis (aw″tol′ĭ-sis) Process of autodigestion (self-digestion) of cells, especially dead or degenerate cells.

Autonomic ganglion Collection of sympathetic or parasympathetic postganglionic neuronal cell bodies.

Autonomic nervous system (ANS) Efferent division of the peripheral nervous system that innervates cardiac and smooth muscles and glands; also called the involuntary or visceral motor system.

Autonomic (visceral) reflexes Reflexes that activate smooth or cardiac muscle and/or glands.

Autoregulation The automatic local adjustment of blood flow to a particular body area in response to its current requirements.

Autosomes Chromosomes number 1 to 22; do not include the sex chromosomes.

Avogadro's number (av″o-gad′rōz) The number of molecules in one mole of any substance, 6.02×10^{23}.

Axial Relating to the head, neck, and trunk; one of the two major divisions of the skeleton.

Axon Neuron process that carries action potentials away from the nerve cell body; efferent process; the conducting portion of a nerve cell.

Axon terminals The bulbous distal endings of the terminal branches of an axon where neurotransmitters are released.

B cells Also called B lymphocytes; oversee humoral immunity; their descendants differentiate into antibody-producing plasma cells.

Baroreceptor (bayr″o-re-sep′tor) A sensory nerve ending in the wall of the carotid sinus or aortic arch sensitive to vessel stretching.

Basal body (ba′sal) An organelle structurally identical to a centriole and forming the base of a cilium or flagellum.

Basal ganglia *See* Basal nuclei.

Basal lamina (lam′ĭ-nah) Noncellular, adhesive supporting sheet consisting largely of glycoproteins secreted by epithelial cells.

Basal metabolic rate (BMR) Rate at which energy is expended (heat produced) by the body per unit time under controlled (basal) conditions: 12 hours after a meal, at rest.

Basal nuclei (basal ganglia) Specific gray matter areas located deep within the white matter of the cerebral hemispheres.

Basal surface The surface near the base or interior of a structure; nearest the lower side or bottom of a structure.

Base A substance capable of binding with hydrogen ions; a proton acceptor.

Basement membrane Extracellular material consisting of a basal lamina secreted by epithelial cells and a reticular lamina secreted by underlying connective tissue cells.

Basophil (ba′zo-fil) White blood cell whose granules stain purplish-black with basic dye.

Benign (be-nīn′) Not malignant.

Bicarbonate buffer system Chemical system that helps maintain pH homeostasis of the blood. Also called carbonic acid–bicarbonate buffer system.

Bile Greenish-yellow or brownish fluid produced in and secreted by the liver, stored in the gallbladder, and released into the small intestine.

Bilirubin (bil″ĭ-roo′bin) Yellow pigment of bile.

Bipolar neuron Neuron with axon and dendrite that extend from opposite sides of the cell body.

Blastocyst (blas′to-sist) Stage of early embryonic development; the product of cleavage.

Blood brain barrier Mechanism that inhibits passage of materials from the blood into brain tissues; reflects relative impermeability of brain capillaries.

Blood pressure (BP) Force exerted by blood against a unit area of the blood vessel walls; differences in blood pressure between different areas of the circulation provide the driving force for blood circulation.

Bolus (bo′lus) A rounded mass of food prepared by the mouth for swallowing; any soft round mass.

Bone marrow Adipose- or blood-forming tissue found within bone cavities; called yellow and red bone marrow, respectively.

Bone (osseous tissue) (os′e-us) A connective tissue that forms the bony skeleton.

Bone remodeling Process involving bone formation and destruction in response to hormonal and mechanical factors.

Bone resorption The removal of osseous tissue; part of the continuous bone remodeling process.

Bowman's capsule (bo-manz) *See* Glomerular capsule.

Boyle's law States that when the temperature is constant, the pressure of a gas varies inversely with its volume.

Bradycardia (brad″e-kar′de-ah) A heart rate below 60 beats per minute.

Brain death State of irreversible coma, even though life-support measures may have restored other body organs.

Brain stem Collectively the midbrain, pons, and medulla of the brain.

Brain ventricle Fluid-filled cavity of the brain.

Bronchioles Smaller (<1 mm in diameter) branching air passageways inside the lungs.

Bronchus (brong′kus) One of the two large branches of the trachea that lead to the lungs, or any of its smaller branches leading to bronchioles.

Buffer Chemical substance or system that minimizes changes in pH by releasing or binding hydrogen ions.

Burn Tissue damage inflicted by intense heat, electricity, radiation, or certain chemicals, all of which denature cell proteins and kill cells in the affected areas.

Bursa (ber′sa) A fibrous sac lined with synovial membrane and containing synovial fluid; occurs between bones and muscle tendons (or other structures), where it acts to decrease friction during movement.

Bursitis Inflammation of a bursa.

Calcaneal tendon (kal-ka′ne-al) Tendon that attaches the calf muscles to the heelbone (calcaneus); also called the Achilles tendon.

Calcitonin (kal″sĭ-to′nin) Hormone released by the thyroid. Lowers blood calcium levels only when present at high (therapeutic) levels.

Calculus (kal′ku-lus) A stone formed within various body parts.

Callus (kal′lus) (1) Localized thickening of skin epidermis resulting from physical trauma; (2) repair tissue (fibrous or bony) formed at a fracture site.

Calorie (cal) Amount of energy needed to raise the temperature of 1 gram of water 1° Celsius. Energy exchanges associated with biochemical reactions are usually reported in kilocalories (1 kcal = 1000 cal), also called large calories (Cal).

Calyx (ka′liks) A cuplike extension of the pelvis of the kidney.

Canaliculus (kan″ah-lik′u-lus) Extremely small tubular passage or channel.

Cancer A malignant, invasive cellular neoplasm that has the capability of spreading throughout the body or body parts.

Capillaries (kap′il-layr″ēs) The smallest of the blood vessels and the sites of exchange between the blood and tissue cells.

Carbohydrate (kar″bo-hi′drāt) Organic compound composed of carbon, hydrogen, and oxygen; includes starches, sugars, cellulose.

Carbonic anhydrase (kar-bon′ik an-hi′drās) Enzyme that reversibly facilitates the joining of carbon dioxide with water to form carbonic acid.

Carcinogen (kar″sĭ′no-jin) Cancer-causing agent.

Cardiac cycle Sequence of events encompassing one complete contraction and relaxation of the atria and ventricles of the heart.

Cardiac muscle Specialized muscle of the heart.

Cardiac output (CO) Amount of blood pumped out of a ventricle in one minute.

Cardiac reserve The difference between resting and maximal cardiac output.

Cardiogenic shock Pump failure; the heart is so inefficient that it cannot sustain adequate circulation.

Cardiovascular system Organ system that distributes the blood to deliver nutrients and remove wastes.

Carotene (kar′o-tēn) Yellow to orange pigment that accumulates in the stratum corneum epidermal layer and in fatty tissue of the subcutaneous tissue.

Carotid body (kah-rot′id) A receptor in the common carotid artery sensitive to changing oxygen, carbon dioxide, and pH levels of the blood.

Carotid sinus (si′nus) A dilation of a common carotid artery; involved in regulation of systemic blood pressure.

Carrier A transmembrane protein that changes shape to envelop and transport a polar substance across the cell membrane.

Cartilage (kar′tĭ-lij) One of four types of connective tissue—avascular and not innervated.

Cartilage bone (endochondral bone) Bone formed by using hyaline cartilage structures as models for ossification.

Cartilaginous joints (kar″tĭ-laj′ĭ-nus) Bones united by cartilage; no joint cavity is present.

Catabolism (ka-tab′o-lizm) Process in which living cells break down substances into simpler substances.

Catalyst (kat′ah-list) Substance that increases the rate of a chemical reaction without itself becoming chemically changed or part of the product.

Cataract Clouding of the eye's lens; often congenital or age-related.

Catecholamines (kat″ĕ-kol′ah-mēnz) Epinephrine, norepinephrine, and dopamine; a class of amines that act as chemical messengers.

Cation (kat′i-on) An ion with a positive charge.

Caudal (kaw′dul) Literally, toward the tail; in humans, the inferior portion of the anatomy.

Cecum (se′kum) The blind-ended pouch at the beginning of the large intestine.

Cell Structural unit of all living things.

Cell cycle Series of changes a cell goes through from the time it is formed until it has reproduced itself.

Cell differentiation The development of specific and distinctive features in cells, from a single cell (the fertilized egg) to all the specialized cells of adulthood.

Cell membrane *See* Plasma membrane.

Cellular immunity Immunity conferred by activated T cells, which directly kill infected or cancerous body cells or cells of foreign grafts and release chemicals that regulate the immune response. Also called cell-mediated immunity.

Cellular respiration Metabolic processes in which ATP is produced.

Cellulose (sel′u-lōs) A fibrous carbohydrate (polysaccharide) that is the main structural component of plant tissues.

Central (Haversian) canal (hah-ver′zhan) The canal in the center of each osteon that contains minute blood vessels and nerve fibers that serve the needs of the osteocytes.

Central nervous system (CNS) Brain and spinal cord.

Centriole (sen′tre-ōl) Minute body found in pairs near the nucleus of the cell; active in cell division.

Centrosome (cell center) A region near the nucleus that contains paired organelles called centrioles.

Cerebellum (ser″ĕ-bel′um) Brain region most involved in producing smooth, coordinated skeletal muscle activity.

Cerebral aqueduct (ser′ĕ-bral, sĕ-re′bral) The slender cavity of the midbrain that connects the third and fourth ventricles.

Cerebral arterial circle (*circle of Willis*) An arterial anastomosis at the base of the brain.

Cerebral cortex The outer gray matter region of the cerebral hemispheres.

Cerebral dominance Designates the hemisphere that is dominant for language.

Cerebral palsy Neuromuscular disability in which voluntary muscles are poorly controlled or paralyzed as a result of brain damage.

Cerebral white matter Consists largely of myelinated fibers bundled into large tracts; provides for communication between cerebral areas and lower CNS centers.

Cerebrospinal fluid (CSF) (ser″ĕ-bro-spi′nal) Plasmalike fluid that fills the cavities of the CNS and surrounds the CNS externally; protects the brain and spinal cord.

Cerebrovascular accident (CVA) (ser″ĕ-bro-vas′ku-lar) Condition in which brain tissue is deprived of a blood supply, as in blockage of a cerebral blood vessel; a stroke.

Cerebrum (ser′ĕ-brum) The cerebral hemispheres (including the cerebral cortex, white matter, and basal nuclei).

Cervical vertebrae The seven vertebrae of the vertebral column that are located in the neck.

Cervix Lower outlet of the uterus extending into the vagina.

Channel A transmembrane protein that forms an aqueous pore, allowing substances to move from one side of the membrane to the other.

Chemical bond An energy relationship holding atoms together; involves the interaction of electrons.

Chemical energy Energy stored in the bonds of chemical substances.

Chemical equilibrium A state of apparent repose created by two reactions proceeding in opposite directions at equal speed.

Chemical reaction Process in which molecules are formed, changed, or broken down.

Chemically gated channel An ion channel that opens when a molecule (ligand) binds to it. Also called a ligand-gated channel.

Chemoreceptor (ke″mo-re-sep′ter) Receptor sensitive to various chemicals in solution.

Chemotaxis (ke″mo-tak′sis) Movement of a cell, organism, or part of an organism toward or away from a chemical substance.

Cholecystokinin (CCK) (ko″le-sis″to-ki′nin) An intestinal hormone that stimulates gallbladder contraction and pancreatic juice release.

Cholesterol (ko-les′ter-ol″) Steroid found in animal fats as well as in most body tissues; made by the liver.

Cholinergic fibers (ko″lin-er′jik) Axons with axon terminals that, upon stimulation, release acetylcholine.

Chondroblast (kon′dro-blast) Actively mitotic cell of cartilage.

Chondrocyte (kon′dro-sīt) Mature cell of cartilage.

Chorion (kor′e-on) Outermost fetal membrane; helps form the placenta.

Choroid (ko′roid) The vascular middle layer of the eye.

Choroid plexus (ko′roid plex′sus) A capillary knot that protrudes into a brain ventricle; produces cerebrospinal fluid.

Chromatin (kro′mah-tin) Strands of DNA (genes) and associated proteins; forms chromosomes when condensed (tightly coiled).

Chromosomes (kro′mo-somz) Barlike bodies of tightly coiled chromatin; visible during cell division.

Chronic obstructive pulmonary disease (COPD) Collective term for progressive, obstructive respiratory disorders; includes emphysema, chronic bronchitis.

Chyme (kīm) Semifluid, creamy mass consisting of partially digested food and gastric juice.

Cilia (sil′e-ah) Tiny, hairlike projections of a cell; may move in a wavelike manner to propel substances across the exposed cell surface.

Circumduction (ser″kum-duk′shun) Movement of a body part so that it outlines a cone in space.

Cirrhosis (sĭ-ro′sis) Chronic disease of the liver, characterized by an overgrowth of connective tissue, or fibrosis.

Cistern (sis′tern) Any cavity or enclosed space serving as a reservoir.

Cisterna chyli (sis-ter′nah ki′li) An enlarged sac at the base of the thoracic duct; the origin of the thoracic duct.

Citric acid cycle Aerobic metabolic pathway occurring within mitochondria, in which food metabolites are oxidized and CO_2 is liberated, and coenzymes are reduced. Also called the Krebs cycle.

Cleavage An early embryonic phase consisting of rapid mitotic cell divisions without intervening growth periods; product is a blastocyst.

Clonal selection (klo′nul) Process during which a B cell or T cell becomes activated by binding with an antigen.

Clone Descendants of a single cell.

Coagulation Process in which blood is transformed from a liquid to a gel; blood clotting.

Cochlea (kok′le-ah) Snail-shaped chamber of the bony labyrinth that houses the receptor for hearing [the spiral organ (organ of Corti)].

Codon (ko′don) The three-base sequence on a messenger RNA molecule that provides the genetic information used in protein synthesis; codes for a given amino acid.

Coenzyme (ko-en′zīm) Nonprotein substance associated with and activating an enzyme; typically a vitamin.

Cofactor Metal ion or organic molecule that is required for enzyme activity.

Collagen fiber The most abundant of the three types of protein fibers found in the extracellular matrix of connective tissue.

Colloid (kol′oid) (1) A mixture in which the solute particles (usually proteins) do not settle out readily. (2) Substance in the thyroid gland containing thyroglobulin protein.

Colloid osmotic pressure (kol′oid ahz-mah′tik) Pressure created in a fluid by large nondiffusible molecules, such as plasma proteins that are prevented from moving across a capillary wall. Such substances tend to draw water to them.

Colon Region of the large intestine; includes ascending, transverse, descending, and sigmoid portions.

Combination (synthesis) reaction Chemical reaction in which larger, more complex molecules are formed from simpler molecules or atoms.

Complement A group of bloodborne proteins, which, when activated, enhance the inflammatory and immune responses and may lead to cell lysis.

Complementarity of structure and function The relationship between a structure and its function; i.e., structure determines function.

Complementary base Refers to how a given nitrogenous base of DNA or RNA bonds to another nitrogenous base. For example, adenine (A) is the complementary base of thymine (T). The result is base pairing.

Complete blood count (CBC) Clinical test that includes counts of all formed elements, a hematocrit, and measurements of erythrocyte size and hemoglobin content.

Compound Substance composed of two or more different elements, the atoms of which are chemically united.

Concentration gradient The difference in the concentration of a particular substance between two different areas.

Conducting zone Includes all respiratory passageways that provide conduits for air to reach the sites of gas exchange (the respiratory zone).

Conductivity A measure of the ability of a substance to transmit an electrical current.

Cones One of the two types of photoreceptor cells in the retina of the eye; provide for color vision.

Congenital (kun-jeh′nĭ-tul) Existing at birth.

Congestive heart failure (CHF) Condition in which the pumping efficiency of the heart is depressed so that circulation is inadequate to meet tissue needs.

Conjunctiva (kon″junk-ti′vah) Thin, protective mucous membrane lining the eyelids and covering the anterior surface of the eye itself.

Connective tissue A primary tissue; form and function vary extensively. Functions include support, storage, and protection.

Consciousness The ability to perceive, communicate, remember, understand, appreciate, and initiate voluntary movements.

Contraception The prevention of conception; birth control.

Contractility Muscle cell's ability to move by shortening.

Contraction To shorten or develop tension, an ability highly developed in muscle cells.

Contralateral Relating to the opposite side.

Cornea (kor′ne-ah) Transparent anterior portion of the eyeball; part of the fibrous layer.

Corona radiata (kor-o′nah ra-de-ah′tah) (1) Arrangement of elongated follicle cells around a mature ovum; (2) crownlike arrangement of nerve fibers radiating from the internal capsule of the brain to every part of the cerebral cortex.

Coronary circulation The functional blood supply of the heart; shortest circulation in the body.

Cortex (kor′teks) Outer surface layer of an organ.

Corticosteroids (kor″tĭ-ko-stě′roidz) Steroid hormones released by the adrenal cortex.

Cortisol (hydrocortisone) (kor′tĭ-sol) Glucocorticoid produced by the adrenal cortex.

Covalent bond (ko-va′lent) Chemical bond created by electron sharing between atoms.

Cranial nerves The 12 nerve pairs that are associated with the brain.

Craniosacral division Another name for the parasympathetic division of the autonomic nervous system.

Cranium (cranial bones) (kra′ne-um) Bony protective encasement of the brain and organs of hearing and equilibrium.

Creatine kinase (kre′ah-tin) Enzyme that catalyzes the transfer of phosphate from creatine phosphate to ADP, forming creatine and ATP; helps power muscle contraction.

Creatine phosphate (CP) (fos-fāt) Compound that serves as an alternative energy source for muscle tissue.

Creatinine (kre-at′ĭ-nin) A nitrogenous waste molecule that is not reabsorbed by the kidney; this characteristic makes it useful for measurement of the GFR and glomerular function.

Crista ampullaris Sensory receptor organ for rotational acceleration and deceleration housed within the ampulla of each semicircular canal of the inner ear.

Cross section A cut running horizontally from right to left, dividing the body or an organ into superior and inferior parts.

Cutaneous (ku-ta′ne-us) Pertaining to the skin.

Cutaneous sensory receptors Receptors located throughout the skin that respond to stimuli arising outside the body; part of the nervous system.

Cyclic AMP (cAMP) Intracellular second messenger that mediates the effects of the first (extracellular) messenger (hormone or neurotransmitter); formed from ATP by a plasma membrane enzyme (adenylate cyclase).

Cystic fibrosis (CF) Genetic disorder in which secretion of overly viscous mucus clogs the respiratory passages, predisposes to fatal respiratory infections.

Cytochromes (si′to-krōmz) Brightly colored iron-containing proteins that form part of the inner mitochondrial membrane and function as electron carriers in oxidative phosphorylation.

Cytokines Small proteins that act as chemical messengers between various parts of the immune system.

Cytokinesis (si″to-kĭ-ne′sis) The division of cytoplasm that occurs after the cell nucleus has divided.

Cytoplasm (si′to-plazm) The cellular material surrounding the nucleus and enclosed by the plasma membrane.

Cytosine (C) (si′to-sēn) Nitrogen-containing base that is part of a nucleotide structure.

Cytoskeleton Literally, cell skeleton. An elaborate series of structural proteins running through the cytosol, supporting cellular structures and providing the machinery to generate various cell movements.

Cytosol Viscous, semitransparent fluid substance of cytoplasm in which other elements are suspended.

Cytotoxic T cell (T_C cell) Effector T cell that directly kills foreign cells, cancer cells, or virus-infected body cells by inducing apoptosis (cell suicide).

Deamination (de″am-ĭ-na′shun) Removal of an amine group from an organic compound.

Decomposition reaction Chemical reaction in which a molecule is broken down into smaller molecules or its constituent atoms.

Defecation (def″ĭ-ka′shun) Elimination of the contents of the bowels (feces).

Deglutition (deg″loo-tish′un) Swallowing.

Dehydration (de″hi-dra′shun) Condition of excessive water loss.

Dehydration synthesis Process by which a large molecule is synthesized by removing water and covalently bonding smaller molecules together.

Dendrite (den′drīt) Branching neuron process that serves as a receptive, or input, region; transmits an electrical signal toward the cell body.

Dendritic cells Protective cells that engulf antigens, migrate to lymph nodes, and present the antigen to T cells, causing them to activate and mount an immune response; those in the skin are sometimes called Langerhans cells.

Depolarization (de-po″ler-ah-za′shun) Loss of a state of polarity; loss or reduction of negative membrane potential.

Dermatome (der′mah-tōm) Portion of somite mesoderm that forms the dermis of the skin; also the area of skin innervated by the cutaneous branches of a single spinal nerve.

Dermis Layer of skin deep to the epidermis; composed mostly of dense irregular connective tissue.

Desmosome (dez′muh-sōm) Cell junction composed of thickened plasma membranes joined by filaments.

Diabetes insipidus (di″ah-be′tēz in-sĭ′pĭ-dus) Disease characterized by passage of a large quantity of dilute urine plus intense thirst and dehydration caused by inadequate release of antidiuretic hormone (ADH).

Diabetes mellitus (DM) (meh′lĭ-tus) Disease caused by deficient insulin release or by insulin resistance, leading to inability of the body cells to use carbohydrates.

Dialysis (di-al′ah-sis) Diffusion of solute(s) through a semipermeable membrane.

Diapedesis (di″ah-pĕ-de′sis) Passage of white blood cells through intact vessel walls into tissue.

Diaphragm (di′ah-fram) (1) Any partition or wall separating one area from another; (2) a muscle that separates the thoracic cavity from the lower abdominopelvic cavity.

Diaphysis (di-af″ĭ-sis) Elongated shaft of a long bone.

Diarthrosis (di″ar-thro′sis) Freely movable joint.

Diastole (di-as′to-le) Period of the cardiac cycle when either the ventricles or the atria are relaxing.

Diastolic pressure (di-as-tah′lik) Arterial blood pressure reached during or as a result of diastole; lowest level of any given cardiac cycle.

Diencephalon (di″en-seh′fuh-lon) That part of the forebrain between the cerebral hemispheres and the midbrain including the thalamus, the epithalamus, and the hypothalamus.

Differential white blood cell count Diagnostic test to determine relative proportion of individual leukocyte types.

Diffusion (dĭ-fu′zhun) The spreading of particles in a gas or solution with a movement toward uniform distribution of particles; driven by kinetic energy.

Digestion A series of catabolic steps in which complex food molecules are broken down to their building blocks by enzymes.

Digestive system System that processes food into absorbable units and eliminates indigestible wastes.

Dipeptide A combination of two amino acids united by means of a peptide bond.

Diploë (dip′lo-e) The internal layer of spongy bone in flat bones.

Diploid chromosomal number The chromosomal number characteristic of an organism, symbolized as $2n$; twice the chromosomal number (n) of the gamete; in humans, $2n = 46$.

Diplopia (dǐ-plo′pe-ah) Double vision.

Dipole (polar molecule) Nonsymmetrical molecules that contain electrically unbalanced atoms.

Disaccharide (di-sak′ah-rīd, di-sak′ah-rid) Literally, double sugar; e.g., sucrose, lactose.

Dislocation (luxation) Occurs when bones are forced out of their normal alignment at a joint.

Displacement (exchange) reaction Chemical reaction in which bonds are both made and broken; atoms become combined with different atoms.

Distal (dis′tul) Away from the attached end of a limb or the origin of a structure.

Diuretics (di″u-ret′iks) Chemicals that enhance urinary output.

Diverticulum (di″ver-tik′u-lum) A pouch or sac in the walls of a hollow organ or structure.

DNA (deoxyribonucleic acid) (de-ok″sǐ-ri″bo-nu-klā′ik) A nucleic acid found in all living cells; it carries the organism's hereditary information.

DNA replication Process that occurs before cell division; ensures that all daughter cells have identical genes.

Dominant trait Occurs when one allele masks or suppresses the expression of its partner.

Dominant-recessive inheritance Reflects the interaction of dominant and recessive alleles.

Dorsal (dor′sul) Pertaining to the back; posterior.

Dorsal root ganglion Peripheral collection of cell bodies of first-order afferent neurons whose central axons enter the spinal cord.

Double helix The secondary structure assumed by two strands of DNA, held together throughout their length by hydrogen bonds between bases on opposite strands.

Duct (dukt) A canal or passageway; a tubular structure that provides an exit for the secretions of a gland, or for conducting any fluid.

Ductus (vas) deferens Extends from the epididymis to the urethra; propels sperm into the urethra by peristalsis during ejaculation.

Duodenum (du″o-de′num) First part of the small intestine.

Dura mater (du′rah ma′ter) Outermost and toughest of the three membranes (meninges) covering the brain and spinal cord.

Dynamic equilibrium Sense that reports on angular (rotational) acceleration or deceleration of the head in space.

Dyspnea (disp-ne′ah) Difficult or labored breathing; air hunger.

Eccrine sweat glands (ek′rin) Sweat glands abundant on the palms, soles of feet, and the forehead. Also called merocrine sweat glands.

Ectoderm (ek′to-derm) Embryonic germ layer; forms the epidermis of the skin and its derivatives, and nervous tissues.

Edema (ě-de′mah) Abnormal increase in the amount of interstitial fluid; causes swelling.

Effector (ef-ek′ter) Muscle or gland (or other organ) capable of being activated by nerve endings.

Efferent (ef′er-ent) Carrying away or away from, especially a nerve fiber that carries action potentials away from the central nervous system.

Elastic cartilage Cartilage with abundant elastic fibers; more flexible than hyaline cartilage.

Elastic fiber Fiber formed from the protein elastin, which gives a rubbery and resilient quality to the matrix of connective tissue.

Electrical energy Energy formed by the movement of charged particles, e.g., across or along cell membranes.

Electrocardiogram (ECG or EKG) (e-lek″tro-car′de-o-gram″) Graphic record of the electrical activity of the heart.

Electrochemical gradient The combined difference in concentration and charge; influences the distribution and direction of diffusion of ions.

Electroencephalogram (EEG) (e-lek″tro-en-sef′ah-lo-gram″) Graphic record of the electrical activity of nerve cells in the brain.

Electrolyte (e-lek′tro-līt) Chemical substances, such as salts, acids, and bases, that ionize and dissociate in water and are capable of conducting an electrical current.

Electrolyte balance Refers to the balance between input and output of salts (sodium, potassium, calcium, magnesium) in the body.

Electromagnetic radiation Emitted photons (wave packets) of energy, e.g., light, X ray, infrared.

Electron Negatively charged subatomic particle; orbits the atom's nucleus.

Electron shells (energy levels) Regions of space that consecutively surround the nucleus of an atom; the atom's electrons are most likely to be found in these regions.

Element One of a limited number of unique varieties of matter that composes substances of all kinds; e.g., carbon, hydrogen, oxygen.

Embolism (em′bo-lizm) Obstruction of a blood vessel by an embolus (blood clot, fatty mass, bubble of air, or other debris) floating in the blood.

Embryo (em′bre-o) Developmental stage extending from fertilization to the end of the eighth week.

Embryoblast Accumulation of cells in the blastocyst from which the embryo develops. Also called the inner cell mass.

Emesis Reflexive emptying of the stomach through the esophagus and pharynx; also known as vomiting.

Endergonic reaction Chemical reaction that absorbs energy, e.g., an anabolic reaction.

Endocardium (en″do-kar′de-um) Endothelial membrane that lines the interior of the heart.

Endochondral ossification (en″do-kon′dral) Embryonic formation of bone by the replacement of calcified cartilage; most skeletal bones are formed by this process.

Endocrine glands (en′do-krin) Ductless glands that empty their hormonal products directly into the blood.

Endocrine system Body system that includes internal organs that secrete hormones.

Endocytosis (en″do-si-to′sis) Means by which fairly large extracellular molecules or particles enter cells, e.g., phagocytosis, pinocytosis, receptor-mediated endocytosis.

Endoderm (en′do-derm) Embryonic germ layer; forms the lining of the digestive tube and its associated structures.

Endogenous (en-doj′ě-nus) Originating or produced within the organism or one of its parts.

Endometrium (en″do-me′tre-um) Mucous membrane lining of the uterus.

Endomysium (en″do-mis′e-um) Thin connective tissue surrounding each muscle cell.

Endoplasmic reticulum (ER) (en″do-plaz′mik rě-tik′u-lum) Membranous network of tubular or saclike channels in the cytoplasm of a cell.

Endosteum (en-dos′te-um) Connective tissue membrane covering internal bone surfaces.

Endothelium (en″do-the′le-um) Single layer of simple squamous cells that line the walls of the heart, blood vessels, and lymphatic vessels.

Energy The capacity to do work; may be stored (potential energy) or in action (kinetic energy).

Energy intake Energy liberated during food oxidation.

Energy output Sum of energy lost as heat, as work, and as fat or glycogen storage.

Enzyme (en′zīm) A protein that acts as a biological catalyst to speed up a chemical reaction.

Eosinophil (e″o-sin′o-fil) Granular white blood cell whose granules readily take up an acid stain called eosin.

Ependymal cell (ě-pen′dǐ-mul) A type of CNS supporting cell; lines the central cavities of the brain and spinal cord.

Epidermis (ep″ĭ-der′mis) Superficial layer of the skin; composed of keratinized stratified squamous epithelium.

Epididymis (ep″ĭ-dĭ′dĭ-mis) That portion of the male duct system in which sperm mature. Empties into the ductus (or vas) deferens.

Epidural space Area between the bony vertebrae and the dura mater of the spinal cord.

Epiglottis (ep″ĭ-glot′is) Elastic cartilage at the back of the throat; covers the opening of the larynx during swallowing.

Epileptic seizures Abnormal electrical discharges of groups of brain neurons, during which no other messages can get through.

Epimysium (ep″ĭ-mis′e-um) Sheath of fibrous connective tissue surrounding a muscle.

Epinephrine (ep″ĭ-nef′rin) Chief hormone produced by the adrenal medulla. Also called adrenaline.

Epiphyseal plate (e″pĭ-fis′e-ul) Plate of hyaline cartilage at the junction of the diaphysis and epiphysis that provides for growth in length of a long bone.

Epiphysis (e-pif′ĭ-sis) The end of a long bone, attached to the shaft.

Epithalamus Most dorsal portion of the diencephalon; forms the roof of the third ventricle with the pineal gland extending from its posterior border.

Epithelium (epithelial tissue) (ep″ĭ-the′le-ul) Pertaining to a primary tissue that covers the body surface, lines its internal cavities, and forms glands.

Erythrocytes (ĕ-rith′ro-sīts) Red blood cells.

Erythropoiesis (ĕ-rith″ro-poy-e′sis) Process of erythrocyte formation.

Erythropoietin (EPO) (ĕ-rith″ro-poy′ĕ-tin) Hormone that stimulates production of red blood cells.

Esophagus (ĕ-sof′ah-gus) Muscular tube extending from the laryngopharynx through the diaphragm to join the stomach; collapses when not involved in food propulsion.

Estrogens (es′tro-jenz) Hormones that stimulate female secondary sex characteristics; female sex hormones.

Eupnea (ūp-ne′ah) Normal respiratory rate and rhythm.

Excess postexercise oxygen consumption (EPOC) The volume of oxygen required after exercise to replenish stores of O_2, ATP, creatine phosphate, and glycogen and oxidize the lactic acid formed during exercise. Also called oxygen debt.

Exchange (displacement) reaction Chemical reaction in which bonds are both made and broken; atoms are combined with different atoms.

Excitability (responsiveness) Ability to respond to stimuli.

Excitation-contraction (E-C) coupling Sequence of events by which transmission of an action potential along the sarcolemma leads to the sliding of myofilaments.

Excitatory postsynaptic potential (EPSP) Depolarizing graded potential in a postsynaptic neuron.

Excretion (ek-skre′shun) Elimination of waste products from the body.

Exergonic reaction Chemical reaction that releases energy, e.g., a catabolic or oxidative reaction.

Exocrine glands (ek′so-krin) Glands that have ducts through which their secretions are carried to a particular site.

Exocytosis (ek″so-si-to′sis) Mechanism by which substances are moved from the cell interior to the extracellular space as a secretory vesicle fuses with the plasma membrane.

Exons Amino acid–specifying informational sequences (separated by introns) in the genes of higher organisms.

Extension Movement that increases the angle of a joint, e.g., straightening a flexed knee.

Exteroceptor (ek″ster-o-sep′tor) Sensory receptor that responds to stimuli from the external world.

Extracellular fluid (ECF) Internal fluid located outside cells; includes interstitial fluid, blood plasma, and cerebrospinal fluid.

Extracellular matrix Nonliving material in connective tissue consisting of ground substance and fibers; separates the living cells.

Extrasystole (ek″strah-sis′to-le) Premature heart contraction.

Extrinsic (ek-strin′sik) Of external origin.

Extrinsic eye muscles The six skeletal muscles that attach to and move each eye.

Facilitated diffusion Passive transport process used by certain large or charged molecules (e.g., glucose, Na^+) that are unable to pass through the plasma membrane unaided. Involves movement through channels or movement facilitated by a membrane carrier.

Fallopian tube (fah-lō′pe-un) *See* Uterine tube.

Fascia (fash′e-ah) Layers of fibrous tissue covering and separating muscle.

Fascicle (fas′ĭ-kl) Bundle of nerve or muscle fibers bound together by connective tissue.

Fatty acids Linear chains of carbon and hydrogen atoms (hydrocarbon chains) with an organic acid group at one end. A constituent of fat.

Feces (fe′sēz) Material discharged from the bowel; composed of food residue, secretions, bacteria.

Fenestrated (fen′es-tra-tid) Pierced with one or more small openings.

Fertilization Fusion of the sperm and egg nuclei.

Fetus Developmental stage extending from the ninth week of development to birth.

Fiber A slender threadlike structure or filament. May refer to a cell, part of a cell, or a protein. *See also* Muscle fiber, Nerve fiber, Collagen fiber.

Fibrillation Condition of rapid and irregular or out-of-phase heart contractions.

Fibrin (fi′brin) Fibrous insoluble protein formed during blood clotting.

Fibrinogen (fi-brin′o-jin) A soluble blood protein that is converted to insoluble fibrin during blood clotting.

Fibrinolysis Process that removes unneeded blood clots when healing has occurred.

Fibroblast (fi′bro-blast) Young, actively mitotic cell that forms the fibers of connective tissue.

Fibrocartilage The cartilage most resistant to compression and stretch. Forms vertebral discs and knee joint cartilages.

Fibrocyte (fi′bro-sīt) Mature fibroblast; maintains the matrix of fibrous types of connective tissue.

Fibrosis Proliferation of fibrous connective tissue called scar tissue.

Fibrous joints Bones joined by dense fibrous connective tissue; no joint cavity is present.

Filtrate A plasma-derived fluid that is processed by the renal tubules to form urine.

Filtration Passage of a solvent and dissolved substances through a membrane or filter.

First-degree burn A burn in which only the epidermis is damaged.

Fissure (fĭ′sher) (1) A groove or cleft; (2) the deepest depressions or inward folds on the brain.

Fixator (fix′a-ter) Muscle that immobilizes one or more bones, allowing other muscles to act from a stable base.

Flagellum (flah-jel′lum) Long, whiplike cellular extension containing microtubules; propels sperm and some single-celled eukaryotes.

Flexion (flek′shun) Movement that decreases the angle of the joint, e.g., bending the knee from a straight to an angled position.

Flexor (withdrawal) reflex Reflex initiated by a painful stimulus (actual or perceived); causes automatic withdrawal of the threatened body part from the stimulus.

Fluid mosaic model A depiction of the structure of the membranes of a cell as phospholipid bilayers in which proteins are dispersed.

Follicle (fah′lĭ-kul) (1) Ovarian structure consisting of a developing egg surrounded by one or more layers of follicle cells; (2) colloid-containing structure of the thyroid gland; (3) B cell–rich region in lymphoid tissue.

Follicle-stimulating hormone (FSH) Hormone produced by the anterior pituitary that stimulates ovarian follicle production in females and sperm production in males.

Fontanelles (fon″tah-nelz′) Fibrous membranes at the angles of cranial bones that accommodate brain growth in the fetus and infant.

Foramen (fo-ra′men) Hole or opening in a bone or between body cavities.

Forebrain (prosencephalon) Anterior portion of the brain consisting of the telencephalon and the diencephalon.

Formed elements Cellular portion of blood.

Fossa (fos′ah) A depression, often an articular surface.

Fovea (fo′ve-ah) A pit.

Fracture A break in a bone.

Free radicals Highly reactive chemicals with unpaired electrons that can scramble the structure of proteins, lipids, and nucleic acids.

Frontal (coronal) plane Longitudinal (vertical) plane that divides the body or an organ into anterior and posterior parts.

Fulcrum The fixed point on which a lever moves when a force is applied.

Fundus (fun′dus) Base of an organ; part farthest from the opening of the organ. For example, the posterior wall of the eye.

G protein Protein that relays signals between extracellular first messengers (such as hormones or neurotransmitters) and intracellular second messengers (such as cyclic AMP) via an effector enzyme.

Gallbladder Sac beneath the right lobe of the liver used for bile storage.

Gallstones (biliary calculi) Crystallized cholesterol that obstructs the flow of bile from the gallbladder.

Gamete (gam′ēt) Sex or germ cell.

Gametogenesis (gam″e-to-jen′eh-sis) Formation of gametes.

Ganglion (gang′gle-on) Collection of nerve cell bodies outside the CNS.

Gap junction A passageway between two adjacent cells; formed by transmembrane proteins called connexons.

Gastrin Hormone secreted in the stomach; regulates gastric juice secretion by stimulating HCl production.

Gastroenteritis Inflammation of the gastrointestinal tract.

Gastrulation (gas″troo-la′shun) Developmental process that produces the three primary germ layers (ectoderm, mesoderm, and endoderm).

Gene One of the biological units of heredity located in DNA; transmits hereditary information.

Genetic code Refers to the rules by which the base sequence of a DNA gene is translated into protein structures (amino acid sequences).

Genitalia (jen″ĭ-ta′le-ă) The internal and external reproductive organs.

Genome The complete set of chromosomes derived from one parent (the haploid genome); or the two sets of chromosomes, i.e., one set from the egg, the other from the sperm (the diploid genome).

Genotype (jēn′o-tīp) One's genetic makeup or genes.

Germ layers Three cellular layers (ectoderm, mesoderm, and endoderm) that represent the initial specialization of cells in the embryonic body and from which all body tissues arise.

Gestation period (jes-ta′shun) The period of pregnancy; about 280 days for humans.

Gland Organ specialized to secrete substances for further use in the body or excrete substances for elimination.

Glaucoma (glaw-ko′mah) Condition in which intraocular pressure increases to levels that cause compression of the retina and optic nerve; results in blindness unless detected early.

Glial cells (gle′al) *See* Neuroglia.

Glomerular capsule (glo-mer′yoo-ler) Double-walled cup at end of a renal tubule; encloses a glomerulus. Also called Bowman's capsule.

Glomerular filtration rate (GFR) Rate of filtrate formation by the kidneys.

Glomerulus (glo-mer′u-lus) (1) Cluster of capillaries forming part of the nephron; forms filtrate; (2) odor-specific processing unit in olfactory bulb.

Glottis (glah′tis) Opening between the vocal cords in the larynx.

Glucagon (gloo′kah-gon) Hormone formed by alpha cells of pancreatic islets; raises the glucose level of blood.

Glucocorticoids (gloo″ko-kor′tĭ-koidz) Adrenal cortex hormones that increase blood glucose levels and aid the body in resisting long-term stressors.

Gluconeogenesis (gloo″ko-ne″o-jen′ĕ-sis) Formation of glucose from noncarbohydrate molecules.

Glucose (gloo′kōs) Principal blood sugar; a hexose.

Glycerol (glis′er-ol) A modified simple sugar (a sugar alcohol); a building block of fats.

Glycocalyx (gli″ko-ka′liks) A layer of externally facing glycoproteins and glycolipids (a "cell coat") on or near a cell's plasma membrane; its components determine blood type and are involved in cellular interactions.

Glycogen (gli′ko-jin) Main carbohydrate stored in animal cells; a polysaccharide.

Glycogenesis (gli″ko-jen′ĕ-sis) Formation of glycogen from glucose.

Glycogenolysis (gli″ko-jĕ-nol′ĭ-sis) Breakdown of glycogen to glucose.

Glycolipid (gli″ko-lip′id) A lipid with one or more covalently attached sugars.

Glycolysis (gli-kol′ĭ-sis) Breakdown of glucose to pyruvic acid—an anaerobic process.

Goblet cells Individual cells (unicellular glands) that produce mucus.

Golgi apparatus (gol′je) Membranous system close to the cell nucleus that packages protein secretions for export, packages enzymes into lysosomes for cellular use, and modifies proteins destined to become part of cellular membranes.

Gonad (go′nad) Primary reproductive organ; i.e., the testis of the male or the ovary of the female.

Gonadocorticoids (gon″ah-do-kor′tĭ-koidz) Sex hormones, primarily androgens, secreted by the adrenal cortex.

Gonadotropins (gon″ah-do-trōp′inz) Gonad-stimulating hormones produced by the anterior pituitary.

Graded muscle contractions Variations in the degree of muscle contraction by changing either the frequency or strength of the stimulus.

Graded potential A local change in membrane potential that varies directly with the strength of the stimulus, declines with distance.

Graves' disease Autoimmune disorder in which the thyroid gland is directly stimulated by antibodies; causes markedly increased secretion of thyroid hormone.

Gray matter Gray area of the central nervous system; contains neuronal cell bodies and their dendrites.

Growth hormone (GH) Hormone that stimulates growth in general; produced in the anterior pituitary; also called somatotropin.

Guanine (G) (gwah′nēn) One of two major purines occurring in all nucleic acids.

Gustation (gus-ta′shun) Sense of taste.

Gyrus (ji′rus) An outward fold of the surface of the cerebral cortex.

Hair follicle Structure with outer and inner root sheaths extending from the epidermal surface into the dermis and from which new hair develops.

Hapten (hap′ten) An incomplete antigen; has reactivity but not immunogenicity.

Haversian system (hah-ver'zhen) *See* Osteon.

Heart attack (coronary) *See* Myocardial infarction.

Heart block Impaired transmission of impulses from atrium to ventricle resulting in abnormally slow heart rhythms.

Heart murmur Abnormal heart sound (usually resulting from valve problems).

Heimlich maneuver Procedure in which the air in a person's own lungs is used to expel an obstructing piece of food.

Helper T cell (T$_H$ cell) Type of T lymphocyte that orchestrates the entire adaptive immune response (both cellular and humoral immunity); also activates macrophages.

Hematocrit (he-mat'o-krit) The percentage of total blood volume occupied by erythrocytes.

Hematoma (he″mah-to′mah) Mass of clotted blood that forms at an injured site.

Hematopoiesis (hem″ah-to-poy-e′sis) Blood cell formation; hemopoiesis.

Hematopoietic stem cell (hem″ah-to-poy-et′ik) Bone marrow cell that gives rise to all the formed elements of blood; hemocytoblast.

Heme (hēm) Iron-containing pigment that is essential to oxygen transport by hemoglobin.

Hemoglobin (he′mo-glo″bin) Oxygen-transporting protein of erythrocytes.

Hemolysis (he-mah'lĕ-sis) Rupture of erythrocytes.

Hemophilia (he″mo-fil′e-ah) A term loosely applied to several different hereditary bleeding disorders that exhibit similar signs and symptoms.

Hemopoiesis (he″mo-poy-e′sis) *See* Hematopoiesis.

Hemorrhage (hem'or-ij) Loss of blood from the vessels by flow through ruptured walls; bleeding.

Hemostasis (he″mo-sta′sis) Stoppage of bleeding.

Heparin Natural anticoagulant secreted into blood plasma.

Hepatic portal system (hĕ-pat′ik) Circulation in which the hepatic portal vein carries dissolved nutrients to the liver tissues for processing.

Hepatitis (hep″ah-ti′tis) Inflammation of the liver.

Hernia (her'ne-ah) Abnormal protrusion of an organ or a body part through the containing wall of its cavity.

Heterozygous (het″er-o-zi′gus) Having different allelic genes at a given locus or (by extension) many loci.

Hilton's law Any nerve serving a muscle that produces movement at a joint also innervates the joint and the skin over the joint.

Hilum (hi'lum) The indented region of an organ from which blood and/or lymphatic vessels and nerves enter and exit.

Hippocampus Limbic system structure that plays a role in converting new information into long-term memories.

Histamine (his'tuh-mēn) A chemical messenger (neurotransmitter or paracrine); causes vasodilation and increased capillary permeability; in stomach causes acid secretion.

Histology (his-tol'o-je) Branch of anatomy dealing with the microscopic structure of tissues.

HIV (human immunodeficiency virus) Virus that destroys helper T cells, thus depressing adaptive immunity; symptomatic AIDS gradually appears when lymph nodes can no longer contain the virus.

Holocrine glands (hol'o-krin) Glands that accumulate their secretions within their cells; secretions are discharged only upon rupture and death of the cell.

Homeostasis (ho″me-o-sta′sis) A state of body equilibrium or stable internal environment of the body.

Homologous (ho-mol'ŏ-gus) (1) Parts or organs corresponding in embryonic origin and structure but not necessarily in function. (2) In chromosomes, a pair of chromosomes (one from each parent) that carry genes for the same traits but may differ in the details of their genetic instructions.

Homozygous (ho-mo-zi'gus) Having identical genes at one or more loci.

Hormones Steroidal or amino acid–based molecules released to the blood that act as chemical messengers to regulate specific body functions.

Humoral immunity (hu'mer-ul) Immunity conferred by antibodies present in blood plasma and other body fluids.

Huntington's disease Hereditary disorder leading to degeneration of the basal nuclei and the cerebral cortex.

Hyaline cartilage (hi′ah-līn) The most abundant cartilage type in the body; provides firm support with some pliability.

Hydrochloric acid (HCl) (hi″dro-klor′ik) Acid that aids protein digestion in the stomach; produced by parietal cells.

Hydrogen bond Weak bond in which a hydrogen atom forms a bridge between two electron-hungry atoms. An important intramolecular bond.

Hydrogen ion (H$^+$) A hydrogen atom minus its electron and therefore carrying a positive charge (i.e., a proton).

Hydrolysis (hi″drah′lă-sis) Process in which water is used to split a substance into smaller parts.

Hydrophilic (hi″dro-fil′ik) Refers to molecules, or portions of molecules, that interact with water and charged particles.

Hydrophobic (hi″dro-fo′bik) Refers to molecules, or portions of molecules, that interact only with nonpolar molecules.

Hydrostatic pressure (hi″dro-stă′tic) Pressure of fluid in a system.

Hydroxyl ion (OH$^-$) (hi-drok′sil) An ion liberated when a hydroxide (a common inorganic base) is dissolved in water.

Hyperalgesia Greater than normal sensitivity to pain.

Hypercapnia (hi″per-kap′ne-ah) High carbon dioxide levels in the blood.

Hyperemia An increase in blood flow into a tissue or organ; congested with blood.

Hyperglycemic (hi″per-gli-se′mik) Term used to describe hormones such as glucagon that elevate blood glucose level.

Hyperopia (hi″per-o′pe-ah) A condition in which visual images are routinely focused behind rather than on the retina; commonly known as farsightedness.

Hyperplasia (hi″per-pla′ze-ah) Accelerated cell division, e.g., in anemia, the bone marrow produces red blood cells at a faster rate.

Hyperpnea (hi″perp-ne′ah) An increase in ventilation in response to metabolic need (e.g., during exercise).

Hyperpolarization An increase in membrane potential in which the membrane becomes more negative than resting membrane potential.

Hypersensitivity Overzealous immune response to an otherwise harmless antigen.

Hypertension (hi″per-ten′shun) High blood pressure.

Hypertonic (hi″per-ton′ik) Excessive, above normal, tone or tension.

Hypertonic solution A solution that has a higher concentration of nonpenetrating solutes than the reference cell; having greater osmotic pressure than the reference solution (blood plasma or interstitial fluid).

Hypertrophy (hi-per'trah-fe) Growth of an organ or tissue due to an increase in the size of its cells; differs from hyperplasia (an increase in size due to increased cell numbers).

Hyperventilation An increase in the depth and rate of breathing that is in excess of the body's need for removal of carbon dioxide.

Hypocapnia Low carbon dioxide levels in the blood.

Hypoglycemia Abnormally low concentration of glucose in extracellular fluids.

Hypoglycemic (hi″po-gli-se′mik) Term used to describe hormones such as insulin that decrease blood glucose level.

Hyponatremia Abnormally low concentrations of sodium ions in extracellular fluid.

Hypoproteinemia (hi″po-pro″te-ĭ-ne′me-ah) A condition of unusually low levels of plasma proteins causing a reduction in colloid osmotic pressure; results in tissue edema.

Hypotension Low blood pressure.

Hypothalamic-hypophyseal tract (hi″po-thah-lam′ik hi″po-fiz′e-al) Bundle of axons that runs through the infundibulum and connects the posterior pituitary to the hypothalamus.

Hypothalamus (hi″po-thal′ah-mus) Region of the diencephalon forming the floor of the third ventricle of the brain.

Hypotonic (hi″po-ton′ik) Below normal tone or tension.

Hypotonic solution A solution that is more dilute (containing fewer nonpenetrating solutes) than the reference cell. Cells placed in hypotonic solutions plump up rapidly as water diffuses into them.

Hypoventilation A decrease in the depth and rate of breathing; characterized by an increase in blood carbon dioxide.

Hypovolemic shock (hi″po-vo-le′mik) Most common form of shock; results from extreme blood loss.

Hypoxia (hi-pok′se-ah) Condition in which inadequate oxygen is available to tissues.

Ileocecal valve (il″e-o-se′kal) Site where the small intestine joins the large intestine.

Ileum (il′e-um) Terminal part of the small intestine; between the jejunum and the cecum of the large intestine.

Immune system A functional system whose components attack foreign substances or prevent their entry into the body.

Immunity (im″ūn′ĭ-te) Ability of the body to resist many agents (both living and nonliving) that can cause disease; resistance to disease.

Immunocompetence Ability of the body's immune cells to recognize (by binding) specific antigens; reflects the presence of plasma membrane–bound receptors.

Immunodeficiency Any congenital or acquired condition causing a deficiency in the production or function of immune cells or certain molecules (complement, antibodies, etc.) required for normal immunity.

In vitro (in ve′tro) In a test tube, glass, or artificial environment.

In vivo (in ve′vo) In the living body.

Incompetent valve Valve that does not close properly.

Incontinence Inability to control micturition or defecation voluntarily.

Infarct (in′farkt) Region of dead, deteriorating tissue resulting from a lack of blood supply.

Infectious mononucleosis Highly contagious viral disease; marked by excessive agranulocytes.

Inferior (caudal) Pertaining to a position toward the lower or tail end of the long axis of the body.

Inferior vena cava Vein that returns blood from body areas below the diaphragm.

Inflammation (in″flah-ma′shun) An innate (nonspecific) defensive response of the body to tissue injury; includes dilation of arterioles and an increase in capillary permeability; cardinal signs are redness, heat, swelling, and pain.

Infundibulum (in″fun-dib′u-lum) (1) A stalk of tissue that connects the pituitary gland to the hypothalamus; (2) the distal end of the uterine (fallopian) tube.

Inguinal (ing′wĭ-nal) Pertaining to the groin region.

Inhibitory postsynaptic potential (IPSP) A graded potential in a postsynaptic neuron that inhibits action potential generation; usually hyperpolarizing.

Innervation (in″er-va′shun) Supply of nerves to a body part.

Inorganic compound Chemical substances that do not contain carbon, including water, salts, and many acids and bases.

Insertion Movable attachment of a muscle.

Insula Lobe of the cerebral cortex that is buried in the lateral sulcus beneath portions of the parietal, frontal, and temporal lobes.

Insulin A hormone that enhances the carrier-mediated diffusion of glucose into tissue cells, thus lowering blood glucose levels.

Insulin resistance State in which a greater than normal amount of insulin is required to maintain normal glucose blood levels.

Integration The process by which the nervous system processes and interprets sensory input and makes decisions about what should be done at each moment.

Integumentary system (in-teg″u-men′tar-e) Skin and its derivatives; provides the external protective covering of the body.

Intercalated discs (in-ter′kah-la″ted) Specialized connections between myocardial cells containing gap junctions and desmosomes.

Interferons (IFNs) (in-ter-fēr′ons) Proteins released from virus-infected (and other) cells that protect uninfected cells from viral takeover. Also inhibit some cancers.

Internal capsule Band of projection fibers that runs between the basal nuclei and the thalamus.

Internal respiration Exchange of gases between blood and tissue fluid and between tissue fluid and cells.

Interneuron (association neuron) Nerve cell located between motor and sensory neurons that shuttles signals through CNS pathways where integration occurs.

Interoceptor (in″ter-o-sep′tor) Sensory receptor in the viscera that is sensitive to changes and stimuli within the body's internal environment; also called visceroceptor.

Interphase One of two major periods in the cell life cycle; includes the period from cell formation to cell division.

Interstitial endocrine cells Cells located in the loose connective tissue surrounding the seminiferous tubules of the testes; they produce androgens (most importantly testosterone), which are secreted into the surrounding interstitial fluid.

Interstitial fluid (IF) (in″ter-stish′al) Fluid between the cells.

Interstitial lamellae Incomplete lamellae that lie between intact osteons, filling the gaps between forming osteons, or representing the remnants of an osteon that has been cut through by bone remodeling.

Intervertebral discs (in″ter-ver′teh-brul) Discs of fibrocartilage between vertebrae.

Intracapsular ligament Ligament located within and separate from the articular capsule of a synovial joint.

Intracellular fluid (ICF) (in″trah-sel′u-ler) Fluid within a cell.

Intrinsic factor Substance produced by the stomach that is required for vitamin B_{12} absorption.

Intron Noncoding segment or portion of DNA that ranges from 60 to 100,000 nucleotides long.

Involuntary muscle Muscle that cannot ordinarily be controlled voluntarily (e.g., smooth and cardiac muscle).

Involuntary nervous system The autonomic nervous system.

Ion (ī′on) Atom or molecule with a positive or negative electric charge.

Ion channel Transmembrane protein with a central pore that allows ions to pass through.

Ionic bond (ī-ah′nik) Chemical bond formed by electron transfer between atoms.

Ipsilateral (ip″sĭ-lă′ter-ul) Situated on the same side.

Ischemia (is-ke′me-ah) Local decrease in blood supply.

Isograft Tissue graft donated by an identical twin.

Isomer (i′so-mer) One of two or more substances that has the same molecular formula but with its atoms arranged differently.

Isometric contraction (i″so-mĕ′trik) Contraction in which the muscle does not shorten (the load is too heavy) but its internal tension increases.

Isotonic contraction (i″so-tah′nik) Contraction in which muscle tension remains constant at a given load, and the muscle shortens.

Isotonic solution A solution with a concentration of nonpenetrating solutes equal to that found in the reference cell.

Isotopes (i′so-tōps) Different atomic forms of the same element, which vary only in the number of neutrons they contain; the heavier species tend to be radioactive.

Jejunum (jĕ-joo′num) The part of the small intestine between the duodenum and the ileum.

Joint (articulation) The junction of two or more bones.

Joint kinesthetic receptor (kin″es-thet′ik) Receptor that provides information on joint position and motion.

Juxtaglomerular complex (JGC) (juks″tah-glo-mer′u-lar) Cells of the distal part of the ascending limb of the nephron loop and afferent arteriole located close to the glomerulus; involved in blood pressure regulation (via release of the hormone renin) and autoregulation of GFR.

Karyotype (kar′e-o-tīp) The diploid chromosomal complement, typically shown as homologous chromosome pairs arranged from longest to shortest (X and Y are arranged by size rather than paired).

Keratin (ker′ah-tin) Fibrous protein found in the epidermis, hair, and nails that makes those structures hard and water resistant; precursor is keratohyaline.

Ketones (ketone bodies) (ke′tōnz) Fatty acid metabolites; strong organic acids.

Ketosis (kē-tō′sis) Excess levels of ketone bodies in blood. Called ketoacidosis if blood pH is low.

Kilocalories (kcal) *See* Calorie.

Kinetic energy (ki-net′ik) The energy of motion or movement, e.g., the constant movement of atoms, or the push given to a swinging door that sets it into motion.

Labia (la′be-ah) (1) Lips; (2) skin folds of the female external genitalia; singular: labium.

Labor Collective term for the series of events that expel the fetus from the uterus.

Labyrinth (lab′ĭ-rinth″) Bony cavities and membranes of the inner ear.

Lacrimal (lak′rĭ-mal) Pertaining to tears.

Lactation (lak-ta′shun) Production and secretion of milk.

Lacteal (lak′te-al) Special lymphatic capillaries of the small intestine that take up lipids.

Lactic acid (lak′tik) Product of anaerobic metabolism, especially in muscle.

Lacuna (lah-ku′nah) A small space, cavity, or depression; lacunae in bone or cartilage are occupied by cells.

Lamella (lah-mel′ah) A layer, such as of bone matrix in an osteon of compact bone.

Lamina (lam′ĭ-nah) (1) A thin layer or flat plate; (2) the portion of a vertebra between the transverse process and the spinous process.

Lamina propria Loose connective tissue supporting an epithelium; part of a mucous membrane (mucosa).

Large intestine Portion of the digestive tract extending from the ileocecal valve to the anus; includes the cecum, appendix, colon, rectum, and anal canal.

Larynx (lar′ingks) Cartilaginous organ located between the trachea and the pharynx; voice box.

Latent period Period of time between stimulation and the onset of a response (such as muscle contraction).

Lateral Away from the midline of the body.

Leakage channel An ion channel that is always open (a nongated channel).

Leptin Hormone released by fat cells that signals satiety.

Leukemia Refers to a group of cancerous conditions of white blood cells.

Leukocytes (loo′ko-sīts) White blood cells; formed elements involved in body protection that take part in inflammatory and immune responses.

Leukocytosis An increase in the number of leukocytes (white blood cells); usually the result of a pathogen attack on the body.

Leukopenia (loo″ko-pe′ne-ah) Abnormally low white blood cell count.

Leukopoiesis The production of white blood cells.

Lever system Consists of a lever (bone), effort (muscle action), resistance (weight of object to be moved), and fulcrum (joint).

Ligament (lig′ah-ment) Band of dense regular connective tissue that connects bones.

Ligands Signaling chemicals that bind specifically to membrane receptors.

Limbic system (lim′bik) Functional brain system involved in emotional response and memory formation.

Lipid (lĭ′pid) Hydrophobic organic compound formed of carbon, hydrogen, and oxygen; examples are fats and cholesterol.

Lipolysis (lĭ-pol′ĭ-sis) The breakdown of stored fats into glycerol and fatty acids.

Liver Lobed accessory organ that overlies the stomach; produces bile to help digest fat, and serves other metabolic and regulatory functions.

Lumbar (lum′bar) Portion of the back between the thorax and the pelvis.

Lumbar vertebrae The five vertebrae of the lumbar region of the vertebral column, commonly called the small of the back.

Lumen (loo′min) Cavity inside a tube, blood vessel, or hollow organ.

Luteinizing hormone (LH) (lu′te-in-īz″ing) Anterior pituitary hormone that aids maturation of cells in the ovary and triggers ovulation in females. In males, causes the interstitial endocrine cells of the testis to produce testosterone.

Lymph (limf) Protein-containing fluid transported by lymphatic vessels.

Lymph node Small lymphoid organ that filters lymph; contains macrophages and lymphocytes.

Lymphatic system (lim-fat′ik) System consisting of lymphatic vessels, lymph nodes, and lymph; drains excess tissue fluid from the extracellular space. The nodes provide sites for immune surveillance.

Lymphatics General term used to designate the lymphatic vessels that collect and transport lymph.

Lymphocyte Agranular white blood cell that arises from bone marrow and becomes functionally mature in the lymphoid organs of the body.

Lysosomes (li′so-sōmz) Organelles that originate from the Golgi apparatus and contain strong digestive enzymes.

Lysozyme (li′so-zīm) Enzyme in sweat, saliva, and tears that is capable of destroying certain kinds of bacteria.

M (mitotic) phase One of two major periods in the cell life cycle; involves the division of the nucleus (mitosis) and the division of the cytoplasm (cytokinesis).

Macromolecules Large, complex molecules containing from 100 to over 10,000 subunits.

Macrophage (mak′ro-fāj″) Immune cell type common in connective tissue, lymphoid tissue, and many body organs; phagocytizes tissue cells, bacteria, and other foreign debris; presents antigens to T cells in the immune response.

Macula (mak′u-lah) (1) Receptor for linear acceleration, deceleration, and gravity, located within the vestibule of the inner ear; (2) a colored area or spot (as on the retina).

Malignant (muh-lig′nent) Life threatening; pertains to neoplasms that spread and lead to death, such as cancer.

Malignant melanoma (mel″ah-no′mah) Cancer of the melanocytes; can begin wherever there is pigment.

MALT (mucosa-associated lymphoid tissue) Diffusely distributed collections of lymphoid tissue in mucous membranes.

Mammary glands (mam′mer-e) Milk-producing glands of the breast.

Mandible (man′dĭ-bl) Lower jawbone; U shaped, largest bone of the face.

Mass number Sum of the number of protons and neutrons in the nucleus of an atom.

Mast cells Immune cells that function to detect foreign substances in the tissue spaces and initiate local inflammatory responses against them; typically found clustered deep to an epithelium or along blood vessels.

Mastication (mas″tĭ-ka′shun) Chewing.

Meatus (me-a′tus) External opening of a canal.

Mean arterial pressure A weighted average of the arterial pressure calculated from the systolic and diastolic pressures; determined by cardiac output and total peripheral resistance.

Mechanical advantage (power lever) Condition that occurs when the load is close to the fulcrum and the effort is applied far from the fulcrum; allows a small effort exerted over a relatively large distance to move a large load over a small distance.

Mechanical disadvantage (speed lever) Condition that occurs when the load is far from the fulcrum and the effort is applied near the fulcrum; the effort applied must be greater than the load to be moved.

Mechanical energy The energy directly involved in moving matter; e.g., in bicycle riding, the legs provide the mechanical energy that moves the pedals.

Mechanoreceptor (meh″kĕ-no-re-sep′tor) Receptor sensitive to mechanical pressure such as touch, sound, or that exerted by muscle contraction.

Medial (me′de-ahl) Toward the midline of the body.

Median (midsagittal) plane Specific sagittal plane that lies exactly in the midline.

Mediastinum (me″de-ah-sti′num) The medial cavity of the thorax containing the heart, great vessels, thymus, and parts of the trachea, bronchi, and esophagus.

Medulla (mĕ-dul′ah) Central portion of certain organs.

Medulla oblongata (mĕ-dul′ah ob″long-gah′tah) Inferiormost part of the brain stem.

Medullary cavity Central cavity of a long bone. Contains yellow or red (bone) marrow.

Meiosis (mi-o′sis) Nuclear division process that reduces the chromosomal number by half and results in the formation of four haploid (*n*) cells; occurs only in certain cells of the reproductive organs.

Melanin (mel′ah-nin) Dark pigment formed by cells called melanocytes; imparts color to skin and hair.

Melatonin (mel″ah-to′nin) A hormone secreted by the pineal gland; secretion peaks at night and helps set sleep-wake cycles; also a powerful antioxidant.

Membrane potential Voltage across the plasma membrane.

Membrane receptors A large, diverse group of integral proteins that serve as binding sites for signaling molecules.

Memory cells Members of T cell and B cell clones that provide for immunological memory.

Menarche (mĕ-nar′ke) Establishment of menstrual function; the first menstrual period.

Meninges (mĕ-nin′jēz) Protective coverings of the central nervous system; from the most external to the most internal, the dura mater, arachnoid mater, and pia mater.

Meningitis (mĕ-nin-ji′tis) Inflammation of the meninges.

Menopause Period of life when, prompted by hormonal changes, ovulation and menstruation cease.

Menstruation (men″stroo-a′shun) The periodic, cyclic discharge of blood, secretions, and tissue from the mature female uterus in the absence of pregnancy.

Merocrine glands (mer′o-krin) Glands that produce secretions intermittently; secretions do not accumulate in the gland.

Mesencephalon (mes″en-sef′ah-lon) One of the three primary vesicles of the developing brain; becomes the midbrain.

Mesenchyme (meh′zin-kīm) Common embryonic tissue from which all connective tissues arise.

Mesenteries (mes″en-ter′ēz) Double-layered extensions of the peritoneum that support most organs in the abdominal cavity.

Mesoderm (mez′o-derm) Primary germ layer that forms the skeleton and muscles of the body.

Mesothelium (mez″o-the′le-um) The epithelium found in serous membranes lining the ventral body cavity and covering its organs.

Messenger RNA (mRNA) Long nucleotide strands that reflect the exact nucleotide sequences of the genetically active DNA and carry the DNA's message.

Metabolic rate (met″ah-bol′ik) Energy expended by the body per unit time.

Metabolic water (water of oxidation) Water produced from cellular metabolism (about 10% of our body's water).

Metabolism (mĕ-tab′o-lizm) Sum total of the chemical reactions occurring in the body cells.

Metaphase Second stage of mitosis, meiosis I, and meiosis II.

Metastasis (mĕ-tas′tah-sis) The spread of cancer from one body part or organ into another not directly connected to it.

Metencephalon (met″en-sef′ah-lon) A secondary brain vesicle; anterior portion of the rhombencephalon of the developing brain; becomes the pons and the cerebellum.

MHC (major histocompatibility complex) proteins Molecules on the outer surface of the plasma membrane of all cells; help the immune system distinguish self from nonself. T cells recognize antigens only when combined with these proteins.

Microfilaments (mi″kro-fil′ah-ments) One of three types of cytoskeletal elements; thin strands of the protein actin.

Microglial cells (mi-kro′gle-al) A type of CNS supporting cell; can transform into phagocytes in areas of neural damage or inflammation.

Microtubules (mi″kro-tu′būlz) One of three types of cytoskeletal elements; hollow tubes made of the spherical protein tubulin.

Microvilli (mi″kro-vil′i) Tiny projections on the free surfaces of some epithelial cells; increase surface area for absorption.

Micturition (mik″tu-rish′un) Urination, or voiding; emptying the bladder.

Midbrain (mesencephalon) Region of the brain stem between the diencephalon and the pons.

Midsagittal (median) plane Specific sagittal plane that lies exactly in the midline.

Milliequivalents per liter (mEq/L) The units used to measure electrolyte concentrations of body fluids; a measure of the number of electrical charges in 1 liter of solution.

Mineralocorticoid (min″er-al″ō-kor′tĭ-koyd) Steroid hormone of the adrenal cortex that regulates Na^+ and K^+ metabolism and fluid balance.

Minerals Inorganic chemical compounds found in nature; salts.

Mitochondria (mi″to-kon′dre-ah) Cytoplasmic organelles responsible for ATP generation for cellular activities.

Mitosis Process during which the chromosomes are redistributed to two daughter nuclei; nuclear division. Consists of prophase, metaphase, anaphase, and telophase.

Mitral (bicuspid) valve (mi′tral) The left atrioventricular valve.

Mixed nerves Nerves containing the axons of motor and sensory neurons; their action potentials travel to and from the central nervous system.

Molar (mo′lar) (1) A solution concentration determined by mass of solute in a given volume of solution—1 liter of a 1 molar solution contains an amount of solute equal to its molecular weight in grams. (2) Broad back teeth that grind and crush.

Molarity (mo-lar′ĭ-te) A way to express the concentration of a solution; moles per liter of solution.

Mole (mōl) A mole of any element or compound is equal to its atomic weight or its molecular weight (sum of atomic weights) measured in grams.

Molecule Particle consisting of two or more atoms joined together by chemical bonds.

Monoclonal antibodies (mon″o-klo′nal) Pure preparations of identical antibodies that exhibit specificity for a single antigen.

Monocyte (mon′o-sīt) Large single-nucleus white blood cell; agranular leukocyte.

Monosaccharide (mon″o-sak′ah-rīd) Literally, one sugar; building block of carbohydrates; e.g., glucose.

Motor areas Functional areas in the cerebral cortex that control voluntary motor functions.

Motor (efferent) nerves Nerves that carry action potentials leaving the brain and spinal cord, and destined for effectors.

Motor unit A motor neuron and all the muscle cells it stimulates.

Mucous membranes (mucosae) Membranes that form the linings of body cavities open to the exterior (digestive, respiratory, urinary, and reproductive tracts).

Mucus (myoo′kus) A sticky, thick fluid secreted by mucous glands and mucous membranes; keeps the free surface of membranes moist.

Multinucleate cell (mul″tĭ-nu′kle-āt) Cell with more than one nucleus, e.g., skeletal muscle cells, osteoclasts.

Multiple sclerosis (MS) Demyelinating disorder of the CNS; causes hardened patches (sclerosis) in the brain and spinal cord.

Multipolar neurons Neurons with three or more processes; most common neuron type in the CNS.

Muscarinic receptors (mus″kah-rin′ik) Acetylcholine-binding receptors of the autonomic nervous system's target organs; named for activation by the mushroom poison muscarine.

Muscle fiber A muscle cell.

Muscle spindle Encapsulated receptor found in skeletal muscle that is sensitive to stretch.

Muscle tension The force exerted by a contracting muscle on some object.

Muscle tone Low levels of contractile activity in relaxed muscle; keeps the muscle healthy and ready to act.

Muscle twitch The response of a muscle to a single brief threshold stimulus.

Muscular dystrophy A group of inherited muscle-destroying diseases.

Muscular system The organ system consisting of the skeletal muscles of the body and their connective tissue attachments.

Myelencephalon A secondary brain vesicle; lower part of the developing hindbrain, especially the medulla oblongata.

Myelin sheath (mi′ĕ-lin) Fatty insulating sheath that surrounds all but the smallest nerve fibers.

Myoblasts Embryonic mesoderm cells from which all muscle fibers develop.

Myocardial infarction (MI) (mi″o-kar′de-al in-fark′shun) Condition characterized by dead tissue areas in the myocardium; caused by interruption of blood supply to the area. Commonly called heart attack.

Myocardium (mi″o-kar′de-um) Layer of the heart wall composed of cardiac muscle.

Myofibril (mi″o-fi′bril) Rodlike bundle of contractile filaments (myofilaments) found in muscle fibers (cells).

Myofilament (mi″o-fil′ah-ment) Filament that constitutes myofibrils. Of two types: actin and myosin.

Myoglobin (mi″o-glo′bin) Oxygen-binding pigment in muscle.

Myogram A graphic recording of mechanical contractile activity produced by an apparatus that measures muscle contraction.

Myometrium (mi″o-me′tre-um) Thick uterine musculature.

Myopia (mi-o′pe-ah) A condition in which visual images are focused in front of rather than on the retina; nearsightedness.

Myosin (mi′o-sin) One of the principal contractile proteins found in muscle.

Myxedema (mik″sĕ-de′mah) Severe hypothyroidism.

Nares (na′rez) Nostrils.

Natural killer (NK) cell Defensive cell (a type of lymphocyte) that can kill cancer cells and virus-infected body cells before the adaptive immune system is activated.

Necrosis (nĕ-kro′sis) Death or disintegration of a cell or tissues caused by disease or injury.

Negative feedback mechanisms The most common homeostatic control mechanism. The net effect is that the output of the system shuts off the original stimulus or reduces its intensity.

Neonatal period The four-week period immediately after birth.

Neoplasm (ne′o-plazm) An abnormal mass of proliferating cells. Benign neoplasms remain localized; malignant neoplasms are cancers, which can spread to other organs.

Nephron (nef′ron) Structural and functional unit of the kidney; consists of the renal corpuscle and renal tubule.

Nerve A bundle of axons in the peripheral nervous system.

Nerve fiber Axon of a neuron.

Nerve growth factor (NGF) Protein that promotes survival and development of neurons; secreted by their target cells and many other cell types.

Nerve impulse A self-propagating wave of depolarization; also called an action potential.

Nerve plexuses Interlacing nerve networks that occur in the cervical, brachial, lumbar, and sacral regions and primarily serve the limbs.

Nervous system Fast-acting control system that triggers muscle contraction or gland secretion.

Neural crest cells Embryonic cells of ectodermal origin that migrate away from the forming neural tube to become a wide variety of different cells such as certain adrenal medulla cells, melanocytes, and neurons in various ganglia.

Neural tube Fetal structure that gives rise to the brain, spinal cord, and associated neural structures; formed from ectoderm.

Neuroglia (nu-rog′le-ah) Nonexcitable cells of neural tissue that support, protect, and insulate the neurons; glial cells.

Neurohypophysis (nu″ro-hi-pof′ĭ-sis) Posterior pituitary plus infundibulum; portion of the pituitary gland derived from the brain.

Neuromuscular junction (motor end plate) Region where a motor neuron comes into close contact with a skeletal muscle cell.

Neuron (nerve cell) (nu′ron) Cell of the nervous system specialized to generate and transmit electrical signals (action potentials and graded potentials).

Neuron cell body The biosynthetic center of a neuron; also called the perikaryon, or soma.

Neuronal pools Functional groups of neurons that process and integrate information.

Neuropeptides (nu″ro-pep′tīds) A class of neurotransmitters including beta endorphins and enkephalins (which act as euphorics and reduce perception of pain) and gut-brain peptides.

Neurotransmitter Chemical messenger released by neurons that, upon binding to receptors of neurons or effector cells, stimulate or inhibit those neurons or effector cells.

Neutralization reaction Displacement reaction in which mixing an acid and a base forms water and a salt.

Neutron (nu′tron) Uncharged subatomic particle; found in the atomic nucleus.

Neutrophil (nu′tro-fil) Most abundant type of white blood cell.

Nicotinic receptors (nik″o-tin′ik) Acetylcholine-binding receptors of all autonomic postganglionic neurons and skeletal muscle neuromuscular junctions; named for activation by nicotine.

Nitric oxide (NO) A gaseous chemical messenger; diverse functions include participation in memory formation in the brain, and causing local vasodilation.

Nociceptor (no″se-sep′tor) Receptor sensitive to potentially damaging stimuli that result in pain.

Nondisjunction Failure of sister chromatids to separate during mitosis or failure of homologous pairs to separate during meiosis; results in abnormal numbers of chromosomes in the resulting daughter cells.

Nonmyelinated fibers (non-mi′ĕ-lĭ-nāt″ed) Axons lacking a myelin sheath and therefore conducting action potentials quite slowly.

Nonpolar molecules Electrically symmetrical molecules.

Nonvolatile (fixed) acid Acid generated by cellular metabolism that must be eliminated by the kidneys.

Norepinephrine (NE) (nor″ep-ĭ-nef′rin) A catecholamine neurotransmitter and adrenal medullary hormone, associated with sympathetic nervous system activation.

Nuclear envelope The double membrane barrier of a cell nucleus.

Nucleic acid (nu-kle′ik) Class of organic molecules that includes DNA and RNA.

Nucleoli (nu-kle′o-li) Dense spherical bodies in the cell nucleus involved with ribosomal RNA (rRNA) synthesis and ribosomal subunit assembly.

Nucleosome (nu′kle-o-sōm) Fundamental unit of chromatin; consists of a strand of DNA wound around a cluster of eight histone proteins.

Nucleotide (nu′kle-o-tīd) Building block of nucleic acids; consists of a sugar, a nitrogen-containing base, and a phosphate group.

Nucleus (nu′kle-is) (1) Control center of a cell; contains genetic material; (2) clusters of neuron cell bodies in the CNS; (3) center of an atom; contains protons and neutrons.

Nutrients Chemical substances taken in via the diet that are used for energy and cell building.

Oblique section A cut made diagonally between the horizontal and vertical plane of the body or an organ.

Occlusion (ah-kloo′zhun) Closure or obstruction.

Octet rule (rule of eights) (ok-tet′) The tendency of atoms to interact in such a way that they have eight electrons in their valence shell.

Olfaction (ol-fak′shun) Sense of smell.

Oligodendrocyte (ol″ĭ-go-den′dro-sīt) A type of CNS supporting cell that composes myelin sheaths.

Oocyte (o′o-sīt) Immature female gamete.

Oogenesis (o″o-jen′ĕ-sis) Process of ovum (female gamete) formation.

Ophthalmic (of-thal′mik) Pertaining to the eye.

Optic (op′tik) Pertaining to the eye or vision.

Optic chiasma (op′tik ki-az′muh) The partial crossover of fibers of the optic nerves to form the optic tracts.

Organ A part of the body formed of two or more tissues and adapted to carry out a specific function; e.g., the stomach.

Organ system A group of organs that work together to perform a vital body function; e.g., the nervous system.

Organelles (or″gah-nelz′) Small subcellular structures (ribosomes, mitochondria, and others) that perform specific functions for the cell as a whole.

Organic compound Any compound composed of atoms (some of which are carbon) held together by covalent (shared electron) bonds. Examples are proteins, fats, and carbohydrates.

Organism The living animal (or plant), which represents the sum total of all its organ systems working together to maintain life; also applies to a microorganism.

Origin Attachment of a muscle that remains relatively fixed during muscular contraction.

Osmolality The number of solute particles dissolved in 1 kilogram (1000 g) of water; reflects the solution's ability to cause osmosis.

Osmolarity (oz″mo-lar′ĭ-te) The number of solute particles present in 1 liter of a solution.

Osmoreceptor (oz″mo-re-sep′tor) Structure sensitive to osmotic pressure or concentration of a solution.

Osmosis (oz-mo′sis) Diffusion of a solvent through a selectively permeable membrane from a dilute solution into a more concentrated one.

Osmotic pressure A measure of the tendency of a solvent to move into a more concentrated solution.

Ossicles *See* Auditory ossicles.

Ossification (os″ĭ-fi-ka′shun) *See* Osteogenesis.

Osteoblasts (os′te-o-blasts) Bone-forming cells.

Osteoclasts (os′te-o-klasts) Large cells that resorb or break down bone matrix.

Osteocyte (os′te-o-sīt) Mature bone cell.

Osteogenesis (os″te-o-jen′e-sis) The process of bone formation; also called ossification.

Osteoid (os′te-oid) Unmineralized bone matrix.

Osteomalacia (os″te-o-mah-la′she-ah) Disorder in which bones are inadequately mineralized; soft bones.

Osteon (os′te-on) System of interconnecting canals in the microscopic structure of adult compact bone; unit of bone; also called Haversian system.

Osteoporosis (os″te-o-po-ro′sis) Decreased density and strength of bone resulting from a gradual decrease in rate of bone formation.

Ovarian cycle (o-vayr′e-an) Monthly cycle of follicle development, ovulation, and corpus luteum formation in an ovary.

Ovary (o′var-e) Female reproductive organ in which ova (eggs) are produced; female gonad.

Ovulation (ov″u-la′shun) Ejection of an oocyte (immature egg) from the ovary.

Ovum (o′vum) Female gamete; egg.

Oxidases Enzymes that catalyze the transfer of oxygen in oxidation-reduction reactions.

Oxidation (oks′ĭ-da″shun) Process of substances combining with oxygen or the removal of hydrogen.

Oxidation-reduction (redox) reaction A reaction that couples the oxidation (loss of electrons) of one substance with the reduction (gain of electrons) of another substance.

Oxidative phosphorylation (ok′sĭ-da″tiv fos″for-ĭ-la′shun) Process of ATP synthesis during which an inorganic phosphate group is attached to ADP; occurs via the electron transport chain within the mitochondria.

Oxyhemoglobin (ok″sĭ-he″mo-glo′bin) Oxygen-bound form of hemoglobin.

Oxytocin (ok″sĭ-to′sin) Hormone synthesized in the hypothalamus and secreted by the posterior pituitary; stimulates contraction of the uterus during childbirth and the ejection of milk during nursing.

Pacemaker cells Specialized cells in cardiac or unitary (visceral) smooth muscle that have unstable resting membrane potentials and so periodically fire spontaneous action potentials, triggering contraction of surrounding myocytes.

Paget's disease (paj′ets) Disorder characterized by excessive bone breakdown and abnormal bone formation.

Palate (pal′at) Roof of the mouth.

Pancreas (pan′kre-us) Gland located behind the stomach, between the spleen and the duodenum; produces both endocrine and exocrine secretions.

Pancreatic juice (pan″kre-at′ik) Bicarbonate-rich secretion of the pancreas containing enzymes for digestion of all food categories.

Papilla (pah-pil′ah) Small, nipple-like projection; e.g., dermal papillae are projections of dermal tissue into the epidermis.

Paracrine (par′ah-krin) A chemical messenger that acts locally within the same tissue and is rapidly destroyed. Examples are prostaglandins and nitric oxide.

Parasagittal planes All sagittal planes offset from the midline.

Parasympathetic division The division of the autonomic nervous system that oversees digestion, elimination, and glandular function; the resting and digesting subdivision.

Parasympathetic tone Normal (background) level of parasympathetic output; sustains normal gastrointestinal and urinary tract activity, lowers heart rate.

Parathyroid glands (par″ah-thi′roid) Small endocrine glands located on the posterior aspect of the thyroid gland.

Parathyroid hormone (PTH) Hormone released by the parathyroid glands that regulates blood calcium level.

Parietal (pah-ri′ĕ-tal) Pertaining to the walls of a cavity.

Parietal serosa The part of the double-layered membrane that lines the walls of the ventral body cavity.

Parkinson's disease Neurodegenerative disorder of the basal nuclei due to insufficient secretion of the neurotransmitter dopamine; symptoms include tremor and rigid movement.

Partial pressure The pressure exerted by a single component of a mixture of gases.

Parturition (par″tu-rish′un) Culmination of pregnancy; giving birth.

Passive immunity Short-lived immunity resulting from the introduction of "borrowed antibodies" obtained from an immune animal or human donor; immunological memory is not established.

Passive (transport) processes Membrane transport processes that move substances down their concentration gradients (e.g., diffusion). They are driven by kinetic energy and so do not require cellular energy (such as ATP).

Pathogen (path′o-jen) Disease-causing organism.

Pectoral (pek′tor-al) Pertaining to the chest.

Pectoral (shoulder) girdle Bones that attach the upper limbs to the axial skeleton; includes the clavicle and scapula.

Pedigree Diagram that traces a particular genetic trait through several generations and helps predict the genotype of future offspring.

Pelvic girdle (hip girdle) Consists of the paired coxal bones and sacrum that attach the lower limbs to the axial skeleton.

Pelvis (pel′vis) (1) Basin-shaped bony structure composed of the hip bones, sacrum, and coccyx; (2) funnel-shaped tube within the kidney continuous with the ureter.

Penis (pe′nis) Male organ of copulation and urination.

Pepsin Enzyme capable of digesting proteins in an acid pH.

Peptide bond (pep′tīd) Bond joining the amine group of one amino acid to the acid carboxyl group of a second amino acid with the loss of a water molecule.

Perforating canals Canals that run at right angles to the long axis of the bone, connecting the vascular and nerve supplies of the periosteum to those of the central canals and medullary cavity; also called Volkmann's canals.

Pericardium (per″ĭ-kar′de-um) Double-layered sac enclosing the heart and forming its superficial layer; has fibrous and serous layers.

Perichondrium (per″ĭ-kon′dre-um) Fibrous, connective-tissue membrane covering the external surface of cartilaginous structures.

Perimysium (per″ĭ-mis′e-um) Connective tissue that bundles muscle fibers into fascicles.

Perineum (per″ĭ-ne′um) That region of the body spanning the region between the ischial tuberosities and extending from the pubic arch to the coccyx.

Periosteum (per″e-os′te-um) Double-layered connective tissue that covers and nourishes the bone.

Peripheral congestion Condition caused by failure of the right side of the heart; results in edema in the extremities.

Peripheral nervous system (PNS) Portion of the nervous system consisting of nerves, ganglia, sensory receptors, and motor endings that lie outside of the brain and spinal cord.

Peristalsis (per″ĭ-stal′sis) Progressive, wavelike contractions that move foodstuffs through the alimentary tube organs (or that move other substances through other hollow body organs).

Peritoneum (per″ĭ-to-ne′um) Serous membrane lining the interior of the abdominal cavity and covering the surfaces of abdominal organs.

Peritonitis (per″ĭ-to-ni′tis) Inflammation of the peritoneum.

Permeability A measure of the ability of molecules and ions to pass through a membrane.

Peroxisomes (pĕ-roks′ĭ-sōmz) Membranous sacs in cytoplasm containing powerful oxidase enzymes that use molecular oxygen to detoxify harmful or toxic substances, such as free radicals.

Peyer's patches (pi′erz) Lymphoid organs located in the small intestine; also called aggregated lymphoid nodules.

pH unit (pe-āch) The measure of the relative acidity or alkalinity of a solution.

Phagocytosis (fag″o-si-to′sis) Engulfing of solids by (phagocytic) cells.

Phagosome (fag′o-sōm) Vesicle formed as a result of phagocytosis.

Pharmacological dose A drug dose that is dramatically higher than normal levels of that substance (e.g., hormone) in the body.

Pharyngeal cleft An indentation of the surface ectoderm in the embryo; the external acoustic meatus develops from it. Also called branchial groove.

Pharyngotympanic tube Tube that connects the middle ear and the pharynx. Also called auditory tube, eustachian tube.

Pharynx (fayr′inks) Muscular tube extending from the region posterior to the nasal cavities to the esophagus.

Phenotype (fe′no-tūp) Observable expression of the genotype.

Phospholipid (fos″fo-lip′id) Modified lipid, contains phosphorus.

Phosphorylation A chemical reaction in which a phosphate molecule is added to a molecule; for example, phosphorylation of ADP yields ATP.

Photoreceptor (fo″to-re-sep′tor) Specialized receptor cell that responds to light energy; rods and cones.

Physiological acidosis (as″ĭ-do′sis) Arterial pH lower than 7.35 resulting from any cause.

Physiological dose A drug dose that replicates normal levels of that substance (e.g., hormone) in the body. (Compare with Pharmacological dose.)

Physiology (fiz″e-ol′o-je) Study of the function of living organisms.

Pineal gland (body) (pin′e-al) A hormone-secreting part of the diencephalon of the brain thought to be involved in setting the biological clock and influencing reproductive function.

Pinocytosis (pe″no-si-to′sis) Engulfing of extracellular fluid by cells.

Pituitary gland (pĭ-tu′ĭ-tayr″e) Neuroendocrine gland located beneath the brain that serves a variety of functions including regulation of gonads, thyroid, adrenal cortex, lactation, and water balance.

Placenta (plah-sen′tah) Temporary organ formed from both fetal and maternal tissues that provides nutrients and oxygen to the developing fetus, carries away fetal metabolic wastes, and produces the hormones of pregnancy.

Plasma (plaz′mah) The nonliving fluid component of blood within which formed elements and various solutes are suspended and circulated.

Plasma cells Members of a B cell clone; effector B cells specialized to produce and release antibodies.

Plasma membrane Membrane, composed of phospholipids, cholesterol, and proteins, that encloses cell contents; outer limiting cell membrane.

Platelet (plāt′let) Cell fragment found in blood; involved in clotting.

Pleurae (ploo′re) Two layers of serous membrane that line the thoracic cavity and cover the external surface of the lung.

Pleural cavity (ploo′ral) A potential space between the two layers of pleura; contains a thin film of serous fluid.

Plexus (plek′sus) A network of converging and diverging nerve fibers, blood vessels, or lymphatics.

Polar molecules Electrically asymmetric molecules.

Polarized State of a plasma membrane of an unstimulated neuron or muscle cell in which the inside of the cell is relatively negative in comparison to the outside; the resting state.

Polycythemia (pol″e-si-the′me-ah) An abnormally high number of erythrocytes.

Polymer A substance of high molecular weight with long, chainlike molecules consisting of many similar (repeated) units.

Polypeptide (pol″e-pep′tīd) A chain of amino acids.

Polyps Benign mucosal tumors.

Polysaccharide (pol″e-sak′ah-rīd) Literally, many sugars, a polymer of linked monosaccharides; e.g., starch, glycogen.

Pons (1) Any bridgelike structure or part; (2) the part of the brain stem connecting the medulla with the midbrain, providing linkage between upper and lower levels of the central nervous system.

Pore The surface opening of the duct of a sweat gland.

Positive feedback mechanisms Feedback that tends to cause the level of a variable to change in the same direction as an initial change.

Posterior pituitary Neural part of pituitary gland; part of the neurohypophysis.

Postganglionic neuron (post″gan″gle-ah′nik) Autonomic motor neuron that has its cell body in an autonomic ganglion and projects its axon to an effector.

Potential energy Stored or inactive energy.

Preganglionic neuron Autonomic motor neuron that has its cell body in the central nervous system and projects its axon to an autonomic ganglion.

Presbyopia (pres″be-o′pe-ah) A condition that results in the loss of near focusing ability; typical onset is around age 40.

Pressure gradient Difference in pressure (hydrostatic or osmotic) that drives movement of fluid.

Primary active transport A type of active transport in which the energy needed to drive the transport process is provided directly by hydrolysis of ATP.

Primary lymphoid organs The red bone marrow and thymus; lymphoid organs in which lymphocytes develop and mature.

Prime mover Muscle that bears the major responsibility for effecting a particular movement; an agonist.

Process (1) Prominence or projection; (2) series of actions for a specific purpose.

Progesterone (pro-jes′ter-ōn) Hormone partly responsible for preparing the uterus for the fertilized ovum.

Prolactin (PRL) (pro-lak′tin) Anterior pituitary hormone that stimulates the breasts to produce milk.

Pronation (pro-na′shun) Inward rotation of the forearm causing the radius to cross diagonally over the ulna—palms face posteriorly.

Prophase The first stage of mitosis, meiosis I, and meiosis II. The chromosomes become visible, the nuclear envelope breaks down, and a spindle forms.

Proprioceptor (pro″pre-o-sep′tor) Receptor located in a joint, muscle, or tendon; concerned with locomotion, posture, and muscle tone.

Prostaglandin (pros″tah-glan′din) A lipid-based chemical messenger synthesized by most tissue cells; acts locally as a paracrine.

Prostate Accessory reproductive gland; produces one-third of semen volume, including fluids that activate sperm.

Protein (pro′tēn) Organic compound composed of carbon, oxygen, hydrogen, and nitrogen; types include enzymes, structural components; 10–30% of cell mass.

Protein kinase Enzyme that phosphorylates a protein (transfers an inorganic phosphate from ATP to the protein), resulting in activation or inactivation of the protein.

Prothrombin time Diagnostic test to determine status of the body's hemostasis system.

Proton (pro′ton) Subatomic particle that bears a positive charge; located in the atomic nucleus.

Proton acceptor A substance that takes up hydrogen ions in detectable amounts. Commonly referred to as a base.

Proton donor A substance that releases hydrogen ions in detectable amounts; an acid.

Proximal (prok′si-mul) Toward the attached end of a limb or the origin of a structure.

Pseudounipolar neuron (soo″do-u″nĭ-po′lar) Another term for unipolar neuron.

Puberty Period of life when reproductive maturity is achieved.

Pulmonary (pul′muh-nayr-e) Pertaining to the lungs.

Pulmonary arteries Vessels that deliver blood to the lungs to be oxygenated.

Pulmonary circuit System of blood vessels that serves gas exchange in the lungs; i.e., pulmonary arteries, capillaries, and veins.

Pulmonary edema (ĕ-de′muh) Leakage of fluid into the air sacs and tissue of the lungs.

Pulmonary veins Vessels that deliver freshly oxygenated blood from the respiratory zones of the lungs to the heart.

Pulmonary ventilation Breathing; consists of inspiration and expiration.

Pulse Rhythmic expansion and recoil of arteries resulting from heart contraction; can be felt from outside the body.

Pupil Opening in the center of the iris through which light enters the eye.

Pus Fluid product of inflammation composed of white blood cells, the debris of dead cells, and a thin fluid.

Pyloric sphincter (pi-lor′ik sfink′ter) Valve of the distal end of the stomach that controls food entry into the duodenum.

Pyramidal (corticospinal) tracts Major motor pathways concerned with voluntary movement; descend from pyramidal cells in the frontal lobes of each cerebral hemisphere.

Pyruvic acid An intermediate compound in the metabolism of carbohydrates.

Radioactivity The process of spontaneous decay seen in some of the heavier isotopes, during which particles or energy is emitted from the atomic nucleus; results in the atom becoming more stable.

Radioisotope (ra″de-o-i′so-tōp) Isotope that exhibits radioactive behavior.

Ramus (ra′mus) Branch of a nerve, artery, vein, or bone.

Rapid eye movement (REM) sleep Stage of sleep in which rapid eye movements, an alert EEG pattern, and dreaming occur.

Reactant A substance that is an input to a chemical reaction.

Receptor (re-sep′tor) (1) A cell or nerve ending of a sensory neuron specialized to respond to particular types of stimuli; (2) protein that binds specifically with other molecules, e.g., neurotransmitters, hormones, paracrines, antigens.

Receptor-mediated endocytosis The type of endocytosis in which engulfed particles attach to receptors before endocytosis occurs.

Receptor potential A graded potential that occurs at a sensory receptor membrane.

Recessive traits A trait due to a particular allele that does not manifest itself in the presence of other alleles that generate traits dominant to it; must be present in double dose to be expressed.

Reduction Chemical reaction in which electrons and energy are gained by a molecule (often accompanied by gain of hydrogen ions) or oxygen is lost.

Referred pain Pain felt at a site other than the area of origin.

Reflex Automatic reaction to stimuli.

Refraction The bending of a light ray when it meets a different surface at an oblique rather than right angle.

Regeneration Replacement of destroyed tissue with the same kind of tissue.

Regulatory T cells (T$_{Reg}$ cells) Population of T cells (usually expressing CD4) that suppress the immune response.

Relative refractory period Follows the absolute refractory period; interval when a threshold for action potential stimulation is markedly elevated.

Renal (re′nal) Pertaining to the kidney.

Renal autoregulation Process the kidney uses to maintain a nearly constant glomerular filtration rate despite fluctuations in systemic blood pressure.

Renal clearance The volume of plasma from which a particular substance is completely removed in a given time, usually 1 minute; provides information about renal function.

Renin (re′nin) Enzyme released by the kidneys that raises blood pressure by initiating the renin-angiotensin-aldosterone mechanism.

Renin-angiotensin-aldosterone mechanism Hormone mechanism that acts to increase blood sodium content and blood pressure.

Rennin Stomach-secreted enzyme that acts on milk protein; not produced in adults.

Repolarization Return of the membrane potential to the initial resting (polarized) state.

Reproductive system Organ system that functions to produce offspring.

Respiration The processes involved in supplying the body with oxygen and disposing of carbon dioxide.

Respiratory system Organ system that carries out gas exchange; includes the nose, pharynx, larynx, trachea, bronchi, lungs.

Resting membrane potential The voltage that exists across the plasma membrane during the resting state of an excitable cell; typically ranges from −50 to −90 millivolts depending on cell type.

Reticular activating system (RAS) (re-tik′u-lar) Diffuse brain stem neural network that receives a wide variety of sensory input and maintains wakefulness of the cerebral cortex.

Reticular connective tissue Connective tissue with a fine network of reticular fibers that form the internal supporting framework of lymphoid organs.

Reticular formation Functional system that spans the brain stem; involved in regulating sensory input to the cerebral cortex, cortical arousal, and control of motor behavior.

Reticular lamina A layer of extracellular material containing a fine network of collagen fibers; together with the basal lamina it is a major component of the basement membrane.

Reticulocyte (rĕ-tik′u-lo-sīt) Immature erythrocyte.

Retina (ret′ĭ-nah) Inner layer of the eyeball; contains photoreceptors (rods, cones).

Rhombencephalon (hindbrain) (romb″en-sef′ah-lon) Caudal portion of the developing brain; constricts to form the metencephalon and myelencephalon; includes the pons, cerebellum, and medulla oblongata.

Ribosomal RNA (rRNA) A constituent of ribosomes that assists in protein synthesis.

Ribosomes (ri′bo-sōmz) Cytoplasmic organelles at which proteins are synthesized.

RNA (ribonucleic acid) (ri′bo-nu-kle′ik) Nucleic acid that contains ribose and the bases A, G, C, and U. Carries out DNA's instructions for protein synthesis.

Rods One of the two types of photoreceptor cells in the retina.

Rotation Movement at a synovial joint in which a bone turns around its own long axis.

Rugae (ru′ge) Elevations or ridges, as in stomach mucosa.

Rule of nines Method of computing the extent of burns by dividing the body into a number of areas, each accounting for 9% (or a multiple thereof) of the total body area.

S (synthetic) phase The part of the interphase period of the cell cycle in which DNA replicates itself, ensuring that the two future cells will receive identical copies of genetic material.

Sagittal plane (saj′ĭ-tal) A longitudinal (vertical) plane that divides the body or any of its parts into right and left portions.

Saliva Secretion of the salivary glands; cleanses and moistens the mouth and begins chemical digestion of starchy foods.

Saltatory conduction Transmission of an action potential along a myelinated fiber in which the action potentials appears to leap from gap to gap.

Sarcolemma The plasma membrane of a muscle fiber.

Sarcomere (sar′ko-mēr) The smallest contractile unit of muscle; extends from one Z disc to the next.

Sarcoplasm The cytoplasm of a muscle fiber.

Sarcoplasmic reticulum (SR) (sar″ko-plaz′mik rĕ-tik′u-lum) Specialized endoplasmic reticulum of muscle cells.

Schwann cell A type of supporting cell in the PNS; forms myelin sheaths and is vital to peripheral nerve fiber regeneration. Also called a neurolemmocyte.

Sclera (skler′ah) White opaque portion of the fibrous layer of the eyeball.

Scrotum (skro′tum) External sac enclosing the testes.

Sebaceous glands (oil glands) (se-ba′shus) Epidermal glands that produce an oily secretion called sebum.

Sebum (se′bum) Oily secretion of sebaceous glands.

Second-degree burn A burn in which the epidermis and the upper region of the dermis are damaged.

Second messenger Intracellular molecule generated by the binding of a chemical (e.g., hormone or neurotransmitter) to a receptor protein; mediates intracellular responses to the chemical messenger.

Secondary active transport A type of active transport in which the energy needed to drive the transport process is provided by the electrochemical gradient of another molecule (which moves "downhill" through the transport protein at the same time as another molecule is moved "uphill" against its gradient). Also called cotransport or symport (when the two transported molecules move in the same direction) or antiport (when the two transported molecules move in opposite directions).

Secondary lymphoid organs Lymph nodes, spleen, and mucosa-associated lymphoid tissue (MALT); lymphoid organs in which lymphocytes encounter antigens and are activated.

Secondary sex characteristics Anatomical features, not directly involved in the reproductive process, that develop under the influence of sex hormones, e.g., male or female pattern of muscle development, bone growth, body hair distribution.

Secretion (se-kre′shun) (1) The passage of material formed by a cell to its exterior; (2) cell product that is transported to the exterior of a cell.

Secretory vesicles (granules) Vesicles that migrate to the plasma membrane of a cell and discharge their contents from the cell by exocytosis.

Section A cut through the body (or an organ) that is made along a particular plane; a thin slice of tissue prepared for microscopic study.

Segregation During meiosis, the distribution of the members of the allele pair to different gametes.

Selectively permeable membrane A membrane that allows certain substances to pass while restricting the movement of others; also called differentially permeable membrane.

Semen (se′men) Fluid mixture containing sperm and secretions of the male accessory reproductive glands.

Semilunar (SL) valves (sě″me-loo′ner) Valves that prevent blood return to the ventricles after contraction; aortic and pulmonary valves.

Seminiferous tubules (sem″ ĭ-nif′er-us) Highly convoluted tubes within the testes; form sperm.

Sense organs Localized collections of many types of cells working together to accomplish a specific receptive process.

Sensory (afferent) nerves Nerves that contain processes of sensory neurons and carry action potentials to the central nervous system.

Sensory areas Functional areas of the cerebral cortex that provide for conscious awareness of sensation.

Sensory receptor A cell or part of a cell (e.g., receptive endings of sensory neurons) specialized to respond to a stimulus.

Serosa (serous membrane) (se-ro′sah) The moist membrane found in closed ventral body cavities.

Serous fluid (sēr′us) Clear, watery fluid secreted by cells of a serous membrane.

Serum (sēr′um) Amber-colored fluid that exudes from clotted blood as the clot shrinks; plasma without clotting factors.

Sesamoid bones (ses′ah-moid) Short bones embedded in tendons, variable in size and number, many of which influence the action of muscles; largest is the patella (kneecap).

Severe combined immunodeficiency syndromes (SCIDs) Congenital conditions resulting in little or no protection against disease-causing organisms of any type.

Sex chromosomes The chromosomes, X and Y, that determine genetic sex (XX = female; XY = male); the 23rd pair of chromosomes.

Sex-linked inheritance Inherited traits determined by genes on the sex chromosomes, e.g., X-linked genes are passed from mother to son, Y-linked genes are passed from father to son.

Sexually transmitted infection (STI) Any infectious disease spread through sexual contact.

Signal sequence A short peptide segment present in a protein being synthesized that causes the associated ribosome to attach to the membrane of rough ER.

Simple diffusion The unassisted transport across a plasma membrane of a lipid-soluble or very small particle.

Sinoatrial (SA) node (si″no-a′tre-al) Specialized myocardial cells in the wall of the right atrium; pacemaker of the heart.

Sinus (si′nus) (1) Mucous-membrane-lined, air-filled cavity in certain cranial bones; (2) dilated channel for the passage of blood or lymph.

Skeletal muscle Muscle composed of cylindrical multinucleate cells with obvious striations; the muscle(s) attached to the body's skeleton; voluntary muscle.

Skeletal system System of protection and support composed primarily of bone and cartilage.

Skull Bony protective encasement of the brain and the organs of hearing and equilibrium; includes cranial and facial bones.

Small intestine Convoluted tube extending from the pyloric sphincter to the ileocecal valve where it joins the large intestine; the site where digestion is completed and virtually all absorption occurs.

Smooth muscle Spindle-shaped cells with one centrally located nucleus and no externally visible striations (bands). Found mainly in the walls of hollow organs.

Sodium-potassium (Na $^+$-K$^+$) pump A primary active transport system that simultaneously drives Na$^+$ out of the cell against a steep gradient and pumps K$^+$ back in. Also called Na$^+$-K$^+$ ATPase.

Sol-gel transformation Reversible change of a colloid from a fluid (sol) to a more solid (gel) state.

Solute (sol′yoot) The substance that is dissolved in a solution.

Solute pump Enzyme-like protein carrier that mediates active transport of solutes such as amino acids and ions uphill against their concentration gradients.

Somatic nervous system (so-mă′tik) Division of the peripheral nervous system that provides the motor innervation of skeletal muscles; also called the voluntary nervous system.

Somatic reflexes Reflexes that activate skeletal muscle.

Somatosensory system That part of the sensory system dealing with reception in the body wall and limbs; receives inputs from exteroceptors, proprioceptors, and interoceptors.

Somite (so′mīt) A mesodermal segment of the body of an embryo that contributes to the formation of skeletal muscles, vertebrae, and dermis of skin.

Spatial discrimination The ability of neurons to identify the site or pattern of stimulation.

Special senses The senses of taste, smell, vision, hearing, and equilibrium.

Specific gravity Term used to compare the weight of a substance to the weight of an equal volume of distilled water.

Sperm Male gamete. Also called spermatozoon.

Spermatogenesis (sper″mah-to-jen′ĕ-sis) The process of sperm (male gamete) formation; involves meiosis.

Sphincter (sfink′ter) A circular muscle surrounding an opening; acts as a valve.

Spinal cord The bundle of nervous tissue that runs from the brain to the first to third lumbar vertebrae and provides a conduction pathway to and from the brain.

Spinal nerves The 31 nerve pairs that arise from the spinal cord.

Splanchnic circulation (splangk′nik) The blood vessels serving the digestive system.

Spleen Largest lymphoid organ; provides a site for immune surveillance, lymphocyte activation, and erythrocyte destruction.

Spongy bone Internal layer of skeletal bone. Also called trabecular bone.

Sprain Ligaments reinforcing a joint are stretched or torn.

Static equilibrium Sense of head position in space with respect to gravity.

Stem cells Unspecialized cells that can divide to produce various kinds of more specialized cell types.

Stenosis (stě-no′sis) Abnormal constriction or narrowing.

Steroids (stě′roidz) A class of lipids derived from (and including) cholesterol; act as hormones and as constituents of phospholipid bilayer membranes.

Stimulus (stim′u-lus) An excitant; a change in the environment that evokes a response.

Stomach Temporary reservoir in the gastrointestinal tract where chemical breakdown of proteins begins and food is converted into chyme.

Stressor Any stimulus that directly or indirectly causes the hypothalamus to initiate stress-compensating responses, such as the fight-or-flight response.

Stroke *See* Cerebrovascular accident.

Stroke volume (SV) Amount of blood pumped out of a ventricle during one contraction.

Stroma (stro′mah) The basic internal structural framework of an organ.

Structural (fibrous) proteins Consist of extended, strandlike polypeptide chains forming a strong, ropelike structure that is linear, insoluble in water, and very stable; e.g., collagen.

Subcutaneous (sub″kyu-ta′ne-us) Beneath the skin.

Subcutaneous tissue Tissue just deep to the skin; consists of adipose plus some areolar connective tissue. Also called hypodermis or superficial fascia.

Subendocardial conducting network Modified ventricular muscle fibers of the conduction system of the heart. Also called Purkinje fibers.

Submucosa Connective tissue underlying a mucosa (mucous membrane that lines body cavities open to the exterior).

Substrate A reactant on which an enzyme acts to cause a chemical action to proceed.

Sudoriferous gland (su″do-rif′er-us) *See* Sweat gland.

Sulcus (sul′kus) A furrow on the brain, less deep than a fissure.

Summation Accumulation of effects, especially those of muscular, sensory, or mental stimuli.

Superficial Located toward or on the body surface.

Superior Toward the head or upper body regions.

Superior vena cava Vein that returns blood from body regions superior to the diaphragm.

Supination (soo″pĭ-na′shun) The outward rotation of the forearm causing palms to face anteriorly.

Surfactant (ser-fak′tant) Secretion produced by certain cells of the alveoli that reduces the surface tension of water molecules, thus preventing the collapse of the alveoli after each expiration.

Suspension Heterogeneous mixtures with large, often visible solutes that tend to settle out.

Suture (soo′cher) An immovable fibrous joint; with one exception, all bones of the skull are united by sutures.

Sweat gland Epidermal gland that produces sweat. Also called sudoriferous gland.

Sympathetic division The division of the autonomic nervous system that prepares the body for activity or to cope with some stressor (danger, excitement, etc.); the fight, fright, and flight subdivision.

Sympathetic (vasomotor) tone State of partial vasoconstriction of the blood vessels maintained by sympathetic fibers.

Symphysis (sim′fih-sis) A joint in which the bones are connected by fibrocartilage.

Synapse (sin′aps) Functional junction or point of close contact between two neurons or between a neuron and an effector cell.

Synapsis (sĭ-nap′sis) Pairing of homologous chromosomes during the first meiotic division.

Synaptic cleft (sĭ-nap′tik) Fluid-filled space at a synapse.

Synaptic delay Time required for a signal to cross a synapse between two neurons.

Synaptic vesicles Small membranous sacs containing neurotransmitter.

Synarthrosis (sin″ar-thro′sis) Immovable joint.

Synchondrosis (sin″kon-dro′sis) A joint in which the bones are united by hyaline cartilage.

Syndesmosis (sin″des-mo′sis) A joint in which the bones are united by a ligament or a sheet of fibrous tissue.

Synergist (sin′er-jist) (1) Muscle that aids the action of a prime mover by effecting the same movement or by stabilizing joints across which the prime mover acts, preventing undesirable movements. (2) Hormone that amplifies the effect of another hormone at a target cell.

Synostosis (sin″os-to′sis) A completely ossified joint; a fused joint.

Synovial fluid Fluid secreted by the synovial membrane; lubricates joint surfaces and nourishes articular cartilages.

Synovial joint Freely movable joint exhibiting a joint cavity; also called a diarthrosis.

Synthesis (combination) reaction A chemical reaction in which larger, more complex atoms or molecules are formed from simpler ones.

Systemic (sis-tem′ik) Pertaining to the whole body.

Systemic circuit System of blood vessels that serves gas exchange in the body tissues.

Systole (sis′to-le) Period when either the ventricles or the atria are contracting.

Systolic pressure (sis-tah′lik) Pressure exerted by blood on the blood vessel walls during ventricular contractions.

T cells Lymphocytes that mediate cellular immunity; include helper, cytotoxic, regulatory, and memory cells. Also called T lymphocytes.

T tubule (transverse tubule) Extension of the muscle cell plasma membrane (sarcolemma) that protrudes deeply into the muscle cell.

Tachycardia (tak″e-kar′de-ah) A heart rate over 100 beats per minute.

Taste buds Sensory receptor organs that house gustatory epithelial cells, which respond to dissolved food chemicals.

Telencephalon (tel″en-seh′fuh-lon) Anterior subdivision of the primary forebrain that develops into olfactory lobes, cerebral cortex, and basal nuclei.

Telophase The final phase of mitosis, meiosis I, and meiosis II; begins when migration of chromosomes to the poles of the cell has been completed and ends with the formation of two daughter nuclei.

Tendon (ten′dun) Cord of dense regular connective tissue attaching muscle to bone.

Tendon organ Proprioceptor located in tendon; monitors muscle tension to prevent tearing and help smooth onset and termination of muscle contraction.

Tendonitis Inflammation of tendon sheaths, typically caused by overuse.

Terminal branches Branching ends of an axon that allow it to form many axon terminals; terminal arborization.

Testis (tes′tis) Male primary reproductive organ that produces sperm; male gonad.

Testosterone (tes-tos′tĕ-rōn) Male sex hormone produced by the testes; during puberty promotes virilization, and is necessary for normal sperm production.

Tetanus (tet′ah-nus) (1) A smooth, sustained muscle contraction resulting from high-frequency stimulation; (2) an infectious disease caused by an anaerobic bacterium.

Thalamus (thal′ah-mus) A mass of gray matter in the diencephalon of the brain.

Thermogenesis (ther″mo-jen′ĕ-sis) Heat production.

Thermoreceptor (ther″mo-re-sep′ter) Receptor sensitive to temperature changes.

Third-degree burn A burn that involves the entire thickness of the skin; also called a full-thickness burn. Usually requires skin grafting.

Thoracic cage (bony thorax) Bones and costal cartilages that form the framework of the thorax; includes sternum, ribs, and thoracic vertebrae.

Thoracic duct Large lymphatic duct that receives lymph drained from the entire lower body, the left upper extremity, and the left side of the head and thorax.

Thoracic vertebrae The 12 vertebrae that are in the middle part of the vertebral column and articulate with the ribs.

Thorax (tho′raks) That portion of the body trunk above the diaphragm and below the neck.

Threshold stimulus Weakest stimulus capable of producing a response in an excitable tissue.

Thrombin (throm′bin) Enzyme that induces clotting by converting fibrinogen to fibrin.

Thrombocyte (throm′bo-sīt) Platelet; cell fragment that participates in blood coagulation.

Thrombocytopenia (throm″bo-si″to-pe′ne-ah) A reduction in the number of platelets circulating in the blood.

Thrombus (throm′bus) A clot that develops and persists in an unbroken blood vessel.

Thymine (T) (thi′mēn) Single-ring base (a pyrimidine) in DNA.

Thymus (thi′mus) Lymphoid organ active in immune response; site of maturation of T lymphocytes.

Thyroid gland (thi′roid) One of the largest of the body's endocrine glands; straddles the anterior trachea.

Thyroid hormone (TH) The major hormone secreted by thyroid follicles; stimulates production of enzymes concerned with glucose oxidation. *See also* Thyroxine, Triiodothyronine.

Thyroid-stimulating hormone (TSH) Anterior pituitary hormone that regulates secretion of thyroid hormones.

Thyroxine (T₄) (thi-rok′sin) Iodine-containing hormone secreted by the thyroid gland; accelerates cellular metabolic rate in most body tissues.

Tight junction Area where plasma membranes of adjacent cells are tightly bound together, forming an impermeable barrier.

Tissue A group of similar cells and their intercellular substance specialized to perform a specific function; primary tissue types of the body are epithelial, connective, muscle, and nervous tissue.

Tissue perfusion Blood flow through body tissues or organs.

Tonicity (to-nis′ĭ-te) A measure of the ability of a solution to cause a change in cell shape or tone by promoting osmotic flows of water.

Tonsils A ring of lymphoid tissue around the entrance to the pharynx. *See also* Adenoids.

Total peripheral resistance (TPR) A measure of the amount of friction encountered by blood as it flows through the blood vessels.

Trabecula (trah-bek′u-lah) (1) Any of the fibrous bands extending from the capsule into the interior of an organ; (2) strut or thin plate of bone in spongy bone.

Trachea (tra′ke-ah) Windpipe; cartilage-reinforced tube extending from larynx to bronchi.

Tract (1) A collection of axons in the central nervous system having the same origin, termination, and function; (2) a major anatomical passageway.

Transcription The transfer of genetic code information from a DNA base sequence to the complementary base sequence of an mRNA molecule; the first major step in polypeptide synthesis.

Transduction (trans-duk′shun) The conversion of the energy of a stimulus into an electrical event (a graded potential).

Transepithelial transport (trans-ep″ĭ-the′le-al) Movement of substances through, rather than between, adjacent epithelial cells connected by tight junctions, such as absorption of nutrients in the small intestine.

Transfer RNA (tRNA) Short-chain RNA molecules that transfer amino acids to the ribosome.

Transfusion reaction Agglutination and destruction of red blood cells following transfusion of incompatible blood.

Translation The second major step in the transfer of genetic code information, in which the information carried by mRNA is decoded and used to assemble polypeptides.

Transverse (horizontal) plane A plane running from right to left, dividing the body or an organ into superior and inferior parts.

Tricuspid valve (tri-kus′pid) The right atrioventricular valve.

Triglycerides (tri-glis′er-īdz) Fats and oils composed of fatty acids and glycerol; are the body's most concentrated source of energy fuel.

Triiodothyronine (T₃) (tri″i-o″do-thi′ro-nēn) Thyroid hormone; secretion and function similar to those of thyroxine (T₄).

Trophoblast (tro′fo-blast) Outer sphere of cells of the blastocyst.

Tropic hormone (tropin) (trōp′ik) A hormone that regulates the secretory action of another endocrine organ.

Trypsin Proteolytic enzyme secreted by the pancreas.

Tubular reabsorption The movement of filtrate components from the renal tubules into the blood.

Tubular secretion The movement of substances (such as drugs, urea, excess ions) from blood into renal filtrate.

Tumor An abnormal growth of cells; a swelling; may be cancerous.

Tunica (too′nĭ-kah) A covering or tissue coat; membrane layer.

Tympanic membrane (tim-pan′ik) Eardrum.

Ulcer (ul′ser) Lesion or erosion of the mucous or cutaneous membrane, such as a gastric ulcer of the stomach.

Umbilical cord (um-bĭ′lĭ-kul) Structure bearing arteries and a vein connecting the placenta and the fetus.

Umbilicus (um-bĭ′lĭ-kus) Navel; marks site where umbilical cord was attached in fetal stage.

Unipolar neuron Neuron in which embryological fusion of the two processes leaves only one process extending from the cell body.

Uracil (U) (u′rah-sil) A smaller, single-ring base (a pyrimidine) found in RNA.

Urea (u-re′ah) Main nitrogen-containing waste excreted in urine.

Ureter (u-re′ter) Tube that carries urine from kidney to bladder.

Urethra (u-re′thrah) Canal through which urine passes from the bladder (and semen passes from the ejaculatory duct in the male) to outside the body.

Uric acid The nitrogenous waste product of nucleic acid metabolism; component of urine.

Urinary bladder A smooth, collapsible, muscular sac that stores urine temporarily.

Urinary system System primarily responsible for water, electrolyte, and acid-base balance and removal of nitrogenous wastes.

Uterine tube (u′ter-in) Tube through which the ovum is transported to the uterus. Also called fallopian tube.

Uterus (u′ter-us) Hollow, thick-walled organ that receives, retains, and nourishes fertilized egg; site where embryo/fetus develops.

Uvula (u′vu-lah) Tissue tag hanging from soft palate.

Vaccine Preparation that provides artificially acquired active immunity.

Vagina Thin-walled tube extending from the cervix to the body exterior; often called the birth canal.

Valence shell (va′lens) Outermost electron shell (energy level) of an atom that contains electrons.

Varicosities Knoblike swellings of certain autonomic axons containing mitochondria and synaptic vesicles.

Vasa recta (va′sah rek′tah) Capillary branches that supply nephron loops in the medulla region of the kidney.

Vascular Pertaining to blood vessels or richly supplied with blood vessels.

Vascular spasm Immediate response to blood vessel injury; results in constriction.

Vasoconstriction (vas″o-kon-strik′shun) Narrowing of blood vessels.

Vasodilation (vas″o-di-la′shun) Relaxation of the smooth muscles of the blood vessels, producing dilation.

Vasomotion (vas″o-mo′shun) Oscillating constriction and dilation of blood vessels. In arterioles, this results in intermittent flow through capillary beds and matches perfusion to tissue requirements.

Vasomotor center (vas″o-mo′ter) Brain area concerned with regulation of blood vessel resistance.

Vasomotor fibers Sympathetic nerve fibers that cause the contraction of smooth muscle in the walls of blood vessels, thereby regulating blood vessel diameter.

Veins (vānz″) Blood vessels that return blood toward the heart from the circulation.

Ventral Pertaining to the front; anterior.

Ventricles (1) Paired, inferiorly located heart chambers that function as the major blood pumps; (2) cavities in the brain.

Venule (ven′ūl) A small vein.

Vertebral column (spine) (ver′tĕ-brul) Formed of a number of individual bones called vertebrae and two composite bones (sacrum and coccyx).

Vesicle (vĕ′sĭ-kul) A small liquid-filled sac or bladder.

Vesicular follicle Mature ovarian follicle.

Vesicular transport Transport of large particles and macromolecules into or out of a cell or between its compartments in membrane-bound sacs.

Vestibule An enlarged area at the beginning of a canal, i.e., inner ear, nose, larynx.

Villus (vil′us) One of the fingerlike projections of the small intestinal mucosa that tremendously increase its surface area for absorption.

Visceral (vis′er-al) Pertaining to an internal organ of the body or the inner part of a structure.

Visceral muscle Type of smooth muscle; its cells are electrically coupled by gap junctions, and so they rhythmically contract as a unit. They often exhibit spontaneous action potentials. Also called unitary smooth muscle.

Visceral organs (viscera) A group of internal organs housed in the ventral body cavity.

Visceral serosa (se-ro′sah) The part of the double-layered membrane that lines the outer surfaces of organs within the ventral body cavity.

Viscosity (vis′kos′ĭ-te) A measurement of thickness (ability to flow) of a fluid.

Visual field The field of view seen when the head and eyes are still.

Vital capacity (VC) The volume of air that can be expelled from the lungs by forcible expiration after the deepest inspiration; total exchangeable air.

Vital signs Includes pulse, blood pressure, respiratory rate, and body temperature measurements.

Vitamins Organic compounds required by the body in minute amounts.

Vocal folds Mucosal folds that function in voice production (speech); also called the true vocal cords.

Volatile acid An acid that can be eliminated by the lungs; carbonic acid is converted to CO_2, which diffuses into the alveoli.

Volkmann's canals *See* Perforating canals.

Voltage-gated channel An ion channel that opens or closes when the transmembrane voltage changes.

Voluntary muscle Muscle under strict nervous control; skeletal muscle.

Voluntary nervous system The somatic nervous system.

Vulva (vul′vuh) Female external genitalia.

Wallerian degeneration (wal-er′ē-an) A process of disintegration of an axon that occurs when it is crushed or severed and cannot receive nutrients from the cell body.

White matter White substance of the central nervous system; myelinated nerve fibers.

Xenograft Tissue graft taken from another animal species.

Yolk sac (yōk) One of the extraembryonic membranes; involved in early blood cell formation.

Zygote (zi′gōt) Fertilized egg.

Photo and Illustration Credits

Photo Credits

Visual Walkthrough. Nik Merkulov/Shutterstock, artjazz/Shutterstock, iShift/Shutterstock

Chapter 1. 1.2: Dmytro Aksonov/Getty Images. 1.9 (top): John Wilson White/Pearson Education, Inc. 1.9a: CNRI/Science Source. 1.9b: Scott Camazine/Science Source. 1.9c: James Cavallini/Science Source. 1.13a: John Wilson White/Pearson Education, Inc. A Closer Look (1): eAlisa/Fotolia. (2): Du Cane Medical Imaging Ltd/Science Source. (3, 4): Zephyr/Science Source. (5) Scott Camazine/Science Source (6): Mangelli/Pearson Education, Inc.

Chapter 2. 2.4.1: Stockbyte/Getty Images. 2.4.2: Andreas Poertner/Shutterstock. 2.4.3: Henry Hong/Pearson Education, Inc. 2.6: Peter Hermes Furian/123RF. 2.10b: Hermann Eisenbeiss/Science Source. 2.17a: Uwimages/Fotolia. 2.17b: Angel Simon/Fotolia.

Chapter 3. 3.8, 3.22.1: David M. Phillips/Science Source. 3.12b: Birgit Satir. 3.14c, 3.16b: Professors Pietro M. Motta & Tomonori Naguro/Science Source. 3.15b: Science Source. 3.18, 3.20a: Dr. Gopal Murti/Science Source. 3.20b: Daniel Schroen/Cell Applications Inc/Science Source. 3.20c: Mark S. Ladinsky/California Institute of Technology. 3.21b: Don W. Fawcett/Science Source. 3.22.2: *Cell motility and the cytoskeleton* by John Wiley & Sons, Inc. Reproduced with permission of John Wiley & Sons, Inc. 3.24: Don W. Fawcett/Science Source. 3.25b.1: *Nucleus*, Lelio Orci M.D., Plate 25 on p. 53, Jan 1, 1975, With permission of Springer. 3.25b.2: *The nuclear pore complex: Its free existence and an hypothesis as to its origin*, A. C. Fabergé, 151, Jan 1, 1974, *Cell* Tiss. Res. 151 (1974):403, With permission of Springer. 3.25b.3: Ueli Aebi et al. *Nature* 323 (a996):560–564, figure 1a, inset. Used by permission. 3.26a: Dr. Victoria E. Foe. 3.26b: Armed Forces Institute of Pathology. Focus Figure 3.4: Photos by Dr. Conly L. Rieder. 3.32b: Barbara Hamkalo. 3.35: Haziq Qadri/Barcroft Media/Getty Images.

Chapter 4. 4.2.1: Steve Gschmeissner/Science Source. 4.2.2: Dr G. Moscoso/Science Source. p.117 (bottom right), 4.4a: William Karkow/Pearson Education, Inc. 4.4b, c, f, 4.11c, i, j: Allen Bell, University of New England/Pearson Education, Inc. 4.4d, 4.12b: Steve Downing/Pearson Education, Inc. 4.4e, 4.11b, f, h: Nina Zanetti/Pearson Education, Inc. 4.6a: SPL/Science Source. 4.11d, e, k, 4.12c: Lisa Lee/Pearson Education, Inc. 4.11g: Karen Krabbenhoft/Pearson Education, Inc. 4.13: Science Source. A Closer Look: David Musher/Science Source.

Chapter 5. 5.2a: William Karkow/Pearson Education, Inc. 5.3: Centers for Disease Control and Prevention. 5.4b: John Wilson White/Pearson Education, Inc. 5.4c: Clouds Hill Imaging Ltd./Science Source. 5.5, 5.8, 5.11b: Dr. P. Marazzi/Science Source. 5.6b: Manfred Kage/Science Source. 5.6d: Teve Downing/Pearson Education, Inc. 5.9a, b: Lisa Lee/Pearson Education, Inc. 5.10: Ian Boddy/Science Source. 5.11a, c: Biophoto Associates/Science Source. 5.12a: Scott Camazine/Science Source. 5.12b: Dr M.A. Ansary/Science Source. Clinical Case Study: Ron Rovtar Photography/Alamy.

Chapter 6. 6.3: Larry DeLay/Pearson Education, Inc. 6.4.1: Skulls Unlimited International, Inc./Pearson Education Inc. 6.4.2: Steve Gschmeissner/Science Source. 6.7: Science Source. 6.9.c.1: Andrew Syred/Science Source. 6.9.c.2: William Karkow/Pearson Education, Inc. 6.12a: Living Art Enterprises/Science Source. 6.12b: Deborah W. Vaughan, BUSM. Table 6.3.1: Lester V. Bergman/Corbis Documentary/Getty Images. Table 6.3.2: Centers for Disease Control and Prevention. Table 6.3.3: Brian Tidey/Pearson Education, Inc. Table 6.3.4: Frank Gaillard/Pearson Education, Inc. Table 6.3.5: © 2017 Dr. Ian Bickle. Image courtesy of Dr. Ian Bickle and Radiopaedia. org. Used under license. Table 6.3.6: Hani Alsalam/Pearson Education, Inc. 6.17a, 6.18b: SPL/Science Source. 6.17b: Professor Pietro M. Motta/Science Source. 6.18a: Temet/E+/Getty Images. 6.19: Scott Camazine/Science Source. Clinical Case Study: Ron Rovtar Photography/Alamy.

Chapter 7. 7.10: Michael Wiley/Univ. of Toronto/Imagineering/Pearson Education, Inc. 7.17d: Medical Body Scans/Science Source. 7.18a: Princess Margaret Rose Orthopaedic Hospital/Science Source. 7.18b: Bengt-Göran Carlsson/Nordicphotos/Alamy. 7.18c: Steve Gorton/DK Images. 7.20d, 7.24c: Creative Digital Visions/Pearson Education, Inc. 7.23b, 7.29a: Karen Krabbenhoft/Pearson Education, Inc. 7.28a: Creative Digital Visions, LLC/Pearson Education, Inc. Table 7.4: From *A Stereoscopic Atlas of Human Anatomy*, by David L. Bassett. 7.35c: Scott Camazine/Alamy. 7.35e: Michael Wiley/Univ. of Toronto/Imagineering/Pearson Education, Inc. 7.37b: Elaine N. Marieb. 7.39a: Medicshots/Alamy. 7.39b: Adam Murphy/Alamy. 7.41: Reik/Mauritius Images/AGE Fotostock. Clinical Case Study: BEW Authors/AGE Fotostock.

Chapter 8. 8.5, 8.6: John Wilson White/Phocasso/Pearson Education, Inc. 8.8b: Mark Nielsen/Pearson Education, Inc. 8.8e, 8.9c: Karen Krabbenhoft/Pearson Education, Inc. 8.10b: From *A Stereoscopic Atlas of Human Anatomy*, by David L. Bassett. 8.13: Elaine N. Marieb. A Closer Look (1): Lawrence Livermore National Laboratory/Science Source. (2): Elaine N. Marieb. 8.14: Dr. P. Marazzi/Science Source. Clinical Case Study: BEW Authors/AGE Fotostock.

Chapter 9. p. 281, 9.2a, Table 9.3.2: Marian Rice. 9.1b: Lisa Lee/Pearson Education, Inc. 9.4: MGPhoto/Alamy. 9.6.1: Don W. Fawcett/Science Source. 9.6.2: Biophoto Associates/Science Source. 9.10b: Eric V. Grave/Science Source. Table 9.3.1: Eric Graves/Science Source. Table 9.3.3: SPL/Science Source. A Closer Look: Iulianvalentin/Fotolia. Clinical Case Study: FluxFactory/E+/Getty Images.

Chapter 10. 10.6a, 10.9b, 10.14d, 10.21b: Karen Krabbenhoft/Pearson Education, Inc. 10.7, 10.19a: John Wilson White/Pearson Education, Inc. 10.10c: Mark Nielsen. 10.11c: From *A Stereoscopic Atlas of Human Anatomy*, by David L. Bassett. 10.14b: Dr. Olga Malakhova/Winston Charles Poulton/University of Florida College of Medicine/Pearson Education, Inc. 10.15d: Mark Nielsen and Shawn Miller. p. 386: Dr. P. Marazzi/Science Source. Clinical Case Study: BEW Authors/AGE Fotostock.

Chapter 11. 11.5b: Maryann Martone. 11.6b: Don W. Fawcett/Science Source. 11.15b: Oliver Meckes/Ottawa/Science Source. 11.21: Republished with permission of American Assn for the Advancement of Science, from *Hardwiring the Brain: Endocannabinoids Shape Neuronal Connectivity*, Paul Berghuis et. al., 25 May 2007. A Closer Look: Science Source. Clinical Case Study: Jetta Productions/Blend Images/Getty Images.

Chapter 12. 12.5a, 12.9b, 12.23b: From *A Stereoscopic Atlas of Human Anatomy*, by David L. Bassett. 12.5b, 12.9a, 12.11b, 12.16a, c: Karen Krabbenhoft/Pearson Education, Inc. 12.6: WDCN/Univ. College London/Science Source. 12.20a: Phanie/SuperStock. 12.25: Images & Stories/Alamy. 12.27: Courtesy of Dr. Michael E. Phelps, David Geffen School of Medicine at UCLA. 12.36: Scott Camazine/Science Source. 12.39: Biophoto Associates/Science Source. Clinical Case Study: Tyler Olson/123RF.

Chapter 13. 13.4a: Thomas Deerinck/NCMIR/Science Source. p. 526: Ray Ellis/Science Source. Clinical Case Study: Javier Larrea/AGE Fotostock.

Chapter 14. 14.11: Graham Dunn/Alamy. Clinical Case Study: Blend Images/ERproductions Ltd/Alamy.

Chapter 15. 15.4b: From *A Stereoscopic Atlas of Human Anatomy*, by David L. Bassett. 15.6c: Lisa Lee/Pearson Education, Inc. 15.7: Thalerngsak Mongkolsin/Shutterstock. 15.9: SPL/Science Source. 15.11: Keith Leighton/Alamy. 15.19b: Stephen Spector/Pearson Education, Inc. 15.27d: Prof. P.M. Motta/Univ. "La Sapienza," Rome/Science Source. 15.35b: Image courtesy of Dr. I. Hunter-Duvar, Hospital for Sick Children Toronto. 15.37: Life in View/Science Source. Clinical Case Study: Jim Craigmyle/Corbis/Glow Images.

Chapter 16. 16.6: General Photographic Agency/Hulton Archive/Getty Images. 16.10a: Scott Camazine/Science Source. 16.10b: Ralph C. Eagle, Jr./Science Source. 16.13b: Lisa Lee/Pearson Education, Inc. 16.15: Charles B. Wilson. Clinical Case Study: KatarzynaBialasiewicz/iStock/Getty Images.

Chapter 17. 17.2a: National Cancer Institute/Science Source. 17.8a: Cheryl Power/Science Source. 17.8b: Science Source. 17.10a, c, d: Biophoto Associates/Science Source. 17.10b: Alvin Telser/Science Source. 17.10e: Jarun011/iStock/Getty Images. 17.15: Eye of Science/Science Source. 17.16: Mediscan/Alamy. 17.17: John F. Scanlon/Holyoke Community College. Clinical Case Study: Barry Diomede/Alamy.

Chapter 18. 18.6c: Lennart Nilsson/Tidningarnas Telegrambyra AB. 18.16a: Frédérik Astier/Science Source. 18.20: Eric Gevaert/Shutterstock. Clinical Case Study: ASTIER/BSIP SA/Alamy.

Chapter 19. 19.2a: Jubal Harshaw/Shutterstock. 19.15: Asia Images Group/Getty Images. 19.20: SPL/Science Source. A Closer Look (1): Sheila Terry/Science Source. (2): GJLP/Science Source. Clinical Case Study: Image Source/Getty Images.

Chapter 20. 20.4: Francis Leroy/Biocosmos/Science Source. 20.6b, 20.8, 20.9: Biophoto Associates/Science Source. Clinical Case Study: Javier Larrea/age fotostock/Alamy.

Chapter 21. 21.2a: Eye of Science/Science Source. 21.10: David Scharf/Science Source. 21.14b: Dr. Lawrence W. Haynes. 21.19b: Dr. Andrejs Liepins/Science Source. Clinical Case Study: Judith Haeusler/Cultura/Getty Images.

Chapter 22. 22.4b, 22.6c, 22.12b: From *A Stereoscopic Atlas of Human Anatomy*, by David L. Bassett. 22.7: CNRI/Science Source. 22.8c, 22.11b: Science Source. 22.13: SPL/Science Source. 22.22a: Biophoto Associates/Science Source. 22.22b: Dr. Tony Brain/Science Source. Clinical Case Study: Corepics VOF/Shutterstock.

Chapter 23. 23.6: Naomi Tjaden. 23.10b, 23.13a, 23.32: Science Source. 23.15b, 23.23: From *A Stereoscopic Atlas of Human Anatomy*, by David L. Bassett. 23.17: CNRI/Science Source. 23.18: Eye of Science/Science Source. 23.24b: M. I. Walker/Science Source. 23.31c: Steve Gschmeissner/Science Source. 23.34a: VideoSurgery/Science Source. Clinical Case Study: B BOISSONNET/BSIP/AGE Fotostock.

Index

NOTE: Page numbers in **boldface** indicate a definition. A *t* following a page number indicates tabular material, an *f* indicates an illustration, and a *b* indicates boxed material.

A agglutinins, 664, 664*t*
A agglutinogens, 664, 664*t*, 665*f*
A bands, **284**, 285*f*, 288*f*, 289, 289*f*, 684*f*, 685
ABCD rule, for melanoma recognition, **166**
Abdomen
 arteries of, 744–747*f*, 744–747*t*
 muscles of, 330*f*, 346*t*, 347*f*
 sympathetic pathways to, 540*f*, 541
 veins of, 756–757*f*, 756–757*t*
Abdominal aorta, **738***t*, 738*f*, 739*f*, 744–747*f*, 744–747*t*, 748*f*
Abdominal aortic plexus, 537, **541**
Abdominal cavity, **18**, 18*f*
Abdominal hernia, 346*t*, **386**
Abdominal organs, arterial supply/ perfusion/blood flow of, 726, 726*f*, 744–747*f*, 744–747*t*
Abdominal reflexes, **525**
Abdominal wall, innervation of, 515
Abdominal wall, muscles of, 330*f*, 346*t*, 347*f*
Abdominopelvic cavity, **18**, 18*f*, 19, 19*f*
Abdominopelvic organs, 19, 19*f*
 muscles compressing, 345*t*, 347*f*
 muscles supporting, 348*t*, 349*f*
Abducens nerve (cranial nerve VI), 450*f*, 452, 499*f*, 503*t*, **506**, 556, 557*f*
Abduction (body movement), **259**, 259*f*, 325*f*, 350*t*, 354*t*, 362*t*, 365*t*, 367*t*, 384*t*
Abductor digiti minimi muscle, **364***t*, 365*f*, **380***t*, 381*f*
Abductor hallucis muscle, **380***t*, 381*f*
Abductor muscles, of thigh, 367*t*
Abductor pollicis brevis muscle, **364***t*, 365*f*
Abductor pollicis longus muscle, **360***t*, 361*f*, 362*t*, 363*f*
ABO blood groups, **664**, 664*t*, 665–666, 665*f*, 811, 1129, 1129*t*
Abortion, **1120**
Abruptio placenta, **1120**
Abscess, 787
Absence (petit mal) seizures, 462
Absolute refractory period, **410**, 410*f*
 in cardiac muscle, 685, 689*f*, 690
Absorption (food/nutrient), **869**, 870*f*, **871**, 872, 889, 904*t*, 912
 in large intestine, 904*t*
 in small intestine, 902–903, 902*f*, 904*t*, 905–906, 913*f*
 vitamin/electrolyte/water, 917
Absorptive cells (enterocytes), of small intestine, 902*f*, 903, 903*f*
Absorptive (fed) state, 950–952, 951*f*, 952*t*
Abstinence methods, of contraception, 1118*b*
Acceleration
 crista ampullaris response to, 592, 592*f*, 593
 macular response to, 590–591, 591*f*
Accessory digestive organs, 869, 869*f*, **870**, 893–900. *See also specific structure*
Accessory hemiazygos vein, 754*f*, **755***t*, 755*f*
Accessory nerve (cranial nerve XI), 340*t*, 352*t*, 450*f*, 451*f*, 499*f*, **506**, 506*t*, 509*f*

Accessory oculomotor (Edinger-Westphal) nuclei, 536
Accessory pancreatic duct, 900, 900*f*
Accessory reproductive ducts. *See* Reproductive duct system
Accessory reproductive organs/glands/ ducts, **1042**, 1047, **1050**, **1052**–1053
Acclimatization, **858**
Accommodation, gastric, **892**
Accommodation, ocular, 565, **566**–567
Accommodation pupillary reflex, **567**
ACE. *See* Angiotensin converting enzyme
Acetabular labrum, 268–269, 269*f*
Acetabulum, **236**, 237*f*, 239*t*, 268–269
Acetic acid, 39, 41, 946
Acetylcholine (ACh), 81, 421*t*, **423**, 517, **533**, 543, 544*t*
 in autonomic versus somatic nervous system, 533, 533*f*, 543, 544*t*
 in hydrochloric acid secretion control, 889
 in muscle contraction, **290**, 291, 291*f*, 292*f*, 293, 293*f*, 294*f*, 315
 in neuromuscular junction, **290**, 292*f*, 293*f*, 517
Acetylcholine (ACh) receptors, **290**, 292*f*, 421*t*, 517
Acetylcholinesterase (AChE), 292*f*, 293, **423**, 517
Acetylcholinesterase (AChE) inhibitors, autonomic activity and, 544*t*
Acetyl coenzyme A (acetyl CoA), 934, 935*f*, 939*f*, **940**, 944*f*, 946–947, 946*f*, 947*t*
ACh. *See* Acetylcholine
Achalasia, **919**
AChE. *See* Acetylcholinesterase
Achilles (calcaneal) tendon, 244, 331*f*
 ruptured, **386**
Achondroplasia, **195**
Acid(s), **39**
 amino acid as, 48
 nonvolatile (fixed), **1029**
 pH of, 40, 40*f*
 strong, **41**, 1027, 1027*f*
 volatile, **1029**
 weak, **41**, 1027, 1027*f*
Acid-base balance, 39–40, **1026**–1035
 abnormalities of, 1033–1035, 1033*t*, 1034*b*, 1035. *See also* Acidosis; Alkalosis
 aldosterone and, 623
 blood in regulation of, 643
 chemical buffer systems in, 41, **1027**–1028, 1027*f*
 developmental aspects of, 1035–1037
 physiological buffer systems in, 1027, 1028–1032, 1030*f*, 1031*f*, 1032*f*
 renal regulation of, 992*f*, 995, 1029–1032, 1030*f*, 1031*f*, 1032*f*, 1035
 respiratory and renal compensations and, **1035**
 respiratory regulation of, 852, 856, 1035
Acidemia, **1027**. *See also* Acidosis
Acid group, amino acid, 48, 49*f*
Acid hydrolases, 86
Acidic diet, 1002
Acidic solution, 40, 40*f*
Acid mantle of skin, **163**, 783, 783*t*
Acid mantle of vagina, 783*t*, 1066

Acidosis, **56**, 1026, **1027**, 1034*b*
 in diabetes, 952
 effects of, 1035
 metabolic, 947, 1033*t*, 1034*b*, **1035**
 newborn's first breath and, 1116
 physiological, 1027
 renal tubular, **1037**
 respiratory (hypoventilation), **858**, 1029, **1033**–1034, 1033*t*, 1034*b*
 respiratory and renal compensations and, **1035**
Acinar (alveolar) glands, **125**, 125*f*
Acini/acinar cells, pancreatic, 630, 630*f*, **898**, 899*f*
Acinus (secretory unit), of multicellular exocrine gland, 125, 125*f*
Acne, 163, 169
Acoustic meatus
 external (external auditory/ear canal), 204*f*, 206*f*, **208**, 208*f*, **216***t*, 581*f*, **582**, 596
 internal, 205*f*, 207*f*, **208**, **216***t*, 503*t*
Acoustic nerve. *See* Vestibulocochlear nerve
Acquired immune deficiency syndrome (AIDS), **811**
Acquired immunodeficiencies, 811
Acquired (learned) reflex, 519
Acromegaly, 612*t*, **616**
Acromial end of clavicle, **228**, 228*f*
Acromioclavicular joint, 227*f*, **228**, 256*t*
Acromion, **228**, 229*f*, 267*f*
Acrosomal reaction, **1093**, 1095*f*
Acrosome, 1057*f*, **1058**
ACTH. *See* Adrenocorticotropic hormone
Actin, 49, 88, 88*f*, 91, 95*t*, 283*t*, 286*f*, **287**, 313*t*
 clot retraction and, 660
 cross bridge activity and, 296, 297*f*
 in excitation-contraction coupling, 291*f*, 295*f*
 in sliding filament model, 289
Actin filaments. *See* Thin (actin) filaments
Action potential, 24, **290**, 396, 403, 404
 as all-or-none phenomenon, **409**, 410
 generation of, 396, 405, 406–407*f*, 408
 across sarcolemma, 293–296, 293*f*
 information processing in retina and, 572*f*, 573
 in muscle fiber contraction, 290
 cardiac muscle, 685, 686, 687*f*, 689–690, 689*f*
 skeletal muscle, 290, 291*f*, 292*f*, 293–296, 293*f*, 296*f*
 in neuron, 396, 404, **405**–412, 406–407*f*, 410*f*
 graded potential compared with, 416–417*t*
 information transfer at chemical synapse and, 414, 415*f*
 propagation of, 291, 293, 293*f*, 295*f*, **409**, 409*f*, 411*f*
 impaired, 412
Activated macrophages, 795
Activation energy, **52**, 52*f*, 938
Activation gate, 407*f*, 408
Active humoral immunity, 797–798, 797*f*
Active hyperemia, **728**, 728*f*
Active processes/transport, **68**, 68*t*, 73–79, 74*f*, 75*f*, 76*f*, 77*f*, 78*f*, 79*f*, 912
 active transport, 74*f*, **75**–76, 75*f*, 78*t*

electrochemical gradient/membrane potentials and, 80–81
 in tubular reabsorption, **990**
 vesicular transport, **76**–79, 76*f*, 77*f*, 78*t*, 79*f*
Active sites, **51**, 52, 53*f*, 287
Active tubular reabsorption, **990**
Acute (immediate/type I) hypersensitivity, **812**–813, 813*f*
Acute glomerulonephritis, 1008. *See also* Glomerulonephritis
Acute hypotension, 724
Acute leukemia, 654
Acute mountain sickness, 858
Adam's apple (laryngeal prominence), **824**, 825*f*
Adaptation, **494**
 dark and light, **573**
 olfactory, 577
Adaptive immune system/defenses, **782**, 782*f*, **790**–811, 809*f*, 810*t*
 antigens, 770, **791**–792, 791*f*
 cells of, 792–795, 793*f*, 794*f*, 795*t*, 795*f*, 810*t*
 cellular immune response, 782*f*, **790**, 795*t*, 801–811, 802*f*, 803*t*, 804*f*, 805*t*, 806*f*, 807*f*, 809*f*, 810*t*
 humoral immune response, 782*f*, **790**, 795*t*, 796–801, 796*f*, 797*f*, 798*f*, 799*t*, 800*f*, 809*f*
Addison's disease, **625**–626, 626*t*, 1021, 1022
Adduction (body movement), **259**, 259*f*, 325*f*
 of lower limb, 367*t*, 384*t*, 385*f*
 of upper limb, 259*f*, 350*t*, 354*t*, 365*t*
Adductor(s)/adductor muscles, of thigh, 330*f*, 331*f*, **367***t*, 368*f*, 369*t*, 385*f*
Adductor brevis muscle, 368*f*, **369***t*, 384*t*
Adductor hallucis muscle, **382***t*, 383*f*
Adductor hiatus, 748*t*, 748*f*, 749*f*
Adductor longus muscle, 330*f*, 368*f*, **369***t*, 384*t*
Adductor magnus muscle, 331*f*, 368*f*, **369***t*, 371*f*, 384*t*
Adductor pollicis muscle, **364***t*, 365*f*
Adductor tubercle, 240*f*, **241**
A delta fibers, 495
Adenine (A), 54*f*, **55**
Adenocarcinoma, 147, **860**
Adenohypophysis, **609**–610. *See also* Anterior pituitary gland
Adenoidectomy (adenotonsillectomy), **863**
Adenoids, 823. *See also* Pharyngeal tonsils
Adenoma, 144*b*, **147**
Adenosine, 422*t*, **423**, 989
Adenosine diphosphate. *See* ADP
Adenosine monophosphate. *See* AMP
Adenosine triphosphate. *See* ATP
Adenotonsillectomy (adenoidectomy), **863**
Adenylate cyclase, 577, 577*f*, **604**, 605*f*
ADH. *See* Antidiuretic hormone
Adherence, in phagocytosis, 784, 784*f*
Adhesions, 146
Adipocytes (adipose/fat cells), 127*t*, 128*f*, **129**, 130, 131*f*, 958*b*, 960
Adiponectin, 634, 635*t*
Adipose cells. *See* Adipocytes

Adipose (fat) tissue, 127*t*, **130**–133, 131*f*, 151*f*
　autonomic innervation/effects and, 546*t*
　cholesterol transport to, 955
　hormones produced by, 634, 635*t*
　lipolysis in, 953, 953*f*, 954*f*
　in postabsorptive state, 953, 953*f*, 954, 954*f*
　white and brown, 130–133
Adluminal compartment, of seminiferous tubule, 1056*f*, **1058**
ADP (adenosine diphosphate), 55*f*, 56, **658**
　phosphorylation of, 304, 304*f*, 305*f*, 936, 936*f*
Adrenal cortex, **622**, 623–626, 623*f*, 624*f*, 625*f*, 626*t*
Adrenal (suprarenal) gland(s), 602*f*, **622**–627, 623*f*, 624*f*, 625*f*, 626*t*, 628–629*f*, 976. *See also* Adrenal cortex; Adrenal medulla
　aging affecting, 636
　arteries of, 738*f*, 746*t*, 746*f*
　autonomic innervation/effects and, 534*f*, 540*f*, 542, 542*f*
　in stress response, 622, 624, 627, 628–629*f*
　veins of, 750*f*, 756*t*, 756*f*, 757*f*
Adrenal hormones, 622–627, 626*t*. *See also* Adrenal medullary hormones; Adrenocortical hormones
Adrenaline. *See* Epinephrine
Adrenal insufficiency/Addison's disease, **625**–626, 626*t*, 1021, 1022
Adrenal medulla, **622**, 623*f*, 626*t*, 627
　autonomic pathways with synapses in/ autonomic innervation and, 540*f*, 542, 542*f*, 546*t*
　blood pressure control and, 721
Adrenal medullary hormones, 626*t*, 627, 628*f*
Adrenergic fibers, **543**
Adrenergic receptors, 543–545, 544*t*
Adrenocortical hormones, 623–626, 623*f*, 624*f*, 626*t*
Adrenocorticotropic hormone (ACTH/ corticotropin), **610**, 611*f*, 613*t*, **616**, 620*t*, 624, 624*f*, 625, 629*f*
Adrenogenital syndrome, 626, 626*t*
Adult respiratory distress syndrome (ARDS), **863**
Adventitia, **874**
　of esophagus, 874, 882*f*, 883
　tracheal, **827**, 827*f*
　of ureter, 1002, 1002*f*
　vaginal, 1066
Aerobic cellular respiration, 84, 304, 304*f*, **305**–306, 305*f*, 313*t*, 686. *See also* Cellular respiration; Krebs cycle
Aerobic endurance, **305**
Aerobic exercise, **309**. *See also* Exercise
Aerobic pathways, **938**–939
Afferent arteriole, glomerular, 978, 978*f*, **723**, **981***f*, **982**, 982*f*
Afferent (sensory) division of peripheral nervous system, **392**, 392*f*, 497
Afferent lymphatic vessels, **772**, 772*f*
Afferent nerves/neurons/tracts. *See* Sensory (afferent) nerves; Sensory (afferent) neurons; Sensory (ascending) pathways/tracts
Afferent pathway, in homeostasis, 10, 10*f*, 11*f*
Affinity
　hemoglobin, for oxygen, 847, 849*f*
　hormone-receptor, 608
Afterbirth, **1116**
After-discharge circuit, parallel, 427, 427*f*
Afterload, **698**
Age-related macular degeneration (ARMD), **596**
Ageusia, **596**
Agglutination, 665–666, 665*f*, **800**, 800*f*, 809*f*
Agglutinins, **664**, 664*t*

Agglutinogens, **663**, 664*t*, 665*f*
Aggregated lymphoid nodules (Peyer's patches), 771, 771*f*, **775**, 775*f*, **903**
Aging, theories of cellular, 110. *See also specific structure or system*
Aging pigment (lipofuscin), 394
Agonist (muscle), **324**
Agouti-related peptides, in food intake regulation, 960, 961*f*
Agranulocytes, 652, 652*f*, **653**, 653*f*, 655*f*, 656*t*
AIDS (acquired immune deficiency syndrome), **811**
Air blood barrier. *See* Respiratory membrane
Air cells, mastoid, **209**, 582
Air flow, volume and pressure relationships and, 834–840, 835*f*, 836*f*, 837*f*, 838*f*, 839*f*. *See also* Airway resistance
Air movements, nonrespiratory, **838**
Airway resistance, 838–839, 839*f*
Ala(e)
　of ilium, **236**, 237*f*
　of nose, 821, 821*f*
　of sacrum, **224**, 224*f*
Alar cartilages, 821, 821*f*
Alar plate, **482**, 482*f*
Albinism, **169**, 1128, 1128*f*, 1129
Albumin, 51, **644**, 644*t*
Albuminuria, 1001*t*
Alcohol
　antidiuretic hormone and, 614, 1000
　teratogenic effects of (fetal alcohol syndrome), 1113
Aldosterone, **623**–625, 623*f*, 624*f*, 625, 626*t*, **723**, **1021**–1022, 1022*f*
　blood pressure/volume and, 624, 624*f*, 721, 721*t*, 722–723, 722*f*, 1022
　electrolyte balance and, 1021–1022, 1022*f*
　potassium regulation and, 624, 624*f*, 1022*f*, 1025
　sodium regulation and, 623, 624, 624*f*, 1021–1022, 1022*f*
　tubular reabsorption and, 624*f*, 992*f*, 994
Aldosteronism (hyperaldosteronism/ Conn's disease), 625, 626*t*, **1037**
Alertness, 462
Alimentary canal. *See* Gastrointestinal (GI) tract (alimentary canal)
Alkalemia, **1027**. *See also* Alkalosis
Alkaline diet, 1002
Alkaline phosphatase, in bone remodeling, 188
Alkaline reserve, 643, 852, **1028**, 1030
Alkaline solution, 40, 40*f*. *See also* Base(s)
Alkaline tide, **891**, 891*f*
Alkalosis, 56, **1027**, 1034*b*
　effects of, 1035
　metabolic, 1033*t*, 1034*b*, **1035**
　respiratory (hyperventilation), **855**, 1029, 1033*t*, **1034**, 1034*b*
　respiratory and renal compensations and, **1035**
Allantois, 1100*f*, **1102**
Allele(s), **1125**–1126, 1126*f*, 1127, 1129, 1129*t*
Allergen, **812**–813, 813*f*
Allergic asthma, 813, 859
Allergic contact dermatitis, **813**–814
Allergies, **812**–813, 813*f*
Allografts, 811
All-or-none phenomenon, **409**, 410
All-or-none rule (amino acid), 929
All-*trans*-retinal, **570**, 570*f*, 571, 571*f*
Alopecia, **169**
Alopecia areata, 169
Alpha (α) adrenergic receptors, **543**, 544*t*
Alpha (α) cells, pancreatic, **630**, 630*f*
Alpha (α) efferent fibers, 520*f*, **521**
α-γ coactivation, **521**, 521*f*
Alpha (α) globin chains, 646, 646*f*
Alpha globulin, 644*t*

Alpha (α)-helix, **50**, 50*f*
Alpha (α) interferon, 788, 790*t*, 805*t*
Alpha (α) motor neurons, **521**, 521*f*
Alpha (α) particles, 28
Alpha waves, **461**, 461*f*
ALS. *See* Amyotrophic lateral sclerosis
Alternative pathway, of complement activation, **788**, 789*f*
Altitude
　partial pressure affected by, 843, 843*t*
　respiratory adjustments and, 857–858
Alveolar cells, types I and II, **829**, 830*f*, **831**
Alveolar dead space, **842**
Alveolar ducts, **829**, 829*f*
Alveolar gas, composition of, 843*t*, 844
Alveolar (acinar) glands, **125**, 125*f*
Alveolar macrophages, **831**
Alveolar pores, 829*f*, 830*f*, **831**
Alveolar processes, 204*f*, 205*f*, **211**, 211*f*, 214*f*
Alveolar sacs/saccules, **829**, 829*f*
Alveolar surface tension, 839
Alveolar ventilation rate, **842**, 842*t*, 855, 855*f*, 856, 1029
Alveoli
　dental (tooth sockets). *See under* Teeth
　of mammary glands, **1068**, 1068*f*
　respiratory/lung, **829**, 829*f*, 830*f*, 831*t*, 843*t*, 860–861
Alzheimer's disease, **469**, 469*f*
Amacrine cells, in retina, 560*f*
Amacrine granule cells, in olfactory bulb, 577
Amenorrhea, 1078
Amine group, amino acid, 48, 49*f*
Amino acid(s), **48**, 49*f*, 50*f*, 913*f*, **915**, 915*f*
　in absorptive state, 950–951, 951*f*
　degradation of, 949, 949*f*
　essential, 929, 929*f*
　in food intake regulation, 961, 961*f*
　as neurotransmitters, 421–422*t*, 423
　in postabsorptive state, 953–954, 953*f*, 954*f*
　protein synthesis/metabolism and, 913*f*, 915, 915*f*, 948*f*, 949, 949*f*
　supply of (amino acid pool), 950, 950*f*
　translation in synthesis of, 99, 99*f*, **103**–104, 104*f*, 105*f*, 106–107*f*, 108*f*
　transport of, 915, 915*f*
Amino acid–based hormones, **603**, 604–605, 605*f*, 609*t*. *See also* Hormone(s)
Amino acid pool, 950, 950*f*
Aminoacyl-tRNA, 103–104
Aminoacyl-tRNA synthetase, 103, 106*f*
γ-Aminobutyric acid (GABA), 421*t*, **423**, 469
Aminopeptidases, 915
Amitotic cells, neurons as, 394, 428
Amnesia, 461
Amniocentesis, **1134**, 1134*f*
Amnion, 1100*f*, 1101*f*, **1102**, 1103*f*, 1104*f*, 1105*f*
Amniotic cavity, 1100*f*, 1105*f*
Amniotic fluid, 1101*f*, **1102**
　testing (amniocentesis), **1134**, 1134*f*
Amoeboid motion, **77**, 88, **652**
AMP (adenosine monophosphate), 55*f*, 56. *See also* Cyclic AMP
Amphiarthroses, **251**
Amphoteric molecules, **1028**
Amplifying circuit, 427*f*
Amplitude, sound wave, 586–587, 586*f*
Ampulla
　of ductus deferens, 1047*f*, **1050**, 1051*f*
　hepatopancreatic, **900**, 900*f*
　of semicircular duct, **583**–585, 583*f*, 592
　of uterine tube, **1062**, 1063*f*
Ampullary cupula, 592*f*, **593**
Amygdaloid body/amygdala, 429*b*, **456**, 456*f*, 457, 578

Amylase, pancreatic, **899**, 913*f*, **914**, 914*f*
Amylase, salivary, **877**, 904*t*, 913*f*, **914**
Amyloid precursor protein (APP), in Alzheimer's disease, 469
Amyotrophic lateral sclerosis (ALS/Lou Gehrig's disease), 475–476
Anabolic hormones/steroids, 317*b*, 319, 929
Anabolism, 8, **36**, **934**, 935*f*, 950, 951*f*
Anaerobic glycolysis, **304**–305, 304*f*, 305*f*, 306, 313*t*, 934, 935*f*, **937**–939, 937*f*, 938*f*, 944*f*
Anaerobic threshold, **305**
Anal canal, 869*f*, 907–909, 907*f*
Anal columns, 907*f*, **909**
Analgesia, **526**
Anal sinuses, 907*f*, **909**
Anal sphincters, external, 348*t*, 349*f*, **907**, 907*f*
Anal sphincters, internal, **907**–909, 907*f*
Anal veins, varices of (hemorrhoids), 713, 909
Anandamide, 422*t*
Anaphase, in meiosis, 1044, 1045*f*, 1046*f*
Anaphase, in mitosis, 96*f*, **98**, 101*f*, 1046*f*
Anaphylactic shock, 725, **813**
Anaplasia, **110**
Anastomoses (vascular), **714**
Anatomical dead space, **841**–842
Anatomical direction/directional terms, **13**, 14*t*
Anatomical neck, of humerus, 230*f*, **231**
Anatomical position, **12**–13, 13*f*, 14*t*
Anatomy, 1, **2**
　complementarity principle and, **3**, 61
　terminology used in, 2, **12**–17, 13*f*, 14*t*, 15*f*
Anchoring junctions. *See* Desmosomes
Anconeus muscle, 355*f*, **357***t*, 361*f*, 362*t*
Androgen(s), 159, 162, 623*f*, **626**, 626*t*, 1055, 1060. *See also* Testosterone
Androgen-binding protein, **1058**, 1059*f*
Andropause, 1086
Androstenedione, 626
Anemia(s), **650**–651, 651*f*. *See also specific type*
Anemic hypoxia, **852**
Anencephaly, **483**
Anergy, **804**
Aneurysm, 710, **759**, 760*b*
Angelman syndrome, 1132
Angina pectoris, **683**, 760*b*
Angiogenesis, 727
Angiogram/angiography, **759**
　digital subtraction (DSA), 16*f*, 16*b*
Angioplasty, balloon, 760*b*
Angiotensin I, 722*f*, **723**
Angiotensin II, **624**, 624*f*, **721**, **723**
　blood pressure/glomerular filtration and, 721, 721*t*, 722*f*, 723, 723*f*, 988*f*, 989
　sodium regulation and, 1022, 1022*f*, 1023*f*, 1024*f*
Angiotensin converting enzyme, 722*f*, **723**, 834
Angiotensinogen, **624**, 722*f*, **723**
Angle of rib, 226*f*
Angular body movements, **258**–260, 258–259*f*
Anion(s), **32**, 39, 403, 1026
Ankle joint, 257*t*
　muscles controlling movements of, 374–379*t*, 375*f*, 376*f*, 378–379*f*, 384*t*, 385*f*
Ankyloglossia, 877
Ankylosing spondylitis, **276**
Ankylosis, in rheumatoid arthritis, 275, 275*f*
Annular (common tendinous) ring, **556**, 557*f*
Anopsias, 500*t*
Anosmias, 500*t*, 578
ANP. *See* Atrial natriuretic peptide
ANS. *See* Autonomic nervous system
Ansa cervicalis, 338*t*, 509*f*, 509*t*
Antacid, **1037**

Antagonism, hormone, **609**
Antagonist (muscle), **324**
Antebrachial vein, median, **754***t*, 754*f*, 755*f*
Antebrachium, 232. *See also* Forearm
Anterior (term), 14*t*. *See also under Ventral*
Anterior arch, 221, 221*f*
Anterior association area, 440*f*, **443**, 444. *See also* Prefrontal cortex
Anterior border, of tibia, 242*f*, **243**
Anterior cardiac veins, 675*f*, **683**, 683*f*
Anterior cerebral artery, **740***t*, 740*f*, 741*f*
Anterior chamber, **562**, 562*f*
Anterior circumflex humeral artery, **742***t*, 742*f*, 743*f*
Anterior commissure, **444**, 447*f*, 448*f*, 456*f*
Anterior communicating artery, **740***t*, 740*f*, **741***t*, 741*f*
Anterior cranial fossa, 201, 201*f*, **202**, 207*f*
Anterior cruciate ligament, 270*f*, **271**, 271*f*, 272, 272*f*
Anterior cutaneous branch of intercostal nerve, 508*f*
Anterior division, brachial plexus, **510**, 511*f*
Anterior extensor compartment of leg, muscles of, 374*t*
Anterior femoral cutaneous nerve, 513*f*
Anterior flexor compartment
  of arm, 363*f*
  of forearm, 355–356*f*, 357*t*, 358–361*f*, 359*f*, 362*t*, 363*f*
Anterior fontanelle, 246, 246*f*
Anterior fornix, 1061*f*, 1066
Anterior (ventral) funiculi, 473*f*, **474**
Anterior gluteal line, **236**, 237*f*
Anterior (ventral) horns, **472**, 473*f*, 474*f*, 474*t*
Anterior inferior iliac spine, 236, 237*f*
Anterior intercostal arteries, 742*f*, **743***t*, 743*f*
Anterior interventricular artery (left anterior descending artery), 675*f*, **682**, 683*f*
Anterior interventricular sulcus, **674**
Anterior lobe, of cerebellum, **454**, 455*f*
Anterior longitudinal ligament, 218, 219*f*
Anterior (ventral) median fissure, **472**, 473*f*
Anterior muscles/muscle compartment
  of forearm, 355–356*f*, 357*t*, 358–361*f*, 359*f*, 362*t*, 363*f*
  of hip/thigh, 367*t*, 368*f*, 369–370*f*, 384*t*, 385*f*
  of leg, 374*t*, 375*f*, 384*t*, 385*f*
Anterior nasal spine, 211, 211*f*, 214*f*
Anterior pituitary gland (adenohypophysis), **609**–611
  aging affecting, 636
  hormones of, 612–613*t*, 614–617, 1077*f*
  hypothalamus relationship and, 607*f*, 608, 610–611, 611*f*
Anterior sacral foramina, **224**, 224*f*
Anterior scalene muscle, **340***t*, 341*f*
Anterior segment of eye, 558, 558*f*, **562**, 562*f*
Anterior semicircular canal, 583, 583*f*
Anterior (ventral) spinothalamic tracts, 475*f*, **476**, 480*t*
Anterior superior iliac spine, **236**, 237*f*
Anterior thalamic nuclei, limbic system and, 456, 456*f*
Anterior tibial artery, 739*f*, **748***t*, 748*f*, 749*f*
Anterior tibial vein, 751*f*, **758***t*, 758*f*
Anterior vagal trunk, 537
Anterograde amnesia, 461
Anterograde movement/transport, 396
Anteverted uterus, 1063
Anti-A antibodies, 664, 664*t*, 665*f*
Anti-B antibodies, 664, 664*t*, 665*f*
Antibiotic-associated diarrhea, 909
Antibodies, 49*f*, 51, **653**, 770, 790, **798**–801, 810*t*. *See also* Humoral (antibody-mediated) immunity

in complement activation, 788, 789*f*, 800–801, 800*f*
diversity and, 794
in humoral immune response, **790**, 795*t*, 796–801, 796*f*, 797*f*, 798*f*, 799*t*, 800*f*, 809*f*
maternal, transfer to fetus/infant and, 797*f*, 798
monoclonal, **801**
red blood cell, blood groups and, 664–665, 664*t*
self (autoantibodies), 811
Antibody-mediated (humoral) immunity. *See* Humoral (antibody-mediated) immunity
Antibody monomer, **799**, 799*t*
Antibody titer, 797
Anticancer drugs (chemotherapy), 145*b*
Anticoagulants, **660**, 662
Anticodon(s), **103**, 103*f*, 106–107*f*, 108*f*
Antidiuretic hormone (ADH/vasopressin), 610*f*, 611, 612*t*, **613**–614, **721**
  blood pressure/volume and, 614, 721, 721*t*, 722*f*, 723, 723*f*, 1018, 1018*f*
  syndrome of inappropriate secretion of (SIADH), 612*t*, 614, **1037**
  tubular reabsorption and, 991, 992*f*, 994, 1017–1018, 1018*f*
  urea recycling and, 999*f*, 1000
  urine concentration and, 998, 999*f*, 1017–1018, 1018*f*
  in water balance regulation, 449, 613, 998, 999*f*, 1000, 1017–1018, 1018*f*
Antigen(s), 770, **791**–792, 791*f*, 809*f*, 810*t*. *See also* Immunity/immune response
  autoimmune disease and, 811
  endogenous, **803**, 803*t*
  exogenous, **803**, 803*t*
  naive lymphocyte encountering, 792, 793*f*
    B lymphocyte, 796–797, 796*f*
  presentation of, 792, 794–795, 795*f*, 804*f*, 809*f*
    MHC proteins and, 802–804, 802*f*, 803*t*, 804*f*
    T cell activation and, 794–795, 804*f*, 809*f*
  receptors for, diversity and, 793–794
  red blood cell, blood groups and, 663–664, 664*t*, 811
  self, **791**–792, 803, 803*t*, 811
  T cell-dependent, **806**
  T cell-independent, **806**
  transplantation and, 811
Antigen-antibody (immune) complexes, **800**, 800*f*, 813
Antigen binding, T cell activation and, 804, 804*f*
Antigen-binding site, 798*f*, **799**
Antigenic determinants, **791**, 791*f*
Antigen-presenting cells, 792, **794**–795, 795*f*, 809*f*, 810*t*
  MHC proteins and, 802–804, 802*f*, 803*t*, 804*f*
Antihemophilic factor (AHF/factor VIII), 659*f*, 661*t*, **662**
Anti-insulin effect, of growth hormone, 615
Antimicrobial proteins, 782*f*, **788**–790, 788*f*, 790*t*
Antioxidant reactions, 931
Antiport system, **76**, 78*t*
Anti-Rh antibodies, 664
Antithrombin III, **661**
Antitoxin, passive immunity and, 798
Antivenom, passive immunity and, 798
Antral (tertiary/vesicular) follicles, 1070*f*, **1071**, 1072*f*, 1073
Antral phase of ovarian cycle, 1073
Antrum
  mastoid, 581*f*, **582**
  ovarian follicle, **1071**, 1072*f*, 1073
  pyloric, **885**, 885*f*

Anucleate cells, **91**, 646
Anular ligament, **268**, 268*f*
Anulospiral (primary sensory) endings, 520*f*, **521**
Anulus fibrosus, **219**, 219*f*
Anuria, 989
Anus, 869*f*, **907**–909, 907*f*, 918. *See also under Anal*
Anvil (incus), 581*f*, **582**, 582*f*, 587*f*
Aorta, 674, 676*f*, 677*f*, 681*f*, **738***t*
  abdominal, **738***t*, 738*f*, 739*f*, 744–748*f*
    branches of, 738*f*, 744–747*f*, 744–747*t*
  ascending, 675*f*, 676*f*, **738***t*, 738*f*, 739*f*
  atherosclerosis affecting, 760*b*
  blood flow velocity and, 730, 730*f*
  blood pressure in, 716*f*
  coarctation of, 701, 701*f*
  descending, **738***t*, 743*f*
  thoracic, **738***t*, 738*f*, 742*f*
Aortic arch, 675*f*, 677*f*, **738***t*, 739*f*, 740*f*, 742*f*
  baroreceptors in, 719, 720*f*
Aortic bodies, **720**, **855**, 856*f*
Aortic impression, 832*f*
Aortic reflex, **720**
Aortic sinus, 738*t*
Aortic valve (aortic semilunar valve), 677*f*, 678*f*, **680**, 681*f*, 696*f*
Apex
  of heart, **672**, 672*f*, 675*f*, 676*f*
  of lung, 832*f*, **833**
  of nose, 821, 821*f*
Apgar score, **1116**
Aphasia, 459
Apical-basal polarity, epithelial, 117
Apical foramen, of root canal, **880**, 880*f*
Apical impulse, **672**, 672*f*
Apical membrane
  in absorption/digestion, 912, 914*f*, 915
  in tubular reabsorption, 990–991, 990*f*, 991*f*
Apical surface, epithelial, **117**–118, 118*f*
Aplastic anemia, 650
Apnea, **855**
  sleep, **860**
Apneustic breathing, 854
Apocrine glands, 126
Apocrine sweat glands, 162*f*, **162**–163
Apoenzyme, **51**
Aponeuroses, 133, **282**
Apoptosis, **109**, 428, 785, 794, 794*f*
APP. *See* Amyloid precursor protein
Appendages
  body. *See* Lower limbs; Upper limbs
  epiploic, **906**, 907*f*
  skin, 151*f*, **157**–163, 158*f*, 160*f*, 161*f*. *See also specific structure*
Appendicitis, **906**
Appendicular region, **13**
Appendicular skeleton, 175*f*, **176**, 200*f*, 227–245, 235*t*, 241*t*. *See also specific region and specific bone*
  congenital abnormalities of, 247
  lower limbs, 240–245, 240*f*, 241*t*, 242–243*f*, 244*f*, 245*f*
  pectoral (shoulder) girdle, 200*f*, **227**–228, 227*f*, 228*f*, 229*f*, 235*t*
  pelvic (hip) girdle, 200*f*, **236**–239, 237*f*, 239*f*, 241*t*
  upper limbs, 200*f*, 230–236, 230–235*f*, 234*f*, 235*t*
Appendix, 771, 771*f*, **775**, 869*f*, **906**, 907*f*
Appetite, 960–962, 961*f*, **969**
Appositional growth of bone, 186, 187, 187*f*
Appositional growth of cartilage, **174**
Aquaporins (AQPs), 70*f*, **71**, 73*t*, **991**, 991*f*
Aqueous humor, 558*f*, **562**, 562*f*, 563
Arachidonic acid, 48
2-Arachidonoylglycerol, 422*t*
Arachnoid granulations, 464*f*, **465**, 466*f*

Arachnoid mater
  of brain, 464*f*, **465**, 465*f*, 466*f*
  spinal, 470*f*, 471, 471*f*, 473*f*
Arbor vitae, 447*f*, 448*f*, **454**, 455*f*
ARC. *See* Arcuate nucleus
Arches of foot, 244–245, 245*f*
  muscles supporting, 380–383*t*, 381*f*, 383*f*
Arcuate arteries
  of foot, 739*f*, **748***t*, 748*f*, 749*f*
  of kidney, **978**, 978*f*, 981*f*
  of uterus, **1065**
Arcuate line, **236**, 237*f*
Arcuate nucleus, 960, 961*f*, 962, 449*f*
Arcuate popliteal ligament, 270*f*, **271**
Arcuate veins, **978**, 978*f*, 981*f*
ARDS (adult respiratory distress syndrome), **863**
Areola, **1068**, 1068*f*
Areolar connective tissue, 127*t*, 128*f*, **130**, 131*f*. *See also* Connective tissue
Arm. *See also* Forearm
  bones of, 230–231*f*, 231, 235*t*
  deep artery of, **742***t*, 742*f*, 743*f*
  muscles of/muscles controlling movement of, 330*f*, 331*f*, 350–353*t*, 351*f*, 353*f*, 354–356*t*, 355–356*f*, 357*t*, 362*t*, 363*f*
Arrector pili muscle(s), 151*f*, 158*f*, **159**, 538, 539*f*, 540*f*, 546*t*
Arrhythmias, **688**, 692
ART. *See* Assisted reproductive technology
Arterial anastomoses, **714**, 748*f*
Arterial blood pressure, 716–717, 716*f*. *See also* Blood pressure
Arterial pulse, **716**, 717, 717*f*
Arteries, 642, **707**, 707*f*, 708*f*, 710. *See also specific structure supplied or specific named artery*
  blood flow velocity and, 730, 730*f*
  blood pressure in, 716–717, 716*f*. *See also* Blood pressure
  blood volume in, 713*f*
  of brain and circle of Willis, 740–741*f*, 740–741*t*
  pulmonary, 675–677*f*, 681*f*, 729, **736**–737*t*, 736*f*, 832*f*, **834**
  structure of, 708*f*, 709*t*, 710
  systemic, 735, 737*t*, 737*f*, 738–739*f*, 738*t*
Arteriolar vasodilation, in blood pressure regulation, **720**
Arterioles, 707, 707*f*, 709*t*, **710**
  blood flow velocity and, 730*f*
  blood pressure in, 716*f*
  blood volume in, 713*f*
  glomerular, 978, 978*f*
  oxygen partial pressures affecting, 845–846, 846*f*
  terminal, 707*f*, **712**, 712*f*
Arteriosclerosis. *See* Atherosclerosis
Arteriovenous anastomoses, 707*f*, **714**, 729
Arthritis, **273**–275, 275*f*, 812
Arthrology, **276**
Arthroplasty, **276**
Arthroscopic surgery, for cartilage tear, **273**, 273*f*
Articular (joint) capsule, 133, 133*f*, **254**, 254*f*
  of elbow joint, 268, 268*f*
  of hip joint, 269, 269*f*
  of knee joint, 270*f*, 271
  of shoulder joint, 266, 267*f*
Articular cartilage, 134, 174, 175*f*, 179*f*, 185, 185*f*, 253, 253*f*, 254, 254*f*
Articular cavity. *See* Joint (articular/synovial) cavities
Articular discs. *See* Menisci (articular discs)
Articular facets, vertebral, 220, 220*f*, 221*f*, 222, 223*t*, 224*f*, 226*f*

Articular processes, 219f, **220**, 220f, 221f, 222f, 223t, 224, 224f
Articular surfaces, 237, 257
Articular tubercle, **264**, 265f
Articulations. *See* Joint(s)
Artifacts, in microscopy, **117**
Artificial insemination, 1119
Artificial joints, 274b
Artificially acquired immunity, 797, 797f, 798
Artificial skin, for burn injury, 167
Arytenoid cartilage, **824**–825, 825f
Ascending aorta, 675f, 676f, **738**t, 738f, 739f
Ascending colon, 869f, **906**, 907f
Ascending limb of nephron loop, 979f, **980**, 981f, 982, 992–993, 993t, 997f, 998
Ascending lumbar veins, **754**t, **755**t, 755f, 756f, 757f
Ascending (sensory) pathways/tracts, 474, 475f, 476–478, 477f, 480t
Ascites, **919**
Ascorbic acid (vitamin C), 147, 931, 932t
A site, 104, 106f, 107f
Aspartate, **423**
Aspiration, **863**
Aspiration pneumonia, cleft lip/palate and, 246
Aspirin, **662**, 760b
Assisted reproductive technology, 1119
Association, memory and, 460
Association areas/cortex, 440f, 441
  homeostatic imbalance and, 444
  language and, 459–460
  motor, 440f
  multimodal, 440f, **443**–444
  sensory, 440f, 442–443
Association fibers, **444**, 445f
Association neurons. *See* Interneurons
Asters, 100f
Asthma, 813, 839, **859**
Astigmatism, **567**
Astrocytes, **393**, 393f, 497
Asystole, **702**
Atelectasis, **836**, 836f
Atheroma(s), 760b
Atherosclerosis, 48, 319, 700, 702, 759, 760b, 957, 959
Atherosclerotic (fibrous) plaques, 760b
Atlantoaxial joint, 256t
Atlanto-occipital joint, 256t
Atlas (C$_1$), **221**, 221f, 222f
Atmospheric pressure, **9**, **835**, 835f, 836f, 843, 843t
Atom(s), 4, 4f, **25**–27, 25f, 27f
Atomic mass unit, **25**
Atomic number, 27
Atomic symbols, 25
Atomic weight, **28**
Atonic bladder, **549**
Atopic dermatitis, 814
ATP (adenosine triphosphate), 8, 24, **55**–56, 55f, **927**
  in active transport, 73–79, 74f, 75f, 78t
  chemical energy in form of, 8, 24, 55–56, 56f
  in exchange reactions, 36, 36f
  mitochondrial supply of, 83, 84
  for muscle contraction, 303–306, 304f, 305f
    in cardiac muscle, 685t
    cross bridge cycling and, 296, 297f
    smooth muscle and, 314, 315f
  muscle fatigue and decreased, 306
  as neurotransmitter, 422t, **423**
  synthesis of, 303–306, 304f, 305f, 934, 935f, 936–945, 936f, 937f, 938f, 939f, 941f, 944f
    fat metabolism in, 935f, 946–948, 946f, 947f
    glucose/carbohydrate metabolism in, 36, 36f, 37, 44, 935f, 936–945, 937f, 938f, 939f, 941f, 942f, 943f, 944f
  in vesicular transport, 76, 78t

ATPase, myosin, 307, 308
ATP synthases, 936, 936f, 941f, **943**, 943f
Atresia, of ovarian follicles, 1071
Atria of heart, 671, 671f, **674**, 675–678f, 681f
  development of, 700, 701f
  hormones produced by, 634, 635t. *See also* Atrial natriuretic peptide
Atrial diastole, 693
Atrial natriuretic peptide (ANP), 624, 624f, **634**, 635t, **721**, **1022**, 1023f
  blood pressure/volume and, 624, 624f, 721, 721t, 723f, 1022, 1023f
  in sodium regulation, 624, 1021, 1022, 1023f
  tubular reabsorption and, 624f, 994, 1022, 1023f
Atrial (Bainbridge) reflex, **699**
Atrial systole, 693, 694f, 695f
Atrial tachycardia, paroxysmal, **702**
Atrioventricular (AV) bundle (bundle of His), **687**, 687f
Atrioventricular groove (coronary sulcus), **674**
Atrioventricular (AV) node, 686, **687**, 687f
Atrioventricular (AV) valves, 678f, **679**, 679f, 681f, 694f. *See also* Mitral (bicuspid) valve; Tricuspid valve
Atrium. *See* Atria
Atrophy, **110**, 310
Attenuated pathogens, for vaccines, 797, 797f
Auditory areas, 443
Auditory association area, 440f, **443**
Auditory cortex, primary, 440f, **443**, 589f, **590**
Auditory (acoustic) canal/meatus, external, 204f, 206f, **208**, 208f, **216**t, 581f, **582**, 596
Auditory nerve. *See* Vestibulocochlear nerve
Auditory ossicles, **217**t, 581f, **582**, 582f, 587, 587f, 596
Auditory pathway, 589–590, 589f
Auditory processing, 590
Auditory (pharyngotympanic) tube, 581f, **582**, 582f, 596, 821f, 823, 823f
Auras, **462**, 578
Auricle (atrial), **674**, 675–677f
Auricle (pinna), 581f, **582**
Auricular nerve, greater, 509t, 509f
Auricular surface of ilium, **236**, 237f
Auricular surface of sacrum, **224**, 224f
Auricular vein, posterior, 752f, 753f
Auscultation, 2
Auscultatory method, for blood pressure measurement, **717**
Autism, **484**
Autoantibodies, 811
Autocrines, **603**
Autografts, 167, **814**
Autoimmune diseases, 275, **811**–812
Autoimmunity, **811**
Autologous chondrocyte implantation, 274b
Autologous transfusions, **665**
Autolysis, **87**
Automatic memory, 460
Automaticity/autorhythmicity, **685**
Autonomic dysreflexia, 548
Autonomic ganglia, **532**, 533, 533f, 535, 535t, 538–539, 538t
Autonomic (visceral) motor (VM) neurons, **474**, 474f. *See also* Motor (efferent) neurons
Autonomic nervous system, **392**, 392f, 531–552, **532**, 532f, 534f. *See also specific division*
  anatomy of, 535, 535t
  blood pressure and, 545
  cardiac muscle innervated by, 313t
  case study and, 552b
  control of, 547–548, 547f
  cooperative effects of, 545
  developmental aspects of, 548–549

digestion/digestive system and, 534, 534f, 536f, 540f, 546t, 550b, 920b
divisions of, 534–542, 534f, 535t. *See also. See* Parasympathetic (craniosacral) division of autonomic nervous system; Sympathetic (thoracolumbar) division of autonomic nervous system
  opposing effects of, 545–547, 546t
  drug effects and, 544t
  dual innervation and, **534**, 534f
  heart regulation and, 534f, 536f, 540f, 545, 546t, 688f, 689, 697f, 698–699
  homeostatic imbalances/disorders of, 545, 548
  homeostatic relationships of, 550b
  hypothalamus in control of, 449, 547–548, 547f
  physiology of, 535t
  smooth muscle innervated by, 311, 311f, 313t, 315
  somatic nervous system compared with, 532–533, 533f
Autonomic neuropathy, **545**
Autonomic (visceral) reflex, **519**, 542–543, 543f
Autophagosomes, 109
Autophagy, 87, **109**
Autopsy, **147**
Autoregulation
  blood flow, **725**–730, 725f, 727f
  cerebral, 728
  renal, **988**–989, 988f
  in ventilation-perfusion coupling, 845–846, 846f
Autorhythmicity/automaticity, **685**
Autosomal dominant/recessive PKD, 1007
Autosome(s), **1081**, **1125**
Avascularity of cartilage, 127, 134
Avascularity of epithelial tissue, 118
AV node. *See* Atrioventricular (AV) node
Avogadro's number, **30**
AVR. *See* Alveolar ventilation rate
AV valves. *See* Atrioventricular (AV) valves
Axial region, **13**
Axial skeleton, 175f, **176**, 199–227, **199**–**200**, 200f. *See also specific region and specific bone*
  skull, 200, 200f
  thoracic cage, 200, 200f, **224**–227, 225f, 226f
  vertebral column (spine), 200, 200f, **218**–224, 218f, 219f, 220f, 221f, 222f, 223t, 224f
Axial tomography. *See* Computed tomography
Axillary artery, 738f, 739f, 741f, **742**t, 742f, 743f
Axillary (lateral) border of scapula, 228
Axillary lymph nodes, 769f
Axillary nerve, 354t, 356t, 510t, 511f, **512**
Axillary vein, 750f, 751f, **754**t, 754f, 755f
Axis (C$_2$), **221**–222, 221f, 222f
Axoaxonal synapses, 413, 413f
Axodendritic synapses, **413**, 413f
Axolemma, **396**
Axon(s), 140, 140f, 290, 394, **395**–397, 395f, 398t, 496, 497f
  action potential conduction velocity and, 410–412
  action potential generated in, 405, 408
  developmental aspects of, 482, 482f
  of motor neurons, 298f, 299, 395f
  in motor unit, 298f, 299
  myelinated, **396**, 397f, 411–412, 411f. *See also* Myelin sheath(s)
  nonmyelinated, **396**, 397, 411–412, 411f
  olfactory, 576, 576f, 577
  regeneration of, 497–498, 498f
  transport along, 396
Axon collaterals, **395**
Axon hillock, **395**, 395f, 417t
Axon terminals, **290**, **395**, 395f
  in chemical synapses, 414, 415f

at neuromuscular junction, 290, 291f, 292f, 293f, 298f, 517
Axosomatic synapses, **413**, 413f
Azygos system, 750f, 754–755f, 754–755t, **754**t
Azygos vein, **754**t, 754f, 755f

Babinski's sign, **526**
Baby teeth (deciduous/primary dentition), **879**, 879f
Back
  innervation of, 515–516
  intrinsic (deep) muscles of, 340–343t, 341f, 343f
Bacteria, retrograde transport used by, 396
Bacterial microbiota (flora), **909**–910, 958b
Bactericidal action, of sebum, 163
B agglutinins, 664, 664t
B agglutinogens, 664, 664t, 665f
Bainbridge (atrial) reflex, **699**
Balance. *See* Equilibrium/orientation
Baldness (true/frank), 160
Ball-and-socket joints, 260, 263f, 264
Balloon angioplasty, 760b
Band cells, **654**, 655f
Bare area of liver, 894, 894f, 895f
Baroreceptors/baroreceptor reflexes, 699, **719**–720
  blood pressure and, 719–720, 720f, 722f, 723f, 989, 1023–1024, 1024f
  sodium regulation and, 1023–1024, 1024f
Barrett's esophagus, **919**
Barrier methods, for contraception, 1118b
Basal bodies, **90**, 90f. *See also* Centrioles
Basal cell carcinoma, of skin, **165**, 165f
Basal compartment, of seminiferous tubule, 1056f, **1058**
Basal epithelial cells, 578, **578**, 579f
Basal ganglia. *See* Basal nuclei
Basal lamina, **118**
Basal layer
  of endometrium (stratum basalis), **1065**
  of skin (stratum basale/germinativum), 152f, **153**
Basal metabolic rate, **962**–963, 969
  thyroid hormone/thyroxine affecting, 618, 618f, 955t, **963**
Basal nuclei (basal ganglia), 435f, 439, **444**–447, 445f, 446f, 458t
  in language, 459
  in motor control, 517, **518**–519, 518f
  in Parkinson's disease, 452, 469
Basal plate, **482**, 482f
Basal surface, epithelial, **117**, 118f
Base
  of bone, of metacarpals, **234**, 234f
  of heart, **672**
  of lung, 832f, **833**
Base(s) (chemical), **39**, 40, 40f, 48
  strong and weak, **41**, 1027
Base(s) (nucleotide), 53t, 54f, 55
Basement membrane, **118**, 120f, 121f, 122f, 467
  glomerular, 984–986, 985f
Base pairing, DNA, 54f, 55
Basic electrical rhythm, of stomach, **892**
Basilar artery, 740f, **741**t, 741f
Basilar membrane, 584f, **585**, 587, 587f, 588, 588f, 589f
Basilar region, occipital bone, 206f
Basilic vein, 751f, **754**t, 754f, 755f
Basolateral membrane
  in absorption/digestion, 912, 914f, 915, 915f
  in tubular reabsorption, 990, 990f, 991f
Basophilic erythroblasts, **647**, 647f
Basophils, 652f, **653**, 653f, 655f, 656t, 812, 813, 813f
B cells. *See* B lymphocytes
Beau's lines, 161
Bedsores (decubitus ulcers), **169**

Behavioral methods, for contraception, 1118*b*
Bell's palsy, 503*t*
Benign neoplasm, 144*b*
Benign prostatic hyperplasia (BPH), **1052**
Beriberi, 932*t*
Bernard, Claude, 1013
Beta (β) adrenergic receptors, **543**–545, 544*t*
Beta-amyloid peptide, in Alzheimer's disease, 469
Beta-carotene, 931
Beta (β) cells, pancreatic, **630**, 630*f*
Beta endorphin, 422*t*, **423**
Beta (β) globin chains, 646, 646*f*
Beta globulin, 644*t*
Beta (β) interferon, 788, 790*t*, 805*t*
Beta oxidation, **946**, 946*f*, 947*f*
Beta (β) particles, 28
Beta (β)-pleated sheet, **50**, 50*f*
Beta waves, **461**, 461*f*
Biaxial movement, **258**, 263*f*
Bicarbonate/bicarbonate ion, **39**, **850**–852
   in acid-base balance, 41, 852, 1027–1028, 1029–1032, 1030*f*, 1031*f*, 1032*f*, 1034*b*
   in acidosis/alkalosis, 1033*t*, 1034–1035, 1034*b*
   carbon dioxide transport as, **850**–852, 851*f*
   generation of, 1031–1032, 1031*f*, 1032*f*
   in mucosal barrier, 887
   in plasma/interstitial fluid/intracellular fluid, 1014*f*
   tubular reabsorption and, 992*f*, 1029, 1030–1031, 1030*f*
   tubular secretion and, 994, 995, 1029, 1030*f*, 1032
Bicarbonate buffer system, **41**, **852**, **1027**–1028, 1028*f*
   bicarbonate reabsorption and, 1029, 1030–1031, 1030*f*
   bicarbonate regeneration and, 1031–1032, 1031*f*, 1032*f*
   bicarbonate secretion and, 1029, 1032
Biceps brachii muscle, 330*f*, 355*f*, 356*f*, **357**t, 362*t*, 363*f*
   tendon of long head of, 266, 267*f*
Biceps femoris muscle, 331*f*, 371*f*, **373**t, 384*f*
Bicipital groove (intertubercular sulcus), 230*f*, **231**
Bicuspid (mitral) valve, 677*f*, 678*f*, **679**, 681*f*, 696*f*
Bicuspids (premolars), 879*f*, **880**
Bifid spinous process, 221, 222*f*
Bilayer, plasma membrane, 63, 64–65*f*, 66
Bile/bile salts, 47*t*, 894, 896*f*, **897**, 897*f*, 913*f*, 916, 916*f*
   secretion of, 900, 900*f*, 901*f*
   in urine, 1001*t*
Bile canaliculi, 896*f*, **897**
Bile duct, **894**, 895*f*, 896*f*, 897, 898, 900*f*
Bile pigment, 897. *See also* Bilirubin
Biliary calculi (gallstones), **898**
Bilirubin, **649**, 649*f*, **897**
Bilirubinuria, 1001*t*
Bimalleolar fracture, 242*f*, 243
Biochemistry, **38**–56
   inorganic compounds in, **38**–41
   organic compounds in, **38**, 41–56
Biogenic amines, 421*t*, **423**
Biological colloids, 38
Biopsy, in cancer diagnosis, 145*b*
Biotin (vitamin B₇), 932*t*
Biotransformation, liver in, 956*t*
Bipennate fascicle pattern/muscles, 326*f*, 327
Bipolar neurons/cells, **397**, 398–399*t*
   retinal, 560*f*, **561**, 572*f*, 573
Birth, 1114–1116, 1114*f*, 1115*f*
   adjustment to extrauterine life after, 1116

bone development at, 194
circulation changes at, 1107, 1109*f*, 1116
digestive system after, 919
oxytocin in labor and, 11, 612, **1114**–1115, 1114*f*
Birth canal. *See* Vagina
Birth control, 1118–1119*b*
Birth control pill, 1118*b*
2,3-Bisphosphoglycerate (BPG), hemoglobin saturation and, 847–850
Bisphosphonates, 193
Bitter taste, 578, 579
Blackhead, 163
Black patients, sickle-cell anemia in, 650
Bladder. *See* Urinary bladder
Bladder cancer, **1007**
Blanching, 157
Blastocyst, 1097, 1097*f*, **1098**, 1098*f*, 1099, 1100*f*
Blastomeres, 1097
Bleaching of the pigment, 570*f*, **571**
Bleeding disorders, 650, 662–663, **662**
Blepharitis, **596**
Blindness
   color, 569
   functional, 443
   night, 573
   optic nerve damage causing, 500*t*
   trachoma causing, **596**
Blind spot (optic disc), 558*f*, 560*f*, **561**, 561*f*
Blinking, 555, 556
Blister(s), 156
Blood, 127*t*, **137**–138, 137*f*, 642–669. *See also* Blood cells; Plasma
   case study and, 669*b*
   components of, 643–645, 643*f*, 644*t*, 645*f*
   developmental aspects of, 666
   formed elements of, 643–644, 643*f*, **644**–645, 645*f*, 656*t*. *See also* Blood cells
   functions of, 643
   gas exchange/transport and, 38, 643, 730, 731*f*, 819*f*, 842–847, 843*t*, 844*f*, 846*f*, 847–852, 848–849*f*, 850*f*, 851*f*
   hemostasis and, 12, 12*f*, **657**–663, 658*f*, 659*f*, 660*f*
   homeostatic imbalances and, 648–649, 654, 665
   oxygen content of, 847
   pH of. *See* Acid-base balance; pH, of blood
   physical characteristics of, 644
   transfusion/replacement of, 663–666, 664*t*, 665*f*
Blood air barrier. *See* Respiratory membrane
Blood brain barrier, **467**–468
Blood buffers, acid-base balance and, 41. *See also* Buffers
Blood cells, 137, 137*f*, 643–645, 643*f*, 645*f*. *See also* specific type
   formation of. *See* Hematopoiesis
Blood chemistry tests, 666, **666**, 1133
Blood clot, 143*f*, 144, 658, 658*f*, 659*f*, 660, 660*f*. *See also* Coagulation; Fibrin mesh
Blood doping, **651**
Blood flow/circulation, **714**. *See also* specific organ
   autoregulation of, **725**–730, 725*f*, 727*f*
   through body tissues (tissue perfusion), **725**–730, 725*f*, 726*f*, 727*f*, 728*f*
   distribution of, 719
      exercise and, 726, 726*f*, 728, 728*f*
      pressure/resistance relationship and, 715, 718–719, 719*f*, 720*f*, 723*f*
   velocity of, 730, 730*f*
Blood fraction, **666**
Blood glucose. *See* Glucose

Blood groups, 663–665, 664*t*
   multiple-allele inheritance and, 1129, 1129*f*
   transplantation and, 811
   typing and, 665–666, 665*f*
Blood islands, 666, **759**
Blood plasma. *See* Plasma
Blood poisoning (septicemia), **667**
Blood pressure, **714**, 716–725, 716*f*
   age affecting, 759
   aldosterone regulation and, 624, 624*f*, 721*t*, 722*f*, 1022
   antidiuretic hormone secretion and, 1018, 1018*f*
   arterial, 716–717, 716*f*
   autonomic nervous system in regulation of, 545
   blood volume and, 721, 722, 722*f*, 723*f*, 1023–1024, 1024*f*
   capillary, 716*f*, 717, 731
   flow/resistance relationship and, 715, 718–719, 719, 719*f*, 720*f*, 723*f*
   glomerular filtration rate and, 987, 987*t*, 988*f*
   homeostatic imbalances in, 724–725. *See also* Hypertension; Hypotension
   hormonal control of, 721, 721*t*, 722*f*, 723*f*
   measuring, 717
   neural control of, 719–721, 720*f*, 723*f*
   normal, 724
   pregnancy and, 1113
   regulation/maintenance of, 718–723, 719*f*, 1023–1024, 1024*f*
      long-term, 719, 721–723, 722*f*, 723*f*
      short-term, 719–721, 720*f*, 721*t*
   renal regulation of, 721–723, 722*f*, 723*f*, 1018, 1018*f*, 1024*f*
   thirst mechanism and, 722*f*, 723, 1017, 1017*f*
   thyroid hormone affecting, 618
   venous, 713, 716*f*, 717–718, 718*f*
Blood pressure cuff, 717
Blood reservoir, 165, 713, 713*f*
Blood sinusoids/sinusoid capillaries, 647, 711, 711*f*
   endometrial, 1065, 1065*f*
   in liver, 896*f*, **897**
   in splenic pulp, 773*f*, **774**
Blood sugar. *See* Glucose
Blood supply. *See* specific organ or system and Blood vessels; Circulation
Blood testis barrier, 1056*f*, **1058**
Blood tests. *See* Blood chemistry tests
Blood thymus barrier, 776
Blood typing, 665–666, 665*f*
Blood vessels, 706–765, **707**, 707*f*, 708*f*, 709*t*. *See also* specific type and Circulation
   atherosclerosis affecting, **700**, 702, 759, 760*b*
   autonomic innervation/effects and, 539*f*, 540*f*, 545, 546*t*
   case study and, 765*b*
   circulation physiology and, 714–734
   circulatory pathways and, 734–759
   clot in (thrombus), **662**, 759
   congenital disorders of, 759
   constriction of, **709**. *See also* Vasoconstriction
   developmental aspects of, 759
   dilation of, **709**. *See also* Vasodilation
   embolus/embolism and, **662**
   homeostatic imbalances and, 713, 734
   interconnections (anastomoses) of, 707*f*, **714**
   resistance and
      diameter and, 715, 719, 719, 719*f*, 723*f*
      length and, 715, 719*f*, 723*f*
   structure/function of, 707–714, 707*f*, 708*f*, 709*f*
   walls/tunics of, 708*f*, 709, 709*t*
Blood viscosity, 644, 715, 719*f*, 723*f*

Blood volume, 644, 713*f*
   aldosterone regulation and, 624, 624*f*, 1022
   antidiuretic hormone secretion and, 1018, 1018*f*
   blood pressure and, 721, 722, 722*f*, 723*f*, 1023–1024, 1024*f*
   pregnancy and, 1113
   reduced, hypovolemic shock and, 725, 1018
   restoration of, 663
   thirst mechanism and, 1017, 1017*f*
Blue cones, 564*f*, 569
B lymphocytes (B cells), **653**, 655*f*, **770**, **792**–794, 795*t*, 809*f*, 810*t*
   activation/development/differentiation/maturation of, 792–793, 793*f*, 795*t*, 796–797, 796*f*, 809*f*
      helper T cells in, 806, 806*f*
   antigen presentation by, 795
   clonal selection of, 796, 796*f*
   in humoral (antibody-mediated) immune response, 782*f*, 795*t*, 809*f*. *See also* Humoral (antibody-mediated) immunity
   memory, 796*f*, **797**, 809*f*
   T cell interaction and, 806, 806*f*
B lymphocyte precursor, **654**, 655*f*, 793*f*, 814
BMI. *See* Body mass index
Body
   of bone
      of ilium, **236**
      of ischium, **236**, 237*f*
      of mandible, 210, **211**, 211*f*
      of pubis, 237*f*, **238**
      of sphenoid bone, **209**, 209*f*
      of sternum, 225, 225*f*
      of vertebra, 219*f*, **220**, 220*f*, 223*t*
   of clitoris, 1066*f*, **1067**
   of epididymis, 1049*f*, 1050
   of nail, 160, 160*f*
   of penis, 1051*f*
   of stomach, **885**, 885*f*
   of uterus, **1063**, 1063*f*
Body cavities, 17–20, 18*f*, 19*f*, 20*f*
Body defenses. *See* Immune system
Body fluids, 1013–1015, 1013*f*, 1014*f*, 1015*f*
Body mass index (BMI), **960**
Body movements, 258–260, 258–260*f*
Body planes/sections, 13–17, 15*f*
Body position, muscles in maintenance of, 281
Body proportions, skeletal growth and, 247, 247*f*
Body regions, anatomical terms describing, **13**, 13*f*
Body surface area, basal metabolic rate and, 962
Body temperature
   homeostatic imbalances and, 967, 967, 968
   metabolic rate and, 963, 967
   normal, 9, 964
   regulation of, 10, 11*f*, 164, 546, 643, 963–968, 963*f*, 965*f*, 966*f*
      energy conversions and, 25
      heat generation by muscles and, 281
      hypothalamus in, 449
      skin in, 164, 728, 729
      thyroid hormone/thyroxine in, 618, 618*t*, 620, 967
Body water, 1013, 1013*f*. *See also* Water balance
   developmental changes in, 1035–1037
   pregnancy and, 1113
Body weight, regulation of, 958–959*b*, 960–962, 961*f*
Bohr effect, **850**, 850*f*
Boils, 169
Bolus (food), **877**
Bonds. *See* Chemical bonds

Bone(s)/skeletal tissue, 127*t*, 136*f*, **137**, 173–198, 174*t*. *See also specific named bone and* Skeletal system; Skeleton
  aging affecting, 193, 194–195, 247
  appositional growth of, 186, 187, 187*f*
  in calcium regulation, 1026
  case study and, 198*b*, 250*b*
  chemical composition of, 182–184
  classification of, 176, 177*f*
  compact, 127*t*, **177**, 177*f*, 178*f*, 179*f*
  developmental aspects of, 184–188, 185*f*, 186*f*, 187*f*, 194–195, 194*f*
  flat, **176**, 177*f*
  formation of, 184–188, 185*f*, 186*f*, 187*f*
  fracture of, **190**–192, 191*f*, 192*t*, 193, 195
  functions of, 6*f*, 175–176
  growth of, 184, 186–188, 187*f*
  homeostasis of. *See* Bone remodeling; Bone repair
  homeostatic imbalances/disorders of, 189, 193–194, 193*f*, 194*f*
  homeostatic relationships and, 196*b*
  hormones produced by, 176, 634, 635*t*
  irregular, **176**, 177*f*
  long, **176**, 177*f*, 178, 179*f*, 184–187, 185*f*, 187*f*
  longitudinal growth of, 186–187, 187*f*
  mechanical stress affecting, 189–190, 190*f*
  membranous, **184**
  muscle relationships and (lever systems), **327**–328, 327–328*f*, 329*f*
  ossification of
    endochondral, **184**–185, 185*f*
    intramembranous, 184, **186**, 186*f*
  osteoporotic, **193**–194, 193*f*
  pagetic, 194, 194*f*
  parathyroid hormone affecting, **189**, 189*f*, 1026
  remodeling. *See* Bone remodeling
  sesamoid, **176**
  short, **176**, 177*f*
  spongy, 127*t*, **177**, 177*f*, 178*f*, 179*f*, 182
  structure of
    gross anatomy, 177–180, 177*f*, 178*f*, 179*f*, 180*t*
    microscopic anatomy, 180–182, 181*f*, 182*f*, 183*f*
  surface markings of (bone markings), **179**, 180*t*
  thyroid hormone affecting, 188, 618, 618*t*
Bone cells, 180–181, 181*f*. *See also* Osteoblasts; Osteoclasts; Osteocytes
Bone collar, in endochondral ossification, 184, 185*f*
Bone density, aging/osteoporosis and, **193**–194, 193*f*
Bone deposition, 188, 193
Bone lining cells, **181**
Bone markings, **179**, 180*t*, 216–217*t*
Bone marrow
  B cell maturation in, 654, 771, 793*f*, 795*t*
  cancer of (leukemia), 654
  hematopoiesis in, **176**, 179, 647, 666
  red, **178**–179, **647**, 666, 771, 771*f*
  stimulation of, in joint therapy, 274*b*
  yellow, 178, 179, 179*f*, 666
Bone marrow biopsy, 179, **666**
Bone mass, aging/osteoporosis affecting, 193
Bone matrix, 127*t*, 137, 181
Bone remodeling, 180, 181, **188**–190, 189*f*, 190*f*
  in fracture repair, 191, 191*f*
  in longitudinal bone growth, 187, 187*f*
Bone repair, 190–192, 191*f*
Bone resorption, **188**, 193
Bone (osseous) tissue. *See* Bone(s)/ skeletal tissue
Bony callus, in bone repair, **191**, 191*f*
Bony (osseous) labyrinth, **583**, 585*t*, 596

Bony pelvis, 236, 237*f*, **238**. *See also* Pelvis
Bony spur, **195**
Bony thorax. *See* Thoracic cage/thorax
Booster shots, vaccine, 798
Boundary maintenance, **8**
  epithelia in, 117
  plasma membrane in, 8, 64*f*
  skin in, 5*f*, 6*f*, 8, 163–164
Bowman's (glomerular) capsule, 979*f*, **980**, 981*f*
Boyle's law, **836**
BPG, hemoglobin saturation affected by, 847–850
Brachial artery, 717*f*, 739*f*, **742***t*, 742*f*, 743*f*
Brachialis muscle, 330*f*, 331*f*, 355*f*, 356*f*, **357***t*, 362*t*, 363*f*
Brachial plexus, 507*f*, **510**–512, 510*t*, 511*f*, 515*t*
Brachial vein, 751*f*, **754***t*, 754*f*, 755*f*
Brachiocephalic trunk (artery), 675*f*, 677*f*, **738***t*, 738–743*f*
Brachiocephalic veins, **750***t*, 750–755*f*, **754***t*
Brachioradialis muscle, 330*f*, 331*f*, 355*f*, **357***t*, 359*f*, **360***t*, 361*f*, 362*t*, 363*f*
Brachium. *See* Arm
Bradycardia, **699**
Bradykinin, 785*t*
Brain, 391, 391*f*, 392*f*, **435**–470. *See also specific part*
  arterial supply of, 740–741*f*, 740–741*t*
  degenerative disease of, 469–470, 469*f*
  developmental aspects/embryology of, 435–436, 435*f*, 436*f*, 482–484, 1104–1106, 1105*f*
  functions/functional systems of, 456–458, 456*f*, 457*f*, 458–459*f*
  gray and white matter of, **436**, 437*f*. *See also* Gray matter; White matter
  higher mental functions and, 458–464, 460*f*, 461*f*, 463*f*
  homeostatic imbalances/injuries/ disorders of, 441, 443, 444, 450, 460–461, 462, 463–464, 465, 467, 468–470, 483
  inflammation of (encephalitis), 465
  perfusion (blood flow) of, 726, 726*f*, 728
  protection of, 464–468, 464*f*, 465*f*, 466*f*, 467*f*
  regeneration and, 497
  regions/organization of, 436, 437*f*
  swelling of (cerebral edema), **468**, 728
  traumatic injury of, 468
  veins of, 735
  ventricles of, 435, 435*f*, **436**–438, 437*f*
Brain death, 461, **462**
Brain stem, **435**, 450–454, 450–451*f*, 452*f*, 453*f*, 459*t*
  autonomic regulation by, 547, 547*f*
  development of, 435, 435*f*, 436*f*
  respiratory centers in, 853–857, 853*f*, 854*f*, 855*f*, 1027
Brain vesicles, 1106, **435**, 435*f*
Brain waves, **461**–462, 461*f*
Branchial groove (pharyngeal cleft), 596
Braxton Hicks contractions, 1114
BRCA1/BRCA2 genes, 1069
Breast(s). *See* Mammary gland(s)/breasts
Breast cancer, **1069**, 1069*f*
Breast feeding. *See* Lactation
Breast milk, mammary glands producing, 1068, 1068*f*, 1116–1119, 1117*f*
  prolactin and, 616, 617, **1117**, 1117*f*
Breathing, 819*f*, **835**, 836–840, 837*f*, 838*f*, 839*f*. *See also* Respiration
  altitude affecting, 857–858
  apneustic, 854
  cessation of. *See* Apnea
  Cheyne-Stokes, **863**
  control of, 454, 853–857, 853*f*, 854*f*, 855*f*, 856*f*

  exercise affecting, 838, 857
  expiration and, 819*f*, **835**, 837*f*, 838, 838*f*
  immediately after birth, 860, 1116
  inspiration and, 819*f*, **835**, 836–838, 837*f*, 838*f*
  muscles of, 344*t*, 345*f*, 836–838, 837*f*
  physical factors influencing, 838–840, 839*f*
  rate and depth of, factors affecting, 854–857, 854*f*, 855*f*, 856*f*, 1029
  volume/pressure changes and, 834–840, 835*f*, 836*f*, 837*f*, 838*f*, 839*f*
Breech presentation, 1116
Bridge of nose, 821, 821*f*
Brittle bone disease (osteogenesis imperfecta), **147**, 195
Broad ligament, **1061**, 1063*f*, 1064*f*
Broca's area, 440*f*, **441**, 459
Bronchi/bronchial tree, **828**–831, 828*f*, 829*f*, 830*f*, 831*t*, 833*f*
  airway resistance in, 839, 839*f*
Bronchial arteries, 738*f*, **743***t*, 834
Bronchial circulation, 834
Bronchial veins, 755*t*
Bronchioles, **828**–829, 829*f*, 830*f*, 845, 846
Bronchitis, chronic, 858*f*, **859**
Bronchomediastinal trunks, 768*f*, **769**
Bronchopulmonary dysplasia, 840
Bronchopulmonary segments, **833**, 833*f*
Bronchoscopy, **863**
Bronchus. *See* Bronchi
Bronze diabetes (hemochromatosis), **667**, **921**
Brown fat (brown adipose tissue), 130–133, 959*b*
Brown or black "necklace" (bruises), 157
Browning reactions, 969
Brows (eyebrows), **554**, 555*f*
Bruises, 157
Brush border, 118, 902*f*, **903**, 980. *See also* Microvilli
Brush border enzymes, **903**, 904*t*, 905, 913*f*, 914, 914*f*, 915, 915*f*
Bruxism, **919**
Buboes, 773
Buccal branch of facial nerve, 503*t*, 504*t*
Buccal cavity. *See* Oral cavity (mouth)
Buccal glands (intrinsic salivary glands), 878
Buccal phase of swallowing, **883**, 884*f*, 904*t*
Buccinator muscle, **334***t*, 335*f*, 336*t*, 339*f*
Buffers, **41**
  chemical, 41, **1027**–1028, 1027*f*, 1028*f*
  physiological, 1027, 1028–1032, 1030*f*, 1031*f*, 1032*f*
Buffy coat, **643**–644, 643*f*
Bulb(s)
  hair, 158*f*, **159**
  olfactory, 440*f*, 452*f*, 456*f*, 499, 499*f*, 500*t*, 576*f*, **577**
  of penis, **1051**, 1051*f*
  of vestibule, 1066*f*, **1067**
Bulbar conjunctiva, **555**, 555*f*, 562*f*
Bulbospongiosus muscle, 348*t*, 349*f*
Bulbo-urethral glands, 1047, 1047*f*, 1051*f*, **1053**, 1082*f*
Bulbous corpuscles (Ruffini endings), 492*t*, **493**
Bulbus cordis, 700
Bulimia, **919**
Bulk flow, 731–734, 732–733*f*, 1015
Bundle branches, **687**, 687*f*
Bundle of His (atrioventricular bundle), **687**, 687*f*
Bundles, cardiac muscle, 673, 673*f*
Burn(s), **166**–167, 166*f*, 167*f*
Burn blister, 166
Bursae, **255**, 255*f*
Bursitis, **273**

C. *See* Carbon
Ca. *See* Calcium

Cadherins, 67*f*, 68, 81
Cafe coronary, 828
Cajal, interstitial cells of, 892
Calcaneal nerve, 515*t*
Calcaneal (Achilles) tendon, 244, 331*f*
  ruptured, **386**
Calcaneal tuberosity, **244**, 244*f*
Calcaneus, 241*t*, **244**, 244*f*
Calcarine sulcus, 440*f*, 442
Calcification, 174, 188
Calcification front, 188
Calcification zone (epiphyseal plate), 186, 187*f*
Calcitonin, **189**, 617*f*, 618, **621**, 1026
Calcitriol, 189, 622, **634**–635
Calcium, 9, 26*t*, 931, 933*t*
  absorption of, 917
  balance of, 188–189, 189*f*, 1020*t*, 1025–1026. *See also* Hypercalcemia; Hypocalcemia, 9999
  in blood clotting, 659, 659*f*, 660, 661*t*
  in bone matrix/tissue, 137, 184
  in bone remodeling, 188–189, 189*f*
  calcitonin and, 189, 621
  cardiac function affected by, 699
  cross bridge activity and, 296, 297*f*
  deficiency of, 193, 933*t*
  dietary, 931, 933*t*
  information transfer at chemical synapse and, 414, 415*f*
  in muscle contraction, 288, 291*f*, 292*f*, 296, 313*t*
    cardiac muscle, 313*t*, 685, 685, 685*t*, 689–690, 689*f*
    excitation-contraction coupling and, 291*f*, 294–295*f*, 296
    smooth muscle, 311, 311*f*, 313*t*, 314, 315*f*
  for osteoporosis, 193
  parathyroid hormone relationship and, 188, 189, 189*f*, 607, 607*f*, 622, 622*f*, 917, **1026**
  in plasma/interstitial fluid/intracellular fluid, 1014*f*
  as second messenger, 81, **425**, 604–605
  in taste transduction, 579
  tubular reabsorption of, 1026
  vitamin D in absorption/regulation of, 164, **917**
Calcium channels
  in cardiac muscle contraction, 685, **686**, **689**, 689*f*
  information transfer at chemical synapse and, 414, 415*f*
  male infertility and, 1058
  in muscle contraction, 292*f*, 295*f*, 296, 311, 311*f*, 315*f*
Calculus (dental), **881**
Callus, 153, 191, 191*f*
Calmodulin, 313*t*, **314**, 315*f*, **605**
Calories, 927, 929
Calorigenic effect, of thyroid hormone, **618**, 618*t*
Calorimeter, **962**
Calvaria (cranial vault), 201
Calyces (calyx), renal (major and minor), **977**, 977*f*
cAMP. *See* Cyclic AMP
CAMs. *See* Cell adhesion molecules
Canal(s), bone, 182, 182*f*, 183*f*
Canaliculi
  bile, 896*f*, **897**
  bone, **182**, 183*f*
  lacrimal, **556**, 556*f*
Cancellous bone. *See* Spongy bone
Cancer, 144–145*b*. *See also specific type or organ affected*
  aging and, 146, 814
  basement membrane ignored in, 118
  glycocalyx changes in, 67
  sentinel nodes in, **777**
Cancer chemotherapy, 145*b*
Cancer vaccine, 145*b*
Canines (eyeteeth), 879*f*, **880**, 880*f*

Cannabinoid receptors, 424
Cannon, Walter, 9
Canthi/canthus, medial and lateral, **554**, 555f
Capacitance vessels, veins as, 707f, 713, 713f
Capacitation, **1093**
Capillaries. *See also specific organ or structure supplied*
  blood, 642, 707, 707f, 708f, 709t, **710**–712, 711f, 712f
    blood flow through, 730–734, 730f, 731f, 732–733f
    blood flow velocity and, 730, 730f
    blood pressure in, 716f, 717, 731
    blood volume in, 713f
    true, **712**, 712f
  lymphatic, 707f, 767–769, 767f
Capillary beds, 708f, **712**, 712f
  nephron, 982, 982f
Capillary colloid osmotic pressure (OP$_c$/oncotic pressure), **731**, 733f, 734
Capillary gas exchange, 730, 731f, 846–847
Capillary hydrostatic pressure (HP$_c$), **731**, 733f, 734
Capillary plexus, primary and secondary, **610–611**, 611f
Capillary pores (fenestrations), 711f
  glomerular, 980
Capitate, **234**, 234f, 235t
Capitulum, 230f, **231**, 231f
Capsular ligaments, **254**, 271
Capsular space, **980**, 980f, 985f, 986, 986f
Capsule. *See specific type*
Carbaminohemoglobin, **647**, **850**, 851f
Carbohydrate(s), 9, **43–45**, 43–44f, 55t, 927–928, 930t
  in absorptive state, 950, 951f
  digestion/absorption of, 913f, 914–915, 914f
  interconversion of, 949–950, 950f
  membrane, 64f, 65f, 66–68
  metabolism of, 935f, 936–945, 937f
    gluconeogenesis in, 944f, **945**
    glucose oxidation in, 937–944, 938f, 939f, 941f, 942f, 943f, 944f
    glycogenesis and glycogenolysis in, 944–945, 944f, 945f, 952, 953f, 954f
    liver in, 945, 950, 951f, 952, 956t
    thyroid hormone affecting, 618, 618t
Carbohydrate pool, **950**, 950f
Carbon, 25, 26t
  as chemically reactive element, 31f
  electroneutrality of, **41**
  isotopes of, 28
  in organic compounds, 41–42
Carbon dioxide, 843, 843t
  acid-base balance/blood pH and, 852, 1028–1029, 1033t, 1034b, 1035
  in acidosis/alkalosis, 1029, 1033t, 1034b
  alveolar, 843t, 844, 844f
  in blood, erythrocytes carrying, 850, 851f
  exchange of, 830, 842–847, 843t, 844f, 846f
  external respiration (pulmonary gas exchange) and, 844–845, 844f, 846f
  hemoglobin saturation affected by, 850, 850f
  internal respiration and, 844f, 846–847
  as nonpolar compound, 34, 34f
  partial pressure of (P$_{CO_2}$), 843, 843t, 844f, 855, 855f
  respiratory rate/depth affected by, 855, 855f, 856, 1029
  solubility of, 843, 844f, 845
  transport of, 643, 647, 850–852, 851f
  ventilation affected by, 845, 846, 846f, 855, 855f, 856, 1029
Carbonic acid, 39, 41, 855. *See also* Bicarbonate buffer system
  in acid-base balance, 41, 852, 1027, 1027f

Carbonic anhydrase, **850**–852, 851f, 1030, 1030f
Carbon monoxide (CO), 422t, **423**–424, 852
Carbon monoxide poisoning, **852**
Carboxyl groups, 1028
Carboxypeptidase, **899**, 899f, 913f, **915**
Carbuncles, **169**
Carcinogenesis, 144b, 145f
Carcinoma, **147**. *See also* Cancer
Cardia (cardial part of stomach), **885**, 885f
Cardiac catheterization, **702**
Cardiac center, 454, 547, 719, 720f
Cardiac conduction system. *See* Conduction system of heart
Cardiac cycle, 692–696, **693**, 694–695f
Cardiac impression, 832f
Cardiac muscle/cardiac muscle tissue, **138**–140, 139f, **280**, 312–313t, 683–686, **684**, 684f, 685t. *See also* Heart
  aging affecting, 702
  contraction of, 313t, 689–690, 689f
  development of, 317
  energy sources for contraction and, 685t, 686
  length-tension relationship in, 697
  regeneration of, 317
  skeletal muscle compared with, 685–686, 685t
Cardiac muscle bundles, 673, 673f
Cardiac muscle cells/fibers (myocytes), 139f, 312t, 683–686, 684f, 685t. *See also* Cardiac muscle/cardiac muscle tissue
  contractile, 685, 689–690, 689f
Cardiac notch, 832f, **833**
Cardiac output, **696**–700, 697f, 698f
  blood pressure and, 700, 718–719, 719f, 720, 723f
  homeostatic imbalance of, 700
  pregnancy and, 1113
Cardiac pacemaker cells, 313t, 685, 685t, **686**–689, 687f
Cardiac plexus, 537, 540f
Cardiac reserve, **696**, 702
Cardiac skeleton, **674**, 678f, 684
Cardiac (gastroesophageal) sphincter, **882**
Cardiac tamponade, 673
Cardiac veins, 675f, 676f, **683**, 683f, 751f
Cardial orifice, **882**
Cardial part of stomach (cardia), **885**, 885f
Cardinal (transverse cervical) ligaments, 1063f, **1064**
Cardioacceleratory center, 688f, **689**, 718, 719, 720f, 723f
Cardiogenic shock, **725**
Cardioinhibitory center, 688f, **689**, 719, 720f
Cardiomyopathy, 700
Cardiovascular center, 454, **719**, 720f
Cardiovascular system, 5f, 6f. *See also* Blood; Blood vessels; Circulation; Heart
  aging of, 319
  blood vessels, 706–765, **707**, 707f, 708f, 709t
  case studies and, 705b, 765b
  digestive system and, 761b, 920b
  endocrine system and, 637b, 761b
  fetal (fetal circulation), 759, 1101f, 1107, 1108–1109f, 1116
  heart, 670–705, **671**, 671f
  homeostatic relationships of, 5f, 761b
  integumentary system and, 168b, 761b
  lymphatic system/immunity and, 761b, 778b
  muscular system and, 318b, 761b
  nervous system and, 550b, 761b
  physiology/functions of, **2**
  pregnancy and, 1113
  reproductive system and, 761b, 1085b
  respiratory system and, 761b, 862b
  skeletal system and, 196b, 761b

stress response and, 628f
  thyroid hormone affecting, 618t
  urinary system and, 761b, 1036b
Carfentanil, 429
Caries, dental (cavities), **881**
Carina, 820f, **827**–828, 828f
Carotene, **156**, 931
Carotid arteries
  common, 717f, **738t**, 738f, 739f, 740t, 740f, 741f, 742f, 743f
  external, 738f, 739f, **740t**, 740f, 741f
  internal, 208, 738f, 739f, **740t**, 740f, 741f
Carotid bodies, 720, **740t**, **855**–856, 856f
Carotid canal, 206f, **208**, **216t**
Carotid sinuses, 719, 720f, **740t**
Carotid sinus reflex, **720**
Carpal bones/carpus (wrist), 200f, **234**, 234f, 235t
  muscles controlling movement of, 358–361, 359f, 361f, 362t, 363f
Carpal tunnel, 234
Carpal tunnel syndrome, 234, 512
Carpometacarpal joint, 256t
Carpus. *See* Carpal bones/carpus (wrist)
Carrier-mediated facilitated diffusion, 70, 70f, 73t
Carriers (genetic), 1129, 1133, 1133f
Carriers (protein), 66, 66f, **70**, 70f, 73t
CART. *See* Cocaine-and amphetamine-regulated transcript
Cartilage(s), 127t, **134**–137, 135–136f, 174t. *See also specific type*
  elastic, 127t, **134**–137, 135f, **174**, 175f
  in endochondral ossification, 184–185, 185f
  fibrocartilage, 127t, 136f, **137**, **174**, 175f
  growth of, 174
  hyaline, 127t, **134**, 135f, **174**, 175f
  injury/repair of, 137, 272–273, 273f
  skeletal, **174**–175, 175f
Cartilage plates, bronchial, 828
Cartilage tears, 272–273, 273f
Cartilaginous joints, 251, **253**–254, 253f, 255t
Cascades, positive feedback mechanisms as, 11
Case studies. *See specific system*
Caspases, in apoptosis, 109
CAT (computerized axial tomography). *See* Computed tomography
Catabolic-anabolic balance, 949–950
Catabolism, 8, **36**, **934**, 935f, 952–955, 953f. *See also* Digestion
Catalase(s), 86
Catalysts, **38**, 51, 52. *See also* Enzymes
Cataract, **563**, 563f
Catecholamines, **423**, 627, 628f
Cathelicidins, 163
Catheter, urinary, **1005**
Catheterization, cardiac, **702**
Cation(s), **32**, 402f, 403
Cauda equina, 470f, 471f, **472**, 507f, 508
Caudal (term), 14t. *See also under* Inferior
Caudate lobe, of liver, 894, 895f
Caudate nucleus, 445f, **446**, 446f
Caveolae, in smooth muscle, **311**, 311f, 312t, 314f
Cavernous sinuses, **752t**, 752f, 753f
Cavities (body), 17–20, 18f, 19f, 20f
Cavities (dental), **881**
C (parafollicular) cells, 617f, 618, **621**
CCK. *See* Cholecystokinin
CD4 cells, 801–802, 802f, 809f. *See also* Helper T cells
  HIV infection/AIDS and, 811
  MHC antigen presentation and, 802f, 803–804, 803t, 804f
  MHC restriction and, 804
CD8 cells, 801–802, 802f, 803, 803t, 804, 806, 806f, 809f. *See also* Cytotoxic T cells
Cdks (cyclin-dependent kinases), 98
Cecum, 869f, **906**, 907f
Celiac disease (gluten-sensitive enteropathy), 918

Celiac ganglion/plexus, 540f, **541**
Celiac trunk, 738f, 739f, 744f, **745t**, 745f, 746f, 747f
  branches of, 744f, **745t**, 745f, 746f
Cell(s), 4, 4f, 60–114, **61**, 61f, 62f, 94–96t. *See also under Cellular*
  in connective tissue, 127t, 128f, 129
  cytoplasm of, 62, 62f, 64f, **83**–91
  death of, programmed. *See* Apoptosis
  destruction of, 109–110
  developmental aspects of, 109–110
  differentiation/diversity/specialization of, 61, 61f, **109**
  division of. *See* Cell division
  environmental interactions with, 63–66, 81–83, 82f
  extracellular materials and, **62**–63
  generalized/composite, 62, 62f
  growth/reproduction of
    cell cycle and, **96**–98, 96f, 97f
    protein synthesis and, 98–108, 99f, 102f, 104f, 105f, 106–107f, 108f
  membrane of. *See* Plasma membrane
  nucleus of, 62, 62f, **91**–110, 92f
  polarity of, resting membrane potential and, **79–80**, **403**
  unneeded, removal of, 109
Cell adhesion molecules (CAMs), 66f, **81**, 128, 428, 498f, 786
Cell body
  neuron, 140f, 290, **394**, 395f, 398t
  rod and cone, 568, 568f
Cell-cell recognition, glycocalyx in, 64f, 65f, 66–67, 66f
Cell center. *See* Centrosome
Cell cycle, **96**–98, 96f, 97f
  cell division, 96f, 98, 100–101f
  cytokinesis, 96f, **98**, 101f
  DNA replication, 97–98, 97f
  interphase, **96**–98, 96f, 100f
  mitosis, 96f, **98**, 100–101f
Cell death, programmed. *See* Apoptosis
Cell differentiation glycoproteins, 801–802
Cell division, 96f, 98, 100–101f, 109–110. *See also* Cytokinesis; Mitosis
  in cancer, 144b
Cell junctions, 66f, 67–68, 67f
Cell lysis, in complement activation/antibody mechanism of action, 788, 800f, 801
Cell-mediated immunity. *See* Cellular (cell-mediated) immunity
Cell membrane. *See* Plasma membrane
Cell theory, **61**
Cellular extensions, 90–91, 90f, 91f, 95t
Cellular (cell-mediated) immunity, 782f, **790**, 795f, 801–811, 802f, 803t, 804f, 805t, 806f, 807f, 809f, 810t
  organ transplants/rejection prevention and, 810–811
  primary response, 805, 808–809f
Cellular level, structural organization at, **4**, 4f
Cellular metabolism, autonomic effects and, 546t. *See also* Metabolism
Cellular organelles. *See* Organelles
Cellular respiration, 8, 84, 304, **305**–306, 313t, 819, **934**, 935f, 936, 937f, 943–944, 944f
  ATP for muscle activity and, 304, 304f, 305, 305f, 306
  cardiac muscle and, 686
  redox reactions in, 37
Cellular secretions, 63
Cellular trophoblast (cytotrophoblast), 1098f, **1099**, 1100f
Cellulose, 44, 928
Cement, **880**, 880f
Central artery
  of retina, 558f, 560f, **561**, 561f
  of spleen, 773f
Central (Haversian) canal, 136f, 137, **182**, 182f, 183f

Central canal of spinal cord, 435*f*, **472**, 473*f*
Central chemoreceptors, 854*f*, **855**, 855*f*
Central deafness, vestibulocochlear damage causing, 504*t*
Central nervous system, **391**, 391*f*, 392*f*, **434**–488. *See also* Brain; Spinal cord
  brain, **435**–470
  case study and, 488*b*
  developmental aspects of, 482–484, 482*f*, 483*f*
  diagnostic procedures for dysfunction of, 470
  gray and white matter in, **436**, 437*f*
  higher mental functions and, 458–464, 460*f*, 461*f*, 463*f*
  homeostatic imbalances/injuries/ disorders of brain and, 441, 443, 444, 450, 460–461, 462, 463–464, 465, 467, 468–470, 483
  myelination in, 393*f*, 397
  neuroglia (supporting cells) of, 140, 140*f*, 393–394, 393*f*
  neuronal circuit types and, **427**, 427*f*
  protection of brain and, 464–468, 464*f*, 465*f*, 466*f*, 467*f*
  regeneration and, 144–145, 497
  spinal cord, **470**–481
Central pattern generators, **518**, 518*f*
Central process, **397**, 398*t*
Central sleep apnea, **860**
Central sulcus, 438*f*, **439**, 440*f*
Central tendon of perineum, **348**t, 1067
Central thermoreceptors, **965**
Central vein
  of liver lobule, 896*f*, **897**
  of retina, 558*f*, 560*f*, **561**, 561*f*
Centrioles, 62*f*, **89**, 89*f*, 90, 90*f*, 95*t*, 97, 100*f*
Centromere, 98, 100*f*
Centrosome/centrosome matrix, 62*f*, 88, **89**, 89*f*, 100*f*
Centrum (body) of vertebra, 219*f*, **220**, 220*f*, 223*t*
Cephalic (reflex) phase, of gastric secretion, **890**, 890*f*
Cephalic vein, 751*f*, **754**t, 754*f*, 755*f*
Cephalization, **435**
Cerebellar cortex, 436, 437*f*, 454, 455, 455*f*
Cerebellar gray matter, 436, 437*f*, 454
Cerebellar hemispheres, **454**
Cerebellar peduncles, 450*f*, 451, 451*f*, 452, 453*f*, **454**, 455*f*
Cerebellar processing, 454–455
Cerebellar white matter, 437*f*, 454, 455*f*
Cerebellum, **454**–456, 455*f*, 459*t*. *See also under* Cerebellar
  development of, 435, 435*f*, 436*f*
  in equilibrium pathway, **593**, 593*f*, 594
  in motor control, 454–455, 517, **518**–519, 518*f*
Cerebral aqueduct, 435*f*, **436**, 437*f*, 447*f*, 448*f*, 451, 453*f*
Cerebral arterial circle (circle of Willis), 714, 740*f*, **741**t, 741*f*
Cerebral arteries, **740**t, **741**t, 740*f*, 741*f*
Cerebral blood flow, 726, 726*f*, 728
Cerebral cortex, 435*f*, 436, 437*f*, **439**–444, 439*f*, 440*f*, 442*f*, 445*f*, 446*f*, 458*t*
  association areas of. *See* Association areas/cortex
  autonomic regulation by, 547*f*, 548
  lateralization and, **444**
  motor areas of, 440*f*, 441, **441**, 442*f*
  motor control and, 517
  respiratory control and, 857
  sensory areas of, 440*f*, 441, 442–443, 442*f*
  sensory processing at level of, 494, 494*f*, 495
Cerebral dominance, **444**
Cerebral edema, **468**, 728

Cerebral fissures, 438*f*, **439**
Cerebral gray matter, 435*f*, 436, 437*f*, 439, 445*f*, 458*t*. *See also* Cerebral cortex
Cerebral hemispheres (cerebrum), **435**, 435*f*, 436, 436*f*, 438*f*, **439**–447, 440*f*, 442*f*, 445*f*, 446*f*, 458*t*
Cerebral palsy, **483**
Cerebral peduncles, **451**
Cerebral white matter, 435*f*, 437*f*, 439, **444**, 445*f*, 446*f*
Cerebrospinal fluid (CSF), **465**–467, 466*f*, 471
Cerebrovascular accidents (strokes), 441, **468**–469, 760*b*
Cerebrum. *See* Cerebral hemispheres
Cerumen (earwax), 163, **582**
Ceruminous glands, **162**–163, **582**
Cervical branch of facial nerve, 503*t*, 504*t*
Cervical canal, 1063*f*, **1064**
Cervical cancer, **1064**
Cervical cap, for contraception, 1118*b*
Cervical curvature of spine, **218**, 218*f*, 246
Cervical enlargement, 470*f*, **472**, 507*f*
Cervical flexure, 436, 436*f*
Cervical ganglia, superior/middle/inferior, 540*f*, **541**
Cervical glands, 1064, 1066
Cervical ligaments, transverse (cardinal), 1063*f*, **1064**
Cervical lymph nodes, 769*f*
Cervical mucus, 1064
Cervical nerves, 450*f*, 451*f*, 470–471*f*, 507*f*, 509*t*
  muscles supplied by, 338*t*, 340*t*, 352*t*, 509*t*
Cervical plexus, 507*f*, **509**, 509*t*, 509*f*, 515*t*
Cervical plug, 1076
Cervical vertebrae, **218**, 218*f*, 221–222, 221*f*, 222*f*, 223*t*
Cervix, 1061*f*, **1063**–1064, 1063*f*
  cancer of, **1064**
  dilation of in labor, 1114, 1115*f*
Cesarean section (C-section), 1116
C fibers, 495
CFTR (cystic fibrosis transmembrane conductance regulator) protein, 861
cGMP. *See* Cyclic GMP
Chadwick's sign, 1112
Chain ganglia (sympathetic trunk/ paravertebral ganglia), 508*f*, **538**, 538*t*, 538*f*, 539*f*, 540*f*
Chalazion, 555
Chancre, syphilitic, 1080
Channel-linked (ionotropic) receptors, **424**–425, 424*f*
Channel-mediated facilitated diffusion, 70, 70*f*, 73*t*
Channels (protein), 66, 66*f*, **70**, 400, 401*f*. *See also specific type*
  action potential and, 405, 406–407*f*, 408
  channel-linked (ionotropic) receptors and, **424**–425, 424*f*
  diffusion and, 70, 70*f*, 73*t*
  gated, 70, **400**–401, 401*f*
  information transfer at chemical synapse and, 414, 415*f*
  leakage (nongated), 70, **400**
  mechanically gated, **401**
  muscle contraction and, 290, 292*f*, 293*f*, 296*f*
    cardiac muscle, 685, 686, **689**, 689*f*
Charley horse, **386**
Charnley, John, 274*b*
Checkpoints, in cell division, 96*f*, 98
Cheekbones. *See* Zygomatic bones
Cheeks, **876**
Chemical barriers, skin, 163, 783
Chemical bonds, **31**–35, 31*f*, 32*f*, 33*f*, 34*f*, 35*f*, 35*t*. *See also specific type*
  chemical reactions and, 35, 36, 36*f*, 37
  energy stored in, 24
  sacrificial, in bone, 183
Chemical buffer systems, 41, **1027**–1028, 1027*f*, 1028*f*. *See also specific type*

Chemical digestion. *See* Digestion
Chemical energy, **24**
Chemical equations, 35–36
Chemical equilibrium, **37**
Chemical level, structural organization at, **4**, 4*f*
Chemically-gated (ligand-gated) channel(s), 290, 292*f*, **400**, 401*f*, 417, 418*f*
Chemical mediators, inflammatory, 143*f*, 785, 785*t*, 787*f*
Chemical properties, 25
Chemical reactions, **35**–38, 36*f*
  enzymes affecting, 38, 51–52, 52*f*
Chemical senses, 575–580, 595. *See also* Smell; Taste
Chemical signaling, 64–65*f*, 81, 82*f*, 98
Chemical synapses, **413**–414, 415*f*. *See also* Neurotransmitter(s)
Chemical (nonshivering) thermogenesis, **967**
Chemiosmotic processes/chemiosmosis, **936**, 941*f*, 943
Chemistry, **23**–59
  bonds and, **31**–35, 31*f*, 32*f*, 33*f*, 34*f*, 35*f*, 35*t*
  chemical reactions and, **35**–38, 36*f*
  energy and, **24**–25
  matter and, **24**
    in combination (molecules/mixtures), 28–30, 29*f*
    composition of (atoms and elements), 25–28, 25*f*, 26*t*, 27*f*, 28*f*
    living matter. *See* Biochemistry
Chemoreceptors, **490**, 491*f*, **575**, 720, **855**
  in blood pressure control, 719, 720, 723*f*
  in respiratory rate/rhythm control, 854*f*, **855**–856, 855*f*, 856*f*, 1029
  in salivation control, 879
  for taste and smell, 575–580
Chemotactic agents, **786**
Chemotaxis, 652, **786**, 786*f*, 787*f*
Chemotherapy, 145*b*
Chest trauma, **226**
Chewing (mastication), 330*f*, 336*f*, 337*f*, 870*f*, 879, **881**, 904*t*
Cheyne-Stokes breathing, **863**
Chiasmata (crossovers), **1044**, 1045*f*, **1127**–1128, 1127*f*
Chief cells, 886*f*, **887**
Childbirth. *See* Birth
Chiropractic, **248**
Chlamydia, 596, **1080**
Chlamydia trachomatis, 1080
Chloasma, 1112
Chloride, 931, 1014*f*, 1015, 1020*t*, 1026
Chloride shift, **851**–852, 851*f*
Chlorine, 26*t*, 32, 32*f*, 933*t*
Choanae, nasal (posterior nasal apertures), 821*f*, **822**, 823, 823*f*
Choking (tracheal obstruction), 828
Cholecalciferol, **634**, 635*t*
Cholecystitis, **919**
Cholecystokinin, 422*t*, 423, 635*t*, 888*t*, 890*f*, **891**, 900, 901*f*, 961, 961*f*
Cholelithiasis, 898
Cholesterol, 47*f*, 48*f*, 63, 928, 930*t*, 947–948, **955**–959
  blood levels of
    atherosclerosis and, 48, 760*b*, 957, 959
    factors regulating, 957–958
    recommended, 957
  in membranes, 63, 65*f*
  metabolism of, 947*f*, 955–959, 957*f*
  transport of, 956–957, 957*f*
Cholesterol-lowering drugs, 760*b*, 959
Choline acetyltransferase, 423
Cholinergic fibers, **543**
Cholinergic receptors, 543, 544*t*
Chondroblasts, 127*t*, **129**, 134, 135*f*
Chondrocytes, 127*t*, **129**, 134, 135*f*, 136*f*, 137, 174, 274
Chondroitin sulfate, 128, 134
Chondromalacia patellae, **276**

Chordae tendineae, 677*f*, 678*f*, **679**, 679*f*
Chorda tympani, in gustatory pathway, 580
Chorea, in Huntington's disease, 470
Chorion, 1098, 1099, **1100**, 1100*f*, 1101*f*
Chorionic gonadotropin, human (hCG), **1099**, 1099*f*, 1101–1102
Chorionic somatomammotropin, human (hCS/human placental lactogen), 1102, **1113**
Chorionic villi, **1100**–1101, 1100*f*, 1101*f*
Chorionic villus sampling, **1134**, 1134*f*
Choroid, 558*f*, **559**, 560*f*
Choroid plexus, 447*f*, 451*f*, 452, 455*f*, **465**–466, 466*f*
Chromaffin cells, medullary, **627**
Chromatids
  nonsister, **1043**
  sister, 93*f*, 100*f*, **1043**, 1045*f*, 1046*f*
Chromatin, 62*f*, 92*f*, **93**, 93*f*, 96*t*, 98, 100*f*
Chromatophilic substance (Nissl bodies), **394**, 395*f*
Chromium, 26*t*, 933*t*
Chromosomal number, diploid (2*n*) and haploid (*n*), **1043**, 1044
Chromosome(s), **93**, 93*f*, 100*f*. *See also* Heredity
  homologous (homologues), **1043**, 1044, 1125, 1125*f*
  maternal, 1043
  paternal, 1043
  recombinant, **1127**, 1127*f*
  segregation of, independent assortment and, 1126*f*, **1127**
  sex, **1081**, **1125**, 1130, 1130*f*
Chronic bronchitis, 858*f*, **859**
Chronic hypotension, 724
Chronic leukemia, 654
Chronic obstructive pulmonary diseases (COPD), **858**–859, 858*f*
Chronic renal disease, **1001**
Chronotropic factors, positive and negative, 698
Churning food (stomach), 870*f*, 871
Chyle, **769**
Chylomicrons, **916**–917, 916*f*, 946, 956, 957*f*
Chyme, **885**, 891, 892, 892*f*, 893, 893*f*, 903, 905
Chymotrypsin, 899*f*, 899*f*, 913*f*, **915**
Chymotrypsinogen, 899*f*, 899*f*
Cigarette smoking. *See* Smoking
Cilia, **90**–91, 90*f*, 91*f*, 95*t*, 117*f*, 783*t*
  in body defense, 783, 808*f*
  epithelial, **118**, 120*f*, 121*f*
  nasal, 822
  olfactory, **576**, 576*f*
  in trachea, 827*f*
Ciliary body, 558*f*, **559**, 562*f*
Ciliary ganglia, **536**
Ciliary glands, 555
Ciliary muscles, 546*t*, **559**, 562*f*, 565*f*, 566
Ciliary processes, 558*f*, **559**, 562*f*
Ciliary zonule (suspensory ligament), 558*f*, **559**, 562*f*, 565, 565*f*, 566
Ciliated pseudostratified columnar epithelium, 121, 121*f*
Cingulate gyrus, 440*f*, 444, **456**–457, 456*f*
Circadian pigment (melanopsin), 575
Circadian rhythm, sleep-wake cycle as, 463
Circle of Willis (cerebral arterial circle), 714, 740*f*, **741**t, 741*f*
Circuit(s), **427**, 427*f*
  sensory processing at level of, 494–495, 494*f*
Circular fascicle pattern/muscles, **326**, 326*f*. *See also* Sphincter(s)
Circular folds, **902**, 902*f*
Circular layer of smooth muscle, 310, 310*f*
  in alimentary canal muscularis, 873, 873*f*
  in stomach muscularis, 885*f*, 886, 886*f*

Circulation, 714–734. *See also specific type and* Blood vessels; Cardiovascular system; Lymphatic system
blood pressure maintenance and, 718–723, 719*f*, 720*f*, 721*t*, 722*f*, 723*f*
enterohepatic, 897
fetal, 759, 1101*f*, 1107, 1108–1109*f*, 1116
flow/pressure/resistance definitions/ relationships and, 715, 718, 719, 719*f*, 720*f*, 723*f*
monitoring, 716–717, 717*f*
physiology of, 714–734
pulmonary, 671, 671*f*, 680, 681*f*, 729, 735, 736–737*t*, 736*f*, 834
systemic, 671, 671*f*, 680–682, 681*f*
arteries of, 735, 737*t*, 737*f*, 738–739*f*, 738*t*
veins of, 735, 750–751*f*, 750*t*
tissue perfusion and, 725–730, 725*f*, 726*f*, 727*f*, 728*f*
Circulatory pathways, 734–759
Circulatory shock, 545, 724–725
Circumcision, 1050
Circumduction (body movement), 259–260, 259*f*
Circumferential lamellae, 182, 183*f*
Circumflex artery (coronary), 675*f*, 682, 683*f*
Circumflex femoral arteries, lateral and medial, 748*t*, 748*f*, 749*f*
Circumflex humeral arteries, anterior and posterior, 742*t*, 742*f*, 743*f*
Cirrhosis, 898
Cis face, of Golgi apparatus, 85*f*
11-*Cis*-retinal, 569, 570*f*, 571, 571*f*
Cisterna chyli, 768*f*, 769, 769*f*
Cisterns, 84
rough endoplasmic reticulum, 84, 84*f*, 85*f*, 86*f*, 92*f*, 105*f*
terminal, 288, 288*f*, 291, 294–295*f*, 296
Citric acid, 939*f*, 940
Citric acid cycle (Krebs cycle), 935*f*, 936, 937*f*, 939*f*, 940, 944*f*
Cl. *See* Chlorine
Class I MHC proteins, 802*f*, 803, 803*t*
Class II MHC proteins, 802*f*, 803–804, 803*t*
Classical pathway, of complement activation, 788, 789*f*
Claudication, intermittent, 319
Clavicles (collarbones), 200*f*, 227–228, 227*f*, 228*f*, 235*t*
Clavicular notches, 225, 225*f*
Claw hand, 512
Cleavage, 1097, 1097*f*
Cleavage furrow, 98, 101*f*, 1045*f*
Cleavage (tension) lines, 155, 155*f*
Cleft lip, 246*f*, 919
Cleft palate, 246, 246*f*, 861, 919
Climax (sexual)
in female, 1079
in male, 1054
Clitoris, 536*f*, 546*t*, 1061*f*, 1066*f*, 1067
Cloaca, 1006, 1006*f*, 1007, 1082*f*
Cloacal membrane, 918, 918*f*
Clonal deletion, 794
Clonal selection, 792, 804
B lymphocyte, 796, 796*f*
T lymphocyte, 804, 804*f*
Clone, 654, 793
Closed (simple) fracture, 191
Closed comedone (whitehead), 163
Closed (external) reduction of fracture, 191
Close vision, focusing for, 565–567, 565*f*
Clostridium difficile, diarrhea caused by, 909–910
Clot (blood), 143*f*, 144, 658, 658*f*, 659*f*, 660, 660*f*. *See also* Coagulation; Fibrin mesh
Clot retraction, 660
Clotting. *See* Coagulation

Clotting cascade, 659, 661
Clotting factors, 657, 658–659, 659*f*, 661*t*
in inflammation, 785–786, 787*f*
in semen, 1053
Clubfoot, 248
CMP (comprehensive medical panel), 666
CNS. *See* Central nervous system
CO. *See* Carbon monoxide; Cardiac output
Coagulation (blood clotting), 658–660, 660*f*, 661*t*
autonomic effects and, 546*t*
disorders of, 662–663
homeostatic control in, 12, 12*f*
platelets in, 12, 12*f*, 657, 658, 658*f*, 659*f*
semen, 1053
in tissue repair, 143*f*, 144
undesirable, prevention of, 662
Coagulation cascade, 659, 661
Coarctation of aorta, 701, 701*f*
Coated vesicles (pits)/coat proteins, 76*f*, 77, 77*f*, 105*f*
Cobalt, 26*t*, 933*t*
Cocaine abuse, 429*b*
Cocaine- and amphetamine-regulated transcript (CART), in food intake regulation, 960, 961*f*, 962
Coccygeal nerve, 507*t*
Coccygeus muscle, 348*t*, 349*f*
Coccyx, 218, 218*f*, 224, 224*f*, 237*f*, 239*f*
Cochlea, 581*f*, 584*f*, 585, 585*t*
Cochlear duct (scala media), 583, 583*f*, 584*f*, 585, 585*t*, 587*f*
Cochlear hair cells, 584*f*, 585, 588, 589*f*. *See also* Hair cells
Cochlear implant, 594–595, 594*f*
Cochlear nerve, 504*t*, 581*f*, 583*f*, 584*f*, 585, 587*f*
Cochlear nuclei, 453, 453*f*, 589, 589*f*
Coding strand, 102, 102*f*
Codominant alleles, 1129
Codon(s), 103, 104, 106–107*f*, 108*f*
Coelom, 1100*f*, 1105*f*, 1107
Coenzyme(s), 51, 931, 934–935
Coenzyme A, 940, 946, 946*f*
Cofactor(s), 51, 940
Cognition/cognitive function, 458–464, 460*f*, 461*f*, 463*f*
aging affecting, 483–484
cerebellum in, 456
prefrontal cortex and, 440*f*, 443
Coiled (spiral) arteries of uterus, 1065
Coitus interruptus, 1118*b*
Cold exposure, hypothermia and, 967
Cold sores (fever blisters), 169
Colic arteries, left/middle/right, 744*f*, 746*t*, 747*f*
Colitis, ulcerative, 921
Collagen, 49*f*, 51
Collagenase, 165
Collagen fibers, 127*t*, 128, 128*f*
in bone, 136*f*, 137, 182, 182*f*, 183
in cartilage, 134, 135*f*
in dense irregular connective tissue, 133, 133*f*
in dense regular connective tissue, 132*f*, 133
in dermis, 155, 155*f*
in fibrocartilage, 136*f*, 137
in tissue repair, 143*f*, 145
Collarbones. *See* Clavicles
Collateral channels, 714
Collateral (prevertebral) ganglia, 538*t*, 539, 539*f*, 540*f*, 541–542
Collateral ligaments
of elbow (radial and ulnar), 268, 268*f*
of knee (fibular and tibial), 270*f*, 271, 271*f*, 272, 272*f*
Collecting ducts, 979*f*, 980, 981, 981*f*
osmotic gradient and, 996*f*, 998, 999*f*
potassium regulation and, 1025
reabsorptive capability of, 992*f*, 993–994, 993*t*
Collecting lymphatic vessels, 767*f*, 769, 769*f*
Colles' fracture, 233

Colliculi
inferior, 450*f*, 451, 451*f*, 589*f*, 590
superior, 450*f*, 451, 451*f*, 453*f*, 478, 574*f*, 575
Colloid(s), 29*f*, 30, 38
thyroid gland storage/thyroid hormone synthesis and, 617, 617*f*, 619*f*
Colloid osmotic pressure, 731, 732–733*f*, 734, 986, 986*f*
Colon, 869*f*, 906–907, 907*f*. *See also* Large intestine
Colon cancer, 144*b*, 145*f*
Colony-stimulating factors, 654
Color blindness, 569
Colorectal cancer, 144*b*, 145*f*
Color of light, 564, 564*f*
Color vision, 561, 569, 569*t*, 595
Colostrum, 1117
Columnar epithelium, 118*f*, 119
pseudostratified, 121, 121*f*
simple, 120*f*, 121
stratified, 123
Coma, 458, 462
Combination (synthesis) reactions, 36, 36*f*
Comminuted fracture, 192*t*
Commissural fibers, 444, 445*f*, 474
Commissures, 444
Commissures (canthi), medial and lateral, 554, 555*f*
Common carotid arteries, 717*f*, 738*t*, 738*f*, 739*f*, 740*t*, 740*f*, 741*f*, 742*f*, 743*f*
Common fibular (peroneal) nerve, 373*t*, 514, 514*f*, 515, 515*f*
Common hepatic artery, 744*f*, 745*t*, 745*f*
Common hepatic duct, 894, 895*f*, 900*f*
Common iliac arteries, 738*t*, 738*f*, 739*f*, 744*f*, 746*t*, 746*f*, 747*f*, 748*t*, 748*f*, 749*f*
Common iliac veins, 750*t*, 750*f*, 751*f*, 756*f*, 757*f*, 758*t*, 758*f*
Common interosseous artery, 742*t*, 742*f*, 743*f*
Common tendinous (annular) ring, 556, 557*f*
Commotio cordis, 702
Communicating arteries, anterior/ posterior, 740*t*, 740*f*, 741*t*, 741*f*
Communication proteins, 49*f*
Compact bone, 127*t*, 177, 177*f*, 178*f*, 179*f*, 182, 182*f*, 183*f*, 186*f*
Complement/complement system, 785, 785*t*, 787*f*, 788–789, 789*f*, 790*t*, 800–801, 800*f*, 810*t*
Complement activation, 788–789, 789*f*, 800–801, 800*f*, 808*f*, 809*f*
Complementary bases, 54*f*, 55
Complete antigens, 791
Complete blood count (CBC), 666
Complete fracture, 191
Complete proteins, 928–929, 930*t*
Complete (fused) tetanus, 300*f*, 301
Complex carbohydrates, 928, 930*t*. *See also* Starch
Compliance
arterial, blood pressure and, 716
lung, 840
Complicated plaque, 760*b*
Compound(s), 29. *See also specific type*
inorganic, 38–41
mixtures differentiated from, 30
organic, 38, 41–56
Compound (open) fracture, 191
Compound glands, 125, 125*f*
Comprehensive medical panel (CMP), 666
Compression fracture, 192*t*
Computed tomography (computerized axial tomography/CAT/CT), 16*f*, 16*b*, 470
Concave lens, 564, 566*f*
Concentrating circuit, 427*f*
Concentration, 29–30, 38

Concentration gradient, 68
active transport and, 75, 75*f*
diffusion and, 68, 69, 69*f*, 70
osmosis and, 71
resting membrane potential and, 401, 402*f*, 403
Concentric contractions, 302*t*, 303
Conception. *See* Fertilization (conception)
Conceptus, 1092, 1092*f*
Concha(e), nasal, 576*f*
inferior, 202*f*, 203*f*, 212, 214*f*, 217*t*, 821*f*, 822
middle, 203*f*, 210, 210*f*, 214*f*, 216*t*, 821*f*, 822
superior, 210, 214*f*, 216*t*, 576, 821*f*, 822
Concussion, 468
Condensed chromatin, 92*f*, 93
Condoms, 1080, 1118*b*
Conducting (elastic) arteries, 707*f*, 709*t*, 710
Conducting region of neuron, 395*f*, 396, 398*t*
Conducting zone, 824, 828–829, 828*f*
Conduction
heat loss by, 965, 965*f*
nerve impulse, 410. *See also* Action potential, propagation of
Conduction deafness, 594
Conduction system of heart, 685*f*
extrinsic, 688*f*, 689
intrinsic, 686–689, 687*f*, 688*f*
Conduction velocity of action potential, 408*f*, 410–412
Conductors, 400
Condylar (ellipsoid) joints, 263*f*, 264
Condylar process, of mandible, 204*f*, 211, 211*f*, 217*t*
Condyle (bone marking), 180*t*
of femur, 240*f*, 241
occipital, 202, 203*f*, 206*f*, 216*t*
of tibia, 242*f*, 243, 243*f*
Cone(s), 560*f*, 561, 564*f*, 568–570, 568*f*, 569*t*, 571, 573. *See also* Photoreceptors
rods compared with, 569, 569*t*
Cone discs, 568*f*, 569
Cone inner segments, 568*f*, 569
Cone opsins, 569
Cone outer segments, 560*f*, 568*f*, 569
Congenital hypothyroidism, 620
Congenital immunodeficiencies, 811
Congenital muscular torticollis, 386
Congenital thymic aplasia, 814
Congestive heart failure, 700
Conjunctiva, 555, 555*f*, 562*f*
Conjunctival sac, 555, 555*f*
Conjunctivitis, 555
Connective tissue, 116*f*, 126–138, 127*t*, 128*f*, 129*f*, 131–137*f*
blood. *See* Blood
bone (osseous). *See* Bone(s)/skeletal tissue
in cardiac muscle, 312*t*
cartilage. *See* Cartilage
dense (fibrous), 127*t*, 129, 132–134*f*, 133–134
elastic, 127*t*, 129, 134, 134*f*
epithelia supported by, 118
exocrine glands supported by, 125
hair follicle sheath of, 158*f*, 159
loose, 127*t*, 129–133, 131–132*f*
reticular, 127*t*, 132*f*, 133, 770, 770*f*
skeletal muscle sheath of, 281–282, 282*f*, 283*t*, 312*t*
in smooth muscle, 311, 312*t*
Connective tissue proper, 126, 127*t*, 129–134, 131–134*f*
Connexons, 67*f*, 68, 413
Conn's disease (aldosteronism/ hyperaldosteronism), 625, 626*t*, 1037
Conoid tubercle, 228, 228*f*
Consciousness, 462
Consensual light reflex, 573
Constant (C) region, 798*f*, 799

Constipation, **911**, 1113
Constrictor muscles, pharyngeal, **338**t, **339**t, 339f, 882
Contact dermatitis, **169**
  allergic, **813**–814
Contact inhibition, 98
Contact signaling, 81
Continuous capillaries, 711, 711f
Continuous conduction, **411**, 411f
Continuous positive airway pressure (CPAP) device, 860
Contraception, **1118**–1119b
Contractile cardiac muscle cells, 685, 689–690, 689f
Contractile proteins, 49f, 51
Contractile ring, in cytokinesis, 98, 101f
Contractility, muscle, **8**, **280**. See also Muscle contraction
  cardiac muscle, 685, 689–690, 689f, 697f, **698**, 698f
  movement and, 8
  stomach muscles and, 892–893, 892f
Contraction, **289**. See also Muscle contraction
  period of, muscle twitch, **299**, 299f, 300f
  uterine, 1114, 1114f, 1116
Contralateral (term), **441**
Control center, in homeostasis, **10**, 10f, 11f
Contusion, brain, **468**
Conus medullaris, 470f, **471**, 471f
Convection, heat loss by, **965**, 965f
Convergence, in close vision, **567**
Convergent fascicle pattern/muscles, **326**, 326f
Converging circuits, 427, 427f
Convex lens, 564, 565f
Convolutions, brain, 436, 436f, 439
COPD. See Chronic obstructive pulmonary diseases
Copper, 26t, 933t
Copulation, 1042
Coracoacromial ligament, 267f
Coracobrachialis muscle, 351f, **354**t, 355f, **356**t, 356f, 362t
Coracohumeral ligament, **266**, 267f
Coracoid process, 228, 229f
Cord(s), brachial plexus, 510, 510t, 511f
Cordotomy, **484**
Core temperature, **964**, 967
Cornea, 555f, 558f, **559**, 562f
Corneal endothelium, 559, 562f
Corneal epithelium, 559, 562f
Corneoscleral junction, 562, 562f
Corniculate cartilage, **824**, 825f, 826f
Cornua, hyoid bone, 212, 212f
Coronal (frontal) plane/section, **15**, 15f
Coronal suture, 201, 201f, **202**, 204f, 205f
Corona radiata (cerebral projection fibers), **444**, 445f
Corona radiata (oocyte), 1072f, **1073**, 1093, 1094f, 1096f
Coronary arteries, 675f, 676f, **682**–683, 683f, 700, 702, **738**t, 739f
Coronary bypass surgery, 760b
Coronary circulation, 546t, **682**–683, 683f, 729
Coronary sinus, **674**, 676f, 681f, **683**, 683f, 714
Coronary sulcus (atrioventricular groove), **674**
Coronary veins, 683, 683f. See also Cardiac veins
Coronoid fossa, 230f, **231**, 231f
Coronoid process
  of mandible, 204f, **210**–211, 211f, **217**f
  of ulna, 231f, 232f, **233**, 233f
Corpora cavernosa
  of clitoris, 1066f, 1067
  of penis, 1047f, 1051f, **1052**
Corpora quadrigemina, 447t, 448f, **451**, 451f
Cor pulmonale, **702**
Corpus albicans, 1072f, 1074
Corpus callosum, 440f, **444**, 445f, 446f, 447f, 448f

Corpus hemorrhagicum, 1074
Corpus luteum, 1070f, 1072f, 1073, **1074**, 1075f, 1076, 1077f
Corpus spongiosum, 1047f, **1051**, 1051f
Corrugator supercilii muscle, **333**t, 333f, 335f
Cortex
  adrenal. See Adrenal cortex
  cerebellar, 436, 437f, 454, 455, 455f
  cerebral. See Cerebral cortex
  hair, 158f, 159
  lymph node, **772**, 772f
  ovarian, 1061f, 1062
  renal, 976–977, 977f, 978f, 979f, 980f, 982f, 995f
Cortex-medulla junction (kidney), 978, 981f
Corti, organ of. See Spiral organ
Cortical nephrons, 981f, **982**
Cortical radiate arteries, **978**, 978f, 981f, 982
Cortical radiate veins, **978**, 978f, 981f
Cortical reaction, 47t, 48, **623**
Corticospinal (pyramidal) tracts, 452, 453f, 475f, **478**, 479f, 481t
Corticospinal tracts. See Pyramidal tracts
Corticosteroids, 47t, 48, **623**
Corticosterone, **625**
Corticotropic cells, **616**
Corticotropin. See Adrenocorticotropic hormone
Corticotropin-releasing hormone (CRH), 611f, 613t, **616**, 624, 629f, 960, 961f, **1113**
Cortisol, 623f, **625**, 625f, 626t, 955t, 1023
Cortisone, **625**
Costal cartilages, 174, 175f, 225f, 226f
Costal facets, superior/inferior/transverse, 222, 222f, 223t, 226f
Costal groove, 226, 226f
Costal margin, 225f, **226**
Costal surface, of lung, **833**
Co-stimulatory signals/costimulation, **804**–805, 804f, 812
Costocervical trunk, 741t, 741f, 742f, **743**t, 743f
Costovertebral joint, 256t
Cotransport. See Secondary active transport (cotransport)
Cotransport protein, **75**
Cough/cough reflex, 825
Coumadin. See Warfarin
Countercurrent exchanger, **995**, 996f, 998, 998f
Countercurrent multiplier, **995**, 996–997f, 998
Covalent bonds, 32–34, 33f, 34f
Covering and lining epithelium (cutaneous/serous/mucous membranes), 117, 141–142, 141–142f
Coxal bones (hip bones/os coxae), **236**–239, 237f, 241t
Coxal joint. See Hip (coxal) joint
C proteins, 287
"Cradle cap," 163, 163f
Cranial (term), 14t. See also under Superior
Cranial base, 201
Cranial bones. See Cranium (cranial bones)
Cranial cavity, **18**, 18f, 201, 202f, 207f
Cranial fossae, 201, 201f
  anterior/posterior, 201, 201f, **202**, 207f
  middle, 201, 201f, 207f, **208**
Cranial nerves, 391, 391f, 392f, **498**–507, 499f, 500–506t, 526. See also specific nerve
Cranial part, of parasympathetic nervous system, 536, 536f
Cranial rootlets, accessory nerve and, 506t
Cranial sensory ganglia, **507**
Cranial sutures, **201**, 201f, 202, **252**, 252f, 256t. See also specific type
Cranial vault, 201

Craniosacral division of autonomic nervous system. See Parasympathetic (craniosacral) division of autonomic nervous system
Cranium (cranial bones), 200f, **201**, 201f, 202–210, 203f, 204–205f, 207f, 216t, 246, 247f. See also specific bone and under Cranial
C-reactive protein, 760b
Creatine kinase, in CP-ADP reaction, **304**, 304f
Creatine phosphate (CP), **304**, 304f
Creatinine, **1001**
C (constant) region, 798f, **799**
Cremaster muscle, **1048**, 1048f
Crenation, in hypertonic solution, 72, 72f
Crepitus, 274
Crest (bone marking), 180t
CRH. See Corticotropin-releasing hormone
Cribriform foramina, 207f, 210, 210f
Cribriform plates, 207f, **210**, 210f, 214f, **216**t, 821f
Cricoid cartilage, 175f, 823f, **824**, 825f
Cricothyroid ligament, 825f
Cricotracheal ligament, 825f
Crista(e), **83**–84, 83f
Crista ampullaris, 583f, 585, 585t, **592**–593, 592f
Crista galli, 205f, 207f, **210**, 210f, 214f, **216**t
Crista terminalis, 674, 676f
Crohn's disease, 921
Cross bridges/cross bridge cycling, **287**, 291, 291f, 295f, 296–298, 297f, 685
  force of muscle contraction and, 306–307, 307f
Crossed-extensor reflex, **524**, 525f
Crossing over (chiasmata), 1043, **1044**, 1045f, **1127**–1128, 1127f
Cross matching, blood, 665
Cross section (transverse section), **15**, 15f
Crowning, in labor, 1115
Crown-rump measurement, 1103
Crown of tooth, **880**, 880f
Cruciate ligaments, 270f, 271, 271f, 272, 272f
Crura/crus of clitoris, 1066f, 1067
Crura/crus of penis, 1051f, **1052**
Crus cerebri, 450f, 451, 453f
Cryptorchidism, 1049, 1084
Crypts, intestinal, 902f, **903**
Crypts, tonsillar, 774f, **775**
Crystal(s), **32**, 32f
Crystallins, lens, **563**
C-section (cesarean section), 1116
CT. See Computed tomography
Cubital vein, median, 751f, **754**t, 754f, 755f
Cuboidal epithelium, 118f, **119**
  simple, 120f, **121**
  stratified, **123**
Cuboid bone, 241t, **244**, 244f
Cuneiform bones, 241t, **244**, 244f
Cuneiform cartilage, **824**, 825f, 826f
Cupula (ampullary), 592f, **593**
Current, **400**
  decremental, 405, 405f
Curvatures
  spinal, 218, 218f, 220, 220f, 246–247, 247f
  of stomach, greater and lesser, **885**, 885f
Cushingoid signs, 625, 625f
Cushing's disease/Cushing's syndrome, **625**, 625f, 626t
Cushioning function, of water, 38
Cuspids (canines/eyeteeth), 879f, **880**, 880f
Cutaneous branches of intercostal nerve, lateral and anterior, 508f, 515
Cutaneous glands, 161–163, 161f, 162f
Cutaneous membrane. See Skin
Cutaneous nerves, **509**, 526
  cervical, 509, 509f
  femoral, 512, 513f, 513f, 515f
  medial, of arm/forearm, 511f

Cutaneous sensation (receptors/sense organs), 151f, 152f, 154, 155, **164**
Cuticle, hair, 158f, **159**
Cuticle, nail, **160**–161, 160f
CVS. See Chorionic villus sampling
Cyanide poisoning, 943
Cyanocobalamin (vitamin B_{12}), 649, 649f, 650, 889, 931, 932t
Cyanosis, 156, 852
Cyclic AMP, **81**, **425**, 425f, 577, 577f, **604**, 605f, 698, 698f
Cyclic GMP, 423, **425**, **571**, 571f, 572f, 605
Cyclic slow waves, of stomach, 892
Cyclin-dependent kinases (Cdks), 98
Cyclins, 98
Cystic duct, **894**, 895f, 898, 900f
Cystic fibrosis, **861**, **919**, 1129, 1133
Cystic fibrosis transmembrane conductance regulator (CFTR) protein, 861
Cystic veins, **756**t, 756f
Cystitis, 1004
Cystocele, **1007**
Cystoscopy, **1007**
Cytochromes, **940**
Cytokines, 785, 785t, **805**, 805t, 808f, 810t
Cytokinesis, 96f, **98**, 101f, 1045f
Cytology, **2**
Cytoplasm, 62, 62f, 64f, **83**–91, 94–95t. See also specific structure
  division of. See Cytokinesis
Cytoplasmic inclusions, **83**, 95t
Cytoplasmic organelles, 4, 4f, **83**–89, 109. See also specific type
Cytosine (C), 54f, **55**
Cytoskeleton, 62f, 64f, 66f, **88**–89, 88f
Cytosol, 30, **83**
Cytotoxic (type II) hypersensitivity, **813**
Cytotoxic T cells, 795t, 802, 802f, **807**, 807f, 809f, 810t
Cytotrophoblast (cellular trophoblast), 1098f, **1099**, 1100f

DAG. See Diacylglycerol
Dalton's law of partial pressures, **843**, 843f
D antigen, in Rh blood group, 664
Dark adaptation, **573**
Dark potential, 571
Dartos muscle, **1048**, 1048f
Daughter cells, 97, 101f, 1044, 1045f, 1046f, 1047, 1056f, 1057
Daughter chromosomes, 101f
DCT. See Distal convoluted tubule
Dead space, 841–842, 842t
Deafness, 504t, 587, **594**–595, 596
Deamination/oxidative deamination, 948f, 949, 949f
Decarboxylation, in Krebs cycle, 939f, **940**
Decaying primary teeth, 879
Decibels (dB), **586**
Decidua basalis, 1100f, **1101**, 1101f
Decidua capsularis, 1100f, **1101**, 1101f
Deciduous teeth (primary dentition), **879**, 879f
Declarative (fact) memory, 460
Decomposition reactions, **36**, 36f
Decremental current, 405, 405f
Decubitus ulcers (bedsores), **169**
Decussation, 476
  of pyramids, 450f, **452**
Deep (term), 14t. See also under Internal
Deep artery, **742**t, 742f, 743f, **748**t, 749f
Deep (intrinsic) back muscles, 340–343t, 341f, 343f
Deep brain stimulation, 462, 469
Deep femoral artery (deep artery of thigh), **748**t, 748f, 749f
Deep fibular (peroneal) nerve, 514f
Deep (forced) inspiration, muscles used in, 344t, 838

Deep palmar arch, 739f, **742t**, 742f, 743f, **754t**, 754f, 755f
Deep plantar arch, 758f
Deep transverse perineal muscle, 348t, 349f
Deep veins, 754–755f, 754t, 758t, 758f
Deep venous thrombosis, **759**
Defecation, 8, 870f, **871**, 904t, 910–911, 911f
Defecation reflex, **910**, 911f
Defense system. See Body defenses
Defensins, 163, **652**, 783, 784, 822, 879
Defensive proteins, 49f
Defibrillation (heart), 688
Degenerative brain diseases, 469–470, 469f
Degenerative joint disease (osteoarthritis), 273–275
Deglutition. See Swallowing (deglutition)
Degradation, neurotransmitter, 414, 415f
Dehydration, 721, 723f, 998, 998, 999f, **1018**, 1019f
Dehydration synthesis, **42**, 42f, 44, 45, 45f, 48, 49f
Dehydroepiandrosterone (DHEA), 626, 1079
Dehydrogenases, **934**
Delayed-action gene, 1129
Delayed (type IV) hypersensitivity, **813**–814
Deletion, **1135**
Delta waves, **461**, 461f
Deltoid muscle, 330f, 331f, 351f, 353f, **354t**, 355f, 362t
Deltoid tuberosity, 230f, **231**
Dementia, **469**, 484
Demifacets, vertebral, 222
Demyelinating diseases, 412
Denaturation, protein, **51**
Dendrites, 140, 140f, 394, **395**, 395f, 398t
Dendritic cells, 164, **770**, **794**–795, 795f, 802f
    epidermal (Langerhans cells), 152f, **153**, 164
Dendritic spines, 395, 395f
Dendrodendritic synapses, 413
Denosumab, fracture prevention and, 193
Dens of axis, **221**–222, 221f, 222f
Dense bodies, **314**, 314f
Dense (fibrous) connective tissue, 127t, **129**, 132–134f, **133**–134
Dental alveoli (tooth sockets), 211, **217t**, 879, 880f
    articulation of tooth with (gomphosis), **252**–253, 252f, 880
Dental caries (cavities), **881**
Dental formula, **880**
Dental plaque, **881**
Dentate gyrus, 456, 456f
Dentate nuclei, 454
Denticulate ligaments, **472**
Dentin, **880**, 880f
Dentinal tubules, 880f, 881
Dentition, 879–880, 879f. See also Teeth
Deoxyhemoglobin, **647**, **847**, 852
Deoxyribonuclease, 913f
Deoxyribonucleic acid. See DNA
Deoxyribose, 43, 43f, 53t, 54f, 55
Depolarization, 290
    cardiac muscle, 685, 686, 687, 687f, 688, 689, 689f, 690f, 691, 691f
    neuron, **403**–404, 404f, 405, 405f, 406f, 407f, 408, 417, 418f
    photoreceptor, 571–573, 572f
    skeletal muscle, 291f, 293, 293f, 296f
Depressed fracture, 192t
Depression (body movement), **260**, 261f, 350t
Depressions (bone markings), 179, 180t
Depressor anguli oris muscle, 334t, 335f
Depressor labii inferioris muscle, **334t**, 335f
Depth perception, **575**, 595
Dermal macrophages, 164
Dermal papillae, 151f, **154**–155

Dermal ridges, 154, 155f
Dermal vascular plexus, 151f
Dermatitis, **169**, 813–814
Dermatology, **169**
Dermatome (somite), 1105f, **1106**, 1106f
Dermatomes, **516**–517, 516f, 526
Dermcidin, 162, 783
Dermis, 133, 141, 151, 151f, **154**–156, 154f, 155f, 156f, 167
Descending aorta, **738t**, 743f
Descending colon, 869f, **907**, 907f
Descending limb of nephron loop, 979f, 980, **980**, 981f, 992, 993t, 997f, 998
Descending (motor) pathways/tracts, 474, 475f, 478–481, 479f, 481t
Desmin (intermediate) filaments, 287
Desmosomes (anchoring junctions), 67f, **68**, 118, 152f, 153, 684, 684f
Detached retina, 561
Detrusor muscle, **1003**, 1003f
Deuterium, 28f
Development. See Human development
Developmental anatomy, **2**
Developmental dysplasia of the hip (DDH), 247
Deviated septum, **863**
Dextrinase, 913f, **914**
Diabetes insipidus, 612t, **614**, **1007**
Diabetes mellitus, **632**–634, 632f, 633b, 952, 968
    gestational, 1113
    obesity and, 959b, 960
Diacylglycerol, **425**, **605**
Diapedesis, **651**, **786**, 786f, 787f
Diaphragm (contraceptive), 1118b
Diaphragm
    pelvic, 348t, 349f
    respiratory, 344t, 345f, 509, 509f, 820f, 836, 837f, 853, 853f
    slit, renal, 985f, 986
    urogenital, 348t, 349f, 1003f, 1004
Diaphysis, **178**, 179f, 184–185, 185f
Diarrhea, **911**
Diarthroses, **251**
Diastole, 693, 694f
Diastolic pressure, **716**, 716f
DIC. See Disseminated intravascular coagulation
Dicrotic notch, **693**, 694f
Diencephalon, **435**, 435f, 436f, 447–448f, **447**–450, 450–451f, 456, 456f, 458t
Diet. See also Nutrient(s); Nutrition
    cholesterol levels and, 928, 930t
    erythropoiesis and, 649, 649f
    hypertension and, 724
    immune system and, 814
    pregnancy and, 1112–1113
    weight loss and, 959b
Diet drugs, for obesity, 959b
Differential permeability, 64f, **69**. See also Membrane(s), permeability of; Membrane transport
Differential white blood cell count, 652f, **666**
Differentiation (diversity), cell, 61, 61f, **109**
Diffuse endocrine system, 124. See also Endocrine system/glands
Diffuse junctions, 311, 311f
Diffuse lymphoid tissue, **771**
Diffusion, **68**–73, 69f, 70f, 71f, 72f, 73t, 467, **730**, 916
    in capillary exchange, 730
    membrane permeability affecting, 69, 70–72, 70f, 71f
    membrane polarity/resting membrane potential and, 79–81, 80f
    neurotransmitter, 414, 415f
Diffusion tensor MRI, 480, 481f
Digastric muscle, **338t**, 339f
Digestion, 5f, 7f, **8**, 869, 870f, 871f, 904t, **912**–918, 913–916f. See also specific digestive organ and nutrient and Digestive system

after birth, 919
    case study and, 925b
    in esophagus, 870f, 883–884, 884f, 904t
    hydrolysis in, 44, **912**
    in large intestine, 870f, 904t, 910–911, 911f
    mechanical breakdown and, 870f, **871**, 871f, 904t
    mechanical/chemical stimuli in, 875, 875f, 884f
    in mouth, 870f, 881, 904t, 913f
    nervous system affecting, 534, 534f, 536f, 540f, 546t, 550b, 920b
    neural controls in, 875, 875f, 889, 890, 890f, 891
    processes of, 870–871, 870f, 871f
    regulation of, 875, 889–891, 890f
    in small intestine, 870f, 902f, 904t, 905–906, 905f, 912, 913f, 914f, 915f
    smooth muscle/gland effectors and, 875, 875f
    in stomach, 870f, 889, 904t, 913f
    taste and, 580
Digestive cavity, **20**
Digestive enzymes, 904t, 905. See also Pancreatic juice
Digestive system, 5f, 7f, 8, **868**–925, 869f. See also specific organ or structure and Digestion
    accessory organs of, 869, 869f, **870**, 893–900
    autonomic innervation/effects and, 534, 534f, 536f, 540f, 546t, 920b
    blood supply of, 874
    cardiovascular system and, 761b, 920b
    case study and, 925b
    developmental aspects of, 918–919, 918f, 1104, 1104f
    endocrine system and, 637b, 920b
    enteric nervous system and, **543**, **874**–875, 874f, 875f
    functional anatomy of, 875–911. See also specific organ or structure
    histology of, 872–874, 873f
    homeostatic imbalances and, 872, 877, 878, 879, 880, 883, 887–889, 893, 897–898, 898, 903, 906, 909–910, 910, 911, 914–915, 918, 919
    homeostatic relationships of, 5f, 920b
    integumentary system and, 168b, 920b
    lymphatic system/immunity and, 778b, 920b
    muscular system and, 318b, 920b
    neural signals from, food intake regulation and, 961, 961f
    peritoneal relationships and, 871–872, 872f
    physiology/functions/processes of, 7f, 8, 870–871, 870f, 871f, 875, 904t, 912–918, 913f, 914f, 915f, 916f. See also Absorption; Digestion
    pregnancy and, 1113
    reproductive system and, 920b, 1085b
    respiratory system and, 862b, 920b
    skeletal system and, 196b, 920b
    urinary system and, 920b, 1036b
    venous drainage of, 735
Digital arteries, 739f, **742t**, 742f, 743f
Digital subtraction angiography (DSA), 16f, 16b
Digital veins, 751f, 754f, 755f, 758f
Digits. See Finger(s)
Dihydrotestosterone, 1060
Diiodotyrosine (DIT), **619**, 619f
Dilated cardiomyopathy, 700
Dilation, vascular, **709**. See also Vasodilation
Dilation (first) stage of labor, **1115**, 1115f
Dilator pupillae, 559, 559f
Dimers, 165, 799, 799t
Dipeptidases, 913f, 915
Dipeptides, 48, 49f, 913f, 915
Diploë, 178, 178f
Diploid chromosomal number (2n), **1043**, 1044f

Diplopia (double vision), 501t, **557**
Dipole(s), **34**
Direct (fleshy) muscle attachments, **282**
Directional terms in anatomy, **13**, 14t
Direct (pyramidal) pathways, 478, 478, 479f, 481t, 518
Disaccharidases, 49f
Disaccharides, 43f, **44**, 913f, 914, 914f
Discharge zone, 426, 426f
Discriminative touch, 476
Discs, intervertebral. See Intervertebral discs
Dislocation(s)/luxation(s), 266, **273**
Disorders of sex development (DSD), **1081**
Displaced fracture, 191
Displacement (exchange) reactions, **36**, 36f
Disseminated intravascular coagulation (DIC), **662**, 663
Dissociation
    of salts in water/water polarity, 38, 39, 39f
    strong and weak acids and, 41, 1027, 1027f
Distal (term), 14t
Distal convoluted tubule, 979f, **980**–981, 980f, 981f, 993–994, 993t, 1025. See also Renal tubules
Distal phalanx
    finger, 234f, 235f, 236
    toe, 241t, 244f
Distal radioulnar joint, 232f, 256t
Distant vision, focusing for, 565, 565f
Distensibility (arterial), blood pressure and, 716
Distributing (muscular) arteries, 707f, 709t, **710**
Disuse atrophy, 310
Diuresis, 613, 994, 1000
Diuretics, 700, **762**, 958, **1000**
Diverging circuits, 427, 427f
Diversity
    antigen receptor/antibody, 793–794
    cellular, 61, 61f
Diverticula, **910**
Diverticulitis, **910**
Diverticulosis, **910**
Division(s), brachial plexus, 510, 511f
DMD. See Duchenne muscular dystrophy
DNA (deoxyribonucleic acid), **53**–55, 53t, 54f. See also Genome; Heredity
    in chromatin/chromosomes, 93, 93f
    digestion/absorption of, 913t, 917
    linker, 93f
    miscellaneous roles of, 108
    mitochondrial, 83f, 84, 1133
    mutations in, **110**, 144b
    in protein synthesis, 53–54, 98–108, 99f, 102f, 106f, 108f. See also Transduction; Translation
    replication of, 97–98, 97f
    UV radiation and, 165
DNA fingerprinting, 53
DNA polymerase, **97**, 97f
DNA probes, 1133
DNA-RNA hybrid, 102f
Domains, cortical, 439
Dominance, cerebral, **444**
Dominant allele, **1126**, 1129t
Dominant follicle, 1070f, **1071**, 1074, 1077f
Dominant-recessive inheritance, **1128**–1130, 1128f, 1129t, 1130f
Dominant traits, 1129, 1129t
L-Dopa, for Parkinson's disease, 469
Dopamine, 421t, 423, 429b, 452, 469, 470, 616
Dorsal (term), 14t. See also under Posterior
Dorsal body cavity, 17, **18**, 18f
Dorsal column-medial lemniscal pathway, **476**, 477f, 480t
Dorsal (posterior) funiculi, 473f, **474**, **476**

Dorsal (posterior) horns, **472**, 473f, 474f, 474t
Dorsal interossei muscles
  of foot, 382t, 383f
  of hand, 365f, **366**t
Dorsalis pedis artery, 717f, **748**t, 748f, 749f
Dorsalis pedis vein, **758**t, 758f
Dorsal (posterior) median sulcus, 451f, 471f, **472**, 473f
Dorsal mesentery, 872, 872f
Dorsal metatarsal arteries, **748**t, 748f, 749f
Dorsal metatarsal veins, 751f, 758f
Dorsal motor nuclei, 453f, 537
Dorsal ramus/rami, **508**, 508f, 509, 515–516
Dorsal respiratory group, **853**, 853f. See also Medullary respiratory centers
Dorsal root(s), 473f, **474**, 474f, **508**, 508f
Dorsal root ganglion, 471f, 473f, **474**, 474f, 482, 482f, 497, 508f
Dorsal rootlets, 473f, 474, 508f
Dorsal scapular nerve, 352t, 510t, 511f
Dorsal spinocerebellar tract, 475f, 477f, **478**, 480t
Dorsal venous arch, foot, 751f, **758**t, 758f
Dorsal venous network, hand, 754t
Dorsal white column/funiculus, 473f, **474**, 475f, **476**, 480t
Dorsiflexion (body movement), **260**, 261f, 384t, 385f
Dorsum of foot, muscles on, 375f, 376f, **380**t
Dorsum nasi, 821, 821f
Double covalent bond, 33, 33f
Double helix, DNA, 54f, **55**, 93f
Double recognition, 804, 804f
Double sugars (disaccharides), 43f, **44**, 913f, 914
Doublets (microtubule), 90, 90f
Double vision (diplopia), 501t, **557**
Down-regulation, **608**
Down syndrome, **1135**
DRG. See Dorsal respiratory group
Drowsiness, 462
Drug abuse, 429b
Dry mouth (xerostomia), 878, 1017, 1017f
Dual innervation, **534**, 534f, 545
Duchenne muscular dystrophy, 287f, **287**
Duct(s)
  glandular, 123, 123f, 125, 125f
  reproductive. See Reproductive duct system
Duct of epididymis, 1049f, **1050**
Ductless glands, 123, 602. See also Endocrine system/glands
Ductules, efferent, 1049, 1049f
Ductus arteriosus, **700**, 701, 701f, 759, **1107**, 1107t, 1108f, 1109f, 1116
Ductus (vas) deferens, 1047, 1047f, 1048f, 1049f, **1050**, 1051f, 1082f
  resection of (vasectomy), **1050**, 1118b, 1119b
Ductus venosus, 759, **1107**, 1107t, 1108f, 1109f, 1116
Duodenal glands, 902f, **903**
Duodenal papilla, major, **900**, 900f, 901
Duodenal ulcers, 903
Duodenum, 635t, 869f, 893, 893f, 900f, **901**, 901f
Dural septa, **464**–465, 465f
Dural venous sinuses, **464**, 465f, 714, 735, 750f, 751f, **752**t, 753f
Dura mater
  of brain, **464**–465, 464f, 465f, 466f
  spinal, 470f, **471**, 471f, 473f
Dwarfism, 188, 612t, 615f, **616**
Dyads, 1044, 1045f
Dynamic equilibrium, 590
Dynein(s)/dynein arms, 89, 90, 90f, 396
Dynorphin, 422t, **423**
Dysarthria, **526**
Dyslexia, **484**
Dysmenorrhea, **1086**
Dysphagia, **921**

Dysplasia, **110**, 247
Dyspnea, **858**, 1113
Dysreflexia, autonomic, 548
Dystocia, 1116
Dystonia, **526**
Dystrophin, **287**
Dysuria, 1004

e⁻. See Electron(s)
Ear, 580–595, 581f, 596. See also specific region and Equilibrium/orientation; Hearing
  abnormalities/disorders of, 594–595, 596
  developmental aspects of, 596
  external (outer), 137, 174, 175f, 208, 581f, **582**
  internal (inner), 581f, **583**–585, 583f, 584f, 585t, 587, 587f, 596
  middle, **20**, 208, 581f, **582**–583, 582f, 596
Earache (otalgia), 596
Ear canal, auditory/external (external acoustic meatus), 204f, 206f, **208**, 208f, **216**t, 581f, **582**, 596
Eardrum (tympanic membrane), 581f, **582**, 582f, 587, 587f, 594, 596
Earlobes (lobule), 581f, **582**
Earwax (cerumen), 163, **582**
Eating. See Ingestion (food intake)
Eccentric contraction, **303**
E-C coupling. See Excitation-contraction coupling
Ecchymoses, 157
Eccrine (merocrine) sweat glands, 151f, **161**–162, 161f, 162f
ECF. See Extracellular fluid
ECG. See Electrocardiograph
Eclampsia, **1114**
Ecstasy abuse, 429b
Ectoderm, **146**, 146f, 1100f, **1103**, 1103f, 1105f
  differentiation of, 482, 1104–1106, 1104f, 1105f, 1106f
Ectopic focus, **688**
Ectopic pregnancy, 1063, **1120**
Eczema, **169**, 814
Edema, 71, 733f, **734**, 734f, **1019**
  cerebral, **468**, 728
  in inflammation, 786, 787f
  lymphatic vessels and (lymphedema), 734, 770
Edges, in visual processing, 575
Edinger-Westphal (accessory oculomotor) nuclei, 536
EEG. See Electroencephalogram
Effacement, in labor, 1115
Effector
  in homeostasis, **10**, 10f, 11f
  in reflex arc, 426f, 427, 519, 519f, 543, 543f
Effector cells, 655f, **793**, 793f, 795t, 802, 802f, 805–810, 806f, 807f, 809f
  plasma cells, 653, 655f, **770**, 795t, **796**–797, 796f, 809f, 810t
  T cells, 655f, 802, 802f, 805–810, 806f, 807f, 809f
Effector organs, 391, 532, 533f
Effector regions (antibody), 799
Efferent arteriole, glomerular, 978f, 981f, **982**, 982f
Efferent (motor) division of peripheral nervous system, **392**, 392f, 497
Efferent ductules, 1049, 1049f, 1082f
Efferent fibers, α and γ, 520f, **521**
Efferent lymphatic vessels, **772**, 772f
Efferent nerves/neurons/tracts. See Motor (efferent) nerves; Motor (efferent) neurons; Motor (descending) pathways/tracts
Efferent pathway, 10, 10f, 11f, 532–533, 533f
Effort, in lever system, **327**–328, 327–328f, 329f

Eggs. See Ova (eggs)
Eicosanoid(s), 47t, **48**, **603**
Eights, rule of (octet rule), **31**–32
Ejaculation, 1050, 1053, **1054**, 1093
Ejaculatory duct, 1047, 1047f, **1050**, 1051f
Elastic (conducting) arteries, 707f, 709t, **710**
Elastic cartilage, 127t, **134**–137, 135f, **174**, 175f
Elastic connective tissue, 127t, 131, **134**, 134f
Elastic fibers, 127t, **128**, 128f, 134f
  bronchial, 830f
  in cartilage, 134, 135f, 174
  in dermis, 155, 156
  in vocal folds, 825
Elastic (titin) filaments, 283t, 284, 285f, 287
Elasticity, muscle cell, **280**
Elastic membrane, arterial, 708f, 710
Elastin, 128
Elbow, student's (olecranon bursitis), 273
Elbow joint, 231f, 233, 256t, 268, 268f
  muscles crossing/controlling movement of, 355–356f, 357t, 362t, 363f
Electrical energy, **24**
Electrical gradient, 401, 402f
Electrically coupled neurons, 413
Electrical synapses, **413**
Electrically polarized cells, **79**–80
Electricity, basic principles of, 400–401, 401f
Electrocardiograph/electrocardiogram (ECG), **690**–692, 690f, 691f, 692f, 695f
Electrochemical gradient, 74f, **75**, 80–81, **401**
Electrochemical proton (H₊) gradient, 936f, 941f, **942**–943
Electroencephalogram/electroencephalography (EEG), **461**, 461f, 462–463, 463f
Electrolyte(s), **39**, 644t, 917, **1013**–1015, 1014f
Electrolyte balance, **1019**–1026, 1020t, 1021t, 1022f, 1024f, 1035–1037
Electrolyte imbalances, 699, 1020t, 1021
Electromagnetic radiation (radiant energy), **24**, **563**
Electromagnetic spectrum, 24
Electromyography, **386**
Electron(s), 25f, **26**, 31–32, 31f
Electron acceptor/donor, 32, 37
Electron cloud, 25f, 27
Electronegativity, **34**
Electroneutrality, **41**
Electron microscopy, 117, 117f
Electron sharing, in covalent bonds, 32–34
Electron shells (energy levels), **31**, 31f
Electron transport chain, 935f, 936f, 937f, **940**–943, 942f, 944f
Electropositive atoms, **34**
Elements, **25**–28, 26t
  atomic structure and, 25–27, 25f, 27f
  chemically inert/reactive, 31, 31f
  identifying, 27–28
Elephantiasis, **777**
Elevation (body movement), **260**, 261f, 350t
Ellipsoid (condylar) joints, 263f, 264
Elongation, 102, 102f, 104, 107f
EM. See Electron microscopy
Embolism/embolus/emboli, **662**, 863
Embryo, **1092**, 1092f, 1100f, 1110f
  blood supply for (fetal circulation), 759, 1101f, 1107, 1108–1109f, 1116
  cleavage/blastocyst formation and, 1097–1098, 1097f
  folding of, 1104, 1104f
  germ layers, **146**, 146f
  implantation of, 1097f, 1098f, **1099**
  sexually indifferent stage of, **1083**, 1083f
  testosterone in, 1060, 1060f
Embryoblast, 1097f, **1098**

Embryology, **2**. See also Embryo; Fetus
Embryonic disc, 1098, 1098f, 1100f, **1102**, 1103f, 1104f
Embryonic period, 1092, 1092f, 1097–1110. See also specific aspect
Emergency contraceptive pills, 1118b, 1119b
Emesis (vomiting)/emetic center, **893**
Emetic center, **893**
Emmetropic eye, **565**, 566f
Emotional memory, 460
Emotional-visceral brain, 456, 456f. See also Limbic system
Emotions
  hypothalamus and, 449
  limbic association area/limbic system and, 444, 456–457
  memory processing and state of, 460
Emphysema, 845, 845f, **858**–859, 858f
Emulsion(s), 30, 913f, 916, 916f
Enamel (tooth), **880**, 880f
Encapsulated nerve endings, **492**–493, 492t. See also specific type
Encephalitis, 465
Encephalopathy, **484**
End diastolic volume, 693, 695f, **697**, 697f
Endemic goiter, 620
Endergonic reactions, **37**
Endocannabinoids, 422t, **424**
Endocarditis, **702**
Endocardium, 673f, **674**, 677f
Endochondral bones, **184**
Endochondral ossification, **184**–185, 185f
Endocrine system/glands, 6f, **123**–124, 123f, 601–641, **602**, 602t, 602f. See also specific gland and Hormone(s)
  autonomic nervous system and, 550b
  cardiovascular system and, 637b, 761b
  case study and, 641b
  developmental aspects of, 636
  diffuse, 124
  digestive system and, 637b, 920b
  environmental pollutants affecting, 636
  functions of, 6f
  homeostatic imbalance in, 614, 615–616, 617, 620–621, 622, 625, 626, 627, 632–634
  homeostatic relationships of, 10, 637b
  hypothalamus in regulation of, 449
  integumentary system and, 168b, 637b
  lymphatic system/immunity and, 637b, 778b
  muscular system and, 318b, 637b
  nervous system and, 637b
  reproductive system and, 637b, 1085b
  respiratory system and, 637b, 862b
  skeletal system and, 196b, 637b
  stimuli activating, 607–608, 607f
  urinary system and, 637b, 1036b
Endocrinology, **602**
Endocytosis, 76–77, 76f, 77f, 78t, 619f, 620
Endoderm, **146**, 146f, 1100f, **1102**, 1103, 1103f, 1104, 1104f, 1105f, 1106f
Endogenous antigen, **803**, 803t
Endolymph, **583**, 584f, 585, 592f, 593, 595
Endomembrane system, **87**, 87f
Endometrial cancer, **1086**
Endometriosis, **1086**
Endometrium, 1063f, **1065**, 1065f, 1076–1078, 1077f
Endomysium, **282**, 282f, 283t, 311, 312t, 684
Endoneurium, **496**, 497f
Endoplasmic reticulum, 84–85, 84f
  rough, 62f, 84, 84f, **85**, 85f, 86f, 87f, 94t, 105, 105f
  smooth, 62f, 84f, **85**, 87f, 94t
Endoplasmic reticulum signal sequence, **105**, 105f
Endorphins, 422t, **423**, 495
Endoscopy, **1086**
Endosome, 76f, 77
Endosteum, **178**, 179f, 183f

Endothelial tissue/endothelium, **121**
  blood vessel, 708*f*, **709**, 709*t*, 760*b*
  corneal, 559, 562*f*
Endothelins, **726**, 727*f*
Endotracheal tube, **863**
End plate potential, 291, 291*f*, 292*f*, **293**, 293*f*, 517
End systolic volume, 693, 695*f*, **697**, 697*f*
Endurance exercise, **309**. *See also* Exercise
Energy, **24**–25
  activation, **52**, 52*f*, 938
  for active transport, 73–79, 74*f*, 75*f*, 78*t*
  from ATP, 8, 24, 55–56, 56*f*
  chemical bonds and, 24
  in chemical reactions, 37
  diffusion and, 68, 73*t*
  enzyme action and, 52, 52*f*
  intake-output balance and, 960–968
  for muscle contraction, 303–306, 304*f*, 305*f*, 313*t*
  in cardiac muscle, 685*t*, 686
  in smooth muscle, 314, 315*f*
Energy balance, 960–968
Energy conversions, 24–25
Energy intake, **960**
Energy investment phase of glycolysis, 938, 938*f*
Energy levels (electron shells), **31**, 31*f*
Energy output, **960**
Engagement, in labor, 1115, 1115*f*
Enkephalins, 422*t*, **423**, 495
Enteric bacteria, **909**–910, 958*b*
Enteric (intestinal) gastrin, 888*f*, 890*f*, **891**. *See also* Gastrin
Enteric nervous system/plexuses, **543**, 874–875, 874*f*, 875*f*
Enteric neurons, 543, **874**
Enteric pacemaker cells, 892
Enteritis, **921**
Enterocytes (absorptive cells), 902*f*, **903**, 903*f*
Enteroendocrine cells, 634, 886*f*, **887**, 901*f*, 902*f*, 903
Enterogastric reflex, 890*f*, **891**, 893, 893*f*, 905
Enterogastrones, 890*f*, **891**, 893, 893*f*, 900, 905
Enterohepatic circulation, **897**, 897*f*
Enterokinase. *See* Enteropeptidase
Enteropathy, gluten-sensitive (celiac disease), 918
Enteropeptidase (enterokinase), **899**, 899*f*
Enucleation, **596**
Enuresis, nocturnal, **1008**
Environmental pollutants, effects of, 636, 958*b*
Enzyme(s)/enzyme activity, 38, 49*f*, **51**–53, 52*f*, 53*f*, 66, 66*f*, 414, 415*f*
Enzyme inhibitor, 52
Enzyme-substrate complex, 52, 53*f*
Eosinophils, 645*f*, **652**, 652*f*, 653*f*, 655*f*, 656*t*
Ependymal cells, **393**–394, 393*f*, 436, 466*f*, 467
Ephedra-containing supplements, for weight loss, 959*b*
Ephrin, 428
Epiblast, 1100*f*, 1102, 1103*f*, 1106*f*
Epicanthic fold, 555
Epicardium (visceral layer of serous pericardium), 19, 19*f*, 142*f*, **673**, 673*f*, 677*f*
Epicondyle (bone marking), 180*t*, 230*f*, **231**, 231*f*, 240*f*, **241**
Epicranial aponeurosis, 333*f*, 335*f*
Epicranius (occipitofrontalis) muscle, 330*f*, 331*f*, **333***t*, 333*f*, 335*f*
Epidermal dendritic cells (Langerhans cells), 152*f*, **153**
Epidermal derivatives. *See* Appendages:skin
Epidermal ridges, 154, 155*f*

Epidermis, 123, 141, 151, 151*f*, **152**–154, 152*f*, 154*f*
  in body defense, 164, 783, 783*t*
  development of, 167
  hormones produced by, 635*t*
  metabolic functions of, 164
Epidermolysis bullosa, **169**
Epididymis, 1047, 1047*f*, 1048*f*, 1049, 1049*f*, **1050**, 1051*f*, 1082*f*
Epidural space, **471**, 473*f*
Epigastric region, 20, 20*f*
Epigenetic marks, 1132–1133
Epiglottis, 137, 174, 175*f*, 823*f*, **825**, 825*f*, 826*f*, 876*f*
Epilepsy/epileptic seizures, **462**
Epimolysis bullosa, **169**
Epimysium, **282**, 282*f*, 283*t*, 312*t*
Epinephrine, 423, 544*t*, 626*t*, **627**, 628*f*, **721**
  adrenal secretion of, 542, 542*f*, 544*t*, 623*f*, 626*t*, 627, 628*f*
  blood pressure and, 721, 721*t*, 723*f*
  cardiac contractility affected by, 698
  in food intake regulation, 961, 961*f*
  heart rate affected by, 699
  metabolism and, 955*t*, 963
  stress response/sympathetic stimulation and, 627, 628*f*
Epineurium, **496**, 497*f*
Epiphyseal arteries and veins, 178
Epiphyseal fracture, 192*t*
Epiphyseal line, **178**, 179*f*
Epiphyseal plates, 134, **178**, 185, 185*f*, 186, 187, 187*f*, 253, 253*f*
Epiphyses, **178**, 179*f*, 185, 185*f*
Epiploic appendages, **906**, 907*f*
Episiotomy, 1115
Epistaxis (nosebleed), **863**
Epithalamus, 435, 435*f*, 447*f*, 448*f*, **450**
Epithelial cells, taste bud, **578**, 579*f*. *See also* Epithelial tissue/epithelium
Epithelial root sheath, 158*f*, **159**
Epithelial tactile complexes (Merkel cells and discs), 491*f*, **492**
Epithelial tissue/epithelium, 116*f*, **117**–126, 118*f*, 119–126*f*. *See also specific type*
  of alimentary canal mucosa, **873**, 873*f*
  bronchial, 828
  corneal, 559, 562*f*
  covering/lining (cutaneous/serous/mucous membranes), 117, 141–142, 141–142*f*
  ectoderm and endoderm as, 1103
  germinal, of ovary, 1061*f*, 1062
  of larynx, 826
  of lens of eye, 562*f*, **563**
  olfactory, **575**–576, 576*f*, 821*f*, **822**
  regeneration of, 118
  of small intestine, 903
  of stomach, 886–887, 886*f*
  of ureters, 1002, 1002*f*
Epitympanic recess, **582**, 582*f*
EPO. *See* Erythropoietin
EPOC. *See* Excess postexercise oxygen consumption
Eponychium (nail cuticle), **160**–161, 160*f*
EPSPs. *See* Excitatory postsynaptic potentials
Epstein-Barr virus, mononucleosis caused by, 654
Equational division (meiosis II), **1044**–1047, 1045*f*, 1046*f*
Equator (spindle), 101*f*, 1044, 1045*f*
Equilibrium (chemical), **37**
Equilibrium/orientation, 443, 590–595, 591*f*, 592*f*, 596. *See also* Ear
Equilibrium (vestibular) cortex, 440*f*, 443
Equilibrium pathway, 593–594, 593*f*
ER. *See* Endoplasmic reticulum
Erectile dysfunction (ED), **1054**
Erectile tissue/bodies, 1051, 1051*f*, 1066*f*, 1067
Erection, 1051, 1053, **1054**
Erector spinae (sacrospinalis) muscles, **340***t*, **342***t*, 343*f*

ERV. *See* Expiratory reserve volume
Erythema, 157
Erythroblastosis fetalis, **665**
Erythroblasts, basophilic/orthochromatic/polychromatic, **647**, 647*f*
Erythrocytes (red blood cells), 127*t*, 137, 137*f*, 643, 643*f*, 644–651, **645**, 645–649*f*, 656*t*
  antigens of, blood groups and, 663–664, 811
  disorders of, 650–651, 651*f*
  functions of, 646–647, 646*f*, 656*t*
  hemoglobin in, **646**–647, 847, 851*f*
  life cycle of, 649, 649*f*, 656*t*
  production of (erythropoiesis), **647**–649, 647*f*, 648*f*, 649*f*, 656*t*
  sickled, 650–651, 651*f*, 1129
  structure of, 645–646, 645*f*, 656*t*
  transfusion of, 663–666
  in urine, 1001*t*
Erythropoiesis, **647**–649, 647*f*, 648*f*, 649*f*, 656*t*
Erythropoietin, 634, 635*t*, **648**–649, 648*f*, 649*f*, 650, 975
Eschar, 167
E site, 104, 106*f*, 107*f*
Esophageal arteries, 738*f*, 743*t*
Esophageal glands, 883
Esophageal hiatus, **882**
Esophageal plexus, 537
Esophageal ulcers, 883
Esophagitis, 883
Esophagus, 823*f*, 824, 869*f*, 876*f*, **882**–883, 882*f*
  Barrett's, **919**
  digestive processes occurring in, 870*f*, 883–884, 884*f*, 904*t*
Essential amino acids, 929, 929*f*
Essential fatty acids, 928
Essential (primary) hypertension, **724**
Essential nutrients, **927**
Estradiol, 1060, 1079*t*
Estriol, 1060
Estrogen(s), 626, 626*t*, **634**, **1022**, **1060**, 1078, 1079*t*
  adrenal, 626, 626*t*
  bone growth and, 188
  labor initiation and, 1114, 1114*f*
  in menopause, 1084–1086
  osteoporosis and, 193
  in ovarian cycle, 1074–1076, 1075*f*, 1079*t*
  placental/during pregnancy, 1099*f*, 1102
  prolactin levels and, 617, 1116–1117
  in sodium regulation, **1022**
  in uterine (menstrual) cycle, 1076, 1077*f*, 1078, 1079*t*
Estrogen (hormone) replacement therapy, 1086
Estrone, 1060
Ethmoidal air cells (sinus), 202*f*, **210**, 210*f*, 215*f*
Ethmoid bone, 202*f*, 203*f*, 204*f*, 205*f*, 207*f*, **210**, 210*f*, 214*f*, **216***t*
Eupnea, **853**
Eustachian tube. *See* Pharyngotympanic (auditory) tube
Evaporation, heat loss by, 164, **965**, 965*f*, 967
Eversion (body movement), **260**, 261*f*, 384*t*, 385*f*
Excess postexercise oxygen consumption (EPOC), **306**
Exchange (displacement) reactions, **36**, 36*f*
Exchange transfusions, 665, **666**
Exchange vessels, in capillary bed, 712, 712*f*
Excitability. *See* Responsiveness (excitability)
Excitation-contraction coupling, 291, 291*f*, 294–295*f*, **296**, 314, 315*f*, 685
Excitatory neurotransmitters, 421–422*t*, 424

Excitatory postsynaptic potentials (EPSPs), **417**, 417*t*, 418*f*, 419*f*, 420, 572*f*, 573
Excitatory synapses, 417
Excitotoxin, in cerebrovascular accident, 469
Excreta, 8
Excretion, 7*f*, **8**, 165
Excretory system. *See* Urinary system
Executive area, in prefrontal cortex, 440*f*, 443
Exercise
  active hyperemia and, **728**, 728*f*
  blood flow affected by, 726, 726*f*, 728, 728*f*
  cardiac output affected by, 697, 697*f*
  energy systems used during, 305–306, 305*f*
  heart affected by, 699, 702, 729
  joint health and, 276
  muscles affected by, 309–310, 317
  for osteoporosis, 193, 276
  respiratory adjustments during, 857
Exergonic reactions, 37
Exertion-induced heat exhaustion, 967
Exhalation. *See* Expiration
Exocrine glands, 123, 123*f*, **124**–126, 124*f*, 125*f*, 602
Exocytosis, 62*f*, **77**–79, 78*t*, 79*f*, 86*f*, 414, 415*f*, 517
Exogenous antigens, **803**, 803*t*
Exons, **99**, 103
Exophthalmos, **596**, 620*f*, 621
Exosomes, 145*b*
Expiration, 819*f*, **835**, 837*f*, 838, 838*f*. *See also* Breathing
  muscles used in, 344*t*, 345*f*, 837*f*, 838
Expiratory reserve volume (ERV), **840**, 841*f*
Expiratory volume, forced (FEV), **842**
Expulsion (second) stage of labor, 1115*f*, 1116
Extended chromatin, 93
Extensibility, muscle cell, **280**
Extension (body movement), 258*f*, **259**, 259*f*, 325*f*
  of lower limb, 259*f*, 367*t*, 384*t*, 385*f*
  of upper limb, 259*f*, 354*t*, 362*t*, 363*f*, 365*t*
Extensor carpi radialis brevis muscle, **360***t*, 361*f*, 362*t*
Extensor carpi radialis longus muscle, 331*f*, 359*f*, **360***t*, 361*f*, 362*t*
Extensor carpi ulnaris muscle, 331*f*, **360***t*, 361*f*, 362*t*
Extensor digiti minimi, 361*f*
Extensor digitorum muscles, 331*f*, **360***t*, 361*f*, 362*t*
  brevis, 375*f*, 376*f*, **380***t*
  longus, 330*f*, **374***t*, 375*f*, 376*f*, 384*t*
Extensor hallucis brevis muscle, 375*f*, 376*f*
Extensor hallucis longus muscle, **374***t*, 375*f*, 376*f*, 384*t*
Extensor indicis muscle, **361***t*, 361*f*, 362*t*
Extensor muscles
  of forearm, 357*t*, 360*t*, 362*t*, 363*f*
  of leg, 367*t*, 374*t*
  of thigh, 367*t*
  of trunk/back, 340–343*t*, 341*f*, 343*f*
Extensor pollicis brevis muscle, **360***t*, 361*f*, 362*t*
Extensor pollicis longus muscle, **360***t*, 361*f*, 362*t*
Extensor retinacula, **358***t*, 360*t*, **374***t*, 375*f*, 376*f*
External (term), 14*t*. *See also under* Superficial
External acoustic meatus (external ear/auditory canal), 204*f*, 206*f*, **208**, 208*f*, 216*t*, 581*f*, **582**, 596
External anal sphincter, 348*t*, 349*f*, **907**, 907*f*
External carotid artery, 738*f*, 739*f*, **740***t*, 740*f*, 741*f*

External (outer) ear, 137, 174, 175*f*, 208, 581*f*, **582**
External genitalia
  embryologic/fetal development of, 1083–1084, 1083*f*
  in female, **1060**, 1066*f*, **1067**
  in male (penis/scrotum), 1047, 1047*f*, **1048**, 1048*f*, **1050**–1052, 1051*f*, 1052*f*
External iliac arteries, 739*f*, **748***t*, 748*f*, 749*f*
External iliac vein, 751*f*, 756*f*, 757*f*, **758***t*, 758*f*
External intercostal muscles, **344***t*, 345*f*, 836–837, 837*f*, 853*f*
External jugular vein, 750*f*, 751*f*, **752***t*, 752*f*, 753*f*, 754*f*, 755*f*
External nose, 821, 821*f*, 824*f*
External oblique muscle, 330*f*, **346***t*, 347*f*
External occipital crest, 202–203, 203*f*, 206*f*, **216***t*
External occipital protuberance, **202**, 203*f*, 205*f*, 206*f*, **216***t*
External os, 1063*f*, 1064
External (closed) reduction of fracture, 191
External respiration, 819*f*, 842, 844–846, 844*f*, 845*f*, 846*f*
External root sheath, 159
External spermatic fascia, 1048*f*
External strabismus, 501*t*
External urethral orifice, 1003*f*, **1004**, 1047*f*, 1050, 1051*f*, 1061*f*, 1066*f*
External urethral sphincter, 348*t*, 349*f*, 1003*f*, **1004**
Exteroceptors, 164, **490**, 491*f*, 492*t*
Extracapsular ligaments, **254**, 271
Extracellular fluid, 8, **62**, 64*f*, 1013, 1013*f*
  developmental changes in, 1035
  electrolytes in, 1014*f*, 1015
  gas/nutrient/water/waste exchanges and, 1015, 1015*f*
  intracellular fluid compared with, 1014*f*, 1015
  resting membrane potential and, 401–403, 402*f*
Extracellular fluid compartment, **1013**, 1013*f*
Extracellular materials, **62**–63
Extracellular matrix, **63**, 66*f*, **126**–127, 127, 128*f*, 174
Extracellular pathogens, 801
Extraembryonic membranes, 1100*f*, **1102**
Extraembryonic mesoderm, 1099–1100, 1100*f*
Extrafusal muscle fibers, **520**, 520*f*, 521*f*
Extraglomerular mesangial cells, 983, 983*f*
Extranuclear (mitochondrial) inheritance, 1131, 1133
Extrapyramidal (indirect/multineuronal) pathways, 478–480, 479*f*, 481*t*, 518
Extrapyramidal system, 478
Extrasystole (premature contraction), **688**
Extrauterine life, infant adjustment to, 1116
Extremities. *See* Lower limbs; Upper limbs
Extrinsic conduction system of heart, 688*f*, 689
Extrinsic controls, for blood flow, **725**–726, 725*f*, 727*f*
Extrinsic eye muscles, **556**–557, 557*f*
  nerves supplying, 501*t*, 503*t*, 556, 557*f*, 557*t*
Extrinsic pathway, **659**–660, 659*f*
Extrinsic (major) salivary glands, **878**
Extrinsic shoulder muscles, 350–353*t*, 351*f*, 353*f*
Extrinsic tongue muscles, **336***t*, 337*f*, 877
Exudate, **785**–786, 787*f*
Eye, 554–575, 555*f*. *See also under Visual* and Vision
  accessory structures of, **554**–557, 555–556*f*, 557*f*. *See also specific structure*

autonomic innervation/effects and, 534*f*, 536*f*, 540*f*, 546*t*, 559*f*, 565*f*
  developmental aspects of, 595
  eyeball structure and, **558**–563, 558*f*, 559*f*, 560*f*, 562*f*
  homeostatic imbalances/disorders of, 555, 556, 557, 561, 563, 567, 569, 573, 575, 595
  internal chambers/fluids of, 562–563, 562*f*
  light/optics and, 563–567, 564*f*, 565*f*, 566*f*
  muscles controlling movement of, **556**–557, 557*f*
  photoreceptors/phototransduction and, **561**, 567–573, 568*f*, 569*t*, **570**–573, 570*f*, 571*f*, 572*f*
  visual pathways and, 573–575, 574*f*
Eyeballs, **558**–563, 558*f*, 559*f*, 560*f*, 562*f*, 567, 595. *See also* Eye
Eyebrows, **554**, 555*f*, 555*f*
Eyelashes, **555**, 555*f*, 555*f*
Eyelids, **554**–555, 555*f*
Eyeteeth (canines), 879*f*, **880**, 880*f*

Face. *See also under Facial*
  bones of, 200*f*, 201, 201*f*, 203*f*, 204*f*, 210–212, 217*t*
  muscles of, 330*f*, 333–334*t*, 333*f*, 335*f*
Facet (bone marking), 180*t*, 220, 220*f*
Facial artery, 717*f*, **740***t*, 740*f*, 741*f*
Facial expression, muscles controlling (facial muscles), 330*f*, 333–334*t*, 333*f*, 335*f*
  facial nerve supplying, 333–334*t*, 503–504*t*, **506**
Facial nerve (cranial nerve VII), 333–334*t*, 450*f*, 451*f*, 452, 499*f*, 503–504*t*, **506**, **536**–537, 536*f*
  in gustatory pathway, **580**, 580*f*
  muscles of mastication supplied by, 336*t*
  parasympathetic fibers of, 499*f*, 503*t*, **536**–537, 536*f*
  salivation control and, 879
  scalp muscles supplied by, 333*t*
Facial paralysis, in Bell's palsy, 503*t*
Facial vein, **752***t*, 752*f*, 753*f*
Facilitated diffusion, 69, **70**–71, 70*f*, 73*t*, 75, 467
Facilitated zone, 426, 426*f*
Facilitation, neuron, **420**
Fact (declarative) memory, 460
F (filamentous) actin, 287
Factor I (fibrinogen), 644*t*, 659*f*, **660**, 661*t*
Factor II (prothrombin), 659*f*, **660**, 661*t*
Factor III (tissue factor), 659*f*, **660**, 661*t*
Factor IV (calcium), 659, 659*f*, 661*t*
Factor V (proaccelerin), 659*f*, 661*t*
Factor VII (proconvertin), 659*f*, 661*t*
Factor VIII (antihemophilic factor/AHF), 659*f*, 661*t*, **662**
  deficiency of, hemophilia A caused by, 662
Factor IX (plasma thromboplastin component/PTC), 659*f*, 661*t*
Factor X (Stuart factor), 659, 659*f*, 661*t*
Factor XI (plasma thromboplastin antecedent/PTA), 659*f*, 661*t*
Factor XII (Hageman factor), 659*f*, 661*t*
Factor XIII (fibrin stabilizing factor), 659*f*, **660**, 661*t*
Facultative water reabsorption, **991**
FAD. *See* Flavin adenine dinucleotide
Fainting, **462**, 549, 728
Falciform ligament, **894**, 894*f*, 895*f*, 908*f*
Fallopian tubes. *See* Uterine (fallopian) tubes
Fallot, tetralogy of, 701, 701*f*
False labor, **1114**
False pelvis, 238*f*, **238**
False (vertebrochondral) ribs, 225*f*, **226**
False vocal cords (vestibular folds), 823*f*, 825*f*, **826**, 826*f*
Falx cerebelli, **465**, 465*f*

Falx cerebri, 464*f*, **465**, 465*f*
Familial hypercholesterolemia, **969**
Far point of vision, **565**
Farsightedness (hyperopia), 566*f*, **567**, 595
Fascia(e)
  lata, 367*t*
  lumbar, 347*f*
  renal, **976**, 976*f*
  spermatic, external and internal, 1048*f*
  superficial, 130, **151**, 151*f*, 167
Fascicle(s)
  muscle, **282**, 282*f*, 283*t*, 324, 326–327, 326*f*
  nerve, **496**, 497*f*
Fasciculus cuneatus, 475*f*, **476**, 477*f*, 480*t*
Fasciculus gracilis, 475*f*, **476**, 477*f*, 480*t*
Fast fibers, **307**, 307*f*, 308*t*
Fast glycolytic fibers, 307*f*, **308**, 308*t*
Fasting (postabsorptive) state, 952–955, 953*f*, 954*f*
Fast motility, 588
Fast oxidative fibers, **308**–309, 308*t*
Fast voltage-gated Na+ channels, in cardiac muscle contraction, **689**, 689*f*
Fat(s)/triglyceride(s), 9, **45**–48, 45*f*, 47*t*, 928. *See also* Lipid(s)
  in atherosclerosis, 959
  dietary, 928, 930*t*
  digestion/absorption of, 889, 913*f*, 916–917, 916*f*
  interconversion of, 949–950, 950*f*
  metabolism of, 935*f*, 946–948, 946*f*, 947*f*
    glycerol/fatty acid oxidation in, 946, 946*f*
    lipogenesis, **946**–947, 946*f*, 947*f*
    lipolysis, 946*f*, **947**, 947*f*, 953, 953*f*, 954*f*
    liver in, 947, 947*f*, 951*f*, 956*t*
    thyroid hormone affecting, 618*t*
  saturated and unsaturated, **46**, 46*f*, 957–958
  storage of in bone, 176
  supply of (fat pool), 950, 950*f*
  transport of, 956–957, 957*f*
Fat cells. *See* Adipocytes (adipose/fat cells)
Fatigue, muscle, **306**
Fat pool, **950**, 950*f*
Fat-soluble vitamins, 47*t*, 917, **931**, 932*t*
Fat tissue. *See* Adipose (fat) tissue
Fatty acids, **45**–47, 45*f*, 913*f*, **916**
  essential, 928
  in food intake regulation, 961, 961*f*
  omega–3, **47**, 930*t*, 958
  oxidation of, 946, 946*f*, 947*f*
  saturated and unsaturated, **46**, 46*f*, 957–958
  trans (trans fats), **46**–47, **958**
Fatty liver disease, non-alcoholic (NAFLD), **898**
Fatty pads, **255**, 268*f*, 270*f*
Fauces, **877**
  isthmus of, **823**, 823*f*
Fe. *See* Iron
Feature abstraction, **495**
Fecal incontinence, 911
Fecal transplant, 909–910
Feces, 8, 870*f*, **871**, 904*t*, **906**, 910–911, 911*f*
Fed (absorptive) state, 950–952, 951*f*, 952*f*
Feedback, 10–12, 10*f*, 518. *See also* Negative feedback; Positive feedback
Feet. *See* Foot
Female(s). *See also* Female reproductive system
  pelvis in, 238, 238*f*, 239*t*
  puberty in, 1084
  urethra in, 975*f*, 1003*f*, 1004, 1061*f*
  urinary system in, 975*f*, 1003*f*, 1004
Female pronucleus, **1096**, 1096*f*

Female reproductive system, 7*f*, 8. *See also specific organ or structure*
  anatomy of, 1060–1069, 1061*f*, 1063–1064*f*, 1065*f*, 1066*f*, 1068*f*, 1069*f*
  embryologic/fetal development of, 1081–1086, 1082*f*, 1083*f*
  functions/physiology of, 7*f*, 8, 1069–1079, 1070*f*, 1072*f*, 1075*f*, 1077*f*, 1079*t*
  homeostatic imbalances and, 1063, 1064, 1066, 1078, 1084
  lactation and, 1068, 1068*f*, **1116**–1119, 1117*f*
  mammary glands (breasts) and, **163**, **1068**–1069, 1068*f*
  parturition (birth) and, **1114**–1116, 1115*f*
  pregnancy and, 1091–1123, **1093**, 1112*f*
Female sex hormones. *See* Estrogen(s); Progesterone
Female sexual response, **1078**–1079
Femoral artery, 717*f*, 739*f*, **748***t*, 748*f*, 749*f*
Femoral cutaneous nerves, 512, 513*t*, 513*f*, 514*f*, 515*f*
Femoral nerve, 367*t*, 369*t*, 370*t*, **512**, 513*t*, 513*f*
Femoral vein, 751*f*, **758***t*, 758*f*
Femoropatellar joint, 257*t*, **270**–271. *See also* Knee joint
Femur, 200*f*, **240**–241, 240*f*, 241*t*
  in hip fracture, 193, 241
  in hip joint, 268, 269, 269*f*
  in knee joint, 270, 270*f*
  ligament of head of (ligamentum teres), **269**, 269*f*
Fenestrated capillaries, 711, 711*f*
Fenestrations (capillary pores), 711*f*, 980, 984, 985*f*
Fentanyl, 429
Fermentation, by gut bacteria, 909
Ferritin, **649**, 649*f*, **917**
Fertilization (conception), 1042, 1092–1096, 1092*f*, **1093**, 1094–1095*f*, 1096*f*, 1097*f*, 1119, 1128
Fetal alcohol syndrome (FAS), 1113
Fetal circulation, 759, 1101*f*, 1107, 1107*t*, 1108–1109*f*, 1116
Fetal fibronectin, 1115
Fetal heart rate, 699
Fetal hemoglobin (hemoglobin F), **666**
Fetal period, 1092, 1092*f*, 1107–1110, 1110*f*, 1111*t*
Fetal testing, 1134–1135, 1134*f*
Fetus, 759, **1092**, 1092*f*, 1100*f*, 1101*f*, 1107–1110, 1110*f*, 1111*t*. *See also under Fetal*
FEV. *See* Forced expiratory volume
Fever, 782*f*, **789**, 790*t*, 968
Fever blisters (cold sores), **169**
Fiber (dietary), 910, 928, 930*t*
Fibers
  connective tissue, 127*t*, 128–129, 128*f*
  muscle. *See* Muscle cells/fibers
Fibrillation, cardiac, **688**, 692*f*
Fibrin, 659*f*, **660**
Fibrin mesh (blood clot), 143*f*, 144, 658, 658*f*, 659*f*, 660, 660*f*. *See also* Coagulation
  factors limiting growth/formation of, 661
  factors preventing undesirable formation of, 662
  in inflammation, 786, 787*f*
  in unbroken vessel (thrombus), **662**, 759
Fibrinogen, 644*t*, 659*f*, **660**, 661*t*
Fibrinolysis, **660**
Fibrin stabilizing factor (FSF), 659*f*, **660**, 661*t*
Fibroblasts, 127*t*, 128*f*, **129**, 130, 131*f*, 132*f*, 133, 133*f*, 143*f*
Fibrocartilage, 127*t*, 136*f*, **137**, **174**, 175*f*, 253, 253*f*
Fibrocartilaginous callus, in bone repair, **191**, 191*f*

Fibrocytes, 127*t*, **129**
Fibromyalgia (fibromyositis), **319**
Fibronectin, fetal, 1115
Fibrosis, **143**, 143*f*, 145, 840
Fibrous capsule
 of kidney, **976**, 976*f*, 977*f*
 of lymph node, **771**, 772*f*
 of multicellular exocrine gland, 125
Fibrous (dense) connective tissue, 127*t*,
 129, 132–134*f*, **133–134**
Fibrous joints, 251, **252–253**, 252*f*, 255*t*
Fibrous layer
 of eye, **558–559**
 of joint capsule, 254, 254*f*
 of periosteum, 178
Fibrous pericardium, **673**, 673*f*
Fibrous (atherosclerotic) plaques, 760*b*
Fibrous (structural) proteins, 49*f*, **51**
Fibrous sheath, hair follicle, 158*f*, **159**
Fibula, 200*f*, 241*t*, 242–243*f*, **243**,
 270*f*, 271*f*
Fibular (peroneal) artery, 748*f*, 749*t*, 749*f*
Fibular collateral ligament, 270*f*, **271**, 271*f*
Fibularis (peroneus) brevis muscle, 376*f*,
 377*t*, 378*f*, 379*f*, 384*t*
Fibularis (peroneus) longus muscle, 330*f*,
 331*f*, 375*f*, 376*f*, **377***t*, 378*f*, 379*f*,
 384*t*
Fibularis (peroneus) tertius muscle, **374***t*,
 375*f*, 376*f*, 384*t*
Fibular (peroneal) muscles, **374***t*, 385*f*.
 See also specific muscle
Fibular (peroneal) nerve
 common, 511*t*, 514, 514*f*, **515**
 deep, 514*f*
 muscles supplied by, 373*t*, 374*t*, 377*t*,
 380*t*, 515, 515*t*
 superficial, 514*f*
Fibular notch, **243**
Fibular (peroneal) retinacula, **374***t*, 376*f*
Fibular (peroneal) vein, 758*t*, 758*f*
"Fight-or-flight" response, 534–535, 626*t*,
 627, 628*f*, 721
Filamentous (F) actin, 287
Filaments. *See* Myofilaments
Filaments, olfactory, 499, 499*f*, 500*t*,
 576, 576*f*
Filiform papillae, **877**, 877*f*
Filling in, 561
Filopodia, 428
Filtrate, **980**, 984
Filtration
 in capillaries, 731, 732–733*f*
 glomerular, 982, **984–989**, 984*f*, 985*f*,
 986*f*, 988*f*
Filtration membrane, 980, **984–986**, 985*f*,
 987
Filtration slits, **980**, 985*f*
Filum terminale, 470*f*, **471**, 471*f*
Fimbriae, 1061*f*, 1062, 1063*f*
Finger(s), **236**
 bones of (phalanges), 200*f*, 234*f*, 235*t*,
 236
 joints of, 256*t*
 muscles controlling movement of,
 358–361*t*, 358*t*, 359*f*, 361*f*, 362*t*,
 363*f*, 364–366*t*, 365*f*
Fingerprints, 155, 155*f*
First breath, after birth, 860, 1116
First line of defense. *See* Mucous
 membranes (mucosae); Skin
 (cutaneous membrane/integument)
First-class lever, **328**, 329*f*
First-degree burns, **166**, 166*f*
First intention, healing by, **147**
First law of thermodynamics, 960
First messenger, 82*f*, **604**, 605*f*
First-order neurons, **476**, 477*f*
First polar body, 1070*f*, **1071**, 1096*f*
First (dilation) stage of labor, **1115**, 1115*f*
Fission, mitochondrial, 84
Fissure (bone marking), 180*t*
Fissure(s)
 cerebral, 438*f*, **439**
 oblique and horizontal, of lung, 832*f*, 833

orbital, inferior and superior, 203*f*, 209*f*,
 210, **212**, 216*t*, **217***t*
Fixation, tissue, for microscopy, **117**
Fixator (muscle), **324**
Fixed (nonvolatile) acid, **1029**
Fixed macrophages, 129, 784
Flaccid paralysis, **475**
Flagella, **90**, 95*t*
Flat bones, **176**, 177*f*, 178*f*, 186, 186*f*
"Flat EEG," 461
Flatfoot (pes planus), 245
Flatus (gas), 909
Flavin(s), **940**
Flavin adenine dinucleotide (FAD), **935**,
 935*f*
Fleshy (direct) muscle attachments, **282**
Flexion (body movement), **258–259**, 258*f*,
 259*f*, 325*f*, 354*t*, 362*f*, 363*f*, 365*t*,
 384*t*, 385*f*
Flexor accessorius (quadratus plantae)
 muscle, **380***t*, 381*f*, 383*f*
Flexor carpi radialis muscle, 330*f*, **358***t*,
 359*f*, 362*t*
Flexor carpi ulnaris muscle, 331*f*, **358***t*,
 359*f*, 361*f*, 362*t*
Flexor digiti minimi brevis muscle, **364***t*,
 365*f*, 381*f*, **382***t*, 383*f*
Flexor digitorum brevis muscle, **380***t*, 381*f*
Flexor digitorum longus muscle, **377***t*,
 379*f*, 384*t*
Flexor digitorum profundus muscle, **359***t*,
 359*f*, 362*t*, 365*f*
Flexor digitorum superficialis muscle,
 359*t*, 359*f*, 362*t*, 365*f*
Flexor hallucis brevis muscle, 381*f*, **382***t*,
 383*f*
Flexor hallucis longus muscle, 376*f*, **377***t*,
 379*f*, 384*t*
Flexor muscles
 of forearm, 355–356*f*, 357*t*, 358–361*t*,
 359*f*, 362*t*, 363*f*
 of leg, 374*t*
 of thigh, 367*t*
Flexor pollicis brevis muscle, **364***t*, 365*f*
Flexor pollicis longus muscle, **359***t*, 359*f*,
 362*t*
Flexor (withdrawal) reflex, 10, **524**, 525*f*
Flexor retinacula, 358*t*, 359*f*, **374***t*
Flexure lines, **155**, 155*f*
Flexures
 colic, right (hepatic) and left (splenic),
 906, 907*f*
 midbrain and cervical, 436, 436*f*
Floating (vertebral) ribs, 225*f*, **226**
Flocculonodular lobe, **454**, 455*f*
Floor
 of nasal cavity, 215
 of orbit, 213*f*
Flora (bacterial), intestinal, **909–910**, 958*b*
Flower spray (secondary sensory) endings,
 520*f*, **521**
Fluid(s), body. *See* Body fluids
Fluid balance. *See* Water balance
Fluid compartments, 731–734, 732–733*f*,
 1013, 1013*f*, 1015, 1015*f*
Fluid mosaic model, **63**, 64–65*f*
Fluid movements (bulk flow), 731–734,
 732–733*f*, 1015
Fluid-phase endocytosis (pinocytosis), **77**,
 77*f*, 78*t*
Fluorine, 26*t*, 933*t*
Foam cells, in atherosclerosis, 760*b*
Focal point(s), **564**, 565*f*, 566*f*
Focusing (light on retina), 565–567, 565*f*
Folacin. *See* Folic acid/folacin
Folding, embryo, 1104, 1104*f*
Folia, **454**, 455*f*
Foliate papillae, **578**, 579*f*, **877**, 877*f*
Folic acid/folacin (vitamin B₉), 483, 649,
 649*f*, 932*t*, 1112
Follicle(s)
 hair, 151*f*, 158*f*, **159**
 lymphoid, **771**, 772*f*, 774*f*
 aggregated (Peyer's patches), 771,
 771*f*, **775**, 775*f*, **903**

ovarian. *See* Ovarian follicles
 thyroid, **617**, 617*f*
Follicle cells, ovarian, 1070*f*
Follicle-stimulating hormone (FSH), 611*f*,
 613*t*, **616**, **1043**
 in female reproductive function, 1043,
 1071, 1073, 1074, 1075, 1075*f*,
 1076, 1077*f*
 in male reproductive function, 1043,
 1059*f*
Follicular cells, thyroid, 617, 617*f*,
 619–620, 619*f*
Follicular phase of ovarian cycle, **1073**,
 1074, 1075*f*, 1077*f*
Fontanelles, **246**, 246*f*
Food. *See also* Nutrient(s); Nutrition
 absorption/digestion of, 5*f*, 7*f*, **8**, **869**,
 870*f*, **871**, 871*f*, 872, 904*t*, **912–918**,
 913–916*f*
 ingestion/intake of, **870**, 870*f*,
 876–881, 904*t*
 metabolic rate and, 963
 regulation of, 449, 958–959*b*,
 960–962, 961*f*
 propulsion of through digestive system,
 870, 870*f*, 892*f*, 904*t*.
 See also Peristalsis
 as survival need, 9
Food guides, 927, 927*f*
Food-induced thermogenesis, **963**
Foot
 arches of, 244–245, 245*f*
 bones of, 241*t*, 244–245, 244*f*
 dorsal venous arch of, 751*f*, 758*t*, 758*f*
 dorsiflexion/plantar flexion of, **260**,
 261*f*, 384*t*, 385*f*
 intrinsic muscles of/muscles acting on,
 380–383*t*, 381*f*, 383*f*
 inversion/eversion of, **260**, 261*f*, 384*t*,
 385*f*
Footdrop, 515
Foot processes, **980**, 983*f*, 985*f*, 986
Foramen (bone marking), 180*t*
Foramen lacerum, 206*f*, 207*f*, **208**
Foramen magnum, **202**, 206*f*, 207*f*,
 216*t*
Foramen ovale (heart), 674, **700**, 701*f*,
 759, **1107**, 1107*t*, 1108*f*, 1109*f*
Foramen ovale (skull), 206*f*, 207*f*, 209*f*,
 210, **216***t*
Foramen rotundum, 207*f*, 209*f*, **210**, **216***t*
Foramen spinosum, 206*f*, 207*f*, 209*f*, **210**,
 216*t*
Forced convection, heat loss by, 965
Forced expiration, 344*t*, 346*t*, 838
Forced expiratory volume (FEV), **842**
Forced (deep) inspiration, muscles used in,
 344*t*, 838
Forced vital capacity (FVC), **842**
Forearm
 bones of, 232–233, 232–233*f*, 235*t*
 muscles of/muscles controlling
 movement of, 330*f*, 331*f*, 355–356*f*,
 357*t*, 358–361*t*, 359*f*, 361*f*, 361*f*,
 362*t*
Forebrain (prosencephalon), **435**, 435*f*,
 1106
Foregut, 918*f*, 1104, 1104*f*
Forehead, 202
Foreskin (prepuce), 1047*f*, **1050**, 1051*f*
Formed elements of blood, 643–645,
 643*f*, **644**, 645*f*, 656*t*. *See also*
 Blood cells
Fornix
 limbic system, 440*f*, 447*f*, 448*f*, **456**,
 456*f*
 vaginal, 1061*f*, **1066**
Fossa(e) (bone marking), 180*t*
Fossa ovalis, 674, 676*f*, 677*f*, **1107**, 1107*t*,
 1109*f*
Fourchette, 1066*f*, **1067**
Fourth ventricle, 435*f*, **436**, 437*f*, 452,
 453*f*
Fovea capitis, 240*f*, **241**
Fovea centralis, 558*f*, **561**

Foveal (hard) focus, 561
Fractures, **190–192**, 192*t*
 bimalleolar, 242*f*, 243
 in osteoporosis, 193
 pathologic, **195**
 repair of, 190–192, 191*f*
Frank-Starling law of heart, **697**
Fraternal twins, 1074
FRC. *See* Functional residual capacity
Freckles, 156
Free edge of nail, 160, 160*f*
Free fatty acids, 916
Free macrophages, 129, 784
Free (nonencapsulated) nerve endings,
 151*f*, 164, **491**, 491*t*
Free radicals, **86**, 110, 931
Free ribosomes, 84
Frenulum
 labial, **876**, 876*f*
 lingual, 876*f*, **877**
Frequency (action potential), stimulus
 coding and, 410, 410*f*
Frequency (sound wave), **586**, 586*f*, 588*f*
Frequency (urinary), infection and, 1004
Friction ridges, **155**, 155*f*
Friction rub, pericardial, 673
Frontal belly of epicranius, 330*f*, **333***t*,
 333*f*, 335*f*
Frontal bone, **202**, 202*f*, 203*f*, 204*f*, 205*f*,
 207*f*, **216***t*, 821, 821*f*
Frontal eye field, 440*f*, **441**
Frontal lobe, 438*f*, **439**
Frontal (coronal) plane/section, **15**, 15*f*
Frontal processes, **211**, 211*f*, **217***t*
Frontal sinuses, **202**, 202*f*, 205*f*, 214*f*,
 215*f*, 821*f*
Frontonasal suture, 202, 203*f*
Frostbite, 967
Fructose, 43, 43*f*, 913*f*, 914
FSH. *See* Follicle-stimulating hormone
Fulcrum, in lever system, **327–328**,
 327–328*f*, 329*f*
Full-thickness burns, 166, 166*f*
Functional blindness, 443
Functional brain systems, 456–458, 456*f*,
 457*f*, 458–459*f*
Functional groups, 43, 48, 49*f*
Functional layer (stratum functionalis) of
 endometrium, **1065**
Functional MRI (fMRI), 17*b*, 439*f*
Functional (globular) proteins, 50, 50*f*, **51**
Functional residual capacity (FRC), **841**,
 841*f*
Functional syncytium, cardiac muscle as,
 684, 685, 685*t*
Fundus
 of eye, **561**, 561*f*
 of stomach, **885**, 885*f*
 of uterus, **1063**, 1063*f*
Fungiform papillae, **578**, 579*f*, **877**, 877*f*
Funiculi (columns), white, 473*f*, 474, 475*f*,
 476, 480*t*
Furosemide, 1000
Fused (complete) tetanus, 300*f*, **301**
Fusiform muscles, 326*f*, **327**
FVC. *See* Forced vital capacity

G₀ phase, **97**
G₁ checkpoint (restriction point), 96*f*, 98
G₁ subphase, **96–97**, 96*f*, 100*f*
G₂/M checkpoint, 96*f*, 98
G₂ subphase, 96*f*, **97**, 100*f*
GABA. *See* Gamma (γ)-aminobutyric acid
G (globular) actin, 287
Galactorrhea, **636**
Galactose, 43, 43*f*, 913*f*, 914, 914*f*
Galactosemia, 968
Gallbladder, 869*f*, 894*f*, 895*f*, **898**, 900*f*,
 901*f*, 904*t*
 autonomic innervation/effects and,
 534*f*, 536*f*, 540*f*, 546*t*
Gallstones, **898**
Gamete(s), 98, **1042**, 1071*t*. *See also*
 Ova; Sperm

Gamete intrafallopian transfer (GIFT), 1119
Gamete variability, 1126f, 1127–1128, 1127f
Gamma (γ)-aminobutyric acid (GABA), 421t, **423**, 469
Gamma (γ) efferent fibers, 520f, **521**
Gamma globulin, 644t, 797f, **798**. See also Antibodies
Gamma (γ) interferon, 788, 790t, 805, 805t
Gamma (γ) motor neurons, 521, 523–524
Gamma (γ) rays, 28, 563, 564f
Ganglia, 391, 391f, **394**, **497**. See also specific type
    autonomic, **532**, 533, 533f, 535, 535t, 538–539, 538t
Ganglion cells, retinal, 560f, **561**, 572f, 573, 575
Ganglionic (postganglionic) neuron, **532**, 533f
Gap 0 (G₀) phase, **97**
Gap 1 (G₁) checkpoint (restriction point), 96f, 98
Gap 1 (G₁) subphase, **96**–97, 96f, 100f
Gap 2 (G₂)/M checkpoint, 96f, 98
Gap 2 (G₂) subphase, 96f, **97**, 100f
Gap junction(s), 67f, **68**
    in cardiac muscle, 313t, 317, 684, 684f, 685, 685t
    in nervous tissue (electrical synapses), 413
    in smooth muscle, 311, 313t, 314, 314f, 316
Gas (flatus), 909
Gases, 24
    basic properties/partial pressure of, **843**–845, 843t, 844f
    as neurotransmitters, 422t, 423–424
    solubility of, 843, 844–845, 844f
Gas exchange, 845f. See also Gas transport; Respiratory gases
    among blood/lungs/tissues, 824, 829, 830, 842–847, 843t, 844f, 846f
    among fluid compartments, 1015, 1015f
    internal respiration and, 844f, 846–847
    pulmonary (external respiration), 819f, 844–846, 844f, 846f
Gas flow, resistance to. See Airway resistance
Gas (air) movements, nonrespiratory, **838**
Gasotransmitters, 422t, 423–424
Gas transport, 643, 730, 731f, 819f, 847–852, 848–849f, 850f, 851f. See also Gas exchange
Gastric accommodation, **892**
Gastric arteries, 744t, **745**t, 745f
Gastric bypass surgery, 959b
Gastric contractile activity, 892–893, 892f
Gastric emptying, 893, 893f
Gastric glands, 546t, 886f, **887**
Gastric inhibitory peptide (GIP/glucose-dependent insulinotropic peptide), 635t, 888t, 952
Gastric juice, 783t, **887**, 889–891, 890f, 904t
Gastric lipase, 887, 889, 913f
Gastric motility, 892–893, 892f, 893f, 905–906, 905t
Gastric phase, of gastric secretion, **890**, 890f
Gastric pits, **886**–887, 886f
Gastric ulcers, **887**–889, 887f, 889f
Gastric veins, 757f
Gastrin, 635t, **887**, 888t, 889, 890, 890f, 891, 905t, 906
Gastritis, 887
Gastrocnemius muscle, 330f, 331f, 375f, 376f, **377**t, 378f, 379f, 384t
Gastrocolic reflex, **910**
Gastroduodenal artery, 744t, **745**t, 745f
Gastroenteritis, **919**
Gastroepiploic arteries, 744f, **745**t, 745f
Gastroepiploic vein, right, 757f

Gastroesophageal reflux disease (GERD), 883
Gastroesophageal (cardiac) sphincter, **882**
Gastroileal reflex, 905t, **906**
Gastrointestinal (GI) tract (alimentary canal), **869**, 869f, 872f. See also specific structure and Digestive system
    autonomic innervation/effects and, 536f, 539f, 540f
    development of, 918–919, 918f, 1104, 1104f
    endocrine (enteroendocrine) cells in, 634, 886f, **887**, 901f, 902f, 903
    enteric nervous system of, **543**, **874**–875, 874f, 875f
    histology of, 872–874, 873f
    hormones produced by, 634, 635t
    neural signals from, food intake regulation and, 961, 961f
    pregnancy and, 1113
    processes occurring in, 870–871, 870f, 871f, 875, 904t, 912–918, 913f, 914f, 915f, 916f
    thyroid hormone affecting, 618t
Gastrula, **1102**
Gastrulation, 1100f, **1102**–1103, 1103f
Gated channels, 70, 400–401, 401f. See also specific type
G cells, **890**, 890f
GDP (guanosine diphosphate), 82f, 604, 605f
Gemellus muscles, 372f, **373**t, 384t
Gender
    basal metabolic rate and, 963
    CNS development and, 482–483
    genetic, determination of, 1081, 1081f
Gene(s), 53, **99**, 1124. See also Heredity
    activation of, in hormone mechanism of action, 606, 606f
    carcinogenesis and, 144b
    diabetes mellitus and, 633b
    expression of, 1131–1133
    hypertension and, 724
    linked, **1127**
    obesity and, 958b
    pairs of (alleles), **1125**–1126, 1126f, 1127, 1129, 1129t
    recombination of, 1127–1128, 1127f
General senses, receptors for, 491–493, **491**. See also Sensory receptors
General visceral motor system, **532**. See also Autonomic nervous system
Generator potential, 404, **494**
Gene therapy, 469, 1135
Genetic code, 103f, **103**
Genetic disorders, dominant and recessive, 1129
Genetic recombination, 1127–1128, 1127f
Genetics, 1124. See also Gene(s); Inheritance
Genetic screening/counseling, 1133–1135, 1133f, 1134f
Genetic sex, determination of, 1081, 1081f
Genetic theory of aging, 110
Genetic variation, 1047, 1126–1128, 1126f, 1127f
Geniculate bodies/nuclei
    lateral, 448, 449f, **574**, 574f
    medial, 448, 449f, 589f, **590**
Geniculate ganglion, 503f
Genioglossus muscle, **336**t, 337f
Geniohyoid muscle, 337f, **338**t, 339f
Genital(s)/genitalia
    autonomic innervation/effects and, 534f, 536f, 540f, 545
    embryologic/fetal development of, 1083–1084, 1083f
    female
        external, **1060**, 1066f, 1067
        internal, **1060**, 1063–1064f
    male, external (penis/scrotum), 1047, 1047f, **1048**, 1048f, **1050**–1052, 1051f, 1052f
Genital herpes, **1080**–1081

Genital tubercle, 1083f, **1084**
Genital warts, **1080**
Genitofemoral nerve, 513t, 513f
Genome, 53, **1125**. See also DNA; Gene(s); Heredity
Genomic imprinting, **1132**
Genotype, **1126**, 1128–1129, 1128f
Germ cells, primordial, **1083**
Germinal centers, **771**, 772f
Germinal epithelium, of ovary, 1061f, 1062
Germ layers, 1102–1107
    differentiation of, 1103–1107, 1104f, 1105f, 1106f
    formation of, 146, 146f, 1100f, 1102–1103, 1103f
    primary, **146**, 146f, 1100f
Germinal centers in lymphoid follicles, 774f, 775
Gestational diabetes mellitus, 1113
Gestation period, 1042, **1092**
GFR. See Glomerular filtration rate
GH. See Growth hormone
GHIH. See Growth hormone-inhibiting hormone
Ghrelin (Ghr), 635t, 961, 961f
GHRH. See Growth hormone-releasing hormone
GIFT. See Gamete intrafallopian transfer
Gigantism, 188, 612t, **615**, 615f
Gingiva (gum), 876f, **880**, 880f
Gingival disease, 881
Gingival sulcus, 880, 880f
Gingivitis, **881**
GIP. See Gastric inhibitory peptide
GI tract. See Gastrointestinal (GI) tract
Glabella, **202**, 203f
Gland(s)/glandular epithelia, 117, **123**–126, 124f, 125f. See also specific type
    autonomic innervation/effects and, 546t
    cutaneous, 161–163, 161f
    endocrine (ductless). See Endocrine system/glands
    exocrine, 123, 123f, **124**–126, 124f, 125f, 126f
    innervation of, 517
Glans of clitoris, 1066f, **1067**, 1083f
Glans penis, 1047f, **1050**, 1051f, 1083f
Glassy membrane, 158f, **159**
Glaucoma, **563**
Gleevec. See Imatinib
Glenohumeral joint. See Shoulder (glenohumeral) joint
Glenohumeral ligaments, **266**, 267f
Glenoid cavity, **228**, 229f, 266, 267f
Glenoid labrum, **266**, 267f
Glial cells (neuroglia), 140, 140f, 392–394, **393**, 393f
Gliding body movements, **258**, 258f, 265
Globin, 646, 646f
Globin chains, 646, 646f, 650
Globular (G) actin, 287
Globular (functional) proteins, 50, 50f, **51**
Globulin(s), 644t
Globus pallidus, 445f, **446**, 446f
Glomerular capillaries. See Glomerulus/glomerular capillaries
Glomerular capsular space. See Capsular space
Glomerular (Bowman's) capsule, 979f, **980**, 981f
Glomerular filtration, 982, **984**–989, 984f, 985f, 986f, 988f
Glomerular filtration rate, **986**–989, 987t, 988f
Glomerular mesangial cells, 983f, 986
Glomeruli (olfactory bulb), 576f, **577**
Glomerulonephritis, 812, **1008**
Glomerulus/glomerular capillaries, 978f, 979f, **980**, 980f, 981f, 982, 982f, 985f, 986, 986f

Glottis, 825f, 826f
Glucagon, **630**, 630f, 631f, **954**, 954f, 955t, 961, 961f
Glucoamylase, 913f, **914**
Glucocorticoids, **625**–626, 626t, 629f, **1023**
Gluconeogenesis, 625, 630, 944f, **945**, 953f
Glucose, 43, 43f, 44, 55, 913f, 914, 914f, **928**
    in absorptive state, 950, 951f, 952f
    aging and, 969
    in ATP formation, 304, 304f, 305, 305f
    in diabetes mellitus, 632, 952
    in food intake regulation, 961, 961f
    insulin/glucagon in regulation of, 10, 630–631, 631f
    metabolism of, 935f, 936–945, 937f, 938f, 939f, 941f, 942f, 943f, 944f
    glycogenesis and glycogenolysis in, 630, 631f, 944–945, 944f, 945f, 952, 953f, 954f
    liver in, 950, 951f, 952
    oxidation in, 937–944, 938f, 939f, 941f, 942f, 943f, 944f
    pregnancy and, 1113
    molecular weight of, 30
    in postabsorptive state, 952–954, 953f, 954f
    sources of, 952–954, 953f, 954f
    transport of, 70, 75f, 914, 914f
    in urine, 632, 632f, 1001t
Glucose-1-phosphate, 945, 945f
Glucose-6-phosphatase, 945, 945f
Glucose-6-phosphate, 937, 938–939, 945, 945f
Glucose-dependent insulinotropic peptide (gastric inhibitory peptide/GIP), 635t, 888t, 952
Glucose sparing, 615, **954**
Glucose tolerance, aging affecting, 636
Glutamate, 421t, **423**, 429b, 469, 495
Glutamic acid, 948f, 949, 949f
Glutamine, renal metabolism of, ammonium ion excretion and, 1032, 1032f
Gluteal arteries, 748t, 748f
Gluteal lines, posterior/anterior/inferior, **236**, 237f
Gluteal nerves (superior and inferior), 370f, 514f, **515**, 515t
Gluteal surface, of ilium, **236**
Gluteal tuberosity, 240f, **241**
Gluten-sensitive enteropathy (celiac disease), 918
Gluteus maximus muscle, 331f, **367**t, **370**t, 371f, 372f, 384t
Gluteus medius muscle, 331f, **367**t, **370**t, 371f, 372f, 384t
Gluteus minimus muscle, **367**t, **370**t, 372f, 384t
Glycerol, **45**, 45f, 46f, 946, 946f, 947f
Glycine, 422t, **423**
Glycocalyx, 64f, 65f, **66**–67, 66f
Glycogen, 44, 44f, 914, 914f
    breakdown of (glycogenolysis), 630, 631f, 944f, **945**, 945f, 952, 953f, 954f
    formation of (glycogenesis), 631, 631f, 944f, **945**, 945f
    muscle fatigue and decreased, 306
Glycogenesis, 631, 631f, 944f, **945**, 945f
Glycogenolysis, 630, 631f, 944f, **945**, 945f, 952, 953f, 954f
Glycogen phosphorylase, 945, 945f
Glycogen storage disease, 968
Glycogen synthase, 945, 945f
Glycolipids, 47t, 65f, **66**, 87
Glycolysis/glycolytic pathway, **304**–305, 304f, 305f, 306, 313f, 934, 935f, **937**–939, 937f, 938f, 944f
Glycolytic fibers, 307f, **308**, 308t
Glycoproteins, 65f, **66**, 66f, 81
Glycosaminoglycans (GAGs), 128, 134
Glycosomes, **284**

Glycosuria, 632f, 1001t
GN. See Glomerulonephritis
GnRH. See Gonadotropin-releasing hormone
Goblet cells, **124**, 124f, 822, 902f, 903, 909
Goiter, **620**, 620f, 933t
G$_{olf}$, **577**, 577f
Golgi apparatus, 62f, **85–86**, 85f, 86f, 87f, 94t
Gomphoses, **252–253**, 252f, 880
Gonad(s), 602f, **634**, **1042**. See also Ovaries; Testes
  aging affecting, 636
  descent of, 1084
  differentiation of, 1082f, 1083
  female, **1060**, 1061f
  male, 1042, 1047, 1047f
Gonadal arteries, 738f, 739f, 744f, **746t**, 746f, 1049
Gonadal hormones, 616
Gonadal ridges, 1082f, **1083**
Gonadal veins, 750f, **756t**, 756f, 757f
Gonadocorticoids, **626**, 626t. See also Sex hormones
Gonadotropic cells, **616**
Gonadotropin(s), **616**. See also Follicle-stimulating hormone; Luteinizing hormone
  in female reproductive function, 1074–1076, 1075f, 1077f
  human chorionic (hCG), **1099**, 1099f, 1101–1102
  in male reproductive function, 1042–1043, 1059f
Gonadotropin-releasing hormone (GnRH), 611f, 613f, **616**
  in female reproductive function, 1074, 1075, 1075f
  in male reproductive function, **1042–1043**, 1059f
Gonorrhea, 1063, **1080**
Gout/gouty arthritis, **275**
G protein(s), **81**, 82f, 424, 425, 425f, 571, 571f, 577, 577f, 579, **604**, 605f
G protein-linked (metabotropic) receptors, **81**, 82f, 315, **425**, 425f, 577, 577f
Gracilis muscle, 330f, 331f, 368f, **369t**, 371f, 384t
Graded muscle contractions, 299–302, **300**, 300f
Graded potentials, 395, 403, **404–405**, 405f
  action potentials compared with, 416–417t
  information processing in retina and, 573
  information transfer at chemical synapse and, 414, 415f
  postsynaptic potentials as, 416, 417, 417t
  sensory processing and, 494
Grand mal (tonic-clonic) seizures, 462
Granular (juxtaglomerular/JG) cells, **983**, 983f, 988f, 989
Granulation tissue, **143–144**, 143f
Granule cells, amacrine, 577
Granulocyte(s), **652–653**, 652f, 653f, 655f, 656t. See also specific type
Granulocyte-colony-stimulating factor (G-CSF), 654
Granulomas, 787
Granulosa cells, **1070**, 1073
Granzymes, **807**, 807f, 810t
Graves' disease, **620**–621, 620f, 812
Gray commissure, **472**, 473f
Gray matter, **436**, 437f
  of brain
    brain stem, 436, 437f
    cerebellar, 436, 437f, 454
    cerebral, 435f, 437f, 439, 445f, 458t
    periaqueductal, 451, 453f
  of spinal cord, 437f, 472–474, 473f, 474f
    development of, 482, 482f

Gray ramus communicans/rami communicantes, 535t, 538f, **539**, 539f
Great cardiac vein, 675f, 676f, **683**, 683f, 751f
Greater auricular nerve, 509t, 509f
Greater curvature of stomach, **885**, 885f
Greater omentum, **885**, 908f
Greater sciatic notch, **236**, 237f, 239t
Greater sphenoid wings, 203f, 204f, 205f, 206f, 207f, **209**, 209f
Greater splanchnic nerve, 540f, **541**
Greater trochanter, 240f, **241**
Greater tubercle, 230f, **231**
Greater vestibular glands, 1061f, 1066f, **1067**
Great saphenous vein, 751f, **758t**, 758f
Great toe (hallux), 241t, **244**
Green cones, 564f, 569
Greenstick fracture, 192t
Gristle, 134. See also Hyaline cartilage
Groove (bone marking), 180t
Gross (macroscopic) anatomy, 2
Ground substance, 127t, **128**, 128f, 130, 131f, 134
Group A, B, and C fibers, **412**
Growth, 8
  cell division and, 96f, 98, 100–101f
  growth hormone affecting, 614f, 615
Growth cone, **428**, 428f
Growth cycles, hair, 159
Growth hormone (GH/somatotropin), 188, 611f, 612t, 614f, **615–616**, 615f, 955t
Growth hormone-inhibiting hormone (GHIH/somatostatin), 422t, 423, 611f, 613t, 614f, **615**, 630, **887**, 888t
Growth hormone-releasing hormone (GHRH), 611f, 613t, 614f, **615**
Growth (proliferation) zone (epiphyseal plate), 186, 187f
GTP (guanosine triphosphate), 76, 82f, **604**, 605f
Guanine (G), 54f, **55**
Guanosine diphosphate (GDP), 82f, 604, 605f
Guanosine monophosphate, cyclic. See Cyclic GMP
Guanosine triphosphate, (GTP), 76, 82f, **604**, 605f
Guanylate cyclase, 423
Gubernaculum, **1084**
Gum (gingiva), 876f, **880**, 880f
Gum disease, 881
Gustatory (taste) cortex, 440f, **443**, 580, 580f
Gustatory epithelial cells, 578, **578**, 579, 579f
Gustatory hairs, **578**, 579f
Gustatory pathway, 580, 580f
Gustducin, 579
Gut bacteria (bacterial flora), **909**–910, 958b
Gut brain, 874, 875f. See also Enteric nervous system/plexuses
Gut-brain peptides, **423**
Gynecology, **1086**
Gynecomastia, **1086**
Gyri (gyrus), 438f, **439**

H. See Hydrogen
H$_2$S. See Hydrogen sulfide
Hageman factor (factor XII), 659f, 661t
Hair(s), **157–160**, 158f, 167
Hair bulb, 158f, **159**
Hair bulge, 159
Hair cells
  cochlear, 584f, 585, 588, 589f
  in crista ampullaris, 592f, 593
  in equilibrium, **590**–594, 591f, 592f
  in hearing, 584f, 585, 588, 589f, 594, 596
  in maculae, **590**, 591, 591f
Hair follicle(s), 151f, 158f, **159**

Hair follicle receptor (root hair plexus), 151f, **159**, 164, 491t, **492**
Hair loss, 160
Hair matrix, 158f, **159**
Hair papilla, 158f, 159
Hair root, 151f, 158, 158f
Hair shaft, 151f, 158, 158f
Hair thinning, 159–160
Haldane effect, **852**
Half-life, 28, **608–609**
Hallucinations, 462, 578
Hallux (great toe), 241t, **244**
Hamate, 234, 234f, 235t
Hammer (malleus), 581f, **582**, 582f, 587t
Hamstring muscles, 331f, **367t**, 371f, **373t**, 385f
Hamstring strain, **386**
Hand
  bones of, 234–236, 234f, 235t
  muscles controlling movement of, 358–361t, 359f, 361f, 364–366t, 365f
Handedness, 190, 444
Haploid chromosomal number (n), **1043**, 1044, 1044f
Hapten(s), **791**
Hard (bony) callus, in bone repair, 191, 191f
Hard (foveal) focus, 561
Hard keratin, 157, 160. See also Keratin
Hard palate, 206f, 211, 214f, 821f, **822**, 823f, **876–877**, 876f
Hashimoto's thyroiditis, **814**
Haustra (haustrum), **906**, 907f
Haustral contractions, **910**
Haversian (central) canal, 136f, 137, **182**, 182f, 183f
Haversian system (osteon), 137, **181–182**, 182f, 183f
Hb. See Hemoglobin
HbO$_2$. See Oxyhemoglobin
hCG. See Human chorionic gonadotropin
HCO$_3^-$. See Bicarbonate/bicarbonate ion
hCS. See Human chorionic somatomammotropin
HDLs. See High-density lipoproteins
He. See Helium
Head. See also Skull
  arteries supplying, 739f, 740–741f, 740–741t
  muscles of/muscles controlling movement of, 330f, 333–334t, 333f, 335f, 336t, 337f, 340t, 341f, 877
  receptors sensing movement of, 590–593, 591f, 592f
  sympathetic pathways to, 540f, 541
  trauma to, brain injury and, 468
  veins of, 750f, 751f, 752–753f, 752t
Headache, sinus, 822
Head (bone marking), 180t
  of femur, 240f, **241**, 269, 269f
  of fibula, 242f, **243**
  of humerus, 230f, **231**
  of metacarpals, **234**, 234f
  of radius, 231f, 232f, **233**
  of rib, 226, 226f
  of ulna, 232f, **233**, 233f
Head of caudate nucleus, 446f
Head of epididymis, 1049f, 1050
Head injury, traumatic brain injury and, 468
Head of pancreas, 898, 900f
Head of sperm, 1057f, **1058**
Healing, 142–146, 143f
  by first and second intention, **147**
  fracture, 190–192, 191f
Hearing, 585–590, 586f, 587f, 588f, 589f, 596. See also Ear
  cortical areas in, 440f, 443, 589f, 590
Hearing loss, 504t, 587, **594–595**, 596
Heart, 670–705, **671**, 671f, 675–678f. See also under Cardiac
  age-related changes affecting, 702
  apex of, **672**, 672f, 675f

autonomic innervation/effects and, 534f, 536f, 540f, 545, 546f, 688f, 689, 697f, 698–699
base of, **672**
blood flow through/perfusion of, 680–683, 681f, 726, 726f, 729
blood supply of (coronary circulation), **682**–683, 683f
blood volume in, 713f
cardiac cycle and, 692–696, **693**, 694–695f
cardiac output and, **696–700**, 697f, 698f
case study and, 705b
chambers of, 674, 675–677f, 700, 701f
conduction system of, 685f, 686–689, 687f, 688f, 689
congenital defects of, **700–701**, 701f
coverings of, 673, 673f
developmental aspects of, 700–702, 701f
Frank-Starling law of, **697**
great vessels and, 674
heart sounds and, **693–696**, 694f, 696f
homeostatic imbalances and, 673, 680, 683, 688, 692, 696, 699, 700–701
hormones produced by, 634, 635t. See also Atrial natriuretic peptide
layers of walls of, 673–674, 673f
muscle tissue of. See Cardiac muscle
pulmonary circuit of, 671, 671f, 680, 681f
size/location/orientation of, 672, 672f
systemic circuit of, **671**, 671f, 680–682, 681f
Heart attack (myocardial infarction), 683, 700, 760b
Heart block, **688–689**, 692f
Heartburn, **883**, 1113
Heart failure, **700**, 845
Heart murmurs, **696**
Heart palpitation, **702**
Heart rate, 698–699
  cardiac output and, 696, 697f, 718–719, 719f, 723f
Heart sounds, **693–696**, 694f, 696f
Heart tube, 700, 701f
Heart valves, 677f, 678–680f, 679–680, 681f, 696, 702
Heat. See also Body temperature
  heart rate affected by, 699
  in inflammation, 785, 787f
  loss of, 164, 963f, 965, 966f, 967
  muscles in generation of, 281
  production of, 963f, 966–967, 966f
  thermoregulatory responses to, 546, 963f, 964–965
Heat capacity, of water, 38
Heat exchange, 964–965, 965f
Heat exhaustion, 967
Heat-loss center, **965**, 966f
Heat-promoting center, **965**, 966, 966f, 967
Heat stroke, **967**
Heat of vaporization, of water, 38, **965**
Heavy (H) chains, 798f, **799**
Heavy metals, **56**
Heavy metal salts, skin penetrated by, 164
Heimlich maneuver, **828**
Helicobacter pylori, peptic ulcers and, 889, 889f
Helicotrema, 584f, **585**, 587, 587f
Helium, 25f, 27, 27f, 31, 31f
Helix, of auricle, 581f, **582**
Helper T cells, 795t, 802, 802f, **806**, 806f, 809f, 810t, 811
Hematocrit, 643f, **644**, 648
Hematology, **666**
Hematoma, 157, 191, 191f, **666**
Hematopoiesis, **176**, 179, **647**, 666
Hematopoietic stem cells, 129, **647**, 647f, 654, 655f, 657f
Hematopoietic tissue, location of in bones, 178–179
Hematuria, 976, 1001t
Heme, **646**, 646f
Hemiazygos vein, 754f, **755t**, 755f

Hemiplegia, 468, 475
Hemochromatosis, **667**, **921**
Hemocytoblast (hematopoietic stem cell),
    647, 647*f*, 654, 655*f*, 657*f*
Hemodialysis, **1001**
Hemoglobin, 49*f*, **156**, **646**–647, 646*f*
    abnormal, 650, 651*f*
    as buffer, 1028
    carbon dioxide transport in, 647, **850**,
        851*f*
    fetal, **666**
    oxygen transport in, 646–647, 847–850,
        848–849*f*, 850*f*
    reduced (deoxyhemoglobin), **647**, **847**,
        852
    skin color affected by, 156
    in urine, 1001*t*
Hemoglobin F (fetal hemoglobin), **666**
Hemoglobin S (HbS), 650, 651*f*
Hemoglobin saturation, 847–850,
    848–849*f*, 850*f*
Hemoglobinuria, 1001*t*
Hemolytic anemia, 650
Hemolytic disease of newborn, **665**
Hemophilias (hemophilia A/B and C),
    662–663
Hemorrhagic anemia, 650
Hemorrhoidal veins, 907*f*, 909
Hemorrhoids, 713, 909
Hemosiderin, **649**, 649*f*
Hemostasis, 12, 12*f*, **657**–663, 658*f*, 659*f*,
    660*f*, 662–663
Henle's loop. *See* Nephron loop
Henry's law, **843**
Heparin, 129, **661**, 662
Hepatic artery, 744*f*, **745***t*, 745*f*
Hepatic artery proper, 744*f*, **745***t*, 745*f*,
    894, 895*f*, 897
Hepatic ducts, 894, 895*f*, 900*f*
Hepatic (right colic) flexure, **906**, 907*f*
Hepatic (stellate) macrophages, 784, 896*f*,
    897
Hepatic portal system, 735, **756***t*, 756*f*,
    757*f*, 874
Hepatic portal vein, 751*f*, **756***t*, 756*f*, 757*f*,
    894, 895*f*, 897
Hepatic veins, 750*f*, 751*f*, **756***t*, 756*f*, 757*f*,
    895*f*, 896*f*
Hepatitis, **897**–898
Hepatocyte(s), **897**, 945, 950, 951*f*, 953,
    954, 955–959, 956*t*
Hepatopancreatic ampulla, **900**, 900*f*
Hepatopancreatic sphincter, **900**, 900*f*, 901*f*
Herceptin. *See* Trastuzumab
Heredity, 1124–1137. *See also*
    Chromosome(s); Gene(s)
    case study and, 1137*b*
    gene expression and, 1131–1133
    genetic screening/counseling/therapy
        and, 1133–1135, 1133*f*, 1134*f*
    genetic variation and, 1047,
        1126–1128, 1126*f*, 1127*f*
    hypertension and, 724
    inheritance patterns and, 1128–1131,
        1128*f*, 1129*t*, 1130*f*
Hering-Breuer (inflation) reflex, **857**
Hernia, 319
    abdominal, 346*t*, **386**
    hiatal, **883**
    inguinal, **1086**
Herniated (prolapsed/slipped) disc,
    219–220, 219*f*, 247, 276
Herpes simplex virus, 169, 1080–1081
Herpes zoster (shingles), **430**
Hertz (Hz), 586
Heterogeneous mixture, 30
Heterozygous (term), **1125**
Hexose sugars, 43, 43*f*
HHb. *See* Deoxyhemoglobin;
    Reduced hemoglobin
Hiatal hernia, 18, **883**
Hiccups, 510
High blood pressure. *See* Hypertension
High-density lipoproteins (HDLs), 760*b*,
    956, 957, 957*f*, 959

High-energy phosphate bonds, 55, 55*f*
Higher mental functions. *See* Cognition/
    cognitive function
Hilton's law, **517**
Hilum
    lymph node, **772**, 772*f*
    pulmonary, 832*f*, **833**
    renal, 975*f*, **976**, 977*f*
    splenic, 773
Hindbrain (rhombencephalon), **435**, 435*f*,
    1106
Hindgut, 918, 918*f*, 1104*f*
Hinge joints, 262*f*, 264
Hinge region (antibody), 798*f*, 799
Hip bones (coxal bones/os coxae),
    236–239, 237*f*, 241*t*
Hip dysplasia, 247
Hip fracture, 193, 241
Hip (pelvic) girdle, 200*f*, **236**–239, 237*f*,
    239*t*, 241*t*
Hip (coxal) joint, 236, 257*t*, **268**–269,
    269*f*
    artificial/replacement, 274*b*
    muscles crossing/muscles acting on,
        331*f*, 367–373*t*, 368*f*, 371–372*f*,
        384*t*
Hippocampus, 444, 456, 456*f*, **457**,
    460–461
Hirsutism, **159**, **636**
His bundle (atrioventricular bundle), **687**,
    687*f*
Histamine, 129, 421*t*, 423, **785**, **887**
    in basophils, 653
    digestion/gastric secretion and, 887, 888*t*
    in immediate hypersensitivity, 812, 813,
        813*f*
    in inflammation, **785**, 785*t*, 787*f*
    in mast cells, 129, 653
Histology, **2**, **116**, 117
Histones/histone proteins, 93, 93*f*, 98,
    102, 1132
Histotoxic hypoxia, **852**
HIV (human immunodeficiency virus),
    811
Hoarseness, 505*t*, 826
Hodgkin's lymphoma, **777**, 811
Holocrine glands, **126**, 126*f*, 162*f*, 163
Holoenzyme, **51**
Homeostasis, **9**–12, 10*f*, 11*f*
Homeostatic imbalance, **12**. *See also*
    *specific type or organ/system
    affected*
Homeostatic relationships. *See specific
    system*
Homogeneous mixture, 29
Homologous chromosomes (homologues),
    1042, **1043**, 1044, 1044*f*, 1046*f*,
    1125, 1125*f*, 1127–1128
Homologous structures, **1042**, 1083*f*
Homozygous (term), **1125**
Horizontal cells, in retina, 560*f*
Horizontal fissure, 832*f*, 833
Horizontal (transverse) plane/section,
    15, 15*f*
Horizontal plates, 206*f*, **212**, 214*f*
Hormonal stimuli, 607–608, 607*f*, 727*f*
Hormone(s), 81, **123**–124, 123*f*, 602,
    603–617, 605*f*, 606*f*, 607*f*, 609*t*. *See
    also specific type*
    absorptive state regulated by, 952, 952*f*
    anabolic, 317, 317*b*, 929
    blood pressure control and, 721, 721*t*,
        722*f*
    in bone growth/remodeling, 188–189,
        189*f*
    bone production of, 176
    chemical structure of, 603
    in contraception, **1118**–1119*b*
    control of release of, 449, 607–608,
        607*f*, 610–611*f*, 611
    in digestion, 888*t*, 889, 890, 890*f*
    in erythropoiesis, 648–649, 648*f*
    food intake regulation and, 961, 961*f*
    glomerular filtration rate control and,
        988*f*, 989

half-life/onset/duration of action of,
    608–609
    heart rate affected by, 699
    hyperglycemic, glucagon as, 630, 631*f*,
        954, 954*f*
    hypoglycemic, insulin as, 630, 631*f*,
        952
    labor induction and, 1114–1115, 1114*f*
    mechanisms of action of, 603–606,
        605*f*, 606*f*
    metabolism affected by, 952, 952*f*, 954,
        954*f*, 955*t*
    non-endocrine tissues/organs producing,
        634–635, 635*t*
    in plasma, 644*t*
    postabsorptive state regulated by, 954,
        954*f*
    protein-based, 49*f*, 51
    protein synthesis regulated by, 606, 606*f*,
        614*f*, 929
    receptors for, 603, 604–605, 605*f*, 606,
        606*f*, 608–609
    reproductive function regulation and,
        1042–1043, 1059*f*, 1074–1076,
        1075*f*, 1077*f*, 1078, 1079*t*
    smooth muscle contraction regulated
        by, 315
    in sodium regulation, 1022–1023
    target cells of, **603**, 608, 609
    tropic, **615**
Hormone (estrogen) replacement therapy,
    1086
Horner's syndrome, **549**
Horns, hyoid bone, 212, 212*f*
Horseshoe kidney, 1007
Housemaid's knee, 273
HP. *See* Hydrostatic pressure
HP$_c$. *See* Capillary hydrostatic pressure
HP$_{cs}$. *See* Hydrostatic pressure, in capsular
    space
HPG axis. *See* Hypothalamic-
    pituitarygonadal axis
HP$_{gc}$. *See* Hydrostatic pressure, in
    glomerular capillaries
HP$_{if}$. *See* Interstitial fluid hydrostatic
    pressure
hPL. *See* Human placental lactogen
HPV. *See* Human papillomavirus
HRT. *See* Hormone (estrogen) replacement
    therapy
5-HT. *See* Serotonin
Human body. *See also under Body*
    maintenance of life and, 5–9, 5*f*
    organization of, 4–5, 4, 4*f*
    orientation to, 1–22
Human chorionic gonadotropin (hCG),
    1099, 1099*f*, 1101–1102
Human chorionic somatomammotropin
    (hCS/human placental lactogen),
    1102, **1113**
Human development, 1091–1123. *See also
    specific aspect and* Pregnancy
    embryonic, 1092, 1092*f*, 1097–1110
    fertilization (egg to zygote), 1092–1096,
        1092*f*, **1093**, 1094–1095*f*, 1096*f*,
        1097*f*
    fetal, 1092, 1092*f*, 1107–1110, 1110*f*,
        1111*t*
    infant adjustment to extrauterine life
        and, 1116
    lactation and, **1116**–1119, 1117*f*
    parturition (birth) and, **1114**–1116,
        1115*f*
Human gene therapy. *See* Gene therapy
Human immunodeficiency virus (HIV),
    811
Human papillomavirus, 1064, **1080**
Human placental lactogen, 1102, **1113**
Humeral artery, anterior and posterior
    circumflex, **742***t*, 742*f*, 743*f*
Humerus, 177*f*, 200*f*, 230*f*, **231**, 231*f*, 235*t*
    in elbow joint, 268, 268*f*
    muscles controlling movement of, 354–
        356*t*, 355–356*f*, 362*t*, 363*f*
    in shoulder joint, 266, 267*f*

Humoral (antibody-mediated) immunity,
    782*f*, **790**, 795*t*, 796–801, 796*f*,
    797*f*, 798*f*, 799*t*, 800*f*, 809*f*. *See also*
    Antibodies
    active, **797**–798, 797*f*
    B cell activation/differentiation and,
        795*t*, 796–797, 796*f*
    helper T cells and, 806, 806*f*
    immunological memory and, 791, **797**
    passive, 797*f*, **798**
    primary response, 796*f*, **797**, 797*f*, 809*f*
    secondary response, 796*f*, **797**, 797*f*
Humoral stimuli, for hormone release,
    607, 607*f*
Humors of eye, 558. *See also* Aqueous
    humor; Vitreous humor
Hunger, food intake regulation and, 449,
    960–962, 961*f*
Huntingtin protein, 469
Huntington's disease, **469**–470, 1129,
    1133
Hyaline cartilage, 127*t*, **134**, 135*f*, **174**,
    175*f*
    in endochondral ossification, 184–185,
        185*f*
    skeletal, **174**, 175*f*
    in symphyses, 253, 253*f*
    in synchondroses, 253, 253*f*
    in synovial joints, 175*f*
    tracheal, 175*f*, 825*f*, 827, 827*f*
Hyaluronic acid, 128, 130, 134
Hydatid (hydatidiform) mole, **1120**
Hydration, hypotonic, **1019**, 1019*f*
Hydration layers, **38**
Hydrocarbon chains, 45, 45*f*
Hydrocephalus, 467, 467*f*
Hydrochloric acid, 39, 41, 886*f*, 887,
    889–891, 891*f*
Hydrocortisone (cortisol), 623*f*, **625**, 625*f*,
    626*t*, 955*t*, 1023
Hydrogen, 25, 26, 26*t*, 27–28, 27*f*, 28*f*,
    31*f*, 579
Hydrogen bonds, 34–35, 35*f*
Hydrogen ions, 39
    in acid-base balance/pH, 39, 40, 40*f*,
        41, 855, 1027. *See also* Acid-base
        balance
    renal regulation of, 992*f*, 994, 995,
        1029–1032, 1030*f*, 1031*f*, 1032*f*
    respiratory regulation of, 852, 856,
        1028–1029
    acids releasing, 39
    bases taking up, 39
Hydrogen peroxide, 86
Hydrogen phosphate, in plasma/interstitial
    fluid/intracellular fluid, 1014*f*, 1015
Hydrogen sulfide (H_2S), **424**
Hydrolase(s), 52, 86
Hydrolysis reactions, 38, **42**, 42*f*, 44, 49*f*,
    56, **912**
Hydrophilic polar head, of phospholipid,
    46*f*, **47**, 63, 65*f*
Hydrophobic nonpolar tail, of
    phospholipid, 46*f*, **47**, 63, 65*f*
Hydrostatic-osmotic pressure interactions,
    71–72, 731–734, 732–733*f*
Hydrostatic pressure, **71**–72, **731**,
    732–733*f*
    in capsular space, **986**, 986*f*
    edema and, 734
    fluid movement among compartments
        and, **71**–72, 731–734, 732–733*f*,
        1015
    in glomerular capillaries, **986**, 986*f*
Hydroxide(s), 39, 41
Hydroxyapatites, 183–184, 188
Hydroxurea, 651
Hydroxyl ion(s), **39**–40, 40*f*
Hymen, **1066**, 1066*f*, 1082*f*
Hyoglossus muscle, **336***t*, 337*f*, 339*f*
Hyoid bone, **212**, 212*f*, 823*f*, 825*f*, 876*f*
Hyperaldosteronism (aldosteronism/
    Conn's disease), 625, 626*t*, **1037**
Hyperalgesia, **496**
Hyperbaric oxygen chambers, 843

Hypercalcemia, 189, 622, 699, 1020*t*, 1026
Hypercapnia, **855**
Hyperchloremia, 1020*t*
Hypercholesterolemia, familial, **969**
Hyperemia, **726**, 728, 728*f*, **785**, 787*f*
Hyperextension (body movement), 258*f*, **259**, 259*f*
Hyperglycemia, 633*b*
Hyperglycemic hormone, glucagon as, 630, 631*f*, **954**, 954*f*
Hyperinsulinism, 634
Hyperkalemia, 699, 1020*t*, 1025
Hypermagnesemia, 1020*t*
Hypernatremia, 1020*t*
Hyperopia (farsightedness), 566*f*, **567**, 595
Hyperparathyroidism, 622
Hyperphosphatemia, 1020*t*
Hyperplasia, **110**
Hyperpnea, **857**
Hyperpolarization, 403, **404**, 404*f*
    action potential and, 406*f*, 407*f*, 408
    IPSPs and, 418*f*, 420
    photoreceptor, 571–573, 571*f*, 572*f*
Hyperprolactinemia, 617
Hypersensitivities, 791, **812**–814, 813*f*
Hypersomnia, **484**
Hypertension, 548, 700, **724**, 760*b*
    portal, **898**
Hyperthermia, **967**, 968. *See also* Fever
Hyperthyroidism, 620–621, 620*f*, 963
Hypertonic solutions, **72**, 72*f*
Hypertrophic cardiomyopathy, **702**
Hypertrophic zone (epiphyseal plate), 186, 187*f*
Hypertrophy, **110**
    muscle, 307, 309
Hyperventilation (respiratory alkalosis), **855**, 1029, 1033*t*, **1034**, 1034*b*, 1035
Hypoaldosteronism (Addison's disease), **625**–626, 626*t*, 1021, 1022
Hypoblast, 1100*f*, 1102, 1103*f*
Hypocalcemia, 189, 622, 622*f*, 699, 1020*t*, 1026
Hypocapnia, **855**
Hypochloremia, 1020*t*
Hypochondriac regions, **20**, 20*f*
Hypodermis (superficial fascia/subcutaneous tissue), 130, **151**, 151*f*, 167
Hypogastric plexus, inferior (pelvic), 536*f*, **537**
Hypogastric (pubic) region, **20**, 20*f*
Hypoglossal canal, **202**, 207*f*, **216***t*, 506*t*
Hypoglossal nerve (cranial nerve XII), 336*t*, 338*t*, 450*f*, 452, 453*f*, 499*f*, 506*t*, **507**, 509*f*
Hypoglycemia, **634**, 945
Hypoglycemic hormone, insulin as, 630, 631*f*, **952**
Hypokalemia, 699, 1020*t*, 1025
Hypomagnesemia, 1020*t*
Hyponatremia, **1019**, 1020*t*
Hyponychium (nail), 160*f*, **161**
Hypoparathyroidism, 622
Hypophosphatemia, 1020*t*
Hypophyseal arteries, inferior and superior, 610*f*, 611*f*
Hypophyseal fossa, 207*f*, **209**, 209*f*
Hypophyseal portal system, 611, 611*f*
Hypophyseal portal vein, **610**–611, 611*f*
Hypophysectomy, **636**
Hypophysis. *See* Pituitary gland (hypophysis)
Hypoproteinemia, **734**
Hypospadias, 1007
Hypotension, **724**
    orthostatic, **549**, 724
Hypothalamic-hypophyseal tract, **610**, 610*f*
Hypothalamic nuclei, 449, 449*f*
Hypothalamic-pituitary-gonadal axis, **1042**–1043, 1059*f*

Hypothalamic thirst center, 449, 1016–1017, 1017*f*
Hypothalamus, 447*f*, **448**–450, 448*f*, 450*f*, 454, 458*t*, 467–468
    autonomic regulation by, 449, 547–548, 547*f*
    development of, 435, 435*f*, 483
    food intake regulation and, 449, 960, 961*f*
    homeostatic imbalances and, 450
    hormone secretion and, 449, 602, 602*f*, 607*f*, 608, 610–611*f*, 611
    in labor, 1114
    limbic system and, 449, 456, 456*f*
    pituitary relationship and, 607*f*, 608, 610–611, 610–611*f*
    respiratory regulation and, 856–857
    sleep and, 449, 463
    suprachiasmatic nucleus of, 627
    stress response and, 628–629*f*
    in temperature regulation, 449, 965–966, 966*f*
    thirst regulated by, 449, 1016–1017, 1017*f*
Hypothenar eminence, muscles of, 364*t*
Hypothermia, **967**
Hypothyroidism, 620, 963
Hypotonic hydration, **1019**, 1019*f*
Hypotonic solutions, **72**, 72*f*
Hypoventilation (respiratory acidosis), **858**, 858*f*, 1029, **1033**–1034, 1033*t*, 1034*b*, 1035
Hypovolemic shock, **725**, 989, 1018
Hypoxemic hypoxia, **852**
Hypoxia, 648, 648*f*, **852**
Hypoxia-inducible factor (HIF), 648
Hysterectomy, **1086**
H zones, **284**, 285*f*, 288, 288*f*, 289, 289*f*

I. *See* Iodine
I bands, **284**, 285*f*, 288*f*, 289, 289*f*, 684*f*, 685
IC. *See* Inspiratory capacity
ICF. *See* Intracellular fluid
Identical twins, 1074
IF. *See* Interstitial (tissue) fluid
IFNs. *See* Interferons
Ig(s) (immunoglobulins), 653, **798**–801, 798*f*, 799*t*, 800*f*, 810*t*. *See also* Antibodies
IgA, 799, 799*t*, 1117
IgA antibodies, in saliva, 879
IgD, 799, 799*t*
IgE, 799, 799*t*, 812, 813*f*
IGFs. *See* Insulin-like growth factor(s)
IgG, 799, 799*t*, 813
IgM, 799, 799*t*, 813
IL. *See under* Interleukin
Ileocecal valve/sphincter, **901**, 901*f*, 905*t*, 906, 907*f*
Ileocolic artery, 744*f*, **746***t*, 747*f*
Ileostomy, 910
Ileum, 869*f*, **901**, 901*f*, 907*f*
Ileus, 921
Iliac arteries
    common, **738***t*, 738*f*, 739*f*, 744*f*, **746***t*, 746*f*, 747*f*, **748***t*, 748*f*, 749*f*
    external and internal, 739*f*, **748***t*, 748*f*, 749*f*
Iliac crest, **236**, 237*f*
Iliac fossa, **236**, 237*f*
Iliac (inguinal) regions, **20**, 20*f*
Iliac spines, **236**, 237*f*
Iliacus muscle, **367***t*, 368*f*
Iliac veins
    common, **750***t*, 750*f*, 751*f*, 756*f*, 757*f*, **758***t*, 758*f*
    external and internal, 751*f*, 756*f*, **757***f*, **758***t*, 758*f*
Iliococcygeus muscle, 349*f*
Iliocostalis cervicis/lumborum/thoracis muscles, **342***t*, 343*f*
Iliofemoral ligament, **269**, 269*f*
Iliohypogastric nerve, 513*f*, 513*t*

Ilioinguinal nerve, 513*t*, 513*f*
Iliopsoas muscle, 330*f*, **367***t*, 368*f*, 384*t*
Iliotibial tract, 331*f*, 370*t*, 371*f*
Ilium, **236**, 237*f*
Imaging (medical), 16–17*b*
Imatinib, 145*b*
Immediate (acute/type I) hypersensitivity, **812**–813, 813*f*
Immune (antigen-antibody) complexes, **800**, 800*f*, 813
Immune complex (type III) hypersensitivity, **813**
Immune interferon, 788
Immune response. *See* Immunity/immune response
Immune surveillance, **807**
Immune system, 781–817, **782**, 782*f*. *See also* Immunity/immune response; Lymphatic system
    adaptive defenses and, **782**, 782*f*, **790**–811, 809*f*, 810*t*
    cancer immunotherapy to activate, 145*b*
    cardiovascular system and, 761*b*, 778*b*
    case study and, 780*b*, 817*b*
    cells of, 792–795, 793*f*, 794*f*, 795*t*, 795*f*, 810*t*. *See also specific type*
    developmental aspects of, 814
    digestive system and, 778*b*, 920*b*
    endocrine system and, 637*b*, 778*b*
    homeostatic imbalances/disorders and, 786–787, 801, 811–814
    innate defenses and, **782**–790, 782*f*, 783*t*, 790*t*, 808*f*. *See also* Innate defenses
    integumentary system and, 168*b*, 778*b*
    lymph nodes in activation of, 771
    muscular system and, 318*b*, 778*b*
    nervous system and, 550*b*, 778*b*, 814
    reproductive system and, 778*b*, 1085*b*
    respiratory system and, 778*b*, 862*b*
    skeletal system and, 196*b*, 778*b*
    stem cells of, 814
    urinary system and, 778*b*, 1036*b*
Immune theory of aging, 110
Immunity/immune response, 7*f*, **782**–817, 782*f*, 808–809*f*. *See also* Immune system
    active, **797**–798, 797*f*
    artificially acquired, 797, 797*f*, 798
    cellular. *See* Cellular (cell-mediated) immunity
    homeostatic imbalances/disorders and, 786–787, 801, 811–814
    humoral. *See* Humoral (antibody-mediated) immunity
    naturally acquired, 797, 797*f*, 798
    passive, 797*f*, **798**
    primary, 796*f*, **797**, 797*f*, 805, 809*f*
    secondary, 796*f*, **797**, 797*f*
    in tissue repair, 142
Immunization, 798, 814
Immunocompetence, **792**
Immunodeficiency, **811**
Immunogenicity, **791**
Immunoglobulin(s), 653, **798**–801, 798*f*, 799*t*, 800*f*, 810*t*. *See also specific type under Ig and* Antibodies
Immunological memory, 791, **797**. *See also* Memory cells
Immunology, **814**
Immunopathology, **814**
Immunosuppressive therapy, after transplant, 811
Immunotherapy, cancer, 145*b*
Impacted teeth, 880
Impaired function, in inflammation, 785, 787*f*
Impetigo, **169**
Implantable cardioverter defibrillators (ICDs), 688
Implantation, 1097*f*, 1098*f*, **1099**
Imprinting, genomic, **1132**
Inactivation gate, 407*f*, 408
Inborn errors of metabolism, 968
Inborn (intrinsic) reflex, 519

Incisive canal, 205*f*, **211**, **217***t*
Incisive fossa, 206*f*, **211**, **217***t*
Incisors, 879*f*, **880**
Inclusions, **83**, 95*t*
Incompetent heart valve, 680, 696
Incomplete antigen, **791**
Incomplete dominance, **1129**
Incomplete fracture, 191
Incomplete proteins, 929, 930*t*
Incomplete (unfused) tetanus, 300*f*, **301**
Incontinence, 911, **1005**, 1007
Incus (anvil), 581*f*, **582**, 582*f*, 587*f*
Independent assortment, 1126*f*, **1127**
Indifferent stage, of gonadal development, **1083**, 1083*f*
Indirect muscle attachments, **282**–284
Indirect (extrapyramidal/multineuronal) pathways, 478–480, 479*f*, 481*t*, 518
Indolamines, **423**
Infant. *See* Newborn
Infant respiratory distress syndrome (IRDS), **839**
Infarct
    myocardial, **683**, 700, 760*b*
    renal, **1008**
Infection, 144, 167
Infectious mononucleosis, 654
Inferior (term), 14*t*
Inferior angle of scapula, 228, 229*f*
Inferior articular process, 219*f*, **220**, 221*f*, 222*f*, 223*t*
Inferior cerebellar peduncles, 450*f*, 451*f*, 452, 453*f*, **454**, 455*f*
Inferior cervical ganglion, 540*f*, **541**
Inferior colliculus, 450*f*, **451**, 451*f*, 589*f*, **590**
Inferior costal facet, 222, 222*f*, 223*t*
Inferior ganglion, 505*t*
Inferior gemellus muscle, 372*f*, **373***t*
Inferior gluteal artery, **748***t*, 748*f*
Inferior gluteal line, **236**, 237*f*
Inferior gluteal nerve, 514*f*, **515**, 515*t*
Inferior horns, 437*f*, 446*f*
Inferior hypogastric (pelvic) plexus, 536*f*, **537**
Inferior hypophyseal artery, 610*f*
Inferior iliac spines, 236, 237*f*
Inferior mesenteric artery, 738*f*, 739*f*, 744*f*, **746***t*, 746*f*, 747*f*
Inferior mesenteric ganglion/plexus, 540*f*, **541**
Inferior mesenteric vein, 751*f*, **756***t*, 756*f*, 757*f*
Inferior nasal conchae, 202*f*, 203*f*, **212**, 214*f*, **217***t*, 821*f*, 822
Inferior nasal meatus, 214*f*, 215, 821*f*
Inferior nuchal line, 203, 203*f*, 206*f*
Inferior oblique muscle, **556**, 557*f*
Inferior olivary nuclei, 452, 453*f*
Inferior orbital fissure, 203*f*, **212**, **217***t*
Inferior pharyngeal constrictor muscle, **339***t*, 339*f*
Inferior phrenic arteries, 738*f*, 744*f*, 746*f*
Inferior phrenic veins, **756***t*, 756*f*, 757*f*
Inferior pubic ramus, 237*f*, **238**
Inferior rectus muscle, **556**, 557*f*
Inferior sagittal sinus, **752***t*, 752*f*, 753*f*
Inferior salivatory nuclei, 537
Inferior suprarenal artery, 746*t*
Inferior thyroid artery, 617, 617*f*
Inferior tibiofibular joint, 242*f*, 243, 257*t*
Inferior vena cava, **674**, 675–677*f*, 681*f*, **750***t*, 750*f*, 751*f*, 755*f*, **756***t*, 756*f*, 757*f*, 758*f*
Inferior vestibular ganglion, 583*f*, **591**
Infertility, male, 1058
Inflammation/inflammatory response, 142, 143*f*, 782*f*, **785**–787, 785*t*, 787*f*, 790*t*, 800*f*, 801, 808*f*, 814
Inflammatory bowel disease, **921**
Inflammatory chemicals. *See* Chemical mediators
Inflammatory joint disease, 273–275, 275*f*
Inflation (Hering-Breuer) reflex, **857**

Influenza virus, immune response to, 808–809*f*
Infraglenoid tubercle, 229*f*
Infrahyoid muscles, **338***t*
Infraorbital foramen, 203*f*, 206*f*, 211*f*, **212**, **217***t*
Infrapatellar bursa, 270*f*
Infrapatellar fat pad, 270*f*
Infraspinatus muscle, 266, 331*f*, 353*f*, **354***t*, 355*f*, **356***t*, 362*t*
Infraspinous fossa, 228, 229*f*
Infraspinous process, 228, 229*f*
Infratemporal fossa, 265*f*, 266
Infundibulum, **448**, **609**
   hypothalamic-pituitary gland, 448, 449*f*, 450*f*, 609, 610*f*
   uterine tube, 1061*f*, **1062**, 1063*f*
Ingestion. *See under* Food
Inguinal hernia, **1086**
Inguinal ligament, 238, 347*f*
Inguinal lymph nodes, 769*f*
Inguinal (iliac) regions, **20**, 20*f*
Inhalation. *See* Inspiration
Inheritance, 1128–1131, 1128*f*, 1129*t*, 1130*f*. *See also specific type*
Inhibin, **1043**, 1058, 1059*f*, 1075, 1075*f*
Inhibiting hormones, hypothalamic, 449, 608, **611**, 611*f*, 613*t*
Inhibitory neurotransmitters, 421–422*t*, 424
Inhibitory postsynaptic potentials (IPSPs), 417*t*, 418*f*, 419*f*, **420**, 572*f*, 573
Inhibitory synapses, 417–420
Initial segment of axon, action potential at, 405, 420
Initiation in transcription, 102, 102*f*
Initiation in translation, 104, 106*f*
Initiator tRNA, **104**, 106*f*
Inlet, pelvic, **238**, 239*t*
Innate (nonspecific) defenses, 782–790, 782*f*, 783*t*, 790*t*, 806, 808*f*
   internal (cells/chemicals), 782*f*, 783–790, 784*f*, 785*t*, 786*f*, 787*f*, 788*f*, 789*f*, 790*t*, 808*f*
   surface barriers (skin/mucosae), 163–164, 782–783, 782*f*, 783*t*, 808*f*
Inner cell mass (embryoblast), 1097*f*, **1098**
Inner (internal) ear. *See* Internal (inner) ear
Inner fiber, rod and cone, 568, 568*f*
Inner hair cells, 584*f*, **585**, 588, 589*f*
Inner layer of eye. *See* Retina
Inner mitochondrial membrane, 83–84, 83*f*
Inner segments, rod and cone, **568–569**, 568*f*
Innervation, of epithelial tissue, 118
Inorganic components, of bone, 183–184
Inorganic compounds, **38–41**. *See also specific type*
Inorganic phosphate, muscle fatigue and increased, 306
Inositol trisphosphate, **605**
Inotropic agents, positive and negative, 698
Input (receptive) regions, **395**, 395*f*
Insensible heat loss/perspiration, 164, **965**
Insensible water loss, **965**, **1016**
Insertion (muscle), **258**, **282**, 324
Insoluble fiber, 928
Insomnia, **463–464**
Inspiration, 819*f*, **835**, 836–838, 837*f*, 838*f*, 860, 1116. *See also* Breathing
Inspiratory capacity (IC), **841**, 841*f*
Inspiratory muscles, 344*t*, 345*f*, 346*t*, **836–838**, 837*f*
Inspiratory reserve volume (IRV), **840**, 841*f*
Insufficient heart valve, 680, 696
Insula, 438*f*, **439**, 443
Insulators, 400
Insulin, 10, 49*f*, **630**–634, 631*f*
   in absorptive state regulation, 952, 952*f*
   β cells producing, **630**, 630*f*
   deficit of/diabetes mellitus and, **632**–634, 632*f*, 633*b*, 952
   excess (hyperinsulinism), 634
   in food intake regulation, 961, 961*f*

mechanism of action of, 605
   metabolism and, 952, 952*f*, 954, 954*f*, 955*t*
   in postabsorptive state regulation, 954, 954*f*
Insulin-dependent (type 1) diabetes mellitus (IDDM), 632, 633*b*, 812, 968
Insulin-like growth factor(s) (IGFs), 614*t*, **615**
Insulin resistance, 633*b*
Integral proteins, 65*f*, **66**, 81
Integration, **391**, 391*f*, **426**–427, 426*f*, 427*f*
   motor, 517–519, 518*f*
   synaptic/postsynaptic, 419*f*, 420
Integration center, in reflex arc, 426*f*, 427, 519, 519*f*, 543*f*
Integrins, 81, 428
Integument. *See* Skin
Integumentary system, 5*f*, 6*f*, **150–172**. *See also* Skin
   autonomic nervous system and, 550*b*
   cardiovascular system and, 168*b*, 761*b*
   case study and, 172*b*
   developmental aspects of, 167–169
   digestive system and, 168*b*, 920*b*
   endocrine system and, 168*b*, 637*b*
   functions of, 5*f*, 6*f*, 8, 163–165
   homeostatic relationships of, 5*f*, 168*b*
   lymphatic system/immunity and, 168*b*, 778*b*
   muscular system and, 168*b*, 318*b*
   nervous system and, 168*b*, 550*b*
   reproductive system and, 168*b*, 1085*b*
   respiratory system and, 168*b*, 862*b*
   skeletal system and, 168*b*, 196*b*
   thyroid hormone affecting, 618*t*
   urinary system and, 168*b*, 1036*b*
Intensity of sound waves, relative, 590
Interatrial septum, **674**, 676*f*
Intercalated cells, 979*f*, 981, 1025, 1030, 1031, 1031*f*, 1032
Intercalated discs, 139*f*, **140**, **684**, 684*f*
Intercarpal joint, 256*t*
Intercellular clefts, **711**, 711*f*
Intercondylar eminence, 242*f*, **243**
Intercondylar fossa, 240*f*, **241**
Intercostal arteries, anterior and posterior, 738*f*, 742*f*, **743***t*, 743*f*
Intercostal muscles, 330*f*, 344*t*, 345*f*, 832*f*, 836–838, 837*f*, 853*f*
Intercostal nerves, 344*t*, 346*t*, 507*f*, 508*f*, **515**, **853**, 853*f*
Intercostal spaces, 225, 225*f*
Intercostal veins, posterior, 754*t*, 754*f*, **755***t*, 755*f*
Interferons (IFNs), **788**, 788*f*, 790*t*, 805*t*, 808*f*
Interleukin(s), 654, 805, 805*t*
Interleukins 1 and 2, **805**, 805*t*
Interleukins 4, 5, 10, 12, and 17, 805*t*
Interlobar arteries, **978**, 978*f*
Interlobar veins, **978**, 978*f*
Interlobular (cortical radiate) arteries, renal, 978, 978*f*, 981*f*, 982
Interlobular veins, hepatic, 896*f*
Intermediate (term), 14*t*
Intermediate cuneiform bone, 241*t*, **244**, 244*f*
Intermediate filaments, 62*f*, **88**, 88*f*, 95*t*
   in desmosomes, 67*f*, 68
   in muscles, 287, 314, 314*f*
   in neurons (neurofibrils), **394**
Intermediate mass (interthalamic adhesion), **447**, 447*f*
Intermediate mesoderm, 1105*f*, **1106**, 1106*f*
Intermediate part (membranous) urethra, 1003*f*, **1004**, 1050, 1051*f*
Intermittent claudication, 319
Internal (term), 14*t*. *See also under* Deep
Internal acoustic meatus, 205*f*, 207*f*, **208**, **216***t*, 503*t*
Internal anal sphincter, **907**–909, 907*f*
Internal capsule, **444**, 445*f*

Internal carotid artery, 208, 738*f*, 739*f*, **740***t*, 740*f*, 741*f*
Internal (inner) ear, 581*f*, **583**–585, 583*f*, 584*f*, 585*t*, 587, 587*f*, 596
   Ménière's syndrome and, **595**
Internal ear cavity, 208
Internal feedback, in motor control, 518
Internal genitalia, in female, **1060**, 1063–1064*f*
Internal iliac arteries, 739*f*, **748***t*, 748*f*, 749*f*, 1065
Internal iliac vein, 751*f*, 756*f*, 757*f*, **758***t*, 758*f*
Internal intercostal muscles, **344***t*, 345*f*, 838
Internal jugular veins, **750***t*, 750*f*, 751*f*, **752***t*, 752*f*, 753*f*, 754*f*, 755*f*
Internal oblique muscles, 330*f*, **346***t*, 347*f*
Internal os, 1063*f*, 1064
Internal pudendal artery, **748***t*, 748*f*, 1054
Internal respiration, 819*f*, 843, 844*f*, 846–847
Internal root sheath, 159
Internal spermatic fascia, 1048*f*
Internal strabismus, 503*t*
Internal thoracic (internal mammary) artery, 741*f*, 742*f*, **743***t*, 743*f*
Internal urethral sphincter, 1003*f*, **1004**
Interneurons (association neurons), **399**, **399***t*, 472, 474*f*, 474*t*, 482, 482*f*
Internodal pathway, of cardiac conduction system, 687
Interoceptors (visceroceptors), **490**, 491*t*, 492*t*
Interossei muscles, **364***t*, 365*f*, **366***t*, **382***t*, 383*f*
Interosseous artery, common, **742***t*, 742*f*, 743*f*
Interosseous membrane, **232**, 232*f*, 242, 242*f*
Interosseous nerve, forearm muscles supplied by, 360*t*
Interphalangeal joints, 256*t*, 257*t*
Interphase, **96**–98, 96*f*, 100*f*, 1044*f*, 1045*f*
Intersex, 1084
Interspinales muscle, 343*f*
Interspinous ligament, 219*f*
Interstitial cells of Cajal, 892
Interstitial endocrine cells (Leydig cells), **1055**, 1055*f*
Interstitial (tissue) fluid, 5*f*, 8, **62**, 128, **1013**, 1013*f*, 1014*f*, 1015, 1015*f*
Interstitial fluid colloid osmotic pressure (OP_{if}), **731**, 733*f*, 734
Interstitial fluid hydrostatic pressure (HP_{if}), **731**, 733*f*, 734
Interstitial growth, **174**, 186
Interstitial lamellae, **182**, 183*f*
Intertarsal joint, 257*t*
Interthalamic adhesion (intermediate mass), **447**, 447*f*
Intertransversarius muscle, 343*f*
Intertrochanteric crest, 240*f*, **241**
Intertrochanteric line, 240*f*, **241**
Intertubercular sulcus (bicipital groove), 230*f*, **231**
Interventricular artery, anterior and posterior, 675*f*, 676*f*, **682**, 683*f*
Interventricular foramen, **436**, 437*f*, 447*f*
Interventricular septum, **674**, 677*f*
Interventricular sulcus, anterior and posterior, 674, 676*f*
Intervertebral discs, 136*f*, 137, 175*f*, 218*f*, **219**, 219*f*, 222*f*, 256*t*
   herniated (prolapsed/slipped), **219**–220, 219*f*, 247, 276
   as symphyses, 253*f*, 254
Intervertebral foramen, 218*f*, 219*f*, **220**
Intervillous spaces, in stratum functionalis, 1100*f*, **1101**, 1101*f*
Intestinal arteries, 744*f*, **746***t*, 747*f*
Intestinal crypts, 902*f*, **903**
Intestinal (enteric) gastrin, 888*t*, 890*f*, **891**. *See also* Gastrin

Intestinal juice, 903–905
Intestinal microbiota, **909**–910, 958*b*
Intestinal motility, 905–906, 905*t*, 910
Intestinal phase, of gastric secretion, 890*f*, **891**
Intestinal trunk, 768*f*, **769**
Intestine(s). *See also* Large intestine; Small intestine
   arteries supplying, 744*f*, **746***t*, 747*f*
   autonomic innervation/effects and, 536*f*, 539*f*, 540*f*
   mucosa-associated lymphoid tissue (MALT) in, **873**, 873*f*, 902*f*, 903
   parathyroid hormone affecting, 1026
Intra-alveolar (intrapulmonary) pressure, **835**, 835*f*, 836*f*, 837*f*, 838*f*
Intracapsular ligaments, **254**, 271–272, 272*f*. *See also* Cruciate ligaments
Intracellular fluid, 8, **63**, 402*f*, 403, 1013, 1013*f*, 1014*f*, 1015, 1015*f*
Intracellular fluid compartment, **1013**, 1013*f*
Intracellular receptors, in hormone mechanism of action, 604, 606, 606*f*
Intracranial dural sinuses. *See* Dural venous sinuses
Intracranial hemorrhage, 468
Intrafusal muscle fibers, 493, **520**–521, 520*f*, 521*f*
Intramembranous ossification, 184, **186**, 186*f*
Intramolecular bonds, hydrogen bonds as, 35
Intramural ganglia, 537
Intraocular pressure, in glaucoma, 563
Intraperitoneal (peritoneal) organs, **872**, 872*f*
Intrapleural pressure, **835**, 835*f*, 836*f*, 838*f*
Intrapulmonary pressure, **835**, 835*f*, 836*f*, 837*f*, 838*f*
Intrauterine device, 1118*b*, 1119*b*
Intravenous pyelogram (IVP), 1004, 1004*f*, **1008**
Intrinsic (deep) back muscles, 340–343*t*, 341*f*, 343*f*
Intrinsic cardiac conduction system, 685, **686**–689, 687*f*, 688, 688*f*, 692*f*
Intrinsic controls (autoregulation), for blood flow, **725**–730, 725*f*, 727*f*
Intrinsic eye muscles, **557**
Intrinsic factor, **650**, 887, **889**, 917, 931
Intrinsic foot muscles, 380–383*t*, 381*f*, 383*f*
Intrinsic nerve plexuses, of alimentary canal, 873*f*, 874, 875*f*
Intrinsic palm muscles, 358*t*, 359*f*, 361*f*, 364–366*t*, 365*f*
Intrinsic pathway, **659**, 659*f*
Intrinsic (inborn) reflex, 519
Intrinsic (minor) salivary glands, **878**
Intrinsic tongue muscles, 336*t*, **877**
Introns, 99, 102–103
Inulin, glomerular filtration rate estimated by clearance rate of, 1000–1001
Inversion (body movement), **260**, 261*f*, 384*t*
In vitro fertilization, 1119
Invagination, 123*f*
Involuntary muscle, **138**, 280. *See also* Cardiac muscle; Smooth muscle
Involuntary nervous system. *See* Autonomic nervous system
Iodide, in thyroid hormone synthesis, 619, 619*f*
Iodine, 26*t*, 619, 619*f*, 620, 620*f*, 932, 933*t*
Iodine–123, 28
Ion(s), **32**, 32*f*, 39, 992, 992*f*
Ion channels, 66, 70, 290, 400, 401*f*. *See also specific type and* Channels
Ionic bonds, **32**, 32*f*, 34*f*
Ionic imbalances, muscle fatigue and, 306
Ionizing radiation, 28, **56**
Ionotropic (channel-linked) receptors, **424**–425, 424*f*
IP_3 (inositol trisphosphate), **605**

Ipsilateral fibers, cerebellar, **454**
Ipsilateral reflex, **523**
IPSPs. *See* Inhibitory postsynaptic
    potentials
Iris, 546*t*, 555*f*, 558*f*, **559**, 559*f*, 562*f*
Iron, 26*t*, 646, 646*f*, 647, 649, 649*f*, 917,
    931, 933*t*
Iron-deficiency anemia, 650, 933*t*
Iron storage disease (hemochromatosis),
    667, 921
Irregular bones, **176**, 177*f*, 185
Irregular dense connective tissue, 127*t*,
    129, **133**, 133*f*
Irreversible denaturation, 51
Irritability. *See* Excitability
Irritable bowel syndrome, **910**
Irritant receptors/reflexes (pulmonary),
    854*f*, 857
IRV. *See* Inspiratory reserve volume
Ischemia, **468**
Ischemic (stagnant) hypoxia, **852**
Ischemic (luteolytic) phase of ovarian
    cycle, 1074
Ischial ramus, 236, 237*f*
Ischial spine, **236**, 237*f*
Ischial tuberosity, **236**–238, 237*f*
Ischiocavernosus muscle, 348*t*, 349*f*
Ischiofemoral ligament, **269**, 269*f*
Ischium, **236**–238, 237*f*
Islets of Langerhans (pancreatic islets),
    630, 630*f*, 899
Isografts, **814**
Isomers (glucose), **43**–44, 43*f*
Isometric contraction, 299, 302*f*, **303**
Isometric exercises, muscles affected by,
    309
Isotonic contraction, 302*f*, **303**
Isotonic solutions, **72**, 72*f*
Isotopes, **27**–28, 28*f*
Isovolumetric contraction/relaxation
    phases, **693**, 694*f*, 695*f*
Isthmus
    of fauces, **823**, 823*f*
    of thyroid gland, 617, 617*f*
    of uterine tube, **1062**, 1063*f*
    of uterus, 1063, 1063*f*
Itch receptor, 491

Jaundice, 157, 898, 1117, **1120**
Jaw/jawbone. *See* Mandible; Maxillary
    bones/maxilla(e)
Jaw joint. *See* Temporomandibular joint
Jejunum, 869*f*, 900*f*, 901*f*
Joint(s), **251**–278, 255*t*, 256–257*t*.
    See also specific joint
    arthritis/osteoarthritis affecting,
        273–275, 275*f*
    artificial (prosthetic), 274*b*
    blood supply of, 255
    cartilaginous, 251, **253**–254, 253*f*, 255*t*
    case study and, 278*b*
    classification of, **251**–252, 255*t*
    common injuries of, 272–273, 273*f*
    developmental aspects of, 275–276
    fibrous, 251, **252**–253, 252*f*, 255*t*
    homeostatic imbalances and, 266, 272
    inflammatory/degenerative conditions
        of, 273–275, 275*f*
    innervation of, 255, 517
    muscles in stabilization of, 258, 281
    synovial, 251, **254**–272, 254*f*, 255*t*,
        256–257*t*, 262–263*f*, 275
Joint (articular) capsule. *See* Articular
    (joint) capsule
Joint (articular/synovial) cavities, **20, 254,**
    254*f*, 267*f*, 268*f*, 270*f*, 271
Joint kinesthetic receptors, 492*t*, **493**
Jugular foramen, 206*f*, 207*f*, **208, 216***t*,
    505*t*, 752*t*, 753*f*
Jugular (suprasternal) notch, **225**, 225*f*
Jugular trunks, 768*f*, **769**
Jugular veins, external and internal, **750***t*,
    750*f*, 751*f*, **752***t*, 752*f*, 753*f*, 754*f*,
    755*f*

Junctional folds, **290**, 291*f*, 292*f*
Junctional rhythm, **688**, 692*f*
Juxtaglomerular (JG/granular) cells, **983**,
    983*f*, 988*f*, 989
Juxtaglomerular complex (JGC), 624,
    982–983, 983*f*, 988, 988*f*, 1022
Juxtamedullary nephrons, 981*f*, **982**, 995,
    996–997*f*

K. *See* Potassium
Karyotype, **1125**, 1125*f*
Keloid, **147**
Keratin, 67*f*, 68, 123, **152**, 153, 157, 160,
    782, 783*t*
Keratin filaments, 67*f*, 68
Keratinization, 123, **153**
Keratinized epithelia, 122*f*, 123
Keratinocytes, **152**–153, 152*f*
Keratohyaline granules, 153
Ketoacidosis, **632**–634
Keto acids, 939*f*, **940**, 947, 948*f*, 949, 949*f*
Ketogenesis, **947**, 947*f*
Ketone(s)/ketone bodies, **632**–634, 632*f*,
    947, 947*f*, 1001*t*, 1027
Ketonuria, 632, 632*f*, 1001*t*
Ketosis, 947
Kidney(s), 974–1002, 975*f*. *See also under
    Renal and* Nephron(s)
    in acid-base balance, 992*f*, 995, 1027,
        1029–1032, 1030*f*, 1031*f*, 1032*f*,
        1035
    arteries supplying, 739*f*, 744*f*, **746***t*, 746*f*
    autonomic innervation/effects and, 540*f*,
        546*t*
    blood pressure regulation and,
        721–723, 722*f*, 723*f*, 1022, 1024*f*
    blood supply/blood flow/perfusion of,
        726, 726*f*, 977–978, 978*f*, 982*f*
    in calcium regulation, 1026
    developmental aspects of, 1006–1007,
        1006*f*
    erythropoietin produced by, 648, 648*f*
    in fluid balance, 974, 995, 998, 999*f*,
        1024*f*
    function, evaluation of, 1000–1002
    homeostatic imbalances and, 976
    hormones produced by, 634, 635*t*
    horseshoe, 1007
    infection of (pyelonephritis), 977, 1004
    internal anatomy of, 976–977, 977*f*
    location/external anatomy of,
        975–976, 976*f*
    nerve supply of, 978
    parathyroid hormone affecting, 1026
    polycystic, 1007
    potassium regulation and, 624*f*, 1025
    pregnancy and, 1113
    sodium balance regulation and, 624,
        624*f*, 1022, 1022*f*, 1024*f*
    stress response and, 629*f*
    urine formation by, 983–1000, 984*f*
    urine transport/storage/elimination and,
        1002–1006, 1002*f*, 1003*f*, 1004*f*,
        1005*f*
    veins of, 750*f*, 751*f*, **756***t*, 756*f*, 757*f*
Kidney stones (renal calculi), **1002**
Killer cells, natural (NK), 782*f*, **785**,
    790*t*, 807
Kilocalories, **927**
Kinases, 81, 82*f*, 98
Kinesins, 89, 396
Kinesthetic receptors, joint, 492*t*, **493**
Kinetic energy, **24**, 52, 52*f*, 68, 69*f*, 73*t*, 75
Kinetochore(s), 100*f*
Kinetochore microtubules, 100*f*
Kinins, **785**, 785*t*, 787*f*
Kinocilium, **590**, 591, 591*f*, 592*f*
Klinefelter's syndrome, 1081
Kneecap (patella), 200*f*, 240*f*, 241, 241*t*,
    242, 270, 270*f*
Knee-jerk reflex, 271, 522*f*, **523**
Knee joint, 257*t*, 270–272, 270–271*f*
    artificial/replacement, 274*b*
    femoropatellar, 257*t*, **270**–271

injuries of, **272**, 272*f*
muscles crossing/muscles acting on,
    330*f*, 331*f*, 367–373*t*, 368*f*,
    371–372*f*, 384*t*
tibiofemoral, 257*t*, **270**
Knob, ATP synthase, 943*f*
Knuckle (metacarpophalangeal joint),
    234, 256*t*
Koilonychia, 161, 161*f*
Korotkoff sounds, 717
Krebs (citric acid) cycle, 935*f*, 936, 937*f*,
    939–940, 939*f*, 944*f*
Kwashiorkor, **969**
Kyphosis, 220, 220*f*

Labia of mouth. *See* Lip(s) (labia)
Labial frenulum, **876**, 876*f*
Labia majora (labium majus), 1061*f*,
    1066*f*, **1067**, 1083*f*
Labia minora (labium minus), 1061*f*,
    1066*f*, **1067**, 1083*f*
Labioscrotal swellings, 1083*f*, **1084**
Labor, 11, 612, **1114**–1116, 1114*f*, 1115*f*
Labyrinth (internal ear), 581*f*, **583**–585,
    583*f*, 584*f*, 585*t*, 587, 587*f*, 596
    Ménière's syndrome and, **595**
Labyrinthitis, **596**
Lacrimal apparatus, **556**, 556*f*
Lacrimal bones, 203*f*, 204*f*, **212, 217***t*
Lacrimal canaliculi, **556**, 556*f*
Lacrimal caruncle, **554**, 555*f*
Lacrimal fossa, 204*f*, **212, 217***t*
Lacrimal gland, 536*f*, 540*f*, 546*t*, **556**, 556*f*
Lacrimal nuclei, 536
Lacrimal puncta, **556**, 556*f*
Lacrimal sac, 212, **556**, 556*f*
Lacrimal secretion (tears), **556**, 783*t*
Lactase, 913*f*, **914**
Lactation, 612, 616, 617, 1068, 1068*f*,
    1116–1119, 1117*f*
Lacteals, 769, 902*f*, **903**
Lactic acid, **304**–305, 304*f*, **938**, 938*f*,
    1027
Lactiferous ducts, **1068**, 1068*f*
Lactiferous sinus, **1068**, 1068*f*
Lactogen, human placental, 1102, **1113**
Lactose, 43*f*, 44, 913*f*, 914, 915
Lactose intolerance, 914–915
Lacunae
    bone, 136*f*, 137, **182**, 183*f*
    cartilage, 134, 135*f*, 136*f*, 174
    in stratum functionalis (intervillous
        spaces), **1100**–1101, 1100*f*, 1101*f*
Lagging strand, 97, 97*f*
Lambdoid suture, 201, 201*f*, **202**, 203*f*,
    204*f*, 205*f*
Lamellae (bone), 136*f*, 137, **182**, 182*f*,
    183*f*
Lamellar bone, 182. *See also* Compact
    bone
Lamellar (Pacinian) corpuscles, 151*f*, 155,
    164, 492*t*, **493**
Lamellar granules, 153
Lamina(e)
    basal, **118**
    nuclear, 92, 92*f*
    osseous spiral, 584*f*, **585**
    reticular, **118**
    vertebral, **220**, 220*f*
Lamina propria, **141**
    of alimentary canal mucosa, **873**, 873*f*
    of areolar connective tissue, 130, 131*f*
    of stomach, 886*f*
Laminar flow, 715
Laminectomy, **248**
Laminin, 428
Lamins, 92, 92*f*
Langerhans, islets of (pancreatic islets),
    630, 630*f*, 899
Langerhans cells (epidermal dendritic
    cells), 152*f*, **153**
Language, 459–460
Lanugo coat, 167, 1111*t*
Laparoscopy, **921, 1086**

Large cell (medial) nuclear group, 453*f*,
    457
Large granular lymphocytes, 785
Large intestine, 869*f*, **906**–911, 907*f*, 908*f*
    autonomic innervation/effects and, 536*f*,
        540*f*
    digestive processes occurring in, 870*f*,
        904*t*, 910–911, 911*f*
Laryngeal cartilages, 174, 175*f*,
    824–825, 825*f*
Laryngeal prominence (Adam's apple),
    824, 825*f*
Laryngitis, **826**
Laryngopharynx, **823**–824, 823*f*, 876*f*, **882**
Laryngotracheal bud, **860**, 861*f*
Larynx (voice box), 174, 175*f*, 820*f*, 823*f*,
    824–826, 825*f*, 826*f*, 831*t*
Lasix. *See* Furosemide
Latent period, muscle twitch, **299**, 299*f*
Latent period after orgasm, 1054
Latent syphilis, 1080
Lateral (term), 14*t*
Lateral angle of scapula, 228
Lateral apertures, **436**–438, 437*f*
Lateral (axillary) border of scapula, 228,
    229*f*
Lateral circumflex femoral artery, 748*t*,
    748*f*, 749*f*
Lateral commissure/canthus, 554, 555*f*
Lateral condyle, 240*f*, **241**, 242*f*, **243**, 243*f*
Lateral cords, brachial plexus, 510*t*, 510*f*,
    511*f*
Lateral corticospinal tract, 475*f*, 479*f*, 481*t*
Lateral cuneiform bone, 241*t*, **244**, 244*f*
Lateral cutaneous branch of intercostal
    nerve, 508*f*
Lateral epicondyle, 230*f*, **231**, 231*f*, 240*f*,
    241
Lateral excursion, 265, 265*f*
Lateral femoral cutaneous nerve, 513*t*,
    513*f*
Lateral fornix, 1063*f*, 1066
Lateral funiculi, 473*f*, **474**
Lateral geniculate bodies/nuclei, 448,
    449*f*, **574**, 574*f*
Lateral horns, **472**, 473, 473*f*, 474*f*, **538**,
    539*f*
Lateral hypothalamic area, 449*f*, 960, 961*f*
Lateral inhibition, 575
Lateralization, **444**
Lateral lemniscus, 589*f*, **590**
Lateral ligaments, of temporomandibular
    joint, **264**
Lateral longitudinal arch of foot, 244–245,
    245, 245*f*
Lateral (lumbar) regions, **20**, 20*f*
Lateral malleolus, 242*f*, **243**, 244*f*
Lateral mass of atlas, 221, 221*f*
Lateral mass of ethmoid bone, **210**, 210*f*
Lateral meniscus of knee, 270*f*, 271*f*
Lateral muscles/muscle compartment, of
    leg, 374*t*, 376*f*, 377*t*, 384*t*, 385*f*
Lateral (small cell) nuclear group, 453*f*,
    457
Lateral patellar retinaculum, 270*f*, **271**
Lateral pectoral nerve, 510*t*, 511*f*
Lateral plantar artery, 748*f*, **749***t*, 749*f*
Lateral plantar nerve, 380*t*, 382*t*, **515**, 515*t*
Lateral plantar vein, 758*t*
Lateral plate mesoderm, 1105*f*, **1106**,
    1106*f*
Lateral pterygoid muscle, **336***t*, 337*f*
Lateral rectus muscle, 556, 557*f*
Lateral reticulospinal tract, 475*f*, 481*t*
Lateral rotation of limbs, 260, 260*f*, 362*t*,
    372–373*t*, 384*t*
Lateral sacral crest, **224**, 224*f*
Lateral semicircular canal, 583, 583*f*
Lateral spinothalamic tract, 475*f*, **476**,
    477*f*, 480*t*
Lateral sulcus, 438*f*, **439**
Lateral supracondylar line, 240*f*, **241**
Lateral supracondylar ridge, 230*f*, **231**
Lateral thoracic artery, **742***t*, 742*f*, 743*f*
Lateral ventricles, 435*f*, **436**, 437*f*

Latissimus dorsi muscle, 331*f*, **350***t*, 353*f*, **354***t*, **355***t*, 355*f*, 362*t*
LDLs. *See* Low-density lipoproteins
L-dopa, for Parkinson's disease, 469
Leading strand, 97, 97*f*
Lead salts, skin penetrated by, 164
Leakage channels, 70, **400**, 402*f*, 403
Learned (acquired) reflex, 519
Least splanchnic nerve, **541**
Lecithin, 916
Lectin(s), 788, 789*f*
Lectin pathway, of complement activation, **788**, 789*f*
Left anterior descending artery (anterior interventricular artery), 675*f*, **682**, 683*f*
Left ascending lumbar vein, **755***t*, 756*f*, 757*f*
Left atrium, 671, 671*f*, 674, 681*f*
Left brachiocephalic vein, **750***t*, 750*f*, 751*f*
Left bundle branch, **687**
Left colic artery, 744*f*, **746***t*, 747*f*
Left colic (splenic) flexure, **906**–907, 907*f*
Left common carotid artery, **738***t*, 738*f*, 740*f*, 742*f*
Left common iliac artery, **738***t*, 738*f*
Left coronary artery, 675*f*, **682**, 683*f*, **738***t*
Left gastric artery, 744*f*, **745***t*, 745*f*
Left gastroepiploic artery, 744*f*, **745***t*, 745*f*
Left hypochondriac region, **20**, 20*f*
Left inguinal (iliac) region, **20**, 20*f*
Left lateral (lumbar) region, **20**, 20*f*
Left lobe, of liver, 894*f*, 895*f*
Left lower quadrant (LLQ), **19**, 19*f*
Left main (primary) bronchus, 820*f*, **828**, 828*f*, 832*f*
Left marginal artery, 682
Left pulmonary artery, 675–677*f*, **736***t*, 736*f*
Left pulmonary vein, 675–677*f*, 736*f*
Left subclavian artery, **738***t*, 738*f*, 742*f*, 743*f*
Left suprarenal vein, 750*f*, 757*f*
Left upper quadrant (LUQ), **19**, 19*f*
Left ventricle, 671, 671*f*, 674, 681*f*, 682, 682*f*, 683*f*
Leg
  bones of, 241*t*, 242–243, 242–243*f*
  muscles of/muscles controlling movement of, 330*f*, 331*f*, 367–373*t*, 368*f*, 371–372*f*, 374–379*t*, 375*f*, 376*f*, 378–379*f*, 384*t*, 385*f*
Length-tension relationship, **307**, 307*f*, 316, 697
Lens (of eye), 558*f*, 562*f*, **563**, 563*f*, 565–567, 565*f*, 595
  suspensory ligament of (ciliary zonule), 558*f*, **559**, 562*f*
Lens crystallins, **563**
Lens epithelium, 562*f*, **563**
Lenses, refraction and, 564, 565*f*
Lens fibers, **563**
Lens vesicle, **595**
Lentiform nucleus, 446
Leptin, **634**, 635*t*, 961*f*, **962**, 1074, 1078
Lesion, 147
Lesser curvature of stomach, **885**, 885*f*
Lesser occipital nerve, 509*t*, 509*f*
Lesser omentum, **885**, 908*f*
Lesser sciatic notch, **236**, 237*f*
Lesser sphenoid wings, 205*f*, 207*f*, **209**, 209*f*
Lesser splanchnic nerve, 540*f*, **541**
Lesser trochanter, 240*f*, **241**
Lesser tubercle, 230*f*, **231**
Let-down reflex, **1117**, 1117*f*
Lethal dominant genes, 1129
Lethal hit, cytotoxic T cell, 807, 807*f*
Lethargy, 462
Leukemias, 654
Leukocyte(s) (white blood cells), 127*t*, **129**, 137, 137*f*, 643–645, 643*f*, 645*f*, **651**–656, 652*f*, 653*f*, 656*t*. *See also specific type*
  agranulocytes, 652, 652*f*, **653**, 653*f*, 655*f*, 656*t*

disorders of, 654
  granulocytes, **652**–653, 652*f*, 653*f*, 655*f*, 656*t*
  in inflammation, 786, 787*f*
  production/life span of, 653–654, 655*f*, 656*t*
  structure/function and, 651–652, 652*f*, 656*t*
  in urine, 1001*t*
Leukocyte count, 652, 652*f*, 666
Leukocytosis, **652**, 786, 786*f*, 787*f*
Leukocytosis-inducing factors, 786, 787*f*
Leukopenia, **654**
Leukoplakia, 144*b*
Leukopoiesis, **653**–654, 655*f*
Leukotrienes, 603
Levator ani muscle, 348*t*, 349*f*, 907*f*, 1004
Levator labii superioris muscle, **334***t*, 335*f*
Levator palpebrae superioris muscle, 555, 555*f*
Levator scapulae muscle, 331*f*, **350***t*, **352***t*, 353*f*
Levator veli palatini, 338*t*, 339*f*
Lever(s)/lever systems, **327**–328, 327–328*f*, 329*f*
Leydig (interstitial endocrine) cells, **1055**, 1055*f*
LH. *See* Luteinizing hormone
Li. *See* Lithium
Lids (eyelids), 554–555, 555*f*
Life
  cellular basis of, 61–63, 61*f*, 62*f*
  maintenance of/necessary functions and, 5–9, 5*f*
Ligament(s), 132*f*, 133. *See also specific named ligament*
  sprains of, **273**
  syndesmoses connected by, 252, 252*f*
  of synovial joints, 254–255, 254*f*, 257
  of vertebral column, 218–219, 219*f*
Ligamentum arteriosum, 675*f*, 677*f*, **1107**, 1107*t*, 1109*f*
Ligamentum flavum, **218**–219, 219*f*
Ligamentum nuchae, 203, 340*t*, 343*f*
Ligamentum teres (ligament of head of femur), **269**, 269*f*
Ligamentum teres (round ligament), 894, 894*f*, 895*f*, **1107**, 1109*f*
Ligamentum venosum, **1107**, 1107*t*, 1109*f*
Ligand(s), **81**, 82*f*
Ligand-gated (chemically-gated) channel, 290, 292*f*, **400**, 401*f*, 417, 418*f*
Ligase, 97
Light, 563–567, 564*f*, 565*f*
  capture of by visual pigments, 570–571
  focusing of on retina, 565–567, 565*f*
  pupillary reflex and, 559, 573
  refraction of, **564**, 564*f*, 565*f*, 566*f*, 567
  wavelength/color of, 564, 564*f*
Light adaptation, **573**
Light (L) chains, 798*f*, **799**
Light chain kinase (myosin), **314**, 315*f*
Light energy, 24
Light microscopy, tissue stains for, 117
Light transduction (phototransduction), **570**–573, 571*f*
Limb(s) (appendages/extremities). *See* Lower limbs; Upper limbs
Limb buds, 1105*f*, **1106**
Limbic association area, **444**
Limbic system, 443, 444, 449, **456**–457, 456*f*, 458*t*, 547*f*, 578
Line (bone marking), 180*t*
Linea alba, 346*t*, 347*f*
Linea aspera, 240*f*, **241**
Linear acceleration, 590, 591
Lingual artery, 740*t*, 740*f*, 741*f*
Lingual frenulum, 876*f*, **877**
Lingual lipase, 889, 913*f*
Lingual tonsils, 774*f*, **775**, **823**, 823*f*, 876*f*, 877, 877*f*
Linked genes, **1127**
Linker DNA, 93, 93*f*
Linker proteins, 67*f*, 68

Linoleic acid, 928, 930*t*
Linolenic acid, 930*t*
Lip(s) (labia), **876**, 876*f*
  cleft, 246*f*, **919**
Lipase(s), 913*f*, **916**
  chief cell secretion of, 887
  gastric, 887, 889, 913*f*
  lingual, 889, 913*f*
  lipoprotein, **917**, 950
  pancreatic, **899**, 913*f*, 916, 916*f*
Lipid(s), **45**–48, 45*f*, 47*t*. *See also* Fat(s)/triglyceride(s)
  in atherosclerosis, 760*b*
  dietary, 928, 930*t*
  digestion/absorption of, 916–917, 916*f*
  metabolism of, 618*t*, 946–948, 946*f*, 947*f*
  as neurotransmitters, 422*t*, 423–424
  in plasma membrane, 63, 65*f*
Lipid bilayer, plasma membrane, 63, 64–65*f*, 66
Lipid solubility, diffusion and, 69, 467
Lipidemia, 666
  diabetic, 632, 633*b*
Lipid-soluble hormones, 604, 606, 606*f*, 608, 609*f*
Lipid-soluble substances, skin penetrated by, 164
Lipofuscin (aging pigment), 394
Lipogenesis, **946**–947, 946*f*, 947*f*
Lipolysis, 946*f*, **947**, 947*f*, 953, 953*f*, 954*f*
Lipoprotein(s), 47*t*, 760, **956**–957, 957*f*, 959. *See also specific type*
Lipoprotein (a), 760*b*
Lipoprotein lipase, **917**, 950
Liposomes, **110**
Liposuction, 958–959*b*
Liquid(s), 24
Lithium, atomic structure of, 27, 27*f*
Lithotripsy, 898, 1002
Liver, 869*f*, 894–896*f*, **894**–898, 896*f*, 897*f*, 900*f*. *See also under* Hepatic
  autonomic innervation/effects and, 534*f*, 536*f*, 540*f*, 546*t*
  bleeding disorders and, 662
  digestive function of, 904*t*
  glycogenolysis in, 952, 953*f*, 954*f*
  homeostatic imbalances of, 897–898
  lipolysis in, 953, 953*f*, 954*f*
  metabolic functions of, 950–959, 951*f*, 953*f*, 954*f*, 956*t*
  round ligament of (ligamentum teres), **894**, 894*f*, 895*f*, **1107**, 1107*t*, 1109*f*
Liver cells (hepatocytes), **897**, 945, 950, 951*f*, 953, 954, 955–959, 956*t*
Liver lobules, 896*f*, **897**
Liver sinusoids, 896*f*, **897**
Load (bone), growth/remodeling and, 189–190, 190*f*
Load (muscle), **299**, 307*f*, 309, 309*f*, 327, 327–328*f*, 329*f*
Lobar arteries, **736***t*, 736*f*
Lobar (secondary) bronchi, **828**, 828*f*
Lobe(s)
  of cerebral hemispheres, 438*f*
  of kidney, **977**
  of liver, 894, 894*f*, 895*f*
  of lungs, 828*f*, 832*f*, **833**, 833*f*
  mammary gland, **1068**, 1068*f*
  of multicellular exocrine gland, 125
  thyroid gland, 617, 617*f*
Lobule(s)
  of ear (earlobes), 581*f*, **582**
  of liver, 896*f*, **897**
  of lung, 832*f*, **833**
  mammary gland, **1068**, 1068*f*
  of testis, 1048–1049, 1049*f*
  thymic, 776

Linoleic acid... (4th column)
Longitudinal arches of foot, 244–245, **245**, 245*f*
Longitudinal fissure, 438*f*, **439**
Longitudinal layer of smooth muscle, 310, 310*f*, 873*f*, 885*f*, 886, 886*f*
Longitudinal ligaments, **218**, 219*f*
Long reflexes, **875**, 875*f*, 905*t*
Long-term memory (LTM), **460**, 460*f*
Long-term potentiation (LTP), 420
Long thoracic nerve, 350*t*, 510*t*, 511*f*
Loop diuretics, 1000
Loop of Henle. *See* Nephron loop
Loose connective tissue, 127*t*, **129**–133, 131–132*f*
Lordosis, 220, 220*f*, 247, 1112
Loudness, **586**, 587, 590, 594, 596
Lou Gehrig's disease (amyotrophic lateral sclerosis/ALS), 475–476
Low-density lipoproteins (LDLs), 760*b*, **956**–957, 957*f*, 959
Lower limbs. *See also* Foot; Leg; Thigh
  arteries of, 739*f*, 748–749*f*, 748–749*f*
  bones of, 200*f*, 240–245, 240*f*, 241*t*, 242–243*f*, 244*f*, 245*f*
  innervation of, 512–515, 513*t*, 513*f*, 514*f*, 515*t*
  veins of, 750*f*, 751*f*, 758*t*, 758*f*
Lower motor neurons, 478, 479*f*
Lower respiratory system, 820, 824–831, 831*t*. *See also specific structure*
Lower trunk, brachial plexus, **510**, 511*f*
Lumbar arteries, 738*t*, 744*f*, **746***t*, 746*f*
Lumbar curvature of spine, **218**, 218*f*, 220, 220*f*, 246–247
Lumbar enlargement, 470*f*, **472**, 507*f*
Lumbar fascia, 347*f*
Lumbar nerves, 369*t*, 370*t*, 372–373*t*, 470–471*t*, 507*f*
Lumbar plexus, 507*f*, **512**, 513*t*, 513*f*, 515*t*
Lumbar puncture, **471**, 471*f*
Lumbar regions, **20**, 20*f*
Lumbar splanchnic nerves, 540*f*, **541**
Lumbar trunks, 768*f*, **769**
Lumbar veins, 750*f*, 755*t*, 755*f*, **756***t*, 756*f*, 757*f*
Lumbar vertebrae, **218**, 218*f*, 222, 222*f*, 223*t*
Lumbosacral plexus, **512**–515, 513*t*, 513*f*, 514*f*, 515*t*
Lumbosacral trunk, **512**, 513*f*, 514, 514*f*
Lumbrical muscles, 359*f*, **364***t*, 365*f*, **366***t*, **380***t*, 381*f*
Lumen, blood vessel, 708*f*, **709**
Lumpectomy, 1069
Lunate, **234**, 234*f*, 235*f*
Lung(s), 820*f*, 828*f*, 831*t*, 832*f*, **833**–834, 833*f*. *See also under* Pulmonary and Respiratory
  autonomic innervation/effects and, 534*f*, 536*f*, 540*f*, 546*t*
  blood supply/blood flow/perfusion of, 729, 832*f*, 834
  collapse of (atelectasis), **836**, 836*f*
  developmental aspects of, 860–861, 861*f*
  innervation of (pulmonary plexus), 537, 540*f*, **834**
Lung (respiratory) alveoli, **829**, 829*f*, 830*f*, 831*t*, 843*t*, 860–861
Lung cancer, 859–860
Lung compliance, **840**
Lung disease, 858–860, 858*f*
Lung recoil, 835, 837*f*, 838
Lung root, 832*f*, **833**
Lunule, 160, 160*f*
Lupus erythematosus, systemic (SLE), 812, **814**
Luteal phase of ovarian cycle, **1073**, 1074, 1075*f*, 1077*f*
Luteinizing hormone (LH), 611*f*, 613*t*, **616**, **1043**, 1059*f*, 1073, 1074–1076, 1075*f*, 1077*f*
LH surge, **1076**, 1077*f*
Luteolytic (ischemic) phase of ovarian cycle, 1074
Luxation(s)/dislocation(s), 266, **273**

Lye. *See* Sodium hydroxide
Lyme disease, **275**
Lymph, **767**, 769, 771
Lymphadenopathy, **777**
Lymphangiography, **777**
Lymphangitis, 769
Lymphatic capillaries, 707*f*, **767**–769, 767*f*
Lymphatic capillary valves, 767*f*, 768
Lymphatic collecting vessels, 767*f*, **769**, 769*f*
Lymphatic ducts, 767*f*, 768*f*, 769, 769*f*
Lymphatic system, 7*f*, 707*f*, **766**–773, 767*f*, 768*f*, 769*f*
 cardiovascular system and, 761*b*, 778*b*
 case study and, 780*b*
 cells and tissues of, **767**, **770**–771, 770*f*
 developmental aspects of, 776
 digestive system and, 778*b*, 920*b*
 endocrine system and, 637*b*, 778*b*
 functions of, 7*f*
 homeostatic imbalances and, 769–770, 772–773, 774
 homeostatic relationships of, 778*b*
 integumentary system and, 168*b*, 778*b*
 lymph nodes in, 707*f*, 767, 767*f*, 769*f*
 muscular system and, 318*b*, 778*b*
 nervous system and, 550*b*, 778*b*
 organs of, **767**, **771**, 771*f*
 reproductive system and, 778*b*, 1085*b*
 respiratory system and, 778*b*, 862*b*
 skeletal system and, 196*b*, 778*b*
 urinary system and, 778*b*, 1036*b*
 vessels of. *See* Lymphatic vessels (lymphatics)
Lymphatic tissues. *See* Lymphoid tissues
Lymphatic trunks, 767*f*, **769**
Lymphatic vessels (lymphatics), 707*f*, 734, 770, 772, 772*f*, **767**–769, 767–769*f*, 772, 772*f*, 776. *See also specific type and* Lymphatic system
Lymphedema, 734, 770
Lymph node(s), 707*f*, 767, 767*f*, 769*f*, **771**–773, 771*f*, 772*f*, 777*t*. *See also specific type*
 in cancer, 773, 777
 developmental aspects of, 776
 in infection, 772–773
 reticular tissue in, 770, 770*f*
 sentinel, **777**
Lymphocytes, 128*f*, 137*f*, 645*f*, 652*f*, **653**, 653*f*, 656*t*, **770**, 770*f*, 785, 792–794, 793*f*, 794*f*, 795*t*. *See also* B lymphocytes; T lymphocytes
Lymphocytic leukemia, 654
Lymphoid cells, 770, 770*f*
Lymphoid follicles/nodules, **771**, 772*f*, 774*f*
 aggregated (Peyer's patches), 771, 771*f*, **775**, 775*f*, **903**
Lymphoid organs, **767**, **771**, 771*f*, 776, 777*t*, 792. *See also specific type and* Lymph node(s)
Lymphoid stem cells, **654**, 655*f*
Lymphoid tissues, **767**, **770**–771, 770*f*, 777*t*
 cancer of, 777, 811
 developmental aspects of, 776
 mucosa-associated (MALT), 771, **774**–775, 774*f*, 775*f*, 777*t*, **873**, 873*f*, 902*f*, 903
Lymphoma, **777**
Lymph sacs, **776**
Lymph sinuses, **772**
Lysis, cell, 72, 72*f*, 788, 800*f*, 801
Lysosomal enzymes, 86, 188, 619*f*, 620
Lysosomes, 62*f*, 76*f*, 77, 77*f*, **86**–87, 86*f*, 87*f*, 95*t*, 652, 784, 784*f*
Lysozyme, **556**, 783, 822, 879

MAC (membrane attack complex), **789**, 789*f*
Macrocytes, **650**
Macromolecules, **42**

Macronutrients, **927**
Macrophage(s), 77, 128*f*, **129**, **653**, 655*f*, **770**, 770*f*, **784**, 784*f*, 785, 786. *See also* Phagocytosis/phagocytes
 activated, 795
 alveolar, **831**
 antigen presentation by, 795
 dermal, 164
 in PNS nerve fiber regeneration, 498*f*
 stellate (hepatic), 784, 896*f*, **897**
 in tissue repair, 143*f*
Macroscopic (gross) anatomy, **2**
Macula(e), of utricle and saccule, 583, 583*f*, 585*t*, **590**–591, 591*f*
Macula densa, **983**, 983*f*, 988–989, 988*f*
Macula lutea, 558*f*, **561**, 561*f*
Macular degeneration, age-related (ARMD), **596**
Magnesium, 26*t*, 306, 931, 933*t*, 1014*f*, 1020*t*
Magnesium hydroxide, 39
Magnetic resonance imaging (MRI), 15, 15*f*, 17*f*, 17*b*, 439*f*, 470, 480, 481*f*
Magnitude estimation, **495**
Main (primary) bronchi, 820*f*, 827, 828*f*, 832*f*
Main pancreatic duct, **900**, 900*f*
Major calyces, **977**, 977*f*
Major duodenal papilla, **900**, 900*f*, 901
Major histocompatibility complex (MHC)/MHC proteins, **791**–792, 802*f*, 804*f*, 811
Major histocompatibility complex (MHC) restriction, 804
Major (extrinsic) salivary glands, **878**
Malabsorption, **918**
Malaria, sickle-cell anemia and, 650
Male(s). *See also* Male reproductive system
 breast tissue in (gynecomastia), **1086**
 pelvis in, 238, 239*t*
 puberty in, 1084
 urethra in, 1003*f*, 1004, 1047, 1047*f*, **1050**, 1051*f*
 urinary system in, 975*f*, 1003*f*, 1004
Male infertility, 1058
Male pattern baldness, **160**
Male pronucleus, **1096**, 1096*f*
Male reproductive system, 7*f*, 8. *See also specific organ or structure*
 anatomy of, 1047–1053, 1047*f*, 1048*f*, 1049*f*, 1051*f*, 1052*f*
 embryologic/fetal development of, 1081–1086, 1082*f*, 1083*f*
 functions/physiology of, 7*f*, 8, 1053–1060, 1056*f*, 1057*f*, 1059*f*
 homeostatic imbalances and, 1049, 1054, 1058, 1084
Male sexual response, 1053–1054
Malignant neoplasm, 144*b*. *See also* Cancer
Malleoli, medial and lateral, 242*f*, **243**, 244*f*
Malleus (hammer), 581*f*, **582**, 582*f*, 587*f*
Malnutrition, protein energy, **969**
MALT. *See* Mucosa-associated lymphoid tissue
Maltase, 913*f*, **914**
Maltose (grain sugar), 43*f*, 44, 913*f*, 914
Mammary artery, internal (internal thoracic artery), 741*f*, 742*f*, **743***t*, 743*f*
Mammary glands/breasts, **163**, **1068**–1069, 1068*f*
 cancer of, **1069**, 1069*f*
 lactation and, 1068, 1068*f*, 1116–1119, 1117*f*
 pregnancy affecting, 1112
 prolactin affecting, 616, 617, 1116–1119, **1117**, 1117*f*
Mammillary bodies, 447*f*, **448**, 448*f*, 449*f*, 450*f*, 452*f*, 456*f*
Mammography, 145*b*, **1069**, 1069*f*
Mandible, 202*f*, 203*f*, 204*f*, 205*f*, **210**–211, 211*f*, **217***t*, 264–265, 265*f*

Mandibular angle, 204*f*, **210**, 211*f*
Mandibular branch of facial nerve, 503*t*, 504*t*
Mandibular division of trigeminal nerve (cranial nerve V₃), 336*t*, 338*t*, 502*t*
Mandibular foramina, 205*f*, **211**, 211*f*, **217***t*
Mandibular fossa, 206*f*, **208**, 208*f*, 211*f*, **216***t*, **264**, 265*f*
Mandibular notch, 204*f*, **210**, 211*f*
Mandibular ramus/rami, 204*f*
Mandibular symphysis, 203*f*, **211**, **217***t*
Manganese, 26*t*, 933*t*
Manubrium, 225, 225*f*
MAP. *See* Mean arterial pressure
Marasmus, **969**
Marfan's syndrome, **147**
Marginal artery, left and right, 675*f*, **682**, 683*f*
Margination, **786**, 786*f*, 787*f*
Marrow (bone). *See* Bone marrow
Marrow (medullary) cavity, **178**, 179*f*, 185, 185*f*
Masculinization, 159, 626, 626*t*
Mass, 24
Masseter muscle, 330*f*, 335*f*, **336***t*, 337*f*
Mass movements/peristalsis, 904*t*, **910**
Mass number, **27**
Mast cells, 128*f*, **129**, 653, **785**, 812–813, 813*f*
Mastication (chewing), 330*f*, 336*t*, 337*f*, 870*f*, 879, **881**, 904*t*
Mastoid air cells, **209**, 582
Mastoid antrum, 581*f*, **582**
Mastoid fontanelle, 246*f*
Mastoiditis, 209
Mastoid process, 203*f*, 204*f*, 206*f*, **208**, 208*f*, 209, **216***t*
Maternal antibodies, passive immunity and, 797*f*, 798
Maternal blood sample, for fetal testing, 1135
Maternal chromosome, 1043
Matrix
 blood (plasma), 127*t*, 137, 137*f*
 bone, 127*t*, 136*f*, 137, 181
 cartilage, 127*t*, 134, 135*f*
 centrosome, 62*f*, 88, 89*f*
 extracellular, **63**, 66*f*, **126**–127, 127*t*, 128*f*, 174
 hair, 158*f*, **159**
 mitochondrial, 83–84, 83*f*
 nail, **160**, 160*f*
Matrix vesicles, 188
Matter, **24**
 in combination (molecules/mixtures), 28–30, 29*f*
 composition of (atoms and elements), 25–28, 25*f*, 26*t*, 27*f*, 28*f*
 living. *See* Biochemistry
Maxillary artery, **740***t*, 740*f*, 741*f*
Maxillary bones/maxilla(e), 202*f*, 203*f*, 204*f*, 205*f*, 206*f*, **211**–212, 211*f*, 214*f*, **217***t*, 821, 821*f*
Maxillary division of trigeminal nerve (cranial nerve V₂), 502*t*
Maxillary sinuses, 202*f*, **211**, 215*f*
Maximal stimulus, **301**
Mean arterial pressure, **716**, 716*f*, 719*f*. *See also* Blood pressure
Meatus (bone marking), 180*t*
Mechanical advantage, **327**, 327*f*
Mechanical breakdown (food), 870*f*, **871**, 871*f*, 889, 904*t*
Mechanical disadvantage, **328**, 328*f*
Mechanical energy, **24**
Mechanically gated channel(s), **401**
Mechanical stress, bone remodeling and, 189–190, 190*f*
Mechanoreceptors, **490**, 491*t*, 492*t*, 1117
Meconium, 1111*t*, **1117**
Medial (term), 14*t*
Medial (vertebral) border of scapula, 228, 229*f*

Medial circumflex femoral artery, **748***t*, 748*f*, 749*f*
Medial commissure/canthus, **554**, 555*f*
Medial condyle, 240*f*, **241**, 242*f*, **243**
Medial cords, brachial plexus, **510**, 510*t*, 511*f*
Medial cuneiform bone, 241*t*, **244**, 244*f*
Medial epicondyle, 230*f*, **231**, 231*f*, 240*f*, **241**
Medial geniculate bodies/nuclei, 448, 449*f*, 589*f*, **590**
Medial lemniscus, 453, 453*f*, **476**, 477*f*
Medial longitudinal arch of foot, 244–245, **245**, 245*f*
Medial malleolus, 242*f*, **243**, 244*f*
Medial meniscus of knee, 270*f*, 271*f*
Medial muscles/muscle compartment, of hip/leg/thigh, 367*t*, 368*f*, 369*t*, 384*t*, 385*f*
Medial (large cell) nuclear group, 453*f*, **457**
Medial patellar retinaculum, 270*f*, **271**
Medial pectoral nerve, 510*t*, 511*f*
Medial plantar artery, 748*t*, **749***t*, 749*f*
Medial plantar nerve, 380*t*, 382*t*, **515**, 515*f*
Medial plantar vein, **758***t*
Medial pterygoid muscle, **336***t*, 337*f*
Medial rectus muscle, 556, 557*f*
Medial reticulospinal tract, 475*f*, 481*t*
Medial rotation, 260, 260*f*, 362*t*, 384*t*
Medial supracondylar line, 240*f*, **241**
Medial supracondylar ridge, 230*f*, **231**
Median antebrachial vein, **754***t*, 754*f*, 755*f*
Median aperture, **436**–438, 437*f*
Median cubital vein, 751*f*, **754***t*, 754*f*, 755*f*
Median nerve, 234, 358*t*, 359*t*, 364*t*, 366*t*, 510*t*, 511*f*, **512**
Median palatine suture, 206*f*
Median (midsagittal) plane/section, **15**, 15*f*
Median sacral artery, 738*f*, 744*f*, **746***t*, 746*f*
Median sacral crest, **224**, 224*f*
Median umbilical ligaments, **1107**, 1107*t*, 1109*f*
Mediastinal arteries, 738*f*, **743***t*
Mediastinum, **18**, 18*f*, **672**, 672*f*
Mediators, chemical, inflammatory, 143*f*, 785, 785*t*, 787*f*
Medical imaging, 16–17*b*
Medications, skin penetrated by, 164
Medulla
 adrenal. *See* Adrenal medulla
 hair, 158, 158*f*
 lymph node, 772*f*, **772**
 ovarian, 1061, 1062*f*
 renal, **977**, 977*f*, 978, 979*f*, 995*f*
Medulla oblongata, 435, 435*f*, 436, 436*f*, 447, 448*f*, 450–451*f*, **452**–454, 452*f*, 453*f*, 458*f*
 blood pressure regulation and, 721
 cardiovascular center in, 454, 719, **721**
 respiratory centers in, 454, 853, 853*f*, 854, 854*f*, 1013
Medullary (marrow) cavity, **178**, 179*f*, 185, 185*f*
Medullary chromaffin cells, **627**
Medullary cords, lymph node, 772*f*, **772**
Medullary osmotic gradient, **995**, 996–997*f*, 999, 1000
Medullary (renal) pyramids, **977**, 977*f*
Medullary respiratory centers, 454, 853, 853*f*, 854, 855*f*, 1027, 1029
Medullary sinuses, 770*f*, **772**, 772*f*
Megakaryoblasts, **657**, 657*f*
Megakaryocytes, **657**, 657*f*. *See also* Platelet(s)
Meiosis, 61, 98, **1043**–1047, 1044*f*, 1045*f*, 1046*f*
 chromosome segregation in, 1126*f*, **1127**
 crossing over and genetic recombination and, **1044**, 1045*f*, **1127**–1128, 1127*f*
 fertilization and, 1070*f*, 1096, 1096*f*
 mitosis compared with, 1046*f*, 1047
 in oogenesis, 1069, 1070*f*, 1071–1073, 1071*t*
 in spermatogenesis, 1043–1047, 1056*f*, 1057, 1071–1073, 1071*t*

Meiosis I, **1044**, 1044f, 1045f, 1046f, 1056f, 1070f, 1126f, 1127–1128, 1127f
Meiosis II, **1044**–1047, 1044f, 1045f, 1046f, 1056f, 1070f, 1096, 1096f
Meissner's (tactile) corpuscles, 154, 164, **492**, 492t
Melanin, 152f, **153**, 156, 159, 163
Melanocytes, 152f, **153**, 156, 158f, 159
Melanoma, 165f, **166**
Melanopsin, 575
Melanosomes, 153, 156
Melatonin, 450, **627**
Membrane(s). See also specific type
    basement, 118, **118**, 120f, 121f, 122f
    of bone, 178, 179f
    boundary maintenance and, 8, 64f
    cell. See Plasma membrane
    covering and lining (cutaneous/serous/mucous), 117, 141–142, 141–142f
    cutaneous. See Skin
    mucous. See Mucous membranes (mucosae)
    nuclear, 62f, 87f, **92**–93, 92f, 96t, 100f, 101f
    permeability of. See Permeability, membrane
    plasma. See Plasma membrane
    serous. See Serous membranes (serosae)
    synovial, **254**, 254f
Membrane attack complex (MAC), **789**, 789f
Membrane-bound ribosomes, 84, 84f, 85
Membrane lipids, 63, 65f
Membrane potentials, **79**–81, 80f. See also Action potential; Graded potentials
    electricity principles and, 400–401, 401f
    in muscle contraction, 290, 292f, 293–296, 293f
    resting, **79**–81, 80f, 400–404, 401f, 402f, **403**, 404f, 686
Membrane proteins, 49f, 63–66, 66f
    in exocytosis, 79, 79f
    in receptor-mediated endocytosis, 77, 77f, 78t
    as receptors, 66, **81**, 82f
Membrane receptors, 66, 66f, **81**, 82f, 604–605, 605f
Membrane transport
    active, **68**, 73–79, 74f, 75f, 76f, 77f, 78t, 79f
    passive, **68**–73, 69f, 70f, 71f, 72f, 73t
Membranous bones, **184**
Membranous (intermediate part) urethra, 1003f, **1004**, 1050, 1051f
Membranous labyrinth, **583**, 583f, 585t, 592f, 596
Membranous organelles, 83
Memory, 440f, 443, **460**–461, 460f, 483
    immunological, 791, **797**
Memory cells, **793**, 793f, 795t, **797**, 810t
    B cells, 796f, **797**, 809f
    T cells, 802, 802f, 809f
Memory consolidation, **460**
Memory loss, in Alzheimer's disease, 469
Menarche, **1074**, 1078, 1084
Mendel, Gregor, 1124
Ménière's syndrome, **595**
Meningeal artery, middle, 210, 740t
Meningeal branch, **508**
Meningeal layer of dura mater, 464, 464f, 466f
Meninges, brain, **464**–465, 464f, 465f
Meninges, spinal, 470–471f, 471, 473f
Meningitis, 465
Meningocele, 483
Meninx. See Meninges
Menisci (articular discs), **255**
    of knee, 175f, **270**, 270f, 271f, 272–273, 272f, 273f
    of temporomandibular joint, 264, 265f
Menopause, **1084**–1086
Menses, 917, **1076**, 1077f, 1084
Menstrual (uterine) cycle, **1076**–1078, 1077f

Menstrual phase of uterine (menstrual) cycle, 1076, 1077f
Menstruation, 917, **1076**, 1077f, 1084
Mental foramina, 203f, 204f, **211**, 211f, **217**t
Mental functioning. See Cognition/cognitive function
Mentalis muscle, 334t, 335f
Mercury salts, skin penetrated by, 164
Merkel (tactile) cells, 152f, **153**, **492**
Merkel (tactile) discs, 153, 491t, **492**
Merocrine glands, **125**–126, 126f
    apocrine sweat glands, 162f, **162**–163
    eccrine sweat glands, 151f, **161**–162, 161f, 162f
Mesangial cells, 983, 983f
Mesencephalon. See Midbrain (mesencephalon)
Mesenchymal cells, 129, 184, 186, 186f
Mesenchymal stem cell regeneration, in joint therapy, 274b
Mesenchyme, 127, **129**, 1103
Mesenteric arteries, inferior and superior, 738f, 739f, 744f, 745f, **746**t, 746f, 747f, 907f
Mesenteric ganglia/plexuses, inferior and superior, 540f, **541**
Mesenteric veins, inferior and superior, 751f, **756**t, 756f, 757f
Mesentery (mesenteries), **872**, 872f, 873f, 901, 907f, 908f
Mesocolons, 885, 907f, 908f, **909**
Mesoderm, **146**, 146f, 1100f, **1103**, 1103f, 1105f
    differentiation of, 1104f, 1105f, 1106–1107, 1106f
    extraembryonic, 1099–1100, 1100f
Mesodermal aggregates, 1105f
Mesometrium, 1063f, **1064**, 1064f
Mesonephric (Wolffian) ducts, **1006**, 1006f, 1082f, **1083**
Mesonephros, **1006**, 1006f, 1082f
Mesosalpinx, **1063**, 1063f, 1064f
Mesothelium, **121**, 142, 873, 873f
Mesovarium, **1061**, 1063f
Messenger RNA (mRNA), 54, **99**, 99f, 102–105, 102f, 105f, 106–107f, 108f
Metabolic acidosis, 947, 1033t, 1034–1035, 1034b, 1035
Metabolic alkalosis, 1033t, 1034–1035, 1034b, 1035
Metabolic controls, of blood flow, 726, 727f
Metabolic rate, 394, 618, 618t, 955t, **962**, 963, 967, 968, 969
Metabolic syndrome, 959b, **968**–969, 968f
Metabolic water (water of oxidation), **1016**
Metabolism, 6f, 7f, **8**, **933**–959. See also specific aspect
    absorptive state and, 950–952, 951f, 952f
    anabolism and catabolism, **934**, 935f
    ATP synthesis, 936, 936f, 937f
    autonomic effects and, 546, 546t
    case study and, 973b
    developmental aspects of, 968–969, 968f
    homeostatic imbalances and, 943, 947, 952, 959, 963, 968–969
    hormone influences and, 614f, 615, 952, 952f, 954, 954f, 955t
    inborn errors of, 968
    liver in, 950–959, 951f, 953f, 954f, 956t
    muscle, 303–306, 304f, 305f, 313t, 952, 953f
    oxidation-reduction reactions/coenzymes, 36–37, **934**–935
    postabsorptive state and, 952–955, 954f
    pregnancy affecting, 1113
    stress response and, 628f, 629f
    thermoregulation and, 963–968, 963f, 965f, 966f
    thyroid hormone/thyroxine affecting, 618, 618t, 955t, **963**
Metabotropic (G protein-linked) receptors, **81**, 82f, 315, **425**, 425f, 577, 577f

Metacarpal(s), 200f, **234**–236, 234f, 235t
Metacarpal arteries, **742**t, 742f
Metacarpal veins, 754f
Metacarpophalangeal joint (knuckle), 234, 256t
Metacarpus (palm of hand), **234**–236
    bones of (metacarpals), 200f, **234**–236, 234f, 235t
    muscles of, 358t, 359f, 361f, 364–366t, 365f
Metalloproteinases, in osteoarthritis, 273
Metanephros, **1006**, 1006f, 1082f
Metaphase in meiosis, 1044, 1045f, 1046f
Metaphase in mitosis, 96f, **98**, 101f, 1046f
Metaphase plate, 101f
Metaphysis, 178
Metarteriole, 712, 712f
Metastasis, 144b
Metatarsal arteries, **748**t, 748f, **749**t, 749f
Metatarsals, 200f, 241t, **244**, 244f
Metatarsal veins, dorsal, 751f, 758f
Metatarsophalangeal joints, 257t
Metatarsus, bones of (metatarsals), 200f, 241t, **244**, 244f
Metencephalon, **435**, 435f, 436f
Methamphetamine abuse, 429b
Methane gas, formation of, 33, 33f
Metencephalon, **435**, 435f, 436f
Mg. See Magnesium
MHC proteins, 791–792, 802f, 811
MHC restriction, 804
Micelles, **916**, 916f
Microbiota, bacterial, **909**–910
Microcephaly, 484
Microcirculation, **712**, 712f
Microcytes, **650**
Microfilaments, **88**, 88f, 95t
Microglial cells, **393**, 393f
Micronutrients, **927**
MicroRNAs (miRNAs), 54, **108**
Microscopic anatomy, 2
Microscopy, preparing tissue for, 117. See also Histology
Microtubule(s), 62f, **88**–89, 88f, 89f, 90, 90f, 95t, 100f
Microtubule organizing center, 89
Microvilli, 67f, **91**, 91f, 95t, 118, 120f, 124f
    in proximal convoluted tubule, 979f, 980, 980f
    in small intestine, 902f, **903**, 903f
Micturition, 1002, **1004**–1005, 1005f, 1007
Micturition center, pontine, 1005, 1005f
Midbrain (mesencephalon), **435**, 447f, 448f, 450–451f, **451**–452, 452f, 453f, 459f
    development of, 435, 435f, 436f, 1106
Midbrain flexure, 436, 436f
Middle cardiac vein, 676f, **683**, 683f
Middle cerebellar peduncles, 450f, 451f, 452, 453f, **454**, 455f
Middle cerebral arteries, **740**t, 740f, 741f
Middle cervical ganglion, 540f, **541**
Middle colic artery, 744f, **746**t, 747f
Middle cranial fossa, 201, 201f, 207f, **208**
Middle ear (tympanic cavity), **20**, 208, 581f, **582**–583, 582f, 596
Middle meningeal artery, 210, 740t
Middle nasal concha, 203f, **210**, 210f, 214f, **216**t, 821f, 822
Middle nasal meatus, 214f, 215, 821f
Middle phalanx, 234f, 235t, 236, 241t, 244f
Middle pharyngeal constrictor muscle, **339**t, 339f
Middle scalene muscle, **340**t, 341f
Middle suprarenal arteries, 744f, **746**t, 746f
Middle thyroid vein, 752f, 753f
Middle trunk, brachial plexus, 510, 511f
Midgut, 918f
Midpalmar muscles, 364t, 365f, 366t
Midpiece of sperm, 1057f, **1058**
Midsagittal (median) plane/section, **15**, 15f
Migrating motor complex (MMC), 905, 905f, **906**t
Milia, 167

Milk ejection, oxytocin and, 612, 612t
Milk intolerance, 915
Milk of magnesia. See Magnesium hydroxide
Milk production. See Lactation
Milk teeth (deciduous/primary dentition), **879**, 879f
Milliequivalents per liter, **1014**
Milligrams per deciliter, 30
Millivolt(s), 400
Mineral(s), 9, **931**–932, 933t
    storage of, 176, 183–184, 956t
Mineralocorticoids, **623**–625, 624f, 626t, 629f. See also Aldosterone
Mineral salts, in bone tissue, 176, 183–184
Minipill, for contraception, 1118b
Minivalves, lymphatic capillary, 767f, 768
Minor calyces, **977**, 977f
Minor (intrinsic) salivary glands, **878**
Minute ventilation, **842**, 842t
miRNAs. See MicroRNAs
MIT. See Monoiodotyrosine
Mitochondria, 62f, **83**–84, 83f, 94t
    ATP production in, 934, 935f, 937f, 944f
    in cardiac muscle, 684f, 685
    free radicals in, 110
    sarcoplasmic reticulum, 288, 288f
Mitochondrial DNA/genes, 83f, 84, 1133
Mitochondrial (extranuclear) inheritance, 1131, 1133
Mitochondrial theory of aging, 110
Mitosis, 61, 96f, **98**, 100–101f
    meiosis compared with, 1046f, 1047
    in oogenesis, 1069, 1070f
    in spermatogenesis, 1047, 1056f, 1057
Mitotic (M) phase of cell cycle, 96f, **98**, 100–101f. See also Cell division
Mitotic spindle, 100f, 101f
Mitral cells, olfactory, 576f, **577**
Mitral (bicuspid) valve, 677f, 678f, **679**, 681f, 696f
Mitral valve prolapse, **702**
Mittelschmerz, 1074
Mixed nerves, **497**
Mixture(s), **29**–30, 29f, 30
M lines, **284**, 285f, 288f, 289, 289f
MMC. See Migrating motor complex
Modiolus, 584f, **585**
Molarity, **30**, 40
Molars, 879f, **880**
Mole (concentration unit), **30**
Mole(s) (skin), 156, 166
Molecular formula, **36**
Molecular mimicry, in diabetes mellitus, 633b
Molecular weight, **30**
Molecule(s), 4, 4f, **28**–29
Molybdenum, 26t, 933t
Monoblasts, **654**, 655f
Monoclonal antibodies, **801**
Monocytes, 645f, 652f, **653**, 653f, 655f, 656t, **784**, 786
Monoglycerides, 913f, **916**
Monohydrogen phosphate, 1031, 1031f
Monoiodotyrosine (MIT), **619**, 619f
Monomer(s), **42**, 42f, 43f, 55t, 799, 799t, 912
Mononucleosis, 654
Monosaccharides, **43**–44, 43f, 914
Monospermy, **1093**
Monosynaptic reflex, **519**, **523**
Monounsaturated fats, 46
Mons pubis, 1061f, 1066f, **1067**
Morning-after pill, 1118b, 1119b
Morning sickness, 1113
Morula, **1097**, 1097f
Motilin, 888t, 905t, **906**
Motility. See Gastric motility; Intestinal motility
Motion sickness, **594**
Motor areas, 440f, **441**, 442f
Motor association cortex, 440f
Motor control, levels of, 517–519, 518f
Motor cortex, primary (somatic), 440f, **441**, 442f

Motor (efferent) division of peripheral nervous system, **392**, 392*f*, 497
Motor endings, **517**
Motor end plate, **290**. *See also* End plate potential; Neuromuscular junction
Motor fibers, somatic and visceral, 392, 392*f*
Motor homunculus, **441**, 442*f*
Motor integration, 517–519, 518*f*
Motor memory, 460
Motor molecules/proteins, **89**, 90, 90*f*, 396
Motor (efferent) nerves, 298*f*, 299, **497**
Motor (efferent) neurons, 298*f*, 299, 395*f*, **399**, 399*t*
  α, **521**, 521*f*
  in autonomic versus somatic nervous system, 532–533, 533*f*
  development of, 482, 482*f*
  in muscle contraction, 290–298, 291*f*, 292*f*, 298*f*, 299
    stimulus frequency changes and, 300, 300*f*, 306–307, 307*f*
    stimulus strength changes and, 300, 301–302, 301*f*
  in neuromuscular junction, 290, 292*f*
  in reflex arc, 426*f*, 427, 519, 519*f*, 543, 543*f*
  somatic (SM), 290, 291*f*, 292*f*, **474**, 474*f*, 479*f*
  upper and lower, 478, 479*f*
  visceral (autonomic), **474**, 474*f*
Motor nuclei, 453*f*, 457*f*, 458
Motor output, **391**, 391*f*
Motor (descending) pathways/tracts, 474, 475*f*, 478–481, 479*f*, 481*t*
Motor proteins/molecules, **89**, 90, 90*f*, 396
Motor speech area (Broca's area), 440*f*, **441**
Motor unit(s), 298*f*, **299**, 307, 307*f*
Mountain sickness, acute, 858
Mouth. *See* Oral cavity (mouth)
Movement, **8**. *See also specific type*
  cerebellar processing and, 454–455
  muscle tissue/system and, 6*f*, 8, 116*f*, 138–139*f*, **138**–140, 280–281
  skeletal system/bones and, 6*f*, 8
  at synovial joints, 258–260, 258–260*f*
M (mitotic) phase of cell cycle, 96*f*, **98**, 100–101*f*
MRI. *See* Magnetic resonance imaging
mRNA. *See* Messenger RNA
Mucin, **124**, 124*f*, 783, 878
Mucosa(e). *See* Mucous membranes (mucosae)
Mucosa-associated lymphoid tissue (MALT), 771, **774**–775, 774*f*, 775*f*, 777*t*, **873**, 873*f*, 902*f*, 903
Mucosal barrier, **887**, 909
Mucous cells, 124, **124**, **878**, 878*f*, 886, 886*f*
Mucous membranes (mucosae), **141**–142, 141*f*
  of alimentary canal, **872**–873, 873*f*
  of anal canal, 909
  in body defense, 782–783, 782*f*, 783*t*, 808*f*
  of esophagus, 882*f*, 883
  of large intestine, 909
  olfactory, 536*f*, 540*f*, **822**
  of pharynx, 882
  respiratory, 783, **822**, 863
  of small intestine, 903, 903*f*
  of stomach, 885*f*, 886, 886*f*
  tracheal, **827**, 827*f*
  of ureters, 1002, 1002*f*
  vaginal, 1066
Mucous neck cells, 886*f*, **887**
Mucus
  in body defense, 783, 783*t*, 808*f*
  cervical glands producing, 1064, 1066
  in cystic fibrosis, 861
  duodenal glands secreting, 903
  in gastric mucosal barrier, 887
  mucosa of alimentary canal producing, 872, 873

mucous neck cells producing (acidic), 886*f*, **887**
  nasal glands producing, 576, 576*f*, 822
  salivary glands producing, **878**
  submucosa of trachea producing, 827
Müllerian (paramesonephric) ducts, 1082*f*, **1083**
Multiaxial movement, **258**, 263*f*
Multicellular glands, 123, 123*f*
  exocrine, 123*f*, **125**–126, 125*f*, 126*f*
Multifidus muscle, 343*f*
Multimodal association areas, 440*f*, **443**
Multineuronal pathways, 478–480, 479*f*, 481*t*
Multinucleate cells, **91**
Multiple-allele inheritance, **1129**, 1129*t*
Multiple electrolyte solution, for volume restoration, 663
Multiple motor unit summation (recruitment), **301**–302, 301*f*, 309
Multiple sclerosis, **412**, 812
Multipolar neurons, **397**, 398–399*t*
Multi unit smooth muscle, **316**
Mumps, **921**
Murmurs, heart, **696**
Muscarinic agents, autonomic activity and, 544*t*
Muscarinic receptors, 421*t*, **543**, 544*t*
Muscle(s)/muscle tissue, 116*f*, 138–139*f*, **138**–140, 279–322, 283*t*, 312–313*t*. *See also* Cardiac muscle/cardiac muscle tissue; Muscular system; Skeletal muscle(s)/skeletal muscle tissue; Smooth muscle/smooth muscle tissue
  characteristics of, 280
  developmental aspects of, 316–319, 316*f*, 317*b*
  functional classification of, 324, 325*f*
  functions of, 280–281
  homeostatic imbalances and, 287, 293, 296–298, 310
  regeneration of, 317
  types of, 280, 312–313*t*
Muscle action, 324, 325*f*, 326, 332
Muscle atrophy, disuse causing, 310
Muscle attachments, 282–284, 326
Muscle cells/fibers, 138, 138*f*, **280**, 283*t*, 312*t*. *See also* Cardiac muscle cells; Skeletal muscle fibers/cells; Smooth muscle fibers/cells
  contraction of. *See* Muscle contraction
  development of, 316–319, 316*f*
  in muscle spindle, 520–521, 520*f*, 521*f*
  naming muscles and direction of, 324
  types/classification of, 307–309, 307*f*, 308*t*
Muscle contraction, 280, 313*t*
  cardiac muscle, 313*t*, 689–690, 689*f*
  force of, 306–307, 307*f*
  graded, 299–302, **300**, 300*f*
  heat production during, 281
  muscle fatigue and, **306**
  skeletal muscle. *See under* Skeletal muscle(s)/skeletal muscle tissue
  sliding filament model of, **289**, 289*f*
  smooth muscle, 313*t*, 314–316, 315*f*
Muscle fascicles, **282**, 282*f*, 283*t*, 324, 326–327, 326*f*
Muscle fatigue, **306**
Muscle fibers/cells. *See* Muscle cells/fibers
Muscle metabolism, 303–306, 304*f*, 305*f*, 313*t*, 952, 953*f*
Muscle spasm, **319**
Muscle spindle, 477*f*, 492*t*, **493**, 520–521, 520*f*, 521*f*, 522*f*, 523–524
Muscle strain, **319**
Muscle stretch, 307, 307*f*, 313*t*, 315–316, 697–698
Muscle tension, **299**, 307, 307*f*, 316
Muscle tone, 258, 281, 302–303, 310, 314
Muscle twitch, **299**, 299*f*, 300*f*

Muscular (distributing) arteries, 707*f*, 709*t*, **710**
Muscular dystrophy, **287**, 317
Muscularis externa
  of alimentary canal, **873**, 873*f*
  of esophagus, 882*f*, 883
  of small intestine, 903
  of stomach, 885*f*, 886, 886*f*
  of ureter, 1002, 1002*f*
  of vagina, 1066
Muscularis mucosae
  of alimentary canal, **873**, 873*f*
  of small intestine, 902*f*
  of stomach, 886*f*
Muscular pump, venous return and, **717**–718, 718*f*, 723*f*
Muscular system, 6*f*, 323–389, 330*f*, 331*f*. *See also* Muscle(s)/muscle tissue; Skeletal muscle(s)
  actions of, 324, 325*f*, 326
  cardiovascular system and, 318*b*, 761*b*
  case study and, 322*b*, 388–389*b*
  digestive system and, 318*b*, 920*b*
  endocrine system and, 318*b*, 637*b*
  functions of, 6*f*, 8, 280–281
  homeostatic imbalances and, 287, 293, 296–298, 310
  homeostatic relationships of, 318*b*
  integumentary system and, 168*b*, 318*b*
  lymphatic system/immunity and, 318*b*, 778*b*
  movement and, 6*f*, 8, 280–281
  naming muscles and, 324–326
  nervous system and, 318*b*, 550*b*
  reproductive system and, 318*b*, 1085*b*
  respiratory system and, 318*b*, 862*b*
  skeletal system and, 196*b*, 318*b*
    lever systems and, **327**–328, 327–328*t*, 329*f*
  thyroid hormone affecting, 618*t*
  urinary system and, 318*b*, 1036*b*
Musculocutaneous nerve, 356*t*, 357*t*, 510*t*, 511*f*, **512**
Mutation(s), **110**, 144*b*, 1132, 1133, **1135**
Myalgia, **319**
Myasthenia gravis, 293, 812
Myelencephalon, **435**, 435*f*, 436*f*
Myelinated axons/fibers, **396**, 397*f*, 411–412, 411*f*
Myelination, 393*f*, 396–397, 397*f*, 483, 411–412, 411*f*
  multiple sclerosis and, **412**, 812
Myelin sheath(s), 393*f*, 394, **396**, 397*f*, 411–412, 411*f*, 812
Myelin sheath gaps (nodes of Ranvier), 395*f*, **396**, 397
Myelitis, **484**
Myeloblasts, **654**, 655*f*
Myelocytes, **654**, 655*f*
Myelogram, **484**
Myeloid leukemia, 654
Myeloid stem cells, **647**, **654**, 655*f*
Myelomeningocele, 483, 483*f*
Myeloproliferative disorder, 667
Myenteric nerve plexus, 873*f*, **874**
Mylohyoid muscle, 338*t*, 339*f*
Myo (prefix), **280**
Myoblasts, **316**, 316*f*
Myocardial infarction (heart attack), **683**, 700, 760*b*
Myocarditis, **702**
Myocardium, **673**, 673*f*, 677*f*
Myoclonus, **484**
Myocytes. *See* Cardiac muscle cells/fibers
Myoepithelial cells, 161
Myofascial pain syndrome, **319**
Myofibrils, 283*t*, **284**–288, 285*f*, 286*f*, 288*f*, 312*t*, 685
Myofilaments, **138**, 283*t*, **284**–287, 285*f*, 286*f*, 311–314, 312*t*
Myogenic controls, of blood flow, 726, **727**, 727*f*, 987*t*, **988**, 988*f*
Myoglobin, **284**
Myogram, **299**, 299*f*
Myoid cells, **1055**, 1055*f*

Myomesin, 287
Myometrium, 1063*f*, **1065**, 1114
Myopathy, **319**
Myopia (nearsightedness), 566*f*, **567**
Myosin, 49*f*, 88, 283*t*, **286**–287, 286*f*
  clot retraction and, 660
  cross bridge activity and, **296**, 297*f*, 298, 314*f*
  in excitation-contraction coupling, 291*f*, 295*f*
  in sliding filament model, 289
  in smooth muscle, 311, 313*t*, 314, 315*f*
Myosin ATPase, 307, 308
Myosin filaments. *See* Thick (myosin) filaments
Myosin heads, 283*t*, **286**–287, 286*f*, 287, 289, 291*f*, 296, 297*f*, 311
Myosin kinase (myosin light chain kinase), **314**, 315*f*
Myosin tail, 286, 286*f*, 287
Myotome, 1105*f*, **1106**, 1106*f*
Myotonic dystrophy, **319**
Myotubes, 316, 316*f*
MyPlate, 927*f*
Mys (prefix), **280**
Myxedema, **620**
Myxovirus, mumps caused by, **921**

N. *See* Nitrogen
n (haploid chromosomal number), **1043**, 1044, 1044*f*
2*n* (diploid chromosomal number), **1043**, 1044*f*
Na. *See* Sodium
Na⁺-glucose cotransporter, 75*f*
Na⁺-K⁺ ATPase, 74*f*, **75**, 75*f*, 402*f*, **403**, 408, 914*f*, 915*f*, 990–991, 991*f*
Nail(s), **160**–161, 160*f*
Nail bed, 160, 160*f*
Nail folds, **160**, 160*f*
Nail matrix, **160**, 160*f*
Nail plate/nail body, 160, 160*f*
Nail root, 160, 160*f*
Naive lymphocytes, antigen encounter and, **792**, 793*f*, 796–797, 796*f*
Narcolepsy, **463**
Nares (nostrils), 820*f*, 821, **821**, 821*f*, 821*f*
Nasal apertures, posterior (choanae), 821*f*, **822**, 823, 823*f*
Nasal bones, 203*f*, 204*f*, 205*f*, **212**, 214*f*, **217**t, 821, 821*f*
Nasal cartilages, 174, 175*f*, 821, 821*f*
Nasal cavity, **20**, 202*f*, 214*f*, **215**, 820*f*, **821**–822, 821*f*, 824*t*
Nasal cilia, 822
Nasal conchae, 576, 576*f*, 821*f*, 822
  inferior, 202*f*, 203*f*, 214*f*, **217**t, 821*f*, 822
  middle, 203*f*, **210**, 214*f*, **216**t, 821*f*, 822
  superior, **210**, 214*f*, **216**t, 576, 821*f*, 822
Nasal glands, 546*t*, 822
Nasal hairs (vibrissae), 783, 783*t*, **822**
Nasal meatuses, 214*f*, 215, 821*f*, 822
Nasal mucosa, 536*f*, 540*f*, 822
Nasal polyps, **863**
Nasal septum, 215, **821**
  deviated, **863**
Nasal spine, anterior, **211**, 211*f*, 214*f*
Nasal vestibule, 821*f*, **822**
Nasolacrimal duct, **556**, 556*f*
Nasopharynx, **823**, 823*f*
Natriuretic peptide, atrial. *See* Atrial natriuretic peptide
Natural killer (NK) cells, 782*f*, **785**, 790*t*, 807, 808*f*
Naturally acquired immunity, 797, 797*f*, 798
Nausea and vomiting, **893**, 1113
Navel (umbilicus), 20, 1109*f*
Navicular, 241*t*, **244**, 244*f*
Near point of vision, **566**–567
Nearsightedness (myopia), 566*f*, **567**
Nebulin, 287

Neck
  arteries supplying, 740–741*f*, 740–741*t*
  innervation of, 509–510, 509*f*, 509*t*. *See also* Cervical nerves
  muscles of, 330*f*, 331*f*, 333*f*, 338–339*t*, 339*f*, 340–343*t*, 341*f*, 343*f*
  veins of, 750*f*, 751*f*, 752–753*f*, 752*t*
Neck of bone
  of femur, 240*f*
  of humerus, 230*f*, **231**
  of radius, 231*f*, 232*f*
  of rib, 226, 226*f*
Neck of tooth, **880**, 880*f*
Necropsy, 147
Necrosis, 110
Negative chronotropic factors, 698
Negative feedback, **10–11**, 11*f*
  hormone secretion and, **607**, 607*f*
  ovarian function and, 1075, 1075*f*, 1076
Negative inotropic agents, 698
Negative nitrogen balance, 929–931
Negative selection, **794**, 794*f*
Neisseria gonorrhoeae, 1080
Neon, as chemically inert element, 31, 31*f*
Neonatal period, **1116**
Neonate. *See* Newborn
Neoplasm, 144*b*. *See also* Cancer
Nephrolithiasis, 1002
Nephron(s), **978–983**, 979*f*, 980*f*, 981*f*, 982, 982*f*, 983*f*
  cortical, 981*f*, **982**
  juxtamedullary, 981*f*, **982**, 995, 996–997
Nephron loop (loop of Henle), 979*f*, **980**, 981*f*, 992–993, 993*t*, 996–997*f*, 998
Nephrotoxin, **1008**
Nerve(s), 391, 391*f*, **395**, **496–517**, 497*f*, 498*f*. *See also specific nerves and* Cranial nerve(s); Spinal nerve(s)
Nerve cell(s). *See* Neuron(s)
Nerve cell adhesion molecule (N-CAM), 428
Nerve conduction studies, **526**
Nerve deafness, vestibulocochlear damage causing, 504*t*
Nerve endings, 151*f*, 164, 491, 491*t*, 492–493. *See also specific type and* Sensory receptors
Nerve fascicles, **496**, 497*f*
Nerve fibers, **395**, 412, 496, 497*f*. *See also* Axon
  regeneration of, 144–145, **497–498**, 498*f*
Nerve growth factor, 428, **548**
Nerve impulse(s), 24, 396, 404, **405–412**, 406–407*f*, 410*f*, 416–417*t*. *See also* Action potential
Nerve plexuses, 507*f*, **509**, 515*t*. *See also specific type*
Nerve roots. *See* Dorsal root(s); Spinal root(s); Ventral root(s)
Nervous system, 6*f*, **390–433**, 391, 391*f*, 392*f*. *See also specific division and* Nerve cells; Nervous tissue
  autonomic. *See* Autonomic nervous system
  cardiovascular system and, 550*b*, 761*b*
  case study and, 433*b*
  central. *See* Central nervous system
  digestive system and, 550*b*, 920*b*
  endocrine system and, 550*b*, 602*t*, 637*b*
  functions of, 6*f*, 8, 391–392, 391*f*
  homeostatic imbalances and, 396, 412
  hormone release controlled by, 607, 607*f*, 608
  integumentary system and, 168*b*, 550*b*
  lymphatic system/immunity and, 550*b*, 778*b*, 814
  muscular system and, 318*b*, 550*b*
  organization of, 391–392, 392*f*
  peripheral. *See* Peripheral nervous system
  reproductive system and, 550*b*, 1085*b*
  respiratory system and, 550*b*, 862*b*
  responsiveness/excitability and, 6*f*, 8, 392, 405

skeletal system and, 196*b*, 550*b*
thyroid hormone affecting, 618*t*
urinary system and, 550*b*, 1036*b*
Nervous tissue, 116*f*, 140–141, 140*f*, 392–400. *See also* Neuron(s)
  case study and, 433*b*
  membrane potentials and, 400–404, 401*f*, 402*f*, 404*f*
  neurotransmitters/receptors and, 396, 414, 415*f*, 420–425, 421–422*t*, 429*b*
  supporting cells (neuroglia), 140, 140*f*, 392–394, **393**, 393*f*
  synapses and, **412–415**, 413*f*, 415*f*
Net filtration pressure (NFP), **731**, 733*f*, **986**, 986*f*
Net osmosis, 71, 918
Netrin, 428
Neural crest, 428, **482**, 482*f*, 548, 1105*f*, **1106**
Neural folds, **482**, 482*f*, **1104**, 1105*f*
Neuralgia, **526**
Neural groove, **482**, 482*f*, **1104**, 1105*f*
Neural integration, **391**, 391*f*, **426–427**, 426*f*, 427*f*, 494–495, 494*f*
Neural integrators, axon's initial segment as, 420
Neural layer of retina, 560*f*, **561**
Neural plate, **482**, 482*f*, **1104**, 1105*f*
Neural processing, 426–427, 426*f*
Neural stimuli
  blood flow/perfusion regulation and, 727*f*
  blood pressure regulation and, 719–721, 720*f*
  digestion control and, 889, 890, 890*f*
  food intake regulation and, 960–961, 961*f*
  glomerular filtration rate control and, 987*t*, 988*f*, 989
  for hormone release, 607, 607*f*, 608
  respiratory control and, 454, 853–854, 853*f*, 854*f*
Neural tube, 428, **435**, 435*f*, **482**, 482*f*, **1104**, 1104*f*, 1105*f*
Neurilemma (outer collar of perinuclear cytoplasm), 396, 397*f*
Neuroblastoma, **430**
Neuroblasts, **428**
Neurocardiogenic (vasovagal) syncope, **549**
Neuroendocrine organ, hypothalamus as, **602**
Neuroepithelial cells, 428
Neurofibrillary tangles, 469
Neurofibrils/neurofilaments, **394**
Neurogenic shock, 725
Neuroglia (glial cells), 140, 140*f*, 392–394, **393**, 393*f*
Neurohormones, **609**
Neurohypophysis, **609**. *See also* Posterior pituitary gland
Neuroimaging, 17*b*, 439*f*, 470, 575
Neurokinin A, 422*t*
Neurolemmocytes (Schwann cells), 393*f*, **394**, 395*f*, 396, 397*f*, 498*f*
Neurologist, **430**
Neuromodulators, 421–422*t*, **424**
Neuromuscular junction(s), **290**, 291*f*, 292*f*, 293*f*, 298*f*, 313*t*, **517**
Neuron(s) (nerve cells), **140**, 391, 392, **394–400**
  association (interneurons), **399**, 399*t*, 472, 474*f*, 474*t*, 482, 482*f*
  in autonomic versus somatic nervous system, 532–533, 533*f*
  case study and, 433*b*
  cell body (perikaryon/soma) of, 140*f*, 290, **394**, 395*f*
  circuit types and, **427**, 427*f*
  classification of, 397–400, 398–399*t*
  developmental aspects of, 428, 428*f*, 482, 482*f*, 483
  electrically coupled, 413
  enteric, 543, **874**. *See also* Enteric nervous system

motor. *See* Motor (efferent) neurons
  organization of (neuronal pools), **426**, 426*f*
  postsynaptic, **412**, 413*f*, 415*f*, 418*f*
  presynaptic, **412**, 413*f*, 415*f*, 419*f*
  processes of, 140, 140*f*, **394–397**, 395*f*
  regeneration of, 144–145, 394
  responsiveness/excitability and, 6*f*, 8, 392, 405
  sensory. *See* Sensory (afferent) neurons
  somatic motor, 290, 292*f*, **474**, 474*f*, 479*f*
  somatic sensory, **474**, 474*f*
  synaptic communication and, **412–415**, 413*f*, 415*f*
  visceral (autonomic) motor (VM), **474**, 474*f*
  visceral sensory, **474**, 474*f*, 542
Neuronal circuits, **427**, 427*f*, 494–495, 494*f*
Neuronal growth cone, **428**, 428*f*
Neuronal pathways, 475*f*, 476–481, 477*f*, 479*f*, 480–481*t*, 481*f*
Neuronal pools, **426**, 426*f*
Neuropathy, **430**, 526, 545
Neuropeptide(s), **423**, 960, 961, 961*f*, 962
Neuropeptide Y, in food intake regulation, 960, 961, 961*f*, 962
Neuropharmacology, **430**
Neurophysiology, **2**
Neuroses, **484**
Neurotoxin, **430**
Neurotransmitter(s), 81, 396, 413, 414, 415*f*, 420–425, 421–422*t*, 429*b*
  in autonomic nervous system, 533, 533*f*, 535*t*, 543–545, 544*t*
  in axonal terminals, 396, 414, 415*f*, 517
  classification of, 421–422*t*, 423–424
  direct versus indirect acting, 421–422*t*, 424
  excitatory and inhibitory, 421–422*t*, 424
  information processing in retina and, 572*f*, 573
  information transfer at chemical synapse and, 414, 415*f*
  in neuromuscular junction, 290, 292*f*
  receptors for, 424–425, 424*f*, 425*f*
  in smooth muscle contraction, 315
  in taste physiology, 579
Neurotransmitter receptor region, 414
Neurotropins, 428
Neurulation, **1104**, 1105*f*
Neutralization, by antibodies, **800**, 800*f*
Neutralization reaction, **40**
Neutron(s), **25**, 25*f*
Neutrophils, 128*f*, 137*f*, 645*f*, **652**, 652*f*, 653*f*, 655*f*, 656*t*, **784**, 785, 786. *See also* Phagocytosis/phagocytes
Nevi, pigmented (moles), 156
Newborn
  adjustment to extrauterine life by, 1116
  circulation in, 1107, 1109*f*, 1116
  digestive system in, 919
  first breath by, 860, 1116
  fluid/electrolyte/acid-base balances and, 1035–1037
  genetic screening of, 1133
  hemolytic disease of, **665**
  hydrocephalus in, 467, 467*f*
  immune system in, 814
  physiological jaundice in, **1120**
  skin in, 167
  skull in, 246, 246*f*
  STIs transmitted to, 1080
Nexus. *See* Gap junction(s)
NH_3. *See* Ammonia
NH_4^+. *See* Ammonium ion
Niacin (vitamin B_3), 932*t*, 935
Nicotinamide adenine dinucleotide (NAD⁺), **935**
Nicotine, 724, 1113
Nicotinic agents, autonomic activity and, 544*t*
Nicotinic receptors, 421*t*, **543**, 544*t*
Night blindness, 573

Nipple(s), **1068**, 1068*f*
Nissl bodies (chromatophilic substance), **394**, 395*f*
Nitric oxide (NO), 422*t*, **423**–424, 658, **726**, 727*f*, 1054
Nitrogen, 25, 26*t*, 843, 843*t*, 844
Nitrogen balance, **929**–931
Nitrogen gas, formation of, 33*f*
Nitrogenous wastes, 1000, **1001**, 1015*f*
NMDA receptors, pain activating, 496
NO. *See* Nitric oxide
Noble gases, 31
Nociceptors (pain receptors), 477*f*, **490**, 491, 491*t*, 492*t*, 495
Nocturia, 1007
Nocturnal enuresis (NE), **1008**
Nodes of Ranvier (myelin sheath gaps), 395*f*, **396**, 397
Nodule(s), lymphoid. *See* Lymphoid follicles/nodules
Noise, hearing loss and, 587, 594, 596. *See also* Loudness
Non-alcoholic fatty liver disease (NAFLD), **898**
Nonaxial movement, **258**, 262*f*
Nondisjunction, **1081**, **1135**
Nondisplaced fracture, 191
Nonelectrolytes, **1013**–1014
Nonencapsulated (free) nerve endings, 151*f*, 164, 491, 491*t*
Nongated (leakage) channels, 70, **400**
Non-Hodgkin's lymphoma, **777**
Non-insulin-dependent (type 2) diabetes mellitus (NIDDM), 632, 633*b*, 968
Nonkeratinized epithelia, 122*t*, 123
Nonkinetochore microtubules, 100*f*
Nonmembranous organelles, 83
Nonmyelinated axons/fibers, **396**, 397, 411–412, 411*f*
Nonpolar molecules, **34**, 34*f*
Nonpolar tails, of membrane phospholipids, 46*f*, 47, 63, 65*f*
Non-rapid eye movement (NREM) sleep, **462–463**, 463*f*
Nonrespiratory air (gas) movements, **838**
Nonself antigens, **791**
Nonshivering (chemical) thermogenesis, **967**
Nonsister chromatids, **1043**
Nonspecific defenses. *See* Innate (nonspecific) defenses
Nonsteroidal antiinflammatory drugs (NSAIDs), peptic ulcers and, 887, 889
Nonstriated muscle. *See* Smooth muscle
Nonvolatile (fixed) acid, **1029**
Noradrenaline. *See* Norepinephrine
Norepinephrine, 421*t*, 423, **533**, 543, 544*t*, 626*t*, **627**, 628*f*, **721**
  adrenal secretion of, 542, 542*f*, 544*t*, 623*f*, 626*t*, 627, 628*f*
  in autonomic versus somatic nervous system, **533**, 533*f*, 543, 544*t*
  basal metabolic rate and, 963
  blood pressure and, 721, 721*t*, 723*f*
  cardiac contractility affected by, 698, 698*f*
  heart rate affected by, 699
  memory processing and, 460
  smooth muscle contraction and, 315
  stress response/sympathetic stimulation and, 627, 628*f*
Normal saline, for volume restoration, 663
Nose, 820–822, 821*f*, 821*f*, 824*t*. *See also under* Nasal
Nosebleed (epistaxis), **863**
Nostrils (nares), 820*f*, 821, **821**, 821*f*, 821*f*
Notch (bone marking), 180*t*
Notochord, **1103**, 1104, 1104*f*, 1105*f*, 1106, 1106*f*
NREM sleep. *See* Non-rapid eye movement (NREM) sleep
Nuchal lines, 203, 203*f*, 206*f*, **216***t*
Nuclear envelope/membrane, 62*f*, 87*f*, **92–93**, 92*f*, 96*t*, 100*f*, 101*f*

Nuclear lamina, 92, 92*f*
Nuclear pores/nuclear pore complexes, **92**, 92*f*
Nucleases, pancreatic, **899**, **917**
Nuclei. *See* Nucleus
Nucleic acids, **53–55**, 53*t*, 54*f*, 55*t*, 913*f*, 917. *See also* DNA; RNA
Nucleoli (nucleolus), 62*f*, 92*f*, **93**, 96*t*, 100*f*, 101*f*
Nucleoplasm, 92–93
Nucleosidases, 913*f*, **917**
Nucleosome(s), **93**, 93*f*
Nucleotides/nucleotide bases, **54–55**, 54*f*, **917**
Nucleus (atomic), **25**, 25*f*
Nucleus (cell), 62, 62*f*, **91–110**, 92*f*, 95
  division of. *See* Meiosis; Mitosis
Nucleus (neural), **394**, 497
  basal. *See* Basal nuclei
  hypothalamic, 449, 449*f*
  reticular, 457*f*, 458
  septal, 456, 456*f*
  thalamic, 448, 449*f*, 456, 456*f*
Nucleus accumbens, 429*b*
Nucleus ambiguus, 453*f*
Nucleus cuneatus, **453**, 477*f*
Nucleus gracilis, **453**, 477*f*
Nucleus pulposus, **219**, 219*f*, 1106, 1106*f*
Nursing (breast feeding). *See* Lactation
Nutrient(s), **9**, **927–933**, 927*f*, 929*f*, 930*t*, 932–933*t*. *See also specific type*
  absorption of. *See* Absorption
  essential, **927**
  exchanges of among fluid compartments, 1015, 1015*f*
  malabsorption of, **918**
  in plasma, 644*t*
  transport of, 643, 730, 731*f*
  tubular reabsorption of, 991–992, 991*f*
Nutrient arteries, 178, 179*f*
Nutrient foramina, **178**, 179*f*
Nutrient pools, 949–950, 950*f*
Nutrient signals, food intake regulation and, 961, 961*f*
Nutrient vein, 178
Nutrition, 167, 927, 968–969, 973*b*, 1112–1113
Nyctalopia (night blindness), 573
Nystagmus, vestibular, **593**

O. *See* Oxygen
Obesity, 958–959*b*, 960, 968, 724, 962
Obligatory water losses, **1017**
Obligatory water reabsorption, **991**
Oblique fissure, 832*f*, 833
Oblique layer, of stomach muscularis, 885*f*, 886, 886*f*
Oblique muscles, 330*f*, **346***t*, 347*f*, 556, 557*f*
Oblique popliteal ligament, 270*f*, **271**
Oblique sections, **15**
Obstructive jaundice, 898
Obstructive pulmonary diseases, 842, 858–859, 858*f*
Obstructive sleep apnea, **860**
Obturator artery, 748*t*, 748*f*, 749*f*
Obturator externus muscle, 372*t*, 372*f*, 384*t*
Obturator foramen, 237*f*, **238**
Obturator internus muscle, 349*f*, **372***t*, 372*f*, 384*t*
Obturator nerve, 369*t*, 372*t*, **512**, 513*t*, 513*f*
Occipital artery, 740*t*, 740*f*, 741*f*
Occipital belly, of epicranius, 331*f*, **333***t*, 333*f*, 335*f*
Occipital bone, **202–203**, 203*f*, 204*f*, 205*f*, 206*f*, 207*f*, **216***t*
Occipital condyles, **202**, 203*f*, 206*f*, **216***t*
Occipital crest, external, 202–203, 203*f*, 206*f*, **216***t*
Occipital lobe, 438*f*, **439**, 440*f*
Occipital nerve, lesser, 509*t*, 509*f*
Occipital protuberance, external, **202**, 203*f*, 205*f*, 206*f*, **216***t*
Occipital vein, 752*f*, 753*f*

Occipitofrontalis (epicranius) muscle, 330*f*, 331*f*, **333***t*, 335*f*
Occipitomastoid suture, 202, 203*f*, 204*f*, 205*f*
Octet rule (rule of eights), **31–32**
Oculomotor nerve (cranial nerve III), 450*f*, **499**, 499*f*, 501*t*, **536**, 536*f*, 556, 557*f*
Oculomotor nuclei, 451, 453*f*, 536, 547
Odontoblast, 881
Odorants, 576–577, 577*f*
Odor sensation. *See* Olfaction
Ohm's law, **400**
Oil(s), triglycerides as, 45
Oil (sebaceous) glands, 151*f*, 158*f*, 161*f*, 162*f*, **163**, 167, 169
Oleoresins, skin penetrated by, 164
Olecranon /olecranon process, 231*f*, 232*f*, **233**, 233*f*
Olecranon bursitis, 273
Olecranon fossa, 230*f*, **231**, 231*f*
Olfaction (sense of smell), 440*f*, 443, 575–578, 576*f*, 577*f*
  developmental aspects of, 595
  disorders/homeostatic imbalances and, 578
  limbic system and, 457, 578
  physiology of, 577
  taste sensation and, 580
Olfactory adaptation, 577
Olfactory aura, 578
Olfactory axons, 576, 576*f*, 577
Olfactory bulbs, 440*f*, 452*f*, 456*f*, 499, 499*f*, 500*t*, 576*f*, **577**
Olfactory cilia, 576, 576*f*
Olfactory (smell) cortex, primary, 440*f*, **443**, 577
Olfactory epithelium, **575–576**, 576*f*, 821*f*, **822**
Olfactory filaments, 499, 499*f*, 500*t*, **576**, 576*f*
Olfactory gland, 576*f*
Olfactory hallucinations, 578
Olfactory mucosa, **822**. *See also* Olfactory epithelium
Olfactory nerve (cranial nerve I), **499**, 499*f*, 500*t*, 576, 576*f*, 821*f*
Olfactory pathway, 577–578
Olfactory pits, **860**
Olfactory placodes, **860**, 861*f*
Olfactory sensory neurons/olfactory receptors, 575–577, **576**, 576*f*, 1093
Olfactory stem cells, **576**, 576*f*
Olfactory tract, 440*f*, 443, 499, 499*f*, 576*f*, **577**
Olfactory transduction, 577, 577*f*
Oligodendrocytes, 393*f*, **394**, 397, 497
Oligosaccharides, 913*f*, 914, 914*f*
Oliguria, 1018
Olivary nuclei, **452**, 453*f*, 589, 589*f*
Olive(s), 450*f*, **452**
Omega-3 fatty acids, **47**, 930*t*, 958
Omega-6 fatty acids, 930*t*
Omentum (omenta), greater and lesser, 872, 885, 908*f*
Omohyoid muscle, **338***t*, 339*f*, 341*f*
Oncogenes, 144*b*
Oncotic pressure (capillary colloid osmotic pressure/OP$_c$), **731**, 733*f*, 734
One-molar solution, 30
Oocyte(s), **1062**, 1070*f*. *See also* Ova
  fertilization of, 1042, 1092–1096, 1092*f*, **1093**, 1094–1095*f*, 1096*f*, 1097*f*
  primary, **1069**, 1070*f*, 1072*f*
  secondary, 1070*f*, **1071**, 1072*f*
  sperm-binding receptors on, 1093, 1095*f*
Oocyte membrane block, **1093**, 1095*f*
Oogenesis, **1069–1073**, 1070*f*, 1071*t*
Oogonia, **1069**, 1070*f*
Oophorectomy, **1086**
OP. *See* Colloid osmotic pressure
OP$_c$. *See* Capillary colloid osmotic pressure
Open (compound) fracture, 191
Open (internal) reduction of fracture, 191
Open comedone (blackhead), 163

Openings (bone markings), 179, 180*t*
OP$_{gc}$. *See* Colloid osmotic pressure, in glomerular capillaries
Ophthalmic artery, **740***t*, 740*f*, 741*f*
Ophthalmic division of trigeminal nerve (cranial nerve V$_1$), 502*t*
Ophthalmic vein, **752***t*, 752*f*, 753*f*
Ophthalmology, 596
OP$_{if}$. *See* Interstitial fluid colloid osmotic pressure
Opponens digiti minimi muscle, **364***t*, 365*f*
Opponens pollicis muscle, **364***t*, 365*f*
Opposition (body movement), 236, **260**, 261*f*, 365*t*
Opsins, 568*f*, **569**
Opsonins, **784**, 789, 789*f*
Opsonization, **784**, 789, 789*f*, 801
Optic canals, 203*f*, 207*f*, **209–210**, 209*f*, **216***t*
Optic chiasma, 447*f*, 448*f*, 450*f*, 452*f*, 499*f*, 500*t*, **573–574**, 574*f*
Optic cups, **595**
Optic disc (blind spot), 558*f*, 560*f*, **561**, 561*f*
Optic nerve (cranial nerve II), 450*f*, 452*f*, **499**, 499*f*, 500*t*, 558*f*, 560*f*, **573**, 574*f*
Optic radiation, 500*t*, **574–575**, 574*f*
Optics, **563–564**, 564*f*, 565*f*
Optic tracts, 450*f*, 452*f*, 499*f*, 500*t*, **574**, 574*f*
Optic vesicles, **595**
Optometrist, **596**
Oral cavity (mouth), **20**, 202*f*, 820*f*, 869*f*, **876–877**, 876*f*. *See also specific structure*
  development of, 918
  digestive processes occurring in, 870*f*, 881, 904*t*, 913*f*
Oral cavity proper, **876**
Oral contraceptive pill, 1118*b*
Oral membrane, **918**, 918*f*
Oral orifice, **876**
Oral vestibule, **876**, 876*f*
Ora serrata, 558*f*, **561**
Orbicularis oculi muscle, 330*f*, **334***t*, 335*f*, **555**, 555*f*
Orbicularis oris muscle, 330*f*, **334***t*, 335*f*, 337*f*
Orbit(s) (orbital cavity), **20**, 202, 202*f*, **212**, 213*f*
Orbital fissure, inferior and superior, 203*f*, 209*f*, **210**, 212, **216***t*, **217***t*
Orbital model of atom/orbitals, 25*f*, **27**
Orbital plates, **210**, 210*f*
Orbitals, **27**
Orbitofrontal cortex, 440*f*
Orchiectomy, 1049
Orchitis, **1086**
Orexins, **463**, 960, 961*f*
Organ(s), 4*f*, **5**. *See also specific organ*
Organ of Corti. *See* Spiral organ
Organelles, 4, 4*f*, **83–89**, 94–95*t*, 109*t*. *See also specific type*
Organic acid groups, 45, 48
Organic components, in bone, 182–183
Organic compounds, **38**, **41–56**. *See also specific type*
Organic solvents, skin penetrated by, 164
Organism/organismal level, structural organization at, 4*f*, **5**
Organization, in tissue repair, 143*f*
Organ level, structural organization at, 4*f*, **5**
Organogenesis (germ layer differentiation), **1103–1107**, 1104*f*, 1105*f*, 1106*f*
Organ systems, 4*f*, 5, 5*f*. *See also specific system*
  structural organization at level of, 4*f*, **5**
Organ transplants, 810–811
Orgasm, **1054**, 1079
Orientation. *See* Equilibrium/orientation
Origin (muscle), **258**, **282**, 324, 326
Origins of replication, 97

Orlistat, 959*b*
Oropharynx, **823**, 823*f*, 876, 876*f*, **882**
Orthochromatic erythroblast, **647**, 647*f*
Orthodontics, **921**
Orthopedist/orthopedic surgeon, **248**
Orthopnea, 863
Orthostatic hypotension, **549**, 724
Os coxae (hip bones/coxal bones), **236–239**, 237*f*, 241*t*
Osmolality, 72, 974, 995, 995*f*, 998, 999*f*, 1016–1019, 1017*f*, 1018*f*. *See also* Water balance
Osmolarity, **71**, 71*f*, 72–73
Osmoreceptors, 449, 613, 1016, 1017*f*
Osmosis, 69, 70*f*, **71–73**, 71*f*, 73*t*
Osmotic diuretics, 1000
Osmotic gradient, medullary, **995**, 995*f*, 996–997*f*, 999*f*, 1000
Osmotic pressure, **72**, 732–733*f*
  colloid, **731**, 732–733*f*, 986, 986*f*
  edema and, 734
  fluid movement among compartments and, **72**, 731–734, 732–733*f*, 1015
Osseous spiral lamina, 584*f*, **585**
Osseous tissue. *See* Bone(s)/skeletal tissue
Ossicles, auditory, **217***t*, 581*f*, **582**, 582*f*, 587, 587*f*, 596
Ossification (osteogenesis), **184–188**, 185*f*, 186*f*, 194–195
  endochondral, **184–185**, 185*f*
  intramembranous, 184, **186**, 186*f*
  of skull bones, 186, 246
Ossification centers, 184, 185, 185*f*, 186*f*, 246, 246*f*
  primary and secondary, **184**, 185*f*, 194, 194*f*
Ossification (osteogenic) zone (epiphyseal plate), 186, 187*f*
Ostealgia, **195**
Osteitis, **195**
Osteitis fibrosa cystica, 622
Osteoarthritis (degenerative joint disease), **273–274**, 276
Osteoblasts, 127*t*, **129**, 137, **181**, 181*f*, 634
Osteocalcin, 176, 634, 635*t*
Osteochondral grafting, in joint therapy, 274*b*
Osteoclasts, **181**, 181*f*, **188**, 189, 189*f*, 622, 623*f*
Osteocytes, 127*t*, **129**, 136*f*, 137, **181**, 181*f*
Osteogenesis. *See* Ossification (osteogenesis)
Osteogenesis imperfecta (brittle bone disease), **147**, **195**
Osteogenic cells, 178, **180–181**, 181*f*
Osteogenic layer, of periosteum, 178
Osteoid, 181, **182–183**, 186*f*
Osteoid seam, 188
Osteomalacia, **193**
Osteomyelitis, **195**
Osteon (Haversian system), 137, **181–182**, 182*f*, 183*f*
Osteophytes, 274
Osteoporosis, 192*t*, **193**, 193*f*
Osteoprogenitor (osteogenic) cells, 178, **180–181**, 181*f*
Osteoprotegerin, 189
Osteosarcoma, **195**
Otalgia, **596**
Otic ganglia, 505*t*, **537**
Otic pit, **596**
Otic placode, **596**
Otic vesicle, **596**
Otitis externa, **596**
Otitis media, **582**, 594
Otoacoustic emissions, **596**
Otolith membrane, **590**, 591*f*
Otoliths, **590**, 591*f*
Otorhinolaryngology, **863**
Otosclerosis, **594**
Outer collar of perinuclear cytoplasm, 396, 397*f*
Outer ear. *See* External (outer) ear
Outer fiber, rod, 568, 568*f*

Outer hair cells, 584f, **585**, 588–589
Outer mitochondrial membrane, 83, 83f
Outer segments, rod and cone, 560f, **568–569**, 568f
Outlet, pelvic, **238**, 238f, 239t
Ova (eggs), 1042, 1070f, **1071**
  fertilization of, 1042, 1092–1096, 1092f, **1093**, 1094–1095f, 1096f, 1097f, 1119, 1128
  production of (oogenesis), **1069**–1073, 1070f, 1070f, 1071t
Oval window, 581f, **582**, 583f, 587, 587f
Ovarian arteries, 746t, 746f, **1061**
Ovarian branch, of uterine arteries, 1061
Ovarian cancer, **1086**
Ovarian cortex, 1061f, 1062
Ovarian cycle, 1072f, **1073**–1076, 1075f, 1077f
Ovarian cysts, **1086**
Ovarian follicles, 1061f, **1062**, **1070**, 1072f
  dominant, **1071**, 1074, 1077f
  preantral, 1073
  primary, 1070f, **1071**, 1072f, 1073
  primordial, **1070**, 1070f, 1071, 1072f, 1073
  secondary, 1070f, **1071**, 1072f, 1073
  vesicular (antral/tertiary), 1070f, **1071**, 1072f
Ovarian ligament, **1061**, 1063f
Ovarian medulla, 1061f, 1062
Ovarian vein, 756t
Ovaries, 602f, 634, 1042, **1060**, 1061–1062, 1061f, 1063f, 1064f
  aging affecting, 636
  cancer of, **1086**
  descent of, 1084
  embryologic/fetal development of, 1082f, 1084
  polycystic, 159, **1086**
  surgical removal of (oophorectomy), **1086**
Overflow incontinence, 1005
Overhydration, 998, 999f, 1019, 1019f
Overload principle, in exercise, 309
Overweight, 960
Oviducts. See Uterine (fallopian) tubes
Ovulation, 1062, 1070f, 1071, 1073, **1074**, 1075f, 1076, 1077f, 1118b, 1119
Ovum. See Ova
Oxaloacetic acid, 939f, **940**
Oxidase(s), 49f, 52, 86, **934**
Oxidation, **934**
  beta, **946**, 946f, 947f
  in Krebs cycle, 939f, 940
  water of (metabolic water), **1016**
Oxidation-reduction (redox) reactions, 36–37, **934**–935
Oxidative deamination, 948f, 949, 949f
Oxidative fibers, fast and slow, 307f, **308**–309, 308t
Oxidative phosphorylation, 935f, **936**, 937f, 940–943, 941f, 944, 944f
Oxidized reactant, in redox reaction, **37**
OxyContin (oxycodone) abuse, 429b
Oxygen, **9**, 25, 26t, 843
  alveolar, 843t, 844, 844f
  in blood, erythrocytes carrying, 646–647
  as chemically reactive element, 31f
  exchange of, 830, 842–847, 843t, 844f, 846f
  hyperbaric, 843
  maximum amount used in aerobic metabolism, aging affecting, 861
  partial pressure of (P_{O_2}), 843, 843t
    external respiration (pulmonary gas exchange) and, 844–846, 844f, 845f, 846f
    hemoglobin saturation affected by, 847, 848–849f
    internal respiration and, 844f, 846–847
    perfusion affected by, 845, 846
    respiratory rate and depth affected by, 855–856
    ventilation affected by, 846, 846f, 856
  solubility of, 843, 844–845, 844f

  supplemental, for COPD, 859
  as survival requirement, **9**
  transport of, 643, 646–647, 847–850, 848–849f, 850f, 851f
Oxygen debt (excess postexercise oxygen consumption), **306**
Oxygen gas, formation of, 33f
Oxygen-hemoglobin dissociation curve, **847**, 848–849f, 850f
Oxygen toxicity, **844**
Oxyhemoglobin, **647**, **847**
Oxyphil cells, 621, 621f
Oxytocin, 11, 610f, 611, **612**–613, 612t, **1114**–1115, 1114f, 1117, 1117f

P. See Phosphorus
Pacemaker, 313t, **687**
Pacemaker cells, 313t, 314, 685, 685t, **686**–689, 687f, 854, 892
Pacemaker potentials (prepotentials), **686**, 687f
Pacinian (lamellar) corpuscles, 151f, 155, 164, 492t, **493**
Packed red cell transfusions, **663**
Pagetic bone, 194, 194f
Paget's disease, **194**, 194f
Pain
  in inflammation, 785, 786, 787f
  perception of, 495–496, 496f
  referred, **496**, 496f
  visceral, 496
Pain receptors (nociceptors), 477f, **490**, 491, 491t, 492t, 495
Pain threshold, 495
Pain tolerance, 495–496
Palate, 822, **876**–877, 876f
  cleft, 246, 246f, 861
  hard, 206f, 211, 214f, 821f, **822**, 823f, **876**–877, 876f
  soft, 821f, **822**, 823, 823f, 876f, **877**
Palatine bones, 205f, 206f, **212**, 214f, 217t
Palatine processes, 205f, 206f, **211**, 214f, **217t**
Palatine raphe, 876f
Palatine suture, median, 206f
Palatine tonsils, 774f, **775**, **823**, 823f, 876f, 877f
Palatoglossal arches, 876f, **877**, 877f
Palatopharyngeal arches, 876f, **877**, 877f
Pallor, 157
Palmar aponeurosis, 359f
Palmar arches, 739f, 742t, 742f, 743f, **754t**, 754f, 755f
Palmar interossei muscle, 365f, **366t**
Palmaris longus muscle, 330f, **358t**, 359f, 362t
Palm of hand. See Metacarpus (palm of hand)
Palpation, 2
Palpebrae (eyelids), **554**–555, 555f
Palpebral conjunctiva, **555**, 555f
Palpebral fissure, **554**, 555f, 555f
Palpitation, heart, **702**
Pampiniform venous plexus, 1048f, **1049**
Pancreas, 602f, **630**–634, 631f
  autonomic innervation/effects and, 534f, 536f, 546t
  in digestion (exocrine), 630, 630f, 869f, **898**–899, 899f, 900, 900f, 901f, 904t
  endocrine, **630**–634, 630f, 631f
  transplantation of for diabetes, 633b
Pancreatic amylase, **899**, 913f, **914**, 914f
Pancreatic ducts, 898, 899f, 900, 900f
Pancreatic enzymes. See Pancreatic juice
Pancreatic islets (islets of Langerhans), **630**, 630f, 899
Pancreatic juice/pancreatic enzymes, **898**, 899, 899f, 900, 900f, 901f, 904t, 913f
Pancreatic lipase, **899**, 913f, 916, 916f
Pancreatic nucleases, **899**, **917**
Pancreatic polypeptide, 630
Pancreatic proteases, **899**, 899f, 915, 915f
Pancreatitis, **921**

Paneth cells, 902f, 903
Pannus, in rheumatoid arthritis, **275**
Panoramic vision, 575
Pantothenic acid (vitamin B_5), 932t
Papanicolaou (Pap) smear, 1064
Papilla(e)
  dermal, 151f, **154**–155
  duodenal, major, **900**, 900f, 901
  hair, 158f, 159
  renal medullary, 977, 977f
  of tongue, **578**, 579f, 877, 877f
Papillary dermis, 151f, **154**–155, 154f
Papillary muscles, **674**, 677f, 678f, 679, 679f
Papilledema, **596**
Papillomavirus, human, 1064, **1080**
Pap smear, 1064
Paracellular route, of tubular reabsorption, 990, 990f
Paracrines, 81, **603**, 888t, 989
Parafollicular (C) cells, 617f, 618, **621**
Parahippocampal gyrus, 440f, 444, 456, 456f
Parallel after-discharge circuits, 427, 427f
Parallel fascicle pattern/muscles, **326**–327, 326f
Parallel processing, 426, **427**
Paralysis, 441, **474**–475
Paramesonephric (Müllerian) ducts, 1082f, **1083**
Paranasal sinuses, **215**, 215f, 822, 824t
Paraneuron(s), 634
Paraplegia, **475**
Parasagittal plane/section, **15**
Parasitic infections, 801
Parasympathetic (craniosacral) division of autonomic nervous system, 392, 392f, 532t, **534**, 535t, **536**–537
  heart regulation and, 534f, 536f, 545, 546t, 688t, 689, 697f, 699
  salivation control and, 534f, 879
Parasympathetic tone, **545**
Parasympathomimetic agents, autonomic activity and, 544t
Parathyroid cells, 621f, **621**–622
Parathyroid glands, **621**–622, 621f, 622f, 636
Parathyroid hormone (PTH/parathormone), 621f, **622**, **1026**
  bone and, **189**, 189f, 1026
  calcium relationship and, 188, 189, 189f, 607, 607f, 622, 622f, 917, 1026
  in tubular reabsorption, 992f, 994
Paraventricular nuclei, 449, 449f, **610**, 610f, 611
Paravertebral (sympathetic trunk/chain) ganglia, 508f, **538**, 538f, 538f, 539f, 540f
Paresthesia, 412, **474**, **526**
Parietal bones, **202**, 203f, 204f, 205f, 206f, 207f, **216t**
Parietal cells, 886f, **887**, 889, 891, 891f
Parietal layer of glomerular capsule, 979f, 980, 983f, 985f
Parietal lobe, 438f, **439**, 440f
Parietal pericardium, 19, 19f, 142f, **673**, 673f
Parietal peritoneum, 19, 142f, **872**, 872f, 908f
Parietal pleura, 19, 142f, 831t, 832f, **834**
Parietal serosa, **19**, 19f
Parieto-occipital sulcus, 438f, **439**, 440f
Parkinson's disease, 452, **469**
Parotid duct, 878f
Parotid glands, 536f, 869f, **878**, 878f, 921
Paroxysmal atrial tachycardia, **702**
Partial pressures/partial pressure gradients, **843**, 843t, 844–847, 844f. See also specific gas
Partial-thickness burns, 166, 166f
Particle size, chemical reaction rate affected by, 38
Parturition, 1042, **1114**–1116, 1115f. See also Birth
Passive humoral immunity, 797f, **798**

Passive processes/transport, **68**–73, 69f, 70f, 71f, 72f, 73t, 75. See also Diffusion; Filtration
Passive tubular reabsorption, **990**, 991–992
Patella, 200f, 240f, 241, 241t, **242**, 270, 270f, 271f
Patellar ligament, 270f, **271**
Patellar (knee-jerk) reflex, 271, 522f, **523**
Patellar retinacula, medial and lateral, 270f, **271**
Patellar surface, 240f, **241**
Patent ductus arteriosus, 701
Paternal chromosome, 1043
Pathogens, **782**, 797, 797f, 801
Pathological anatomy, 2
Pathologic fracture, **195**
Pathology, 147
Pattern generators, central, **518**, 518f
Pattern recognition, **495**
Pattern recognition receptors, **783**–784
P_{CO_2}. See Carbon dioxide, partial pressure of
Pectinate muscles, **674**, 676f, 677f
Pectineus muscle, 330f, 368f, **369t**, 384t
Pectoral (shoulder) girdle, **227**, 227f
  bones of, 200f, 227–228, 227f, 228f, 229f, 235t
  muscles of, 330f, 331f, 350–353t, 351f, 353f, 354–356t, 355–356f, 362t
Pectoralis major muscle, 330f, **350t**, 351f, **354t**, 355f, 362t
Pectoralis minor muscle, 330f, **350t**, 351f
Pectoral nerves, 350t, 354t, 510t, 511f
Pedicles, vertebral, **220**, 220f
Pedigree, **1133**, 1133f
Pelvic brim, 236, 237f, **238**, 238f, 239t
Pelvic cavity, **18**, 18f
Pelvic diaphragm, **348t**, 349f
Pelvic floor, muscles of, 348t, 349f, 1064
Pelvic (hip) girdle, 200f, 227, **236**–239, 237t, 239t, 241t. See also under Hip
Pelvic inflammatory disease (PID), **1063**, 1080
Pelvic inlet, **238**, 239t
Pelvic (pelvic splanchnic) nerves, 536f, **537**
Pelvic outlet, **238**, 238f, 239t
Pelvic (inferior hypogastric) plexus, 536f, **537**
Pelvimetry, **248**
Pelvis, 236, 237f, **238**, 239t
  arteries of, 748–749f, 748–749t
  muscles of, 330f
  sympathetic pathways to, 540f, 541–542
  veins of, 758t, 758f
Pelvis (renal), **977**, 977f, 978f, 979f
Penetration enhancers (drug agents), 164
Penis, 1047, 1047f, 1048f, **1050**–1052, 1051f, 1052f
  autonomic innervation/effects and, 536f, 540f, 546t
  embryologic/fetal development of, 1082f, 1083f, 1084
Pennate fascicle pattern/muscles, 326f, **327**
Pentamer, 799, 799t
Pentose sugars, 43, 43f, 913f
Pepsin, 886f, **887**, 889, 904t, 913f, **915**
Pepsinogen, chief cell secretion of, 886f, 887, 915
Peptic ulcers, **887**–889, 887f, 889f, **921**
Peptide(s), 422t, 423, 913f
Peptide bond(s), **48**, 49f, 104, 107f
Percent, solution concentration, **29**
Perception, 490, **493**, 494, 494f, 495
Perceptual detection, **495**
Perforated eardrum, 594
Perforating (Volkmann's) canals, **182**, 183f
Perforating fibers, 178, 179f, 183f
Perforins, **807**, 807f, 810t
Perfusion, 845–846. See also Blood flow
Periaqueductal gray matter, 451, 453f
Pericardial arteries, 738f, **743t**
Pericardial cavity, **18**, 18f, 673, 673f
Pericardial friction rub, 673

Pericarditis, 673
Pericardium, 19, 19f, 142f, **142**, **673**, 673f, 677f
Perichondrium, 134, 174
Pericytes, 467, **710**, 711f
Perikaryon (cell body), of neurons, 140f, **394**, 395f
Perilymph, **583**, 584f, 585
Perimetrium, 1061f, 1063f, **1065**
Perimysium, **282**, 282f, 283t, 312t
Perineal body (central tendon of perineum), 348t
Perineal muscle, deep and superficial transverse, 348t, 349f
Perineal space, superficial, 348t, 349f
Perineum, 348t, 349f, **1052**, 1052f, 1066f, **1067**
Perineurium, **496**, 497f
Perinuclear cytoplasm, outer collar of, 396, 397f
Period of contraction, muscle twitch, 299, 299f, 300f
Periodic table, **25**
Periodontal disease/periodontitis, **881**
Periodontal ligament, 252f, **253**, **880**, 880f
Period of relaxation, muscle twitch, 299, 299f, 300f
Periosteal bud, **185**, 185f
Periosteal layer of dura mater, 464, 464f, 466f
Periosteum (bone), **178**, 179f, 183f, 186f
Peripheral chemoreceptors, 854f, **855**–856, 855f, 858
Peripheral congestion, in heart failure, **700**
Peripheral connective tissue sheath, hair follicle, 158f, **159**
Peripheral nervous system, **391**–392, 391f, 392f, 489–530, **490**, 490f
    case study and, 530b
    developmental aspects of, 526
    homeostatic imbalances and, 496, 500t, 501t, 502t, 503t, 504t, 505t, 506t, 510, 512, 515, 523, 526
    motor endings/motor activity and, **517**–519, 518f
    myelination in, 393f, 396, 397f
    nerves (nerve structure/repair) and, 391, 391f, **395**, **496**–517, 497f, 498f
        cranial nerves, 391, 391f, 392f, **498**–507, 499f, 500–507t
        spinal nerves and, 391, 391f, 392f, 472, 473f, 474, **474**, 474f, **507**–517, 507f, 508f, 509f
    neuroglia (supporting cells) of, 140, 140f, 393f, **394**
    reflex activity and, **426**–427, 426f, 519–526, 519f, 520f, 521f, 522f
    regeneration and, 498, 498f
    sensory processing and, 493–496, 494f
    sensory receptors/sensation and, **490**–496, 491–492t, 494f, 496f
Peripheral neuropathy, **526**
Peripheral process, 397, 398f, 399
Peripheral proteins, 65f, **66**, 66f
Peripheral (vascular) resistance, total, **714**–715
    blood pressure and, 718–719, 719f, 720, 720f, 723f
Peripheral thermoreceptors, **965**
Peripheral tissues, cholesterol transport to, 956–957
Perirenal fat capsule, **976**, 976f
Peristalsis, **870**, 870f, 871f, 904t
    in intestines, 904t, 905–906, 905t
        mass, 904t, **910**
    in stomach, 892–893, 892f, 904t
    swallowing and, 884f, 904t
Peritoneal cavity, **872**, 872f
Peritoneal (intraperitoneal) organs, **872**, 872f
Peritoneum(s), **142**, 142f, 871–872, 908f, 909
    parietal, 19, 142f, **872**, 872f, 908f
    visceral, 19, 142f, **872**, 872f, 873, 873f, 908f

Peritonitis, 19, **872**, 906
Peritubular capillaries, 978f, 981f, **982**, 982f
Permanent teeth/permanent dentition, **879**–880, 879f
Permeability, membrane, 64f, **69**.
    See also Membrane transport
    action potential and, 408, 409
    diffusion and osmosis and, 69, 70–72, 70f, 71f, 73
    membrane potential and, 79–81, 80f, 402f, 403
    vascular, inflammation and, 785–786, 787f
Permissiveness, hormone, **609**
Pernicious anemia, 650, 889
Peroneal (fibular) artery, 748f, **749t**, 749f
Peroneal (fibular) muscles, **374t**, 385f
Peroneal (fibular) nerves, 373t, 380t, 514, 514f, 515, 515t
Peroneal (fibular) retinacula, **374t**
Peroneal (fibular) vein, **758t**, 758f
Peroneus (fibularis) brevis muscle, 376f, **377t**, 378f, 379f, 384t
Peroneus (fibularis) longus muscle, 330f, 331f, 375f, 376f, **377t**, 378f, 379f, 384t
Peroneus (fibularis) tertius muscle, **374t**, 375f, 376f, 384t
Peroxisomes, 62f, **86**, 94t
Perpendicular plate, 203, 205f, **210**, 210f, **212**, 214f
Perspiration, 164, 965. *See also* Sweat; Sweating
Pes planus (flatfoot), 245
Petechiae, 662
Petit mal (absence) seizures, 462
Petrous part, temporal bone, 205f, 206f, 207f, **208**, 208f
Peyer's patches, 771, 771f, **775**, 775f, **903**
PF₃ (platelet factor 3/phosphatidylserine), 659
pH, 39–40, 40f
    of blood, 40f, 41, 644. *See also*
        Acid-base balance
        in acidosis/alkalosis, 1033, 1033t, 1034b, 1035
        carbon dioxide affecting, 852, 1029
        chemical control and, 41
        hemoglobin saturation affected by, 850, 850f
        normal, 40f, 41, 1026–1027
        renal control and, 992f, 995, 1029–1032, 1030f, 1031f, 1032f, 1035
        respiratory control and, 852, 856, 1035
    blood in regulation of, 643
    of semen, 1053
    of urine, 1002
    of vagina, 1066
Phagocytosis/phagocytes, 77, 77f, 78t, 652, 782f, 784–785, 784f, 786, 786f, 787f, 790t, 800, 800f, 808f
Phagolysosome, **784**, 784f
Phagosome, 77, 77f, 78t, 86f, **784**, 784f
Phalanges, 200f, 234f, 235t, **236**, 241t, 244, 244f
Phantom limb pain, **496**
Pharyngeal cleft, **596**
Pharyngeal constrictor muscles, 338t, 339t, 339f, 882
Pharyngeal-esophageal phase of swallowing, **883**, 884f, 904t
Pharyngeal pouches, **596**
Pharyngeal tonsils, 774f, **775**, **823**, 823f
Pharyngotympanic (auditory) tube, 581f, **582**, 582f, 596, 821f, 823, 823f
Pharynx, 820f, **822**–824, 823f, 824t, 869f, 882, 904t
    muscles of, swallowing control and, 338–339t, 339f, 882
Phasic receptors, **494**
Phenocopies, **1131**
Phenotype, **1126**, 1128, 1128f, 1131

Phentermine, 959b
Phenylketonuria (PKU), 968, 1133
Pheochromocytoma, 627
Pheomelanin, 159
Phlebitis, **762**
Phlebotomy, **762**
Phosphatases, 913f, **917**
Phosphate, 188, 1020t, 1026
Phosphate bonds, high-energy, 55, 55f
Phosphate buffer system, **1028**, 1028f, 1031, 1031f
Phosphatidyl inositol bisphosphate (PIP₂), **605**
Phosphatidylserine (PF₃/platelet factor 3), 659
Phosphodiesterase (PDE), 571, 571f, **604**
Phospholipase C, **605**
Phospholipids, 46f, **47**, 47t, 63, 64–65f, 947–948
Phosphoric acid, 1027
Phosphorus, 26t, 931, 933t
Phosphorylation, 55, 56f, 604, **934**
    direct, ATP for muscle contraction and, 304, 304f, 305f
    oxidative, 935f, **936**, 937f, 940–943, 941f, 944, 944f
    substrate level, **936**, 937f, 938, 944, 944f
Photodamage, melanin and, 153, 156, 163
Photons, 564
Photopigments (visual pigments), 568f, **569**–571, 570f, 571f, 573
Photoreceptors, **490**, 560f, **561**, 564f, 568–569, 568f, 569t, 572f, 573
Photosensitivity, 156
Phototransduction, **570**–573, 571f
Phrenic arteries, inferior and superior, 738f, **743t**, 744f, 746f
Phrenic nerve, 344t, **509**–510, 509t, 509f, **853**, 853f
Phrenic veins, inferior, **756t**, 756f, 757f
pH units, **40**, 40f
Phylloquinone (vitamin K), 47t, 658, 662, 931, 932t
Physical properties, 25
Physiological acidosis, 1027
Physiological buffering systems, 1027, 1028–1032, 1030f, 1031f, 1032f
    renal, 992f, 994–995, 1029–1032, 1030f, 1031f, 1032f
    respiratory, 1028–1029
Physiological jaundice, 1117, **1120**
Physiology, 1, **2**–3, 61
Pia mater
    of brain, 464f, **465**
    spinal, 471, 473f
Pica, 969, 1021
Pickup molecule, oxaloacetic acid as, 939f, 940
Pigment(s)
    inside neuron cell bodies, 394
    skin, 156
    visual (photopigments), 568f, **569**–571, 570f, 571f, 573
Pigmented layer of retina, 560f, **561**
Pigmented nevi (moles), 156, 166
Pili, **157**. *See also* Hair(s)
Pineal gland/body, 447f, 448f, **450**, 451f, 602f, **627**
Pinealocytes, **627**
Pinkeye, 555
Pinna (auricle), 581f, **582**
Pinocytosis (fluid-phase endocytosis), 77, 77f, 78t
*P*ip. *See* Intrapleural pressure
PIP₂ (phosphatidyl inositol bisphosphate), **605**
PIP₂-calcium signaling mechanism, 604–605
Piriformis muscle, 349f, **372t**, 372t, 384t
Piriform lobe, 443, 577
Pisiform, **234**, 234f, 235t
Pit(s), gastric, **886**–887, 886f
Pitch, **586**, 586f, 590
Pitocin. *See* Oxytocin
Pitting edema, 734, 734f

Pituicytes, 609
Pituitary dwarfism, 612t, 615f, **616**
Pituitary gland (hypophysis), **448**, 449f, 450f, 602f, **609**–617, 607f, 610–611f, 612–613t, 614f, 615f, 616, 616f. *See also* Anterior pituitary gland; Posterior pituitary gland
Pituitary growth hormone. *See* Growth hormone
Pivot joints, 262f, 264
Placenta, 634, **1099**–1102, 1100f, 1101f
    delivery of, 1115f, 1116
    in fetal circulation, 1101f, 1108f, 1109f
    formation of, **1099**–1102, 1100f, 1101f
    hormones secreted by, 634, 1113
Placental growth hormone, **1113**
Placental (third) stage of labor, 1115f, **1116**
Placenta previa, **1120**
Placentation, **1099**–1102, 1100f, 1101f
Plane joints, 262f, 264
Planes (body), 13–17, 15f
Planetary model of atom, 25f, **27**
Plantar arch, 748t, 748f, 749f
Plantar arteries, medial and lateral, 748f, **749t**, 749f
Plantar flexion (body movement), 260, 261f, 384t, 385f
Plantar interossei muscles, **382t**, 383f
Plantaris muscle, **377t**, 378f, 379f, 384t
Plantar metatarsal arteries, 748f, **749t**
Plantar nerves, medial and lateral, 380t, 382t, **515**, 515t
Plantar reflex, **525**
Plantar veins, **758t**, 758f
Plaque (dental), **881**
Plaques (desmosome), 67f, 68
Plaques (fibrous/atherosclerotic), 760b
Plasma, 8, 127t, 137, 137f, 643, 643f, **644**, 644t, **1013**, 1013f
    carbon dioxide transport in, 850–852, 851f
    electrolytes in, 1014, 1014f, 1015
    gas/nutrient/water/waste exchanges and, 1015, 1015f
    oxygen transport in, 847, 848–849f, 851f
Plasma cells, 653, 655f, **770**, 795t, **796**–797, 796f, 809f, 810t
Plasmalemma. *See* Plasma membrane
Plasma membrane, 62, 62f, **63**–83, 64–65f, 94t
    boundary maintenance and, 8, 64f
    cell-cell communication/interactions and, 64f, 65f
    cell-environment interactions and, 63–66, 81–83, 82f
    cell junctions in, 66f, **67**–68, 67f
    endomembrane system and, 87, 87f
    fluid mosaic model of, **63**–68, 64–65f
    functions of, 64f
        active and passive transport, **68**–79, 69f, 70f, 71f, 72f, 73t, 74f, 75f, 76f, 77f, 78t, 79f
        glycocalyx of, 64f, 65f, **66**–67, 66f
        lipids associated with, 63, 65f
        proteins associated with. *See* Membrane proteins
        resting membrane potential and, **79**–81, 80f, 400–404, 401f, 402f, **403**, 404f
        selective/differential permeability of. *See* Permeability, membrane
Plasma membrane receptors. *See* Membrane receptors
Plasmapheresis, 667
Plasma proteins, 49f, 608, **644**, 644t
Plasma thromboplastin antecedent (PTA/factor XI), 659f, 661t
Plasma thromboplastin component (PTC/factor IX), 659f, 661t
Plasmin, **660**
Plasminogen, **660**
Plasminogen activator. *See* Tissue plasminogen activator
Plateau, in cardiac muscle contraction, 689, **689**, 689f

Platelet(s), 127t, 137, 643f, 644–645, 645f, 652f, **657**
  deficiency of (thrombocytopenia), **662**
  formation of, 656t, 657, 657f
  in hemostasis, 12, 12f, 658, 658f, 659f
Platelet count, **666**
Platelet-derived growth factor (PDGF), **660**
Platelet factor 3 (PF₃/phosphatidylserine), 659
Platelet plug, 12, 12f, 658, 658f
Platysma muscle, 330f, 333f, **334t**, 335f
Pleura(e), 142, 831t, 832f, **834**
  parietal and visceral, 19, 142f, 831t, 832f, **834**
Pleural cavity, **18**, 18f, 832f, **834**. See also Pleura(e)
Pleural effusion, 834
Pleural fluid, **834**
Pleurisy, 19, **834**
Plexuses
  nerve, 507f, **509**. See also specific type
  venous, 728
P_{N₂}, 843, 843t
Pneumonia, 246, 845, **863**
Pneumothorax, **836**, 836f
PNS. See Peripheral nervous system
P_{O₂}. See Oxygen, partial pressure of
Podocytes, **980**, 983f, 985f, 986
Poisons, chemiosmotic model of oxidative phosphorylation and, 943
Polar bodies, 1070f, 1071, 1094f, 1096, 1096f
Polar heads, of membrane phospholipids, 46f, 47, 63, 65f
Polarity
  of epithelial tissue, 117–118
  of membrane, 79–80, **403**
  phospholipid, 46f, 47, 63, 65f
  of sarcolemma, 293, 293f
  of water, 34, 34f, 38
Polar molecule(s), **34**, 34f
Poliomyelitis, 475
Pollex. See Thumb
Pollutants, environmental, endocrine system affected by, 636
Polyatomic ions, 39
Polychromatic erythroblast, **647**, 647f
Polycystic kidney disease, 1007
Polycystic ovary syndrome (PCOS), 159, **1086**
Polycythemia, **651**
Polycythemia vera, 651
Polydipsia, in diabetes, **632**, 632f
Polygene inheritance, 1130f, **1131**
Polymer(s), **42**, 55t
Polymerases, DNA, **97**, 97f
Polymerases, RNA, **102**, 102f
Polymorphonuclear leukocytes (PMNs/polys), 652. See also Neutrophils
Polyp, colon, 144b, 145f
Polyp, nasal, **863**
Polypeptide(s), **48**, 98–108, 99f, 102f, 104f, 105f, 106–107f, 108f, 913f. See also Protein synthesis
Polyphagia, in diabetes, **632**, 632f
Polyribosome, 104, 104f
Polys (polymorphonuclear leukocytes), **652**. See also Neutrophils
Polysaccharides, **44**, 44f, 914, 928
Polyspermy, **1093**–1096, 1095f
Polysynaptic reflexes, 519
Polyunsaturated fats, **46**
Polyuria, in diabetes, **632**, 632f
POMC. See Pro-opiomelanocortin
Pons, 435, 435f, 436f, 447f, 448f, 450–451f, **452**, 452f, 453f, 459t. See also under Pontine
Pontine micturition center, 1005, 1005f
Pontine nuclei, 452, 453f
Pontine respiratory centers/group, 853f, **854**
Pontine storage center, 1005, 1005f
Popliteal artery, 717f, 739f, **748t**, 748f, 749f
Popliteal ligaments, arcuate and oblique, 270f, **271**

Popliteal vein, 751f, **758t**, 758f
Popliteus muscle, 374t, **377t**, 378f, 379f, 384t
Pores
  alveolar, 829f, 830f, **831**
  in fenestrated capillaries, 711f
  nuclear, 92, 92f
  in plasma membrane. See Channels (protein)
  sweat, 151f, 161–162, 161f
  taste, **578**, 579f
Porphyria, 169
Porta hepatis, 894, 895f
Portal hypertension, **898**
Portal system, hepatic, 735, **756t**, 756f, 757f, 874
Portal system, hypophyseal, 611, 611f
Portal triad (portal tract region), 896f, **897**
Portal veins, 896f
  hepatic portal vein, 751f, **756t**, 756f, 757f, 894, 895f, 897
  hypophyseal portal vein, **610**–611, 611f
Positional terms/anatomical position, 12–13, 13f, 14t
Positive chemotaxis, **652**
Positive chronotropic factors, 698
Positive feedback, **11**–12, 12f
  action potential generation and, 408, 416f
  labor and, 612, 1115
  lactation and, 612, 1117, 1117f
  ovarian function and, 1075–1076, 1075f
Positive inotropic agents, 698
Positive nitrogen balance, 929
Positive selection, **794**, 794f
Positron emission tomography (PET), 16f, 16b
Postabsorptive (fasting) state, 952–955, 953f, 954f
Postcapillary venules, 707f, **712**, 712f, 713
Postcentral gyrus, 438f, **439**, 442f
Postcoital contraception, 1119b
Posterior (term), 14t. See also under Dorsal
Posterior arch, 221, 221f
Posterior association area, 440f, **444**
Posterior auricular vein, 752f, 753f
Posterior cerebral arteries, 740f, **741t**, 741f
Posterior chamber, **562**, 562f
Posterior circumflex humeral artery, **742t**, 742f, 743f
Posterior commissure, **444**, 447f, 448f, 450
Posterior communicating arteries, 740f, **741t**, 741f
Posterior cords, brachial plexus, **510**, 510f, 511f
Posterior cranial fossa, 201, 201f, **202**, 207f
Posterior cruciate ligament, 270f, **271**
Posterior division, brachial plexus, **510**, 511f
Posterior extensor compartment, of forearm, 357t, 360–361t, 362t, 363f
Posterior femoral cutaneous nerve, 514f, 515t
Posterior flexor compartment, of leg, muscles of, 374t
Posterior fontanelle, 246f
Posterior fornix, 1061f, 1066
Posterior (dorsal) funiculi, 473f, **474**, **476**
Posterior gluteal line, **236**, 237f
Posterior (dorsal) horns, **472**, 473f, 474f, 474t
Posterior inferior iliac spine, 236, 237f
Posterior intercostal arteries, 738f, 742f, **743t**, 743f
Posterior intercostal veins, **754t**, 754f, **755t**, 755f
Posterior interventricular artery, 676f, **682**, 683f
Posterior interventricular sulcus, **674**
Posterior lobe, of cerebellum, **454**, 455f
Posterior longitudinal ligament, 218, 219f
Posterior (dorsal) median sulcus, 451f, **472**, 473f

Posterior muscles/muscle compartment
  of arm, 363f
  of forearm, 357t, **358t**, 360–361t, 362t, 363f
  of hip and thigh, 367t, 372f, 373t, 384t, 385f
  of leg, 374t, 377t, 378–379f, 384t, 385f
Posterior nasal apertures (choanae), 821f, **822**, 823, 823f
Posterior pituitary gland, **609**, 610, 610f, 611–614, 612t
Posterior sacral foramina, 224, 224f
Posterior scalene muscle, 340t, 341f
Posterior segment of eye, 558, 558f, **562**, 562f
Posterior semicircular canal, 583, 583f
Posterior superior iliac spine, 236, 237f
Posterior tibial artery, 717f, 739f, 748f, **749t**, 749f
Posterior tibial vein, 751f, **758t**, 758f
Posterior vagal trunk, 537
Postganglionic axons/fibers, **532**, 533, 533f, 535, 535t, 536, 536f, 540f
Postganglionic neurons, **532**, 533f
Postmortem examination, 147
Postovulatory (secretory) phase of uterine (menstrual) cycle, 1076–1078, 1077f
Postpolio syndrome, 475
Postsynaptic neuron, **412**, 413f, 415f, 416t, 418f, 419f, 420
Postsynaptic potentials, 404, 416–420, 417t, 418–419f, 420, 572f, 573
Posture, skeletal muscles in maintenance of, 281
Potassium, 26t, 933t
  absorption of, 917
  balance/imbalance of, 623, 624, 624f, 1020t, 1024–1025
  dietary, 933t, 1024
  heart function and, 699, 1025
  in plasma/interstitial fluid/intracellular fluid, 1014f, 1015
  plasma levels of, 623, 624, 624f, 626t, 1022f, 1025
  resting membrane potential and, 80, 80f, 402f, 403
  tubular reabsorption of, 1025
  tubular secretion of, 992f, 994–995
Potassium channels, 401f
  action potential and, 406f, 407f, 408
  muscle contraction and, 292f, 293f, 296f, 686, 689–690, 689f
Potential/potential difference, **400**
Potential energy, 24, 31
Potentiation, synaptic, **420**
Power lever, 327, 327f
Power stroke, of cilia, 90–91, 91f
P_{pul}. See Intrapulmonary pressure
Prader-Willi syndrome, 1132
Pre-granulosa cells
Preantral follicles, 1073
Preantral phase of ovarian cycle, 1073
Precapillary sphincter, **712**, 712f
Precentral gyrus, 438f, **439**, 442f
Precipitation, by antibodies, **800**, 800f
Precommand level of motor control, **518**–519, 518f
Preeclampsia, **1114**
Prefix in chemical equation, 36
Preformed antibodies (agglutinins), **664**, 664f
Prefrontal cortex (anterior association area), 440f, **443**
Preganglionic axons/fibers, **532**, 533, 533f, 535, 535t, 536, 536f, 538–542, 539f, 540f
Preganglionic neurons, **532**, 533f
Pregnancy, 1091–1123, **1092**. See also Human development
  anatomical changes of, 1112–1113, 1112f
  case study and, 1123b
  ectopic, 1063, **1120**
  metabolic/hormonal changes during, 1099f, 1113

parturition (birth) and, **1114**–1116, 1115f
  physiologic changes during, 1113–1114
  prevention of (contraception), **1118**–1119b
  Rh incompatibility and, 665
Pregnancy tests, hCG-based, 1099
Pre-granulosa cells, **1070**, 1073
Prehypertension, 724
Pre-keratin, 153
Preload, **697**–698
Premature birth, infant respiratory distress syndrome and, 839
Premature contraction (extrasystole), **688**
Premolars (bicuspids), 879f, **880**
Premotor cortex, 440f, **441**
Pre-mRNA, 99f, 102, 106f
Prenatal testing, 1134–1135, 1134f
Preoptic nucleus, 449f, 463
Preovulatory (proliferative) phase of uterine (menstrual) cycle, 1076, 1077f
Prepatellar bursa, subcutaneous, 270f, 271
Prepatellar bursitis, 273
Prepotentials (pacemaker potentials), **686**, 687f
Prepuce of clitoris, 1066f, **1067**
Prepuce of penis (foreskin), 1047f, **1050**, 1051f
Presbycusis, **596**
Presbyopia, **567**, 595
Pressure gradient (blood pressure), 714, 716
Pressure gradient (breathing), 838
Pressure points, 717, 717f
Pressure receptors, 151f, 155, 164, 492t, **493**
Pressure reservoirs, elastic arteries as, 707f, 710
Pressure-volume relationships, in respiratory physiology, 834–840, 835f, 836f, 837f, 838f, 839f
Presynaptic inhibition, **420**
Presynaptic neuron, **412**, 413f, 415f, 419f
Pretectal nuclei, 574f, **575**
Prevertebral (collateral) ganglia, 538t, **539**, 539f, 540f, 541–542
Primary active transport, 74f, **75**, 78t, 990, 991f
Primary auditory cortex, 440f, **443**, 589f, **590**
Primary brain vesicles, **435**, 435f
Primary (main) bronchi, 820f, 827, 828f, 832f
Primary capillary plexus, **610**–611, 611f
Primary curvatures, **246**, 247f
Primary dentition (deciduous teeth), 879, 879f
Primary follicle, 1070f, **1071**, 1072f, 1073
Primary germ layers, 146, 146f, 1100f, **1102**–1103, 1103f, 1106f
Primary (essential) hypertension, **724**
Primary immune response, 796f, **797**, 797f, 805, 808–809f
Primary lymphoid organs, 771, 771f, **792**
Primary (somatic) motor cortex, 440f, **441**, 442f
Primary olfactory (smell) cortex, 440f, **443**, 577
Primary oocytes, **1069**, 1070f, 1072f
Primary ossification centers, **184**, 185f, 194, 194f
Primary sensory (anulospiral) endings, 520f, **521**
Primary sex organs, **1042**. See also Gonad(s); Ovaries; Testes
Primary somatosensory cortex, 440f, **442**, 442f
Primary spermatocyte, 1056f, **1057**
Primary structure of protein, 48–50, 50f
Primary visual (striate) cortex, 440f, **442**–443, 574f, **575**
Prime mover, **324**
Primitive gut, **918**, **1104**, 1104f

Primitive streak, **1102**, 1103*f*
Primitive tissues, 1103
Primordial follicles, **1070**, 1070*f*, 1071, 1072*f*, 1073
Primordial germ cells, **1083**
Principal cells, 979*f*, 981, 1025
Principle of complementarity of structure and function, **3**, 61
P-R interval, 690*f*, **691**
PRL. *See* Prolactin
Proaccelerin (factor V), 659*f*, 661*t*
Probabilities, in inheritance, 1128–1129, 1128*f*
Procarboxypeptidase, 899, 899*f*
Procedural (skills) memory, 460
Process (bone marking), 180*t*
Processes (neuron), 140, 140*f*, **394**–399, 395*f*, 398–399*t*
Procoagulants (clotting factors), 657, **658**–659, 659*f*, 661*t*
Proconvertin (factor VII), 659*f*, 661*t*
Proctodeum, **918**, 918*f*
Proctology, **921**
Product(s), in chemical reaction, **36**
Proerythroblasts, **647**, 647*f*
Progeria, 110
Progerin, 110
Progesterone, **634**, **1060**
    in ovarian cycle, 1075*f*, 1076, 1079*t*
    placental/during pregnancy, 1099*f*, 1102
    in sodium regulation, **1022**–1023
    in uterine (menstrual) cycle, 1076–1078, 1077*f*, 1079*t*
Progestins, for contraception, 1118*b*, 1119*b*
Programmed cell death. *See* Apoptosis
Prohormone, 616
Proinsulin, **630**
Projection, **495**
Projection fibers, **444**, 445*f*
Projection level of motor control, **518**, 518*f*
Projections (bone markings), 179, 180*t*
Prolactin (PRL), 611*f*, 613*t*, **616**–617, **1117**, 1117*f*
Prolactin cells, 616
Prolactin-inhibiting hormone (PIH), 611*f*, 613*t*, **616**
Prolactinoma, **636**
Prolactin-releasing factors, 616, 617, 1117, 1117*f*
Prolapsed (herniated/slipped) disc, **219**–220, 219*f*, 247, 276
Prolapse of uterus, **1064**
Proliferation (growth) zone (epiphyseal plate), 186, 187*f*
Proliferative (preovulatory) phase of uterine (menstrual) cycle, 1076, 1077*f*
Promonocytes, **654**, 655*f*
Promoter, in transcription, **102**, 102*f*
Promyelocytes, **654**, 655*f*
Pronation (body movement), 232, **260**, 261*f*, 362*t*
Pronator quadratus muscle, **358***t*, 359*f*, **360***t*, 362*t*
Pronator teres muscle, 330*f*, **358***t*, 359*f*, 362*t*, 363*f*
Pronephric duct, **1006**, 1006*f*
Pronephros, **1006**, 1006*f*
Pronuclei, male and female, **1096**, 1096*f*
Pro-opiomelanocortin (POMC), **616**, 960, 961*f*
Propagation of action potential, 291, 293, 293*f*, 295*f*, **409**, 409*f*, 411*f*, 412
Propecia. *See* Finasteride
Prophase in meiosis, 1044, 1045*f*, 1046*f*
Prophase in mitosis, 96*f*, **98**, 100*f*, 1046*f*
Proprioception, conscious, 476
Proprioceptors, **442**, 455, 477*f*, **490**, 491*t*, 492*t*, 520
Propulsion (food), **870**, 870*f*, 889, 892*f*, 904*t*. *See also* Peristalsis
Prosencephalon (forebrain), **435**, 435*f*, 1106
Prostacyclin, **658**

Prostaglandin(s), 47*t*, 48, 603, **785**, 785*t*, 787*f*, 928, **1114**, 1114*f*, 1053
Prostaglandin E₂, in glomerular filtration rate regulation, 989
Prostate cancer, **1053**
Prostate gland, 1003, 1003*f*, 1047, 1047*f*, 1051*f*, **1052**–1053, 1082*f*
Prostate-specific antigen (PSA), 1052, 1053
Prostatic hyperplasia, benign (BPH), **1052**
Prostatic secretion, 1052
Prostatic urethra, 1003*f*, **1004**, 1050, 1051*f*
Prostatitis, **1052**, **1086**
Prostheses (artificial joints), 274*b*
Protease(s)
    in mast cells, 129
    pancreatic, **899**, 899*f*, 915, 915*f*
Proteasomes, **109**
Protective function
    of alimentary canal mucosa, 872
    of blood, 643
    of bones, 175
    of skin, 5*f*, 163–164
    of water, 38
Protein(s), 9, **48**–53, 49*f*, 50*f*, 55*t*, 928–931, 929*f*, 930*t*. *See also specific type*
    amino acids/peptide bonds and, **48**, 49*f*, 49*f*, 50*f*
    complete, **928**–929, 930*t*
    contractile, 49*f*, 51
    denaturation of, **51**
    dietary, 928–931, 929*f*, 930*t*
    digestion/absorption of, 889, 913*f*, 915, 915*f*
    enzymes/enzyme activity and, 49*f*, **51**–53
    fibrous (structural), 49*f*, **51**
    globular (functional), 50, 50*f*, **51**
    incomplete, 929, 930*t*
    interconversion of, 949–950, 950*f*
    membrane. *See* Membrane proteins
    metabolism of, 618*t*, 935*f*, 948–949, 948*f*, 949*f*, 953–954, 953*f*, 956*t*
    plasma, 49*f*, 608, 644, 644*t*
    structural levels of, 48–51, 50*f*
    synthesis of. *See* Protein synthesis
    transport, 49*f*, 51, 66, 66*f*, 70
    in urine, 1001*t*
Protein buffer system, **1028**, 1028*f*
Protein C, **661**
Protein channels. *See* Channels (protein)
Protein-coated vesicles/pits, 76*f*, 77, 77*f*, 105*f*
Protein energy malnutrition, **969**
Protein hormones, 49*f*, 51
Protein kinases, 81, 82*f*, **604**, 605*f*
Protein synthesis, 394, 949
    DNA in, 53–54, 98–108, 99*f*, 102*f*, 106*f*, 108*f*
    in ribosomes, 84, 85, 85*f*, 99, 99*f*, 102, 103–105, 105*f*, 106–107*f*
    RNA in, 54, 98–108, 99*f*, 102*f*, 105*f*, 106–107*f*, 108*f*
    transcription, 99, 99*f*, **102**–103, 102*f*, 108*f*
    translation, 99, 99*f*, **103**–105, 104*f*, 105*f*, 106–107*f*, 108*f*
Proteinuria, 1001*t*
Proteoglycans, 128
Proteolysis, 109
Proteome. *See* Protein(s)
Prothrombin, 659*f*, **660**, 661*t*
Prothrombin activator, **659**–660, 659*f*
Prothrombin time, **666**
Proton(s), **25**, 25*f*, 188
Proton acceptors, **39**, 48, 1027
Proton donors, **39**, 48, 1027
Proton (H⁺) gradient, 936*f*, 941*f*, **942**–943
Proton pumps, 891
Proto-oncogenes, 144*b*
Protraction (body movement), **260**, 261*f*, 350*t*
Provirus, HIV, 811
Provitamin(s), 931

Provitamin D₃ (cholecalciferol), **634**, 635*t*
Proximal (term), 14*t*
Proximal convoluted tubule, 979*f*, **980**, 980*f*, 981*f*, 992, 993*t*. *See also* Renal tubules
Proximal phalanx, 234*f*, 235*t*, 236, 241*t*, 244*f*
Proximal radioulnar joint, 232*f*, 256*t*
Pseudomembranous colitis, 909
Pseudopods, 77, 77*f*
Pseudostratified columnar epithelium, **121**, 121*f*
Pseudounipolar neurons, **397**, 398–399*t*
P site, 104, 106*f*, 107*f*
Psoas major muscle, **367***t*, 368*f*
Psoas minor muscle, 368*f*
Psoriasis, **169**
Psychoneuroimmunology, 814
Psychoses, **484**
Psychosomatic illnesses, **457**
Pterygoid muscles, **336***t*, 337*f*
Pterygoid processes, 205*f*, **209**, 209*f*, 214*f*
Pterygopalatine ganglia, 503*t*, **536**
PTH. *See* Parathyroid hormone
Ptosis, eyelid, 501*t*
Puberty, **1043**, **1084**
    in females, 1043, 1070*f*, 1071, 1079*t*, 1084
    in males, 1043, 1060, 1060*f*, 1079*t*, 1084
Pubic arch, 237*f*, **238**, 239*t*
Pubic bone (pubis), 237*f*, **238**
Pubic crest, 237*f*, **238**
Pubic rami, superior and inferior, 237*f*, **238**
Pubic (hypogastric) region, **20**, 20*f*
Pubic symphysis, 175*f*, 237*f*, **238**, 253*f*, 254, 257*t*
Pubic tubercle, 237*f*, **238**
Pubis (pubic bone), 237*f*, **238**
Pubococcygeus muscle, 349*f*
Pubofemoral ligament, **269**, 269*f*
Pudendal artery, internal, **748***t*, 748*f*, 1054
Pudendal nerve, 514*f*, **515**, 515*t*
Pudendum (external genitalia), **1060**, 1066*f*, **1067**, 1083–1084, 1083*f*
Pulmonary arteries, 675–677*f*, 681*f*, 729, **736**–737*t*, 736*f*, 832*f*, **834**
Pulmonary capillaries, **736***t*, 736*f*, 830, 830*f*, **834**, 844–845, 845*f*
Pulmonary circuit/circulation, **671**, 671*f*, 680, 681*f*, 713*f*, 729, 735, 736–737*t*, 736*f*, 834
Pulmonary congestion, in heart failure, **700**
Pulmonary disease, 858–860, 858*f*
Pulmonary embolism, 662, **863**
Pulmonary function tests, 842
Pulmonary gas exchange (external respiration), 819*f*, 844–846, 844*f*, 845*f*, 846*f*
Pulmonary irritant receptors/reflexes, 854*f*, 857
Pulmonary plexus, 537, 540*f*, **834**
Pulmonary trunk, **674**, 675*f*, 677*f*, 681*f*, **736***t*, 736*f*, 832*f*
Pulmonary valve (pulmonary semilunar valve), 677*f*, 678*f*, **680**, 681*f*, 696*f*
Pulmonary veins, **674**, 675–677*f*, 681*f*, **736***t*, 736*f*, 832*f*, **834**
Pulmonary ventilation, 819*f*, **835**, 836–840, 837*f*, 838*f*, 839*f*. *See also* Breathing; Respiration
Pulp, dental, **880**
Pulp, splenic, 773*f*, 774
Pulp cavity, **880**, 880*f*
Pulse (arterial), **716**, 717, 717*f*
Pulse (pressure) points, **717**, 717*f*
Pulse pressure, **716**
Pumps (transport proteins), **75**, 78*t*
Puncta, lacrimal, **556**, 556*f*
Punnett square, **1128**, 1128*f*
Pupil, 555*f*, 558*f*, **559**, 559*f*, 567, 573
Pupillary reflex, accommodation, 567
Pupillary reflex, light, 559, 573
Purines, 55, 422*t*, 423
Purkinje cells, **454**

Purkinje fibers (subendocardial conducting network), **687**–688, 687*f*
Pus, **147**, 786, 1001*t*
Putamen, 445*f*, **446**, 446*f*
P wave, 690*f*, **691**, 695*f*
Pyelogram, 1004, 1004*f*, **1008**
Pyelonephritis, 977, 1004
Pyloric antrum, **885**, 885*f*
Pyloric canal, **885**, 885*f*
Pyloric part of stomach, **885**, 892, 892*f*
Pyloric stenosis, **921**
Pyloric valve/sphincter, **885**, 885*f*, 892, 892*f*, 901*f*
Pylorus, **885**, 885*f*
Pyramidal cells, **441**, 479*f*
Pyramidal (direct) pathways, 478, 478, 479*f*, 481*t*, 518
Pyramidal (corticospinal) tracts, 441, 452, 453*f*, 475*f*, **478**, 479*f*, 481*t*
Pyramids, medulla oblongata, 450*f*, **452**, 453*f*
Pyramids, renal (medullary), **977**, 977*f*
Pyridoxine (vitamin B₆), 932*t*
Pyrimidines, 55
Pyrogens, **789**, 968
Pyruvic acid, 304, 934, 935*f*, 937, 937*f*, **938**, 938*f*, 939–940, 939*f*
Pyuria, 1001*t*

QRS complex, 690*f*, **691**, 695*f*
Q-T interval, 690*f*, **691**
Quadrants, abdominopelvic, 19, 19*f*
Quadrate lobe, of liver, 894, 895*f*
Quadratus femoris muscle, 372*f*, **373***t*, 384*t*
Quadratus lumborum muscle, **342***t*, 343*f*, 368*f*
Quadratus plantae (flexor accessorius) muscle, **380***t*, 381*f*, 383*f*
Quadriceps femoris muscle, 270*f*, 271*f*, 272, **367***t*, 368*f*, **369***t*
Quadriceps strain, **386**
Quadriplegia, **475**
Qualities, **495**
    sound, 586
Quality discrimination, **495**
Quarks, 28
Quaternary structure of protein, 50*f*, 51
Quiescent period, cardiac cycle, **693**
Q wave, 691

Rabies, **430**
Radial artery, 717, 717*f*, 739*f*, **742***t*, 742*f*, 743*f*, **1065**
Radial collateral ligament, **268**, 268*f*
Radial fossa, 230*f*, 231*f*, **231**
Radial groove, 230*f*, **231**
Radial nerve, 357*t*, 360*t*, 510*f*, 511*f*, **512**
Radial notch, 231*f*, 232*f*, **233**, 233*f*
Radial pulse, 717, 717*f*
Radial tuberosity, 231*f*, 232*f*, **233**
Radial vein, 751*f*, **754***t*, 754*f*, 755*f*
Radiant (electromagnetic) energy, 24, **563**
Radiate arteries, cortical, **978**, 978*f*, 981*f*, 982
Radiate veins, cortical, **978**, 978*f*, 981*f*
Radiation
    heat loss by, **964**–965, 965*f*
    ionizing, 28, **56**
    ultraviolet. *See* Ultraviolet radiation
Radiation sickness, **56**
Radiation therapy, for cancer, 145*b*
Radioactivity, 28
Radiographic anatomy, 2
Radioisotopes, **28**
Radioulnar joints, distal and proximal, **232**, 232*f*, 256*t*
Radius, 200*f*, 231*f*, 232, 232–233*f*, **233**, 235*t*, 268, 268*f*
Ramus/rami (bone marking)
    of ischium, **236**, 237*f*
    of mandible, **210**, 211*f*
    of pubis, 237*f*, **238**

Ramus/rami (nerve), 509
    communicantes, **508**, 508*f*, 535*t*, 538*f*,
        539–541, 539*f*, 540*f*
    dorsal, **508**, 508*f*, 509
    ventral, 367*t*, **508**, 508*f*, 509
Random fertilization, 1128
Range of motion, 256–257*t*, 258
RANK-L protein, 189
Ranvier, nodes of (myelin sheath gaps),
    395*f*, **396**, 397
Raphe, hard palate, 877
Raphe nuclei, 453*f*, **457**
Rapid eye movement (REM) sleep,
    462–463, 463*f*
RAS. *See* Reticular activating system
Raynaud's disease, 548, 548*f*
RBCs (red blood cells). *See* Erythrocytes
Reabsorption, 731, 733*f*
    tubular. *See* Tubular reabsorption
Reactant(s), in chemical reaction, **36**, 38
Reactive hyperemia, **726**
Reactivity, of antigens, **791**
Real image, 564, 565*f*
Receptive endings, 398*t*, 399
Receptive field, 494
Receptive regions, **395**, 395*f*, 398*t*
Receptive relaxation, of stomach muscles,
    892
Receptor(s). *See also* specific type
    in homeostasis, **10**, 10*f*, 11*f*
    in hormone mechanism of action, 603,
        604–605, 605*f*, 606, 606*f*, 608–609
    membrane, 66, 66*f*, **81**, 82*f*, 604–605,
        605*f*
    phasic, **494**
    in reflex arc, 426*f*, 427, 519, 519*f*
    tonic, **494**
    up- and down-regulation of, **608**
Receptor binding, in chemical signaling,
    81, 82*f*
Receptor cells, in special senses, 554
Receptor-mediated endocytosis, **77**, 77*f*,
    78*t*
Receptor membrane, gustatory epithelial
    cell (gustatory hair), 578
Receptor potential, 404, **494**
Receptor region, neurotransmitter, 414
Recessive allele, **1126**, 1129*t*
Recessive traits, 1129*t*
Reciprocal activation, **524**, 525*f*
Reciprocal inhibition, **523**
Recoil (lung), 837*f*, 835, 838
Recombinant chromosomes, **1127**, 1127*f*
Recombination, gene, 1127–1128, 1127*f*
Recombination, antigen receptor diversity
    and somatic, 794
Recovery stroke, of cilia, 90–91, 91*f*
Recruitment (multiple motor unit
    summation), 300, **301**–302, 301*f*,
    309
Rectal artery, superior, 744*f*, **746***t*, 747*f*
Rectal valves, **907**, 907*f*
Rectouterine pouch, 1061*f*, 1064, 1064*f*
Rectum, 536*f*, 540*f*, 869*f*, **907**, 907*f*. *See
    also under Rectal*
Rectus abdominis muscles, 330*f*, **346***t*,
    347*f*
Rectus femoris muscle, 330*f*, **367***t*, 368*f*,
    369*t*, 384*t*, 385*f*
Rectus muscles, of eye, 556, 557*f*
Rectus sheath, 346*t*
Red blood cells. *See* Erythrocytes (red
    blood cells)
Red cones, 564*f*, 569
Red marrow, **178**–179, 186*f*, **647**, 666,
    771, 771*f*
Redness in inflammation, 785, 787*f*
Redness of skin, 157
Red nucleus, **452**, 453*f*
Redox (oxidation-reduction) reactions,
    36–37, **934**–935
Red pulp, 773*f*, **774**
Reduced hemoglobin (deoxyhemoglobin),
    647, 847, 852
Reduced reactant, in redox reaction, **37**

Reduction, dislocation, 273
Reduction, fracture, 191
Reduction division (meiosis I), **1044**,
    1045*f*, 1046*f*
Reference man/woman, 2
Referred pain, **496**, 496*f*
Reflection (light), 564
Reflex(es), 8, **426**–427, 426*f*, 519–526,
    519*f*, 520*f*, 521*f*, 522*f*
    enteric nervous system (short and long),
        874–875, 875*f*
    spinal, 520–526. *See also* specific type
        testing, 470
Reflex arcs, 426*f*, **427**, 517, 519–520, 519*f*
    visceral, **542**–543, 543*f*
Reflex (cephalic) phase, of gastric
    secretion, **890**, 890*f*
Refraction (light), **564**, 564*f*, 565*f*,
    566*f*, 567
Refractory periods
    muscle fiber, **296**
    neuron, **410**, 410*f*
    after orgasm, 1054
Regeneration, **143**, 143*f*, 144–146
    cardiac tissue, 144
    epithelial tissue, 118, **143**, 143*f*, 144
    muscle, 317
    nerve fiber/axon, 144–145, 497–498,
        498*f*
    neuronal, 144–145, 394
    visual pigment, 570*f*, 571
Regeneration tube, 498*f*
Regional anatomy, 2
Regional terms in anatomy, **13**, 13*f*
Regular dense connective tissue, 127*t*,
    129, 132*f*, **133**
Regulatory functions of blood, 643
Regulatory T cells, 776, 795*t*, 802, 802*f*,
    807–810, 810*t*
Rehearsal, memory and, 460
Rejection (transplant), 810–811
Relative intensity/relative timing (sound
    wave), 590
Relative refractory period, **410**, 410*f*
Relaxation
    period of, muscle twitch, 299, 299*f*, 300*f*
    receptive, of stomach muscles, **892**
Relaxin, 1053, 1102, **1112**
Relay(s), in neuronal pathways, 476
Relay proteins, 605
Release factor, 107*f*
Releasing hormones, hypothalamic, 449,
    608, **611**, 611*f*, 613*t*
Remodeling (bone). *See* Bone remodeling
REM sleep, **462**–463, 463*f*
Renal anemia, 650
Renal arteries, 739*f*, 744*f*, **746***t*, 746*f*, 975*f*,
    976*f*, **977**–978, 978*f*
Renal autoregulation, **987**–989, 987*t*, 988*f*
Renal calculi, **1002**
Renal calyces (major and minor), **977**, 977*f*
Renal clearance, **1000**–1001
Renal columns, **977**, 977*f*
Renal corpuscle, 979*f*, **980**, 980*f*, 981*f*,
    985*f*
Renal cortex, **976**–977, 977*f*, 978*f*, 979*f*,
    980*f*, 982*f*, 995*f*
Renal failure, **1001**
Renal fascia, **976**, 976*f*
Renal function, evaluating, 1000–1002
Renal hilum, 975*f*, **976**, 977*f*
Renal infarct, **1008**
Renal medulla, **977**, 977*f*, 978*f*, 979*f*, 995*f*
Renal pelvis, **977**, 977*f*, 978*f*, 979*f*
Renal physiology, 2. *See also* Kidney(s);
    Urine, formation of
Renal plexus, **978**
Renal (medullary) pyramids, **977**, 977*f*
Renal sinus, 976, 978
Renal tubular acidosis, **1037**
Renal tubules, 979*f*, **980**–981,
    980*f*, 981*f*
    reabsorption in. *See* Tubular
        reabsorption
    secretion by. *See* Tubular secretion

Renal veins, 750*f*, 751*f*, **756***t*, 756*f*, 757*f*,
    975*f*, 976*f*, **978**, 978*f*
Renin, 546, **624**, 624*f*, **634**, 722–723, 722*f*,
    975, 983, 988*f*, **989**, 1021, 1022
Renin-angiotensin-aldosterone
    mechanism, 624, 624*f*, 626*t*,
    722–723, 722*f*, 987*t*, 988*f*, **989**,
    1021, 1022
Rennin, 889
Repair, bone fracture, 190–192, 191*f*
Repair, tissue, 142–146, 143*f*
    healing by first/second intention, **147**
Replication bubble, **97**, 97*f*
Replication fork, **97**, 97*f*
Repolarization
    cardiac muscle, 686, 687*f*, 689–690,
        689*f*, 690*f*, 691*f*, 692
    neuron, action potential and, 406*f*, 407*f*,
        408, 416*t*
    skeletal muscle, 293*f*, 296, 296*f*
Reproduction, 6*f*, 7*f*, **8**
Reproductive duct system
    embryologic/fetal development of,
        1081–1084, 1082*f*
    in females, 1062–1066, 1063–1064*f*,
        1065*f*, 1066*f*
    in males, 1047, 1047*f*, **1050**, 1052
Reproductive system, 1041–1090, **1042**.
    See also Female reproductive
    system; Male reproductive system
    accessory organs of, **1042**
    cardiovascular system and, 761*b*, 1085*b*
    case study and, 1090*b*
    developmental aspects of, 1081–1086,
        1082*f*, 1083*f*
    digestive system and, 920*b*, 1085*b*
    endocrine system and, 637*b*, 1085*b*
    homeostatic relationships of, 1085*b*
    integumentary system and, 168*b*, 1085*b*
    lymphatic system/immunity and, 778*b*,
        1085*b*
    muscular system and, 318*b*, 1085*b*
    nervous system and, 550*b*, 1085*b*
    respiratory system and, 862*b*, 1085*b*
    sexually transmitted infections and,
        1080–1081
    shared male and female features,
        1042–1047
    skeletal system and, 196*b*, 1085*b*
    thyroid hormone affecting, 618*t*
    urinary system and, 1036*b*, 1085*b*
Residual volume (RV), **841**, 841*f*
Resistance (airway), 838–839, 839*f*
Resistance (electrical), **400**
Resistance (vascular), total peripheral,
    714–715, 718–719, 719*f*, 720,
    720*f*, 723*f*
Resistance exercise, **309**–310
Resistance vessels, arterioles as, 710
Resistin, 633*b*, 634, 635*t*
Resolution phase of sexual response, in
    male, **1054**
Resonance, 588, 588*f*
Resorption, bone, **188**
Respiration, **819**. *See also* Breathing
    cellular. *See* Cellular respiration
    external, 819*f*, 842, 844–846, 844*f*,
        845*f*, 846*f*
    gas exchange and, 824, 829, 830,
        842–847, 843*t*, 844*f*, 846*f*
    gas transport and, 643, 730, 731*f*, 819*f*,
        847–852, 848–849*f*, 850*f*, 851*f*
    internal, 819*f*, 843, 844*f*, 846–847
    muscles of, 344*t*, 345*f*, 836–838, 837*f*
Respiratory acidosis (hypoventilation),
    858, 1029, **1033**–1034, 1033*t*,
    1034*b*, 1035
Respiratory alkalosis (hyperventilation),
    855, 1029, 1033*t*, **1034**, 1034*b*,
    1035
Respiratory/lung alveoli, 829, 829*f*, 830*f*,
    831*t*, 843*t*, 860–861
Respiratory bronchioles, **829**, 829*f*, 830*f*
Respiratory burst, **652**, 784
Respiratory capacities, **841**, 841*f*

Respiratory cartilages, 174, 175*f*
Respiratory centers
    medullary, 454, 853, 853*f*, 854*f*, 1027,
        1029
    pontine, 853*f*, **854**
Respiratory distress syndrome, adult
    (ARDS), **863**
Respiratory distress syndrome, infant
    (IRDS), **839**
Respiratory enzyme complexes, **940**, 941*f*
Respiratory gases, 643, 644*t*, 730, 731*f*,
    819*f*, 847–852, 848–849*f*, 850*f*,
    851*f*. *See also* under Gas
Respiratory membrane (blood air barrier),
    829–831, **830**, 830*f*, 844–846, 844*f*
Respiratory mucosa, 783, **822**
Respiratory pump, venous return and,
    718, 723*f*
Respiratory rate/depth
    alveolar ventilation and, **842**, 842*t*
    factors affecting, 854–857, 854*f*, 855*f*,
        856*f*, 861, 1029
    normal (eupnea), **853**
Respiratory rhythm, 454, 853, 854
Respiratory system, 5*f*, 7*f*, 818–867, **819**,
    820*f*. *See also* specific organ or
    structure and under Pulmonary
    in acid-base balance, 852, 856, 1035
    adjustments and, 857–858
    in body defense, 783
    cardiovascular system and, 761*b*, 862*b*
    case study and, 867*b*
    developmental aspects of, 860–863,
        861*f*, 1104, 1104*f*
    digestive system and, 862*b*, 920*b*
    endocrine system and, 637*b*, 862*b*
    functional anatomy of, 820–834, 824*t*,
        831*t*
    functions/physiology of, 7*f*, 8, 834–861
    gas exchange and, 824, 829, 830,
        842–847, 843*t*, 844*f*, 846*f*
    gas transport and, 643, 730, 731*f*, 819*f*,
        847–852, 848–849*f*, 850*f*, 851*f*
    homeostatic imbalances/diseases and,
        822, 823, 826, 827, 828, 834, 836,
        839–840, 844, 845, 852, 855,
        858–860, 858*f*, 861
    homeostatic relationships of, 5*f*, 862*b*
    integumentary system and, 168*b*, 862*b*
    lower, 820, 824–831, 831*t*
    lymphatic system/immunity and, 778*b*,
        862*b*
    muscular system and, 318*b*, 862*b*
    nervous system and, 550*b*, 862*b*
    nonrespiratory air movements and, **838**
    pregnancy and, 1113
    reproductive system and, 862*b*, 1085*b*
    respiration control and, 853–857, 853*f*,
        854*f*, 855*f*, 856*f*, 1028–1029
    skeletal system and, 196*b*, 862*b*
    stress response and, 628*f*
    upper, 820–824, 824*t*
    urinary system and, 862*b*, 1036*b*
    volume/pressure changes and, 834–840,
        835*f*, 836*f*, 837*f*, 838*f*, 839*f*
Respiratory volumes, 834–841, **840**,
    835–839*f*, 841*f*
Respiratory zone, **824**, 829–831, 829*f*
Respirometer, **962**
Responsiveness (excitability), 8, 290
    muscle cell, **280**
    neuron, 6*f*, 8, 392, 405. *See also* Nerve
        impulse(s)
Restenoses, after angioplasty, 760*b*
Resting heart rate, 699
Resting membrane potential, **79**–81, 80*f*,
    400–404, 401*f*, 402*f*, **403**, 404*f*, 686
Resting state, of action potential, 406*f*,
    407*f*, 408
Resting (quiescent) zone (epiphyseal
    plate), 186, 187*f*
Restriction points, in cell cycle, 96*f*, 98
Restrictive pulmonary disorders,
    pulmonary function testing in, 842
Rete testis, **1049**, 1049*f*

Reticular activating system (RAS), **457**–458, 457*f*, 463
Reticular cells, 132*f*, **133**, 647, **770**, 770*f*
Reticular connective tissue, 127*t*, 132*f*, **133**, **770**, 770*f*
Reticular dermis, 154*f*, **155**
Reticular fibers, 127*t*, 128*f*, **129**, 132*f*, 133, 770, 770*f*
Reticular formation, 452, 453*f*, **457**–458, 457*f*, 459*t*, 547, 547*f*
Reticular lamina, **118**
Reticular nuclei, 457*f*, 458
Reticulocyte(s), **647**–648, 647*f*
Reticulocyte count, **648**
Reticulospinal tracts, 458, 475*f*, **478**, 481*t*
Reticulum, 647
Retina, 435, 435*f*, 558*f*, **559**–561, 560*f*, 561*f*
  focusing light on, 565–567, 565*f*
  information/visual processing in, 572*f*, 573, 575
  photoreceptors in. *See* Photoreceptors
Retinal (visual pigment), 568*f*, **569**–570, 571, 571*f*
Retinal artery, central, 558*f*, 560*f*, **561**, 561*f*
Retinal detachment, 561
Retinal vein, central, 558*f*, 560*f*, **561**, 561*f*
Retinitis pigmentosa, 573
Retinol. *See* Vitamin A
Retraction (body movement), **260**, 261*f*, 350*t*
Retrograde amnesia, 461
Retrograde movement/transport, 396
Retroperitoneal organs, **872**, 872*f*
Retropulsion, 892, 892*f*
Retroverted uterus, 1063
Reuptake, neurotransmitter, 414, 415*f*
Reverberating circuits, 427, 427*f*
Reverse transcriptase, HIV and, 811
Reversibility, of chemical reactions, 37
R group, amino acid, 48, 49*f*, 50
Rh blood groups, **664**–665
Rheumatic fever, 812
Rheumatism, **276**
Rheumatoid arthritis (RA), **275**, 275*f*, 812
Rh factor, **664**–665
Rhinencephalon, **443**, 457
Rhinitis, 822
Rhodopsin, 568*f*, 570*f*, **571**–572, 573
RhoGAM, 665
Rhombencephalon (hindbrain), **435**, 435*f*, 1106
Rhomboid major muscle, 331*f*, **352***t*, 353*f*
Rhomboid minor muscle, 331*f*, **352***t*, 353*f*
Rhomboid muscles, 331*f*, **350***t*, **352***t*, 353*f*
Rhythm methods, of contraception, 1118*b*
Rib(s), 200*f*, 225*f*, **226**–227, 226*f*
Riboflavin (vitamin B₂), 932*t*, 935
Ribonuclease, 913*f*
Ribonucleic acid. *See* RNA
Ribose, 43, 43*f*, 53*t*, 55
Ribosomal RNA (rRNA), 54, 84, **99**
Ribosomes, 62*f*, 83*f*, **84**, 84*f*, 85, 94*t*
  mitochondrial, 84
  in protein processing, 84, 85, 85*f*, 99, 99*f*, 102, 103–105, 105*f*, 106–107*f*
RICE, **319**
Rickets, 193, 932*t*
Right ascending lumbar vein, **754***t*, 756*f*. *See also* Lumbar veins
Right atrium, 671, 671*f*, 674, 675–677*f*, 681*f*
Right brachiocephalic vein, **750***t*, 750*f*, 751*f*. *See also* Brachiocephalic veins
Right bundle branch, **687**
Right colic artery, 744*f*, **746***t*, 747*f*
Right colic (hepatic) flexure, **906**, 907*f*
Right common carotid artery, **738***t*, 738*f*, 740*f*, 742*f*
Right common iliac artery, **738***t*, 738*f*, 747*f*
Right coronary artery, 675*f*, 676*f*, **682**, 683*f*, **738***t*
Right gastric artery, 744*f*, **745***t*, 745*f*

Right gastroepiploic artery, 744*f*, **745***t*, 745*f*
Right gastroepiploic vein, 757*f*
Right hypochondriac region, **20**, 20*f*
Right inguinal (iliac) region, **20**, 20*f*
Right lateral (lumbar) region, **20**, 20*f*
Right lobe, of liver, 894, 894*f*, 895*f*
Right lower quadrant (RLQ), **19**, 19*f*
Right lymphatic duct, 768*f*, **769**, 769*f*
Right main bronchus, 820*f*, **828**
Right marginal artery, 675*f*, **682**, 683*f*
Right posterior intercostal veins, 754*f*, **754***t*
Right pulmonary artery, 675–677*f*, **736***t*, 736*f*
Right pulmonary vein, 675–677*f*, 736*f*
Right subclavian artery, **738***t*, 738*f*, 742*f*, 743*f*
Right suprarenal vein, 750*f*, 757*f*. *See also* Suprarenal veins
Right upper quadrant (RUQ), **19**, 19*f*
Right ventricle, 671, 671*f*, 674, 681*f*, 682, 682*f*, 683*f*
Rigor mortis, 296–298, 297*f*
Ringer's solution, for volume restoration, 663
Risorius muscle, 334*t*, 335*f*
RNA (ribonucleic acid), **53**–55, 53*t*
  in chromatin, 93
  digestion/absorption of, 913*f*, 917
  messenger (mRNA), 54, **99**, 99*f*, 102–105, 102*f*, 105–108*f*
  micro (miRNAs), 54, **108**
  in protein synthesis, 54, 98–108, 99*f*, 102*f*, 105*f*, 106–107*f*, 108*f*
  ribosomal (rRNA), 54, 84, **99**
  small noncoding, 1131–1132
  small/small interfering, **108**, 1132
  transfer (tRNA), 54, **99**, 103–104, 103*f*, 106–107*f*, 108*f*
RNA polymerase, **102**, 102*f*
Rod(s), 560*f*, **561**, 564*f*, 568–571, 568*f*, 569*t*, 573, 943*f*. *See also* Photoreceptors
Rod discs, 568*f*, 569
Rod inner segments, 568, 568*f*
Rod outer segments, 560*f*, 568, 568, 568*f*, 569
Roof of nasal cavity, 215
Roof of orbit, 213*f*
Root
  of brachial plexus, **510**, 511*f*
  hair, 151*f*, 158, 158*f*
  of lung, 832*f*, **833**
  nail, 160, 160*f*
  nerve, 509. *See also* Dorsal root(s); Spinal root(s); Ventral root(s)
  of nose, 821, 821*f*
  of penis, 1051*f*
  of tongue, 877
  of tooth, **880**, 880*f*
Root canal, **880**, 880*f*
Root hair plexus (hair follicle receptor), 151*f*, **159**, 164, 491*t*, **492**
Rooting reflex, 919
Rootlets, spinal nerve, 471*f*, **508**, 508*f*
Root sheath, epithelial, 158*f*, **159**
Rosacea, **170**
Rotation (body movement), **260**, 260*f*, 350*t*, 354*t*, 362*t*, 367*t*, 384*t*
Rotational acceleration, 592, 592*f*, 593
Rotator cuff, **266**, 267*f*, 354*t*, 355–356*f*
Rotatores muscle, 343*f*
Rotator muscles of thigh, lateral, 372–373*t*
Rotor, ATP synthase, 941*f*, 943*f*
Rough endoplasmic reticulum, 62*f*, 84, 84*f*, **85**, 85*f*, 86*f*, 87*f*, 94*t*, 105, 105*f*
Round ligament (ligamentum teres), **894**, 894*f*, 895*f*, **1107**, 1107*t*, 1109*f*
Round ligament of uterus, 1061*f*, 1063*f*, **1064**, 1064*f*
Round window, 581*f*, **582**, 583*f*, 587*f*
rRNA. *See* Ribosomal RNA
Rubrospinal tract, 475*f*, **478**, 479*f*, 481*t*
Ruffini endings (bulbous corpuscles), 492*t*, **493**

Ruffled border, osteoclast, 181, 188
Rugae
  of bladder, 1003, 1003*f*
  gastric, **885**, 885*f*
  vaginal, 1066
Rule of eights (octet rule), **31**–32
Rule of nines, **167**, 167*f*
Ruptured eardrum, 594
RV. *See* Residual volume

S. *See* Sulfur
Saccule, **583**, 583*f*, 585*t*, 590, 591*f*
Sacral artery, median, 738*f*, 744*f*, **746***t*, 746*f*
Sacral canal, **224**, 224*f*
Sacral crests, lateral and median, **224**, 224*f*
Sacral curvature of spine, **218**, 218*f*, 246
Sacral foramina, anterior and posterior, **224**, 224*f*
Sacral hiatus, **224**, 224*f*
Sacral nerves, 370*t*, 372–373*t*, 470*f*, 507*t*
Sacral part, of parasympathetic nervous system, 536*f*
Sacral plexus, 507*f*, **514**–515, 514*f*, 515*t*
Sacral promontory, **224**, 224*f*, 237*f*
Sacral splanchnic nerves, 540*f*, **541**
Sacrificial bonds, in bone, 183
Sacroiliac joint, 224, 236, 237*f*, 257*t*
Sacrospinalis (erector spinae) muscles, **340***t*, **342***t*, 343*f*
Sacrospinous ligament, 236
Sacrotuberous ligament, 238
Sacrum, 200*f*, **218**, 218*f*, 224, 224*f*, 237*f*, 239*f*
Saddle joints, 260, 263*f*, 264
Sagittal plane/section, **15**, 15*f*
Sagittal sinuses
  inferior, **752***t*, 752*f*, 753*f*
  superior, 464*f*, 465*f*, 466*f*, **752***t*, 752*f*, 753*f*
Sagittal suture, 201, **202**, 203*f*
Saline, for volume restoration, 663
Saliva, 580, 783*t*, **877**, 878–879
Salivary amylase, 49*t*, **877**, 904*t*, 913*f*, **914**
Salivary glands, 534*f*, 536*f*, 540*f*, 546*t*, 869*f*, 877–879, **878**, 878*f*
Salivation, control of, 879
Salivatory nuclei, 536–537, **879**
Salpingitis, **1086**
Salt(s), 32, **39**, 39*f*
Saltatory conduction, 411–412, 411*f*
Salt substitutes, potassium in, 1025
Salty taste, 578, 579
SA node. *See* Sinoatrial (SA) node
Saphenous nerve, 513*f*
Saphenous veins, great and small, 751*f*, **758***t*, 758*f*
Sarco (prefix), **280**
Sarcolemma, 280, **284**
  cardiac muscle, 684*f*
  skeletal muscle, 283*t*, **284**, 285*f*, 288*f*, 290, 292*f*, 293–296, 293*f*
  smooth muscle, 311
Sarcoma, **147**. *See also* Cancer
Sarcomeres, **284**
  cardiac muscle, 312*t*, 684*f*, 685
  skeletal muscle, 283*t*, **284**, 285*f*, 286*f*, 289, 289*f*, 307, 307*f*, 312*t*
Sarcopenia, 317–319
Sarcoplasm, 280, **284**, 292*f*
Sarcoplasmic reticulum, 85, **288**, 288*f*, 311, 313*t*, 684*f*, 685, 685*t*
Sartorius muscle, 330*f*, **367***t*, 368*f*, 384*t*
Satellite cells, muscle, 317
Satellite cells, neuroglia, 393*f*, **394**
Satiety signals, food intake regulation and, 449
Saturability of transport process, 73
Saturated fats/fatty acids, **46**, 46*f*, 957–958
Saturation, hemoglobin, 847–850, 848–849*f*, 850*f*
Saturation of carrier proteins, 70
"Saturday night paralysis," 512

Ruffled border, osteoclast, 181, 188
Scab, 143*f*, 144
Scala(e), 584*f*, **585**
Scala media (cochlear duct), 583, 583*f*, 584*f*, **585**, 585*t*, 587*f*
Scala tympani, 584*f*, **585**, 587*f*
Scala vestibuli, 584*f*, **585**, 587, 587*f*
Scalene muscles, **340***t*, 341*f*
Scalp, muscles of, 333*t*, 333*f*, 335*f*
Scanning electron microscopy (SEM), 117, 117*f*
Scaphoid, **234**, 234*f*, 235*t*
Scapulae (shoulder blades), 200*f*, 227, 227*f*, **228**, 229*f*, 235*t*, 266, 267*f*
  muscles controlling movement of, 350–353*t*, 351*f*, 353*f*
Scapular nerve, dorsal, 352*t*, 510*t*, 511*f*
Scapular spine, **228**, 229*f*
Scar tissue, **143**, 143*f*, 145–146
  keloid, **147**
Schwann cells (neurolemmocytes), 393*f*, **394**, 395*f*, 396, 397*f*, 498*f*
Sciatica, 515
Sciatic nerve, 369*t*, 373*t*, **514**, 514*f*, 515, 515*t*
Sciatic notch, greater, **236**, 237*f*, 239*t*
Sciatic notch, lesser, **236**, 237*f*
SCID (severe combined immunodeficiency) syndromes, **811**
Sclera, 555*f*, **558**–559, 558*f*, 560*f*, 562*f*
Scleral venous sinus, 558*f*, **562**, 562*f*
Scleroderma, **170**
Scleroses, in multiple sclerosis, 412
Sclerotherapy, **762**
Sclerotome, 1105*f*, **1106**, 1106*f*
Scoliosis, 220, 220*f*, 247
Scotoma, **596**
Scrotal septum, 1048, 1048*f*
Scrotum, 1047, 1047*f*, **1048**, 1048*f*, 1052*f*, 1083*f*, 1084
Scurvy, **147**, 932*t*
Sebaceous (oil) glands, 151*f*, 158*f*, 161*f*, 162*f*, **163**, 167, 169
Seborrhea, 163, 163*f*
Sebum, **163**
Second line of defense. *See* Innate (nonspecific) defenses
Secondary active transport (cotransport), **75**–76, 75*f*, 78*t*
  in tubular reabsorption, 990–991, 991, 991*f*
Secondary brain vesicles, **435**, 435*f*
Secondary (lobar) bronchi, **828**, 828*f*
Secondary cancer, 144*b*
Secondary capillary plexus, **611**, 611*f*
Secondary curvatures, **246**
Secondary follicle, 1070*f*, **1071**, 1072*f*, 1073
Secondary hypertension, **724**
Secondary immune response, 796*f*, **797**, 797*f*
Secondary lymphoid organs, **771**, 771*f*, **792**
Secondary oocyte, 1070*f*, **1071**, 1072*f*. *See also* Oocyte(s)
Secondary ossification centers, **185**, 185*f*, 194
Secondary polycythemias, 651
Secondary sensory (flower spray) endings, 520*f*, **521**
Secondary sex characteristics, **1060**, 1079*t*
Secondary spermatocytes, 1056*f*, **1057**
Secondary structure of protein, 50, 50*f*
Second-class lever, **328**, 329*f*
Second-degree burns, 166, 166*f*
Second-degree heart block, 692*f*
Second intention, healing by, **147**
Second messengers, **81**, 82*f*, 424, 425, 425*f*, **604**–605, 605*f*
Second-order neurons, **476**, 477*f*
Second polar body, 1070*f*, **1071**, 1096, 1096*f*
Second (expulsion) stage of labor, **1115**–1116, 1115*f*

Secretin, 635t, 888t, 890f, **891**, 900, 901f
Secretion, glandular, **123**
Secretion, renal tubular. *See* Tubular secretion
Secretory IgA, 799t
Secretory (postovulatory) phase of uterine (menstrual) cycle, 1076–1078, 1077f
Secretory region of neuron, **396**, 398t
Secretory unit (acinus), of multicellular exocrine gland, 125, 125f
Secretory vesicles/granules, 79, 79f, 86f, 87, 87f
Sectioning, tissue, for microscopy, **117**
Sections (body), 13–17, 15f
Segmental arteries, **978**, 978f
Segmental (tertiary) bronchi, **828**, 828f
Segmental level of motor control, **518**, 518f
Segmentation (digestive), 870f, **871**, 871f, 904t, 905, 905t, 906
Segregation (chromosome), independent assortment and, 1126f, **1127**
Seizures, epileptic, **462**
Selective permeability, 64f, **69**. *See also* Permeability, membrane; Membrane transport
Selenium, 26t, 933t
Self-antigens, **791**–792, 803, 803t, 811, 812
Self-propagating action potential, 409
Self-tolerance, **792**, 812. *See also* Self-antigens
Sella turcica, 205f, 207f, **209**, 209f, 214f, **216**t
Semaphorin, 428
Semen, **1053**
Semicircular canals, 581f, **583**–585, 583f, 585t, 590, 592, 592f, 593
Semicircular duct, **583**–585, 583f, 585t
Semiconservative replication (DNA), **97**, 97f
Semilunar cartilages (menisci of knee), 175f, **270**, 270f, 271f, 272–273, 272f
Semilunar valves, 678f, **680**, 680f, 681f, 694f. *See also* Aortic valve; Pulmonary valve
Semimembranosus muscle, 331f, 371f, **373**t, 384t
Seminal glands/vesicles, 1047, 1047f, 1051f, **1052**, 1082f
Seminiferous tubules, **1048**, 1049f, 1055f, 1056f, 1083
    spermatogenesis in, 1044f, 1045f, 1046f, **1054**–1058, 1056f, 1057f
Semispinalis capitis/cervicis/thoracis muscle, **342**t, 343f
Semitendinosus muscle, 331f, 371f, **373**t, 384t
Sensation, 490, **493**. *See also under* Sensory
Sense organs, **491**. *See also* Sensory receptors
Sensible heat loss/perspiration, 164
Sensible water loss, **1016**, 1017
Sensorineural deafness, **594**, 596
Sensory areas, 440f, 441, 442–443, 442f
Sensory association cortex, 440f
Sensory (afferent) division of peripheral nervous system, **392**, 392f, 497
Sensory endings, 151f, 152f. *See also* Sensory receptors
    primary and secondary, 520f, **521**
Sensory fibers, 392, 392f, 521f
Sensory ganglia, cranial, **507**
Sensory input, **391**, 391f, 493–496, 494f
Sensory (afferent) nerves, **497**
Sensory (afferent) neurons, **399**, 399t
    in reflex arc, 426f, 427, 519, 519f, 542, 543f
    somatic (SS), **474**, 474f
    visceral (VS), **474**, 474f, **542**
Sensory (ascending) pathways/tracts, 474, 475f, 476–478, 477f, 480t

Sensory receptors, **490**–496, 491–492t, 494f, 496f. *See also specific type*
    classification of, 490, 491–493, 491–492t
    cutaneous, 151f, 152f, 154, 155, **164**
    peripheral, 490–496, 491–492t, 494f, 496f
    in reflex arc, 426f, 427, 519, 519f, 543f
    sensory processing at level of, 494, 494f
    for special senses, **491**, **554**. *See also* Special senses
Sensory terminals (receptive endings), 398t, 399
Sentinel node, **777**
Sepsis (septic shock), 167, 725, **814**
Septal cartilage, 214f, 215, 821, 821f
Septal defects of heart, 701, 701f
Septal nuclei, 456, 456f
Septicemia, 667, 725
Septic shock (sepsis), 167, 725, **814**
Septum
    interatrial, **674**, 676f
    interventricular, **674**
    nasal, 215, **821**
        deviated, **863**
    scrotal, 1048, 1048f
    testicular, 1048, 1049f
Septum pellucidum, **436**, 437f, 447f
Serial processing, **426**–427, 426f
Seromucous glands, 822, 827, 827f
Serosae. *See* Serous membranes (serosae)
Serotonin, 421t, 423, 429b, **658**, **887**, 888t
Serous cells, 878, 878f
Serous fluid, **19**, 19f, 142
Serous membranes (serosae), **18**–19, 19f, **142**, 142f
    of alimentary canal (peritoneum), 19, **872**, 872f, **873**–874, 873f, 908f
    of stomach, 885f, 886f
Serous pericardium, **673**
Serratus anterior muscle, 330f, **350**t, 351f
Sertoli cells (sustentocytes), **1055**, 1055f, 1056f, **1058**
Serum, **660**
Sesamoid bones, 176, 234f
Set point, in homeostasis, 10
Severe combined immunodeficiency (SCID) syndromes, **811**
Sex (gender). *See* Gender
Sex characteristics, secondary, **1060**, 1079t
Sex chromosomes, **1081**, **1125**, 1130, 1130f
Sex development disorders, **1081**
Sex hormones, 47t, 48, **1042**, 1043, 1075. *See also specific hormone*
    adrenal, **626**, 626t
    bone growth and, 188
    gonadal secretion of, 634
    osteoporosis and decreased, 193
    terminal hair growth and, 159
Sex-linked inheritance, **1130**, 1130f
Sexual differentiation, 1081–1084, 1082f
Sexual intercourse (copulation), 1042
Sexually indifferent stage, **1083**, 1083f
Sexually transmitted infections/diseases (STIs/STDs), 1007, **1080**–1081
Sexual response, female, **1078**–1079
Sexual response, male, 1053–1054
Shaft of bone, 226, 226f, 234f
Shaft of hair, 151f, 158, 158f
Shell temperature, **964**
Shinbone. *See* Tibia
Shingles (herpes zoster), **430**
Shin splints, **386**
Shivering, 966f, 967, 968
Shock
    anaphylactic, 725, **813**
    cardiogenic, **725**
    circulatory, 545, **724**–725
    hypovolemic, **725**, 989, 1018
    neurogenic, 725
    septic (sepsis), 167, 725, **814**
    spinal, **475**, 520
    vascular, **725**
Shock wave lithotripsy, 898, 1002

Short bones, **176**, 177f, 185
Short reflexes, **874**–875, 875f, 905, 905t
Short-term (working) memory (STM), **460**, 460f
Shoulder blades. *See* Scapulae
Shoulder dislocation, **266**
Shoulder girdle. *See* Pectoral (shoulder) girdle
Shoulder (glenohumeral) joint, 256t, 266, 267f
    muscles crossing/acting on, 330f, 331f, 350–353t, 351f, 353f, 354–356t, 355–356f, 362t
Shuttle molecule, 944
Sickle-cell anemia, **650**–651, 651f, **1129**
Sickle-cell crisis, 1129
Sickle-cell trait, 650, **1129**
Sickling gene(s), 650, 1129, 1133
SIDS (sudden infant death syndrome), **863**
Sigmoidal artery, 744f, **746**t, 747f
Sigmoid colon, 869f, **907**, 907f
Sigmoid mesocolon, 908f
Sigmoid sinus, **752**t, 752f, 753f
Signaling chemicals, 81
Signaling pathways, in cancer, drugs targeting, 145b
Signal recognition particle (SRP), 105, 105f
Signal sequence, 105, 105f
Signal transduction, membrane proteins in, 66, 66f
Silicon, 26t
Simple carbohydrates, 928, 930t. *See also* Sugars
Simple diffusion, 69, **70**, 70f, 73, 73t, 467
Simple epithelium, **118**, 118f, 119–121, 119–121f
    columnar, 120f, **121**
    cuboidal, 120f, **121**
    pseudostratified columnar, **121**, 121f
    squamous, **119**–121, 119f
Simple (closed) fracture, 191
Simple glands, **125**, 125f
Simple sugars (monosaccharides), 43f, **44**–45
Sine wave, 586, 586f
Single covalent bond, 33, 33f
Sinoatrial (SA) node, **686**–687, 687f, 688
Sinus (bone marking), 180t
Sinuses, paranasal, **215**, 215f, **822**, 824t
Sinus headache, 822
Sinusitis, **822**
Sinusoids/sinusoid capillaries, 647, 711, 711f, **773**f, **774**, 896f, 897, 1065, 1065f
Sinus rhythm, **687**, 692f
Sinus venosus, 700
siRNAs. *See* Small/small interfering RNAs
Sister chromatids, 93f, 100f, **1043**, 1044, 1045f, 1046f, 1047
Size principle of recruitment, 301, 301f, 307, 307f
Sjögren's syndrome, 878
Skeletal cartilages, **174**–175, 175f
Skeletal muscle(s)/skeletal muscle tissue, 6f, 116f, **138**, 138f, **280**, **281**–310, 283t, 312–313t, **323**–389, 330f, 331f. *See also specific named muscle and* Muscular system; Skeletal muscle fibers/cells
    actions of, 324, 325f
        naming muscles and, 326
    anabolic steroids affecting, 317, 317b
    attachments of, 282–284
        naming muscles and, 326
    blood supply/perfusion of, 281, 726, 726f, 728, 728f
    cardiac muscle compared with, 685–686, 685t
    case study and, 322b, 388–389b
    connective tissue sheaths of, 281–282, 282f, 283t, 312t

contraction of, 280, 290–309, 291f, 313t
    action potential generation/propagation and, **290**, 291f, 293–296, 293f
    channels in, 290, 292f, 293f, 296f
    cross bridge activity/cycling and, **287**, 296–298, 297f
    energy (ATP) for, 303–306, 304f, 305f, 313t
    excess postexercise oxygen consumption (EPOC) and, **306**
    excitation-contraction coupling and, 291f, 294–295f, **296**
    fiber types affecting velocity/duration of, 307–309, 307f, 308t
    force of, 306–307, 307f
    frequency of stimulation affecting force and, 300–301, 300f, 306–307, 307f
    graded, 299–302, 300f, **300**, 300f
    heat production and, 281
    isometric, 299, 302f, **303**
    isotonic, 302f, **303**
    load affecting, 307f, 309, 309f
    motor neuron stimulation and, 290–298, 291f, 298f, 299
    motor unit and, 298f, **299**
    muscle fatigue and, **306**
    muscle stretch affecting force of, 307, 307f, 313t
    muscle tone and, 302–303
    muscle twitch and, **299**, 299f, 300f
    myofilaments in, 286–287
    nerve stimulus/neuromuscular junction events and, 291, 292f, 293, 313t
    stimulus frequency and, 300–301, 300f, 306–307, 307f
    stimulus strength and, 301–302, 301f
    number of stimulated fibers affecting, 307, 307f, 309
    recruitment affecting, **301**–302, 301f, 309
    sarcoplasmic reticulum and T tubules in, 288–289, 288f, 312t, 313t
    size of fibers affecting force of, 301, 301f, 307, 307f
    sliding filament model of, **289**, 289f
    velocity/duration of, 307–309, 307f, 308t, 309f
    wave summation and, **300**–301
    development of, 316, 316f, 317–319
    exercise affecting, 309–310
    homeostatic imbalances and, 287, 293, 296–298, 310
    innervation of, 281, 290, 291, 292f, 293, 298f, 299, 517
    metabolism in, 303–306, 304f, 305f, 313t, 952, 953f
    movement and, 6f, 8
    naming, 324–326
    regeneration of, 317
    skeletal system (bones) and, 196b, 318b
        lever systems and, **327**–328, 327–328f, 329f
Skeletal muscle fibers/cells, **138**, 138f, 280, 283t, 284–298, 285f, 286f, 288f, 289f, 312t
    contraction of, 290–298, 291f, 301, 301f, 307, 307f. *See also under* Skeletal muscle(s)/skeletal muscle tissue, contraction of
    differences between smooth muscle fibers and, 311–314
    in motor unit, 298f, 299
    naming muscles and direction of, 324
    types of, 307–309, 307f, 308t
Skeleton/skeletal system, 6f, 175f, **199**–250, 200f. *See also* Bone(s)/skeletal tissue; Cartilage(s); Joint(s); Ligament(s)
    appendicular, 175f, **176**, 200f, 227–245, 235t
    autonomic nervous system and, 196b, 550b

axial, 175*f*, **176**, 199–227, **199–200**, 200*f*
cardiovascular system and, 196*b*, 761*b*
case study and, 198*b*, 250*b*
digestive system and, 196*b*, 920*b*
endocrine system and, 196*b*, 637*b*
formation/developmental aspects of, 184–188, 185*f*, 186*f*, 246–247, 246*f*, 247*f*
functions of, 6*f*, 8
homeostatic relationships of, 196*b*
hormones produced by, 176, 634, 635*t*
integumentary system and, 168*b*, 196*b*
lymphatic system/immunity and, 196*b*, 778*b*
movement and, 6*f*, 8
muscular system and, 196*b*, 318*b*
lever systems and, **327**–328, 327–328*f*, 329*f*
parathyroid hormone affecting, **189**, 189*f*
reproductive system and, 196*b*, 1085*b*
respiratory system and, 196*b*, 862*b*
thyroid hormone affecting, 188, 618, 618*t*
urinary system and, 196*b*, 1036*b*
Skills (procedural) memory, 460
Skin (cutaneous membrane/integument), 5*f*, 123, **141**, 141*f*, 150–157, 151*f*, 152*f*, 154*f*, 155*f*, 156*f*
appendages of, 151*f*, **157**–163, 158*f*, 160*f*, 161*f*. *See also specific structure*
artificial/synthetic, for burn injury, 167
autonomic innervation/effects and, 534*f*, 539*f*, 540*f*
burn injury and, **166**–167, 166*f*, 167*f*
case study and, 172*b*
color of, 156–157, 1130*f*, 1131
cutaneous sensation (receptors/sense organs) and, 151*f*, 152*f*, 154, 155, **164**
developmental aspects of, 167–169
excretion via, 165
functions of, 5*f*, 6*f*, 8, 163–165
biological barriers, 164
as blood reservoir, 165
body defense, 5*f*, 6*f*, 8, 163–164, 782–783, 782*f*, 783*t*
boundary maintenance, 5*f*, 6*f*, 8, 163–164
chemical barriers/protective secretions and, 163, 783
metabolic, 164–165
as physical barrier, 163–164
protective, 5*f*, 6*f*, 8, 163–164, 782–783, 782*f*, 783*t*
temperature regulation, 164, 728, 729
homeostatic imbalances/disorders of, 156–157, 163, 164, 165–167
homeostatic relationships of, 5*f*, 168*b*
hormones produced by, 634–635, 635*t*
innervation of (dermatomes), **516**–517, 516*f*
perfusion (blood flow) of, 164, 726*f*, 728–729
sun damage and, 153, 156, 163, 165, 169
thick, **153**
thin, 153
Skin cancer, 156, 165–166, 165*f*
Skin-fold test, **969**
Skin grafts, for burn injury, 167
Skin patch, for contraception, 1118*b*
Skin pigments, 156
Skin (decubitus) ulcers, **169**
Skull, 200, 200*f*, **201**–217, 201*f*, 202*f*, 204–207*f*, 206*f*, 207*f*, 216–217*t*, 246, 246*f*
Sleep, 461, 461*f*, 462–464, 463*f*
Sleep apnea, **860**
Sleep-wake cycles, 449, 458, 462–464
Sliding filament model of contraction, **289**, 289*f*, 314
Slipped (herniated/prolapsed) disc, **219**–220, 219*f*

Slit diaphragm, 985*f*, 986
Slow block to polyspermy (zona reaction), **1096**
Slow Ca^{2+} channels, in cardiac muscle contraction, 685, **689**, 689*f*
Slow fibers, **307**, 307*f*, 308*t*
Slow oxidative fibers, 307*f*, **308**, 308*t*
Slow waves, cyclic, of stomach, 892
Slow-wave sleep, **463**
Small cardiac vein, 675*f*, **683**, 683*f*
Small cell carcinoma, of lung, **860**
Small cell (lateral) nuclear group, 453*f*, **457**
Small intestine, 869*f*, 900–906, **901**, 901*f*, 902*f*, 903*f*
autonomic innervation/effects and, 536*f*, 540*f*
in calcium regulation, 1026
digestive processes occurring in, 870*f*, 902*f*, 904*t*, 905–906, 905*t*, 912, 913*f*, 914*f*, 915*f*
hormones produced by, 635*t*
motility of, 905–906, 905*t*
Small noncoding RNAs, 1131–1132
Small saphenous vein, 751*f*, **758***f*, 758*f*
Small/small interfering RNAs, **108**, 1132
Smell, 440*f*, 443, 457, 575–578, 576*f*, 577*f*, 580, 595. *See also under Olfactory*
Smell transduction, 577, 577*f*
Smoking, effects of, 724, 858, 858*f*, 860
Smooth endoplasmic reticulum, 62*f*, 84*f*, **85**, 87*f*, 94*t*
Smooth muscle/smooth muscle tissue, 116*f*, 139*f*, **140**, 280, 310–316, 310*f*, 312–313*t*
bronchial, 828–829, 830*f*
contraction of, 313*t*, 314–316, 315*f*
development of, 317, 319
innervation of, 311, 311*f*, 313*t*, 517, 538, 548
microscopic structure of, 310, 310*f*, 311*f*, 312–313*t*
multi unit, **316**
regeneration of, 317
unitary (visceral), **316**
Smooth muscle fibers/cells, 139*f*, 140, 310–314, 310*f*, 311*f*, 312*t*
Smooth muscle tone, 314
SNARE proteins, in exocytosis, 79, 79*f*, 414
Sneeze, 822
Sodium, 26*t*, 931, 932, 933*t*
absorption of, 917
aldosterone and, 623, 624, 624*f*, **1021**–1022, 1022*f*
balance of, 623, 1020*t*, 1021–1024, 1022*f*, 1023*f*, 1024*f*
as chemically reactive element, 31*f*
concentration versus content and, 1021, 1021*t*
dietary, 931, 932, 933*t*
in fluid and electrolyte balance, 1020*t*, 1021
in plasma/interstitial fluid/intracellular fluid, 1014*f*, 1015
resting membrane potential and, 80, 80*f*, 81, 403
sodium chloride formation and, 32, 32*f*
in taste transduction, 579
tubular reabsorption of, 990–991, 991*f*, 992*f*, 1022, 1022*f*, 1023
Sodium bicarbonate, 1027. *See also Bicarbonate*
Sodium channels, 401*f*
action potential and, 406*f*, 407*f*, 408
muscle contraction and, 292*f*, 293*f*, 296*f*, **689**, 689*f*
Sodium chloride, formation of, 32, 32*f*
Sodium-glucose cotransporter (Na⁺-glucose cotransporter), 75*f*
Sodium hydroxide, 39
Sodium-potassium pump, 74*f*, **75**, **403**
action potential and, 408
carbohydrate absorption/digestion and, 914*f*

in primary active transport, 74*f*, 75, 990
protein absorption/digestion and, 915*f*
resting membrane potential and, 80, 80*f*, 402*f*, **403**
in secondary active transport, 75, 75*f*, 990–991, 991*f*
Soft callus, in bone repair, 191, 191*f*
Soft keratin, 157, 160. *See also Keratin*
Soft palate, 821*f*, **822**, 823, 823*f*, 876*f*, **877**
Sole of foot, muscles on, 380–383*t*, 381*f*, 383*f*
Soleus muscle, 330*f*, 331*f*, 375*f*, 376*f*, 377*t*, 378*f*, 379*f*, 384*t*
Sol-gel transformation, 29*f*, **30**
Solid(s), 24
Solitary nucleus, 453*f*, **580**
Soluble fiber, 928
Solute(s), **29**, 29*f*, 991–992, 991*f*, 1015, 1015*f*
Solution(s), **29**–30, 29*f*
Solvent(s), **29**, 164
universal (water), 29, **38**, 1013
Soma (cell body), of neurons, 140*f*, **394**, 395*f*
Somatic afferent fibers/nerves, 392, 392*f*, 497
Somatic efferent nerves, 392, 392*f*, 497
Somatic mesoderm, 1106, 1106*f*
Somatic (primary) motor cortex, 440*f*, **441**, 442*f*
Somatic motor fibers, 392, 392*f*, 497
Somatic motor (SM) neurons, 290, 291*f*, 292*f*, **474**, 474*f*, 479*f*. *See also Motor (efferent) neurons*
Somatic nervous system, **392**, 392*f*
autonomic nervous system compared with, 532–533, 533*f*
motor neurons of, 290, 291*f*, 292*f*, **474**, 474*f*, 479*f*
sensory neurons of, **474**, 474*f*
Somatic recombination, antigen receptor diversity and, 794
Somatic reflexes, **519**
Somatic sensation, cortical areas for, 440*f*
Somatic sensory fibers, 392, 392*f*, 497
Somatic sensory (SS) neurons, **474**, 474*f*. *See also Sensory (afferent) neurons*
Somatodendritic synapses, 413
Somatomammotropin, human chorionic (hCS/human placental lactogen), 1102, **1113**
Somatosensory association cortex, 440*f*, **442**
Somatosensory cortex, primary, 440*f*, **442**, 442*f*
Somatosensory homunculus, **442**, 442*f*
Somatosensory system, **493**–495, 494*f*
Somatostatin (growth hormone-inhibiting hormone/GHIH), 422*t*, 423, 611*f*, 614*f*, **615**, 630, **887**, 888*t*
Somatotopy, **441**, 442*f*, 476
Somatotropic cells, **615**
Somatotropin/somatotropic hormone. *See Growth hormone*
Somites, 1104*f*, 1105*f*, **1106**, 1106*f*
Sonography (ultrasound), 17*f*, 17*b*, **1120**
Sörensen, Sören, 40
Sound quality, **586**
Sounds of Korotkoff, 717
Sound/sound waves, 582, **585**–589, 586*f*, 587*f*, 590
Sound transduction, 588–589, 589*f*
Sour taste, 578, 579
Spasm (muscle), **319**
Spasmodic torticollis, 386
Spastic paralysis, **475**
Spatial discrimination, **442**, **495**
Spatial summation, by postsynaptic neuron, 416*t*, 419*f*, 420
Special senses, **491**, 553–600, **554**. *See also specific sense and structure*
case study and, 600*b*
chemical (taste and smell), 575–580, 595
developmental aspects of, 595–596

ear (hearing and balance), 580–595, 581*f*, 596
eye and vision, **554**–575, 555*f*, 595
Special sensory receptors, 491, **554**. *See also specific type*
Specific gravity, of urine, **1002**
Specific immune system/defenses. *See Adaptive (specific) immune system/defenses*
Specificity of transport process, 73
Spectrin, 646
Speech, laryngeal voice production and, 826
Speech area, motor (Broca's area), 440*f*, **441**
Speed lever, 328, 328*f*
Sperm (spermatozoa), 90, 1042, 1055*f*, **1057**–1058, 1057*f*
epididymal maturation/storage/ejaculation of, 1050
formation of. *See Spermatogenesis*
infertility and, 1058
penetration of oocyte by (fertilization), 1042, **1093**, 1094–1095*f*, 1096*f*, 1097*f*
transport of, 1050, 1093
Spermatic cord, 1048*f*, **1049**, 1049*f*, 1084
Spermatic fascia, external and internal, 1048*f*
Spermatids, 1056*f*, **1057**, 1057*f*
Spermatocytes, 1056*f*, **1057**
Spermatogenesis, 1044*f*, 1045*f*, 1046*f*, **1054**–1058, 1056*f*, 1057*f*, 1060*f*
oogenesis compared with, 1071–1073, 1071*t*
sustentocytes in, **1055**, 1055*f*, 1056*f*, **1058**
Spermatogenic cells, **1055**, 1055*f*, 1056*f*
Spermatogonia, 1056*f*, **1057**
Spermatozoa. *See Sperm*
Sperm-binding receptors, 1093, 1095*f*
Spermicides, for contraception, 1118*b*
Spermiogenesis, 1056*f*, **1057**–1058, 1057*f*
S (synthetic) phase of cell cycle, 96*f*, **97**–98, 97*f*, 100*f*
Sphenoidal fontanelle, 246*f*
Sphenoidal sinuses, 205*f*, **209**, 214*f*, 215*f*, 821*f*
Sphenoid bone, 203*f*, 204*f*, 205*f*, 206*f*, 207*f*, **209**–210, 209*f*, 214*f*, **216***t*
Sphincter(s), 326, 326*f*. *See also specific type*
in alimentary canal, 873, **907**–909, 907*f*
laryngeal structures as, 826
Sphincter pupillae muscles, 559, 559*f*
Sphygmomanometer, 717
Spina bifida, **248**, **483**
Spina bifida occulta, 483
Spinal accessory nerve. *See Accessory nerve*
Spinal (vertebral) cavity, **18**, 18*f*
Spinal column. *See Vertebral column (spine/spinal column)*
Spinal cord, 391, 391*f*, 392*f*, **470**–481, 472*f*, 473*f*, 474, 474*f*, 475*f*, 477*f*, 479*f*, 480*t*, 481*t*
autonomic regulation by, 547, 547*f*
cross-sectional anatomy of, 472–474, 473*f*, 474*f*, 475*f*
development of, 435, 435*f*, 436*f*, 482–484, 482*f*, 483*f*
gray matter of, 437*f*, 472–474, 473*f*, 474*f*, 482, 482*f*
gross anatomy/protection of, 470–472, 470–471*f*
regeneration and, 497
segments, 472, 472*f*
trauma/disorders of, 474–475
white matter of, 437*f*, 473*f*, 474, 475*f*, 482, 482*f*
Spinal curvatures, 218, 218*f*, 220, 220*f*, 246–247, 247*f*
Spinal dura mater, 470*f*, **471**, 471*f*, 473*f*
Spinal fusion, **248**

Spinal (dorsal root) ganglion, 471*f*, 473*f*, **474**, 474*f*, 497
Spinalis cervicis/thoracis muscle, **342***t*, 343*f*
Spinal meninges, 470–471*f*, 471, 473*f*
Spinal nerve(s), 391, 391*f*, 392*f*, 472, 472*f*, 473*f*, 474, **474**, 474*f*, **507–517**, 507*f*, 508*f*, 509*f*, 516*f*, 526
Spinal nerve plexuses, 507*f*, **509**, 515*t*
Spinal reflexes, **520–526**. See also specific type
Spinal root(s), 472–474, 473*f*, 474*f*, 508
Spinal rootlets, 471*f*, **508**, 508*f*
Spinal shock, **475**, 520
Spindle, 100*f*, 101*f*, 1044, 1045*f*
    muscle, 477*f*, 492*t*, **493**, 520–521, 520*f*, 522*f*, 523–524
Spindle equator, 101*f*, 1044, 1045*f*
Spindle pole, 100*f*
Spine (bone marking), 180*t*
Spine (spinal column). See Vertebral column (spine/spinal column)
Spinocerebellar pathways/tracts, 475*f*, 476, 477*f*, **478**, 480*t*
Spinothalamic pathways/tracts, 475*f*, **476**–478, 477*f*, 480*t*
Spinous process, vertebral, 218*f*, 219*f*, **220**, 220*f*, 221, 221*f*, 222*f*, 223*t*
Spiral (coiled) arteries, of uterus, **1065**
Spiral fracture, 192*t*
Spiral ganglion, 504*t*, 584*f*, **589**, 589*f*
Spiral lamina, osseous, 584*f*, **585**
Spiral organ (organ of Corti), 583*f*, 584*f*, **585**, 585*t*, 589*f*, 596
Spirometer/spirometry, **840**, 842
Splanchnic circulation, **874**
Splanchnic mesoderm, 1106, 1106*f*
Splanchnic nerves, 539, 539*f*, 540*f*, **541**
    lumbar, 540*f*, **541**
    pelvic, 536*f*, **537**
    sacral, 540*f*, **541**
    thoracic (greater/lesser/least), 538*f*, 540*f*, **541**
Spleen, 540*f*, 771, 771*f*, **773–774**, 773*f*, 777, 777*t*, 869*f*
Splenectomy, 774
Splenic artery, 744*f*, **745***t*, 745*f*, 773, 773*f*
Splenic cords, 773*f*, **774**
Splenic (left colic) flexure, **906**–907, 907*f*
Splenic sinusoids, 773*f*, **774**
Splenic vein, 751*f*, **756***f*, 756*f*, 757*f*, 773, 773*f*
Splenius capitis/cervicis muscles, 331*f*, 335*f*, **340***t*, 341*f*
Splenomegaly, **777**
Spliceosomes, 103
Spondylitis, ankylosing, **276**
Spongy bone, 127*t*, **177**, 177*f*, 178*f*, 179*f*, 182, 185, 185*f*
Spongy urethra, 1003*f*, **1004**, 1050, 1051*f*
Spoon nail, 161, 161*f*
Sprain(s), **273**
Squamous cell carcinoma of lung, **860**
Squamous cell carcinoma of skin, **165**, 165*f*
Squamous epithelium, 118*f*, **119**
    simple, **119**–121, 119*f*
    stratified, 122*f*, **123**
Squamous part of frontal bone, 202, 203*f*
Squamous part of temporal bone, **208**, 208*f*
Squamous suture, 201, 201*f*, **202**, 204*f*, 205*f*, 208
SRY gene, **1081**, 1082*f*
Stage/staging systems, cancer, 145*b*
Stagnant (ischemic) hypoxia, **852**
Staining tissue, for microscopy, **117**
Stapedius muscle, 582*f*, **583**
Stapes (stirrup), 581*f*, **582**, 582*f*, 583*f*, 587*f*
Starch, 44, 913*f*, 914, 914*f*, 928, 930*t*. See also Carbohydrate(s)
Start codon, 104, 106*f*
Startle reflex, 451
Start point, in transcription, 102
Static equilibrium, 590

Statins, 760*b*, 959
Stator, ATP synthase, 943*f*
Steady state, 80
Stellate (hepatic) macrophages, 784, 896*f*, **897**
Stem cell(s), 146, 814
    bone (osteogenic cells), 178, **180**–181, 181*f*
    damaged gastric epithelial mucosal cells replaced by, 887
    hematopoietic, 129, **647**, 647*f*, 654, 655*f*, 657*f*
    of immune system, 814
    olfactory, **576**, 576*f*
    in small intestine, 903
Stem region (antibody), 798*f*, 799
Stenosis, heart valve, 680, 696
Stents, coronary angioplasty and, 760*b*
Stercobilin, 649, 649*f*, 897
Stereocilia of epididymis, 1050
Stereocilia of hair cells, 584*f*, **588**, 589*f*, 591*f*, 592*f*
Sterilization, for contraception, 1118*b*, 1119*b*
Sternal angle, **225**, 225*f*
Sternal end of clavicle, **228**, 228*f*
Sternoclavicular joint, 256*t*
Sternocleidomastoid muscle, 330*f*, 331*f*, 333*f*, 335*f*, 339*f*, **340***t*, 341*f*
Sternocostal joint, 256*t*
Sternohyoid muscle, 330*f*, **338***t*, 339*f*, 341*f*
Sternothyroid muscle, **338***t*, 339*f*, 341*f*
Sternum, 177*f*, 200*f*, **225**–226, 225*f*
Steroid(s)/steroid hormones, 47–48, 47*t*, 48*f*, **603**, 609*t*, 623. See also Hormone(s); Sex hormones
    anabolic, 317, 317*b*, 929
    mechanism of action of, 606, 606*f*
Steroid diabetes, 625
Stimuli, 10, 10*f*, **490**
    action potential and, 416–417*t*
        coding, 410, 410*f*
        threshold/subthreshold, **301**, 406*f*, **408–409**, 494
Stirrup (stapes), 581*f*, **582**, 582*f*, 583*f*, 587*f*
Stomach, 869*f*, 884–893, **885**, 885*f*, 886*f*. See also under Gastric
    autonomic innervation/effects and, 534*f*, 536*f*, 540*f*
    digestive processes occurring in, 870*f*, 889, 904*t*, 913*f*
    filling/emptying of, 892–893, 892*f*, 893*f*
    gross anatomy of, 885–886, 885*f*
    hormones produced by, 635*t*
    microscopic anatomy of, 886–889, 886*f*
Stomodeum, 918, 918*f*
Stool, 906. See also Feces
Stop codon, 103, 104, 107*f*, 108*f*
Storage center, pontine, 1005, 1005*f*
Storage function, of bone, 176
Strabismus, 501*t*, 503*t*, **557**
Straight arteries, of uterus, **1065**
Straight sinus, 465*f*, **752***t*, 752*f*, 753*f*
Straight tubule, **1049**, 1049*f*
Strain (muscle), **319**
Stratified epithelium, 118*f*, **119**, 122*f*, 123
    columnar, **123**
    cuboidal, **123**
    squamous, 122*f*, **123**
    transitional, 122*f*, **123**
Stratum basale (stratum germinativum), 152*f*, **153**
Stratum basalis of endometrium, **1065**
Stratum corneum, 152*f*, **154**, 163
Stratum functionalis of endometrium, **1065**
    lacunae (intervillous spaces) in, **1100**–1101, 1100*f*, 1101*f*
Stratum granulosum, 152*f*, **153**–154
Stratum lucidum, **154**
Stratum spinosum, 152*f*, **153**
Streamlining (laminar flow), 715
Stress (mechanical), bone remodeling and, 189–190, 190*f*
Stress incontinence, 1005, 1113

Stress-relaxation response, 313*t*, **315**–316, 892
Stress/stress response, 534–535, 622, 624, 625, 627, 628–629*f*, 724, 963
Stretch (muscle). See Muscle stretch
Stretch marks (striae), 155–156, 156*f*
Stretch receptors, 477*f*
    bladder, micturition and, 1004, 1005, 1005*f*
    in digestion/gastric motility, 890, 890*f*, 892–893
    in food intake regulation, 961, 961*f*
    renin production and, 989
    in respiration, 854*f*
Stretch reflex, 520–524, 522*f*, **523**
Stria(e), 155–156, 156*f*
Striate (visual) cortex, primary, 440*f*, **442**–443, 574*f*, **575**
Striations, **284**
    cardiac muscle, 139*f*, 280, 684
    skeletal muscle, 138, 138*f*, 280, **284**
Striatum, 446, 446*f*
Stria vascularis, 584*f*, **585**
Stroke (cerebrovascular accident), 441, **468**–469, 760*b*
Stroke volume, 695*f*, **696**, 697–698, 697*f*, 698*f*, 718–719, 719*f*, 723*f*
Stroma, 132*f*, **133**, 770
    of lung, 833
    reticular tissue forming, 132*f*, **133**, 770, 777*t*
    of thymus, 776, 777*t*
Strong acids, 41, 1027, 1027*f*
Strong bases, 41, 1027
Structural (fibrous) proteins, 49*f*, **51**
S-T segment, 690*f*, **691**
Stuart factor (factor X), 659, 659*f*, 661*t*
Student's elbow, 273
Stupor, 462
Sty, 555
Styloglossus muscle, 336*t*, 337*f*
Stylohyoid ligaments, 212
Stylohyoid muscle, 337*f*, **338***t*, 339*f*
Styloid process
    of radius and ulna, 232*f*, **233**, 233*f*
    temporal bone, 204*f*, 206*f*, **208**, 208*f*, **216***t*
Stylomastoid foramen, 206*f*, **208**, 216*t*, 503*t*, 503*t*
Stylopharyngeus, glossopharyngeal nerve supplying, 505*t*
Subacromial bursa, 255*f*, 267*f*
Subacute hypersensitivities, **813**
Subarachnoid hemorrhage, **468**
Subarachnoid space, spinal, 471, 471*f*, 473*f*
Subarachnoid space of brain, 438, 464*f*, **465**, 466*f*
Subcapsular sinus, **772**, 772*f*
Subclavian arteries, **738***t*, 738*f*, 739*f*, 741*f*, **742***t*, 742*f*, 743*f*
Subclavian trunks, 768*f*, **769**
Subclavian vein, **750***t*, 750*f*, 751*f*, 752*f*, 753*f*, **754***t*, 754*f*, 755*f*
Subclavius muscle, **350***t*, 351*f*
Subcostal arteries, 743*t*
Subcostal nerve, **515**
Subcutaneous prepatellar bursa, 271
Subcutaneous tissue (hypodermis/superficial fascia), 130, **151**, 151*f*, 167
Subdural hemorrhage, **468**
Subdural space, spinal, 473*f*
Subdural space of brain, 464*f*, **465**
Subendocardial conducting network (Purkinje fibers), **687**–688, 687*f*, 709
Subendothelial layer, blood vessel, 708*f*, **709**
Sublingual duct, 876*f*, 878*f*
Sublingual glands, 536*f*, 869*f*, **878**, 878*f*
Subluxation, 273
Submandibular duct, 876*f*, 878*f*
Submandibular ganglia, 503*t*, **537**
Submandibular gland, 536*f*, 869*f*, **878**, 878*f*

Submucosa
    of alimentary canal, **873**, 873*f*
    of esophagus, 882*f*, 883
    of small intestine, 902*f*, 903
    of stomach, 886*f*
    tracheal, **827**, 827*f*
Submucosal nerve plexus, 873*f*, **874**
Subpubic angle (pubic arch), 237*f*, **238**, 239*t*
Subscapular artery, **742***t*, 742*f*, 743*f*
Subscapular fossa, 228, 229*f*
Subscapularis muscle, 266, 351*f*, **354***t*, **356***t*, 356*f*, 362*t*
Subscapular nerves, 356*t*, 510*t*, 511*f*
Subscript in chemical equation, 36
Substance abuse, 429*b*
Substance P, 422*t*, **423**, 495
Substantia nigra, 446, **451**–452, 453*f*, 469
Substrate, **51**, 52, 53*f*
Substrate level phosphorylation, **936**, 937*f*, 938, 944, 944*f*
Subthalamic nuclei, 446
Subthreshold stimuli, **301**, 408, 410*f*
Sucking reflex, 919
Sucrase, 913*f*, **914**
Sucrose, 42*f*, 43*f*, 44, 913*f*, 914
Sudden infant death syndrome (SIDS), **863**
Sudoriferous glands. See Sweat (sudoriferous) glands
Sugars, 927–928, 930*t*. See also Carbohydrate(s); Fructose; Galactose; Glucose; Sucrose
    double (disaccharides), 43*f*, **44**, 913*f*, 914, 914*f*
    simple (monosaccharides), **43**–44, 43*f*, 914
Sulci (sulcus), 438*f*, **439**, 442
Sulfate, in plasma/interstitial fluid/intracellular fluid, 1014*f*
Sulfur, 26*t*, 933*t*
Sulfuric acid, 41
Summation, **301**–302, 301*f*, 309, 416*t*, 419*f*, **420**
Sun exposure/sunlight. See Ultraviolet radiation
Superficial (term), 14*t*. See also under External
Superficial fascia (hypodermis/subcutaneous tissue), 130, **151**, 151*f*, 167
Superficial fibular (peroneal) nerve, 514*f*
Superficial palmar arch
    arterial, 739*f*, **742***t*, 742*f*, 743*f*
    venous, **754***t*, 754*f*, 755*f*
Superficial perineal space, 348*t*, 349*f*
Superficial reflexes, **525**–526
Superficial temporal artery, 717*f*, **740***t*, 740*f*, 741*f*
Superficial temporal vein, **752***t*, 752*f*, 753*f*
Superficial thrombophlebitis, 762
Superficial transverse perineal muscle, 348*t*, 349*f*
Superficial veins, 754–755*f*, 754*t*, 758*t*, 758*f*
Superior (term), 14*t*
Superior angle of scapula, 228, 229*f*
Superior articular process, **220**, 220*f*, 222*f*, 223*t*
    of sacrum, 224, 224*f*
Superior border of scapula, 228, 229*f*
Superior cerebellar peduncles, 450*f*, 451, 451*f*, 453*f*, **454**, 455, 455*f*
Superior cervical ganglion, 540*f*, **541**
Superior colliculi, 450*f*, **451**, 451*f*, 453*f*, 478, 574*f*, **575**
Superior costal facet, 222, 223*t*, 226*f*
Superior ganglion, 505*t*
Superior gemellus muscle, 372*f*, **373***t*
Superior gluteal artery, **748***t*, 748*f*
Superior gluteal nerve, 514*f*, **515**, 515*t*
Superior hypophyseal artery, 611*f*
Superior iliac spines, **236**, 237*f*
Superior mesenteric artery, 738*f*, 739*f*, 744*f*, 745*f*, **746***t*, 746*f*, 747*f*, 907*f*

Superior mesenteric ganglion/plexus, 540*f*, **541**

Superior mesenteric vein, 751*f*, **756***t*, 756*f*, 757*f*

Superior nasal conchae, 210, 214*f*, **216***t*, 576, 821*f*, 822

Superior nasal meatus, 214*f*, 215, 821*f*

Superior nuchal line, 203, 203*f*, 206*f*

Superior oblique muscle, **556**, 557*f*

Superior olivary nuclei, **589**, 589*f*

Superior orbital fissure, 203*f*, 209*f*, **210**, **216***t*

Superior pharyngeal constrictor muscle, **339***t*, 339*f*

Superior phrenic arteries, 738*f*, **743***t*

Superior pubic ramus, 237*f*, **238**

Superior rectal artery, 744*f*, **746***t*, 747*f*

Superior rectus muscle, **556**, 557*f*

Superior sagittal sinus, 464*f*, 465*f*, 466*f*, **752***t*, 752*f*, 753*f*

Superior salivatory nuclei, 536–537

Superior suprarenal artery, 746*t*

Superior thyroid artery, 617, 617*f*, **740***t*, 740*f*, 741*f*

Superior thyroid vein, 752*f*, 753*f*

Superior tibiofibular joint, 242*f*, 243, 257*t*

Superior vena cava, **674**, 675–677*f*, 681*f*, **750***t*, 750*f*, 751*f*, **752***t*, 752*f*, 753*f*, 754*f*, 755*f*

Superior vesical arteries, 1107

Superior vestibular ganglion, 583*f*, **591**

Supination (body movement), **260**, 261*f*, 362*t*

Supinator muscle, **358***t*, 359*f*, **360***t*, 361*f*, 362*t*

Support function, of bone, 175

Supporting cells, **576**, **590**

in cochlea, 584*f*, 585

in crista ampullaris, 592*f*, 593

in maculae, **590**, 591*f*

in nervous tissue (neuroglia), 140, 140*f*

olfactory, **576**, 576*f*

Suppressor factors, 805*t*

Suprachiasmatic nucleus, 449, 449*f*, 463, 574*f*, **575**, 627

Supraclavicular nerves, 509*t*, 509*f*

Supracondylar lines, medial and lateral, 240*f*, **241**

Supracondylar ridges, medial and lateral, 230*f*, **231**

Supraglenoid tubercle, 229*f*, 231

Suprahyoid muscles, **338***t*

Supraoptic nuclei, 449, 449*f*, **610**, 610*f*, 611

Supraorbital foramen/notch, **202**, 203*f*, **216***t*

Supraorbital margins, **202**, 203*f*

Suprapatellar bursa, 270*f*

Suprarenal arteries, 738*f*, 744*f*, 746*t*, 746*f*

Suprarenal gland(s). See Adrenal (suprarenal) gland(s)

Suprarenal veins, 750*f*, **756***t*, 756*f*, 757*f*

Suprascapular artery, 742*f*, 743*f*

Suprascapular nerve, 356*t*, 510*t*, 511*f*

Suprascapular notch, **228**, 229*f*

Supraspinatus muscle, 266, 353*f*, **354***t*, 355*f*, **356***t*, 362*t*

Supraspinous fossa, 228, 229*f*

Supraspinous ligament, 219*f*

Suprasternal (jugular) notch, **225**, 225*f*

Sural nerve, 514*f*, **515**, 515*t*

Surface anatomy, **2**

Surface area. See Body surface area

Surface (germinal) epithelium, of ovary, 1061*f*, 1062

Surface tension, 35, 35*f*, 835, **839**

Surface-volume relationships, cell division control and, 98

Surfaces (bone markings), 179, 180*t*

Surfactant, 831, **839**, 1116

Surfactant protein A, 1114

Surgery, for obesity, 959*b*

Surgical neck of humerus, 230*f*, **231**

Survival needs, 8–9

Suspension(s), 29*f*, **30**

Suspensory ligament

of breast, **1068**, 1068*f*

of lens of eye (ciliary zonule), 558*f*, **559**, 562*f*, 565, 565*f*, 566

of ovary, **1061**, 1061*f*, 1063*f*

Sustentaculum tali, **244**, 244*f*

Sustentocytes (Sertoli cells), 1049, **1055**, 1055*f*, 1056*f*, **1058**

Sutural bones, 203*f*, **210**

Sutures (cranial), **201**, 201*f*, 202, **252**, 252*f*, 256*t*. See also specific type

Swallowing (deglutition), 870, 870*f*, **883**–884, 884*f*, 904*t*

difficulty with, 505*t*

achalasia, **919**

dysphagia, **921**

epiglottis in, 825

muscles controlling, 338–339*t*, 339*f*

Sweat, 162, 164, 965

Sweat (sudoriferous) glands, **161**–163, 161*f*, 169

apocrine, 162*f*, **162**–163

autonomic innervation/effects and, 161, 538, 539*f*, 540*f*, 546*t*

eccrine (merocrine), 151*f*, **161**–162, 161*f*, 162*f*

Sweating, 161–163, 164, 165, 965, 966*f*, 967

Sweat pores, 151*f*, 161–162, 161*f*

Sweet taste, 578, 579

Swelling in inflammation, 785, 786, 787*f*. See also Edema

Symmetry, of neuronal pathways, 476

Sympathetic (thoracolumbar) division of autonomic nervous system, **392**, 392*f*, 532*f*, **534**–535, 535*t*, **538**–542, 538*t*, 538*f*, 539*f*, 545–546

adrenal medulla and, 540*f*, 542, 542*f*, 546*t*

glomerular filtration regulation and, 988*f*, 989

heart regulation and, 534*f*, 540*f*, 545, 546*t*, 688*f*, 689, 697*f*, 699

postabsorptive state regulation and, 954–955

salivation control and, 534*f*, 540*f*, 879

stress response and, 534–535, 627, 628*f*

Sympathetic pathways, 539–542, 539*f*, 540*f*, 542*f*

with synapses in adrenal medulla, 540*f*, 542, 542*f*

with synapses in collateral ganglia, 539*f*, 540*f*, 541–542

with synapses in trunk ganglia, 539–541, 539*f*, 540*f*

Sympathetic (vasomotor) tone, **545**

Sympathetic trunk (chain), **538**, 538*f*, 539*f*

Sympathetic trunk (paravertebral) ganglia, 508*f*, **538**, 538*t*, 538*f*, 539*f*, 540*f*

Sympathetic venoconstriction, venous return and, 718

Sympatholytic agents, autonomic activity and, 544*t*

Sympathomimetic agents, autonomic activity and, 544*t*

Symphyses, 253–254, 253*f*

Symport system, 75–76, 75*f*, 78*t*

Synapse(s), **412**–415, 413*f*, 415*f*

chemical, **413**–414, 415*f*. See also Neurotransmitter(s)

electrical, **413**

en passant, 517

postsynaptic potential/synaptic integration and, 416–420, 417*t*, 418*f*, 419*f*

Synapsis, 1044

Synaptic cleft, **290**, 292*f*, **414**, 415*f*

Synaptic delay, **414**

Synaptic integration, 419*f*, 420

Synaptic potentiation, **420**

Synaptic terminals, rod and cone, 568, 568*f*

Synaptic vesicles, **290**, 292*f*, 293*f*, **414**, 415*f*

Synaptotagmin, 414

Synarthroses, **251**

Synchondroses, **253**, 253*f*

Syncope, **462**, 549, 728

Syncytiotrophoblast (syncytial trophoblast), 1098*f*, **1099**, 1100*f*

Syncytium, cardiac muscle as, **684**, 685, 685*t*

Syndesmoses, **252**, 252*f*

Syndrome of inappropriate ADH secretion (SIADH), 612*t*, 614, **1037**

Synergism, hormone, **609**

Synergist (muscle), **324**

Synostoses, **252**

Synovial cavities. See Joint (articular/synovial) cavities

Synovial fluid, **254**, 254*f*

Synovial joints, 251, **254**–272, 254*f*, 255*t*, 256–257*t*, 262–263*f*. See also specific joint

aging/development and, 275

blood supply of, 255

bursae and tendon sheaths and, **255**, 255*f*

diversity of, 264–272

factors affecting stability of, 257–258

general structure of, 254–255, 254*f*

innervation of, 255

movements allowed by, 258–260, 258–260*f*

types of, 262–263*f*, 264

Synovial membrane, 141, **254**, 254*f*

Synovitis, 275, **276**

Synthesis reactions, **36**, 36*f*

dehydration, **42**, 42*f*, 44, 45, 45*f*, 48, 49*f*

Synthetic growth hormone, 616

Synthetic skin, for burn injury, 167

Syphilis, **1080**

System(s), organ. See Organ systems

Systemic anatomy, **2**

Systemic blood pressure. See Blood pressure

Systemic circuit/circulation, **671**, 671*f*, 680–682, 681*f*, **735**, 737*t*, 737*f*

arteries of, 735, 737*t*, 737*f*, 738–739*f*, 738*t*

veins of, 735, 750–751*f*, 750*t*

Systemic lupus erythematosus (SLE), 812, **814**

Systole, **693**, 694*f*

Systolic pressure, **716**, 716*f*

T

T₃. See Triiodothyronine

T₄. See Thyroxine

Tabes dorsalis, **526**

Tachycardia, **699**, 702

Tachykinin(s), 422*t*

Tactile epithelial (Merkel) cells, 152*f*, **153**, **492**

Tactile (Meissner's) corpuscles, 154, 164, **492**, 492*t*

Tactile (Merkel) discs, 153, 491*t*, **492**

Tailbone. See Coccyx

Tail of caudate nucleus, 446*f*

Tail of epididymis, 1049*f*, 1050

Tail of pancreas, 898, 900*f*

Tail of sperm, 1057*f*

Talar shelf (sustentaculum tali), **244**, 244*f*

Talus, 177*f*, 241*t*, **244**, 244*f*

Tamponade, cardiac, 673

Target cells, hormone, **603**, 608, 609

Target-SNAREs, in exocytosis, 79, 79*f*

Tarsal bones/tarsus, 200*f*, 241*t*, **244**, 244*f*

Tarsal glands, **555**, 555*f*

Tarsal plates, **555**, 555*f*

Tarsometatarsal joint, 257*t*

Tarsus, bones of (tarsals), 200*f*, 241*t*, **244**, 244*f*

Tastant, 579

Taste, 440*f*, 443, 578–580, 579*f*, 580*f*, 595

Taste buds, **578**, 579, 579*f*, 580, 580*f*, 877

Taste cells, receptors for (gustatory epithelial cells), **578**, 579, 579*f*

Taste pore, **578**, 579*f*

Taste receptors. See Taste buds

Taste sensations, 578–579

Taste transduction, 579

Tau protein, in Alzheimer's disease, 469

Tay-Sachs disease, 87, 1129, 1133

T₍C₎ cells. See Cytotoxic T cells

T cell antigen receptors (TCRs), 802*f*, **804**

T cell-dependent antigens, **806**

T cell-independent antigens, **806**

T cells. See T lymphocytes

TCRs. See T cell antigen receptors

Tears, 556, 783*t*

Tectorial membrane, 584*f*, **588**, 589*f*

Tectospinal tract, 475*f*, **478**, 481*t*

Tectum, 451, 451*f*, 453*f*

Teeth, **879**–881, 879*f*, 880*f*

sockets (alveoli), 211, **217***t*, 879, 880, 880*f*

articulation of tooth with (gomphosis), **252**–253, 252*f*, 880

Telencephalon, **435**, 435*f*, 436*f*

Telogen effluvium (TE), 160

Telomerase, 110

Telomere(s), 110

Telophase in meiosis, 1045*f*, 1046*f*

Telophase in mitosis, 96*f*, **98**, 101*f*, 1046*f*

Temperature

body. See Body temperature

chemical reaction rate affected by, 38

diffusion speed affected by, 69

gas solubility and, 843

hemoglobin saturation affected by, 850, 850*f*

Temperature receptors. See Thermoreceptors

Template strands, 97, 97*f*, 102, 102*f*, 106*f*

Temporal artery, superficial, 717*f*, **740***t*, 740*f*, 741*f*

Temporal bones, **203**–209, 203*f*, 204*f*, 205*f*, 206*f*, 207*f*, 208*f*, **216***t*

Temporal branch of facial nerve, 503*t*, 504*t*

Temporalis muscle, 330*f*, 335*f*, **336***t*, 337*f*

Temporal lobe, 438*f*, **439**, 440*f*, 460–461

Temporal (wave) summation, 300*f*, **300**–301, 416*t*, 419*f*, 420

Temporal vein, superficial, **752***t*, 752*f*, 753*f*

Temporary abstinence methods, of contraception, 1118*b*

Temporomandibular joint (TMJ), 208, 211, 211*f*, 256*t*, **264**–266, 265*f*

Tendon(s), 132*f*, 133, 258, **282**–284, 282*f*. See also specific named tendon

of long head of biceps brachii muscle, 266, 267*f*

Tendonitis, **273**

Tendon organs, 492*t*, **493**, 520, 520*f*, 524, 524*f*

Tendon reflex, 493, 520–524, **524**, 524*f*

Tendon sheaths, **255**, 255*f*

Teniae coli, **906**, 907*f*

Tennis elbow, **386**

Tension (muscle), **299**, 307, 307*f*, 316

Tension (cleavage) lines, **155**, 155*f*

Tensor fasciae latae muscle, 330*f*, **367***t*, 368*f*, **370***t*, 384*t*

Tensor tympani muscle, **582**–583, 582*f*

Tensor veli palatini, 338*t*, 339*f*

Tentorium cerebelli, **465**, 465*f*

Teratogens, **1113**

Teres major muscle, 331*f*, 353*f*, **354***t*, 355*f*, **356***t*, 362*t*

Teres minor muscle, 266, 353*f*, **354***t*, 355*f*, **356***t*, 362*t*

Terminal arteriole, 707*f*, **712**, 712*f*

Terminal branches/arborizations, **395**, 395*f*

Terminal bronchioles, **828**, 829*f*, 830*f*

Terminal cisterns, **288**, 288*f*, 294–295*f*, 296

Terminal ganglia, **536**, 538*t*

Terminal hair, **159**

Terminal sulcus, **877**, 877*f*

Terminal web, 91, 91*f*

Termination in transcription, 102, 102*f*

Termination in translation, 104, 107*f*
Termination signal, 102*f*
Terminology (anatomical), 2, 12–17, 13*f*, 14*t*, 15*f*
Tertiary (segmental) bronchi, **828**, 828*f*
Tertiary (antral/vesicular) follicles, 1070*f*, **1071**, 1072*f*, 1073
Tertiary structure of protein, 50, 50*f*
Tertiary syphilis, 1080
Testes, 602*f*, 634, 1042, **1047**, 1047*f*, 1048–1049, 1051*f*
  aging affecting, 636
  descent of, 1049, 1084
  embryologic/fetal development of, 1082*f*, 1084
  rete, **1049**, 1049*f*
Testicular arteries, **746***t*, 746*f*, 1048*f*, **1049**
Testicular cancer, 1049
Testicular fluid, **1058**
Testicular veins, **756***t*, **1049**
Testis. *See* Testes
Testosterone, 626, 626*t*, **634**, 1060, 1079*t*
  adrenal, 626, 626*t*, 1060
  aging and production of, 636, 1060, 1060*f*
  anabolic steroid use and, 317*b*, 319
  bone growth and, 188
  in embryos, 1060, 1060*f*
  erythropoietin production and, 649
  gonadal development and, 1084
  interstitial endocrine (Leydig) cell production of, 1055
  in male reproductive function, 1043, 1059*f*, 1060, 1060*f*, 1079*t*
  mechanism/effects of activity of, 1060, 1079*t*
  metabolism and, 955*t*, 1060, 1079*t*
  muscle strength/growth and, 317
  somatic effects of, 1060, 1060*f*, 1079*t*
  terminal hair growth affected by, 159
Tetanus (infection), **319**
Tetanus (tetanic muscle contraction), 300*f*, **301**, 319
  absence of in cardiac muscle, 685–686, 685*t*
Tetany, 622
Tetrads, **1044**, 1045*f*, 1046*f*, 1126*f*, 1127, 1128
Tetrahydrocannabinol (THC), 424
Tetralogy of Fallot, 701, 701*f*
Tetramer fibrils, 88, 88*f*
T$_H$1 cells, 806. *See also* Helper T cells
T$_H$2 cells, 806. *See also* Helper T cells
T$_H$17 cells, 806. *See also* Helper T cells
Thalamic nuclei, 448, 449*f*, 456, 456*f*
Thalamus, 435, 435*f*, 445*f*, 446*f*, **447**–448, 447*f*, 448*f*, 450–451*f*, 458*t*
Thalassemias, **650**
Thalidomide, 1131
Theca folliculi, 1072*f*, **1073**
Thecal cells, **1073**
Thenar eminence, muscles of, 359*f*, 364*t*
Thermodynamics, first law of, 960
Thermogenesis, chemical (nonshivering), **967**
Thermogenesis, food-induced, **963**
Thermoreceptors, 477*f*, **490**, 491, 491*f*, **965**
Thermoregulation. *See* Body temperature, regulation of
Thermoregulatory centers, **965**
Theta waves, **461**, 461*f*
Thiamine (vitamin B$_1$), 932*t*
Thiazide diuretics, 1000
Thick ascending limb/segment of nephron loop, 979*f*, 980. *See also* Nephron loop
Thick (myosin) filaments, 138, **284**
  in cardiac muscle, 685
  cross bridge cycling and, 296, 297*f*
  in skeletal muscle, 283*t*, **284**, 285*f*, 286–287, 286*f*
  in sliding filament model, 289, 289*f*
  in smooth muscle, 311, 312*t*, 314, 315*f*
Thick skin, **153**

Thigh
  bones of, 240–242, 240*f*, 241*t*
  deep artery of, **748***t*, 749*f*
  muscles of/muscles controlling movement of, 330*f*, 331*f*, 367–373*t*, 368*f*, 371–372*f*, 384*t*, 385*f*
Thin ascending limb/segment of nephron loop, 980. *See also* Nephron loop
Thin descending limb/segment of nephron loop, 979*f*. *See also* Nephron loop
Thin (actin) filaments, 88, 88*f*, 95*t*, 138, **284**
  in cardiac muscle, 313*t*, 685
  cross bridge cycling and, 296, 297*f*
  in cytokinesis, 98, 101*f*
  in microvilli, 91, 91*f*
  in skeletal muscle, 283*t*, **284**, 285*f*, 286*f*, 287, 313*t*
  in sliding filament model, 289, 289*f*
  in smooth muscle, 311, 312*t*, 314, 314, 315*f*
Thin skin, 153
Third-class lever, **328**, 329*f*
Third-degree burns, 166, 166*f*
Third-order neurons, **476**, 477*f*
Third (placental) stage of labor, 1115*f*, **1116**
Third line of defense. *See* Adaptive (specific) immunity/defenses
Third ventricle, 435*t*, **436**, 437*f*
Thirst/thirst mechanism, 449, **1016–1017**, 1017*f*
  blood pressure and, 722*f*, 723, 1017, 1017*f*
Thirst center, 449, 1016–1017, 1017*f*
Thoracic aorta, **738***t*, 738*f*, 742*f*
Thoracic arteries
  internal (internal mammary artery), 741*f*, 742*f*, **743***t*, 743*f*
  lateral, **742***t*, 742*f*, 743*f*
Thoracic cage/thorax, 200, 200*f*, **224–227**, 225*f*, 226*f*
  aging affecting, 247
  anatomical relationships of organs in, 832*f*
  arteries of, 741*f*, 742–743*f*, 743*t*
  bones of/bony, 200, 200*f*, 224–227, 225*f*, 226*f*
  compliance of wall of, 840
  innervation of, 470–471*f*, 507*f*, 515. *See also* Thoracic nerves
  muscles of, 330*f*, 344*t*, 345*f*
    deep (breathing), 344*t*, 345*f*, 836–838, 837*f*
    superficial (arm/scapular movements), 350–353*t*, 351*f*, 353*f*
  sympathetic pathways to, 540*f*, 541
  veins of, 754–755*f*, 754–755*t*
Thoracic cavity, **18**, 18*f*
  organs in, 832*f*
  pressure relationships in, 835–838, 835*f*, 836*f*, 837*f*, 838*f*
Thoracic curvature of spine, **218**, 218*f*, 220, 220*f*, 246
Thoracic duct, 768*f*, **769**, 769*f*
Thoracic nerves, 470–471*f*, 507*f*, 515
  long, 350*t*, 510*t*, 511*f*
Thoracic (greater/lesser/least) splanchnic nerves, 538*f*, 540*f*, **541**
Thoracic vertebrae, **218**, 218*f*, 222, 222*f*, 223*t*
Thoracic visceral organs, arteries of, 743*t*
Thoracoacromial artery, **742***t*, 742*f*, 743*f*
Thoracodorsal nerve, 355*t*, 511*f*
Thoracolumbar division of autonomic nervous system. *See* Sympathetic (thoracolumbar) division of autonomic nervous system
Thorax. *See* Thoracic cage/thorax
Thoroughfare channel, 712, 712*f*
Three-dimensional vision (depth perception), **575**
Threshold (anaerobic), **305**
Threshold stimulus, **301**, 406*f*, **408–409**, 494

Throat. *See* Pharynx
Thrombin, 659*f*, **660**
Thrombocytopenia, **662**
Thromboembolic disorders, 662
Thrombolytic agents, 760*b*
Thrombophlebitis, superficial, **762**
Thrombopoietin, **657**
Thromboxane(s), 47*t*
Thromboxane A$_2$, platelet aggregation and, **658**
Thrombus (thrombi), **662**, 760*b*, 759
Thumb (pollex), **236**
  carpometacarpal joint of, 256*t*
  muscles controlling movement of, 359*f*, 364*t*, 365*f*
  opposition of, **260**, 261*f*
Thymic aplasia, congenital, **814**
Thymic corpuscles, **776**, 776*f*
Thymic lobules, 776
Thymine (T), 54*f*, **55**
Thymopoietin(s), 635, 635*t*
Thymosin(s), 635, 635*t*
Thymulin, 635, 635*t*
Thymus, **635**, 771, 771*f*, **776**, 776*f*, 777*t*
  failure of development of (congenital thymic aplasia), **814**
  hormones produced by, 635, 635*t*
  T-cell differentiation/education/ maturation and, 635, 654, 771, 776, 793*f*, 794, 794*f*, 795*t*, 802*f*
Thyrocervical trunk, 741*t*, 741*f*, 742*f*, 743*f*
Thyroglobulin, **617**, 619, 619*f*, 620
Thyrohyoid membrane, 825*f*
Thyrohyoid muscle, 337*f*, **338***t*, 339*f*
Thyroid arteries, inferior, 617, 617*f*
Thyroid arteries, superior, 617, 617*f*, **740***t*, 740*f*, 741*f*
Thyroid cartilage, 175*f*, 823*f*, **824**, 825*f*
Thyroid crisis/storm, **636**
Thyroid follicles, **617**, 617*f*
Thyroid gland, 602*f*, **617–621**, 617*f*, 618*t*, 619*f*, 620*f*, 963
Thyroid hormone (TH), 616*f*, 617, 617*f*, **618**, 618*t*, 619*f*
  growth/bone growth and, 188, 618, 618*t*
  hyper-/hyposecretion of, 620, 963
  metabolism and, 618, 618*t*, 963
  synthesis of, 618–620, 619*f*, 636
  thermoregulation and, 618, 618*t*, 620, 967
  transport and regulation of, 616*f*, 620–621
Thyroiditis, Hashimoto's, **814**
Thyroid-stimulating hormone (thyrotropin/ TSH), 611*f*, 613*t*, **616**, 616*f*, 618–619, 620, 967
Thyroid storm/crisis, **636**
Thyroid vein, middle and superior, 752*f*, 753*f*
Thyrotropic cells, **616**
Thyrotropin (TSH), 611*f*, 613*t*, **616**, 616*f*, 618–619, 620, 967
Thyrotropin-releasing hormone (TRH), 611*f*, 613*t*, **616**, 616*f*, 620, 967
Thyroxine (T$_4$), **618**, 618*t*, 619, 619*f*, 620, 699, 955*t*, **963**, 967. *See also* Thyroid hormone
Thyroxine-binding globulins (TBGs), 620
Tibia, 200*f*, 241*t*, 242–243*f*, **243**, 270, 270*f*, 271*f*
Tibial arteries
  anterior, 739*f*, **748***t*, 748*f*, 749*f*
  posterior, 717*f*, 739*f*, 748*f*, **749***t*, 749*f*
Tibial collateral ligament, 270*f*, **271**, 272, 272*f*
Tibialis anterior muscle, 330*f*, **374***t*, 375*f*, 376*f*, 384*t*, 385*f*
Tibialis posterior muscle, **377***t*, 379*f*, 384*t*
Tibial nerve, 373*t*, 377*t*, **514–515**, 514*f*, 515*t*
Tibial tuberosity, 242*f*, **243**, 243*f*
Tibial veins, anterior and posterior, 751*f*, **758***t*, 758*f*
Tibiofemoral joint, 257*t*, **270**. *See also* Knee joint

Tibiofibular joints, 242, 242*f*, 243, 257*t*
Tic douloureux (trigeminal neuralgia), 502*t*
Tidal volume (TV), **840**, 841*f*, 842, 842*t*, 1113
Tight junctions, **67–68**, 67*f*, 118
  in blood brain barrier, 467
  in capillaries, 711, 711*f*
  epithelial, 118
  in gastric mucosal barrier, 887
  of sustentocytes (blood testis barrier), 1056*f*, **1058**
  in tubular reabsorption, 990, 990*f*
Timing of sound waves, relative, 590
Tin, 26*t*
Tinnitus, **595**
Tip links, 588, 589*f*
Tissue(s), 4–5, 4*f*, 115–149, **116**, 116*f*
  blood flow through (tissue perfusion), **725–730**, 725*f*, 726*f*, 727*f*, 728*f*
  connective. *See* Connective tissue
  covering and lining (cutaneous/serous/ mucous) membranes, 117, 141–142, 141–142*f*
  developmental aspects of, 146, 146*f*
  epithelial. *See* Epithelial tissue/ epithelium
  lymphoid. *See* Lymphoid tissues
  muscle. *See* Muscle(s)/muscle tissue
  nervous. *See* Nervous tissue
  preparation of for microscopy, 117
  regenerative capacity of. *See* Regeneration
  structural organization at level of, 4–5, 4*f*
Tissue factor (TF/factor III), 659*f*, **660**, 661*t*
Tissue fluid. *See* Interstitial (tissue) fluid
Tissue perfusion, **725–730**, 725*f*, 726*f*, 727*f*, 728*f*
Tissue plasminogen activator (tPA), 469, **660**, 760*b*
Tissue repair, 142–146, 143*f*
  first/second intention, **147**
Titin (elastic) filaments, 283*t*, 284, 285*f*, **287**
TLC. *See* Total lung capacity
TLRs. *See* Toll-like receptors
T lymphocytes (T cells), **653**, 655*f*, **770**, **792–794**, 795*t*, 801–802, 802*f*, 809*f*
  activation/development/differentiation/ maturation of, 792–793, 793*f*, 795*t*, 804–805, 804*f*, 805*t*, 809*f*
    MHC proteins and, 802*f*
    thymus in, 635, 654, 771, 776, 793*f*, 794, 794*f*, 802*f*
  antigen presentation and, 794–795, 804, 804*f*, 809*f*
  in cellular immunity, 782*f*, 795*t*, 809*f*
  clonal selection and, 804, 804*f*
  cytotoxic, 795*t*, 802, 802*f*, **807**, 807*f*, 809*f*, 810*t*
  effector roles of, 655*f*, 809*f*
  helper, 795*t*, 802, 802*f*, **806**, 806*f*, 809*f*, 810*t*, 811
  memory, 802, 802*f*, 809*f*
  receptors on, 802*f*, **804**
  regulatory, 776, 795*t*, 802, 802*f*, **807–810**, 810*t*
  selection/education of, 794, 794*f*, 804, 804*f*
T lymphocyte precursor, **654**, 655*f*, 793*f*, 814
Tocopherol (vitamin E), 47*t*, 931, 932*t*
Toe(s)
  bones of (phalanges), 200*f*, 241*t*, 244, 244*f*
  interphalangeal joints of, 257*t*
  muscles controlling movements of, 374–379*t*, 375*f*, 376*f*, 378–379*f*, 380–383*t*, 381*f*, 383*f*, 384*t*, 385*f*
Tolerance
  to pain, 495–496
  self-tolerance, **792**, 812
Toll-like receptors (TLRs), **783**–784
Tone (sound), 586

Tongue, 869f, 876f, **877**, 877f
  hypoglossal nerve supplying, 336t, 506t
  muscles of, extrinsic and intrinsic, 336t, 337f, **877**
  papillae of, **578**, 579f, 877, 877f
  taste buds on, **578**, 579f, 877
"Tongue-tied," 877
Tonic-clonic (grand mal) seizures, 462
Tonicity, **72–73**, 72f
Tonic receptors, **494**
Tonsil(s), 771, 771f, 774f, **775**
  inflammation of (tonsillitis), **777**
  lingual, 774f, **775**, **823**, 823f, 876f, 877, 877f
  palatine, 774f, **775**, **823**, 823f, 876f, 877f
  pharyngeal, 774f, **775**, **823**, 823f
  tubal, **775**, 821f
Tonsillar crypts, 774f, **775**
Tonsillitis, **777**
Tooth. *See* Teeth
Torticollis, **386**
Total body water, 1013, 1013f, 1113
Total dead space, **842**
Total lung capacity (TLC), **841**, 841f
Total metabolic rate, **963**
Total peripheral resistance, **714–715**
Touch, discriminative, 476
Touch receptors, 153, 154, 164, 477f, 491t, **492**, 492t
TP53 gene, 144b
Trabeculae
  carneae, **674**, 677f
  in intramembranous ossification, 186f
  lymph node, **771–772**, 772f
  in ossification, 185, 186f
  splenic, 773f, 774
  in spongy bone, **177**, 177f, 178f, 182, 183f
Trabecular bone. *See* Spongy bone
Trace elements, 26t
Trace minerals, 931, 933t
Trachea, 820f, **826–828**, 827f, 828f, 831t
Tracheal cartilages, 175f, 825f, 827, 827f
Trachealis muscle, **827**, 827f
Tracheoesophageal fistula, 919
Tracheotomy, **863**
Trachoma, **596**
Tracings, myogram, 299
Tract(s), **395**. *See also specific type*
Traction, **195**
Transamination, 948f, 949, 949f
Transcellular route, of tubular reabsorption, 990, 990f
Transcription, 99, 99f, **102**–103, 102f, 108f
Transcription factors, 102
Transcytosis, **76**, 76f, 77, 78t, 467
Transducin, **571**, 571f, 573
Transduction, **494**
  light (phototransduction), **570**–573, 571f
  in sensory processing, 494
  signal, membrane proteins in, 66, 66f
  smell, 577, 577f
  sound, 588–589, 589f
  taste, 579
  vestibular, 589f, 591, 591f
Transepithelial process, tubular reabsorption as selective, 990
Trans face, of Golgi apparatus, 85f, 86f
Trans fats/trans fatty acids, **46–47**, 958
Transferrin, 649, 649f, 917
Transfer RNA (tRNA), 54, **99**, 103f, 108f
  aminoacyl, 103–104
  initiator, **104**, 106f
  in translation, 103–104, 106–107f
Transforming growth factor beta (TGF-β), 805t
Transfusion (blood), 663–666, 664t, 665f
Transfusion reactions, **665**
Transient ischemic attacks (TIAs), **468**
Transitional epithelium, 122f, **123**, 1002, 1002f
Transitional period, of newborn, **1116**
Transitional phase, Krebs cycle, **939**, 939f
Translation, 99, 99f, **103**–105, 104f, 105f, 106–107f, 108f

Translocation, 104, 107f
Transmembrane proteins, 65f, 66, 66f, 68, 70. *See also* Membrane proteins
Transmission electron microscopy (TEM), 117, 117f
Transplants (organ), 810–811
Transport, membrane. *See* Membrane transport
Transport maximum, **992**
Transport medium, water as, 38
Transport proteins, 49f, 66, 66f, 70
Transport vesicles, 76, 76f, 77, 85f, 87, 87f
Transposons, 1132
Transpulmonary pressure, **835–836**, 835f, 838f
Transthyretin, 50f, 51
Transurethral resection of prostate, 1052
Transverse arch of foot, 244–245, **245**, 245f
Transverse cerebral fissure, 438f, **439**
Transverse cervical (cardinal) ligament, 1063f, **1064**
Transverse cervical nerve, 509t, 509f
Transverse colon, 869f, **906**, 907f
Transverse costal facets, 222, 222f, 223t, 226f
Transverse foramen, **221**, 221f, 223t
Transverse mesocolon, 907f, 908f
Transverse pathways (commissural fibers), **444**, 445f, 474
Transverse perineal muscle, deep and superficial, 348t, 349f
Transverse (horizontal) plane/section, **15**, 15f
Transverse processes, vertebral, 218f, 219f, **220**, 220f, 221f, 222f, 223t
Transverse ridges, in sacrum, 224, 224f
Transverse sinuses, 465f, **752**t, 752f, 753f
Transversus abdominis muscle, 330f, **346**t, 347f
Trapezium, **234**, 234f, 235t
Trapezius muscle, 330f, 331f, **350**t, **352**t, 353f
Trapezoid, **234**, 234f, 235t
Trapezoid line, 228
T_{Reg} cells. *See* Regulatory T cells
Treponema pallidum, 1080
TRH. *See* Thyrotropin-releasing hormone
Triad(s), **288**–289, 288f, 294–295f, 296
Triceps brachii muscle, 330f, 331f, **355**f, **357**t, 362t, 363f
Triceps surae muscle, **377**t, 385f
Trichomoniasis, **1080**
Tricuspid valve, 677f, 678f, **679**, 681f, 696f
Trigeminal ganglion, 502t
Trigeminal nerve (cranial nerve V), 450f, 452, 453f, 499f, 502t, **506**, 536
  mandibular division of (cranial nerve V₃), 336t, 338t, 502t
  maxillary division of (cranial nerve V₂), 502t
  ophthalmic division of (cranial nerve V₁), 502t
Trigeminal neuralgia (tic douloureux), 502t
Trigger zone, 396, 398t
Triglycerides, **45**–47, 45f, 47t, 913f, 916–917, 928
  in absorptive state, 950, 951f
  metabolism of, 946, 946f, 947f
  storage of in bone, 176
Trigone (bladder), **1003**, 1003f
Triiodothyronine (T₃), **618**, 618t, 619, 619f. *See also* Thyroid hormone
Tripeptide, 48, 913f
Triple covalent bond, 33, 33f
Triplet (nucleotide base), **99**, 108f
Triplets (microtubule), 89, 90, 90f
Triquetrum, **234**, 234f, 235t
Tritium, 28f
tRNA. *See* Transfer RNA
Trochanter (bone marking), 180t
  greater and lesser, 240f, **241**

Trochlea
  of eye, **556**, 557f
  of humerus, 230f, **231**, 231f
  of talus, 244, 244f
Trochlear nerve (cranial nerve IV), 450f, 451f, **499**, 499f, 501t, 556, 557f
Trochlear notch, 232f, **233**, 233f
Trochlear nuclei, 451
Trophoblast/trophoblast cells, 1097f, **1098**, 1098f, 1099, 1100f
Tropic hormones/tropins, **615**
Tropomyosin, 286f, **287**, 295f, 296
Troponin, 286f, **287**, 291f, 295f, 296, 297f, 313t
True capillaries, **712**, 712f
True pelvis, 236, 238f, **238**
True (vertebrosternal) ribs, 225f, **226**, 226f
True solutions, 29–30
True vocal cords (vocal folds), 823f, 825–826, 825f, 826f
Truncus arteriosus, 700
Trunk(s), brachial plexus, 510, 511f
Trunk movements, muscles controlling, 346t, 347f
  extension, 340–343t, 341f, 343f
Trunk region
  arteries supplying, 738–739t
  veins of, 750–751t, 751f
Trypsin, 899, 899f, 913f, **915**
Trypsinogen, 899, 899f
TSH. *See* Thyroid-stimulating hormone
t-SNAREs, in exocytosis, 79, 79f
T tubules, **288**
  in cardiac muscle, 312t, 684f, 685, 685t
  in excitation-contraction coupling, 291f, 294–295f, 296
  in skeletal muscle, **288**, 288f, 312t
Tubal ligation, 1118b, 1119b
Tubal tonsils, **775**, 821f
Tubercle(s) (bone markings), 180t, 226, 226f. *See also specific type*
Tuberculosis (TB), 814, **859**
Tuberosity (bone marking), 180t
Tubular glands, **125**, 125f
Tubular reabsorption, 984, 984f, **989**–994, 990f, 991f, 992f, 993t
  of bicarbonate, 1029, 1030–1031, 1030f
  of calcium, 1026
  of nutrients/water/ions, 991–992, 991f
  of potassium, 1025
  reabsorptive capabilities of tubules/collecting ducts and, 992–994, 992f, 993t
  of sodium, 990–991, 991f, 1022, 1022, 1022f, 1023
Tubular secretion, 984, 984f, 992f, **994**–995
  of bicarbonate, 1029, 1030f, 1032
  of potassium, 992f
Tubulin(s), 88, 88f
Tubuloalveolar glands, **125**, 125f
Tubuloglomerular feedback mechanism, 987–989, **988**–989, 988f
Tumor. *See* Neoplasm
Tumor necrosis factor (TNF), 275, 633b, 805t
Tumor suppressor genes, 144b, 165
Tunic(s)
  of alimentary canal organs, 872–874, 873f
  of blood vessels, 708f, 709, 709t
Tunica albuginea
  of ovary, 1061f, **1062**
  of testis, **1048**, 1048f, 1049f, 1051f
Tunica externa (tunica adventitia), 708f, **709**
Tunica intima, 708f, **709**, 760b
Tunica media, 708f, **709**
Tunica vaginalis, **1048**, 1048f, 1049f, 1084
Turbulent flow, 715
Turner's syndrome, 1081
TV. *See* Tidal volume
T wave, 690f, **691**, 695f
Twins, fraternal/identical, 1074

Twitch, muscle, **299**, 299f, 300f
2n (diploid chromosomal number), **1036**
Two-neuron chain, 532, 533f, 543, 543f
Two-point discrimination test, **495**
Tympanic cavity (middle ear), **20**, 208, 581f, **582**–583, 582f, 596
Tympanic membrane (eardrum), 581f, **582**, 582f, 587, 587f, 594, 596
Tympanic part, temporal bone, 208, 208f
Type I cells, alveolar, **829**, 830f, 831
Type 1 (insulin-dependent) diabetes mellitus, 632, 633b, 812, 968
Type I (immediate/acute) hypersensitivity, **812**–813, 813f
Type II cells, alveolar, 830f, **831**
Type 2 (non-insulin-dependent) diabetes mellitus, 632, 633b, 968
Type II (cytotoxic) hypersensitivity, **813**
Type III (immune complex) hypersensitivity, **813**
Type IV (delayed) hypersensitivity, **813**–814
Type A blood, 664, 664t, 665f
Type AB blood, 664, 664t, 665f
Type A daughter cells, 1056f, **1057**
Type A intercalated cells, 981, 1025, 1030, 1031, 1031f
Type B blood, 664, 664t, 665f
Type B daughter cells, 1056f, **1057**
Type B intercalated cells, 981, 1032
Type O blood, 664, 664t, 665f
Tyrosine/tyrosinase, 156, 619, 619f
Tyrosine kinase, as insulin receptor, 605

Ubiquitin-proteasome pathway, **109**
Ubiquitins, **109**
Ulcer(s)
  decubitus (skin), **169**
  duodenal, 903
  esophageal, 883
  gastric/peptic, **887–889**, 887f, 889f
Ulcerative colitis, 921
Ulna, 200f, 231f, 232, 232–233f, **233**, 235t, 268, 268f
Ulnar artery, 739f, **742**t, 742f, 743f
Ulnar collateral ligament, **268**, 268f
Ulnar nerve, 358t, 359t, 364t, 366t, 510t, 511f, **512**
Ulnar notch, 232f, 233f, **233**
Ulnar vein, 751f, **754**t, 754f, 755f
Ultrasound imaging (ultrasonography/sonography), 17f, 17b, **1120**
Ultraviolet radiation, 156, 564f
  melanocyte/melanin response to, 156
  metabolic functions of skin and, 164
  skin damage/cancer caused by, 153, 156, 163, 165, 169
  vitamin D and, 634
Umami, 578, 579
Umbilical arteries, 759, 1101f, **1107**, 1107t, 1108f, 1109f, 1116
Umbilical cord, 1100f, 1101f, **1102**, 1109f
Umbilical ligaments, median, **1107**, 1107t, 1109f
Umbilical region, **20**, 20f
Umbilical vein, 759, 1101f, **1107**, 1107t, 1108f, 1109f, 1116
Umbilicus (navel), 20, 1109f
Unconsciousness, 462
"Uncouplers," poisons as, 943
Uncus, 440f, 443
Undescended testis (cryptorchidism), 1049, 1084
Unfused (incomplete) tetanus, 300f, **301**
Uniaxial movement, **258**, 262f
Unicellular (one-cell) organisms, **116**
Unicellular glands, 123, **124**, 124f
Unipennate fascicle pattern/muscles, 326f, 327
Unipolar neurons, **397**, 398–399t, 399
Unitary (visceral) smooth muscle, **316**
Universal donor, **665**
Universal recipient, **665**
Universal solvent (water), 29, **38**, 1013

Unsaturated fats/fatty acids, **46**, 46*f*, 957–958
Upper limbs. *See also* Arm; Forearm; Hand
  arteries of, 739*f*, 742–743*f*, 742*t*
  bones of, 200*f*, 230–236, 230–235*f*, 234*f*, 235*t*
  innervation of, 507*f*, 510–512, 510*t*, 511*f*
  veins of, 750*f*, 751*f*, 754–755*f*, 754*t*
Upper motor neurons, 478, 479*f*, 518
Upper respiratory system, 820–824, 824*t*. *See also specific structure*
Upper trunk, brachial plexus, 510, 511*f*
Up-regulation, **608**
Uracil (U), **55**
Urea, 948*f*, **949**, 992, 992*f*, 999*f*, 1000, **1001**
Urea cycle, **949**
Uremia, **1001**
Ureter(s), 536*f*, 975, 975*f*, 977*f*, 978*f*, **1002**, 1002*f*, 1003*f*
Ureteric buds, **1006**, 1006*f*
Ureteric ducts, **1006**
Ureterolithiasis, 1002
Urethra, 975, 975*f*, 1003*f*, **1004**
  autonomic innervation/effects and, 546*t*
  embryologic/fetal development of, 1006*f*, 1007, 1082*f*
  in female, 975*f*, 1003*f*, 1004, 1061*f*
  hypospadias and, 1007
  in male, 1003*f*, 1004, 1047, 1047*f*, **1050**, 1051*f*
Urethral folds, 1083*f*, **1084**
Urethral glands, 1050
Urethral groove, 1083*f*, **1084**
Urethral orifice, 1003*f*, **1004**, 1047*f*, 1050, 1051*f*, 1061*f*, 1066*f*
Urethral sphincters
  external, **348***t*, 349*f*, 1003*f*, **1004**
  internal, 1003*f*, **1004**
Urethritis, 1080
Uric acid, 275, **1001**
Urinalysis, **1000**
Urinary bladder, 975, 975*f*, **1003**–1004, 1003*f*, 1004*f*
  aging affecting, 1007
  atonic, **549**
  autonomic innervation/effects and, 534*f*, 536*f*, 540*f*, 546*t*
  cancer of, **1007**
  embryologic/fetal development of, 1006*f*, 1007
  emptying (micturition), **1004**–1005, 1005*f*, 1007
  infection of (cystitis), 1004
Urinary incontinence, **1005**, 1007, 1113
Urinary retention, **1005**
Urinary system, 5*f*, 7*f*, 974–1011, **975**, 975*f*. *See also specific structure and* Kidneys
  cardiovascular system and, 761*b*, 1036*b*
  case study and, 1011*b*, 1040*b*
  congenital abnormalities of, 1007
  developmental aspects of, 1006–1007, 1006*f*
  digestive system and, 920*b*, 1036*b*
  endocrine system and, 637*b*, 1036*b*
  homeostatic imbalances and, 976, 977, 989, 1001, 1002, 1004, 1005, 1007
  homeostatic relationships of, 5*f*, 1036*b*
  integumentary system and, 168*b*, 1036*b*
  lymphatic system/immunity and, 778*b*, 1036*b*
  muscular system and, 318*b*, 1036*b*
  nervous system and, 550*b*, 1036*b*
  physiology/functions of, 7*f*, 8
  pregnancy and, 1113
  reproductive system and, 1036*b*, 1085*b*
  respiratory system and, 862*b*, 1036*b*
  skeletal system and, 196*b*, 1036*b*
Urinary tract infections, 1004, 1007
Urination (micturition), 1002, **1004**–1005, 1005*f*, 1007

Urine, **984**, 1001–1002, 1001*t*
  blood in (hematuria), 976
  in body defense, 783*t*
  chemical composition of, 1001, 1001*t*
  concentration/volume regulation and, 995–1000, 996–997*f*, 999*f*
  formation of, 983–1000, 984*f*
    dilute/concentrated urine and, 998, 999*f*
    fetal, 1007
    glomerular filtration in, 982, **984**–989, 984*f*, 985*f*, 986*f*, 988*f*
    tubular reabsorption in. *See* Tubular reabsorption
    tubular secretion in. *See* Tubular secretion
  output, 989, 1016, 1016*f*, 1017, 1018
  physical characteristics of, 1001–1002
  transport/storage/elimination of, 1002–1006, 1002*f*, 1003*f*, 1004*f*, 1005*f*
Urobilinogen, 649
Urochrome, **1001**
Urogenital diaphragm, 348*t*, 349*f*, 1003*f*, 1004, 1047*f*, 1051*f*, 1061*f*
Urogenital ridges, 1006, 1006*f*
Urogenital sinus, 1006*f*, **1007**, 1082*f*, 1083*f*
Urologist, **1008**
Usnic acid, for weight loss, 959*b*
Uterine arteries, 1061, **1065**
Uterine contractions, 1114, 1114*f*, 1116
Uterine (menstrual) cycle, **1076**–1078, 1077*f*
Uterine glands, 1065, 1065*f*
Uterine (fallopian) tubes, 1060, 1061*f*, **1062**–1063, 1063*f*, 1064*f*
  embryologic/fetal development of, 1082*f*
  inflammation of (salpingitis), **1086**
  resection of (tubal ligation), 1118*b*, 1119*b*
Uterosacral ligaments, 1061, 1063*f*, **1064**
Uterus, 1060, 1061*f*, **1063**–1065, 1063*f*, 1064*f*, 1065*f*, 1100*f*, 1101*f*
  autonomic innervation/effects and, 540*f*
  cancer of (endometrial cancer), **1086**
  embryologic/fetal development of, 1082*f*
  pregnancy affecting, 1112, 1112*f*
  prolapse of, **1064**
  surgical removal of (hysterectomy), **1086**
Utricle, **583**, 583*f*, 585*t*, 590–591, 591*f*
Uvea, 558*f*, **559**
Uvula, 821*f*, 823, 876*f*, **877**

VAC (vacuum-assisted closure), **147**
Vaccines, 145*b*, **797**–798, 797*f*, 814
Vagal tone (heart), **699**
Vagal trunks, anterior and posterior, 537
Vagina, 1060, 1061*f*, 1063*f*, **1066**, 1066*f*, 1082*f*
  acid mantle of, 783*t*, 1066
  autonomic innervation/effects and, 536*f*, 540*f*, 546*t*
Vaginal fornix, 1061*f*, **1066**
Vaginal orifice, **1066**, 1066*f*
Vaginal process, 1084
Vaginal ring, for contraception, 1118*b*
Vagotomy, **549**, **921**
Vagus nerve (cranial nerve X), 450*f*, 451*f*, 452, **506**, 537
  in food intake regulation, 961, 961*f*
  in gustatory pathway, **580**, 580*f*
  parasympathetic fibers of, 499*f*, 505*t*, 536*f*, 537
  resection of (vagotomy), **549**, **921**
Vagus nerve stimulator, in seizure control, 462
Valence shell, **31**, 31*f*
Vallate papillae, **578**, 579*f*, **877**, 877*f*
Valsalva maneuver, 346*t*, **826**, 911
Valves. *See* Heart valves; Lymphatic capillary valves; Venous valves
Valvular heart disease, 680, 696, 702

Vanadium, 26*t*
Vanilloid receptor, 491
Vaporization, heat of, of water, 38, **965**
Variable, in homeostasis, **10**, 10*f*
Variable (V) region, 798*f*, **799**
Variation, genetic, 1047, 1126–1128, 1126*f*, 1127*f*
Varicose veins, **713**, 1113
Varicosities, autonomic nerve fiber, in smooth muscle, **311**, 311*f*, **517**
Vasa recta, 978*f*, 981*f*, **982**, 996*f*, 998, 998*f*
Vasa vasorum, 708*f*, **709**
Vascular anastomoses, **714**
Vascular dilation. *See* Vasodilation
Vascular endothelial growth factor (VEGF), 759
Vascularity/avascularity
  of bone, 136*f*, 137
  of cartilage, 127, 134
  of connective tissue, 127
  of epithelial tissue, 118
  of skeletal muscle, 281
  of synovial joints, 255
Vascular layer of eye, 558*f*, **559**
Vascular permeability, inflammation and, 785–786, 787*f*
Vascular plexus, dermal, 151*f*
Vascular (total peripheral) resistance, **714**, 718–719, 719*f*, 720, 720*f*, 723*f*
Vascular shock, **725**
Vascular shunts
  in capillary bed, 712, 712*f*
  in fetal circulation, 759, 1107, 1108*f*, 1109*f*, 1116
Vascular spasm, in hemostasis, **658**, 658*f*
Vascular system, **735**–759. *See also* Blood vessels; Circulation
Vascular tissue. *See* Blood
Vas deferens. *See* Ductus (vas) deferens
Vasectomy, **1050**, 1118*b*, 1119*b*
Vasoactive intestinal peptide, 888*t*
Vasoconstriction, **709**, 727*f*
  blood pressure and, 719, 720, 722*f*, 723, 724
  heat promotion and, 164, 966, 966*f*
  in hemostasis, 658, 658*f*
  Raynaud's disease and, 548
  venous return and, 718
Vasodilation, 164, **709**, 720, 727*f*, 785, 787*f*, 966*f*, 967
Vasomotion, **730**
Vasomotor center, 454, 547, **719**, 723*f*
Vasomotor fibers, **545**, 709, **719**
Vasomotor (sympathetic) tone, **545**, **719**, 723*f*
Vasopressin. *See* Antidiuretic hormone (ADH/vasopressin)
Vasovagal (neurocardiogenic) syncope, **549**
Vastus muscles, 368*f*, 384*t*, 385*f*
  intermedius, 368*f*, **370***t*, 385*f*
  lateralis, 330*f*, 368*f*, **369***t*, 385*f*
  medialis, 330*f*, 368*f*, **369***t*, 385*f*
Vault, cranial, 201
Veins, 642, **707**, 707*f*, 708*f*, 709*t*, 712–714. *See also specific named vein*
  blood flow velocity and, 730, 730*f*
  blood pressure in, 716*f*, 717–718
  blood volume in, 713, 713*f*
  constriction of, venous return and, 718
  dilation of, blood pressure regulation and, 720
  pulmonary, **674**, 675–677*f*, 681*f*, **736***t*, 736*f*, 832*f*, **834**
  structure of, 708*f*, 709*t*, 712–714
  systemic, 735, 750–751, 750*t*
  valves in, 708*f*, **713**
  varicose, **713**, 1113
Vellus hair, **159**
Venae cavae, 674, 675–677*f*, 750*f*, 750*t*, 751*f*
  blood flow velocity and, 730*f*

blood pressure in, 716*f*
  inferior, **674**, 675–677*f*, 681*f*, **750***t*, 750*f*, 751*f*, 755*f*, **756***t*, 756*f*, 757*f*, 758*f*
  superior, **674**, 675–677*f*, 681*f*, **750***t*, 750*f*, 751*f*, **752***t*, 752*f*, 753*f*, 754*f*, 755*f*
Venoconstriction, venous return and, 718
Venodilation, blood pressure regulation and, **720**
Venous anastomoses, **714**
Venous blood pressure, 713, 716*f*, 717–718, 718*f*
Venous palmar arches, deep and superficial, **754***t*, 754*f*, 755*f*
Venous plexuses, 728
Venous reserve, 847
Venous return, **697**–698, 697*f*, 717–718, 718*f*, 720, 723*f*
Venous sinuses, **713**–714
  dural, **464**, 465*f*, 714, 735, 750*f*, 751*f*, **752***t*
  scleral, 558*f*, **562**, 562*f*
  splenic (splenic sinusoids), 773*f*, **774**
Venous system, 707*f*, 708*f*, 709*t*, 712–714. *See also* Veins
Venous valves, 708*f*, **713**
Ventilation, 845
  alveolar, 842, 842*t*, 855, 855*f*, 856, 1029
  minute, **842**, 842*t*
  pulmonary, 819*f*, **835**, 836–840, 837*f*, 838*f*. *See also* Breathing; Respiration
Ventilation-perfusion coupling, 845–846, 846*f*
Ventral (term), 14*t*. *See also under* Anterior
Ventral body cavity, 17, **18**–20, 18*f*, 19*f*
Ventral corticospinal tract, 475*f*, 479*f*, 481*t*
Ventral (anterior) funiculi, 473*f*, **474**
Ventral (anterior) horns, **472**, 473*f*, 474*f*, 474*t*
Ventral (anterior) median fissure, **472**, 473*f*
Ventral mesentery, 872, 872*f*
Ventral posterolateral nuclei, 448, 449*f*
Ventral ramus/rami, 367*t*, **508**, 508*f*, 509, 515*t*
  brachial plexus, 510, 510*t*, 511*f*
  cervical plexus, 509, 509*f*, 509*t*
  lumbar plexus, 513*f*, 513*t*
  sacral plexus, 514*f*, 515*t*
  in thorax, 515
Ventral respiratory group, **853**, 853*f*. *See also* Medullary respiratory centers
Ventral root(s), **472**, 473*f*, 474*f*, **508**, 508*f*. *See also* Ventral ramus/rami
Ventral rootlets, 472, 473*f*, 508*f*
Ventral spinocerebellar tract, 475*f*, **478**, 480*t*
Ventral (anterior) spinothalamic tracts, 475*f*, **476**, 480*t*
Ventral tegmental area (VTA), 429*b*
Ventricles
  of brain, 435, 435*f*, **436**–438, 437*f*
  of heart, 671, 671*f*, **674**, 675–677*f*, 681*f*, 700, 701*f*
Ventricular diastole, 693
Ventricular ejection phase, 693, 695*f*
Ventricular fibrillation, **688**, 692*f*
Ventricular filling, 693, 695*f*
Ventricular septal defect, 701*f*
Ventricular systole, 693, 694*f*
Ventricular tachycardia, **702**
Ventromedial nucleus, 449*f*, 960
Venules, 707, 707*f*, 709*t*, 712, 712*f*, **713**, 713*f*, 716*f*, 730*f*
Vermis, **454**, 455*f*
Vernix caseosa, 167, 1111*t*
Vertebrae, 177*f*, 200*f*, **218**, 218*f*, 219*f*, 220*f*, 221*f*, 222*f*, 223*t*
  cervical, **218**, 218*f*, 221–222, 221*f*, 223*t*
  compression fractures of, in osteoporosis, 192*t*, 193
  lumbar, **218**, 218*f*, 222, 222*f*, 223*t*

regional characteristics of, 221–224, 221*f*, 222*f*, 223*t*, 224*f*
structure of, 220, 220*f*
thoracic, **218**, 218*f*, 222, 222*f*, 223*t*
Vertebral arch, **220**, 220*f*
Vertebral arteries, 738*f*, 739*f*, **740**–741*t*, 740*f*, 741*f*, 742*f*, 743*f*
Vertebral (medial) border of scapula, 228
Vertebral canal, **220**
Vertebral (spinal) cavity, **18**, 18*f*
Vertebral column (spine/spinal column), 200, 200*f*, **218**–224, 218*f*, 219*f*, 220*f*, 221*f*, 222*f*, 223*t*, 224*f*
developmental aspects of, 246–247, 247*f*
muscles of, 340–343*t*, 341*f*, 343*f*
osteoporotic fractures of, 192*t*, 193
Vertebral foramen, **220**, 220*f*, 223*t*
Vertebral (floating) ribs, 225*f*, **226**
Vertebral vein, 750*f*, 751*f*, **752***t*, 752*f*, 753*f*
Vertebra prominens, **221**, 222*f*
Vertebrochondral (false) ribs, 225*f*, **226**
Vertebrosternal (true) ribs, 225*f*, **226**, 226*f*
Vertex presentation, 1116
Vertical (perpendicular) plate, 203*f*, 205*f*, **210**, 210*f*, **212**, 214*f*
Very low density lipoproteins (VLDLs), 950, **956**–957, 957*f*
Vesical arteries, superior, 1107
Vesicle(s), 76–77
brain, 1106
primary and secondary, **435**, 435*f*
in exocytosis, 79, 79*f*
lens, **595**
matrix, 188
optic, **595**
otic, **596**
secretory, 79, 79*f*, 86*f*, 87, 87*f*
synaptic, **290**, 292*f*, 293*f*, **414**, 415*f*, 415*f*
transport, 76, 76*f*, 77, 85*f*, 87, 87*f*
Vesicle SNAREs, in exocytosis, 79, 79*f*
Vesicouterine pouch, 1061*f*, 1064, 1064*f*
Vesicular (antral/tertiary) follicles, 1070*f*, **1071**, 1072*f*, 1073
Vesicular trafficking, 76, 78*t*
Vesicular transport, **76**–79, 76*f*, 77*f*, 78*t*, 79*f*
Vestibular apparatus, **590**
Vestibular (equilibrium) cortex, 440*f*, 443
Vestibular folds (false vocal cords), 823*f*, 825*f*, **826**, 826*f*
Vestibular ganglia, 504*t*, 583*f*, **591**
Vestibular glands, greater, 1061*f*, 1066*f*, **1067**
Vestibular membrane, 584*f*, **585**
Vestibular nerve, 504*t*, 581*f*, 583*f*, **591**, 591*f*, 592*f*
Vestibular nuclear complex/vestibular nuclei, 453, 453*f*, **593**–594, 593*f*
Vestibular nystagmus, **593**
Vestibule
of bony labyrinth, 581*f*, **583**, 583*f*, 585*t*, 590
nasal, 821*f*, **822**
oral, **876**, 876*f*
of vagina, 1066*f*, **1067**, 1082*f*
Vestibulocochlear nerve (cranial nerve VIII), 450*f*, 451*f*, 453, 499*f*, 504*t*, **506**, 584*f*, 585, 589*f*. *See also* Cochlear nerve; Vestibular nerve
Vestibulo-ocular reflex, 594
Vestibulospinal tract, 475*f*, **478**, 481*t*
Viagra, 1054
Vibrissae (nose hairs), 783, 783*t*, **822**
Villi
chorionic, **1100**–1101, 1100*f*, 1101*f*
testing (chronic villus sampling), **1134**, *1134f*
in small intestine, **902**–903, 902*f*
VIP. *See* Vasoactive intestinal peptide
Viruses, 396, 788, 788*f*
Viscera/visceral organs, **18**, 18*f*, 346*t*, 347*f*, 743*t*. *See also specific organ*

Visceral afferent fibers/nerves, 392, 392*f*, 497
Visceral efferent fibers/nerves, 392, 392*f*, 497
Visceral layer
of glomerular capsule, 979*f*, 980, 983*f*, 985*f*
of serous pericardium (visceral pericardium/epicardium), 19, 19*f*, 142*f*, **673**, 673*f*, 677*f*
Visceral motor fibers, 392, 392*f*, 497
Visceral (autonomic) motor (VM) neurons, **474**, 474*f*. *See also* Autonomic nervous system; Motor (efferent) neurons
Visceral motor system, general, **532**. *See also* Autonomic nervous system
Visceral motor zones (lateral horns), **472**, 473*f*, **538**, 539*f*
Visceral muscle, 280, **316**, 517. *See also* Smooth muscle
Visceral pain, 496
Visceral pericardium, 19, 19*f*, 142*f*, **673**, 673*f*, 677*f*
Visceral peritoneum, 19, 142*f*, **872**, 872*f*, 873, 873*f*, 908*f*
Visceral (pulmonary) pleura, 19, 142*f*, 831*t*, 832*f*, **834**
Visceral reflex arc, **542**–543, 543*f*
Visceral (autonomic) reflexes, **519**, 542–543, 543*f*
Visceral sensory area, 443
Visceral sensory fibers, 392, 392*f*, 497
Visceral sensory (VS) neurons, **474**, 474*f*, **542**. *See also* Sensory (afferent) neurons
Visceral serosa, **19**, 19*f*
Visceroceptors (interoceptors), **490**, 491*t*, 492*t*
Viscosity (blood), 644, 715, 719*f*, 723*f*
Visible light, **564**, 564*f*
Visible spectrum, **564**, 564*f*
Vision, 554–575. *See also under Visual and Eye*
cortical areas in, 440*f*, 442–443, 574*f*, 575
depth perception and, **575**
developmental aspects of, 595
focusing of light on retina in, 565–567, 565*f*
homeostatic imbalances/disorders of, 557, 561, 563, 567, 569, 573, 575
information/visual processing in retina and, 572*f*, 573, 575
light/optics and, 563–567, 564*f*, 565*f*, 566*f*
pathway to brain and, 573–575, 574*f*
photoreceptors/phototransduction in, **561**, 567–573, 568*f*, 569*t*, **570**–573, 570*f*, 571*f*, 572*f*
Visual acuity, developmental aspects of, 595
Visual areas, 442–443
Visual association area, 440*f*, **443**
Visual (striate) cortex, primary, 440*f*, **442**–443, 574*f*, **575**
Visual fields, 574, 574*f*, 575
Visual pathways, 573–575, 574*f*
Visual pigments (photopigments), 568*f*, **569**–571, 570*f*, 571*f*, 573
Visual processing, 575
Visual receptor cells. *See* Photoreceptors
Vital capacity (VC), **841**, 841*f*
forced (FVC), **842**
Vital signs, **716**–717
Vitamin(s), 9, **931**, 932*t*
absorption of, 917
fat-soluble, 47*t*, 917, **931**, 932*t*
gut bacteria in synthesis of, 909
immune system and, 814
storage of in liver, 956*t*
water-soluble, 917, **931**, 932*t*
Vitamin A (retinol), 47*t*, 156, 569, 570*f*, 571, 573, 931, 932*t*

Vitamin B$_1$ (thiamine), 932*t*
Vitamin B$_2$ (riboflavin), 932*t*, 935
Vitamin B$_3$ (niacin), 932*t*, 935
Vitamin B$_5$ (pantothenic acid), 932*t*
Vitamin B$_6$ (pyridoxine), 932*t*
Vitamin B$_7$ (biotin), 932*t*
Vitamin B$_9$ (folic acid/folacin), 483, 649, 649*f*, 932*t*, 1112
Vitamin B$_{12}$ (cyanocobalamin), 649, 649*f*, 932*t*
intrinsic factor and, 650, 889, 931
Vitamin C (ascorbic acid), 147, 931, 932*t*
Vitamin D, 47*t*, **917**, 931, 932*t*
calcium absorption/regulation and, 164, 917
deficiency of, osteomalacia and rickets and, 193, 932*t*
immune system and, 814
for osteoporosis, 193
parathyroid hormone affecting activation of, 622, 622*f*
synthesis of in skin, 164
Vitamin E (tocopherol), 47*t*, 931, 932*t*
Vitamin K (phylloquinone), 47*t*, 658, 662, 931, 932*t*
Vitiligo, **170**
Vitreous humor, 558*f*, **562**, 562*f*
VLDLs. *See* Very low density lipoproteins
VMN. *See* Ventromedial nucleus
V$_{O_2max}$, aging affecting, 861
Vocal cords
false (vestibular folds), 823*f*, 825*f*, **826**, 826*f*
true. *See* Vocal folds
Vocal folds (true vocal cords), 823*f*, 825–826, 825*f*, 826*f*
Vocal ligaments, **825**
Voice box. *See* Larynx
Voice production, 826
Voiding (micturition), **1004**–1005, 1005*f*, 1007
Volatile acid, **1029**
Volkmann's (perforating) canals, **182**, 183*f*
Volt(s), 400
Voltage, 79, **400**
Voltage-gated channel(s), **290**, **400**, 401*f*
action potential and, 406*f*, 407*f*, 408
information transfer at chemical synapse and, 414, 415*f*, 572*f*
muscle contraction and, **290**, 292*f*, 293*f*, 315*f*, **689**, 689*f*
Volume-pressure relationships, in respiratory physiology, 834–840, 835*f*, 836*f*, 837*f*, 838*f*, 839*f*
Volume replacement, 663
Voluntary muscle, **138**, **280**. *See also* Skeletal muscle(s)
Voluntary nervous system, **392**, 392*f*. *See also* Somatic nervous system
Vomer, 202*f*, 203*f*, 205*f*, 206*f*, **212**, 214*f*, **217***t*
Vomiting (emesis), **893**
von Willebrand factor, 658
V (variable) region, 798*f*, **799**
VRG. *See* Ventral respiratory group
v-SNAREs, in exocytosis, 79, 79*f*
VTA. *See* Ventral tegmental area
VT/V-tac. *See* Ventricular tachycardia
Vulva (external genitalia), **1060**, 1066*f*, **1067**, 1083–1084, 1083*f*

Wallerian degeneration, 498*f*
Walls of nasal cavity, 215
Walls of orbit, 212, 213*f*
Warfarin, **662**
Warts, genital, **1080**
Water, 9, **38**, 39*f*, 843, 843*t*
absorption of, in digestion, 917–918
alveolar, 843*t*, 844
body, 1013, 1013*f*, 1035–1037, 1113. *See also* Water balance
exchanges of among fluid compartments, 1015, 1015*f*

hydrogen bonding and, 34–35, 35*f*
intake of, 1016–1017, 1016*f*, 1017*f*
loss of
insensible, **965**, **1016**
sensible, **1016**, 1017
metabolic (water of oxidation), **1016**
output of, 1016, 1016*f*, 1017
partial pressure of, 843, 843*t*
polarity of, 34, 34*f*, 38
as solvent, 29, **38**, 1013
as survival requirement, **9**, 38
tubular reabsorption of, 991, 991*f*, 992*f*
Water balance, 1016–1019, 1016*f*, 1018*f*, 1019*f*
antidiuretic hormone in, 449, 613, 998, 999*f*, 1000, 1017–1018, 1018*f*
developmental aspects of, 1035–1037
disorders of, 1018–1019, 1019*f*
hypothalamus in, 449
intake/intake regulation and, 1016–1017, 1017*f*
kidneys in, 974, 995
output/output regulation and, 1016, 1016*f*, 1017
sodium in, 1020*t*, 1021–1024, 1021*t*, 1022*f*, 1023*f*, 1024*f*
Water intake, 1016–1017, 1017*f*
Water on the knee, 273
Water output, 1016, 1016*f*, 1017
Water-soluble hormones, 604–605, 605*f*, 609*t*
Water-soluble vitamins, 917, **931**, 932*t*
Wavelength, 563–564
light, 564, 564*f*
sound, 586
Wave (temporal) summation, **300**–301, 416*t*, 419*f*, 420
WBCs (white blood cells). *See* Leukocyte(s)
Weak acids, 41, 1027, 1027*f*
Weak bases, 41, 1027
"Wear-and-tear arthritis" (osteoarthritis/ degenerative joint disease), 273–275
Wear-and-tear theory, of aging, 110
Weber's test, **596**
Weeping lubrication, 254
Weight, regulation of, 958–959*b*, 960–962, 961*f*
Weight gain, during pregnancy, 1112
Weight-loss supplements, 959*b*
Weight training, anabolic steroid abuse and, 317*b*
Wernicke's area, 440*f*, 443, **459**
White blood cell(s). *See* Leukocyte(s)
White blood cell count, 652
differential, 652*f*, **666**
White columns/funiculi, 473*f*, **474**, 475*f*, **476**, 480*t*
White fat (white adipose tissue), 130
Whitehead, 163
White matter, **436**, 437*f*
of brain
brain stem, 437*f*
cerebellar, 437*f*, 454, 455*f*
cerebral, 435*f*, 437*f*, 439, **444**, 445*f*, 446*f*
of spinal cord, 437*f*, 473*f*, 474, 475*f*, 482, 482*f*
White pulp, 773*f*, **774**
White ramus communicans/rami communicantes, 535*f*, **538**, 538*f*, 539–541, 539*f*, 540*f*
Whole blood transfusions, **663**
Willis, circle of (cerebral arterial circle), 714, 740*f*, **741***t*, 741*f*
Window of implantation, 1099
Windpipe. *See* Trachea
Wisdom teeth, 879–880
Withdrawal (coitus interruptus), 1118*b*
Withdrawal (flexor) reflex, 8, 10, **524**, 525*f*
Wolffian (mesonephric) ducts, **1006**, 1006*f*, 1082*f*, **1083**

Wolff's law, 189–190
Working (short-term) memory, **460**, 460*f*
Wound healing/repair, 142–146, 143*f*
  by first/second intention, **147**
Wrist (carpus). *See* Carpal bones/carpus
  (wrist)
Wrist drop, 512
Wrist joint, 256*t*
Wrong site surgery, 20

X chromosome, **1081**, 1125, 1130, 1130*f*
Xenical. *See* Orlistat
Xenografts, **814**

Xerostomia (dry mouth), 878, 1017, 1017*f*
Xiphisternal joint, 225*f*, **226**
Xiphoid process, 225, 225*f*
X-linked inheritance, **1130**
X-ray imaging, 16*f*, 16*b*
X rays, 16*b*, 563, 564*f*

Y chromosome, **1081**, 1125, 1130, 1130*f*
Yellow bone marrow, 178, 179, 179*f*, 666
Yellow bone marrow cavity, 178
Yellowness of skin, 157
Yolk sac, 666, 1100*f*, 1101*f*, **1102**, 1103*f*,
  1104*f*, 1105*f*

Z discs/lines, **284**
  in cardiac muscle, 684*f*, 685
  in skeletal muscle, **284**, 285*f*, 286*f*, 288*f*,
    289, 289*f*
ZIFT. *See* Zygote intrafallopian transfer
Zinc, 26*t*, 933*t*
Zona fasciculata, **623**, 623*f*
Zona glomerulosa, **623**, 623*f*
Zona pellucida, 1072*f*, **1073**, 1093, 1094*f*,
  1095*f*, 1096, 1096*f*
Zona reaction, 1095*f*, **1096**
Zona reticularis, **623**, 623*f*
Zygomatic arch, 206*f*, **208**

Zygomatic bones, 202*f*, 203*f*, 204*f*, 206*f*,
  212, **217***t*
Zygomatic branch of facial nerve, 503*t*,
  504*t*
Zygomatic process, 204*f*, 206*f*, **208**, 208*f*,
  211*f*, **212**, **216***t*, **217***t*
Zygomaticus muscles, 330*f*, **334***t*, 335*f*
Zygote, **1042**, **1093**, 1096, 1096*f*, 1097*f*
  cleavage of, **1097**, 1097*f*
Zygote intrafallopian transfer (ZIFT),
  1119
Zymogen granules, **898**, 899*f*

Word Roots, Prefixes, Suffixes, and Combining Forms

Prefixes and Combining Forms

a-, an- *absence, lack* acardia, lack of a heart; anaerobic, in the absence of oxygen

ab- *departing from, away from* abnormal, departing from normal

acou- *hearing* acoustics, the science of sound

ac-, acro- *extreme or extremity, peak* acrodermatitis, inflammation of the skin of the extremities

ad- *to, toward* adorbital, toward the orbit

aden-, adeno- *gland* adeniform, resembling a gland in shape

adren- *toward the kidney* adrenal gland, adjacent to the kidney

aero- *air* aerobic respiration, oxygen-requiring metabolism

af- *toward* afferent neurons, which carry impulses to the central nervous system

agon- *contest* agonistic and antagonistic muscles, which oppose each other

alb- *white* corpus albicans of the ovary, a white scar tissue

aliment- *nourish* alimentary canal, or digestive tract

allel- *of one another* alleles, alternative expressions of a gene

amphi- *on both sides, of both kinds* amphibian, an organism capable of living in water and on land

ana- *apart, up, again* anaphase of mitosis, when the chromosomes separate

anastomos- *come together* arteriovenous anastomosis, a connection between an artery and a vein

aneurysm *a widening* aortic aneurysm, a weak spot that causes enlargement of the blood vessel

angi- *vessel* angiitis, inflammation of a lymph vessel or blood vessel

angin- *choked* angina pectoris, a choked feeling in the chest due to dysfunction of the heart

ant-, anti- *opposed to, preventing, inhibiting* anticoagulant, a substance that prevents blood coagulation

ante- *preceding, before* antecubital, in front of the elbow

aort- *great artery* aorta

ap-, api- *tip, extremity* apex of the heart

append- *hang to* appendicular skeleton

aqua-, aque- *water* aqueous solutions

arbor *tree* arbor vitae of the cerebellum, the treelike pattern of white matter

areola- *open space* areolar connective tissue, a loose connective tissue

arrect- *upright* arrector pili muscles of the skin, which make the hairs stand erect

arthr-, arthro- *joint* arthropathy, any joint disease

artic- *joint* articular surfaces of bones, the points of connection

atri- *vestibule* atria, upper chambers of the heart

auscult- *listen* auscultatory method for measuring blood pressure

aut-, auto- *self* autogenous, self-generated

ax-, axi-, axo- *axis, axle* axial skeleton, axis of vertebral column

azyg- *unpaired* azygous vein, an unpaired vessel

baro- *pressure* baroreceptors for monitoring blood pressure

basal *base* basal lamina of epithelial basement membrane

bi- *two* bicuspid, having two cusps

bili- *bile* bilirubin, a bile pigment

bio- *life* biology, the study of life and living organisms

blast- *bud, germ* blastocyte, undifferentiated embryonic cell

brachi- *arm* brachial plexus of peripheral nervous system supplies the arm

brady- *slow* bradycardia, abnormally slow heart rate

brev- *short* fibularis brevis, a short leg muscle

broncho- *bronchus* bronchospasm, spasmodic contraction of bronchial muscle

bucco- *cheek* buccolabial, pertaining to the cheek and lip

calor- *heat* calories, a measure of energy

capill- *hair* blood and lymphatic capillaries

caput- *head* decapitate, remove the head

carcin- *cancer* carcinogen, a cancer-causing agent

cardi-, cardio- *heart* cardiotoxic, harmful to the heart

carneo- *flesh* trabeculae carneae, ridges of muscle in the ventricles of the heart

carot- *(1) carrot, (2) stupor* (1) carotene, an orange pigment; (2) carotid arteries in the neck, blockage of which causes fainting

cata- *down* catabolism, chemical breakdown

caud- *tail* caudal (directional term)

cec- *blind* cecum of large intestine, a blind-ended pouch

cele- *abdominal* celiac artery, in the abdomen

cephal- *head* cephalometer, an instrument for measuring the head

cerebro- *brain, especially the cerebrum* cerebrospinal, pertaining to the brain and spinal cord

cervic-, cervix *neck* cervix of the uterus

chiasm- *crossing* optic chiasma, where optic nerves cross

chole- *bile* cholesterol; cholecystokinin, a bile-secreting hormone

chondr- *cartilage* chondrogenic, giving rise to cartilage

chrom- *colored* chromosomes, so named because they stain darkly

cili- *small hair* ciliated epithelium

circum- *around* circumnuclear, surrounding the nucleus

clavic- *key* clavicle, a "skeleton key"

co-, con- *together* concentric, common center, together in the center

coccy- *cuckoo* coccyx, which is beak-shaped

cochlea *snail shell* the cochlea of the inner ear, which is coiled like a snail shell

coel- *hollow* coelom, the ventral body cavity

commis- *united* gray commissure of the spinal cord connects the two columns of gray matter

concha *shell* nasal conchae, coiled shelves of bone in the nasal cavity

contra- *against, opposite* contraceptive, agent preventing conception

corn-, cornu- *horn* stratum corneum, outer layer of the skin composed of (horny) cells

corona *crown* coronal suture of the skull

corp- *body* corpse; corpus luteum, hormone-secreting body in the ovary

cort- *bark* cortex, the outer layer of the brain, kidney, adrenal glands, and lymph nodes

cost- *rib* intercostal, between the ribs

crani- *skull* craniotomy, a skull operation

crypt- *hidden* cryptomenorrhea, a condition in which menstrual symptoms are experienced but no external loss of blood occurs

cusp- *pointed* bicuspid, tricuspid valves of the heart

cutic- *skin* cuticle of the nail

cyan- *blue* cyanosis, blue color of the skin due to lack of oxygen

cyst- *sac, bladder* cystitis, inflammation of the urinary bladder

cyt- *cell* cytology, the study of cells

de- *undoing, reversal, loss, removal* deactivation, becoming inactive

decid- *falling off* deciduous (milk) teeth

delta *triangular* deltoid muscle, roughly triangular in shape

den-, dent- *tooth* dentin of the tooth

dendr- *tree, branch* dendrites, branches of a neuron

derm- *skin* dermis, deep layer of the skin

desm- *bond* desmosome, which binds adjacent epithelial cells

di- *twice, double* dimorphism, having two forms

dia- *through, between* diaphragm, the wall through or between two areas

dialys- *separate, break apart* kidney dialysis, in which waste products are removed from the blood

diastol- *stand apart* cardiac diastole, between successive contractions of the heart

diure- *urinate* diuretic, a drug that increases urine output

dors- *the back* dorsal, dorsum, dorsiflexion

duc-, duct *lead, draw* ductus deferens, tube which carries sperm from the epididymis into the urethra during ejaculation

dura *hard* dura mater, tough outer meninx

dys- *difficult, faulty, painful* dyspepsia, disturbed digestion

ec-, ex-, ecto- *out, outside, away from* excrete, to remove materials from the body

ectop- *displaced* ectopic pregnancy; ectopic focus for initiation of heart contraction

edem- *swelling* edema, accumulation of water in body tissues

ef- *away* efferent nerve fibers, which carry impulses away from the central nervous system

ejac- *to shoot forth* ejaculation of semen

embol- *wedge* embolus, an obstructive object traveling in the bloodstream

en-, em- *in, inside* encysted, enclosed in a cyst or capsule

enceph- *brain* encephalitis, inflammation of the brain

endo- *within, inner* endocytosis, taking particles into a cell

entero- *intestine* enterologist, one who specializes in the study of intestinal disorders

epi- *over, above* epidermis, outer layer of skin

erythr- *red* erythema, redness of the skin; erythrocyte, red blood cell

eso- *within* esophagus

eu- *well* euesthesia, a normal state of the senses

excret- *separate* excretory system

exo- *outside, outer layer* exophthalmos, an abnormal protrusion of the eye from the orbit

extra- *outside, beyond* extracellular, outside the body cells of an organism

extrins- *from the outside* extrinsic regulation of the heart

fasci-, fascia- *bundle, band* superficial and deep fascia

fenestr- *window* fenestrated capillaries

ferr- *iron* transferrin, ferritin, both iron-storage proteins

flagell- *whip* flagellum, the tail of a sperm cell

flat- *blow, blown* flatulence

folli- *bag, bellows* hair follicle

fontan- *fountain* fontanelles of the fetal skull

foram- *opening* foramen magnum of the skull

foss- *ditch* fossa ovalis of the heart; mandibular fossa of the skull

gam-, gamet- *married, spouse* gametes, the sex cells

gangli- *swelling, knot* dorsal root ganglia of the spinal nerves

gastr- *stomach* gastrin, a hormone that influences gastric acid secretion

gene *beginning, origin* genetics

germin- *grow* germinal epithelium of the gonads

gero-, geront- *old man* gerontology, the study of aging

gest- *carried* gestation, the period from conception to birth

glauc- *gray* glaucoma, which causes gradual blindness

glom- *ball* glomeruli, clusters of capillaries in the kidneys

glosso- *tongue* glossopathy, any disease of the tongue

gluco-, glyco- *sweet* gluconeogenesis, the production of glucose from noncarbohydrate molecules

glute- *buttock* gluteus maximus, largest muscle of the buttock

gnost- *knowing* the gnostic sense, a sense of awareness of self

gompho- *nail* gomphosis, the term applied to the joint between tooth and jaw

gon-, gono- *seed, offspring* gonads, the sex organs

gust- *taste* gustatory sense, the sense of taste

hapt- *fasten, grasp* hapten, a partial antigen

hema-, hemato-, hemo- *blood* hematocyst, a cyst containing blood

hemi- *half* hemiglossal, pertaining to one-half of the tongue

hepat- *liver* hepatitis, inflammation of the liver

hetero- *different or other* heterosexuality, sexual desire for a person of the opposite sex

hiat- *gap* the hiatus of the diaphragm, the opening through which the esophagus passes

hippo- *horse* hippocampus of the brain, shaped like a seahorse

hirsut- *hairy* hirsutism, excessive body hair

hist- *tissue* histology, the study of tissues

holo- *whole* holocrine glands, whose secretions are whole cells

hom-, homo- *same* homeoplasia, formation of tissue similar to normal tissue; homocentric, having the same center

hormon- *to excite* hormones

humor- *a fluid* humoral immunity, which involves antibodies circulating in the blood

hyal- *glass, clear* hyaline cartilage, which has no visible fibers

hydr-, hydro- *water* dehydration, loss of body water

hyper- *excess* hypertension, excessive tension

hypno- *sleep* hypnosis, a sleeplike state

hypo- *below, deficient* hypodermic, beneath the skin; hypokalemia, deficiency of potassium

hyster-, hystero- *uterus, womb* hysterectomy, removal of the uterus; hysterodynia, pain in the womb

ile- *intestine* ileum, the last portion of the small intestine

im- *not* impermeable, not permitting passage, not permeable

inter- *between* intercellular, between the cells

intercal- *insert* intercalated discs, the end membranes between adjacent cardiac muscle cells

intra- *within, inside* intracellular, inside the cell

iso- *equal, same* isothermal, equal, or same, temperature

jugul- *throat* jugular veins, prominent vessels in the neck

juxta- *near, close to* juxtaglomerular complex, a cell cluster next to a glomerulus in the kidneys

karyo- *kernel, nucleus* karyotype, the assemblage of the nuclear chromosomes

kera- *horn* keratin, the water-repellent protein of the skin

kilo- *thousand* kilocalories, equal to 1000 calories

kin-, kines- *move* kinetic energy, the energy of motion

labi-, labri- *lip* labial frenulum, the membrane which joins the lip to the gum

lact- *milk* lactose, milk sugar

lacun- *space, cavity, lake* lacunae, the spaces occupied by cells of cartilage and bone tissue

lamell- *small plate* concentric lamellae, rings of bone matrix in compact bone

lamina *layer, sheet* basal lamina, part of the epithelial basement membrane

lat- *wide* latissimus dorsi, a broad muscle of the back

laten- *hidden* latent period of a muscle twitch

later- *side* lateral (directional term)

leuko- *white* leukocyte, white blood cell

leva- *raise, elevate* levator labii superioris, muscle that elevates upper lip

lingua- *tongue* lingual tonsil, adjacent to the tongue

lip-, lipo- *fat, lipid* lipophage, a cell that has taken up fat in its cytoplasm

lith- *stone* cholelithiasis, gallstones

luci- *clear* stratum lucidum, clear layer of the epidermis

lumen *light* lumen, center of a hollow structure

lut- *yellow* corpus luteum, a yellow, hormone-secreting structure in the ovary

lymph *water* lymphatic circulation, return of clear fluid to the bloodstream

macro- *large* macromolecule, large molecule

macula *spot* macula lutea, yellow spot on the retina

magn- *large* foramen magnum, largest opening of the skull

mal- *bad, abnormal* malfunction, abnormal functioning of an organ

mamm- *breast* mammary gland, breast

mast- *breast* mastectomy, removal of a mammary gland

mater *mother* dura mater, pia mater, membranes that envelop the brain

meat- *passage* external acoustic meatus, the ear canal

medi- *middle* medial (directional term)

medull- *marrow* medulla, the middle portion of the kidney, adrenal gland, and lymph node

mega- *large* megakaryocyte, large precursor cell of platelets

meio- *less* meiosis, nuclear division that halves the chromosome number

melan- *black* melanocytes, which secrete the black pigment melanin

men-, menstru- *month* menses, the cyclic menstrual flow

meningo- *membrane* meningitis, inflammation of the membranes of the brain

mer-, mero- *a part* merocrine glands, the secretions of which do not include the cell

meso- *middle* mesoderm, middle germ layer

meta- *beyond, between, transition* metatarsus, the part of the foot between the tarsus and the phalanges

metro- *uterus* endometrium, the lining of the uterus

micro- *small* microscope, an instrument used to make small objects appear larger

mictur- *urinate* micturition, the act of voiding the bladder

mito- *thread, filament* mitochondria, small, filament-like structures located in cells

mnem- *memory* amnesia

mono- *single* monoglyceride, a carbohydrate consisting of a single sugar molecule

morpho- *form* morphology, the study of form and structure or organisms

multi- *many* multinuclear, having several nuclei

mur- *wall* intramural ganglion, a collection of neuron cell bodies within an organ

muta- *change* mutation, change in the base sequence of DNA

myelo- *spinal cord, marrow* myeloblasts, cells of the bone marrow

myo- *muscle* myocardium, heart muscle

nano- *dwarf* nanometer, one-billionth of a meter

narco- *numbness* narcotic, a drug producing stupor or numbed sensations

natri- *sodium* atrial natriuretic peptide, a sodium-regulating hormone

necro- *death* necrosis, tissue death

neo- *new* neoplasm, an abnormal growth

nephro- *kidney* nephritis, inflammation of the kidney

neuro- *nerve* neurophysiology, the physiology of the nervous system

noci- *harmful* nociceptors, receptors for pain

nom- *name* innominate artery; innominate bone

noto- *back* notochord, the embryonic structure that precedes the vertebral column

nucle- *pit, kernel, little nut* nucleus

nutri- *feed, nourish* nutrition

ob- *before, against* obstruction, impeding or blocking up

oculo- *eye* monocular, pertaining to one eye

odonto- *teeth* orthodontist, one who specializes in proper positioning of the teeth in relation to each other

olfact- *smell* olfactory nerves

oligo- *few* oligodendrocytes, neuroglial cells with few branches

onco- *a mass* oncology, study of cancer

oo- *egg* oocyte, precursor of female gamete

ophthalmo- *eye* ophthalmology, the study of the eyes and related disease

orb- *circular* orbicularis oculi, muscle that encircles the eye

orchi- *testis* cryptorchidism, failure of the testes to descend into the scrotum

org- *living* organism

ortho- *straight, direct* orthopedic, correction of deformities of the musculoskeletal system

osm- *smell* anosmia, loss of sense of smell

osmo- *pushing* osmosis

osteo- *bone* osteodermia, bony formations in the skin

oto- *ear* otoscope, a device for examining the ear

ov-, ovi- *egg* ovum; oviduct

oxy- *oxygen* oxygenation, the saturation of a substance with oxygen

pan- *all, universal* panacea, a cure-all

papill- *nipple* dermal papillae, projections of the dermis into the epidermal area

para- *beside, near* paranuclear, beside the nucleus

pect-, pectus *breast* pectoralis major, a large chest muscle

pelv- *a basin* pelvic girdle, which cradles the pelvic organs

peni- *a tail* penis; penile arteriole

penna- *feather* unipennate, bipennate muscles, whose fascicles have a feathered appearance

pent- *five* pentose, a 5-carbon sugar

pep-, peps-, pept- *digest* pepsin, a digestive enzyme of the stomach; peptic ulcer

per-, permea- *through* permeate; permeable

peri- *around* perianal, situated around the anus

phago- *eat* phagocyte, a cell that engulfs and digests particles or cells

pheno- *show, appear* phenotype, the physical appearance of an individual

phleb- *vein* phlebitis, inflammation of the veins

pia *tender, gentle* pia mater, delicate inner membrane around the brain and spinal cord

pili *hair* arrector pili muscles of the skin, which make the hairs stand erect

pin-, pino- *drink* pinocytosis, the engulfing of small particles by a cell

platy- *flat, broad* platysma, broad, flat muscle of the neck

pleur- *side, rib* pleural serosa, the membrane that lines the thoracic cavity and covers the lungs

plex-, plexus *net, network* brachial plexus, the network of nerves that supplies the arm

pneumo- *air, wind* pneumothorax, air in the thoracic cavity

pod- *foot* podiatry, the treatment of foot disorders

poly- *multiple* polymorphism, multiple forms

post- *after, behind* posterior, places behind (a specific) part

pre-, pro- *before, ahead of* prenatal, before birth

procto- *rectum, anus* proctoscope, an instrument for examining the rectum

pron- *bent forward* prone, pronate

propri- *one's own* proprioception, awareness of body parts and movement

pseudo- *false* pseudotumor, a false tumor

psycho- *mind, psyche* psychogram, a chart of personality traits

ptos- *fall* renal ptosis, a condition in which the kidneys drift below their normal position

pub- *of the pubis* puberty

pulmo- *lung* pulmonary artery, which brings blood to the lungs

pyo- *pus* pyocyst, a cyst that contains pus

pyro- *fire* pyrogen, a substance that induces fever

quad-, quadr- *four-sided* quadratus lumborum, a muscle with a square shape

re- *back, again* reinfect

rect- *straight* rectus abdominis, rectum

ren- *kidney* renal; renin, an enzyme secreted by the kidney

retin-, retic- *net, network* endoplasmic reticulum, a network of membranous sacs within a cell

retro- *backward, behind* retrogression, to move backward in development

rheum- *watery flow, change, flux* rheumatoid arthritis; rheumatic fever

rhin-, rhino- *nose* rhinitis, inflammation of the nose

ruga- *fold, wrinkle* rugae, the folds of the stomach, gallbladder, and urinary bladder

sagitt- *arrow* sagittal (directional term)

salta- *leap* saltatory conduction, the rapid conduction of impulses along myelinated neurons

sanguin- *blood* consanguineous, indicative of a genetic relationship between individuals

sarco- *flesh* sarcomere, unit of contraction in skeletal muscle

saphen- *visible, clear* great saphenous vein, superficial vein of the thigh and leg

sclero- *hard* sclerodermatitis, inflammatory thickening and hardening of the skin

seb- *grease* sebum, the oil of the skin

semen *seed, sperm* semen, the discharge of the male reproductive system

semi- *half* semicircular, having the form of half a circle

sens- *feeling* sensation, sensory

septi- *rotten* sepsis, infection; antiseptic

septum *fence* nasal septum

sero- *serum-related* serological tests, which assess blood conditions

serrat- *saw* serratus anterior, a muscle of the chest wall that has a jagged edge

sin-, sino- *a hollow* sinuses of the skull

soma- *body* somatic nervous system

somn- *sleep* insomnia, inability to sleep

sphin- *squeeze* sphincter

splanchn- *organ* splanchnic nerve, autonomic supply to abdominal viscera

spondyl- *vertebra* ankylosing spondylitis, rheumatoid arthritis affecting the spine

squam- *scale, flat* squamous epithelium, squamous suture of the skull

steno- *narrow* stenosis, abnormal narrowing of a blood vessel or valve

strat- *layer* strata of the epidermis; stratified epithelium

stria- *furrow, streak* striations of skeletal and cardiac muscle tissue

stroma *mattress* stroma, the connective tissue framework of some organs

sub- *beneath, under* sublingual, beneath the tongue

sucr- *sweet* sucrose, table sugar

sudor- *sweat* sudoriferous glands, the sweat glands

super- *above, upon* superior, quality or state of being above other parts

supra- *above, upon* supracondylar, above a condyle

sym-, syn- *together, with* synapse, the region of communication between two neurons

synerg- *work together* synergism

systol- *contraction* systole, contraction of the heart

tachy- *rapid* tachycardia, abnormally rapid heartbeat

tact- *touch* tactile sense

telo- *the end* telophase, the end of mitosis

templ-, tempo- *time* temporal summation of nerve impulses

tens- *stretched* muscle tension

terti- *third* fibularis tertius, one of three fibularis muscles

tetan- *rigid, tense* tetanus of muscles

therm- *heat* thermometer, an instrument used to measure heat

thromb- *clot* thrombocytopenia, thrombus

thyro- *a shield* thyroid gland

tissu- *woven* tissue

tono- *tension* tonicity, hypertonic

tox- *poison* toxicology, study of poisons

trab- *beam, timber* trabeculae, spicules of bone in spongy bone tissue

trans- *across, through* transpleural, through the pleura

trapez- *table* trapezius, the four-sided muscle of the upper back

tri- *three* trifurcation, division into three branches

trop- *turn, change* tropic hormones, whose targets are endocrine glands

troph- *nourish* trophoblast, from which the fetal portion of the placenta develops

tuber- *swelling* tuberosity, a bump on a bone

tunic- *covering* tunica albuginea, the covering of the testis

tympan- *drum* tympanic membrane, the eardrum

ultra- *beyond* ultraviolet radiation, beyond the band of visible light

vacc- *cow* vaccine (first made from cowpox)

vagin- *a sheath* vagina

vagus *wanderer* the vagus nerve, which starts at the brain and travels into the abdominopelvic cavity

valen- *strength* valence shells of atoms

vas *vessel, duct* vasoconstriction, vas deferens

venter, ventr- *abdomen, belly* ventral (directional term), ventricle

vent- *the wind* pulmonary ventilation

vert- *turn* vertebral column

vestibul- *a porch* vestibule, the anterior entryway to the mouth and nose

vibr- *shake, quiver* vibrissae, hairs of the nasal vestibule

villus *shaggy hair* microvilli, which have the appearance of hair in light microscopy

viscero- *organ, viscera* visceroinhibitory, inhibiting the movements of the viscera

viscos- *sticky* viscosity, resistance to flow

vita- *life* vitamin

vitre- *glass* vitreous humor, the clear jelly of the eye

viv- *live* in vivo

vulv- *a covering* vulva, the female external genitalia

zyg- *a yoke, twin* zygote

Suffixes

-able *able to, capable of* viable, ability to live or exist

-ac *referring to* cardiac, referring to the heart

-algia *pain in a certain part* neuralgia, pain along the course of a nerve

-apsi *juncture* synapse, where two neurons communicate

-ary *associated with, relating to* coronary, associated with the heart

-asthen *weakness* myasthenia gravis, a disease involving paralysis

-blast *bud, germ* osteoblast

-bryo *swollen* embryo

-cide *destroy or kill* germicide, an agent that kills germs

-cipit *head* occipital

-clast *break* osteoclast, a cell that dissolves bone matrix

-crine *separate* endocrine organs, which secrete hormones into the blood

-cyte *cell* osteocyte, adipocyte

-dips *thirst, dry* polydipsia, excessive thirst associated with diabetes

-ectomy *cutting out, surgical removal* appendectomy, cutting out of the appendix

-ell, -elle *small* organelle

-emia *condition of the blood* anemia, deficiency of red blood cells

-esthesi *sensation* anesthesia, lack of sensation

-ferent *carry* efferent nerves, nerves carrying impulses away from the CNS

-form, -forma *shape* cribriform plate of the ethmoid bone

-fuge *driving out* vermifuge, a substance that expels worms of the intestine

-gen *an agent that initiates* pathogen, any agent that produces disease

-glea, -glia *glue* neuroglia, the connective tissue of the nervous system

-gram *data that are systematically recorded, a record* electrocardiogram, a recording showing action of the heart

-graph *an instrument used for recording data or writing* electrocardiograph, an instrument used to make an electrocardiogram

-ia *condition* insomnia, condition of not being able to sleep

-iatrics *medical specialty* geriatrics, the branch of medicine dealing with disease associated with old age

-ism *condition* hyperthyroidism

-itis *inflammation* gastritis, inflammation of the stomach

-lemma *sheath, husk* sarcolemma, the plasma membrane of a muscle cell

-logy *the study of* pathology, the study of changes in structure and function brought on by disease

-lysis *loosening or breaking down* hydrolysis, chemical decomposition of a compound into other compounds as a result of taking up water

-malacia *soft* osteomalacia, a process leading to bone softening

-mania *obsession, compulsion* erotomania, exaggeration of the sexual passions

-nata *birth* prenatal development

-nom *govern* autonomic nervous system

-odyn *pain* coccygodynia, pain in the region of the coccyx

-oid *like, resembling* cuboid, shaped as a cube

-oma *tumor* lymphoma, a tumor of the lymphatic tissues

-opia *of the eye* myopia, nearsightedness

-ory *referring to, of* auditory, referring to hearing

-pathy *disease* osteopathy, any disease of the bone

-phasia *speech* aphasia, lack of ability to speak

-phil, -philo *like, love* hydrophilic, water-attracting molecules

-phobia *fear* acrophobia, fear of heights

-phragm *partition* diaphragm, which separates the thoracic and abdominal cavities

-phylax *guard, preserve* prophylaxis, to guard in advance, as in preventive treatment

-plas *grow* neoplasia, an abnormal growth

-plasm *form, shape* cytoplasm

-plasty *reconstruction of a part, plastic surgery* rhinoplasty, reconstruction of the nose through surgery

-plegia *paralysis* paraplegia, paralysis of the lower half of the body or lower limbs

-rrhagia *abnormal or excessive discharge* metrorrhagia, uterine hemorrhage

-rrhea *flow or discharge* diarrhea, abnormal emptying of the bowels

-scope *instrument used for examination* stethoscope, instrument used to listen to sounds of parts of the body

-some *body* chromosome

-sorb *suck in* absorb

-stalsis *compression, constriction* peristalsis, muscular contractions that propel food along the digestive tract

-stasis *arrest, fixation* hemostasis, arrest of bleeding

-stitia *come to stand* interstitial fluid, between the cells

-stomy *establishment of an artificial opening* enterostomy, the formation of an artificial opening into the intestine through the abdominal wall

-tegm *cover* integument

-tomy *to cut* appendectomy, surgical removal of the appendix

-trud *thrust* protrude, detrusor muscle

-ty *condition of, state* immunity, condition of being resistant to infection or disease

-uria *urine* polyuria, passage of an excessive amount of urine

-zyme *ferment* enzyme